11th International Conference on Damage Assessment of Structures

(DAMAS 2015)

Journal of Physics: Conference Series Volume 628

Ghent, Belgium
24 – 26 August 2015

Part 1 of 2

ISBN: 978-1-5108-0994-9
ISSN: 1742-6588

Printed from e-media with permission by:

Curran Associates, Inc.
57 Morehouse Lane
Red Hook, NY 12571

Some format issues inherent in the e-media version may also appear in this print version.

Printed by Curran Associates, Inc. (2015)

For permission requests, please contact the Institute of Physics at the address below.

Institute of Physics
Dirac House, Temple Back
Bristol BS1 6BE UK

Phone: 44 1 17 929 7481
Fax: 44 1 17 920 0979

techtracking@iop.org

Additional copies of this publication are available from:

Curran Associates, Inc.
57 Morehouse Lane
Red Hook, NY 12571 USA
Phone: 845-758-0400
Fax: 845-758-2634
Email: curran@proceedings.com
Web: www.proceedings.com

TABLE OF CONTENTS

PART 1

STRUCTURAL HEALTH AND CONDITION MONITORING

FATIGUE DAMAGE OF STEAM TURBINE SHAFT AT ASYNCHRONOUS CONNECTIONS OF TURBINE GENERATOR TO ELECTRICAL NETWORK 1
A P Bovsunovsky

DAMAGE IDENTIFICATION BASED ON INCOMPLETE MODAL DATA AND CONSTRAINED NONLINEAR MULTIVARIABLE FUNCTION 9
S S Kourehli

STATISTICAL EVALUATION OF CHARACTERISTIC SDDLV-INDUCED STRESS RESULTANTS TO DISCRIMINATE BETWEEN UNDAMAGED AND DAMAGED ELEMENTS 15
L M Hansen, R J Johansen, M D Ulriksen, D Tcherniak, L Damkilde

MULTI-DAMAGE LOCALIZATION IN PLATE STRUCTURE USING FREQUENCY RESPONSE FUNCTION-BASED INDICES 23
Hai-Yang Gao, Xing-Lin Guo, Huajiang Ouyang, Xiu-Ming Yang

LOCATION OF THE CORROSION DAMAGE IN RECTANGULAR PLATES 31
M Tufoi, G R Gillich, I C Mituletu, V Iancu

CRACK DIAGNOSIS OF METALLIC PROFILES BASED ON STRUCTURAL DAMAGE INDICATORS 39
A Preisler, C Steenbock, K-U Schröder

HOW WOODPECKER AVOIDS BRAIN INJURY? 47
C W Wu, Z D Zhu, W Zhang

A GENERALIZED FUNCTIONAL MODEL BASED METHOD FOR VIBRATION-BASED DAMAGE PRECISE LOCALIZATION IN 3D STRUCTURES 58
Christos S Sakaris, John S Sakellariou, Spilios D Fassois

ASSESSMENT OF A MULTIPLE MODEL BASED PARAMETRIC METHOD FOR OUTPUT-ONLY VIBRATION-BASED DAMAGE DETECTION FOR A POPULATION OF LIKE STRUCTURES 66
Kyriakos J Vamvoudakis-Stefanou, John S Sakellariou, Spilios D Fassois

IMPROVED STOCHASTIC SUBSPACE SYSTEM IDENTIFICATION FOR STRUCTURAL HEALTH MONITORING 74
Chia-Ming Chang, Chin-Hsiung Loh

MULTI-OBJECTIVE SHAPE OPTIMIZATION OF PLATE STRUCTURE UNDER STRESS CRITERIA BASED ON SUB-STRUCTURED MIXED FEM AND GENETIC ALGORITHMS 82
Pierre Garambois, Sebastien Besset, Louis Jézéquel

EVALUATION OF DAMAGE INDEXES FOR STRUCTURAL HEALTH MONITORING IN NONLINEAR MECHANICAL SYSTEMS BASED ON VOLTERRA SERIES 90
Sidney Bruce Shiki, Samuel Da Silva

NOVEL SHM METHOD TO LOCATE DAMAGES IN SUBSTRUCTURES BASED ON VARX MODELS 98
U Ugalde, J Anduaga, F Martínez, A Iturrospe

ANALYSIS OF STATIONARY ROVING MASS EFFECT FOR DAMAGE DETECTION IN BEAMS USING WAVELET ANALYSIS OF MODE SHAPES 106
Mario Solís, Antonio J Benjumea, Mario Algaba, Pedro Galvín

A VERSATILE INSPECTION SYSTEM FOR PIPE STRUCTURE USING ULTRASONIC WAVES PROPAGATION IMAGER 114
T C Truong, J R Lee

DAMAGE CHARACTERIZATION ON HUMAN FEMUR BONE BY MEANS OF ULTRASONICS AND ACOUSTIC EMISSION 122
M Strantza, D Polyzos, O Louis, F Boulpaep, D Van Hemelrijck, D G Aggelis

COMPARATIVE STUDY OF PERFORMANCE OF NEUTRAL AXIS TRACKING BASED DAMAGE DETECTION 129
R Soman, P Malinowski, W Ostachowicz

INTEGRATED NON-DESTRUCTIVE ASSESSMENT OF RELEVANT STRUCTURAL ELEMENTS OF AN ITALIAN HERITAGE SITE: THE CARTHUSIAN MONASTERY OF TRISULTI 139
C Rainieri, A Marra, G M Rainieri, D Gargaro, M Pepe, G Fabbrocino

DAMAGE ASSESSMENT THROUGH CHANGES IN MODE SHAPES DUE TO NON-PROPORTIONAL DAMPING 147
F Iezzi, D Spina, C Valente

A MULTI-RESOLUTION BASED METHOD TO PRECISE IDENTIFY THE NATURAL FREQUENCIES OF BEAMS WITH APPLICATION IN DAMAGE DETECTION 155
I C Mituletu, N Gillich, C N Nitescu, C P Chioncel

THE INFLUENCE OF OPERATIONAL AND ENVIRONMENTAL LOADS ON THE PROCESS OF ASSESSING DAMAGES IN BEAMS 163
H Furdui, F Muntean, A A Minda, Z I Praisach, N Gillich

EXPERIMENTAL FAULT DIAGNOSIS IN SYSTEMS CONTAINING FINITE ELEMENTS OF PLATE OF KIRCHOFF BY USING STATE OBSERVERS METHODOLOGY 171
D M Alegre, E H Koroishi, G P Melo

INTEGRATION OF ACOUSTIC EMISSION SYSTEMS WITHIN INTEGRI-TECHTM ANALYSIS SYSTEM FOR STRUCTURAL HEALTH MONITORING OF PRESSURISED ENGINEERING PLANT 179
A A Ghouri, Steven Rafferty, Andy Pickwell, Walter Galbraith, S Gareth Pierce, Anthony Gachagan

A WAVELET-BASED SCHEME FOR OPTIMUM MEASUREMENT/MONITORING OF DYNAMIC RESPONSES OF STRUCTURES 187
S H Mahdavi, H A Razak

RESEARCH ON ASSESSMENT OF BOLTED JOINT STATE USING ELASTIC WAVE PROPAGATION 195
R Kedra, M Rucka

WAVE PROPAGATION IN DAMAGE ASSESSMENT OF GROUND ANCHORS 203
B Zima, M Rucka

MAGNETIC FLUX LEAKAGE AND PRINCIPAL COMPONENT ANALYSIS FOR METAL LOSS APPROXIMATION IN A PIPELINE 211
M Ruiz, L E Mujica, M Quintero, J Florez, S Quintero

DAMAGE LOCALIZATION IN PLATES WITH USE OF THE PROCEDURE BASED ON MODAL FILTRATION 219
Ziemowit Dworakowski, Lukasz Ambrozinski, Krzysztof Mendrok

STRUCTURAL DAMAGE DETECTION USING AUTO CORRELATION FUNCTIONS OF VIBRATION RESPONSE UNDER SINUSOIDAL EXCITATION 226
Muyu Zhang, Rüdiger Schmidt, Bernd Markert

LOAD IDENTIFICATION APPROACH BASED ON BASIS PURSUIT DENOISING ALGORITHM 234
D Ginsberg, M Ruby, C P Fritzen

DAMAGE ASSESSMENT OF CEMENT-BASED GEOMATERIAL DURING LOADING BY ULTRASONIC TOMOGRAPHY 242
Duc-Phi Do, Ndao Baba, Dashnor Hoxha, Truong-Son Bui

OPTIMAL SAMPLING IN DAMAGE DETECTION OF FLEXURAL BEAMS BY CONTINUOUS WAVELET TRANSFORM 250
B Basu, B M Broderick, L Montanari, A Spagnoli

MILLING DAMAGE ON CARBON FIBRE REINFORCED POLYMER USING TIALN COATED END MILLS 258
Mohamed Konneh, Sudin Izman, Abdullah Abdul Rahman Kassim

DAMAGE DETECTION BASED ON THE NATURAL FREQUENCY SHIFTING OF A CLAMPED RECTANGULAR PLATE MODEL 266
Irfan Hilmy, M M Abdel Wahab, Erry Yullian T Adesta, Tasnim Firdaus

GENETIC ALGORITHM BASED OBJECTIVE FUNCTIONS COMPARATIVE STUDY FOR DAMAGE DETECTION AND LOCALIZATION IN BEAM STRUCTURES 273
S Khatir, I Belaidi, R Serra, B Benaissa, A Ait Saada

SINGLE SIDE DAMAGE SIMULATIONS AND DETECTION IN BEAM-LIKE STRUCTURES 281
Yun-Lai Zhou, M Abdel Wahab, R Perera, N Maia, R Sampaio, E Figueiredo

DAMAGE IN CIVIL ENGINEERING

LONG TERM APPLICATION OF BUS MONITORING SYSTEM TO SHORT AND MEDIUM SPAN BRIDGES AND DAMAGE DETECTION ... 291
Ayaho Miyamoto, Válter J G Lúcio, Hisao Emoto, Hideaki Tanaka

PROBABILITY DISTRIBUTION FUNCTIONS FOR COVER USED IN 3-D MODEL SIMULATING CONCRETE DETERIORATION IN PORT ASSETS ... 305
Rachel M Homer, David W Law, Thomas C K Molyneaux

FINITE ELEMENT SIMULATION OF GUIDED WAVES IN PIPELINES FOR LONG RANGE MONITORING AGAINST THIRD PARTY ATTACKS ... 313
Salisu El-Hussein, John J Harrigan, Andrew Starkey

BUILDING DAMAGE DUE TO STRUCTURAL POUNDING DURING EARTHQUAKES ... 322
B Soltysik, R Jankowski

NONLINEAR DYNAMIC RESPONSES OF SUSPENSION BRIDGE TO SUDDEN BREAKAGE OF HANGERS ... 332
W L Qiu, M Jiang, Z Zhang

CEMENTITIOUS COMPOSITES ENGINEERED WITH EMBEDDED CARBON NANOTUBE THIN FILMS FOR ENHANCED SENSING PERFORMANCE ... 339
Kenneth J Loh, Jesus Gonzalez

NUCLEATION OF SQUAT CRACKS IN RAIL, CALCULATION OF CRACK INITIATION ANGLES IN THREE DIMENSIONS ... 347
Meysam Naeimi, Zili Li, Rolf Dollevoet

DAMAGE DETECTION IN BRIDGE STRUCTURE USING VIBRATION DATA UNDER RANDOM TRAVELLING VEHICLE LOADS ... 360
C H Loh, T Y Hung, S F Chen, W T Hsu

EXPERIMENTAL VERIFICATION OF THE INTERPOLATION METHOD ON A REAL DAMAGED BRIDGE ... 368
M P Limongelli, A Morassi

LIMIT STATES FOR THE DAMAGE ASSESSMENT OF BRIDGES SUPPORTED ON LRB BEARINGS ... 376
M Jara, C Gómez, J M Jara, B A Olmos

THE EFFECT OF POST-TENSIONING FORCE MAGNITUDE AND ECCENTRICITY ON THE NATURAL BENDING FREQUENCY OF CRACKED POST-TENSIONED CONCRETE BEAMS ... 384
D Noble, M Nogal, A J O'Connor, V Pakrashi

EXPECTED DAMAGES OF RETROFITTED BRIDGES WITH RC JACKETING ... 392
O Montes, J M Jara, M Jara, B A Olmos

OVERTOPPING FAILURE ANALYSIS OF COASTAL FLOOD DEFENCES AFFECTED BY CLIMATE CHANGE ... 400
Mehrdad Bahari Mehrabani, Hua-Peng Chen, Morris W Stevenson

ASSESSMENT OF CONCRETE DAMAGE AND STRENGTH DEGRADATION CAUSED BY REINFORCEMENT CORROSION ... 408
Jaya Nepal, Hua-Peng Chen

EXPERIMENTAL STUDY ON THE EFFECTIVENESS OF POLYMER DAMPER IN DAMAGE REDUCTION OF TEMPORARY STEEL GRANDSTAND ... 416
N Lasowicz, A Kwiecien, R Jankowski

DAMAGE ASSESSMENT OF CURTAIN WALL GLASS ... 423
H Puga, Ba Olmos, L Olmos, J M Jara, M Jara

ACQUISITION AND ANALYSIS OF DYNAMIC RESPONSES OF A HISTORIC PEDESTRIAN BRIDGE USING VIDEO IMAGE PROCESSING ... 431
Michael O'Byrne, Bidisha Ghosh, Franck Schoefs, Deirdre O'Donnell, Robert Wright, Vikram Pakrashi

LONG-TERM BRIDGE HEALTH MONITORING FOCUSING ON THE MAHALANOBIS DISTANCE OF MODAL PARAMETERS ... 439
Chul-Woo Kim, Tomoaki Morita, Ziran Wang, Kunitomo Sugiura

EXPERIMENTAL COMPARISON OF DYNAMIC RESPONSES OF A TENSION MOORED FLOATING WIND TURBINE PLATFORM WITH AND WITHOUT SPRING DAMPERS ... 447
C Wright, K O'Sullivan, J Murphy, V Pakrashi

THE HURST EXPONENT AS AN INDICATOR OF THE BEHAVIOUR OF A MODEL MONOPILE IN AN OCEAN WAVE TESTING BASIN ... 455
Vikram Pakrashi, Richard O'Shea, Vesna Jaksic, Jimmy Murphy

PERFORMANCE OF A SINGLE LIQUID COLUMN DAMPER FOR THE CONTROL OF DYNAMIC RESPONSES OF A TENSION LEG PLATFORM ... 463
V Jaksic, C Wright, Afeef Chanayil, Shaikh Faruque Ali, Jimmy Murphy, Vikram Pakrashi

A DELAY VECTOR VARIANCE BASED MARKER FOR AN OUTPUT-ONLY ASSESSMENT OF STRUCTURAL CHANGES IN TENSION LEG PLATFORMS 471
V Jaksic, C Wright, D P Mandic, J Murphy, V Pakrashi

PRINCIPAL COMPONENT ANALYSIS FOR CONDITION MONITORING OF A NETWORK OF BRIDGE STRUCTURES 480
Ciarán Hanley, Denis Kelliher, Vikram Pakrashi

DAMAGE ASSESSMENT OF RC BUILDINGS SUBJECTED TO THE DIFFERENT STRONG MOTION DURATION 488
Alireza Mortezaei, Mohsen Mohajer Tabrizi

ANALYSIS ABOUT FACTORS AFFECTING THE DEGREE OF DAMAGE OF BUILDINGS IN EARTHQUAKE 495
Jing Jia, Jinghong Yan

CHARACTERISING THE EFFECT OF GLOBAL AND LOCAL GEOMETRIC IMPERFECTIONS ON THE NUMERICAL PERFORMANCE OF A BRACE MEMBER 503
M S Hassan, J Goggins, S Salawdeh

FOOTPRINT CAUSED BY A VEHICLE CONFIGURATION ON THE DYNAMIC AMPLIFICATION OF THE BRIDGE RESPONSE 513
Omar Mohammed, Arturo González

PART 2

NUMERICAL MODELLING OF REINFORCED CONCRETE SLABS UNDER BLAST LOADS OF CLOSE-IN DETONATIONS USING THE LAGRANGIAN APPROACH 522
M Shuaib, O Daoud

THE EVALUATION OF DAMAGE MECHANISM OF UNREINFORCED MASONRY BUILDINGS AFTER VAN (2011) AND ELAZIG (2010) EARTHQUAKES 530
D Güney, E Aydin, B Öztürk

DAMAGE DETECTION IN WIND TURBINE TOWERS USING A FINITE ELEMENT MODEL AND DISCRETE WAVELET TRANSFORM OF STRAIN SIGNALS 543
A Kenna, B Basu

DAMAGE IN MACHINERIES

EXPERIMENTAL DAMAGE DETECTION IN A WIND TURBINE BLADE MODEL USING PRINCIPAL COMPONENTS OF RESPONSE CORRELATION FUNCTIONS 551
S Hoell, P Omenzetter

DAMAGE LOCALIZATION IN A RESIDENTIAL-SIZED WIND TURBINE BLADE BY USE OF THE SDDLV METHOD 559
R J Johansen, L M Hansen, M D Ulriksen, D Tcherniak, L Damkilde

BLIND SOURCE SEPARATION AND DYNAMIC FUZZY NEURAL NETWORK FOR FAULT DIAGNOSIS IN MACHINES 567
Haifeng Huang, Huajiang Ouyang, Hongli Gao

VIBRATION DAMAGE MECHANISM ANALYSIS ON ROTOR OF DIESEL GENERATING SET WITH RIGID COUPLING 578
Bing Yan, Weizhen Shi, Chunrong Hua, Jingming Liu, Dawei Dong, Jun Chen

VIBRATION-BASED SHM SYSTEM: APPLICATION TO WIND TURBINE BLADES 586
D Tcherniak, L L Mølgaard

NATURAL VIBRATION RESPONSE BASED DAMAGE DETECTION FOR AN OPERATING WIND TURBINE VIA RANDOM COEFFICIENT LINEAR PARAMETER VARYING AR MODELLING 594
L D Avendaño-Valencia, S D Fassois

USE OF HIGH FREQUENCY ANALYSIS OF ACOUSTIC EMISSION SIGNALS TO DETERMINE ROLLING ELEMENT BEARING CONDITION 602
A Cockerill, K M Holford, T Bradshaw, P Cole, R Pullin, A Clarke

SURROGATE MODELS FOR EFFICIENT STABILITY ANALYSIS OF BRAKE SYSTEMS 610
Lyes Nechak, Frédéric Gillot, Sébastien Besset, Jean-Jacques Sinou

NUMERICAL STUDIES ON TIME-VARYING STIFFNESS OF DISK-DRUM TYPE ROTOR WITH BOLT LOOSENING 618
Zhaoye Qin, Fulei Chu

SELECT ASPECTS OF FEM ANALYSIS FOR BONDED JOINTS OF POLYMER COMPOSITE MATERIALS 624
A Rudawska
THE DETECTION OF LUBRICATING OIL VISCOSITY CHANGES IN GEARBOX TRANSMISSION SYSTEMS DRIVEN BY SENSORLESS VARIABLE SPEED DRIVES USING ELECTRICAL SUPPLY PARAMETERS 633
S Abusaad, K Brethee, M Assaeh, R Zhang, F Gu, A D Ball
A FEATURE EXTRACTION METHOD FOR FAULT CLASSIFICATION OF ROLLING BEARING BASED ON PCA 641
Fengtao Wang, Jian Sun, Dawen Yan, Shenghua Zhang, Liming Cui, Yong Xu
PROJECTILES IMPACT ASSESSMENT OF AIRCRAFT WING STRUCTURES WITH REAL DYNAMIC LOAD 648
Lu Han, Qing Han, Changlin Wang
STRUCTURAL HEALTH MONITORING CHALLENGES ON THE 10-MW OFFSHORE WIND TURBINE MODEL 656
E Di Lorenzo, G Kosova, U Musella, S Manzato, B Peeters, F Marulo, W Desmet
AUTOMATIC WELDING DETECTION BY AN INTELLIGENT TOOL PIPE INSPECTION 664
C J Arizmendi, W L Garcia, M A Quintero
A THEORETICAL MODEL FOR FAULT DIAGNOSIS OF LOCALIZED BEARING DEFECTS UNDER NON-WEIGHT-DOMINANT CONDITIONS 673
Q K Han, F L Chu
A NEW ANGULAR RESAMPLING ALGORITHM FOR THE BEARING FAULT DIAGNOSIS UNDER THE TIME-VARYING ROTATIONAL SPEED 679
Tianyang Wang, Weidong Cheng, Jianvong Li, Fulei Chu
IMPACT DAMAGE EVALUATION METHOD OF FRICTION DISC BASED ON HIGH-SPEED PHOTOGRAPHY AND TOOTH-ROOT STRESS COUPLING 688
L Yin, Y M Shao, J Liu, H L Zheng
DAMAGE ASSESSMENT FOR WIND TURBINE BLADES BASED ON A MULTIVARIATE STATISTICAL APPROACH 694
David García, Dmitri Tcherniak, Irina Trendafilova
DETECTION AND CLASSIFICATION OF ALARM THRESHOLD VIOLATIONS IN CONDITION MONITORING SYSTEMS WORKING IN HIGHLY VARYING OPERATIONAL CONDITIONS 702
M Straczkiewicz, T Barszcz, A Jablonski
AUTOREGRESSIVE MODELLING FOR ROLLING ELEMENT BEARING FAULT DIAGNOSIS 710
H Al-Bugharbee, I Trendafilova
A CONTROL STRATEGY FOR ADAPTIVE ABSORBER BASED ON VARIABLE MASS 718
Qiang Gao, Ning Han, Yanqing Zhao, Chendong Duan, Wanqin Wang

DAMAGE IN COMPOSITE MATERIALS

FUZZY MODAL CONTROL APPLIED TO SMART COMPOSITE STRUCTURE 724
E H Koroishi, A W Faria, F A Lara-Molina, V Steffen Jr
NATURAL FREQUENCY CHANGES DUE TO DAMAGE IN COMPOSITE BEAMS 732
I Negru, G R Gillich, Z I Praisach, M Tufoi, N Gillich
BALLISTIC DAMAGE IN HYBRID COMPOSITE LAMINATES 740
Vaibhav A Phadnis, Kedar S Pandya, Niranjan K Naik, Anish Roy, Vadim V Silberschmidt
DAMAGE MECHANISMS OF RANDOM FIBROUS NETWORKS 748
E Sozumert, F Farukh, E Demirci, M Acar, B Pourdeyhimi, V V Silberschmidt
DYNAMIC RESPONSE OF RC BEAMS STRENGTHENED WITH NEAR SURFACE MOUNTED CARBON-FRP RODS SUBJECTED TO DAMAGE 756
R Capozucca, M G Blasi, V Corina
DAMAGE LOCALIZATION IN A GLASS FIBER REINFORCED COMPOSITE PLATE VIA THE SURFACE INTERPOLATION METHOD 764
M P Limongelli, V Carvelli
ON THE MONITORING AND IMPLICATIONS OF GROWING DAMAGES CAUSED BY MANUFACTURING DEFECTS IN COMPOSITE STRUCTURES 772
M Schagerl, C Viechtbauer, S Hörrmann
NONLINEAR DYNAMIC RESPONSE ANALYSIS OF LOCALIZED DAMAGED LAMINATED COMPOSITE STRUCTURES IN THE CONTEXT OF COMPONENT MODE SYNTHESIS 780
S Mahmoudi, F Trivaudey, N Bouhaddi

A COMPARATIVE STUDY OF A SELF STRAIN-MONITORING CARBON NANOTUBE FILM AND CARBON FIBERS UNDER FLEXURAL LOADING BY ELECTRICAL RESISTANCE CHANGES 789
B Pinto, S Kern, J J Ku-Herrera, J Yasui, V La Saponara, K J Loh

IMPACT DAMAGE RESISTANCE OF THIN STITCHED CARBON/EPOXY LAMINATES 796
L Francesconi, F Aymerich

MOISTURE DETECTION IN COMPOSITES BY TERAHERTZ SPECTROSCOPY 804
Pawel Malinowski, Norbert Palka, Szymon Opoka, Tomasz Wandowski, Wieslaw Ostachowicz

ASSESSMENT OF DAMAGE DETECTION IN COMPOSITE STRUCTURES USING 3D VIBROMETRY 812
S Grigg, M Pearson, R Marks, C Featherston, R Pullin

NON-DESTRUCTIVE TESTING ON ARAMID FIBRES FOR THE LONG-TERM ASSESSMENT OF INTERVENTIONS ON HERITAGE STRUCTURES 820
R Ceravolo, A De Marchi, E Pinotti, C Surace, L Zanotti Fragonara

ANALYSIS OF ELASTIC NONLINEARITY FOR IMPACT DAMAGE DETECTION IN COMPOSITE LAMINATES 828
A Frau, L Pieczonka, M C Porcu, W J Staszewski, F Aymerich

NON-DESTRUCTIVE VISUALIZATION OF LINEAR EXPLOSIVE-INDUCED PYROSHOCK USING PHASE ARRAYED LASER-INDUCED SHOCK IN A SPACE LAUNCHER COMPOSITE 836
Jae Kyeong Jang, Jung Ryul Lee

ACCURATE DAMAGE LOCATION IN COMPLEX COMPOSITE STRUCTURES AND INDUSTRIAL ENVIRONMENTS USING ACOUSTIC EMISSION 842
M Eaton, M Pearson, W Lee, R Pullin

FINITE ELEMENT BASED DAMAGE ASSESSMENT OF COMPOSITE TIDAL TURBINE BLADES 850
Edward M Fagan, Sean B Leen, Ciaran R Kennedy, Jamie Goggins

AN INVESTIGATION INTO THE IMPACT OF CRYOGENIC ENVIRONMENT ON MECHANICAL STRESSES IN FRP COMPOSITES 859
O Fifo, B Basu

SENSING AND SENSORS

PREDICTION OF DYNAMIC STRAINS ON A MONOPILE OFFSHORE WIND TURBINE USING VIRTUAL SENSORS 866
A N Iliopoulos, W Weijtjens, D Van Hemelrijck, C Devriendt

LOCAL INTERROGATIONS FOR DETECTION OF MULTIPLE DEFECTS USING AN ACTIVE SENSOR NETWORK 874
F Gao, J Lin, L Zeng, Z Luo

DAMAGE DETECTION MONITORING APPLICATIONS IN SELF-HEALING CONCRETE STRUCTURES USING EMBEDDED PIEZOELECTRIC TRANSDUCERS AND RECOVERY 882
G Karaiskos, E Tsangouri, D G Aggelis, A Deraemaeker, D Van Hemelrijck

FREQUENCY DOMAIN DECOMPOSITION PERFORMED ON THE STRAIN DATA OBTAINED FROM THE ALUMINIUM MODEL OF AN OFFSHORE SUPPORT STRUCTURE 890
M Mieloszyk, S Opoka, W Ostachowicz

INFLUENCE OF SUPPORT CONDITIONS AND TEMPERATURE ON THE EMI CHARACTERISTICS 897
A Stancalie, D Sporea, P Malinowski, M Mieloszyk, S Opoka, T Wandowski, W Ostachowicz

COLLECTING DATA FROM A SENSOR NETWORK IN A SINGLE-BOARD COMPUTER 904
F Casciati, S Casciati, Z-C. Chen, L Faravelli, M Vece

ASSESSMENT OF PIEZOELECTRIC SENSOR ADHESIVE BONDING 912
T Wandowski, J Moll, P Malinowski, S Opoka, W Ostachowicz

EMBEDDED FIBRE BRAGG GRATING SENSOR RESPONSE MODEL: CRACK GROWING DETECTION IN FIBRE REINFORCED PLASTIC MATERIALS 921
G Pereira, L P Mikkelsen, M McGugan

USING GENETIC ALGORITHMS TO OPTIMIZE AN ACTIVE SENSOR NETWORK ON A STIFFENED AEROSPACE PANEL WITH 3D SCANNING LASER VIBROMETRY DATA 929
R Marks, A Clarke, C Featherston, L Kawashita, C Paget, R Pullin

SENSOR MEASUREMENT STRATEGIES FOR MONITORING OFFSHORE WIND AND WAVE ENERGY DEVICES 937
Deirdre O'Donnell, Bruno Srbinovsky, Jimmy Murphy, Emanuel Popovici, Vikram Pakrashi

NOVEL FBG ROSETTE FOR DETERMINING IMPACT LOCATION IN THIN PLATE-LIKE STRUCTURE 945
K Majewska, S Opoka, P Kudela, W Ostachowicz
THE USE OF STRAIN GAUGES IN VIBRATION-BASED DAMAGE DETECTION 953
Fabio Luis Marques Dos Santos, Bart Peeters, Jenny Lau, Wim Desmet, Luiz Carlos Sandoval Goes
IDENTIFICATION OF FAULTY SENSOR USING RELATIVE PARTIAL DECOMPOSITION VIA INDEPENDENT COMPONENT ANALYSIS 961
Z Wang, S T Quek

SIGNAL PROCESSING

DAMAGE LOCALIZATION BY STATISTICAL EVALUATION OF SIGNAL-PROCESSED MODE SHAPES 969
M D Ulriksen, L Damkilde
PULSE COMPRESSION AND DISPERSION COMPENSATION FOR HIGH- RESOLUTION LAMB WAVE INSPECTION 978
J Hua, J Lin, L Zeng
MODE PURIFICATION FOR ULTRASONIC GUIDED WAVES UNDER PSEUDOPULSE EXCITATION 986
Z Luo, J Lin, L Zeng, F Gao
MACHINERY VIBRATION SIGNAL DENOISING BASED ON LEARNED DICTIONARY AND SPARSE REPRESENTATION 994
Liang Guo, Hongli Gao, Jun Li, Haifeng Huang, Xiaochen Zhang
IDENTIFICATION AND STOCHASTIC MODELLING OF SOURCES IN COPPER ORE CRUSHER VIBRATIONS 1002
Agnieszka Wylomanska, Radoslaw Zimroz, Joanna Janczura
DETECTION OF STRUCTURAL DAMAGE USING NOVELTY DETECTION ALGORITHM UNDER VARIATIONAL ENVIRONMENTAL AND OPERATIONAL CONDITIONS 1010
M El Mountassir, S Yaacoubi, F Dahmene
INVESTIGATION ON THE USE OF ARTIFICIAL NEURAL NETWORKS TO OVERCOME THE EFFECTS OF ENVIRONMENTAL AND OPERATIONAL CHANGES ON GUIDED WAVES MONITORING 1017
M El Mountassir, S Yaacoubi, F Dahmene
PIPE LEAKS CLASSIFICATION BY USING A DATA-DRIVEN APPROACH BASED ON FEATURES FROM CROSS-CORRELATED PIEZO-VIBRATION SIGNALS 1023
Jhonatan Camacho-Navarro, Magda Ruiz, Oscar Perez, Rodolfo Villamizar, Luis Mujica
STUDY OF CROSS-CORRELATION SIGNALS IN A DATA-DRIVEN APPROACH FOR DAMAGE CLASSIFICATION IN AIRCRAFT WINGS 1032
Jhonatan Camacho-Navarro, Magda Ruiz, Rodolfo Villamizar, Luis Mujica, Alfredo Güemes, Ignacio González-Requema

REPLACEMENT ARTICLE

PROTOCOLS FOR IMAGE PROCESSING BASED UNDERWATER INSPECTION OF INFRASTRUCTURE ELEMENTS 1040
Michael O'Byrne, Bidisha Ghosh, Franck Schoefs, Vikram Pakrashi
Author Index

Proceedings of the 11th International Conference on Damage Assessment of Structures, DAMAS 2015, Ghent University, Belgium, 24-26 August 2015

Chairman

Prof. Magd Abdel Wahab,
Ghent University, Belgium

International organisation committee

Prof. Biswajit Basu, Trinity College Dublin, Ireland
Prof. Nao-Aki Noda, Kyushu Institute of Technology, Japan
Prof. Huajiang Ouyang, University of Liverpool, UK

International Scientific Committee

F Aymerich, University of Cagliari, Italy
J Barton, University of Southampton, UK
B Basu, Trinity College Dublin, Ireland
S K Bhalla, Indian Institute of Technology Delhi, India
F Casciati, University of Pavia, Italy
R Ceravolo, Politechnico di Torino, Italy
F Chu, Tsinghua University, China
G De Roeck, KU Leuven, Belgium
L Faravelli, University of Pavia, Italy
S Fassois, Patras University, Greece
M Friswell, University of Swansea, UK
L Garibaldi, Politecnico di Torino, Italy
G-R Gillich, Eftimie Murgu Univ Resita, Romania
J-C Golinval, University of Liege, Belgium
A Gonzalez, University College Dublin, Ireland
K Holford, University of Cardiff, UK
G Manson, University of Sheffield, UK
T McCarthy, University of Wollongong, Australia
N-A Noda, Kyushu Institute of Technology, Japan
K Oda, Oita University, Japan
P Omenzetter, University of Aberdeen, UK
W Ostachowicz, Polish Academy of Science, Poland
H Ouyang, University of Liverpool, UK
B Peeters, LMS International, Belgium
R Pullin, Cardiff University, UK
A Rudawska, Lublin University of Technology, Poland
V Silberschmidt, University of Loughborough, UK
J J Sinou, Ecole Centrale de Lyon, France
W Staszewski, AGH University of Science and Technology, Poland
C Surace, Politecnico di Torino, Italy
K Worden, University of Sheffield, UK
W Zhu, University of Maryland, Baltimore, USA
X Zhu, University of Western Sidney, Australia

Editorial

This volume contains the proceedings of the 11th International Conference on Damage Assessment of Structures (DAMAS) 2015. DAMAS has a long history of almost 20 years. The first DAMAS conference took place in 1995 (Pescara, Italy), followed by a biannual meeting in 1997 (Sheffield, UK), 1999 (Dublin, Ireland), 2001 (Cardiff, UK), 2003 (Southampton, UK), 2005 (Gdansk, Poland), 2007 (Torino, Italy), 2009 (Beijing, China), 2011 (Oxford, UK) and 2013 (Dublin, Ireland). The eleventh edition of DAMAS conference series, DAMAS 2015, is hosted by Ghent University, Belgium, and is held at the congress center Het Pand in Ghent city. Ghent is the capital and the largest city of the East Flanders province of the Flemish region of Belgium. Het Pand is the culture and congress center of Ghent University and is a historical monument.

The conference is established as a major international forum for research topics relevant to damage assessment of engineering structures and systems including numerical simulations, signal processing of sensor measurements and theoretical techniques as well as experimental case studies. The presentations of DAMAS 2015 are divided into 6 main sessions, namely 1) Structural Health and Condition Monitoring, 2) Damage in Civil Engineering, 3) Damage in Machineries, 4) Damage in Composite Materials, 5) Sensing and Sensors and 6) Signal Processing.

The organising committee is grateful to keynote speakers; Professor Guido De Roeck, Head of Structural Mechanics Division, KULeuven, Belgium, for his keynote lecture entitled ‘Structural Health Monitoring: highlights and challenges‘, Professor Weidong Zhu, Department of Mechanical Engineering, University of Maryland, USA, for his keynote lecture entitled ‘Vibration-based Structural Damage Detection: Theory and Applications’ and Professor Wieslaw Ostachowicz, Head of the Laboratory of Active Materials and Smart Structures, Polish Academy of Sciences, Poland, for his keynote lecture entitled ‘Damage Assessment and Reliability in Offshore Wind Turbines Technology’.

Special thanks go to members of the Scientific Committee of DAMAS 2015 for reviewing the articles published in this volume and for judging their scientific merits. Based on the comments of reviewers and the scientific merits of the submitted manuscripts, the articles were accepted for publication in the conference proceedings and for presentation at the conference venue. The accepted papers are of a very high scientific quality and contribute to advancement of knowledge in all research topics relevant to DAMAS conference.

The organising committee would like to thank prestigious research groups, who made a great contribution to DAMAS 2015: the group of Professor Lars Damkilde, Aalborg University, Denmark; the group of professor Gilbert-Rainer Gillich, Eftimie Murgu" University of resita, Romania, the group of Professor Wieslaw Ostachowicz, Polish Academy of Sciences, Poland and the group of Dr Vikram Pakrashi, University College Cork, Ireland. Special thanks go to Dr Vikram Pakrashi for organizing the mini-symposium ‘Damage Detection, System Identification and Health Monitoring for Offshore Wind and Wave Energy Devices’.

Finally, the organising committee would like to thank all authors, who have contributed to this volume and presented their research work at DAMAS 2015.

Professor Magd Abdel Wahab
Chairman of DAMAS 2015

Journal of Physics: Conference Series **628** (2015) 011002 doi:10.1088/1742-6596/628/1/011002

Peer review statement

All papers published in this volume of *Journal of Physics: Conference Series* have been peer reviewed through processes administered by the proceedings Editors. Reviews were conducted by expert referees to the professional and scientific standards expected of a proceedings journal published by IOP Publishing.

Fatigue damage of steam turbine shaft at asynchronous connections of turbine generator to electrical network

A P Bovsunovsky

National University of Food Technologies, Department of Mechanical Engineering, Standardization and Certification of Equipment, 68 Volodymyrska Str., 03033, Kyiv, Ukraine

E-mail: apbovsunovsky@gmail.com

Abstract. The investigations of cracks growth in the fractured turbine rotors point out at theirs fatigue nature. The main reason of turbine shafts fatigue damage is theirs periodical startups which are typical for steam turbines. Each startup of a turbine is accompanied by the connection of turbine generator to electrical network. During the connection because of the phase shift between the vector of electromotive force of turbine generator and the vector of supply-line voltage the short-term but powerful reactive shaft torque arises. This torque causes torsional vibrations and fatigue damage of turbine shafts of different intensity. Based on the 3D finite element model of turbine shaft of the steam turbine K-200-130 and the mechanical properties of rotor steel there was estimated the fatigue damage of the shaft at its torsional vibrations arising as a result of connection of turbine generator to electric network.

1. Introduction.

The analysis of vibration of steam turbine shaft usually focuses upon their transverse vibrations, the intensity of which is especially significant when passing through the critical speeds of shaft rotation. At the same time, in certain cases namely torsional vibrations of turbine shafts are much more dangerous. The evidences of the latter are the catastrophic failure of the rotor of medium pressure steam turbine at the power plant in Gallatin (USA) in 1974, caused by the growth of a crack for several years under the cyclic torsion [1], as well as the destruction of the steam turbine in Kashirskaia thermal station (Russian Federation) in 2002 [2].

The main reasons of the turbine shaft torsional vibrations are the short circuit of turbine generator, the connection of turbine generator to electric network, and the dynamic instability of the system turbine generator – electric network [1-4].

Short circuits of turbine generators happen several times during the operation time. Under certain conditions a short circuit can cause the instant destruction of turbine [2], but the accumulation of fatigue damage of a turbine shaft as a result of short circuit is unlikely. At the same time, the analysis of crack propagation in the destroyed or decommissioned turbine shafts makes it possible to conclude that cracks developed over a long period of time [1, 5]. The reasons of accumulation of fatigue damage of turbine shafts should be sought in theirs multiple startups, which are typical for steam turbine units operating in a flexible manner. The predetermined number of such startups during the operation time is 2000, but in some cases it can be prolonged to 2500 and more. At each startup at least one connection of turbine generator to electric network occurs. At this because of the difference of the phase shift angle between the vector of electromotive force of turbine generator and the vector of electric network voltage the short-term reactive electromagnetic shaft torque arises, resulting in the turbine shaft torsional vibrations of various intensity. The intensity of torsional vibrations may be so high that they cause the fatigue damage of the shaft.

According to the established practice, the evaluation of residual life of steam turbines is mainly based on the analysis of thermal stress state of rotors and on the study of degradation of mechanical properties of materials of turbine structural elements. Thus the fatigue damage of turbine shaft due to the torsional vibrations is not taken into account, which may be the reason of significant errors of estimation of actual damage of turbines structural elements. The importance of the problem of torsional vibrations of turbine shafts is emphasized in ref. [6].

The aim of the present research was to investigate the factors which affect the fatigue damage of turbine shaft, as well as to assess the level of this damage caused by torsional vibrations arising at the connections of turbine generator to electrical network.

2. Model of steam turbine shaft.

The steam turbine K-200-130 is a complex mechanical system consisting of a high pressure rotor (HPR), middle pressure rotor (MPR), low pressure rotor (LPR) and turbine generator (figure 1; the generator is not shown in this figure). The numbers 1, 2 and 3 in figure 1 point out the most stressed areas of turbine shaft the fatigue damage of which has been the subject of research.

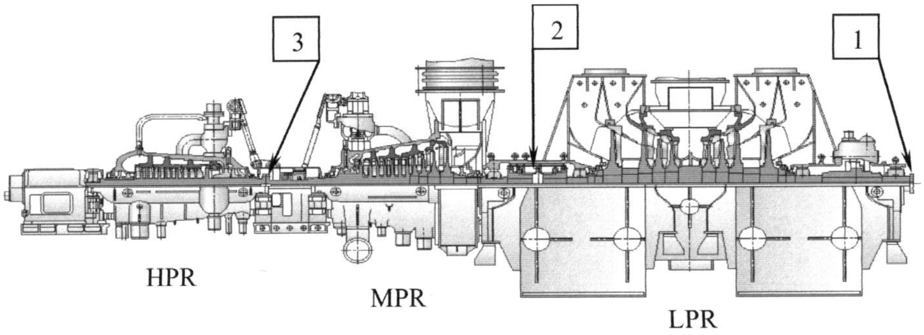

Figure 1. Steam turbine K-200-130.

The study on forced vibrations of such a mechanical system was performed numerically with a 3-D finite element model (figure 2). As an object for the research was selected the shaft of turbine K-200-130 (200 MW). The model was a mesh of 50000 eight-node elements. The detailed description of the model and the numerical methods used for the solution of the problem can be found in ref. [7].

The scheme of the shaft loading is presented in figure 2, namely the torques arising in HPR (M_{hp}), MPR (M_{mp}), LPR (M_{lp}) and the reactive torque (M_r) which arises at connection of turbine generator (TG) to electrical network. Total operating shaft torque is determined as follows

$$M_{oper} = M_{hp} + M_{mp} + M_{lp} \,. \tag{1}$$

Calculations of the electromagnetic torque at the connection of the turbine generator TGV-200 to the electric network of voltage 110 KV showed that the reactive torque strongly depends on the phase shift angle θ (the so called synchronizing angle) between the vector of electromotive force of the turbine generator and the vector of electrical network voltage [8]. The results of calculation of time dependence of the relative value of reactive shaft torque at different values of synchronizing angle are shown in figure 3. As can be seen, the reactive shaft torque increases in direct proportion to the value of the synchronizing angle and may exceed up to 16 times the operating shaft torque. Since these functions are damped, the almost complete attenuation of the electromagnetic processes in turbine generator occurs in about 4 seconds.

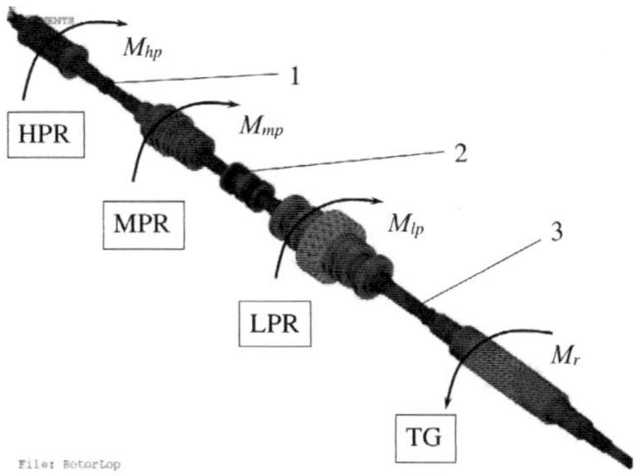

Figure 2. 3-D finite element model of the turbine K-200-130 shaft.

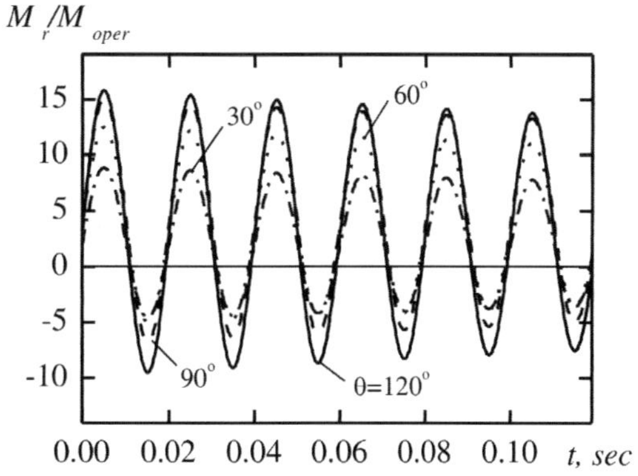

Figure 3. Time dependences of the reactive shaft torque at different values of angle θ.

3. Torsional vibrations of turbine shaft at asynchronous connection of turbine generator to electrical network.

Calculations of forced torsional vibrations of turbine shaft were performed for the synchronizing angles θ=30, 60, 90 and 120°. The level of damping of the mechanical system was estimated by the logarithmic decrement of vibration δ. As far as its actual value is unknown it was accepted in calculations that δ=0.3. Such a high level of damping was justified by the presence of at least four mechanisms of energy dissipation, namely the energy dissipation in the material, structural energy dissipation, the resistance of the air-steam medium and the resistance of turbo generator itself.

The connection of a turbine generator to electrical network can be successful and unsuccessful. In the case of successful connection the turbine generator is not disconnected from the electrical network and the reactive shaft torque gradually vanishes. But if in the process of the connection the intense vibrations of a turbine arise, the turbine generator is disconnected from electrical network, the result of which is the almost immediate disappearance of the reactive torque and of the vibration of the turbine as well. The latter is the case of the so-called unsuccessful connection. The attempt of connection is repeated for as long as it would be successful one.

Torsional vibrations of the turbine shaft at successful and unsuccessful connection of turbine generator to electrical network have some specific features.

Thus at successful connection of turbine generator torsional vibrations of the shaft are maintained at relatively high level for almost the entire duration of the reactive torque action (figure 4). The most intense vibrations are excited in the case of the angle θ=120°, as far as the reactive torque reaches maximum value. With the decrease of the synchronizing angle from 120° to 30° the amplitude of torsional vibrations of the shaft drops in half.

The most intense stresses are observed in areas 2 and 3. The fatigue limit of R2MA rotor steel at normal temperature is τ_{-1}=230 MPa (it is shown in figure 4, *c* and *d* by a horizontal line). As can be seen, at certain conditions the stress level may exceed the fatigue limit of rotor steel, that is the reason of fatigue damage of the shaft. It is evident that the higher is the synchronizing angle the more intense is the fatigue damage of the shaft.

As the reactive electromagnetic torque attenuates the torsional vibrations of the shaft gradually reduce. At high level of damping in the mechanical system fatigue damage occurs only at first few cycles of vibration as far as the intensity of stress quickly drops to the safe level.

In the case of unsuccessful connection of a turbine generator to electrical network the intensity of torsional vibrations of the shaft depends not only on the synchronizing angle, but also on the duration of the reactive shaft torque t_r.

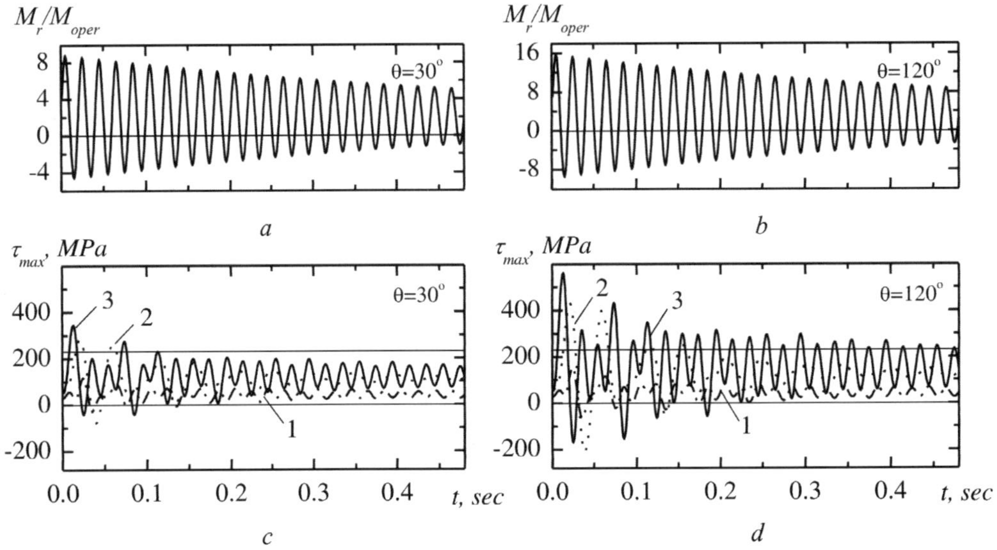

Figure 4. Time dependences of the reactive torsional moment (*a, b*) and shear stresses (*c, d*) in areas 1, 2 and 3 of the shaft at successful connection of turbine generator.

The development of torsional vibrations of the shaft when trying to connect the turbine generator to electrical network and subsequent its emergency disconnection at different duration of the reactive shaft torque is demonstrated in figure 5. As can be seen, the intensity of forced vibrations in both cases (t_r=0.271 and 0.291 sec) is almost the same. However, the difference between the intensity of free vibrations reaches several times. So the duration of the reactive torque strongly affects free vibrations of the shaft.

4. Calculations of fatigue damage of turbine shaft.

The fatigue damage of turbine shaft as a result of torsional vibrations was evaluated based on the hypothesis of Palmgren-Miner [9]

$$D = \sum_{i=1}^{n} \frac{1}{N_{id}}, \qquad (2)$$

where n is the number of damaging cycles of vibration, in which shear stresses exceed the fatigue limit of rotor steel τ_{-1}; N_{id} is the number of cycles till the fracture under the stress τ_{imax}. In 2000 connections of a turbine generator the total fatigue damage was determined as D_{total}=2000D. The limiting state of the rotor steel is attained in condition D_{total}=1.

Calculations of cyclic damage of the shaft were performed with the fatigue curve for the rotor steel R2MA, which was obtained at symmetric cycle of loading and for normal temperature [10]. The influence of scale effect, temperature and asymmetry of cyclic deformation on the fatigue properties of the steel was taken into account by means of correction factors

$$\tau_{\max} = K_s K_a K_t \left[\eta_0 + \eta_v (N)^c + \eta_u (N)^b\right], \qquad (3)$$

where η_0=208,3; η_v=3176,2; η_u=1,86·10^5; c=–0,3114; b=–0,8348; N is the number of loading cycles;

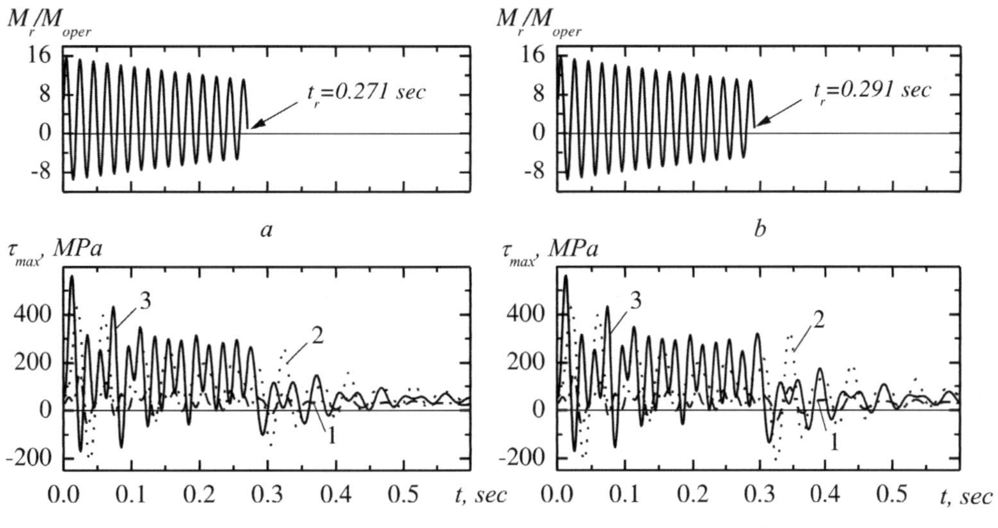

Figure 5. Time dependences of the reactive torque (*a*, *b*) and shear stresses (*c*, *d*) in areas 1, 2 and 3 of the shaft at unsuccessful connection of turbine generator (θ=120°).

K_s is the scale factor; K_a is the asymmetry factor; K_t is the temperature factor. Based on the experimental research [11] the values of scale and temperature factors were excepted to be K_s=0.58 and K_t=0.78. The asymmetry factor was determined by the formula

$$K_a = \frac{\sqrt{\tau_{-1}^2 - \kappa^2 \psi_\tau^2 \tau_m^2}}{\tau_{-1}}, \tag{4}$$

where τ_{-1}=230 MPa; κ=0.922; ψ_τ=0.505 [10]; τ_m is the mean stress.

The results of calculation of the shaft's fatigue damage as a result of torsional vibrations for 2000 successful connections of turbine generator to electrical network are shown in the table 1. As can be seen, in the case of synchronizing angle θ=30° the total fatigue damage of the shaft does not exceed 16%, so there is a significant reserve of residual durability. The increase of synchronizing angle causes the growth of damaging cycles and, consequently, the spread of fatigue damage. Most dangerous situation is observed in the case θ=120° when even high level of damping is not able to prevent the rotor steel to attain the limiting state. In all cases being considered the area 1 of the shaft practically is not subjected to the fatigue damage.

At unsuccessful connection of turbine generator to electrical network it is necessary to take into account the duration of the reactive shaft torque. In this case the vibrations consist of forced vibration under the action of reactive torque and of free vibration arising after the disconnection of turbine generator from electrical network.

The dependencies of total damage, as well as damage due to forced and free vibrations of the shaft on the duration of the reactive torque are shown in figure 6. As can be seen, the curves are quasi-periodic with the trend of gradual reduction of the maximal damage. At small damping (δ=0.02) the damage of area 2 is higher than of area 3, but at high damping (δ=0.3) the damage of area 3 exceeds that one of area 2. The data on damage of area 1 is not shown because it does not exceed 15%.

It is also evident from figure 6 that if the level of damping would be low, the damage accumulated mainly at free vibrations (figure 6, *a, b*). High damping rapidly reduces theirs intensity to the safe level and damage accumulates mainly at forced vibrations (figure 6, *c, d*). It is necessary to note that each point of the curves in figure 6 was obtained under the assumption that the duration of reactive torque at every of 2000 connection was the same. Obviously, in practice this value is different at each connection.

Table 1. Fatigue damage of turbine shaft at successful connection of turbine generator to electrical network

θ, °.	Areas of the shaft	n	D_{total}
30	2	2	0.113
	3	3	0.167
60	2	3	0.414
	3	4	0.669
90	2	3	0.762
	3	4	1.343
120	2	3	0.969
	3	10	1.776

Therefore, a more realistic assessment of damage should be performed based on the assumption that the duration t_r is distributed equiprobably in a certain range $\Delta t_r = t_{r,K} - t_{r,1}$ ($t_{r,1}$ and $t_{r,K}$ are the initial and end value of the range, respectively). The initial value of the range was taken to be $t_{r,1}$=0.25 sec. The step of change of duration t_r was 0.001 sec. The averaged over the range damage was determined by the formula

$$D_{av} = \frac{\sum_{i=1}^{K} D_{total,i}(t_{r,i})}{\Delta t_r}, \quad (5)$$

where $D_{total,i}$ is the total damage at the duration of the reactive torque $t_{r,i}$.

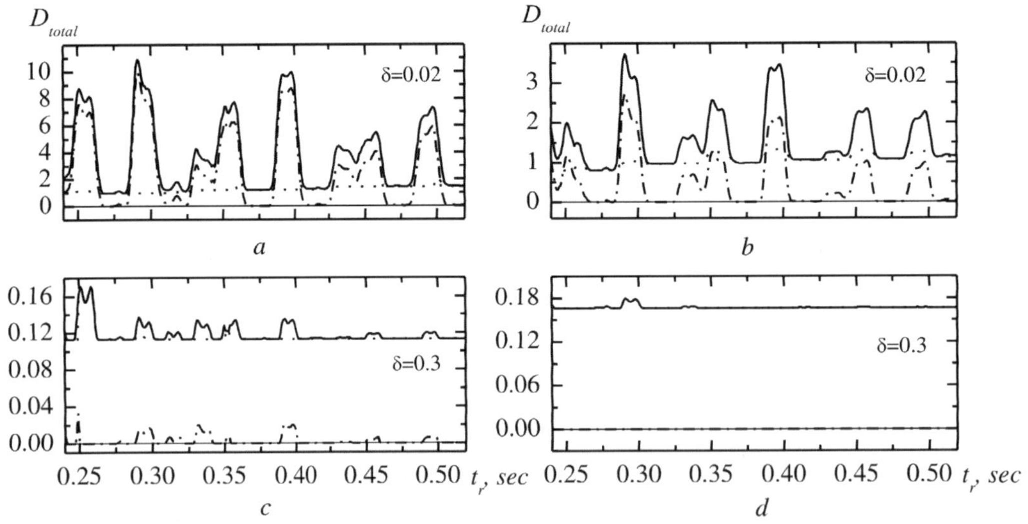

Figure 6. Dependences of total fatigue damage of turbine shaft in area 2 (*a, c*) and 3 (*b, d*) on the duration of the reactive torque at unsuccessful connection of turbine generator (θ=30°): free vibration (dash dot line); forced vibration (dot line); sum (solid line).

As can be seen from figure 7, the averaged damage over the lifetime of the turbine is quite smooth characteristic and depends weakly on the averaging range. This characteristic can be used to assess the fatigue damage of the shaft as a result of torsional vibrations through the turbine lifetime.

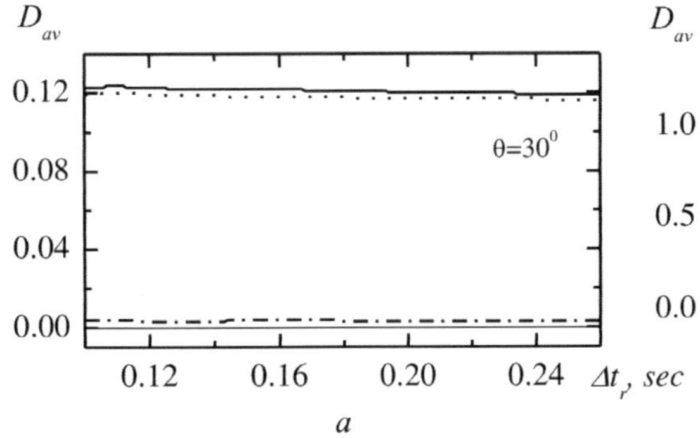

Figure 7. Dependences of averaged fatigue damage of turbine shaft in area 2 on the range of duration of the reactive torque at unsuccessful connection of a turbine generator: free vibration (dash dot line); forced vibration (dot line); sum (solid line).

It is practically unlikely that all 2000 connections of turbine generator to electrical network would be successful. The more realistic scenario should count for possibility of certain number of unsuccessful connections which is different for various steam turbines. Figure 8 presents the results of averaged damage assessment for two hypothetical cases: first case considers 2000 successful connections and second one – 2000 unsuccessful connections. It should be noted that the attempts to connect the turbine generator to electrical network continues until it will be successful. Therefore, in the calculations there was taken into account that 2000 unsuccessful connections were accompanied by the same amount of successful ones.

In all cases presented in figure 8 the averaged damage is strongly dependent on the synchronizing angle and on the number of unsuccessful connections. The scenario of damage accumulation presented in figure 8 by solid line is optimistic in term of the number of unsuccessful connections. The actual number could be much higher than 2000.

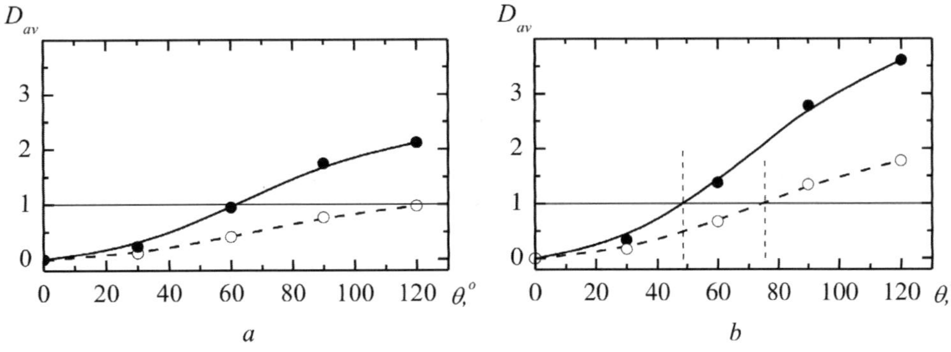

Figure 8. Dependences of averaged fatigue damage of turbine shaft in area 2 (*a*) and 3 (*b*) on the angle *θ*: 2000 successful connections (dash line); 2000 unsuccessful and 2000 successful connections (solid line).

Comparison of figure 8, *a* and *b* demonstrates that the fatigue damage of area 3 is more intense than of area 2 of the shaft. Therefore, the conclusion about the possibility of safe operation of the turbine should be done based on the analysis of fatigue damage accumulated in area 3. In such a way the limiting state of the rotor steel can be achieved in condition $\theta \geq 75°$ if all connections would be successful and in condition $\theta \geq 50°$ if the number of unsuccessful connections would be 2000 (vertical dash lines in figure 8, *b*). It is evident that the higher is the number of unsuccessful connections the lower is the synchronizing angle.

Consequently the safety of steam turbines in operation to a great extent is affected by the number of unsuccessful connections and by the synchronizing angle. The value of synchronizing angle should be controlled at every connection of the turbine generator to electrical network.

5. Conclusions.

The results of research have shown that the evaluation of fatigue damage of turbine shaft should be based on the account not only of the transverse but also of the torsional vibrations occurring at the connection of turbine generator to electrical network. The intensity of the latter depends essentially on the synchronizing angle, duration of the reactive shaft torque and damping level.

The torsional vibrations of turbine shaft are the reason of turbine shaft fatigue damage. In general case the fatigue damage is a function of three parameters: the synchronizing angle, the number of unsuccessful connections and the duration of the reactive shaft torque at unsuccessful connections.

The assessment of fatigue damage of turbine shaft at torsional vibrations for 2000 successful connections of turbine generator to power network showed that at high level of damping in the mechanical system the limiting state of the rotor steel can be achieved in condition $\theta \geq 75°$. Each unsuccessful connection decreases this value by about 0.013°.

The efficient way to control the extent of fatigue damage of turbine shaft is the on-line measurement of the shaft vibrations and the use this information to assess the damage based on the model proposed in the present work.

References

[1] Kramer L D and Randolph D D 1976 Analysis of the Tennessee valley authority, Gallatin unit N2 turbine rotor burst *ASME-MPC Symp. on Creep-Fatigue Interaction* 1

[2] Zagretdinov I Sh, Kostyuk A G, Trukhnii A D and Dolzhanskii P R 2004 Failure of the 300 MW turbine unit of the state district power station at Kashira: causes, consequences and conclusions *Teploénergetika* **5** 5–15

[3] Dorfman L S and Trubelja M 1999 Torsional monitoring of turbine-generators for incipient failure detection *Proc. of the sixth EPRI steam turbine/generator workshop, St. Luis, Missouri* (August 17–20) 1–6

[4] Stein J and Fick H 1980 The torsional stress analyzer for continuously monitoring turbine generators *IEEE Trans. PAS-99(2) Mar.-Apr.* 703–710

[5] Barella S, Bellogini M, Boniardi M and Cincera S 2011 Failure analysis of a steam turbine rotor *Eng. Fail. Anal.* **18** 1511–19

[6] Tetsuro Kakinoki, Ryuichi Yokoyama, Goro Fujita, Kaoru Koyanagc, Toshihisa Funabashi and Kwang Y Lee 2004 Shaft torque observer and excitation control for turbine–generator torsional oscillation *Elect. Pow. Syst. Res.* **68** 248–57

[7] Bovsunovskii A P, Chernousenko O Yu, Shtefan E V and Bashta D A 2010 Fatigue damage and failure of steam turbine rotors by torsional vibrations *Strength Mater.* **42** 108–113

[8] Bovsunovsky A P, Kuevda V P, Kuevda Yu V and Shtefan E V 2013 Fatigue damage of steam turbine shaft at asynchronous switching on of turbo generator to power network *Vibr. in Eng. Technol.* **4** 51–7

[9] Palmgren A 1924 Life of ball bearings *Zeitschrift des Vereines Deutscher Ingen.* **68** 339–41

[10] Bovsunovskii A P 2011 Assessment of fatigue damage in steam turbine shafting due to torsional vibrations *Strength Mater.* **43** 487–97

[11] Kudryavtsev I V and Naumchenkov N E 1978 Fatigue strength characteristics of steel 25KhNZMFA in relation to absolute dimensions and stress concentrations *Strength Mater.* **10** 386–392

Damage identification based on incomplete modal data and constrained nonlinear multivariable function

S. S. Kourehli
Department of Civil Engineering, Ahar Branch, Islamic Azad University, Ahar, Iran

E-mail: s-kourehli@iau-ahar.ac.ir

Abstract. Damage detection and estimation in structures using incomplete modal data is presented. In the proposed approach, damage location and severity is determined by solving an optimization problem using the constrained nonlinear multivariable function of Matlab (so-called fmincon) to perform constrained minimization. The feasibility of the presented method is validated with a three-story plane frame as numerical example containing one or several damages with different value of damage severity. The obtained results indicated that proposed method is effective and robust in detection and estimation of damage.

1. Introduction

Damage detection is one of the branches of structural health monitoring (SHM) which has recently attracted many scientific efforts. Health monitoring refers to a process of measuring and interpreting data from a system of sensors distributed about a structural system to objectively quantify the condition of the structure [1, 2]. Damage detection techniques have been successfully applied to several real-world problems. Based on the performance of structures, damage detection methods can be categorized into four levels [3]. The first level is devoted to detection of existing damages in a structure, and the second and third levels focus on the determination of the location and severity of damage in structures, respectively. The last level is a complete study that includes the estimation of the residual life of a structure, reaching a point that requires more information from fracture mechanics and structural reliability.

Among numerous methods, approaches that are based on the observation of the dynamic behavior of a structure heve been developed [4-7]. Many of these techniques use the identified modal parameters like mode shapes and natural frequencies for structural damage detection and estimation. The identification of alteration in mode shapes and natural frequencies at the damaged system in comparison with the undamaged system is one of the popular methods in the structural damage detection. Carvalho et al. [8] presented a direct method for model updating with incomplete modal data. The proposed method uses an algorithmic way without requiring any model reduction or modal expansion techniques. Huajun et al. [9] extended the CMCM method to simultaneously update the mass, damping and stiffness matrices of a finite element model when only few spatially incomplete, complex-valued modes are available. The results reveal that applying the CMCM method, along with an iterative Guyan reduction scheme can yield good damage detection in general. Also, Chen [10] presented an approach for detecting local damage in large scale frame structures by utilizing regularization methods with incomplete noisy data. A system of linear basic equations for determining the damage indicators has been developed by directly adopting the measured incomplete modal data.

In this research, a new method for localizing and estimating the severity of structural damage is introduced. The damage identification is carried out through fmincon to minimize an objective function derived from incomplete dynamic characteristics of damaged structure. Numerical example shows that the proposed method can be considered as a flexible and robust approach in damage identification of structures.

2. Problem formulation

The modal characteristics of a structure without damage are described by the following equations:

$$([\mathbf{K}^{ud}]-\lambda_i[\mathbf{M}])\{\mathbf{\Phi}_i\}=0 \quad i=1,2,\ldots,n \tag{1}$$

where, $\mathbf{K}^{ud}$ and $\mathbf{M}$ are undamaged stiffness and mass matrices, respectively; λ_i is the square of the natural frequency corresponding to the mode shape $\mathbf{\Phi}_i$; and n is the total number of obtained mode shapes.

One of the simplest techniques to determine damage-induced alteration stiffness is the degradation in Young's modulus of an element as follows:

$$E_j^d = E_j^{ud}(1-d_j) \tag{2}$$

where, E_j^d and E_j^{ud} are the damaged and undamaged Young's modulus of the jth element in the finite element model, respectively; and d_j indicates the damage severity at the jth element in the finite element model whose values are between 0 for an element without damage and 1 for a ruptured element.

Moreover, it is assumed that no change would occur after damage in the mass matrix, which seems to be reasonable in most real problems.

Thus, as it was mentioned above, the eigenvalue equations for a damaged structure became:

$$([\mathbf{K}^d]-\lambda_i^d[\mathbf{M}])\{\mathbf{\Phi}_i^d\}=0 \quad i=1,2,\ldots,n \tag{3}$$

where, $\mathbf{K}^{ud}$ is the damaged stiffness matrix; λ_i^d and $\mathbf{\Phi}_i^d$ are the square of the ith natural frequency and the ith mode shape of the damaged structure, respectively.

As the number of sensors used to measure modal data is normally limited and usually is less than the number of DOFs in the finite element model, either the model reduction method should be used to match with incomplete measured mode shapes or the measured mode shapes must be expanded to the dimension of the analytical mode shapes. Because of no convergence in the proposed optimization method using the modal expansion, the first option has been adopted using the Guyan [11] static reduction method. This method is employed to condense the mass and stiffness matrices. In this method, the mass and stiffness matrices, and the displacement and acceleration vectors in Eq. (1) are partitioned into a set of master and slave DOFs:

$$\begin{bmatrix}\mathbf{M}_{mm} & \mathbf{M}_{ms}\\ \mathbf{M}_{sm} & \mathbf{M}_{ss}\end{bmatrix}\begin{Bmatrix}\ddot{\mathbf{x}}_m\\ \ddot{\mathbf{x}}_s\end{Bmatrix}+\begin{bmatrix}\mathbf{K}_{mm} & \mathbf{K}_{ms}\\ \mathbf{K}_{sm} & \mathbf{K}_{ss}\end{bmatrix}\begin{Bmatrix}\mathbf{x}_m\\ \mathbf{x}_s\end{Bmatrix}=\begin{Bmatrix}0\\ 0\end{Bmatrix} \tag{4}$$

In which, $\mathbf{x}$ and $\ddot{\mathbf{x}}$ are the displacement and acceleration vectors, respectively and the subscript m and s are the master and slave coordinates, respectively. To eliminate the slave DOFs, the inertia terms for the second set of equations are neglected that leads to:

$$\begin{Bmatrix} \mathbf{x}_m \\ \mathbf{x}_s \end{Bmatrix} = [\mathbf{T}]\{\mathbf{x}_m\} \tag{5}$$

where:

$$[\mathbf{T}] = \begin{bmatrix} \mathbf{I} \\ -\mathbf{K}_{ss}^{-1}\mathbf{K}_{sm} \end{bmatrix} \tag{6}$$

Substitution of Eq. (5) into Eq. (4), followed by permultiplication by $[\mathbf{T}]^{\mathrm{T}}$ and using incomplete mode shapes, yields the reduced eigenproblem as:

$$([\mathbf{K}_r^d] - \lambda_{i,r}^d [\mathbf{M}_r])\{\mathbf{\Phi}_{i,r}^d\} = 0 \quad i = 1,2,\ldots,n \tag{7}$$

where:

$$[\mathbf{K}_r^d] = [\mathbf{T}]^T [\mathbf{K}^d][\mathbf{T}] \tag{8}$$

$$[\mathbf{M}_r] = [\mathbf{T}]^T [\mathbf{M}][\mathbf{T}] \tag{9}$$

in which, $\mathbf{K}^d{}_r$ and $\mathbf{M}_r$ are the reduced stiffness and mass matrices of damaged structure, respectively; $\lambda_{i,r}{}^d$ and $\mathbf{\Phi}_{i,r}{}^d$ are the square of the ith natural frequency and the ith incomplete mode shape of the damaged structure in the reduced state, respectively.

Applying the incomplete mode shapes and natural frequencies of damaged structure to Eq. (7) leads to form the inverse problem of determining the damage severity parameter. The definition of a local damage severity parameter d in the finite element model allows estimating damage quantity and location together, since damage identification is then carried out at the element level. The problem can be formulated as optimization problem of objective function, while using some transforms as a direct inversion to obtain solution is impossible most of the time.

The general statement for the objective function is:

$$F = f(d_1, d_2, \ldots, d_{N_e}) \tag{10}$$

In the process of substituting the incomplete measured modal parameters of the damaged structure in Eq. (10), a dynamic residue vector can be defined over each measured mode as follows:

$$[\mathbf{R}_i(d)] = ([\mathbf{K}_r^d] - \lambda_{i,r}^m [\mathbf{M}_r])\{\mathbf{\Phi}_{i,r}^m\} \quad i = 1,2,\ldots,m \tag{11}$$

where, $\lambda_{i,r}{}^m$ and $\mathbf{\Phi}_{i,r}{}^m$ are the square of the ith natural frequency and the ith incomplete mode shape from measurements, respectively; and m is the number of available mode shape for damage detection.

Then, if structural damages are determined correctly, the residue vector would be next to 0 in Eq. (11). Therefore, the problem of damage detection can be formulated as an optimization problem. So, the first objective function can be formulated as follows:

$$f_1(d) = \sum_{i=1}^{m} \left\| [R_i(d)] \right\|^2 \tag{12}$$

$$0 \le d_1 \le 1, \quad 0 \le d_2 \le 1, \quad \ldots, \quad 0 \le d_{N_e} \le 1$$

where, $\| \; \|$ represents the Euclidean length of $\mathbf{R}_i(d)$.

The optimization technique used in this paper was a constrained nonlinear minimization, fmincon, which is available in the MATLAB Optimization Toolbox. This routinely implement Sequential Quadratic Programming (SQP) to minimize the nonlinear cost function was subjected to linear and nonlinear equality and inequality constraints. SQP converts a nonlinear minimization to a linear minimization using a Hessian matrix of cost function and gradient of nonlinear constraints. In addition to the use of fmincon, multiple start points were used to ensure that the global minimum was reached. The global minimum can be seen with repeated answers in the final output.

3. Numerical study

Consider a three-story plane steel frame for which the finite-element model consists of nine elements (six columns and three beams) and six free nodes, as shown in Fig. 1. For the steel frame considered, the material properties of the steel include Young's modulus E=200 GPa, mass density ρ=7850 kg/m3. The mass per unit length, moment of inertia, and cross-sectional area of the columns are: m=117.75 kg/m, I=3.3× 10−4 m4 and A=1.5× 10−2 m2, respectively; for the beams are: m=119.32 kg/m, I=3.69× 10−4 m4 and A=1.52× 10−2 m2. Also, the damage severity in each element is given by the reduction factor listed in Table 1.

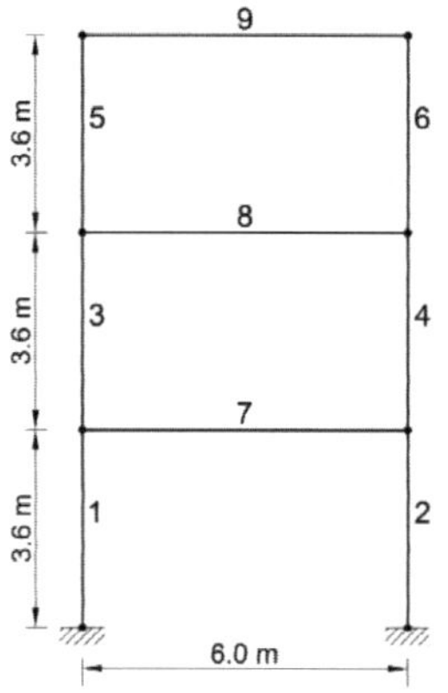

Fig. 1 Three-story plane frame with the finite element model

In this example, two damage scenarios are represented as the elements with reduction in Young's modulus. The damage severity in each element is given by the reduction factor listed in Table 4.1. In this case, only 6 translational DOFs are selected as the measured DOFs in the process of damage detection and quantification.

Table 1. Damage scenarios for three story plane frame

Scenario 1		Scenario 2	
Element 2	20%	Element 2	20%
		Element 4	35%
		Element 9	25%

Different initial values of damage severities in the proposed method have been tested to check its convergence. Figures 2 and 3 show the results of damage identification in the three story plane frame for two damage patterns with zero and 50% initial values, respectively. It depicts that the proposed method is a robust and effective method in detecting and quantifying various damage patterns with different initial values of the damage severities.

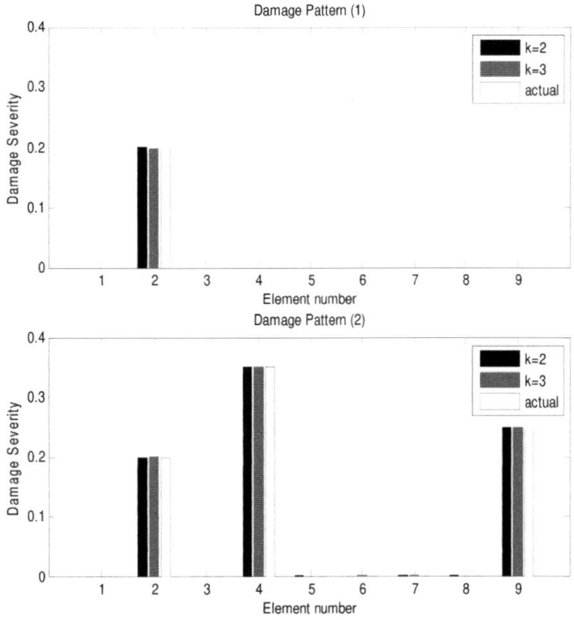

Fig. 2 The obtained results for two damage patterns of the three story frame with 0% initial values

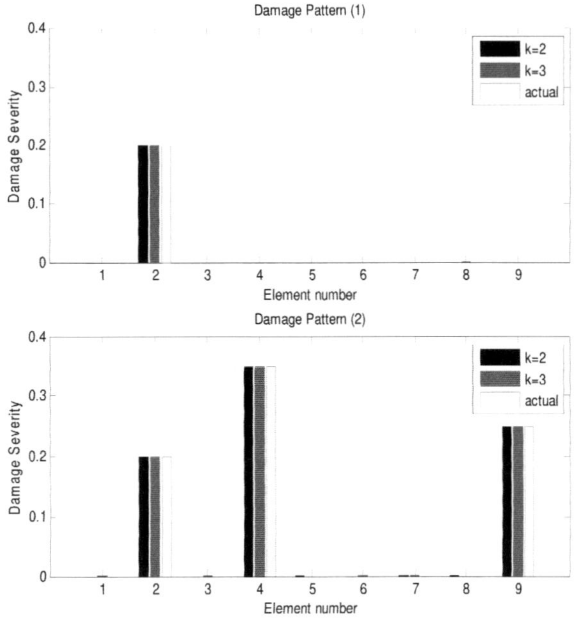

Fig. 3 The obtained results for two damage patterns of the three story frame with 50% initial values

4. CONCLUSIONS

In this paper, a method has been developed for detection and estimation of damage in structures based on the incomplete modal data of the damaged structure using an optimization problem. In this method, constrained nonlinear minimization, fmincon, is used to determine the damage in structures by optimizing a cost function. For damage detection and estimation, this proposed method was applied to a three story plane frame with one or several damage patterns. The obtained results indicated that the proposed method is a strong and viable method to the problem of detection and estimation of damage in the structures. The results revealed high sensitivity of the proposed method to the damage in spite of incomplete measurements.

References

[1] Johnson E A, Lam H F, Katafygiotis L S and Beck J L 2004 *Phase I IASC-ASCE structural health monitoring benchmark problem using simulated data* J. Eng. Mech. 130(1) 3-15.

[2] Zingoni A 2005 *Structural health monitoring, damage detection and long-term performance* J. Eng. Struct. 27(12) 1713-1714.

[3] Rytter, A 1993 *Vibration based inspection of civil engineering structures* PhD Dissertation Aalborg University, Aalborg.

[4] Doebling SW, Farrar CR, Prime MB, Schevitz DW 1996 *Damage identification and health monitoring of structural and mechanical systems from changes in their vibration characteristics: A literature review* Technical Report LA-13070-MS, Los Alamos National Laboratory, Los Alamos.

[5] Kourehli, S S 2015 *Damage assessment in structures using incomplete modal data and artificial neural network* International Journal of Structural Stability and Dynamics Vol. 15 No. 6 1450087. DOI: 10.1142/S0219455414500874.

[6] Kourehli S S, Ghodrati Amiri G, Ghafory-Ashtiany M., Bagheri A 2013 *Structural damage detection based on incomplete modal data using pattern search algorithm* Journal of vibration and control Vol.19 No.6 pp. 821-833.

[7] Kourehli S S, Bagheri A, Ghodrati Amiri, G, Ghafory-Ashtiany, M 2013 *Structural damage detection using incomplete modal data and incomplete static response* KSCE journal of civil engineering Vol.17 No.1 pp. 216-223.

[8] Carvalho J, Datta BN, Gupta, A and Lagadapati M *2007A direct method for model updating with incomplete measured data and without spurious modes Mechanical System and Signal Processing*, Vol. 21 No. 7 pp. 2715-2731.

[9] Huajun L, Fushun L, James HS 2008 *Employing incomplete complex modes for model updating and damage detection of damped structures Sci China Ser E-Tech Sci*. Vol. 51 No. 12 pp. 2254-2268.

[10] Chen H 2008 *Application of regularization methods to damage detection in large scale plane frame structures using incomplete noisy modal data Engineering Structures* Vol. 30 No. 11 pp. 3219-3227.

[11] Guyan RJ 1965 *Reduction of stiffness and mass matrices AIAA Journal*, Vol. 3, No. 2, pp. 380-387.

Statistical evaluation of characteristic SDDLV-induced stress resultants to discriminate between undamaged and damaged elements

L.M. Hansen[1], R.J. Johansen[1], M.D. Ulriksen[1], D. Tcherniak[2] and L. Damkilde[1]

[1]Dept. of Civil Engineering, Aalborg University, Niels Bohrs Vej 8, Esbjerg, Denmark
[2]Brüel & Kjær Sound & Vibration Measurement A/S, Skodsborgvej 307, Nærum, Denmark

E-mail: lhanse10@student.aau.dk, rjohan10@student.aau.dk, mdu@civil.aau.dk, dmitri.tcherniak@bksv.com, lda@civil.aau.dk

Abstract. The stochastic dynamic damage location vector (SDDLV) method utilizes the vectors from the kernel of a damaged-induced transfer function matrix change to localize damages in a structure. The kernel vectors associated with the lowest singular values are converted into static pseudo-loads and applied alternately to an undamaged reference model with known stiffness matrix, hereby, theoretically, yielding characteristic stress resultants approaching zero in the damaged elements. At present, the discrimination between potentially damaged elements and undamaged ones is typically conducted on the basis of modified characteristic stress resultants, which are compared to a pre-defined tolerance value, without any thorough statistical evaluation. In the present paper, it is tested whether three widely-used statistical pattern-recognition-based damage-detection methods can provide an effective statistical evaluation of the characteristic stress resultants, hence facilitating general discrimination between damaged and undamaged elements. The three detection methods in question enable outlier analysis on the basis of, respectively, Euclidian distance, Hotelling's T^2 statistics, and Mahalanobis distance. The study of the applicability of these methods is based on experimentally obtained accelerations of a cantilevered residential-sized wind turbine blade subjected to an unmeasured multi-impulse load. The characteristic stress resultants are derived by applying the static pseudo-loads to a representative finite element (FE) model of the actual blade.

1. Introduction

Structural health monitoring (SHM) is widely employed within many systems, and the importance of its application for wind turbine blades is growing due to the increasing size and number of operating wind turbines, see, e.g., [1]. Vibration-based approaches are commonly applied in such a way that the dynamic response from the current state is compared to the representative response from the healthy reference state. In principle, the structure is potentially damaged if the current state differs significantly from the reference state namely due to changes in physical properties, typically a stiffness reduction.

The present paper concentrates on damage localization for which stochastic dynamic damage location vectors (SDDLVs) are employed. The basis of the SDDLV method is the damage location vector (DLV) method [4], in which the null space of the changes in the flexibility matrix from a pre- and post-damaged structure is utilized to locate the damage. The vectors that form the basis of the null space are designated as DLVs, since that particular null vector associated with the lowest singular value is applied to an undamaged reference model as static-pseudo loads. The stresses in the model will, theoretically, approach zero in the damaged area(s).

The SDDLV method is an extension of the DLV method that treats output-only measurements and includes the dynamics of the system by applying the null space of the changes in the transfer function matrix instead of the changes in the flexibility matrix. In [2], D. Bernal has proved that the method is applicable for localizing structural damages in numerical models of truss systems. Additionally, the applicability of the SDDLV method has, by the authors of the present paper, been demonstrated in an experimental context with the residential-sized wind turbine blade shown in figure 1 [3]. Here, it was suggested that the implementation of a statistical procedure for discrimination between actual damage-induced irregularities and noise-induced ones may significantly improve the method. This is explored in the present paper where different statistical approaches to discriminate between damage and other irregularities in the aforementioned residential-sized wind turbine blade are tested. Thus, the study is based on the same experimental setup as the one presented in [3].

2. Experimental setup

The approximately 800 mm polymer blade is reinforced with carbon fibre and composed of two separable shells by means of 25 bolts along the leading and trailing edges. Different damage conditions can be examined by untightening one or more bolts. In the present study, the damage indicated in figure 2 is analysed.

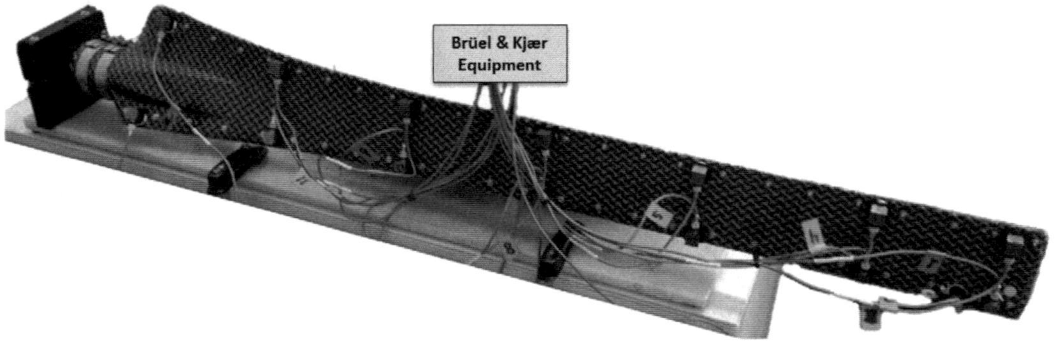

Figure 1. Experimental setup for OMA of the residential-sized wind turbine blade.

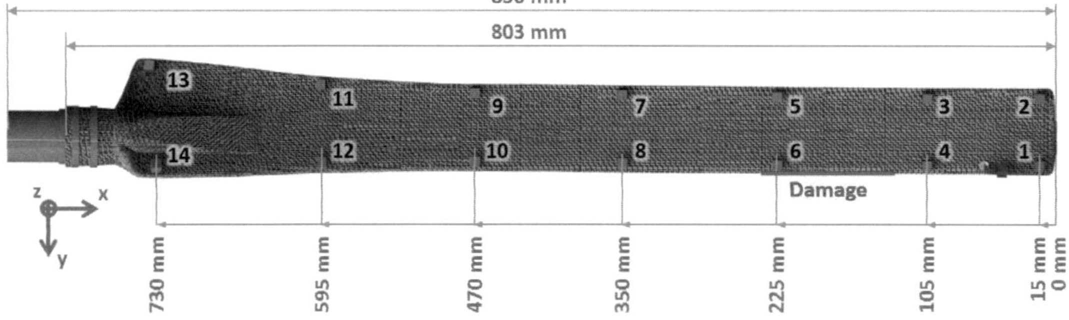

Figure 2. Dimensions of the blade, and location of the simulated damage.

The blade was subjected to unmeasured multi-impulse loads that were applied by hitting the structure randomly over the surface with a pencil. The response was captured by use of 14 equally spaced Brüel & Kjær uniaxial accelerometers, as illustrated in figure 2, which were measuring

Journal of Physics: Conference Series **628** (2015) 012003 doi:10.1088/1742-6596/628/1/012003

perpendicularly to the surface. A sufficiently high sampling frequency of 8192 Hz was utilized throughout the experiments. The recordings were uninterrupted time series that were divided into smaller partitions in order to obtain more than one experiment from the different states.

3. Damage localization using the SDDLV method

The recordings from the reference state and the current state are mathematically described by means of system identification techniques. Here, the dynamic properties of the system are defined in a state matrix, A_c, and the output measurements are defined in an output matrix, C_c.

Generally, the state matrix and the output matrix contribute to establish the kernel of the transfer function matrix which, as previously declared, is utilized for locating potential damages from changes between the two states. The basic form of the transfer function matrix is

$$G(s) = C_c(sI - A_c)^{-1}B_c + D_c\,, \tag{1}$$

where the input matrix, B_c, and the direct transmission matrix, D_c, are unknown since the system is purely stochastic with output-only measurements. An approach for estimating a fictive input is implemented from [2], hereby yielding

$$G(s) = R(s)D_c \tag{2}$$

where

$$R(s) = C_cA_c^{-2}(sI - A_c)^{-1}\begin{bmatrix} C_cA_c \\ C_cI \end{bmatrix}^{\dagger}\begin{bmatrix} I \\ 0 \end{bmatrix}, \tag{3}$$

with the dagger sign designating the Moore-Penrose pseudo-inverse. The direct transmission term, D_c, is assumed to be constant, because it is unaffected by the introduction of damage. The change in the transfer function matrix is then proportional to the change in $R(s)$, i.e.

$$\Delta G(s) \propto \Delta R(s) = R_d(s) - R_u(s)\,. \tag{4}$$

The SDDLVs, used as pseudo-loads, are estimated from the quasi-null space of $\Delta R(s)^T$ by singular value decomposition (SVD). In particular, the right singular vector associated with the smallest singular value is chosen to constitute the SDDLV.

The pseudo-loads are applied to a finite element (FE) model of the blade, which is based on an approximated 3D CAD geometry, discretized by equally sized first order shell elements. The model is calibrated against four experimentally-obtained natural eigenfrequencies and modes shapes of the undamaged blade. The pseudo-loads are applied at the same position and direction as the accelerometers.

4. Selection of s-values

The damage localization is not efficient for all s-values of the transfer function matrix. The guidance in [2] is to evaluate close, but not too close, to the poles of the system. Previous research by the authors of the present paper, see [3], reveals that a slight increase of an s-value corresponding to an identified system pole is applicable for damage localization, see figure 3.

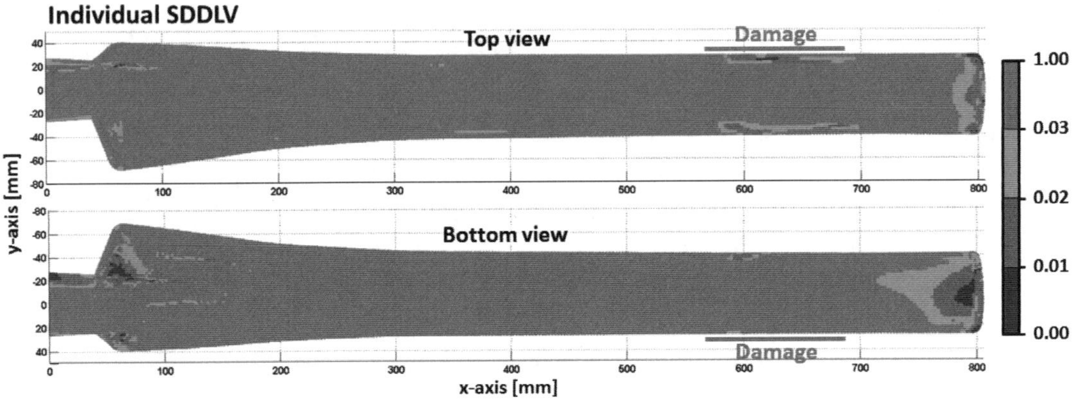

Figure 3. Stress field illustrating the normalized elemental mean von Mises stress for one SDDLV and showing the appearance of the damage.

As shown in figure 3 and concluded in [3], the results of applying individual SDDLVs imply that statistical evaluation can be included in the methodology to improve the damage localization. Such evaluation can be based on several experiments and/or several s-values. The s-value selection process is still a quite unexplored area of the SDDLV method, hence no strictly generalized guidance exists. The s-values can be chosen from infinite combinations of increments/decrements of the real and imaginary parts of an identified pole. In the present study, the two selection approaches depicted in figure 4 are tested, i.e. 10 s-values arranged in intervals of 2 to 20 % increment of the imaginary part and 28 s-values arranged in four branches of seven equally-spaced points from combinations of increments /decrements of the real and imaginary parts. It is noticed that the third flapwise bending mode is chosen, since it was most consistently excited during the experiments.

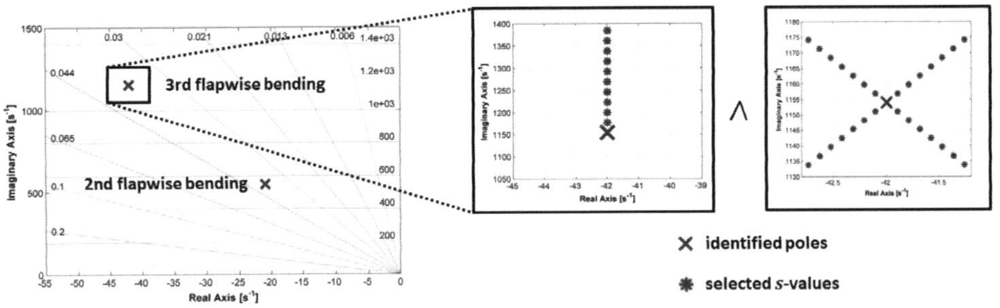

Figure 4. Selection of s-values based on the identified poles.

All the examined s-values are applicable for damage localization. However, the disturbance in the individual SDDLVs is larger close to the pole.

The stresses evaluated along the four branches are not found to vary in any specific manner with regards to direction. Consequently, all experiments are evaluated for the 10 s-values to establish the basis of the statistical evaluation.

5. Methodology of statistical evaluation

Statistical evaluation using pattern recognition works by training a statistical baseline model from several healthy states followed by testing the current state against the baseline in order to capture any significant changes. For multivariate data, several outlier analysis methods based on, e.g., T^2 statistics, Q statistics and the Mahalanobis distance have been successfully applied for damage detection, see e.g., [6] and [7]. The methods have the following of steps in common with the process of declaring the health of the structure:

- Train a baseline from the data measured in the healthy state.
- Determine a threshold defined by an appropriate quantile of the baseline.
- Test the current data using a pattern recognition technique for estimation of outliers.

In the present paper, it is tested whether some of these damage detection methods can be implemented in the SDDLV method to provide an effective statistical evaluation of the characteristic stress resultants obtained by applying the SDDLVs to the FE model of the undamaged blade. The elements of the FE model are evaluated one by one, and the results are gathered in a three-dimensional matrix, with the structure depicted in figure 5.

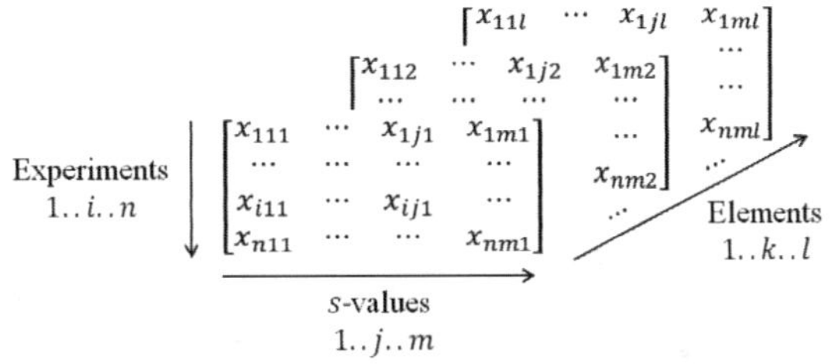

Figure 5. Suggestion for organizing the three-dimensional SDDLV data.

The outlier analysis methods used are based on Euclidean distance, T^2 statistics, and the Mahalanobis distance. The Euclidean distance is the distance between the current data from the experiment and the mean data from the experiments used for training, i.e.

$$d(x_{ik}) = \sqrt{\sum_{j=1}^{m} \left(x_{ijk} - \mu_{ik}\right)^2}, \tag{5}$$

where $x_{ijk} \in \mathbb{R}$ is the stress resultant from the current experiment and $\mu_{ik} \in \mathbb{R}$ is the mean of the training data.

In the T^2 statistics-based method, the dimensionality is reduced to dimension r by use of principal component analysis (PCA). Hereby, the similarity between m characteristic stress resultants from the current experiment and the training data is derived through

$$T^2(x_{ik}) = \left(x_{ik} - \mu_{jk}\right) P \Lambda^{-1} P^T \left(x_{ik} - \mu_{jk}\right)^T \tag{6}$$

in which $P \in \mathbb{R}^{m\times r}$ and $\Lambda \in \mathbb{R}^{r\times r}$ contain, respectively, the eigenvectors and eigenvalues of the covariance matrix computed from the trained data. In this context, it is found that the dimensionality can be reduced to one-tenth of original size without losing any significant information.

The Mahalanobis-squared distance employs the inverse covariance matrix of the data, i.e.

$$D^2(x_{ik}) = \left(x_{ik} - \mu_{jk}\right) \Sigma^{-1} \left(x_{ik} - \mu_{jk}\right)^T, \tag{7}$$

where $\Sigma \in \mathbb{R}^{m\times m}$ is the covariance matrix of the training data.

The similarity measures calculated from any of the three above-mentioned methods need to be compared to a threshold calculated from data from the healthy structure. It has been chosen to calculate the threshold for each element matrix l based on the following steps:

- Test the first row of the undamaged data matrix and calculate the similarity using one of the three methods while applying the rest of the undamaged data matrix as the base.
- Repeat this for all vectors in the training matrix and sort the distances in descending order.
- The threshold is then equal to the value exceeded by 5 % of the similarity measures.

6. Results

The following results are based on 280 undamaged and 200 damaged experiments that are based on evaluation of 10 SDDLVs in different points along the imaginary axis, as illustrated in figure 4. The variances of the experiments are reduced by taking the mean of 10 experiments and using these new vectors for the statistical evaluation. In this way, 28 undamaged and 20 damaged experiments are obtained, where the first 20 undamaged experiments are utilized for training the baseline. The previous research in [3] states that noise often appears in the area around the tip and root of the blade. Consequently, it is chosen to demonstrate the proposed statistical evaluation procedures for the locations shown in figure 6.

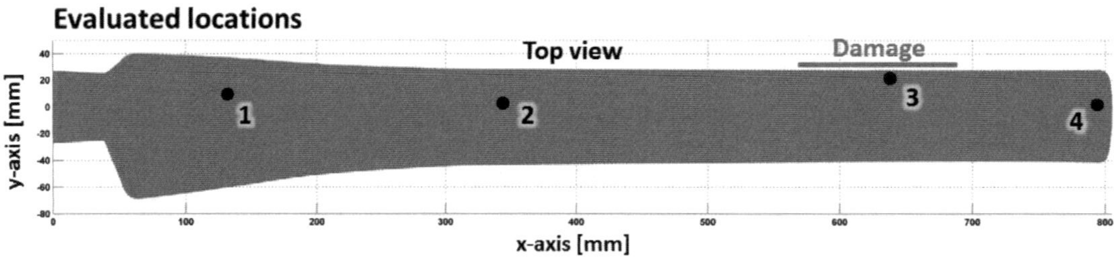

Figure 6. Location of elements utilized for inspection of the statistical evaluation.

Location 3 should, theoretically, be the only one to have distances higher than the calculated threshold in the damaged experiment, while distances at the other locations should be below the threshold. The results of the locations are presented in figure 7.

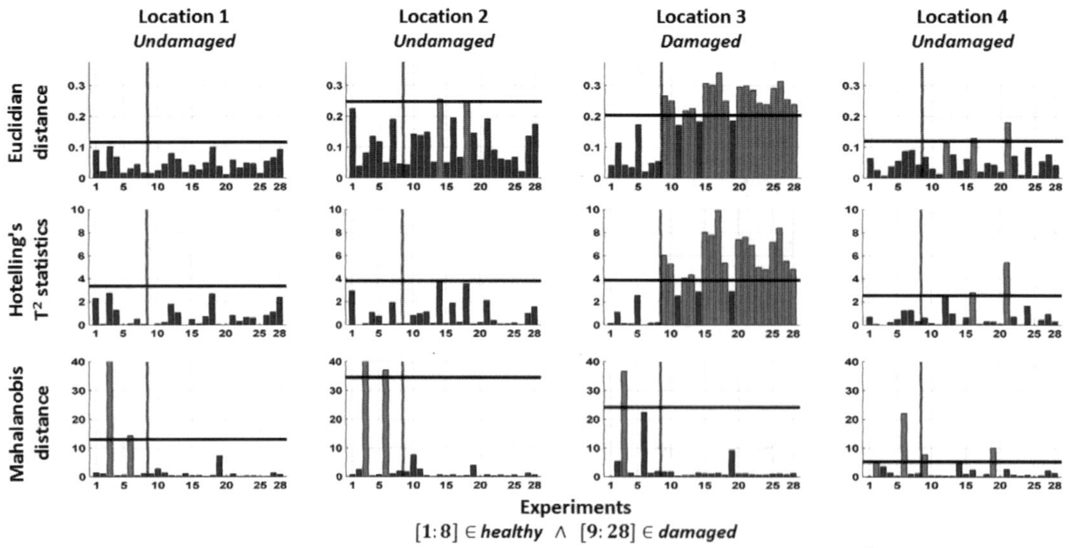

Figure 7. Statistical results from four locations by use of Euclidean distance, T^2 statistics, and Mahalanobis distance. The horizontal lines indicate the threshold and the vertical lines discriminate between the healthy experiments to the left and the damaged experiments to the right.

It appears that the two methods employing Euclidean distance and T^2 statistics detect the damage in location 3, without too many false alarms at any of the other locations. It is also noticed that the T^2 statistics-based method is slightly better than the Euclidean distance-based method. On the contrary, the Mahalanobis distance does not appear to be applicable to the measured data. This has been shown to improve if the data is taken as the mean of many more experiments, but that is a problem due to the limited amount of available data.

The Euclidean distance has been calculated for all elements using all of the damaged experiments, and in figure 8, the amount of experiments that yield a distance exceeding the threshold is presented as a percentage of all experiments. Evidently, the damage is clearly located and the amount of noise present in the plot is reduced significantly compared to the results from individual SDDLVs, see figure 3. The same procedure has been applied for the results of T^2 statistics. The results, which can be seen in figure 9, show how the damage is located practically perfect without any disturbances. It is noticed how the exact size of the damage is assessed, which is important information with respect to commercial applications.

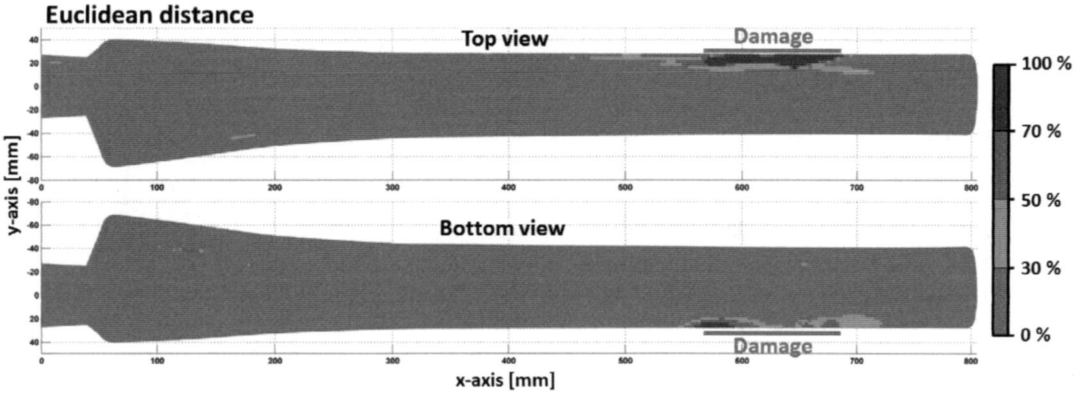

Figure 8. Statistical evaluation for SDDLV-induced stress resultants using the Euclidian distance.

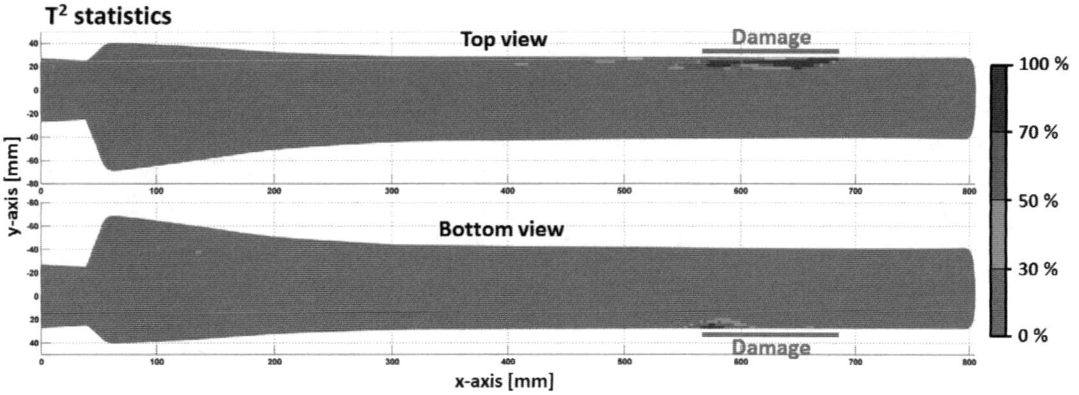

Figure 9. Statistical evaluation for SDDLV-induced stress resultants using T^2 statistics.

7. Conclusion

The presented paper deals with the localization of damage in a residential-sized wind turbine blade by use of the SDDLV method. The SDDLV method has previously been demonstrated, by the authors, as capable of locating different damages in the aforementioned blade, although noise disturbances made it difficult to obtain unambiguous localization when using individual SDDLVs. This is handled in the present study by extending the SDDLV method to contain a statistical evaluation yielding the final discrimination between damaged and undamaged areas.

The statistical evaluation is based on outlier analysis of the characteristic stress resultants, for a total of 10 s-values, in each element. The statistical patterns, i.e. mean vector and covariance matrix, are established on the basis of several experiments from the healthy state. Subsequently, data from the current state are tested against this pattern by alternate use of three different similarity measures, namely the Euclidian distance, T^2 statistics, and the Mahalanobis distance. While the Mahalanobis distance proves completely inapplicable, it is found that by employing either Euclidian distance or T^2 statistics, the damage is located unambiguously. Of the latter two approaches, T^2 statistics is preferable as it not only provides unambiguous localization but also assesses the size of the damage practically perfect.

Future research activities will deal with further improvement of the s-value selection process in order to establish a particular guidance. Moreover, a sensitivity study, in which the influence of damage size and location is examined, will be conducted.

8. Acknowledgement

The experimental part of the project was supported by the EUDP (Danish Energy Technology Development and Demonstration Programme) grant 64011-0084 "Predictive Structure Health monitoring of Wind Turbines". The authors also acknowledge Robert Flemming Mikkelsen who provided the blade for the experiments.

9. References

[1] Ciang C C *et al* 2008 Structural health monitoring for a wind turbine system a review of damage detection methods *Meas. Sci. Technol.* **19** 122001

[2] Bernal D 2010 Load vectors for damage location in systems identified from operational loads *J. Eng. Mech.* **136** 31-39

[3] Johansen R J *el al* 2015 Damage localization in residential-sized wind turbine blade by use of SDDLV method *Submitted to 11th International Conference on Damage Assessment of Structures*

[4] Bernal D 2002 Load vectors for damage localization *J. Eng. Mech.* **128** 7-14

[5] Van Overschee P and Moor B L R 1996 Subspace identification for linear systems theory implementation applications Kluwer academic publishers

[6] Mujica LE *et al* 2010 Q-statistic and T^2-statistic PCA-based measures for damage assessment in structures *Structural Health Monitoring* **10** 539-553

[7] Balsamo L *et al* 2014 Damage Detection Using Large-Scale Covariance Matrix *Structural Health Monitoring Volume 5 Proceedings of the 32nd IMAC* **10** 89-97

Journal of Physics: Conference Series **628** (2015) 012004 doi:10.1088/1742-6596/628/1/012004

Multi-damage localization in plate structure using frequency response function-based indices

Hai-yang Gao[1], Xing-lin Guo[2], Huajiang Ouyang[2,3], Xiu-ming Yang[2]
[1]Beijing Institute of Spacecraft Environment Engineering, Beijing, P. R. China
[2]State Key Laboratory of Structural Analysis for Industrial Equipment, Dalian University of Technology, Dalian, P. R. China
[3]University of Liverpool, Liverpool L69 3BX, United Kingdom

E-mail: eric415@126.com

Abstract. Vibration signal and its derivative have shown some promise in structural damage detection in previous research. However, the theoretical and practical difficulties of multi-damage detection in plate structures based on dynamic responses remain. In this paper, an efficient damage localization index based on frequency response function (FRF) is presented. The imaginary part of FRF (IFRF) is extracted to derive the new localization index due to its relation to modal flexibility. For avoiding the finite element model error, two-dimensional gapped smoothing method (GSM) is employed without the need for baseline data from a presumably undamaged structure. Experimental studies on a steel plate with two localized defects in different boundary conditions are performed. The results are compared with some typical damage indices in the literature, such as mode shapes, uniform load surface and IFRF. In order to mitigate the inherent disadvantages of GSM in anti-noise ability, a simple statistical treatment based on Thompson outlier analysis is finally used for noise suppression. The effect of damage level and boundary condition on the detection results is also investigated.

1. Introduction

A great deal of numerical methods for structural damage detection have been proposed and developed during recent years. However, few methods can successfully solve this problem with satisfactory accuracy in practical engineering. In this sense, a promising method, not requiring measurements on the healthy structure, for use in practice would have the ability to identify slight damage with limited sensor placement, more importantly, to have strong noise immunity.

Mode shapes are widely used for damage localization, which can be easily extracted from modal analysis, and possible to describe the local features of structures [1]. Then their derivatives (e.g. modal flexibility and uniform load surface) were found to have better performance in high-noise cases. Some studies indicated that only a few pieces of information can be obtained from mode shapes [2]. To construct a direct damage localization index without modal identification, FRF was increasingly used by many researchers [3-5] which can be obtained directly from vibration measurement. Another advantage is that FRF provides a wide range of frequencies and gives more information related to structural dynamical characteristics.

Liu et al. [6] proposed a non-modal scheme including some significant modifications such as using the imaginary parts of FRF shapes (IFS) and normalized FRF shapes (NFS) before comparison to detect the fault in a single damaged beam. The damage can be easily found by using IFS index. To validate the proposed method, they demonstrated the close relation between residual imaginary FRF shapes and residual mode shapes at the same modal frequencies. To the authors' knowledge, only this work has explored the feasibility of IFS for damage detection. This motivated the authors to further explore the potential of IFS and its derivatives for identifying damage in plate-like structures.

On the base of the authors' previous research [7], a novel efficient damage localization index from FRFs is formulated in this paper. For carrying out multi-damage localization in plate structures, two-dimensional gapped smoothing method (GSM) is employed. To further obtain a more explicit detection result, a simple statistical treatment based on outlier analysis is finally performed. By comparing with some traditional indices (e.g. modal shape, uniform load surface, and IFS), the effectiveness of this proposed method is verified by the results of experimental studies on a multi-damage steel plate.

2. Principle of Two-dimensional GSM

The basic concept of GSM for damage detection in a two-dimensional structure was proposed by Yoon [8], and demonstrated to be robust by many researchers. Here a brief description of the fundamental theory is given. Firstly, a two-dimensional grid is meshed on the plate structure in the x and y directions. The curvature of the damage localization index ψ can be calculated at grid point (s, t) by

$$\nabla^2\psi_{s,t} = \psi_{s,t}^{xx} + \psi_{s,t}^{yy} \tag{1}$$

in which

$$\psi_{s,t}^{xx} = (\psi_{s+1,t} + \psi_{s-1,t} - 2\psi_{s,t})/h_x^2 \tag{2a}$$

$$\psi_{s,t}^{yy} = (\psi_{s,t+1} + \psi_{s,t-1} - 2\psi_{s,t})/h_y^2 \tag{2b}$$

where h_x and h_y are the distances between a point and its neighbors in the x and y directions, respectively. Assume that the healthy structure has a smooth deformed curvature, it can be approximated by the polynomial function constructed as

$$C_{s,t} = a_0 + a_1 x + a_2 y \tag{3}$$

where the coefficients a_i can be determined explicitly by the curvatures of neighboring points from the damaged plate structure.

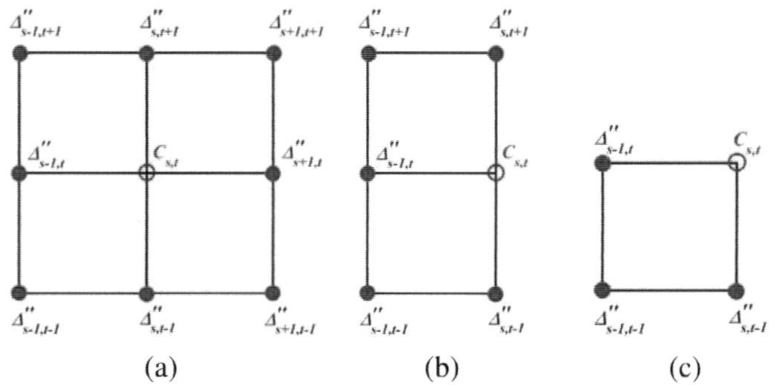

Figure 1. Interpolation for 2-D GSM: (a) inner point; (b) boundary point; (c) corner point

The least squares method is employed to solve this curve-fitting problem. More details can be found in [8]. Consequently, the value of function $C_{s,t}$ at grid point (s, t) can be calculated by using Eq. 3. It should be noted that there are three cases for interpolation: (a) The value of the inner point can be estimated by using the neighboring eight points around it, and hence Eq.3 can be extended as $C_{s,t}=a_0+a_1x+a_2y+a_3xy+a_4x^2+a_5y^2+a_6xy^2+a_7x^2y$ to obtain a higher-order surface; (b) if the point is on a boundary line, its value should be evaluated by using the neighboring five points. In this case, Eq. 3 can be extended as $C_{s,t}=a_0+a_1x+a_2y+ a_3x^2+a_4y^2$; (c) particularly, only three points can be used when the point is at a borderline intersection, as shown in figure 1. The damage index map is then calculated by using the absolute difference between the estimated value and the curvature

$$\delta_{s,t} = \left|\nabla^2 \psi_{s,t} - C_{s,t}\right| \tag{4}$$

Finally the outlier of the damage localization index indicates the damage region.

3. Construction of damage localization index

In this section, some previous indices (e.g. modal shape, uniform load surface, and IFS) are introduced here for performance comparison by making an experimental study reported in the next section. Mode shape index was first used in GSM by Ratcliffe [9] for locating structural damage in beams, which can be written as

$$\mathbf{MS} = \{\phi_{ik}\} \quad i = 1.2...N \ , \ k = 1.2...m \tag{5}$$

where ϕ_{ik} is the kth mode shape obtained at the ith point when m lower modes are available for an N-DOFs structure. Since the MS index may be easily contaminated by measurement noise, uniform load surface index (ULS) [10] was proposed as

$$\mathbf{ULS} = \{u(i)\} = \sum_{k=1}^{m} \frac{\phi_{ik} \sum_{l=1}^{N} \phi_{lk}}{\omega_k^2} \tag{6}$$

where $u(i)$ represents the modal deflection at point i under uniform unit load all over the structure. Some researchers extended mode shape-based index to broadband frequency ranges using FRF shape information. Liu et al [6] presented what was possibly the first use of imaginary part of FRF shape (IFS) because of the close connection between residual curvatures of IFS and MS. The IFS can be easily extracted out of FRF. Each element in FRF matrix can be written as

$$\alpha_{ij}(\omega)=\sum_{k=1}^{m}\frac{\phi_{ik}\phi_{jk}}{(\omega_k^2-\omega^2)+2\mathrm{j}\xi_k\omega_k\omega}=\sum_{k=1}^{m}\frac{\phi_{ik}\phi_{jk}(\omega_k^2-\omega^2)}{(\omega_k^2-\omega^2)^2+(2\xi_k\omega_k\omega)^2}-\mathrm{j}\sum_{k=1}^{m}\frac{\phi_{ik}\phi_{jk}2\xi_k\omega_k\omega}{(\omega_k^2-\omega^2)^2+(2\xi_k\omega_k\omega)^2} \quad (7)$$

where α_{ij} is the receptance FRF measured at location i for a force input at location j, ϕ_{ik} and ϕ_{jk} are the ith and jth elements of the kth mode shape. ω_k *and* ξ_k denote the k-th frequency and damping ratio, respectively. The l-th column of the IFS is denoted as

$$IFS_l(\omega)=\left[-\sum_{k=1}^{m}\frac{2\phi_{ik}\phi_{1k}\xi_k\omega_k\omega}{(\omega_k^2-\omega^2)^2+(2\xi_k\omega_k\omega)^2},\ -\sum_{k=1}^{m}\frac{2\phi_{ik}\phi_{2k}\xi_k\omega_k\omega}{(\omega_k^2-\omega^2)^2+(2\xi_k\omega_k\omega)^2},\ \mathrm{L}\ ,\ -\sum_{k=1}^{m}\frac{2\phi_{ik}\phi_{mk}\xi_k\omega_k\omega}{(\omega_k^2-\omega^2)^2+(2\xi_k\omega_k\omega)^2}\right]^{\mathrm{T}} \quad (8)$$

It can be seen that when $\omega=\omega_k$, each element in Eq.8 can be reduced to

$$\sum_{k=1}^{m}\frac{\phi_{ik}\phi_{jk}2\xi_k\omega_k\omega}{(\omega_k^2-\omega^2)^2+(2\xi_k\omega_k\omega)^2}=\sum_{k=1}^{m}\frac{1}{2\xi_k}\frac{\phi_{ik}\phi_{jk}}{\omega_k^2} \quad (9)$$

where $\phi_{ik}\phi_{jk}/\omega_k^2$ is the kth modal flexibility. This means that the IFS index at each natural frequency is closely related to its modal flexibility form. Therefore the damage location can be described by IFS index, which in turn supports the validity of the method in [6]. Inspired by ULS and IFS index, here a novel index is proposed as

$$\textbf{ULS - IFS}=\{w(i)\}=\sum_{\omega=\omega_q}^{\omega_s}\frac{IFS_i(\omega)\sum_{l=1}^{N}IFS_l(\omega)}{\omega^2} \quad (10)$$

in which ω_q and ω_s limited the frequency range, consisting of m natural frequencies. It is noteworthy that the definition of ULS-IFS here includes more shape-based information in considered frequency broadband, compared with conventional ULS. Meanwhile, it inherits the merit of IFS index that the random error can be averaged out in each measuring point.

4. Experimental study on Multi-damage detection of a steel plate

To assess the detection performance and the advantage of the proposed index, a rectangular plate with an area of 300×300mm and 5mm thickness is considered. To simulate the damage, two defects a and b were manufactured both with area of 30×30 mm. Their position is shown as dotted box in figure 4. The material properties of the steel plate are as follows: Young's modulus E=208.5GPa, Poisson's ratio μ=0.27, density ρ=7830kg/m^3. For studying the effects of damage magnitude and structural boundary condition on the detection results, three damage cases are studied, as defined in table 1.

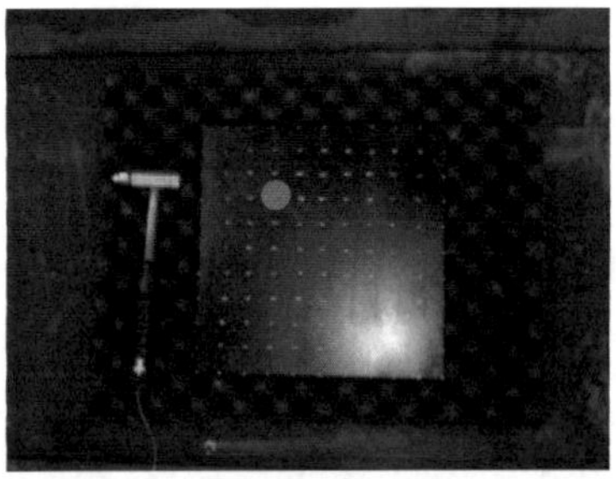

Figure 2. Rectangle steel plate for experiment

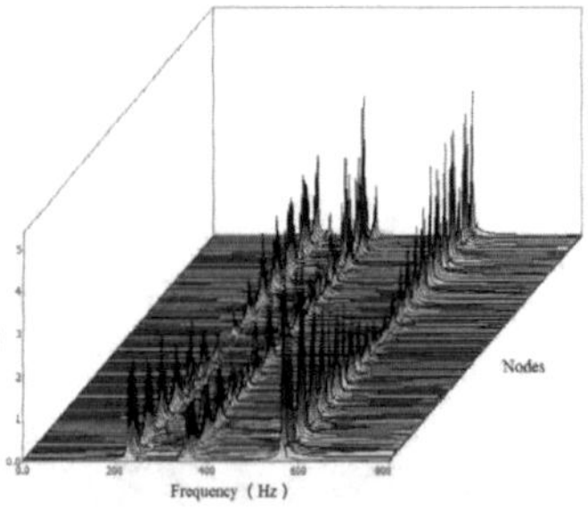

Figure 3. The measured FRF waterfall map

Table 1. Damage cases of the steel plate

Case	Damage level (mm)	Boundary condition	Measuring point
1	*a*: 1.0 *b*:1.2 (slight)	resilient polyurethane foam	121
2	*a*: 2.0 *b*:3.0 (severe)	resilient polyurethane foam	121
3	*a*: 2.0 *b*:3.0 (severe)	One edge clamped	110

The grid map consists of 121 points as shown in figure 2. An acceleration transducer is mounted at the position marked by a red solid dot. All points are excited by roving hammer impact. Test data are collected and processed using a 6-channel intelligent signal acquisition and analysis system B&K 3050. The FRFs of the damaged plate are shown in figure 3. Then the natural frequencies and corresponding mode shapes of the damaged plate can be easily identified to calculate MS index and ULS index for the performance comparison.

Table 2. First three modal frequencies for all cases

Case	1	2	3
1st frequency (Hz)	231	228	59.5
2nd frequency (Hz)	336	336	128
3rd frequency (Hz)	359	356	313

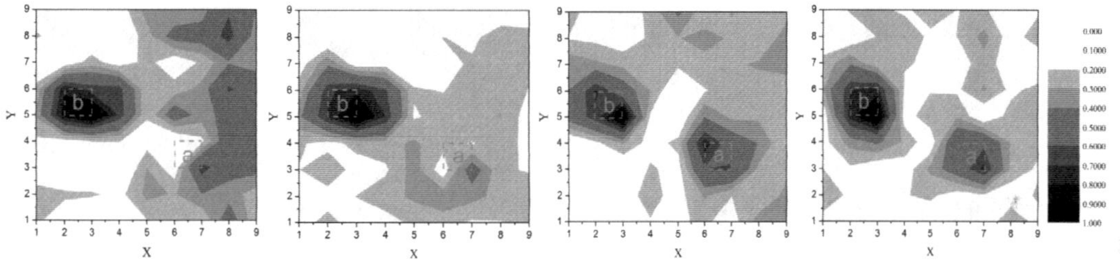

(a) (b) (c) (d)

Figure 4. Damage localization results of Case 1: (a) MS; (b) ULS; (c) IFS; (d) ULS-IFS

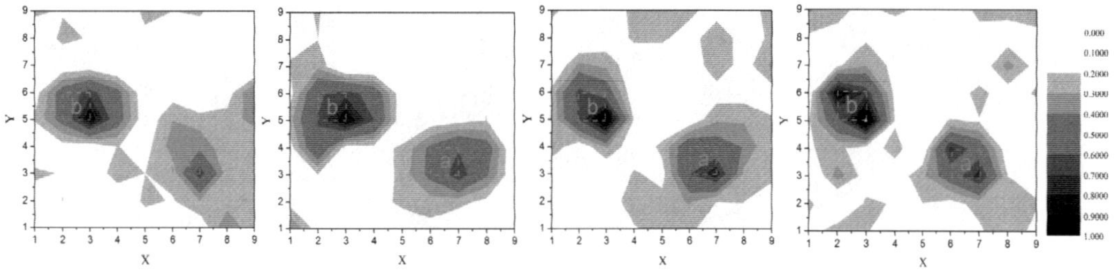

(a) (b) (c) (d)

Figure 5. Damage localization results of Case 2: (a) MS; (b) ULS; (c) IFS; (d) ULS-IFS

Figures 4(a)-(d) compare the normalized values of each kind of indices at the same damage level. It is found that MS and ULS indices can approximately detect defect *b*. However, a shortcoming is that they give a poor prediction of the relatively slightly damaged region *a*. It can be observed from figures 4(c) and 4(d) that both damaged regions are successfully detected and the peak values explicitly show the relative levels of two damaged regions. And one can find large gradient of the IFS index surface in undamaged area, while the ULS-IFS index surface is relatively smooth.

Case 1 and Case 2 shows the effects of damage magnitude on detection performance of all the indices. It can be noted that the values of all the indices in the undamaged region are lower than the damaged ones in Case 1 if the damage magnitude increases.

To assess the effectiveness and reliability of the proposed index when changing structural boundary conditions, the right edge of the steel plate is clamped so that only 110 measuring points on the plate are available for Case 2. The detection results of all four indices can be calculated in a similar way, as shown in figure 6. It is found that MS and ULS indices in the undamaged region are somewhat sensitive to the boundary change, while IFS and ULS-IFS indices still show stable performance. It can be seen that ULS-IFS index provides a clearer damage localization result of both damaged regions than IFS index based on the measured points, as shown in figures 6(c) and (d). In the respect of detection performance in the undamaged area, most of ULS-IFS index values are less than 0.1, which is exactly what we expect.

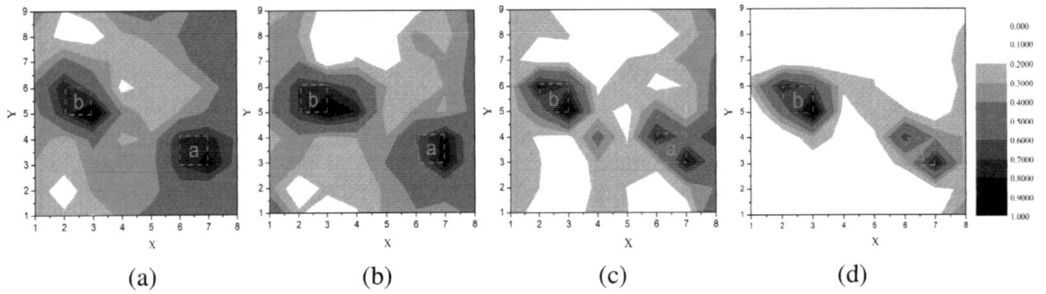

Figure 6. Damage localization results of Case 2: (a) MS; (b) ULS; (c) IFS; (d) ULS-IFS

5. Noise suppression method based on outlier analysis

GSM is based on curvature which can amplify outliers as a result of the presence of damage. However, GSM makes outliers caused by noise more obvious on the smooth curve simultaneously. In this aspect, GSM seems to be less capable in noise suppression. To obtain more explicit results of damage detection, a statistical treatment is performed in this section. The basic idea is from [8] but some refinement is made.

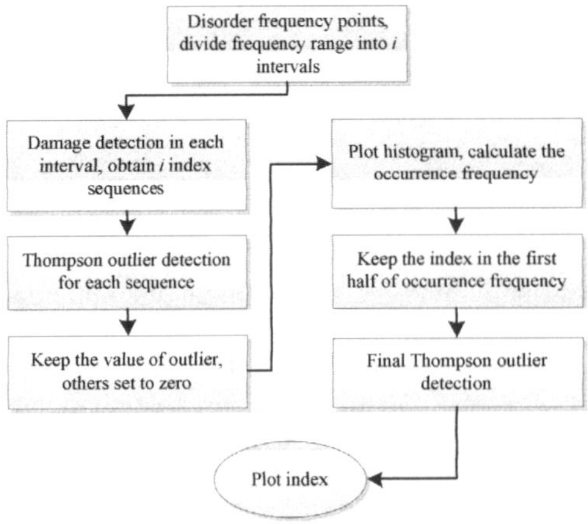

Figure 7. Process of damage detection with noise suppression method based on outlier detection

It is assumed that damage region is very small compared with the entire structure, and more importantly, the mean and standard deviation of damage indices of the undamaged area of a damage structure are similar to those from the repeated tests of an undamaged structure. Therefore, outliers of averaged damage indices correspond to the irregularity caused by damage. The process of damage detection with noise suppression method using Thompson outlier detection [11] is shown as following steps. To show the effectiveness of proposed noise suppression method, ULS-IFS index is used to recalculate all damage cases. The considered frequency points was randomly ordered and divided into eight intervals to carry out the outlier detection. The damage detection results are shown in figures 8 and 9.

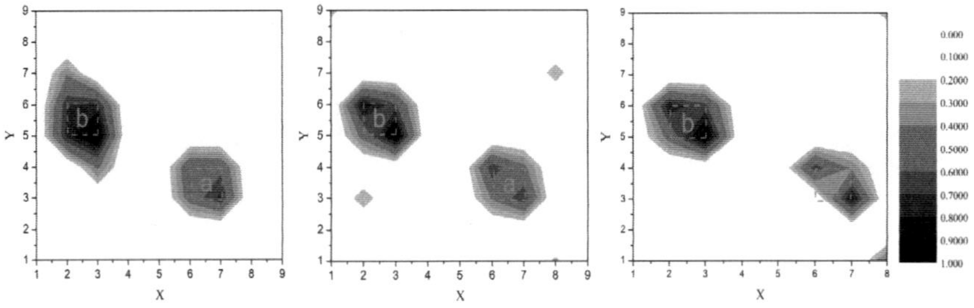

(a) (b) (c)

Figure 8. Localization results based on ULS-IFS index in: (a) Case 1; (b) Case 2; (c) Case 3

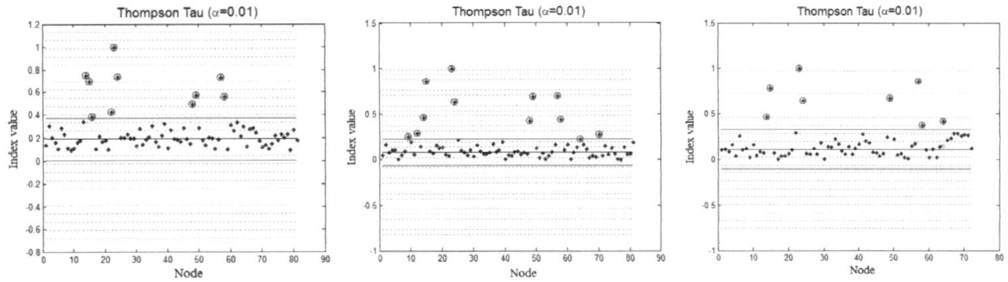

(a) (b) (c)

Figure 9. Final outlier analysis based on ULS-IFS index in: (a) Case 1; (b) Case 2; (c) Case 3. The red circles are outliers, and the blue dots between red parallel lines represent values caused by noise.

It can be seen that the irregular values in the undamaged area have been filtered so that the damaged area is highlighted. But the identified damage area is somewhat larger than the true area because a plate is a continuum on which the measurement data at different points are probably not independent, that is, the identified boundary between slightly damage areas and undamaged areas seems vague. Figure 9(b) shows that the bandwidth of non-outliers decreases with increasing damage severity. Also one can see that different boundary conditions can affect the identified results. It is important to note that, the accuracy of the GSM depends on density of the measurement mesh on a structure. Laser vibrometers can be used to obtain more measurement points and give a relatively smooth interpolating surface for damage detection.

6. Conclusions

In this paper, a novel damage localization index from FRFs is firstly formulated for damage detection. To avoid using a finite element model, two-dimensional gapped smoothing method with a statistical noise suppression technique is employed. An experimental study on a steel plate demonstrates the effectiveness of the proposed index. Some conclusions can be obtained as follows:

1. For the same boundary conditions, FRF-based index has a better anti-noise ability than mode shape-based index, since it can be averaged over the frequency bandwidth. Also it can be seen that the ULS-based index shows superior anti-noise ability over those basic indices (MS and IFS index) because the summations $\sum_{l=1}^{N}\phi_{lk}$ and $\sum_{l=1}^{N}IFS_l(\omega)$ average out the random error at each measuring point.
2. ULS-IFS index gives a better performance than other three indices, which combines the advantages of IFS and ULS in noise suppression.
3. Boundary conditions and damage level do affect the damage localization results.
4. Outlier detection is an easy and effective technique for noise suppression, which makes damage localization results of GSM more explicit.

Reference

[1] Pandey A K, Biswas M, Samman M M, Damage detection from changes in curvature mode shapes, Journal of Sound and Vibration, 1991, **145** (2):321-332

[2] Srinivas V, Sasmal S, Ramanjaneyulu K, Studies on methodological developments in structural damage identification, Structural Durability and Health Monitoring, 2009, **5**(2): 133-160

[3] Maia N M M, Silva J M M, Sampaio R P C, Localization of Damage using Curvature of the Frequency Response Functions, Proceedings of the 15th International Modal Analysis Conference, Society of Experimental Mechanics, 1997, **1**: 942-946

[4] Sampaio R P C, Maia N M M, Silva J M M, Damage detection using the frequency response function curvature method, Journal of Sound and Vibration, 1999, **226**(5):1029-1042

[5] Sampaio R P C, Maia N M M, Strategies for an efficient indicator of structural damage, Mechanical Systems and Signal Processing, 2009, **23**:1855-1869

[6] Liu X, Lieven N A J, Escamilla-Ambrosio, P.J., Frequency response function shape based methods for structural damage localization, Mechanical Systems and Signal Processing, 2009, **23**:1243-1259

[7] Gao H, Guo X, Zhao Y, Experimental Study of Multi-damage Detection in a Plate Based on Non-modal Method. Experimental Techniques, 2014, **38**: 6-15

[8] Yoon M K, Heider D, Gillespie J W, Ratcliffe C P, Crane, R.M., Local damage detection using the two-dimensional gapped smoothing method, Journal of Sound and Vibration, 2005, **279**:119-139

[9] Ratcliffe C P, Damage detection using a modified Laplacian operator on mode shape data, Journal of Sound and Vibration, 1997, **204** (3):505–517

[10] Wu D, Law S S, Damage localization in plate structures from uniform load surface curvature, Journal of Sound and Vibration, 2004, **267**:227-244

[11] Thompson W R, On a criterion for the rejection of observations and the distribution of the ratio of deviation to sample standard deviation. The Annals of Mathematical Statistics, 1935, **6**(4): 214-219

Location of the corrosion damage in rectangular plates

M Tufoi[1*], **G R Gillich**[1], **I C Mituletu**[1], **V Iancu**[1]
[1] Department of Mechanical Engineering, Eftimie Murgu University of Resita, P-ta Traian Vuia 1-4, 320085, Resita, Romania

E-mail: m.tufoi@uem.ro

Abstract. The presence of corrosion influences the dynamic behaviour of plates, by modifying its natural frequencies and mode shapes. Many methods to assess damages in structural elements are known, but they mainly address to the geometric discontinuities, where mass loss is insignificant. In some of the prior research, we have got success in achieving a mathematical relation between damage geometry and frequency shifts. Since the mass decrease is also associated with numerous corrosion mechanisms, the idea of this paper is to establish the influence of mass changes in the natural frequencies of plates and including this information in a baseline, to apply it in processes for corrosion assessment. Thus, by research the evaluation of frequency changes due to mass loss and stiffness decrease has been followed. The finite element analysis was performed, considering corroded area as being simulated by a constantly decreasing plate thickness. In this way, is highlighted the correlation between frequency changes in the transversal vibration modes and mass loss. The dependency between mass loss and frequencies is accomplished, as a mathematical relation as well. This relation is associated with another relation, who predicts the frequency decreases due to stiffness reduction, in order to find the curves of frequency shifts and promote them as a baseline for corrosion assessment. Also, a corrosion assessment method, based on the contrived frequency shift curves, has been developed and proved as reliable and easily applicable.

1. Introduction

Degradation of structural elements due to wear or corrosion affects the safety of the structures by reducing their cross-section or/and mechanical strength. In literature numerous methods developed to assessing damages using vibration signals are presented: these methods base on changes of modal parameters, most common being the frequency shifts; comprehensive reviews are provided for instance in [1]. Most methods dedicated to assess the damage location and severity are model-based. The main disadvantage of these methods is that it's difficult to distinguish between model errors and changes of the modal parameters [2]. Consequently, the process of finding the modal parameters associated with the damaged system in order to fit its answers to that obtained by measurement is complex and often inadequate results are obtained [3, 4].

In our previous research we contrived mathematical relations that precise predict, for slender beams, the frequency decrease due to a transversal discontinuity [5-7] or change due to mass variation [8]. These relations are the base of an original damage detection method applicable for breathing and open cracks as well. It consists of comparing a sequence of normalized frequency shifts for the weak-axis bending vibration modes, measured at different time points, with patterns derived from the mode shapes and their derivatives.

In this paper we present investigations, performed on rectangular plates clamped at all ends, in which both the stiffness changes and loss of mass due to corrosion are considered, aiming to fit the method for the case of plates.

2. Frequency changes in beam-type structures due to mass and rigidity variation

The vibration of beams with geometrical discontinuities and/or not uniformly distributed mass was earlier studied and permitted the enunciation of following inferences:
- if the stiffness reduction is placed on slices not subjected to bending, i.e. regions 1 and 4 in figure 1, the frequency drop is insignificant;
- if the stiffness reduction is placed on slices achieving important bending, i.e. regions 2 and 3 in figure 1, the frequency drop is important;
- if the mass loss is placed on slices subjected large displacement, i.e. regions 1 and 3 in figure 1, the frequency increase is important;
- if the mass loss is placed on slices achieving no displacement, i.e. regions 2 and 4 in figure 1, the frequency change is insignificant.

As a consequence, the kinetic and potential energy distribution for the bending modes are altered from the healthy state with respect to the modal curvature and mode shape respectively. Moreover, the natural frequencies of the bending modes change in relation to the stored energy alteration. By occurrence of a discontinuity at location x, the frequencies drop $\Delta f_{i-S}(x)$ is:

$$\Delta f_{i-S}(x) = f_{i-U} - f_{i-S}(x) = f_{i-U} \cdot \frac{\sqrt{\delta_D} - \sqrt{\delta_U}}{\sqrt{\delta_D}} \cdot \left(\bar{\phi}_i''(x) \right)^2 \quad (1)$$

where f_{i-U} is the frequency of the undamaged beam and $f_{i-S}(x)$ for the ith transverse mode, and $\bar{\phi}_i''(x)$ is the value of the dimensionless mode shape curvature at location x. The terms δ_U and δ_D are the static deflections at the free end of the similar cantilever beam for the healthy state and if the discontinuity is located at the fixed end respectively. These deflections, being in direct relation with the beam's global rigidity EI and EI_{eq}, reflect the stored energy in the two mentioned states, thus the fraction in equation (1) represent the damage severity.

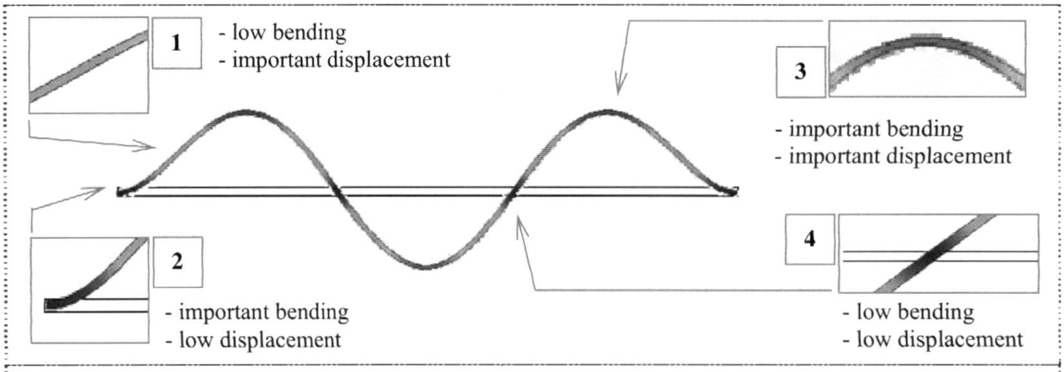

Figure 1. Three typical regions for the vibrating fixed-fixed beam.

If a loss of mass ΔM at location x is observed, the bending frequency increase $\Delta f_{i-M}(x)$ is:

$$\Delta f_{i-M}(x) = f_{i-M}(x) - f_{i-U} = f_{i-U} \cdot \frac{\sqrt{M} - \sqrt{M - 4\Delta M}}{\sqrt{M - 4\Delta M}} \cdot \left(\bar{\phi}_i(x) \right)^2 \quad (2)$$

where M is the healthy beam's mass and $\bar{\phi}_i(x)$ is the dimensionless mode shape value at location x.

If a discontinuity associated with a mass loss is present, the natural frequency shift becomes:

$$\Delta f_{i-D}(x) = f_{i-U} \cdot \left[\frac{\sqrt{M} - \sqrt{M - 4\Delta M}}{\sqrt{M - 4\Delta M}} \cdot \left(\bar{\phi}_i(x) \right)^2 - \frac{\sqrt{\delta_D} - \sqrt{\delta_U}}{\sqrt{\delta_D}} \cdot \left(\bar{\phi}_i''(x) \right)^2 \right] \quad (3)$$

By dividing the equations (1) and (2) with f_{i-U} relative frequency shifts are accomplished. A sequence of relative frequency shifts, lets say $i = 1 \ldots n$, constitute patterns characterizing the damage location. In next sections the possibility to contrive patterns for plate-type structures is studied.

3. Model description

In our concern to demonstrate that stiffness and mass loss influence the plate frequencies with respect to the locally stored energy (i.e. the squared of the mode shape curvatures) and achieved accelerations (or displacements), we performed numerical analysis using the *SolidWorks* software. A plate, presented in figure 2, with dimensions a = 1 m, b = 0.5 m and h = 0.002 m was the subject of our analysis, both in healthy state and with a simulated corroded area. The corroded region has a constant thickness h_C = 0.001 m and a square shape with edges d; herein we present the case of damage extension d_1 = 0.03 m, d_2 = 0.04 m and d_3 = 0.05 m. For simplicity, we consider the corrosion in the plate centre, as depicted in figures 2 and 3.

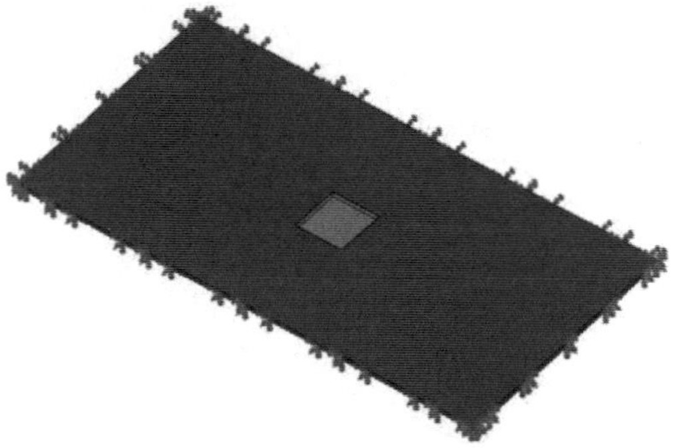

Figure 2. Corroded plate with fixed edges.

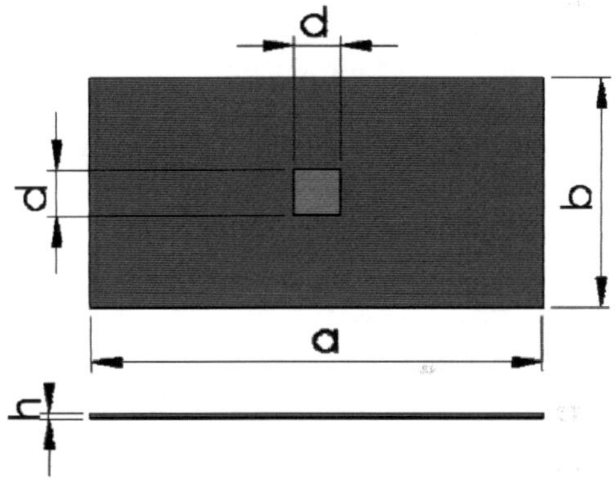

Figure 3. Plate and corroded area dimensions.

The material related from the *SolidWorks* software library is the **AISI 1045 Steel**, cold drawn, having the mechanical properties indicated in Table 1.

Table 1. Mechanical properties of the plate material

Yield strength [N/mm^2]	Tensile strength [N/mm^2]	Mass density [kg/m^3]	Elastic modulus [N/mm^2]	Poisson's ratio [-]	Volumetric mass density [kg/m^3]
630	580	7,850	200,000	0.3	7850

In this analysis, the plates are fixed on all four contour edges and the mesh was generated by tetrahedral elements with 21 characteristic points and the maximum size of 2 mm.

4. Results and discussions

The natural frequencies for both the healthy and corroded plates were evaluated by means of FEA simulations; in all cases 30 vibration modes are considered. Table 2 present these frequencies grouped after the *i-j* indices of the vibration modes, this providing a clear overview of the plate behavior.

To find out if it can be made a parallelism between the dynamic behavior of corroded beams and plates, we derived the relative frequency shift Δf_{ij}^* of the plate's mode *i-j*, after the relation used for beams [9]. This relation can be written in the form presented below:

$$\Delta f_{ij\_U}^* = \frac{\Delta f_{ij}}{f_{ij-U}} = \frac{f_{ij-U} - f_{ij-D}}{f_{ij-U}} \tag{4}$$

In the relation above, f_{ij-U} is the natural frequencies of mode *i-j* of the healthy plate and f_{ij-U} is the natural frequencies of mode *i-j* for the damaged plate. To express the relative frequency shift in percents, equation (4) has to be multiplied by 100.

Table 2. Frequency values for four type of plate grouped into m-n and relative shift for three scenarios damage

Mode *i-j*	Without defect	Corroded area 30x30x 1 mm	Corroded area 40x40x1 mm	Corroded area 50x50x 1 mm	Corroded area 30x30x 1 mm	Corroded area 40x40x1 mm	Corroded area 50x50x 1 mm
	Frequency [Hz]				Relative frequency shift [%]		
1-1	47.803	47.770	47.749	47.731	0.208	0.317	0.396
1-2	61.898	61.903	61.899	61.892	-0.006	0.008	0.037
1-3	87.066	86.957	86.884	86.807	0.332	0.511	0.661
1-4	123.154	123.160	123.133	123.076	0.013	0.072	0.192
1-5	169.660	169.398	169.230	169.059	0.352	0.508	0.605
1-6	226.232	226.232	226.122	225.903	0.037	0.164	0.407
1-7	292.687	292.268	292.008	291.762	0.293	0.386	0.392
1-8	368.883	368.858	368.568	368.000	0.065	0.267	0.639
1-9	454.768	454.214	453.897	453.614	0.226	0.246	0.181
2-1	124.429	124.436	124.400	124.329	0.010	0.089	0.238
2-2	138.218	138.149	138.076	137.984	0.082	0.168	0.274
2-3	161.926	161.946	161.910	161.835	0.002	0.062	0.177
2-4	195.980	195.794	195.619	195.400	0.160	0.300	0.476
2-5	240.470	240.521	240.472	240.370	-0.009	0.041	0.136
2-6	295.290	295.083	294.859	294.577	0.127	0.249	0.402
2-7	360.278	360.393	360.324	360.169	-0.022	0.024	0.114
2-8	435.287	435.138	434.894	434.565	0.076	0.171	0.297
2-9	520.223	520.466	520.396	520.209	-0.043	-0.011	0.044
3-1	239.634	239.221	238.943	238.671	0.361	0.524	0.611
3-2	253.438	253.518	253.508	253.482	-0.039	-0.025	-0.002
3-3	276.787	276.481	276.290	276.125	0.248	0.312	0.291
3-4	310.031	310.131	310.073	309.954	-0.025	0.018	0.099
3-5	353.362	353.170	353.041	352.959	0.141	0.134	0.031
3-6	406.869	406.997	406.852	406.571	-0.006	0.077	0.219
3-7	470.537	470.396	470.295	470.251	0.082	0.029	-0.124
4-1	393.100	393.101	392.738	392.055	0.057	0.296	0.733
4-2	406.970	407.084	407.028	406.955	-0.033	0.000	0.037
4-3	430.281	430.317	429.982	429.361	0.046	0.238	0.555
4-4	463.239	463.211	463.008	462.757	0.034	0.106	0.195
4-5	505.995	506.116	505.761	505.113	0.029	0.199	0.476

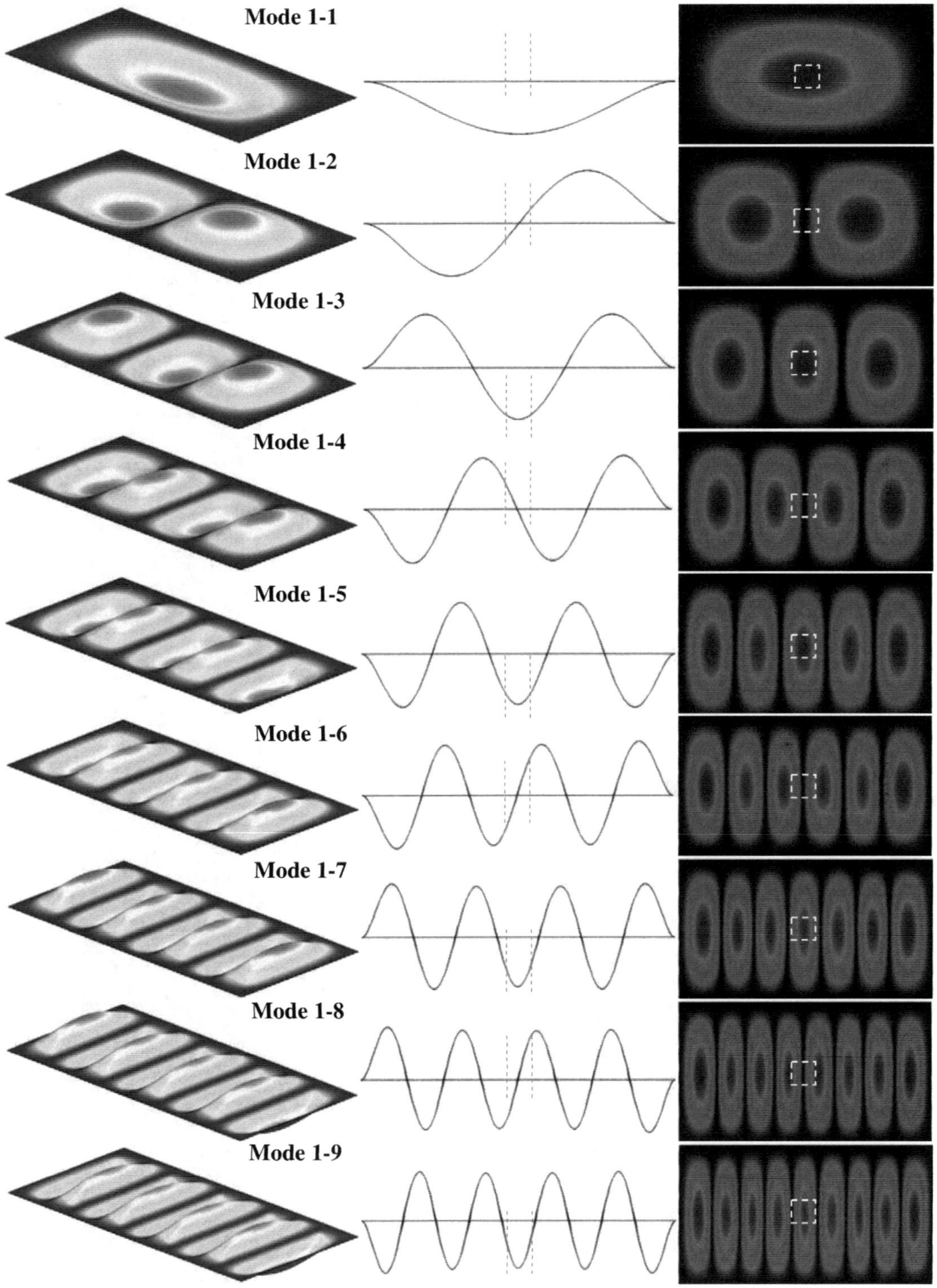

Figure 4. Vibration modes 1-1 to 1-9 (isometric and profile view) and detail of the central region.

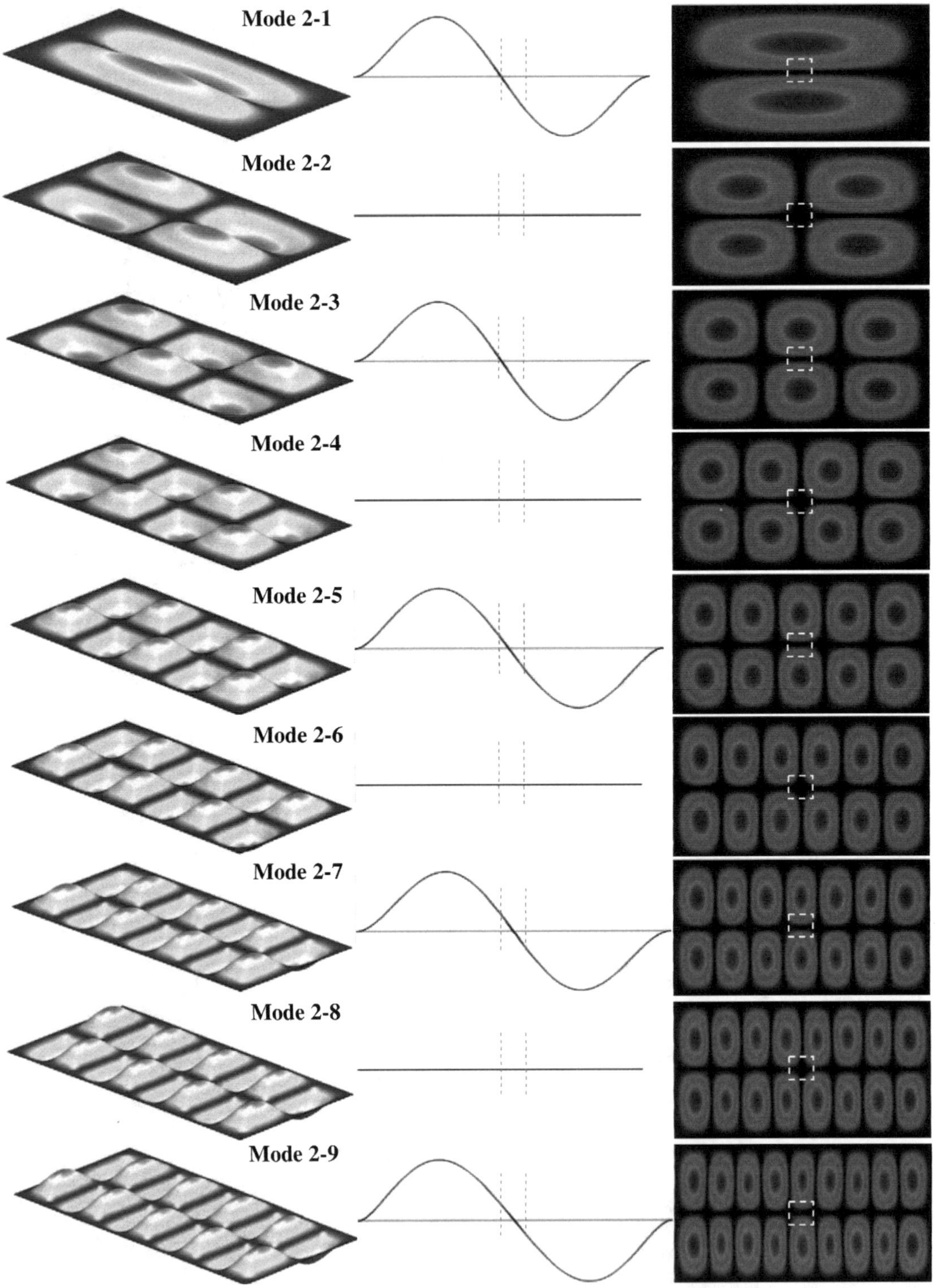

Figure 5. Vibration modes 2-1 to 2-9 (isometric and profile view) and detail of the central region.

Typical mode shapes for modes $i = 1$ and $j = 1\ldots9$ are presented in figure 4. On the left column, the axonometric views are presented, while in the central column the profile view is depicted. It is remarkable that, for the even modes j, the displacement of the centrally located corrosion area is inexistent. Opposite, for the odd modes j the displacement is significant. Similar is the plate behavior for all combinations including odd modes i.

In contrary, for the combination of even modes i, the odd modes j induce important displacement, while for the even modes j the plate's central area displacement is null. This is depicted in figure 5 for the modes $i = 2$ and $j = 1\ldots4$.

The right column in figure 4 illustrates the displacements achieved by the plate's central area. Analyzing the displacements in corroboration with the rotations induced in this region, one can qualitatively evaluate the attained bending moment and consequently the stored energy for the healthy plate. By occurrence of corrosion, thus thickness reduction, the frequencies will change in concordance with this locally stored energy. The relative frequency shifts, shown in figure 6, reflect this phenomenon.

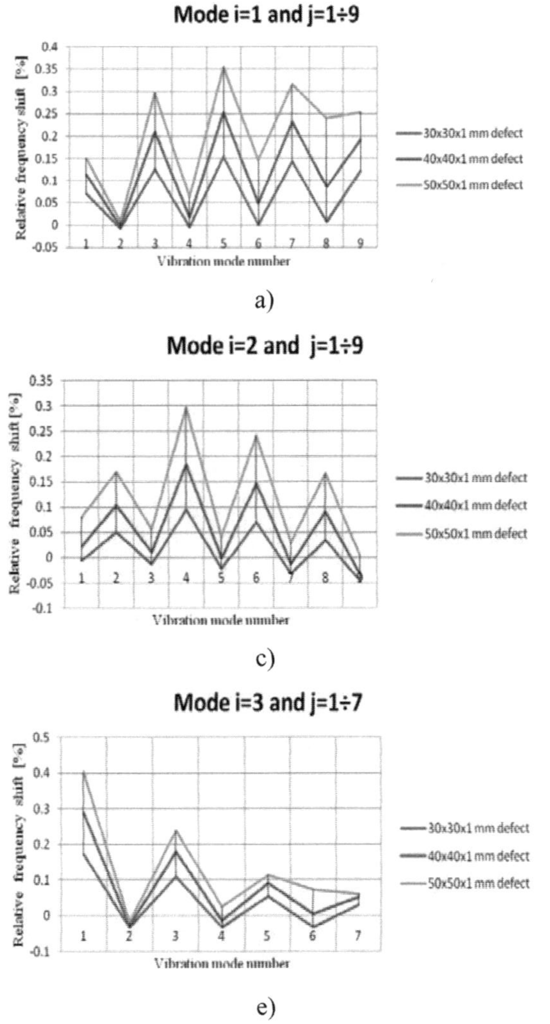

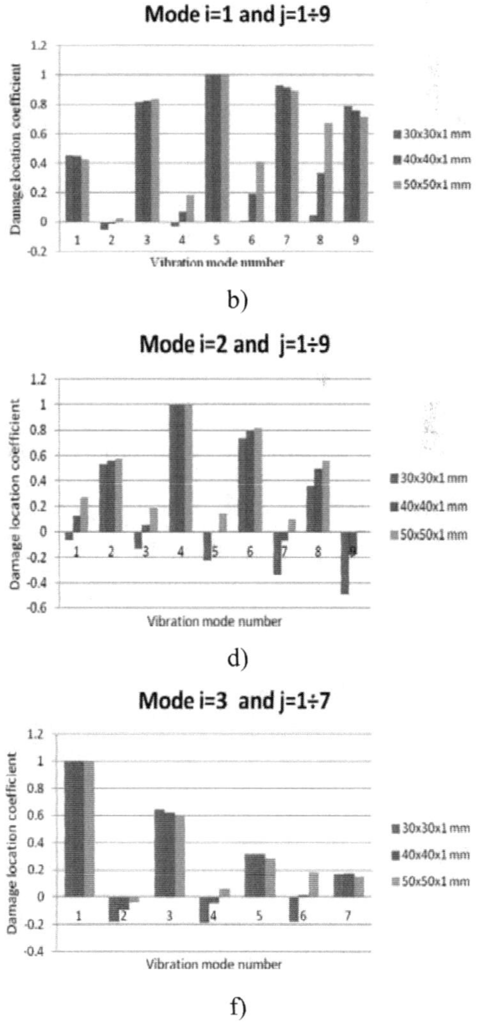

Figure 6. Relative frequency shifts and subsequent damage location coefficients.

Figure 7. Relative frequency shifts and subsequent damage location coefficients.

It is obvious that the frequencies admit changes with respecting to the rules imposed by the physical laws. Each corroded region produce different combination of frequency changes [10], thus these can be used as a benchmark in damage detection. One advanced solution is to normalize the sequence of relative frequency shifts, in order to obtain Damage Location Coefficients (see figure 7) that in a manner are similar with those designed for beams.

5. Conclusions

This paper analyses the frequency changes occurring in corroded rectangular plates, with a special focus on the case of a central damaged area. Interpretation of the results emphasizes two divergent tendencies: (i) the frequency decreasing due to stiffness (thickness) reduction and (ii) the frequency increasing due to loss of mass. The four definitory parameters of these changes are damage location, damaged surface, damage shape (i.e. the ratio between edges of the rectangular area) and thickness reduction. By normalization, the variables can be taken off and the number of parameters is consequently reduced, as well as in the case of defining damage location coefficients by suppressing the influence of corroded thickness. Also, it is possible to define valid patterns for any damage scenario by performing a set of simulations, and they can be assumed as reference in damage detection processes.

Acknowledgement

The work has been funded by the Sectoral Operational Programme Human Resources Development 2007-2013 of the Romanian Ministry for European Funds through the Financial Agreement POSDRU/159/1.5/S/132395.

References

[1] Doebling S W, Farrar C R, Prime M B and Shevitz D 1996 *Damage Identification and Health Monitoring of Structural and Mechanical Systems From Changes in their Vibration Characteristics: A literature Review* Los Alamos National Laboratory Report LA-13070-MS

[2] Morassi A and Vestroni F 2008 *Dynamic Methods for Damage Detection in Structures* (Vienna: Springer)

[3] Araújo dos Santos J V, Maia N M M, Mota Soares C M and Mota Soares C A 2008 *Structural Damage Identification: A Survey* in: B.H.V. Topping, M. Papadrakakis (Eds.) Trends in Computational Structures Technology (Stirlingshire: Saxe-Coburg Publications) 1-24

[4] Friswell M I 2007 Damage identification using inverse methods *Phil. Trans. R. Soc. A* **365** 393-410

[5] Gillich G R and Praisach Z I 2014 Modal identification and damage detection in beam-like structures using the power spectrum and time–frequency analysis *Signal Process.* **96**(Part A) 29–44

[6] Gillich G R, Praisach Z I and Negru I 2012 Damages influence on dynamic behaviour of composite structures reinforced with continuous fibers *Mater. Plast.* **49**(3) 186-191

[7] Praisach Z I, Gillich G R and Birdeanu D E 2010 Considerations on natural frequency changes in damaged cantilever beams using FEM *Latest trends on engineering mechanics, structures, engineering geology* 214-9

[8] Gillich G R, Praisach Z I, Wahab M A and Vasile O 2014 Localization of transversal cracks in sandwich beams and evaluation of their severity *Shock Vib.* **9**(4-5) 607125

[9] Gillich G R and Praisach Z I 2013 Detection and Quantitative Assessment of Damages in Beam Structures Using Frequency and Stiffness Changes *Key Eng. Mat.* **569** 1013-20

[10] Tufoi M, Hatiegan C, Vasile O and Gillich G R 2013 Dynamic Analysis of Thin Plates with Defects by Experimental and FEM Methods *Rom. J. Acoust. Vib.* **10**(2) 83-8

Crack diagnosis of metallic profiles based on structural damage indicators

A Preisler[1], C Steenbock[2] and K-U Schröder[1]

[1] RWTH Aachen University, Institute of Structural Mechanics and Lightweight Design, Wüllnerstr. 7, 52062 Aachen, Germany

[2] Altair Engineering GmbH, Hildesheimerstr. 98b, 30173 Hannover, Germany

E-mail: andreas.preisler@sla.rwth-aachen.de

Abstract. Structural Health Monitoring (SHM) faces several challenges before large-scale industrial application. First of all damage diagnosis has to be reliable. Therefore, common SHM approaches use highly advanced sensor techniques to monitor the whole structure on all possible failures. This results in an enormous amount of data gathered during service. The general effort can be drastically reduced, if the knowledge achieved during the sizing process is used. During sizing, potential failure modes and critical locations, so called hot spots, are already evaluated. A very sensitive SHM system can be developed, when the monitoring effort shifts from the damage to its impact on the structural behaviour and the so called damage indicators. These are the two main components of the SmartSHM approach, which reduces the monitoring effort significantly. Not only the amount of data is minimized, but also reliability and robustness are ensured by the SmartSHM approach.

This contribution demonstrates the SmartSHM approach by a cracked four point bending beam. To show general applicability a parametric study considering different profiles (bar, box, I, C, T, L, Z), crack positions and lengths has been performed. Questions of sensitivity and minimum size of the sensor network are discussed based on the results of the parametric study.

1. SmartSHM

In order to ensure structural integrity, Structural Health Monitoring (SHM) must be capable of monitoring the damage state of a structure in service during the whole lifetime. Potential damages have to be detected before they become critical to the structural integrity. In addition to the information whether the structure is damaged or not, informations about the location of damage and the corresponding mode and size has to be evaluated. This leads to a full Level 4 Diagnosis [1]. But a Level 4 Diagnosis does not enable to assess a damage. For damage assessment the loading state also has to be known. With that set of information the remaining lifetime of the damaged part can be predicted in order to define and schedule repair actions.

A common SHM approach is to design a system that is able to monitor the entire structure. The target of such a system is to have a full understanding of the current state of the structure in order to detect every potential failure on all possible positions. In most cases, this approach leads to a rather complex sensor network and an enormous amount of data which has to be recorded and analysed. However, most structures which are in use nowadays are analysed in detail during sizing process. Due to the structural analysis, stress and strain state of the structure under given load cases are already known. Critical areas of the structure, so called hot spots, and the corresponding predominant failure modes are investigated while sizing. This knowledge can be used in order to set up a so called SmartSHM system [2]. Within this approach, the monitoring focuses on the detection of the expected damages at certain hot spots of the structures.

The second important element of SmartSHM, besides focussing on hot spot areas with their corresponding failure modes, is the application of structural damage indicators. From sizing process, stress and strain state of the theoretically intact structure are known. In general, a damage results into a characteristic change of the structural behaviour, e. g. a crack causes a redistribution in stress at tensile loading while a delamination causes a stiffness reduction. Relating the current structural behaviour to a reference state, a damage can be detected by evaluating the change. That change of the structure's response can be described with the structural damage indicator

$$\iota_{\mathrm{d}} = \frac{X(x,t) - X_0(x)}{X_0(x)}, \tag{1}$$

where $X(x,t)$ describes the structural parameter at the current structural state and $X_0(x)$ at the reference state, which is in general the intact structure at service launch. That structural damage indicator is the monitored object. So SmartSHM does not focus on the direct measurement of damage itself, but on its implication on the structure's response.

The third element, which makes the SmartSHM approach smart, is the way the reference state is chosen. Equation (1) indicates that the damage indicator ι_{d} is very sensitive to a structural parameter with zero signal at reference state. In that case even a weak signal, i. e. a small change in the structural parameter $X(x,t)$ due to a damage, leads to a high damage indicator ι_{d}. Therefore it is more efficient to monitor the location with the most severe effect on the damage indicator rather than near to the damage. A further advantage of the presented approach is that all sensors will measure a zero signal, if the structure is undamaged. Therefore, sending and processing of data is not necessary and the amount of monitoring data is reduced significantly. With some further investigation of the characteristics of the damage indicator, a damage can be localized, qualified and quantified [3].

SmartSHM is not designed to detect uncertainties like misuses or impacts. However, during the sizing process uncertainties are already taken into account by the implementation of design load and design values along with safety factors. Still those damages can be detected by an additional global monitoring. Since the probability of those accidental events is rather small and as they represent extreme events, a level 1 diagnosis (i. e. the information whether the structure is damaged or not) is feasible. In this case, a damage signal within the global monitoring would lead to an immediate inspection.

2. Application of the SmarthSHM approach on a four point bending beam

The cracked four point bending beam is used as striking example for SmartSHM. The general test setup is shown in Figure 1. The maximum bending moment is reached between both load application points. This area is identified as hot spot and therefore, SmartSHM focuses on monitoring the middle span. Considering linear-elastic material response, the stress distribution is linear over cross-section as shown in Figure 2 (left). The upper flange is compression loaded. Therefore, the corresponding failure mode is buckling. The lower flange is tension loaded. Its corresponding failure mode is cracking. This contribution focuses on a potential crack in the lower flange.

At this point the hot spot and the corresponding failure mode of the structure are known. Hence, the crack in the lower flange of the mid span is the most likely damage which can occur at standard conditions of service. The next step following the SmartSHM approach is to evaluate the change of the structure's response due to that damage and to derive the most suitable damage indicator. For that reason a crack is introduced in the lower flange of the mid span

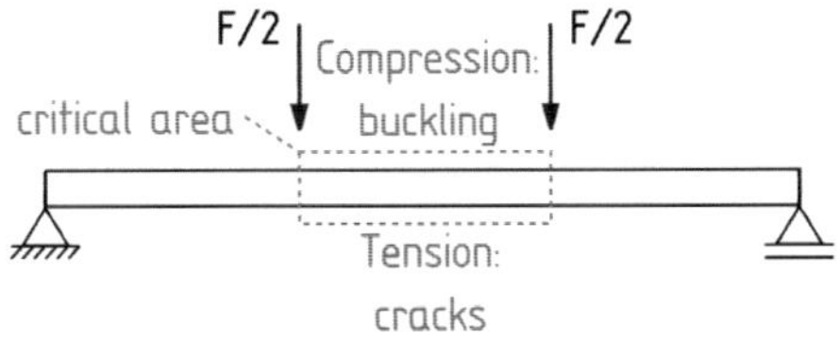

Figure 1. General test set-up for four point bending specimens

and the resulting stress redistribution is analysed. Figure 2 (right) shows the expected stress distribution.

A possible damage indicator is the decrease of strain in the lower flange directly adjacent to the crack. Applying a strain gauge at this position, comparable high change in the absolute value will occur. However, the reference value is also maximal. Being aware of the conclusions drawn from Equation (1), it is recommended to look for a structural parameter which is (almost) zero in the undamaged (reference) state. By definition, the neutral axis is free of stress and strain under pure bending. If a crack occurs, the neutral axis shifts away from the crack and the former neutral axis is exposed to stress and hence to strain. Therefore, longitudinal strain at the original neutral axis is chosen to be the damage indicator within this example. The general suitability of this damage indicator was demonstrated numerically end experimentally in [3]. However, this approach raises a lot of additional questions. First of all, does the approach also work for other cross-section types than shown in [3]? Second, within which distance is the damage indicator influenced by the crack? And are the characteristics of the damage indicator constant with varying distance to the crack? These questions are answered in this contribution on the base of extensive numerical investigations, considering different profiles (bar, box, I, C, T, L, Z) and crack positions.

3. FEM model of the parametric study

The above mentioned four point bending beam with crack is computed with the Finite Element Method. The beam is meshed with second order solid elements. The mesh is highly refined in close proximity to the crack in order to reproduce stress peaks at the crack tip. The crack itself is modelled by node separation. The model includes plasticity. However, crack growth is not considered. The material data is summarized in Tab. 1. A static load F, smaller than failure load, is applied in the shear centre of the cross-section by distributing coupling constraints. Weighting factors are chosen to be linear. The supports are modelled with kinematic coupling constraints, ensuring a uniform support load distribution.

Within the parametric study, the applied moment is assumed to be constant. The cross-section properties are derived under the condition, that in every case the longitudinal stress of the undamaged cross-section is equal to +100 MPa at the potential crack position. Therefore, the section modulus is constant. In addition, the length-to-height-ratio $\frac{l}{h}$ of the beam is assumed

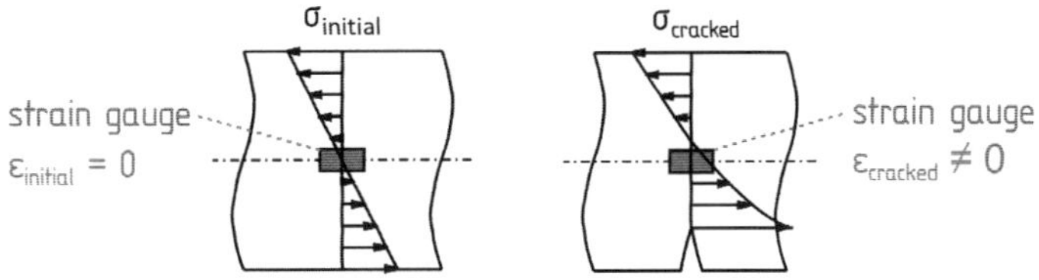

Figure 2. Stress state for a non-damaged (left) and a damaged (right) cross-section [3]

E - Young's modulus	74500 MPa
ν - Poisson's ratio	0.33
$R_{p0,2}$ - Yield strength	290 MPa
R_m - Ultimate tensile strength	395 MPa
A - Ultimate tensile strain	12%

Table 1. Material data for Aluminum 2024-T3 (extrusion) [4]

to be constant. This leads to a beam length l which is dependent on the height h. Since the moment is assumed to be constant, the force F is dependent on the height h as well. The selected profiles and the corresponding geometries are shown in Figure 3. The derived values are summarized in Tab. 2.

In the first step of the parametric study, the crack is placed in the middle section. Afterwards, two additional crack positions are introduced. The second crack is shifted towards one load application point. The last crack is applied outside of the critical area where bending moment is decreasing linearly. A distance of h to the load application is chosen in order to reduce the influence of the disturbance. The considered crack positions are shown in Figure 3.

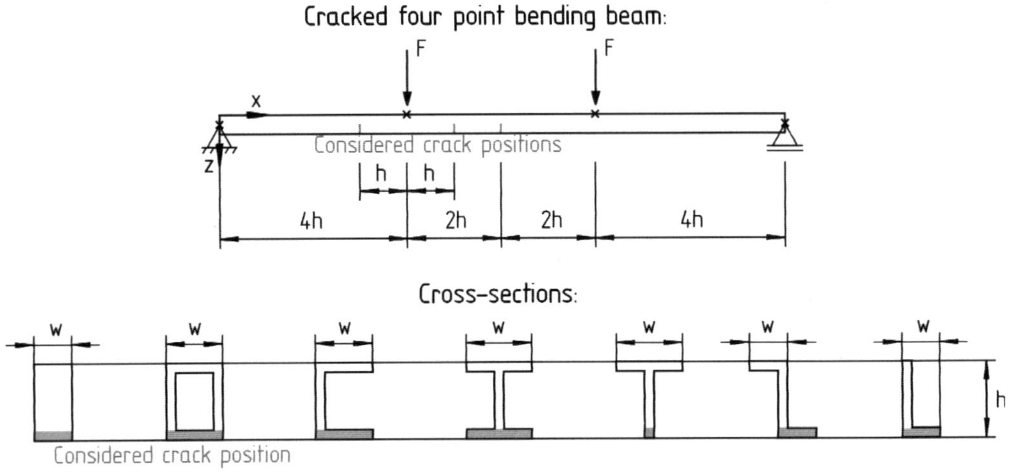

Figure 3. Geometry of the four point bending beam and considered crack positions

Profile	h - Height	w - Width	t - Thickness	l - Total length	F - Force
Bar	35 mm	17.5 mm	-	420 mm	2620 N
Box1	40 mm	40 mm	2 mm	480 mm	2293 N
Box2	50 mm	25 mm	2 mm	600 mm	1834 N
C	42 mm	42 mm	2 mm	504 mm	2183 N
I	42 mm	42 mm	2 mm	504 mm	2183 N
T	68 mm	68 mm	3 mm	816 mm	1349 N
Z	60 mm	30 mm	3 mm	720 mm	1528 N
L	84 mm	28 mm	4 mm	1008 mm	1092 N

Table 2. Variation of parameters used in the study

Journal of Physics: Conference Series **628** (2015) 012006 doi:10.1088/1742-6596/628/1/012006

4. Results and evaluation

The presentation and evaluation of the results are divided into three parts. Since the I-profile is the most suitable cross-section type for bending problems, it is analysed in detail in the first part. In the second part, the selected cross-sections are compared to each other. Afterwards the influence of the crack position is discussed.

4.1. I-profile

Figure 4 shows the strain at the former neutral axis within a distance of $\pm h$ to the crack in the middle section. As explained earlier, a crack leads to a stress redistribution which affects the former neutral axis over a certain distance. A change of the damage indicator can be observed within a distance of $\pm h$ to the cracked cross-section. However it becomes smaller for higher distances. This trend complies Saint-Venant's Principle. In addition the extend of redistribution shows a direct correlation to the damage size.

Multiple effects are responsible for the extend of redistribution. A crack in general leads to a separation and therefore to a reduced effective cross-section. This influences the effective centroid and the section modulus. However this approach is only valid for an undisturbed cross-section. In case of a sudden change of cross-section due to crack, stress concentration at the crack tip occurs. These two effects are counteracting and have to be superimposed. For larger cracks the effect of the shifting centroid is predominant. This can be easily seen for crack lengths in range of 2 mm and 4 mm. There is a major jump in the level of stress redistribution at a crack length of 2 mm. This curve represents a complete cut-through of the flange. Therefore, the crack introduces a severe change of effective cross-section and hence a severe change of centroid position. In addition, plasticity occurs at the crack tip. However, for smaller crack lengths the influence of plasticity on the strain at the former neutral axis is low. For large crack lengths plasticity shows a significant impact on the results. Stress concentration at the crack tip is limited and a large area with high stresses and strains occurs, resulting in high strain at the former neutral axis within the cracked cross section. This effect can be easily seen for a crack length of 5 mm.

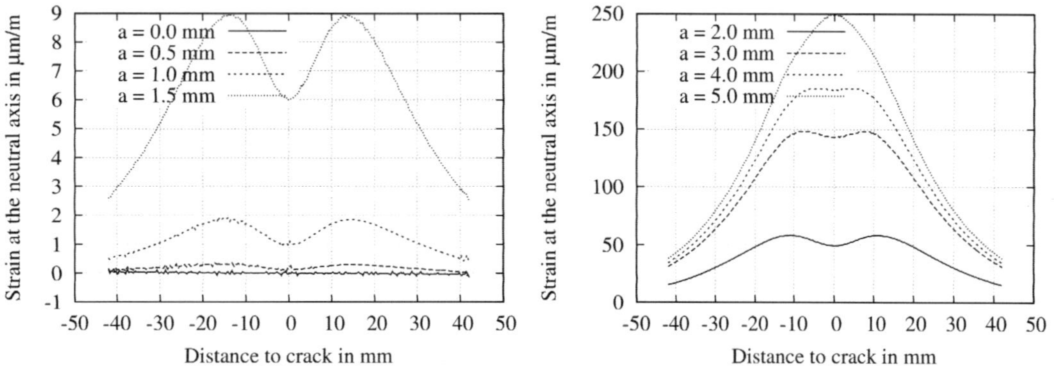

Figure 4. Strain at the former neutral axis for different crack lengths (I-profile)

4.2. Comparison of cross-section types

The change of strains of all considered cross-sections is compared in Figure 5, where the change of strain relative to each maximum value is plotted for a crack length of 4 mm. The maximum absolute values of strain are shown in Tab. 3. The curves are derived at the z-coordinate of

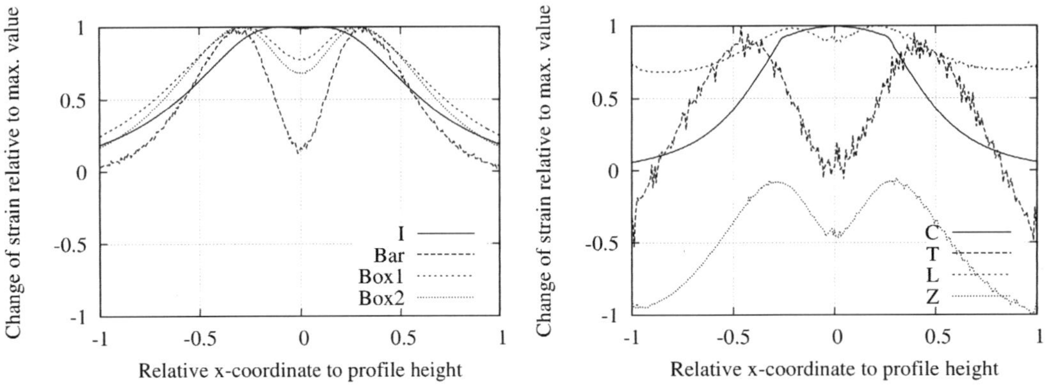

Figure 5. Strain relative to each maximum value for a crack length of 4 mm

Profile	Max. change of strain	Profile	Max. change of strain
I	185,4 μm/m	C	336,4 μm/m
Bar	5,0 μm/m	T	0,5 μm/m
Box1	51,9 μm/m	L	6,7 μm/m
Box2	18,2 μm/m	Z	8,1 μm/m

Table 3. Maximum absolute value of strain at the monitored positions

centroid. In most cases this point coincides with the neutral axis. However, for L and Z-profile oblique bending is present due to the rotation of principal axis of inertia. Therefore the measuring points are not strain-free in the undamaged case. Still deviations in proximity of crack are easily visible. This attests feasible and robust application of the general approach, even if a strain free reference state is not present (e.g. due to sensor misplacement or interfering loads).

Most curves shown in Figure 5 have comparable behaviour. The maximum deviation is reached within certain distance to the cracked cross-section. Within the cracked cross-section high stress concentrations occurs at the crack tip. Therefore, the amount of measured strain is reduced. However, for I and C-profile stress concentration leads to significant plasticity. Those two curves reach their maximum deviation within the cracked cross-section. In case of flange cut-through a large amount of stress is redistributed. For I and C-profile this stress is transferred to the web, leading to high stress concentration and plasticity. Those two profiles show very high sensitivity to crack (see Tab. 3). Sensitivity for T-profile is rather poor. Therefore some numerical scatter is visible within the curve. However the influence of a crack within a T-profile is very low. In this case a crack just cuts through a very small area and is less critical. The change of centroid and section modulus is small.

4.3. Influence of the crack position

The influence of the crack position is investigated for three different positions: $x_{Crack} = 3h$, $5h$ and $6h$ (reference crack). Strain at the former neutral axis for those three models is plotted in Figure 6. The cracks at $5h$ and $6h$ are within the hot spot area. Therefore stress and strain redistribution is similar. The curve for $5h$ indicates larger deviations at the left end due to boundary conditions. At this point the load F is applied to the cross-section with distributing coupling. A comparable behaviour is present for a crack position of $3h$. However the level of the absolute change of strains is lower due to a reduced acting bending moment by a factor of

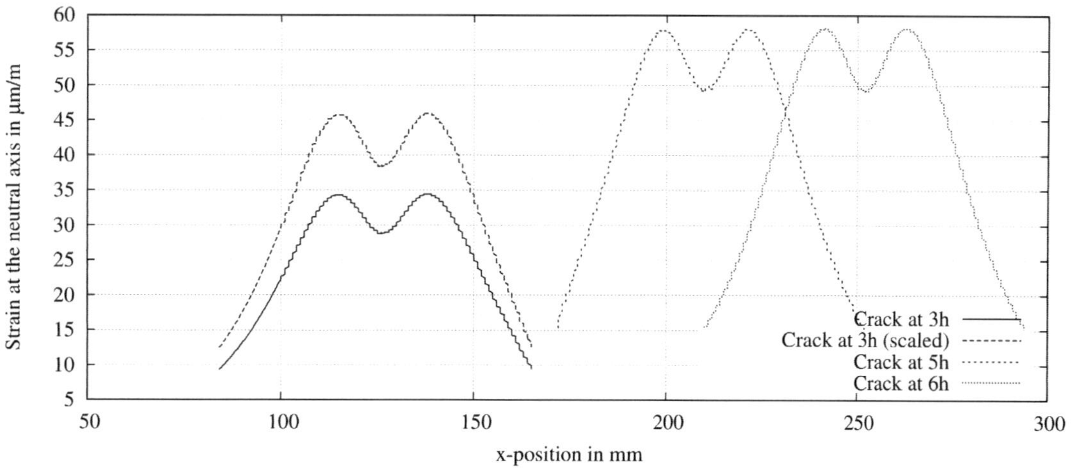

Figure 6. Strain at the former neutral axis for a crack length of 2 mm for chosen crack positions

$\frac{3}{4}$. Therefore, the curve is scaled with a factor of $\frac{4}{3}$ in order to compensate the difference of bending moment. The remaining deviations indicate the effect of plasticity. Due to non-linear material, the relationship between load and strain is non-linear and larger strains occur.

5. Level 4 damage diagnosis and sensor layout

As shown in the previous section, a crack introduces a characteristic disturbance of strain on the former neutral axis. There is certain distance from cracked cross-section, where this damage indicator is sensitive to the crack. The value of the damage indicator is directly correlating to the crack length, the distance to the cracked cross-section and the acting bending moment. Therefore, only one monitoring spot is insufficient to determine a distinct crack length and position. The same level of change in strain under given load might be achieved at a location which is close to a small crack and at a location which has a high distance to a huge crack. If two monitoring locations with a known distance to each other are used within the sensitivity range of the damage, a distinct result of crack length and position can be obtained [3]. In addition, a Loads Monitoring is needed. For the example presented within this contribution, a feasible Loads Monitoring can be achieved via two strain gauges within one cross-section. The gradient between both measurements is in direct relationship to the applied bending moment.

In earlier studies, strain at the former neutral axis was approximated by a second order polynomial, which was superimposed with a logistic function. The coefficients of the approximation were derived by numerical studies, which were verified by test results. With that approach multiple functions with similar shape for certain crack lengths were derived [3]. They can be modified by the effect of plasticity.

To perform a Level 4 Diagnosis, data of at least two sensors with given distance between both sensors is required alongside Loads Monitoring. A rather simple script can be created which perform a loop over the whole set of approximation curves and determine the best fitting curve. This curve refers to a certain crack length and is used in a second step to evaluate the distance of the crack to both sensor positions. This algorithm allows a very efficient evaluation of the sensor signal. Since critical crack length is known from sizing, it contains an intrinsic damage assessment. The algorithm is illustrated in Figure 7.

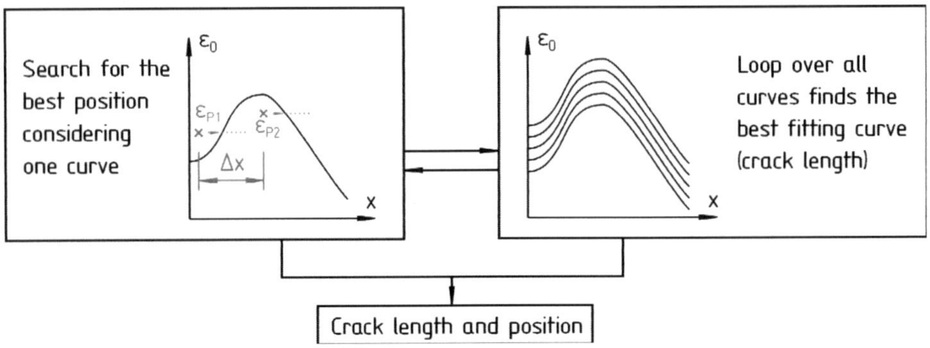

Figure 7. Illustration of the algorithm to determine crack length and position

6. Conclusion

In this contribution, the SmartSHM approach is presented on the basis of a cracked four point bending beam. SmartSHM uses the knowledge obtained during sizing process and within structural analyses. The monitoring effort is focussed on critical areas and the corresponding predominant failure modes. Afterwards, the investigation of the impact of the damage on the structural behaviour leads to the most suitable and sensitive damage indicator, which can be used to design an efficient and robust sensor network. In order to cover accidental damages, the presented SmartSHM can be augmented with a simple global monitoring.

Applicability and sensitivity of the SmartSHM approach is discussed on basis of a parametric study considering varying metallic profiles. For most cases, high sensitivity is achieved by the use of strain at the former neutral axis as damage indicator under pure bending. Due to a zero reference signal, even small deviations are easy to determine. The results of the parametric study show a direct correlation of the chosen damage indicator to the length and distance of the cracked cross-section. In addition, the applied bending moment needs to be derived with Loads Monitoring. For severe stress redistribution, the effect of plasticity has to be considered.

In the presented example, two sensor positions within the influenced range are needed alongside loads monitoring. The data can be processed by a set of curves. A rather simple script is able to identify the curve matching sensor signals and conclude length and position of the crack. Since critical damage size is known from sizing, this approach enables an intrinsic damage assessment.

Within future investigations, SmartSHM will be applied on more complex structures. Still, hot spots, the corresponding failure modes and the most suitable structural damage indicator can be derived by the use of structural analyses.

References

[1] A. Rytter. *Vibration Based Inspection of Civil Engineering Structures.* PhD thesis, Department of Building Technology and Structural Engineering, Aalborg University, Denmark, 1993.

[2] K.-U. Schröder, C. Viechtbauer, and M. Schagerl. Identification and Monitoring of Structural Parameters as Damage Indicators for Plates in the Post-Buckling Regime. In *Structural Health Monitoring 2013 - A Roadmap to Intelligent Structures*, volume 1, pages 1137–1143, September 2013.

[3] K.-U. Schröder, A. Preisler, C. Viechtbauer, and M. Schagerl. On the Damage Diagnosis based on Structural Analysis Data. In *EWSHM - 7th European Workshop on Structural Health Monitoring (2014)*, July 2014.

[4] *Handbook Structure Analysis.* Industry Committee for Structure Analysis, 2009.

How woodpecker avoids brain injury?

C W Wu, Z D Zhu and W Zhang[1]
[1]State Key Lab of Structural Analysis for Industrial Equipment, Department of Engineering Mechanics, Dalian University of Technology, Dalian 116024, China

Email: wei.zhang@dlut.edu.cn

Abstract. It has long been recognized that woodpecker is an excellent anti-shock organism, as its head and brain can bear high deceleration up to 1500 g under fast pecking. To investigate the mechanism of brain protection of woodpecker, we built a finite element model of a whole woodpecker using computed topography scanning technique and geometry modeling. Numerical results show that the periodical changing Young's modulus around the skull affects the stress wave propagation in head and makes the stress lowest at the position of the brain. Modal analysis reveals the application of pre-tension force to the hyoid bone can increase the natural frequency of woodpecker's head. The large gap between the natural and working frequencies enable the woodpecker to effectively protect its brain from the resonance injury. Energy analyses indicate the majority of the impact energy (99.7%) is stored in the bulk of body and is utilized in the next pecking. There is only a small fraction of it enters into the head (0.3%). The whole body of the woodpecker gets involved in the energy conversion and forms an efficient anti-shock protection system for the brain.

1. Introduction

Woodpeckers have excellent anti-shock ability to adapt high-speed and high-acceleration pecking activities. Many observations have showed that the woodpecker pecks at a frequency of 18–28 Hz with a maximum speed 5–7 m/s, and the maximum deceleration of a collision in less than one millisecond can reach 600–1500 g [1-3]. The special anti-shock and energy-absorption function have received much attention, and numerous efforts have been devoted to understand the mechanism.
Gibson [4] compared the woodpecker's head structure with human being's, and found that small sized head, short impact duration and large contact area between brain and skull can protect the brain from damage of shock. Zhou et al. [5] dissected a grey woodpecker and found that the hyoid, which is the specific structure of woodpecker, has excellent mechanical properties with Young's modulus of 3.72 GPa and tensile strength of 131 MPa. The hyoid, with one end fixing on the right nostril and the other end binding the skull tightly and stretching out from the mouth, is four times longer than the beak, and

the binding of the hyoid to the skull reduces the impact stress in the head [6].

To study the impact response of the woodpecker head, numerical simulations of finite element (FE) were introduced. Oda et al. [7] set up a stereo lithography head model which is three times larger than the actual head, and measured the strain in different positions in the head. Then a 2D FE model was built and proved that little cerebral liquid and the existence of hyoid bone can effectively protect its brain from shock damage. Wang et al. [8] experimentally obtained the pecking force of the woodpecker was about 8 N, which is useful for numerical simulation. They built a 3D FE model of the woodpecker head based on micro-computed topography (CT) scanning and compared the results of three models with different lengths of beaks when impacting the rigid wall. They explained that the majority of the pecking force was carried by the low beak to the neck and the brain avoided being hurt. Zhu et al. [9] observed that the upper beak is longer than the lower one and interpreted the mechanism of anti-shock in terms of stress wave propagation. They established a 3D FE model of the woodpecker head including the contact relationships among the inner structures, and studied the dynamic response of the head on pecking trees. They concluded that the stress wave mainly propagated from the upper beak and the head structure and the hyoid helps spreading the stress wave, the viscoelasticity of biomaterials is responsible for the decrease of the stress value.

To date, however, the papers published in the literatures merely focus on the stress and strain analyses of a sole woodpecker head under a single impact. Little is known on the frequency response of the head and dynamic response under successive process of pecking. In light of these we herein built a whole woodpecker model including the body based on the CT scanning technology. Different from the existing single head model, the whole model has the advantages as follows: (i) Constraints can be applied at the claws and tail, where are far away from the brain. Thereby the effect of localized stress concentration initiated by constraints could be reduced dramatically; (ii) The successive pecking process is more close to the actual pecking of woodpecker, which could truly reflect the distribution and accumulation of stress. Based on this model, the anti-shock characteristic of the woodpecker was studied from the aspects of modal analysis, stress spectrum analysis and energy analysis.

2. Material property

The woodpecker skull is a small and spheroid-like structure and it is hard to obtain the mechanical property by conventional tension test. In this paper we used nano-indentation equipment (Triboindenter, Hysitron TI-950, Hysitron, USA) to obtain the property of skull [10-12]. Twenty-eight samples with the average size of 2.5 mm×2.5 mm were cut from the skull and washed with deionized water. Then the sample was indented with a trapezoidal load function from 50 μ N to 1,250 μ N. Figure 1 shows the positions of the samples on the sagittal section of the woodpecker head and the local Young's modulus measured by nano-indentation. The results showed that the Young's modulus changed ups and downs with the average value of 6.6 GPa and the coefficient of variation of measured Young's modulus of each specimen ranges from 16% to 40% [13, 14].

The Young's modulus of woodpecker skull has large standard deviation and changes in sinusoidal form. How the changing modulus will affect the stress wave propagation in the skull when woodpecker pecks? Limited by the huge amount of the meshes in the head, it is difficult to discuss the effect in the actual head model, thus we built a simplified model of a viscoelastic bar. Figure 2 shows the model of stress wave propagation in the viscoelastic bar. The bar is 2,000 mm long and divided into three parts including incident part, material changing part and ending part. A projectile strikes at the start of the bar with the velocity of 7 m/s, and the boundary condition of the bar is complete free. We suppose the bar has similar material property with the bone and the material changing of the middle part is in sinusoidal form. Equations (1) and (2) give the viscoelastic property of the bar.

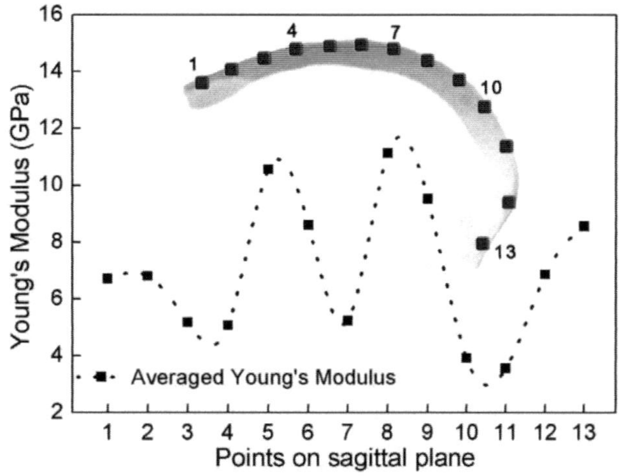

Figure 1. Positions of the samples on the sagittal section of the woodpecker head and the local Young's modulus measured by nano-indentation, the average distance between the adjacent samples is 2.5 mm.

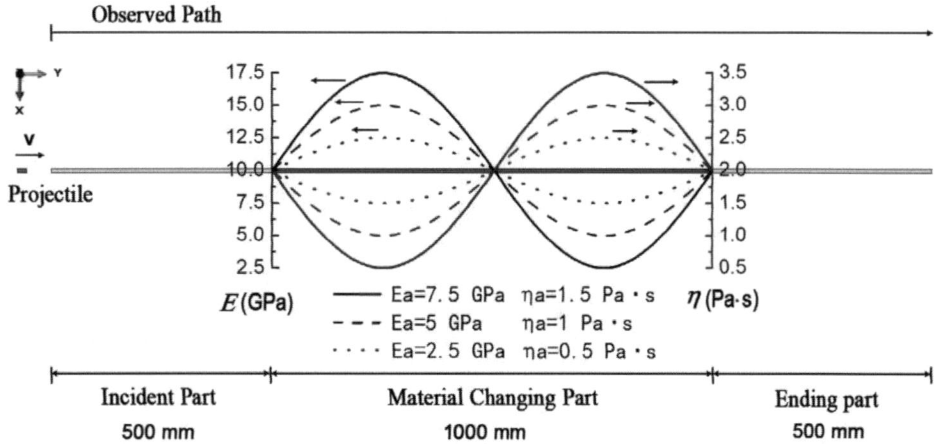

Figure 2. The model of stress wave propagation in the viscoelastic bar. The bar is impacted by a projectile with the velocity of 7 m/s in one end, the mass of the projectile is 50 g and the mass of the bar is about 600 g. The bar is divided into three parts including incident part, material changing part and ending part. The curves in the middle shows three situation of different Young's modulus and viscosity assigned in the material changing part.

$$E = E_0 + E_a \sin(2\pi \frac{kx}{l} + \varphi_E) \tag{1}$$

$$\eta = \eta_0 - \eta_a \sin(2\pi \frac{kx}{l} + \varphi_\eta) \tag{2}$$

E_0 and η_0 represent the average values of the Young's modulus and viscosity, which are 10 GPa and 2 Pa·s respectively. E_a and η_a are the changing amplitude of them, with the maximum value of 7.5 GPa and 1.5 Pa·s. According to the amplitudes of the material parameter assigned to the middle part, we set three computational situations totally. In addition, k and φ represent the cycle count in the specify distance (l=1000 mm) and phase delay. When the cycle count is large enough, the effect of phase delay could be negligible. Bone with higher Young's modulus usually contains less water. As a result, the viscosity variation of skull should demonstrate the opposite trend with respect to the Young's modulus.

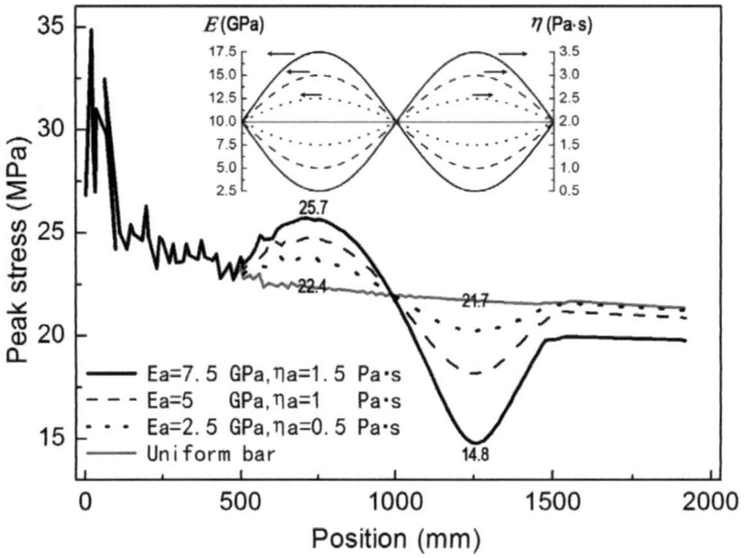

Figure 3. Material parameters assigned in the middle bar and the peak stress curves of the whole bar. The peak stress curve is composed of the maximum stress in the impact direction at different moment along the bar. The red line represents stress curve of the bar when uniform material property assigned in the middle bar. The curves have the same variation in the incident segment but change fiercely in the middle segment.

Then we discussed the effect of the changing amplitude of material parameters on the stress wave propagation and obtained the following results. Figure 3 shows material parameters assigned in the middle bar and the peak stress curves of the whole bar. Compared with the uniform distribution of material property, peak stress curves of the bar under three situations have the same variation in the incident segment but change fiercely in the middle segment. From Fig. 3 we could find that the peak stress changes in agreement with the changing of the Young's modulus. The stress is large in the high-modulus zone and small in the low-modulus zone. In Figure 3, take the maximum amplitude of the material parameter as an example, the peak stress is 25.7 MPa in the increasing segment which is 14.7% higher than the uniform distribution of material property and 14.8 MPa in the decreasing segment which is 31.8% lower than the uniform one. Therefore, the variation of the material parameter indeed affects the propagation of the stress wave and makes the peak stress smaller in the low-modulus zone than the other positions. The same conclusion can be deduced under multi-periodic conditions (where k is greater than one), and the peak stress in the final period of the material property will be the smallest in the whole bar.

For the woodpecker head, stress wave is generated by pecking at the tip of beak, and is propagated into the brain through the skull. In the direction of stress wave passing, as measured, the Young's modulus of the skull changed periodically as shown in Figure 1. Here we provide an analogy between the woodpecker head and the viscoelastic bar model as shown in Figure 4. The beak works as the incident segment of the head to make the impact stress steady, and the skull reacts as the material changing part to reduce the stress in the brain. Figure 4a shows the peak stress curve of the whole bar when material parameter changes in two periods and Figure 4b shows the changing of measured modulus of the skull on the sagittal section. According to the conclusion of the former analysis, we deduce that the peak stress of skull is smallest at the position of the brain, which is the final low-modulus zone of the material property in the skull.

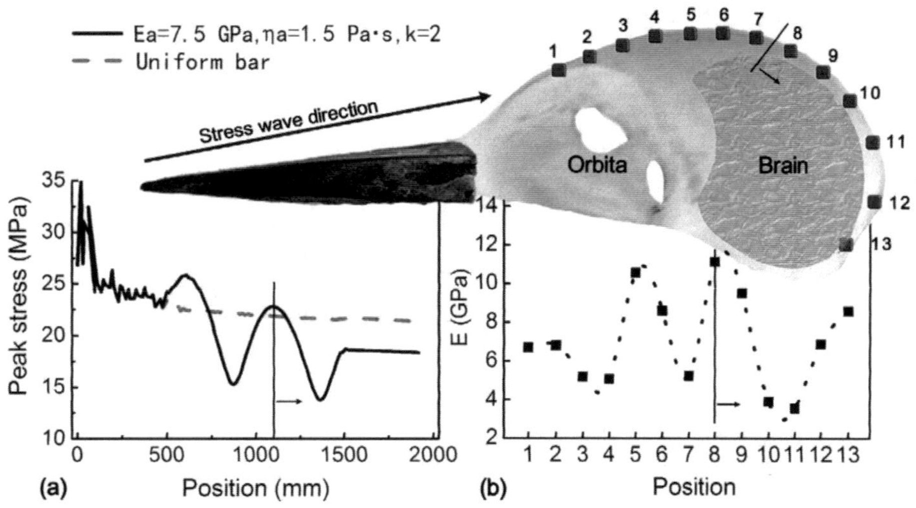

Figure 4. An analogy between the woodpecker head and the viscoelastic bar model. (a) The peak stress curve of the whole bar when material changes in two periods (E_a=7.5 GPa, η_a=1.5 Pa·s and k=2), and the arrow points is the position of the smallest peak stress in the bar; (b) The changing of the measured modulus on the sagittal section, there are two cycles of the modulus in the curve and brain is located at the final low-modulus zone of the material property in the skull.

3. FE model

The woodpecker investigated here is a female grey-faced one widely living in the Northern China. Woodpecker usually has small head, little amount of cerebrospinal fluid, big eyeballs, hard beaks and long flexible hyoid bone [15]. For this kind of bird, the hyoid bone is three or four times the length of the beak with one end fixing on the right nostril and the other end binding the skull tightly and stretching out from the mouth to reinforce the head. In order to get an accurate finite element (FE) model including these features, the procedure is as follows. The woodpecker was put on a CT scanning device (CT scanner, LightSpeed VCT XT, GE, USA) and the inner structure of head can be discerned through the CT images without breaking their outer parts [16-18]. Then Mimics and ProE software were used to establish 3D geometric configurations by integrating the scattered points of the images. After the geometric repairs, the FE model meshed by tetrahedron elements was established using Abaqus software.

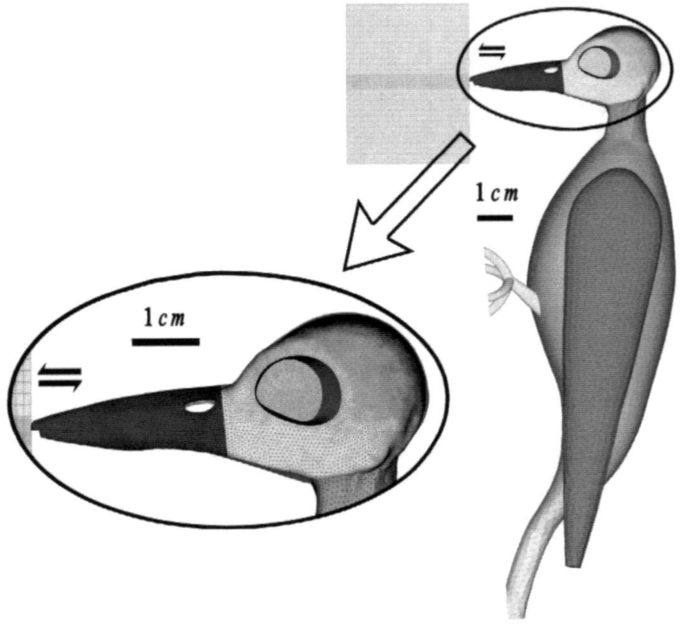

Figure 5. FE model of a whole woodpecker, the woodpecker has 1,030,000 elements in total, and the head has more than 940,000 elements which accounts for 91% of the total elements, the arrows show the moving direction of the head.

The final FE model contains more than 940,000 elements with a minimum size of 0.07 mm in the head, 70,000 coarse elements with a maximum size of 3.5 mm in the body and 20,000 elements with a minimum size of 0.16 mm in the trunk. Figure 5 shows the FE model of a whole woodpecker, as can be observed, the element size increases from the top of the head to the body gradually. The whole model could be divided into ten parts, the body is composed of parts named torso, wing, neck, claw and tail, and the woodpecker head is composed of parts named beak, hyoid, skull, dura and brain. The body parts are mainly composed of muscles and other elastic tissues, and thus are considered as elastic materials. The head is set as viscoelastic material to reflect the effect of brain on the stress propagation and the energy absorption. Table 1 lists the physical and mechanical parameters of the ten parts [19-26]. In the FE model there are totally eighteen contact pairs and tie constraints which are set as surface to surface contact, and the sliding formulation was finite sliding to ensure the convergence and the accuracy of simulation. For a woodpecker, the pores of the skull bone are always filled with tissue fluids. It has been verified that the biological fluids and soft tissue in the pores of the bone does not affect the bone's modulus significantly. Thus we build a continuum model with the ignorance of pores and then assign the material properties to the whole model.

Table 1. Physical and mechanical parameters of the ten parts

Parts	Torso	Wing	Tail	Neck	Claw	*Skull*	*Brain*	*Beak*	*Dura*	*Hyoid*
Volume (mm^3)	51961	14924	822	705	321	1797	1230	180	143	31
Density (mg/mm^3)	1.3	1.2	1.2	1.2	3	1.2	1.04	1.46	1.04	1.2
Mass (g)	67.6	17.9	0.99	0.85	0.96	2.16	1.28	0.26	0.15	0.04
Relative Mass (%) [a]	73.3	19.4	1.07	0.92	1.04	2.34	1.39	0.28	0.16	0.04
Young's Modulus (Pa)	2.5×10^9	4×10^9	4×10^9	5×10^9	2×10^{10}	6×10^9	3.1×10^3	7×10^9	3.7×10^8	3.7×10^9
Possion's Ratio	0.4	0.4	0.4	0.4	0.4	0.4	0.499	0.3	0.3	0.3
Coefficient of Viscosity (Pa·s)	—	—	—	—	—	0.086	3.85	0.086	3.85	0.086
References	[19,20]	[21]	[21]	[19,22]	[19,23]	[19]	[24,25]	[26]	[9,25]	[5]

[a] Relative mass is calculated on the mass of the whole woodpecker.

4. Modal analysis

The natural frequency is highly dependent on the material properties and boundary conditions. Herein we focus on the frequency characteristic of the head, and fixed the woodpecker's head at the bottom of neck to simulate the actual conditions. To examine the effect of pre-tension force on hyoid, we applied different pre-tension forces to the hyoid of the model. As the maximum tension strength and the across-section area of the hyoid are reported to be about 131 MPa and 0.2 mm^2 respectively, we estimated the maximum pull force of the hyoid bone is around 26 N [5, 6]. As such, in the model, we assumed the upper limit of pre-tension force to be 25 N. With one end of the hyoid fixed on the nostril, the other end was pulled with a force of 0 N to 25 N. The first thirty modes of woodpecker head are obtained. It is found that they are the local vibration occurring within the brain due to its low Young's modulus. The whole vibration frequency of woodpecker head must be higher than the local one.

Figure 6 shows that the first ten natural frequencies of woodpecker head are in the range of 50-100 Hz. These frequencies are more than twice the working frequency (18-28 Hz) of the woodpecker. Apparently, such natural frequencies can protect the woodpecker head from resonance damage when pecking the trees. To imitate the contract of hyoid bone of a woodpecker on pecking, we investigate the effect of the application of pre-tension forces on the natural frequencies of woodpecker's head. As illustrated in Figure 6, the larger the pre-tension force of hyoid, the higher the natural frequency of head. Take the first natural frequency as an example, when the applied pre-tension is increased from 0

N to 25 N, the natural frequency increases from 47 Hz to 57 Hz, see the inset to Figure 6, an increment of 21.3%.

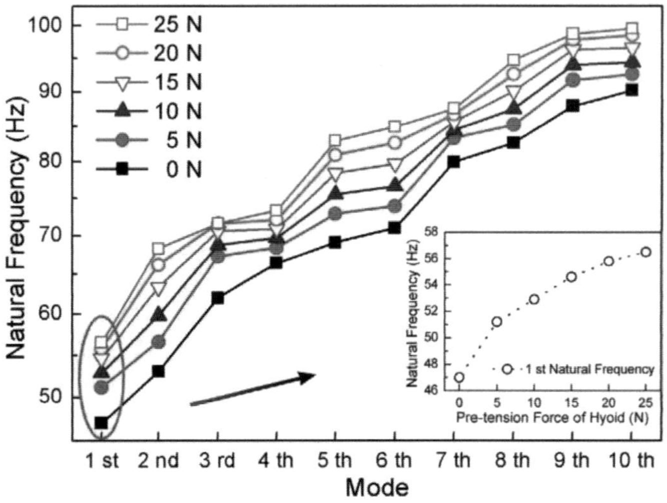

Figure 6. The first ten natural frequencies of the woodpecker head under 0 N to 25 N pre-tension force on hyoid. Illustration in the bottom right corner shows the effect of the pre-tension force on the first natural frequency of the woodpecker head, which raises the frequency from 47 Hz to 57 Hz.

5. Energy analysis

Next we will study the distribution and conversion of energy in a whole woodpecker during the successive peckings in details. When woodpecker pecks, the muscles of the neck control the motion of head towards and away from the trunk directly, and the energy stored in the body is converted from SE into kinetic energy (KE) and then is converted reversely. The total SE in the woodpecker is mainly composed of recoverable strain energy (RSE) and dissipated energy (DE) which reflect the degree of deformation and heat generation respectively.

Figure 7 shows the total strain energy of the different parts of the woodpecker. Computational results show that the strain energy stored in parts of body is much higher than that in the head. When the woodpecker head returns to the initial position after three successive peckings, the total SE stored in the body accounts for 99.7% of the whole bird, and there is merely 0.3% in the head. The stored energy will be utilized in the next pecking, and the energy enters into the head is only a small fraction of it. The circulation of energy conversion repeats itself while the head pecks trunk successively.

The strain energy in woodpecker head increases periodically due to the dissipated energy of the viscosity of materials. For the brain part, the RSE density is the lowest and the DE density is the largest which is much higher than the other parts. High dissipated energy causes the temperature increment of the brain, and the value can be calculated by Equation (3).

$$Q = C_p m \Delta T \tag{3}$$

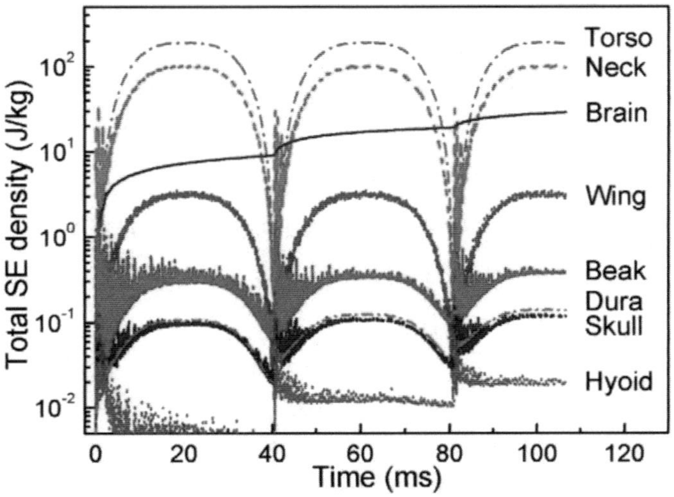

Figure 7. The total strain energy of the main parts of the woodpecker. The claws and tail are fixed on the trunk during the whole pecking process, and the energy distribution and conversion in them are affected by the constraints readily. The mechanism of energy generation is quite different from the other parts of the woodpecker, thus we do not compare the energy in them with the other parts.

Herein, m presents the mass and C_p is the specific heat, which was approximated as 2.18×10^3 J/kg·℃ [27]. The average DE density (Q/m) of brain during one pecking is about 12.5 J/kg. Thus, the temperature increment ΔT is found to be 0.0057℃. After one successive pecking process (10 to 30 peckings), the temperature increment of the brain will be 0.06 to 0.17℃. This theoretically explains why the woodpecker cannot conduct successive peckings for a long time, but has to take a break for heat dissipation after certain peckings. The other structures of the head such as beak, skull, hyoid and dura matter, provide energy absorption and physical protection for the brain to lower the stress and deformation of the brain further.

6. Conclusions

The woodpecker has the unique ability to protect its brain from impact damage. In this paper we studied the anti-shock mechanism of the woodpecker and got the conclusions as follows:

(a) The Young's modulus around the skull is non-uniform but changes periodically. The results of stress wave propagation in the viscoelastic bar shows that the peak stress changes accord with the changing of the Young's modulus and the stress is lowest in the final low-modulus zone. The material property distribution of the skull makes the peak stress smallest at the position of the brain.

(b) The application of pre-tension force to hyoid bone can increase the natural frequency of woodpecker head with the maximum increment of 21.3%. The large gap among the natural and working frequencies enable the woodpecker to effectively protect its brain from the resonance injury.

(c) Most of the impact energy is converted into the strain energy stored in the body and is utilized in the next pecking. There is only a small fraction of it enters into the head. The energy dissipated in the brain finally raises the temperature of the brain. This may explain why the woodpecker has to take a break after certain successive peckings.

Acknowledgments
This work was supported by the National Natural Science Foundation of China (11272080) and the Doctoral Education Foundation of China Education Ministry (20110041110021).

References

[1] Stark R D *et al* 1998 A quantitative analysis of woodpecker drumming *Condor* vol 100 pp 350–356
[2] May P R *et al* 1979 Woodpecker drilling behavior *Arch. Neurol.* vol 36 pp 370–373
[3] Ono K *et al* 1980 Human head tolerance to sagittal impact *Stapp Car Crash Conf. Proceedings* pp 101–161
[4] Gibson L J 2006 Woodpecker pecking: how woodpeckers avoid brain injury *J. Zool.* vol 270 pp 462–465
[5] Zhou P *et al* 2009 The novel mechanical property of tongue of a woodpecker *J. Bionic. Eng.* vol 6 pp 214–218
[6] Yoon S H *et al* 2011 A mechanical analysis of woodpecker drumming and its application to shock-absorbing systems *Bioinspir. Biomim.* **6** 016003
[7] Oda J *et al* 2006 Mechanical evaluation of the skeletal structure and tissue of the woodpecker and its shock absorbing system *JSME Int. J. Ser. A* vol 49 pp 390–396
[8] Wang L Z *et al* 2011 Why do woodpeckers resist head impact injury: a biomechanical investigation *Plos One* **6** e26490
[9] Zhu Z D *et al* 2012 Numerical study of the impact response of woodpecker's head *Aip. Adv.* **2** 042173
[10] Oliver W C *et al* 1992 An improved technique for determining hardness and elastic modulus using load and displacement sensing indentation experiments *J. Mat. Res.* vol 7 pp 1564–1583
[11] Pharr G M *et al* 1992 On the generality of the relationship among contact stiffness, contact area, and elastic modulus during indentation *J. Mat. Res.* vol 7 pp 613–617
[12] Oliver W C *et al* 2004 Measurement of hardness and elastic modulus by instrumented indentation: advances in understanding and refinements to methodology *J. Mat. Res.* vol 19 pp 3–20
[13] Stefan U *et al* 2010 Effects of three different preservation methods on the mechanical properties of human and bovine cortical bone *Bone* vol 47 pp 1048–1053
[14] Chen P Y *et al* 2009 Comparison of the structure and mechanical properties of bovine femur bone and antler of the North American elk *Acta Biomat.* vol 5 pp 693–706
[15] Villard P *et al* 2004 How do woodpeckers extract grubs with their tongues *The Auk* vol 121 pp 509–514
[16] Agam G *et al* 2005 Vessel tree reconstruction in thoracic CT scans with application to nodule detection *IEEE Trans. Med. Imag.* vol 24 pp 486–499
[17] Wan S Y *et al* 2002 Multi-generational analysis and visualization of the vascular tree in 3D micro-CT images *Comput. Biol. Med.* vol 32 pp 55–71
[18] Rydberg J *et al* 2000 Multi-section CT: scanning techniques and clinical applications. *Radiographics*

vol 20 pp 1787–1806

[19] Wegst U G K *et al* 2004 The mechanical efficiency of natural materials *Philos. Mag.* vol 84 pp 2167–2181

[20] Chen P Y *et al* 2008 Structure and mechanical properties of selected biological materials *J. Mech. Behav. Biomed. Mater.* vol 1 pp 208–226

[21] Cheng S *et al* 2009 Mechanical and thermal properties of chicken feather fiber/PLA green composites *Compos. Part B Eng.* vol 40 pp 650–654

[22] Meyers M A *et al* 2008 Biological materials: Structure and mechanical properties *Prog. Mater. Sci.* vol 53 pp 1–206

[23] Li B W *et al* 2010 Experimental study on the mechanical properties of the horn sheaths from cattle *J. Exp. Biol.* vol 213 pp 479–486

[24] Taylor Z *et al* 2004 Reassessment of brain elasticity for analysis of biomechanisms of hydrocephalus *J. Biomech.* vol 37 pp 1263–1269

[25] Sack I *et al* 2009 The impact of aging and gender on brain viscoelasticity *Neuroimage* vol 46 pp 652–657

[26] Seki Y *et al* 2005 Structure and mechanical behavior of a toucan beak *Acta Materialia* vol 53 pp 5281-5296

[27] Cooper T E *et al* 1971 Correlation of thermal properties of some human tissue with water content *Aerosp. Med.* vol 42 pp 24–27

A Generalized Functional Model Based Method for Vibration–Based Damage Precise Localization in 3D Structures

Christos S Sakaris, John S Sakellariou and Spilios D Fassois

Stochastic Mechanical Systems & Automation (SMSA) Laboratory
Department of Mechanical & Aeronautical Engineering
University of Patras, GR 265 04 Patras, Greece
URL: http://www.smsa.upatras.gr

E-mail: sakaris@mech.upatras.gr, sakj@mech.upatras.gr, fassois@mech.upatras.gr

Abstract.
A Generalized Functional Model Based Method for vibration–based damage precise localization on structures consisting of 1D, 2D, or 3D elements is introduced. The method generalizes previous versions applicable to structures consisting of 1D elements, thus allowing for 2D and 3D elements as well. It is based on scalar (single sensor) or vector (multiple sensor) Functional Models which – in the inspection phase – incorporate the mathematical form of the specific structural topology. Precise localization is then based on coordinate estimation within this model structure, and confidence bounds are also obtained. The effectiveness of the method is demonstrated through experiments on a 3D truss structure where damage corresponds to single bolt loosening. Both the scalar and vector versions of the method are shown to be effective even within a very limited, low frequency, bandwidth of 3–59 Hz. The improvement achieved through the use of multiple sensors is also demonstrated.

1. Introduction

Vibration-based damage localization has been tackled mainly as a classification problem, where a damage is classified as belonging to one out of a number of specific pre-determined locations/areas [1,2]. Although such an approach is often useful, it does not solve the problem of damage *precise localization*, that is the estimation of the precise damage coordinates.

From the broad family of vibration based methods, those based on Finite Element Models (FEMs) may achieve damage *precise* localization on structural topologies provided that an accurate large-scale FEM, proper updating techniques, and a sufficiently high number of vibration signals are available [2,3]. Yet, such an approach has its difficulties when confronted with large and complicated structures. On the other hand, *data–based* methods often have the capability to use partial models of the dynamics (often of reduced order too), along with fewer measured signals. Such approaches have been mainly restricted to approximate localization based on the aforementioned damage classification concept [1,4].

The problem of formulating data–based methods for damage precise localization on continuous structural topologies has been tackled by the present authors and co–workers within a statistical time series framework through a novel *Functional Model Based Method (FMBM)* [5, 6]. The method's cornerstone is the special family of Functional Models (FMs) which may represent a structure under damage of any magnitude or location. The method also uses appropriate data pooling techniques and pertinent estimation procedures. The FMBM was initially developed for structures consisting of 1D structural elements. In this context, the damaged element is determined in a first step, and, a dedicated Functional Model – specific to the determined substructure – is then employed for precise localization [6]. The method thus becomes elaborate for structures consisting of many elements, as both decision making mechanisms and several individual FMs are required.

In recent developments, scalar [7] or vector [8] (for a single or multiple sensors, respectively) versions of the FMBM have been presented that do not require determination of the damaged element in a separate step and may use a single FM for the structure. Yet, these are still limited to structures consisting of 1D elements.

In this study a generalized form of the FMBM which allows for damage precise localization in structures consisting of not only 1D elements, also 2D and 3D elements using a single or multiple sensors, is introduced. This is made possible through the use of scalar or vector Functional Models which – in the inspection phase – incorporate the mathematical form of the specific structural element topology. Precise localization is then based on coordinates estimation through non–linear schemes within this model structure, and confidence bounds are also obtained. The method's effectiveness is experimentally assessed using a laboratory 3D truss structure (also used in [7], [8]) and damage scenarios corresponding to the loosening of a single bolt at each time. Either one or two vibration response sensors are used. In all cases, a limited, low frequency, bandwidth of 3–59 Hz is used, so that the applicability of the method under natural, limited bandwidth, excitation is explored.

2. Overview of the damage localization methodology

The Generalized Functional Model Based Method is equipped with scalar or vector FMs within a statistical time series framework consisting of two phases [9, 10]: (a) The baseline (training) phase, which is performed once based on signals from known structural health states, and (b) the inspection phase, which runs periodically or continuously during the structures service cycle based on current (fresh) signals with the structure being in unknown health state. In all cases damage localization is activated once a damage has been properly detected [7, 8].

2.1. Baseline (training) Phase

The general form of an FM capable of representing the considered structural topology under damage is identified in this phase of the methodology based on signals obtained from experiments corresponding to different damage locations on the structure. A total number of $M = M_1 \times M_2 \times M_3$ experiments are performed for a sample of potential damage locations that cover the considered structural topology, with each experiment being characterized by a specific damage location with coordinates x_l, y_m, z_n on a selected Cartesian system. The complete series of experiments covers $[x_{min},\ x_{max}]$, $[y_{min},\ y_{max}]$ and $[z_{min},\ z_{max}]$ via the discretizations:

$$x_l \in \{x_1, x_2, \ldots, x_{M_1}\}, \qquad y_m \in \{y_1, y_2, \ldots, y_{M_2}\}, \qquad z_n \in \{z_1, z_2, \ldots, z_{M_3}\} \tag{1}$$

with each subscript designating the discretization index. Thus for a single damage located at a specific point on the 3D structural topology, the following *operating parameter vector* $\boldsymbol{k}$ is defined:

$$\boldsymbol{k} = [x_l\ y_m\ z_n]^T \iff \boldsymbol{k}_{l,m,n} \text{ with } l = 1, \ldots, M_1, \qquad m = 1, \ldots, M_2, \qquad n = 1, \ldots, M_3 \tag{2}$$

and the complete set of baseline experiments yields a pool of M sets of nx-excitation and ny-response signals, each of length N:

$$\boldsymbol{x}_{\boldsymbol{k}}[t] = [x_{1,\boldsymbol{k}}[t] \cdots x_{nx,\boldsymbol{k}}[t]]^T_{[nx\times 1]}, \qquad \boldsymbol{y}_{\boldsymbol{k}}[t] = [y_{1,\boldsymbol{k}}[t] \cdots y_{ny,\boldsymbol{k}}[t]]^T_{[ny\times 1]} \tag{3}$$

$$\text{with } t = 1, \ldots, N, \qquad x_l \in \{x_1, \ldots, x_{M_1}\}, \qquad y_m \in \{y_1, \ldots, y_{M_2}\}, \qquad z_n \in \{z_1, \ldots, z_{M_3}\}$$

A mathematical representation of the structural dynamics under *any potential damage location* in the three axis of the Cartesian system, is obtained in the form of a Vector dependent Functionally Pooled AutoRegressive with eXogenous excitation model, in short VFP-VARX(na, nb), of the form:

$$\boldsymbol{y_k}[t]+\sum_{i=1}^{na}\boldsymbol{A}_i(\boldsymbol{k})\cdot\boldsymbol{y_k}[t-i]=\sum_{i=0}^{nb}\boldsymbol{B}_i(\boldsymbol{k})\cdot\boldsymbol{x_k}[t-i]+\boldsymbol{e_k}[t],\ \ \boldsymbol{e_k}[t]\sim \text{iid}\,\mathcal{N}\big(\boldsymbol{0},\boldsymbol{\Sigma_e}(\boldsymbol{k})\big)\ \ \boldsymbol{k}\in\mathbb{R}^3 \quad \text{(4a)}$$

$$\boldsymbol{A}_i(\boldsymbol{k})=\sum_{j=1}^{p}\boldsymbol{A}_{i,j}\cdot G_j(\boldsymbol{k}),\qquad \boldsymbol{B}_i(\boldsymbol{k})=\sum_{j=1}^{p}\boldsymbol{B}_{i,j}\cdot G_j(\boldsymbol{k}) \quad \text{(4b)}$$

The fully parametrized AR and X matrices $\boldsymbol{A}_i(\boldsymbol{k}),\ \boldsymbol{B}_i(\boldsymbol{k})$ of equation (4b) have dimensions $[ny \times ny]$ and $[ny \times nx]$, respectively, and are expressed as explicit functions of the operating parameter vector $\boldsymbol{k}$ belonging to p-dimensional functional subspaces spanned by the (mutually independent) functions $G_1(\boldsymbol{k}), G_2(\boldsymbol{k}), \ldots, G_p(\boldsymbol{k})$. These form a functional basis that consists of polynomials of three variables (vector polynomials) obtained as tensor products from typical univariate polynomials such as Chebyshev, Legendre, Jacobi and so on [11, pp. 771-802]) (also see [7] for a detailed description of the functional basis). The matrices $\boldsymbol{A}_{i,j},\ \boldsymbol{B}_{i,j}$ include the corresponding AR and X, coefficients of projection to the basis functions. If a single pair of excitation-response signals is used, the scalar version of the FMBM may be employed where the representation of the continuous 3D structural topology is obtained through a simpler VFP-ARX model [7].

The VFP–VARX$(na, nb)_p$ model of equations (4a)-(4b) may be written in linear regression form as follows:

$$\boldsymbol{y_k}[t]=[\boldsymbol{I}_{ny}\otimes\boldsymbol{\varphi}^T_{\boldsymbol{k}}[t]\otimes\boldsymbol{g}^T(\boldsymbol{k})]\cdot\boldsymbol{\theta}+\boldsymbol{e_k}[t]=\boldsymbol{\Phi}^T_{\boldsymbol{k}}[t]\cdot\boldsymbol{\theta}+\boldsymbol{e_k}[t] \quad (5)$$

with $\boldsymbol{I}_{ny}$ designating the $[ny \times ny]$ identity matrix, $\boldsymbol{\varphi_k}[t] = [-\boldsymbol{y_k}[t-1]\ldots-\boldsymbol{y_k}[t-na]\vdots\boldsymbol{x_k}[t]\ldots\boldsymbol{x_k}[t-nb]]^T$ the regression vector with dimension $[(na\,ny+(nb+1)nx)\times 1]$, $\boldsymbol{g}(\boldsymbol{k}) = [G_1(\boldsymbol{k})\ldots G_p(\boldsymbol{k})]^T_{[p\times 1]}$ a vector containing the functional basis, $\boldsymbol{\theta}$ the vector that includes the projection coefficients with dimension $[(na\,ny^2\,p+(nb+1)nx\;ny\,p)\times 1]$, and $\otimes$ the Kronecker product [12, pp. 660–661].

Pooling together the VFP–VARX model of equation (5) for all operating parameter vectors $\boldsymbol{k}$ ($\boldsymbol{k}_{1,1,1}$, $\boldsymbol{k}_{1,1,2},\ldots,\boldsymbol{k}_{M_1,M_2,M_3}$) considered in the experiments for a single value of t, leads to:

$$\underbrace{\begin{bmatrix}\boldsymbol{y}_{\boldsymbol{k}_{1,1,1}}[t]\\ \vdots \\ \boldsymbol{y}_{\boldsymbol{k}_{M_1,M_2,M_3}}[t]\end{bmatrix}}_{[nyM\times 1]}=\underbrace{\begin{bmatrix}\boldsymbol{\Phi}^T_{\boldsymbol{k}_{1,1,1}}[t]\\ \vdots \\ \boldsymbol{\Phi}^T_{\boldsymbol{k}_{M_1,M_2,M_3}}[t]\end{bmatrix}}_{[nyM\times(na\,ny^2\,p+(nb+1)nx\;ny\,p)]}\cdot\boldsymbol{\theta}+\underbrace{\begin{bmatrix}\boldsymbol{e}_{\boldsymbol{k}_{1,1,1}}[t]\\ \vdots \\ \boldsymbol{e}_{\boldsymbol{k}_{M_1,M_2,M_3}}[t]\end{bmatrix}}_{[nyM\times 1]}\implies \boldsymbol{y}[t]=\boldsymbol{\Phi}[t]\cdot\boldsymbol{\theta}+\boldsymbol{e}[t] \quad (6)$$

Then, following substitution of the data for $t = 1,\ldots,N$ yields:

$$\boldsymbol{y}=\boldsymbol{\Phi}\cdot\boldsymbol{\theta}+\boldsymbol{e} \quad (7)$$

with:

$$\boldsymbol{y}=\begin{bmatrix}\boldsymbol{y}[1]\\ \boldsymbol{y}[2]\\ \vdots \\ \boldsymbol{y}[N]\end{bmatrix}_{[NnyM\times 1]},\qquad \boldsymbol{\Phi}=\begin{bmatrix}\boldsymbol{\Phi}[1]\\ \boldsymbol{\Phi}[2]\\ \vdots \\ \boldsymbol{\Phi}[N]\end{bmatrix}_{[NnyM\times(na\,ny^2\,p+(nb+1)nx\;ny\,p)]},\qquad \boldsymbol{e}=\begin{bmatrix}\boldsymbol{e}[1]\\ \boldsymbol{e}[2]\\ \vdots \\ \boldsymbol{e}[N]\end{bmatrix}_{[NnyM\times 1]}$$

The parameter vector $\boldsymbol{\theta}$ may be thus estimated based on the Ordinary Least Squares estimator [8]:

$$\widehat{\boldsymbol{\theta}} = \left[\boldsymbol{\Phi}^T\boldsymbol{\Phi}\right]^{-1} \cdot \left[\boldsymbol{\Phi}^T\boldsymbol{y}\right] = \left[\sum_{t=1}^{N} \boldsymbol{\Phi}^T[t]\,\boldsymbol{\Phi}[t]\right]^{-1} \cdot \left[\sum_{t=1}^{N} \boldsymbol{\Phi}^T[t]\,\boldsymbol{y}[t]\right] \tag{8}$$

and the residual covariance matrix estimate is obtained as:

$$\widehat{\boldsymbol{\Sigma}}_{\boldsymbol{e}}(\boldsymbol{k}, \widehat{\boldsymbol{\theta}}) = \frac{1}{N}\sum_{t=1}^{N} \boldsymbol{e}_{\boldsymbol{k}}[t, \widehat{\boldsymbol{\theta}}]\boldsymbol{e}_{\boldsymbol{k}}^T[t, \widehat{\boldsymbol{\theta}}] \quad \text{for} \quad \boldsymbol{k} = \boldsymbol{k}_{1,1,1}, \boldsymbol{k}_{1,1,2}, \ldots, \boldsymbol{k}_{M_1,M_2,M_3} \tag{9}$$

The AR and X orders of the VFP-VARX(na, nb) model is selected based on typical procedures with the determination of its functional subspace dimensionalities for a given basis function family to be achieved via Genetic Algorithms (GAs) [6–8] based on the minimization of a proper version of the BIC [13]. Finally model validation is based on formal verification of the residual sequence uncorrelatedness (whiteness) hypothesis [12, pp. 157–174] corresponding to the M experiments used in model estimation.

2.2. *Inspection Phase*

A pair of excitation $\boldsymbol{x}_u[t]$ and response $\boldsymbol{y}_u[t]$ signals is obtained from the current (unknown) state of the structure. Once the presence of a damage is detected based on appropriate method [7, 8], damage precise localization is activated through the generalized FMBM.

The current pair of signals is driven through the VFP-VARX$(na, nb)_p$ model of the baseline phase, which is now re-parametrized with corresponding residual series $\boldsymbol{e}_u[t, \boldsymbol{k}]$ with covariance matrix $\boldsymbol{\Sigma}_{\boldsymbol{e}_u}$ [8]:

$$\boldsymbol{y}_u[t] + \sum_{i=1}^{na} \boldsymbol{A}_i(\boldsymbol{k}) \cdot \boldsymbol{y}_u[t-i] = \sum_{i=0}^{nb} \boldsymbol{B}_i(\boldsymbol{k}) \cdot \boldsymbol{x}_u[t-i] + \boldsymbol{e}_u[t, \boldsymbol{k}] \tag{10}$$

The procedure is then specialized as follows according to the type (1D, 2D, or 3D) of structural elements involved.

2.2.1. *1D structural elements*

For structures consisting of *1D structural elements*, each element may be expressed in 3D space by a set of functions of an independent variable s [8]:

$$x = \mathrm{x}(s), \quad y = \mathrm{y}(s), \quad z = \mathrm{z}(s) \tag{11}$$

with $s \in S = [s_{lb}, s_{ub}] \subset \mathbb{R}$ defined by the structural topology. Thus the damage location $\boldsymbol{k}$ is a function of s, $\boldsymbol{k}(s) = [\mathrm{x}(s)\,\mathrm{y}(s)\,\mathrm{z}(s)]^T$. The estimation of s that leads to minimum Residual Sum of Squares (RSS) is obtained through Nonlinear Least Squares (NLS) over the considered structural topology (optimization realized via golden search and parabolic interpolation; Matlab function: *fminbnd.m*) [8]:

$$\widehat{s} = \arg\min_{s\in S} \mathrm{RSS}(s) = \arg\min_{s\in S} \sum_{t=1}^{N} \boldsymbol{e}_u^T[t, \boldsymbol{k}(s)]\boldsymbol{e}_u[t, \boldsymbol{k}(s)], \ \widehat{\boldsymbol{\Sigma}}_{\boldsymbol{e}_u} = \frac{1}{N}\sum_{t=1}^{N} \boldsymbol{e}_u[t, \boldsymbol{k}(\widehat{s}\,)]\boldsymbol{e}_u^T[t, \boldsymbol{k}(\widehat{s}\,)] \tag{12}$$

with $\boldsymbol{e}_u[t, \boldsymbol{k}(s)]$ given by equation (10).

The estimate $\widehat{s}$ may be shown to be asymptotically ($N \to \infty$) a Gaussian distributed random variable, with mean equal to the true μ_s and variance σ_s^2 $\left(\widehat{s} \sim \mathcal{N}(\mu_s, \sigma_s^2)\right)$ which is statistically optimal as it coincides with the Cramer-Rao lower bound (details in [8]):

$$\widehat{\sigma}_s^2 = \left[\sum_{t=1}^{N} \boldsymbol{\xi}[t, \widehat{s}]\cdot\widehat{\boldsymbol{\Sigma}}_{\boldsymbol{e}_u}^{-1}\cdot\boldsymbol{\xi}^T[t, \widehat{s}]\right]^{-1}, \quad \boldsymbol{\xi}[t, \widehat{s}] = \left.\frac{\partial \boldsymbol{e}_u^T[t, \boldsymbol{k}(s)]}{\partial s}\right|_{s=\widehat{s}} = -\widehat{\boldsymbol{\theta}}^T\cdot\left[\boldsymbol{I}_{ny}\otimes\boldsymbol{\varphi}[t]\otimes\left.\frac{\partial \boldsymbol{g}(\boldsymbol{k}(s))}{\partial s}\right|_{s=\widehat{s}}\right] \tag{13}$$

with $\varphi[t] = [-\boldsymbol{y}_u[t-1] \ \ldots \ -\boldsymbol{y}_u[t-na] \vdots \boldsymbol{x}_u[t] \ \ldots \ \boldsymbol{x}_u[t-nb]]^T_{[(na\,ny+(nb+1)nx)\times 1]}$ being the regression vector that corresponds to the unknown structural state and $\widehat{\boldsymbol{\theta}}$ the projection coefficient vector estimated in the baseline phase. Thus each damage coordinate estimate included in $\widehat{\boldsymbol{k}} = \boldsymbol{k}(\widehat{s})$ is a random variable with probability density function (pdf) that may be obtained from that of $\widehat{s}$ (also see [8]).

In the present study where each rod/element of the truss (see section 3) is one-dimensional (negligible thickness), it is modelled as a straight line connecting two nodes with coordinates $\{x_0\, y_0\, z_0\}$ and $\{x_1\, y_1\, z_1\}$ and equation (11) may be expressed as:

$$x = x_0 + s\,(x_1 - x_0) \quad \text{and likewise for } y,\ z \tag{14}$$

Based on them, estimates of the damage Cartesian coordinates are constructed using the estimate of s. Indeed, the random variable $\widehat{x}$ follows Gaussian distribution with mean and variance:

$$\mu_x = x_0 + \mu_s(x_1 - x_0), \quad \sigma_x^2 = (x_1 - x_0)^2 \sigma_s^2 \tag{15}$$

and the interval estimate of x then is (at the α risk level):

$$\left[\widehat{x} + Z_{\frac{\alpha}{2}} \cdot \widehat{\sigma}_x,\ \widehat{x} + Z_{1-\frac{\alpha}{2}} \cdot \widehat{\sigma}_x\right] \tag{16}$$

with $Z_{\frac{\alpha}{2}}, Z_{1-\frac{\alpha}{2}}$ designating the normal distribution's critical points at the α risk level, and $\widehat{\sigma}_x$ the estimated standard deviation of $\widehat{x}$ obtained from that of s via equation (15).

2.2.2. 2D structural elements For structures consisting of *2D structural elements*, as for instance shell type elements, a set of functions of two independent variables $\boldsymbol{s} = [s_1\ s_2]^T$ may be used in equation (11) for their expression in 3D space with $\boldsymbol{s} \in S : (s_1, s_2) \in [s_{1,lb}\ s_{1,ub}] \times [s_{2,lb}\ s_{2,ub}] \subset \mathbb{R}^2$ defined by the structural topology. As in the previous case the estimation of $\boldsymbol{s}$ that leads to minimum RSS is achieved via NLS over the considered structural topology as in equation (12). In this case, where the minimization of the left part of equation (12) shall be be achieved for more than a single variable, well known Sequential Quadratic Programming (SQP) techniques (Matlab function: *fmincon.m*) are employed.

Like before, the estimate $\widehat{\boldsymbol{s}}$ may be shown to be asymptotically a Gaussian distributed random vector, with mean $\boldsymbol{\mu_s}$ and covariance matrix $\boldsymbol{\Sigma_s}$ $\left(\widehat{\boldsymbol{s}} \sim \mathcal{N}(\boldsymbol{\mu_s}, \boldsymbol{\Sigma_s})\right)$ being statistically optimal as coinciding with that of the Cramer-Rao lower bound:

$$\widehat{\boldsymbol{\Sigma}}_{\boldsymbol{s}} = \left[\sum_{t=1}^{N} \Xi[t,\widehat{\boldsymbol{s}}] \cdot \widehat{\boldsymbol{\Sigma}}_{\boldsymbol{e}_u}^{-1} \cdot \Xi^T[t,\widehat{\boldsymbol{s}}]\right]^{-1}, \ \Xi[t,\widehat{\boldsymbol{s}}] = -\left[\boldsymbol{I}_n \otimes \widehat{\boldsymbol{\theta}}^T\right] \cdot \left[\boldsymbol{I}_{ny} \otimes \text{vec}\left(\varphi[t] \otimes \left.\frac{\partial \boldsymbol{g}(\boldsymbol{k}(\boldsymbol{s}))}{\partial \boldsymbol{s}^T}\right|_{\boldsymbol{s}=\widehat{\boldsymbol{s}}}\right)\right] \tag{17}$$

where $n = 2$ for 2D structural elements and the vec$(\cdot)$ operator transforms a matrix into a vector by stacking its columns.

Each damage coordinate in $\widehat{\boldsymbol{k}} = \boldsymbol{k}(\widehat{\boldsymbol{s}})$ is a random variable with pdf that may be obtained from that of $\widehat{\boldsymbol{s}}$. For example, the pdf $f_x(x)$ of the coordinate x may be obtained based on the joint distribution $f_{xy}(x,y)$ of coordinates x, y which is constructed by using the joint normal pdf of $\boldsymbol{s}$, $f_{\boldsymbol{s}}(s_1, s_2)$, as follows (see forthcoming journal version of [7]):

$$f_x(x) = \int_{y_{lb}}^{y_{ub}} f_{xy}(x,y)\ dy \tag{18}$$

$$\text{with} \quad f_{xy}(x,y) = f_{\boldsymbol{s}}(s_1,s_2) \cdot \left|\det\begin{bmatrix} \frac{\partial s_1}{\partial x} & \frac{\partial s_1}{\partial y} \\ \frac{\partial s_2}{\partial x} & \frac{\partial s_2}{\partial y} \end{bmatrix}\right|, \ s_1 = \mathrm{x}^{-1}(x,y),\ s_2 = \mathrm{y}^{-1}(x,y)$$

provided that $\mathrm{x}(\boldsymbol{s}), \mathrm{y}(\boldsymbol{s})$ are one-to-one. y_{lb}, y_{ub} are the lower and upper bounds of coordinate y obtained by equation (11) and det denotes determinant of a matrix. Thus the mean and variance of $\widehat{x}$, based on

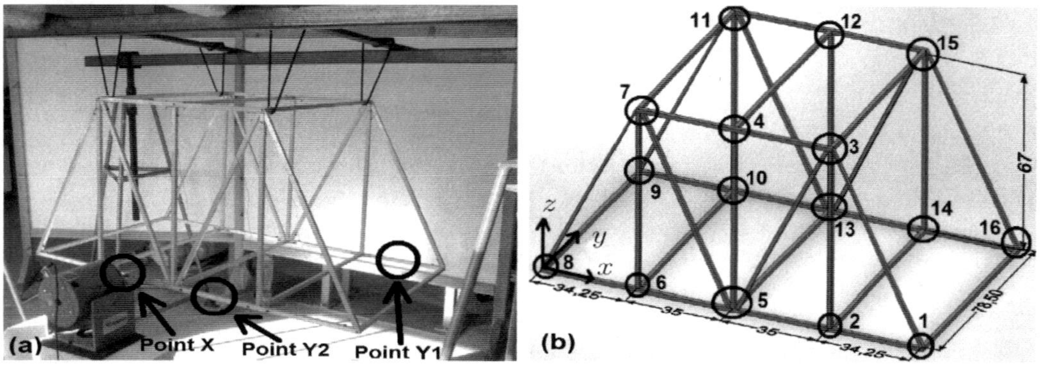

Figure 1. (a) The truss structure and the experimental setup: The force excitation (Point X) and the vibration acceleration measurement positions (Point Y1 & Y2), (b) Schematic diagram of the truss with the considered 3D Cartesian coordinate system with origin at node 8. The 16 nodes are indicated by circles (dimensions are in cm).

Table 1. Experimental details.

Structural State	Description	No of experiments
Damaged	loosening of a bolt at a single location	390

Sampling frequency: $f_s = 128$ Hz; Signal length in samples (s): $N = 9758$ samples (76.23 s)
Bandpass filter: Chebyshev Type II; 12th order (Matlab function: *filter.m*)
Signal Bandwidth: $[3 - 59]$ Hz

which corresponding confidence intervals (uncertainty bounds) are constructed, may be then obtained. The pdfs of the other coordinates are similarly obtained.

The construction of the precise damage confidence region (uncertainty bound) on the 2D structural element is based on the Gaussianity of s_1, s_2. If the transformations in equation (11) lead to known pdfs (such as Gaussian, χ^2 and others) for coordinates x, y, z then their damage confidence region may be constructed based on typical expressions. In the case where the transformations lead to unknown pdfs, the damage confidence region may be arithmetically constructed through equation (11) based on $\widehat{\boldsymbol{s}}$ and $\widehat{\boldsymbol{\Sigma}}_{\boldsymbol{s}}$.

Similarly for structures consisting of *3D structural elements* the above procedure with a set of functions of three independent variables in vector $\boldsymbol{s} = [s_1 \; s_2 \; s_3]^T$ is employed.

3. The Structure, the damage scenarios, and the experiments

The truss structure is suspended from metallic beams with a set of elastic cords and hooks (figure 1(a)) and consists of 26 rods with rectangular cross sections (1.5 $\times$ 1.5 cm) jointed together via steel elbow plates and 74 bolts. All parts are constructed from standard aluminum with the overall dimensions being 140 $\times$ 80 $\times$ 70 cm.

Each considered damage corresponds to the complete loosening of a single bolt at a node. The bolts are distributed at 16 nodes as shown in figure 1(b) and each node has certain coordinates defined based on the origin point ($x = 0, y = 0, z = 0$) taken at node 8 (see figure 1(b)). Thus, a damage is designated as $F_{x,y,z}$, with x, y, z the coordinates along the corresponding Cartesian axis. Furthermore each rod of the truss is considered as a line without thickness.

A single-excitation and two vibration acceleration response signals measured via two lightweight accelerometers along y direction at Points Y1 & Y2 (figure 1(a)). The excitation is a random Gaussian force applied horizontally along the y direction at Point X (figure 1(a)) via an electromechanical shaker equipped with a stinger, and measured via an impedance head. A number of experiments are carried out for each damaged state (table 1). Full details about the equipment and the experiments are provided in [8].

4. Experimental results

4.1. Baseline phase

The identification procedure [7, 8] is based on $M = 16$ experiments and leads to the estimation of a scalar VFP-ARX(84,84)$_{16}$ (based on response at Point Y1) and a VFP-VARX(62,62)$_{16}$ model (based on responses at Points Y1 and Y2) that may represent the 3D truss structure under the considered damage type. The first model is used for the scalar version of the FMBM while the second for the vector one. For both models the functional subspace consists of 16 trivariate Shifted Legendre polynomials (details in [7, 8]).

4.2. Inspection phase

Once a damage is detected both VFP-ARX(84,84)$_{16}$ and VFP-VARX(62,62)$_{16}$ models are re-parametrized based on the current signals corresponding to each test case of the inspection phase and the NLS estimator in the left part of equation (12) (realized via golden search and parabolic interpolation [7, 8]) leads to estimates of the unknown damage coordinates.

Indicative damage localization results using the nonlinear optimization framework which searches for the damage coordinates over the truss structure through the minimization of equation (12) with respect to variable s based on the scalar and the vector version of the FMBM, are shown in figure 2. The color code depicts the level of the cost function at each location on the structure, with the darkest

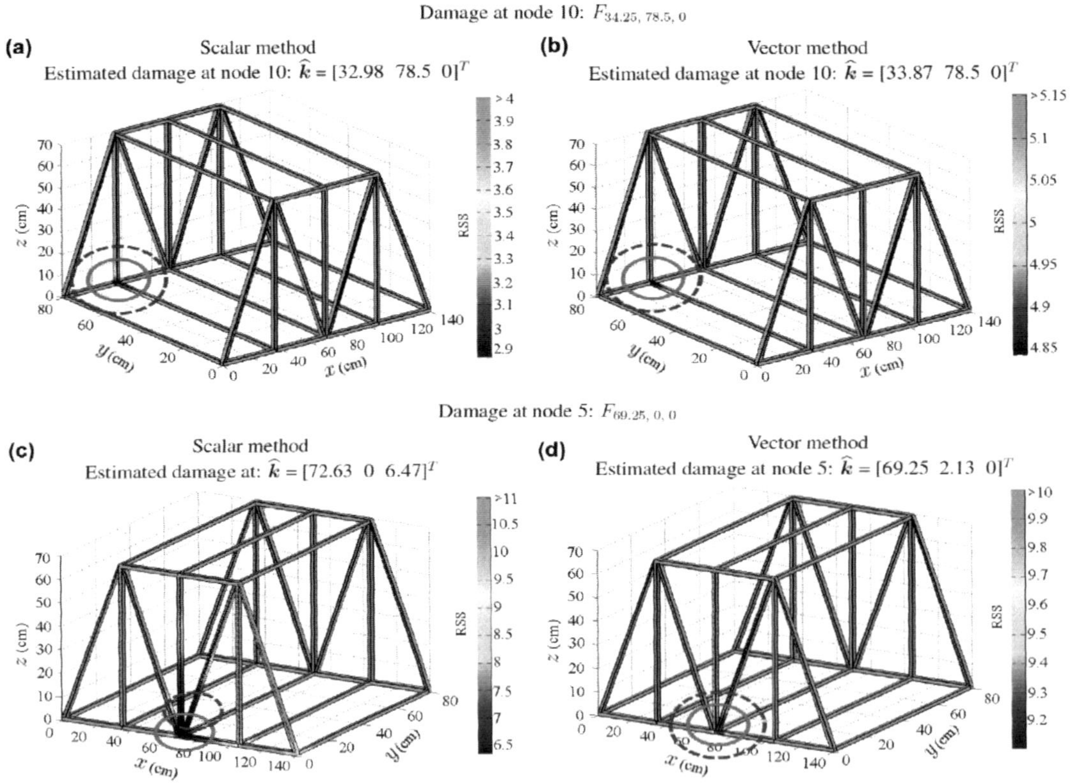

Figure 2. Indicative damage localization results for two damage scenarios based on: (a) & (c) the scalar and, (b) & (d) the vector FMBM. The values of the RSS are color coded as on the right vertical scale and the darkest value indicates the minimum. The actual and estimated damage coordinates are shown above the plots (actual damage location: —; estimated damage location: – –).

Damage location 1, $F_{104.25,\,78.5,\,0}$
Scalar method: $\widehat{\boldsymbol{k}} = [105.34 \;\; 78.5 \;\; 0]^T$
Vector method: $\widehat{\boldsymbol{k}} = [104.62 \;\; 78.5 \;\; 0]^T$

Damage location 2, $F_{69.25,\,78.5,\,0}$
Scalar method: $\widehat{\boldsymbol{k}} = [74.70 \;\; 78.5 \;\; 0]^T$
Vector method: $\widehat{\boldsymbol{k}} = [69.87 \;\; 78.5 \;\; 0]^T$

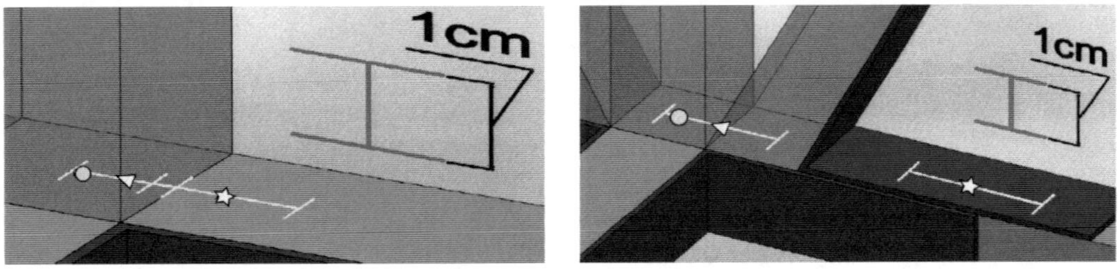

Figure 3. Indicative damage localization results and corresponding comparisons between the scalar and the vector version of the FMBM for four damage scenarios (•: true damage location; : point estimate based on the vector FMBM; : point estimate based on the scalar FMBM; : confidence intervals at the 0.0001 risk level; the true and estimated damage location coordinates are numerically provided above each plot as well).

value corresponding to the minimum point. Overall, the vector method leads to more accurate damage localization results, and additionally manages to successfully tackle certain "pathological" cases that could not be effectively handled via the scalar version. Point estimates for the damage coordinates, along with corresponding confidence intervals, are for two indicative damage scenarios presented in figure 3 for both the scalar and vector forms of the FMBM. The localization improvement achieved by the vector version is evident.

5. Concluding remarks

A Generalized FMBM for damage precise localization on structures consisting of 1D, 2D, or 3D elements was introduced. The method generalizes previous versions applicable to structures consisting of 1D structural elements. It is based on scalar or vector Functional Models (FMs) which – in the inspection phase – incorporate the mathematical form of the specific structural topology. Precise localization is then based on damage coordinate estimation within this model structure, and confidence bounds are also obtained. The method's effectiveness was assessed via experiments on a truss structure, where damage corresponds to the loosening of a single bolt. The results demonstrated that both the scalar (single sensor) and vector (multiple sensor) versions of the method may achieve damage precise localization even within the very limited and low frequency bandwidth of 3–59 Hz, with the vector version achieving somewhat higher accuracy in certain cases.

References

[1] Kopsaftopoulos F P and Fassois S D 2011 *Journal of Theoretical and Applied Mechanics* **49** 727–756
[2] Perera R and Ruiz A 2008 *Mechanical Systems and Signal Processing* **22** 970–991
[3] Teughels A and Roeck G D 2005 *Archives of Computational Methods in Engineering* **12** 123–164
[4] Lautour O R D and Omenzetter P 2010 *Mechanical Systems and Signal Processing* **24** 1556–1569
[5] Sakellariou J S and Fassois S D 2008 *Mechanical Systems and Signal Processing* **22** 557–573
[6] Kopsaftopoulos F P and Fassois S D 2013 *Mechanical Systems and Signal Processing* **39** 143–161
[7] Sakaris C S, Sakellariou J S and Fassois S D 2012 *Proceedings of the 6th European Workshop on Structural Health Monitoring* (Dresden, Germany) Journal version submitted for publication
[8] Sakaris C S, Sakellariou J S and Fassois S D 2015 *Structural Health Monitoring* **in press** doi:10.1177/1475921714568407
[9] Fassois S D and Sakellariou J S 2007 *The Royal Society - Philosophical Transactions: Mathematical, Physical and Engineering Sciences* **365** 411–448
[10] Fassois S D and Sakellariou J S 2009 *Encyclopedia of structural health monitoring* ed Boller C, Chang F K and Fujino Y (Chichester: John Wiley & Sons) pp 443–472
[11] Abramowitz M and Stegun I 1972 *Handbook of mathematical functions with formulas, graphs and mathematical tables* 10th ed (New York: Dover)
[12] Lütkepohl H 2005 *New introduction to multiple time series analysis* (Berlin: Springer-Verlag)
[13] Hios J D and Fassois S D 2014 *Mechanical Systems and Signal Processing* **49** 77–94

Assessment of a Multiple Model Based Parametric Method for Output–Only Vibration–Based Damage Detection for a Population of Like Structures

Kyriakos J. Vamvoudakis-Stefanou, John S. Sakellariou and Spilios D. Fassois
Stochastic Mechanical Systems & Automation (SMSA) Laboratory, Department of Mechanical & Aeronautical Engineering, University of Patras, Patras, Greece

E-mail: vamvoudakisk@mech.upatras.gr, sakj@mech.upatras.gr, fassois@mech.upatras.gr

Abstract. This study focuses on the problem of vibration-based damage detection for a population of like structures. Although nominally identical, like structures exhibit variability in their characteristics due to variability in the materials and manufacturing. This inevitably leads to variability in the dynamics, which may be so significant as to mask deviations due to damage. Damage detection via conventional vibration-based methods, using a common threshold in the decision making mechanism thus becomes highly challenging. The study presents a detailed assessment of a recently introduced Multiple Model (MM) based AutoRegressive (AR) model parameter method aiming at addressing this problem. The assessment is based on high numbers of experimental test/inspection cases using composite beams damaged via impact, as well as comparisons with the corresponding conventional (single model based) method. The results confirm significant improvement over the method's conventional counterpart. A sensitivity analysis additionally indicates that the method is relatively insensitive to the model order, but sensitive to the specific beams selected as baseline (training) ones; in fact their selection may lead to excellent results.

1. Introduction

When dealing with a population of like structures there are two possibilities for vibration-based Structural Health Monitoring: Either a *common* threshold is adopted in the decision making mechanism [1] for all structures, or an *individual* threshold is established for each individual structure. The former approach has some obvious advantages in terms of set-up and cost, and is of high interest. Yet, the problem is that like structures are never identical due to inevitable variability associated with materials and manufacturing, plus of course the fact that they may operate under somewhat different conditions. This inevitably leads to variability in the dynamics, which may be so significant as to mask deviations due to damage.

Unlike other uncertainty factors, such as those associated with environmental and operational variability [2], the handling of the variability associated with a population of like structures and its effects on vibration-based SHM is, thus far, effectively unexplored. In two recent studies [3,4] by the present authors, a Multiple Model (MM) vibration-based framework for SHM for a population of like structures (composite beams subject to delamination) was introduced, and very interesting results and favourable comparisons with corresponding conventional methods were obtained. The main idea behind this framework is on the use of vibration data obtained from *several* - instead of a single - like structures for modelling the healthy dynamics during the methods' training (baseline phase) and then decision making based on the simultaneous use of all baseline (healthy) models based on a minimal distance type criterion. The results of these studies have illustrated the poor performance of conventional methods and the significantly improved performance of the Multiple Model methods. Yet, the assessments presented are only preliminary and *not* comprehensive, in the sense that neither the sensitivity of the methods with respect to the specific structures selected for inclusion in the baseline, nor that referring to the selected model order have been studied.

Thus the aim of the present study is exactly to address these issues and provide a comprehensive assessment of one of the methods - the MM based AutoRegressive (AR) model parameter method. The laboratory-scale composite beams subject to delamination and other types of damage induced by impact, used in the previous studies [3,4], are presently employed as well, along with a *high number* of test/inspection cases. This is made possible by "rotating" the baseline beams within the set of all healthy beams. Moreover, the sensitivity of the method with respect to the *number* and *specific* structures selected for inclusion in the baseline phase, as well as with respect to *model order* employed, are studied in detail.

The rest of this article is organized as follows: The experimental set-up and the damage scenarios are briefly reviewed in section 2, the MM based AutoRegressive (AR) model parameter method is elaborated upon in section 3, the assessment of the method is performed in section 4, and the conclusions are summarized in section 5.

2. The experimental set-up and the damage scenarios

The study is based on 26 nominally identical, composite beams consisting of several layers of woven and unidirectional (UD) fabric. They have square hollow cross section, uniform along their length with dimensions 600x65x65 mm (LxWxH), wall thickness of 3mm and corner radius of 8mm (figure 1). Each beam represents the topology of the main part of a commercial UAV boom. In the experimental setup each beam is tightly clamped at one end, simulating its connection to the fuselage, while its free end is attached to an aluminum mass representing the aircraft tail (figure 1). The beam is excited at its free end with a random, approximately white, Gaussian force applied vertically at Point X via an electromechanical shaker, while vibration acceleration response signals are measured at point Y1 via a lightweight accelerometer.

Figure 1. Photo of the experimental set-up: Point x represents the force excitation position, Point Y1 the vibration acceleration measurement position, Point D the damage position.

Three damage scenarios (table 1) that may include a combination of delamination, small cracks and broken fibres are induced in 3 of the 26 beams at point D (figure 1) by using a pendulum type impact hammer with distinct energy level for each damage. Welch-based FRF magnitude estimates [5, pp 329, 338] for the healthy and damaged beams, revealing significant variability in the dynamics of the healthy beams, as well as the damages effects, are presented in figure 2. For a detailed presentation of the experimental set-up, the employed equipment and the damage scenarios the reader is referred to [3].

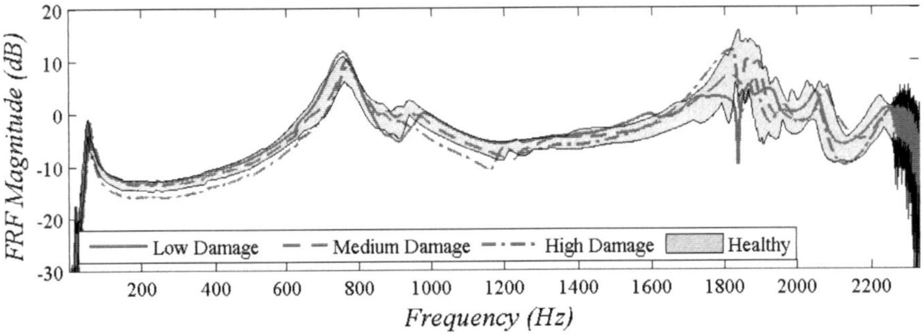

Figure 2. Welch-based FRF magnitude estimates for 23 healthy and 3 damaged beams (7 experiments per beam).

Table 1. The damage scenarios and experimental details.

Structural State	Impact Energy (J)	Number of beams	No of experiments per beam	Total number of experiments
Healthy	---	23	7	161
Damage A	5	1	7	7
Damage B	10	1	7	7
Damage C	15	1	7	7

Sampling frequency fs = 4 654.5 Hz; signal bandwidth [3 – 2 327.25] Hz.

3. The Multiple Model (MM) based AR parameter method

The method is of the *response-only type*, based on the parametric AR modelling [6] of the random vibration response signal(s) - the excitation is not required. A scalar AR model is of the form:

$$y[t] + \sum_{i=1}^{n_a} a_i y[t-i] = e[t]\,, \qquad e[t] \sim NID(0, \sigma_e^2) \tag{3}$$

with $y[t]$ designating the random vibration response signal, n_a the model order, a_i the i-th AR parameter, and $e[t]$ the model residual that is a white Gaussian zero-mean with variance σ_e^2 sequence. NID stands for Normally Independently Distributed. The employed characteristic quantity Q is the AR model parameter vector ($Q = \boldsymbol{\theta}$) [7].

The main idea behind the method is that AR models corresponding to *several* healthy structures, instead of just one (as in conventional methods), are used to effectively represent the healthy structural dynamics while taking variability into account. As a vibration-based method [7-9], the MM based AR parameter method is applied in two phases: (a) The baseline (training) phase, and (b) the inspection (damage detection) phase.

Baseline phase. In the baseline phase vibration signals are obtained from a number of like structures in their healthy state. Let v be the number of employed healthy structures and $Y_o = \{\mathrm{y}_{\mathrm{o,k}}[\mathrm{t}]\}$ the respective response signals with $t = 1, \ldots, N$ and $k = 1, \ldots, v$ (the subscript o is used to designate the healthy state). Using the available signals (presently one per healthy structure) a set of AR models $m_o = \{\mathrm{m}_{\mathrm{o},1}, \ldots, \mathrm{m}_{\mathrm{o},\mathrm{v}}\}$ are obtained, representing the population's healthy dynamics.

Inspection phase. In this phase vibration response signal(s) are obtained from the current structure under inspection (which is in unknown state). The objective is to decide whether the current structure's dynamics are represented by the set of healthy AR models obtained in the baseline phase, in which case the current structure is declared as healthy - if this is not the case, then the structure is declared as damaged.

In doing so, let $Y_u[t]$ designate the current response signal(s) and m_u the respective AR model (the subscript u is used to designate unknown state). Then a proper distance metric in the method's characteristic quantity space (AR model parameters) is employed, and the distances between Q_u and each one of the $Q_{o,1}, \dots, Q_{o,v}$ used as baseline are obtained.

The KL divergence [10] is the distance employed, which for Q_u, $Q_{o,k}$, each following l-dimensional Gaussian distribution with means $\boldsymbol{\mu}_{\mathrm{u}}, \boldsymbol{\mu}_{\mathrm{o}}$ and covariance matrices $\boldsymbol{\Sigma}_{Q_u}, \boldsymbol{\Sigma}_{Q_{o,k}}$, respectively, is defined as (capital/lower case bold face symbols represent matrix/vector quantities):

$$\begin{aligned} d(Q_u, Q_{o,k}) &= d_{KL}(Q_u, Q_{o,k}) \\ &= \frac{1}{2}\left(tr\left(\boldsymbol{\Sigma}_{Q_u}^{-1} \boldsymbol{\Sigma}_{Q_{o,k}} \right) + (\boldsymbol{\mu}_{\mathrm{u}} - \boldsymbol{\mu}_{\mathrm{o}})^{\mathrm{T}} \boldsymbol{\Sigma}_{Q_u}^{-1} (\boldsymbol{\mu}_{\mathrm{u}} - \boldsymbol{\mu}_{\mathrm{o}}) - 1 - \ln\left(\frac{\det \boldsymbol{\Sigma}_{Q_{o,k}}}{\det \boldsymbol{\Sigma}_{Q_u}} \right) \right) \end{aligned} \tag{1}$$

with $\mathrm{tr}(\cdot)$ designating trace and $\det(\cdot)$ determinant of the indicated quantity.

Once the distances between the current AR model and those of the baseline are available, damage detection is based on the following decision making mechanism that takes *all* distances to the baseline models into account through an *overall distance metric* D:

$$\begin{array}{lcl} D = \sum_{k=1}^{v} d(Q_u, Q_{o,k}) \leq l_{\lim} & \rightarrow & \text{Healthy structure} \\ \text{Otherwise} & \rightarrow & \text{Damaged structure} \end{array} \tag{2}$$

with $l_{\lim}$ designating a user defined (based on the baseline signals) threshold.

4. Assessment of the MM based AR parameter method

The assessment of the MM based AR parametric method is based on damage detection tests with an increasing number of beams used in the baseline phase ($v = \{1,4,8,12,16\}$). Note $v = 1$ designates the corresponding conventional method. The testing procedure for each number of beams in the baseline phase is based on the following algorithm:

Step 1: All possible combinations among the 23 healthy beams in sets of v beams are considered.
Step 2: If the total number of combinations is very large (for $v = \{4,8,12,16\}$), a random subset of combinations is employed, so that each one of the 23 healthy beams is used in at least one combination. Otherwise, all possible combinations are employed (these are referred to as the <u>baseline combinations</u>).
Step 3: The first baseline combination and one experiment per beam (<u>baseline experiments</u>) are used in the baseline phase of the method. At present the remaining experiments with the baseline beams are not subsequently used.
Step 4: The remaining healthy beams ($23 - v$) are used in the inspection phase, and along with the damaged beams, formulate the set of <u>inspection beams</u>. Damage detection is performed for all the experiments associated with the inspection beams (<u>inspection experiments</u>).
Step 5: Steps 3 and 4 are repeated until all baseline combinations are exhausted.

At the end, a total number of damage detection tests for all the inspection experiments and all the baseline combinations is obtained, forming the set of inspection cases.

This algorithm is a modified cross validation eliminating the effects of baseline beam selection, plus leading to a drastically higher number of inspection cases to assess the method with. For each number of baseline beams considered, the number of inspection beams, baseline experiments, inspection experiments, baseline combinations, and inspection cases are presented in table 2.

Table 2. Inspection beams, inspection experiments, & inspection cases for various values of v.

No. of baseline beams (v)	No. of inspection beams (i)	Baseline experiments	Inspection experiments ($z = 7 \cdot i$)	Baseline combinations (h)	Inspection cases ($h \cdot z$)
1	25	1	175	23	4 025
4	22	4	154	500	77 000
8	18	8	126	500	63 000
12	14	12	98	500	49 000
16	10	16	70	500	35 000

4.1. Baseline phase
The estimation of the parametric AR models is based on $N = 10\ 000$ sample long ($\approx$ 2.15s) data records using Ordinary Least Squares with QR implementation [11, pp 318-320]. The model order selection is based on the Bayesian Information Criterion (BIC) and the Residual Sum of Squares / Signal Sum of Squares (RSS/SSS) [11, pp 498-514]. The selected model order for all models is presented in table 3, along with a summary of estimation details (the performance characteristics correspond to an individual baseline model).

Table 3. AR model estimation details

Model	Measur. Point	Selected model	No. of estimated parameters	SPP*	BIC**	RSS/SSS(%)**
AR	Y2	AR(47)	47	~212	-1.271	28.17

Estimation method: Ordinary Least Squares (OLS), QR implementation (MATLAB function arx.m)
Signals normalized via sample mean removal and division by own sample standard deviation.
Signal length used: N = 10 000 samples (2.15 s)
* Samples Per Parameter ** Refers to a single baseline (healthy) model

4.2. Inspection phase
Damage detection results for $v = \{1,12\}$ and the corresponding number of inspection cases (table 2) are presented in figure 3. The distance metric D from the current model to the baseline model set is provided per inspection case. As depicted, for $v = 1$ (conventional approach) several distance metrics related to damaged beams are lower than those related to their healthy counterparts, thus rendering damage detection ineffective. Improved results are obtained by the MM based method as v is increased to 12.

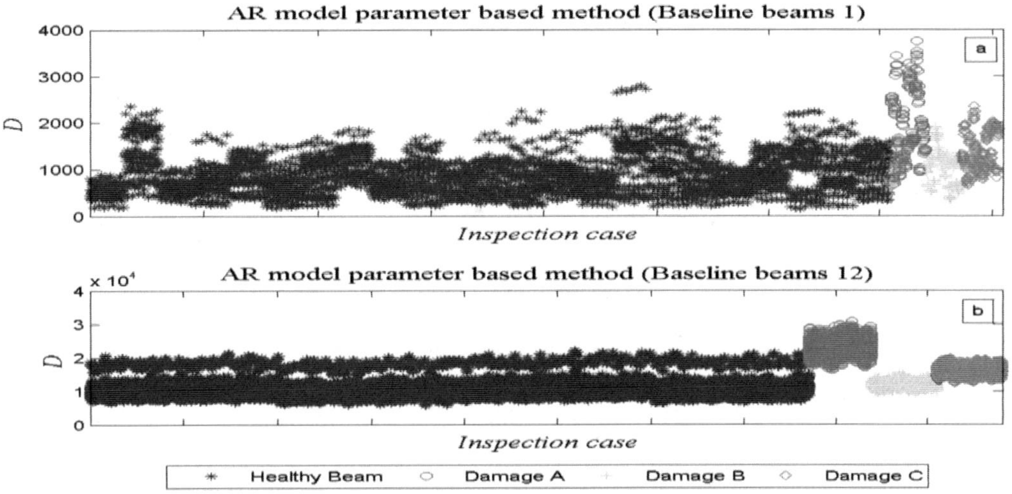

Figure 3. Damage detection by the MM based AR model parameter method: Overall distance metric D for each inspection case. (a) Single-model baseline (4 025 inspection cases); (b) 12-model baseline (49 000 inspection cases).

The remarks made above are also depicted by figure 4 in terms of Receiver Operating Characteristic (ROC) curves [12, pp 34-35], for different values of v. These curves represent the true positive rate (percentage of correct damage detections) versus the false positive rate (percentage of false alarms) for varying decision thresholds. The potential improvement in damage detection is evident when moving from the conventional ($v = 1$) to the Multiple Model ($v > 1$) based method.

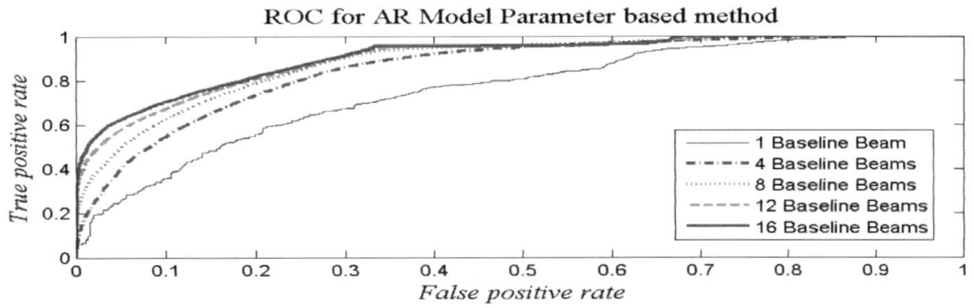

Figure 4. ROC curves for different values of ν ({1,4,8,12,16}) and 500 combinations of baseline beams.

Despite the significant improvement, the method is still not effective as shown in figure 4, which is reasonable as the fundamental difficulty stemming from the high variability in the healthy beam dynamics *cannot* be fully overcome. Then a sensitivity analysis is performed based first on the AR model order and secondly on the selection of specific baseline beam combinations. Thus, tests with 5 different AR orders ($n_a = \{27,37,47,57,67\}$) and 12 baseline beams are made, examining whether the selection of a different model order affects detection performance. The results of these tests are presented in figure 5 in terms of ROC curves, depicting limited sensitivity of the MM based method.

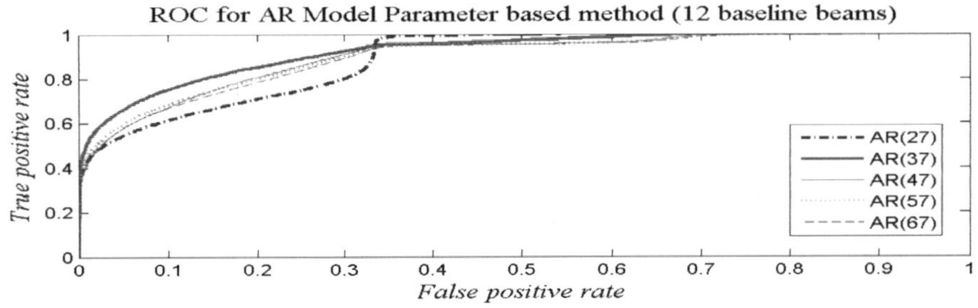

Figure 5. ROC curves for different AR model orders, 12 baseline beams and 500 combinations of baseline beams.

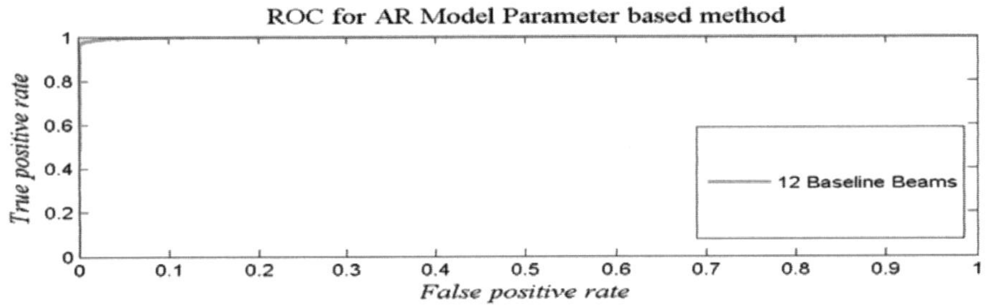

Figure 6. ROC curve for $v = 12$ and 50 properly selected combinations of baseline beams for achieving effective damage detection.

Subsequently, several baseline beam combinations are tested in order to examine whether some specific combinations may lead to effective detection. Indicative results for 50 properly selected combinations with $v = 12$, are presented in figure 6 in terms of the ROC curve. It is evident that with proper selection of the baseline beams the method's performance improves significantly; in fact reaching excellence.

5. Concluding remarks

A detailed assessment of a recently introduced Multiple Model (MM) based AutoRegressive (AR) model parameter method for Structural Health Monitoring of a population of like structures using a *single*, common, threshold in the decision making mechanism has been presented. The assessment has been based on high numbers of experimental test/inspection cases using composite beams subject to damage via impact. Comparisons with the corresponding conventional (single model based) method have been also made. The results achieved demonstrate the MM method's significant improvement over that of the corresponding conventional method, but also that it is not possible to generally (that is in all cases, *irrespectively* of the beams selected as baseline) and fully overcome the fundamental difficulty associated with significant variations in the healthy population dynamics. A sensitivity analysis has additionally indicated that the method is relatively insensitive to the selected model order, but sensitive to the specific beams selected as baseline (training) ones, with proper selection potentially leading to excellent results.

Acknowledgements
The support of this study by the European Commission (FP7 Project 605549 on "Global in Flight Healthy Monitoring Platform for Composite Aerostructures Based on Advanced Vibration Based

Methods" – VIBRATION) is gratefully acknowledged. The authors wish to thank all project partners for their support and contributions.

References

[1] Fassois S and Sakellariou J 2009 Statistical time series methods for structural health monitoring, in: *Encyclopedia of Structural Health Monitoring*, ed C Boller, F Chang, Y Fujino (John Wiley & Sons Ltd.) pp 443-472

[2] Sohn H 2007 Effects of environmental and operational variability on structural health monitoring *The Royal Society - Philosophical Transactions: Mathematical, Physical and Engineering Sciences* **365** 539-560

[3] Vamvoudakis-Stefanou K, Sakellariou J and Fassois S Random vibration response-only damage detection for a set of composite beams: *Proceedings of the ISMA International Conference on Noise and Vibration Engineering (Leuven, Belgium 2014)*

[4] Vamvoudakis-Stefanou K, Fassois S, Vibration-based damage detection for a population of like structures via a multiple model framework: *Proceedings of the 6th International Symposium on NDT in Aerospace (Madrid, Spain 2014)*

[5] Bendat J and Piersol A 2000 *Random Data: Analysis and Measurement Procedures* 3rd edition (New York: Wiley–Interscience)

[6] Fassois S 2001 Parametric identification of vibrating structures, in: *the Encyclopedia of Vibration*, ed S Braun, D Ewins, S Rao (Academic Press) pp 673-685

[7] Fassois S and Sakellariou J 2007 Time series methods for fault detection and identification in vibrating structures *The Royal Society - Philosophical Transactions: Mathematical, Physical and Engineering Sciences* **365** 411–448

[8] Avendaño-Valencia L and Fassois S 2014 Robust fault detection based on multiple functional series tar models for structures with time-dependent dynamics: *Proceedings of the 7th European Workshop on Structural Health Monitoring (EWSHM) (Nantes, France 2014)* pp 804-811

[9] Fassois S and Kopsaftopoulos F 2013 Statistical time series methods for vibration based structural health monitoring, in: *New Trends in Structural Health Monitoring*, ed W Ostachowicz and A Guemes (Springer) pp 209-264

[10] Press W, Teukolsky S, Vetterling W and Flannery B 2007 Kullback–Leibler distance, in: *Numerical Recipes: The Art of Scientific Computing* (3rd ed.) (New York: Cambridge University Press)

[11] Ljung L 1999 *System Identification: Theory for the User* 2nd Edition (Prentice-Hall)

[12] Duda R, Hart P and Stork D 2000 *Pattern Classification* 2nd Edition (Wiley)

Improved Stochastic Subspace System Identification for Structural Health Monitoring

Chia-Ming Chang[1] and Chin-Hsiung Loh[2*]
[1,2] Department of Civil Engineering, National Taiwan University, Taipei 10617, Taiwan

[*]E-mail: lohc0220@ntu.edu.tw

Abstract. Structural health monitoring acquires structural information through numerous sensor measurements. Vibrational measurement data render the dynamic characteristics of structures to be extracted, in particular of the modal properties such as natural frequencies, damping, and mode shapes. The stochastic subspace system identification has been recognized as a power tool which can present a structure in the modal coordinates. To obtain qualitative identified data, this tool needs to spend computational expense on a large set of measurements. In study, a stochastic system identification framework is proposed to improve the efficiency and quality of the conventional stochastic subspace system identification. This framework includes 1) measured signal processing, 2) efficient space projection, 3) system order selection, and 4) modal property derivation. The measured signal processing employs the singular spectrum analysis algorithm to lower the noise components as well as to present a data set in a reduced dimension. The subspace is subsequently derived from the data set presented in a delayed coordinate. With the proposed order selection criteria, the number of structural modes is determined, resulting in the modal properties. This system identification framework is applied to a real-world bridge for exploring the feasibility in real-time applications. The results show that this improved system identification method significantly decreases computational time, while qualitative modal parameters are still attained.

1. Introduction

Performance of civil infrastructure directly affects public safety and society cost. Civil infrastructure refers to the integration of various systems such as buildings, bridges, transportation networks, lifeline, etc. Components in such systems need to be functional; otherwise, a huge amount of economic loss and dead lives would occur and impact the entire society. For example, the I-35W Mississippi River Bridge collapsed on August 1, 2007, resulting in 13 people killed and 145 people injured. A replacement bridge, the I-35W Saint Anthony Falls Bridge, was then constructed and opened on September 18, 2008. The estimated users' economic loss was US$71,000 to US$220,000 a day, while more than 50,000 users needed to reroute [1]. This example demonstrates that deficient and aging infrastructure systems require diagnosis of present conditions to prevent catastrophic failures. Thus, structural health monitoring (SHM) is of need to early identify these problems in structures [2, 3]. Operational modal analysis is to extract the dynamic characteristics of structures based on the structural vibration responses. These dynamic characteristics are composed of natural frequencies,

[1] Research Associate, Dept. of Civil Engineering, National Taiwan University, Taipei, Taiwan.
[2*] Professor (corresponding author), Dept. of Civil Engineering, National Taiwan University, Taipei, Taiwan.

damping ratios, and mode shapes. Deviations in these dynamic characteristics reflect the changed properties of structures. Detailed inspection may be required for the structures. A number of researchers applied system identification to real-world structures for operational modal analysis such as Farrar and James [4], Brownjohn [5], Lynch *et al.* [6], Siringoringo and Fujino [7], Weng *et al.* [8] and Jang *et al.* [9]. With growth of sensing technology, structural health monitoring has drawn a lot of attentions to researchers in order to assure structures of their serviceability and safety. Structural integrity can be then studied through operational modal analysis to the sensor measurements of structures.

Due to unavailability of input excitation, the operational modal analysis is directed to the stochastic system identification. This type of system identification methods is emphasized on performance assessment of structures using measured outputs. In terms of time-domain approaches, one of the popular methods is stochastic subspace system identification (SSI) proposed by Van Overschee and De Moor in 1991 [10]. This method utilizes the extended observability to derive modal parameters Moreover, Peeters and De Roeck in 1999 further extended the SSI method to improve computational efficiency [11]. Peeters and De Roeck in 2001 also proposed the stabilization diagram for SSI to enhance the quality of identification results [12]. Even though such nice methods are developed for operational modal analysis, both identification efficiency and quality are still a concern in practice.

One of common seen problems in SSI is the noise modes obtained in results. This noise modes may be reduced or removed by means of time series analysis prior to system identification such as the Singular Spectrum Analysis (SSA). Numerous fields has employed SSA to investigate the trend and periodic components to various components [13-16]. Moreover, SSA can be extended to multi-channel time series. For structural health monitoring applications, a large amount of data are measured from structures. When applying SVD to the Hankel matrix, the loading in computation becomes a problem. Further reductions on the dimension of the Hankel matrix is a solution to expedite the computational rate as well as to be suitable for structural health monitoring problems.

Another possible reason to result in noise modes in identification results is the orthogonal projection. Yang and Nagarajaiah in 2014 found that the outliers in measure signals can contaminate identification results and introduce noise modes to be obtained [17]. Moreover, the low-amplitude signals in measurements can also induce the difficulty in the calculation of a subspace through the orthogonal projection. To effectively eliminate the noise modes in identification results, these unfavourable components need to be removed in advance.

As introduced in [12], a stabilization diagram is helpful for determining true modes in identification. However, establishing a stabilization diagram may take a large amount of time, especially state-space realization. Many studies developed a series of methods that allow determining the number of states [18-20]. The resulting number of states from these methods are tended to be overestimated. Thus, these methods can be employed in a stabilization diagram to limit the numbers of states used. To accelerate the determination of true modes, modifications should be made to a stabilization diagram.

In this study, an improved stochastic subspace system identification method is developed. This method consists of 1) signal preprocessing using SSA, 2) forming a modified Hankel matrix with a reduced dimension, 3) effectively orthogonal projection to obtain multiple subspaces, and 4) extraction of modal parameters from a given range of numbers of states. The signal preprocessing using SSA is employed to reduce the noise effect in system identification as well as to retain the principle components in measured signals. A modified Hankel matrix is developed to eliminate the low-energy components in this matrix. Then, multiple subspaces are calculated using the proposed effective projection method. Each subspace can render a set of modal parameters. By analysing these sets of modal parameters, the dynamic characteristics of a structure can be obtained. Moreover, the proposed method is evaluated by field measurements from a cable-stayed bridge. The identification results illustrate that the efficiency and quality of the proposed stochastic subspace system identification method suppress those of the conventional SSI.

2. Framework of Proposed System Identification Method

Figure 1 illustrates the improved stochastic subspace system identification method. This improved method contains four sequential components: 1) preprocessing, 2) forming Hankel matrix, 3) obtaining subspaces through projection, and 4) determining stable modes. The signal preprocessing eliminates signal trends and offsets and prepares denoised signals. The Hankel matrix is then formed, and only those columns with higher norms are retained. In comparison to the convectional stochastic subspace system identification, the improved method calculates several subspaces instead of one. The state-space system and measurement matrices are derived based on the extended observability matrix that is calculated from a subspace. A range of orders (or a range of numbers of states) are used to examine different sets of identified modes, and the range is determined by proposed criteria. Finally, a number of modes are obtained with respect to subspaces and orders. The stable modes are determined using a similar approach to the stabilization diagram.

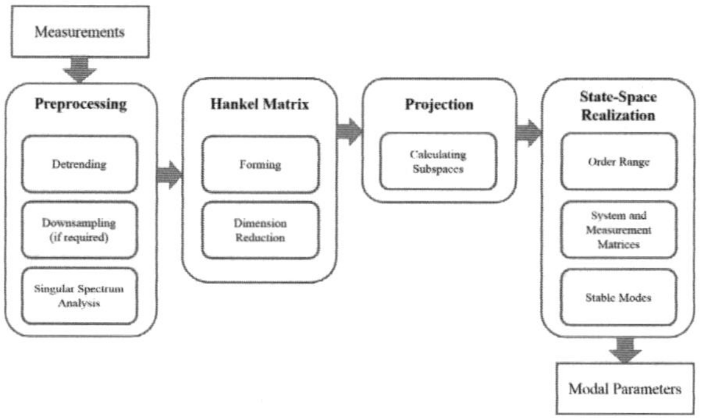

Figure 1. Framework of improved stochastic subspace system identification.

In this study, the proposed system identification method is developed in accordance to the stochastic state-space model for a structure. The stochastic discrete-time state-space model is defined by

$$\begin{aligned} &\mathbf{x}[\mathrm{k}+1]=\mathbf{A}_d x[\mathrm{k}]+\mathbf{w}[\mathrm{k}] \\ &\mathbf{y}[\mathrm{k}]=\mathbf{C}_d[\mathrm{k}]+\mathbf{v}[\mathrm{k}] \\ &\mathbf{x}\in\Re^{n_S\times 1},\ \mathbf{y}\in\Re^{n\times 1},\ \mathbf{w}\in\Re^{n_S\times 1},\ \mathbf{v}\in\Re^{n\times 1} \end{aligned} \tag{1}$$

where **x** and **y** are the state and output measurement vectors at time step, k; $\mathbf{A}_d$ and $\mathbf{C}_d$ are the system and measurement matrices in the stochastic state-space representation; and **w** and **v** are the input and output noise. The time span is defined as $k\in[1,N]$ where N is the total number of samples.

2.1 Pre-processing

Measured signals from structures are sometimes distorted or contaminated by drift, offset, and noise. A drift and offset can be corrected by a simple function such as low-order polynomial function and piecewise linear functions. Noise can be reduced or eliminated by filtering or time series analysis. Downsampling signals to the frequency range of interest is an approach to limit the frequency content in low frequencies, and then noise in high frequencies can be reduced. Another method is to utilize the Multi-channel Singular Spectrum Analysis (MSSA). MSSA consists of three steps: embedding, singular value decomposition, and reconstruction [16]. The basic concept of MSSA is to find a transformation matrix that satisfies

$$\mathbf{Y}=\begin{bmatrix} \mathbf{y}[1] & \mathbf{y}[1+h] & \cdots & \mathbf{y}[1+(n_c-1)h] \\ \mathbf{y}[2] & \mathbf{y}[2+h] & \cdots & \mathbf{y}[2+(n_c-1)h] \\ \vdots & \vdots & \ddots & \vdots \\ \mathbf{y}[l] & \mathbf{y}[l+h] & \cdots & \mathbf{y}[l+(n_c-1)h] \end{bmatrix}=\mathbf{TY}+(\mathbf{I}-\mathbf{T})\mathbf{Y} \tag{2}$$

where **Y** is the extended Hankel matrix (or the delay coordinates) with a specific delay, h; l is the window length each column vector; n_c is the number of delays in the row; **T** is the transformation matrix; and (**I-T**) indicates the noise or less significant components in the **Y** space. The transformation matrix represents the mapping of most significant components in the space of the measurements, $\mathbf{y}[k]$.

Embedding in SSA is to form an extended Hankel matrix such as **Y** in Eq. (2). In the SSA or MSSA theory, h in Eq. (2) is typically equal to 1; l should be in a range of 2~N/2; and n_c is equal to N-l+1. To reduce the dimension in embedding, an additional parameter, h, is introduced. A smaller size of the extended Hankel matrix can expedite the rest of the SSA process (e.g., singular value decomposition and reconstruction). When using h, some rules should be satisfied such as

$$\begin{gathered} 1 \le h \le l \\ \left(1+\left(n_c-1\right)h\right) > nl \\ 2 < l < N/2 \end{gathered} \tag{3}$$

Singular value decomposition in SSA is to obtain a transformation matrix. When **Y** in Eq. (2) is decomposed by SVD, the transformation matrix can be represented by

$$\begin{gathered} \mathbf{Y} = \mathbf{U}\boldsymbol{\Lambda}\mathbf{V}^T = \mathbf{U}_m\mathbf{U}_m^{\mathrm{T}}\mathbf{Y} + \left(\mathbf{I} - \mathbf{U}_m\mathbf{U}_m^{\mathrm{T}}\right)\mathbf{Y} \\ \mathbf{T} = \mathbf{U}_m\mathbf{U}_m^{\mathrm{T}} \end{gathered} \tag{4}$$

where $\boldsymbol{\Lambda}$ is a matrix which contains nonnegative singular values of **Y** in diagonal terms; **U** and **V** are unitary matrices. The singular values in $\boldsymbol{\Lambda}$ are in a decreasing order so that the first few column vectors in $\mathbf{U}_{\mathrm{m}}$ indicate the most significant components in the space of **Y**. Assume that the number of modes in noise-free structural measurements is known, and the number of significant components are equal to two times of number of modes; the rest of singular values are very close to zero. Consider structural measurements with noise, and the rest of singular values will be nonzero. For a single-channel time series (e.g., n = 1), the variance of noise can be estimated through SVD to the covariance matrix of **Y**, given by

$$\mathbf{M} = \frac{1}{\left(n_c-1\right)h}\mathbf{Y}\mathbf{Y}^T = \mathbf{U}\bar{\boldsymbol{\Lambda}}\mathbf{U}^T \tag{5}$$

where

$$\begin{gathered} diag\left(\boldsymbol{\Lambda}\right) = \left[\lambda_1 \quad \cdots \quad \lambda_{no} \quad \lambda_{no+1} \quad \cdots \quad \lambda_l\right] \\ \lambda_1 \ge \lambda_2 \ge \cdots \ge \lambda_{no} > \lambda_{no+1} \ge \cdots \ge \lambda_l \end{gathered} \tag{6}$$

Because **Y** has a maximum rank equal to the row dimension, the left unitary matrix, **U**, in both Eqs. (4-5) is the same. no is assumed to be two times of number of modes, and the measurement noise is assumed to be Gaussian white noise. The variance of measurement noise, σ , in this single-channel time series is estimated by

$$\sigma \cong \frac{1}{\left(l-no\right)}\sum_{i=no+1}^{l}\lambda_i \tag{7}$$

To extend the same idea in multi-channel structural measurements, Eq. (7) is modified by

$$\frac{1}{n}\sum_{j=1}^{n}\sigma_j \cong \frac{1}{\left(nl-no\right)}\sum_{i=no+1}^{nl}\lambda_i \tag{8}$$

where σ_j is the variance of measurement noise with respect to channel. When the number of modes or no is unknown, Eq. (8) can be treated as a criterion to examine no. Because no ranges from 2 to l, only a few trial-and-error attempts are required to approach no using Eq. (8). Consequently, the transformation matrix is determined.

The final step in MSSA is reconstruction that averages out the processed measurements (e.g., **TY**). When single-channel measurements are considered and h equals 1, the reconstruction cam be completed by the diagonal averaging of **TY** [16]. For MSSA, a block-diagonal averaging method can be applied if h is equal to 1. When h is greater than 1, the reconstruction can be completed by a similar approach, counting the number of occurrences per step in **Y** and then averaging the signals at a step by this number. The reconstructed signals contains the most dominate components in a structure.

2.2 Forming Hankel Matrix

As described in [10], a Hankel matrix is formed in the beginning of the SSI method. Likewise to Eq. (2), the Hankel matrix, **H**, is composed of the processed measurements, given by

$$\mathbf{H} = \begin{bmatrix} \mathbf{Y}_p \\ \hline \mathbf{Y}_f \end{bmatrix} = \left[\begin{array}{cccc} \bar{\mathbf{y}}[1] & \bar{\mathbf{y}}[2] & \cdots & \bar{\mathbf{y}}[n_h] \\ \bar{\mathbf{y}}[2] & \bar{\mathbf{y}}[3] & \cdots & \bar{\mathbf{y}}[n_h+1] \\ \vdots & \vdots & \ddots & \vdots \\ \bar{\mathbf{y}}[l] & \bar{\mathbf{y}}[l+1] & \cdots & \bar{\mathbf{y}}[n_h+l] \\ \hline \bar{\mathbf{y}}[l+1] & \bar{\mathbf{y}}[l+2] & \cdots & \bar{\mathbf{y}}[n_h+l+1] \\ \bar{\mathbf{y}}[l+2] & \bar{\mathbf{y}}[l+3] & \cdots & \bar{\mathbf{y}}[n_h+l+2] \\ \vdots & \vdots & \ddots & \vdots \\ \bar{\mathbf{y}}[2l] & \bar{\mathbf{y}}[2l] & \cdots & \bar{\mathbf{y}}[n_h+2l] \end{array}\right] \tag{9}$$

where the subscripts, "p" and "f", denote the "past" and "future" delay coordinates, and $\bar{\mathbf{y}}$ is the processed measurements using the preprocessing methods. The purpose of forming the Hankel matrix is to obtain a column subspace through projection. However, the subspace may be derived a low-resolution projection if the column vectors in **H** has some small numbers. Thus, the improved method introduces a criterion that eliminates "ill-conditioned" column vectors in **H**. These ill-conditioned column vectors can be viewed as those have lower norms. Define that the column vector has the maximum norm in **H** as h_{max}, and then only those column vectors having a norm greater than $\alpha_h h_{\max}$ are retained, given by

$$\bar{\mathbf{H}} = \begin{bmatrix} \bar{\mathbf{Y}}_p \\ \hline \bar{\mathbf{Y}}_f \end{bmatrix} = \begin{bmatrix} \mathbf{h}_1 & \mathbf{h}_2 & \cdots & \mathbf{h}_{m_h} \end{bmatrix}$$
$$\|\mathbf{h}_i\| \geq \alpha_h h_{\max},\ i \in [1, m_h] \tag{10}$$

where m_h is the number of columns in $\bar{\mathbf{H}}$. Utilizing Eq. (10), $\bar{\mathbf{H}}$ is derived with a reduced dimension, as well as the column vectors which may induce the projection errors are deleted.

2.3 Projection

The second step in the proposed SSI method is projection that allows one or multiple subspace(s) to be obtained. In Eq. (9), the Hankel matrix is divided into the past and future portions. As given in [10], the projected subspace is calculated using the past and future matrices in Eq. (10) and defined as

$$\mathbf{O} = \bar{\mathbf{Y}}_f / \bar{\mathbf{Y}}_p = \bar{\mathbf{Y}}_f \bar{\mathbf{Y}}_p^T \left(\bar{\mathbf{Y}}_p \bar{\mathbf{Y}}_p^T\right)^{-1} \bar{\mathbf{Y}}_p \tag{11}$$

where **O** is the projected subspace. Consider using SVD to represent $\mathbf{Y}_p$ and assume that $\mathbf{Y}_p$ is a low-rank matrix, and then Eq. (11) can be simplified as

$$\bar{\mathbf{Y}}_f / \bar{\mathbf{Y}}_p \cong \bar{\mathbf{Y}}_f \mathbf{V}_{p,m} \mathbf{V}_{p,m}^T \to \bar{\mathbf{Y}}_f \mathbf{V}_{p,m}$$
$$\bar{\mathbf{Y}}_p = \mathbf{U}_{p,m} \mathbf{\Lambda}_{p,m} \mathbf{V}_{p,m}^T + \mathbf{U}_{p,0} \mathbf{\Lambda}_{p,0} \mathbf{V}_{p,0}^T \tag{12}$$

where the subscripts, "m" and "0", denote the main and null spaces. All singular values in $\mathbf{\Lambda}_{p,0}$ are assumed to be close to zero. $\mathbf{V}_{p,m}^T$ in Eq. (12) is neglected because this matrix can be a similarity matrix to both extended observability and controllability matrices in the SSI theory. Therefore, a projected subspace is derived from $\bar{\mathbf{Y}}_f \mathbf{V}_{p,m}$.

In the SSI theory, a Hankel matrix in Eq. (9) only supports to calculate one subspace. The separation line in this Hankel matrix let an identical dimension of the past and future submatrices to be obtained. Thus, the two submatrices result in the extended observability and controllability matrices with an equal number of rows and columns, respectively. However, the separation line can be shifted as long as the resulting past and future submatrices have a number of time lags greater than the number of system states. This approach allows multiple subspaces to be attained and multiple sets of modal parameters to be compared.

2.4 State-space Realization

These dynamic characteristics, such as natural frequencies, damping, and mode shapes, are implied in the stochastic state-space representation of the structure. Because the projected subspace represents the product of the extended observability and controllability matrices, SVD is utilized to separate these two matrices. Then, the system and measurement matrices, $\mathbf{A}_d$ and $\mathbf{C}_d$, are calculated by

$$\bar{\mathbf{Y}}_f \mathbf{V}_{p,m} = \mathbf{U}_{id,m} \mathbf{\Lambda}_{id,m} \mathbf{V}_{id,m} + \mathbf{U}_{id,0} \mathbf{\Lambda}_{id,0} \mathbf{V}_{id,0}$$
$$\mathbf{\Gamma} = \mathbf{U}_{id,m} = \begin{bmatrix} \mathbf{C}_d \\ \mathbf{C}_d\mathbf{A}_d \\ \mathbf{C}_d\mathbf{A}_d^2 \\ \vdots \\ \mathbf{C}_d\mathbf{A}_d^{n_s} \end{bmatrix}, \quad \mathbf{A}_d = \begin{bmatrix} \mathbf{C}_d \\ \mathbf{C}_d\mathbf{A}_d \\ \mathbf{C}_d\mathbf{A}_d^2 \\ \vdots \\ \mathbf{C}_d\mathbf{A}_d^{n_s-1} \end{bmatrix}^{\dagger} \begin{bmatrix} \mathbf{C}_d\mathbf{A}_d \\ \mathbf{C}_d\mathbf{A}_d^2 \\ \mathbf{C}_d\mathbf{A}_d^3 \\ \vdots \\ \mathbf{C}_d\mathbf{A}_d^{n_s} \end{bmatrix} \tag{13}$$

where the subscripts, "m" and "0", denote the main and redundant components in the product of the extended observability and controllability matrices; $\mathbf{\Gamma}$ represents the extended observability matrix; n_s is a given number of states; and $(\ )^{\dagger}$ denotes the pseudo inverse. By applying the eigen analysis to $\mathbf{A}_d$, the eigenvalues are the complex-valued natural frequencies, and the eigenvectors represents the mode shapes with respect to states. In each mode, these modal parameters can be obtained by

$$\mathbf{A}_d \boldsymbol{\eta} = \lambda_d \boldsymbol{\eta}$$
$$\frac{1}{\Delta t} \ln\left(\lambda_d, \lambda_d^*\right) = -\xi\omega_n \pm j\omega_n\sqrt{1-\xi^2} \tag{14}$$
$$\boldsymbol{\varphi} = \mathbf{C}_d \boldsymbol{\eta}$$

where λ_d is one of the eigenvalues in $\mathbf{A}_d$; λ_d^* is the complex conjugated λ_d; $\boldsymbol{\eta}$ is the eigenvector with respect to the state vector; ω_n and ξ are the natural frequency and damping ratio; Δt is the sample time; and $\boldsymbol{\varphi}$ is the mode shapes with respect to the output. By calculating all eigenvalues and eigenvectors from $\mathbf{A}_d$, the modal parameters can be extracted.

3. Application to Field Test Measurements

The last example is to employ the proposed system identification method in a structural health monitoring problem of a cable-stayed bridge in Taiwan. The detailed instrumentation on this bridge are available from [8]. In this study, the focus is to examine the proposed system identification method using the ambient vibration responses, collected from the deck of the cable-stayed bridge in the vertical direction. As compared to the methods in [8], this study centers on the efficiency and qualitative results of system identification to be achieved.

The parameters used in the proposed system identification method are first presented. The structural responses collected are velocities, and the total duration of the measurements is 150 seconds. The sensor locations with respect to the bridge dimension are exhibited in Figure 2. The vertical velocity responses are used in the system identification. These measurements are recorded at 100 Hz. Before employing the recorded data in system identification, the constant offsets in the measured signals are removed, and the data are resampled to 50 Hz. In MSSA, the parameters, l and h, are set to be 50 and 5, respectively. The resulting number of orders by Eq. (8) is 30. In SSI, the window length, l, remains the same, and the locations of separation lines are {320 336 352 368 384 400 416 432 448 464 480}. By exploring the power density functions of all measurements, the preliminary guess of modes is 9; however, the 8th and 9th modes have lower energy so that the adjusted number of modes is 7. Then, the range of numbers of states is 12~30 with an increment of 2. Because of these parameters, the total number of runs using the proposed system identification method is 110.

Figure 3 shows the identified natural frequencies with the range of numbers of states between 12 and 26. Likewise, the magenta curves represent the summation of power spectral density functions, while the "star" markers indicate the identified natural frequencies with respect to a separation line and number of states. The vertical lines are the finalized natural frequencies after cluster analysis. As shown in this figure, the identified modes are concentrated in low frequencies. The first peak at a very frequency in the power spectral density is interpreted as a suspicious mode. In the identification results, only a few runs show this suspicious natural frequency. The number of successfully identified modes is 6. To increase the success number, a higher number of states may be used. In addition, the identified mode shapes are drawn in Figure 4. The results are quite comparable to the study in [8], though two of modes are missing in this study. In comparison of computational time, 110 runs performed using the proposed method is about 40 seconds, while the conventional approach including the stabilization diagram needs at least 10 minutes using the same computer. For system identification, the proposed method is able to attain the computational efficiency and qualitative results.

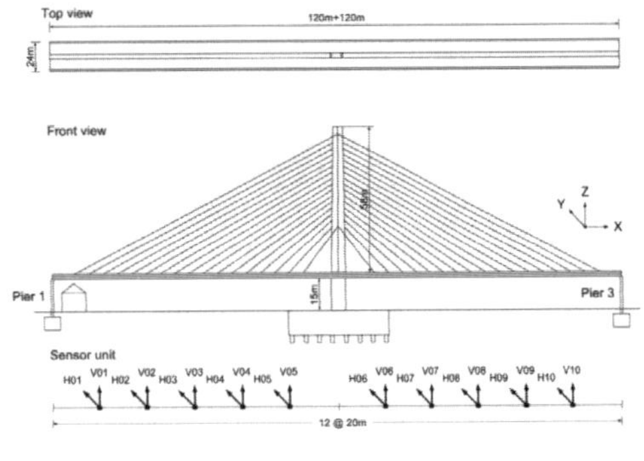

Figure 2. Sensor setup for bridge deck measurements.

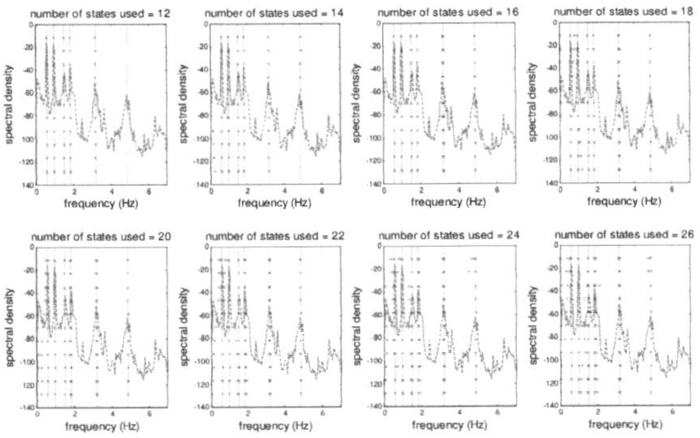

Figure 3. Identified natural frequencies from the field measurements of a cable-stayed bridge.

4. Conclusions

This study proposed an improved stochastic subspace system identification method that can enhance the efficiency and quality in operational modal analysis. In this method, multi-channel singular spectrum analysis was employed to remove noise and uncertainties in measurements, and to estimate the number of states that can be used in the stochastic state-space realization. When forming a Hankel matrix for the stochastic subspace system identification, the low-energy components were removed by a proposed criterion. An efficient projection method was also developed to expedite the process of generating a subspace. By shifting separation lines, multiple sets of the past and future matrices can be obtained, and multiple subspaces were then derived. These subspaces can yield different sets of modal parameters to be obtained. The finalized modal parameters were determined from these sets of results with respect to separation line and number of states.

An example provided was the proposed method applied to the field measurements of a cable-stayed bridge. The results were quite comparable to the previous study, and the computation efficiency was significantly improved in comparison to the conventional stochastic subspace system identification method. Because of these assessments, the proposed method was applicable to be used in most structural health monitoring problems in terms of operational modal analysis.

Acknowledgement
The authors are grateful for the support from the Monistry of Science & Technology, Taiwan, under the grant number MOST 103-2625-M-002 -006.

References

[1] Xie F and Levinson D 2011 Evaluating the effects of the I-35W bridge collapse on road-users in the twin cities metropolitan region Transportation Planning and Technology. 34(7) 691-703
[2] Farrar CR and Worden K 2007 An introduction to structural health monitoring Philosophical Transactions of the Royal Society of London A. 373 303-315
[3] Sim SH and Spencer BF Jr. 2009 Decentralized strategies for monitoring structures using wireless smart sensor networks Newmark Structural Engineering Laboratory (NSEL) Report Series, No. 19, University of Illinois at Urbana-Champaign, Urbana, Illinois (http://hdl.handle.net/2142/14280)
[4] Farrar CR and James GH 1997 System identification from ambient vibration measurements on a bridge Journal of Sound and Vibration 205(1) 1-18
[5] Brownjohn JMW 2003 Ambient vibration studies for system identification of tall buildings Earthquake Engineering and Structural Dynamics 32 71-95
[6] Lynch JP, Wang Y, Loh KJ, Yi J-H and Yun C-B 2006 Performance monitoring of the Geumdang Bridge using a dense network of high-resolution wireless sensors Smart Materials and Structures 15 1561-1575
[7] Siringoringo DM and Fujino Y 2008 System identification of a suspension bridge from ambient vibration response Engineering Structures 30 (2) 462-477
[8] Weng J-H, Loh C-H, Lynch JP, Lu K-C, Lin P-Y, and Wang Y 2008 Output-only modal identification of a cable-stayed bridge using wireless monitoring systems Engineering Structures 30 1820-1830
[9] Jang SA, Jo H, Cho S, Mechitov KA, Rice JA, Sim SJ, Jung HJ, Yun CB, Spencer BF and Agha G 2010 Structural health monitoring of a cable-stayed bridge using smart sensor technology: deployment and evaluation Journal of Smart Structures and Systems 6(5-6) 439-459
[10] Van Overschee P and De Moor B 1991 Subspace algorithms for the stochastic identification problem Proceedings of the 30th IEEE Conference on Decision and Control, Brighton, UK, 1321-1326
[11] Peeters B and De Roeck G 1999 Reference-based stochastic subspace identification for output-only modal analysis Mechanical Systems and Signal Processing 13(6) 855-878
[12] Peeters B and De Roeck G 2001 Stochastic system identification for operational modal analysis: a review Journal of Dynamic Systems, Measurement, and Control 123(4) 659-667
[13] Broomhead DS and King GP 1986 Extracting qualitative dynamics from experimental data Physica D 20 217-236
[14] Broomhead DS, Jones R and King GP 1987 Singular system analysis with application to dynamical systems In E.R. Pike & L.A. Lugaito (Eds.) Chaos, noise, and fractals 15-27
[15] Fraedrich K 1986 Estimating dimensions of weather and climate attractors Journal of the Atmospheric Sciences 43 419-432
[16] Elsner JB and Tsonis AA 1996 Singular spectrum analysis: a new tool in time series analysis Plenum Press, New York
[17] Yang Y and Nagarajaiah S 2014 Blind denoising of structural vibration responses with outlier via principal component pursuit Journal of Structural Control and Health Monitoring 21(6) 962-978
[18] Akaike H 1974 Anew look at the statistical identification model IEEE Trans. Autom. Control 716-723
[19] Schwarz G 1978 Estimating the dimension of a model Ann. Stat. 6 461-464
[20] Koehler AB and Murphree ES 1988 A comparison of the Akaike and Schwarz of criteria selecting model order J. R. Stat. Soc. Ser. C: Appl. Stat. 37(2) 187-195

Multi-objective shape optimization of plate structure under stress criteria based on sub-structured mixed FEM and genetic algorithms

Pierre Garambois, Sebastien Besset, Louis Jézéquel
Laboratoire de Tribologie et Dynamique des Systèmes, UMR CNRS 5513,
Ecole Centrale de Lyon, 36 avenue Guy de Collongue 69134 Ecully Cedex, France
E-mail: [pierre.garambois,sebastien.besset,louis.jezequel]@ec-lyon.fr

Abstract.
This paper presents a methodology for the multi-objective (MO) shape optimization of plate structure under stress criteria, based on a mixed Finite Element Model (FEM) enhanced with a sub-structuring method. The optimization is performed with a classical Genetic Algorithm (GA) method based on Pareto-optimal solutions and considers thickness distributions parameters and antagonist objectives among them stress criteria. We implement a displacement-stress Dynamic Mixed FEM (DM-FEM) for plate structure vibrations analysis. Such a model gives a privileged access to the stress within the plate structure compared to primal classical FEM, and features a linear dependence to the thickness parameters. A sub-structuring reduction method is also computed in order to reduce the size of the mixed FEM and split the given structure into smaller ones with their own thickness parameters. Those methods combined enable a fast and stress-wise efficient structure analysis, and improve the performance of the repetitive GA. A few cases of minimizing the mass and the maximum Von Mises stress within a plate structure under a dynamic load put forward the relevance of our method with promising results. It is able to satisfy multiple damage criteria with different thickness distributions, and use a smaller FEM.

1. Introduction

The goal of this paper is to implement a DM-FEM for plate structure enhanced with a sub-structuring reduction method for MO structural optimizations with thickness repartition parameters and mass/stress criteria. This reduced model provides a fast and efficient analysis of complex plate structures vibrations and facilitates the work of the GA used for the optimization.

The optimization of a structure's parameters in order to prevent damages is a major concern in mechanical engineering but can only be achieved at the expense of others essential criteria. Even though often carried out manually, MO-optimization methods also naturally appear in the litterature, such as gradient-based method [1, 7] and MO-GA-based methods [6, 8]. In this study, the optimization is performed with a classical GA-based NSGA-II method [4], as it permits to dispense with any weighting of criteria. This kind of method enables to find a set of Pareto-optimal solutions, which are all optimal compared to each other, for at least one criterion. Nevertheless, an inconvenient lies in the repetitions of the criteria's evaluation, depending on the chosen method and theory.

In the literature, some of the previous works on thin structures optimization focus on laminated composites [1, 8] with thickness and orientation layer parameters, as well as piezoelectric smart structures [6] with input energy and deflection variables. Shell structures optimization has also been studied [7, 3] and make the integration between FEM and CAD modeling softwares. In all these cases, the calculation of stresses is often necessary as a constraint of the optimization, and more rarely as a proper objective. Furthermore, they use a primal FEM that discretizes the displacements and requires extra calculation to get to the stress, and eventually needs rebuilding of the meshing, which increase the CPU time. The originality of the present work is the use of a different FEM formulation giving stresses as a primary result of the structural analysis, and avoid rebuilding as well, in order to improve the repetitive GA.

Many different formulations for mechanical problems permit to access to various parameters in the same model. Washizu's book [10] gives a good insight of all the variational methods with different fields. Among these, the Hellinger-Reissner's [9] describes both displacements and stresses, and the application to FEM is well explained by Wriggers' book [11]. In this work, we program a mixed displacement-generalized stress dynamic mixed FEM (DM-FEM), based on Kirchoff-Love thin plate theory, which allows a direct access to the stress field within the plate, and save operations, each iteration. Moreover, the plate theory implementation permits to act on the thickness of the plate without building a whole new assembly. This also improves the performance of the optimization techniques as the thicknesses of the different zone of the structure constitute parameters for the optimization.

In addition to the DM-EM, a sub-structuring reduction method imagined by the authors [5] and adapted to mixed models is used in order to reduce the numerical size of the structure. That method allows to compute the stress criteria on a mixed reduced FEM, even smaller than the primal corresponding FEM (that would only give the displacement). Moreover, that method also provides the advantage of being linearly dependent from the thickness, and turn the sub-structures into small super-elements defined by its thicknesses. Thus, it allows to quickly work on a smaller model and improves the technique all the more.

A few investigations are made, using a GA-based MO optimization method improved by the DM-FEM and the sub-structuring method. We consider stress criteria and thickness parameters. The idea is to split the test structure into sub-structures with their own thicknesses (parameters) so as to minimize the mass and the Von Mises (VM) stress within the whole plate (objectives) under a dynamic load. We put forward the principal differences between the regular method and ours, and show interesting results satisfying multiple damage criteria with different thickness distributions.

In the two first parts, we present the optimization problem and give a quick description of the GA. We continue with the main point of the paper: the implementation of the mixed FEM and the associated sub-structuring method. Finally we present the principal relevant examples.

2. Optimization problem

The "example" structure we want to optimize is built up with 768 thin plate elements (see figure 1). It is composed of 3 flat sub-structures, with their own thicknesses, which are the parameters of the optimization (range defined by the user).

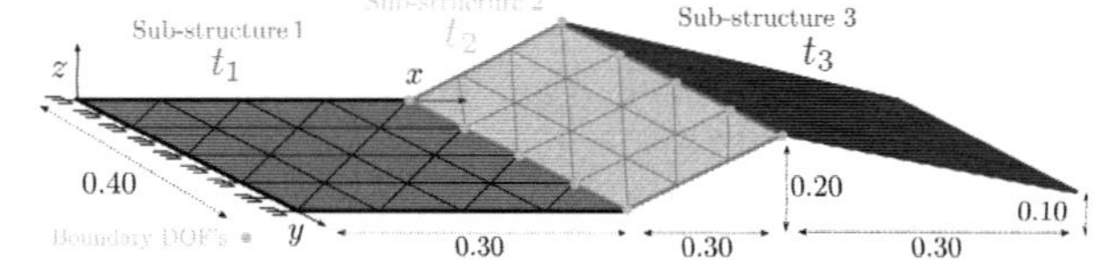

Figure 1: Plate structure composed of 3 flat sub-structures with their own thicknesses

The problem we try to solve is the following: we want to minimize both the maximum VM stress within the whole plate structure for a given n^{th} mode, and the mass in a bid of costs. These two objectives being antagonistic, the goal is to find some thickness distributions that are good compromises between the mass and the maximum stress undergone by the structure. As such, the MO-GA-based method quite naturally appears.

3. Genetic Algorithm method

The GA methods are based on the evolution of species in their natural environment. It consists in making evolve a population, whose individuals, solutions of the problem, tend to improve for the purpose of our objectives as the generations follow. In this context, the word used to explain its functioning are taken from biology, namely:

- an "individual" is a solution to the problem (here, a combination of 3 thicknesses (one each sub-structure) that gives a compromises between the objectives)
- a "population" is a group of "individuals" (here, a set of thickness combinations, that forms a Pareto front with one compromises per thickness distribution)
- a "generation" is an iteration of the algorithm and correspond to a population

The functioning of the algorithm is summarized in figure 2). It is composed of 5 different stages ("evaluation" stages 2 being the most costly).

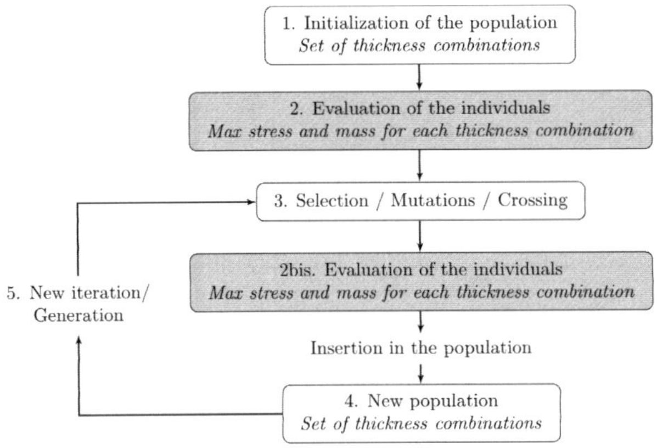

Figure 2: Principle of the genetic algorithm

4. Mixed Finite Element Model for thin Kirchhoff-Love plate

The "evaluation step" of the GA is by far the most costly because we need to evaluate the stress within the whole plate structure, and identify the maximum. The classical method that consists in building a primal FEM to access the displacement and use extra calculations to reach the stress is the most common and intuitive method, but it appears to be heavy going because of the extra calculations. The originality of our method consists in the construction of a displacement-generalized stress DM plate FEM that gives a direct access to both fields in the response of the structure. Another important benefit of this method, when choosing a plate theory, is the possibility of modifying the thickness as a parameters without rebuilding a whole assembly.

This type of DM-FEM is based on the HR mixed functional [9, 11] expressed for dynamics problems. It may correspond to the regular Lagrangian used in dynamics, but computed with mixed component, both functions of displacements and stresses, as follows:

$$\Pi_{HRD} = \iiint_V -\boldsymbol{\sigma}_{ij}\boldsymbol{e}_{ij}(\boldsymbol{u}_i) + \frac{1}{2}\boldsymbol{\sigma}_{ij}\boldsymbol{S}_{ijkl}\boldsymbol{\sigma}_{kl} + \boldsymbol{b}_i\boldsymbol{u}_i + \frac{1}{2}\rho\dot{\boldsymbol{u}}_i^2 \, \mathrm{dV} \tag{1}$$

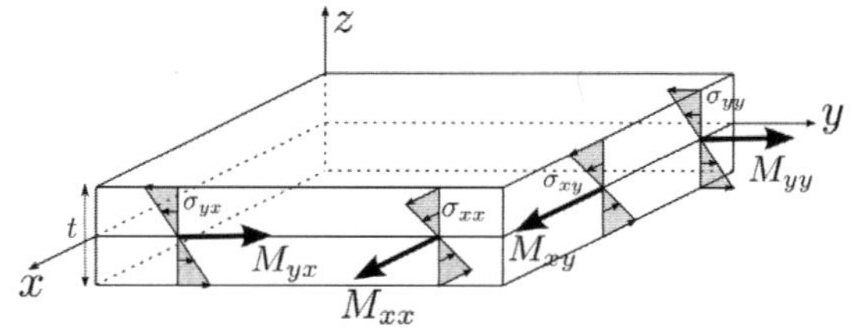

Figure 3: Bending and twisting moment and transverse shear force

considering $\boldsymbol{\sigma}_{ij}$ the stress, $\boldsymbol{u}_i$ the displacement, $\boldsymbol{e}_{ij}(\boldsymbol{u}_i)$ the strain function of the displacement $\boldsymbol{u}_i$, $\boldsymbol{b}_i$ the body force, ρ the density and $\boldsymbol{S}_{ijkl}$ the elastic compliance matrix. The stationary condition or Euler-Lagrange equations can then be applied to the functional so as to conventionally solve a dynamic structure problem.

The discretization of the generalized stresses and the displacements (according to figure 3) with plate elements using Kirchhoff-Love (KL) thin plate is as follows:

$$\boldsymbol{\sigma}_{ij} = \left\{M_x, M_y, M_{xy}\right\}^T = \boldsymbol{P\beta} \tag{2}$$

$$\boldsymbol{u}_i = \left\{w, \theta_x, \theta_y\right\}^T = \boldsymbol{NU} \tag{3}$$

$$\boldsymbol{e}_{ij} = \left\{\epsilon_{xx}, \epsilon_{yy}, \gamma_{xy}\right\}^T = \boldsymbol{D}\,\boldsymbol{u}_i = \boldsymbol{DNU} \tag{4}$$

with $\left\{M_x, M_y, M_{xy}\right\}^T$ respectively the bending moments in the x, and y direction, and the twisting moment within the plate (thickness t), $\left\{w, \theta_x, \theta_y\right\}^T$ respectively the transverse displacement of the plate, and the section rotations around x and y. $\boldsymbol{P}$ is the generalized stresses shape function matrix, $\boldsymbol{\beta}$ the generalized stress parameters vector within the plate, $\boldsymbol{N} = \left\{\boldsymbol{N}_w, \boldsymbol{N}_\theta\right\}^T$ the displacements shape function matrix, $\boldsymbol{U} = \left\{\boldsymbol{U}_w, \boldsymbol{U}_\theta\right\}^T$ the displacements parameters vector et $\boldsymbol{D}$ the displacement-strain tensor-operator.

When carefully splitting the matrix in order to separate the matrices dependent on t, t^3 and $\frac{1}{t^3}$, the application of the Euler-Lagrange equations leads us to the following matrix formulation of the dynamic mixed FEM (for one plate element of surface S):

$$\begin{Bmatrix} t\tilde{\boldsymbol{M}}_{F_w} & \boldsymbol{0} & \boldsymbol{0} \\ \boldsymbol{0} & t^3\tilde{\boldsymbol{M}}_{F_\theta} & \boldsymbol{0} \\ \boldsymbol{0} & \boldsymbol{0} & \boldsymbol{0} \end{Bmatrix} \begin{Bmatrix} \ddot{\boldsymbol{U}}_{F_w} \\ \ddot{\boldsymbol{U}}_{F_\theta} \\ \ddot{\boldsymbol{\beta}} \end{Bmatrix} + \begin{Bmatrix} \boldsymbol{0} & \boldsymbol{0} & \boldsymbol{G}_w^T \\ \boldsymbol{0} & \boldsymbol{0} & \boldsymbol{G}_\theta^T \\ \boldsymbol{G}_w & \boldsymbol{G}_\theta & \frac{1}{t^3}\tilde{\boldsymbol{H}} \end{Bmatrix} \begin{Bmatrix} \boldsymbol{U}_{F_w} \\ \boldsymbol{U}_{F_\theta} \\ \boldsymbol{\beta} \end{Bmatrix} = \begin{Bmatrix} \boldsymbol{F}_w \\ \boldsymbol{F}_\theta \\ \boldsymbol{0} \end{Bmatrix} \tag{5}$$

with $\tilde{\boldsymbol{M}}_{F_w} = \iint_S \boldsymbol{N_w}^T \rho \boldsymbol{N_w} \mathrm{dS}$, $\tilde{\boldsymbol{M}}_{F_\theta} = \iint_S \boldsymbol{N_\theta}^T \frac{\rho}{12} \boldsymbol{N_\theta} \mathrm{dS}$, $\boldsymbol{G_w} = \iint_S \boldsymbol{P}^T \boldsymbol{D} \boldsymbol{N_w} \mathrm{dS}$, $\boldsymbol{G_\theta} = \iint_S \boldsymbol{P}^T \boldsymbol{D} \boldsymbol{N_\theta} \mathrm{dS}$, $\boldsymbol{G} = \left\{\boldsymbol{G}_w, \boldsymbol{G}_\theta\right\}$ $\tilde{\boldsymbol{H}} = \iint_S -\boldsymbol{P}^T \tilde{\boldsymbol{S}}_{ijkl} \boldsymbol{P} \mathrm{dS}$, $\tilde{\boldsymbol{S}}_{ijkl} = t^3 \boldsymbol{S}_{ijkl}$ and $\boldsymbol{F} = \left\{\boldsymbol{F}_w, \boldsymbol{F}_\theta\right\}^T$ being the nodal forces (splitted for w and θ parameters).

The advantages of this method are the direct access to the stress (doesn't requires to rebuild the stress field from the displacements) and the thickness parameters easily mutable for the same meshing (doesn't require to rebuild $\tilde{\boldsymbol{M}}_{F_w}$, $\tilde{\boldsymbol{M}}_{F_\theta}$, $\boldsymbol{G}$ and $\tilde{\boldsymbol{H}}$). Nevertheless, it has the inconvenient of being numerically bigger than a primal FEM (see Table 1). A sub-structuring reduction method is thus implemented, in order to decrease the size of the DM-FEM, taking care of the thickness independence of the method for each sub-structure.

5. Sub-structuring method

The method used in this paper has been explained in [5]. The idea is to split the structure into few sub-structures, and reduce each of them separately using the primal corresponding FEM. The initial Degrees Of Freedom (DOF) of a given sub-strucures a are separated in internal DOF $\boldsymbol{U}_i^a$, junction DOF $\boldsymbol{U}_j^a$ (permitting to link sub-structures between them) and stress DOF $\boldsymbol{\beta}^a$. The displacements are projected on a basis composed of truncated primal "fixed" modes (see Craig & Bampton method [2]) and the stresses on a basis composed of truncated primal "fixed" modes as well, but projected on the stresses. Thus, the new reduced parameters are: modal components $\boldsymbol{\eta}^a_{FI\,(U,\beta)}$ representing both the displacements and the stresses, and the junction DOF $\boldsymbol{U}_j^a$ remaining unchanged. The reduction of the whole sub-structure a is given by:

$$\begin{Bmatrix} \boldsymbol{U}_i^a \\ \boldsymbol{U}_j^a \\ \boldsymbol{\beta}^a \end{Bmatrix} = \begin{Bmatrix} \boldsymbol{\Phi}^a_{FIi\,(U,\beta)} & \boldsymbol{\Psi}_i^a \\ \boldsymbol{0} & \boldsymbol{I}_{ij} \\ t^3\tilde{\boldsymbol{P}}^a \begin{Bmatrix} \boldsymbol{\Phi}^a_{FIi\,(U,\beta)} \\ \boldsymbol{0} \end{Bmatrix} & t^3\tilde{\boldsymbol{P}}^a \begin{Bmatrix} \boldsymbol{\Psi}_i^a \\ \boldsymbol{I}_{ij} \end{Bmatrix} \end{Bmatrix} \begin{Bmatrix} \boldsymbol{\eta}^a_{FI\,(U,\beta)} \\ \boldsymbol{U}_j^a \end{Bmatrix} \tag{6}$$

with $\tilde{\boldsymbol{P}}^a = -(\tilde{\boldsymbol{H}}^a)^{-1}\boldsymbol{G}^a$, $\boldsymbol{\Phi}^a_{FIi\,(U,\beta)}$ being a truncated basis of the "fixed" modes of the sub-structures a (with the junction considered fixed) and $\boldsymbol{\Psi}_i^a = -\boldsymbol{K_{ii}}^{-1}\boldsymbol{K_{ij}}$ being the matrix of the constraint static modes, taken from the primal FEM. We assemble two sub-structures a and b considering:

$$\boldsymbol{U}_j^a = \boldsymbol{U}_j^b = \boldsymbol{U}_j \tag{7}$$

The formulation of the reduction basis gives us another advantage: in fact, the "fixed" modes of the sub-structures a contained in the basis $\boldsymbol{\Phi}^a_{FIi\,(U,\beta)}$ remains the same for a flat sub-structure, even when the thickness change. In the same way, the basis $\left\{\boldsymbol{\Psi}_i^a, \boldsymbol{I}_{ij}\right\}^T$ representing the reaction of the sub-structure when its junction is moving doesn't depend on the thickness of the plate, as long as it remains flat. In this way, our reduction method is all the more robust that it is easily transposable to different thickness distributions (doesn't requires to rebuild $\tilde{\boldsymbol{P}}^a$, $\boldsymbol{\Psi}_i^a$ and $\tilde{\boldsymbol{P}}^a$).

Basically, each sub-structure works as a super-element, defined by its thickness only. Each individual of each generation, an almost instantaneous simple operation is made to modify the thickness (colored in red) in the matrices of equation 5 and 6. Thus, the "evaluation step" (see figure 2) to reach the stress simply comes down to the computation of the structure response, using mixed reduced matrices smaller than the primal ones classically used (see Table 1).

Table 1: Characteristics of the FEMs and DOFs for a 768 elements structure

FEM	DOFs	Stress field
Primal (displacement)	2448	Reconstruction needed
Mixed (displacement and stress)	9360	Direct
Mixed reduced (20 truncated modes each sub-structure)	168	Direct

6. Example: optimization of the thickness combination to minimize the mass & the maximum stress for one mode

The example we use is schematized in figure 1 in section 2. We minimize the maximum VM stress within the whole structure for a given mode n (see shape of mode 1 in figure 4c and mode 7 in figure 5c), and the mass. The parameters are the thicknesses of each "single plate" or "sub-structure" (thickness zones in figures 4b and 5b). Thickness parameters vary in a range between 1 and 2 mm. The plate is made of $S210$ steel. The load applied is a harmonic force in the z direction, whose pulsation corresponds to the n mode, and is considered normalized. A hysteretic damping is also added to the model for each mode to keep a realistic system. The FEM is composed of 768 elements. Table 1 shows the improvement DOF-wise for this example.

The results are presented as "Pareto Front" (black points in 4a and 5a). These points are defined by the right ordinate axis and the abscissa axis representing respectively the mass of the structure and the maximum VM stress. Each of these optimal points are better than the others in the sense of at least one of the two objectives. In that way they are all compromises we can select to design our structure, depending on the objective we want privilege. Once a Pareto point chosen, the colored curves plotted in addition to the Pareto Front in 4a and 5a leads to the parameters corresponding to the point. They are defined by the left ordinate axis and the abscissa axis representing respectively the thickness of each of the sub-structures for the chosen Pareto point (with the same abscissa) and the maximum VM stress in the whole structure for this point. Basically, when selecting a point of the Pareto Front, the corresponding thicknesses to design the structure correspond to ordinate of the colored three points with the same abscissa.

Those results were obtained with NSGA-II with a population of 100 individuals and 1000 generations. We used the DM-FEM reduced with the sub-structuring method to improve the CPU time, and easily changed the t thicknesses of each sub-structure for each evaluation step. The GA has been compiled several times so as to check the convergence of the results.

The first subsection deals with the optimization for the mode 1, a simple example whose sketch of results could be intuitive. The second one treats of the optimization for the mode 7 and shows interesting results with "parametric typing".

6.1. Simple example: mode 1

The results are presented in figure 4. This example shows some thickness combinations that allow to have a VM stress between 7 and 140 MPa, with a mass of the structure between 2.9 and 4.7 Kgs. In this case, more than the Pareto-Front itself, the most relevant is the evolution of the thicknesses in function of VM stress (colored curves in figure 4a). In fact, we observe that the most important zone to "reinforce" so as to decrease damages is the first one, and then the second one. When taking a look at the stress distribution for the first mode (figure 4c), this evolution of thicknesses in function of the stress seems coherent. Intuitively, we would decrease the thickness where the stress is low, and increase it where it is high. This test puts forward the validity of the method qualitatively, and brings a quantitative point in term of stress.

6.2. Test example: mode 7

The results are presented in figure 5. Considering the stress distribution for the mode 7 (figure 5c), the results are not intuitive. This example shows some thickness combinations that allows us to have a VM stress between 7 and 150 MPa, with a mass of the structure between 2.9 and 4.7 Kgs. The evolution of the 3 parameters in function of VM stress (colored curves in figure 5a) is very specific. In fact, we distinguish 3 main parametric typing:

- from 7 until 56 MPa: it is the lowest stress range. It appears that the thickest sub-structures of this range are the first one near the housing, and the second one in the middle of the whole structure. The evolution of each of them is not identical, but the first zone remains thicker than the second one. On the other hand, that range contains a particularity: the thickness of the third sub-structure remains equal to the range minimum 1 mm.
- from 57 until 72 MPa: it is the middle stress range. It presents some significant differences with the first one. In fact, the second sub-structure becomes the thickest, and the third one is no longer minimum. Plus, that range still contains a particularity: the first sub-structure which was the thickest in the first stress range is now the thinest with a minimal thickness.

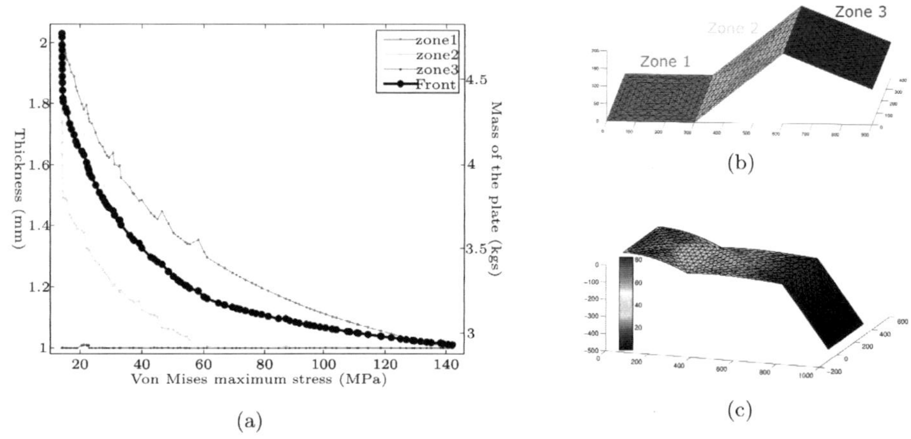

Figure 4: Optimization results for mode 1: a/ Pareto Front (black) & thicknesses for each point (colored), b/ Independent thickness zones, c/ Stress distribution on mode 1 (Mpa)

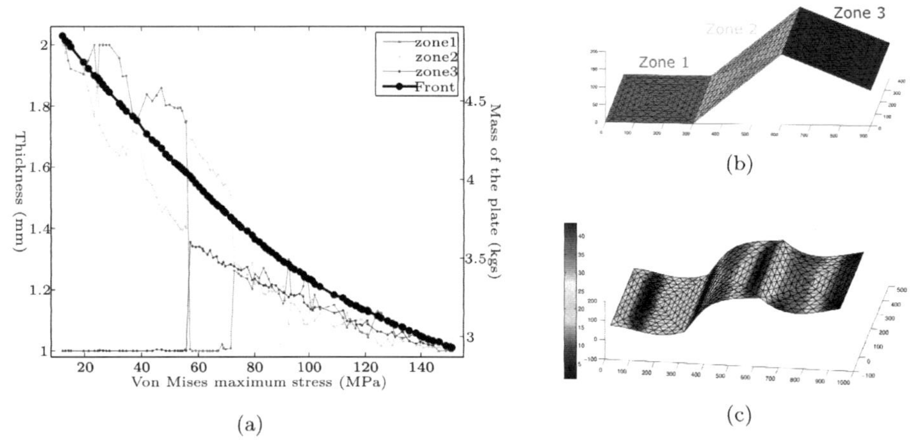

Figure 5: Optimization results for mode 7: a/ Pareto Front (black) & thicknesses for each point (colored), b/ Independent thickness zones, c/ Stress distribution on mode 7 (Mpa)

- from 73 until 150 MPa: it is the highest stress range. The evolution of the thicknesses and their respective order of magnitude is completely different. In fact, there is no "minimal value" zone compared to the 2 first ranges. The three parameters have close values (between 1 and 1.3 mm) and their evolution is close as well, since they all decrease the same from 1.3 till 1 mm as the VM stress increase until its maximum.

Those results define a phenomenon we call "parametric typing". Although the Pareto Front has no singularities, the parameters' evolution do not follow any specific or logical rule. Nevertheless, they can be divided into three main stress domains which correspond to significantly different thickness distributions. That presents an interesting advantage in terms of design possibilities: it means you can find two different thickness distributions (two designs) that leads to almost the same Pareto point (same property in terms of the objectives we minimize). That example shows promising results that could be interesting in a more complex industrial case, and puts forward the interest of such a method, in addition to the efficiency previously highlighted.

7. Conclusion

This paper introduces a new methodology for MO optimization technique of structural dynamic plate problems with stress criteria and thickness parameters. The optimization is made with a classical NSGA-II method that finds a set of Pareto-optimal solutions, but presents the originality of using a displacement-stress DM plate FEM. Not only the latter permits a direct access to the stress field compared to a classical FEM, but it is also featured by a linearly dependence to the thickness parameters which dispense with rebuilding the meshing each iteration. Furthermore, in order to offset the inconvenient of a bigger numerical size, a fast sub-structuring reduction method is implemented, each sub-structure with its own thickness. It turns the DM-FEM into a mixed reduced DM-FEM, smaller than the classical FEM and requires less calculation to get the stress criterion. These two methods combined permits to improve the costly "evaluation step" of the GA and make this MO optimization method more efficient.

This study shows two practical cases, using a plate structure composed of 3 sub-structures and 768 elements. Their goals are to minimize the maximum VM stress and the mass of the plate structure in function of the proper thickness of each sub-structure, under a dynamic load (harmonic modal pulsation). Our optimization technique permits compute the response on a reduced FEM and save CPU time, especially with high density meshing. We find a whole group of compromises between the maximum VM stress and the mass for different thickness distributions. The first example shows intuitive results and puts forward the relevance of the techniques. The second example, less obvious, presents specific results such as parametric typing, namely some zone in the Pareto front characterized by significantly different thickness parameters evolution in function of the maximum stress. This feature leads to interesting design possibility as two antagonist thickness distributions prove a close stress response. Beyond those findings, we have implemented an efficient MO optimization method for dynamic plate structural problems, that gives the user an opportunity to have a whole set of design possibilities in order to satisfy multiple antagonist damage criteria. We could now consider the perspective of testing industrial structures and study the response on a wide frequency band.

References

[1] A.L Araujo, P Martins, C.M Mota Soares, Mota Soares C.A, and J Herskovits. Damping optimization of viscoelastic laminated sandwich composite structures. Struct. Mult. Opt., 39:569–579, 2009.

[2] D Brizard, L Jezequel, S Besset, and B Troclet. Determinental method for locally modified structures. application to the vibration damping of a space launcher. Computational Mechanics, 20:631–644, 2012.

[3] L Dai, Z.Q Guan, B.S Chen, and H.W Zhang. An open platform of shape design optimization for shell structures. Structural and Multidisciplinary Optimization, 35:609–622, 2008.

[4] K Deb, S Agrawal, A Pratap, and T Meyarivan. A fast elitist non-dominated sorting genetic algorithm formulti-objective optimization: Nsga-ii. Notes in Computer Science, 2000.

[5] P Garambois, S Besset, and L Jézéquel. Modal synthesis applied to a reissner mixed plate finite element dynamic model. Proc. of the 9th Int. Conference on Structural Dynamics, EURODYN 2014, 2014.

[6] R Kudikala, K Deb, and B Bhattacharya. Multi-objective optimization of piezoelectric actuator placement for shape control of plates using genetic algorithms. Journal of Mechanical Design, 131:091007–1–11, 2009.

[7] N.D Lagaros and V Papadopoulos. Optimum design of shell structures with random geometric, material and thickness imperfections. International Journal of Solids and Structures, 43:6948–6964, 2006.

[8] B Paluch, M Grédiac, and A Faye. Combining a finite element programme and a genetic algorithm to optimize composite structures with variable thickness. Composite Structures, 83:284–294, 2008.

[9] E Reissner. On a mixed variational theorem and on shear feformable plate theory. International Journal For Numerical Methods In Engineering, 23(2):193–198, 1986.

[10] Washizu. Variational methods in elasticity and plasticity, 2nd edition. Pergamon Press, 1975.

[11] P Wriggers and C Cartsensen. Mixed Finite Element Technologies. App. Math./Comp. Meth. of Eng., 2009.

Evaluation of damage indexes for structural health monitoring in nonlinear mechanical systems based on Volterra series

Sidney Bruce Shiki, Samuel da Silva

UNESP - Univ Estadual Paulista, Faculdade de Engenharia de Ilha Solteira, Departamento de Engenharia Mecânica, Av. Brasil 56, 15385-000, Ilha Solteira, SP, Brasil

E-mail: sbshiki@gmail.com, samuel@dem.feis.unesp.br

Abstract. Nonlinear behavior can be important for the monitoring of the structural state of mechanical systems since it can be mistakenly classified as a damage in the systems. However, nonlinear tools are still not consolidated and need further research for applications in mechanical system. Discrete-time Volterra series is an interesting mathematical framework to deal with nonlinear dynamics. It is a clear generalization of the linear convolution for weakly nonlinear systems. In the present work two different damage indicators are proposed based on Volterra models by considering linear and nonlinear contributions of the total response. The goal is to evaluate these indicators using an experimental test rig where the damages are simulated and considering the nonlinear regime of motion. The main advantages and drawbacks of the proposed methodology are highlighted in the final remarks.

1. Introduction

There is an increasing importance on considering nonlinear phenomena in structural dynamics due to the need of better performance in terms of comfort, weight, noise, durability, among others features [1]. Nonlinearities can bring complex effects as jumps, limit cycle oscillations, harmonic distortion, chaos and others [2]. These effects can be strong in structures which are subjected to clearance, impacts, dry friction and bolted connections [3]. All these features shows that the study of the effect of nonlinearities in the monitoring of mechanical systems are highly important for the development of structural health monitoring (SHM) methodologies since these effects can mask or be mistakenly viewed as structural changes [4]. Many techniques as harmonic balance [5], nonlinear autoregressive models [6], nonlinear normal modes [7], among others, have been already presented in the literature. However, there is still no general technique to deal with this kind of system.

In this sense, the Volterra series expansion can be an interesting framework since it is a generalization of the linear convolution and the impulse response function [8]. Because of this, methodologies that are useful in the linear domain can be extended, until certain point, to nonlinear structures. Another advantage is that this model can separate the linear and nonlinear part of the response enabling to have an idea of the degree of nonlinearity of the system.

A few papers have already applied similar formulations in the damage detection problem. Chatterjee [9] identified a Volterra model representing a clamped beam with a breathing crack modeled by a bilinear oscillator. The model was identified using the harmonic probing method and the variations in the harmonics in the response signal were related to the opening of the crack. Tang *et al* [10] applied the Volterra expansion for the diagnosis of a rotor-bearing system. In this paper the Volterra kernels were monitored and compared to a baseline model in order to detect unbalance and rubbing conditions. Rébillat *et al* [11] used a Hammerstein model to calculate damage indicators based on the degree of nonlinearity of the response of the system. The technique was tested in initially linear systems subjected to breathing cracks that allows the system to behave in nonlinear way.

The goal of the present paper is to apply the Volterra series to identify the system using the input and output data. Numerical issues of this model are avoided using orthonormal Kautz functions. The identified model of the structure in the healthy condition is used to monitor the structural state of the system. In previous papers of the authors of the present work the Volterra model was already applied in the problem of damage detection in simulated [12] and experimental examples [13]. The current paper tries to improve the previous results in a benchmark structure with a better control on the application of the damage and also with a more detailed characterization of the behavior of the proposed damage index. A few analysis of the quality of the damage indicators are presented with the results. Finally, the main advantages and drawbacks of the technique are highlighted.

2. Application of discrete-time Volterra series for damage detection

The total response of the system in discrete-time, $y(k)$, can be separated in a linear component, $y_1(k)$, and a sum of nonlinear components $y_2(k) + y_3(k) + \ldots$ using the Volterra series by:

$$y(k) = \sum_{\eta=1}^{+\infty} y_\eta(k) = y_1(k) + y_2(k) + y_3(k) + \ldots \tag{1}$$

Despite being a infinite expansion [14], this expression can usually be truncated until a low-order component (i.e. order 2 or 3) for the case of smooth nonlinearities [15, 16] .

Each $\eta-$th order component of the expansion, $y_\eta(k)$, can be represented as a multidimensional convolution between the $\eta-$th Volterra kernel, $\mathcal{H}_\eta(n_1, \ldots, n_\eta)$, and the input signal $u(k)$:

$$y_\eta(k) = \sum_{n_1=0}^{N_1} \cdots \sum_{n_m=0}^{N_\eta} \mathcal{H}_\eta(n_1, \ldots, n_\eta) \prod_{i=1}^{\eta} u(k - n_i) \tag{2}$$

where $N_1, ..., N_\eta$ is the number of terms considered to represent the $\eta-$th Volterra kernel. The main problem of this approach is that this number of terms can be high, specially in higher-order kernels. This can make it difficult to calculate the kernels since it is an ill-posed numerical problem with severe problems of convergence [17]. This issue can be minimized by using orthonormal functions $\psi_{i_j}(n_j)$ to represent the $\eta-$th Volterra kernel $\mathcal{H}_\eta(n_1, \ldots, n_\eta)$:

$$\mathcal{H}_\eta(n_1, \ldots, n_\eta) \approx \sum_{i_1=1}^{J_1} \cdots \sum_{i_\eta=1}^{J_\eta} \mathcal{B}_\eta\,(i_1, \ldots, i_\eta) \prod_{j=1}^{\eta} \psi_{i_j}(n_j) \tag{3}$$

where $J_1, ..., J_\eta$ is the number of samples in each $\eta-$th orthonormal projection of the kernel $\mathcal{B}_\eta\,(i_1, \ldots, i_\eta)$. With this reduced representation of the Volterra kernels, it is possible to write the Volterra model in a matrix form by grouping the input signal in a matrix $\mathbf{\Gamma}$ and the output signal in the vector $\mathbf{y}$. In this way the vector with the orthonormal projections of the kernels, $\mathbf{\Phi}$, can be calculated by a least-squares estimation:

$$\mathbf{\Phi} = (\mathbf{\Gamma}^T\mathbf{\Gamma})^{-1}\mathbf{\Gamma}^T\mathbf{y} \tag{4}$$

With the identified Volterra kernels it is possible to make the multidimensional convolution as depicted in the eq. 2 to calculate each $\eta-$th component of the response. In this work, this model is considered to represent the studied system in its reference undamaged condition. Therefore, this model can be used to monitor the state of an initially nonlinear system.

The damage index was based on the prediction error of the reference Volterra model [12]. The main idea of this metric is that the response of the model should deviate from the measured response if a change in the structure happens. The prediction error, e_η, is defined considering η components in the Volterra series expansion:

$$e_\eta = y_{exp} - \sum_{m=1}^{\eta} y_m \tag{5}$$

where y_{exp} is the measured response, and y_m is the $m-$th component of the series expansion. An $\eta-$th order metric, named λ_η, was defined as the ratio between the standard deviation of the prediction error in the unknown condition, $e_{\eta,unk}$, and in the reference condition, $e_{\eta,ref}$:

$$\lambda_\eta = \frac{\sigma\left(e_{\eta,unk}\right)}{\sigma\left(e_{\eta,ref}\right)} \tag{6}$$

In the cases where the structure is undamaged, it is expected that $\lambda_\eta \approx 1$ since the prediction errors in the reference and unknown conditions should have similar statistical distributions. Otherwise, any other effect modified the structure making the index to deviate from 1.

The effects of structural variations and input levels will be showed in the next section.

3. Experimental study of damage detection using Volterra-based metrics

The experimental setup is composed by a clamped-free aluminum beam with dimensions of $300 \times 19 \times 3.2$ mm. The beam have a small steel mass attached to the tip of the free end of the beam and a neodymium magnet positioned 2 mm distant to the mass as seen in figure 1. A shaker is attached 50 mm distant from the clamped end instrumented with a load cell. A laser vibrometer is used to measure the velocity of the oscillation in the free end.

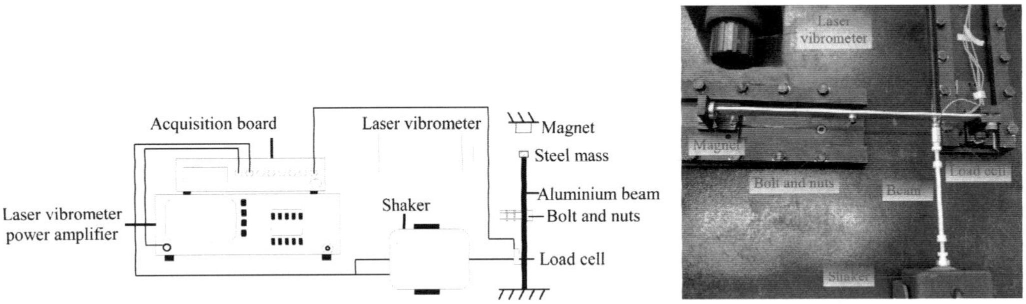

(a) Schematic diagram. (b) General view.

Figure 1. Experimental setup of the magneto-elastic system.

The structure was excited by a sine sweep signal from 10 to 50 Hz in 4 amplitude levels applied in the shaker: 0.01, 0.05, 0.10 and 0.15 V. For each level 40 blocks were collected in a sample rate of 1024 Hz with 8192 samples for each block. A bolt with 4 nuts was placed in a hole in the center of the beam. The reference condition was considered to be the structure with 4 nuts (state 1) and the damage was simulated by removing these nuts one-by-one until only 1 nut remains. After this the same kind of tests were performed again by putting back the nuts in the bolt one-by-one in order to simulate a gradual repair in the structure. This was done in order to investigate if the damage index shows the structure returning to the reference condition. A summary of the tests is illustrated in the table 1.

Table 1. Structural states simulated in the nonlinear system.

State	Condition	State	Condition
1	4 nuts (reference)	5	1 nut (repair)
2	3 nuts (damaged)	6	2 nuts (repair)
3	2 nuts (damaged)	7	3 nuts (repair)
4	1 nut (damaged)	8	4 nuts (repair)

A stepped sine test was also performed in the beam mapping the frequency range from 10 to 40 Hz in 4 different input amplitudes. The frequency response function (FRF) calculated with the sweep sine input and the stepped sine response are showed in the figure 2.

In the sweep sine FRF it is possible to observe the hardening behavior of the system and many distortions in the FRF with the increase of the excitation. The hardening effect can be more clearly seen in the stepped sine response where it is also possible to observe the jump phenomenon characterized by the sudden fall in the response level [3]. The main issue with these effects is that they can be mistakenly viewed as a damage if a proper nonlinear model is not used. Another possibility is that the nonlinearity can also hide effects of structural damage. These facts make it important to identify a model that take in account this kind of phenomenon.

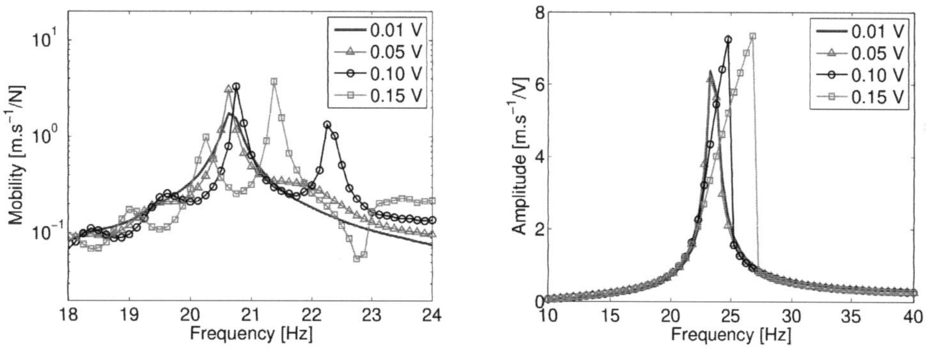

(a) FRF with 4 nuts in the beam.

(b) Stepped sine response for the beam with 4 nuts.

Figure 2. Nonlinearity detection of the magneto-elastic system for different input amplitudes.

In order to model the nonlinearity of the system a 3rd order Volterra model was employed to identify the benchmark structure. The Kautz orthonormal functions were defined through an optimization procedure aiming to minimize the prediction error of the model. The frequencies and damping ratios representing the orthonormal functions for each kernel are presented in table 2. The high-level input (0.15 V) was used to estimate the first three kernels that are illustrated in figure 3. The 1st kernel is analogous to the impulse response function while the other kernels have additional dimensions in the representation. The 3rd kernel can not be fully represented and only the main diagonal is showed for illustration.

With the identified model it is possible to estimate the linear and nonlinear components of the response of the system using the multidimensional convolutions showed in the eq. 2 . The performance of the model in the high-level input is illustrated in figure 4 where it is showed a direct comparison between the responses of the experimental system and of the model, and also the linear and nonlinear components of this response. The response of the model to a sine excitation with 21 Hz and 0.15 V of amplitude is illustrated by the power spectral densities (PSD) in a direct comparison with the experimental response and the components of the response in figure 5. The harmonics are clearly showed in the figure 5 (b) where it is seen that the 1st, 2nd and 3rd kernels reproduces the first three harmonics of the input frequency.

Table 2. Information for the identification of the Kautz functions.

f_1 [Hz]	ζ_1[%]	f_2 [Hz]	ζ_3[%]	f_3 [Hz]	ζ_3[%]
20.65	0.65	19.03	0.27	19.91	1.38

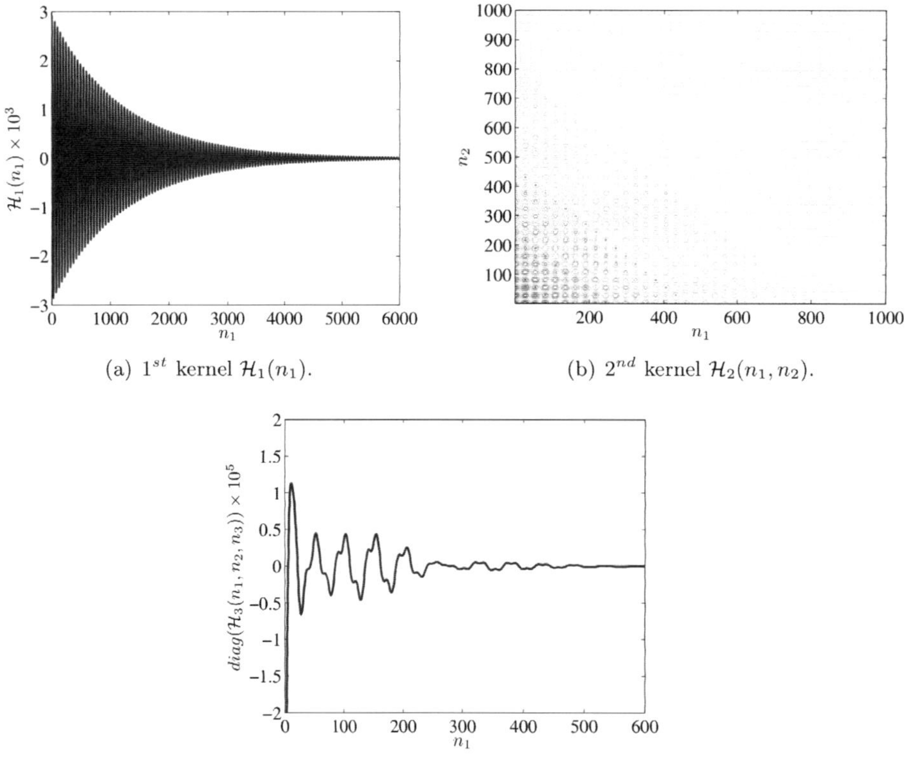

(a) 1st kernel $\mathcal{H}_1(n_1)$.

(b) 2nd kernel $\mathcal{H}_2(n_1, n_2)$.

(c) Main diagonal of the 3rd kernel $\mathcal{H}_3(n_1, n_2, n_3)$

Figure 3. Volterra kernels of the magneto-elastic system in the physical basis.

The damage index based on the prediction error of the reference Volterra model showed in eq. 6 was applied in the acquired dataset for each one of the structural states showed in the table 1. The indexes were calculated with the last 39 blocks in each combination of amplitude levels and structural states. The first block was rejected due to the presence of transients related to the start in the testing. The results for the 39 realizations considering the error bars with $2 \times \sigma$, where σ is the standard deviation of the samples, are depicted in the figure 6 for the low-level input (0.01 V) and in the figure 7 for the high-level input (0.15 V). These figures show a linear

(λ_1) and a nonlinear metric (λ_3) while the 2nd-order index λ_2 was not taken in account since the contributions of y_2 does not showed to be relevant to the response. In the figure 6, for the low-input level (0.01 V), both indexes are sensitive to structural variations since it is possible to observe the increase in the index from the states 1 to 4, where the damage is applied, and the decrease in λ_η from the states 5 to 8, where the repair of the bolts was done.

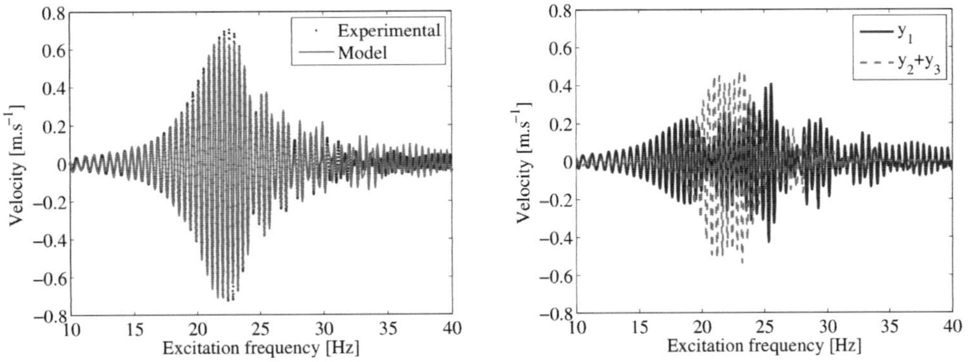

(a) Direct comparison. (b) Components of the response.

Figure 4. Output estimated by Volterra model in healthy condition.

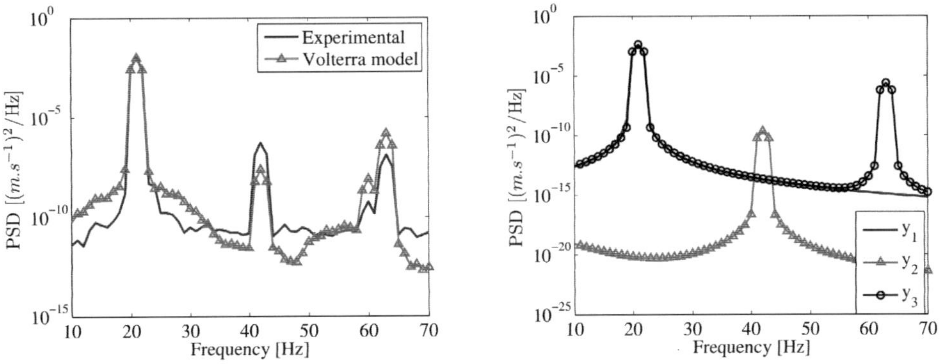

(a) Direct comparison between the PSDs. (b) PSDs of the components of the response.

Figure 5. Response of the Volterra model to a sine excitation.

However, in the figure 7 with the high-level input (0.15 V) only the nonlinear index λ_3 points to the real structural condition while λ_1 detailed in the figure 7(b) does not have a trend consistent with the evolution of the structural state of the system. This happens due to the fact that at this level of displacement of the beam the nonlinearity is very relevant in the response of the structure and so the linear index is not able to distinguish between the nonlinear effects and the damage. This fact illustrates what would happen if one chooses to model the system by assuming only the linear behavior of this kind of structure.

Figure 8 shows the indexes λ_1 against λ_3 for all the realizations in each structural state. In the diagram for the input of 0.01 V both indexes separate well the states of the structure while for the 0.15 V input only λ_3 really characterize the damage. Also a lot of dispersion is observed mainly in the case with 4 nuts and input level of 0.15 V during the application of the damage (blue circles) which was probably caused by the loosening of the nuts during the test cycles. This can also explain why index do not really return to the reference value during the repair.

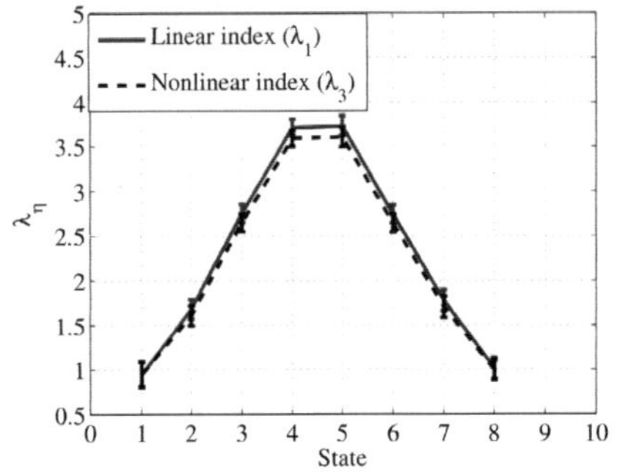

Figure 6. Damage index based on the prediction error (λ_η) for an input level of 0.01 V.

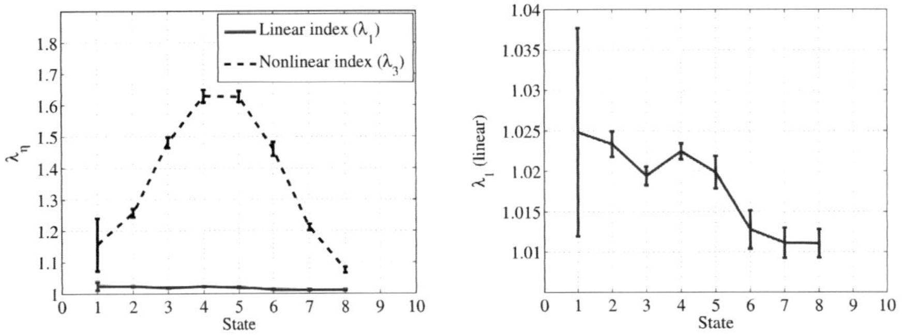

(a) Damage indexes λ_η for an input of 0.15 V. (b) Linear index λ_1 for an input of 0.15 V.

Figure 7. Damage index based on the prediction error (λ_η) for an input level of 0.15 V.

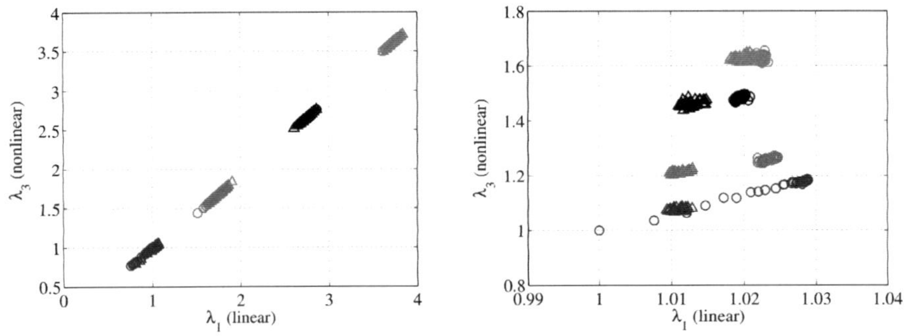

(a) λ_1 *versus* λ_3 for an input of 0.01 V. (b) λ_1 *versus* λ_3 for an input of 0.15 V.

Figure 8. Comparison between the linear index λ_1 and the nonlinear index λ_3. Blue is the reference state with 4 nuts, Red is the case with 3 nuts, Black represents the case with 2 nuts and Purple is the case with 1 nut. Circles ○ represent the case where the damage is being applied and triangles △ represent the repaired cases.

4. Final remarks

This work proposed the application of the Volterra series in the problem of damage detection in an initially nonlinear structure. The structure showed to behave with a hardening type of nonlinearity. If linearity is assumed this kind of behavior can be mistakenly viewed as a damage in the system. Orthonormal Kautz functions were applied in order to minimize numerical problems of the identification of the kernels. An index based on the prediction error of the reference model was tested to detect structural variations. The metric showed to be sensitive to structural variations even under the nonlinear regime of motion. The index also illustrated a clear difference between considering or not the nonlinear behavior of the system since the linear version of the index can fail to represent the structural state.

The results used the interesting property of the Volterra expansion in separating the linear and nonlinear parts of the response of a nonlinear system. This was useful to show the possible drawbacks of the linearity assumption. However, one should keep in mind that the identified model has a limited range of operation since for higher levels of nonlinearity the model could fail since higher-order kernels may be necessary. This is still a problem in most of the nonlinear tools since there is still not a general tool to apply in this kind of system.

5. Acknowledgments

The authors acknowledge the financial support provided by São Paulo Research Foundation (FAPESP, Brasil) by the grant number 12/09135-3, and the National Council for Scientific and Technological Development (CNPq, Brasil) by the grant number 470582/2012-0. The first author is thankful to FAPESP for his scholarship grant number 13/25148-0.

References

[1] Kerschen G, Worden K, Vakakis A F 2006 Past, present and future of nonlinear system identification in structural dynamics *Mech. Syst. Signal Pr.* **20** 502-592.

[2] Virgin L N 2000 *Introduction to Experimental Nonlinear Dynamics: A Case Study in Mechanical Vibration* (Cambridge University Press)

[3] Worden K and Tomlinson G R 2001 *Nonlinearity in Structural Dynamics* (London : Institute of Physics Publishing)

[4] Bornn L, Farrar C R and Park G 2010 Damage detection in initially nonlinear systems *Int. J. Eng. Sci.* **48** 909-920

[5] Peng Z K, Lang ZQ, Billings S A and Tomlinson G R 2008 Comparisons between harmonic balance and nonlinear output frequency response function in nonlinear system analysis *J. Sound Vib.* **311** 56-73

[6] Fassois S D and Sakellariou J S 2007 Time-series methods for fault detection and identification in vibrating structures *Philos. T. R. Soc. A* **365** 411-448

[7] Peeters M, Kerschen G and Golinval J C 2011 Modal testing of nonlinear vibrating structures based on nonlinear normal modes: Experimental demonstration *Mech. Syst. Signal Pr.* **25** 1227-1247

[8] Schetzen M 1980 *The Volterra and Wiener theories of nonlinear systems* (New York: Wiley)

[9] Chatterjee A 2010 Structural damage assessment in a cantilever beam with a breathing crack using higher order frequency response functions *J. Sound Vib.* **329** 3325-3334

[10] Tang H, Liao Y H, Cao J Y and Hang X 2010 Fault diagnosis approach based on Volterra models *Mech. Syst. Signal Pr.* **24** 1099-1113

[11] Rébillat M, Hajrya R and Mehbal N 2014 Nonlinear structural damage detection based on cascade of Hammerstein models *Mech. Syst. Signal Pr.* **48** 247-259

[12] Shiki S B, Lopes Jr V and da Silva S 2013 Damage Detection in Nonlinear Structures using Discrete-Time Volterra Series *Key Eng. Mat.* **569** 876-883

[13] Shiki S B, Hansen C and da Silva S 2014 Nonlinear features identified by Volterra series for damage detection in a buckled beam *International Conference on Structural Nonlinear Dynamics and Diagnosis*

[14] Rugh W J 1991 *Nonlinear System Theory – The Volterra/Wiener Approach* (The Johns Hopkins University Press)

[15] da Silva S 2011 Non-linear model updating of a three-dimensional portal frame based on Wiener series *Int. J. Nonlin. Mech* **46** 312-320

[16] Shiki S B, Lopes Jr V and da Silva S 2013 Identification of nonlinear structures using discrete-time Volterra series *J. Braz. Soc. Mech. Sci.* **36** 523-532

[17] da Rosa A, Campello R J G B and Amaral W C 2007 Choice of free parameters in expansions of discrete-time Volterra models using Kautz functions *Automatica* **43** 1084-1091

Journal of Physics: Conference Series **628** (2015) 012013 doi:10.1088/1742-6596/628/1/012013

Novel SHM method to locate damages in substructures based on VARX models

U Ugalde[1], J Anduaga[1], F Martínez[1], A Iturrospe[2]

[1]Department of Sensors, IK4-Ikerlan, Mondragon, Spain
[2]Department of Electronics and Computer Sciences, Mondragon Goi Eskola Politeknikoa, Mondragon, Spain

E-mail: uugalde@ikerlan.es

Abstract. A novel damage localization method is proposed, which is based on a substructuring approach and makes use of Vector Auto-Regressive with eXogenous input (VARX) models. The substructuring approach aims to divide the monitored structure into several multi-DOF isolated substructures. Later, each individual substructure is modelled as a VARX model, and the health of each substructure is determined analyzing the variation of the VARX model. The method allows to detect whether the isolated substructure is damaged, and besides allows to locate and quantify the damage within the substructure. It is not necessary to have a theoretical model of the structure and only the measured displacement data is required to estimate the isolated substructure's VARX model. The proposed method is validated by simulations of a two-dimensional lattice structure.

1. Introduction

Structural Health Monitoring (SHM) is the process of implementing a damage detection and characterization strategy for engineering structures [1]. SHM is regarded as a very important engineering field in order to secure structural and operational safety; issuing early warnings on damage or deterioration, avoiding costly repairs or even catastrophic collapses [2].

Most of the existing vibration based SHM methods could be classified into two different approaches: global approaches and local approaches [3]. In the global approaches, the goal is to monitor the health of the entire structure. These global methods have been tested and implemented in different types of structures during the last 30 years [4]. However, for many large systems, global monitoring is not practical due to the lack of sensitivity of global features regarding local damages, inaccuracies of developed models or the high cost of sensing, cabling and computational operations [5]. On the other hand, local SHM methods are focused on evaluating the state of reduced parts within the entire structures, based on substructuring methods. This approach aims to overcome global method's problems, dividing the whole structure into substructures and analyzing each one individually.

Several research works have proposed substructuring methods for large-scale structures. Koh [6] presented a "divide and conquer" strategy to monitor large structures based on the division of the whole structure into isolated substructures. For each substructure, structural parameters are identified using the Extended Kalman filter (EKF). However, the EKF usually require knowledge of the system and its dynamics [7]. Yun and Lee [8] detected damage in structures combining a substructuring method and experimental modelling [9]. Most recently, Xing [10] presented another damage detection method based on a substructuring method and Auto-Regressive Moving Average with eXogenous input (ARMAX) models. Damage indicators were obtained for each estimated substructural model calculating the difference between the squared natural frequencies in the healthy state and the squared natural frequencies during the structure lifetime. All the natural frequencies were computed from their respective estimated ARMAX models. This damage detection method was validated through simulations and experimental test. The method proposed by Xing [10] doesn't require any theoretical model of the structure [9]. Nevertheless, as only one internal DOF is measured in each substructure, is not possible to give information about the damage location within the substructures.

In this paper, a damage localization method based on the combination of a substructuring method and experimental modelling is proposed. The substructuring method is used to isolate a multi-DOF substructure from the rest of the structure, and each isolated substructure is modelled as a Vector Auto-Regressive with eXogenous input (VARX) model. VARX models incorporate data measured in different internal DOFs and their coefficient matrices describe the relationship between the measured internal DOFs through some structural characteristics (mass, stiffness, damping...). Therefore, the proposed method could potentially locate the damage within the substructure by analyzing variations on the VARX model over the time. Furthermore, the proposed method doesn't require any theoretical model of the structure.

The rest of the paper is organized as follows. First, the proposed method is presented in section 2. Secondly, the proposed method is evaluated by series of simulations. In section 3, simulation results are discussed and finally the concluding remarks are presented in section 4.

2. The proposed method

The behavior of the structure is described by a lumped parameter model, where we assume that all objects are rigid bodies and all interactions between the rigid bodies take place via springs and dampers. We assume the structure consists of bars that are connected together by rigid joints. The point in which two or more bars are joined is called node and the number of structural nodes will depend on the topology of the structure. The forces could only be transmitted along the axial direction of the bars and the load could only be applied at the two ends of each bar. We assume that the mass of each bar is distributed equally between its two nodes, a half in the first one and the other half in the second one. On the other hand, the structure could be subjected to arbitrary external loading that is assumed to be known and could act on any node.

The structure is divided into different substructures and these substructures are isolated from the remaining structure. The substructures could contain several internal (i) and interface (j) nodes. The interface nodes are located in the border between the selected substructure and the remaining structure. On the other hand, internal nodes are located within the substructure and they are not connected to the nodes of the remaining structure.

The dynamic equations for an internal node i are formulated as follows:

$$\begin{aligned} m_i\ddot{\mathbf{z}}_{\mathbf{xi}} &= \sum_{k=1}^{N}(f_{xk}(\mathbf{z_i},\dot{\mathbf{z}}_\mathbf{i},\mathbf{z_k},\dot{\mathbf{z}}_\mathbf{k}))+\mathbf{F_{xi}} \\ m_i\ddot{\mathbf{z}}_{\mathbf{yi}} &= \sum_{k=1}^{N}(f_{yk}(\mathbf{z_i},\dot{\mathbf{z}}_\mathbf{i},\mathbf{z_k},\dot{\mathbf{z}}_\mathbf{k}))+\mathbf{F_{yi}} \\ m_i\ddot{\mathbf{z}}_{\mathbf{zi}} &= \sum_{k=1}^{N}(f_{zk}(\mathbf{z_i},\dot{\mathbf{z}}_\mathbf{i},\mathbf{z_k},\dot{\mathbf{z}}_\mathbf{k}))+\mathbf{F_{zi}} \end{aligned} \tag{1}$$

where m_i is the lumped mass of the internal node i and $\ddot{\mathbf{z}}_{\mathbf{xi}}$, $\ddot{\mathbf{z}}_{\mathbf{yi}}$ and $\ddot{\mathbf{z}}_{\mathbf{zi}}$ are the absolute accelerations of the internal node i in x, y and z axes respectively. f_{xk}, f_{yk} and f_{zk} are linear or non-linear functions used to calculate the total internal force applied in the node i by the N nodes connected to him. These functions depend on the value of $\mathbf{z_i},\dot{\mathbf{z}}_\mathbf{i},\mathbf{z_k}$ and $\dot{\mathbf{z}}_\mathbf{k}$, where $\mathbf{z_i}$ and $\dot{\mathbf{z}}_\mathbf{i}$ represent the absolute displacement and the velocity of the node i and $\mathbf{z_k}$ and $\dot{\mathbf{z}}_\mathbf{k}$ are the absolute displacements and velocities of the N nodes that are connected to the node i, all of them in x, y and z axes. Furthermore, $\mathbf{F_{xi}}$, $\mathbf{F_{yi}}$ and $\mathbf{F_{zi}}$ are the external forces that are acting in the internal node i.

Expanding f_{xk}, f_{yk} and f_{zk} functions as Taylor series [11] and selecting only the first term, the dynamic equations for an internal node i are stated as:

$$\begin{aligned} m_i\ddot{z}_{xi} &= \sum_{k=1}^{N}(k_{xxki}(z_{xi}-z_{xk})+k_{xyki}(z_{yi}-z_{yk})+k_{xzki}(z_{zi}-z_{zk})+c_{xxki}(\dot{z}_{xi}-\dot{z}_{xk})+c_{xyki}(\dot{z}_{yi}-\dot{z}_{yk})+c_{xzki}(\dot{z}_{zi}-\dot{z}_{zk}))+F_{xi} \\ m_i\ddot{z}_{yi} &= \sum_{k=1}^{N}(k_{yxki}(z_{xi}-z_{xk})+k_{yyki}(z_{yi}-z_{yk})+k_{yzki}(z_{zi}-z_{zk})+c_{yxki}(\dot{z}_{xi}-\dot{z}_{xk})+c_{yyki}(\dot{z}_{yi}-\dot{z}_{yk})+c_{yzki}(\dot{z}_{zi}-\dot{z}_{zk}))+F_{yi} \\ m_i\ddot{z}_{zi} &= \sum_{k=1}^{N}(k_{zxki}(z_{xi}-z_{xk})+k_{zyki}(z_{yi}-z_{yk})+k_{zzki}(z_{zi}-z_{zk})+c_{zxki}(\dot{z}_{xi}-\dot{z}_{xk})+c_{zyki}(\dot{z}_{yi}-\dot{z}_{yk})+c_{zzki}(\dot{z}_{zi}-\dot{z}_{zk}))+F_{zi} \end{aligned} \tag{2}$$

where z_{xi}, z_{yi}, z_{zi} and $\dot{z}_{xi}, \dot{z}_{yi}, \dot{z}_{zi}$ are the absolute displacements and velocities of the node i and z_{xk}, z_{yk}, z_{zk} and $\dot{z}_{xk}, \dot{z}_{yk}, \dot{z}_{zk}$ are the absolute displacements and velocities of the N nodes that are connected to the node i, all of them in x, y and z axes. On the other hand, k and c are coefficients related to the stiffness and damping values of the N bars that are in contact with node i.

The finite central difference method [10] is used to obtain the approximation of the displacement's first and second derivatives. Repeating the explained process (equation 2) for the other internal nodes and representing the expressions in matrix form, the substructural dynamic equation is stated as:

$$\begin{bmatrix} z_{i1x}(n) \\ z_{i1y}(n) \\ z_{i1z}(n) \\ \dots \\ z_{inx}(n) \\ z_{iny}(n) \\ z_{inz}(n) \end{bmatrix} = -A_1 \begin{bmatrix} z_{i1x}(n-1) \\ z_{i1y}(n-1) \\ z_{i1z}(n-1) \\ \dots \\ z_{inx}(n-1) \\ z_{iny}(n-1) \\ z_{inz}(n-1) \end{bmatrix} - A_2 \begin{bmatrix} z_{i1x}(n-2) \\ z_{i1y}(n-2) \\ z_{i1z}(n-2) \\ \dots \\ z_{inx}(n-2) \\ z_{iny}(n-2) \\ z_{inz}(n-2) \end{bmatrix} \quad (3)$$

$$+ B_{10} \begin{bmatrix} z_{ij1x}(n) \\ z_{ij1y}(n) \\ z_{ij1z}(n) \\ \dots \\ z_{ijnx}(n) \\ z_{ijny}(n) \\ z_{ijnz}(n) \end{bmatrix} + B_{11} \begin{bmatrix} z_{ij1x}(n-1) \\ z_{ij1y}(n-1) \\ z_{ij1z}(n-1) \\ \dots \\ z_{ijnx}(n-1) \\ z_{ijny}(n-1) \\ z_{ijnz}(n-1) \end{bmatrix} + B_{12} \begin{bmatrix} z_{ij1x}(n-2) \\ z_{ij1y}(n-2) \\ z_{ij1z}(n-2) \\ \dots \\ z_{ijnx}(n-2) \\ z_{ijny}(n-2) \\ z_{ijnz}(n-2) \end{bmatrix} + B_{21} \begin{bmatrix} F_{i1x}(n-1) \\ F_{i1y}(n-1) \\ F_{i1z}(n-1) \\ \dots \\ F_{inx}(n-1) \\ F_{iny}(n-1) \\ F_{inz}(n-1) \end{bmatrix}$$

where $z_{i1x}, \dots, z_{inz}$ and $z_{ij1x}, \dots, z_{ijnz}$ are the absolute displacements of all internal and interface nodes in x, y and z axes.

Equation (3) could be regarded as a VARX model [12], where $z_{i1x}, \dots, z_{inz}$ corresponds to the endogenous variables and $z_{ij1x}, \dots, z_{ijnz}$ and $F_{i1x}, \dots, F_{inz}$ correspond to the exogenous variables. A_k is a n x n endogenous coefficient matrix, where n denotes the number of internal DOFs of the substructure. The elements of A_k matrix are related to the physical properties of the bars that connect the internal nodes between them. On the other hand, B_k is a n x m exogenous coefficient matrix, where m denotes the number of interface DOFs of the substructure. The elements of B_k are related to the physical properties of the bars that connect one internal node to another interface node.

In the proposed method, the substructures are modelled as a VARX models. The state of each substructure is evaluated analyzing deviations in its estimated coefficient matrices (A_k and B_k) respect to its estimated coefficient matrices (A_k and B_k) in the healthy condition. Firstly substructural damages are detected and secondly the damages are located within the substructure.

3. Numerical results

A linear and time invariant two-dimensional lattice structure is studied in this section. The structure consists of stainless steel bars that are connected together by rigid joints and we assume that the forces could only be transmitted along the axial direction of the bars and the load could only be applied at the two ends of each bar. The structural behaviour is described by a lumped parameter model, where we assume that all object are rigid bodies and all interactions between the rigid bodies take place via springs.

In the studied case, the structure is modelled as a sixteen DOF mass-spring model (see figure 1). Furthermore, a ten DOF substructure is isolated from the general structure, where absolute displacement z_{6x}, z_{6y}, z_{7x}, z_{7y}, z_{8x} and z_{8y} correspond to internal DOFs and absolute displacement z_{4x},

Journal of Physics: Conference Series **628** (2015) 012013 doi:10.1088/1742-6596/628/1/012013

z_{4y}, z_{5x} and z_{5y} correspond to interface DOFs. As shown in figure 1, none external force is applied within the substructure.

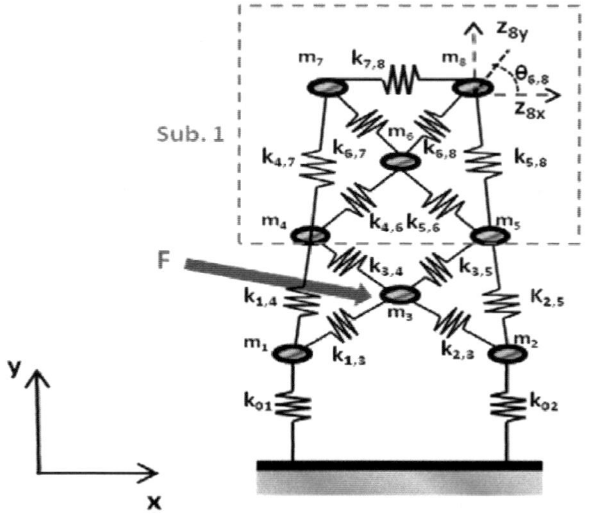

Figure 1. Isolated substructure in the structural model

Below, the dynamic equations for the internal node *i* are formulated:

$$m_i \ddot{z}_{xi} = \sum_{k=1}^{N} (k_{k,i} \cos^2 \theta_{k,i} (z_{xi} - z_{xk}) + k_{k,i} \cos \theta_{k,i} \sin \theta_{k,i} (z_{yi} - z_{yk})) \tag{4}$$

$$m_i \ddot{z}_{yi} = \sum_{k=1}^{N} (k_{k,i} \cos \theta_{k,i} \sin \theta_{k,i} (z_{xi} - z_{xk}) + k_{k,i} \sin^2 \theta_{k,i} (z_{yi} - z_{yk}))$$

where m_i represent the lumped mass of the internal node i and $\ddot{z}_{xi}$ and $\ddot{z}_{yi}$ are the absolute accelerations of the internal node *i* in x and y axes. The internal node *i* is supporting *N* internal forces, one force for each node connected to him. These forces depend on the stiffness and the angle respect to the x axis of the springs located between the internal node *i* and the *N* nodes that are connected to him, as well as the displacements of these nodes in x and y axes (z_{xi}, z_{yi}, z_{xk}, z_{yk}).

Following the procedure described in section 2, we get the VARX model. Equation (5) could be regarded as a four exogenous and six endogenous variables VARX model [12]. The exogenous variables are the measured absolute displacements in z_{4x}, z_{4y}, z_{5x} and z_{5y} and the endogenous variables are the measured absolute displacement in z_{6x}, z_{6y}, z_{7x}, z_{7y}, z_{8x} and z_{8y}.

$$\begin{bmatrix} z_{6x}(n) \\ z_{7x}(n) \\ z_{8x}(n) \\ z_{6y}(n) \\ z_{7y}(n) \\ z_{8y}(n) \end{bmatrix} = -A_1 \begin{bmatrix} z_{6x}(n-1) \\ z_{7x}(n-1) \\ z_{8x}(n-1) \\ z_{6y}(n-1) \\ z_{7y}(n-1) \\ z_{8y}(n-1) \end{bmatrix} - A_2 \begin{bmatrix} z_{6x}(n-2) \\ z_{7x}(n-2) \\ z_{8x}(n-2) \\ z_{6y}(n-2) \\ z_{7y}(n-2) \\ z_{8y}(n-2) \end{bmatrix} + B_1 \begin{bmatrix} z_{4x}(n-1) \\ z_{5x}(n-1) \\ z_{4y}(n-1) \\ z_{5y}(n-1) \end{bmatrix} \tag{5}$$

A_1 and A_2 are 6 x 6 endogenous coefficient matrices and B_1 is a 6 x 4 exogenous coefficient matrix. These matrices are related to the physical properties of the substructural bars (mass, stiffness and angle) and also depend on the used sampling period. In addition to this, equation 6 shows the dependence between each matrix element and the substructural stiffness values.

$$A_1 = \begin{bmatrix} f_{11}(k_{4,6},k_{5,6},k_{6,7},k_{6,8}) & f_{12}(k_{6,7}) & f_{13}(k_{6,8}) & f_{14}(k_{4,6},k_{5,6},k_{6,7},k_{6,8}) & f_{15}(k_{6,7}) & f_{16}(k_{6,8}) \\ f_{21}(k_{6,7}) & f_{22}(k_{4,7},k_{6,7},k_{7,8}) & f_{23}(k_{7,8}) & f_{24}(k_{6,7}) & f_{25}(k_{4,7},k_{6,7},k_{7,8}) & f_{26}(k_{7,8}) \\ f_{31}(k_{6,8}) & f_{32}(k_{7,8}) & f_{33}(k_{5,8},k_{6,8},k_{7,8}) & f_{34}(k_{6,8}) & f_{35}(k_{7,8}) & f_{36}(k_{5,8},k_{6,8},k_{7,8}) \\ f_{41}(k_{4,6},k_{5,6},k_{6,7},k_{6,8}) & f_{42}(k_{6,7}) & f_{43}(k_{6,8}) & f_{44}(k_{4,6},k_{5,6},k_{6,7},k_{6,8}) & f_{45}(k_{6,7}) & f_{46}(k_{6,8}) \\ f_{51}(k_{6,7}) & f_{52}(k_{4,7},k_{6,7},k_{7,8}) & f_{53}(k_{7,8}) & f_{54}(k_{6,7}) & f_{55}(k_{4,7},k_{6,7},k_{7,8}) & f_{56}(k_{7,8}) \\ f_{61}(k_{6,8}) & f_{62}(k_{7,8}) & f_{63}(k_{5,8},k_{6,8},k_{7,8}) & f_{64}(k_{6,8}) & f_{65}(k_{7,8}) & f_{66}(k_{5,8},k_{6,8},k_{7,8}) \end{bmatrix} \quad (6)$$

$$A_2 = I$$

$$B_1 = \begin{bmatrix} g_{11}(k_{4,6}) & g_{12}(k_{5,6}) & g_{13}(k_{4,6}) & g_{14}(k_{5,6}) \\ g_{21}(k_{4,7}) & 0 & g_{23}(k_{4,7}) & 0 \\ 0 & g_{32}(k_{5,8}) & 0 & g_{34}(k_{5,8}) \\ g_{41}(k_{4,6}) & g_{42}(k_{5,6}) & g_{43}(k_{4,6}) & g_{44}(k_{5,6}) \\ g_{51}(k_{4,7}) & 0 & g_{53}(k_{4,7}) & 0 \\ 0 & g_{62}(k_{5,8}) & 0 & g_{64}(k_{5,8}) \end{bmatrix}$$

In this work, the structure is excited in the third mass (outside the substructure) by a Gaussian white noise and the displacements are recorded for each substructural DOF using a data sampling frequency of 1000 Hz. Later, the substructural VARX model is estimated by the Multivariable Least-Square estimator (MLS) method [12] for a healthy state and for the damaged scenarios. All considered damages are stiffness losses of a specific spring within the structure. Three different damage severities (5%, 10% and 20%) and six different damage locations are evaluated. In some of them, the damaged springs are within the substructure ($k_{4,6}$, $k_{4,7}$, $k_{6,7}$, $k_{6,8}$) and in the others, they correspond to external spring ($k_{1,3}$, $k_{2,5}$).

As we could see in equation (6), the elements of matrices A_1 and B_1 are function, among other things, of the stiffness of the substructural springs. Matrix A_1 depends on the state of the springs that are located between the internal nodes ($k_{6,7}$, $k_{6,8}$, $k_{7,8}$) and matrix B_1 depends on those springs located between the internal and interface nodes ($k_{4,6}$, $k_{4,7}$, $k_{5,6}$, $k_{5,8}$).

In this work we evaluate the state of the whole substructure. For this purpose, firstly the substructural VARX model of the healthy state is estimated. Later, the substructural VARX model is reestimated in each new scenario and the state of the substructure is evaluated in these new scenarios comparing these updated A_1 and B_1 matrices and the healthy ones. In the case of substructural damages, the damages will be firstly detected and later located in one of the substructural spring depending on the varied elements within A_1 and B_1 matrices. On the other hand, if external springs are damaged, the elements of matrices A_1 and B_1 will not change, so these external damages will not be detected and located within the substructure.

Table 1 shows which elements of A_I and B_I matrices depend exclusively on the properties of the substructural springs. In this work, only the variations of these elements (see table 1) are analyzed and the state of each substructural spring is evaluated depending on this analysis. For example, the stiffness of the spring that joins nodes 5 and 6 ($k_{5,6}$) affects exclusively in four elements within the matrix B_I ($B_{I(1,2)}$, $B_{I(1,4)}$, $B_{I(4,2)}$ and $B_{I(4,4)}$). For this reason, the variations of these four elements are analyzed to determine if the spring $k_{5,6}$ is damaged or not. In the present study, the damage severity for each spring is given by the mean variation of the elements that should be analyzed (see table 1).

Table 1. Analyzed elements within A_I and B_I matrices

Internal spring	A_I elements	B_I elements
$k_{4,6}$	-	(1,1),(1,3).(4,1),(4,3)
$k_{4,7}$	-	(2,1),(2,3),(5,1),(5,3)
$k_{5,6}$	-	(1,2),(1,4),(4,2),(4,4)
$k_{5,8}$	-	(3,2),(3,4),(6,2),(6,4)
$k_{6,7}$	(1,2),(1,5),(2,1),(2,4) (4,2),(4,5),(5,1),(5,4)	-
$k_{6,8}$	(1,3),(1,6),(3,1),(3,4) (4,3),(4,6),(6,1),(6,4)	-
$k_{7,8}$	(2,3),(2,6),(3,2),(3,5) (5,3),(5,6),(6,2),(6,5)	-

Regarding to the results, for external damages (reducing $k_{1,3}$ and $k_{2,5}$ values), the estimated stiffness modification for all substructural springs are almost zero, so the method determines that the substructure is healthy. For internal damages (reducing $k_{4,6}$, $k_{4,7}$, $k_{6,7}$ and $k_{6,8}$ values), the estimated stiffness modifications are shown in figure 2. In these four scenarios, the substructural damages are firstly detected and later located in the proper spring. Furthermore, the results show that the method estimates the severity of the damage.

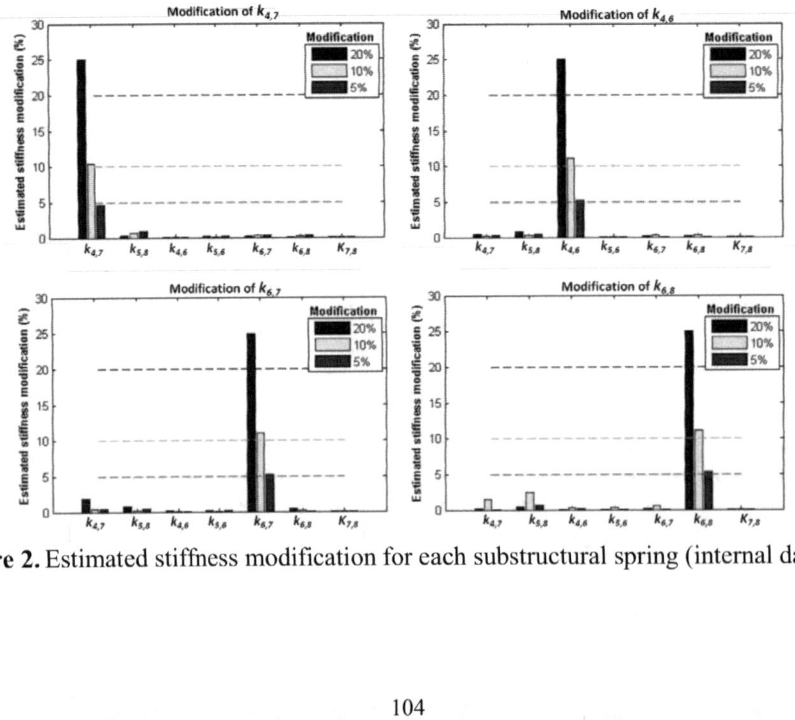

Figure 2. Estimated stiffness modification for each substructural spring (internal damages)

4. Conclusions

This paper proposes a novel SHM method to locate damages in multidimensional structures. A substructure of interest is isolated by a substructuring method and a VARX model of the isolated substructure is obtained. The analysis of the estimated VARX model is carried out in order to evaluate the health of the isolated substructure. It is not necessary to have any theoretical model and only the measured displacement data is required to estimate the isolated substructure's VARX model.

A linear and time invariant model of a two-dimensional lattice structure is simulated to evaluate the proposed method. The results show that the method not only allows detecting damages within the substructure, because it also allows estimating their location and their severity.

The proposed method is also suited for three dimensional lattice structures, where the number of element's connections increases. Our research group is already applying this method in a laboratory lattice structure and the results will be published soon.

References

[1] J. Ko and Y. Ni, "Technology developments in structural health monitoring of large-scale bridges," Engineering structures, vol. 27, pp. 1715-1725, 2005.
[2] A. Mita, "Structural dynamics for health monitoring," Sankeisha Co., Ltd, Nagoya, vol. 114, 2003.
[3] L. Jankowski, "Dynamic load identification for structural health monitoring," 2013.
[4] S. W. Doebling, C. R. Farrar, M. B. Prime, and others, "A summary review of vibration-based damage identification methods," Shock and vibration digest, vol. 30, pp. 91-105, 1998.
[5] J. Hou, L. Jankowski, and J. Ou, "Structural damage identification by adding virtual masses," Structural and Multidisciplinary Optimization, vol. 48, pp. 59-72, 2013.
[6] C. G. Koh, L. M. See, and T. Balendra, "Estimation of structural parameters in time domain: a substructure approach," Earthquake Engineering \& Structural Dynamics, vol. 20, pp. 787-801, 1991.
[7] A. W. Oreta and T.-a. Tanabe, "Element identification of member properties of framed structures," Journal of Structural Engineering, vol. 120, pp. 1961-1976, 1994.
[8] C. Yun and H. Lee, "Substructural identification for damage estimation of structures," Structural Safety, vol. 19, pp. 121-140, 1997.
[9] R. Isermann and M. Munchhof, "Identification of dynamic systems: an introduction with applications", Springer Science \& Business Media, 2010.
[10] Z. Xing and A. Mita, "A substructure approach to local damage detection of shear structure," Structural Control and Health Monitoring, vol. 19, pp. 309-318, 2012.
[11] L. B. Zhao, J. Y. Zhang, and Y. X. Zhao, "Taylor Series Numerical Method in Structural Dynamics," Advanced Materials Research, vol. 33, pp. 1213-1220, 2008.
[12] H. Lutkepohl, "New introduction to multiple time series analysis: Springer, 2007".

Journal of Physics: Conference Series **628** (2015) 012014 doi:10.1088/1742-6596/628/1/012014

Analysis of stationary roving mass effect for damage detection in beams using wavelet analysis of mode shapes

Mario Solís, Antonio J. Benjumea, Mario Algaba, Pedro Galvín

Escuela Técnica Superior de Ingeniería Universidad de Sevilla 41092 Sevilla (Spain)

E-mail: msolis@us.es

Abstract. One of the main challenges in damage detection techniques is sensitivity to damage. During the last years, a large number of papers have used wavelet analysis as a sensitive mathematical tool for identifying changes in mode shapes induced by damage. This paper analyzes the effect of adding a mass to the structure at different positions. Depending on the location and severity of damage, the presence of the mass affects the natural frequencies and mode shapes in a different way. The paper applies a damage detection methodology proposed by the authors, although it has been modified in order to consider the addition of the mas. This methodology is based on a wavelet analysis of the difference of mode shapes of a damaged and a reference state. The singular behavior of a normalized weighted addition of wavelet coefficients is used as an indicator of damage. The presence of damage is detected by combining all the information provided by mode shapes and natural frequencies for different positions of the roving mass. A continuous wavelet transform is used to detect the difference between the response of a healthy state and a damaged one. The paper shows the results obtained for a beam with different cracks. The paper analyzes the sensitivity to damage of the proposed methodology by considering some practical issues such as the size of the crack, the number of measuring points and the effect of experimental noise.

1. Introduction

It is well known that the presence of damage (cracks) in a beam implies a change in its dynamic properties. Thus, vibration based damage detection techniques try to detect the presence of damage by analyzing the change on natural frequencies, mode shapes and/or damping ratios. Some pioneering damage detection techniques (1) were based on the analysis of changes in natural frequencies, which are the most simple dynamic parameters to measure. However, natural frequencies lack of sensitivity to damage. Only a significant damage would induce a significant change in natural frequency. Moreover, the effect of damage may be masked by the effect of environmental changes, experimental noise and uncertainty, etc. On the other hand, natural frequencies are a global parameter of the structure, and therefore it can provide information only about the presence of damage but not about its location. In order to locate damage and to be able to detect more little damage, the mode shapes of the structure may be used. From a experimental point of view, the identification of mode shapes requires a larger amount of sensors (more complex and expensive experimental setups) as well as more sophisticated system identification methods. Despite of these experimental and mathematical efforts, the changes in mode shapes induced by damage are usually subtle (unless severe damage is present) so damage can not be identified from mode shapes in a straightforward way. There is a significant number of papers that propose different techniques and damage detection parameters to analyze the information provided by mode shapes (1).

The wavelet transform is a rather new mathematical tool that has been developed from the 90s for signal processing and information encoding. After some pioneering works that extended the use of wavelet transform to damage detection in structures (2), a number of authors have made different proposals for the same purpose by applying wavelet transform to mode shapes, time response, static deflection, etc (3; 4).

The wavelet transform is sensitive to local changes in the original signal. Wavelet coefficients show a singular behavior, ridges or peaks when some discontinuity or a sudden change occurs. Thus, they can be used as an indicator of damage when applied to mode shapes, assuming that damage may lead to some kind of discontinuity in mode shapes.

The authors have recently proposed a simple damage detection technique based on the wavelet analysis of difference in mode shapes from a healthy and a damaged state (5). The main idea is to combine all the information provided from all identified mode shapes and natural frequencies by a weighted addition of the wavelet coefficients according to changes in natural frequencies for each mode. The methodology has been successfully applied to cracked steel beams. This paper includes a new idea for making the proposed damage detection method more robust and sensitive to little damage. A non-structural mass is attached to the structure, and modal analysis is performed for different positions of the mass. The mass is at a fixed position for each experimental test (it is not a moving load) but it changes its position along the experimental campaign, so it is called a stationary roving mass.

The effect of the mass on the dynamic response of the structure will depend on the position and severity of the damage. As a result, additional information is available for damage detection and the methodology becomes more sensitive to damage. The information of all available mode shapes and frequencies is combined through the addition of wavelet coefficients of difference in mode shapes from a healthy and a damage state. The paper shows numerical results for a steel cantilever beam with different damage scenarios.

2. Combined modal-wavelet methodology for damage detection with a stationary roving mass

In this section, the proposed methodology for damage detection in beams is described step by step.

2.1. Modal analysis

The first task of the proposed methodology is to obtain the mode shapes and natural frequencies of the structure. As any vibration based damage detection method, it requires modal information about the undamaged and damaged state. The modal parameters for each position of the added mass for the undamaged beam as well as for the damaged one must be obtained.

2.2. Extension of mode shapes

The next step of the methodology deals with a very important issue in wavelet analysis, especially when applied to space based damage detection. The wavelet transform is defined for an infinite integration interval, whereas the original signal is defined over a finite interval. When the wavelet transform is performed, there is a singular behavior at the beginning and at the end of the signal. The signal starts and finishes at those points, so there is a significant local change there, unless the signal trends softly in an asymptomatic way to a constant value, which will never be the case for a mode shape. This unstable behavior of the wavelet coefficients in the vicinity of the beginning and the end of the analyzed signal is known as the edge effect, and it is a serious drawback of the wavelet transform when the damage is close to the beginning or the end of the signal. Moreover, the high values of wavelet coefficients near those regions of the signal can mask the structural damage effect on the wavelet transform along the structure.

This paper applies a simple method to avoid edge effects. It consists of an anti-symmetric extension of the signal of the same length of the original signal at both ends (5; 6).

2.3. Smoothing, interpolation and noise reduction of mode shapes

Another important issue when analyzing mode shapes is the reduction of experimental noise effect. The experimental mode shapes will always be affected by noise, so they will always show some kind of irregularities. This undesirable behavior will affect the wavelet analysis and it could eventually mask the effect of damage. In order to reduce this effect, a smoothing technique may be introduced, so local peaks induced by experimental noise are eliminated without affecting any local trend that could have been induced by damage, since damage is not expected to produce only a local peak as it is the case of experimental noise. In this paper a softening technique already proposed by the authors is applied (5).It is based on a windowed quadratic regression.

2.4. Wavelet transform of difference of modes

Once the extended and smoothed mode shapes have been obtained, the wavelet analysis is applied. Firstly, the extended difference of each mode shapes (i) for each position of the mass (j) is obtained $(\Phi^{ij}_{diff,ext})$ by computing the difference between the smoothed extended damaged $(\Phi^{ij}_{s,ext,d})$ and undamaged $(\Phi^{ij}_{s,ext,u})$ mode shapes:

$$\Phi^{ij}_{diff,ext}(x) = \left(\Phi^{ij}_{s,ext,d}(x) - \Phi^{ij}_{s,ext,u}(x)\right) \tag{1}$$

Then, a CWT of each extended mode shape difference is done to give information about changes in mode shapes. The CWT for the *ith* mode shape and for the *jth* position of the mass can be written as:

$$CWT_{\Phi^{ij}_{diff,ext}}(u,s) = \frac{1}{\sqrt{s}} \int_{-\infty}^{+\infty} \Phi^{ij}_{diff,ext}(x)\Psi^*\left(\frac{x-u}{s}\right) dx \tag{2}$$

This paper uses the well-known Daubechies wavelet function with two vanishing moments. This function has already provided successful results in damage detection (5). From this point, only the CWT coefficients that corresponds to the original signal $(CWT_{\Phi^{ij}_{diff}})$, and therefore to the real structure, will be considered.

2.5. Normalized weighted addition of wavelet results based on frequency changes

In order to simplify the analysis of the CWT for each mode shape and to draw an overall result for damage detection in one single picture, the values of CWT coefficients of each mode shape and for each position of the mass are added up to obtain a global result for damage detection (Equation 3). On the other hand, this combination of the results mitigates the effect of experimental noise.

For a more precise and clear detection of singularities, the proposed methodology combines the absolute values of the wavelet coefficients. In addition, the coefficients for each mode shape are weighted according to its corresponding change in natural frequencies:

$$CWT_{sum}(u,s) = \sum_{j=1}^{M}\sum_{i=1}^{N} \left|CWT_{\Phi^{ij}_{diff}}(u,s)\right| \cdot \left(1 - \frac{\omega^{ij}_u}{\omega^{ij}_d}\right)^2 \tag{3}$$

where ω^{ij}_u and ω^{ij}_d stand for the natural frequencies of mode shape i for the position M of the added mass, for the undamaged and the damaged state, respectively.

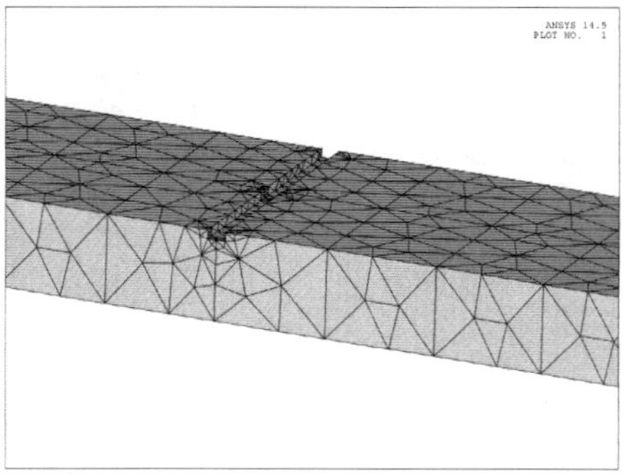

Figure 1. Detail of the finite element model around the notch (1mm depth).

The weighting of the difference of the damaged and undamaged mode shapes with their natural frequencies relations is used to emphasize the most sensitive mode shapes to damage. It is assumed that those modes that exhibit a higher frequency change are more sensitive to damage and therefore changes in those mode shapes are more significant. On the other hand, the mode shapes that do not change their natural frequencies are almost disregarded. They are likely to introduce mainly noise in the final result when all the mode shapes are combined.

Finally, the resulting weighted addition of CWT coefficients is normalized to unity for each scale:

$$CWT_{sum-mass-norm}(u,s) = \frac{CWT_{sum-mass}(u,s)}{max[CWT_{sum-mass}(u,s)]_s} \tag{4}$$

Normalized coefficients give a more clear final result since the information for all scales can be analyzed altogether. The idea of the addition and normalization of wavelet coefficients has also been applied by other authors (7).

2.6. Analysis of the results for damage detection and location

The normalized weighted addition of absolute values of CWT coefficients of mode shapes differences ($CWT_{sum-mass-norm}(u,s)$) is finally plotted so information for all scales is available along the beam in just one final figure. A singular behavior of wavelet coefficients may indicate the presence of damage. This singular behavior is typically shown as high values (unity values because of the normalization of coefficients) for all scales around damage location.

3. Numerical model

A finite element model was built to simulate the effect of a crack in a beam. The length (L) of the beam is 800mm and its cross section is 300mm wide and 10mm high. The beam was modeled with 4 node tetrahedral elements using Ansys software. Figure 1 shows a picture of the finite element model around the crack, which is modeled as a notch of 2mm width. The severity of the damage is defined by the depth of the crack and it will be referred as a percentage of the height of the beam. Thus, a 10% damage will mean a crack of 1mm depth. The mesh of the model is different depending on the depth of the crack since it has to be adapted to the irregular geometry around the crack. The added mass is modeled with a point mass element. The beam is made of steel, with an elastic modulus $E = 210GPa$, Poisson ratio $\nu = 0.3$ and density $7850kg/m^3$. The beam is fixed at one end and free at the other end. All displacements are constrained for all nodes at the fixed end. For damage detection purposes, only bending modes in the plane parallel to the shortest dimension of the beam (vertical plane) are considered. Vertical modal displacements of nodes along the axis on the upper side of the beam are considered as

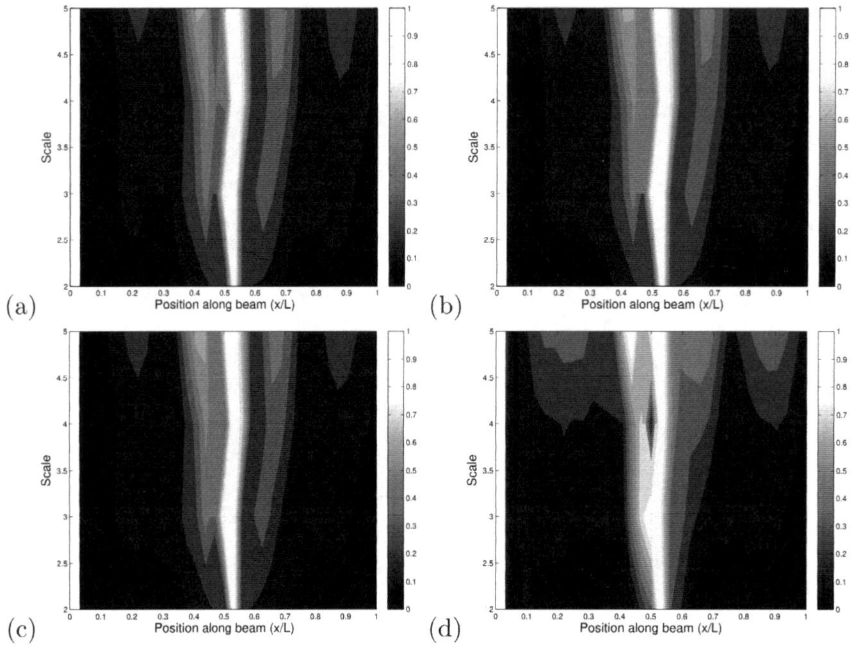

Figure 2. Normalized weighted addition of wavelet coefficients using 32 measuring points, no added noise and crack depth 20% (a), 10% (b), 5% (c) and 1% (d).

virtual measuring points, simulating the modal information that could be obtained by attaching accelerometers to that side of the beam. The influence of the amount of sensors along the beam on the final results is analyzed by considering different number of nodes along the beam. These nodes must be uniformly spaced, since the wavelet transform assumed that the samples of the signal to be transformed are equally spaced. In real applications, if the accelerometers are not uniformly distributed then some kind of interpolation is required (5).

4. Results

This section analyzes the sensitivity to damage of the proposed methodology considering the effect of the number of measuring points, the presence of experimental noise and the use of the added roving mass. The damage is located at the middle section of the beam ($0.5L$). The result for each case is presented by a coloured 2D picture of the normalized weighted addition of wavelet coefficients of mode shapes differences. The vertical axis is the scale of the wavelet transform whereas the horizontal axis is the position along the beam. The values of the wavelet coefficient are represented by a color map. The results are obtained considering the first five bending modes of the beam.

Figure 2 shows the results for several depths of the crack, considering 32 measuring points along the beam. The added mass is 20% of the total mass of the beam and it is located at 10 different positions along the beam (from the position at $0.1L$ from the fixed end of the beam to its free end, in steps of $0.1L$). It can be observed that the damage can be clearly identified for all the damage scenarios presented: high values (about unity) of normalized wavelet coefficients are located at damage location for every scale. The methodology is able to identify the presence of a damage even when the notch is only 1% deep ($0.1mm$).

The results of Figure 2 shows that the methodology is very sensitive to damage from a

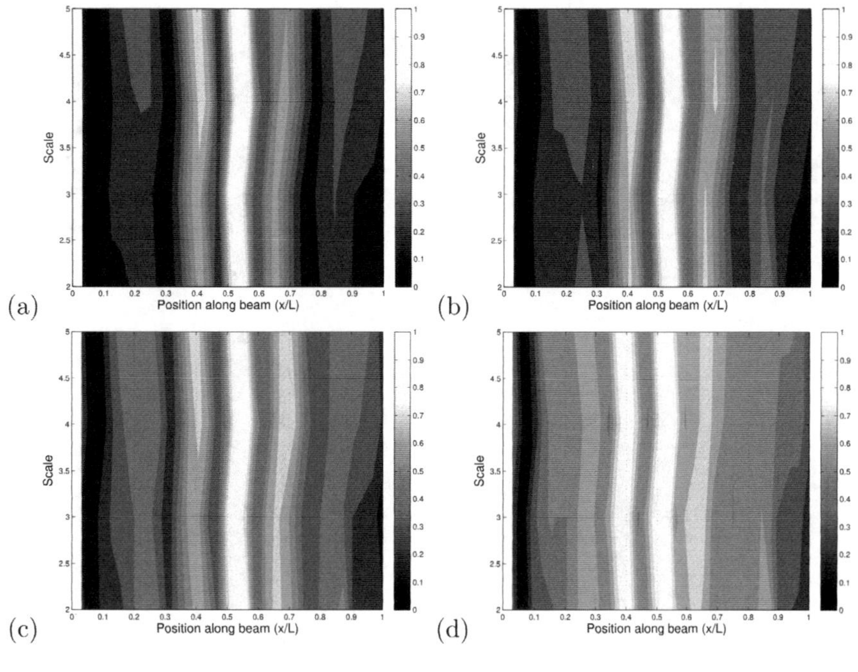

Figure 3. Normalized weighted addition of wavelet coefficients using 32 measuring points, 40db gaussian noise and crack depth 50% (a), 40% (b), 30% (c) and 20% (d).

mathematical point of view, when noise-free mode shapes are obtained from the numerical model of the beam. However, experimental noise will always be present in real applications. Experimental errors will affect the modal displacements values and noisy mode shapes will be obtained. In order to analyze the effect of noise in the numerical simulations, a white gaussian noise is added to the mode shapes using the 'awgn' built-in function of Matlab software. The intensity of this synthetic noise is defined by the Signal to Noise Ratio expressed in terms of the average power of the original signal (mode shape vector) and the added noise.

Figure 3 shows the results obtained when the mode shapes are contaminated with high level of noise (40db). The sensitivity to damage is significantly affected by the presence of noise. The effect of the presence of the crack is present for a crack depth higher than 20%, but it can not be not clearly identified, even for crack depths up to 50%.

Figure 4 shows the results obtained when 64 measuring points are considered. It is clear from figures 3 amd 4 that the undesirable effect of noise is reduced by increasing the number of measuring points. When using 64 measuring points the results are satisfactory from crack depth values of 20%, even with a 40db noise level.

It should be noted here that, according to the changes in natural frequencies, all the considered damage scenarios are very demanding for damage detection purposes. Even when the crack is 50% deep, the drift in natural frequencies from the healthy state is less than 4%. When the crack depth is 20%, the drift is less than 0.6%, and for a crack of 10% the drift is less than 0.2%. The use of the added roving mass enhance the sensitivity to damage, as it can be seen by comparing figures 4 and 5. When no mass is added, the results are much more affected by noise. If the level of noise is reduced to 60db, then up to a 10% deep crack can be clearly identified with 32 measuring points, whereas even a very small crack of 5% depth can be identified with 64 measuring points (Figure 6).

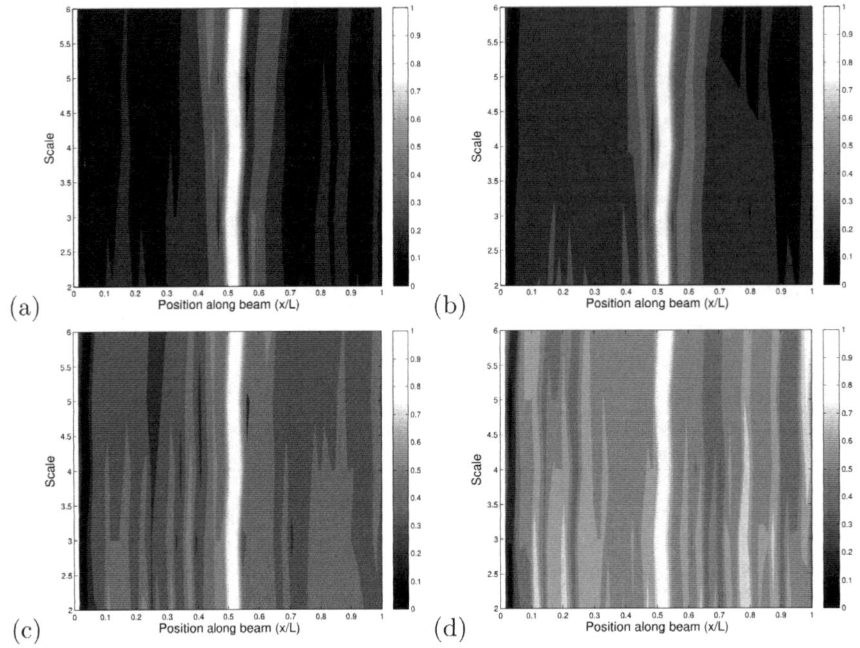

Figure 4. Normalized weighted addition of wavelet coefficients using 64 measuring points, 40db gaussian noise and crack depth 50% (a), 40% (b), 30% (c) and 20% (d).

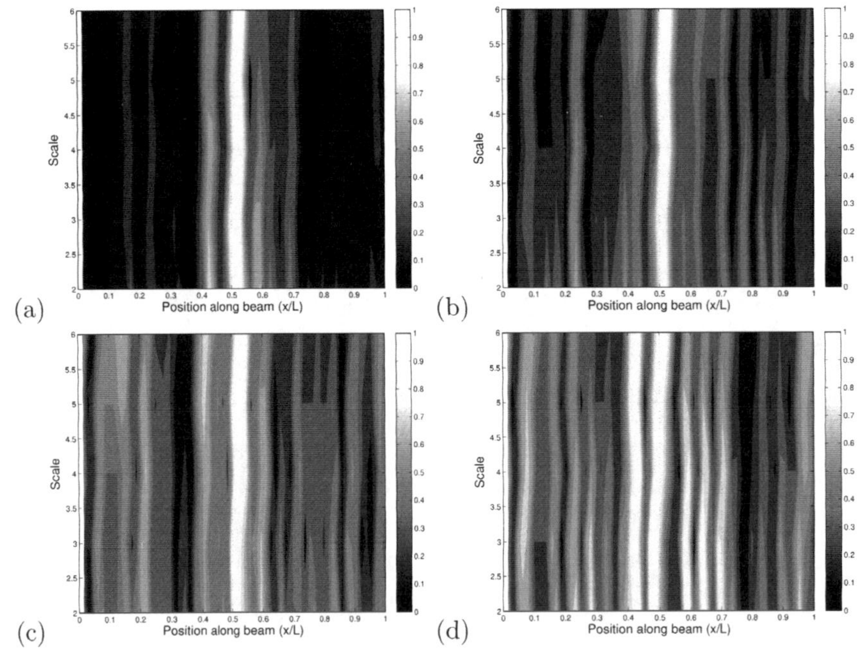

Figure 5. Normalized weighted addition of wavelet coefficients with no added mass,using 64 measuring points, 40db gaussian noise and crack depth 50% (a), 40% (b), 30% (c) and 20% (d).

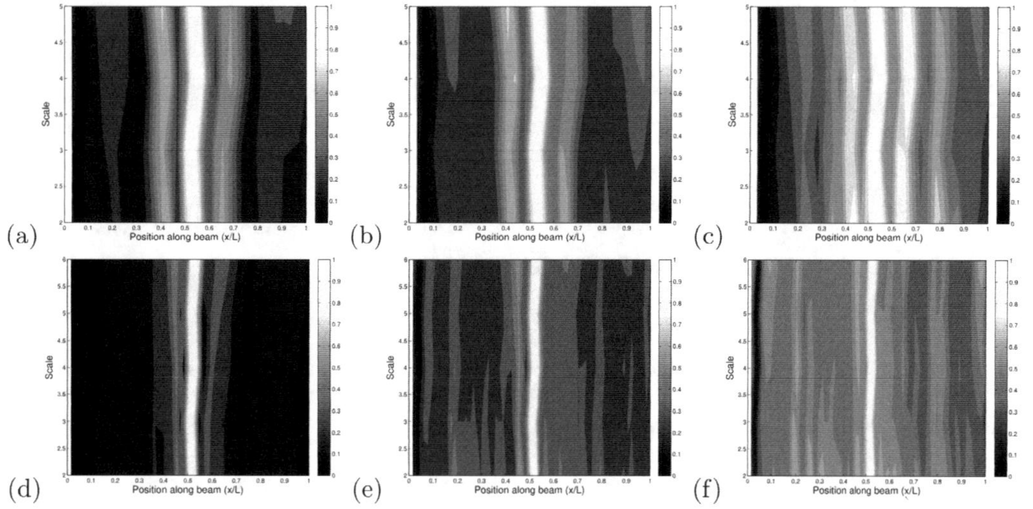

Figure 6. Normalized weighted addition of wavelet coefficients with 60db gaussian noise, using 32 measuring points (a-c) and 64 measuring points (d-f), for crack depth 20% (a,d), 10% (b,e) and 5% (c,f).

5. Conclusions

This paper has presented a novel damage detection methodology based on the wavelet analysis of changes in mode shapes due to the presence of damage. A stationary roving mass is added in order to emphasize the effect of damage as well as to reduce the effect of experimental noise. The influence of experimental noise, size of the crack and number of measuring points has been addressed in the paper. The obtained results show that the proposed methodology is a promising tool for detecting small cracks in beams. The actual sensitivity in real applications will depend on the real experimental noise and the available number of sensors.

6. Acknowledgments

This work was supported by the Consejería de Economía, Innovación, Ciencia y Empleo of Andalucía (Spain) under project P12-TEP-2546. The financial support is gratefully acknowledged.

References

[1] Fan W and Qiao P 2011 *Structural Health Monitoring* **10** 83–111

[2] Surace C and Ruotolo R 1994 *Proceedings of the 12th International Modal Analysis Conference* pp 1141–1147

[3] Taha M M R, Noureldin A, Lucero J L and Baca T J 2006 *Structural Health Monitoring* **5** 267–295

[4] Katunin A 2013 *International Journal of Composite Materials* **3** 1–9

[5] Solís M, Algaba M and Galvín P 2013 *Mechanical Systems and Signal Processing* **40** 645–666

[6] Strang G and Nguyen T 1996 *Wavelets and Filter Banks* (Wellesley- Cambridge Press)

[7] Katunin A 2015 *Archives of Civil and Mechanical Engineering* **15** 251–261

A Versatile Inspection System for Pipe Structure Using Ultrasonic Waves Propagation Imager

TC Truong[1] and JR Lee[2]

[1] PhD student, Department of Aerospace Engineering
[2] Associate Professor, Department of Aerospace Engineering
Korea Advanced Institute of Science and Technology, Republic of Korea

E-mail: leejrr@kaist.ac.kr

Abstract. Pipe structure is vulnerable to many types of damage, such as flow-accelerated corrosion, crack, and in the case of multi-layers pipe, debonding damage. A versatile damage inspection system is needed, where it must be easily used for variety types of pipeline, must have the capability to detect many types of damage, as well as must be able to carry out inspection in the working condition of the pipe system. In this paper, we present the Ultrasonic Propagation Imager (UPI) that demonstrated to meet those demands. The UPI system consists of a high speed Q-switched laser, a high speed scanning mirror, a DAQ system, and a changeable sensing system depends on applications. Advanced signal processing using ultrasonic wavenumber imaging algorithm and energy mapping were applied for damage detection of the pipe structures.

1. Introduction

There are millions of kilometres of pipes around the world, transport various types of material, from water to oil and gas. Each application often required a specialized type of pipe, which different with each other in sizes, materials, and designs. From the perspective of an inspection engineer, such as from a pipeline manufacturer, this complexity poses a big challenging. An inspection system that designed to work well for one type of pipe does not necessary work for another. Therefore there is a need of a versatile inspection system that can be easily used for a variety types of pipeline, must have the capability to detect many types of damage, as well as must be able to carry out inspection in the working condition of the pipe system, and ideally obtain the inspection result in a short amount of processing time.

The system introduced in this paper answers some of those very demanding aspects. It is a damage inspection system based on guided wave technique. The research on guided waves using for damage detection has been around since the 60s and 70s, with a high expectation of becoming the paradigm shift of the structural health monitoring (SHM) for various structures, including the pipes structure. The high hope is mainly due to the capability of guided waves to propagate over a long distance from an excitation point [1]. Unfortunately that expectation has not yet been truly delivered in the sense that the guided waves techniques have not yet became the standard of the industry. The main limitation is that most guided wave SHM system relies on a very limited set of sensing and excitation points, which severely limits the ability to diagnose of the structure's condition. To fully investigate the structure,

Journal of Physics: Conference Series **628** (2015) 012015 doi:10.1088/1742-6596/628/1/012015

we need a vast amount of data recorded the propagation of the ultrasonic waves of a very dense grid points on the structure. The technique that able to do that is so-called the full wavefield measurement technique. The ultrasonic propagation imager (UPI) system as showing here is one of currently three techniques that enable the measurement of full wavefield. The other two techniques are the scanning air-coupled transducer and the scanning laser vibrometer. Reader who interested in these techniques may refer to the recent review of Ostachowicz et al. [2].

This paper is organized as follows. We start with the introduction of working principle of the UPI system. Next, the algorithms for damage detection using the full wavefield are very briefly presented. Finally we show the applications of UPI for damage detection of a variety of pipes, with many types of damages and in at different working conditions of the pipes.

2. The Ultrasonic Propagation Imager (UPI) system

The principle of the UPI system relies on the reciprocity property of ultrasonic waves. The UPI uses one or many fixed sensing points and controls a Q-switched laser beam for scanning. Through reciprocity, a virtual laser ultrasonic wavefield is generated from the sensing point(s). This technique is applicable to realistic surfaces, in the sense that there is no need to prepare the structure's surface, unlike the scanning laser vibrometer technique. Also it allows for the manufacture of very high speed scanning systems. The latest generation of the Q-switched laser scanning system [3] has reached to scan speed of 20,000 points per second, which means that the inspection time for an area of 1 m × 1 m with a spatial resolution of 1 mm is less than 1 minute. The inspected area can be extended using a set of electric-controlled mirrors to alter the direction of the laser beam.

Figure 1 shows the UPI system for measurement the full wavefield of a pipe specimen. The data of this test is one among several examples as we will show later in this paper. The distance between the laser scan mirrors to the specimen is about 2 meters, however it can be easily extended up to tens of meters using a laser beam collimator.

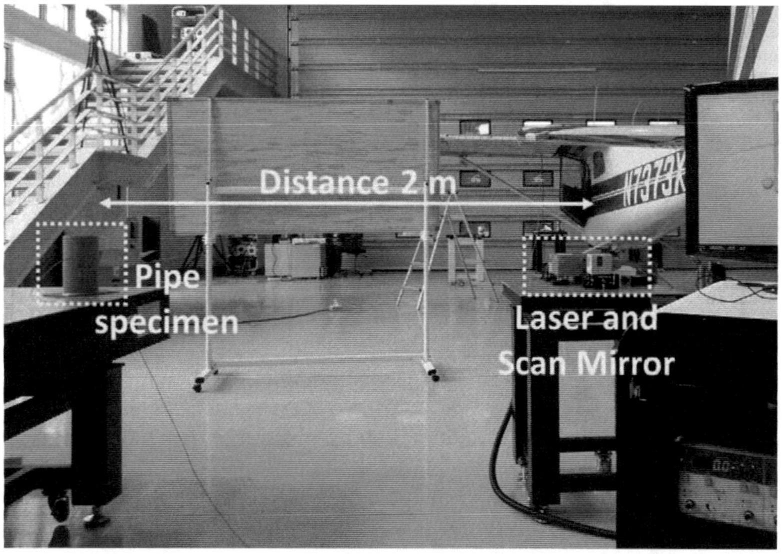

Figure 1. The Ultrasonic Propagation Imager (UPI) system

3. Algorithms for damage detection with full wavefield data

The essence of several algorithms for the processing of the full wavefield data used in this paper are presented. Readers who interests in detail may refer to the provided references.

3.1. Ultrasonic Wave Propagation Imaging (UWPI)

This is the basic algorithm that suitable for a preliminary inspection since it take very little computation and the result is in real time. The full scanning data forms a three-dimensional matrix $V \times H \times N$, where V and H represent the number of scanning points in the vertical and horizontal directions of the scanning area, while N is the sampling length of the measurement data for each scanning point. A slide perpendicular to the N axis maps the ultrasonic wave amplitude of the scan area at a particular time point. Plotting successive slides along the time axis generates a movie of ultrasonic waves propagation. A technician with some minimum training can analyse the wave propagation and able to give the assessment about the integrity of structure. If certain anomaly in amplitude or direction of the ultrasonic waves appears, further analysis with more complicated algorithms may needed for a stronger conclusive assessment.

3.2. Ultrasonic wavenumber imaging (UWI)

The main idea of the UWI algorithm [5] is as follows. The Lamb waves propagate across a thin plate at various modes. In the case when the plate has uniform stiffness, at one particular frequency of one particular mode, the guided waves have one corresponding dominant wavenumber. When the waves propagates across a damaged area, which has lower stiffness, the dominant wavenumber at that damaged area alters: increasing in the case of the A_0 mode and decreasing in the case of the S_0 mode. By mapping the dominant wavenumber at one particular frequency, it is able to detect the damages as well as estimate the damages size.

3.3. Waves energy mapping

The UWI algorithm only works with area-like damage, which means that it only can detect the damages with relatively large area: the damage size must be in the order of the ultrasonic wavelength. For the type of damage has at least one dimensions too small (slit-type or point-type), other algorithm such as waves energy mapping [6] can be used. The basic assumption of the waves energy mapping algorithm is that the wavenumber of the ultrasonic waves changes when the waves propagate through structural inhomogeneity such as geometry discontinuities or debonding. Therefore, by applying a narrowband wavenumber filter, and then calculating the waves energy of every spatial point in the scan area, it is able to visualize the damage areas.

4. Test specimens and analysis results

In this section, we will show the damage detection results of several pipe specimens. Since the target of this paper is to introduce a versatile damage inspection system, various types of pipes with various type of damages were chosen: plastic pipes with area-like damage, slit-like damage, and crack damage; steel pipe with slit-like damage; composite pipe with the debonding damage between two layers; and finally joint damage between two plastic pipes. Also, we will show that the UPI system is able to detect damage of the pipe even in working condition: with and without water inside the pipe. First, all specimens used for the test are shown in Figure 2 to 5, including description of structures and damages. Notice that figures 2a, 2b, and 2c show three plastic pipes with different kinds of damages: the flow-accelerated corrosion (FAC) damage, slit-type damages, and crack damages. Figure 3a shows the steel pipe before welding (which reveals the slit-like damages) and after welding. Figure 4a shows the picture and figure 4b shows the diagram of two plastic pipes joining together, but with an artificial damages, as shown more clearly using two plane cut AA and BB. Finally, figure 5b shows the pipe with two layers: steel outside and plastic inside, with the debonding damage.

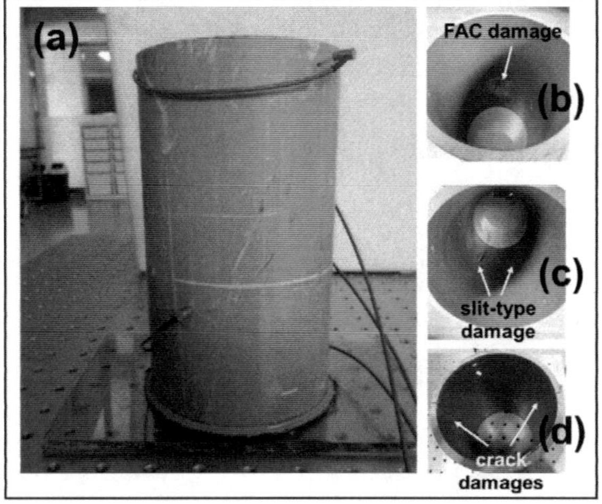

Figure 2. Plastic pipe with thickness 5.5 mm

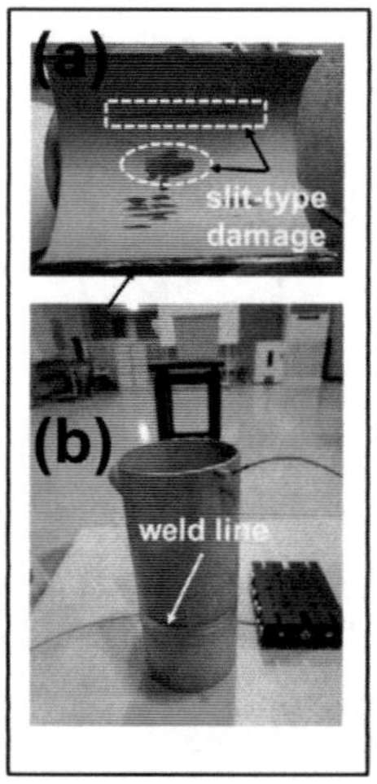

Figure 3. Steel pipe with thickness 2 mm

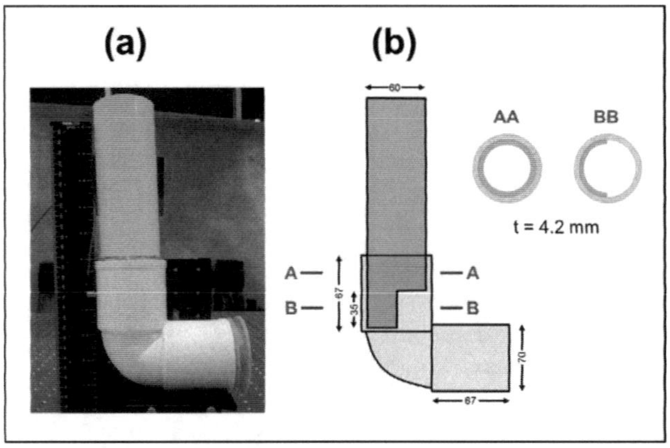

Figure 4. Plastic pipes joint with damage

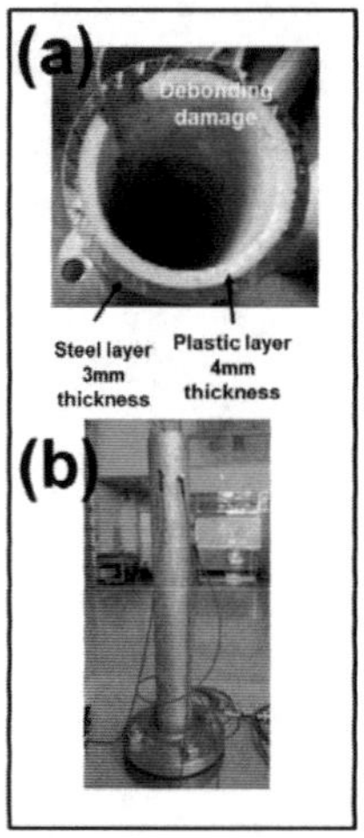

Figure 5. Composite pipe with debonding damage

4.1. Plastic pipe with flow-accelerated corrosion damage

The damage was made to simulate the damage that happens due to flow of liquid. Two cases were tested, with and without the water inside the pipe. Figure 6 and 8 show the damage detection result using UWPI and UWI, respectively. The presence of the water in the pipe can be easily visualize in the waves propagation in time domain with the appearance of the Scholte waves. Figure 7 shows the wavenumber frequency curves. Interestingly, even with the presence water, at the chosen frequency 150 kHz, the dominant wavenumber is still follows the same A_0 curve of the case without water The size of the damage was estimated as in Figure 7. The damage size identified using UWI algorithm for two case: with and without water are approximately equivalent with the different of two dimensions is less than 3 mm.

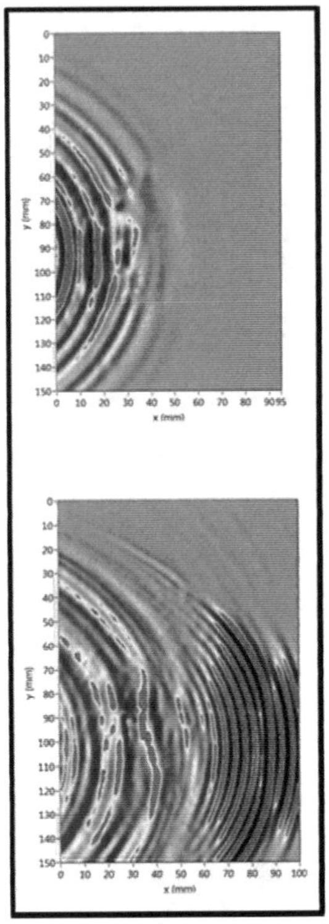

Figure 6. UWPI. Top: no water in the pipe. Bottom: with water

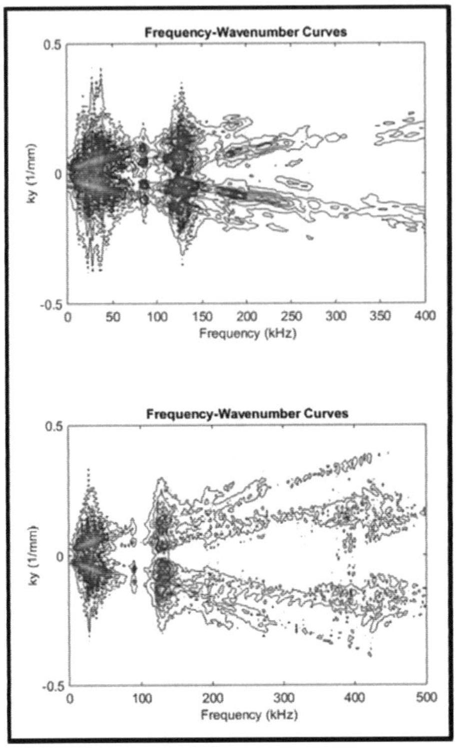

Figure 7. k-f curves. Top: case no water inside the pipe. Bottom: with water

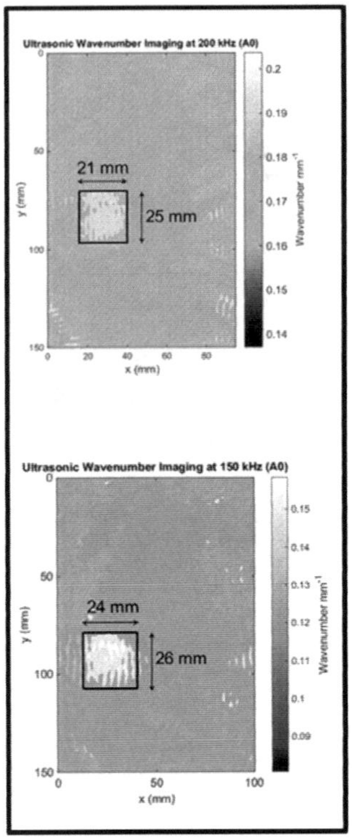

Figure 8. UWI. Top: no water in the pipe. Bottom: with water

4.2. Plastic pipe with slit-type and crack damages

For the case of damage the width of the damages are too small, the UWI algorithm cannot be used as explained in previous section. Figure 9 and 11 show the snap shot of the waves propagation, which indicates the location of the slit damages and the crack damage, respectively. The damage locations were confirmed using the Energy Mapping algorithm as shown in figures 10 and 12.

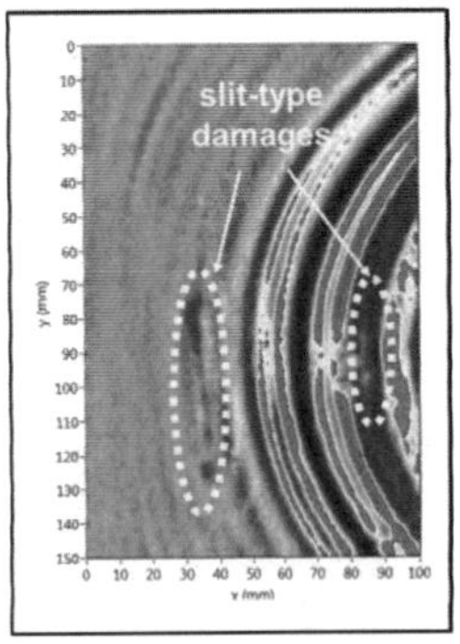

Figure 9. UWPI. Plastic pipe with two slit-type damages

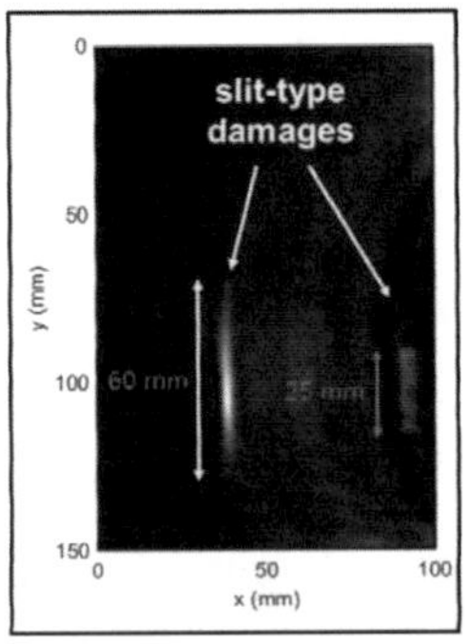

Figure 10. Energy Mapping. Plastic pipe with two slit-type damages

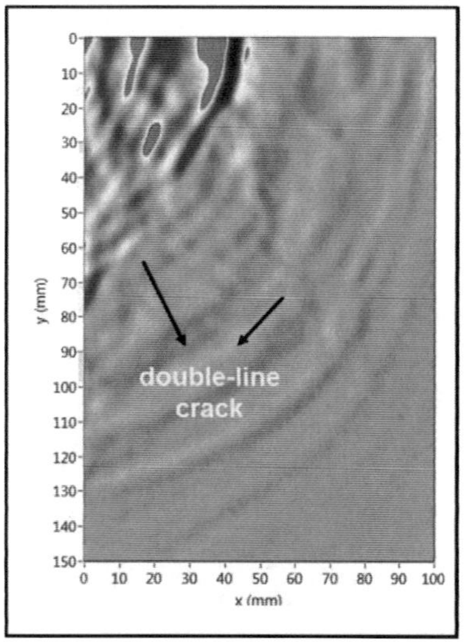

Figure 11. UWPI. Plastic pipe with crack damages

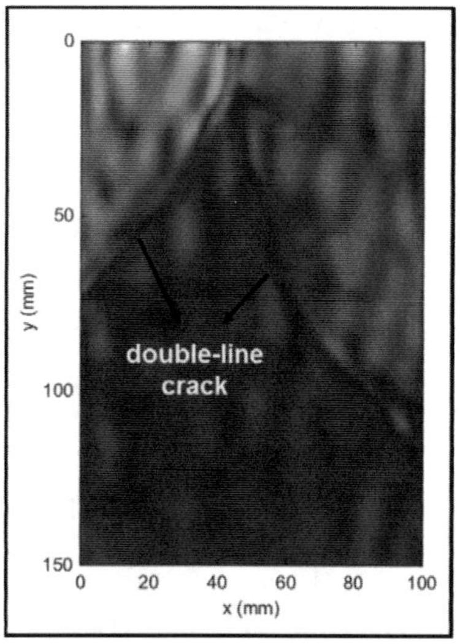

Figure 12. Energy Mapping. Plastic pipe with crack damages

4.3. Composite pipe

The total thickness of this pipe is approximately 7 mm, which potentially make the detection of the damage difficult. However, the actual inspection result in both the UWPI (figure 13) and the waves energy mapping (figure 14) were successfully detected the debonding damage. Also, two small damages were detected at the right side of the debond line (figure 14). This is an unexpected results since these two damages were unnoticed before actual doing the inspection test.

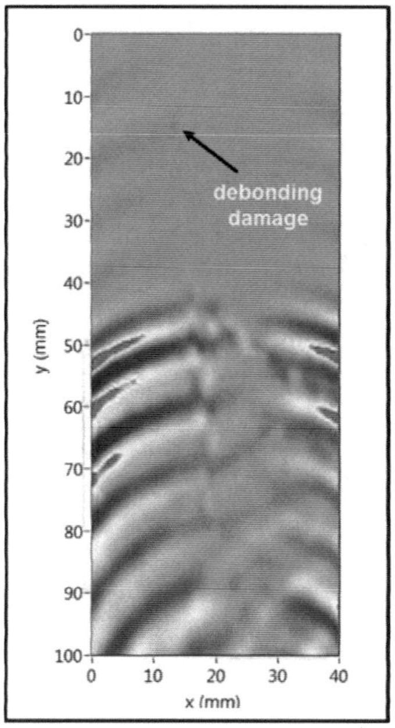

Figure 13. UWPI

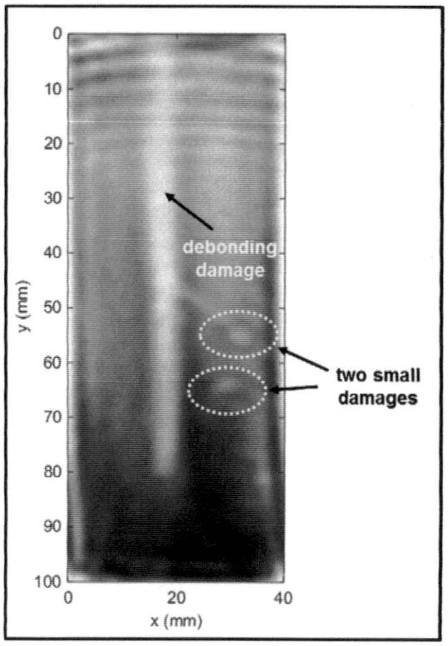

Figure 14. Energy Mapping

4.4. Steel pipe

The large slit-type damage on the left and the weld line were detected using both the UWPI and the energy mapping algorithm. However, the long slit at right side is only found using the energy mapping algorithm. Also, this test poses an interesting problem to the UPI system for future development, which is how to distinguish the damages from the structural features, which is in this case the weld line.

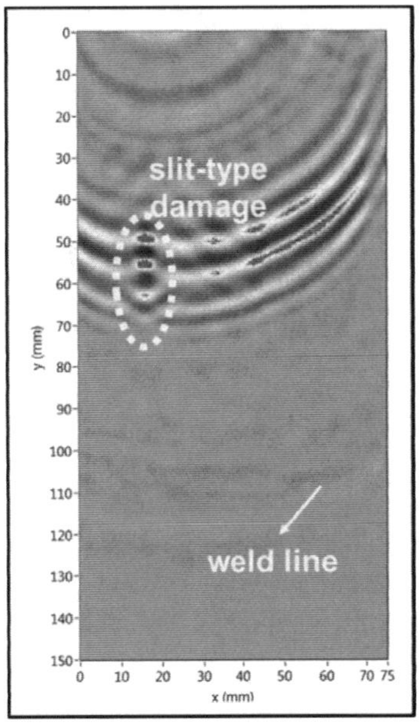

Figure 15. UWPI

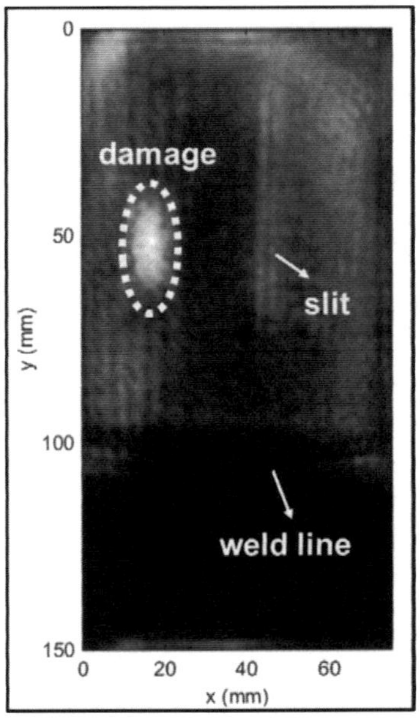

Figure 16. Energy Mapping

4.5. Pipes with a damaged joint

This is the test case where the use of the UWI algorithm shows a superior performance as compare the simple UWPI algorithm. We found that even observing the complete UWPI movie, it is still unable to recognize any significant indication of the damage, whereas the UWI algorithm is able to find quite clearly the area where the damage occurs, which is corresponding to the area with lower stiffness, and as the results: a higher wavenumber of A_0 mode.

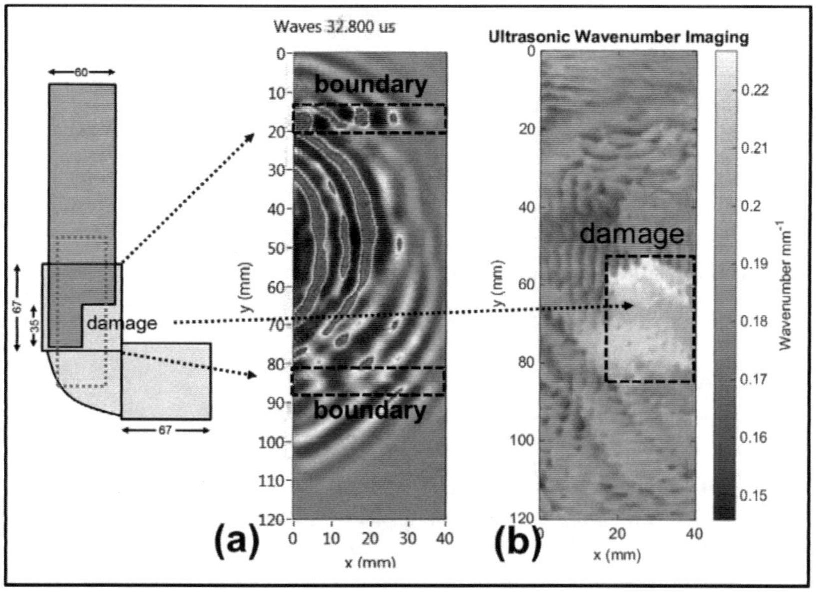

Figure 17. Damage detection of joint damage between two plastic pipes

Conclusions

In this paper, the application of the Ultrasonic Propagation Imager (UPI) for damage inspection of the pipeline were presented. We showed that the UPI can be used effortlessly to detect various types of damages for various types of pipes in different sizes and materials. The developed algorithm has overcome the difficulties posed by the working condition of the pipe structure with water inside the pipe. Those results have proved the feasibility of the UPI system as an effective structural health management system for the real-world pipeline inspection system.

Acknowledgments

This research was supported by the Space Core Technology Development (2013-042548) and Leading Foreign Research Institute Recruitment Program (2011-0030065), through the National Research Foundation of Korea, funded by the Ministry of Science, ICT and Future Planning.

Reference

[1] Rose J L 2004 *Key. Eng. Mat.* **270-273** 14
[2] Ostachowicz W, Radzienski M and Kudela P 2014 *Strain.* **50**, 27
[3] Chong S Y and Lee J R 2014 *Procs. of Advances in Structural Health Management and Composite Structures* (Jeonju, Korea) vol I, p18-1
[4] Chia C C, Lee J R, Park C Y and Jeong H M 2012 *Opt Laser Technol.* **44** 1507
[5] Flynn E B, Chong S Y, Jarmer G J and Lee J R 2013 *NDT&E Int* **59** 1
[6] Lee J R, Cho C M, Park C Y, Shin H J, Jeong H M and Truong T C 2014 *Procs. of Advances in Structural Health Management and Composite Structures* (Jeonju, Korea) vol I, p20-1

Damage characterization on human femur bone by means of ultrasonics and acoustic emission

M Strantza[1], D Polyzos[2], O Louis[3], F Boulpaep[1], D Van Hemelrijck[1], D G Aggelis[1]

[1]Department of Mechanics of Materials and Constructions, Vrije Universiteit Brussel, Pleinlaan 2, 1050 Brussels, Belgium

[2]Department of Mechanical Engineering and Aeronautics, University of Patras, University of Patras, Panepistimioupolis Rion, 26500 Patra, Greece

[3]Universitair Ziekenhuis Brussel, Department of Radiology, Avenue du Laerbeek 101, 1090 Brussels, Belgium

E-mail: Dimitrios.Aggelis@vub.ac.be

Abstract. Human bone tissue is characterized as a material with high brittleness. Due to this nature, visible signs of cracking are not easy to be detected before final failure. The main objective of this work is to investigate if the acoustic emission (AE) technique can offer valuable insight to the fracture process of human femur specimens as in other engineering materials characterization. This study describes the AE activity during fracture of whole femur bones under flexural load. Before fracture, broadband AE sensors were used in order to measure parameters like wave velocity dispersion and attenuation. Waveform parameters like the duration, rise time and average frequency, were also examined relatively to the propagation distance as a preparation for the AE monitoring during fracture. After the ultrasonic study, the samples were partly cast in concrete and fixed as cantilevers. A point load was applied on the femur head, which due to the test geometry resulted in a combination of two different patterns of fracture, bending and torsion. Two AE broadband sensors were placed in different points of the sample, one near the fixing end and the other near the femur head. Preliminary analysis shows that parameters like the number of acquired AE signals and their amplitude are well correlated with the load history. Furthermore, the parameters of rise time and frequency can differentiate the two fracture patterns. Additionally, AE allows the detection of the load at the onset of fracture from the micro-cracking events that occur at the early loading stages, allowing monitoring of the whole fracture process. Parameters that have been used extensively for monitoring and characterization of fracture modes of engineering materials seem to poses characterization power in the case of bone tissue monitoring as well.

1. Introduction

Acoustic emission (AE) is a technique used in several occasions for the monitoring of the fracture behaviour of materials. Usually piezoelectric sensors are attached on the surface of the material to record the elastic waves generated by cracking incidents in the material. This provides valuable input on the failure process from early stages, certainly before fracture is apparent by visible macro-cracks [1]. Study of the rate and amount of recorded activity is correlated to the applied load and the damage condition of the material [2-6]. Additionally, indices based on the energy or amplitude of the recorded waveforms are used to characterize the intensity of fracture and possibly make projections for the future life [7,8]. In different types of engineering materials AE has demonstrated the capability to characterize the damage mode. This has been demonstrated in different materials like rock, concrete, metal, ceramics and composites [2-9]. The application of AE in human tissue studies though, reveals specific difficulties. A basic one is that due to the nature of specimens (excised from cadavers), limited number of experiments can be conducted. However, there is an extensive literature study by Browne et al. [10] and Shrivastava et al. [11] regarding the use of AE technique in the general biomedical field.

Apart from the limited number of experimental works, another difficulty is the geometry of the specimens and the interpretation of the results. It should me mentioned that considerable effort is given in the ultrasonic assessment of properties, which can be applied for diagnosis purposes of fracture or healing of bones [12-16]. Cortical bone is one of the most complex materials. Due to the microstructure, the properties of porosity and stiffness are varying from the periosteum to endosteum. It appears that the thin and curved geometry of cortical bone endows additional plate wave dispersion. In that kind of heterogeneous media, the elastic waves have the tendency to separate into different modes due to the different velocities of the several frequency components. This complexity and dispersive nature may lead to false AE characterization during fracture. There are extensive studies in AE monitoring on structural materials with respect to the crack characterization with sufficient results [17,18]. Nevertheless, this is not the case for bone tissue where the AE waveform character cannot linked with a fracture mode under a certain load level. Due to the aforementioned irregularities, it is essential to make an ultrasonic study on the same specimens as a preparation for the AE characterization during fracture experiments. Crucial parameters of the AE waveforms can provide information which will exhibit differences due to the propagation distance through the volume of the bone.

In the present study results on fracture experiments in human femur specimens with concurrent monitoring of AE are reported. Waveform parameters like the duration, rise time and average frequency, are also examined relatively to the propagation distance as a preparation for the AE monitoring. The setup applies a mixed bending-torsion monotonic loading on the head until fracture. The AE activity shows the point of micro-cracking onset as well as its development. AE parameters like frequency and rise time exhibit strong shifts with the increase of load, showing that the fracture mechanisms are not stable throughout loading. Discussion is made on the possible correlations between AE parameters and maximum load, thickness and ultrasonic parameters that have been investigated prior to failure [19].

2. Experimental details

This study was performed on eleven femur specimens excised from cadavers. The specimens were supplied by the Anatomy Department of the School of Medicine of the Vrije Universiteit Brussel.

Initially, a preliminary study was conducted on five specimens with the purpose to study the separating burst of the waveforms. In order to obtain these results, four sensors were recording the changes during wave propagating. The sensors were placed in a way to capture the propagation in axial transmission and not through the thickness of the bone. The distance between the sensors was set on 10 mm and the pulse was excited by fractures of mechanical pencil leads in a distance of 5 mm in front of the first sensor. The sensors were of Pico type with a broadband response and maximum sensitivity at 450 kHz. The electric waveforms were pre-amplified by 40 dB and recorded to the acquisition board with sampling rate of 10 MHz in the acquisition board (PAC micro-II, 8 channels). Vaseline grease was used as acoustic coupling and the sensors were located on flat areas of the femur specimens in order to avoid contact problems.

After the ultrasonic study, the samples were partly cast in concrete and fixed as cantilevers (see Figure 1). The "head" of the femur was 120 mm away from the fix point in all specimens. A support was provided in the main body of the specimens (point of minimum elevation) by a metal bolt in order to avoid the fracture at the fix point due to bending moments. The load was applied by a piston resulting in a vertical (nearly) point force. The geometry of the setup resulted in a combination of bending and torsion and different fracture patterns as will be mentioned in the results section.

During AE monitoring two AE broadband sensors (of the same specifications) were used. The first was placed near the fix point and the second at the bottom of the head as it is depicted from Figure 1. While the general placement was similar, their exact position could not be identical in all specimens due to local differences in geometry and curvature. The sensor position was such that AE signals could be potentially captured from the fix point of maximum bending moment as well as from the head which is usually the most exposed part in hip fracture. The threshold was set at 30 dB and acoustic coupling was improved by vaseline grease between the sensors and the surface of the femur, while the sensors were secured by means of tape. Despite the geometry and heterogeneity of the medium, event location was enabled and it resulted in satisfactory localization since pencil lead breaks could correctly be identified in three areas (head, middle of the specimen and close to the fix point). The pulse velocity was measured in an earlier study on the level of 3500 m/s [19].

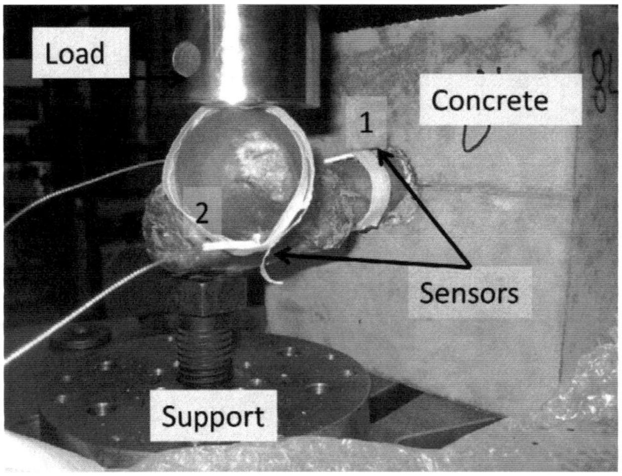

Figure 1. Acoustic emission sensors placed on the femur diaphysis.

3. Results

3.1 General AE activity

Figure 2 shows two typical cases of AE activity history. The load is also depicted in both graphs. In Figure 2a, it is seen that AE starts to accumulate shortly after application of load. Specifically it started at around 15% of the maximum load indicating that this load level was necessary to activate microcracking. The load at the onset of AE was very repeatable for all tested specimens. The activity of sensor 1 (placed near the fixing end) was almost ten times higher than sensor 2 something justified by the fact that fracture took place near the fix point, while the area of the femur head of this specimen was seemingly undamaged (see Figure 4a). On the other hand Figure 2b shows the activity of a specimen fractured at the head (photograph in Figure 4b). Although sensor 1 still recorded the highest amount of activity at the end, sensor 2 accumulated a much higher proportion of AE activity (close to 1/3) showing that significant fracture mechanisms were taking place in the head as well. It is indicative that sensor 2 recorded the first hits and up to 50% of the load it had received most of the activity before fracture was widely spread to the whole specimen. It is worth to mention that the AE activity from concrete is considered minor. The modulus of elasticity of the bone is lower than the corresponded of the concrete. Due to this reason, the damage was concentrated either on the diaphysis or on the head of the femur bone.

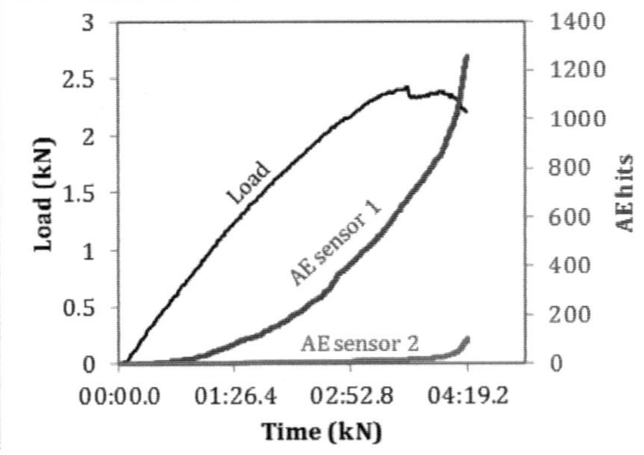

(a)

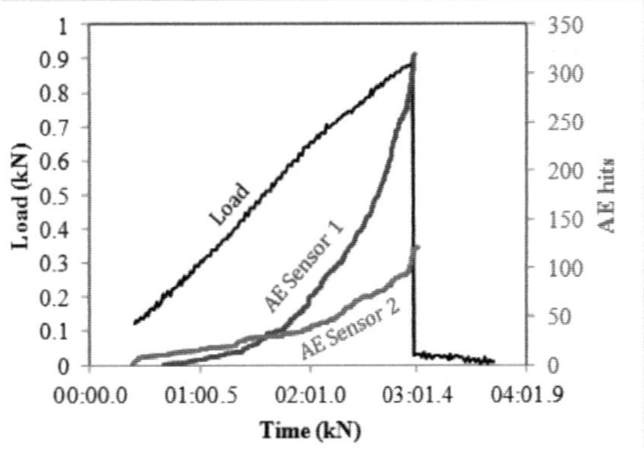

(b)

Figure 2. Load history and cumulative AE activity of different sensors for (a) specimen 1, and (b) specimen 6

The maximum load ranged from less than 1 kN up to 3.6 kN depending on the geometry and thickness of the specimens. Although the number of specimens cannot be considered sufficient for establishing robust correlations, some trends between AE parameters and mechanical data seem promising and are worth discussing in this text.

Figure 3 shows the correlation plot between the maximum load and the accumulated energy received by both sensors. A general increasing trend is seen. The two specimens with the lowest load exhibited also the lowest amount of AE energy, while as the maximum load increased, so did the AE accumulated energy. The coefficient R is of the order of 0.8 while in the expense of a single point (that of the higher energy) the coefficient R rises higher than 0.9, as shown by the blue dash line in Figure 3. These correlations cannot be regarded as global with the limited amount of points, but still they are encouraging in the sense that parameters from acoustic monitoring techniques show some trend with the mechanical results and specifically the load, something that has not been examined thoroughly so far.

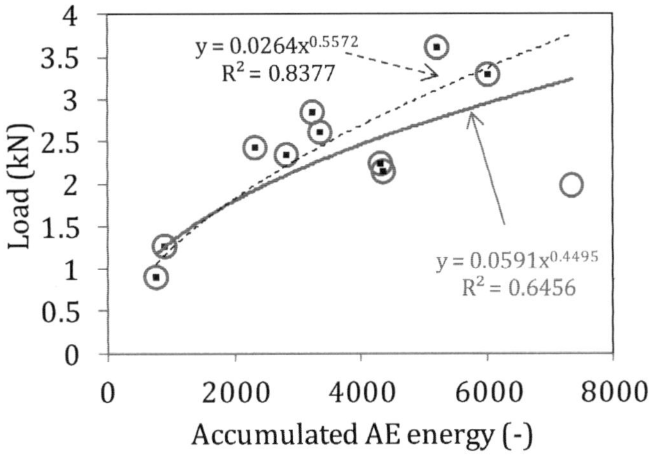

Figure 3. Maximum load and accumulated AE energy for both sensors

(a) (b)

Figure 4. Photograph of femur specimen after the fracture (a) fracture on the diaphysis of the femur bone and (b) on the head of the femur bone

3.2 Ultrasonic assessment

The ultrasonic investigation, offers some other correlations. Specifically, pulse velocity was calculated by the first threshold crossing of the waves. It is mentioned that the pulse velocity corresponds to the fastest mode (which has been identified as S0, symmetric), while a few μs later another burst is identified as A0 (antisymmetric) [20]. An indicative waveform of the last sensor (approx. 40 mm from the excitation) is seen in Figure 5. There, a fast small burst can be identified followed by a larger wave packet. In the same graph the wavelet transformation of the waveform is depicted based on the "Gabor" wavelet [21].

Most of the energy is included in the time window between 120 and 160 μs that corresponds to the second high amplitude mode. By following the maxima of each frequency component of this mode it becomes evident that the wave packet is dispersive. The higher frequencies arrive earlier something typical of plate wave propagation. This behavior may be further highlighted by the microstructure of the tissue that imposes further dispersion (frequency dependence of velocity). It is mentioned that the maxima of the wavelet intensity shown by cross symbols coincide with the theoretical dispersion curves validating the correspondence to the S0 and A0 modes. The curves were developed for material with longitudinal wave velocity 3800 m/s and shear wave velocity of 1800 m/s, close to the velocities of the two modes of this study [20].

Attenuation coefficient was calculated by the exponential decay of the maximum amplitude of the four successive sensors after each excitation meaning that it corresponds to the 2nd burst (A0) which was the strongest packet. In total 5 specimens were ultrasonically investigated and later they were cut for microscopy assessment of the cross sections at the area of ultrasonic measurements.

Pulse velocity, as measured by the propagation distance the first threshold crossing ranged from 3100 m/s to 3700 m/s indicating a quite high stiffness (this value is close to concrete longitudinal wave velocity). The velocity of the second mode measured by a characteristic peak was always between 1000 m/s to 1300 m/s, values that coincide with values in literature. The velocity values showed weak correlation to the maximum sustained load. However, the attenuation exhibited a much better correlation as shown in Figure 6.

The specimens with the highest measured attenuation (between 75 dB/m to 80 dB/m), exhibited the lowest mechanical capacity, below 2.5 kN while as the attenuation coefficient decreased to 50 dB/m maximum load reached 3.6 kN. This correlation between an ultrasonic parameter and strength should first be further validated by more specimens and then investigated as to its origin. However, higher attenuation is connected to lower stiffness and higher amount of heterogeneity in engineering materials [22, 23].

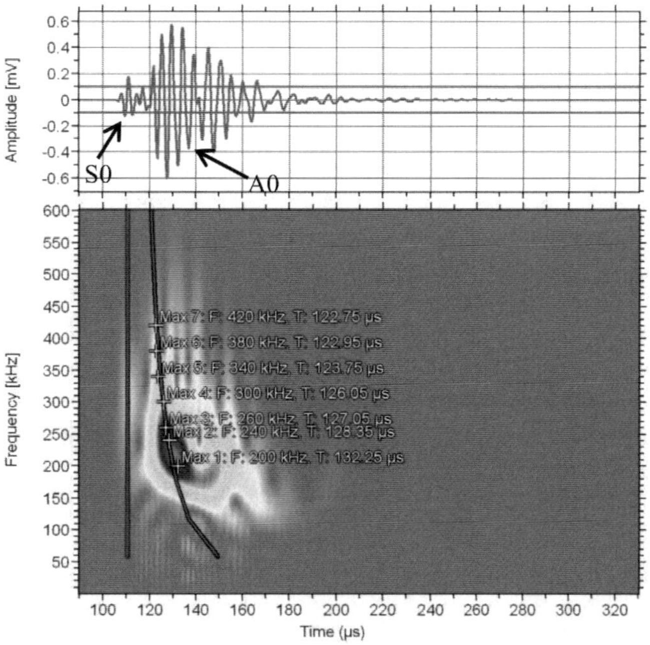

Figure 5. Typical waveform of the 4th sensor (top) and its wavelet transformation

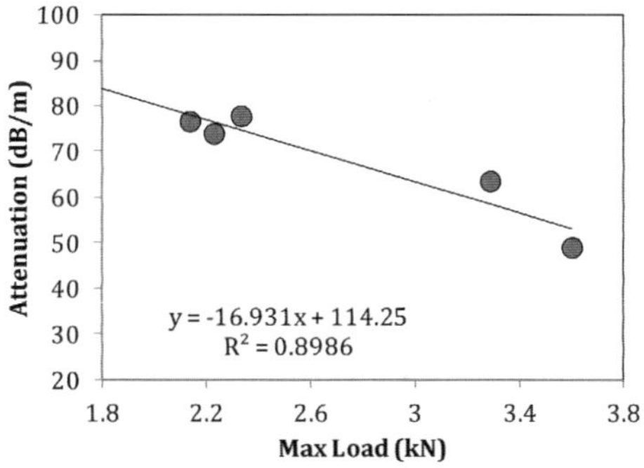

Figure 6. Attenuation coefficient vs. maximum load for five femur specimens.

4. Conclusions

In the present paper, the AE activity during fracture of human bone tissue is discussed. The specimens were femurs the fracture of which is very common especially in aged people (hip fracture) [24-25]. Results show that AE activity can be used for identification of the onset of cracking which occurs much earlier than macroscopic fracture or visible cracks. Additionally, the increase of rate of incoming signals is a precursor of serious fracture phenomena, while the parameters of the obtained waveforms reveal more information as to the shift between fracture mechanisms. During ultrasonic investigations, strong dispersive and attenuative trends are observed with high frequencies propagating faster for the band up to about 420 kHz. Also, it is highlighted from the results that the specimens with the highest measured attenuation exhibited the lowest mechanical capacity. Mechanical properties are also taken into account in an effort to examine the possibility of applying AE indices to interpret the fracture behaviour of bones based on the experience from other material fields.

References

[1] Grosse C U and Ohtsu M 2008 Acoustic emission testing Heidelberg Springer
[2] Shanyavskiy A and Banov M 2010 The twisting mechanism of subsurface fatigue cracking in Ti–6Al–2Sn–4Zr–2Mo–0.1Si alloy, Engineering Fracture Mechanics **77** 1896–1906
[3] Han Z, Luo H, Cao J, Wang H 2011 Acoustic emission during fatigue crack propagation in a micro-alloyed steel and welds, Materials Science and Engineering A **528** 7751– 7756
[4] Zárate B, Caicedo J, Yu J, Ziehl P 2012 Probabilistic Prognosis of Fatigue Crack Growth Using Acoustic Emission Data J. Eng. Mech. **138(9)** 1101–1111
[5] Maillet E, Godin N, R'Mili M, Reynaud P, Fantozzi G, Lamon J 2014 Damage monitoring and identification in SiC/SiC minicomposites using combined acousto-ultrasonics and acoustic emission, Composites A **57** 8–15
[6] Iwamoto M, Ni Q-Q, Fujiwara T, Kurashiki K 1999 Intralaminar fracture mechanism in unidirectional CFRP composites Part I: Intralaminar toughness and AE characteristics Engineering Fracture Mechanics **64** 721-745
[7] Shiotani T, Fujii K, Aoki T, Amou K 1994 Evaluation of progressive failure using AE sources and improved b-Value on slope model tests Progress in Acoustic Emission **7** 529– 534
[8] Kurz J H, Finck F, Grosse C U , Reinhardt H W 2006 Stress drop and stress redistribution in concrete quantified over time by the b-value analysis Structural Health Monitoring **5** 69
[9] Shiotani T, Ohtsu M, Ikeda K. 2001 Detection and evaluation of AE waves due to rock deformation Construction and Building Materials **15** 235–246
[10] Browne M, Roques A, Taylor A. 2005 The acoustic emission technique in orthopaedics-a review The Journal of Strain Analysis for Engineering Design **40** 59–79
[11] Shrivastava S and Prakash R 2009 Assessment of bone condition by acoustic emission technique: A review J. Biomedical Science and Engineering **2** 144-154
[12] Ossi Z, Abdou W, Reuben R L, Ibbetson R J 2013 Transmission of acoustic emission in bones, implants and dental materials Proc. Inst. Mech. Eng. H. 227(**11**) 1237–45
[13] Agcaoglu S and Akkus O 2013 Acoustic emission based monitoring of the microdamage evolution during fatigue of human cortical bone J. Biomech. Eng. 135(**8**) 81005
[14] Shark L-K, Chen H, Goodacre J 2011 Acoustic emission: a potential biomarker for quantitative assessment of joint ageing and degeneration Medical Engineering & Physics **33** 534–545
[15] http://www.ultrasonics.org/Proceedings_2007_UIA/Laugier_2007_UIA.pdf
[16] Vavva M G, Protopappas V C, Gergidis L N, Charalambopoulos A, Fotiadis D I, Polyzos D 2009 Velocity dispersion of guided waves propagating in a free gradient elastic plate: Application to cortical bone J. Acoust. Soc. Am. 125(**5**) 3414-3427
[17] Shahidan S, Pulin R, Muhamad Bunnori N, Holford K M 2013 Damage classification in reinforced concrete beam by acoustic emission signal analysis. Construction and Building Materials **45** 78–86
[18] Farhidzadeh A, Dehghan-Niri E, Salamone S, Luna B, 2013 Whittaker A Monitoring Crack Propagation in Reinforced Concrete Shear Walls by Acoustic Emission Journal of Structural Engineering doi:10.1061/(ASCE)ST.1943-541X.0000781
[19] Strantza M, Louis O, Polyzos D, Boulpaep F, Van Hemelrijck D, Aggelis D G 2014 Elastic wave propagation on human femur tissue, Proc. SPIE 9062, Smart Sensor Phenomena, Technology, Networks, and Systems Integration 2014, 90620T doi: 10.1117/12.2048393 ISBN 9780819499882
[20] Strantza M, Louis O, Polyzos D, Boulpaep F, van Hemelrijck D, Aggelis DG. 2014 Wave Dispersion and Attenuation on Human Femur Tissue. *Sensors* 14(**8**) 15067-15083
[21] Hamstad M A An Illustrated Overview of the Use and Value of a Wavelet Transformation to Acoustic Emission Technology 2005 Wavelet R 1121 User's Guide AGU-Vallen: Icking Germany
[22] Aggelis D G, Shiotani T 2008 Effect of Inhomogeneity Parameters on Wave Propagation in Cementitious Material, ACI Materials Journal 105(**2**) 187-193
[23] Shah S P, Popovics J S, Subramanian K V, Aldea C M 2000 New Directions in Concrete Health Monitoring Technology Journal of Engineering Mechanics ASCE **126** 754-760
[24] Feik S A, Thomas C D L, Clement J G 2002 Age-related changes in cortical porosity of the midshaft of the human femur, Journal of Anatomy **191** 407–416
[25] Stein M S, Thomas C D L, Feik S A, Wark J D, Clement J G 1998 Bone size and mechanics at the femoral diaphysis across age and sex, Journal of Biomechanics 31(**12**) 1101-1110

Comparative study of performance of neutral axis tracking based damage detection

R Soman[1], P Malinowski[1], W Ostachowicz[1,2]

[1] Institute of Fluid Flow Machinery, Polish Academy of Science
14 Fiszera St., 80-231 Gdansk, Poland

[2] Faculty of Automotive and Construction Machinery, Warsaw University of Technology
84 Narbutta St., 02-524 Warsaw, Poland

e-mail: rsoman@imp.gda.pl

Abstract. This paper presents a comparative study of a novel SHM technique for damage isolation. The performance of the Neutral Axis (NA) tracking based damage detection strategy is compared to other popularly used vibration based damage detection methods viz. ECOMAC, Mode Shape Curvature Method and Strain Flexibility Index Method. The sensitivity of the novel method is compared under changing ambient temperature conditions and in the presence of measurement noise. Finite Element Analysis (FEA) of the DTU 10MW Wind Turbine was conducted to compare the local damage identification capability of each method and the results are presented. Under the conditions examined, the proposed method was found to be robust to ambient condition changes and measurement noise. The damage identification in some is either at par with the methods mentioned in the literature or better under the investigated damage scenarios.

1. Introduction

Over the last few years, there has been an increasing awareness about the environmental hazards posed by the extensive use of fossil fuels. As a result a search for clean source of energy has intensified. Wind Energy appears to be one of the leading sources of clean energy. The main hindrance for more widespread deployment of the wind turbines is the high initial costs for setting up of wind farms and its subsequent maintenance. These high initial costs make the energy more expensive than the conventional energy sources. The cost of generation being the biggest drawback of wind energy, there is a concerted effort to reduce it. This can be achieved by increasing the life-time of the wind turbines; reducing maintenance costs and ensuring high availability. The lifetime may be increased by ensuring a more robust design while the maintenance cost may be lowered and the high availability ensured through the use of condition monitoring (CM) and structural health monitoring (SHM), [1].

Traditionally, the evaluation of the condition of civil infrastructure has been carried out through visual inspection mainly due to its low cost and simplicity. However, visual inspection is prone to human errors, as it depends on the perception and the experience of the inspector. Furthermore, visual inspection can be carried out periodically and therefore cannot provide information of the condition of the system in between two inspections. The realization of these shortcomings along with the increasing dependence of the society on the civil engineering infrastructure has made the Structural

Health Monitoring (SHM) systems an imperative. The process of damage can be broadly classified into 4 levels, namely detection of the presence of damage (level 1), identification of damage location (level 2), determining the extent of damage (level 3), and estimating the remaining life of the structure (level 4), [2].

The SHM systems should be able to perform continuous monitoring, be low cost for setting up and maintenance and able to raise an alarm in case the system performance undergoes a sudden change. The common SHM strategy thus is the study of vibration characteristics like the natural frequencies and mode shapes is an inexpensive way for continuous monitoring. These techniques are based on the concept that, the change in mechanical properties of the structure will be captured by a change in its dynamic characteristics [3]. The SHM process involves the observation of a system over time using periodically sampled dynamic response measurements from an array of sensors, followed by the extraction of damage-sensitive features from these measurements, and the statistical analysis of these features to determine the current state of the system's health. The SHM process requires use of sensors for data collection, filters for data cleansing, and central data processing units for feature extraction and post processing. Doebling et al. [4] give an extensive overview of the vibration based detection methods. Many different parameters for the damage detection have been suggested in literature [5-8]. These parameters yield promising results under controlled conditions, but the damage detection is not accurate in in-service conditions. Thus there is a need for a robust methodology, which can detect damage in the presence of measurement noise and ambient temperature changes.

The present paper proposes a comparative study of a new methodology, making use of strain sensors for tracking the Neutral Axis (NA) through the use of Kalman Filter (KF). The position of the neutral axis can be assessed by measuring the strains on opposite surfaces of the tower in bending. The NA is the property of the cross section of the tower independent of the bulk temperature effects, and the ambient wind loading. The methodology is applied to a finite element (FE) model of the DTU 10MW RWT and its performance is compared with some widely used methods namely the Enhanced Co-ordinate Modal Assurance Criterion (ECOMAC) [8], the Mode Shape Curvature Method [5], and the Strain Flexibility Index Method [9]. The performance of these established methods is compared in simulated operating conditions, like, error in yaw angle measurement, measurement noise and ambient condition changes.

2. Proposed Methodology

2.1. Neutral axis

The primary function of the tower structure is to support the hub, and the nacelle of the wind turbine. The nacelle and the hub are axial loads which are eccentrically loaded on the tower. This eccentric loading gives rise to axial compressive loads as well as bending loads as shown in Figure 1. The axial compression is uniform over the entire cross section while the bending loads will be tensile at one end and compressive at the other. Furthermore, the tower experiences wind loads which result in bending strains in the tower. The wind turbine tower is a tall slender structure and hence, Euler-Bernoulli beam theory may be used with limited approximation errors [10]. The bending strains are given by (1)

$$\epsilon = \frac{M_b y}{EI} \tag{1}$$

where, ε is the longitudinal strain in bending, M_b is the net bending moment at the cross section due to wind loading and eccentricity, E is the Young's modulus and I is the area moment of Inertia, y is the distance from the NA [11]. The Figure 1, explains the abbreviations used and the concept. The NA can thus be estimated based on the strain measurements.

Thus, one surface of the tower experiences, a combination of two axial compressions, (right side in Figure 1) while the other end experiences a combination of compressive load due to the weight and tensile load due to the bending, (left side in Figure 1). If the line connecting the two strain levels is extended, there will be a point where the strain experienced will be zero, which is identified as the Neutral Axis point. The neutral axis of the section is a function of the flexural rigidity of the structure, and does not depend on the applied bending loads, thus by, measuring the strains at the opposite edges of the beam, the neutral axis can be located, which in turn may be an indicator of the damage. The figure 1 shows a solid beam-type structure for ease of visualization of the NA. For wind turbine applications, the tower is generally hollow but the response of the structure is more or less similar. A more in depth discussion and sensitivity studies about the methodology may be found in [12-15]. The method is able to detect damage even when the damage is not in the vicinity of the sensor. Although for accurate damage isolation, (along the longitudinal direction as well as along the circumference) bi-axial neutral axis tracking is necessary.

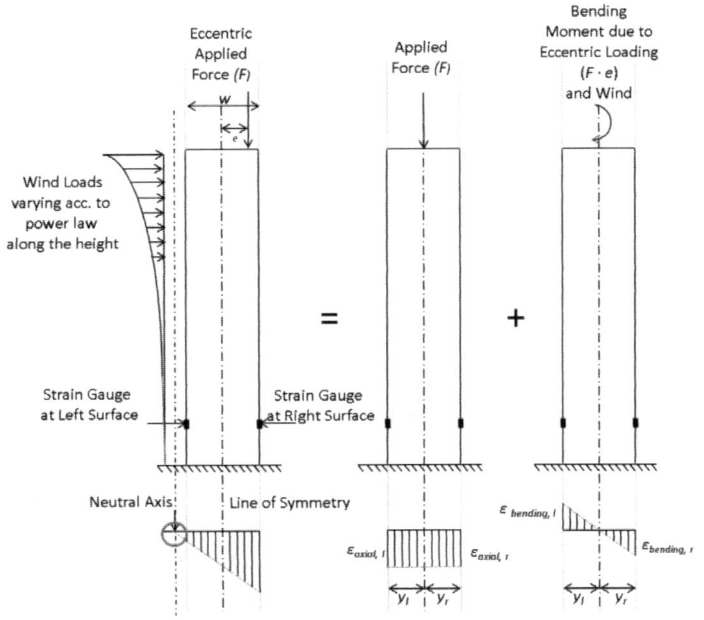

Figure 1: Flexural Strain Distribution over the tower cross-section subject to loading

2.2. *Kalman Filter*

The KF is a set of mathematical equations that provides an efficient computational (recursive) solution of the least-squares method. Theoretically, KF combines a system's dynamic model (physical laws of motion) and measurements (sensor readings) to form an estimate of the systems varying quantities (system state) that is better than the estimate of the system obtained by measurement alone, [16]. Figure 2 concisely explains the implementation of the KF.

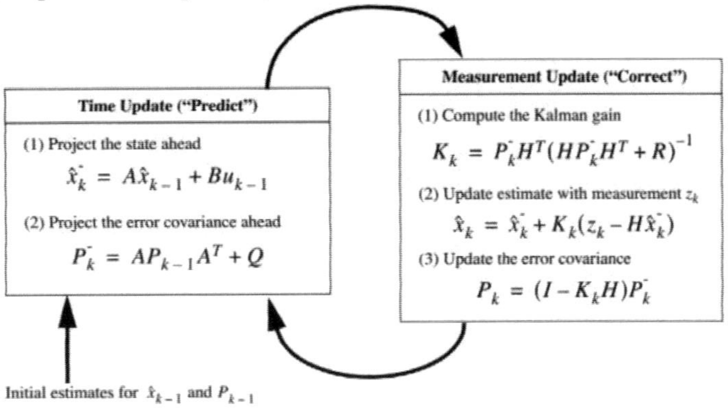

Figure 2: Flow Chart for the implementation of the KF [16]

where, x is the estimate of the state, A is the state transition matrix, B is the control matrix, u is the control variable, P is the state variance matrix, Q is the process variance matrix, K is the Kalman gain, H is the measurement matrix, z is the measurement variable, the 'super minus' indicates a priori estimate while the subscripted k indicates the time step.

In the present application, the state estimate variable is a 3×1 vector consisting of the ratio, $\frac{2x}{w}$ (referred to as Neutral Axis Estimate - NAE) which in undamaged condition should remain constant independent of the applied loads. x is the location of the neutral axis, the second variable tracks constant 1. This constant for tracking 1 is incorporated to ensure, accurate system depiction, and formation of square measurement matrix which allows faster computations. The third variable is θ for the yaw angle. It is a linear estimate of the measurement from the sensor. The measurement update matrix takes into consideration the observability of the neutral axis based on the locations of the sensors.

3. Numerical Modelling

The proposed methodology was verified on a simulated FE model of the DTU 10 MW Reference Wind Turbine. The model was simulated in commercial FE modelling software ABAQUS [17] based on the design data in the reference [18].

The tower is a 115.63 m tall hollow steel structure. The outer diameter varies linearly from 8.3 m at the base to 5.5 m at the top of the tower. The tower is divided in 10 sections where the wall thickness is constant in each section, but gradually decreasing from the bottom to the top. The tower is encastred at the bottom (Figure 3). The tower is made from steel S355 with a Young's modulus of 210 GPa, Poisson's ratio 0.3 and the density 8500 kg/m3 (8% increase of the density to account for the secondary structural components).

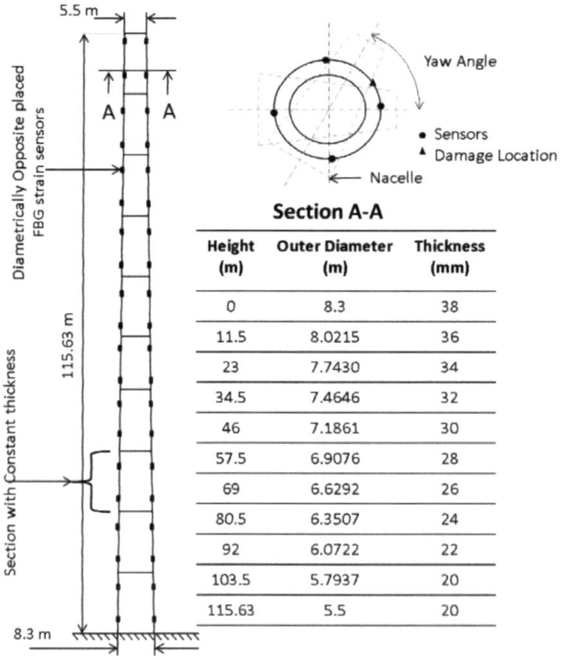

Height (m)	Outer Diameter (m)	Thickness (mm)
0	8.3	38
11.5	8.0215	36
23	7.7430	34
34.5	7.4646	32
46	7.1861	30
57.5	6.9076	28
69	6.6292	26
80.5	6.3507	24
92	6.0722	22
103.5	5.7937	20
115.63	5.5	20

Figure 3: FE modelling details of the Tower.

The nacelle and hub loads were applied as point loads, at specified eccentricity and height indicated from the design specifications. The wind loads were simulated as random loads using Euro-codes, [19]. A peak wind pressure was selected and applied on the surface area facing the wind, in order to compute the force. The force increases according to the power law along the height of the tower. The blades, however, were assumed to be pitched into a full aerodynamic brake position to ensure minimal rotor motion and consequent change in mass distribution, which may affect the NA [20].

3.1. Modelling the effect of temperature

For a steel structure like the tower, the ambient temperature changes affect the performance in two primary ways [21], the bulk effects due to the change in the temperature of the entire structure and the gradient effects, caused due to the direction of the sun. Tower being made of steel is a good conductor of heat, and as such the resultant gradient of the temperature will be quite small. Thus, only the bulk effects have been modelled. The bulk effect of the change in temperature manifests in many different ways, but two of these effects are considered to be most significant namely, changes in the geometry of the structure, and the change in the material properties.

3.1.1. Change in geometry of the structure due to temperature. Changes in the dimensions of the components of a structure are directly proportional to the change in temperature. The rate of change in dimension depends on the co-efficient of expansion α, of the material. For the S355 steel, the expansion coefficient is assumed to be 1.7×10^{-5} m/C° [22]. In studies carried out on civil structures, the effect of these changes alone is significant. A temperature difference of 40°C causes a 1.3% variation in the natural frequencies [23].

3.1.2. Change in the material properties. The material properties have a linear relationship with the change in temperature. The Modulus of Elasticity (E) varies with temperature, the modulus thermal coefficients for steel is -3.6×10^{-4} Pa/C° [21]. The constants are higher by an order of magnitude than the coefficient of expansion but there contribution to the change in natural frequencies is of the order 1.72% for a variation of 40°C [23].

4. Vibration based damage identification (VBDI)

4.1. Vibration parameters

The vibration based damage identification methods use vibration characteristics to monitor the performance of the structure. It is an effective strategy for low-cost continuous monitoring. The methods are distinguished based on the basis of the structure property used for monitoring. The commonly used parameters are: change in natural frequencies, change in mode shapes, change in mode shape derivatives, change in modal strains etc. The properties monitored are acceleration data and strain data for the structure subject to unknown ambient excitation.

The modal properties are a function of the condition of the structure. Hence any change in the structure will be reflected by a change in these parameters. These parameters have a varying sensitivity to the physical changes and have their own merits and limitations. Some of the commonly used VBDI methods are mentioned below.

4.2. Enhanced coordinate modal assurance criterion method (ECOMAC)

The first systematic use of mode shapes for damage detection was carried out by West (1984) [24]. In the future, several other methods were proposed in order to overcome the insensitivity of mode shapes. In 1992, Hunt [25] proposed the ECOMAC method which correlates the two mode shapes. The ECOMAC [8] metric can be computed by

$$ECOMAC(j) = \frac{1}{2N}\sum_{i=1}^{N}\left|\psi_i{}^*(j) - \psi_i(j)\right| \qquad (2)$$

where, $\psi_i{}^*$, and ψ_i, denote the ith damaged and undamaged mode shape respectively, N, is the number of mode shapes and j, is the joint number. The highest peak in the ECOMAC plot indicates damage. The derivation of ECOMAC is given in detail in [8].

4.3. Mode shape curvature method (MSC)

The sensitivity of the mode shape changes to damage was low. Thus the mode shape curvature was proposed by Pandey et al. [5] as a damage indicator. The modal curvature is the second derivative of the mode shape and is highly sensitive to damage. This metric is, hence ideal for the detection and isolation of damage. The central difference operator may be used to get the mode shape curvature (MSC) from the measured mode shapes [5]. It is given by:

$$\phi_i'' = \frac{\phi_{i+1} - 2\phi_i + \phi_{i-1}}{l^2} \tag{3}$$

where, ϕ is the mode shape curvature at i^{th} location, ϕ is the mode shape at particular location, and l is the distance between the sensor nodes.

4.4. Strain modal flexibility index method (MFI)

There is a recent trend to use strain mode shapes for damage detection [9]. The strain mode shapes are more sensitive to damage, and the numerical error in the computation of MSC as a result of the central difference operator can be overcome. Strain Energy metrics are commonly used for damage detection [3]. An improvement to these methods is the use of strain based flexibility matrix. The flexibility matrix of the structure is an inverse of the stiffness matrix and hence can be calculated precisely even with a few lower modes of vibration [8]. The flexibility can be computed by:

$$F = \sum_{i=1}^{N} \frac{1}{\omega^2} \psi_i \psi_i^T \tag{4}$$

where, ψ_i, denotes the ith mode shape, N, is the number of mode shapes, and ω, is the diagonal matrix with natural frequencies of vibration.

The change in flexibility matrix computed from the first few modes of vibration for the damaged and undamaged structure can be a used for damage detection and isolation as shown in equation 5.

$$\boldsymbol{MFI(j)} = \frac{|\Delta \boldsymbol{F_{jj}}|}{\boldsymbol{F_{jj}}} \tag{5}$$

where, $MFI(j)$ is the flexibility index, j denotes the jth degree of freedom, and ΔF is the change in flexibility, F is the flexibility of the undamaged structure.

5. Analysis results and discussion

The performance of the four methods introduced in the earlier sections can be assessed in terms of their ability to isolate damage successfully. The isolation method is considered to be successful, if the value of the corresponding parameter is highest at the damage location. The damage was introduced at various locations and the ability of the methods was validated. Figure 4 indicates one such analysis carried out. It is noted that in these analyses 42 sensor nodes (21 each for acceleration measurement along x and y axis) assumed without taking into account measurement noise and temperature variation. For this case the damage was introduced by reducing the Flexural Rigidity of an element (size-0.38m × 0.25m) by 25%. Reduction of flexural rigidity is a valid damage simulation strategy as indicated by [26]. It may be treated equivalent to loss of material thickness due to corrosion or cracking and is a commonly used strategy for global level damage simulation in bridge structures, [12]. Analyses have demonstrated that the NA based method is able to detect low damage extents (10% reduction) [13-15]. However, a higher degree of damage is assumed, to facilitate the comparison of the four methods under varying conditions of measurement noise or temperature, for which the performance of the above methods has already been, validated [8]. As can be seen in Figure 4, all the four methods were able to isolate the damage successfully.

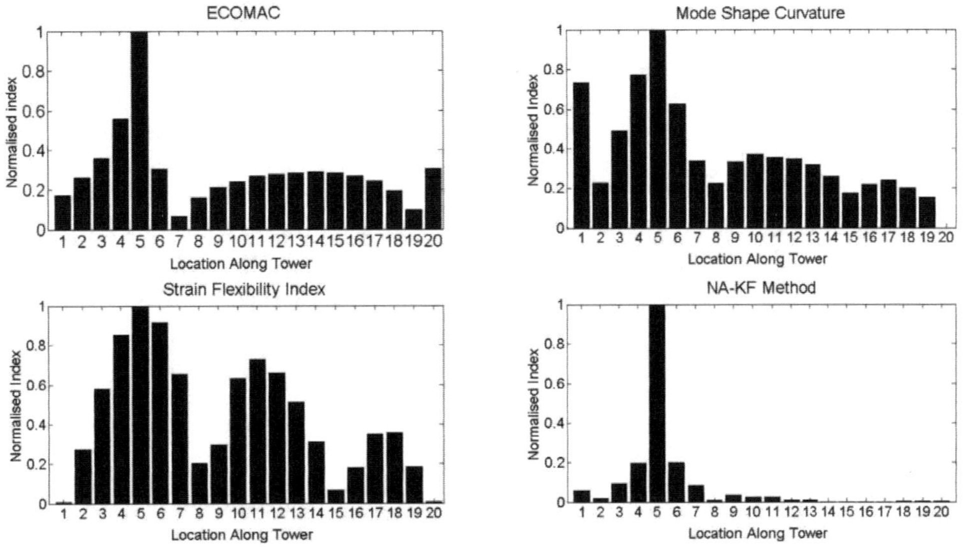

Figure 4: Comparative performance of damage isolation methods

The relative difference in the peak corresponding to the damage location and the other locations is highest for the NA tracking based method, while it is the least for the MFI method. Similar results were obtained for different damage locations along the circumference of the tower. The damage was not detected in the NA based method only when the damage was at a location perpendicular to the plane of the sensors, and as a result bi-axial neutral tracking is necessary. But for other locations, the NA method is able to detect damage even when the sensors are not in the vicinity of the damage.

5.1. Noise

The experimental readings are usually altered by measurement noise thus making the selected parameters less sensitive to damage isolation. The effectiveness of the proposed method was therefore checked through analyses under various levels of measurement noise. The noise component was generated randomly and added to the base measurement [3]. The damage isolation algorithm was run 1000 times to determine the probability of damage isolation and is given in Table 3. Strain and modal curvature based methods like MFI and MSC are more sensitive to damage, than the mode shape based methods, and hence, are expected to perform better in the presence of noise. But, due to the numerical differentiation and round-off errors associated with the computation of MSC, its performance is affected significantly at higher noise levels. On the other hand, the MFI method does not need any post-processing, and hence demonstrates a better performance. The use of KF as indicated in [13] yields robustness to the NA tracking methodology and the performance of the NA-KF method is better than the other methods. It must be noted that the noise was added to the measurement of strain and displacement mode shapes only, the yaw angle measurement is assumed to be accurate. The discussion on the effect of noise on yaw angle measurement is included in a later section.

Table 1: % Probability of Damage Isolation with varying Noise.

	Noise Level				
Methods	0%	1%	2%	5%	10%
ECOMAC	100	100	78	58	19
MSC	100	100	58	46	15
MFI	100	100	95	90	76
KF-NA	100	100	99	91	82

5.2. Temperature
The effect of ambient condition changes has been discussed in Section 3.1. The performance of the proposed method was compared with that of the other methods in varying ambient temperature conditions. The ECOMAC and MFI methods have a comparable performance in changing temperature conditions. The MSC being more sensitive to damage performs better than the methods based on mode shapes, while KF-NA method performs better than MSC for the reasons explained in Section 5.1. Only the effects of temperature change mentioned in section 3.1 on the structure were considered. The acquisition equipment was assumed to be unaffected by the temperature, as the sensors are almost always, temperature compensated, and the rest of the acquisition equipment is designed for the temperature range.

Table 2: Damage Isolation for changing temperature conditions

	Temperature Variation				
Methods	-30°C.	-15°C.	0°C.	15°C.	30°C.
ECOMAC	×	✓	✓	✓	×
MSC	×	✓	✓	✓	✓
MFI	×	✓	✓	✓	×
KF-NA	✓	✓	✓	✓	✓

5.3. Yaw
Yawing of the nacelle for alignment with the direction of the wind improves the efficiency and as a result the power output of the wind turbine. Unfortunately, due to the yawing of the nacelle, the load distribution around the tower changes and as a result the mode shapes of vibration of the tower change as well. So, a damage detection system which does not consider the yaw of the nacelle will not be able to successfully detect damage in the tower structure. Thus it is of utmost importance to track the yaw of the nacelle. Unfortunately the 3 methods namely ECOMAC, MSC and MFI have not been formulated specifically for such an application and do not fuse the information in the prediction. Thus, this information which is easily available has to be incorporated artificially, which may lead to systematic errors if the estimate of yaw angle is wrong. This estimation may be due to system faults or just measurement noise. A small change of 2° significantly affects the resultant mode shapes and the damage detection. On the other hand, the KF based method can easily incorporate the Yaw angle through data fusion as shown by Soman et al. [13].Furthermore; the yaw angle tracking based method is robust to measurement noise and hence yields accurate damage detection in presence of noise in the yaw angle measurement

6. Conclusion
A novel damage detection method using KF for NA tracking is compared to other popular damage detection techniques. The performance of the methodology is examined in conditions which simulate the different temperature conditions and different levels of contamination of measurements because of noise. The study indicates that the KF based methodology performs better than the other methods. It combines the benefits of the use of a sensitive damage detection parameter, use of strain sensors in the dynamic domain and powerful numerical tool for post processing, and allows us to have reliable damage isolation and also damage diagnosis. However, an experimental validation is required for this method. Furthermore, more realistic damage scenarios need to be introduced like fatigue cracks, and the effectiveness of the methodology should be validated.

Acknowledgement
The authors would like to acknowledge the European Commission for their research grant under the project FP7-PEOPLE-20120ITN 309395 “MARE-WINT” (new Materials and REliablity in offshore WINd Turbines technology). The authors would also like to thank DTU Wind Energy, for providing

valuable information for the modeling of the 10 MW RWT tower for the purpose of this study. The authors are also grateful to TASK-CI for allowing the use of their computational resources. The opinions expressed in this paper do not necessarily reflect those of the sponsors.

References

[1] Cho S, Park J, Jung H J, Yun C B, Jang S, Jo H, Spencer B F Jr, Nagayama T, Seo J W, Structural health monitoring of cable-stayed bridge using acceleration data via wireless smart sensor network, *Bridge Maintenance, Safety, Management and Life Cycle Optimization* (2010), pp. 158-164.

[2] Rytter A, PhD Dissertation: Vibrational Based Inspection of Civil Engineering Structures, Aalborg, University of Aalborg (1993).

[3] Abedwuyi A, Wu Z, Serker N H K M, Assessment of Vibration-based Damage Identification Methods using Displacement and Distributed Strain Measurement, Structural Health Monitoring *Structural Health Monitoring, 8(6), 443-461 (2009)*

[4] Doebling S, Farrar C, Prime M, A Summary Review of Vibration based Damage Identification Techniques, *Shock and Vibration Digest*, 30(2) (1998) 91-105.

[5] Pandey A K, Biswas M, Samman M M, Damage Detection From Changes In Curvature Mode Shapes, *Journal of Sound and Vibration,* Vol. **145**, No. 2, pp. 321–332, (1997)

[6] Pandey A K, Biswas M, Damage Detection In Structures Using Changes In Flexibility, *Journal of Sound and Vibration,* (1994) Vol. **169**, No. 1, pp. 3–17.

[7] Stubbs N, Kim J T, Farrar C R, Field verification of a non-destructive damage localization and severity estimation algorithm, *Proceedings of 13th International Modal Analysis Conference, vol. 1*, (1995).

[8] Talebinejad I, Fischer C, Ansari F, Numerical Evaluation of Vibration-Base Methods for Damage Assessment of Cable Stayed Bridges, *Computer-Aided Civil and Infrastructural Engineering 26* (2011).

[9] Zonta D, Bernal D, Strain-based approaches to damage localization in civil structures, *Proceedings of XXIV international modal analysis conference, Saint Louis* (2006).

[10] Labuschagne, A., Janse van Rensburg, NF., and Van der Merwe, A., Comparison of linear beam theories. *Mathematical and Computer Modelling* 49, no. 1 (2009): pp. 20-30

[11] Xia H W, Ni Y Q, Ye X W, Neutral-axis position based damage detection of bridge deck using strain measurement: formulation of a Kalman filter estimator, *Proceedings of the 6th European Workshop on Structural Health Monitoring, Dresden, Germany.* (2012).

[12] Soman R, Malinowski P, Ostachowicz W, Neutral Axis Tracking for damage detection in wind turbine towers, *Proceedings of the EWEA 2014, Barcelona*, Spain. 10-14 March 2014.

[13] Soman R, Malinowski P, Ostachowicz W, Kalman-Filter Based Data Fusion for Neutral Axis Tracking for Damage Detection in Wind-Turbine Towers, *Proceedings of 7th European Workshop of Structural Health Monitoring Vol.20 No.02 - The e-Journal of Nondestructive Testing pp-245-252.*

[14] Soman, R, Malinowski, P and Ostachowicz, W. "Threshold determination for neutral axis tracking based damage detection in wind turbine towers", In: *Proceedings of the EWEA Offshore2015, Copenhagen, Denmark.* 10-12 March 2015.

[15] Soman, R, Malinowski, P and Ostachowicz, W., Paulsen, U "Kalman filter based data fusion for neutral axis tracking in wind turbine towers", In: *Proceedings of the SPIE smart Structures.NDE, San Diego,US.* 8-12 March 2015.

[16] Brown R G, Hwang P Y C, Introduction to Random Signals and Applied Kalman Filtering, 3rd Edition, *John Wiley & Sons, New York, (1997).*

[17] ABAQUS, Analysis User's manual, version 6.12-3 ed., (2013).

[18] DTU Wind Energy Report-I-0092. "Description of the DTU 10 MW Reference Wind Turbine",

July (2013)

[19] Eurocode NS-EN 1991-1-4; "General actions - Wind actions". Standards Nor-way, 2005+NA: (2009).

[20] Bas J, Carriveau R, Cheng S, Newson T, Strain Response of a Wind Turbine Tower as a Function of Nacelle Orientation, *In BIONATURE 2012, The Third International Conference on Bioenvironment, Biodiversity and Renewable Energies (pp. 12-18)*, (2012)

[21] Soman, R, Onoufriou T, Kyriakides, M, Damage assessment of bridges using compound SHM-signal processing and communication challenges, *(EUSIPCO), 2013 Proceedings of the 21st European Signal Processing Conference*, pp. 1-5. IEEE, 2013.

[22] Xu Z D, Wu Z, Simulation of the Effect of Temperature Variation on Damage Detection in a Long Span Cable Stayed Bridge, *Journal of Structural Health Monitoring* (2007), 6: pp. 177-189.

[23] Xia Y, Chen B, Wong S, Ni Q, Xu Y, Temperature effect on vibration properties of civil structures: A literature reviews and case studies, *Journal of Structural Health Monitoring,* 2012, 2 pp. 29-46.

[24] West W M, Illustration of the Use of Modal Detection in Steel Framed Structures by Vibration, *Proceedings of the AirForce Conference on Aircraft Structural Integrity, Los Angles* (1984).

[25] Hunt D L, Application of an enhanced coordinate modal assurance criterion, *Proceedings of the 10th International Modal Analysis Conference,* San Diego, CA, 1, 66–71 (1992).

[26] Jang, S, Jo, H, Cho, S, et al. Structural health monitoring of a cable-stayed bridge using smart sensor technology: deployment and evaluation. *Smart Structures and Systems*, 2010 *6*(5-6), 439-459.

Integrated non-destructive assessment of relevant structural elements of an Italian heritage site: the Carthusian monastery of Trisulti

C Rainieri[1], A Marra[1], G M Rainieri[2], D Gargaro[1], M Pepe[2] and G Fabbrocino[1]
[1]Structural and Geotechnical Dynamics Laboratory, University of Molise, Via Manzoni, 86100 Campobasso – Italy
email: carlo.rainieri@unimol.it, adriana.marra@unimol.it, danilo.gargaro@unimol.it, giovanni.fabbrocino@unimol.it,
[2]S2X s.r.l., Via Duca degli Abruzzi, 86039 Termoli – Italy,
email: gianmaria.rainieri@s2x.it, michele.pepe@s2x.it

Abstract: The analysis of historical structures in need of preservation and restoration interventions is a very complex task due to the large uncertainties in the characterization of structural properties and detailing in view of the structural response. Moreover, the predictive performance of numerical analyses and simulations depend on the availability of information about the constructional properties of the architectural complex, crack patterns and active degradation phenomena. In particular, local changes in material properties or damage due to past events (such as earthquakes) can affect individual structural elements. They can be hardly detected as a result of the maintenance interventions carried out over the centuries and the possibility to carry out limited or even no destructive investigations due to the historical relevance of the structure. Thus, non-destructive investigations play a fundamental role in the assessment of historical structures minimizing, at the same time, the invasiveness of interventions. The present paper deals with an explanatory case study concerning the structural investigations carried out in view of the seismic assessment of an Italian historical monument, the Carthusian monastery of Trisulti in Collepardo, erected in 1204 under Pope Innocenzo III. The relevance of the case study is due to the application, in combination, of different NDT methods, such as sonic tests, and active and passive infrared thermography, in order to characterize relevant masonry elements. Moreover, an advanced system for the in-situ non-destructive vibration-based estimation of the tensile loads in ancient tie-rods is described and the main results obtained from its application for the characterization of the tie-rods of the cloister are presented.

1. Introduction

The seismic vulnerability assessment of historical masonry structures is a challenging task due to the uncertainties affecting material properties and structural schemes, as well as to the structural modifications and interventions occurred over the centuries and hidden defects or damage. Type of masonry (e.g., regular brick or rubble stone), material properties, construction techniques and effectiveness of connections among structural elements play a fundamental role in defining the structural response. As a result, the seismic vulnerability assessment of historical masonry buildings requires a detailed knowledge of the traditional construction techniques to set appropriate models and the identification of existing damage and degradation phenomena [1].

Non Destructive Testing (NDT) can give a relevant contribution to the analysis of the constructional properties of the architectural complex as well as of individual structural components to identify, for instance, masonry texture, connections among different elements, mechanical properties of materials or stresses in operational conditions, eventual damage or degradation phenomena. In particular, attention is focused on the integration between historical and iconographic research and the on-site experimental assessment of the actual conditions of the architectural complex. It is a crucial phase of the knowledge path of the historical asset that is sometimes adopted to support questionable invasive techniques aimed at evaluating local mechanical parameters, i.e. elastic moduli or compressive/shear strength of selected masonry panels, even though the latter could be not really relevant for the results of the analysis. Herein, these aspects are analyzed with reference to a very interesting explanatory case study: the Library and the Cloister of the Carthusian Monastery of Trisulti in Collepardo [2]. It represents a unique site for the experimentation of the integrated application of different NDT techniques to collect relevant information for the setting of an appropriate model, representative of the actual structural behavior. The Carthusian monastery of Trisulti (Figure 1) is a monumental site lying on the Ernici mountains in the Lazio Apennines (Central Italy). The erection of the primitive monastery is attributed to Saint Dominic by Sora, but the risk of landslides and the need for a rearrangement of the existing complex to fit the rules of the different monastic organization (in 1204 Pope Innocenzo III entrusted to the Carthusian order what remains of the former Benedictine monastery) led to the erection of another Carthusian monastery in a more suitable place. The construction of the new monastery was completed in 1211; thereafter, the complex was further enlarged and remodeled in the Baroque Age and later in the Eighteenth Century [3]. In order to minimize the uncertainties in the evaluation of the structural behavior of the most relevant buildings in the Monastery and identify past restoration interventions [3], extensive NDT has been carried out by means of different and often complementary methods. The results of investigations have been also compared with information collected from historical archives and detailed analysis of bibliographic sources such as historical documents and manuals on the construction techniques in the 13th century.

Figure 1. Overview of the Carthusian Monastery of Trisulti

2. Integrated NDT of the Library of the Carthusian Monastery of Trisulti

Extensive non-destructive investigations by means of a number of complementary NDT techniques have been carried out at the Library of the Carthusian Monastery to identify texture and hidden features of the masonry, presence of voids and other internal defects, degree of connection of the walls. Merging the results of experimental tests with information collected from a number of bibliographical sources, an accurate knowledge of the structure was achieved, including information and details about past interventions and hidden defects. The use of different, complementary in nature, NDT techniques allowed the accurate investigation of the structural elements without damaging paintings and decorations on the walls.

The following tests, alone or in combination, were performed and they are briefly reviewed in the following:

- *Passive and active thermography*, in order to gain information about the origin and historical evolution of the building, identify the type of masonry and investigate the connections between the walls;
- *Endoscopic tests*, in order to check and complement the basic information obtained from thermography at locations characterized by absence of artistic heritage;
- *Sonic tests*, in order to confirm the similarity between different masonry panels and to detect macroscopic voids, cracks or portions of masonry with different characteristics.

2.1. Infrared thermography

Infrared thermography can support geometric survey and investigations on materials and structural details, as well as the analysis of the historical evolution of the building [4]. By means of a thermal imaging camera the temperature distribution on the investigated objects surface can be detected. In fact this device provides a color map of the temperature field by exploiting the correlation between the intensity of the infrared rays, that a body emits, and its temperature.

Depending on the environmental conditions (outdoor, indoor) and the test objectives, passive as well as active thermography has been performed. Passive tests took advantage of the effect of the daily temperature variations. The main feature of the thermal imaging camera used for tests are: 320x240 pixels IR image resolution; 7.5-13 μm spectral range; thermal sensitivity <0.045 °C; 60 Hz frame rate.

The infrared thermography has provided interesting information about the evolution of the building. In fact, it showed that some openings (Figure 2) were closed after interventions (probably in 1958, as resulting from the analysis of the literature) aimed at restoring the original configuration of the building. Such interventions could not be easily recognized by exterior visual inspections.

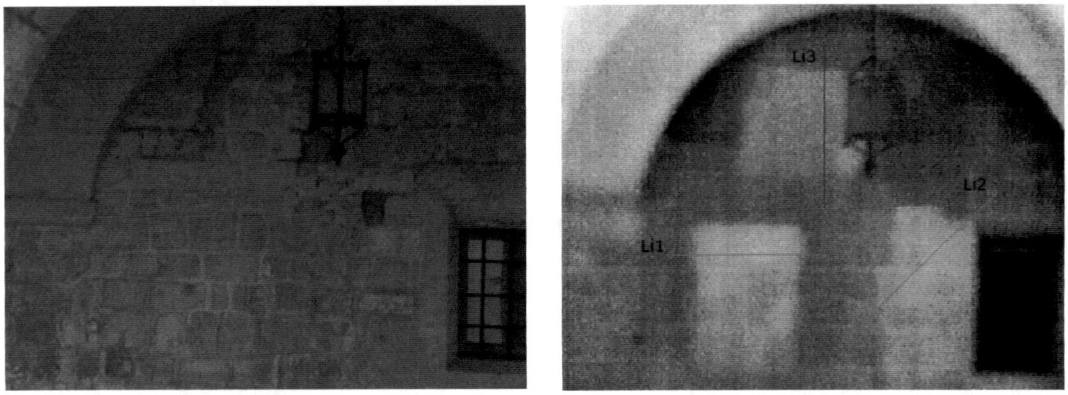

(a) (b)

Figure 2. Digital (a) and infrared (b) image of the outer wall of the Library

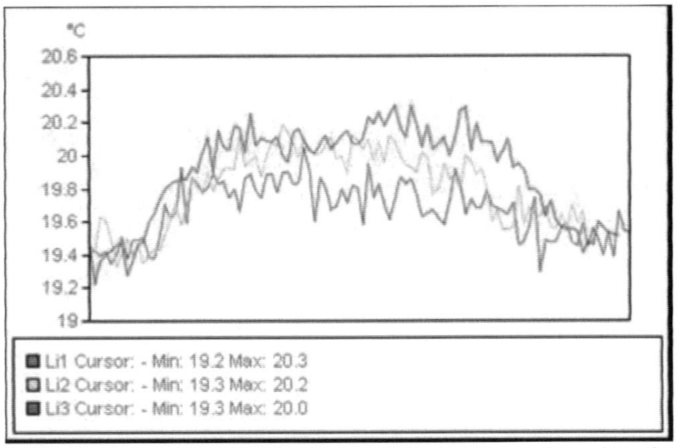

Figure 3. Temperature profile of the segments Li1, Li2, Li3 of Figure 2

The analysis of the infrared image highlights three portions of the wall characterized by a slightly higher temperature with respect to the rest. The analysis of the thermal profiles along three segments intercepting these areas shows that they are comparable (Figure 3), denoting common characteristics of the material in these areas.

2.2. *Endoscopic tests*

Endoscopy has been used, alone and in combination with thermography, to identify the types of masonry constituting the structure of the Library. By direct visual inspection of cavities the inside of walls has been analyzed, verifying the absence of voids or a core made of different and poor materials. Inspections have been carried out by means of a miniaturized digital camera with crystal clear output at the end of a 8.2 mm diameter probe with six adjustable LED lights; the system was completed by a 3.5" color LCD screen.

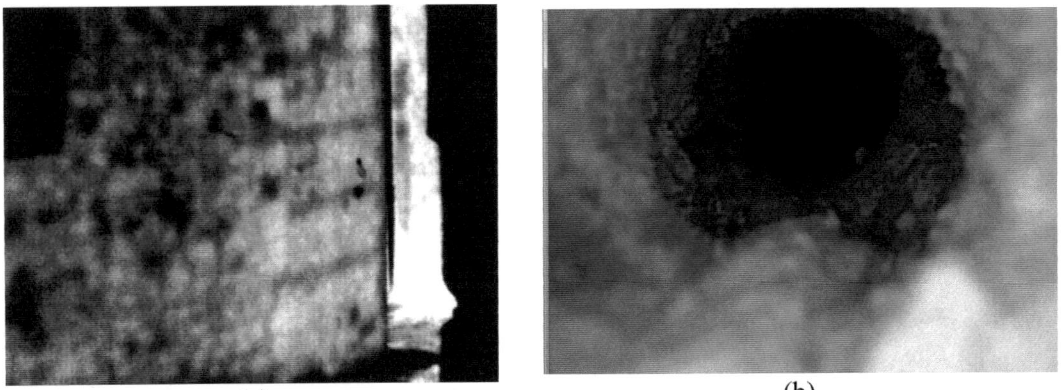

(a) (b)

Figure 4. Masonry Class A1/II: infrared image (a), endoscopic image (b)

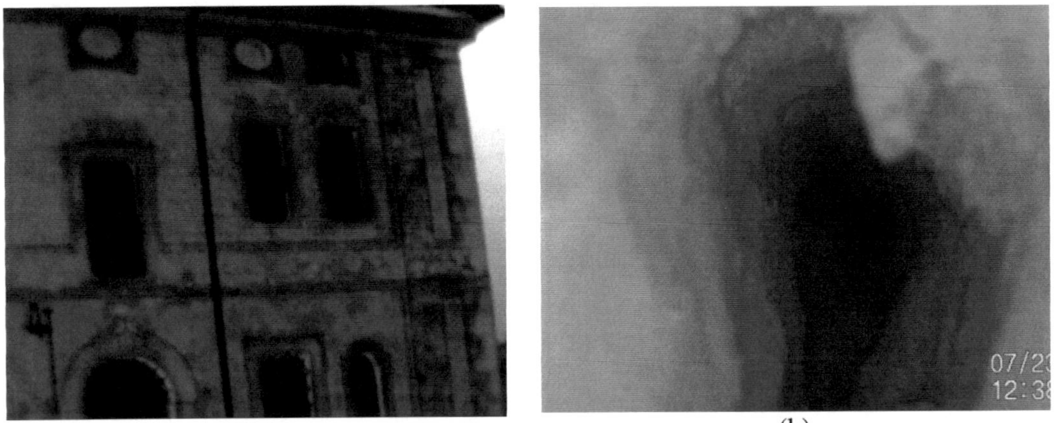

(a) (b)

Figure 5. Masonry Class A3/I: infrared image (a), endoscopic image (b)

Endoscopic tests, in combination with thermography, allowed the identification of the following masonry typologies [5]:

- *Class A1 - Group II*: irregular masonry characterized by the presence of compact limestone and other materials (travertine and bricks), and corners made by large stone blocks (Figure 4).
- *Class A3 - Group I*: masonry made by compact limestone with lime mortar and aggregates; the texture is homogeneous and undifferentiated; this class of masonry is also characterized by the presence of adjustments of horizontality along the height (Figure 5).

The combination of thermography and endoscopy has revealed its potential in the analysis of the vaults (Figure 6) confirming some assumptions based on the historical research [3] and literature reviews [5].

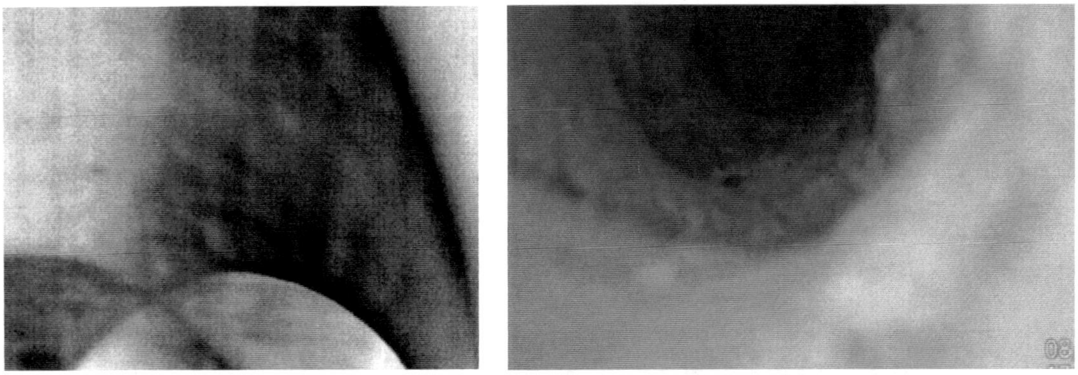

(a) (b)

Figure 6. Vault in the portico of the Library: infrared image (a) and endoscopy (b)

The vaults were made of limestone with elements arranged in form of shell ("*in foglio*"). It was also recognized that the stone slabs have been put with the broader face along the intrados of the vault, then cemented with mortar.

2.3. Sonic tests

Sonic tests have been also carried out for the nondestructive characterization of the masonry walls of the Library. They are based on the analysis of wave propagation in the material. The time for the emitted impulse to cover the distance between the emitter and the receiver is related to the dynamic elastic modulus, the Poisson's ratio and the density in the case of homogeneous and isotropic material. However, in the case of masonry, the inherent heterogeneity and anisotropy prevents the correlation of wave velocity with the elastic properties of the material [6].

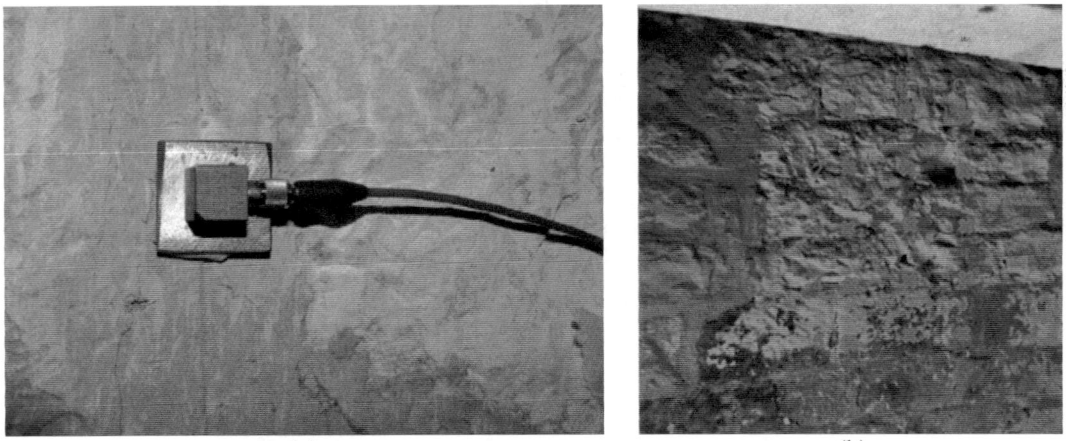

(a) (b)

Figure 7. Sonic tests: sensor (a) and measurement mesh (b)

Direct sonic tests, aimed at obtaining qualitative information about texture and homogeneity of the walls, have been performed at the Library of the Carthusian Monastery of Trisulti.
The sensors were arranged according to a regular mesh grid (20 cm spacing in both x and y direction) on the investigated wall surfaces (Figure 7).

The test equipment consisted of an impact hammer (22.240 kN full scale range, 0.23 mV/N sensitivity), IEPE accelerometers (1 V/g sensitivity, ±5 g pk full scale range, 0.5-3000 Hz frequency range, 0,00005 g rms resolution), a data acquisition unit (24 bit resolution).

The sonic tests highlighted a discontinuity in the response of the masonry between the upper and lower part of one of the outer walls of the Library (Figure 8). It was probably due to the reconstruction of the upper part of the wall during the past. The discontinuity of material was visually detectable and also confirmed by the infrared thermography.

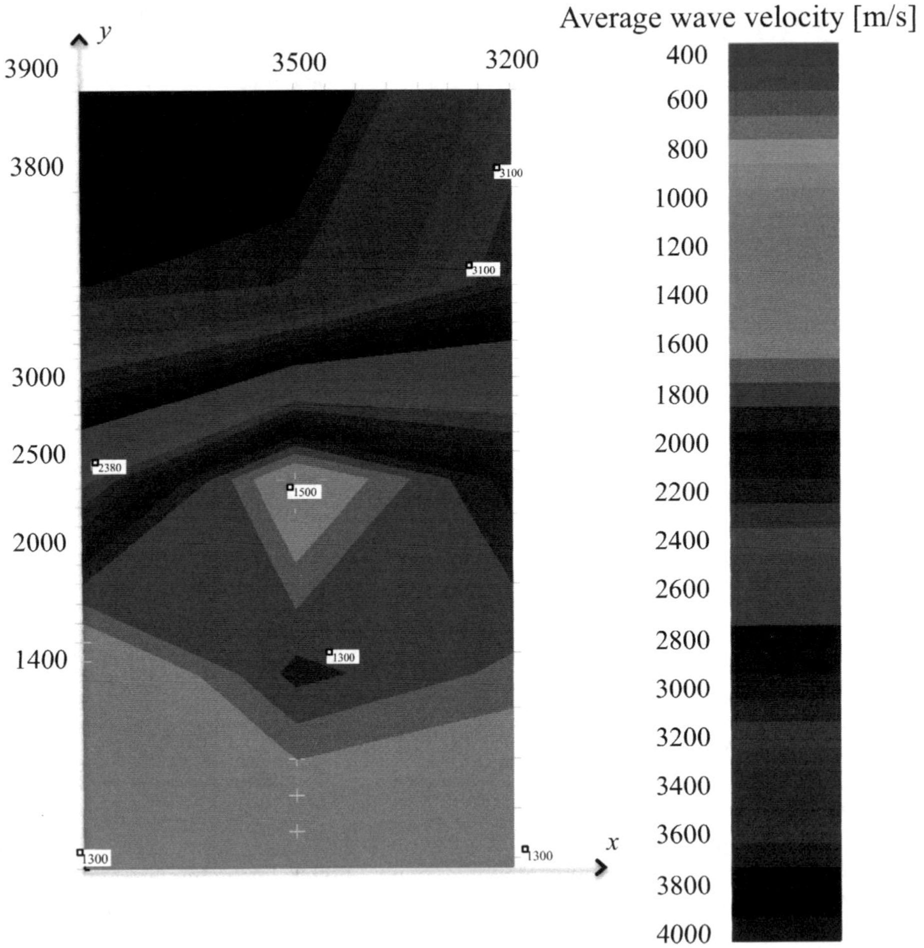

Figure 8. Mapping of average wave velocity

3. Vibration-based estimation of tensile loads in tie-rods

The analysis of documents collected in the Public Archive in Frosinone highlighted that, after the restoration interventions carried out in 1936, the tie-rods in the cloister of the Monastery were not tensioned, so that they could be activated by external loads such as the earthquake. Non-destructive vibration-based tests have been therefore carried out to estimate the tensile load [7] in four tie-rods of the cloister of the Carthusian Monastery (Figure 9) and assess their role in the current equilibrium condition of the structure.

For the vibration-based tensile load estimation, an innovative system developed by S2X s.r.l., a spin off company of the University of Molise, has been used. It is a compact system able to acquire and automatically process the ambient vibration response of cables and tie-rods, providing estimates of the fundamental modal properties and the tensile load in operational conditions [7]. The portable system (Figure 10a) for in-situ evaluation of tensile loads in tie-rods consists of two boards working in parallel. The first board is dedicated to data acquisition and storage, while the second carries out the automated modal parameter identification and the estimation of the tensile load.

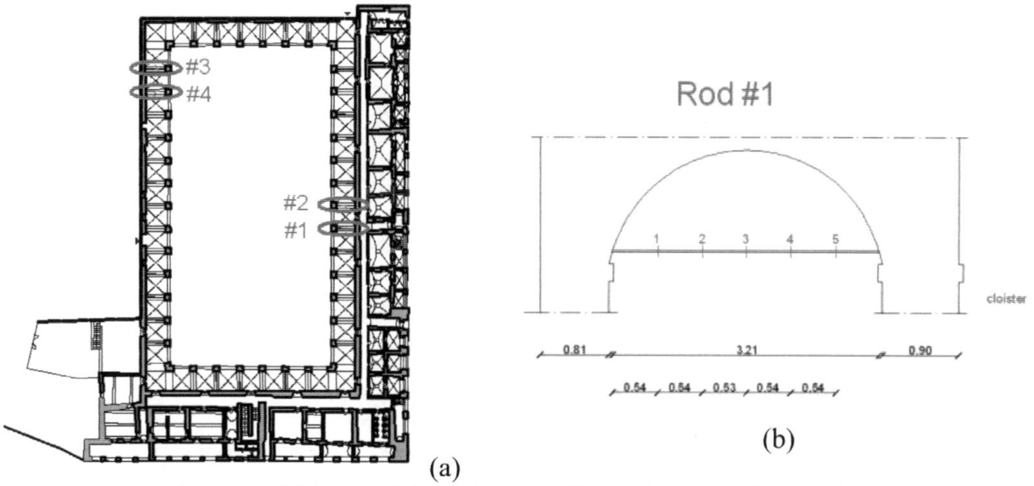

(a) (b)

Figure 9. The tested tie-rods (a) and sample sensor layout (b)

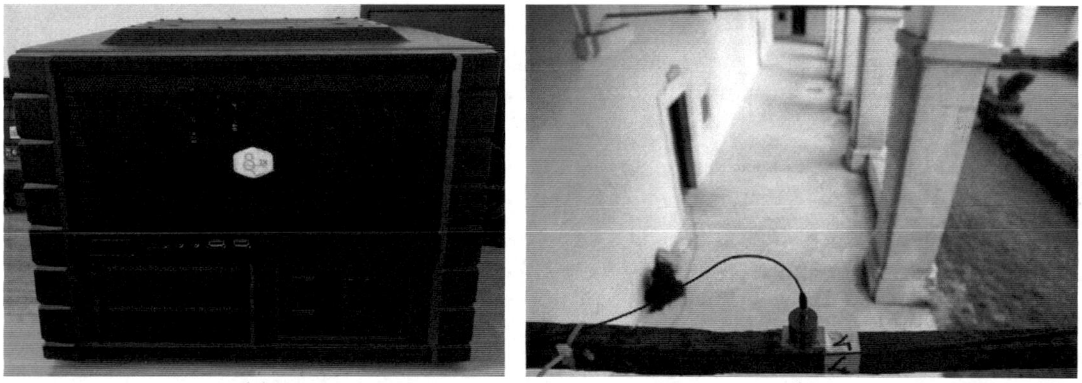

(a) (b)

Figure 10. Vibration-based testing of tie-rods: measurement system (a), installed sensor (b)

The system can acquire data from a maximum of eight IEPE accelerometers by means of the integrated NI PCI 4472b board. The system in complemented by a set of seismic, miniature (50 gm) IEPE accelerometers (Figure 10b). The characteristics of the sensors are: 10 V/g sensitivity, 0.7-450 Hz bandwidth, 0.5 g pk full-scale range. The data acquisition system is characterized by 24-bit Sigma-Delta ADCs for simultaneous sampling of the eight channels, analog antialiasing filters, 110 dB dynamic range, 102.4 kHz maximum sampling rate.

Reference values of 210 GPa and 7800 kg/m^3 have been adopted for the Young's modulus and the density of the steel, respectively. Stable estimates of the tensile loads have been obtained, as reported in Table 1. The results point out that the investigated tie-rods are subjected to tensile loads in operational conditions.

Table 1. Results from in-situ tests at the Carthusian monastery of Trisulti.

Rod #	Base (mm)	Height (mm)	Net span (m)	Number of estimates	f_{av} (Hz)	N_{av} (kN)	σ_N (kN)	γ_N (%)
1	28.8	29.3	3.21	31	13.58	30.3	0.9	3.0
2	29.4	29.0	3.24	12	15.72	43.0	0.6	1.4
3	29.3	29.7	3.17	35	14.46	38.5	0.6	1.6
4	29.5	29.4	3.17	12	16.17	39.4	0.2	0.5

4. Conclusions

The application of different NDT methods, complementary in nature, allowed the accurate investigation of the characteristics of masonry elements (walls and vaults) of the Library of the Carthusian Monastery of Trisulti. The paper has shown how an integrated approach to NDT can provide relevant information for the structural assessment and makes the process more robust as a result of the cross validation of the results provided by the different testing methodologies. Moreover, an innovative vibration-based system for the estimation of the tensile load in tie-rods has been introduced and applied to evaluate the load in the tie-rods of the vaults belonging to the central cloister of the Monastery. Most of the test results confirmed the information obtained from extensive literature review and analysis of documents collected into historical archives. NDT, therefore, provides a valuable support to the enhancement of knowledge about the structural behavior of historical structures, when destructive tests have to be limited. Furthermore, the experience of the conducted research confirmed the effectiveness and reliability of the integrated use of historical and iconographic research and experimental evidences taken on site by means of different advanced NDT techniques.

5. Acknowledgments

The present work was carried out in the context of the Agreement Arcus – DG PaBAAC Rep. n. 113/2011 del 30/09/2011 issued by MiBac for the seismic assessment of the Carthusian monastery of Trisulti in Collepardo. Arcus-MiBac contribution is gratefully acknowledged.

6. References

[1] Presidenza del Consiglio dei Ministri 2011 *Valutazione e riduzione del rischio sismico del patrimonio culturale con riferimento alle norme tecniche per le costruzioni di cui al DM 14 Gennaio 2008*, Direttiva P.C.M. 9 febbraio 2011 (In Italian).

[2] Taglienti A 1985 *Il Monastero di Trisulti e il Castello di Collepardo*. Storia e documenti, Roma.

[3] Savorra M, Belli G, Marra A, Fabbrocino G 2014 Analisi storico-critica della Certosa di Trisulti. *Technical Report B01*, Arcus-MiBac-Reluis (in Italian).

[4] Avdelidis N P and Moropoulou A 2004 Applications of infrared thermography for the investigation of historic structures. *J Cult Herit* **5** 119–127.

[5] Fiorani D 1996 *Tecniche costruttive murarie medievali - Il Lazio meridionale*, Ed. L'Erma di Bretschneider (in Italian).

[6] Binda L, Saisi A and Tiraboschi C 2001 Application of sonic tests to the diagnosis of damaged and repaired structures. *NDT&E Int*, **34** 123-138.

[7] Rainieri C and Fabbrocino G 2015 Development and validation of an automated operational modal analysis algorithm for vibration-based monitoring and tensile load estimation. *Mech Syst Signal Pr* (In press, available online: doi:10.1016/j.ymssp.2015.01.019).

Damage assessment through changes in mode shapes due to non-proportional damping

F Iezzi[1], D Spina[2] and C Valente[1]

[1]Department of Engineering and Geology, University "G. d'Annunzio" of Chieti-Pescara, Viale Pindaro, 42 – 65127 Pescara, Italy
[2]Department of Civil Protection, Seismic and Volcanic Risk Office, Via Vitorchiano, 2 – 00189 Roma, Italy

E-mail: fabrizio.iezzi@unich.it

Abstract. Modal parameters are often used for the structural damage assessment in the dynamic field. Usually, the changes in the modal parameters between different states are assumed as measures of damage. The frequencies are easy to identify, but in some circumstances they are not sensitive to damage and moreover they are mainly a global measure. On the contrary, the mode shapes are more suited for damage localization, but they are generally hard to identify accurately. The energy dissipation, and hence the damping, increases with damage. This feature is stable and has a monotonic behaviour, therefore, damping can be confidently used as an alternative or complementary measure for damage assessment in spite of the accuracy of its identification. However, the damping by itself suffers of the same drawbacks as the frequencies. The joint use of damping and mode shapes is an effective procedure for the damage identification. In the real world the damping is of non-proportional type and the measured mode shapes are complex. It is assumed that an increase of damage causes a modification of non-proportional damping and a variation of the modal complexity. The extent of modal complexity between two different structural states can be used to identify the damage through appropriate indicators. A number of such indicators is introduced and discussed. The effectiveness and sensitivity of the damage indicators are tested on theoretical and pseudo-experimental data.

1. Introduction

Damage assessment is an important research area in structural engineering. In the past, it has received considerable attention, especially in the dynamic field, from the pioneering work [1]. This subject, although well established, still rises interest [2] due to continuous developments in signal analysis and experimental identification techniques. The basic idea arises from the notion that the modal parameters are functions of the physical properties of the structure. Therefore, changes in the physical properties cause changes in the modal parameters. The changes in the modal parameters are assumed as damage measures.

Changes in natural frequencies have been the topic of numerous research studies due to the ease of their identification [3]. These studies have revealed that frequency changes can suffer of practical limitations in several applications, especially in the case of large structure. It is highlighted that detectable frequency changes could require high damage level or global damage.

Mode shapes are more suited for damage localization than frequencies [4], but did not receive as much consideration because their determination requires the simultaneous measurement of vibrations in a high number of points together with the use of more refined identification techniques.

Damping has seldom been used for damage assessment. This is due to the existence of multiple damping mechanisms and the large scatter in estimating damping values. A review of the literature, however, suggests that damping may prove, in some cases, to be more advantageous than detection schemes based on frequencies and mode shapes alone. Among the others, use of damping for damage detection is mentioned in several applications such as: cracked elements [5], cyclic fatigue [6], composite materials [7] and structures [8] and [9]. In the last paper damage assessment is provided by proper damage indicators based on the effects of damping on the mode shapes rather than damping itself.

This paper aims to review and deepen the approach followed in [9] by testing the effectiveness and sensitivity of these indicators. The basic assumption consists of a direct relationship among the damage increase, the energy dissipation increase and the damping increase. No account is made for the actual damping mechanism and an equivalent non-proportional viscous damping is considered. Although simplified, this model allows to capture the complex nature of the mode shapes of real structures, provided the assumption of proportional damping is relaxed. In the proposed approach the complexity level of the mode shapes (i.e. the non-proportional entity) is assumed as a measure of damage and the damage is quantified by comparing different structural states. The mode shapes are not compared directly, but through appropriate synthetic measures of the modal complexity: the damage indicators.

The paper is organized in four sections. In the first section the damage modelling and its relations to modal complexity are discussed together with the damage indicators. In the second section the analysis methodologies are introduced along with a case study. In the third section sensitivity analyses of the damage indicators are carried out using theoretical data obtained through standard modal analysis. In the fourth section the most sensitive indicator is used to evaluate the damage through the analysis of pseudo-experimental data.

2. Damage modelling and indicators

A general equivalent viscous damping capable to encompass different damping mechanisms is considered. This type of damping can be proportional (real mode shapes) or non-proportional (complex mode shapes) depending whether the system matrices satisfy the conditions expressed in [10]. In the case of framed civil structures, damage is mainly due to the formation of plastic hinges. The local constitutive relations show energy dissipation because of material hysteresis. The energy loss can be predicted by an equivalent dissipation mechanism based on viscous damping. The three quantities (damage, hysteresis and equivalent damping) are related by one to one relations at least for moderate strains and ductility. In this sense, an increase of damage causes an increase of hysteresis and consequently an increase of equivalent damping. However, the hysteretic behaviour does not lead only to energy loss, but involves complex interactions between loss of stiffness and energy dissipation. In order to make clear the effects of these two contributions they are idealized and studied by means of two elementary and independent contributions. The two elementary contributions are modeled through changes in the coefficients of the stiffness $\mathbf{K}$ or the damping $\mathbf{C}$ system matrices (constant mass matrix $\mathbf{M}$). When the damage causes a reduction of stiffness then it is modeled assuming a reduced stiffness matrix $\mathbf{K}_r$ with $\mathbf{C}$ kept constant. On the contrary, when the damage causes an energy dissipation then it is modeled assuming an increased damping matrix $\mathbf{C}_r$ with $\mathbf{K}$ kept constant. In either case the final effect is an increase of the fraction of non-proportional damping compared to the initial one (undamaged structure) and hence of the extension of the modal complexity.

In the literature several indices have been proposed to quantify the extension of the modal complexity [9], [11] and [12]. The difference between the values assumed by a particular damage indicator in two different structural states quantifies the damage occurred. In this paper five of the above indices are assumed as damage indicators:

$$I_1 = \frac{\|\mathrm{Im}(X_i)\|}{\|\mathrm{Abs}(X_i)\|}; \quad I_2 = \frac{\mathrm{Abs}\left(\mathrm{Re}(X_i)^T\,\mathrm{Im}(X_i)\right)}{\sqrt{\left(\mathrm{Re}(X_i)^T\,\mathrm{Re}(X_i)\right)\left(\mathrm{Im}(X_i)^T\,\mathrm{Im}(X_i)\right)}}; \quad I_3 = \frac{\sum_{j=1}^{N}\mathrm{Abs}\left(\mathrm{Im}(X_{ij})\right)}{N};$$

$$I_4 = \frac{\mathrm{Abs}(\phi_{i,\max}) - \mathrm{Abs}(\phi_{i,\min})}{\pi}; \quad I_5 = \frac{A_i}{A_{i,\max}}.$$

The I_1 (modal imaginary ratio) indicator is taken from [9], the I_2 (modal collinearity) and I_3 (modal dispersity) indicators are taken from [11] whereas the I_4 (modal phase difference) and I_5 (modal polygon area) indicators derive from [12]. The symbols have the following meaning: X_i is the i-th complex mode shape; X_{ij} is the j-th component of X_i; $\phi_{i,max}$ and $\phi_{i,min}$ are the maximum and minimum phase angle of the i-th mode shape; A_i is the modal polygon area of the i-th mode shape and $A_{i,max}$ the maximum modal polygon area; $\|\cdot\|$ is the Euclidean 2-norm operator and Abs(·) is intended as the componentwise absolute value; N is the number of degrees of freedom (dofs) of the system.

The damage is essentially related to the imaginary part of the mode shapes, it is hence advisable to normalize, in some sense, the mode shapes before their insertion into the formula of the damage indicators. The normalization is carried out using the procedure proposed in [11], and it is equivalent to minimize the imaginary part of the mode shape. Geometrically, it is equivalent to rotate the complex mode shape such that it optimally aligns with the real axis in the complex plane.

The meaning of the indicators is as follows: the I_1 indicator weighs the importance of the imaginary part with respect to the overall length of the complex mode shape. The I_2 indicator is based on the fact that the degree of interdependence of the real and imaginary parts of a complex mode shape is directly affected by the damping proportionality: the higher the damping proportionality, the more correlated the real and imaginary parts of the mode shapes are; if the imaginary parts of a complex mode shapes are completely dependent on the real parts, the damping is proportional. The I_3 indicator measures the degree of scatter of the complex mode shape that is directly related to the amplitude of the imaginary parts. The idea behind the I_4 and I_5 indicators relies on a geometric interpretation. If the mode shape components are plotted in the complex plane, the effects of the non-proportional damping become apparent. The I_4 indicator considers the phase differences between the dofs of a mode shape as a consequence of the non-proportionality of damping. In fact, each component of a mode shape of a system endowed of proportional damping lies on a straight line, whereas those of system endowed of non-proportional damping do not; in effect, these latter exhibit an angular dispersion equivalent to the phase differences. If the individual components of a mode shape are connected by straight lines, an N-side polygon is formed. If the damping is proportional the components of a mode shape lie on a straight line and the polygon has zero area. As the non-proportionality of the damping increases, the area of this polygon also increases. The I_5 indicator measures the area of this polygon.

3. Case study and methodology

The effectiveness and sensitivity of the five damage indicators are analysed via a 3dofs spring-dashpot system representative of basic models of framed structures. The structure is a single span three floors shear-type frame (Figure 1a). Linear degrading behaviour and viscous damping are assumed. The inter-storey height is 3 m and a mass of 20 t is condensed at each floor. The columns have a constant cross-section 30 by 30 cm^2 and a Young modulus $E = 3\text{x}10^7$ kN/m^2. The system, in its initial undamaged state, is characterized by a tridiagonal stiffness matrix **K**, a diagonal mass matrix **M** and a damping matrix $\mathbf{C} = \beta\mathbf{K}$ (β is a proportionality constant).

The initial frequencies and mode shapes are: $f_1 = 2.13$ Hz, $f_2 = 5.96$ Hz, $f_3 = 8.61$ Hz and $X_1 = \{0.45, 0.80, 1.00\}$, $X_2 = \{1.00, 0.45, -0.80\}$, $X_3 = \{-0.80, 1.00, -0.45\}$; β is equal to 0.0075 so that the damping ratios are 5%, 14%, 21% respectively for the first, second and third mode (Figure 1b).

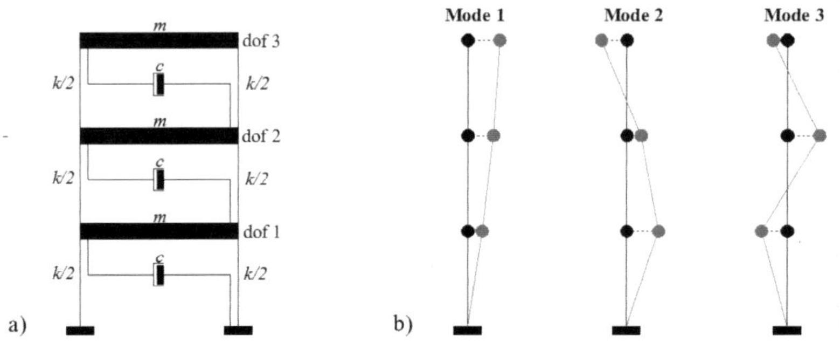

Figure 1. Structural model (a) and its initial mode shapes (b).

For simplicity, the initial reference state coincides with the undamaged structure characterized by proportional damping and real mode shapes, so that the damage indicators are zero.

According to the damage modelling of Section 2, two elementary damage cases are studied: (a) inter-storey stiffness degradation and (b) inter-storey energy dissipation (damping) increase. Further for each case two damage scenarios are studied: localized or diffused. Localized damage involves damage alternatively at the first, second or third floor, whereas diffused damage involves a simultaneous and identical damage at each floor. In the case (a) the inter-storey stiffness k is reduced stepwise up to 50% of the undamaged structure, with the damping matrix fixed at the initial (undamaged) state; similarly, in the case (b) the inter-storey dashpot constant c is increased stepwise up to a maximum damping ratio of 30%, with the stiffness matrix fixed at the initial (undamaged) state. In both cases a damage factor, $\lambda_k \leq 1$ or $\lambda_c \geq 1$, has been applied respectively to k ($\lambda_k k$) or c ($\lambda_c c$) as appropriate.

The main difference between the case (a) and (b) is that in the second case there exists a direct relation between energy loss, damping non-proportionality and mode shapes complexity; whereas in the first case the non-proportionality arises only as an indirect effect of the stiffness loss. It is hence expected that damage related mainly to stiffness loss, case (a), is more difficult to be detected than damage related mainly to energy dissipation, case (b).

In order to work in a controlled environment the damage indicators are tested on theoretical and pseudo-experimental data. The theoretical mode shapes are derived using the standard modal analysis according to the state space method to deal with the non-proportional damping [13]. The pseudo-experimental mode shapes can be derived in principle by any identification technique. In the present context the Complex Plane Representation (CPR) method [14] is used. Briefly speaking, the CPR method is an output-only time domain technique in which the original signal is represented in the complex plane by computing its imaginary counterpart via the Hilbert transform [15]. This new representation makes it very simple to identify the phase shift of the motion between the different measurement points, and therefore it is particularly effective for the identification of complex mode shapes of general viscously damped systems. The method processes the harmonic response close to resonance and it is therefore suited with field applications based on the use of electromechanical actuators, but without the necessity of knowing the applied force, which is rarely measured in the case of civil structures.

4. Sensitivity analysis using theoretical data

4.1. Localized damage

Because of the limited room available only the results relative to damage at the columns of the lower floor are discussed. Broadly speaking, the damage indicators trend depends on the type of structure and damage scenario considered. However, the features shown below by the damage indicators are shared by the different situations that structures can actually exhibit. In this respect, the results can be considered to have generality. In Figure 2 the damage indicators are reported against the damage factors affecting k (λ_k from 1 to 0.5; case a) or c (λ_c from 1 to 4; case b) for each of the three structural modes.

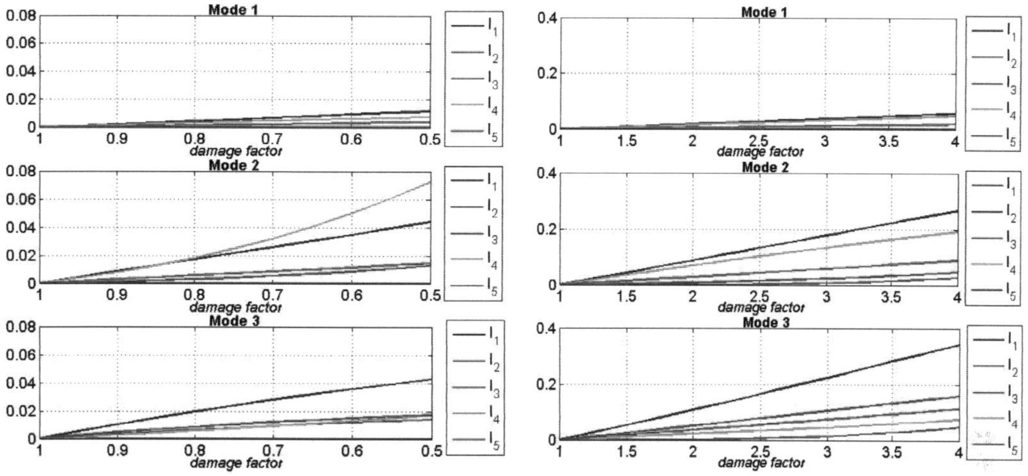

Figure 2. Damage indicators. Localized damage: case a (left) and b (right).

4.2. Uniform damage

Uniformly distributed damage is hardly exhibited by structures and damage is more or less always localized. However, it is interesting to analyze such a limit case. In this particular situation the non-proportionality between the damping and stiffness matrix is spatially uniform and equal to the damage factor; in this condition, the mode shapes are very close to the real ones obtained using the proportional damping (i.e. modal complexity is very small) and the damage indicators are practically zero. Thus, in the case of uniform damage the indicators are useless and it is not possible to distinguish between a healthy structure and a very damaged one. Different quantities sensitive to global damage can help to indicate whether damage is present or not. Plots of the damping ratios against the natural frequencies can help to answer the problem, Figure 3.

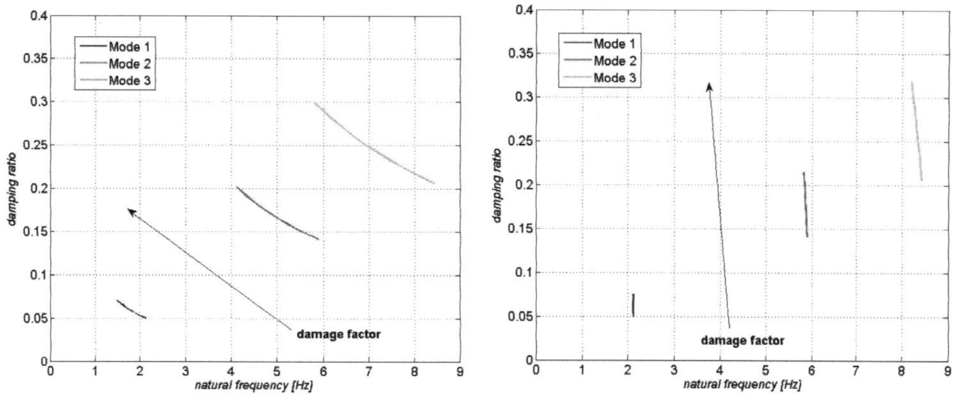

Figure 3. Damping ratio vs. natural frequency. Diffused damage: case a (left) and b (right).

4.3. Comments of the results

By the inspection of Figure 2 some conclusion can be drawn. Initially and as expected, it must be observed that the indicators are more sensitive to damage related to energy dissipation than damage related to stiffness degradation. This feature preserves regardless the mode order; however, higher order modes are generally more sensitive than lower order modes. This is partly due to the type of damping assumed and partly due to the raise of strain energy involved along with the increase of the mode order. Another note concerns the absolute values of the indicators that are contained also for important damages.

Further, as concerns the damage identifiability and quantification, the main interesting feature shared by all the indicators is their monotonical trend along with the damage increase. This allows to set univocal and unambiguous relationships between the damage level and the indicator value.

Among all, the I_1 indicator, that is essentially a direct measure of the modal complexity, is generally and largely the most sensitive. This is a positive result in view of the robustness of the results considered the ease of computation of I_1. I_4 is the second more sensitive indicator and I_2 is the less sensitive one. I_1 is therefore the best candidate to deal with modal complexity (damage detection) and is therefore the only one used in the following.

One particular problem arises in the case of uniform damage. In this instance any damage indicator is practically zero, due to the non-proportionality spatially uniform of the damping matrix, and cannot be used to answer the question whether damage is present or not. The representation of Figure 3 can be useful on purpose. If the damage is caused by stiffness reduction (case a) a simultaneous change of the structure frequencies (reduction) and energy dissipation (increase) takes place; whereas if it is caused by energy loss (case b) a significant addition of the energy dissipation is observed with virtually no changes of the frequencies.

5. Pseudo-experimental results

The CPR method [14] is used to identify the complex modal parameters in operating conditions. Two different excitation types are considered: distributed and point excitation. The distributed excitation consists of a forcing function whose profile coincides with mode shape of the undamped and undamaged structure. This configuration is ideal and is considered for reference. The point excitation corresponds to the current practice in civil engineering to shake the structure by means of a electromechanical actuator. For definiteness the point excitation is here assumed located at the top floor. Damage is assumed localized at the columns of the first floor. The results are presented according to the first mode (less sensitive to damage) and to the I_1 indicator (most sensitive indicator). The response of the structure is generated in terms of acceleration time histories with a 15% rms added white noise to account for real field conditions. The responses to both excitation types are processed via the CPR method. The identified complex mode shapes are used to compute the I_1 indicator.

5.1. Distributed excitation

The Figure 4 compares the theoretical and pseudo-experimental (noise free) results in the presence of distributed excitation for both damage cases (a) and (b). The comparison is carried out via the I_1 indicator plotted against the damage factor. In any case no differences are appreciated for the first mode. On the contrary, some deviations are observed for modes two and three. These deviations are lesser in case (a) than in case (b) where some under estimation is observed. Further, apart mode three of case (b), where the differences between theoretical and pseudo-experimental data increase with the damage factor, no such feature is detectable in the other cases. In these cases the deviation is almost constant regardless the damage level so that damage can be reliably detected.

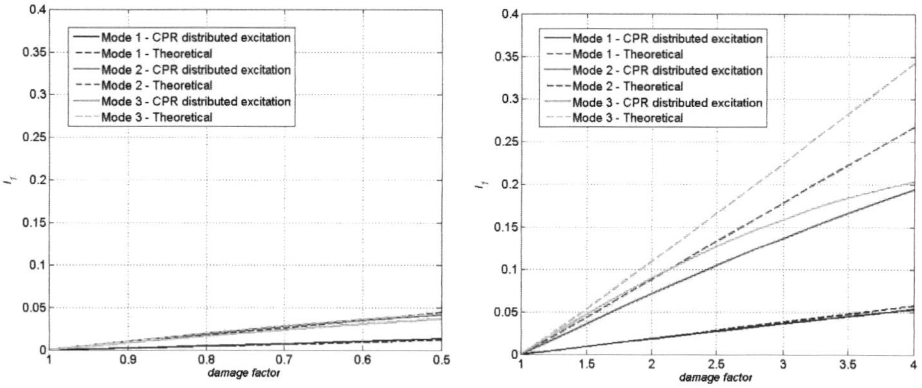

Figure 4. Distributed excitation, I_1 indicator: pseudo-experimental vs. theoretical. Case (a): stiffness reduction (left), case (b) energy dissipation (right).

5.2. Point excitation

As for the previous case, the Figure 5 compares the theoretical and pseudo-experimental results in the presence of point excitation for both damage cases (a) and (b). Here again, the comparison is carried out via the I_1 indicator plotted against the damage factor. For reference, the results of the distributed excitation are also reported. Only the results relevant to the first mode shape are shown. This is why the first mode proves to be the less sensitive in the present context and hence the one more difficult to deal with. Unlike the previous case, noisy pseudo-experimental data have been now considered. Because of that the pseudo-experimental curves are not straight lines in the Figure 5 as it happens for the theoretical data. Despite of that a good agreement is observed. Therefore experimental dynamic data collected via conventional excitation schemes of civil structures can be confidently used for damage detection purposes.

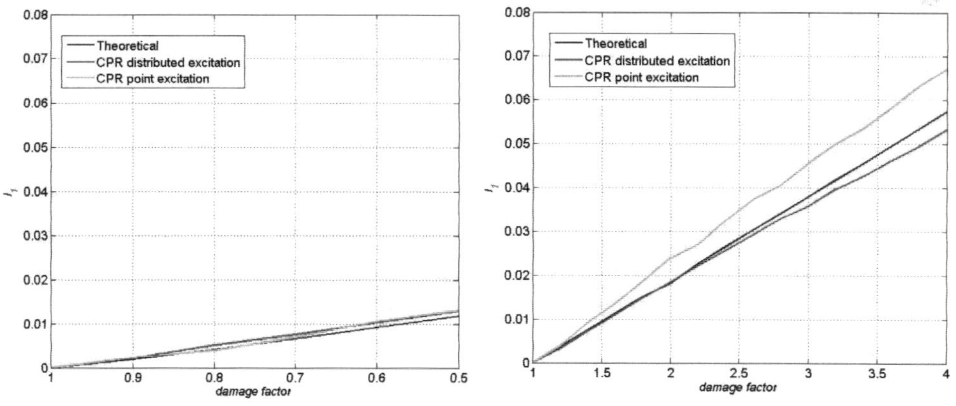

Figure 5. Point excitation, I_1 indicator: pseudo-experimental vs. theoretical. Case (a): stiffness reduction (left), case (b) energy dissipation (right).

6. Conclusions

The paper is concerned with the use of the mode shapes complexity for damage assessment purposes. The proposed method is based on the observation that the damage, either due to stiffness degradation or to energy dissipation, causes an increase of the non-proportionality of the system damping matrix. In the above context, complex modes arise naturally and the amount of modal complexity is assumed as a quantification of the structural damage. In order to get a measure of the modal complexity, several indicators derived from the literature are compared. These indicators are used to transform the identified complex mode shapes into scalar quantities readily usable in one to one relationships with the damage factor. In principle, the identification of the mode shapes can be carried out through any available signal processing procedure. Actually, the CPR method has been used because of two nice properties: it suits well with typical excitation schemes of civil structures, it proves highly robust and reliable. A simple shear-type frame structure is used as a case study. Two distinct damage cases have been considered: stiffness reduction and energy dissipation increase. In both cases all the damage indicators show a quasi-linear and monotonic behavior in the range of feasible damage. The indicators show effective for localized damage, they fail instead in the case of uniformly distributed damage because the mode shapes do not vary despite the damage level. In this case a different strategy should be followed. The best indicator happens to be the simplest one based on the ratio of the amount of the imaginary part to the whole mode vector. This indicator is used with sample examples to show its effectiveness. Theoretical and pseudo-experimental data (either noise free or noise corrupted) are compared along with the different modes excited for different excitation profile (either distributed and mode shape proportional or localized and point applied).

References

[1] Cawley P and Adams R D 1979 The location of defects in structures from measurements of natural frequencies *Journal of Strain Analysis* **14** 49-57.

[2] Gunes B and Gunes O 2013 Structural health monitoring and damage assessment Part I: A critical review of approaches and methods *International Journal of Physical Sciences* **8**(34) 1694-1702.

[3] Salawu O S 1997 Detection of structural damage through changes in frequency: a review *Engineering Structures* **19**(9) 718-723.

[4] Shi Z Y, Law S S and Zhang L M 2000a Damage localization by directly using incomplete mode shapes *Journal of Engineering Mechanics* **126**(6) 656-660.

[5] Modena C, Sonda D and Zonta D 1999 Damage localisation in reinforced concrete structures by using damping measurements *Key Engineering Materials* **167-168** 132-141.

[6] Lage Y, Cachão H, Reis L, Fonte M, de Freitas M and Ribeiro A 2014 A damage parameter for HCF and VHCF based on hysteretic damping *International Journal of Fatigue* **62** 2-9.

[7] Montalvão D, Ribeiro A M R and Duarte-Silva J 2009 A method for the localization of damage in CFR plate using damping *Mechanical Systems and Signal Processing* **23** 1846-1854.

[8] Curadelli R O, Riera J D, Ambrosini D and Amani M G 2008 Damage detection by means of structural damping identification *Engineering Structures* **30** 3497-3504.

[9] Valente C, Spina D, Gabriele S and De Leonardis A M 2010 The complex plane representation method for structural damage detection *Proceedings of the Thenth International Conference on Computational Structures Technology*, Valencia, Spain.

[10] Adhikari S 2014 *Structural dynamic analysis with generalized damping models - Analysis* (London: ISTE Ltd).

[11] Liu K, Kujath M R and Zheng W 2000 Quantification on non-proportionality of damping in discrete vibratory systems *Computers and Structures* **77** 557-569.

[12] Prater G, Singh Jr and Singh R 1986 Quantification of the extent of non-proportional viscous damping in discrete vibratory systems *Journal of Sound and Vibration* **104**(1) 109-125.

[13] Meirovitch L 1980 *Computational methods in structural dynamics* (The Netherlands: Sijthoff & Noordhoff).

[14] Gabriele S, Iezzi F, Spina D and Valente C 2014 Effects of modal density in system identification using the Hilbert transform *Proceedings of 2014 IEEE Workshop on Environmental, Energy and Structural Monitoring Systems (EESMS)*, Naples, Italy.

[15] Feldman M 2011 *Hilbert transform applications in mechanical vibrations* (India: Johon Wiley & Sons ltd).

A multi-resolution based method to precise identify the natural frequencies of beams with application in damage detection

I C Mituletu[1*], N Gillich[1], C N Nitescu[1], C P Chioncel[1]
[1]Department of Mechanical Engineering, "Eftimie Murgu" University of Resita, P-ta Traian Vuia 1-4, 320085 Resita, Romania

E-mail: mituic@yahoo.com

Abstract. This paper presents an enhanced method that accurately identifies natural frequencies of structures, via an important improvement of spectral lines density. Spectral lines density increases due to the superposition of many spectra, each spectrum being accomplished by subtracting certain number of samples from the original signal and using the power spectral density analysis. The newly achieved overlapped-spectrum shows important additions to spectral line number and makes more accurately identification of frequencies in the peak position of relevant amplitudes. In this way, modal analysis becomes more reliable and efficient. Method application assumes small frequency changes and the advantage of establishing more precisely the position and dimension of damage. The analysis of all results explicitly indicates an important assessment in the evaluation of frequency. Considering relevant aspects, it can be appreciated that the approached method offers a new development way in the vibration-based damage detection, allowing de possibility to perform more accurate evaluation of frequencies and introducing new features into the domain.

1. Introduction

In time, whatever physical structure, under the application of a large variety of strength types, gets damage. By detecting damages in their very early stage, the keeping of structure integrity and functionality can be ensured via an adequate maintenance. The impact of early damage detection is significant for big and complex structures, causing money and work-time saving and also decreasing the possibility to occur accidents. There is a great area of methods that can perform non-destructive tests as regard the damage detection. Dynamic methods offer an import advantage they do not require access to the damaged place [1].

Within damage detection domain, the vibration-based methods are very popular and also widely spread. Herein natural frequency shifts are determined and afterwards attentively analyzed [2]-[3]. These methods can be commonly divided into two categories those having limitations at the damage detection or designed to predict detection, location and quantification of damage [4]. Into the second category, all methods based on the natural frequency shifts are defined as model-based, [5] and [6], and use finite element models. The readability of frequency shifts determines the accuracy of damage detection. Also, a great precision in frequency reading and clear measurements lead to better results in detection. Small frequency changes caused by early structural damage or even small damage cannot be distinguished from those of the healthy structure, and so that the mathematical model is under the impossibility to work well. Thus, mostly it is not possible to make a difference between improperly models and fine changes caused by the occurrence of small or early damage [7] and [8]. And these are not all the issues regarding the frequency shifts, the temperature action or/and structural loads application induce more other interferences.

Also, other methods available as well, besides a comprehensive classification and description are presented in [3]. Each method fits a particular application, due to its specific advantage. Until now, at least one method does not meet all the requirements supposed by the high variety of analyzed structures and the diversity of imposed conditions.

In this paper, a new method to enhance the frequency evaluation is presented. It is based on an extensive study regarding the transformation of acquired time-domain signal in a frequency spectrum. The achieved spectrum has to provide many possibilities to find a better fit for a certain frequency. Herein, an overlapped-spectrum has been accomplished via an iterative algorithm, performing a modulation from the right-side of acquired signal window. A step-by-step windowing of time signal is applied to an algorithmic structure implemented in LabView. Each step of iteration consists in subtracting from the signal right-side of a certain number of time samples. Also, after subtracting the samples, an analysis applying Power Spectral Density (PSD) feature of LabView is made. The amplitude values of each frequency are stored and at the end of last iteration step, they are displayed all in one single spectrum, achieving in this way a much denser frequency overlapped-spectrum.

2. Description of frequency identification method

Method consists of three steps: signal acquisition, reference analysis and algorithm setup. Considering the spectrogram and Wigner-Vile distribution, as could be seen in figure 1, it is observed that certain frequencies are damping very fast, providing in this way a very short time length [9]-[10]. As known, time length gives the resolution of the frequency spectrum after Fourier analysis is performed. Thus, spectrum resulted by applying ordinary Fourier Transform provides poor density and in case of small damage, determining fine frequency shifts, this is impossible to distinguish. So that, even if a faster acquisition device is taken to achieve much more samples in a short time interval, any enhancement of the frequency spectrum can be accomplished.

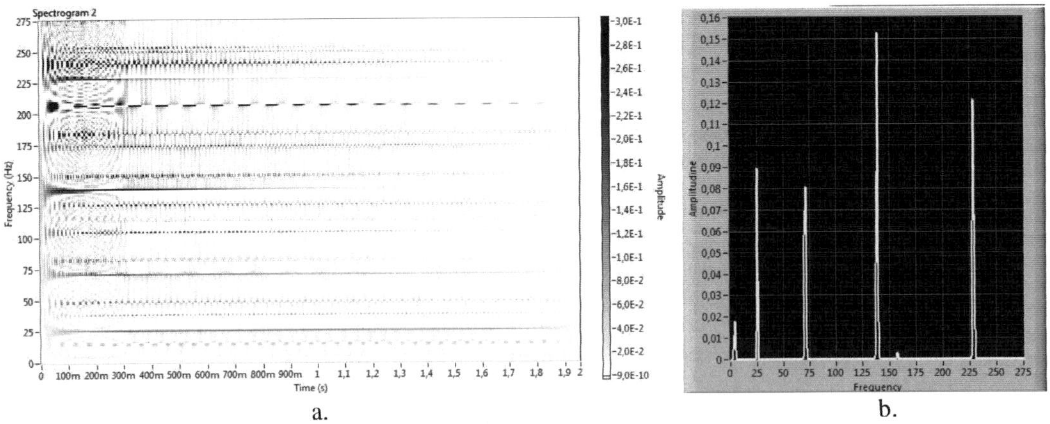

Figure 1. Wigner-Vile distribution (a) and power spectrum (b) of vibration signal

To significantly enhance the spectrum density, the following argument can be assumed: by decreasing the length of signal, from the side of low amplitude in the time history, the number of spectral lines is decreasing as well, but also their position changes. From this last point of view, cropping the signal length become important and can be iteratively applied to achieve a multi-spectrum. If an iterative step-by-step signal cropping is imagined, which is performed by an algorithm, and also the obtained spectral lines are afterwards displayed in one single graphical spectrum, we finally can achieve a much denser frequency spectrum.

Algorithm to analysis and iteratively crop the signal is implemented in LabView under a logical loop structure. Loop considers a start point and a stop one, within these points performing a number of iteration. From the beginning have to be established some functional parameters, as initial length of time signal, entire cropping portion of signal and number of samples cropped at the iteration step.

2.1. Signal acquisition and pre-processing

In order to prove the method performance, make a right comparison of results and have a better flexibility for input signals, we provide as time signals those generated by LabView. Of course, they have been generated to meet few important requirements: short length, are mixed and have different amplitudes. Figure 2 shows an auto-generated time signal, different cropped and mixed from two signals having the following characteristics: entire time length *t*, total number of samples *N*, frequency *f* and amplitude *A*. First signal S_1 has $f_1 = 4.2$ Hz and $A_1 = 1$ mm/s^2, the second one S_2 is at $f_2 = 23.7$ Hz and $A_2 = 0.5$ mm/s^2.

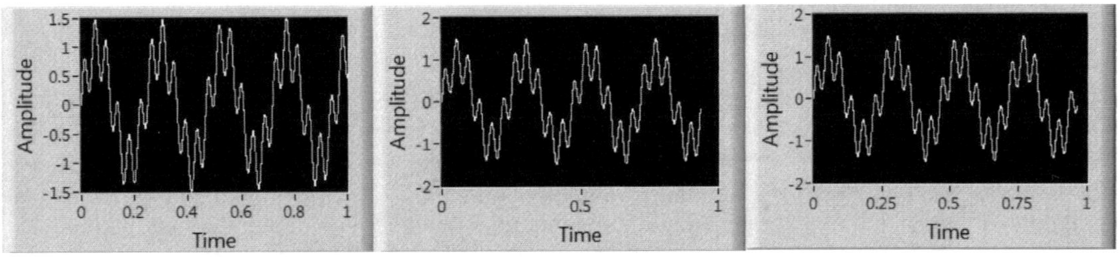

Figure 2. Mixed sine signal with two variable amplitudes and cropped.

Left image of figure 2 presents the entire signal t = 1 s, middle image shows the signal cropped with a period $T_1 = 0.0(6)$ s and the right-side one has a cropped period $T_2 = 0.037$. The reason of cropping is assumed by algorithm setup and will be later detailed.

2.2. The reason of applying reference analysis

Actually, reference analysis is performed to acquire an initial spectrum and establish who the frequencies of interests in a certain case are. At this time, the cropping period is chosen taking into account the lowest frequency considered as expected from experience reasons. It is very important to establish the right frequencies of interest, due to the fact that in correlation with them the algorithm must be set. In the presented case, this second step is redundant, the frequencies are well known being generated by ourselves. In real cases, the result of this test can be assumed as being similar to that shown in figure 1 b.

2.3. Explaining the algorithm implementation

Algorithm is a logic structure implemented in a virtual environment employed to ensure an explicitly work and satisfy certain requirements. This algorithm is formulated to get the signal, compute some parameters, iteratively crop and analyze the signal, and display all the results in one graphical frequency spectrum. First consideration is about the input signal, which can be auto-generated or real acquisitioned and stored as a file with a given extension. In LabView, the "Block Diagram" (*BD*) window ensures the placement and interconnectivity of functional blocks. Auto-generated signal can be applied by using the functional block "Simulate Signal" (*SS*), where the waveform, amplitude, frequency, length and number of samples were set in this case. Main part of algorithm block diagram (*ABD*) is the logic structure placed into the limited loop, where "Extract Portion of Signal" (*EPS*) and "Spectral Measurements" (*SM*) blocks perform signal cropping and spectral analysis. At the end of loop execution, all data achieved from the iterative analysis is sending to the "Graph" (*G*) block. In a different window "Front Panel" (*FP*), data is show as graphical spectrum. The parameters chosen to be set or visualize are also shown in *FP*.

Journal of Physics: Conference Series **628** (2015) 012020 doi:10.1088/1742-6596/628/1/012020

Figure 3 presents the *BD* image that is meant to comprehensively clear up all the aspects associated with the algorithm implementation.

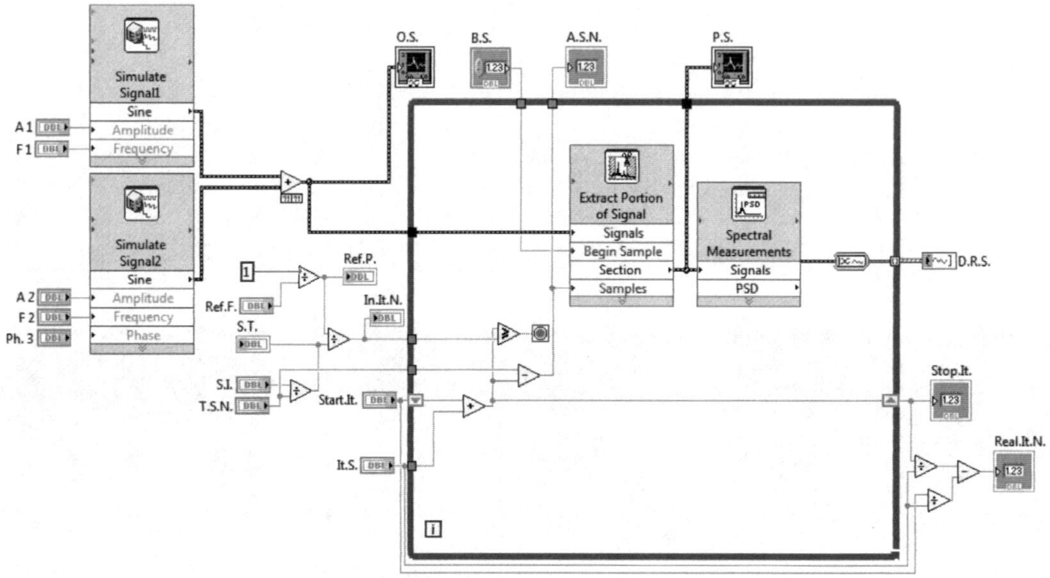

Figure 3. Software implementation of frequency identification algorithm.

In table 1 parameter abbreviations included in the figure 3 are presented and detailed as well. Also, for each of them the sense of consideration (as input or output) is presented, regarding the displayed values meaning that can be divided in two categories, set by hand or resulted after computing.

Table 1. Explanation of parameters involved in the identification algorithm.

Parameter	Sense	Detail
$\mathbf{A_{1,2}}$	Input	Signal amplitude
$\mathbf{F_{1,2}}$	Input	Signal frequency
$\mathbf{Ph_3}$	Input	Signal phase - not used in these tests
Ref.F.	Input	Reference frequency - expected to reach by analysis
Ref.P.	Output	Reference period - corresponds to Ref.F.
T.S.N.	Input	Total samples number - all signal samples considered in test, equal to N
S.I.	Input	Sampling interval - whole signal length in time, equal to t
S.T.	Output	Sampling time
B.S.	Input	Begin sample - first sample from the signal left-side
A.S.N.	Output	All samples number - remaining samples after last iteration is done
Start.It.	Input	Start iteration - sample whence algorithm begins the iterative cropping
It.S.	Input	Iteration step - number of samples cropped at iteration
Stop.It.	Output	Stop iteration - number of samples cropped at the iteration end
In.It.N.	Output	Initial iteration number - resulted by computing
Real.It.N.	Output	Real iteration number - resulted by applying It.S. to In.It.N. and rounding
O.S.	Output	Original signal - graphically displayed
P.S.	Output	Portioned signal - graphically displayed when the iteration ends
D.R.S.	Output	Displaying of resulted data - data displayed as overlapped-spectrum

Above parameters have not been considered only for simple signal cropping and iteration, but also to create many possibilities of data and results management from whatever section of the algorithmic structure. Therefore, considering important the way in which identification algorithm works and can be set, in order to get suitable results more details are provided in the next chapter.

3. Easy frequency evaluation by adequate algorithm setup

From the beginning has to be mentioned the importance of following few steps to properly setup the identification algorithm and easy achieve clear results:

- Chose the lowest expected frequency value, decrease its value to at least 10% and use it as Ref.F.
- Perform reference analysis and make the choice regarding the frequencies of interest.
- For each of these frequencies a new algorithm setup has to be applied and also a new analysis with the identification algorithm must be taken.

Let us consider figure 4 as example for how algorithm parameters have to be managed and what for. Firstly, image from figure 4 has been accomplished by writing the following values for parameters: T.S.N. = 50000, S.I. = 1 s, Ref.F. = 3.5 and It.S. = 50. Ref.F. was deduced by assuming a lowest expected frequency in about of 4 Hz, subtracting 10% and rounding to 3.5 Hz. Afterwards, overlapped-spectrum achieved is zoomed in around of 4 Hz, so as the image offers a good vision.

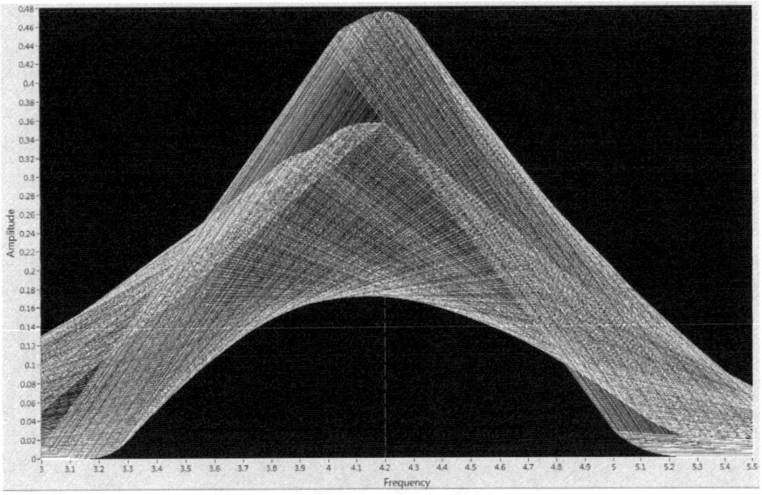

Figure 4. Overlapped-spectrum achieved at Ref.F. = 4 Hz.

Overlapped-spectrum is realized from 286 graphical lines, being resulted from the same number of iteration performed by the algorithm. Image can trick us, due to its apparent complexity, and make enough difficult the evaluation of frequency because of the presence of some superposed small lobes.

To clear it up, values Ref.F. as well as Start.It. are recommended to be modified. By increasing the value of Ref.F., smaller lobes disappear and right-side of the main lobe is going to clear as well. Also, increasing the value of Start.It., left-side of the main lobe become more definite.

But in the overlapped-spectrum two values are distinguished, as well as in the auto-generated time signal provided to the input of identification algorithm. Magnifying area in around of 23 Hz, the image of the other main lobe is depicted in figure 5.

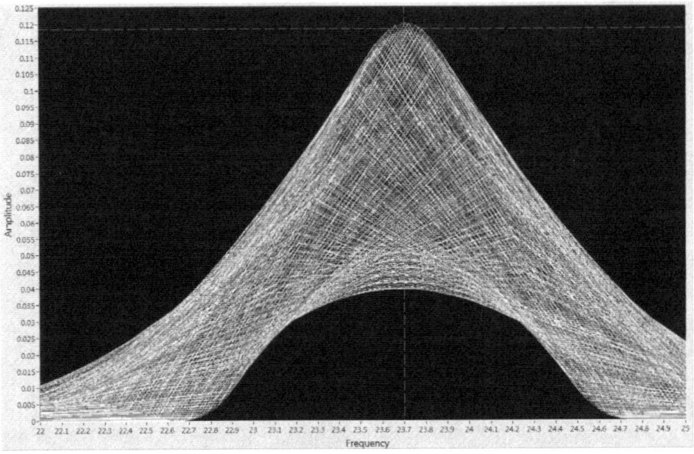

Figure 5. Overlapped-spectrum achieved at Ref.F. = 23 Hz.

As easily can be seen, more superposed lobes are affecting the image of the main lobe at about of 23 Hz. To refine its area, values Ref.F. and Start.It. have to be managed differently than before. First modification is addressed to Ref.F., which must be significantly increased at least up to the value of expected frequency or even more, if the superposed lobes do not disappear. As for lobe's left-side, Start.It. value will be set much lower, otherwise lobe will entirely vanish.

4. Results analysis and discussion

Taking into account the previously presented technique for managing the parameters of frequency identification algorithm, we offer the values of those strictly involved in clearing lobe areas. In this way, making easy reading frequencies within the overlapped-spectrum resulted after iteration completes.

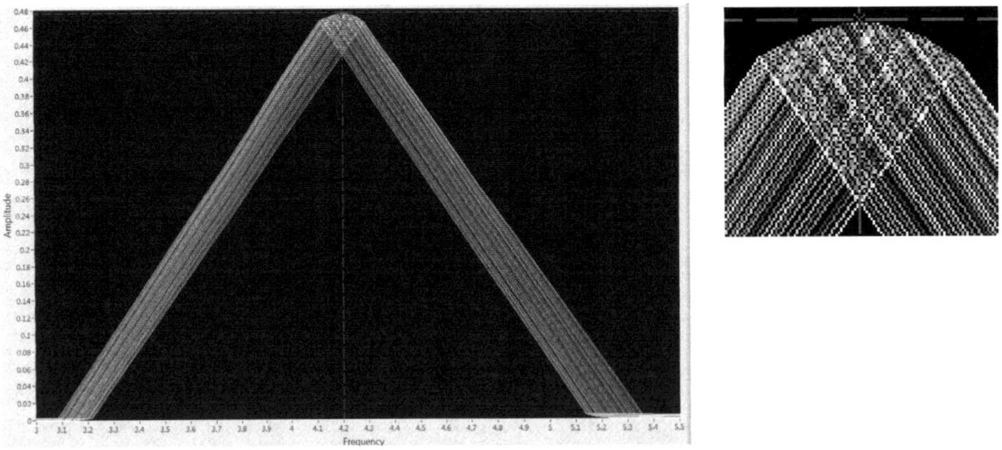

Figure 6. Frequency identification at 4.2 Hz, after performing an adequate algorithm setup.

Figure 6 shows how easy is the identification of frequency after performing adequate setting of algorithm parameters, frequency matching area being magnified and placed at the up-right side in the same image. Due to the clearness of the lobe area, simply identify the peak of lobe, and there must be the best frequency matching. Being in LabView, the cursor numerically shows values for frequency and amplitude in the pointed location.

The values of input parameters that suitable characterize the first frequency identification at 4.2 Hz are the following: Ref.F. = 15, Start.It. = 1300 and It.S. = 50.

For the other frequency value of 23.7 Hz that is also provided into the mixed signal, a new parameterization and analysis have been performed, and the identification was made using a graphical evolution as presented in figure 7.

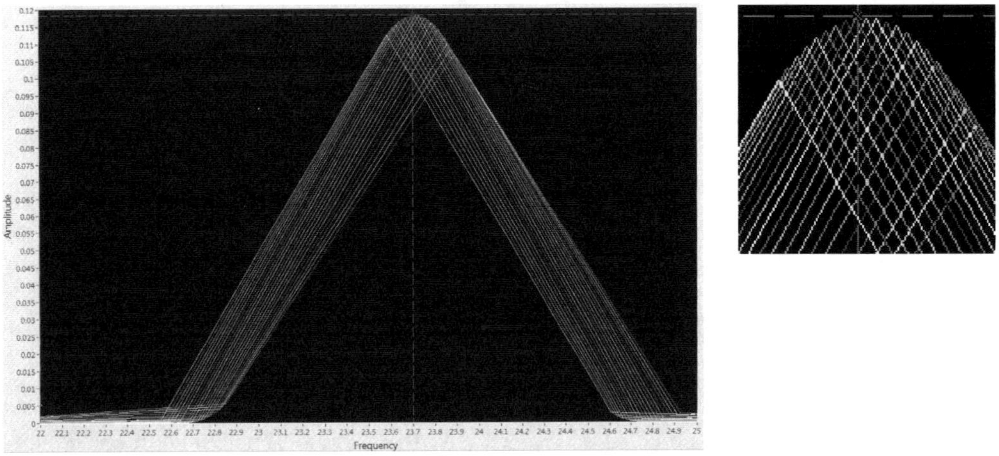

Figure 7. Second frequency is clearly identified at 23.7 Hz, after performing recommended setup.

Suitable values for parameters setting, ensuring a clear area of the lobe, in this case, are Ref.F. = 27, Start.It. = 1200 and It.S. = 30. By these settings, a number of 22 spectra have been accomplished, which formed the overlapped-spectrum.

Method limits can be defined as belonging to three groups: signal structure, acquisition device capability and computing power. First group refers to number of signal samples, in case of short signal length, at low-frequency value and for small frequency shifts; we have to achieve as many samples as possible. Also, signal length has to stay bigger than one period of expected lowest frequency. Second group considers the characteristics of acquisition device. As for the sensitivity, a higher number of bits for analog-to-digital conversion are requested, and for short signals faster devices have to be provided.

Third group treats computing power, regarding the number of iteration. Once frequency shift value decreases, much iteration must be performed, and greater computing power to diminish the analysing time is recommended.

In the end, it has to be specified, these frequency values actually were chosen as being nearby to those from real cases [11]-[14], in order to give significance and weighty to the results of performed tests. So that, results confirm the important improvement introduced by this method, and also highlights the great advantage in using the described new technique and algorithm, in the frequency identification field.

5. Conclusion

The paper presents an enhance method to accurately identify frequencies within a multi-resolution spectrum, here introduced as overlapped-spectrum. The method consists of a software algorithm, for iteratively cropping of the signal and generating the overlapped-spectrum, and a technique to adequate setup the algorithm, which clears the spectrum area in around of desired frequency.

Performing analyses on signals similar to real once, method confirms its effectiveness in improving the readability of frequencies, especially in the case of a short time signal.

The clarity and easiness of frequency identification, strongly suggests this method for performing in cases as rapid vibration damping or early damage occurrence.

The maximum precision in frequency reading depends only on the accuracy of the acquisition signal and the number of samples, when short time signals are analyzed.

Acknowledgments
The work has been funded by the Sectoral Operational Programme Human Resources Development 2007-2013 of the Romanian Ministry for European Funds through the Financial Agreement POSDRU/159/1.5/S/132395.

References

[1] Balageas D, Fritzen C P and Güemes A 2006 Structural health monitoring, ISTE Ltd, London
[2] Cawley P and Adams R D 1979 The location of defects in structures from measurements of natural frequencies *Journal of Strain Analysis* **14** 2 pp 49-57
[3] Doebling S W, Farrar C R, Prime M B and Shevitz D W 1996 Damage identification and health monitoring of structural and mechanical systems from changes in their vibration characteristics: a literature review Rep. No. LA 13070-MS *Los Alamos National Laboratory (Los Alamos, NM)*
[4] Salawu O S 1997 Detection of structural damage through changes in frequency: a review *Engineering Structures* **19** 9 pp 718-723.
[5] Rytter A 1993 Vibration based inspection of civil engineering structures *Ph.D. Thesis (Aalborg University, Denmark)*
[6] Morassi A and Vestroni F 2008 Dynamic Methods for Damage Detection in Structures *(CISM Courses and Lectures vol 499) (Springer Wien New York)*
[7] Friswell M I 2007 Damage identification using inverse methods *Phil. Trans. R. Soc. A* 365 pp 393-410
[8] Ewins D J 2000 *Model Testing: Theory, Practice and Applications (2nd Edition) Research Studies (Press Ltd., Hertfordshire, England)*
[9] Gillich G R and Praisach Z I 2014 Modal identification and damage detection in beam-like structures using the power spectrum and time–frequency analysis *Signal Process.* **96**(Part A) 29–44
[10] Onchis-Moaca D, Gillich G R and Frunza R 2012 Gradually improving the readability of the time-frequency spectra for natural frequency identification in cantilever beams *Proceedings of the 20th European Signal Processing Conference - EUSIPCO 2012 (Bucharest)*
[11] Gillich G R, Praisach Z I and Negru I 2012 Damages influence on dynamic behaviour of composite structures reinforced with continuous fibers *Mater. Plast.* **49**(3) 186-191
[12] Gillich G R, Praisach Z I, Wahab M A and Vasile O 2014 Localization of transversal cracks in sandwich beams and evaluation of their severity *Shock Vib.* **9**(4-5) 607125
[13] Praisach Z I, Gillich G R and Birdeanu D E 2010 Considerations on natural frequency changes in damaged cantilever beams using FEM *Latest trends on engineering mechanics, structures, engineering geology* 214-9
[14] Gillich G R and Praisach Z I 2013 Detection and Quantitative Assessment of Damages in Beam Structures Using Frequency and Stiffness Changes *Key Eng. Mat.* **569** 1013-20

The influence of operational and environmental loads on the process of assessing damages in beams

H Furdui[1], F Muntean[1], A A Minda[2 *], Z I Praisach[1], N Gillich[2]
[1] Department of Mechanical Engineering, "Eftimie Murgu" University of Resita, P-ta Traian Vuia 1-4, 320085 Resita, Romania
[2] Department of Economical Engineering, "Eftimie Murgu" University of Resita, P-ta Traian Vuia 1-4, 320085 Resita, Romania

E-mail: a.minda@uem.ro

Abstract. Damage detection methods based on vibration analysis make use of the modal parameter changes. Natural frequencies are the features that can be acquired most simply and inexpensively. But this parameter is influenced by environmental conditions, e.g. temperature and operational loads as additional masses or axial loads induced by restraint displacements. The effect of these factors is not completely known, but in the numerous actual research it is considered that they affect negatively the damage assessment process. This is justified by the small frequency changes occurring due to damage, which can be masked by the frequency shifts due to external loads. The paper intends to clarify the effect of external loads on the natural frequencies of beams and truss elements, and to show in which manner the damage detection process is affected by these loads. The finite element analysis, performed on diverse structures for a large range of temperature values, has shown that the temperature itself has a very limited effect on the frequency changes. Thus, axial forces resulted due to obstructed displacements can influence more substantially the frequency changes. These facts are demonstrated by experimental and theoretical studies. Finally, we succeed to adapt a prior contrived relation providing the frequency changes due to damage in order to fit the case of known external loads. Whereas a new baseline for damage detection was found, considering the effect of temperature and external loads, this process can be performed without other complication.

1. Introduction

In the scientific literature different models for obtaining modal parameters are presented, that rely on beam slenderness [1-3], cross-section variation due to constructive reasons [4] or damage [5-8]. Among them, those that take into account the operational loads are much fewer [9]. The temperature effect on natural frequent changes of the healthy and damaged structures is shown in [10-11].

In previous research a method for detecting the damage position was developed; is bases on an original mathematical relation that allows the calculation of natural frequencies of the damaged beam [12-16]. In studies presented in this paper we researched the effect that the temperature changes have on the frequency changes and, as a result, we succeed to develop a damage detection method which takes into account these changes of temperature.

2. Buckling of beams

In vibration analysis performed to detect damages in beams under changing conditions (temperature variations for instance) we must take into account the compressive loads that may occur. In reality, in practical cases, embedding at both ends of the beams is avoided. If at one end a relative displacement is allowed, for instance hinge with friction, then the change in temperature will have a lower influence, depending on the compression lodes that occurs due to friction. Thus, in this paper the case of clamped-clamped beam is analyzed.

For a beam of length L, width b and height h, mass density ρ at an initial temperature T_0, clamped at both ends, if the temperature changes with ΔT , an internal load P occurs, that is:

$$P(T) = \alpha \cdot E \cdot A \cdot (T - T_0) = \alpha \cdot E \cdot A \cdot \Delta T \tag{1}$$

Here A is the cross-section area, E is the Young's modulus and α is the thermal expansion coefficient. When the first critical temperature T_{cr-1} is reached, the first critical buckling load P_{cr-1} appears and the structure will suffer a vertical displacement, as can be seen in figures 1 and 2.

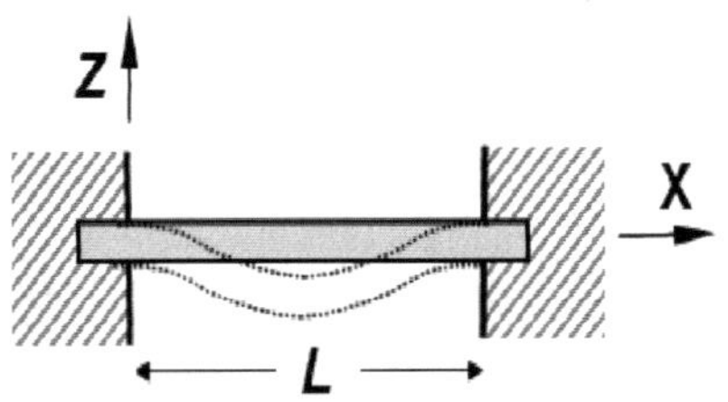

Figure 1. Double-clamped beam in initial state and subjected to buckling (first mode)

Figure 2. Beam in buckling mode one, after passing the critical load

The equation of the deformed beam, due to buckling, is given by:

$$EI\frac{d^4v}{dx^4} + P\frac{d^2v}{dx^2} = \frac{d^4v}{dx^4} + \varsigma^2\frac{d^2v}{dx^2} = 0 \tag{2}$$

where v is the lateral displacement and ς is the buckling eigenvalue.

The solution of this equation is of the form [17]:

$$v(x) = c_1 \sin(\varsigma x) + c_2 \cos(\varsigma x) + c_3 x + c_4 \tag{3}$$

with $c_1 \ldots c_4$ coefficients to be derived for the specific boundary conditions. For the double clamped beam we obtain the characteristic equation:

$$\varsigma \sin\varsigma + 2\cos\varsigma - 2 = 0 \tag{4}$$

with the first six buckling eigenvalues presented in table 1

Table 1. The first six buckling eigenvalues

Symmetric buckling modes	Eigenvalues	Asymmetric buckling modes	Eigenvalues
1	6.2831853071	2	8.9868189158
3	12.566370614	4	15.450503673
5	18.849555921	6	21.808243318

The mode shape value, for any location x along the beam, is given by the following function:

$$v_i(x) = c\left\{1-\cos\left(\varsigma_i \frac{x}{L}\right) - \left[\sin\left(\varsigma_i \frac{x}{L}\right) - \varsigma_i \frac{x}{L}\right]\frac{1-\cos\varsigma_i}{\sin\varsigma_i - \varsigma_i}\right\} \tag{5}$$

The critical loads are obtained from:

$$P_{cr-i} = \frac{\varsigma_i^2 E\cdot I}{L^2} \tag{6}$$

so, the critical temperatures are:

$$T_{cr-i} = T_0 + \Delta T_{cr} = T_0 + \frac{P_{cr-i}}{\alpha\cdot E\cdot A} \tag{7}$$

These temperatures are critical because at these points the frequency becomes zero and then, from here on, the apparent stiffness increases. If we have negative temperatures, P decreases and the frequency increase. This frequency is derived as:

$$f_i = \frac{\lambda_i^2}{2\pi L^2}\sqrt{\frac{EI}{\rho A} - \frac{P}{\rho A}\left(\frac{L^2}{\varsigma_i^2}\right)} \tag{8}$$

where λ_i is the dimensionless wave number of bending vibration mode i.

We made an analytical calculation and determined the critical load values and the critical temperatures for a clamped steel beam with the length L = 1200 mm, width b = 50 mm and the thickness h = 5 mm and the yield strength of 530 N/mm^2, elastic modulus $E = 2\cdot 10^5$ N/mm^2, Poisson's ratio μ = 0.3, mass density $\rho = 7850\cdot 10^{-9}$ kg/mm^3 and the thermal coefficient $\alpha = 1.25\cdot 10^{-5}/{}^\circ\mathrm{C}$. The results of calculation are given in table 2.

3. FEM analysis

For finite element analysis was used *SolidWorks* simulation software with which was performed an analysis over a clamped steel beams using the features stated in the previous paragraph. In *Thermal analysis* module we perform a thermal analysis for temperatures in the range of 0…100 °C, with reference temperature of 24.85°C. Next, we test whether the critical temperature obtained from the analysis in *Buckling module* corresponds to the one analytically obtained.

Table 2. Critical loads and temperatures for the analyzed double-clamped beam

Mode	Critical load		Critical temperature	
	Analytic [N]	FEM [N]	Analytic [°C]	FEM [°C]
1	2855.79	2857.47	29.8	29.8
2	5842.22	5846.48	35.01	35.01
3	11423.1	11440.6	44.71	44.74
4	17268.3	17301.3	54.88	54.94
5	25702.1	25776.3	69.54	69.68
6	34403.9	34523.2	84.68	84.89

Analyzing the data in table 2 is seen that the critical buckling load values and the critical temperature values are very similar both for analytical and FEM analysis.

In figure 3 we graphically represented the buckling shapes for both, symmetric and asymmetric modes analytically obtained and in figure 4 the buckling shapes for the same modes obtained from the simulation program.

Comparing figures 3 and 4 is observed similarity between the shapes of vibration modes analytically obtained and those obtained from finite element analysis.

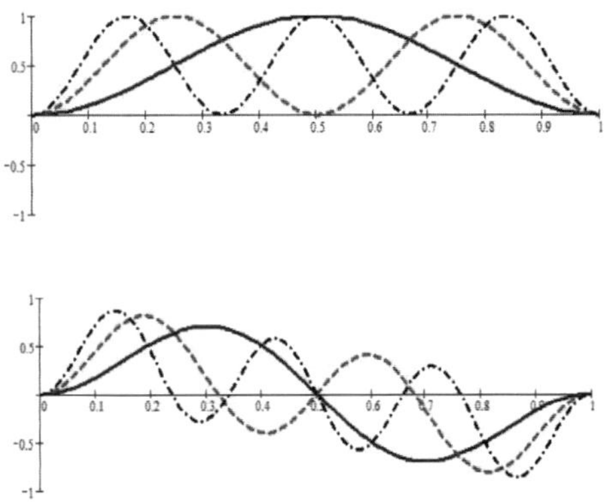

Figure 3. Buckling shapes for the symmetric and asymmetric modes analytically obtained

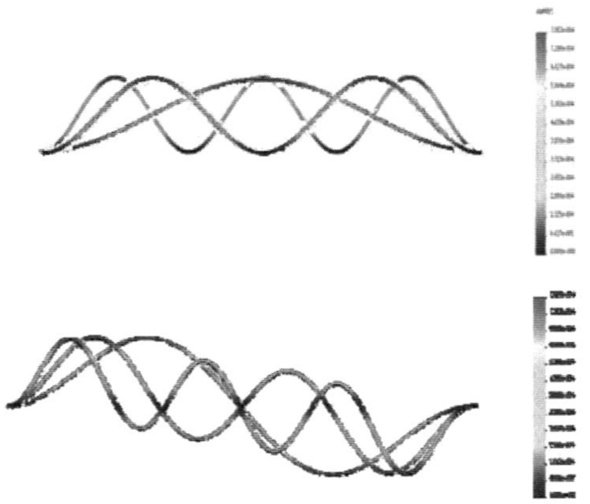

Figure 4. Buckling shapes for the symmetric and asymmetric modes from FEM analysis

Further analysis will be done using the *Modal analysis* tool in which we will consider temperatures between 0 and 100 °C, ranging from 5 to 5 °C in the interval 0…24.85 °C and then, for temperatures higher than the reference temperature, will go with a step of 1 °C.

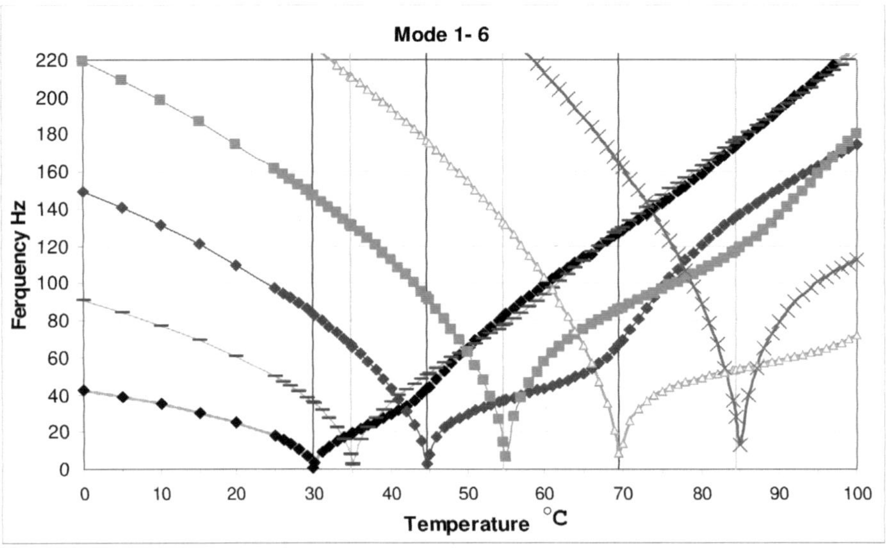

Figure 5. Frequency changes for transversal modes 1-6

The frequencies obtained by modal analysis for the first six weak-axis transversal vibration modes are presented in figure 5, presenting good similarity with that presented in [18].

In figure 5 we observed that the frequency decreases with increasing temperature until reaching the buckling in that mode. After this point, the frequency increases due to the important contribution of the apparent stiffness increase.

Figure 6 presents the frequency values calculated analytically for the vibration mode three in the state before buckling and with continuous line the natural frequency values obtained from simulation program for temperatures between 25 °C and 45 °C.

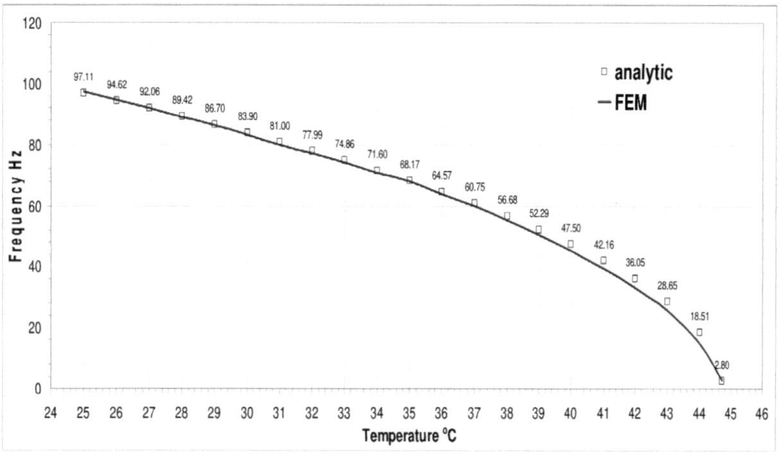

Figure 6. Frequency decrease due to compression for vibration mode three

It is important to mention that the vibration mode shapes deviate from the classical look due to thermal loads and expansion.

4. Testing the damage location indicator for buckled beams

Analyzing the frequency distribution depending on the position of the damage, and depending on the depth of the damage, we conclude that the reduction of stiffness in a transverse segment does not automatically changes frequencies. With other words, damages located in two different positions produce different effects on their frequencies for a given mode of vibration. On the other hand, if we place it in the same position the damage produces different effects for different vibration modes. All the above observations help us to understand the phenomena occurring and provide the basis for developing a mathematical model and an algorithm by which it is possible to identify the position of the damage which occurs in a beam.

The natural frequency of the beam with transverse open is [12]

$$f_{i\_D}(x_D,a) = f_{i\_U}(1-\gamma(a))(\bar{\phi}_i''(x_D))^2 \quad (9)$$

where f_{i_D} is the damaged beam frequency, f_{i_U} is the frequency for the undamaged beam, $\gamma(a)$ reflect the damage severity, $i=1..n$ the mode number and n the number of the weak-axes bending vibration modes considered. The normalized mode shape curvature at location x_D is denoted $\bar{\phi}_i''(x_D)$.

If we have a set of frequencies measurements for an undamaged beam, we can calculate the relative changes in frequencies using the following equation:

$$\delta f_i(x,a) = \frac{f_{i\_U} - f_{i\_D}(x,a)}{f_{i\_U}} \quad (10)$$

If these relative changes in frequencies will be normalized by dividing each value with the maximum of the sequence

$$\Psi_1 = \frac{\delta f_1}{\max(\delta f_i)},\ \Psi_2 = \frac{\delta f_2}{\max(\delta f_i)},\ \ldots,\ \Psi_n = \frac{\delta f_n}{\max(\delta f_i)},\quad i = 1..n \tag{11}$$

we get a sequence of sub unitary positive numbers, only one element having the value 1, that is:

$$\Psi : \{\Psi_1, \Psi_2, \ldots, \Psi_n\} \tag{12}$$

On the other hand, introducing equation (9) in equation (10) results:

$$\delta f_i(x,a) = \frac{f_{i\_U} - f_{i\_D}(x,a)}{f_{i\_U}} = \gamma(a)\cdot(\bar{\phi}_i''(x))^2 \tag{13}$$

By normalization we obtain a sequence which is independent of severity:

$$\Phi_1 = \frac{\delta f_1}{\max(\delta f_i)} = \frac{\gamma(a)\left(\bar{\phi}_1''(x)\right)^2}{\max\left\{\gamma(a)\left(\bar{\phi}_i''(x)\right)^2\right\}} = \frac{\left(\bar{\phi}_1''(x)\right)^2}{\max\left\{\left(\bar{\phi}_i''(x)\right)^2\right\}}$$

$$\cdots\cdots\cdots \tag{14}$$

$$\Phi_n = \frac{\delta f_n}{\max(\delta f_i)} = \frac{\gamma(a)\left(\bar{\phi}_n''(x)\right)^2}{\max\left\{\gamma(a)\left(\bar{\phi}_i''(x)\right)^2\right\}} = \frac{\left(\bar{\phi}_n''(x)\right)^2}{\max\left\{\left(\bar{\phi}_i''(x)\right)^2\right\}}$$

If we want to generalize, we consider k positions of the damage on the beam. For each position $x_1, x_2, \ldots, x_k$ on the beam we determine the squared of the mode shape curvature at that point $\left(\bar{\phi}_i''(x_j)\right)^2$.

Normalizing these values for each position $x_1, x_2, \ldots, x_k$ considered of the beam, by dividing with the maximum values on a line we obtain the influence coefficients obtained defect position:

$$\Phi_1(x_1) = \frac{(\bar{\phi}_1''(x_1))^2}{\max\left\{(\bar{\phi}_i''(x_1))^2\right\}},\ \Phi_2(x_1) = \frac{(\bar{\phi}_2''(x_1))^2}{\max\left\{(\bar{\phi}_i''(x_1))^2\right\}},\ \ldots,\ \Phi_n(x_1) = \frac{(\bar{\phi}_n''(x_1))^2}{\max\left\{(\bar{\phi}_i''(x_1))^2\right\}}$$

$$\Phi_1(x_2) = \frac{(\bar{\phi}_1''(x_2))^2}{\max\left\{(\bar{\phi}_i''(x_2))^2\right\}},\ \Phi_2(x_2) = \frac{(\bar{\phi}_2''(x_2))^2}{\max\left\{(\bar{\phi}_i''(x_2))^2\right\}},\ \ldots,\ \Phi_n(x_2) = \frac{(\bar{\phi}_n''(x_2))^2}{\max\left\{(\bar{\phi}_i''(x_2))^2\right\}} \tag{15}$$

$$\cdots\cdots\cdots$$

$$\Phi_1(x_k) = \frac{(\bar{\phi}_1''(x_k))^2}{\max\left\{(\bar{\phi}_i''(x_k))^2\right\}},\ \Phi_2(x_k) = \frac{(\bar{\phi}_2''(x_k))^2}{\max\left\{(\bar{\phi}_i''(x_k))^2\right\}},\ \ldots,\ \Phi_n(x_k) = \frac{(\bar{\phi}_n''(x_k))^2}{\max\left\{(\bar{\phi}_i''(x_k))^2\right\}}$$

So, for the position x_j on the beam, we obtain the damage location indicator DLI as

$$\Phi_j : \{\Phi_1(x_j), \Phi_2(x_j), \ldots, \Phi_n(x_j)\} \tag{16}$$

Considering k positions equidistantly distributed along the beam, i.e. $j = 1..k$ we produce a matrix which has k rows and n columns. Each matrix row, or set Φ_j, uniquely characterize a certain location on the beam [12]. The sequence Ψ, as a single row matrix, is compared successively with each row of the matrix derived using equation (16). The value of distance x_j, for which the elements of row j of the matrix Φ are closest to the values of the matrix Ψ, indicate the damage location.

In this study we want to see if we can use the damage location indicators (DLI) without considering the influence of temperature. For this, we consider the double-clamped beam presented in paragraph 2, with a transverse damage with a width of 0.1 mm, situated at a distance of 400 mm to one end and reducing the cross-sectional area by 50%. The analysis was made for the temperature of 0 °C, 10 °C, 15 °C, 20 °C, 26 °C having a reference temperature by 24.85 °C. Using the frequency values obtained the graphs in figure 6 have been traced.

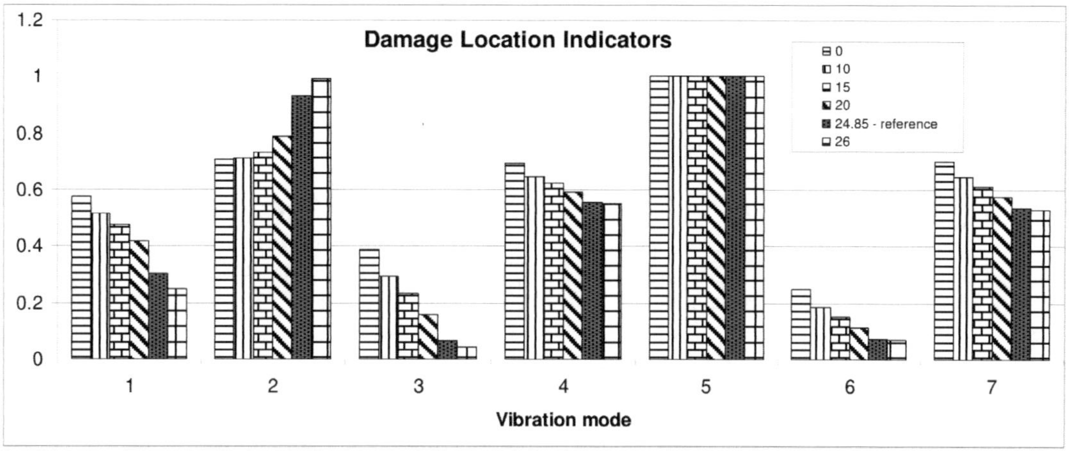

Figure 6. Damage location indicators for damage located at 400 mm

Analyzing the results, it can be seen that there are quite large deviations of frequency values obtained at different temperatures. From this, we can conclude that either we have to analyze the frequencies at the same temperature, or we have to find a formula to compensate the temperature change. This is possible by introducing the explicit relation of the buckling load given in equation (1) in the frequency equation (8). It results

$$f_i(T) = \frac{\lambda_i^2}{2\pi L^2}\sqrt{\frac{EI}{\rho A} - \frac{P(T)}{\rho A}\left(\frac{L^2}{\varsigma_i^2}\right)} = \frac{\lambda_i^2}{2\pi L^2}\sqrt{\frac{EI}{\rho A} - \frac{\alpha EA \cdot \Delta T}{\rho A}\left(\frac{L^2}{\varsigma_i^2}\right)} \tag{17}$$

For practical applications it is sufficient to study the frequency until the first buckling mode occurs.

5. Conclusion

The paper presents the influence of axial loads induced by temperature on the damage detection process. It is shown that, if in the structure monitoring the vibration measurements are made at different temperatures, the initially acquired frequencies of the healthy clamped-clamped beam cannot be considered a reference as they are. This happens because the frequencies of different modes are affected in different ways by the same temperature change. A solution of the problem was tested and presented in this paper: the adjustment of frequencies by considering the loads induced by temperature changes. Thus, a new reference is contrived and the Damage Location Indicators introduced by the authors can be used to assess the damage position. This approach is valid also if compression or tensile loads are applied on the beam; in this case the induced stress has to be evaluated and introduced in the frequency relation in order to make the monitored frequencies compare. Since in real cases the boundaries are not perfectly fixed, but small displacements are possible, it is preferable to apply this second approach also if temperature changes produce the internal loads.

Acknowledgment
The work has been funded by the Sectoral Operational Programme Human Resources Development 2007-2013 of the Romanian Ministry for European Funds through the Financial Agreement POSDRU/159/1.5/S/132395.

References

[1] Han S M, Benaroya H and Wei T 1999 Dynamics of transversely vibrating beams using four engineering theories *J. Sound Vib.* **225**(5) 935-988
[2] Timoshenko S P 1921 On the correction for shear of the differential equation for transverse vibrations of bars of uniform cross-section *Philos. Mag. A* **41** 744-6
[3] Cowper G R 1966 The Shear Coefficient in Timoshenko's Beam Theory *J. Appl. Mech.-T ASME* **33**(2) 335-40
[4] Ece M C, Aydogdu M and Taskin V 2007 Vibration of a variable cross-section beam *Mech. Res. Commun.* 34(1) 78–84
[5] Salawu O S 1997 Detection of structural damage through changes in frequency: a review *Eng. Struct.* **19**(9) 718-723
[6] Gillich G R, Praisach Z I and Iavornic C M 2012 Reliable method to detect and assess damages in beams based on frequency changes, *Proceedings of the ASME 2012 International Design Engineering Technical Conferences (Chicago)* paper DETC2012-70094
[7] Al-Ansari L, Al-Waily M and Al-Hajjar Y 2012 Experimental and Numerical Study of Crack Effect on Frequency of Simple Supported Beam *Al-Khwarizmi Eng. J.* **8**(2) 30-41
[8] Chondros T G, Dimarogonas A D and Yao J 2001 Vibration of a beam with a breathing crack, *J. Sound Vib.* **239**(1) 57-67
[9] Azadi M 2011 Free and forced vibration analysis of FG beam considering temperature dependency of material properties *J. Mech. Science Tech.* 25(1) 69-80
[10] Zapico J L, Worden K and Molina F J 2001 Vibration-based damage assessment in steel frames using neural networks *Smart Mater. Struct.* **10**(3) 553–9
[11] Hios J D and Fassois S D 2014 A global statistical model based approach for vibration response-only damage detection under various temperatures: A proof-of-concept study *Mech. Syst. Signal Process.* **49**(1-2) 77-94
[12] Gillich G R and Praisach Z I 2013 Detection and Quantitative Assessment of Damages in Beam Structures Using Frequency and Stiffness Changes *Key Eng. Mat.* **569** 1013-20
[13] Gillich G R and Praisach Z I 2014 Modal identification and damage detection in beam-like structures using the power spectrum and time–frequency analysis *Signal Process.* **96**(Part A) 29–44
[14] Praisach Z I, Gillich G R and Birdeanu D E 2010 Considerations on natural frequency changes in damaged cantilever beams using FEM *Latest trends on engineering mechanics, structures, engineering geology* 214-9
[15] Gillich G R, Praisach Z I and Negru I 2012 Damages influence on dynamic behaviour of composite structures reinforced with continuous fibers *Mater. Plast.* **49**(3) 186-191
[16] Gillich G R, Birdeanu E D, Gillich N, Amariei D, Iancu V and Jurcau C S 2009 Detection of damages in simple elements *Annals of DAAAM and Proceedings of the International DAAAM Symposium* 623-624
[17] Reddy J N 2006 *Theory and Analysis of Elastic Plates and Shells* Second edition (Boca Raton: CRC Press)
[18] Tufoi M, Gillich G R, Praisach Z I, Iancu V and Furdui H, About the Influence of Temperature Changes on the Natural Frequencies of Clamped-Clamped Euler-Bernoulli Beams *Rom. J. Accoust. Vib.* **11**(2) 84-7

Experimental Fault Diagnosis in Systems Containing Finite Elements of Plate of Kirchoff by Using State Observers Methodology

D M Alegre[1], E H Koroishi[2], G P Melo[3]

[1] Federal Fluminense University, Department of Mechanical Engineering, Rua Passos da Pátria, 156, Sala 213, Bloco E, Niterói – RJ – 24210-240, Brazil
[2] Federal Technological University of Paraná, Department of Mechanical Engineering, Av. Alberto Carazzai, 1640 - Campus Cornélio Procópio - Cornélio Procópio – PR - 86300-000, Brazil
[3] São Paulo State University, Department of Mechanical Engineering, Avenida Brasil Centro, 56, Ilha Solteira - SP – 15385-000, Brazil

E-mail: dmalegre@gmail.com, edsonh@utfpr.edu.br, gilberto@dem.feis.unesp.br

Abstract. This paper presents a methodology for detection and localization of faults by using state observers. State Observers can rebuild the states not measured or values from points of difficult access in the system. So faults can be detected in these points without the knowledge of its measures, and can be track by the reconstructions of their states. In this paper this methodology will be applied in a system which represents a simplified model of a vehicle. In this model the chassis of the car was represented by a flat plate, which was divided in finite elements of plate (plate of Kirchoff), in addition, was considered the car suspension (springs and dampers). A test rig was built and the developed methodology was used to detect and locate faults on this system. In analyses done, the idea is to use a system with a specific fault, and then use the state observers to locate it, checking on a quantitative variation of the parameter of the system which caused this crash. For the computational simulations the software MATLAB was used.

1. Introduction

The state observer's methodology consists on developing a model for the analyzed system and compares the estimated output with the measured output. The mathematical models which represent the behavior of the systems are not free of unknown perturbations and variations of the parameters. In the most of the state observer's designs, the parameters of the system are known or they can be identified by specific methods presents in the literature. In the cases in which the parameters are not known or they are subject to changes while the system working, the output of the observer may provide a incorrect estimation of the rebuild states, leading to false alarms in the detection and location of faults. In the last few years the problem of variation of the parameters in design of state observers have been studied by countless researchers [1-4].

According to Luenberger [5], the state observers can rebuild the states not measured or values from points of difficult access in the system. So, faults can be detected in these points without knowing its values. The existing methodologies of state observers are designed to solve problems of control and detection of possible faults in sensors and instruments [6].

This work presents a methodology of fault detection and location in a system, which is similar to a simplified vehicular platform, by using of state observers.

2. Methodology

In the project of many control systems, it is considered that all the state variables are available for retroaction. However, in practices, there are situations that this is not real, so, it's necessary to estimate the state variables that are not available [7].

A state observer estimates the variables using the measurement of the outputs and control variables. According to Ogata [7], the state observers can be projected, if and only if, the observability condition is satisfied.

Considering the system defined by

$$\dot{x}(t) = A \cdot x(t) + B \cdot u(t) \tag{1}$$

$$y(t) = C \cdot x(t) \tag{2}$$

where A, of size $n \times n$, is the dynamical matrix, B, of size $n \times r$, is the input matrix and C, of size $n \times n$, is the output matrix, where n is the dimension of the system and r is the dimension of the inputs vector.

Admit that the state vector $x(t)$ must approximate by the state vector $\hat{x}(t)$ of the dynamic model

$$\dot{\hat{x}}(t) = A \cdot \hat{x}(t) + B \cdot u(t) + L \cdot (y - C \cdot \hat{x}(t)) \tag{3}$$

$$\hat{y}(t) = C \cdot \hat{x}(t) \tag{4}$$

which represents the state observer. The gain matrix L was determined using the Linear Quadratic Regulator (LQR).

3. Fault Detection By Using State Observers

The aim of this paper is the detection and location of faults in a system that is similar to a simplified vehicle platform. According to many recent papers [3-4,8], by For this purpose the methodology of state observers was used. For the fault detection in the system, it was projected a Global Observer. This Global Observer has the function of indicate a fault in the system. It was also projected a bank of Robust Observers, which has the objective of locate the positions of the fault in the system and its intensity. Figure 1 illustrates the mount scheme, using state observers.

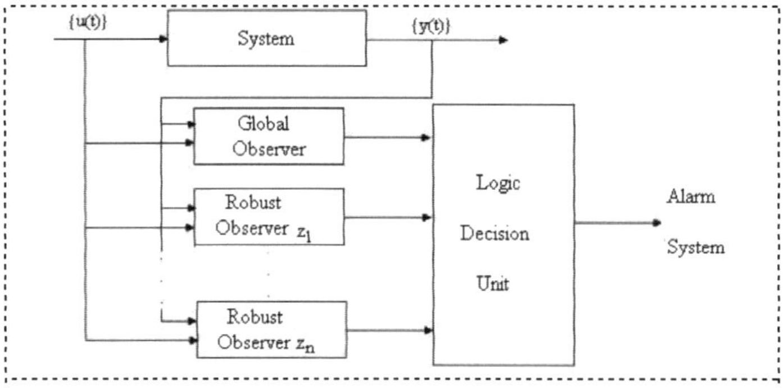

Figure 1. Simplified Scheme of a System with State Observers.

The State Observers can rebuild the non-measured states of the system, so the Global Observer was used to simulate the dynamic behaviour of the system without any faults. The signal produced by the global observer is exactly the same of the output signal of the real system with no faults. In the case of a fault occur in any component of the system, the output signal produced by the global observer and the real system will not be equal anymore, in this way the Logical Unit Decision will indicate the presence of a fault in the system, by comparing these two signals.

In the presence of a fault in the system, the global observer detects it, so becomes necessary the location of the component of the system in which the fault occurred. For this, a bank o Robust Observers is mounted for the system. In this bank, the observers are projected to estimate the outputs of the system for the case of specific faults in the many components of the system. There are utilized many observers, where each one represents a system with a fault in a determined component and a given percentage of fault. By this bank is possible to locate the fault in the system and its intensity.

4. Finite Elements Method

The Finite Elements Method was used to determine the system´s model. By using this method, mathematical models can be obtained for complex mechanical systems. The equations of this section are according to [9-11].

The motion equations of the system are obtained using the mass and stiffness matrices of the system. These matrices are obtained from the mass and stiffness matrices of the finite elements of the system. Now it will be shown the way that these matrices are obtained for the elements used in this paper, then these matrices will be used to obtain the mass and stiffness matrices of the entire system and these matrices will be used to determine the movement equations for the system.

Figure 2, shows the system utilized in this paper. The plate will be divided as the plate shown in figure 3.

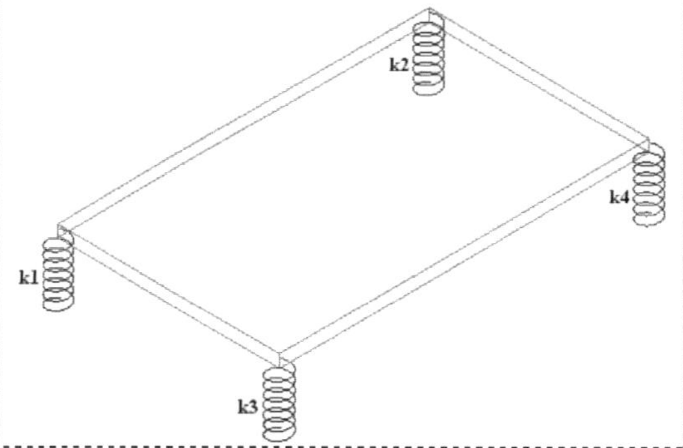

Figure 2. System considered.

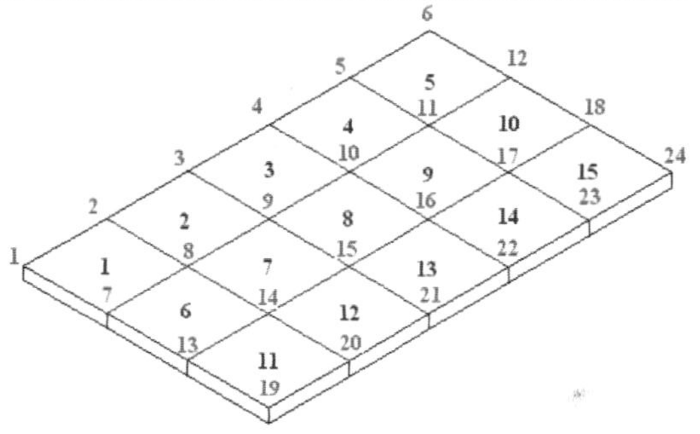

Figure 3. Plate composed by finite elements.

The element that composed the plate presented in figure 3 is known as plate of Kirchoff. Some considerations will be adopted in the following theory:

1. The plate's thickness is small compared with the lateral dimension;
2. The plate's deflection is small comparing with its thickness and the inclination of the medium deflected plane is small compared with the unit;
3. The plate's deformation is such that, straight lines that are initially perpendicular to the medium surface, stay straight and perpendicular to the medium surface in the structure with the loading. This is the Kirchoff's Hypothesis;
4. There are no tensions in the medium surface of the plate.

The theory considered here, for which the above considerations are applied, is know as the Kirchoff-Love's Plate Theory, or classic plate theory. It will also be assumed that the material is homogeneous and isotropic. Using these consideration, the system presentd in figure 2 can be represented by the equation 5.

$$M \cdot \ddot{D}(t) + K \cdot D(t) = Q \tag{5}$$

where Q is a vector of dimensions 75X1, which represents the forces and bendings applied to the system; M is the mass matrix; K is the stiffness matrix and D(t) is displacement vector.

5. Experimental Results

The studied system is shown in figure 4.

Figure 4. Experimental Apparatus.

The plate utilized was built with aluminum and its dimensions were 0.5m of length, 0.3m of width and 0.015m of thickness (how the thickness is equal to 5% of the smaller side of the plate, it was considered that this plate is a thin plate, so the plates of Kirchoff can be used). It was considered a density of 2700kg/m^3 for the aluminum. It was utilized an electric motor of Weg manufacturer, with a maximum rotation of 3380 rpm and a mass of 4 kg. This motor had a frequency regulator, which permits the control of the velocity of rotation.

There were supports for the springs screwed to the plate, where the springs were fixed. They were used springs with two different values of stiffness.

Table 1. Dimensions of the used springs.

d(m)	R(m)	k (N/m)	Fault (%)
0.003	0.0015	5437.884	0
0.0028	0.0014	4207.363	25

The springs in table 1 present high percentage of faults (25%), but in practices the faults are usually smalls, less then 10%. However, to build a spring with a loss of 10% of stiffness, for example, it would be necessary to made very small variations in its dimensions, so it was no possible to find a manufacturer of springs which provides this precision. Therefore it was chosen springs with a higher percentage of faults, so the springs could be made. For the computational simulation of the vertical movement of the apparatus shown in figure 6, it was used the model shown in figure 2 and the plate was divided in finite elements how is shown in figure 3. This configuration was chosen for a better distribution of motor weight and the forces of the springs in the dots.

It was considered that the springs k_1, k_2, k_3 e k_4 were exactly beneath the center of elements 1, 5, 11 and 15, respectively. The motor was positioned on the center of the element number 8, so its mass was considered as a set of discrete masses distributed into the dots 9, 10, 15 and 16. The same was done to the masses of the supports of the springs, distributing then equally into the dots of the elements 1, 5, 11 and 15. For the forces applied by the springs, it was adopted a similar procedure of the distribution of the masses of the supports and the motor. The forces were equally distributed into the dots of the elements 1, 5, 11 and 15.

Although there were not any dampers in the experimental apparatus, the springs present a natural damping, so this damping was determined by the logarithmic decrement. For this purpose, it was applied impacts in the center of element number 8 and it was measured the decay of the amplitude of the vertical movement of the plate. It was obtained a value of 0.022 for the logarithmic decrement and, consequently a value of 3.685 N.s/m for the equivalent damping of the springs. How the springs were in parallel, the value of the damping for each spring was 0.921 N.s/m. This value was used in the computational simulations, considering the dampers together with the springs.

For the movement of the plate, it was used a rotation of 1127 rpm (frequency of 20Hz) of the axis of the motor. It was also used an unbalanced mass of 0.05kg in a distance of 0.05m of the center of the axis. This force was distributed in the dots 10 and 16. How the plate was made of aluminum, the accelerometer was fixed by pasting a bronze support in the plate and then fixing the accelerometer by a screw. The accelerometer was positioned in the dot 14, so all the results presented in this section are for the vertical displacement of this dot.

For the signal acquisition, the accelerometer was connected to an amplifier, which was connected to an acquisition board DaqBook 112. The signal from DaqBook was sent to a computer that, together with software Dasylab, produced the visualization of the signal in the computer screen. The sample frequency was 500Hz. The software MATLAB was used for the design of the state observers that simulates the movement of the plate. A signal for the vertical displacement of dot 14 was generated by the global observer and compared with the signal from the accelerometer. The results are shown in figure 5.

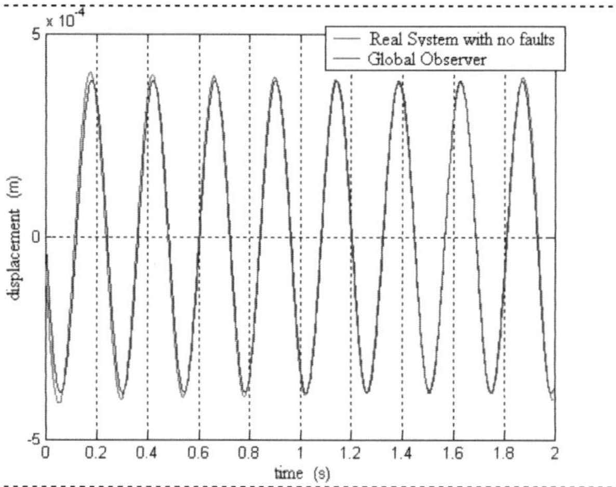

Figure 5. Vertical displacement of dot 14 for the system with no faults and the signal of global obsever.

By figure 5, it can be seen that the signal from the global observer represents very well the signal from the experimental apparatus.

For the simulation of the system with faults, they were used the springs shown in table 1. These springs were positioned in specific positions and the vertical displacement of dot 14 was measured. The measured signal was compared with the signals from the bank of state observers. Initially the spring k_1 was replaced by a spring with a loss of 25% of stiffness. The signal measured for the vertical displacement of the dot 14 and the signal of the global observer for the vertical displacement of this same dot are shown in figure 6.

In figure 6, it is possible to see a difference between the two signals, showing a fault in the system. Figure 7 below shows the inverse of the difference, between the RMS values, of the signal of the movement of dot 14 and the signals from the robust observers.

Figure 7 showed that the higher value occurs in the position 25% of robust observer k_1. This indicates that the signal from the robust observer to k_1, for a fault of 25%, is closer of the signal from the experimental apparatus, then the signal from any other robust observer. Therefore the difference between the RMS values of these two signals is smaller then the difference between any other RMS values from the experimental apparatus and the other observers. So the inverse of these differences produces a better visualization of the position of the fault (spring k_1) and its intensity (25%).

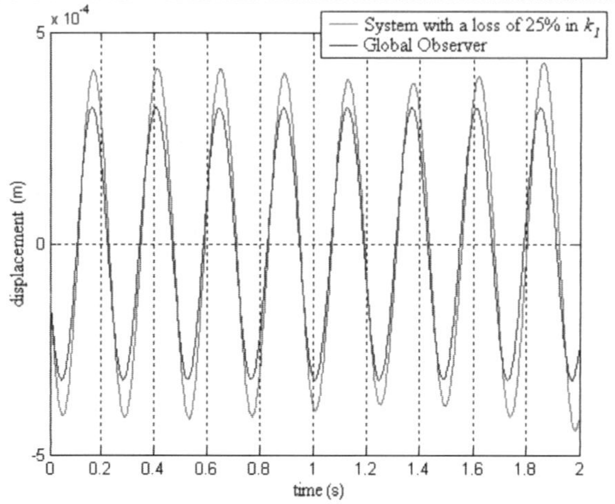

Figure 6. Vertical displacement of dot 14 for the system with a fault of 25% in k_1 and the signal of global observer.

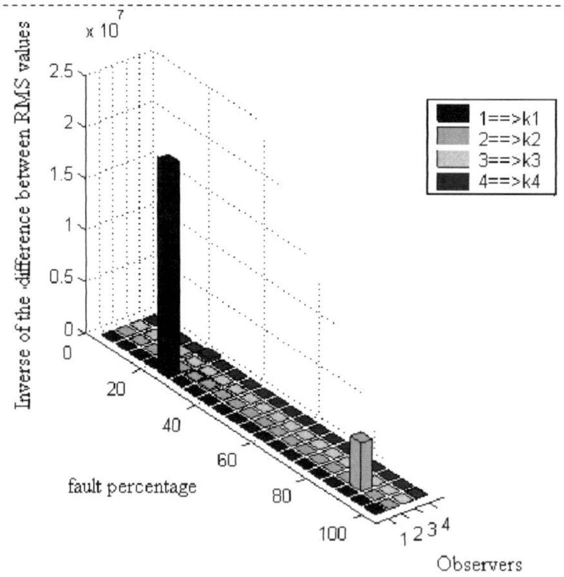

Figure 7. Inverse of the difference, between RMS values, of the signals of the vertical displacement of dot 14, from robust observers and form the experimental apparatus, for the system with a fault of 15% in k_1.

In sequence, both the springs k_1 and k_3 were replaced by springs with a loss of 25% in the stiffness. With the system functioning with these springs, the vertical displacement of the dot 14 was measured. The measured signal was compared with the signal form the global observer. These two signals are shown in figure 8.

Figure 8 showed a large difference between the two signals, showing a fault in the system. Figure 9, below, presents the inverse of the difference, between RMS values, of the signals from this same system and from the robust observers.

By the figure 9, it was noticed that was necessary the introduction of a new observer, which represents the system with simultaneous faults on springs k_1 and k_3. It's also verified that the higher value occurs in the position 25% of this observer. This means that the signal from the robust observer to k_1 and k_3, for a fault of 25%, is the signal that is closer to the signal from the accelerometer, so the difference between the RMS values of these two signals is very smaller then the difference between the signal from the accelerometer and the other observers. The inverse of these values was done for a better visualization of the position of the fault. Therefore, is possible to locate the position of the fault (in springs k_1 and k_3 simultaneously), and its intensity (25%).

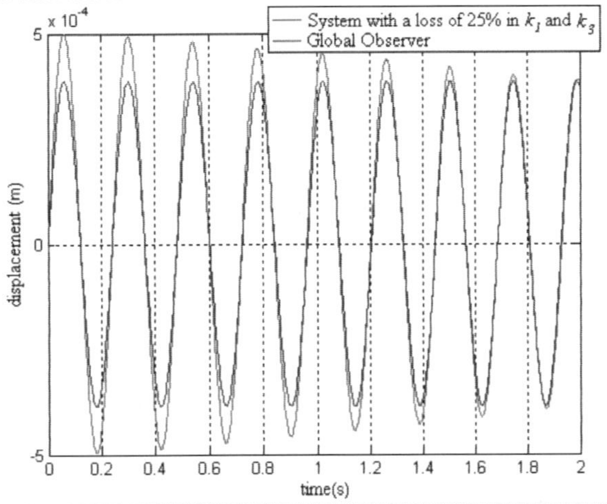

Figure 8. Vertical displacement of dot 14 for the system with 25% of fault in springs k_1 and k_3, simultaneously, and from the global observer.

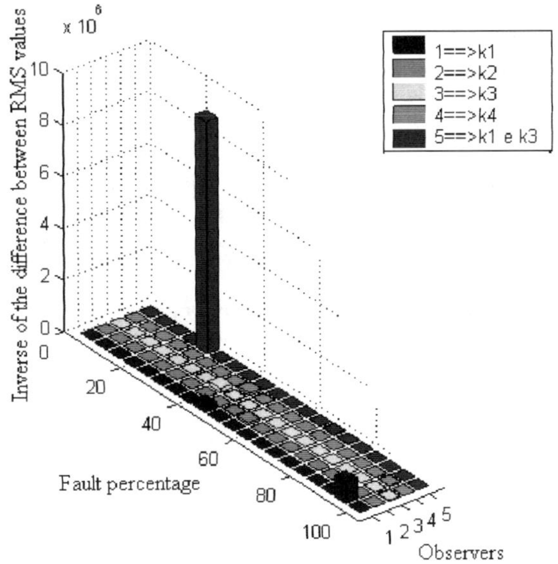

Figure 9. Inverse of the difference, between RMS values, of the vertical displacement of dot 14 from the system with faults of 25% in springs k_1 and k_3, and the robust observers.

6. Conclusions

In this paper the methodology of the state observers was applied for the detection and location of faults in a model of platform that had a suspension system. The mathematical equations for the system used were determined by using the finite elements method.

By the presented results, it is possible to conclude that the state observer, projected by the Linear Quadratic Regulator (LQR) method, were efficient in the detection and location of faults in the system considered. Was verified that, for the system considered, the observers can detect and locate faults even that the measurement point is not close to the point where the faults occurs. It was also verified that the observers can detect and locate simultaneous faults in the system, that's the case when two springs presented faults at the same time.

Computational programs were developed, by the usage of the finite elements method, which allowed the simulation of the movement of the system and the detection and location of faults in a continuous system. The system used presented a very simplified suspension system. Although this system didn't present any parameter based on the parameters of a real vehicle, the computational programs can be expanded, in future works, to simulate the movement of a real vehicle.

The number of finite elements present on the plate could not be higher, because this would compromise the observability of the system.

7. References

[1] Marano J H 2002 *Localização de falhas via observadores de estado em sistemas com variação de parâmetros* (Universidade Estadual Paulista, Faculdade de Engenharia de Ilha Solteira: Ilha Solteira).

[2] Melo G P, Morais T S and Daniel G B 2005 Diagnosys of faults in mechanical systems with unknown inputs using proportional and integral state observers *Proceedings of International Congress of Mechanical Engineering - COBEM.*

[3] Koroishi E H, Melo G P and Assunção E 2010 State Observer Using Decay Rate LMI Constraints for Fault Detection in Mechanical Systems Ciência & Engenharia 12-21

[4] Koroishi E H and Melo G P 2012 Experimental Fault Detection in Rotation System Using State Observers by LMIs *Journal of Mechanics Engineering and Automation* 470-475.

[5] Luenberger D G 1964 Observing the state of a linear system *IEEE Military Electronics* 74-80.

[6] Koroishi E H, Borges A S, Cavalini Jr A A and Steffen Jr V 2014 Numerical and Experimental Modal Control of Flexible Rotor Using Electromagnetic Actuator *Mathematical Problems in Engineering (Print).*

[7] Ogata K 1998 *Engenharia de controle moderno* (Prentice-Hall do Brasil: Rio de Janeiro).

[8] Melo G P and Lemos G F 2004 Fault Diagnosis in Rotation System Using well Conditioned State Observer *Proceedings of Conference on Structural Dynamics - IMAC.*

[9] Azevedo A F M 1976 *Método dos elementos finitos* (Faculdade de Engenharia da Universidade do Porto)

[10] Bathe K J 1976 *Numerical methods in finite elements analysis* (Prentice Hall: Englewood Cliffs).

[11] Dawe D J 1984 *Matrix and finite element displacement analysis of structures* (Oxford: Clarendon).

Integration of acoustic emission systems within Integri-Tech™ analysis system for structural health monitoring of pressurised engineering plant

A.A. Ghouri[1, 2], Steven Rafferty[2], Andy Pickwell[2], Walter Galbraith[1], S. Gareth Pierce[1], Anthony Gachagan[1]

1- Department of Electronic and Electrical Engineering, University of Strathclyde, Glasgow, G1 1XW, UK

2- Aker Solutions 11 Marine Crescent, Glasgow, G51 1HD, Scotland, UK

E-mail: aamir.ghouri@strath.ac.uk

Abstract. The aim of this Acoustic Emission (AE) based Structural Health Monitoring project is to enable accurate location of AE sources in pressurised engineering plant and to use AE source location data to establish defect locations for use within Integri-Tech™; a finite element based analysis, monitoring and fitness for service assessment system. Integri-Tech™ is a windows based system which carries out combined analysis and assessment providing fatigue life and remnant life calculations and inspection priorities presenting the results in an accessible web portal format. The software uses finite element stress models created in the companion software Model Wizard. The AE monitoring system that has been developed can be used with an array of up to four AE broad band sensor channels with associated signal processing. Using a flexible approach in MATLAB, the authors have developed algorithms which were used for analysing the received AE signals to extract information about the nature and location of the source. The ability to carry out source location and possibly perform real time monitoring (detecting cracking as it occurs) is attractive feature of the AE system developed for this project. The time of arrival (TOA) data was used by Integri-Tech™ software to calculate source location using its own built-in algorithm, and this was verified independently using a MATLAB approach.

1. Introduction

Acoustic emission testing (AET) is a non-destructive technique that monitors defect creation and failure within a material through the detection and analysis of acoustic signals under load condition. AET is fundamentally different from other non- destructive testing methods because it passively detects energy that is released from a material, whereas other methods require an energy input for the defect to be detected (i.e., x-rays, ultrasound, thermal energy and microwaves). The technique is largely acceptable for the detection of possible AE events in pressure vessels, piping system and tanks. An important ability of this technique is to use the arrival times of the signals to estimate the AE source location. The arrival times from a single event at three or more sensors can be used to calculate the location of the source in the test object.

In this paper we cover the measurement aspects of our research work which include: monitoring of AE events in plate and pressure vessel samples, post-processing of AE raw data for calculating time of arrival at each single node in an array of transducers and source location on the object under test using a Trilateration algorithm implemented in MATLAB. The finite element stress models for the specimen geometries have been designed into Integri-Tech™ using its companion software Modal Wizard and the calculated arrival time value of the wave forms were used to estimate source locations on the designed models using the Integri-Tech™ built in source location algorithm.

2. Experimental Setup

Wide band sensors UT-1000 from Physical Acoustics (working frequency range from 100-1000 kHz) were used throughout this research work. Plate and pressure vessel specimen geometries were used which are representative of real structures encountered in pressurised engineering plant. The plate was manufactured from aluminium with dimensions of 2 mm thick, 550 mm long and 450 mm width. The vessel was manufactured from mild steel with dimensions of 5 mm thick, 1000 mm length and 675 mm circumference. For the plate structure, three transducers were coupled onto the plate surface in a triangular pattern, silicon grease was used as couplant to provide good acoustic coupling between the transducers and the surface of the geometry. For the vessel structure, four sensors were coupled on it. The transducers were arranged on the vessel surface in a fashion that they made four triangles for detecting possible AE events as shown in Figure 1. A four channel Tie-Pie pico-scope was used for AE waveform acquisition with the setting of 12-bit data resolution, 5000 data points and selected first transducer as trigger. Detected events were pre-amplified by four matched 40 dB preamplifiers. The acquired waveforms were stored on to the computer for further analysis.

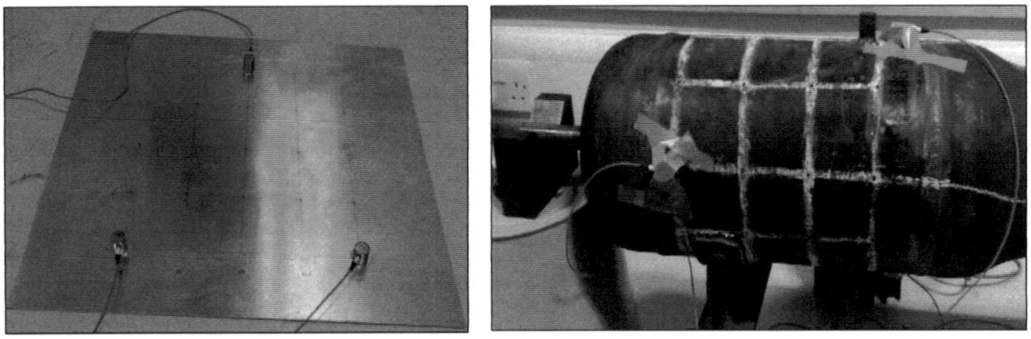

Figure 1. Experimental specimens for source location experiment

For the plate sample, an area of 300 mm by 300 mm was selected on the surface and a grid with size of (75x75) mm^2 was drawn (Figure 1). Out of 25 points on a grid, Hsu-Nielson (H-N) sources [1] were conducted at 22 points. AE signals were generated by breaking 0.7 mm pencil leads at selected locations on the plate. To standardise the tests, the pencil was held at the same angle to the surface using a bespoke of aluminium shoe and the same length of leads were broken for each test. A grid with a spacing of (75x75) mm^2 was drawn on the vessel surface to cover 675 mm circumference and 300 mm length (Figure 1). Out of 30 points on a grid, Hsu-Nielson (H-N) sources were excited at 23 points in the same manner as conducted for plate geometry. To verify the reproducibility of lead break tests for both geometries the tip of a pencil was broken 8 times on a randomly selected point on each sample.

3. On-Set Time Calculation for Source estimation

The ability to locate the source is one of the advantageous characteristic of the acoustic emission technique, but in complex material and structures it can be difficult to locate an AE event accurately. A significant volume of research has been carried out to improve the accuracy of AE source location. Recent advances in signal analysis using a statistical approach. The objective of this technique is to statistically differentiate the noise and meaningful data from the raw AE signal [2]. This approach called the *AIC* method [3-5] and can be applied to AE signals coming from a structural monitoring.

Zhang et al. [6] and Akaike [3] have shown that a time series can be divided into locally stationary segments, each modelled as an autoregressive (AR) process [3,4].

The approach described in this paper has focused on using an autoregressive process first proposed by Akaike [3] also known as Akaike information criterion (AIC). According to this approach, the AE signal can be divided into two intervals. Each segment is modelled as an autoregressive process and represents the part of the AE signal mostly dominated by the noise and the original signal, respectively. Then the two series are separated by the point at which the AIC is minimised and this point corresponds to the onset time of the signal of interest [7].

The minimum of the AIC function can be calculated using Equation1 for defining the arrival time of an AE, the function can be derived directly from the raw AE signal according to Yokota et al. [7] and Maeda [8] as:

$$AIC\,(t_w) = t_w \,.\, \log\,(\mathrm{var}\,(R_w\,(t_w\,,1))) + (T_w - t_w - 1)\,.\, \log\,(\mathrm{var}\,(R_w\,(1 + t_w\,, T_w\,))) \quad (1)$$

Where $\boldsymbol{R_w}$ is the selected window, $\boldsymbol{T_w}$ is the last sample of the selected window, $\boldsymbol{t_w}$ ranges through all samples of $\boldsymbol{R_w}$ and **var** is the variance function. $\boldsymbol{R_w\,(t_w\,,1)}$ expresses the variance that is measured from the beginning of the window until the current value $\boldsymbol{t_w}$, while $\boldsymbol{R_w\,(1 + t_w\,,\, T_w\,)}$ expresses the variance that is taken for all samples ranging from $\boldsymbol{1 + t_w}$ to $\boldsymbol{T_w}$.

For this research work the AIC algorithm was coded using MATLAB. Considering the nature of an artificial AE source, asymmetric group velocity values for plate and pipe (2400 m/s and 3100 m/s) respectively were extracted theoretically using dispersion curves and were used in the algorithm for arrival time estimation. The threshold defined for the AIC algorithm was 0.4 for the waveforms in the higher frequency range (40 kHz-250 kHz), and it could be increased up to 0.8 for lower frequency signals as the observed level of noise in these frequencies was considerably higher. The window around the prearranged on-set was selected to be from the starting point of the waveform until 100 sampling points after the prearranged onset point. However in the lower frequency range signals, this interval can be adjusted starting from the beginning of the waveform until 450-650 sampling points after the prearranged onset point. This ensured that the onset point would not be `buried' in the high levels of noise.

4. Source Estimation using Trilateration Algorithm

One-dimensional geometry problems are rare in AE source location. Although, a good understanding of the principles of source location in one-dimension provides a solid grounding for higher dimensional problems, two-dimensional problems are more prevalent in AE source location than those in one-dimensional geometries. An illustration of two dimensional source location geometry can be seen in Figure 2. Three sensors **S1**, **S2** and **S3** are placed on the surface of a structure at locations $\mathbf{(x_1, y_1)}$, $\mathbf{(x_2, y_2)}$ and $\mathbf{(x_3, y_3)}$. The location of the source, to be determined, is $\mathbf{(x_s, y_s)}$. The distance between the sensors are $\mathbf{D_1}$, $\mathbf{D_2}$ and $\mathbf{D_3}$ (which are known), and the distance between the source and the sensors are $\mathbf{d_1}$, $\mathbf{d_2}$ and $\mathbf{d_3}$ (to be solved). $\mathbf{t_1}$, $\mathbf{t_2}$ and $\mathbf{t_3}$ are the times the signal takes to reach the sensors **S1**, **S2** and **S3** respectively.

Using the derivations as made in [9], the distance $\mathbf{d_1}$ can be written to be:

$$d_1 = D_1^{\,2} - \Delta t_1^{\,2}\,.\,c^2\,[2(\Delta t_1\,.\,c + D_1\cos\,(\theta - \theta_1))]^{-1} \quad (2)$$

$$d_1 = D_2^{\,2} - \Delta t_2^{\,2}\,.\,c^2\,[2(\Delta t_2\,.\,c + D_2\cos\,(\theta_3 - \theta))]^{-1} \quad (3)$$

Where angles $\boldsymbol{\theta}$, $\boldsymbol{\theta_1}$ and $\boldsymbol{\theta_3}$ are marked in Figure 4 and

$$d_2 - d_1 = c\,(t_2 - t_1) = \Delta t_1.c \quad (4)$$

$$d_3 - d_1 = c\,(t_3 - t_1) = \Delta t_2.c \quad (5)$$

$\mathbf{\Delta t_1}$ and $\mathbf{\Delta t_2}$ are time differences $\mathbf{(t_2 - t_1)}$ and $\mathbf{(t_3 - t_1)}$ and **c** is the wave speed.

The location of the source is given by

$$x_s = x_1 + d_1\cos\theta \quad (6)$$

$$y_s = y_1 + d_1\sin\theta \quad (7)$$

For this research work MATLAB codes were used for writing the trilateration algorithm. For the calculated arrival times of the signals the time differences were calculated in the trilateration algorithm. Using a suitable iteration scheme, the values of **θ** are varied and those that minimise the error between two calculated source locations were used to identify the AE source location.

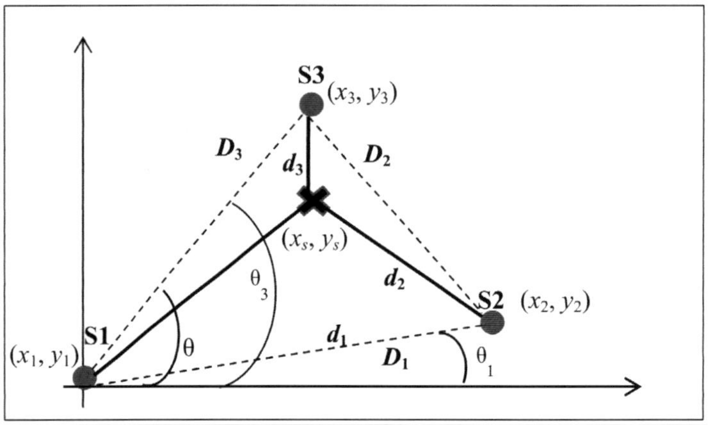

Figure 2. Two Dimensional source location [9]

For the plate sample, because of the 2-D nature, this algorithm was applied directly and calculations were done for all emission sources. For the pipe, the region of interest was divided into four non-equilateral triangles in such a way that all of four triangles cover the complete region. Four sensors were placed on a pipe to cover all possible AE events within the area of interest. Division of the pipe surface in four triangles made easy implementation of the trilateration algorithm. The pipe sample was unfolded and 2-D coordinates were used in calculations for source estimations. For both geometries the results are presented graphically in Figure 3. In this figure the actual source locations were indicated by circle (black in colour) and the estimated source locations were indicated using cross (blue in colour), the lines joining the calculated source locations with their respective actual locations.

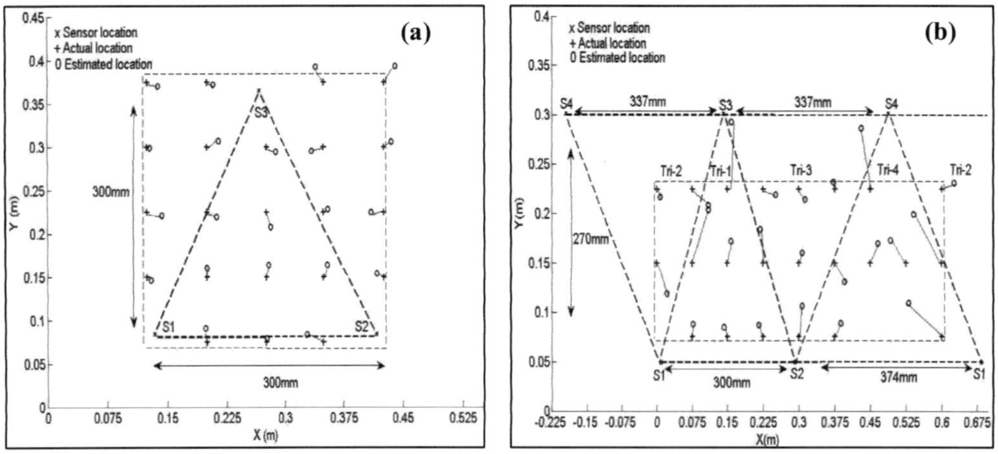

Figure 3. Trilateration Source locations for plate and unwrapped vessel in 2D plane

5. Source Estimation using Integri-Tech^{TM} Software

Integri-Tech^{TM} is a Windows application which carries out a combined finite element analysis and fatigue life calculation, presenting the results in an easy to understand web portal format. The software uses finite element models created in the companion software Modal Wizard. For this research work this software was used for the first time to estimate AE events location on the specimen under test. The software utilises its own built-in proprietary algorithm for source location. Initially the specimens were modelled using Modal Wizard as per experimental geometries measurements. The sensors were placed in the same fashion on the models and at same locations as in experiments. Figure 4 shows both plate and vessel models designed in Modal Wizard with AE sensors placed on them. The on-set time values for each of four sensors in response to the AE events were calculated using AIC algorithm and given as input AE primary parameter to Integri-Tech^{TM} for estimating source location. Depending on the given arrival time of wave forms on each node, the estimated source locations were calculated by Integri-Tech^{TM}. Figure 5 presents 3D geometry models with estimated source locations calculated by Integri-Tech^{TM} source location algorithm and its 2D graphical presentation.

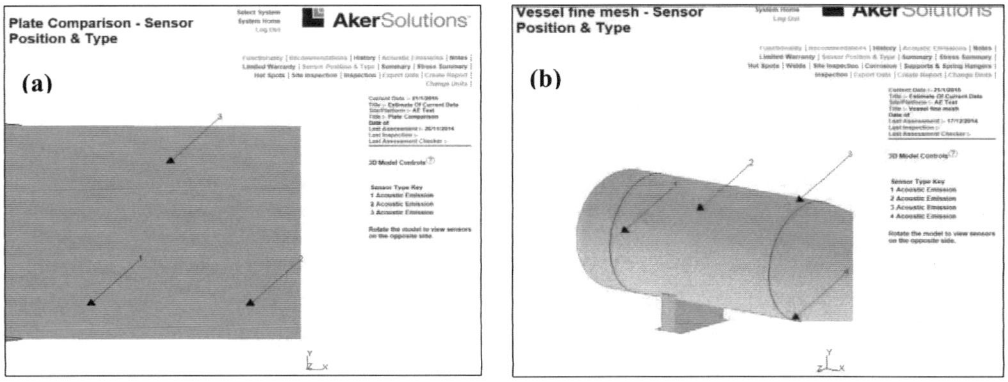

Figure 4. Integri-Tech^{TM} 3D model for each specimen with sensors positions

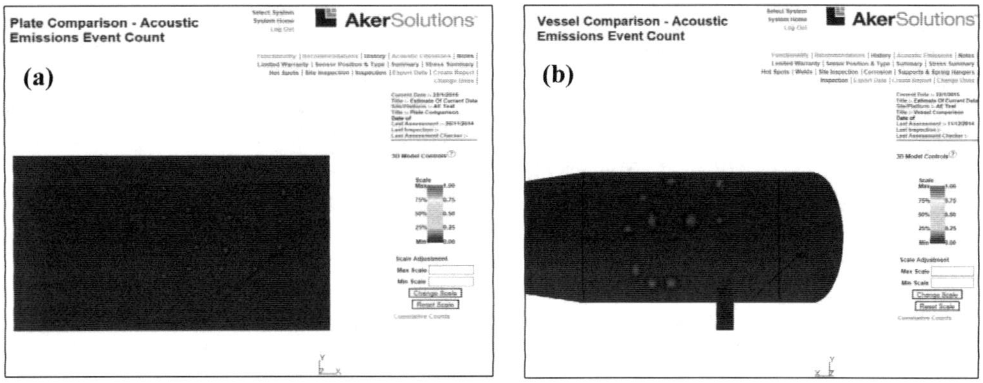

Figure 5. Estimated source locations on 3D model of each geometry using Integri-Tech^{TM}

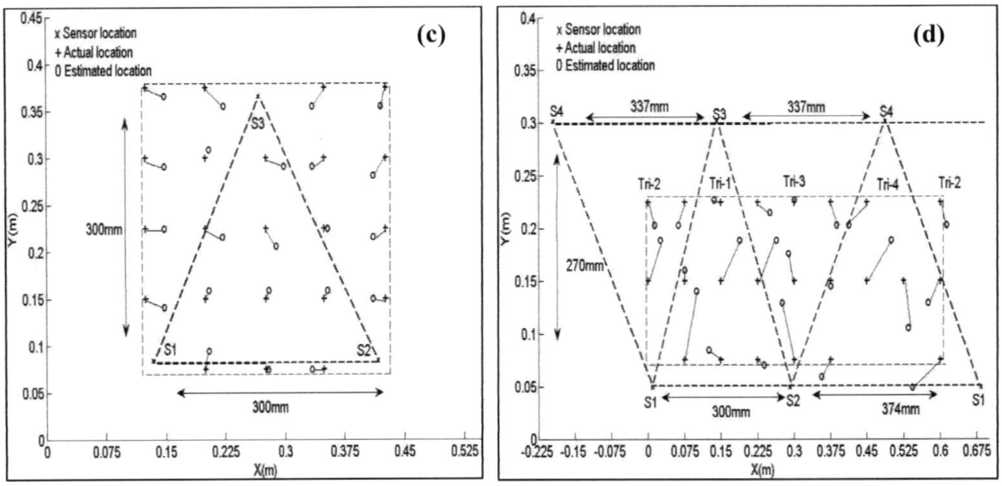

Figure 5. Integri-Tech™ Source locations for plate and unwrapped vessel in 2D plane

6. Results and Discussion

Figure 6 shows the plate and vessel samples in a 2D plane, the graphical presentation compares calculated source positions with actual source locations. The figure indicates estimated source locations calculated using two algorithms: (i) Trilateration MATLAB & (ii) Integri-Tech™ multilateration in comparison to actual source locations. Figure 6(a) shows 22 AE events on a selected grid area of a plate which represent 66 AE signals. The location analysis was done using Trilateration technique in MATLAB, a mean of distance error for all 22 AE locations was found to be 14mm (RMSE). Using the Integri-Tech™ source location algorithm for plate, a mean of the distance error for all 22 AE locations was found to be 18mm (RMSE). For the vessel, the geometry was unfolded and divided into four triangles to perform trilateration technique in MATLAB and cover whole area of interest as shown in Figure 5(d) total 23 AE events were made on a selected grid area of vessel which represent 92 AE signals. The graphical presentation in Figure 6(b) indicated four sensors positions with actual source locations and estimated source locations calculated using both source location algorithms. From trilateration technique, a mean distance error for all 23AE locations was found to be 34mm (RMSE) on the other hand with Integri-Tech™ multilateration technique the mean distance error for all 23 AE locations was found to be 38mm (RMSE).

The errors between the calculated and real source locations may be due to variability of the artificial AE sources, error in the measurement of actual source locations on a specimen, sound attenuation, and uncertainty of the wave propagation path. Additionally, the Integri-Tech™ introduced another source of error which is called mesh error. Since the model is built from a finite element mesh with a mesh size of (55 x 55) mm^2, the discretization error associated with the mesh density provides a limit to the source location accuracy as the AE event can only be located at a mesh node. The Integri-Tech™ reports the *Mesh Error* for each AE event. This is the difference between the measured time of arrivals, and the expected times of arrival for an event at the reported location calculated from the model geometry and the input velocity. Events are assigned to the node with the lowest error. Improved accuracy can be obtained by using a finer mesh, at the cost of increasing analysis times.

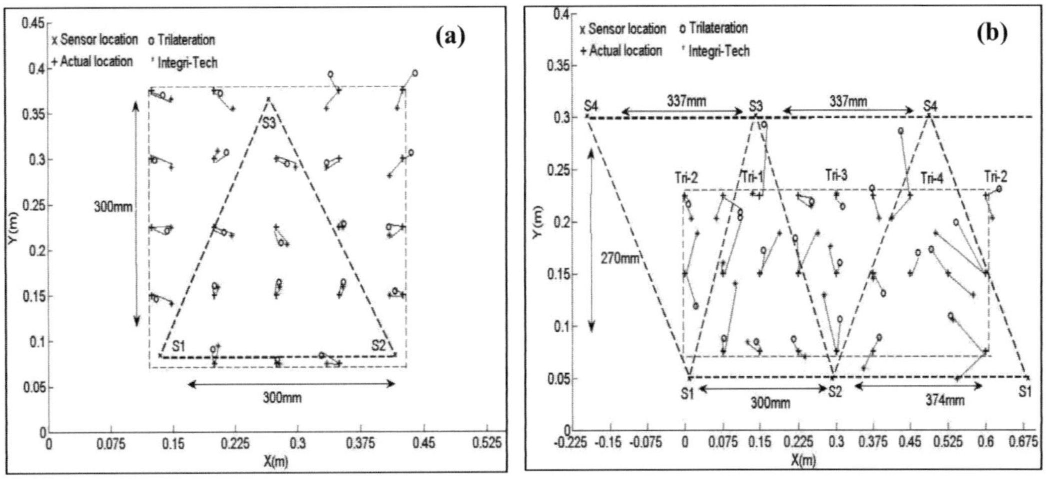

Figure 6. 2D Plate and Vessel models with estimated source locations using both algorithms

For both geometries, a final comparison table with selected AE events was then produced (Table 1 - 4). It is concluded that the Integri-Tech^{TM} has good capability for locating AE events. AE source location can be estimated more accurately using the Integri-Tech^{TM} by making reasonable changes in mesh setting.

Table 1. Comparison of source estimation using both algorithms for Vessel

Actual source location		Trilateration (MATLAB)		Integri-Tech^{TM}	
x(mm) ±10mm	y(mm) ±10mm	x(mm)	y(mm)	x(mm)	y(mm)
0	150	22	119	25	189
300	150	308	160	288	176
375	150	396	131	375	145

Table 2. Location error as a percentage of the distance from the source for Vessel

Actual source location		Trilateration (MATLAB)	Integri-Tech^{TM}
x(mm) ±10mm	y(mm) ±10mm	% of Error	% of Error
0	150	14	17
300	150	7	16
375	150	15	3

Table 3. Comparison of source estimation using both algorithms for plate

Actual source location		Trilateration (MATLAB)		Integri-Tech^{TM}	
x(mm) ±7mm	y(mm) ±7mm	x(mm)	y(mm)	x(mm)	y(mm)
0	150	18	146	23	150
150	150	156	133	163	131
225	150	231	154	229	150

Table 4. Location error as a percentage of the distance from the source for plate

Actual source location		Trilateration (MATLAB)	Integri-Tech™
x(mm) ±7mm	y(mm) ±7mm	% of Error	% of Error
0	150	8	10
150	150	9	12
225	150	3	2

7. Conclusion

Originally the Integri-Tech™ software was designed for remote integrity and fitness for service monitoring (vibration, strain and process), remaining life and inspection priorities. This work has demonstrated that the software can also be used to make reliable estimates of acoustic emission source location. This paper has presented the results of experiments aimed to verify source locations calculated by Integri-Tech™ multilateration with a standard trilateration technique coded in MATLAB based on Akaike information criterion (AIC) for on-set time calculation. Two different geometries were selected for this research work to locate AE events. A plate specimen was used in experiments with an array of three transducers to detect possible AE events. Similarly a vessel specimen was used with an array of four transducers to detect possible AE events. Recorded waveforms from H-N sources at each known location were processed in MATLAB using the AIC algorithm to find on-set time on each node. For the current work all tests were carried out in a lab environment, however for future use of AE system a high level of noise would be encountered in the resulting signals due to the environmental condition of pressurised engineering plant. For the plate sample RMS errors of 14mm were obtained in MATLAB vs 18mm in the Integri-Tech™. For the vessel sample RMS errors of 34mm were obtained in MATLAB vs 38mm in the Integri-Tech™. This close level of agreement provides confidence in the basic AE source location capabilities of the Integri-Tech™ software; however future work is essential to focus on testing the robustness of the technique for accurate operation in noisy environments.

8. Acknowledgements

The authors would like to acknowledge support for this project through the UK Knowledge Transfer Partnership scheme (Project reference number 8967).

9. References

[1] Holford, K. M. and Carter, D. C. Acoustic emission source location. *Key Eng. Mater.* , 1999, **167-168**, 162-171.
[2] Kurz, J., Grosse, C., Reinhardt, H., "Strategies for reliable automatic onset time picking of acoustic emissions and of ultrasound signals in concrete". *Ultrasonics* **43**, 538–546, (2005).
[3] Akaike H. Markovian "Representation of stochastic processes and its application to the analysis of autoregressive moving average processes". *Ann Inst Stat Math* 1974; **26**: 363–387.
[4] Kitagawa G and Akaike H. "A procedure for the modelling of non-stationary time series". *Ann Inst Stat Math* 1978; **30**: 351–363.
[5] Maeda, N. "A method for reading and checking phase times in auto-processing system of seismic wave data", *Zisin,* **38**, 365–379, (1985).
[6] Zhang, H., Thurber, C., Rowe, C., "Automatic p-wave arrival detection and picking with multiscale wavelet analysis for single-component recordings", *Bull. Seismol. Soc. Am.* **93**, 1904–1912, (2003).
[7] Yokota, T., Zhou, S., Mizoue, M., Nakamura, I., "An automatic measurement of arrival time of seismic waves and its application to an on-line processing system", *Bull. Earthq. Res. Inst.*,**55**, 449-484, (1981).
[8] Maeda, N. "A method for reading and checking phase times in auto-processing system of seismic wave data", *Zisin,* **38**, 365–379, (1985).
[9] P. Nivesrangsan, J.A. Steel, R.L. Reuben. "Source location of acoustic emission in diesel engines", *Mechanical system and signal processing.* 2007; **21**(2): 1103-14

A wavelet-based scheme for optimum measurement /monitoring of dynamic responses of structures

S H Mahdavi[1] and H A Razak[1]

[1] StruHMRS Group, Department of Civil Engineering, University of Malaya, 50603 Kuala Lumpur, Malaysia

Email: s.h.mahdavi@siswa.um.edu.my

Abstract. This paper presents an optimum measurement for the dynamic responses of structures from the operation of integration to respective derivation of acceleration's data using free-scaled wavelet functions. For this purpose, the numerical approach of integration and derivation has been developed for displacement measurement or determination of the third-order derivative (known as quantity of the jerk) from acquired accelerations. A simple improved algorithm is developed in order to optimally measure the dynamic quantities particularly, using Chebyshev and Haar wavelet functions. It is deduced that, stable measurement of dynamic quantities is independently achieved from the structural materials through a satisfactory optimum algorithm; that is capable of online monitoring, while emphasizing on maximum accuracy of the measurement with less computational time.

1. Introduction

In general, the design of mechanical systems is widely influenced by significant innovations in design technology. One of the most affected structural systems has emerged in civil engineering structures, which are increasingly being designed for optimized behaviours and optimum materials to achieve aesthetic and economic benefits. In fact, most of the time, various natural and environmental situations such as earthquakes, wind and tsunamis govern on these structures. As a result, the aforementioned structures are often subjected to abnormal ultimate loadings and may even collapse under such unexpected loadings that were not anticipated at the design stage. Therefore, it is indispensable to evaluate and monitor the structural reliability, integrity and safety [1].

Fundamentally, the main parameter being evaluated for the integrity analysis of a structure is displacement, however, several items of dynamic responses can be monitored. This is because the most important indicators of structural behaviour are short-term and long-term displacements, thus the integrity of structures can be examined for diverse options using displacement quantities. Accordingly, a sufficiently accurate displacement measurement is required to ensure the reliable monitoring of a structure's situation [2,3]. In addition, for many small-scaled mechanical systems, the precise measurement of other dynamic responses, e.g., velocity or the derivative of acceleration in respect to time (jerk) is essential for an engineering reliability analysis. Practically, the accelerometer is a widely utilized tool for measuring structural accelerations and displacements [4]. For instance, while the displacement is acquired by processing the acceleration data from the accelerometer, and the dynamic displacement and velocity seem precise enough, the process of filtering of pure accelerations can be implemented one step before the displacement measurement [5].

Recently, wavelet solution techniques have attracted much attention in different branches of science and technology. Basically, multi-resolution analysis and preserving time information during time-scaled-frequency analysis constitute the value of wavelet transforms in contrast to the well-known Fourier analysis [6-10]. Accordingly, wavelet functions have been utilized to numerically solve either ordinary or partial differential equations (ODEs, PDEs). Therefore, it is introduced as a powerful tool in structural dynamics. Moreover, free scales of wavelet functions have been implemented on the direct time history analysis of 2-dimensional 2D frames to calculate rotational and transitional displacements [11]. Furthermore, the direct approach (not inverse scheme) to structural dynamics of framed structures involving 3-dimensional (3D) transitional degrees of freedom has benefitted from using wavelet functions [12]. It was concluded that, as an indirect time integration method the dynamic analysis was optimally accomplished with the maximum accuracy of responses. The emphasis was on de-noising the input data and numerical integration scheme, simultaneously [12]. So far, the implementation of wavelet functions for the solution of invers problems through the aforementioned procedure is not received in the literature.

This paper is organized as follows. The next section provides a brief review of the background of wavelet, which includes 3D and complex Chebyshev wavelet functions and 2D and simple Haar wavelet basis. In Section 3, implementation of the proposed approach is considered for two main operations. For this aim, the proposed scheme is firstly employed for operation of integration to compute velocities and displacements from the acceleration data. Secondly, it is explored for operation of the derivation to calculate the third-ordered derivative of displacement in respect to time (namely, jerk). Finally, in Section 4, the effectiveness of the proposed scheme is experimentally evaluated.

2. Fundamentals of Chebyshev and Haar wavelet functions

As a prefix to this section, the basic notations and definitions of wavelet theory, continuous wavelet transforms (CWT) and characteristics of Haar and Chebyshev wavelet may be found in Refs. [6-16]. In this study, 2M is presented for the number of segmentations (wavelet collocation points) in each global time interval regarding to the proposed scales of wavelet (refers to the segmentation method, SM). Each function f(x) can be decomposed with wavelet transforms as [6-16]:

$$f(x) \approx \sum_{n=1}^{2M} \sum_{m=0}^{M-1} c_{n,m}\psi_{n,m}(x) = C^T\psi(x) \tag{1}$$

where, vectors C and $\psi(x)$, indicate expansion coefficients and wavelet coefficients, respectively. m is the degree of polynomials and n denotes the considered scale of wavelet. Subsequently, 2M squared matrix of Wavelet coefficients ($\phi_{(x)}$) are calculated for 2M local nodes x_l defined as follows [11,12]:

$$x_l = (2M^{-1})(l - 0.5), \;\; l = 1,2,3, \dots ,2M \tag{2}$$

It should be noted that, for 2D wavelets, the second expansion in equation (1) is ignored (m). Eventually, to calculate integration of f(x), the product matrix of integration (P) is operated on right side of equation (1) [6-11].

3. Implementation of the proposed scheme

3.1. Measurement of displacements from accelerations

A linear dynamic equilibrium governing on corresponding mass (M*), damping (C*) and stiffness (K*) of a structure can be expressed as [1,12]:

$$M^*\ddot{u}(t) + C^*\dot{u}(t) + K^*u(t) = F(t) \tag{3}$$

where, $\ddot{u}(t)$, $\dot{u}(t)$ and $u(t)$ represent the vectors of acceleration, velocity and displacement of each degree of freedom (DOF) existing in the structure. Furthermore, F(t) indicates the vector of nodal lateral loads applied to the degrees of freedom. In this study, converse to Mahdavi and Razak [11,12], where the direct dynamic analysis was considered, the $\ddot{u}(t)$ is assumed to be the known accelerometer's data (e.g., acquired with piezoelectric sensors). Thus, the unit of input data follows acceleration (m/sec^2). To numerically approximate velocities and displacements we first start to decompose the input data ($\ddot{u}(t)$) on the 2M local points of the wavelet basis, with emphasis on the scale of wavelet corresponding to the required accuracy of frequency decomposition of the input data. According to equation (1), the known vector of the accelerometer's data is numerically approximated using wavelet functions corresponding to the local fixed 2M points as follows (by any basis function):

$$\ddot{u}(t) = C^T \psi(t) \tag{4}$$

Note that, $\psi(t)$ indicates the corresponding wavelet coefficients matrix. For instance, for Haar wavelet the h(t) will be replaced in equations. Accordingly, the coefficient vector corresponding to the decomposed accelerations is obtained as:

$$C^T_{1\times 2M} = \ddot{u}(t)_{1\times 2M}(\phi_{2M\times 2M})^{-1} \tag{5}$$

The vector of velocity is therefore approximated on the global time, multiplying by the operation of integration (P), as follows ($d_t = t_{i+1} - t_i$):

$$\dot{u}(t) = d_t C^T P\psi(t) + v_n \tag{6}$$

Eventually, displacements are numerically expanded as follows:

$$u(t) = d_t^2 C^T P^2 \psi(t) + u_n \tag{7}$$

To transfer numerical calculations from 2M local fixed points of the wavelet window to the 2M global points of the considered time interval, $d_t$is operated as a coefficient of the aforesaid transformation. In other words, quantities of the dynamic system are modified corresponding to local times as:

$$\dot{u}(t) = d_t . v \tag{8}$$

$$\ddot{u}(t) = d_t . F(t_n + d_t . \tau , u , v) \tag{9}$$

Moreover, in equations (6) and (7), v_n and u_n represent the initial conditions of each global time, approximated by the wavelet as a constant value. For this aim, the unity is being approximated by the Chebyshev wavelet as:

$$1 \cong I^* \psi(t) \cong \left(\frac{\sqrt{\pi}}{2}\right) [1,0,0, \dots ,1,0,0, \dots]\psi(t) \tag{10}$$

Or alternatively is obtained with Haar wavelet as:

$$1 \cong I^* h(t) \cong [1,0,0, \dots ,0,0,0, \dots]h(t) \tag{11}$$

Therefore, the initial quantities are numerically developed as:

$$v_n = S_1^T \psi(t) \tag{12}$$

$$u_n = S_2^T \psi(t) \tag{13}$$

where, S_1^T and S_2^T are $2M \times 1$ dimension vectors that may be obtained by Chebyshev wavelet as follows:

$$S_1^T \cong v_{n(0)} \left(\frac{\sqrt{\pi}}{2}\right) [1,0,0, \dots ,1,0,0, \dots]^{\,T} \tag{14}$$

$$S_2^T \cong u_{n(0)} \left(\frac{\sqrt{\pi}}{2}\right) [1,0,0, \dots ,1,0,0, \dots]^{\,T} \tag{15}$$

Substituting equations (14) and (15) into equations (6) and (7) the vectors of velocity and displacement are defined as:

$$\dot{u}(t) = d_t C^T P\psi(t) + S_1^T \psi(t) \tag{16}$$

$$u(t) = d_t^2 C^T P^2 \psi(t) + d_t S_1^T P\psi(t) + S_2^T \psi(t) \tag{17}$$

The robustness of the proposed method is demonstrated through the two steps of computation; firstly, the quantities of displacements and velocities have been independently approximated from the mass, damping, stiffness and applied loadings of the considered system. Secondly, the accelerometers' data containing wide-frequency contents may be decomposed on the specific range of frequencies.

3.2. Measurement of derivative of accelerations with respect to time

Physically, the derivative of acceleration in respect to time introduces the quantity of jerk in the corresponding DOF. In following section, the proposed scheme is improved for approximation of the derivatives using free-scaled wavelet functions. For this purpose, integral functions are numerically developed from local coordinates to the global system. For a differentiable function of $\mathrm{F}(t)$: $t \in [t_n, t_{n+1}]$, calculation of integral on the interval of integration is considered based on Newton's theorem as follows:

$$f(t) = f(t_n) + \int_{t_n}^{t} f'(t)dt \tag{18}$$

Equivalently, from our point of consideration we have:

$$\ddot{u}(t) = \ddot{u}(t_n) + \int_{t_n}^{t} \dddot{u}(t)dt \tag{19}$$

For the first step $\ddot{u}(t) = \ddot{u}(t_{n+1})$ is initialized for a given t_{n+1}; Using equation (1) to approximate derivative function on global coordinates yields:

$$\dddot{u}(t) = C^T \psi(t) \tag{20}$$

Substituting into equation (19) yields:

$$\ddot{u}(t_{n+1}) = \ddot{u}(t_n) + \int_{t_n}^{t_{n+1}} C^T \psi(t)\ dt \tag{21}$$

Multiplying by operational matrices and adding initial constant of integration, we have:

$$\ddot{u}(t_{n+1}) = \ddot{u}(t_n) + d_t C^T P \psi(t) + \ddot{u}(t_0) \tag{22}$$

Using the idea proposed in equations (10), (14) and (15), constant quantities are also approximated by the corresponding wavelet in each step (Haar wavelet or Chebyshev wavelet). After substituting the initial quantities, equation (22) is developed as follows:

$$\ddot{u}(t_{n+1}) I^* \psi(t) \ = \ddot{u}(t_n) I^* \psi(t) + d_t C^T P \psi(t) + \ddot{u}(t_0) I^* \psi(t) \tag{23}$$

Subsequently, eliminating $\psi(x)$ from the both sides of equation (23) and after a set of algebraic calculations C^T is calculated. Accordingly, using equation (20) the 3rd ordered derivative of displacement (known as the jerk quantity) is numerically calculated.

4. Experimental verification

A simple SDOF model evaluated as an actual experiment under a regular hammer test. For the considered system, a comprehensive program code was codified in MATLAB, to measure online the dynamic responses. Furthermore, g= 9.81 m/sec^2 is presumed as the ground acceleration. As shown in Figure 1, the considered system including the two narrow columns are made of aluminum and a rigid Plexiglas, which are only vibrating in the x direction. To invoke only one DOF, the thickness of the columns was chosen to be much narrower than its width ($\cong 1/37$). The geometry and details of the experiment setup are shown in the figure. As illustrated in the figure, a non-contact laser device was installed on the base in order to identify the accurate time history of displacements (sampling rate 102.4).

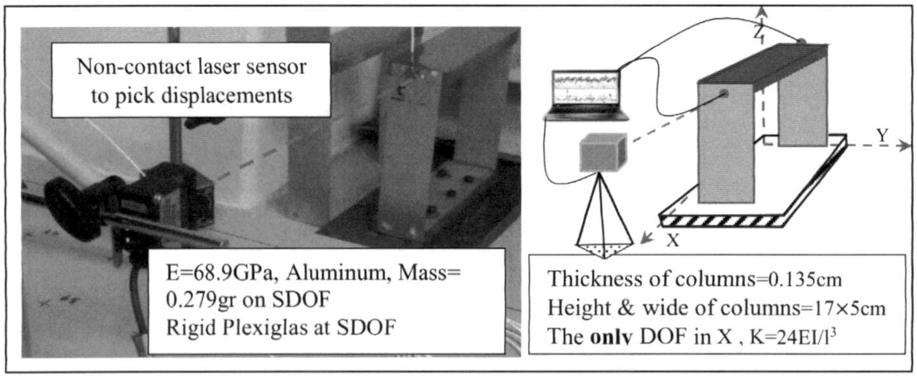

Figure 1. The ideal SDOF experiment setup and the schematic view, measuring displacements by laser sensor.

Moreover, two piezoelectric sensors (accelerometers) were embedded on the DOF of the system. To evaluate the proposed method, a hammer strikes the structure from 1sec after its initial condition.

Accordingly, the time history of the velocity corresponding to the SDOF has been computed for the 4th scale of Haar and Chebyshev wavelet (designated by Cal-Vel-Cheby and Haar) for the time interval of d_t=0.05 sec as depicted in Figure 2. The results were compared against those picked by the non-contact laser sensor for d_t=0.009766 sec (designated by Orig-Vel).

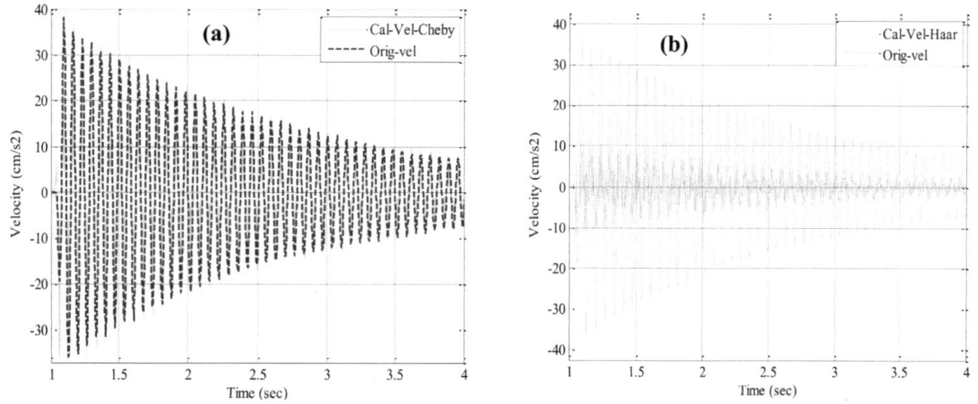

Figure 2. Time history of velocity for SDOF system shown in Figure 1. (a) Calculated velocity using the 4th scale of Chebyshev wavelet (Cal-Vel-Cheby). (b) Calculated velocity using the 4th scale of Haar wavelet (Cal-Vel-Haar).

To clarify the discrepancy between the computed results through the proposed method, approximation of the original acceleration (accelerometer data) was plotted corresponding to the 4th scale of Haar and Chebyshev wavelet by illustrating the wavelet coefficients for 1-1.2 sec.

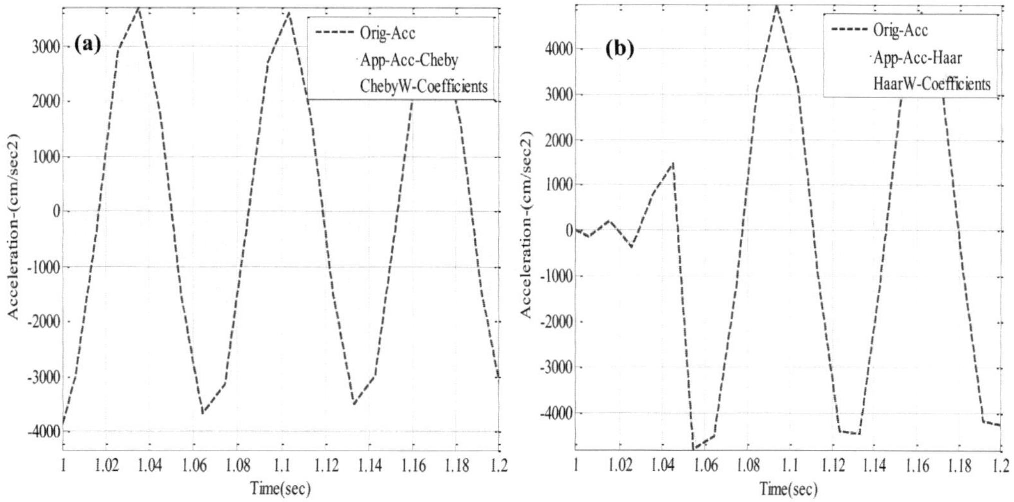

Figure 3. Coefficients of wavelet for the 4th scale of (a) ChebyW-Coefficients=Chebyshev wavelet (b) HaarW-Coefficients=Haar wavelet (App-Acc= approximation of original acceleration).

Figures 2 and 3 clearly demonstrate the source of significant errors of Haar wavelet (2M=4) for d_t=0.05 sec; while it is shown that details of the acceleration are accurately captured by the coefficient of Chebyshev wavelet for the same conditions, therefore, a sufficiently precise approximation is satisfied at the first step of the proposed scheme. Furthermore, to evaluate the practice of various scale of Haar wavelet in detail, computed responses for the same d_t=0.05 sec and 8th, 32nd and 64th scales of simple

Haar wavelet (designated by HA(2M)) were compared with the 4th scale of complex Chebyshev wavelet. The results depicted in Figure 4 are from the computational time involved and percental total average error (PTAE) point of view.

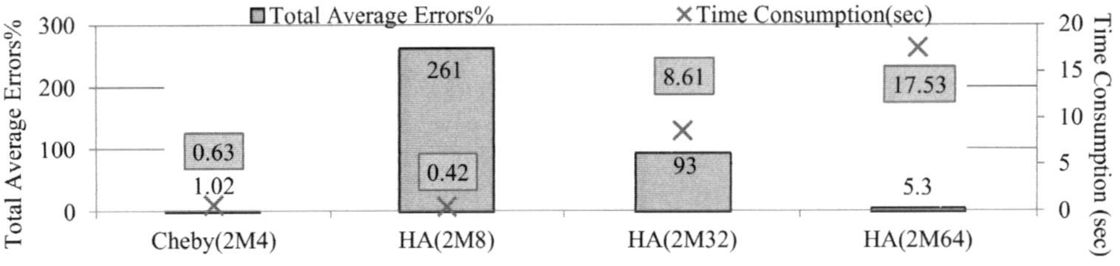

Figure 4. Computational time and PTAE for various scale of Haar (HA) and the 4th scale of Chebyshev wavelet (Cheby).

It is clearly distinguishable from Figure 4 that for the accurate scales of 2D and simple Haar wavelet (2M64) the cost of analysis is dramatically increased; however, the optimum results can be computed even using a low scale of 3D and complex Chebyshev wavelet. Consequently, it is shown that practice of Haar wavelet is not applicable for the online measurement of dynamic responses, however, for an initial evaluation it may be considered where the fastest computation is satisfied by implementation of this basis function. Eventually, the time history of displacement of the only DOF is optimally calculated through the proposed optimal measurement scheme, particularly, using a long time interval of d_t=0.05 sec by the 4th scale of Chebyshev wavelet (designated by Cal-Disp-Cheby) and is compared with those measured by the laser sensor (Orig-Disp) for the sampling rate of 102.4 (shown in Figure 5).

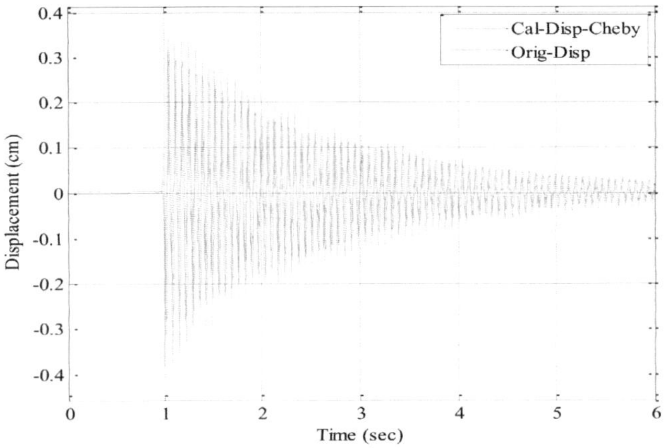

Figure 5. The first 6 sec time history displacement of SDOF system shown in Figure 1 (Cal-Disp-Cheby = Calculated displacement using Chebyshev wavelet, Orig-Disp=Original displacement).

The capability of the proposed algorithm of measurement is overtly illustrated in Figure 5, using low scales of Chebyshev wavelet. It should be pointed out again that details of the broad-frequency component acceleration are completely captured by the diverse scale of this comprehensive wavelet function in contrast to the simple Haar wavelet. Finally, the third derivative of displacement in respect to time namely, Jerk ($\ddot{u}(t)$) is considered using the 8th scale of Chebyshev wavelet (designated by computed jerk), as depicted in Figure 6 and compared with a normal incremental calculation (d(Acc)/dt) for the long interval of d_t=0.5 sec. Note that the quantity of jerk is often considered for very small scaled or flexible systems. Moreover, measurement of this value is very helpful to make an engineering judgment on the behaviour of a dynamic system.

Figure 6 shows that the proposed method optimally and exactly detects the time of the strike at the first second in contrast with the normal incremental calculation. This is because the time interval of d_t=0.5 sec (d_n=d_t for incremental computations) does not allow an incremental approach to detect an abrupt derivative of acceleration in respect to time. On the other hand, the 8th scale of Chebyshev wavelet, firstly, saves an accurate decomposition of the broad-frequency of acceleration even for d_t=0.5 sec for the collaborative collocation points, and, secondly, approximates the sharp incremental acceleration. Overall, it is demonstrated that the quantity of jerk as one of the reliable criteria of dynamic behaviour is optimally measured through the proposed which is capable of optimal and online measurement of dynamic responses.

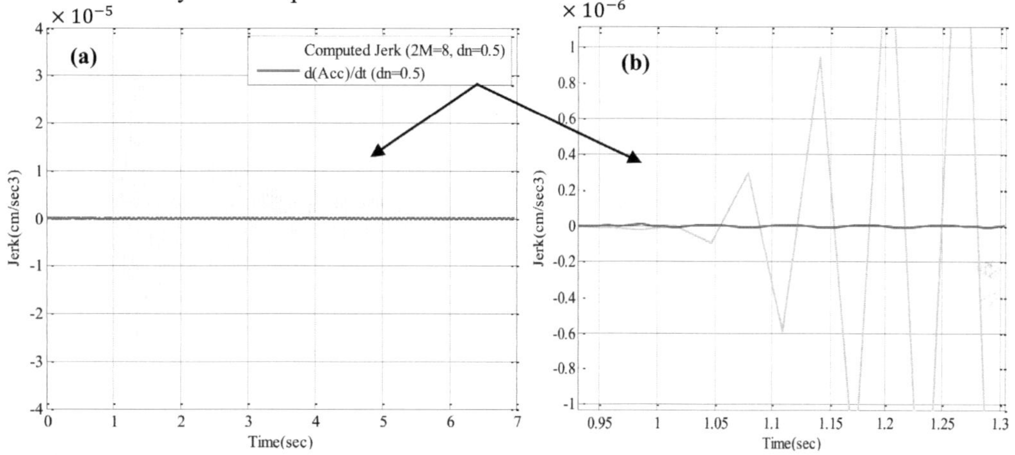

Figure 6. The first 7 sec analysis of the third derivative of displacement (jerk). (a) 0-7 sec Computed jerk using the 8th scale of Chenyshev wavelet vs. normal calculation of the derivative with respect to time (d(Acc)/dt). (b) Zooming plane on 0.9-1.3 sec.

5. Conclusions

This paper introduces a new numerical integration and derivation scheme to optimally measure the dynamic responses of structures in order to monitor structural integrity and reliability. A straightforward algorithm is implemented to measure particularly, displacements with high precision by tracking the existing acceleration data on the structure without requiring the installation of special and extra devices of displacement measurement with very expensive costs and difficult access (i.e., laser sensors or displacement transducers). It is shown that, the dynamic responses are independently obtained from the behavior of the materials, e.g., mass, damping or stiffness. Furthermore, it is demonstrated that this method is also capable of online monitoring in which the optimum and stable calculation have been conducted from de-noising and numerical integration or derivation in a similar sense of analysis for small sampling rates. However, in this study, the practice of free-scaled and simple 2D Haar wavelet and complex and 3D Chebyshev wavelet have been comparatively considered; the implementation of several basis functions is recommended for further study to achieve an adaptive measurement scheme.

Acknowledgement

The authors wish to acknowledge financial supports by the University of Malaya (UM) and Ministry of Education Malaysia (Grant No.: PG078/2013B and UM.C/625/1/HIR /MOHE/ENG/55).

References

[1] Balageas D, Fritzen C P and Guemes A 2010 *Structural Health Monitoring* (New York: John Willy Sons).
[2] Chan W S, Xu Y L, Ding X L and Dai W J 2006 An integrated GPS-accelerometer data processing

technique for structural deformation monitoring *J. Geodesy* **80** 705-19.

[3] Tzou H S and Tseng C I 1990 Distributed piezoelectric sensor/actuator design for dynamic measurement/control of distributed parameter systems: a piezoelectric finite element approach *J. Sound Vib.* **138** 17-34.

[4] Muenchhof M and Singh T 2003 Jerk limited time optimal control of flexible structures *J. Dyn. Syst. Measure. Control.* (Transactions-American Society of Mechanical Engineers) **125** 139-42.

[5] Basu B, Nagarajaiah S and Chakraborty A 2008 Online identification of linear time-varying stiffness of structural systems by wavelet analysis *Structural Health Monitoring* **7** 21-36.

[6] Mahdavi S H and Shojaee S 2013 Optimum time history analysis of SDOF systems using free scale of Haar wavelet *Struct. Eng. Mech. Int. Journal* **45** 95-110.

[7] Babolian B and Fatahzadeh F 2010 Numerical solution of differential equations by using Chebyshev wavelet operational matrix of integration *Appl. Math. Comput.* **188** 417-426.

[8] Cattani C 2004 Haar wavelets based technique in evolution problems *Proc. Estonian Academy of Sciences Phys. Math.* **53** 45-63.

[9] Lepik U 2008 Haar wavelet method for solving higher order differential equations *Int. J. Math. Comput.* **1** 84-94.

[10] Lepik U 2009 Haar wavelet method for solving stiff differential equations *Int. J. Math. Comput.* **14** 467-81.

[11] Mahdavi S H and Razak H A 2014 Optimum dynamic analysis of 2D frames using free-scaled wavelet functions Latin American J. Solids Struct. **11** 1036-48.

[12] Mahdavi S H and Razak H A 2013 A wavelet-based approach for vibration analysis of framed structures *Appl. Math. Comput.* **220** 414-28.

[13] Mason J C and Handscomb D C 2003 *Chebyshev Polynomials* (Chapman and Hall/CRC).

[14] Mahdavi S H and Razak H A 2015 An indirect time integration scheme for dynamic analysis of space structures using wavelet functions *J. Eng. Mech. (ASCE)* In Press (DOI: 10.1061/(ASCE) 13 EM.1943-7889.0000914).

[15] Mahdavi S H and Razak H A 2015 A comparative study on optimal structural dynamics using wavelet functions *Mathematical Problems in Eng.*

[16] Mahdavi S H and Razak H A 2015 An efficient iterative scheme using family of Chebyshev's operations *Mathematical Problems in Eng.*

Research on assessment of bolted joint state using elastic wave propagation

R. Kędra[a], **M. Rucka**[1,b]
[a]Department of Metal Structures and Management in Civil Engineering, Faculty of Civil and Environmental Engineering, Gdańsk University of Technology, Poland
[b]Department of Structural Mechanics, Faculty of Civil and Environmental Engineering, Gdańsk University of Technology, Poland

E-mail: mrucka@pg.gda.pl

Abstract. The work contains results of experimental investigation of elastic wave propagation in a bolted single-lap joint. Tests were carried out for the excitation perpendicular to the connection plane. In experimental studies, PZT transducers were used for both excitation and registration of ultrasonic waves. The analyses took into account varying contact conditions between the elements of the connection depending on the value of the prestressing force. The influence of loosening/tightening of bolts on the energy dissipation was analysed. The experimental results proved the influence of bolt torque on quantitative characteristics of the signals. To improve the diagnostic possibilities only the initial parts of signals were analysed.

1. Introduction

Joints are integral parts of steel structures, which allow formation of subtle and unconventional structures of buildings. However, due to abrupt stiffness changes, stress concentrations, technical difficulties and susceptibility to atmospheric corrosion, connection points are the places most vulnerable to damage and they are relatively frequently subjected to failures. In building industry welded and bolted joints are the most commonly used forms of connections. Less prevalent are riveted and glued joints. In the group of bolted connections, two types of joint can be distinguished, namely standard and prestressed joints. The second of those subgroups is recently becoming more popular. The main reasons, and also the biggest advantages of prestressed bolted joints, are durability and high load capacity. Besides, they are easy to assemble and do not require expensive equipment during mounting. However, prestressed bolted joints require thorough preparation of connected surfaces and providing a controlled bolt torque value. Additionally bolts may be subjected to loosening under time-varying external loads. This results in the need for continuous monitoring of joints, which are especially important for structural reliability. Therefore in recent years bolted joints have become an area of intensive research associated with diagnostics, in particular in the context of ultrasonic waves. Methods based on the propagation of elastic waves are successfully used to detect defects in many types of structural components, especially in plate-like structures [1–4]. They allow for non-destructive assessment of structural condition. Moreover, they are relatively inexpensive and allow testing of inaccessible areas of structures.

[1] To whom any correspondence should be addressed

2. Diagnostics of bolted joints using elastic waves

One of the first approaches to diagnosis of bolted connections using elastic waves was developed by Mita and Fujimoto [5]. They developed an algorithm using the support vector machine which allows to detect loosening of bolts and damage localisation. Similar methodology is presented in [6]. Park et al. described the technique to detect defects simulated as loosening of selected connectors by the means of the time of flight and coefficients based on the wavelet transform. Additionally they used neutron probabilistic networks and support vector machines to assess the state of the connection. Experimental validation of the proposed method showed its high effectiveness in detecting failures of bolts. Yang and Chang [7, 8] proposed attenuation-based system of monitoring of bolted joint in composite thermal protection panels. Doyle et al. [9] showed changes in the phase of registered signals for differing values of a bolt torque in the context of monitoring of space structures. An and Sohn [10] presented a combination of methods using impedance and elastic wave techniques. A completely different approaches based on nonlinear acoustic phenomena are presented by Amerini and Meo [11]. The authors developed three indicators for three different ranges of excitation frequency. These indicators were based on the energy of the signal in the frequency domain, represented by the acoustic moment, sidebands in frequency spectrum and second harmonic generation. Wang et al. [12] investigated the effect of tightening torque on the signal energy in the time domain.

The aim of this study is to analyse the possibilities of using quantitative indicator based on signal energy of propagating elastic waves for assessment of the state of a steel lap bolted joint. The effects of sensor positions, excitation frequency and duration of the analysed time interval were examined.

3. Experimental case studies

3.1. The bolted joint model and experimental set-up

The object of the research was a single lap bolted joint (Figure 1). The model consisted of two steel plates with dimensions 6.2 mm × 59.8 mm × 404 mm which were assembled using two steel bolts of class 5.6. and diameter of 12 mm. The joint was mounted in a steel frame to simulate fixed-fixed boundary conditions and ensure a stable position during tightening and loosening process. The excitation and measurements of elastic waves were carried by the device PAQ-16000D and piezoelectric transducers Noliac NAC2024 with dimensions of 3 mm × 3 mm × 2 mm attached to the structure at selected points. The excitation had forms of five-peak sine modulated by the Hanning window. The output voltage signals were measured with a sampling frequency of 2 MHz. In order to reduce the influence of noise, each measurement was repeated 100 times and averaged.

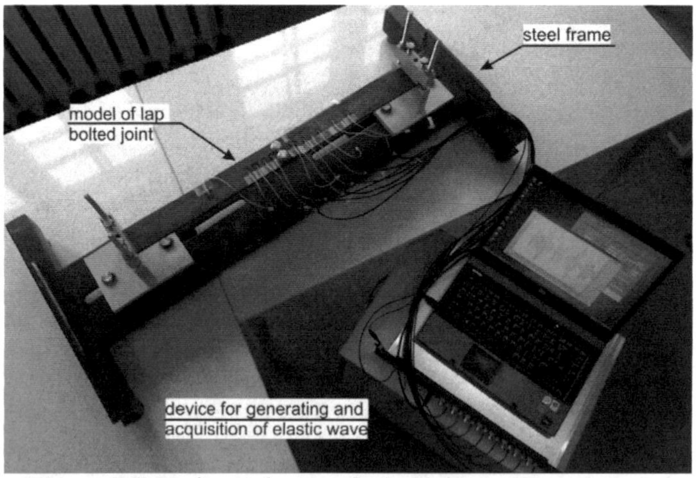

Figure 1. Experimental set-up for lap-joint measurements

3.2. Research program

Ultrasonic waves were excited at one point, at the distance of 8 cm from the axis of connection. Measurements were performed at 8 points located both before and behind the connection area. The arrangement of sensors is shown in Figure 2. Research was carried out for different values of the wave frequency and the bolt torque. The central frequency of excitation varied from 60 kHz to 120 kHz with step of 20 kHz. The controlled value of the bolt torque was ensured by using a torque wrench. The study included variation of the bolt torque in the range from 0 Nm to 100 Nm.

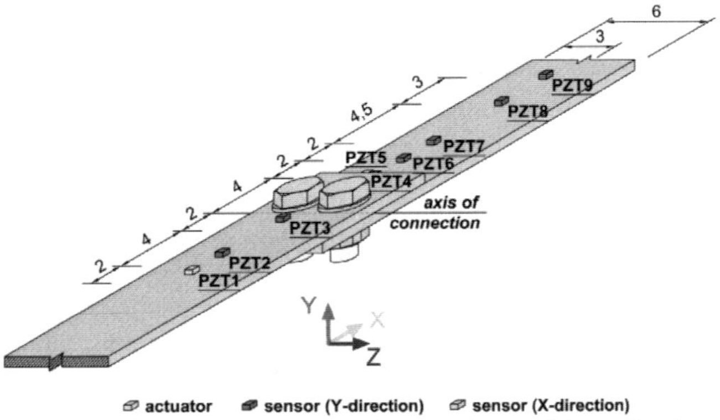

Figure 2. Sensors arrangement in experimental investigation

3.3. Results

The experimental results showed a significant impact of the bolt torque value on the characteristics of ultrasonic waves propagating in the tested joint. Differences between wave propagation signals were caused mainly by the changes of the contact area between the plates in the micro and macro scale. On macro level, this effect is caused by the variations of the radius of area around the bolt where two plates interact with each other [13]. In the micro-scale this phenomenon is associated with the change of the true contact area in dependence on the value of the contact stress and the surface roughness. The relationship between the nominal and true contact area for sinusoidal wavy surface using Hertz contact theory was discussed in paper [7]. Illustration of the changes of the contact surface is shown in Figure 3.The increase of the true contact surface also reduces the dissipation of energy at the interface between the connected plates. Less influence on variability of registered waveforms has the stress and tension state due to bolt tightening process.

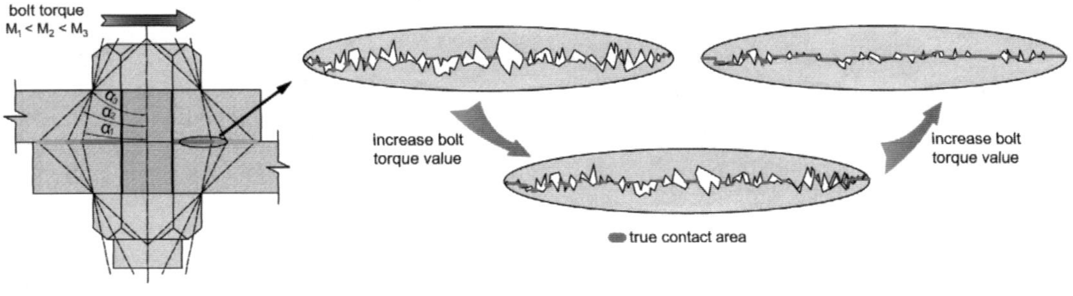

Figure 3. Changes of nominal and true contact area with increasing bolt torque value

The examples of ultrasonic signals acquired by sensors PZT5 and PZT9, for the excitation with the central frequency 60 kHz and 100 kHz, are presented in Figures 4, 5 and 6. It can be seen that there are significant differences between signals registered for different values of bolts pre-load. The registered waveform signals are similar only at the initial time interval, namely 0–0.11 ms for PZT5 and

0–0.15 ms for PZT9. However, despite of resemble in the shape, a slight shift of the registered signals occurred with the increase of the torque value (cf. Figure 5, 6). Analogical results were obtained for the other piezoelectric sensors.

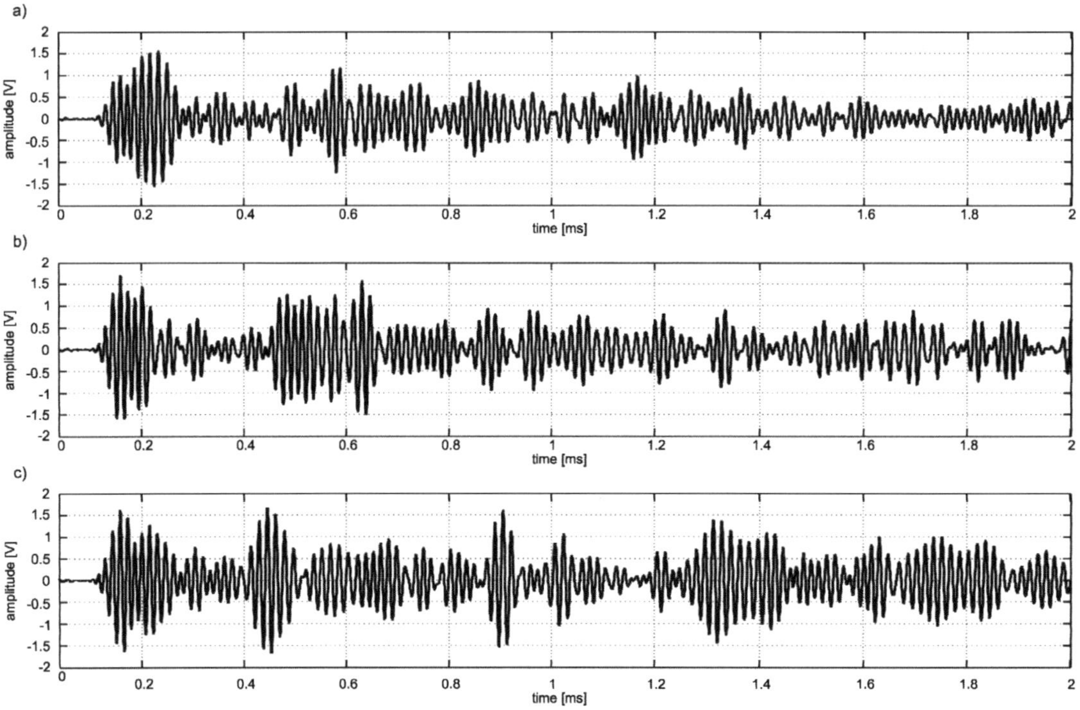

Figure 4 Output voltage signal registered in the case of excitation frequency 60 kHz by PZT9 for different value of bolt torque: a) 20 Nm; b) 60 Nm; c) 100 Nm

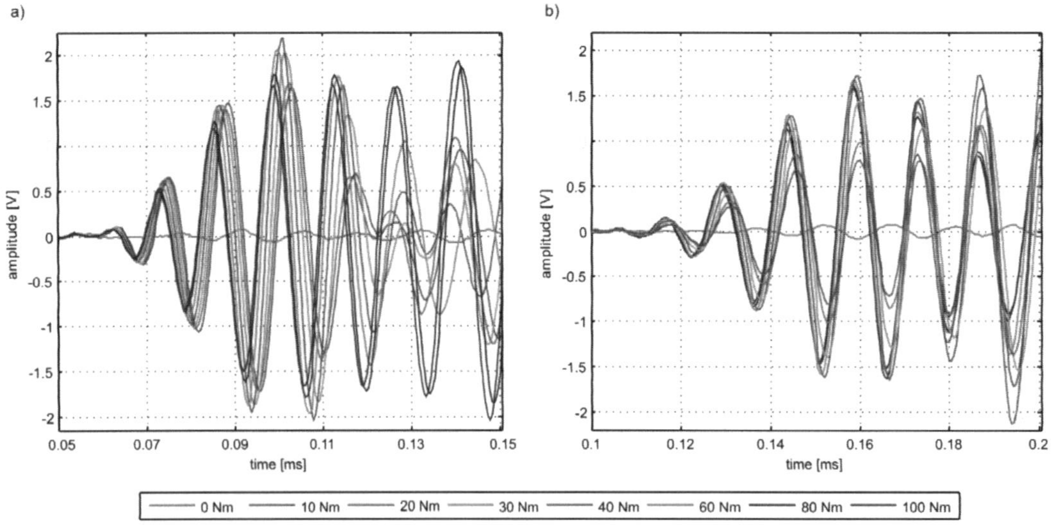

Figure 5. Wave propagation signals for different values of bolt torque in the case of excitation frequency 60 kHz registered by: a) PZT5; b) PZT9

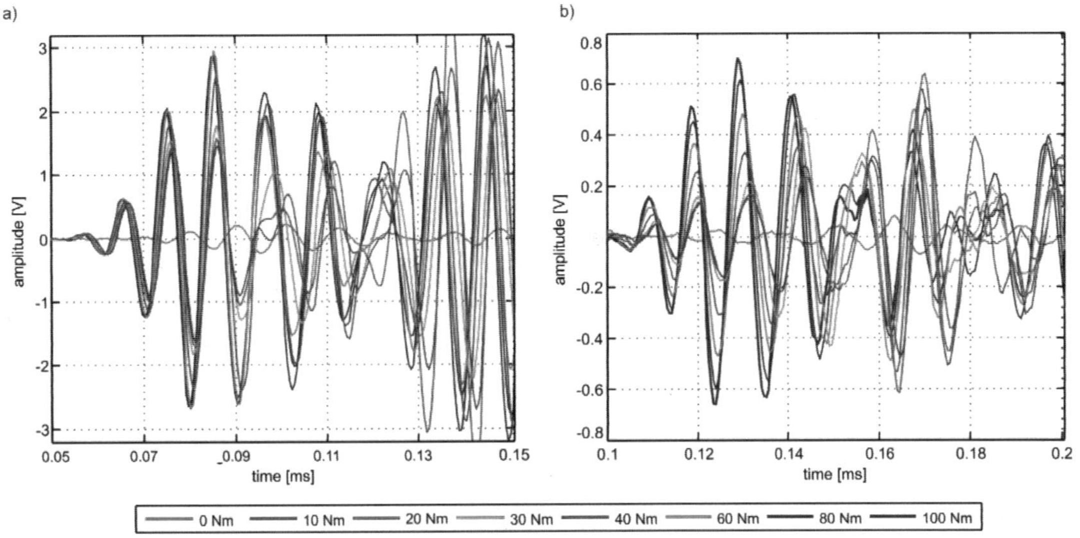

Figure 6. Wave propagation signals for different values of bolt torque in the case of excitation frequency 100 kHz registered by: a) PZT5; b) PZT9

Figure 7 shows the time-frequency characteristics of the signal registered by sensor PZT9 for excitation with the central frequency of 60 kHz. The values of the signal spectrum increased with increasing bolt torque value. Simultaneously, the spectrum was flattened, and few side lobes with similar amplitude appeared. In addition, registered waveforms were highly contaminated by reflections from the edges of the bolted joint model. As a result, the qualitative comparison of signals in wide time range was extremely difficult. Therefore, in this study the energy of signals has been proposed as a quantitative characteristic directed to diagnostic purposes of bolted joints.

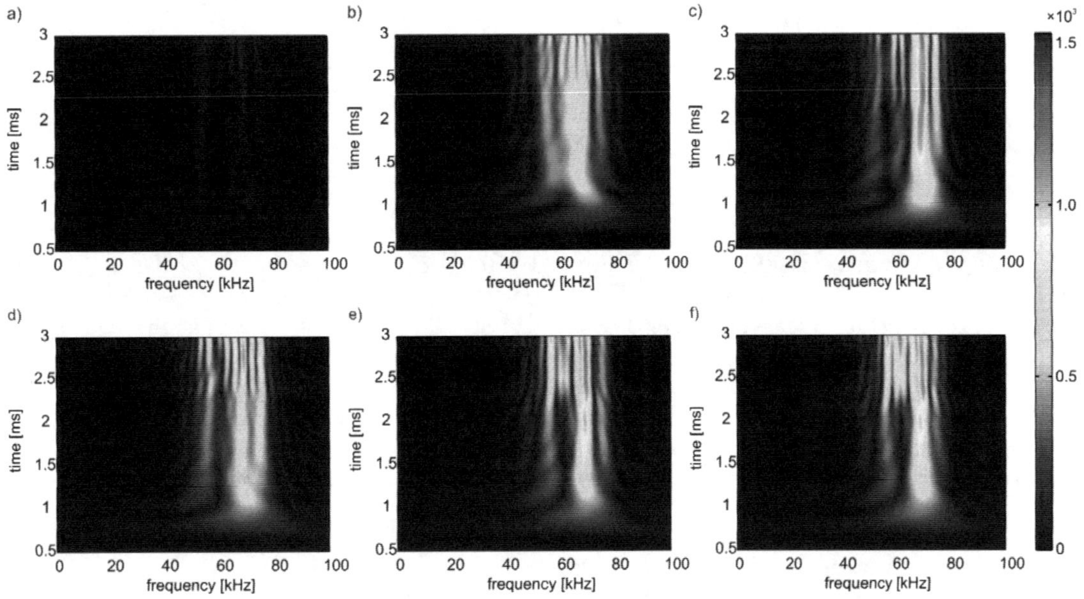

Figure 7. Time-frequency characteristics of signals registered by PZT9 in case of excitation frequency 60 kHz and bolt torque value: a) 0 Nm; b) 20 Nm; c) 40 Nm; d) 60 Nm; e) 80 Nm; f) 100 Nm

The energy of a signal can be defined for continuous and discrete signals in both time and frequency domains. In the theory of signals, the energy of discrete signal can be written as [14]:

$$E_x = \sum_{n=n_0}^{\infty} x_n^{\ 2}, \tag{1}$$

where x_n denotes the value of discrete signal at a given time point. In Equation (1) summation occurs in the infinite range, which is impossible to carry out for experimentally acquired signals. The problem disappears if the recorded signal is periodic. However, it does not occur in this case. To overcome this limitation, we must assume that outside the specified range $\langle n_0;n_1 \rangle$ the signal has only a zero or very close to zero values, so the signal energy in the range $\langle n_1;+\infty \rangle$ is negligible. It is obvious that due to the physical nature of a waveform induced by a single wave packet its amplitude after a certain time significantly decreases. An additional difficulty is the noise in the signal, whose influence can be reduced by using averaging or filtering techniques. After determination of the suitable registration time, the analysed wave signal becomes finite, hence its energy is finite and calculable.

Figure 8 shows the changes of the selected signal energies depending on the time range of the measurement. It may be noted that after significant initial increase, the energy of all signals was stabilized. In the time range of 5–10 ms energy values rose slightly and further measurement was therefore unnecessary.

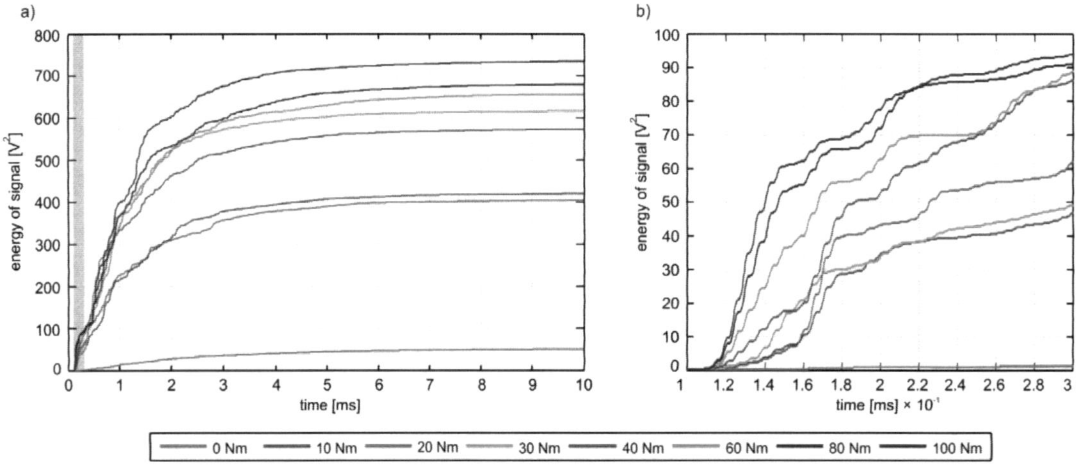

Figure 8. Signal energy changes depending on the measurement time in the case of excitation frequency 100 kHz for sensor PZT9 in the range: a) 0–10 ms; b) 0.1–0.3 ms

Table 1 presents the values of the signal energy for all PZT-sensors at the excitation frequency of 100 kHz normalized with respect to the maximum value obtained for relevant sensor. Only in the case of sensor PZT9 a stable trend between the values of the signal energy and the bolt torque can be observed. For sensors located in the lap area, the decline of the maximum signal energy with the increasing value of prestressing force can be observed. In the cases of excitation frequencies of 60 kHz, 80 kHz and 120 kHz, there was no clear relationship between the signals energy and the bolt torques for any of sensors. It demonstrates that in the case of evaluation of the joint state based on the signal energy over the entire range of its variability, it is necessary to determine accurately both the position of the sensor and the excitation frequency.

Another approach is to assess the state of the connection only on the basis of the energy of the part of the signal, at the certain time internal. In this case, the greatest difficulty was to determine the end of the analysed range. Based on the analysis of the variability of the recorded waveform it was found that analysed time should correspond to half duration of the excitation time. For this purpose, a cross-correlation function of Hilbert spectrums of the excitation signal and waveforms recorded by sensor PZT9 was used. Results obtained in this manner are summarized in Table 2. In this case, the relationship between the increase of the bolt torque and the signal energy in the analysed time range can be observed for all sensors situated outside the connection area except PZT2.

Table 1. Energy of signals registered by various PZT sensors for excitation frequency of 100 kHz

	normalised signal energy [-]							
bolt torque [Nm]	0	10	20	30	40	60	80	100
PZT2	1,0	0,832	0,852	0,892	0,921	0,887	0,926	0,968
PZT3	1,0	0,717	0,703	0,770	0,644	0,651	0,743	0,710
PZT4	1,0	0,613	0,511	0,417	0,466	0,564	0,515	0,499
PZT5	0,059	0,703	0,631	1,0	0,983	0,790	0,861	0,914
PZT6	0,076	0,658	0,752	0,755	1,0	0,832	0,899	0,843
PZT7	0,042	0,768	0,644	0,901	1,0	0,903	0,900	0,824
PZT8	0,072	0,631	0,557	0,675	1,0	0,924	0,933	0,869
PZT9	0,069	0,572	0,551	0,839	0,780	0,893	0,925	1,0

Table 2. Energy of initial part of signals registered by various PZT sensors for excitation frequency of 100 kHz

	normalised signal energy [-]							
bolt torque [Nm]	0	10	20	30	40	60	80	100
PZT2	0,992	0,994	0,995	0,999	1,0	0,999	0,996	0,997
PZT3	0,508	1,0	0,975	0,827	0,858	0,934	0,927	0,891
PZT4	0,658	0,795	0,810	0,574	0,655	1,0	0,841	0,721
PZT5	0,001	0,370	0,397	0,450	0,628	0,883	0,970	1,0
PZT6	0,001	0,425	0,410	0,510	0,705	0,855	0,959	1,0
PZT7	0,001	0,638	0,611	0,622	0,815	0,938	0,982	1,0
PZT8	0,002	0,166	0,161	0,222	0,394	0,716	0,910	1,0
PZT9	0,004	0,048	0,049	0,075	0,198	0,451	0,758	1,0

4. Conclusions

The paper contains experimental investigations of elastic wave propagation in a laboratory model of the single lap bolted joint. The research focused on the changes in the waveforms registered by piezoelectric sensors attached to the structure at several locations. Quantitative signal changes were analysed. It has been shown that based on the signal energy over the entire range of its variability only the full loosening state can be detected. In order to improve the diagnostic possibilities the initial part of signals was analysed. To find the limits of the range, which should be considered, a cross-correlation function of Hilbert spectrums of the excitation signal and waveforms was used. In this case, for sensors located on the opposite side of the joint with respect to the actuator, the regular increase of the signal energy with increasing bolt torque was observed.

References

[1] Mallet L., Lee B. C., Staszewski W. J., Scarpa F. Structural health monitoring using scanning laser vibrometry: II: Lamb waves for damage detection. Smart Materials and Structures, vol. 13, 2004, pp 261–269.

[2] Żak A., Radzieński M., Krawczuk M., Ostachowicz W. Damage detection strategies based on propagation of guided elastic waves. Smart Materials and Structures, vol. 21, no. 3, 2012, Article ID 035024.

[3] Malinowski P., Wandowski T., Ostachowicz W. Damage detection potential of a triangular piezoelectric configuration, Mechanical Systems and Signal Processing, vol. 25, issue 7, 2011, pp. 2722–2732.

[4] Rucka M. Modelling of in-plane wave propagation in a plate using spectral element method and Kane-Mindlin theory with application to damage detection. Archive of Applied Mechanics, vol. 81, 2011, pp. 1877–1888.

[5] Mita A., Fujimoto A. Active detection of loosened bolts using ultrasonic waves and support vector machines. Proceeding of the 5th International Workshop on Structural Health Monitoring Stanford University Stanford, 2005, pp. 1017–1024.

[6] Park S.-H., Yun C.-B., Roh Y. PZT-induced Lamb waves and pattern recognitions for on-line health monitoring of jointed steel plates. Proceedings of SPIE-Smart Structures and Materials vol. 5765, 2005, pp. 364–375.

[7] Yang J., Chang F.-K. Detection of bolt loosening in C-C composite thermal protection panels: I. Diagnostic principle. Smart Materials and Structures, vol. 15, 2006, pp. 581–590.

[8] Yang J., Chang F.-K. Detection of bolt loosening in C-C composite thermal protection panels: II. Experimental verification Smart Materials and Structures, vol. 15, 2006, pp. 591–599.

[9] Doyle D., Zagrai A., Arritt B., Çakan H. Damage detection in bolted space structures. Journal of Intelligent Material Systems and Structures, vol. 21(3), 2010, pp. 251–264.

[10] An Y.-K., Sohn H. Integrated impedance and guided wave based damage detection. Mechanical Systems and Signal Processing, vol. 28, 2012, pp. 50–62.

[11] Amerini F., Meo M. Structural health monitoring of bolted joints using linear and nonlinear acoustic/ultrasound methods. Structural Health Monitoring, vol. 10, 2011, pp. 659–672.

[12] Wang T., Song G., Wang Z., Li Y. Proof-of-concept study of monitoring bolt connection status using a piezoelectric based active sensing method. Smart Materials and Structures, vol. 22, 2013, Article ID 087001.

[13] Huda F., Kajiwara I., Hosoya N., Kawamura S. Bolt loosening analysis and diagnosis by non-contact laser excitation vibration tests. Mechanical Systems and Signal Processing, vol. 40, 2013, pp. 589–604.

[14] Mandal M., Asif A. Continuous and Discrete Time Signals and Systems. Cambridge University Press, 2007.

Wave propagation in damage assessment of ground anchors

B. Zima[1] **and M. Rucka**
Department of Structural Mechanics, Faculty of Civil and Environmental Engineering, Gdańsk University of Technology, Poland

E-mail: beazima@pg.gda.pl

Abstract. The inspection possibilities of ground anchors are limited to destructive test such as pull-out test. Guided wave propagation gives an opportunity to develop an inspection system dedicated to determine the condition of inspected element without violation of their integrity. In this paper the experimental study on wave propagation in laboratory models of ground anchors are presented. Experiments were conducted for different bonding lengths and different frequencies of excitation. Waves were generated by a piezoelectric actuator and the laser vibrometry technique was used to register velocity signals. For all tested anchors it was possible to identify the boundary between steel and concrete based on the registered reflections in wave propagation signals.

1. Introduction

The primary task of reinforced concrete elements, which are widely used in civil engineering is a significant load transfer achieved by the cooperation of concrete and reinforcing steel. Adhesion of concrete to steel is the most important component of load-bearing capacity of the element. The cooperation of steel and concrete in reinforced concrete structures is possible due to the binding, which consists of adhesion between steel and concrete, friction of steel on concrete clamping around due to shrinkage and mechanical resistances associated with the shaping surface of the bar and its configuration. The main problems of reinforced structures are slip or separation of two materials under load and in consequence, decrease of the load capacity. Moreover, it is an important issue because the visual assessment of the state is very often impossible after performing the item. Load capacity is determined during destructive testing which is based on applying gradually increasing load until destruction of tested structure. This method is very expensive, time consuming and requires performing elements, to be destroyed. Another drawback is the lack of information on other elements. Load capacity and workmanship of the group of elements is deduced on the base of investigation for one or few of them. For these reasons non-destructive diagnostic methods based on elastic waves propagation phenomenon have become the subject of intensive study over several last decades. In recent year, a large number of papers have been published on the use of guided waves for non-destructive evaluation (NDE) and structural health monitoring (SHM) and many of investigators have studied the guided wave propagation in various types of structures and materials, e.g. [1–10]. Wave propagation techniques offer the advantage of fast diagnostics of structural elements without violation of their integrity. They are totally adapted to automated inspection of large areas with a small number of sensors. That is the reason why methods based on wave propagation are very promising solutions for advanced monitoring systems or diagnostics of structures which cannot be assessed visually.

[1] To whom any correspondence should be addressed.

Very important group of relatively simply engineering structures made of a steel rod or tendon and concrete are ground anchors, which are used to support geotechnical structures. The tendon is a long element transferring only tensile force into deeper layers of the ground. The main element of the ground anchor is the anchor body. The compression of the soil causes resistance to movement of the whole element. Most common failures of the anchor systems are failures of the steel tendon caused by corrosion or failures of anchor body caused by weak condition of performed injected element [11]. Guided wave propagation gives an opportunity to assess the state of these elements without performing destructive tests called pull-out tests.

The study focuses on the identification and recognition of the phenomena, which occur during wave propagation in laboratory models of ground anchors with different bonding lengths. Different excitation frequencies have been tested with the aim to select the most sensitive frequency for detection of border of media and determination of the bonding length of the anchors.

2. Wave propagation phenomena in ground anchor

2.1. Wave modes in a free bolt

Waves propagating in a free rod are dispersive waves. The relationship between the frequency and the group velocity for guided waves propagating in a rod with a circular cross section can be obtained by solving the frequency equation known as the Pochhammer equation [12]. The equation has the following form [13]:

$$\frac{2\alpha}{r}\left(\beta^2+k^2\right)J_1(\alpha r)J_1(\beta r)-\left(\beta^2-k^2\right)^2 J_0(\alpha r)J_1(\beta r)-4k^2\alpha\beta J_1(\alpha r)J_0(\beta r)=0, \quad (1)$$

where:

$$\alpha^2=\frac{\omega^2}{c_L^2}-k^2, \quad (2)$$

$$\beta^2=\frac{\omega^2}{c_s^2}-k^2, \quad (3)$$

$$k=\frac{\omega}{c}, \quad (4)$$

and r denotes a radius of the rod, k is the wavenumber, ω is a circular frequency. J_0 and J_1 are the Bessel functions of the first kind. Velocities of the bulk waves, i.e. the velocity of the pressure wave c_L and the velocity of the shear wave c_S are defined as:

$$c_L=\left(\frac{E(1-v)}{\rho(1+v)(1-2v)}\right)^{1/2}, \quad (5)$$

$$c_S=\left(\frac{E}{2\rho(1+v)}\right)^{1/2}, \quad (6)$$

where ρ denotes the density, E is the Young modulus and v is the Poisson coefficient.

2.2. Wave propagation on the border of the media

At the interface between media with different physical properties two phenomena may occur: the wave can be reflected from the boundary or it can be transmitted into the second medium (Figure 1). The intensity of the reflection and the transmission depends on the difference between values of the acoustic impedance of two media. In the case of two media with different values of longitudinal wave velocity, only the part of the wave energy penetrates into the second medium, while the remaining part is reflected. A large difference in the impedance can result in complete reflection of wave, while a small difference in the impedance may be the cause of the dominance of the wave transmission into the surrounding medium.

The intensity of the leakage of the wave energy not only depends on material parameters of two media but also on the bonding length and the quality of connection. The difference in the quality of the connection between steel and concrete can be noticed when the intensity of the leakage for two different cases (destroyed and undestroyed anchor) is compared. When the bonding quality is high, the transfer of the energy to the surrounding medium is more effective and results in greater leakage of the energy. When the bonding is weak, the phenomenon of wave reflection from the border of media dominates (Figure 2).

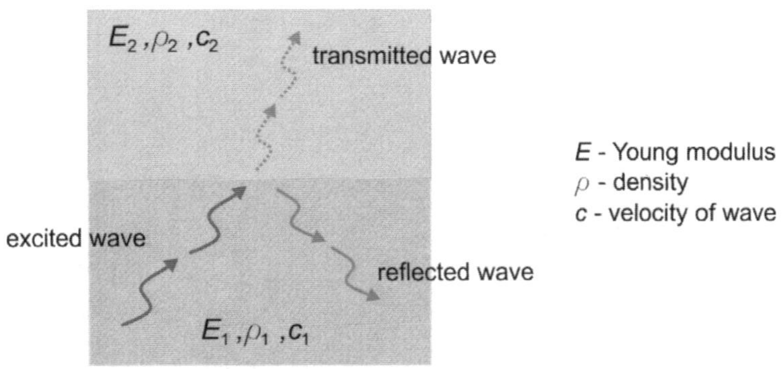

Figure 1. Wave reflecting and passing through the boundary between media

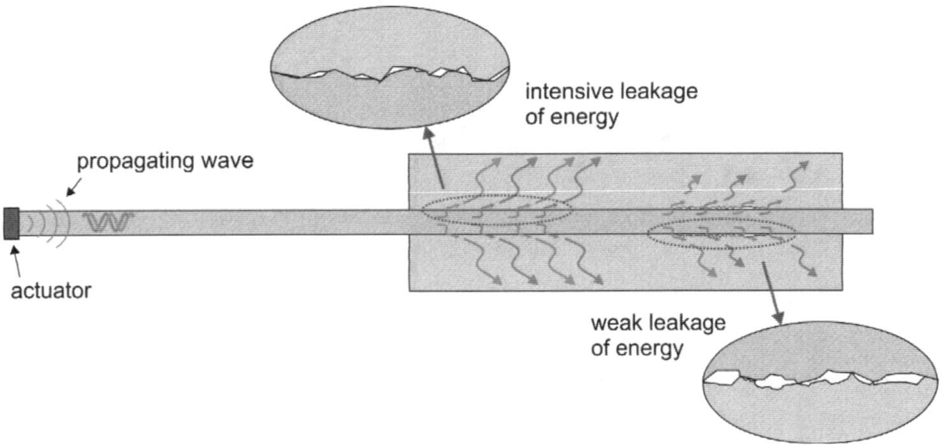

Figure 2. Wave leakage on the steel-concrete interface

3. Experimental investigation

3.1. Ground anchor specimens

Laboratory models of ground anchors were considered as testing structures. The ground anchor consists of a steel circular rod having diameter of 2 cm and 150 cm long embedded in concrete cylinder having outer diameter of 10 cm and the length *a*. The geometry of the model is presented in Figure 3. Nine different cases of bonding length were considered, namely a = 0, 10, 20, 30, 40, 50, 60, 80, 100 cm. The material parameters were for steel: E =210 GPa; ν = 0,3; ρ = 7820 kg/m^3 and for concrete: E =26 GPa; ν = 0,2; ρ = 2084 kg/m^3.

3.2. Experimental set-up
Experimental set-up is shown in Figure 4. Guided waves were excited by the arbitrary waveform generator (Tektronix AFG 3022) with the high voltage amplifier (EC Electronics PPA 2000). The PZT actuator was bonded on the left end of the steel rod. By the use scanning head of the laser vibrometer Polytec PSV-3D-400-M propagating velocity signals $v(t)$ were sensed and recorded parallel to the rod axis on the same end where the excitation was realized.

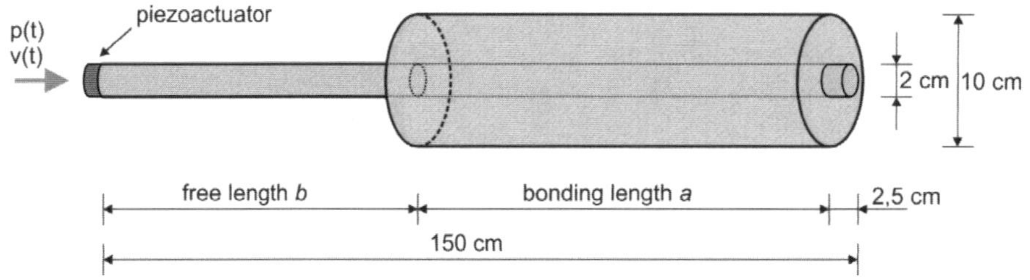

Figure 3. Geometry of the tested specimen

Figure 4. Experimental set-up for wave propagation measurements of ground anchors

As the excitation signal a wave packet was used (Figure 5). It was obtained by multiplying a 5-cycle sinusoidal function of carrier frequency f and a window function $w(t)$:

$$p(t) = \begin{cases} p_o \sin(2\pi f t) \cdot w(t) & t \in [0, T_w], \\ 0 & t > T_w, \end{cases} \quad (7)$$

where T_w is the length of a window and p_o denotes an amplitude of a sinusoidal function. As a window function, the Hanning window was chosen:

$$w(t) = 0.5\left(1 - \cos\left(2\pi f t / n_w\right)\right), \quad t \in [0, T_w], \quad (8)$$

where n_w is the number of counts in the tone burst. The carrier frequency f of the wave packet varied from 30 kHz to 90 kHz with the step of 10 kHz.

Figure 5. Wave excitation signal used in experimental investigations

3.3. Dispersion curves for a free rod

The solution of the Pochhammer equation can be presented graphically as dispersion curves. The lowest longitudinal mode L(0,1) is the most important from the practical point of view in the context of non-destructive diagnostics. In Figure 6 numerical and experimental results are compared. Numerical curves were calculated based on Eq. (1) in the frequency range up to 300 kHz. In this range, besides the fundamental mode, two higher longitudinal modes appear above their cut-off frequencies: L(0,2) and L(0,3) modes. Experimental part of the curve was determined by dividing the distance equal to twice the length of the rod by the time interval between registration peaks of incident and reflected wave packet.

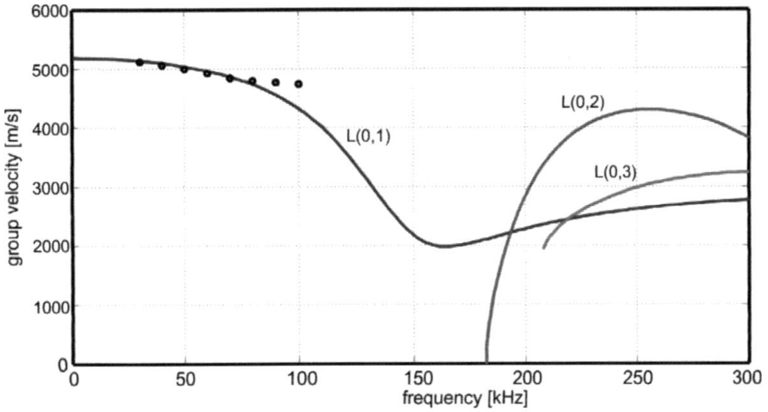

Figure 6. Group velocity dispersion curves: (—) numerical results (solutions of Pochhammer equation); (**o**) experimental results

4. Results of experimental investigations

4.1. Selection of excitation frequency

Before the assessment of condition of ground anchors, the selection of appropriate frequency of excitation was performed. At first, experimental investigation was performed for the free rod in the frequency range 30–100 kHz. Figure 7 shows envelopes of wave propagation signals created using the Hilbert transform. The envelopes of the vibration velocity signals were normalized with respect to the input wave amplitude by the following formula:

$$\hat{v}_n(t) = \frac{\hat{v}(t)}{\max\left(\hat{v}(t)\right)}, \tag{9}$$

where $\hat{v}(t)$ is the spectrum of the Hilbert transform of velocity signal $v(t)$. In the time interval form 0 ms to 1.5 ms the incident wave and two reflections are visible. It can be observed that the amplitude of the first and second reflections from the end of the rod varies with the change of frequency of incident wave. Even though reflections for all considered frequencies are clear, a significant difference between amplitude values for different carrier frequencies can be observed. With the increase of the excitation frequency, the increase of the first and second reflection can be observed. The maximum value was archived for the frequency of 90 kHz (Figure 7).

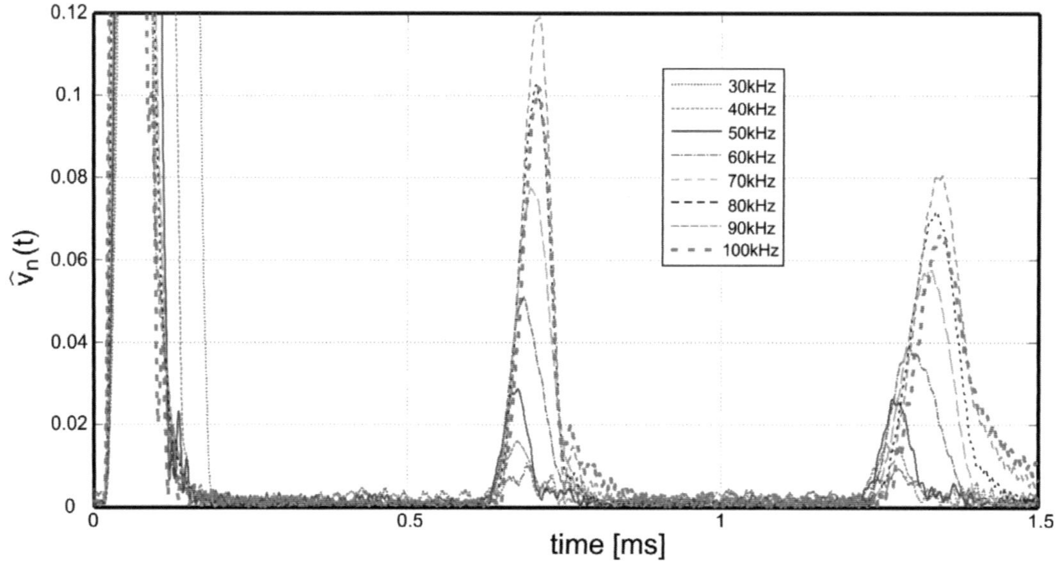

Figure 7. Set of wave propagation signals (in the form of signal envelopes) collected during experiments for different carrier frequency of excitation for the free rod

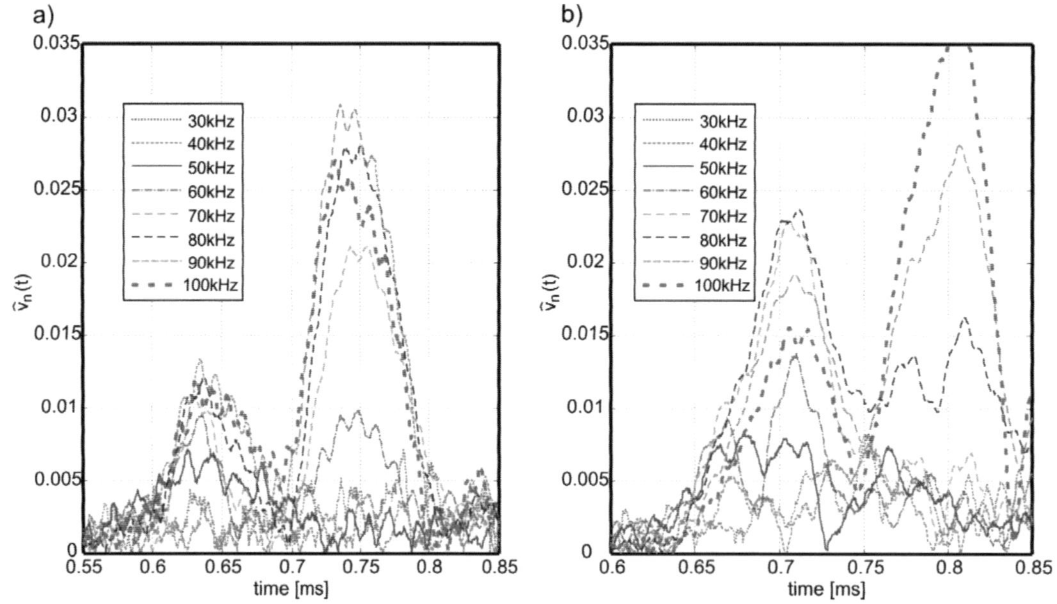

Figure 8. Set of reflections from anchorage collected during experiments for different carrier frequency of excitation for anchor with bonding length equal to: a) 20 cm, b) 40 cm

The increase of the amplitude proportional to the increase of the excitation frequency has been also observed for the rest of the anchors. An example of results for two different bonding lengths (a = 20 cm and a = 40 cm) are shown in Figure 8. This comparison allowed establishing that further analysis should include the propagation of waves with a carrier frequency equal to 90 kHz, because the energy of reflection was the highest for this frequency.

4.2. Identification of bonding lengths

In Figure 9, a set of wave propagation signals for all tested bonding lengths and for chosen carrier frequency equal to 90 kHz are presented. Due to the leakage of the energy caused by surrounding concrete, the reflections from the end of the steel rod are not visible. The wave packets highlighted in Figure 9 result from the wave reflection form the anchorage, i.e. the boundary between steel and concrete. One can see that the length of the propagation path is proportional to the time interval between the registered incident wave and the reflected wave. In consequence this time interval is in inverse proportion to the bonding length.

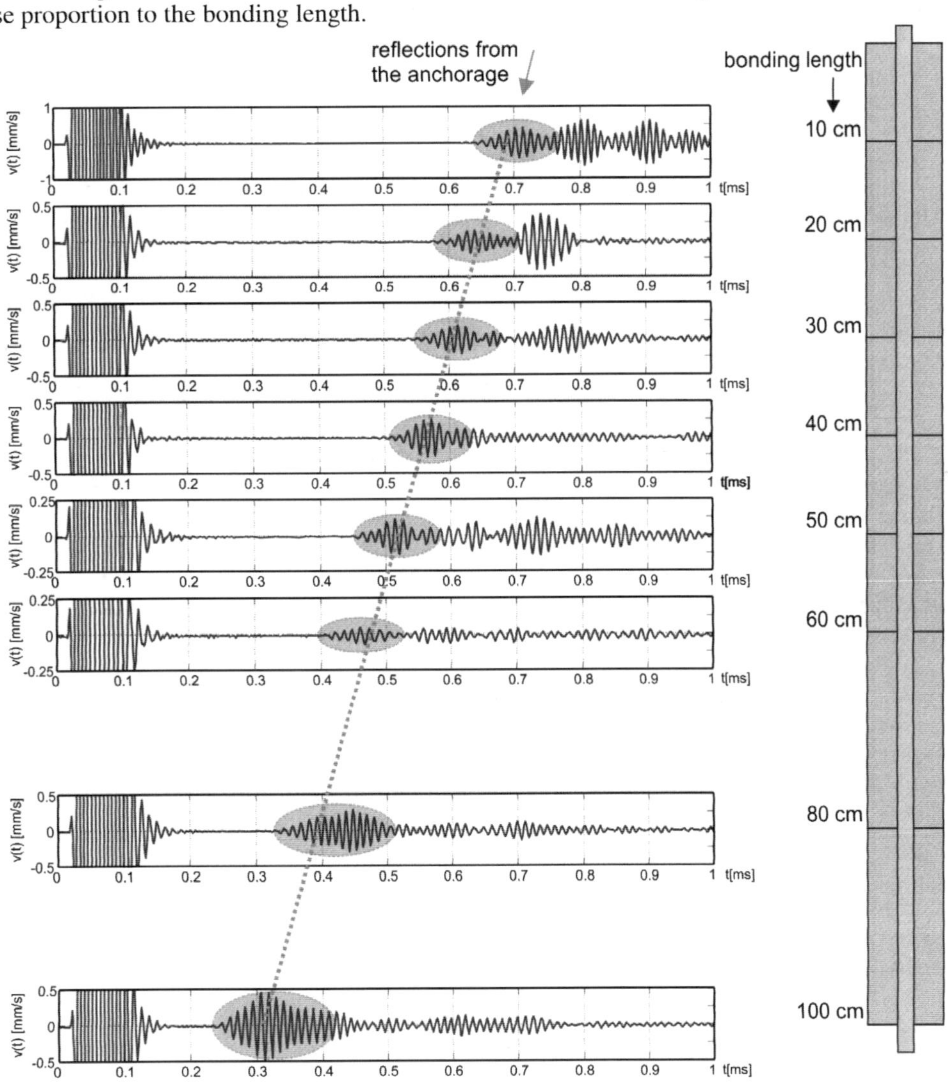

Figure 9. Set of signals collected during experiments for carrier frequency of excitation equal to 90 kHz and for different bonding lengths with indicated reflections from the anchorage

5. Conclusions

This paper presents an experimental investigation of guided wave propagation in laboratory models of ground anchors. The specimens with different bonding lengths were tested by the wave packet excitation with different carrier frequencies. The main conclusions are connected with the fact that properties of guided waves, such as attenuation or velocity, are functions of the input wave frequency. As the bonding length increases and the quality of the connection between steel rod and concrete gets better, the more energy of wave is transferred into concrete and the amplitude quickly attenuates. In investigated anchors, even for relatively short bonding lengths, the dissipation of energy was enough intensive to prevent extraction of information about reflection from the end of the rod. For all tested anchors it was possible to identify the boundary between steel and concrete based on reflections in the registered wave propagation signals.

6. References

[1] Park H. W., Sohn H., Law K. H., Farrar C. R. Time reversal active sensing for health monitoring of a composite plate. Journal of Sound and Vibration, vol. 302, 2007, pp. 1907–1914.

[2] Yu L., Giurgiutiu V. In situ 2D piezoelectric wafer active sensors arrays for guided wave damage detection. Ultrasonics, vol. 48, 2008, pp. 117–134.

[3] Mallet L., Lee B., Staszewski W. J., Scarpa F. Structural health monitoring using scanning laser vibrometry: II: Lamb waves for damage detection. Start Materials and Structures, vol. 13, 2004, pp. 261–269.

[4] Żak A., Radzieński M., Krawczuk M., Ostachowicz W. Damage detection strategies based on propagation of guided elastic waves. Smart Materials and Structures, vol. 21, 2012, 035024.

[5] Malinowski P., Wandowski T., Ostachowicz W. Damage detection potential of a triangular piezoelectric configuration. Mechanical Systems and Signal Processing, vol. 25(7), 2011, pp. 2722–2732.

[6] Rucka M. Experimental and numerical study on damage detection in an L-joint using guided wave propagation. Journal of Sound and Vibration, vol. 329, 2010, pp. 1760–1779.

[7] Rucka M., Wilde K. Experimental study on ultrasonic monitoring of splitting failure in reinforced concrete. Journal of Nondestructive Evaluation, vol. 32, 2013, pp. 372–383.

[8] Alleyne D. N., Pavlakovic B., Lowe M. J. S., Cawley P. Rapid, long range inspection of chemical plant pipework using guided waves. Key Engineering Materials, 270–273, 2004, pp. 434–441.

[9] Ciang C. C., Lee J. R., Bang H. J. Structural health monitoring for a wind turbine system: a review of damage detection methods Measurements science and Technology vol. 19(12), 2008, 122001.

[10] Beard M. D., Lowe M. J. S. 2003 Non-destructive testing of rock bolts using guided ultrasonic waves International Journal of Rock Mechanics and Mining Sciences vol. 40, 2003, pp. 527–536.

[11] Sabatini P.J., Pass D.G., Bachus R.C. Ground anchors and anchored systems – Technical Report FHWA-IF-99-015, Office of Bridge Technology, Washington, 1999.

[12] Pochhammer L. Beitrag zur Theorie der Biegung des Kreiscylinders. Journal für die reine und angewandte Mathematik 81, 1876, pp. 33–61.

[13] Achenbach J.D. Wave propagation in solids. North-Holland Publishing Company, 1975.

Magnetic Flux Leakage and Principal Component Analysis for metal loss approximation in a pipeline

M. Ruiz[1], L.E. Mujica[1], M. Quintero[2], J. Florez[2] and S. Quintero[2]

[1]Departament of Applied Mathematics III. Escola Universitària d'Enginyeria Tècnica Industrial de Barcelona. Consorci Escola Industrial de Barcelona. Universitat Politècnica de Catalunya. BARCELONATECH. Spain

[2]Department of Electronics. Research Institute of Corrosion CIC. Colombia. Km 2 via Refugio Sede UIS Guatiguará, Piedecuesta, Colombia

E-mail: magda.ruiz@upc.edu / luis.eduardo.mujica@upc.edu / mquintero@corrosion.uis.edu.co / jflorez@corrosion.uis.edu.co / squintero@corrosion.uis.edu.co

Abstract.

Safety and reliability of hydrocarbon transportation pipelines represent a critical aspect for the Oil an Gas industry. Pipeline failures caused by corrosion, external agents, among others, can develop leaks or even rupture, which can negatively impact on population, natural environment, infrastructure and economy. It is imperative to have accurate inspection tools traveling through the pipeline to diagnose the integrity. In this way, over the last few years, different techniques under the concept of structural health monitoring (SHM) have continuously been in development.

This work is based on a hybrid methodology that combines the Magnetic Flux Leakage (MFL) and Principal Components Analysis (PCA) approaches. The MFL technique induces a magnetic field in the pipeline's walls. The data are recorded by sensors measuring leakage magnetic field in segments with loss of metal, such as cracking, corrosion, among others. The data provide information of a pipeline with 15 years of operation approximately, which transports gas, has a diameter of 20 inches and a total length of 110 km (with several changes in the topography). On the other hand, PCA is a well-known technique that compresses the information and extracts the most relevant information facilitating the detection of damage in several structures. At this point, the goal of this work is to detect and localize critical loss of metal of a pipeline that are currently working.

1. INTRODUCTION

In Colombia, the petrochemical industry facilities have structures with more than 30 years in service. The ferrous pipe structures of oil and gas production and, the transmission pipelines are, in majority, buried. Nowadays, phenomena like corrosion, mechanical stress, soil erosion, worker mistakes and damages caused by third parts (for instance, excavation machinery that can strike the pipe causing scratches or dents) have generated several problems on pipelines.

The need to manage and maintain pipeline system has become increasingly in an important priority to operators. Thus, major investment on integrity programs with In-Line Inspection Tools - smart pigs - have been improved in order to examine the pipelines and avoid environmental, financial and social disasters. At the same time, international regulations have raised levels of requirements on reliable hydrocarbon and gas transmission.

Journal of Physics: Conference Series **628** (2015) 012027 doi:10.1088/1742-6596/628/1/012027

The Colombian government is developing a regulation which contains the integrity concept of oil and gas transport in accordance with the international standards. These rules also should fulfill the national requirements. As a result of this process, operators will require the establishment of the physical condition of oil and gas pipelines, as well as the generation of correction and prevention strategies to integrity management assurances hold in the pipeline. Smart inspection is an effective technique to consider perform detailed inspections on pipelines.

Internal inspection of pipelines are currently carried out in Colombia by foreigner companies. However, the cost of the service is excessively high without negotiation opportunity, limitation in the analysis of results and without monitoring of action plans. All these disadvantages restrict the regular use of the In-Line Inspection tools.

Recently, Research Institute of Corrosion - CIC (Corporación para la Investigación de la Corrosión) run their own smart pig ILI tool in pipelines. This is the first device for this purpose developed completely in Colombia. The inspection technology is based on inertial and operational trends, ITION (Inertial Technology Inspection and Operational Trends). Up to date, the technology has been tested several times inside of pipelines providing valuable information along of thousand kilometers. These records contain a huge amount of data. An univariate statistical method can be used to determine the thresholds for each observation variable. However, it does not analyze the correlated information between variables. In this way, the main contribution of Principal Component Analysis (PCA) in this work is to monitor the structure by using the whole available variables to detect statistically significant events or damages since the information is compressed and a pattern recognition in the signal is performed for structural monitoring [1].

For this work, The CIC has provided the first measurement made with the ITION technology -smart pig ILI- pilot test. A brief summary of this technology is described in sections 2 and 3. PCA approach is explained in section 4. In section 5, the raw data analysis and PCA results are presented. Finally, discussion and conclusions are drawn.

2. ITION Technology Description

2.1. Generalities

One way to monitor pipelines is the In-Line Inspection pilot Tool - smart pig - which is a vehicle that travels inside the pipeline. The CIC has developed the Inertial Technology Inspection and Operational Trends -ITION-. This tool is composed by electronic system capable of acquiring, processing and storing data signals (Figure 1). The sensor system has mainly an Inertial Measurement Unit. It consists by accelerometers and gyroscopes with high precision and sensitivity. As the tool travels along the pipe, the inertial system records information on all components of accelerations and angular velocities associated with the movement dynamics of the device within the pipeline.

Figure 1: Inertial Technology Inspection and Operational Trends -ITION- developed by CIC

A second set of accelerometers is installed on the tool in order to filter and remove characteristic noise of inertial systems. In addition, the tool contains pressure and temperature sensors to study the phenomena present in the conveying pipeline. The system has been expanded to incorporate a prototype array of linear transducers that varies its output voltage in response to magnetic fields, providing a constant driving current to the sensors and amplifying the output signal. Odometers systems measures the distance traveled by the tool and allows the calculation of the instantaneous speed, information that is used to compensate the error associated with the nature of the inertial measurement ROA2012. All electronics are protected by a mechanical housing and designed to fit most of the conventional scraper routinely used in pipeline cleaning processes. The flexibility and adaptation of the ITION technology allow its application using high frequency with low cost [3]. In consequence, an effective monitoring and diagnosis of pipelines changes due to the passing of time can be achieved.

2.2. Signals description

After running the ITION, the following signals are recorded:

- Signal 1: Intensity of axial movement (the direction of flow of the transported product)
- Signal 2: Intensity rate of axial rotation
- Signal 3: Intensity resulting from the rotational sensors (three-axis)
- Signal 4: Intensity sensor remanent fields
- Signal 5: Propulsive force experienced by the tool
- Signal 6: Temperature of the transported product

Each signal has more than 7 millons of samples. These signals are depicted in Figure 2. For the purpose of this article, the signals are processed without emphasis on the characteristics of the phenomenon that describes or the behavior associated with any technique.

3. Magnetic flux leakage technique

Magnetic Flux Leakage technique - MFL - is used to detect metal loss defects and flaws in steel pipelines. It is implemented on In-Line inspection tools with powerful neodymium magnets, that requires a saturated field through the ferrous wall to achieve desired performance [4].

When the ITION goes through the pipe structure, a remanence effect is presented. The remanence effect is the magnetization left in the pipe wall by a pair of high energy permanent magnets employed in the inspection vehicle. Laboratory tests show that some kind of remanence are undesirable before running MFL inspection. Remanence is a magnetic distortion that affects the offset of the signal position in homogeneous fields on pipe walls without metal loss or corrosion.There is some array of magnetometers transducers installed inside and outside of the body of the tool; the external magnetometers register the residual field over the internal pipe wall surface.

The main objective is to understand the residual magnetization for improving the ability to reliably detect this kind of damage. In this way, identifying characteristics of the residual field signals is possible through the application of computational techniques [5]. In consequence, this work is focused on identifying pattern of defects, pipe accessories and reference grouped tags, which are important in the study and post-processing analysis of the data collected by ITION.

4. Principal Component Analysis

PCA is widely used in this kind of problems since it allows represent graphically as effectively as possible observations belongs to a general m-dimensional space in a small dimensional space (r) [6]. Besides, PCA allows transform original variables, usually correlated, to new uncorrelated variables, making easier its interpretation. The goal of PCA is to find a subspace with dimension

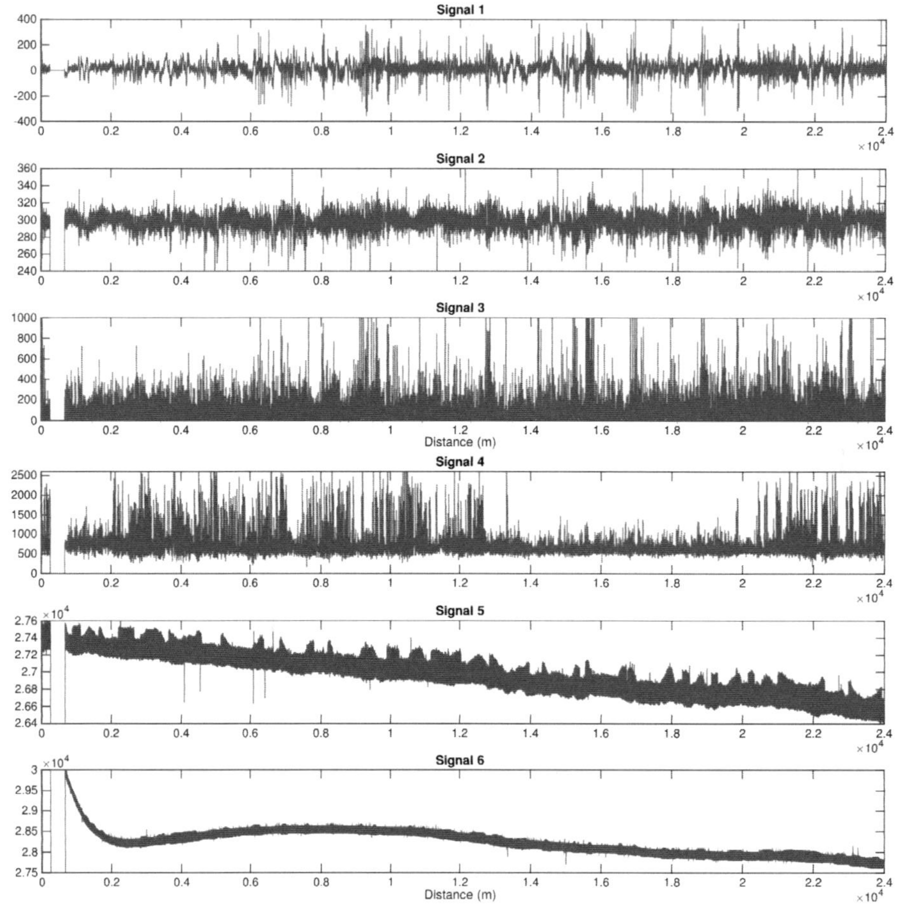

Figure 2: Original measurements detected by ITION

lesser than m such that projecting into it, the new variables keep its structure and minimize the distortion. In other words, a linear transformation orthogonal matrix $\mathbf{P}$, which is used to transform the original data matrix $\mathbf{X}$ into the form

$$\mathbf{T} = \mathbf{X}\mathbf{P}. \tag{1}$$

In the literature, it can be found that the r-dimensional space ($r \leq m$) that represents better the original data is defined by the eigenvectors associated with the highest eigenvalues of the covariance matrix of the observations as follows:

$$\mathbf{C}_X = \frac{1}{n}\mathbf{X}^T\mathbf{X}, \tag{2}$$

$$\mathbf{C}_X\mathbf{P} = \mathbf{P}\Lambda, \tag{3}$$

where $\mathbf{C}_X$ is the covariance matrix of the original data $\mathbf{X}$, the eigenvectors of $\mathbf{C}_X$ are the columns of $\mathbf{P}$, and the eigenvalues are the diagonal terms of Λ (the off-diagonal terms are zero). The eigenvectors p_j forming the transformation matrix $\mathbf{P}$ (its columns) are sorted according to the eigenvalues by descending order, the eigenvector with the highest eigenvalue represents the most important pattern in the data with the largest quantity of information. In Equation 1, Columns of $\mathbf{P}$ are called the Principal Components (or loading matrix in other references) of the data set and $\mathbf{T}$ the projected or transformed matrix to the principal component space (or score matrix in other articles).

In the full dimension case (using all the n principal components), this projection is invertible (since $\mathbf{PP}^T = \mathbf{I}$) and the original data can be recovered as $\mathbf{X} = \mathbf{TP}^T$. But, PCA also seeks to reduce the dimensionality of the data set $\mathbf{X}$ by choosing only a reduced number r of principal components ($r < n$). Now, with $\mathbf{T}$ given by the reduced matrix $\mathbf{P}$, it is not possible to fully recover $\mathbf{X}$, but $\mathbf{T}$ can be projected back onto the original m-dimensional space and obtain another data matrix as follows:

$$\hat{\mathbf{X}} = \mathbf{TP}^T. \tag{4}$$

Therefore, the original data matrix $\mathbf{X}$ can be decomposed by the projected back data $\hat{\mathbf{X}}$ and the residual error matrix $\mathbf{E}$, which describes the variability not described by the model as follows:

$$\mathbf{X} = \mathbf{TP}^T + \mathbf{E}. \tag{5}$$

Two well-known statistics are commonly used to this aim: the Q-statistic (or SPE-statistic) and the Hotelling's T^2-statistic (D-statistic). Q-statistic is based on analyzing the residual data matrix $\mathbf{E}$ to represent the variability of the data projection in the residual subspace. It denotes the change of the events that are not explained by the model of principal components. The Q-statistic of the i-th sample or experiment (row vector x_i of data matrix $\mathbf{X}$) is defined as follows:

$$Q_i = e_i e_i^T = x_i \left(\mathbf{I}\text{-}\mathbf{PP}^T\right) x_i^T. \tag{6}$$

where e_i is its projection into the residual subspace (row vector of residual data matrix $\mathbf{X}$).

T^2-statistic is based in analyzing the score matrix $\mathbf{T}$ to check the variability of the projected data in the new space of the principal components. The T^2-statistic of the i-th sample (or experiment) is defined in the form:

$$T_i^2 = t_{si}\Lambda^{-1}t_{si}^T = x_i \left(\mathbf{P}\Lambda^{-1}\mathbf{P}^T\right) x_i^T. \tag{7}$$

where t_{si} is its projection into the new space (row vector of the score matrix $\mathbf{T}$) [7][9].

5. Results

5.1. Initial analysis and organization of the collected data

The six recorded signals, previously detailed in section 2.2 (See Figure 2), belong to a test in a buried pipeline of 20 inches diameter and 110 Km length. This Pipeline is made of sections of thin-walled steel tubing of 12 m length, which are welded together using a circumferential weld. It has been in service for more than 15 years and nowadays it transports gas. The owner of the structure provided the location of 58 tags across the pipeline (See Figure 3). These tags include elements of the pipeline (e.g. VA belongs to valves) and damages among others. Unfortunately no complete explanation can be given (due to confidence reasons) of what these events are. An initial analysis of the measurements is performed. The test is conducted in 24 Km of the structure, this means that the "smart pig" (ITION) traveled inside the pipeline 24 Km collecting data from 6 sensors. A total of 7′426.500 measurements (samples) were collected by each sensor. At the first 10 meters of the pipeline, during the tuning of the ITION, the sample frequency is higher than the rest of the structure (1′600.000 samples). On the other hand, in these profiles any event or damage can be directly observed. Since it is not possible to give some diagnosis of the structure observing directly the measurements, PCA is applied to carry out a multivariable analysis, in other words, to analyze all measurements and its correlations as a whole [8]. In this way, possible patterns of strange events can be recognized (elements of the pipeline, damages, welds, etc). For this purpose, the original data are organized in a matrix. This $n \times m$ matrix contains information from m sensors and n experimental trials (samples). Consequently, each row vector represents measurements from all the sensors at a specific time instant or experiment trial. In the same way, each column vector represents measurements from one sensor in the whole set of experiment trials.

Figure 3: Tags provided by the owner of the pipeline

5.2. Multivariable analysis by using PCA

The original data ($7'426.500$ samples $\times$ 6 sensors) is used to build the PCA model and compare with the tags provided by the pipeline owner. The data matrix is scaled, the loading and score matrices are calculated (**P** and **T** in Equation 1). Besides, statistical indices (Q and T^2) are calculated for each experiment (see Equations 6 and 7). However, the results for detecting tags are not entirely satisfactory. Going back to the initial analysis, the oversampling of the first 10 meters and, the profile of the signal 6 should be considered. The temperature of the transported product (signal 6) is irrelevant for the goal of the analysis: detection of tags. Therefore, signal 6 is removed from the original data matrix and three PCA models are built:

- Model 1 which uses all measurements ($7'426.500 \times 5$)
- Model 2 which uses the measurements of the first 10 meters of the pipeline ($1'600.000 \times 5$)
- Model 3 which uses the measurements of the rest of the pipeline ($5'826.500 \times 5$)

Each model is built with 4 principal components since around 90% of the cumulative variance is retained. Anew, scores and statistical indices (Q and T^2) are calculated and depicted for each model. Figure 4 shows scores 1 to 3 and the statistical indices for model 1. Apparently, the scores and statistical indices exhibit a local maximum (peak) in some tags. But a deeper validation confirmed that the best approximation is achieved by using Q and T^2-statistics.

Model 2 describes the first 10 meters of the pipeline and the tuning of the ITION. 14 from the 58 tags (24%) are located in this section. In Figure 5, where the indices are depicted, it can be seen that some maximum match with some tags and others tags are nearby to be localized. 3 from the 9 tag named TA are well matched. VA, TE and GR are detected without any problem. In contrast, BR is not matched.

Model 3 describes the rest of the pipeline, 44 tags are localized in this section. According to Figure 6, a better tags localization is observed. First and third PM tag are detected, the second PM is detected only by score 1 (no shown). The set of tags around the same point are detected by at least one score or index: E.g. around the spot 0.2 x 10^4 meters (2 Km), tags are remarked by T^2 index; tags in 0.8 x 10^4 meters (8 Km) are detected by all scores and indices. In the interval from 14 Km to 18 Km it is observed some maximum values, but any tag is located there. Finally, in the last kilometers, all tags are correctly detected.

On the other hand, analyzing more in detail plots in Figure 6, it can be seen that indices increases every 12 meters (approximately). For instance, in Figure 7, that presents a zoomed section of Figure 6, a pattern can be identified every 12 meters. The reason of this "out of control" of these indices can be attributed to the weld that join the sections of the pipeline.

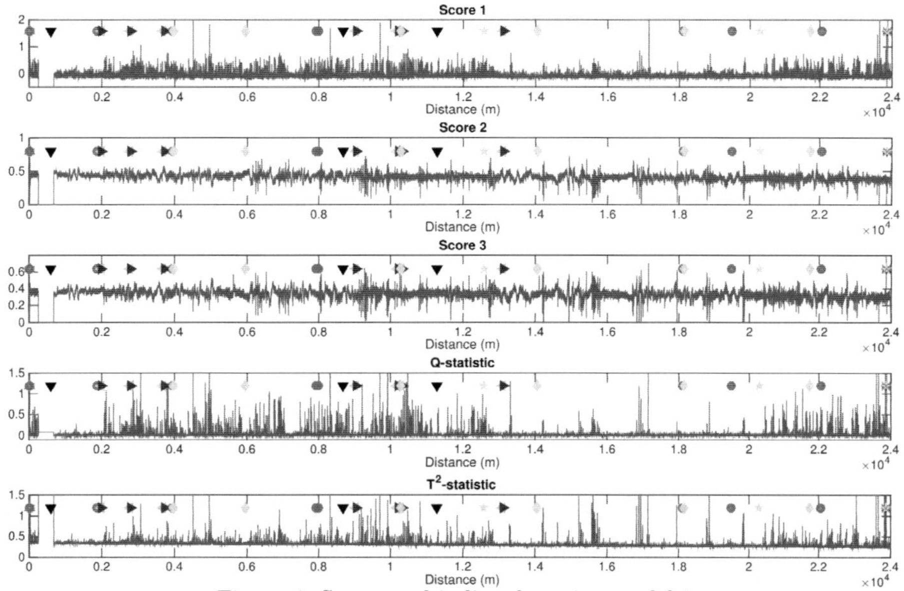

Figure 4: Scores and indices by using model 1

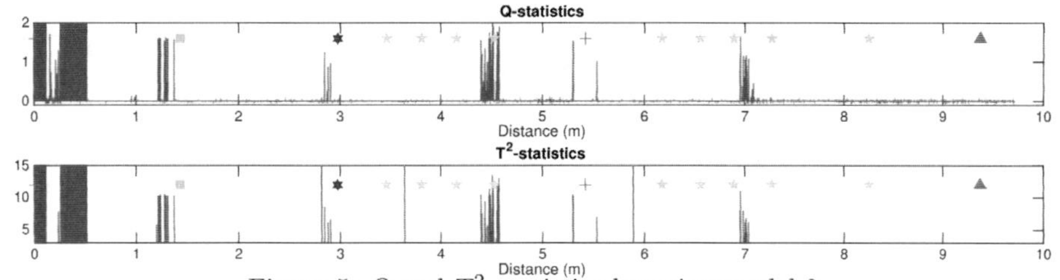

Figure 5: Q and T^2 statistics by using model 2

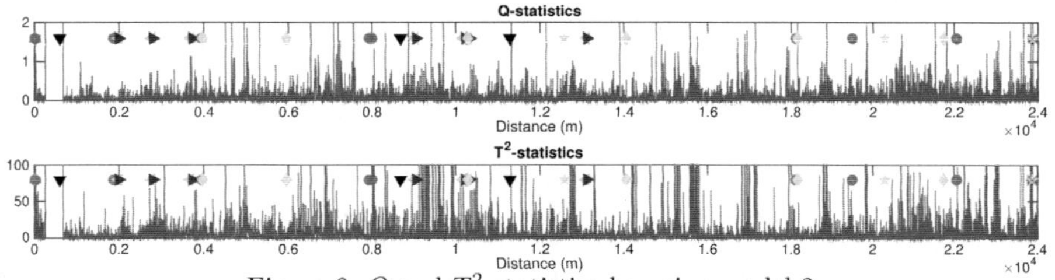

Figure 6: Q and T^2 statistics by using model 3

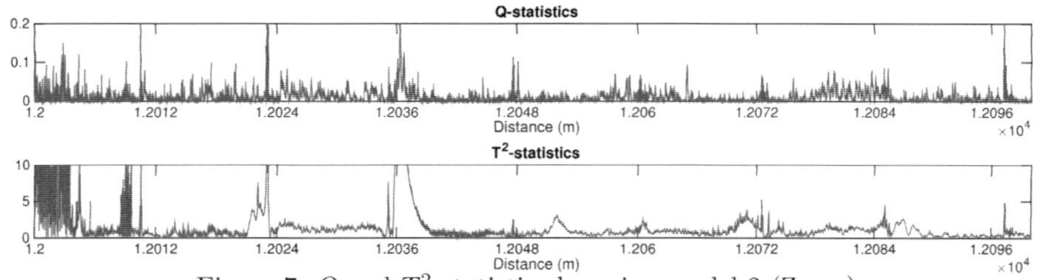

Figure 7: Q and T^2 statistics by using model 3 (Zoom)

6. Discussion and Conclusions

The main goal of this work is to localize critical loss of metal in a pipeline. The smart pig tool (ITION) and the methodology (based on PCA) have been validated in a pipeline (currently in service) made of tubing sections welded every 12 meters. The owner of the structure provided the location of 58 tags belong to different operational elements and damages of the pipeline, however any explanation of the meaning was given. A huge amount of measurements are gathered in the first run of the tool through 24 Km of the pipeline. This information is processed by means of PCA and some indices are calculated for every location of measurement. The validation of the methodology is carried out by comparing the location of the "alarms" or values out of control of the mentioned indices and, the location of the tags. From results, it is concluded that the localization of abnormal events (operational elements or damages) are improved considerably when data are arranged in two models (model 2 and 3). These results can be considered as successful (despite that it is a novelty detection real application) due to the pipeline complexity. Even though several possible false alarms are presented, it is inferred that the pipeline welds could be the responsibilities. In the near future, it is expected to run the second version of the smart pig (ITION) which includes 18 sensors and the sample frequency is higher. On the other hand, the methodology must be improved if the owner is not interested in detecting welds.

Acknowledgments

The authors would like to thank the support from the "Ministerio de Economía y Competitividad" in Spain through the Coordinated Project DPI2011-28033-C03-01, to the Escola Universitària d'Enginyeria Tècnica Industrial de Barcelona, Consorci Escola Industrial de Barcelona and, the Colombian Research Institute of Corrosion - CIC-.

References

[1] Russell E L, Chiang L H and Braatz R D 2000 *Data-driven techniques for fault detection and diagnosis in chemical processes "Advances in Industrial Control". London: Springer*

[2] Roa S, Parra J, Garnica H, Quintero S and Quintero M 2012 Design Optimization of a Magnetic Circuit and Sensor System for MFL Testing of Pipelines. *The 2012 Review of Progress in Quantitative Nondestructive Evaluation*, 2012.

[3] Perez I and Dobmann G 2013 Surface Open Corrosive Wall Thinning Effects. *International Workshop on Smart Materials, Structures & SHM*

[4] Jansen H J M, Van de Camp P B J and Geerdink M 1994 Magnetisation as a key parameter of magnetic flux leakage pig for pipeline inspection. *INSIGHT*, 36(9) pp 672-673

[5] Zuoying H, Peiwen Q and Liang C 2006 3D FEM analysys in Magnetic Flux Leakage method. *NDT&E International*, 39 pp 61-66

[6] Jolliffe I T 2002 *Principal Component Analysis.* Springer Series is Statistics. Springer-Verlag

[7] Mujica L E, Rodellar J, Fernández A and Güemes A 2011 Q-statistic and T2-statistic PCA-based measures for damage assessment in structures. *Struc. Health Monitoring* 10(5) pp 539-553

[8] Mujica L E, Vehí J, Ruiz M, Verleysen M, Staszewski W J and Worden K 2008 Multivariate Statistics Process Control for Dimensionality Reduction in Structural Assessment. *Mechanical Systems and Signal Processing* 22 pp 155-171

[9] Ruiz M, Mujica L E, Berjaga X and Rodellar J 2013 Partial least square projection to latent structures (PLS) regression to estimate impact localization in structures. *Smart materials ans structures* 22(2) pp 025028

Damage localization in plates with use of the procedure based on Modal Filtration

Ziemowit Dworakowski[1],Lukasz Ambrozinski[2],Krzysztof Mendrok[3]
AGH University of Science and Technology, al. Mickiewicza 30, 30-059 Krakow, Poland
E-mail: [1]zdw@agh.edu.pl, [2]ambrozin@agh.edu.pl, [3]mendrok@agh.edu.pl

Abstract. Modal filtering has numerous applications in the analysis of object dynamics. One possible use of this technique, recently presented in the literature, is damage detection. The main idea behind it, is to compare system characteristics filtered with modal filter for damaged and undamaged state. If a structural change appears in the system the filter does not work perfectly and its output significantly differs from one obtained for the healthy structure. Method is based on the modal parameter variation due to damage but does not require the modal analysis for every diagnosis. Another advantages are: computational simplicity and robustness for ambient temperature changes. The method was extended in 2008 with the possibility of damage localization. The idea here was based on the fact that local damage disturbs mode shapes locally. Instead of one modal filter, set of local modal filters were identified and used for further processing. Area of the structure connected with the filter with the worst performance is suspected of containing the damage. The method was already presented and positively verified but only for the beam-like structures. In this paper, application for plate-like structures is shown.

1. Introduction

Structural Health Monitoring (SHM) is a field of growing importance. Autonomous detection and localization of damage in structures allows for considerable reduction of expenses related to their operation as well as increase in safety. Large amount of methods were proposed in this scope, including ultrasonics [1, 2], vision-based methods [3], acoustic emission [4] and many more. Among those, vibration-based methods play an important role, due to their global character and relative insensitivity to environmental changes [5]. Damage can be detected by observation of changes in natural frequencies [6], features calculated from signal spectrum or obtained by time-frequency methods [7]. In Modal Analysis, a modal model (MM) of an object under investigation is build out of vibration signals acquired in distinct points of its structure. Signals are usually acquired with sensors attached to the structure. However, several non-contact techniques were also proposed [8, 9]. MM consist of natural frequencies associated with damping factors and Modal Vector (MV). Damage detection with use of MM is performed by comparing model calculated for a known, intact state of a structure with one acquired in unknown conditions. Modal analysis has been successfully applied in many cases of SHM [9–11]. One of approaches employ *Modal Filtration* (MF) introduced by Meirovitch and Baruh [12]. It consists of calculation of a set of *Reciprocal modal vectors* (RMVs), out of which each is orthogonal to all but one Modal Vector. That feature can be used in decomposition of a system response to components related to natural frequencies. Occurrence of damage results usually in local drop of stiffness, which, in turn, cause changes in modal parameters. Signal acquired in such a case cannot properly be decomposed with use of previously calculated RMVs. In contrast, global change of parameters does not modify the modal vectors and thus does not alter the output of decomposition with RMVs [13]. Modal filtration can successfully be used in damage detection under variable environmental conditions. However, its main disadvantage lies in inability to locate damage. That issue can be solved by dividing a structure into areas, for which modal filtration is performed separately. The extent of mode shapes change is dependent to the proximity of damage. This solution introduced in [14] was successfully tested for truss structures [15]. In this paper, the authors verified its efficiency on a plate-like structure.

2. Local modal filtration

The idea to extend the MF method by adding damage localization [13, 14] is based on the fact that damage, in most cases, only disturbs the mode shapes locally. That is why many methods of damage localization use mode shapes as input data [16–18]. It is then possible to divide an object into areas measured with the use of several sensors and build separate local modal filters for data coming only from these sensors. In areas without damage, the shape of modes does not change and the modal filter keeps working there are no additional peaks on the filter output. When a group of sensors placed near the damage is considered, mode shape is disturbed locally due to damage and the MF does not perfectly filter the characteristics measured by these sensors.

It has been said [19] that the minimum number of sensors required to build an effective MF is equal to or greater than the number of modes in the frequency range of interest. Thus, by limiting the frequency range of analysis to the first two modes, it would be theoretically possible to construct MFs for object areas measured with only 2 sensors. This would significantly increase the damage location accuracy. In practice, however, to construct the modal filter for two modes, data from at least 4 sensors has to be used. The accuracy of damage localization depends on measuring net density. The graphical presentation of this idea is presented in Figure 1.

For each section tuning of local modal filter (LMF) is performed according to following procedure: The construction of the r-th modal filter, which corresponds to r-th pole of the transfer function $H(\omega)$, starts with the determination of the theoretical output of the considered modal filter $G_r(\omega)$. The function is calculated basing on the modal parameters of the system:

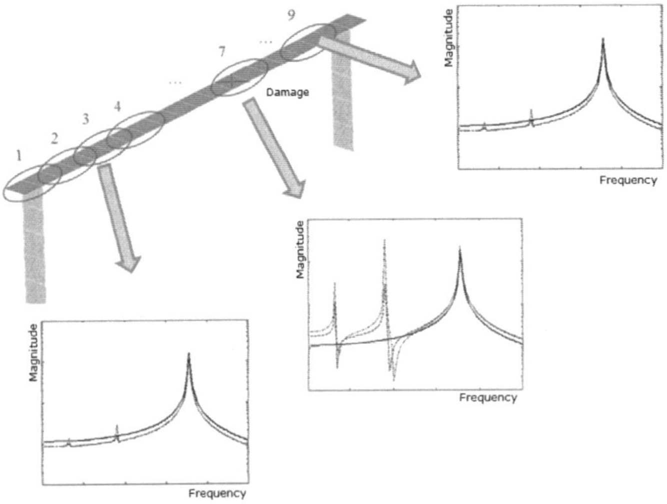

Figure 1. Scheme of the proposed method of damage localization

$$G_r(\omega) = \frac{2\xi_r\omega_r^2}{\omega_r^2 - \omega^2 + 2j\xi_r\omega_r^2\omega^2} \tag{1}$$

Where ξ_r denotes r-th modal damping coefficient while ω_r represents r-th natural frequency. For the given frequency range, the filter output defined by eq. (1) is determined in the k values: Assuming that the single excitation was used and response signals were measured in N points, the experimental FRFs can be presented as the k x N matrix:

$$H_{kN}(\omega) = \begin{bmatrix} H_1(\omega_1) & H_2(\omega_1) & \dots & H_N(\omega_1) \\ H_1(\omega_2) & H_2(\omega_2) & & H_N(\omega_2) \\ \vdots & & & \\ H_1(\omega_k) & H_2(\omega_k) & & H_N(\omega_k) \end{bmatrix} \tag{2}$$

The FRF matrix formed in this way is used to deremine the r-th reciprocal modal vector Ψ_r:

$$\Psi_r = H_{kN}^+ \cdot G_r \tag{3}$$

Which can be used to decompose acquired system response $\{x(\omega)\}$ to modal coordinates η_r:

$$\eta_r(\omega) = \Psi_p^T \cdot \{x(\omega)\} \tag{4}$$

In the monitoring phase, some unknown signal $\{x_u(\omega)\}$ is filtered by tuned RMVs according to the equation

$$\eta_{ru}(\omega) = \Psi_p^T \cdot \{x_u(\omega)\} \tag{5}$$

which, in turn, lead to FRFs. These can be compared with respective FRFs associated with a reference state. A measure of dissimilarity between both FRFs can be used as a damage index (DI). The authors decided to employ for this purpose a simple RMS of the signal calculated according to the equation:

$$DI_{RMS} = 1 - \frac{\int_{t_1}^{t_2} [y(t) - x(t)]^2}{\int_{t_1}^{t_2} x(t)^2}, \tag{6}$$

It is expected, that occurrence of damage would influence mainly those LMFs, which are placed in damage neighborhood. For practical applications it would be useful to set the threshold for the difference between damage index value for the regions (groups of sensors) with and without damage. Such a threshold would make it possible to detect damaged region more easily and even introduce some automation of diagnostic procedure. However, the threshold needs to be assessed individually depending on the object type and minimum damage size to be detected.

3. Experimental setup

The experiments were performed using a laser scanning Doppler vibrometer PSV-400, provided by Polytec, Germany. The device enabled contact-less measurement of out-of-plane plate velocity vibrations. The plate was fixed to the stand at one extreme. The signals were captured in 209 points distributed over the sample in rectangular-shape lattice (19x11 points), with horizontal spacing of 11 mm and vertical spacing of 16 mm. The acquisition was performed with sensitivity of the device was 50 mm/s/V and sampling frequency of 512Hz. Time duration of signal captured at each point was 32 s. The test sample was excited using white noise generated by a shaker K2004E01, provided by ModalShop. The excitation was transferred using a steel stinger to an impedance head 288D01, provided by PCB, capable of force and acceleration measurements, however, only force signal was acquired for further processing.

Five measurements were performed: three for intact specimen, two for two damage cases. Two notches were introduced to the sample. Their location and size, along with specimen's dimensions and measurement points placement, is depicted in Fig. 2(a). An image of the back side of the specimen after damage introduction is given in Fig. 2(b).

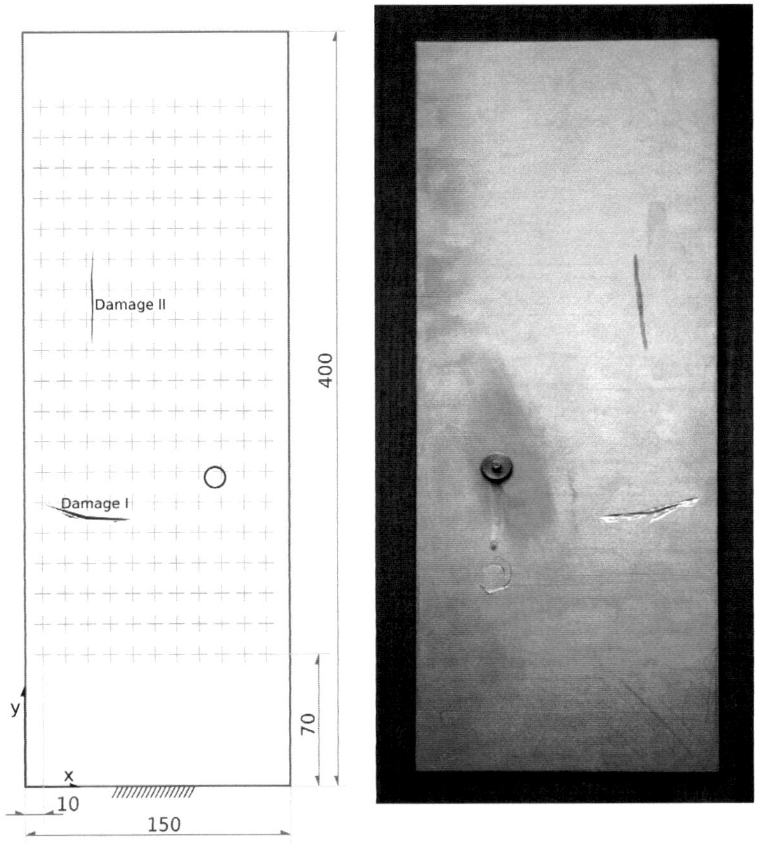

(a) (b)

Figure 2. A scheme of specimen with measurement points' locations marked by crosses. Circle refers to location of an excitation source (a). Image of the specimen is given in (b). Since frame of reference was defined on the face side of the beam, damage locations are shifted in X direction

4. Results

Signals acquired in first measurement step were used to perform modal analysis of a specimen and construct LMFs. For each measurement point its 3x3 neighborhood was used for creation of a local modal filter and corresponding FRFs. In following measurement steps, these filters were used for filtration of newly acquired data. FRFs obtained in that fashion were compared with corresponding reference FRFs with use of DI_{RMS}. Maximal DI acquired for all points under monitoring can be treated as a damage index for the whole specimen, thus damage is detected

Table 1. Results of damage localization with use of LMF. Percentage error given with respect to longest specimen dimension

Damage location measured	Damage location detected	Error [mm]	Error [%]
(30,140)	(88,150)	58.85	14.71
(35,250)	(36,246)	4.12	1.03

by comparing value of DI with a given threshold. Once the threshold is exceeded, location of damage can be obtained by observation of DIs calculated for all LMFs.

Results of this procedure for $5th$ natural frequency are provided in Fig. 3. Detection and localization steps are both performed successfully: For undamaged state (Fig. 3(a) and Fig. 3(b)) maximum value of DI_{RMS} is significantly lower than for damaged states. Measured and detected damage location is given in table 1. Damage is not only located close to its actual position, but also correctly oriented and properly sized.

5. Conclusions

The method developed by authors was successfully applied in detection and localization of damage in plates. Good agreement between measured and detected location, size and orientation of damage reveals as well yet undiscovered potential for damage quantification. Since method has been tested both for truss structures and for plates, its wide potential of applicability include large amount structures, including bridges, cranes or high-voltage transmission lines. The main drawback of the developed solution is its vulnerability to selected mode shape. Selection of a proper mode is not easily achieved beforehand, since modes differ in their sensitiveness to damage in distinct areas of a specimen. Additional work should be performed to create a system that is able to perform diagnosis based on several modes at once.

6. Acknowledgements

This research was conducted under a GEKON1/02/214108/19/2014 project funded by NCBiR.

References

[1] Su Z, Ye L and Lu Y 2006 *Journal of Sound and Vibration* **295** 753–780 ISSN 0022460X URL `http://linkinghub.elsevier.com/retrieve/pii/S0022460X0600109X`
[2] Dworakowski Z, Ambrozinski L, Packo P, Dragan K and Stepinski T 2014 *Structural Control and Health Monitoring, in press*
[3] Colaco D F, de Alexandria A R, Cortez P C, Frota J B B, Lima J N D and de Albuquerque V H C 2010 *Energy Conversion and Management* **51** 2279–2284 ISSN 01968904
[4] Behnia A, Chai H K and Shiotani T 2014 *Construction and Building Materials* **65** 282–302 ISSN 09500618 URL `http://dx.doi.org/10.1016/j.conbuildmat.2014.04.103`
[5] Cury A, Cremona C and Dumoulin J 2012 *Mechanical Systems and Signal Processing* **33** 13–37 ISSN 08883270 URL `http://dx.doi.org/10.1016/j.ymssp.2012.07.005`
[6] Ihesiulor O K, Shankar K, Zhang Z and Ray T 2014 *Composites Part B: Engineering* **56** 906–925 ISSN 13598368 URL `http://dx.doi.org/10.1016/j.compositesb.2013.09.032`
[7] Omar F K and Gaouda a M 2012 *Mechanical Systems and Signal Processing* **26** 190–204 ISSN 08883270 URL `http://dx.doi.org/10.1016/j.ymssp.2011.06.021`
[8] Kohut P and Kurowski P 2009 *Journal of Theoretical and Applied Mechanics* **47** pp. 855–870 ISSN 14292955 URL `http://www.warminski.pollub.plwww.ptmts.org.pl/Kohut-Kur-4-09.pdf`
[9] Xu Y F, Zhu W D, Liu J and Shao Y M 2014 *Journal of Sound and Vibration* **333** 6273–6294
[10] Herman a P, Orifici a C and Mouritz a P 2013 *Composite Structures* **104** 34–42 ISSN 02638223 URL `http://dx.doi.org/10.1016/j.compstruct.2013.04.012`
[11] Hu W H, Caetano E and Cunha A 2013 *Engineering Structures* **57** 578–593 ISSN 01410296 URL `http://dx.doi.org/10.1016/j.engstruct.2012.06.051`
[12] Meirovitch L and Baruh H 1982 *Journal of Guidance, Control, and Dynamics* **8** 60–66

[13] Mendrok K and Uhl T 2008 *Structural Health Monitoring 2008: Proceedings of the Fourth European Workshop* ed Uhl T, Ostachowicz W and Holnicki-Szulc J (Krakow) pp 929–936
[14] Mendrok K and Uhl T 2011 *Structural Health Monitoring* **10** 157–171 ISSN 1475-9217
[15] Mendrok K, Uhl T and Maj W 2013 *Research and Applications in Structural Engineering, Mechanics and Computation - Proceedings of the 5th International Conference on Structural Engineering, Mechanics and Computation, SEMC 2013* pp 2263–2267 ISBN 9781138000612 URL `http://www.scopus.com/inward/record.url?eid=2-s2.0-84889004365&partnerID=tZOtx3y1`
[16] Carrasco C, Osequeda R, Ferregut C and Grygier M 1997 *Proceedings of the 15th International Modal Analysis Conference (IMAC)* (Orlando, Florida) pp 1786–1792
[17] Kohut P and Kurowski P 2006 *Proceedings of the 24th International Modal Analysis Conference (IMAC)* (St. Louis, Missouri)
[18] Ho Y and Ewins D 2000 Numerical Evaluation of Damage Index
[19] Zhang Q, Allemang R and Brown D 1990 *Proceedings of 8th IMAC* (Orlando, Florida) pp 487–496

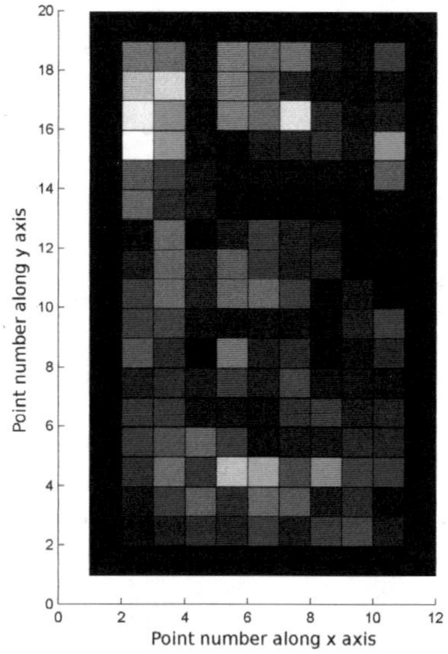

(a) Intact, $DI_{RMS} = 0.60$

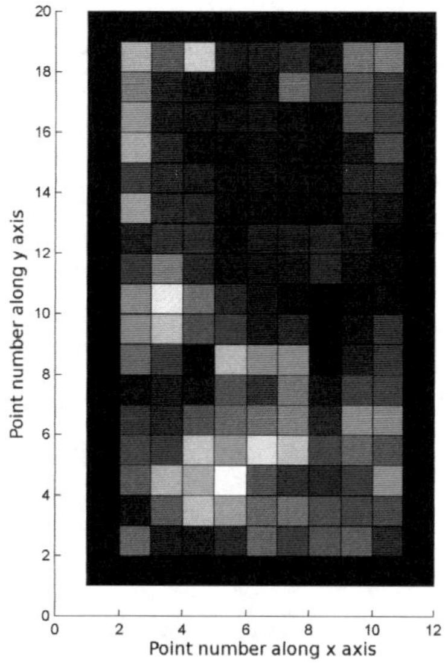

(b) Intact, $DI_{RMS} = 0.64$

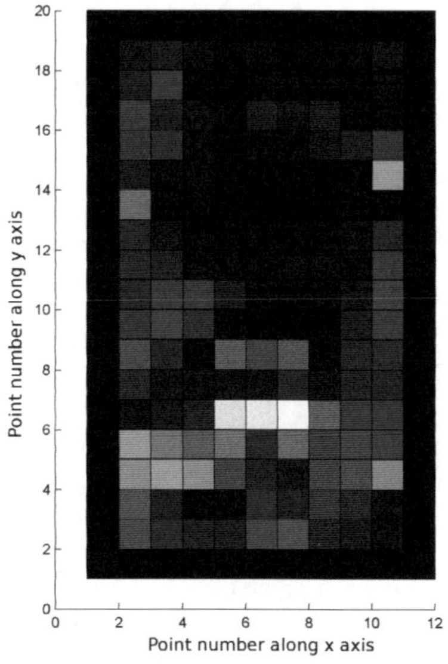

(c) Damage I, $DI_{RMS} = 0.91$

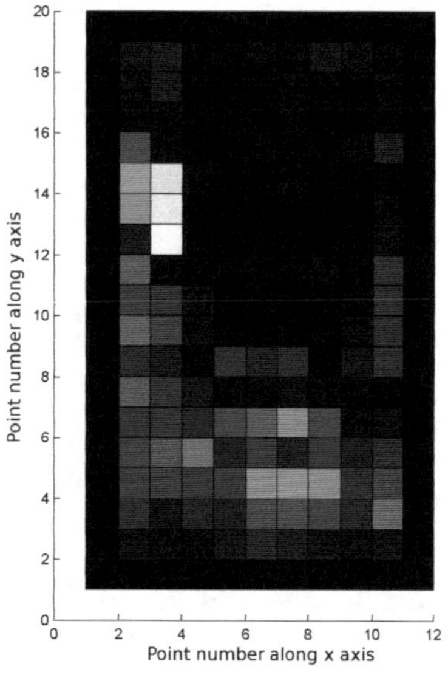

(d) Damage II, $DI_{RMS} = 1.30$

Figure 3. Results of local modal filtration for intact state (a,b) and for damaged specimen (c,d) along with corresponding values of the DI_{RMS}. Since LMFs are constructed from 3x3 group of points, borders of the sample are excluded from the damage search. The light intensity refer to the value of DI_{RMS} calculated for given LMF

Structural damage detection using auto correlation functions of vibration response under sinusoidal excitation

Muyu Zhang[1], Rüdiger Schmidt[1], Bernd Markert[1,2]

[1]Institute of General Mechanics, RWTH Aachen University, Templergraben 64, 52062 Aachen, Germany
[2]Faculty of Engineering and Science, University of Agder, Televeien 9, 4879 Grimstad, Norway

E-mail: muyu@iam.rwth-aachen.de

Abstract. Structural damage detection using time domain vibration responses has attracted more and more researchers in recent years because of its simplicity in calculation and no requirement of a finite element model. This paper proposes a new approach to locate the damage using the auto correlation function of vibration response signals under sinusoidal excitation from different measurement points of the structure, based on which a vector named Auto Correlation Function at Maximum Point Value Vector (AMV) is formulated. A sensitivity analysis of the normalized AMV with respect to the local stiffness shows that under several specific frequency excitations, the normalized AMV has a sharp change around the local stiffness change location, which means that even when the damage is very small, the normalized AMV is a good indicator for the damage. In order to locate the damage, a damage index is defined as the relative change of the normalized AMV before and after damage. Several example cases in stiffness reduction detection of a frame structure valid the results of the sensitivity analysis, demonstrate the efficiency of the normalized AMV in damage localization and the effect of the excitation frequency on its detectability.

1. Introduction

Structural health monitoring and damage detection using time domain vibration response is appealing, because the important health information of the structure is involved in the vibration response signals, which can be easily measured by conventional techniques. Moreover, it has advantages of simplicity in calculation and no finite element analysis is required that can be performed in real-time and online.

In the literature, several approach than use the auto/cross correlation functions of the vibration responses for damage detection are reported. Nichols and Seaver [1] proposed the correlation-function-based transfer entropy, which is used for detecting the presence of damage-induced nonlinearities in structures. Overbuy and Todd [2] modified this transfer entropy and used it for the damage identification of an experimental 8-degrees-of-freedom oscillator. Trendafilova [3] developed a damage indicator from the normalized cross correlation to detect the presence and location of the different kinds of delamination damage in a composite laminate beam that is subjected to random excitation. Li and Law [4] used the matrix of the covariance of covariance of acceleration response signals under white noise excitation to successfully detect the damages of a simply supported plane truss structure. Yang and Yu [5] proposed a vector named Cross Correlation Function Amplitude Vector (CorV) for damage detection. Based on the concept of CorV and Natural Excitation Technique (NExT) [6], Yang and Wang [7] proposed a damage detection method using the Inner Product Vector

(IPV) of the vibration responses. Similar to the IPV, Zhang and Schmidt [8] introduced a damage index named Auto Correlation Function at Maximum Point Value Vector (AMV). The stiffness reduction detection of a 12-story frame structure showed that the AMV-based damage detection method can locate the damage effectively, even in the presence of noise and has a better detectability compared to the other correlation-function-based damage detection methods.

In this paper, the effectiveness of AMV is studied when the sinusoidal excitation is applied on the structure. Firstly, the theory of auto correlation function of vibration responses under sinusoidal excitation is presented, based on which the damage index is formulated. Secondly, sensitivity analysis is used to substantiate the choice of this damage index for damage detection. As an example, damage detection of a 12-story frame structure is used to verify the proposed method, followed by a comparison of the excitation frequency on the detectability. Finally, some conclusions are made.

2. Theory of auto correlation function and damage index

2.1. Auto correlation function

The standard matrix equations of motion is given by

$$\mathbf{M}\ddot{\mathbf{x}}(t)+\mathbf{C}\dot{\mathbf{x}}(t)+\mathbf{K}\mathbf{x}(t)=\mathbf{f}(t) \tag{1}$$

where $\mathbf{M}$ is the mass matrix, $\mathbf{K}$ is the stiffness matrix, $\mathbf{C}=\alpha\mathbf{M}+\beta\mathbf{K}$ is the damping matrix, $\mathbf{f}$ is a vector of the force and $\mathbf{x}(t)$ is the vector of displacement.

Assume real normal modes and the excitation is set at point k, the auto correlation function of the displacement response under sinusoidal excitation at point i when $T=0$ can be expressed as

$$R_{i,i}=\sum_{r=1}^{n}\psi_{ir}[\sum_{s=1}^{n}\psi_{is}\mu_k^{rs}] \tag{2}$$

where ψ_{ir} and ψ_{is} represent the ith component of mode shape $\mathbf{\Psi}_r$ and $\mathbf{\Psi}_s$, respectively, and

$$\mu_k^{rs}=\frac{X_k^2}{2}\frac{\psi_{kr}\psi_{ks}}{m^r m^s L_r L_s}P_{rs} \tag{3}$$

where X_k is the amplitude of the sinusoidal excitation, m^r is the r th modal mass and

$$L_i=[(\omega_n^i)^2+(2\pi f_{0,k})^2]^2-4(2\pi f_{0,k}\omega_d^i)^2 \tag{4}$$

$$P_{rs}=D_r D_s+4(2\pi f_{0,k})^2\zeta^r\omega_n^r\zeta^s\omega_n^s \tag{5}$$

$$D_i=(\omega_n^i)^2-(2\pi f_{0,k})^2 \tag{6}$$

Herein, ω_n^r is the r th modal frequency, $\zeta^r=\frac{1}{2}(\frac{\alpha}{\omega_n^r}+\beta\omega_n^r)$ is the r th modal damping ratio, ω_d^r is the r th damped modal frequency and $f_{0,k}$ is the frequency of the sinusoidal excitation.

2.2. Damage index

Set the values of the auto correlation function of the responses from different measurement points at $T=0$ as a vector

$$\mathbf{R}_{AMV}=[R_{1,1},R_{2,2},\ldots,R_{n,n}]^T \tag{7}$$

where $R_{i,i}$ is the value of the auto correlation function of the response from measurement point i at $T=0$. Since the auto correlation function value at $T=0$ is also its maximum value, we define this vector as Auto Correlation Function at Maximum Point Value Vector (AMV).

In order to eliminate the influence of the excitation, $\mathbf{R}_{AMV}$ is normalized by its root mean square value as follows

$$\overline{\mathbf{R}}_{AMV} = \frac{\mathbf{R}_{AMV}}{rms(\mathbf{R}_{AMV})} = [\overline{R}_{1,1}, \overline{R}_{2,2}, ..., \overline{R}_{n,n}]^T \tag{8}$$

where $rms(\mathbf{R}_{AMV}) = \sqrt{\frac{1}{n}\sum_{i=1}^{n} R_{i,i}^2}$.

Then the relative change of the ith element in $\overline{\mathbf{R}}_{AMV}$ is defined as

$$D_{AMV,i} = \frac{\overline{R}_{i,i}^d - \overline{R}_{i,i}^u}{\overline{R}_{i,i}^u} \tag{9}$$

where the superscript d denotes the damaged structural state and u indicates the intact structural state. Thus, the damage index of the AMV method can be defined as $\mathbf{D}_{AMV} = [D_{AMV,1}, D_{AMV,2}, ... D_{AMV,n}]^T$.

3. Sensitivity analysis and damage detection

The sensitivity is defined as the relative change of a variable to another variable z , so the sensitivity of the ith element of the vector $\overline{\mathbf{R}}_{AMV}$ to the local stiffness k_j can be given by [8]

$$\eta(\overline{R}_{i,i} / k_j) = k_j\left(\frac{\partial R_{i,i} / \partial k_j}{R_{i,i}} - \frac{\partial rms(\mathbf{R}_{AMV}) / \partial k_j}{rms(\mathbf{R}_{AMV})}\right) \tag{10}$$

which describes how the value $\overline{R}_{i,i}$ will change due to changes of the local stiffness k_j .

Deduced from the basic vibration theory and assume the proportional damping $\mathbf{C} = \alpha\mathbf{M} + \beta\mathbf{K}$, the partial derivative of the frequency to the local stiffness $\partial\omega_n^r / \partial k_j$ can be obtained from the real part of

$$\frac{\partial \lambda_r}{\partial k_j} = -\frac{1}{2(\lambda_r + \zeta^r \omega_n^r)} \mathbf{\Psi}_r^T \left(\lambda_r^2 \frac{\partial \mathbf{M}}{\partial k_j} + \lambda_r \frac{\partial \mathbf{C}}{\partial k_j} + \frac{\partial \mathbf{K}}{\partial k_j}\right) \mathbf{\Psi}_r \tag{11}$$

where $\lambda_r = -\zeta^r \omega_n^r \pm i\omega_d^r$. The partial derivative of the mode shape to the local stiffness $\partial\mathbf{\Psi}_r / \partial k_j$ is given by

$$\frac{\partial \mathbf{\Psi}_r}{\partial k_j} = \sum_{u=1}^{n} a_u \mathbf{\Psi}_u \tag{12}$$

where $a_u = \begin{cases} -\dfrac{1}{(\lambda_r^2 - \lambda_u^2) + (\lambda_r - \lambda_u)[\alpha + \beta(\omega_n^u)^2]} \mathbf{\Psi}_u^T \left(\lambda_r^2 \dfrac{\partial \mathbf{M}}{\partial k_j} + \lambda_r \dfrac{\partial \mathbf{C}}{\partial k_j} + \dfrac{\partial \mathbf{K}}{\partial k_j}\right) \mathbf{\Psi}_r, & u \neq r \\ -\dfrac{1}{2} \mathbf{\Psi}_r^T \dfrac{\partial \mathbf{M}}{\partial k_j} \mathbf{\Psi}_r, & u = r \end{cases}$

According to equation (2), $R_{i,i}$ is only related to the frequency and mode shape of the system, so that the partial derivative of $R_{i,i}$ to the local stiffness k_j can be expressed by

$$\frac{\partial R_{i,i}}{\partial k_j} = \frac{\partial f(\psi_{kr}, \psi_{ks}, \psi_{ir}, \psi_{is}, \omega_n^r, \omega_n^s)}{\partial k_j} \tag{13}$$

By substitution of equations (11, 12) into equation (13), the partial derivate $\partial R_{i,i} / \partial k_j$ can be obtained. Then by substituting equation (13) into equation (10), the sensitivity $\eta(\overline{R}_{i,i} / k_j)$ of the normalized correlation function value at $T = 0$ to the local stiffness k_j can be obtained.

3.1. Sensitivity analysis results

As shown in figure 1, a 12-story shear frame structure is adopted as the numerical simulation model. The centred mass of each floor m_i is 1 kg, which is centralized on its floor. The stiffness of each floor is supplied by the braces, and the stiffness in the out-of-the-plane direction y that is perpendicular to the x-z plane is much larger than that in x direction. As a result, only the movement in x direction is considered. The stiffness coefficient k_i of each floor is 20,000 N/m. Based on these assumptions, the mechanical model of the frame structure can be expressed as a 12 degree of freedom (DOF) discrete system, the mass matrix $\mathbf{M}$ and stiffness matrix $\mathbf{K}$ can be obtained easily from the basic vibration theory. Besides, proportional damping $\mathbf{C} = \alpha\mathbf{M} + \beta\mathbf{K}$ is adopted.

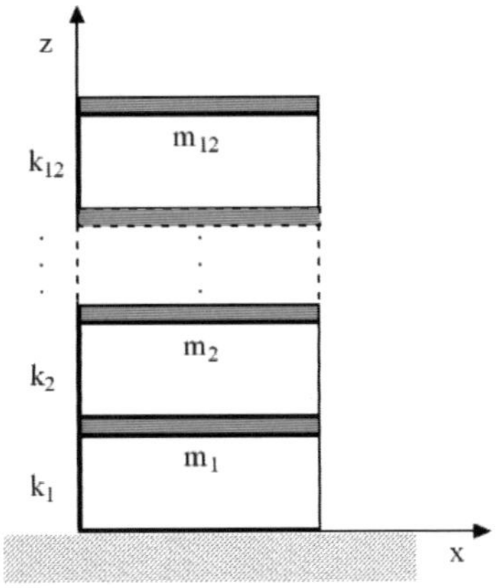

Figure 1. 12-story frame structure

Substituting $[\partial\mathbf{M}/\partial k_j]_{pq} = \partial m_{pq} / \partial k_j$ and $[\partial\mathbf{K}/\partial k_j]_{pq} = \partial k_{pq} / \partial k_j$, where m_{pq} and k_{pq} are the p th row, q th column elements of mass matrix $\mathbf{M}$ and stiffness matrix $\mathbf{K}$, respectively, into equation (11) and equation (12), the sensitivity of the normalized AMV with respect to the local stiffness changes can be calculated by equation (10).

Figure 2 to 4 depict the sensitivity of the normalized AMV to the local stiffness k_j, where the local stiffness change location j is chosen from measurement point 1 to measurement point 12. In each figure, the red solid line, the green dotted line, the yellow dash-dotted line and the blue dotted line are the sensitivity of the normalized AMV calculated using the auto correlation function of vibration responses under 1 Hz, 2.5 Hz, 3 Hz and 10 Hz sinusoidal excitation, respectively.

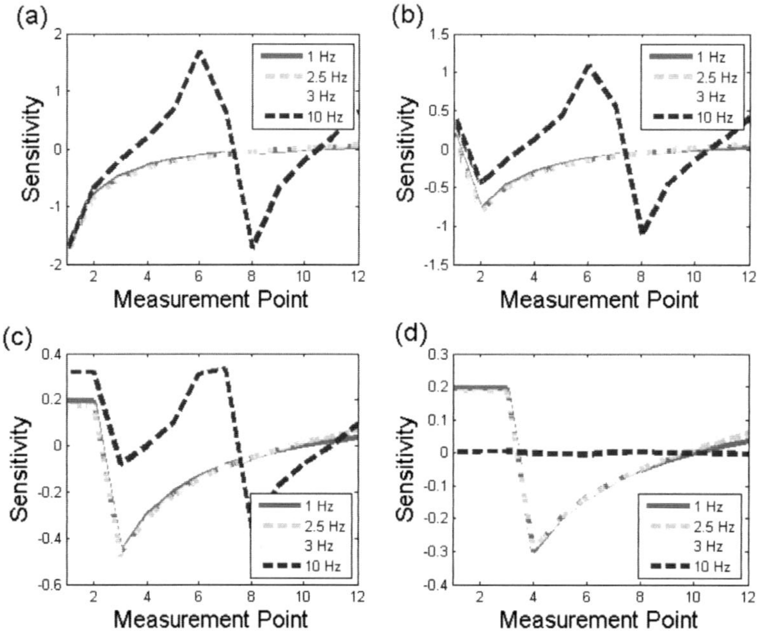

Figure 2. Sensitivity of normalized AMV to the local stiffness k_j:
(a) $j=1$; (b) $j=2$; (c) $j=3$; (d) $j=4$.

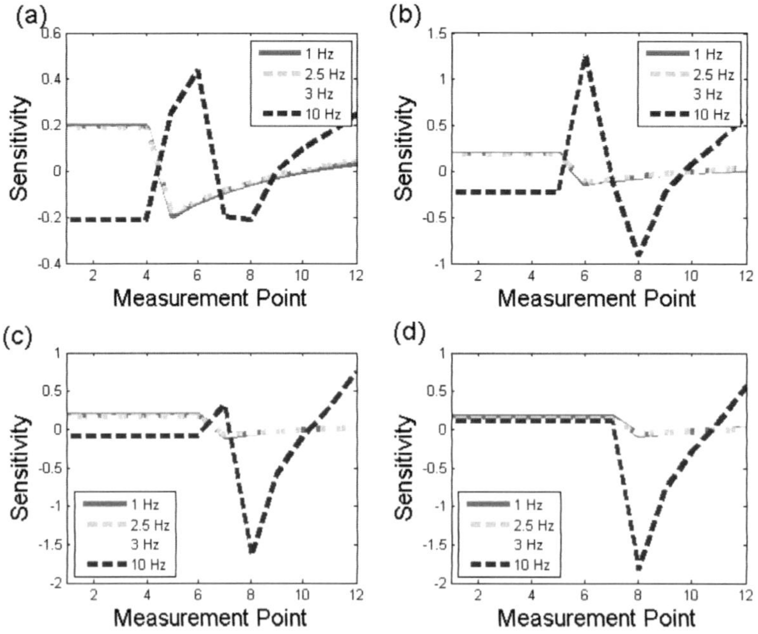

Figure 3. Sensitivity of normalized AMV to the local stiffness k_j:
(a) $j=5$; (b) $j=6$; (c) $j=7$; (d) $j=8$.

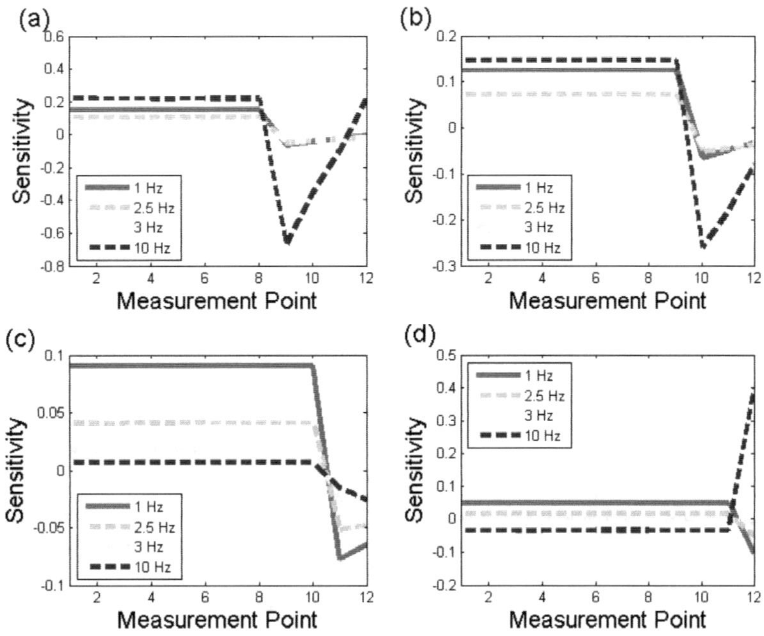

Figure 4. Sensitivity of normalized AMV to the local stiffness k_j:
(a) $j=9$; (b) $j=10$; (c) $j=11$; (d) $j=12$.

From figure 2 to 4, we can observe one common phenomenon for the sensitivities calculated from the responses under 1 Hz, 2.5 Hz and 3 Hz sinusoidal excitation. As the 1st natural frequency of the structure is 2.8 Hz, the excitation frequency is below or a little higher than the 1st natural frequency. The sensitivity of the normalized AMV to the local stiffness k_j has a sharp change at the measurement point j, before point j, it has a positive value while after j, it has a negative value. Consequently, when the local stiffness k_j decreases, the value of $\overline{R}_{i,i}$ also decreases before the measurement point j, but it increases after the measurement point j. Clearly, the relative change of $\overline{R}_{i,i}$ before and after k_j decreases will have a sharp change around the local stiffness change location, i.e. the measurement point j. Therefore, using different measurement point values of $\overline{R}_{i,i}$, local stiffness change locations can be observed, and the damage index calculated from equation (9) can be used for the damage localization.

As for the sensitivity calculated from the response under 10 Hz sinusoidal excitation, the analysis does not reveal the phenomenon as the previous three excitation frequencies. The method just result in random values from measurement point 1 to measurement point 12. Consequently, the damage index calculated from equation (9) cannot be used for damage localization if the excitation frequency is 10Hz.

3.2. Damage detection results

The damage of the frame structure is simulated by reducing the stiffness coefficient of 2nd, 5th, 8th and 11th floor by 5%, respectively. The different damage cases are listed in Table 1.

The excitation is applied on the first floor of the structure. It has duration of 20 s with a sample frequency of 1024 Hz and a magnitude of 5 N. The *Wilson-θ* method is utilized to get the vibration responses of the intact structure and the different damage cases.

Table 1. Damage cases of the frame structure

Damage case	D2	D5	D8	D11
Damage location	2	5	8	11
Damage extent/%		5		

The value of the auto correlation function can be obtained by the inner product of the vibration response. After getting the auto correlation function of the response before and after damage, the damage index can be calculated using the equations (7-9). The results of the different damage cases are shown in figure 5. Therein, the red solid line, the green dotted line, the yellow dash-dotted line and the blue dotted line represent the damage indexes calculated using the auto correlation function of vibration responses under 1 Hz, 2.5 Hz, 3 Hz and 10 Hz, respectively.

The damage indexes coincide well with the sensitivity analysis in section 3.1. Consider the red solid lines in figure 3(a) and figure 5(b) for instance. The red solid line in figure 3(a) is the sensitivity of the normalized AMV to the local stiffness, when the local stiffness change location is set at the 5th floor and the excitation frequency is 1 Hz. The red solid line in figure 5(b) is the damage index value obtained by the vibration responses under 1Hz sinusoidal excitation when the 5th floor has a 5% stiffness reduction. In figure 3(a), the value of $\eta(\overline{R}_{i,i}/k_5)$ has a positive value from measurement point 1 to measurement point 4 and they are almost identical. The value drops sharply at measurement point 5, where it has a negative value. After measurement point 5, it slowly increases to its initial level again. $\eta(\overline{R}_{i,i}/k_5)$ has a minimal value at measurement point 5. Thus, when k_5 decreases, $\overline{R}_{4,4}$ will decrease and $\overline{R}_{5,5}$ will increase. Thereby, the decrease and increase rates are very large. As a consequence, before and after k_5 decreases, the relative change of $\overline{R}_{4,4}$, i.e. $D_{AMV,4}$ is negative while the relative change of $\overline{R}_{5,5}$, i.e. $D_{AMV,5}$ is positive, such that there is a large difference between $D_{AMV,4}$ and $D_{AMV,5}$. This can be seen in figure 5(b). Moreover, for $\eta(\overline{R}_{i,i}/k_5)$, the values at measurement point 1 to measurement point 4 are almost the same and the values at measurement point 5 to measurement point 12 just have a slight difference. So the relative change of $\overline{R}_{1,1} \sim \overline{R}_{4,4}$ before and after k_5 decreases just show a very small change and likewise the relative change of $\overline{R}_{5,5} \sim \overline{R}_{12,12}$. Accordingly, the values of $D_{AMV,1} \sim D_{AMV,4}$ and $D_{AMV,5} \sim D_{AMV,12}$ do not change much, which can also be seen in figure 5(b). Other measurement points shon the same phenomenon. While the sensitivity analysis uses only the modal parameters of the structure and the damage detection uses only the time domain vibration responses of the structure, the good agreement of the results reveals that the damage index proposed in this paper is a good indicator for damage detection.

Similar results can be obtained using the sinusoidal excitation of 2.5 Hz and 3 Hz. For the damage index when the auto correlation function is calculated by the responses under 10 Hz sinusoidal excitation, they also coincide well with the sensitivity results in section 3.1 that the AMV-based damage index cannot locate the damage.

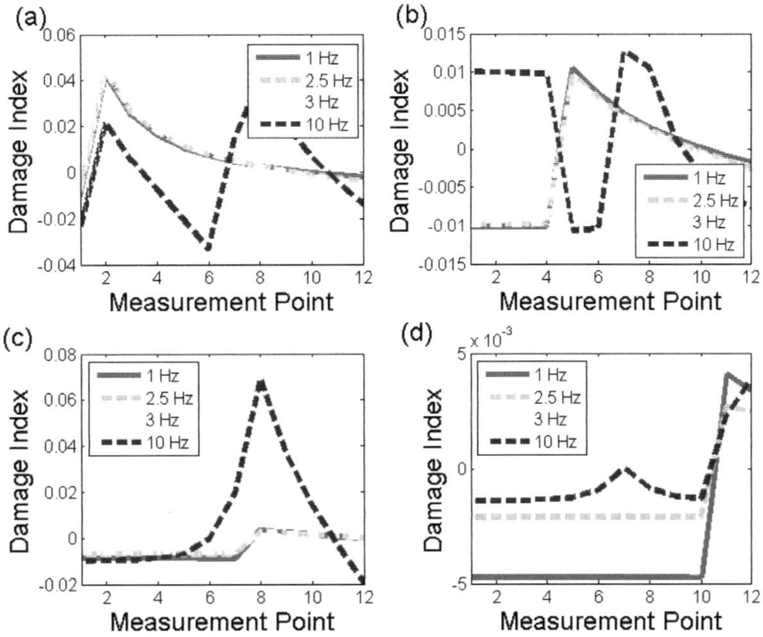

Figure 5. Damage index of different damage cases:
(a) Damage case D2; (b) Damage case D5; (c) Damage case D8; (d) Damage case D11.

4. Conclusions

An AMV-based damage detection method is proposed and studied in this paper. Sensitivity analysis shows that the normalized AMV can be used to locate the damage when the frequency of sinusoidal excitation is below or a little larger than the 1st natural frequency of the structure. Damage detection results of a 12-floor story frame structure prove the results of the sensitivity analysis, showing that the AMV is a good indicator for damage detection.

However, the AMV-based damage index cannot locate the damage for all the frequencies of the sinusoidal excitation. So it is important to choose the suitable excitation frequency when the AMV is applied for damage detection.

References

[1] Nichols J M, Seaver M, Trickey S T, Todd M D, Olson C and Overbey L 2005 *Phys. Rev.* E **72** 046217
[2] Overbey L and Todd M D 2009 *J. Sound. Vib* **322** 438-53
[3] Trendafilova I, Palazzetti R and Zucchelli A 2015 *European Journal of Mechanics-A/SolidsEur. J. Mech* A-Solid **49** 197-204
[4] Li X Y and Law S S 2010 *Mech. Syst. Signal. Pr* **24** 945-56
[5] Yang Z C, Yu Z F and Sun H 2007 *Mech. Syst. Signal. Pr* **21** 2918-32
[6] James G H III and Farrar C R 1997 *J. Sound. Vib* **205** 1-18
[7] Wang L, Yang Z C and Waters T P 2010 *J. Sound. Vib* **329** 5070-86
[8] Zhang M Y and Schmidt R 2014 *J. Sound. Vib* **333** 7352-63

Load identification approach based on basis pursuit denoising algorithm

D Ginsberg[1], M Ruby[2], C P Fritzen[3]
[1]Ph.D.-Student, University of Siegen, Department of Mechanical Engineering, Siegen, Germany
[2]Master-Student, University of Siegen, Department of Mechanical Engineering, Siegen, Germany
[3]Head of Workgroup, University of Siegen, Department of Mechanical Engineering, Siegen, Germany

E-mail: daniel.ginsberg@uni-siegen.de

Abstract. The information of the external loads is of great interest in many fields of structural analysis, such as structural health monitoring (SHM) systems or assessment of damage after extreme events. However, in most cases it is not possible to measure the external forces directly, so they need to be reconstructed. Load reconstruction refers to the problem of estimating an input to a dynamic system when the system output and the impulse response functions are usually the knowns. Generally, this leads to a so called ill-posed inverse problem, which involves solving an underdetermined linear system of equations. For most practical applications it can be assumed that the applied loads are not arbitrarily distributed in time and space, at least some specific characteristics about the external excitation are known a priori. In this contribution this knowledge was used to develop a more suitable force reconstruction method, which allows identifying the time history and the force location simultaneously by employing significantly fewer sensors compared to other reconstruction approaches. The properties of the external force are used to transform the ill-posed problem into a sparse recovery task. The sparse solution is acquired by solving a minimization problem known as basis pursuit denoising (BPDN). The possibility of reconstructing loads based on noisy structural measurement signals will be demonstrated by considering two frequently occurring loading conditions: harmonic excitation and impact events, separately and combined. First a simulation study of a simple plate structure is carried out and thereafter an experimental investigation of a real beam is performed.

1. Introduction

Online monitoring of external loads is a very important technique in the field of structural health monitoring (SHM). For SHM-Systems, the information of the time history of acting forces in combination with the location where these forces are applied, is useful to make a statement about the occurrence of damage and its extent after adverse loading conditions. In reality, external loads are reconstructed using measured structural responses, such as acceleration, velocity, position or strain. A direct measurement is mostly difficult or sometimes even not possible, e.g. when the force application position is unknown. The reconstruction of external forces from dynamic measurements results in an inverse problem, which can be considered as mathematically ill-posed in most cases, which means that either the existence, the uniqueness or the stability of the solution is violated [1][2][3]. If this inversion can be done, the structure itself becomes its own force sensor [4]. In order to overcome the problem of ill-posedness several studies have been carried out and a considerable amount of literature has been published on this topic, e.g. [3][4][5][6][7].

In this contribution the inverse problem is tackled by the so called basis pursuit denoising algorithm. Some characteristics of the applied force are employed to transform the force reconstruction problem in a sparse recovery task. This paper is structured as follows: Section 2 briefly discusses the general problem of force reconstruction and explains how it can be transformed into a sparse identification problem. In Section 3, the applied approach for solving the sparse reconstruction problem is introduced. A simulation study and an experimental investigation in Section 4 and 5 show the capability of the proposed identification method. Finally, concluding remarks are presented in Section 6.

2. Problem

Dynamic external forces applied on a mechanical structure will stimulate the system to vibrate. In general these structural responses can easily be measured by appropriate sensors. For linear systems the structural responses $\mathbf{y}(t) \in \mathbb{R}^r$ are linked to the external forces input $\mathbf{u}(t) \in \mathbb{R}^f$ in the time continuous case by the linear convolution integral

$$\mathbf{y}(t) = \int_0^t \mathbf{h}(\tau)\,\mathbf{u}(t-\tau)\,\mathrm{d}\tau \qquad (1)$$

where $\mathbf{h}(t) \in \mathbb{R}^{r\times f}$ denotes a matrix containing the impulse response functions (IRF) $h_{i,j}$. It is assumed that $\mathbf{u}(t) = \mathbf{h}(t) = \mathbf{y}(t) = \mathbf{0}$ for $t < 0$. The IRF $h_{i,j}$ describes the transmission of an external force applied at a discrete input position $j = 1, 2, \ldots, f$, to the structural responses at output position $i = 1, 2, \ldots, r$. In a discrete time domain ($t = n\Delta t,\ n \in \mathbb{N}$), the convolution integral becomes an algebraic equation:

$$\mathbf{y}_k = \sum_{i=0}^{k} \mathbf{h}_i \mathbf{u}_{k-i}\ , \quad (k = 0, 1, 2, \ldots, n) \qquad (2)$$

Here, $\mathbf{h}_i \in \mathbb{R}^{r\times f}$ are the so-called Markov parameters. Rewriting Equation (2) in vector matrix form leads to

$$\mathbf{Y} = \mathbf{HU}\ , \qquad (3)$$

with the transfer matrix $\mathbf{H} \in \mathbb{R}^{r\cdot n\times f\cdot n}$, the measurement vector $\mathbf{Y} \in \mathbb{R}^{r\cdot n}$ and the input vector $\mathbf{U} \in \mathbb{R}^{f\cdot n}$:

$$\mathbf{H} = \begin{bmatrix} \mathbf{h}_0 & \mathbf{0} & \cdots & \mathbf{0} \\ \mathbf{h}_1 & \mathbf{h}_0 & \ddots & \vdots \\ \vdots & \vdots & \ddots & \mathbf{0} \\ \mathbf{h}_n & \mathbf{h}_{n-1} & \cdots & \mathbf{h}_0 \end{bmatrix}, \quad \mathbf{Y} = \begin{bmatrix} \mathbf{y}_0 \\ \mathbf{y}_1 \\ \vdots \\ \mathbf{y}_n \end{bmatrix}, \quad \mathbf{U} = \begin{bmatrix} \mathbf{u}_0 \\ \mathbf{u}_1 \\ \vdots \\ \mathbf{u}_n \end{bmatrix} \qquad (4)$$

In case of a forward problem the IRFs or the Markov parameters and the external forces are known, so that the desired system response can be calculated by Equation (1), (2) or (3). However, generally the external force respectively the input to the system is unknown and the structural responses are determined by (noisy) measurements. Thus force identification becomes a deconvolution problem. But this also requires the determination of the system dynamics in advance. If the input position of a single force is known a priori, the number of inputs f is reduced to 1 and the force reconstruction problem is a bit more simplified. Since in general the

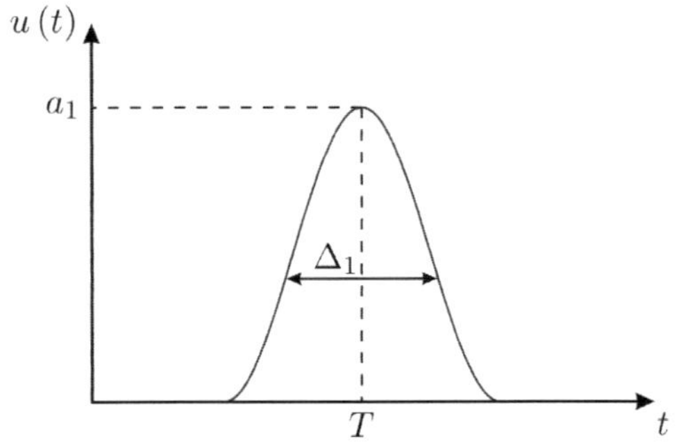

Figure 1. General representation of an impact force history

input position is unknown, every location on the surface becomes a potential input position. Depending on the type of structural model, these positions are usually coinciding with the degrees of Freedom (DOF). For real applications, the number of measurement sensors r is normally smaller than the number of DOFs, so that $r < f$ and the deconvolution problem becomes even more challenging. Nevertheless, in most practical applications at least some characteristics about the acting forces are known and can be taken into account in the estimation process. Two frequently occurring loading conditions are harmonic forces and impact events. These loading types will be further investigated in this paper.

Impact forces are characterized by a short time duration and a spatial concentration. The profile of the force history looks approximately like the pulse function shown in Figure 1. This pulse-shape is commonly modeled as half or quarter cycle-sine pulses function (see e.g. [8] and [9]). The force history can be described by three parameters: the pulse width, indicated by the parameter $\Delta \in [\Delta_{\min}, \Delta_{\max}]$, the time $T \in [0, T_{\max}]$ at which the magnitude is reached and the magnitude value of this pulse $a_{T,\Delta} \in \mathbb{R}$ itself. However, if the impact is intended to be reconstructed using measured structural responses these parameters are unknown and need to be identified. Additionally the location where the load is applied is unknown as well.

But, employing these parameters an impact dictionary $\tilde{\mathbf{U}}_{\mathrm{I}}$ can be created which contains the force time history of all potential parameter combinations at all potential force input locations. To this end, the interval $[0, T_{\max}]$ needs to be discretized according to the sampling time (see Equation (2)) with $T = l\Delta t$ $(l = 0, 1, 2, \ldots, m)$ and $T_{\max} = m\Delta t$ and also the interval $[\Delta_{\min}, \Delta_{\max}]$ describing the range of potential impact durations, has to be discretized appropriate, with $\Delta = \Delta_{\min} + (p-1)\,\Delta_p$ and $(p = 1, 2, \ldots, q)$. By multiplying this dictionary $\tilde{\mathbf{U}}_{\mathrm{I}} \in \mathbb{R}^{f \cdot n \times f \cdot m \cdot q}$ by a vector $\mathbf{A}_{\mathrm{I}} \in \mathbb{R}^{f \cdot m \cdot q}$

$$\hat{\mathbf{U}}_{\mathrm{I}} = \tilde{\mathbf{U}}_{\mathrm{I}} \cdot \mathbf{A}_{\mathrm{I}} \ , \tag{5}$$

$\hat{\mathbf{U}}_{\mathrm{I}}$ equals now the actual force input of a single impact, if all entries in $\mathbf{A}_{\mathrm{I}}$ are zeros, except the one which belongs to the actual impact. Here $\mathbf{A}_{\mathrm{I}}$ can be considered as magnitude vector, if the force histories in $\tilde{\mathbf{U}}_{\mathrm{I}}$ are normalized to unit.

A similar dictionary matrix can be created for harmonic loads. It is well known that harmonic signals can also be described by a small number of parameters: the frequency of the force and its cosine and sine amplitude. If the applied load is in a known frequency bandwidth $\omega \in [\omega_{\min}, \omega_{\max}]$ the harmonic force dictionary $\tilde{\mathbf{U}}_{\mathrm{H}} \in \mathbb{R}^{f \cdot n \times f \cdot 2 \cdot s}$ reads as:

$$\tilde{\mathbf{U}}_{\mathrm{H}} = \begin{bmatrix} \tilde{\mathbf{U}}_{\mathrm{HC}} & \tilde{\mathbf{U}}_{\mathrm{HS}} \end{bmatrix} \tag{6}$$

It is combined from a dictionary $\tilde{\mathbf{U}}_{\mathrm{HC}}$ containing normalized cosine force histories of different frequencies in the interval $[\omega_{\min}, \omega_{\max}]$ and a corresponding sine dictionary $\tilde{\mathbf{U}}_{\mathrm{HS}}$. The discretization of the frequency interval depends on the length of the measured response time

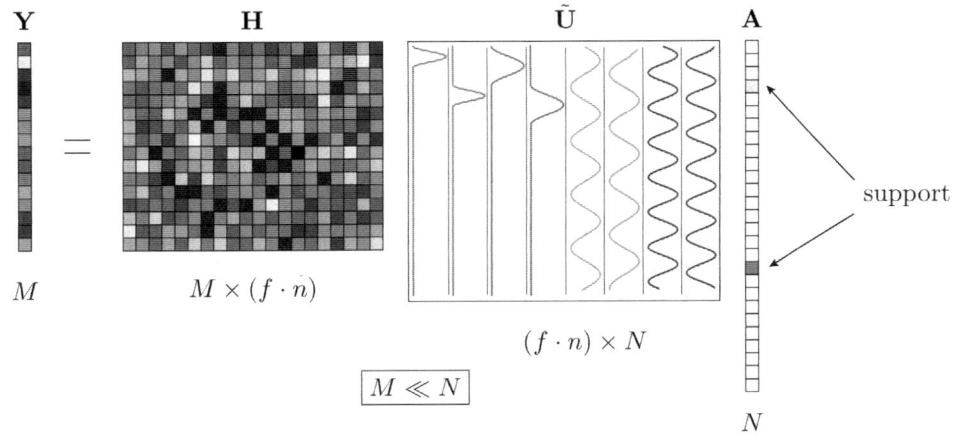

Figure 2. By pre-multiplying a force dictionary, identification becomes a sparse recovery problem

history, $\omega = \omega_{\min} + (v-1)\,\Delta\omega$ with $(v = 1, 2, \ldots, s)$. Again, if the dictionary $\tilde{\mathbf{U}}_{\mathrm{H}}$ is multiplyed by a magnitude vector $\mathbf{A}_{\mathrm{H}} = \begin{bmatrix}\mathbf{A}_{\mathrm{HC}}^T & \mathbf{A}_{\mathrm{HS}}^T\end{bmatrix}^T$

$$\hat{\mathbf{U}}_{\mathrm{H}} = \tilde{\mathbf{U}}_{\mathrm{H}} \cdot \mathbf{A}_{\mathrm{H}} \tag{7}$$

$\hat{\mathbf{U}}_{\mathrm{H}}$ equals the actual harmonic load (with the same frequency, phase and amplitude), if all elements of $\mathbf{A}_{\mathrm{H}}$ are zero, except cosine and sine magnitudes corresponding to the actual frequency of the harmonic force. The impact and the harmonic force dictionary are united to entire input dictionary $\tilde{\mathbf{U}}$ with an associated entire magnitude vector $\mathbf{A}$:

$$\tilde{\mathbf{U}} = \begin{bmatrix}\tilde{\mathbf{U}}_{\mathrm{H}} & \tilde{\mathbf{U}}_{\mathrm{I}}\end{bmatrix} \in \mathbb{R}^{f \cdot n \times f(m \cdot q + 2 \cdot s)} \text{ and } \mathbf{A} = \begin{bmatrix}\mathbf{A}_{\mathrm{H}}^T & \mathbf{A}_{\mathrm{I}}^T\end{bmatrix}^T \in \mathbb{R}^{f(m \cdot q + 2 \cdot s)} \tag{8}$$

If this entire dictionary is used as an input to the system, it follows from Equation (3):

$$\mathbf{Y} = \mathbf{H} \cdot \tilde{\mathbf{U}} \cdot \mathbf{A} = \tilde{\mathbf{H}} \cdot \mathbf{A} \tag{9}$$

Thus, a new transition matrix $\tilde{\mathbf{H}} = \mathbf{H} \cdot \tilde{\mathbf{U}} \in \mathbb{R}^{r \cdot n \times f(m \cdot q + 2 \cdot s)}$ can be obtained (see Figure 2).

Now force reconstruction becomes a problem of solving Equation (9) for $\mathbf{A}$. However, it can be easily seen that $\tilde{\mathbf{H}}$ tends to have a lot more columns $(N := f\,(m \cdot q + 2 \cdot s))$ than rows $(M := r \cdot n)$. Hence the linear system of equations in Equation (9) has a lot more unknowns than knowns and thus solving for $\mathbf{A}$ is in general not possible. But it is known that $\mathbf{A}$ has just a very few nonzero entries and can be considered as sparse vector. The property of sparsity of the desired magnitude vector $\mathbf{A}$ is used in the following to obtain an optimal estimate.

3. Sparse solution

The sparse solution of a high-dimensional underdetermined problem is an application of Occam's Razor: in face of many possibilities, all of which are plausible, favor the simplest candidate solutions [10]. Here, it can also been regarded as a problem of finding the columns in $\tilde{\mathbf{H}}$ which correlate the most with the measurement vector $\mathbf{Y}$.

A possibility to obtain a sparse solution for the linear system in Equation (9) is to minimize the l_0-norm of the solution vector $\mathbf{A}$:

$$\hat{\mathbf{A}} = \underset{\mathbf{A} \in \mathbb{R}^N}{\arg\min} \|\mathbf{A}\|_0 \text{ subject to } \mathbf{Y} = \tilde{\mathbf{H}}\mathbf{A} \tag{10}$$

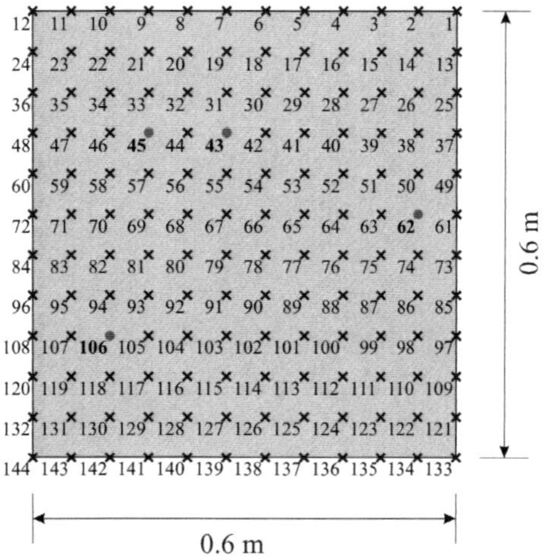

Figure 3. Node numbering of the finite element plate model; node 43, 45, 62 and 106 are acceleration measurement positions

The l_0-norm specifies the number of nonzero entries of $\mathbf{A}$. Equation (10) leads to the sparsest solution which agrees with the measurement vector $\mathbf{Y}$. Unfortunately, solving Equation (10) requires a combinatorial search, which makes it nearly impossible to solve computationally. Under the assumption of a sparse solution, e.g. [11] and [12] have shown that by replacing the l_0-norm with the l_1-norm almost the same solution can be obtained as solving Equation (10):

$$\hat{\mathbf{A}} = \underset{\mathbf{A}\in\mathbb{R}^N}{\arg\min} \|\mathbf{A}\|_1 \text{ subject to } \mathbf{Y} = \tilde{\mathbf{H}}\mathbf{A} \tag{11}$$

Now, the solution to equation (11) can be found by applying linear programming techniques, since l_1-regularized optimization is a convex problem.

For a measured output vector $\hat{\mathbf{Y}}$, the measurement data are usually polluted by some measurement noise $\boldsymbol{\omega}$, so Equation (9) becomes:

$$\hat{\mathbf{Y}} = \tilde{\mathbf{H}}\mathbf{A} + \boldsymbol{\omega} \tag{12}$$

To obtain a sparse solution for Equation (12) the minimization problem (11) needs to be modified, since it is not capable to deliver a proper estimate of $\mathbf{A}$ in the presence of noise:

$$\hat{\mathbf{A}} = \underset{\mathbf{A}\in\mathbb{R}^N}{\arg\min} \left\|\hat{\mathbf{Y}} - \tilde{\mathbf{H}}\mathbf{A}\right\|_2^2 + \lambda \|\mathbf{A}\|_1 \tag{13}$$

This problem is known as basis pursuit denoising (BPDN) [13]. In this case, a least-squares minimization is combined with a l_1-norm which penalizes solutions with many nonzero elements. The parameter λ regularizes the trade-off between sparsity of the solution and congruence of $\hat{\mathbf{Y}}$ and $\tilde{\mathbf{H}}\mathbf{A}$. λ can be adjusted according to the noise level $\boldsymbol{\omega}$ [14].

In recent years, the possibility of finding a sparse solution via l_1-minimization has been attracted a lot of attention, particularly in the field of Compressive Sensing (CS). Hence, a number of algorithms have been developed to solve Equation (11) and Equation (13) (e.g. [15] and [16]). In this contribution the so called *In-Crowd Algorithm* developed by Gill et al. in 2011 [10] is employed. This algorithm is one of the fastest solvers for very large sparse problems. A detailed description of the *In-Crowd* Algorithm, including a precise prescription of each calculation step, can be found in [10].

The so obtained magnitude vector $\hat{\mathbf{A}}$ has just a few nonzero elements, called the support of $\hat{\mathbf{A}}$. By considering this values and its corresponding columns in the dictionary $\tilde{\mathbf{U}}$ the

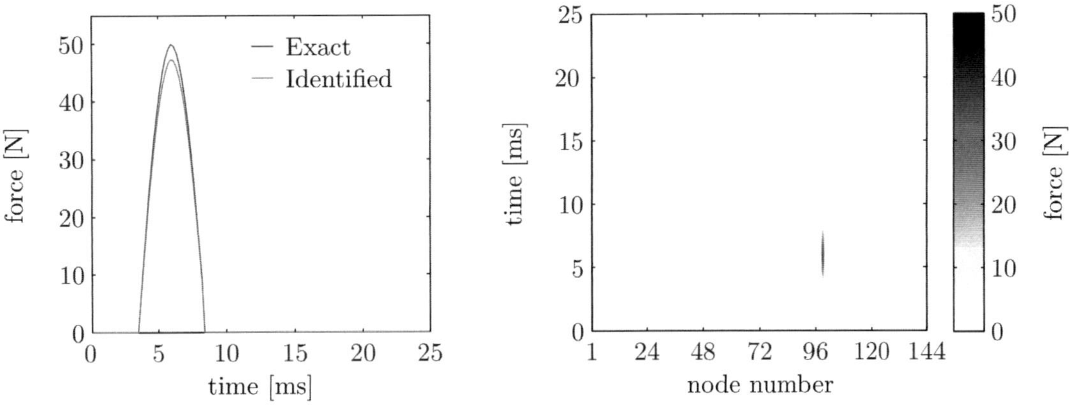

Figure 4. Impact reconstruction result; force history at node 99 (left), which was correctly identified as impact location (right)

force application points and the associated parameters of impact force and harmonic load are identified. The estimated force history at each potential input location $\hat{\mathbf{U}} \in \mathbb{R}^{f \cdot n}$ can be calculated by:

$$\hat{\mathbf{U}} = \tilde{\mathbf{U}}\hat{\mathbf{A}} \tag{14}$$

4. Simulation study

The capability to identify the force history and the force input location of impacts and harmonic loads from noisy measurements by solving the BPDN problem will be shown first by means of simulation studies. For illustration purposes, a completely free uniform 0.6 m by 0.6 m square steel plate of 5 mm thickness is investigated. The structural dynamics of the plate due to external forces are described by a finite element model. The plate is modelled by quadratic shell elements and 144 nodes (each with 6 degrees of freedom), see figure 3. The employed structural responses used in the simulation studies as well as in the experimental investigations are acceleration measurements perpendicular to the plate plane. White Gaussian noise, with a standard deviation of three percent of the maximum measurement value, is added to the simulated outputs to imitate real measurement data. It should be mentioned that the shown identification strategy is also applicable for the use of other structural models, e.g. analytical models and other types of measurement data, e.g. strain [17].

In a first study an impact load is applied perpendicular to the plate at Node 99. The acceleration responses due to this impact are recorded only at four output positions (Node: 43, 45, 62 and 106). The impact dictionary is composed of 13 different impact durations ($q = 13$). In order to investigate the effect of deviations in the employed reconstruction dictionary, the actual applied impact duration Δ_{true} is not included. The relative deviation to the best fitting impact Δ_{dic} amounts $(\Delta_{\text{dic}} - \Delta_{\text{true}})/\Delta_{\text{true}} = 3\%$. The reconstructed force history at Node 99 is shown in the left part of figure 4, the estimated force history at other nodes is zero (or nearly zero), thus the impact location is identified as well (figure 4 right). The magnitude estimation error amounts 5%.

In the next simulation a harmonic load with a frequency of 158 Hz is applied at node 79. The harmonic force dictionary is composed of frequencies from 38.5 Hz up to 275.5 Hz, the step size is (according to the recorded time history) 3 Hz. Again, the harmonic force dictionary does not contain the true frequency. Figure 5 displays the identification results. In the left plot the estimated and the true force frequency spectra are compared. The right plot illustrates once

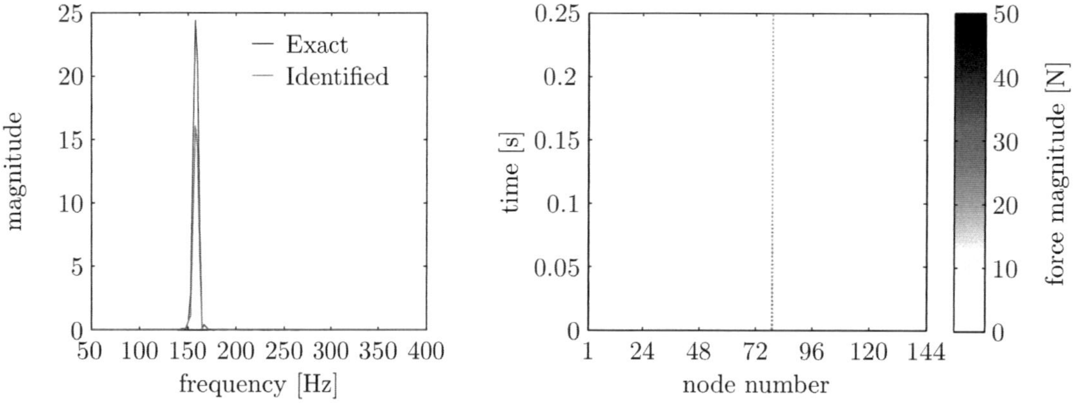

Figure 5. Reconstructed harmonic load; left: comparison of force spectra, right: identified force application at node 79

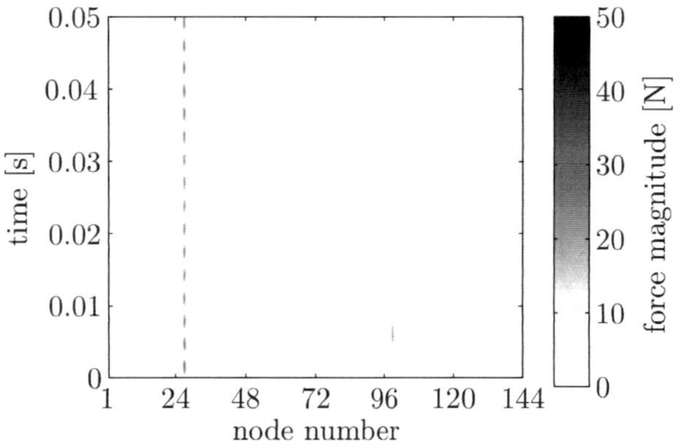

Figure 6. Reconstruction of superimposed impact force (node 99) and harmonic load (node 27); both application points are identified correctly

more the amount of force history for all nodes. It can be seen that the estimated force is assigned to the correct input node. However, the amplitude of the signal is slightly attenuated.

An estimation result for a superposition of impact and harmonic loads is shown in figure 6. The proposed load reconstruction strategy identifies the input location for both loading types correctly (see figure 6). The estimated magnitude of the impact and the identified frequency is slightly attenuated as well in this case.

5. Experimental results

For experimental validation of the presented load identification strategy an experimental test bench is set up. A beam of $L = 3\,\mathrm{m}$ length is placed onto two bearing blocks. Two piezoelectric accelerometers are mounted on the beam bottom side. The sensors are placed at positions $x_{s1}/L = 0.5$ and $x_{s2}/L = 0.73$ of the normalized beam length. The beam structure is excited by an impact of an impulse force hammer. The built-in force sensor is able to measure the force history applied by the hammer, which is used later to confirm the reconstructed load. Due to the simplicity of the simply supported beam structure an analytical model is used as structural model [17]. Figure 7 show the reconstruction results for an impact applied at the normalized beam position $x/L = 2/3$. The deviation of the calculated impact location amounts 1.94 cm. In assessing the localization accuracy, the hammer tip with a width of 2 cm needs to be considered. The relative magnitude estimation error amounts in this case 8.26%.

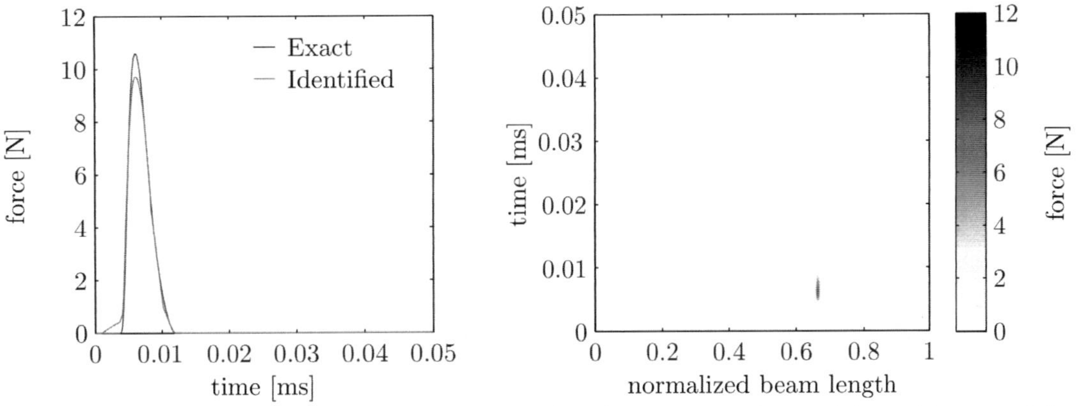

Figure 7. Experimental reconstruction results; left: comparison of force history, right: identified force over normalized beam length and time

6. Conclusion

In this paper the problem of reconstructing an impact from structural responses is addressed. It has been shown that by employing prior knowledge of the external force characteristics the proposed load identification method is able to reconstruct the force location and the force history, simultaneously. Generally it can be observed that this identification method estimates the input location very reliably. Also the force history deviations of the shown reconstruction studies are within reasonable limits.

Compared to other identification methods, significantly fewer sensors are required to estimate simultaneously the location and the force history. For the investigated beam structure two accelerometers are sufficient to obtain satisfying estimates. The proposed sparse identification method opens up a lot of new opportunity in the field of load reconstruction. The idea of employing prior knowledge of the force characteristics to create a sparse problem is transferable to other load reconstruction problem, e.g. spatial distributed loads.

References

[1] Stevens K 1987 *Proceedings of SEM* (Houston) pp 838–844
[2] Jacquelin E, Bennani A and Hamelin P 2003 *Journal of Sound and Vibration* **265** 81–107
[3] Inoue H, Harrigan J J and Reid S R 2001 *Applied Mechanics Reviews* **54** 503–524
[4] Klinkov M and Fritzen C P 2007 *Key Engineering Materials* **347** 461–466
[5] Park J, Ha S and Chang F K 2009 *AIAA Journal* **47** 2011–2021 ISSN 0001-1452
[6] Torres-Arredondo M A and Fritzen C P 2012 *4th International Symposium on NDT in Aerospace*
[7] Gillijns S and Moor B D 2007 *Automatica* **43** 934 – 937 ISSN 0005-1098
[8] Grady J 1988 *Contact Force History and Dynamic Response Due to the Impact of a Soft Projectile*
[9] Yan G and Zhou L 2009 *Journal of Sound and Vibration* **319** 869–884 ISSN 0022-460X
[10] Gill P, Wang A and Molnar A 2011 *IEEE Transactions on Signal Processing* **59** 4595–4605 ISSN 1053-587X
[11] Donoho D L and Elad M 2003 *Proceedings of the National Academy of Sciences* **100** 2197–2202
[12] Donoho D L 2006 *IEEE Transactions on Information Theory* **52** 1289–1306 ISSN 0018-9448
[13] Chen S, Donoho D and Saunders M 1998 *SIAM Journal on Scientific Computing* **20** 33–61
[14] Selesnick I 2012 Sparse deconvolution (an mm algorithm) Tech. rep. Polytechnic Institute of New York University
[15] van den Berg E and Friedlander M P 2008 *SIAM J. Sci. Comput.* **31** 890–912 ISSN 1064-8275
[16] Kim S J, Koh K, Lustig M, Boyd S and Gorinevsky D 2007 *IEEE Journal of Selected Topics in Signal Processing* **1** 606–617 ISSN 1932-4553
[17] Ginsberg D and Fritzen C P 2014 *Proceedings of the International Conference on Noise and Vibration Engineering* ed Sas P, Moens D and Denayer H pp 2051–2064

Damage assessment of cement-based geomaterial during loading by ultrasonic tomography

Duc-Phi Do[1], Ndao Baba[1], Dashnor Hoxha[1], Truong-Son Bui[2]

[1] Laboratory PRISME, Polytech'Orleans, 8 rue Léonard de Vinci, 45072, Orleans Cedex 2

[2]Department of Engineering Geology, Ha Noi University of Mining and Geology

E-mail: duc-phi.do@univ-orleans.fr

Abstract. Damage assessment of cement-based geomaterials during loading was conducted in this work by using the through-transmission ultrasound. For this purpose a built up system of ultrasound consisting of 96 channels and the specific sensors allowing to measure at the same time three types of waves (a bulk wave and two shear waves) were used. The continuous measurements enable to assess the damage of material through the constructed image of ultrasonic velocity as well as the attenuation of each wave during loading. The difference tomography method using the differential arrival times or relative amplitudes with respect to the initial stage confirms its efficacy through this work. The results show that all three types of wave can be used to capture the progressive damage in material but the bulk wave seems to be more sensitive than the shear waves.

1. Introduction

Investigation of the damage phenomenon in geomaterial (rock, concrete, mortar) which plays an essential role on the sustainability of many structures such as buildings, tunnels, nuclear reactors or radioactive waste disposal galleries as mentioned a few is always a grand challenge not only on the theoretical aspect but also on the experimental and numerical points of view. Experimentally, this phenomenon is usually characterized in laboratory through some classical experiments like the uniaxial or triaxial compression test thanks to their possibility to reproduce the loading state that the material could be undergone during their life of service. During these tests, the specimen will be loaded monotonously or cyclically until failure while the damage can be interpreted from the result of the stress and deformation curve [1, 2]. It is widely accepted that damage in geomaterial is the consequence of the nucleation and propagation of a microcrack network which can coalesce to produce macrofracture at failure of specimen. These explanations are verified and confirmed from the observation by microscope realized on the specimen's surface or on some cross sections obtained from the thin cuts. This efficient method owes however some disadvantages notably due to its destructive characteristic and hence less adopted in situ. Otherwise, the observation using thin cuts after loading cannot reflect exactly the process evolving during loading due to the change of sample's statement.
Recent tendency is focused on the application of the non-destructive technique as an independent method or associated with the destructive method to assess damage in geomaterial. The principal idea of these methods consists of measuring one or some physical properties (thermal or electrical conductivity, permeability, acoustic waves etc.) of material which could be sensitive with the

evolution of damage in geomaterial. Among different techniques, the measurement based on the ultrasonic wave has showed its efficacy as demonstrated in some contributions (see [1, 2] with references cited therein). A high sensitivity of ultrasonic waves with respect to the variation of the microstructure of material due to the presence of heterogeneity like inclusions, pores and eventually the evolution of the microcrack network was highlighted in these works. Particularly, by increasing the number of measure at different position on the surface of sample, the image of some acoustic parameter (velocity, attenuation) can be reconstructed through an inversion procedure. As for instance, this technique was used in [3] to image the progressive damage of granite subjected to uniaxial cyclic loading while in [4] it allows to investigate the process of shear and compaction band formation. It is noted however that in these contributions, usually only one type of wave (longitudinal or one shear wave) was chosen.

In this study, ultrasonic tomography was conducted to characterize damage in a cement-based geomaterial. By using a specific transducer which allows measuring at the same time three types of waves (a bulk wave and two shear waves), we will discuss the feasibility of this non destructive technique and also the efficacy of each type of wave to assess damage in geomaterial.

2. Experimental setup and tomographic reconstruction procedure

In this work, damage assessment by ultrasound of a cement-based geomaterials like mortar was conducted in associated with the uniaxial compression experiment. The considered material was fabricated using Portland cement CEMII 32.5 and standard sand with maximum grain size 2mm. The $40\times40\times80mm^3$ specimens were produced from a mixing sand, cement and water with the ratio $s/c=1/1$ and $w/c=0.4$. During the axial loading, to assess the continuous process evolving in sample, the ultrasonic measurement was carried out continually thanks to a system comprising 96 channels developed by Diagnostic Sonar Ltd. This system including a generator, a group of multiplexers, preamplifiers for each channel and the card of acquisition allows recording the wave on all the receivers in a short chosen time (one second in this work). More precisely, three analogue hardware (NI-5752) 32 channel adapter modules connected with FlexRIO FPGA module to produce 96 channel 12 bit 50 MSPS acquisition module with digitally controlled swept gain. This digitizer is combined with other multi-channel preamplifier and pulser boards and three multiplexers 32 channel module. The multiplexers were used to gate the excitation 150V pulse to each of the transmitting transducer. The operation of this system ensures at a specific moment one transducer was chosen as transmitter while the remaining transducers were used as receivers. It means that any transducers in our system can be used as emitter or receiver. In this work, we decide to utilize the piezoceramic transducer (P-143.01 of PI Ceramic GmbH) with 150 kHz of central frequency. This piezoceramic transducer ($10mm\times10mm\times7mm$ of size) composes three layers: the first layer is sensitive and can produce the longitudinal wave (qP-wave) while the second and the third layer are sensitive and can produce the shear waves (qSV and qSH waves respectively). Thus each of these sensors allows us to measure three modes of wave instead of using three separate transducers. The signals, preamplified with 30dB gain, were acquired at a sampling frequency of 50MHz with the resolution 12 bits and were stored in compressed form in the *.png file. From this database, the recorded waveforms can be extracted to realize the data processing. It is important to note also that, to maximize the acoustic contact, it requires a coupling material between the sample and sensors. During this work, honey is chosen due to its ability to support both longitudinal and shear waves. In figure 1 is captured the schema of the uniaxial compression test associated by a system of ultrasonic measurement. Twenty two acoustic sensors mounted on the surface of sample aim to image two vertical planes which pass the center vertical axis of specimen. More precisely, five sensors were glued on each lateral surface while on the top or bottom surface, only one sensor was placed due to the presence of the steel end platens. Before loading, all surface of sample were carefully prepared to ensure the good contact between sample and sensors. The measurements at the initial state were selected as standard references for successive measurements.

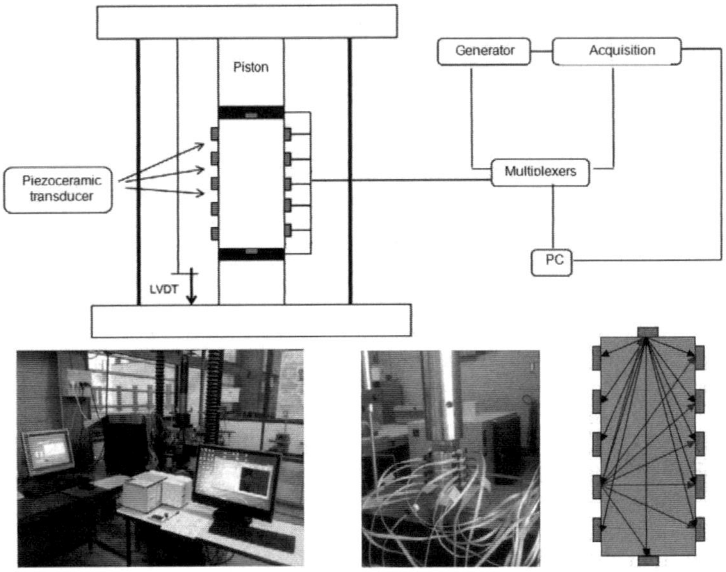

Figure 1. Disposition of piezoceramic transducers on sample and ultrasonic measurement during uniaxial loading.

The data processing procedure consist of determining the time of flight (TOF) of recorded waveforms and particularly the differential traveltime for rays traversing identical paths at two states before and after loading ($\delta\tau = \tau^{\sigma} - \tau^{\sigma=0}$). In this work, these traveltimes were determined automatically. Concretely, the TOF was evaluated by using the automatic picking procedure based on the AIC method [5]. For the calculation of the relative arrival time, the recorded signals during loading were cross-correlated with those of the same raypaths measured at the initial state (referenced signals) by windowing the first portion of the arrival [6]. However, due to the fact that the accuracy of the determined TOF and differential traveltime affect importantly the result of tomographic image, potential inaccurate arrival time was checked and corrected manually during this work. Concerning the attenuation evaluation, the fast Fourier transformation was used to determine the magnitude of signals.
The traveltime data of ultrasonic waves were used as input in the next step to reconstruct the tomograhic images through an inversion procedure. In effect, these images present a set of the studied parameter (slowness, attenuation,…) which is considered to be constant in each grid cell (called pixel) of the discretized media. The principal idea of this discretization step is to replace the continuous model to the discrete model which allows establishing a system of equations for an inversion procedure. Indeed, following the well-known theory of traveltime in inhomogeneous, for a ray-path R_i, the time τ_i of wave propagating in the general media is defined as:

$$\tau_i = \int_{R_i} \frac{1}{U} ds = \int_{R_i} V ds; \qquad (1)$$

where V is the slowness (inverse of the velocity U) at each location on the ray-path and ds is the incremental length along the ray-path R_i.
Using the discretization step, the integral of traveltime along a ray-path, measured by each pair of emitting and receiving transducers, can be reduced to the sum of the traveltime within each grid cell through which the ray pass. Thus we can rewrite the equation (1) as:

$$\tau_i = \sum_j \tau_{ij} = \sum_j R_{ij} V_j; \qquad (2)$$

where R_{ij} is the ray length of the ith ray in the jth cell, with slowness V_j.

Similarly, if the interest focuses on the tomographic image of the slowness variation with respect to the initial state ($\delta V = V^{\sigma} - V^{\sigma=0}$), from the equation (2), we have these following expressions:

$$\delta\tau_i = \sum_j \delta\tau_{ij} = \sum_j R_{ij}.\delta V_j; \tag{3}$$

This equation was deduced by ignoring the influence of the ray-path perturbation, thus the different traveltime is considered as a function of slowness variation (the perturbation theory as defined in [7]). Gathering the result obtained from all ray-path yields the following system of equations:

$$t = L.\chi; \tag{4}$$

where t denotes the data vector representing the measured traveltime (or the relative traveltime), χ defines the vector containing the unknown parameter of each grid cell and the matrix L is function of the ray-path segment R_{ij}.

The inversion procedure consists to calculate the solution χ of this system of equations which minimizes the error between the data, t, and the theoretical traveltime given by $L.\chi$. In general the matrix L (of size $M \times N$) is not square due to the fact that in most case there are many more data than unknowns (M>N). Otherwise this system of equations can be non-linear due to the dependence of matrix L on the model parameter. It is usually the case even in the isotropic media where the contrast of the model parameter of grid cells is high and hence the ray-path of each source-receiver can take the curve form instead of the straight ray. The methodology to reconstruct the ray propagating in media called ray-tracing method has been extensively discussed in the literature [8]. All the strategies based on the Fermat's principle in which the traveltime from a source to a receiver must be the minima over all the possible ray-paths.

Consequently, the solution of the nonlinear system of equations (4) can be obtained through an iterative procedure:

$$\chi_{k+1} = \chi_k + (J^T J + \lambda.I)^{-1} J^T (t - L.\chi_k); \tag{5}$$

where J is the sensitivity (Jacobien) matrix representing the derivative of the theoretical traveltime with respect to the unknown parameter χ. Furthers, in equation (5) because the inverse problem is ill-posed, a small positive damping parameter λ ($\lambda \in 0.001 \sim 10$) is introduced with I the identity matrix. The procedure is repeated until the convergence is reached.

Note that in this work, we suppose that the initial media is isotropic and homogeneous and hence we can use the straight ray hypothesis between each pair source-receiver to reconstruct the tomography.

The same procedure can be used to construct the tomography of the attenuation due to its similarity from mathematical point of view. Indeed, the attenuation phenomenon is characterized by the attenuation coefficient α which represents the relationship of the amplitude measured at the receiver A_i decayed exponentially with respect to the amplitude at the source A_0 and the distance of the ray-path l_{R_i} as follow:

$$A_i = \frac{1}{l_{R_i}} A_0 \exp(-\int_{R_i} \alpha.ds); \tag{6}$$

This equation (6) can be written in a similar form of equation (1) as:

$$\tau_i = \int_{R_i} \alpha.ds; \tag{7}$$

where $\tau_i = \ln(A_0) - \ln(A_i l_{R_i})$.

By applying the perturbation theory in equation (7), we can obtain without difficulty the following expression which has the analogue form as equation (3):

$$\delta\tau_i = \sum_j \delta\tau_{ij} = \sum_j R_{ij}.\delta\alpha_j; \tag{8}$$

where the amplitude difference of a ray-path ($\delta\tau_i = \ln(A_i^{\sigma=0}) - \ln(A_i^{\sigma})$) is function of relative variation of the attenuation coefficient $\delta\alpha = \alpha^{\sigma} - \alpha^{\sigma=0}$.

3. Result and discussions

As mentioned previously, the tomography can be presented in two forms: the first form image the absolute value of the studied parameter while the second form aims to capture the difference tomography with respect to the referenced state. In this study, the authors used the difference data (of slowness and attenuation coefficient) versus the initial state so that the evolution damage during loading could be observed.

We presented in figure 2 tomographic images constructed from the slowness difference of longitudinal wave measured at different loading stages. These tomograms show that even at low stress (about 20% of maximum stress at failure), a microfracture appears at the left high corner of sample where the slowness difference increases (velocity decreases correspondingly). This damaged zone becomes visible from 40% of maximum loading stage represented by a considerable increase of slowness difference. Moreover, this latter parameter changes significantly in the loading range from 40% to 80% of peak stress. Beyond this latter stage load, the maximum value of slowness difference increases slightly but the extension of damaged zone becomes remarkable. Particularly, when the stress approaches the material's strength, the results show that a fracture ranging from the left high corner to the right bottom corner was created in the sample.

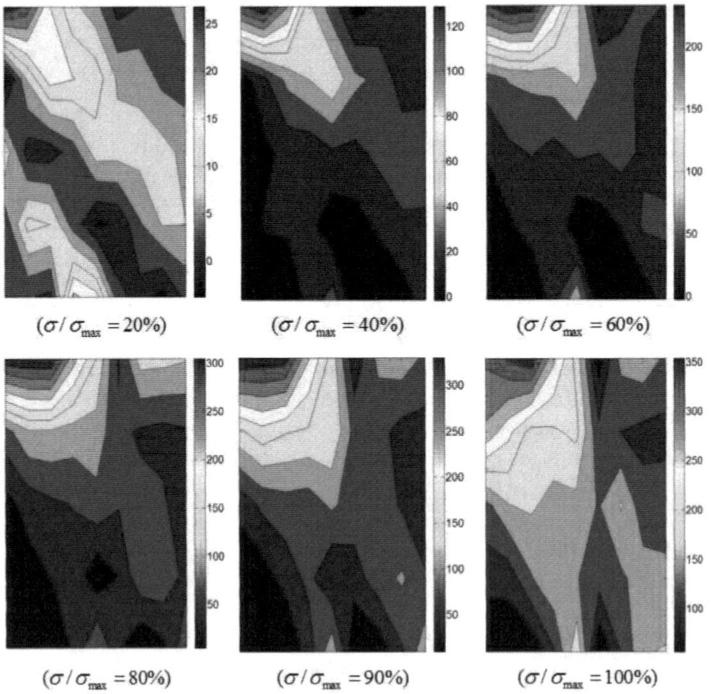

Figure 2. Slowness difference tomography ($s/\mu m$) of longitudinal wave at different stages of loading with respect to the initial state.

The same remarks can be noted for the images obtained from the shear wave as illustrated in figure 3. However, more artefacts were observed in these images in comparison with the tomograms of bulk wave and the damaged zone may be distinguished only from the loading stage corresponding to 60% of stress at failure. The results obtained at different stages of loading are less consistent than those of longitudinal wave. In effect, taking insight the waveform of bulk and shear waves recorded during loading, we observed more noise and eventually greater reduction of amplitude in the shear wave represented by smaller signal to noise ratio in comparison with one of the bulk wave.

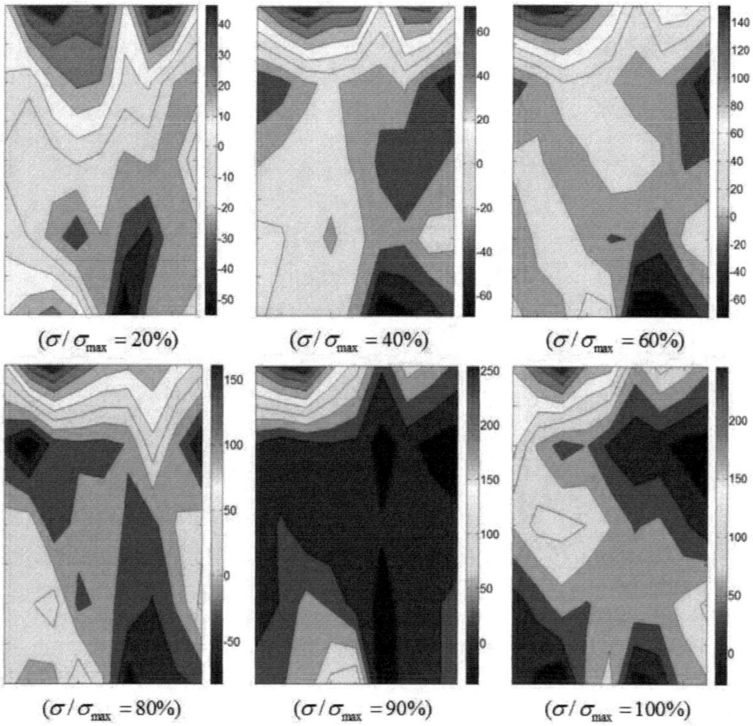

Figure 3. Slowness difference tomography ($s/\mu m$) of shear wave at different stages of loading with respect to the initial state.

In figures 4 and 5, we presented images of the attenuation difference of bulk and shear waves during loading of sample. In this case, no significant difference between bulk and shear waves can be observed but higher value of the relative attenuation coefficient was noted for the shear wave. As in the previous study case of slowness difference, a weakened zone on the left of the top surface of sample begins to develop at the loading stage $\sigma/\sigma_{max} = 40\%$ illustrated by a sharply increased value of relative attenuation coefficient. This zone tends to extend toward the bottom surface near the right corner and accelerate when the stress pass 80% maximum stress. At failure, the results showed a fracture created with an inclined angle with the axial axis of speciment. This type of failure is usually observed in uniaxially loaded samples and the damage phenomena is known as consequence of the nucleation, propagation of microcracks which coalesce at high density to create a fracture. The position of the damaged zone observed from the reconstructed images of the slowness and attenuation difference agree quite well with the real fracture created in sample as captured in figure 6. A small offset of the damaged zone from the real position in space can be explained by the potential artefacts included by the tomography technique.

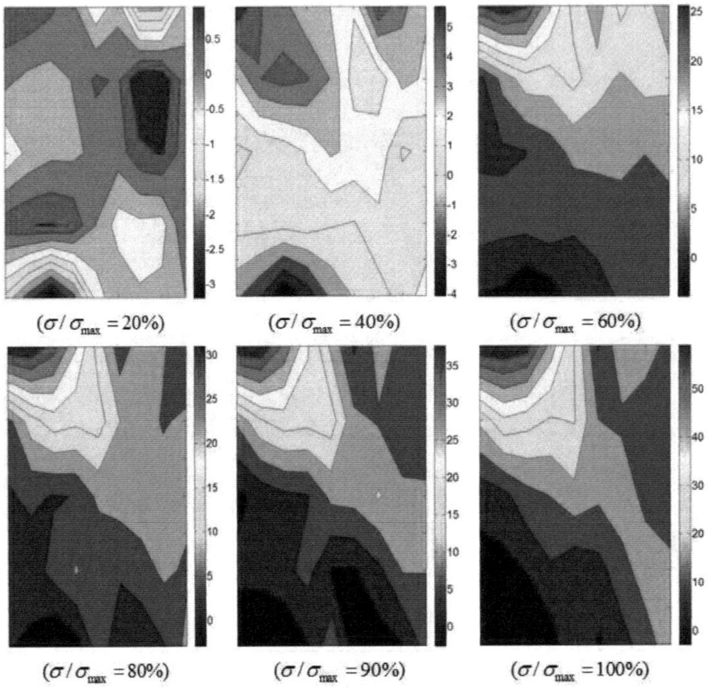

Figure 4. Relative attenuation coefficient (dB/m) tomography of longitudinal wave at different stages of loading with respect to the initial state.

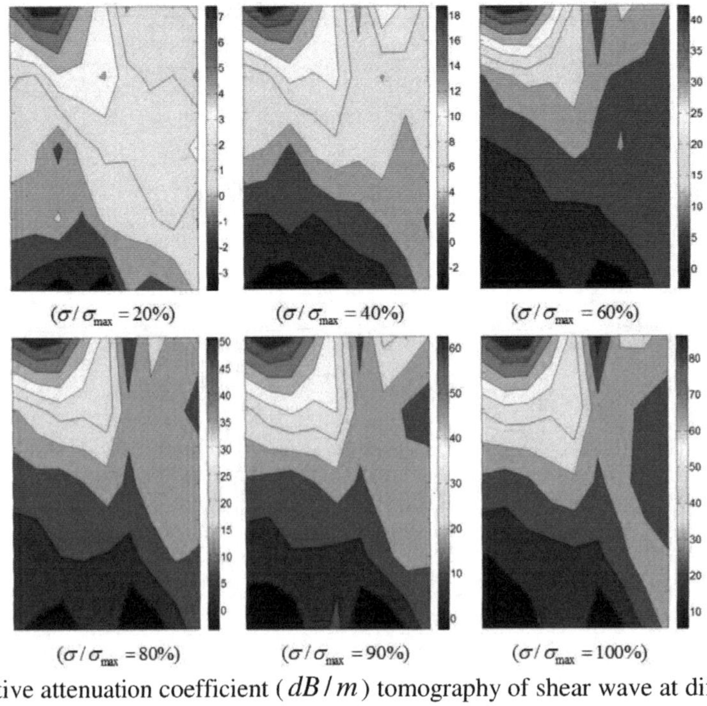

Figure 5. Relative attenuation coefficient (dB/m) tomography of shear wave at different stages of loading with respect to the initial state.

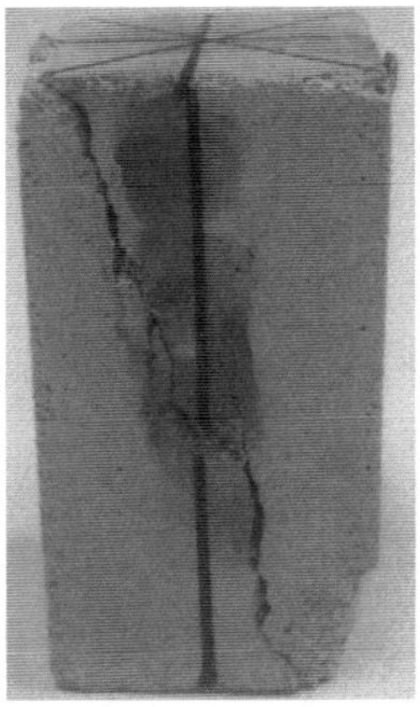

Figure 6. Fracture created in sample after failure.

4. Conclusions

In this work, we present an application of the non-destructive method like ultrasound to assess the damage during loading of the cement-based geomaterial. Using a system consists of 96 channels with the specific piezoceramic transducer, we can capture the tomographic image of all types of waves and hence the continuous process evolving in the uniaxially loaded sample. The quite good agreement between the reconstructed tomograms and the visible fractured specimen after failure confirms the accuracy of the utilized inversion procedure and the efficacy of the ultrasound method to assess damage in geomaterial. However, the slowness tomography of longitudinal wave presents a clearer image with respect to the shear waves. Contrary, the tomograms of relative attenuation seem more stable with no significant difference for all types of wave. The question concerning the presence of artefacts particularly in the slowness images of the shear waves require more investigation notably to find a more robust arrival picking. Otherwise, the combination of the longitudinal and shear wave data could provide a complete answer of damage process in material and will be discussed in our near future work.

References

[1] Eslami J, Grgic D and Hoxha D 2010 *Geo. Jour. Int.* **183** 1362–75
[2] Fortin J, Guéguen Y and Schubnel A 2007 *Jour. Geo. Res.* **112** 1–16
[3] Chow T M, Meglis I L and Young R P 1995 *Int. Jour. Rock Mec. Min. Sci. Geo.* **112** 1–16
[4] Charalampidou E M, Hall S A, Stanchits S, Lewis H and Viggiani H 2011 *Techtonophysics* **503** 8–17
[5] Kurz J H, Grosse C U and Reinhardt H W 2011 *Ultrasonics* **43** 538–46
[6] Peterson M 1997 *Ultrasonics* **35** 17–29
[7] Chapman C H and Pratt R G 1992 *Geo. Jour. Int.* **109** 1–19
[8] Zhou B and Greenhalgh S A 2005 *Geo. Prospec.* **53** 19-41

Optimal sampling in damage detection of flexural beams by continuous wavelet transform

B Basu[1], B M Broderick[1], L Montanari[2] and A Spagnoli[2]

[1]Dept. Civil, Structural and Environmental Engineering, Trinity College, Dublin 2, Ireland
[2]DICATeA, University of Parma, Parco Area delle Scienze 181/A, Parma, Italy

E-mail: basub@tcd.ie

Abstract. Modern measurement techniques are improving in capability to capture spatial displacement fields occurring in deformed structures with high precision and in a quasi-continuous manner. This in turn has made the use of vibration-based damage identification methods more effective and reliable for real applications. However, practical measurement and data processing issues still present barriers to the application of these methods in identifying several types of structural damage. This paper deals with spatial Continuous Wavelet Transform (CWT) damage identification methods in beam structures with the aim of addressing the following key questions: (i) can the cost of damage detection be reduced by down-sampling? (ii) what is the minimum number of sampling intervals required for optimal damage detection ? The first three free vibration modes of a cantilever and a simple supported beam with an edge open crack are numerically simulated. A thorough parametric study is carried out by taking into account the key parameters governing the problem, including level of noise, crack depth and location, mechanical and geometrical parameters of the beam. The results are employed to assess the optimal number of sampling intervals for effective damage detection.

1. Modelling of the cracked beam

A cracked Euler-Bernoulli beam characterized by an open edge crack under different boundary conditions at the two ends, i.e. clamped-free (cantilever beam) and supported-supported (simply supported beam), is considered (Figure 1). The free vibration responses are analytically evaluated by solving the free vibration equations of the two uncracked sub-beams connected by a rotational spring (representing the local stiffness k_c of the cracked cross-section of the beam) at the crack location x_c. The beam has a rectangular cross-section with height h and width b; the crack depth is a and L is the beam length. The symbols I, A, E and ρ represent, respectively, moment of inertia and area of the cross-section and Young's modulus and density of the material.

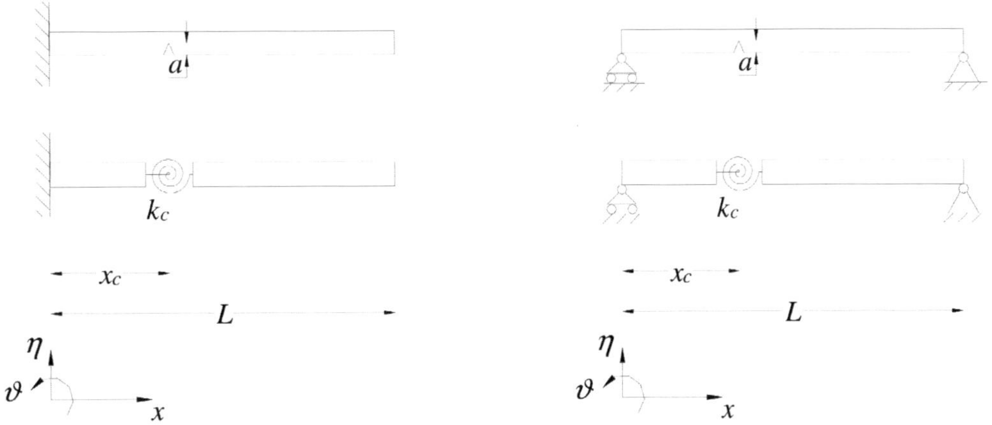

Figure 1. Sketch of the two cracked beams under study.

The free vibration equation related to the two uncracked parts of the beam can be written as

$$EI\frac{\partial^4 v(x,t)}{\partial x^4}+\rho A\frac{\partial^2 v(x,t)}{\partial t^2}=0, \quad (1)$$

where $v(x, t)$ is the transversal displacement of the beam from its static equilibrium position at a distance x from the left end at the time t. Separating the variables in Eq. (1) ($v(x,t)=\eta(x)\,g(t)$) and solving the characteristic equation function of x, the mode shapes η_L and η_R of the left and right sub-beam, respectively, are as follows

$$\eta_L = C_{1L}\mathrm{sen}(\alpha x)+C_{2L}\cos(\alpha x)+C_{3L}\mathrm{senh}(\alpha x)+C_{4L}\cosh(\alpha x) \quad 0\le x\le x_c \quad (2.1)$$

and

$$\eta_R = C_{1R}\mathrm{sen}(\alpha x)+C_{2R}\cos(\alpha x)+C_{3R}\mathrm{senh}(\alpha x)+C_{4R}\cosh(\alpha x) \quad x_c\le x\le L, \quad (2.2)$$

where $\alpha=\left(\frac{\rho A\,\omega^2}{EI}\right)^{1/4}$ and ω is the natural frequency of the cracked beam.

The $C_{(.)}$ terms are integration constants arising from the solution of the fourth order differential equation in space. By imposing the boundary conditions (see Eq. (3) below), a system of eight linear equations is formed. The natural frequencies of the cracked beam are found by setting the determinant of the matrix of the linear system to zero and solving it numerically for the roots of α. The coefficient C_{1L} is imposed to be equal to unity.

The boundary conditions of the cantilever beam at the clamped and at the free end are, respectively:

$$\eta_L(0)=0 \text{ and } \eta_L'(0)=0,\ \eta_R''(L)=0 \text{ and } \eta_R'''(L)=0. \quad (3.1)$$

The boundary conditions of the simple supported beam at the two ends are:

$$\eta_L(0)=0 \text{ and } \eta_L''(0)=0,\ \eta_R(L)=0 \text{ and } \eta_R''(L)=0 \quad (3.2)$$

For both the beams, the conditions of continuity of displacement, moment and shear at the crack location can be expressed as

$$\eta_L(x_c)=\eta_R(x_c),\ \ \eta_L''(x_c)=\eta_R''(x_c) \text{ and } \eta_L'''(x_c)=\eta_R'''(x_c). \quad (3.3)$$

The rotational spring at the cracked section introduces a discontinuity of the rotation which can be written as

$$\eta_R'(x_c)-\eta_L'(x_c)=\frac{EI}{k_c}\eta_R''(x_c). \quad (3.4)$$

The local stiffness k_c is evaluated, according to linear elastic fracture mechanics concepts, through the following expression [1]

$$k_c = \frac{bh^2 E}{24(\delta/(1-\delta))^2(5.93-19.69\delta+37.14\delta^2-35.84\delta^3+13.12\delta^4)} \quad (4)$$

where $\delta = a/h$ is the relative crack depth.

2. Damage detection by spatial CWT

Due to its multi-resolution properties, wavelet analysis, acting as a signal microscope, appears to have a better ability to analyze the details of non-stationary signals in comparison to traditional analysis tools, such as Fourier Transform or Short-Time Fourier Transform.

A wavelet function $\psi(x)$ is a zero mean local wave-like function which decays rapidly and satisfies particular conditions [2]. A family of wavelet functions can be obtained by considering:

$$\psi_{k,s}(x) = \frac{1}{\sqrt{s}}\psi\left(\frac{x-k}{s}\right), \quad (5)$$

where s and k are, respectively, the scale and the translation parameter. The continuous wavelet transform of a signal $\eta(x)$ with respect to the wavelet function $\psi(x)$ is defined as

$$W(k,s) = \int_{-\infty}^{+\infty} \eta(x)\frac{1}{\sqrt{s}}\psi^*\left(\frac{x-k}{s}\right)dx \quad (6)$$

where ψ^* is the complex conjugate of ψ.

The issue of the sampling interval impact in CWT damage detection is dealt with considering the 4th order Coiflets wavelet ('Coif4'), with 8 vanishing moments, which, exhibit better performances in comparison to other wavelet functions [3].

Furthermore, a MATLAB routine to perform the CWT for the aforementioned wavelets is implemented by the authors to improve the accuracy of the existing built-in one (the original routine approximates the signal through a constant piecewise function, while the implemented one considers a linear piecewise trend) [4].

Realistic situations are simulated by superimposing synthetic Gaussian white noise to the signal corresponding to the beam deflection obtained from the mechanical model described in section 1. In the following the noise level is quantified through the signal to noise ratio (SNR), defined in decibels as

$$SNR = 10\log_{10}\left(\frac{P_{signal}}{P_{noise}}\right). \quad (7)$$

The term P with the subscripts in Eq. (7) denotes power and is computed as

$$P_{(.)} = \frac{1}{N_z}\sum_{i=1}^{N_z}|z(x_i)|^2, \quad (8)$$

where N_z denotes the number of discrete points of a generic sampled function $z(x)$.

As above-mentioned, the CWT is defined by the convolution of the input signal $\eta(x)$ with a wavelet function generated by scaling and translating its mother wavelet $\psi(x)$. For a finite-length signal, when the convolution operation is executed close to the signal ends, the wavelet window extends into a region with no available data, so that the transform values close to the borders of the signal are tainted by the nonexistence of data. Consequently, the values of the CWT coefficients very close to the signal extrema arise abnormally (border distortions) and the real signal features of that region are consequently corrupted by the transform. As a result, edge effects can provoke the masking of the damage and yield false indicators. To handle edge effects is used the polynomial extension method suggested by Montanari et al. [5], which pads the original signal by means of two high degree polynomial functions obtained through a fitting procedure. The extension functions satisfy continuity conditions and extend the trend of the noisy signal and its derivatives in an average sense.

3. Parametric study

The effect of sampling interval of the cracked beam mode shape in damage detection through the spatial CWT is numerically analysed varying the relevant parameters of the problem, that is: the normalized sampling interval $\Delta x/L$ ($\Delta x/L$ is considered equal to 0.025, 0.01, 0.005, 0.0025, 0.001, 0.0005 and 0.00025), the noise level (SNR value is assumed equal to 130 dB, 100 dB or 70 dB), the deformed shape of the beam (i.e. the first three mode shapes of the cantilever and of the simply supported beam), the relative crack position along the beam (x_c/L is considered equal to 0.1, 0.3, 0.5, 0.7 or 0.9) and the beam parameters (ρ, h, b, L and E). Initially the parametric study is carried out assuming fixed values of the beam parameters; subsequently, their influence is analyzed to generalize the results.

In order to investigate the impact of sampling interval in CWT based damage detection, the minimum detectable (threshold) crack size for a given spatial sampling interval is obtained according to the following criterion based on an iterative procedure. While increasing gradually the relative crack depth δ from a value of 0.0001 to 0.95, the wavelet transform is executed at fixed values of the scale s ($s = 2,4,6,\mathrm{K},200$) and the maximum absolute value of the transform is determined for each scale. If for a given relative crack depth δ this maximum is attained for all scales at the crack position for each of an arbitrary number of 20 different random Gaussian white noise distributions, this value of δ can be regarded to be detectable, otherwise a larger value of δ has to be assessed. To allow for the numerical approximation of the CWT, the crack depth is also regarded as detectable even if the CWT absolute maximum is attained in either of the two nearest points (i.e. the preceding or the following point) to the actual damage location. For very small sampling intervals ($\Delta x / L < 0.005$), the criterion is relaxed to any of the four nearest points to the damage location.

A cracked beam of length L = 1 m with a rectangular cross-section of height $h = 0.05L$ and width $b = 0.5h$, constituted by an elastic linear isotropic material of Young modulus E = 200 GPa and density $\rho = 7850$ kg/m^3, is considered.

Figure 2 shows in bilogarithmic graphs the minimum detectable crack size as a function of the pseudo-frequency, $f_a = f_c / (\Delta x \cdot s)$, where f_c is the center frequency of the mother wavelet (i.e. the frequency maximizing the Fourier transform of the mother wavelet modulus [6]); for 'Coif4' f_c = 0.6957. The damage identification turns out to be, with good approximation, a function of the pseudo-frequency only (in the simulations of figure 2 the fitting parameters of the polynomial padding method are assumed to be $\beta_1 = \beta_2 = 1$, $H_1 = \overline{H}_1$ and $H_2 = \overline{H}_2$ [5]; in the other simulations, whereas the values of β_1 and β_2 vary, in order to reduce the computational cost $H_1 = \overline{H}_1$ and $H_2 = \overline{H}_2$ is assumed). The curves of figure 2 decrease monotonically from high to low pseudo-frequencies until, at a specific frequency, a sudden jump occurs toward the upper limit of δ (say, equal to 0.95). This specific value of pseudo-frequency represents, with good approximation, the *lowest* and *optimal* value of pseudo-frequency for damage identification (the optimal value is that allowing the minimum crack size to be detected as the pseudo-frequency is made to vary). It needs to be underlined that this lowest/optimal value of f_a is the same for all levels of noise.

The jump in the curves of figure 2 at the lowest/optimal value of f_a occurs because, by analyzing the cracked beam shape with wavelet functions characterized by lower pseudo-frequencies, even large cracks cannot be localized due to the influence of the edge effects. It can be noted that the curves for $\Delta x/L$ = 0.00025 exhibit jumps to the upper limit of δ at a higher value of pseudo-frequency than those for larger sampling intervals. This trend is due to the fact that, when the crack is near the signal extremum, the wavelet analysis is tainted by edge distortions, and, particularly for very low sampling intervals, the maximum CWT coefficient value falls not exactly at the crack location (or at its two four nearest points, as permitted by the detection criterion), but at sampling points other from that of the crack location.

Hereafter, the $f_{a,opt}$ is used to indicate the optimal value of pseudo-frequency f_a that gives the minimum detectable crack size (in figure 2 $f_{a,opt}$ is equal to about 12 m^{-1}). The value $f_{a,opt}$ of pseudo-frequency approximately coincides, as shown above, with the lowest pseudo-frequency for performing damage detection.

(a)

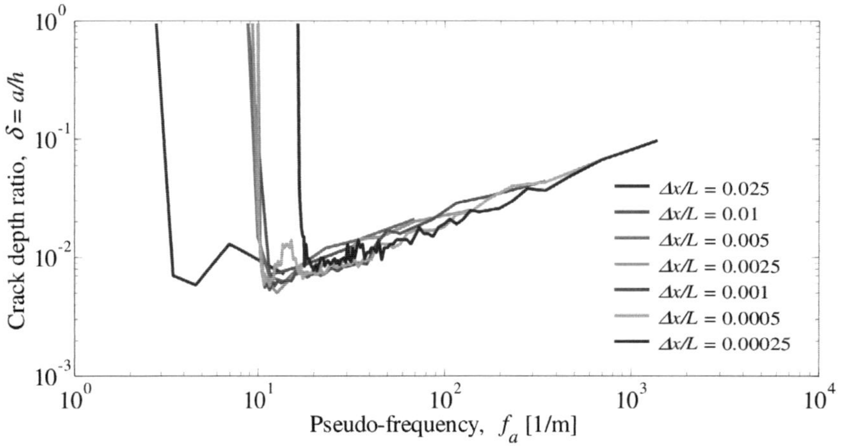

(b)

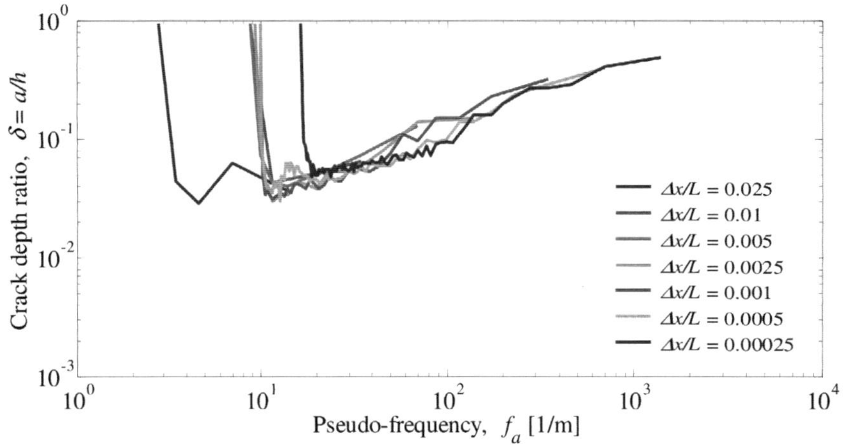

(c)

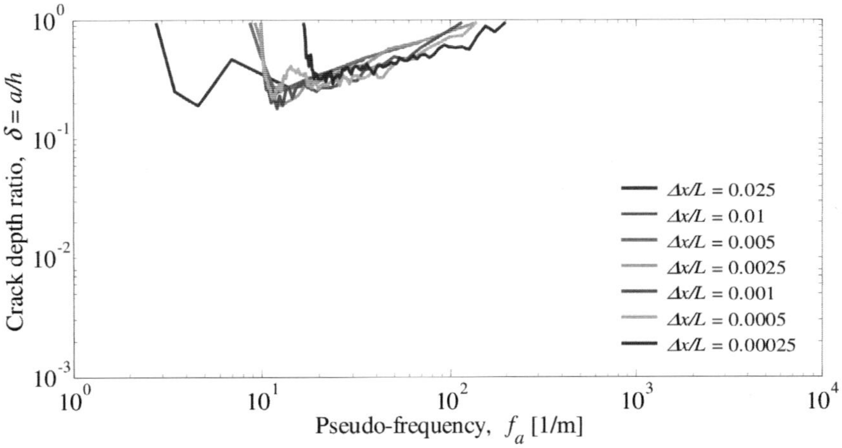

Figure 2. Impact of the sampling interval in damage detection by spatial CWT with 'Coif4'. The polynomial padding method is used. The first mode shape of the cantilever beam with crack at x_c/L = 0.1 is analysed. (a) SNR = 130 dB; (b) SNR = 100 dB; (c) SNR = 70 dB.

Considering that CWT based damage detection depends on the pseudo-frequency only, it is numerically illustrated how the use of the proper scale range is essential in detecting damage when different sampling intervals are considered [3]. The first mode shape of the cantilever beam with a crack of $\delta = 0.2$ at $x_c/L = 0.1$ is analysed. The noise level is assumed to be equal to 70 dB. The same mode shape is sampled at $\Delta x/L = 0.01$ and $\Delta x/L = 0.0004$.

Figures 3(a,b) report the contour plots of the CWT (for the polynomial padding method $\beta_1 = \beta_2 = 1$ are used) when the signal is sampled at $\Delta x/L = 0.01$, and the scale ranges $s = [4-24]$ (i.e. $s = 4,5,6\ldots,24$) and $s = [4-6]$ are respectively used. The CWT coefficients, considering the broad scale range, completely mask the crack location (figure 3a), which on the contrary can be identified using the narrow scale range (figure 3b).

Figures 3(c-d) highlight that even if the signal is sampled at an extremely small sampling interval ($\Delta x/L = 0.0004$), the CWT damage detection can be achieved only considering a proper scale range. The damage localization at the scale range $s = [2-40]$ fails (figure 3c), while it succeeds for $s = [96-158]$ (figure 3d).

In this Section, the influence of these beam parameters ρ, h, b, L and E on the optimal pseudo-frequency in CWT based damage detection is analysed and a further generalization of the parametric study is obtained.

According to Eq. (2), the functions describing the general mode shape depend on the boundary conditions and on the parameter , and in turns the corresponding natural frequency is a function of the material density ρ and of α. Hence, the material density ρ affects the natural frequencies of the cracked beam but not its mode shapes and it is not considered in the following.

Then, by substituting the LEFM expression of the local rotational stiffness k_c due to the crack (Eq. (4)) in the boundary condition representing the rotation discontinuity at the crack section (Eq. (3.4)), we obtain:

$$\eta_2'(x_c) - \eta_1'(x_c) = 2h\,\tilde{f}(\delta)\,\eta_2''(x_c), \tag{9}$$

where $\tilde{f}(\delta) = (\delta/(1-\delta))^2(5.93 - 19.69\delta + 37.14\delta^2 - 35.84\delta^3 + 13.12\delta^4)$ is a function of the relative crack depth δ only. For a given beam deflection shape (and, hence, for a given value of the curvature in x_c, $\eta_1''(x_c) = \eta_2''(x_c)$), Eq. (9) demonstrates that the rotation discontinuity due to the crack is a function of h and δ, but not of the other beam parameters b and E. Therefore, the CWT based damage detection results are invariant for constant values of h, δ, and L.

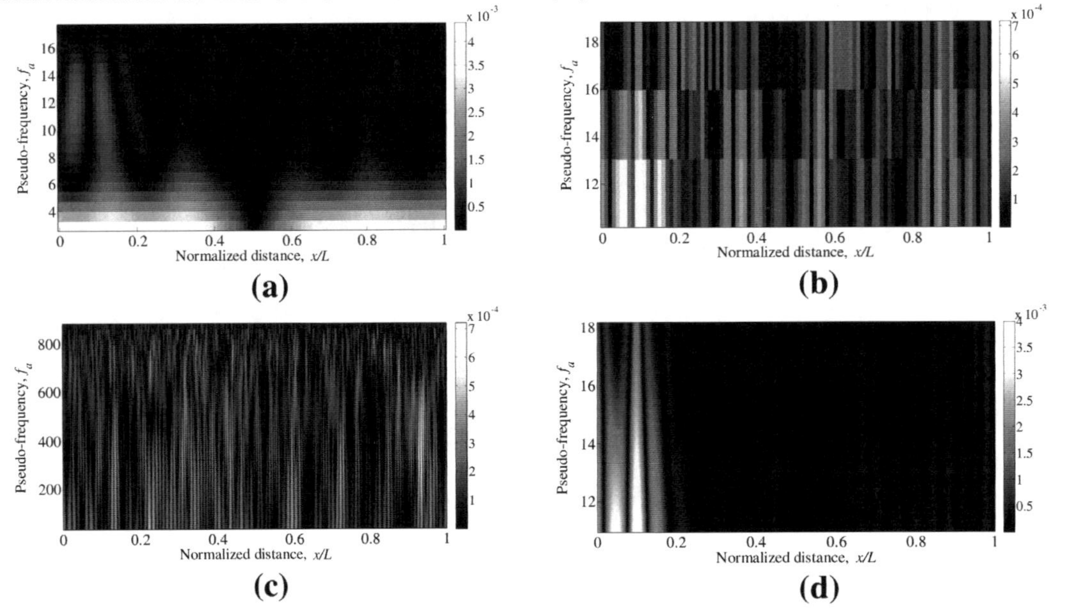

Figure 3. Contour plots of the spatial CWT using 'Coif4' and the polynomial padding method (SNR = 70 dB). The first mode shape of the cantilever beam with crack of δ = 0.2 at x_c/L = 0.1 is analysed. Sampling interval $\Delta x/L$ and scale range are varied: (a) $\Delta x/L$ = 0.01, $s=[4-24]$; (b) $\Delta x/L$ = 0.01, $s=[4-6]$; (c) $\Delta x/L$ = 0.0004, $s=[2-40]$; (d) $\Delta x/L$ = 0.0004, $s=[96-158]$.

As for the beam parameter h, it can be demonstrated [4] that its variation influences the detectable crack depth but not the optimal pseudo-frequency value of CWT damage detection. In summary, the optimal pseudo-frequency $f_{a,opt}$ is independent of the beam parameters b, h, E and ρ and hence, using the polynomial padding method and a given wavelet function, it is a function of the deformed beam shape and the crack position x_c. Formally, this can be written as:

$$f_{a,opt} = \text{f (beam deflection shape}, x_c, L). \tag{10}$$

Through the dimensional analysis and Buckingham's theorem [7], Eq. (10) can be expressed in the following dimensionless form:

$$f^*_{a,opt} = f_{a,opt}\, L = \text{f (beam deflection}, x_c / L), \tag{11}$$

Equation (11) states that the optimal pseudo-frequency $f_{a,opt}$ multiplied by the beam length L is a function of the beam deflection shape and the relative crack position x_c/L only.

Figure 4 illustrates the $f^*_{a,opt}$ against x_c/L curves pertaining to the first three mode shapes of the cantilever and simply supported beams. It can be noted that, mainly due to the edge effects, $f^*_{a,opt}$ turns out to be higher near the beam ends, especially for the third mode shapes.

According to the curves of figure 4, for a given wavelet scale, it is possible to calculate the optimal number of sampling intervals needed to detect the smallest crack located at x_c/L using a given beam vibration mode shape.

Finally, from the definition of $f^*_{a,opt}$, the optimal number of sampling intervals yielding the best CWT damage detection at the scale s for a given crack location x_c/L, is equal to $\left.\frac{L}{\Delta x}\right|_{opt} = \frac{f^*_{a,opt}\, s}{f_c}$.

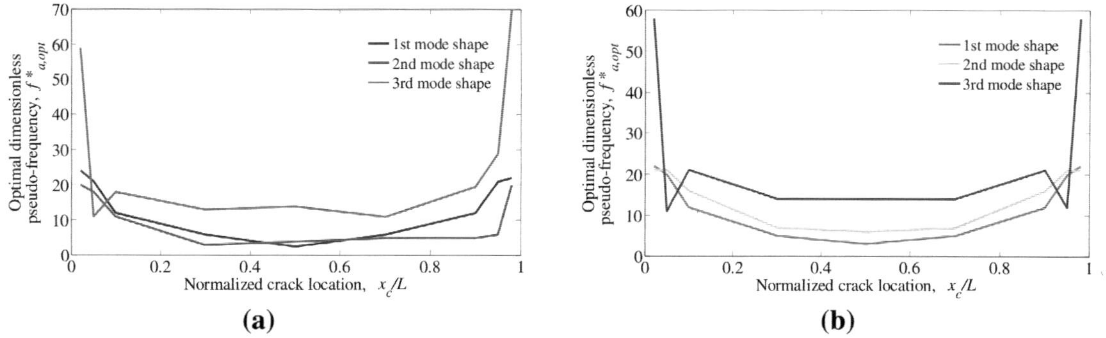

Figure 4. Variation of $f^*_{a,opt}$ for the CWT damage detection with relative crack location for the first three mode shapes of: (a) cantilever beams; (b) simply supported beams.

4. Conclusions

The present investigation examined the effect of the spatial sampling interval in damage detection by CWT, with the aim of answering the following key questions: Can the cost of damage detection be reduced by down-sampling? What is the minimum number of sampling intervals required for optimal damage detection? An in-depth parametric study has been carried out by analyzing the first three mode shapes of a cantilever and a simply-supported beam varying the relevant parameters of the problem, that is: the sampling interval, the noise level, the crack depth and position along the beam, and the mechanical and geometrical beam parameters. The effect of the sampling interval in CWT based damage detection is investigated in terms of the minimum detectable crack size (which is defined on the basis of an ad-hoc criterion) as a function of the key parameter, pseudo-frequency. The curves of minimum detectable crack size against pseudo-frequency for different relative sampling intervals coincide and decrease monotonically with decreasing pseudo-frequency down to a lower bound value of pseudo-frequency f_a below which no damage detection is possible. This lower bound of f_a can be regarded as the optimal pseudo-frequency $f_{a,opt}$ capable of detecting the smallest crack size by CWT. The parametric study shows that such an optimal pseudo-frequency depends on the mode shape and on the crack location along the beam. In addition, it has been observed that, for a given mode shape and crack position, different beam structures, characterized by the same value of x_c/L, attain the same optimal value of the dimensionless pseudo-frequency $f_{a,opt}L$.

References

[1] Tada H, Paris P C and Irwin G R 1985 *The Stress Analysis of Crack Handbook* Del Research Corporation, St. Louis

[2] Mallat S 2001 *A Wavelet Tour on Signal Processing* Academic Press NewYork

[3] Montanari L, Spagnoli A, Basu B and Broderick B M 2015 On the effect of spatial sampling in damage detection of cracked beams by continuous wavelet transform *Journal of Sound and Vibration* (in press)

[4] Montanari L 2014 Vibration-based damage identification in beam structures through wavelet analysis PhD Thesis Univerity of Parma

[5] Montanari L, Basu B, Spagnoli A and Broderick B M 2015 A padding method to reduce edge effects for enhanced damage identification using wavelet analysis *Mechanical Systems and Signal Processing* **52** 264-277

[6] Misiti M, Misiti Y, Oppenheim G and Poggi J 2000 *Wavelet Toolbox* The MathWorks Inc.

[7] Langhaar H L 1951 *Dimensional Analysis and Theory of Models* Wiley New York

Milling damage on Carbon Fibre Reinforced Polymer using TiAlN coated End mills

Mohamed Konneh[1], Sudin Izman[2] and Abdullah Abdul Rahman Kassim[3]

[1]International Islamic University Malaysia, Faculty of Engineering, Department of Manufacturing and Materials Engineering, P. O. Box 10, 50728 Kuala Lumpur, Malaysia
[2]Universiti Teknologi Malaysia, Faculty of Mechanical Engineering, 81310 UTM Skudai, Johor, Malaysia.
[3]International Islamic University Malaysia, Faculty of Engineering, Department of Manufacturing and Materials Engineering, P. O. Box 10, 50728 Kuala Lumpur, Malaysia

Email: mkonneh@iium.edu.my

Abstract. This paper reports on the damage caused by milling Carbon Fibre Reinforced Composite (CFRP) with 2-flute 4 mm-diameter solid carbide end mills, coated with titanium aluminium nitride. The machining parameters considered in work are, rotation speed, feed rate and depth of cut. Experiments were designed based on Box-Behnken design and the experiments conducted on a Mikrotool DT-110 CNC micro machine. A laser tachometer was used to ascertain a rotational speed for conducting any machining trial. Optical microscopy examination reveals minimum delamination value of 4.05 mm at the spindle speed of 25,000 rpm, depth of cut of 50μm and feed rate of 3 mm/min and the maximum delamination value of 5.04 mm at the spindle speed of 35000 rpm, depth of cut of 150μm and feed rate of 9 mm/min A mathematical model relating the milling parameters and delamination has been established.

1. Introduction

The desire to fabricate materials for light-weight applications has gained renewed interest in composite materials research. Research in the processing of reinforced composites in the recent years seems a gateway to producing light-weight materials. Global applications of fibre reinforced plastic (FRP) for example, include but not limited to marine transportation components, architectural cladding components, aerospace transportation and weapons components, automotive components, energy sector components (wind turbines) and static structural components (buildings/bridges). It well documented in literature that fiberglass, a type of fibre reinforced plastic, is the first modern composite materials. Nevertheless, although glass is a very strong material, it is however brittle and will break easily if bend sharply. Because of this disadvantage, advanced composites made of made carbon fibers are now used instead of glass. Carolin, A. [1] has referred to Fibre Reinforced Polymers (FRP) as composites, materials being formed from two or more separated parts with a distinguished phase between them. It has been reported [2, 3] that Fibre Reinforced Polymers (FRP) composites are lighter and stronger than fiberglass but it costs more to produce them. However, their light weights and high strength characteristics offer them possibilities to be used for making lighter cars and aircraft for less fuel consumption when compared with the heavier vehicles that are currently in use and for sports equipment such as golf clubs. It has also been reported [4] that Boeing 787 Dreamliner was the first world major commercial airliner that used composite materials for the construction of most of its parts.

Although Carbon Fibre Reinforced Plastics (CFRP) are lighter, yet stronger than those typically used, they pose challenges when it comes to machining such as milling and drilling. The materials are difficult to cut because they are very abrasive and tough, these characteristics obviously will affect both the work-piece and cutting tools. The damage on the machined CFRP referred here as delamination, is the separation of individual carbon fiber plies and it is difficult to prevent from occurring while machining. However these challenges have been met with some success with the use of hard, sharp solid-carbide milling cutters that employ special surface coatings. The work of Konneh et al [5] has demonstrated that CFRP can hardly be machined without inducing delamination.

This experimental investigation was done to study how machining parameters influence the production of delamination on milled CFRP using TiAlN coated end mills on a CNC micro machine with maximum rotational speed limited to 2500 rpm. Box-Behnken design of experiments was employed for the designing and analysis of the experiments. A mathematical model relating the milling parameters and delamination response has been established.

2. Experimental details

This section describes the work-piece and cutting tool materials, the experimental conditions, setup and procedure.

2.1 Work and Tool Materials, and experimental conditions

Locally prepared Carbon Fiber Reinforced Polymer (CFRP) by hand lay-up process, with dimensions of 50 mm x 50 mm x 3 mm, was used as the work-piece, Figure 1

The tools chosen for this experimental work are 2-flute 4 mm-diameter solid carbide end mills, coated with titanium aluminium nitride (TiAlN), Figure 2. The cutting tool was selected based on the recommendation from HPMT Cutting Tool Catalogue.

Figure 1. CFRP work-piece material

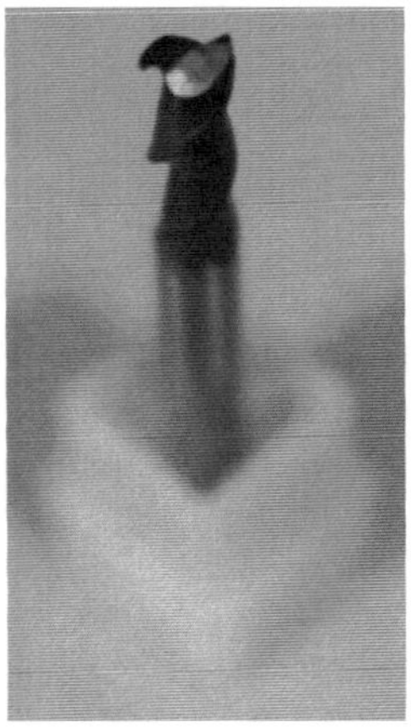

Figure 2. TiAlN used as cutting tools

2.2 Experimental conditions

The machining variable for conducting the experiments are summarised in Table 1.

Table 1. Factors and levels in the design of experiments

Factors	Levels		
	Low	Medium	High
Spindle Speed (rpm)	15 000	25 000	35 000
Feed (mm/min)	3.00	9.00	15.00
Depth of Cut (μm)	50.00	100.00	150.00

2.3 Experimental Setup and Procedure

The Figure 3 depicts the setup for the experiments on a CNC Micro-machines having maximum rotational speed of 2500 rpm. To facilitate a high speeding machining, an air turbine was attached to the spindle of the machine to augment the speed of the Mikrotool machine. This could enable rotational speeds up to 50,000 RPM to be obtained. The locally prepared CFRP composite work-piece, consisted of 10 layers of carbon fiber, epoxy and hardener. The layers were compressed by a Universal Testing Machine with a compression force of 55 kPa and left to dry for 6 hours. Seventeen experiments were conducted based on the Box-Behnken experiment design. Straight cuts analogous to end milling were taken along the length of the work-piece. A laser tachometer (A DT 2336B) was used to ascertain a rotational speed for conducting any machining trial. 4 mm-diameter 2-fluted carbides coated with titanium aluminium nitride (TiAlN) were used to cut the work-piece based on the predetermined machining conditions. Delamination of the milled CFRP was determined in terms of the profile path of the surface integrity. Measurements of delamination were done on a Nikon Epiphot 200 Optical Microscope and the average value of the widths of the traversed path by the tool for each trial run noted. Some three points noted during a measurement are shown in the Figure 4.

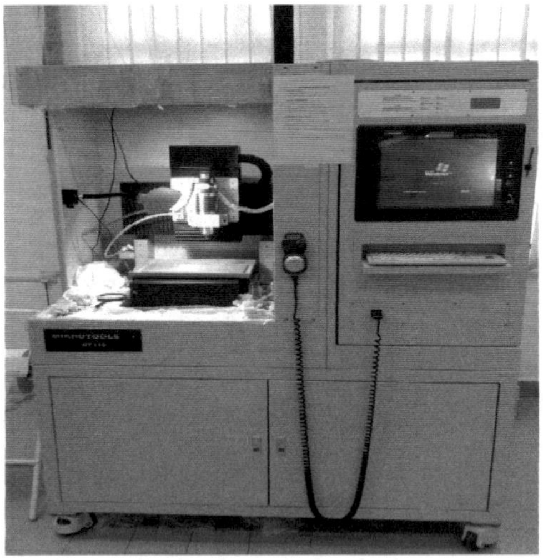

Figure 3. Micro-machine (DT-110) that was used for experiments

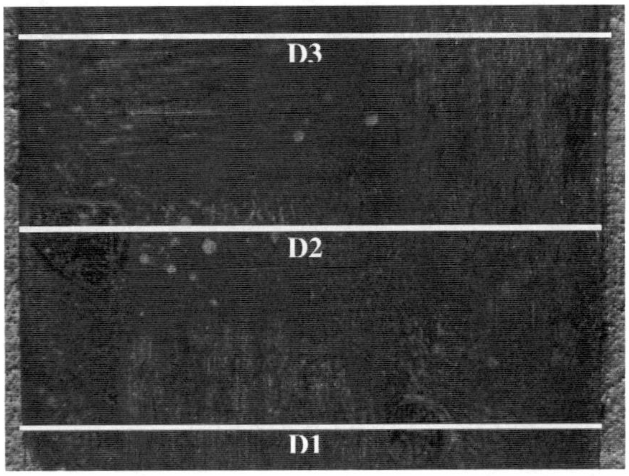

Figure 4. Delamination measurement method along the traversed path on the work-piece; D1, D2 and D3 denote the path widths after milling.

3. Results and discussion

The milling experiments that were conducted to study the effect of machining parameters on delamination are listed in Table 2. It is clearly seen in Table 3 depicting the analysis of variance (ANOVA) that, for the Model F-value of 12.86, the model is significant, thus there is only a 0.14% chance that such a "Model F-Value" could occur due to noise. Statistically, "Prob > F" values less than 0.0500 indicate model terms are significant. This being the case, it is clear that A, C, C^2, BC are the significant model terms. This means that there are some insignificant model terms, not considering those required to support hierarchy.In the light of this, a model reduction became a necessity for improvement, the improved (reduced model) shown in Table 4

In the reduced model (Table 4), the Model F-value of 16.69 implies that the model is significant and there is only a 0.01% chance that a "Model F-Value" this large could occur due to noise. This F-value is better than that obtained (12.86) in the original ANOVA Table 3. In the reduced ANOVA (Table 4), it is evdent that A, C, C^2, BC are the significant model terms the fact that values of "Prob > F" less than 0.0500 statistically imply model terms are significant and values greater than 0.1000 indicating the model terms are not significant. The 1.42 "Lack of Fit F-value" implies the is not significant relative to the pure error. With this Lack of Fit value, is a 38.53% chance that it occur due to noise. A non-significant lack of fit is obviously preferred since it desirable for a model to fit. This is (38.53%) preferrable compared with the 49.78% in Table 3.

Based on statistics, the 0.6836 "Pred R-Squared" seems to be in a reasonable agreement with the 0.8306 "Adj R-Squared" in the Table 4. Also since the "Adeq Precision" measures the signal to noise ratio and a ratio greater than 4 is desirable, it can be concluded that the ratio of 13.640 here indicates an adequate signal. The fact that these statistical indicators in (Table 4) have proven better than the corresponding indicators in Table 3, was necessary to explore the design space for further analysis.

Normality of residuals is evident in the normal probability plot of the studentized residuals, Figure 4. It is also obvious in the Outlier t versus run order (Figure 5) that there are no outliers i.e., influential values. The model statistics and diagnostic plots being adequate, the Model Graphs, Figures 6 and 7 were examined.

Table 2. Box-Behnken designs showing 17 trial runs and response delamination.

Std	Run	Block	Factor 1 A:Rotational sp RPM	Factor 2 B:Feed rate mm/min	Factor 3 C:Depth of cut Microns	Response 1 Delamination mm
7	1	Block 1	15000.00	9.00	150.00	4.209
14	2	Block 1	25000.00	9.00	100.00	4.792
12	3	Block 1	25000.00	15.00	150.00	4.159
13	4	Block 1	25000.00	9.00	100.00	4.526
1	5	Block 1	15000.00	3.00	100.00	4.171
10	6	Block 1	25000.00	15.00	50.00	4.341
11	7	Block 1	25000.00	3.00	150.00	4.753
15	8	Block 1	25000.00	9.00	100.00	4.558
2	9	Block 1	35000.00	3.00	100.00	5.02
4	10	Block 1	35000.00	15.00	100.00	5.024
6	11	Block 1	35000.00	9.00	50.00	4.505
3	12	Block 1	15000.00	15.00	100.00	4.146
17	13	Block 1	25000.00	9.00	100.00	4.792
9	14	Block 1	25000.00	3.00	50.00	4.051
5	15	Block 1	15000.00	9.00	50.00	4.137
8	16	Block 1	35000.00	9.00	150.00	5.049
16	17	Block 1	25000.00	9.00	100.00	4.654

Table 3. Experimental results diagnosed by Analysis of Variance (ANOVA)

ANOVA for Response Surface Quadratic Model Response: Delamination
Analysis of variance table [Partial sum of squares]

Source	Sum of Squares	DF	Mean Square	F Value	Prob > F	
Model	1.79	9	0.20	12.86	0.0014	significant
A	1.08	1	1.08	69.80	< 0.0001	
B	0.013	1	0.013	0.86	0.3857	
C	0.16	1	0.16	10.46	0.0144	
A^2	5.897E-003	1	5.897E-003	0.38	0.5559	
B^2	0.052	1	0.052	3.40	0.1078	
C^2	0.22	1	0.22	14.04	0.0072	
AB	2.103E-004	1	2.103E-004	0.014	0.9103	
AC	0.056	1	0.056	3.61	0.0992	
BC	0.20	1	0.20	12.66	0.0092	
Residual	0.11	7	0.015			
Lack of Fit	0.045	3	0.015	0.95	0.4978	not significant
Pure Error	0.063	4	0.016			
Cor Total	1.89	16				

Std. Dev.	0.12	R-Squared	0.9430
Mean	4.52	Adj R-Squared	0.8696
C.V.	2.75	Pred R-Squared	0.5690
PRESS	0.82	Adeq Precision	11.529

Verification of the adequacy of the model is shown in the ANOVA Table 4 where the Model F-value of 16.69 implies the model is significant.

It can be observed in from the model graph in the Figure 6 showing the individual effects of the process machining parameters that the depth of cut and the rotational speed a considerable effect on delamination under investigation, while the feed rate virtually showing little effect on the same

Table 4. Reduced ANOVA for delamination

Response: Delamination ANOVA for Response Surface Reduced Quadratic Model
Analysis of variance table [Partial sum of squares]

Source	Sum of Squares	DF	Mean Square	F Value	Prob > F	
Model	1.67	5	0.33	16.69	< 0.0001	significant
A	*1.08*	*1*	*1.08*	*53.72*	*< 0.0001*	
B	*0.013*	*1*	*0.013*	*0.66*	*0.4343*	
C	*0.16*	*1*	*0.16*	*8.05*	*0.0162*	
C²	*0.23*	*1*	*0.23*	*11.27*	*0.0064*	
BC	*0.20*	*1*	*0.20*	*9.75*	*0.0097*	
Residual	0.22	11	0.020			
Lack of Fit	*0.16*	*7*	*0.022*	*1.42*	*0.3853*	*not significant*
Pure Error	*0.063*	*4*	*0.016*			
Cor Total	1.89	16				

Std. Dev.	0.14	R-Squared	0.8835
Mean	4.52	Adj R-Squared	0.8306
C.V.	3.13	Pred R-Squared	0.6836
PRESS	0.60	Adeq Precision	13.640

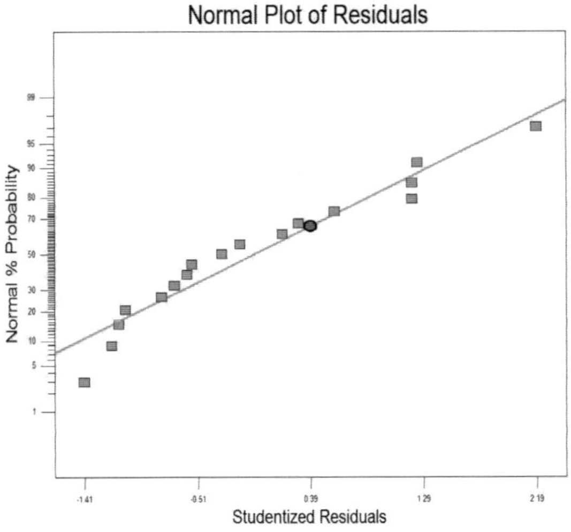

Figure 5. The normal plot of residuals

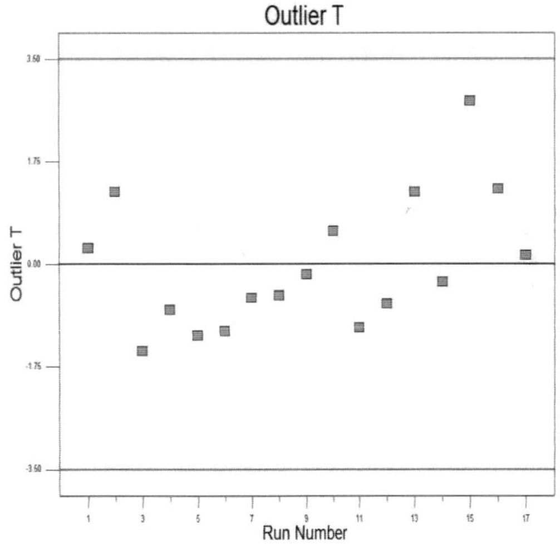

Figure 6. The outlier plot indicating the trial runs are within range.

response variable. Also, taking the cutting tool diameter into consideration, it is evident from the graph that high rotational speed is not a necessary condition for minimising delamination defects at a machined work-piece surface, while taking the tool diameter into consideration. The ANOVA (Table 4) clearly reveals interaction effect between depth of cut (C) and feed rate (B) parameters, the effect which is depicted in both 2D and 3D in the Figures 7 and 8 respectively. Such interaction effect seemed to have caused the feed rate to almost has negligible effect on the machining process.

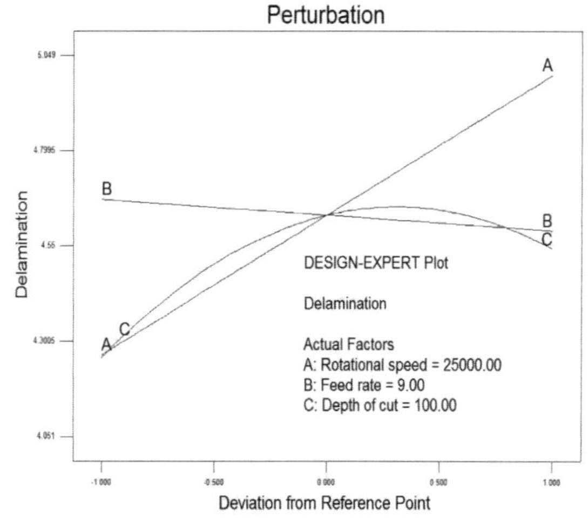

Figure 7. A plot of individual effects of the machining parameters

It seems that the level of depth of cut has an effect of feed rate, when the depth of cut is high, delamination becomes smaller. When the depth of cut is low delamination becomes relatively higher. Such a behaviour has likely hindered a meaningful contribution of the feed rate to the response variable during the machining process. Figure 8 is the 3D plot exemplifying the factor reaction.

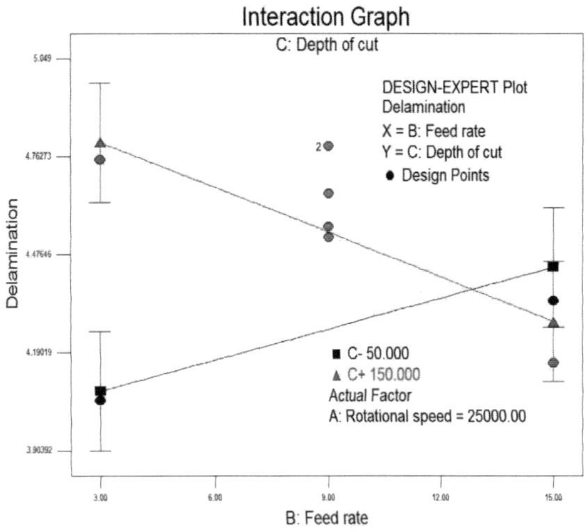

Figure 8. A plot of individual effects of feed rate and depth of cut.

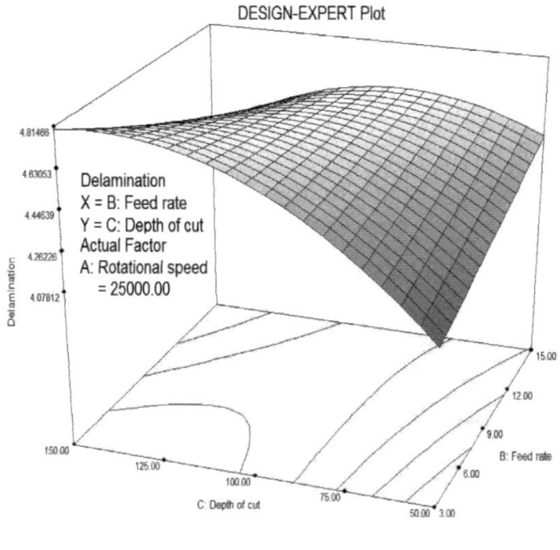

Figure 9. The 3D plot of the interaction effect of feed rate and grit size

A mathematical model is used to relate the response variable (D) and the milling parameters represented by:

$$D = f(A, B, C) \qquad (1)$$

where D is Delamination; f function of, A rotational speed; B feed rate and C depth of cut. It is observed that a second order model suitable to relate delamination and the milling parameters tested under study. In the light well established procedure [6] for predicting first and second order models the model equation (2) is predicted based on the experimental results obtained in this study.

$$\text{Delamination (D)} = + 4.63+0.37A - 0.041B +0.14C - 0.23C^2 -0.22BC. \qquad (2)$$

where the variables A, B and C are defined above.

4. Conclusion

The following conclusions can be drawn from the experiments that were conducted when CFRP was milled using2-flute 4 mm diameter carbide end mills coated with TiAlN:

- It has been observed that rotational speed and depth of cut have marked effect on delamination, when the feed rate has little effect on the response variable.
- Optical microscopy examination reveals minimum delamination value of 4.05 mm at moderate spindle speed of 25000 rpm, low depth of cut (50µm) and low feed rate (3 mm/min); the maximum delamination value of 5.04 mm at the spindle speed (35000 rpm), depth of cut (150µm) and feed rate (9 mm/min).
- A mathematical model predicted for delamination for the range of machining parameters used in the study is presented by:

  $$\text{Delamination (D)} = + 4.63+0.37A - 0.041B +0.14C - 0.23C^2 -0.22BC$$

  where D is Delamination, A rotational speed; B feed rate and C depth of cut
- High rotational speed and moderate depth of cut combination is likely to yield a favourable delamination when a Carbon Fibre Reinforced Plastic is milled with a TiAlN coated tool.

5. Acknowledgement

The authors are indebted to the Kulliyyah of Engineering, International Islamic University Malaysia (IIUM) and the Research Management Centre for moral support while undertaking this research work. They would also like to express their gratitude and thanks to the Ministry of Higher Education (MOHE), Malaysia for their financial support (FRGS).

6. References

[1] Carolin, A. (2003). Carbon Fiber Reinforced Polymer for Strengthening of Structural Elements. Doctorial Thesis, Lulea University of Technology, 13-18.
[2] Kindo, S. (2010). Study on mechanical behavior of coir fibre reinforced polymer composites, B. Tech Thesis, Department of Mechanical Engineering, NITR, 1-5
[3] Babu, G. D., Babu, K. S. and Gowd, U. M. (2013). Effect of machining parameters on milled natural fifer-renfocred plastic composites, Journal of Advanced Mechanical Engineering, 1: 1-12
[4] Hota, V.S., Rao, G., Vijay, P. E. P. V. (2010). Feasibility review of FRP materials for structural applications. http://www.cemr.wvu.edu/cfc/research/projects/usacereport.pdf
[5] Konneh, M., Sudin, I., Padil, M. E. D. and Roszat , R. (2014).Surface Roughness Study Of Milled Carbon Fiber Reinforced Polymer (CFRP) Composite Using 4 mm 2-Flute Titanium Aluminum Nitride (TiAlN) Coated Carbide End Mills. Journal of Advanced Materials Research Vols. 887-888, 1101-1106
[6] Alao, A. R. (2007). Precision Micro-scaled Partial Ductile Mode Machining of Silicon. MSc Thesis, International Islamic University, Malaysia.

Damage Detection based on the Natural Frequency shifting of a clamped rectangular plate model

Irfan Hilmy[1], MM. Abdel Wahab[2], Erry Yullian T. Adesta[1], Tasnim Firdaus[1]

[1]International Islamic University Malaysia, Faculty of Engineering, Department of Manufacturing and Materials Engineering, P. O. Box 10, 50728 Kuala Lumpur, Malaysia

[2]Department of Mechanical Construction and Production, Ghent University, Belgium

E-mail: ihilmy@iium.edu.my

Abstract. Damage detection of any structure becomes the main concern in a failure analysis. Early failure detection is very important as it can prevent any catastrophic failure by replacing or repairing the damage part at early stage. One of the non-destructive methods of damage detection is using frequency based vibration analysis. Identification and comparison of a set of natural frequencies before and after damage is the main concern of this research. A rectangular plate clamped at all edges represented an initial undamaged structure. Based on Kachanov's definition, damage existence in a structure is introduced in the presence of some circular voids. The voids are generated randomly at different level of damage value. To obtain the Natural Frequencies, a Finite Element Model (FEM) of a clamped plate with the updated value of Young's Modulus is analyzed. From the FEM analysis result, it is found that the Natural Frequencies are shifted as the void existence increase. Using curve fitting, the model of Natural Frequency shifting as a function of damage evolution has been generated. It is found that the shifting of the Natural Frequency is greater at higher frequency value as indicated by the higher absolute gradient.

1. Introduction

The eigenvalue analysis technique is a fundamental tool and the most well-known for matrix computations. In vibration analysis, eigenvalue resemble natural frequency of an engineering structure. It is well known that any normal/perfect/undamaged structure has a set of natural frequency that will change due to the degradation of its mechanical properties. This status is called as damage or abnormal structure. The main aim of the damage detection is to avoid any catastrophic failure by replacing the damage part before it propagates to other main/vital part. To detect or identify the damage level of the structure based on the change of the natural frequency becomes the objective of this present paper.

The detection to determine a degree of damage in existing engineering structures is of great importance from the point of view of their safety and serviceability. An extensive testing and a visual inspection can be employed to measure and locate the degradation of structure by non-destructive techniques such as ultrasonic methods, thermography, acoustic emission, or the vibration testing. In vibration analysis, many methods have been developed, but the common method is Frequency Response Function (FRF) analysis. Using FRF, a shifting of natural frequency, mode shape, FRF or the existence of the new natural frequency can become an indication of the occurrence of damage evolution [1]. The damage term in this paper relates with the damage evolution law theory as initiated by Kachanov [2]. To determine a failure of a structure, damage mechanics is considered a relatively new tool compare to the fracture mechanics that already established for quite some times. Fracture mechanics uses strain energy release rate and crack propagation phenomenon for its calculation while damage mechanics use the presence of large number of randomly distributed micro-cracks and voids. Damage mechanics deal with the crack initiation phase in a structural failure.

Kwak & Han [3] has developed a free vibration analysis of a rectangular plate with a rectangular or a circular hole based on the Independent Coordinate Coupling Method (ICCM). The latest publication related to the Eigen analysis of a plate with circular cut-out was generated by Hossain et al. [4]. The explicit closed form solution to natural frequencies and its mode shapes has been derived by Kang & Kim [5]. As far as the author's knowledge, none of the model as stated in the publications above incorporating the updated Young's Modulus as a function of material degradation in their numerical models. Figure 1 shows a model of structure/material contain damage in term of the existence of void/air bubble as proposed by Kachanov.

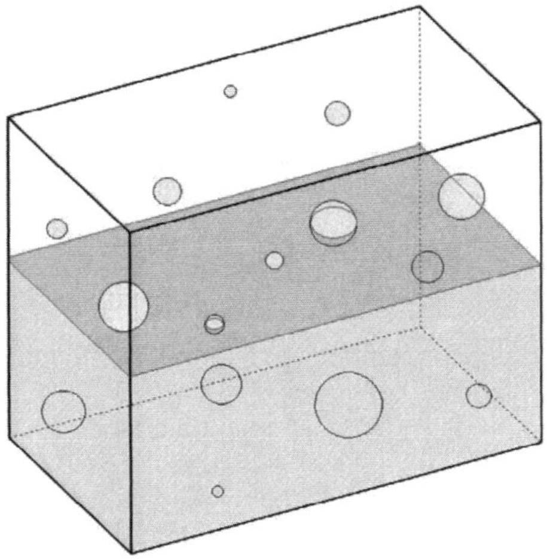

Figure 1. Damage definition, adapted from Kachanov [2]

In reality, this is the case where any real material always contains defect no matter how sophisticated any material forming process is. If from Figure 1, any imaginary plane was taken, then it can be modelled as a rectangular plane with any circles that represent void/air bubble as shown in Figure 2.

Analytical solutions for flexible plate case with some boundary conditions were developed by Brethee [6], Hashemi et al. [7] and Safizadeh & Mat [8]. Determination of the natural frequencies of flexible rectangular plates was developed by Geradin [9] using approximation method, which follows from the properties of deflection of the mid-plane. Some researchers have developed methods to improve and validate the analytical solution. Using asymptotic theory as proposed by Geveci & Walker [10], numerical solutions for plates with all-clamped edges are obtained and validated. Mochida & Ilanko [11] used the finite difference and superposition methods to obtain the natural frequencies of a free rectangular plate.

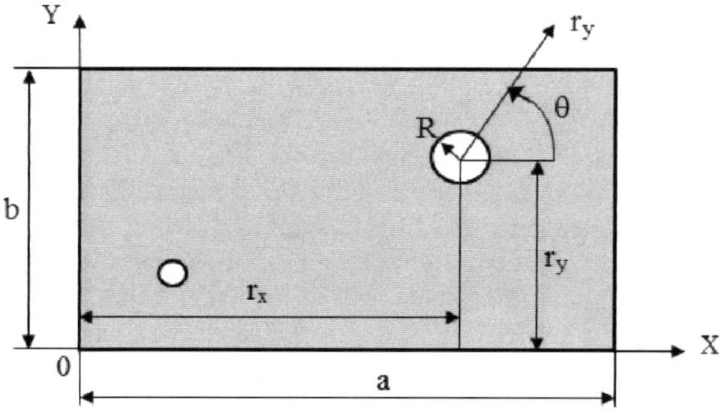

Figure 2. Rectangular plate with a circular hole as a model of damage, adapted from Moon [3]

2. Clamped plate model

Based on reference [11], the analytical formula of natural frequencies of all-clamped edges plate are obtained using the following formula:

$$f_{ij} = \frac{\lambda_{ij}^2}{2\pi\emptyset^2}\left[\frac{Eh^2}{12\rho(1-v^2)}\right]^{1/2} \qquad \text{Eq. 1}$$

where the natural frequency (f_{ij}) depends upon Young modulus (E), Poisson ratio (v), mass density (ρ) and the dimensionless natural frequency parameter (λ).

Refer to the notations stated in Figure 2, a fictitious plate model with the dimensions a = 6 inch, b = 5 inch and thickness=0.063 inch with all edges clamped was developed using ABAQUS Standard/Explicit model. A perfect model whereby no voids exist was validated with Moon's analytical equation. The overall flowchart is shown in Figure 3.

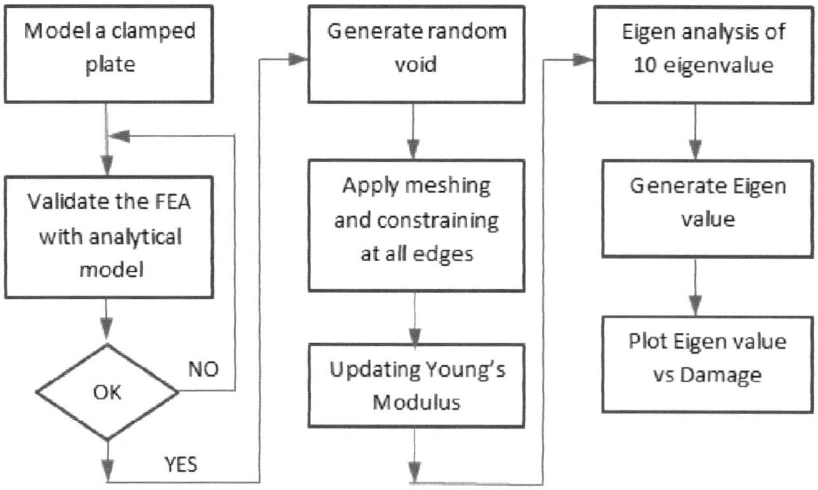

Figure 3. Flowchart of the research

Using Matlab, the coordinate of the voids were randomly generated, while maintaining the size of the void constant. The models were generated near two conditions: very low damage value (0 until 0.034) and near critical damage value (0.1 until 0.134). The geometry with the holes generation at different level of damage is shown in Figure 4.

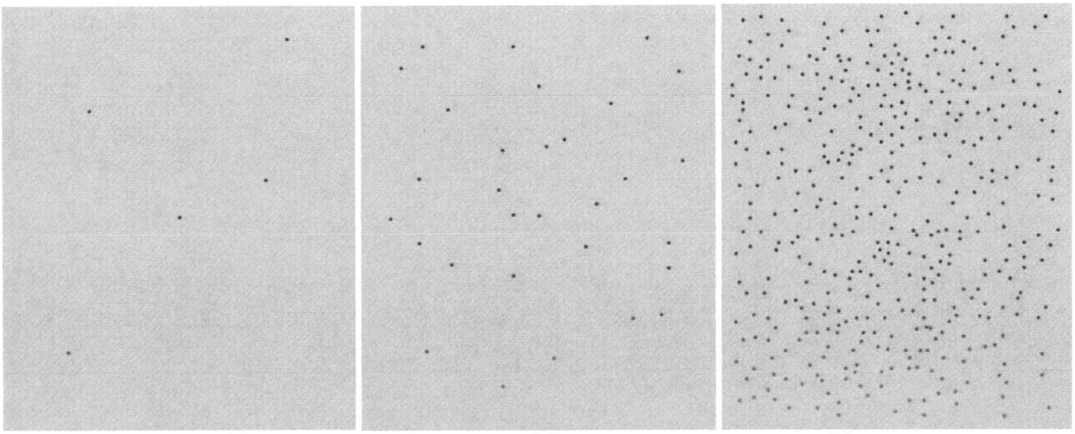

(a) (b) (c)

Figure 4. Generation of void model with (a) 5, (b) 30, and (c) 100 holes

Figure 5 shows meshing status for different damage level. It can be seen that the number of elements rising rapidly to accommodate finer meshing near hole.

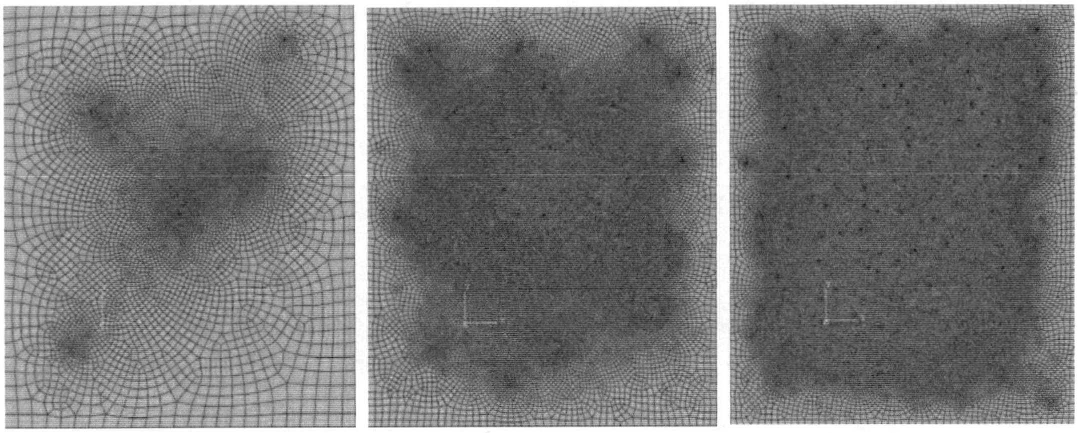

(a) (b) (c)

Figure 5. Generation of meshing for different hole number respective to the Figure 3
(a) 7,118 (b) 39,065, and (c) 77,496 elements

During the development of Damage, the Young's Modulus value degrades. The relation of Damage as a function of Young's Modulus degradation has been developed by Wahab et al. [12] and shown in Figure 6. Based on this phenomenon, the value of Young's Modulus was updated respective to the damage value. It can also be seen that after the damage reach the critical value (beyond 200 number of cycles), damage value increases rapidly that eventually it leads to a catastrophic failure.

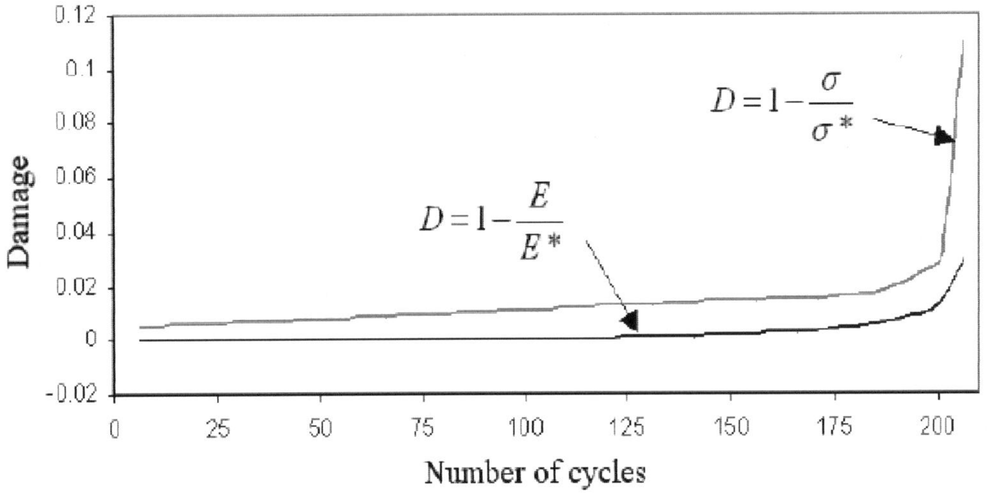

Figure 6. Damage calculation using different methods, adapted from Wahab et al. [12]

3. Numerical results

Typical result of the Eigen analysis is shown in Figure 7, where the mode shapes of the first three natural frequencies are plotted.

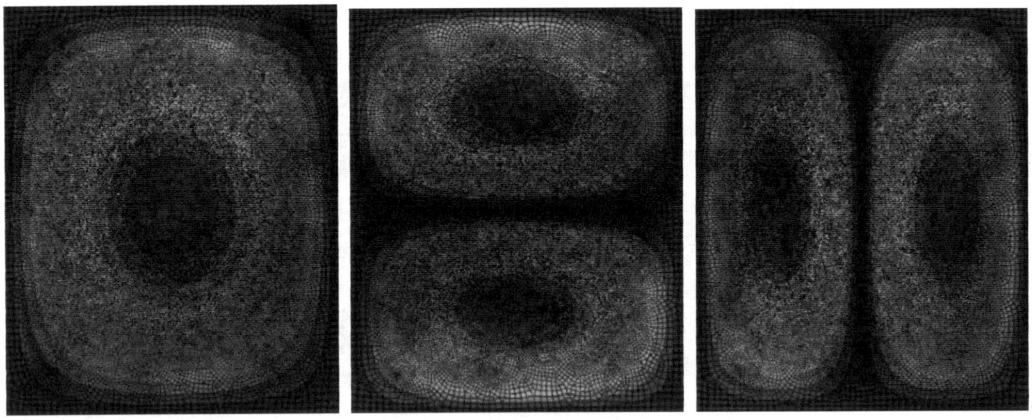

(a) (b) (c)

Figure 7. Mode shape of the first four natural frequencies (case damage value = 0.034): (a) 19.83, (b) 44.15, and (c) 54.97 (Hz)

Figure 8 shows the result of the Eigen analysis. Ten natural frequencies were extracted for different damage level up to Damage value = 0.134. The reason behind this maximum value is based on the assumption that at that level, the damage already reaches its critical value. In theory, above critical value, the damage will increase rapidly that eventually it will lead to a catastrophic failure.

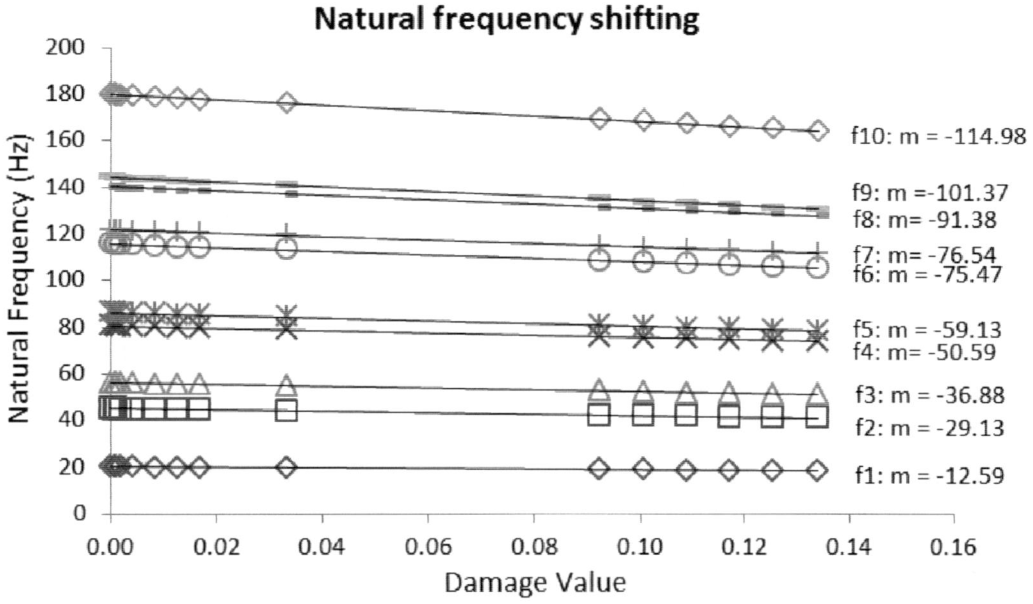

Figure 8. Natural frequency shifting vs Damage evolution

It also can be seen from Figure 8 that the absolute value of the gradient of the trend line increases (from 12.59 to 114.98) as the Natural frequency rises (from f_1 up to f_{10}). It means that the reduction of Natural frequency due to material's property degradation is higher at high Natural frequency number.

4. Conclusions

The present paper is concerned with the shifting of Natural frequency as a function of damage evolution. Based on this finding, damage value is treated as the value of the whole structure and not refers to any particular place in the structure. This is due to the assumption that the void exist in a randomly manner. For future works, the location of void will be determined as a function of space. The FEA model also will be upgraded into a solid model with the void randomly scatter in a 3D space. Therefore, the shifting of the natural frequency can be determined also as a function of the location of the void.

5. Acknowledgement

This work was supported by The Fundamental Research Grant Scheme (FRGS), FRGS14-128-0369, Ministry of Higher Education (MOHE), Malaysia.

6. Reference

[1] K. Dems, J. Turant 2011 Structural damage identification using frequency and modal changes, *Bulletin Of The Polish Academy Of Sciences Technical Sciences*, Vol. 59, No. 1

[2] L. M. Kachanov 1986 Introduction to continuum damage mechanics *Mechanics of Elastic Stability* Volume 10 ISBN: 978-90-481-8296-1

[3] Moon K. Kwak, Sangbo Han 2007, Free vibration analysis of rectangular plate with a hole by means of independent coordinate coupling method, *Journal of Sound and Vibration*, 306:12-30

[4] Nazmul Hossain, KH. Nazmul Ahshan, Md. Zahid Hossain and Md. Shahriar Islam 2015 Effect on Natural Frequency of a Simply Supported Plate Due to Circular Cutouts, *Proceedings of 10th Global Engineering, Science and Technology* ISBN: 978-1-922069-69-6

[5] Sang Wook Kang & Sang-Hyun Kim 2008 Vibration analysis of simply supported rectangular plates with unidirectionally, arbitrarily varying thickness, *Jour. of Sound & Vibr.* 312,551-562.

[6] Khaldoon F. Brethee 2009 Free Vibration Analysis of A Symmetric and Anti-Symmetric Laminated Composite Plates With a Cutout at The Center *Al-Qadisiya Journal For Engineering Sciences* Vol. 2 No. 2,

[7] Shahrokh Hosseini-Hashemi, Mahmoud Karimi, Hossein Rokni 2012 Natural frequencies of rectangular Mindlin plates coupled with stationary fluid *Applied Mathematical Modelling* 36 764–778.

[8] Safizadeh, M. R.; Darus, I. Z. Mat May 2010 Natural Frequency Analysis of All Edges Clamped Flexible Thin Plate, *International Review of Mechanical Engineering* Vol. 4 Issue 4, p433

[9] Michel Geradin 2006, *Mechanical Vibrations: Theory and Application to Structural Dynamics*, Sons Inc & John Wiley, USA,

[10] Berk Geveci, Walker, J. D. A. May 2009 Nonlinear Resonance of Rectangular Plates *Mathematical, Physical and Engineering Sciences*. 457, pp. 1215-1240, ,

[11] Mochida, Y, Ilanko, S 2008 Bounded natural frequencies of completely free rectangular plates *Journal of Sound and Vibration* 311 pp. 1–8,

[12] MM Abdel Wahab, I Hilmy, IA Ashcroft and AD Crocombe, Damage Modelling of Adhesive Bonding, *4th International Conference on Fracture and Damage Mechanics*, Mallorca, Spain, 2005

Genetic Algorithm Based Objective Functions Comparative Study for Damage Detection and Localization in Beam Structures

K Samir[1], B Idir[1], R Serra[2], B Brahim[1] and A Aicha[1]
[1]University M'hamed Bougara Boumerdes, LEMI Laboratory Research Team Modelling and Simulation in Mechanical Engineering, 35000 Boumerdes, ALGERIA
[2]Laboratoire de Mécanique et Rhéologie INSA Centre Val de Loire, LMR, 3 Rue de la chocolaterie, 41000 Blois, France

E-mail: khatir_samir@hotmail.fr

Abstract. The detection techniques based on non-destructive testing (NDT) defects are preferable because of their low cost and operational aspects related to the use of the analyzed structure. In this study, we used the genetic algorithm (GA) for detecting and locating damage. The finite element was used for diagnostic beams. Different structures considered may incur damage to be modelled by a loss of rigidity supposed to represent a defect in the structure element. Identification of damage is formulated as an optimization problem using three objective functions (change of natural frequencies, Modal Assurance Criterion MAC and MAC natural frequency). The results show that the best objective function is based on the natural frequency and MAC while the method of the genetic algorithm present its efficiencies in indicating and quantifying multiple damage with great accuracy . Three defects have been created to enhance damage depending on the elements 2, 5 and 8 with a percentage allocation of 50% in the beam structure which has been discretized into 10 elements. Finally the defect with noise was introduced to test the stability of the method against uncertainty.

1. Introduction

In recent years, genetic algorithms have been recognized as promising techniques for intelligent search difficult optimization problems. A method of the genetic algorithm is very attractive compared to conventional methods because it does not require a search of solution within the entire space of solution. Iturrioz et al [1] used a separate index: COMAC criterion (Coordinate Modal Assurance Criterion). Numerical and experimental vibration modes were determined the index position and the magnitude of the damage.

Messina et al. [2] proposed an approach to uncertainty for the detection of damage that was later extended in [3] to identify the extent of damage in several sites. Data validation was performed by free digital noise tests. This approach, however, may involve significant computational effort when it comes to large structures.

For structures with higher degree of freedom, MI Friswell et al. (1998) [4] had developed a technique which is based on the combined use of the sensitivity of eigenvalues and genetic algorithms to identify the location and extent of damage from the measured vibration data. They used a genetic algorithm to minimize a square value of the frequency error.

J. H. Chou and J. Ghaboussi (2001) [5] used a GA to solve an optimization problem formulated for the detection and identification of structural damage. The "Output error" indicating the difference between the measured and calculated responses under static loads and the equation error indicating the residual force in the equilibrium equations were used to formulate the objective function to be optimized. The proposed method can successfully detect the location and magnitude of damage and determine the correct unmeasured nodal displacement, while avoiding the complete finite element analyzes. Cawley and Adams [6] used the variations in the natural frequency finite element models to find the damage. H. M. Gomes and N.R.S. Silva [7] used two methods, one based on the frequency sensitivity to damage and the second is based on optimization and parametric finite element modelling techniques. [8] have Used a substructure FE-based approach to identify and localize damages in a complex mechanical systems using the subspace fitting method.

1.1 Aim and scope

In this paper the damage identification problem is addressed by means of a Genetic Algorithm (GA) approach. It has utilized in order to achieve the optimal results in the damage detection problem. The objective functions used in the optimization process are based on the dynamic analysis data in the structure (frequencies and mode shape). Various numerical finite element methods will be performed on clamped free beams discretized into 10 elements.

2. Modeling and Applied stresses identification

We consider a clamped free beam of pure unidirectional composite of AS4/3501-6 graphite-epoxy materials with symmetrical order of layers flexion. The finite element model is discretized in 10 elements as shown in Figure 1. Each node of the finite element has three degrees of freedom, normal displacement w along the z axis, a rotation γ around the y axis and longitudinal displacement u. since the beam is macroscopically considered homogeneous, i.e. there is only one layer, the shear correction coefficient is the same as for isotropic beam ,i.e. K Correction =5/6 [9].

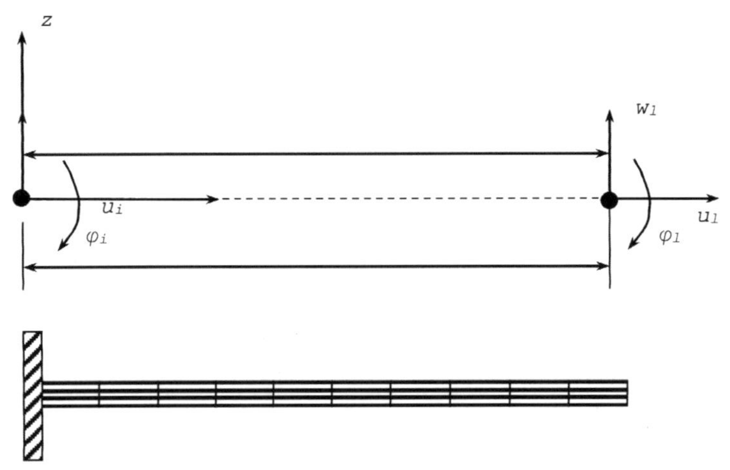

Figure.1 beam structure of unidirectionally reinforced graphite-epoxy beam

The material properties and beam dimensions of AS4/3501-6 graphite epoxy [10] are given in Table 1.

Table 1. Dimension and material of beam

Ply property	Mean value
Length (m)	0.75
Width (m)	0.03
Thickness (m)	0.005
Longitudinal modulus (GPa)	141.96
Transverse Shear modulus (GPa)	6
Density (kg.m-3)	1600
Poisson's ratio v	0.42

3. Modal Assurance Criterion (MAC)

The Modal Assurance Criterion (MAC) value indicates the degree of correlation between the two modes and ranges from 0 to 1, with 1 for any correlation and 0 for no correlation. The deviation from 1 can be interpreted as a damage indicator of in structures. This is based on comparison between the changes in the mode shapes obtained from both tests and calculations, the MAC is defined by Ostachowicz WM et al. 1996 [11]:

$$\text{MAC}\,(\varphi_i, \varphi_j) = \frac{(\varphi_i^T \varphi_j)^2}{\varphi_i^T \varphi_i \varphi_j^T \varphi_j} \tag{1}$$

Where: φ_i is the tested Eigen vector and $\varphi_j$the calculated eigen vector.

The objective function to be minimized is defined as follows:

$$\text{Fitness} = \Sigma\,(1 - \text{MAC})^2 \tag{2}$$

4. Natural frequencies variations

The natural frequency is used as a diagnostic parameter in the procedures for structural assessment using vibration monitoring. A great advantage to use only specific to the damage assessment structures values that can easily obtained and cheaply acquired that make the approach provide an assessment of the economic structure technique. The objective function to be minimized is defined as follows according to MTV Baghmisheh et al 2008 [12]:

$$\Delta\omega = \sum_i^n((\omega_i^m - \omega_i^a)^2 / (\omega_i^m)^2) \tag{3}$$

i = mode number (i=1, 2, 3,….,n), ω_i^m = tested natural frequencies and ω_i^a = Calculated natural frequencies.

5. Application of the Genetic Algorithm

The Genetic Algorithm (GA) is an evolutionary optimization method, used efficiently for different kinds of optimization problems in last decades [13]. In our study 10 individual, also called chromosomes, represent the two damage parameters of position and rate, are converted to binary encoding. The population evolves toward better solution iteratively in a process inspired from the natural evolution, where they are allowed to cross among themselves in order to obtain favorable solutions. The fitness is the objective function value, calculated in Eqs (2-3). The best feasible solutions have higher probability to be chosen as parent of new individuals, where the properties of the parents are combined by exchanging chromosomes parts, to produce two new designs. After, the mutation is performed by randomly replacing the digits inside of a randomly selected chromosome.

These basic operators are repeated to create the next generations until the maximum number of iterations is reached [14].the methodological approach to the damage detection and localization problems using FEM and GA according to the following:

1- Modeling beam structures using FEM
2- Expressing the damage by stiffness reduction at a global level by identification of dynamic measure (frequencies and eigenvector).
3- Calculation the frequencies and eigenvectors proposed by GA
4- Solving the inverse problem after the providing given frequencies and eigenvector to objective function of optimization algorithm in order to calculate a local damage using GA.

5.1 *optimization parameter*

In GA method, each of the 100 individuals contains two chromosomes representing the wanted damage parameters using a total number of iterations of 200.After several applications, a crossover coefficient of 0.8 and mutation of 0.1 were used in the GA parameters.

6. Results and discussions

The approach presented in this study where the natural frequency is used as a diagnostic parameter in the procedures for structural assessment using vibration monitoring allows drawing the following interpretation:

6.1 Objective function based on the changes in the natural frequency – Figure 2

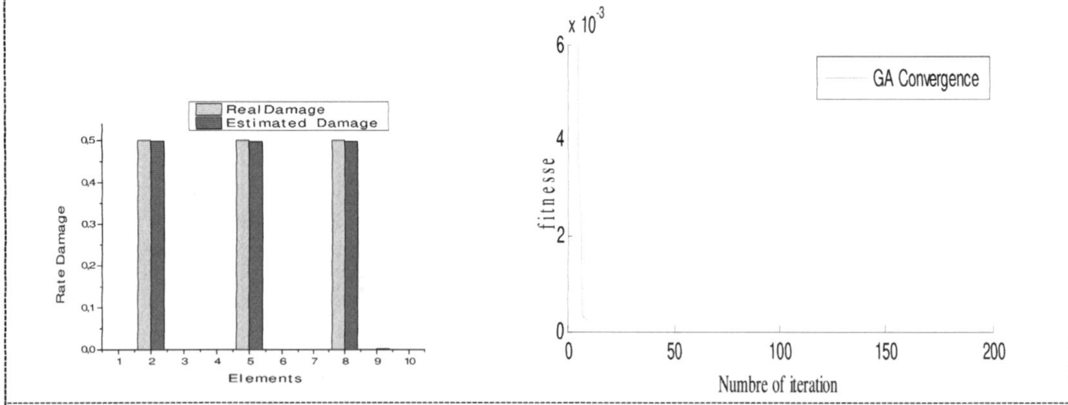

Figure 2. Objective function based on the change of the natural frequency.

The objective function based on the change of the natural frequency gives good results because the reduction in rigidity has a relatively large effect on the natural frequency in relation to the eigenmodes.

6.2 Objective function based on Modal Assurance Criterion (MAC) – Figure 3

From the results it can be observed that the objective function based on Modal Assurance Criterion (MAC) provides good results of the damage position in the beam elements. However, errors are estimated between actual and estimated damage with disturbances in other elements.

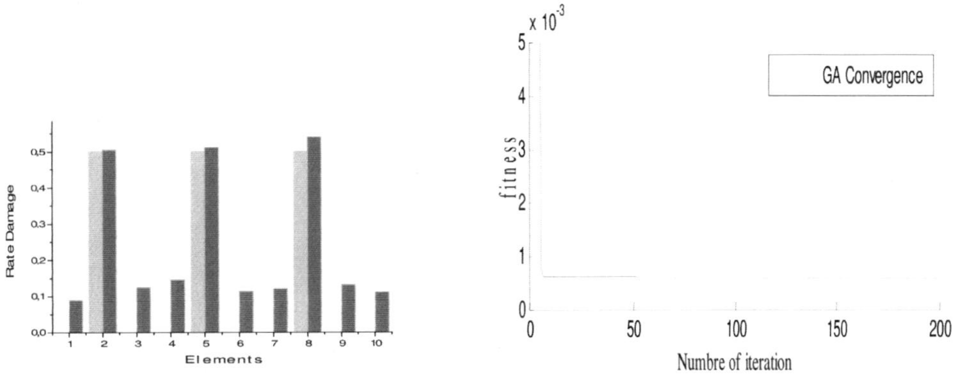

Figure 3. Objective function based based on MAC

6.3 Objective function based on Modal Assurance Criterion (MAC) and natural frequency – Figure 4.

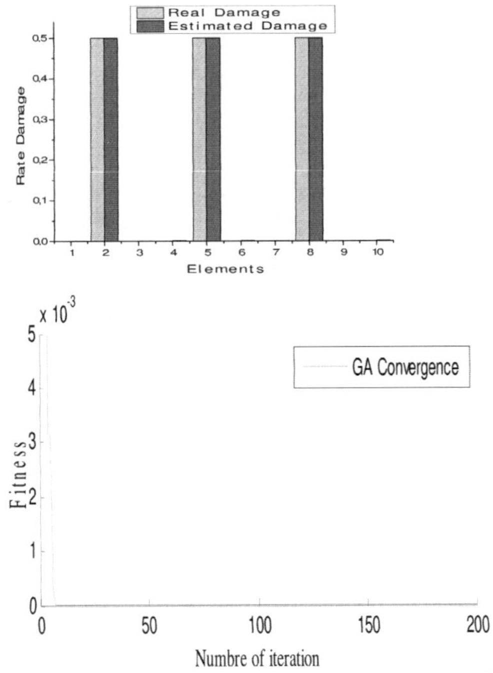

Figure 4. Objective function based on the change on MAC and natural frequency

The objective function based on the change of the natural frequency and Modal Assurance Criterion (MAC) is the best objective function, because the reduction in stiffness has a relatively large effect on the natural frequencies and the Modal Assurance Criterion (MAC).

6.4 Defect detection with noise 5 % Level – Figure 5

A white Gaussian noise was introduced using MATLAB code to the input data to test the precision of the method with presence of uncertainty, the figure Fig.5 shows that the algorithm could estimate the damaged element precisely, the damage rate was successfully guessed for a high noise level of 5%, it is noted that other damaged element with a weak damage rate was proposed.

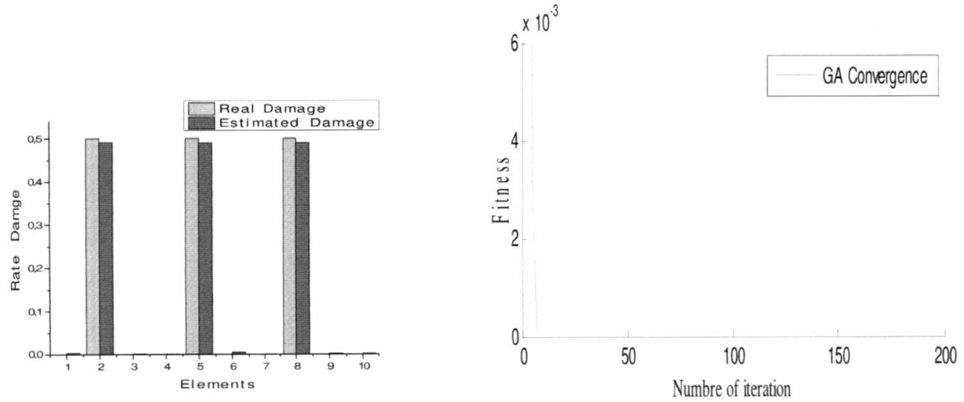

Figure 4. Objective function based on the change on MAC and natural frequency with noise 5% level

7. Conclusion

In this work an optimization approach has been developed and used for vibration analysis of beam structure to detect and locate multi damages using genetic algorithm. Three objective functions were considered. The main conclusions of this work can be formulated as follows:

- The study shows that the genetic algorithm is effective to identify the positions and assignment of multiple degree of damage. The objective function based on the change of the natural frequency gives good results because the reduction in rigidity has a relatively large effect on the natural frequency in relation to the eigen modes.

- The objective function based on the change of the natural frequency and Modal Assurance Criterion (MAC) is the best objective function since the reduction in stiffness has a relatively large effect on the natural frequencies and Modal Assurance Criterion (MAC) compared to own modes and eigenvectors.

- The noise level was introduced to test the stability of the approach. The algorithm proved its accuracy even after a high level of perturbation was imposed.

References

[1] Gomes, H. M., & Silva, N. R. S. (2008). Some comparisons for damage detection on structures using genetic algorithms and modal sensitivity method.Applied Mathematical Modelling, 32(11), 2216-2232.

[2] I.A. Jones Messina, E.J. Williams, Damage detection and localization using natural frequency changes, in: Proceedings of Conference on Identification in Engineering Systems, Swansea, UK, 1996, pp. 67–76.

[3] A. Messina, E.J. Williams, T. Contursi, Structural damage detection by a sensitivity and statistical-based method, J. Sound Vib. 216 (5) (1998) 791–808.

[4] M. I. Friswell, J. E. T. Penny, S. D. Garvey (1998). "A combined genetic and eigen sensitivity algorithm for the location of damage in structures". Journal of Computers and Structures. Vol. 69, Pages 547-556.

[5] J. H. Chou and J. Ghaboussi (2001). " Genetic algorithm in structural damage detection". Journal of Computers and Structures". Vol. 79, Pages 1335-1353.

[6] P. Cawley, R.D. Adams, The location of defects in structures from measurements of natural frequencies, J. Strain Anal. 14 (1979) 49–57.

[7] Gomes, H. M., & Silva, N. R. S. (2008). Some comparisons for damage detection on structures using genetic algorithms and modal sensitivity method.Applied Mathematical Modelling, 32(11), 2216-2232.

[8] Gautier, G., Serra, R., Mencik, J.-M. A substructure based approach for detecting And locating damages in mechanical systems.

[9] Rikards, R. (1991). Analysis of Laminated Structures: Course of Lectures. Riga: Riga Technical University.

[10] Gayathri, P., Umesh, K. & Ganguli, R. (2010) 'Effect of matrix cracking and material uncertainty on composite plates', Reliability Engineering & System Safety, 95, 716-28.

[11] Abdulkareem, A. A., & Hadi, A. P. D. N. H. Damage Detection and Assessment of Stiffness and Mass Matrices in Curved Simply Supported Beam Using Genetic Algorithm.

[12] Baghmisheh, M. V., Madani, K., & Navarbaf, A. (2011). A discrete shuffled frog optimization algorithm. Artificial Intelligence Review, 36(4), 267-284.

[13] Abraham, A., & Jain, L. (2005). Evolutionary multiobjective optimization (pp. 1-6). Springer London

[14] Gen, M., & Cheng, R. (2000). Genetic algorithms and engineering optimization (Vol. 7). John Wiley & Sons.

Post-Publication Change: Genetic Algorithm Based Objective Functions Comparative Study for Damage Detection and Localization in Beam Structures

The correct author details for the article are as follows:

S Khatir[1], I Belaidi[1], R Serra[2], B Benaissa[1] and A Ait Saada[1]

Single side damage simulations and detection in beam-like structures

Yun-Lai Zhou[1], M. Abdel Wahab[2], R. Perera[1], N. Maia[3], R. Sampaio[3, 4], E. Figueiredo[5]

[1] Department of Mechanical Engineering, Technical University of Madrid
Jose Gutierrez Abascal 2, 28006 Madrid, Spain
[2] Faculty of Engineering and Architecture, Ghent University
Technologiepark Zwijnaarde 903, B-9052 Zwijnaarde, Belgium
[3] LAETA, IDMEC, Instituto Superior Técnico, University of Lisbon
Av. Rovisco Pais, 1049-001 Lisbon, Portugal
[4] Instituto Superior de Engenharia de Lisboa, Polytechnical Institute of Lisbon
Rua Conselheiro Emídio Navarro, 1, 1959-007 Lisbon, Portugal
[5] Faculdade de Engenharia da Universidade Lusófona
Campo Grande 376, 1749-024 Lisbon, Portugal

E-mail: Magd.AbdelWahab@UGent.be

Abstract. Beam-like structures are the most common components in real engineering, while single side damage is often encountered. In this study, a numerical analysis of single side damage in a free-free beam is analysed with three different finite element models; namely solid, shell and beam models for demonstrating their performance in simulating real structures. Similar to experiment, damage is introduced into one side of the beam, and natural frequencies are extracted from the simulations and compared with experimental and analytical results. Mode shapes are also analysed with modal assurance criterion. The results from simulations reveal a good performance of the three models in extracting natural frequencies, and solid model performs better than shell while shell model performs better than beam model under intact state. For damaged states, the natural frequencies captured from solid model show more sensitivity to damage severity than shell model and shell model performs similar to the beam model in distinguishing damage. The main contribution of this paper is to perform a comparison between three finite element models and experimental data as well as analytical solutions. The finite element results show a relatively well performance.

1. Introduction

Structural health monitoring (*SHM*) has been a research focus in the past decades since more and more structures are sophisticated with the development of science and technology. As failure of the structure functionality can be caused by wear, corrosion, delamination and so on, various *SHM* techniques are developed, for instance, vibration based techniques [1-5].

Vibration based methods have been processed a booming developing period since the appearance of computer in 1970s. Modal analysis has become an important tool in *SHM*. Detailed theoretical and experimental modal analysis can be referred to [6]. Besides, techniques such as acoustic emission, ultrasonic, X-ray and so on are also developed with the high developing technology. However, experiments may confront two problems in real engineering, one is the availability in testing experiments, and another one is that the higher cost required for the experiments.

Due to the high cost of experiment in real engineering structures, numerical analysis has been exploited as another useful but affordable tool in *SHM*, and many other directions. Various commercial programs have been developed and used. Numerical analysis might be divided into three stages: preprocessing, simulation and postprocessing. In scientific field, real engineering structures are basically studied with some representative structures, such as beams, frames, and so on. As for *SHM*, numerical analysis and experiment verification are the commonly used approaches.

Even lots of research work have been conducted on beam-like structures, difficulty in detecting the damage still emerges, and double side saw cut in beam has been more often studied. However, single side damage may be more frequently confronted. And although a large amount of beam-like structure simulations have been carried out, few of them discussed the comparison of different finite element models, for instance, using solid element, shell element and beam element.

In this study, in order to show the performance of solid, shell and beam elements in beam-like structures, a real steel beam is studied using ABAQUS [7]. Afterwards, a saw cut is introduced into the side of the beam with varying depths for simulating damage severity, henceforth, natural frequencies are extracted and analyzed under each damage scenario and compared to the experimental results.

2. Theoretical background

Considering an elastic system, for example, a loaded beam, the dynamic equilibrium equation of damped vibration can be expressed as

$$M\ddot{x} + C\dot{x} + Kx = f(t) \tag{1}$$

where M, C and K are the system's mass, damping and stiffness matrices, which are assumed to be symmetric and positive definite, and $f(t)$ includes all possible types of time dependent loading [6].

As to modal analysis determining the dynamic characteristics of the systems, damping and external loading forces are assumed to be zero in order to achieve a simplest harmonically vibrating model. And therefore to vibration system, characteristic equation becomes

$$\det(K - \lambda M) = 0 \tag{2}$$

where each eigenvalue λ_i corresponds to a resonant frequency ω_i with $\lambda_i = \omega_i^2$, $i = 1,...,n$.

Then mode shapes can be calculated by substituting the resonant frequencies into the following equation

$$(K - \lambda_i M)\phi_i = 0 \tag{3}$$

In regard to the sensitive to structural damage, natural frequency change is firstly taken into account for detecting damage. Herein, an indicator for describing the natural frequency decrease due to the damage effect can be expressed as

$$FD_i = \frac{\omega_i^u / 2\pi - \omega_i^d / 2\pi}{\omega_i^u / 2\pi} \times 100\% \tag{4}$$

where ω_i^u means frequency under intact state, and ω_i^d frequency under damaged state.

In addition, modal assurance criterion (*MAC*) is also frequently used; for mode shape before and after damage, it can be expressed as

$$MAC(\phi_i^d, \phi_i^u) = \frac{\left|(\phi_i^d)^T(\phi_i^u)\right|^2}{\left((\phi_i^d)^T(\phi_i^d)\right)\left((\phi_i^u)^T(\phi_i^u)\right)} \tag{5}$$

where ϕ_i^u represents the *ith* mode shape under intact state, ϕ_i^d means the *ith* mode shape under damaged state.

3. Test specimen and experimental setup

A steel beam with length, height, and thickness of 1004 × 35 × 6 mm was selected to analyze, and is shown in Figure 1. The physical properties of the beam are, density 7917 kg/m^3 and Young's modulus 185.2 GPa. And the beam with rectangular cross-section was analyzed under free-free conditions (transverse bending). For the convenience in analyzing the beam, 23 equally spaced nodes (1-23) for translation response measurements were considered. For each node, one accelerometer is used for capturing the dynamic response.

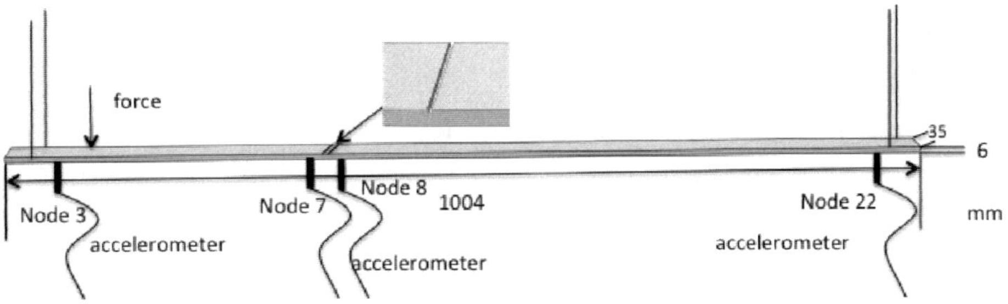

Figure 1. Schematic representation of the beam.

Afterwards, the damage was introduced into the beam by a saw cut, split into eight scenarios. Basically, a single cut, with varying depths, between node 7 and 8 (Figure 1) was introduced into the beam in order to create eight damage scenarios, as summarized in Table 1.

Two inextensible cables simulating "free-free" support conditions were used to sustain the test beam. A fan was used to generate wind to simulate the varying operational and environment conditions. On the other hand, a Brüel & Kjaer 4809 shaker was used to excite the beam at node 3, pseudo-randomly, and a Brüel & Kjaer 2706 power amplifier was used to amplify the excitation force signal. The force was transmitted through a stinger and measured by a Brüel & Kjaer 8200 force transducer. In terms of data acquisition, the responses were measured by 23 piezoelectric CCLD accelerometers at each node. The signals were fed into the Multi-channel Data Acquisition Unit Brüel & Kjaer 2816 (PULSE) and analyzed directly with the Labshop 6.1 Pulse software from the attached laptop (Dell series 400).

For each measurement, time domain responses were obtained, as well as frequency response functions (*FRFs*) and coherence functions from 30 averages, in order to get rid of some operational influences. In the original undamaged structural condition, 20 measurements were conducted, namely ten with and ten without artificial wind. For each damage scenario, five measurements were obtained with and without artificial wind. Finally, the frequency analysis of the beam was carried out in a frequency range of 0-800 Hz (3200 lines) with Hanning windows applied upon force time series as well as response acceleration time series.

Table 1. Cut properties of each damage scenario.

Damage scenario	Width (mm)	Depth (mm)
1	1.5	0.8
2	1.5	1.0
3	1.5	1.3
4	1.5	1.6
5	1.5	2.2
6	1.5	3.0
7	1.5	4.0
8	1.5	4.8

4. Numerical analysis

In this section, test specimen is illustrated in the first step, and later the numerical models for simulating the experimented specimen will be given, finally simulation for modeling the cut will be also addressed.

4.1 Model description

For simulating the experimental beam, three models: solid (C3D20R), shell (S4R) and beam (B31) models are used for the analysis. Figures 2 and 3 show solid model and shell model, respectively.

For solid model, C3D20R element is used, which is a general-purpose quadratic brick element, with reduced integration (2x2x2 integration points). It takes reduced integration by using a lower-order integration (eight integration points) to form the element stiffness matrix that will reduce running time. The mass matrix and distributed loadings use full integration. And for shell model, a general-purpose linear four-sided shell element - S4R is used. Finally, for beam model, a general 2-node beam model B31 is utilized.

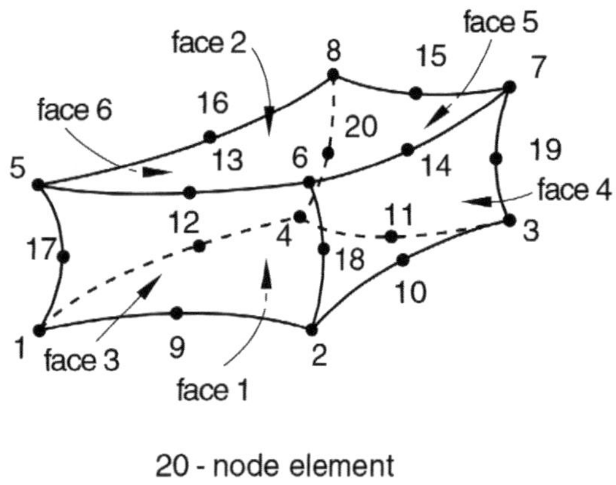

Figure 2. Solid model.

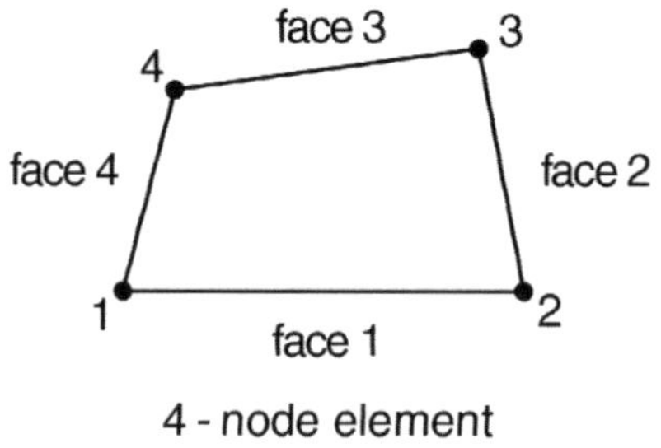

Figure 3. Shell model.

4.2 Saw cut damage simulation
For simulating the saw cut damage, the aforementioned three models are also used.

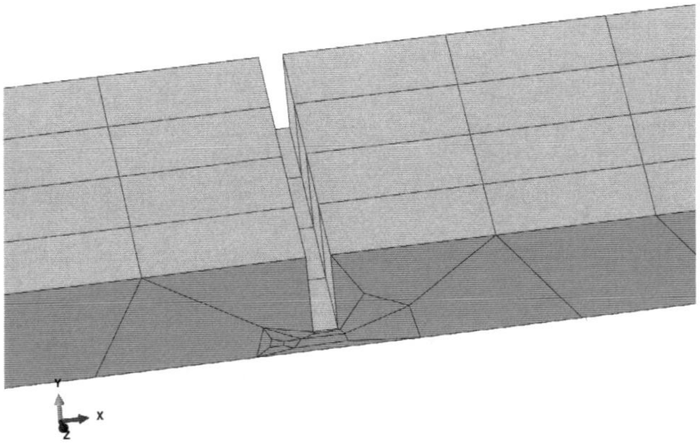

Figure 4. Zooming saw cut with solid model.

For solid model, it exactly describes the cut showing in Figure 4 in a local zooming scale. One can find that cut is indicated with solid element in exact physical saw cut geometry.

For shell model, in this study damage was simulated with stiffness reduction corresponding to the saw cut in section, with ignoring the mass reduction influence, as it is found that the thickness change will greatly affect the simulation results.

For beam model, Figure 5 shows the section under intact state, and Figure 6 shows the section under damaged state, from which one can find the difference clearly. And after occurrence of saw cut, the horizontal symmetrical line is moved downward. In order to overcome this difficulty, a generalized section is used as indicated in Figure 7. From the characteristic properties: area, the moment of inertia and so on, the generalized section can regenerate a section for the simulation.

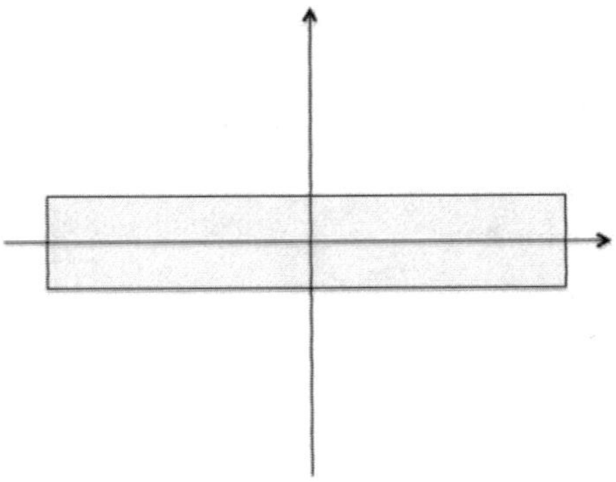

Figure 5. Section under intact state.

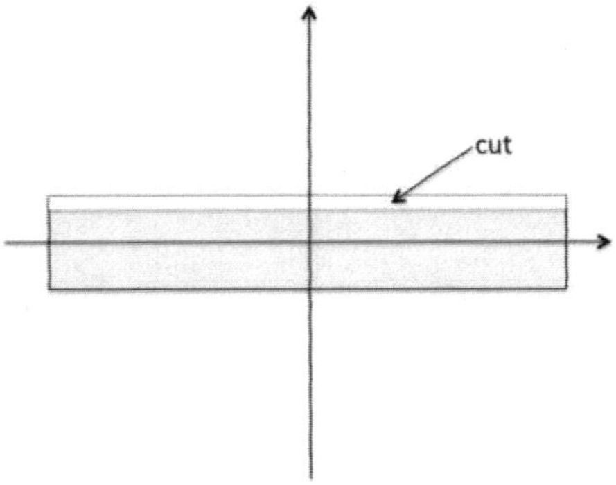

Figure 6. Section under D1 state.

5. Results and discussions
In order to illustrate the applicability of the performance of the three models, the aforementioned simulation results are analyzed as follows:

5.1 Natural frequencies

In this section, a captured natural frequencies comparison is proposed Table 2 for different models and experimental results as well as analytical values.

From Table 2, firstly, one can find that all the captured frequencies from analytical values and three models are close to the experimental results. The error percentages are all less than 1.375%, and this suggests that the experiment and simulations are well conducted. Secondly, Solid model shows better performance than shell model and beam model, as the captured natural frequencies are more close to experimental and analytical values as well. And similarly, shell model performs better than beam model. Note that solid model will consume much more time than shell model and beam model.

Table 2. Natural frequencies (Hz) extraction under intact state.

Mode order	1	2	3	4	5	6
Experiment	29.616	82.250	161.250	267.250	397.500	553.000
Analytical	29.593	81.573	159.916	264.349	394.892	551.544
(%)	0.078	0.823	0.827	1.178	0.656	0.263
Solid model	29.591	81.564	159.890	264.290	394.760	551.290
(%)	0.084	0.834	0.843	1.200	0.689	0.309
Shell model	29.591	81.561	159.880	264.250	394.680	551.130
(%)	0.084	0.838	0.850	1.215	0.709	0.338
Beam model	29.580	81.503	159.700	263.820	393.830	549.620
(%)	0.122	0.908	0.961	1.375	0.923	0.611

Figure 7 shows the captured natural frequencies difference and analytical values compared with the experimental results. It is clear that all are under 1.4%, which means that in this case, the experiment is well conducted, and the three models can all well describe the beam-structure. And one can also find that from analytical solution, solid model, shell model to beam model, the errors of each mode increase, which confirms that analytical solution considers the beam-like structure to be continuous elastic system, and finite element models discrete the structure and approximate the results.

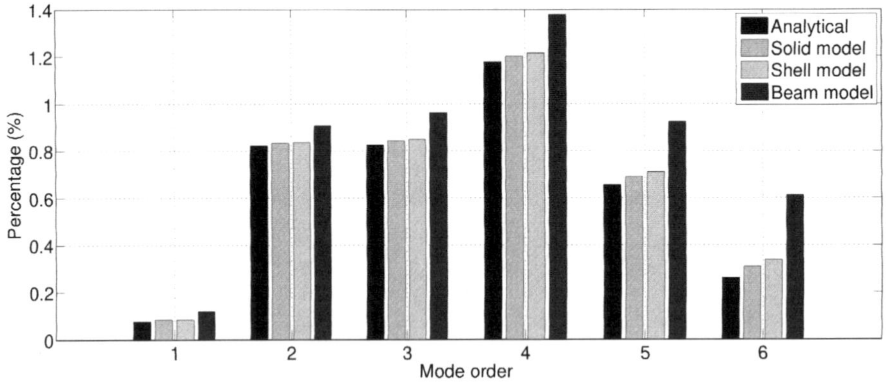

Figure 7. Frequency difference to the experimental value.

5.2 Frequency decrease

During damage detection process, the natural frequency identification before and after deterioration is the most commonly used indicator, which can reveal the characteristic change caused by the deterioration. Figure 8 shows the first six natural frequencies of the three aforementioned models and experiment results with respecting to each damage scenario and intact condition, respectively. For comparison reason, the vertical axis value means the natural frequencies divided by the experimental value under intact state.

From Figure 8, one can find that (i). all the three models show clear decrease as the damage increases, while for light damage scenarios, the natural frequencies vary not too much; (ii). in all three models, the natural frequencies in mode three and six show little differences in comparison with other modes, this is resulted from that the saw cut is close to the node of third and sixth mode shape, then the node value is almost zero, and recalling Equation (3), it will be clear that the corresponding natural frequencies will be relatively stable with little change. (iii). apart from mode three and mode six, to the other four modes, one can see that the natural frequencies captured by solid model are more sensitive to saw cut, as in damage scenario five (D5), the natural frequency captured by solid model shows more difference than that of shell model and beam model. And this may confirm that solid model performs better than shell model and beam model. (iv). for shell model and beam model, the difference between captured natural frequencies is little, this means that for beam-like structures herein presented, shell model and beam model show the same capability in characterizing the structure.

5.3 MAC value decrease

Figure 9 shows the *MAC* value of the three aforementioned models from mode one to six. From Figure 9, it is clear that (i). as damage increases, the *MAC* value decreases; (ii). from D0-D5, the *MAC* value decrease little, this suggests for small damage cases, the *MAC* can endure more challenge in detecting the damage; (iii). comparing the three models, one can find that solid model decreases earlier and greater than other two models. Besides, comparing Figure 9 with Figure 8, one can find that (i). similar to natural frequency change, to mode three and six, unlike other modes, the *MAC* value also change very small. This can be explained by Equation (3), each natural frequency is corresponding to the mode shape; (ii). generally, the natural frequencies change much more than the *MAC* value, this suggests that the natural frequency is more sensitive than *MAC* of mode shapes.

6. Conclusions

A steel experimented beam is simulated and analysed with three models: solid model, shell model, and beam model. Firstly, under intact state, the solid model performs better than shell model and shell model performs better than beam model in capturing the natural frequencies; while under damaged states, solid model performs better than shell model and shell model shows similar performance as beam model in distinguishing the saw cut by natural frequencies change. Secondly, to saw cut simulation, in shell model it can be interpreted by equivalent stiffness reduction in the saw cut section, however, this might reduce the capability in detecting the saw cut; and in beam model, the saw cut can be represented by generalized section.

Acknowledgement

The first author thanks CWO (Commissie Wetenschappelijk Onderzoek), Faculty of Engineering and Architecture, Ghent University for providing financial suypport for a research stay at Soete Laboratory.

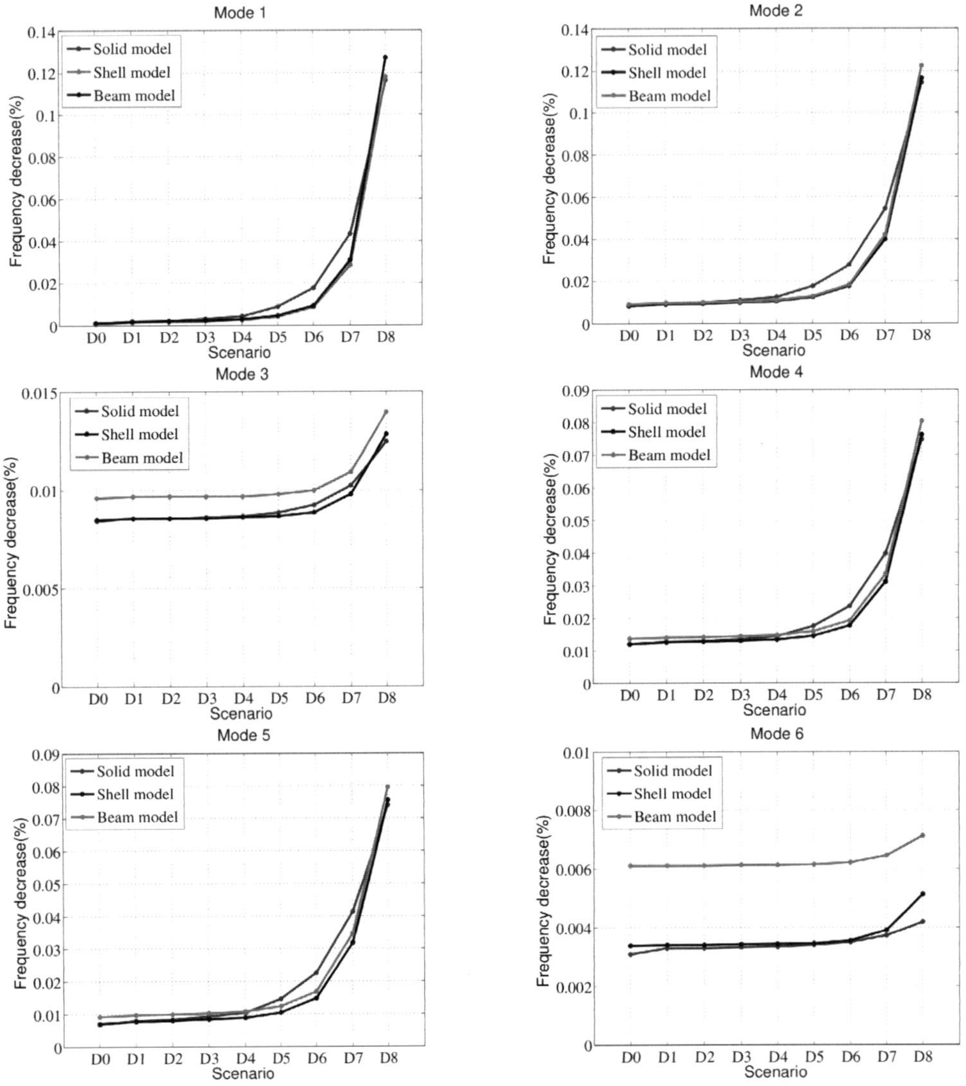

Figure 8. Natural frequency change of mode one to six for the three models.

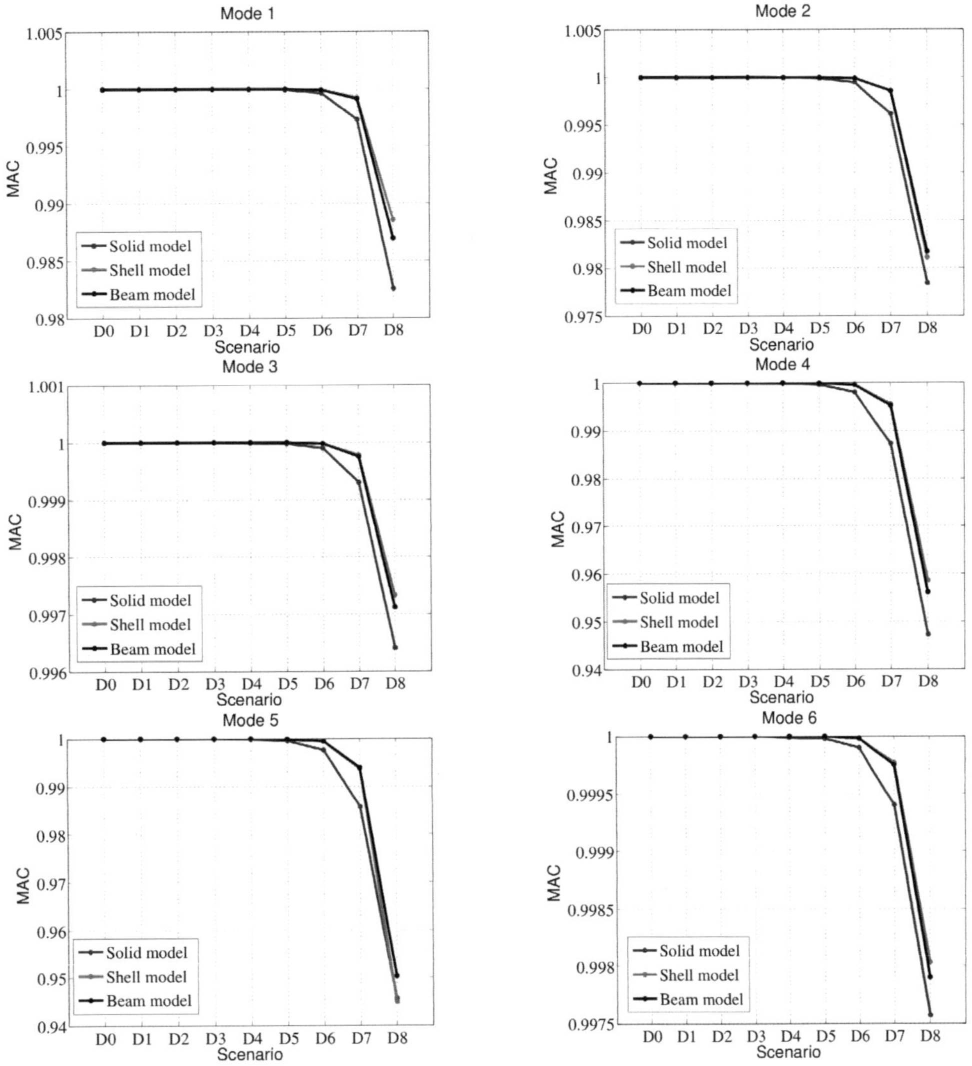

Figure 9. MAC value of mode shape from mode one to six for the three models.

Reference

[1] Farrar CR, Doebling SW, Nix DA. Vibration-based structural damage identification. *Philosophical Transactions of the Royal Society*; A 2001: 356(1778): 131-149.

[2] M. M. Abdel Wahab, G. De Roeck (1999). Damage detection in bridges using modal curvatures: application to a real damage scenario. *Journal of Sound and vibration*; 226 (2): 217-235.

[3] Montalvao D., Maia N.M.M., Ribeiro A.M.R. (2006). A review of vibration-based structural health monitoring with special emphasis on composite materials. *The Shock and Vibration Digest*; 38 (4): 295-324.

[4] Yun-Lai Zhou, E. Figueiredo, Nuno M. Maia, R. Perera (2015). Damage detection and quantification using transmissibility coherence analysis. *Shock and Vibration*; Volume 2015, Article ID 290714.

[5] Yun-Lai Zhou, E. Figueiredo, N. Maia, R. Sampaio, R. Perera (2015). Damage detection in structures using a transmissibility-based Mahalanobis distance. *Structural Control and Health Monitoring*; DOI: 10.1002/stc.1743

[6] Maia, N. M. M. and Silva, J. M. M. (1997). *Theoretical and Experimental Modal Analysis*. Research Studies Press, Ltd.

[7] ABAQUS/CAE, (2004), User's Manuals, Version 6.4.1, Hibbitt Karlsson & Sorensen, Inc., USA.

Long term application of bus monitoring system to short and medium span bridges and damage detection

Ayaho Miyamoto[1,a], **Válter J. G. Lúcio**[2,a], **Hisao Emoto**[1,b] and **Hideaki Tanaka**[1,c]
[1] Graduate School of Science & Engineering, Yamaguchi Univ., Ube, Japan
[a]E-mail: miya818@yamaguchi-u.ac.jp
[2] Faculty of Science and Engineering, NOVA de Lisboa Univ., Lisbon, Portugal
[a]E-mail: vlucio@unl.pt

Abstract. In this study, as one solution to the problem for condition assessment of existing short and medium span reinforced/prestressed concrete bridges, a new monitoring method using a public bus as part of a public transit system (bus monitoring system) is proposed, along with safety indices, namely, characteristic deflection, which is relatively free from the influence of dynamic disturbances due to such factors as the roughness of the road surface, and a structural anomaly parameter. A basic study was conducted by using the results of technical verification experiments and numerical analysis simulation. This paper describes the details of not only how to assess the bridge condition by public bus vibration measured in operating on Ube City bus network as a specific example for verify the system but also what kind of consideration we need to apply the system to existing bridges in overseas country.

1. Introduction

In order to properly maintain numerous short and medium span bridges in Japan by conducting periodic inspections, it is necessary to develop an easy-to-use and efficient bridge monitoring system. The authors are developing a short and medium span bridge monitoring system for detecting deterioration in safety and other performance of existing bridges by use of under-rear-wheel-spring acceleration sensors installed on fixed-route buses (hereinafter referred to as the "bus monitoring system") (Miyamoto and Yabe 2012; Yabe, Isoda and Miyamoto 2011). The proposed monitoring system represents a monitoring system, to be applied mainly to short and medium span bridges, making use of a fixed-route bus operated as part of a public transit system. The aim of the proposed system is to detect the transition from the "acceleration stage," in which the safety performance of a bridge sharply declines because of aging, to the "deterioration stage" (see Fig. 1). This paper presents the result of a three-year field test of the bus monitoring system conducted by using an in-service fixed-route bus in Ube City. For the bus monitoring system, "characteristic deflection" was defined as an indicator that may be useful in efficiently detecting a structural anomaly of a bridge. For the purpose of examining the sensitivity of characteristic deflection, the effect of artificial damage (bridge guardrail removal) on characteristic deflection was evaluated by using an aged (out-of-service) bridge being removed. Another field test was conducted jointly with the Universidade Nova de Lisboa (New University of Lisbon) in Portugal as an overseas application of the bus monitoring system. By analyzing the field test results thus obtained, possible steps in realizing practical application of the bus monitoring system were considered systematically.

2. Principles and Overview of the Bus Monitoring System

The bus monitoring system has been developed by the method of monitoring short and medium span bridges by using fixed-route buses operated as part of public transport systems. Fig. 2 illustrates the bus monitoring system, and Fig. 3 shows the analysis flow of the monitoring method. Main reasons why the use of fixed-route buses, which are large and heavy vehicles, was considered are as follows:

1) If a large vehicle about 10 m in length is used for on-the-move measurement on a short or medium span bridge, that vehicle is likely to be the only vehicle passing over the bridge at point in time.
2) In order to make a short or medium span bridge, whose stiffness is relatively high, it is necessary to use a reasonably heavyweight vehicle.

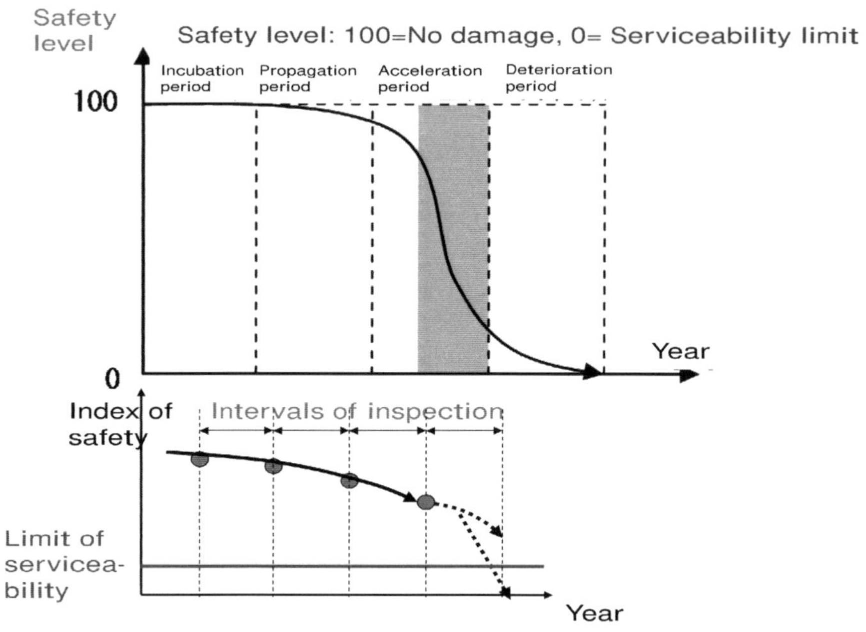

Figure 1 Reduction path of safety level and coverage of proposed system

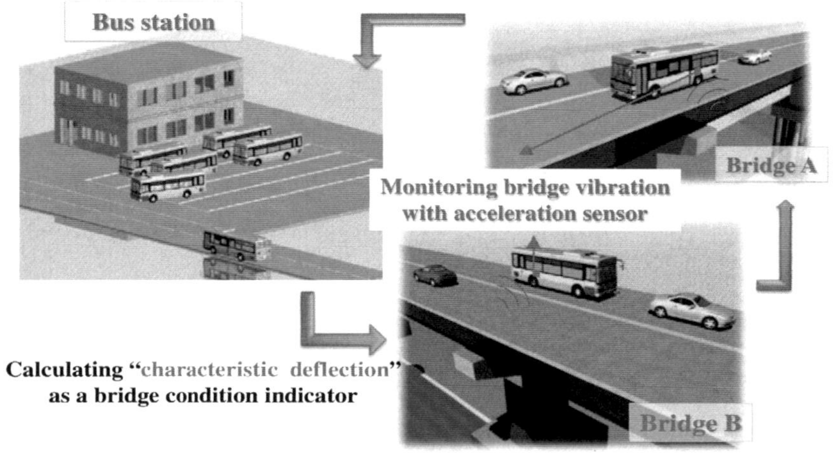

Figure 2 Bus monitoring system

3) Large vehicles used to vibrate bridges make it easier to reproduce measuring conditions such as travel time, route, frequency of passages and vehicle speed.
4) By using a fixed-route bus equipped with acceleration sensors, main short and medium span bridges in the area of interest can be monitored on a regular basis. By so doing, monitoring cost would be substantially less than in the case where sensors were installed at all bridges.
5) Power for the measuring devices can be supplied from the power supply available in the bus.

The vertical displacement, $\delta_a(t)$ of the bridge and the bus under its rear wheel springs is an estimated deflection obtainable by integrating the vertical acceleration waveform twice. That displacement can be expressed, as the sum of (1) the static displacement, $\delta_{sa}(t)$ dependent on bridge stiffness and bus weight, (2) a non-steady-state vibration component having a stochastic nature characterized as a Gaussian process attributable to a rough road surface and having a mean value of 0, (3) vibration components governed by the equation of motion such as the bridge and the vehicle, and (4) external disturbance factors, $\delta_x(t)$ due to differences in the operating conditions of the bus and differences in data extraction and processing methods, as follows:

$$\delta_a(t) = \delta_{sa}(t) + S_r(\Omega, t) + \int_{-\infty}^{+\infty} X(f) \cdot e^{2\pi fti} df + \delta_x(t) \tag{1}$$

where,

$S_r(\Omega, t)$: density function of road roughness,
Ω: spatial frequency of road surface,
$X(f)$: Fourier coefficient.

For the purposes of this study, characteristic deflection μ_a (Yabe 2006) is defined as the estimated deflection obtained from Eq. (1) averaged in the bridge section of the road. Studies have shown that characteristic deflection can be made to converge to the extent that anomaly evaluation results are not affected by increasing the number of samples, N, by using the central limit theorem, in order to allow for the influence of various external disturbance factors, $\delta_x(t)$, including the operating conditions of the bus. In this study, the authors used the moving average method to process characteristic deflection data and see what happens. In the moving average method, averages are calculated for incrementally shifted data sections. Fig. 4 shows standard deviations of characteristic deflection values obtained by calculating characteristic deflection averages for

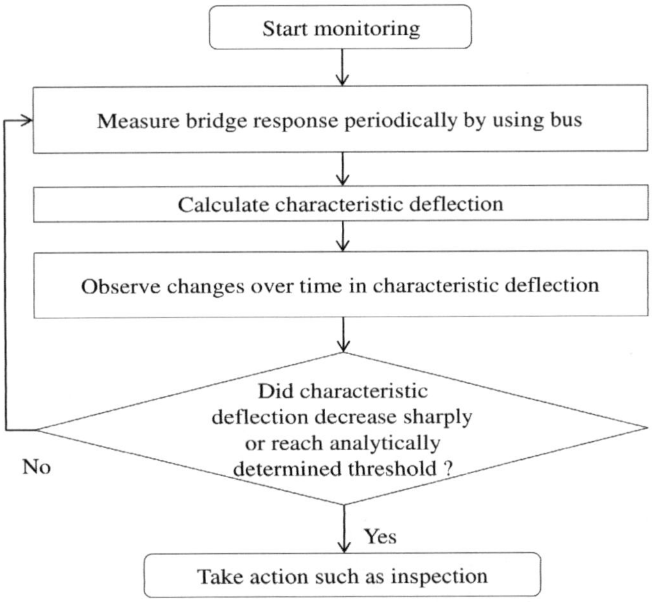

Figure 3 Analysis flow for bus monitoring system

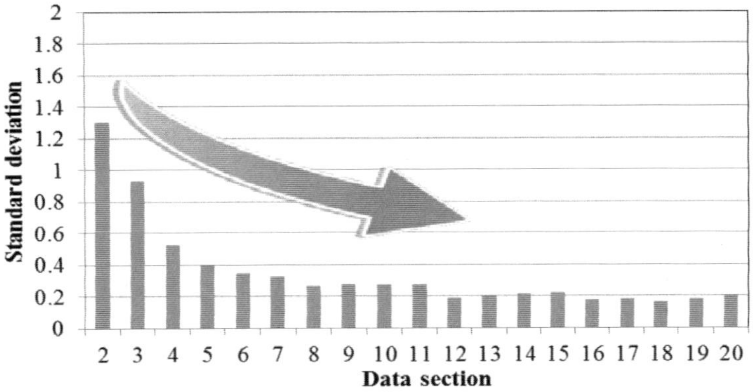

Figure 4 Changes in standard deviation estimated by the moving average method

incrementally shifted data sections. As shown in Fig. 4, as data sections become larger, standard deviations tend to decrease. It can also be seen that after data section size exceeds a certain limit, standard deviations do not change significantly thereafter. This indicates that by the moving average method mentioned above, the variability of characteristic deflection values decreased and converged so that anomalies can be detected by observing such changes.

3. Long-term Monitoring of Short and Medium Span Bridges along Bus Routes in Ube City

3.1 Field test

As a long-term application of the bus monitoring method of measurement, a field test utilizing Ube City's bus routes has been underway. The long-term experiment, which uses fixed-route buses being operated in the city, has been continued for about three years since December 2010 and is supposed to be continued in the years to come. Fig. 5 shows the fixed-route bus used in the experiment and the under-rear-wheel-spring acceleration sensors installed to the bus. The acceleration sensors were bonded to the under-rear-wheel-spring section of the bus and was coated with epoxy resin for protection over a long period of time. The sensors used consisted of a triaxial acceleration sensor and uniaxial acceleration sensors (backups). These sensors were oriented along the direction of travel of the bus (X-axis), the transverse direction (Y-axis) and the vertical direction (Z-axis). The acceleration sensor cables were passed through the floor drain of the bus. Thus, the acceleration sensors are connected to the data logger and the computer in the bus so that the vibration of the bus crossing the bridge can be monitored in real time. Although the acceleration sensors, the data logger and the computer are designed to be powered by the power source of the bus, in the experiment a portable rechargeable battery was used. Table 1 lists main bridges of interest located on the bus routes and summarizes data on those bridges. As an example of measurement data, this paper shows the measurements results obtained from the Shingondai Bridge (see Fig. 6), one of the bridges (prestressed concrete(PC) girder bridge) managed by Ube City on the bus routes.

3.2 Characteristic deflection calculation method

This section describes how to calculate characteristic deflection, an indicator of bridge performance used by the bus monitoring system. A flow diagram for the calculation of characteristic deflection following measurement data acquisition is shown in Fig. 7, and each step is described in detail below. The steps (1) through (5) shown below correspond to the numbers shown in Fig. 7.

Figure 5 Fixed-route bus and sensor used for long-term field testing

Table 1 Bridge data

Bridge name	Completed in	Type of superstructure			Span length (m)	Bridge length (m)
Jase Bridge	1976	Span No.	Start point side 1	Prestressed concrete slab bridge (pretensioned slab)	18.0	85.0
			2	Prestressed concrete slab bridge (pretensioned slab)	16.0	
			3	Prestressed concrete slab bridge (pretensioned slab)	18.0	
			4	Prestressed concrete slab bridge (pretensioned slab)	14.0	
			End point side 5	Prestressed concrete slab bridge (pretensioned slab)	19.0	
Shiratsuchi Daini Bridge	1933 (estimated)	Span No.	Start point side 1	Reinforced concrete (T-girder)	7.0	15.8
			End point side 5	Reinforced concrete (T-girder)	7.0	
Shingondai Bridge	June 1998	Girder bridge (other)			22.4	23.6

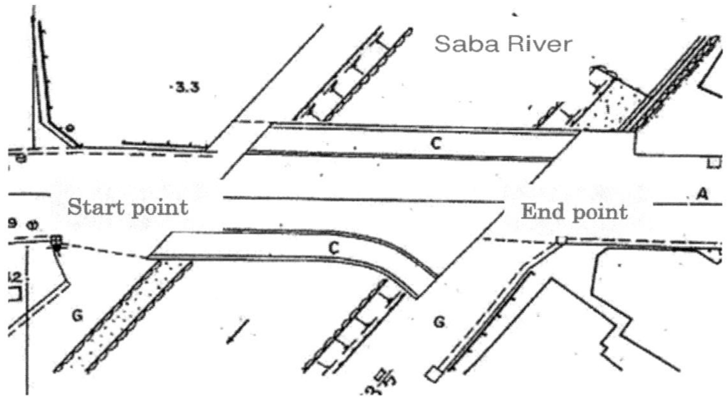

Figure 6 Plan view of Shingondai Bridge

1) Extract data on vertical acceleration during bridge crossing

Extract data on vertical acceleration during the crossing of the bridge from the acceleration measurement data by referring to the combination of the time at which the bus crossed the bridge and GPS data.

2) Estimate the time when the bus passed midspan points

The extracted data on acceleration during the crossing of the bridge include data on strong vibrations recorded at bridge joints. It is therefore necessary to use girder midspan acceleration data that do not include such joint vibration data. The time when the midspan point of a girder is passed can be estimated by extracting data for between-joint data meeting certain criteria such as time intervals(sec) and the number of waves. Since, however, acceleration data vary considerably in magnitude depending on such factors as the operating conditions of the bus, it may be difficult to identify joint-zone data. In order to accurately estimate the time when the midspan point of a girder is passed, attention was paid to deflection estimates that can be obtained by integrating vertical acceleration data twice. As an example, Fig. 8 shows measured acceleration waveforms and estimated deflection values obtained by integrating bridge-crossing acceleration data twice. As shown in Fig. 8, distinctively characteristic waveforms appear when the bus passes joints. Hence, on the basis of characteristic waveforms appearing on estimated deflection diagrams when the bus passes the joints of the bridge of interest, girder midspan waveforms are estimated and determined by synchronizing different data sets.

3) Extract midspan vertical acceleration data

Extract the midspan vertical acceleration data determined as described in Step (2). Fig. 9 shows examples of joint and midspan portions of vertical acceleration data. In the example shown in Fig. 9, a portion of midspan region data corresponding to about 0.8 sec (about 10 waves) was identified as data to be extracted. The most important thing in calculating characteristics deflection is to determine the range of data to be extracted according to such criteria as the number of waves and time(sec) and extract acceleration waveforms from the same regions every time.

4) Integrate extracted acceleration data twice

Extracted acceleration data are converted to velocity by the first integration and to displacement by the second integration. At this step, the vertical displacement obtained by integrating vertical acceleration twice is taken as an estimated deflection.

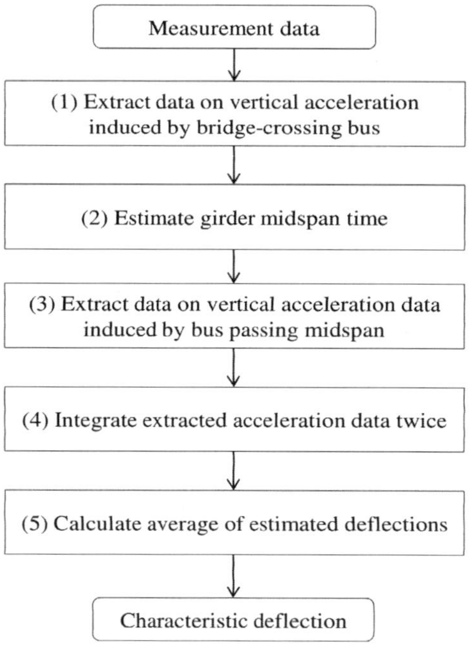

Figure 7 Flow of characteristic deflection calculation following measurement data acquisition

An example of a deflection estimation result is shown in Fig. 10.

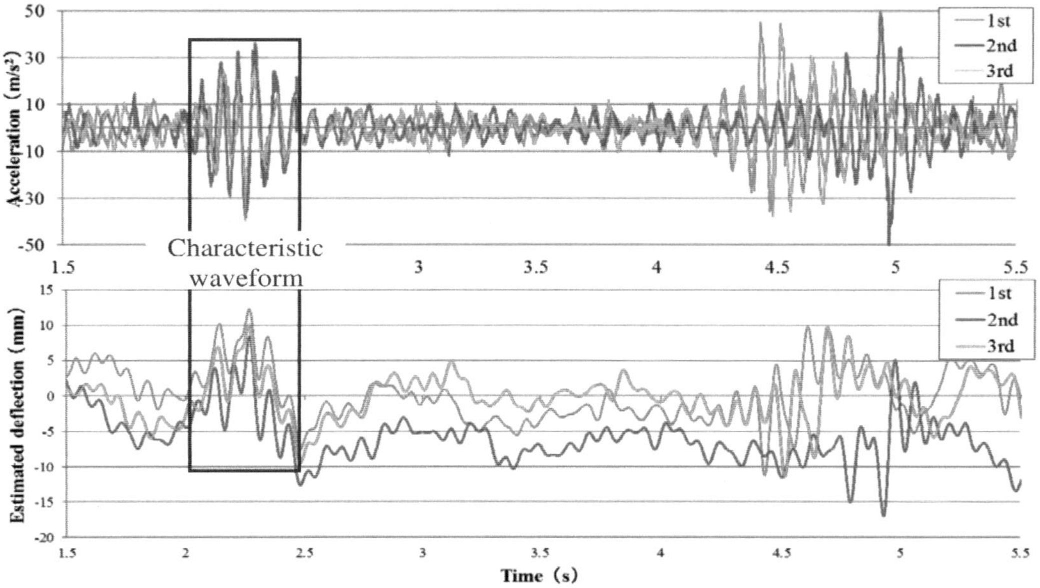

Figure 8 Examples of characteristic waveforms of estimated deflection

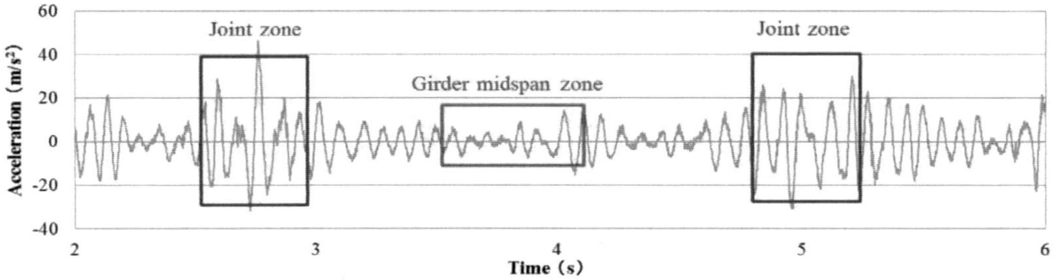

Figure 9 Example of determination of acceleration data extraction range

5) Average estimated deflection
Calculated characteristic deflection by averaging the deflection estimates shown in the graph obtained at Step (4) in the direction of time (Fig. 10).

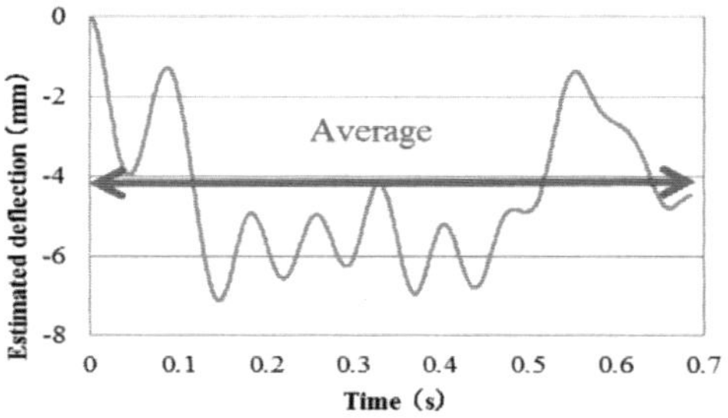

Figure 10 Example of deflection calculation

3.3 Experiment results
First of all, let us consider whether or not it is possible, in the event of a structural anomaly of the bridge of interest, to detect the anomaly from the under-rear-wheel-spring acceleration response of the bus. In a previous study, vibration analysis using FSWT (Frequency Slice Wavelet Transform), a time–frequency analysis method, was conducted to analyze acceleration response (Yan, Miyamoto and Jiang 2009). The FSWT analysis of the girder-midspan and under-rear-wheel-spring acceleration responses conducted in this study indicated that the under-rear-wheel-spring region of the bus was always vibrating at around 12 Hz, regardless of whether or not the bus was crossing the bridge. On the other hand, with respect to the acceleration response of the bridge, it has been confirmed that the girder-midspan region of the bridge was actually vibrating at around 12 Hz (Yabe *et al.* 2012). Fig. 11 shows the acceleration responses of the under-rear-wheel-spring region of the bus and the girder midspan region of the bridge observed when the bus passed the girder midspan section of the bridge. It can be seen that in terms

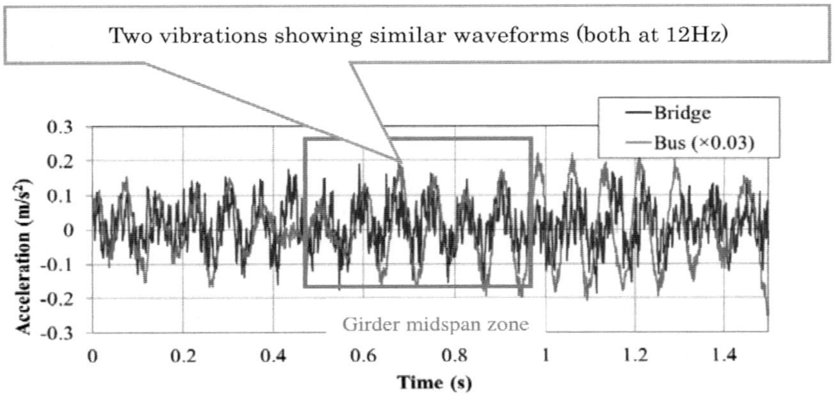

Figure 11 Comparison of girder-midspan and under-rear-wheel-spring acceleration responses

of the magnitude of acceleration response, the two waveforms show a fair degree of similarity although the under-rear-wheel-spring response values tend to be larger. From these results, it can be concluded that the vibration properties of the bridge of interest can be identified according to the under-rear-wheel-spring vibration of the bus.

Next, let us consider characteristic deflection calculation results. Fig. 12(a) and (b) show changes in characteristic deflection at the Shingondai Bridge during the initial and return trips, respectively, of the bus calculated from the measurement data obtained during a total of 56 days in 2010 to 2013. There are differences in the number of data sets between initial trip data and return trip data because there were some invalid cases such as a case in which the bus stopped on a bridge. The data shown in Fig. 12(a) and (b) were obtained by applying a serious deterioration threshold specifying the limit of the degree of seriousness of bridge anomalies.

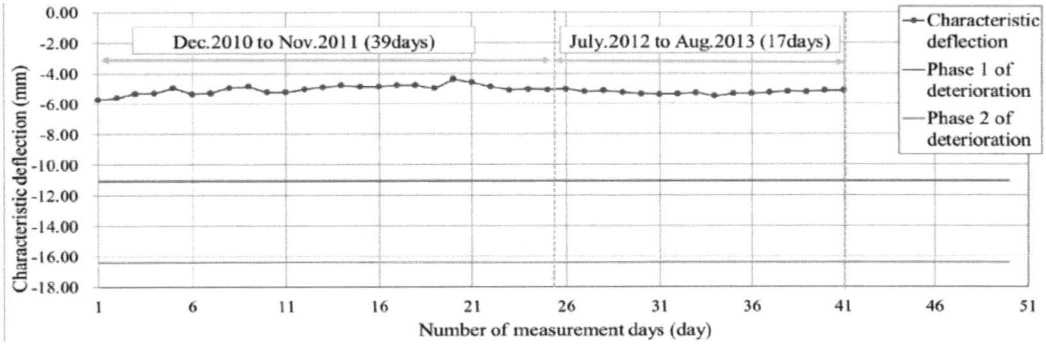

(a) Tokonami→Nishikiwa Gakkomae

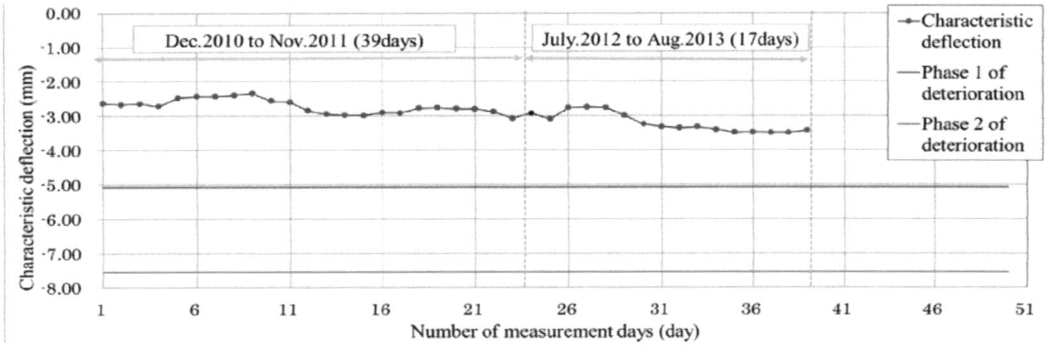

(b) Nishikiwa Gakkomae→Tokonami

Figure 12 Long-term changes in characteristic deflection (Shingondai Bridge)

In this connection, it has been found that characteristic deflection increases to 193% in the event of deterioration that causes the introduced prestressing force to decrease by 50% and to 286% in the event of deterioration that causes the introduced prestressing force to decrease by 90% (NILIM 2010). It was therefore decided to calculate the average of the characteristic deflection values obtained from the first 15 measurements and use the characteristic deflection value equal to 1.93 times the average value as the threshold level for Phase 1 deterioration (red line) and the characteristic deflection value equal to 2.86 times the average value as the threshold level for Phase 2 deterioration (green line). The actual procedure is to measure characteristic deflection on a continual basis and, if the threshold level is reached or exceeded, suspect the occurrence of some kind of serious damage to the bridge of interest and conduct visual inspection immediately. Inspection needs to be conducted when characteristic deflection has reached or exceeded the threshold level on the initial or return trip. Characteristic deflections in "Data Section 15" obtained by the moving average method are shown in Fig. 12, where a single characteristic deflection point represents the average of 15 measurement results. Comparison of Fig. 12(a) and (b) reveals that characteristic deflection values differ between the initial trip and the return trip. This is thought to be because the Shingondai Bridge is not symmetrical as shown in Fig. 6. Long-term changes in characteristic deflection showed little change although Fig. 12(a) shows slight changes. In contrast, Fig. 12(b) shows that the lowest value of characteristic deflection decreased gradually. Nevertheless, in both cases, the serious deterioration threshold was not reached or exceeded. It can therefore be concluded that serious damage has not yet occurred on the initial trip route or on the return trip route.

4. Effect of artificial damage on characteristic deflection

This section reports on the result of a study on the influence of artificial damage (bridge guardrail removal) conducted, by using an aged (out-of-service) bridge being removed, to investigate the sensitivity of the bus monitoring system to characteristic deflection.

In the experiment conducted for the purpose mentioned above, acceleration sensors were installed to a sightseeing bus weighing about 15 tons as shown in Fig. 13 by using a bridge being removed for the construction of a new bridge. The bridge is a 168.29-meter-long, eight-span reinforced concrete(RC) cantilever T-girder bridge as shown in Fig. 14. In consideration of such factors as seasonal changes and the presence or absence of bridge railings, the measurement was conducted four times, namely, in September, 2012, and January, February and March, 2013. Fig. 15 shows how the bridge looks before and after guardrail removal.

Figure 13 Sightseeing bus used for the field test

Fig. 16 shows the average values of 15 sets of characteristic deflection measurements calculated from the total of 4 sets of measurements. As can be seen from Fig. 16, the concrete guardrails were removed between the second measurement and the third measurement. The characteristic deflection values before and after the guardrail removal, therefore, were

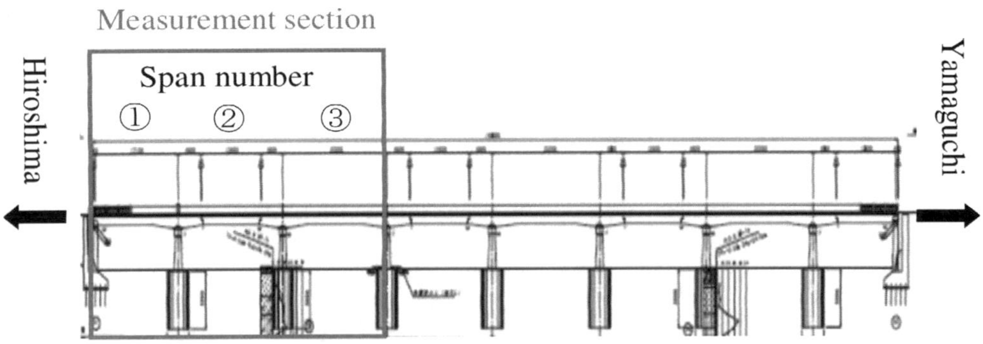

Figure 14 Side view of the out-of-service bridge used for field testing

(a) Before guardrail removal (b) After guardrail removal

Figure 15 Bridge view before and after guardrail removal

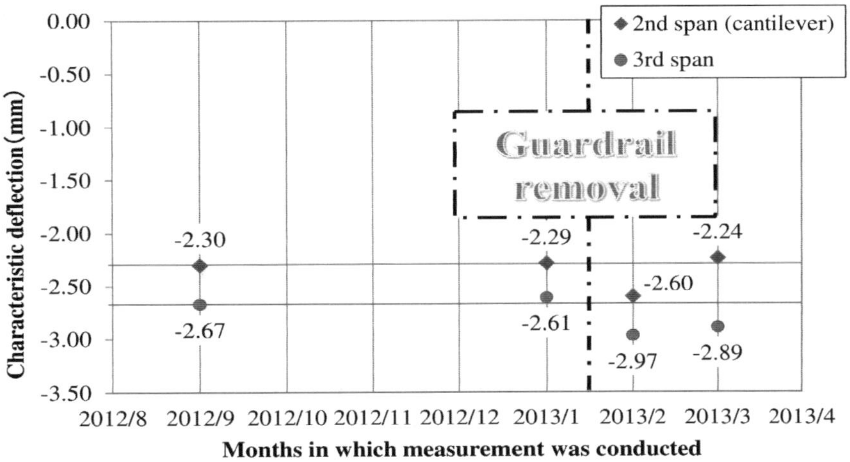

Figure 16 Changes in characteristic deflection due to guardrail removal

compared. The amount of change in the flexural stiffness of the bridge structure due to the guardrail removal was a decrease of about 5%. Examination of Span 3 (see Fig. 14) reveals that the two measured values of characteristic deflection after the guardrail removal (−2.97, −2.89) were about 10% smaller than the values of characteristic deflection after the guardrail removal (−2.67, −2.61). This is thought to be because the decrease in the flexural stiffness of the bridge structure influenced characteristic deflection. However, examination of Span 2, which includes the cantilever structure, reveals that the measured value of characteristic deflection obtained from the fourth measurement (−2.24) is slightly greater than the characteristic deflection before the guardrail removal. The reason for this is thought to be that since Span 2 includes the cantilever structure, it had some kind of influence on characteristic deflection. It is necessary, therefore, to continue various studies, including simulation analyses, on how to deal with spans that include cantilever structures.

5. Overview of Overseas Field Test

This section reports on the field test conducted in the city of Lisbon in Portugal as an overseas application of the bus monitoring system.

As part of a joint study made by Yamaguchi University and the Universidade Nova de Lisboa (NOVA), the field test was conducted with the cooperation of a group led by Prof. Válter J. G. Lúcio of NOVA. Fig. 17 shows the fixed-route bus used in the experiment. The bus used in the field test is a fixed-route bus actually operating in Lisbon. The total weight is about 14 tons, and an under-rear-wheel-spring triaxial acceleration sensor was installed to the bus. Two bridges located near the university were chosen for the experiment. In this paper, the two bridges are referred to as "FCT-in" and "FCT-out." Table 2 and Fig. 18(a) and (b) show the structural types, dimensions and appearances of the bridge, along with other general information. For the purpose of checking the coupling between the bus and the bridge of interest, a velocity sensor was installed in the middle section of each bridge to measure the velocity of the bus passing by on the bridge. The velocity sensor actually used in the field test is shown in Fig. 19. At each bridge, measurement was conducted 15 times at a travel speed of 40 km/h.

Figure 17 Fixed-route bus used and sensor location

Fig. 20(a) and (b) compare the under-rear-wheel-spring acceleration of the bus and the girder midspan acceleration of the bridge measured when the bus crossed the two bridges. The under-rear-wheel-spring acceleration response of the bus here is the vertical acceleration data obtained from a triaxial acceleration sensor, and the center-of-bridge acceleration response of the bridge was obtained by converting the velocity data obtained from a velocity sensor by differentiating it once. At FCT-in, which is a continuous slab girder bridge as shown in Fig. 18(a), the under-rear-wheel-spring acceleration response of the bus and the middle section acceleration response of the bridge are fairly similar as shown in Fig. 20(a). At FCT-out, which is a cantilever bridge as shown in Fig. 18(b), the under-rear-wheel -spring acceleration response and the bridge acceleration response were fairly similar when the bus was passing the center-of-cantilevered-span zone at the left as shown in Fig. 20(b). When the bus was passing the center-of-bridge joint zone, however, the bridge showed a larger acceleration response to the extent that similarity with the under-rear-wheel-spring acceleration of the bus was no longer discernible. This is thought to be because the bus practically climbed over the central joint (i.e. cantilever joint) so that the vibration caused by the weight of the bus had a great influence on the bridge. Thus, it was found that the bus monitoring system can be applied to a continuous girder or simple girder bridge but cannot be applied to a cantilever bridge with a midspan joint like FCT-out.

This section shows characteristic deflection calculation results for FCT-in, which was thought to be a bridge of the type to which the bus monitoring system can be applied. As the first step, to determine the characteristic deflection of FCT-in, data for a period of about 1.8 seconds corresponding to the time the bus took to pass the middle section of the slab girder bridge (the encircled area in Fig. 20(a)) were extracted, and deflection was calculated by integrating acceleration twice. Fig. 21 shows examples of deflection estimates thus obtained. Showing estimated deflection on the vertical axis and time on the horizontal axis, Fig. 21 shows estimated deflection values obtained from the 15 measurements. As shown in Fig. 21, data were extracted from the same section, and the obtained waveforms show similarity. Characteristic deflection is calculated by averaging estimated deflection values in this extraction section. Table 3 lists the characteristic deflection values obtained from the 15 measurements. Thus, the average of the results of the 15 measurements that make characteristic deflection converge to a median is −4.10 mm (standard deviation: 0.63). It is important to continue to conduct more measurements by using this average as a criterion value and monitor changes in characteristic deflection over a long period of time.

Table 2 Bridge data

Bridge name	Type of superstructure				Bridge length (m)	
FCT-in	Span number	Start point	1	3-span continuous prestressed concrete slab girder bridge	16.00	68.13
			2		36.13	
		End point	3		16.00	
FCT-out	Span number	Start point	1	Reinforced concrete T-girder cantilever bridge	10.65	49.04
			2		13.87	
			3		13.87	
		End point	4		10.65	

(a)FCT-in

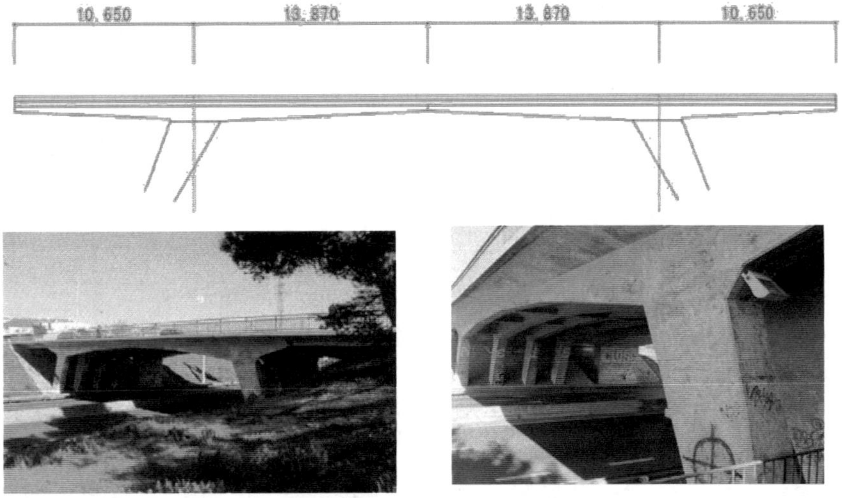

(b)FCT-out

Figure 18 Fixed-route bus used and sensor location

Figure 19 Velocity sensor

As mentioned above, the field test results that have been obtained indicate that the proposed system can be applied to a continuous or simple girder bridge but cannot be applied to a cantilever bridge with a midspan joint like FCT-out. It is therefore necessary to conduct more field tests on other types of bridges. In view of the result of the test conducted in a country other than Japan, it has also been found that the proposed system can be used in other countries where conditions such as the instrumentation environment and vehicles used for measurement differ from those in Japan. The next task is to apply the proposed system to more bridges in other countries, conducting not only short-term but also long-term behavior measurements.

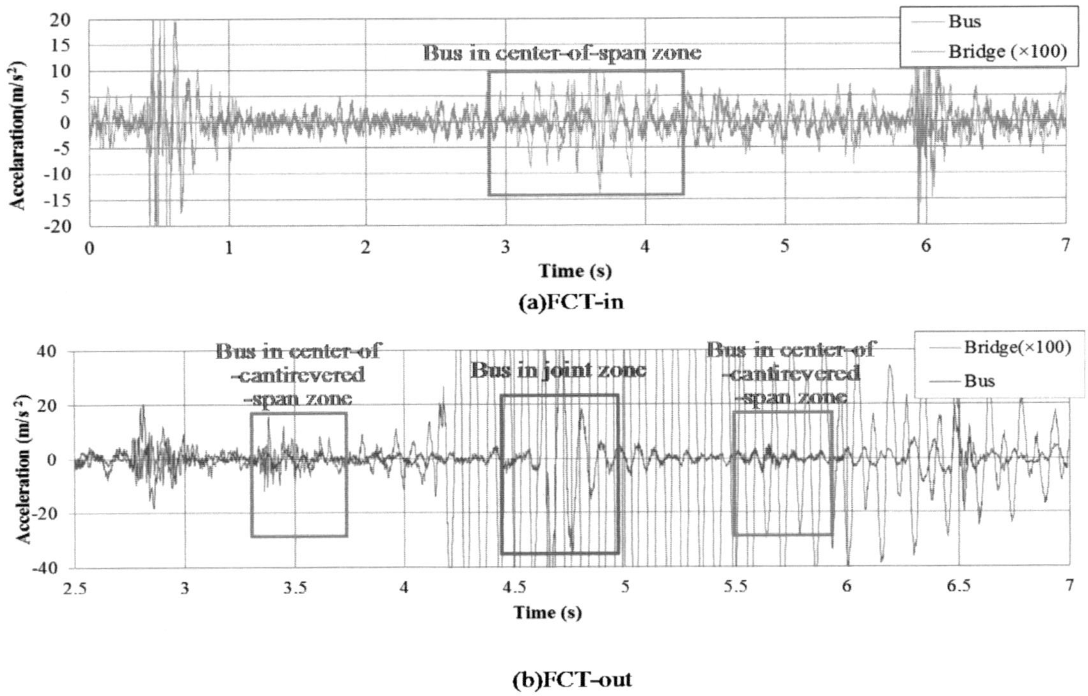

Figure 20 Comparison of under-rear-wheel-spring and girder-midspan acceleration responses during bridge crossing

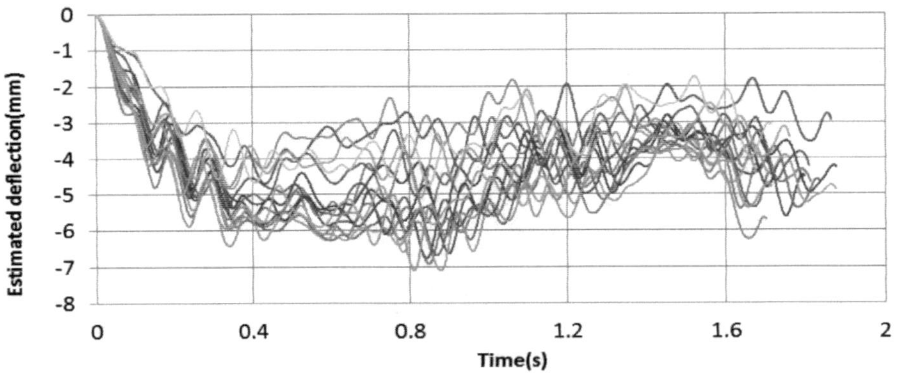

Figure 21 Deflection estimated from under-rear-wheel-spring acceleration response of bus (FCT-in)

6. Conclusions

This paper has reported on the development of a new short and long span bridge monitoring method using a fixed-route bus, presented field test results and discussed evaluation methods. Then, as a result of this study, this paper has shown that it may be possible to detect structural anomalies of bridges efficiently from changes in "characteristic deflection." On the other hand, it has also been found that characteristic deflection values calculated from measurement data varied widely. This is thought to be due to differences in the operating conditions of the bus. As the next step, it is necessary to conduct more measurements and investigate the effects of long-term operating conditions on results obtained, with the aim of improving characteristic deflection so as to reduce its variability and putting the proposed system to practical use as a long-term monitoring system.

Table 3 Characteristic deflection (FCT-in)

No.	Characteristic deflection (mm)
1	−4.51
2	−4.06
3	−4.37
4	−4.26
5	−4.65
6	−4.67
7	−3.72
8	−2.97
9	−4.87
10	−4.23
11	−3.33
12	−4.83
13	−4.44
14	−3.43
15	−3.11
AVERAGE	−4.10

The findings of this study are summarized below:

1) The development of a new short and medium span bridge monitoring system utilizing a publicly operated fixed-route bus and field test results have been reported, and evaluation methods have been discussed. As a result, it has been shown that it may be possible to detect structural anomalies of bridges through long-term measurement of characteristic deflection. It has also been found, however, that characteristic deflection values calculated from measurement data varied considerably. This is thought to be attributable to differences in the operating conditions of the bus. Changes in characteristic deflection have been observed for about three years through field testing. Since no sharp declining trend has been observed, it is necessary to continue long-term measurement.

2) In the field test carried out by using an aged (out-of-service) bridge being removed, a study was conducted on the influence (sensitivity) of artificial damage (bridge railing removal) on characteristic deflection. The experimental results for Span 3 of the bridge indicate that characteristic deflection is reasonably sensitive to declines in the flexural stiffness of the bridge structure. Since the characteristic deflection results for Span 2 show a considerable degree of variability, various studies including simulation analyses need to be conducted.

3) The field test of the bus monitoring system in an overseas environment showed that the system can be used without being affected by the measuring environment and the vehicle used. It has also been found, however, that it may be difficult to apply the system to certain types of bridges such as cantilever bridges.

7. References

[1] A. Miyamoto and A. Yabe: Development of Practical Health Monitoring System for Short-and Medium-Span Bridges based on Vibration Responses of City Bus, Journal of Civil Structural Health Monitoring, International Society for Structural Health Monitoring of Intelligent Infrastructure (ISHMII,), Vol.2, No.1, pp.47-63, May 2012.

[2] Yabe, A., Isoda, S. and Miyamoto, A.: "Development of bridge anomaly detection using characteristic deflection: analytical study" (in Japanese), Proceedings of the 66th Annual Conference of the Japan Society of Civil Engineers, I-275, Sept. 2012.

[3] Japan Society of Civil Engineers: Standard specifications for concrete structures: maintenance, Dec. 2011.

[4] Yabe, A.: "Development of a method for analyzing moving body–structure dynamic interaction using the substructure method" (in Japanese), Proceedings of the 61st Annual Conference of the Japan Society of Civil Engineers, pp. 845–846, Sept. 2006.

[5] Z. Yan, A. Miyamoto and Z. Jiang: "Frequency Slice Wavelet Transform for Transient Vibration Response Analysis", Mechanical Systems and Signal Processing, Vol. 23, Issue 5, pp. 1474-1489, July, 2009.

[6] Yabe, A., Isoda, S., Nagata, M. and Miyamoto, A.: "Field test on short/medium span bridge monitoring system based on public bus vibration" (in Japanese), Proceedings of the 67th Annual Conference of the Japan Society of Civil Engineers, VI-232, 2012.

[7] National Institute for Land and Infrastructure Management and Japan Prestressed Concrete Contractors Association: "Joint study on refinement of soundness evaluation of prestressed concrete highway bridges, chapter 4" (in Japanese), ISSN1346-7328, Technical Note of NILIM No. 613, 2010.

Probability distribution functions for cover used in 3-D model simulating concrete deterioration in port assets.

Rachel M Homer[1], David W Law[2] & Thomas C K Molyneaux[2]

[1] School of Geography, Leeds University, Leeds, LS2 2JT, UK
[2] School of Civil, Environmental & Chemical Engineering, RMIT University, Melbourne, Vic, 3001, Australia

Email: r.m.homer@leeds.ac.uk

Abstract. In previous studies, a 1-D numerical predictive tool to simulate the salt induced corrosion of port assets in Australia has been developed into a 2-D and 3-D model based on current predictive probabilistic models. These studies use a probability distribution function based on the mean and standard deviation of the parameters for a structure incorporating surface chloride concentration, diffusion coefficient and cover. In this paper, this previous work is extended through an investigation of the distribution of actual cover by specified cover, element type and method of construction. Significant differences are found for the measured cover within structures, by method of construction, element type and specified cover. The data are not normally distributed and extreme values, usually low, are found in a number of locations. Elements cast insitu are less likely to meet the specified cover and the measured cover is more dispersed than those in elements which are precast. Individual probability distribution functions are available and are tested against the original function. Methods of combining results so that one distribution is available for a structure are formulated and evaluated. The ability to utilise the model for structures where no measurement have been taken is achieved by transposing results based on the specified cover.

1. Introduction

The corrosion of reinforcing steel in concrete structures is one of the primary causes of concern for owners worldwide. The reinforcing bars in concrete are normally in a passive condition due to the alkaline nature of the cementitious matrix. However corrosion can be initiated when this passive layer is destroyed, either due to a drop in pH caused by carbonation, or by the build-up of chloride ions to a level sufficient to break down the passive layer. The chloride ions can come from three principal sources: inclusion in the initial mix, ingress from de-icing salts and from exposure in a marine environment. In Europe and North America, de-icing salts are the most common cause, while in Australasia it is from marine exposure. The most common models in the literature are 2-stage models based on Tuutti [1-4].

The service life in the 2-stage approach comprises initiation followed by propagation stages [2, 5-7]. The initiation stage refers to the time required for the chloride concentration at the steel surface to reach a critical threshold concentration to initiate corrosion [2]. The propagation stage is the time between initiation of active corrosion until the structural integrity is compromised, where major repair or in the worst case structural failure occurs [1, 8].

The most commonly used solution to model the initiation phase of the service life model is based on Fick's 2nd Law of diffusion, and was developed in the early 1970's[9]. The solution is given in Eqn 1:

$$C_{(x,t)} = C_s\left[1 - erf\left(\frac{x}{2\sqrt{D_C t}}\right)\right] \tag{1}$$

where x is the distance from concrete surface (m); t is time (s); D_C is the diffusion coefficient (m^2/s); C_s is the chloride concentration at the concrete surface; $C_{(x,t)}$ is chloride concentration at the depth of x from the surface at time t; and *erf* is the error function.

This model has been modified as a greater understanding of the process has been developed [10-13] including the DuraCrete model which involves a limit state formulation for chloride induced corrosion initiation [14], Eqn 2.

$$\frac{C}{Cs} = 1 - erf\left(\frac{x}{\sqrt{\left(k_c \cdot k_e \cdot D_{RCM} \cdot \left(\frac{t'_0}{t + t'_{ex}}\right)^n \cdot t\right)}}\right) \tag{2}$$

where C is chloride content at x depth; Cs is surface chloride concentration k_c is the curing factor; k_e is the environmental factor; D_{RCM} is the diffusion coefficient determined by the rapid chloride migration test; t'_0 is age D_{RCM} is measured at; t is duration of exposure; t'_{ex} is age of concrete; n is age factor.

Further developments have been undertaken and incorporated in the current fip model[15], Eqn 3;

$$C_{(x,t)} = (C_0 + (C_{s,\Delta x} - C_0)\left[1 - erf\left(\frac{x - \Delta x}{2\sqrt{D_{app,c} \cdot t}}\right)\right] \tag{3}$$

where x is the distance from concrete surface (m); t is time (s); $D_{app,c}$ is the apparent diffusion coefficient (m^2/s); $C_{s,\Delta x}$ is the chloride concentration at the depth of x from the surface; C_0 is the initial chloride concentration; Δx depth of the convection zone (m); and *erf* is the error function.

In order to be able to solve these equations a number of variables must be known, eg the depth of the rebar, the chloride diffusion coefficient of the material, the age factor and the depth of the convection zone. Each of these variables has a level of uncertainty, ie while the specified cover may be given the actual cover will be a distribution function, which will affect the reliability of the solution to Fick's equation.

This paper reports the analysis of data taken for a number of elements from a number of structures, both precast and cast insitu, to determine their statistical distributions and any statistically significant differences. The intention is to utilise this knowledge to create a set of more tailored probability distribution functions (pdfs) of cover which may be used in the solution of Fick's equation. Systematic sampling was used to select the 30 data points which form the pdf.

2. Site data

The depth of cover has been measured on various elements of four port structures and a bridge in Australia. The port structures and the bridge have concrete elements some of which were manufactured insitu and some of which were precast. The date of construction of the structures ranges from 1968 to 2008, their age at date of measurement has been used to report the analysis. The depth of cover was measured using a covermeter. The sampling strategy was to systematically sample within the accessible parts of the structure. The original plans were consulted to establish the specified cover.

The two oldest structures had specified cover of 44.45mm and 50.8mm (1.75 and 2 inches), with the later structures having specified covers of 65mm and 75mm. CQS (48 years) and LS (44 years) are referred to as "old" structures and DQC (21 years) and CQ4 (2 years) as "new" structures. These variations in cover reflect the changes in design practices and material innovations over the last 50 years.

Due to access restriction only the piers of the bridge were accessible and measurements were taken of the vertical and horizontal elements from 5 of the 18 piers. The specified cover is believed to be 65mm.

3. Analysis of the data

3.1 The port structures.

The port structures had 12 distinct data sets of measured cover from different elements. This included 5 elements, using both methods of construction. The 12 data sets all had a negative skew and did not follow a normal distribution. Figure 1 illustrates the disparity in homogeneity for different element types. Extremely low values were observed in several data sets. Square pile caps manufactured insitu have the smallest dispersion.

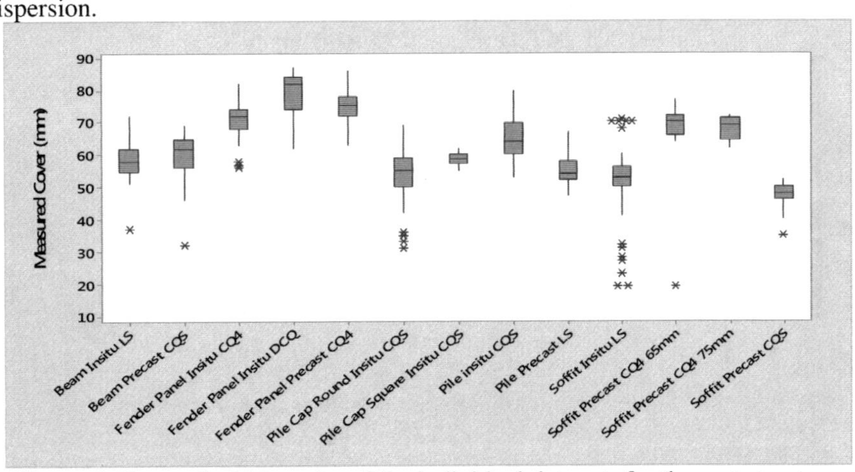

Figure 1. Box and whisker plot of the individual data sets for the port structures

In addition to the presence of outliers, usually small values, there was variation in the proportion of measurements that exceeded specification. Table 1 summarises the size of the spread of the data and the proportion of measurements that exceed specification.

Table 1. Summary of the element types, with specified cover and % exceeding specified cover

	Structure	Sample size (n)	Specified Cover (mm)	Median Measured Cover (mm)	Spread (mm)	Proportion above specified cover
Beam insitu	LS	30	50.8	58	35	97%
Beam Precast	CQS	52	44.45	62	37	98%
Fender Panel Insitu	DCQ	70	75	82	25	100%
Fender Panel Insitu	CQ4	81	75	72	26	23%
Fender Panel Precast	CQ4	104	65	75	25	92%
Pile Insitu	CQS	96	50.8	64	27	100%
Pile Precast	LS	158	50.8	54	20	91%
Round pile cap insitu	CQS	47	50.8	55	38	70%
Square Pile cap insitu	CQS	32	50.8	58.5	7	100%
Soffit Insitu	LS	121	44.45	53	54	91%
Soffit Precast	CQS	47	44.45	48	17	96%
Soffit Precast	CQ4	27	65	70	58	81%
Soffit Precast	CQ4	6	75	69	10	0%

3.1.2 The construction method

The change in specification of cover requires the analysis of elements constructed insitu and those precast, for the older and younger structures. There was a statistically significant difference between the measured cover for all the elements constructed insitu or precast for both old and new structures, illustrated in Figure 2 ((W=23301, p=0.0358) for new structures and (W=106799, p<0.001)) for old structures. However, in each case the difference in the median between precast and insitu was less than 3mm. The cover for elements constructed insitu for old structures was greater than for precast elements, however in CQ4 cover for elements cast insitu was lower than for those precast.

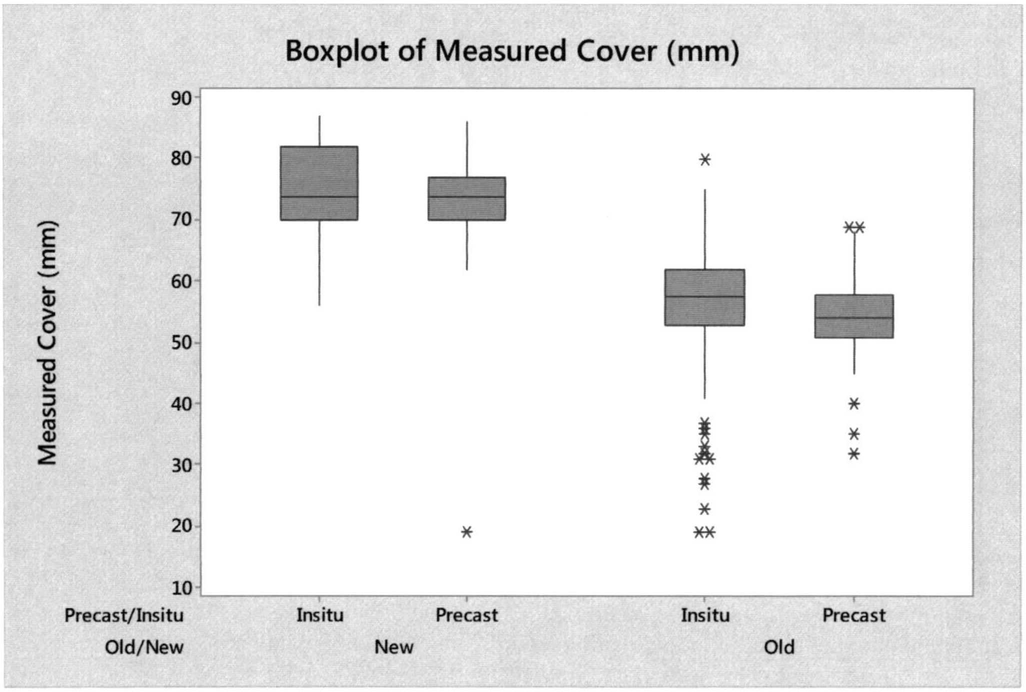

Figure 2. Box and whisker plot comparing the construction method within new and old port structures

In addition to the actual dispersion decreasing between the old and new structures, the prevalence of outliers has decreased with only one outlier present in the data for the new structure. This indicates an improvement in the consistency of construction with fewer instances of insufficient cover achieved.

3.1.3 Port structures

The nature of the sampling mechanism used to take the measurements and the change in cover specification results in no opportunity to compare like for like results. However, some comparisons are possible. Table 2 summarises the results of individual elements manufactured insitu or precast. With the exception of beams there was a statistically significant difference between elements constructed using different methods. Table 3 contains the findings from comparisons where one or more characteristics differs, for example the comparison of the distribution of cover of fender panel was made for the same specified cover, and method of construction, but different locations (and hence different construction teams). Mann-Whitney (Non-parametric) tests were used to test for differences between distributions from different sites or using different methods.

Table 2. Results for tests for difference between the distributions of elements from different structures, different construction methods or different specifications

Element type	Construction method	Specified cover	Location	Results of Mann Whitney test. W=Test statistic, N=sample size	
Pile	Insitu Precast	50.8mm	CQS LS	Significant difference ($p<0.05$,) W=18814, (N96,158)	Insitu (64mm) >Precast (54mm) Difference 10mm
Fender panel	Insitu	75mm	DQC CQ4	Significant difference ($p<0.05$), W=6963.5, (N70,81)	Age 21 (82mm) >Age 2(72mm) Difference 10mm
Fender panel	Insitu Precast	75mm 65mm	CQ4	Significant difference in distribution. ($p<0.05$) W=1113.5 (N=81,104)	Insitu (72) <Precast (75mm) Difference 3mm
Soffits	Insitu Precast	44.45mm	LS CQS	Significant difference in distribution. ($p<0.05$) W=2132.5 (N=121,47)	Insitu (53) >Precast (48mm) Difference 5mm
Beam	Insitu Precast	50.8mm 44.45mm	LS CQS	No significant difference p=0.13, W=1087 (N=52,30)	

The second set of tests was performed to establish if differences exist between different elements with the same specification in the same structure. The results in Table 3 show the significant differences; however these are small. This validates the decision to combine the measurements of different elements to produce a probability distribution function (pdf) to use in the model based on specified cover.

Table 3. Results for tests for difference between the distributions of elements in CQS with different construction methods or different specifications

Element type	Construction method	Specified cover	Location	Results of Mann Whitney or Kruskall Wallis tests	
Round pile caps Square pile caps	Insitu	50.8mm	CQS	Significant difference in distribution. ($p<0.05$) W=1605 (N47,32)	Round cap (55mm) <Square cap (58.5mm) Difference 3.5mm
Beam Soffit	Precast	44.45mm	CQS	Significant difference in distribution. ($p<0.05$) W=1261 (N52,47)	Beam (62mm) > Soffit (48mm) Difference 14mm
Pile Round pile cap	Insitu	50.8mm	CQS	Significant difference in distribution. ($p<0.05$) W=8537 (N=96,47)	Pile (64mm)> Round cap (55mm) Difference 9mm
Pile Square pile cap	Insitu	50.8mm	CQS	Significant difference in distribution. ($p<0.05$) W=7205 (N=96,32)	Pile (64mm)> Square cap (58.5mm) Difference 5.5mm

3.2 Bridge

A total of 162 measurements were made, 44 of horizontal and 118 of vertical elements. The data was normally distributed, with one low value outlier measured. This data is normally distributed, but does contain one very low value. The distribution of the data is much less dispersed than for the port structures. Horizontal and vertical elements were measured and a significant difference in the distribution found ($p=0.008$, $T= 2.71$, N 44,118), with the mean measured cover for horizontal elements less than that for vertical elements. The mean for horizontal measurements (66.0mm) was only 2.2mm smaller than the mean for vertical measurements (68.2mm).

4. Creating probability distribution functions (pdf)

4.1 Port structures

The development of the model from 1D to 3D involved the inclusion of surface chloride concentration, diffusion coefficient and cover. The input for the cover used the mean and the standard deviation to produce a distribution function based on a Gaussian distribution. The use of such a distribution is more representative of reality than the use of a single value. The hypothesis under test is that creating a distribution function based on actual measurements will produce output that will produce a more realistic model.

The statistical analysis revealed that the data has a negative skew. The mean cover is smaller than the median cover and the implication of using a pdf based on the mean rather than median is that the predicted time to onset will be shorter than the actual time to onset. The distributions of the individual elements were compared to a number of known distributions¸ including Weibull, Gamma and Poisson and none of the elements could be represented by one of these distributions. Additionally the cover data was transformed using log, exponential and Box-Cox transformation to establish if this would result in a normal distribution. None of the transformations was successful.

Some exploratory work was undertaken looking at transforming the data with different specified cover. Precast soffits are present in two structures, with three different specified covers. The data for the cover was transformed to produce a set with a notional specified cover of 44.45mm. There is no significant difference between the distributions of the two normalised precast situations, but there is a significant difference between the distribution of the soffits manufactured insitu and those precast. This result confirms that transformation of data is an appropriate technique to employ. Hence pdfs created using transformed data will be evaluated.

A series of bespoke pdfs was created. These were tested by using them as the input for the model, with the value for Sodium Chloride concentration (C_s) held constant at 2.9% and the Diffusion Coefficient (D_c) at $1.0\text{x}10^{-12}$ ms^{-2} in Eqn 1. It is important to evaluate the differences in the pdfs created using different sampling strategies. For every pdf 30 values were drawn from either the entire data set for structures or for specific subsets. The data was ordered and a systematic sampling method (1 in n) was used to select 30 values from the data set. Pdfs were created with different starting places for the selection process and also on data sets which excluded the lowest 5%, highest 5% or lowest and highest 5% of readings.

4.2 Bridge

The data sets for vertical and horizontal measurements are normally distributed as is the data set of combined values. The creation of a pdf is less complex than for the port structures as there are only two variants of the element and the distributions have a low dispersion. The pdfs were created using data from the vertical, horizontal and combined data set. The inclusion of the low value was also investigated.

5. Analysis of the probability distribution functions

The bespoke pdfs created for port structures and bridge were analysed using the model. Constant values for the mean were set at 2.9% for the Sodium Chloride concentration (C_s) and $1.0\text{x}10^{-12}$ m^2/s for the diffusion coefficient (D_c). The evaluation of the output will identify the impact of the created pdfs with the original input based on mean and standard deviation. The evaluation will also evaluate the method of selecting the sample used to create the pdf, from the set of measured cover. The characteristics determining the different pdfs for port structures include:

- Different starting places when systematically sampling from the ordered data
- Data normalised on 44.45mm, 50.8mm, 96mm and 75mm
- Data excluding the lowest 5% of values, the highest 5% of values and both the lowest and highest 5% of values
- Data sets for old structures, in total and by method of construction
- Data sets for new structures, in total and by method of construction

Those for bridges include:

- Including and excluding the observed outlier
- Considering the vertical and horizontal measurements separately and combined

The impact of the pdf is evaluated using the 10% and 90% time to onset values. A 10% time to onset of 17.2 years means that there is a 10% chance that onset will have occurred by 17.2 years after construction. The associated 90% value of 31.9 years means that there is a 90% chance that onset will have occurred by 31.9 years after construction.

5.1 Findings

The pdfs using different starting values and the exclusion of the highest and lowest 5% were created to investigate the impact of including extremely low and high measured cover data points. The impact was very small, indicating that the model is robust and does not respond to extreme values. The non-parametric Kruskal-Wallis test was used to test for difference between the pdfs created with different starting positions. No significant differences were found at ether the 10% or 90% years to impact values.

Pdfs were created excluding the lowest 5%, highest 5% and both the lowest and highest 5% of values. Kruskal-Wallis tests established that there were no significant differences for either the 10% or 90% years to impact values.

Significant differences were found comparing method of construction for the old structures (those with a specified cover of 44.45mm or 50.8mm) for the 90% to onset value. The time to onset for precast elements was 39.1 years compared to 47.2 years for insitu (W=15). There was no significant difference at the 10% years to onset between precast and insitu.

There were no significant differences between time to onset (either 10% or 90%) between precast and insitu for the "new" structures (65mm and 75mm).

Pdfs were created to evaluate the scenario that the older structures had been built to the new specifications and the new structures had been built to the old specification. The outcome is shown in table 4.

Table 4. Results for 10% and 90% time to onset for the four port structures.

	CQS	**CQS built as new**	**LS**	**LS built as new**	**DCQ**	**DCQ built as old**	**CQ4**	**CQ4 built as old**
10% years to impact	21	46	24	40	40	18	42	19
90% years to impact	47	103	36	67	73	33	64	29

It can therefore be concluded that the impact of changing the specification has had a dramatic increase in the time to onset between the older and younger structures.

Small differences were found for vertical and horizontal elements of the bridge as shown in table 5.

Table 5. Results for 10% and 90% time to onset for vertical and horizontal elements of the bridge

	Vertical	Horizontal	Vertical and Horizontal
10% years to onset	35.8 years	33.2 years	36.8 years
90% years to onset	58.6 years	51.3 years	54.1 years

6. Conclusions and recommendations

The analysis of the data reveals substantial differences between the old structures and the new structures. The change in specified thickness has resulted in an increase of time to onset at 10% from 23 years to 42 years and at 90% from 42 years to 68 years. The impact of manufacturing method varies by element and hence is more complex. The use of the data, without weighting for the prevalence of the different element types, indicates that the manufacture method within new structures only impacts the 90% years to onset, with elements manufactured insitu lasting 6 years longer than precast elements.

Further analysis of pdfs created for individual elements is required to refine the method of creating a generic model to be used for structures where measurements have not been taken.

7. References

[1] Liu, Y.P. and R.E. Weyers, *Modeling the time-to-corrosion cracking in chloride contaminated reinforced concrete structures.* Aci Materials Journal, 1998. **95**(6): p. 675-681.

[2] El Maaddawy, T. and K. Soudki, *A model for prediction of time from corrosion initiation to corrosion cracking.* Cement & Concrete Composites, 2007. **29**: p. 168-175.

[3] Gulikers, J. and M. Raupach, *Numerical models for the propagation period of reinforcement corrosion - Comparison of a case study calculated by different researchers.* Materials and Corrosion-Werkstoffe Und Korrosion, 2006. **57**(8): p. 618-627.

[4] Tuuti, K., *Report No 4, Corrosion of steel in concrete*, 1982: Sweden.

[5] Melchers, R.E., C.Q. Li, and W. Lawanwisut, *Modelling deterioration of structural behaviour of reinforced concrete beams under saline environment corrosion.* Magazine of Concrete Research, 2006. **58**(9): p. 575-587.

[6] Peng, J.X., et al., *Modeling of Carbon Dioxide Measurement on cement plants.* Progress in Environmental Science and Engineering, Pts 1-4, 2013. **610-613**: p. 2120-2128.

[7] Cairns, J.J. and D.W. Law, *Prediction of the ultimate limit state of degrading concrete structures*, in *Integrated Lifetime Engineering of Buildings and Civil Infrastructure*2003: Helsinki.

[8] Raupach, M., *Chloride-induced macrocell corrosion of steel in concrete-theoretical background and practical consequences.* Construction and Building Materials, 1996. **10**(5): p. 329-338.

[9] Crank, J., *The Mathematics of Diffusion.* 1975, Oxford, England: Clarendon Press.

[10] Collepardi.M, Turrizia.R, and Marciali.A, *Penetration of Chloride-Ions into Cement Pastes and Concretes.* Journal of the American Ceramic Society, 1972. **55**(10): p. 534-542.

[11] Mejlbro, L., *The complete solution of Fick's second law of diffusion with time dependent diffusion coefficient and surface concentration*, in *Durability of Concrete in Saline Environment,*1996, Cementa AB: Sweden. p. 127-158.

[12] Tang, L. and L.O. Nilsson, *A New Approach to the Determination of Pore Distribution by Penetrating Chlorides into Concrete.* Cement and Concrete Research, 1995. **25**(4): p. 695-701.

[13] Tang, L.P. and J. Gulikers, *On the mathematics of time-dependent apparent chloride diffusion coefficient in concrete.* Cement and Concrete Research, 2007. **37**(4): p. 589-595.

[14] Duracrete, *Modelling of Degradation, Probabilistic Performance Based Durability Design of Concrete Structures Report*, 1998, EU-Project(Brite Euram III) No. BE95-1347.

[15] fip, *The fib Model Code for Concrete Structures 2010* 2013: Ernst & Sohn.

Finite element simulation of guided waves in pipelines for long range monitoring against third party attacks

Salisu El-Hussein, John J. Harrigan and Andrew Starkey

Lloyd's Register Foundation Centre for Safety and Reliability Engineering, School of Engineering, University of Aberdeen, Aberdeen AB24 3UE, United Kingdom

Email:r02se13@abdn.ac.uk

Abstract. Guided waves (GW) are finding more applications for structural health monitoring (SHM) of pipelines and other long, slender structures, particularly in the areas of corrosion and crack detection. Third party impact, both accidental and intentional, is also a major cause of pipeline failure. The use of low frequency (below 10 kHz) GW to detect damage caused by a third-party is investigated. Field test data on a 1 km long pipeline are compared with finite element (FE) predictions to illustrate the potential of low frequency GW to travel long distances along a pipeline. An FE study indicates the type and frequency of GW that can propagate long distances (low attenuation) without significant change in shape (low dispersion). The FE analysis is conducted on a typical 10 in (255 mm) diameter steel pipe with 7.8 mm wall thickness. The effects of pipe diameter and thickness on the GW propagation characteristics are illustrated. It is shown that certain frequencies for certain pipe geometries produce a very dispersive signal and should be avoided for GW SHM and the reasons for this are discussed.

1. Introduction and Background

Vandalism and stealing from petroleum product pipeline has been a major problem in the Niger-Delta area of Nigeria. The consequence of this is huge in terms of repair cost, environmental damage and fatalities. It was reported that a total of 15,796 cases of pipeline vandalism was recorded in Nigeria between 2000-2010 with 2,800 fatalities, about $1.2bn cost of repairs and daily revenue loss to the government of about $10.4million [1]. This research is investigating the potential use of GW for pipeline monitoring to detect third party interference such as act of vandalism and theft. GW propagates in a thin material thickness of the waveguide in similar fashion to Rayleigh and Lamb waves. Plates, rods and hollow cylinders are the common example of natural waveguides. Unlike the longitudinal and shear waves their velocity is not only dependent on the material properties but also on the thickness of the material and propagation frequency. Its propagation is in one dimension which allows the retention of energy thereby ensuring long distance propagation with potentially low energy loss. GWs are also natural properties of the waveguide similar to a natural vibration; hence, changes in the waveguide in the form of damage can be used to detect and locate the damage. The major challenges of GW application are the energy dissipation (material attenuation), dispersion (phase velocity dependence frequency) and attenuation due to reflections from features such as welds and supports.

The interest in the use of GW in SHM is stimulated by its ability to propagate long distances along slender structures allowing 100 percent circumferential monitoring [2]. Other advantages of GW over the conventional methods include: cost effectiveness (of about 5%) and inspection speed [3]. GW can also propagate over curved and straight surfaces and is applicable for both ferromagnetic and non-ferromagnetic materials [4]. These advantages provide the incentive for the choice of GW in this research work. The behaviour of GW in the ultrasonic frequency regime is well researched and documented [5-8] and has been successfully employed as a pipe inspection tool [9, 10] over short distances. This is evidenced by their use in through-thickness Non-destructive testing (NDT) and short range in-plane Non-destructive Evaluation (NDE) of structures. The current research is focussed on low frequency GW (below 10 kHz). Numerical analysis of GW propagation at low frequencies was carried out using Abaqus/Explicit Commercial FE analysis software. It is shown that the use of selected low frequency-mean diameter product can achieve long distance monitoring. To verify the FE result a field test data on 1 km pipeline was analysed. This paper reports the potential of GW for long distance propagation along a pipeline. A brief introduction and background of the research is given in section one followed by a short discussion on GW in cylindrical structures in section 2. Current GW long range application is discussed in section 3. The potential of low frequency GW for long range monitoring is discussed in section 4; followed by long range propagation potential from field test in section 5. Discussion and conclusion are contained in section 6.

2. Guided Waves in Cylindrical Structures

Cylindrical waveguides support 3 modes of GW vibrations: longitudinal, torsional and flexural while for plate like structures only longitudinal and shear horizontal modes are supported. The governing equation for bulk wave and GW are the same; only that a traction free boundary condition needs to be satisfied for GW. The main difference is that GWs are multi-modal and more of these modes are produced in cylindrical waveguides. Fig.1 shows the schematic of a traction-free hollow cylinder with inner and outer radius ‘a’ and ‘b’ respectively. Under each of the 3 modes, several other modes are produced depending on the frequency-thickness product. According to the convention by Silk and Bainton [11] these modes are labelled as *L(0,m), T(0,m)* and *F(n,m)* for longitudinal, torsional and flexural modes respectively. The letter ‘*m*’ represents the GW vibration within the pipe wall thickness while ‘*n*’ describes the number of spiral vibrations of the pipe as a whole. Each of the 3 modes is distinguished by its particles displacement fields in the radial, circumferential and axial directions, as described by the following equations:

$$\begin{aligned} u_r &= U_r(r)\cos n\theta \cos(\omega t + kz) \\ u_\theta &= U_\theta(r)\sin n\theta \cos(\omega t + kz) \\ u_z &= U_z(r)\cos n\theta \sin(\omega t + kz) \end{aligned} \quad (1)$$

where ω is the angular frequency, kz is the component of wave vector in the axial direction and $n =$ 0,1,2,3,... which describes the spiral propagation of GW around the circumference. The axisymmetric modes correspond to the situation where the displacement field has no angular component (i.e. $n = 0$). Non-axisymmetric modes have ‘n’ sinusoidal variation along the circumference (i.e. $n \neq 0$). The quantities U_r, U_θ and U_z are the displacement amplitudes composed of Bessel functions. The displacement fields in the material are obtained by the summation of partial bulk waves in the form of matrix, called ‘characteristics function’.

$$|c_{ij}| = 0, \quad (i, j = 1 \ to \ 6) \quad (2)$$

Eq. (2) is the dispersion equation for GW propagating along a pipe as a function of inner '*a*' and outer diameter '*b*' of the pipe, Lame's constants λ and μ, angular frequency ω and wavenumber [3]. When the wave number $n = 0$, the determinant matrix in eq. (2) can be expressed as product of sub determinant $D_1 D_2 = 0$.

$$D_1 = \begin{vmatrix} c_{11} & c_{12} & c_{14} & c_{15} \\ c_{31} & c_{32} & c_{34} & c_{35} \\ c_{41} & c_{42} & c_{44} & c_{45} \\ c_{61} & c_{62} & c_{64} & c_{65} \end{vmatrix} \text{ and } D_2 = \begin{vmatrix} c_{23} & c_{26} \\ c_{53} & c_{56} \end{vmatrix} \qquad (3)$$

Where $D_1 = 0$, corresponds to the longitudinal wave propagation in which the particle's displacement is independent of angular coordinate while $D_2 = 0$, refers to the torsional wave propagation in which the particle's displacement is dependent on angular coordinate only. The detailed derivation of this equation can be found in [4].

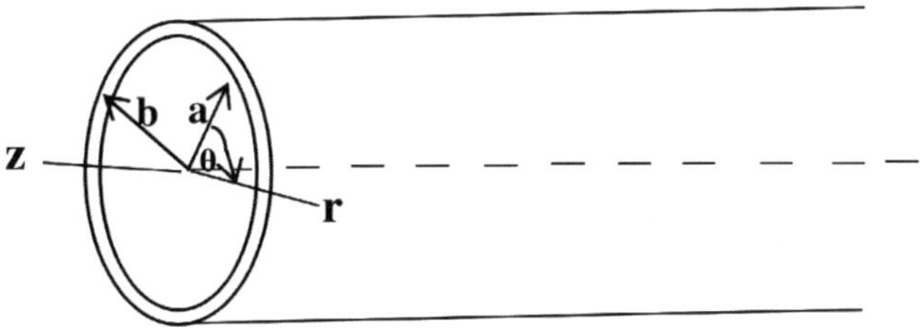

Figure 1. Schematic of hollow cylinder with inner radius 'a' and outer radius 'b' and cylindrical coordinate

An important aspect in the use of GW in SHM is the excitation of the required mode (selective excitation). Multiple modes excitation generates signals which are difficult to interpret. Single mode excitation within its non-dispersive region is therefore a prerequisite for a successful long range GW application.

3. Long range guided wave application

GW has potential to achieve the objective of SHM of providing real-time data on the state of a structure to be monitored. Its application involves sending a wave pulse through a probe attached to a structure to be examined and recording the reflected signals from defects and features along the structure as shown in fig. 2. However, there is a trade-off between the potential propagation distance and the sensitivity to detect defects. The sensitivity of GW is frequency dependent and as the frequency reduces, the sensitivity to defect also reduces. Hence, GW is currently being applied in short (< 5 m), medium (up to 5m) and long (< 10m) range depending on the frequency used [6]. The predominant application areas have been on elongated structures such as pipeline and rail line networks. A GW rail inspection system with the potential to inspect 100 m of rail from a single position was jointly developed by Imperial College and Guided Ultrasonics Ltd [12]. A 1 km GW monitoring range has been reported for the detection of breaks in the railway line [13]. A portable GW rail inspection system was also developed at Pennsylvania State University using electromagnetic acoustics transducer (EMATs) [14]. GW has also being used for defect detection in bones [15], plate [16] and rock bolt testing [17] for mine roof support. In the area of pipeline inspections there are many commercial vendors such as Plant integrity Ltd, Wavesinsolids LLC (WINS), MISTRAS Asset Property Solution, etc., with varying reported inspection distance ranging from 35 m to 300 m. All these operate at higher frequencies above 200 kHz. To date GW application in the low frequency region is yet to be fully exploited. As human ingenuity in the area of inspection and monitoring continues to grow, understanding the behaviour of GW at low frequency will further expand its application boundary especially in elongated structures. FE analysis of GW propagation in pipes can provide the understanding that is required to extend the inspection distance.

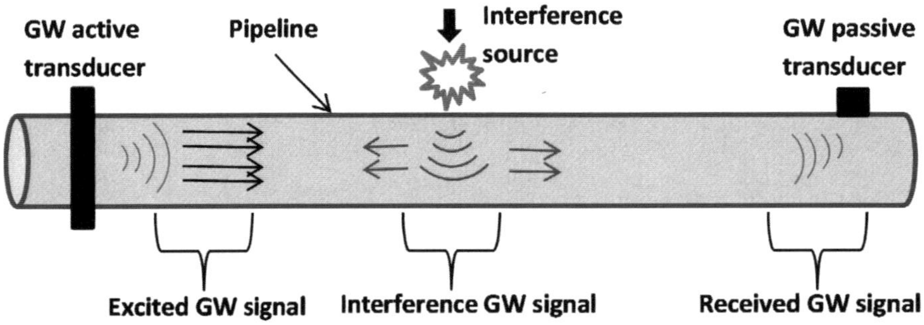

Figure 2. Illustration of guided wave inspection/monitoring techniques

4. Long range guided wave finite element simulation

GW propagation along 10 in pipe 1000 m long, with 7.8 mm wall thickness was simulated in Abaqus/Explicit on a High Performance computer Cluster (HPC). A L(0,1) GW mode was excited by applying a uniform pulse load from one end of the pipe as shown in fig. 3. The pulse load was excited with 5 cycles tone burst modified by Hanning window with a centre frequency of 2.5 kHz. Stresses, displacements and acceleration signals were recorded at 3 nodal positions, 125 m, 250 m and 500 m away from excited end. The nodes positions were chosen so that they correspond to the same nodes positions used in a field test discussed in section 5. Each of these (stress, displacement and acceleration) signals can be used to determine the characteristics of the propagated wave. Here displacement signals were used. Fig. 4 (a) and (b) show the time domain displacements at the 3 nodal positions and their corresponding frequency spectrum respectively. The dispersion and attenuation of the signals were low at this centre frequency as observed from the fig.4 (a) and (b). The propagation characteristics (wavenumber, attenuation and phase velocity) were determined using the transfer function between the nodes 1 and 2 signals and shown in fig.5. Up to 5 kHz (where the magnitude of the signal spectrum is significant), the attenuation is low and dependence of phase velocity on frequency is nearly constant. This shows the potential of GW long range propagation at this region.

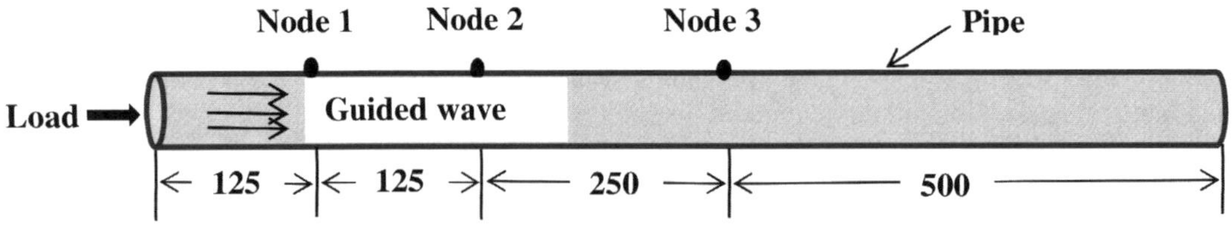

Figure 3. Schematic of the pipe used in FE model and the nodes positions

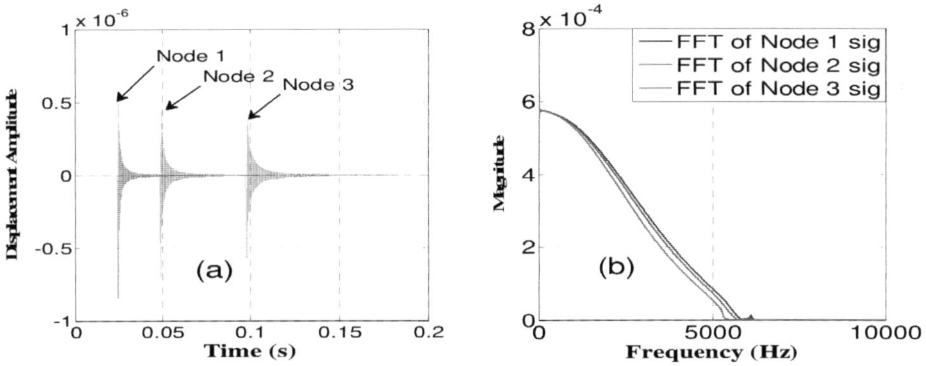

Figure 4. Displacements signals recorded at the nodal positions (a) time domain and (b) its corresponding frequency spectrum

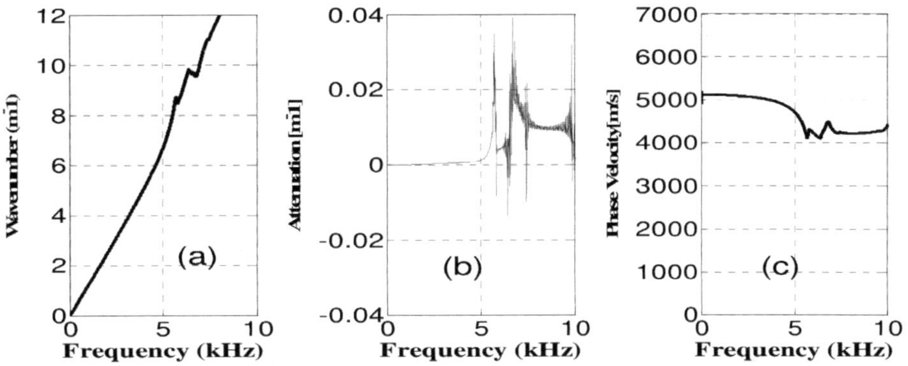

Figure 5. propagation characteristics between nodes 1 and 2 signals

4.1. Effect of frequency-diameter on guided wave propagation

A frequency sweep of 1 – 10 kHz was carried out to investigate the relationship between the centre frequency of the input signal and external diameter of the pipe. It was discovered that at certain frequencies the generated GW was excessively dispersive and does not decay to zero. The reason for this is that GW in pipe has wavenumber in both axial and circumferential direction. Along the length of the circumference, there are 0,1,2…n wavelength λ. Where the λ of the input signal is comparable to the circumferential dimension of the pipe, this type of dispersion is observed. At this region, other frequencies were excited which make the signal processing difficult. For a 12 in diameter pipe, this happens when the centre frequency of the input signal was 5 kHz. Fig.7 (a – f) illustrates the effect of the external diameter and the wavelength of the input signal. At centre frequency of 2.5 kHz, the λ is about 2.064 m which is not comparable to the pipe circumference of 0.958 m and hence, a good signal was recorded as shown fig.7 (a) and (b). When the frequency of the input signal was changed to 5 kHz (λ now 1.032 m), the signal becomes dispersive as shown fig. 7 (c) and (d). The external diameter was then increase to 25 in (635mm) and 2.5 kHz centre frequency was used, the signal was again dispersive as was the case with 5 kHz centre frequency as shown in fig.7 (e) and (f). Therefore, the selection of the right input signal frequency would enhance signal purity, reduce dispersion and ensure long propagation with retention of sufficient signal energy. The results show the need for frequency selection for different pipe dimension in order to achieve long range potential of GW propagation along a pipeline.

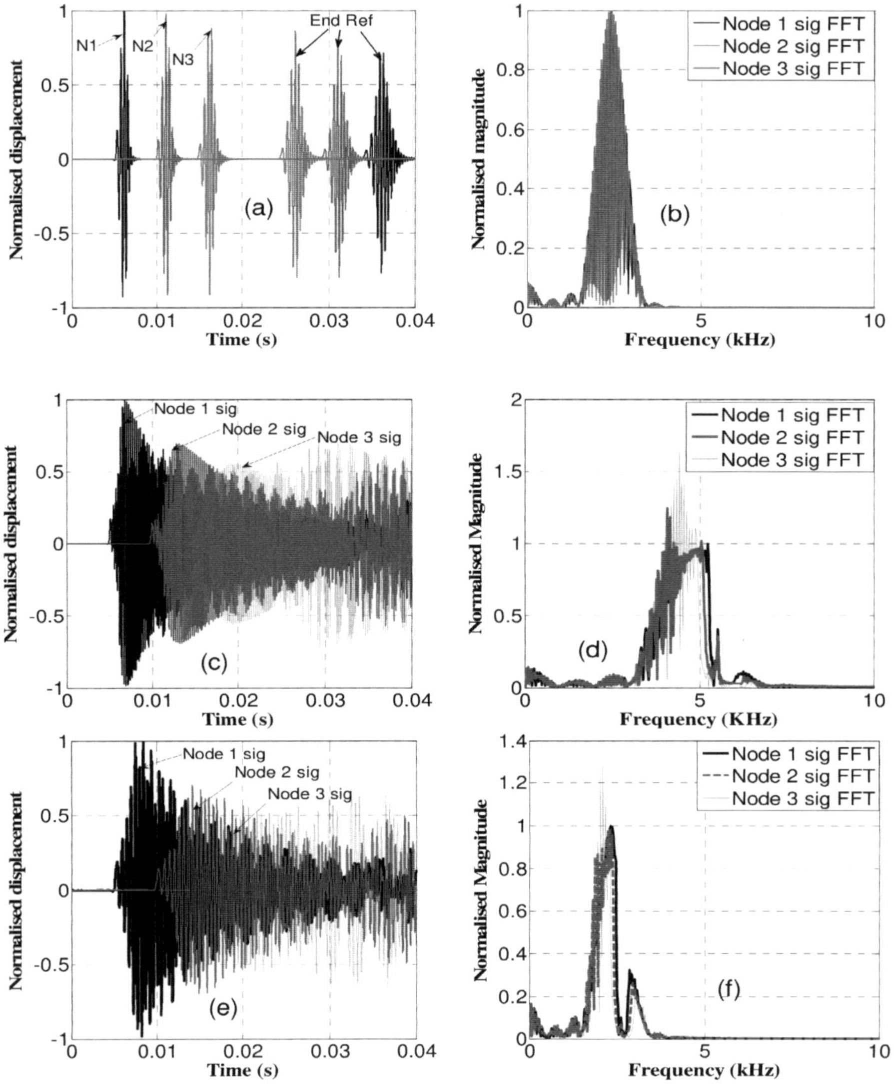

Figure 6. Time domain and frequency spectrum (a, b) 12 in pipe with 2.5 kHz Centre frequency (c, d) 12 in pipe with 5 kHz Centre frequency and (e, f) 25 in pipe with Centre frequency of 2.5 kHz

5. Potential guided wave long distance monitoring from field tests

A field test data conducted on a 1 km long pipeline looped with a 5D bends (i.e. bend radius of 5 times the nominal diameter) was analysed for long range propagation of GWs. The test was carried out by TBG Solution Ltd, a specialist control and automation company which is developing technology for automatic detection of pipeline theft (ADEPT). The test pipe is coated with 5 mm bitumen and has flange connections at approximately every 20 m. PCB Piezotronics 3-wire TO-8 accelerometer (1000mV/g, range 0.13 -800 Hz), PCB Piezotronics 2-wireT0-8 accelerometers (1000mV/g, 0.32 – 8000 Hz) and Hansford HS-100 series accelerometers (100mV/g, 2- 10000 Hz) were used to record the vibration signals. Fig.7 shows the test pipeline loop and position of the sensors. The pipeline has 10 in (254 mm) outer diameter and 7.8 mm wall thickness coated with about 5 mm thick bitumen. Drop tests with a 534 g ball bearing were conducted at heights 2 cm – 50 cm above the pipeline. An

FE model of a similar pipe (but without the bitumen coating) was also created with a simulated ball bearing drop similar to the field trials, and Fig.8 shows the frequency domain analysis comparison of the field tests with the FE results. The field signals were recorded at about 500 m away from the initial impact point where the ball bearing was dropped onto the pipeline while FE signals were taken at a nodal point 25 m away from the ball bearing drop point due to memory and processing constraints on the FE model. The field test result confirms the low attenuation of GW discussed in section 4, with the figure showing the presence in the signal of low frequencies that match the profile of the simulated ball bearing drop (particularly centred on 1.5 kHz). It can be observed that the energy of the propagated frequency in the field tests shows no signal above approximately 2.5 kHz, whereas the FE results do exhibit a frequency response in these higher frequency regions. These results show how the frequency profile over longer distances has changed and that the higher frequencies have higher attenuation and effectively disappear from the analysis, whereas the lower frequencies particularly in the main dominant frequency range produced by the ball bearing drop continue to propagate through the pipeline. This result confirms the potential for long distance propagation of GW through pipelines, at defined frequency ranges which will be determined by the physical attributes of the pipeline as explained in section 4.

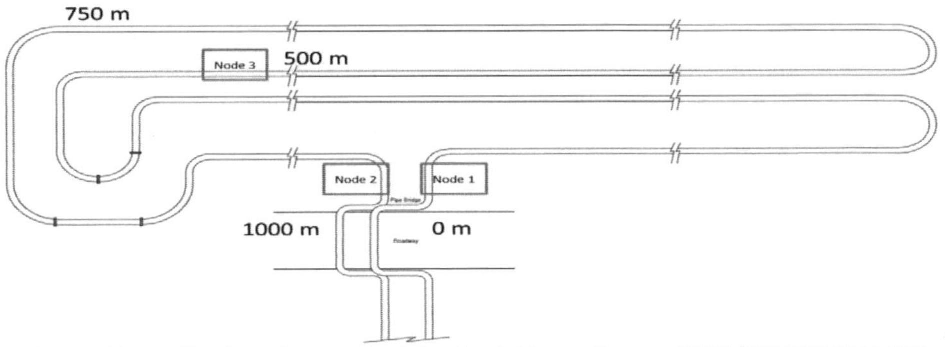

Figure 7. Pipeline loop Layout used for the field test (Source: TBG SOLUTION LTD)

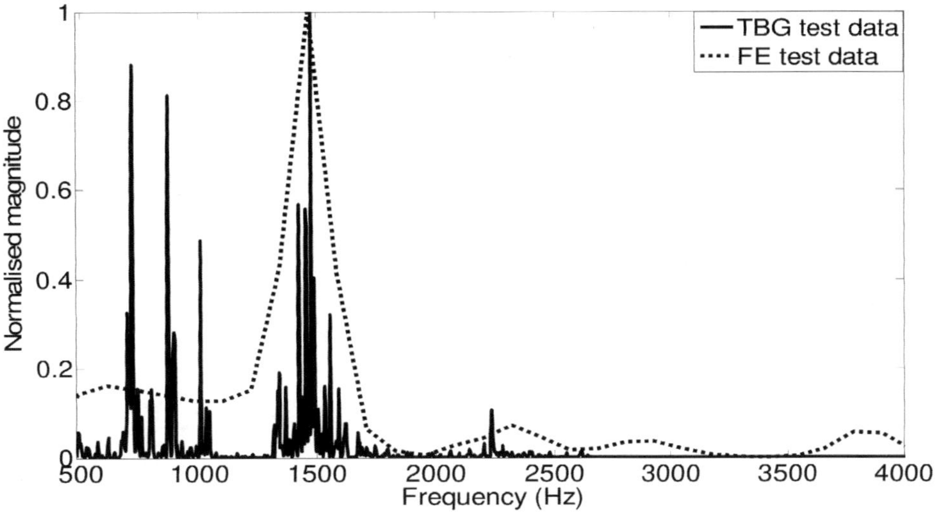

Figure 8. Comparison of frequency content of TBG field test and FE results

6. Discussion and Conclusion

Conventional ultrasonic waves propagate with constant velocity whereas GW velocity varies with waveguide geometry and frequency. Low frequency GW has been shown to have potential to propagate long distance and still retain sufficient energy for detection. In GW propagation, the waveguide plays a more significant role in determining the characteristics of the transmitted signal than the input signal. This is evident in the input frequency and external diameter relationship. However, the input signal has significant effect on the quality of the GW propagation along a pipeline when its wavelength (which has inverse relationship with frequency) is comparable in dimension to the length of the pipe circumference. This region should be avoided for long range GW monitoring and inspection. The good news is that, at low frequency, the wavelength is large; hence, a wide range of pipelines can be excited without approaching the unwanted frequency-diameter region. For example at 2.5 kHz, the wavelength is about 2 m and corresponding unwanted external diameter with about 2 m circumferential length is for about 25 in diameter pipe. The frequency spectrum of the generated GW from ball bearing drop field tests on a 1 km pipeline shows that the energy of the signal concentrated below 3 kHz and can propagate through long distances, and the results at these lower frequencies agree with a scaled FE ball bearing drop simulation results. The results from FE and field tests indicate the potential of low frequency GW for long range monitoring applications. The multimodal nature of GW in particular poses difficulty in exciting, reception and processing of the required signals. Despite these shortcomings, GW application has advantages over the conventional methods such as cost effectiveness, inspection speed and ability to propagate over curved and straight surfaces. Further work and field trials is required to prove the effectiveness of GW for SHM purposes.

Acknowledgement

The authors wish to acknowledge the TBG Sensing Solution LTD for sharing thèir field data and permission to use it in this paper. We also like to appreciate the contribution of Faculty of Engineering, University of Manchester for conducting a preliminary report on the test data which was made available to authors.

References

[1] Udofia O O and Joel OF 2012 Pipeline vandalism in Nigeria: Recommended best of practice of checking the menace. Society of Petroleum Engineers, SPE Nigerian Annual International Conference and Exhibition Abuja-Nigeria 6-8 August 2012.

[2] Raghavan A and Cesnik CE 2007 Review of guided based structural health monitoring Shock Vibration Digest **39** 91-114

[3] Rose J L 2002 A Baseline and vision of ultrasonic guided wave inspection potential Journal of pressure vessel technology **124** (273).

[4] Rose J L Cho Y and Avioli M J 2009 Next generation guided wave health monitoring for long range inspection of pipes. Journal of Loss Prevention in the Process Industries **22** (6), pp. 1010-1015.

[5] Gaziz D C Three dimensional investigation of the propagation of waves in hollow circular cylinders I. Analytical foundation J Accoust. Soc. Am. **31** 568-73

[6] Graff KL Waves motion in elastic solids Dover publication Inc 1975

[7] Rose J L Ultrasonic Waves in Solid Media (Cambridge: Cambridge University Press)

[8] Lowe M J S and Cawley P 2006. Guided waves long inspection usage – current commercial capabilities and research directions Department of Mechanical Engineering Imperial College London.

[9] Lowe M J S, Alleyne P and Cawley P 1998 Defect Detection in Pipes using guided waves Ultrasonic **36** 147-154

[10] Loveday PW 2012 Guided wave Inspection and monitoring of rail way track J Nondestructive Evaluation **31** 303-309 DOI 10.1007/s10921-012-0145-9

[11] Silk MG and Bainton K F 1979 The propagation in metal tubing of ultrasonic wave modes equivalent to Lamb waves Ultrasonics **17** 11 – 19

[12] Wilcox P, Pavlakovic B, Evans M, Vine K, Cawley P, Lowe M and Alleyne DN 2003 Long range inspection of rail using guided wave. In review of progress ion Quantitative Nondestructive Evaluation **22** 236-243.

[13] Burger F A A practical continuous operating rail breaks detection system using guided waves In: 18th World Conference on Destructive Testing, 16 -20 April 2012 Durban, South Africa.

[14] Rose JL Avioli M 2004 Guided wave inspection potential of defects in rail NDT E Int. **37** 174-180.

[15]. Xu K Ta D and Wang W 2010 Multiridge-based analysis for separating individual modes from multimodal guided wave signals in long bones IEEE transactions on ultrasonic, ferroelectrics, and frequency control, **57(11)** 2480-2490.

[16]. Davies J Simonetti F Lowe M and Cawley P 2006 Review of the synthetically focused guided wave imaging techniques with application to defect sizing, American Institute of Physics, Review of NDE Vol. 25, ed. By Dr Thompson and D.E. Chimenti.

[17]. Yan F Royer JR and Rose J L 2010 Ultrasonic guided wave imaging techniques in structural health monitoring, Journal of intelligent material Systems and Structures, Vol. 21, pp. 377 – 384.

Building damage due to structural pounding during earthquakes

B Sołtysik[1] and R Jankowski[2]
[1,2]Department of Metal Structures and Construction Management, Faculty of Civil and Environmental Engineering, Gdansk University of Technology, ul. Gabriela Narutowicza 11/12, 80-233 Gdańsk, Poland

E-mail: [1]barwiech@pg.gda.pl, [2]jankowr@pg.gda.pl

Abstract. Earthquake-induced pounding between adjacent buildings has been identified as one of the reasons for substantial damage or even total collapse of colliding structures. A major reason leading to interactions in buildings results from the differences in their dynamic parameters and also from insufficient distance between the structures. Although the research on structural pounding has been much advanced, the studies have mainly been conducted for concrete structures. The aim of this paper is to show the results of the non-linear numerical analysis focuses on damage due to pounding between two steel buildings under earthquake excitation. The numerical analysis has been performed using models of steel asymmetric structures with different number of storeys which makes them vibrate out-of-phase. Pounding between buildings has been controlled using three-dimensional gap-friction elements which become active when contact is detected. In order to identify the dynamic characteristics of analyzed structures, the modal analysis has been first conducted. Then, the detailed non-linear dynamic analysis of colliding structures has been performed. The acceleration time histories of the El Centro earthquake have been used in the numerical analysis. The results of the study clearly indicate that pounding may substantially influence the response of steel buildings intensifying their damage during earthquakes.

1. Introduction

As a result of high urbanization, the need to build closely-spaced buildings forces the designers to take under consideration earthquake-induced interaction between insufficiently separated structures. Observations after earthquakes confirm that structural pounding may cause some local damages to the elements, like it happened during moderate ground motion in the north-eastern Poland in 2004 [1]. The phenomenon may also lead to the total collapse of structures, as for example during the Mexico earthquake in 1985 [2] or the Loma Prieta earthquake in 1989 [3]. The main reason of collisions between adjacent buildings under earthquake excitation, besides insufficient distance between structures, is the difference in stiffness and/or mass. This difference leads to the out-of-phase vibrations and finally may cause structural interactions [4,5]. The phenomenon of earthquake-induced pounding between buildings has been intensively studied using various structural models and applying different models of collisions (see, for example [6-13]). However, most of the analyses concerned

[1] To whom any correspondence should be addressed.

collisions between reinforced concrete buildings and the studies on pounding between steel structures are very limited (see, for example, [14]).

The aim of this paper is to show the results of the numerical study on colliding two asymmetric steel structures with different dynamic parameters under earthquake excitation. In the first part of the work, the dynamic properties of both structures have been identified. Afterwards, the detailed nonlinear dynamic analysis on earthquake-induced pounding between the structures has been conducted.

2. Numerical model of colliding structures

Two adjacent steel buildings with different geometry and different number of storeys have been considered in the numerical analysis (see Figure 1). The structural members of 3-storey and 4-storey buildings have been modelled by 10560 and 9376 four-node quadrilateral shell elements, respectively. I-shaped columns, made of steel S275, have been rigidly fixed to the ground (soil-structure interaction has not been taken into account). The initial separation gap between structures equal to 2 cm has been considered. To investigate the pounding-involved response of the buildings, six three-dimensional gap-friction elements have been used (two for each storey, placed between the corner nodes of the buildings). These elements assure frictional and gapping connection - when contact is detected, the nodes become fixed in the longitudinal direction and friction forces are imposed in the transverse and vertical directions [15]. The friction coefficient of 0.5 has been applied in the analysis.

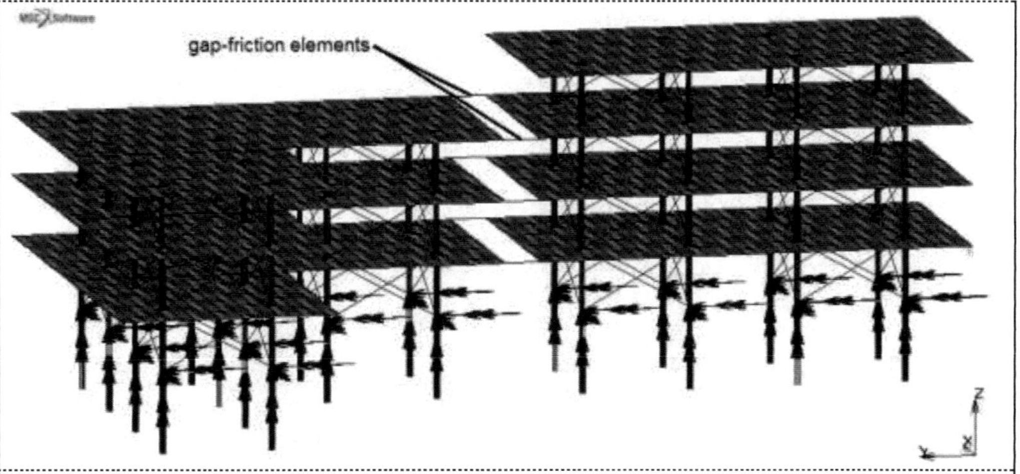

Figure 1. FE model of colliding steel buildings.

3. Modal analysis

In order to identify the dynamic properties of the structural models created, the modal analysis has been first conducted. The examples of the first three natural vibration modes of both buildings, as the result of the analysis, are presented in Figure 2 and Figure 3. The corresponding natural frequencies for the modes are also summarized in Table 1.

Table 1. Natural frequencies for free vibrations modes.

	Transverse (Hz)	**Longitudinal (Hz)**	**Torsional (Hz)**
3-storey building	2.636	2.588	2.721
4-storey building	1.134	1.234	1.151

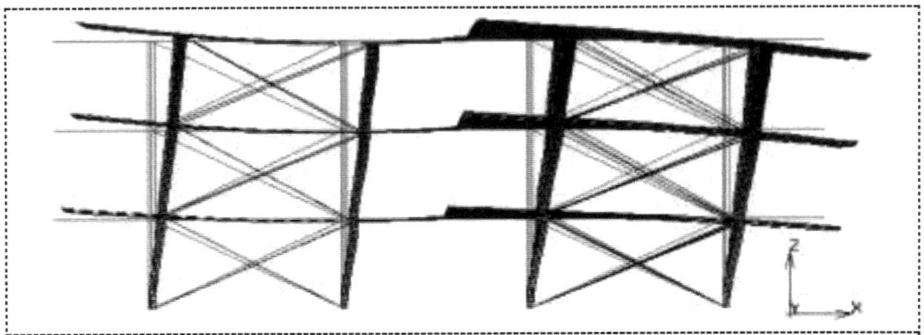

(a) 1st mode

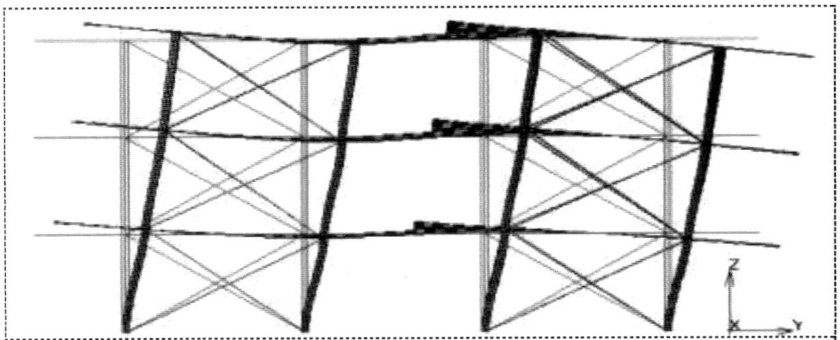

(b) 2nd mode

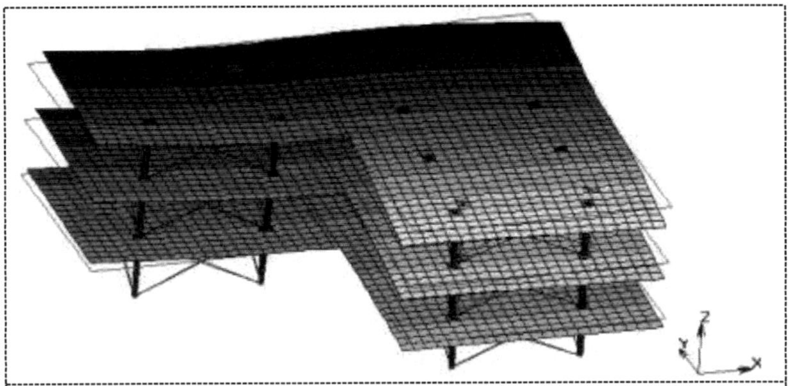

(c) 3rd mode

Figure 2. Modes of free vibrations for 3-storey building.

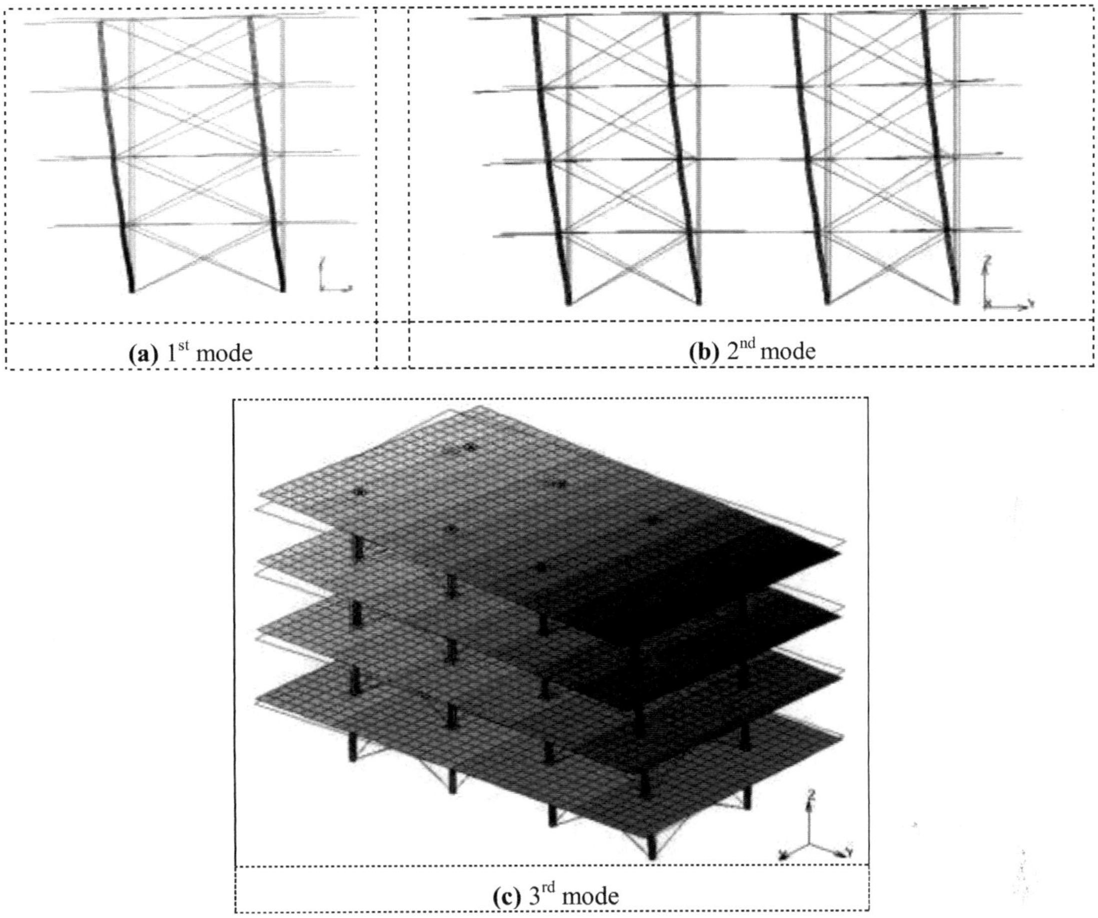

(a) 1st mode **(b)** 2nd mode

(c) 3rd mode

Figure 3. Modes of free vibrations for 4-storey building.

4. Dynamic analysis of colliding buildings

The detailed dynamic analysis, focused on the response of interacting structures under seismic excitation, has been carried out in the second part of the study. The El Centro earthquake (18.05.1940) records have been used in the numerical analysis (see Figure 4). The NS and EW components of the ground motion have been applied in the longitudinal (Y) and transverse (X) direction, respectively. Elastic-perfectly plastic behaviour of steel has been taken into account (material damage model). The examples of the results of the analysis are shown in Tables 2-3 and Figures 5-10. In particular, Table 2 summarizes the peak values of displacement in the longitudinal and transverse direction for both structures. Figure 5 and Figure 6 present a comparison between the displacement time histories in the longitudinal and transverse direction for the 3-storey building (node no. 34668 at the corner of the third storey) with and without pounding (large separation gap); whereas Figure 7 and Figure 8 show the corresponding results for the 4-storey building (node no. 30364 at the corner of the third storey). Moreover, Figure 9 and Figure 10 demonstrate the maps of plastic strains for chosen columns of 3-storey building and 4-storey building with and without pounding. Finally, Table 3 summarizes the peak values of plastic strain for both structures.

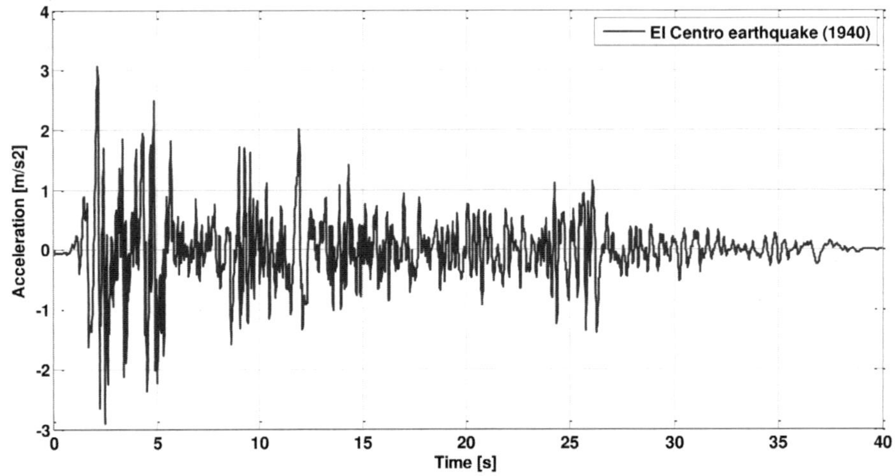

a) NS component (longitudinal direction)

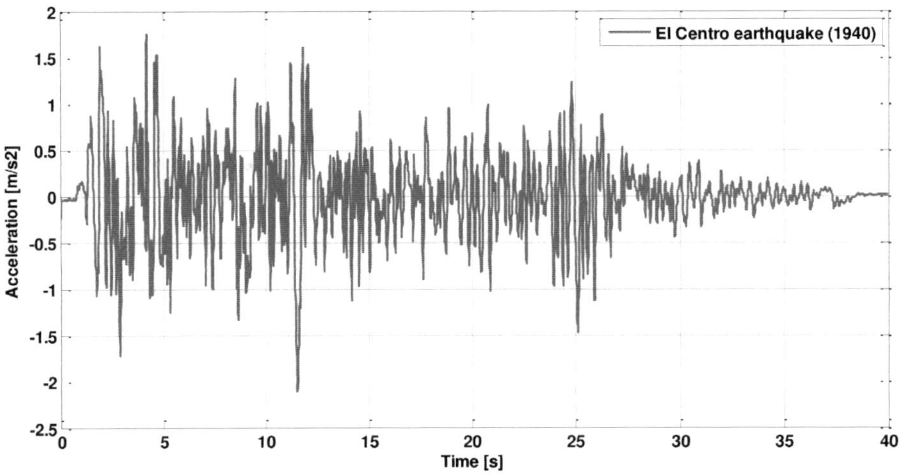

b)

b) EW component (transverse direction)

Figure 4. Acceleration time histories for the El Centro earthquake (1940)

Table 2. Peak values of displacement.

		Without pounding (m)	With pounding (m)	Difference (%)
3-storey building	Longitudinal (Y) direction	0.0263	0.0245	Decrease by 6.8%
	Transverse (X) direction	0.0411	0.0322	Decrease by 21.7%
4-storey building	Longitudinal (Y) direction	0.0353	0.0659	Increase by 86.7%
	Transverse (X) direction	0.0535	0.1046	Increase by 95.5%

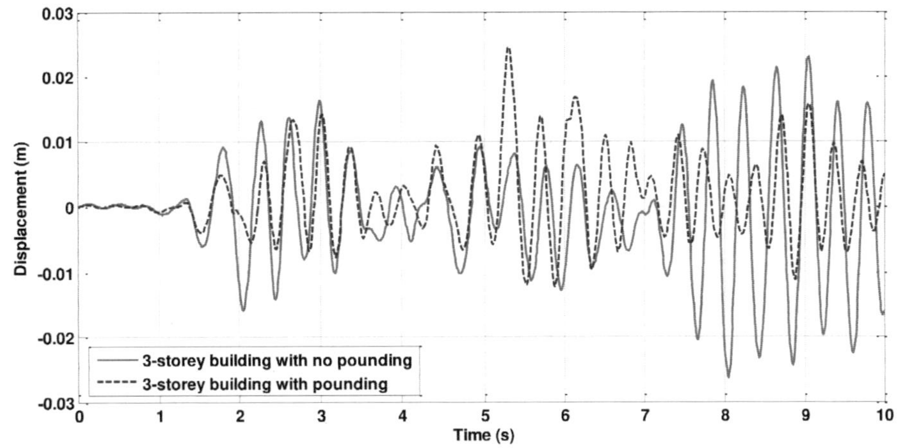

Figure 5. Displacement time histories in the longitudinal (Y) direction for the 3-storey building with and without pounding.

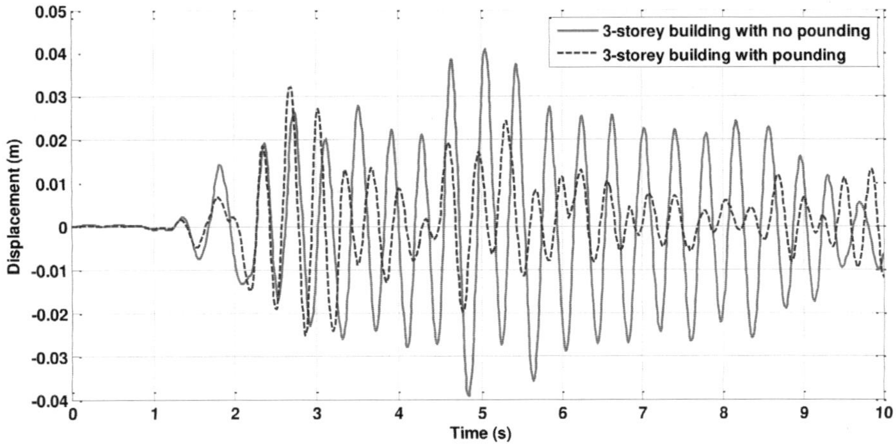

Figure 6. Displacement time histories in the transverse (X) direction for the 3-storey building with and without pounding.

Journal of Physics: Conference Series **628** (2015) 012040 doi:10.1088/1742-6596/628/1/012040

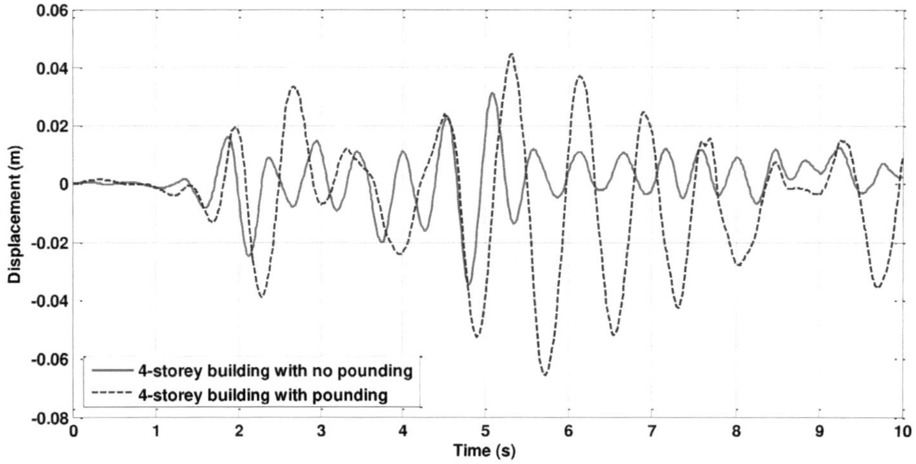

Figure 7. Displacement time histories in the longitudinal (Y) direction for the 4-storey building with and without pounding.

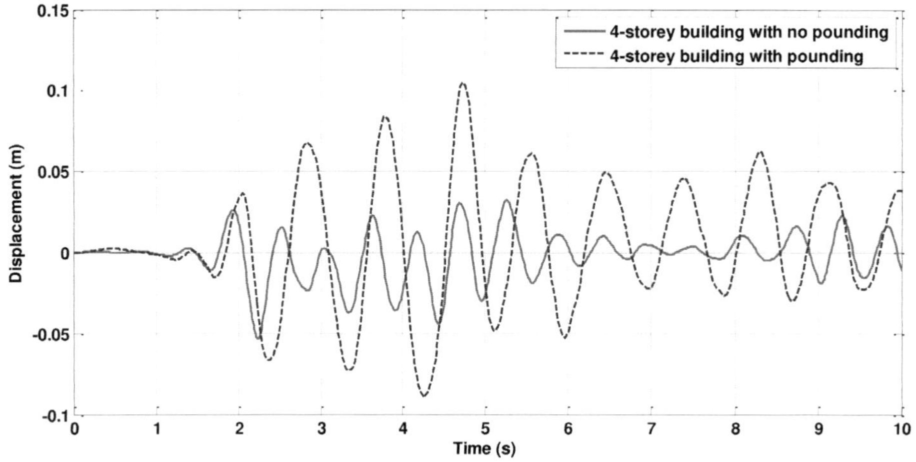

Figure 8. Displacement time histories in the transverse (X) direction for the 4-storey building with and without pounding.

Figures 5-8 clearly show that structural pounding may substantially influence the response of two adjacent steel buildings in the longitudinal and transverse direction under earthquake excitation. It can be seen from Table 2 as well as from Figure 5 and Figure 6 that the peak response of the 3-storey structure decreases due to collisions by 6.8% and 21.7% in the longitudinal and transverse direction, respectively. On the other hand, Figure 7 and Figure 8 indicate that collisions between two steel structures lead to the increase in the peak displacement of the 4-storey building, and this increase is as large as 86.7% and 95.5% for the longitudinal and transverse direction, respectively (see Table 2).

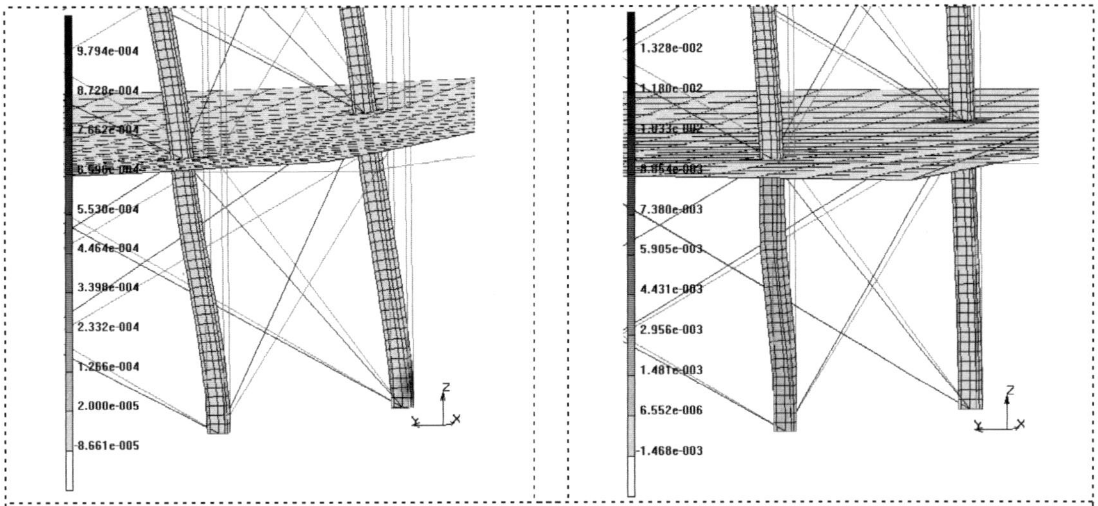

Figure 9. Maps of plastic strains for ground columns of 3-storey building without pounding (on the left) and with pounding (on the right).

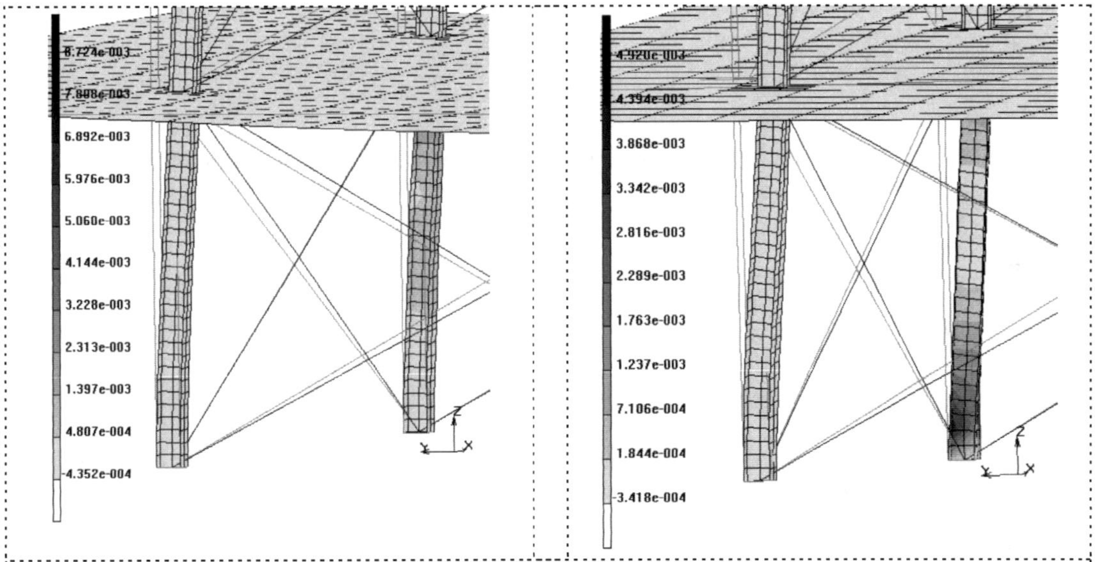

Figure 10. Maps of plastic strains for ground columns of 4-storey building without pounding (on the left) and with pounding (on the right).

Table 3. Peak values of plastic strain.

	Without pounding	**With pounding**	**Difference (%)**
3-storey building	3.924×10^{-2}	1.214×10^{-2}	Decrease by 69.1%
4-storey building	4.938×10^{-2}	7.289×10^{-2}	Increase by 47.6%

The results of the study also indicate that the ground columns of both buildings experience considerable inelastic behaviour at the bases, what results in the permanent deformation of structures. Moreover, it can be seen from Figure 9 and Figure 10 that the size of yielding zone is substantially different for the cases with and without pounding. Also, the peak values of the plastic strains from Table 3 are substantially different for the responses with and without structural interactions. In the case of 4-storey building, pounding results in substantial increase in the size and intensity of yielding zone (increase in the peak plastic strains by 47.6%). On the other hand, Figure 9 and Table 3 show that the earthquake-induced collisions have positive effect on the response of the 3-storey building leading to the decrease in the peak plastic strains by 69.1%.

5. Concluding remarks

The results of the non-linear numerical analysis, focused on damage due to earthquake-induced pounding between two steel buildings, have been presented in this paper. The elastic-perfectly plastic material properties have been implemented in the numerical model. The results of the study clearly indicate that pounding may substantially influence the response of steel buildings during earthquakes, both in the longitudinal and transverse directions. Collisions resulted in the substantial increase in the response of one of the structures, intensifying its damage, while had a positive role in the case of the second building. The use of the FEM with a detailed representation of the geometry, as well as the non-linear material behaviour, allows us to investigate the earthquake-induced pounding-involved response of steel structures quite precisely. However, further research is still needed so as to study in more details the behaviour of colliding steel buildings with different numbers of storeys under different earthquake excitations.

Literature

[1] Zembaty Z, Cholewicki A, Jankowski R, Szulc J 2005 Trzęsienia ziemi 21 września 2004 r. w Polsce północno-wschodniej oraz ich wpływ na obiekty budowlane (Earthquakes of September 21, 2004 in north eastern Poland and their effects on structures) *Inżynieria i Budownictwo* **61** 3-9 (in Polish)

[2] Rosenblueth E and Meli R 1986 The 1985 earthquake: causes and effects in Mexico City *Concrete International* **8** 23-4

[3] Kasai K, Maison B 1997 Building pounding damage during the 1989 Loma Prieta earthquake *Engineering Structures* **19** 195-207

[4] Jankowski R 2007 Assessment of damage due to earthquake-induced pounding between the main building and the stairway tower *Key Engineering Materials* **347** 339-44

[5] Mahmoud S, Abd-Elhamed A, Jankowski R 2013 Earthquake-induced pounding between equal height multi-storey buildings considering soil-structure interaction *Bulletin of Earthquake Engineering* **11** 1021-48

[6] Anagnostopoulos 1988 Pounding of buildings in series during earthquakes *Earthquake Engineering and Structural Dynamics* **16** 443-56

[7] Maison B and Kasai K 1992 Dynamics of pounding when two building collide *Earthquake Engineering and Structural Dynamics* **21** 771-86

[8] Anagnostopoulos SA, Spiliopoulos KV 1992 An investigation of earthquake induced pounding between adjacent buildings *Earthquake Engineering and Structural Dynamics* **21** pp 289-302

[9] Jankowski R 2005 Impact force spectrum for damage assessment of earthquake-induced structural pounding *Key Engineering Materials* **293-294** 711-8

[10] Karayannis CG, Favvata MJ 2005 Earthquake-induced interaction between adjacent reinforced concrete structures with non-equal heights *Earthquake Engineering and Structural Dynamics* **34** 1-20

[11] Jankowski R 2007 Theoretical and experimental assessment of parameters for the non-linear viscoelastic model of structural pounding *Journal of Theoretical and Applied Mechanics* **45** 931-42
[12] Komodromos P, Polycarpou PC, Papaloizou L, Phocas MC 2007 Response of seismically isolated buildings considering poundings *Earthquake Engineering and Structural Dynamics* **36** 1605-22
[13] Mahmoud S, Jankowski R 2011 Modified linear viscoelastic model of earthquake-induced structural pounding *Iranian Journal of Science and Technology* **35** C1 51-62
[14] Sołtysik B, Jankowski R 2013 Non-linear strain rate analysis of earthquake-induced pounding between steel buildings *International Journal of Earth Sciences and Engineering* **6** 429-33
[15] Wriggers P 2002 *Computational Contact Mechanics* (UK: John Wiley & Sons)

Journal of Physics: Conference Series **628** (2015) 012041 doi:10.1088/1742-6596/628/1/012041

Nonlinear dynamic responses of suspension bridge to sudden breakage of hangers

W L Qiu[1], M Jiang[2] and Z Zhang[1]

[1] School of Civil Engineering, Dalian University of Technology, Lingong Road, Dalian 116024, China

[2] School of Hydraulic Engineering, Dalian University of Technology, Lingong Road, Dalian 116024, China

E-mail: qwl@dlut.edu.cn

Abstract. It is found that the hangers in many suspension bridges in China faced serious corrosion. The resistance capacities of the hangers decrease for corrosion, and the corroded hangers may break suddenly under tension forces. The sudden breakage of a hanger can induce violent vibrations and large changes of internal forces of the whole bridge. The safety of suspension bridges is endangered by corrosion of hangers. Using nonlinear dynamic analysis method and adopting 3D finite element model, the nonlinear dynamic responses of an actual suspension bridge to sudden breakage of a hanger are studied in this paper. The influences of sudden breakage of hanger on the bridge are analyzed in detail. Based on the analysis results, some useful suggestions about design and maintain of hangers are proposed.

1. Introduction

In recent years, many accidents of breakage of cables, such stay cables in cable-stay bridges and hangers in arch bridges, have happened in the world for the reasons of corrosion, fatigue or traffic accident. In China, serious corrosion and breakage of cables occurred in many bridges[1]. On September 7th in 2002, one stay cable of Guangzhou Haiyin Bridge that was only used for 7 years broke, and it hit an oil tank truck. On April 12th in 2011, two hangers of Xinjiang Kongque Bridge broke, and the deck supported by these hangers fell in the river. On November 11th in 2011, eight hangers of Yibin Southgate Bridge in China broke, and the deck fell in the river and there two men died in this accident.

Because the breakage of cable usually occurs suddenly, and it can cause strong vibration and large change of internal forces of the structure, the sudden breakage may endanger the safety of the bridge. Because one hanger is broken, Mahakam II suspension Bridge in Indonesia collapsed in 2011. It happened very fast, only about 30 seconds. This bridge was completed in 2002, and it was less than 10 years old. At least 11 people are dead and 30 people are missing in this accident.

The effects of breakage of stay cables on cable-stayed bridge were studied by many researchers. Considering the influences of different layout of stays, number of planes of stays, and stiffness of deck, Mozos and Aparicio studied the effects of breakage of stays on deck, tower and stays in detail[2][3]. The study showed that sudden breakage of hanger controls the design of deck and tower. Because the time that the breakage occurs (breakage time) has significant effects on the responses of structure, Mozos and Aparicio studied the breakage time through experiments and found that the breakage time was 0.00375s for damaged cables, and 0.0085s for undamaged cables[4].

Considering the geometric and material nonlinearity, Cai et al. studied the nonlinear responses and progressive collapses of cable-stayed bridge due to sudden breakage of stays[5]. When a single stay broke, the tension of other stays changed little. When two adjacent stays broke simultaneously, the bridge collapse did not occur, but the deck had large plastic deformation and the stays yielded.

Ruiz-Teran and Aparicio studied the effects of breakage of stays on under-deck cable-stayed bridge[6]. The study results showed that the responses produced by breakage of stays must be calculated using dynamic analysis method.

Using nonlinear dynamic analysis methods and considering the corrosion of hangers, the responses of a suspension bridge with main span of 200m due to the breakage of hangers are studied in this paper.

2. Structure of the suspension bridge and analysis method

2.1. Structure of the suspension bridge

A concrete self-anchored suspension bridge with main span of 200m and side span of 70m, as shown in figure 1, is used to study the structural responses caused by sudden breakage of hanger. Its stiffening girder is reinforced concrete box girder, as shown in figure 2. The tower is reinforced concrete with box section. The bridge has two main cables and each cable is made of 3937 paralleled, 5mm in diameter, high strength galvanized steel wires. There are 65 hangers on each side of the bridge, and the distance of the hangers along the girder is 5m. The hangers are made of 97 paralleled, 7mm in diameter, steel wires with strength of 1.67×10^6 kPa. The hangers are numbered from #1 to #65.

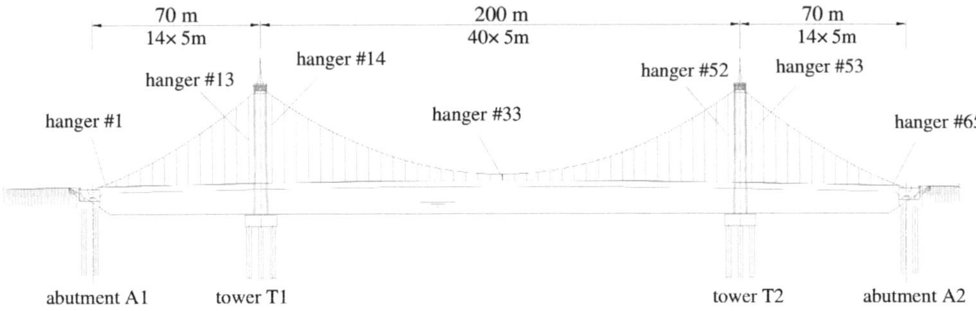

Figure 1. Layout of the Bridge.

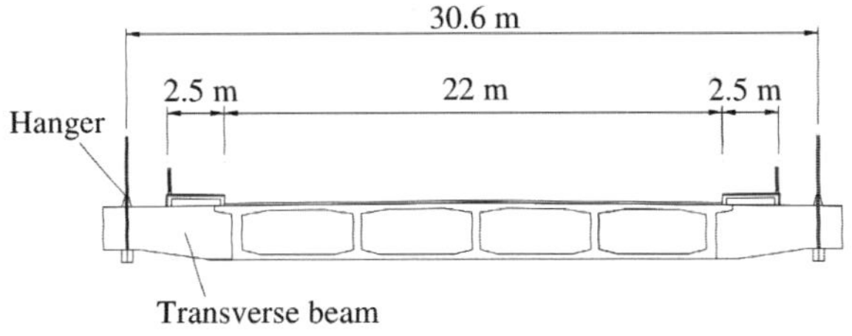

Figure 2. Cross section of girder.

2.2. Analysis model

To modulate the nonlinear vibration process caused by the sudden breakage of hanger, the software ABAQUS V.6.8 is used to establish a three dimension (3D) finite element (FE) model of the bridge, as shown in figure 3. In the FE model, the girder, towers and transverse beams are modelled using 3D beam elements, and their torsion stiffness and torsion mass are considered. The main cables are also modelled using beam elements with very small flexible stiffness, and contribution of the tension to the flexible stiffness of the cables is considered. The hangers are modelled using truss elements. The main cables between two hangers are modelled using 5 elements to consider geometry and vibration of the main cables more precisely. The masses of clamps are concentrated on the nodes where the clamps are located. Considering the piles of the bridge had little effects on the analysis of sudden breakage of hanger, they are not modelled in the FE model.

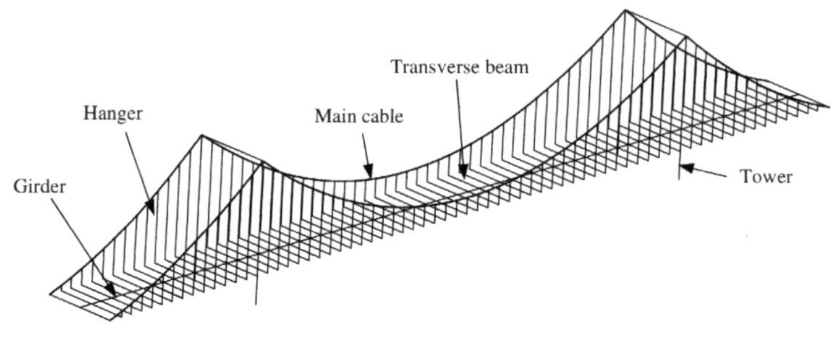

Figure 3. FE model of the bridge.

2.3. Nonlinear behavior of corroded hanger

To modulate the effects of corrosion of steel wires on strength of hanger, the model of corroded hanger are established as shown in figure 4. The cross section area of the hanger is A, and the length of the hanger is L. The steel wires of the hanger are divided into two parts named as part A and part B. The wires in part A are not corroded, but the wires in part B are corroded. The cross section areas of the two parts are $(1\text{-}\alpha)A$ and αA respectively. To simplify the analysis, it is assumed that the corrosions of every wire in part B are same. The length of corroded part (part B-2) of every wire is L_c, and the rest area of corroded part of every wire is β times of the original area. To consider the material nonlinearity of hanger, the nonlinear relationship between stress and strain of steel wire is adopted in the analysis, as shown in figure 5. The steel wires break when their strains reach 0.04.

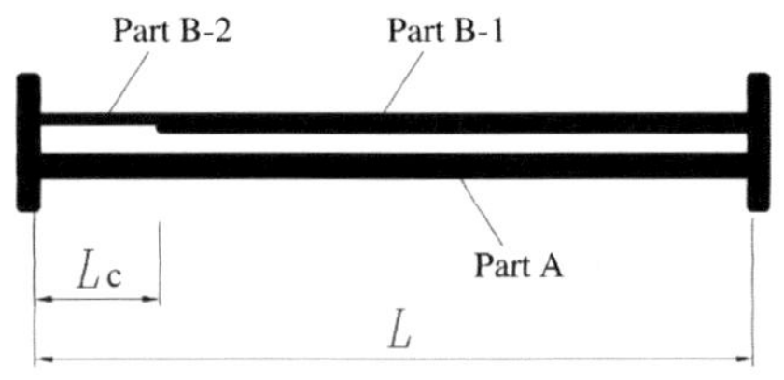

Figure 4. Model of corroded hanger.

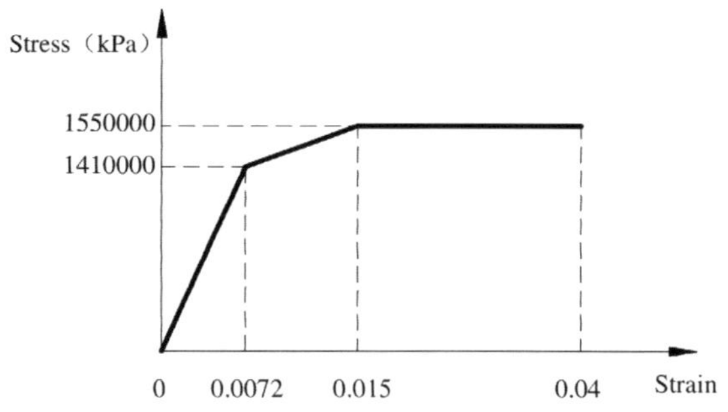

Figure 5. Relationship between stress and strain of steel wire.

2.4. Analysis method

Responses of the bridge due to sudden breakage of hangers on the bridge are analyzed by means of nonlinear dynamic analysis using the finite element software ABAQUS V.6.8. In dynamic analysis, the direct time integration method is used. First, the static state of the bridge under design load is determined, then the broken hanger is removed suddenly and the tension of the broken hanger is unloaded in breakage time $\triangle t$. The breakage time $\triangle t$ takes a value of 0.005s. A Rayleigh damping of 2% is used in the dynamic analysis.

3. Analysis results

In order to study the responses of the bridge caused by breakage of hanger, breakage of the left hanger #23 at the position of 1/4 of mid-span is taken as an example. The study includes nonlinear analysis and linear analysis. The nonlinear analysis considers both geometric nonlinearity and nonlinear behavior of corroded hanger, but the linear analysis considers only geometric nonlinearity. In the nonlinear analysis, for the hangers except the broken hanger, the values of α, β and L_c are taken as 0.5, 0.7 and 0.1m. The stresses of the hangers are about 5.56×10^5kPa under design loads before hanger #23 breaks in the two analysis cases. In the dynamic analysis, the hanger #23 broke at the time of 25s.

3.1. Responses of hangers

From the analysis results, it is can be seen that the breakage of a hanger has large effects on stresses of the hangers near the broken hanger, and has little effects on stresses of the hangers far away from the broken hanger. When the corrosion is considered, the stress of part B-2 in hanger #22 is 7.94×10^5 kPa before hanger #23 breaks, which is larger than the stresses of part B-1 and part A, as shown in figure 6. After hanger #23 breaks, the stress of part B-2 in hanger #22 reaches 1.55×10^6 kPa rapidly, and the steel wires in part B-2 break when their strains reach 0.04. The steel wires in part B withdraw from the structure, and the steel wires in part A bear the tension of hanger #22 alone. In the vibration, the stresses of steel wires in part A exceed the yield stress of 1.41×10^6 kPa, but they are less than the ultimate stress of 1.55×10^6 kPa.

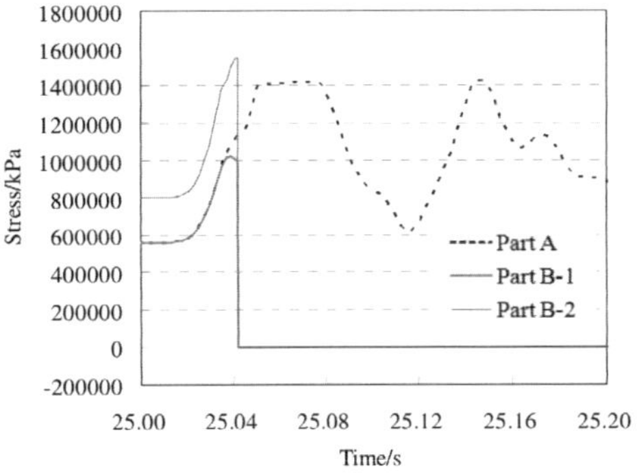

Figure 6. Stresses of the 3 parts in hanger #22.

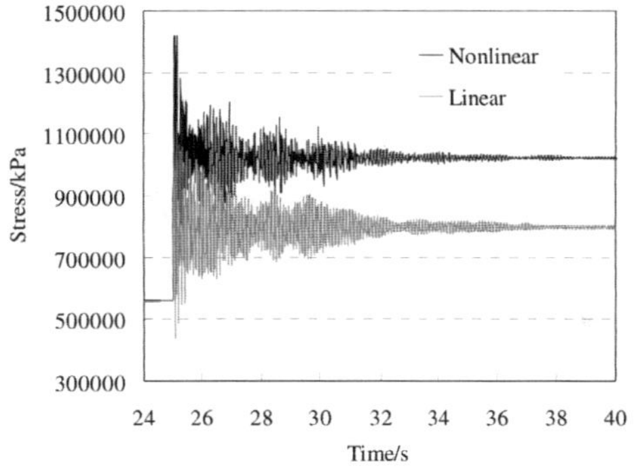

Figure 7. Stresses of part A in hanger #22.

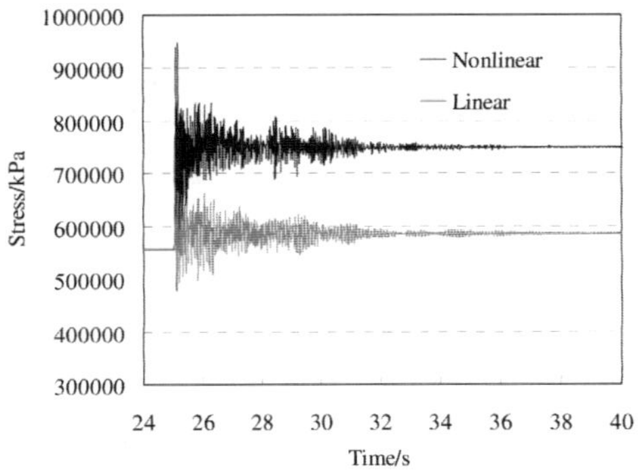

Figure 8. Stresses of part A in hanger #21

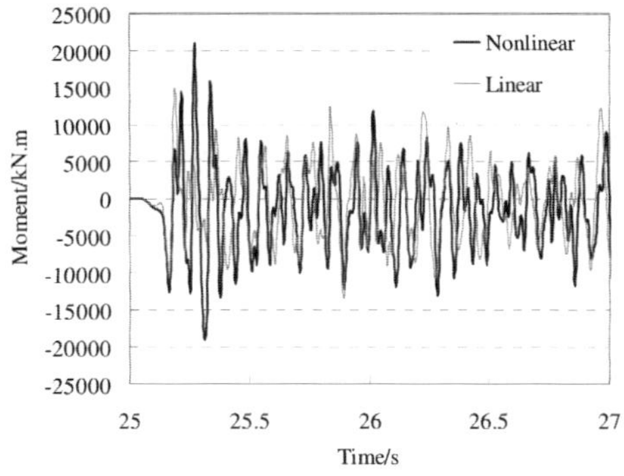

Figure 9. Torsion moments of main girder near tower T1.

Figure 7 and figure 8 show the time histories of stresses of part A in hangers #21 and #22. It can be seen from the figures that stresses of hangers #21 and #22 change very largely after breakage of hanger #23, and the stresses of hanger #22 vary more markedly than those of hanger #21. When the corrosion is considered, the maximum stresses of hanger #22 and #21 are 1.42×10^{6} kPa and 9.48×10^{5} kPa respectively. When the corrosion is not considered, the maximum stresses of hanger #22 and #21 are 1.11×10^{6} kPa and 7.27×10^{5} kPa respectively. So, the corrosion has very large effects on the stresses of the hangers.

3.2. Responses of main girder

Figure 9, figure 10 and figure11 show the torsion moment of main girder near tower T1, vertical bending moments of main girder near hanger #23 and vertical displacement of main girder near hanger #23 caused by breakage of hanger #23. When the corrosion is considered, the maximum torsion moment of main girder is 21014kN.m, the maximum vertical bending moment of main girder is 1365kN.m, and the maximum vertical displacement of main girder is 0.011m. When the corrosion is not considered, the maximum torsion moment, maximum vertical bending moment and the maximum vertical displacement of main girder are 15038kN.m, 2282kN.m and 0.006m respectively. So, the torsion moments of main girder caused breakage of a hanger are very large, and the vertical bending moment and displacement of main girder are small relatively. The maximum torsion moment and displacement become larger, and vertical bending moment become less when the material nonlinearity of hanger is considered in the dynamic analysis.

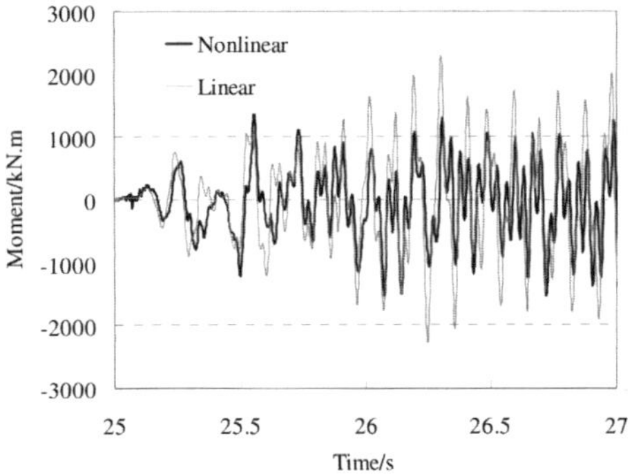

Figure 10. Vertical bending moments of main girder near hanger #23.

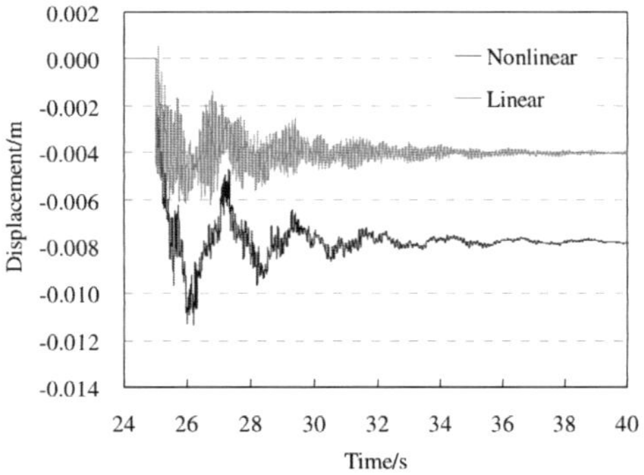

Figure 11. Vertical displacement of main girder near hanger #23.

3.3. Responses of main cable

Figure 12 shows the time histories of stresses of main cable after breakage of hanger #23. The maximum stress of main cable is only larger than the initial value by 1.13×10^{4} kPa, and the results obtained by the nonlinear and linear analysis are same nearly. Figure 14 and figure 15 show the time histories of displacement of main cable at the positions of the hangers #22 and #23. The displacements of main cable obtained by nonlinear analysis are very different from those obtained by linear analysis. Especially, when the vibration ceases, the displacement of main cable at the position hanger #22 obtained by nonlinear analysis is 0.072m, which is much larger than the value of 0.009m obtained by linear analysis. This is because hangers #22 and #24 yields in the nonlinear analysis. The maximum displacement of main cable at the position of hanger #23 is 0.167m, which shows that the amplitude of the vibration is very large. So, the breakage of a hanger causes little change of stress and large displacement of main cable.

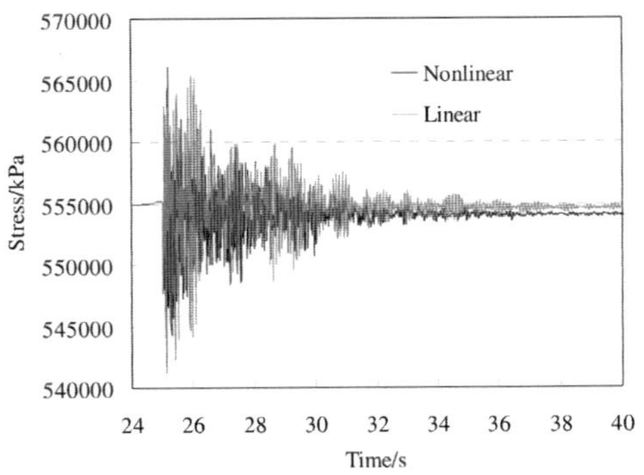

Figure 12. Stresses of main cable near tower T1.

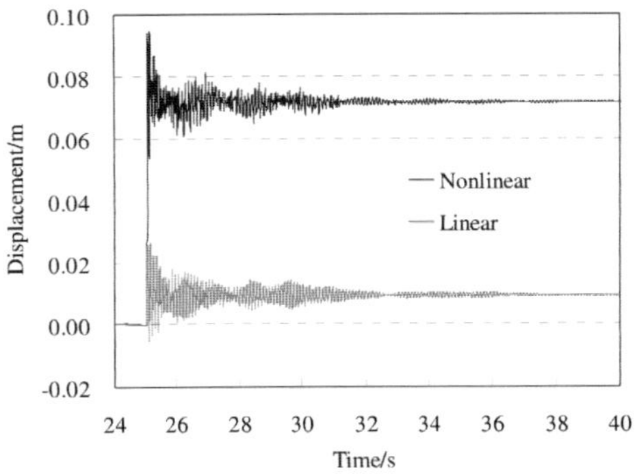

Figure 13. Vertical displacement of main cable at the position of hanger #22.

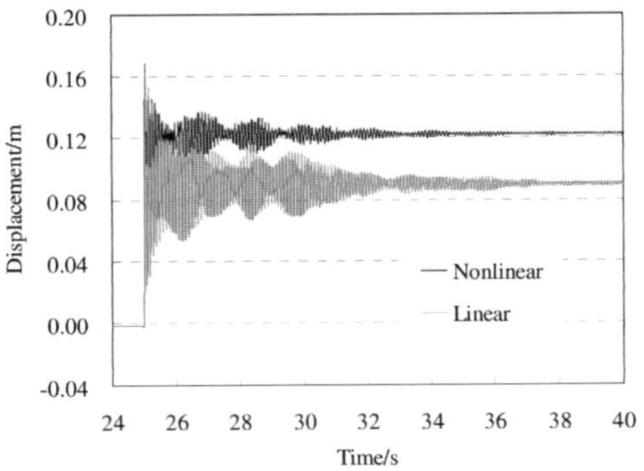

Figure 14. Vertical displacement of main cable at the position of hanger #23.

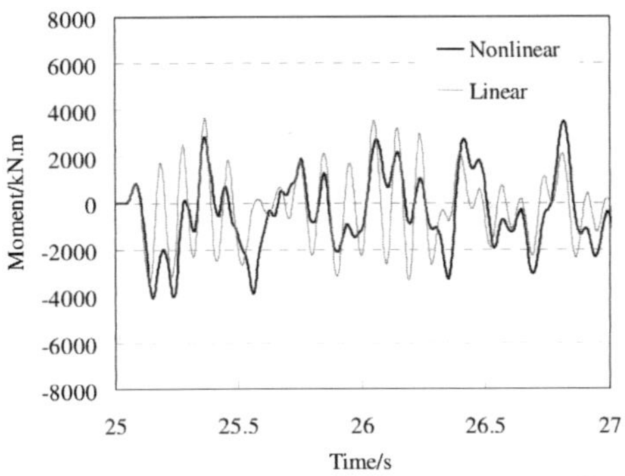

Figure 15. Moments of the bottom of tower T1.

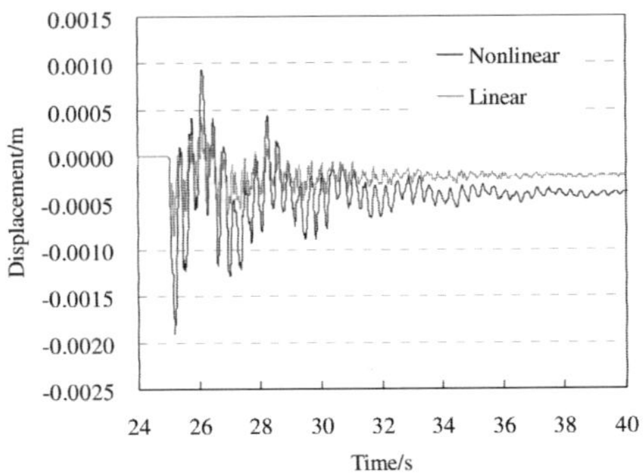

Figure 16. Horizontal displacement of the top of tower T1.

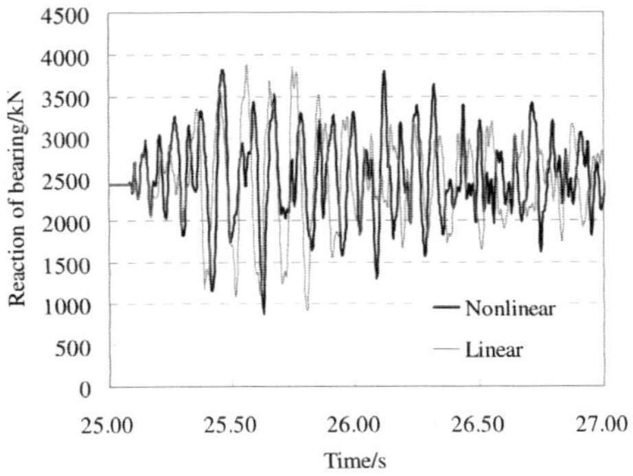

Figure 17. Reaction forces of a bearing at tower T1.

3.4. Responses of tower

Figure 15 shows the time histories of moment of the bottom of tower T1 after breakage of hanger #23. The maximum moment caused by breakage of hanger is 4041kN.m. Figure 16 shows the time history of displacement of the top of tower T1 after breakage of hanger #23. The maximum horizontal displacement caused by breakage of hanger is only 0.0019m. The moments and displacement obtained by the nonlinear are larger than those obtained by linear analysis, but the values are all very small.

3.5. Responses of bearing

Figure 17 shows the time histories of reaction forces of one bearing at tower T1 after breakage of hangers #23. The maximum compression force subjected by the bearings is 3875kN, which is 1.59 times of the initial value. The maximum compression forces obtained by the nonlinear and linear analysis are same nearly.

4. Conclusions

The following conclusions can be obtained through analyzing the responses of a suspension bridge with corroded hanger caused by sudden breakage of a hanger.

(1) The breakage of a hanger causes very large stresses of the hangers near to the broken hanger. If the adjacent hangers are corroded, the corroded part of steel wires in the hangers may break, and the part of steel wires without corrosion may yield during the vibration caused by the breakage of hanger.

(2) Through comparison of the results obtained by nonlinear analysis and linear analysis, when corrosion of hangers are considered, the changes of hanger stresses, torsion moments of main girder, displacement of main cable and displacement of tower top are much larger.

(3) The corrosion of hangers has marked effects on the responses of the bridge caused by breakage of hanger, so the study on the effects of hanger breakage should consider the corrosion of hanger. The corroded hanger should be replaced in time to avoid serious damage of the bridge.

Acknowledgments

This study is funded by the National Natural Science Foundation of China (NSFC-51178080).

References

[1] Wang L L and Yi W J 2007 Cases analysis on cable corrosion of cable-stayed bridges *Central South Highway Engineering* **32** 94-32

[2] Mozos C M and Aparicio A C 2010 Parametric study on the dynamic response of cable stayed bridges to the sudden failure of a stay, Part I: Bending moment acting on the deck *Engineering Structures* **32** 3288-3300

[3] Mozos C M and Aparicio A C 2010 Parametric study on the dynamic response of cable stayed bridges to the sudden failure of a stay, Part II: Bending moment acting on the pylons and stress on the stays *Engineering Structures* **32** 3301-12

[4] Mozos C M and Aparicio A C 2011 Numerical and experimental study on the interaction cable structure during the failure of a stay in a cable stayed bridge *Engineering Structures* **33** 1330-41

[5] Cai J G, Xu YX Zhuang L P, Peng J and Zhang J 2012 Comparison of various procedures for progressive collapse analysis of cable-stayed bridges *Zhejiang Univ-Sci A (Appl Phys & Eng)* **13** 323-334

[6] Ruiz-Teran A M and Aparicio A C 2009 Response of under-deck cable-stayed bridges to the accidental breakage of stay cables *Engineering Structures* **31** 1425-34

Cementitious Composites Engineered with Embedded Carbon Nanotube Thin Films for Enhanced Sensing Performance

Kenneth J. Loh* **and Jesus Gonzalez**
Department of Civil & Environmental Engineering, University of California, Davis
One Shields Avenue, Davis, CA 95616-5294, USA

E-mail: kjloh@ucdavis.edu

Abstract. Cementitious composites such as concrete pavements are susceptible to different damage modes, which are primarily caused by repeated loading and long-term deterioration. There is even greater concern that damage could worsen and occur more frequently with the use of heavier vehicles or new aircraft carrying greater payloads. Thus, the objective of this research is to engineer cementitious composites with capabilities of self-sensing or detecting damage. The approach was to enhance the damage sensitivity of cementitious composites by incorporating multi-walled carbon nanotubes (MWNT) as part of the mix design and during casting. However, as opposed to directly dispersing MWNTs in the cement matrix, which is the current state-of-art, MWNT-based thin films were airbrushed and coated onto sand particles. The film-coated sand was then used as part of the mix design for casting mortar specimens. Mortar specimens were subjected to compressive cyclic loading tests while their electrical properties were recorded simultaneously. The results showed that the electrical properties of these cementitious composites designed with film-coated sand exhibited extremely high strain sensitivities. The electrical response was also stable and consistent between specimens.

1. Introduction

Multifunctional materials have gained considerable attention over the last few decades. In particular, a material, whether it is homogeneous or a composite, is multifunctional when it has been specifically encoded with unique properties that allow it to perform multiple engineering functionalities [1]. Such multifunctional materials are particularly desirable for structural applications. For instance, while infrastructure materials (*e.g.*, steel, concrete, and fiber-reinforced polymer (FRP) composites) are designed to carry loads, these structures can also experience damage during their operational service lifetime. As an example, FRPs are known to suffer from matrix cracking, debonding, and delamination that are not only invisible to the naked eye but can also propagate to cause catastrophic structural failure. Thus, multifunctional FRP composites that can bear loads and sense damage are currently being developed [2, 3].

Another even more widely used infrastructure material is cementitious composites, in which they are often employed for civil infrastructure applications such as for bridges, buildings, foundations, and pavements. While the use of concrete as load-bearing materials can be traced back several centuries ago, engineering them to be multifunctional occurred more recently. In fact, in the early 1990s, Chen and Chung [4] showed that the electrical properties of cementitious composites could be engineered to be sensitive to damage by adding electrically conductive, short, carbon fibers to mortar during casting.

Because the electrical resistivity (or conductivity) of the composite was calibrated and varied in response to applied stress/strain or damage, one could use the cementitious composite as both a load-bearing structure and a sensor.

A variety of research aimed at developing *self-sensing* or *smart* concrete was investigated over the past few decades. The conductive additives embedded in the cementitious matrix ranged from using carbon fibers (as mentioned earlier) [5], steel fibers [6], carbon black [7], to, more recently, carbon nanotubes (CNT) [8-10]. While different advantages and unique attributes were reported for each of these cement composites enhanced with different types of conductive additives, their general approach to designing and casting remained the same. The procedure was to disperse the conductive fibers during mixing (*i.e.*, either by stirring the fibers with other components in a dry state or by pre-dispersing the additives in a solution that was then added to the mix later). For both scenarios, fiber agglomeration, material uniformity, and consistency could not be easily controlled or guaranteed. In addition, even if one were able to disperse 0.1% (by weight of cement) of multi-walled carbon nanotubes (MWNT) in cement (as was used by Han *et al.* [8]), the increase in raw material costs would be significant (*i.e.*, an order of magnitude or greater), as compared to the plain concrete.

Thus, the goal of this research is to demonstrate a new paradigm of nanomaterial-enhanced smart concrete. Instead of undertaking the challenging task of dispersing low concentrations of CNTs in the cement matrix, nanotubes were successfully dispersed in polymer thin films in previous studies [11, 3]. These nanocomposites exhibit high electrical conductivity, are ultra-thin (*i.e.*, on the order of 1 μm), and efficiently uses the smallest quantities of CNTs possible. In that regard, MWNT-films were deposited and coated onto sand particles, prior to using them for casting mortar specimens. The rationale was to modify the cement-particle interface to create a more effective electrically conductive network within the cement matrix. With this technique, MWNTs would remain well dispersed within the thin film, and yet, the films coated on sand could still make contact with other films/particles to enhance the bulk cementitious composite's electromechanical (or sensing) response. To the authors' best knowledge, this approach of designing self-sensing concrete using embedded particles pre-coated with nanocomposites has not been reported before. This paper begins with a presentation of the mix design, casting procedures, and testing protocol. Then, the nanocomposite-enhanced smart concrete's sensing results are presented, discussed, and compared to other results reported in the literature.

2. Experimental Details

The aim of this work was to establish a new, effective, and low-cost method for incorporating nano-materials (*i.e.*, MWNTs) in mortar specimens so that they could be used as multifunctional materials for load bearing and self-sensing. First, section 2.1 begins with a discussion of the mortar mix design and casting procedure. Second, since the mortar mix employed sand pre-coated with an MWNT-based thin film, section 2.2 describes the film preparation procedure. Finally, a description of the cyclic compressive tests and the data acquisition strategy are discussed in section 2.3.

2.1. Mix Design and Casting Procedure

Mortar specimens were cast following U.S. Federal Aviation Administration (FAA) rigid concrete pavement specifications [12], as well as those provided by ASTM C109 for strength testing [13]. Here, Type I/II Portland cement mixed with 25 *wt.%* (of total cementitious material) Type F ground granulated blast furnace slag (GGBFS) was used. A 2.75:1 sand-to-cement ratio was employed, as stated in ASTM C109 [13]. The sand was crushed granite with a density of 2.71 g/cm^3 and a minimum particle size of 0.149 mm. The water-to-cement (w/c) ratio was 0.52, and 2.9 mL/kg of cement of Advacast superplasticizer was also incorporated as a water-reducing agent.

(a) (b)

Figure 1. Pictures of the (a) pristine and (b) MWNT-based nanocomposite-coated sand are shown.

The procedure for casting the mortar specimens began with mixing all the various constituents using a Hobart N-50 mixer. First, the water, superplasticizer, and cementitious materials were placed in a stainless steel mixing bowl, and the contents were mixed at low speed (60 rpm) for 30 s. Second, the nanocomposite-coated sand (as will be discussed later in section 2.2) was gradually poured into the bowl over the course of 30 s and while the mixer was operating at 60 rpm. Upon doing so, the contents were mixed at medium speed (124 rpm) for 1 min to obtain an even mixture. Third, after mixing was complete, any materials adhered to the walls of the bowl were scrapped off, and the entire mixture was set aside for ~15 s prior to pouring them into the molds for casting.

The molds that were used in this study were 5×5×5 cm^3 (2×2×2 in^3) cubes. The molds were lubricated using WD-40, and the seams were sealed with petrolatum. Since the objective of this study was to measure how the electrical properties of these specimens varied in response to applied strains and stresses, electrodes were also incorporated during specimen casting. Here, 4.5×6.5 cm^2 (1.75×2.25 in^2) copper mesh electrodes were cut and placed close to two opposite ends of the mold. They were inserted such that the mesh electrodes covered the entire specimen's cross-section with a portion of it protruding out for electrical measurement purposes during load testing. Then, with the electrodes in place and the molds ready, the mortar mixture was scooped into each mold until it was half-full. The mold was gently vibrated for 10 s before it was filled completely and then vibrated for another 10 s. Next, the filled molds were sealed with plastic wrap and left for 24 h in a sealed container at room temperature. Finally, the specimens were de-molded, submerged in a limewater solution, and cured for 28 days prior to testing. A total of five mortar specimens were casted as part of this study.

2.2. Nanocomposite Fabrication and Sand Coating

As mentioned earlier, the sand used for mortar casting was pre-coated with an MWNT-based thin film. Film deposition onto the sand was accomplished via airbrushing and by following procedures developed in a previous study reported in Mortensen *et al.* [11]. In general, MWNTs (SouthWest NanoTechnologies) were dispersed via sonication in a 2 *wt.%* poly(sodium 4-styrenesulfonate) (PSS) solution. N-methyl-2-pyrrolidinone (NMP) was added to the MWNT-PSS solution and then mixed with a latex solution. This procedure yielded MWNT-based inks that could then be spray coated onto virtually any surface using a Paasche airbrush [3].

Prior to spray coating the sand (figure 1a), sand was dried in a vacuum oven for 24 h. The gradation of sand used was reported in Wong *et al.* [9]. Upon drying, sand was poured onto a flat surface and placed in a fume hood. It should be mentioned that the sand was spread to form a thin layer no thicker

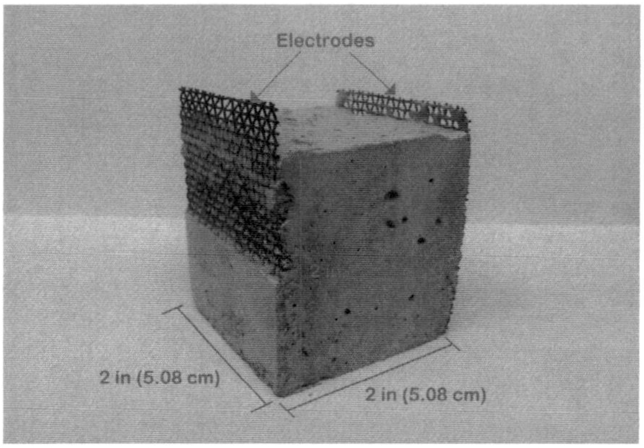

Figure 2. A dried mortar specimen that incorporated nanocomposite-coated sand is shown. The specimen also had two copper mesh electrodes embedded on its two opposite ends so that its electrical resistance could be measured during load testing.

than ~6 mm (~0.25 in). Then, using the MWNT-based ink solution and the airbrush, the thin film was sprayed onto the surface of the sand by using a sweeping motion while holding the spray nozzle ~30 cm away from the target. Once one coat was complete, the sand was then mixed in place using a plastic spatula so as to reveal any underlying sand that had not been coated with the MWNT-based ink. This procedure was repeated for three times. Upon doing so, the sand was left to air-dry for 1 h, followed by additional drying in a vacuum oven set to 50 °C for 30 min. A picture of the coated sand is shown in figure 1b; note that the color of the sand became dark-gray after film deposition.

2.3. Electromechanical Characterization
The 5×5×5 cm^3 (2×2×2 in^3) mortar cubes were subjected to compressive cyclic load tests so that their electromechanical (sensing) response could be characterized. An MTS load frame equipped with a 407 controller was employed for load testing. Each load test was conducted using the following procedure. Upon mounting the specimen in the load frame, a 5,000 N preload was applied. Then, the load frame was commanded to execute a 0.1 Hz cyclic-compressive load pattern (*i.e.*, five to 10 cycles) to a peak strain of ~0.45% during the loading ramp, followed by unloading until the preload was released. The tests were displacement-controlled. Applied load and displacement were sampled at 55 Hz and were recorded using a customized *LabVIEW* program. In addition to measuring the specimen's load-displacement response, an Agilent 34401A digital multimeter (DMM) also recorded the electrical resistance of each specimen (sampling rate: 2.8 Hz). Resistance data was also recorded and time-synchronized using the same *LabVIEW* client. It should be mentioned that the Agilent DMM was controlled to record ~20 s of baseline resistance prior to the start of each load test. This was conducted to minimize any electrical polarization (*i.e.*, resistance drift) inherent to these cementitious composites.

However, immediately prior to test, each mortar specimen was removed from the limewater solution after 28 days of curing. The saturated specimen was pat dried and then dried in a vacuum oven operating at 50 °C for 1 h. Figure 2 shows a picture of a pristine mortar specimen. Then, the specimen was subjected to three consecutive multi-cycle compressive tests. The time for testing was limited to be between 2 to 3 min for minimizing specimen drying during each test. Then, the specimen was placed back in the vacuum oven for an additional 5 h of drying so that the total drying time was 6 h. Upon drying, the specimen was subjected to three more consecutive cyclic compressive tests. This test procedure was repeated for all of the five different mortar specimens casted for this study.

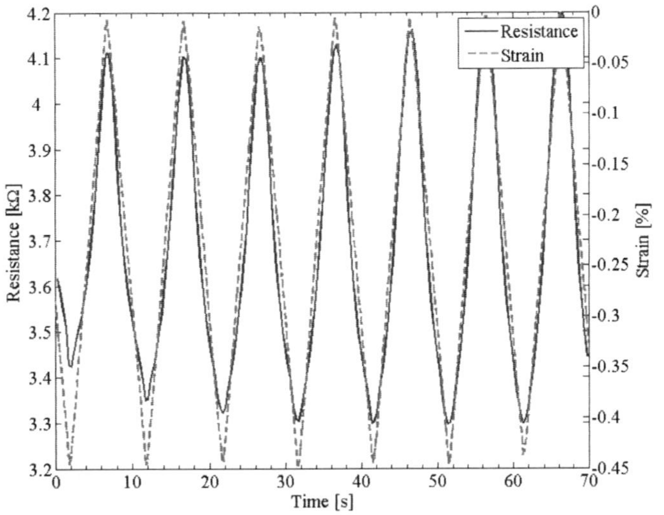

Figure 3. A mortar specimen subjected to 1 h of vacuum oven drying was subjected to compressive cyclic loading. Its resistance time history was overlaid with the applied strain pattern.

3. Results and Discussions

As mentioned in section 2.3, each of the five mortar specimens was subjected to a total of six multi-cycle compressive load tests. The first three test results corresponded to the case when the mortar was dried for 1 h. These tests considered the response of the specimens in a more saturated or damp state. In contrast, the latter three tests investigated the electromechanical response of mortar specimens dried for a total time of 6 h, in which the specimens were presumed to be in a much drier state.

Figure 3 shows a representative result of a nanocomposite-enhanced mortar specimen subjected to seven cycles of compressive loads to ~0.45%. The plot in figure 3 includes the measured resistance time history overlaid on top of the applied cyclic compressive strain pattern. From this result, it is clear that the mortar specimen's electrical resistance decreased in tandem with increasing compressive strains. Its response was linear, as can be observed from the overlay in figure 3. In addition, its nominal resistance did not appear to drift or vary with time significantly. Overall, the response was repeatable and stable.

On the other hand, figure 4 shows two representative results for a mortar specimen that was dried for 6 h in the vacuum oven. Similar trends can be observed when comparing figures 3 and 4, such that in both cases, resistance decreased with greater compressive strains applied. The main difference for these specimens is that specimens dried for 6 h exhibited more prominent nominal resistance drifts, that fortunately and typically only occurred during the first of three consecutive load tests (figure 4a). Afterward, its electromechanical response varied closely with that of the applied strain (figure 4b). It should be mentioned that such nominal resistance drifts are expected, since electrical polarization in cementitious composites had been documented in the literature [14].

In order to quantify the nanocomposite-enhanced mortar specimen's sensitivity to strains or, equivalently, to calculate its gage factor, the raw data from figure 4b is used to plot figure 5. Figure 5 shows the normalized change in resistance of the specimen plotted as a function of the applied strains; note that the *x*-axis shows negative strains because the applied loading was in compression (similar to figures 3 and 4). In general, strain sensitivity (S) or gage factor can be calculated using (1):

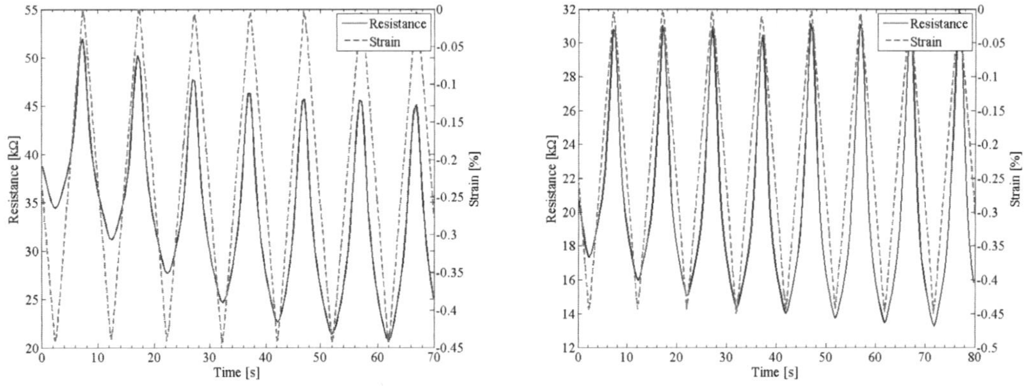

(a) (b)

Figure 4. Nanocomposite-enhanced mortar specimens dried for 6 h in a vacuum oven were subjected to cyclic compressive tests. (a) While the specimen's resistance exhibited sensitivity to applied strains, nominal resistance drift was observed. (b) However, the nominal resistance drift plateaued after a short period of time to yield stable electromechanical response.

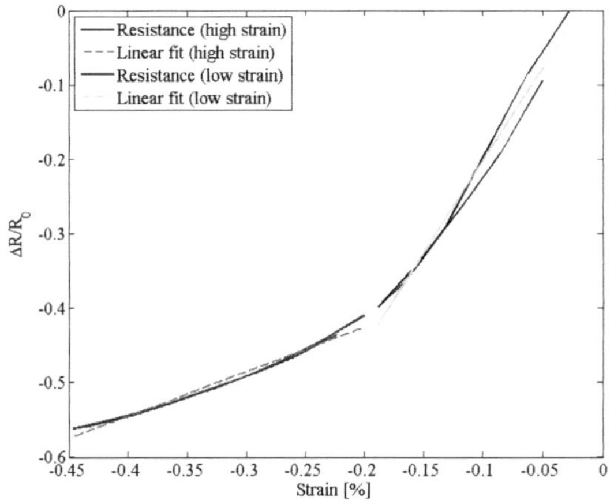

Figure 5. The normalized change in resistance of the raw data plotted in figure 4b is plotted with respect to the applied strain.

$$S = \frac{\Delta R / R_0}{\varepsilon} \tag{1}$$

where ε is strain, R_0 is the mortar's nominal or unstrained resistance, and ΔR is the instantaneous change in resistance of the mortar relative to its nominal resistance. Using figure 5, the specimen's strain sensitivity can be obtained from the slope of the linear least-squares regression line fitted to the data. Upon doing so, the average strain sensitivity computed from all the tests and specimens was found to be ~118. For reference, the typical gage factor for a strain gage is 2. As compared to another previous study in which 0.025 *wt.%* of MWNTs were directly dispersed in the cement matrix of mortar specimens, the gage factor obtained ranged between 12.2 and 13.5 [9].

Nevertheless, while an average gage factor was obtained and reported here, closer examination of figure 5 shows that these specimens exhibited bilinear sensitivity to applied compressive strains. At low strains (*i.e.*, up to -0.2%), the sensitivity was extremely high; S is 248 for the data shown in figure 5. When compressive strain exceeded -0.2%, another linear least-squares best-fit line was fitted to the 'high strain' region. The corresponding gage factor obtained for the dataset shown in figure 5 was 60.

Given that the mortar specimens were subjected to a peak strain of -0.45%, it was unlikely that these results were due to non-linear response or damage occurring in the specimens, although additional tests need to be conducted to verify this assumption. The current working hypothesis is that bilinear strain sensitivity occurred due to a change in the percolated network formed by the MWNT-based thin film-coated sand particles. It is possible that compression up to -0.25% strains allowed thin film-coated sand to begin to come into contact with one another, thereby facilitating significantly better electrical current transport with each additional direct contact formed. After this point, additional compression would yield less significant changes in particle contacts, thereby reducing the relative change in resistance to result in lower sensitivities. Previous studies by Mortensen *et al.* [11] and Loyola *et al.* [3] revealed that these MWNT-latex films exhibited piezoresistivity. Another possibility is that the inherent piezoresistivity of these thin films contributed to the bilinear strain sensitivity observed in figure 5. Future studies will focus on an in-depth characterization of how these thin films affected the cement-sand interface from not only an electrical perspective but also on the mechanical performance of the bulk cementitious composite.

4. Conclusions

In this study, a novel technique for incorporating carbon nanotube fillers for designing multifunctional cementitious composites was presented. Instead of directly dispersing nanotubes in the cement matrix, the approach was to pre-coat sand with conductive, multi-walled carbon nanotube-latex thin films. The nanocomposite-coated sand was then incorporated during mixing and casting of mortar specimens. Upon casting 5×5×5 cm^3 (2×2×2 in^3) nanocomposite-enhanced mortar cube specimens, each specimen was subjected to multiple cycles of compressive load tests while their electrical resistance was measured. Specimens were tested after two different drying times, namely after 1 h and 6 h of total drying at 50 °C in a vacuum oven.

Both sets of results showed that the electrical properties of these cementitious composites exhibited sensitivity to applied compressive strains. Specifically, its electrical resistance decreased in tandem with increasingly applied compressive strains. For specimens that were dried for 6 h, a slight nominal resistance drift was observed, but this effect quickly plateaued to produce stable electromechanical or sensing response. While the average gage factor or strain sensitivity was found to be 118, closer examination of the results showed that bilinear strain sensitivity was observed. At low strains, the sensitivity was higher; one specimen yielded a gage factor of 248 up to -0.2% strains. Subsequent application of compressive loading caused an immediate decrease in gage factor to 60. Future work will focus on characterizing the effects of the films on the cement-sand interface. This casting technique will also be scaled up for fabricating larger specimens, as well as for casting concrete specimens that incorporate large aggregates.

5. Acknowledgements

The U.S. Federal Aviation Administration (FAA) supported this research under cooperative agreement no. 13-G-017. The authors also thank Prof. John Bolander, Mr. Brandon Dashwood, and Ms. Jennifer Yasui for their assistance with equipment training/use, specimen casting, and testing. Dr. Rongzong Wu and Dr. Navneet Garg are also acknowledged for their collaboration and support for this project.

6. References

[1] Momoda L A 2004 The future of engineering materials: Multifunction for performance-tailored structures *The Bridge* **34** 18-21

[2] Boger L, Wichmann M H G, Meyer L O and Schulte K 2008 Load and health monitoring in glass fibre reinforced composites with an electrically conductive nanocomposite epoxy matrix *Composites Science and Technology* **68** 1886-1894

[3] Loyola B R, Briggs T M, Arronche L, Loh K J, Saponara V L, O'Bryan G and Skinner J L 2013 Detection of spatially distributed damage in fiber-reinforced polymer composites *Structural Health Monitoring* **12** 225-240

[4] Chen P-W and Chung D D L 1993 Carbon fiber reinforced concrete for smart structures capable of non-destructive flaw detection *Smart Materials and Structures* **2** 22-30

[5] Fu X and Chung D D L 1996 Self-monitoring of fatigue damage in carbon fiber reinforced cement *Cement and Concrete Research* **26** 15-20

[6] Teomete E and Kocyigit O I 2013 Tensile strain sensitivity of steel fiber reinforced cement matrix composites tested by split tensile test *Construction and Building Materials* **47** 962-968

[7] Ou J and Han B 2009 Piezoresistive cement-based strain sensors and self-sensing concrete components *Journal of Intelligent Material Systems and Structures* **20** 329-336

[8] Han B, Yu X and Kwon E 2009 A self-sensing carbon nanotube/cement composite for traffic monitoring *Nanotechnology* **20** 1-5

[9] Wong I, Loh K J, Wu R and Garg N 2014 Sensing properties of mortar incorporating ultra-low concentrations of carbon-based conductive additives. In: *Proceedings of the 5th Asia-Pacific Workshop on Structural Health Monitoring,* (Shenzhen, China pp 1-10

[10] Wille K and Loh K J 2010 Nano-engineering ultra-high performance concrete with multi-walled carbon nanotubes *Journal of the Transportation Research Board* **2142** 119-126

[11] Mortensen L P, Ryu D, Zhao Y and Loh K J 2013 Rapid assembly of multifunctional thin film sensors for wind turbine blade monitoring *Key Engineering Materials* **569-570** 515-522

[12] O'Donnell M J 2011 Standards for specifying construction of airports. In: *Advisory Circular*: U.S. Department of Transportation, Federal Aviation Administration)

[13] ASTM standard c 109/c 109m - 02. In: *Standard Test Method for Compressive Strength of Hydraulic Concrete Mortars (Using 2-in. (or 50-mm) Cube Specimens),*

[14] Wen S and Chung D D L 2001 Electric polarization in carbon fiber-reinforced cement *Cement and Concrete Research* **31** 141-147

Nucleation of squat cracks in rail, calculation of crack initiation angles in three dimensions

Meysam Naeimi[1], Zili Li[1] and Rolf Dollevoet[1]
[1] Section of Railway Engineering, Faculty of Civil Engineering and Geoscience, Delft University of Technology, Delft, the Netherlands

E-mail: m.naeimi@tudelft.nl

Abstract. A numerical model of wheel-track system is developed for nucleation of squat-type fatigue cracks in rail material. The model is used for estimating the angles of squat cracks in three dimensions. Contact mechanics and multi-axial fatigue analysis are combined to study the crack initiation mechanism in rails. Nonlinear material properties, actual wheel-rail geometries and realistic loading conditions are considered in the modelling process. Using a 3D explicit finite element analysis the transient rolling contact behaviour of wheel on rail is simulated. Employing the critical plane concept, the material points with the largest possibility of crack initiation are determined; based on which, the 3D orientations/angles of the possible squat cracks are estimated. Numerical estimations are compared with sample results of experimental observations on a rail specimen with squat from the site. The findings suggest a proper agreement between results of modelling and experiment. It is observed that squat cracks initiate at an in-plane angle around 13°-22° relative to the rail surface. The initiation angle seen on surface plane is calculated around 29°-48°, while the crack tend to initiate in angles around 25°-31° in the rail cross-section.

1. Introduction

Squat is a type of rolling contact fatigue (RFC) that occurs in the railhead and can develop into the rail fracture, if it is not detected and treated with maintenance measures. Squats are one of the most prevalent forms of RCF defects in the Netherlands. These defects usually appear along with surface and subsurface cracks in the railhead. The squat cracks after initiation, continue to grow during the railway service. A large and growing body of literature has investigated different stages of squat crack life in rails; see for example [1, 2]. Among various stages of a squat crack development, the initial phase is technically a crucial issue that has considerable importance in determining the life time of rail materials. Some researchers [3-5] concentrated on microstructural features related to fatigue crack initiation in rail materials. In this category of works, squat initiation is treated with various laboratory techniques and field studies, mainly to provide information at the micron level and study of the material microstructure. A number of numerical models [6-8] tried to estimate the stress-strain levels inside rail material and to predict RCF initiation, mainly based on finite element method (FEM) and wheel-rail contact mechanics. In spite of a considerable amount of work on crack initiation studies in rails, there has been little discussion about the angles of squat cracks in the initiation phase. Crack initiation angles in rails under rolling contact were predicted in number of studies using 2D models [9-11], which did not take into account the real geometry and loading conditions of the wheel-rail system. In addition, there has been few examples of 3D models [12, 13] for RCF crack initiation; however, they were not dedicated to rail squats, as they predicted cracks in gauge corner (GC) rolling contact conditions (an indication for another sort of RCF cracks called as head checks). The present study seeks to remedy this lack of knowledge, by developing a 3D model of wheel-track contact using the actual geometries, material properties and contact loading conditions. The nucleation orientations of the squat crack in rail are presented using the results of numerical simulations. Moreover, some experimental observations, carried out on an example of rail with squat are presented to assess the effectiveness of the numerical predictions.

Figure 1 gives various phases (A to F) of development of a squat crack in the railhead based on the observations on the majority of the squat cracks [14, 15]. Using this figure, stage A (shown with the dashed-line rectangle) is considered as the objective phase of crack initiation in the present study.

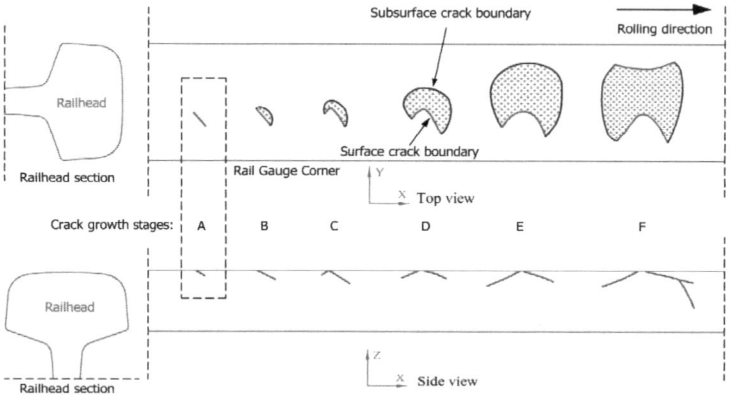

Figure 1. Searching for the critical plane with the maximum *FP* in the first critical point.

2. Numerical modelling of squat crack initiation

A wide range of FE modelling approaches is available for wheel-rail contact problem. Considering the effect of vehicle-track dynamic interaction on RCF phenomenon, a 3D model is developed in this research to study the initiation process of squat cracks. As shown in Figure 2, one loading cycle of a single wheel component, running at limited length of a straight railway track is modelled. The general methodology of the FE simulation and the process of wheel-rail contact are taken from [16, 17].

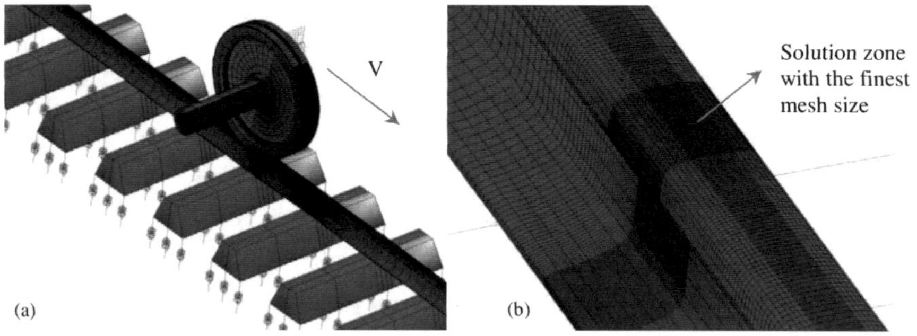

Figure 2. (a) The 3D FE model of the wheel-track system, (b) magnification on rail surface in the solution zone with the detailed meshing style.

2.1. Multi-axial fatigue criteria of crack initiation

A multi-axial fatigue model of the rail material under rolling contact is considered in the present study based on [18]. The fatigue parameter in this model is expressed by:

$$FP_{\max} = (\langle \sigma^{\max} \rangle \frac{\Delta\varepsilon}{2} + J\Delta\tau\Delta\gamma)_{\max} \tag{1}$$

where in each rolling cycle, $\Delta\varepsilon$ is the normal strain range, σ_{max} is the maximum normal stress, $\Delta\gamma$ is the shear strain range, $\Delta\tau$ is the shear stress range, J is a material-dependent constant and $\langle\ \rangle$ denotes the McCauley bracket $\langle x \rangle = (|x| + x)/2$. The constant J is obtained from tension/torsion tests. This model considers the influence of normal and shear loadings on damage occurrence. The first term of (1) considers the mean stress effect in normal direction, while the second term incorporates the shear stress-strain effects. The proposed multi-axial fatigue model has the capability of capturing the synergism between the shear and normal stress components. Stress and strain components are the integral parts of the multi-axial fatigue analysis. The histories of loads are obtained from the results of FE analysis. This fatigue criterion has been recently used in [19, 20], where the crack initiation life of rail material was governed by the cyclic fatigue, rather than the ratchtting failure. Therefore, the ratchetting-based criterion (see for instance [21]) is not used in this research.

2.2. Critical plane approach

All the stress and strain quantities of (1) need to be obtained at the critical plane where the largest fatigue parameter (FP_{max}) occurs. Critical plane approach incorporates the effects of all stress components in three dimensions to find out the angles situation that produces the maximum possibility of fatigue damage. Through a tensor rotation for the stress and strain, the maximum fatigue parameter and the critical plane angles are determined by surveying all the possible planes at a material point. In tensor analysis, various mathematical formulations exist to express the stress components in three dimensions, when the coordinate system is subjected to rotations. In the present research, the spatial rotations of the stress matrices are carried out by the application of Euler angles and mathematical transformations. The 3D transformation matrix can be viewed as a series of three successive rotations about coordinate axes. As shown in Figure 3(a), the first rotation occurs with the angle α about the z-axis and then the second rotation of angle β takes place about the new x-axis (which has already rotated itself due to the first rotation). Finally, the last rotation of γ happens about the new-tilted y-axis to generate the new-rotated coordinate system x′y′z′.

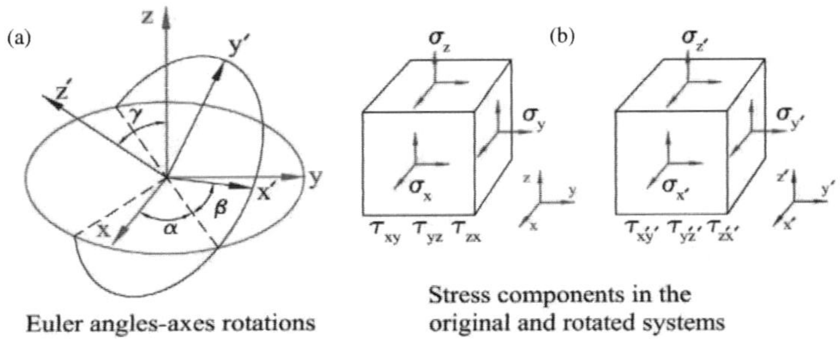

Figure 3. (a) Original and rotated coordinate systems, demonstration of the Euler angles, (b) transforms of stress components.

When a Cartesian coordinate system xyz (the original state, in which the stresses are obtained) is subjected to three sequential rotations demonstrated in Figure 3, the transformation matrix of the stress tensor can be related to the Euler angles α, β, γ:

$$\sigma=\begin{pmatrix}\sigma_x & \tau_{xy} & \tau_{xz}\\ \tau_{xy} & \sigma_y & \tau_{yz}\\ \tau_{xz} & \tau_{yz} & \sigma_z\end{pmatrix} \quad , \quad \sigma'=\begin{pmatrix}\sigma'_x & \tau'_{xy} & \tau'_{xz}\\ \tau'_{xy} & \sigma'_y & \tau'_{yz}\\ \tau'_{xz} & \tau'_{yz} & \sigma'_z\end{pmatrix} \tag{2}$$

$$\sigma'=\mathrm{Q}\cdot\sigma\cdot\mathrm{Q}^{\mathrm{T}} \quad , \quad \sigma'_{\mathrm{mn}}=\lambda_{\mathrm{mi}}\lambda_{\mathrm{nj}}\sigma_{\mathrm{ij}} \quad , \quad \lambda_{\mathrm{ij}}=\cos(\mathrm{x}'_{\mathrm{i}},\mathrm{x}_{\mathrm{j}})$$

$$\sigma'=\begin{pmatrix}\cos(x',x) & \cos(x',y) & \cos(x',z)\\ \cos(y',x) & \cos(y',y) & \cos(y',z)\\ \cos(z',x) & \cos(z',y) & \cos(z',z)\end{pmatrix}.\begin{pmatrix}\sigma_x & \tau_{xy} & \tau_{xz}\\ \tau_{xy} & \sigma_y & \tau_{yz}\\ \tau_{xz} & \tau_{yz} & \sigma_z\end{pmatrix}.\begin{pmatrix}\cos(x',x) & \cos(y',x) & \cos(z',x)\\ \cos(x',y) & \cos(y',y) & \cos(z',y)\\ \cos(x',z) & \cos(y',z) & \cos(z',z)\end{pmatrix} \tag{3}$$

$$\boldsymbol{Q}_x=\begin{pmatrix}1 & 0 & 0\\ 0 & \cos\alpha & \sin\alpha\\ 0 & -\sin\alpha & \cos\alpha\end{pmatrix} \boldsymbol{Q}_y=\begin{pmatrix}\cos\beta & 0 & -\sin\beta\\ 0 & 1 & 0\\ \sin\beta & 0 & \cos\beta\end{pmatrix} \boldsymbol{Q}_z=\begin{pmatrix}\cos\gamma & \sin\gamma & 0\\ -\sin\gamma & \cos\gamma & 0\\ 0 & 0 & 1\end{pmatrix} \tag{4}$$

$$\boldsymbol{Q}=\boldsymbol{Q}_z.\boldsymbol{Q}_y.\boldsymbol{Q}_x=\begin{pmatrix}\cos\beta\cos\alpha & \cos\gamma\cos\alpha+\sin\beta\cos\gamma\sin\alpha & \sin\gamma\sin\alpha-\cos\gamma\sin\beta\cos\alpha\\ \cos\beta\sin\gamma & \cos\gamma\cos\alpha-\sin\gamma\sin\theta\sin\alpha & \sin\gamma\sin\alpha-\sin\gamma\sin\beta\cos\alpha\\ \sin\beta & -\sin\gamma\cos\beta & \cos\beta\cos\alpha\end{pmatrix} \tag{5}$$

2.3. Combining fatigue analysis and critical plane approach

Based on the prescribed formulations, there must be dozens of variations with respect to the spatial angles, since any combination of axes can be considered to perform stress transformation. To determine the location and angles of crack in rail, the material points that are experiencing the largest von Mises stress (criterion for determining the critical damage point) under wheel passage are determined. Then the critical plane approach is employed to determine the angles in which the maximum fatigue parameter is achieved (multi-axial fatigue criterion). These angles provide the situation by which the rail is highly susceptible to crack initiation. Finally, the critical angles are represented by in-plane angles of crack initiation for rail material. As shown in Figure 4(a), there are two forms of demonstrations available for crack plane angles. The in-plane angles θ, φ, ψ are used in the present study to show the orientations of crack initiation. This angles are given in Figure 4(b) with details of traffic direction, gauge corner side and axes definitions.

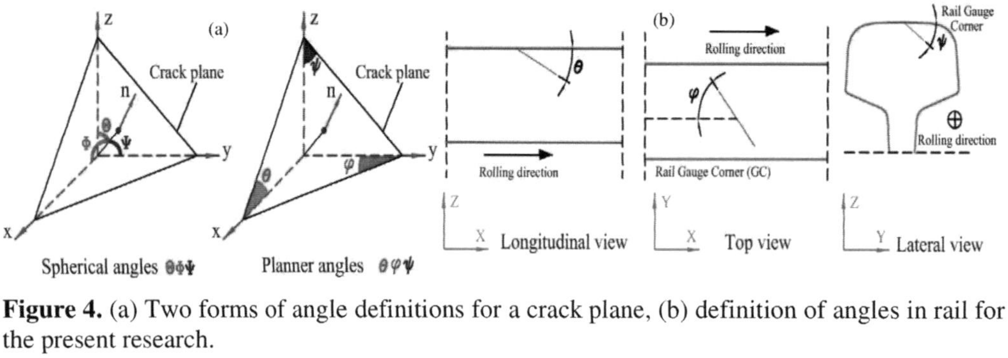

Figure 4. (a) Two forms of angle definitions for a crack plane, (b) definition of angles in rail for the present research.

2.4. Material and loading parameters

The consecutive model of the rail material in the contact interface are defined by the combined nonlinear isotropic and kinematic hardening model, which is developed by Lemaitre and Chaboche [22, 23]. This model can be used for simulation of cyclic plasticity and ratchetting material response with decaying ratchetting rate [11]. The yield stress in this model is defined by a Von Mises yield surface, where the yield function ϕ is expressed by:

$$\phi(\boldsymbol{\sigma}, \boldsymbol{X}, K) = \sqrt{\frac{3}{2}} \, | \boldsymbol{\tau}_{dev} | - K - \sigma_y$$
$$\boldsymbol{\tau}_{dev} = \boldsymbol{\sigma}_{dev} - \boldsymbol{X}, \; | \boldsymbol{\tau}_{dev} | = \sqrt{\boldsymbol{\tau}_{dev} : \boldsymbol{\tau}_{dev}} \tag{6}$$

Where $\boldsymbol{\sigma}$ is the Cauchy stress tensor, X is the backstress tensor, K is the drag stress, σ_y is the yield stress, $\boldsymbol{\tau}_{dev}$ is the deviator stress tensor, $\boldsymbol{\sigma}_{dev}$ is the deviator stress tensor of $\boldsymbol{\sigma}$, and the operator ":" defines the contraction $a : b = a_{ij} \, b_{ij}$. The material is elastic when $\phi(\boldsymbol{\sigma}, \boldsymbol{X}, K) < 0$ and plastic, when $\phi(\boldsymbol{\sigma}, \boldsymbol{X}, K) = 0$. The isotropic hardening law (K) and X tensors are defined by:

$$\dot{K} = \dot{\lambda} \underline{b}(1 - \frac{K}{K_\infty}), \quad \dot{\boldsymbol{X}} = \dot{\lambda} \underline{c}(\sqrt{\frac{2}{3}} \boldsymbol{n}_{dev} - \underline{\gamma} \boldsymbol{X}) \quad , \quad \boldsymbol{n}_{dev} = \frac{\boldsymbol{\tau}_{dev}}{| \boldsymbol{\tau}_{dev} |} \tag{7}$$

where $\underline{b}$ is the initial rate of the isotropic hardening, $\dot{\lambda}$ is the plastic multiplier, K is the drag stress, and K_∞ is the saturated drag stress due to isotropic hardening. The kinematic hardening law follows the nonlinear Armstrong and Frederick law, where $\underline{c}$ and $\underline{\gamma}$ are material parameters. The rail steel that is generally being used in the Netherlands is the typical pearlitic steel. The material constants that are used in the failure models can be obtained by normal fatigue tests like tension-compression and torsion fatigue tests. In the present study, however, these constants are taken from [24]. The material and load dependent constant J was set to 0.2, based on results in the literature [18]. The plasticity parameters of the material model used in the current study are extracted from [11]. The parameters of the FE model in the present research are listed in Table 1. This table also gives the properties of the pearlitic rail steel employed in this work.

Table 1: Mechanical/loading parameters of the wheel-track system and rail material

Parameters of the model and linear materials	Values	Pearlitic rail steel (nonlinear)	Values
Static wheel load, P_0 *(kN)*	120	σ_y (MPa)	456
Wheel weight *(kg)*	900	k_e (MPa)	234
Sleeper mass M_s *(kg)*	280	E (GPa)	210
Stiffness of ballast, K_b *(kN/m)*	45000	v	0.3
Damping of ballast, C_b *(N.s/m)*	32000	ε_f' (%)	10.3
Stiffness of rail pad, K_p *(kN/m)*	1300000	σ_f' (MPa)	936
Damping of rail pad, C_p *(N.s/m)*	45000	γ_f' (%)	15.45
Stiffness of primary suspension, K_c *(kN/m)*	880	τ_f' (MPa)	468
Damping of primary suspension, C_c *(N.s/m)*	4000	$\underline{b}$ (MPa)	10.7
Young's modulus of wheel-rail, E_r *(GP)*	210	$\underline{c}$ (GPa)	6.49
Poisson's ratio of wheel-rail material, v_r	0.3	$\underline{\gamma}$	14.4
Density of wheel-rail material, ρ_r *(kg/m3)*	7800	K_∞ (MPa)	22.8
Young's modulus of concrete, E_c *(GP)*	38.4	ε_c (%)	11.5
Poisson's ratio of concrete material, v_c	0.2	b	−0.089
Density of sleeper material, ρ_c *(kg/m3)*	2520	c	−0.559
Rolling speed *(km/h)*	140		
Friction coefficient of wheel-rail contact	0.6		
Traction coefficient (ratio of tangential load)	0.25		

3. Results of numerical investigations

The histories of stresses and strains under rolling contact were computed from FE analysis. The rail material points with the largest von Mises stresses were determined as the critical damage points. The fatigue parameters (FP) of these points were calculated in a range of 3D transformations. The results of FP calculations for six critical points (the first six points with largest von Mises stress) are given in Figure 5,6, with respect to angles variations. The angles in these figures are the ones defined in Figure 4(b). The dimensionless *FP* values were obtained by normalizing fatigue parameters over the peak *FP* in the entire loading cycle. It should be noted that for the search of the critical plane, only two of three angles need to be calculated. The third angle is dependent open the first two angles by spherical law of cosines.

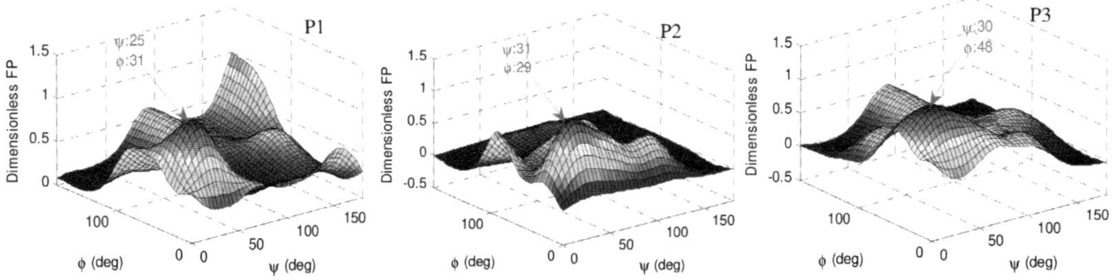

Figure 5. Searching for the critical plane with the maximum *FP* in material points P1-P3.

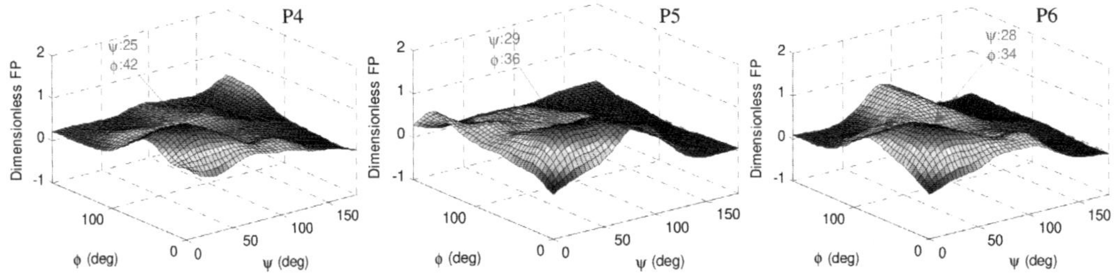

Figure 6. Searching for the critical plane with the maximum *FP* in material points P4-P6.

Based on these results, angles of crack initiation were estimated and listed in Table 2. Three approximate limits are given in the last column of the table, as tentative ranges for the angles. Note that, from the proposed numerical calculation it is not possible to expect exact values for the angles; however suggesting the obtained ranges seem to be rational.

Table 2: Crack initiation angles by numerical observations

Critical point	P1	P2	P3	P4	P5	P6	**Range**
Angle θ	19	20	18	22	15	13	**13-22**
Angle φ	31	29	48	42	36	34	**29-48**
Angle ψ	25	31	30	25	29	28	**25-31**

4. Experimental observations on a sample squat

So far, orientations of the initiated cracks were numerically studied. To assess the effectiveness of the proposed model, a rail sample with relatively severe squat (squat class B according to [25]) is taken from the site. The overall view of the rail sample with squat is demonstrated in Figure 7(a), representing the local plastic deformation of the surface and a typical lung-shape pattern. Traffic direction was from left to right and the rail was taken from the left side of a straight track between Meppel and Steenwijk in the Netherlands. The chosen rail is an R260Mn type Dutch rail made of steel with a chemical composition of 0.64%C, 1.51%Mn, 0.25%Si, 0.02%P and 0.02%S (mass %). The rail was manufactured in 1989 and was loaded with an annual gross tonnage of 4.1 million tons. Experimental observations were made to assess the locations and orientations of the squat cracks for the chosen rail. Owing to this, a cubic sample was cut from the rail top in the area by which the squat cracks can be relatively captured. Figure 7(b, c) gives the location of the selected cube on rail surface.

The sectioning pattern and dimensions of the cutting samples are also shown in the figure. The size of the cube was 14 mm × 15 mm in the longitudinal and lateral directions, respectively. The cube was 10 mm deep in the vertical direction. According to Figure 7, the cube was divided into seven thin samples with 15*10*2 mm dimensions. This was done to reproduce the shape/geometry of the cracks with serial sectioning. Using this method, the actual geometry of cracks was observed. As demonstrated in Figure 7, direction of cutting was perpendicular to the traffic direction and the interval length between sections was 2 mm. Therefore, seven slices were obtained in total, namely S1-S7.

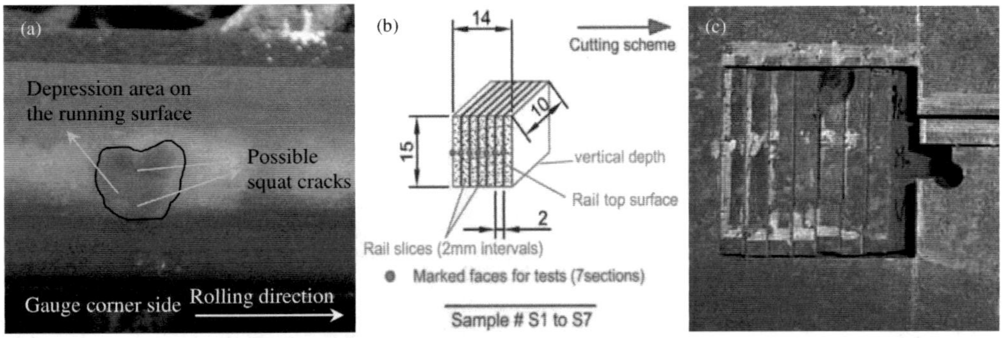

Figure 7. (a) The rail sample with relatively severe squat, (b) rail sectioning pattern, dimensions of cutting samples, (c) sliced samples after the serial cutting process.

All rail slices were cut and the faces of the cuts were ground and roughly polished. After sample preparation, the microstructure of the rail was observed with optical microscopes in transverse directions. The geometries of squat cracks in all sequential sections were traced. The microstructures of the sections with cracks were mapped with 2D images including details of crack tip locations, angles, sizes and depth of propagation. Figure 8 shows the microstructure of the cracked rail in sections S1, S3, S5 as an example of such observations.

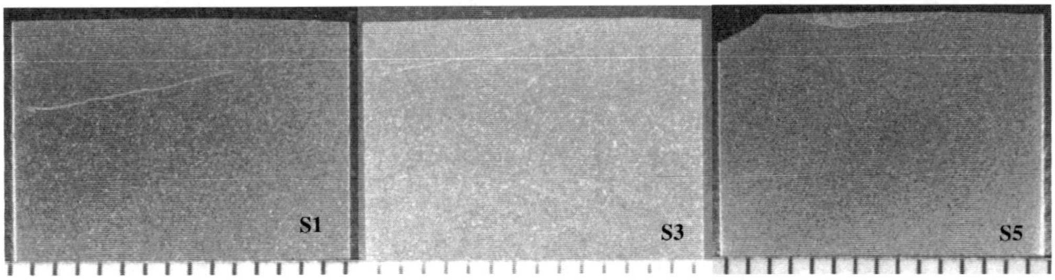

Figure 8. Microstructure of the cracked rail in sections S1, S3, S5, ruler indentations are in mm.

Geometries of the cracks in 2D images were incorporated to generate the 3D configurations of the squat cracks. Figure 9 gives the 3D representation of the cracks, constructed using the prescribed method. This figure shows the full morphology of the cracks in 3 dimensions.

A survey was done about locations and geometries of squat cracks on the rail surface. Based on the 3D geometry of cracks in Figure 8, the intersection lines (boundary curves) of the crack with the rail surface were determined as shown in Figure 10. This figure shows the geometric shape of squat cracks from the top view. Note that two different views of the same sample are shown in Figure 10(a, b). According to these observations, there were two separate lines (crack mouths) on rail surface. Each crack mouth had a turning point somewhere in the middle of the curve. The turning points (marked

Journal of Physics: Conference Series **628** (2015) 012043 doi:10.1088/1742-6596/628/1/012043

with 1 and 2 in Figure 10) can be possible points of crack initiation at rail surface. By the way, the experimental observations cannot confirm that the cracks are initiated at such turning points.

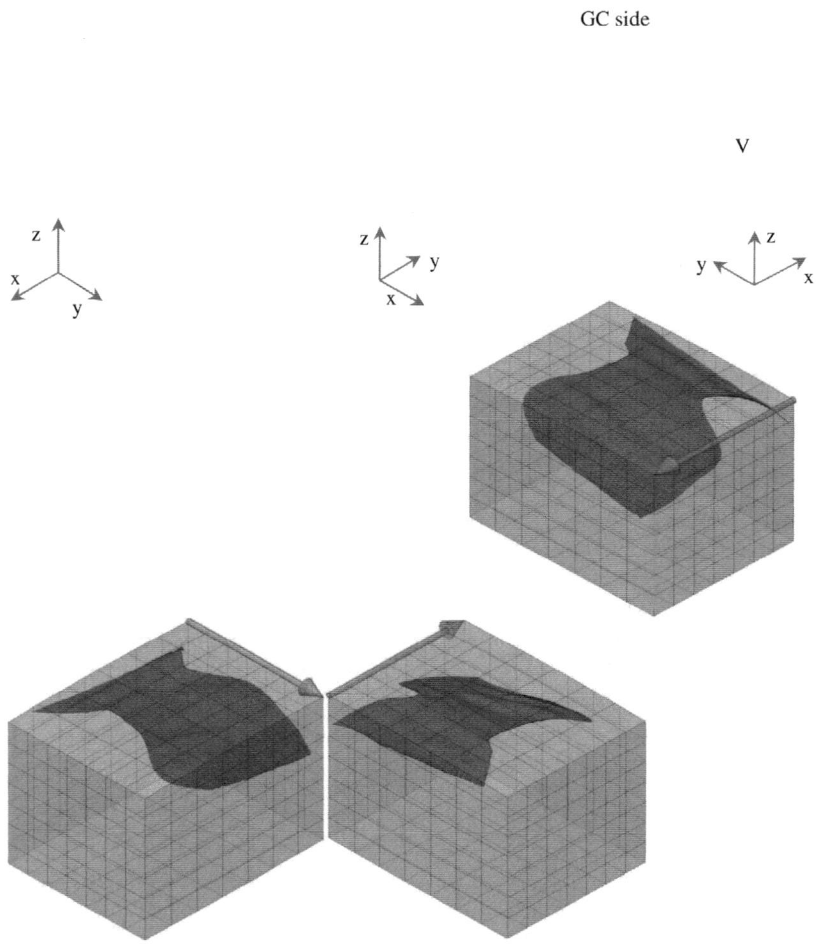

Figure 9. 3D views of the cracked rail; vectors at cube corners show the rolling direction.

Based on results of numerical simulations in this paper, the critical damage points were located at the rail surface. Therefore, it can be hypothesized that the crack initiation point is located somewhere in the surface (along the cracks boundaries in Figure 10, but not necessarily on turning points). Based on this hypothesis, four sides of crack faces (C1-C4) are defined in Figure 10 as the possible faces for crack initiation. Any of two crack mouths on rail surface was divided into two individual curves to generate such faces. Now, it can be hypothesized that the squat is initiated from each of the C1-C4 faces, but still at the surface.

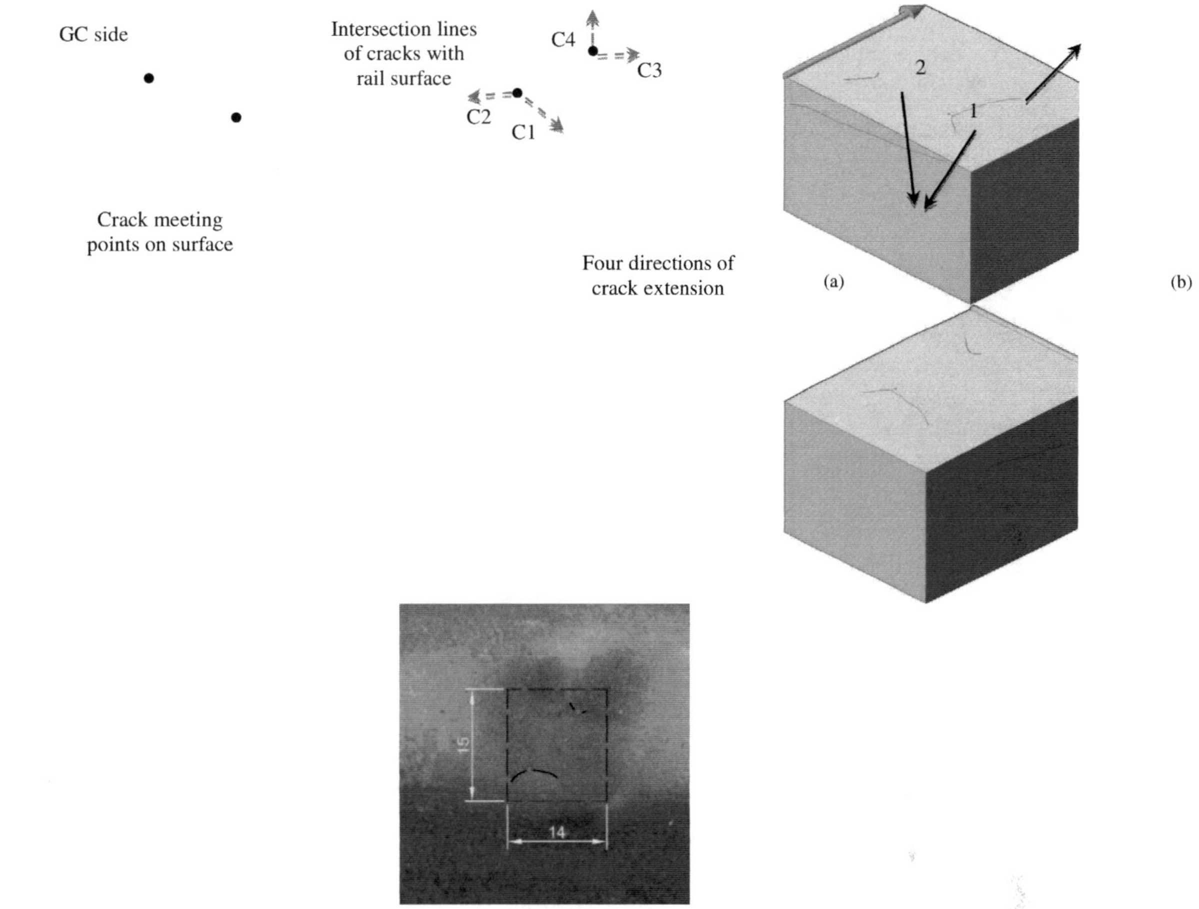

Figure 10. Intersection lines of cracks with rail surface and crack meeting points on rail, (a,b) 3D views, (c) surface view with nomination of crack-surface meeting points and cube dimensions.

Considering C1-C4 as the crack faces carrying the crack initiation point, it is possible to determine the angles of the initiated crack. By mapping each of these faces on different planes of the Cartesian coordinate system, the angels of crack initiations were calculated. The results of angle measurements are shown in Figure 11, in which the orientation angles of each individual face were separately demonstrated. The corresponding angles of crack initiation for C1-C4 crack faces are listed in Table 3. Note that all angles were measured from 0 to 180° with respect to the original axes in Figure 3.

Comparing the results in tables 2 and 3, the crack angles from numerical studies relatively matches with the results of C1 crack face in experimental investigation. This means that among the four crack faces, only one face offers initiation angles, having proper agreement with the numerical predictions. This evidence suggests that the C1 face is the possible face, in which the squat crack has initiated in the rail sample. Although the squat in this sample was a kind of severe squat with well-developed cracks, only situations of cracks at locations very close to the surface was studied. Therefore, it is possible to consider such results for the squat at its initiation phase. By the way, it was assumed that the initial situation of crack angles has not been changed during rail service (after crack initiation).

Looking at results of θ, φ, ψ in table 3, it can be seen that the values of θ and ψ are in the range that is suggested in Table 2 (θ: 14°-23° and ψ: 29°-52°). However, experimental observation on the chosen squat reported a value of 25.7° for φ, which is not in the range 29°-52° suggested by the numerical work (the experimental value of φ was slightly lower than numerical one). This is mainly due to the fact that, many sorts of uncertainties may play a role in the complex orientation of the cracks that cannot be seen in the numerical models. Apart from this discrepancy, the findings of experimental observations appropriately supported the numerical predictions.

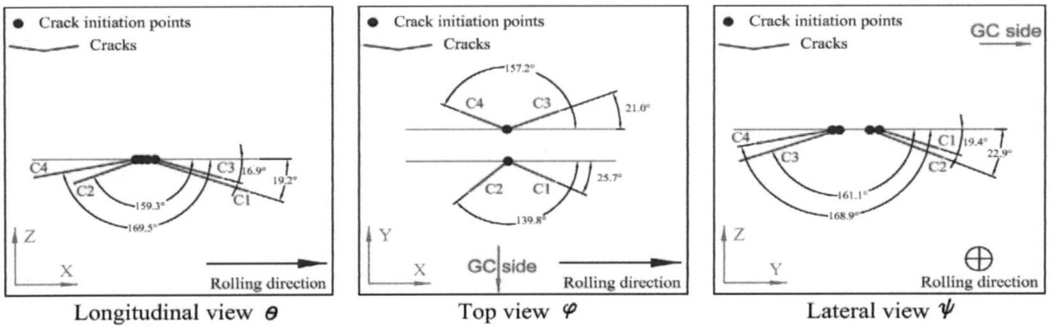

Figure 11. Crack initiation angles in different 2D views for C1-C4 crack faces

Table 3: Crack initiation angles by experimental observations

Crack mouth	θ	φ	ψ
C1	**19.20**	**25.70**	**19.40**
C2	159.30	139.80	22.90
C3	16.90	21.10	161.10
C4	169.50	157.20	168.90

5. Conclusions

This paper presented the results of crack initiation angles in rail squats using both numerical investigations and experimental observations. The 3D finite element model of the wheel-track system was built, capable of simulating rolling contact conditions. Applying the multi-axial fatigue analysis and critical plane approach, the material points and angles that are mostly susceptible to fatigue damage were determined. A rail specimen with squat taken from the site was cut with serial sectioning and the shape/geometry of the cracks was reproduced. Comparing the results of experimental work with numerical predictions, a positive correlation was observed. Based on results of analyses, it can be concluded that squat cracks initiate at an angle 13°-22°, relative to the rail surface (θ). The initiation angle seen on surface plane (φ) was around 29°-48°. Looking at rail cross-section, the angle of crack initiation toward rail surface (ψ) was in the range of 25°-31°. Although the chosen squat in the experimental work was relatively severe with well-developed cracks, it was possible to trace the initiation point and angles of the squat crack, using the prescribed method. As an attempt to quantify the angles of initiated fatigue crack in rails, this study contributed to ascertain the mechanism by which rail squats are generated.

6. Acknowledgment

This research is part of an ExploRail project namely as Development of High–Performance Rail through Intelligent Metallurgy and Engineering (PRIME), in Delft University of Technology. This project (Code: 11247/ C38A07) is funded by Dutch rail infra manager ProRail and the Netherlands organization for scientific research (STW/NWO).

7. References

[1] Bogdanski, S., Olzak, M., & Stupnicki, J., *Numerical modelling of a 3D rail RCF 'squat'-type crack under operating load.* Fatigue & Fracture of Engineering Materials & Structures, 1998. **21**(8): p. 923-935.

[2] Farjoo, M., Daniel, W., & Meehan, P.A., *Modelling a squat form crack on a rail laid on an elastic foundation.* Engineering Fracture Mechanics, 2012. **85**: p. 47-58.

[3] Pal, S., Daniel, W.J.T., & Farjoo, M., *Early stages of rail squat formation and the role of a white etching layer.* International Journal of Fatigue, 2013. **52**: p. 144-156.
[4] Pal, S., Valente, C., Daniel, W., & Farjoo, M., *Metallurgical and physical understanding of rail squat initiation and propagation.* Wear, 2012. **284**: p. 30-42.
[5] Cheng, W., Cheng, H., Mura, T., & Keer, L., *Micromechanics modeling of crack initiation under contact fatigue.* Journal of tribology, 1994. **116**(1): p. 2-8.
[6] Ringsberg, J., Loo-Morrey, M., Josefson, B., et al., *Prediction of fatigue crack initiation for rolling contact fatigue.* International Journal of Fatigue, 2000. **22**(3): p. 205-215.
[7] Šraml, M., Flašker, J., & Potrč, I., *Numerical procedure for predicting the rolling contact fatigue crack initiation.* International journal of fatigue, 2003. **25**(7): p. 585-595.
[8] Franklin, F. & Kapoor, A., *Modelling wear and crack initiation in rails.* Proceedings of the Institution of Mechanical Engineers, Part F: Journal of rail and rapid transit, 2007. **221**(1): p. 23-33.
[9] Akama, M., *Development of finite element model for analysis of rolling contact fatigue cracks in wheel/rail systems.* Railway Technical Research Institute, Quarterly Reports, 2007. **48**(1).
[10] Deng, T.S., Zhao, X., Wu, B., et al., *Prediction of crack initiation of rail rolling contact fatigue.* Applied Mechanics and Materials, 2013. **344**: p. 75-82.
[11] Ringsberg, J.W., *Cyclic ratchetting and failure of a pearlitic rail steel.* Fatigue & Fracture of Engineering Materials & Structures, 2000. **23**(9): p. 747-758.
[12] Akama, M., Matsuda, H., Doi, H., & Tsujie, M., *Fatigue crack initiation life prediction of rails using theory of critical distance and critical plane approach.* Journal of Computational Science and Technology, 2012. **6**: p. 54-69.
[13] Ringsberg, J. & Josefson, B., *Finite element analyses of rolling contact fatigue crack initiation in railheads.* Proceedings of the Institution of Mechanical Engineers, Part F: Journal of Rail and Rapid Transit, 2001. **215**(4): p. 243-259.
[14] Bogdański, S., *Predicting the growth of RCF crack with the use of the 3D multi size finite element model.* Tribology and Interface Engineering Series, 2005. **48**: p. 709-721.
[15] Bogdański, S. & Lewicki, P., *3D model of liquid entrapment mechanism for rolling contact fatigue cracks in rails.* Wear, 2008. **265**(9): p. 1356-1362.
[16] Zhao, X. & Li, Z., *The solution of frictional wheel–rail rolling contact with a 3D transient finite element model: Validation and error analysis.* Wear, 2011. **271**(1–2): p. 444-452.
[17] Zhao, X., Li, Z., & Dollevoet, R., *The vertical and the longitudinal dynamic responses of the vehicle–track system to squat-type short wavelength irregularity.* Vehicle System Dynamics, 2013. **51**(12): p. 1918-1937.
[18] Jiang, Y. & Sehitoglu, H., *A model for rolling contact failure.* Wear, 1999. **224**(1): p. 38-49.
[19] Naeimi, M., Li, Z., Petrov, R., et al., *Substantial fatigue similarity of a new small-scale test rig to actual wheel-rail system.* International Journal of Mechanical, Aerospace, Industrial and Mechatronics Engineering, 2014. **8**(12): p. 1795-1803.
[20] Deng, X., Naeimi, M., Li, Z., et al. *Residual fatigue life evaluation of rail at squats seeds using 3D explicit finite element analysis.* in *AMS 14: Proceedings of the 1st Ageing of Materials & Structures Conference.* 2014. Delft, The Netherlands: DCMat Ageing Centre, Delft University of Technology.
[21] Kapoor, A., *A re-evaluation of the life to rupture of ductile metals by cyclic plastic strain.* Fatigue & fracture of engineering materials & structures, 1994. **17**(2): p. 201-219.
[22] Chaboche, J.-L., *Constitutive equations for cyclic plasticity and cyclic viscoplasticity.* International Journal of Plasticity, 1989. **5**(3): p. 247-302.
[23] Lemaitre, J. & Chaboche, J.-L., *Mechanics of solid materials.* 1990: Cambridge University Press.
[24] Tyfour, W., Beynon, J., & Kapoor, A., *The steady state wear behaviour of pearlitic rail steel under dry rolling-sliding contact conditions.* Wear, 1995. **180**(1): p. 79-89.
[25] Li, Z., Zhao, X., Esveld, C., et al., *An investigation into the causes of squats—correlation analysis and numerical modeling.* Wear, 2008. **265**(9): p. 1349-1355.

Corrigendum: Nucleation of squat cracks in rail, calculation of crack initiation angles in three dimensions

The following are the corrected versions of equation 7 and figures 9 and 10:

$$\dot{K} = \dot{\lambda}\underline{b}(1-\frac{K}{K_{\infty}}), \quad \dot{X} = \dot{\lambda}(\underline{c}\sqrt{\frac{2}{3}}\boldsymbol{n}_{dev} - \underline{\gamma}X) \quad , \quad \boldsymbol{n}_{dev} = \frac{\boldsymbol{\tau}_{dev}}{|\boldsymbol{\tau}_{dev}|} \tag{7}$$

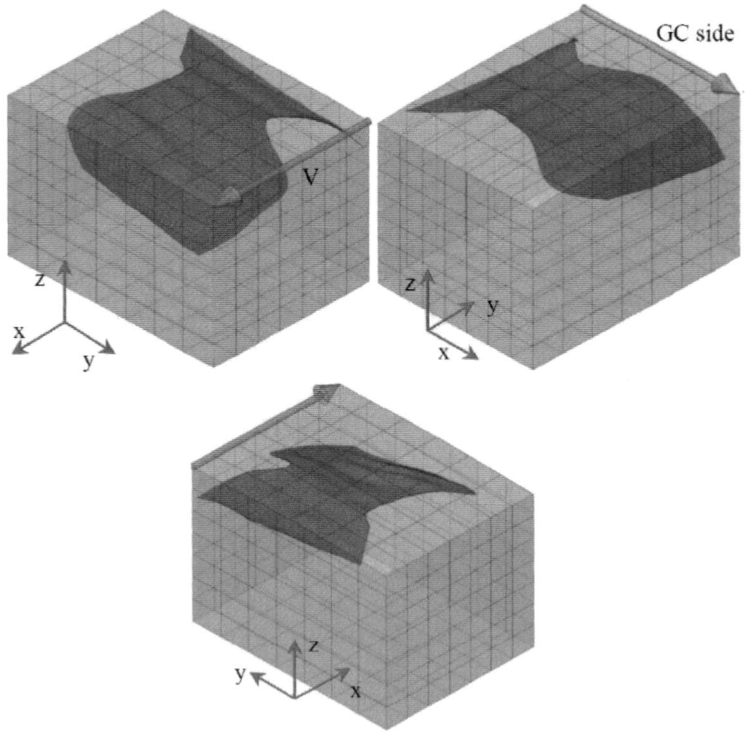

Figure 9. 3D views of the cracked rail; vectors at cube corners show the rolling direction.

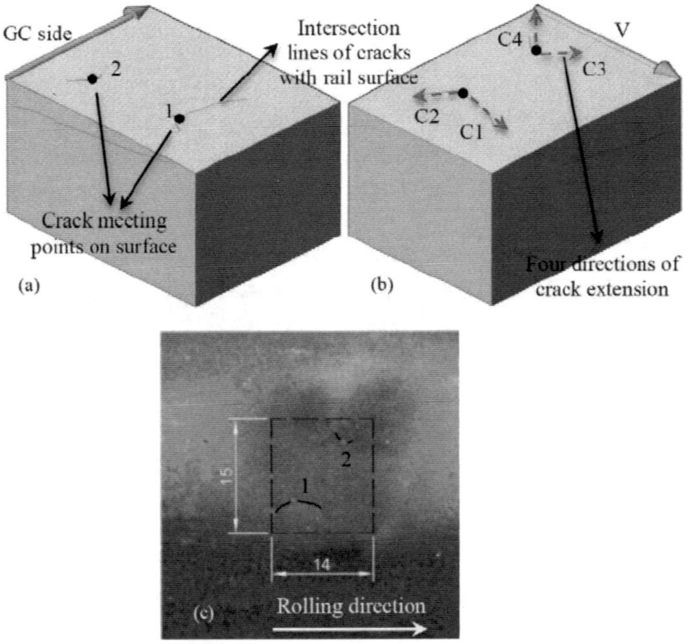

Figure 10. Intersection lines of cracks with rail surface and crack meeting points on rail, (a,b) 3D views, (c) surface view with nomination of crack-surface meeting points and cube dimensions.

Damage Detection in Bridge Structure Using Vibration Data under Random Travelling Vehicle Loads

C H Loh[1], T Y Hung[1], S F Chen[1] and W T Hsu[1]
Department of Civil Engineering, National Taiwan University
Taipei 10617, Taiwan

E-mail: lohc0220@ccms.ntu.edu.tw

Abstract. Due to the random nature of the road excitation and the inherent uncertainties in bridge-vehicle system, damage identification of bridge structure through continuous monitoring under operating situation become a challenge problem. Methods for system identification and damage detection of a continuous two-span concrete bridge structure in time domain is presented using interaction forces from random moving vehicles as excitation. The signals recorded in different locations of the instrumented bridge are mixed with signals from different internal and external (road roughness) vibration sources. The damage structure is also modelled as the stiffness reduction in one of the beam element. For the purpose of system identification and damage detection three different output-only modal analysis techniques are proposed: The covariance-driven stochastic subspace identification (SSI-COV), the blind source separation algorithms (called Second Order Blind Identification) and the multivariate AR model. The advantages and disadvantages of the three algorithms are discussed. Finally, the null-space damage index, subspace damage indices and mode shape slope change are used to detect and locate the damage. The proposed approaches has been tested in simulation and proved to be effective for structural health monitoring.

1. Introduction

One of questions in research attention is the use of structural response from operational dynamic loads in the damage detection procedures. In the long service life and continuous operation process, bridge may suffer resistance and damage accumulation under the comprehensive effect of random vehicle loads and environmental effects such as wind load and rain load. By placing sensors on bridges, the bridge health status can be monitored online through health monitoring system in order to determine the degree of damage. For bridge monitoring the vehicle-bridge structure interaction effect needs to be considered in the modal parameter identification. Generally, the modal parameters of the vehicle-bridge system are changing when the random vehicles are moving on top of the bridge (include the road roughness). For such a system, system identification tools must be developed and tested using numerically generated time histories so as to verify the methodologies. For identification, using time-frequency tools, which are non-parametric methods for processing responses of non-linear or non-stationary system, had been studied [1, 2]. A linear approach derived from the stochastic subspace identification (SSI) method, together with the moving window SSI to examine the time-varying system, has also been adopted for vibration-based system identification [3].

[1] CH Loh, Professor, Dept. of Civil Engineering, National Taiwan University, Taipei 10617, Taiwan
TY Hong, SF Chen, WT Hsu, graduate student, Civil Eng. Dept., National Taiwan University

In this study, comparison of three different time-domain identification and damage detection methods are proposed to identify the bridge model properties by using the simulation data of a vehicle-bridge structural interaction under random travelling loads and with consideration of surface roughness. First, the automatic covariance-drive stochastic subspace identification (SSI-COV) is proposed to extract the system poles through stability diagram. In cooperated with the wavelet packet transform (WTP) sifting process for modal phase consistency, spurious modes can be removed from using SSI. For well-separated modes, the blind source separation (called Second-order Blind Identification, SOBI) algorithm can also be used for identification and damage detection. Advantages and disadvantages of the two methods are discussed. Finally, a parametric identification method, using multivariate AR-model, is adopted to enhance the damage detection and compare the result with other methods.

2. Simulation of bridge-vehicle interaction system

A two span bridge model is proposed in this study, as shown in Fig.1. It shows a bridge carrying a series of random moving vehicles which are modelled as a sprung-mass system. The bridge is modelled as a Bernoulli-Euler beam by finite element method. The information for the bridge and vehicles are provided in Table 1. The traffic loads are assumed as random and to be moving as a group from both sides (two ways) of the bridge with random velocity. The prescribed velocity of each car is random generated and shown in the travel time across the bridge with span length of total 200 m, as shown in Fig.2. A total of 16 cars are on the bridge during the measurement. From Fig.2 one can see that during the recorded time window (i.e. 30. sec) some cars are still on the bridge and some cars may just enter/leave the bridge. In this paper, the simulation of road roughness in vehicle–bridge interaction is also considered [4, 5]. The random road surface roughness of the bridge can be described by a kind of zero-mean, real-valued, stationary Gaussian process defined by ISO 8608 [6]. The moving vehicle is also modelled as a moving mass with spring-mass system. Based on the finite element method a total of 10 beam elements (Bernoulli-Euler beam) are used and the nodal points are considered as measurements. First, the system equation of motion is formed, then the equations of motion of the bridge-vehicle interaction system can be solved by Newmark-β method. Acceleration responses from each nodal point (similar to measurement point) are collected with duration of 30. sec. Both intact and damaged situation of the bridge structure are simulated. For damaged case a 50% of stiffness reduction at element 7 (between measurement point 5 and 6) is assumed.

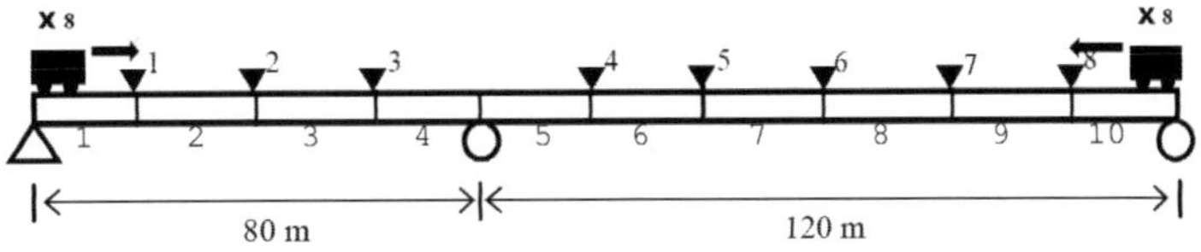

Figure 1. Bridge model with two-way travelling vehicles and the location of measurement points.

Table 1. Natural frequencies of bridge model and vehicle information.

Bridge Data	Bridge Freq. (Hz)	1st mode	2nd mode	3rd mode	4th mode	5th mode	6th mode	7th mode	8th mode
		2.12	5.29	8.37	16.42	20.50	30.31	41.55	48.65
Vehicle Data	Vehicle system natural frequency: 1.96 Hz ~ 6.5 Hz								
	Vehicle Weight (ton) : 1.5 ~ 9.0								

3. Identification of bridge-vehicle interaction system

To identify the dynamic characteristics of the bridge directly from the vibration measurement during random travelling vehicles, two different approaches are employed: the covariance-drive stochastic subspace identification (SSI-COV) and the second-order blind identification (SOBI). Damage detection will also be studied from the extracted vibration features and system modal properties.

Discussion on the identification of damage location using parametric Multivariate-AR model is also presented for comparison.

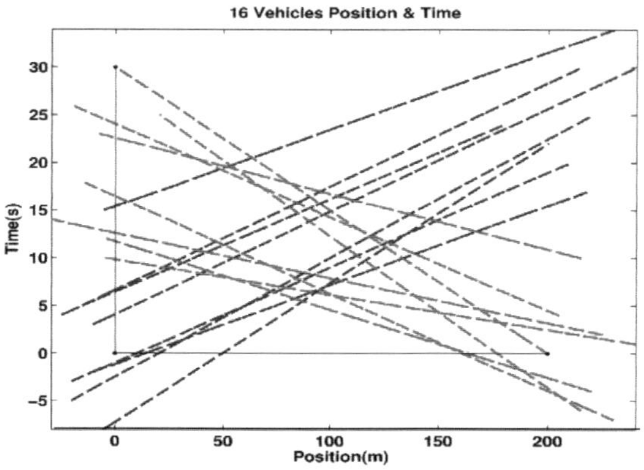

Figure 2. Plot the relationship between the recorded time window (30. sec) and the location of vehicles (8 cars from right and another 8 cars from left) travelling with different speed.

3.1. Automatic covariance-drive stochastic subspace identification (SSI-COV)
A fully automatic identification scheme of reference-base SSI-COV algorithm is developed to extract the dynamic characteristics of the monitoring structure. Technique to remove the spurious modes is included in the automatic identification scheme and the calculation of uncertainty on the identified modal parameters is to confirm the accuracy of the identified modes. The following 4 stages are developed to provide a good quality of the stability diagram from using SSI-COV [13]:

a. Conduct down-sampling on the recorded data (from 200. Hz to 50 Hz), and then adopt singular spectrum analysis (SSA) on the data Toeplitz matrix so as to construct the stability diagram (fix number of rows in data Hankel matrix and change number of order in Singular Value Decomposition).
b. Select output modal amplitude criteria (OMAC) and mode phase co-linearity (MPC) to remove the spurious modes from stability diagram. The hard criteria, such as threshold values on mode assurance criteria, system natural frequency and damping ratio are also employed.
c. Use the 4-means clustering algorithm to minimize the sum of the squared Euclidean distance between the extracted poles, and determine the suitable poles satisfy the criteria.
d. Grouping of poles with each individual mode and identify the physical modal parameters of each mode by using uncertainty bounds of modal frequency and damping ratio as weighting factor for each pole in each specific group of poles.

Through the proposed 4 stages in SSI-COV analysis, one can significantly increase the autonomous ability to remove the spurious modes and create a clear and identifiable stability diagram. The system dynamic characteristics can also be identified.

3.2. Second-order blind identification (SOBI)
In applying SSI-COV some parameters need to be determined, such as the number of row in data Hankel matrix and the number of order, which depends on the dynamic characteristics of the system. Different from the SSI-COV algorithm, blind source separation (BSS) is also a signal processing technique, which attempts to recover individual unknown but statistically independent components with only knowledge of measurements [7]. There is no need to provide parameters in using BSS algorithm. The simple model for BSS assumes the existence of n independent signals $\mathbf{S}^T = [s_1(t), \cdots, s_n(t)]$ and m measurements $\mathbf{X}^T = [x_1(t), \cdots, x_m(t)]$. This is completely represented by the mixing equation

$$\mathbf{X}(t) = \mathbf{AS}(t) \tag{1}$$

where $A \in R^{mxn}$. If the number of measurements and the excitation sources are remain the same, then the source vector **S** can be obtained once the mixing matrix **A** is estimated. But the physical meaning of the mixing equation needs to be established. Now consider a linear time-invariant system with n degree-of-freedom (DOF). The system response can be formulated as the product of mode shapes $[\mathbf{\Phi}] = [\phi_1, \phi_2, \cdots, \phi_n]$ and modal coordinates **U**(t):

$$\mathbf{X}(t) = [\mathbf{\Phi}]\mathbf{U}(t) \tag{2}$$

Compare Eqs.(1) and (2), under the conditions where the modal coordinates are mutually uncorrelated with non-similar spectra for dynamic system which is similar to the assumption of BSS that provided a one-to-one mapping between the mixing matrices (sources) and the vibration modes (modal responses) of the system, therefore, the BSS can be interpreted as the modal expansion.

Since most BSS techniques are based on a model in which the sources are mutually independent, however, the second order blind identification (SOBI) belongs to a BSS technique that makes use of the temporal structure. It means that the sources have different autocorrelation function (or power spectra) and are mutually uncorrelated [8, 9]. To perform SOBI two computation steps are required: whiting process and determining the matrix **U** that approximately diagonal the whitened covariance matrix. For whiting process, first, conduct the eigenvalue decomposition of the signal covariance matrix:

$$[\mathbf{R_X}(0)] = \mathbf{EDE}^{\mathrm{T}} \tag{3}$$

From which the weighting matrix is defined:

$$\mathbf{W} = \mathbf{D}^{-\frac{1}{2}}\mathbf{E}^{T} \tag{4}$$

Additionally, the whitening observed data **Z** can be expressed as:

$$\mathbf{Z} = \mathbf{WX} \tag{5}$$

The covariance matrix of the whitening data becomes:

$$E[\mathbf{ZZ}^{\mathrm{T}}] = \mathbf{WA}E[\mathbf{SS}^{\mathrm{T}}]\mathbf{A}^{\mathrm{T}}\mathbf{W}^{\mathrm{T}} = \mathbf{WAA}^{\mathrm{T}}\mathbf{W}^{\mathrm{T}} = \mathbf{UU}^{\mathrm{T}} = \mathbf{I} \tag{6}$$

where $E[\mathbf{SS}^{\mathrm{T}}] = I$ and $\mathbf{U} = \mathbf{WA}$. Because sources are stationary and uncorrelated, therefore, the time-delayed covariance matrix of sources is diagonal. From this one can easily determine the **U** matrix through an eigenvalue decomposition of the time-delayed covariance matrix of whitened data. For different time-delayed τ_i, the whitened covariance matrix is also different. SOBI algorithm employs the joint approximation diagonalization (JAD) technique to determine the matrix **U** that approximately diagonalized the whitened covariance matrix with different time-delayed value τ_i [10]. Procedures of using SOBI for modal identification are listed as follows:

1. Pre-Processing : Zero-mean can remove the offset and slope of the measured signals. Whitening makes the measured signals uncorrelated and have unit variance.
2. JAD Analysis : After applying joint approximation diagonalization, the **U** matrix which diagonalized the whitened covariance matrices with different time-delayed value can be obtained.
3. Modal Analysis : When mixing matrix **A** is obtained; the source vector **S** can be determined and apply Fourier transform to the source vector to determine the system natural frequencies.

4. Vibration-based system identification of bridge under travelling vehicle loads

Acceleration response data (vertical vibration) of the vehicle-bridge structure interaction through numerical simulation was used for analysis. The original sampling rate was 500 Hz. Before adopting the SSI-COV algorithm to extract the system dynamic characteristics, the data was down-sampled to 125 Hz for the analysis (through Butterworth low-pass filter). In this simulation a total of 16 cars (both ways) travelled on the bridge with different speed simultaneously. Figs. 3a and 3b shows the stability diagram developed from SSI-COV. The size of the data Hankel matrix is (500x500). Fig. 3a used the OMAC criteria of 99.99% to remove the spurious mode while Fig. 3b used both OMAC (99.99%) and MPC (99.9%) criteria. It is clear that with adding extra MPC criteria some spurious modes near frequency 4.5 Hz and 6.4 Hz can be removed from the stability diagram, but still has some poles which still stay in the stability diagram. This is the phenomenon of complex modes (the effect of travelling waves). To check poles under such ambiguity, wavelet packet transform (WPT) is applied to extract signal near the frequency where ambiguity poles are located. As shown in Fig. 4, the response (from WPT) from each measurement is plotted at frequency of 4.28 Hz and 5.26 Hz (from the location of ambiguity poles) [14]. It is observed that the phase difference among the measurement at frequency of 4.28 is not consistent even though the identified poles are satisfy both the OMAC and MPC criteria. Through such an observation these ambiguity poles can be also be removed from the stability diagram.

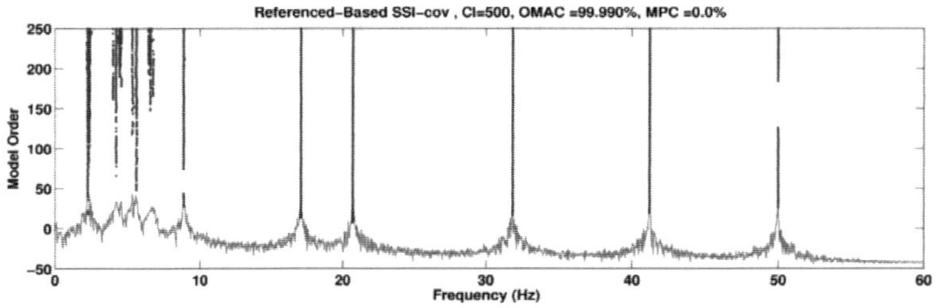

. b

Figure 3a. Stability diagram with using OMPC criteria to remove the spurious modes.

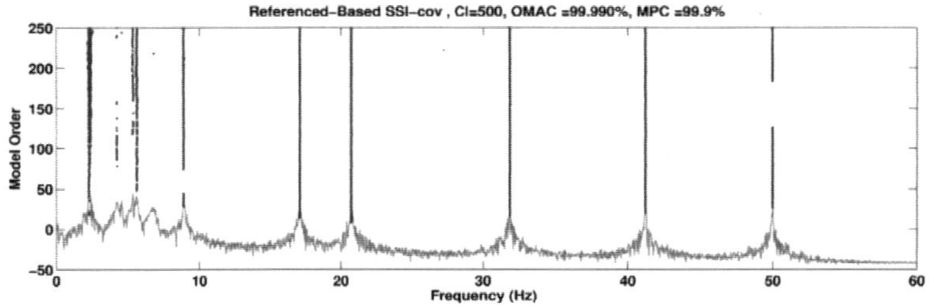

Figure 3b. Stability diagram with using both OMPC and MPC criteria to remove the spurious modes.

(a) (b)

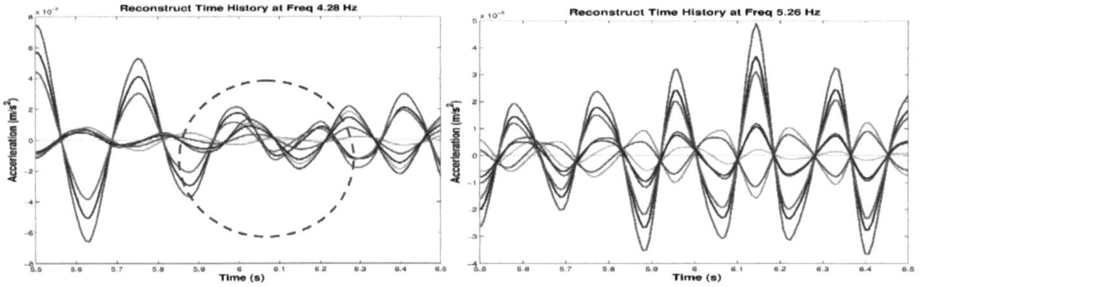

Figure 4: (a) Extract vibration signal at frequency of 4.28 Hz,
(b) Extract vibration signal at frequency of 5.26 Hz.

As discuss before, in using the SSI-COV to identify the system dynamic characteristics there is still have some hard criteria, such as the size of data Hankel matrix or threshold of damping ratio, that need to be determined (cannot automatically setup). On the contrary, in applying BSS less pre-determined parameters is required, therefore, the BBS technique is used to minimize of using model parametersfor system identification. Fig. 5 shows the Fourier amplitude spectrum of the identified source signals (a total of 8 modes which are equivalent to the system modal responses). There is no spurious mode needs to be removed. It is confirmed that the modal frequency from each identified source is consistent with the result from SSI-COV, as shown in Table 2. The result is very consistence. Results of identification from SSI-COV and BSS-SOBI algorithms can also be also applied to the damage case (a 50% reduction of flexible stiffness between measurment pont 5~6). Table 2 also lists the identified system natural frequencies for comparison. It is observed that the change of natural frequencies of the damage case is obvious and particulary focus on the higher modes. The identified mode shapes from the mixing matric **A** of BSS for both damaged and undamaged case is shown in Fig. 6. It is observed that there is no significant difference among them, so the identified mode shapes are not a good indicator for damage detection even though there have obvious chang on the identified modal frequencies. To perform damage detection further analysis is required.

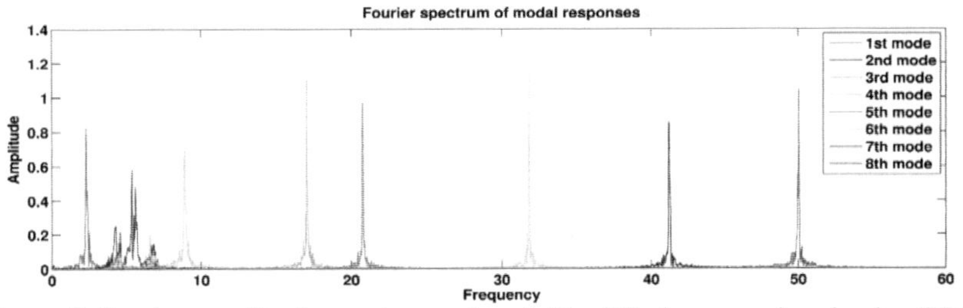

Figure 5. fourier amplitude spectrum of each identified source signal using SOBI

Table 2. Comparison on the identified natural frequencies between damaged and udamafged case.

Bridge Model		Undamaged case		Damaged case	
Mode	Theory (Hz)	SSI-COV (Hz)	SSOBI (Hz)	SSI-COV (Hz)	SOBI (Hz)
1st	2.305	2.262	2.300	2.078 (-0,184)	2.133 (-0.167)
2nd	5.300	5.613	5.367	5.596 (-0.017)	5.367 (-0.0)
3rd	8.829	8.840	8.867	8.366 (-0,474)	8.367 (-0.500)
4th	17.071	17.002	17.033	16.377 (***-0.625***)	16.430 (***-0.603***)
5th	20.815	20.714	20.733	20.386 (-0.346)	20.430 (-0.303)
6th	32.251	31.823	31.867	29.941 (***-1.882***)	29.970 (***-1.897***)
7th	42.134	41.196	41.233	40.643 (***-0.553***)	40.070 (***-1.163***)
8th	51.692	49.981	50.033	47.193 (***-2.788***)	47.230 (***-2.803***)

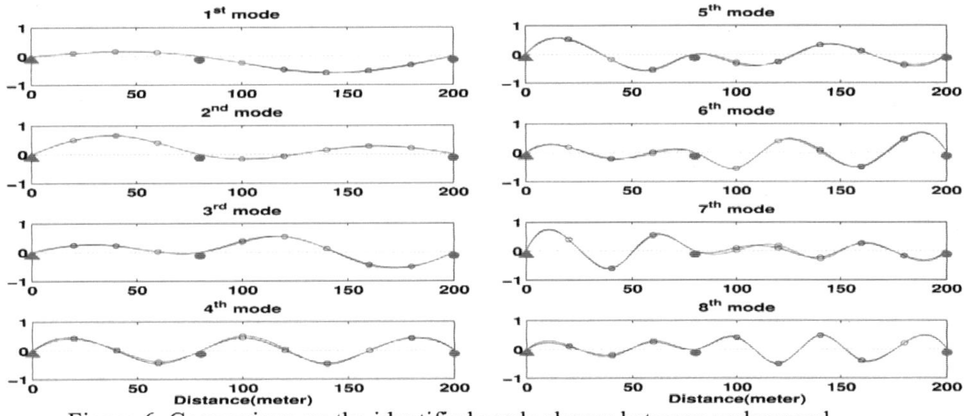

Figure 6. Comparison on the identified mode shapes between undamaged and damaged cases(SOBI method)

5. Damage detection using reponse measurement

To identify the damage location of the beam-like structure, three different approaches are used. First, taking the advantage of constructing the data Hankel matrix, a method based on null subspace and subspace damage detection is used [11]. It is defined the null-space damage index (DI_n) as:

$$DI_n = mean\left\{\left|\mathbf{U}_s \mathbf{U}_{n0}^{\mathbf{T}}\right|\right\} \tag{7}$$

where $\mathbf{U}_s$ is the subspace of the target analysis matrix $\mathbf{Y}$ (multiplication of the transpose of data Hankel matrix) and $\mathbf{U}_{n0}$ is the null-space of the reference analysis matrix $\mathbf{Y}_0$. Different from the null-space damage index, the subspace damage index, DIs, is also defined as [11]:

$$DI_S = \frac{\sum_{i=1}^{K}(\mathbf{y}_i^T\mathbf{y}_i - \mathbf{y}_i^T\mathbf{U}_{s0}\mathbf{U}_{s0}^T\mathbf{y}_i)}{\sum_{i=1}^{K}\mathbf{y}_i^T\mathbf{y}_i} \quad (8)$$

where $\mathbf{y}_i$ is the ith column of matrix $\mathbf{Y}$. The term in the denominator is used to normalize the damage index DI_s so as to have DI_s ranges from 0 to 1. The reference data (undamaged data) is needed to calculate the damage indices. Larger DI_s or DI_n means the monitor structure is different from the reference structure.

It is also believed that data collected from near damage location may record significant response signals due to the occurrence of damage in the structural member. Damage index using the difference of Hilbert amplitude spectrum of the damaged data and the reference (undamaged) data can be used for locating the damage. It is now defined the Hilbert amplitude assurance criteria (HAAC) between the damage case and reference case of the Hilbert amplitude spectrum at sensor location i as $\mathbf{e}^i_{t,f}$:

$$\mathbf{e}^j_{t,f} = \frac{\sum_t\sum_f(\mathbf{H}^{d_j}(t,f)\times\mathbf{H}^{Ref}(t,f))}{(\sum_t\sum_f\mathbf{H}^{Ref}(t,f)^2\cdot\sum_t\sum_f\mathbf{H}^{d_j}(t,f)^2)^{1/2}} \quad (9)$$

Large value of $\mathbf{e}^i_{t,f}$ indicated the similarity is higher at meaurement location i. As shown in Fig. 7a, at node 5 and node 6 the HAAC value is much lower which indicate the damage location.
Finally, to verify the accuracy of the proposed method in damage location identification, the the multivariate AR model with order of 3 was used to identify the model parameters [12]. The analysis of multivariate AR model is conducted using all the measurement data. By using the identified model parameter from undamaged data, and adopt the damaged data for one step prediction for all measurement node, predict error can determined. Fig. 8 shows the error norm increases significanrtly near the location where element stiffness reduced. Comparison with the estimated damage location using the change of mode shape curvature is als shown in the same plot. The result shows the identified damage location is almost the same from three different approaches.

(a) (b)

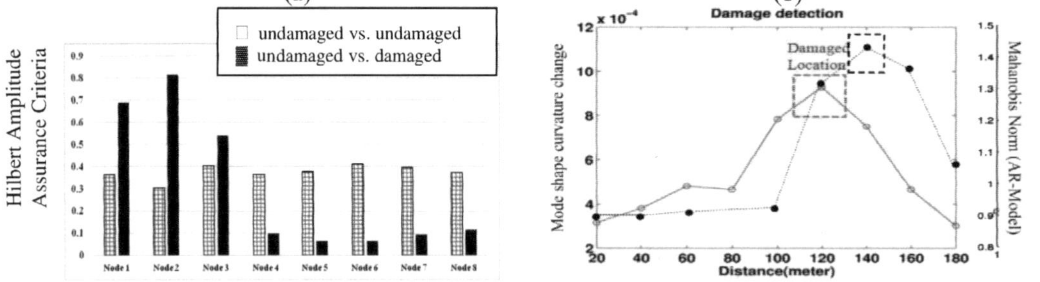

Figure 7. (a) Calculate HAAC between undamaged and damaged cases to estimate the damage location; (b) Plot the change of mode shape curvature change (open circle) and the prediction error norm (close circle) with respect to measurement location.

6. Conclusions

This study presented three different system identification methods to extract the dynamic characteristics and damage detection by using output-only measurement of the bridge-vehicle interaction system. Through numerical simulation the dynamic response of a 2-span bridge subject to random travelling vehicle loads and with the consideration of surface roughness was generated for this study. A new developed automatic covariance-drive stochastic subspace identification (SSI-COV) algorithm, the second-order blind identification (SOBI) and multivariate AR-model are used to extract the dynamic characteristics of the bridge system. Technique to remove the spurious modes in developing the stability diagram of using SSI-COV is discussed. It is demonstrated that both methods can exactly identify the system natual frequencies and mode shapes. With the implemeation of reduction in element stiffness to simulate the damage of the bridge, through the proposed identification merthods, one can also detect the damage and identify the damage location by considering the mode shape curvature change, norm of the prediction error and Hilbert amplitude assurance criteria. It demonstrated that each proposed identyification method has it goodness as well as the drawback in system identifiction. For system with well separate modes either one of the method can detect the damage location well.

Acknowlefgement
The support from Monistry of Science & Technology under grant No. MOST 103-2625-M-002 -006 is acknowledged.

References

[1] A.G.Poulimenos, S.D.Fassois, *Parametric time-domain methods for non-stationary random vibration modeling and analysis—a critical survey and comparison*, Mechanical Systems and Signal Processing, 20(4) (2006)763–816.

[2] P.Argoul,T.-P.Le, *Instantaneous indicators of structural behavior based on the continuous Cauchy wavelet analysis*, Mechanical Systems and Signal Processing,17(1) (2003)243–250.

[3] Bart Peeters and Guido De Roeck, *Reference-based stochastic subspace identification for output-only modal analysis. Mechanical Systems and Signal Processing* 13(6) (1999) 855-878.

[4] Yang, Y. B., and Lin, B. H. (1995). *Vehicle-bridge interaction analysis by dynamic condensation method*, J. Struct. Engrg., ASCE. 121(11) (1995) 1636-1643.

[5] Y.B. Yang, Y.S. Wu, *A versatile element for analyzing vehicle–bridge interaction response,* Engineering Structures 23 (2001) 452–469.

[6] International Organization for Standardization (ISO), Mechanical vibration–road surface profiles, reporting of measured data, ISO8608:1995(E).

[7] Chaojun Huang and Satish Nagarajaiah, *Experimental study on bridge structural health monitoring using blind source separation method: arch bridge*, Structural Monitoring and Maintenance, Vol. 1, No. 1 (2014) 069-087

[8] B. Swaminathan, B. Sharma, S. Chauhan, *Utilization of blind source separation techniques for modal analysis,* Proceedings of the IMAC-XXVIII February 1–4, Jacksonville, Florida USA ©2010 Society for Experimental Mechanics Inc, 2010.

[9] F. Poncelet, G. Kerschen, J.-C. Golinval, D. Verhelst, *Output-only modal analysis using blind source separation techniques*, Mechanical Systems and Signal Processing 21 (2007) 2335–2358.

[10] Adel Belouchrani, *A Blind Source Separation Technique Using Second-Order Statistics*, IEEE transaction on Signal Proceedings, Vol. 45, No. 2, February 1997.

[11] Yan, Ai-Min and Jean-Claude Golinval, *Null subspace-based damage detection of structures using vibration measurements*, Mechanical Systems and Signal Processing, Vol: 20, (2006) 611–626.

[12] Vu, V. H., et al.. *Operational modal analysis by updating autoregressive model. M*echanical Systems and Signal Processing, 25(3), (2011) 1028-1044.

[13] Wei-Ting Hsu, Chin-Hsiung Loh and Shu-Hsien Chao,· *Uncertainty Calculation for Modal Parameters Used with Stochastic Subspace Identification: An Application to a Bridge Structure*, Paper ID 9435-25, SPIE Smart Structures/NDE 2015 conference, San Diego, March 9-12, 2015.

[14] Sheng-Fu Chen, Tzu-Yun Hung and Chin-Hsiung Loh, Analysis of traffic-induced vibration and damage detection by blind source separation with application to bridge *monitoring,* Paper ID 9435-11, SPIE Smart Structures/NDE 2015 conference, San Diego, March 9-12, 2015.

Experimental verification of the Interpolation Method on a real damaged bridge

M P Limongelli [1], **A Morassi**[2]
[1]Politecnico di Milano, Department of Architecture, Built Environment and Construction Engineering, Milano, Italy.
[2]Università degli Studi di Udine, Department of Civil Engineering and Architecture, Udine, Italy.

E-mail: mariagiuseppina.limongelli@polimi.it

Abstract. The identification of damage in a bridge from changes in its vibrational behavior is an inverse problem of important practical value. Significant advances have been obtained on this topic in the last two-three decades, both from the theoretical and applied point of view. One of the main problems when dealing with the assessment of vibration based damage identification methods is the lack of experimental data recorded on real damaged structures. Due to this, a large number of damage identification algorithms are tested using data simulated by numerical models. The availability of data recorded on a damaged bridge before its demolition gave the authors the uncommon chance to verify the sensitivity and reliability of the IDDM basing on data recorded on a real structure. Specifically data recorded on a reinforced concrete single-span supported bridge in the Municipality of Dogna (Friuli, Italy) were used to apply the damage localization algorithm. Harmonically forced tests were conducted after imposing artificial, increasing levels of localized damage. In this paper the sensitivity of the method is discussed with respect to the number of instrumented locations and to the severity of the damage scenarios considered

1. Introduction

The interest in evaluating the structural health condition basing on non-destructive vibrational methods has significantly increased over the past few decades. Changes in the dynamic characteristics of the structure can indicate the emergence of possible damage occurring during the structure lifetime and provide quantitative estimates of the level of residual safety. Among vibration-based damage-detection methods [1], [2], those based on changes in modal shapes or their derivatives, and on frequency response functions (FRFs), are particularly effective allowing the localization of damage in addition to its detection, when variations of the deformed shape of the structure are taken into account in the definition of the damage feature. One of the main problems when dealing with the assessment of vibration based damage identification methods is the lack of experimental data recorded on real damaged structures. Due to this, a large number of algorithms have been tested using data simulated by numerical models that in many cases are affected by simplifications and assumptions that somehow alter the correct representation of reality. For example the effect of nonlinearities, soil-structure interaction, non-structural components but also experimental errors or environmental and operational changes, is often neglected when working on numerical models. On the contrary full scale testing such as for example those reported at references [3]-[5] allows taking into account all the phenomena contributing to the real behaviour of the structure thus providing an effective mean to check the performances of methods and algorithms for damage identification.

The availability of data recorded on a damaged bridge before its demolition gave the authors the uncommon chance to verify, using data recorded on a real structure, the sensitivity and reliability of the

[1] corresponding author

recently proposed Interpolation Damage Detection Method. Specifically, data recorded on a reinforced concrete single-span supported bridge in the Municipality of Dogna (Friuli, Italy), were used to apply the damage localization algorithm. The bridge consists of a slab supported by three longitudinal beams simply supported at the ends. Harmonically forced tests were conducted after imposing artificial, increasing levels of localized damage. In this paper the sensitivity of the IDDM is discussed with respect to the number of instrumented locations and to the frequency range considered for the estimation of the damage index. A couple a variants in the definition of the damage index are investigated as well with reference to the experimental data recorded on the bridge.

2. The basis of the IDDM for damage localization

The method applied herein to identify the location of damage is the Interpolation Damage Detection Method previously presented [6], [7] for the case of seismically excited structures. The damage detecting feature in the IDDM is defined in terms of the error related to the use of spline functions in approximating the *displacement profile* of the bridge. Specifically, the modeling accuracy at a given location of the bridge is defined as the difference between the displacement profile actually measured and the displacement profile computed at that same location by interpolating the measured displacements at all the other locations. The possible increase of the interpolation error at one instrumented location between a reference (undamaged) state and the inspection state (possibly, a damaged state) is assumed to be a symptom of the existence of damage close to the location where the change is detected. The displacement profile, either in the reference and in the inspection phase, is defined in terms of the frequency response functions (FRF) of the acceleration. Namely, for each frequency value, the values of the transfer functions between the acceleration at the measurement points and the input acceleration provide the "*Operational Deformed Shape*" (ODS) at that frequency. At the l-th location z_l and at the frequency f_i, the spline interpolation error is defined as the difference of the transfer function $H_R(z_l,f_i)$ calculated from the recorded signals, and the function $H_S(z_l,f_i)$ calculated through interpolation of the transfer functions $H_R(z_k,f_i)$ at all the other instrumented locations, with $k \neq l$, that is

$$H_S(z_l, f_i) = \sum_{j=0}^{3} c_{j,l}(f_i)(z_l - z_{l-1})^j , \tag{1}$$

where the coefficients (c_{0l}, c_{1l} c_{2l} c_{3l}) are functions of the values $H_R(z_k, f_i)$ at locations $\{z_k\}_{k=1}^{n}$, e.g., $c_{j,l}(f_i) = g(H_R(z_k, f_i))$, $k \neq l$. In terms of FRFs the interpolation error at location z (in the following the index l will be dropped for clarity of notation) at the i-th frequency value f_i, is defined as the difference between the magnitudes of recorded and interpolated FRFs:

$$E(z, f_i) = |H_R(z, f_i) - H_S(z, f_i)| , \tag{2}$$

where H_R is the FRF of the response recorded at location z and H_S is the spline interpolation of the FRF at z. In order to characterize each location z with a scalar-valued error parameter, the norm of the error on the whole range of frequencies has been considered:

$$E(z) = \sqrt{\sum_{i=n_o}^{n_o+N} E(z, f_i)^2} , \tag{3}$$

where N is the number of frequency lines correspondent to the frequency range starting at line n_o, where the signal to noise ratio is high enough to allow a correct definition of the FRF.

The difference between the two values, respectively E_0 and E_d of E calculated at location z respectively in the baseline (undamaged) and in the inspection (potentially damaged) phases, provides an indication about the existence of a degradation at the considered location:

$$\Delta E(z) = E_d(z) - E_0(z) . \tag{4}$$

An increase in the interpolation error, i.e. $\Delta E(z) > 0$, highlights a localized variation in the operational deformed shape and it is assumed to be a symptom of a local variation of stiffness at z associated with the occurrence of damage. In order to remove the effect of random variations of $\Delta E(z)$

a threshold is calculated in terms of the mean $\mu_{\Delta E}$ and variance $\sigma_{\Delta E}$ of $\Delta E(z)$ on the population of available values (that is calculated at all the instrumented locations).

$$\Delta E(z) > \mu_{\Delta E} + \nu\sigma_{\Delta E} \ . \tag{5}$$

The threshold is assumed as a minimum value beyond which a damage is considered to exist at locations where the damage index exceeds the threshold. In our simulations ν takes the values 1 or 2 that, assuming a Normal distribution of this parameter, correspond to accept respectively the 15% and the 2% probability of having a false alarm (threshold exceeded by chance).

Therefore, the damage index $D(z_l)$ at a given location z_l is defined by the relation

$$D(z_l) = \Delta E(z_l) - (\mu_{\Delta E} + \nu\sigma_{\Delta E}), \quad l=1,...,n, \tag{6}$$

Positive values of $D(z_l)$ are assumed to indicate a possible damage occurred at z_l. It should be recalled that if the right hand side of equation (6) is negative, then the damage index is assumed to be equal to zero.

3. A case study: the Dogna Bridge

Dogna Bridge is the four-span, one-lane concrete bridge shown in Figure 1. The length of each span is 16.0 m and the lane is about 4 m width. The bridge deck is formed by a reinforced concrete (RC) slab of 0.18 m thickness, supported by three longitudinal RC beams of rectangular cross-section 0.35×1.20 m. Beams are simply supported at their ends and are connected at the supports, at mid-span and at span-quarters by transverse RC diaphragms. Dynamic tests were performed from April 2 to April 11, 2008 on the end span highlighted in **Figure 2** before the demolition of the bridge. This span was made independent of the adjacent span by removing the deck-joint in correspondence of the pier. Moreover, the asphalt overlay of about 0.1 m thickness was also removed before testing.

Figure 1. General view of Dogna Bridge (left) and detail of damages D1-D6 (right).

We refer to the reference [8] for a complete report on the experiment. Dynamic tests were performed on the bridge in its actual condition (undamaged configuration, indicated by U in what follows) and in seven damaged configurations (D1-D7), see Figure 2. The first six damage states were obtained by cutting a lateral RC beam. The seventh level of damage was obtained by removing the concrete near the mid-span cross-section of the same beam. The vertical motions of the deck structure were produced by means of a vibration generator . The experimental layout is shown in Figure 3. Based on this experimental setup, deck's inertance of the bridge was measured by means of zoom analyses within narrow neighborhoods of natural frequencies. Frequency resolution ranged from 0.02 for the lower modes (up to 15 Hz) to 0.04 Hz for higher modes. The above procedure has been applied for the characterization of all the damaged configurations D1-D7. Figure 4 reports the magnitudes of the FRFs in the undamaged and in three damaged configurations at locations A10 (downstream beam), A11 (central beam) and A14 (upstream beam) considering a linear interpolation for the frequency ranges not measured during the forced vibration tests.

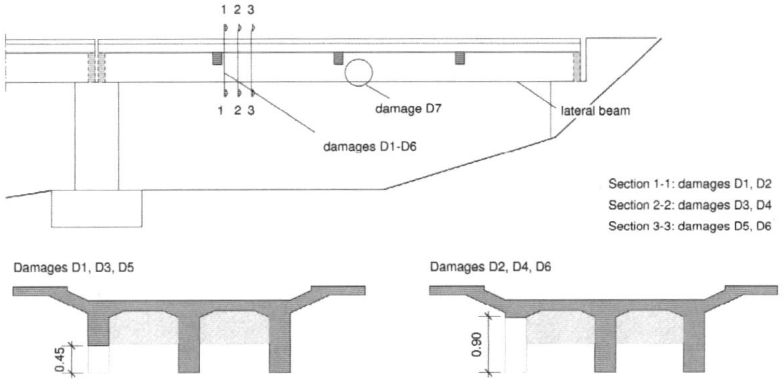

Figure 2. Damaged configurations.

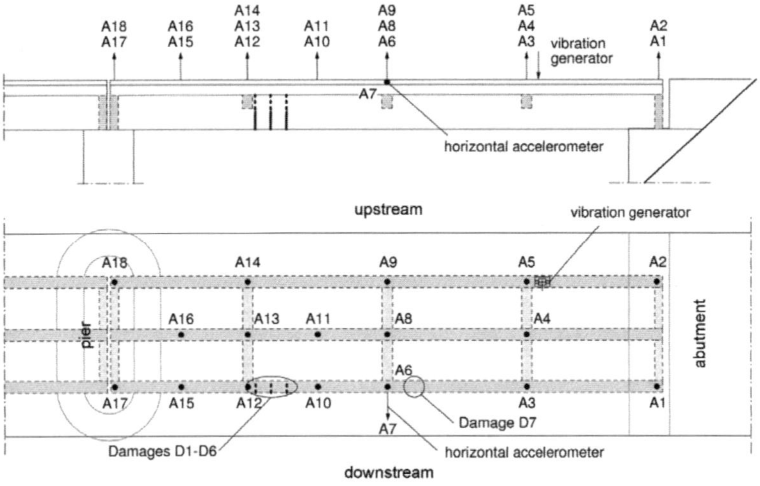

Figure 3. Instrumental layout.

The variations of the first two natural frequencies (around 9 Hz and 13 Hz) are evident from the FRFs at locations A10 and A14 (Figure 4a-c); the evolution of the frequency of the fourth mode (around 35 Hz) can be easily recovered from the FRF associated to location A14 (Figure 4c). The FRFs at location A11 (Figure 4b) are particularly useful to follow the evolution of the frequency of the fifth mode (about 47 Hz). From the undamaged configuration D0 up to configuration D3 a gradual reduction of the frequencies of all the modes 1 to 5 occurs. Starting from configuration D4 and up to D5 the frequencies of all the modes slightly increase. Finally a reduction of all the modal frequencies is exhibited for configurations D6 and D7. This behavior has already been pointed out in previous papers. Particularly, we refer to reference [9] for an interpretation of the dynamic tests and for structural identification analysis on Dogna Bridge. All the measured vibration modes have dominant vertical components.

4. Detecting damage on Dogna Bridge

As recalled in Section 3, the bridge was harmonically excited by means of a shaker located at one fourth of the upstream beam. The main aim of the testing was to carry out a modal characterization of the structure; hence responses were measured only within narrow neighborhoods of the expected natural frequencies values and the measured frequency range was adjusted at each test in order to match the current values of natural frequencies. The FRFs at instrumented locations are thus available only in narrow frequency ranges that change from one test to the other. The application of the IDDM requires through equation (3) and (6), requires that the FRFs are measured in the same frequency range in the reference and in the damaged configurations. In order to calculate the values of the FRF in the frequency ranges where they were not available, a linear interpolation of the measured values was carried out in the range 8-50 Hz. Even if this is a rather crude estimation of the actual values of the FRFs, it allows to correctly detecting the damaged locations, as will be shown in the following.

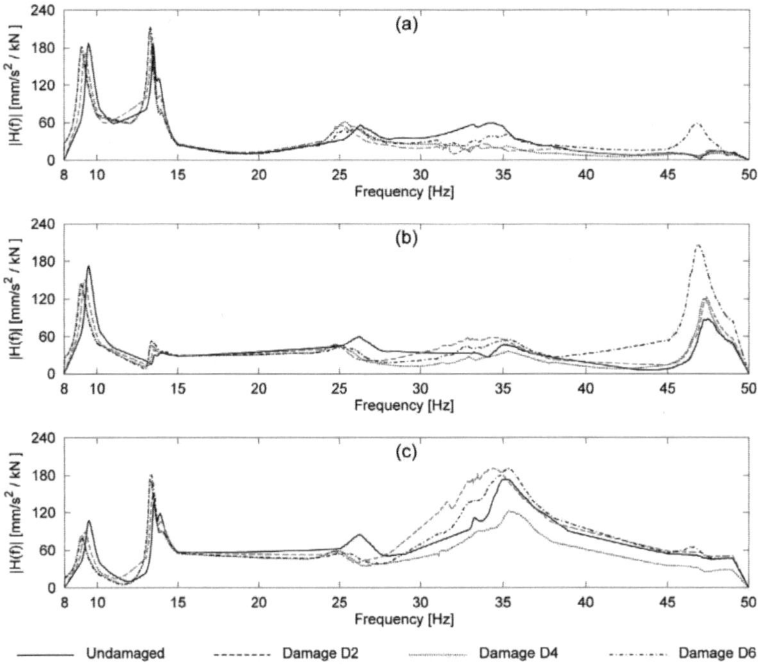

Figure 4. FRF at locations A10 (a), A11 (b) and A14 (c) for configurations U, D2, D4, D6 in the frequency range 8-50 Hz.

In order to investigate the sensitivity of the diagnostic method to the parameters on which the damage feature depends (number of instrumented locations, frequency range for the evaluation of the interpolation error, accepted probability of missing alarm) several analysis where carried out considering different values of these parameters. Referring to the paper [9] for more details, the main findings of the analysis are as follows:

- *Selection of the accepted probability of false alarm P_f through the parameter v*: Simulations show that the choice v=2 ($P_f \cong 2\%$) allows for a correct localization of severe damage only and, in addition, the probability of missing alarms turns out to be very high. Conversely, the choice v=1 ($P_f \cong 35\%$) leads to a better compromise between the probability of having false and missing alarms, compare Figure 5with Figure 6.
- *Number of instrumented locations*: the IDDM was initially applied considering separately the three alignments of sensors at downstream, central axis and upstream. More precisely, the damage index at an instrumented point was evaluated by considering only the FRFs measured at sensors belonging to the same alignment, and neglecting the responses recorded at all the other instrumented locations. Results are reported in Figure 5 and Figure 6 respectively for the two considered values (ν=1 and ν=2) of the probability of false alarm.

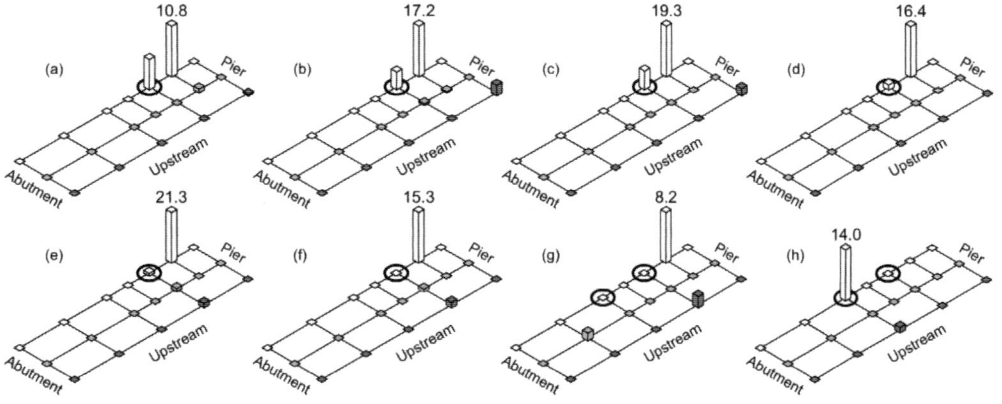

Figure 5. Influence of the probability of the accepted probability of false alarm. **Threshold ν=1**. Damage index D evaluated in the interval 8-38 Hz (first four vibrating modes) from the reference configuration (U) to actual damage configuration: (a) D1; (b) D2; (c) D3; (d) D4; (e) D5; (f) D6; (g) D7; (h) from D6 to D7. Circles denote the actual damage locations.

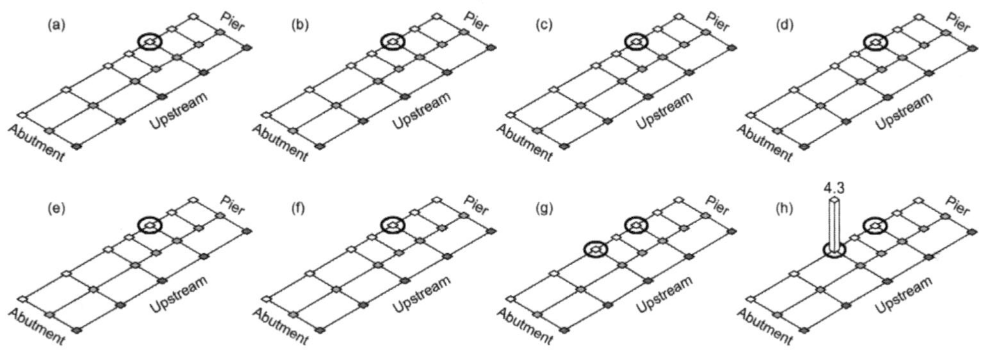

Figure 6. Influence of the probability of the accepted probability of false alarm. **Threshold ν=2.** Damage index D evaluated in the interval 8-38 Hz (first four vibrating modes) from the reference configuration (U) to actual damage configuration: (a) D1; (b) D2; (c) D3; (d) D4; (e) D5; (f) D6; (g) D7; (h) from D6 to D7.Circles denote the actual damage locations.

The above analyses were repeated by including all the instrumented points in the evaluation of the damage index. In Figure 7 and Figure 8 results obtained considering all the instrumented locations for the estimation of the damage feature are reported.
In general terms, the larger amount of data improves the identification and damage localization turns out to be more reliable, whatever the value chosen for the probability of false alarm P_f(ν=1 or ν=2). In particular, false alarms located far from the actual damage position are reduced in number and the performance of the choice ν=2 is now comparable to that resulting from taking ν=1 that is results (reported in Figure 7 and Figure 8) appear less sensitive to the assumption about the accepted probability of false alarm.

A couple of variants in the estimation of the damage parameter $E(z_l, f_i)$ were also investigated. A first variant consists in neglecting the phase information by assuming that the interpolation error $E(z_l, f_i)$ is given as the absolute value of the difference between the modulus of the recorded and interpolated FRFs, namely

$$E(z_l, f_i) = \big\| H_R(z_l, f_i) \big| - \big| H_S(z_l, f_i) \big\| \tag{7}$$

where z_l is the location considered and f_i the ith frequency value. In spite of the reduction in input information, the quality of the identification is comparable to that obtained previously. The identification improves when all the instrumented points are considered and the parameter ν is larger, see, for example, Figure 8 reporting results obtained under the assumption ν=2.

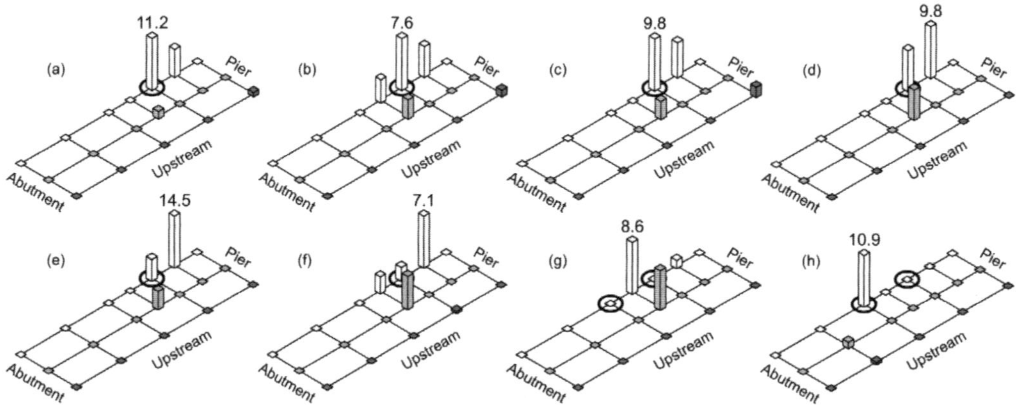

Figure 7. Damage index D evaluated according to equation (8) in the interval 8-38 Hz (first four vibrating modes) from the reference configuration (U) to actual damage configuration: (a) D1; (b) D2; (c) D3; (d) D4; (e) D5; (f) D6; (g) D7; (h) from D6 to D7. **Threshold ν=1** and all the instrumented points are considered in the evaluation of the damage index.. Circles denote the actual damage locations.

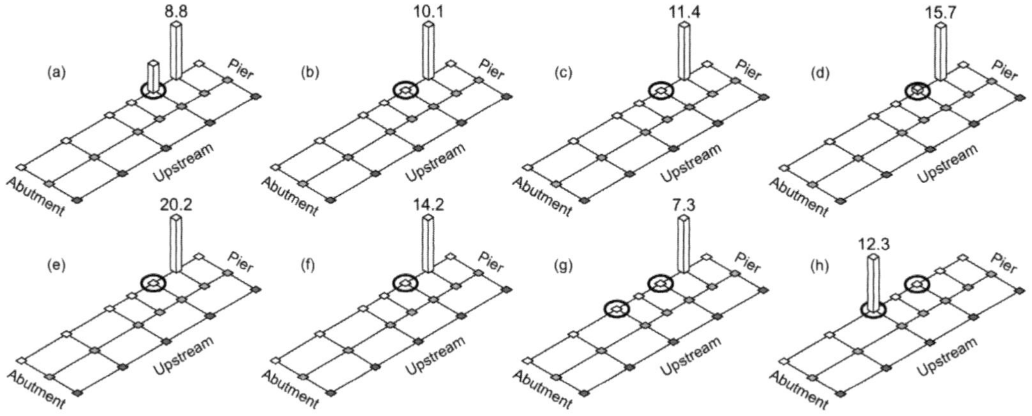

Figure 8. Damage index D evaluated according to equation (8) in the interval 8-38 Hz (first four vibrating modes) from the reference configuration (U) to actual damage configuration: (a) D1; (b) D2; (c) D3; (d) D4; (e) D5; (f) D6; (g) D7; (h) from D6 to D7. **Threshold ν=2** and all the instrumented points are considered in the evaluation of the damage index. Circles denote the actual damage locations

In a second variant, the sensitivity of the IDDM to different extension of experimental FRFs outside the measuring ranges was investigated. In particular, the values of the FRFs were reconstructed in the whole range 0-50 Hz was basing on their analytical expression and using the values of the modal parameters of the first five modes recovered from the Experimental Modal Analysis of the bridge.

Despite the high accuracy of the local reconstruction of experimental FRFs, see reference [10], significant discontinuities in the magnitude of the FRFs occur at frequencies corresponding to the boundary between measured and reconstructed FRFs. Overall, the quality of the damage localization worsens with respect to using a linear interpolation to estimate the FRFs. This can be attributed to discontinuities in the reconstructed FRFs due to the extension of the local approximation of the FRFs from the narrow neighborhoods of the natural frequencies to the whole range 0-50 Hz. Furthermore, as previously recalled, the measurement frequency ranges (hence their boundaries) changed from one test to the following leading to a shift of the discontinuities in the frequency domain that makes the comparison of the FRF corresponding to different damaged configurations quite unreliable.

5. Conclusions

In the paper the sensitivity of the Interpolation Damage Detection Method to several parameters is investigated using experimental data recorded on the Dogna bridge, a four-span, one-lane, reinforced concrete real damaged structure. The influence of both experimental conditions (number of instrumented locations, frequency range of the measurements) and of managing decisions (value of the accepted probability of false alarm) were studied with reference to data relevant to the several damaged scenarios inflicted to the bridge.

Results show that the accuracy of the damage localization increases with the amount of the available experimental data, both in terms of instrumented locations and in terms of frequency range of the measured responses. Furthermore, the increase in the amount of available experimental make results less sensitive to the choice of the accepted probability of false alarm.

6. Acknowledgments

The present work was partially supported by the ReLuis-DPC Executive Project 2014-2016, Special Project "Monitoring", whose contribution is gratefully acknowledged.

7. References

[1] Fan W, Qiao P 2008. Vibration-based damage identification methods: a review and comparative study. *Structural Health Monitoring*; 10(1) p.83–29.

[2] Yan Y J, Cheng L, Wu Z Y, Yam L H 2007 Development in vibration-based structural damage detection technique. *Mechanical Systems and Signal Processing*; 21(5) p.2198–2211.

[3] Cruz P J S , Salgado R 2008 Performance of vibration-based damage detection methods in bridges. *Comput-Aided Civ. Inf.* 24 p. 62–79.

[4] Teughels A, De Roeck G 2004 Structural damage identification of the highway bridge Z24 by FE model updating. *J. Sound Vibr.* 278 p. 589–610.

[5] Toksoy T, Aktan A E 1994 Bridge-condition assessment by modal flexibility. *Exp. Mech.* 34(3) p. 271–278.

[6] Limongelli M P 2010 Frequency response function interpolation for damage detection under changing environment. *Mech. Syst. Signal Pr.* 24(8) p. 2898–2913.

[7] Limongelli M P 2011 The interpolation damage detection method for frames under seismic excitation. *J. Sound Vibr.* 330 p. 5474–5489.

[8] Dilena M, Morassi A 2011 Dynamic testing of a damaged bridge. *Mech. Syst. Signal Pr.* 25(5) p. 1485–1507.

[9] Dilena M, Limongelli M P, Morassi A 2015 Damage localization in bridges via FRF interpolation method, *Mech. Syst. Signal Pr.* 52-53 p. 162-180.

[10] Dilena M, Morassi A, Perin M 2011 Dynamic identification of a reinforced concrete damaged bridge. *Mech. Syst. Signal Pr.* 25(8) p. 2990–3009.

Limit states for the damage assessment of bridges supported on LRB bearings

M Jara, C Gómez, JM Jara and BA Olmos
Professor, Dept. of Civil Engineering, University of Michoacan,
Morelia, México, CP 58030

E-mail: mjarad10@gmail.com

Abstract. There is now a common awareness that excessive strength is neither essential nor a synonymous of good performance in strong earthquakes. It is widely recognized that the design process should put more emphasis on predicting structural damage, according to the performance –based design philosophy. This paper presents an approach for assessing the damage level of lead rubber isolation devices and reinforced concrete piers, according to the performance-based design framework. The proposed performance limit states are based on the control of the materials strain, by means of the shear displacement of the bearing and the hinge rotation of the pier. The performance of the piers of different types of bridges is evaluated with the proposed methodology considering four different limit states.

1. Introduction

The emphasis in seismic design of bridge structures has shifted from resistance of large seismic forces to the control of deformations, rotations or displacements. There is now a common awareness that excessive strength is neither essential nor a synonymous of good performance in strong earthquakes. There are numerous cases of structures that were designed for a life safety performance according to current codes, but they were not operational after moderate earthquakes [1]. It means, that present code approach of designing only for the life safety performance, is not satisfactory.

Today, it is widely recognized that the design process should put more emphasis on predicting structural performance, and several methodologies have been proposed for performance-based design. A discussion about the characteristics and capabilities of each method can be found in [1].

Given the many sources of uncertainty that are inherent in earthquake design, a true performance based design procedure must be based on the reliability-design principles. An important task in performance-based design is the association of earthquake probability of exceedance with suitable measures of structural performance. In this context, limit states relating the seismic intensity with the damage level and structural performance are needed. This paper presents an approach for assessing the performance levels of lead rubber isolation devices and reinforced concrete piers, according to the performance-based design framework by means of four limit states. The proposed performance limit states are based on the control of the materials strain, by means of the shear displacement of the bearing and the hinge rotation of the pier.

2. Limit states for lead rubber bearings

Common earthquake protection systems for bridges are seismic isolators in association with passive energy dissipaters. In particular, lead rubber bearing (LRB) is probably one of the systems more cited in literature as an example of a single isolation device that combines flexibility and damping, providing also an economic and simple solution for protecting bridges. Rubber bearings provide the vertical stiffness needed for supporting gravitational loads and lead bars supply horizontal stiffness to control the lateral bridge displacements under service loads while giving energy dissipating capability under strong motion movements. If the bearing displacements are large enough the lead yields in shear as it is illustrated in figure 1. When coupled with the linear properties of the elastomeric bearing, the combined unit showed a bilinear force – shear deformation relationship. Typically, for computer modelling purposes, a bilinear hysteresis loops are assumed.

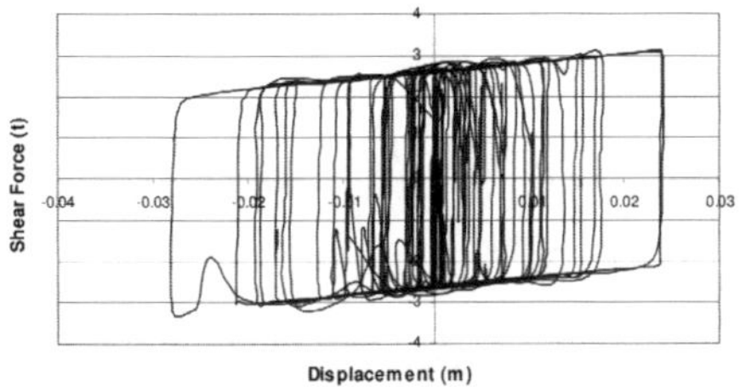

Figure 1. Hysteresis loops for lead elements

Since the shear deformation is the principal displacement in a lead rubber bearing, the shear strain γ_b is selected as the damage index for this type of isolators. Shear strain is defined as the lateral displacement x_b divided by the total thickness of rubber, excluding the total thickness of internal steel plates (figure 2).

$$\gamma_b = \frac{x_b}{h_r} \; ; \; h_r = h_t - \sum_{i=1}^{n} t_i$$

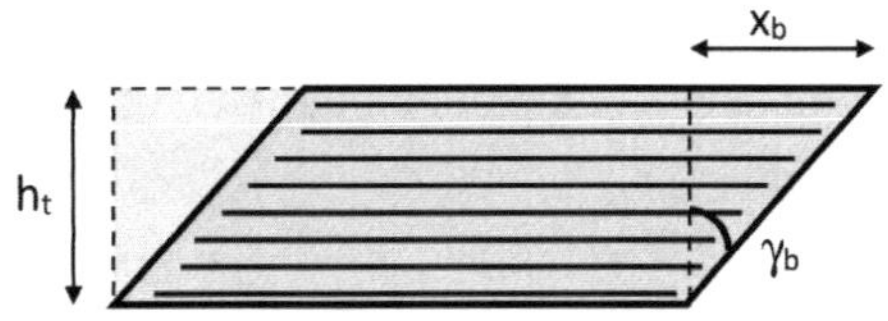

Figure 2. Shear strain as damage index for lead rubber bearings

2.1 Limit state for service earthquakes

For frequent or service earthquakes it is desirable that the bridge continues fully operational without damage; residual displacements are not allowed for the isolator bearings in order to avoid the necessity of application of restoring forces, or the substitution of one or more devices. Thanks to the recovery of the lead mechanical properties after shear inelastic deformations, the yield strength of lead can be exceeded during frequent earthquakes without residual displacements.

As suggested by Chen [2], the response of the idealized bilinear model for the lead rubber bearing shown in figure 3 can be obtained by solving a sequence of linear problems. Then, the solution of the equation of motion can be divided in the steps: 0-1, 1-2, 2-3, 3-4 and 4-0 shown in figure 3. The necessary conditions for obtaining a non-residual displacement are twofold: a) the displacement at the beginning of the last branch t = tx4 must be: xt4 = –xy, and b) the velocity at t = tx4 must be: 0. Then, the maximum displacement of the bearing for non-residual displacements occurs when xt4/xy = -1.0. Figure 4 shows the relationship between the bearing displacement ductility (μ_b = $x_{máx}/x_y$) and the displacement ratio xt4/xy, for the inelastic to elastic stiffness ratio k2 / k1= 0.1, typical ratio for lead rubber bearings. The lower curve corresponds to ξ = 0%; the middle curve is for a 5% equivalent damping; and the upper curve corresponds to a variable damping for the LRB isolators, obtained as a function of the displacement ductility developed by the bearing, according to the following equation [3]:

$$\xi_{eq} = 0.05 + \eta \ \ln(\mu_b) \leq 0.30 \tag{1}$$

where the coefficient η depended on the characteristics of the record: η = 0.065 for earthquakes recorded on firm soils and 0.085 for earthquakes recorded on soft soils.

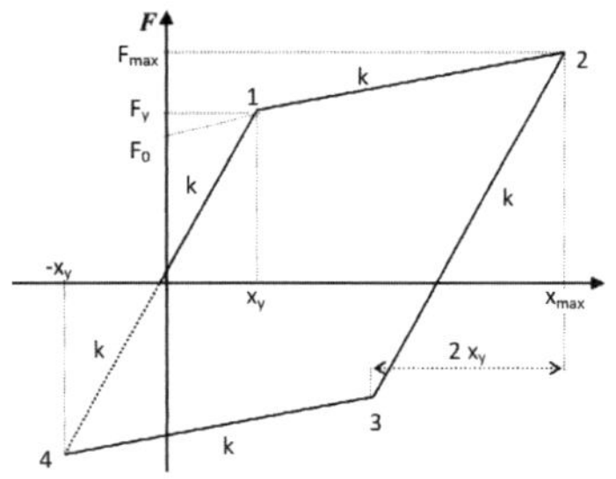

Figure 3. Idealized bilinear model for the lead rubber bearing

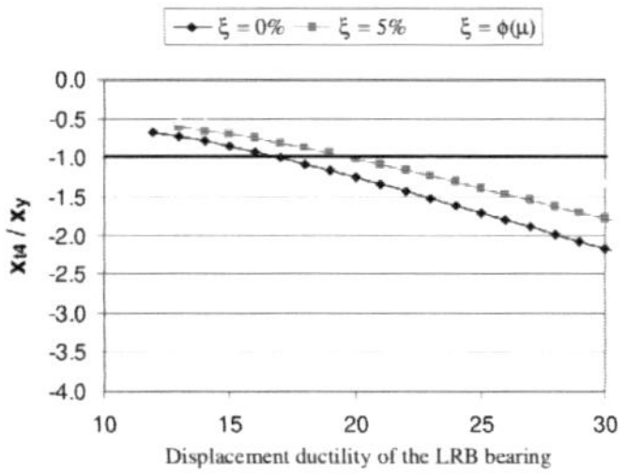

Figure 4. Displacement at time xt4 for different displacement ductility of the bearing

As the maximum displacement of the bearing for non-residual displacements occurs when xt4 / xy = -1.0, the corresponding allowable displacement ductility of the bearing (figure 4) is:

$(\mu_b)_{máx} = 17$ for ξ = 0% $(\mu_b)_{máx} = 20$ for ξ = 5% $(\mu_b)_{máx} = 24$ for ξeq = $\phi(\mu_b)$

Based on the mechanical properties of the LRB bearings, the relationship between the shear deformation and displacement ductility is [3]:

$$\gamma_{xr=0} = 17 \frac{k_2}{k_1} \frac{A_l \tau_y}{G_r A_r} \left(\frac{1}{1 + 10 \frac{A_l}{A_r}} \right) \tag{2}$$

where, τ_y is the lead shear yield stress, Gr the shear modulus of the rubber, A_l the lead core area and A_r the plan area of the rubber. After substituting typical values for the stiffness ratio, A_l / A_r ratio, shear yield stress and shear modulus of the bearing, the maximum distortion expression may be simplified [3] as:

$$\gamma_{x\,r=0} = 17(0.1)(0.05)(10)\left(\frac{1}{1+10(0.05)}\right) = 0.57 \tag{3}$$

An initial shear deformation of the bearing can be present due to tolerances and imperfections during construction; then, a reduction coefficient of 0.7 is considered for accounting the potential initial distortion of the device. The proposed maximum shear deformation for non-residual displacement of the LRB bearing (service limit state) is:

$$\gamma_{\max} \leq 0.7(0.57) = 0.4 \tag{4}$$

2.2 Limit state for service earthquakes

It is expected that the bridge remains in almost total operation condition after the earthquake, and only minor limitations to vehicle circulation, if any, can be admitted. Therefore the maximum shear deformation for non-residual displacement of the bearing is also desirable for this limit state, without any consideration of possible initial distortion of the bearing, and the proposed maximum deformation is:

$$\gamma_{\max} \leq 0.6 \tag{5}$$

2.3 Limit state for rare earthquakes

The life safety is the principal objective for this limit state and the structural damage and rehabilitation works are limited to easy access zones. Restricted traffic on the bridge can be allowed after the earthquake. For this condition it is expected that significant amount of energy is dissipated by the bearing without excessive displacements. To achieve this objective it is proposed that the maximum force transmitted by the bearing to the piers is limited to four times the yield strength of the device [3]. Then, the maximum shear distortion of the bearing for a maximum force $F_{máx} = 4\ F_y$ is equal to:

$$\gamma_{\max} \leq \mu_b \frac{k_2}{k_1}\frac{A_l\,\tau_y}{A_r\,G_r}\frac{1}{1+10\lambda} = \left[1+\frac{k_1}{k_2}\left(\frac{F_{\max}}{F_y}-1\right)\right]\frac{k_2}{k_1}\frac{A_l\,\tau_y}{A_r\,G_r}\frac{1}{1+10\lambda} \tag{6}$$

Based on the LRB´s properties the previous expression is simplified to:

$$\gamma_{\max} \leq \left[1+10\left(\frac{F_{\max}}{F_y}-1\right)\right]\frac{0.05}{1.5} = 1.03 \tag{7}$$

2.4 Limit state for rare earthquakes

The purpose is to maintain the stability of the bridge after the maximum credible earthquake in the region. The allowable deformation of the bearing must be below the rupture strain of the rubber (Ramberger [4] reports a rupture strain γ_u between 3.0 and 5.0). The maximum deformation of the bearing must consider the seismic action and the additional deformations produced by rotation of the bearing, compressive stresses, creep, shrinkage, postension and temperature. The aspects of stability, overturning, and resistance of connection should also be carefully checked. Taking into account all these aspects, the proposed maximum deformation of the isolator is:

$$\gamma_{\max} \leq 2.0 \tag{8}$$

3. Limit states for concrete piers

Based on the experience of past earthquakes and laboratory tests, damage level of reinforced concrete elements can be associated for practical purposes to maximum strain. Then, the proposed performance limit states derived in this study are based on the control of the materials strain, by means of the hinge rotation of the pier. The following expressions for estimating the allowable rotations for each limit state were obtained from the mechanical properties and usual hypothesis adopted in the design of reinforced concrete structures and were favorable compared to the experimental results reported on [5-6]. Initially, the yield rotation and ultimate compression strain of the concrete should be defined.

3.1 Yield Rotation θ_y

Rotation of the plastic hinge is due to the contribution of three effects: flexural deformations, shear deformations and bond slip of the steel reinforcement. The yield rotation caused by these factors can be assessed with the following equation proposed by [7]:

$$\theta_y = \frac{\phi_y L}{2}\left(1 - \frac{H}{3L}\right) + 0.0025 + a_{sl}\frac{0.25\,\varepsilon_{sy}\, d_b\, f_y}{(d-d')\sqrt{f_c}} \tag{9}$$

In the above expression L is the free length of the pier and H is the distance between the pier top and the center of the plastic hinge, usually H ≈ L. The second term is the shear deformation contribution to the plastic rotation and the last term is the effect of bond slip. According to the analyses of the experimental results obtained by [7], inclined cracks and shear deformations can be estimated by a constant factor of 0.0025. The third term in the above expression is proportional to the yield force of the steel in tension ($\pi d^2_b f_y/4$), and inverse to the perimeter of the bar (πd_b) and the bond strength ($\sqrt{f_c}$). If bond slip can be neglected, the last term in the equation can be dropped.

3.2 Ultimate Compression Strain for Concrete εcu

According to the strain energy balance equation [8], the total stored energy equals the strain energy in the stirrups before the rupture plus the strain energy for the unconfined concrete and the required strain energy for maintaining the plastic strain in the longitudinal steel until the stirrups failure. After solving this energy balance equation, the ultimate compression strain for concrete is [3]:

$$\varepsilon_{cu} = \frac{110\rho_s + 3.4\rho_{sl} + 0.017\sqrt{f'_{co}} - 0.07}{\rho_{sl}\left(0.94 f_{yl} + 302\right) - 0.015 f_c'^2 + 1.1 f_c' + 8} \tag{10}$$

3.3 Limit State for Service Earthquakes

The bridge must be fully operational after the earthquake for this limit state, then, if some minor cracks are formed during the earthquake, they have to be closed after the event. For typical pier elements, the first cracks are horizontal as a consequence of flexural strains, and shear deformations and bond slip can be neglected. According to experimental data, the strains before the initiation of cracking are $\varepsilon_c = 0.0015$ and $\varepsilon_s = 0.7\varepsilon_{sy}$, for the concrete and longitudinal steel respectively, and the allowable rotation for this limit state is:

$$\theta_{ser} = \frac{0.7\dfrac{f_y}{E_s} + 0.0015}{3d} L \tag{11}$$

3.4 Limit State for Occasional Earthquakes

Traffic on the bridge after the earthquake is not disrupted and only unimportant limitations to vehicle circulation can be admitted during the repairing works. Minimum repairing works such as epoxy injections in minor cracks are allowed, width cracks should be less than 2 mm without loss of cover concrete, but yield strain can be exceeded. Inclined cracks might be present and the second term of Eqn. 3.1 should be included in the estimation of θ_y. Bond slip has a small contribution at this level of strain and can be neglected. Two non-linear conditions may occur: one is the result of steel yielding, and the other is present when the concrete strain is clearly in the non linear range ($\varepsilon_{cu} \geq 1.2\, f_c / E_c$). Applying the usual hypothesis of reinforced concrete design, the rotations for these non-linear conditions are:

$$\theta_{cd,s} = \frac{L f_{sy}}{3E_s\left(d - c_y\right)} + 0.0025 \qquad \theta_{cd,c} = \frac{1.2 L f_c}{3E_c c_y} + 0.0025 \tag{12}$$

In the above expressions c_y is the distance of the outer compression fiber at the beginning of yield.

3.5 Limit State for Rare Earthquakes

For this limit state the bridge stability must be assured and restricted traffic over the bridge is admitted after the seismic event. Loss of the concrete cover is allowed but damage of the concrete core is limited to minor cracks. As spalling of concrete cover is admitted, the rotation is the result of θ_y and some additional plastic rotation:

$$\theta_{pv} = \theta_y + \left(\phi_{pv} - \phi_y\right) L_p \left(1 - \frac{L_p}{2L}\right) \tag{13}$$

where the limit curvature ϕ_{pv} depends on the maximum strain in concrete $\varepsilon_{pv,c}$ or steel $\varepsilon_{pv,s}$ for the expected damage level. The maximum values of strain can be estimated by [5]:

$$\varepsilon_{pv,s} = \frac{\varepsilon_{cc} f_{cc} b}{1.25N}\left(d - \frac{1.25N}{f_{cc} b}\right) \qquad \varepsilon_{pv,c} = \frac{1.25N\,(0.5\varepsilon_{su})}{f_c b_{co} d_{co} - 1.25N} \tag{14}$$

In the above expressions N is the axial force on the pier, f_{cc} is the confined compression resistance of the concrete, ε_{cc} the strain corresponding to the confined resistance and b_{co} and d_{co} are the width an effective depth of the confined or core section respectively.

3.6 Limit State for Very Rare Earthquakes

The bridge stability for the maximum earthquake expected in the region must be assured and emergency traffic is allowed over the bridge after the shock. Moderate damage of the core concrete is permitted but the stability of the element is not compromised neither its vertical load capacity. Rupture of stirrups must not happen and vertical rebars do not buckle for this condition. Ultimate rotation θ_u is obtained using the ultimate curvature $\phi_{u,c}$ instead of ϕ_{pv}. For assessing θ_u, bond slip should be added. ϕ_u can be originated in the rupture of the transverse bars, rupture of longitudinal steel or the loss of axial load carrying capacity of the pier. Eqn. 13 gives the limit state for very rare earthquakes, after the substitution of Eqn. 15.

$$\phi_{u,c} = \frac{\varepsilon_{cu}}{c_c} = \frac{0.68 b_{co} f_{cu} \varepsilon_{cu}}{N} \tag{15}$$

4. Displacement demand

The nonlinear static analysis procedures proposed in some current seismic design recommendations are a reasonable tool for the evaluation of the displacement capacity of new and existing bridges. According to these procedures the seismic demand is represented in an acceleration-displacement response spectra space (ADRS). Considering the limit states proposed in this study, the seismic demand can be represented in an acceleration-rotation response spectra space as illustrated in figure 5. Using the capacity spectrum method for a reinforced concrete bridge, the rotation demand and the capacity for the four limit states proposed in this study are summarised in table 1. According to the capacity/demand ratio, the bridge is adequate for the very rare earthquake with a rotation capacity twice the rotation demand; however, for the other limit states, the bridge has not adequate capacity for the seismic demand imposed by the earthquake.

Table 1. Seismic demand and rotation capacity for the bridge of figure 7.

Limit state	Rotation capacity (rad)	Equivalent damping (%)	Effective period (s)	Rotation demand	Capacity/demand
Service	0.0076	9	1.58	0.0143	0.53
Occasional	0.0106	16	1.87	0.0182	0.58
Rare	0.0227	18	2.22	0.0292	0.78
Very rare	0.0629	22	2.41	0.0316	2.00

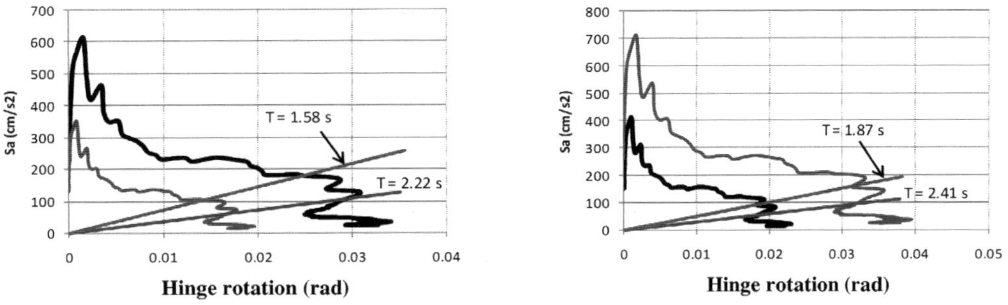

Figure 5. Seismic demand in an acceleration-rotation response spectra

5. Conclusions

This paper presents expressions for determining limit states for concrete piers and LRB isolators according to the performance-based design framework. The rotations of piers and shear deformations are selected as damage indexes and the expressions proposed were favorable compared to experimental data bases.

An expression for estimating the ultimate compression strain in concrete is also proposed. The maximum deformation predicted with this equation, improves the results obtained with other expressions proposed previously.

Considering the four limit states proposed in this study, the seismic demand can be represented in an acceleration-rotation response spectra space, and the capacity/demand ratio can be obtained for isolated bridges, with some of the nonlinear static procedures proposed in the latest version of seismic design codes.

References

[1] Bertero R D and Bertero V V 2002 Performance-Based Seismic Engineering : The Need for a Reliable Conceptual Comprehensive Approach *Earthquake Engineering and Structural Dynamics* **31** 627-652

[2] Chen Q, Douglas B M, Maragakis E A and Buckle I G 2002 Extraction of hysteretic properties of seismically isolated bridges from quick-release field tests *Earthquake Engineering and Structural Dynamics* **31** 333-351

[3] Jara M, Jara J M, Olmos B A and Casas J R 2012 Improved procedure for equivalent linearization of bridges supported on hysteretic isolators *Engineering Structures* 35 99-106

[4] Ramberger G 2002 Structural bearings and expansion joints for bridges *6th International Association for Bridge and Structural Engineering,* Sweden

[5] PEER 2004 Structural Performance Database Pacific Earthquake Engineering Research Center, University of California at Berkeley

[6] UCSD 1997 Performance/Capacity Catalog Components, Sub-Assemblages and Systems Department of Structural Engineering University of California at San Diego

[7] FIB State of the art report 2004 Displacement based-seismic design of reinforced concrete buildings Task Group 7.2, 25 ISBN 2-88394-065-7 196

[8] Mander J B, Priestley M J N y Park R, 1988 Theoretical stress-strain model for confined concrete *Journal of Structural Engineering* 114 1804-1826

The effect of post-tensioning force magnitude and eccentricity on the natural bending frequency of cracked post-tensioned concrete beams

D Noble[1], M Nogal[1], A J O'Connor[1] and V Pakrashi[2]

[1] Dept. of Civil, Structural & Environ. Eng., Trinity College Dublin, Dublin 2, Ireland.
[2] Dynamical Systems & Risk Laboratory, Civil & Environ. Eng., School of Eng., University College Cork, Cork, Ireland.

E-mail: `nobleda@tcd.ie`, `nogalm@tcd.ie`, `oconnoaj@tcd.ie`, `V.Pakrashi@ucc.ie`

Abstract. The effect of prestress force magnitude on the dynamic properties of uncracked prestressed concrete structures is something that has been widely debated among researchers to date. The effect of pre- and post-tensioning force magnitude on the natural bending frequencies of cracked prestressed concrete structures is something that is more established, and widely agreed upon. This paper describes the results of dynamic impact testing on damaged post-tensioned concrete beams. The natural bending frequency of the cracked beams were determined through experimental modal analysis. Dynamic impact response signals were obtained at different levels of post-tensioning force for the cracked beams. The Fast Fourier Transform was implemented and a peak picking algorithm was subsequently used to determine the natural bending frequencies of the beams. The relationship between prestressing force and natural frequency for both the cracked and uncracked beam sections was determined. The results for the cracked beams were compared to the results for the same uncracked beam sections. A marked difference in vibration behaviour was observed for the cracked beams between the non-fully prestressed and the fully prestressed case. Conclusions from the study are drawn and have profound implications in the fields of system identification and structural health monitoring in pre- and post-tensioned concrete structures.

1. Introduction

The effect of prestress force magnitude on the modal properties (frequency, damping and mode-shape) of uncracked prestressed concrete structures is something that has been widely debated among researchers to date [1]. Some researchers suggest that *"compression softening"* theory governs for uncracked pre- and post-tensioned concrete structures, therefore the natural frequency decreases with increasing prestress force magnitude [2, 3]. Other researchers [4] have produced non-linear mathematical models showing there is no relationship between natural bending frequency and prestress force magnitude, while others [5, 6] primarily through experimental approaches, have suggested that there is actually an increase in natural bending frequency with increasing pre- and post-tensioning force magnitude.

The effect of pre- and post-tensioning force magnitude on the natural bending frequencies of cracked prestressed concrete structures is something that is more established, and widely agreed upon. Saiidi et al. [5] report an increase in natural frequency with increasing post-tensioning force. As pointed out by Bruggi et al. [7] the tests carried out by Saiidi et al. [5] were conducted on cracked beam sections only. Uncracked sections were not tested. Williams & Falati [8] present a formula to calculate the average effective second moment of area of a cracked concrete cross section. The effect of crack closure in accordance to this method is that it increases the effective second moment of area of the cross section, and subsequently the natural bending frequencies. Hop [9] agrees, reporting a decrease in natural bending frequencies with increased cracking, and states that increasing the prestressing force acts as to close the cracks, stiffen the section and increase the natural bending frequencies of the beam sections. Grace & Ross [10] also report a decrease in girder stiffness leading to a decrease in natural frequency also attributing it to cracking in the cross section. Unger et al. [11] state that a loss in post-tensioning increases the appearance of cracks which reduces the bending stiffness and subsequent natural frequencies of the system. DeRoeck concurs [12] that prestress loss results only in measurable changes in frequency if accompanied by originating cracks. Hamed & Frostig also report that large cracking damage yields drastic reduction in the natural frequencies of cracked prestressed concrete beams [13].

Pavic et al. [14] agree that prestressing is used to overcome excessive cracking and static deflection but argue that prestressing does not significantly improve dynamic behaviour, as that is governed by stiffness, mass and damping, on which prestress force has little influence. Dall'Asta & Dezi [15] consider it is possible to determine the prestressing force by measuring the natural frequency of a PSC structure in its cracked state only. Rodriguez et al. [16] acknowledges this fact in relation to post-tensioned concrete wind turbine towers, stating that uncracked towers maintain their original stiffness and frequency, but once the towers are cracked and the cracks have been decompressed, any vibrations in the tower will mobilise smaller stiffness, which will be shown by the vibration frequencies.

This paper will describe the outcome of dynamic impact testing conducted on nine cracked post-tensioned concrete beam sections in the laboratory. The results for one beam will be discussed in detail. The relationship between the natural bending frequency and the post-tensioning force for cracked concrete sections will be determined through experimental modal analysis of the cracked beams. Comparisons will be made between the behaviour of the uncracked beams and the cracked beams. The conclusions have profound implications in the fields of system identification, damage detection and structural health monitoring.

2. Experimental Set-Up

Following dynamic impact tests in the lab on nine uncracked post-tensioned concrete beam sections, which have been described previously [17], 4-point bending was applied to each beam as outlined in Figure 1. The individual beams differed in that they had different straight profiled post-tensioning strand eccentricities. The transverse load was applied incrementally and the crack patterns were observed at different values of load application. The loading increments and yield moment for Beam 1 (e=0mm) are outlined in Table 1. First, second and third cracks were identified by visual inspection only, and were not determined scientifically.

Table 1: The incremental damage states of Beam 1 (e=0) until yielding.

Beam	**1st crack (kN)**	**2nd crack (kN)**	**3rd crack (kN)**	**Yield (kN)**	**M_{yield} (kN-m)**
C1	24.6	41.3	49.8	70.4	21.12

Yielding was identified as the point where the beams could no longer hold any additional load and deflections increased significantly. Visible structural cracks formed in the section. Vertical

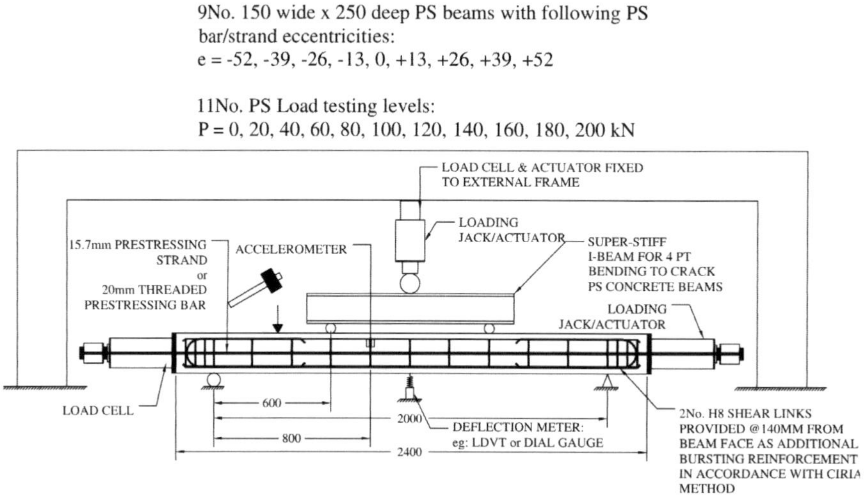

Figure 1: Experimental set-up.

flexural cracks formed in the area of high moment, between the span of the spreader beam, whereas diagonal shear cracks formed in the areas between the end of the spreader beam and the supports. Following cracking, dynamic impact testing was conducted on the beams. Figure 1 shows the experimental set-up. An accelerometer was placed 800mm from one of the supports as shown, enabling the first three modes of vibration to be captured. Three strain gauges were placed equi-distant between the support and mid-span, in order to capture the mode shapes of vibration. The beam was struck ten times at each strain gauge location using an impact rig assembled in the laboratory. This ensured repeatability of the experiment and enabled the error in the estimation of the frequencies to be calculated. This experimental procedure was then repeated at incremental values of post-tensioning load, as outlined in Figure 1. The post-tensioning load was applied using a 15.7mm Freyssinet post-tensioning strand, secured either side of the 300 ton loading jacks as shown. The different straight profiled strand eccentricities are also outlined in Figure 1. A similar testing procedure was applied to the uncracked beam sections and has been outlined in previous works by the same authors [17].

3. Experimental Analysis

Each beam was struck ten times at 3 different impact locations, for 11 different post-tensioning load levels. The response was measured by one accelerometer and three strain gauges. As such, this method of dynamic testing is known as the multiple input, multiple output (MIMO) method. The Fast Fourier Transform (FFT) was then applied to the response signals to convert into the frequency domain. No data was collected on the impact signal, and as such, the frequency response functions (FRFs) of the beams could not be determined. The FRFs are defined as the transfer function, or the ratio between the input and output signals in the frequency domain. Since the input signal was not quantified, the FRFs were not obtained. In this way, the study is an output-only study. However, FFTs of the output signal were sufficient to determine the natural vibration frequencies of the structural system.

As outlined in previous work [17, 18], having calculated the FFTs of the signals, the fundamental vibration mode was intelligible, however, due to the presence of significant noise in the data, higher modes were not. In order to eliminate this noise and identify higher modes, a signal processing technique was established by the authors and applied to the raw data.

Figure 2 and Figure 3 show the results of two processed signals. The higher modes are now intelligible. A peak picking algorithm is invoked to determine the natural frequencies of the system. Figure 2 shows an accelerometer response signal of a cracked beam for zero post-tensioning load, indicating a structural dynamic response of the cracked beam for a non-fully prestressed condition, in which the cracks are open. It is evident that the structural response of the non-fully prestressed condition is complex. The fundamental mode, in this case contributes most to the response, but its contribution is not dominant. The contribution of the other modes is relatively similar.

Figure 3 shows an accelerometer response signal of the same cracked beam for a post-tensioning load level of 180kN, indicating a structural dynamic response of the cracked beam for a fully prestressed condition. Here, the dominance of the first mode of vibration is evident. The beam, when fully prestressed, behaves monolithically and vibrates as one entity rather than a series of individual cracked entities, and hence the dominance of the first vibration mode returns. This response is in line with the response of the uncracked 'virgin' beams.

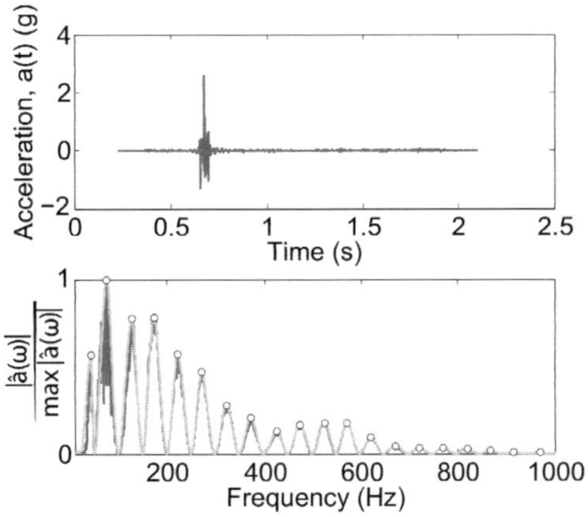

Figure 2: Structural dynamic response of cracked beam for non-fully prestressed case, Beam 1, e=0mm

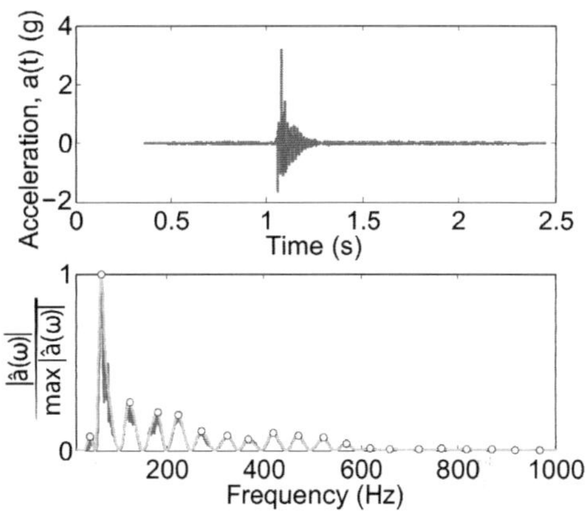

Figure 3: Structural dynamic response of cracked beam for fully prestressed case, Beam 1, e=0mm

4. Experimental Results

Figure 4 shows the structural dynamic response of Beam 1 (e=0) in its uncracked condition. The dominance of the first mode of vibration is evident in this case, since the relative modal amplitude of the fundamental bending mode accounts for a large proportion of the modal mass of the structure. Figure 5 shows the structural dynamic response of Beam 1 in its cracked condition. In this case, the dominance of the first mode of vibration is not clear. The structural dynamic response is much more complex and the contribution of the higher modes of vibration are clearly far more significant in the cracked case. This is evident up to a certain threshold value of post-tensioning force. The dominance of the first vibration mode again becomes apparent in the cracked case when the post-tensioning force causes the structure to behave as one monolithic structure rather than a series of cracked entities. This is the fully prestressed condition. In this case, for Beam 1, e=0mm, it was identified to be at a value of 160kN.

Figure 6 and Figure 7 show regression analysis of fundamental frequency versus post-tensioning load for the same beam in its uncracked and cracked states. From the regression analysis of the uncracked case in Figure 6, no statistically significant change in fundamental

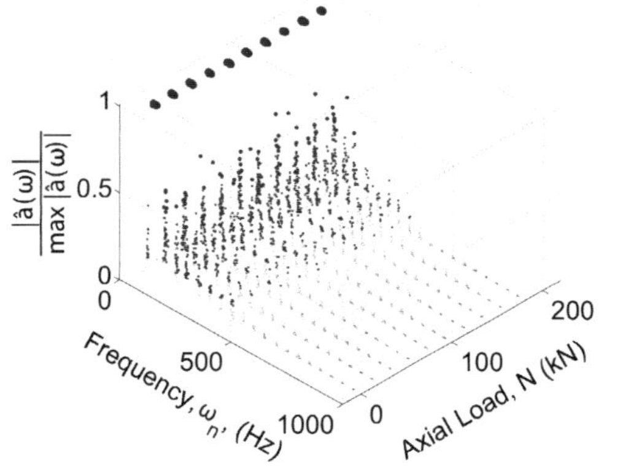

Figure 4: Rel. Modal Amp. Uncracked Beam 1, e=0mm

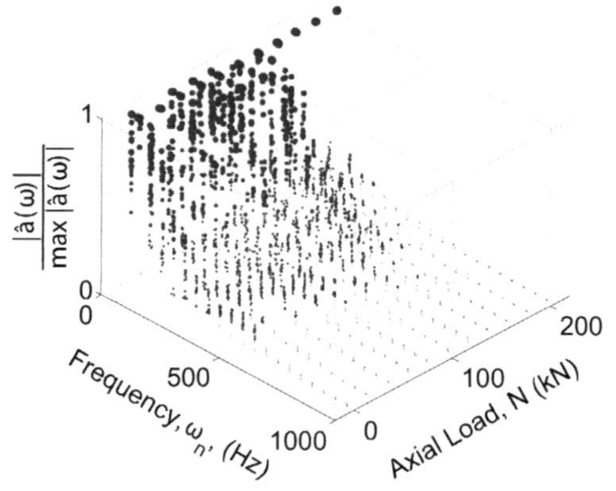

Figure 5: Rel. Modal Amp. Cracked Beam 1, e=0mm

frequency with increasing post-tensioning force can be observed, as previously reported [17]. From first glance at the simple linear regression analysis of the cracked beam case, an overall decreasing trend in fundamental frequency versus post-tensioning load is observed. However, when analysed in conjunction with Figure 5, it becomes apparent that there are two distinct trends in the data. For the non-fully post-tensioned case (0-160kN), the structural dynamic response is complex, the error in the estimation of the natural frequency is high, the structure behaves dynamically as a series of independent vibrating entities and a decreasing trend in fundamental frequency is observed. However, at a threshold post-tensioning load level and upward (160-200kN), the dominance of the first vibration mode returns, the error in the estimation of the fundamental frequency decreases, the structure begins to behave monolithically again, and an increasing trend can be observed for the final 3 data points. This increasing trend, has been widely reported for cracked beam sections [5, 8, 10, 11, 12, 13, 14, 15], however, is not observed for uncracked beam sections [4, 15, 17]. The stiffness of the section is increased by increasing the prestressing force due to crack closure, and the second moment of area moves from the cracked to the uncracked value.

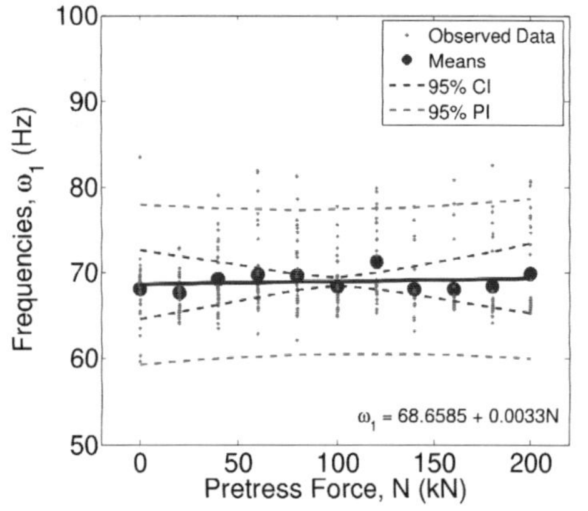

Figure 6: Regression analysis of fundamental frequency vs. post-tensioning load; uncracked case.

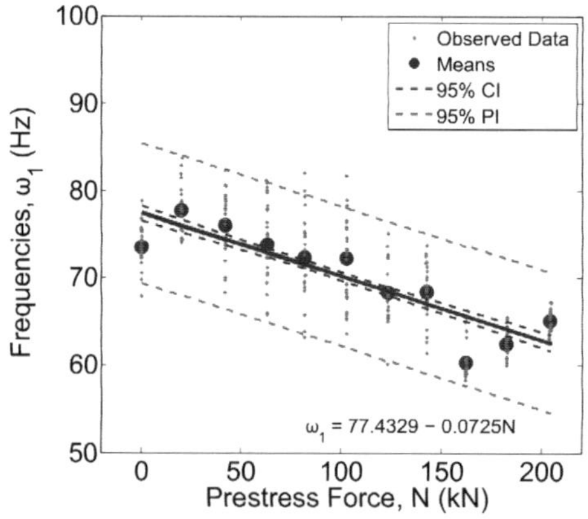

Figure 7: Regression analysis of fundamental frequency vs. post-tensioning load; cracked case.

Figure 8 shows the comparison between the simple linear regression models for the cracked and uncracked beam cases. As discussed previously, this is not strictly a reasonable comparison as two different trends can be observed in the cracked beam case. The initial fundamental frequency for the cracked beam is higher to begin with, which is counter-intuitive. If it is considered that the beam in its damaged state behaves as a series of different vibrating cracked entities rather than as one monolithic structure, then the span length of each individual vibration entity is significantly reduced. Since the span length is a denominator in the equation for the prediction of natural frequency for simply supported beams, this makes intuitive sense. This span length increases as the post-tensioning force increases and more and more structural cracks are closed, meaning the measured frequency decreases. On the other hand, when the beam begins to vibrate again monolithically, the structural dynamic response becomes less complex and the fundamental mode dominates, an increasing trend is observed in the frequency with increasing post-tensioning force. At the threshold value of post-tensioning force, the span length reaches its original value. From this point onwards, as shown in Figure 9 an increasing trend in fundamental frequency is observed, and this is attributed to the gain in geometric stiffness due to crack closure, in accordance with previous research works [5, 8, 10, 11, 12, 13, 14, 15].

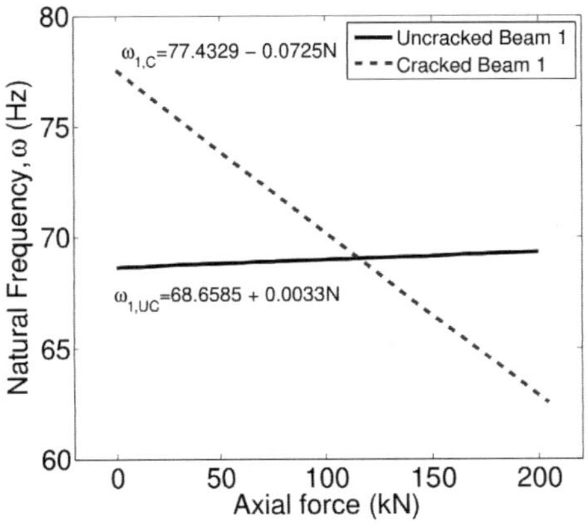

Figure 8: Regression comparison between cracked and uncracked beam cases.

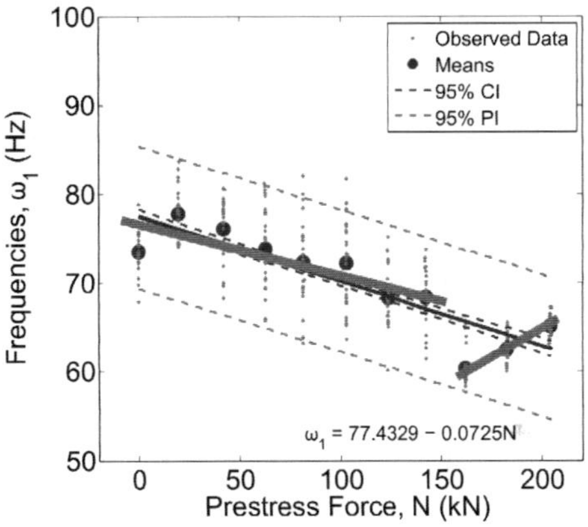

Figure 9: Evidence of the two trends present in the data.

Table 2 shows the statistical regression parameters - intercept parameter ($\alpha_{0,iuc/c}$), slope parameter ($\alpha_{1,iuc/c}$), standard error, t-values and 95% confidence intervals when regressing ω_1 on N for Beam 1, e=0 ($i = 1$) in its cracked (c) and uncracked (uc) case. The corresponding linear regression equations are obtained by substituting into Equation 1. Table 2 should be read in conjunction with Figure 8 for completeness.

$$\omega_1 = \alpha_{0,i} + \alpha_{1,i}N \tag{1}$$

5. Conclusions

Whereas there is significant disagreement among researchers as to the effect of prestress force magnitude on the natural vibration frequency of uncracked prestressed and post-tensioned concrete sections [1], the effect of prestressing force on the bending frequency of cracked pre- and post-tensioned sections is more established. Researchers agree that, for cracked pre- and post-tensioned concrete beams, increasing the post-tensioning force acts as to close the cracks, increase the second moment of area of the cross section, resulting in an increase in the natural bending frequencies [5, 8, 10, 11, 12, 13, 14, 15].

Table 2: Statistical analysis on regression parameters for ω_1 on N

Beam	e (mm)	Reg. P.	Value	SE	t-value	t-crit.	p	95% CI
B1	0	$\alpha_{0,1uc}$	68.6585	2.0515	33.4680	1.9672	0.0000	(64.6228,72.6942)
		$\alpha_{1,1uc}$	0.0033	0.0204	0.1635	1.9672	0.8702	(-0.0368,0.0434)
C1	0	$\alpha_{0,1c}$	77.4329	0.4465	173.4027	1.9680	0.0000	(76.5541,78.3117)
		$\alpha_{1,1c}$	-0.0725	0.0038	-18.8934	1.9680	0.0000	(-0.0801,-0.0650)

For the uncracked beam sections, the fundamental mode of vibration dominates the structural dynamic response and no statistically significant change in the fundamental vibration frequency can be observed. As such, for uncracked pre- and post-tensioned concrete beams, the prestressing force does not change the dynamic stiffness significantly. Pavic et al. [14] concluded similar, stating that the dynamic properties are dependent on mass, stiffness and damping properties, on which prestressing does not have a major influence. It should also be noted that the error in the estimation of the frequency is high. Changes in frequency are affected by temperature effects, lack of ideal support conditions and material variability, which is reported to be in the order of 6-18% [19, 20], therefore any potential effect of prestressing force on dynamic properties is lost within this error.

For the cracked beam sections, the implications of prestressing force on dynamic properties is more apparent. For the cracked beam sections, it was found that the structural dynamic response was far more complex than for the uncracked beams. There is no clearly dominant vibration mode and the response comprised of all modes of system. However, at threshold values of post-tensioning force there is evidence of dominance of first vibration mode. This threshold post-tensioning load value is the load value at which the structure begins to vibrate monolithically again. The trend from this point onwards is easily identified. There is an increase in fundamental frequency with increasing post-tensioning force. This is attributed to crack closure and the subsequently increase the flexural stiffness of the member, as reported by previous authors [5, 8, 10, 11, 12, 13, 14, 15].

The implications that this has for system identification, and structural health monitoring are profound. The above suggests that the identification of the existing prestressing force in uncracked prestressed concrete structures is not feasible through measurement of the natural frequency of the structure only, due to the confounding variables (temperature effects, support conditions, material properties). However, the existence of the relationship between the frequency and the level of cracking in the structure does allow for the calculation of the damage (cracking) state of the structure from the measurement of the natural frequency [12, 21, 22] and application of the formula presented in [8].

Acknowledgments

The authors would like to gratefully acknowledge the financial support donated by the Irish Research Council (IRC) under its Embark initiative. The authors would also like to sincerely thank Banagher Concrete, Heitons Steel, Roadstone Ireland, Fairyhouse Steel, and Freyssinet Ireland for their support in supplying testing materials throughout the duration of the project to date.

References

[1] Quilligan A, O'Connor A and Pakrashi V 2012 *Engineering Structures* **36** 270 – 282

[2] Tse F, Morse I and Hinkle R 1978 *Mechanical vibrations: theory and applications* Allyn and Bacon series in mechanical engineering and applied mechanics (Allyn and Bacon)

[3] Miyamoto A, Tei K, Nakamura H and Bull J 2000 *Journal of Structural Engineering* **126** 1033–1044

[4] Hamed E and Frostig Y 2006 *Journal of Sound and Vibration* **295** 28 – 39

[5] Saiidi M, Douglas B and Feng S 1994 *Journal of Structural Engineering* **120** 2233–2241

[6] Kim J T, Yun C B, Ryu Y S and Cho H M 2004 *Structural Engineering and Mechanics* **17** 467–482

[7] Bruggi M, Caprioli A, Vanali M and Venini P 2008 Investigation of the pre-stress loss in the dynamical behaviour of concrete beams *Conference: 2008 IMAC-XXVI: Conference & Exposition on Structural Dynamics* (Society for Experimental Mechanics (SEM))

[8] Williams M S and Falati S 1999 Modal testing of a post-tensioned concrete model floor slab vol 1 pp 14–20

[9] Hop T 1991 *Materials and Structures* **24**(3) 210–220

[10] Grace N and Ross B 1996 *Journal of Structural Engineering* **122** 643–650

[11] Unger J F, Teughels A and De Roeck G 2006 *Journal of structural engineering* **132** 1691–1698

[12] De Roeck G 2003 *Journal of Structural Control* **10** 127–134

[13] Hamed E and Frostig Y 2004 *Engineering Structures* **26** 1611–1621

[14] Pavic A, Reynolds P, Waldron P and Bennett K 2001 *Finite elements in analysis and design* **37** 305–323

[15] Dall'Asta A and Dezi L 1996 *Journal of Structural Engineering* **122** 458–458

[16] Rodríguez J, Martí J and Martínez F 2010 Post-cracking behaviour of a wind turbine concrete tower *Proceedings of SIMULIA Customer Conference* (Providence, Rhode Island) pp 632–646

[17] Noble D, Nogal M, O'Connor A and Pakrashi V 2014 The effect of prestress force magnitude on the natural frequencies of prestressed concrete structures *Proceedings of the* 23rd *Australasian Conference on the Mechanics of Structures and Materials (ACMSM23)* vol I (Southern Cross University) pp 333–338

[18] Noble D, Nogal M, O'Connor A and Pakrashi V 2014 Impact hammer testing on post-tensioned steel RHS sections; an investigation of the "Compression Softening" effect *Proceedings of Civil Engineering Research in Ireland Conference* vol I (Civil Engineering Research Association of Ireland) pp 427–432

[19] Peeters B and De Roeck G 2001 *Earthquake Engineering and Structural Dynamics* **30** 149–171

[20] Cornwell P, Farrar C, Doebling S and Sohn H 1999 *Experimental Techniques* **23** 45–48

[21] Kim J T and Stubbs N 2002 *Journal of Sound and Vibration* **252** 223 – 238

[22] Li J, Law S and Hao H 2013 *International Journal of Mechanical Sciences* **74** 99 – 111

Expected damages of retrofitted bridges with RC jacketing

O Montes[1], JM Jara[2], M Jara[2], BA Olmos[2]

[1] Graduate student, UMSNH, 58030 Morelia, Michoacán, México
[2] Professor UMSNH, 58030 Morelia, Michoacán, México

E-mail: omalvarez01@gmail.com, jmjara70@gmail.com, mjard10@gmail.com, ba.olmos@gmail.com

Abstract. The bridge infrastructure in many countries of the world consists of medium span length structures built several decades ago and designed for very low seismic forces. Many of them are reinforced concrete structures that according to the current code regulations have to be rehabilitated to increase their seismic capacity. One way to reduce the vulnerability of the bridges is by using retrofitting techniques that increase the strength of the structure or by incorporating devices to reduce the seismic demand. One of the most common retrofit techniques of the bridges substructures is the use of RC jacketing; this research assesses the expected damages of seismically deficient medium length highway bridges retrofitted with reinforced concrete jacketing, by conducting a parametric study. We select a suite of twenty accelerograms of subduction earthquakes recorded close to the Pacific Coast in Mexico. The original structures consist of five 30 m span simple supported bridges with five pier heights of 5 m, 10 m, 15 m 20 and 25 m and the analyses include three different jacket thickness and three steel ratios. The bridges were subjected to the seismic records and non-linear time history analyses were carried out by using the OpenSEEs Plataform. Results allow selecting the reinforced concrete jacketing that better improves the expected seismic behavior of the bridge models.

1. Introduction

Many of the medium span length bridges constructed around the world are old structures with more than 40 years. Most of them are non-seismically designed bridges and they have to fulfill the current seismic regulation codes. Moreover, the continuous increments in load live amplitudes and dead loads derive in cracks formation in the structural elements that conduct to the necessity of evaluate the safety of these structures.

Although one of the main challenges in earthquake engineering is to mitigate the risk of existing structures, most of the regulation codes in the world are oriented to the project of new structures, without explicit mention of the criteria to be taken on the retrofit techniques of existing bridges. The rehabilitation and seismic retrofitting of a bridge involve a set of parameters in addition to the purely technical aspects that influence the decision to intervene structure. It must be evaluated at least the structural capacity level, the seismic demand on site, the importance of the bridge structure in the context of the road network, the path redundancy, environmental restrictions, the expected life service, the importance of the structure and the viability of access among other variables.

The interruption of a bridge service has serious effects in the transportation systems. Immediately after a disaster, the closure of roadway can prevent emergency operations. In addition to the direct costs among the different possibilities of rehabilitation and strengthening, there are indirect costs, which often exceed the direct costs, caused by the economic and social impact of transit disruption. Indirect costs depend on the time the bridge remains unused, the economic importance of traffic on the road, the traffic delay caused by the use of alternative routes, among other parameters. Even though there are many rehabilitation techniques, this study deals with the reinforced concrete jacketing because of its spread use around the world to retrofit bridge piers.

2. RC jacketing

Bridges located in seismic prone areas have usually important displacement demands in piers. The seismic design of old bridges considered low seismic coefficients that conduct to small lateral forces. Additionally the small lap splice length in longitudinal reinforcement of the piers and the small amount of transverse reinforcement conducted to several failures [1]. One column pier bridges are particularly affected by insufficient lap spice length in the zone of plastic hinge formation (Figure 1).

Figure 1. Fukae viaduct collapse during the Kobe earthquake in Japan.

The traditional techniques for the rehabilitation of the bridge substructures increase their resistance and involve Reinforced Concrete (RC) or steel jacketing (Figure 2). Several studies analyze the most important parameters related to RC jacketing in analytical approaches and experimental tests [2-5].

Figure 2. RC jacketing in a column and shear wall

The jacket, in addition to provide lateral confinement, improves the bending and shear capacity of the original column. An important aspect to be considered in all jackets is that this capacity increase at the base of the column causes additional demands on the existing foundation. The bending overstrength at the column ends can induce failures in other regions of the element with insufficient shear or bending capacity and a brittle failure can occur. For this reason, it is common to leave a gap between the jacket and the foundation.

3. Seismic demand

The seismic records to conduct the nonlinear analyses were selected from the Mexican base of strong motion data that contains earthquakes recorded in the period of 1960-1999. The database provides information of 527 strong motion stations in Mexico. We chose a suite of 20 seismic records from the subduction zone, one of the most important seismic sources in the country. Table 1 shows some characteristics of the earthquakes, including the epicenter location, the magnitude and the focal depth.

Table 1. Earthquakes' epicenter from the subduction seismic source.

Date	**Earthquake ID**	**Lat. N**	**Long. W**	**Focal depth (km)**	**Magnitude (Mw)**
09/19/1985	850016	18.08	102.94	15	8
09/21/1985	850018	17.62	101.82	22	7.5
02/08/1988	880004	17.49	101.16	19.2	5.8
04/25/1989	890024	16.60	99.4	19	6.9
05/15/1993	930005	16.47	98.72	15	6
10/24/1993	930009	16.54	98.98	19	6.6
09/14/1995	950001	16.31	98.88	22	7.3

Table 2 presents the information of the 20 seismic records selected to perform the analyses of the bridges. The table displays the station name, the record ID, the station location, the peak ground acceleration (PGA) and the distance between the epicenter and the seismic station.

Table 2. Seismic records selected to perform the analyses.

Earthquake	**Station ID**	**Lat. N**	**Long. W**	**PGA (gals)**	**Distance (km)**
850016	AZIH8509.191	17.60	101.46	153.93	166.17
	FICA8509.191	17.65	99.84	69.18	331.62
	PAPN8509.191	17.33	101.04	154.95	218.24
	PARS8509.191	17.34	100.21	109.82	300.44
	SUCH8509.191	17.23	100.64	103.12	261.66
	UNIO8509.191	17.98	101.81	165.29	120.76
	VILE8509.191	17.65	99.84	69.18	331.62
850018	ATYC8509.211	17.21	100.43	79.66	154.27
	PAPN8509.211	17.33	101.04	242.69	88.9
	PARS8509.211	17.34	100.21	625.78	173.12
	SUCH8509.211	17.23	100.64	85.98	132.47
880004	MAGY8802.081	17.38	100.58	102.09	62.91
	PARS8802.081	17.34	100.21	246.91	101.46
890024	ACAP8904.251	16.84	99.91	104.39	60.58
	COYC8904.251	16.97	100.08	85.08	83.39
	OCTT8904.251	17.25	99.51	201.16	72.93
	PARS8904.251	17.34	100.21	117.11	119.55
930005	VIGA9305.152	16.76	99.234	67.31	63.59
930009	MSAS9310.241	17.01	99.46	119.05	72.58
950001	VIGA9509.141	16.76	99.24	100.35	62.55

The PGA of the seismic records was scaled with the results of a seismic hazard assessment of the country that included the contribution of all seismic sources. The uniform hazard spectra for sites close to the pacific coast and five return periods (Figure 3) were obtained. The scale factor was the required value to match the maximum amplitude of the response spectra of each accelerogram to the maximum amplitude of the uniform hazard spectrum for a 2500 return period.

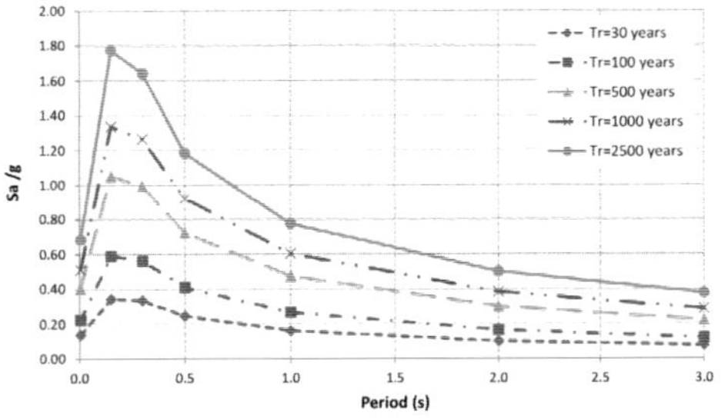

Figure 3. Uniform hazard spectra for a site close to the pacific coast in Mexico.

4. RC bridge models

We select typical RC bridges of medium span length that can be typically found in many countries. In México, this type of structures is distributed all around the country; all of them have similar superstructure and substructures as described in the next section.

4.1. Bridges' characteristics

The bridges are composed of five simple supported 30 m long spans with four possible pier heights, 5 m, 10 m, 15 m, 20 m and 25 m. Most of the bridges designed in the 70's have piers with small longitudinal reinforcement ratios product of gravitational designs. In this study, the longitudinal reinforcement ratio for columns was of 0.5%.

The models are composed of reinforced concrete slab resting on prestressed concrete AASTHO type IV girders (Figure 4). At each span end and at intermediate span length, there are diaphragms to provide lateral stiffness to the superstructure. The substructure consists of frame type piers with four circular columns of constant transverse section and wall type RC abutments. The girders are supported on elastomeric bearings located on top of the bent cap and over the abutments.

The circular columns were strengthened by using three possible jacket thicknesses of 10 cm, 15 cm and 20 cm and three reinforcement ratios of 0.5%, 1% and 1.5%. The parameter combination (heights, jacket thickness and reinforcement ratio) produced 50 bridge models, five original structures and 45 retrofitted models. The bridges were subjected to the 20 seismic records in longitudinal and transverse directions, which conducted to 2000 nonlinear time history analyses.

4.2. Seismic analysis

3D models of the bridges were created using the SAP2000 software [6]. The girders, diaphragms, pier columns, and cap beams were modeled using frame type elements with six degrees of freedom in each node. The 0.20 m thick slab was formed using shell type finite elements and link-type elements were used to model the neoprene bearings and gap elements for the expansion joints.

The frame elements were divided in sub-elements to distribute the mass in several nodes. The columns diameter and the cap beam width depend on the pier height (from 5m to 25m) and they are in the range of 0.85 m - 1.50 m and 1.25 m - 1.90 m, respectively. These cross sections fulfill the AASHTO regulation requirements for a load combination design for gravitational loads.

The nomenclature used to describe the bridge models identifies the pier height, the jacketing thickness and the jacketing reinforcement ratio, e.g. P25_E10_R0.5% corresponds to the 25 meters high bridge with a 0.10m jacket and 0.5% of reinforcement ratio.

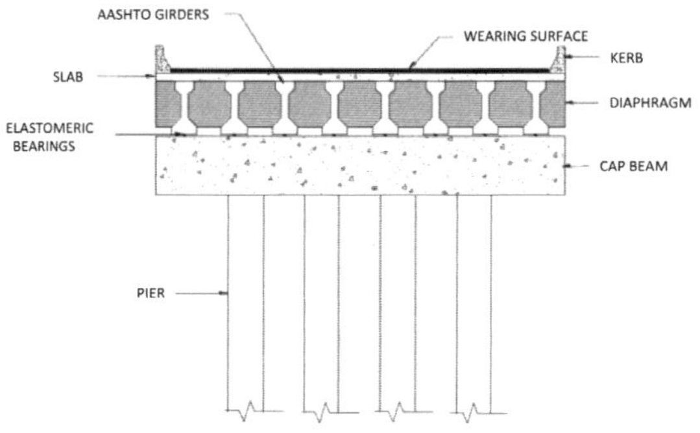

Figure 4. Pier elevation of the bridge models.

The nonlinear analysis were performed with the OpenSEES program. We use a concentrated plasticity model assigning plastic hinges at both ends of the bridge columns. Figure 5 displays typical moment-curvature relationships for an exterior and interior columns of the 10 m and 25 m high models. The curves correspond to the 0.10 m thick jacket and 0.5% longitudinal reinforcement ratio.

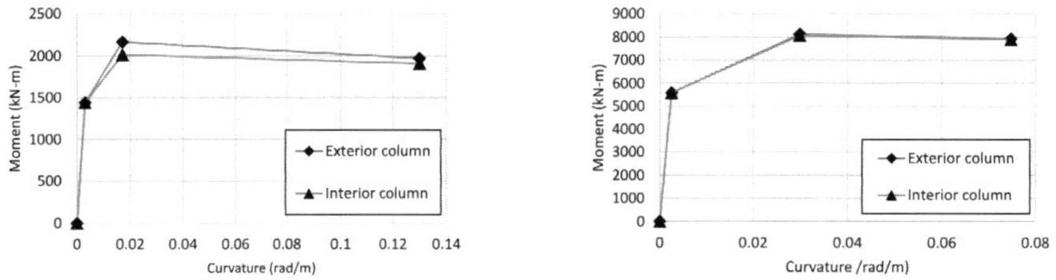

Figure 5. Moment-curvature relationship for columns of the 10 m (left) and 25 m (right) high bridge models.

Figure 6 shows the analytical model of one of the bridges created with the OpenSees navigator platform [7]. In the left a general view of the bridge and in the right the discretization of one of the piers. The columns are divided in several elements to distribute the mass along the length of the bars.

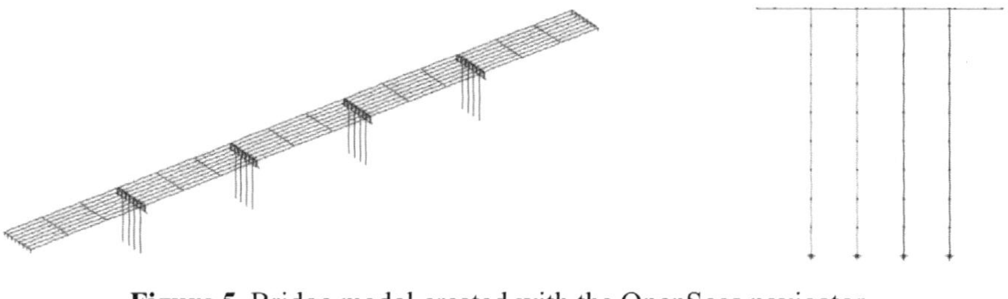

Figure 5. Bridge model created with the OpenSees navigator.

The nonlinear analyses allow determining the pier demands for each of the seismic records. The parameters of interest in columns to correlate them with the expected damages were: rotations, curvatures and lateral displacements. The proposal of Dutta and Mander [8] relates drift ratios with the column expected damages for a non-seismically and seismically designed bridges (Table 3). To identify the damage state, we use the drift limits of the non-seismically case.

A maximum column response for each seismic record was obtained. The mean values of rotation and drift demands for the 5 m high retrofitted bridges is presented in Table 4. The table displays the longitudinal and the transverse demands of the bridge models; the first row presents the demands in the original bridges and the following rows are the results of the retrofitted models.

Table 3. Drift ratios and expected column damages [4].

Damage state	**Description**	**Drift limit***	**Drift limit****
No damage	First yield	0.005	0.008
Slight damage	Cracking, spalling	0.007	0.010
Moderate damage	Loss of anchorage	0.015	0.025
Extensive damage	Incipient column collapse	0.025	0.050
Complete damage	Column collapse	0.050	0.075

* Non-seismically designed
** Seismically designed

Table 4. Rotation and drift ratio demands in longitudinal and transverse directions of the 5m high bridges.

Bridge model	θ_{long}	**Drift long**	θ_{transv}	**Drift transv**
P05	0.0589	0.0641	0.0232	0.0253
P05_E10_R0.5%	0.0395	0.0448	0.0278	0.0311
P05_E10_R1.0%	0.0319	0.0415	0.0179	0.0205
P05_E10_R1.5%	0.0335	0.0394	0.0237	0.0282
P05_E15_R0.5%	0.0360	0.0411	0.0180	0.0198
P05_E15_R1.0%	0.0326	0.0381	0.0221	0.0266
P05_E15_R1.5%	0.0339	0.0389	0.0219	0.0234
P05_E20_R0.5%	0.0404	0.0411	0.0211	0.0198
P05_E20_R1.0%	0.0285	0.0328	0.0144	0.0156
P05_E20_R1.5%	0.0279	0.0324	0.0141	0.0153

As expected, the longitudinal direction had larger demands than those demands in transverse direction of analysis. The pier deformation in longitudinal direction is single curvature whereas in transverse direction a frame type behavior is presented with a double curvature shape. The rotation ratio between the original models and the retrofitted models is in the range of 1.46 to 2.11 in longitudinal direction and in the range of 0.83 to 1.65 in transverse direction. In general, the rotation demands decrease faster when the reinforcement ratio changes from 0.5% to 1.0% than they change with an increase of the reinforcement ratio from 1.0% to 1.5%. It is also evident that the change of the jacket thickness (from 0.10 to 0.20 m) is less important than the change in the longitudinal reinforcement (from 05% to 1.5%) for the rotation and drift demands in both directions of analysis. Figure 6 shows the rotation demands of the 5 m and 15 m high bridges in longitudinal direction.

The horizontal dot line is the rotation demand of the non-retrofitted bridge models and the other three lines are the rotation demands of the retrofitted models. The three jacket thicknesses in the 5 m high bridges reduce notably the pier rotation demands and the results are similar regardless the thickness value. As mentioned before, changing the reinforcement ratio from 0.5% to 1.0% decreases more the demand than in the range of 1.0% to 1.5%.

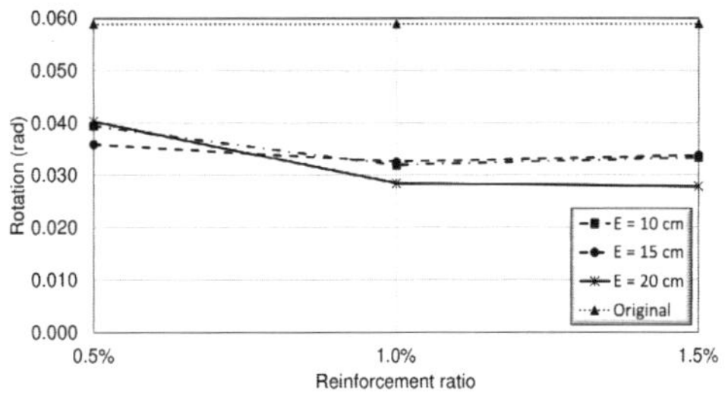

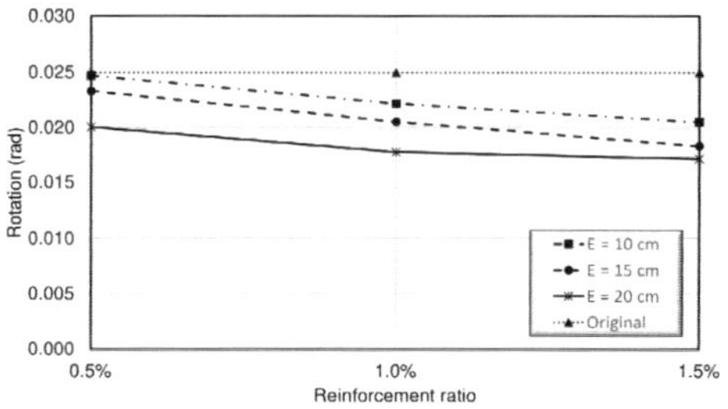

Figure 6. Rotation demands in longitudinal direction of the 5 m (left) and 15 m (right) high bridges.

Higher bridges reduce the efficiency of the RC jacketing. The 15 m high model (Figure 6) displays similar rotation demands of the original and the retrofitted models with the smallest reinforcement ratio and 0.10m jacket thickness. However, slightly smaller rotation demands are presented with the other two jacket thickness. Even though the influence of the reinforcement ratio is similar to the results of 5 m high bridge, the drop of the rotation demands in the range of 0.5% to 1.0% is not now quite different than the drop between 1.0% and 1.5%.

Table 5 displays the column rotation demands for the 15 m and 25 m high pier bridges. The rotation demands continue decreasing with the height increase of the bridges. Now, the rotation ratio between the original models and the retrofitted models is in the range of 1.01 to 1.45 in longitudinal direction and in the range of 0.79 to 1.01 in transverse direction of the 15 m high bridges. These values are in the range of 0.87 to 1.24 and 0.70 to 1.01 in the 25 m high models. Even though the general trends mentioned in the former paragraph are similar, the contribution of the RC jacketing to reduce the seismic response of the bridges is reduced.

Finally, Figure 7 displays the drift demands in the longitudinal direction and the limit states of behavior of the 5 m, 15 m, 20 m and 25 m high bridges. The horizontal discontinuous lines are the limit states of behavior, the dot line is the response of the original models, whereas the rest of the lines are the rotation demands of the retrofitted models. According to the drift demands, the original 5 m high model is located in the complete damage zone while the rehabilitated bridges are in the moderate limit state of behavior. The bridges with higher piers present less expected damages reducing the limit state of behavior to the zone of moderate and in a few cases extensive damages.

Table 5. Rotation demands in both directions of the 15 m and 25 m high bridges.

Bridge model	P15		P25	
	θ_{long}	θ_{transv}	θ_{long}	θ_{transv}
Original	0.0249	0.0104	0.0168	0.0079
E10_R0.5%	0.0247	0.0132	0.0176	0.0089
E10_R1.0%	0.0222	0.0103	0.0161	0.0085
E10_R1.5%	0.0205	0.0122	0.0161	0.0085
E15_R0.5%	0.0233	0.0108	0.0186	0.0115
E15_R1.0%	0.0205	0.0115	0.0191	0.0109
E15_R1.5%	0.0183	0.0131	0.0194	0.0110
E20_R0.5%	0.0200	0.0119	0.0172	0.0078
E20_R1.0%	0.0178	0.0132	0.0140	0.0092
E20_R1.5%	0.0172	0.0128	0.0136	0.0090

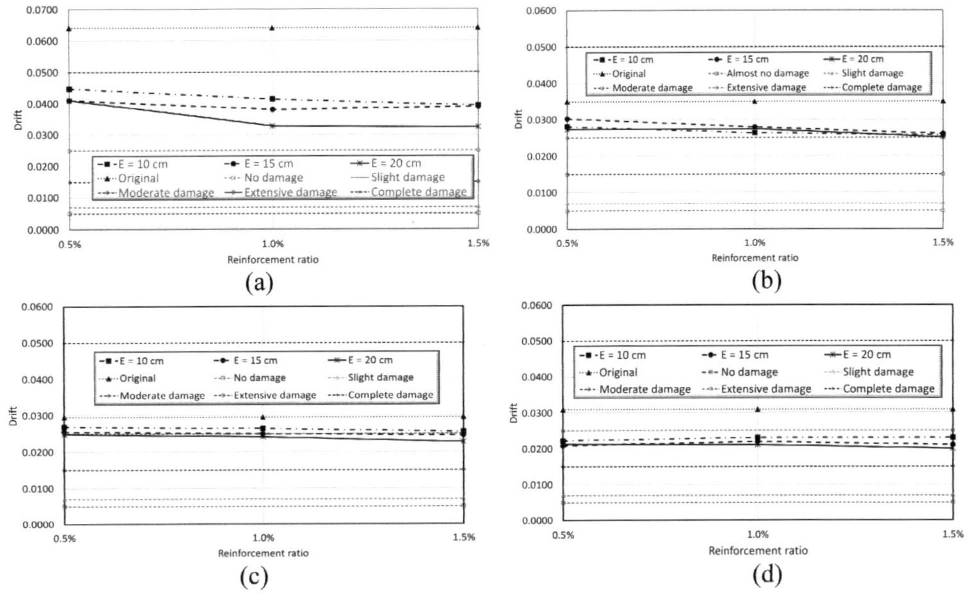

Figure 7. Drift demand and limit states of behavior. (a) 5 m high bridge, (b) 15 m high bridge, (c) 20 m high bridge, (b) 25 m high bridge

5. Conclusions

The results of the analyses allow concluding that small reinforcement ratios combined with jacket thickness in the range of 0.10 m to 0.15 m are the most efficient way to reduce the seismic response of the bridges. The height of the pier resulted also an important parameter to be considered to select the best retrofit technique. The nonlinear analyses showed that the efficiency of the RC jacketing is reduced with the pier height increase, showing important benefits in bridges with short length piers.

The seismic responses of the retrofitted models with piers higher than 5 m were similar regardless the jacket thickness and the reinforcement ratio. It was also observed that changes in the reinforcement ratio modify more the seismic response than the variations of the jacket thickness.

References

[1] Priestley M J N, Seible F and Calvi G M 1996 *Seismic design and retrofit of bridges*, Wiley-Interscience Publication. Jonh Wiley & Sons, Inc.

[2] Júlio, E S, Branco, F and Silva, D 2003 *Structural rehabilitation of columns with reinforced concrete jacketing*, Prog. Struct. Engng. Mater., 5, pp 29-37.

[3] Marlapalle V C, Salunke P J and Gore N G 2014 *Analysis & design of RCC jacketing for buildings*, International Journal of Recent Technology and Engineering, 3, pp 62-63

[4] Breña, S and Alcocer, S 2009 *Seismic performance evaluation of rehabilitated reinforced concrete columns through jacketing*, Proc Improving the seismic Performance of Existing Buildings and Other Structures, San Francisco, Cal, pp 572-583.

[5] Thermou, G E, Pantazopoulou, S J and Elnashai, A S 2007 *Flexural behavior of brittle RC members rehabilitated with concrete jacketing*, 133, pp 1373-1384.

[6] CSI SAP2000 V14.0 2009 *Computers and Structures Inc.*, California, USA.

[7] Schellenberg A, Yang T, Kohama, E 2011 *OpenSees Navigator,* Pacific Earthquake Engineering Research Center, University of California, Berkeley, CA.

[8] Dutta, A, Mander J B 1998 *Seismic Fragility Analysis of Highway Bridges*, Workshop on Earthquake Engineering Frontiers in Transportation Systems, International Center for Disaster-Mitigation Engineering (INCEDE), Tokyo, Japan, pp. 311-325.

Overtopping failure analysis of coastal flood defences affected by climate change

Mehrdad Bahari Mehrabani[1], Hua-Peng Chen[*1] and Morris W Stevenson[2]
[1] School of Engineering, University of Greenwich, Chatham Maritime, Kent, ME4 4TB, UK
[2] Southern Testing, Keeble House, Stuart Way, East Grinstead, West Sussex RH19 4QA, UK

Abstract. Sea defence structures are expected to protect coasts for a long period, hence requiring reliable performance assessment strategies, in order to ensure their integrity and functionality. It has been demonstrated that rising sea level together with changing wave height can lead to increase risks of the failure to coastal defence structures. This paper presents a method for assessing the risk of wave overtopping failure, analysing the joint probability of sea water level and significant wave height under future hydraulic conditions due to climate change. Monte Carlo simulations are utilised to analyse the time-dependant overtopping failure probability of a seawall in the UK subjected to sea level rise. The numerical results for the flood defence example show that the seawall subjected to the sea level rise with high emission scenario could face to a significant increase of the frequency and the rate of overtopping discharge in comparison with the present date conditions without consideration of seawall crest settlement.

1. Introduction

The change of future climate and performance deterioration of coastal flood defences will pose significant risks to human life, homes and infrastructure along the coastlines. Although research has been undertaken on flood protection schemes for controlling coastal flood risk in the future, the risk of overtopping over the crest of coastal defences still needs to be assessed accurately with consideration of uncertainties. From existing investigations, sea level rise and deterioration performance are considered as the major threat to coastal flood defences due to the effect of climate change. This will lead to the requirement of the time dependent reliability analysis of flood defence systems by using a probabilistic approach, especially for those existing structures which are have been designed without considering the impacts of environmental changes. Research show that environmental change leads to sea level rise [1], and a higher sea level probably induces a higher wave height. It is thus important to balance between the cost and reliability of structures.

Some studies have been carried recently in order to improve the assessment of integrity and reliability of coastal defence affected by future climate change especially sea level rise. Chini and Stansby [2] introduced a method for extreme values of coastal wave overtopping to estimate the projected failure due to the joint probability of sea level rise and wave height change. Chen and Alani [3] presented a reliability based approach to analyse and evaluate the failure probability of sea defences by

* Corresponding author, E-mail: h.chen@gre.ac.uk

considering future climate change and to provide cost-effective maintenance operation plan for the structures. Wave overtopping is still a main failure parameter due to sea level rise. Recently, UK Climate Projection 2009 (UKCP09) [4] updated hydrological sea conditions till 2095 under various scenarios. Rakonczai & Zempléni [5] developed a joint probability method to evaluate bivariate or multivariate extreme values. The method is compatible with the nature of rise in sea level, which is useful to probabilistic approaches. Hawkes [6] introduced joint probability methods in flood and coastal defences by applying into several case studies. More studies on time dependant reliability analysis have undertaken to develop the integrity assessment methods [7-10]. Wave and water level are main factors to induce coastal flooding, lead to overtopping and damage at coastal locations. In the design of coastal structures, joint exceedance of significant wave height and still water level should be considered as the critical design factors.

This paper presents a method for overtopping failure analysis and a framework of joint probability estimation for still water level and wave height based on UKCP09 climate projection scenarios. Overtopping often occurs when extreme sea levels and high near-shore wave heights happen at the same time, however sometimes extreme wave height can lead to overtopping independently. Finally, a case study is adopted to demonstrate the proposed methodology to estimate the probability of wave overtopping failure in a coastal defence structure affected by hydrological change in environment.

2. Joint probabilistic approach

Still water level (SWL) and significant wave height (SWH) are considered as two main factors to estimate wave overtopping probability failure. These data are used to produce an estimated SWL-SWH joint probability in the period of 2001-2095 based on UKCP09 projection scenarios. Recorded recent local SWL and SWH are considered as datum. For example, data during 2001-2010 could be a useful datum to predict overtopping in the next 100 years.

2.1. Impact of sea level and near-shore wave height change

A summarised estimation for change in SWL and SWH is provided in UKCP09 second version briefing report. Sea level is projected to increase between +48 to +75 cm with 95th percentile confidence level for various emission scenarios by 2095 [4]. Significant wave height is also estimated to have a change from -150 to +100 cm, as shown in Figure 1. Predictions of other emission scenarios for both SWL and SWH are also provided in the range with various percentile confidence levels such as 5th and 95th in the briefing report.

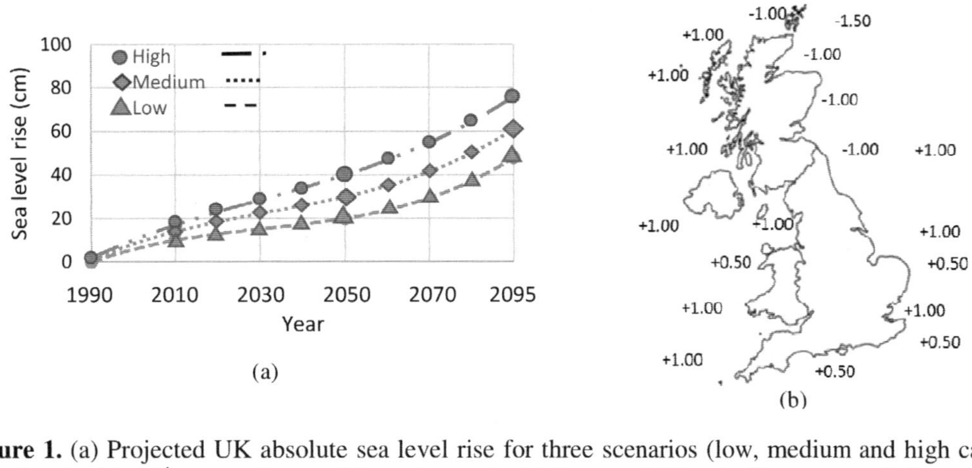

(a) (b)

Figure 1. (a) Projected UK absolute sea level rise for three scenarios (low, medium and high carbon emissions) with 95th percentile confidence interval; (b) Projected UK significant wave height annual changes for various locations from 1980–1999 to 2080–2095 in metre at the 95th percentile level.

These relative changes should be captured based on location and year (e.g. North Wales, UK, 2095, high emission scenario), and the absolute level in 1990. The local water depth $H_w(t)$ at time (t) due the future sea level rise could be calculated as

$$H_w(t) = H_w(0) + \Delta H_w(t) \tag{1}$$

where $H_w(0)$ represents initial local sea water level (1990), and $\Delta H_w(t)$ is the change in local water level from initial time. Similarly, the significant wave height $H_s(t)$ at time (t) is expressed as

$$H_s(t) = H_s(0) + \Delta H_s(t) \tag{2}$$

where $H_s(t)$ represents significant wave height at initial time (1990), $\Delta H_s(t)$ is the change in significant wave height from the initial time. Based on the extreme value analysis methods, e.g. generalised extreme value (GEV) and generalised pareto distribution (GPD), the parameters of the distributions, μ, σ and ξ, can be estimated with statistic methods provided in literatures. Chini and Stansby [2] considered changes in time on μ parameter, e.g. linear trend μ_t and initial value μ_0, to evaluate new distribution parameter for the desired year, expressed as

$$\mu = \mu_0 + \mu_t t \tag{3}$$

2.2. Overtopping analysis

For the overtopping rate of existing coastal defences subjected to sea level rise and crest settlement, the freeboard of a structure at specific time, $R_c(t)$, in the future equation can be expressed here as

$$R_c(t) = R_c(0) - \Delta h_w(t) - \sum \Delta H_d(\Delta t) \tag{4}$$

where $R_c(0)$ is initial freeboard at present day, $\Delta h_w(t)$ is sea level rise, $\Delta H_d(\Delta t)$ is the time-dependant deterioration of dyke crest height due to settlement. Due to the decrees in freeboard, a significant increase will occur in wave run-up and wave overtopping discharge. An average overtopping discharge over time $q(t)$ is calculated from the methods in Pullen *et al.* [1], TAW [11], and CIRIA [12], expressed here as

$$\frac{q}{gH_{m0}{}^{3/2}} = \frac{A}{\tan\alpha^{1/2}}\gamma_f \varepsilon_{m-1,0} \exp(-B\frac{R_c}{H_{m0}}\frac{1}{\varepsilon_{m-1,0}\,\gamma_b\gamma_f\gamma_\beta}) \tag{5}$$

In TAW [11] overtopping is described by two formulae, one is for breaking waves ($\gamma_b\varepsilon_{m-1,0} < \cong 2$) where wave overtopping increases for increasing breaker parameter, and another for non-breaking waves ($\gamma_b\varepsilon_{m-1,0} > 2$) where maximum overtopping is achieved. In a probabilistic analysis the values of A and B are taken as 0.067 and 4.75, respectively. The critical overtopping rate in Eq. (6) can be illustrated as the given line in Figure 2 for distribution analysis.

2.3. Joint probability

Joint probability estimation for various factors in coastal defences has been investigated in many studies [6, 13-16]. Plausible damage to a structure and maximum allowable overtopping rate are two main factors in performance of the structure, which can be assessed by using joint probability distributions. Hence, while a wave height is near to crest level, it is expected to have maximum damage to a structure [13].

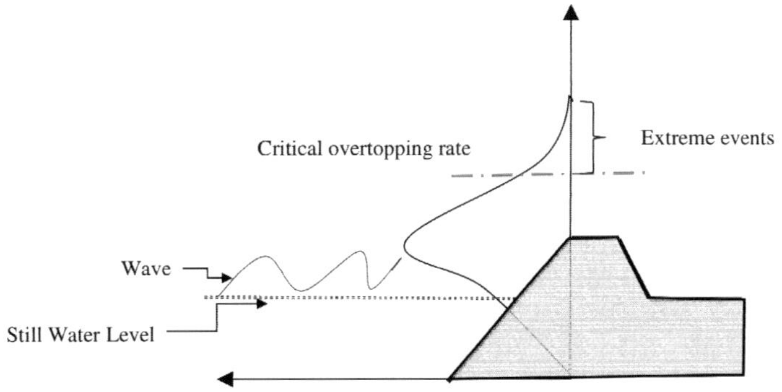

Figure 2. The distribution of wave overtopping discharge in the coastal defence indicated by critical overtopping rate(q_{cr}).

Chini and Stansby [2] proposed the following equation in case of joint extreme value estimation for water level and significant wave height

$$\frac{1}{T} = 1 - F(x) - G(y) + H(x,y) \tag{6}$$

where *F* and *G* are the cumulative distributions for each single variable, and *H* is the joint cumulative distribution based on the Gumble generator function. The marginal distributions should be determined by fitting a generalised extreme value (GEV) or a generalised pareto distribution (GPD). The cumulative and probability density functions of GPD distribution are given in [5], respectively, as

$$F(x) = 1 - (1 + \xi x/\psi)^{-1/\xi} \tag{7}$$

$$f(x;\sigma,\xi) = \frac{1}{\sigma|\xi|}(1 + s\frac{x}{\sigma})^{-(1+\xi)/\xi} \tag{8}$$

where $\psi > 0$ and ξ are scale and shape parameters. For $\xi > 0$ GPD is type I, if $\xi = 0$ it is the exponential distribution, and for $\xi < 0$ it becomes GPD type II, s = *sign* (ξ). There are several methods to estimate the distribution parameters such as moment based method, maximum likelihood, and likelihood-moment estimation in Jonathan, *et al.*, [14] and Armagan, *et al.*, [1].

2.4. Probabilistic analysis

From the sea dyke overtopping probability analysis with consideration of sea level rise discussed above, the summation of the probability density of all combinations of load and resistant can be computed [6] as

$$\text{Pr}_{ot}(t) = \iint_{z<0}^{\infty} f(\text{H}_w(\text{t}), \text{H}_s(\text{t}))\text{dH}_w\text{dH}_s \tag{9}$$

where $Pr_{ot}(t)$ is dyke overtopping probability, and $f(\text{H}_w(\text{t}), \text{H}_s(\text{t}))$ is the joint GPD of two variables by assuming that they are conditionally dependent. Under the condition of statistical independence, the probability value of the exceedance sub-domain can be calculated as

$$\text{Pr}(H_{exceedance}) = \text{Pr}(\text{H}_w > t_w, \text{H}_s > t_s) = (1 - p_w).(1 - p_s) \tag{10}$$

where t_w and t_s are the threshold points on the two probability distributions, in this study for water level and significant wave height, respectively.

3. Numerical Example

A numerical example for a sea wall situated in North Wales is used in this paper for analysing the overtopping failure of a structure due to climate change. The data applied in numerical analysis are as follow: crest level = 14.5 m @Charted Datum, seaside slope = 1 : 4.5 m, roughness factor for the structure = 1 and crest width = 2 m. Joint data are generated based on case studies in [6]. In this paper, North Wales's dataset in the UK including sea water level and significant wave height within 14 years period is chosen as indicated in Figure 3. In this study, mean wave period T_m is used as an average value. The influence of sea level rise and significant wave height change in the calculations is based on the UK climate projection briefing report by [4,17], where the high emission scenario at 95th percentile interval confidence is considered.

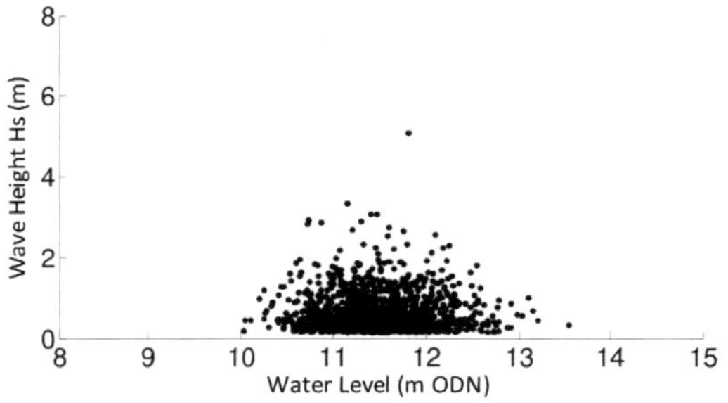

Figure 3. Original data for North Wales, wave height and water level

Based on the data shown in Figure 1, the projected sea level rise of +0.76 m is estimated by the end of 21st century, and the significant wave height increases approximately +1.00 m at the same time in North Wales. From these data, Monte Carlo simulations are utilised to give the estimation for the cumulative probability and joint probability of extreme SWH and SWL. In order to estimate the joint extreme value, cumulative probability distributions fitted by GEV distribution are plotted in Figure 4, for sea water level denoted as *F(x)* and for wave height denoted as *G(x)*, respectively. The fitted GPD cumulative distributions for the same input are shown in Figure 5. It is found that the results by using GPD distribution may be lower than those from the GEV distributions. For instance, sea level for 100 years joint return period in GPD and GEV fitted data are 13.68 m ODN and 14.12 m ODN.

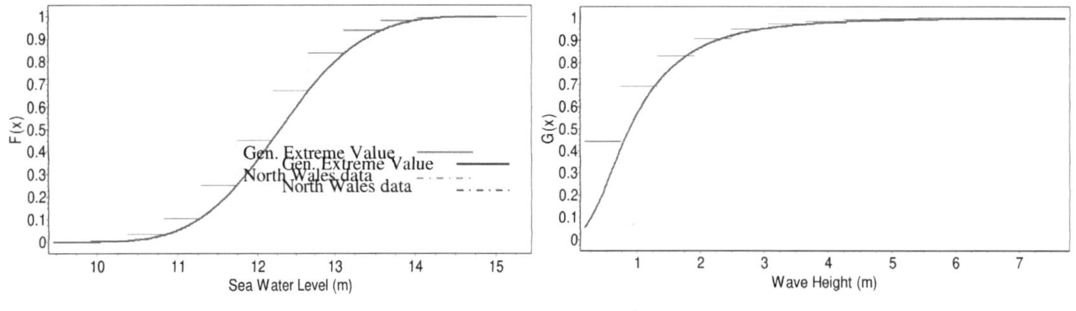

Figure 4. Cumulative probability plots for significant wave height and sea water level, respectively, by considering high emission scenario, fitted by Generalised Extreme Distribution

The frequency analysis is often used to evaluate the hydraulic parameter values for various return periods. The return period here refers to the average period of time between occurrences of a particular high value of the variable. Recurrence interval plots for sea water level and wave height are estimated for 2, 5, 10, 25, 50, 100 and 200 years by considering sea level rise, as shown in Figure 6. By assuming independency between sea water level and wave height, the joint probability counter lines can be estimated from the values of return period. For example, for 50 years return period values the joint return periods will be 15.3 m ODN and 6.9 m for sea water level and wave height, respectively.

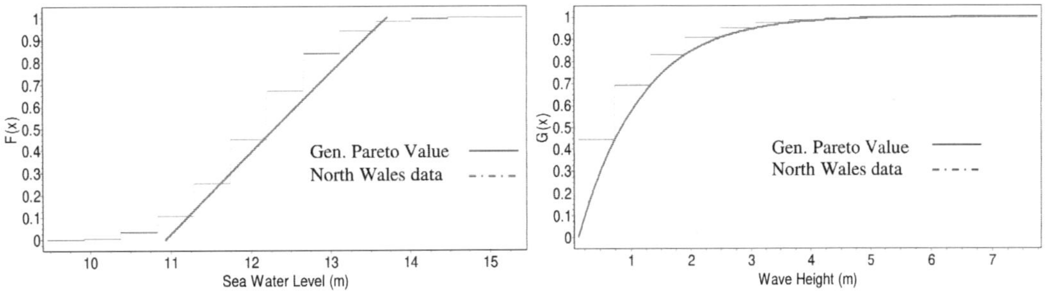

Figure 5. Cumulative probability plots for significant wave height and sea water level, respectively, by considering high emission scenario, fitted by Generalised Pareto Distribution

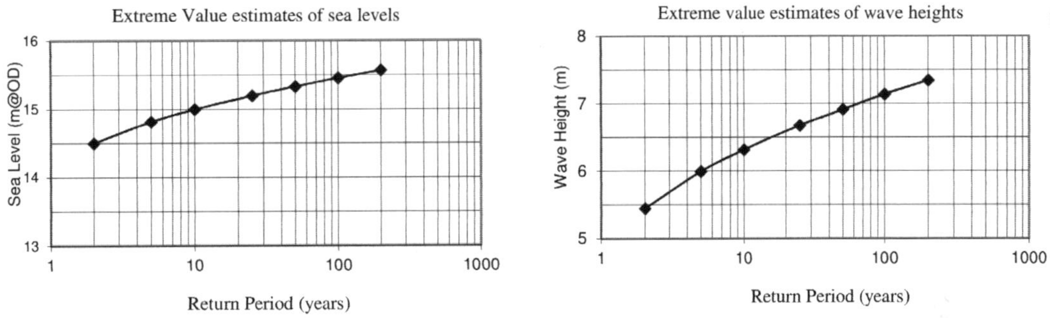

Figure 6. Recurrence interval plots by considering high emission scenario sea level

Sea water level and wave height are usually not independent. By applying dependency based on Eq. (6) and by using cumulative values, the counter lines of return periods of 5, 10, 50 and 100 years are given in Figure 7, by utilising GEV cumulative probability approach. Here, return period curves were estimated by 'counting back' through the ranked data in descending order above individual thresholds.

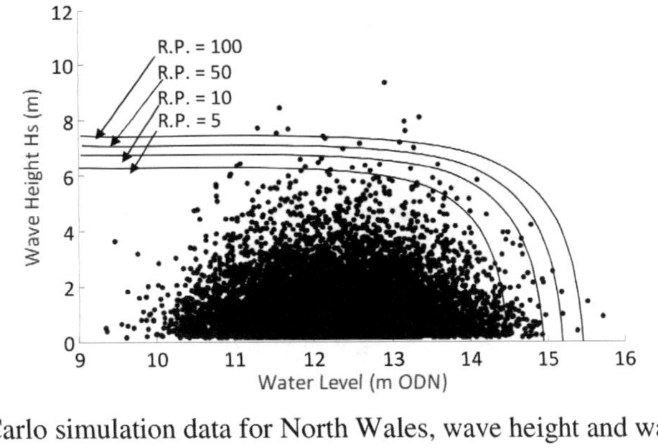

Figure 7. Monte Carlo simulation data for North Wales, wave height and water level.

The counter lines in Figure 7 provides data for extreme conditions for the combination of wave height and sea water level. For example, the exceedance probability in 50 years return period for sea level and significant wave height are the same for case (14.5 m ODN, 6.8 m) and case (15.2 m ODN, 2.0 m). The linear trend μ_t is not significant for SWH at 2095 [2]. The dependency of water level and significant wave height is considered at good level ($0.12 < \rho < 0.36$). The return period estimated from joint return period shown in Figure 7 is lower than single return periods given in Figure 6.

Figure 8 shows the histograms of wave overtopping rates from Monte Carlo simulations. The rate and frequency of wave overtopping for the given structure are calculated over the period of almost 100 years from 2001 to 2095 in two cases, i.e. without and with consideration of Sea Level Rise (SLR), plotted in Figure 8(a) and Figure 8(b), respectively. In the case of SLR, the changes of sea water level and significant wave height are added to the original data yearly between +0.01 m to +0.76 m for sea water level and between +0.01 m to +1.00 m for significant wave height from 2002 to 2095. The results indicate environmental change has severe effects on the probability of failure in overtopping. Hence, the crest of the sea wall should be increased accordingly.

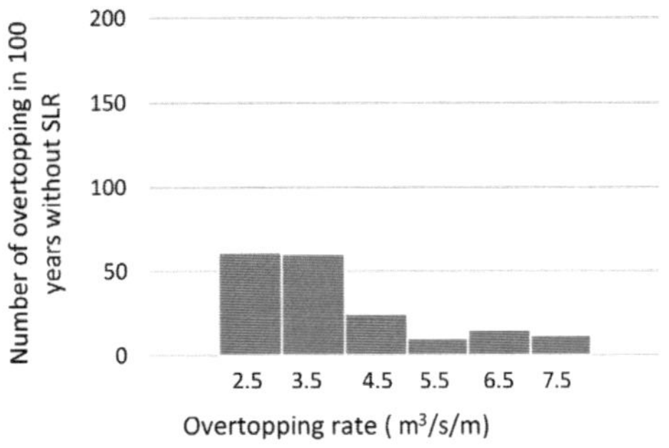

(a)

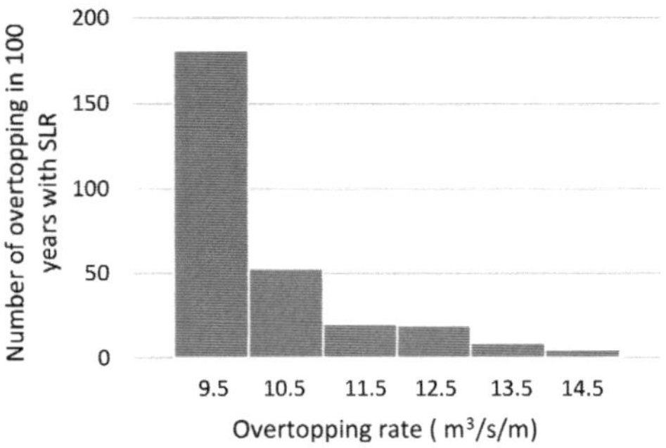

(b)

Figure 8. Histogram of overtopping rate ($m^3/s/m$) between 2001 and 2095: (a) without considering Sea Level Rise (SLR), (b) with considering Sea Level Rise (SLR)

4. Conclusion

This study presents an approach for assessing overtopping failure probability for coastal defences by considering joint probability failure affected by sea level rise. Different approaches of numerical calculations are adopted in this study, i.e. GPD and GEV distributions, in order to provide a better understanding of probability failure analysis. A method by using joint probability distribution model has been proposed to analyse the hydrological sea conditions to evaluate failure risk in the future. This method can be more accurate and flexible to allow a more specific investigation of various structural parameters such as overtopping, by considering joint of hydrological elements and their changes over the time.

From the numerical studies, rising of sea level and increasing significant wave height are important factors which generate significantly higher risks of the current coastal defences around the UK over the next 80 years. The frequency and the rate of overtopping discharge will be increased almost 3 times, without consideration of seawall crest settlement. Environmental change, particularly the increases in sea water level and significant wave height, has substantial effects on the future performance of the coastal defences. As a result, the frequency and rate of overtopping discharge will be escalated dramatically for the existing flood defence structures. Potential research for next steps includes investigation on the effect depth-limited significant wave height in the future climate change and the influence of performance deterioration on the structural reliability of coastal defences by using limit sate analysis.

Acknowledgements
The authors are grateful for the financial support from the Institution of Civil Engineers through the R & D Enabling Fund (Project No. 1305). The findings and opinions expressed in this study are those of the authors alone and are unnecessarily the views of the sponsors.

References

[1]. Pullen T, Allsop NWH, Bruce T, Kortenhaus A, Schüttrumpf H and van der Meer JW 2007 EurOtop: Wave Overtopping of Sea Defences and Related Structure: Assessment Manual. London, UK
[2]. Chini N and Stansby P 2012 Extreme Values of Coastal Wave Overtopping Accounting for Climate Change And Sea Level Rise *Coastal Engineering* 65 pp 27–37
[3]. Chen HP and Alani AM 2012 Reliability and Optimised Maintenance for Sea Defences *Maritime Engineering* 163(2) pp 51–63
[4]. Jenkins G, Murphy J, Sexton D and Lowe J 2010 UK Climate Projections: Briefing Report Version 2 Newcastle *Met Office Hadley Centre*
[5]. Rakonczai P and Zempléni A 2011 Bivariate Generalized Pareto Distribution in Practice: Models And Estimation *Environmetrics* **3** 23 pp 219–227
[6]. Hawkes P 2008 Joint Probability Analysis for Estimation of Extremes *Journal of Hydraulic Research* **2** 46 pp 246–256
[7]. Chen HP and Alani AM 2013 Optimized Maintenance Strategy for Concrete Structures Affected by Cracking Due To Reinforcement Corrosion *ACI Structural Journal* 110 pp 229–238
[8]. Buijs FA, Hall JW, Sayers P and Van Gelder HAJM 2009 Time-Dependent Reliability Analysis of Flood Defences *Reliability Engineering and System Safety* 94 pp 1942–1953
[9]. Burcharth HF, Andersen TL and Lara JL 2014 Upgrade of Coastal Defence Structures Against Increased Loadings Caused by Climate Change: A First Methodological Approach *Coastal Engineering* 87 pp 112–121
[10]. Chen HP and Bicanic N 2010 Identification of Structural Damage in Buildings Using Iterative Procedure and Regularisation Method *Engineering Computations* **8** 27 pp 930–950
[11]. TAW 2002 Wave Run-Up and Wave Overtopping at Dikes *Technical Report, Technical Advisory Committee for Flood Defence in the Netherlands (TAW)* Delft
[12]. CIRIA 2007 The Rock Manual. The Use of Rock in Hydraulic Engineering *Report C683* London UK *CIRIA(Construction Industry Research and Information Association)*
[13]. Hames D and Reeve D 2007 The Joint Probability of Waves and High Sea Levels in Coastal Defence *Cardiff, Institute of Mathematics and its Applications (IMA)*
[14]. Jonathan P, Ewans K and Randell D 2013 Joint Modelling of Extreme Ocean Environments Incorporating Covariate Effects *Coastal Engineering* 79 pp 22–31.
[15]. Armagan A, Dunson DB and Lee J 2013 Generalized Double Pareto Shrinkage *Statistica Sinica* **1** 23 pp 119–143
[16]. Gouldby B, Sayers P, Mulet-Marti J, Hassan MAAM and Benwell D 2008 A Methodology For Regional-Scale Flood Risk Assessment *Proceedings of the Institution of Civil Engineers Water Management* **3** 161 pp 169–182
[17]. Chen HP 2014 Maintenance Of Earth Sea Dykes in Changing Environmental and Structural Conditions *Structural Safety & Reliability* IALCCE 2014 CRC Press

Assessment of concrete damage and strength degradation caused by reinforcement corrosion

Jaya Nepal and Hua-Peng Chen[1]
Department of Civil Engineering, University of Greenwich, Chatham Maritime, Kent, ME4 4TB, UK.

Email: h.chen@gre.ac.uk

Abstract. Structural performance deterioration of reinforced concrete structures has been extensively investigated, but very limited studies have been carried out to investigate the effect of reinforcement corrosion on time-dependent reliability with consideration of the influence of mechanical characteristics of the bond interface due to corrosion. This paper deals with how corrosion in reinforcement creates different types of defects in concrete structure and how they are responsible for the structural capacity deterioration of corrosion affected reinforced concrete structures during their service life. Cracking in cover concrete due to reinforcement corrosion is investigated by using rebar-concrete model and realistic concrete properties. The flexural strength deterioration is analytically predicted on the basis of bond strength evolution due to reinforcement corrosion, which is examined by the experimental data available. The time–dependent reliability analysis is undertaken to calculate the life time structural reliability of corrosion damaged concrete structures by stochastic deterioration modelling of reinforced concrete. The results from the numerical example show that the proposed approach is capable of evaluating the damage caused by reinforcement corrosion and also predicting the structural reliability of concrete structures during their lifecycle.

1. Introduction

Deterioration of reinforced concrete (RC) structures due to reinforcement corrosion is a growing problem worldwide. It typically involves cracking and spallling of concrete cover and reduction in area of steel reinforcement and loss of bond between concrete and corroded steel. This eventually affects the service life of the concrete structures and also increases the resources required for the maintenance and rehabilitation over time [1-5]. Managing corrosion damaged RC structure has become a great challenge both financially and technically. RC structures damaged by reinforcement corrosion compromises structural safety and durability by affecting their performance. The cost associated with managing these corrosion damaged RC structure (repair, rehabilitation, demolition) is in billion dollars [6-9]. For optimum and cost effective infrastructure management, time-dependent reliability analysis is considered as the effective tool. In time-dependent reliability analysis of deteriorating RC structures, the quantification of the damage caused by reinforcement corrosion is essential. Among these types of damage, only cracking in the concrete cover is visible and can be measured without affecting the functionality of the RC structures in operation.

Cracking in concrete cover is an important parameter which helps in condition monitoring of the RC structures. It is necessary to predict the internal damages such as residual strength deterioration

1. Corresponding author, E-mail: h.chen@gre.ac.uk

from the observable surface condition during the routine inspection or maintenance process. Therefore, it is always beneficial to establish a prediction method to quantitatively assess the structural performance by assessing cracking in the concrete cover. In order to evaluate the residual strength of corroded RC structures, considerable investigations have been undertaken in the past two decades, mainly focusing on the relationship between rebar mass loss and residual strength [1-4]. However, limited research has focused on the relationship between cracking in the concrete cover surface and residual strength and its effect on the structural reliability of these corrosion damaged RC structures.

This paper presents a time-dependent reliability analysis of corrosion affected RC structures. At first, the evaluation of the damages caused by reinforcement corrosion such as rebar mass loss, cracking in concrete cover and residual strength loss is presented analytically. Then, a stochastic deterioration model based on a gamma process is adopted to assess the structural reliability. The results from the numerical example show that the proposed approach is capable of evaluating the damage caused by reinforcement corrosion and the structural reliability of corroded RC structures.

2. Damages caused by reinforcement corrosion

The progress of corrosion directly affects the performance and hence the remaining service life of a corroding RC structure [7]. The quantitative description of these damages associated with performance deterioration of corrosion damaged RC structures is the first step in structural reliability analysis of these structures. Therefore, in this section quantitative assessment of the damages caused by reinforcement corrosion is discussed.

2.1. Corrosion induced cover cracking

The corrosion products formed during corrosion process are expansive in nature, which causes two to six times volume increase as compared with the original steel [10]. The increase of volume per unit length due to bar corrosion ΔV can be obtained from the volume of rust minus the volume of the original rebar of a diameter R_b consumed. This increment of volume per unit length of rebar creates a radial displacement at the rebar-concrete interface u_{bx} which can be estimated from

$$u_{bx} = \frac{\Delta V}{2\pi R_b} = \frac{1}{2}(\gamma_{vol} - 1) R_b X_p \tag{1}$$

where X_p is the corrosion level defined by the ratio of the mass loss of the corroded rebar to the original mass of the rebar and γ_{vol} is the volume ratio of the corrosion product formed to its parent metal lies between 1.8 to 6.4 [10]. In this paper corrosion has been considered as uniform, therefore, reduction in rebar radius from the initial state R_b when uniform attack penetration x occurs can be evaluated from $R_{bx} = R_b - x$. The evolution of cracks in concrete cover is discussed in the analytical investigations by Chen and Alani [6], where the equivalent cracks width over the time was defined as the cumulated crack width over the cover surface. The intact cover concrete is treated as elastic material and the cracked concrete is considered as anisotropic in nature [6, 11]. From the anisotropic property and the bilinear softening law of the cracked concrete, normalized cumulative crack width over the concrete cover surface is obtained by considering boundary conditions and by ignoring the Poisson's effect associated with the hoop strain of the completely cracked concrete, expressed here as

$$W_{cx} = \frac{\frac{E}{f_t} u_{bx} - a\left[R_b + R_c(l_o - R_c)(l_o - R_b).\delta(R_c, R_b)\right]}{b(l_o - R_b)\left[1 - R_c(l_o - R_c).\delta(R_c, R_b)\right]} \tag{2}$$

where l_o is the material constant given by $l_o = n_c l_{ch}/2\pi b$ in which n_c is the number of cracks taken as 3 or 4 for the concrete around the rebar and l_{ch} is the characteristic length defined as $l_{ch} = EG_f/f_t^2$ by Hillerborg et al. [12] ; and $\delta(R_c, R_b)$ is the crack factor associated with the material properties and radial distance r between rebar surface R_b and concrete cover surface R_c, defined as.

$$\delta(R_c, R_b) = \frac{R_c - R_b}{l_o(l_o - R_c)(l_o - R_b)} + \frac{1}{l_o^2}\ln\frac{R_c}{R_b}\frac{|l_o - R_b|}{|l_o - R_c|} \tag{3}$$

2.2. Residual strength deterioration

The flexural strength deterioration due to reinforcement corrosion was investigated by Nepal and Chen [7], where the analytical method was proposed to evaluate the residual strength of RC beam with corroded reinforcement by considering different failure modes of concrete and steel. In case of un-corroded perfectly bonded RC beam the strain compatibility condition exists, as given by design codes. Therefore, the initial flexural resistance of RC beams can be evaluated by using design codes. For the corroded RC beam when ultimate bond strength is insufficient to prevent anchorage failure, the tensile force generated in the corroded tensile steel can be obtained from

$$f_{stx} = 2n_b \pi R_{bx} l_d T_{ubx} \tag{4}$$

where n_b is the number of the bottom tensile steel and l_d is the development length which can be evaluated from design code. T_{ubx} is the ultimate bond strength of corroded rebar and is obtained from Nepal and Chen [9]. The strain compatibility of a RC beam with corroded reinforcement can be considered between un-bonded and bonded condition [13]. Assuming the deformation of concrete is mainly due to plastic deformation occurring within the plastic equivalent region, new strain compatibility of the corroded beam can be expressed as

$$\frac{\varepsilon_{stx}}{\varepsilon_{ccx}} = g_x \frac{d_x - Y_x}{Y_x}, \quad \frac{\varepsilon_{scx}}{\varepsilon_{ccx}} = g_x \frac{Y_x - d_x'}{Y_x} \tag{5}$$

where the plastic equivalent region is defined as $L_{eq} = 9.3Y_x$ [14]. Parameters in equation (5) are defined as: ε_{ccx} is ultimate strain of concrete; ε_{stx} and ε_{scx} are strains of tensile steel and compression steel, respectively; Y_x is the neutral axis depth from the edge of compression zone; d_x is the effective depth of beam and d_x' is the distance from the centroid of the compression steel to edge of the compressive fibre corresponding to corrosion level X_p; and g_x is the interpolation factor which can be obtained by considering the bond strength value of perfectly bonded and un-bonded condition of the RC beam. By utilizing the concept given by Cairns and Zhao [15] that the corroded RC beam still follows the condition of equilibrium of resultant tensile and compressive forces acting at the beam section, the residual flexural strength can be evaluated by

$$M_{ux} = f_{ccx}(d_x - 0.4Y_x) + f_{scx}(d_x - d_x') \tag{6}$$

where f_{ccx} and f_{scx} are the compressive forces acting at the centroid of compression zone and the centroid of the compressive steel of the corroded beam, respectively.

2.3. *Structural reliability analysis*

The gamma process has been often adopted for structural deterioration modelling [6, 16, 17]. The gamma process is a stochastic process with independent non-negative increments having a gamma distribution with a given average of deterioration rate. Structural resistance degradation caused by reinforcement corrosion is a continuous and non-negative phenomenon [5, 7]. Therefore, the gamma process is suitable for the stochastic modelling of structural resistance deterioration in corrosion affected RC structures during their lifecycle. In this gamma process deterioration model, cumulative resistance deterioration J is considered as a random quantity with the gamma distribution, and has the shape parameter $\eta_x > 0$ and scale parameter $\lambda > 0$. Then, the probability density function of this random quantity J, i.e. the structural resistance during the lifecycle at time t and corrosion level X_p $(X_p > 0)$, can be formulated as

$$f_{J_x}(J) = Ga(J, \eta_x, \lambda) = \begin{cases} \dfrac{\lambda^{\eta_x}}{\Gamma \eta_x} J^{\eta_x - 1} e^{-\lambda J}, & for\, J \geq 0 \\ 0, & elsewhere \end{cases} \quad (7)$$

where $\Gamma(\eta_x) = \int_0^\infty v^{\eta_x - 1} e^{-v} dv$ is the gamma function for shape parameter $\eta_x > 0$. The scale parameter λ could be estimated from statistical estimation methods such as a Maximum Likelihood Method by maximizing the logarithm of the likelihood function of the increment of the parameter [17] and the shape function η_x can be obtained from $\eta_x = \lambda J_x$ in which J_x indicates the average deterioration rate associated with the reinforcement corrosion such as flexural strength deterioration in ultimate limit state. Assuming J_L as the maximum allowable limit of the structural deterioration, from the definition of probability of failure and by integrating probability density function given in equation (7), the structural reliability associated with structural resistance deterioration is given by

$$R(t) = 1 - F(t) = 1 - P_r[J_x \geq J_L] = 1 - \int_{J=J_L}^{\infty} f_{J_x}(J)dJ = \frac{\Gamma(\eta_x) - \Gamma(\eta_x, J_L \lambda)}{\Gamma(\eta_x)} \quad (8)$$

where $\Gamma(\eta, z) = \int_{v=z}^{\infty} v^{\eta-1} e^{-v} dv$ is the incomplete gamma function for $z \geq 0$ and $\eta > 0$.

3. Numerical example

A simply supported RC beam of 5m span of a bridge exposed to an aggressive environment as defined by Eurocode 2 is now utilised to demonstrate the applicability of the proposed method for assessing the structural performance and a time-dependent reliability analysis during its service life. The beam is doubly reinforced with the cross-sectional width b =300 mm and effective depth d =560 mm Four steel rebar with diameter D_b =20 mm are provided as the tensile reinforcement and two rebar of diameter D_{sc} =16 mm are provided as the compressive steel with clear cover thickness C =40 mm along with the stirrup of diameter D_{st} =6 mm at spacing of 100 mm. The concrete has a characteristic compressive strength f_{ck} =40 MPa, the yield strength of original reinforcing steel f_{yk} = 460 MPa with modulus of elasticity E_{st} =200 GPa. The characteristic compressive strength of concrete is used for estimating other relevant properties of concrete i.e. tensile strength f_t =4.6 MPa; modulus of elasticity E_c=37 MPa [18, 19].The concrete fracture energy G_F=200 N/m is adopted and ultimate cohesive crack width w_u =1.48 mm and critical crack width w_{cr} =0.23 mm are estimated from CEB-FIP [19] for given compressive strength and assumed maximum aggregate size of 20 mm. The total number of crack n_c = 4 is adopted here. The concrete creep coefficient θ_c = 2.0 and Poisson's ratio ϑ = 0.18 and the volume ratio γ_{vol} of the corrosion products is taken as 2.0 [10]. The RC beam has minimum service life of 60 years and is operated in aggressive environments with mean annual corrosion current per unit length i_{corr} =1 μA/cm^2.

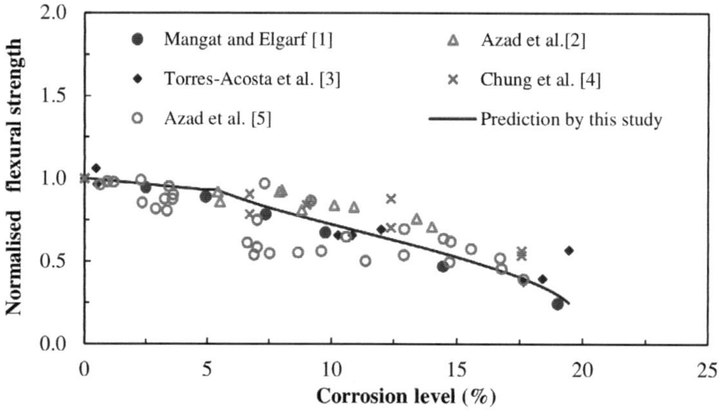

Figure 1. Analytical prediction of residual flexural strength versus corrosion level.

The results in figure 1 show the predicted results of residual flexural strength as a function of corrosion level. The predicted results are then compared with previous experimental investigations obtained from various sources. In figure 1, the residual load capacity is represented by the normalised flexural capacity, which is calculated by dividing the flexural capacity of corroded element by the capacity of the non-corroded element. It is found that the trend of predicted flexural strength deterioration match well with the published experimental data available from various sources. At the initial stage, the flexural strength of the corroded element deteriorates slowly almost in linear trend. When corrosion level reaches about 5%, considerable deterioration occurs due to the reduction in bond strength and corresponding anchorage failure, which occurs before yielding of the steel and the surrounding concrete.

Structural damage assessment and performance predictions using monitored data is critical to determine cost-effective infrastructure management strategies [20,21]. During the routine inspections of concrete bridges, cracking in concrete cover is the most important information recorded for condition rating. Based on the condition ratings collected during inspections, Bridge Management Systems (BMSs) are developed for optimum allocation of limited resources available [22]. Depending on the size of the cracks, these defects in concrete cover due to corrosion can be classified in different categories such as spalling; minor and major cracking etc. The defects in concrete and corresponding rebar loss are described in figure 2. Here the hair line crack is represented by crack ≤0.05 mm; minor cracking 0.05-0.1 mm; major cracking 0.1-0.4 mm and spalling 0.4-1.0 mm. When hair line crack appears at the concrete cover there is about 2% loss in rebar and as the crack size progress the reduction in rebar continuously increases, reaching approximately 13% when the spalling of the concrete cover takes place.

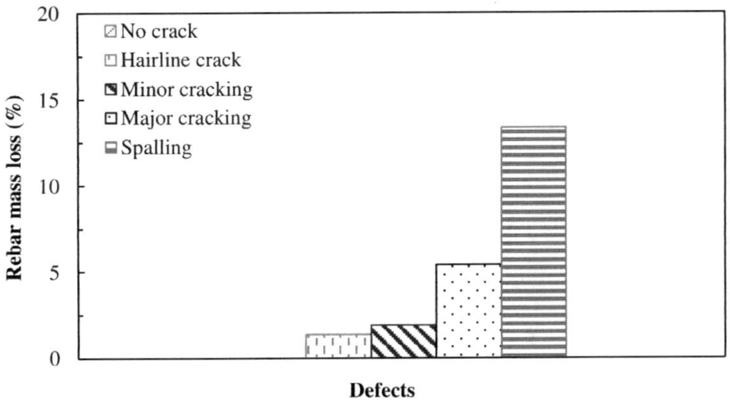

Figure 2. Reduction in rebar mass versus defects.

Influence of different types of aforementioned defects in concrete cover on the structural behaviour of corroded RC structure is presented in figure 3. From the results, till minor cracking in the concrete cover there is no significance change in residual flexural strength. As the defects reach to spalling stage, flexural strength decreases significantly. This clearly shows that, defects in concrete cover have significant effect on residual strength of corroded RC structures. In comparing these results of residual strength, the reduction in residual strength in unconfined concrete is relatively higher than in confined concrete. For instance when the defect in the concrete cover is spalling, the residual flexural strength of the confined element maintains about 60% of its initial strength, whereas in unconfined element it only maintains 40% of its initial strength. This is due to the absence of transverse reinforcement (stirrups) in unconfined concrete. Hence, the results from figure 3 and figure 4 show that, at the same stage of defects in the concrete cover, unconfined concrete is more susceptible than confined concrete.

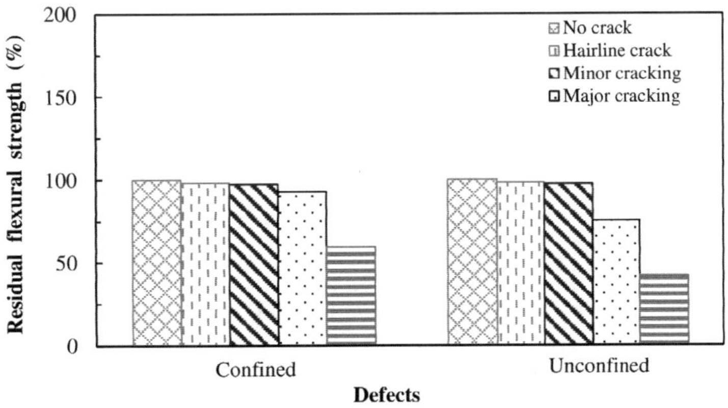

Figure 3. Normalised residual flexural strength versus defects for confined and unconfined concrete.

The structural reliability of confined and unconfined concrete in terms of flexural strength deterioration is given in figure 4. Here, different allowable flexural strength deterioration limits, i.e. J_L = 20%; 25% and 30%, respectively, have been considered during the analysis. Here again, at any stage of cover cracking structural reliability continuously decreases for both unconfined and confined concrete, showing higher probability of failure for a lower allowable deterioration limit. Furthermore from the time-dependent reliability analysis shown in figure 4, it is clear that the unconfined concrete has considerably lower structural reliability than the confined concrete when the same predefined allowable limit and concrete cover crack width are considered.

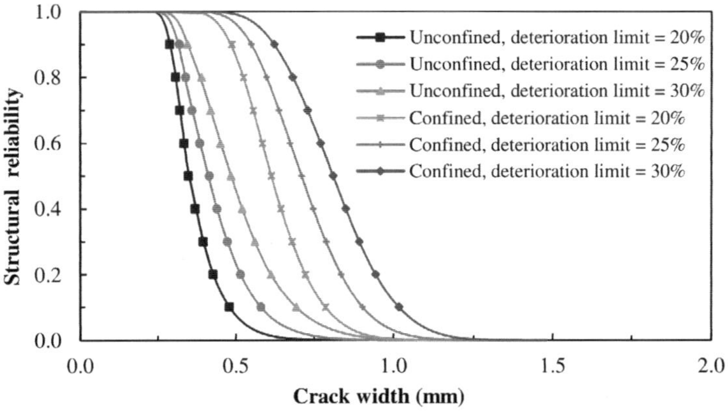

Figure 4. Structural reliability versus surface crack width for various allowable flexural strength deterioration limits of unconfined and confined concrete.

4. Conclusions

This paper presents a new approach for evaluating the damages caused by reinforcement corrosion together with its effect on structural reliability. On the basis of the results obtained from the numerical example, following conclusions are drawn: a) The proposed approach is capable of evaluating structural behaviour and defects of corrosion damaged RC structures; b) Flexural strength decreases significantly after 5% mass loss due to significant reduction in bond strength loss; c) Further progress of corrosion causes significant reduction in rebar size which in turn widens the crack in concrete cover, and consequently reduces residual strength of bond and flexural strength; d) The reliability of the corroded structure decreases with progress of defects in concrete. Further investigations are required to include the effect of external loading on the performance of corroded RC structure serving in aggressive environment.

Acknowledgements

The authors are grateful for the financial support from The Royal Academy of Engineering through Newton Fund (Project No. NRCP/1415/14). The findings and opinions expressed in this study are those of the authors alone and are unnecessarily the views of the sponsors.

References

[1] Mangat P S and Elgarf M S 1999. Flexural strength of concrete beams with corroding reinforcement *ACI Structural Journal* **96** 149-158.

[2] Azad A K, Ahmad S and Azher S A 2007. Residual strength of corrosion-damaged reinforced concrete beams *ACI Materials Journal* **104** 40-47.

[3] Torres-Acosta A A, Navarro-Gutierreza S, and Teran-Guillen J 2007. Residual flexure capacity of corroded reinforced concrete beams *Engineering Structures* **29** 1145-1152.

[4] Chung L, Najm H and Balaguru P 2008. Flexural behavior of concrete slabs with corroded bars *Cement and Concrete Composites* **30** 184-193.

[5] Azad A K, Ahmad S and Al-Gohi B H A 2010. Flexural strength of corroded reinforced concrete beams *Magazine of Concrete Research* **62** 405-414.

[6] Chen H P and Alani A M 2013. Optimized maintenance strategy for concrete structures affected by cracking due to reinforcement corrosion *ACI Structural Journal* **110** 229-238.

[7] Nepal J and Chen H P. Evaluation of residual strength of corrosion damaged reinforced concrete structures *Fourth International Symposium on Life-Cycle Civil Engineering* 2014, Tokyo, Japan.

[8] Nepal J, Chen H P and Alani A M 2013. Analytical modelling of bond strength degradation due to reinforcement corrosion *Key Engineering Materials* **569** 1060-1067.

[9] Nepal J and Chen H P. Gamma process modelling for lifecycle performance assessment of corrosion affected concrete structures *World Congress on Advances in Civil, Environmental, and Materials Research* 2014, Busan, Korea.

[10] Papakonstantinou K G and Shinozuka M 2013. Probabilistic model for steel corrosion in reinforced concrete structures of large dimensions considering crack effects. Engineering Structures **57** 306-326.

[11] Chen H P and Xiao N 2012. Analytical solutions for corrosion-induced cohesive concrete cracking *Journal of Applied Mathematics* Article ID 769132.

[12] Hillerborg A, Modeer M and Petersson P E 1976. Analysis of crack formation and crack growth in concrete by means of fracture mechanics and finite elements *Cement and Concrete Research* **6** 773-781.

[13] Wang X H and Liu X L 2010. Simplified methodology for the evaluation of residual strength of corroded reinforced concrete beams *Journal of Performance of Constructed Facilities ASCE* **24** 108-119.

[14] Au F T K and Du J S 2004. Prediction of ultimate stress in un-bonded prestressed tendons *Magazine of Concrete Research* **56** 1-11.

[15] Cairns J and Zhao Z 1993. Behaviour of concrete beams with exposed reinforcement *Proceedings of the ICE - Structures and Buildings* **99** 141-154.

[16] Van Noortwijk J M and Frangopol D M 2004. Two probabilistic lifecycle maintenance models for deteriorating civil infrastructures *Probabilistic Engineering Mechanics* **19** 345-359.

[17] Van Noortwijk J M 2009. A survey of the application of gamma processes in maintenance *Reliability Engineering and System Safety* **94** 2-21.

[18] EC2 2004. *Eurocode 2: Design of concrete structures - Part 1-1: General rules and rules for buildings (BS EN 1992-1-1:2004)*, European comitee for standardization, Brussels.

[19] Comité Euro-international du Béton - Fédération International de la Précontrainte (CEB-FIP) 1993. *CEB-FIP Model Code 1990 Design Code*, Thomas Telford, London.

[20] Chen H P and Bicanic N 2010. Identification of structural damage in buildings using iterative procedure and regularisation method *Engineering Computations* **27**(8) 930-950.

[21] Chen H P, Alani AM 2012. Reliability and optimised maintenance for sea defences *Proceedings of the ICE-Maritime Engineering* **165**(2) 51-64.

[22] Liu M and Frangopol D M 2005. Time-dependent bridge network reliability: Novel Approach *Journal of Structural Engineering ASCE* **131** 329-337.

Experimental study on the effectiveness of polymer damper in damage reduction of temporary steel grandstand

N Lasowicz[1], **A Kwiecień**[2] **and R Jankowski**[3]
[1,3]Faculty of Civil and Environmental Engineering, Gdansk University of Technology, ul. Narutowicza 11/12, 80-233 Gdańsk, Poland
[2]Institute of Structural Mechanics, Cracow University of Technology, ul. Warszawska 24, 31-155 Cracow, Poland

E-mail: natmajew@pg.gda.pl[1], akwiecie@pk.edu.pl[2], jankowr@pg.gda.pl[3]

Abstract. A large number of accidents, involving collapses of temporary grandstands during different types of events, were observed in the past. If the synchronized movement of people is tuned with the natural frequency of the affected part of the structure, resonance might occur. It may lead to severe damages of grandstands, their collapse or panic among the spectators. The aim of the paper is to assess, through preliminary experimental tests, the effectiveness of a polymer damper in damage reduction of a temporary steel grandstand. The damper considered in the study has been constructed out of two L-shape steel members bonded with polymer mass of high damping properties. The element has been installed as a diagonal one at the back part of the structure. The method has been compared with the typical solution of strengthening the grandstand with the diagonal stiffener of tubular cross section. The results of the study show that the responses of a temporary steel grandstand equipped with the polymer damper as well as with the typical stiffener are substantially different. The application of the polymer damper leads to the substantial increase in the level of structural damping ratio allowing to minimize the possibility of structural damage.

1. Introduction

There are a number of reasons for structural vibrations. Some of them are easily assessed, while others are purely random. The most common causes include: earthquakes (see, for example, [1-4]), large mechanical vibrating entities, air pressure, vibrations resulting from vehicles/trains using the nearby road/track or movement of people. The last one, if acting on relatively flexible structure, such as temporary steel grandstand, may cause serious problems [5, 6]. If the synchronised movement of people excites a natural frequency of the affected part of the structure, resonance will occur [7]. A large number of accidents, involving collapses of temporary grandstands during different types of events were observed in the past. In 1992, a collapse of such a structure took place in Corsica resulting in several deaths and hurting thousands of people [8]. Detailed analysis concerning disasters of temporary grandstands, that took place between 1889 and 2008, with 60 persons killed and around 6000 injured, has been presented in [8]. A lot of collapses were caused by wrong connections or anchors, difficulties during assembly, not properly fitted bracing system and pour quality of material.

[1] natmajew@pg.gda.pl

The main problems were related to unexpected excessive loads and more than 60% of disasters involved dynamic loads as the result of behaviour of spectators, mostly jumping.

Temporary steel grandstands are these types of structures that are commonly erected for a particular sporting or entertainment event to fit enlarged amount of spectators and they are removed when the event is finished. Temporary steel grandstands are characterized by many advantages, such as quick assembly, relatively low costs and possibility of using the same structures many times and locate them on uneven terrain. Most of them are erected using scaffolding system that gives an opportunity to use these structures during events that can fit small group of people or even thousands of spectators. Temporary grandstands are erected using different approaches that should be accurate and appropriate to the considered situation. Different configurations of structural members (mostly bracing systems), types of connection and anchors have a significant influence on changes in dynamic parameters of a structure. Dynamic loads have to be taken into account at the design stage of temporary grandstands, especially if light and slender structural members are used since they are more easily excitable to vibrations [9, 10].

In this paper, an idea of using polymer damper, in order to reduce steel grandstand vibrations and therefore minimize damage under human-induced excitations, is proposed. The polymer mass, applied in the damper, is a specially designed flexible two-component grout, which is based on polyurethane resin [11, 12]. It has been shown, based on the experimental studies, that this kind of material has very good damping properties [11]. Two different members have been analysed during the experimental study. One of them concerns a typical diagonal stiffener of tubular cross section as a solution of strengthening the grandstand, while the second member has been constructed out of two L-shape steel members bonded with polymer mass.

2. Experimental study

A typical repeatable part of a three-storey temporary steel grandstand has been considered in the study. The main dimensions of a structure are: length: 2.1 m, width: 2.7 m and height: 2.42 m (see Figure 1). The structure has been erected using scaffolding system that consists of tubular members. Additional structural members used to built the temporary grandstand are wooden benches, so as to provide seats for people, and steel platforms. Typical connections for scaffolding system have been used. All horizontal members are pushed in one to another allowing for some rotation at the connections what, actually, results in some additional damping in the structure.

Figure 1. Steel temporary grandstand with typical diagonal stiffener.

Two different members have been examined and analysed during the experimental study. One of them concerns a typical diagonal stiffener of tubular cross section as a solution of strengthening the grandstand (see Figure 1). The second one has been constructed out of two L-shape steel members (50x50x5 mm) bonded with polymer mass of thickness 5 mm of high damping properties (see Figure 2). This element, which can be called as polymer damper, has also been installed as a diagonal one at the back part of the structure (see Figure 3).

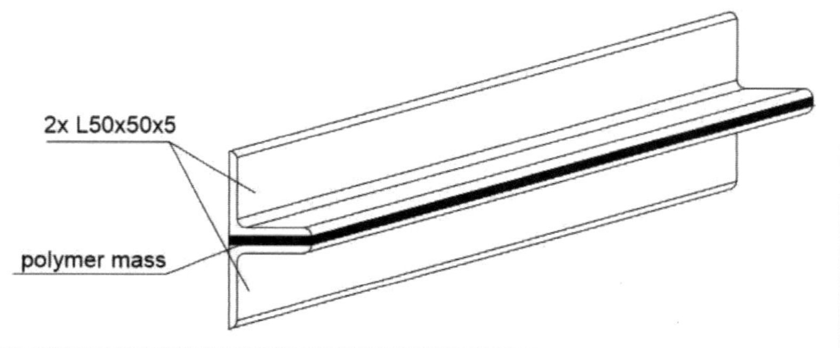

Figure 2. A scheme of polymer damper.

Figure 3. Steel temporary grandstand with diagonal polymer damper.

The aim of the experimental study was to assess the effectiveness of the polymer damper in reduction of vibrations of the temporary steel grandstand under two different load cases. One of them describes empty structure, while the second one takes into account mass of the spectators which, due to safety reasons, has been simulated by blocks of concrete. The weight of each spectator has been assumed to be equal to 104 kg. The mass of twelve spectators has been placed on the structure, according to existing standards, so as to allow four persons to seat on each bench (see Figure 4).

Figure 4. Front view of occupied steel temporary grandstand.

The temporary steel grandstand has been excited to free vibrations by applying an initial impact at the top of vertical tubular members. The response of the structure has been recordered by four accelerometers, placed on each vertical member, till structural damping brought the system to stop. Behaviour of empty and occupied temporary steel grandstand with typical diagonal stiffener is presented in Figure 5 and Figure 6, respectively.

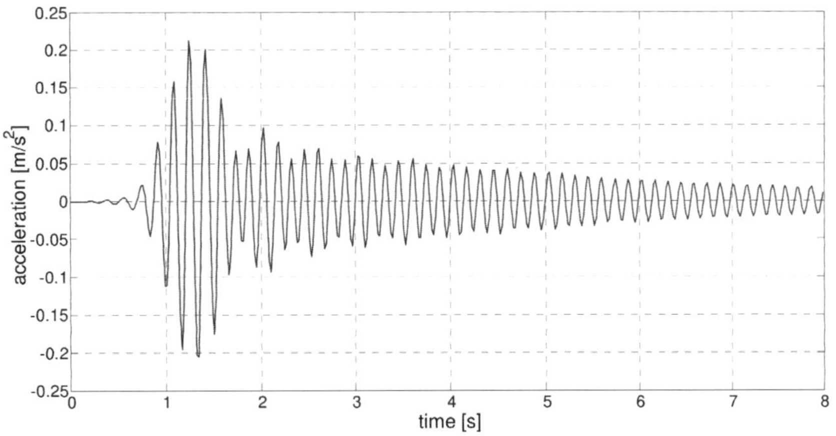

Figure 5. Acceleration time history for empty structure with typical diagonal stiffener.

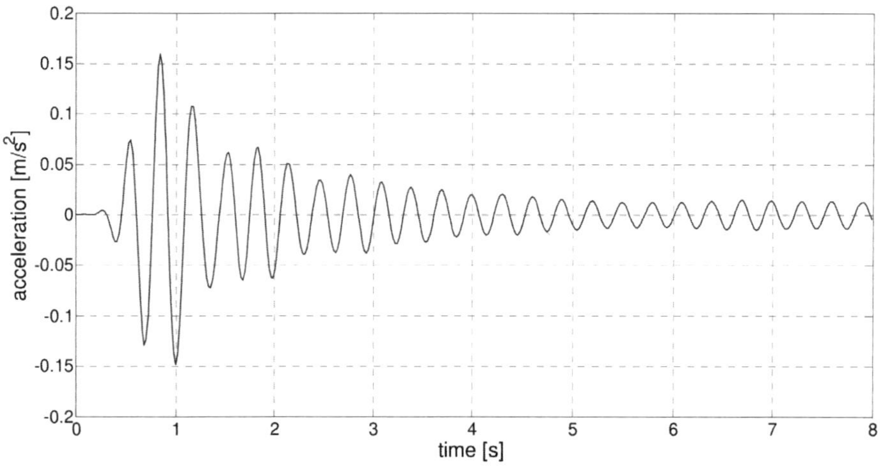

Figure 6. Acceleration time history for occupied structure with typical diagonal stiffener.

The corresponding time histories obtained for the temporary steel grandstand equipped with the polymer damper are shown in Figure 7 and Figure 8. Based on the results, dynamic parameters, such as natural frequencies and damping ratios, have been calculated. These parameters, for the empty and occupied structures with typical stiffener of tubular cross section, are summarized in Table 1. The corresponding dynamic parameters for the empty and occupied structures equipped with the polymer damper are also presented in Table 2.

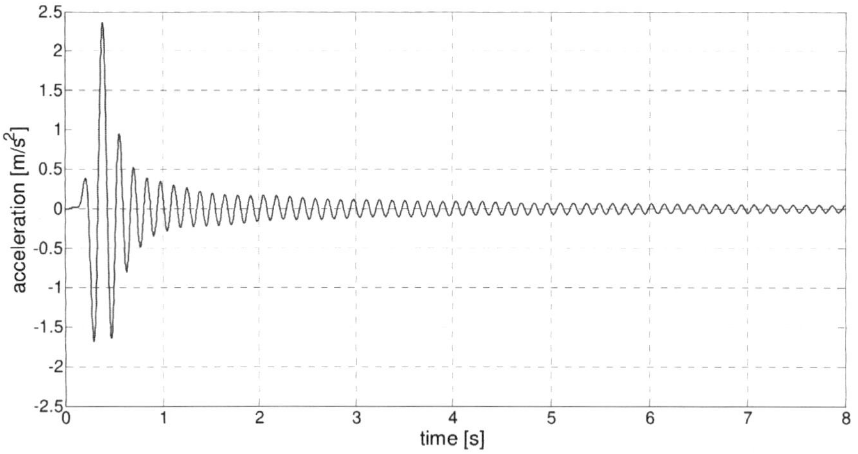

Figure 7. Acceleration time history for empty structure with polymer damper.

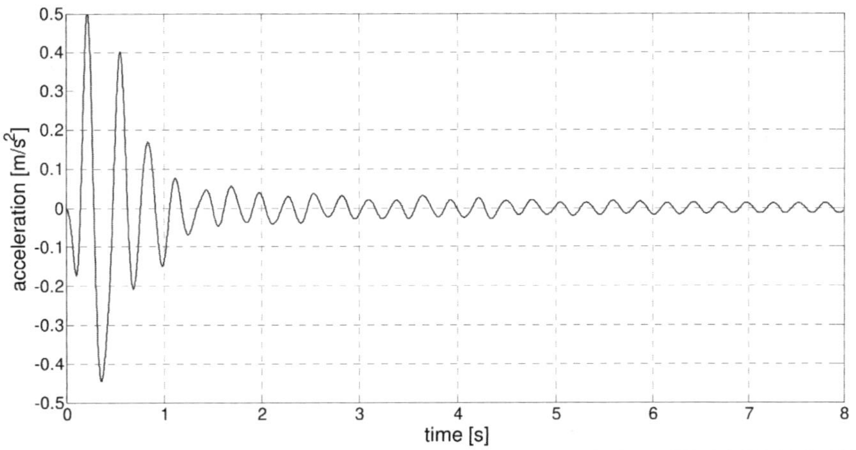

Figure 8. Acceleration time history for occupied structure with polymer damper.

Table 1. Natural frequencies and damping ratios of structure with typical diagonal stiffener member.

Type of load case	f (Hz)	ζ (%)
Empty structure	6.80	0.54
Occupied structure	3.12	0.98

Table 2. Natural frequencies and damping ratios of structure with polymer damper.

Type of load case	f (Hz)	ζ (%)
Empty structure	5.88	1.05
Occupied structure	3.38	3.05

It can be seen from Table 1 and Table 2 that mass of spectators leads to the significant decrease in the natural frequencies of the structure analyzed. The value of the natural frequency calculated for the occupied structure is about two times lower than the value obtained for the empty structure. What is more important, the results shown in Tables 1-2 and Figures 5-8 indicate that the application of polymer damper leads to much higher values of structural damping ratio comparing to the case when the typical diagonal stiffener is used. Actually, the increase in the damping ratio is as large as 95% and 211% for the empty and occupied structure, respectively.

3. Concluding remarks

The effectiveness of a polymer damper in reduction of vibrations of a temporary steel grandstand has been assessed in this paper based on the results of the preliminary experimental tests. The damper considered in the study has been constructed out of two L-shape steel members bonded with polymer mass of high damping properties. The method has been compared with the typical solution of strengthening the grandstand with the diagonal stiffener of tubular cross section. The results of the

study show that the response of the temporary steel grandstand equipped with the polymer damper as well as with the typical stiffener is substantially different. The application of the polymer damper leads to the substantial increase in the level of structural damping ratio.
The analysis described in this paper is a first stage of a wider study. The preliminary results are very promising. They show that the method of installation the polymer damper can be considered as an effective one in reduction of vibrations and therefore preventing from structural damage. However, further detailed experimental and numerical study is required so as to determine the behaviour of the steel grandstand under dynamic load induced by spectators due to their jumping. The results of further investigation will allow us to verify the effectiveness of the method.

Acknowledgments
The authors would like to acknowledge the company RUBO-Serwis for renting, transportation and permission to undertake the experimental work on temporary grandstand using scaffolding system. We would also like to thank Mr. Maciej Lasowicz and Mr. Piotr Brzeziński for help in conducting the experimental study as well as Mr. Bogusław Zając for help in preparing the polymer damper.

4. References

[1] Jankowski R 2005 Impact force spectrum for damage assessment of earthquake-induced structural pounding *Key Engineering Materials* **293-294** 711-8
[2] Jankowski R 2007 Assessment of damage due to earthquake-induced pounding between the main building and the stairway tower *Key Engineering Materials* **347** 339-44
[3] Mahmoud S, Austrell P-E, Jankowski R 2012 Simulation of the response of base-isolated buildings under earthquake excitations considering soil flexibility *Earthquake Engineering and Engineering Vibration* **11** 359-74
[4] Mahmoud S, Abd-Elhamed A, Jankowski R 2013 Earthquake-induced pounding between equal height multi-storey buildings considering soil-structure interaction *Bulletin of Earthquake Engineering* **11** 1021-48
[5] Jones C A, Reynolds P, Pavic A 2010 Vibration service-ability of stadia structures subjected to dynamic crowd loads *Journal of Sound and Vibration* **330** 1531-66
[6] Ellis B R, Little, J D 2000 The response of grandstands to dynamic crowd loads, *Proc. of Institution of Civil Engineering Structures and Buildings* **140** 355-66
[7] Salyards K, Hanagan L 2005 Evaluation of finite element model for dynamic characteristic prediction of a stadium facility, *Proc. of Conference and Exposition on Structural Dynamics (Orlando, USA, 31 January–3 February 2005)*
[8] Valkisfran L, Pimentel R 2009 Cases of collapse of demountable grandstands *Jornal of Performance of Constructed Facilities* **23** 151-9
[9] Reynolds P, Pavic A, Ibrahim Z 2004 Changes of modal properties of a stadium structure occupied by a crowd, *Proc. of Conference and Exposition on Structural Dynamics (Dearborn, USA, 26–29 January 2004)*
[10] Nhleko S P, Williams M S, Blakeborough A 2009 Vibration perception and comfort level for an audience occupying a grandstand with perceivable motion, *Proc. of Conference and Exposition on Structural Dynamics (Orlando, USA, 9–12 February 2009)*
[11] Falborski T, Jankowski R, Kwiecień A 2012 Experimental study on polymer mass used to repair damaged structures *Key Engineering Materials* **488-489** 347-50
[12] Lasowicz N, Kwiecień R, Jankowski R 2015 Enhancing the seismic resistance of columns by GFRP confinement using flexible adhesive - experimental study *Key Engineering Materials* **624** 478–85

Damage assessment of curtain wall glass

H Puga[1], BA Olmos[2], L Olmos[2], JM Jara[2] and M Jara[2]

[1] Graduate student, UMSNH, 58030 Morelia, Michoacán, México
[2] Professor UMSNH, 58030 Morelia, Michoacán, México

E-mail: bolmos@umich.mx

Abstract. The failure prediction of simply supported annealed glass plates subjected to uniform loads is one of the main purposes in the design codes of the United States, Canada and the European Community. The methodologies and codes available in the literature are based on concepts and criteria applicable to the glass failure prediction; they evaluate the load associated to a specific probability of failure. The aim of this work is to estimate fragility curves for curtain glass under different uniform loads representative of the wind loads that they can be subjected, using the lifetime prediction model. The capacity of the structural elements was determined experimentally considering as-received annealed soda lime silica glass; this material is used in structural elements although the material is brittle and random. The capacity and demand are associated with the life time prediction model. The results let us understand the glass failure mechanisms of glass panels with different thickness, as well as assess their probability of failure by estimating fragility curves.

1. Introduction

The glass is a material with an important development on the techniques used for its manufacturing. The float glass process was developed in the 1050s by Alastair Pilkington. This technique allowed producing float glass elements with good qualities to be used in the engineering practice. However, the pioneering work of Griffith [1] showed that the glass surface flawed by having micro-cracks, which represented the main causes of their fragility and resistance randomness. These and other factors are accounted in the methodologies presented in the design codes of North America and Europe to estimate the glass panels' thickness for a design load. The structural design of curtain wall glass can be achieved using the ASTM E 1300-12a [2] in the United States, the CAN/CGSB 12.20-M89 [3] in Canada, and with the EN 13474 [4] and Haldimann's life time prediction model [5] in Europe. All of these methodologies relay on the estimation of the failure probability of the glass panels.

The standard ASTME 1300-12a offers a wide variety of graphics to estimate the glass thickness required for a specific problem. The fundamental concepts used for the development of those curves is the glass failure prediction model proposed by Beason in 1980 and the Vallavhan's research [6], to calculate deflections accounting for the non-geometric linearity's. The glass failure prediction model was developed based on two flaw surface parameters, characterizing the surface condition and a parameter dependent of the crack velocity. The design aids correspond to glasses characterized with a surface condition of $m=7$, $k=2.86\times10^{-53}N^{-7}m^{12}$, elastic modulus of 71.7 GPa and by the glass thickness [5]. Basically, the standard is related to define the load resistance for different glass types with rectangular geometry supported continuously in one, two, three or four sides. The resistance load corresponds to a probability of failure of 0.8% for load duration of 3 seconds. The design load can be one or a combination of the effects of wind, snow, earthquake and glass' self-weight whereas the load is assumed to be not larger than 15 kPa. Two factors define the load resistant: the glass type factor and the non-factored load. The first one accounts for glass type and load duration whereas the second is for the non-factorized load with duration of 3 seconds.

The Canadian standard, CAN/CGSB 12.20-M89 has a similar methodology to the ASTM E1300-12a, both based on the glass failure prediction model [6], and considering glass elements under uniform lateral loads as well as a failure probability of 0.8%. The main difference between them is the load duration assumed, 60 seconds in this case vs. 3 seconds, to estimate the load resistance.

In 1999 the European Committee for Standardization (CEN) published the standard EN 13474-1 that states the general bases for the design of glass elements. The second edition, prEN 13474-2 presents a design method for curtain walls. The third edition of the standard is under revision, prEN 13474-3, where the glass' strength can be estimated with a general methodology or with a detailed methodology based on experimental tests. The parameters used in this standard relay on two crack velocity parameters estimated based on coaxial double ring tests with constant stress rates; with these parameters, the fundamental surface parameters can be calculated using the expression defined in the standard.

The life time prediction model [5] is a generalized probabilistic methodology that can be used for any type of structural glass elements (girders, columns and plates) and load condition. The linear elastic fracture mechanics and the theory of probability are the foundations of the model. A thoroughly consideration of the variables dictating the glass strength behavior (phenomenon of stress corrosion causing subcritical growth of surface flaw, the time-dependent behavior of flaws with random depth, location and orientation) is considered to establish a risk integral approach. The methodology is powerful because of the consideration of many random variables that are simplified in design codes that allows to estimate closely, the real glass behavior for different types of structural elements. However, it is necessary to extend the knowledge of probability and glass mechanics, and also conduct experimental studies which increase its implementation cost and its generalization for the practitioner practice. In spite of these, there is a simplification of the model [5] under research to be considered as standard in a close future.

The aim of this work is to develop general design aids, as the ones presented on the standard ASTME 1300-12a, applying the methodology for the use of the practitioner engineer. First we lead an experimental research program that characterizes the mechanical properties of silica glass for the estimation of the probabilistic parameters [7] used to estimate the fragility curves presented in this work, and later the fragility curves are the data base needed to create the design aids, results that will be reported in a following work. The results presented in the present study show increases on the load resistance by curtain glass panels as a function of their thickness, and the design loads are about 25% lower when considering the probability of failure associated to the European code than the ones related to the probability of failure of the American and Canadian codes.

2. The lifetime prediction model

The life time prediction model [5] is based on concepts of the linear elastic fracture mechanics and the theory of probability where it is assumed a random crack distribution in the glass surface with a random length and number. The main hypotheses of the method are: the glass surface has a random number of cracks with random depth; the cracks' depth can be represented with a Pareto probability density function; the cracks' behavior is independent of the one for the neighbor cracks; the cracks' location and orientation have the same probability to be in any location on the glass surface, so they can be defined with a uniform probability density function; the failure mode I defined for the linear fracture mechanics is assumed to be accurate; and the probability of failure accounts only for tensile strengths, neglecting the compression effects. The theory underneath is:

1. The relationship between crack velocity and stress intensity is defined with an equivalent formulation based on the linear crack velocity parameter v_0, a dimensionless crack velocity parameter n, the stress intensity factor, K_I, and the material constant, K_{Ic}, that is well known. The relationship between v_0 and K_I is simplified into three regions defining the environment influence. Region I defined linear behavior with a small slope since the subcritical crack growth is slow and depends on the environmental conditions; Region II defined with a constant value where v is independent of K_I but depends on the amount of humidity in the environment; and Region III defined linear behavior with a steep slope where v is independent of the environment. It is assumed region I because of the ratio of the size glass elements in buildings and the surface flaws expected for the service life time.
$$v = v_0 \left(\frac{K_I}{K_{Ic}}\right)^n \quad (1)$$
2. The basic rule of glass failure: "A glass element fails, if the stress intensity factor K_I due to tensile stress at the tip of one crack reaches its critical value K_{Ic}", [2]. It is expressed by the relationship among the stress intensity factor K_I, the nominal tensile stress normal to the crack's plane σ_n, a correction factor Y, and the crack depth a, as:
$$\mathrm{K}_I = Y\sigma_n\sqrt{\pi a} \geq \mathrm{K}_{Ic} \quad (2)$$
which is the base to define implicitly the critical strength σ_c and critical crack depth a_c considered for the development of the integral risk.
3. A subcritical crack growth and lifetime is estimated by differentiating with respect time the crack depth that after substitution, integration and algebraic manipulation led to the risk integral expressed in equation (3) for a time failure or lifetime t_f.
$$\int_0^{t_f} \sigma_n^n(\tau)d\tau = \frac{2}{(n-2)v_0 K_{Ic}^{-n}\left(Y\sqrt{\pi}\right)^n a_i^{(n-2)/2}} \quad (3)$$
4. The previous concepts are extended to a random surface flaw population assuming a constant stress and no subcritical crack growth, being the failure probability of a crack the probability that its random size a is larger than the critical crack depth a_c:
$$P_f = \int_{a_c}^{\infty} f_a(a)da = 1 - F_a(a_c) \quad (4)$$

By defining the adequate probability density function to each of the random variables defined and extending the concepts to a multimodal failure criterion, to a time-dependent loading and to account for subcritical crack growth, the lifetime prediction model is defined with [2],
$$P_f(t) = 1 - exp\left[-\frac{1}{A_0}\int_A \frac{2}{\pi}\int_{\varphi=0}^{\frac{\pi}{2}} \left[\max_{\tau\in[0,t]}\left[\left\{\left(\frac{\sigma_n(\tau,\vec{r},\varphi)}{\theta_0}\right)^{n-2} + \frac{\int_0^\tau \sigma_n^n(\tilde{\tau},\vec{r},\varphi)d\tilde{\tau}}{U\cdot\theta_0^{n-2}}\right\}^{\frac{1}{n-2}}\right]\right]^{m_0} dAd\varphi\right] \quad (5)$$

where P_f is the failure probability, A_0 is a unit surface area (A_0=1 m^2), A is a surface area of the glass element, $\sigma_n(\tau,\vec{r},\varphi)$ is an in-plane surface stress component normal to a crack orientation φ at the point $\vec{r}(x,y)$ on the surface and at a time τ, t is the point time, K_{Ic} is the fracture toughness, Y is the geometry factor, v_0 and n are the velocity parameters, θ_0 and m_0 are the surface condition parameters, U is a coefficient containing parameters related to fracture mechanics and subcritical crack growth, defined by:
$$U = \frac{2K_{Ic}^2}{(n-2)\cdot v_0\cdot Y^2\pi} \quad \text{(units of stress}^2 \times \text{time)} \quad (6)$$

To implement this methodology, we conducted dynamic fatigue test trough a suitable coaxial double ring test setup for new silica glass specimens, with the aim to determine the failure load and the maximum deflection associated. The experimental results were correlated with an analytical model of the tested plates under uniform load, to validate the probabilistic parameters determined during this step. Finite element models for 672 glass plates where developed to calculate the tensile stresses on the plates' surface. The lifetime prediction model was applied to the data generated to get the fragility curves; from them, it was estimated the load associated to two failure probabilities, 0.8 and 0.12%, commonly used on the design codes, it was estimated in order to analyze the structural behavior of curtain glass.

2.1. Silica glass characterization

The fatigue dynamic tests were based on the Standard Test Method for Monotonic Equibiaxial Flexural Strength of Advanced Ceramics at Ambient Temperature [8], also known as coaxial double ring test setup. The specimens correspond to new silica glass with quadrilateral section of 200mmx200mm and thicknesses of 3, 4, 5, 6, 8, 10, 12 and 19 mm. Ten specimens for each of the geometries were tested up to failure on an Instron-5500R machine under a load failure rate of 79 kN/min and ambient conditions, the details of the experimental work can be found in [7]. Here only the generalities and the final results are reported. The principal stresses that lead to the specimen failure were determined from the experimental test. This information was used to estimate the parameters related to the silica glass surface under ambient conditions (m_0=5, θ_0=59.78 MPa). Figure 1 shows a load-deflection curve for some of the specimens tested and a picture of one of the failure modes presented by the specimens.

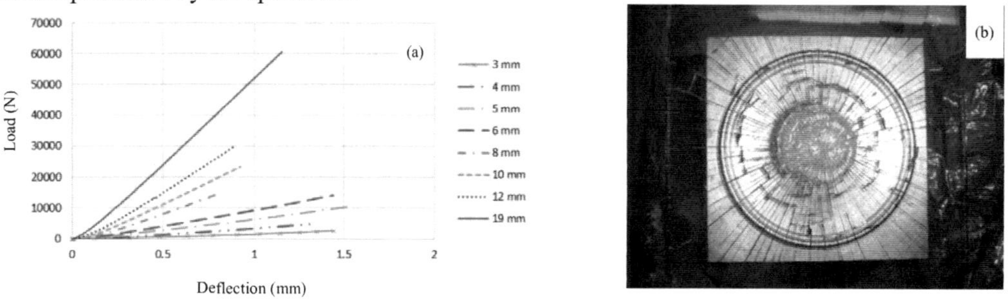

Figure 1. Load vs. deflection curve for some of the specimens (a), and failure mode of a glass specimen with 5 mm of thickness (b).

2.2. Numerical model

For an accurate representation of glass plates, it is required to account for the geometric nonlinearity described by the theory of large deflections when the deflections are larger than one third of the plate's thickness. This problem must be considered on glass plates because the aspect ratio led to inaccurate results when applying the linear plate's theory of Kirchoff that neglects the strains at the middle of the edges and the membrane stresses. The ANSYS software was implemented to create the numerical models with the "Static Structural" module considering the glass as an isotropic material, and accounting for the geometric non linearity that could be present in the plates. Previously to build all them to estimate the fragility curves, there were developed in ANSYS some of the specimens tested to validate the numerical models. After the validation, 672 curtain glass models were built varying the glass plates' size and thickness and assuming them simple supported on their four sides. The cases studied consider width to depth ratio (a/b where a>b) of 1.0, 1.5, 2.0, 2.5, 3.0, 3.5, 4.0, 4.5, 5.0, 5.5, 6.0, 6.5 and 7.0, and the curtain glass thicknesses of 2.5, 3.0, 4.0, 5.0, 6.0, 8.0, 10, 12, 16, 19, 22 and 25 mm. The cases are limited by the maximum glass curtain size offered by the industry, 2.5x3.5 m, as the commercial thicknesses as well, NOM-146-SCFI-2001. Based on these, the glass plates' width and depth selected cover surface plates' areas in the range of 0.375 to 9.375 m^2.

To optimize the time required for the analyses and taking advantage of the plates' symmetry, each of the curtain plates was represented with a quarter of the plate modeled with an isotropic elastic shell finite elements type Shell 181. The finite element sizes vary from 10x10 mm to 30x30 mm. The material properties for the results presented in this work are: elastic modulus of 74000 MPa, Poisson ratio of 0.23 and specific density of 2500 kg/m^3. A fixed condition was considered in the nodes located at the plate edges with an exception of the degree of freedom related to the displacement running out of the glass plate plane. The simple supported condition, recommended in [3] and [4], led to the maximum deflection that can reach a glass plate, representing also the large effects of geometric nonlinearities.

The numerical models were subjected to a uniform pressure under 20 load stages with increments of 0.001 MPa. Since the uniform load value required to cause the failure of the glass plates was unknown, the maximum intensity considered in each of the analyses was 0.02 MPa. Figure 2 shows the results for one of the models under a uniform load of 0.02 MPa.

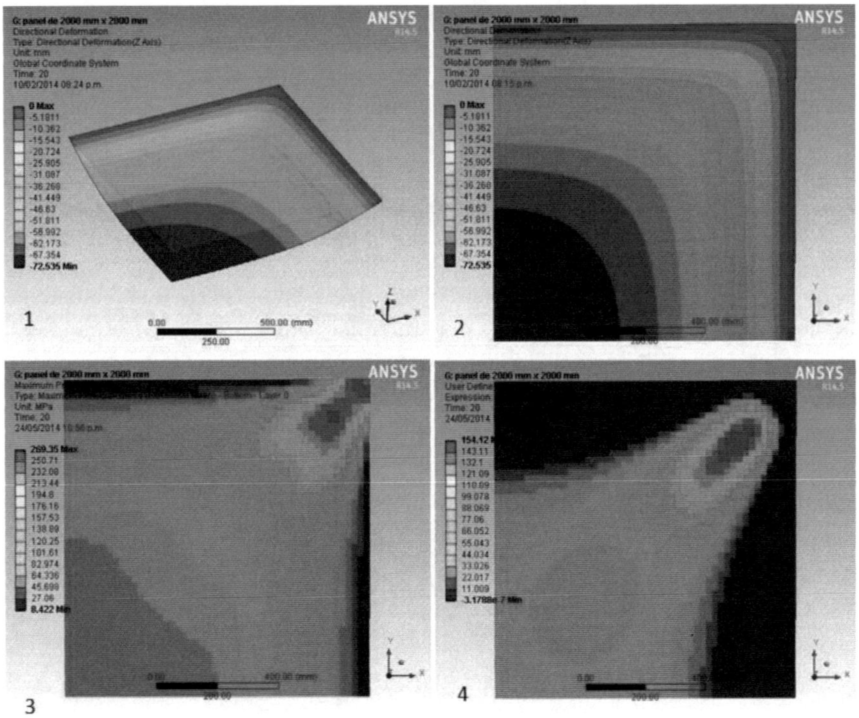

Figure 2. Finite element model for a curtain glass under a uniform pressure of 0.02 MPa: 1) Deformed plate shape, 2) Maximum deflection at the center of the plate (72.5 mm), 3) Maximum principal stress plate distribution, and 4) minimum principal stress plate distribution.

The estimation of the fragility curves based on the lifetime prediction model needs a great number of data defining the stresses on the plates' surface subjected to the entire possible uniform loads. Because this implies an infinite number of analyses, regression analyses were used to extrapolate the stresses associated to loads not considered in our models. This is possible, due to the elastic behavior of the glass associated to a linear load or uniform pressure. Figure 3 shows the regressions associated to the principal stresses.

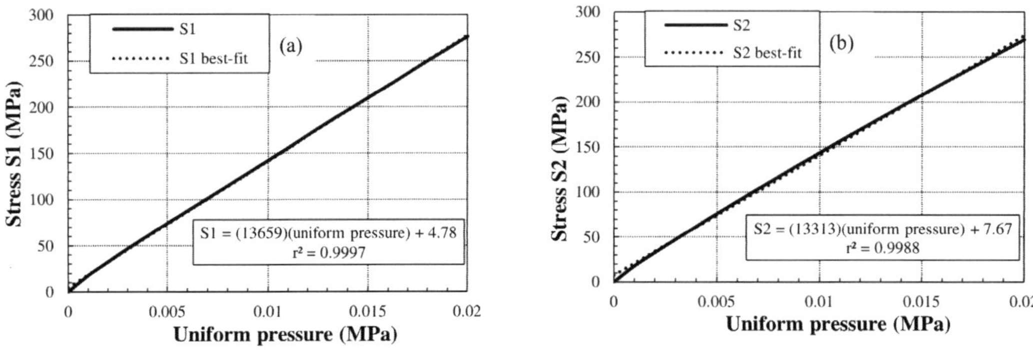

Figure 3. Uniform pressure vs. principal stresses and adjusted with a linear regression, (a) S1 and (b) S2.

3. Fragility curves

Fragility is the probability to reach or to exceed a limit state of behavior. It is defined with equation (5) and programmed using the Matlab software that allows generating a data base relating glass plate sizes with a uniform pressure, associated with a probability of failure. Table 1 shows the parameters considered in the lifetime prediction model.

Table 1. Glass curtains' mechanical properties and parameters considered for the fragility estimates.

Parameter(s)	**Value**
Critical stress intensity factor (K_{IC})	23.7172 MPaxmm$^{0.5}$
Correction factor (Y)	1.12 (dimensionless)
Crack velocity parameters (v_0, n)	v_0 = 6 mm/s
θ_0	n = 16 (dimensionless)
m_0	59.88 MPa
Young's modulus (E)	74 000 MPa
Uniform load time duration (t)	3 s
Sub-area (A_i)	Function of shell size (mm^2)
In-plane principal stresses $S1$ and $S2$	From the finite element (MPa)

The graphs plotted on figure 4 show representative fragility curves for curtain glass panels with sizes of 2500mmx2500mm and 3500mmx500mm, and thicknesses of 2.5, 8 and 25 mm respectively in rows 1 to 3. The results correspond to the uniform pressure on the plates associated with a probability of failure, and it is also reported the uniform pressure for the probabilities considered on the design codes of North America (0.008) and Europe (0.0012). The results presented in figure 4 show that for a defined probability of failure the load needed to reach a limit state increases as the plate thickness increases, and the quadrilateral plates have less strength than the rectangular ones. To confirm these results, table 2 resumes the failure probability for all the thicknesses considered in these sets of plates.

Figure 5 summarizes the fragility curves for all the thicknesses (h) considered for curtains with size of 2000mmx2000mm, plotting only the failure probability interval used with design purposes. The results confirm that the increase of the plate thickness increases the load capacity. It is noticeable that the curtain glass strength associated with the probability of failure adopted by the European code

(0.0012) is in general 25% lower than the one assumed with the American code (0.008), when the lifetime prediction model is implemented.

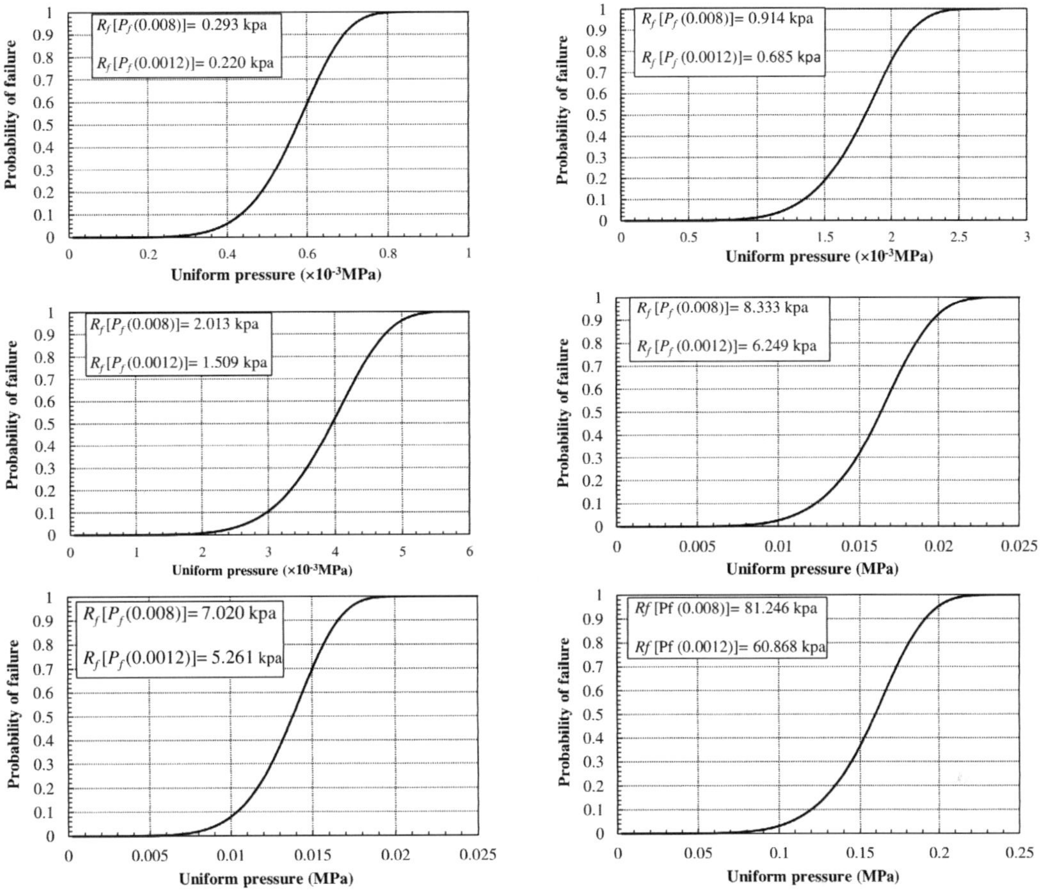

Figure 4. Fragility curves for a uniform pressure for curtain glass plates of 2500 mm x 2500 mm (first column), 3500 mm x 500 mm (second column), and thicknesses of 2.5 (first row), 8 (second row) and 25 mm (third row).

Table 2. Uniform pressure, R_f (kPa), on the curtain glass panels to rich a design failure probability.

Curtain size	R_f [P_f(0.008)]	R_f [P_f(0.0012)]	Curtain size	R_f [P_f(0.008)]	R_f [P_f(0.0012)]
2500x2500x3	0.375	0.281	3500x500x10	13.016	9.751
3500x500x3	1.207	0.905	2500x2500x12	3.138	2.351
2500x2500x4	0.723	0.541	3500x500x12	18.743	14.038
3500x500x4	2.096	1.569	2500x2500x16	4.888	3.661
2500x2500x5	0.974	0.730	3500x500x16	33.309	24.977
3500x500x5	3.259	2.441	2500x2500x19	5.397	4.074
2500x2500x6	1.398	1.047	3500x500x19	46.984	35.202
3500x500x6	4.693	3.515	2500x2500x22	6.156	4.614
2500x2500x10	2.607	1.953	3500x500x22	62.933	47.143

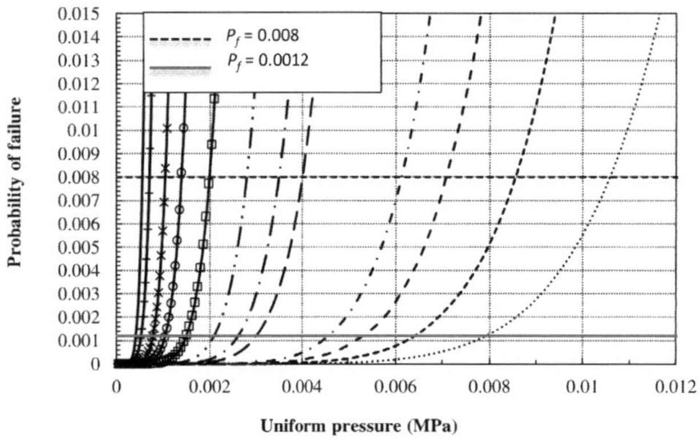

h (mm)	R_f (kPa)	
	P_f = 0.0012	P_f = 0.008
2.50	0.42	0.56
3.00	0.53	0.71
4.00	0.78	1.04
5.00	1.04	1.39
6.00	1.49	1.99
8.00	2.09	2.79
10.00	2.61	3.48
12.00	3.00	4.00
16.00	4.59	6.12
19.00	5.21	7.06
22.00	6.40	8.55
25.00	7.92	10.57

Figure 5. Fragility curves for a uniform pressure of glass plates' 2000 mm x 2000 mm x h mm.

4. Conclusions

672 curtain glass finite element models were developed using the ANSYS software. The models are representative of plates' sizes built by the glass industry. We developed functions to interpolate and extrapolate the stresses on the shell plates for different pressures with an extended data base to estimate the fragility curves. We also programed the lifetime prediction model offering a tool to evaluate any fragility curve as a function of the glass' surface parameters [7]. The load resistance of curtain glass panels increases with the glass thickness increase. We also found that the failure probability proposed by the European code conducts to design loads 25% lower than the loads obtained using the failure probability stated in the American and Canadian codes. The next step is to develop curve design aids using the glass probability of failure estimated with the lifetime prediction model, research that will be published posteriorly.

5. References

[1] Griffith A. A. (1920) The Phenomena of Rupture and Flow in Solids *Philosiphical Transactions, Series A.* **221** 163–198

[2] ASTM E 1300-12a (2012) *Standard practice for determining load resistance of glass in building.* American Society for Testing Materials, EUA, 1–59

[3] CAN/CGSB-12.20-M89 (1989) *Structural design of glass for buildings.* Canidian General Standards Board, Canada, 55

[4] EN 1374-1 (1999) *Vidrio para la identificacion-diseno de acristalamientos-Parte 1: Base general de calculo.* Comite Europeo de Normalizacion, enero, 1–29

[5] Haldiman M. (2006) *Fracture strength of structural glass elements-analytical and numerical modelling, testing and design.* PhD Dissertation No. 3671, Swiss Federal Institute of Technology Lusanne

[6] Beason W.L. and Morgan J.R. (1998) Glass Failure Prediction Model *Journal of Structural Engineering, ASCE.* **110 (2)** 197–212

[7] Puga H. (2014) *Modelo de prediccion de falla de cortinas de vidrio.* Thesis, Universidad Michoacana de San Nicolas de Hidalgo

[8] ASTM C 1499-05 (2005) *Standard test method for monotonic equibiaxial flexural strength of advanced ceramics at ambient temperature.* American Society for Testing Materials, EUA, 1–11

Acquisition and Analysis of Dynamic Responses of a Historic Pedestrian Bridge using Video Image Processing

Michael O'Byrne[1], Bidisha Ghosh[1], Franck Schoefs[2], Deirdre O'Donnell[3], Robert Wright[3] and Vikram Pakrashi[3]

[1]Department of Civil, Structural and Environmental Engineering, Trinity College Dublin, Ireland
[2]LUNAM Universite´, Universite´ de Nantes-Ecole Centrale de Nantes, CNRS, GeM, Institute for Research in Civil and Mechanical Engineering, University of Nantes, Nantes, France; and IXEAD/CAPACITES Society, Nantes, France[1]Dynamical Systems and Risk Laboratory, Civil and Environmental Engineering, School of Engineering, University College Cork, Ireland
[3]Dynamical Systems and Risk Laboratory, Civil and Environmental Engineering, School of Engineering, University College Cork, Ireland

v.pakrashi@ucc.ie Ph: 00353214903862 Fax: 00353214276648

Abstract. Video based tracking is capable of analysing bridge vibrations that are characterised by large amplitudes and low frequencies. This paper presents the use of video images and associated image processing techniques to obtain the dynamic response of a pedestrian suspension bridge in Cork, Ireland. This historic structure is one of the four suspension bridges in Ireland and is notable for its dynamic nature. A video camera is mounted on the river-bank and the dynamic responses of the bridge have been measured from the video images. The dynamic response is assessed without the need of a reflector on the bridge and in the presence of various forms of luminous complexities in the video image scenes. Vertical deformations of the bridge were measured in this regard. The video image tracking for the measurement of dynamic responses of the bridge were based on correlating patches in time-lagged scenes in video images and utilising a zero mean normalised cross correlation (ZNCC) metric. The bridge was excited by designed pedestrian movement and by individual cyclists traversing the bridge. The time series data of dynamic displacement responses of the bridge were analysed to obtain the frequency domain response. Frequencies obtained from video analysis were checked against accelerometer data from the bridge obtained while carrying out the same set of experiments used for video image based recognition.

1. Introduction

Pedestrian bridges have an important role in societies worldwide. Effective monitoring of these structures is vital to ensure that they remain safe and fit for service. Vibration-based assessments are often used in this regard as they provide a way of characterising the bridge in terms of its dynamic response. The dynamic nature of a bridge provides an indication of its serviceability condition and the level of dynamic acceleration can be related to the comfort level experienced by bridge users [1]. Additionally, a knowledge of the bridge parameters, such as the natural frequency and mode shapes, is useful for numerous applications, including for verifying or strengthening Finite Element (FE) models so that they more accurately reflect the true behaviour of the structure. This aspect has attracted a growing interest in recent years as FE modelling is increasingly being utilised as a powerful and effective assessment tool [2-3].

Vibration based assessments are carried out using either direct or indirect instrumentation techniques. Monitoring via direct instrumentation usually requires the bridge to be outfitted with a range of sensors placed directly on the bridge, which is generally expensive and time consuming. The development of wireless sensor networks in recent times has partially alleviated this issue by adding a degree flexibility and value compared to permanent wired systems [4]. Indirect vibration based assessments typically involve vehicles equipped with on-board sensors traversing a bridge and recording the induced vibrations [5-6]. Such an approach is attractive in cases where the cost associated with permanent instrumentation and routine inspections cannot be justified.

This paper presents an alternative to these conventional approaches by proposing a video tracking based technique. Video tracking is suitable for analysing bridge vibrations that are characterised by large amplitudes and low frequencies. Video tracking has the advantage of being a low-cost and easily accessible option as the only hardware required is a standard digital camera capable of recording video. The onsite set-up and video acquisition is straightforward and does not require lengthy configuration or calibration procedures. Furthermore, the acquired data is easy to interpret as the excitation sources can be identified in the video itself. The proposed video tracking technique involves focusing a camera at the mid-span of a bridge and tracking the precise location of a designated point on the bridge in each frame of the video as it is undergoes user-induced vibrations. The time series data related to the location of this tracked point is then analysed to obtain the frequency domain response.

The technique is demonstrated as a proof-of-concept on a historic steel pedestrian suspension bridge in Ireland, called Daly's Bridge, which is known locally as the 'Shaky Bridge' due to its notable, and for some, unnerving swaying motion under pedestrian loading. This bridge is an ideal candidate for video based analysis as the dynamic displacement responses of the bridge are large enough to be detected in the video. The frequencies obtained from video analysis are checked against accelerometer data obtained from sensors on the bridge. The following section describes the video tracking technique. The technique is applied to Daly's Bridge in Section 3 and the results are presented and interpreted. Section 4 concludes the paper.

2. Methodology

This section consists of three parts. The first part describes the relevant camera settings and the set-up. The second part details the video tracking technique. Finally, the third part outlines the limitations and challenges associated with video tracking as a tool for vibration assessment.

2.1. Equipment and set-up

There are a number of equipment related factors that have a crucial impact on the performance of video tracking based vibrations assessments. Firstly, the maximum frequency of vibration that can be encoded is controlled by the sampling rate, or the video frame rate. As an example, the digital video camera used in this case study was a Canon 600D, which can record video at a rate of 60 Frames per Second (FPS) with a moderate resolution of 1280 pixels x 720 pixels, or at a slower rate of 35 FPS with a higher resolution of 1920 pixels x 1080 pixels. The 35 FPS option was chosen as the need for the highest resolution was greatly felt in order to compensate for the extensive distance between the video device and the mid-span of the bridge. This extensive distance was due to the fact that the video camera had to be set-up and positioned relatively far away on the bank of the river as shown in Figure 1. By Nyquist's criterion, the highest frequency that can be coded is half of the sampling rate, which for a frame rate of 35 FPS is 17.5 Hz.

Secondly, the minimum amplitude of the vibration that can be detected in the video is largely affected by the resolution, as previously mentioned, as well as the optical power of the lens. Long-focus lenses are able to make distant appear closer. This effect means that a long-focus lens would magnify bridge displacements and make them more apparent in the acquired video. A 55 mm lens was used in this case study, which was sufficient as the bridge under consideration was characterised by large amplitude vibrations.

Figure 1. Daly's Suspension Bridge in Cork, Ireland, and the location of a mounted video camera on the river bank.

2.2. Video tracking technique

The video tracking technique operates by tracking the movement of a point on the bridge whilst in an excited state. Tracking is done by picking a small patch, or window, *W*, which is centred on the chosen point in the first frame of a video sequence and following it throughout the duration of the video clip. For every successive frame, the point is located by finding the patch which best correlates with the patch in the previous frame. The best correlation was determined using the Zero-mean Normalised Cross Correlation (ZNCC) metric. For video tracking, the window centred on the tracked point in a frame f_n, is matched to the corresponding window in the next frame, f_{n+1}. The search space for locating the corresponding window is confined to a stationary region in the video, *R*, which must be carefully chosen such that it encloses the tracked point throughout the entire video. Confining the search space to the local neighbourhood allows for greater computational parsimony and minimises the risk of false matches. For this task, the ZNCC is defined as:

$$C = \frac{\sum\limits_{u,z\in W} (f_{n+1}(u,z)_{ab} - \bar{f}_{n+1,ab})(f_n(u,z) - \bar{f}_n)}{\sqrt{\sum\limits_{u,z\in W} (f_{n+1}(u,z)_{ab} - \bar{f}_{n+1,ab})^2 (f_n(u,z) - \bar{f}_n)^2}} \qquad a,b \in R \tag{1}$$

where $f_n(u,z)$ and $f_{n+1}(u,z)$ denote the intensity values of the pixels in the window, *W*, centred on the point being tracked in the nth and n+1th frame respectively, *u* and *z* are the horizontal and vertical spatial indices of *W*, *a* and *b* are the horizontal and vertical indices of the stationary region *R*, and $\bar{f}_n$

and $\bar{f}_{n+1}$ are the mean pixel intensity values within the area of the *W* for the nth frame, and within *R* for the $n+1^{th}$ frame. The zero mean normal cross correlation coefficient, *C*, has a high value when there is a high degree of similarity between two patches in each frame. The new location of the tracked point in frame $n+1$ is taken as the one which corresponds to the highest value of *C*. This procedure is repeated until the final frame in the video sequence. The pixel locations for the tracked point are recorded at each frame, which may be viewed as a time series data of dynamic displacement responses. Spectral analysis on the signal may then be performed in order to identify the major frequency components.

2.3. *Tracking challenges*

There are a number of challenges associated with tracking a moving point in a video sequence. Most notably, the tracked point may drift or become completely lost as a result of temporary occlusion or luminous complexities in the scene such as shadows or light reflections. To mitigate these issues, a prominent and distinct point (i.e. a patch in the image with high local contrast) should be selected for tracking. If needed, multiple trials can be carried out by selecting various high-contrast regions near the mid-span of the bridge and assessing how well they are tracked over the duration of the video.

Another challenge relates to locating the position of the tracked region as precisely as possible. Even at the highest video resolution, the motion of the bridge can translates to minute changes of only a few pixels in the video frames. In order to address this, a quadratic is fitted to surrounding values for sub-pixel and sub-scale interpolation as described by Brown and Lowe [7].

3. Experimental study: dynamic response estimation using video tracking

A full-scale experiment is carried out on a pedestrian suspension bridge to validate the video tracking method. This section describes the pedestrian bridge, the experiment overview including details of the designed pedestrian movements, and the results obtained from application of the video tracking method.

3.1. *Bridge description*

Daly's Bridge is a steel pedestrian suspension bridge that spans the River Lee in Cork City, Ireland, as shown in Figure 1. The bridge provides a link from the northside of the city to the southside. It has a regular flow of traffic throughout the day that includes leisurely walkers and commuters. The main span has a length of 51 m and a width of 1.45 m. Two steel lattice towers at each side of the river carry a total of four high tension steel cables, with two cables on the upstream side and two cables on the downstream side. Each tower consists of two piers that are braced together and are cast into concrete abutments on the river bank. The walkway consists of vertical steel lattice parapets and timber decking.

The bridge was built in 1926 and partially refurbished in the 1980s. With this in mind, it is no surprise that there are many instances of corrosion of varying degrees of severity. The four suspension cables are in a good state, showing only mild surface corrosion, however, fatigue damage is present in the free anchor length on the south side of the structure.

3.2. *Experiment overview*

Controlled pedestrian and cyclist movements were used to excite the fundamental vibration modes of the bridge. The pedestrian movements were carried out by a female and male who naturally had different walking signatures. All other traffic around the bridge was stopped during each controlled crossing. The camera began recording just before the controlled pedestrian or cyclist was about to begin the crossing, and continued until a short time after they reached the other side and the bridge returned to a state of rest.

In the acquired video, a distinctive point (e.g. a corroded spot that was easily discernible from the background or some other region of discolouration) was chosen to be tracked at the mid-span of the bridge, as shown in Figure 2. This allowed for more robust and effective tracking from frame to frame. As previously mentioned, tracking can be performed on any number of different points, which is particularly useful if some of the tracked points are prone to drifting from their true position over the course of the video.

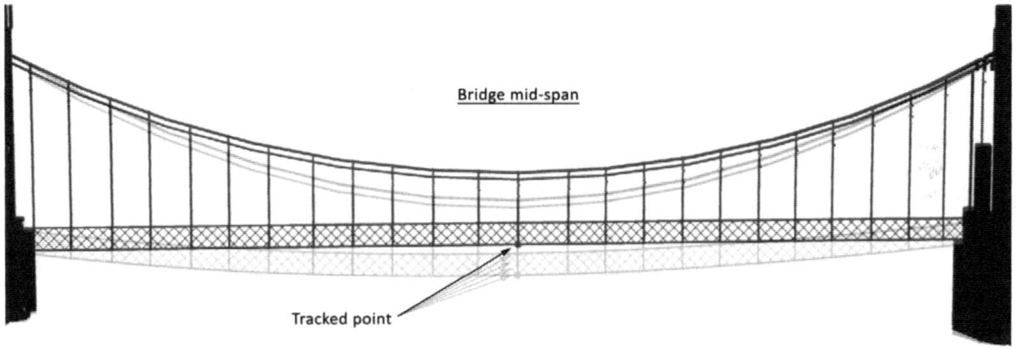

Figure 2. Tracked point at mid-span of the bridge.

As a further step, pixels units can be related to real world metric units by establishing a scale factor. The scale factor is computed by measuring the length of an object in a video frame in pixels and then comparing it with the known real world length of the corresponding object in the scene. In this case of Daly's Bridge, the hanger length at the mid-span was used as the object of known dimensions. This approach is valid as long as the bridge movement is roughly in the same plane as the measured object and the length of the object in the image in pixels is linearly related to its real length (i.e. there is negligible distortion such as perspective change between the top and bottom of the object). The step of relating pixel units to real world units is optional as the scale of the oscillations do not have a bearing on the estimated fundamental frequency, however, it is useful for other applications and conversion to real world units gives engineers a sense of the bridge displacements in a more meaningful light.

3.3. Results

This section presents the results obtained from video tracking of Daly's Bridge. Three excitation sources are considered for this case study: i) pedestrian induced vibration - female, ii) pedestrian induced vibration - male, and iii) cyclist induced vibrations. Two trials are shown for each excitation source. The displacement versus time graphs corresponding to each scenario are shown in Figures 3. Each of the graphs relate to a time period of about 30 seconds, depending on how long it took the pedestrian/cyclist to complete the bridge crossing.

Journal of Physics: Conference Series **628** (2015) 012053 doi:10.1088/1742-6596/628/1/012053

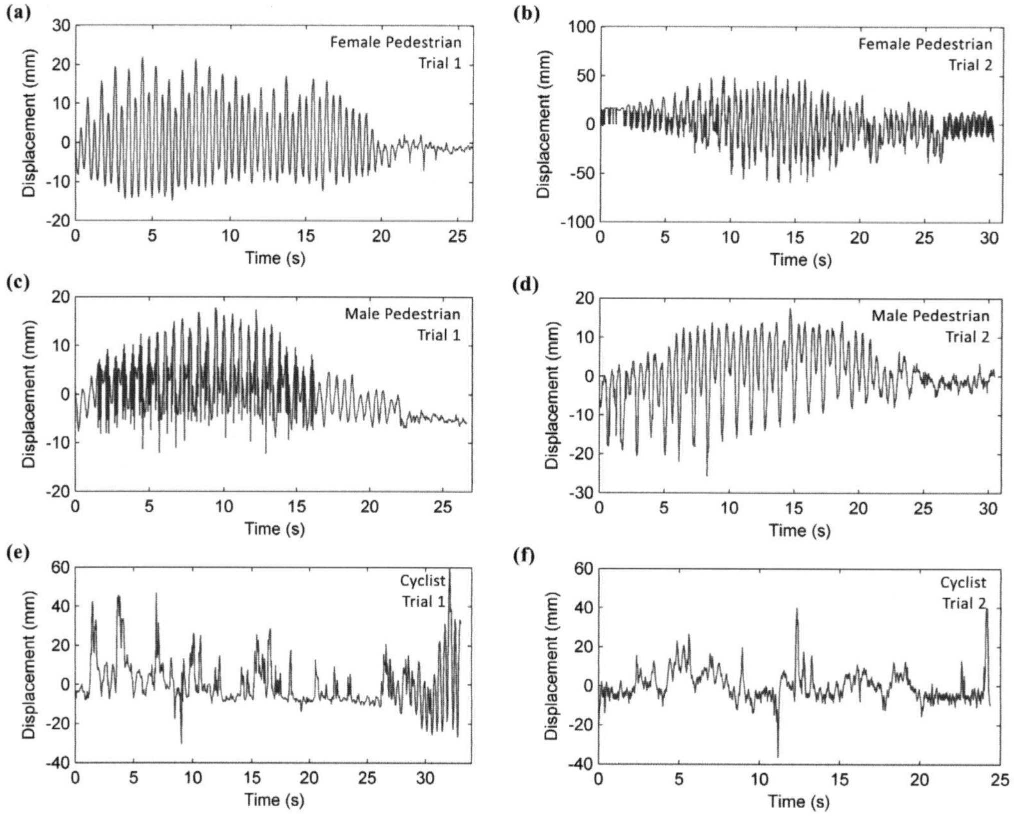

Figure 3. Vertical displacement versus time graph measured at mid-span of the bridge for: (a) female pedestrian - first trial; (b) female pedestrian - second trial; (c) male pedestrian - first trial; (d) male pedestrian - second trial; (e) cyclist - first trial; and (d) cyclist - second trial.

Fourier spectrum of these dynamic displacement responses are shown in the Power Spectral Density (PSD) plot in Figure 4.

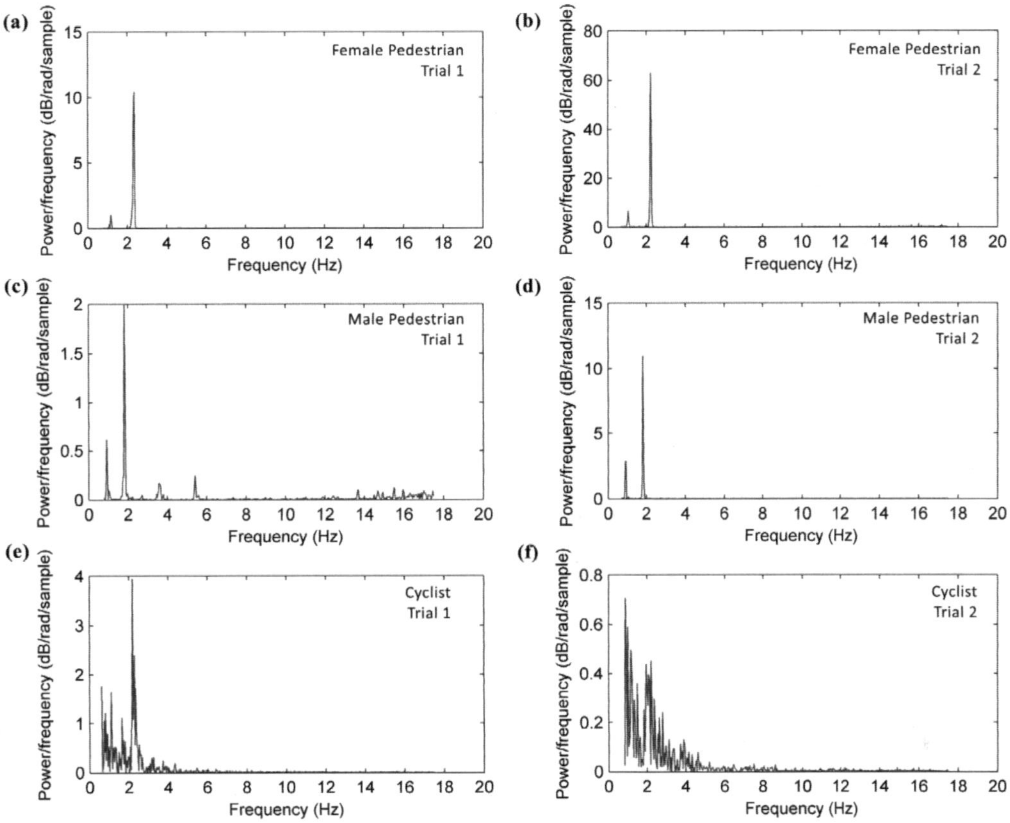

Figure 4. Fourier spectrum of vibration data for: (a) female pedestrian - first trial; (b) female pedestrian - second trial; (c) male pedestrian - first trial; (d) male pedestrian - second trial; (e) cyclist - first trial; and (f) cyclist - second trial.

From the results of these six trials, it may be observed that the values of the estimated first natural frequency of the bridge are relatively consistent. Prominent peaks appear in each PSD plot at around 2 Hz, indicating high repeatability of the results. An exception to this is in the case of the second trial for the cyclist (Figure 4f), which shows prominent peaks in the range 1-1.5 Hz. This is likely to correspond to the forced frequency of the bicycle. Nevertheless, a prominent peak is still identifiable at the estimated natural frequency of 2 Hz. Accelerometers placed at the mid-span of the bridge indicate similar values for the bridge's natural frequency. Table 1 provides a summary of the results for both sensor types.

Table 1. Natural frequency results from image analysis method and from accelerometers on the bridge.

Sensor type	Mean Natural Frequency	Standard Deviation from Mean, %
Accelerometers	2.27 Hz	4.41
Video Analysis	2.11 Hz	1.56

These results from this full scale experiment validate the proposed video tracking technique as a convenient tool for identifying the natural frequency of bridges characterised by large amplitudes and low frequencies to a reasonable degree of accuracy.

4. Conclusion

This paper presents a video tracking technique for the purpose of identifying the fundamental frequency of bridges. The technique is successfully applied and validated on a pedestrian suspension bridge in Cork, Ireland. The bridge was excited by designed pedestrian movements and by individual cyclists traversing the bridge. In each scenario, the video tracking method consistently obtained similar values for the fundamental frequency. The mean natural frequency from all of the trials was 2.11 Hz, which showed good agreement with data from accelerometers placed directly on the bridge, which reported a mean natural frequency of 2.27 Hz.

Overall, the positive results of the video tracking technique suggest that it is a viable low-cost method for carrying out quick vibration based assessments of bridges that exhibit strong dynamic behaviour.

Acknowledgements

The authors wish to thank the Irish Research Council for Science, Engineering and Technology (IRCSET) for providing grant to support this research and CAPACITES/IXEAD society for the practical assistance. The authors acknowledge the help of Fiona Edwards-Murphy and Emanuel Popovici of University College Cork for comparison with accelerometers.

References

[1] Van Nimmen K, Lombaert G, De Roeck G and Van den Broeck P 2014 Vibration serviceability of footbridges: Evaluation of the current codes of practice *Engineering Structures*. **59** 448-61

[2] Ren W X and Peng XL 2005 Baseline finite element modeling of a large span cable-stayed bridge through field ambient vibration tests. *Computers & Structures* **83** 536–50

[3] Živanović S, Pavic A and Reynolds P 2006 Modal testing and FE model tuning of a lively footbridge structure. *Engineering Structures*. **28** 857–68.

[4] Pakzad S N, Fenves G L, Kim S and Culler D E 2008 Design and implementation of scalable wireless sensor network for structural monitoring *Journal of Infrastructure Systems*. **14**, 89–101

[5] Siringoringo D M and Fujino Y 2012 Estimating bridge fundamental frequency from vibration response of instrumented passing vehicle: Analytical and experimental study *Advances in Structural Engineering*. **15** 417-34

[6] González A, O'Brien E J and McGetrick P J 2012 Identification of damping in a bridge using a moving instrumented vehicle *Journal of Sound and Vibration*. **331** 4115-31

[7] Brown M and Lowe D 2002 *Invariant features from interest point groups* Proc., BMVC2002: British Machine Vision Conference 2002, 2-5 Sept. 2002, British Machine Vision Assoc., 253-262.

Long-term bridge health monitoring focusing on the Mahalanobis Distance of modal parameters

Chul-Woo Kim[1], Tomoaki Morita[2], Ziran Wang[3] and Kunitomo Sugiura[4]

[1, 4] Professor, Dept. of Civil and Earth Resources Eng., Graduate School of Eng., Kyoto University, Kyoto, JAPAN
[2] Graduate student, Dept. of Civil and Earth Resources Eng., Graduate School of Eng., Kyoto University, Kyoto, JAPAN
[3] Undergraduate student, Dept. of Global Engineering, Kyoto University, Kyoto, JAPAN

E-mail: kim.chulwoo.5u@kyoto-u.ac.jp

Abstract. Maintaining civil infrastructure, including bridges, has been a keen technical issue in developed countries and will surely be one in developing countries in the near future. An effective maintenance strategy strongly depends on timely decisions on the health condition of the structure. Bridge health monitoring (BHM) using vibration data is widely recognized to be one of the effective technologies that aid decision-making on bridge maintenance. This research focuses on long-term BHM expecting that changes in physical properties of the bridge subject to aged-deterioration progress slowly. In the practical application of the long-term BHM, one of the difficulties is that the observed vibration data includes environmental influences such as temperature change. In order to achieve high accuracy in evaluating modal parameters of the bridge, other influencing variables have to be taken into consideration. In this study, temperature is considered as the main environmental factor by means of a regression analysis. The Mahalanobis distance (MD), a multivariate statistical distance, is adopted to emphasize potential changes in the identified modal parameters. The validity of the proposed approach is investigated utilizing vibration data measured at a real bridge in service.

[1] Corresponding author: kim.chulwoo.5u@kyoto-u.ac.jp

1. Introduction

Damage in bridges has been reported in many countries due to deterioration, aging, heavy loads, etc. How to establish an economical and efficient structural soundness evaluation system has been a keen issue in the field of civil engineering for years. As a strong tool, long-term bridge health monitoring by observing change in bridge vibration characteristics is attracting wide attention.

Theoretically, damage in the bridge is supposed to change the way the bridge vibrates and thus be quantitatively reflected in the change in modal parameters of the bridge. However in the practice of long-term bridge health monitoring, environmental factors also exert influences on modal parameters. Consequently, how to remove the environmental disturbances and extract the change in modal parameters due to damage in the bridge is a major challenge. In this study, regression method is adopted to eliminate the environmental effect.

There have been attempts for damage detection focusing on changes in modal parameters in the past as well [e.g. 1]. One of the difficulties is that before damage occurs, it is difficult to predict which modal parameter is the most sensitive to change in the dynamic characteristic of the bridge. Moreover, even if there appear changes in some of the modal parameters, the level of change in each single modal parameter is usually too small for judging whether they are due to the occurrence of damages in the bridge or just regular fluctuation. Therefore to overcome these difficulties, it is attempted in this study to integrate the information of multiple modal parameters into a single indicator, the Mahalanobis distance, for damage detection.

This study, at first, is intended to identify the modal parameters of a real bridge from monitoring data utilizing single-variate auto-regressive model. Then, a linear regression method is adopted to eliminate influence of environmental factors from modal parameters. After transforming modal parameters, the Mahalanobis distance is calculated and utilized as the damage-detecting indicator for structural diagnoses of bridges. In order to verify validity of the proposed approach, the monitoring data taken from a real bridge, including the data recorded during a traffic accident are utilized.

2. Damage-detecting indicator

The Mahalanobis distance (hereafter MD) is a generalized distance, which can be considered as a single-dimensional quantity that expresses the degree of divergence of a multivariate sample with respect to the central point of the set of normally distributed multivariate samples it belongs in considering the correlations between the variables. MD is a quite useful way for determining the similarity of a set of values from an unknown sample to a set of values measured from a collection of known samples [2]. In this study, the unknown sample is the modal parameters of the target span of the bridge of interest identified when whether damages have occurred is unknown and the known samples are those identified when the bridge was assumed to be healthy. By estimating MD of a certain observation point, the similarity of the vibrational properties at that time to the intact ones can be quantitatively expressed, which is focused at for damage-detecting in this study. Another advantage of adopting MD as the indicator is that it contains the information of multiple variables, the modal parameters, which vary in a level that is not obvious enough to draw conclusion about the existence of damages, and by accumulation, minimal changes in modal parameters due to change in vibration characteristics during long-term monitoring become noticeable difference in MD, which makes it easier for decision-making. Detailed mathematical derivation is interpreted in **4.1**.

3. Real bridge monitoring

3.1. Target bridge

The monitoring data used in this study was taken from the first span from the west of a seven-span plate-Gerber bridge, which was constructed in March, 1960 and has been in service for over 50 year. The photographs of the target bridge and monitored span are shown in **Figure 1** and **Figure 2** respectively and the main information of the bridge are summarized in **Table 1**. In 2004, as a preventive measure against material fatigue, the beam end was strengthened and bridge slab was reinforced by increasing the thickness. Furthermore, considering the bridge is located in a highly occupied national road where around 60% of passing vehicles are heavy trucks, structural monitoring was thought to be necessary.

3.2. Sensors

Monitoring data used in this study are ambient vibration data and temperature data, which are measured by four accelerometers (UA1, UA2, DA1 and DA2) and two thermometers (T-5 and T6). The locations of the sensors are shown in **Figure 3**. UA1 and DA1 were deployed near mid-span, where information of global vibration including 1st mode are relatively easy to extract, while UA2 and DA2 were deployed near the hinge, where deterioration and fatigue damages occurred.

Figure 1 The target bridge

Figure 2 The monitored span

Table 1 The main information of the bridge

Construction year		1960
Bridge length (m)		187.0
Span length (m)	Hanging girder	16.0
	Anchorage girder	6.2+28.4+6.2
Width (m)		8.0

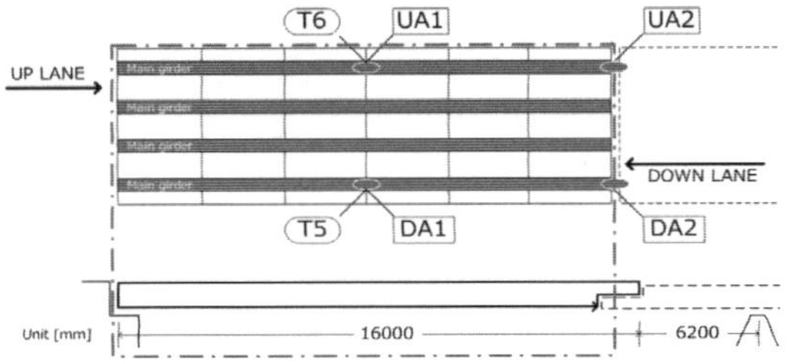

Figure 3 Sensor deploying map

3.3. Observation period

The monitoring started from August 2008 on a full-scale and is still continuing now. Due to hardware problem and operational mistakes, there exist several discontinuities during the six years so far. Though not continuous, available data are from August 10, 2008 to July 13, 2014 and they are all utilized in this study. Besides those, the data of October 18, 2007, on the morning of which a traffic accident took place on the monitored span of the bridge, is used to test the sensitivity of MD.

4. Long-term health monitoring focusing on the Mahalanobis distance of modal parameters

4.1. Derivation of damage detecting indicator
Detailed procedures for calculating MD are described in this subsection.

4.1.1. Identification of frequencies and damping constants. Identifying the system parameters from vibration data by means of linear time-series models such as auto-regressive (hereafter AR) model is a promising approach [3, 4]. Frequencies and damping constants were identified by means of a single-variate AR model from ambient vibration data measured at each sensor [5].

In view of huge amount of data, this study, by means of moving time windows shown in **Figure 4**, identifies modal parameters 200 times for each one hour of vibration data at each sensor. Each time, up to ten frequencies are generated as well as their respective damping constants, and their average values of 200 times are the output of the data of each hour. As it turned out from the results, frequencies near 4Hz and 10Hz are the dominant domains. Therefore, they, as well as their damping constants, were chosen to focus on.

4.1.2. Measures to remove environmental influences. Undoubtedly, the modal parameters identified in this way include environmental influences. In order to propose an indicator that only changes along with the bridge's structural soundness, disturbances of these exterior factors have to be removed.

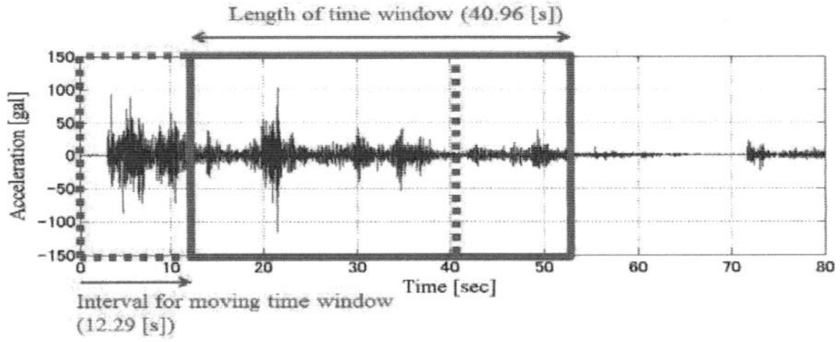

Figure 4 Time windows for identifying modal parameters

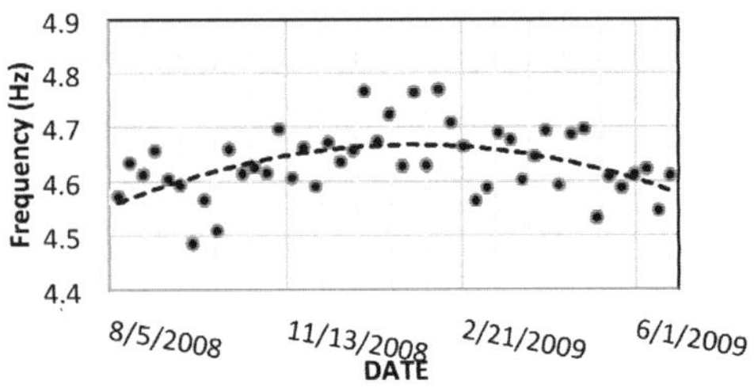

Figure 5 Frequency around 4Hz at DA1

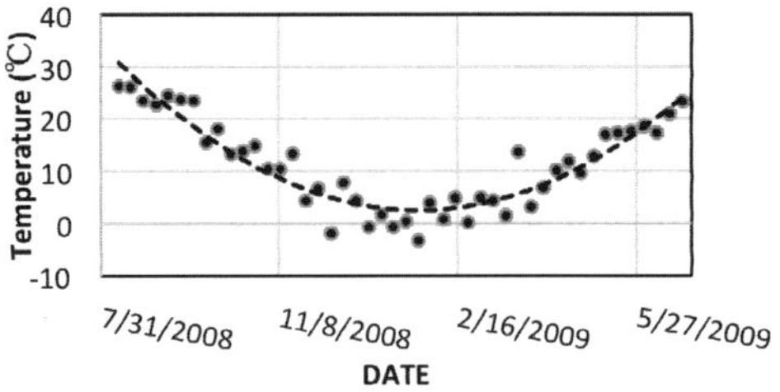

Figure 6 Yearly temperature of the bridge

Clear seasonal fluctuations were noticed in most of frequencies and damping constants. As an example, frequencies around 4Hz identified at DA1 from August 10, 2008 to June 21, 2009 are plotted in **Figure 5**. These seasonal changes were thought to be due to environmental factors, of which temperature was considered to be the most influential one in this case since temperature shows extremely similar but opposite rise and fall during the year, as shown in **Figure 6**. One of the possible reasons of this correlation could be the temperature-dependence of young's modulus of materials, as is also suspected as one of the main reasons in other studies [6]. Based on experiment results, both concrete and steel, which are the main materials of the target bridge, show minimal changes when temperature change [7, 8]. In order to verify whether these changes quantitatively match with those in natural frequencies, rough calculation was done. Theoretically, for the same system, natural frequencies are proportional to the Young's modulus of the system. As can be approximately estimated from **Figure 7** [7] and **Figure 8**, [8] when temperature increases from 0 °C to 30 °C, where temperature of the target bridge varies, Young's modules of steel and concrete decrease with about 0.9 percent and 1.6 percent respectively. Therefore, change in the equivalent Young's modulus should be between 0.9 percent and 1.6 percent. In the same temperature range, identified natural frequencies drop with around 2.5 percent in average. Considering the inaccurateness of the experiment result, it can be concluded that the temperature-dependences of young's modules of materials is one of the possible explanations for the correlation between natural frequencies and temperature. Further investigation is necessary before making more definite conclusion. As for damping constants, which slightly linearly increases as temperature increases, physical mechanism about the correlation remains unknown and future research ought to be done to understand it.

Due to the difficulty of accurately modeling temperature's impact, statistical approach was adopted to remove its influence. Since most modal parameters show linear relation with temperature, linear regression model was utilized in this study. Considering data from August 10, 2008 to June 21, 2009 are the closest to intact condition, regression coefficients were estimated with data in this period. Again, the procedure of this manipulation is illustrated taking frequency around 4Hz identified at DA1 as the example. Firstly, a linear regression model is established as below,

$$f = kT + C \tag{1}$$

where f and T stand respectively for frequency and temperature, and k and C indicate the regression coefficient and constant in respect.

By subtracting the temperature related term, kT, from the frequency data, it is transformed into a temperature-independent variable, as shown in **Figure 9**. The same procedure was done to all other modal parameters before progressing to calculating MD.

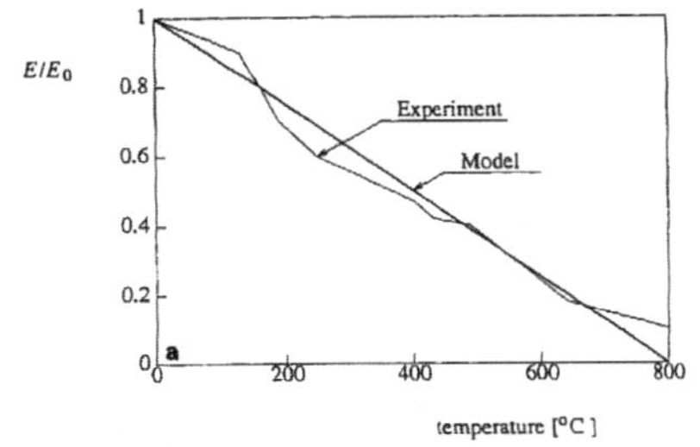

Figure 7 Young's modulus of concrete [7]

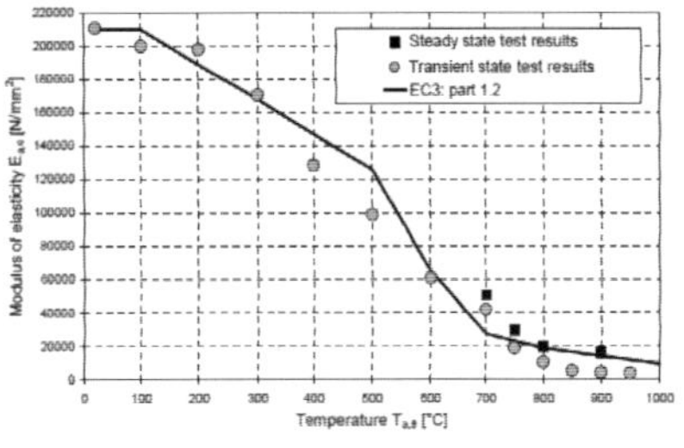

Figure 8 Young's modulus of steel [8]

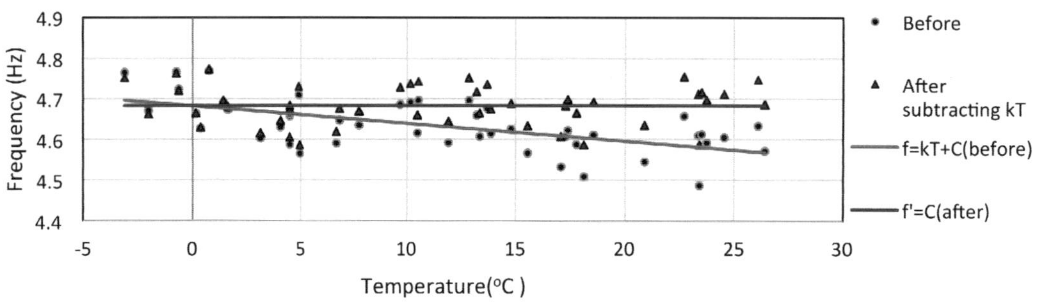

Figure 9 Frequency before and after removing temperature's influence

Another major environmental factor is the variation of passing vehicles. The correlation between natural frequencies of the target bridge and properties, such as volume, speeds and weights, of passing vehicles was proved by the result of data analysis in past study by Heng [9]. Since this study is the first attempt to utilize MD as the damage-detecting indicator, in order to simplify the problem, it is proposed to consider only one environmental factor, change in bridge temperature, during data analysis and minimize the influence of variation of passing vehicles by narrowing down data period. From the monitoring data, it was noticed that at 7 a.m. on Sunday mornings, amount of passing vehicles stabilizes in a relatively low level and shows small variation. On account of this, only data of this hour every week were chosen to be the objects for data analysis.

4.1.3. Estimation of Mahalanobis distance. Since it is generally known that accuracy of identification of frequencies from vibration data is higher than that of damping constants, frequencies were believed to be more suitable for detecting structural damage. Therefore, it was thought that MD calculated with frequencies only would be optimal. For comparison, analyses were also done for MD calculated with only damping constants and with both frequencies and damping constants.

This study attempts to detect damage by observing change in the vibration characteristics, therefore the reference data ought to be the intact modal parameters of the monitored span, which are data from August 10, 2008 to June 21, 2009 in this study for they are the closest to intact condition among available data. We here denote these reference data as a matrix, $A \in R^{k \times l}$, where l and k stand for number of reference data and number of variables respectively.

Firstly, all elements are normalized with,

$$B_{ij} = \frac{A_{ij}-m_i}{s_i}, (i=1,\dots,l; j=1,\dots,k) \quad (2)$$

where m_i and s_i are the average value and standard deviation of the i-th column. Then, the correlation matrix, denoted as $C \in R^{l \times l}$, are formed with each element calculated with the following expression,

$$C_{ij} = \frac{1}{l}\sum_{n=1}^{l} B_{in}B_{jn}, \qquad (i = 1, \dots, l;\ j = 1, \dots, l) \quad (3)$$

Finally, for a normalized sample, $X \in R^k$, its MD can be calculated with the expression shown below[2],

$$MD = \left(\frac{1}{k}\right) X^T C^{-1} X \quad (4)$$

Journal of Physics: Conference Series **628** (2015) 012055 doi:10.1088/1742-6596/628/1/012055

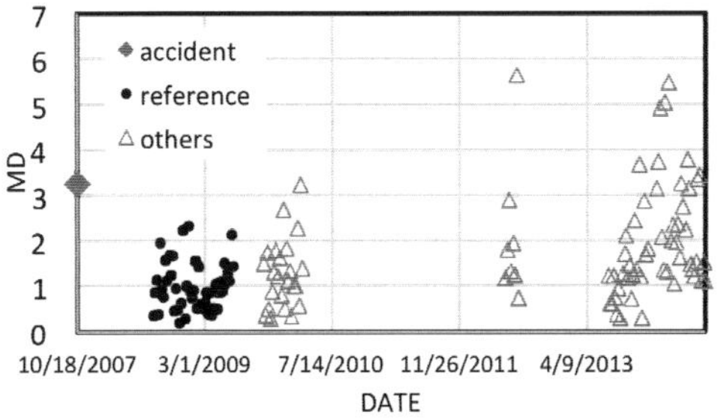

Figure 10 Case 1: *MD (f)*

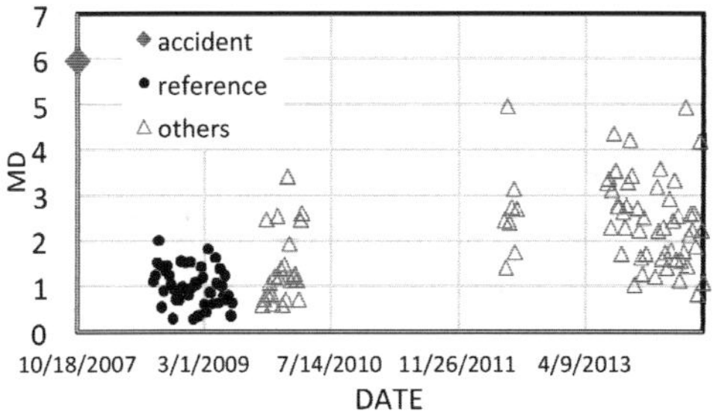

Figure 11 Case 2: *MD (d)*

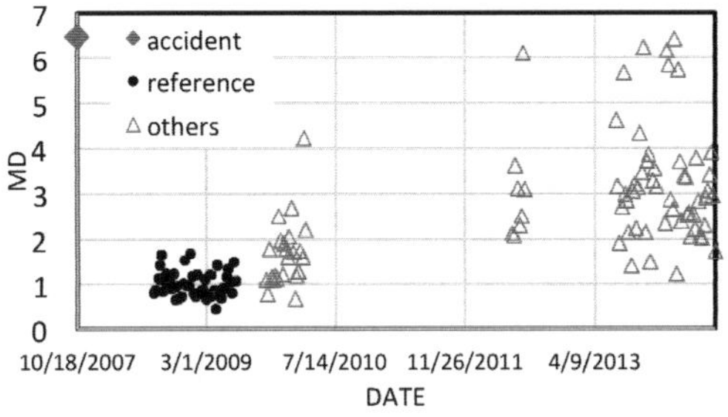

Figure 12 Case 3: *MD (f & d)*

Table 2 Results summary

Sensitivity to gradual change in vibrational characteristics	*MD (f&d)>MD(d)>MD(f)*
Sensitivity to abnormal (accident) data	*MD(d)>MD(f&d)>MD(f)*
Scale of fluctuation	*MD(f)>MD(f&d)>MD(d)*

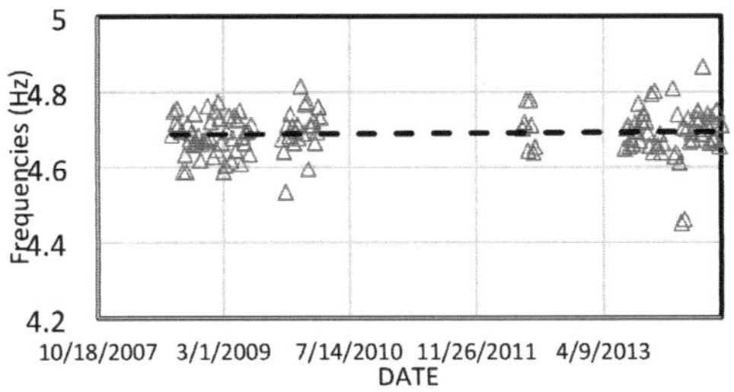

Figure 13 F4 at DA1

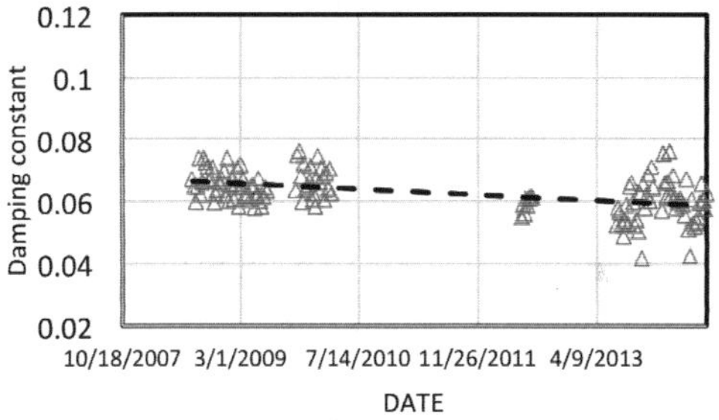

Figure 14 D4 at DA1

4.2. Application to long-term monitoring data

With MD as the indicator, data analysis is conducted with long-term monitoring data, including an abnormal data recorded in a traffic accident. With data from August 10, 2008 to June 21, 2009 as reference data, *MD*s of Case1, Case2 and Case3 are calculated for all data and plotted in **Figure 10**, **Figure 11** and **Figure 12** respectively. Therein Case1 considers MDs from the observed frequencies of all sensors near 4Hz and 10Hz, Case 2 considers MDs from the observed damping ratios of all sensors corresponding to the frequencies near 4Hz and 10Hz, and Case 3 considers MDs from both frequencies and damping constants of all sensors. In all three graphs, black points represent reference data. Red solid squares on the left of the graphs stand for accident data, and how large they are represent the sensitivity of each MD to abnormal data. And red triangles on the right represent unknown data, which are the rest of the monitoring data besides reference data. How much they deviate from reference data shows how sensitive each MD is to gradual structural change.

All three *MD*s showed noticeable increasing tendencies through the six years of monitoring.

Whereas, as shown by two examples of results of modal parameters in **Figure 13** and **Figure 14**, not all of them show obvious deviation from intact data.

5. Discussion

This study investigates the validity of utilizing Mahalanobis distance (*MD*) of modal parameters as the damage-detecting indicator for long-term bridge monitoring. Temperature, as the most influential environmental factor, was taken into consideration by means of regression method.

The proposed approach was applied to the long-term monitoring data of a seven-span plate-Gerber bridge. Conclusion drawn based on the results of this study could be summarized as follows: 1) all three *MD*s showed noticeable increasing tendencies through the six years of monitoring, which could imply the probability of occurrence of damage in the monitored span; 2) MD is useful for integrating information of multiple variables and amplifying minimal changes in modal parameters into more noticeable changes; 3) different from the prediction stated in 4.1.3, *MD* calculated with both frequencies and damping constants seemed to have the best quality to be used as the indicator, considering sensitivity to gradual change in vibrational characteristics, sensitivity to abnormal data, scale of fluctuation and accuracy of identification; and 4) with current results, *MD* showed potential usefulness for damage detection and thus is worthy of further investigation.

For improvement of accuracy, data of passing vehicles ought to be utilized as well in future research. Also, shortage in continuous monitoring data is one of the reasons why more definite conclusion couldn't be reached. Therefore, now that the bridge is still under long-term monitoring, this study will make further analyses with the long-term monitoring data to come. Last but not least, for the realization of practical application, concrete managing standards is necessary to be established for decision-making.

References

[1] M. Dilena and A. Morassi, "Dynamic testing of damaged bridge", *Mech. Systems Signal Processing*, *25*, pp. 1485-1507, 2011.
[2] G. Taguchi and R. Jugulum, The Mahalanobis-Taguchi Strategy: A Pattern Technology System, John Wiley & Sons, Inc., 2002.
[3] M. Hoshiya, and E. Saito, "Structural identification by extended Kalman filter," *J. Engg. Mech. ASCE, 110(12)*, pp. 1757-1770, 1984.
[4] L. Liung, System identification-Theory for the user 2nd ed., *PTR Prentice Hall, Upper Saddle River, MJ.*, 1999.
[5] C. W. Kim, R. Isemoto, K. Sugiura and M. Kawatani, "Structural fault detection og bridges based on linear system parameter and MTS method," *J. of JSCE*, *1*, pp. 32-43, 2013.
[6] J. Kullaa, "Benchmark Data for Structural Health Monitoring," Le Cam, Vincent and Mevel, Laurent and Schoefs, Franck. *EWSHM - 7th European Workshop on Structural Health Monitoring,* 2014.
[7] Rene de Borst and P. J. M. Peeters, "Analysis of concrete structures under thermal loading," *Computer Methods in Applied Mechanics and Engineering, 77(3)*, pp. 293–310, 1989.
[8] J. Outinen, "Mechanical properties of structural steel at elevated temperatures and after cooling down," *Proc. of the 10th Int. Fire and Materials Conference*, San Francisco, USA, 29-31 January 2007.
[9] S. Heng, Y. Oshima, and H. Kawano, "One Year Monitoring of Bridge Frequency and Traffic Load on a Road Bridge," *Proc. of 24th KKCNN Symposium on Civil Engineering*, 2011.

Experimental Comparison of Dynamic Responses of a Tension Moored Floating Wind Turbine Platform with and without Spring Dampers

C Wright[1], K O'Sullivan[2], J Murphy[1] and V Pakrashi[3]

[1]MaREI/Beaufort Research, Environmental Research Institute, University CollegeCork, Pouladuff Road, Cork, Ireland

[2]Black & Veatch Ltd, Grosvenor House, 69 London Road, Redhill, Surrey, RH1 1LQ, United Kingdom

[3]Dynamical Systems and Risk Laboratory, Civil and Environmental Engineering, School of Engineering, University College Cork, Ireland

E-mail: c.s.wright@umail.ucc.ie

Abstract. The offshore wind industry is rapidly maturing and is now expanding to more extreme environments in deeper water and farther from shore. To date fixed foundation types (i.e. monopoles, jackets) have been primarily used but become uneconomical in water depths greater than 50m. Floating foundations have more complex dynamics but at the moment no design has reached commercialization, although a number of devices are being tested at prototype stage. The development of concepts is carried out through physical model testing of scaled devices such that to better understand the dynamics of the system and validate numerical models. This paper investigates the testing of a scale model of a tension moored wind turbine at two different scales and in the presence and absence of a spring damper controlling its dynamic response. The models were tested under combined wave and wind thrust loading conditions. The analysis compares the motions of the platform at different scales and structural conditions through RAO, testing a mooring spring damper for load reductions.

1. Introduction

The first offshore wind farm was commissioned in 1991 and since then the total European offshore installed capacity has risen to greater than 7.3GW [1]. Offshore wind offers a number of advantages over onshore wind, including, higher wind speeds, larger turbines possible, no site access restrictions, larger sites available, reduced turbulence and lower wind shear [2]. As shallow water, nearshore sites are depleted or countries with no shallow continental shelf seek to develop offshore wind, the industry must look to deeper water sites. In water depths greater than 60m floating platforms begin to become economical [3]. In Ireland there is the potential for 25-27GW of offshore floating wind [4]. Tension moored platforms are one floater type generating a great deal of interest. The platform is restrained due to excess buoyancy providing tension in the mooring lines. Part of the development process for such type platforms involves scaled model tests in wave basins. Testing of small scale devices typically 25-100th scale is carried out in an ocean wave basin. This testing is vital to understand the interactions between hydrodynamic and aerodynamic loading. Varied structural settings can be recreated relating to both the operational and survivability conditions. In this paper, a novel floating platform is tested in a deep-water wave basin where adjustments are made to the mooring stiffness in order to reduce peak mooring loads.

2. Scaled Tests

Model testing of tension leg platforms (TLPs) have been carried out at scales of 1:100 [5] and 1:50 [6, 7]. These tests involve Froude scaling of both the hydrodynamics of the floater and aerodynamics of the wind turbine. These tests produce Response Amplitude Operators (RAOs) which help benchmark the performances and often act as a guiding factor for more detailed modelling or testing at a larger scale or for an actual deployment in offshore conditions.

There are four main methods for modelling wind turbine loading on a floating platform; **1) Mechanical Pulley System:** Constant horizontal load applied through a mass hanging via a pulley from the nacelle [8]. **2) Ducted Fan:** A ducted fan with high revs/min that is attached to the top of the tower. A constant or variable wind thrust can me modelled, and with the inclusion of a feed-back loop, which includes the motions of the platform, the relative (between incoming wind and platform motion) wind thrust can be applied [9]. **3) Applied Wind on Disk:** Wind is generated using an array of fans and applied to a thin disk. A drag coefficient of 1.2 is used to model the thrust force [10]. **4) Applied Wind on Model Turbine:** Wind is generated using an array of fans, and applied to a model turbine. As Froude scaling and Reynolds scaling (required for wind blade lift modelling) are not compatible (either thrust force of aerodynamic damping being incorrect) turbine blade geometry must be modified to include longer chord lengths in order to correctly model all effects [11]. Advantages of this method over the previously mentioned include correct modelling of aerodynamic torque and the gyroscopic effect of the rotor [12]. The gyroscopic effect can be compensated for in the first three modelling methods by attaching a motor and simple mass rotor that turns at the correct revolutions per minute.

3. Monitoring

As offshore wind turbines move to more severe environments the need to monitor loadings and platform performance become more important. Environmental observations required include metoocean data such as wind and wave climate. Platform and turbine conditional monitoring includes Platform motions in degrees of freedom, mooring loads, turbine and blade parameters such as pitch and yaw. These parameters are relevant to the testing of platforms at model scale and in the case of this study it is the motions and mooring loads that are most relevant.

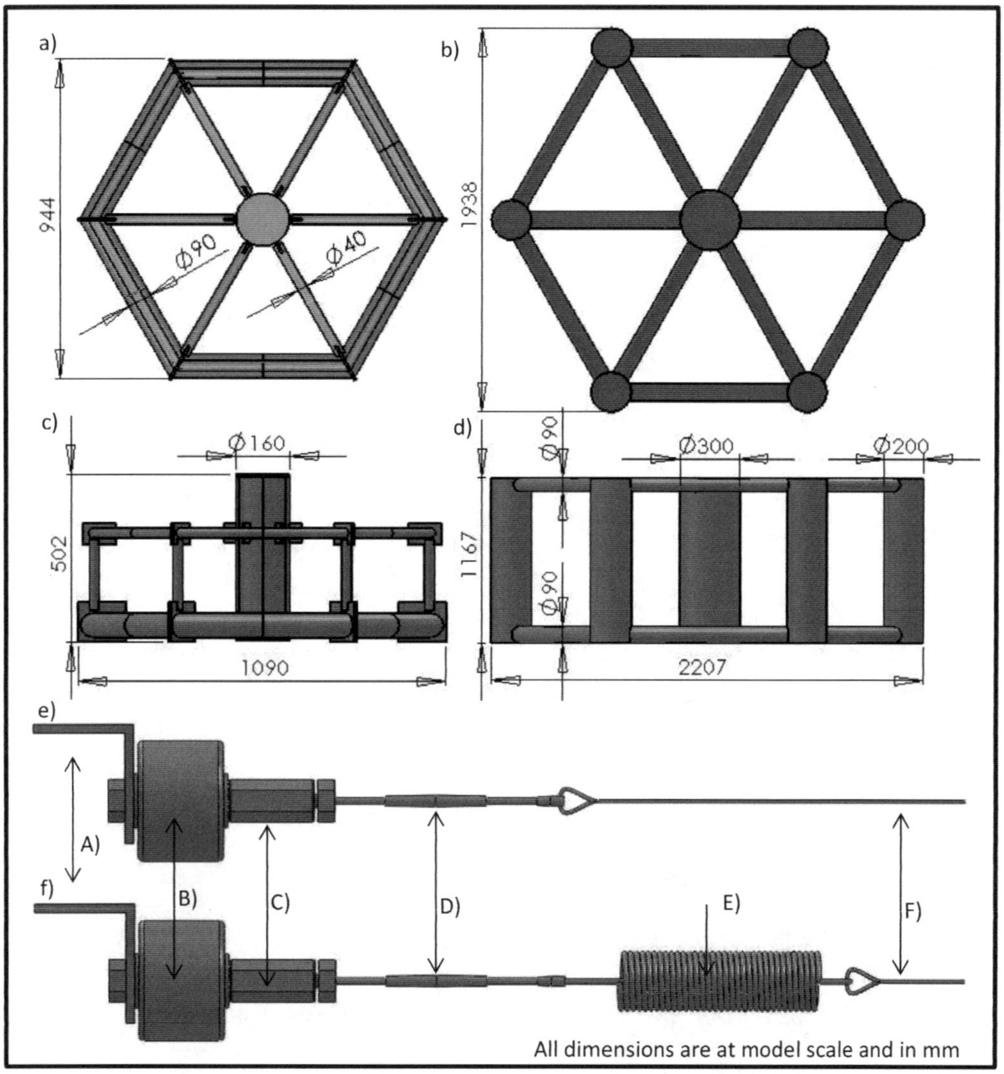

Fig 1.a) Platform 1 Plan λ =50. b) Platform 2 Plan λ = 30 c) Platform 1 Elevation λ =50. d) Platform 2 Elevation λ = 30. e) Mooring Configuration without Spring Damper. f) Mooring Configuration with Spring Damper. A) Fixing Bracket, B) Load Cell, C) Extension Nut, D) Tie Rod, E) Damper Spring, F) Mooring Line.

4. Experimental Setup

Testing was undertaken on two TLPs of similar design at different scales. Both floating platform has the characteristics of a semi-submersible but are moored with vertically tensioned lines and support NREL's 5MW reference turbine [13]. The first platform has a scale of 1:50 (figure 1a& c) and is described in detail in [14]. The second TLP tested (figure 1b& d) is a truss-type structure with a floating hexagonal platform which in the model tests was connected by six mooring tethers to a rigid base located at a depth of 5m (basin depth was 10m). The model was scaled according to Froudian scaling laws and has scale of 1:30 and the floating hexagonal platform consists of six outer and one central buoyancy column(s). The outer buoyancy columns consists of six 0.2m diameter polyvinyl chloride (PVC) pipes, joined to the central column by twelve (six top and six bottom) 0.09m diameter PVC pipes. The outer buoyancy columns are also connected to each other by twelve (six top six bottom 0.09m diameter PVC pipes. The central column was fabricated from 0.3m diameter PVC pipe and provides sufficient buoyancy to counteract the weight of the tower and nacelle. The excess buoyancy force was passed to the six mooring lines made of 3mm diameter stainless steel wire to ensure that they remain in tension at all times. The weight of the TLP was 121.1kg. The wind turbine tower was 2.58m high and was fabricated with a 0.07m diameter stainless steel pipe (7.4 kg) with the thrust load applied using a high revolutions per minute ducted fan (0.75kg), according to the wind speed thrust load curve given in [13]. As the thruster was not heavy enough to model the turbine, additional lead weights are also placed at the top of the tower (8.5 kg). Each wind load case involved a constant wind speed which was verified by a bending load cell connecting the fan to the tower, as can be seen in fig 2.c.

Fig 2.a) Platform 1 λ =50. b) Platform 2λ = 30 *Photo Credit IFREMER*. c) Thrust Generation. d) Mooring Spring Damper

A second mooring case was modelled, where a high stiffness extension spring (6N/mm) was added to each of the 6 mooring lines (fig 1.f). The spring has a pre-tension of 130N meaning the change in stiffness will only be noticed in extreme wave conditions. It was designed such that the motion of wind turbine will be constrained during operational wave conditions and that during extreme wave conditions this constraint will be relaxed allowing the platform to move and thus reducing the peak mooring loads. The TLP was tested in the deep water basin, in IFREMER, Brest, France. The tank is 50m long, 12.5m wide and 10m deep. Panchromatic waves of a Bretschneider (BS) spectrum were generated using a single hydraulic v-shaped wave plunger, and reflections absorbed at the opposite end by a porous parabolic beach. Waves were measured using two servo-controlled wave gauges, motions were captured using three Qualisys motion tracking captures picking up four UV markers, mooring loads were measured with six 500N load cell.

5. Results

All results are shown and compared at prototype scale for standardisation purposes. Figure 3 shows the comparison between surge RAO's for wave loading only for the two different scaled floaters (λ =30 and λ =50) It must be noted that a number of differences exist between the models. This can be attributed to the change in shape of floater, main buoyancy has moved from horizontal ring in λ =50, to vertical ring in λ =30, which greatly increases the water plane area. Also there is a change in initial tension between platforms, 1.55 MN (λ =50), and 2.44 MN (λ =30), a change in floater diameter, 47.2m (λ =50), and 58 m (λ =30), and also a change in water depth, 50m (λ =50), and 150 m (λ =30). 150m being the design water depth and 50m from shallow tank constraints. These changes explains the difference in peak surge frequency from 0.035 Hz (λ =50) to 0.021 Hz (λ =30). As the scale increases it can be seen that the consistency of the results improve.

Figure 4 compares the second mooring case, including the high stiffness extension spring to each mooring line. It shows how the addition of the spring dampers to the mooring lines affects the surge RAO and mooring tension RAO for three wind cases (Constant wind speeds of $0ms^{-1}$ (No thrust), $25ms^{-1}$ (Medium thrust), and $11.5ms^{-1}$ (Max thrust). It can be seen that the springs had the intended effect of reducing the surge stiffness, thus increasing the surge displacement, and decreasing the mooring tension RAO. As surge and pitch are coupled in TLPs a large increase in the pitch RAO was also noted as expected.

Figure 5 shows the surge set-back time series and surge energy of the platform under wind loading of $25ms^{-1}$ and $11.5ms^{-1}$. The natural surge frequency of the platform is seen to increase as wind thrust loading increases as follows: 0.0195 Hz in no wind, 0.0197 Hz in $25ms^{-1}$ and 0.0209Hz in $11.5ms^{-1}$. Differences in the time-series and spectral energy for the same wind thrust loading were deemed to come from transverse waves which had not completely dissipated before testing had begun.

Journal of Physics: Conference Series **628** (2015) 012056 doi:10.1088/1742-6596/628/1/012056

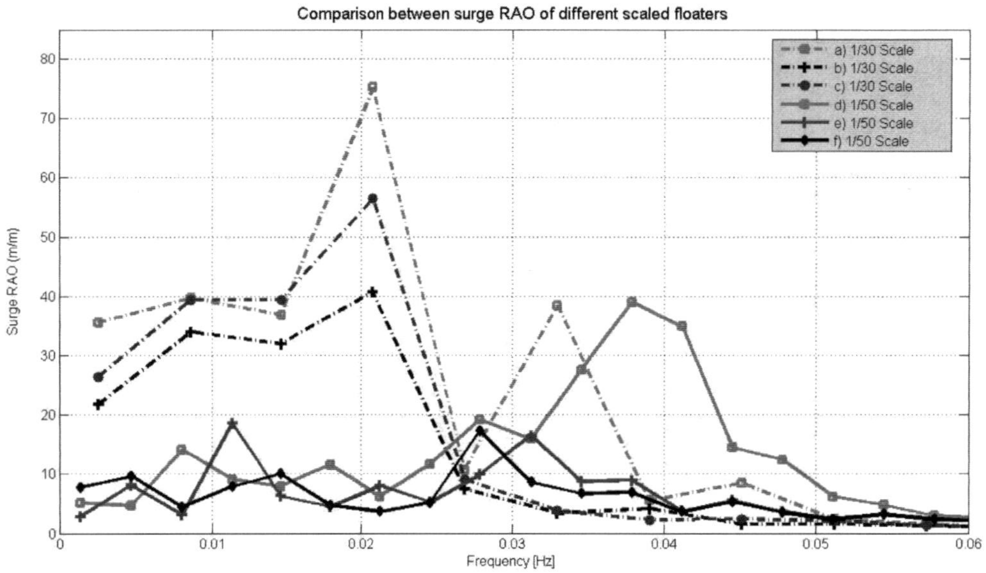

Fig 3. Comparison Between Model Scale λ =30 and λ =50. a) BS Hs 6.6m Tp 11.5 b) BS Hs 8.7m Tp 13s c) BS Hs 6m Tp 12s d) BS Hs 3m Tp 7.6s, e) BS Hs 3m Tp 17.7s, f) BS Hs 3m Tp 19.2s

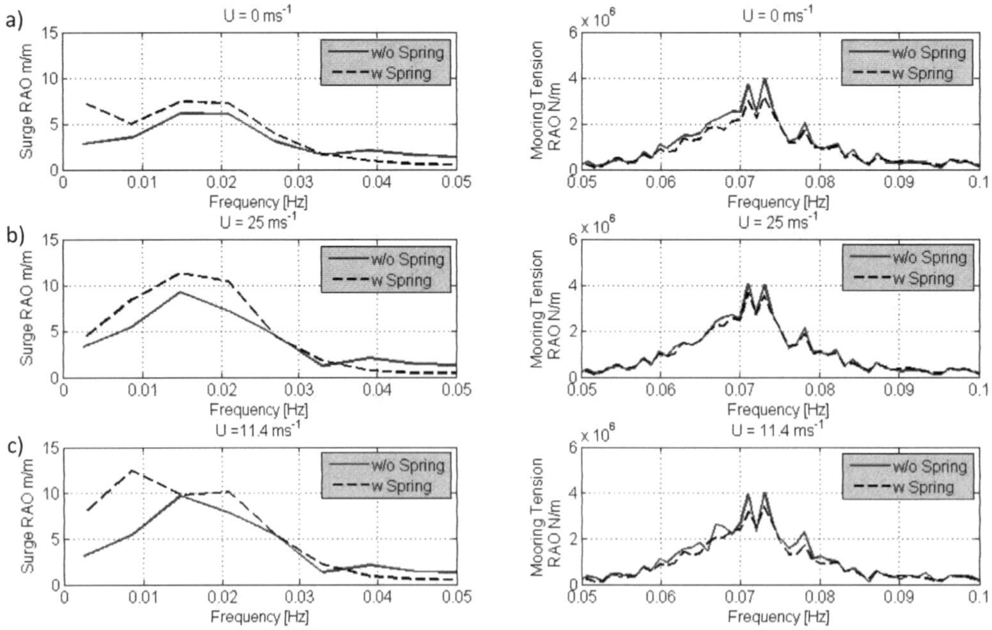

Fig 4. Comparison of Surge RAO and Mooring Tension RAO without Spring and with Spring Damper, for Wind Speeds of 0 (a), 25 (b), 11.4 (c) ms^{-1}. Hs = 9m, Tp = 18s, Model Scale λ =30

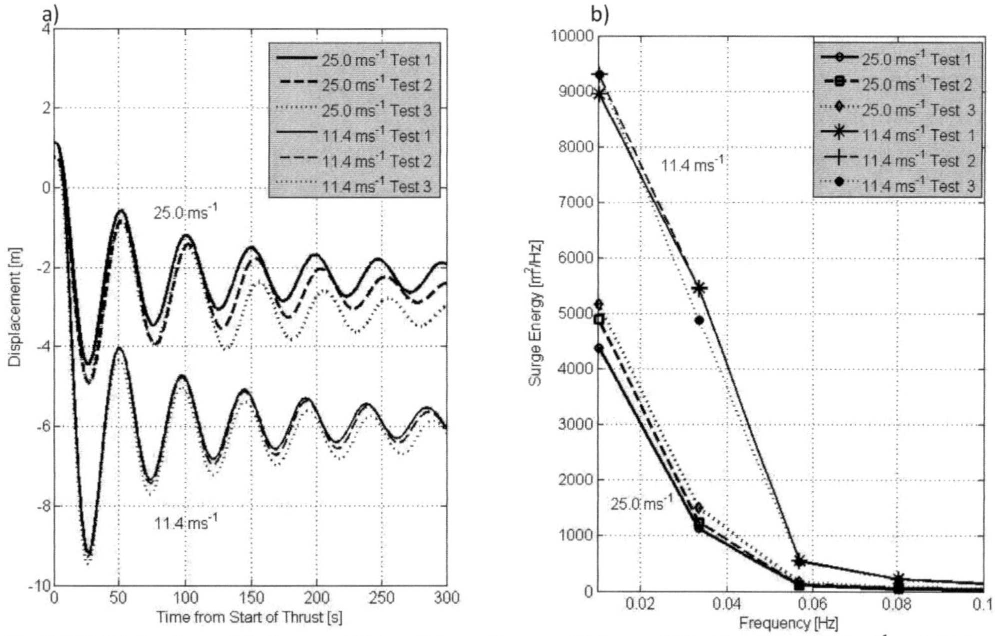

Fig 5. Still Water Surge Displacement (a) and Surge Energy (b) from a Wind Loading of 25 ms^{-1} and 11.4 ms^{-1}, Model Scale λ =30

6. Conclusion

This paper compared the RAO's of two floating platforms as the design evolves through the development proves. It shows the importance in terms of reliability in testing at larger scales. This testing successfully showed and benchmarked the capabilities of the platform in operational and survivability states. There was a large increase in the maximum dynamic response of the system when the spring damper was implemented (25-33%), this had the intended effect of reducing the maximum mooring load (11-25%). A slight increase in higher frequency surge motion of the platform was detected and was deemed to be an effect of the increased pitch motion. The spring damper was demonstrated at a proof of concept level for reducing mooring line tension through softening the restoring stiffness. More testing of this is envisaged in the future. The maximum mooring line force holds a great deal of importance in determining the capital cost of the floater, mooring line and mooring foundation (piles or gravity base) and thus any reduction would be extremely important to the future of deploying wind turbines on these kinds of platforms. Determining natural frequencies and there variance due to wind loading are vital for validating numerical models, and has been demonstrated here.

7. Acknowledgements

The testing at IFREMER was funded by MARINET, a European Community research infrastructure initiative which funds periods of access to marine renewables test facilities. The authors would also like to acknowledge Marine Research Energy Ireland (MaREI), a Science Foundation Ireland (SFI) supported project, grant no. 12/RC/2302 who funded this research.

8. References

[1] Corbetta G, Pineda I and Wilkes J 2014 The European offshore wind industry - key trends and statistics 1st half 2014, EWEA.

[2] Henderson AR, Morgan C, Smith B, Sørensen HC, Barthelmie RJ and Boesmans B 2003 Offshore wind energy in Europe -a review of the state-of-the-art *Wind Energy* **6** pp 35–52

[3] Musial W, Butterfield S and Ram B 2006 Energy from offshore wind *Offshore Technology Conference (May)* pp 1888-1898

[4] Department of Communications, Energy & Natural Resources, 2014 Offshore Renewable Energy Development Plan

[5] Nihei Y and Fujioka H 2010 Motion characteristics of TLP type offshore wind turbine in waves and wind *29th International Conference on Ocean, Offshore and Arctic Engineering* **3**

[6] Stewart G, Lackner M, Robertson A, Jonkman J and Goupee A 2012 Calibration and validation of a FAST floating wind turbine model of the DeepCwind scaled tension-leg platform *International Offshore and Polar Engineering Conference (Greece)* **22**

[7] Koo B, Goupee AJ, Lambrakos K and Kimball RW 2012 Model tests for a floating wind turbine on three different floaters. *31st International Conference on Offshore Mechanics and Arctic Engineering (Brazil)* **7**

[8] Matha D, Sandner F, Molins C, Campos A and Cheng PW 2015 Efficient preliminary floating offshore wind turbine design and testing methodologies and application to a concrete spar design. *Phil. Trans. R. Soc. A* **373** **20140350.** (doi:10.1098/rsta.2014.0350).

[9] Azcona J, Bouchotouch F, González M, Garciandía J, Munduate X, Keberlau F and Nygaard TA 2014 Aerodynamic thrust modelling in wave tank tests of offshore floating wind turbines using a ducted fan *J. Phys.: Conf. Ser* **524**

[10] Wan L, Gao Z and Moan T 2014 Model Test of the STC Concept in Survival Mode *OMAE2014*

[11] Martin H R 2011 Development of a scale model wind turbine for testing of offshore floating wind turbine systems *M.S. thesis (University of Maine)*

[12] Müller K Sandner F, Bredmose H and Azcona J 2014 Improved tank test procedures for scaled floating offshore wind turbines *International Wind Engineering Conference*

[13] Jonkman J, Butterfield S, Musial W and Scott G 2009 Definition of a 5-MW reference wind turbine for offshore system development *National Renewable Energy Laboratory*

[14] Jaksic V, Wright CS, Murphy J, Afeef C, Ali SF, Mandic DP and Pakrashi V. 2015 Dynamic response mitigation of floating wind turbine platforms using tuned liquid column dampers. *Phil. Trans. R. Soc. A* **373**(2035) **20140079**-1-9. (doi:10.1098/rsta.2014.0079).

Journal of Physics: Conference Series **628** (2015) 012057 doi:10.1088/1742-6596/628/1/012057

The Hurst Exponent as an Indicator of the Behaviour of a Model Monopile in an Ocean Wave Testing Basin

Vikram Pakrashi[1,2], Richard O'Shea[1,2], Vesna Jaksic[1,2] and Jimmy Murphy[2]

[1]Dynamical Systems and Risk Laboratory, Civil and Environmental Engineering, School of Engineering, University College Cork, Ireland

[2]Beaufort, Environmental Research Institute, University College Cork, Ireland

v.pakrashi@ucc.ie

Abstract. With the importance of renewable energy well-established worldwide, and targets of such energy quantified in many cases, there exists a considerable interest in the assessment of wind and wave devices. While the individual components of these devices are often relatively well understood and the aspects of energy generation well researched, there seems to be a gap in the understanding of these devices as a whole and especially in the field of their dynamic responses under operational conditions. The mathematical modelling and estimation of their dynamic responses are more evolved but research directed towards testing of these devices still requires significant attention. Model-free indicators of the dynamic responses of these devices are important since it reflects the as-deployed behaviour of the devices when the exposure conditions are scaled reasonably correctly, along with the structural dimensions. This paper demonstrates how the Hurst exponent of the dynamic responses of a monopile exposed to different exposure conditions in an ocean wave basin can be used as a model-free indicator of various responses. The scaled model is exposed to Froude scaled waves and tested under different exposure conditions. The analysis and interpretation is carried out in a model-free and output-only environment, with only some preliminary ideas regarding the input of the system. The analysis indicates how the Hurst exponent can be an interesting descriptor to compare and contrast various scenarios of dynamic response conditions.

1. Introduction

Popularity of offshore wind and wave energy devices is growing steadily with time [1, 2] and such devices are increasingly being considered in deeper waters [3]. While the offshore devices are getting popular and their co-location in high wind and wave [4] conditions ideally leads to a more efficient generation of energy from these natural resources, it also creates problems related to monitoring for maintenance and management since there is limited access to these structures. While relatively calm periods of 'weather-windows' may be employed for operations and maintenance [5] of these devices, continuous online monitoring is preferable for reasons of accessibility, safety and quality of data.

Monitoring of such devices pose a challenge since the devices are most appropriately tested only under deployed condition and not enough deployment has taken place globally to understand the full-scale dynamic behavior of these structures. Consequently, the developed models of the devices have some uncertainty or error associated to them. On the other hand, forces from wind and wave are difficult to measure for these devices to accessibility, cost and logistics issues. While efficient monitoring of these devices are important and instrumented structures have become more commonplace in different fields of engineering [6-8], the uncertainty in models and the non-availability or poor quality of input force data for these structures lead towards development of performance indicators of monitoring devices that are dependent of their dynamic responses under operational conditions. This output-only approach has significant potential to be used and a number of applications have recently illustrated this [9, 10]. In this regard, applications in the field of offshore renewables have been scarce. This paper presents the Hurst exponent of the dynamic responses of such structures as a possible output-only indicator for their monitoring. In this connection, several tests have been carried out on a model monopile structure. The paper is expected to illustrate how Hurst exponents can be important for monitoring these offshore wind and wave renewable energy devices and also highlight the importance of developing and using output-only markers of monitoring in this field.

2. Experimental Setup

2.1. Description of the scaled monopile model

The Froude scaled [11] model monopile device was tested as a model of a point absorber mounted on a mono pile support which was free to move in heave. While some of the details of the model may be accessed from [11], this section briefly reports the setup for the ease of understanding and for completeness. The monopile consisted of an inner steel frame of square tubing of 30mm outside dimension and 3mm thickness. The inner steel frame was bolted to a base comprised of four metal legs. This was then attached to a large circular gravity base which sat on the bottom of the wave basin

Figure 1. Photographs of ocean wave testing basin and model monopile.

during testing. A plastic pipe, 110mm outside diameter, 3mm thickness and 2688mm in length was then attached to the steel frame. The plastic pipe was attached to the inner steel frame at four points along the pipe where load cells were mounted and was also attached to the base of the frame. Two air columns or pipes were contained within the plastic tubing. These air columns or pipes were constructed from plastic tubing 25mm outside diameter and 1.5mm thickness. At one end of the pipes were pistons which were subsequently attached to the point absorber during testing to compress the air in the pipes. At the other end of each pipe was an orifice which could be varied in diameter in order to adjust the damping of the motion of the point absorber when attached to the pistons. These orifices were initially situated above the surface of the water but were moved to a subsurface position during testing. The point absorber was a buoyant cylinder, 380mm outside diameter with a 120mm bore, which was able to move in heave (vertically) against the mono pile. The point absorber was connected to the pistons during testing. The mass of the point absorber was varied by the addition of strips of lead to the outer rim of the point absorber. The entire model was placed in a wave basin with a constant water depth of 1m for testing. Figure 1 presents the photograph of the wave basin where the tests were carried out along with the photo of the model monopile.

2.2. Instrumentation during testing of the scaled monopile model

Readings (in N) were obtained from four load cells located at different heights of the scaled monopile model. The measured force was the force that was exerted between the inner steel frame of the monopile and the outer plastic tubing (sewage pipe). The load cells were attached to the inner metal frame using two bolts and were attached to the outer casing using one bolt. The nomenclature of load cells and the heights of different bolts are presented in Table 1. Table 2 outlines the exposure conditions of the difference load sensors during testing.

Table 1. Details of load-cell locations

Load Cell Name	Height of bottom-most bolt above the tank bottom (mm)	Height of top-most bolt above the tank bottom
Yellow	333	433
Green	651	751
Red	971	1071
Brown	1171	1271

Table 2. Exposure conditions of load cells during testing.

Load cell	Conditions
Yellow	Constantly submerged
Green	Constantly submerged
Red	Submerged and above water at all wave amplitudes.
Brown	Constantly above water

The red load cell is alternately submerged and exposed to air for all the wave heights. Figure 2a provides the indicative locations of load cells and Figure 2b annotates the various parts of the model.

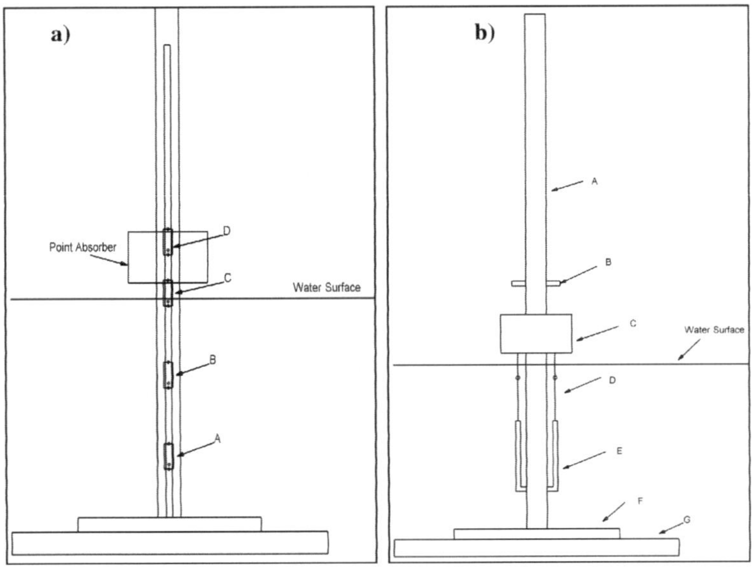

Figure 2. a) P Load Cell Location. A: Yellow, B: Green, C: Red, D: Brown. b) A: Mono pile, B: Orifice, C: Point Absorber, D: Connecting Rods, E: Pistons, F: Frame Base, G: Gravity Base

Each load cell was offset by a given amount so as to be able to distinguish them from one another during testing. Before each new configuration of each setup, a still water reading of all sensors was obtained. The average still water reading of each load cell during testing is presented in Table 3. In order to determine the actual readings from the subsequent tests, the average value of the still water readings were subtracted from the experimentally obtained readings in terms of their magnitude.

Table 3. Still-water readings corresponding to various load cells.

Load Cell	Average Still Water Reading (including offset) (N)
Yellow	0
Green	-10
Red	-20
Brown	-30

Two water level probes were situated on either side of the point absorber and monopile assembly measuring the height of the water surface (mm) during testing. Two pressure sensors, one at the top of each piston, were used in the testing of the monopile and point absorber measuring the pressure in the pistons (mmH_2O) during the tests. Adjustments related to still water readings were carried out as presented in this section.

2.3. Details of experiments

Five different configurations of experiments were carried out in the wave basin. These differing set ups were designed to determine the effect of: wave amplitude, wave period, point absorber mass, the resistance to motion of the pistons, and the effect of damping on the model.

*2.3.1. Set-up 1.*This consisted of the mono pile alone without the point absorber attached. The mono pile was placed in the wave basin and was exposed to Froude scaled waves of 60, 90 and 120 mm height. The period of the waves were varied for all different wave amplitudes. Additionally, the monopile was subject to wave samples using the Bretschneider wave spectrum in order to simulate a more realistic set of sea condition.

2.3.2. Set-up 2. This test configuration consisted of the mono pile and the point absorber which was free to move in heave against the mono pile. The point absorber was not attached to the pistons in order to eliminate the effect of the piston friction on the movement of the point absorber. Tests conducted here involved varying the mass of the point absorber and the period of the waves while keeping the amplitude of the waves constant.

2.3.3. Set-up 3. Tests conducted involved attaching the pistons to the point absorber and varying the period of the waves for constant wave amplitude of 60mm. There was no damping of the point absorber motion. The damping ratio is the ratio of the orifice to the diameter of the piston. A damping ratio of 1 implies that the piston is essentially open at one end. The effect of damping on the motion on the point absorber is absent in this set-up and only the effect of the piston friction is investigated in this set-up.

2.3.4. Set-up 4. Testing consisted of fixing the value of the mass of the point absorber to 6.1kg, attaching the pistons to the point absorber and varying the damping ratio to values of 100, 44, and 400. Each configuration was exposed to waves of amplitude 60mm and varying period. The purpose of this set-up was to determine the effect of damping on the mono pile and the point absorber. During set-up 4 the orifice was exposed to air, i.e. it was above the surface of the water.

2.3.5. Set-up 5. The model was changed more significantly in set up 5. The orifice was moved to an underwater position in order to increase the damping on the motion of the point absorber and to better model the effect of a load on the point absorber. The first third of tests carried out involved varying the mass of the point absorber between 6.1kg and 10.1kg, while it was connected to the pistons, with a constant damping ratio of 100 and was exposed to waves of varying periods and constant amplitude of 60mm.Following this, the damping was removed (DR =1). In this configuration, the mass of the point absorber was varied between 6.1kg and 10.1kg, DR was kept constant at 1 and the model was exposed to waves of constant amplitude (60mm) and varying period. The final portion of testing carried out consisted of varying the mass of the point absorber between 6.1kg and 10.1kg while maintaining a set damping ratio of 400 and exposing the device to waves of constant amplitude (60mm) and varying period.

3. Output-Only Analysis and Hurst Exponents

This section outlines the idea behind Hurst exponent for completeness and discusses some advantages related to using this marker for monitoring. Details of this approach can be found in various literature [12, 13] and applications for infrastructure elements have recently been considered [14].

The *q*-order moments of the distribution of the increments of a time series *y(t)* obtained from the monitoring of an offshore renewable energy device with a value at t^{th} time instant with $t=\nu$, $2\nu,...k\nu,...T$ is considered, where T is the time period of observation, ν is the time resolution of observation and k is an arbitrary positive integer value. The stochastic evolution of the series is characterised and related to the generalised Hurst exponent *H(q)* through

$$K_q(\tau) = \frac{\langle |y_j(t+\tau) - y_j(t)|^q \rangle}{\langle |y_j(t)|^q \rangle} \sim (\tau/\nu)^{qH(q)} \tag{1}$$

The slope of the graph of *qH(q)* against *q* was used to estimate the Hurst exponent. A Hurst exponent based approach is extremely useful when a wide range of underlying activities exist within a system over a measured length of time, but the measurement from the system is not suitable for Fourier analysis or time-frequency analysis based methods [15, 16]. The method can be quite useful for monitoring offshore renewable energy devices since an approximate model or only the output responses can be used to obtain Hurst exponent as a marker for such monitoring.

4. Results

Around 150 tests were carried out for the five set-up conditions described in a previous section in this paper and the Hurst exponents were estimated. The different values of estimated Hurst exponents at different load cells are presented in Figure 3. The horizontal axis represents the index of the test carried out, while the vertical axis represents the estimated Hurst exponent. The scatter of the exponents is quite similar, with the Yellow load cell, submerged deepest in water, showing a larger scatter. Table 4 summarizes the values of the estimated Hurst exponents for different load cells, excluding the stillwater readings. The submerged load cells seem to distinguish themselves from those which are not. The estimated Hurst exponents seem to increase with the height from the basin, with the Red cell showing a small mismatch in this increase when the topmost Brown cell is considered due to the fact that it was partially submerged and partially in air during the tests. The standard deviations of the tests indicate individual tests may not be representative for a good estimate for monitoring and consequently the use of a larger number of tests is recommended to obtain a consistent marker for monitoring that is representative of the location of the sensors, while retaining the fundamental characteristic of what the system is undergoing.

Figure 4 presents the estimated Hurst exponent average values for the five different set-ups considered in this paper. It is observed that the submerged load cell at the bottom (Yellow) is the most representative in tracking the different set-ups considered, giving rise to different types of dynamic responses. While the other submerged cell closer to the free water level is almost constant or only slightly sensitive to such changes, the ones in air are not representative of these changes. Consequently, the location of instrumentation is quite important for tracking changes in structures or monitoring them. On the other hand, the average and consistent values of estimated Hurst exponents at different locations of sensors and for different set-ups may be representative of the device itself.

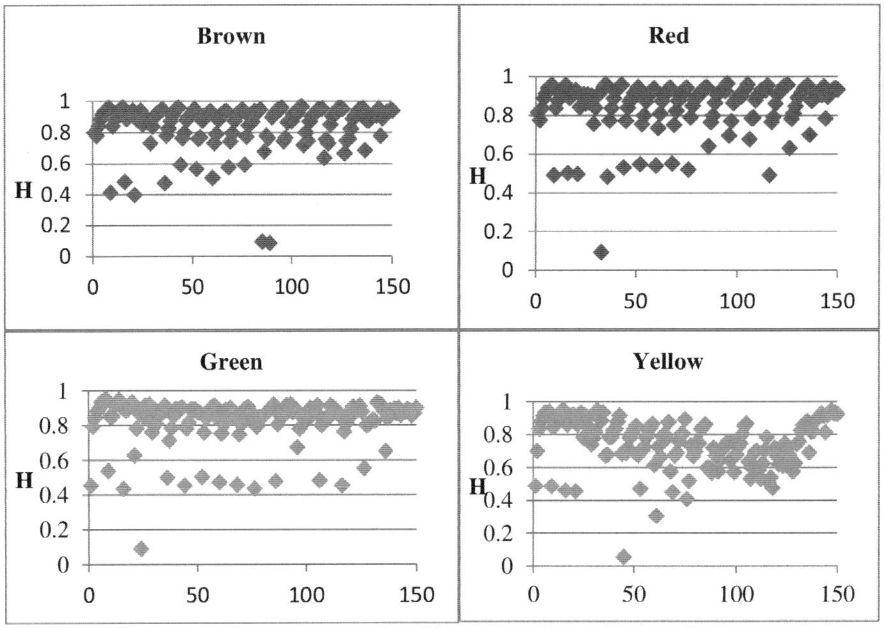

Figure 3. Estimated Hurst exponents for different locations of load cells for a range of experiments.

Table 4. Estimated Hurst exponents for different load cells excluding stillwater values.

	Brown	Red	Green	Yellow
Mean	0.877	0.884	0.859	0.768
Standard Deviation	0.116	0.089	0.089	0.143

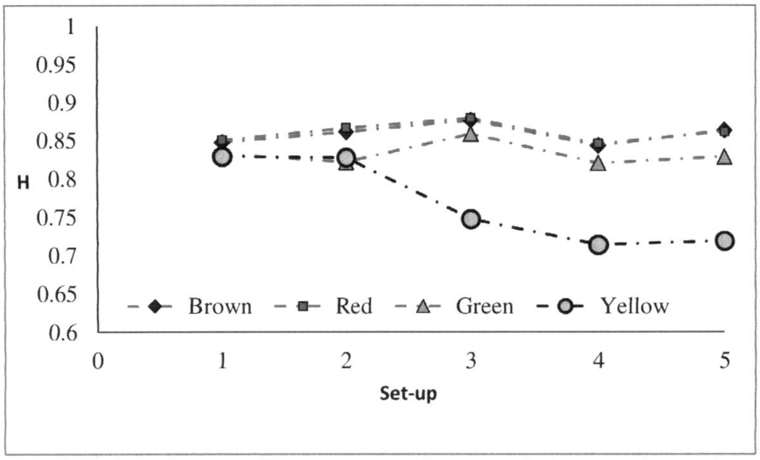

Figure 4. Estimated average Hurst exponents for set-up during experimentation.

5. Discussion and Conclusions

This paper has experimentally presented the potential of Hurst exponent as a possible output-only marker for offshore renewable energy devices. Tests have been carried out in an ocean wave basin and a number of different set-ups have been considered for a Froude scaled monopile model. It is observed that Hurst exponent can be an effective marker for monitoring such devices or platforms when an output from the device is available. In the current paper, the action of waves and equivalent wind forces have been considered at different locations of load cells and it appears that the location of the cell most representative of different loads from water waves is the appropriate location for estimating Hurst exponents. The average estimation of Hurst exponents may be related to the signature of the device itself. Additionally, for the method to be effective, average values over a relatively large number of tests should be carried out.

Previous studies [17, 18] have already shown that such devices typically exhibit nonlinear behaviour although a calibrated theoretical model is not always available. However, estimation of nonlinearity using the structure as a filter is not a new concept [19] and this approach may be extended to the problem under consideration. The requirement of carrying out a relatively larger number of tests is not ideal for rapid assessments, but is often considered as a part of best practice guidelines in normative documents [20]. However, the experiments are usually carried out at a relatively larger scale but in a laboratory environment and consequently it may not be extremely difficult to carry them out. For full-scale deployment, an initial time should be spent to calibrate and benchmark the monitoring markers for any of such structure when the responses are available from a sensor or sensors as a time series. Finally, to obtain better confidence on whether the average values of the estimated Hurst exponents are representative of the signature of the device or not, detailed experiments should be carried out in a wave basin with other offshore renewable energy device solutions. The development of a repository of such device testing can be extremely helpful in this regard. Additionally, such exponents add to the growing interest in developing markers for assessing structural control [21].

Acknowledgements. The authors acknowledge Marine Research Energy Ireland (MaREI), grant no. 12/RC/2302, a Science Foundation Ireland (SFI) supported project and SFI Advance award 14/ADV/RC3022.

References

[1] Kaldellis K, Kavadias K and Christinakis E 2001 Evaluation of the Wind–Hydro Energy Solution for Remote Islands *Energy Conversion and Management* **9**(42) 1105-20

[2] Lakkoju V 1996 Combined Power Generation with Wind and Ocean Waves *Renewable Energy* **9**(42) 870-74

[3] de Vries W and Krolis V 2007 *Effects of Deep Water on Monopile Support Structures for Offshore Wind Turbines* EWEC, Milan

[4] Agarwal P and Manuel L 2009 Simulation of Offshore Wind Turbine Response for Long-Term Extreme Load Prediction *Engineering Structures* **3** 2236-2246

[5] Standard: Design of Offshore Wind Turbines 2007 Federal Maritime and Hydrographic Agency

[6] Swartz RA, Lynch JP, Zerbst S, Sweetman B and Rolfes R 2010 Structural Monitoring of Wind Turbines using Wireless Sensor Networks *Smart Structures and Systems* **6**(3) 1-14

[7] Pakzad SN, Fenves GL, Kim S and Culler DE 2008 Design and Implementation of Scalable Wireless Sensor Network for Structural Monitoring. *ASCE Journal of Infrastructure Systems* **14**(1) 89-101

[8] Pakrashi V, Harkin J, Kelly J, Farrell A and Nanukuttan S 2013 Monitoring and Repair of an Impact Damaged Prestressed Concrete Bridge *Proceedings of the Institute of Civil Engineers, Journal of Bridge Engineering* **166**(1) 16-29

[9] Parloo E, Vanlanduit S, Guillaume P and Verboven P 2004 Increased reliability of reference-

based damage identification techniques by using output-only data *Journal of Sound and Vibration* **270** 813–832

[10] Pakrashi V, Basu B and O' Connor A 2009 A Statistical Measure for Wavelet Based Singularity Detection *ASME Journal of Vibrations and Acoustics***131**(4) 041015-1 – 041015-6

[11] Pakrashi V, O'Sullivan K, O'Shea R and Murphy J 2013 *Experimental Responses of a Monopile Foundation with a Wave Energy Converter Attached*, Proceedings of the Hydro 2013 International, Chennai, India

[12] Di Matteo T, Aste T and Dacorogna MM 2003 Scaling behaviour in differently developed market *Physica A*, **324**, 183 – 188

[13] Ihlen EAF 2012 Introduction to multifractal detrended fluctuation analysis in Matlab. *Frontiers in Physiology* **3**(141) 1-18

[14] Pakrashi V, Kelly J, Harkin J and Farrell A 2013 Hurst Exponent Footprints from Activities on a Large Structural System *Physica A* **392**(8) 1803 – 1817

[15] Melhem H and Kim H 2003 Damage Detection in Concrete by Fourier and Wavelet analyses. *ASCE Journal of Engineering Mechanics* **129**(5) 571-577

[16] Pakrashi V, Basu B and O' Connor A 2007 Structural Damage Detection and Calibration using Wavelet-Kurtosis Technique *Engineering Structures* **29**(9) 2097- 2108

[17] Murphy J, O'Shea R, O'Sullivan K and Pakrashi V 2013 Dynamic Responses of a Scaled Tension Leg Platform, Wind Turbine Support Structure in a Wave Tank *Key Engineering Materials* **569-570** 563-570.

[18] Jaksic V, O'Shea R, Cahill P, Murphy J, Mandic DP and Pakrashi V 2015 Dynamic Response Signatures of a Scaled Model Platform for Floating Wind Turbines in an Ocean Wave Basin *Philosophical Transactions of the Royal Society A: Mathematical, Physical and Engineering Sciences* **373**(2035) 20140078:1-18

[19] Nichols JM, Todd MD, Seaver M and Virgin LN 2003 Use of chaotic excitation and attractor property analysis in structural health monitoring *Physical Review E* **67** 016209:1-8.

[20] BS 6349-7 (1991). Maritime structures part 7: Guide to the design and construction of breakwaters.

[21] Jaksic V, Wright CS, Murphy J, Afeef C, Ali SF and Pakrashi V 2015 Dynamic response mitigation of floating wind turbine platforms using tuned liquid column dampers *Philosophical Transactions of the Royal Society A: Mathematical, Physical and Engineering Sciences* **373**(2035) 20140079:1-20140079:9.

Performance of a Single Liquid Column Damper for the Control of Dynamic Responses of a Tension Leg Platform

V Jaksic[1,2], C Wright[1], Afeef Chanayil[3], Shaikh Faruque Ali[4], Jimmy Murphy[1] and Vikram Pakrashi[1,2]

[1]MaREI/Beaufort Research, Environmental Research Institute, University College Cork, Pouladuff Road, Cork, Ireland
[2]Dynamical Systems and Risk Laboratory, Civil and Environmental Engineering, School of Engineering, University College Cork, College Road, Cork, Ireland
[3]Department of Ocean Engineering, Indian Institute of Technology Madras, India
[4]Department of Applied Mechanics, Indian Institute of Technology Madras, India

v.pakrashi@ucc.ie Ph: 00353214903862 Fax: 00353214276648

Abstract. Tuned liquid column dampers have been proved to be successful in mitigating the dynamic responses of civil infrastructure. There have been some recent applications of this concept on wind turbines and this passive control system can help to mitigate responses of offshore floating platforms and wave devices. The control of dynamic responses of these devices is important for reducing loads on structural elements and facilitating operations and maintenance (O&M) activities. This paper outlines the use of a tuned single liquid column damper for the control of a tension leg platform supported wind turbine. Theoretical studies were carried out and a scaled model was tested in a wave basin to assess the performance of the damper. The tests on the model presented in this paper correspond to a platform with a very low natural frequency for surge, sway and yaw motions. For practical purposes, it was not possible to tune the liquid damper exactly to this frequency. The consequent approach taken and the efficiency of such approach are presented in this paper. Responses to waves of a single frequency are investigated along with responses obtained from wave spectra characterising typical sea states. The extent of control is quantified using peak and root mean squared dynamic responses respectively. The tests present some guidelines and challenges for testing scaled devices in relation to including response control mechanisms. Additionally, the results provide a basis for dictating future research on tuned liquid column damper based control on floating platforms.

1. Introduction

The offshore wind energy industry is moving towards sites which are further from shore in deeper waters and have higher capacity factors and [1, 2]. In such sites where the water depth is greater than 50m floating platforms are considered to be cost competitive with typical fixed foundations [2, 3]. Whilst currently there is no commercial floating wind farm, significant work has been done developing different supporting platform types. They can be categorized into three main types which have adapted from the offshore oil and gas industry: Spar (ballast stabilized), Tension Leg Platform (TLP) (mooring line stabilized)and Semi-submersible (buoyancy stabilized)[4]. Stabilizing a floating wind turbine platform is a major challenge because turbines inherently have a large mass in the nacelle and rotor located up to 100m above the sea surface.

TLP platforms are very buoyant and are kept on station by tensioned mooring lines, called tendons, that usually extend vertically to the seabed [4]. The stability of a TLP in the vertical direction is regulated by these tendons which are kept under tension by the buoyancy of the platform, thus reducing the vertical motions [5].Therefore, TLPs are most suited to deep water sites without significant tidal fluctuation and currents [6] in order to maintain the tension in mooring lines. However, TLP drift motions (surge–sway–yaw motion), due to the action of coupled wind–wave forces, can be significant during extreme weather conditions [7, 8]. These motions could affect the wind turbine performance, accessibility during O&M and ultimately the LCOE (levelised cost of energy [9].

Damping devices integrated into the TLP structure could improve the response characteristics [10], as they have been proved to be beneficial in reducing the vibrations in civil engineering structures [11-13]. Tuned Liquid Dampers (TLDs) are beneficial in offshore wind energy applications for their high performance and low cost [13, 14] whilst Tuned Liquid Column Dampers (TLCD) is the type of TLD which work on the principle of motion of a liquid column in a U-shaped tube counterbalancing external forces imposed onto the structure. The TLCD is found to be successful way of vibration control when structure is exposed to the wind and earthquake loading [15-17].

Lee et al. [18] studied a pontoon type offshore floating platform which incorporated with a TLCD both numerically and experimentally. This paper extends this work to an experimental structure with a much lower frequency (around 0.33 Hz). However, it is practically difficult to match this frequency due to the very large size of upstands necessary for the TLCD. Consequently, the TLCD is tuned to a close enough frequency of 0.5Hz so that the frequency response in the range of the natural frequency of the structure and that of the TLCD overlap significantly. A smaller structure would perhaps allow for testing with exact tuning to a model with a higher natural frequency and consequently more reasonable size of the upstand. However, for such conditions it is difficult to produce scaled wave conditions at that size in a flume [19]. The TLCD design in this paper is based on that described in Yalla and Kareem [20]. Additionally, variations in damper characteristics are also considered. The results of this study are encouraging and form the benchmark for further investigation of TLCD application in offshore wind energy substructure motion mitigation. The results of the study inspired further investigation of the application of the multiple TLCD (MTLCD) for TLP motion mitigation [10]. Jaksic et al. [10] investigated capacity of MTLCD to mitigate motions of TLCD. They found that the additional weight of the dampers on structure causes the tension of the tendons to decrease. In order to keep the tension in tendons additional buoyancy was added to the system. The consequence was that the TLP with MTLCD could be tested only for the low wave height (e.g. Froude-scaled wave amplitudes, Hs, 0.015, 0.02, 0.025, 0.03, and 0.035m) with single period Tp=2.4Hz in order to remain with in load cell capacity. This study shows the results for the wider range of the wave height and period which is in better agreement with the real sea states (e.g. Froude scaled wave amplitudes, Hs, 0.05, 0.075, and 0.1m and T_p, 1.2, 2.4, and 3.6s). To understand the interaction between TLCD and wave energy devices with low natural frequency, a single TLCD-tension leg platform interaction was considered to be an appropriate investigation and this has been addressed in the current paper.

2. Theoretical Modelling

The surge motion of TLP is considered for the simulation studies. Uncoupled linear equation of this motion is given as [18]

$$M\ddot{X} + C_X\dot{X} + K_X X = F_w(t) \quad (1)$$

where M represents of the TLP structure, C_X and K_X are the damping and stiffness associated with the TLP. The wave force from the Pierson Moskowitz (PM) spectra is represented as F_w. X is the motion along the surge direction.

The equation of motion of the system, i.e., TLP with TLCD is given as [18],

$$\rho_d A_d L_d \ddot{H} + \frac{1}{2}\rho_d A_d h_d |\dot{H}|\dot{H} + 2\rho_d A_d g H = -\rho_d A_d B_d \ddot{X} \quad (2)$$

$$M\ddot{X} + C_X\dot{X} + K_X X = F_w - \rho_d A_d B_d \ddot{H} - \rho_d A_d L_d \ddot{X} \quad (3)$$

Where h_d is the factor of the head loss of the liquid, H, $\dot{H}$ and $\ddot{H}$ are the displacement, velocity and the acceleration of the liquid in the vertical tube, respectively. ρ_d is the density of the liquid. The geometric parameters of the damper are inner cross sectional area, A_d, length of horizontal section, B_d and L_d being the total length.
The random force F_w is obtained from Pierson–Moskowitz (PM) spectrum using the following equation.

$$F_w(t) = \sum_{i=1}^{k} A_i \cos(2\pi f_i - \phi_i)$$

where, $A_i = \sqrt{S(f_i)\Delta f_i}$, A_i is the ith amplitude, f_i is the ith frequency, ϕ_i is the ith phase angle. The value of $\phi_i \in [0, 2\pi]$ are randomly generated uing uniform distribution. $S(f_i)$ is the spectral density value of the wave force spectrum at the frequency f_i and $\Delta f_i = 0.1$ Hz is the frequency step considered. A swept sine simulation is carried out to obtain the frequency response of the TLP and TLP with TLCD with different values of the liquid density of the TLCD (m_d*). Different fluid density is considered to present different fluid within TLCD while keeping all other parameters of the damper constant.

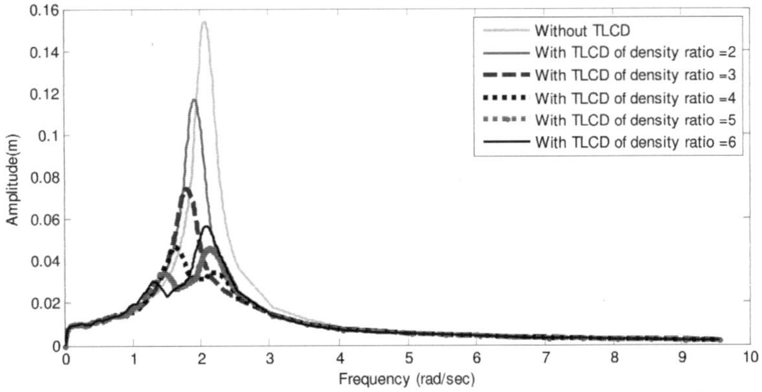

Figure 1: Frequency response results of TLP without and with TLCD.

Figure 1 shows comparative results of the simulation. The frequency sweep is carried out till 10 rad/s. The liquid density ratios were varied from 2 to 6. Responses in the frequency domain were compared with responses which were obtained without the use of TLCD dampers. The responses decrease as m_d* increases up to m_d*=4 and then again it increases for m_d*>4. The minimum response was observed when m_d* is between 3 and 4. The displacement responses for TLP with and without TLCD are compared and reported in Figure 2 for different values of density ratio. Forced oscillation was simulated for 150s and then the system is allowed to vibrate freely.

Figure 2 shows the response with TLCD has reduced compared with only TLP. A reduction of 48.6% in peak displacement is obtained using TLCD. Norm value also was calculated, and obtained a reduction from 1.465 to 0.33 (77% reduction)

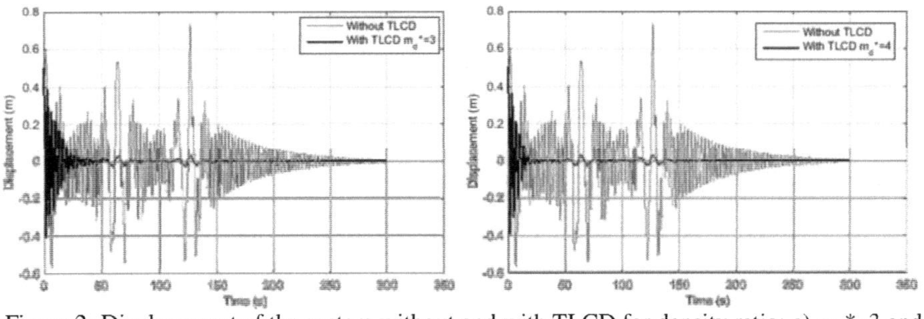

Figure 2: Displacement of the system without and with TLCD for density ratio: a) m_d*=3 and b) m_d*=4.

3. Experiments

3.1. TLP Model with Tuned Liquid Column Damper (TLCD)

The model TLP is a truss type structure with a floating hexagonal platform connected with six mooring tethers to a large circular gravity base located on the floor of the wave basin (Figure 3). The floating hexagonal platform is made up of a buoyancy ring and the upper structure. The buoyancy ring consists of six 90mm diameter Polyvinyl chloride (PVC) pipes, joined to the central column by six 40mm diameter PVC pipes. Situated above the buoyancy ring is the upper structure, fabricated from 40mm diameter PVC pipe, which is joined to the buoyancy ring by six 40mm diameter sections of pipe, and to the central column by six 40mm diameter PVC pipes. The upper structure provides no buoyancy as it is not submerged. The central column is fabricated from 160mm diameter PVC pipe and provides sufficient buoyancy to counteract the weight of the tower and nacelle. The excess buoyancy force ensures that the six mooring lines, made of 2mm diameter stainless steel wire, remain in tension at all times. The weight of the TLP is 16.8kg. The wind turbine tower is 1.15m high 50mm diameter PVC pipe (0.8kg) with the 2.2kg thrust load and a nacelle weight of 0.8kg.

A U-shape TLCD device was attached to the upper structure at the level of the center of the gravity (CG). It was designed following Yalla and Kareem [20] and is tuned to the average requency of the longest Bredschneider waves the basin can generate (0.59Hz). The effects of two TLCD designs on the behaviour of the structure were studied. In the first case the mass ratio, μ (mass of liquid in the tube (m_d) to the mass of the primary system represented by the platform with the turbine tower (M_s)) was 5%, and in the second case 10%, with pipe diameter 30mm and 40mm, respectively. The horizontal part of the TLCD was 1m long, and the vertical part was 0.6m long (filled with water up to 0.2m) to allow for water movement during testing. The experimental set up of TLP is shown in Figure 3a, while the gravity base with load cell arrangement and position of TLCD in relation to the incident wave direction is shown in Figure 3b.

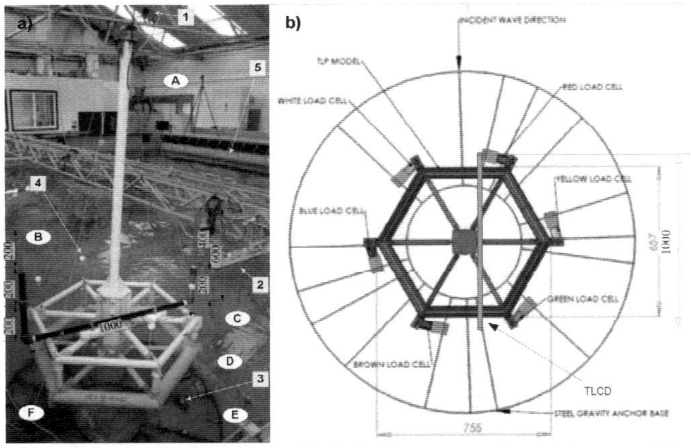

Figure 3: a) 1:50 Scale Model of Truss type TLP Platform experimental setup: (A) mast, (B) central column, (C) upper structure, (D) buoyancy ring, (E) gravity base, and (F) TLCD. The locations of devices used: (1) motion camera, (2) wave probes, (3) load cells, (4) reflective motion markers, and (5) flap type wave-maker; b) TLP view from above: position of TLCD and gravity base with load cell arrangement in relation to the incident wave direction.

3.2. Instrumentation
The TLP equipped with a TLCD device was tested using specified wave conditions and the performance was recorded using six load cells, two conductivity probes, and four motion capture cameras. The Tedea-Huntleigh stainless steel single ended bending beam load cells measured the tension in mooring tethers in Newtons (with a maximum load of ~50N) and were bolted to the gravity base. Each load cell was given a colour code (name) during the testing, i.e. White, Red, Blue, Yellow, Brown, and Green were located at Bow Port, Bow Starboard, Amidships Port, Amidships Starboard, Stern Port, and Stern Starboard, respectively. Two conductivity probes measured the surface elevations during testing and allowed the incident wave conditions to be determined. In order to measure the motions of the TLP during the testing, four reflective markers were attached to the corners of the hexagonal base. The instantaneous positions of the markers were monitored by the Qualisys 3-Series Oqus Marker Tracking Cameras with a sampling frequency of 32Hz. The load cells and wave probes were triggered by the National Instruments Labview 2011 Version 11.0 software. The Qualisys Marker Tracking system was time synchronised with the other instrumentation using Labview.

3.3. Experimental Procedure
The model testing was carried out in wave basin (25m x 18m x 1m) equipped with 40 flap type paddles capable of generating sinusoidal wave profiles as well as 2 and 3-D wave spectra. The TLP was tested for Bretschneider (BS) and Joint North Sea Wave Observation Project (JONSWAP or JS) spectra, wave periods, Tp, 1.2, 2.4, and 3.6s, considering Froude 1:50 scaled (structural models) and waves with heights, Hs as 0.05, 0.075, and 0.1m respectively.

4. Results and Discussion
The TLP was excited by different waves for about one minute in each test and the dynamic responses were recorded. The results of this experimental work are presented in Figure 4.The comparison of the surge energy measured by yellow load cell between completely undamped (without TLCD) and damped (with TLCD) case for H_s 50 and 75mm are presented in Figure 4a. The results show that the surge energy decreases when a TLCD is installed onto the structure. This observation is in agreement with the numerical modeling results presented in this paper and those obtained by and Lee et al. [18].

Figures 4b-h show the comparison of results for TLP with TLCD for BS and JS spectra for μ = 5% and 10%, with and without thrust loading. Figure 4b shows H_s vs. percentage reduction (in relation to undamped case) in tension (+ve means %age reductions in all figures) in yellow cell measurements. The tension reduces with increasing μ for both types of spectra. The presence of the thrust reduces the tension visibly more for BS than JS spectra. The results for BS spectra show that for μ = 5% the tension increases with the wave height. However, the results for μ = 10% are inconclusive since the tension is reduced for wave height 75mm in comparison with 50mm and 100mm waves. When thrust is present, the tension decreases with wave height. The results for JS spectra show almost the same level of reduction for μ = 5% with or without thrust when H_s increases. On the other hand, in the case of μ = 10% the tension decreases with wave height increase. Figures 4c-e shows the H_s vs. the percentage reduction in displacement (X), velocity (V), and acceleration (A) RMS, respectively, in relation to the undamped case. For μ = 5 % the percentage displacement RMS decrease for both wave spectra up to 0.36% for BS wave height 50mm, while for μ = 10% they increase up to 0.2% for wave height 75mm. When thrust is present, displacement response increases for both wave spectra were observed with the maximum being -1% for JS. This phenomena was observed by Roderick [14] where numerical analysis for three different offshore wind turbine configurations, including various locations for TLCD installations were carried out and the results indicated that in most cases the TLCD could successfully reduce overall tower displacement and platform pitch. However, in some cases the use of TLCD actually increased the overall tower and platform motions.

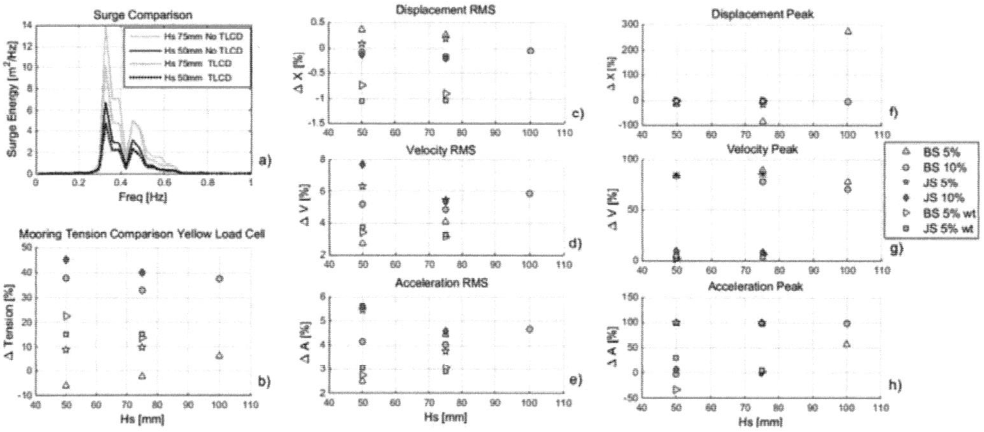

Figure 4: a) Yellow load cell: frequency vs. surge energy for damped and undapend case for 50 and 75mm wave amplitude. b-h) Results for TLP with TLCD for BS and JS spectra when μ = 5 and 10% without and with thrust in comparison with undamped case: b) wave amplitude vs. percentage reduction in tension measured by yellow load cell; c) wave amplitude vs. percentage reduction in displacement RMS; d) wave amplitude vs. percentage reduction in velocity RMS; e) wave amplitude vs. percentage reduction in acceleration RMS; f) wave amplitude vs. percentage reduction in displacement peak; g) wave amplitude vs. percentage reduction in velocity peak; h) wave amplitude vs. percentage reduction in acceleration peak.

The velocity and acceleration RMS is reduced more significantly for higher μ and lower H_s. The presence of the thrust in experiments gives the lowest reduction in the RMS values of velocity and acceleration. Figures 4f-h show the Hs vs. the percentage reduction in peak displacement (X), velocity (V), and acceleration (A). The highest reduction in displacement peak was observed for μ = 5%, when the platform was exposed to 100mm BS waves. On the other hand, the lowest reduction in displacement peak was observed for μ = 5% when the platform was exposed to 75mm BS waves.

However, these two extremes are much different from other results and can be attributed to some leakage of water from TLCD at pipe connections. The displacement peak varies negligibly between remaining measurements, with slightly increased values for μ = 10%. The velocity peak decreases generally around 80% for BS and JS when μ = 5%, while it decreases negligibly in other cases. The acceleration peak reduction figures show the same trend with higher reduction in acceleration for BS and JS spectra related to μ = 5% as compared to μ = 10%. The only exception of this is that the contribution of the thrust reduces the peak in the case of JS, while it increases the peak in the case of BS.

5. Conclusions

This paper numerically and experimentally studies the effectiveness of two designs of a single tuned liquid column damper (TLCD) on the dynamic performance of an offshore wind floating tension leg platform (TLP). The numerical studies showed that the reduction in the displacement of the structure is achieved by incorporation of the TLCD. Moreover, the density of the liquid is a factor that can control the level of the reduction of the TLP motion to a certain extent. Testing of the floating offshore wind structures in a wave tank with actual ocean wave spectra confirms that TLCD can reduce the overall motions of the TLP substructures when exposed to such wave conditions. A reduction in the load cell tension measurement has been observed due to installation of TLCD but further studies are required to make a more conclusive observation. The reduction in peak and RMS values for displacement, velocity, and acceleration of the platform has been reported in this paper. This study forms the basis for further investigation of application of TLCD for mitigating dynamic responses of TLPs. The findings of this research encouraged further investigation into employment of multiple dampers for reducing the motions of the TLPs in the extreme weather conditions [10].

The paper emphasizes the importance of using larger scaled model testing in more realistic conditions to assess control or monitoring strategies and designs for full scale deployment. Small scale models do not necessarily capture certain complexities and challenges related to the mitigation of dynamic responses of offshore renewable energy device platforms. These experiments also show that to achieve an optimal arrangement for the control of dynamic responses of TLP, a number of adjustments is required to be undertaken frequently over the lifespan of structure. For structures with low natural frequency, exact tuning of TLCD can be difficult due to geometric constraints but a tuning close to that low frequency can still achieve an adequate level of vibration mitigation.

Acknowledgement. The authors would like to express their gratitude to Mr Florent Thiebaut, Mr Niall O'Sullivan, and Mr Thomas Walsh, for their help with setting up the experiments reported in this paper. The authors acknowledge Marine Research Energy Ireland (MaREI), grant no. 12/RC/2302, a Science Foundation Ireland (SFI) supported project and SFI Advance award 14/ADV/RC3022.

References

[1] INWORK. 2013 Floating Offshore Wind Foundations: Industry Consortia and Projects in the United States, Europe and Japan. An Overview., pp. 1-44, Marine International Consulting LLC.

[2] Sclavounos, P.D., Tracy, C. & Lee, S. 2008 Floating Offshore Wind Turbines: Responses in a Seastate, Pareto Optimal Designs and Economic Assessment. In *Offshore Mechanics and Arctic Engineering Conference OMAE 2008*. Lisbon, Portugal.

[3] Bulder, B.H., Henderson, A., Huijsmans, R.H.M., Peeringa, J.M., Pierik, J.T.G., Snijders, E.J.B., van Hees, M.T., Wijnants, G.H. & Wolf, M.J. 2003 Floating offshore Wind Turbines for Shallow waters. In *The European Wind Energy Association - EWEC 2003*. Madrid, Spain.

[4] Arapogianni, A., Genachte, A.-B., Ochagavia, R.M., Vergara, J.P., Castell, D., Tsouroukdissian, A.R., Korbijn, J., Bolleman, N.C.F., Huera-Huarte, F.J., Schuon, F., et al. 2013 Deep Water The next step for offshore wind energy. The European Wind Energy Association.

[5] Wang, C.M., Utsunomiya, T., Wee, S.C. & Choo, Y.S. 2010 Research on floating wind turbines: a literature survey. *The IES Journal Part A: Civil & Structural Engineering***3**, 267-277. (doi:10.1080/19373260.2010.517395).

[6] Roddier, D. & Weinstein, J. 2010 Floating Wind Turbines. *ASME Mechanical Engineering*.

[7] Lefebvre, S. & Collu, M. 2012 Preliminary design of a floating support structure for a 5MW offshore wind turbine. *Ocean Engineering***40**, 15-26. (doi:10.1016/j.oceaneng.2011.12.009).

[8] Emmerhoff, O.J. 1994 The slow drift motions of offshore structures. Massachusetts, USA, Massachusetts Institute of Technology.

[9] Butterfield, S., Musial, W., Jonkman, J., Sclavounos, P. & Wayman, L. 2005 Engineering Challenges for Floating Offshore Wind Turbines. In *Proceedings Copenhagen Offshore*. Copenhagen, Denmark.

[10] Jaksic, V., Wright, C.S., Murphy, J., Afeef, C., Ali, S.F., Mandic, D.P. & Pakrashi, V. 2015 Dynamic response mitigation of floating wind turbine platforms using tuned liquid column dampers. *Phil. Trans. R. Soc. A***373: 20140079**. (doi:10.1098/rsta.2014.0079).

[11] Gao, H., Kwok, K.S.C. & Samali, B. 1999 Characteristics of multiple tuned liquid column dampers in suppressing structural vibration. *Engineering Structures***21**, 316-331. (doi:10.1016/S0141-0296(97)00183-1).

[12] Diana, G., Resta, F., Sabato, D. & Tomasini, G. 2013 Development of a methodology for damping of tall buildings motion using TLCD devices. *Wind Struct.***17**, 629-646.

[13] Symans, M.D. & Constantinou, M.C. 1999 Semi-active control systems for seismic protection of structures: a state-of-the-art review. *Engineering Structures***21**, 469-487. (doi:10.1016/S0141-0296(97)00225-3).

[14] Roderick, C. 2012 Vibration Reduction of Offshore Wind Turbines Using Tuned Liquid Column Dampers. Massachusetts, US, University of Massachusetts Amherst.

[15] Yalla, S.K., Kareem, A. & Kantor, J.C. 2001 Semi-active tuned liquid column dampers for vibration control of structures. *Engineering Structures***23**, 1469-1479. (doi:10.1016/S0141-0296(01)00047-5).

[16] Xu, Y., Samali, B. & Kwok, K. 1992 Control of Along-Wind Response of Structures by Mass and Liquid Dampers. *Journal of Engineering Mechanics***118**, 20-39. (doi:10.1061/(ASCE)0733-9399(1992)118:1(20)).

[17] Won, A.Y.J., Pires, J.A. & Haroun, M.A. 1996 Stochastic seismic performance evaluation of tuned liquid column dampers. *Earthquake Engineering & Structural Dynamics***25**, 1259-1274.

[18] Lee, H.H., Wong, S.H. & Lee, R.S. 2006 Response mitigation on the offshore floating platform system with tuned liquid column damper. *Ocean Engineering***33**, 1118-1142. (doi:10.1016/j.oceaneng.2005.06.008).

[19] Stive, M.J.F. 1985 A scale comparison of waves breaking on a beach. *Coastal Engineering***9**, 151-158. (doi:http://dx.doi.org/10.1016/0378-3839(85)90003-1).

[20] Yalla, S. & Kareem, A. 2000 Optimum Absorber Parameters for Tuned Liquid Column Dampers. *Journal of Structural Engineering***126**, 906-915. (doi:10.1061/(ASCE)0733-9445(2000)126:8(906)).

[21] Farshidianfar, A., Oliazadeh, P. & Farivar, H.R. 2009 Optimal Parameter's Design in Tuned Liquid Column Damperv. In *17th. Annual (International) Conference on Mechanical Engineering-ISME2009*. University of Tehran, Iran.

[22] MathWorks. 2014 MATLAB 8.4 (R2014b). Natick, MA, USA, The MathWorks, Inc. Software.

A Delay Vector Variance based Marker for an Output-Only Assessment of Structural Changes in Tension Leg Platforms

V Jaksic[1,2], C Wright[1], D P Mandic[3], J Murphy[1] and V Pakrashi[1,2]

[1]Beaufort, Environmental Research Institute, University College Cork, Cork, Ireland
[2]Dynamical Systems and Risk Laboratory, Civil and Environmental Engineering, School of Engineering, University College Cork, College Road, Cork, Ireland
[3]Department of Electrical and Electronic Engineering, Imperial College London, London SW7 2BT, UK

v.pakrashi@ucc.ie Ph: 00353214903862 Fax: 00353214276648

Abstract. Although aspects of power generation of many offshore renewable devices are well understood, their dynamic responses under high wind and wave conditions are still to be investigated to a great detail. Output only statistical markers are important for these offshore devices, since access to the device is limited and information about the exposure conditions and the true behaviour of the devices are generally partial, limited, and vague or even absent. The markers can summarise and characterise the behaviour of these devices from their dynamic response available as time series data. The behaviour may be linear or nonlinear and consequently a marker that can track the changes in structural situations can be quite important. These markers can then be helpful in assessing the current condition of the structure and can indicate possible intervention, monitoring or assessment. This paper considers a Delay Vector Variance based marker for changes in a tension leg platform tested in an ocean wave basin for structural changes brought about by single column dampers. The approach is based on dynamic outputs of the device alone and is based on the estimation of the nonlinearity of the output signal. The advantages of the selected marker and its response with changing structural properties are discussed. The marker is observed to be important for monitoring the as-deployed structural condition and is sensitive to changes in such conditions. Influence of exposure conditions of wave loading is also discussed in this study based only on experimental data.

1. Introduction

The wind industry is moving toward bigger turbines, larger blades, taller towers, and into deeper ocean waters in order to use the potential of the strong winds. The studies show that the deep offshore designs, floating foundations, are competitive in terms of the levelised cost of energy (LCOE) with bottom fixed foundations in water exceeding 50m in depth [1-3].

However, with moving offshore structures into deeper water the maintenance and monitoring of the structures are becoming critical problem [4]. Further offshore structures supporting energy devices are exposed to often unpredictable and extreme weather conditions, which could influence their performance, stability during maintenance and operations, and overall cost of the offshore device floater [5-7]. Therefore, continuous observation of a floating supporting structure for offshore devices exposed to the highly variable met-ocean conditions is one of the ways to monitor performance and health of the structure [8, 9]. Hence, the remote monitoring of the complex dynamics of the floaters [10, 11] caused by the wind, wave, and wind-wave loading combined with statistical analysis of its responses can be used for structural health monitoring (SHM). The markers identified through the statistical analysis of the structure output only signal can help in the pattern forming recognition system of the behavior changes of the offshore structures. However, in the early stage of development, the quantitative estimates of the stochastic dynamic responses and assessment of the corresponding markers are only possible using scaled testing in laboratory setting [6, 12]. Further, measurement of the dynamic responses of scaled floating foundation model in wave tank is the first step in assessment and understanding of structural dynamic responses, since the prototype in sea testing can be difficult and expensive during at the design stage [13, 14].

This study presents the results of the wave tank testing of a 1:50 Froude scaled Tension Leg Platform (TLP) supporting wind turbine structure. The platform is equipped with the Tuned Liquid Colum Damper (TLCD) which is used to mitigate horizontal motions of the platform [12, 15]. The design of the TLCD is based on the Yalla and Kareem [16] study. Therefore, the overall length of the water column is calculated by equalising irregular wave frequency with water column frequency.

Structural response of the floater exposed to sea states characterized by the Bretschneider (BS) and Joint North Sea Wave Observation Project – JONSWAP (JS) spectra was monitored for two different damper–TLP mass ratio (μ = 5% and 10%) and for three different experimental setups of the damper (when TLCD active /open, inactive/closed, and no TLCD case). The dynamic response was recorded using load cells and camera based motion recognition system. The structural behaviour of the platform was observed for when there is and there is no thrust loading on the mast representing wind effects, for various wave periods and amplitudes, in order to establish the correlation between imposed changes and the selected statistical marker.

This study examines the possibility of employing the Delay Vector Variance (DVV) method for analysis of output only signal, i.e. responses of the TLP, in order to investigate this marker for assessing change in variation of structural parameters brought about by intervention in the form of control [8]. DVV method is used in quantification of linearity or nonlinearity of the structural system response signals as indicator of the nature of the structural behaviour due to changes in the excitation force and TLCD design. DVV method was previously used successfully in different scientific disciplines where outputs of the system response were analyzed for the change in its linearity due to the variation of the system parameters [8, 17-21]. The change in the degree of nonlinearity in the signal is characterized by a Root Mean Square Error, (RMSE) as a marker [22]. The study uses obtained RMSE to discuss TLP tendons behavior changes caused by different sea states and two TLCD designs. The results of this study are encouraging and form the basis in further investigation of TLCD application in offshore wind energy substructure behaviour.

2. Experimentation

2.1. TLP Model equipped with Tuned Liquid Column Damper (TLCD)

The TLP platform tested is a truss structure with hexagonal floating platform. The platform is connected to the gravity base, located at the bottom of an ocean wave basin, with the six mooring tethers. The floating platform consist of the buoyancy ring and the upper structure. The buoyancy ring structure is made of six 90mm diameter Polyvinyl chloride (PVC) pipes, joined to the central column by six 40mm diameter PVC pipes. The upper structure, located above the buoyancy ring, is built from 40mm diameter PVC pipe, which is joined to the buoyancy ring by six 40mm diameter sections of pipe, and to the central column by six 40mm diameter PVC pipes. The upper structure is not submerged. The central column is fabricated from 160mm diameter PVC pipe and provides sufficient buoyancy to counterbalance the tower and nacelle weights. The additional buoyancy force is passed on the six mooring lines made of 2mm diameter stainless steel wire which are in tension at all times. The weight of the TLP is 16.8kg. The wind turbine tower is made of 50mm diameter 1.15m high PVC pipe (0.8kg). The thrust load at the top of the tower is 2.2kg.

A U-shape TLCD device, designed following Yalla and Kareem [16], is attached to the TLP upper structure at the level of the center of the gravity (CG). TLCD is tuned to the average frequency of the longest Bredschneider waves that the basin can generate (0.59Hz). The TLP response was monitored using two TLCD designs. The mass ratio, μ (mass of liquid in the tube, m_d, vs. mass of the primary system, Ms) was in the first 5%, and in the second case 10%, having pipe diameter 30mm and 40mm, respectively. The horizontal part of the TLCD was 1m long while the vertical part was 0.6m long (filled with water up to 0.2m) to allow for water sloshing and prevent its loss. Figure 1a shows the experimental set up of TLP.

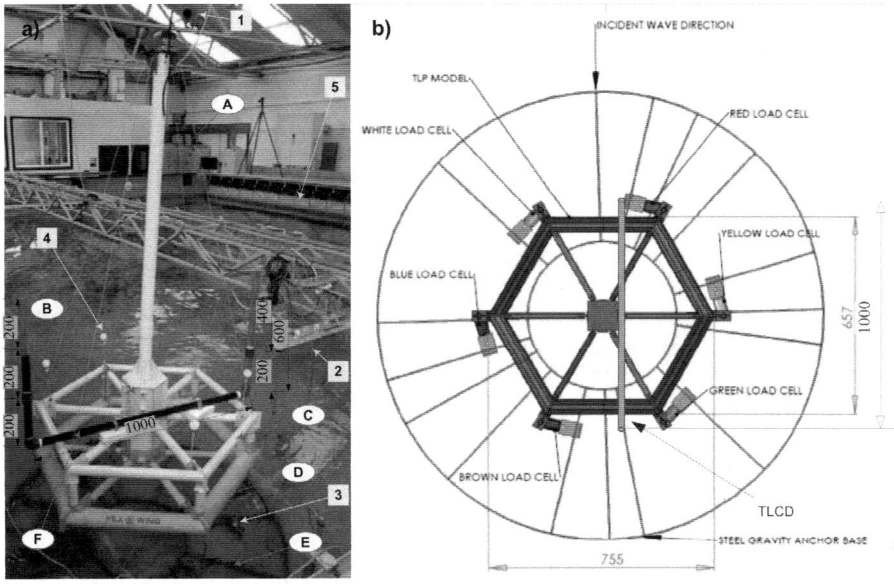

Figure 1: a) 1:50 Scale Model of Truss type TLP Platform experimental setup: (A) mast, (B) central column, (C) upper structure, (D) buoyancy ring, (E) gravity base, and (F) TLCD. The locations of devices used: (1) motion camera, (2) wave probes, (3) load cells, (4) reflective motion markers, and (5) flap type wave-maker; b) TLP view from above: position of TLCD and gravity base with load cell arrangement in relation to the incident wave direction.

2.2. Instrumentation

The response of the TLP structure and the effects of TLCD device on its performance, while exposed to the known characteristic waves for a period of time, was recorded in the laboratory environment using six load cells, four motion capture cameras, and two water level probes. Stainless steel single ended bending beam load cells (Tedea-Huntleigh) measured the tension in mooring lines in Newton (maximum load ~ 50N). The gravity base with load cell arrangement and position of TLCD in relation to the incident wave direction is shown in Figure 1b. Each load cell was given a name during the testing, i.e. White (Bow Port), Red (Bow Starboard), Blue (Amidships Port), Yellow (Amidships Starboard), Brown (Stern Port), and Green (Stern Starboard). Two water level probes were measuring the height of water in mm during testing and could therefore measure the profile of the waves generated in the wave basin which the model was exposed to. Four reflective markers were attached to the corners of the hexagonal base for assisting in measuring the motions of the TLP during the experiment (Figure 1a). Qualisys 3-Series Oqus Marker Tracking Cameras, with 32Hz sampling frequency, were used to determine the momentary positions of the markers. The load cells and wave probes were triggered by the National Instruments Labview 2011 Version 11.0 software. The Qualisys Marker Tracking system was time synchronised using Labview.

2.3. *Experimental strategy*

The TLP model was tested in an ocean wave basin equipped with 40 flap type paddles able of generating sinusoidal wave profiles as well as 2 and 3-D wave spectra. The still water depth is constant at 1.0m. The TLP was tested for BS and JS spectra, wave period, T_p= 2.4sec, wave frequency, f_p=0.42Hz, and Froude scaled (model) wave amplitudes, Hs_m, 0.05 and 0.075m (H_s = 2.5 and 3.75m respectively), with and without thrust loading. Additionally, TLP platform was tested for: BS spectra when T_p=1.2 and 3.6s corresponding to f_p=0.83 and 0.28Hz, respectively, for wave amplitude Hs_m, 0.05m without thrust loading and BS spectra when T_p=2.4s, f_p=0.42Hz for wave amplitude Hs_m, 0.1m (H_s = 5m) without thrust loading. A mass (0.8kg) was attached to the top of the mast to act as the loading of a wind turbine nacelle in no wind conditions. The TLP was provided with the TLCD and four different setups of the damping device were tested, i.e. with μ = 5% and 10% mass ratio when active and inactive. Effects of reflected waves at the boundaries of the basin were removed by dissipating the energy in an immersed barrier made of randomly oriented, rigid objects.

3. Delay Vector Variance (DVV) Methodology

The load cell measurements were analysed using the Delay Vector Variance (DVV) method [23] in order to quantify the extent of nonlinearity of the structural system response to the variable sea conditions and to determine to what extent the nonlinearity of the response signal is linked to changes of the observed TLP system and its exposure to different forces. DVV method, based on surrogate data methodology [24], detects the determinism and nonlinearity in a time series. The explanation and testing of the DVV methodology is given in Mandic et al. [25] and Gautama et al. [26]. DVV was successfully employed as statistical marker to track the response changes due to the changes in the system using the output only signal in different research fields [18, 20-22, 27]. The potential of using DVV method to analyze and track changes in structural properties of the floating platform due to its exposure to different wave types is investigated in Jaksic et al. [8]. The advantages of the proposed method are related to the facts that it does not require any prior knowledge about the system or the excitation force, it is robust to the presence of noise, straight forward to interpret and typically has improved performance over traditionally used methods [20].

Numerical analyses were performed using the DVV Toolbox [27]. The output of the method is one number for each response signal recorded represented by the Root Mean Square Error (RMSE) and represents the degree of nonlinearity of the response [20]. Following DVV parameters were kept constant during analysis: the embedding dimension m=3, time lag τ= 1, the maximal span parameter is n_d=3, the number of standardized distances for which target variances are computed is N_{tv}=50, number of surrogates considered is N_s=50, and the number of reference DVs considered is N_{sub}=200. Discussions related to the choice of these parameters and computation of DVV is reported by Jaksic et al. [8].

DVV scatter diagram can be produced by performing DVV analysis on the original and a number of surrogate time series. If the surrogate time series yields DVV plots similar to the original time series, in which case the DVV scatter diagram coincides with the bisector line, then the original time series is considered to be linear. The deviation from the bisector line, quantified by an RMSE is an indicator of nonlinearity of the original time series [26]. As the degree of nonlinearity increases, the deviation from bisector line grows. The example of the load cell measurements recorded and DVV scattered plots obtained are given in Figure 2.

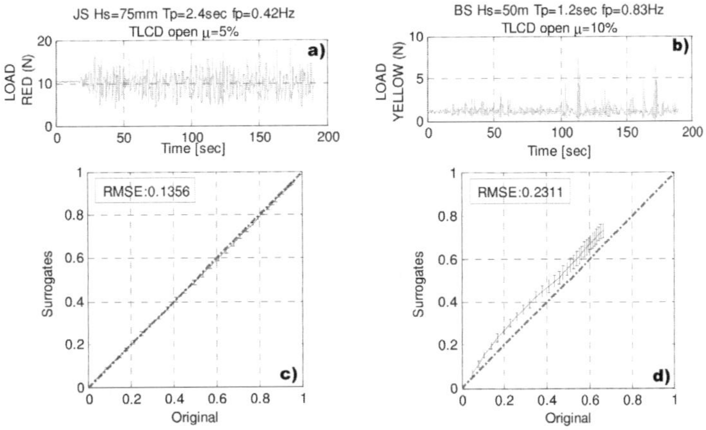

Figure 2: a) Red load cell measurements for TLP with no Thrust excited by JONSWAP waves Hs=75mm high, period Tp=2.4sec, and frequency fp=0.42Hz; b) Yellow load cell measurements for TLP with no Thrust excited by Bretschneider waves Hs=50mm high, period Tp=1.2sec, and frequency fp=0.83Hz; c) DVV scatter diagram corresponding to a) measurements, and d) DVV scatter diagram corresponding to b) measurements.

4. Results and Discussion

The TLP was excited by different waves for about three minute in each test. The total number of test performed is 42. The load cell measurements were recorded and DVV analysed. The degree of the nonlinearity (RMSE) of forces in tendons due to the changing wave characteristics, presence of two different TLCD designs, and thrust loading is estimated and compared.

Figure 3 shows results of DVV analysis when TLP is exposed to BS and JS waves (T_p=2.4s and f_p=0.42Hz). The presented results indicate that there is the change in the degree of linearity of recorded load cell measurements for both types of loading. When TLP with no TLCD is exposed to BS wave spectra the highest RMSE in loading is experienced by Blue and Yellow load cell. However the difference in RMSE (ΔRMSE) due to variation in wave heights is negligible being greatest for the Brown (ΔRMSE=0.006). For TLP exposed to BS with TLCD (μ=5%) installed, the RMSE graph becomes distorted in such way that the highest variation in degree of nonlinearity is recorded by White cell when TLCD is closed and Hs_m=100mm (Figure 3a). However, due to the presence of the damper active and inactive ΔRMSE is highest for Brown and Blue load cells. This could be due to the eccentricity in TLP loading caused by the TLCD location since ΔRMSE is lower for Red, Yellow, and Green load cells which are on the side where TLCD is installed. The results obtained for TLP exposed to BS when there is thrust loading show in general small reduction in RMSE (Figure 3b) in comparison with results observed for when there is no thrust loading (Figure 3a). However, it appears that when trust present level of nonlinearity in load cells measurements depend on the H_s rather than on the activity of the TLCD in such way that with increased H_s RMSE is reduced for all load cells but for Brown. However, this could be due to combined effect of the position of the TLCD and the thrust loading hanging in between Green and Brown cell from the top of the tower. Thus, when TLCD and thrust loading combined they become responsible for the distribution of the loading in the load cells causing different degree of the nonlinearity in the signal measured. Figure 3c shows the results of the TLP exposed to BS spectra for TLCD with μ=10%. In this case the variation in RMSE due to the different in Hs and TLCD two setups is greater than when TLCD μ=5% for all load cells. However, the highest variation ΔRMSE obtained on the White load cell measurements. The reason could be the change in the load cells pretension loading due to the heavier TLCD. The results of the TLP exposed to JS with varying Hs with closed and opened TLCD show that the difference in the level of nonlinearity depends on wave height but also on the activity of TLCD (Figure 3d). When thrust loading introduced to the TLP system exposed to JS (Figure3e) the RMSE decrease in comparison with the cases without the thrust. Also it appears that the level of nonlinearity between load cells is distributed almost equally. However, the RMSE is higher for inactive TLCD being the highest for the tendons on the side of (below) the damper, i.e. Read, Yellow and Green. The web figure obtained for RMSE of TLP exposed to JS equipped with TLCD (μ=10%) has the same shape footprint as the one obtained for BS wave spectra (Figure 3f). However, RMSE is higher for JS than for BS, which proves that the degree of linearity depend on the types of waves as well as on the location of the TLCD.

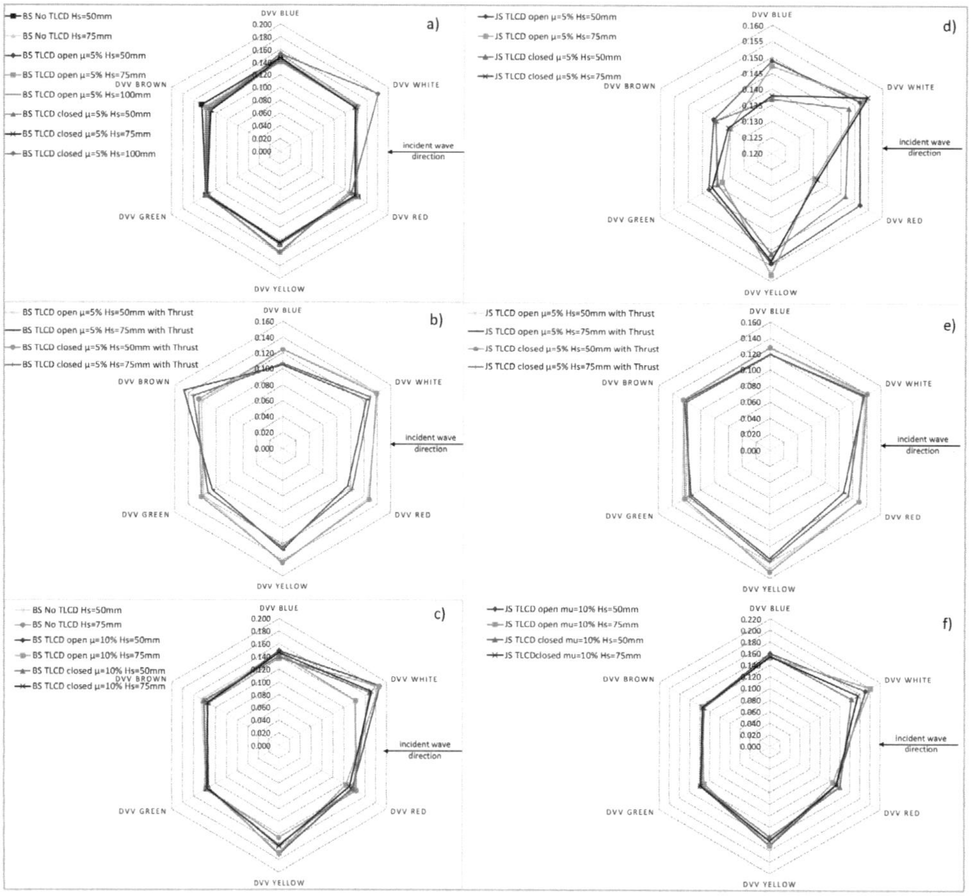

Figure 3: DVV analysis results (RMSE) for waves with Tp=2.4s and fp=0.42Hz a-c) Bretschneider spectra: a) for Hs_m= 50, 75, and 100mm without TLCD and with open and closed TLCD (μ=5%); b) for Hs_m= 50 and 75mm with open and closed TLCD (μ=5%), and with thrust loading; c) for Hs_m= 50, and 75mm without TLCD and with open and closed TLCD (μ=10%), and d-f) JONSWAP spectra: d) for Hs_m= 50 and 75mm with open and closed TLCD (μ=5%); e) for Hs_m= 50 and 75mm with open and closed TLCD (μ=5%), and with thrust loading; f) for Hs_m= 50, and 75mm with open and closed TLCD (μ=10%).

The trend in DVV results of tendon loading when TLP subjected to BS wave spectra of T_p=1.2, 2.4, and 3.6s, corresponding to f_p=0.83, 0.42, and 0.28Hz is shown in Figure 4. In majority observed cases the degree of the nonlinearity of the load cell response is increasing with wave period increase (i.e. wave frequency decrease. The exceptions are the results for open TLCD (μ=10%) in case of Yellow, White, and to less extent Brown load cell and open TLCD (μ=5%) in case of White load cell for lower period / higher frequency waves. This also proves that the location of the TLCD contributes to the change in the system behavior. Further, it shows that the change in the damper–TLP mass ratio and the tuning of TLCD design are contributing the system change as well. This finding is in agreement with findings by Lee et al. [15] and Jaksic et al. [8].

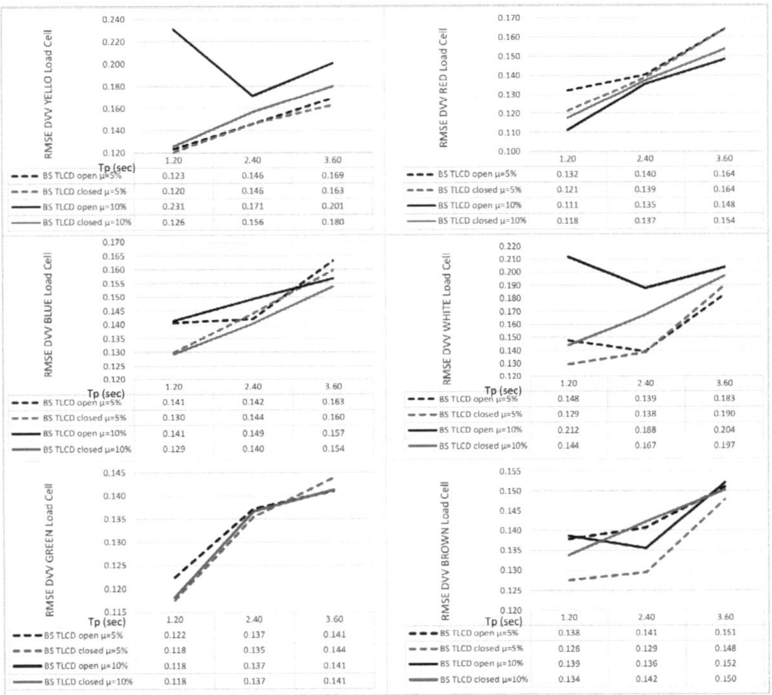

Figure 4: DVV analysis results (RMSE) for Bretschneider spectra waves in function of wave period Tp=1.2, 2.4, and 3.6s and frequency fp=0.83, 0.42, and 0.28Hzfor TLP with open and closed TLCD (μ=5% and 10%).

5. Conclusion

The Delay Vector Variance (DVV) method has been employed to characterize the behavior of the tendons of the scaled floating foundation model in the wave tank. The quantification of the change in degree of nonlinearity of the loadcell measurements as indicators of the nature of the structure and changes within are discussed. The results from different measurement suggests that the variability in estimated nonlinearity is observed to be representative of the different sea state condition and activities due to presence of the TLCD and the relationship of various components are also found. The analysis of the DVV results obtained from the variation in the sea states indicate that the level of the responses depend on the tendon locations in relation to the incident wave direction, but also depends the damper–TLP mass ratio (TLCD design) and its location, i.e. mass eccentricity relative to the TLP center of the gravity. The influential factor for the degree of nonlinearity is also thrust loading.

The DVV method in combination with online structural monitoring can be used for the fast and inexpensive structure assessment. Hence, the finding of the research can be the part of the early warning system for potential malfunctioning or structural failure which can in return minimize the costs related to the operation and maintenance of the offshore (floating) structure. Additionally the evolution of the estimated nonlinearities of different components over time may serve as an output-only marker for the assessment of the structural health.

Acknowledgement. The authors would like to express their gratitude to Mr Florent Thiebaut, Mr Niall O'Sullivan, and Mr Thomas Walsh, for their help with setting up the experiments reported in this paper. The authors acknowledge Marine Research Energy Ireland (MaREI), grant no. 12/RC/2302, a Science Foundation Ireland (SFI) supported project and SFI Advance award 14/ADV/RC3022.

References

[1] Arapogianni A 2007 Deep water - The next step for offshore wind energy. The European Wind Energy Association (EWEA).

[2] IEA. 2011 Offshore foundation technology and knowledge, for shallow, middle and deep water. In *66nd IEA Topical Expert Meeting*. Esbjerg, Denmark, International Energy Agency

[3] Hobohm J, Krampe L, Peter F, Gerken A, Heinrich P and Richter M 2013 Cost Reduction Potentials of Offshore Wind Power in Germany (long version). In *For the The German Offshore Wind Energy Foundation and partners*. Berlin, Germany.

[4] Martínez I, López JL and Auer G. 2013 Report on risk assessment of deployment (D5.1).

[5] Butterfield S, Musial W, Jonkman J, Sclavounos P and Wayman, L 2005 Engineering Challenges for Floating Offshore Wind Turbines. In *Proceedings Copenhagen Offshore*. Copenhagen, Denmark.

[6] Murphy J, O'Shea R, O'Sullivan K and Pakrashi V 2013 Dynamic Responses of a Scaled Tension Leg Platform, Wind Turbine Support Structure in a Wave Tank *Key Engineering Materials* **569-570**, 563. (doi:10.4028/www.scientific.net/KEM.569-570.563).

[7] Waris MB and Ishihara T 2012 Dynamic response analysis of floating offshore wind turbine with different types of heave plates and mooring systems by using a fully nonlinear model. *Coupled Systems Mechanics* **1**, 247-268.

[8] Jaksic V, O'Shea R, Cahill P, Murphy J, Mandic DP and Pakrashi V 2015 Dynamic response signatures of a scaled model platform for floating wind turbines in an ocean wave basin *Phil. Trans. R. Soc. A* **373**(2035) **20140078-**1-18

[9] Jaksic V, Pakrashi V, Basu B and Ryan K Experimental Detection of Sudden Stiffness Change in a Structural System Employing Laser Doppler Vibrometry. In *EACS 5th European Conference on Structural Control* (eds. A.E.D. Grosso & P. Basso). Genoa, Italy.

[10] Godø SN 2013 Dynamic Response of Floating Wind Turbines. Trondheim, Norway, Norwegian University of Science and Technology.

[11] Jonkman JM 2007 Dynamics Modeling and Loads Analysis of an Offshore Floating Wind Turbine. Colorado, USA, National Renewable Energy Laboratory, U.S. Department of Energy Office of Energy Efficiency and Renewable Energy by MRI.

[12] Jaksic V, Wright CS, Murphy J, Afeef C, Ali SF, Mandic DP and Pakrashi V. 2015 Dynamic response mitigation of floating wind turbine platforms using tuned liquid column dampers. *Phil. Trans. R. Soc. A* **373**(2035) **20140079-**1-9. (doi:10.1098/rsta.2014.0079).

[13] Azcona J, Bouchotrouch F, Gonzalez M, Garciandıa J, Munduate X, Kelberlau F and Nygaard, TA 2014 Aerodynamic Thrust Modelling in Wave Tank Tests of Offshore Floating Wind Turbines Using a Ducted Fan. In *The Science of Making Torque from Wind 2014 (TORQUE 2014); Journal of Physics.*

[14] Recommended Procedures and Guidelines: Testing and Extrapolation Methods Loads and Responses, Ocean Engineering Floating Offshore Platform Experiments 2005 In *International Towing Tank Conference (ITTC).*

[15] Lee HH, Wong SH and Lee RS 2006 Response mitigation on the offshore floating platform system with tuned liquid column damper *Ocean Engineering* **33** 1118-1142.

[16] Yalla S Kareem A 2000 Optimum Absorber Parameters for Tuned Liquid Column Dampers. *Journal of Structural Engineering* **126**, 906-915.

[17] Hou S, Li Y and Zhao S 2007 Detecting the Nonlinearity in Time Series from Continuous Dynamic Systems based on Delay Vector Variance Method *International Journal ofv Engineering and Applied Sciences* **3** 258-263.

[18] Hongying H and Fuliang Y 2009 Diesel Engine Fault Information Acquisition Based on Delay Vector Variance Method In *Second International Symposium on Knowledge Acquisition and Modeling (KAM '09)*, 199-202. Wuhan, China, IEEE.

[19] Kuntamalla S and Reddy RGL 2011 The Effect of Aging on Nonlinearity and Stochastic Nature of Heart Rate Variability Signal Computed using Delay Vector Variance Method. *International Journal of Computer Applications* **14** 40-44

[20] Gautama T, Mandic DP and Hulle MMV 2003 Indications of nonlinear structures in brain electrical activity *Physical Review E* **67**, 046204-1-5. (046205).

[21] JianjunY, Haijun W, Chunming X, Yiqin W, Fufeng L, Rui G and Tiancai M 2008 Nonlinear Analysis in TCM Acoustic Diagnosis Using Delay Vector Variance In *The 2nd International Conference on Bioinformatics and Biomedical Engineering (ICBBE 2008)*, pp. 2099-2102. Shanghai, China.

[22] Gautama T, Mandic DP and Van Hulle MV 2003 Signal Nonlinearity in fMRI: A Comparison Between BOLD and MION *IEEE Transactions on Medical Imagining* **22** 636 - 644.

[23] Mandic DP and Chambers JA 2001 *Recurrent Neural Networks for Prediction: Learning Algorithms, Architectures and Stability*, John Wiley & Sons, Ltd.

[24] Schreiber T and Schmitz A 2000 Surrogate time series *Physica D: Nonlinear Phenomena* **142**, 346-382.

[25] Mandic DP, Chen M, Gautama T, Van Hulle MM and Constantinides A 2008 On the characterization of the deterministic/stochastic and linear/nonlinear nature of time series *Proceedings of the Royal Society A: Mathematical, Physical and Engineering Science* **464**, 1141-1160.

[26] Gautama T, Mandic DP and Van Hulle MV 2004 The delay vector variance method for detecting determinism and nonlinearity in time series *Physica D: Nonlinear Phenomena* **190** 167–176.

[27] Mandic DP 2010 Delay Vector Variance MATLAB Toolbox.

Principal Component Analysis for Condition Monitoring of a Network of Bridge Structures

Ciarán Hanley[1], Denis Kelliher[2], & Vikram Pakrashi[1]

[1]Dynamical Systems & Risk Laboratory (DSRL), School of Engineering, University College Cork, Ireland

[2]Research Unit for Structures & Optimisation (RUSO), School of Engineering, University College Cork, Ireland

E-mail: c.m.hanley@umail.ucc.ie, d.kelliher@ucc.ie, v.pakrashi@ucc.ie

Abstract. The use of visual inspections as the primary data gathering tool for modern bridge management systems is widespread, and thus leads to the collection and storage of large amounts of data points. Consequently, there exists an opportunity to use multivariate techniques to analyse large scale data sets as a descriptive and predictive tool. One such technique for analysing large data sets is principal component analysis (PCA), which can reduce the dimensionality of a data set into its most important components, while retaining as much variation as possible. An example is applied to a network of bridges in order to demonstrate the utility of the technique as applied to bridge management systems.

1. Introduction

Bridge maintenance and management systems (BMS) are usually populated by bridge inspectors and the indicators of the condition of different components of the bridge may be used to comment on the condition of the entire network, to a certain extent. Often, the interpretation of the condition of the network, based on summary statistics using the condition ratings become the guiding factor behind possible intervention or investment of individual bridges, including prioritization of investment on one bridge as compared to another. While there exist studies on the choice of appropriate indicators and related modelling techniques, studies over actual networks are still scarce.

This paper presents a principal component analysis (PCA) of the ratings in a regional road network of bridge structures in Ireland. The network consists of ninety-four bridges and the locations overlap urban and rural zones, with varied exposure conditions. The system of ratings follows the same definitions as that of the national road network [1]. This statistical analysis method, first formalised by Pearson [2] and developed by Hotelling [3], identifies the component ratings that are more important than the others in terms of explained variability and reassesses these relative importance ratings of different factors from an engineering point of view, based on additional data. The method is widely used in a variety of different science disciplines [4], but its use in infrastructure management applications has not been explored. This paper attempts to address this issue by investigating the suitability of the approach on a BMS and its limitations in relation to extracting network-wide information.

The study creates a helpful benchmark in the regional road network of Ireland for a larger and more comprehensive study involving the entire national road network. The benchmarking also allows for comparison and contrast with regional road networks of other European Union (EU) countries and countries with an established bridge maintenance management system [5]. The results can be used for the specific network for its evolution with time at a later stage to identify the change of relative importance of different parameters that define the performance of individual bridges based on inspection. The results can also be useful for comparing relative importance of structures, networks of structures or components when different indicators are proposed for such purpose.

2. Bridge Management Systems

Most modern bridge management systems (BMS) are based on the condition assessment of the network through visual inspection as a first step with which more detailed assessment can be planned for [6]. Each individual bridge element is visually assessed and assigned a condition rating based on a prescribed scale, of which is varied across different BMS. A popular method of processing the results of these visual inspections is the Markov chain method; used to predict the future condition states of elements if intervention actions are carried out or not. Such a method is used in the United States (US) systems PONTIS [7] and BRIDGIT [8]. Other times, visual inspections are merely used as a means of identifying areas for further assessment [9]; either through destructive/non-destructive testing or structural assessment, such as deterministic and probabilistic analysis [10].

In the BMS used for this analysis, visual inspection results are recorded for 13 elements in the bridge structure, with an additional result recorded for the condition of the overall structure in general (Table 1). This result is typically a summation of the bridge inspector's impression of the bridge, and is not based on any linear combination of the other 13 results. Thus, this summary observation has been excluded from the input data, and the analysis will be conducted on the 13 elements only.

Table 1. BMS inspection variables

Variable	Description	Variable	Description
1	Bridge surface	8	Piers
2	Expansion joints	9	Bearings
3	Footways/median	10	Deck/slab
4	Parapets/guardrails	11	Beams/girders/transverse beams
5	Embankments/slopes	12	Riverbed
6	Wing walls/retaining walls	13	Other elements
7	Abutments	14	Structure in general

The condition ratings are evaluated and assigned for each variable present or accessible in the bridge structure, and is assigned on a scale of 0-5 in increasing order of level of deterioration/damage. The condition ratings are defined to be a certain state and the inspector assigns these ratings on an observational basis (Table 2).

3. Principal Component Analysis

Principal component analysis (PCA) is a multivariate analysis technique [12], the primary purpose of which is to reduce the dimensionality of a set of data [13]. A background to the theory is presented here for the sake of completeness, and further information on the method can be found in the referenced texts.

The desired outcome of the analysis is to redefine the input variables as principal components (PC), being a linear combination of the original variables, but having a magnitude less than the original data set, but while preserving most of the information. This is accomplished by

Table 2. Condition rating descriptions [11]

Rating	Description
0	No or insignificant damage.
1	Minor damage but no need of repair.
2	Some damage, repair needed when convenient. Component is still functioning as originally designed. Observe the condition development.
3	Significant damage, repair needed very soon. i.e. within next financial year
4	Damage is critical and it is necessary to execute repair works at once, or to carry out a detailed inspection to determine whether any rehabilitation works are required.
5	Ultimate damage. The component has failed or is in danger of total failure, possibly affecting the safety of traffic. It is necessary to implement emergency temporary repair work immediately or rehabilitation work without delay after the introduction of load limitation measures.

highlighting the variables that demonstrate the most variance in the data set. The first principal component Y_1 is defined as:

$$Y_1 = \boldsymbol{\alpha}_1'\mathbf{x} = \alpha_{11}x_1 + \alpha_{12}x_2 + \cdots + \alpha_{1p}x_p = \sum_{j=1}^{p} \alpha_{1j}x_j \tag{1}$$

Where $\boldsymbol{\alpha}_1'\mathbf{x}$ is a linear function of the elements $\mathbf{x}$ having maximum variance, and $\boldsymbol{\alpha}$ is a vector of p coefficients α. The sum of the square of the coefficients α_i is equal to unity, and is a better indicator of the influence the coefficient has than the raw value:

$$\sum_{i=1}^{p} \alpha_i^2 = \alpha'\alpha = 1 \tag{2}$$

The first principal component is the direction along which the data set shows the largest variation [14], and the second component is determined under the constraint of being orthogonal to the first component and to have the largest variance [15]. The second principal component $Y_2 = \boldsymbol{\alpha}_2'\mathbf{x}$ is found in a similar manner to the first principal component, and so on for the subsequent principal components, up to p PCs. It is, however, desired that most of the variance in the data set is accounted for in the PCs $\ll p$. In order to locate the principal components, it is necessary to determine the covariance matrix $\boldsymbol{\Sigma}$ of the vector of random variables $\mathbf{x}$. It can then be shown that $\boldsymbol{\alpha}_k$ is an eigenvector of $\boldsymbol{\Sigma}$ corresponding to its kth largest eigenvalue λ_k [13].

The above can be discussed in matrix terms where a PCA can be conducted through an eigenvalue decomposition (EVD) or a more robust and generalized singular value decomposition (SVD) [16]. For a data matrix $\mathbf{X}$ of n observations on p variables measured about their means:

$$\mathbf{X} = \mathbf{ULA}' \tag{3}$$

Where $\mathbf{L}$ is an $(r \times r)$ diagonal matrix, and $\mathbf{U}$ and $\mathbf{A}$ are $(n \times r)$ and $(p \times r)$ matrices, respectively, with orthonormal columns, and r is the rank of $\mathbf{X}$. It has been observed that SVD approach to PCA is a computationally efficient and generalised method to determining the PCs.

It has been suggested that PCA should only be conducted on continuous variables that conform to a Gaussian distribution [17], and that its application to discrete data, such as condition ratings of a BMS, is inappropriate. However, so long as inferential techniques that require the assumption of multivariate normality are not invoked, there is no necessity for the variables in the data set to have any associated probability distribution [4].

The practical applications of PCA are many, and include [18]:

- the examination of correlations between variables;
- the reduction of the basic dimensions of the variability in the measured set to the smallest number of meaningful dimensions;
- the elimination of variables which contribute relatively little extra information;
- the examination of the grouping of individuals in n-dimensional space;
- determination of the objective weighting of measured variables in the construction of meaningful indices;
- the allocation of individuals to previously demarcated groups;
- the recognition of misidentified individuals;
- orthogonalization of regression calculations.

It is often considered wise to use the correlation matrix for a PCA, as the standardized variates are dimensionless and can be more readily compared [13]. However, when the variables are measured in the same units and have a low variance, using the covariance matrix is sometimes appropriate, and can be beneficial when statistical inference is important. In this case, when the condition ratings are already dimensionless, it is not entirely necessary to standardise the variables. A biplot [19] is most often used to visualise the output of a PCA, as it can handle a matrix of a higher rank than two by approximating it as a matrix of rank two. The biplot displays the orthogonal component coefficients for *each variable* and the principal component scores for *each observation*. An aspect of the biplot is the plotting of variable vectors, the direction and length of each indicate how each variable contributes to the two principal components in the plot. Further discussion of the method can be seen in depth with the additional references [20, 21, 22, 23, 24, 25].

4. Results

A network of 94 bridges was assessed overlapping rural and urban areas. These bridges were predominantly observed to be masonry arch bridges (rural) and reinforced concrete (RC) slab bridges (urban), and so the results are presented for these two bridge types, as the sample size for the remaining bridges in the network was insufficient. As the condition ratings for each of the variables are dimensionless and assessed under the same ratings scale, the PCA was conducted using the correlation matrix for the data.

The distribution of the condition ratings for the individual elements for both bridge types is shown in a box plot (Figure 1). In this plot, the edges of the box represent the 25th and 75th percentiles of the data, with the central marker showing the median. The whiskers extend from the boxes to the extreme values of the data, and outliers are plotted individually. It can be seen that the condition ratings for the masonry arch bridges tend to deviate between 0 and 2, with extreme values reaching a rating of 3 (Figure 1a), whereas the RC bridges are much more bunched around 0-1 with an extreme rating of 2, with the exception of a few outliers (Figure 1b). It can be seen that the widest distribution of condition ratings for a particular element are for the parapets/guardrails (4). This is not unexpected as these are typically the elements that are most likely to suffer vehicular impact, or made from lower grade materials than used for the structural elements.

The number of important components can be visualised with a scree plot of the eigenvalues [26]. This can be a useful tool for determining what components are important and what components can be discarded from the data set. As the slope of the scree plot begins to flatten, the components become less important as they have retain less variance then the previous components. Another method of determining which components to retain is by eliminating any components with an eigenvalue less than the average. For this analysis, it can be seen that there is a progressive decrease in the influence of the higher-order eigenvalues λ, and it has been

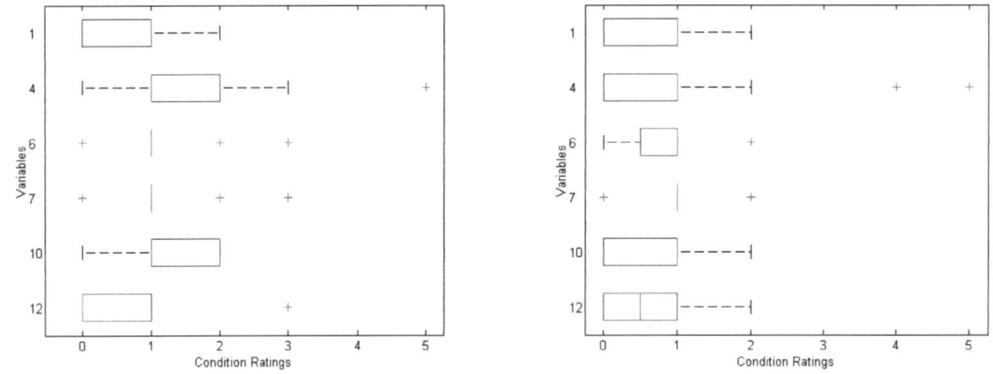

Figure 1. Boxplot of condition ratings for masonry (a) and RC (b) bridges in network.

determined that the first three components retain most of the variance for both the masonry arch and RC bridges; accounting for 75.19% and 70.8% of the data, respectively (Figure 2). The inclusion of a fourth PC would have retained 86.1% and 83.4% of the variance, however its inclusion, in both cases, would've violated the practice of only retaining eigenvalues greater than the average.

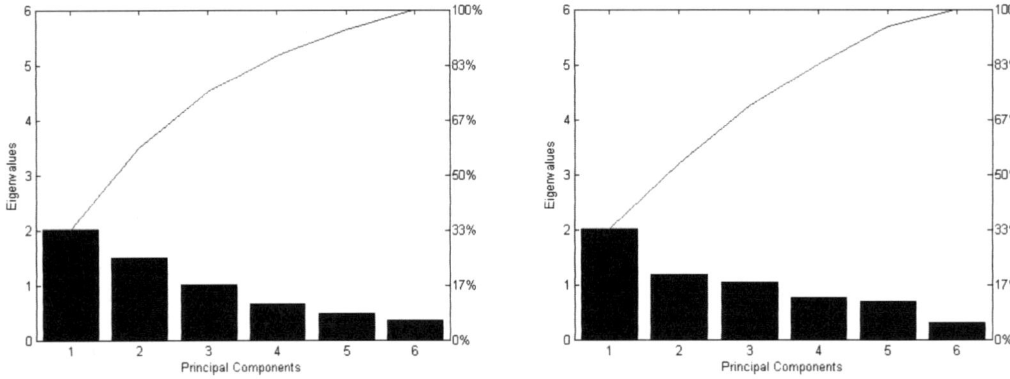

Figure 2. Scree plot of principal components for masonry (a) and RC (b) bridges in network.

The six bridge variables/elements for the masonry and RC bridges are represented by vectors on a bi-plot of the principal component space (Figure 3). The bi-plot displays vectors of the correlation coefficients α (Table 3) between the bridge elements and the PCs, as well as a scatter plot representing the level of correlation between two PCs for each bridge structure. It can be seen that the six relevant bridge variables are all positively correlated for the first PC, with the exception being the road surface (1) for masonry bridges being negatively correlated for this PC (Figure 3). The correlation coefficients with the highest absolute values indicate how much the PC represents certain variables, and the sign describes the relationship between the PC to the actual state. For the masonry bridges in the network, it can be seen that first PC is most positively correlated with the state of the abutments (7), the deck/slab (10), and the parapets/guardrails (4), and negatively correlated with the bridge surface (1) (Figure 3a). It should be noted that because a higher condition rating corresponds to a deteriorated/failed state, the positive PC coefficients represent a negative state, and thus a negative coefficient corresponds to a more favourable state. Thus, it can be said that an individual masonry bridge structure in the network with a high score for the first PC is one that has a bridge surface in a favourable state and whose abutments, deck/slab, and parapets/guardrails are in a deteriorated state. Similarly, it can be seen that the second PC is most highly correlated with positive values for coefficients relating to wing walls/retaining walls (6), the bridge surface, and the riverbed (12). Thus, bridges with a high score for the second PC can be described as having these three variables in a poor state. It should be noted that the bridge surface is described in contradictory terms for the first two PCs, and so it expectantly observed that there is little correlation between the first two PC scores for masonry bridges.

Table 3. Principal component coefficients α and squared coefficients α^2 for network

	Masonry Arch						**RC**					
	PC 1		**PC 2**		**PC 3**		**PC 1**		**PC 2**		**PC 3**	
x	α	α^2	α	α^2	α	α^2	α	α^2	α	α^2	α	α^2
1	-0.350	0.122	0.556	0.309	-0.239	0.057	-0.354	0.126	0.363	0.132	0.504	0.254
4	0.455	0.207	0.173	0.030	-0.477	0.228	-0.584	0.341	-0.359	0.129	-0.254	0.064
6	0.250	0.063	0.561	0.314	-0.311	0.097	0.619	0.383	0.214	0.046	-0.312	0.097
7	0.521	0.272	-0.005	0.000	0.417	0.173	-0.294	0.086	0.673	0.453	-0.126	0.016
10	0.473	0.224	-0.370	0.137	-0.393	0.154	0.238	0.057	0.099	0.010	0.647	0.418
12	0.334	0.112	0.457	0.209	0.539	0.291	0.083	0.007	-0.481	0.231	0.388	0.150
Σ		1.000		1.000		1.000		1.000		1.000		1.000

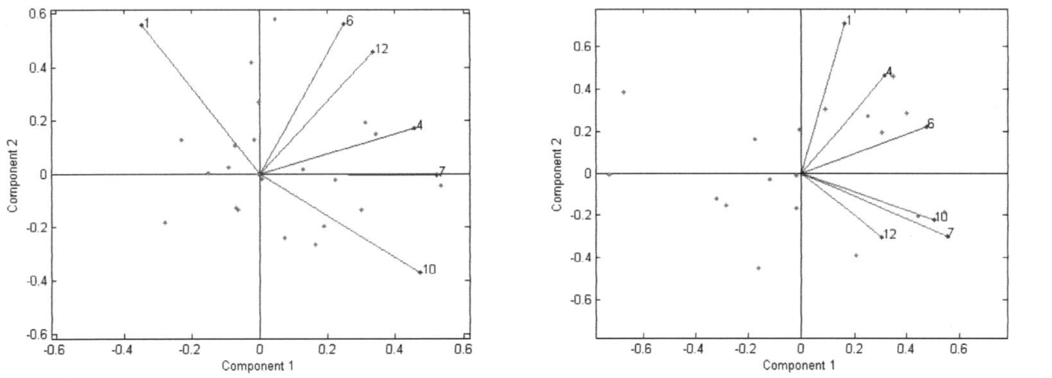

Figure 3. Correlation coefficients for PCs 1 & 2 for masonry (a) and RC (b) bridges in network.

For the RC bridges in the network (Figure 3b), the first PC is most highly correlated with the abutments (7), the deck/slab (10), and the wing walls/retaining walls. As these all have positive correlation coefficients for the first PC, this component can be said to describe how deteriorated these elements are in a RC bridge structure, again with a high PC score describing a more deteriorated state than a lower PC score. The second PC is correlated mostly with the bridge surface (1) and to an extent the parapets/guardrails (4). However, the bridge surface is the most dominant variable in this PC and so the second PC can be described as a measure of the state of deterioration of the bridge surface and, to an extent, the parapets/guardrails. Alternatively, this PC can be described as a measure of the "topside" of the RC bridge structures in the network.

However, as has been noted previously, the third PC also retains a notable amount of variance (Figure 2), and so it cannot be expected that the first two PCs satisfactorily define the data alone. It is possible to display the relationship between the coefficients for three PCs in a three-dimensional bi-plot, however it cannot be optimally interpreted. Thus, additional bi-plots for the second and third PCs have been presented (Figure 4). When compared to the plots for the first two PCs (Figure 3), it can be seen that the new plot confirms the description of the second PC, as the most correlated vectors are demonstrating a consistent relationship with the PC coefficients α. For the masonry bridges in the network, it can be seen that the third PC is highly correlated (positively) to the deteriorating condition of the riverbed (12) and the abutments (7), and to the favourable condition of the parapets/guardrails (7) and the deck/slab (10). Thus, this PC can be said to describe a discrepancy between the conditions of the substructure and the superstructure of the masonry bridges; where a bridge with a high positive PC score describing a bridge with a good superstructure and a poor substructure, and a "high" negative score describing a bridge with a poor superstructure and good substructure. For the RC bridges, it is evident that the third PC correlates mostly to the unfavourable condition of the riverbed (12), and to the favourable condition of the deck/slab (10), and bridges that exhibit corresponding condition ratings will be possess a high score for this PC.

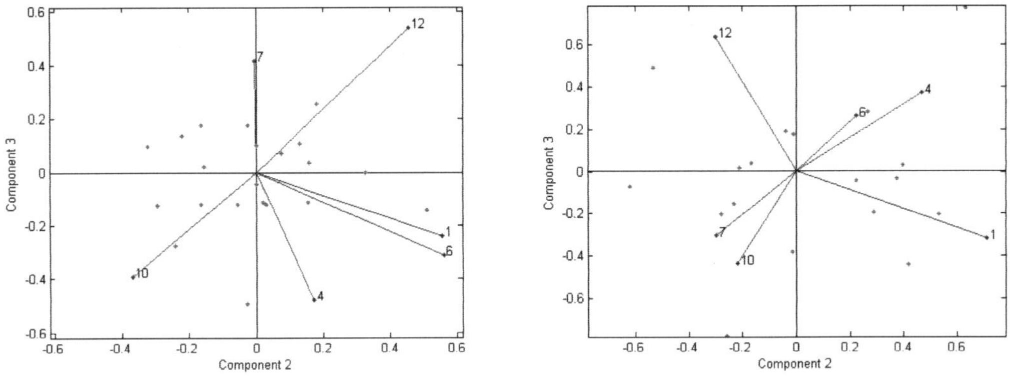

Figure 4. Correlation coefficients for PCs 2 & 3 for masonry (a) and RC (b) bridges in network.

The descriptions of the PCs were compared against the original condition rating data, and the masonry arch and RC bridges with upper and lower bound PC scores are presented below (Table 4). It is evident that there is a clear difference between the elements important to each PC for the upper and lower limits of the PC score. This indicates that the use of the PCA method is a viable tool for the assessment of a BMS based on condition ratings.

5. Conclusion

A principal component analysis was conducted on a BMS for a network of bridges in an urban and rural environment. The principal components obtained allowed for the bunching of bridge types based on their state of disrepair, and this indicates that the use of PCA can be a viable tool in the assessment of large data sets relating to engineering applications. An examination of a larger data-set over a bigger network would afford the opportunity to additionally assess meta-data for the bridges in a PCA; such as age, exposure, time since last inspection/repair, etc. This, coupled with the relative ease with which it can be applied to inspection-based systems, show its potential utility for effective application to an existing BMS.

Table 4. Condition ratings for upper and lower limit PC scores

		Masonry						RC					
PC	**Limits**	**1**	**4**	**6**	**7**	**10**	**12**	**1**	**4**	**6**	**7**	**10**	**12**
1	Upper	**0**	**5**	1	**2**	**2**	1	1	1	**1**	**2**	**2**	0
	Lower	**2**	**0**	0	**0**	**0**	0	0	1	**0**	**0**	**0**	0
2	Upper	**2**	2	**3**	1	1	**1**	**1**	**5**	2	1	0	1
	Lower	**0**	0	**1**	2	2	**0**	**0**	**0**	0	1	1	1
3	Upper	0	**1**	1	**2**	**1**	**3**	1	5	2	1	**0**	**1**
	Lower	1	**5**	1	**1**	**2**	**0**	1	1	1	2	**2**	**0**

Acknowledgements
The authors would like to gratefully acknowledge the financial support from the *John Sisk Postgraduate Research Scholarship in Civil Engineering.*

References
[1] Duffy L 2004 *Proceedings of the ICE - Bridge Engineering* **157** 139–146 ISSN 1478-4637
[2] Pearson K 1901 *Philosophical Magazine Series 6* **2** 559–572 ISSN 1941-5982
[3] Hotelling H 1933 *Journal of Educational Psychology* **24** 417–441
[4] Jackson J E 2003 *A User's Guide to Principal Components* (Wiley)
[5] Žnidarič A, Pakrashi V, O'Brien E J and O'Connor A 2011 A review of road structure data in six European countries
[6] Das P C 1998 *Structural Engineering International: Journal of the International Association for Bridge and Structural Engineering (IABSE)* **8** 299–302 ISSN 10168664
[7] Thompson P D, Small E P, Johnson M and Marshall A R 1998 *Structural Engineering International: Journal of the International Association for Bridge and Structural Engineering (IABSE)* **8** 303–308 ISSN 10168664
[8] Hawk H and Small E P 1998 *Structural Engineering International: Journal of the International Association for Bridge and Structural Engineering (IABSE)* **8** 309–314 ISSN 10168664
[9] Schellenberg K, Vogel T, Chèvre M and Alvarez M 2013 *Structural Engineering International: Journal of the International Association for Bridge and Structural Engineering (IABSE)* **23** 402–410 ISSN 10168664
[10] Pakrashi V, O'Connor A and Salta M 2012 Assessment and Maintenance Planning of Infrastructure Networks *Transportation Research Board (TRB) 91st Annual Meeting* (Washington, D.C.)
[11] NRA 2008 EIRSPAN System Manual No. 3, Principal Inspection National Roads Authority, Ireland
[12] Mardia K, Kent J and Bibby J 1979 *Multivariate Analysis* (Academic Press) ISBN 9780124712522
[13] Jolliffe I 2002 *Prinicpal Component Analysis* 2nd ed (Springer)
[14] Ringnér M 2008 *Nature biotechnology* **26** 303–304 ISSN 1546-1696
[15] Abdi H and Williams L J 2010 *Wiley Interdisciplinary Reviews: Computational Statistics* **2** 433–459 ISSN 19395108
[16] Chambers J 1977 *Computational Methods fo Data Analysis* (New York: Wiley)
[17] Qian G, Qian G, Gabor G, Gabor G, Gupta R and Gupta R 1994 *Journal of Multivariate Analysis* **49** 55–75 ISSN 0047259X
[18] Jeffers J N R 1967 *Applied Statistics* **16** 225–236 ISSN 00359254
[19] Gabriel K R 1971 *Biometrika* **58** 453–467
[20] Gower J C 1966 *Biometrika* **53** 325–338 ISSN 00063444
[21] Anderson T W 1963 *The Annals of Mathematical Statistics* **34** 122–148
[22] Wold S, Esbensen K and Geladi P 1987 *Chemometrics and Intelligent Laboratory Systems* **2** 37–52 ISSN 01697439
[23] Hui Zou T, Hastie R and Tibshirani R 2006 *Journal of Computational and Graphical Statistics* **15** 265–286
[24] Rao C R 1964 *Sankhyā: The Indian Journal of Statistics, Series A* **26** 329–358
[25] Tipping M E and Bishop C M 1999 *Journal of the Royal Statistical Society: Series B (Statistical Methodology)* **61** 611–622 ISSN 1369-7412
[26] Cattell R B 1966 *Multivariate Behavioral Research* **1** 245–276 ISSN 0027-3171

Damage assessment of RC buildings subjected to the different strong motion duration

Alireza Mortezaei[1], Mohsen mohajer Tabrizi[2]

[1] Assistant Professor, Civil Engineering Dep., Engineering Faculty, Semnan Branch, Islamic Azad University, Semnan, Iran
[2] M.Sc. Student, Civil Engineering Dep., Engineering Faculty, Semnan Branch, Islamic Azad University, Semnan, Iran

a.mortezaei@semnaniau.ac.ir

Abstract. An earthquake has three important characteristics; namely, amplitude, frequency content and duration. Amplitude and frequency content have a direct impact but not necessarily the sole cause of structural damage. Regarding the duration, some researchers show a high correlation between strong motion duration and structural damage whereas some others find no relation. This paper focuses on the ground motion durations characterized by Arias Intensity (AI). High duration may increase the damage state of structure for the damage accumulation. This paper investigates the response time histories (acceleration, velocity and displacement) of RC buildings under the different strong motion durations. Generally, eight earthquake records were selected from different soil type, and these records were grouped according to their PGA and frequency ranges. Maximum plastic rotation and drift response was chosen as damage indicator. In general, there was a positive correlation between strong motion duration and damage; however, in some PGA and frequency ranges input motions with shorter durations might cause more damage than the input motions with longer durations. In soft soils, input motions with longer durations caused more damage than the input motions with shorter durations.

1. Introduction

One of the most seismically active regions, that have 5–6% of earthquakes and 17% of the world's largest earthquakes, is the Alpide belt. Iran is one of the most seismically active countries in this belt, being crossed by several major fault lines that cover at least 90% of the country [1]. Accordingly, earthquakes in Iran are inevitable [2]. In order to accurately describe any earthquake, one must be able to describe the three major characteristics of the motion: a) amplitude, b) frequency content and c) duration. These parameters describe characteristics of the recorded ground motion time history, which fully describes the ground motion at a particular site.

The duration of strong ground motion is an important characteristic affecting the response of the structures. It is widely recognized that strong motion duration is one of the challenging characteristics of ground motion, which affects the damage level of structures.

The duration of strong ground motion, rather than the duration of the entire time history, is the main interest of engineers. It is important to recognize the strongest part of an accelerogram and its characteristics, by identifying which part of the ground motion have most of energy and therefore the highest damage potential. For example, as it has been known, for equal accelerations greater duration is generally more damaging and for equal energy shorter duration presents a greater damage.

There are more than 30 different definitions of strong motion duration [3]. While there is no common view regarding which of the definitions of strong motion duration is to be chosen. Although a large number of definitions of strong motion duration have been presented in the literature, the available definitions can be grouped into four different categories: (a) bracketed duration [4, 5]; (b) uniform duration [6]; (c) effective duration [3], and; (d) significant duration [7]. In recent years, Montejo and Kowalsky [8] proposed a procedure for estimation of frequency dependent strong motion duration based on the continuous wavelet transform and the decomposition of the earthquake record. Yaghmaei-Sabegh et al. [9] presented a simple and effective empirical model for predicting the significant duration of ground motions based on recorded earthquake events in Iran.

Bracketed duration is defined as the interval between the first and last acceleration excursions greater than certain absolute amplitude of an accelerogram (Figure 1).

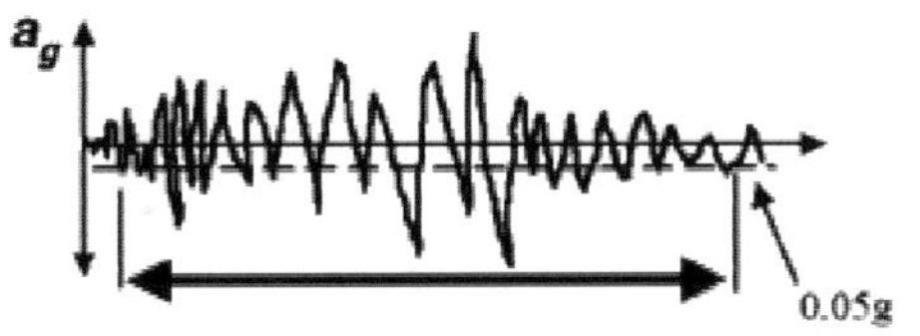

Figure 1. Definition of bracketed duration

Uniform duration is the total sum of time intervals, during which a particular level of acceleration is exceeded.

Trifunac and Brady [7] defined the duration of the strong motion as the time interval in which a significant contribution to the integral of the square of the acceleration, referred to as the accelerogram intensity, takes place. They selected the time interval between the 5% and the 95% contributions as the duration of strong motion. Hence, *significant duration* is based on the accumulation of energy in the accelerogram represented by the integral of the square of the ground acceleration, velocity or displacement (Figure 2).

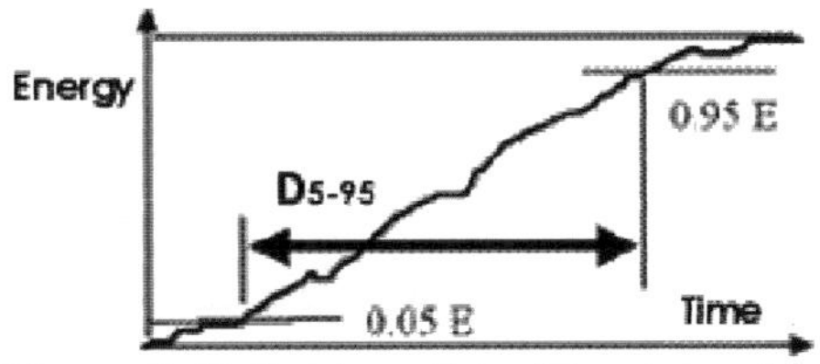

Figure 2. Definition of significant duration

Response based duration is related to the energy input in structures as well as structural frequency and damping. Since there is no standard definition of strong-motion duration, the selection of a procedure for computing it for a study depends on the purpose of the study.

2. Characteristics of selected records

Ground motions often lead to stiffness and strength weakening of structures. For two earthquake ground motions with similar amplitude but of different duration, the motion of longer duration would be anticipated to be more damaging. Hence, the duration of earthquake ground motion should be considered an important parameter in addition to the maximum amplitude and frequency content for adequately characterizing the effect of earthquake ground motion on seismic damage of structures.

In this paper, a total of 8 records were selected to cover a range of frequency content, duration and amplitude. These records come from earthquakes having a magnitude (M_W) range of 6.2 to 7.3, and were recorded at closest fault distance of 0.0 to 7 km. Information pertinent to the ground motion data sets, including station, components of earthquake and peak ground acceleration (PGA) of vertical and horizontal components are presented in Table 1. The comparison of response spectrum of selected records has been shown in Figure 3.

Utilized in this study is a data processing technique proposed by Iwan et al. [10] and refined by Iwan and Chen [11].

Table 1. Near-fault ground motion database

	Earthquake	Year	Station	Distance (km)	M_w	Component	PGA (g)
1	Tabas	1978	9101 Tabas	10	7.4	TAB-TR	0.85
2	Imperial Valley	1979	Bonds Corner	2.68	6.4	0	0.588
3	Morgan Hill	1984	Coyote Lake Dam	0.30	6.2	90	0.711
4	Erzican (Turkey)	1992	Erzincan	4.38	6.8	111	0.496
5	Landers	1992	Lucerne	2.19	7.3	90	0.721
6	Northridge	1994	Rinaldi Rec Stn	6.50	6.7	180	0.472
7	Kobe (Japan)	1995	KJMA	0.96	6.9	360	0.599
8	Chi-Chi	1999	TCU072	7.9	7.6	360	0.37

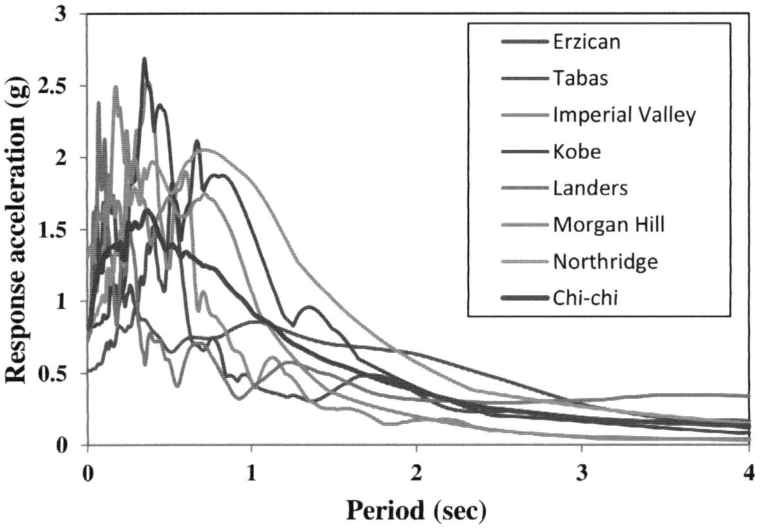

Figure 3. Comparison of response spectrum of selected records

3. Description of Buildings used for Evaluation

Six existing reinforced concrete special moment-resisting frame buildings of 3, 6, 10, 14, 16 and 19 stories were selected as representative case studies to evaluate their seismic demands when subjected to the different ground motions. These buildings were designed in compliance to the Iranian Code of Practice for Seismic Resistant Design of Buildings [12]. The rectangular plan of all buildings measures

30 m × 25 m. The floor plans view of the buildings is shown in Figure 4. The buildings are assumed to be fixed at the base with a damping ratio of 5% in all modes, and the floors as rigid diaphragms with infinite in-plane stiffness.

The modulus of elasticity (Young's modulus) E = 30 kN/mm^2, Poisson's ratio $\nu = 0.20$ and the mass density ρ = 24 kN/m^3 are assumed in all models. The uniaxial strength for nonlinear modeling of the concrete is considered to be 35 MPa. The rebar is modeled as steel with yield strength of 400 MPa and an ultimate strength of 600 MPa.

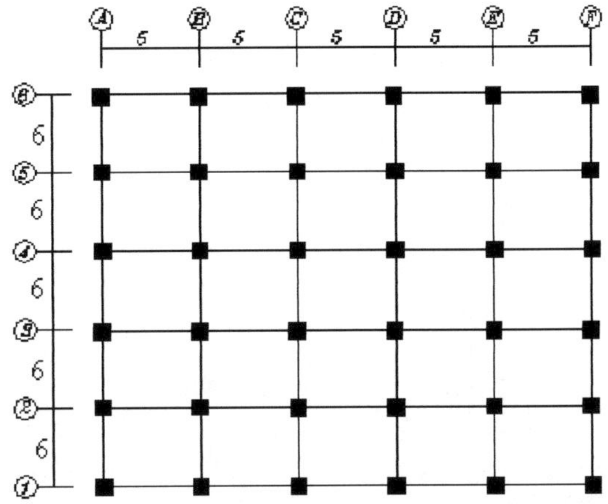

Figure 4. Structural configuration of selected buildings

4. Analytical results

All buildings were analysed under the selected ground motions. The shear strength of a column tended to degrade faster than its flexural strength with cycling of the lateral load. For beam members, flexural cracks were observed. In some cases, flexural cracks occurred on the bottom face of the beams. Shear cracking was observed in the columns and beam-column joints during the earthquakes.

For the sake of clarity, results of the demands on the interior frame of the 10-story building experiencing the largest demand among each ground motion category are presented. For the 10-story building, plastic hinges occurred at the upper and lower levels of the building (Fig. 5). The results show that the failure mechanism is formed of two main damage zones, one at the base and one in the top part of the structures. In almost all the analyses, the cracking is always initiated at the second story which may be due to stress concentration.

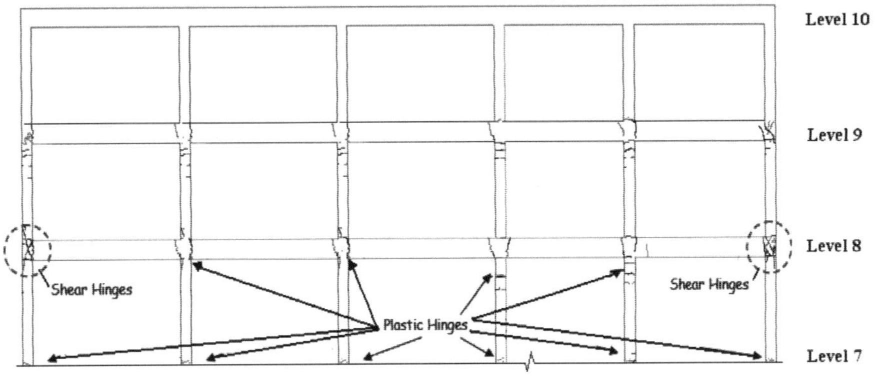

Figure 5. Formation of plastic hinge in upper levels of 10-story RC buildings

For the seismic motions associated with a short duration, the actual response of the low-rise buildings will exhibit some tensile cracking, and some small damage is identified. But it will not severely influence the analytical results of the building. The buildings subjected to the moderate duration, some moderate damage is found. However, the results corresponding to the input motions with a long duration show major strength degradation in the buildings, with a cracking pattern that extends across the upper level.

Figure 6 is the surface plot for Displacement-Duration-PGA. Based on this figure, there is a relative relation between maximum nonlinear displacement (damage) and duration due to large nonlinear displacements for different duration motions. For example, larger nonlinear displacements occur at short durations and low frequencies and larger nonlinear displacements are at short duration and moderate PGA levels. Based on this figure, it can be concluded that maximum nonlinear displacement has a positive relation with strong motion duration.

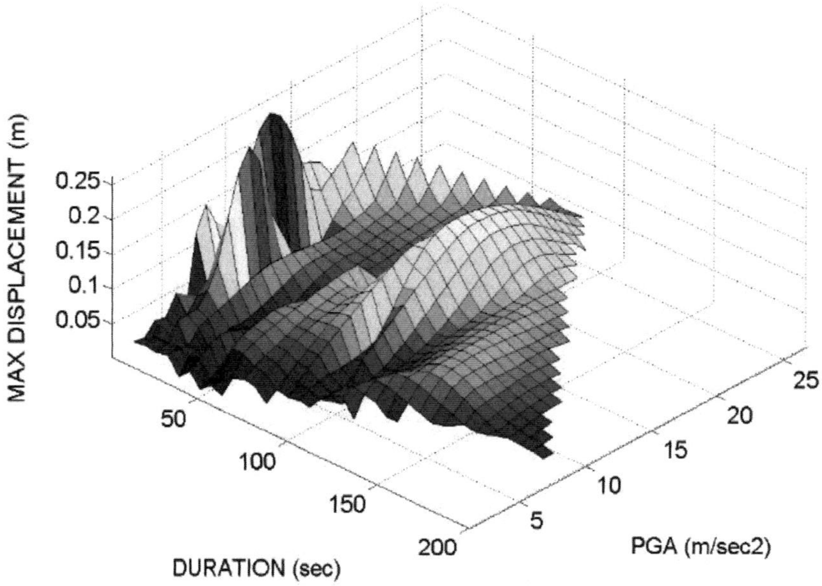

Figure 6. Displacement-Duration-PGA relationship

After the nonlinear dynamic analyses of the building structures for all the selected records, the peak displacement and damage energy dissipation are found, and the local and global damage indices are computed for each seismic excitation. The damage index values for each used record are presented in Table 2. To analyze the effects of strong motion duration on the damage of building structures, damage of the buildings subjected to the 8 records for a given level of intensity in terms of local and global damage indices are determined. Trend lines for each level of intensity are determined with the aim of identifying a general tendency. The results show that S-form trend lines are fitted to the plots to show the general trends for the relationship between strong motion duration and local damage index. The trend lines show that, as would be expected, the local and global damage indices of the building structures during the records with longer duration are generally greater than those under shorter events for the same level of spectral acceleration.

Table 2. Damage index values for each used record

	Earthquake	Local damage index				Global damage index			
		PGA= 0.2g	PGA= 0.25g	PGA= 0.3g	PGA= 0.35g	PGA= 0.2g	PGA= 0.25g	PGA= 0.3g	PGA= 0.35g
1	Tabas	0.20	0.23	0.27	0.30	0.18	0.22	0.24	0.27
2	Imperial Valley	0.17	0.18	0.21	0.24	0.20	0.23	0.27	0.30
3	Morgan Hill	0.15	0.19	0.21	0.25	0.17	0.19	0.21	0.24
4	Erzican (Turkey)	0.17	0.19	0.21	0.24	0.18	0.21	0.24	0.28
5	Landers	0.10	0.14	0.16	0.20	0.13	0.15	0.18	0.22
6	Northridge	0.13	0.15	0.18	0.22	0.17	0.19	0.21	0.24
7	Kobe (Japan)	0.15	0.18	0.21	0.25	0.20	0.23	0.27	0.30
8	Chi-Chi	0.18	0.22	0.24	0.27	0.22	0.26	0.31	0.34

5. Conclusion

Different definitions of strong-motion duration are considered in the study including uniform duration, bracketed duration, significant duration, and effective duration. The structural damage is represented by interstorey drift, column ductility, beam ductility, and base shear. Based on the results from the study, it was found that for building structures, the effect of strong motion duration, which measured by different definitions, depends on both the peak ground acceleration and the using damage indices. Although measure using peak response is mostly used in design and evaluation purposes because of simplicity, the damage measure based on the peak displacement response has a poor relationship with the duration of the ground motion. Strong motion duration has been found to be important to the damage of the building structures. Damage measures such as local damage index, global damage index and damage energy dissipation are consistently greater for ground motions with longer duration. Among the examined definitions of strong motion duration, the significant duration has the better relationship with the damage. The influence of duration depends on a number of different factors including the definition of strong motion duration, the damage measure, the strong motion record and the structural model. These results are the starting point for further exploration, of more seismic records and of other types of building structures including regular or irregular.

6. References

[1] Probabilistic seismic hazard analysis, phase 1: greater Tehran regions. Final report. Vice-presidency for Strategic Planning and Supervision, 2005.

[2] Tavakoli, B., and Ghafory-Ashtiany, M. 1999. Seismic hazard assessment of Iran. Annali Di Geofisica 42, The Global Seismic Hazard Assessment Program (GSHAP) 1992-1999, 1013-1021.

[3] Bommer JJ, Martínez-Pereira A. 1999. The effective duration of earthquake strong motion. J Earthq Eng, 3(2):127–72.

[4] Ambraseys NN, 1967. Sarma SK. The response of earth dams to strong earthquakes. Geotechnique, 17(3):181–213.

[5] Bolt BA. 1973. Duration of strong ground motion. In: Proceedings 5th World conference on earthquake engineering. Rome, p. 1304–13.

[6] Bommer JJ, Stafford PJ, Alarcon JE. 2009. Empirical equations for the prediction of the significant, bracketed, and uniform duration of earthquake ground motion. Bull Seismol Soc Am, 99(6):3217.
[7] Trifunac MD, Brady AG. 1975. A study on the duration of strong earthquake ground motion. Bull Seismol Soc Am, 65(3):581–626.
[8] Montejo LA, Kowalsky MJ. 2008. Estimation of frequency-dependent strong motion duration via wavelets and its influence on nonlinear seismic response. Comput Aided Civil Infrastruct Eng, 23(4):253–64.
[9] Yaghmaei-Sabegh S, Shoghian Z, Neaz Sheikh M. 2014. A new model for the prediction of earthquake ground-motion duration in Iran. Nat Hazards, 70(1):69–92.
[10] Iwan, W. D., Moser, M. A. & Peng, C. Y. 1985. Some observations on strong-motion earthquake measurements using a digital accelerograph. Bulletin of the Seismological Society of America, Vol. 75, pp. 1225–1246.
[11] Iwan, W. D. & Chen, X. D. 1994. Important near-field ground motion data from the Landers earthquake. Proceedings of the 10th European Conference on Earthquake Engineering, Vienna, Austria.
[12] Iranian Code of Practice for Seismic Resistant Design of Buildings. 2005. Standard NO. 2800-05, Building and Housing Research Center.

Analysis about factors affecting the degree of damage of buildings in earthquake

Jing JIA[1] and Jinghong YAN[2]

[1] Room1406 , Building 1 of College of Engineering , Ocean University of China, No.238 Songling Road, Qingdao, 266100 P.R.China

[2] Room A726, Building 25, Tianjin University, No.92 Weijin Road, Tianjin, 300072 P.R.China

E-mail:yjh_flyatdawn@163.com; TEL:86-15122520952

Abstract: Earthquakes have been affecting human's safety through human's history. Previous studies on earthquake, mostly, focused on the performance of buildings or evaluating damages. This paper, however, compares different factors that have influence on the damage of buildings with a case study in Wenchuan earthquake, using multiple linear regression methodology, so as to identify to what extend this factors influence the buildings' damages, then give the rank of importance of these factors. In this process, authors take the type of structure as a dummy variable to compare the degree of damages caused by different types of structure, which were barely studied before. Besides, Factor Analysis Methodology(FA) will be adapted to classify factors, the results of which will simplify later study. The outcome of this study would make a big difference in optimizing the seismic design and improving residential seismic quality.

1. Introduction

There is no need to emphasis damages caused by earthquakes, especially by buildings' falling, which cause almost 95% of casualties. Therefore, it is weighty to study buildings' damages in earthquakes. Scholars have been studying in buildings' damages in earthquakes in terms of destruction mechanism of building or structure. Young-Ji Park and Alfredo H.-S. Ang(1985) proposed a method for evaluating structural damage of reinforced concrete buildings in which the damage can be expressed as a linear combination of the maximum deformation and absorbed hysteretic energy under random earthquake excitation[1]. They also proposed a model in which available monotonic and cyclic test data were analyzed to evaluate the statistics of the appropriate parameters[2]. Peter

Fajfar and Peter Gaspersic(1996) used N2 method to analyze seismic damage of reinforced concrete buildings, which characterized by the use of two separate mathematical models, application of the response spectrum approach and of the nonlinear static analysis, and the choice of a damage model which includes cumulative damage[3]. A.Ghobarah and H. Abou-Elfath(1999) discussed a number of available response-based damage indices and proposed a practical method based on the static pushover analysis to estimate the expected damage to structures[4]. Rob Shepherd(2000) reviewed the acceptability criteria, for vertical flexibility of residential building floor systems and concluded the typical timber frame residence is vulnerable to being loosened up by earthquake shaking[5]. Brendon A.Bradley(2010) deemed that conventionally fragility functions, defining the probability of incurring at least a specified level of damage for a given level of seismic demand, did not take epistemic uncertainties into consideration. Based on the situation, he presented and discussed the sources of epistemic uncertainty in fragility functions, their consideration, combination, and propagation[6].

Studies mentioned above are concerned with damage of particular structure, component and failure mechanism. There are few studies on factors and its' influence degree affecting buildings' damage from perspective of Multivariate Regression Analysis. By extracting information from Report of safety appraisal of Houses in Mianyang earthquake as data resource, this research try to rate factors that affect houses' falling using Multivariate Regression Analysis and classify them into several groups according to results of Factor Analysis. In the end, suggestions are given based on results of regression coefficients and factor analysis, which may provide directional guidance to damage prevention in the future.

2. Data Collection and Processing

After Wenchuan earthquake in SiChuan province of China, faculties in department of civil engineering of Ocean University of China went to stricken areas to conduct houses' safety appraisal and damage evaluation, thus got 231 Safety Appraisal Reports(SAR), including photograph of damage and digital data. 125 of them satisfying this research was selected as original samples.

2.1 Extraction of Independent Variables

First of all, independent variables are extracted from SAR, including types of structure, number of layer, year of construction. Secondly, Anti-seismic Design code was referred and three independent variables was selected: the degree of regularity of house, roof integrity and seismic fortification intensity. Besides, houses' maintenance status can be determined by observing photographs and site grade by address of the house. Finally, eight independent variables affecting damage of houses in earthquake are selected. They are: Structure Type(ST), Number of Layer(NL), Years of Construction(YC), Degree of Regularity of House(DRH), Roof Integrity(RI), Seismic Fortification Intensity(SFI), Maintenance Status(MS) and Site Grade(SG).

2.2 Quantification of data

Original information should be quantified for the use of multiple regression analysis and factor analysis.

First of all, the value of independent variables "Structure Type " is of no order and restricted in three types in this research: brick-concrete structure, bottom-frame structure and frame structure. Namely, there are three value of independent variables ST. In order to analyzing difference of the three structure types, dummy variables D1, D2 are proposed in the study.

$$\text{Set: } D_1 = \begin{cases} 1 & \text{Brick} - \text{concrete structure} \\ 0 & \text{Non} - \text{Brick} - \text{concrete structure} \end{cases}$$

$$D_2 = \begin{cases} 1 & \text{Bottom} - \text{frame structure} \\ 0 & \text{Non} - \text{Bottom} - \text{frame structure} \end{cases}$$

Secondly, for independent variable "Years of Construction", "Seismic Fortification Intensity", "Number of Layer" and 'Site Grade", original data is adopted for convenience and authenticity.

For the value of independent variable "Maintenance Status", "Roof Integrity", "Degree of Regularity of House", quantification is conducted according its original grade. Results of quantification are seen in table 1.

Table 1 Quantification of Variables of Grade

Grade	A	B	C	D
Value	1	2	3	4

Similarly, the independent variable "Degree of Damage of Houses" is quantified according to status of damage, as shown in table 2.(Noted: status of damage is recorded in the report.)

Table 2 Quantification of Degree of Damage of Houses

Status of damage	No damage	Slight damage	Moderate damage	Heavy damage	Collapse
quantification	1	2	3	4	5

3. Multiple Regression Analysis

3.1 Model's Building

Before building the multiple regression model, a series of variables should be set.

Set:

Structure Type(ST)=X1, represented by dummy variable "D1" and "D2".

Years of Construction(YC)=X2

Seismic Fortification Intensity(SFI)=X3

Number of Layer(NL)=X4

Maintenance Status(MS)=X5

Roof Integrity(RI)=X6

Degree of Regularity of House(DRH)=X7

Site Grade(SG)=X8

Degree of Damage of Houses(DDH)=Y

Then, the multiple regression model in this study is built as follows:

Journal of Physics: Conference Series **628** (2015) 012062 doi:10.1088/1742-6596/628/1/012062

$$Y = \beta_0 + \beta_{d1} D_1 + \beta_{d2} D_2 + \beta_3 X_3 + \beta_4 X_4 + \beta_5 X_5 + \beta_6 X_6 + \beta_7 X_7 + \beta_8 X_8$$

3.2 Regression Coefficients and Analysis

Quantified data of independent variables is the value of Xi and Y. After inputting them into SPSS and running, parameters are got as table 3.

Table 3 Regression Coefficient

Model	Non-standard Coefficient		Standard coefficient	t	Sig.
	B	Standard error			
(constant)	45.830	22.009		2.082	.040
D1	.321	.198	.133	1.619	.108
D2	.642	.250	.194	2.564	.012
YC	-.021	.011	-.197	-1.796	.075
SFI	-.120	.156	-.082	-.770	.443
NL	.054	.025	.133	2.139	.035
MS	-.206	.078	-.177	-2.639	.009
RI	-.415	.075	-.328	-5.546	.000
DRH	-.425	.090	-.306	-4.707	.000
SG	.782	.187	.252	4.188	.000

A primary regression model is got as follow:

$$Y = 45.830 + 0.321D_1 + 0.642D_2 - 0.021X_2 - 0.120X_3 + 0.054X_4 - 0.206X_5 - 0.415X_6 - 0.425X_7 + 0.782X_8$$

Firstly, an analysis of type of structure is conducted.

When D1=0, D2=0, the value of Y is the smallest when the the value of other variables are unchanged, which means that frame structure cause to the least damage to house. When D1=1, D2=0, the value of Y is 0.321(0.642-0.321) more than that when D1=0 and D2=1. Therefore, conclusion could be reached that frame structure's capacity of resisting earthquake is the best, followed by brick-concrete structure and bottom-frame structure, which is in consistence with the fact in theory and practice.

For independent variable "YC" and "SFI", we could observe that the earlier the house was constructed and the higher SFI of a house, the slighter the damage of a house was. Similarly, the better the maintenance status was and the higher the RI and DRH was, the lighter the damage was; the higher the SG is (the first is the best), the heavier the damage is; the more the layers are, the heavier the damage is. Those deduction is correspond with common sense, therefore this model is authentic in experience.

3.3 Testing of The Model

First of all, goodness of fit is tested according to R^2. R^2 of this model is 0.655>0.5, suggesting a good fitness. Seen in table 4. In spite of this, the R^2 is far from 0.9 and indicates that some of the independent variables are sightly related to Y, so a T-test is conducted next.

Table 4 Information of The Model

R	R^2	Adjusted R^2	error
.809	.655	.628	.465

The last volume of table 3 shows that when the significance level is 0.05, significance of variable "D1", "YC" and "SFI" is higher than the level. That is, these variables in the model may be collinear. The reality implies the impossibility of colinearity between structure type(D1 and D2) and other variables. The reason why significance of D1 is lower than 0.05 may be the shortage of the data. Collinearity does exist between year of construction and seismic fortification intensity in reality because SFI is 5 before 1993 and 6 after that. Under this condition, this study still chose variable "YC" and "SFI" to explore the impact of year of construction on the buildings' destruction when the SFI is the same.

Next, colinearity is analyzed again by the value of tolerance and VIF(Variance Inflation Factor). Seen in table 5.

Table 5 Statistics of Model's Colinearity

Statistics	D1	D2	YC	SFI	NL	MS	RI	DRH	SG
Tolerance[a]	.442	.521	.250	.267	.780	.664	.856	.709	.826
VIF[b]	2.263	1.918	4.007	3.741	1.282	1.506	1.168	1.410	1.210

a: The value of tolerance is between 0 and 1. The closer to 1 the tolerance is, the less colinear the variable is.

b: The value of VIF is between 1 and 10. The closer to 1 the tolerance is, the less colinear the variable is.

Table5 demonstrates that all variables' VIF is close to 1, except variable "YC" and "SFI". That is, significant colinearity exists between the two variables. This may because the scale of sample is small and most of the value of Y is "2" or "3", being unable to reflect the impact of YC on buildings' destruction, which leads to a significant correlation between YC and SFI.

3.4 Modification of The Model

The result of tests above suggest a strong correlation between YC and SFI. Considering that SFI is a crucial factor affecting damage, variable "YC" is deleted in the modification.

After the modification, a series of tests are conducted the same as before. All the variables' VIF are close to 1. Significance of all variables are less than 0.05, excluding "NL". On the condition that the number of houses' layer is less than 7 in the study, variable "NL" has no direct relation with the damage of house[7]. Therefore, "NL" is deleted in the following factor analysis. And, regression function after deleting "YC" and "NL" is shown below.

$$Y = 6.633 + 0.349D_1 + 0.671D_2 - 0.350X_3 - 0.186X_5 - 0.434X_6 - 0.429X_7 + 0.606X_8$$

Normalized function is shown next.

$$Y = 0.145D_1 + 0.203D_2 - 0.238X_3 - 0.161X_5 - 0.343X_6 - 0.309X_7 + 0.195X_8$$

4. Factor Analysis

Factors should reflect sample information comprehensively, so all variables except "ST" and "NL" are qualified to factor analysis. The reason why "ST" is excluded is that "ST" is a classified variable, which makes no sense to be used to conduct factor analysis. Given the fact that it is an important variable to affect houses' damages, we take it as an individual factor in later analysis. "NL" has no direct relationship with the damage of house according to analysis above, so it is excluded.

4.1 Feasibility Testing

At the beginning, correlation test is conducted and correlation coefficient matrix is shown in table 6.

Table 6 Correlation Coefficient Matrix

	SFI	MS	RI	DRH	SG	YC
SFI	1.000	.413	.274	.264	-.126	.836
MS	.413	1.000	.320	.390	-.184	.418
RI	.274	.320	1.000	.210	-.105	.269
DRH	.264	.390	.210	1.000	-.119	.288
SG	-.126	-.184	-.105	-.119	1.000	-.022
YC	.836	.418	.269	.288	-.022	1.000

The coefficient is lower than 0.5 excluding coefficient between"YC" and "SFI", suggesting a slight correlation of the variables.

Next, test of KMO and Bartlett is conducted, as shown in table 7.

Table 7 KMO and Bartlett Test

KMO	Bartlett sphericity test		
	Approximate Chi-square	df	Sig.
.661	220.026	15	.000

The value of KMO is higher than 0.5 and significance of Bartlett test is 0.000, all indicating a feasibility of factor analysis.

4.2 Identification of Factor

First of all, by observing the sample scree plot (Seen in fig.1, the horizontal ordinate represents the number of factors, the longitudinal ordinate represents the factors' characteristic value. It is a plot reflecting to which degree the factor could explain the variables), the number of factors could be determined. Normally, factors whose characteristic value is higher than 1 will be retained. But according to research of Nyaradzo H.Mvududu and Christopher A.Sink. The point at which the line begins to show a clear bend indicates the actual number of factors that should be retained[8]. Therefore, three factors could be extracted in the study.

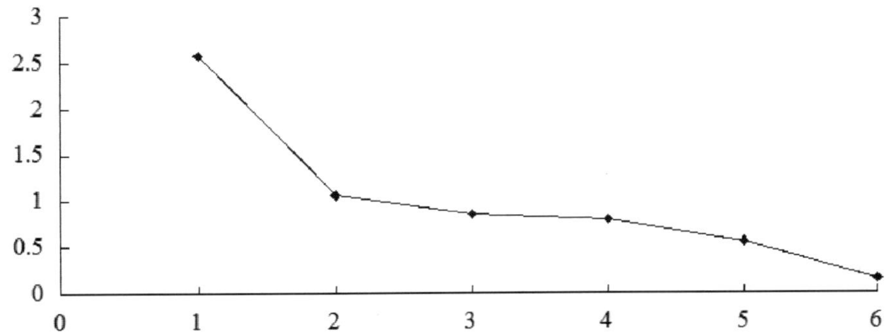

Fig. 1 Sample Scree Plot

Next, total variance explained by the former three factors is 74.819% and higher than 0.7, suggesting a good reflecting of all factors. Seen in table 8.

Table 8 Variance Explained

Component	Extraction Sums of Squared Loadings		
	Eigenvalue	% of Variance	Total Variance Explained: Cumulative %
1	2.571	42.847	42.847
2	1.061	17.688	60.536
3	.857	14.284	**74.819**
4	.796	13.261	88.081
5	.559	9.311	97.391
6	.157	2.609	100.000

In the end, the six variables are classified into three factors according to their factor loading to every factor (That is,component in table 9).

Table 9 Rotated Component Matrix

	Component		
	1	2	3
SFI	**.924**	.234	-.072
YC	**.919**	.260	.055
MS	.059	**.771**	-.012
DRH	.096	**.687**	.024
RI	.330	**.678**	-.163
SG	-.029	-.121	**.986**

From table 9, the first component has a bigger loading on variable "SFI" and "YC", so it could be named as "Time Factor"; the second component has a bigger loading on variable "MS", "DRH", "RI", and it could be named as "House status Factor"; the third component has a bigger loading on variable "SG" and be named using variable name "Site grade Factor".

5. Conclusion

According to the normalized regression function got in the second part in the paper, conclusion could be reached that the rank of degree of variables affecting houses' damage is "Roof Integrity", "Degree of Regularity of House", "Seismic Fortification Intensity", "Site Grade", "Maintenance Status". As for "Structure Type", frame structure causes the least damages to house, followed by brick-concrete structure. Bottom-frame structure causes the biggest damages, so this type of structure should be avoided especially in the town in the future.

In the process of factor analysis, there factors are extracted: "Time Factor", "House status Factor", "Site grade Factor". Cumulative variance explained by the there factors is up to 74.819%, suggesting a comprehensive reflection of variables studied. The outcome of factor analysis could be used in damage prediction and study in the future.

The research has limitations. Restricted by difficulties in data collection and the need for integrity of data, samples in the study are not enough and limited in urban district of Mianyang, so most of the degrees of damage are "slight" or "moderate", which fails to represent all the degrees of damage greatly.

6. Acknowledgments

This research is supported by Safety Appraisal Reports, which are the outcome of teachers' hard-working. We thank their work. We also appreciate the anonymous reviewers for taking the effort and time to review an earlier version of this paper, which will enhance it significantly.

Reference

[1] Young-Ji P. Alfredo H.S 1985 Seismic damage analysis of reinforced concrete buildings *J. Struc. Eng.* **111** pp 740-757

[2] Young-Ji P. Alfredo H.S. 1985 Mechanistic seismic damage model for reinforced concrete *J. Struc. Eng.* **111** pp 722-739

[3] Peter F. Peter G. 1996 The N2 method for the seismic damage analysis of reinforced concrete buildings *Earthquake Eng. and Struc.* **25** pp 31-46

[4] Rob S. 2000 Unacceptable floor flexibility resulting from earthquake damage *Forensic Eng.* pp 200-208

[5] Brendon A. Bradley. 2010 Epistemic uncertainties in component fragility functions *Earthquake Spectra.* **26** pp 41-62

[6] Guoqing L. Qingshan Y. 2009 Detailed investigation and analysis of damage of structures in Dujiangyan city in Wenchuan Earthquake *J. Bldg. Struc.* **39** pp227-231(in Chinese)

[7] Nyaradzo H. Christopher A. 2013 Factor analysis in counseling research and practice *CORE(Counseling Outcome Research. and Evaluation).* **4.2** pp 75-98

Characterising the effect of global and local geometric imperfections on the numerical performance of a brace member

M S Hassan[1], J Goggins[1] and S Salawdeh[2]
[1]College of Engineering & Informatics, National University of Ireland, Galway, Ireland
[2]Department of Civil Engineering, An-Najah National University, Nablus, Palestine

Email: m.hassan3@nuigalway.ie

Abstract. A numerical imperfection study is carried out on a hot rolled tubular brace member under displacement controlled amplitudes. An appropriate range of global and local imperfections is used in the finite element analyses to evaluate the initial-post buckling compressive strength, lateral storey drift, energy dissipation and mid-length lateral deformation of the brace member. The purpose of this study is to assess the impact of the geometrical imperfection on the numerical performance, and to determine an amplitude range that can be used unequivocally for numerical modelling of brace members. It is shown that the amplitude of global imperfections has an effect on the initial response, whereas the amplitude of local imperfections has influence on the resistance capacity of the brace member at higher ductility level. Based on the results, a refined range of amplitude of global and local imperfections is proposed. This range is found to have a good agreement with design standards. In addition, an already established equation to find lateral deformation is compared to results from the analyses and found that the equation with some modification can be used accurately in design. In this paper, a modification factor is proposed in the equation to find the lateral deformation to account for the imperfection amplitude in the numerical analyses of brace members.

1. Introduction

The current knowledge on both the distribution and magnitude of cross-sectional and member imperfection is significantly insufficient for replication of physical configuration into a numerical prototype. Also, the guidelines in the design standards and in the product certificates only cover a conservative range of the upper limit for amplitudes of imperfection to be used in the analysis. One of the reasons of such physical absence is the complexity of the imperfection associated with the rolling and fabrication process which have excluded their definite characterisation [1]. Consequently, numerical techniques have adopted a variety of forms and amplitudes to account for imperfections. However, those forms are entirely based either on numerical studies or on the mathematical formulation rather than physical tests. An alternative solution to the problem is the probabilistic treatment of imperfection proposed by Schafer and Pekož [2].

In advanced modelling of braces, such as physical theory based models, geometrical imperfections categorised longitudinally into two or more elements to form half-sine wave along the member length. Such models were developed by Salawdeh and Goggins [3] and Uriz et al [4]. However, imperfection in a similar manner cannot be modelled in the finite element (FE), as they do not represent an explicit approach to model imperfection. In such models, the behaviour of imperfection is characterised by very small spaced Eigenvectors representing the deformed values of the buckling mode. Alternatively, these deformations could be applied to the nodes directly into the desired form of imperfection to the model. A model of this latter type was developed by Nip et al [5].

Thus, this paper attempts to highlight the way different imperfection types and their interaction affect the compressive response of the hollow member by means of finite element analyses. This is carried out to find a suitable range of amplitudes to be used for the forthcoming research. This study is part of a larger research project concentrating on the inelastic behaviour of the tubular bracing members in concentrically braced frames by Broderick et al. [6]. The FE analyses are carried out on a series of geometrical imperfections incorporated in an axially compressed brace member. These imperfections are categorised into (i) global imperfection consisting of half-sine wave and named as Method 1, (ii) equivalent lateral load-global imperfection as Method 2, and (iii) local imperfection as Method 3.

2. Finite element modelling of SHS brace member

A commercial finite element package ABAQUS [7] is used to construct the model of a brace member subjected to a displacement controlled axial loading. The displacement amplitudes are implemented into a nonlinear static solver that accounts for large geometrical nonlinearity during the analysis based on the Newton-Raphson technique. The symmetry of the cross section has been considered herein, such that only half of the section was modelled.

2.1 Brace properties

The studied brace member is a 40x40x3-1250 mm SHS formed from hot rolled carbon steel of grade S355 J2H, which is classified as Class 1 according to the slenderness criteria set by EC3 Part 1-1 [8]. The normalised global slenderness, $\bar{\lambda}$ is 0.6, which is defined by EC3 Part 1-1 [8] as $\bar{\lambda} = \sqrt{f_y A / N_{cr}}$, where f_y is the yield strength, A is the gross cross sectional area, and N_{cr} is the elastic critical buckling load. This value is less than the upper limit on the global slenderness $\bar{\lambda}$ specified in EC8 [9] of 2.0 for bracing member in the concentrically braced frames. The monotonic tensile mechanical properties of the material are given in Table 1.

2.2 Element and mesh size

The 4 nodes doubly curve shell element with reduced integration, designated S4R, is used to mesh the model. This element deforms explicitly in shear and provides less computational cost compared to the other quadratic and cubic elements. In addition, this element has been used in previous similar studies for modelling of tubular members [5, 10].

Nip et al [5] carried out a detailed mesh study prior to simulation, and found that a variation of strain output in the highly localised strain region was small beyond a mesh size of 32 elements per face. In addition, this size of element possesses an aspect ratio close to unity. Thus, it was found suitable to be used at the mid-length and ends of the models, where the local buckling was likely to occur. The extend of the refined mesh at the regions of the local buckling was set to 1.5 times the larger dimension of the cross section. In other regions, coarser mesh was used to reduce computational cost.

2.3 Boundary conditions

The boundary conditions are applied to both ends of the brace, such that they are restrained in all directions (six degree of freedom at each node) except the loaded end. In addition, restraint in the y-direction was imposed along the length of the member due to the symmetry of the cross-section as shown in Figure 1.

Table 1. Monotonic tensile engineering material properties [5].

Model	Young's Modulus, E (N/mm^2)	Yield Strength, f_y (N/mm^2)	Ultimate tensile strength, f_u (N/mm^2)
40x40x3-CS-HR	219610	478	555

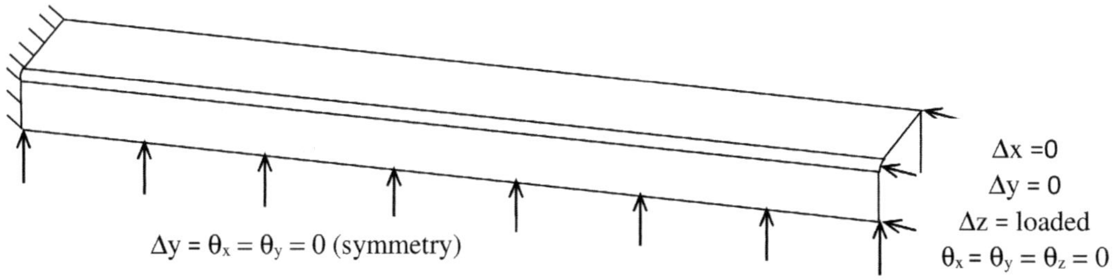

Figure 1. Symmetry condition and end restraints of the brace model

2.4 Geometrical imperfection

A detailed literature review was carried out on the types and amplitudes of imperfection that have been used previously for the prediction of load-displacement behaviour of brace members. For global imperfection, two approaches are widely used in numerical models, (i) a half-sine wave profile is used to derive the initial geometric shape of the brace to capture its out-of-straightness (henceforth, known as Method 1) (see, for example, [3,4,11-13]) and (ii) the application of an equivalent notional lateral load normal to the plane of the buckling at the mid length of the un-deformed member (henceforth, known as Method 2) (see, for example, [5]). In the first approach, the nodal coordinates of the brace represent the unloaded imperfect brace member, using a half sine wave imperfection over its length (Figure 2a). However, in the latter approach, the nodal point loads at the mid-length as shown in Figure 2b, represents an initial imperfection in the brace member.

Long, thin plate buckles locally into half sine waves under uniform axial compression, therefore, it was assumed appropriate to model the brace with local imperfection by using a sine function, which is given as:

$$\omega_2 \cdot \sin\left(\frac{x \cdot \pi}{L}\right) \quad (1)$$

Where, ω_2 represents the amplitudes of local imperfection, x's are the nodal coordinates of the models and L is the overall length of the brace model. The equated imperfections are incorporated into the numerical model by defining the nodal coordinates into their imperfect configuration, which results in a 'flowing fish' pattern of imperfections over the brace length as in the Figure 2c.

In order to characterise the varying effect of global imperfections across a range of amplitudes, a suitable range of global amplitudes, ω_1, from L/100 to L/2000 for initial mid-length out-of-straightness was selected. This range represents a broad scale of amplitudes compared to the tolerance limit of L/500 prescribed in the European (EN 10210-2 2006) [14], North American (ASTM A501 2005) [15] and Australian (AS 1163 1991) [16] standards for hot-rolled sections. Furthermore, the wall thickness, t, tolerance in those standards for hot rolled section is given as $\pm 10\%$ of the section thickness. Thus, a range of t/5 to t/200 was chosen for local amplitudes, ω_2, to account local imperfection in brace models.

3. Discussion of results

3.1 Initial and post buckling compressive strength

The initial buckling loads (F_{cr}) correspond to the types of imperfections with varying amplitudes are presented in Table 2. These values are normalised to the un-factored design strengths calculated according with EC3 [8] using buckling curve *a* and AISC [17] design standards. The results are tabulated in the order of decreasing geometrical imperfection. Similar arrangements are used in the figures throughout this study.

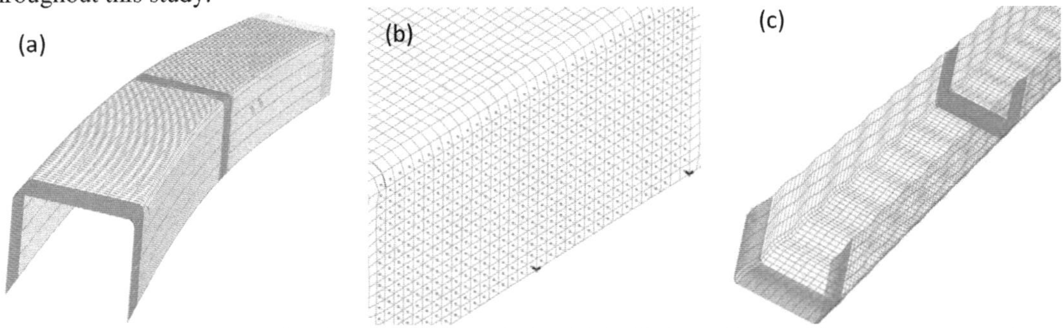

Figure 2. Brace model derived using three methods: (a) Half sinusoidal profile used to derive the initial geometric shape (Method 1), (b) Concentrated nodal point load at mid-length corresponding to the equivalent lateral load imperfection (Method 2), (c) Initial sinusoidal local imperfection profile (flowing fish pattern imperfection) (Method 3).

When the initial global out-of-straightness of the brace member has a magnitude $\omega_l \leq L/500$, the initial buckling loads obtained from the FE models that employ Method 1 to capture initial global imperfections is between 2 and 8% higher than estimates from EC3 [8] and 5-11% higher than predictions from AISC [17]. On the other hand, the estimates from the codes overestimated the initial buckling capacity of between 5 and 15% for models where higher imperfection is used. As expected, the buckling capacity of the member reduces with increase in initial out-of-straightness. From Table 2, it is evident that lower initial buckling capacities are obtained for a given amplitude of the initial out-of-straightness when the imperfections are applied using a notional lateral load at mid-length (i.e. Method 2), rather than creating a geometric model of the unloaded member incorporating its out-of-straightness (i.e. Method 1).

The initial buckling capacity of the braces obtained using Method 2 is affected by both the residual stresses and deformed geometry induced by the equivalent lateral load, whereas no residual stresses are present in Method 1. Consequently, the local deformation induced by the equivalent lateral load applied locally to the brace surface at mid-length has a magnificent effect on overall brace buckling capacity, especially, when $\omega_l \geq L/250$. Based on the FE results, the tolerance of L/500 specified by international standards for the initial out-of-straightness seems appropriate when considering the initial buckling capacity of brace members. However, engineers should be aware of the impact that various methods of applying global imperfection have on the buckling capacity of the brace. In reality, the real physical behaviour of brace lies between Method 1 and Method 2.

In figures 3 and 4, the buckling resistance (F_{cr}) normalised by the yield strength of the section (F_yA) is plotted against normalised axial displacement (δ_c) to yield displacement (δ_y). This facilitates the direct comparison of the response at buckling strength during the entire loading history. Models with initial lower out-of straightness shows a clear interaction of global-local effect by a bell-shaped transition curve in the post-buckling range, which indicates that the local buckling was initiated at higher strain rates. However, such transition effect is suppressed in braces where higher imperfection is present. This is due to the inelastic strain being concentrated over a large critical region, that affect the strain rate in the critical region of the models with higher amplitudes of imperfection. This was proven by monitoring the strain rate at locally buckled region around the mid-length of the brace models.

Table 2. Comparison of the initial buckling load for braces subject to axial compression loading and with various amplitudes of out-of-straightness, ω_1.

Amplitude ω_1^*	Initial imperfection (mm)	Method-1			Method-2		
		F_{cr}	F_{cr} / $F_{EC3\text{-}a^*}$	F_{cr} / F_{AISC}	F_{cr}	F_{cr} / $F_{EC3\text{-}a^*}$	F_{cr} / F_{AISC}
L/100	12.5	160	0.85	0.87	68	0.36	0.37
L/175	7.14	172	0.91	0.94	121	0.64	0.66
L/250	5	180	0.95	0.98	160	0.84	0.87
L/500	2.5	193	1.02	1.05	170	0.90	0.92
L/1000	1.2	200	1.05	1.09	184	0.97	1.00
L/1500	0.833	203	1.07	1.11	186	0.98	1.01
L/2000	0.625	204	1.08	1.11	194	1.02	1.06

Methods of applying global imperfection.

a^* represents buckling curve *a* of EC3 used for calculation of buckling strength
ω_1^* imperfection assumed at the mid-length from the face of an equivalent perfect straight member

Upon local buckling, the local post-buckling capacities of the models with method-1 reduce further regardless of the influence of global amplitudes of imperfection. However, a fluctuation is observed in results of models with method-2, due to the same reason outline in this section previously.

From the above observation, we can say that the predicted response corresponding to the lower amplitudes of imperfection, i.e. when $\omega_1 \leq L/500$, represents the true buckling behaviour of brace under axial compression loading that meet the limits specified by international standards for initial out-of-straightness.

The initial elastic compressive strength, F_e, obtained from the local analyses of the models consisting of sinusoidal wave imperfection is presented in Table 3. The elastic response of the imperfect models under axial compression is similar to that for a member under tensile loading and is independent of the influence of initial local imperfection. In those locally imperfect models, local buckling occurs after the yield strength is exceeded.

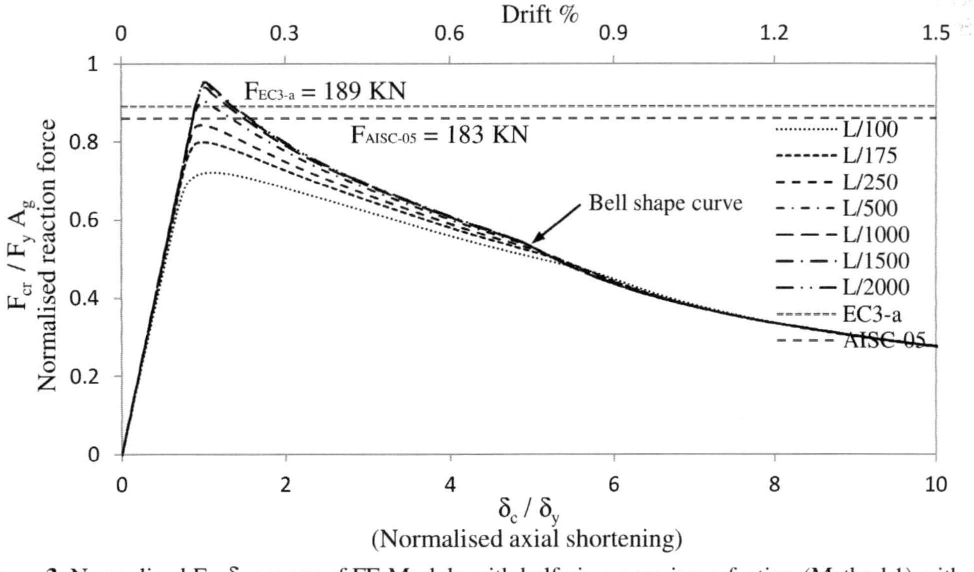

Figure 3. Normalised F_{cr}-δ_c curves of FE Models with half-sine wave imperfection (Method 1) with drift capacity.

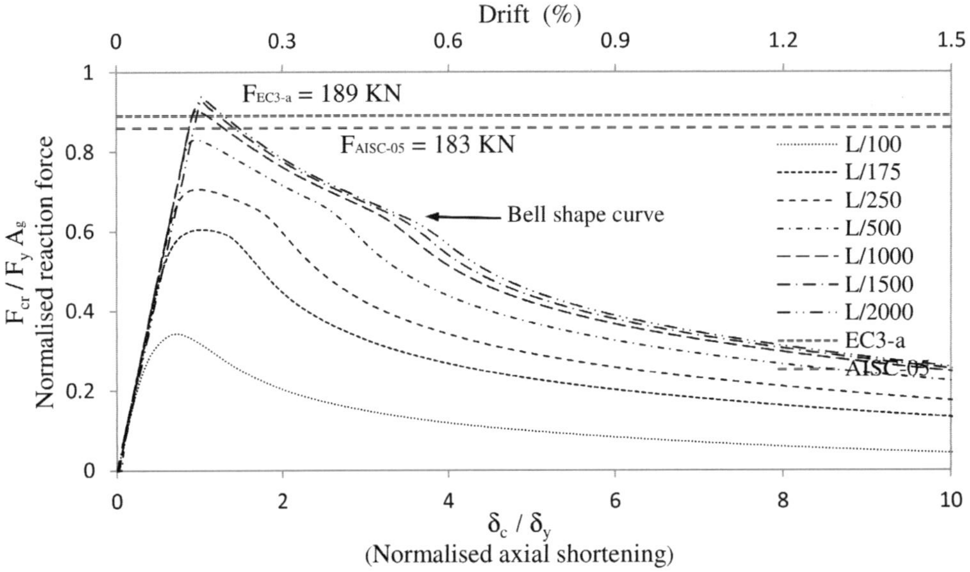

Figure 4. Normalised F_{cr}-δ_c curves of FE Models with sinusoidal wave imperfection (Method 2) with drift capacity.

In figure 5, at ductility level $\mu \geq 3$ the compressive resistance of the brace models with local imperfection $\omega_2 \leq t/100$ is increased relative to a perfectly straight member. However, at the same ductility level the compressive strength of the models is reduced when $\omega_2 \geq t/50$. It may be due to the difference of the numbers of local buckling occurred in the critical regions in the models when, $t/5 > \omega_2 \geq t/50$ (i.e. simultaneous five local buckling) and $\omega_2 \leq t/100$ (seven local buckling occurs) respectively. In general, we can say that amplitudes of initial local imperfection has negligible effect on elastic response, F_e, but significant effect on the resistance capacity at higher ductility level.

3.2 Energy dissipation capacity

The area under the force-displacement curves represent the energy dissipation capacity of the brace members and are presented in tables 4 and 5. These values are calculated at a ductility level of 4 and 10, and indicated as $W_{\mu=4}$ and $W_{\mu=10}$. The obtained values are normalised with the elastic strain energy of the member, which is a function of the cross sectional area and yield strength.

When the dissipated energy corresponding to the braces with higher amplitude of imperfection (i.e. ω=L/100) are compared with those having lower amplitude of imperfection (i.e. ω=L/2000), an average of 35% less energy is dissipated by the braces with a higher amplitude of imperfection at two ductility levels, i.e. at μ= 4 and μ=10, when modelled using Method 1. At the same ductility levels and with same amplitudes of imperfection, the brace modelled using method-2 (i.e. with an equivalent lateral load imperfection), dissipated 90% lesser energy when it contained a high level of imperfection in comparison with the model having lower amplitude of imperfection. The reasons for such a markedly varying dissipated capacity of the models in two methods are the same outlined in the previous section. It should be noted that despite two different imperfection methods, almost same dissipated energy capacity is achieved when lower amplitude of imperfections are employed in the models (Table 4).

The models with higher amplitudes of local imperfection, i.e. ω≥ t/50, show on average 15% reduction in energy dissipation capacity relative to a model with no imperfection at a higher ductility level, i.e. μ=10. However, at the same ductility, an increase of 12% in energy dissipation capacity is observed in models where lower amplitudes of imperfection were used. At lower ductility level, μ=4, the variation in energy dissipation capacity is minimal due to the difference in the numbers of occurrences of local buckling in the models, as discussed earlier.

3.3 Lateral deformation

The results of the FE models in relation to the out-of-plane deformation are plotted against axial displacement in Figure 6. The values are normalised with the brace length, L and yield displacement, δ_y in order to facilitate the comparison between the lateral deformation and the axial shortening. The figure shows curves that lateral deformations of braces with larger amplitudes of global imperfection commences during the elastic loading. However, for braces with smaller amplitudes (i.e. $\omega_1 \leq L/1000$), such excessive deformation was initiated upon the onset of the global buckling. This response agrees with the response of the brace member concluded by Tremblay [18].

The curves are compared to the predictive equations proposed by Tremblay [18] and given in Eqs (2) and (3) for higher and lower ductility levels. These equations are a function of the applied axial displacement, δ_c, cross sectional depth, h, and brace length, L. having a coefficient at the front that depend on the support conditions.

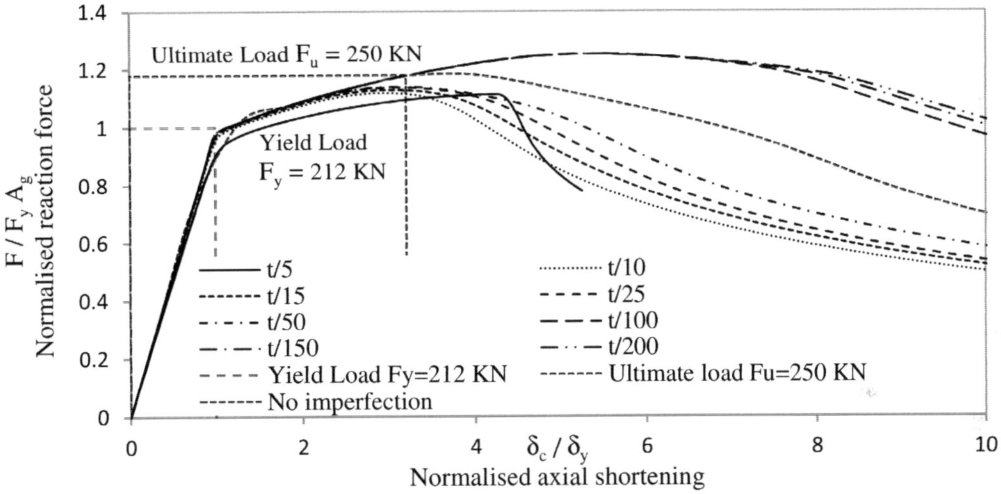

Figure 5. Normalised load-axial displacement response of models incorporated with sinusoidal local imperfection (Method 3).

Table 3. Summary of the numerical results for local analyses of braces with local imperfections over length as sinusoidal waves (flowing fish pattern imperfection).

Amplitude ω_2^*	Initial imperfection (mm)	F_e (KN)	$F_e/F_{EC3\text{-}a}$	F_e/F_{AISC}
t/5	0.6	192	1.01	1.05
t/10	0.3	203	1.07	1.11
t/15	0.2	205	1.08	1.12
t/25	0.12	205	1.08	1.12
t/50	0.06	205	1.08	1.12
t/100	0.03	206	1.09	1.13
t/150	0.02	206	1.09	1.13
t/200	0.015	206	1.09	1.13
No imperfection	0	204	1.08	1.12

ω_2^* imperfection assumed from the face of an equivalent perfect straight member

Table 4. Summary of the dissipation capacity of the brace models with global imperfection.

Amplitude	Method-1		Method-2	
ω_1	$W_{\mu=4}$	$W_{\mu=10}$	$W_{\mu=4}$	$W_{\mu=10}$
L/100	3.3	6.6	0.5	0.8
L/175	3.7	7.7	1.8	3.1
L/250	4.2	8.0	2.6	4.6
L/500	4.6	9.0	3.9	6.8
L/1000	5.1	9.4	4.6	8.3
L/1500	5.3	10.0	4.7	8.8
L/2000	5.5	10.1	5.0	9.0

Table 5. Summary of the dissipation capacity of the brace models with local imperfection.

Amplitude	Method-3	
ω_2	$W_{\mu=4}$	$W_{\mu=10}$
t/5	5.8	N/A
t/10	6.5	15.1
t/15	6.5	15.6
t/25	6.5	16.4
t/50	6.8	17.0
No imperfection	7.3	19.3
t/100	7.4	21.4
t/150	7.4	21.5
t/200	7.4	21.6

Three different responses of the lateral deformation can be observed. Firstly, the curves running close to the higher predictive equation are those with a larger global imperfection corresponding to Method 2. Secondly, the curves running furthest away from the same equation. Thirdly, the curves running between the boundaries of the two predictive equations, and are similar to each other in shape.

$$\Delta = 0.7 \cdot \sqrt{\delta_c \cdot (L - 2h)} \qquad \text{For higher ductility level} \qquad (2)$$

$$\Delta = 0.16 \cdot \sqrt{\delta_c \cdot L} \qquad \text{For lower ductility level} \qquad (3)$$

The predictive equation for higher ductility level tends to give conservative results. The same conclusion was reported by Nip et al. [5]. This can be expected because it accounts for the lateral deformation of buckled slender braces in the post buckling region, which show larger lateral deformations at larger axial displacements compared to less slender braces.

Based on the FE results and the predictive equation, a dimension-less factor is introduced to the equation proposed by Tremblay [18]. This factor consists of a Γ factor that depends on the adjacent ratio of the imperfection amplitude, ω_1, and the brace length, L. The imperfection amplitudes ω_1 can take global amplitudes, ω_1, up to a maximum and minimum of L/5 to L/2000 respectively.

$$\Delta = \Gamma \cdot \frac{\omega_1}{L} \cdot 0.7\sqrt{\delta_c \cdot (L - 2h)} \qquad (4)$$

The values for Γ can be obtained using Equation (5). This equation is derived from a curve fitting analysis, with a correlation coefficient (R^2) = 1. The ratio of imperfection amplitude, ω_1, and the brace length, L, has been replaced with variable α for equation convenience, such that:

$$\Gamma = \frac{0.97}{\alpha^{0.96}} \qquad (5)$$

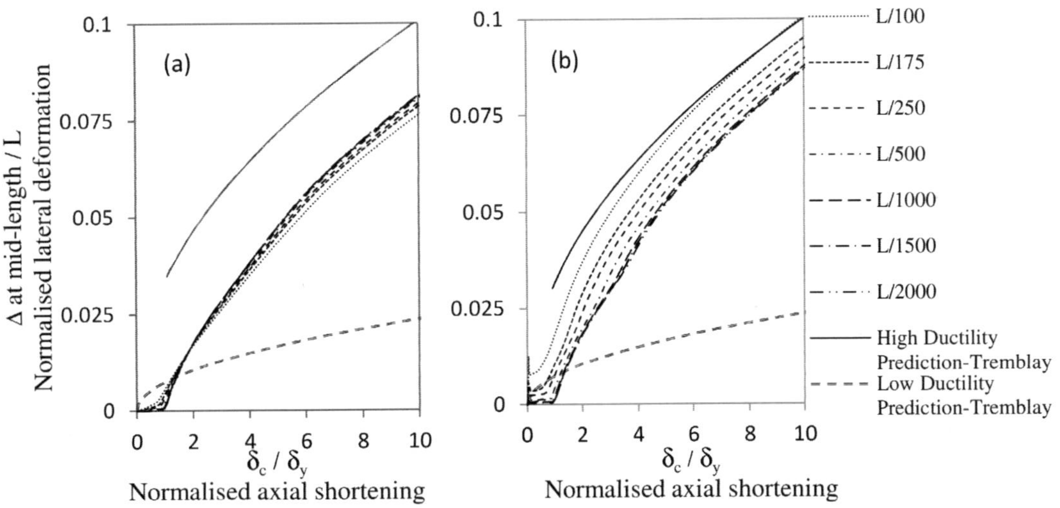

Figure 6. Normalised lateral deformation at mid-length corresponds to (a) half-sine wave imperfection and (b) equivalent load imperfection.

The proposed factor significantly accounts the variation of the lateral deformation across the imperfection amplitudes for short brace member. At higher amplitudes it will produce the magnitude close to the unity and the equated results will be approximately in line with the results of the Tremblay Equation (2). However, the factor will reduce the magnitude of the lateral deformation when lower imperfection amplitudes are used in the equation.

4. Recommendation for geometrical imperfection for less-slender hot rolled sections

The summary of the imperfection study with the view of finding a suitable range of imperfection amplitudes for the numerical modelling of hot-rolled tubular brace member is:

- When using global half sine wave imperfection, a minimum amplitude of L/2000 is used but not greater than L/500, which is the maximum tolerance specified by the international standards.
- For notional lateral load imperfection, amplitudes L/1000 to L/2000 are recommended to be used at the mid length of the critical region, provided that the computational cost is of significance interest.
- An amplitude range of t/10 to t/50 is recommended for local imperfection in order to avoid the enhancement of the forcible resistance at higher ductility levels.
- If physically measured imperfection is available these should be utilised in the numerical model to more accurately capture the actual performance of brace member.

Acknowledgment

This paper has emanated from the research conducted through the financial support scheme of the National University of Ireland, Galway under the COEI fellowship program. The second author would like to acknowledge the support of Science Foundation Ireland through the Career Development Award programme (Grant No. 13/CDA/2200).

References

[1] Gardner L and Nethercot D A 2004 Numerical modelling of stainless steel structural components —A consistent approach *J. Struct Div ASCE.* **130** 1586-01.

[2] Schafer B W and Pekož T 1998 Computational modelling of cold formed steel: characterizing geometrical imperfection and residual stresses *J. Construct Steel Res.* **47** 193-10.

[3] Salawdeh S and Goggins J 2013 Numerical simulation for steel braces incorporating a fatigue model *Eng Struct.* **46** 332-49.
[4] Uriz P, Filippou F C and Mahin S A 2008 Model for cyclic inelastic buckling of steel braces *J. Struct Div ASCE.* **34** 619-28.
[5] Nip K H, Gardner L and Elghazouli A Y 2010 Cyclic testing and numerical modelling of carbon steel and stainless steel tubular bracing members *Eng Struct.* **32** 424-41.
[6] Broderick BM, Hunt A, Mongabure P, Goggins JM, Salawdeh S, O'Reilly G, Beg D, Moze P, Sinur F and Elghazouli AY 2015 'Assessment of the seismic response of concentrically-braced steel frames' In: Taucer, Fabio, Apostolska, Roberta (Eds). *Experimental Research in Earthquake Engineering, Geotechnical, Geological and Earthquake Engineering,* Vol. 35. London.
[7] ABAQUS 2013 User manual, version 6.13. Hibbitt, Karlsson and Sorensen Inc.
[8] CEN 2005 Eurocode 3: Design of steel structures - Part 1-1: General rules and rules for buildings, EN 1993-1-1:2005/AC: 2009.
[9] EN 1998-1:2004 Design of structures for earthquake resistance – Part 1: General rules, seismic actions and rules for building. European Standard EN 1998-1:2004.
[10] Haddad M 2015 Concentric tubular steel braces subjected to seismic loading: Finite element modelling *J.Construct Steel Res.* **104** 155-66.
[11] Mamghani IHP, Usami T and Mizuno E 1996 Inelastic large deflection analysis of structural steel members under cyclic loading *Eng Struct.* **89** 659-68.
[12] Wijesundra KK 2009 Design of Concentrically braced steel frames with RHS shape braces PhD thesis. Pavia: European Centre for Training and Research in Earthquake Engineering (EUCENTRE).
[13] Fell B V 2008 Large-Scale Testing and Simulation of Earthquake-Induced Ultra Low Cycle fatigue in Bracing Members Subjected to Cyclic Inelastic Buckling PhD thesis. Davis, UC.
[14] EN 10210-1, CEN; 2006 Hot finished structural hollow section of non-alloy and fine grain steels-Pat 1: Technical delivery conditions, European Standard.
[15] ASTM A501 2005 standard specification for hot-formed, welded and seamless carbon steel structural tubing in round and shapes.
[16] AS 1163 1991 Australian Standard Structural steel hollow sections.
[17] ANSI/AISC 360-05 An American National Standard Specification for Structural Steel Building.
[18] Tremblay R 2002 Inelastic seismic response of steel bracing members *J. Construct Steel Res.* **58** 665-01.

Footprint caused by a vehicle configuration on the dynamic amplification of the bridge response

Omar Mohammed and Arturo González
School of Civil, Structural and Environmental Engineering; Newstead; University College Dublin; Belfield, Dublin 4; Ireland

E-mail: arturo.gonzalez@ucd.ie

Abstract: The passage of a vehicle over a bridge leaves a unique footprint in the form of measured strains (or displacements) across the structure. This paper proposes a new level I damage detection method for short-span bridges using footprints of Dynamic Amplification Factor (DAF) versus vehicle speed. The total response of a bridge to a moving load is time-varying, and it can be assumed to be made of two components: 'static' and 'dynamic'. Here, DAF is defined as the ratio of the maximum total response to the maximum 'static' component. For a given bridge, DAF patterns will vary with vehicle configuration. However, for a vehicle configuration (or a number of them), the mean DAF pattern measured on the bridge will remain unaltered unless the conditions of the bridge changed. The latter is the subject of investigation in this paper. In order to test the feasibility of using these patterns for monitoring purposes, damage is simulated within a bridge model as stiffness losses of 10% and 30 % at mid-span. Changes in stiffness are identified by differences between DAF patterns corresponding to the healthy and damaged bridges. Results show to be more sensitive to damage than a traditional level I damage detection method based on variation of natural frequencies.

1. Introduction

Bridge damage detection techniques have gained increasing consideration from the engineering communities to prevent sudden structural failure leading to catastrophic, economic, and human life losses. Local damage detection techniques, such as ultrasonic methods and X-ray methods, entail that there is damage and that the vicinity of damage is known a priori and readily accessible for testing. Therefore, vibration-based damage detection methods are typically used first to establish if there has been damage (level I technique), the location (level II), its quantification (level III) and finally the remaining life of the structure (level IV) [1]. The essential idea behind vibration-based damage identification is that damage induces changes in the physical properties (mass, damping, and stiffness) that will be noticeable in modal properties (i.e., natural frequencies, damping and mode shapes). Clearly, a drop in stiffness in a narrow localised portion of the bridge will go unnoticed to the 'static' component of a strain measurement (i.e., in a static loading test), unless the transducer would be located at the damaged location [2]. However, the total strain due to a moving load will be affected by a localised loss in stiffness, as a result of the changes in the 'dynamic' component which is related to the modal properties of the bridge.

As a vehicle traverses a bridge, strain at a given section varies with the vehicle position. If the same vehicle crossed the bridge along the same transverse path with the same speed a second time, the bridge strain would not vary except for a possible alteration in the initial conditions. If the vehicle

Journal of Physics: Conference Series **628** (2015) 012064 doi:10.1088/1742-6596/628/1/012064

'static' configuration (vehicle 'static' configuration refers here to number of axles, their weights and spacing) or speed changes, then the response of the bridge will also change. Similarly, the measured strain due to a given vehicle 'static' configuration will change with an alteration of the structural bridge parameters (boundary conditions, stiffness distribution). Although 'dynamic' mechanical properties of the vehicle (i.e., stiffness and damping of tires and suspension, rotationary moments of inertia associated to the sprung masses, etc…) will also affect the bridge response, these are not considered here, as it is impracticable to assume that these properties could be identified on the field. However, 'static' mechanical properties and speed of a vehicle can be readily captured via a Weigh-In-Motion pavement-based system near the bridge or a Bridge Weigh-In-Motion installed on the bridge. Therefore, it is assumed that the impact of a vehicle fleet with a given 'static' configuration on a bridge can be characterized by a mean response, except for deviations due to vehicular 'dynamic' parameters. The road profile will also have a significant impact on the vehicle forces. While rough profiles will typically lead to a high variability in the bridge response, smooth profiles will produce more consistent (less variable) bridge responses. In fact, the response of a bridge to a sprung vehicle model running on a smooth (flat) profile is practically identical to that of a simpler model made of constant forces equal to the static axle weights of the sprung model. Therefore, this paper analyses the Dynamic Amplification Factor (DAF) patterns resulting from different vehicle 'static' configurations and how the "vehicle configuration – DAF" footprint on the bridge response can be used to monitor not only the passing traffic, but also the structural health of the bridge.

The impact of vehicle speed and 'static' configuration, namely number of axles, axle weights and their spacings, on the bridge response is now well understood in terms of their incremental static load effect as shown by the literature. For example, vehicle speed is considered a factor of supreme importance affecting bridge dynamics [3-7]. General speaking, the bridge dynamic response tends to: (i) increase with an increase in vehicle speed; (2) decrease as the vehicle weight increases [8- 11] although [12] and [13] find that both static and dynamic components increase as the mass of the vehicle increases. The influence of the number of axles on the bridge response to a moving vehicle are specified by [14] (OMT 1991) and [15] (CSA 2006), however, their conclusions are related to particular scenarios and need to be interpreted with caution.

In order to test the use of changes in DAF patterns as a monitoring tool, a bridge is modelled as simply supported finite element discretized beam and vehicles are modelled as series of moving constant loads. This relatively simple analysis ignores interaction between vehicle and bridge, but it is used to capture and understand the underlying pattern (i.e., a contour plot of measured load effect versus time and section location) associated to different 'static' vehicle configurations that could be hindered by more complex models. This 'healthy' pattern is characterized by matching effects between the natural period of the bridge, the vehicle speed and configuration. Then, DAF is employed to detect changes in structural stiffness. In this process, the identification of the correct vehicle 'static' configuration, i.e., via a Weigh-In-Motion installed on the site, is key to the successful application of this method.

2. Simulation models

2.1 Bridge model

The bridge model consists of 150 discretised 1D beam elements (each element being 0.1 m long with two degrees of freedom at each end node). The beam section is assumed to have an inertia of 0.5273 m^4 and a mass per unit length of 28125 m^{-1} kg. The beam material has a modulus of elasticity of 35 GPa. The bridge in its healthy form is assumed to have constant properties throughout the bridge length. The values adopted for these properties are typical of a simply supported short-span solid slab deck, and they lead to main natural frequencies of 5.65 and 22.62 Hz. Damping ratio is assumed to be 0.03 for all modes.

2.2 Vehicle model

A model consisting of a series of constant point loads is used to represent a single vehicle crossing a simply supported bridge at uniform speed. This simulation model is obviously different from the reality in many aspects, as it does not take into account vehicle dynamics, road roughness and the interaction between vehicle and bridge. However, it does provide some insight into the underlying patterns in more complex simulation models. Three types of vehicle configurations based on the Eurocode 1 part 2 [16] are investigated (Table 1).

Table 1. EN 1991-2:2003+NA: 2010

	Axle spacing (m)	Frequent axle loads (kN)
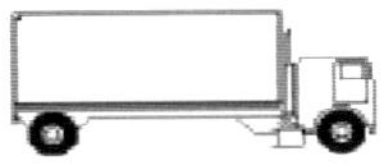	4.5	90 190
	4.2 1.3	80 140 140
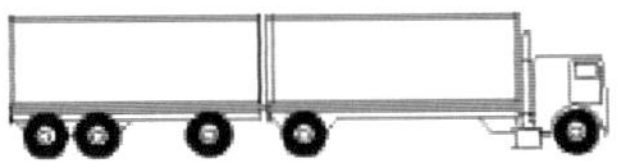	4.8 3.6 4.4 1.3	90 180 120 110 110

2.3 Calculation of the Bridge Response

The response of the discretized beam model to a series of moving time varying forces is given by the system of equations in Equation (1).

$$M\ddot{y} + C\dot{y} + Ky = N f \tag{1}$$

where M, C and K are the mass, damping and stiffness matrices respectively of the beam model and $\ddot{y}$, $\dot{y}$ and y are vectors of nodal bridge acceleration, velocity and displacement respectively. f is the time varying external forces vector. N is a $(n \times nf)$ matrix for distributing nf forces to the n nodal degrees of freedom of the finite element beam. Rayleigh damping is used here to model the damping and is given by:

$$C = \alpha M + \beta K \tag{2}$$

where α and β are constants. These constants are obtained from $\alpha = 2\zeta\omega_1\omega_2/(\omega_1 + \omega_2)$ and $\beta = 2\zeta/(\omega_1 + \omega_2)$, where ω_1 and ω_2 are the first two natural frequencies of the bridge in rad/s. The damping ratio ζ is assumed to be 0.03, the same for all modes [17].

Equation (1) is implemented in MATLAB and solved using the Wilson-Theta integration scheme [18] with a time interval of 0.002 s. The optimal value of the parameter θ=1.420815 is used for unconditional stability in the integration scheme. Initial conditions are assumed to be zero for all displacements, velocities and accelerations.

3. Definition and Testing of the Proposed Level 1 Damage Detection Algorithm

DAF has been used to assess the dynamic bridge response due to the crossing of vehicles by many researchers [19-22]. In this paper, DAF will be applied to damage detection. For this purpose, nine sensors are assumed to be uniformly spaced along the beam described in Section 2 and DAF is calculated for each sensor location using simulations of the response to a passing vehicle. The first and last sensors are located at 1.5 m and 14.5 m respectively from the first support and sensors in-between are spaced by 1.5 m each. Here, DAF refers to the ratio of maximum total strain ('static' + 'dynamic') to maximum 'static' strain for a given sensor location. Strain is calculated at the bottom of these nine locations based on the assumption that the neutral axis is located at 0.375 m of the measurement point. The damage scenario is represented as a localised reduction in stiffness at 10 elements at and around the mid-span section (covering a full length of 1 m). Two different damage levels are simulated, i.e., stiffness reductions of 10% and 30%. The first natural frequencies of the bridge for each of the three states and the frequency ratios of the damaged states to the healthy one are shown in Table 2 for reference purposes. It can be noticed that the difference in first natural frequency amongst the different states is relatively small.

Table 2. Main frequencies of the three bridge states under investigation

Bridge type	First natural frequency (Hz)	Frequency ratio of damaged to healthy
healthy	5.65	1
10 % stiffness reduction at mid-span	5.61	0.99
30 % stiffness reduction at mid-span	5.48	0.97

The damage detection algorithm relies on a prior knowledge of the DAF pattern for the 'healthy' structure (or structure at a time T), and by comparison to this DAF at time T, it establishes if the structural conditions at time "$T + \Delta T$" have changed. Hence, the first step of the algorithm is the calculation of DAF for the 'healthy' bridge due to the three vehicle configurations under investigation.

3.1 DAF pattern at a point in time T

For the calculations in this section, the bridge at time T is assumed to be in a 'healthy' state (i.e., as defined in Section 2.1). Figure 1 shows a 3D plot of DAF versus velocity (km/h) due to the crossing of the 2-axle truck defined in Table 1. Velocity increments of 1.08 km/h are employed in the generation of this graph.

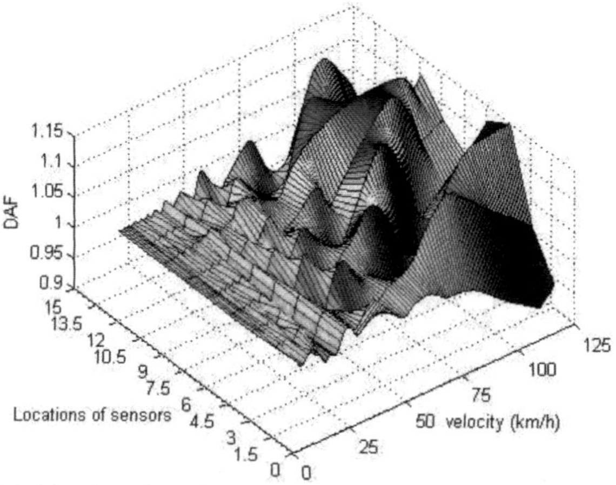

Figure 1. DAF at 9 bridge locations due to a 2-axle truck versus velocity in a healthy bridge

Then, the plot is divided into slices to work out a damage index. For example, Figures 2(a) and (b) show slices at two consecutive velocities, 42.12 km/h and 43.2 km/h. The area under each slice is calculated using the trapezoidal rule. These areas are used to calculate the volume of each strip between two slices using Equation (3).

$$Vol_{strip(i)} = \frac{A_{\mathrm{slice(j)}}+A_{\mathrm{slice(j+1)}}}{2} \times 1.08 \quad (3)$$

where, i and j are the number of the strips and slices respectively. The total volume of the plot is calculated by the sum of the volumes of all individual strips between consecutive slices.

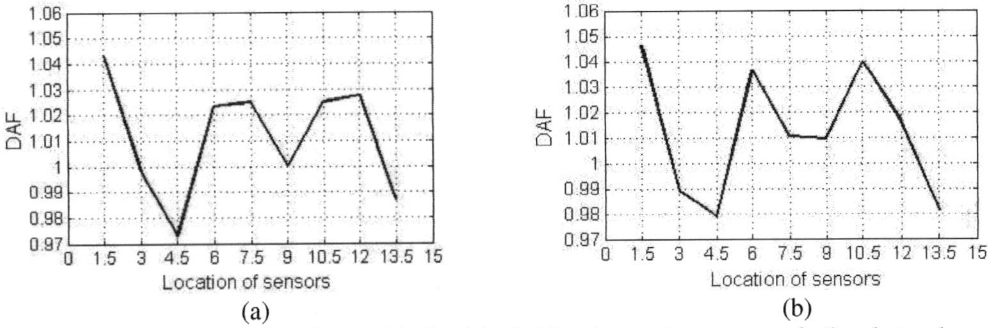

Figure 2. DAF at nine locations of the healthy bridge due to the passage of a 2-axle truck: (a) velocity of 42.12 km/h, (b) velocity of 43.2 km/h

3.2 DAF pattern at a point in time T +ΔT

Figure 3 shows the 3D plot of DAF versus velocity (km/h) for a 2-axle vehicle passing over bridge with a 30% stiffness loss localised at mid-span. The total volume of this plot is calculated using the same method defined in Section 3.1. Figures 4(a) and (b) show slices at two subsequent velocities 42.12 km/h and 43.2 km/h.

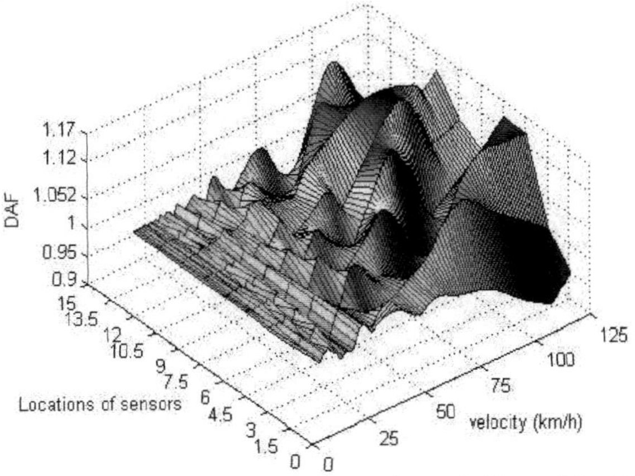

Figure 3. DAF versus velocity due to a 2-axle truck crossing a bridge with a 30% stiffness loss at mid-span

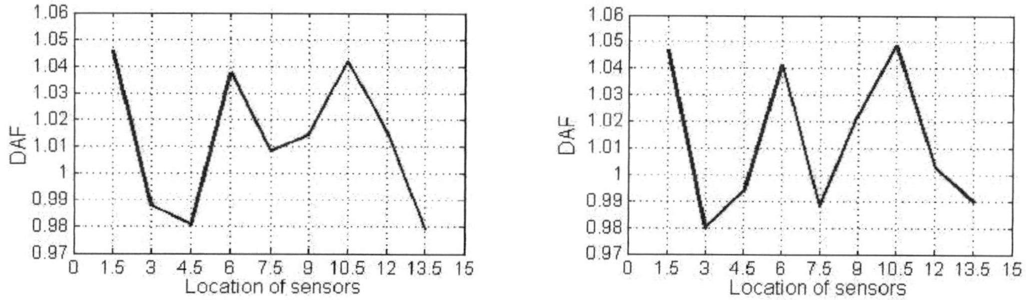

(a) (b)

Figure 4. DAF at nine locations versus two subsequent velocities of a 2-axle truck crossing damaged bridge with 30 % stiffness loss at mid-span: (a) velocity of 42.12 km/h, (b) velocity of 43.2 km/h

3.3 Comparison of DAF patterns at different points in time

DAF corresponding to two points in time are subtracted for depicting if significant change had occurred. To quantify the damage, the resultant DAF ("damaged-healthy") is shown in the contour plots of Figures 5-7 for the two percentages of stiffness loss and the three vehicle configurations.

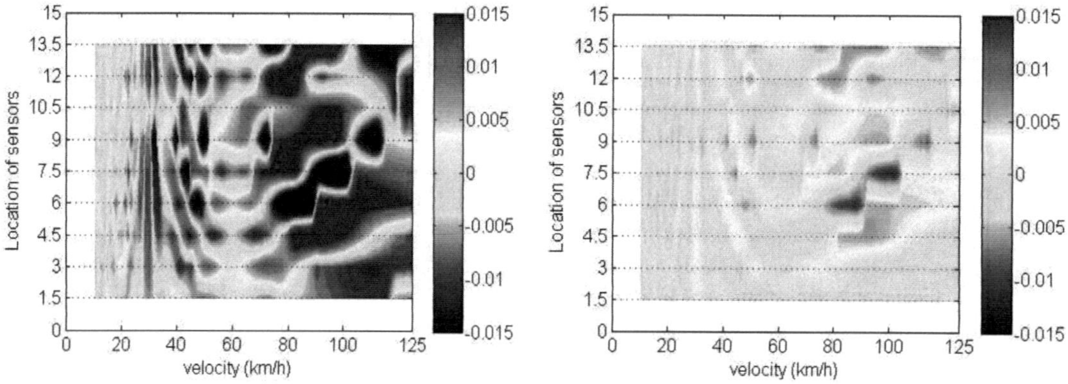

(a) (b)

Figure 5. Resultant DAF using a 2-axle truck: (a) 30% stiffness loss, (b) 10% stiffness loss

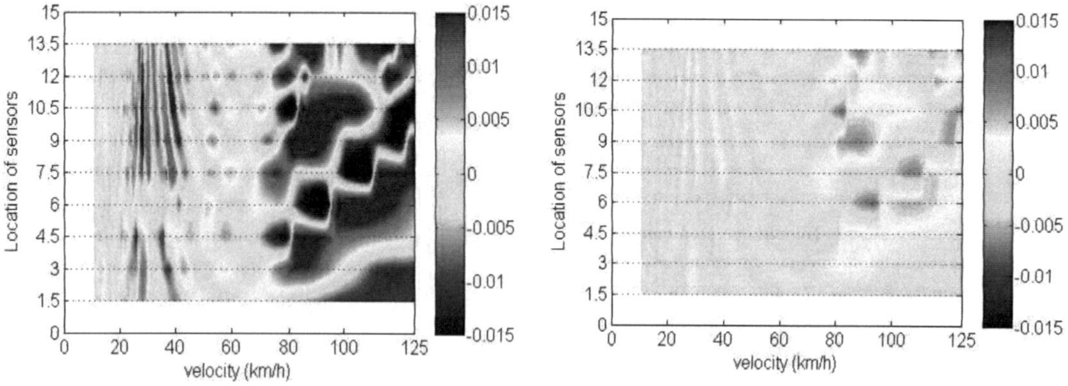

(a) (b)

Figure 6. Resultant DAF using a 3-axle truck: (a) 30% stiffness loss, (b) 10% stiffness loss

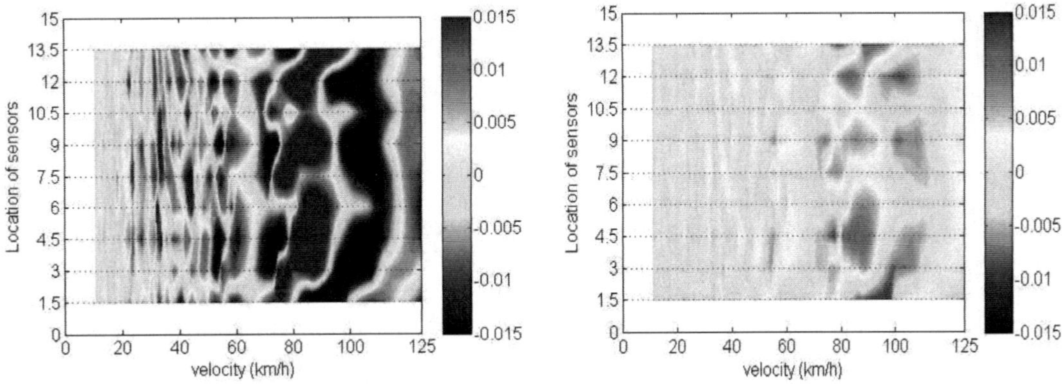

(a) (b)

Figure 7. Resultant DAF using a 5-axle truck: (a) 30% stiffness loss, (b) 10% stiffness loss

As expected, changes in DAF for 10% stiffness loss at mid-span are not as clear as for 30% stiffness loss. Table 3 shows the values of the DAF damage indexes. The resultant DAF damage indexes are the subtraction between the two volumes of the damaged and 'healthy' bridges for each of the vehicle configurations (the volumes have been calculated using Equation (3)).

Table 3: DAF damage indexes for the six scenarios under investigation

	Percentage of stiffness loss	
Truck Configuration	10%	30%
2-axle truck	0.663	2.621
3-axle truck	0.615	2.320
5-axle truck	0.265	1.173

From Table 3, it can be noticed that, the bridge under the impact of a 2-axle truck is more sensitive to the mid-span damage than the other two vehicles (3-axle and 5-axle trucks). In addition, the results show that the DAF damage index is comparatively more sensitive to changes in stiffness than those changes in frequencies shown in Table 2.

4. Conclusions

This paper has proposed a new level I damage detection method based on differences in DAF at different points in time. The method assumes that the vehicle speed, number of axles and their spacing and weight distribution can be obtained on-site, i.e., via a weigh-in-motion system. It has been theoretically tested with the simulated response of a 15 m simply supported bridge to a series of moving constant load resembling three types of vehicle configurations (2-axle, 3-axle and 5-axle trucks). The responses have been obtained at nine equally spaced locations that allowed gathering a full picture of the DAF patterns throughout the length. Damage has been modelled as a localised reduction in stiffness of 10% and 30 % at bridge mid-span. The DAF damage index that has resulted from subtracting the DAF of a 'healthy' bridge from that of damaged ones have shown to be potentially able to detect damage. The resultant 3D plots have been shown to be more sensitive to the crossing of the 2-axle truck than the other vehicle configurations. Despite the natural frequency of the damaged bridge having changed very slightly for the 10% stiffness loss, the new method has still been able to successfully detect changes in the structural response. However, a number of issues still need to be addressed in the future which include: influence of variations in vehicle dynamics and road profile on the pattern, inaccuracies in obtaining the static response via filtering, etc...

Acknowledgement
The authors are grateful for the financial support received from the Iraqi Ministry of Higher Education and Scientific Research (MoHESR) towards this investigation.

References:
[1] Wei Fan and Pizhong Qiao. Vibration-based Damage Identification Methods: A Review and Comparative Study. *Struct Heal Monitoring*. 2010;**10**(1):83–111.

[2] Cantero D, González A. Bridge Damage Detection Using Weigh-in-Motion Technology. *J Bridg Eng*. 2014 ;(1):1–10.

[3] Dongil Chang and Heehyun Lee. Impact factors for simple-span highway girder bridges. *J Struct Eng*. 1994 ;**120**(0733-9445):704–15.

[4] Sennah K, Zhang X, Kennedy J. Impact factors for horizontally curved composite box girder bridges. *J Bridg Eng*. 2004; **9** (6):512–20.

[5] Calcada R, Cunha A, Delgado R. Analysis of traffic-induced vibrations in a cable-stayed bridge. Part II: Numerical modeling and stochastic simulation. *J Bridg Eng*. 2005; **10** (4): 386–97.

[6] Kwasniewski L, Wekezer J, Roufa G, Li H, Ducher J, Malachowski J. Experimental Evaluation of Dynamic Effects for a Selected Highway Bridge. *J Perform Constr Facil*. 2006;**20**(3):253–60.

[7] Ding L, Hao H, Zhu X. Evaluation of dynamic vehicle axle loads on bridges with different surface conditions. *J Sound Vib*. 2009 Jun 2014; **323**(3-5):826–48.

[8] Huang D, Wang T, Shahawy M. Impact studies of multigirder concrete bridges. *J Struct Eng* 1993;**119**(8):2387–402.

[9] Ashebo DB, Chan THT, Yu L. Evaluation of dynamic loads on a skew box girder continuous bridge Part II: Parametric study and dynamic load factor. *Eng Struct*. 2007; **29** (6):1064–73.

[10] Kim S, Nowak A. Load distribution and impact factors for I-girder bridges. *J Bridg Eng* . 1997; **2** (3):97–104.

[11] Laman J, Pechar J, Boothby T. Dynamic load allowance for through-truss bridges. *J Bridg Eng*. 1999;**4** (4):231–41.

[12] Broquet C, Bailey S. Dynamic behavior of deck slabs of concrete road bridges. *J Bridg Eng* 2004 ;**9** (2):137–46.

[13] Tin Chan TH, O'Connor C. Vehicle Model for Highway Bridge Impact. *J Struct Eng*. 1990;**116** (7):1772–93.

[14] Ontario Ministry of Transportation (OMT). (1991). Ontario highway bridge design code, *3rd Ed., Downsview, ON, Canada.*

[15] Canadian Standards Association (CSA). (2006). Canadian highway bridge design code, *Mississauga, ON, Canada.*

[16] EN 1991-2:2003+NA:2010. Eurocode 1 – Actions on structures – Part 2: Traffic loads on bridges. (Including Irish National Annex).

[17] Clough RW, Penzien J. Dynamics of structures. *McGraw-Hill*; 1993.

[18] Tedesco JW, McDougal WG, Ross CA. Structural dynamics: theory and applications. *California, USA: Addison- Wesley Longman*; 1999.

[19] OBrien EJ, Cantero D, Enright B, González A. Characteristic Dynamic Increment for extreme *traffic* loading events on short and medium span highway bridges. *Eng Struct*.2010 ;**32**(12):3827–35

[20] González A, OBrien EJ, Cantero D, Li Y, Dowling J, Žnidarič A. Critical speed for the dynamics of truck events on bridges with a smooth road surface. *J Sound Vib*. 2010 ;**329**(11):2127–46.

[21] Brady S, O'Brien E. Effect of vehicle velocity on the dynamic amplification of two vehicles crossing a simply supported bridge. *J Bridg Eng* . 2006; **11** (2) :250–6.

[22] Cantero D, González A, Obrien EJ. Comparison of bridge dynamic amplifications due to articulated 5-axle trucks and large cranes . *Balt J ROAD Bridg Eng*. 2011;**6** (1):39–47.

AUTHOR INDEX

Abusaad, S 633
Acar, M 748
Adesta, Erry Yullian T 266
Aggelis, D G 122, 882
Al-Bugharbee, H 710
Alegre, D M 171
Algaba, Mario 106
Ali, Shaikh Faruque 463
Ambrozinski, Lukasz 219
Anduaga, J 98
Arizmendi, C J 664
Assaeh, M 633
Avendaño-Valencia, L D 594
Aydin, E 530
Aymerich, F 796, 828
Baba, Ndao 242
Ball, A D 633
Barszcz, T 702
Basu, B 250, 543, 859
Belaidi, I 273
Benaissa, B 273
Benjumea, Antonio J 106
Besset, Sebastien 82, 610
Blasi, M G 756
Bouhaddi, N 780
Boulpaep, F 122
Bovsunovsky, A P 1
Bradshaw, T 602
Brethee, K 633
Broderick, B M 250
Bui, Truong-Son 242
Camacho-Navarro, Jhonatan 1023, 1032
Capozucca, R 756
Carvelli, V 764
Casciati, F 904
Casciati, S 904
Ceravolo, R 820
Chanayil, Afeef 463
Chang, Chia-Ming 74
Chen, Hua-Peng 400, 408
Chen, Jun 578
Chen, S F 360
Chen, Z-C. 904
Cheng, Weidong 679
Chioncel, C P 155
Chu, F L 673
Chu, Fulei 618, 679
Clarke, A 602, 929
Cockerill, A 602
Cole, P 602
Corina, V 756
Cui, Liming 641
Da Silva, Samuel 90
Dahmene, F 1010, 1017
Damkilde, L 15, 559, 969
Daoud, O 522
De Marchi, A 820
Demirci, E 748
Deraemaeker, A 882
Desmet, W 656
Desmet, Wim 953
Devriendt, C 866
Di Lorenzo, E 656
Do, Duc-Phi 242
Dollevoet, Rolf 347
Dong, Dawei 578
Dos Santos, Fabio Luis Marques 953
Duan, Chendong 718
Dworakowski, Ziemowit 219
Eaton, M 842
El Mountassir, M 1010, 1017
El-Hussein, Salisu 313
Emoto, Hisao 291
Fabbrocino, G 139
Fagan, Edward M 850
Faravelli, L 904
Faria, A W 724
Farukh, F 748
Fassois, S D 594
Fassois, Spilios D 58, 66
Featherston, C 812, 929
Fifo, O 859
Figueiredo, E 281
Firdaus, Tasnim 266
Florez, J 211
Fragonara, L Zanotti 820
Francesconi, L 796
Frau, A 828
Fritzen, C P 234
Furdui, H 163
Gachagan, Anthony 179
Galbraith, Walter 179
Galvín, Pedro 106
Gao, F 874, 986
Gao, Hai-Yang 23
Gao, Hongli 567, 994
Gao, Qiang 718
Garambois, Pierre 82
García, David 694
Garcia, W L 664
Gargaro, D 139
Ghosh, Bidisha 431, 1040
Ghouri, A A 179
Gillich, G R 31, 732
Gillich, N 155, 163, 732
Gillot, Frédéric 610
Ginsberg, D 234
Goes, Luiz Carlos Sandoval 953
Goggins, J 503
Goggins, Jamie 850

AUTHOR INDEX

Gómez, C 376
González, Arturo 513
Gonzalez, Jesus 339
González-Requema, Ignacio 1032
Grigg, S 812
Gu, F 633
Güemes, Alfredo 1032
Güney, D 530
Guo, Liang 994
Guo, Xing-Lin 23
Han, Lu 648
Han, Ning 718
Han, Q K 673
Han, Qing 648
Hanley, Ciarán 480
Hansen, L M 15, 559
Harrigan, John J 313
Hassan, M S 503
Hilmy, Irfan 266
Hoell, S 551
Holford, K M 602
Homer, Rachel M 305
Hörrmann, S 772
Hoxha, Dashnor 242
Hsu, W T 360
Hua, Chunrong 578
Hua, J 978
Huang, Haifeng 567, 994
Hung, T Y 360
Iancu, V 31
Iezzi, F 147
Iliopoulos, A N 866
Iturrospe, A 98
Izman, Sudin 258
Jablonski, A 702
Jaksic, V 463, 471
Jaksic, Vesna 455
Janczura, Joanna 1002
Jang, Jae Kyeong 836
Jankowski, R 322, 416
Jara, J M 376, 392, 423
Jara, M 376, 392, 423
Jézéquel, Louis 82
Jia, Jing 495
Jiang, M 332
Johansen, R J 15, 559
Karaiskos, G 882
Kassim, Abdullah Abdul Rahman 258
Kawashita, L 929
Kedra, R 195
Kelliher, Denis 480
Kenna, A 543
Kennedy, Ciaran R 850
Kern, S 789
Khatir, S 273
Kim, Chul-Woo 439
Konneh, Mohamed 258
Koroishi, E H 171, 724
Kosova, G 656
Kourehli, S S 9
Kudela, P 945
Ku-Herrera, J J 789
Kwiecien, A 416
La Saponara, V 789
Lara-Molina, F A 724
Lasowicz, N 416
Lau, Jenny 953
Law, David W 305
Lee, J R 114
Lee, Jung Ryul 836
Lee, W 842
Leen, Sean B 850
Li, Jianvong 679
Li, Jun 994
Li, Zili 347
Limongelli, M P 368, 764
Lin, J 874, 978, 986
Liu, J 688
Liu, Jingming 578
Loh, C H 360
Loh, Chin-Hsiung 74
Loh, K J 789
Loh, Kenneth J 339
Louis, O 122
Lúcio, Válter J G 291
Luo, Z 874, 986
Mahdavi, S H 187
Mahmoudi, S 780
Maia, N 281
Majewska, K 945
Malinowski, P 129, 897, 912
Malinowski, Pawel 804
Mandic, D P 471
Manzato, S 656
Markert, Bernd 226
Marks, R 812, 929
Marra, A 139
Martínez, F 98
Marulo, F 656
McGugan, M 921
Mehrabani, Mehrdad Bahari 400
Melo, G P 171
Mendrok, Krzysztof 219
Mieloszyk, M 890, 897
Mikkelsen, L P 921
Minda, A A 163
Mituletu, I C 31, 155
Miyamoto, Ayaho 291
Mohammed, Omar 513
Mølgaard, L L 586

AUTHOR INDEX

Moll, J 912
Molyneaux, Thomas C K 305
Montanari, L 250
Montes, O 392
Morassi, A 368
Morita, Tomoaki 439
Mortezaei, Alireza 488
Mujica, L E 211
Mujica, Luis 1023, 1032
Muntean, F 163
Murphy, J 447, 471
Murphy, Jimmy 455, 463, 937
Musella, U 656
Naeimi, Meysam 347
Naik, Niranjan K 740
Nechak, Lyes 610
Negru, I 732
Nepal, Jaya 408
Nitescu, C N 155
Noble, D 384
Nogal, M 384
O'Byrne, Michael 431, 1040
O'Connor, A J 384
O'Donnell, Deirdre 431, 937
Olmos, B A 376, 392
Olmos, Ba 423
Olmos, L 423
Omenzetter, P 551
Opoka, S 890, 897, 912, 945
Opoka, Szymon 804
O'Shea, Richard 455
Ostachowicz, W 129, 890, 897, 912, 945
Ostachowicz, Wieslaw 804
O'Sullivan, K 447
Ouyang, Huajiang 23, 567
Öztürk, B 530
Paget, C 929
Pakrashi, V 384, 447, 471
Pakrashi, Vikram 431, 455, 463, 480, 937, 1040
Palka, Norbert 804
Pandya, Kedar S 740
Pearson, M 812, 842
Peeters, B 656
Peeters, Bart 953
Pepe, M 139
Pereira, G 921
Perera, R 281
Perez, Oscar 1023
Phadnis, Vaibhav A 740
Pickwell, Andy 179
Pieczonka, L 828
Pierce, S Gareth 179
Pinotti, E 820
Pinto, B 789
Polyzos, D 122
Popovici, Emanuel 937
Porcu, M C 828
Pourdeyhimi, B 748
Praisach, Z I 163, 732
Preisler, A 39
Puga, H 423
Pullin, R 602, 812, 842, 929
Qin, Zhaoye 618
Qiu, W L 332
Quek, S T 961
Quintero, M 211
Quintero, M A 664
Quintero, S 211
Rafferty, Steven 179
Rainieri, C 139
Rainieri, G M 139
Razak, H A 187
Roy, Anish 740
Ruby, M 234
Rucka, M 195, 203
Rudawska, A 624
Ruiz, M 211
Ruiz, Magda 1023, 1032
Saada, A Ait 273
Sakaris, Christos S 58
Sakellariou, John S 58, 66
Salawdeh, S 503
Sampaio, R 281
Schagerl, M 772
Schmidt, Rüdiger 226
Schoefs, Franck 431, 1040
Schröder, K-U 39
Serra, R 273
Shao, Y M 688
Shi, Weizhen 578
Shiki, Sidney Bruce 90
Shuaib, M 522
Silberschmidt, V V 748
Silberschmidt, Vadim V 740
Sinou, Jean-Jacques 610
Solís, Mario 106
Soltysik, B 322
Soman, R 129
Sozumert, E 748
Spagnoli, A 250
Spina, D 147
Sporea, D 897
Srbinovsky, B 937
Stancalie, A 897
Starkey, Andrew 313
Staszewski, W J 828
Steenbock, C 39
Steffen Jr, V 724
Stevenson, Morris W 400
Straczkiewicz, M 702

AUTHOR INDEX

Strantza, M 122
Sugiura, Kunitomo 439
Sun, Jian 641
Surace, C 820
Tabrizi, Mohsen Mohajer 488
Tanaka, Hideaki 291
Tcherniak, D 15, 559, 586
Tcherniak, Dmitri 694
Trendafilova, I 710
Trendafilova, Irina 694
Trivaudey, F 780
Truong, T C 114
Tsangouri, E 882
Tufoi, M 31, 732
Ugalde, U 98
Ulriksen, M D 15, 559, 969
Valente, C 147
Vamvoudakis-Stefanou, Kyriakos J 66
Van Hemelrijck, D 122, 866, 882
Vece, M 904
Viechtbauer, C 772
Villamizar, Rodolfo 1023, 1032
Wahab, M Abdel 281
Wahab, M M Abdel 266
Wandowski, T 897, 912
Wandowski, Tomasz 804
Wang, Changlin 648
Wang, Fengtao 641
Wang, Tianyang 679
Wang, Wanqin 718
Wang, Z 961
Wang, Ziran 439
Weijtjens, W 866
Wright, C 447, 463, 471
Wright, Robert 431
Wu, C W 47
Wylomanska, Agnieszka 1002
Xu, Yong 641
Yaacoubi, S 1010, 1017
Yan, Bing 578
Yan, Dawen 641
Yan, Jinghong 495
Yang, Xiu-Ming 23
Yasui, J 789
Yin, L 688
Zeng, L 874, 978, 986
Zhang, Muyu 226
Zhang, R 633
Zhang, Shenghua 641
Zhang, W 47
Zhang, Xiaochen 994
Zhang, Z 332
Zhao, Yanqing 718
Zheng, H L 688
Zhou, Yun-Lai 281
Zhu, Z D 47
Zima, B 203
Zimroz, Radoslaw 1002

11th International Conference on Damage Assessment of Structures

(DAMAS 2015)

Journal of Physics: Conference Series Volume 628

Ghent, Belgium
24 – 26 August 2015

Part 2 of 2

ISBN: 978-1-5108-0994-9
ISSN: 1742-6588

Printed from e-media with permission by:

Curran Associates, Inc.
57 Morehouse Lane
Red Hook, NY 12571

Some format issues inherent in the e-media version may also appear in this print version.

Printed by Curran Associates, Inc. (2015)

For permission requests, please contact the Institute of Physics at the address below.

Institute of Physics
Dirac House, Temple Back
Bristol BS1 6BE UK

Phone: 44 1 17 929 7481
Fax: 44 1 17 920 0979

techtracking@iop.org

Additional copies of this publication are available from:

Curran Associates, Inc.
57 Morehouse Lane
Red Hook, NY 12571 USA
Phone: 845-758-0400
Fax: 845-758-2634
Email: curran@proceedings.com
Web: www.proceedings.com

TABLE OF CONTENTS

PART 1

STRUCTURAL HEALTH AND CONDITION MONITORING

FATIGUE DAMAGE OF STEAM TURBINE SHAFT AT ASYNCHRONOUS CONNECTIONS OF TURBINE GENERATOR TO ELECTRICAL NETWORK 1
A P Bovsunovsky

DAMAGE IDENTIFICATION BASED ON INCOMPLETE MODAL DATA AND CONSTRAINED NONLINEAR MULTIVARIABLE FUNCTION 9
S S Kourehli

STATISTICAL EVALUATION OF CHARACTERISTIC SDDLV-INDUCED STRESS RESULTANTS TO DISCRIMINATE BETWEEN UNDAMAGED AND DAMAGED ELEMENTS 15
L M Hansen, R J Johansen, M D Ulriksen, D Tcherniak, L Damkilde

MULTI-DAMAGE LOCALIZATION IN PLATE STRUCTURE USING FREQUENCY RESPONSE FUNCTION-BASED INDICES 23
Hai-Yang Gao, Xing-Lin Guo, Huajiang Ouyang, Xiu-Ming Yang

LOCATION OF THE CORROSION DAMAGE IN RECTANGULAR PLATES 31
M Tufoi, G R Gillich, I C Mituletu, V Iancu

CRACK DIAGNOSIS OF METALLIC PROFILES BASED ON STRUCTURAL DAMAGE INDICATORS 39
A Preisler, C Steenbock, K-U Schröder

HOW WOODPECKER AVOIDS BRAIN INJURY? 47
C W Wu, Z D Zhu, W Zhang

A GENERALIZED FUNCTIONAL MODEL BASED METHOD FOR VIBRATION-BASED DAMAGE PRECISE LOCALIZATION IN 3D STRUCTURES 58
Christos S Sakaris, John S Sakellariou, Spilios D Fassois

ASSESSMENT OF A MULTIPLE MODEL BASED PARAMETRIC METHOD FOR OUTPUT-ONLY VIBRATION-BASED DAMAGE DETECTION FOR A POPULATION OF LIKE STRUCTURES 66
Kyriakos J Vamvoudakis-Stefanou, John S Sakellariou, Spilios D Fassois

IMPROVED STOCHASTIC SUBSPACE SYSTEM IDENTIFICATION FOR STRUCTURAL HEALTH MONITORING 74
Chia-Ming Chang, Chin-Hsiung Loh

MULTI-OBJECTIVE SHAPE OPTIMIZATION OF PLATE STRUCTURE UNDER STRESS CRITERIA BASED ON SUB-STRUCTURED MIXED FEM AND GENETIC ALGORITHMS 82
Pierre Garambois, Sebastien Besset, Louis Jézéquel

EVALUATION OF DAMAGE INDEXES FOR STRUCTURAL HEALTH MONITORING IN NONLINEAR MECHANICAL SYSTEMS BASED ON VOLTERRA SERIES 90
Sidney Bruce Shiki, Samuel Da Silva

NOVEL SHM METHOD TO LOCATE DAMAGES IN SUBSTRUCTURES BASED ON VARX MODELS 98
U Ugalde, J Anduaga, F Martínez, A Iturrospe

ANALYSIS OF STATIONARY ROVING MASS EFFECT FOR DAMAGE DETECTION IN BEAMS USING WAVELET ANALYSIS OF MODE SHAPES 106
Mario Solís, Antonio J Benjumea, Mario Algaba, Pedro Galvín

A VERSATILE INSPECTION SYSTEM FOR PIPE STRUCTURE USING ULTRASONIC WAVES PROPAGATION IMAGER 114
T C Truong, J R Lee

DAMAGE CHARACTERIZATION ON HUMAN FEMUR BONE BY MEANS OF ULTRASONICS AND ACOUSTIC EMISSION 122
M Strantza, D Polyzos, O Louis, F Boulpaep, D Van Hemelrijck, D G Aggelis

COMPARATIVE STUDY OF PERFORMANCE OF NEUTRAL AXIS TRACKING BASED DAMAGE DETECTION 129
R Soman, P Malinowski, W Ostachowicz

INTEGRATED NON-DESTRUCTIVE ASSESSMENT OF RELEVANT STRUCTURAL ELEMENTS OF AN ITALIAN HERITAGE SITE: THE CARTHUSIAN MONASTERY OF TRISULTI 139
C Rainieri, A Marra, G M Rainieri, D Gargaro, M Pepe, G Fabbrocino

DAMAGE ASSESSMENT THROUGH CHANGES IN MODE SHAPES DUE TO NON-PROPORTIONAL DAMPING 147
F Iezzi, D Spina, C Valente

A MULTI-RESOLUTION BASED METHOD TO PRECISE IDENTIFY THE NATURAL FREQUENCIES OF BEAMS WITH APPLICATION IN DAMAGE DETECTION 155
I C Mituletu, N Gillich, C N Nitescu, C P Chioncel

THE INFLUENCE OF OPERATIONAL AND ENVIRONMENTAL LOADS ON THE PROCESS OF ASSESSING DAMAGES IN BEAMS 163
H Furdui, F Muntean, A A Minda, Z I Praisach, N Gillich

EXPERIMENTAL FAULT DIAGNOSIS IN SYSTEMS CONTAINING FINITE ELEMENTS OF PLATE OF KIRCHOFF BY USING STATE OBSERVERS METHODOLOGY 171
D M Alegre, E H Koroishi, G P Melo

INTEGRATION OF ACOUSTIC EMISSION SYSTEMS WITHIN INTEGRI-TECHTM ANALYSIS SYSTEM FOR STRUCTURAL HEALTH MONITORING OF PRESSURISED ENGINEERING PLANT 179
A A Ghouri, Steven Rafferty, Andy Pickwell, Walter Galbraith, S Gareth Pierce, Anthony Gachagan

A WAVELET-BASED SCHEME FOR OPTIMUM MEASUREMENT/MONITORING OF DYNAMIC RESPONSES OF STRUCTURES 187
S H Mahdavi, H A Razak

RESEARCH ON ASSESSMENT OF BOLTED JOINT STATE USING ELASTIC WAVE PROPAGATION 195
R Kedra, M Rucka

WAVE PROPAGATION IN DAMAGE ASSESSMENT OF GROUND ANCHORS 203
B Zima, M Rucka

MAGNETIC FLUX LEAKAGE AND PRINCIPAL COMPONENT ANALYSIS FOR METAL LOSS APPROXIMATION IN A PIPELINE 211
M Ruiz, L E Mujica, M Quintero, J Florez, S Quintero

DAMAGE LOCALIZATION IN PLATES WITH USE OF THE PROCEDURE BASED ON MODAL FILTRATION 219
Ziemowit Dworakowski, Lukasz Ambrozinski, Krzysztof Mendrok

STRUCTURAL DAMAGE DETECTION USING AUTO CORRELATION FUNCTIONS OF VIBRATION RESPONSE UNDER SINUSOIDAL EXCITATION 226
Muyu Zhang, Rüdiger Schmidt, Bernd Markert

LOAD IDENTIFICATION APPROACH BASED ON BASIS PURSUIT DENOISING ALGORITHM 234
D Ginsberg, M Ruby, C P Fritzen

DAMAGE ASSESSMENT OF CEMENT-BASED GEOMATERIAL DURING LOADING BY ULTRASONIC TOMOGRAPHY 242
Duc-Phi Do, Ndao Baba, Dashnor Hoxha, Truong-Son Bui

OPTIMAL SAMPLING IN DAMAGE DETECTION OF FLEXURAL BEAMS BY CONTINUOUS WAVELET TRANSFORM 250
B Basu, B M Broderick, L Montanari, A Spagnoli

MILLING DAMAGE ON CARBON FIBRE REINFORCED POLYMER USING TIALN COATED END MILLS 258
Mohamed Konneh, Sudin Izman, Abdullah Abdul Rahman Kassim

DAMAGE DETECTION BASED ON THE NATURAL FREQUENCY SHIFTING OF A CLAMPED RECTANGULAR PLATE MODEL 266
Irfan Hilmy, M M Abdel Wahab, Erry Yullian T Adesta, Tasnim Firdaus

GENETIC ALGORITHM BASED OBJECTIVE FUNCTIONS COMPARATIVE STUDY FOR DAMAGE DETECTION AND LOCALIZATION IN BEAM STRUCTURES 273
S Khatir, I Belaidi, R Serra, B Benaissa, A Ait Saada

SINGLE SIDE DAMAGE SIMULATIONS AND DETECTION IN BEAM-LIKE STRUCTURES 281
Yun-Lai Zhou, M Abdel Wahab, R Perera, N Maia, R Sampaio, E Figueiredo

DAMAGE IN CIVIL ENGINEERING

LONG TERM APPLICATION OF BUS MONITORING SYSTEM TO SHORT AND MEDIUM SPAN BRIDGES AND DAMAGE DETECTION ... 291
Ayaho Miyamoto, Válter J G Lúcio, Hisao Emoto, Hideaki Tanaka

PROBABILITY DISTRIBUTION FUNCTIONS FOR COVER USED IN 3-D MODEL SIMULATING CONCRETE DETERIORATION IN PORT ASSETS ... 305
Rachel M Homer, David W Law, Thomas C K Molyneaux

FINITE ELEMENT SIMULATION OF GUIDED WAVES IN PIPELINES FOR LONG RANGE MONITORING AGAINST THIRD PARTY ATTACKS ... 313
Salisu El-Hussein, John J Harrigan, Andrew Starkey

BUILDING DAMAGE DUE TO STRUCTURAL POUNDING DURING EARTHQUAKES ... 322
B Soltysik, R Jankowski

NONLINEAR DYNAMIC RESPONSES OF SUSPENSION BRIDGE TO SUDDEN BREAKAGE OF HANGERS ... 332
W L Qiu, M Jiang, Z Zhang

CEMENTITIOUS COMPOSITES ENGINEERED WITH EMBEDDED CARBON NANOTUBE THIN FILMS FOR ENHANCED SENSING PERFORMANCE ... 339
Kenneth J Loh, Jesus Gonzalez

NUCLEATION OF SQUAT CRACKS IN RAIL, CALCULATION OF CRACK INITIATION ANGLES IN THREE DIMENSIONS ... 347
Meysam Naeimi, Zili Li, Rolf Dollevoet

DAMAGE DETECTION IN BRIDGE STRUCTURE USING VIBRATION DATA UNDER RANDOM TRAVELLING VEHICLE LOADS ... 360
C H Loh, T Y Hung, S F Chen, W T Hsu

EXPERIMENTAL VERIFICATION OF THE INTERPOLATION METHOD ON A REAL DAMAGED BRIDGE ... 368
M P Limongelli, A Morassi

LIMIT STATES FOR THE DAMAGE ASSESSMENT OF BRIDGES SUPPORTED ON LRB BEARINGS ... 376
M Jara, C Gómez, J M Jara, B A Olmos

THE EFFECT OF POST-TENSIONING FORCE MAGNITUDE AND ECCENTRICITY ON THE NATURAL BENDING FREQUENCY OF CRACKED POST-TENSIONED CONCRETE BEAMS ... 384
D Noble, M Nogal, A J O'Connor, V Pakrashi

EXPECTED DAMAGES OF RETROFITTED BRIDGES WITH RC JACKETING ... 392
O Montes, J M Jara, M Jara, B A Olmos

OVERTOPPING FAILURE ANALYSIS OF COASTAL FLOOD DEFENCES AFFECTED BY CLIMATE CHANGE ... 400
Mehrdad Bahari Mehrabani, Hua-Peng Chen, Morris W Stevenson

ASSESSMENT OF CONCRETE DAMAGE AND STRENGTH DEGRADATION CAUSED BY REINFORCEMENT CORROSION ... 408
Jaya Nepal, Hua-Peng Chen

EXPERIMENTAL STUDY ON THE EFFECTIVENESS OF POLYMER DAMPER IN DAMAGE REDUCTION OF TEMPORARY STEEL GRANDSTAND ... 416
N Lasowicz, A Kwiecien, R Jankowski

DAMAGE ASSESSMENT OF CURTAIN WALL GLASS ... 423
H Puga, Ba Olmos, L Olmos, J M Jara, M Jara

ACQUISITION AND ANALYSIS OF DYNAMIC RESPONSES OF A HISTORIC PEDESTRIAN BRIDGE USING VIDEO IMAGE PROCESSING ... 431
Michael O'Byrne, Bidisha Ghosh, Franck Schoefs, Deirdre O'Donnell, Robert Wright, Vikram Pakrashi

LONG-TERM BRIDGE HEALTH MONITORING FOCUSING ON THE MAHALANOBIS DISTANCE OF MODAL PARAMETERS ... 439
Chul-Woo Kim, Tomoaki Morita, Ziran Wang, Kunitomo Sugiura

EXPERIMENTAL COMPARISON OF DYNAMIC RESPONSES OF A TENSION MOORED FLOATING WIND TURBINE PLATFORM WITH AND WITHOUT SPRING DAMPERS ... 447
C Wright, K O'Sullivan, J Murphy, V Pakrashi

THE HURST EXPONENT AS AN INDICATOR OF THE BEHAVIOUR OF A MODEL MONOPILE IN AN OCEAN WAVE TESTING BASIN ... 455
Vikram Pakrashi, Richard O'Shea, Vesna Jaksic, Jimmy Murphy

PERFORMANCE OF A SINGLE LIQUID COLUMN DAMPER FOR THE CONTROL OF DYNAMIC RESPONSES OF A TENSION LEG PLATFORM ... 463
V Jaksic, C Wright, Afeef Chanayil, Shaikh Faruque Ali, Jimmy Murphy, Vikram Pakrashi

A DELAY VECTOR VARIANCE BASED MARKER FOR AN OUTPUT-ONLY ASSESSMENT OF STRUCTURAL CHANGES IN TENSION LEG PLATFORMS 471
V Jaksic, C Wright, D P Mandic, J Murphy, V Pakrashi

PRINCIPAL COMPONENT ANALYSIS FOR CONDITION MONITORING OF A NETWORK OF BRIDGE STRUCTURES 480
Ciarán Hanley, Denis Kelliher, Vikram Pakrashi

DAMAGE ASSESSMENT OF RC BUILDINGS SUBJECTED TO THE DIFFERENT STRONG MOTION DURATION 488
Alireza Mortezaei, Mohsen Mohajer Tabrizi

ANALYSIS ABOUT FACTORS AFFECTING THE DEGREE OF DAMAGE OF BUILDINGS IN EARTHQUAKE 495
Jing Jia, Jinghong Yan

CHARACTERISING THE EFFECT OF GLOBAL AND LOCAL GEOMETRIC IMPERFECTIONS ON THE NUMERICAL PERFORMANCE OF A BRACE MEMBER 503
M S Hassan, J Goggins, S Salawdeh

FOOTPRINT CAUSED BY A VEHICLE CONFIGURATION ON THE DYNAMIC AMPLIFICATION OF THE BRIDGE RESPONSE 513
Omar Mohammed, Arturo González

PART 2

NUMERICAL MODELLING OF REINFORCED CONCRETE SLABS UNDER BLAST LOADS OF CLOSE-IN DETONATIONS USING THE LAGRANGIAN APPROACH 522
M Shuaib, O Daoud

THE EVALUATION OF DAMAGE MECHANISM OF UNREINFORCED MASONRY BUILDINGS AFTER VAN (2011) AND ELAZIG (2010) EARTHQUAKES 530
D Güney, E Aydin, B Öztürk

DAMAGE DETECTION IN WIND TURBINE TOWERS USING A FINITE ELEMENT MODEL AND DISCRETE WAVELET TRANSFORM OF STRAIN SIGNALS 543
A Kenna, B Basu

DAMAGE IN MACHINERIES

EXPERIMENTAL DAMAGE DETECTION IN A WIND TURBINE BLADE MODEL USING PRINCIPAL COMPONENTS OF RESPONSE CORRELATION FUNCTIONS 551
S Hoell, P Omenzetter

DAMAGE LOCALIZATION IN A RESIDENTIAL-SIZED WIND TURBINE BLADE BY USE OF THE SDDLV METHOD 559
R J Johansen, L M Hansen, M D Ulriksen, D Tcherniak, L Damkilde

BLIND SOURCE SEPARATION AND DYNAMIC FUZZY NEURAL NETWORK FOR FAULT DIAGNOSIS IN MACHINES 567
Haifeng Huang, Huajiang Ouyang, Hongli Gao

VIBRATION DAMAGE MECHANISM ANALYSIS ON ROTOR OF DIESEL GENERATING SET WITH RIGID COUPLING 578
Bing Yan, Weizhen Shi, Chunrong Hua, Jingming Liu, Dawei Dong, Jun Chen

VIBRATION-BASED SHM SYSTEM: APPLICATION TO WIND TURBINE BLADES 586
D Tcherniak, L L Mølgaard

NATURAL VIBRATION RESPONSE BASED DAMAGE DETECTION FOR AN OPERATING WIND TURBINE VIA RANDOM COEFFICIENT LINEAR PARAMETER VARYING AR MODELLING 594
L D Avendaño-Valencia, S D Fassois

USE OF HIGH FREQUENCY ANALYSIS OF ACOUSTIC EMISSION SIGNALS TO DETERMINE ROLLING ELEMENT BEARING CONDITION 602
A Cockerill, K M Holford, T Bradshaw, P Cole, R Pullin, A Clarke

SURROGATE MODELS FOR EFFICIENT STABILITY ANALYSIS OF BRAKE SYSTEMS 610
Lyes Nechak, Frédéric Gillot, Sébastien Besset, Jean-Jacques Sinou

NUMERICAL STUDIES ON TIME-VARYING STIFFNESS OF DISK-DRUM TYPE ROTOR WITH BOLT LOOSENING 618
Zhaoye Qin, Fulei Chu

SELECT ASPECTS OF FEM ANALYSIS FOR BONDED JOINTS OF POLYMER COMPOSITE MATERIALS 624
A Rudawska
THE DETECTION OF LUBRICATING OIL VISCOSITY CHANGES IN GEARBOX TRANSMISSION SYSTEMS DRIVEN BY SENSORLESS VARIABLE SPEED DRIVES USING ELECTRICAL SUPPLY PARAMETERS 633
S Abusaad, K Brethee, M Assaeh, R Zhang, F Gu, A D Ball
A FEATURE EXTRACTION METHOD FOR FAULT CLASSIFICATION OF ROLLING BEARING BASED ON PCA 641
Fengtao Wang, Jian Sun, Dawen Yan, Shenghua Zhang, Liming Cui, Yong Xu
PROJECTILES IMPACT ASSESSMENT OF AIRCRAFT WING STRUCTURES WITH REAL DYNAMIC LOAD 648
Lu Han, Qing Han, Changlin Wang
STRUCTURAL HEALTH MONITORING CHALLENGES ON THE 10-MW OFFSHORE WIND TURBINE MODEL 656
E Di Lorenzo, G Kosova, U Musella, S Manzato, B Peeters, F Marulo, W Desmet
AUTOMATIC WELDING DETECTION BY AN INTELLIGENT TOOL PIPE INSPECTION 664
C J Arizmendi, W L Garcia, M A Quintero
A THEORETICAL MODEL FOR FAULT DIAGNOSIS OF LOCALIZED BEARING DEFECTS UNDER NON-WEIGHT-DOMINANT CONDITIONS 673
Q K Han, F L Chu
A NEW ANGULAR RESAMPLING ALGORITHM FOR THE BEARING FAULT DIAGNOSIS UNDER THE TIME-VARYING ROTATIONAL SPEED 679
Tianyang Wang, Weidong Cheng, Jianvong Li, Fulei Chu
IMPACT DAMAGE EVALUATION METHOD OF FRICTION DISC BASED ON HIGH-SPEED PHOTOGRAPHY AND TOOTH-ROOT STRESS COUPLING 688
L Yin, Y M Shao, J Liu, H L Zheng
DAMAGE ASSESSMENT FOR WIND TURBINE BLADES BASED ON A MULTIVARIATE STATISTICAL APPROACH 694
David García, Dmitri Tcherniak, Irina Trendafilova
DETECTION AND CLASSIFICATION OF ALARM THRESHOLD VIOLATIONS IN CONDITION MONITORING SYSTEMS WORKING IN HIGHLY VARYING OPERATIONAL CONDITIONS 702
M Straczkiewicz, T Barszcz, A Jablonski
AUTOREGRESSIVE MODELLING FOR ROLLING ELEMENT BEARING FAULT DIAGNOSIS 710
H Al-Bugharbee, I Trendafilova
A CONTROL STRATEGY FOR ADAPTIVE ABSORBER BASED ON VARIABLE MASS 718
Qiang Gao, Ning Han, Yanqing Zhao, Chendong Duan, Wanqin Wang

DAMAGE IN COMPOSITE MATERIALS

FUZZY MODAL CONTROL APPLIED TO SMART COMPOSITE STRUCTURE 724
E H Koroishi, A W Faria, F A Lara-Molina, V Steffen Jr
NATURAL FREQUENCY CHANGES DUE TO DAMAGE IN COMPOSITE BEAMS 732
I Negru, G R Gillich, Z I Praisach, M Tufoi, N Gillich
BALLISTIC DAMAGE IN HYBRID COMPOSITE LAMINATES 740
Vaibhav A Phadnis, Kedar S Pandya, Niranjan K Naik, Anish Roy, Vadim V Silberschmidt
DAMAGE MECHANISMS OF RANDOM FIBROUS NETWORKS 748
E Sozumert, F Farukh, E Demirci, M Acar, B Pourdeyhimi, V V Silberschmidt
DYNAMIC RESPONSE OF RC BEAMS STRENGTHENED WITH NEAR SURFACE MOUNTED CARBON-FRP RODS SUBJECTED TO DAMAGE 756
R Capozucca, M G Blasi, V Corina
DAMAGE LOCALIZATION IN A GLASS FIBER REINFORCED COMPOSITE PLATE VIA THE SURFACE INTERPOLATION METHOD 764
M P Limongelli, V Carvelli
ON THE MONITORING AND IMPLICATIONS OF GROWING DAMAGES CAUSED BY MANUFACTURING DEFECTS IN COMPOSITE STRUCTURES 772
M Schagerl, C Viechtbauer, S Hörrmann
NONLINEAR DYNAMIC RESPONSE ANALYSIS OF LOCALIZED DAMAGED LAMINATED COMPOSITE STRUCTURES IN THE CONTEXT OF COMPONENT MODE SYNTHESIS 780
S Mahmoudi, F Trivaudey, N Bouhaddi

A COMPARATIVE STUDY OF A SELF STRAIN-MONITORING CARBON NANOTUBE FILM AND CARBON FIBERS UNDER FLEXURAL LOADING BY ELECTRICAL RESISTANCE CHANGES 789
B Pinto, S Kern, J J Ku-Herrera, J Yasui, V La Saponara, K J Loh

IMPACT DAMAGE RESISTANCE OF THIN STITCHED CARBON/EPOXY LAMINATES 796
L Francesconi, F Aymerich

MOISTURE DETECTION IN COMPOSITES BY TERAHERTZ SPECTROSCOPY 804
Pawel Malinowski, Norbert Palka, Szymon Opoka, Tomasz Wandowski, Wieslaw Ostachowicz

ASSESSMENT OF DAMAGE DETECTION IN COMPOSITE STRUCTURES USING 3D VIBROMETRY 812
S Grigg, M Pearson, R Marks, C Featherston, R Pullin

NON-DESTRUCTIVE TESTING ON ARAMID FIBRES FOR THE LONG-TERM ASSESSMENT OF INTERVENTIONS ON HERITAGE STRUCTURES 820
R Ceravolo, A De Marchi, E Pinotti, C Surace, L Zanotti Fragonara

ANALYSIS OF ELASTIC NONLINEARITY FOR IMPACT DAMAGE DETECTION IN COMPOSITE LAMINATES 828
A Frau, L Pieczonka, M C Porcu, W J Staszewski, F Aymerich

NON-DESTRUCTIVE VISUALIZATION OF LINEAR EXPLOSIVE-INDUCED PYROSHOCK USING PHASE ARRAYED LASER-INDUCED SHOCK IN A SPACE LAUNCHER COMPOSITE 836
Jae Kyeong Jang, Jung Ryul Lee

ACCURATE DAMAGE LOCATION IN COMPLEX COMPOSITE STRUCTURES AND INDUSTRIAL ENVIRONMENTS USING ACOUSTIC EMISSION 842
M Eaton, M Pearson, W Lee, R Pullin

FINITE ELEMENT BASED DAMAGE ASSESSMENT OF COMPOSITE TIDAL TURBINE BLADES 850
Edward M Fagan, Sean B Leen, Ciaran R Kennedy, Jamie Goggins

AN INVESTIGATION INTO THE IMPACT OF CRYOGENIC ENVIRONMENT ON MECHANICAL STRESSES IN FRP COMPOSITES 859
O Fifo, B Basu

SENSING AND SENSORS

PREDICTION OF DYNAMIC STRAINS ON A MONOPILE OFFSHORE WIND TURBINE USING VIRTUAL SENSORS 866
A N Iliopoulos, W Weijtjens, D Van Hemelrijck, C Devriendt

LOCAL INTERROGATIONS FOR DETECTION OF MULTIPLE DEFECTS USING AN ACTIVE SENSOR NETWORK 874
F Gao, J Lin, L Zeng, Z Luo

DAMAGE DETECTION MONITORING APPLICATIONS IN SELF-HEALING CONCRETE STRUCTURES USING EMBEDDED PIEZOELECTRIC TRANSDUCERS AND RECOVERY 882
G Karaiskos, E Tsangouri, D G Aggelis, A Deraemaeker, D Van Hemelrijck

FREQUENCY DOMAIN DECOMPOSITION PERFORMED ON THE STRAIN DATA OBTAINED FROM THE ALUMINIUM MODEL OF AN OFFSHORE SUPPORT STRUCTURE 890
M Mieloszyk, S Opoka, W Ostachowicz

INFLUENCE OF SUPPORT CONDITIONS AND TEMPERATURE ON THE EMI CHARACTERISTICS 897
A Stancalie, D Sporea, P Malinowski, M Mieloszyk, S Opoka, T Wandowski, W Ostachowicz

COLLECTING DATA FROM A SENSOR NETWORK IN A SINGLE-BOARD COMPUTER 904
F Casciati, S Casciati, Z-C. Chen, L Faravelli, M Vece

ASSESSMENT OF PIEZOELECTRIC SENSOR ADHESIVE BONDING 912
T Wandowski, J Moll, P Malinowski, S Opoka, W Ostachowicz

EMBEDDED FIBRE BRAGG GRATING SENSOR RESPONSE MODEL: CRACK GROWING DETECTION IN FIBRE REINFORCED PLASTIC MATERIALS 921
G Pereira, L P Mikkelsen, M McGugan

USING GENETIC ALGORITHMS TO OPTIMIZE AN ACTIVE SENSOR NETWORK ON A STIFFENED AEROSPACE PANEL WITH 3D SCANNING LASER VIBROMETRY DATA 929
R Marks, A Clarke, C Featherston, L Kawashita, C Paget, R Pullin

SENSOR MEASUREMENT STRATEGIES FOR MONITORING OFFSHORE WIND AND WAVE ENERGY DEVICES 937
Deirdre O'Donnell, Bruno Srbinovsky, Jimmy Murphy, Emanuel Popovici, Vikram Pakrashi

NOVEL FBG ROSETTE FOR DETERMINING IMPACT LOCATION IN THIN PLATE-LIKE STRUCTURE 945
K Majewska, S Opoka, P Kudela, W Ostachowicz
THE USE OF STRAIN GAUGES IN VIBRATION-BASED DAMAGE DETECTION 953
Fabio Luis Marques Dos Santos, Bart Peeters, Jenny Lau, Wim Desmet, Luiz Carlos Sandoval Goes
IDENTIFICATION OF FAULTY SENSOR USING RELATIVE PARTIAL DECOMPOSITION VIA INDEPENDENT COMPONENT ANALYSIS 961
Z Wang, S T Quek

SIGNAL PROCESSING

DAMAGE LOCALIZATION BY STATISTICAL EVALUATION OF SIGNAL-PROCESSED MODE SHAPES 969
M D Ulriksen, L Damkilde
PULSE COMPRESSION AND DISPERSION COMPENSATION FOR HIGH- RESOLUTION LAMB WAVE INSPECTION 978
J Hua, J Lin, L Zeng
MODE PURIFICATION FOR ULTRASONIC GUIDED WAVES UNDER PSEUDOPULSE EXCITATION 986
Z Luo, J Lin, L Zeng, F Gao
MACHINERY VIBRATION SIGNAL DENOISING BASED ON LEARNED DICTIONARY AND SPARSE REPRESENTATION 994
Liang Guo, Hongli Gao, Jun Li, Haifeng Huang, Xiaochen Zhang
IDENTIFICATION AND STOCHASTIC MODELLING OF SOURCES IN COPPER ORE CRUSHER VIBRATIONS 1002
Agnieszka Wylomanska, Radoslaw Zimroz, Joanna Janczura
DETECTION OF STRUCTURAL DAMAGE USING NOVELTY DETECTION ALGORITHM UNDER VARIATIONAL ENVIRONMENTAL AND OPERATIONAL CONDITIONS 1010
M El Mountassir, S Yaacoubi, F Dahmene
INVESTIGATION ON THE USE OF ARTIFICIAL NEURAL NETWORKS TO OVERCOME THE EFFECTS OF ENVIRONMENTAL AND OPERATIONAL CHANGES ON GUIDED WAVES MONITORING 1017
M El Mountassir, S Yaacoubi, F Dahmene
PIPE LEAKS CLASSIFICATION BY USING A DATA-DRIVEN APPROACH BASED ON FEATURES FROM CROSS-CORRELATED PIEZO-VIBRATION SIGNALS 1023
Jhonatan Camacho-Navarro, Magda Ruiz, Oscar Perez, Rodolfo Villamizar, Luis Mujica
STUDY OF CROSS-CORRELATION SIGNALS IN A DATA-DRIVEN APPROACH FOR DAMAGE CLASSIFICATION IN AIRCRAFT WINGS 1032
Jhonatan Camacho-Navarro, Magda Ruiz, Rodolfo Villamizar, Luis Mujica, Alfredo Güemes, Ignacio González-Requema

REPLACEMENT ARTICLE

PROTOCOLS FOR IMAGE PROCESSING BASED UNDERWATER INSPECTION OF INFRASTRUCTURE ELEMENTS 1040
Michael O'Byrne, Bidisha Ghosh, Franck Schoefs, Vikram Pakrashi
Author Index

Numerical Modelling of Reinforced Concrete Slabs under Blast Loads of Close-in Detonations Using the Lagrangian Approach

M Shuaib[1] and O Daoud[2]

[1] Postgraduate Student, Department of Civil Engineering, University of Khartoum, Khartoum, Sudan.

[2] Associate Professor, Building and Road Research Institute, University of Khartoum, Khartoum, Sudan.

Email: mmnshuaib@gmail.com

Abstract. This paper includes an investigation for the deformations, including deflections and damage modes, which occur in reinforced concrete (RC) slabs when subjected to blast loads of explosions. The slab considered for the investigation is a one-way square RC slab with the dimensions of 1000 x 1000 x 40 mm, fixed supported at two opposite sides. It was subjected to close-in detonations of three different charge weights for a constant standoff distance. For the study, the slab was analysed using the numerical method by means of nonlinear finite element analysis. The slab was modelled as 3-D structural continuum using LS-DYNA software. For concrete modelling, two constitutive models were selected, namely the KCC and Winfrith concrete models. Blast loads were applied to the slab through the Lagrangian approach, and the blast command available in the software, namely LOAD_BLAST_ENHANCED, was selected for the application. The deflections and damage modes results obtained were compared to those from a previously published experiment. From the study, both the KCC and Winfrith concrete models effectively and satisfactorily estimated the actual slab maximum deflection. For damage modes, the KCC model appeared to be capable to capture satisfactorily the general damage mode including flexural cracks. However, the model could not capture the local shear mode at the middle of slab (spallation) because the Lagrangian approach does not simulate the interaction between the ambient air and the solid slab.

1. Introduction

Nowadays, the need for the protection against the effects of blast loads has increased as the number of explosion incidents that caused serious damages have greatly increased. Explosions can be caused either accidentally, as those occur by handling or storing explosives, or intentionally as in terrorist attacks. The principal material used for blast mitigation is reinforced concrete (RC) as it encompasses good characteristics in blast resistance. Among the RC structural elements, slabs are important elements and they are used widely for both structural and non-structural purposes.

Blast wave generated in an explosion imposes a short duration dynamic loads on structures. In general, materials including concrete and steel behave differently under dynamic loads. They experience an increase in strength under rapidly applied loads. The deformations which occur in a structural member due to blast can be assessed by using either the experimental approach or the analytical approach which includes simplified empirical and advanced numerical techniques. The finite element (FE) method is an advanced technique which is used to solve complex structural analysis problems including blast problems. Many works such as those carried out by Hao and Zhongxian [1], Wang et al [2], Zhao and Chen [3], Tai et al [4] and others have proved that by using the FE method of analysis, the deformation which occurs in concrete due to blast loads can be effectively and satisfactory assessed.

In the FE method, blast loads can be simulated using many approaches. The Lagrangian approach is one of them in which blast loads are applied directly in the form of load-time history on a designated surface. The use of this approach is advantageous as many software packages, such as the LS-DYNA, offer the automatic generation of blast loads by defining the basic parameters of the explosion. In addition, it does not require the modelling of air as in the Multi-Material Aribtrary Lagrangian Eulerian (MM-ALE) approach [5]. However, the Lagrangian does not take into account the interaction between air and structure [6].

RC elements possess different general behaviours depending on the intensity and standoff distance of the explosive. In the case of close-in detonations, the blast load produced is non-uniform with high intensity [7]. Thus, extremely high pressure detonations are developed which can produce local (punching) failure (including spalling and scabbing) of an element which is regarded to the brittle nature of concrete. The deformation is associated with considerable deflection values as well. The effect of close-in detonations on one-way concrete slabs was investigated by Wang et al [2] by using the experimental and numerical approaches. In this paper, this experiment was simulated numerically using the Lagrangian approach with nonlinear finite element analysis software, namely LS-DYNA. The obtained results were verified and compared with the experimental results.

2. Model Description

The model was for one-way RC slab supported at two opposite sides and subjected to close-in detonation of TNT explosives, and it similar to the slab in reference [2]. The charge weights considered for the detonations were 0.2, 0.31 and 0.46kg, and their configuration is as shown in Figure 1. The software used for modelling was LS-DYNA software, and the program library was used for the selection of element types and constitutive models for concrete and steel. The simulation of blast loading was carried out using the blast command available in the software.

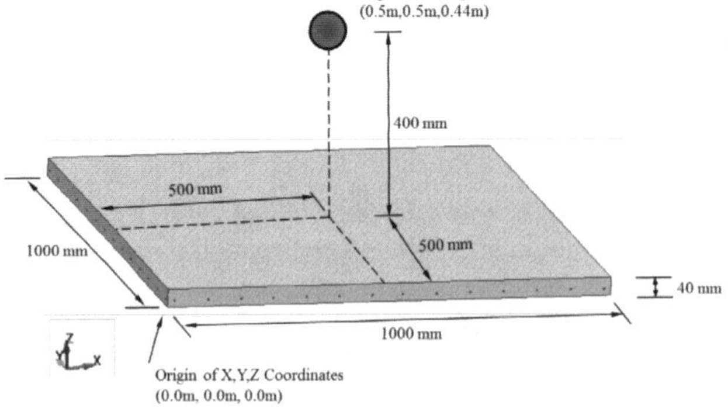

Figure 1. Configuration of the explosive charge

The slab dimensions were 1000 mm x 1000 mm x 40 mm. It was reinforced at its both directions by 6 mm steel bar reinforcement mesh with 75 mm spacing and 20 mm concrete cover. It was assumed to be fixed at the two supporting sides, although it was partially fixed in the experiment [2]. The dimensions of the slab in addition to the reinforcement detailing and supporting conditions are shown in Figure 2.

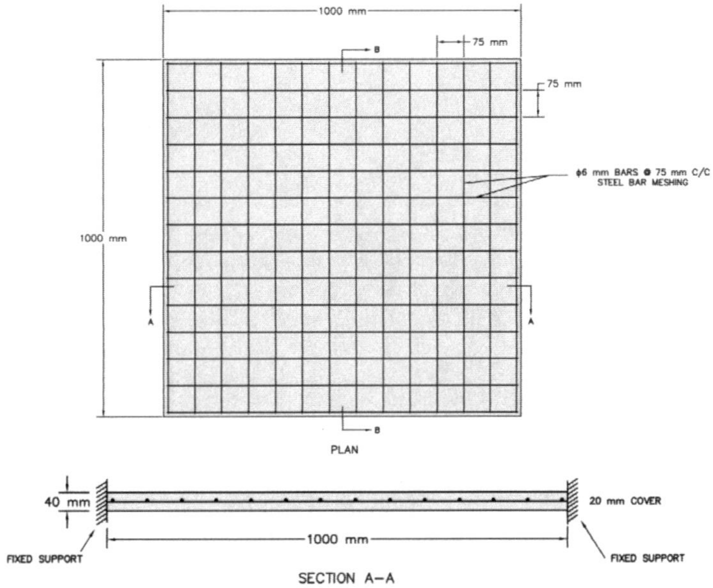

Figure 2. RC slab geometry and reinforcement details

The approach used for slab modelling was the three dimensional (3-D) approach in which the concrete was modelled as a 3-D block. This approach allows capturing the local deflections and damage. In the model, the concrete was discretized into solid elements, and the steel reinforcement into beam elements. Both meshed parts are connected using discrete technique which would insure the displacement compatibility of concrete and steel at the shared nodes. In addition, it would simulate the effect of links between steel bar mesh. The fixity at supports is applied by assigning fixed boundary conditions to support nodes. A schematic representation for slab finite element details is shown in Figure 3.

Figure 3. Schematic representation for finite element scheme details of RC slab

2.1. RC Slab Meshing

2.1.1. Concrete Meshing

The concrete of the RC slab was modelled as a solid box of dimensions 1000mm x 1000mm x 40mm which was composed of 8-node hexahedron solid elements with constant stress solid element formulation. The size of each element is 5mm x 5mm x 5mm which results in a number of 8 elements through the slab thickness. In the model, the total number of solid elements is 320,000 consisting of 363,639 nodes.

Journal of Physics: Conference Series **628** (2015) 012065 doi:10.1088/1742-6596/628/1/012065

2.1.2. Steel Reinforcement Meshing

The steel reinforcement mesh was modelled as one layer of 6 mm diameter reinforcement bars and spaced at 75 mm. The type of element assigned to bars was Hughes-Liu beam element with mesh size of 5 mm, knowing that either truss or beam element can be used to model reinforcement bars. The total number of beam elements is 6,000 consisting of 5,835 nodes.

2.2. Material Constitutive Models

The RC used for slabs in the experiment had the properties of 39.5 MPa compressive strength (f_c), 4.2 MPa tensile strength (f_t'), and 28.3 GPa Young's modulus (E_c). For concrete modelling, two constitutive models were selected. These models are namely Karagozian & Case Concrete (KCC) (MAT072R3) and the Winfrith concrete (MAT084) models. Both KCC and the Winfrith concrete models are invariant isotropic plasticity models, and they are suitable for blast problems, as they can capture the keys of concrete behaviours such as strain rate enhancement effect; however, the former proved to have better prediction for the localized shear. Both models take relatively simple input [8][9].

2.2.1. Concrete Constitutive Models

2.2.1.1 The KCC Concrete Model

The KCC (MAT072R3), is one of the material models that commonly used for concrete modelling. The main advantage offered by the KCC model that it allows the automatic generation of all the parameters by inputting only the unconfined compressive strength and the density of concrete [9]. For modelling, the default model parameter generation feature was used. This required the specification of the unconfined compression strength and density of concrete. For 39.5MPa concrete, the inputs in the model cards were as follows: concrete density (R0) 2400 kg/m^3, negative of the unconfined compressive strength (A0) -39.5 MPa, conversion factors for length factors for length (inches-to-meters) and pressure (psi-to-MPa) (RSIZE & UCF) 39.72 and 145 for respectively. The remaining parameters were generated and determined automatically by the software.

2.2.1.2 The Winfrith Concrete Model

The Winfrith concrete model (MAT084) is a smeared crack model implemented in the 8-node single integration point continuum element. Similar to MAT072R3, it also allows the automatic generation of all the parameters by inputting certain parameters [10]. Those imported to the software were as follows: mass density of concrete (R0) 2400 kg/m^3, initial tangent modulus of concrete (TM) 28.3 GPa, Possion's ratio (PR) 0.29, uniaxial compressive strength (UCS) 39.5 MPa, uniaxial tensile strength (UTS) 4.2 MPa and aggregate size (ASIZE)16 mm (assumed). The remaining parameters were generated and determined automatically by the software.

2.2.2. Steel Constitutive Models

The steel material properties in the slab reinforcement used in the experiment were as follows: 600 MPa yield stress (f_y), 200 GPa Young's modulus (E_s) and 0.29 Poisson's ratio (υ). MAT003, which suited to model isotropic and kinematic hardening, was used for steel modelling.

2.3. Blast Modelling

The detonations considered in this study were for three different TNT charge weights of 0.2, 0.31 and 0.46 kg which had 40 cm standoff distance. For blast simulation, the Lagrangian approach was used for applying the blast loads on the RC slab surface. LOAD_BLAST_ENHANCED (LBE) keyword provided by LS-DYNA was selected. In LBE, the air blast pressure is computed empirically with ConWep [10] data which is a collection of conventional weapons effects calculation from the equations and curves of TM 5-855-1 manual (similar to UFC 03-340-02 [7]) [11] [12].

3. Results and Discussions

3.1. Blast Pressure-Time History Results

The general blast wave characteristics provided by LBE were verified by observing the pressure-time histories which proved to follow the Friedlander curve as shown in Figure 4. The scaled distance Z for the 0.2, 0.31 and 0.46 kg TNT charge weights and 0.4 m standoff distance was 0.68, 0.59 and 0.52 m/kg$^{1/3}$ respectively. Since Z values are less than 0.71 m/kg$^{1/3}$, the detonation type for all charges is close-in [7]. The resulting pressures obtained from LBE for the TNT charge weights were relatively high, and their values in addition to other blast parameters, shown in Table 1, exhibited similar results as in UFC 03-340-02[7]. Pressure contours on slab for W_{TNT}=0.2 kg at different time steps Figure 5.

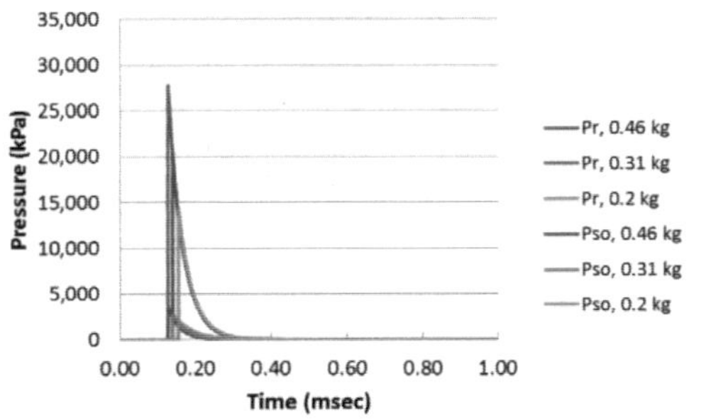

Figure 4. Pressure-time histories of reflected (P_r) and incident (P_{so}) pressures for 0.2, 0.31 and 0.46 kg TNT charge weights and 0.4 m standoff distance

Table 1. Positive phase blast parameters for different charge weights obtained from LBE

	Charge Weights (W_{TNT})		
	0.2 kg	0.31 kg	0.46 kg
Standoff distance, R (m)	0.40	0.40	0.40
Scaled distance, Z=R/W$^{1/3}$ (m/kg$^{1/3}$)	0.68	0.59	0.52
Overpressure, (refleceted pressure), P_r (kPa)	14216.00	20406.00	27813.00
Incident pressure, P_{so} (kPa)	2108.87	2804.48	3595.09
P_r/P_{so} ratio	6.74	7.28	7.74
Refleceted impulse, i_r (kPa.msec)	548.75	774.28	1065.95
Incident impulse, i_s (kPa.msec)	100.85	104.07	110.64
Arrival time, t_a (msec)	0.16	0.14	0.13
Positive phase duration, t_o (msec)	0.46	0.33	0.27

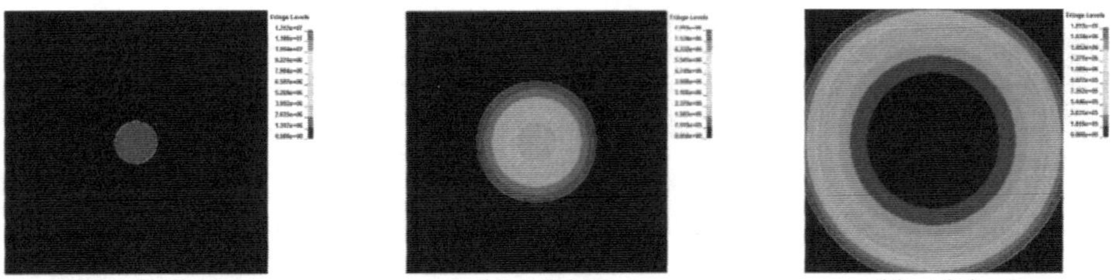

(a) at 0.16 msec (b) at 0.2 msec (c) at 0.4msec

Figure 5. Pressure contours for the applied reflected pressure on slab surface. W_{TNT} = 0.2 kg and R = 0.4 m (Z = 0.68 m/kg$^{1/3}$)

3.2. Deflection of Slab

The FE analysis indicated the pattern of deflection for the one-way square slab under blast loading through time. Before RC slab reached the one-way slab failure pattern, the square slab went through different stages. Since the blast loading was applied at the centre of slab and expanded circularly on the blast application surface, the slab correspondingly responded and started to deflect circularly when the load was applied. This deflection pattern continued until the unsupported edges started to respond. Then, the deflection was redistributed until the slab reached the one-way slab pattern. Both the KCC and Winfrith Concrete material models exhibited a similar behaviour. Graphical presentations for the slab deflection contours at different time instants for 0.2 kg TNT charge weight for the KCC material model is shown in Figure 6, and cross section of the slab at maximum deflection is shown in Figure 7.

Deflections were measured at the slab surface mid-point, and the displacement-time histories at this point for slabs with different concrete models are shown in Figure 8. In general, the maximum deflection, shown in Table 2, was noticed to be increased when the TNT charge weight was increased.

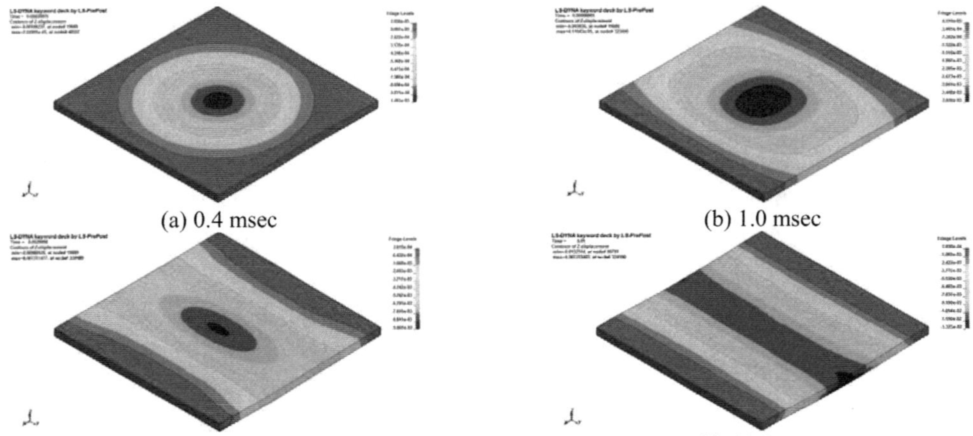

(a) 0.4 msec (b) 1.0 msec

(c) 3.0 msec (d) 10.0 msec

Figure 6. Pressure contours for the applied reflected pressure on slab surface. $W_{TNT} = 0.2$ kg and R = 0.4 m (Z = 0.68 m/kg$^{1/3}$)

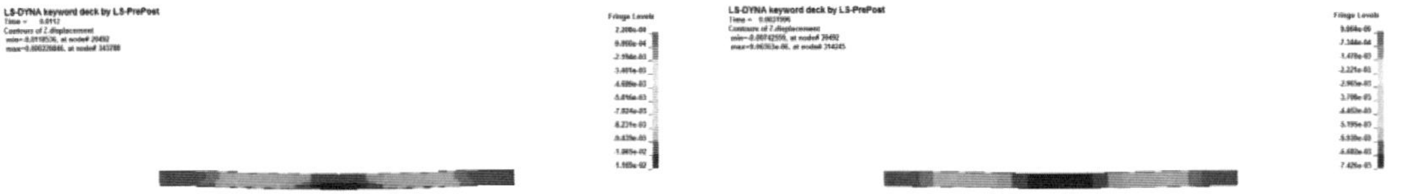

(a) KCC Model (MAT072R3) (b) The Winfrith Model (MAT084)

Figure 7. Cross sections at the middle of slab model in LS-DYNA along supports direction for maximum deflection. $W_{TNT} = 0.2$ kg, R = 0.4 m (Z = 0.68 m/kg$^{1/3}$)

Comparing results obtained from KCC and the Winfrith models with those form the experiment (as shown in Figure 9 and Table 2), it was noticed that both constitutive models estimated the maximum slab deflection effectively. While the former appeared to estimate it more accurately for 0.2 and 0.46 kg TNT charge weights.

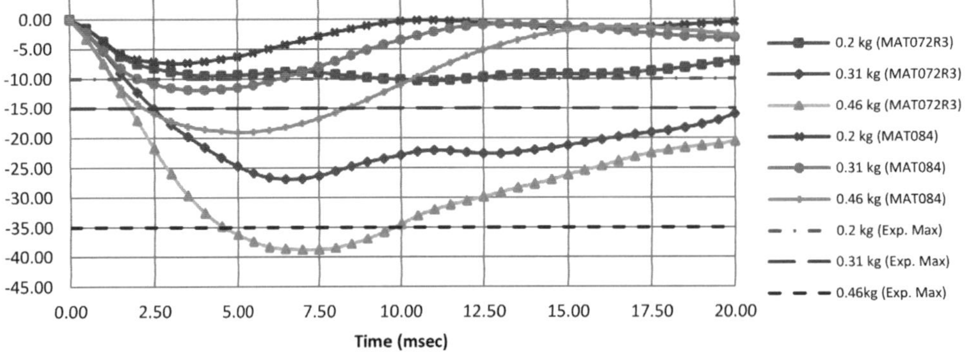

Figure 8. Displacement-time histories at middle of slab for the KCC (MAT072R3), the Winfrith concrete models (Exp. Max: maximum displacement in the experiment)

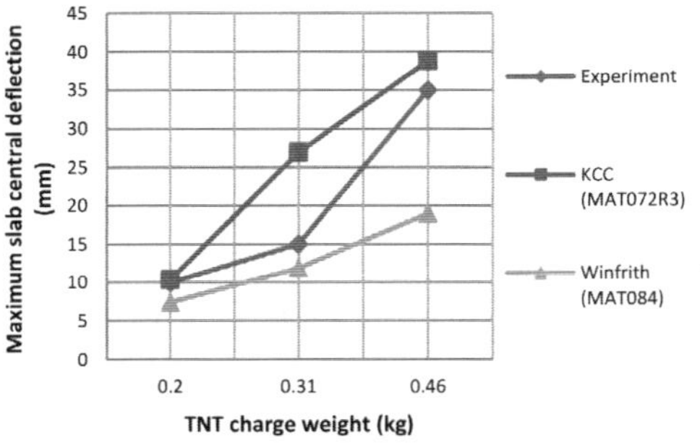

Figure 9. Variation in maximum slab central deflection

Table 2. Summary of slab maximum deflection for charge weights

Slab No.	W_{TNT} (kg)	R (m)	Z (m/kg$^{1/3}$)	Max. Deflection (mm)		
				Experiment	FEM (LS-DYNA)	
					KCC (MAT072R3)	Winfrith (MAT084)
1	0.2	0.40	0.68	10.00	10.36	7.41
2	0.31	0.40	0.59	15.00	26.92	11.87
3	0.46	0.40	0.52	35.00	38.84	18.99

3.3. Damage Modes

Damage modes obtained from models at maximum deflection were verified by observing the plastic strain in slabs. In the experiment [2], the actual slab damage modes were varied from low damage associated with cracks for W_{TNT}=0.2 kg (0.68 m/kg$^{1/3}$) and moderate damage associated with spallation, scabbing and cracks for W_{TNT}=0.31 and 0.46 kg (0.59 and 0.52 m/kg$^{1/3}$). These modes were compared with those in the numerical models [13]. As example, the results for W_{TNT}=0.31 (0.59 m/kg$^{1/3}$) are as shown in Figure 10 and Figure 11.

For the KCC model, the captured damage at slab top was the yielding at the supports. The model could not assess the spallation effectively because the LBE command applies blast loading using the Lagrangian approach which does not consider interaction between the ambient air and solid slabs is not considered [6]. While at the slab bottom the flexural cracks in addition to localised shear regions were appeared at supports and middle of slab, and they agreed approximately with the actual damage. On the other hand, the Winfrith concrete model appeared to behave more rigidly as the damage mode obtained was composed of a damage strip that expanded along the middle of the slab at top, and support yielding at slab bottom.

Therefore, the damage modes can be better estimated using the KCC model, and the interaction between the air and solid slab can be included using the ALE approach which will consequently result in the effective estimation of spallation such as the obtained in reference [2].

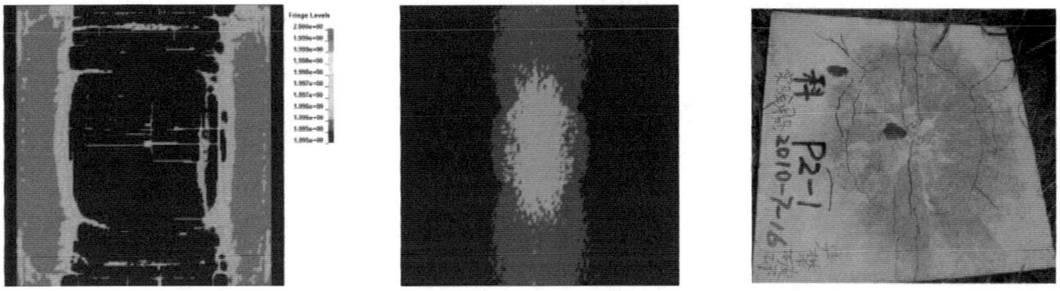

Figure 10. Damage mode of top of slab for 0.31 kg, and R = 0.4 m (Z = 0.59 m/kg1/3). KCC (left), the Winfrith (middle) the experiment [2] (right)

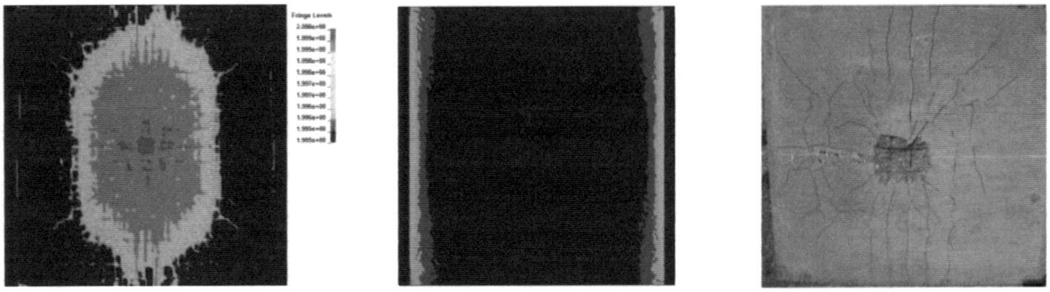

Figure 11. Damage mode of bottom of slab for 0.31 kg, and R = 0.4 m (Z = 0.59 m/kg1/3). KCC (left), the Winfrith (middle) the experiment [2] (right)

4. Conclusions and Recommendations

The numerical models of the one-way RC slab subjected to blast loads, which is applied using Lagrangian approach, for close-in detonations indicated that the slab deflection can be estimated satisfactorily. Both KCC and the Winfrith concrete models can indicate the slab deflection effectively, while the former demonstrated more accurate results. Also, the KCC appeared to be capable of capturing the flexural damage on slab and yielding at supports. While the spallation occur in the tension zone at the top of slab could not be captured. This effect can be achieved using the Arbitrary-Lagrangian Eulerian (ALE) approach which is capable to simulate the interaction between the ambient air and the solid slab.

References

[1] D. Hao and L. Zhongxian, "Numerical analysis of dynamic behavior of RC slabs under blast loading," *Transactions of Tianjin University,* vol. 15, no. 1, pp. 61-64, February 2009.

[2] W. Wang, D. Zhang, L. Fangyun, S.-C. Wang and F. Tang, "Experimental study and numerical simulation of the damage mode of a square reinforced concrete slab under close-in explosion," *Engineering Failure Analysis,* vol. 27, p. 41–51, 2013 .

[3] C. F. Zhao and J. Y. Chen, "Damage mechanism and mode of square reinforced concrete slab subjected to blast loading," *Theoretical and Applied Fracture Mechanics* , vol. 63–64, p. 54–62, (2013.

[4] Y. S. Tai, T. L. Chu, H. T. Hu and J. Y. Wu, "Dynamic response of a reinforced concrete slab subjected to air blast load," *Theoretical and Applied Fracture Mechanics,* vol. 56, p. 140–147, 2011.

[5] Z. S. Tabatabaei and J. S. Volz, "A Comparison between Three Different Blast Methods in LS-DYNA: LBE, MM-ALE, Coupling of LBE and MM-ALE," in *12th International LS-DYNA Users Conference*, Dearborn, Michigan, USA, 2007.

[6] Y. Lin, G. Shunfeng and J. Weiliang, "Spallation Mechanism of RC Slabs Under Contact Detonation," *Trans. Tianjin Univ.,* vol. 14, pp. 464-469, 2008.

[7] USA Department of Defence, "Structures to Resist The Effect of Accidental Explosions - Unified Facilities Criteria (UFC) 3-340-02", Department of Defence, 2008.

[8] L. Schwer and L. Malvar, "Simplified Concrete Modeling with *MAT_CONCRETE_DAMAGE_REL3," in *JRI LS-DYNA User Week 2005*, Nagoya, Japan, 2005.

[9] L. Schwer, "The Winfrith Concrete Model : Beauty or Beast ? Insights into the Winfrith Concrete Model," in *8th European LS-DYNA Users Conference*, Strasbourg, 2011.

[10] D. W. Hyde, "User's Guide for Computer Programs CONWEP and FUNPRO. Applications of TM 5-855-1, "Fundamentals of Protective Design for Conventional Weapons"," Washington, DC, 1988.

[11] G. Randers-Pehrson and K. A. Bannister, "Airblast Loading Model for DYNA2D and DYNA3D," Adelphi, Maryland, 1997.

[12] L.S.T.C., LS-DYNA Keyword User's Manual, Livermore: Livermore Software Technology Corporation (LSTC), 2007.

[13] M. M. N. Shuaib, "Numerical Modeling of Reinforced Concrete Slab under Blast Loads", M.Sc. Thesis, University of Khartoum, Khartoum, Sudan, 2014.

The evaluation of damage mechanism of unreinforced masonry buildings after Van (2011) and Elazig (2010) Earthquakes

D Güney[1], **E Aydın**[2] , **B Öztürk**[3]
[1]Yildiz Technical University, Faculty of Architecture, Dep. of Arch., Istanbul, Turkey
[2]Nigde University, Faculty of Engineering, Civil Eng. Dep., Nigde, Turkey
[3]Hacettepe University, Faculty of Engineering, Civil Eng. Dep., Ankara, Turkey

E-mail: deguney@yildiz.edu.tr

Abstract. On March 8th, 2010 Karakocan-Elazig earthquake of magnitude 6.0 occurred at a region where masonry and adobe construction is very common. Karakocan-Elazig is located in a high seismicity region on Eastern Anatolian Fault System (EAFS). Due to the earthquake, 42 people were killed and 14'113 buildings were damaged. Another city, Van located at South east of Turkey is hit by earthquakes with M = 7.2 occurred on October 23rd, 2011 at 13:41 (local time), whose epicenter was about 16 km north of Van (Tabanlı village) and M = 5.6 on November 9th, 2011 with an epicenter near the town of Edremit, south of Van and caused the loss of life and heavy damages. Both earthquakes killed 644 people and 2608 people were injured. Approximately 10'000 buildings were seriously damaged. There are many traditional types of structures existing in the region hit by earthquakes (both Van and Elazig). These buildings were built as adobe, unreinforced masonry or mixed type. These types of buildings are very common in rural areas (especially south and east) of Turkey because of easy workmanship and cheap construction cost. Many of those traditional type structures experienced serious damages. The use of masonry is very common in some of the world's most hazard-prone regions, such as in Latin America, Africa, the Indian subcontinent and other parts of Asia, the Middle East, and southern Europe. Based on damage and failure mechanism of those buildings, the parameters affecting the seismic performance of those traditional buildings are analyzed in this paper. The foundation type, soil conditions, production method of the masonry blocks, construction method, the geometry of the masonry walls, workmanship quality, existence of wooden beams, type of roof, mortar between adobe blocks are studied in order to understand the reason of damage for these types of buildings.

1. Introduction

On October 23rd, 2011 Van city in eastern Turkey was hit by a large earthquake at 13:41 (10:41 GMT), on Sunday afternoon of magnitude 7.2. This earthquake had a shallow hypocenter depth of about 10 km. Effective duration of the earthquake according to Muradiye station records, was 20 sec, while it was 18 sec. according to Bitlis station records. The Van earthquake, where epicenter was about 16 km north of Van province, between Ercis county with a population of about 77 000 and Van

1 Correspondence author: Deniz Güney, Yildiz Technical University, Istanbul, Turkey
2 Ersin Aydın: Nigde University, Nigde, Turkey.
3 Baki Öztürk: Hacettepe University, Ankara, Turkey

city with population of about 370 000, has devastated the area, demolished many buildings with hundreds of people dead and thousands injured under the ruins. The location of epicenter can be seen in Fig. 1. The earthquake mainly affected Ercis County that is 90 km away from Van city. Hundreds of buildings totally collapsed; thousands of them were heavily damaged and 644 people died (Table 1). The 604 deaths during the first earthquake are 61 in the center, 66 in villages in the vicinity and 477 in Ercis; 40 people died because of second earthquake. The total economic loss is about 1 billion Turkish Lira (TL) to 4 billion TL (approx. 555 million–2.2 billion USD). This would represent around 17 to 66% of the provincial GDP (Gross Domestic Product) of Van [1].

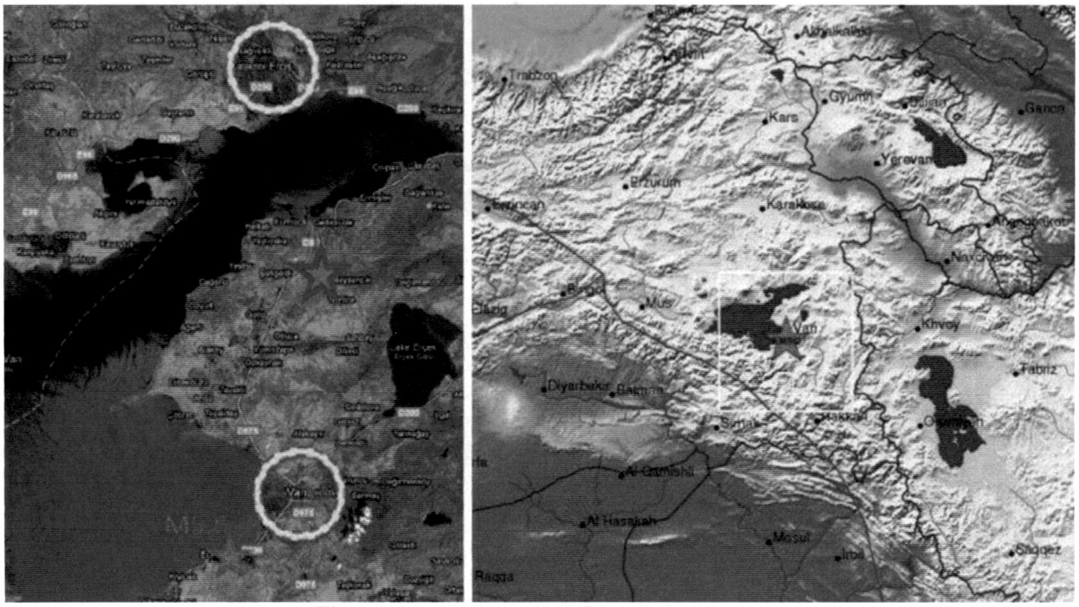

Figure 1. Location of earthquake epicenter

Table 1. Human losses and collapsed buildings.

Location	Life Losses	Injured	Total Collapsed Buildings
Center, Van	101	1150	10
Erciş, Van	477	1058	100
Villages	66	400	2197
Total	644	2608	2307

According to the information given by AFAD (The Disaster and Emergency Management Presidency), 644 people lost their lives and 252 people were saved alive from the debris. AFAD informed that, by December 9th, 2011, 17'005 dwelling units were determined as collapsed and/or heavily damaged in Van City Center, Erciş and villages. AFAD has reported deaths, Injured, Medical Personnel and Search and Rescue trends between 23.10.2011 and 1.11.2011 as shown in Figure 2.

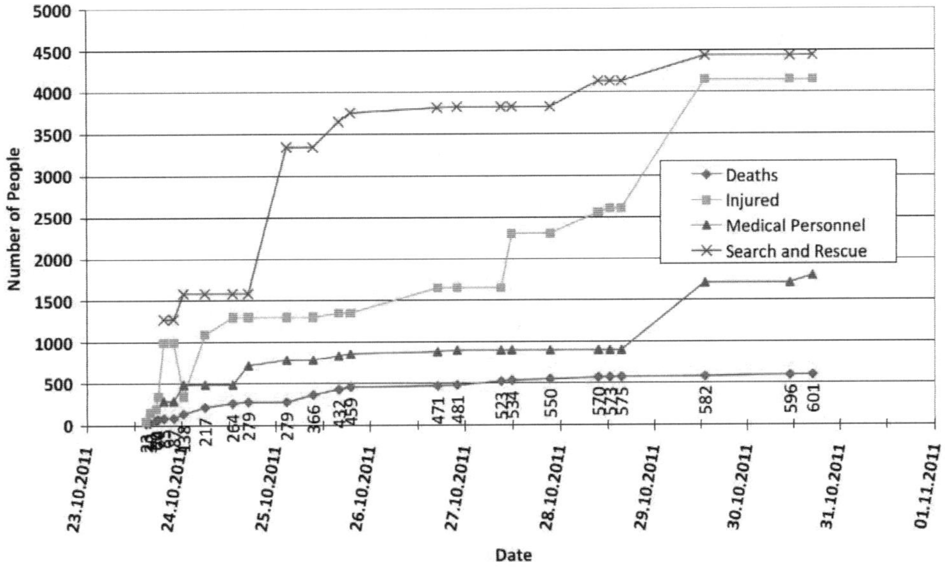

Figure 2. Deaths, Injured, Medical Personnel and Search and Rescue trends as of November 1st, 2011 (AFAD)

On November 9th, 2011, the second Van-Edremit centered earthquake occurred at 21.23 (18.23 GMT) of magnitude 5.6. The epicenter of the earthquake was near the Edremit town, south of Van (Figure 1). Effective duration of the earthquake according to Van station records was 18 sec, according to Van-Edremit station records was 23 sec. During the second earthquake 40 people lost their lives. This earthquake caused collapse or heavy damage of the buildings having slight or medium damage in Van city center and death of 40 people [1, 2]. On October 23rd, 2011, Van city in eastern Turkey was hit by a large earthquake at 13:41 (10:41 GMT), on Sunday afternoon of magnitude 7.2. This earthquake had a shallow hypocenter with a depth of about 10 km. Effective duration of the earthquake according to Muradiye station records, was 20 sec, according to Bitlis station records was 18 sec. The Van earthquake, where epicenter was about 16 km north of Van province, between Ercis county with population of about 77 000 and Van city with population of about 370 000, has devastated the area, demolished many buildings with hundreds of people dead and thousands injured under the ruins. The location of epicenter can be seen in Fig. 1. The earthquake mainly affected Ercis County that is 90 km away from Van city. Hundreds of buildings totally collapsed; thousands of them were heavily damaged and 644 people died (Table 1). The 604 deaths during the first earthquake are 61 in the center, 66 in villages in the vicinity and 477 in Ercis; 40 people died because of second earthquake. The total economic loss is about 1 billion Turkish Lira (TL) to 4 billion TL (approx. 555 million–2.2 billion USD). This would represent around 17 to 66% of the provincial GDP (Gross Domestic Product) of Van [1].

On November 9th, 2011, the second Van-Edremit centered earthquake occurred at 21.23 (18.23 GMT) of magnitude 5.6. The epicenter of the earthquake was near the Edremit town, south of Van (Figure 1). Effective duration of the earthquake according to Van station records was 18 sec, according to Van-Edremit station records was 23 sec. During the second earthquake 40 people were killed. This earthquake caused collapse or heavy damage of the buildings having slight or medium damage in Van city center and death of 40 people [1, 2].

2. Seismic Properties of the Region

In the area of Lake Van and further east, tectonics is dominated by the Bitlis suture zone (in eastern Turkey) and Zagros fold and thrust belt (towards Iran). The October 23rd, 2011 earthquake occurred in a broad region of convergence beyond the eastern extent of Anatolian strike-slip tectonics. The focal mechanism of recent earthquakes is consistent with oblique-thrust faulting similar to mapped faults in the region. Given its tectonic history, a major earthquake in Anatolia is by no means an unusual event and other major earthquake events are to be expected in the region as the central block continues to be squeezed westwards and lateral movement occurs along the fault complexes of both North and East Anatolian Faults (Figure 3).

Major earthquakes such as this one have occurred in the year 1111 causing major damage and having a magnitude around 6.5-7. In the year 1646 or 1648, Van was again struck by a M 6.7 quake killing around 2000 people. In 1881, a M6.3 earthquake near Van killed 95 people. Again, in 1941, a M5.9 earthquake affected Ercis and Van killing between 190 and 430 people. 1945-1946 as well as 1972 brought again damaging and casualty-bearing earthquakes to the Van province.

In 1976, the Van-Muradiye earthquake struck the border region with a M7, killing around 3'840 people and causing around 51'000 people to become homeless. In recent years (according to historical records from 1900), there were 10 earthquakes which happened with a magnitude in between 5-6, three earthquakes with a magnitude in between 6-7, two earthquakes with a magnitude in between 7-7.5. The damages and human loss can be seen in Figure 4.

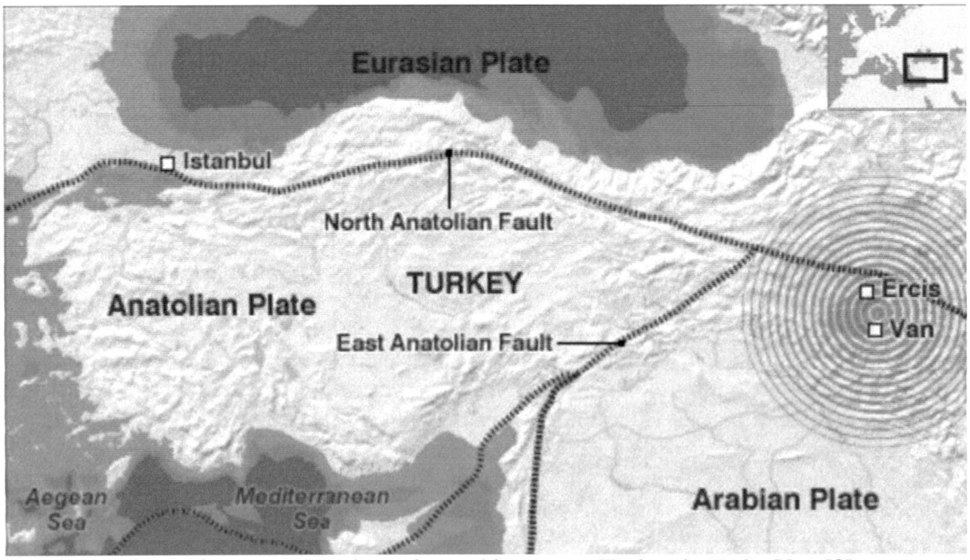

Figure 3. Van and Erciş are located in eastern Turkey by Lake Van [3]

The damages and human loss can be seen in Figure 3. The energy released by earthquake is calculated as 2.09x1015 Joule. This energy is 33.2 times more than Hiroshima atomic bomb explosion energy [4].

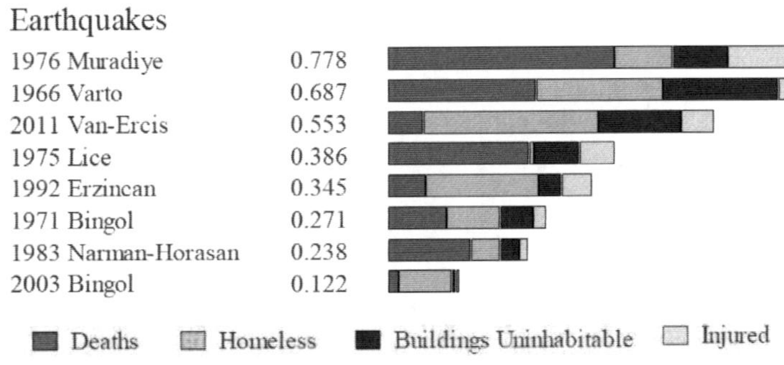

Figure 4. Loss score for historic earthquakes in eastern Turkey

As of the October 31st, 2011 reported from Turkish Red Crescent, of the 43'548 buildings that have been examined, 2'309 have collapsed, 11'847 have been severely damaged or are uninhabitable. 17'923 houses have been slightly or moderately damaged. In addition, 11'469 buildings have been undamaged as shown in Figure 5. A group of almost 200 staff have been undertaking preliminary assessment of the damaged buildings. In the Van City, the natural gas system (building collapse on a system regulator), water supply system (pipeline damage), the power and communication systems (general interruptions) were all affected, however were reported to be functional again within 24 hours after the earthquake. The Van-Ercis road was also reported to have been damaged in the form of road collapse and cracking.

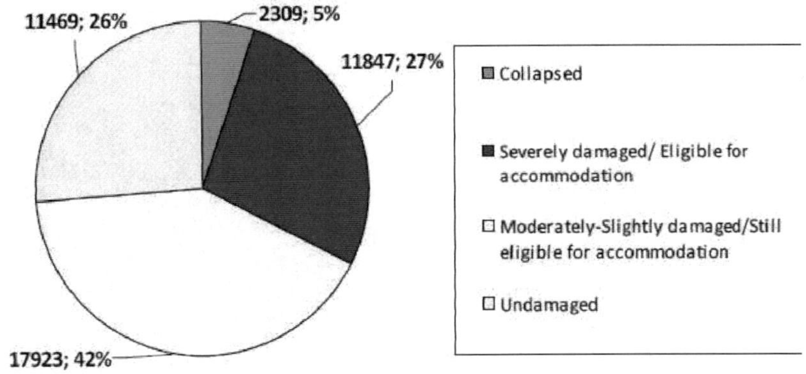

Figure 5. The current statistics of building damage as determined by different damage classes [5]

Unreinforced masonry has traditionally been the primary construction method of rural areas in Turkey. Though reinforced concrete was introduced during the first decade of 1900, the adobe, stone and later concrete in rural houses using local materials and unskilled labour continued to be constructed. In the region, there are many houses and public buildings constructed with stone, brick, adobe and concrete blocks. Some of them were heavily affected where some had slight damages. Most of the buildings in the region were built without any design, control or supervision. In addition to lack of design and control of mixed buildings, the material and workmanship quality was much less than desired level [6]. This earthquake caused extensive damages not only to the reinforced concrete structures but also unreinforced masonry buildings. Many of the damaged building types were stone, brick or briquette masonry or adobe with low construction and material quality. In this paper, the results of the site surveys are presented and the lessons learned from the earthquake and structural damages are discussed.

3. Earthquakes and Consequences

On October 23rd, 2011 Van-Merkez earthquake is unique from several aspects. Very high number of aftershocks within short period after the event was not experienced previously. Within the first week of the earthquake, there happened 114 earthquakes with magnitudes between 4.0 and 4.9 and 7 earthquakes with magnitudes bigger then M l:5.0. Within the first month after the event daily average aftershock number is around 180 earthquakes. By December 9th, 2011, the number of aftershocks reached to 6284. Focal depths of aftershocks vary between 2.5 km to 25 km (AFAD). The area between Van and Erciş is tectonically complex and there are several faults with different characteristics. The reason for such big amount of aftershocks and diversity of the focal mechanism solutions are due to this tectonic complexity. Very generally, earthquake with Mw:7.0 at 19 km. depth activated this systems and small scaled faults triggered one and each other within this period and increased the earthquake activity. It was observed, as a result of the field studies, the fault (N60,70E) that caused earthquake was a buried fault. Morphological trace of fault is observed on the shaded relief map and 3D digital elevation model map as shown in Figure 6.

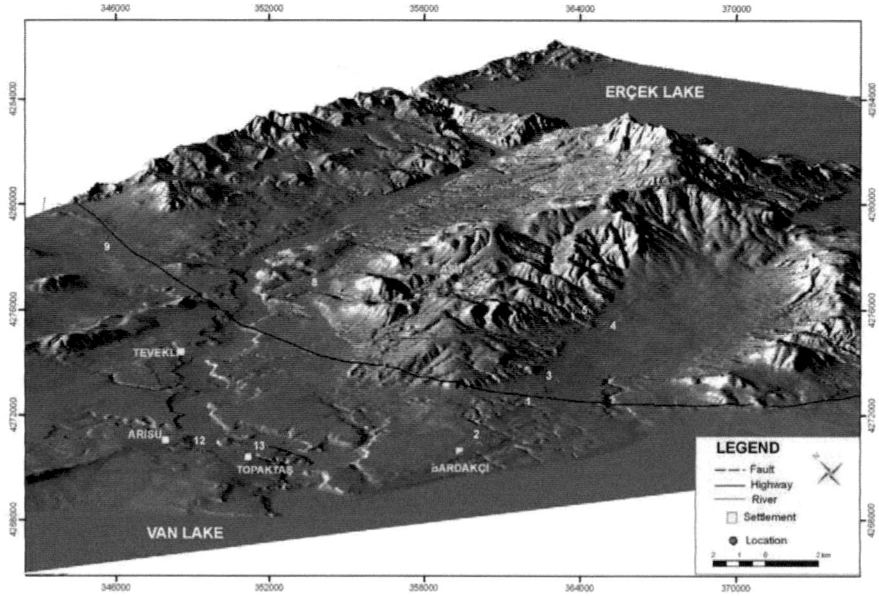

Figure 6. 3D Digital elevation model map (AFAD)

To describe earthquake force; Acceleration, velocity and displacement response spectrum of acceleration records are calculated by AFAD. In order to Ml=6.7 Van-Merkez earthquake and Ml=5.6 Van-Edremit earthquake, response spectrum that is obtained from acceleration records were calculated for 5%, 10% and 15% damping ratios. The results can be followed from Figure 7a and 7b.

Compare with acceleration response spectrum and design spectrum (according to Turkish Earthquake Resistant Code (TDY) 2007) of October 23rd, 2011 Van earthquake Ml=6.7 and November 9th, 2011 Van-Edremit earthquake Ml=5.6 are given in Figs. 8a and 8b. When examined at the calculated response spectrum curve, it is seen that each one of two ground motions under the design spectrum that identified for first degree earthquake zone.

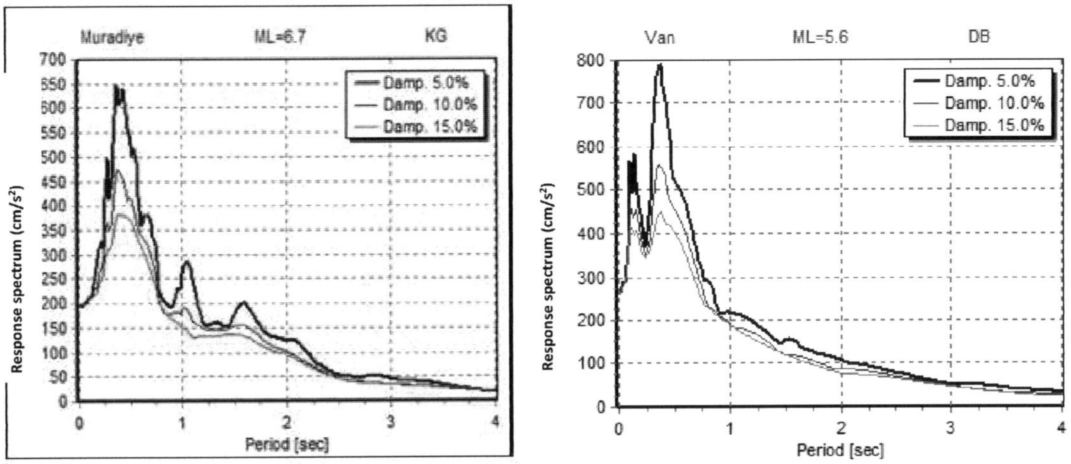

a) b)

Figure 7. Response spectra of October 23rd, 2011, a) Ml=6.7 Van-Merkez earthquake Muradiye record NS component, b) Ml=5.6 Van-Edremit earthquake Van-Merkez record EW component (AFAD)

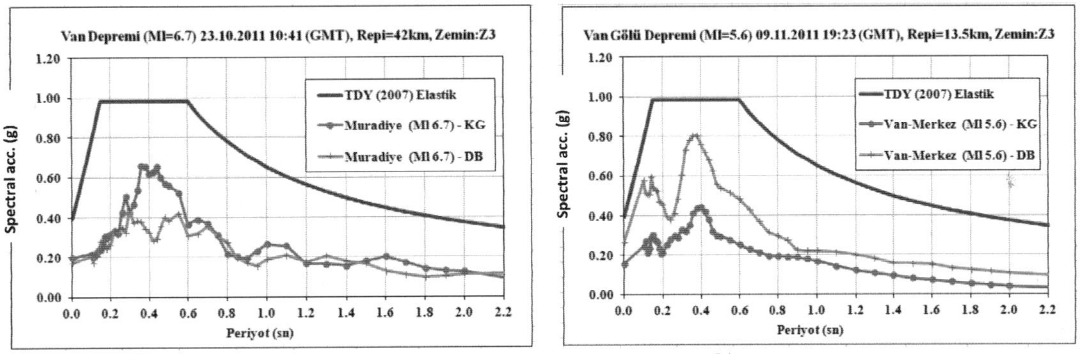

a) b)

Figure 8. Comparison of NS and EW component response spectrum and TDY 2007 design spectrum a) October 23rd, 2011 Van earthquake Ml=6.7 and b) November 9th, 2011 Van-Edremit earthquake (AFAD)

Building stock in Van and Erciş Center generally consists of 4-8 storey reinforced concrete structures, which is very common in Turkey. In most of the buildings asmolen slab (infilled joist slab) is used. Especially in collapsed buildings, commercial parts such as shops or markets having almost two times normal floor height has been determined. In rural areas, most of the existing building stock is comprised of adobe, stone and brick masonry buildings with ages longer than their service life. They are constructed as one or two-storey by local people without taking into consideration any regulation, standard and earthquake resistant design rules. It is observed in the masonry structures at this region that horizontal and vertical supporting members, used to distribute loads safely, are made from wood, number of these members is inadequate and they are placed irregularly. Also, it is determined that lengths of their connections to load carrying walls are very short and weak. Briefly, poor quality construction material, structures with non-conforming earthquake code and lack of inspection are the main reasons of damage in the region.

Journal of Physics: Conference Series **628** (2015) 012066 doi:10.1088/1742-6596/628/1/012066

4. Damage Mechanism of Unreinforced Masonry Buildings

According to site investigations, unreinforced masonry building houses are more common building type in the region hit by earthquake. The distribution of the buildings according to construction types and location is given in Table 2. As shown in the table, 75% of the buildings in Van (towns and villages included) is unreinforced masonry. The unreinforced masonry building ratio in Ercis is 63%, in Muradiye is 81%, in the city center of Van is 82%.

Table 2. Distribution of investigated buildings

	Van (Total)	Ercis	Muradiye	Center
Number of buildings	78.000	10.700	3.600	35.200
RC Buildings	12.7 %	27 %	5 %	5 %
Unreinforced masonry	75 %	63 %	81 %	82 %
Adobe	9.5 %	8 %	12 %	9 %
Rubble stone	2.8 %	2 %	2 %	4 %

Unreinforced masonry buildings mean that masonry infills in non-ductile reinforced concrete frames can be found in many places around the world. Masonry infills have been widely used because of their good thermal and acoustic insulation properties both for aesthetic reasons and fire resistance. Although in design masonry infills are considered as non-structural elements, they can develop a strong interaction with the bounding frames when subjected to earthquake loads and, therefore, contribute significantly to the lateral stiffness and load resistance of the structure.

Architectural characteristics of unreinforced masonry buildings are similar in the region: the rectangular plan, single door, and small lateral windows are predominant. Quality of construction in urban areas is generally better than construction in rural areas. The foundation, if present, is made of medium to-large stones joined with mud or coarse mortar. Walls are made with briquette or hollow brick with mud or cement mortar. The size of briquette varies from region to region. Generally the size of briquette is about 12x15x34 cm and the weight is about 10-18 kg. Average size of hollow brick is about 19x13.5x19 cm and the weight is about 2.6-3.0 kg [8]. Average size of adobe block size varies between 30x15x10 cm and 35x17x12 cm and the weight is about 6-8 kg.

Figure 9. Briquette, hollow brick, adobe, stone

In addition to briquette and hollow brick buildings, adobe, brick, stone masonry or timber houses are also common in the region. Generally the masonry houses are built without wooden confinements. When horizontal and vertical ties do not exist, horizontal forces cannot be taken by thin adobe walls and damage occurs.

A typical masonry wall consists of piers between openings plus a portion below openings (sill masonry) and above openings (spandrel masonry), as shown in Figure 10a. When subjected to in-plane earthquake shaking, masonry walls demonstrate either rocking or diagonal cracking. Rocking is illustrated in Figure 10b, and is characterized by the rotation of an entire pier, which results in the crushing of pier end zones. Alternatively, masonry piers subjected to shear forces can experience diagonal shear cracking (also known as X-cracking), as shown in Figure 10c. Diagonal cracks develop when tensile stresses in the pier exceed the masonry tensile strength, which is inherently very low. This type of damage is typically observed in the bottom story of a building. Several factors influence the in-plane failure mechanism of stone masonry buildings, including pier dimensions; wall thickness, building height, and masonry shear strength. Rocking behavior is more desirable than diagonal shear cracking. [9]

Out-of-plane wall collapse is one of the major causes of destruction in stone masonry buildings, particularly in buildings with flexible floors and roofs. The connections between structural components are important for maintaining building integrity [10]. Integrity is absent or inadequate when the walls are not connected at their intersections and there are no ties or ring beams at the floor and roof levels. As a result, each wall vibrates on its own when subjected to earthquake ground shaking and is therefore likely to total or partial collapse. The typical failure mode in these structures was out-of-plane wall failure due to the absence of diaphragms tying the walls together. In multi-story buildings, this type of collapse usually takes place at the top floor level due to the significant earthquake accelerations there.

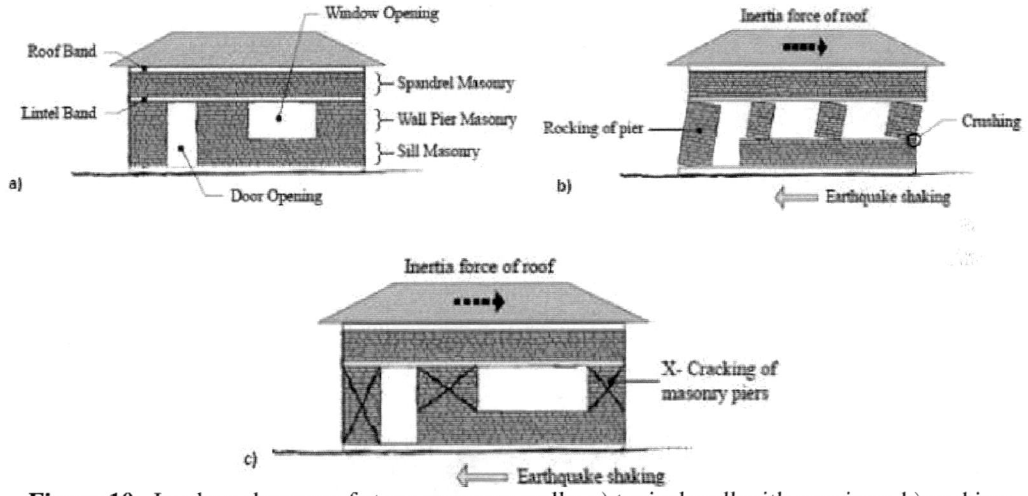

Figure 10. In-plane damage of stone masonry walls: a) typical wall with openings; b) rocking failure, and c) diagonal shear cracking [10]

Unreinforced masonry buildings are very common in the region because of socio-economic reasons. Annual income average in this region is far lower than the other regions of Turkey. In addition to this, education level is too low. Therefore people do go to the engineers or architects for preparing their projects. Generally they prefer local uneducated or uncertified constructors for construction of their buildings. Unfortunately local authorities do not properly control and supervise construction activities in the region. Damage observed in masonry infilled frame structures varied from small cracking to severe damage (diagonal cracks, out-of-plane failure etc) and collapse.

The masonry buildings in the region hit by the earthquakes were constructed by mud bricks, stones/pebbles taken from the river banks, soft natural stone blocks and lime hollow/solid blocks.

Adobe buildings were constructed from adobe bricks made by formation of mud with wooden bracings. These materials have been used as structural material; however, they do not fulfill code or any provision requirements [11].

After October 23rd, 2011 and November 9th, 2011 earthquakes, a detailed site investigation was performed in the region hit by earthquakes. Especially damaged or collapsed unreinforced masonry buildings were observed to discover their damage reasons.

One of the most important reasons was the use of improper construction material. Rubble stone, briquette and soft stone blocks were the masonry materials widely used in the damaged or collapsed buildings. Their quality is low and production is out of standard. Therefore low quality materials caused serious damages as shown in Figure 11.

Figure 11. Low quality briquette and adobe

The use of mud mortar for adobe and briquette blocks cause low lateral force resistance. Mud as mortar has low strength and has poor bonding characteristics. As a result of this, masonry buildings performed very poor lateral resistance to seismic forces and damaged or collapsed easily as shown in Figure 12.

Figure 12. Improper mortar use such as mud

In addition to previous reasons, construction method is also against basic structural design criteria. Wall-to-wall and wall-to-floor connections are very poor. Therefore infill walls collapsed easily during the earthquake as shown in Figure 13.

Figure 13. Poor wall-to-wall and wall-to-floor connections

Many damaged buildings have flexible floor diaphragm, which prevents the transfer and distribution of lateral forces in a uniform manner as shown in Figure 14.

Figure 14. Flexible floor diaphragm

Improper placement of door and window openings in walls caused serious damages in the region, which creates vulnerable and weak zones in the structure as shown in Figure 15.

Figure 15. Improper placement of door and window openings

5. Conclusions

Unreinforced masonry is a widely used building type all over the world and also in the investigated area. In rural areas, it is preferred because of economic, easy, simple workmanship construction type. However; seismic performance of the unreinforced masonry buildings is generally less than the desired level. National Construction Control and Supervising Law was not a mandatory regulation in the region until 2011. In rural areas of Turkey, seismic codes for buildings were not strictly enforced as much as they were enforced in urban areas. Therefore; the buildings in the region didn't have adequate engineering assistance. Most of the houses used different construction materials and systems in the same building.

Unreinforced masonry houses in the villages of the Van city were seriously affected from the earthquake. Similar to adobe buildings, improper low quality reinforced concrete or unreinforced masonry houses experienced serious damages in the region.

The serious damage of masonry infill walls subjected to out of-plane accelerations makes load-bearing masonry buildings very vulnerable to seismic damage and possible collapse. Due to the dynamic interaction between the vibrating structure, slab diaphragms and the infill wall loaded out-of-plane, enormously increased accelerations occur on the face of loaded infill wall, resulting in greatly increased inertia forces.

Traditional masonry buildings, regardless of their important architectural or cultural value, are also prone to suffering damage during strong earthquakes. Thus, it is important to provide adequate preservation methods to ensure safety and their authenticity. Using simple and economic confinement techniques, seismic performance of the unreinforced houses can be provided. In order to reach to this target, education of local workers and contractors is very important. Skilled workmanship will increase seismic resistance of unreinforced masonry buildings. Construction of the houses should be controlled and supervised by experienced engineers and architects assigned by the government. It is necessary to develop guidelines for the construction of unreinforced masonry buildings in rural areas.

References

[1] Disaster and Emergency Management Directorate (AFAD) Report: *Press Bulletin on Oct.23rd, and Nov. 9th, 2011 Van EQ. (in Turkish)*, November 24th, 2011. http://www.deprem.gov.tr/sarbis/Shared/WebBelge.aspx?param=103, November 24th, 2011.

[2] Güney, D., Technical Report of October 23rd, 2011 Van Earthquake, (in Turkish), YTÜ, No:YTÜ.MF.KP-2013.0872, *Yildiz Technical University Publication*, İstanbul, 2013.

[3] British Broadcasting Corporation (BBC): Turkey earthquake: Survivors outdoors on freezing night, http://www.bbc.co.uk/news/world-europe-15437014, September 28th, 2011.

[4] Özkaymak, Ç., Sağlam, A., Köse, O.: Van Gölü doğusu aktif tektonik özellikleri, Aktif Tektonik Araştırma Grubu 7. Toplantısı – ATAG VII, Van, Turkey, 2003.

[5] Turkish Red Crescent Society (TRCS) (2011) "Van Earthquake Humanitarian Aid Operation, Operations Update 1-10", available from URL: http://reliefweb.int/node/456744.

[6] Güney, D., Kuruşcu, A., O., Arun, G., Damage Evaluation of Adobe Houses after Van Earthquakes (October 23rd, 2011 and November 9th, 2011), *International Conference Kerpiç'13*, 11-15 September, IAU Istanbul, Turkey.

[7] AFAD Report, Disaster and Emergency Management Presidency, http://www.afad.gov.tr, 12.12.2011.

[8] Güney, D., Kuruscu, A., O., Arun, G., Damage Evaluatıon of Masonry Houses after Van Earthquakes (October 23rd, 2011 and November 9th, 2011), *9th International Conference on Structural Analysis of Historical Constructions*, 14-17 October 2014, Mexico City, Mexico.

[9] Murty, C. V. R., 2005. IITK-BMTPC Earthquake Tips: Learning Earthquake Design and Construction, National Information Centre of Earthquake Engineering, IIT Kanpur, India, www.nicee.org.

[10] Bothara, J., Brzev, S., A tutorial: Improving the Seismic Performance of Stone Masonry Buildings, July 2011, EERI, Publication Number WHE-2011-01.

[11] Güney, D.: Van Earthquakes (23.10.2011 and 9.11.2011) and Performance of Masonry and Adobe Structures, *Nat. Hazards Earth Syst. Sci.*, Vol. **12(11)**, pp. 3337-3342, November 14th , 2012.

Damage Detection in Wind Turbine Towers using a Finite Element Model and Discrete Wavelet Transform of Strain Signals

A. Kenna[1], B. Basu[2*]

[1] Dept. Civil Structural & Environmental Engineering, Trinity College Dublin, Ireland
[2] Dept. Civil Structural & Environmental Engineering, Trinity College Dublin, Ireland

Abstract. Wind turbine support towers at heights in excess of 90m are nowadays being formed in steel, concrete and hybrid concrete and steel structures. As is the case for all towers of this height, the towers will be assembled using a number of segments, which will be connected in some way. These local connections are to be viewed as areas of potential local weakness in the overall tower assembly and require care in terms of design and construction. This work concentrates on identifying local damage which can occur at an interface connection by either material or bolt/tendon failure. Spatial strain patterns will be used to try to identify local damage areas around a 3-dimensional tower shell. A Finite Element (FE) model will be assembled which will describe a hybrid tower as a continuum of four-noded, two-dimensional Reisser-Mindlin shell elements. In order to simulate local damage, an element around the circumference of the tower interface will be subjected to a reduced stiffness. Strain patterns will be observed both in the undamaged and damaged states and these signals will be processed using a Discrete Wavelet Transform (DWT) algorithm to investigate if the damaged element can be identified.

1. Modelling of the Wind Turbine Tower

A well-established means of generating the elemental stiffness of finite elements in engineering is to follow a variational functional approach commonly used in structural mechanics based on the potential energy functional [1]. What results is an integral expression which implicitly contains the differential equations and is known as the weak form of the problem statement. This method was employed by Kenna and Basu [2] to generate local stiffness matrices for shell elements with in-plane and out-of plane degrees of freedom which were then assembled into a three-dimensional wind turbine FE model. Each model elemental stiffness matrix, $[k]$ is expressed as

$$k = \int_v [B]^T [C][B] dV \quad (1)$$

* basub@tcd.ie

The term $[C]$ here is the material constitutive description. The element form used in defining the tower continuum in this paper is an isoparametric bilinear quadrilateral, with membrane and plate bending properties to give the element shell capabilities. The membrane stiffness matrix is to be mapped to local co-ordinates and integrated numerically.

$$k_m = t\iint \{B_m\}^T [C_m]\{B_m\}\det[J]d\varepsilon d\eta \tag{2}$$

The term t is the thickness of the shell element. The subscript 'm' here denotes the application to the membrane degrees of freedom of the element. The term $[J]$ denotes the Jacobian matrix, used to transform derivatives of displacements with respect to x and y to those with respect of ε and η. For the out-of-plane bending and transverse displacement degrees of freedom, a separate portion of the element stiffness matrix must be set up as outlined in a text by Zienkiewicz [3]. The Mindlin plate stiffness matrix is defined as follows

$$k_p = \iint [B_b]^T [C_b][B_b]\det[J]d\varepsilon d\eta + \iint [B_s]^T [C_s][B_s]\det[J]d\varepsilon d\eta \tag{3}$$

where the subscripts '*b*' and '*s*' relate to the bending and shear strain-displacement respectively. Once the membrane and plate co-efficients are in place for each element, they are then combined to form the shell element elastic stiffness matrix as follows.

$$k_{elastic} = \begin{bmatrix} [k_m]_{8\times8} & [0]_{8\times12} \\ [0]_{12\times8} & [k_p]_{12\times12} \end{bmatrix} \tag{4}$$

Note that there are no coupling terms between the membrane and bending co-efficients in this matrix. The material elasticity matrices used in defining the membrane and plate co-efficients are as follows

$$C_m = \frac{E_c}{(1-v^2)}\begin{bmatrix} 1 & v & 0 \\ v & 1 & 0 \\ 0 & 0 & \frac{1-v}{2} \end{bmatrix} \quad C_b = \frac{E_c t^3}{12(1-v^2)}\begin{bmatrix} 1 & v & 0 \\ v & 1 & 0 \\ 0 & 0 & \frac{1-v}{2} \end{bmatrix} \quad C_s = \frac{E_c t 5/6}{2(1+v)}\begin{bmatrix} 1 & 0 \\ 0 & 1 \end{bmatrix} \tag{5}$$

E_c is the modulus of concrete and v denotes the poissons ratio. It is noted that the same matrix assemblies are used in the case of steel elements, where an appropriate modulus and poissons ratio for steel are used. The above local element quantities are defined for each element in the tower model before being transformed to the global structural co-ordinate system and finally assembled to a global stiffness matrix.The global matrix [K] is simply the summation of all individual element matrices which have been transformed as above.

$$[K] = \sum_{i=1}^{numelements} [K_G]_i \tag{6}$$

2. Load Application

The load to be applied to the tower will be a lateral tower top load which will simulate the loading applied by the rotor. The Blade Element Momentum (BEM) method is used to determine a suitable load for this thrust, whereby the aerodynamic properties of an airfoil (rotor blade) are utilised in tandem with the prevailing wind speeds and blade rotational speed to generate nodal drag and lift loadings acting on elements of the blades. Hansen [4] describes this method in detail in his text on the subject. The NREL 5mw baseline wind turbine [5] was used in developing the BEM loading information in terms of the airfoil aerodynamic properties and blade geometry and material properties. The speed of rotation was held to a constant 12.1rpm/0.201Hz, which is the normal operating rotation speed for this rotor. The algorithm was run a number of times to assess the variation of load at different normal operating wind speeds. The results are provided in Table 1.

Table 1. Rotor Thrust Loads

Wind Speed (mps)	Thrust (kN)
7.000	370.927
8.000	450.605
9.000	534.139
10.000	617.821
11.000	699.506
12.000	771.966
13.000	833.871

3. Damage Detection using Wavelet Transformation

Wavelet transformation can be seen as being an extension of the Fourier transform, being a modern means of decomposing a temporal or spatial signal. Chang and Chen [6] presented a paper proposing the use of a spatial based wavelet approach in order to detect the location of a crack in a rotating blade. They found that the distributions of wavelet coefficients can identify the crack location by showing a peak at the position of the cracking. Liew and Wang [7] presented an earlier work whereby they showed that wavelets can be used to identify a crack location in an Euler-Bernoulli beam element. Law et al [8] studied the use of wavelets from the point of view of assessing the sensitivity of the wavelet coefficient from structural responses with respect to system parameters. Their work concluded that the wavelet co-efficients are more sensitive than the structural responses to local structural changes.

A continuous wavelet transform of a function $f(x)$ is defined as

$$W_f(a,b) = \frac{1}{\sqrt{|a|}} \int f(x)\psi^*\left(\frac{x-b}{a}\right)dx \tag{7}$$

Where b is a translation parameter, a is a scale parameter, $f(x)$ is the input signal expressed spatially and which is to be transformed, $\psi^*(\cdot)$ is the complex conjugate of the basis wavelet function $\psi(\cdot)$. Finally, W_f is the wavelet transform which is to be found. A discrete version of this transformation is to be employed in this paper given the discretised finite element results and thus signal. The scale and translation parameters are therefore discretized in a binary format as

$$a = 2^j \qquad b = 2j^j k \qquad j,k \in Z \tag{8}$$

In the above, Z is a defined set of positive integers. The signal can now be decomposed into a series as below

$$d_{j,k} = \int f(x)\psi_{j,k}(x)dx \tag{9}$$

Where $d_{j,k}$ is the wavelet co-efficient. In this paper, the Daubechies wavelet of order 10 and Level 1 is used to decompose the strain signal.

A wind turbine tower is subjected here to a lateral load of 500kN as noted above. The geometry of this tower is approximately based upon a conceptual hybrid tower put forward by the Concrete Centre [9] for a 100m hub height. The said tower has a base and top diameter of 12m and 3m respectively. The concrete-steel interface height is taken as 70m above ground level. The shell thickness taken for the lower concrete segment is 350mm and 24mm for the upper steel shell. The concrete compressive strength is given as 50MPa. Although not provided in the literature, a density of 2450 kg/m^3 and a poissons ratio of 0.15 is assumed for the concrete. A density of 7850 kg/m^3 and a poissons ratio of 0.30 is assumed for the steel segment. After having applied the lateral load of 500kN, a series of loads increasing from 370kN to 833kN (representing prevalent wind speeds of 7mps to 13mps) were applied to the tower top. This was carried out to investigate any change in the strain pattern as a result of the increase in load for the undamaged tower. It was found that after normalising the strain at each load level that there was no change in pattern and so this pattern is independent of the load intensity.

Figure 1 shows the deflected shape of the overall tower after having imposed the tower top lateral load. Damaged circumferential locations were next simulated by reducing a given finite shell elements' planar stiffness. This was carried out at three locations. On each occasion, the strain signal was extracted from the interface circumference and this was processed with a discrete wavelet transformation as discussed above.

Figure 2 (a, b and c) shows the strain signals of the undamaged circumference overlaid with the damaged circumference appropriate for each damaging event. It can be seen that the change in strain profile is marginal but nonetheless is present in the region around the damaged finite element. Figure 3 presents a close-up view of the strain signals around the damaged element whose overall strain is shown in Figure 2b.

It was found that after transforming the strain signal of the damaged towers that this lead to a visible change in the pattern of the detail component of the wavelet transform. In order to better show where a change in the peaks has occurred, the component quantities for the damaged and undamaged states were combined. The results are presented in Figure 4 (a, b and c). Here the damaged locations are clearly evident by the new peaks which represent the arithmetic difference between the damaged and undamaged coefficients.

Finally, the tower model which depicted damage at the ½ circumference length position was taken and strain quantities were extracted for all shell elements in a vertical line from the tower base to the tower top. A DWT was carried out on these spatial signals in the same way as the circumferential signal to observe if the damaged location could be identified. It was found that even though a sudden change in strain profile exists given the change in material at this location, this does not prohibit the ability of the DWT to locate damage in this immediate area. In this case a stiffness reduction from the original 100% to a damaged stiffness of 50% was imposed.

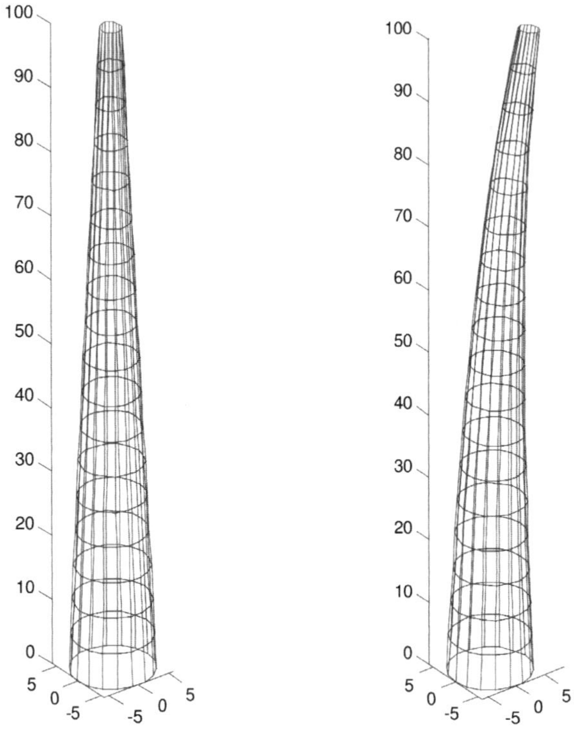

Figure a - Tower Mesh, Before and After Loading

4. Conclusions

It was proposed to use a spatial strain signal extracted from the interface circumference as a means of identifying damage at this location. A first step was to assess if this method would be independent of the size of loading imposed at the tower top due to rotor thrust. After applying a number of loads and normalising the strains, it was found that the strain pattern was not affected by loading intensity. As such, this means damage could be detected at any level of tower response.

After having extracted the normalised strain signal for the undamaged tower, a number of signals for various damaged towers were found. These spatial signals were all transformed using the DWT technique and co-efficients were found for low and high frequency components. The high frequency components of these signals identified the change in slope from one finite element to the next. By combining the coefficients for this high frequency component in terms of taking the undamaged and relevant damaged co-efficients, it can be seen that the region of damage can be clearly identified. Although absolute changes in the strain signal were shown to be minute and possibly undetectable without some form of signal processing, the DWT carried out was well suited to detecting the changes. This method of using Spatial DWT of the strain signal could be a powerful means of locating damage at wind turbine tower interface connections. Strain profiles were taken circumferentially in the majority of cases in this work, with an advantage of this approach being that the strain pattern is independent of the load level.

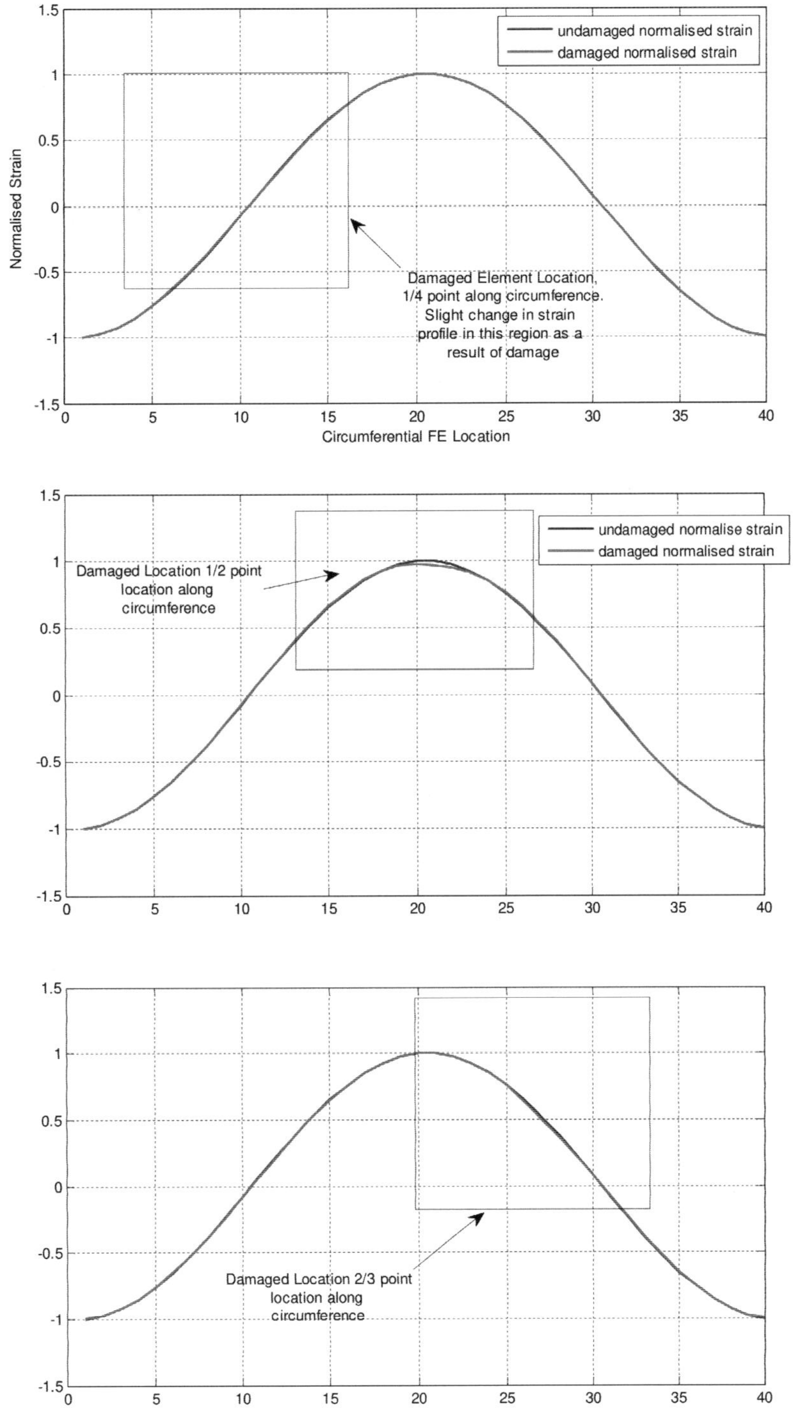

Figure 2 (a, b and c) -Damaged vs Undamaged Strain Signals

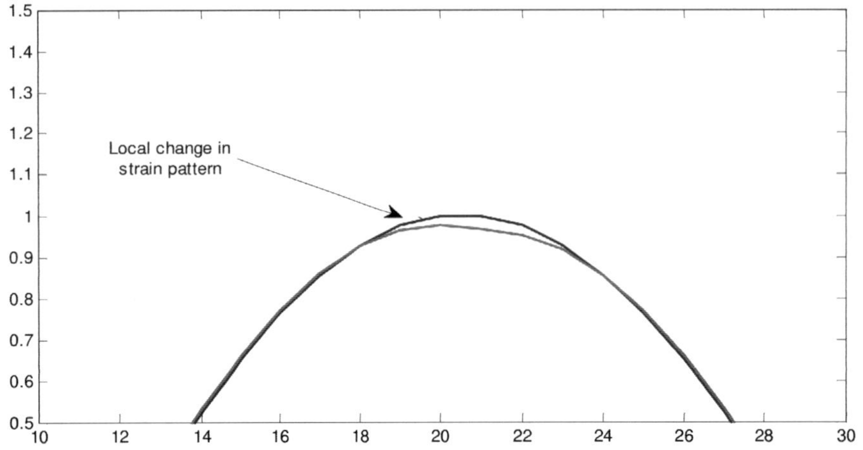

Figure 3 – View of Local Strain Profile in Damaged Region at 1/2 circumference length

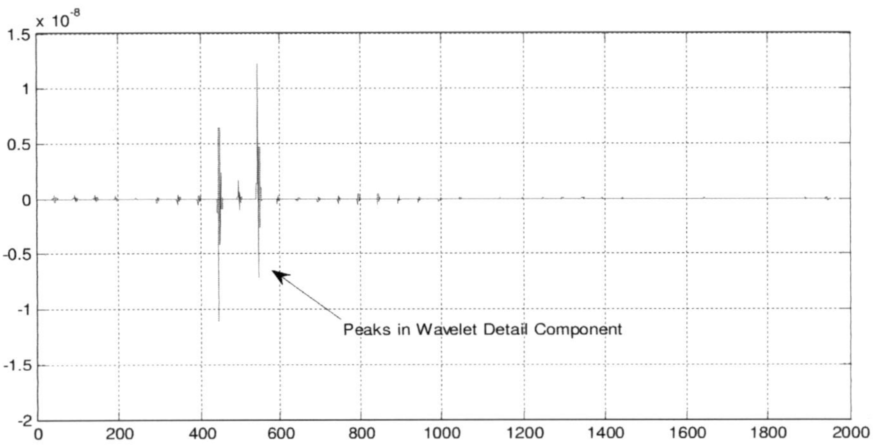

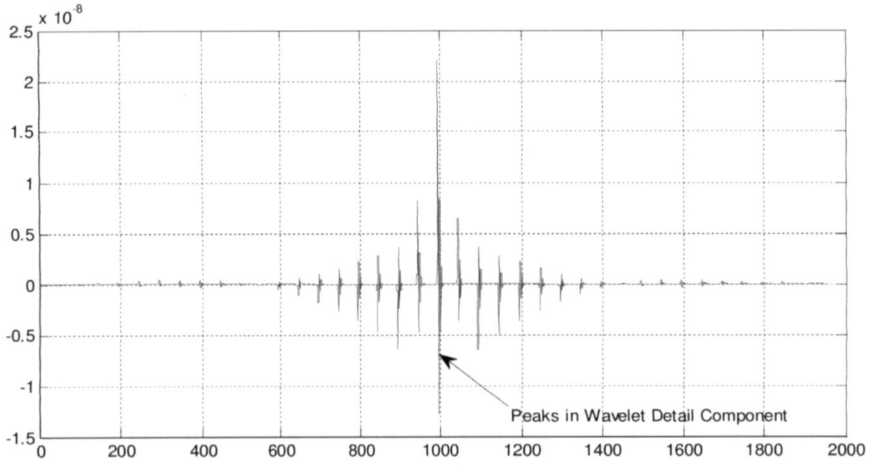

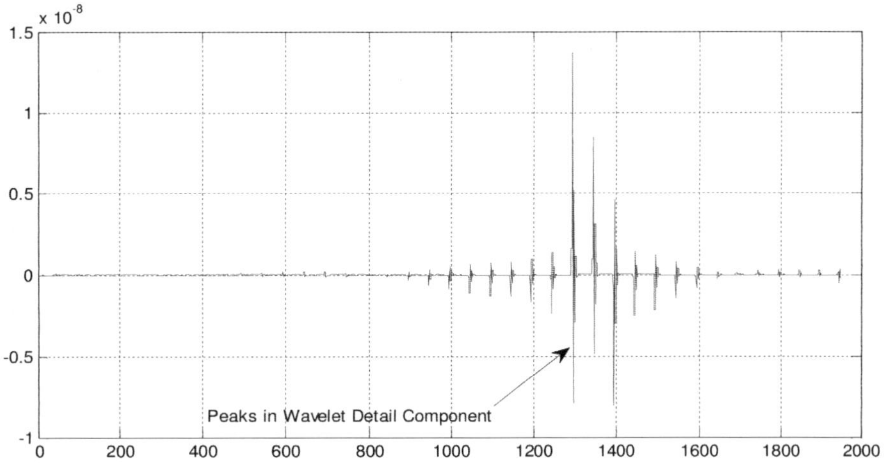

Figure 4 (a, b and c) - Wavelet Decomposition, Detail Component Differences

References

[1] R. Cook, D. Malkus, M. Plesha, and R. Witt, "Concepts and applications of finite element analysis. 2002," ed: Wiley, New York.

[2] A. Kenna and B. Basu, "A finite element model for pre-stressed or post-tensioned concrete wind turbine towers," *Wind Energy,* pp. n/a-n/a, 2014.

[3] O. C. Zienkiewicz, R. L. Taylor, and J. Z. Zhu, *The Finite Element Method: Its Basis and Fundamentals: Its Basis and Fundamentals*: Butterworth-Heinemann, 2005.

[4] M. O. Hansen, *Aerodynamics of wind turbines*: Earthscan, 2008.

[5] J. M. Jonkman, S. Butterfield, W. Musial, and G. Scott, *Definition of a 5-MW reference wind turbine for offshore system development*: National Renewable Energy Laboratory Colorado, 2009.

[6] C.-C. Chang and L.-W. Chen, "Damage detection of cracked thick rotating blades by a spatial wavelet based approach," *Applied Acoustics,* vol. 65, pp. 1095-1111, 2004.

[7] K. Liew and Q. Wang, "Application of wavelet theory for crack identification in structures," *Journal of engineering mechanics,* vol. 124, pp. 152-157, 1998.

[8] S. Law, X. Li, and Z. Lu, "Structural damage detection from wavelet coefficient sensitivity with model errors," *Journal of engineering mechanics,* vol. 132, pp. 1077-1087, 2006.

[9] A. Tricklebank, B. Magee, and P. Halberstadt, "Concrete Towers for Onshore and Offshore Wind Farms," 2007.

Experimental damage detection in a wind turbine blade model using principal components of response correlation functions

S Hoell, P Omenzetter
The LRF Centre for Safety and Reliability Engineering, The University of Aberdeen, Fraser Noble Building, Aberdeen, AB24 3UE, Scotland, UK

E-mail: r01sh13@abdn.ac.uk; piotr.omenzetter@abdn.ac.uk

Abstract. The utilization of vibration signals for structural damage detection (SDD) is appealing due to the strong theoretical foundation of such approaches, ease of data acquisition and processing efficiency. Different methods are available for defining damage sensitive features (DSFs) based on vibrations, such as modal analysis or time series methods. The present paper proposes the use of partial autocorrelation coefficients of acceleration responses as DSFs. Principal component (PC) analysis is used to transform the initial DSFs to scores. The resulting scores from the healthy and damaged states are used to select the PCs which are most sensitive to damage. These are then used for making decisions about the structural state by means of statistical hypothesis testing conducted on the scores. The approach is applied to experiments with a laboratory scale wind turbine blade (WTB) made of glass-fibre reinforced epoxy composite. Damage is non-destructively simulated by attaching small masses and the WTB is excited with the help of an electrodynamic shaker using band-limited white noise. The SDD results for the selected subsets of PCs show a clear improvement of the detectability of early damages compared to other DSF selections.

1. Introduction

Vibration-based structural health monitoring (SHM) is the process of acquiring vibrational measurements, extracting information and knowledge from these observations and determining the current structural performance [1]. It is based on the premise that damage leads to changes in stiffness, mass or energy dissipation mechanisms of a structure and, thus, vibrational response are affected. This enables to define damage sensitive features (DSFs) which describe the current structural state by processing these signals.

Although traditional structural damage detection (SDD) with the help of modal parameters, i.e. natural frequencies, mode shapes and their spatial derivatives, is a mature approach [2], the estimation of these parameters entails practical difficulties due to computational demands and hindrances preventing wider automation of the process [3]. Parametric and non-parametric time series representations of vibration signals, as data-driven approaches, avoid this estimation of modal characteristics. Autoregressive (AR) models are one example of parametric models, for which the theoretical relationship between the structural stiffness and AR coefficients (ARCs) was demonstrated by Nair et al. [4]. However, the use of such models for SHM purposes requires a priori model structure selection, identification and validation. Non-parametric time series representations are less demanding and computationally efficient. They can be classified according to their working domain as frequency, mixed time-frequency and time domain approaches.

Decisions about the current structural state can be made with the help of statistical hypothesis tests of distance measures, e.g. the Mahalanobis distance, between DSF vectors from the healthy and current state [5], where statistical assumptions about the DSF vectors are used to define thresholds for these distances. Even though this is a structured approach with minimal user intervention, the performance with respect to the detectability of early damages depends significantly on the DSF vector dimension and the damage sensitivity of the individual DSF vector entries. The use of principal component analysis (PCA) received considerable attention in the field of SHM in the past for several purposes. The general aim of PCA is to identify a projection defined by orthogonal basis vectors which maximizes the variation in the data for the first few dimensions [6]. This property has been used to reduce the effects of changes in environmental and operational conditions on DSFs [7], where it is generally assumed that the highest variations are a result of these changes and removing the a few first coefficients counteracts the adverse effects. Bandara et al. [8] used PCA for the reduction of frequency response function data for SDD by means of artificial neural networks in a three storey bookshelf structure. Statistical hypothesis testing and PCA of mechanical response data obtained from piezoelectric transducers attached to a small aluminium plate under Lamb wave excitation was applied by Mujica et al. [9]. They used univariate statistical tests of the transformed data, the so called scores, for making decisions about the current structural state.

The present paper discusses the application of partial autocorrelation coefficient (PACC) estimates from acceleration time series as DSF vectors. Due to the high dimensionality of this non-parametric time series representation, dimensionality reduction is achieved by means of PCA, where only selected principal component (PC) estimates from the healthy system are used to transform the DSF vectors to scores. The most damage sensitive PCs for the considered damage scenarios are selected with respect to the corresponding statistical threshold. Then, decisions about the structural state are made by multivariate statistical tests of scores from the current state against the healthy state. The approach is applied to data gathered from dynamic experiments of a laboratory scale wind turbine blade (WTB) under band limited white noise shaker excitation. Damage is simulated non-destructively by attaching small magnets as additional masses.

The paper is structured as follows. First, the proposed methodology is introduced. Second, the structure under study and the experimental setup is presented. Then, the SDD results are discussed. Finally, the paper is rounded up with a set of conclusions and prospects for the future work.

2. Methodology

For the proposed SDD methodology, it is assumed that the utilized vibrational responses are stationary. A discrete time series $x[t]$ can be divided into n segments $x_i[t]$ with i=1, 2, …, n and a common number of samples. Then, removing the estimated means, $\hat{\mu}_{x_i}$, and dividing by the estimated standard deviations, $\hat{\sigma}_{x_i}$, gives the normalized time series segments $z_i[t]$. PACCs, $\alpha_i[\tau]$, are defined as the correlations between time series segments $z_i[t]$ and $z_i[t-\tau]$ shifted by a time lag, τ, without the effects of intermediate variables $z_i[t-1]$, $z_i[t-2]$,…, $z_i[t-\tau+1]$. The sample PACC, $\hat{\alpha}_i[\tau]$, can be efficiently computed by the recursion [10]:

$$\hat{\alpha}_i[\tau] = \frac{\hat{r}_i[\tau] - \sum_{k=1}^{\tau-1} \hat{a}_{i,\tau-1,k}\hat{r}_i[\tau-k]}{1 - \sum_{k=1}^{\tau-1} \hat{a}_{i,\tau-1,k}\hat{r}_i[k]} \tag{1}$$

where $\hat{r}_i$ is the sample autocorrelation function of $z_i[t]$, and $\hat{a}_{i,l,k}$ denotes the estimated l-th coefficient of an AR model of order k of the i-th segment. This underlines the close relationship between autocorrelations, AR models and PACCs, which can be estimated from each other [11]. In the present study, consecutive PACCs are used to construct a DSF matrix, $\mathbf{A}$, comprising DSF vectors, $\boldsymbol{\alpha}_i$, estimated from different time series segments, i, as:

$$\mathbf{A}^T = \left[(\boldsymbol{\alpha}_1 - \hat{\boldsymbol{\mu}}_\alpha) \quad (\boldsymbol{\alpha}_2 - \hat{\boldsymbol{\mu}}_\alpha) \quad \cdots \quad (\boldsymbol{\alpha}_n - \hat{\boldsymbol{\mu}}_\alpha)\right] \text{ with } \boldsymbol{\alpha}_i = \left[\hat{\alpha}_i[1] \quad \hat{\alpha}_i[2] \quad \cdots \quad \hat{\alpha}_i[m]\right]^T \tag{2}$$

where $\hat{\boldsymbol{\mu}}_{\alpha}$ is the estimated mean DSF vector. The feature matrix, **A**, of dimension $n \times m$ can be linearly transformed into a new matrix, **S**, with the help of a linear transformation matrix, **T**, as:

$$\mathbf{S} = \mathbf{AT} \tag{3}$$

PCA enables to identify a transformation matrix such that the new features are linearly independent and the variances are maximized for the first few dimensions. where **S** is in this case called the score matrix. This is done by performing singular value decomposition (SVD) of the sample variance-covariance matrix, $\hat{\boldsymbol{\Sigma}}$, as

$$\hat{\boldsymbol{\Sigma}} = \mathbf{T}\boldsymbol{\Lambda}\mathbf{T}^T \text{ with } \boldsymbol{\Lambda} = diag(\lambda_1, \lambda_2, \ldots, \lambda_m) \text{ and } \mathbf{T} = \begin{bmatrix} \mathbf{t}_1 & \mathbf{t}_2 & \cdots & \mathbf{t}_m \end{bmatrix} \tag{4}$$

where λ_k are the eigenvalues sorted in descending order and $\mathbf{t}_k$ are the corresponding normalized eigenvectors, or PCs. In the present study, the transformation matrix, $\mathbf{T}_h$, estimated from the PCACs of the healthy state, $\mathbf{A}_h$, is used. Then, it is proposed to make decisions about the structural state by statistical hypothesis testing of the scores in the current structural state, $\mathbf{S}_c$, with respect to the healthy state, $\mathbf{S}_h$. In order to improve the detectability of early damages, only a subset of all PCs is used, which includes only these PCs that are the most sensitive to damage. For the selection of this subset, the scores for the healthy and a selected damage state, $\mathbf{S}_h$ and $\mathbf{S}_d$, are computed as:

$$\mathbf{S}_h = \mathbf{A}_h\mathbf{T}_h \text{ and } \mathbf{S}_d = \mathbf{A}_d\mathbf{T}_h \tag{5}$$

where both feature matrices, $\mathbf{A}_h$ and $\mathbf{A}_d$, are obtained by removing the estimated DSF mean of the healthy state only.

It can be assumed that the covariance structure for the initial features in the healthy state and states with small damage extent is almost equal, thus the computed scores $\mathbf{S}_d$ are nearly linearly independent. This enables to estimate the contribution of k-th PC to a multivariate distance between healthy and damaged score features in terms of Fisher's criterion, FC_k, as:

$$FC_k = \frac{n_d n_h (n_d + n_h - 2)}{n_d + n_h} \frac{(\hat{\mu}_{k,d} - \hat{\mu}_{k,h})^2}{(n_d - 1)\hat{\sigma}^2_{k,d} + (n_h - 1)\hat{\sigma}^2_{k,h}} \text{ where } \hat{\mu}_k = n^{-1}\sum_{i=1}^{n} \mathbf{s}_{i,k} \text{ and } \hat{\sigma}_k^2 = n^{-1}\sum_{i=1}^{n}(\mathbf{s}_{i,k} - \hat{\mu}_k)^2 \tag{6}$$

The subscripts h and d refer to the healthy and damage state, respectively. The estimated mean, $\hat{\mu}_k$, and the estimated variance, $\hat{\sigma}_k^2$, are calculated from the score vectors, $\mathbf{s}_{i,k}$, of the corresponding k-th PC computed using n samples. Then, the cumulative sum of Fisher's criteria sorted in descending order, FC_k^{sort}, divided by a statistical threshold with the number of degrees-of-freedom (DOFs) corresponding to the number of coefficients can be used to define the relative Fisher's criteria, FC_k^{rel}, as:

$$FC_k^{rel} = \sum_{j=1}^{k} FC_k^{sort} \Big/ F_{T^2_{k,n_h+n_d-2}}(1-\alpha) \tag{7}$$

Here, it is assumed that a single score, $\mathbf{s}_i$, follows a multivariate Gaussian distribution $\mathcal{N}(\boldsymbol{\mu}, \boldsymbol{\Sigma})$ with true mean, $\boldsymbol{\mu}$, and true variance-covariance matrix, $\boldsymbol{\Sigma}$. The statistical threshold is therefore selected using the cumulative distribution function, $F_{T^2_{k,n_d+n_h-2}}(1-\alpha)$, of Hotelling's probability distribution, T^2_{k,n_d+n_h-2}, at a selected level of significance, α, and with k and n_d+n_h-2 DOFs. The number of considered PCs is k, and the number of samples used for estimating the mean and variance in the healthy and damage state is n_d and n_h, respectively.

Note it requires $FC_k^{rel} > 1$ for damage detection and that the selection of m_{select} PCs for optimal damage detection requires maximising FC_k^{rel} with respect to k:

$$m_{select} = \max_k FC_k^{rel} \tag{8}$$

The assumption that the covariance structure of PACCs for the damaged structure is the same as for the healthy one is leading to the diagonalization of scores from the damage state and makes it easy to determine which scores contribute the most to FC_k^{rel}. However, the during the simulated SDD stage, this assumption can be relaxed. The multivariate mean of DSF vector estimated from a system in the current state, $\boldsymbol{\mu}_c$, can then be tested against the healthy state, $\boldsymbol{\mu}_h$, by the following statistical hypothesis:

$$\begin{array}{ll} H_0: \ \boldsymbol{\mu}_c = \boldsymbol{\mu}_h & \text{tested with} \quad T^2 \le F_{T^2_{m_{select}, n_c+n_h-2}}(1-\alpha) \ \Rightarrow \ H_0 \text{ is accepted} \\ H_1: \ \boldsymbol{\mu}_c \ne \boldsymbol{\mu}_h & \qquad\qquad\qquad Else \qquad\qquad\qquad \Rightarrow \ H_0 \text{ is rejected} \end{array} \quad (9)$$

where the healthy state is represented by the null hypothesis, H_0, and the alternative hypothesis, H_1, corresponds to the damage state. In practical applications these true statistical properties are generally not available. Thus, the hypothesis is tested by means of the estimated quantities and the T^2 statistic [12]:

$$T^2 = \frac{n_c n_h}{n_c + n_h}(\hat{\boldsymbol{\mu}}_c - \hat{\boldsymbol{\mu}}_h)^T \hat{\boldsymbol{\Sigma}}_{pl}^{-1}(\hat{\boldsymbol{\mu}}_c - \hat{\boldsymbol{\mu}}_h) \sim T^2_{m_{select}, n_c+n_h-2} \quad (10)$$

where the subscript c refers to the current state. The T^2 statistic is a standardized distance between two estimated means, $\hat{\boldsymbol{\mu}}_c$ and $\hat{\boldsymbol{\mu}}_h$. The statistical threshold is given at a selected level of significance, α, by the cumulative distribution function, $F_{T^2_{m_{select}, n_c+n_h-2}}$, because the T^2 statistic follows Hotelling's probability distribution, $T^2_{m_{select}, n_c+n_h-2}$, with m_{select} and n_c+n_h-2 DOFs. The estimated pooled variance-covariance matrix, $\hat{\boldsymbol{\Sigma}}_{pl}$, can be calculated with

$$\hat{\boldsymbol{\Sigma}}_{pl} = \frac{(n_c - 1)\hat{\boldsymbol{\Sigma}}_c + (n_h - 1)\hat{\boldsymbol{\Sigma}}_h}{n_c + n_h - 2} \quad (11)$$

where $\hat{\boldsymbol{\Sigma}}_c$ and $\hat{\boldsymbol{\Sigma}}_h$ are the unbiased estimates of the variance-covariance matrices in the current and the healthy state. The relative distance, a counterpart of Eq. (7) becomes:

$$T^2_{rel} = T^2 \Big/ F_{T^2_{m_{select}, n_h+n_d-2}}(1-\alpha) \quad (12)$$

3. Experiment

The structure under study is the blade of a small wind turbine with 5 kW rated power output and a rotor diameter of 5 m. The WTB has a length of 2.36 m and a width of 150 mm. Defined by the aerofoil E387, the solid cross-section is constant along the blade, see Figure 1. The WTB is made of a glass-fibre reinforced epoxy composite. The total mass and mass density are estimated from measurements as 7,110 g and 2.30 g/cm^3, respectively.

Figure 1. Cross section of experimental wind turbine blade.

A vertical configuration is chosen for the dynamic experiments, as shown in Figure 2. In order to create cantilever-type boundary conditions, the WTB is clamped to a massive steel base sitting on a concrete floor. The WTB is excited with the help of an electrodynamic shaker LDS V406, which is rigidly connected to a stiff vertical column made of a steel quadratic hollow section. The column is supported on the same steel base as the WTB, thus only negligible uncontrolled excitations sources are allowed. The shaker is attached to the WTB with the help of a modal stringer, where only normal forces are applied because of the thin stringer flexibility. The high-pressure side of the WTB profile is flatter then the low-pressure side, therefore the stringer is attached to the flatter side. The location is eccentric to the WTB's centre of mass in order to excite flapwise and torsional responses.

Acceleration responses are measured with four Metra uniaxial miniature piezoelectric accelerometers KS94B-100 with sensitivities of approximately 100 mV/g. They can operate in a frequency range from 0.5 Hz to 28 kHz. The accelerometers are attached to the high-pressure side of the WTB with the help of adhesive wax. The sensor and stringer attachment locations are given in Figure 3. The sensor positions are close the selected damage location, which is also indicated in Figure 3. The response accelerations are read with a National Instruments NI-9234 data acquisition card connected to the NI cDAQ-9174 chassis and a laptop. The National Instruments software LabView is used to process the signals.

Although, damage is simulated non-destructively by adding small masses, the actual location is selected based on WTB inspection reports and damage studies of larger WTBs. The trailing-edge bondline is prone to disbonding due to the production process of large WTBs, where the upper and lower shells are bonded together. Additionally, trailing-edge buckling affects adversely the structural resistance [13]. Inspections of 99 WTs with 100 kW and 300 kW rated power output in Egypt showed frequent failures in trailing-edge bonding [14]. The maximum chord location was found to be exposed for this damage type because of the sharp geometrical changes. The simulated damage is therefore located on the trailing-edge at approximately 33% of the WTB length from the root, which corresponds to the typical maximum chord location in large WTBs. To be able to attach magnets to the WTB, a small steel washer is glued onto the surface. It is assumed that this small mass of 5 g does not affect the global WTB dynamics, thus baseline measurements are done with this mass attached. Then, damage with increasing extent is introduced by attaching small magnets and steel plates of know masses.

Figure 2. Configuration for dynamic experiments.

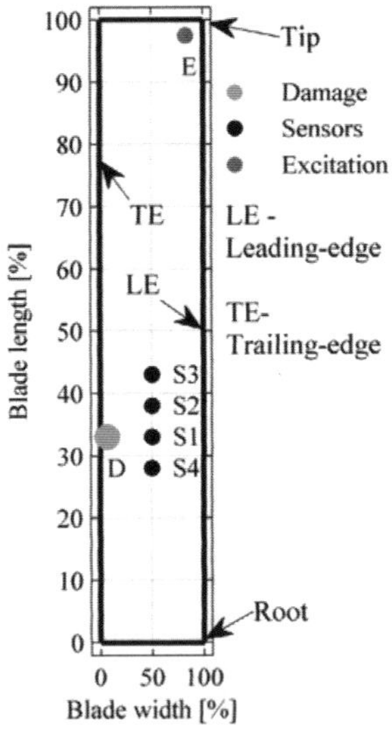

Figure 3. Schematic representation of experimental setup (not to scale).

The excitation characteristics are chosen to be band-limited white noise. Therefore, the National Instruments LabView software is used to generate zero-mean normally distributed time series with constant standard deviation. This broadband noise is then low-pass filtered with a Chebyshev filter of order eight, 100 Hz cut-off frequency and 0.6 dB band-pass ripple. The shaker is controlled by the National Instruments NI 9263 voltage output card and an amplifier with constant gain. This card is also connected to the NI cDAQ-9174 chassis and the laptop. Measurements are taken for 900 s at a sampling rate of 2048 Hz.

4. Structural Damage Detection

For the following discussion of vibration-based SDD with the help of PACCs as time series representation and statistical hypothesis testing of selected PCA scores, the initial acceleration response signals of 900 s are divided into 100 segments of 15 s with an overlap of 40%. Although, the time series segments are directly used without subsampling or filtering, normalization is performed to account for possible variations of the shaker excitation. Therefore, the estimated means of segments are removed before dividing the time series by the estimated standard deviations.

The PACC estimates of acceleration responses from sensor 'S1' (see Figure 3) are shown for lags 1 to 100 in Figure 4. These estimates are given in terms of the mean and standard deviations obtained from the 100 time series segments of the healthy baseline structure. It can be seen that with increasing numbers of lags the mean values decrease while the standard deviations are nearly constant over the whole range. For the following analysis, DSF vectors, $\boldsymbol{\alpha}_i$, are created with PACCs with consecutive lags from 1 to 100 in order to captured sufficient information.

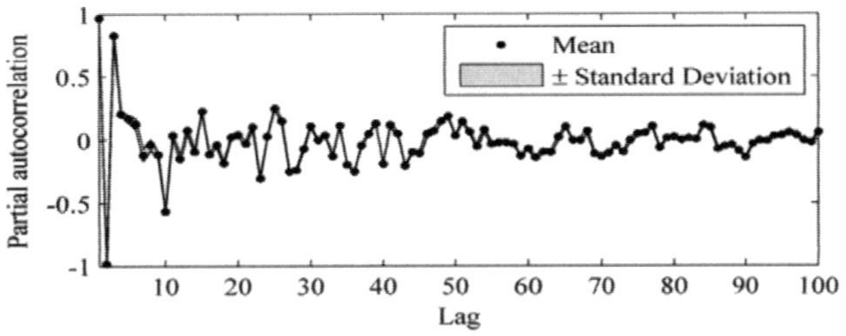

Figure 4. Mean and standard deviations of PACCs estimated from 100 samples of healthy state data.

These DSF vectors enable to construct the initial feature matrix, $\mathbf{A}_h$, for the healthy structure. Since the proposed SDD methodology utilizes PCA, SVD of $\mathbf{A}_h$ is performed. This gives the transformation matrix, $\mathbf{T}_h$, as the eigenvectors or PCs.

The next step is the selection of PCs in order to improve the damage detectability. This requires data from the healthy and one of the simulated damage states, where masses of 1 g, 3 g and 5 g are used resulting in the maximum overall change in the WTB's mass of less than 0.1%. Next, the corresponding score matrices, $\mathbf{S}_h$ and $\mathbf{S}_d$, are calculated with the transformation matrix from the healthy state, $\mathbf{T}_h$. To select PCs for SDD, the relative Fisher's criteria are calculated according to Eq. (7), where mean values and variances are estimated from the 100 samples of the healthy and the reference damage states and the significance level is chosen as 5%. The results are shown in Figure 5. For small damage extents of 1 g and 3 g the maxima are achieved by using only PCs 6 and 20. However, the larger reference damage of 5 g leads to selection of PCs 95, 92, 65 and 74. It should be noted that for all the PC selections and damage extents the relative Fisher's criteria, FC_k^{rel}, are larger than one, thus the corresponding statistical threshold for damage detectability is always exceeded and damage successfully detected.

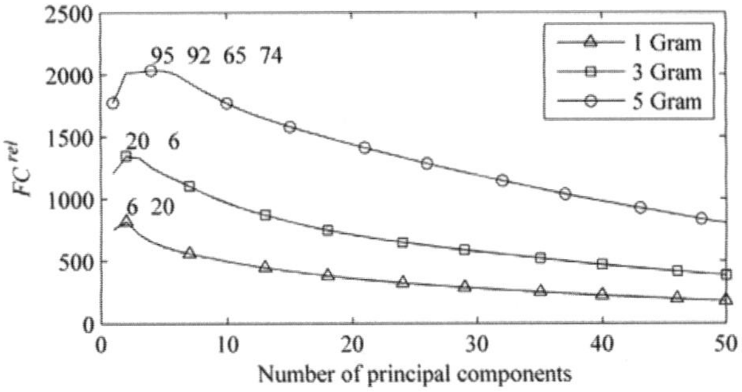

Figure 5. Relative Fisher's criterion for various selections of PCs (selected PCs are given for the maxima).

For the preceding identification of the most damage sensitive PCs, a large number of samples is used for estimating the statistical model in the damaged state. This may not be optimal in practical applications because the need for a large number of samples will lead to delays in making a decision about the structural state. Therefore, the SDD phase discussed now uses only two sample estimates from the current structural state. To reduce the dependencies due to the overlap of time series segments, these samples are randomly selected from each state.

The detection results using two sample estimates are shown in Figure 6 in terms of the relative statistic, T^2_{rel}, at 5% significance level in order to allow for a comparison of results obtained with different numbers of PCs. Furthermore, the damage detection threshold for T^2_{rel} equal to one is shown, where exceedance indicates the detection of damage. The results are shown for the two previously identified optimal subsets, i.e. (6, 20) and (95, 92, 65, 74), and the full set of PCs. The healthy state results are obtained from another 900 s measurement without attached masses but using the initially obtained PCs from the first 900 s of healthy state data. Here, it can be seen that the selection of four PCs leads to a significant number, in fact 100%, of false positive alarms in contrast to the two PC subset and the full set of PCs. PC selection (95, 92, 65, 74) shows then a monotonic increase in T^2_{rel} with increasing damage but with a high variation. The full set of PCs has a smaller variation, and for damage extents of 1 g and 2 g occasional false negatives are present. The selection with two PCs leads to a significant increase in T^2_{rel} from the healthy state to the damage state with 1 g, which later increases slowly up to the 4 g damage. The variation for all damage extents is here the smallest of all selections. Even though there is a drop for the highest damage of 5 g compared to the smaller ones, in all cases damage is clearly indicated, while low false positive alarm rates are maintained. This selection can therefore be considered optimal for small damage extents.

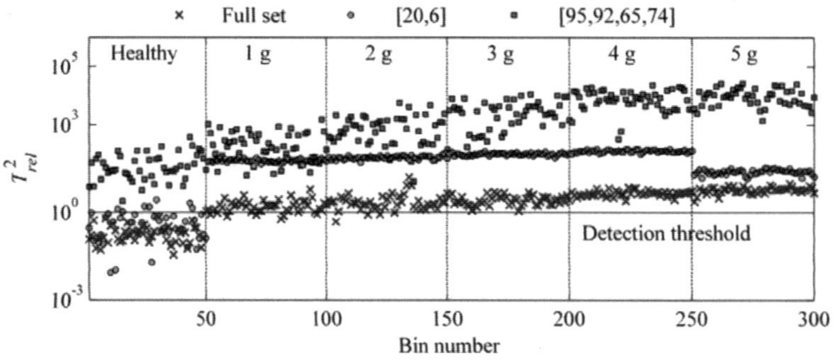

Figure 6. Relative detection rates for increasing damage extents using a subset and the full set of PCs.

5. Conclusions

This paper explored the application of PC selection of acceleration time series correlation-based DSFs for improving the detectability of structural damage. Damage decision making by means of statistical hypothesis testing of scores obtained by transforming the initial DSFs with PCA is discussed. The approach is applied to experimental data gathered from a small scale WTB under band-limited white noise excitation, where damage is non-destructively introduced by attaching small masses at the WTB's trailing edge.

The advantages of PC selection have been demonstrated with the proposed methodology, namely that optimal selections of PC subsets can be achieved for enhanced early damage detectability. Nevertheless, further studies are required to increase the understanding of the effect of different damage extents for the selection. The presented results suggest that there might not be one optimal selection for all damage extents, thus each state may require its own selection in order to achieve best detectability. Here, the orthogonality of PCs is advantages because the contribution of individual scores is directly available without the need for considering the cross-correlations of DSF. Furthermore, only one sensor close to the actual damage location is used, and the future design of sensor setups with respect to the economic constraints and achievable damage detectability is another important aspect to be studied. However, the SDD results are promising for future developments, which can lead to beneficial applications under realistic conditions.

Acknowledgments
Piotr Omenzetter and Simon Hoell's work within the Lloyd's Register Foundation Centre for Safety and Reliability Engineering at the University of Aberdeen is supported by Lloyd's Register Foundation. The Foundation helps to protect life and property by supporting engineering-related education, public engagement and the application of research.

References

[1] Farrar, C R, Doebling, S W and Nix, D A 2001 Vibration–based structural damage identification *Philosophical Transactions of the Royal Society A: Mathematical, Physical and Engineering Sciences,* **359**(1778), 131-149
[2] Carden, E P and Fanning, P 2004 Vibration based condition monitoring: A review *Structural Health Monitoring,* **3**(4), 355-377
[3] Reynders, E 2012 System identification methods for (operational) modal analysis: Review and comparison *Archives of Computational Methods in Engineering,* **19**(1), 51-124
[4] Nair, K K, Kiremidjian, A S and Law, K H 2006 Time series-based damage detection and localization algorithm with application to the ASCE benchmark structure *Journal of Sound Vibration,* **291**(1-2), 349-368
[5] Fassois, S D and Sakellariou, J S 2007 Time-series methods for fault detection and identification in vibrating structures *Philosophical Transactions of the Royal Society A: Mathematical, Physical and Engineering Sciences,* **365**(1851), 411-448
[6] Jolliffe, I T 2002 Principal component analysis, 2nd ed., New York, USA, Springer
[7] Yan, A M, Kerschen, G, De Boe, P and Golinval, J C 2005 Structural damage diagnosis under varying environmental conditions - Part I: A linear analysis *Mechanical Systems and Signal Processing,* **19**(4), 847-864
[8] Bandara, R P, Chan, T H and Thambiratnam, D P 2014 Structural damage detection method using frequency response functions *Structural. Health Monitoring,* **13**(4), 418-429
[9] Mujica, L E, Ruiz, M, Pozo, F, Rodellar, J and Güemes, A 2014 A structural damage detection indicator based on principal component analysis and statistical hypothesis testing *Smart Materials and Structures,* **23**(2), 1-12
[10] Cryer, J D and Chan, K 2008 In Casella G., Fienberg S. and Okin I. (Eds.), Time series analysis with applications in R, New York, NY, USA, Springer Science+Business Media, LLC
[11] Kay, S M 1988 Modern spectral estimation: Theory and application, Upper Saddle River, USA, Prentice Hall
[12] Rencher, A C 1998 Multivariate statistical inference and applications, New York, USA, Wiley
[13] Jensen, F M, Sørensen, J D, Nielsen, P H, Berring, P and Flores, S 2011 Failures in trailing edge bondlines of wind turbine blades *Proceedings of the 32nd Risø International Symposium on Materials Science,* Roskilde, Denmark
[14] Ataya, S and Ahmed, M M Z 2013 Damages of wind turbine blade trailing edge: Forms, location, and root causes *Engineering Failure Analisis,* **35**, 480-488

Damage localization in a residential-sized wind turbine blade by use of the SDDLV method

R.J. Johansen[1], L.M. Hansen[1], M.D. Ulriksen[1], D. Tcherniak[2] and L. Damkilde[1]

[1]Dept. of Civil Engineering, Aalborg University, Niels Bohrs Vej 8, Esbjerg, Denmark
[2]Brüel & Kjær Sound & Vibration Measurement A/S, Skodsborgvej 307, Nærum, Denmark

E-mail: rjohan10@student.aau.dk, lhanse10@student.aau.dk, mdu@civil.aau.dk, dmitri.tcherniak@bksv.com, lda@civil.aau.dk

Abstract. The stochastic dynamic damage location vector (SDDLV) method has previously proved to facilitate effective damage localization in truss- and plate-like structures. The method is based on interrogating damage-induced changes in transfer function matrices in cases where these matrices cannot be derived explicitly due to unknown input. Instead, vectors from the kernel of the transfer function matrix change are utilized; vectors which are derived on the basis of the system and state-to-output mapping matrices from output-only state-space realizations. The idea is then to convert the kernel vectors associated with the lowest singular values into static pseudo-loads and apply these alternately to an undamaged reference model with known stiffness matrix. By doing so, the stresses in the potentially damaged elements will, theoretically, approach zero. The present paper demonstrates an application of the SDDLV method for localization of structural damages in a cantilevered residential-sized wind turbine blade. The blade was excited by an unmeasured multi-impulse load and the resulting dynamic response was captured through accelerometers mounted along the blade. The static pseudo-loads were applied to a finite element (FE) blade model, which was tuned against the modal parameters of the actual blade. In the experiments, an undamaged blade configuration was analysed along with different damage scenarios, hereby testing the applicability of the SDDLV method.

1. Introduction

Research activities on vibration-based structural health monitoring (SHM) systems for wind turbine blades have been growing rapidly over the last two decades, see, e.g., [1]. The most common approach is to compare collected data from a reference state, which is typically a healthy one, and the current state. The current state is potentially damaged if it differs significantly from the reference state. A typical partition of the damage identification process was suggested in [2] and contains the following four steps: 1) detection, 2) localization, 3) assessment and 4) consequence. There are plenty of well-documented methods for damage detection, see, e.g., [3], thus the present paper concentrates only on the damage localization process.

In [4], Bernal presented the damage location vector (DLV) method, which utilizes the null space of the changes in the flexibility matrix from a pre- and post-damaged structure to locate the damage. The method assumes that the system behaves linearly in both the pre- and post-damaged states. The vectors that form the basis of the null space are designated as DLVs, and they contain usable information about the location of the damage. It is proved in [4] that by applying a DLV as loads to the undamaged structure, the stresses in the damaged elements approach zero.

The DLV method only includes static properties of the system. However, it was lately extended to include the dynamics of the system, namely the dynamic damage location vector (DDLV) method [5]. Here, the dynamics of the system is included by applying the changes in transfer function matrix, instead of changes in flexibility matrix, to obtain the static pseudo-loads [4].

The dynamic version is further extended to an output-only measurements scenario, which is designated as the stochastic dynamic damage location vector (SDDLV) method [6]. Thus far, the SDDLV method has primarily been tested in the context of numerical models of simple truss and frame structures, see, e.g., [6].

In the present paper, the SDDLV method is applied to locate damages in a residential-sized wind turbine blade tested experimentally.

2. Test setup

To demonstrate the applicability of the SDDLV method, experiments are performed on a cantilevered residential-sized wind turbine blade, see figure 1. Specifically, the structural vibration responses are measured for an undamaged and two damaged cases subjected to unmeasured multi-impulse loading conducted by tapping the blade with a pencil. For each case, a state space model, which forms the basis for determination of the pseudo-loads, i.e., SDDLVs, is estimated from the collected acceleration data.

The blade depicted in figure 1 is about 800 mm long and constructed from a composite material, namely carbon-fibre-reinforced polymer. The blade was originally constructed for another project [7], and later modified for the purpose of testing different SHM methods. The blade is separable into two shells, which are assembled by means of 25 bolts along the leading and the trailing edges. Different damage conditions can be examined by untightening one or more bolt(s).

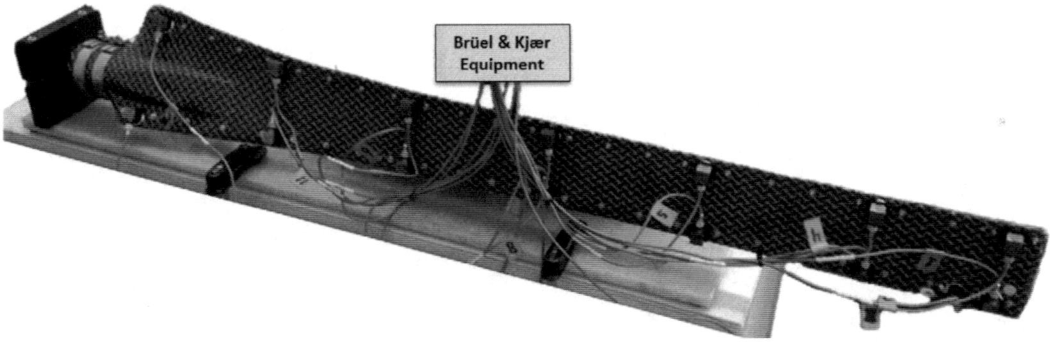

Figure 1. Experimental setup for test on the residential-sized wind turbine blade.

The blade is excited by hitting the structure randomly over the surface in order to simulate operational conditions, i.e., only the output vibrations are collected. The vibrations are captured by use of 14 equally spaced Brüel & Kjær Type 4507-B-004 uniaxial accelerometers along each edge of the blade, as illustrated in figure 1 and figure 2. The accelerometers are mounted, such that they measure perpendicularly to the surface: thus they do not measure in exactly the same direction.

For each experiment, the sampling frequency was set to 8192 Hz, since a sufficiently high sampling frequency is required to ensure that the dynamics of the system is captured properly. The recordings have duration of 200 seconds and are later divided into smaller partitions in order to obtain more than one experiment.

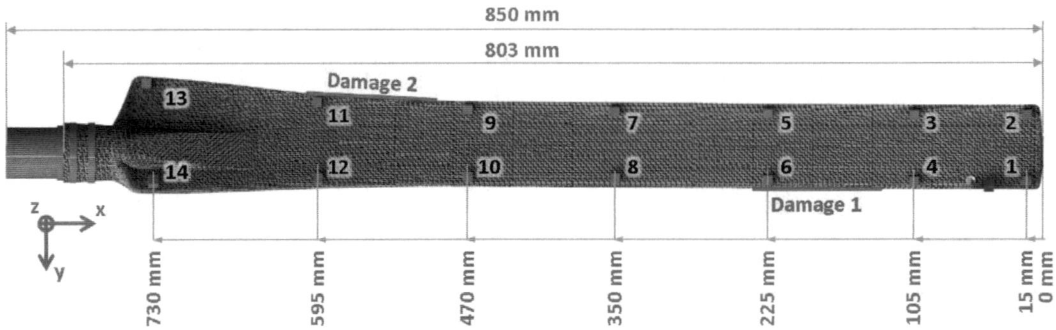

Figure 2. Dimensions of the blade and the locations of the two simulated damages.

The location and size of the two separately simulated areas of damages are shown in figure 2. Both of the damage areas are simulated with three bolts untightened, but kept in the blade to avoid mass changes between the experiments.

3. System identification

System identification techniques are used to mathematically describe the captured acceleration data. In this context, subspace identification is found to be applicable and the fundamental principle of this method is a state-space representation based on the output-only continuous time state-space model, see, e.g., [8].

$$\dot{x}(t) = A_c x(t) + w_k \, , \tag{1}$$

$$y(t) = C_c x(t) + v_k \, , \tag{2}$$

where $A_c \in \mathbb{R}^{n \times n}$ is the state/system matrix for the system containing the dynamic properties, $C_c \in \mathbb{R}^{m \times n}$ is the output matrix, while $x(t) \in \mathbb{R}^n$ and $y(t) \in \mathbb{R}^m$ are the state vector and the output vector. The two last vectors, $w_k \in \mathbb{R}^n$ and $v_k \in \mathbb{R}^m$, are unmeasured stationary noise/disturbances related to the process and the output, respectively. The sizes of the matrices and vectors depends on the order, n, of the state-space model and the number of outputs, m. The system identification is performed using MATLAB System Identification Toolbox, namely using N4SID. For each of the three system states, several state-space models are derived on the basis of different segments of the data. A total of 25 models are derived for each of the two areas of damage and used for estimating a corresponding SDDLV.

4. SDDLV method

The SDDLV method is, as previously declared, based on the change in transfer function matrix for systems where the input is unknown and, as such, the transfer function matrix is inaccessible. Instead, the estimated state matrix and the output matrix are applied to estimate vectors from the kernel of the change in transfer function matrix The transfer function is basically the relation between the output, $Y(s)$, and input, $F(s)$, in the Laplace domain, i.e.,

$$Y(s) = G(s)F(s) \tag{3}$$

where

$$G(s) = C_c(sI - A_c)^{-1}B_c + D_c \, . \tag{4}$$

The input matrix, B_c, and the direct transmission matrix, D_c, are not directly used for the estimation of the SDDLVs, as clarified in the following.

This basic form of the transfer function (4) is not applicable for stochastic systems with output only. In [5], an approach to estimate a fictive input from the state matrix and the output matrix is documented. The idea is to use the fact that there should always be a correlation between the input and the output. The approach ends up in the following:

$$G(s) = R(s)D_c \tag{5}$$

where

$$R(s) = C_c A_c^{-p}(sI - A_c)^{-1} H_p^{\dagger} L \quad , \tag{6}$$

with the value of the exponent $p = 0, 1, 2$, depending on whether the measurements are displacements, velocities or accelerations. The terms estimating a fictive input to the system are defined by

$$H_p = \begin{bmatrix} C_c A_c^{(1-p)} \\ C_c A_c^{(-p)} \end{bmatrix} \tag{7}$$

and

$$L = \begin{bmatrix} I \\ 0 \end{bmatrix} \; , \tag{8}$$

where the dagger sign in equation (6) designates, that the Moore-Penrose pseudo-inverse is applied to H_p. The direct transmission term D_c is assumed to be a constant, since the 'feedthrough' is assumed to be non-changing and thus not affected from the system properties if damage occurs. The change in transfer function is then proportional to the change in $R(s)$, i.e.,

$$\Delta G(s) \propto \Delta R(s) = R_d(s) - R_u(s) \, . \tag{9}$$

The SDDLVs are found from the quasi-null space of $\Delta R(s)^T$ by singular value decomposition (SVD), hence yielding

$$\Delta R(s)^T = U \Sigma \mathrm{V}^{\mathrm{T}} \, , \tag{10}$$

in which each of the singular values contained in Σ has a corresponding left singular vector in U and a right singular vector in V. The right singular vector associated with the smallest singular value is used as pseudo-loads.

The damage localization is not efficient for all s-values of the quasi-null space of $\Delta R(s)$. Proper ones are selected on the basis of the response characteristics of the system. This information is available when solving an eigenvalue problem of the state matrix. Only the s-value(s) near the poles of the system are selected for the $\Delta R(s)$, hereby introducing a modal truncation of the system. The applied s-values are increased by 1% since studies in context of this paper confirm that the value must be slightly different from the poles of the system, as stated in [5]. A more robust selection may exist and the selection is further discussed by the authors in [9].

5. Finite element model

The pseudo-loads obtained in section 4 are applied to the finite element (FE) model of the blade, in which elements containing stresses approaching zero are potentially damaged according to the (-)DLV approach. The model is based on an approximated 3D CAD geometry and is discretized by equally sized first order shell elements. The FE model is fixed at the blade root, i.e., all displacements and rotations are equal to zero, in order to simulate the clamping mechanism from the experimental setup shown in figure 1. It is calibrated against the first four experimental natural eigenfrequencies and mode shapes obtained from operational modal analysis (OMA) of the undamaged structure, see table 1. The bolts and accelerometers are not included in the model; however, the increased mass is taken into consideration by calibrating the density of the material.

Table 1. Comparison of eigenfrequencies between OMA and calibrated FE model

	Description	OMA [Hz]	Calibrated FE model [Hz]
Mode 1	1st flapwise bending	15.9	16.0
Mode 2	1st edgewise bending	none [a]	48.3
Mode 3	2nd flapwise bending	87.9	88.2
Mode 4	1st torsional	109.0	109.2
Mode 5	3rd flapwise bending	183.3	182.8

[a] *was not found during the experiment.*

As previously mentioned, the accelerations are measured perpendicular to the surface, which consequently defines the direction of the applied pseudo-loads. The local coordinate systems depicted in figure 3 illustrate the perpendicular direction, namely the z-axis, for each of the 14 accelerometer positions. The pseudo-loads are applied over an area corresponding to the size of the accelerometers in order to avoid stress disturbance at the specific positions.

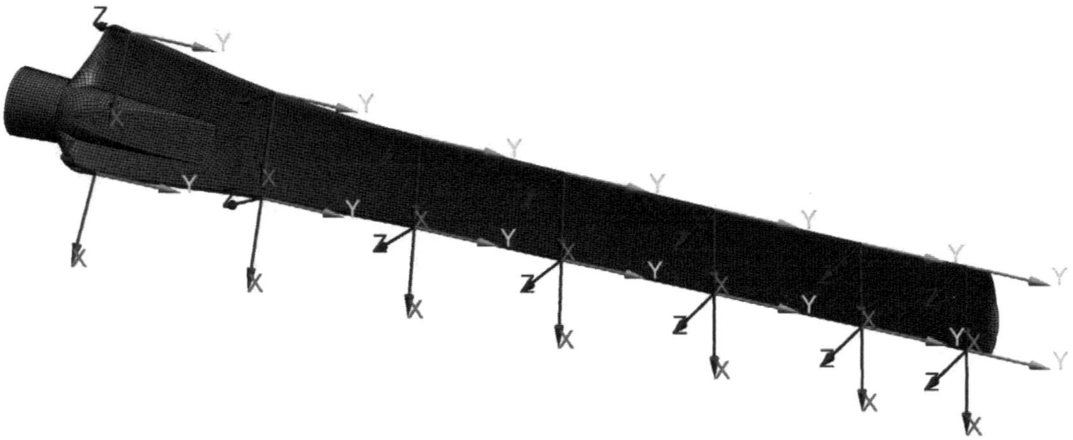

Figure 3. Discretized FE model of the blade and local coordinate systems corresponding to the orientation of the 14 accelerometers.

In the post-processing, the elemental mean von Mises stresses are chosen for locating the damaged area, since this particular stress type contains information from all stress components and is thereby applicable for different load cases, e.g., shearing and bending.

6. Summary of methodological process

A summary of the steps in the damage localization process presented in this paper is outlined in order to make a clear overview of the method before presenting the results.

- Preparation:
    - Collect reference measurements from a healthy state.
    - Perform a system identification of the reference measurements in order to estimate $R_u(s)$.
    - Calibrate an FE model based on experimental modal parameters obtained using OMA.

- Damage localization:
  - Perform system identification of the current measurements in order to estimate $R_d(s)$.
  - Perform SVD on $\Delta R(s)^T$ for a proper s-value in order to estimate the SDDLV.
  - Apply the SDDLV as static pseudo-loads to the calibrated FE model and compute stresses, e.g., von Mises stresses.
  - The stresses approaching zero are identifying damaged location(s).

7. Results

By examining the SDDLV-induced stress fields for the two analysed cases, it is generally found that the global x-directional location of the areas of damage are estimated accurately and consistently, whereas the y-directional location can vary depending on the specific SDDLV applied.

The system identification of the experiments reveals that the poles corresponding to the 2nd and 3rd flapwise bending modes are the most excited and consequently only poles associated with these modes are used to estimate the SDDLV. The s-value used for determination of the SDDLV is taken as the value of the current pole increased by 1 %, as described in section 4.

In figure 4, the stress field from one SDDLV is presented for Damage 1. Here, the damage is located between 560-680 mm from the root at the leading edge, see figure 2. The stress field in figure 4 approaches zero stress in the damaged area, but also at the root and at the tip.

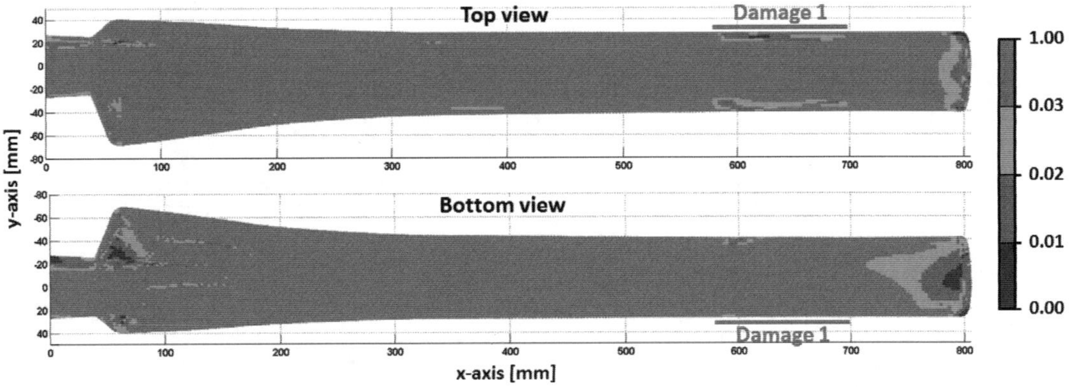

Figure 4. Stress field illustrating the normalized elemental mean von Mises stress for one SDDLV showing the appearance of Damage 1.

These disturbances and lack of precision for localization in the y-direction would most likely disappear if more sensors were used for monitoring the vibrations. The amount and location of this noise are varying for all SDDLVs, but the noise often appears at the tip and root of the blade. The disturbance at the root is understandable, as the geometry is varying more in this area and the number of sensors is small. At the tip, it is clear that noise will appear since the area from sensor 1 and 2 to the end of the blade is only affected a little by the loads applied to the structure.

Examination of SDDLVs for Damage 2 has confirmed many of the observations from Damage 1. Some SDDLVs locate the damage clearly, while others contain mostly noise. Damage 2 is located 180-300 mm from the root at the trailing edge, which defines the maximum damage size based on the same conditions as described for Damage 1. One stress field for Damage 2 is illustrated in figure 5, where the damage is clearly localized, albeit with disturbances at the root and at the tip. In some situations it is hard to distinguish between actual damage and disturbances when observing the stress field for a single SDDLV. However, examination of several SDDLVs has revealed that the damaged location is the only area in which the stresses always approach zero. An informative overview of examined SDDLVs from different operational experiments is listed in table 2. The table, which is based on 25 experiments for each of the two damage scenarios, shows consistency in localization of the damage, even when the disturbance is changing.

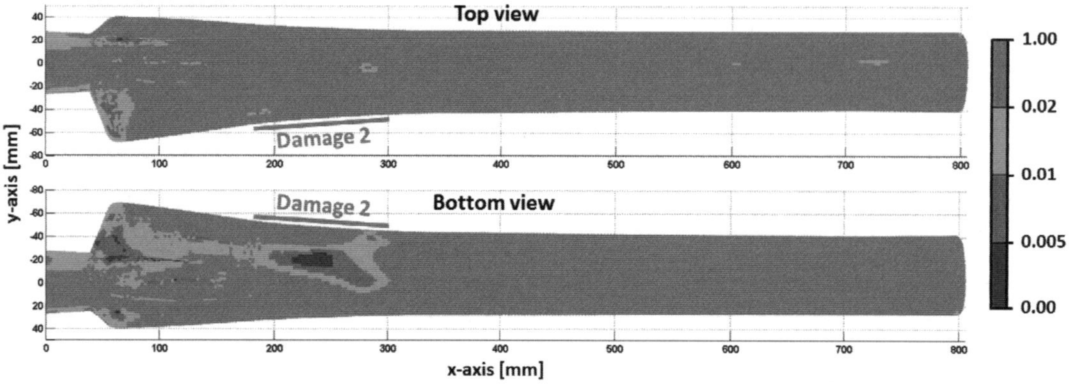

Figure 5. Stress field illustrating the normalized elemental mean von Mises stress for one SDDLV showing the appearance of Damage 2.

Table 2. SDDLV-based damage localization results for two damage types.

	Experiments	**Localized**		**Noise root**			**Noise tip**		
		Yes	No	High	Low	None	High	Low	None
Damage 1	25	23	2	10	8	7	10	3	12
Damage 2	25	25	0	8	10	7	13	0	12

A clear explanation of the varying precision of the SDDLVs throughout the examination of the two areas of damage has not been observed. However, it has been observed that SDDLVs based on *s*-values corresponding to the second flapwise bending mode and the third flapwise bending mode yield less stress disturbance than those for the remaining identified blade modes for Damage 1 and Damage 2, respectively. It has been noticed during OMA, that these two modes are excited significantly better than other higher modes.

From the experiments visualized in figure 4 and figure 5 and the results listed in table 2, it is clear that the stresses in the damaged areas approach zero, while the zero stresses elsewhere occur rather randomly. This observation suggests that one way of reducing the level of disturbance, in order to separate damages from noise, is to apply a statistical evaluation on a large set of stress fields obtained from different SDDLVs. This is treated in [9].

8. Conclusion

The presented paper deals with localization of damage in a residential-sized wind turbine blade by use of the SDDLV method. The method employs the vectors from the quasi-null space of the damage-induced change in transfer function matrix as static pseudo-loads, which are applied to a known model of the undamaged system. The damage is then, suggestively, found at locations where the stresses are approaching zero.

The SDDLV method is demonstrated as capable of locating different areas of damage on both the leading and trailing edges of the blade. The stress field generally approaches zero in the location of the damage. However, noise is, in most cases, present, making it difficult to obtain unambiguous localization. The disturbances primarily appear at the tip and at the root of the blade; thus a way to reduce these noise contributions could be to apply a statistical approach in which several SDDLV-induced stress fields are evaluated in a combined manner. This is a part of future research activities.

Further future research activities will deal with the selection of s-value(s). In the present study, it has been set to be a 1 % increment of the s-value corresponding to the consistently identified pole. This procedure has been chosen somewhat randomly, and it could therefore be an interesting study to apply different s-values around the pole of interest in order to see, if any general approach can be established for the choice of s-values, hereby increasing the reliability and robustness of the SDDLV method.

9. Acknowledgement

The experimental part of the project was supported by the EUDP (Danish Energy Technology Development and Demonstration Programme) grant 64011-0084 "Predictive Structure Health monitoring of Wind Turbines". The authors also acknowledge Robert Flemming Mikkelsen who provided the blade for the experiments.

10. References

[1] Ciang C C *et al* 2008 Structural health monitoring for a wind turbine system a review of damage detection methods *Meas. Sci. Technol.* **19** 122001

[2] Rytter A 1993 Vibration based inspection of civil engineering structures Ph.D. Thesis Aalborg University Denmark

[3] Farrar C R *et al* 2001 Vibration-based structural damage identification *Philosophical Transactions: Mathematical, Physical and Engineering Sciences* **359** 131-149

[4] Bernal D 2002 Load vectors for damage localization *J. Eng. Mech.* **128** 7-14

[5] Bernal D 2007 Damage localization from the null space of the transfer matrix *AIAA Journal* **45** 374-381

[6] Bernal D 2010 Load vectors for damage location in systems identified from operational loads *J. Eng. Mech.* **136** 31-39

[7] Bottoli F and Pignatti L 2001 Design and processing of structural components in biocomposite materials rotor blade for wind turbine cars M.Sc. Thesis Risø DTU

[8] Van Overschee P and Moor B L R 1996 Subspace identification for linear systems theory implementation applications Kluwer academic publishers

[9] Hansen L M *el al* 2015 Statistical evaluation of characteristic SDDLV-induced stress resultants to discriminate between damaged and undamaged elements *Submitted to 11th International Conference on Damage Assessment of Structures*

Blind Source Separation and Dynamic Fuzzy Neural Network for Fault Diagnosis in Machines

[1*]Haifeng Huang, [1,2]Huajiang Ouyang and [1]Hongli Gao
[1] School of Mechanical Engineering, Southwest Jiaotong University, Chengdu, China
[2] School of Engineering, University of Liverpool, Liverpool, U.K.

E-mail: hlonghaifeng@163.com

Abstract. Many assessment and detection methods are used to diagnose faults in machines. High accuracy in fault detection and diagnosis can be achieved by using numerical methods with noise-resistant properties. However, to some extent, noise always exists in measured data on real machines, which affects the identification results, especially in the diagnosis of early-stage faults. In view of this situation, a damage assessment method based on blind source separation and dynamic fuzzy neural network (DFNN) is presented to diagnose the early-stage machinery faults in this paper. In the processing of measurement signals, blind source separation is adopted to reduce noise. Then sensitive features of these faults are obtained by extracting low dimensional manifold characteristics from the signals. The model for fault diagnosis is established based on DFNN. Furthermore, on-line computation is accelerated by means of compressed sensing. Numerical vibration signals of ball screw fault modes are processed on the model for mechanical fault diagnosis and the results are in good agreement with the actual condition even at the early stage of fault development. This detection method is very useful in practice and feasible for early-stage fault diagnosis.

1. Introduction

Mechanical equipment plays a very important role in modern industry and its quality, performance and intelligence reflects the level of production capability and development of a country [1, 2]. However, once mechanical faults occur in machines they may cause financial losses and even heavy casualties. It is often reported that machinery faults bring about catastrophic accidents because of damage of key parts. Therefore, the problems of how to ensure mechanical equipment working safely have been the objective of many researchers. Thus, it is very important to study fault diagnosis in machines for machinery reliability.

According to statistics, unrepairable faults of aircraft reduce by 82% due to fault diagnosis technology (FDT) and fault prognostic technology (FPT). Maintenance manpower required reduces by 20%-40%, and logistics incurred reduces by 50%. As a result, aircraft will work more reliably and maintenance and repairing cost is now half of the amount without FDT and FPT. Furthermore, service life of aircraft can be extended to 8000 hours [3]. Above all, research on diagnosis of early faults not only is beneficial for cost effective maintenance but also increases overall efficiency of operation and life-span of machinery. It can also reduce risk of loss of human lives due to unexpected mechanical equipment failures.

In the case of a ball screw, during its service, there will be wear, cracking, pitting, gaps with other components and so on. Most of these faults evolve over a long operating time and the fault information can be obtained from the mass of fault data so that these mechanical faults can be diagnosed and predicted correctly. However, installation positions of sensors are often limited by the structure of the machine involved, the working environment and the operating conditions and it is often hard to put sensors in the best locations. Usually, important fault data are buried in the large amount of fault data collected and going through them takes much calculation time. On the other hand, even a large amount of data may still miss some important ones because of the poor sensor locations and operating conditions and hence faults developed at their early stages can go unnoticed. Furthermore, it is very difficult to obtain features of an early fault and potential ones in the early fault stage for the feature information of early-stage faults is very weak. Moreover, noise to signal ratio of early fault is very high. All of these make it very difficult to obtain accurate fault diagnosis from the complex signals of measurement [4, 5].

One of the key problems of diagnosis is how to extract useful information and sensitive fault features from weak fault signals subjected to all kinds of interference.

Nowadays, fault detection methods based on measured vibration data have been widely utilised for assessing common machinery faults and many methods on signal processing and feature extraction have been used in the field of machine fault diagnosis including spectrum analysis, correlation analysis, fine spectral analysis, time series analysis, cepstrum analysis, envelopment analysis, holo-spectrum, wavelet analysis, principal component analysis, Empirical Mode Decomposition (EMD) and so on [6]. Most of the methods mentioned above are shown to be successful in dealing with faults in rotating machinery.

Most of the afore-mentioned signal processing methods cannot separate a mechanical fault as the vibration source from other vibration sources. As for the mixed signal, especially for an early-stage fault, generated by signal sources whose number is unknown, the vibration wave of each signal source is not clear yet. However, the vibration of a machine part itself includes the most important information of the fault, while other vibration sources are interferences that should be excluded. Exacting the vibration wave of the faulty part itself correctly is the key step to predict the performance of the machine part.

Nowadays, there are few studies in which fault features are obtained successfully from weak fault signals. In view of this difficulty, blind source separation is adopted to separate all the vibration sources apart, and then fault information from vibration signals of key machine parts is extracted. In this article, the method combining blind source separation and neural network is used to diagnose the early-stage faults of machines. The results of simulation show blind source separation is suitable for weak fault signals of a ball screw.

2. Signal processing

A ball screw is one of the key parts of a Numerical Control (NC) machine, and its performance is most associated with machining performance of an NC machine because the accurate positioning of the cutter in relation to the workbench is the main factor for machining accuracy. And the accurate position depends on transmission accuracy of the ball screw. If its performance declines or a fault occurs the positioning accuracy of the cutter will get worse rapidly. In order to improve an NC machine's reliability, it is very important to reveal early fault mechanism of the ball screw and how much any fault affects machining performance, especially the early-stage ones.

2.1. Signal pretreatment

Most real vibration signals contain zero drift, which needs mean value processing to remove it. On the other hand, mechanical fault diagnosis usually needs multi-dimensional data to allow correlation among them to be assessed. Concerning multidimensional data, a whitening process is an essential step before machine diagnosis and hence will be used here.

2.1.1. Blind Source Separation

Blind source separation (BSS) is a typical method to separate source signals from several different observations provided by sensors. Due to lack of prior knowledge of the source signals, generally, this method assumes that these source signals are independent.

Many examples show that BSS has been successfully used in many fields, such as biomedicine, telecommunications, and so on. However, BSS methods own its special predominance to process mechanical vibration signal for fault diagnosis especially for weak signals [7].

2.1.2. BSS Introduction

Blind source separation is a signal processing method, by which mixed signals, from more than one vibration sources, are separated from the observation of several mixtures without any knowledge of source signals or mixing process. BSS methods include Principal Component Analysis (PCA), Independent Component Analysis (ICA) and Second-Order Blind Identification (SOBI) [8].

Independent Component Analysis (ICA) is adopted in this article to process early-stage fault vibration signals.

2.1.3. Mathematical mode of BSS

The mathematical mode of liner BSS can be expressed as

$$x(t) = \mathrm{A}s(t) + n(t) \tag{1}$$

where $x(t)$ is a vector of sensor signals, and $s(t)$ is an independent signal source vector. A is an unknown mixing matrix, $n(t)$ is a white noise signal vector, as shown in Figure 1. Equation (1) shows that any signal vector can be expressed by the liner combination of $s(t)$ with unknown matrix A .

The formula of observation signals and blind separation sources can be expressed in equation (2),

$$y(t) = \mathrm{M}x(t) \tag{2}$$

where $y(t)$ is a vector separated source signals, and M is the separation matrix.

BSS separation model can be shown in Figure 1.

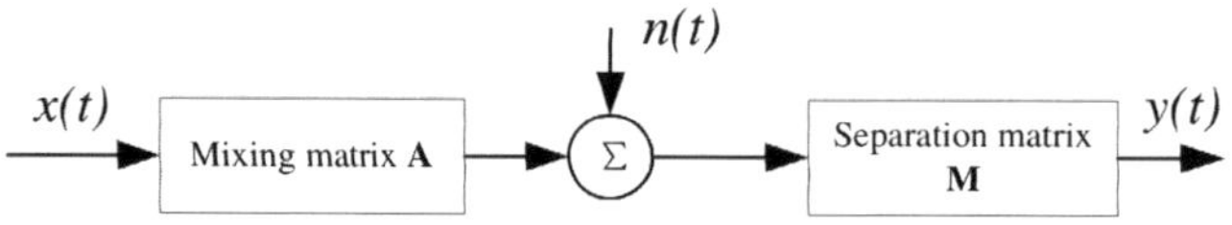

Figure 1. Basic theory figure of BSS

2.1.4. Processing and algorithm of BSS

Independent component analysis (ICA) is one of the most common BSS methods, whose advantages include the versatility and the simplicity in practice. By ICA method, it is easy to separate independent signal sources from mixed signals. However, a main drawback of BSS techniques is the need of several sensors. FastICA method is ICA combined with fixed point algorithm, and FastICA method can effectively evaluate source signals when observation signals are independent from each other. Many researchers focus on FastICA method because of fast calculating speed, excellent separated results and so on. Therefore, FastICA is widely used in fields of image processing, telecommunications, speech processing, underwater acoustics. The calculation flow chart is shown in Figure 2.

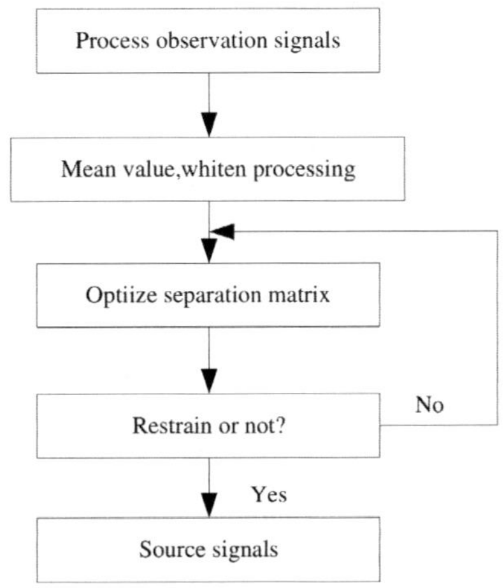

Figure 2. FastICA flow chart

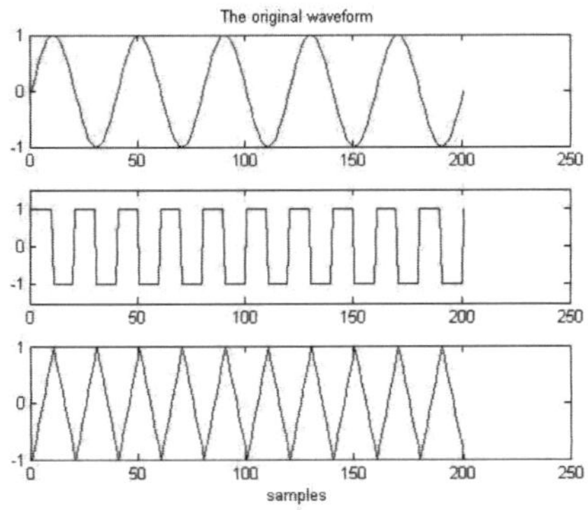

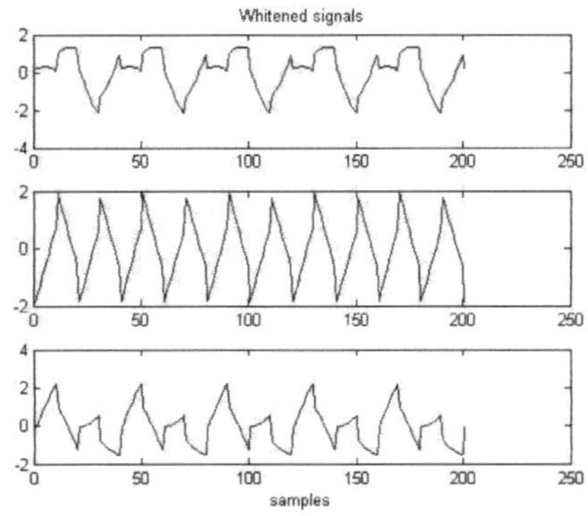

Figure 3. Source signals **Figure 4.** Mixed signals

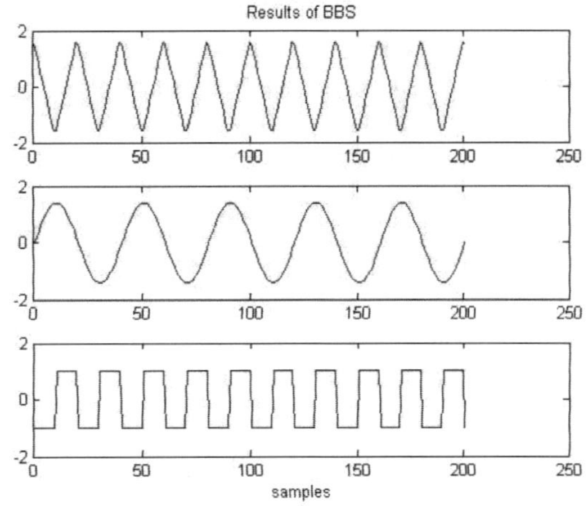

Figure 5. Results of BSS

In order to test the BSS method, a triangular wave, a sine wave and a pulse in different proportions are added together to get different mixed signals, as shown in Figure 4. Then the mixed signals are separated with FastICA method to recover individual signal sources shown in Figure 5.

It can be seen that by comparing Figure 3 with Figure 5, FastICA can extract source signals from mixed ones successfully. Therefore, FastICA method is used to separate early-stage fault signals from mixed signals in this paper.

3. Diagnosis technique

A diagnosis model plays an important role in machine fault detection. The main idea in conventional approaches is to build a mapping function between signals and faults based on mathematical tools. However, these methods have been found to be difficult in coping with ill-defined and uncertain systems, especially for uncertain vibration sources.

Many neural network methods are focused on solving uncertain systems too. Their main idea is to set up a relationship between inputs and outputs, through training neural networks from inputs and outputs.

Dynamic Fuzzy Neural Network (DFNN) is one of neural networks. This kind of neural network fuses fuzzy theory and neural network theory. In many fields, it is widely used for system prediction, realization, on-line control, communications and pattern recognition. Its computational efficiency and dynamic characteristic are better than those of other typical neural networks and DFNN also show excellent adjustment characteristics by uncertain topology structure. A DFNN system starts with no hidden units and neurons can be recruited or deleted dynamically based on their significance in the system so that not only the system parameters can be adjusted, but also the structure can be self-adapted simultaneously. A simple hierarchical learning approach has been proposed for DFNN by pruning technology and significant neurons are selected so that a parsimonious structure with excellent performance can be achieved [9].

The diagnosis model based on DFNN is very suitable to deal with uncertain systems for its dynamic performance. That means changing inputs and DFNN structure, and the network will expand or contract along with inputs. The dynamic performance of DFNN especially meets the process of detecting mechanical faults. In this article, dynamic mechanical fault diagnosis model is built based on DFNN and BSS to predict mechanical faults.

Using DFNN with a Gaussian function to be activation function, each fuzzy rule represents a neural network node that realizes dynamic response to external conditions.

The topological structure of DFNN is shown in Figure 6, where x_1, x_2, $\cdots$, x_r are inputs, and y is output. MF_{ij} represents the jth activation function for the ith input. R_j is the jth fuzzy rule. N_j as the jth normalized node, w_j as weight values, u as the total number of rules.

Just as shown in Figure 5, the first layer is input where each node refers to one input. The second layer is activation function layer where each node means one activation function (a Gaussian function). It can be expressed in equation (3).

$$\mu_{ij}(x_i) = \exp\left[-\frac{(x_i - c_{ij})^2}{\sigma_j^2}\right] \tag{3}$$

In (3), μ_{ij} represents the jth activation function for x_i, c_{ij} is the jth Gaussian function center for x, and σ_j is the jth Gaussian function width for x_i.

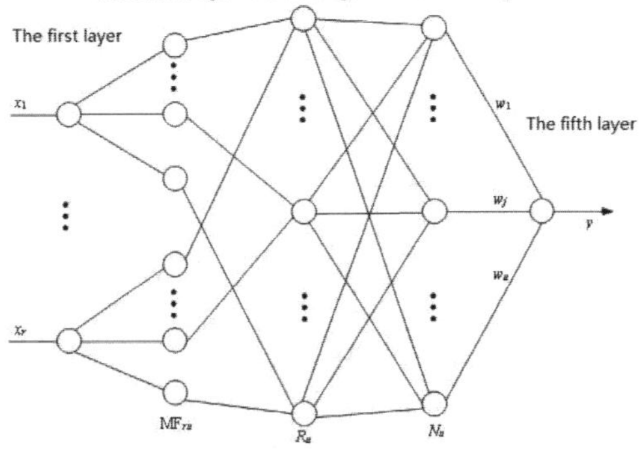

Figure 6. Topology of DFNN

The third layer of DFNN topology is T-norm, where every node may become a fuzzy rule and the rule number is the node number. The jth rule for R_j can be expressed as

$$\phi_j = \exp\left[-\frac{\sum_{i=1}^{r}(x_i - c_{ij})^2}{\sigma_j^2}\right] = \exp\left[-\frac{\left\|X - C_j\right\|^2}{\sigma_j^2}\right], \quad j=1, 2, \cdots, u \tag{4}$$

where $X=(x_1, x_2, \cdots, x_r) \in R_r$ is input vector, and $C_j=(c_{1j}, c_{2j}, \cdots, c_{rj}) \in R_r$ is the jth neural network node centre. In a word, each node is one unit of neural network in the third layer

The forth layer is a normalized layer, where the node number N is equal to number of fuzzy rule nodes. The jth node N_j is expressed as formula (5).

$$\phi_j = \frac{\phi_j}{\sum_{k=1}^{\mu}\phi_k}, \quad j=1, 2, \cdots, u \tag{5}$$

The fifth layer is the output layer and the output vector is expressed by $y(X)$, which can be calculated from (6).

$$y(x) = \sum_{k=1}^{u} w_k \bullet \varphi_k \tag{6}$$

where w_k is weighting of the kth rule.

The weight model can be expressed as below

$$w_k = \alpha_{k0} + \alpha_{k1}x_1 + \ldots + \alpha_{kr}x_r, \quad k=1, 2, \cdots, u$$

This means [10, 11]

$$y(X) = \frac{\sum_{i=1}^{u}\left[(\alpha_{i0} + \alpha_{i1}x_1 + \ldots + \alpha_{ir}x_r)\exp\left(-\frac{\left\|X - C_i\right\|^2}{\sigma_i^2}\right)\right]}{\sum_{i=1}^{u}\exp\left(-\frac{\left\|X - C_i\right\|^2}{\sigma_i^2}\right)} \tag{7}$$

The algorithm of diagnose model is shown in Figure 7.

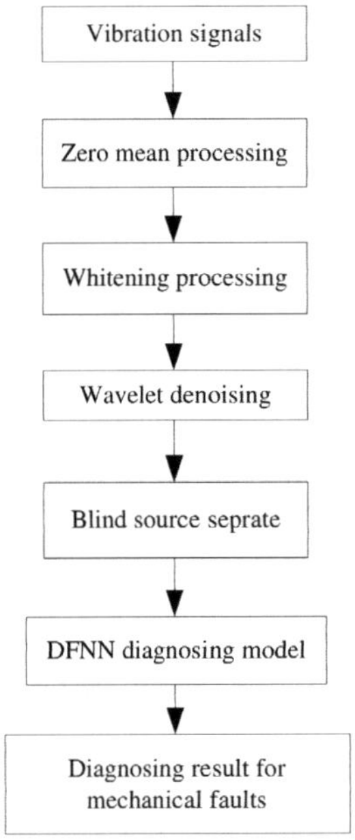

Figure 7. Fluent chart of diagnosis model based on DFNN

4. Example

4.1. 1 Test for a ball screw

The early-stage fault signals of a ball screw are collected in the performance degradation test, which is shown in Figure 8.

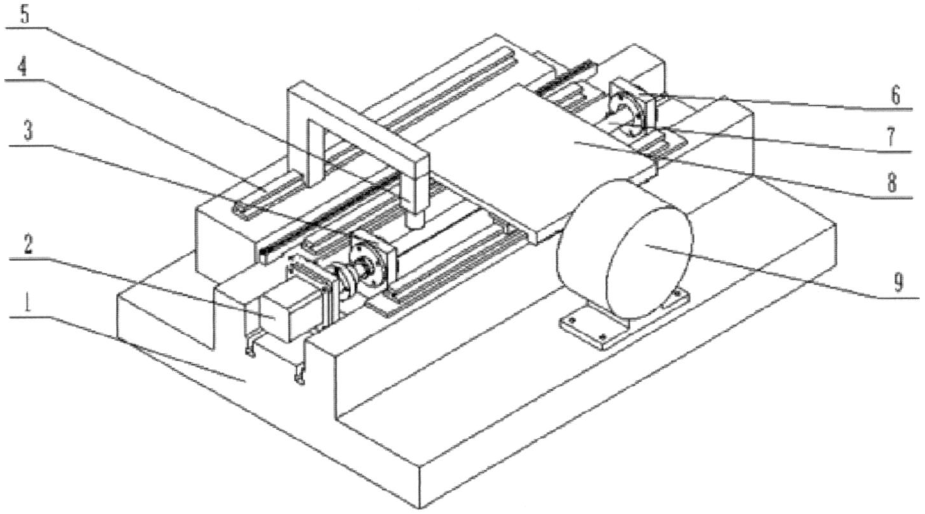

Figure 8. Performance degradation test-bed

1-Bed 2-Motor 3-Rolling bearing 4- Ball slide 5- Image acquisition apparatus
6- Rolling bearing 7- Ball screw 8- Workbench 9- Magnetic powder brake

On the test-bed, a ball screw driven by motor is supported by two pieces of rolling bearings. The nut of the ball screw pair is connected with the workbench, which is supported by two ball slides. The load of the ball screw pair is loaded by a magnetic powder brake. Vibration sensors are located on both of the rolling bearing seats and the nut seat to collect vibration signals of the ball screw in the X, Y, and Z directions, once for every 30 minutes during acceleration test. Acquisition frequency is 5 kHz, and acquisition time is 10 seconds. The vibration and noise of ball screw increases in the same situation. The nut of ball screw vibrates violently and when adjust pre-tightening force, the ball screw can work normally again, which is the early-stage wearing. The information of sensors is shown in table 1.

Table 1. Type and accuracy of sensors

Type of sensor	Model	Purpose	Installation position
Torque sensor	AKC-215	Load detection	Magnetic powder brake
Three acceleration sensors	14530	Vibration of nut holder	Nut holder
Acceleration sensors	14200	Vibration of ball screw	Fixed end of ball screw
Acceleration sensors	14200	Vibration of ball screw	Support end of ball screw

Three sensor signals are collected on the nut and both ends of the ball screw. All signals are separated by BSS method and the separated results are shown in Figure 9 to Figure 14. Then, early-stage fault signals and normal signals are processed to test the fault diagnosis method based on BSS and DFNN. Compared with normal vibration signal characteristics, most of differences of same kind characteristic of early fault signals are getting larger when use BSS method to deal with vibration data, which can be seen from Table 2 to Table 4.

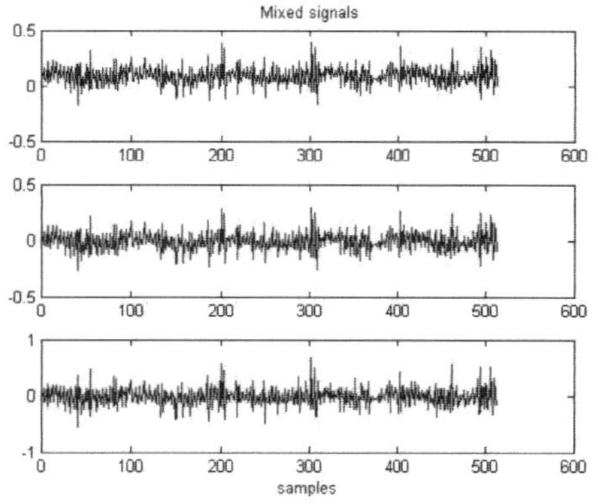

Figure 9. Mixed signals of normal signal

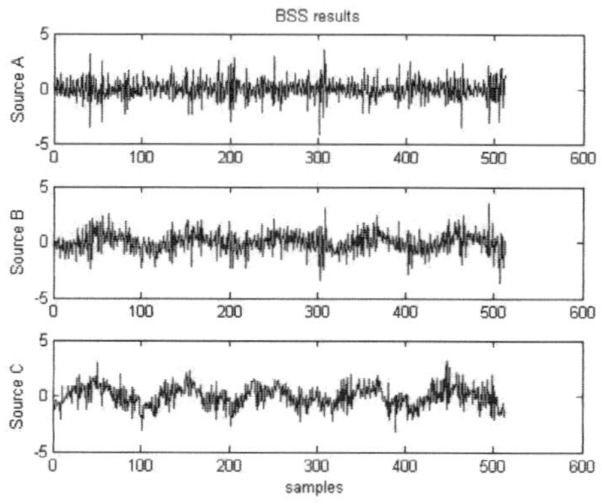

Figure 10. BSS of normal signals

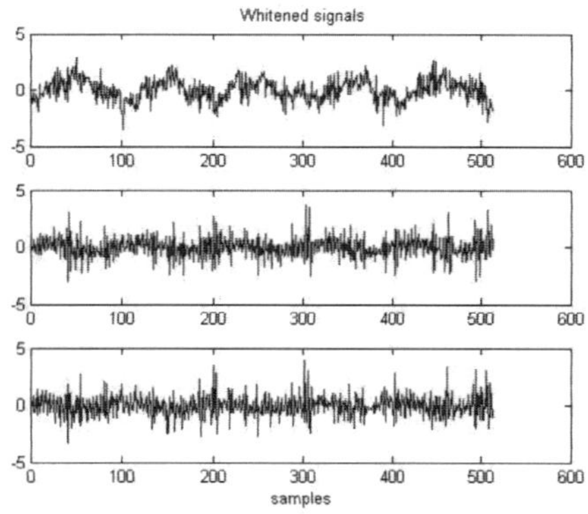

Figure 11. Whitened signals of normal signals

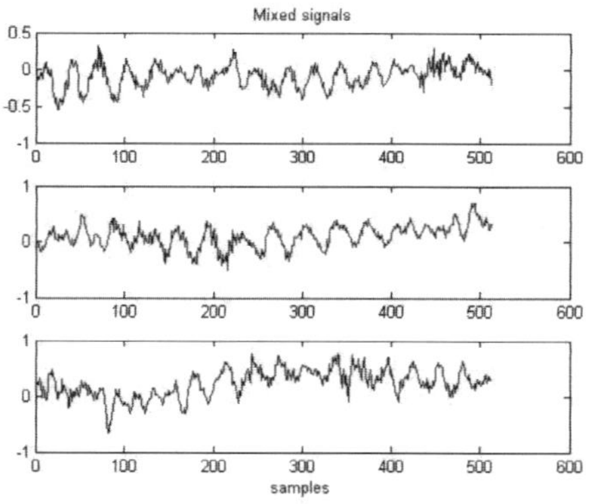

Figure12. Mixed signals of early fault signals

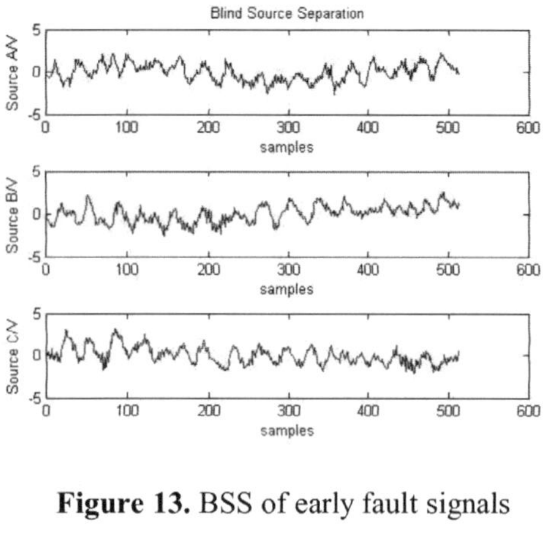

Figure 13. BSS of early fault signals

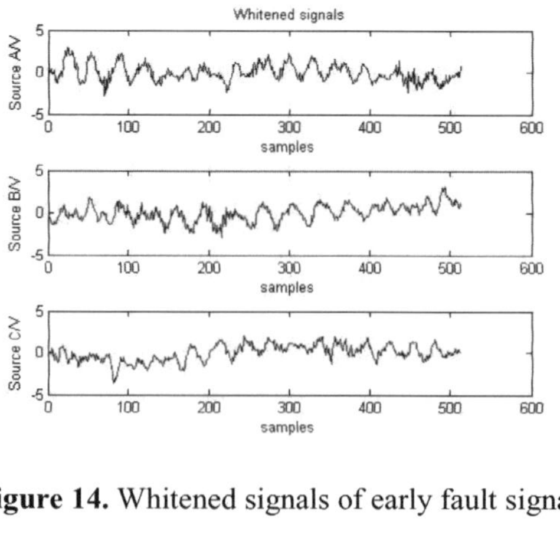

Figure 14. Whitened signals of early fault signals

As a result, characteristics of signals separated by BSS can be recognized by DFNN more easily than those of early-stage fault vibration signals. 25 groups of normal data and 25 groups of early-stage fault data are used to train the DFNN model, and another 25 groups of normal data and early-stage fault data are used to test diagnosis accuracy. It is found that the early-stage fault of 25 groups data is recognized by the combined method of DFNN and BSS. However, only 23 early fault groups are recognized without BSS method, which may be affected by other vibration signals that have nothing to do with ball screw faults.

Table 2 Characteristics of fault signal measured on the ball screw nut

Mean value	Peak	Peak factor	Kurtosis factor	Pulse factor	Margin coefficient	Centroid frequency	Frequency variance	Mean–square frequency
0.08	0.36	4.84	3.65	6.19	7.35	0.98	0.46	1.42
0.09	0.42	4.46	3.25	5.63	6.66	1.07	0.67	1.81
0.11	0.67	6.14	4.40	7.99	9.60	1.31	0.59	2.30
……	……	……	……	……	……	……	……	……

Table 3 Characteristics of normal signal on the ball screw nut

Mean value	Peak	Peak factor	Kurtosis factor	Pulse factor	Margin coefficient	Centroid frequency	Frequency variance	Mean–square frequency
0.15	0.39	2.58	2.86	3.19	3.71	0.38	0.35	0.50
0.20	0.51	2.54	3.02	3.16	3.69	0.32	0.29	0.39
0.25	0.57	2.30	3.04	2.84	3.29	0.21	0.15	0.20
……	……	……	……	……	……	……	……	……

Table 4 Characteristics of fault signals on the ball screw nut processed by BSS

Mean value	Peak	Peak factor	Kurtosis factor	Pulse factor	Margin coefficient	Centroid frequency	Frequency variance	Mean–square frequency
1.00	2.17	3.25	3.25	3.11	3.56	0.29	0.25	0.33
1.00	2.32	3.30	3.35	3.80	3.39	0.28	0.24	0.36
1.00	2.22	3.15	4.11	3.77	3.23	0.31	0.26	0.39
……	……	……	……	……	……	……	……	……

5. Conclusion

The fault diagnosis of machines, especially for early-stage faults, is practised to reduce maintenance cost and risk to people. However, it is difficult to diagnose early-stage faults based on vibration signals in practical environment. In this article BSS and DFNN methods are combined to get rid of interference from other sources and provide a more accurate diagnosis for key parts of machines, such as ball screws. Firstly, Fast ICA, one of BSS methods, is tested and found to be suitable to deal with mixed signals from different vibration signals. Then, the diagnosis model is built based on BSS and DFNN. Through an acceleration test, the vibration signals of the ball screw in both normal state and early-stage fault state are collected. The results of numerical simulation by Matlab show that the diagnosis accuracy with BSS method is higher than that without BSS method.

Reference

[1] Jordan M and Markus T 2011 Feature extraction for novelty detection as applied to fault detection in machinery *Pattern Recognition Letters* 32 pp1054-1061

[2] Lee J et al 2006 Intelligent prognostics tools and e-maintenance *Computers in Indu*stry vol 56 pp 476-489

[3] MA N and LU C. 2009 Research on PHM architecture of aircraft *Journal of Huazhong University of Science and Technology (Natural Science Edition)* vol 39 pp 207-209

[4] TEO C Y and GOOI H B 1997 Application of knowledge-based systems for fault diagnosis and supply restoration *Engineering Applications of Artificial Intelligence* vol 10 pp 631-638

[5] MAK B M T 2002 Rule extraction from expert heuristics：a comparative study of rough sets with neural networks and ID3 *European Journal of Operational Research* vol 136 pp 212-229

[6] Du Y et al 2008 The Application of Information Fusion in Fault Diagnosis and Equipment Monitoring *Proceedings of the 27th Chinese Control Conference* (July Kunming Yunnan China) pp 16-18

[7] Huang W G et al 2009 Research on Blind Source Separation for Machine Vibrations *Wireless Sensor Network* 1 pp 453-457

[8] Nguyen V H et al 2012 Fault diagnosis in industrial systems based on blind source separation techniques using one single vibration sensor *Shock and Vibration* 19 pp 795-801

[9] Wu S Q and Meng J E 2000 Dynamic Fuzzy Neural Networks-A Novel Approach to Function Approximation *Systems, Man, and Cybernetics (Part B: Cybernetics, IEEE Transactions on)* vol 30 pp 358-364

[10] Wu S Y and Xu J 2008 *Dynamic Fuzzy Neural Network-Design and Application* (Beijing: Tsinghua University press) pp27-38

[11] Huang H F et al 2013 Study on evaluation of performance degradation for NC machine tool spindle system *Journal of Vibration, Measurement and Diagnosis* **33** 646-652.

Acknowledgments

This study is supported by the Fundamental Research Funds for the Central Universities (No. 2682014CX034).

Vibration damage mechanism analysis on rotor of diesel generating set with rigid coupling

Bing Yan[1], Weizhen Shi, Chunrong Hua, Jingming Liu, Dawei Dong and Jun Chen

College of Mechanical Engineering, Southwest Jiaotong University, China

E-mail: yanbingwd@163.com

Abstract. The crankshaft output end is generally connected with generator rotor through the coupling in diesel generating set. When using rigid coupling, the attachments and connecting parts of generator rotor (especially at larger gyration radius) are vulnerable to fatigue damage even if the vibration level of the generating set does not exceed the acceptable "usual value". In order to investigate the reasons, the torsional vibration of the rotor in the diesel generating set was calculated and measured in this paper, which shows that using high rigidity coupling would result in large torsional vibration on the generator rotor, and that the linear vibration (the tangential vibration) value induced by torsional vibration at larger gyration radius of generator motor is almost the same as the vibration level of the generating set. Then, the vibration level of generating set was obtained, and the maximum vibration velocities of the generator are below the permissible value regulated by ISO 8528-9. But the velocities of synthetic vibration of the generating set vibration and the linear vibration induced by torsional vibration at larger gyration radius are much higher than permissible value 2(28mm/s) regulated by ISO 8528-9, which may be the reason of the mechanical damage of the attachments and connecting parts at larger gyration radius of generator motor caused by exceeded vibration.

1. Introduction

As the power unit of much mechanical equipment, diesel generating set is widely used in industrial production, traffic and other fields. Therefore the failure of diesel generating set threatens the normal operation of equipment and personnel safety significantly.

There are various generator faults including stator fault, rotor fault , bearing fault and so on [1][2],

[1] No.111, North 1st Section of Second Ring Road, Chengdu, Sichuan province, China

among which the rotor fault is the main fault type[3]. Rotor broken bar and the fracture of end ring are general rotor faults, which will make the generator emitting heat during operation, and lead the guide bar and end ring to suffer cyclic thermal stress and deformation, resulting in further expansion of the fault. The reasons of the generator rotor faults include unbalanced rotor mass, uneven temperature and unbalanced electromagnetic force, which will cause rotor's bending vibration and generator vibration; there are many studies in these areas [4-6]. Diesel generating set has its particularity of uneven tangential torque acting on the engine crankshaft, which may produce obvious torsional vibration on rotor. This can be the reason that the vibration level of a generating set does not exceed the acceptable "usual value", but the fatigue failure of the attachments and connecting parts on generator rotor (especially at larger gyration radius) often occurs, and especially when the output end of the crankshaft is coupled rigidly with the generator rotor in the generating set, generator rotor is more vulnerable to fatigue damage. Therefore in this paper the torsional vibration of the generator rotor in the generating set and its linear vibration characteristics were studied, and its effect on structure vibration of the rotor was investigated combining with examples.

2. Dynamic characteristics analysis of generator rotor torsional vibration for typical generating set

Generally speaking, with the increasing of the number of the cylinders and the length of the crankshaft in diesel engine, the crankshaft stiffness decreases, thereby the crankshaft modal (the nodes are in the crankshaft) frequency decreases, and vice versa. The above two kinds of crankshaft torsional vibration modal characteristics are distinctly different, and the modal natural characteristics and vibration response characteristic of shafting torsional vibration are also different when different stiffness couplings are used to connect the generator, but they all have the generality of larger rotor torsional vibration induced by high rigidity coupling (such as metal diaphragm coupling) in generating set shafting.

2.1. Larger crankshaft rigidity

With the decreasing of the number of the cylinders and the length of the crankshaft in diesel engine, the crankshaft stiffness increases, thereby the crankshaft modal frequency increases. Taking a generating set with a 6-cylinder diesel used in a diesel multiple units of railway locomotive as an example, the engine speed ranges from 900r/min to 1800r/min. The equivalent system diagram of the generating set shafting is shown in Figure 1.

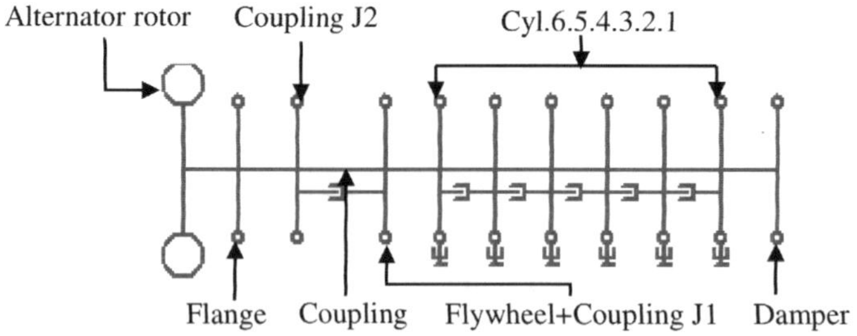

Figure 1. The equivalent system diagram of 6-cylinder engine shafting.

Figure 2 is the curves of the maximum synthesized torsional vibration displacement amplitude (hereinafter referred to as Syn) at crankshaft free end, generator rotor and crankshaft flywheel changing against coupling stiffness. It shows that when using highly elastic coupling (for example, below 0.031 MNm/rad), the torsional vibration amplitude of the generator is small (below 0.066°), and the torsional vibration amplitude of the flywheel and the crankshaft free end is large (about 0.6°).

When using large stiffness coupling, the generator rotor is rigidly connected with the crankshaft, and the torsional vibration response characteristics are mainly determined by roll-vibration [7]. In low speed condition, compared with highly elastic coupling, the main harmonic torsional vibration amplitude of the crankshaft is reduced obviously and that of the generator rotor is increased evidently. It is shown from Figure 2 that the maximum Syn at crankshaft free end decreases significantly (slightly larger than 0.2°) and that of the generator rotor increases greatly (about 0.19°), when using larger stiffness coupling (Supposing the stiffness of the diaphragm coupling is 92.7 MNm/rad), therefore the ability to bear torsional vibration and the induced tangential vibration of the generator rotor should be taken into consideration. The maximum Syn of the generator rotor reaches 0.19°, which means that the larger tangential vibration amplitude may exist at larger gyration radius of the generator rotor. For example, the tangential vibration amplitude induced by torsional vibration may reach 0.83mm at gyration radius of 250mm of the generator rotor, and the effective value of tangential vibration velocity is approximately 166 mm/s, which is significantly larger than usual vibration velocity permissible value of the generator. Therefore the vibration velocity of the generator rotor caused by torsional vibration is relatively larger in this circumstance, and may result in strong excitation on rotor parts, which may possibly make the generator rotor generate strong structural vibration.

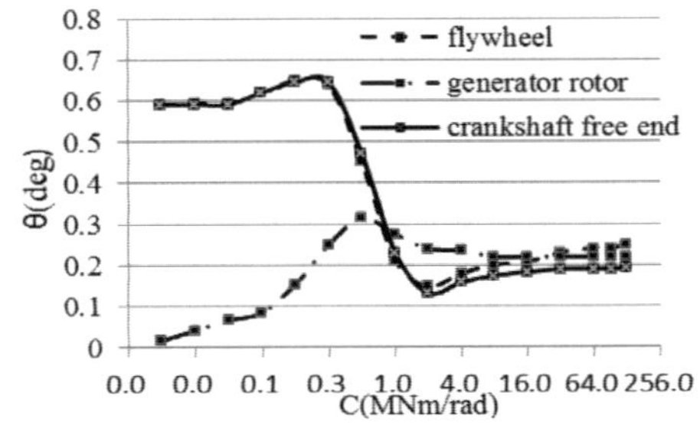

Figure 2. The curves of maximum Syn changing against coupling stiffness.

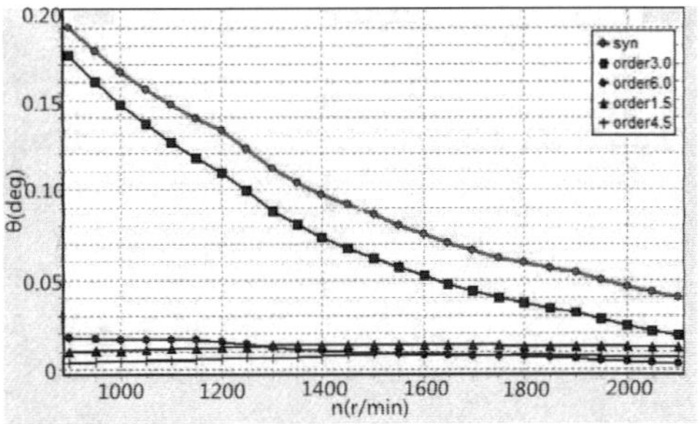

Figure 3. The torsional vibration amplitude of generator rotor changing against speed.

2.2. Smaller stiffness crankshaft

For diesel engine with more cylinders and longer crankshaft, the crankshaft stiffness is relatively small, and the effect of using the above two kinds of coupling on torsional vibration characteristic of the shafting is not as obvious as the larger rigidity crankshaft mentioned above, but the natural frequency and response characteristic of torsional vibration coupled modal for crankshaft and generator rotor are affected apparently.

Taking the 3150kW diesel generating set used in railway locomotive as an example, the shafting mass-elastic system model is shown in Figure 4. A measurement and calculation is carried out for the torsional vibration of the shafting.

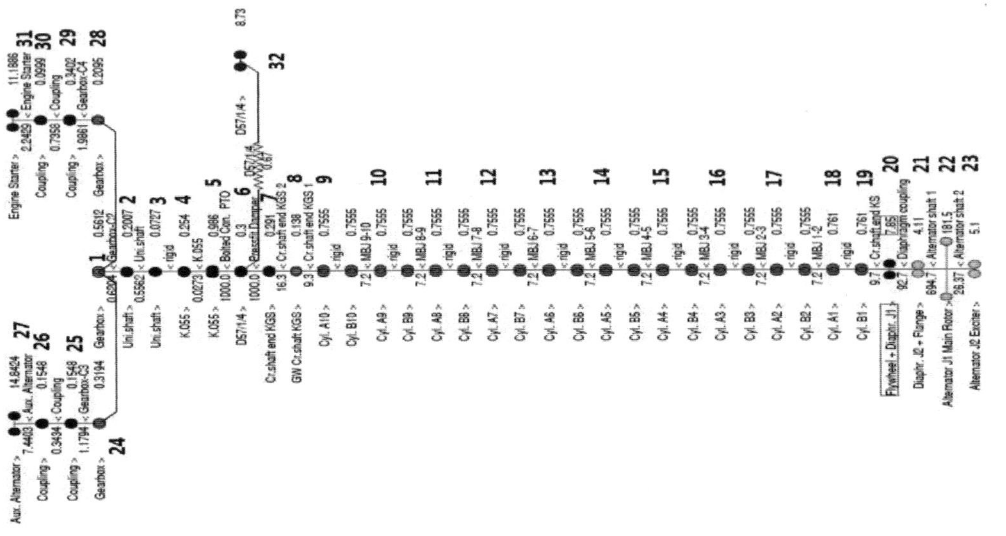

Figure 4. The mass-elastic system diagram of crankshaft for a locomotive generating set.

Figure 5 is the resonance speed diagram of the shafting using higher stiffness diaphragm coupling (the stiffness of the diaphragm coupling is 92.7 MNm/rad). Calculations and analysis for the shafting show that the third order modal is the coupled modal between the crankshaft and the generator rotor through diaphragm coupling. The resonance speed of the 2.5 order and 1.5 order engine excitations corresponding to the third order modal (1757/60=29.3Hz) of the shafting is 703r/min and 1171r/min respectively. Figure 6 is the measurement results of torsional vibration at the generator end, which shows that the resonance arising from the interaction of the 2.5 and 1.5 order engine excitations and the third order modal of the shafting appears apparently on the generator rotor.

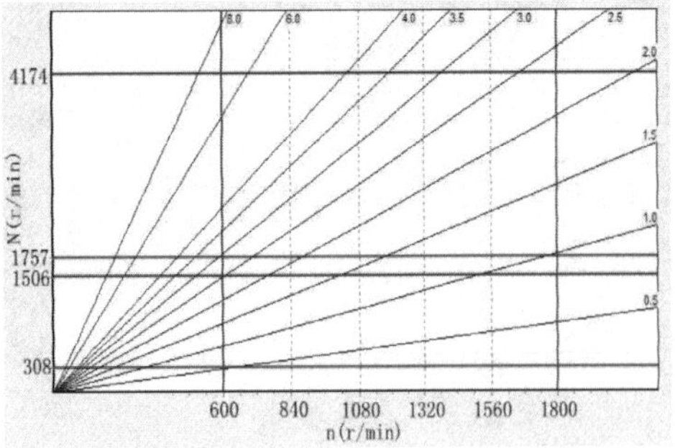

Figure 5. The resonance speed diagram connected with diaphragm coupling.

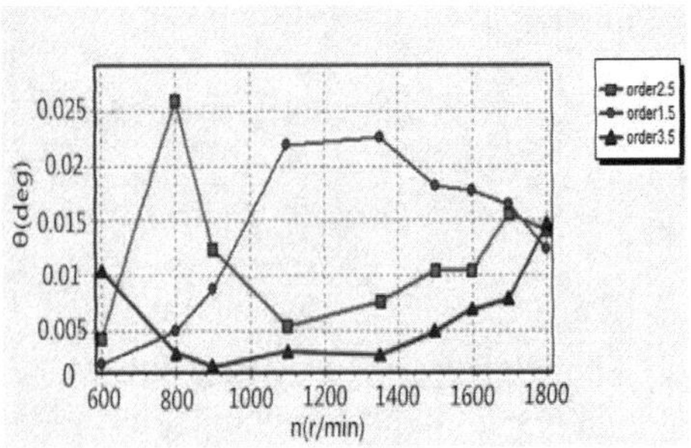

Figure 6. The measurement results of torsional vibration at the generator.

The maximum Syn at crankshaft free end, generator rotor and crankshaft flywheel changing against the coupling stiffness could be obtained from further calculation for the torsional vibration (as shown in Figure 7). The figure shows that the vibration amplitude of the generator rotor is large when using high stiffness coupling (above 1.44 MNm/rad). The torsional vibration amplitude of the generator rotor will reduce along with the decreasing of coupling stiffness (below 1.44 MNm/rad).

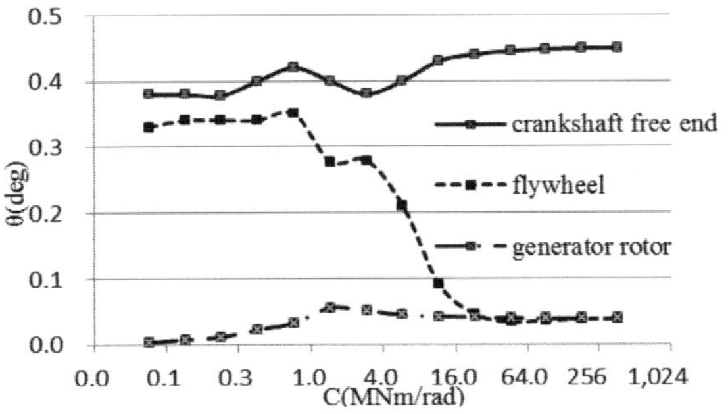

Figure 7. The curves of maximum Syn changing against coupling stiffness.

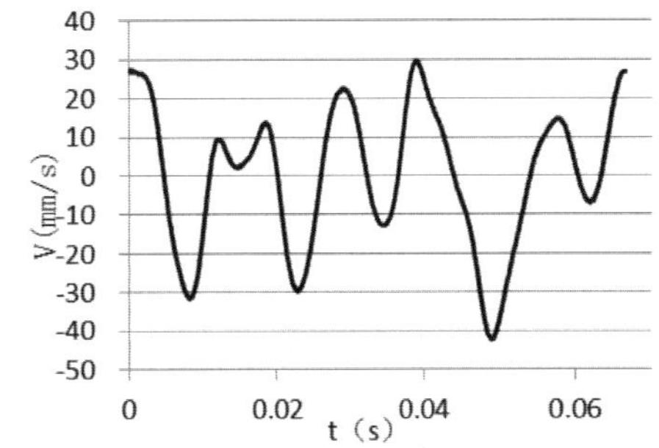

Figure 8. The linear vibration curve induced by torsional vibration in radial 250mm position of generator rotor.

Selecting the maximum torsional vibration working condition from Figure 6 and calculating the generator vibration, the linear vibration induced by torsional vibration in radial 250mm position of generator rotor is shown in Figure 8, in which the effective value is 18.0mm/s, almost the same vibration level with the generating set.

In summary, although the torsional vibration modal characteristics of the shafting is different, it is a general character that using large stiffness coupling may lead to large torsional vibration on rotor and the linear vibration (tangential vibration) induced by the torsional vibration at larger gyration radius of generator rotor reaches the same level with the vibration of the generating set.

3. Vibration analysis for the generating set

3.1. The measured results

The 3150kW diesel generating set used in diesel locomotive comprises a 20-cylinder diesel engine (20V4000R53) and a generator (CDJF201L), and the generating set is connected with the locomotive body through eight rubber isolators, among which six isolators are on the diesel engine and two are on the generator. The structure of the generating set is shown in Figure 9,

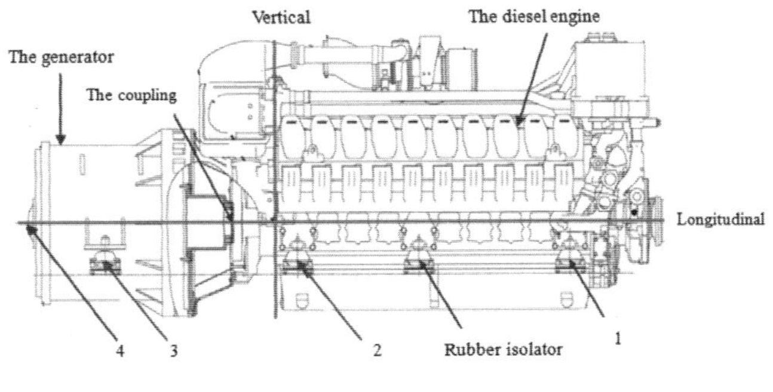

Figure 9. The structure schematic of the generating set.

The experimental results show that the vibration value of the generating set reaches the maximum in the condition of 1800r/min and 3150kW. The number and the position of measuring points are shown in Figure 9 and the effective values of three-dimensional vibration velocities are shown in Table 1. The experimental results above are provided by the MTU Company.

Table 1. The measured vibration velocity of the generating set.

NO.	Measuring Point	Effective Value of Vibration Velocity (mm/s)		
		Vertical	Transverse	Lateral
1	Upper Edge of Crankshaft Free End Bearing	21	15	21
2	Upper Edge of Drive End Bearing	15	15	17
3	Upper Edge of Generator Bearing	21	15	18
4	Back End of Generator Stator	21	16	13

It can be seen from Table 1 that the effective values of vibration velocity of measuring points on the generator are below the permissible value 2 (28mm/s) regulated by ISO 8528-9.

3.2. The calculated vibration results of the generating set

The calculated vibration results of the generating set in the condition of 1800r/min and 3150kW through a six degree-of-freedom model of the generating set with eight isolators are in consistent with the measurement results. The calculated effective values of lateral and vertical vibration velocity of the measuring points on the generator are below the permissible value 2 (28mm/s) regulated by ISO 8528-9.

4. The synthesized results of the generating set vibration and linear vibration induced by torsional vibration

From above measurement and calculation results, we know that the vibration level of the generating set meets the standards, but the attachments and connecting parts at larger gyration radius of generator rotor are locations where structural damage initiates in practical operation, which may be caused by the combination action of the generating set vibration and the linear vibration induced by torsional vibration. Therefore the following paper synthesizes the vibration of the generating set and the linear vibration induced by torsional vibration in radial 250mm of generator rotor, and analyzes its actual vibration condition(as shown in Figure 10).

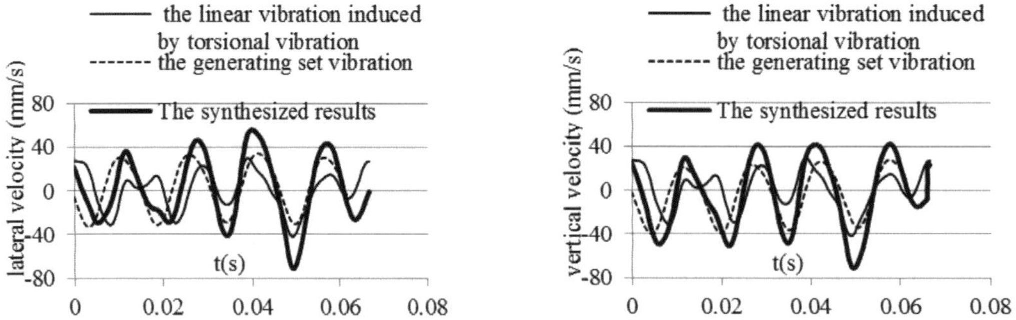

Figure 10. The synthesized results of the generating set vibration and the linear vibration induced by torsional vibration in in radial 250mm of generator rotor.

The calculation results show that the effective values of lateral vibration velocity at gyration radius of 250mm of the rotor is 30.98mm/s, the effective value of vertical vibration velocity is 32.05mm/s, exceeding the maximum vibration velocity value 2(28mm/s) regulated by ISO 8528-9, which will make the generator rotor and connecting parts be subjected to excessive vibration and appear structural failure; the normal operation of generator will be affected, accordingly.

5. Torsional vibration control of the generating set rotor

From the analysis mentioned above, we know that the structural damage commonly initiates at the attachments and connecting parts where the gyration radius of generator rotor is larger, which is possibly caused by the combined action of the generating set vibration and the linear vibration induced by torsional vibration. Under the condition that the generating set vibration is difficult to be reduced further, it is another method to reduce the structural damage of generator rotor through controlling torsional vibration on generator rotor of the generating set for the purpose of reducing the influence of torsional vibration on the rotor.

From Figure 7, we can see the reduction of coupling stiffness (below 1.44 MNm/rad) is apparently favorable for the decreasing torsional vibration amplitude of the generator rotor. On the premise of ensuring transmitting required torque for the generating set shafting, the coupling is replaced with a Geislinger elastic coupling (its stiffness is 0.51MNm/rad). Results show that the natural frequency of coupled modal of the crankshaft and generator rotor is obviously decreased with the high elastic coupling; the resonance speed diagram of the first four modal is shown in Figure 11. Further calculations and analysis show that the second order modal is the coupled modal between the crankshaft and generator rotor through diaphragm coupling. The resonance speed of the 1.5 and 1.0 order engine excitations corresponding to the second order modal (1104/60=18.4Hz) of the crankshaft is 736r/min and 1104r/min respectively. The measurement results of torsional vibration are in accordance with the above calculation on the aspect of general rules. The Figure 12 shows that the resonance arising from the interaction of the 1.5 and 1.0 order engine excitations and the second order modal of the shafting appears on the generator rotor.

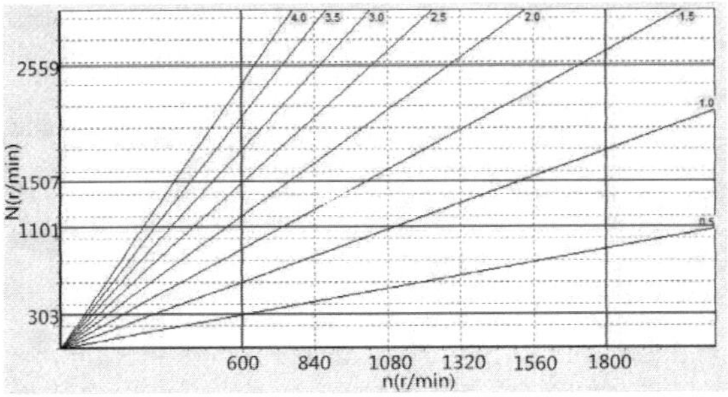

Figure 11. The resonance speed diagram with high elastic coupling.

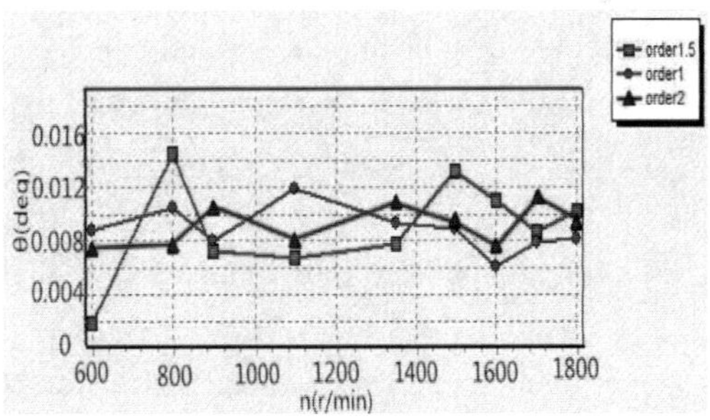

Figure 12. The measurement results of torsional vibration at the generator.

Selecting the condition of maximum torsional vibration from Figure 12 to calculate, the effective value of the linear vibration caused by torsional vibration in radial 250mm position of generator rotor is decreased to 7.5mm/s, and by 58.3% compared to the unimproved structure.

When the shafting are connected by highly elastic coupling, through synthesizing the linear vibration caused by torsional vibration (tangential vibration) at radial 250mm position of generator rotor and the generating set vibration, it can be calculated that the effective value of lateral vibration velocity at gyration radius of 250mm is 23.21mm/s, and the effective value of vertical vibration velocity is 24.24mm/s, falling in between the value 1 (20mm/s) and 2 (28mm/s) regulated by the ISO

8528-9, satisfying the standard demands. When using highly elastic coupling, the fault rate of the generator rotor decreases in practical engineering application.

6. Conclusion

Combining with the examples, this paper studied the characteristic of the torsional vibration and linear vibration of generator rotor when the generating set uses rigid coupling, and discussed the influence on the structure vibration of the rotor. Conclusions which have a certain guidance and reference value for coupling selection for the generating set and the strength design for accessories and the connecting parts on generator rotor, are as follows:

1) Although different types of diesel engine crankshafts have markedly different torsional vibration modal characteristics, but they all have the generality that using high rigidity coupling would result in large torsional vibration on the generator rotor.
2) When using high rigidity coupling, the linear vibration (tangential vibration) value caused by torsional vibration at larger gyration radius of the generator rotor is equivalent to the vibration level of the generating set. The synthesized vibration of the generating set vibration and the linear vibration caused by torsional vibration maybe the mechanical damage reason of the attachments and connecting parts at larger gyration radius of generator motor because of the exceeded vibration.
3) Under the condition that the vibration of the generating set is difficult to further reduced, through controlling the torsional vibration of the generator rotor to reduce the influence of the torsional vibration on the generator rotor, it is a feasible method to reduce the structure damage to the generator rotor.

References

[1] Iasmotor Reliabitity Working Group. 1985 Report of Large Motor Reliability Survey of Industrial and Commercail Installations-Part I [J].IEEE Trans on Ind.App I. 21 (1) :853-64

[2] Subhasis N, Hamid A. T and Xiaodong Li. 2005 Condition Monitoring and Fault Diagnosis of Electrical Motors—A Review[C]. IEEE Transactions On Energy Conversion, 719-729

[3] Tongguang Yang and Xinhua Jiang. 2010 Research and Development Trends of Induction Motor Fault Diagnosis[J]. *Micromotors Servo Technique*,43 (4) 68-69

[4] Lijuan Zhang,Xinhua Yang,Chengxia Ma and Hongwei Wang. 2009 Remedy on Vibration of Turbogenerator Due to Thermal-caused Bending of Rotor[J]. *Large Electric Machine and Hydraulic Turbine*, 4 (4)

[5] Xueshi Yao and Chunlong Zheng. 2014 Research on Bending Vibration of a Rotor Based on Spin Softening[J]. *Turbine Technology*, (2)

[6] Xiongqing Li,Shi Liu,Yongxin Feng,Xiaowen Deng,Fei Wang and Weimin Que. 2011 Diagnosison on Vibration Problem of Generator with Bending Rotor[J]. *Journal of Vibration,Measurement & Diagnosis*, 31 (2)

[7] Bing Yan,Meiyun Sun,Yumei Li,Xiao Li and Chunrong Hua. 2006 The Single Order Quasi Rigid Body Model of Torsional Vibration Diagnosis for ICE Crankshaft[J]. *Chinese Internal Combustion Engine Engineering*, 27 (4) : 62-65

Vibration-based SHM System: Application to Wind Turbine Blades

D Tcherniak L L Mølgaard[2]

[1] Research Engineer, Brüel & Kjær Sound & Vibration Measurement A/S, Nærum, Denmark
[2] Senior Researcher, Department of Applied Mathematics and Computer Science, Technical University of Denmark, Lyngby, Denmark

E-mail: `dmitri.tcherniak@bksv.com`

Abstract. This study presents an vibration-based system designed for structural health monitoring of wind turbine blades. Mechanical energy is introduced by means of an electromechanical actuator mounted inside the blade. The actuator's plunger periodically hits the blade structure; the induced vibrations propagate along the blade and are measured by an array of accelerometers. Unsupervised learning is applied to the data: the vibration patterns corresponding to the undamaged blade are used to create a statistical model of the reference state. During the detection stage, the current vibration pattern is compared with the reference state, and the novelties can be associated with damage. The vibration pattern is described by the covariance matrix between the accelerometer signals. The mid-range frequencies are used: this range is above the frequencies excited by blade-wind interaction, thus ensuring a good signal-to-noise ratio. Simultaneously, the frequencies are low enough to be able to propagate the entire blade length, so good results can be obtained even using only one actuator. The system is demonstrated on a real 34m blade mounted on a test rig. Using the suggested approach, the system enables detection of, e.g., a 20cm long trailing edge opening under realistic noise conditions. It is also demonstrated that the system provides rough information about damage location. Progression of damage, if any, can also be detected.

1. Introduction

Wind turbine blades today are extremely complex and expensive structures; from the cost perspective, they constitute a significant asset for wind turbine owners. Therefore, monitoring of their structural health becomes economically rational, especially for big, remote (offshore) turbines.

There are many different approaches to SHM of wind turbine blades, varying by utilizing different physical phenomena, sensor types and signal processing; a detailed review can be found, for example, in [1]. The present study introduces a vibration-based, active system. *Vibration-based* means that changes in mechanical vibrations of the blade serve as a feature indicating blade damage. Active means that the vibrations are introduced artificially, by using a dedicated *actuator* mounted inside the blade; this contrasts with other vibration-based approaches which rely on ambient vibration of the blades due to wind. In authors knowledge, so far there is no evidence that the latter can provide damage detection resolution satisfying industry's demand. The suggested technique differs from another well-known active approach, *guided waves*, in the frequency range used. If, in the guided waves case, the tens of kHz range is typical, the vibration frequencies excited by the suggested electromechanical actuator are normally below 1 kHz.

The vibrations induced by the actuator are picked by a number of accelerometers distributed over the monitored structure. The recorded signals are processed by the set of algorithms that determine whether or not the blade is damaged and find the location of the damage. A video demonstrating the SHM in action can be found in [2].

The *medium* frequency range which utilizes the proposed SHM is a compromise between the propagation range and detection resolution. Low-frequency vibrations can propagate long distances but are very insensitive to small areas of damage. It is well documented, see, for example, [3], that the modal parameters of the lowest blade modes are quite insensitive to blade damage, thus SHM systems utilizing changes in the modal parameters can only detect very significant damages. Ultrasonic range frequencies employed in the guided waves approach have good damage-detection resolution but the vibrations at such high frequencies quickly decay and cannot propagate long distances. This means the monitored blade has to be equipped with a large number of actuators and sensors, making such a system economically infeasible. The proposed approach utilizes the medium frequency range. Being well above the very low frequencies, where the blade is mainly excited by the wind, ensures satisfactory signal-to-noise ratio of the acceleration signals picked up by the accelerometers. At the same time, the frequency range is low enough for the vibrations to propagate long distances, and the experiments showed that a single actuator is sufficient to excite even a long modern blade. This also provides a good signal-to-noise ratio of the measured acceleration signals. On the other hand, the utilized frequency is high enough to ensure that the level of detection resolution is satisfactory for wind turbine owners; as it will be shown later, e.g., the system can detect a 20cm long trailing edge opening of a 34m blade.

Data-driven models for damage detection have been employed in numerous studies, for example, [4], with positive results. The experiments presented are often performed in laboratory conditions with models trained in a supervised setting where measurements from undamaged and damaged structures were available for the fitting. In contrast, the approach presented here is based on unsupervised anomaly detection, and thus does not require the learning of the damaged states. The SHM system is to be installed on an operational wind turbine continually monitoring the state so the algorithm must be robust to changing conditions such as wind, precipitation, and the operational modes such as changing rotational speeds and blade orientation. These phenomena may result in false alarms and the influence must therefore be investigated to implement a practical SHM system.

2. SHM system implementation on an SSP34m blade

Instead of making a general description of the proposed SHM system, this section uses one example of its implementation (namely, in application to an SSP34m blade). It is thought, however, that it should be easy to get a general idea about the concept of the proposed system and its usage. Before installing the SHM system on an operating wind turbine, its concept was

Figure 1. SSP34m blade mounted on the test rig

validated in laboratory conditions. An SSP34m blade, mounted on a test rig at DTU Wind

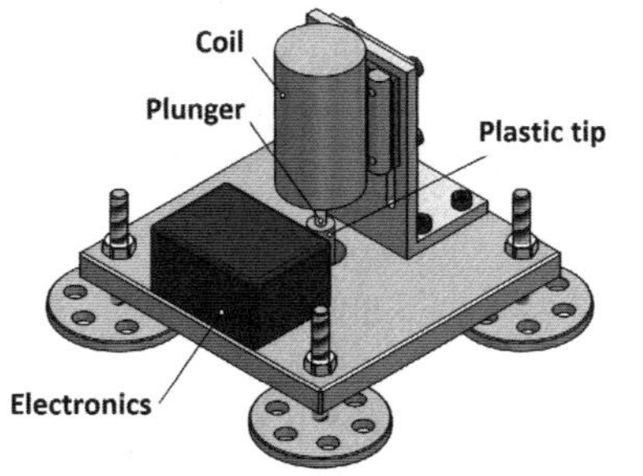

Figure 2. Design of electro-mechanical actuator.

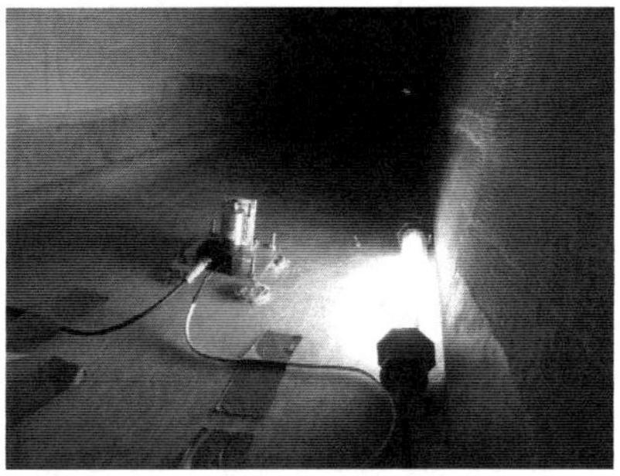

Figure 3. Mounting inside an SSP34m blade

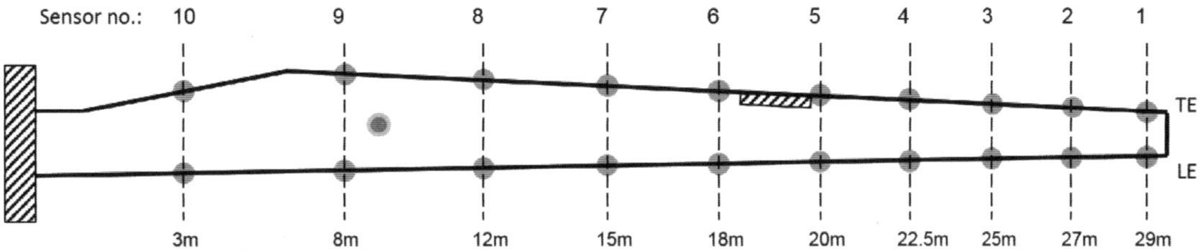

Figure 4. Test setup. Location of the actuator (●), accelerometers (●) and the damage (▨).

Energy facilities in Roskilde, Denmark (figure 1), was instrumented, and a series of experiments were conducted on the blade in undamaged and damaged states.

2.1. SHM system hardware

The hardware of the SHM system consists of an electromechanical actuator and array of accelerometers.

The actuator is an electromechanical device consisting of two main parts (figure 2): a steel plunger and a coil. Driven by an electrical pulse, the coil 'shoots' the plunger towards the structure; after the hit, the plunger returns to its initial position by spring. Though, there is an advantage of measuring the injected force, the practical implementation of such sensors would be quite difficult and thus omitted in this study.

The actuator is meant to be mounted inside the blade, on the shear web or the spar cap (figure 3), in the root section of the blade, i.e., at the first third of the blade length. This area is typically easily accessible in most modern blades. From a few pretests, it became obvious that even one actuator can excite the entire blade. The chosen location of the actuator is shown in figure 4.

The blade was also instrumented with 20 Brüel & Kjær accelerometers Type 4524-B, ten on the leading edge (LE) and ten on the trailing edge (TE) (figure 4). The same accelerometer setup was used for modal analysis, therefore triaxial accelerometers were used [3]. However, in the experiment described, only the data from one axis, perpendicular to the blade surface, was used.We acknowledge that mounting sensors on wind turbine blades is a difficult task, especially in a retrofitting scenario. However, the last years advances in sensing technology, e.g. the fiber optic sensing, can make blade instrumentation feasible in the nearest future.

It should be mentioned that we did not use any systematic approach for selecting the number of accelerometers, their placement and the placement of the actuator. As mentioned before, the accelerometers were reused from another modal analysis experiment, and the placement of the actuator was mainly driven by ease of access and a few pretests. A systematic approach to sensor/actuator placement may significantly improve the effectiveness of the SHM system and contribute to its cost reduction by reducing the number of sensors (and hence, data acquisition channels) by placing them at most optimal points. The first steps in this direction were taken by Parker, who used a genetic algorithm [5], and Lagerbon, who utilized a topology optimization approach [6].

A Brüel & Kjær data acquisition unit (see [3] for more details) was used for collecting the acceleration data. The unit also contains a signal generator, which was set to generate a rectangular pulse, which, after amplification, was fed to the actuator, resulting in an actuator hit. The data acquisition was triggered by this pulse, set to collect vibration data for a few milliseconds before and a few seconds after the pulse. Using the trigger significantly reduces the amount of data collected for the analysis.

The damage-detection algorithms are based on statistics to make detection more robust. In order to collect the necessary statistics, the blade in undamaged and damaged states was subjected to about fifty hits at each state. To speed up the experiment, the time between the successive hits were selected quite small, varying from one to five minutes. In real life application, the time between hits could be, for example, an hour or more.

2.2. Artificial damage

Only one type of damage was tested in the presented study: a trailing edge opening (figure 5). This damage is typical for many types of wind turbine blades, caused by de-bounding of the shells forming the pressure and suction sides of the blade. The damage was introduced into the blade gradually; we started with a 20cm trailing edge opening, drilling a series of holes through the glue between the pressure and suction sections of the blade. Then, using a saw, the holes were merged into one opening. After that, using a chisel and saw, the crack was gradually extended to 120cm; thus we had four damaged states: 20cm, 60cm, 90cm and 120cm. The data acquisition were conducted for each state, including the initial undamaged state, using about fifty actuator hits for each state.

Figure 5. Blade damage in the trailing edge, with the chisel used to introduce the damage

Figure 6. Bolts and metal plates used to change damage size in the second experiment.

When the damage was introduced, it was realized that operating with a heavy hammer and chisel can introduce some other unwanted changes into the blade, which can be erroneously associated with the effect of the damage. To avoid this, the experiment was repeated: the de-bounded parts of the blades were connected by bolts, placed 10cm from each other. The glue removed between the pressure and suction parts was replaced by thin metal plates (figure 6).

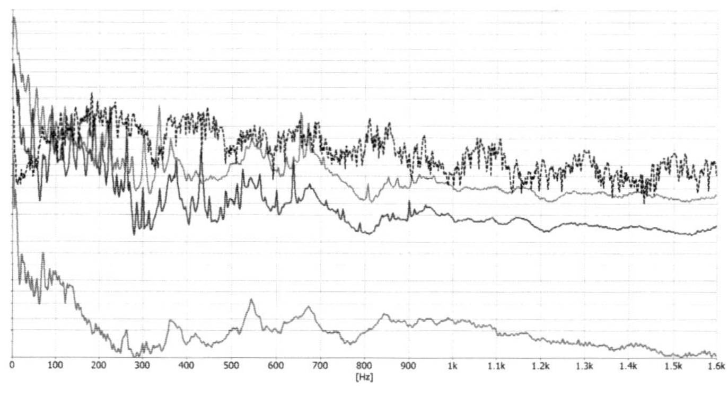

Figure 7. Power Spectral Densities (PSD) of the acceleration measured at the blade tip. V27 wind turbine: —— idle condition, moderate wind; —— operating at 32RPM, moderate wind; —— operating at 43RPM, strong wind, ······ response from actuator measured on SSP34m.

By tightening and loosening the bolts, it was easy to control the size and location of the opening without using the chisel.

2.3. Modelling of noise

Noise in the data is known to be a factor that can significantly reduce the robustness of damage-detection algorithms. In the case of operating wind turbines, the acceleration responses due to the actuator hits will overlap with the vibrations due to blade-wind interaction, and the vibrations due to the numerous mechanisms of the wind turbine, such as gearbox, generator, yaw and pitch mechanisms, cooling system, etc. To mimic such noise, vibration data measured on the blades of an operating Vestas V27 was used. This data was obtained during another measurement campaign, which is described in detail in [7]. Though the V27 and SSP34m blades are quite different (for example, the first one is 13m long, while the second is 34m), using this noise data was considered the best among other alternatives. Figure 7 shows the PSD of the acceleration measured at the tip of an operating V27 wind turbine under the different operating conditions. It should be mentioned that the presented study did not model the possible effects of varying weather conditions, such as temperature, wind speed and direction, etc. We also did not model the effects of different rotor speed and blade orientation (azimuth and pitch angles), which may strongly affect the robustness of the damage-detection algorithms.

3. Damage-detection method

The proposed damage-detection algorithm is based on an unsupervised anomaly detection method [8]. The method is based solely on modelling data obtained from the blade in a healthy state. This is in contrast to many earlier data-driven approaches that use supervised methods to detect a limited number of areas of damages introduced in experiments [4]. The supervised approach is consequently too restrictive for an operational system such as a wind turbine where it is impossible to perform experiments for all imaginable damage scenarios.

The damage-detection method is therefore based on a *training phase*, where we establish a model of the normal state exclusively from healthy state data, and subsequently a *detection phase* where a new sample is compared to the normal-state model. The modelling of the normal state takes as input the multivariate accelerometer signal recorded from the sensors. The processing for the model can be split into three parts; preprocessing, feature extraction and statistical modelling.

The set of accelerometer signals for each hit requires some preprocessing steps to enhance the damage detection. The first step is *windowing*, in which the signal is truncated to start at the actuator trigger signal and end when the impulse from the strike has died out, i.e., after 1s.

Figure 7 compares the spectra of the blade tip response due to wind and due to the actuator hit. As mentioned before, the wind mainly excites low-frequency vibrations, while the actuator excites a wider range. If at low frequencies, the signal due to the actuator is buried in the noise due to wind, at higher frequencies the situation is the opposite, and one can expect a good signal-to-noise ratio if the low frequencies are filtered out. This observation helps select the cut-off frequencies of the applied high-pass or bandpass filter that is applied in the filtering step of the preprocessing.

3.1. Damage indicator

The data-driven modelling approach employed in this study requires the definition of a compressed representation of the state of the structure. A damage metric based on the covariance matrix between the time histories, similar to the one introduced in [5], is employed. Structural damage will change the energy propagation paths from the actuator to the accelerometers, and the vibration pattern due to actuator hit. Since the cross-covariance function is a measure of the similarity between two signals, the change in vibration pattern can be very well characterized by the change of the covariance matrix. The cross-covariance between signals x_1 and x_2 of length N is calculated as:

$$c_{x_1x_2}(m) = \begin{cases} \sum_n^{N-m-1} \left(x_1(n+m) - \frac{1}{N}\sum_{i=0}^{N-1} x_1(i) \right) \left(x_2(n) - \frac{1}{N}\sum_{i=0}^{N-1} x_2(i) \right) & , m \geq 0 \\ c(-m) & , m < 0 \end{cases} . \quad (1)$$

In this application, only the cross-covariance for lag 0, $m = 0$, is used for the damage indicator. Calculating the covariance for a set of M sensors, produces $M(M-1)/2$ distinct values for each hit that constitutes the damage indicator. Figure 8a and 8b show the covariance features for the SSP34m blade in an undamaged state and with an opening at 120cm. It is seen that the feature values change for the covariance between sensor $c_{x_2x_{10}}$, $c_{x_4,x_{10}}$, and $c_{x_5,x_{10}}$, depicted in the lower left corner of the plots. Inspection of the covariance-feature vectors reveal that

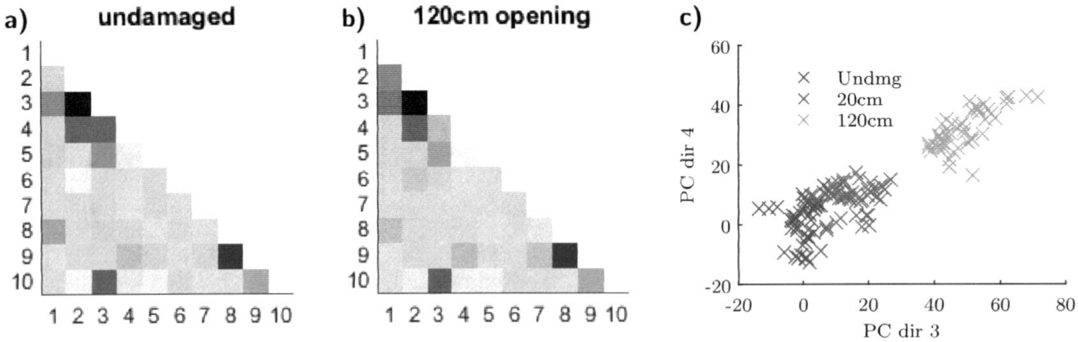

Figure 8. Covariance-based damage indicator using the 10 trailing edge accelerometers for the undamaged blade and the blade with 1.2m opening in trailing edge. The third panel shows the 3rd and 4th principal component dimensions for undamaged and two damage sizes

many of the dimensions are correlated, and in addition the dimensionality of the feature vectors grows quadratically with the number of sensors which could impede a statistical model of the samples. The dimensionality of the covariance-feature vectors is reduced using principal component analysis (PCA) [9]. The PCA is estimated using the samples from the undamaged case which produces a principal component space, that all samples are projected into. This projection is seen to separate damaged and undamaged samples quite well as illustrated in figure 8c.The dimensionality, i.e., the number of principal components, used for the PCA was chosen using a heuristic method. The method chooses the minimal number of principal components such that 99% of the variance in the data is accounted for.

3.2. Damage-detection metric
To evaluate whether a sample vector in the PCA-space, $\mathbf{y}$, is normal or damaged we use the Mahalanobis distance. The Mahalanobis distance is calculated as the distance between the normal-state samples $\mathbf{X}_n$, summarised by the mean $\boldsymbol{\mu}_{X_n}$ and the covariance matrix $\boldsymbol{\Sigma}$, and the new sample:

$$d(\mathbf{y}, \mathbf{X}_n) = \sqrt{(\mathbf{y} - \boldsymbol{\mu}_{X_n})^\top \boldsymbol{\Sigma}^{-1} (\mathbf{y} - \boldsymbol{\mu}_{X_n})}. \qquad (2)$$

To make the decision whether a new sample $\mathbf{y}$ is anomalous/damaged, a threshold must be chosen. This choice of threshold must either be based on observations of data from normal and from damaged blades, or chosen based on the normal-state data only. The former approach of supervised tuning of the threshold will have superior performance but requires that the system is trained on all damages seen. The current system uses the latter approach by calculating the distance for each normal-state sample $\mathbf{x}_{n,i}$ and then choosing the threshold as the maximum among the $d(\mathbf{x}_{n,i}, \mathbf{X}_n)$. This choice of threshold will thus classify all data in the training as normal.

3.3. Damage localization
The proposed SHM system also provides a rough damage localization. The localization is based on evaluation which elements of the covariance matrix are most affected. The damage is likely to be located between the sensors whose signals cross-covariance significantly changed (cf. figure 8a and 8b). Typically, several most affected elements of the covariance matrix indicate the damage position with precision, sufficient for damage severity evaluation and facilitation of its particular location for the repair.

4. Results
The damage-detection method was tested using data for the undamaged state and six different damage sizes of 20, 40, 60, 80,100, and 120cm. The performance was evaluated using tenfold cross-validation to get unbiased estimates of the error rates shown in table 1. The results show a very low error rate already with the use of the 5 sensors at the trailing edge (TE) and has approximately similar performance for all damage sizes and as we add more sensors. A realistic instrumentation that uses both leading-edge (LE) and trailing-edge (TE) sensors also reaches the lowest observed error rates. As mentioned earlier, the algorithm was also tested with added

Table 1. Results of damaged vs. undamaged classification with different sensor configurations. The table shows the mean error rate (in %) from the tenfold cross-validation followed by the 95% confidence interval of the mean in brackets.

State	20cm	40cm	60cm	80cm	100cm	120cm
No. samples	53	52	52	48	49	53
TE1-5	2.0 [0.2;7.2]	2.1 [0.3;7.3]	2.1 [0.3;7.3]	2.2 [0.3;7.6]	2.1 [0.3;7.5]	2.0 [0.2;7.2]
TE1-6	3.1 [0.6;8.7]	3.1 [0.6;8.8]	3.1 [0.6;8.8]	3.2 [0.7;9.1]	3.2 [0.7;9.0]	3.1 [0.6 8.7]
TE1-7	5.1 [1.7;11.5]	5.2 [1.7;11.6]	5.2 [1.7;11.6]	5.4 [1.8;12.1]	5.3 [1.7;12.0]	5.1 [1.7;11.5]
TE1-10	2.0 [0.2;7.2]	2.1 [0.3;7.3]	2.1 [0.3;7.3]	2.2 [0.3;7.6]	2.1 [0.3;7.5]	2.0 [0.2;7.2]
TE1-5,LE1-5	2.0 [0.2;7.2]	2.1 [0.3;7.3]	2.1 [0.3;7.3]	2.2 [0.3;7.6]	2.1 [0.3;7.5]	2.0 [0.2;7.2]

artificial wind noise from an operational turbine to test the robustness of the damage-detection method in real operation. The results of these experiments are shown in figure 9. The wind noise is measured in terms of signal-to-noise ratio (SNR) between the clean signal in trailing edge sensor 1 from the arrival of the strike and 0.25s forward and the generated wind noise signal. The performance is measured using the well-established Area Under the (ROC) Curve (AUC) [10] that scores a binary classifier from perfect classification with a score of 1 to random guessing at a score of 0.5 independently of the chosen detection threshold. It is clear that all damage can be detected down to an SNR level of 0dB while larger areas of damage are detectable at even lower SNR levels.

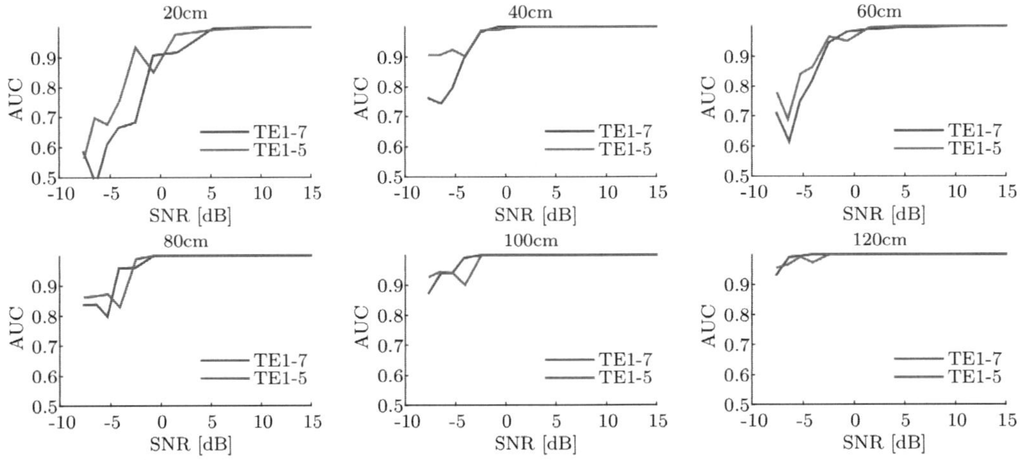

Figure 9. AUC for experiments with added wind noise. The legend in the figure shows the numbers of the trailing edge sensors used.

5. Conclusion

The paper presented a prototype of a vibration-based SHM system for wind turbine blades. The proposed approach is based on active excitation of the blade by an electro-mechanical actuator and measurements of the exited vibration by an array of accelerometers. The unsupervised abnormality detection method is employed for the analysis of the response signals for damage detection and localization. The method is demonstrated on a 34m wind turbine blade mounted on a test rig. It is shown that the proposed system is able to detect a 20cm length of trailing edge damage in presence of realistic (though artificial) noise.

Acknowledgments

The work was partly supported by EUDP (Danish Energy Technology Development and Demonstration Programme), grant number 64011-0084 Predictive Structure Health monitoring of Wind Turbines. The authors would like to thank DTU Wind Energy for giving access to the test object and research engineer Per Hørlyk Nielsen for his great assistance in setting up and conducting the experiment.

References

[1] Hameed Z, Hong Y S, Cho Y M, Ahn S H and Song C K 2009 Condition monitoring and fault detection of wind turbines and related algorithms: A review *Renewable and Sustainable Energy Reviews* **13** 1–39 ISSN 13640321

[2] 2014 Predictive structural health monitoring - Brüel & Kjær URL `https://www.youtube.com/watch?v=A7UYwV-bc2w\&feature=youtu.be`

[3] Larsen G C, Berring P, Tcherniak D, Nielsen P H r and Branner K 2014 Effect of a Damage to Modal Parameters of a Wind Turbine Blade *EWSHM - 7th European Workshop on Structural Health Monitoring* (Nantes, France: IFFSTTAR, Inria, Université de Nantes) URL `https://hal.inria.fr/hal-01020347`

[4] Fassois S D and Sakellariou J S 2007 Time-series methods for fault detection and identification in vibrating structures. *Phil Trans R S A* **365** 411–448 ISSN 1364-503X

[5] Parker D L 2011 *Multi-objective design optimization framework for structural health monitoring* Ph.D. thesis Mississippi State University

[6] Lagerbon M 2014 Optimal Sensor Positioning for Structural Health Monitoring Tech. rep. Technical University of Denmark

[7] Tcherniak D and Larsen G C 2013 Application of OMA To an Operating Wind Turbine: Now Including Vibration Data *5th International Operational Modal Analysis Conference*

[8] Tarassenko L, Clifton D, Bannister P, King S and King D 2009 Novelty detection *Encyclopaedia of Structural Health Monitoring* (John Wiley & Sons, Ltd) pp 653–675

[9] Jolliffe I T 2002 *Principal Component Analysis* vol 98 (Springer) ISBN 0-387-95442-2

[10] Fawcett T 2006 An introduction to ROC analysis *Pattern Recognition Letters* **27** 861–874 ISSN 01678655

Natural vibration response based damage detection for an operating wind turbine via Random Coefficient Linear Parameter Varying AR modelling

L D Avendaño-Valencia and S D Fassois

Stochastic Mechanical Systems and Automation (SMSA) Laboratory, Department of Mechanical and Aeronautical Engineering, University of Patras, Greece

E-mail: `ldavendanov@upatras.gr`, `fassois@mech.upatras.gr`

Abstract. The problem of damage detection in an operating wind turbine under normal operating conditions is addressed. This is characterized by difficulties associated with the lack of measurable excitation(s), the vibration response non-stationary nature, and its dependence on various types of uncertainties. To overcome these difficulties a stochastic approach based on Random Coefficient (RC) Linear Parameter Varying (LPV) AutoRegressive (AR) models is postulated. These models may effectively represent the non-stationary random vibration response under healthy conditions and subsequently used for damage detection through hypothesis testing. The performance of the method for damage and fault detection in an operating wind turbine is subsequently assessed via Monte Carlo simulations using the FAST simulation package.

1. Introduction

Structural Health Monitoring (SHM) is of primary importance for wind turbine structures as damage may have catastrophic consequences. SHM systems are indeed appealing for reducing maintenance and repair costs and guaranteeing the safety of the infrastructure [1, 2]. Vibration based SHM is of particular interest, as it has the potential of global monitoring at minimal cost [3, 4]. Within this context, statistical time series methods are particularly attractive [5]. The effectiveness of such a method is strongly affected by the efficacy of the statistical time series model employed in representing the structural random vibration response. Thus, obtaining an accurate model is a main concern in the process of developing a vibration–based SHM system. In particular for operating wind turbines, the vibration response is random non-stationary, and also subject to various types of uncertainties [6, 7]. Thus achieving robust SHM is a challenging task [4]. Naturally, the proper handling of the vibration signal properties – including non-stationarity – is vital for effective SHM [8, 9]. In conventional methods, baseline time series models are identified from a single vibration signal under particular environmental and operational conditions, assuming that the dynamics observed in the vibration response would remain constant even under the effects of uncertainties [4, 5]. Yet, in order to effectively treat uncertainties it is necessary to monitor the behavior of the healthy structure during prolonged time periods in order to account for different operational and environmental conditions. More importantly, it is necessary to devise a methodology capable of combining the information conveyed in the baseline set, so as to yield a robust decision making mechanism.

In this sense, two different approaches are frequently used: a feature extraction/selection approach and a global model-based approach. In the *feature extraction/selection* approach, the characteristics that are least affected by uncertainties are extracted from the whole set of characteristics estimated through the models. Data transformation techniques, like principal component analysis, factor analysis, and others are often used in this context [10, 11]. The *global model-based* approach attempts to build a (normally parametric) stochastic model capable of explicitly taking into account the effects of environmental and operational conditions. For this purpose, it is possible to construct a model whose parameters are functions of one or more environmental/operational variables. Functionally Pooled AR/ARMA/ARX models [12], Linear Parameter Varying models [13], or Polynomial Chaos Basis AR models [14] are main examples. The identification of this type of models requires estimating the parameters and selecting a set of multivariate polynomial functions defined over a space spanned by the considered environmental and operational variables. The models may become complex, and computational complexity problems may arise.

As an alternative, Random Coefficient (RC) stochastic models may be used to describe the effects of uncertainties on the dynamics [15, 16]. Model identification in this case typically involves a Bayesian framework.

The present study addresses the problem of damage and fault detection in structures with non-stationary random vibration response subject to important environmental and operational variability through the use of *Random Coefficient Linear Parameter Varying AutoRegressive* (RC–LPV–AR) models. The work continues a previous study by the authors [17], in which Functional Series Time-dependent AutoRegressive (FS-TAR) models were used for representing the wind turbine's non–stationary random vibration response. The two key differences of the approach postulad in this study is the random nature of the model parameters, better accommodating operating and other uncertainties, as well as the Linear Parameter Varying nature (instead of the direct Time Varying) nature of the model that may lead to a better representation of the non–stationary dynamics itself. The obtained model parameters and corresponding modal quantities (frozen natural frequencies and damping ratios) are studied for healthy and damaged scenarios, while damage detection is performed via proper hypothesis testing.

2. The method

2.1. Non–stationary random vibration response representation via RC–LPV–AR

Let $y[t]$ be a non-stationary signal defined over the normalized discrete time $t \in \mathbb{Q}$, corresponding to a single realization of the *RC–LPV–AR* model:

$$y[t] = -\sum_{i=1}^{n_a} a_i(\beta[t]) \cdot y[t-i] + w[t], \quad w[t] \sim \texttt{NID}\left(0, \sigma_w^2(\beta[t])\right) \tag{1a}$$

$$a_i(\beta[t]) = \sum_{k=1}^{p_a} a_{i,k} \cdot G_{b_{a(k)}}(\beta[t]), \quad \boldsymbol{a} \sim \mathcal{N}\left(\bar{\boldsymbol{a}}, \boldsymbol{\Sigma_a}\right), \quad \sigma_w^2(\beta[t]) = \sum_{k=1}^{p_s} s_k \cdot G_{b_{s(k)}}(\beta[t]) \tag{1b}$$

where:

$a_i(\beta[t]), \sigma_w^2(\beta[t])$	: AR parameters and innovations variance
$w[t]$	: Non-stationary innovations sequence
$\beta[t]$	: Scheduling variable
$G_{b_{a(k)}}(\beta[t]), G_{b_{s(k)}}(\beta[t])$	: AR and innovations variance functional basis

and

n_a, p_a, p_s	: Model order and basis subspace dimensionalities
$\boldsymbol{a} = \begin{bmatrix} a_{1,1} & a_{1,2} & \dots & a_{n_a,p_a} \end{bmatrix}^T$	: AR parameter vector (random)
$\bar{\boldsymbol{a}} = \mathrm{E}\{\boldsymbol{a}\}$, $\boldsymbol{\Sigma_a} = \mathrm{E}\{(\boldsymbol{a}-\bar{\boldsymbol{a}})\cdot(\boldsymbol{a}-\bar{\boldsymbol{a}})^T\}$	: AR parameter vector mean and covariance
$\boldsymbol{s} = \begin{bmatrix} s_1 & \dots & s_{p_s} \end{bmatrix}^T$	: Innovations parameter vector (deterministic)

Each realization of the parameter vector $\boldsymbol{a}$ describes the dynamics at a particular interval of analysis where the environmental and operational conditions may be assumed to be constant. At the same time, the parameter variability under changing operational and environmental conditions is described via a Gaussian density with hyperparameters $\bar{\boldsymbol{a}}$ and $\boldsymbol{\Sigma_a}$. Therefore, the identification of the RC–LPV–AR model from a single realization is equivalent to the identification of a conventional LPV-AR model, and consists of estimating the coefficients of projection and identification of the model order and subspace dimensionalities [18, 19]. Thus the estimation of the RC–LPV–AR model requires the estimation of the hyperparameters associated to the Gaussian distribution of the parameter vector.

The joint probability of observing the vibration response $\boldsymbol{y} = \{y[1], \dots, y[N]\}$ with an associated parameter vector $\boldsymbol{a}$, for the particular RC–LPV–AR model $\bar{\mathcal{M}} = \{\mathcal{P}, \mathcal{M}\}$, with hyperparameters $\mathcal{P} = \{\bar{\boldsymbol{a}}, \boldsymbol{\Sigma_a}, \boldsymbol{s}\}$ and structure $\mathcal{M} = \{n_a, p_a, p_c, G_{b_{a(k)}}[t], G_{b_{s(k)}}[t]\}$, is as follows:

$$\begin{aligned} p(\boldsymbol{y}, \boldsymbol{a}|\bar{\mathcal{M}}) &= p(\boldsymbol{y}|\boldsymbol{a}, \bar{\mathcal{M}}) \cdot p(\boldsymbol{a}|\bar{\mathcal{M}}) \\ &= p(y[0]) \cdot \prod_{t=1}^{N} \mathcal{N}\left(\sum_{i=1}^{n_a} a_i(\beta[t]) \cdot y[t-i], \sigma_w^2(\beta[t])\right) \cdot \mathcal{N}(\bar{\boldsymbol{a}}, \boldsymbol{\Sigma_a}) \end{aligned} \tag{2}$$

where $p(\boldsymbol{y}|\boldsymbol{a}, \bar{\mathcal{M}})$ is the likelihood of the observed time series given the parameter vector $\boldsymbol{a}$ and model $\mathfrak{m}$, $p(\boldsymbol{a}|\bar{\mathcal{M}})$ is the PDF of the parameters under model $\bar{\mathcal{M}}$, and $p(y[0])$ is the initial condition. Model estimation may be performed by optimizing Equation (2) with respect to the parameters of interest, as presented in [20, Ch. 5]. For the present work, a simple approach using sample averages is adopted. Thus, given M estimates of the parameter vector, each one being referred to as $\boldsymbol{a}_m$, the sample mean and covariance matrix are obtained as:

$$\bar{\boldsymbol{a}} = \frac{1}{M}\sum_{m=1}^{M} \boldsymbol{a}_m, \qquad \boldsymbol{\Sigma_a} = \frac{1}{M-1}\sum_{m=1}^{M} (\boldsymbol{a}_m - \bar{\boldsymbol{a}}) \cdot (\boldsymbol{a}_m - \bar{\boldsymbol{a}})^T \tag{3}$$

It is well known that these estimators suffer when M is small, especially for the covariance matrix, and may yield inaccurate results and/or ill-conditioned covariance. Diagonal covariance, PCA-based, shrinkage, or maximum entropy estimators may be used to alleviate this problem and obtain robust estimates [21, Ch. 18].

2.2. Damage detection method using RC-LPV-AR models

Damage detection consists of determining whether a given test signal $\boldsymbol{y}_u$ with corresponding parameter vector $\boldsymbol{a}_u$ constitutes a realization of the model corresponding to the healthy state of the structure. In practice the parameter vector $\boldsymbol{a}_u$ is not available, but its replaced by its corresponding estimate. Let $\bar{\mathcal{M}}_o = \{\mathcal{P}_o, \mathcal{M}\}$ be the identified model of the vibration response of the structure at its baseline (healthy) state, with hyperparameters $\mathcal{P}_o = \{\bar{\boldsymbol{a}}_o, \boldsymbol{\Sigma}_{\boldsymbol{a}_o}, \boldsymbol{s}_o\}$, and structure $\mathcal{M}$. Let $\mathfrak{m}_u$ designate the conventional LPV–AR model from the current state of the structure, then the damage detection problem consists on determining whether or not the event $\mathfrak{m}_u \in \bar{\mathcal{M}}_o$ is correct or not. With no extra information, the probability that the current model is from the healthy state is equal to the prior probability of that event, but once a vibration response signal $\boldsymbol{y}_u$ (and thus the corresponding parameter vector $\boldsymbol{a}_u$) is available, one may use the Bayes theorem [21, Ch.4] as follows:

$$P(\mathfrak{m}_u \in \bar{\mathcal{M}}_o|\boldsymbol{y}_u, \boldsymbol{a}_u) \propto p(\boldsymbol{y}_u, \boldsymbol{a}_u|\bar{\mathcal{M}}_o) \cdot P(\mathfrak{m}_u \in \bar{\mathcal{M}}_o) \tag{4}$$

where $p(\boldsymbol{y}_u, \boldsymbol{a}_u|\mathfrak{m}_o)$ is the joint probability from equation (2), and $P(\mathfrak{m}_u \in \bar{\mathcal{M}}_o)$ is the prior probability of the event $\mathfrak{m}_u \in \bar{\mathcal{M}}_o$. Under the Gaussianity assumption for the innovations and the coefficients of projection, the logarithm of $P(\mathfrak{m}_u \in \bar{\mathcal{M}}_o|\boldsymbol{y}_u, \boldsymbol{a}_u)$ may be expressed as:

$$\ln P(\mathfrak{m}_u \in \bar{\mathcal{M}}_o|\boldsymbol{y}_u, \boldsymbol{a}_u) = \ln \mathcal{L}(\boldsymbol{a}_u|\boldsymbol{y}_u) - \frac{1}{2} d_M^2(\boldsymbol{a}_u, \bar{\boldsymbol{a}}_o) + C \tag{5a}$$

$$\ln \mathcal{L}(\boldsymbol{\theta}_u|\boldsymbol{y}_u) = -\frac{N}{2} \cdot \ln 2\pi - \frac{1}{2} \sum_{t=1}^{N} \left(\ln \sigma_w^2(\beta[t]) + \frac{y_u[t] + \sum_{i=1}^{n_a} a_i(\beta[t]) \cdot y_u[t-i]}{\sigma_w^2(\beta[t])} \right) \tag{5b}$$

$$d_M(\boldsymbol{a}_u, \bar{\boldsymbol{a}}_o) = \left((\boldsymbol{a}_u - \bar{\boldsymbol{a}}_o)^T \cdot \boldsymbol{\Sigma}_{\bar{\boldsymbol{a}}_o}^{-1} (\boldsymbol{a}_u - \bar{\boldsymbol{a}}_o) \right)^{1/2} \tag{5c}$$

where $\ln \mathcal{L}(\boldsymbol{a}_u|\boldsymbol{y}_u)$ is the log-likelihood of the parameter vector $\boldsymbol{a}_u$ given the current vibration signal $\boldsymbol{y}_u$, $d_M(\boldsymbol{a}_u, \bar{\boldsymbol{a}}_o)$ is the Mahalanobis distance between the mean parameter vector $\bar{\boldsymbol{a}}_o$ and the parameter vector $\boldsymbol{a}_u$ corresponding to the current signal, and C is a constant gathering the remainder constant terms. Then, for damage detection the following test may be performed:

$$- \ln \mathcal{L}(\boldsymbol{a}_u|\boldsymbol{y}_u) + \frac{1}{2} d_M^2(\boldsymbol{a}_u, \bar{\boldsymbol{a}}) \leq l_{\text{lim}} \quad \Rightarrow \quad \mathfrak{m}_u \in \bar{\mathcal{M}}_o \quad \text{(healthy structure)} \tag{6a}$$

$$\text{otherwise} \quad \Rightarrow \quad \mathfrak{m}_u \notin \bar{\mathcal{M}}_o \quad \text{(damaged structure)} \tag{6b}$$

where l_{lim} is a detection threshold selected by the user. This threshold can also be optimized through a cross-validation approach [17], [21, Ch. 7].

3. Application example

3.1. Data description

The vibration signals used in this application example are obtained from a finite element model of a "NREL offshore 5-MW baseline wind turbine" simulated by means of the FAST wind turbine simulation code [22]. Four accelerometers are set on the structure: two at the tower top in the fore-aft and lateral directions; two around 25% of the blade chord of the third blade in the edgewise and flapwise directions. The rotor position is measured as well. The obtained vibration responses are 200 s long, sampled at 25 Hz. Further details can be found in [17].

3.2. Identification of the healthy state using RC-LPV-AR models

The obtained vibration response signals from the healthy state and from all the sensor locations are represented by means of RC-LPV-AR models. The RC-LPV-AR models use trigonometric basis for the expansion of the parameter evolutions, where the instant phase of the trigonometric basis is the instantaneous value of the rotor position. Specifically, the basis function is defined as follows

$$G_0(\beta[t]) = 1, \qquad G_{2k-1}(\beta[t]) = \sin(2\pi k\beta[t]), \qquad G_{2k}(\beta[t]) = \cos(2\pi k\beta[t])$$

where the scheduling variable $\beta[t] \in (0, 1)$ is the measured rotor position at time t. Notice that the use of this type of functional basis makes the parameters of the RC-LPV-AR model directly dependent on the instant position of the rotor and thus any variation in the rotor speed is tracked by the model. This improves the tracking ability of the postulated RC-LPV-AR models in comparison to the FS-TAR models (as shown in [17]) in the representation of wind turbine vibration, since the rotor speed may have small variations as the control system of the wind turbine responds to the upcoming wind. The models are computed independently for each sensor to achieve simpler models and to compare the sensitivity of damage detection for different types of damage at each sensor location. The settings of the estimation methods and the identified model structures for each one of the sensor locations can be found in Table 1.

Table 1. Identification details of RC-LPV-AR models.

ID step	Description
Estimation	Multi-stage method with WLS parameter estimation and instantaneous innovations variance estimate. Stopping criteria: parameter tolerance 10^{-9}, objective fcn. tolerance 10^{-9}, maximum No. of iterations 100.
Identification	Model order: Estimate LPV-AR$(n)_{[p_a,p_s]}$ models with $n = 6, \ldots, 20$ and $p_a = p_s = 9$. Basis order: Estimate LPV-AR$(n_a)_{[p,p]}$ models with best n_a of previous step and $p_a = p_s = 1, 3, \ldots, 15$. Regularized covariance estimates.
Achieved structures	Tower top – fore-aft: LPV-AR$(9)_{[7,7]}$; Tower top – lateral: LPV-AR$(17)_{[7,7]}$; Blade 3 – flapwise: LPV-AR$(6)_{[7,7]}$; Blade 3 – edgewise: LPV-AR$(17)_{[7,7]}$.

3.3. Simulated damage scenarios

Five types of damage are simulated with four increasing levels corresponding to typical damages or malfunctions in wind turbines [1, 2]. Each scenario is described in Table 2. For each scenario 100 realizations of the wind turbine response are simulated with different seeds for the generation of the turbulence time series. In comparison with the previous work [17], the present work evaluates two more damage scenarios (damages D and E), and utilizes larger number of realizations per structural state (50 realizations were used in the previous work).

Table 2. Simulated damage scenarios

Type	Description
Damage A	Increased mass on the last 10% of the length of the third blade. Four damage levels: 20%, 40%, 60% and 80% blade mass density increment.
Damage B	Stiffness reduction of the root of the third blade in the edgewise direction. Four damage levels: 15%, 30%, 45% and 60% blade edgewise stiffness decrement from the root up to 10% of the blade chord.
Damage C	Stiffness reduction in the lateral direction at the base of the tower. Four damage levels: 5%, 10%, 15% and 20% tower stiffness decrement in the lateral direction, from the base up to 10% of the tower height.
Damage D	Error in the yawing mechanism. Four damage levels: 2, 4, 6 and 8 degree errors from the upwind position in the yawing system.
Damage E	Reduction of the damping coefficient of the slow speed shaft. Four damage levels: 15%, 30%, 45% and 60% damping coefficient decrement.

Figure 1 shows 90% confidence intervals of the RC-LPV-AR parameter trajectories derived from the vibration responses at the blade in the flapwise direction for all realizations at different structural states (healthy and damages A to E). The most obvious difference from the healthy state is evidenced by damage D, where the parameter trajectories travel through almost totally different areas. In the remaining scenarios, the areas covered by parameter trajectories at the different structural states are largely overlapping.

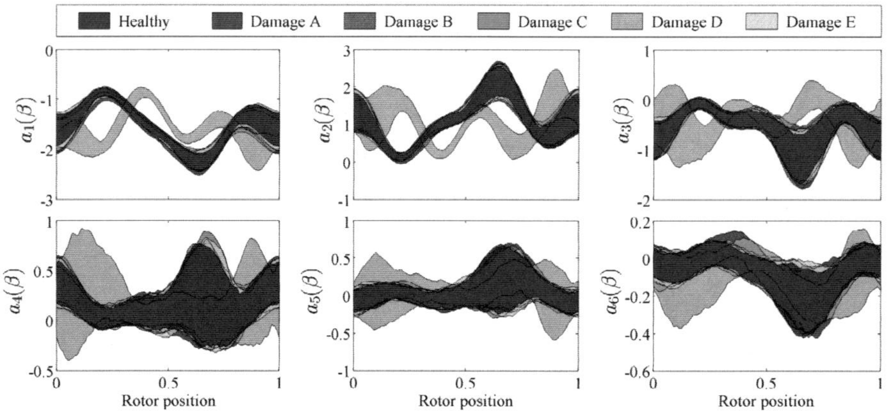

Figure 1. 90% confidence intervals of the RC-LPV-AR parameter trajectories as a function of the rotor position β on 100 realizations at different structural states. Results shown for sensor on the blade – flapwise direction with a RC-LPV-AR$(6)_{[7,7]}$ model. Only highest damage levels are shown.

Figure 2 provides a comparison of the "frozen" modal quantities extracted from the mean parameter vectors $\bar{\boldsymbol{a}}_s$ for the healthy and damage scenarios at their highest level of damage. Certain "frozen" natural frequencies and damping ratios are shown for a single period of rotation of the blades and are computed as explained in [18]. The trajectories followed at the healthy state are shown in black color, whereas the damages are shown in different colors. The trajectories of the modal quantities remain almost the same on the sensors at the tower top fore-aft direction and at the third blade in the edgewise direction. The changes are more obvious for the remaining sensors. Damage D induces the largest deviations in both frequency and damping ratio. In particular a large decrement of the damping ratio is evident in the first blade mode at the edgewise direction, and in the trajectories of the modal quantities on the flapwise direction. The changes are more discrete for the other types of damages.

3.4. Damage detection

Damage assessment is performed via the test described by Equation (6) based on the obtained RC-LPV-AR models for the healthy state. A cross-validation approach is used for the optimization of the detection threshold. Namely, the set of vibration response signals of each sensor is divided into 10 subsets, each one containing 10 vibration response signals from each reference state. Then, 90 signals from each structural state are used for training of the method (RC-LPV-AR model identification), while the remaining signals are used for evaluation.

Figure 3 shows the distribution of the damage detection statistic obtained for the vibration response at the blade on the flapwise direction for each type of damage and for each level of damage. The change in the distribution of the detection statistic is evident especially for damages A and D. Table 3 provides a summary of the performance of the damage detection method in terms of the Receiver Operating Characteristic (ROC) Area Under the Curve (AUC), measuring the discriminative ability of the detector [21, Ch. 9]. The results obtained with

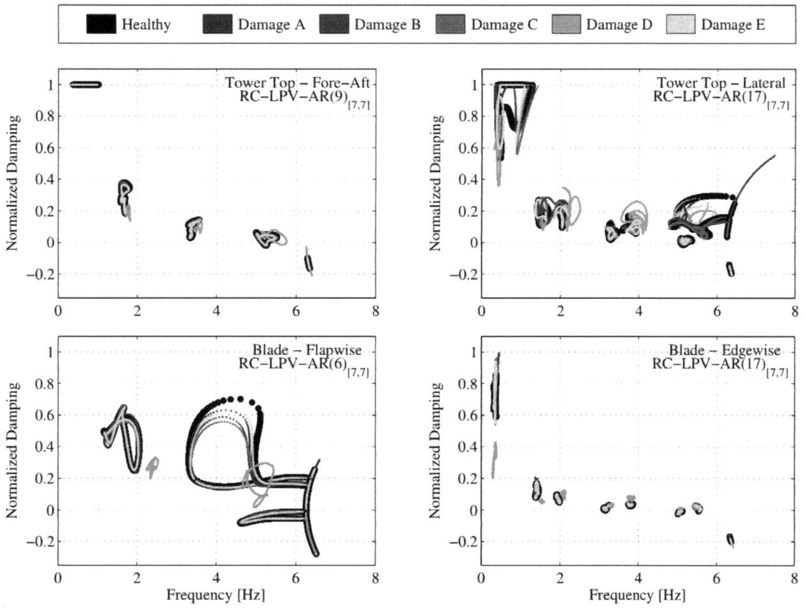

Figure 2. Selected trajectories of the "frozen" natural frequencies and damping ratios derived from the average parameter vectors of the RC-LPV-AR models at different sensor locations and for different structural states. Highest damage levels considered for each damage type.

single and multiple FS-TAR models methods shown in [17] are also provided. The proposed method provides outcomes comparable with those obtained with the multiple model methods from previous works and demonstrate the effectiveness of the method.

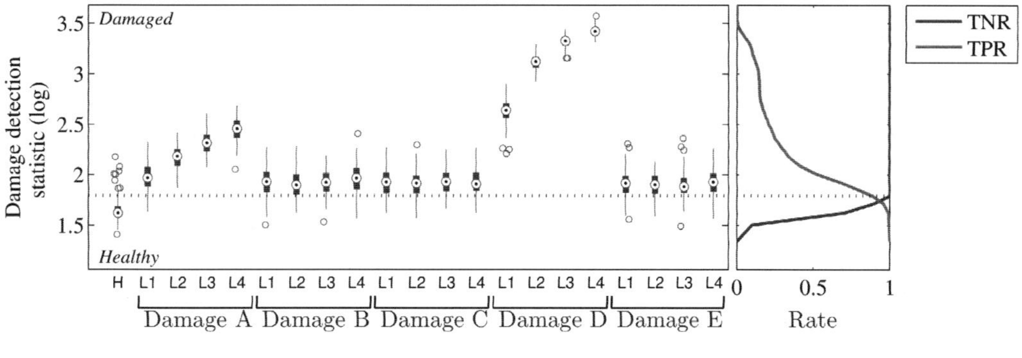

Figure 3. Distribution of the damage detection statistic obtained by the damage detection method based on RC-LPV-AR modeling for each one of the structural states and for each level of damage (indicated as L1,...,L4). Right plot shows the distribution of true negatives and true positives for increasing values of the threshold.

4. Conclusions

In this study the use of Random Coefficient Linear Parameter Varying AutoRegressive (RC–LPV–AR) models was postulated for the modeling and damage detection of structures with

Table 3. Best detection performance per damage type of the proposed RC-LPV-AR model based damage detection method, and the single/multiple model based methods shown in [17].

Method	Damage A	Damage B	Damage C	Damage D	Damage E
Single FS-TAR	0.69	0.85	0.64	0.98	0.66
Multiple FS-TAR - min rule	0.78	0.94	0.95	1.00	0.94
Multiple FS-TAR - sum rule	0.79	0.95	0.92	1.00	0.89
RC-LPV-AR model	0.90	0.75	0.94	1.00	0.94

Performance figures provided in terms of the mean of the ROC–AUC obtained after cross–validation. A ROC–AUC close to one indicates an excellent performance of the detector, whereas a ROC–AUC equal to 0.5 indicates that the detector has no discriminative ability [21, Ch. 9].

non-stationary vibration response and exposed to uncertainties due to changes in environmental and operational conditions. The advantage of the this type of modeling is twofold: the models may provide compact representations of the complex non–stationary dynamics observed on time-dependent systems subject to uncertain operating/environmental conditions and, on the other hand, they may be effectively used for damage/fault detection with superior performance compared to that of previously used approaches.

Acknowledgements

This work was partially supported by the EU FP7 ITN project SYSWIND (Grant 238325). The first author also acknowledges useful discussions with Professors Soren Nielsen (Aalborg Universitet, Denmark) and Theodore Philippidis (University of Patras, Greece).

References

[1] Ciang C, Lee J and H B 2008 *Measurement Science and Technology* **19** 122001
[2] Hameed Z, Hong Y, Cho Y, Ahn S and Song C 2009 *Renewable and Sustainable Energy Reviews* **13** 1–39
[3] Doebling S, Farrar C and Prime M 1998 *Shock Vib. Dig.* **30** 91–105
[4] Farrar C R and Worden K 2007 *Philosophical Transactions of the Royal Society A* **365** 303–315
[5] Fassois S and Sakellariou J 2009 *Encyclopedia of Structural Health Monitoring* ed Boller C, Chang F and Fujino Y (John Wiley & Sons Ltd.) pp 443–472
[6] Murtagh P and Basu B 2007 *Proceedings of the Institution of Mechanical Engineers, Part K: Journal of Multi-body Dynamics* **221** 577–589
[7] Allen M S, Sracic M W, Chauhan S and Hansen M H 2011 *Mechanical Systems and Signal Processing* **25** 1174–1191
[8] Antoni J 2009 *Mechanical Systems and Signal Processing* **23** 987–1036
[9] Skjoldan P and Hansen M 2009 *Journal of Sound and Vibration* **327** 424–439
[10] Sohn H 2007 *Philosophical Transactions of the Royal Society A* **365** 539–560
[11] Yan A M, Kerschen G, Boe P D and Golinval J C 2005 *Mechanical Systems and Signal Processing* **19** 847–864
[12] Hios J and Fassois S 2014 *Mechanical Systems and Signal Processing* **49** 77–94
[13] Bamieh B and Giarré L 2002 *International Journal of Robust and Nonlinear Control* **12** 841–853
[14] Spiridonakos M and Fassois S 2014 *Signal Processing* **96** 63–79
[15] Zao Z W and Wang D H 2012 *Mathematical and Computer Modelling* **56** 152–166
[16] Michaelides P G and Fassois S 2013 *Engineering Structures* **53** 112–121
[17] Avendano-Valencia L and Fassois S 2014 *7th European Workshop on Structural Health Monitoring*
[18] Poulimenos A and Fassois S 2006 *Mechanical Systems and Signal Processing* **20** 763–816
[19] Spiridonakos M and Fassois S 2014 *Mechanical Systems and Signal Processing* **47** 175–224
[20] Rasmussen C and Williams C 2006 *Gaussian Processes for Machine Learning* (MIT Press)
[21] Hastie T, Tibshirani R and Friedman J 2009 *The Elements of Statistical Learning, Data Mining, Inference, and Prediction* Springer Series in Statistics (Springer)
[22] Jonkman J M and Buhl M L 2005 FAST user's guide Tech. rep. National Renewable Energy Laboratory

Use of high frequency analysis of acoustic emission signals to determine rolling element bearing condition

A Cockerill[1], K M Holford[1], T Bradshaw[2], P Cole[2], R Pullin[1] and A Clarke[1]

[1] *School of Engineering, Cardiff University, UK, CF24 3AA*
[2]*Mistras Group Limited, Over, Cambridge, UK, CB24 5QE*

Email: *CockerillA@cardiff.ac.uk*

Abstract: Acoustic Emission (AE) sensors were used to detect signals arising from a cylindrical roller bearing with artificial defects seeded onto the outer raceway. An SKF N204ECP roller bearing was placed between two double row spherical roller bearings, type SKF 22202E, and loaded between 0.29 and 1.79kN. Speed was constant at 5780rpm. High frequency analysis allowed insight into the condition of the bearings through the determination of an increase in the structural resonances of the system as the size of an artificial defect was increased. As higher loads were applied, frequencies around 100kHz were excited, indicating the release of AE possibly attributed to friction and the plastic deformation as peaks, induced through engraving of the raceway, were flattened and worn down. Sensitivity of AE to this level in bearings indicates the potential of the technique to detect the early stages of bearing failure during life tests.

INTRODUCTION

Rolling element bearings are found within machinery in a wide variety of industries including the aerospace, automotive and renewable energy sectors. Failure of such bearings can lead to costly asset downtime or at worst, catastrophic failure and the loss of human life. Although bearings are designed with longevity in mind, it has been calculated that approximately 10% of bearings will fail before reaching their predicted life, L_{10}, as derived by Lundberg and Palmgren [1]. In real world applications, the bearing can be significantly overloaded, vastly reducing the number of cycles that the bearing can sustain. This, coupled with variation in operational temperature, oil cleanliness and lubrication regime, make it very difficult to predict with confidence when or how a bearing will fail due to fatigue. With advancing research in the areas of tribology and condition monitoring, it is now possible to detect incipient damage on bearing raceways and rolling elements before complete bearing failure occurs [2].

Of the condition monitoring techniques available, vibrational analysis has been widely adopted within industry to detect defects within rotating machinery [3]. It is well understood that the repeatable passing of the rolling elements over the point of maximum load produces characteristic frequencies directly associated with the geometry and specifications of the bearing itself [4]. The characteristic frequencies have long been used to determine the location of a bearing defect where an increase in frequency amplitude over time is attributed to an increase in defect severity. Although this provides a good indication of a defects existence, when this method is applied to heavy machinery operating with a large amount of low frequency noise and high inertia, the low amplitude defect frequencies of small defects can often be masked [5][6]. Complex transmission paths also result in a large attenuation of a defect signal making it difficult to detect a defect in its earliest stages. Although weak signal extraction methods can be used, the time between defect detection and machine failure can still be relatively small [5]. This is inherent in the use of vibration techniques that detect the response of a structure to a defect rather than the occurrence and growth of the defect itself. As an alternative to vibrational analysis, Acoustic Emission (AE) is proposed as a more sensitive technique to detect early-stage damage within rotating machinery. AE is the detection of transient elastic waves caused by plastic deformation, crack tip growth, fluid flow, sharp impacts, friction/rubbing and fractures/dislocations of fibres and is commonly used to detect damage in static structures such as bridges and pressure vessels. For bearings it is thought that AE is induced through sub-surface and surface cracking, friction and plastic deformation of surface asperities during the running in phase and rolling fatigue failure [7]. As a crack grows, a short burst of high frequency energy is released and transmitted through the structure until it propagates to the surface and is detected by a sensor [8]. A benefit of using AE over vibration is its signal to noise ratio. Most rotating machinery operates in the low frequency range (sub 20kHz) but AE is sensitive from 50kHz-1MHz. This allows AE sensors to detect frequencies both in the vibration spectrum as well as the high frequency components, currently unable to be detected by traditional accelerometers.

Most analysis of AE data focuses on basic characteristics or parameters that are calculated from the raw signal, such as Root Mean Square (RMS), hits and counts. Although able to determine bearing condition holistically, use of techniques such as RMS presents difficulty when trying to identify the failure mechanism occurring within the bearing. Pullin *et al.* determined the crack growth of a spur gear tooth by monitoring the frequency response from an AE sensor over a period of time [8]. Howard [9] states that the three important steps of condition monitoring are detection, diagnosis and prognosis. The work presented in this paper investigates how RMS, high frequency analysis and characteristic defect frequencies can be used together to detect and diagnose the condition of artificially damaged bearings.

EXPERIMENTAL SET UP

2.1 Test Head

Figure 1 illustrates the bearing test head that was driven by a 0.55kW motor at a speed of approximately 5780rpm through a timing belt pulley. The test bearing (1) is situated between two support bearings (2) and the 10:1 lever arm (3) is used to apply a radial load directly to the test bearing. The two spherical roller support bearings, type SKF 22205/20E, were selected due to their high fatigue load capacity. The cylindrical roller test bearing, type SKF N204ECP, was selected due to the simplicity in the separation of the bearing into its constituent parts. All bearings are 20mm inner raceway bore. Table 1 shows the characteristic frequencies of the bearings used for a shaft frequency of 97.83Hz (5780rpm). Since the support bearings only see half of the load experienced by the test bearing, any damage is highly likely to occur solely in the test bearing.

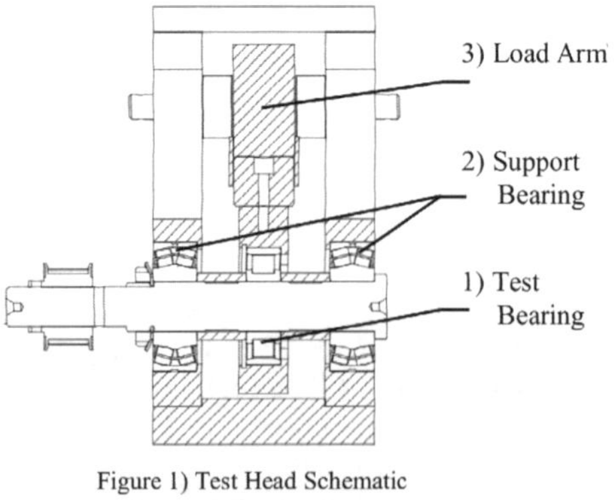

Figure 1) Test Head Schematic

Table 1. Determination of Characteristic Frequencies

Defect Frequencies	N204ECP	22202/20E
Outer (Hz)	419.4	604.5
Inner (Hz)	656.7	863.0
Ball Spin (Hz)	211.0	262.3
Train (Hz)	38.1	40.3

Each bearing was lubricated via individual oil jet feeds with a mineral oil (meeting OEP-80 performance specification) supplied at 50°C. For the healthy bearing, five radial loads were applied: HL1, HL2, HL3, HL4 and HL5 which represent 0.29, 0.61, 1.29, 1.61 and 1.8kN respectively. For the damaged bearing four loads were applied: L1, L2, L3 and L4, representing 0.29, 0.79, 1.29 and 1.8kN respectively.

2.2 Data Acquisition

Mistras Group Limited (MGL) Nano30 AE sensors were placed onto the bearing housings, close to the outer raceway of each bearing. Lithium-based grease was applied as a couplant to improve the transmission of the signal from the test head into the sensor. Magnets were used to secure the sensors into position. Each sensor was connected to an amplifier with 40db gain and a bandpass filter of 20-1500kHz. Nano30 sensors were selected due to their small size and high resonant frequency of 300kHz whilst maintaining reasonable bandwidth. A MGL PCI-2 system was used to collect manually triggered waveforms at a sampling frequency of 2MHz. Before waveforms were recorded, the test head was run until steady state temperatures were achieved for all bearings. For each load, a total of four waveforms were recorded at intervals of 2,000 revolutions. The load was then increased and the next four waveforms were taken after approximately 10,000 revolutions, counted by an optical revolution counter mounted on the test shaft.

2.3 Seeded Defects

To replicate common bearing failure modes, artificial defects were engraved onto the outer raceway of the test bearing. Before and after each test, the defect was measured using a Taylor Hobson Talysurf surface profilometer. Artificial defects are by no means a direct replication of natural damage propagation seen in bearings. It is common for sub-surface cracks to initiate due to the stresses induced as the rolling element passes over the point of maximum load [7]. As these cracks grow, they can propagate to the surface of the raceway causing pieces of material to dislodge. Engraving the surface does not remove material but instead creates an engraved pit surrounded by peaks as the material is forced upwards. Although not a true representation of the failure modes of a bearing, engraving allows for quick validation of the equipment and methodology before conducting life tests as well as demonstrating the sensitivity of AE for use within rotating machinery to detect realistically sized defects. Table 2 quantifies the size of each of the seeded defects through the determination of the volume of the peaks and the troughs created through engraving using the least squares method.

Table 2. Defect Volumes

	Defect 1	Defect 2	Defect 3	Defect 4	Defect 5	Defect 6
Peak (mm^3)	2.62E-004	1.05E-003	5.76E-003	6.56E-003	5.36E-003	1.44E-002
Valley (mm^3)	2.62E004	1.06E-003	5.76E-003	6.56E-003	5.34E-003	1.45E-002

3. RESULTS AND DISCUSSION

Figure 2 is the RMS_{AE} vs. load for each of the seeded defects. Defect 1 and defect 2 are both point defects in the centre of the outer raceway and show a large rise in the RMS_{AE} value compared to the healthy bearing. As the defect is expanded from a point defect to multiple points across the width of the bearing, defect 3, a large reduction in the RMS_{AE} is seen, although the amplitude of the RMS_{AE} remains higher than that of the undamaged bearing. As the depth and width of the defect is increased from defect 3 to defect 6, the RMS_{AE} continued to increase. The significant reduction in RMS value between defect 2 and defect 3 is considered to be due to an increase in contact area as the defect is spread across the raceway. The RMS value indicates that the AE generated by each defect depends on the defect size, but little information can be determined about what is happening within the bearing itself.

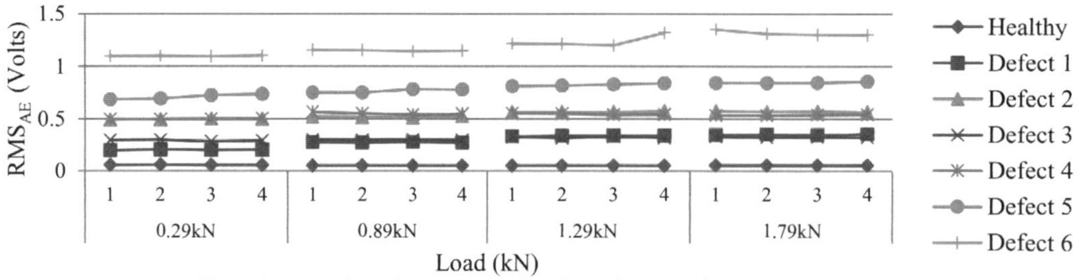

Figure 2) Comparison of RMS values of healthy and damaged bearings

For each collected waveform, a Fast Fourier Transform (FFT) was calculated and plotted against time to produce a 3-dimensional graph. In order to smooth the data for easier interpretation, the x-axis was separated into a number of frequency bands, or bins, where each bin is the sum of all of the frequency amplitudes within. Figure 3a and 3b are stacked FFT's between 0-75kHz and 75-150kHz respectively for a healthy bearing with a bin width of 20Hz. The y-axis is divided into the separate loads applied where each load contains the four collected waveforms, demonstrating amplitude fluctuation over time.

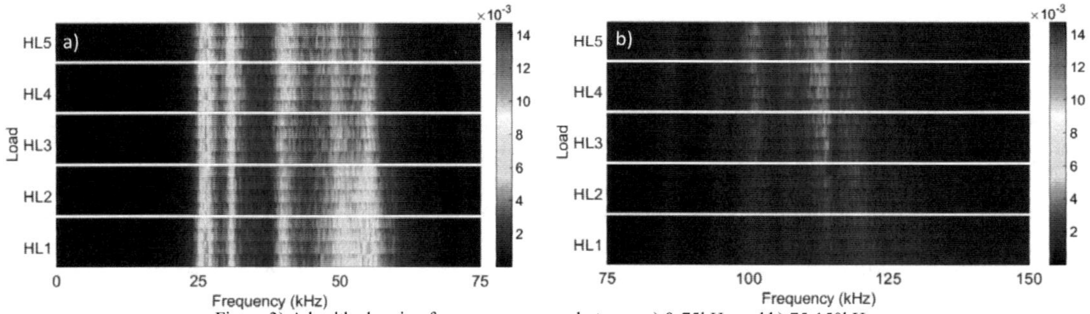

Figure 3) A healthy bearing frequency response between a) 0-75kHz and b) 75-150kHz

It can be seen that the dominant frequency within the healthy bearing signal is around 26kHz. It is thought that the value of 26kHz relates to either the structural resonance of the test housing or of the bearing itself [10][11]. Since the resonant frequency of the Nano30 sensor is 300kHz, to excite the piezoelectric crystal in such a way a significant amount of energy must be present at 26kHz. Figure 3a also shows significant energy at approximately 60 kHz for loads HL1 and HL2. Between HL3 and HL5 the amplitude around 60kHz decreases and the amplitude around 35 kHz increases. This could be due to the system being damped as the load is increased or even evidence of the end of the running in phase. Figure 3b illustrates that higher frequency AE can also be witnessed as the load is increased. Energy in the frequency band of 100-125kHz is indicative of AE sources occurring within the bearing. It is thought that the high frequency emission seen in a healthy bearing is induced through friction between the roller and the races as the bearing completes its 'running in' phase. During the running in phase, surface roughness asperities are smoothed over as the rollers pass over in quick succession. This flattening of the raceways requires plastic deformation of asperities and is therefore a likely source of high frequency emission.

Figures 4(a-f) are low frequency (0-75kHz) stacked FFT's of defects 1-6. The suspected system resonances excited are at similar frequencies to those witnessed within the healthy bearing, albeit at an order of magnitude higher in amplitude. As the defect size increases the magnitude of the resonant frequencies increase dramatically. This increase in amplitude of the resonant frequencies could be used to determine the severity of the defect or be used to demodulate the signals to help ease weak signal extraction. Figure 5(a-f) illustrates the high frequency (75-150kHz) stacked FFT's of defects 1-6. Unlike the low frequency stacked FFT's, an increase in amplitude can be seen as the load increased.

Journal of Physics: Conference Series **628** (2015) 012074 doi:10.1088/1742-6596/628/1/012074

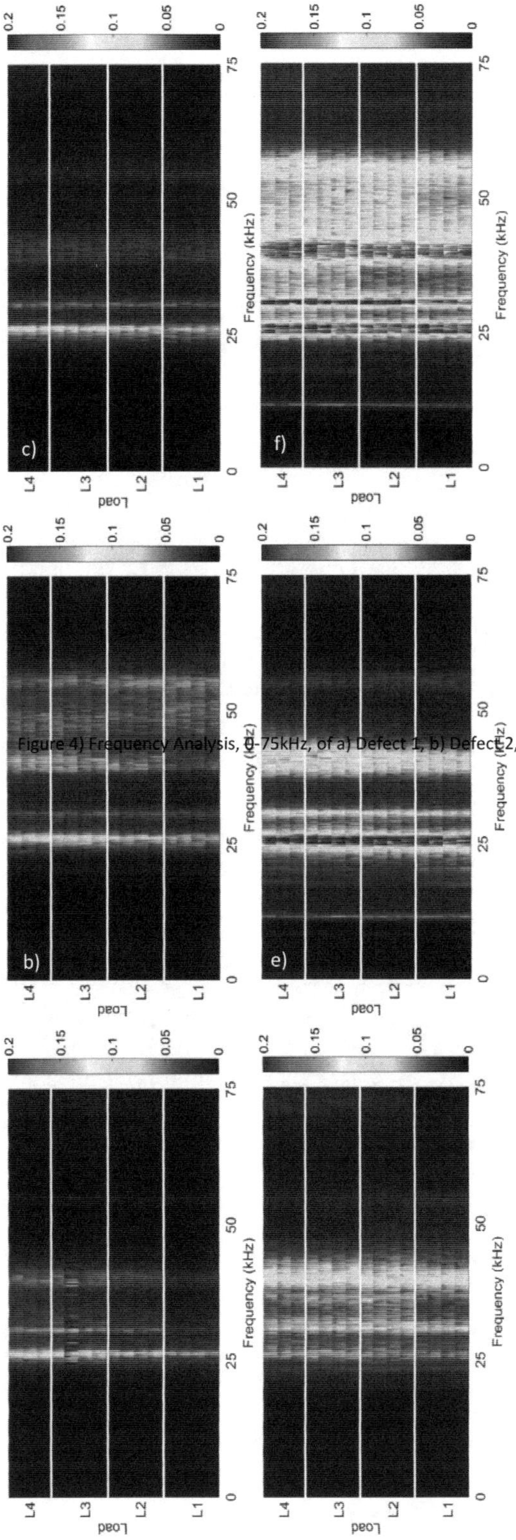

Figure 4) Frequency Analysis, 0-75kHz, of a) Defect 1, b) Defect 2, c) De

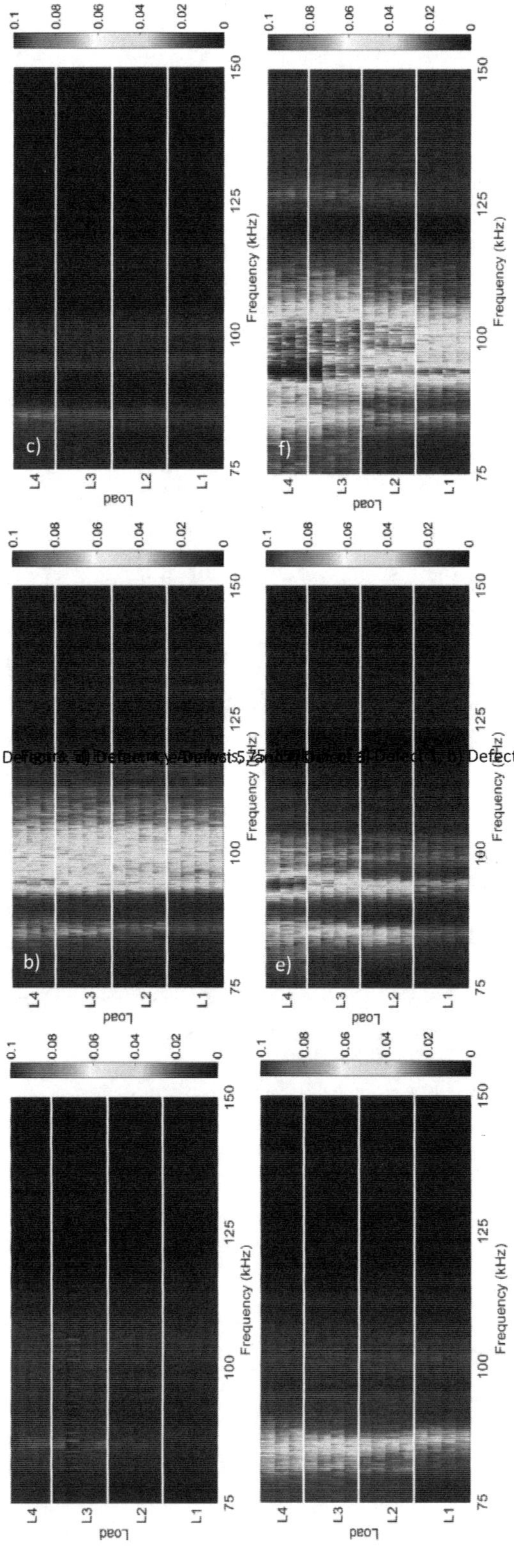

Figure 5) Frequency Analysis, 75-150kHz of a) Defect 1, b) Defect 2, c) Defect 3, d) Defect 4, e De

It is thought that these high frequency components are caused as the rollers pass over the sharp peaks of the seeded defects, as well as friction between the interacting bodies. Similar to the high frequency AE that occurs during crack growth, high localised stresses within the area of contact cause the peaks to become plastically deformed or to fracture. Evidence of this can be seen in Figure 6 where a pre and post-test measurement of defect 2 are compared together. It can be seen that the sharp peaks witnessed in the pre-test measurement are plastically deformed and flattened significantly during the test.

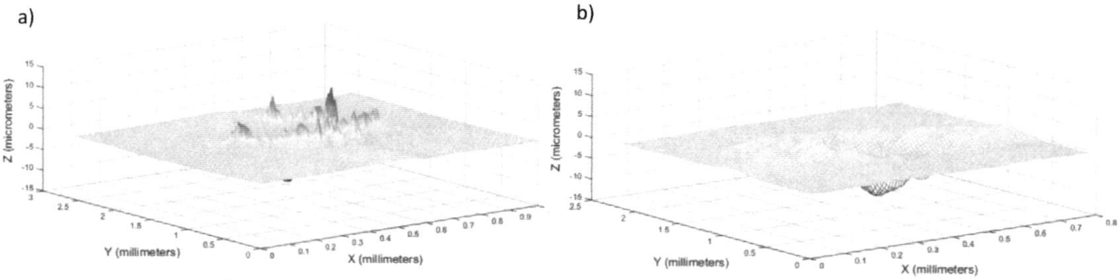

Figure 6) a) Pre-Test surface measurement and b) Post-Test surface measurement

During the natural degradation of bearing raceways, there are rarely any sharp peaks surrounding pits in the surface as material tends to become dislodged from the surface as opposed to being forced upwards as occurs in the artificial defects used in this work. Any sharp edges of a naturally occurring defect would become quickly flattened and smoothed during the repeated over-rolling in the life of a bearing. Figure 5 demonstrates that the high frequency acoustic emission released from the bearing raceway can be detected at the surface of the bearing housing. Therefore, it is hoped that the AE released during the natural fatigue of the bearing surface during a life test to failure will be detectable using similar techniques.

The RMS value indicates gross bearing change and the stacked FFT illustrates the detailed nature of the interaction between rollers and defects within the bearings. With such a large number of possible failure mechanisms within a bearing, the location at which a defect is propagating is of importance as the attenuation from the inner raceway is far more severe than from the outer raceway. For a sensor mounted on the stationary outer race bearing housing, an AE wave from the inner race has to pass through two elastohydrodynamic oil films, and through two additional solid bodies (roller and outer race) before it reaches the sensor. To determine location, the characteristic defect frequencies determined in Table 1 are used. Figure 7 is the raw waveform and FFT of a healthy bearing. The low frequency response has low amplitude and is dominated by orders of shaft rotation with no other dominant frequency spikes. A similar signal and frequency response is witnessed for defect 1 albeit with larger amplitudes (Figure 8).

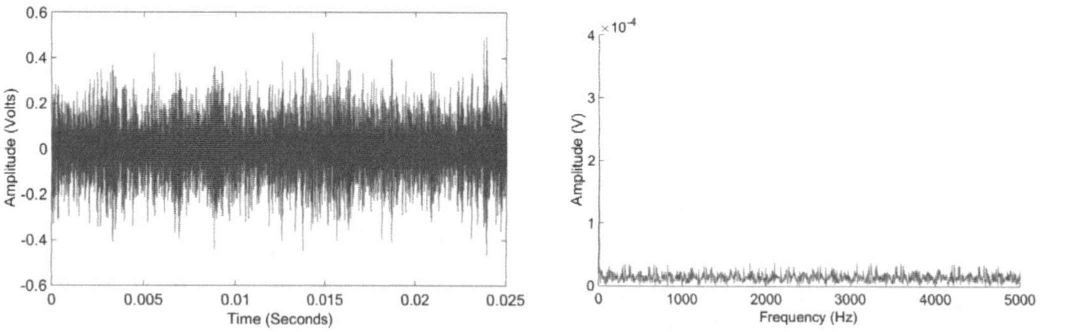

Figure 7) The raw waveform and FFT of a healthy bearing

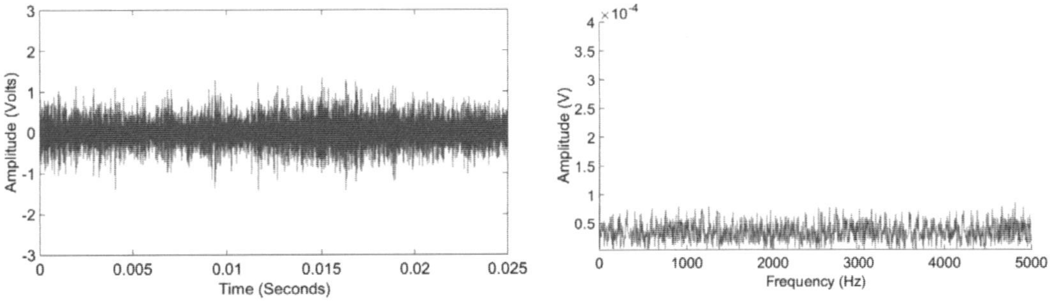

Figure 8) The raw waveform and FFT of defect 1

As the size of the defect is increased from defect 1 to defect 2, repeatable high frequency bursts are witnessed within the raw waveform (Figure 9). These bursts occur close to the outer race defect frequency of 415.8Hz as shown in the FFT. As the damage size is increased to the maximum of defect 6, the amplitude of the outer race defect frequency within the FFT becomes more apparent and multiple orders are witnessed within the FFT (Figure 10).

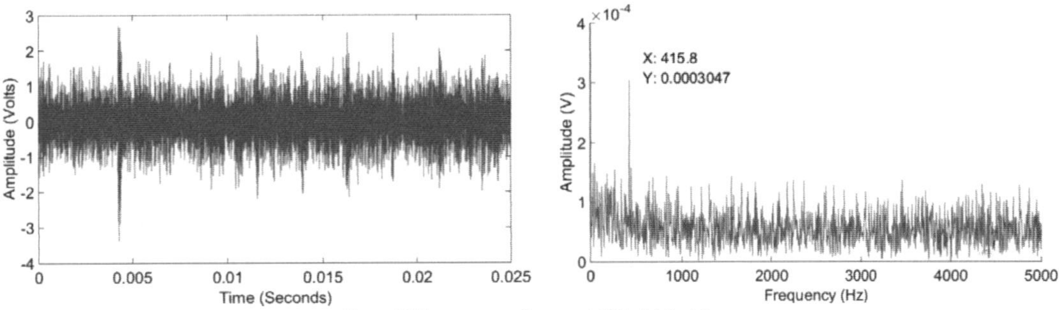

Figure 9) The raw waveform and FFT of defect 2

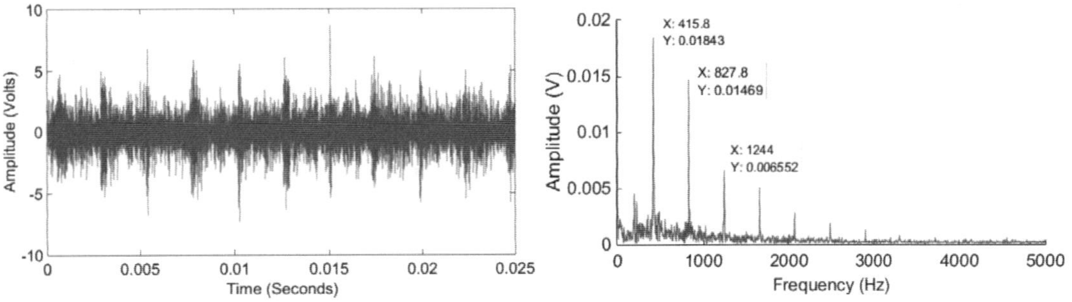

Figure 10) The raw waveform and FFT of defect 6

Although bearing defect frequencies are commonly relied upon in the condition monitoring of bearings with both vibrational and AE signals, Figure 11b is the frequency response from an AE sensor mounted upon a support bearing during the analysis of defect 6. It can be seen that a frequency close to the outer raceway defect frequency of the support bearing, (approx. 605Hz), has increased amplitude when compared with its healthy counterpart Figure 11a. It is thought that the damage of the test bearing is exciting the characteristic frequencies of the support bearing. Current condition monitoring techniques may represent this bearing as being damaged however, post-test inspection of this bearing showed the outer raceway to be unmarked and visually healthy.

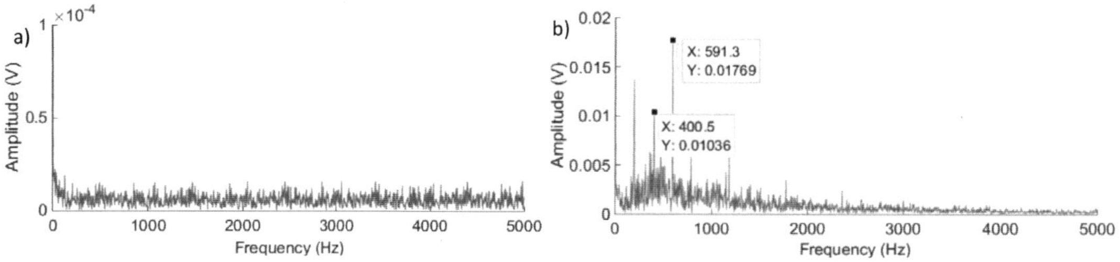

Figure 11) The FFT of the support bearing of a) a healthy bearing and b) defect 6

4. CONCLUSION

Seeded defects placed upon the outer raceway of a rolling element bearing were detectable through RMS and frequency analysis, in both low and high frequency ranges. As the defect spreads across the raceway, the amount of energy released reduces and this can be seen in both the high frequency response and the RMS values. Thus the size and shape of the defect has a large influence on the release of AE from it. The high frequency analysis has determined that as the defect size increases, the structural frequencies are excited without a significant change as the load is increased. The high frequency components, (75-150kHz), do increase in magnitude with an increase of load and therefore it is deduced that the change in this high frequency section is most likely to be due to plastic deformation occurring as the rollers pass over sharp peaks produced by engraving.

Acknowledgements

The authors wish to acknowledge the support of EPSRC Grant references EP/L021757/1, and EP/K031635/1 which facilitated this work in part. The first author acknowledges the financial support of Mistras Group Limited towards his studentship.

References

[1] Lundberg G and Palmgren A 1952 Dynamic capacity of roller bearings Acta Polytech Mechanical Engineering Series, 2.

[2] Halme J and Andersson P 2010 *Proceedings of the Institution of Mech. Eng. Part J: J. Eng. Tribology* **224(4)** 377–393

[3] García Márquez F P, Tobias A M, Pinar Pérez J M and Papaelias M 2012 *Renewable Energy* **46** 169–178.

[4] Harris T A and Kotzalas M N 2006. Essential concepts of bearing technology CRC press

[5] Qiu H, Lee J, Lin J and Yu G 2003 *Advanced Engineering Informatics* **17(3-4)** 127–140

[6] Ho D and Randall R B 2000 Mech. Systems & Sig Processing **14(5)** 763-788.

[7] Lees A W, Quiney Z, Ganji A and Murray B 2011 *J. Phys. Conf. Series* **305** 012074

[8] Pullin R, Clarke A, Eaton M J, Pearson M R and Holford K M 2012 *J. Phys: Conf. Series* **382(1)** 2050

[9] Howard I 1994 *A Review of Rolling Element Bearing Vibration 'Detection, Diagnosis and Prognosis'.* No. DSTO-RR-0013. DEFENCE SCIENCE AND TECHNOLOGY ORGANIZATION CANBERRA (AUSTRALIA)

[10] Tandon N and Choudhury A 1999 *Tribology international* **32** 469-480

[11] Rai V K and Mohanty A R 2007 *Mech. Systems & Sig. Processing* **21(6)** 2607-2615

Surrogate models for efficient stability analysis of brake systems

Lyes Nechak, Frédéric Gillot, Sébastien Besset and Jean-Jacques Sinou

Laboratoire Tribologie et Dynamique de Systémes, Ecole Centrale de Lyon-CNRS, 36 av. Collongue, 69134 Ecully cedex, France

E-mail: `lyes.nechak, frederic.gillot, sebastien.besset, jean-jacques.sinou@ec-lyon.fr`

Abstract. This study assesses capacities of the global sensitivity analysis combined together with the kriging formalism to be useful in the robust stability analysis of brake systems, which is too costly when performed with the classical complex eigenvalues analysis (CEA) based on finite element models (FEMs). By considering a simplified brake system, the global sensitivity analysis is first shown very helpful for understanding the effects of design parameters on the brake system's stability. This is allowed by the so-called Sobol indices which discriminate design parameters with respect to their influence on the stability. Consequently, only uncertainty of influent parameters is taken into account in the following step, namely, the surrogate modelling based on kriging. The latter is then demonstrated to be an interesting alternative to FEMs since it allowed, with a lower cost, an accurate estimation of the system's proportions of instability corresponding to the influent parameters.

1. Introduction

The increasingly stringent requirements concerning tolerable noise levels in agglomerations and stations have made squeal prediction a very important issue for car and train manufacturers [1, 2, 3] since brake squeal is characterized by high and audible frequencies and high intensities that are discomfort for passengers while it does not affect the braking's efficiency [4]. Two main groups of methods can be distinguished in brake squeal prediction's framework. The nonlinear dynamic transient analysis is used to determine the time evolution of mechanical solutions (such as displacements, velocities, etc). However, it remains still prohibitive regarding to its time consuming even when associated with reduced and/or simplified FEMs [5]. So, the complex eigenvalue analysis remains the most used method [6]. Its main principle is to linearize the system's model around its equilibrium state then to analyze the sign of the resulting linear system's eigenvalues according to the Lyapunov theory. The main drawback of the CEA is that the associated stability analysis may lead to an underestimation or an over-estimation of the unstable modes observed in the nonlinear time simulation due to the fact that linear conditions (i.e. the linearized stability around an initial equilibrium point with a defined contact state for each node are not valid during transient oscillations [7]). However, the method remains less expensive than the nonlinear transient dynamic analysis but, because of the complexity of finite element models, it stays costly when robustness and/or optimal objectives are required. Therefore, it is necessary to search for alternative models that can be efficiently

used in stability analysis instead of FEMs. This is the main goal of this paper. Indeed, we propose the combination of the global sensitivity analysis together with kriging modeling to build simpler models helping for a more efficient stability analysis of brake systems. The global sensitivity analysis helps, by means of Sobol indices [8], for the quantification of individual and/or collective impacts of parameters on the brake system's stability. This issue has been treated in other studies but sensitivities was calculated locally (in a relatively small neighborhood of nominal values of parameters, see [9] and references are therein). Based on the global sensitivity analysis, the kriging formalism [10] is then proposed to construct substituting models to FEMs, which can be used efficiently to analyze the stability of brake systems by taking into account the uncertainty of the influent parameters.
This paper is organized as follows. First, a simplified brake system is described in Section 2. The global sensitivity analysis of its stability is discussed in Section 3 while kriging is proposed to represent its instability proportions in Section 4.

2. Finite element model

This study concerns a simplified brake system represented by a finite element model (Fig.1) obtained from the assembling of finite element models of a pad and a disc with a contact surface modeled by nine uniformly spaced contact nodes. The inner surface of the disc is assumed

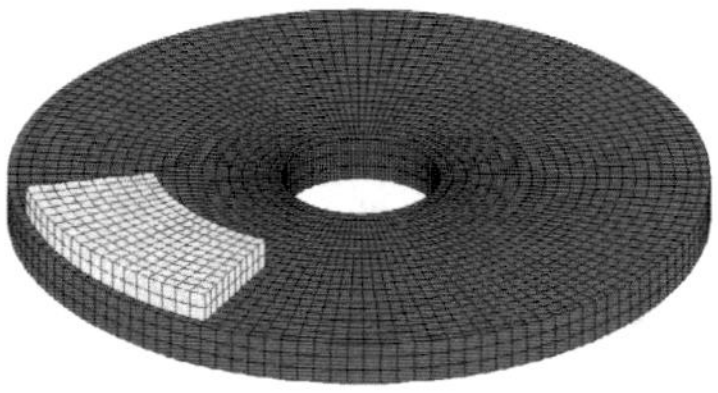

Figure 1: Finite element model of a simplified brake system

clamped while the outline of the upper surface of the pad is just translating along the normal direction [11]. The simplified assembled model has 174 dofs issued from 80 dofs for the disc , 40 dofs for the pad and 54 dofs for the contact interface. As the main objective of this study is to assess the feasibility of the combination of global sensitivity analysis together with kriging, we consider in the following a non-damped dynamic behavior for the brake system. It is then described by the classic second order differential equation given by

$$\mathbf{M}\ddot{\mathbf{X}} + (\mathbf{K} + \mathbf{K_c})\mathbf{X} = \mathbf{F}_{\text{pressure}} \tag{1}$$

where $\mathbf{M}$ is the mass matrix, $\mathbf{K}$ denotes the structural stiffness matrix while $\mathbf{X}$ is the displacement vector with the associated velocity $\dot{\mathbf{X}}$ and acceleration $\ddot{\mathbf{X}}$. Otherwise, $\mathbf{F}_{\text{pressure}}$ represents the force due to the brake pressure applied on the pad. $\mathbf{K_c}$ represents a non-symmetric matrix defined to model the contact interface between the disc and the pad at the selected interface nodes where normal and tangential contact forces are applied by the pad over the disc and conversely by the disc over the pad. The friction contact is assumed to follow a Coulomb friction law with a constant friction coefficient μ and a permanent sliding. Normal and tangential contact forces are expressed for each node as follows [11].

$$F^{\text{d}}_{\text{normal},i} = \begin{cases} k_{\text{c}}(X_{\text{d},i} - X_{\text{p},i}) & \text{if}(X_{\text{p},i} - X_{\text{d},i}) > 0, \\ 0 & \text{otherwise.} \end{cases} \tag{2}$$

$$F^{\text{d}}_{\text{tangent},i} = \mu F^{\text{d}}_{\text{normal},i}\text{sign}(v_{\text{r},i}) \tag{3}$$

where $X_{i,\mathrm{d}}$ and $X_{i,\mathrm{p}}$ define the ith node displacements of the disc and the pad denoted by d and p respectively, $v_{i,\mathrm{r}}$ is the corresponding velocity.

3. Stability and sensitivity analysis

The main question in this section is to quantify how the stability of the considered brake system is impacted by some design parameters $p_i \in \{e_\mathrm{d}, e_\mathrm{p}, \rho_\mathrm{d}, \rho_\mathrm{p}, \mu, k_\mathrm{c}\}$ including Young modulus $\{e_\mathrm{d}, e_\mathrm{p}\}$ and densities $\{\rho_\mathrm{d}, \rho_\mathrm{p}\}$ of the pad and disc, the friction coefficient $\{\mu\}$ and the linear contact stiffness $\{k_\mathrm{c}\}$, see Table 1 where intervals variations I_{p_i} of parameters p_i are given. These parameters are assumed to be driven by independent uniform laws within their respective intervals while the other parameters defining the geometry of the pad and disc are considered as determinist.

Stability of the considered brake system is characterized by its eigenvalues $\lambda = a + jb$ which are obtained from the solution of the following CEA problem,

$$(\lambda^2 \mathbf{M} + (\mathbf{K} + \mathbf{K}_\mathrm{c}))\boldsymbol{\Psi} = \mathbf{0} \tag{4}$$

The Monte Carlo method is often used to analyze parameters effects on the stability properties. It consists of generating parameters samples following their probability density functions and calculating eigenvalues from the corresponding CEA problem. We have used this procedure so, real parts of the resulting eigenvalues are plotted against imaginary parts in Fig. 2. A stable eigenvalue $\lambda = a + jb$ is said to be asymptotically stable if its real part a is strictly negative and unstable if a is strictly positive.

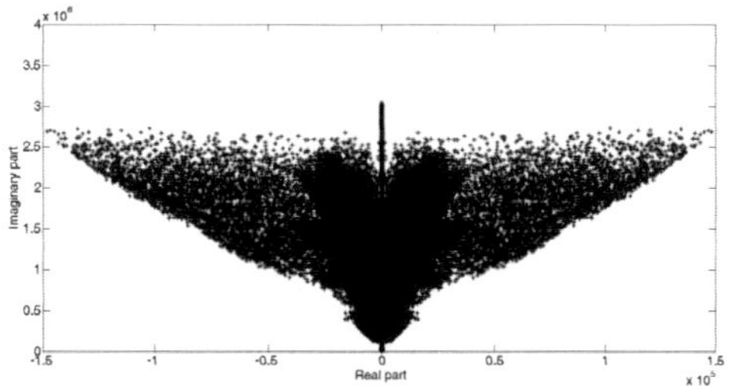

Figure 2: Brake system's eigenvalues in the complex plane obtained from Monte Carlo simulations

From the obtained results, it is so difficult to objectively understand how the considered design parameters impact the distance of eigenvalues from the imaginary axis which defines the stability/instability border. Hence, we propose to carry out the global sensitivity analysis from the Sobol sense of the following distance function in design parameters:

$$\mathrm{D}_{\text{stability}} = \sum_{i=1}^{n} |\mathrm{Re}(\lambda_i(e_\mathrm{d}, e_\mathrm{p}, \rho_\mathrm{d}, \rho_\mathrm{p}, \mu, k_\mathrm{c})| \tag{5}$$

where $\mathrm{Re}(\cdot)$ denotes the real part of the ith eigenvalue $\lambda_i(\cdot)$.

The main idea is to determine Sobol indices which quantify the variability of the distance function induced by the variability of one or several parameters. From sensitivity analysis theory [8]), first order sensitivity indices (also named first order Sobol indices) quantifying the individual influence of parameter p_i on the stability function $\mathrm{D}_{\text{stability}}$ can be defined by

Table 1: Variability intervals of mechanical input parameters

Parameters	Minimum value	Maximum value
Friction coefficient	0.2	0.7
Disc Young modulus (pa)	$14\,10^{10}$	$22\,10^{10}$
Pad Young modulus (pa)	$0.5\,10^{9}$	$2\,10^{9}$
Disc density (kg.m^{-3})	7100	7900
Pad density (kg.m^{-3})	1400	2600
Linear contact stiffness (n.m^{-1})	$1\,10^{7}$	$4\,10^{7}$

$$S_i = \frac{\mathrm{Var}(\mathbf{E}[\mathrm{D_{stability}} \backslash p_i])}{\mathrm{Var}(\mathrm{D_{stability}})} \tag{6}$$

where Var(.) denotes the variance operator while $\mathbf{E}[.\backslash.]$ is the conditional expectation operator.

Similarly, high order sensitivity indices measuring influences of couples or combinations of several parameters on the analyzed function can be defined. For example second order indices are given by:

$$S_{ij} = \frac{\mathrm{Var}(\mathbf{E}[\mathrm{D_{stability}} \backslash p_i, p_j])}{\mathrm{Var}(\mathrm{D_{stability}})} \tag{7}$$

Ultimately, a total index S_{T_i} can be obtained for a single parameter p_i by the summation of all indices including p_i, that is:

$$S_{T_i} = S_i + S_{ij} + S_{ijk} + ... \tag{8}$$

A high value for S_i indicates that the corresponding parameter p_i influences strongly the variability of the stability function (5) while a small value for a total index S_{T_i} shows that the parameter p_i possesses a small impact on the stability function even if it is combined with other parameters.
The main step is then to compute the Sobol indices. The referential Monte carlo method is used in the present study [8, 12]. The sample size ensuring a convergent calculus of Sobol indices is obtained equal to $N = 10000$. Table 2 gives both the computed first order Sobol indices and the total indices of input parameters ($p_i \in \{e_\mathrm{d}, e_\mathrm{p}, \rho_\mathrm{d}, \rho_\mathrm{p}, \mu, k_\mathrm{c}\}$). Two sets of parameters can be

Table 2: First order Sobol indices estimated by Monte Carlo method

Parameters	S_i with $N = 10000$	S_{T_i} with $N = 10000$
Friction coefficient	0.756	0.865
Disc Young modulus	−0.007	0.021
Pad Young modulus	−0.007	0.002
Disc density	−0.007	0.004
Pad density	0.001	0.055
Linear contact stiffness	0.204	0.226

distinguished. The first one includes the friction coefficient and linear contact stiffness while the second contains the remaining parameters. The friction coefficient and the linear contact stiffness present the most significant influences on the stability function (5) since the corresponding first order Sobol indices are significantly high. However, Young modulus and density parameters possess very small first order Sobol indices (negative sign denotes that the corresponding indices are closed to zero,[12]) which denote the non-significant impact of parameters on the considered stability function. Otherwise, relatively small differences between total and first orders Sobol indices especially for friction and linear contact stiffness coefficients show that these parameters influence stability function individually while the collective impacts can be considered as negligible.

4. Kriging based meta-modeling

Based on conclusions of the performed sensitivity analysis, the relevance of using of kriging based models to predict instability proportions of the used brake system is assessed. More accurately, we are interested by the prediction, under the uncertainty of the friction coefficient μ and the linear contact stiffness k_l previously shown as influent parameters, of instability proportions denoted by $\underset{f_\mu, f_{k_c}}{\text{prop}}(e_d, e_d)$ which depend implicitly on μ and k_c and explicitly on deterministic Young modulus (e_d, e_p) of the disc and the pad, while the corresponding densities of the pad and disc are omitted for the sake of simplicity.

The instability proportion corresponds to the pourcentage of instability given by the number of times the brake system is unstable within the sample data set $\{\mu^j, k_c^j\}$ with $j = 1, ..., 1000$ for each couple(e_d^i, e_p^i). Fig. 3 represents $\underset{f_\mu, f_{k_c}}{\text{prop}}(e_d, e_d)$ computed from the solution of 10^7 CEA problems. The final solution is obtained after 21 days. Kriging is then proposed to estimate instability proportions with a lower cost.

4.1. Mathematical formulation of kriging

From kriging theory, it is possible to determine a function at unknown points from some calculus of correlations between couples of sample points generated from the function domain [10]. The correlation is built such as two infinitely distant points be not correlated while two identical points have a unitary correlation. We present in the following main principles of kriging based modeling through its exploiting to represent the instability proportion based function $\underset{f_\mu, f_{k_c}}{\text{prop}}(e_d, e_d)$. So, let consider m sample points $\mathbf{e}_{\text{prop}} = [\mathbf{e}^1, \mathbf{e}^2, ..., \mathbf{e}^m]$ with $\mathbf{e} = [e_d, e_p]$ and the m outputs $\mathbf{Y} = [y^1, y^2, ..., y^m]^T$ defining instability proportions for each input sample such that $y^i = \underset{f_\mu, f_{k_c}}{\text{prop}}(\mathbf{e}^i)$. Thus, from kriging theory [10], the instability proportion $\underset{f_\mu, f_{k_c}}{\text{prop}}(\mathbf{e})$ can be written as follows:

$$\underset{f_\mu, f_{k_c}}{\text{prop}}(\mathbf{e}) = f(\mathbf{e})^T \boldsymbol{\beta} + z(\mathbf{e}) \tag{9}$$

The first part in the summation is a constant, linear or polynomial regression model with q basis polynomial functions $f_i, i = 1, ..., q$ weighted by the regression parameters β_i. It defines the mean structure of the function while the second part z is the realization of a normally distributed random process with a zero mean value and a covariance function given by:

$$\mathbf{E}[z(\mathbf{w})z(\mathbf{e})] = \sigma^2 \Re(\boldsymbol{\theta}, \mathbf{w}, \mathbf{e}) \tag{10}$$

where σ^2 is the process variance and $\Re(\boldsymbol{\theta}, \mathbf{w}, \mathbf{e}) \in [0\,1]$ is a spatial correlation model with the scaling vector $\boldsymbol{\theta} = [\theta_1, \theta_2]$ which has to be estimated later, $\mathbf{E}[.]$ is the expectation operator. This model is a monotone function depending on the distance positive or null between two points $\mathbf{w}$ and $\mathbf{e}$. It is built such as two infinitely distant points be not correlated while two identical points have a unitary correlation. Correlation models (see Table 3) used are of the following form:

$$\Re(\boldsymbol{\theta}, \mathbf{w}, \mathbf{e}) = \prod_{j=1}^{2} \Re_j[-\boldsymbol{\theta}(w_j - e_j)^2] \tag{11}$$

Table 3: Correlation functions

Correlation model	$\Re(\theta, w_j, e_j)$
Exponential	$\exp(-\theta_j \lvert w_j - e_j \rvert)$
Gaussian	$\exp(-\theta_j (w_j - e_j)^2)$
Linear	$\max\{0, 1 - \theta_j \lvert w_j - e_j \rvert\}$

Denoting by $\mathbf{G}$ a matrix whose entries are given by $\mathbf{g}_{i,:} = [f(\mathbf{e}^i)]^T$, the regression parameter vector $\boldsymbol{\beta}$ given by

$$\boldsymbol{\beta}^* = (\mathbf{G}^T \boldsymbol{\Phi}^{-1} \mathbf{G})^{-1} \mathbf{G}^T \boldsymbol{\Phi}^{-1} \mathbf{Y} \tag{12}$$

is the least square solution of the following classical regression problem

$$\mathbf{G}\boldsymbol{\beta} \simeq \mathbf{Y} \tag{13}$$

Otherwise, the associated optimal variance σ^2 is given by

$$\sigma^2 = \frac{1}{m}(\mathbf{Y} - \mathbf{G}\boldsymbol{\beta}^*)^T \boldsymbol{\Phi}^{-1} (\mathbf{Y} - \mathbf{G}\boldsymbol{\beta}^*) \tag{14}$$

where $\boldsymbol{\Phi}$ is the correlation matrix with entries defined by $\phi_{i,j} = \Re(\boldsymbol{\theta}, e^i, e^j)$ and $\mathbf{r}(\mathbf{x}) = [\Re(\boldsymbol{\theta}, \mathbf{e}^1, \mathbf{x}), ..., \Re(\boldsymbol{\theta}, \mathbf{e}^m, \mathbf{x})]^T$ is a vector containing the values of the correlation between each of the m points of the input sample data set and a variable point $\mathbf{x}$ in the design space.
The previous optimal parameters $\boldsymbol{\beta}^*$ and $\boldsymbol{\theta}^*$ depend on the inverse matrix Φ^{-1} which depends in turn on the process correlation parameter vector $\boldsymbol{\theta}$. The optimal choice $\boldsymbol{\theta}^*$ is defined as the solution with respect to $\boldsymbol{\theta}$ of the following optimization problem [13].

$$\min f(\boldsymbol{\theta}) = |\boldsymbol{\Phi}|^{\frac{1}{m}} \sigma^2 \tag{15}$$

where $|\boldsymbol{\Phi}|$ is the determinant of $\boldsymbol{\Phi}$. Based on the decomposition (9) and the obtained $\boldsymbol{\theta}^*$ from the solution of (15), we are interested by the prediction of the unbiased instability proportion at a given sample point $\boldsymbol{e}$ in the design space. The kriging based predictor [10] of instability proportions is defined by

$$\hat{y}(\mathbf{e}) = \mathbf{f}^T(\mathbf{e})\boldsymbol{\beta}^* + \mathbf{r}(\mathbf{e})^{\mathrm{T}} \boldsymbol{\gamma}^* \tag{16}$$

where $\boldsymbol{\beta}^*$ is the solution (12) of the regression problem (13) while $\boldsymbol{\gamma}^* = \boldsymbol{\Phi}^{-1}(\mathbf{Y} - \mathbf{G}\boldsymbol{\beta}^*)$.

4.2. Prediction of instability proportions

In this sub-section kriging models are used to estimate the referential map of instability proportions shown in Fig. 3.

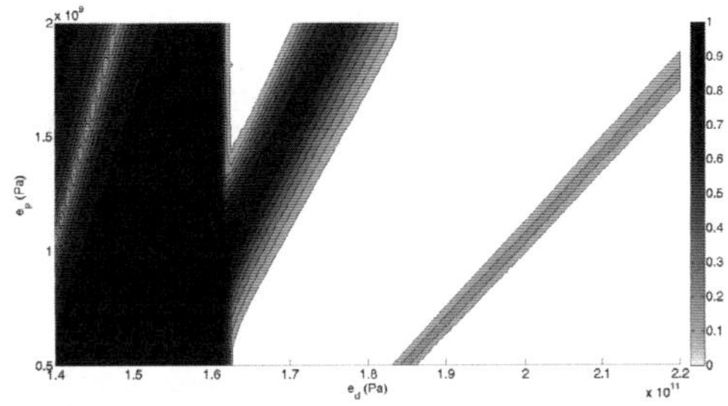

Figure 3: Instability proportions from FEM based computing

The sample data size m is varied to observe its effect on the accuracy of kriging based estimations. Otherwise, the gaussian spatial correlation model is chosen. In fact, no criterion helping for determining the most suitable correlation model is found. A more complete study would consider other correlation models to select the best one. Kriging estimations of instability proportions obtained by using Dace toolbox [13] for different sample data sizes $m = 100 : 100 : 1000$ are shown in Fig .4.

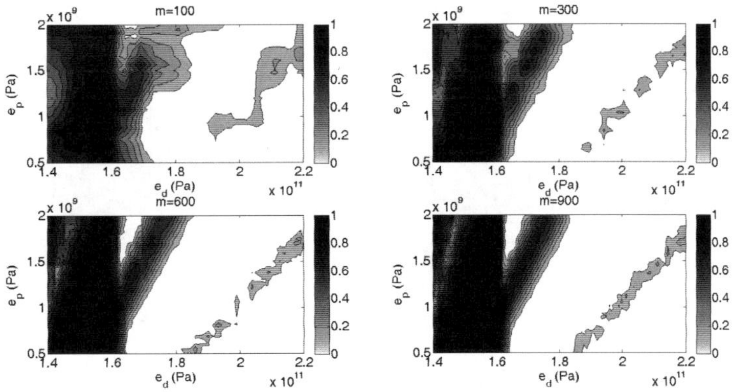

Figure 4: Instability proportions estimated via kriging based models using different sample data sizes $m \in \{100, 300, 600, 900\}$ with exponential correlation model

It can be observed that zones of estimated instability proportions against $(e_\mathrm{d}, e_\mathrm{p})$ presented in Fig.4 globally show the same shape as the referential map plotted in Fig.3. The accuracy of the estimated instability proportions depends on the sample data size m. Indeed, it increases with m. A relatively high sample data size is required for a suitable estimation of stability/instability borders. We could expect this result since zones defining borders between stability and instability proportions are submitted to fast variations that can also be nonlinear. Otherwise, kriging based estimations have required much smaller time that the referential calculus. For example with $m = 300$ the corresponding kriging model has cost 3.10^5 CEA problems solved after 54432 seconds (almost 15hours). The cost difference is logically paid by less accurate predictions than the reference. A suitable compromise between the accuracy of kriging based estimated instability proportions and the time consuming has to be defined in practice.

5. Conclusion

This study focused on capacities of the global sensitivity analysis and kriging modeling to be combined together in order to generate simpler models that can be used instead of finite element models for a more efficient robust stability analysis of brake systems. The feasibility of the proposed strategy is tested on a simplified brake system. Firstly, the global sensitivity analysis is shown to be very important in the understanding of the stability behaviour against parameters variabilities. Secondly, kriging is demonstrated to be an interesting alternative to finite element models in brake systems's robust stability analysis. Indeed, with a fewer number of CEA solutions than the one required by the referential method, a kriging based model can be built and used to suitably estimated instability proportions corresponding to uncertain parameters. Finally, it can be noted that the proposed strategy can be considered for other classes of dynamic systems as rotors.

6. Acknowledgments

The authors thank the financial support provided by the French National Research Agency through the framework of its project ANR-12-JS09-0009.

References

[1] N.M. Kinkaid, O.M. O'Reilly, and P. Papadopoulos. Automotive disc brake squeal. *Journal of Sound and Vibration*, 267:105–166, 2003.

[2] J. J. Sinou, O. Dereure, G. M. Mazet, F. Thouverez, and L. Jézéquel. Friction-induced vibration for an aircraft brake system - part 2, nonlinear dynamics. *Interbational Journal of Mechanical Sciences*, 48(5), 2006.

[3] J.-J. Sinou, A. Loyer, O. Chiello, G. Mogenier, X. Lorang, F. Cocheteux, and S. Bellaj. A global strategy based on experiments and simulations for squeal prediction on industrial railway brakes. *Journal of Sound and Vibration*, 332(20):5068–5085, 2013.

[4] H. Ouyang, W. Nack, Y. Yuan, and F. Cocheteux. Numerical analysis of automotive disc brake squeal: a review. *International Journal of Vehicle Noise and Vibration*, 1(3):207–231, 2005.

[5] A. Loyer, J.-J. Sinou, O. Chiello, and X. Lorang. Study of nonlinear behaviors and modal reductions for friction destabilized systems. application to an elastic layer. *Journal of Sound and Vibration*, 331(5):1011–1041, 2012.

[6] A.R. AbuBakar and H. Ouyang. A prediction methodology of disk brake squeal using complex eigenvalue analysis. *International Journal of Vehicle Design*, 46(4):416–435, 2008.

[7] J.-J. Sinou. Transient non-linear dynamic analysis of automotive disc brake squeal - on the need to consider both stability and non-linear analysis. *Mechanics Research Communications*, 37(1):96–105, 2010.

[8] I.M. Sobol. Global sensitivity indices for nonlinear mathematical models and their monte carlo estimates. *Mathematics and Computers in Simulation*, 55(1-3):271–280, 2001.

[9] T. Bultin and J. Woodhouse. Sensitivity of friction-induced vibration in idealised systems. *Journal of Sound and Vibration*, 319(1-2):182–198, 2009.

[10] J. Sacks, W. J. Welch, T. J. Mitchell, and H. P. Wynn. Design and analysis of computer experiments. *Statistical Science*, 4(4):409–435, 1989.

[11] E. Sarrouy, O. Dessombz, and J. J. Sinou. Piecewise polynomial chaos expansion with an application to brake squeal of linear brake system. *Journal of Sound and Vibration*, 332(3):577–594, 2013.

[12] A. Saltelli, S. Tarantola, F. Campolongo, and M. Ratto. *Sensitivity Analysis in practice- a guide to assessing scientific models*. 2004.

[13] J. Sondergaard S. N. Lophaven, H. B. Nielsen. Dace a Matlab Kriging Toolbox. Technical report, Technical University of Denmark, 12 2002.

Numerical Studies on Time-Varying Stiffness of Disk-Drum Type Rotor with Bolt Loosening

Zhaoye QIN[1], Fulei CHU

State Key Laboratory of Tribology, Department of Mechanical Engineering, Tsinghua University. , Beijing. 100084. P.R. China

qinzy@mail.tsinghua.edu.cn

Abstract. Disk-drum type rotors are widely used in industry for their high stiffness and low weight properties. In disk-drum type rotors, the adjacent disks and drums are commonly connected by bolted joints. Those rotating joint interfaces are subjected to numerous combinations of loads during normal operation, where loosening of the connecting bolts might occur. The bolt loosening will change the local stiffness of the rotor, which in turn affect the rotor dynamics and even result in structural failures. In this paper, the local stiffness of a disk-drum rotor with bolt loosening is investigated numerically. A three-dimensional (3D) finite element (FE) model for the bolted disk-drum joint is established in ANSYS, where the bolt loosening is simulated by reducing the preloads of certain bolts, and removing those bolts as the limiting case. Simulations are performed on the FE model to evaluate the joint behaviour under static loads. Periodic variations of the joint deflections with respect to the rotation angle of the shaft are obtained, which implies the appearance of the time-varying local stiffness in the rotor system. The studies in this paper help accurate prediction of the rotor dynamics and early detection of bolt loosening.

1. Introduction

Disk-drum type rotors are commonly employed in heavy rotating machines, such as aero-engines, gas turbines, and so on. The configuration of a typical disk-drum type rotor is shown in Fig. 1, where the adjacent disks and drums are fastened together through bolts distributing in the circumference of joint interfaces. Preloads are applied to those bolts to prevent joint separation and provide sufficient strength to maintain the integrity of rotor systems. Despite the strict control of preloads, bolt loosening may occur due to the severe loading conditions in the course of normal operation. The bolt loosening will change the local stiffness of the rotor, which in turn affect the rotor dynamics and even result in structural failures. Investigating the joint behavior with bolt loosening may help accurately predicting the rotor dynamics, which assists in the early detection of the bolt loosening before damage occurs.

As one of the most common means to fasten substructures in industry applications, bolted joints have been extensively studied in decades, most of which deal with the joints in non-rotating structures [1-5]. When investigating the dynamics of rotor systems, the bolted joints assembling rotor components are usually assumed to be rigid connections. The few studies on bolted joints in rotor

[1] Corresponding author: Department of Mechanical engineering, Tsinghua University, Haidian District, Beijing, China, 100084. email: qinzy@mail.tsinghua.edu.cn

systems focused on evaluating the nonlinear influence of the joints on the rotor dynamics without taking into account the loosening phenomenon [6-10].

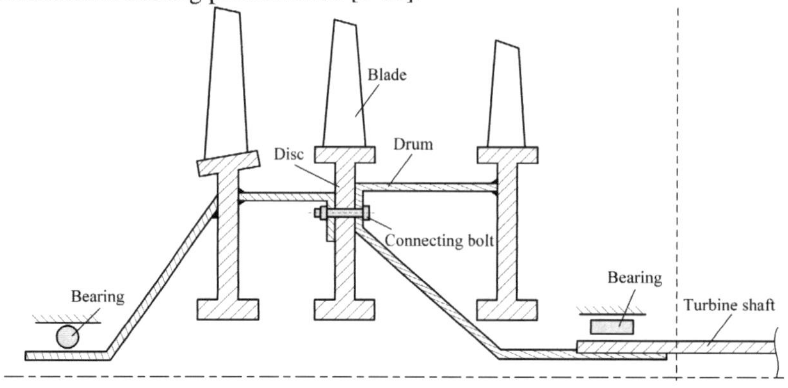

Figure 1 Sketch of fan rotor of a certain type of aero-engine.

Although no report has been found on bolt loosening in rotating structures, some researchers [11, 12] suggested that the dynamics of a rotor with a transverse crack could be simulated by the removing parts of the connecting bolts equally distributing in the circumference of the joint interface of rotating components. Treating the bolt removal as the limiting case of the bolt loosening, the rationality of the bolt-removal method implies that the bolt loosening shares similar signatures with the so called crack breathing subjected to static loads.

FE method is the most commonly used tool in structure design and analysis, especially for those structures involving joints, since it can deal with contact problem accurately and conveniently, which is one of the primary characteristics of the joint structures [4, 13-14]. In this paper, the time-varying stiffness of the disk-drum type rotor with bolt loosening is studied using FE method. Initially, a 3D FE model of the joint interface is constructed in ANSYS. The bolt loosening is taken into account by degrading the preload level of certain bolts or removing those bolts. Nonlinear static simulations are performed to evaluate the joint behavior with bolt loosening. It is revealed that the deflections of the joint interface and the corresponding joint stiffness are varying along with the rotation angle of the shaft periodically. The deflection amplitude is proportional to the preload decrement and the number of loosened bolts.

2. Finite Element Modelling of Disk-Drum Type Rotor with Bolt Loosening

2.1. Rotor description

The disk-drum type rotor studied here consists of three disks, two drums and two shafts supported by a rolling element bearing at each end, as shown in Fig. 1. In the rotor, the connections between the components are assumed to be rigid, except that the middle disk and two drums are connected by inner bolted flanges with 20 M4 bolts distributed uniformly in the circumference of the joint interfaces. The values of the dimensional parameters of the rotor are illustrated in Fig. 2. The materials of the disks, drums and shafts are identical, of which the density, elastic modulus and Poisson ratio are 7800 kg/m3, 210 GPa and 0.3, respectively.

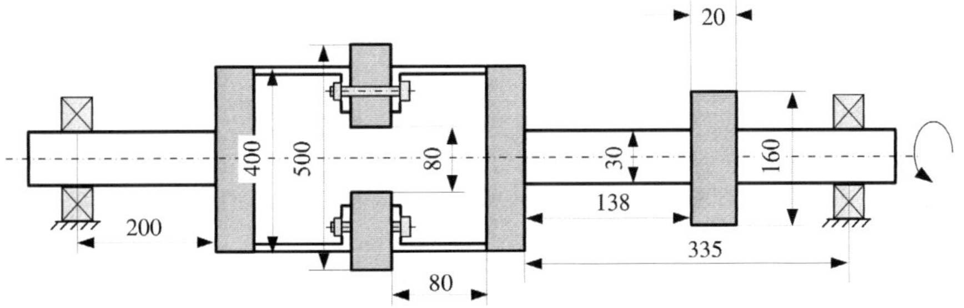

Figure 2 Sketch of the rotor specimen.

2.2. Finite element modelling

A 3D FE model for the local joint interface, involving the middle disk and one drum in each side fastened with bolts, is developed in ANSYS as shown in Fig. 3. The disk, drums and bolts are all modeled using 8-node solid elements SOLID185, and the frictional surface-to-surface contact is defined between joint components using contact elements CONTA173 and target segment elements TARGE170. The model involves 58995 nodes and 55025 elements.

The boundary conditions of the model are defined as follows: all degrees of freedom (DOFs) of the nodes at the left edge surface of the left-side drum are constrained; all DOFs of the nodes at the right edge surface of the right-side drum are coupled together to form a rigid region with an auxiliary node defined at the region center as the master node. Multiple load steps are defined, where preloads are imposed to the bolts using the virtual thermal deformation method in the first load step. In the subsequent load steps, bending moments are applied at the master node.

To simulate the bolt loosening phenomenon in the FE model, certain bolts bear the preload of lower levels; whereas, the preloads applied to other bolts are identical with the desired magnitude. The preload for those normal bolts is 0.5kN, and 2 preload levels of 0.25kN, and 0 are applied to the loosening bolts, respectively. As the limiting case, the joint interface with some bolts removed is also discussed. Note that the deflection at the right side of the FE model is caused by both the elastic deformation of the joint components and the bolt loosening. To eliminate the deflection caused by the elastic deformation, a FE model of the joint interface with all components rigidly connected is also constructed, of which the deflections is subtracted from those of the joint interface with bolt loosening.

As mentioned in Section 1, if the rotor is mainly subjected to static loads, such as the gravity, the joint stiffness at the interface with bolt loosening would vary periodically with the shaft revolution, just like the breathing mechanism of the cracked rotors. To evaluate the varying joint stiffness, relative deflections of the disk and drums at the joint interface are calculated for one full revolution with the increment of 9° with the directions and magnitudes of the external loads remaining unchanged. The initial position of the joint interface and the load directions are illustrated in Fig. 4.

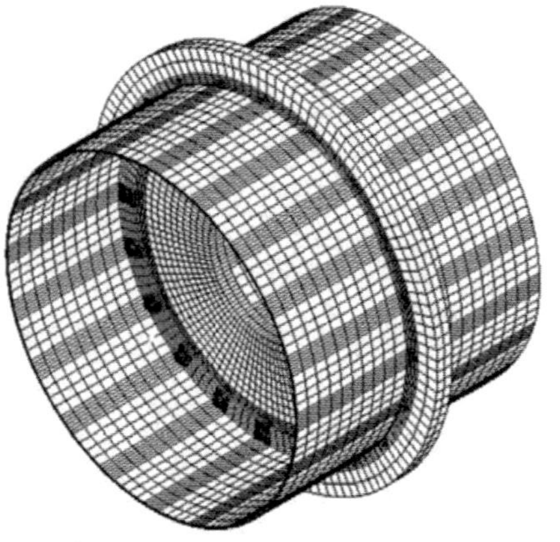

Figure 3 Finite element model for joint interface.

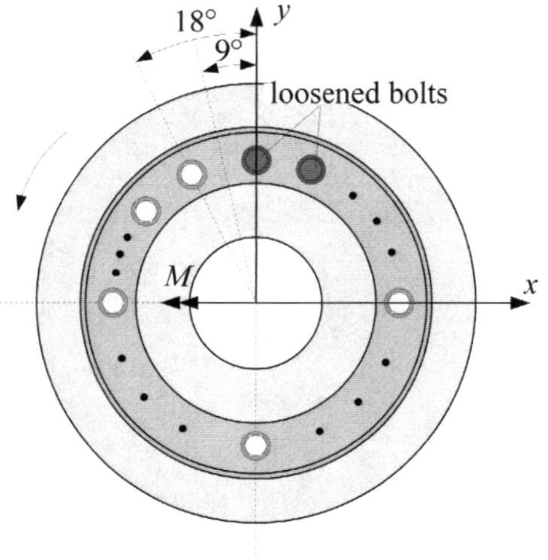

Figure 4 Schematic of joint interface with bolt loosening.

3. Simulations and Discussions

Subjected to the bending moment of 10kNm, the deflections of the joint interface with a single bolt loosening are investigated. The horizontal and vertical deflections versus the rotation angle Φ are shown in Fig. 5, where periodical variations of the deflections along with the rotation angle can be observed.

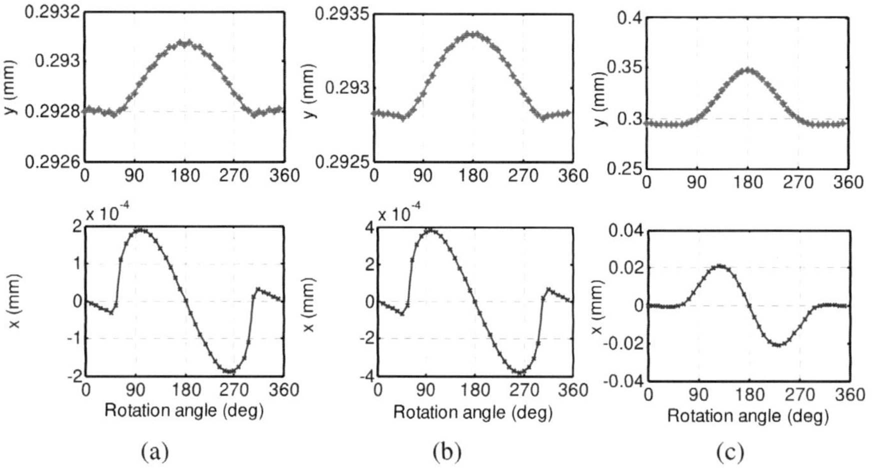

Figure 5 Deflections with a single bolt loosening (a) under the preload of 0.25kN, (b) with no preload and (c) with bolt removed.

Considering that the bending moment is constant as the shaft rotates, the upper portion of the joint structure is always in the compression state, and the lower one is subjected to tension with the application of the bending moment. At the initial stage, the region involving the loosened bolt is in compression. In this case, the bolt loosening has little influence on the joint stiffness, which is equal to the one without bolt loosening. As the shaft rotates, the bolt loosening region will move into the tensile stress region, where the disk and drums tend to open up and the joint stiffness in the vertical direction decreases. The joint stiffness will reach its minimum as the loosened bolt reaches the bottom of the interface. Further rotation enlarges the joint stiffness again. Thus, along with the shaft revolution, the vertical joint stiffness varies periodically, so does the vertical deflection, which quite resembles the breathing mechanism of the crack shafts. The horizontal deflection of the joint interface also changes periodically due to the structural asymmetry resulting from the bolt loosening. The horizontal deflection shifts from one side to the other at $\Phi = 180^\circ$, where the maximum vertical deflection takes place.

Comparing the curves of Fig. 5 (a)-(c), it can be seen that the aforementioned signatures of the deflections appear in all the cases of 50% preload degradation, no preload and bolt removal. However, it should be noted that although the deflections of the joint interface are enlarged with the decreasing of the loosened bolt preload, the deflections without removing the loosened bolt are extremely small. This means that unless severe loosening, that is, the loosened bolt provides no constraint regardless of the joint interface deformations, takes place, the single bolt loosening causes no obvious variations of the joint stiffness.

The deflections at the joint interface with loosening of multiple adjacent bolts are also calculated. The simulation results corresponding to 1, 2, and 3 loosened bolts are compared in Fig. 6. It can be seen that the deflection signatures remain unchanged for the multiple bolt loosening cases. As the number of loosened bolts increases, the deflection amplitudes are amplified. Correspondingly, the variation of the joint stiffness becomes more significant. Due to the definition of the initial position in the numerical simulations, a slight shift of the deflection peaks to the right side for more loosened bolts is observed in Fig. 6.

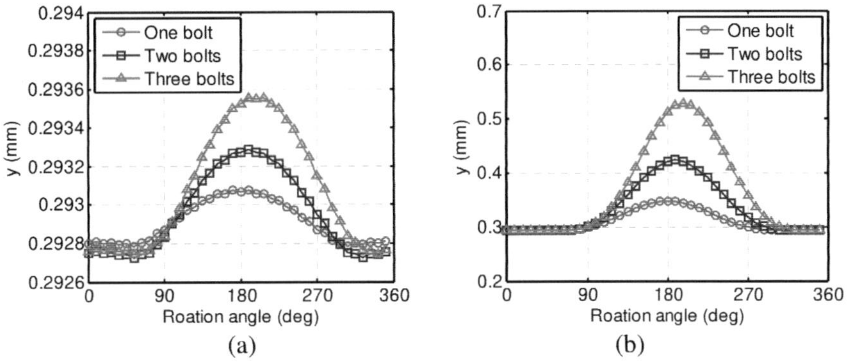

Figure 6 Vertical deflections with multiple bolt loosening (a) under the 50% reduction of preload, and (b) with bolt removed.

4. Conclusions

In this paper, the time-varying stiffness of the joint interface in a disk-drum type rotor with bolt loosening was investigated using FE method. A nonlinear 3D FE model for the joint interface of the rotor was developed in ANSYS. The bolt loosening was simulated by reducing the bolt preload levels or bolt removal. Nonlinear static simulations were performed to calculate the deflections at the joint interface with bolt loosening. An analysis of the simulation results led to the following conclusions.

(1) The deflections of the joint interface varied periodically along with the shaft rotation angle, which implied the time-varying stiffness of the rotor with bolt loosening as the shaft rotates.

(2) The amplitude of the stiffness varying was proportional to the severity of the loosening and the number of loosened bolts.

(3) Only severe loosening, that is, the loosened bolt(s) provided no constraint on the interface deflections, could induce significant varying stiffness.

In further studies, the time-varying stiffness should be incorporated into the rotor model to evaluate the influence of bolt loosening on the rotor dynamics.

Acknowledgements
This research is supported by Beijing Natural Science Foundation (Grant No. 3143024) and Natural Science Foundation of China (Grant No. 11472148).

Reference

[1] Gaul L and Lenz J 1997 Nonlinear dynamics of structures assembled by bolted joints *Acta Mech.* **125** 160-181

[2] Song Y, Hartwigsenb C J, McFarlanda D M, Vakakis A F and Bergmana L A 2004 Simulation of dynamics of beam structures with bolted joints using adjusted Iwan beam elements *J. Sound Vib.* **273** 249-276.

[3] Ouyang H, Oldfield M J and Mottershead J E 2006 Experimental and theoretical studies of a bolted joint excited by a torsional dynamic load *Int. J. Mech. Sci.* **48** 1447-1455

[4] Kim J, Yoon J C and Kang B S 2007 Finite element analysis and modeling of structure with bolted joints *Appl. Math. Model.* **31** 895 - 911

[5] Luan Y, Guan Z Q, Cheng G D and Liu S 2012 A simplified nonlinear dynamic model for the analysis of pipe structures with bolted flange joints *J. Sound Vib.* **331** 325-344

[6] Klompas N 1983 Effects of anomalous rotor joints on turbomachine dynamics *ASME J. Eng. Power* **105** 927-934

[7] Mat Isa A A, Penny J E T and Garvey S D 2000 Dynamics of bolted and laminated rotors. *Proc. of Int. Modal Anal. Conf. – IMAC (San Antonio Texas 2000)* **1** 867-872.

[8] Liu S G, Ma Y H, Zhang D Y and Hong J 2012 Studies on dynamic characteristics of the joint in the aero-engine rotor system *Mech. Syst. Signal Pr.* **29** 120-136.

[9] Qin Z Y, Han Q K and Chu F L 2014 Analytical model of bolted disk-drum joints and its application to dynamic analysis of jointed rotor *P. I. Mech. Eng. C: J. Mech.* **228** 646–663

[10] Yuan Q, Gao J and Li P 2014 Nonlinear dynamics of the rod-fastened Jeffcott rotor. *ASME J. Vibr. Acoust.* **136** 1-10

[11] Tamura A, Iwata Y and Sato H 1988 Unstable vibration of a rotor with a transverse crack *Vibrations in Rotating Machinery* London UK 647-653

[12] Keiner H, Gadala S and Tang W 2004 Validity of bolt-removal and gap-insertion techniques to experimentally simulate the vibration of a cracked rotor *Fatigue Fract. Engng. Mater. Struct.* **27** 449-458

[13] Reid J D and Hiser N R 2005 Detailed modeling of bolted joints with slippage *Finite Elem. Anal. Des.* **41** 547 - 562

[14] Croccolo D, Cuppini R and Vincenzi N 2009 Design improvement of clamped joints in front motorbike suspension based on FEM analysis *Finite Elem. Anal. Des.* **45** 406 - 414

Select aspects of FEM analysis for bonded joints of polymer composite materials

A Rudawska

Department of Engineering Production, Lublin University of Technology, Nadbystrzycka 36 Street, 20-618 Lublin, Poland

E-mail: a.rudawska@pollub.pl

Abstract. The paper presents selected aspects of modelling bonded joints of polymer composite materials by finite element method. The shear-loaded adhesive lap joints made of epoxy-graphite and epoxy-glass composite materials were investigated. The research objective was to determine correct modelling of adhesive layers using cohesive elements and of bonded joints for selected epoxy composite materials with different mechanical properties (e.g. Young's modulus) and geometrical dimensions, using, however, the same type of adhesive. The numerical analysis was performed based on experimental tests. A comparison is made between the distribution of reduced stress in the examined joint models according to the H-M-H hypothesis and that determined according to the maximum principal stress hypothesis. The finite elements analysis was performed in ABAQUS software and the traction-separation failure criterion was used for the damage onset and growth in the adhesive layer.

1. Introduction

Composite materials can be joining in a variety of ways and then used to produce numerous structural components [1,2]. One of such joining methods is adhesive bonding. The bonding technology offers numerous benefits, including the possibility of bonding various structural materials that have different properties and geometrical dimensions. When investigating adhesive joints, it is vital that both design and experimental tests be supplemented by finite element method based numerical analysis [3-5]. Such analysis is very useful at the design stage as it allows for simulating joint behaviour under various types of external load [6-8]. The application of numerical analysis to the investigation of adhesive joints also allows the determination of stress distributions in an adhesive layer [9] as well as indication of crucial structure elements that require special attention at the stage of their design or operation [10]. As a result, it is possible then to introduce some changes, e.g. to dimensions and shape of the given structure as well as the manner and location of applied load [11-14]. The application of FEM to design an adhesive joint allows both prompt verification of this adhesive joint and optimization of the above-mentioned parameters of such joint [15].

Some issues of FEM analysis of bonded composites joints are presented in some articles [16-17]. Pinto et al. [1] investigated the carbon-epoxy structures in repairing aspects used finite element stress analysis. Ashcroft et al. [18] conducted the analysis of the effect of environment on the fatigue of bonded composite joints. Finite element analysis is used to develop fatigue threshold prediction procedure. Madani et al. [19] analysed by finite element method the single- and double-sided composite repairs designed to reduce the concentration of the stresses at circular notches and cracks.

2. Experimental

2.1. Samples preparation
The object of the experiments concerned bonded single-lap joints of epoxy-graphite and epoxy-glass composite materials. A geometrical model of the examined joints is shown in figure 1.

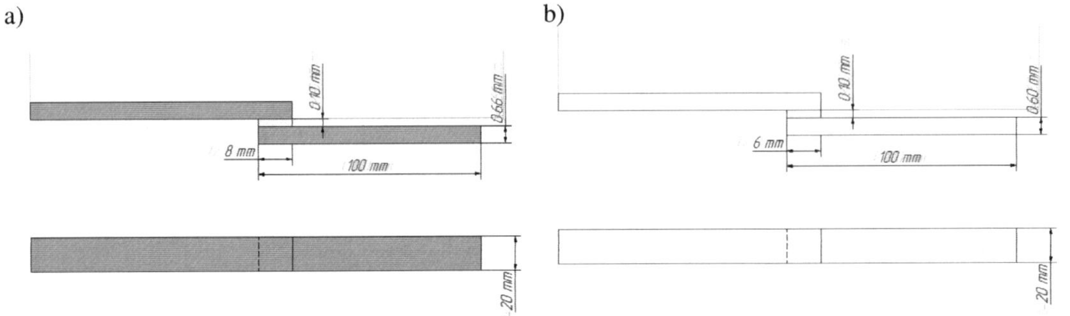

Figure 1. Geometrical model of joints: a) epoxy-graphite composite, b) epoxy-glass composite.

The samples were made of epoxy-graphite and epoxy-glass composite materials. The epoxy-graphite composite material was made of two layers of graphite fabric designated as GR-EP 199-45-003; the layer configuration applied was [0/90]. These were autoclave cured at 175±5°C for 60 min. The thickness of the composite obtained was 0.66±0.04 mm. The lap joint length was set equal to 8 mm (between 7.82 mm and 8.43 mm). The epoxy-glass composite material was made of two layers of glass fabric designated as glass laminate 3200-7781; the layer configuration applied was [0/90]. These were autoclave cured at 132°C for 60 min. The composite had a thickness of 0.60±0.03 mm. The thickness of the adhesive layer was 0.10±0.025. The lap joint length was set equal to 6 mm (between 5.63 mm do 6.51 mm).

2.2. Experimental conditions
Bonded joints were made under the following experimental conditions:

1) surface preparation: degreasing with acetone-based degreasing agent Loctite7063,
2) adhesive: a two-component epoxy Loctite Fast Epoxy 3430,
3) curing conditions: temperature 20±2°C, curing time 48 h, load 0.02 MPa.

The degreasing operation is described in the study [20], while the adhesive characteristic is given in the paper [21]. Strength tests were undertaken to determine both failure force and strength of shear-loaded single-lap joints. The tests were performed using a Zwick Roell 150 testing machine, in compliance with DIN EN 1465 [22].

3. Numerical analysis

3.1. Numerical model
Based on the geometrical model of adhesive joints used in the experiments (Fig. 1), a numerical model of adhesive joints was developed. To model the adherends, 8-noded solid elements were used, with three translational degrees of freedom each. The adhesive layer was modelled with COH3D8 [23] cohesive finite elements that enabled the description of adhesive layer failure in the loading process of the adhesive joints. The mechanical properties of the composites used in the analyses were those supplied by the manufacturer and are shown in table 1.

A schematic of the numerical model is illustrated in figure 2a. Taking into account the data included in the studies, finite element mesh refinement in the vicinity of adhesive layer edges was applied, as it is there where sudden stress gradients can be expected to occur [24].

Table 1. Mechanical properties of tested composites.

Composite	Young modulus, MPa	Poisons ratio
Epoxy-graphite composite	140 000	0.26
Epoxy-glass composite	39 000	0.25

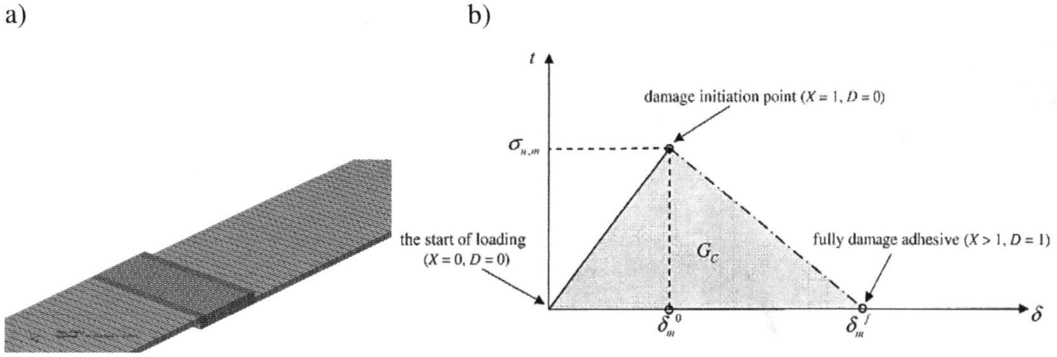

Figure 2. a) The part of the numerical model for examined adhesive joints, b) Traction-separation failure criterion used to describe the adhesive layer [26].

The boundary conditions of the numerical model were defined by means of blocking the translational degrees of freedom of the nodes located on the front surface of the upper sheet and leaving the possibility of displacement for the front surface of the lower steel only in the direction of applied load.

3.2. Failure criterion

The application of cohesive elements in the meshing process for adhesive layers requires that material with failure mechanism be defined. Figure 2b shows a material model of the adhesive used in the FEM analysis whose characteristics with failure initiation description included and evolution until complete loss of element rigidity based on the traction-separation law were defined. The cohesive zone model (figure 2b), combines a strength-based failure criterion to predict the damage initiation and fracture mechanics-based criterion to determine the damage propagation [25]. The same model the fatigue damage in adhesive bonded joints is presented by some research, e.g. Khoramishad et al. [11]. In addition to that, there exist numerous models of the traction-separation failure criterion that are described in the studies: (e.g. Pinto et al [1], De Moura et al. [25], Zadpoor et al. [26], Campilho et al. [27], Abdel Wahab et. Al [28]. Khoramishad et al. [11].

4. Test results

4.1. Experimental results

The conducted experimental tests enabled the determination of failure force considered basic for verifying the numerical analysis performed.

Table 2. Experimental results.

The type of joints	Failure force, N	Standard deviation, N	Strength, MPa
Epoxy-graphite composite	2875	287	17.97
Epoxy-glass composite	1838	331	15.33

Table 2 lists, among others, the results of failure force obtained in the experiments. The results presented are an average value of 10 measurements. The failure force value listed in table 2 was used to model load in a numerical model of adhesive joints for epoxy-graphite and epoxy-glass composite materials.

4.2. Numerical analysis results

4.2.1. Stress distribution. The presented numerical analysis results concern both adhesive joints and adhesive layers. As for the adhesive joints, the values and distributions of reduced stresses were determined according to the Huber-Mises-Hencky (H-M-H) strength hypothesis; while the results obtained for adhesive layers were determined using both the H-M-H strength hypothesis [11] and maximum principal stress hypothesis [18]. In many works [16,29] the maximum principal stresses as well as two principal stress components: S33 (peel stress) and S13 (shear stress) were analysed. Based on the numerical analysis performed, the value of failure load for adhesive joints was determined.

4.2.2. Stress distribution – epoxy/graphite composite. An example of the reduced stress distribution in bonded joints of the epoxy-graphite composite near the lap joint is shown in figure 3 a. In these numerical models there is one adhesive layer which was modelled cohesive finite elements. The nodes along lap joints lengths applied to the adhesive layer which it is joining to the samples.

a) b)

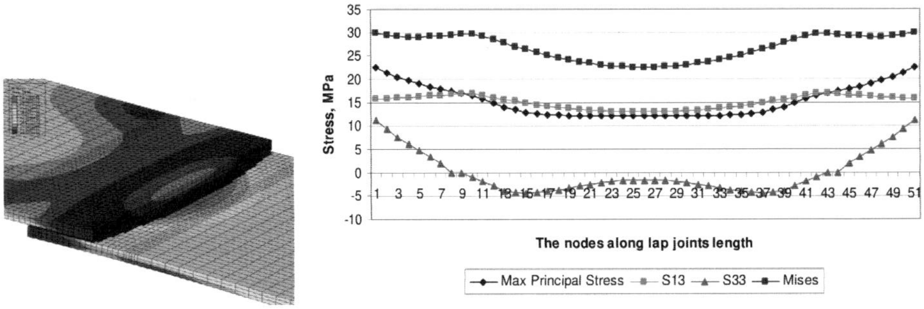

Figure 3. Bonded joints of epoxy-graphite composite: a) the reduced stress distribution, b) the stress distribution along lap joint length

The numerical results obtained for the adhesive layer are presented for one of the stages of adhesive layer failure, i.e. the initial failure process for which the highest value of maximum principal stress was obtained. Employing the maximum principal stress hypothesis, results were obtained for both maximum principal stress and principal stress components: S33 and S13. These components denoting peel stress and shear stress, respectively. The results are listed in Table 3 together with values of the stress components in general adhesive layer effort.

Table 3. Numerically obtained values of stresses in adhesively bonded joints of epoxy-graphite composite.

The type of stress	Maximum principal stress, MPa	S33, MPa	S13, MPa
Value	26.63	17.10	16.94

Examining epoxy graphite composite joints, it was observed that values of the peel stress (S33) and of the shear stress (S13) are similar. For the same stage of adhesive layer failure the value of reduced stress determined according to the Huber-Mises-Hencky hypothesis was 32.46 MPa.

It can be observed that a higher stress value is obtained if the Huber-Mises-Hencky reduced stress hypothesis (σ_z = 32.46 MPa) is employed. Therefore, it can be concluded that the application of the maximum principal stress hypothesis enables better evaluation of the load-bearing capacity of adhesive joints, defined as the maximum load value at which the joint fails. The distributions of the examined stresses along the lap joint length are shown in figure 3b.

According to the Goland and Reissner theory [9,30,31], the maximum principal stress in an adhesive lap joint occurs on the adhesive layer edges, and its value can be calculated using the relationship mentioned, among others, in the study. Shown in figure 3b, the distributions of maximum principal stress for the subsequent stages of adhesive layer loading are confirmed by the results of previous research [9,20,30] undertaken to investigate the character of stress distributions in adhesive joints. The presence of maximum principal stress along the adhesive layer edge and its very low values in the central region of the adhesive lap joint (figure 3b) prove the concentration of stress on the adhesive layer edges. It can also be noted that shear stresses prevail in the central region of the adhesive layer, while peel stresses do not occur there at all (which is proved by the negative values shown in figure 3b).

Based on the charts given in figure 3b, it can be claimed that both peel and shear stresses are vital at the initial stage of adhesive layer failure, these stresses being concentrated on the edges (peel stresses) and near the edges (shear stresses). In the central region of the lap joint, shear stresses play a dominant role over peel stresses.

4.2.3. Stress distribution - epoxy/glass composite. An example of the distributions of reduced stress in adhesively bonded joints of the epoxy-glass composite in the region of the lap joint is presented in figure 4a.

a) b)

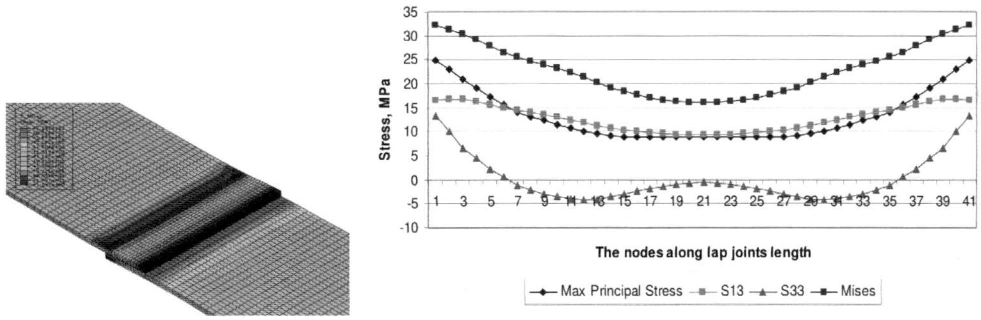

Figure 4. Bonded joints of epoxy-glass composite: a) the reduced stress distribution, b) the stress distribution along lap joint

Table 4 lists the values of maximum principal stress and principal stress components: S33 and S13 that describe peel stress and shear stress, respectively.

Table 4. Numerically obtained values of stresses in bonded joints of epoxy-glass composite.

The type of stress	Maximum principal stress, MPa	S33, MPa	S13, MPa
Value	29.74	20.19	16.93

In the case of the epoxy-glass composite joints, the value of peel stress (S33) is higher than that of shear stress (S13). This stress value can characterize the way in which adhesive joints of these composites fail. During the experimental failure tests, adhesive and cohesive adhesive layer failures were observed. This may result from similar values of the shear and peel stresses.

Comparing the values of reduced stress and maximum principal stress, it can be observed that a higher value stress was obtained using the Huber-Mises-Hencky hypothesis for reduced stress (σ_z = 35.50 MPa). In effect of the application of the hypothesis for maximum principal stress, it is possible to obtain values of components of this stress. As a result, a more thorough analysis of the stress state in adhesive layers can be performed. The distributions of stresses examined along the lap joint length are presented in figure 4b.

With the adhesively bonded joints of the epoxy-glass composite, the distributions of the examined stress types are of a similar character – the highest values occur on the adhesive layer edges, while much lower values are present in the central region of the adhesive layer. Examining the distributions of shear and peel stresses, it can clearly be observed that the values of these stresses on the adhesive layer edges are similar; however, the shear stresses (S13) prevail in the central region of the adhesive layer. This can mean that both the shear and peel stresses are active at the beginning of the failure process, with the shear stresses taking on the leading role later on in the course of the process.

The visualisation of peel stress and its distribution along lap joints length of epoxy-glass composite joints in vector form are presented in figure 8.

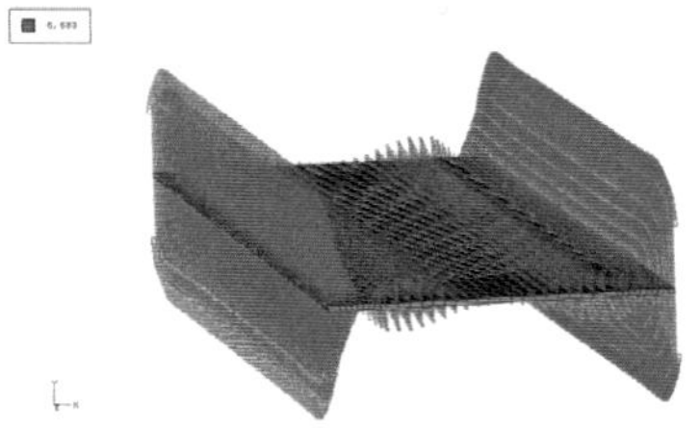

Figure 8. The visualisation of peel stress along lap joints length of epoxy-glass composite joints.

The distribution of the peel stress (figure 8) confirming the character of distribution which is presented in figure 7. In figure 8 it can be observed this character but in figure 7 it can be obtained the values of stress. In many cases the visualisation of stress distribution is very helpful to quick analysis of proper applied of boundary condition.

4.2.4. Comparison of stress distribution. A comparison of the stress distribution along the lap joint length for the epoxy-graphite and epoxy-glass composite is presented in figure 9 and figure 10.

It was observed (figure 9a) that higher maximum principal stress values were obtained for the adhesively bonded joint of the epoxy-glass composite characterized by a shorter lap joint length. These stresses have lower values in the central region along the lap joint length for the epoxy-glass composite material. The principal stress distribution in the adhesive joints of the epoxy-graphite composite is less varied than that for the epoxy-glass composite joints.

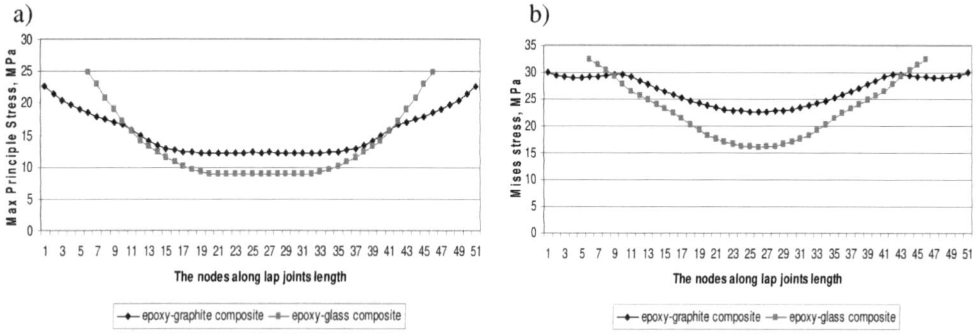

Figure 9. Distribution of: a) maximum principal, b) reduced stresses along lap joint length for epoxy-graphite and epoxy-glass composite bonded joints.

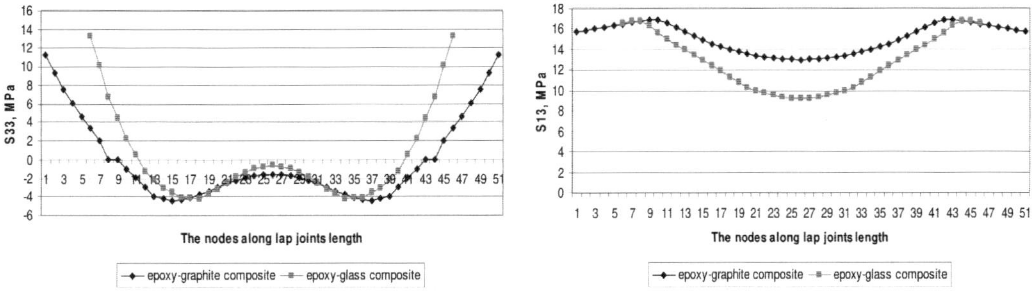

Figure 10. Distribution of: a) peel stresses b) shear stress along lap joint length for epoxy-graphite and epoxy-glass composite bonded joints.

Comparing the distributions of peel stresses (figure 10a) in the adhesive layer for both examined joints, it was observed that these distributions are quite similar in terms of both their values and characteristics. The distribution and values of the shear stress on the adhesive layer edges are the same. Differences occur in the central region of the lap joint length and they pertain to stress values. Lower values of the shear stress were obtained for the epoxy-glass composite joints with a value of Young's modulus lower by three times.

The differences in the distributions of reduced stresses can be observed on both the adhesive layer edges and in the centre of the lap. The differences exist in the values and character of the stress distribution. As for the epoxy-glass composite material, it was observed that maximum values of these stresses occur on the lap joint edge, accompanied by a gradual decrease in the value of this stress progressing from the adhesive layer edge. Furthermore, equal values of reduced stress on much a shorter length of the lap in its central region than is the case with epoxy-graphite composite joints. It can be observed that the decrease in reduced stresses occurs at some distance from the lap edge and the variation in stress is less dramatic than is the case with the epoxy-graphite composite.

4.3. Comparison of the experimental and numerical results

Table 5 lists the values of failure forces obtained in the experiments and numerical analysis for the following polymer composite materials: epoxy-glass and epoxy-graphite.

Table 5. Values of adhesive failure force obtained in the experiments and numerical analysis for bonded joints of polymer composites.

Composite	Failure force, N	
	Experimental test	Numerical analysis
Epoxy-graphite composite	2875	2555
Epoxy-glass composite	1838	1796

Comparing the results for the selected polymer composite materials listed in table 5, it can be observed that the experimental and numerical results compare quite well. The result agreement for the epoxy-glass composite adhesive joint was approx. 98%. In effect of a comparative analysis of the experimental and numerical results for the adhesively bonded joints of the epoxy-graphite composite, a certain difference in failure load can be observed: the failure load value obtained in the experiment is P = 2875 N, while the FEM-calculated load value is P = 2555 N. The agreement of the results in this case amounts to 89%. The difference in the failure load values obtained can result from the assumptions made in the modelling of adhesive joints and also the different properties of the adherends. The higher the tensile rigidity of the adherends, the higher the value of the force that causes adhesive joint failure.

5. Conclusions

Comparing the results obtained in the numerical analysis with those obtained in the experiments, it can be claimed that the numerical model of the tested joint agrees well with the real joint, given the simplified assumptions taken in modelling and confidence intervals for the experiments. The higher the tensile rigidity of the adherends, the higher the value of the force that causes adhesive joint failure, which accords with the Volkersen theory.

Using numerical analysis, it is possible to obtain the distributions of stresses based on various strength hypotheses, which at the same time allows a more detailed evaluation of adhesive joint effort. It is observed that the concentration of stresses occurs on adhesive layer edges. The distributions of shear and peel stresses on the adhesive layer edges are similar; however, it should be noted that the shear stresses (S13) prevail in the central region of the adhesive layer. This can mean that at the beginning of the failure process both the shear and peel stresses are active, with the former taking on the leading role later on in the course of the process.

Owing to different properties of the adherends (including Young's modulus), the obtained distributions of stresses are varied. The material with a higher modulus (epoxy-graphite composite) is exhibited higher values of stresses in the central region along the lap joint length. In contrast, the less rigid material (with a lower Young's modulus) is characterized by more dramatic variations in the stresses along the lap joint length. These results confirm the effect of mechanical properties on the distributions and values of stresses.

References

[1] Pinto A M G, Campilho R D S G, de Moura M F S F and Mendes I R 2010 *Int. J. Fatigue* **30** 329
[2] Gude M R, Prolongo S G, Gomez-del Rio T and Urena A 2011 *Int. J. Adhes. Adhes.* **31** 695
[3] He X 2011 *Int. J. Adhes. Adhes.* **31** 248
[4] Aydin M D, Özel A and Temiz S 2004 *J. Adhes. Sci.Technol.* **18** 1589
[5] Ávila A F and de O. Bueno P 2004 *Int. J. Adhes. Adhes.* **24** 407
[6] Pirondi A and Nicoletto G 2004 *Eng. Frac. Mech.* **71** 859
[7] Liljedahl C D M, Crocomble A D, Wahab M A and Ashcroft I A 2007 *Int. J. Adhes. Adhes.* **27** 505
[8] Khalili S M R, Khalili S, Pirouzhashemi M R, Shokuhfar A and Mittal R K 2008 *Int. J. Adhes. Adhes* **28** 411
[9] Castagnetti D and Dragoni E 2009 *Int. J. Adhes. Adhes.* **29** 125
[10] Diaz Diaz A, Hadj-Ahmed R, Foret G and Ehrlacher A 2009 *Int. J. Adhes. Adhes.* **29** 67
[11] Khoramishad H, Crocombe A D, Katnam K B and Ashcroft I A 2010 *Int. J. Fatigue* **32** 1146
[12] Magalhães A G, de Moura M F S F and Gonçalves J P M 2005 *Int. J. Adhes. Adhes.* **25** 313
[13] Gcoin A, Lestriez P, Assih J, Objois A and Delmas Y 2009 *Int. J. Adhes. Adhes.* **29** 572
[14] Apalak M K and Engin A 2004 *J. Adhes. Sci.Technol.* **18** 529
[15] You M, Yan Z-M, Zheng X-L, Yu H-Z and Li Z 2007 *Int. J. Adhes. Adhes.* **27** 696
[16] Taib A A, Boukhili R, Achiou S, Gordon S and Boukehili H 2006 *Int. J. Adhes. Adhes.* **26** 226
[17] Wahab M A, Ashcroft I A, Crocombe A D and Smith P A 2004 *Composites A* **35** 213
[18] Ashcroft I A,Wahab M A, Crocomble A D, Hughes D J and Shaw S J 2001 *Composites A* **32** 45
[19] Madani K, Touzain S, Feaugas X, Benguediab M and Ratwani M 2009 *Int. J. Adhes. Adhes.* **29** 225
[20] Rudawska A 2010 *Int. J. Adhes. Adhes.* **30** 574
[21] Rudawska A 2013 *Selected issues of constitutional processes of homogeneous and hybrid adhesion joints* (Lublin: Publisher University of Lublin)
[22] Standard DIN EN 1465. Adhesives. Determination of tensile lap-shear strength of bonded joints.
[23] Abaqus 6.9 – Documentation 2009
[24] de Morais A B, Pereira A B, Teixeira J P and Cavaleiro N C 2007 *Int. J. Adhes. Adhes.* **27** 679

[25] de Moura M F S F, Concalves J P M, Chousal J A G and Campilho R D S G 2008 *Int. J. Adhes. Adhes.* **28** 419
[26] Zadpoor A A, Sinke J and Benedictus R 2009 *Int. J. Adhes. Adhes.* **29** 558
[27] Campilho R D S G, de Moura M F S F, Ramantani D A, Morais J J L and Domingues J J M S 2009 *Int. J. Adhes. Adhes.* **29** 678
[28] Wahab M A, Hilmy I, Ashcroft I A and Crocombe A D 2010 *J. Adhes. Sci. Technol.* **24** 325
[29] Wahab M A, Crocomble A D, Beevers A and Ebtehaj K 2002 *Int. J. Adhes. Adhes.* **22** 61
[30] Adams R D, Comyn J and Wake W C 1997 *Structural Adhesive Joints in Engineering* (London: Chapman and Hall)
[31] Godzimirski J 2002 The strength of constructional bonding joints (Warsaw: WNT)

The detection of lubricating oil viscosity changes in gearbox transmission systems driven by sensorless variable speed drives using electrical supply parameters

S. Abusaad[1], K. Brethee[1], M. Assaeh[1], R. Zhang[2], F. Gu[1] and A.D. Ball[1]
[1]Centre for Efficiency and Performance Engineering, University of Huddersfield, Huddersfield, U.K.
[2]School of Mechanical Engineering, Taiyuan University of Technology, Shanxi, China

E-mail: f.gu@hud.ac.uk

Abstract. Lubrication oil plays a decisive role to maintain a reliable and efficient operation of gear transmissions. Many offline methods have been developed to monitor the quality of lubricating oils. This work focus on developing a novel online method to diagnose oil degradation based on the measurements from power supply system to the gearbox. Experimental studies based on an 10kW industrial gearbox fed by a sensorless variable speed drive (VSD) shows that measurable changes in both static power and dynamic behaviour are different with lube oils tested. Therefore, it is feasible to use the static power feature to indicate viscosity changes at low and moderate operating speeds. In the meantime, the dynamic feature can separate viscosity changes for all different tested cases.

1. Introduction

Gearbox lubricants are critical in maintaining efficient and effective operations [1]. Lubricating oils must reduce wear and friction at the contact surfaces and separate them to ensure appropriate operation and avoid failures. Lube oil properties also effects gear dynamics, such as vibration, power characteristics and heating [2]. Therefore, continuously monitoring changes in lubricant properties is crucial for avoiding defects within machines [3] and prevent unexpected breakdowns. Lubricants in gearboxes are degraded mainly due to oxidation, particle and water contaminations [1]. Both oxidation [4] and water [5, 6] generally increase the lube viscosity.

Studies show that oil properties considerably affect gearbox behaviors and performance due to churning and splashing [7]. Oil churning causes significant power losses in the gearbox. churning losses increase with increasing viscosity at low speeds, yet these losses decrease at high speeds [8, 9]. Likewise , as found in [10] viscosity increases temperature due to internal lube friction. However, high temperature motivates chemical activities and good tribological layer formation. Conversely this will decrease the viscosity and hence thinner oil film is formed.

Different offline techniques have been developed for monitoring oil degradation [11-16]. However these techniques are not always applicable, require addition measuring equipment and expensive to implement. The need for an easy to apply, reliable and cost effective condition monitoring technique is becoming more imperative important in order to provide timely assessment of oil quality.

Vibration and current signature analysis are the most widely used schemes for condition monitoring and diagnosis of different equipment and faults detection. Though, the research on developing lubricating degradation detection based on these schemes is very little. Rui and Linilson [17] studied the relationship between the viscosity and vibration signature. They found that vibration features in high frequency ranges can be correlated with the oil viscosity. Nevertheless, the potential of using signatures from electrical measurements for lube oil degradation detection based on viscosity is not yet well discovered [3].

Therefore, this study investigates the effect of varying oil viscosity in a gearbox transmission system on power supply parameters from a variable speed drive (VSD). After an examination of the effect on gear transmission process, an experimental study was performed based on a 10 kW gearbox transmission system with four different oils in turn. The diagnostic capability of both a static feature and dynamic feature which are developed based on the current and voltage measurements are then examined for different operating conditions.

2. Oil Viscosity Related Power Losses in Gearboxes

Power loss sources in a gear transmission relate to various gear parameters [4, 15] such as teeth geometry, specific sliding, and lube properties. Yet the viscosity is a key element in improving the gearbox efficiency and performance. Research in [4, 15] show that the use of a lubricant with a suitable viscosity could save up to 20% of power losses. Power losses P_T in gears can be approximated as follows [15]:

$$P_T = P_{fr} + P_{spl} + P_{M1} + P_{M0} + P_{sl} \tag{1}$$

where P_{fr} denotes gear friction power loss and P_{spl} is gear churning power loss, which are more significant compared with other three losses: P_{M1} denoting load dependent power loss in rolling bearings, P_{M0} denoting power loss in rolling bearings, and P_{sl} denoting seal power loss. The gear frictional power losses are load dependent and also related to gears geometry, number of teeth of both gear and pinion (z_1, $z_2$respectively), and friction coefficient(μ_m), and can be estimated based on [18-20]:

$$P_{fr} = \pi\left(\frac{1}{z_1}+\frac{1}{z_2}\right)\left(1-\left(\frac{g_f+g_a}{p_b}\right)+\left(\frac{g_f}{p_b}\right)^2+\left(\frac{g_a}{p_b}\right)^2\right)P_{in}\mu_m \tag{2}$$

It shows that the average friction coefficient has a significant impact on P_{fr}. According to studies in [18, 19, 21], the friction coefficient can be calculated from the following equation:

$$\mu_m = 0.048\left(\frac{F_{bt}/b}{v\,.\,\rho_c}\right)^{0.2} \eta_{oil}^{-0.05} R_a{}^{-0.25} X_L \tag{3}$$

Equation (3) shows that μ_m is inversely proportional to the lube viscosity η_{oil} and velocity v. On the other hand, the churning loss P_{spl} in a gear are load independent losses; and mainly depend on the velocity, geometry of the moving parts immersed in the oil and oil density ρ and viscosity, which can be estimated from [18, 19, 22]:

$$P_{spl} = \frac{\pi}{30}.n.\left[\frac{1}{2}.\rho\left(\frac{\pi.n}{30}\right)^2 A_i\left(\frac{d}{2}\right)^3\right]\left[\left(\frac{2.h}{d}\right)^{0.45}\left(\frac{V_{oil}}{d}\right)^{0.1} F_R^{-0.6} Re^{-0.21}\right] \tag{4}$$

Equation (4) shows the complex influence of gear geometry and velocity on the power losses due to oil churning and splashing. The Reynolds number R_e depends on the oil density and viscosity and given as:

$$R_e = \frac{\rho UD}{\eta_{oil}} \tag{5}$$

While losses from churning in bearings depend on bearing type and size, bearing arrangement, and lubricant viscosity supply [8], and can be calculated from [21]:

$$P_{M0} = \frac{\pi}{30}.n.G_{Br}.\eta_{oil}{}^{2/3}.10^{-3} \tag{6}$$

where: P_{in} donates input power, g_f is the length of approach, g_a represents the length of recess, F_R Froude number, R_a mean roughness, p_b base pitch, F_{bt} tooth normal force in the transverse section, X_L an adjustable parameter relating to each lubricant, b face width, $v_{\Sigma c}$ sum velocity at pitch point, ρ_c

equivalent curvature radius at pitch point, ρ specific weight, V_{oil} is oil volume, G_{Br} a parameter depends on the bearings geometry , n input speed (rpm), D a characteristic linear dimension, U the mean velocity of the flow and η_{oil} is the dynamic viscosity.

Equations (2) to (6) show that oil viscosity has a significant effect on gearbox power losses. The frictional power loss is inversely proportional to viscosity values and operating speeds, as shown by Equations (2) and (3). Conversely, the power losses from churning increase as the speed and viscosity increase, by Equations (4) to (6). However, the churning induced loss may be more dominant as its fractional power value is higher. Nevertheless, both of these two types of losses will also alter the dynamics of gear transmission. Especially, the churning effect can reduce the effect of damping and increase the moment of inertia of the rotational system. Consequently, it leads to more oscillatory motion of the system.

In theory, these power losses can be reflected by the VSD behavior. Particularly, when the mechanical system has any disturbances such as those caused by viscosity changes, the VSD adjusts the electrical supply parameters in order to maintain the operation at the desired speed. Previous studies [23, 24] show that both dynamic and static effects occurring on downstream machines can be observed in the electrical signals. Specifically, the dynamic effect can be represented by the sideband components at frequencies related to supply and shaft frequencies. The amplitude changes at the supply frequency can be based to examine the static effects. While in systems driven by sensorless VSDs the supply parameters are adjusted by the control system. This may give more chances to observe such effects, rather than just using the feature frequency components based on the high slippage changes under direct voltage/Hz control methods.

3. Test facility and Procedures

The test facility employed for this study consists of a mechanical system and an electrical control system as shown in the schematic of Figure 1. The mechanical system includes a 15kW AC induction motor as the prime driver, two back-to-back two stage helical gearboxes for coupling the AC motor with a load generator using flexible spider rubber couplings. The first gearbox operates as a speed reducer while the other is a speed increaser so that the system maintains sufficient speed for the load generator to produce sufficient load to the AC motor through the two gearboxes. The control system consists of a programmable logic controller (PLC) for setting up different speed-load profiles specified by operator, an AC VSD, Parker 650V, that can be set either to a sensorless flux vector control mode or V/Hz mode for adjusting the speed of the system, a DC VSD, Parker 550c, ensuring a controlled load to the AC motor by regulating the torque of the load generator.

A high speed data acquisition system collects the data from sensors measuring vibration, three phase currents and voltages, gearbox temperatures sends it into a PC for post processing and analysis in the Matlab environment. The dynamic data is used for both evaluating the performance of the conventional analysis methods for detecting and diagnose different oils viscosities, and also for benchmarking the developed scheme from the static data.

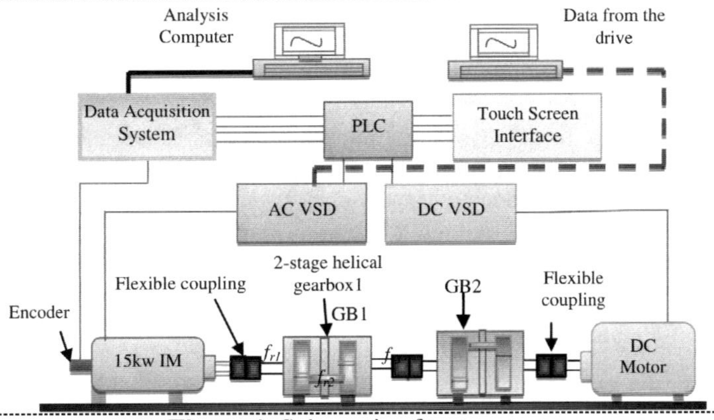

Figure 1. Schematic of test system.

Tests were carried out on the gearbox 1, denoted as GB1, which is a common industrial gearbox with a transmission ratio of 3.6 and a power rate of 10kW at 1460rpm. Four different oils having different viscosities, EP 100, EP 320, EP650 and EP 1000 respectively, have been tested. The gearbox manufacturer recommends the EP 320. The EP 650 have been made in the laboratory by mixing 61% of EP 1000 with 39% of EP 100, while other types have been provided from a supplier that recommended by the gearbox manufacturer, with specifications listed in table 1.

Table 1. Specifications of the oils used for tests.

	Specific Gravity (at 15°C)	Kinematic Viscosity (at 100°C, c.St)	Kinematic Viscosity (at 40°C, c.St)	Viscosity Index	Pour Point (°C)	Flash Point (°C)
EP 100	0.885	10.95	100	93	-9	200
EP 320	0.901	23.5	320	92	-9	200
EP 1000	0.927	71.0	1000	140	-6	200

This allows the variation of different measurements underlying to examine wide range of viscosities for defining the corresponding detection methods. The alignment of the test rig was firstly minimized by tuning installation base carefully for the lowest vibration. Oil was added and removed from GB1 without affecting the alignment condition using the drainage holes on the GB1. The rig is operated under three speed settings: 50%, 75% and 100% of the full motor speed, under four incremental load settings: 0%, 30%, 70%, and 100% of the gearbox rated load for each speed cycle, attempting for examining the detection performance under variable speed and load operations, which are common scenarios in real applications. Each load setting was for a period of two minutes and changed to the next step automatically by the PLC controller incorporated in the rig control system. In total, each load cycle lasts 8 minutes. In addition, the VSD was set under sensorless control mode for evaluating the detection capability under this particular mode.

To ensure the data quality for reliable comparison each speed/load cycles ran consecutively five times for each of oil type. During these repeating operations, the lube temperature in GB1 was observed on-line and reached to around 46°C-47°C from the room temperature when the system operating parameters also became stabilized. By using an automated acquisition procedure based on time advancement, 40 seconds of dynamic data were collected at every load setting. In the meantime, the static data from the VSD were also logged for the entire speed/load cycle. Moreover, oil samples were also taken for measuring their viscosity values under different temperatures.

4. Results and Discussion

Previous work [3] shows that the average spectral amplitudes at the supply frequency and motor speed associated sidebands can represent the influence of viscosity changes on the features of static and dynamic electrical power respectively. Yet, it was difficult to extract consistent changes based on these characteristic frequencies from voltage and current spectra separately. This is due to the fact that the VSD acts on both current and voltage simultaneously to maintain the system speed, and also because of the high noise associated with the drive actions. Therefore, the power signal from the combination of both the current and voltage was based to extract the spectral amplitudes at these characteristic frequencies.

4.1. Viscosity measurements results

The viscosity of each type of oils was measured before test. The measurement was done based on rotary viscometer test method. The results are given as an absolute dynamic viscosity in centipoise (cP), equivalent to Pa•s in SI units. As shown in figure 2 (a), differences in viscosity values are clear between tested oils and become smaller as temperature increases.

4.2. Influence of temperature

To examine the influence of temperature on the system, signals from the lubricant temperature sensors in GB1 and GB2, and speed are processed to obtain their static feature values. Figure 2 (b-d) shows these key measurements against testing run numbers under different loads. It can be seen in figure 2 (b and (c) that temperature in GB1 and GB2 increases gradually and reaches stable status by the 3rd test when the system stabilized. Noticeably, differences in temperature values between different lubricants in GB1 represent viscosity values of lubricants. In addition, the similarity of temperature trends in GB2 shows that tests were conducted with good consistency between different tests, which shows critically that the churning loss is much more significant than the frictional loss.

The temperature stabilization affects not only the losses inside the gearbox but also the circuitry performance of the electric and electronic components in the VSD based system. When it is under the lower temperature during the transient operation, the effect is stronger as the load and speed exhibit wider variation. As shown in figure 2 (d) during the low temperature operation for the 1st test run the VSD has poorer performance in maintaining the speed of the system at the set-point, which leads to higher speed under high load. On the other hand, when the system reaches its stable conditions the VSD is then able to control the speed with higher accuracy under different load settings, as shown by the speed results for the testing runs 3 to 5. However, the test 5 shows a temperature drop for the EP 1000 oil, indicating the instability either due to lubricant or ambient temperature. Based on these observations, it can be concluded that measurements from the 3rd and 4th test runs have less transient effects and more stable for examining the effect of lube viscosity more accurately.

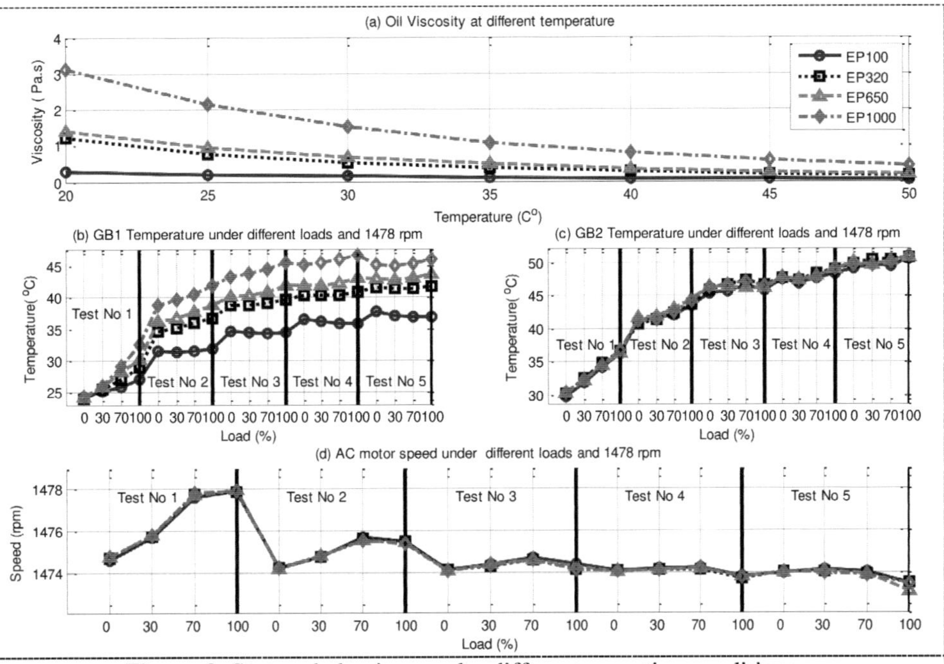

Figure 2. System behaviour under different operating conditions

4.3. *Effect of oil viscosity on the power spectra*

In addition to the power loss oil churning also affects the dynamics of the rotational system. When gear pairs rotate, a large quantity of oil circulates with the gear motion. The amount of oil formed increases with the speed. It means that this effect will decrease the oil level in the gearbox reservoir. Consequently, it lowers the damping effect of the rotating motion. In addition, the effective moment of inertia can also vary as the shape of the rotating oil is not perfectly uniform due to the inherent eccentricity and unbalanced mass. This generates more torsional oscillations at the shaft frequency which then modulate the supply component of power system. The power spectra have been investigated in order to investigate the effect of this phenomenon. Figure 3 displays the power spectra for the EP 100 and EP 1000 representing the highest and the lowest viscosities under different operational conditions. It shows the feature frequencies related to the tested gearbox. Particularly, the sidebands at the shaft frequency exhibit noticeable changes with viscosity, indicating the potential of using them for diagnosis. It also shows that the speed has a significant effect, where the higher speed the more mass circulates with gears and hence the more oscillations.

4.4. *Viscosity change detection*

To analyze the performance of power signals, both static and dynamic parameters from the power supply data have been investigated. The analysis is performed by averaging the results from the 3rd and 4th testing runs and then differences to values of the EP100 test data are made in percentages to obtain more reliable analysis. Figure 4 (a - c), depicts the changes in the static data of power in (%). It shows that at low and moderate speeds i.e. at 50 and 75% shown in figure 4(a) and (b) respectively, changes in the static power represent the differences in viscosity, displaying good performance for oil viscosity changes detection and diagnosis.

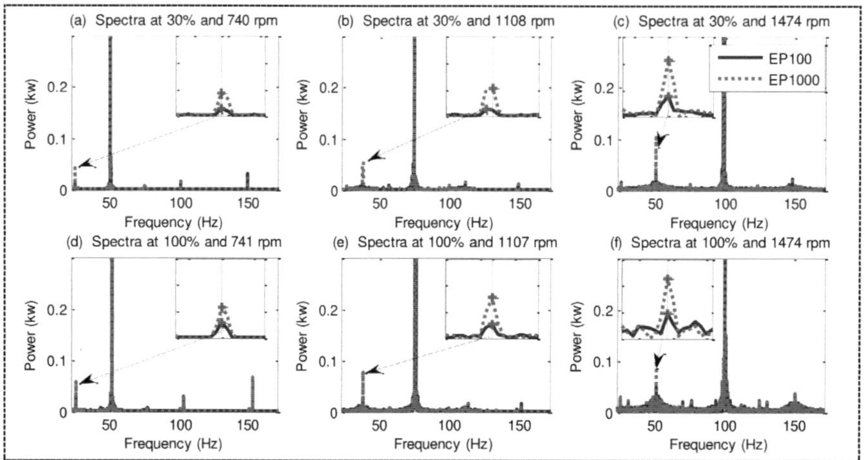

Figure 3. Power spectra under different operating conditions.

However, when the speed is high, i.e. at 100% in figure 4 (c), it shows poor performance in make difference between EP650 and EP1000 oils. This can be explained due to higher disturbances from oil churning and splashing. The drive adjusts the electrical supply parameters to maintain the system stability, and hence more noise generated preventing from separating signals at high speed. On the other hand, the dynamic feature of the sideband amplitudes in figure 4 (d-f) shows more correct performance in indicating different oils under different operating conditions, making it more suitable for monitoring the change of oil quality.

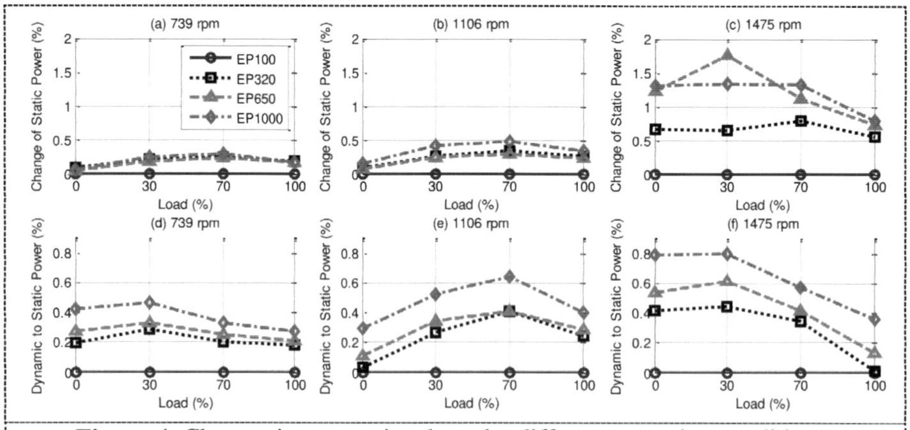

Figure 4. Changes in power signals under different operating conditions.

5. Conclusion

In this work, a new online and cost effective method is developed to diagnose lube oil degradation based on measurements of electrical parameters. The change of oil viscosity due to oil degradation will cause corresponding changes in both static power consumption and dynamic behavior of a gearbox transmission system. These changes have been verified experimentally with a 10 kW gearbox fed by a sensorless VSD. Specifically, it has found that the increase in oil viscosity lead to a measurable increase of power consumption due to the effect of viscous friction and churning.

Simultaneously, the effect also changes the dynamics of the gear transmission system as it reduces damping effect and increases the unbalance of inertia moments. Therefore, it leads to higher oscillation of rotation system, which eventually reflected by increasing the sidebands around supply frequency. Based on these two changes a static power feature and a dynamic power feature can be developed to make differences between different oils under different operating conditions. Results show that the static power feature can show viscosity changes correctly at low and moderate speeds. However the dynamic feature gives the expected diagnostic performance as it can separate viscosity changes for all different cases.

References

[1] Zhu J, He D and Bechhoefer E 2013 *Journal of Chemical Science and Technology* **2** 100-115
[2] Perera A 1986 Effect of lubricating oil characteristics on gear vibrations. *A Thesis submitted to the University of Newcastle Upon Tyne.*
[3] Abusaad S, Benghozzi A, Brethee K, Gu F and Ball A 2014 *3rd International Workshop and Congress on eMaintenance, 17th - 18th June 2014*, Lulea, Sweden..
[4] Ahmed N S and Nassar A Lubricating Oil Additives *Egyptian Petroleum Research Institute* Available from:http://www.intechopen.com/books/tribology-lubricants-and-lubrication/lubricating-oil-additives.
[5] Johnsen E E and Rønningsen H P 2003 *Journal of Petroleum Science and Engineering* **38** 23-36
[6] Jakoby B, and Vellekoop M J 2004 *Sensors and Actuators A: Physical* **110** 28-32
[7] Mehta N S, Parekh N J, and Dayatar R K 2013 *International Journal of Engineering and Advanced Technology (IJEAT)* **2** 120-123
[8] Michaelis K, Höhn B R and Hinterstoißer M 2011 *Industrial lubrication and tribology*,. **63** 46-55
[9] Michaelis K and Winter H 1994 *48th Annual Meeting, Tribology Transactions,* Alberta **37** 161-167
[10] Höhn R and Michaelis K 2004 *Tribology International* **37** 103-109
[11] Musial W, Butterfield S and McNiff B 2007 Improving wind turbine gearbox reliability *in European Wind Energy Conference, Milan, Italy*
[12] Sheng S 2011 *The American Wind Energy Association (AWEA) Project Performance and Reliability Workshop, January 12-13, 2011San Diego, California*
[13] Zhu J, Yoon J, He D, Qu Y and Bechhoefer E 2013 *International Journal of Prognostics and Health Management*, **20** 2153
[14] Greenwood M S and Bamberger J A 2002 *Ultrasonics* **39** 623-630
[15] Agoston A, Ötsch C and Jakoby B 2005 *Sensors and Actuators A: Physical* **121** 327-332
[16] Sheng S and Veers P S 2011 Wind turbine drivetrain condition monitoring-an overview *National Renewable Energy Laboratory*.
[17] Almeida R and Padovese L R 2003 Characterization of oil viscosity alterations in a gearbox through vibration signal analysis *Anais do COBEM*
[18] Höhn R, Michaeliset K and Vollmer T 1996 *American Gear Manufacturers Association*
[19] Magalhães L, Martins R, Locateli C and Seabra J 2010*Tribology International* **43** 1861-1871
[20] Olver A 2002 *Journal of Engineering Tribology* **216** 255-267.
[21] Martins R, Seabra J, Brito A, Ch S, Luther R and Igartua A 2006 *Tribology International* **39** 512-521
[22] Gonçalves D, Fernandes C, Martins R and Seabra J *Lubrication Science*, 2013. **25** 297-311
[23] Tian X, Guojin F, Chen Z, Albraik A, Gu F and Ball A 2014 *Comadem 2014* Brisbane, Australia
[24] Abusaad S, Benghozzi A, Gu F and Ball A 2013 *Engineering Materials* **569** 465-472

A Feature Extraction Method for Fault Classification of Rolling Bearing based on PCA

Fengtao Wang[1], Jian Sun[1], Dawen Yan[2], Shenghua Zhang[1], Liming Cui[1] and **Yong Xu[3]**.

[1]Institute of Vibration Engineering, Dalian University of Technology, Dalian, China
[2]School of Mathematical Sciences, Dalian University of Technology, Dalian, China
[3]West Pacific Petrochemical Company Ltd, Dalian, China

E-mail: wangft@dlut.edu.cn
E-mail: sunjian0804220@163.com

Abstract. This paper discusses the fault feature selection using principal component analysis (PCA) for bearing faults classification. Multiple features selected from the time-frequency domain parameters of vibration signals are analyzed. First, calculate the time domain statistical features, such as root mean square and kurtosis; meanwhile, by Fourier transformation and Hilbert transformation, the frequency statistical features are extracted from the frequency spectrum. Then the PCA is used to reduce the dimension of feature vectors drawn from raw vibration signals, which can improve real time performance and accuracy of the fault diagnosis. Finally, a fuzzy C-means (FCM) model is established to implement the diagnosis of rolling bearing faults. Practical rolling bearing experiment data is used to verify the effectiveness of the proposed method.

1. Introduction

Rolling bearing is widely used in modern rotating machine, but it is also one of the components which are easily damaged. Statistics show that 30% rotating machinery faults are caused by rolling bearings. Therefore, successful fault diagnosis of rolling bearings is of profound significance, which can reduce the incidents and economic loss. Feature extraction of rolling bearing vibration signals is the key of fault diagnosis [1, 2]. When rolling bearing faults occur, the time-frequency statistical features of vibration signal will change. Attempts at analyzing large numbers of original features are motivated by how to effectively extract the potential features responsible for fault characterization.

PCA has been proved to be a useful tool for feature reduction from the original features on fault diagnosis [3].When PCA-based feature extraction is used for machine bearing fault classification, it is believed that reduced dimensional data which is arranged along the principal eigenvectors could represent features effectively. For example, a PCA-based feature selection scheme was presented to choose the most representative features from a multi-domain feature set for bearing defect

[1] Corresponding author. Tel.: +86 411 84709291; Fax: +86 411 84708429-8007
E-mail address: wangft@dlut.edu.cn (Fengtao Wang).

classification [4]. Studies in [5] used the low-dimensional principal component representations from the statistical features of the measured signals to characterize and monitor gearbox conditions. In another study, by adopting the first seven principal components as inputs, a Euclidian distance discriminating approach is utilized to distinguish the bearing fault data [6]. While the frequency statistical features based on Hilbert transformation is not considered in the above studies, this paper summarizes the multiple statistical time-frequency features to construct the high dimensional feature set, then the low-dimensional feature set are obtained by PCA. Experimental data sets which include different fault types and different fault sizes are considered in this study.

The rest of the paper is organized as follows. The theoretical background is briefly introduced in Section 2. The proposed method for fault classification of rolling bearing is described in Section 3.An application case is presented in Section 4. Finally, conclusions are drawn in Section 5.

2. A brief description of PCA

The principal component analysis technique is a statistical analysis approach to mapping multiple characteristic parameters to a few comprehensive features. These PCA-based comprehensive features are not related to each other and can represent original fault features effectively

For a given feature vector set $\boldsymbol{X} = \{\boldsymbol{x}_1,\cdots,\boldsymbol{x}_m\}, \boldsymbol{x}_i \in R^n$ which consists of m feature vectors, each with n-dimensional, the algorithm to extract sensitive features from the defect conditions is expressed as follows:

1) Calculate the average value

$$\boldsymbol{\mu}=\frac{1}{m}\sum_{i=1}^{m}\boldsymbol{x}_i \tag{1}$$

2) Compute the covariance matrix C of eigenvectors

$$C=\frac{1}{m}\sum_{i=1}^{m}(\boldsymbol{x}_i-\boldsymbol{\mu})(\boldsymbol{x}_i-\boldsymbol{\mu})^T \tag{2}$$

3) Compute the eigenvalues λ_i and eigenvectors $\boldsymbol{v}_i (i=1,2,...,n)$ of C

$$C\boldsymbol{v}_i=\lambda_i\boldsymbol{v}_i \tag{3}$$

4) Arrange the eigenvalues in descending order $\lambda_1 \geq \lambda_2 \geq \cdots \geq \lambda_n$, composite the first k eigenvalues $\Delta=(\lambda_1,\lambda_2,\cdots,\lambda_k)$ and corresponding eigenvectors $\boldsymbol{V}=[\boldsymbol{v}_1,\boldsymbol{v}_2,...,\boldsymbol{v}_k]$ .Thus the cumulative contribution rate is defined as

$$R_k=\frac{\sum_{i=1}^{k}\lambda_i}{\sum_{i=1}^{n}\lambda_i} \tag{4}$$

R_k indicates the percentage of the total variance by the first k principal components(PCs), and the threshold value should be carefully selected to choose the most representative PCs .

5) Finally, a k-dimensional feature vector ($k<n$) is obtained

$$\boldsymbol{P}=\boldsymbol{V}^T\boldsymbol{X} \tag{5}$$

P is the first k PCs. Then the goal to reduce feature dimension achieves.

3. The proposed method for fault classification of rolling bearing based on PCA

3.1. Feature extraction

A general feature extraction scheme to classification is divided into two steps. Firstly, select relative statistical features that cover both time and frequency domain, and transform the original data space into feature space. Then, convert global feature space matrix by linear or nonlinear transformation.

In this study, PCA-based feature extraction scheme is introduced for fault classification of rolling bearing. The time-domain statistical features including mean, variance, peak, root mean square, skewness ,kurtosis, crest factor and shape factor etc , totally 16 parameters ($p_1 \sim p_{16}$) as defined in table 1, have been considered for bearing fault diagnosis. Since it is difficult to recognize bearing conditions only with time-domain features, the characteristics of the frequency-domain also have been proposed. Frequency domain feature contains 11 parameters ($p_{17} \sim p_{27}$) as defined in table 1, including the mean frequency, root mean square frequency and the centre frequency and the variance frequency, in which p_{17} reflects the size of the vibration energy in the frequency domain; p_{21} and $p_{23} \sim p_{25}$ reflect changes in the frequency band positions; $p_{18} \sim p_{20}$, p_{22} and $p_{26} \sim p_{27}$represents the degree of concentration or dispersion of the spectrum. Extract the 11 parameters ($p_{28} \sim p_{38}$) in Hilbert envelope spectrum in the same way as $p_{17} \sim p_{27}$. Finally, 38 feature parameters compose the initial feature set [7].

3.2. Feature selection

Apply PCA to the obtained high-dimensional feature set and select the most representative features from the original time-frequency statistics feature set for bearing defect classification based on the defined cumulative contribution rate in equation 4. Finally, adopting the selected representative principal components as inputs, a FCM approach is utilized to classify the bearing fault data. The flowchart of the proposed method is as follows:

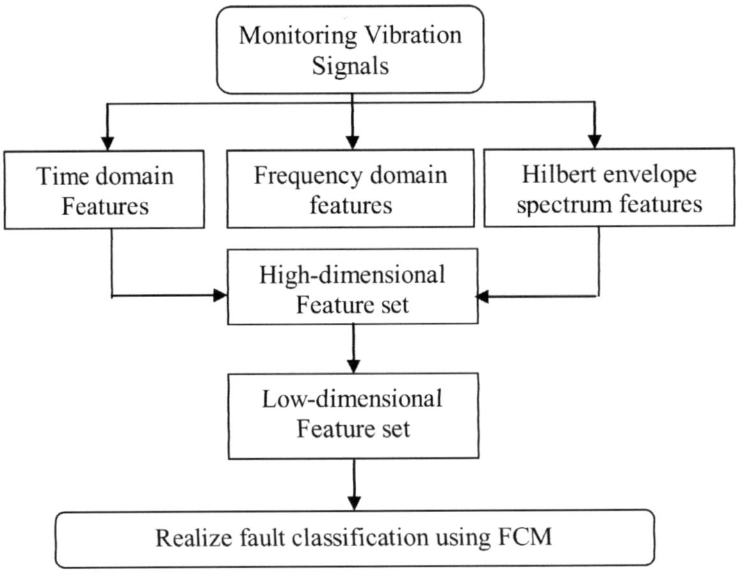

Figure 1. Flowchart of the proposed method.

Journal of Physics: Conference Series **628** (2015) 012079 doi:10.1088/1742-6596/628/1/012079

Table1. The definition of the time-frequency statistical features.

No.	Feature Definition	No.	Feature Definition	No.	Feature Definition
1	$p_1 = \frac{\sum_{n=1}^{N} x(n)^{a}}{N}$	11	$p_{11} = \frac{p_2}{p_4}$	21	$p_{21} = \frac{\sum_{k=1}^{K} (f_k^{\,c} y(k))}{\sum_{k=1}^{K} y(k)}$
2	$p_2 = \sqrt{\frac{\sum_{n=1}^{N} (x(n) - p_1)^2}{N-1}}$	12	$p_{12} = \frac{p_8}{p_2}$	22	$p_{22} = \frac{\sqrt{\sum_{k=1}^{K} [(f_k - p_{21})^2 y(k)]}}{K}$
3	$p_3 = (\frac{\sum_{n=1}^{N} \sqrt{\lvert x(n) \rvert}}{N})^2$	13	$p_{13} = \frac{p_8}{p_4}$	23	$p_{23} = \frac{\sqrt{\sum_{k=1}^{K} (f_k^2 y(k))}}{\sum_{k=1}^{K} y(k)}$
4	$p_4 = \frac{\sum_{n=1}^{N} \lvert x(n) \rvert}{N}$	14	$p_{14} = \frac{p_8}{p_3}$	24	$p_{24} = \frac{\sqrt{\sum_{k=1}^{K} (f_k^4 y(k))}}{\sum_{k=1}^{K} (f_k^2 y(k))}$
5	$p_5 = \frac{\sum_{n=1}^{N} (x(n))^3}{N}$	15	$p_{15} = \frac{p_5}{(\sqrt{p_7})^3}$	25	$p_{25} = \frac{\sum_{k=1}^{K} (f_k^2 y(k))}{\sqrt{[\sum_{k=1}^{K} (f_k^4 y(k))][\sum_{k=1}^{K} y(k)]}}$
6	$p_6 = \frac{\sum_{n=1}^{N} (x(n))^4}{N}$	16	$p_{16} = \frac{p_6}{p_7^2}$	26	$p_{26} = \frac{p_{22}}{p_{21}}$
7	$p_7 = \frac{\sum_{n=1}^{N} (x(n))^2}{N}$	17	$p_{17} = \frac{\sum_{k=1}^{K} y(k)^{b}}{K}$	27	$p_{27} = \frac{\sum_{k=1}^{K} (\sqrt{\lvert f_k - p_{21} \rvert} y(k))}{K\sqrt{p_{22}}}$
8	$p_8 = \max\lvert x(n) \rvert$	18	$p_{18} = \frac{\sum_{k=1}^{K} (y(k) - p_{17})^2}{K-1}$		
9	$p_9 = \min\lvert x(n) \rvert$	19	$p_{19} = \frac{\sum_{k=1}^{K} (y(k) - p_{17})^3}{K(\sqrt{p_{18}})^3}$		
10	$p_{10} = p_8 - p_9$	20	$p_{20} = \frac{\sum_{k=1}^{K} (y(k) - p_{17})^4}{K p_{18}^2}$		

[a] $x(n), n = 1, \cdots, N$ is the time-domain signal sequence, N is the number of samples.

[b] $y(k), k = 1, 2, \cdots, K$ is the spectrum of the signal $x(n)$, K is the number of spectral lines.

[c] f_k is the value of k-th frequency spectrum

4. Case study

To verify the effectiveness of the proposed method, apply it to the Case Western Reserve University Experimental data [8]. The test bearings are 6205-2RS JEM SKF deep groove ball bearings. The driver end bearing is tested under 2 hp loads and speed of 1772 r/min. The vibration data set including four different fault types (normal, inner race fault, outer race fault, and ball fault) is collected at 12K sampling frequency. Single point faults were made on the test bearings using electro-discharge machining with fault diameters of 0.007, 0.014 and 0.021 inches.

The three fault types each with three fault diameters and the normal compose 10 cases. In each case, there are 15 samples with each having 2048 data points. Choose 5 samples randomly from the 15 ones for each case to construct the training data set. The rest ones are used as the testing data set. Data sets in this paper are described in table 2.And we use the D_{07}, D_{14} and D_{21} data sets to identify the defective components of bearing and use the D_B, D_I and D_O data sets to differentiate the defective bearings on the basis of defect size. The time domain acceleration signals for defect classification (four types) and severity classification (four degrees) are depicted in figure 2.

Table 2. Description of experimental data sets.

Dataset	Diameter(mil)	Type	Size
D_{07}	0 7 7 7	N I B O	15 15 15 15
D_{14}	0 14 14 14	N I B O	15 15 15 15
D_{21}	0 21 21 21	N I B O	15 15 15 15
D_B	0 7 14 21	N B B B	15 15 15 15
D_I	0 7 14 21	N I I I	15 15 15 15
D_O	0 7 14 21	N O O O	15 15 15 15

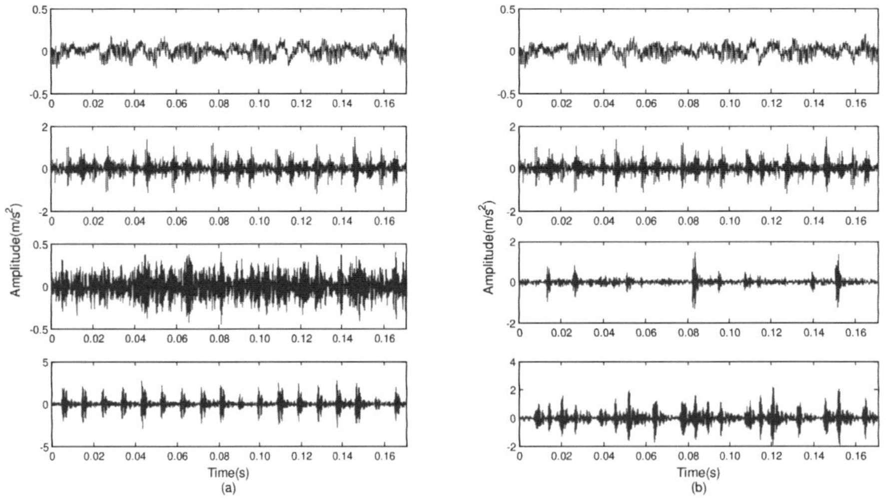

Figure 2. The time-domain vibration signal (a) D_{07} and (b) D_I.

The dimensions of original feature set are 38, including 16 time-domain statistic parameters, 11 FFT-based statistic parameters and 11 Hilbert-based parameters. Totally we get 60*38 feature set for each dataset. Then the PCA is performed on the obtained high dimensional feature set to acquire low dimensional feature set. The results for all datasets show that the accumulate contribution R_k of the first two principal component reaches 99%. So the first 2 PCs are selected to represent the fault. The other PCs are not considered as key contribution of the fault.

Figure 3 shows that 2D PCA-based representation of the D_{07}, D_{14} and D_{21}. X, Y coordinate axes of each figure are represented by first and second principal eigenvector. It can be seen that corresponding to the four types: inner race fault, ball fault, outer race fault and normal condition, the selected first 2 PCs perform well in classifying the three dataset.

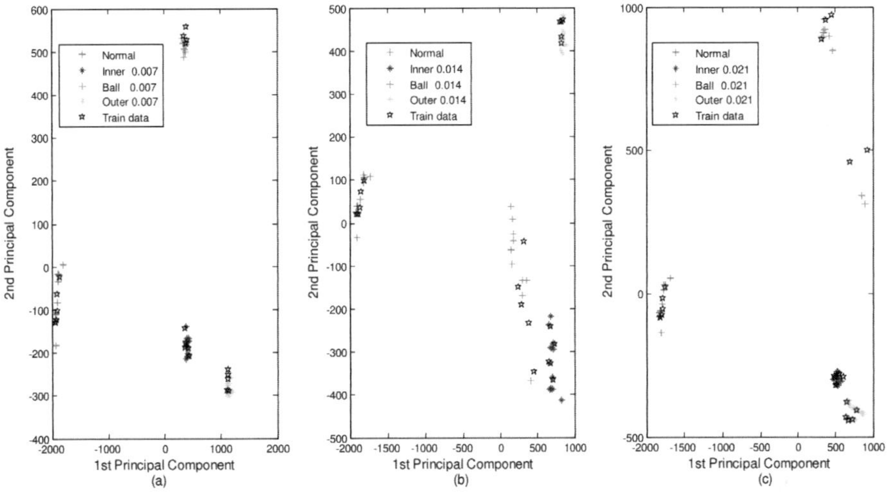

Figure 3. The defect classification with data set (a) D_{07}, (b) D_{14} and (c) D_{21}.

Figure 4 shows that 2D PCA-based representation of the D_B, D_I and D_O .X, Y coordinate axes of each figure are also represented by first and second dominant eigenvector. As depicted in figure 4, corresponding to the four degrees: 0 inches fault, 0.007 inches fault, 0.014 inches fault, 0.021 inches fault, the difference is clearer in the D_I and D_O than in D_B.

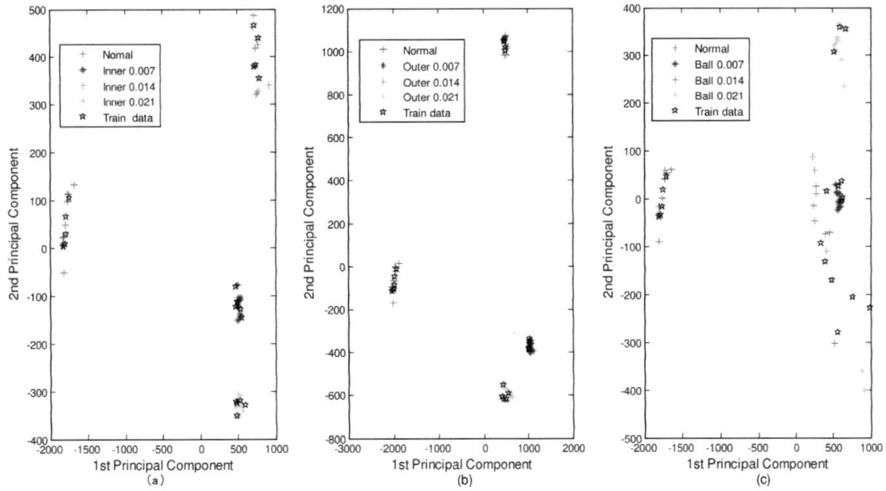

Figure 4. The severity classification with data set (a) D_I, (b) D_O and (c) D_B.

To give quantitative estimation of the classification results, the Fuzzy C-means clustering method, is applied to the first 2 PCs data set. The number of clusters is 4 and other clustering parameters all select the default value. Finally we classify the clusters according to the membership function matrix. The classifier accuracy with FCM for multiclass data sets is shown in Table 3.

Table 3. Accuracy of FCM for each of the first 2 PCs data set.

Dataset	D_{07}	D_{14}	D_{21}
PCA	100%	100%	95%
Dataset	D_B	D_I	D_O
PCA	100%	100%	90%

The result shows the proposed method has a rather good performance for defect classification (D_{07} and D_{14}) and severity classification (D_B and D_I). And the low accuracy of dataset D_{21} and D_O dues to the wrong classification of the ball fault. Overall, the proposed method works well in defect classification and distinguishing inner race fault and outer race fault better than ball fault in severity classification.

5. Conclusion

The primary objective of this paper was to realize defect-classification and severity-classification of rolling element bearings. A PCA-based approach to selecting the most representative features for the classification is developed. Based on the defined time-frequency statistical features, the efficacy of the low-dimensional representations obtained by the proposed method was evaluated via a FCM classifier. The results of the practical rolling bearing experiment data validated the suitability of the PCA-based feature selection scheme.

Acknowledgement

This work is supported by the National Natural Science Foundation of China (No.51375067), the Aviation Science Foundation of China (No. 20132163010) and the Fundamental Research Funds for the Central Universities of China (No. DUT13JS08).

Reference

[1] Zio E, Baraldi P and Gola G 2008 Feature-based classifier ensembles for diagnosing multiple faults in rotating machinery *Applied Soft Computing.* **8** 1365-80

[2] Zhang YZ, Xu GH, Liang L and Wang J 2009 Feature extraction methods for fault classification of rolling element bearing based on nonlinear dimensionality reduction and SVMs *Artificial Intelligence and Computational Intelligence AICI International Conference on.* **3** 228-234

[3] Widodo A, Yang B S and Han T 2007 Combination of independent component analysis and support vector machines for intelligent faults diagnosis of induction motors *Expert Systems with Applications.* **32** 299-312

[4] Malhi A and Gao R X 2004 PCA-based feature selection scheme for machine defect classification *IEEE Transaction on Instrumentation and Measurement.* **53** 1517-25

[5] He QB, Yan RQ, Kong FR and Du RX 2009 Machine condition monitoring using principal component representations *Mechanical Systems and Signal Processing.* **23** 446-66

[6] Su Y, Yuan ZH and Qi XX 2013 A feature extraction method of rolling bearings faults based on PCA and Power Spectrum Analysis *International Journal of Advancements in computing Technology.* **5** 681-83

[7] Ou L and Yu DJ 2014 Rolling bearing fault diagnosis based on supervised laplaian score and principal component analysis *Journal of Mechanical Engineering.* **50** (*Preprint* gr-qc/0303038)

[8] Loparo KA 2013 Bearing vibration data set Case Western Reserve University (http://csegroups.case.edu/bearingdatacenter/home)

Projectiles Impact Assessment of Aircraft Wing Structures with Real Dynamic Load

Lu Han [1] **, Qing Han** [1] **, and Changlin Wang** [1]
[1] Department of Aircraft Design Engineering. P. O. BOX 120, Northwestern Polytechnical University, Xi'an, Shaanxi 710072, P. R. China

E-mail: hanlu1119@mail.nwpu.edu.cn

Abstract. This paper presents an analysis to achieve the impact damage of the wing structure under real dynamic load. MPCCI tools are utilized to convert wing aerodynamic load into structural Finite Element Method (FEM) node load. The ANSYS/LS-DYNA code is also used to simulate the dynamic loading effects of the wing structure hit by several projectiles, including both active damage mechanism and common damage mechanism. In addition, structural node force on the leading edge and the midline is compared to the aerodynamic load separately. Furthermore, the statistical analysis of the penetrating size and the stress concentration around the damage holes indicates that under the same load situation, the structural damage efficiency of active damage mechanism is significantly higher than the one of common damage mechanism.

1. Introduction

Experimental and numerical studies have been carried over the past few years in order to evaluate the impact damage of aircraft structures. Aircraft damage assessment, especially the damage under projectiles impact has become a non-ignorable part of aircraft vulnerability analysis and high survivability design. The aircraft will suffer the aerodynamic load and impacting load during different flight phases of a mission, the assessment of aircraft damage becomes more complicated under this coupling effect. The aerodynamic characteristics analysis and combat process are considered separately in current research: García-Castillo S K[1] evaluated the ballistic limits of aluminium alloy under normal impact of spherical steel projectiles, when they are subjected to uniaxial in-plane. F R Ahad[2] simulated the failure of aluminium alloy under high velocity impact. J Donea[3] provided an arbitrary Lagrangian-Eulerian kinematical description of the fluid domain in which the grid points can be displaced independently of the fluid motion. C Farhat and M Lesoinne[4] presented a new algorithms for converting the fluid pressure and stress fields at the fluid/structure interface into a structural load, and for transferring the structural motion to the fluid system. Therefore, it is meaningful and necessary to do some research to take the aerodynamic loading and impact loading into consideration in wing structure damage assessment.

The different types of threats that an aircraft is faced with during flight mission can be divided into common damage mechanism and active damage mechanism due to their killing mechanisms. The common damage mechanism is non-explosive penetrating projectile while the active mechanism is the multifunctional structural fragment[5] with energetic material which was presented in the last century.

Lu Han hanlu1119@mail.nwpu.edu.cn

When collision happens, rapid chemical reaction occurs in the internal energetic material because of the strong impact, releasing a certain amount of energy, even burning at the same time. Due to the aftereffect explosion and burning, the damage efficiency is significantly enhanced[6].

In this paper, the research of damage efficiency of active damage mechanism is aimed at the explosive active damage mechanism. The damage efficiency is blasting and energy output which is mainly represented by shock wave. Different from common damage mechanism, most of the present active damage mechanism are explosions that coated with inert high strength metal shell due to the launching conditions and terminal damage effect and the active damage mechanism researched in this paper is a cylinder projectile with a hemisphere head.

The analysis of common and active damage mechanisms are conducted and compared respectively, section 2 introduces the material model and the state equations used in the simulation model; section 3 gives a detailed introduction of the modelling method and information; in section 4, the dynamic response is analysed after the aerodynamic load is transformed into structural load by the MPCCI tools; section 5 assumes that wing structure is hit by two rows of ten stars of the projectile, and simulates its injury condition.

2. Calculation Methods

Both the common damage mechanism and the shell of the active damage mechanism are modelled by 45# steel with JOHNSON-COOK [7] material model and GRUNEISEN equation of state. The internal explosive material model and equation state of active damage mechanism are respectively chosen as *ELASTIC-PLASTIC-HYDRO and *IGNITION-GROWTH-OF-REACTION-IN-HE; The wing structure uses the *MAT-PLASTIC-KINEMATIC material model and *EOS-LINEAR-POLYNOMIAL equation of state.

2.1.1. JOHNSON-COOK model

Johnson and Cook presented this suitable constitutive model of large deformation, high strain rate and high temperature conditions for metal material. Due to its simple form and convenient use, this material model is widely used for the study on metal impact mechanics. The yield stress is expressed by equation (1) in this model:

$$\sigma_y = (A + B\bar{\varepsilon}^{pn})(1 + c\ln\varepsilon^*)(1 - T^{*m}) \tag{1}$$

where A、B、c、n、m are material parameters determined by experiment, $\bar{\varepsilon}^{pn}$ is the effective plastic strain, T^{*m} is the homologous temperature.

$$\varepsilon^* = \bar{\varepsilon}^{p}\varepsilon_0^{-1} \tag{2}$$

$$T^{*m} = (T - T_{room})(T_{melt} - T_{room})^{-1} \tag{3}$$

T_{room} is the room temperature, T_{melt} is the melt temperature of the material.

2.1.2. GRNUEISEN equation of state

The Gruneisen equation of state with cubic shock velocity-particle velocity defines pressure for compressed materials as:

$$p = \frac{\rho_0 C^2 \mu[1 + (1 - \frac{\gamma_0}{2})\mu - \frac{a}{2}\mu^2]}{[1 - (S_1 - 1)\mu - S_2\frac{\mu^2}{\mu + 1} - S_3\frac{\mu^2}{(\mu + 1)^2}]^2} + (\gamma_0 + a\mu)E \tag{4}$$

where C is the intercept of the Vs-Vp curve; S_1, S_2, and S_3 are the coefficients of the slope of the Vs-Vp curve; γ_0 is the Gruneisen gamma; a is the first order volume correction to γ_0; and $\mu = \rho\rho_0^{-1} - 1$.

*2.1.3. *IGNITION-GROWTH-OF-REACTION-IN-HE model*

The *IGNITION-GROWTH-OF-REACTION-IN-HE model, presented by Wackerle and Johnson[8] in 1976, was used to simulate the shock ignition (or ignition failure) and the shock wave propagation of solid explosive. This three-ignition growth model based on hot-hole collapse mechanism which was corrected by Lee and Tarver, and then developed into the complete IGNITION-GROWTH-OF-REACTION-IN-HE model, it is the so called Lee-Tarver model. It is mainly used to calculate the explosive shock reaction conditions from start to finish. It assumes that a small amount of explosives is ignited by the shock heating, the reaction rate is controlled both by pressure and surface area.

$$\dot{\lambda}=\mathrm{I}(1-\lambda)^{b}(\rho\rho_0^{-1}-1-a)^{x}+G_1(1-\lambda)^{c}\lambda^{d}P^{y}+G_2(1-\lambda)^{e}\lambda^{g}P^{z} \tag{5}$$

where the first item indicates the hot point form and the ignition of the heating zone; the second item represents the slow growth inward or outward of the hot point; the third item expresses the rapid completion of the reaction where the ρ is the density of explosive under specific pressure.

2.1.4. MAT-PLASTIC-KINEMATIC model

This model is suited to model isotropic and kinematic hardening plasticity with the option of including rate effects into consideration. It is a cost effective model and available to shell elements. Strain rate is accounted for using the Cowper and Symonds model which scales the yield stress with the factor:

$$1+\left(\dot{\varepsilon}C^{-1}\right)^{1/p} \tag{6}$$

where $\dot{\varepsilon}$ is the strain rate.

The *MAT_003 model itself has a failure criterion which is better than *MAT-JOHNSON-COOK for shell elements.

2.1.5. EOS-LINEAR-POLYNOMIAL equation state

The linear polynomial equation of state is linear in internal energy. The pressure is given by:

$$P=C_0+C_1\mu+C_2\mu^2+C_3\mu^3+(C_4+C_5\mu+C_6\mu^2)E \tag{7}$$

where $\mu=\rho\rho_0^{-1}-1$, and $\rho\rho_0^{-1}$ is the ratio of current density to initial density. Equation (7) and (8) have the same meaning because C_0~C_6 can be expressed as related forms of the specific heat ratio:

$$P=(\gamma-1)E\rho\rho_0^{-1} \tag{8}$$

the units of E are the units of pressure.

3. Numerical Models

3.1.1. Wing structure model

LS-DYNA software is used to establish the numerical models which are impacted by two different kinds of projectiles. The finite element model of wing structure consists of three parts: the wing cover, the front and back girder and ten ribs. All of the three parts are modelled with 163 shell elements, for all the contacts between the parts the *CONTACT-ERODING-SURFACE-TO-SURFACE contact algorithm of LS-DYNA is used without friction.

The size of wing bounding box is $1220\mathrm{mm}\times800\mathrm{mm}\times45\mathrm{mm}$, the structure thickness is equivalent to aluminium alloy due to their rigidity. Table 1 indicates the parameters of LY-12cz [9]:

Table 1. The parameters of LY-12cz.

E(GPa)	ρ(kg/mm^{-3})	μ	σ_s(MPa)	E_t(MPa)	h	ε_f
72	2.7800E-6	0.79	0.51	0.26	0.014	1.03

3.1.2. Damage Mechanism

The active damage mechanism is a cylindrical projectile that contains both shell and the insert explosive, while the common projectiles are steel projectiles explosion which have the same shape and dimensions with active damage mechanism. Solid 164 element type is used in both numerical models, figure 1 shows the impact finite element model (the active damage mechanism and the cover of the wing are shown as a 1/2 model in order to express more clearly).

(a) Common damage mechanism model

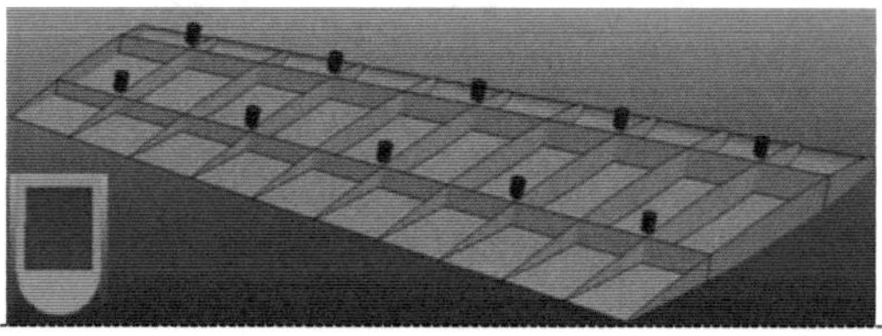

(b) Active damage mechanism model

Figure 1. The impact finite element model of wing structure

• Common Damage Mechanism

The Lagrange grid algorithm and solid 164 element type with *MAT-JOHNSON-COOK mat keyword are used for common damage mechanism, the bounding box size is $26mm \times 26mm \times 39mm$ and the velocity is 1300m/s. The material parameters of the common damage mechanism are shown in table 2:

Table 2. The parameters of 45# steel [10].

ρ(gm^{-3})	G(GPa)	A	B	n	C	m	T_m(K)	γ_0	α
7.8500E-6	78	0.79	0.51	0.26	0.014	1.03	1520	2.17	0.46

• Active Damage Mechanism

Large mesh deformation will happen during the explosion process for which the element type is chosen as solid 164 with ALE options. The thickness of the shell is 3mm, while the explosive size is $20mm \times 20mm \times 26mm$ which is filled in the cavity. The material and equation of state models of projectiles shell are the same as wing structure. Comp B [11] is selected as the internal explosive due to its efficient killing effect. As mentioned in section 2, the material model of the explosive is *ELASTIC-PLASTIC-HYDRO and the equation of state is *IGNITION-GROWTH-OF-REACTION-IN-HE. The Explosion and the shell both use automatic surface to surface contact (ASS2D) while the projectile and wing structure uses eroding surface to surface contact (ESS). The properties of the parameters of explosive are presented in table 3.

Table 3. The parameters of Comp B explosion.

ρ(g/m^3)	G(GPa)	σ_y(GPa)	w	F_{igmax}	a_p(GPa)	b_p(GPa)
1.6300E-6	3.54	0.2	0.912	0.3	557.48	7.83
r_{1p}	r_{2p}	b	$wcvp$	C_v	x	a
4.5	1.2	0.222	0.34	2.48E-3	4	0.01

4. The calculation of structural node load based on MPCCI tools

MPCCI tools can realize the coupling problem between multiple disciplines by proving a single subject simulation program[12]. The data can be exchanged between two or more FEM software by dividing the calculation into a number of step length, simplex data is transformed during each step and exchanged between two steps. The way uses in this paper is that the pressure which imported into MPCCI is picked up ANSYS CFD™ software by calculating the flow field outside of the wing cover, while the force on structural node is the output.

The finite element sizes of aerodynamic grid should be small enough due to the limiting calculation precision demand. It is complicated and unnecessary to use the large number of load for structural loading assessment, therefore, the pneumatic load is transferred to minor structural node for ensuring the reaction of real pneumatic load based on the simplified calculation. In this paper, the pneumatic node load is obtained at 0.83 Maher and 3.06° angle of attack with ANSYS CFD™ software, then it is transferred to structural node load by MPCCI tools. The structural node can well reacting the real pneumatic load of the wing structure. The forces of structural node and aerodynamic node are compared in figure 2.

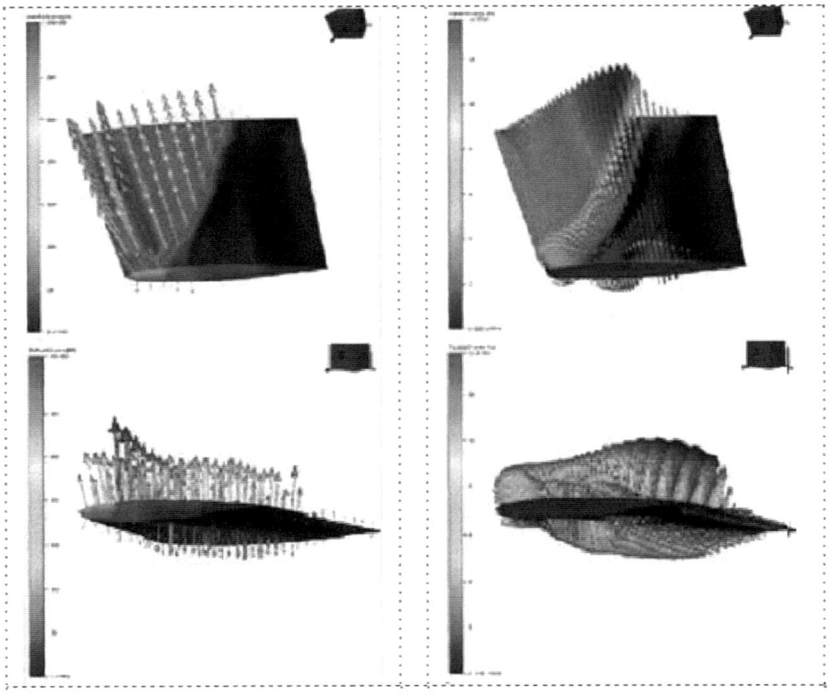

(a) Structural node pressure. (b) Aerodynamic node pressure.

Figure 2. Force comparison of structural node and aerodynamic node

Forces on the 72 aerodynamic nodes and 11 structural nodes distributed along the midline and the leading edge of the wing are compared separately (Node number from the wing root to tip starts from 1). Figure 3 shows the aerodynamic node force on the leading edge and the midline. Aerodynamic force decreases smoothly from root to tip. The one loads on the upper wing surface is more than on the lower surface, and on the leading edge is higher than on the midline.

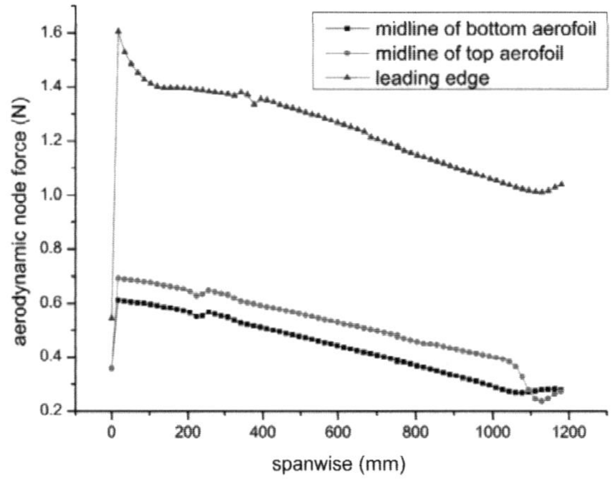

Figure 3. Aerodynamic node force at leading edge and midline.

The structural node force on the leading edge and the midline is shown in figure 4. Since there are far less structural nodes than aerodynamic nodes, more force contributes to each structural node than aerodynamic one. Figure 4 shows that the trend of transformed structural force agrees to aerodynamic characteristics.

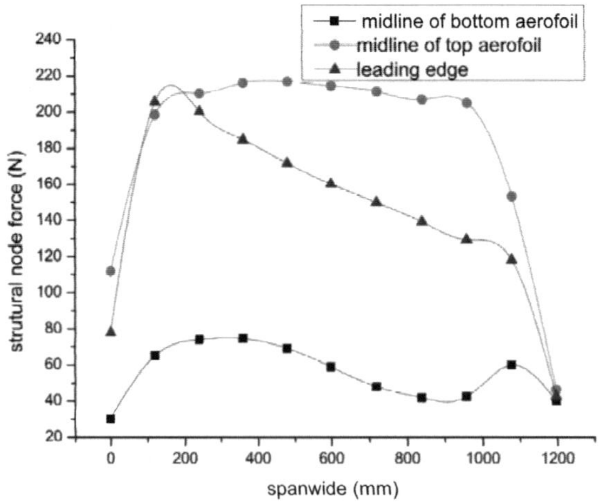

Figure 4. Structural node force at leading edge and midline.

5. The comparison of penetrating size and stress concentration around the damage holes

The structural node load that obtained in section 4 is loaded on the corresponding node of wing cover by the keyword *LOAD-NODE-POINT. Section 2 describes a model that the wing structure is impacted by two rows of ten projectiles at the intersection of the girders and ribs, the ten projectiles are numbered from 1 to 10 based on the position (from the root to tip, from the leading edge to the trailing edge).The penetrating holes are numbered from 1 to 20 according to the same rules as the projectiles which are from 1 to 10 on the bottom aerofoil while from 11 to 20 on the top aerofoil (assuming the aircraft is impacted from the bottom aerofoil).

The penetrating sizes result from no. 1 to 10 are counted at 0.3ms. Figure 5 shows the sizes of the penetrating holes under the two kinds of damage mechanism and the stress concentration around the damage holes are compared in figure 6. The abscissa is the broken holes number corresponded with the projectile.

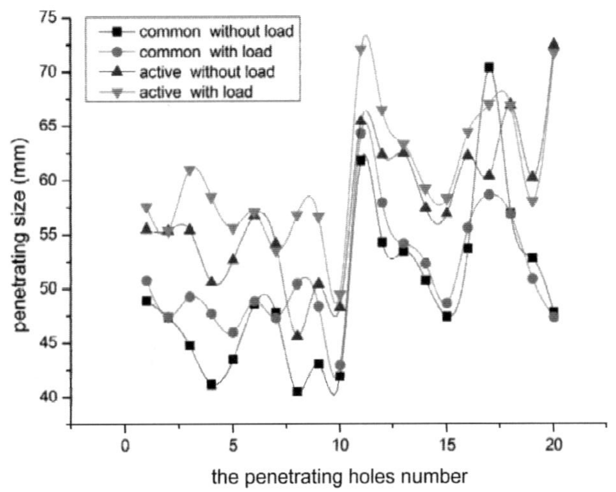

Figure 5. The comparison of penetrating size

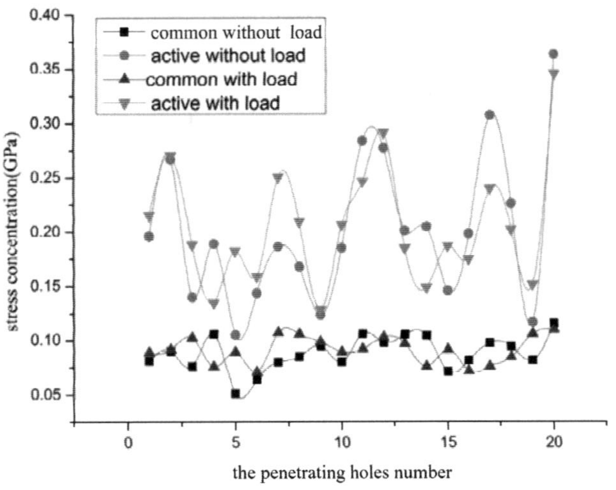

Figure 6. The comparison of penetrating size stress concentration.

6. Conclusions

The penetrating sizes and the stress concentration around the damage holes of metal wing structure are recorded in order to analyse the response under the common damage mechanism and active damage mechanism condition. In both of the numerical simulations, the aerodynamic load is well transformed into structural load by MPCCI tools. And there is no significant differences of penetrating holes size as well as stress concentration between the loading condition and unloading condition, while the type of damage mechanism has a major influence on these two parameters. The damage size of wing structure under active damage mechanism is about 15% more than common damage mechanism and the stress concentration around the holes is almost twice of the latter. The structural damage efficiency of active damage mechanism is significantly better than common damage mechanism.

References

[1] García-Castillo S K, Sánchez-Sáez S and Barbero E 2011 Behaviour of uniaxially preloaded aluminium plates subjected to high-velocity impact *J. Mech. Res. Commun.* **38** p 404-407
[2] Ahad F R, Enakoutsa K, Solanki K N and et al 2014 Nonlocal modeling in high-velocity impact failure of 6061-T6 aluminum *Int. J. Plasticity* **55** p 108-132
[3] Donea J, Giuliani S and Halleux J P 1982 An arbitrary Lagrangian-Eulerian finite element method for transient dynamic fluid-structure interactions *J. Comput. Method. Appl. M.* **33** p 689-723
[4] Farhat C, Lesoinne M and Le Tallec P 1998 Load and motion transfer algorithms for fluid/structure interaction problems with non-matching discrete interfaces: Momentum and energy conservation, optimal discretization and application to aeroelasticity *J. Comput. Method. Appl. M* **157** p 95-114.
[5] Ames R 2005 Vented Chamber Calorimetry for Impact-Initiated Energetic Materials *43rd AIAA Aerospace Sciences Meeting and Exhibit.*
[6] Huang, Heng-jian et al 2007 Preliminary Research on Damage Enhanced Fragment *J.ENERG MATER* **15** p 566.
[7] Liu Y 2008 ANSYS and LS-DYNA used for structural analysis *Int. J. Comput Aided Engineering and Technology* **1** p 31.
[8] Wackerle J, Johnson J O and Halleck P M Shock 1976 *Initiation of High Density PETN 6th Sym. on Detotonation Coronado (California: Office of Naval Research-Department of the Navy 20).*
[9] Lee E L and Tarver C M 1980 Phenomenological model of shock initiation in heterogeneous explosives *J. Phys. Fluids* **23** p 2362.
[10] Zhou P, ZHANG J and HOU R 2004 Study of Battle-Damage Prediction Mode of Aircraft Structure Material LY-12 CZ with Projectile *J. Air Force Engineering University* **1** p 007.
[11] Taylor S, Lever J H, Fadden J and et al 2009 Simulated rainfall-driven dissolution of TNT, Tritonal, Comp B and Octol particles *J. Chemosphere* **75** p 1074-1081.
[12] Joppich W and Kürschner M MpCCI-a tool for the simulation of coupled applications 2005 Concurrency and Computation: Practice and Experience **18** p 183–192.

Structural Health Monitoring challenges on the 10-MW offshore wind turbine model

E Di Lorenzo[1,2,3], G Kosova[2], U Musella[2], S Manzato[1], B Peeters[1], F Marulo[2] and W Desmet[3]

[1] Siemens Industry Software NV, RTD Test Department
Interleuvenlaan 68, 3001, Leuven, Belgium
[2] University of Naples "Federico II", Department of Industrial Engineering, via Claudio 21, 80125, Naples, Italy
[3] KU Leuven, Department of Mechanical Engineering, Division PMA
Celestijnenlaan 300, 3001, Heverlee, Belgium.

E-mail: emilio.dilorenzo@siemens.com

Abstract. The real-time structural damage detection on large slender structures has one of its main application on offshore Horizontal Axis Wind Turbines (HAWT). The renewable energy market is continuously pushing the wind turbine sizes and performances. This is the reason why nowadays offshore wind turbines concepts are going toward a 10 MW reference wind turbine model. The aim of the work is to perform operational analyses on the 10-MW reference wind turbine finite element model using an aeroelastic code in order to obtain long-time-low-cost simulations. The aeroelastic code allows simulating the damages in several ways: by reducing the edgewise/flapwise blades stiffness, by adding lumped masses or considering a progressive mass addiction (i.e. ice on the blades). The damage detection is then performed by means of Operational Modal Analysis (OMA) techniques. Virtual accelerometers are placed in order to simulate real measurements and to estimate the modal parameters. The feasibility of a robust damage detection on the model has been performed on the HAWT model in parked conditions. The situation is much more complicated in case of operating wind turbines because the time periodicity of the structure need to be taken into account. Several algorithms have been implemented and tested in the simulation environment. They are needed in order to carry on a damage detection simulation campaign and develop a feasible real-time damage detection method. In addition to these algorithms, harmonic removal tools are needed in order to dispose of the harmonics due to the rotation.

1. Introduction

The growth of the wind energy market is driven by several factors. One of the most important is the impressive improvements in the wind turbine technologies. The rated power capacity for a single wind turbine has dramatically increased from some kW in the 1980s to 8-MW in 2014. At the same time the rotor diameter has reached values higher than 150 meters.

Historical trends suggest that the on-shore wind turbine size and power capacity will be continuously increasing. On the other hand, since high potential sites on the land have already been taken, in the past years the off-shore wind potential has been exploited taking advantage of the strong and steady wind.

In the offshore sector, the cost associated to Operation and Maintenance (O&M) is extremely high. In order to reduce this cost, efficient Structural Health Monitoring (SHM) techniques need to be developed for such systems. Wind turbine manufacturers, owners and operators would benefit financially from SHM technology, which provides an indication of the reliability of each wind turbine in its lifecycle. SHM will provide actionable information that the maintainers can use to schedule maintenance ahead of time to avoid more costly unforeseen maintenance actions.

A finite element model of a 10-MW reference wind turbine has been built by using an aeroelastic code. The final aim of the work is to combine together Operational Modal Analysis and SHM techniques by developing a feasible damage detection method based on the modal parameters of the structure. Most existing modal parameter estimation techniques assume that the system to be analysed is a Linear Time Invariant (LTI) system. In the case of a wind turbine, this can be considered true if the brake is engaged and the blades are not rotating. When the wind turbine is in operating conditions, the structure cannot be considered anymore as LTI. If the angular speed can be approximated with a constant value, then the wind turbine can be considered as a Linear Time Periodic (LTP) system. Several modal parameter estimation techniques can be adapted and applied to such a case.

2. Operational Modal Analysis for LTP systems

Operational Modal Analysis (OMA) is a suitable tool for estimating the modal parameters of a structure. It has its main application in cases of huge systems such as bridges, stadiums, airplanes, and spacecraft because there are very limited or no practical ways to excite them in controlled manner.

Several pre-processing techniques have been developed in the past years in order to apply standard modal parameter estimation techniques to rotating structures, such as helicopters and wind turbines. The main two techniques which will be analysed in details are the so-called Multi-Blade Coordinate (MBC) transformation and the Harmonic Power Spectrum (HPS) method.

2.1. *Multi-Blade Coordinate transformation*

The MBC approach is also known as Coleman theory because it was developed, in its first formulation, by Coleman [1] for helicopter analysis. The transformation allows converting the rotating degrees of freedom into a non-rotating frame. In [2] a numerical approach has been introduced for wind turbines. It properly models the dynamic interaction between the non-rotating systems such as tower and nacelle, and the spinning rotor. It also offers a better understanding of the interaction between fixed and rotating bodies. One of the main advantages of the technique is that it filters out all the periodic terms except those ones which are integral multiples of NΩ, where Ω is the rotor angular speed and N is the number of blades.

Considering a rotor with 3 blades, we need to assume that the blades are equally spaced around the rotor azimuth. In such case, the azimuth angle of the b-th blade can be obtained by using equation (1). ψ represents the instantaneous azimuth angle of a reference blade by assuming that this blade is vertically up when $\psi = 0$.

$$\psi_b = \psi_1 + (b-1)\frac{2\pi}{3}, \qquad b=1,\ 2,\ 3 \tag{1}$$

The blade coordinates $q_{b,i}$ can be, for example, the accelerations measured by sensors placed along the blades at the location i. They can be transformed into the multi-blade coordinates $q_{0,i}$, $q_{c,i}$ and $q_{s,i}$ by means of equation (2).

$$q_{0,i} = \frac{1}{N}\sum_{b=1}^{N} q_{b,i} \qquad q_{c,i} = \frac{2}{N}\sum_{b=1}^{N} q_{b,i}\cos(\psi_b) \qquad q_{s,i} = \frac{2}{N}\sum_{b=1}^{N} q_{b,i}\sin(\psi_b) \tag{2}$$

The physical interpretation of the new coordinates is not immediate and it depends on the considered degree of freedom. Anyway the first coordinate $q_{0,i}$ represents the collective symmetric behavior of the blades, while the other two coordinates represent the asymmetric one.

Concluding, MBC transformation can be used as a data pre-processing step before an OMA algorithm is applied. The following steps need to be accomplished:

- acquire the accelerations at several locations along the blades and along the tower;
- transform the accelerations of the blade by using equation (2) and by taking into account the azimuth angle, whereas the accelerations acquired on the non-rotating subsystems remain unchanged;
- combine the two sets of accelerations together and perform OMA by using Operational Polymax [3] for estimating the modal parameters (natural frequencies, damping ratios and multi-blade mode shapes);
- apply equation (3), also known as inverse MBC transformation, in order to get the physical mode shapes.

$$q_{b,i} = q_{0,i} + q_{c,i}\cos(\psi_b) + q_{s,i}\sin(\psi_b) \quad (3)$$

Previous studies have shown the applicability of this procedure to operating wind turbines built by using aeroelastic codes [4], [5]. The application of the same technique to real operating data obtained in the experimental field has been addressed only in few works [6], [7].

2.2. *Harmonic Power Spectrum method*

For Linear Time Invariant (LTI) systems, the signal of fundamental importance is the complex exponential signal, also referred as sinusoidal signal. When such a signal of a given frequency is injected into a LTI system, the output response is a complex exponential signal of the same frequency, but with possibly different amplitude and phase. Instead, for a Linear Time Periodic (LTP) system the response to the same input will be at a combination of frequencies, each separated by the fundamental, or rotating, frequency Ω.

In order to extend the concept of transfer function to LTP systems, Wereley [8] introduced the Exponentially Modulated Periodic (EMP) signals which can be expressed as the complex Fourier series of periodic signals of fundamental frequency Ω, modulated by sinusoidal signals. They can be seen as a collection of frequency shifted signals, which are described by equation (4). Theoretically, an infinite number of harmonics should be considered to characterize a LTP system, but most part of the systems is well approximated with a finite, perhaps small number.

$$u(t) = \sum_{n=-\infty}^{\infty} u_n e^{(j\omega + jn\Omega)t} \qquad y(t) = \sum_{n=-\infty}^{\infty} y_n e^{(j\omega + jn\Omega)t} \quad (4)$$

The property for which an EMP input signal $u(t)$ causes an EMP output signal $y(t)$ of the same frequency but with possibly different amplitude and phase allows defining the Harmonic Transfer Function (HTF). The HTF is completely analogous to the commonly known transfer function for LTI systems. The main difference between the two transfer function is that the standard one relates the output at a single frequency to the input at the same frequency, whereas the HTF relates the input at a collection of frequencies each separated by Ω to the output at the same collection of frequencies.

The use of the Harmonic Transfer Function (HTF) has been extended to output-only linear time periodic systems by introducing the so-called Harmonic Power Spectrum (HPS). The method can be seen as an extension of classical Operational Modal Analysis to time periodic systems [9].

The autospectrum $S_{yy}(\omega)$ of an EMP signal can be obtained in the standard way by breaking the modulated time signals into several blocks, applying a Hanning window, computing the Discrete Fourier Transform of each block and then averaging. The autospectrum formulation is shown in equation (5) where E denotes the expectation and $(\ )^H$ denotes the Hermitian, whereas $Y(\omega)$ is the frequency domain representation of the EMP output signal.

$$S_{yy}(\omega) = E\left[Y(\omega)Y(\omega)^H\right] \approx \sum_{r=1}^{N} \sum_{l=-\infty}^{\infty} \frac{C_{r,l} W(\omega)_{r,l} C_{r,l}^H}{\left[j\omega - (\lambda_r - jl\Omega)\right] \left[j\omega - (\lambda_r - jl\Omega)\right]^H} \tag{5}$$

In equation (5), $W(\omega)$ is a function of the input spectrum and the input characteristic of the system. The autospectrum of the output can be approximated by a sum of modal contributions if the excitation is reasonably flat in the frequency domain. The term λ_r corresponds to the complex eigenvalue of the r-th mode of the system and the mode vector $C_{r,l}$ is not the usual mode vector describing the mode shape at different points, but it is comprised of the Fourier coefficients for the r-th mode but with the position of the elements shifted by l in the coefficient vector [10].

Summarizing, the following steps need to be accomplished in order to apply successfully the HPS method:

- acquire the output signal $y(t)$ of a LTP system subjected to a white noise broadband excitation;
- construct the EMP signal by considering shifted copies of the output, as shown by equation (4);
- compute the autospectrum of the EMP signal by averaging multiple blocks of the time history;
- use Operational Polymax system identification technique to estimate the modal parameters from the autospectrum, as shown by equation (5);
- reconstruct the time periodic mode shapes using the identified Fourier coefficients $C_{r,l}$.

3. 10-MW wind turbine model

In the past years, DTU Wind Energy developed a new reference 10-MW offshore wind turbine (DTU 10MW RWT). A detailed report has been provided. It contains a description of the complete aerodynamic, structural and aeroelastic model for the DTU 10MW RWT [11]. The aeroelastic solution has been computed by using LMS Samcef Wind Turbines (SWT). This is an aeroelastic code which integrates all the main aerodynamic, structural, multi-body and control features of the wind turbine in a fully dynamic environment.

The model is shown in figure 1 and its main parameters are listed in table 1. The tower is made from steel and it has been modelled by using 10 beam elements with a constant thickness for each segment. The outer diameter is varying linearly from the tower bottom to the top. The three identical blades have been modelled by specifying the mass, elastic and aerodynamic properties provided in [11]. Finally, the drivetrain has been built by using rigid connection with a gear ratio equal to 50:1.

Figure 1. 3D view of 10-MW wind turbine model

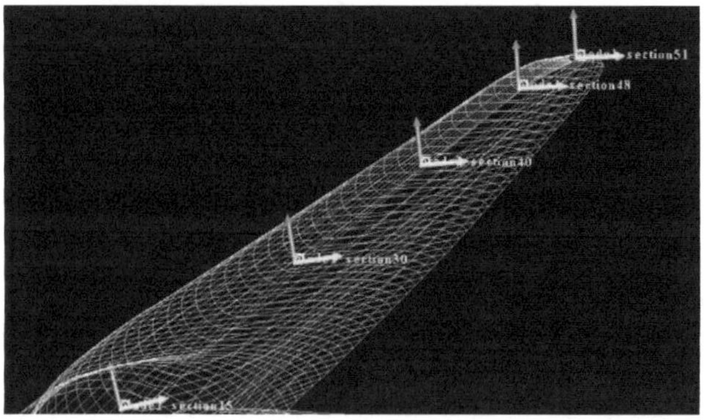

Figure 2. Virtual accelerometers placed along the blade

Journal of Physics: Conference Series **628** (2015) 012081 doi:10.1088/1742-6596/628/1/012081

The Kaimal turbulence model has been selected for performing the analysis [12]. The main wind component is in the direction perpendicular to the rotor plane and the turbulent fluctuations are in the other two directions. The average wind speed has been set equal to 10 m/s.

In order to validate the model, the natural frequencies extracted from the model after the linearization process have been compared to the ones provided by DTU in [11]. A very good agreement has been found and the comparison is shown in table 2.

Table 1. Key parameters of the DTU 10 MW RWT

Main parameters	
Rated power	10 MW
Rated rotor speed	9.6 rpm
Rotor diameter	178.3 m
Tower height	119.0 m
Rotor mass	227962 kg
Tower mass	628422 kg
Gearbox ratio	50:1

Table 2. Natural frequencies comparison (FA = Fore-Aft; S2S = Side-to-Side)

Mode	DTU [Hz]	SWT [Hz]
1st Tower FA	0.249	0.247
1st Tower S2S	0.251	0.251
1st flap with yaw	0.547	0.550
1st flap with tilt	0.590	0.598
1st collective flap	0.634	0.636
1st edge with tilt	0.922	0.940
1st edge with yaw	0.936	0.956
2nd flap with yaw	1.376	1.413
2nd flap with tilt	1.550	1.573
2nd collective flap	1.763	1.812

4. Application of the pre-processing techniques to the wind turbine in operating conditions

The wind turbine has been instrumented by means of twenty virtual accelerometers: five along the tower and five for each blade distributed along the pitch axis as shown in figure 2. The sampling frequency has been set to 100 Hz and the simulations have been running for 1000 seconds. In general, long time histories are necessary for confident modal parameters estimation, but on the other hand a reasonable computational time is also needed. Afterwards, a down-sampling to 10 Hz has been performed since the modes of interests are below 5 Hz.

First of all the DTU 10MW RWT has been analysed in parked conditions, which means by disconnecting the generator and by activating the brake. In this case, the blades are not rotating and standard Operational Modal Analysis techniques can be applied because the system is linear time invariant. The time series can be exported to LMS Test.Lab in which the correlations and the spectra can be computed and Operational Polymax can be applied in order to extract the modal parameters. A very good agreement with the modal parameters listed in table 2 has been found, with a relative error lower than 1%.

While the application of OMA techniques is straightforward in case of parked wind turbines, the same does not apply to operating wind turbines. In fact, when the blades are rotating, the implemented algorithms discussed in Section 2 need to be applied to the time series in order to take into account the time varying nature of the system in operating conditions. In this case, the harmonics due to the rotation mask the dynamic behaviour of the structure, as shown in figure 3 where the comparison between operating and parked conditions has been done in terms of crosspowers. The rotor angular speed has been set to 9.6 rpm, which correspond to 0.16 Hz, and an accelerometer placed on the tower top has been chosen as reference channel for the computation.

By applying the Multi-Blade Coordinate transformation, all the periodic terms except those which are integral multiples of 3p, where p is the rotor angular speed are filtered out, as shown in figure 4. All pairs of asymmetric rotor edgewise modes in parked conditions become pairs of rotor whirling modes owing to the rotation. They can be observed and identified only after that one of the two techniques mentioned in Section 2 has been applied.

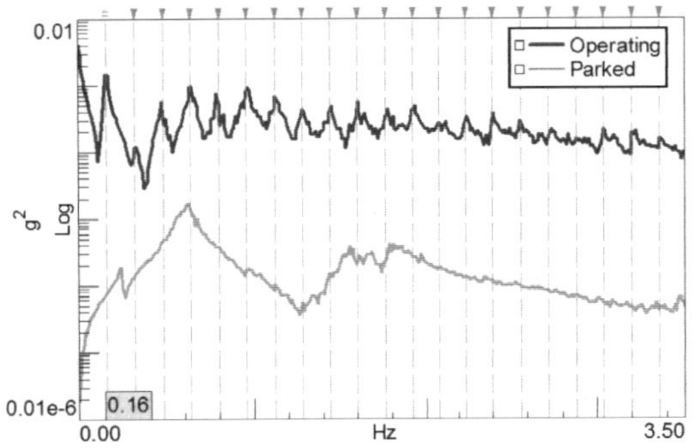

Figure 3. Crosspower comparison in parked configuration (light grey) and in operating configuration (dark grey)

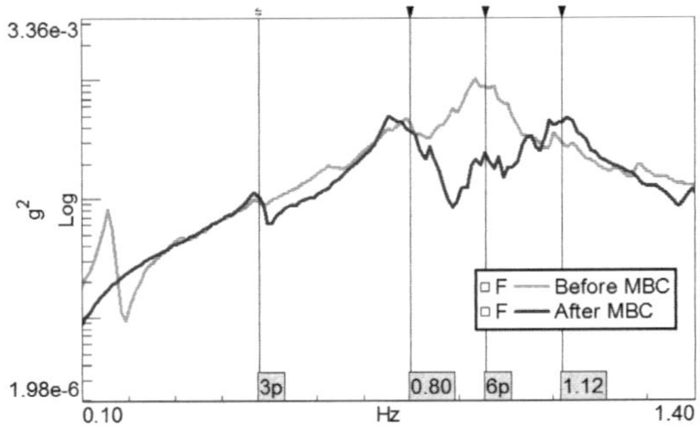

Figure 4. Crosspower comparison before (light grey) and after (dark grey) MBC transformation

The two whirling modes are identified at the frequencies listed in table 3. They are separated by 2p (0.32 Hz) in accordance with the literature. Once that the natural frequencies and damping ratios have been evaluated, the mode shapes need to be converted back to the physical coordinates in order to be animated and compared to the ones obtained by using the Harmonic Power Spectrum method. Table 3 shows the comparison of the results by applying OMA without any pre-processing technique and by applying the two techniques introduced in Section 2. A very good agreement in terms of natural frequencies has been found by applying MBC and HPS.

The whirling modes can also be identified by looking only at the tower accelerometers signals in the lateral direction. In figure 5 the autopowers of such signals are shown for three different conditions: a parked configuration and two rotating configurations with a different rotational speed (8 rpm and 9.6 rpm).

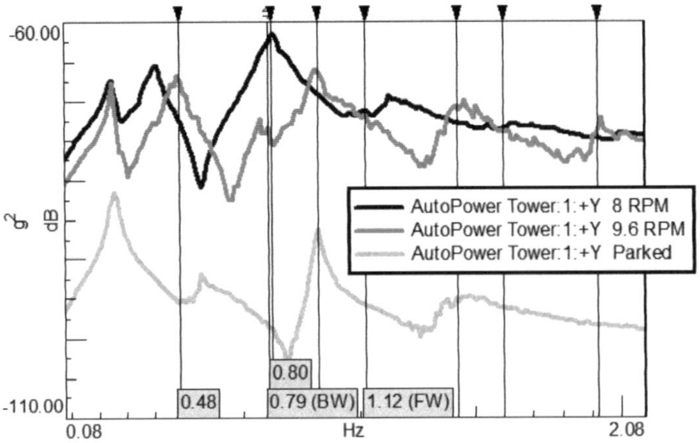

Figure 5. Whirling modes identification from tower accelerometers in three different conditions: parked configuration (light grey), operating configuration with p=8 rpm, operating configuration with p=9.6 rpm

In operating conditions the amplitudes of the signals are much higher than in the parked configuration. The shift toward lower frequencies of the 3p components can be seen in operating conditions by reducing the rotational speed from 9.6 rpm (nominal speed) to 8 rpm. The whirling modes are shifting as well because the difference among them, which is equal to 2p, reduces when the rotational speed decreases.

Table 3. Backward and forward whirling modes identification procedure

Mode	OMA Parked [Hz]	MBC + OMA Rotating [Hz]	HPS + OMA Rotating [Hz]
1st edge with tilt/ BW	0.94	0.80	0.79
1st edge with yaw / FW	0.95	1.12	1.12

5. Structural Health Monitoring considerations

In order to perform Structural Health Monitoring (SHM) studies, the first step is the identification of the most sensitive modal characteristic. In this paper we mainly look at the whirling modes which are the most identifiable modes in operating conditions. Both magnitude and phase of these mode shapes can be considered as damage detection indicators. By looking at the backward whirling mode, but the same considerations can be drawn also for the forward whirling mode, it can be seen that the shapes have constant amplitudes and the phase lag between the blades is constant and equal to 120°.

By considering a simple bi-dimensional model of the DTU 10 MW RWT, it can be seen that reducing the stiffness of one blade by 2%, the two properties of the whirling mode shape are lost. As shown in figure 6 and figure 7, the phase lag between the blades is not anymore equal to 120° and the amplitude is not anymore the same for all the blades.

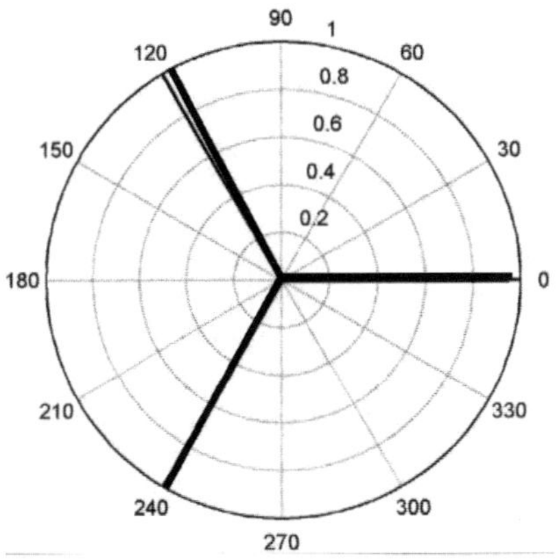

Figure 6. Normalized amplitude and phase of backward whirling mode in healthy conditions

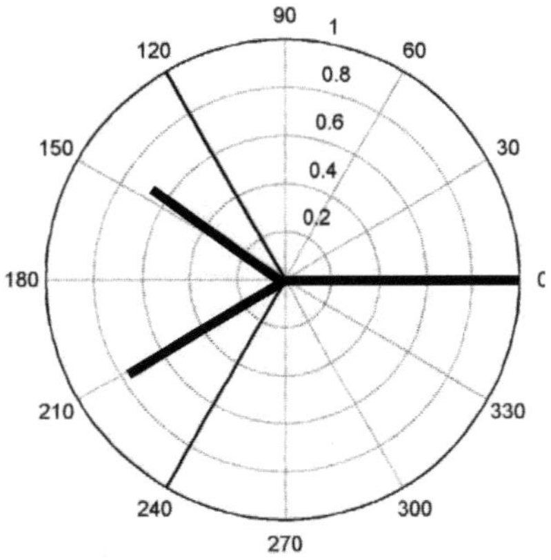

Figure 7. Normalized amplitude and phase of backward whirling mode in damaged conditions (0° blade stiffness reduced by 2%)

Conclusions

The operational data pre-processing techniques, known as Multi-Blade Coordinate transformation and Harmonic Power Spectrum method, seem to be very promising to allow the estimation of modal parameters in case of Linear Time Periodic system such as operating wind turbines and helicopters. Once they are successfully applied, standard techniques for estimating the modal parameters can be used. In this work, Operational Polymax has been used in order to perform the Operational Modal Analysis step.

The so-called whirling modes are always well identified in several conditions and with both the pre-processing techniques combined with OMA. The magnitude and phases between the blades for these modes have demonstrated to be very sensitive to the introduction of small blade damages. Anisotropies can be introduced in the system by modifying the stiffness parameters of a single blade. In a simulation environment, it has been proven that by looking at the in-plane whirling mode shapes it is possible to identify small variations of the stiffness, which means that small damages into the blade structure could be identified.

Acknowledgements

The authors acknowledge the European Commission (EC) for their research grant under the project FP7-PEOPLE-2012-ITN 309395 "MARE-WINT" (new MAterials and REliability in offshore WINd Turbines technology), see: http://marewint.eu/.

References

[1] Coleman R P 1943 Theory of self-excited mechanical oscillations of hinged rotor blades *NACA technical report G29*

[2] Bir G 2008 Multiblade coordinate transformation and its application to wind turbine analysis *Proc. of 2008 ASME Wind Energy Symp.* (Reno, Nevada, USA, 2008)

[3] Peeteers B, Guillaume P, Van der Auweraer H, Cauberghe B, Verboven P and Leuridan J 2004 The PolyMAX frequency-domain method: a new standard for modal parameter estimation? *Shock and Vibration* **11** 395-409

[4] Tcherniak D, Chauhan S, Rossetti M, Font I, Basurko J and Salgado O 2010 Output-only modal analysis on operating wind turbines: application to simulated data *Proc. of European Wind Energy Conf.* (Warsaw, Poland, April 2010)

[5] Di Lorenzo E, Manzato S, Peeters B and Marulo F 2014 Modal parameter estimation for operational wind turbines *Proc. of 7th European Workshop on Structural Health Monitoring* (Nantes, France, July 2014)

[6] Tcherniak D and Larsen G C 2013 Application of OMA to an operating wind turbine: now including vibration data from the blades *Proc. of 5th Int. Operational Modal Analysis Conf.* (Guimaraes, Portugal, May 2013)

[7] Di Lorenzo E, Manzato S, Peeters B and Marulo F 2014 Structural health monitoring techniques applied to operating wind turbines *Proc. of 9th Int. Conf. on Structural Dynamics* (Porto, Portugal, July 2014)

[8] Wereley N M 1991 Analysis and control of linear periodically time varying systems *PhD dissertation: Massachusetts Institute of Technology, Department of Aeronautics and Astronautics*

[9] Yang S and Allen M S 2014 Lifting approach to simplify output-only continuous-scan laser vibrometry *Mech. Syst. and Sign. Proc.* **45** 267-282

[10] Allen M S, Sracic M W, Chauhan S and Hansen M H 2011 Output-only modal analysis of linear time periodic systems with application to wind turbine simulation data *Mech. Systems and Signal Processing* **25** 1174–1191

[11] Bak C, Zahle F, Bitsche R, Kim T, Yde A, Henriksen L C, Natarajan A and Hansen M H 2013 Description of the DTU 10 MW reference wind turbine *DTU Wind Energy Report-I-0092*

[12] Kaimal J C, Wyngaard J C, Izumi Y and Cote R O 1972 Spectral characteristics of surface layer turbulence *Q. J. R. Meteorol. Soc.* **98** 563–598

Automatic welding detection by an intelligent tool pipe inspection

C J Arizmendi[1], W L Garcia [2,3], M A Quintero[3]

[1] Mechatronic Department, Universidad Autónoma de Bucaramanga, Avenue 42 N° 48-11, Bucaramanga, Colombia.
[2] System Engineering Department, Universidad Autónoma de Bucaramanga, Avenue 42 N° 48-11, Bucaramanga, Colombia.
[3] Corrosion Research Institute, Km 2 via Refugio, Piedecuesta, Colombia.

Abstract. This work provide a model based on machine learning techniques in welds recognition, based on signals obtained through in-line inspection tool called "smart pig" in Oil and Gas pipelines . The model uses a signal noise reduction phase by means of pre-processing algorithms and attribute-selection techniques. The noise reduction techniques were selected after a literature review and testing with survey data. Subsequently, the model was trained using recognition and classification algorithms, specifically artificial neural networks and support vector machines. Finally, the trained model was validated with different data sets and the performance was measured with cross validation and ROC analysis. The results show that is possible to identify welding automatically with an efficiency between 90 and 98 percent.

Keywords: Data analysis, In-Line Inspection, Smart Pig, Pattern Recognition, Pipeline, Pipe weld joints.

1. Introduction

The pipelines are considered the most efficient way to carry fluids (oil and gas) over long distances [1]. Over time the integrity of this type of infrastructure is of concern for companies, mainly because much of the piping is nearing the end of its useful life due to the effect of external agents (corrosion, incorrect operations for civil works, landslides, vandalism etc.), which affects its operation. Consequently, it is necessary to monitor, evaluate and maintain pipelines to ensure reliability from a comprehensive analysis and reduce the probability of failure. The effects of a failure usually end up being accidents that impact the environment and population.

In recent decades, industry have developed different technologies to assess the condition of pipeline, of which one of the most used is the inspection with intelligent tools. An intelligent pipeline inspection tool[1] (see figure 1) as defined in [2] is an instrumented vehicle is introduced into a pipeline travels through its extension, driven by the fluid itself and can perform various functions, such as:

- Measure pipeline operational condition.
- Detect and locate mechanical damages, defects and anomalies.

[1] Also known as ILI, In-line inspection, among others.

- Mapping the route of the pipeline, knowing the accurate position identify pipe movement and bending strain.

Figure 1. Smart tool inspection ITION and scraper.
Source: Ition-E project - CIC.

This type of intelligent tools has a number of sensors coupled to the internal and external of the chassis. The signals generated by the sensors are acquired, processed and stored while traveling through the pipe, then, large amount of data is downloaded, and artificial intelligence data analysis is applied to obtain relevant information, identification of discontinuities known as indications like anomaly, imperfection or component. Anomaly which are reported as possible segments of mechanical damage or pipe metal loss defects. Additionally, artificial intelligence detects piping accessories such as welds, coupling flanges, valves, shunts, among others.

Among the possibilities offered in-line inspection services, are implemented with different technologies depending on the purpose of the study of phenomena in pipes, vehicle technologies most commonly used are: MFL[2] [1], caliper [3], X-rays [4], ultrasound [5] and inertial [6]. From the point of view of data analysis, computational techniques used for processing, data analysis and recognition of welds are filtering signals and artificial intelligence.

Filtering techniques have been used in several works in order to improve the training data performance, for example in [1], which seeks to recognize defects in welds in pipes of carbon steel, based on information acquired by through a smart tool MFL inspection technology, tests were performed with Fourier [7] Wavelet [8] and Savitzky-Golay [9] filters to pre-process the input data seeking to eliminate irrelevant features that may hinder the classification defects. Validation results showed an improvement in the efficiency of the algorithm for almost all types of defects passing from 66.7% to 92.5%. Furthermore, it was able to confirm an improvement in training time using techniques such filtering. In [10], the wavelet transform is used to pre-process the data of 71 types of geometric defects in pipes, using data from a smart tool MFL inspection technology. Work results demonstrate that is possible with the aid of the Wavelet transform and other computational techniques to predict the depth and shape of these defects in 3D. In [11] - [12] Wavelet transform and a modified version of this transform is applied to remove noise in the data delivered by an intelligent tool MFL inspection technology to prevent recognition algorithms find defects wrongly. Finally, in [13], the wavelet transform is used to locate leaks in natural gas pipelines.

Different artificial intelligence techniques have been used for the recognition of phenomena in pipelines, in [1], [10], [14] neural networks were used to classify and identify some classes of geometrical defects in welds, in [15] using a genetic algorithm was possible to determine the shape, size and location of defects in about 1000 lines carrying natural gas. In [16] - [17], support vector machines applied to improve recognition performance of the algorithm, for defects in the metal loss and pipeline leaks.

[2] Magnetic Flux Leakage

In the present study, inspection data acquired by the smart tool to inspection trends integrity and operation, called ITION, developed by the Research Institute of Corrosion (CIC), for this work, inertial mapping technique is used. This purpose differs from other work (such as [1], [10], [11] and [16]) mainly because does not consist only of recognizing defects in the weld but also in locating it. In addition, inspection technology used by CIC is different (Inertial vs MFL), it is notable that no references about identification based on inertial welding techniques found.

Presented the prior differences, the aim of this work is to identify a weld using data acquired through an inspection tool with inertial technology from the study of computational intelligence applied to similar phenomena using other technologies inspection.

This paper is organized as follows. In section 2, the method is explained, describing each phase of the methodology and emphasizing the model developed. In section 3, the experiments performed together with the results obtained are shown, in section 4 discusses the results and section 5 the overall conclusions of the study.

2. Methodology

In order to achieve quality in the data mining process, the CRISP-DM methodology (Cross Industry Standard Process for Data Mining) [18] was applied. Next, main stages implemented are presented below:

2.1. Business understanding

First step was to understand the representative features of a weld. A weld defined in [19] as "fixed joining metal parts, made with or without melting the edges to be joined using a filler metal or not". In Oil and Gas transport a welding joint is a point or edge where two metal pipes or another component of the pipeline, valves, tees, flanges, branch, supports, are joined together by fusion bonding or metal to metal according to a particular geometry .An example of a weld is illustrated applied to join two tubes, figure 2.

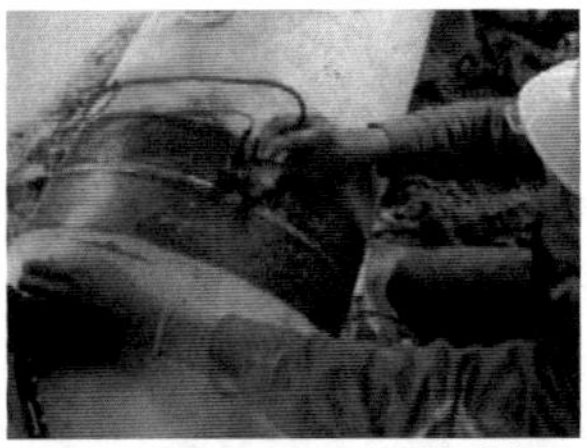

Figure 2. Welding pipe applied in section.
Source: Ition-E project - CIC.

When the intelligent line inspection passes by a welding, weld acts as a barrier and attempts to stop its path due to the excess material that is adhered to the inner surface protruding tube, however, at that moment and thrust forces phenomena occurs of inertia about the tool which keeps the dynamics and certain sensors recorded this tool behaviour.

Find all pipeline welds is an important purpose for the operator. During construction of the pipeline, manufacturer supplies the location of all welds, however, over time, repairs or changes occur due to maintenance which in some cases are not documented or reported, which generates the obsolescence of this information. Additionally, vandalism install accessories as valves in order to extract illegally pipeline products, in this case, apply welding to join the valve to the pipe. By construction conditions, the average distance between welds not exceeding twelve (12) meters. The objective of the analysis of welds from variables defined by experts is to identify sites where the welds are located with less than ten (10) percent error.

2.2. Data understanding

According expert criteria to identify a welding, information input variables shown in Table 1, is required

Table 1. Variables used by the experts to identify a weld

Variable	Notation	Unit	Accepted value	
			Maximum	**Minimum**
Rotation	Gh	lsb	6000	-6000
Acceleration	Ah	lsb	5454	-5454
Magnetization	Mh	lsb	+2500	-2500
Vibration	Vh	lsb	32000	-32000
Position	Od	m	50000	0

Regarding the performance experienced by the tool during its journey through the pipe the three axes of rotation is recorded, which are variables that measures changes in the angular velocity, acceleration also recorded in all three axes, measuring the speed changes linear magnetization measures changes in the magnetic field remaining in the pipe due to the previous step of a magnetic arrangement. The vibration sensor records changes in the intensity of the mechanical vibrations generated by the friction between the tool body and the pipe walls. The position reference variable shows the axial distance (i.e. parallel to the flow direction) to measure the length of travel of the tool through the pipe.

The values delivered by the rotation, acceleration and magnetization are represented in the spatial axes (X, Y, Z), that is, there is data for each axis. In total have 12 input variables, 3 of acceleration, 3 of rotation 3, 3 related to the variation of the magnetic field, 2 of vibration and position. All data are numerical and are in the same unit related to the resolution of digital analog converter called *lsb* except the sensor position, which is measured in meters.

It was requested the experts that be defined an output variable, which welding was called, in the table 2 is shown the characteristics of this variable.

Table 2. Characteristic of the variable output

Variable	Notation	Data type	Accepted value	
Weld	Weld	Text	S	N

Then, data were reviewed to discard values outside the permitted range, based on information provided by experts (see table I). An exploration of a given set of data was performed and found the following assumptions:

- When the tool passes a weld, vibration signals show a higher intensity than in most other pipe segments.
- The acceleration magnitude, this being understood as the square root of the sum of the X, Y, Z squared is less than or equal to the magnitude of the acceleration before reaching the welding possible. This proves that the tool is "slowing" in its initial phase of step welding.

- The magnitude of the acceleration is greater than or equal after going through the possible welding. This means that tool speeds for an instant of time when the solder is overcome.

Finally, developed and implemented an algorithm, which was applied in all data in order to eliminate those records with values outside the permitted ranges for each variable or was unrepresentative values. After performing this task, is determined for example, using only the information recorded by the tool once was moving through the pipe. This decision was made because it was found that when the tool is installed at the point of release, spends a considerable period of time before run into the pipeline. While the tool is stationed at the launch site captures information that is not representative for the study of the inertial behaviour welds.

2.3. Data Preparation

The information provided was delivered in a text file in comma separated csv extension columns. The first row of the file represents the names of each of the variables, both inlet and outlet. The other lines contain the data for each variable. Then, in order twenty constructed data files, the former contains 2.5% of welds and twentieth file contains 50% of the welds present during inspection. That is, each file contains about 2.5% more than the previous welds. However, the results of this work will be presented based on four of the twenty files built, with the following number of records, hereafter levels: 148, 642, 1464 and 1838 to be used in the training phase and a file 600 instances as test data. Each data set contains the same number of instances classified as the output variable as welding and soldering, for example, in the case of assembly 148, there are instances with the solder 74 and 74 instances label as welding. The output variable was validated considering information from a document called Pipe Tally Data by the owner of the pipeline, which accurately describe the location of welds applied during construction.

The 5 files are converted to csv format to format arff through the Weka tool. Weka is one of the most comprehensive data analysis software, using different pre-processing techniques and data modeling tools that implements. Developed with Java technology which allows it to run on almost any platform, is freely available under GNU license and its libraries can be easily integrated into other software applications.

Taking advantage of the benefits offered this software tool and following the recommendation of the work reported in the literature, applies preprocessing to reduce the number of input variables. The attribute selection technique and algorithm CfsSubsetEval [20] is used. In total, 10 files are created in the initial five arff file type with all the input variables and the other five products pre-processing.

2.4. Modeling

In figure 3 the proposed model for the recognition of welds, which consists of two phases, the first called learning is part of the generated files in the pre-processing to train a classification algorithm, which according to illustrated the work reported in the literature may be artificial neural networks [1] or support vector machines [16]. Then, in the evaluation phase, each classification algorithm is evaluated using cross validation technique [21] and a set of tests not used during training. The results of the evaluation are validated by two metrics, the percentage error and the ROC [22] analysis. Whether validation results meet the stated objective process, then process ends, otherwise, it returns to the learning phase to review the causes of the problem and rethink a new learning scenario with the new settings.

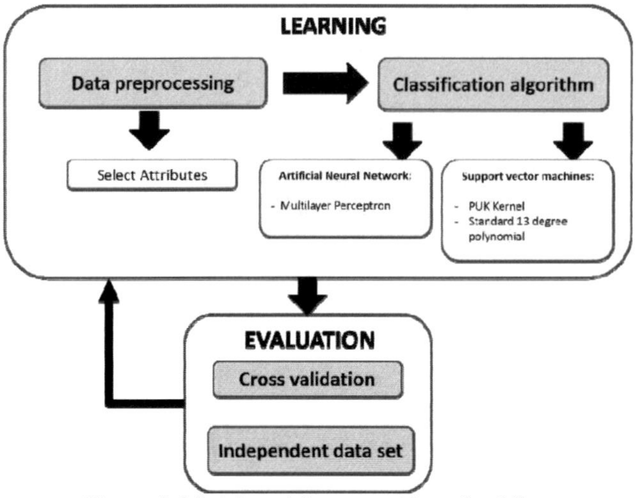

Figure 3. Model used for recognition of welding

3. Experiments and results

The information which the experiments were developed corresponds to an intelligent inspection in a Colombian pipeline in September 2012, with the following specifications:

- 12" nominal diameter.
- 36 km inspection.
- 2'150.835 records with an approximate size of 3GB.
- 3000 welds.

As it were mentioned in section 2 using the Weka software, 10 dataset was constructed, half of them are pre-processed using the technique of selecting attributes and the other half were left with all the input variables defined by the expert. The data for training were taken from a section of the inspection and consecutively, each training file has the same number of records with no welding and welding. Then, these files were used during training model for learning an artificial neural network of the multilayer perceptron and support vector machine with kernel called PUK or Pearson Universal Kernel5 [23] function. The performance of each model is evaluated using cross validation and independent set of data with a total of 600 instances, which, were taken from the final section of the inspection. It is taken into account as a criterion of success that the error rate with two evaluation metrics is less than ten percent. For each of the tests were performed to model eight (8) set of instances experiments, four (4) of the eight (8) experiments twelve (12) defined by the input variables used and the expert is mentioned in the understanding phase of data in section 2. In the other four (4) experiments, a feature selection algorithm, which recommended the use of certain input variables for each case was used.

Model Tests 1838 instances

For tests with this data set, in four (4) of the eight (8) experiments, CfsSubsetEval algorithm recommended using Y-axis acceleration, Y-axis magnetization with two vibration was employed. According to the results obtained when evaluating each model against a set of 600 instances untrained and cross-validation by the metric error rate (see table 3). Shows that two models recognize the location of the weld clearly when the variables acceleration (AHY) magnetic field in the pipe axis (MHY) and vibration (VH2) are used. The trained model that best meets the evaluation data is to support vector machines with PUK kernel with less than two (2%) percent error.

TABLE 3 Error rate models using 1838 instances in training

	Error rate in each test			
	Artificial neural networks		Support vector machines	
Pre-processing technique	Cross validation 10 folds	Independent set 600 instances	Cross validation 10 folds	Independent set 600 instances
None	2.67	12.19	1.41	7.33
Select attributes (Ahy, Mhy,Vh2)	3.67	4.98	2.61	1.33

Results of the tests conducted in set of 600 instances and cross-validation metrics, but using ROC curves (see Table 4) shows that when pre-processing is applied, the false positive rate (number of records evaluated where the algorithm obtained recognition as a false output and was actually true) and true negative (number of records evaluated where recognition algorithm obtained as true output but is actually false) decreases compared with models that do not use the pre-processing. Furthermore the model that implements an algorithm of support vector machines is more accurate than the model with the algorithm of artificial neural networks (the value of the sensitivity and specificity is closer to 100% compared to the model that implements the artificial neural network).

TABLE 4 ROC analysis using models 1838 instances in training

	ROC analysis							
	Artificial neural networks				Support vector machines			
Pre-processing technique	Cross validation 10 folds		Independent set 600 instances		Cross validation 10 folds		Independent set 600 instances	
	TPR	SPC	TPR	SPC	TPR	SPC	TPR	SPC
None	99.2%	99.0%	99.7%	89.3%	99.2%	99.3%	99.7%	93.0%
Select attributes (Ahy, Mhy, Vh2)	98.6%	98.9%	100%	98.0%	98.3%	99.1%	99.7%	99.0%

4. Discussion

With the results, it can be concluded that, using only 3 of the 12 input variables, the two developed models recognize the weld. This benefit by reducing processing time and analysis, considerably as large amount of data is processed.

The attributes that offered better performance for the developed models were acceleration and magnetic field variation on the Y axis and one of the vibration sensors. The pipeline inspection tool is configured such that the axial axis is the axis Y. This is consistent with a priori the experts have identified. However, during testing, as the number of data is increased, the rotation information in all three axes was not classified by the algorithm attribute selection as a representative and was in these assemblies where the best results were obtained model. This finding contradicts the opinion of experts who consider this information during the analysis process. The reason why the model uses only one of the vibration sensors is because one of them is filtered and real time processing which phenomenon intensifies the welding step and reduces the dynamic noise.

Regarding related work in the literature, a similar procedure to that reported in [1] and [16], was employed (i.e., pre-processing, training based on artificial intelligence and data validation) , the results demonstrate the feasibility of use of these computational techniques for identifying welds.

5. Conclusions

The results of this work show that welds can be identified automatically using artificial intelligence techniques and data mining with a percentage error of less than five percent (5%). Furthermore, it is concluded that 3 of 12 input information variables is only necessary to locate a weld: the acceleration and change of magnetic field in the Y axis and the vibration sensors.

It can be concluded that, under the conditions established by each test, it is possible to automatically recognize the position where the welds are located in a pipeline. The foregoing allows experts now have a versatile tool to provide timely reports. . However, as the developed models have a margin of error, in such cases, is at the discretion to the expert determine whether the point has or not welding, according to their experience and pipeline design information.

References

[1] Carvalho A A, Rebello J M A, Sagrilo L V S, Camerini C S, and Miranda I V J 2006 MFL signals and artificial neural networks applied to detection and classification of pipe weld defects. Ndt & E International, Vol. 39 no. 8, pp. 661-667.

[2] Souza R D 2003 Avaliação Estrutural de Dutos com Defeitos de Corrosão Reais. Pós-Graduação em Engenharia Mecânica, PUC-Rio, Rio de Janeiro, Brasil, capítulo 2, sección 2.1, Dissertação de Mestrado, 11p.

[3] Hart J D, Powell G H, Hackney D, and Zulfiqar N 2002 Geometry monitoring of the trans-alaska pipeline, in 11-th International Conference on Cold Region Engineering, Anchorage.

[4] Carneval R, Marinho M G, and Santos J 2006 Flexible line inspection, in European Conference on Nondestructive Testing (ECNDT).

[5] Beller M 2006 Applying ultrasound for in-line inspection: Facts and issues, in PPSA Aberdeen Seminar, UK.

[6] Santana D, Maruyama N, and Furukawa C 2010 Estimation of trajectories of pipeline pigs using inertial measurements and non linear sensor fusion, in Industry Applications (INDUSCON), 2010 9th IEEE/IAS International Conference on. IEEE, pp. 1-6.

[7] Oppenheim A V and Schafer R W 1989 *Discrete-time signal processing*. Engle Wood Cliffs, NJ: Prentice-Hall.

[8] Misiti M, Misiti Y, Oppenheim G and Poggi J M 1996 *Wavelet tool box for use with MATLAB, user's guide,* version 1. Natick, MA: The Math-Works Inc.

[9] Sophocles J O 1999 *Introduction to signal processing*. Engle Wood Cliffs, NJ: Prentice-Hall.

[10] Joshi A, Udpa L, Udpa S and Tamburrino A 2006 Adaptive wavelets for characterizing magnetic flux leakage signals from pipeline inspection. IEEE Transactions on Magnetics, vol. 42, no. 10, pp. 3168–3170.

[11] Afzal M and Udpa S 2002 Advanced signal processing of magnetic flux leakage data obtained from seamless gas pipeline, NDT &E International, vol. 35, no. 7, pp. 449–457.

[12] Han W and Que P 2006 A modified wavelet transform domain adaptive FIR filtering algorithm for removing the SPN in the MFL data, Measurement, vol. 39, no. 7, pp. 621–627.

[13] Yang Z, Xiong Z and Shao M 2010 A new method of leak detection for the natural gas pipeline based on wavelet analysis. Energy, 35, 3814–3820.

[14] Carvalho A A, Silva R R, Rebello J M A and Sagrilo L V S 2010 Pattern recognition techniques applied to the detection and classification of welding defects by magnetic testing. Research in Nondestructive Evaluation, vol. 21 no. 2, pp. 91-111.

[15] Hari K C, Nabi M and Kulkarni S V 2007 Improved FEM model for defect-shape construction from MFL signal by using genetic algorithm, IET Science, Measurement and Technology, vol. 1,no. 4, pp. 196–200.
[16] Khodayari-Rostamabad A, Reilly J P, Nikolova N K, Hare J R and Pasha S 2009 Machine learning techniques for the analysis of magnetic flux leakage images in pipeline inspection. IEEE Transactions on Magnetics, vol. 45, no. 8, pp. 3073-3084.
[17] Mandal S, Chan F and Tiwari M K 2011 Leak detection of pipeline: An integrated approach of rough set theory and artificial bee colony trained SVM. Expert Systems with Applications, vol. 39, pp. 3071-3080.
[18] Chapman P, Clinton J, Kerber R, Khabaza T, Reinartz T, Shearer C and Wirth R 2000 CRISP-DM 1.0 Step-by-step data mining guide.
[19] Jeffus L 2009 *Soldadura: principios y aplicaciones*, vol. 3. Editorial Paraninfo.
[20] Selvakuberan K, Indradevi M and Rajaram R 2008 Combined Feature Selection and classification–A novel approach for the categorization of web pages. Journal of Information and Computing Science, vol. 3 no. 2, pp. 83-89.
[21] Krogh A and Vedelsby J 1995 Neural network ensembles, cross validation, and active learning. Advances in neural information processing systems, pp. 231-238.
[22] Burgueño M J, García-Bastos J L and González-Buitrago J M 1995 Las curvas ROC en la evaluación de las pruebas diagnósticas. Med Clin Barcelona, vol. 104, no. 17, pp. 661-70.
[23] Üstün B, Melssen W J and Buydens L M 2006 Facilitating the application of support vector regression by using a universal Pearson VII function based kernel. Chemometrics and Intelligent Laboratory Systems, vol. 81, no 1, pp. 29-40.

A theoretical model for fault diagnosis of localized bearing defects under non-weight-dominant conditions

Q K Han, F L Chu[1]
The State Key Laboratory of Tribology, Tsinghua University, Beijing 100084, China

E-mail:chufl@tsinghua.edu.cn

Abstract. Fault diagnosis of localized bearing defects under non-weight-dominant conditions is studied in this paper. A theoretical model with eight degrees of freedom is established, considering two transverse vibrations of the rotor and bearing raceway and one high-frequency resonant degree of freedom. Both the Hertzian contact between rolling elements and raceways, bearing clearance, unbalance force and self-weight of rotor are taken into account in the model. The localized defects in both inner and outer raceways are modeled as half sinusoidal waves. Then, the theoretical model is solved numerically and the vibrational responses are obtained. Through envelope analysis, the fault characteristic frequencies of inner/outer raceway defects for various conditions, including the weight-dominant condition and non-weight-dominant condition, are presented and compared with each other.

1. Introduction

Rolling element bearings are important components of rotating machinery. Localized defects are frequently found in inner/outer raceways. The growth of localized defects can cause the fault of rolling bearings and even severe accidents to the rotating machinery if undetected early. For years, extensive efforts have been devoted to the modelling of localized defects and then predict their effects upon the vibrational response of the bearing system [1-3].

Most current studies focused on the bearing systems under the weight-dominant condition [4-6]. The load zone due to the self-weight of the rotor is time-invariant. The rolling elements periodically pass through the load zone and impact with the inner/outer raceway defects. The fault characteristic frequencies for the defects could be predicted accordingly.

However, the self-weight might not be the dominant load in many rotating machinery, such as the vertical rotors, lightly weight rotors, and et.al. In these rotors, the unbalance force is the dominant load, and cause the load zone of the rolling bearing varies with time. In this case, the impacts between the rolling element and raceway defects become more complicate. Obviously, the fault characteristic frequencies would differ distinctly with that of the weight-dominant case.

Therefore, fault diagnosis of localized bearing defects under non-weight-dominant conditions is studied in this paper. A theoretical model with eight degrees of freedom is established, considering two transverse vibrations of the rotor and bearing raceway and one high-frequency resonant degree of freedom. Both the Hertzian contact between rolling elements and raceways, bearing clearance, unbalance force and self-weight of rotor are taken into account in the model. The localized defects in both inner and outer raceways are modelled as half sinusoidal waves. Then, the theoretical model is

[1] To whom any correspondence should be addressed.

solved numerically and the vibrational responses are obtained. Through envelope analysis, the fault characteristic frequencies of inner/outer raceway defects for various conditions, including the weight-dominant condition and non-weight-dominant condition, are presented and compared with each other. Finally, some conclusions are summarized.

2. Theoretical model for rotor-bearing system with localized bearing defects

2.1. Nonlinear restoring force of rolling bearing with localized defects

The schematic diagram for a deep grove ball bearing is shown in Fig. 1. Generally, the outer race is fixed on the pedestal. The inner race, connected with rotor, rotates under constant angular speed ω_s. The coordinate for dynamic analysis is $O-X-Y$. There are two degrees of freedom for the inner race and rotor, i.e. x_s, y_s. The mass and damping are m_s, c_s, and the stiffness is provided by the nonlinear Hertz deformation between the ball and raceways. The outer race and pedestal also have two degrees of freedom, i.e. x_p, y_p. The corresponding mass, stiffness and damping are m_p, k_p, c_p. Besides, in order to simulate the high frequency resonant response of the rolling bearing, a spring-mass-damping system (m_r, k_r, c_r) with high resonant frequency is added to the model [3], as shown in Fig. 1.

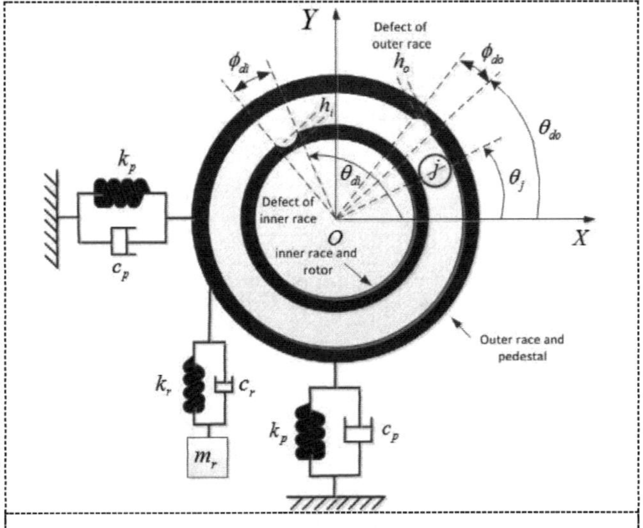

Figure 1. Defective bearing model

The location angle of the j th ball is $\theta_j(t)=\omega_{bc}t+2\pi(j-1)/N_b$, where N_b the number of balls, ω_{bc} the orbital speed of the ball. Under pure rolling conditions, the orbital speed of deep grove rolling bearing is expressed by $\omega_{bc}=\omega_s(1-d_b/D_b)/2$, in which d_b the ball diameter and D_b the pitch diameter of the bearing. According to the geometric relations shown in Fig. 1, the contact deformation for the the j th ball at time t is

$$\delta_j=(x_s-x_p)\cos\theta_j+(y_s-y_p)\sin\theta_j-c_l \tag{1}$$

Where c_l denotes the radial clearance.

A localized defect in the outer race is considered and shown in Fig. 1. The defect is modelled by a half sinusoidal wave, and its location, size and depth are, respectively, expressed by θ_{do}, ϕ_{do} and h_o. Similarly, the location, size and depth for the inner race defect are denoted by θ_{di}, ϕ_{di} and h_i. When the ball is contact with the defect, the contact deformation should be modified to take into account the additional clearance induced by the defect. For the rolling bearing with outer race defect, the contact deformation could be rewritten as

$$\delta_j = \begin{cases} (x_s - x_p)\cos\theta_j + (y_s - y_p)\sin\theta_j - \left(c_l + h_o \sin\left(\frac{\pi}{\phi_{do}}(\theta_j - \theta_{do})\right)\right) & \theta_{do} \le \theta_j < \theta_{do} + \phi_{do} \\ (x_s - x_p)\cos\theta_j + (y_s - y_p)\sin\theta_j - c_l & else \end{cases} \quad (2)$$

As the outer race is fixed, so the location angle of the defect θ_{do} is also constant. However, as the inner race is rotating, so the inner race defect also rotates with time. Its location angle should be $\theta_{di} = \omega_s t + \varphi_i$, in which φ_i the initial angle with respect to the X axis at time $t = 0$. Thus, when the inner race defect is considered, the contact deformation could be expressed similar with Eq. (2), while the parameters θ_{do}, ϕ_{do} and h_o should be substituted by θ_{di}, ϕ_{di} and h_i.

According to the Hertz contact theory, the contact force between the inner race and ball is given by $Q_j = \lambda_j K_c \delta_j^{3/2}$, in which K_c the contact stiffness, λ_j the force coefficient. For $\delta_j > 0$, $\lambda_j = 1$; otherwise, $\lambda_j = 0$. The contact stiffness K_c is related to the geometry and material property of the contact pairs, and could be solved from Harris' monograph. Considering the number of balls in contact, the total restoring forces of the rolling bearing in the two axes are expressed as follows

$$\begin{aligned} F_{bx} &= \sum_{j=1}^{N_b} \lambda_j K_c \delta_j^{3/2} \cos\theta_j \\ F_{by} &= \sum_{j=1}^{N_b} \lambda_j K_c \delta_j^{3/2} \sin\theta_j \end{aligned} \quad (3)$$

2.2. Equations of motion for the rotor-bearing system

A rigid rotor system supported by two deep grove rolling bearings is shown in Fig. 2. Thus, the equations of motion for the system could be derived as follows

$$\begin{aligned}
& m_s\ddot{x}_s + c_s\dot{x}_s + F_{bx1} + F_{bx2} = m_u e_u \omega_s^2 \cos\omega_s t \\
& m_s\ddot{y}_s + c_s\dot{y}_s + F_{by1} + F_{by2} = m_u e_u \omega_s^2 \sin\omega_s t \\
& m_{p1}\ddot{x}_{p1} + c_{p1}\dot{x}_{p1} + k_{p1}x_{p1} - F_{bx1} = 0 \\
& m_{p1}\ddot{y}_{p1} + (c_{p1} + c_{r1})\dot{y}_{p1} - c_{r1}\dot{y}_{r1} + (k_{p1} + k_{r1})y_{p1} - k_{r1}y_{r1} - F_{by1} = 0 \\
& m_{r1}\ddot{y}_{r1} + c_{r1}(\dot{y}_{r1} - \dot{y}_{p1}) + k_{r1}(y_{r1} - y_{p1}) = 0 \\
& m_{p2}\ddot{x}_{p2} + c_{p2}\dot{x}_{p2} + k_{p2}x_{p2} - F_{bx2} = 0 \\
& m_{p2}\ddot{y}_{p2} + (c_{p2} + c_{r2})\dot{y}_{p2} - c_{r2}\dot{y}_{r2} + (k_{p2} + k_{r2})y_{p2} - k_{r2}y_{r2} - F_{by2} = 0 \\
& m_{r2}\ddot{y}_{r2} + c_{r2}(\dot{y}_{r2} - \dot{y}_{p2}) + k_{r2}(y_{r2} - y_{p2}) = 0
\end{aligned} \quad (4)$$

In which x_{pi}, y_{pi}, x_{ri} ($i = 1,2$) represents the degrees of freedom for the bearing 1 and 2, m_{pi}, m_{ri}, c_{pi}, c_{ri} and k_{pi}, k_{ri} ($i = 1,2$), respectively, denote the mass, damping and stiffness of bearing 1 and 2. Obviously, considering the degrees of freedom of the rotor, pedestal and high-frequency spring-mass system, there are eight degrees of freedom for the rotor-bearing system. Due to the nonlinear restoring forces induced by the rolling bearing, Eq. (4) could only be solved by the numerical integration method. In the following section, we will show the fault features of the dynamic response for the system under various load conditions, including the non-weight-dominant condition.

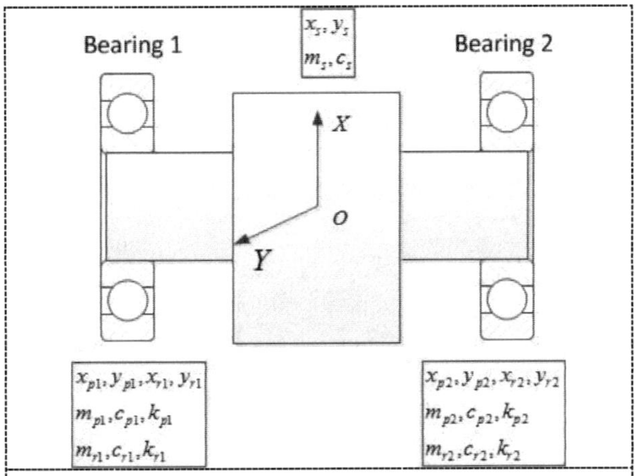

Figure 2. Rotor-bearing model

3. Simulations and discussions

In the simulation, the parameters for the rotor-bearing system are shown in Table. 1. Two types of load conditions are considered: (1) only weight condition (just the weight-dominant condition); (2) only unbalanced force condition (just the non-weight-dominant condition). In the following analysis, the envelope spectra for the rolling bearing without defect, with inner race defect and with outer race defect are obtained to show the values and distributions of fault frequencies.

Table 1. System parameters

Parameters	Values	Parameters	Values
Rotor mass (Kg)	3	Pedestal mass (Kg)	12.638
Rotor damping (N/s)	1200	Pedestal stiffness (N/m)	15e6
Rotor stiffness (N/m)	1e6	Pedestal damping (N/s)	2210
Unbalanced mass (Kg)	0.01	Spring's mass (Kg)	1
Mass eccentricity (m)	5e-4	Spring's damping (N/s)	9500
Bearing pitch diameter (m)	0.0449	Spring's stiffness (N/m)	8e9
Ball diameter (m)	0.0072		
Contact stiffness (N/m$^{3/2}$)	8e9		
Number of balls	8		
Radial clearance (m)	1e-5		

3.1. Only rotor self-weight condition (weight-dominant condition)

The unbalanced force excitation is ignored in this section in order for comparisons with the following analysis. Rotor speed is 20 rad/s, and the localized defect parameters are assumed to be $\theta_{do} = 0.1rad$, $\phi_{do} = 1e-3rad$ and $h_o = 1e-3m$ for the outer race defect; $\theta_{di} = 0.1rad$, $\phi_{di} = 1e-3rad$ and $h_i = 1e-3m$ for the inner race defect. The envelope spectra for the system with outer race defect and inner race defect are, respectively, shown in Fig. 3. It is shown that the fault characteristic frequencies are consistent with current results. For the outer race defect, the characteristic frequencies could be expressed by nf_o ($n = 1, 2, 3, ...$), where $f_o = N_b \omega_{bc}$ denotes the ball passing frequency. When the inner race has a localized defect, one can see from Fig. 3(b) that the characteristic frequencies are the combinations of inner race fault frequency with rotating speed, i.e. $|nf_i + m\omega_s|$ ($n = 0, 1, 2, \ldots, m = \ldots, -2, -1, 0, 1, 2, \ldots$). These frequencies are also identified and marked in Fig. 3.

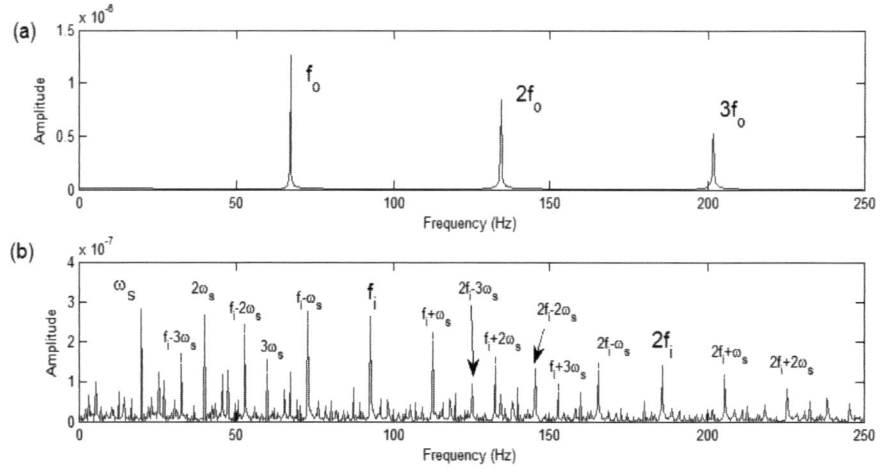

Figure 3. Envelop spectra for the system under only weight condition: (a) outer race defect; (b) inner race defect.

3.2. Only unbalanced force excitation condition (non-weight-dominant condition)

For the lightly rotor, especially the vertical rotor, its weight might be small enough and even could be ignored in the analysis. In this section, only the unbalanced force is considered in the simulation. The envelop spectra for the system with outer race defect and inner race defect are present in Fig. 4. Compared with the weight-dominant condition (Fig. 3), one can see that the fault frequencies differ distinctly. For the outer race defect, it is shown from Fig. 4(a) that the fault frequencies are relatively complex. They could be summarized as: $|nf_o + m\omega_s|$ ($n = 0, 1, 2, \ldots, m = \ldots, -2, -1, 0, 1, 2, \ldots$). When the inner race defect is consider, as shown in Fig. 4(b), it is shown that the fault frequencies are related to the inner race frequency and odd times of rotating speed, i.e. $|nf_i + m\omega_s|$ ($n = 0, 1, 2, \ldots, m = \ldots, -4, -2, 0, 2, 4, \ldots$). Although the fault frequencies are still the combinations of inner race frequency and rotating speed, the difference is that only the odd times of rotating speed are found.

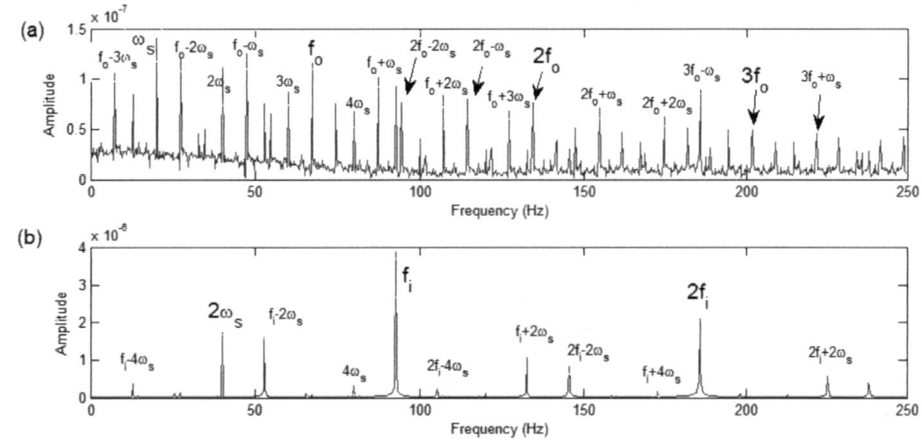

Figure 4. Envelop spectra for the system under only unbalanced force condition: (a) outer race defect; (b) inner race defect.

4. Conclusions

The fault characteristics of a rolling bearing with localized defects under non-weight-dominant conditions are studied. Through envelope analysis, the fault characteristic frequencies of inner/outer raceway defects for various conditions, including the weight-dominant condition and non-weight-dominant condition, are presented and compared with each other. It is shown that the fault frequencies for the non-weight-dominant condition differ distinctly with that of the weight-dominant condition. Experiments on a vertical rotor with defective rolling bearing are being conducted to verify the theoretical results.

References

[1] Sopanen J and Mikkola A 2003 Dynamic model of a deep-groove ball bearing including localized and distributed defects. Part 1: theory *Proc. Instn Mech. Engrs Part K: J. Multi-body Dynamics* 217 201-211.

[2] Zeki K and Hira K 2003 Simulation and analysis of vibration signals generated by rolling element bearing with defects *Tribology International* **36** 667–678.

[3] Sawalhi N and Randall R B 2008 Simulating gear and bearing interactions in the presence of faults Part I. The combined gear bearing dynamic model and the simulation of localised bearing faults *Mechanical Systems and Signal Processing* **22** 1924–1951.

[4] Rafsanjania A, Abbasiona S, Farshidianfarb A 2009 Nonlinear dynamic modeling of surface defects in rolling element bearing systems *Journal of Sound and Vibration* **319** 1150–1174.

[5] Ashtekar A, Sadeghi F and Stacke L-E 2010 Surface defects effects on bearing dynamics *Proc. IMechE Vol. 224 Part J: J. Engineering Tribology* **224** 25-35.

[6] Patel V. N., Tandon N and Pandey R K 2010 A Dynamic Model for Vibration Studies of Deep Groove Ball Bearings Considering Single and Multiple Defects in Races *Journal of Tribology* **132** 041101-1.

A new angular resampling algorithm for the bearing fault diagnosis under the time-varying rotational speed

Tianyang Wang[1], Weidong Cheng[2], Jianyong Li[2], Fulei Chu[1]
[1]Department of Mechanical Engineering, Tsinghua University, Beijing 100084, China
[2]School of Mechanical Electronic and Control Engineering, Beijing Jiaotong University, Beijing 100044, China

E-mail: wty19850925@126.com

Abstract: Order tracking is one of the most effective algorithms to eliminate the effect of time-varying rotational speed on the rotary machines. However, this algorithm is not suitable for the faulty rolling bearing unless the peak time of the fault-induced impulse is set as zero which cannot be met in the real engineering. The traditional resampling process will cause uneven intervals between the adjacent impulse peaks in the angular domain and then affect the envelope analysis-based diagnosis result. To solve this problem, a new resampling algorithm with three parts is proposed: (a) linearly fitting the instantaneous rotational speed measured by the tachometer, (b) resampling the vibration signal from the time domain to the angular domain using the traditional method, (c) calculating the envelope deformation amount and then compensating the resampled result. The effectiveness of the proposed method has been validated by both the simulated and experimental bearing vibration signals.

1. Introduction

Rolling element bearings have been widely used in different rotating equipment. Reliable bearing conditional monitoring is crucial to avoid the failure or degraded performance of the whole machine [1]. In real engineering, the bearings always operate under time-varying speed which will then make the bearing vibration signal non-stationary and the techniques based on the assumption of constant rotating speed inapplicable [2, 3]. To solve this problem, the order tracking algorithm was proposed to remove the effects of speed fluctuation and smearing of the spectrum by resampling the original vibration signal at a constant angle increment and converting the non-stationary signal in time domain into the stationary one in angular domain [4-6].

However, when the resampling algorithm is used to analysis the rolling bearing with time-varying rotational speed, it will cause the envelope deformation [7] which may then effect the final bearing fault diagnosis. To solve this problem, a new resampling algorithm is proposed to compensate the deformation. At first, the distribution character of the peak time under the time-varying rotational speed is deduced. The rules of envelope deformation is then summarized based on the analysis in [7]. And the new resampling algorithm based on the piecewise fitting and envelope deformation compensation is finally put forward.

The rest of the paper is organized as follows: Section 2 deduces the distribution character of peak time of bearing fault-induced impulse and presents the rules of envelope deformation caused by the traditional angular resample algorithm. Section 3 presents the new resampling method. Section 4 and 5 examine the performance of the new resampling algorithm using the simulated and experimental data respectively. The conclusions are drawn in Section 6.

2. The principle of envelope deformation

As proposed in [7], the envelope deformation happens when the traditional angular resampling algorithm is used to analysis the faulty rolling bearing signal under the time-varying rotational speed. And the essential reason of this phenomenon is the existence of peak time and the corresponding effect on the traditional angular resampling algorithm. The distribution character of the peak time under time-varying rotational speed is firstly obtained, and the envelope deformation rule is then presents with the assumption of the constant peak time.

2.1 The distribution character of the peak time

In general, when a rolling element bearing has a local fault, the contact between the defect and its mating surface will generate an impulse with a short duration in which the peak time (t_p) is defined as the time interval between the starting point and the maximum value. In this section, the real response of the bearing fault impulse is simulated with the time response of the undamped spring-mass system to a half sine wave excitation. Eq. (1) represents the impulse and the corresponding response can be divided into two phase: load stage ($0 < t < T=\pi/\omega$) and unload stage ($t > T=\pi/\omega$).

$$F(t)=\begin{cases} P_0 \sin \omega t (0<t<T=\pi/\omega) \\ 0(\text{others}) \end{cases} \tag{1}$$

where T and ω are the duration time and the angular frequency. Using the Duhamel's integral, the peak time can be expressed:

$$t_p=\begin{cases} \frac{\pi}{2}(\frac{2p+1}{\omega_n}-\frac{1}{\omega})(\omega>\omega_n) \\ \frac{2p\pi}{\omega_n+\omega}(\omega<\omega_n) \end{cases} (p=1,2,\text{L}\ ) \tag{2}$$

According to the Eq. (2), if the ω is bigger than the ω_n, the load time is relatively small. So the maximum value happens after the load time. On the contrary, if the ω is smaller than the ω_n, the peak value happens before the load time. To evaluate the distribution character of peak time under time-varying speed, the parameters in Eq. (2) should be given. The ω can be expressed by the rolling bearing parameters:

$$\omega = \frac{\pi}{T} = \frac{\pi}{l_{fault}/v_{line}} = \frac{\pi}{l_{fault}/(\omega_{bearing} \times \pi \times d_{pitch})} \tag{3}$$

where, l_{faul} is width of local fault; v_{line} is the linear velocity fault point; $\omega_{bearing}$ is the rotational frequency of the rolling bearing; d_{pitch} is the corresponding pitch diameter. So, the Eq. (2) can be expressed by:

$$t_p = \begin{cases} \frac{\pi}{2}(\frac{2p+1}{\omega_n} - (\frac{\pi}{l_{fault}/(\omega_{bearing} \times \pi \times d_{pitch})})^{-1})(\omega_n < \frac{\pi}{l_{fault}/(\omega_{bearing} \times \pi \times d_{pitch})}) \\ 2p\pi(\omega_n + \frac{\pi}{l_{fault}/(\omega_{bearing} \times \pi \times d_{pitch})})^{-1}(\omega_n > \frac{\pi}{l_{fault}/(\omega_{bearing} \times \pi \times d_{pitch})}) \end{cases} \quad (p = 1, 2, \mathrm{L}\ ) \tag{4}$$

Without loss of generality, the specific value of all the parameters in Eq. (4) are set as follow: the higher resonance frequency is set as 10000Hz, the varying range of the rotational speed is set as an extremely wide interval of [0 10000Hz], *p*, d_{pitch} and l_{fault} are separately set as 1, 33.5 mm and 1 mm.

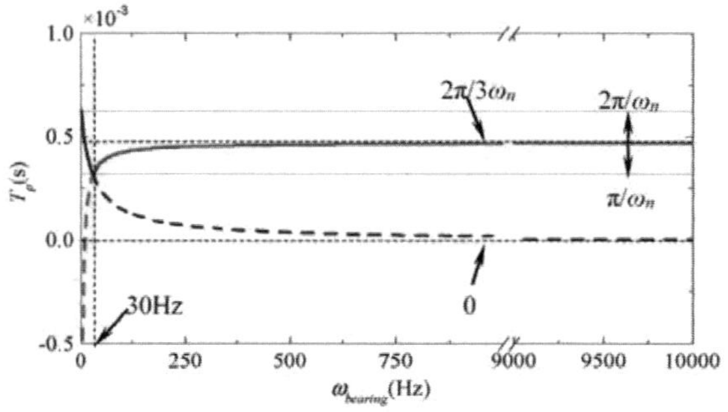

Figure 1. The rule of peak time changing with bearing rotational frequency

In Fig. 1, The blue line and red line separately represent the first and the second expression of Eq.(4). And the solid part represents the real change law of the peak time. In this way, the peak time actually changes in the interval of [π/ω_n, $2\pi/\omega_n$]. Considering the resonance frequency ω_n is relative high, the corresponding value range are usually narrow. As for the simulation mentioned above, the specific value range is [3.14×10^{-4}, 6.28×10^{-4}] which is extremely narrow. Based on the aforementioned analysis, we assume the peak time as a constant value with the time-varying rotational speed.

2.2 The presentation of the envelope deformation phenomenon

If the peak time can be considered as a constant value with the time-varying bearing rotational speed, the peak times at the higher rotational speed will be lengthened more significant than the one at lower rotational speed in the angular domain using the traditional resampling algorithm. This phenomenon of unequally spaced peaks is called envelope deformation which will then affect the envelope spectrum and the fault diagnosis of rolling bearing. In this subsection, an analytical model is constructed to specify this kind of the envelope deformation where the half-sine wave is used to simplify the fault-induced impulse, and the projection of the ascending segment on the base is the peak time. Fig. 2(a) is the schematic diagram of envelope deformation theory using the simulated signal of faulty rolling bearing with a constant accelerator. $t_{1,s}$, $t_{2,s}$ and $t_{3,s}$ are the start moments of the adjacent three impulses, $t_{1,p}$, $t_{2,p}$ and $t_{3,p}$ are the peak moments. Just because the peak time does not change with rotational speed, $t_{1,p}$ - $t_{1,s}$, $t_{2,p}$ - $t_{2,s}$ and $t_{3,s}$ - $t_{3,p}$ should equal to each other. Fig. 2(b) is the resampled angular waveform, in which $\varphi_{1,s}$, $\varphi_{2,s}$ and $\varphi_{3,s}$ are the corresponding start phases, $\varphi_{1,p}$, $\varphi_{2,p}$ and $\varphi_{3,p}$ are the peak phases. In the angular domain, the start phases are equally spaced ($\Delta\varphi_{s,2}$=$\Delta\varphi_{s,3}/2$), however, the peak phases are not.($\varphi_{1,p} - \varphi_{1,s} \neq \varphi_{2,p} - \varphi_{2,s} \neq \varphi_{3,p} - \varphi_{3,s}$).

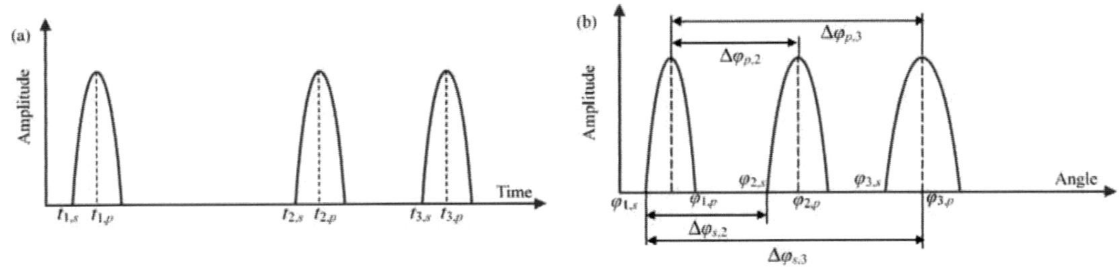

Figure 2. The schematic diagram of envelope deformation theory: (a) time-domain waveform with rotational speed up; (b) outcome of angular resample

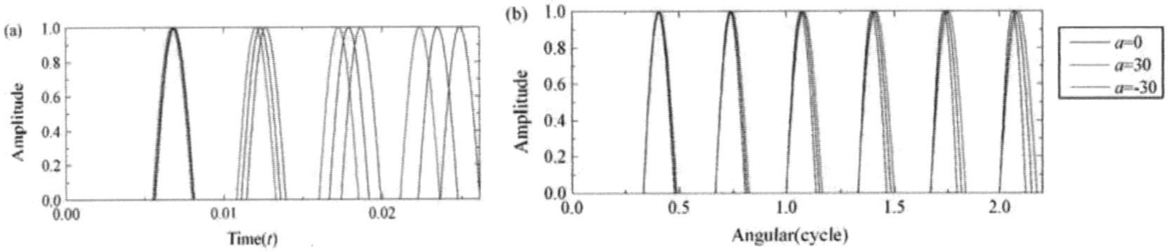

Figure 3. The schematic diagram of the envelope deformation: (a) the waveform of simulated faulty bearing vibration; (b) the angular resampling outcome using traditional order tracking algorithm

Figure 3 gives examples to visualize the envelope deformation. In Fig. 3(a), the black line represents the signal with a constant rotational speed and the red and blue signals are separately speed-up and speed-down signal with the acceleration of 30 and -30. As for the resampled results in the angular domain shown in Fig 3(b), we can see that the start phases of all the impulses are equally spaced and the peak phases are not which can be used to testify the existence of the envelope deformation.

Fig 4(a) and (b) are the corresponding order spectrums of the speed up and speed down situations. The theoretical positions of the fault characteristic frequency and its harmonics are '3', '6' and '9' which are displayed by vertical dash lines. As shown in Fig. 4(a) and Fig. 4(b), the peaks will both deviate from their theoretical positions. This can prove that the tradition angular resampling algorithm will the envelope deformation will then affect the rolling bearing fault diagnosis

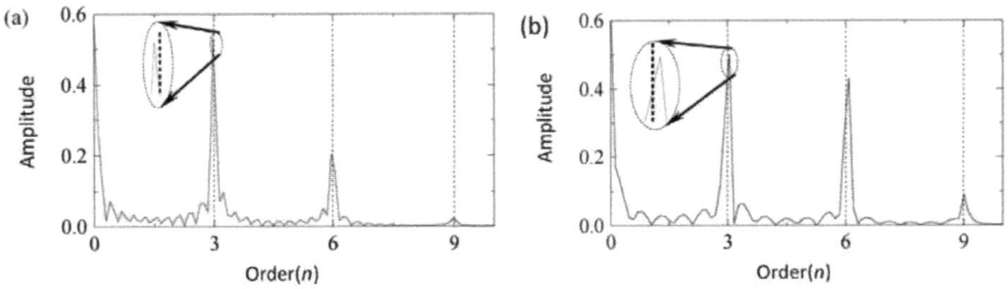

Figure 4. The order spectrums of simulated faulty bearing vibration signal under different laws of time-varying rotational speed: (a) positive acceleration; (b) negative acceleration

3. The revised angular resampling algorithm

As mentioned in [7], the relative deformation value of the nth impulse is listed as follow:

$$\Delta\varphi_{1,n} = (\varphi_{n,p} - \varphi_{n,s}) - (\varphi_{1,p} - \varphi_{1,s}) = ((2an\Delta\varphi_s + n_{1,s}^2)^{1/2} - n_{1,s})\,t_p \qquad (5)$$

where, $\varphi_{n,s}$ and $\varphi_{n,p}$ are separately the start phase and the peak phase of the nth impulse, $\varphi_{1,s}$ and $\varphi_{1,p}$ are the start phase and the peak phase of the first impulse, a is the acceleration , $n_{1,s}$ is the initial rotational speed, t_p is the peak time. Following Eq. (5), Table 1 gives the peak phases and the corresponding relative deformation value:

Table 1: The peak angle and relative deformation of the impulse waveforms

	Peak phase	Relatively deformation
The first impulse	$\varphi_{1,s} + n_{1,s} \cdot t_p + \frac{1}{2}at_p^2$	0
The second impulse	$\varphi_{1,s} + \Delta\varphi_s + (2a\Delta\varphi_s + n_{1,s}^2)^{1/2} t_p + \frac{1}{2}at_p^2$	$((2a\Delta\varphi_s + n_{1,s}^2)^{1/2} - n_{1,s})\,t_p$
The nth impulse	$\varphi_{1,s} + (n-1)\Delta\varphi_s + (2a((n-1)\Delta\varphi_s) + n_{1,s}^2)^{1/2} t_p + \frac{1}{2}at_p^2$	$((2a(n-1)\Delta\varphi_s + n_{1,s}^2)^{1/2} - n_{1,s})\,t_p$
The n+1th impulse	$\varphi_{1,s} + (n)\Delta\varphi_s + (2a(n)\Delta\varphi_s + n_{1,s}^2)^{1/2} t_p + \frac{1}{2}at_p^2$	$((2a(n\Delta\varphi_s) + n_{1,s}^2)^{1/2} - n_{1,s})\,t_p$

According to the peak phases and relatively deformation values listed in Table 1, it is found that if we subtract the relatively deformation value from the corresponding peak time, the adjusted peak time are equally spaced. $\Delta\varphi'_{n+1,n}$ is the phase interval between the adjusted nth phase peak and the n+1th phase peak.

$$\begin{aligned}\Delta\varphi'_{n+1,n} &= \varphi'_{n+1,p} - \varphi'_{n,p} = (\varphi_{n+1,p} - \Delta\varphi_{1,n+1}) - (\varphi_{n,p} - \Delta\varphi_{1,n}) \\ &= (\varphi_{1,s} + n_{1,s} \cdot t_p + \frac{1}{2}at_p^2 + n\Delta\varphi_s) - (\varphi_{1,s} + n_{1,s}t_p + \frac{1}{2}at_p^2 + (n-1)\Delta\varphi_s) = \Delta\varphi_s\end{aligned} \qquad (6)$$

As shown in Eq. (6), the phase interval after compensation become a constant value. In this way, the aforementioned envelope deformation phenomenon can be compensated by subtracting the relative deformation value from the x-axis of resampled angular waveform. In this paper, we call it the ideal compensation value. In real application, this ideal compensation value can be rewritten as follow:

$$\varphi_{c,i} = (\ (2(a(n-1)\Delta\varphi_s) + n_{1,s}^2)^{1/2} - n_{1,s})\, t_p = (\,(2a_i\Delta\varphi_i + n_b^2)^{1/2} - n_b)\, t_p \tag{7}$$

where, $\varphi_{c,i}$ is the compensation value of angular i in angular domain, a_i is the corresponding acceleration, $\Delta\varphi_i$ is angle the rolling bearing has rotated, n_b is the initial rotational speed.

The a_i and t_p can be determined by Eq. (8) and (9)

$$a_i = \frac{IF_{i+1} - IF_i}{\Delta t} \tag{8}$$

$$t_p = \frac{3\pi}{2\omega_n} \tag{9}$$

where, IF_{i+1} and IF_i are the corresponding instantaneous rotational speed measures by the speed sensor. The value of t_p is the mean of interval of $[\pi/\omega_n, 2\pi/\omega_n]$ mentioned in section 2.1. ω_n is the resonance frequency whose value can be determined by the kurtogram-based spectrum kurtosis[8-10].

Above all, the revised resampled algorithm is presented as follow:

STEP 1: Determine the variation trend of IF piecewise.

According to the number of the discrete IFs measured by the speed sensor, the raw signal is divided into n segments. We assume the acceleration of every segment as a constant value a_i ($0 < i < n$-1), the IF in every segment can be expressed by:

$$\text{IF}(t) = k_i t + b_i (\frac{T}{n-1}(i-1) < t < \frac{T}{n-1} i) \tag{10}$$

where, T is the time duration of the signal, n is the number of the segments, k_i and b_i are the coefficients.

STEP 2: Determine the angular axis of every time-domain sampled point in angular domain.

$$\text{Angle}(t) = \sum_{i=1}^{k} \frac{T}{2(n\text{-}1)} (\text{IF}(i) + \text{IF}(i+1)) + \int_{\frac{T}{n\text{-}1}k}^{t-\frac{T}{n\text{-}1}k} a_{k+1} t + b_{k+1} dt \tag{11}$$

where k is the number of the segments before t.

STEP 3: Calculate the compensation value.

$$\begin{aligned}\text{Angle}_c(t) &= (\ (2a_i\Delta\varphi_i + n_b^2)^{1/2} - n_b)\, t_p \\ &= (\,(2\frac{IF_i - IF_{i-1}}{T/(n\text{-}1)}\text{Angle}(t) + IF^2(1))^{1/2} - IF(1))\frac{2\pi}{3\omega_n}((\frac{T}{n\text{-}1}(i-1) < t < \frac{T}{n\text{-}1} i))\end{aligned} \tag{12}$$

STEP 4: Compensate the envelope deformation.

$$\text{Angle}'(t) = \text{Angle}(t) - \text{Angle}_c(t) \tag{13}$$

STEP 5: Determine the sample points in the angular domain.

$$\frac{f_s}{\text{IF}_{base}} : \frac{f_s}{\text{IF}_{base}} : \sum_{i=1}^{n\text{-}1} \frac{T}{2(n\text{-}1)} (\text{IF}(i) + \text{IF}(i+1)) \qquad 14)$$

where f_s/IF_{base} is the sample rate of the angular domain, IF_{base} is minimum of the measured IF.

STEP 6: Complete the resample process using the interpolation algorithm.

4. Simulation analyses

In this section, the simulated signal in Fig. 3(a) are used to testify the effectiveness of the propose algorithm. In specific, the change rules of the red and blue are separately $f(t)$=30t+20 and $f(t)$=-30t+20. Fig. 5 and Fig. 6 show the results using the algorithm described in Section 3. Fig. 5(a) and 6(a) are the compensation values; Fig. 5(b) and 6(b) are the corresponding envelope spectrum. It is obvious that the peaks representing fault characteristic frequency and corresponding harmonics in Fig. 5(b) and 6(b) are all compensated to its theoretical values, '3', '6' and '9'. It is proved that the revised angular resampled algorithm are effective both under speed up and speed down situations.

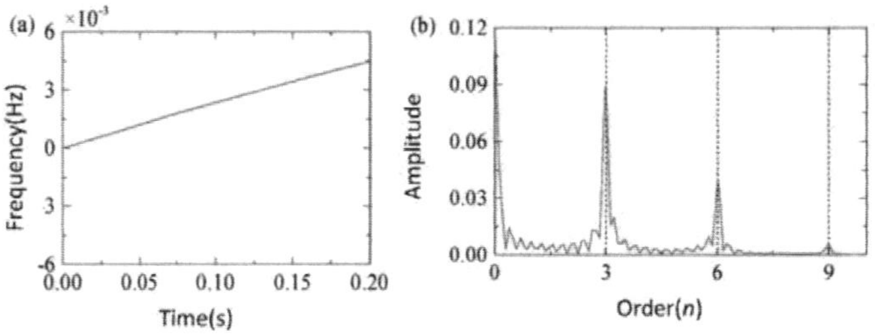

Figure 5. Speed-up model: (a) the compensation values; (b) the order envelope spectrum

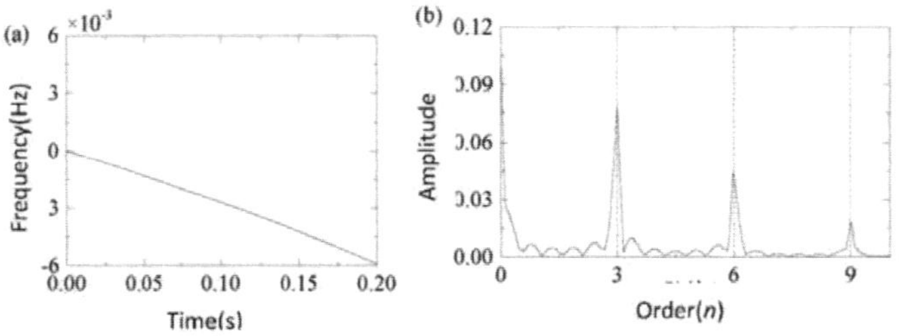

Figure 6. Speed-down model: (a) the compensation values; (b) the order envelope spectrum

5. Experimental tests

The effectiveness of the proposed algorithm is further tested using a set of experimental bearing vibration signals with fault in the outer race.

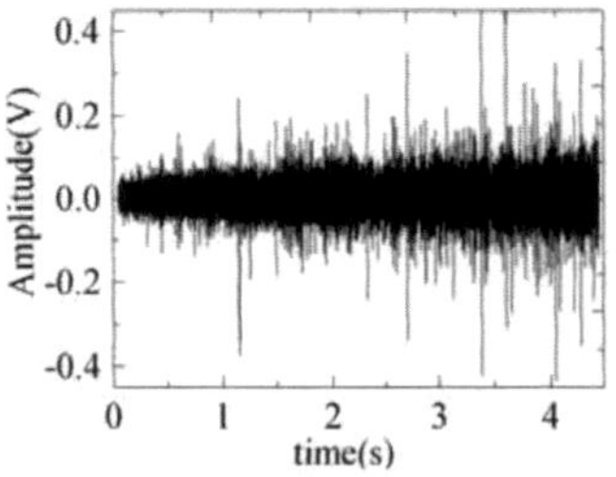

Figure 7. Raw signal with the increasing speed

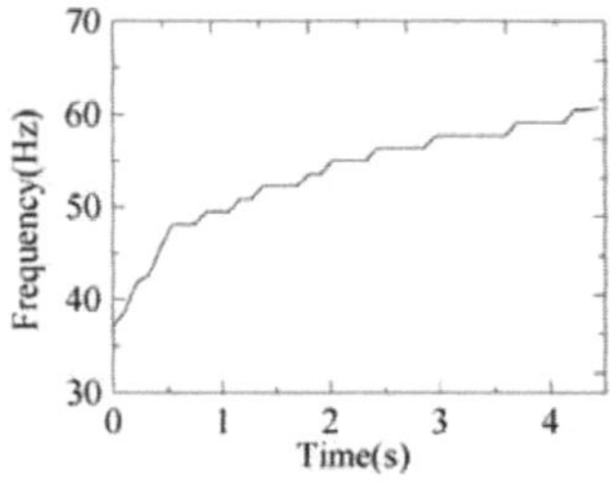

Figure 8. The measured IF of the rolling bearing

The corresponding rotational frequency of the rolling bearing increases from 38 Hz to 60 Hz. The sampling rate is 24,000 samples/s. The type of rolling bearing is ER16K, the fault characteristic frequency equals to 3.45×f_r. The data lasts for 4.3 s. Fig. 7 and Fig. 8 are separately the waveform and the corresponding IF. Fig. 9 shows the resampled waveform using the revised resampling algorithm and the corresponding order envelops spectrum is shown in Fig.10. For comparison, Fig. 11 shows the order envelope spectrum using the traditional resampling algorithm. In Fig. 10 and 11, there both exist three vertical dash lines which represent theoretical fault characteristic frequency and its second and third harmonics. Their x-axis coordinates are 3.45, 6.9 and 10.35.

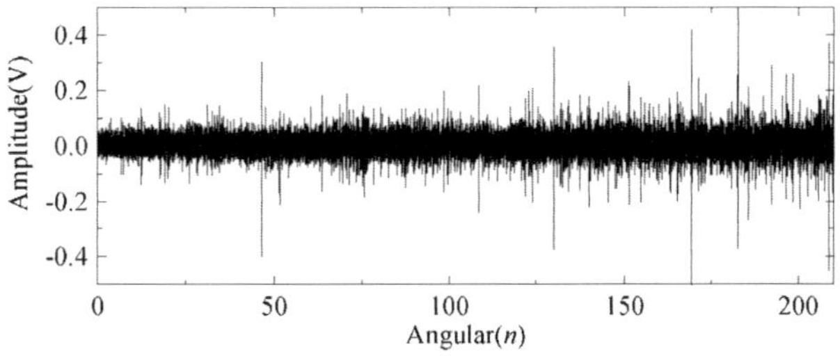

Figure 9. The resampled waveform using the revised resampling algorithm

It is obvious that the peaks representing the fault characteristic frequency and its harmonics in Fig 11 are not on their theoretical value. In Fig. 10, the revised resampling algorithm can compensate the envelope deformation and the corresponding fault characteristic order and harmonics are all adjusted to the theoretical value. This can testify the effectiveness of the proposed algorithm.

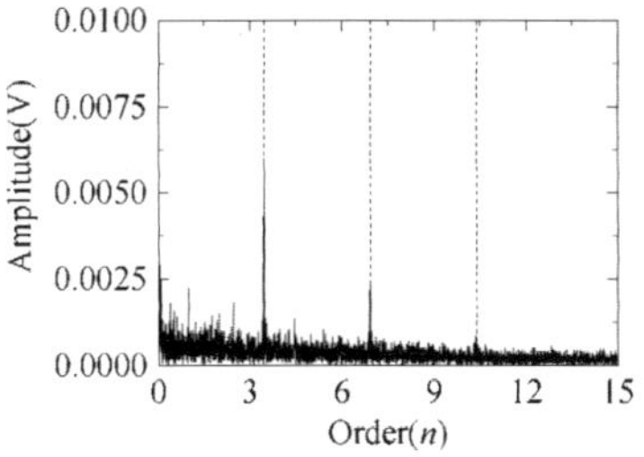

Figure 10. The order envelope spectrum using the revised resampling algorithm

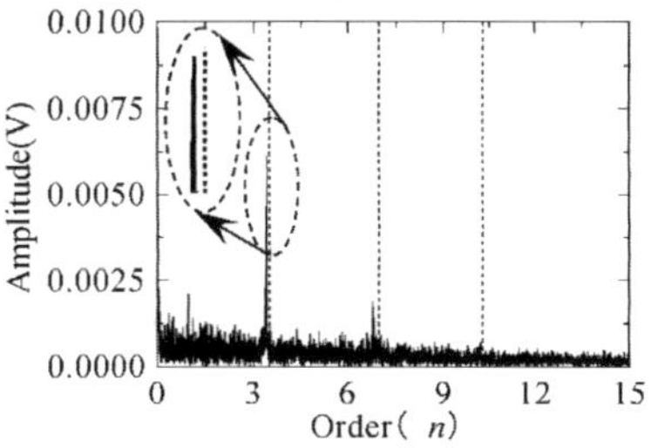

Figure 11. The order envelope spectrum using the traditional resampling algorithm

6. Conclusions

A revised resampling algorithm has been proposed to deal with the envelope deformation phenomenon caused by the traditional resampling algorithm. And the corresponding order envelope spectrums are more accurate than the ones obtained using traditional resampling algorithm. It is proved that this revised resampling algorithm is effectiveness under both the speed up and speed down situations.

Acknowledgments

The research work described in this paper was supported by Natural Science Foundation of China under Grant no.51335006 and Beijing Natural Science Foundation under Grant no.3131002.

References:

[1]. R B Randall and J Antoni 2001 Rolling element bearing diagnostics—a tutorial *Mechanical system signal processing* **25** 485

[2]. McFadden P D and J D Smith 1984 Vibration monitoring of rolling element bearings by the high-frequency resonance technique-a review *Tribology international* **17** 3

[3]. Wang T Y, Liang M, Li J Y and Cheng W D 2014 Rolling element bearing fault diagnosis via fault characteristic order (FCO) analysis *Mechanical system signal processing* **45** 139

[4]. Borghesani P, Ricci R, Chatterton S, and Pennacchi P 2013 A new procedure for using envelope analysis for rolling element bearing diagnostics in variable operating conditions, *Mechanical system signal processing* **38** 23

[5]. Gao Y, Guo Y, Chi Y L, Qin S R 2006 Order Tracking Based on robust peak search instantaneous frequency estimation. *J. Phys.: Conf. Ser* **48** 479

[6]. Fyfe K R and E D S Munck 1997 Analysis of computed order tracking *Mechanical system signal processing* **11** 187

[7]. Cheng W D, Gao R X, Wang J J, Wang T Y, Wen W G and Li J Y 2014 Envelope deformation in computed order tracking and error in order analysis *Mechanical system signal processing* **48** 92

[8]. Antoni J 2007 Fast computation of the kurtogram for the detection of transient faults *Mechanical system signal processing* **21** 108

[9]. Antoni J 2006 The spectral kurtosis: a useful tool for characterizing non-stationary signals, *Mechanical system signal processing* **20** 282

[10]. Antoni J and R B Randall 2006 The spectral kurtosis: application to the vibratory surveillance and diagnostics of rotating machines *Mechanical system signal processing* **20** 308

Impact Damage Evaluation Method of Friction Disc Based on High-Speed Photography and Tooth-Root Stress Coupling

L Yin, Y M Shao, J Liu and H L Zheng
State Key Laboratory of Mechanical Transmission, Chongqing University, PR China

E-mail: ymshao@cqu.edu.cn

Abstract. The stability of friction disc could be seriously affected by the tooth surface damage due to poor working conditions of the wet multi-disc brake in heavy trucks. There are few current works focused on the damage of the friction disc caused by torsion-vibration impacts. Hence, it is necessary to investigate its damage mechanisms and evaluation methods. In this paper, a damage mechanism description and evaluation method of a friction disc based on the high-speed photography and tooth-root stress coupling is proposed. According to the High-Speed Photography, the collision process between the friction disc and hub is recorded, which can be used to determine the contact position and deformation. Combined with the strain-stress data obtained by the strain gauge at the place of the tooth-root, the impact force and property are studied. In order to obtain the evaluation method, the damage surface morphology data of the friction disc extracted by 3D Super Depth Digital Microscope (VH-Z100R) is compared with the impact force and property. The quantitative relationships between the amount of deformation and collision number are obtained using a fitting analysis method. The experimental results show that the damage of the friction disc can be evaluated by the proposed impact damage evaluation method based on the high-speed photography and tooth-root stress coupling.

1. Introduction

Wet multi-disc brakes are widely used in heavy trucks due to the big braking torque, ability to resist thermal recession and pollution. During the work process, the stability of the wet multi-disc brake could be seriously affected by the damage of the friction disc tooth in poor working conditions. Therefore, it is important for researchers to investigate the damage mechanisms of the friction disc tooth to improve the stability and useful life of the wet multi-disc brake.

A significant amount of research works has been presented to investigate the wear and thermal damage of friction pairs. For instant, Zhang et al. [1] studied the wear law of a clutch friction plate by the experimental method. Li et al. [2] used the surface analysis techniques and iron spectrum technology to study the wear failure mechanisms of a clutch friction plate. Chen et al. [3] analyzed the thermal stress of a friction plate, and obtained the temperature and thermal stress distributions under the thermal loads. Przemyslaw et al. [4] investigated the local high temperature regions of a friction plate by the temperature simulation analysis on the combination of clutch. Zagrodzki [5] analyzed the influences of the different pressure distributions on the thermal stress based on a finite element analysis of thermal elastic theory. As for the methods that can be used to evaluate the impact damage, some researchers adopted the finite element analysis [6], nonlinear fatigue damage theory [7] and high-speed digital cameras [8].

However, there are few works focused on the damage of the friction disc tooth caused by the torsion-vibration impact. Hence, in order to improve the stability and useful life of the wet multi-disc brake, it is very useful to study the damage mechanisms and evaluation methods of the friction disc. In this work, an impact damage evaluation method of a friction disc based on the high-speed photography and tooth-root stress coupling is proposed to investigate its damage mechanisms.

2. Experiment and impact damage evaluation method

2.1. Experiment setup

To observe the actual impact damage of the friction disc tooth caused by the torsion-vibration impact, an experiment setup is carried out. A impact damage experimental system consists of the impact test rig of friction disc, strain signal acquisition module, ultra high speed photoelectric framing camera, coupling, drive shaft, strain measuring instrument, 3D super depth digital microscope (VH-Z100R), as shown in figure 1.

The specimen material is 45 steel. By the strain gauges at roots of the teeth, the strains of the friction disc tooth are measured under torsion-vibration impact loads. A repeated impact load is applied to the friction disc, and the impact frequency is determined by the frequency conversion motor to meet the requirements of the impact frequency range. The tooth surface elastic-plastic deformation of the friction disc under a single load is obtained by an ultra high speed photoelectric framing camera. The real-time root strain data of the friction disc is obtained by the strain measuring instrument. The tooth surface cumulative elastic-plastic deformation of the friction disc under the repeated impact load is obtained by the 3D super depth digital microscope (VH-Z100R).

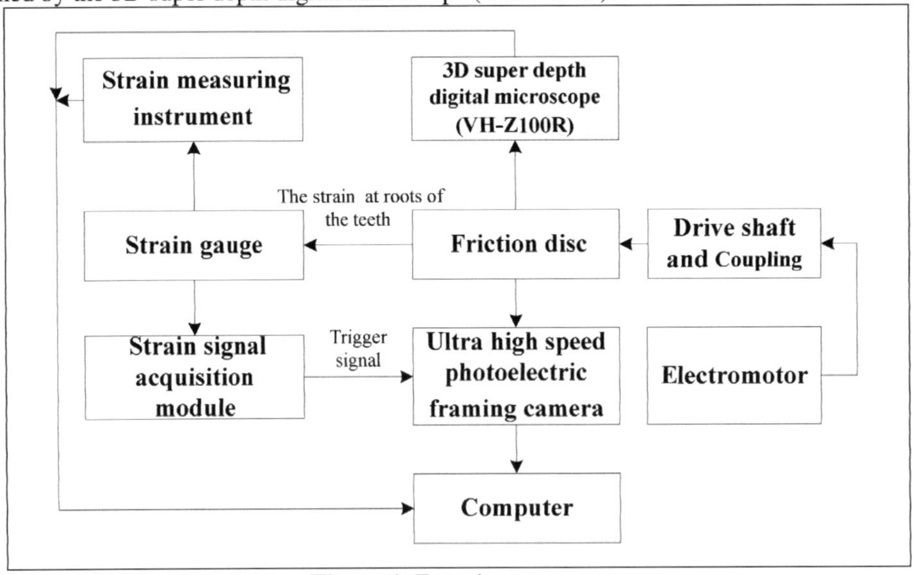

Figure 1. Experiment setup.

2.2. Ultra high speed photoelectric framing photography

The ultra high speed photoelectric framing camera is used to output object by six channel using the prism spectrometer system, as shown in figure 2. Same images can be output from the each channel output by using a fine adjustment light path. The images from the six channels are coupled to the photocathode of the six image-intensifiers. The high-pressure gating pulse from the each channel is controlled by the precise timing circuit, respectively. Therefore, it can control the exposure moment and exposure time of the six image-intensifiers independently. The six CCD cameras are used to record the output image of image-intensifiers. The camera can acquire six images of the target in different time in a shot, the exposure time of the each image is 5 ns, the picture interval is adjustable from 5 ns to 10 ms. Compared with the high-speed camera, using a camera to record the high-speed collision process can greatly shorten the exposure time, which can eliminate the boundary motion blur caused by the movement of the object. Shortening the exposure time does not affect the spatial resolution of the camera.

The target is imaged on the cathode plane of the gating type micro-channel panel image intensifier through the main objective, second objective and spectroscopic prism. The image amplified by the image-intensifier is coupled to the CCD chip. The images from the CCD camera are displayed and processed in a computer. Shooting frequency and single exposure time are set by the synchronization control part and special pulsed power supply of gating type micro-channel panel image intensifier.

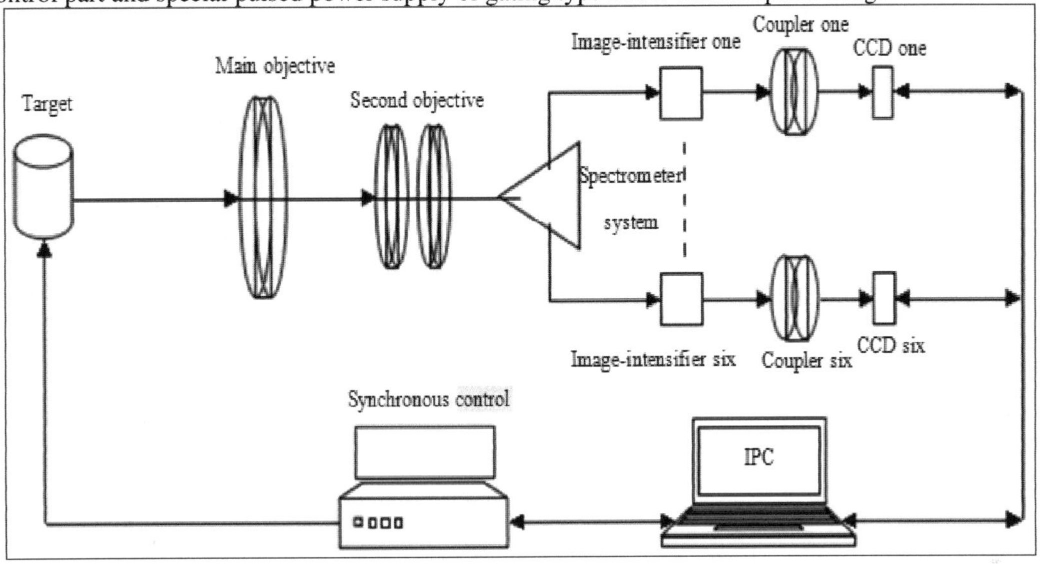

Figure 2. Schematic diagram of the ultra high speed photoelectric framing camera.

2.3. *Impact damage evaluation method*

According to linear cumulative damage theory, a method for estimating the impact damage life for a wet multi-disc brake of heavy trucks under load spectrum is proposed. For a constant stress amplitude σ, the friction disc may be fail when the number of cyclic loading is N .The large plastic deformation occurs on the contact surface between the contact components could cause their damage and failure. If the number of the cyclic loading is n less than N , the damage rate is equal to n/N for the constant stress amplitude σ .The tooth surface impact damage of the friction disc under a single load can be assumed to be same [9].

If a component is given in a series of stress levels, each stress level will cause an amount of impact damage rate. The friction disc will fail when the sum of these damage rates is equal to level one. Therefore, the impact damage evaluation expression can be written as follows:

$$\sum_{i=1}^{i}\frac{n_i}{N_i}=\frac{n_1}{N_1}+\frac{n_2}{N_2}+......+\frac{n_i}{N_i}=1 \tag{1}$$

Where $\sum_{i=1}^{i}\frac{n_i}{N_i}$ can be used to evaluate the impact damage of the friction disc.

The following steps are presented to verify the impact damage evaluation method based on the linear cumulative damage theory.

- d_i is the tooth surface plastic deformation of the friction disc of a single load obtained by the ultra high speed photoelectric framing camera and image processing technology for a stress level σ_i. The photograph from the ultra high speed photoelectric framing camera is shown in figure 3.
- The actual stress σ and number of cyclic impact loading n can be measured with the strain measuring instrument. The stress spectrum of the actual stress σ is obtained by rain flow counting method. n_i should match the following conditions:

$$\sum_{i=1}^{i} n_i = n_1 + n_2 + ...+ n_i = n \tag{2}$$

- According to linear cumulative damage theory, the total plastic deformation of the cumulative impact damage is calculated by the expression as follows:

$$D' = n_1 d_1 + n_2 d_2 + ... + n_i d_i \tag{3}$$

- The actual total plastic deformation D is obtained by the VH-Z100R. A photograph from the VH-Z100R is shown in figure 4. The absolute error is determined by the absolute value of the difference between D' and D. The absolute error and relative error are analyzed to validate the impact damage evaluation method. Therefore, the absolute error can be written as follows:

$$AE = |D' - D| \tag{4}$$

and the relative error can be written as follows:

$$RE = \frac{|D' - D|}{D} \tag{5}$$

Figure 3. Photograph from the ultra high speed photoelectric framing camera.

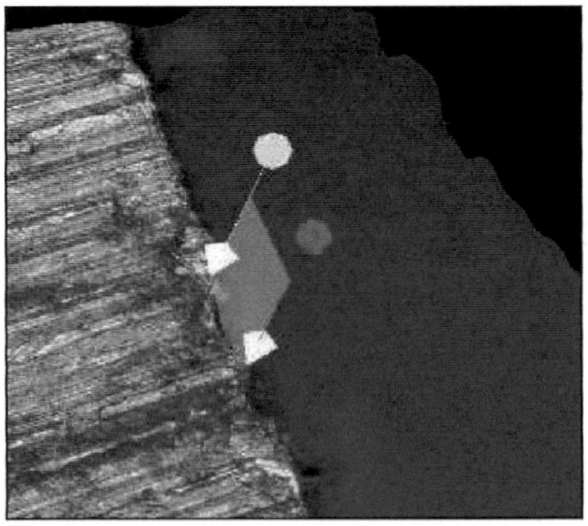

Figure 4. Photograph from the VH-Z100R

For a same stress level, the plastic deformation values d on the tooth surface are 15.6μm, the minimum measurable height is 15.6μm, as shown in figure 5. Therefore, the value between 0 and 15.6μm is displayed as 15.6μm. The plastic deformation d is 15.6μm. The number of the cyclic impact loading n is 10. The total plastic deformation D'=156μm. The actual total plastic deformation D is 24.2μm as shown in figure 6. The absolute error AE=131.8μm, and the elative error: RE=544%.

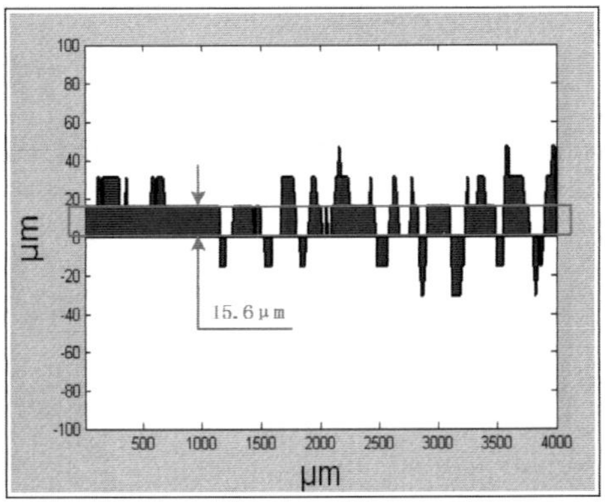

Figure 5. The plastic deformation on the tooth surface of the friction disc for a single load.

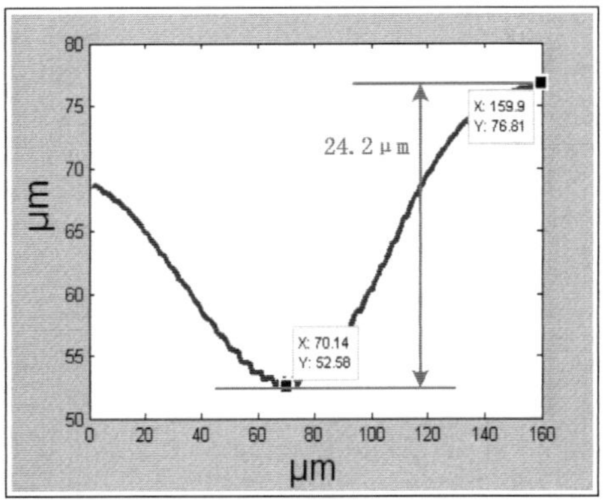

Figure 6. The actual total plastic deformation of the friction disc.

3. Conclusions

- The instantaneous impact state of the friction disc under the impact loading, and the tooth surface impact damage of the friction disc under a single load are obtained by the ultra high speed photoelectric framing camera. date published.
- The cumulative damage value is obtained by the VH-Z100R, and the surface morphology characteristics of damage location are presented.
- According to linear cumulative damage theory, a method for estimating the impact damage based on the high-Speed photography and tooth-root stress coupling is proposed. The results show that the proposed method can provide some valuable conclusions for the impact damage evaluation method for the friction disc.

Acknowledgments

The authors are grateful for the financial support provided by the project of the National Key Scientific Instrument and Equipment Development Projects of China under Contract No. 2011YQ13001906.

References

[1] Zhang X Y, Liu W Y, Tao F H and Wang P 2002 A study on the wear rules of the friction disk *Lubrication Engineering* **5** 44-45
[2] Li L, Zhu Y L, Huang Y L, Xu B S and Chen J M 2005 Failure mechanism of the driving clutch of a heavy loaded pedrail vehicle based on surface analysis and ferrography *Tribology* **25** 270-274
[3] Chen G J, Yin X L, Gong Y P and Yin L P 2008 Coupling heat stress analysis of the friction disk in clutch *Mechanical & Electrical Engineering Magazine* **25** 100-102
[4] Przemyslaw Zagrodzki and Todd D. 1998 Analysis of Temperatures and Stresses in Wet Friction Disks Involving Thermally Induced Changes of Contact Pressure *SAE Transactions* **107** 360-367
[5] Zagrodzki P and Truncone S.A. 2003 Generation of hot spots in a wet multidisk clutch during short-term engagement *Wear* **254** 474-491
[6] Dean G and Wright L 2003 An evaluation of the use of finite element analysis for predicting the deformation of plastics under impact loading *Polymer Testing* **22** 625-631
[7] Zhang W H, Ren T H and Qiu Z H 2005 Analysis of impact fatigue damage in hammer foundation *Journal of Vibration Engineering* **18** 506-511
[8] Timothy E, Schmidt, John Tyson, Konstantin Galanulis, Duane M Revilock and Matthew E

Melis 2005 Full-field dynamic deformation and strain measurements using high-speed digital cameras *26th International Congress on High-Speed Photography and Photonics* (Alexandria, VA, United states, 20-24 September 2004) vol 5580 (SPIE) p 174-185
[9] Zhao S B Study 2000 On the accuracy of fatigue life predictions by the generally used damage accumulation theory *Journal of Mechanical Strength* **22** 206-209.

Damage assessment for wind turbine blades based on a multivariate statistical approach

David García [1], Dmitri Tcherniak [2] and Irina Trendafilova [1]
[1]Mechanical & Aerospace Engineering, University of Strathclyde, 75 Montrose street, G11XJ, Glasgow, UK
[2]Bruel & Kjær Sound & Vibration Measurement A/S, Skodsborgvej 307, Nærum, Denmark
E-mail: david.garcia@strath.ac.uk

Abstract. This paper presents a vibration based structural health monitoring methodology for damage assessment on wind turbine blades made of composite laminates. Normally, wind turbine blades are manufactured by two half shells made by composite laminates which are glued together. This connection must be carefully controlled due to its high probability to disbond which might result in collapse of the whole structure. The delamination between both parts must be monitored not only for detection but also for localisation and severity determination. This investigation consists in a real time monitoring methodology which is based on singular spectrum analysis (SSA) for damage and delamination detection. SSA is able to decompose the vibratory response in a certain number of components based on their covariance distribution. These components, known as Principal Components (PCs), contain information about of the oscillatory patterns of the vibratory response. The PCs are used to create a new space where the data can be projected for better visualization and interpretation. The method suggested is applied herein for a wind turbine blade where the free-vibration responses were recorded and processed by the methodology. Damage for different scenarios *viz* different sizes and locations was introduced on the blade. The results demonstrate a clear damage detection and localization for all damage scenarios and for the different sizes.

1. Introduction

Vibration-based Structural Health Monitoring (VSHM) has been widely used for damage assessment in a multitude of engineering structures based on the features of the vibration response measured along the structure [1, 2]. Nowadays, VSHM became a trend in the future techniques for monitoring the health of modern civil engineering and aerospace engineering among many others sectors. Indeed, the growth of the off-shore wind turbines place VSHM at the forefront of the contemporary research. The visual inspections of these structures are dangerous, expensive and it might requires a tedious planning, which can be particular and different for each case. The idea of develop an on-line remote system to monitor the health of the structure is the great interest for these structures.

Although failure can happen in any structural component of the wind turbine, one of the most likely parts are the turbine blades [3]. One of the most common failure mechanism in turbine occurs in the glued interface between the two shell parts of the blade. This failure is relative small compare with the total dimension of the blade and it can growth until collapse the entire part. Therefore, it is very important to detect, control and localise this kind of failures.

VSHM can be divided in two groups: model base methodologies [4] and non-model based methodologies [5]. Model based methodologies requires the existence of a model which is able to describe the behaviour of the system. However, the second group is based on pure data-driven techniques to create a reference system space where compare the changes that happen in the system.

Thus the aim of this study is to apply a data-driven technique based on Singular Spectrum Analysis (SSA). This technique is able to separate the stationary and non-stationary components from a vibratory response [6] . SSA divides the data signal in blocks with the same mean and variance over the time. Those blocks called Principal Components (PCs) are used to reduce the dimension of the system by considering the most relevant for each case of study. The projection of the reconstructed signals based on the importance of the PCs is used as a feature for damage detection and localization [7].

The papers is organized as follows: the first sections describe the methodology implemented and the following sections describe the basis and results of the application case.

2. Stochastic subspace approach for damage assessment

The stochastic subspace approach presented in this paper is an output-only measurement damage assessment method. The methodology is performed to compare two data sets: one from the undamaged system which is considered as baseline (reference data) and another from the damage system (observation data). The approach is based on Singular Spectrum Analysis (SSA) which is an extension of Principal Component Analysis able to decompose and compress the non-independent values such as time series in their covariance distribution. Stochastic subspace methods are efficient tools for system identification and hence for damage assessment [8, 9]. The projection of the observation data onto the new space reduces the distances between elements from the same system/category and on the same time the distances from different systems/categories increase.

2.1. Step 1: Baseline covariance subspace model

The discrete time-acceleration measurements taken from the undamaged system are considered as a reference data to create the new subspace.

The time-acceleration signals are represented in the frequency domain. The amplitudes values of the frequency spectrum are arranged into a vector as $\mathbf{z} = (z_1, z_2, ..., z_N)$. The new vector is embedding in a Hankel matrix form as bellow

$$\check{\mathbf{Z}}^i = \begin{pmatrix} z_1^i & z_2^i & z_3^i & \cdots & z_w^i & \cdots & z_W^i \\ z_2^i & z_3^i & z_4^i & \cdots & z_{w+1}^i & \cdots & z_{W+1}^i \\ z_3^i & z_4^i & z_5^i & \cdots & z_{w+2}^i & \cdots & z_{W+2}^i \\ z_4^i & z_5^i & z_6^i & \cdots & z_{w+3}^i & \cdots & \vdots \\ z_5^i & z_6^i & \vdots & \cdots & \vdots & \cdots & z_N^i \\ z_6^i & \vdots & \vdots & \cdots & z_N^i & \cdots & 0 \\ \vdots & \vdots & z_N^i & \cdots & 0 & \cdots & 0 \\ \vdots & z_N^i & 0 & \cdots & 0 & \cdots & 0 \\ z_N^i & 0 & 0 & \cdots & 0 & \cdots & 0 \end{pmatrix} \tag{1}$$

where W is the window length and $i = 1...M$ the number of realisations considered to build the new subspace. Each M realisation is performed in the same Hankel matrix form. The group of embedding matrices constructs a large matrix which contains the information about the oscillatory patterns of the undamaged system as $\check{\mathbf{Z}} = (\check{\mathbf{Z}}^1, \check{\mathbf{Z}}^2, ..., \check{\mathbf{Z}}^M)$. The computation of the covariance matrix of $\check{\mathbf{Z}}$ distributes the information contained in the dynamics of the undamaged system in vectors with the same mean and variance over the frequency spectrum. These vectors are namely as Empirical Orthogonal Functions (EOFs).

The EOFs are obtained by the eigen-decomposition of the covariance matrix $\mathbf{C}_{\check{Z}}$. The eigenvalues λ_k are then ordered in the diagonal matrix $\mathbf{\Lambda}_{\check{Z}}$ in decreasing order and the matrix $\mathbf{E}_{\check{Z}}$ contains their corresponding eigenvectors ρ_k written as columns.

$$\mathbf{E}'_{\check{Z}}\mathbf{C}_{\check{Z}}\mathbf{E}_{\check{Z}} = \mathbf{\Lambda}_{\check{Z}} \tag{2}$$

The EOFs are considered as the basis for the new subspace. The projection of the reference data onto the new basis yields the Principal Components (PCs) as

$$\mathbf{A} = \check{\mathbf{Z}}\mathbf{E}_{\check{Z}} \tag{3}$$

The dynamic response is then distributed in WxM number of PCs placed into the matrix $\mathbf{A}$. The PCs are allocated in the matrix $\mathbf{A}$ in decreasing order from the ones which contain more percent of the total variance to the components with the lowest percent of variance.

The dynamic response is then reconstructed by the projection of the PCs onto the EOFs. For a given set of indices $\mathcal{K}$ corresponding to the set of PCs contained in $\mathbf{A}$, the Reconstructed Components (RCs) are obtained as is shown in the equation bellow

$$R^k_{m,n} = \frac{1}{W}\sum_{w=1}^{W} A^k_{n-w}E^k_{m,w} \tag{4}$$

where $k-$eigenvectors give the kth RC at $n-$frequency between $n = 1...N$ for each $m-$realisation ($m = 1...M$) which was embedded in a $w-$lagged vectors with the maximum $W-$length.The RCs are placed in the $\mathbf{R}$ matrix.

The RCs are the new reconstructed signals based on the percent of variance contained in the PCs used in their reconstruction. Each RC contains different variance of the total reference signal (original signal). Therefore, the RCs can be used separately and independently for the damage assessment methodology.

2.2. Step 2: Subspace projection. Damage index

The RCs are well separated signals reconstructed by the components which contain the same mean and variance over the whole frequency spectrum [10]. Therefore, the new subspace allows us to project the data from the damaged systems (observation data) onto the new subspace based on the undamaged system. The representation of the data onto the new space is performed by using an inner-product between the RCs and the observation data. This projection is able to characterise the vibratory response into a single point as

$$T = \langle \mathbf{O}, \mathbf{R} \rangle \tag{5}$$

where $\mathbf{O}$ is the observation data matrix and the matrix $\mathbf{R}$ is the set of RCs, which correspond to the undamaged system (referene data).

2.3. Step 3: Dimensionality magnitude estimation

The data obtained in the projection onto the new space can be utilised as damage index because it contains the information of the dynamic response. The damage index T is a vector which contains the inner product of the observation data in each RCs. As mention in the previous sections, each RC contains a certain percent of variance from the signals of the reference data. The first RCs contains more variance than the rest of RCs.

Computing the Mahalanobis distance (MD) between the projection of the observation data $\mathbf{T}_{\mathcal{O}}$ and the projection of the reference data $\mathbf{T}_{\mathcal{Z}}$ onto the new space as the following equation,

$$\mathbf{D_M}(T_{\mathcal{Z}}) = \sqrt{(\mathbf{T}_{\mathcal{O}} - \mu_{\mathcal{Z}})^T \mathbf{S}_{\mathcal{Z}}^{-1} (\mathbf{T}_{\mathcal{O}} - \mu_{\mathcal{Z}})} \tag{6}$$

provides a magnitude used for system identification and hence for damage assessment. $\mathbf{D_M}(T_{\mathcal{Z}})$ is the Mahalanobis distance to the reference data set $\mathbf{T}_{\mathcal{Z}}$, $\mathbf{T}_{\mathcal{O}}$ is the observation data, $\mu_{\mathcal{Z}}$ is the means of the $\mathbf{T}_{\mathcal{Z}}$ and $\mathbf{S}_{\mathcal{Z}}$ is the standard deviation of $\mathbf{T}_{\mathcal{Z}}$.

Theoretically, the dimension of the $\mathbf{D_M}$ can be as much as the dimension of the projection vector T. However, the dimension of $\mathbf{D_M}$ is best established by inspecting results from the PCs decomposition of the reference system.

3. Application for a wind turbine blade

3.1. Description of the experiment

For testing different damage detection and localization algorithms, a dedicated test setup was created. The test object is a 80 cm long rotor blade, which was created in the frame of DTU Wind Car project at Wind Energy department of Technical University of Denmark. The blade consists of two parts: the pressure and suction sides are manufactured separately from composite materials ([11] details the blade design and implementation). In Wind Car projects, the parts are glued together, which very much resembles the manufacturing process of real wind turbine blades. For the described test setup, the parts were squeezed together by means of a big number of small bolts, placed along the leading and trailing edges of the blade approximately 25 mm from each other. This solution greatly simplifies introducing a damage into the blade: one shall simply loose some of the bolts; this way, it is easy to control damage location and size. For damage "repair", the loosen bolts need to be re-tightened. This approach allows us modelling only one type of damage: leading and trailing edge de-bounding, which are quite common for many types of real blades. However, there are other types of damages, which cannot be modelled with the current setup.

For the experiment, the blade was clamped from the root end, in a similar way as it is supposed to be mounted on the rotor hub (Figure 1(a)). During the tests, the blade was artificially excited by a small electro-mechanical actuator mounted close to the root. The actuator is driven by a signal generator. This setup enables periodic highly repeatable force impulses being introduced into the blade structure. The response due to the actuator strikes was being measured by an array of accelerometers. Fifteen B&K Type 4507 B 4 monoaxial accelerometers were used, seven were mounted along the leading edge, another seven-along the trailing edge, and one in a vicinity of the actuator (Figure1(b)). The data acquisition was conducted using B&K Pulse LAN-Xi modules Type $3053 - B - 120$ and $3160 - A - 042$, the latter also includes the signal generator. In total, 16 channels were recorded: 15 acceleration signals and the driving signal from the signal generator, the latter to facilitate triggering during the post-processing.

3.2. Data collection

The most of damage detection and localization algorithms heavily rely on statistics in order to improve the robustness and detection rate and minimize the number of false alarms. The described algorithm belongs to the class of unsupervised learning, i.e. the algorithm is trained on healthy state (or also called reference state) of the test object, and the damage state is associated with a deviation from the normal state. First, the statistics for the reference state is collected: when all bolts are tightened, a series of approximately hundred actuator hits is recorded. Then a damage was introduced and expanded. The presented study uses the measurements from three damages: Damage 1 leading edge tip section; Damage 2 trailing edge middle section; Damage 3 leading edge root section. All three damages were introduced following the same scenario: first

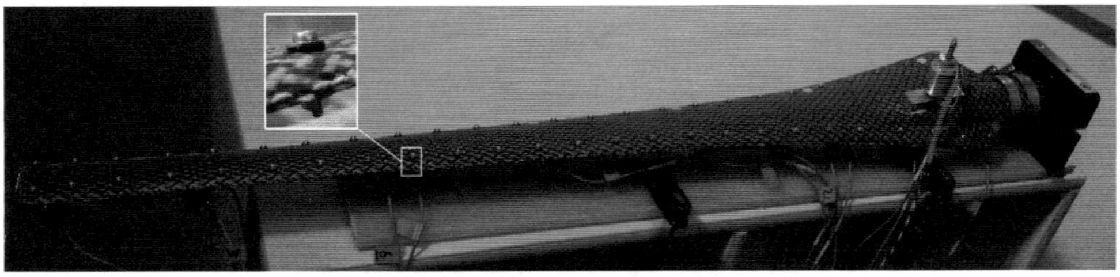

(a) Disposition of the bolts on the blade and blade support

(b) Distribution of the accelerometers along the blade

Figure 1. Experiment test ring for the wind turbine blade

only one bolt was loosened, and a series of new measurements (corresponding to about 60 hits) is conducted. Then a neighboring bolt was loosen, which modeled damage progression. Finally, the third bolt was untightened. After the state corresponding to the biggest damage was measured (namely, three loosened bolts for Damage 1 and Damage 2 cases and four bolts for Damage 3 case), all bolts were re-tightened, thus the structure was brought to the undamaged state, again. We did not tested the cases with more than one damage. It is important to note that, when loosen, the bolts were not removed from the blade, thus the total mass of the structure kept unchanged. The soft rubber washers keep the bolts fixed in the holes, thus preventing the bolt rattling. Doing this, we tried to avoid any possible side effects of loosening the bolts, which the algorithm can confuse with the changes in local structural stiffness, which we are trying to detect.

3.3. Damage detection

The discrete time-acceleration measurements (free-decay responses), from the undamaged blade, were recorded and processed by the methodology explained in §2. The new subspace based on the undamaged blade was used as reference state where the vibratory responses measured from the different damage locations and sizes can be compared. Ten signals were used to build the new subspace and the window length was selected as $W = 10$. The reason to choose this window length follows similar considerations than [12]. The statistical model was created by using only the undamaged responses for one sensor. The signals from the damaged blades were recorded form the same sensor and consequently projected onto the new subspace. This procedure was separately repeated for each sensor.

Figure 2 represents the projection of undamaged and damaged responses characterised in a single point onto the new subspace. Each damage location was studied separately. For the case of Damage 1 (damage located on the tip of the blade) and Damage 2 (damage located on the middle of the leading edge) can clearly be observed that the values cluster within groups of the same

Journal of Physics: Conference Series **628** (2015) 012086 doi:10.1088/1742-6596/628/1/012086

category. The distances from elements of the same category reduces their distances meanwhile elements from different categories increase the distances from the other categories groups (see Figure2(a) and 2(b)). Therefore, the detection between undamaged and damaged blades is achieved as well as the detection between different damage sizes. However, the clustering effect in the case of the Damage 3 (damage located on the trailing edge close to the support) does not clearly differentiate between different damage sizes. Although, the detection between damaged and undamaged is clearly achieved (see Figure2(c)).

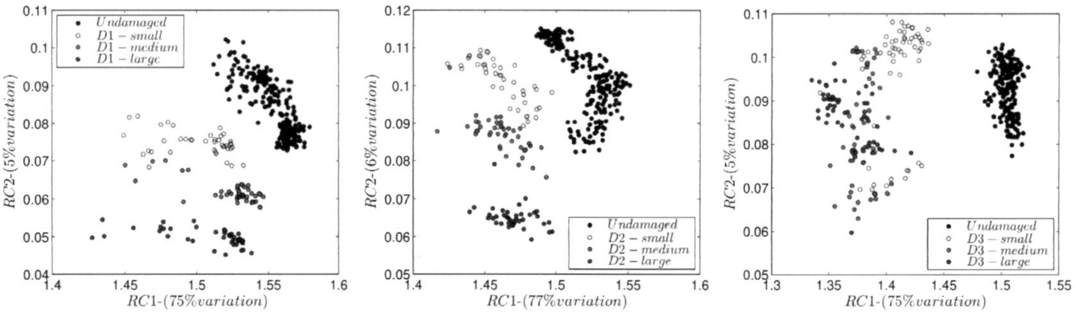

(a) Damage 1 by sensor 1 (b) Damage 2 by sensor 7 (c) Damage 3 by sensor 14

Figure 2. Projection of all damages location and sizes (small, medium and large) onto the new sub space based on undamaged structure.

In order to quantify the results obtained in the projection figures above, the MD was computed to measure the distances between undamaged elements to the different damages locations and sizes. Figure 3 shows separately the MD for each damage location. It can be observed within the three damage locations that the damaged was successfully detected. Also, the progression of the damaged is well observed even for the case of Damage 3 (Figure 3(c)) where the cluster classification in Figure 2(c) was not very well achieved. Figure 3(a) and 3(b) demonstrate that as much as the damage increases the distance from the undamaged state also increase. Moreover, the distance between different sizes of the damages is also distinguishable. For the three cases the damage progression is perfectly described.

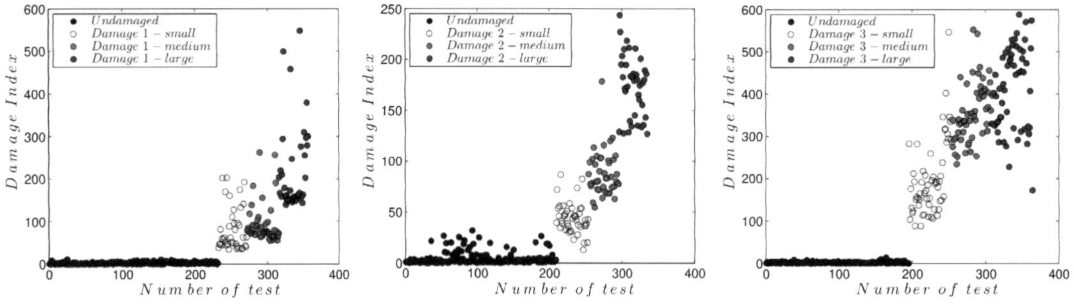

(a) Damage 1 by sensor 1 (b) Damage 2 by sensor 7 (c) Damage 3 by sensor 14

Figure 3. Mahalanobis distance from the undamaged scenario to different damage sizes (small, medium and large) for each damage location.

Journal of Physics: Conference Series **628** (2015) 012086 doi:10.1088/1742-6596/628/1/012086

3.4. Damage localization
Other important aspect to consider for VSHM is the localization of the damage along the blade. The damage index described in the previous sections and the localisation of the sensors along the blade are correlated to determinate a potential location of the damage. The sensors which are located close to the damage are more prone to detect such damage [13]. Therefore, the sensors which obtain larger value of MD index indicates that damage occurs close to the region where such sensor is located.

Figure 4(a) shows the sensitivity of the sensors on the Damage 1. It can be observed that the largest values of the MD occurs between sensors 1-5. These sensors are located around the area close to the tip of the blade which is the region where Damage 1 was performed. Similarly, Figure 4(b) represents the sensitivity to the Damage 2. It can be observed that although damage 2 occurs between sensor 5 and 7, the sensors around the area are more sensitive to damage. Finally, Figure 4(c) shows clearly the sensitivity of the sensor 12 on the Damage 3. The area where Damage 3 occurs is exactly in this position. As a general comment, the growth of the damage, in all the damages scenarios, is clearly observed in all sensor of the structure.

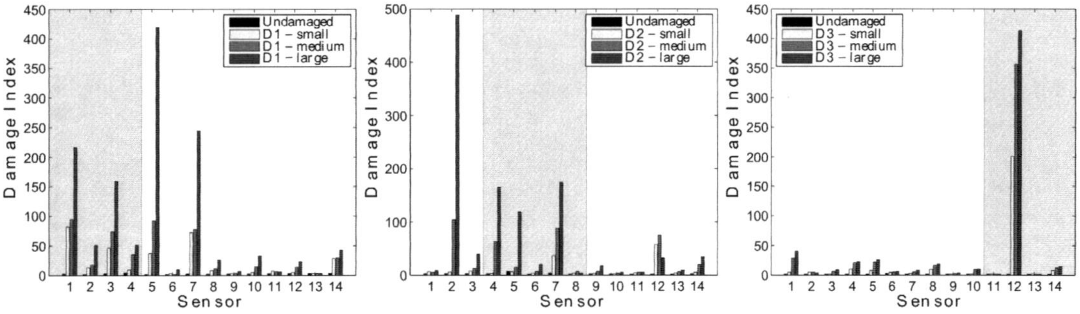

(a) Damage 1 (b) Damage 2 (c) Damage 3

Figure 4. Mahalanobis distance for each damage location and size corresponding to each sensor. The grey region involves the sensors which are closer to the damage.

4. Conclusion
The study presented in this paper is focused of using a multivariate signal processing technique for damage assessment in turbine blades. The experimental lab data on a turbine blade was used to evaluate the effectiveness of the methodology for damage detection and localization. The damage detection is fruitfully achieved for the different damages scenarios. The methodology is also able to monitor the progression of the damage being clearly detected the different sizes introduced during the experimental test procedure. The correlation of the damage index and the location of the sensors along the blade provides a high potential for damage localization. The damages were approximately located in the regions where each damage was performed. Further studies in this direction are extremely recommend due to the successful results obtained. It is also important to pointed out that the test data used in this study is focus on the low frequency damage data. As a conclusion, the methodology clearly present an high potential for on-line damage assessment in wind turbine blades.

Acknowledgments
The experimental part of the project was supported by the EUDP (Danish Energy Technology Development and Demonstration Programme) grant 64011 – 0084 "Predictive Structure Health

monitoring of Wind Turbines". The authors also acknowledge Robert Flemming Mikkelsen who provided the blade for the experiments.

References

[1] Montalvao, Diogo and Maia, Nuno Manuel Mendes and Ribeiro, António Manuel Relógio, 2006. *A review of vibration-based structural health monitoring with special emphasis on composite materials.* The Shock and Vibration Digest, 38(4):295324.

[2] Zou, Y and Tong, LPSG and Steven, GP, 2000. *Vibration-based model-dependent damage (delamination) identification and health monitoring for composite structuresa review.* Journal of Sound and vibration.

[3] M. M. Khan, M. T. Iqbal, and F. Khan, 2005. *Reliability and condition monitoring of a wind turbine.* Proceedings of 18th Annual Canadian Conference on Electrical and Computer Engineering.

[4] Jaishi, Bijaya and Ren, Wei-Xin,2006. *Damage detection by finite element model updating using modal flexibility residual.* Journal of sound and vibration, 290(1):369-387

[5] S.D. Fassois and J.S. Sakellariou, 2007. *Time series methods for fault detection and identification in vibrating structures.* The Royal Society Philosophical Transactions: Mathematical, Physical and Engineering Sciences, 365:411448.

[6] Schölkopf, Bernhard and Smola, Alexander and Müller, Klaus-Robert, 1998. *Nonlinear component analysis as a kernel eigenvalue problem.* Neural computation, 10(5):1299-1319.

[7] Garcia, David and Trendafilova, Irina, 2014. *A multivariate data analysis approach towards vibration analysis and vibration-based damage assessment: Application for delamination detection in a composite beam.* Journal of Sound and Vibration, 333(25):7036-7050.

[8] Marin, Luciano and Döhler, Michael and Bernal, Dionisio and Mevel, Laurent, 2014. *Robust statistical damage localization with stochastic load vectors.* Structural Control and Health Monitoring.

[9] Kopsaftopoulos, FP and Fassois, SD, 2013. *A functional model based statistical time series method for vibration based damage detection, localization, and magnitude estimation.* Mechanical Systems and Signal Processing, 39:143161.

[10] Baktashmotlagh, Mahsa and Harandi, Mehrtash and Bigdeli, Abbas and Lovell, Brian and Salzmann, Mathieu, 2013. *Non-Linear Stationary Subspace Analysis with Application to Video Classification.* Proceedings of The 30th International Conference on Machine Learning, p.450-458.

[11] Bottoli F and Pignatti L, 2001 *Design and processing of structural components in biocomposite materials rotor blade for wind turbine cars M.Sc.*Thesis Ris DTU

[12] Garcia, D and Trendafilova, I, 2014. *Singular Spectrum Analysis for identifying structural nonlinearity using free-decay responses. Application for delamination detection and diagnosis in composite laminates.* 26th International Conference on Noise and Vibration Engineering, Leuven, Belgium.

[13] Liu, Gang and Mao, Zhu and Todd, Michael and Huang, Zongming, 2014. *Localization of nonlinear damage using state-space-based predictions under stochastic excitation.* Smart Materials and Structures, 23(2):25-26.

Detection and classification of alarm threshold violations in condition monitoring systems working in highly varying operational conditions

M Strączkiewicz[1], T Barszcz, A Jabłoński

AGH University of Science and Technology,
Faculty of Mechanical Engineering and Robotics,
30 Mickiewicza Av., 30-059 Krakow, Poland

E-mail: mstr@agh.edu.pl

Abstract. All commonly used condition monitoring systems (CMS) enable defining alarm thresholds that enhance efficient surveillance and maintenance of dynamic state of machinery. The thresholds are imposed on the measured values such as vibration-based indicators, temperature, pressure, etc. For complex machinery such as wind turbine (WT) the total number of thresholds might be counted in hundreds multiplied by the number of operational states. All the parameters vary not only due to possible machinery malfunctions, but also due to changes in operating conditions and these changes are typically much stronger than the former ones. Very often, such a behavior may lead to hundreds of false alarms. Therefore, authors propose a novel approach based on parameterized description of the threshold violation. For this purpose the novelty and severity factors are introduced. The first parameter refers to the time of violation occurrence while the second one describes the impact of the indicator-increase to the entire machine. Such approach increases reliability of the CMS by providing the operator with the most useful information of the system events. The idea of the procedure is presented on a simulated data similar to those from a wind turbine.

1. Introduction

In the last several years we may observe a rapid growth of condition monitoring systems (CMS), employed nowadays in any field, where safety requirements meet a need to maximize the effectiveness of machinery operation. Regardless the area of usage, all CMS gather and process information from machinery residual processes, like vibration, oil or temperature, in order to reveal any sign of fault development. This information is often referred as diagnostic features (estimates), i.e. features that carry diagnostic information for the user of the system. The number of such diagnostic features to monitor is still increasing together with the advancement of signal processing techniques and computing capabilities. As stated in [1], in typically used wind turbine's vibration monitoring process, there is nearly 150 diagnostic features monitored simultaneously and this number could rise if, for instance, additional sensors were installed or temperature

[1] To whom any correspondence should be addressed.

features were taken into consideration. For a machinery park number of features to track reaches thousands [2], for example for new designs of wind farms.

In order to simplify decision-making process on such extensive amount of data, the statistical process control limits are set using various methods, such as presented in [3-5]. Moreover, for machinery working in varying operation conditions, the operational states were employed [6-8] to limit the number of false positive alarms resulting from high fluctuation of process parameters.

Nonetheless, diagnostic engineers responsible for operation of particular group of machines are still forced to deal with overwhelming number of threshold violations. This is due to several causes, besides the actual fault detection, just to name the most severe ones: improper or lack of threshold setup, irregular and unexpected behavior of machinery process parameters, technical problems not related to machinery condition, etc.

Many attempts were made in the field of alarm management starting from the early 80s [9-10], when first scientists pointed out a large number of alarms to be dealt with in rapidly growing power and process industries. They also noticed a need for distinguishing a crucial alarms to the entire system from less significant ones. In [11] the authors emphasize the difficulty with definition of the priority of particular alarms in operational context. Simultaneously they refer to the survey with process plant operators and claim that on average there are about 80 alarms in the first minute of a severe plant breakdown. The number of problems to be checked with no special priority for action is overwhelming to operators. Based on these conclusions the International Society of Automation developed the standard [12] that laid a foundation to the industrial applications, as mentioned in [13]. It assumed identification of necessary alarms and became the groundwork for this paper, in which the authors would like to discuss the idea of violation attributes and introduce the new approach – the violation priority coefficient (VPC) – for simplification of decision-making process. Although the proposed method is mainly focused on condition monitoring systems of wind turbines, it could be employed as well to other groups of rotating machinery.

This article is organized as follows. At first, the violation attributes are described to differentiate significance of various alarm features to the machine's operation. Then, the methodology of reasoning is proposed. The case study on simulated data is presented step-by-step for several threshold violations of three, typically used in CMS, diagnostic features. Finally, the results are discussed and paper is concluded.

2. Violation attributes

Firstly it is essential to note that in the further discussion as a violation one should understand any singular crossing of defined threshold level by selected diagnostic estimator of rotating machine. We do not take into account the industrial practice of delaying the detection of an alarm or other practical modifications of the threshold violation management.

Every violation can be described using few attributes, that inform of a value that estimator reached, time when it happened, or what kind of estimator is above the limit. In authors opinion the listed details of violation may be used as a measure for the priority calculation.

2.1 Severity

The severity factor can be understood as level of the abnormal behavior compared to the acceptable, regular projection of particular phenomena (e.g. vibrations or temperature). In some cases, the distance from the normal operation can be fixed, as suggested in standards [14-15], where recommended vibration levels for rotating machines categorized into different classes based on their size and possible vibrations. Other approaches assume calculation of thresholds after 'referential' period of observation using various methodology [16-17]. For the typically used CMS common practice is to distinguish two basic levels of severity:

- Warning,

- Alarm.

2.2 Time of occurrence
An attribute related to date is an important information for decision making process. Based on authors' experience this parameter should modify its value in descending order. This is due to the fact that violation that appears, e.g. today should have higher priority than the ones from 7 days ago. Therefore authors propose to use gradual time of occurrence justification:

- 1 – old, e.g. older than 7 days,
- 2 – recent, e.g. between 2-7 days,
- 3 – new, e.g. until 2 days ago.

The problem of how to set the limits between particular levels can be solved in many ways. The values proposed above are based on authors' practice and should be treated merely as a starting point. Following this manner will allow user to keep his insight about systems current performance up-to-date.

2.3 Type of feature
In the past few decades, a numerous health indicators (also called features) were proposed for a measure of the machinery dynamic condition. For instance, for vibration-based condition monitoring, these indicators, that could also be referred as 'diagnostic features', consists of two main groups: narrowband and broadband. Narrowband features are calculated from a filtered band of frequency-domain signal, i.e. from the selected row of spectral lines. In the contrary, broadband features does not require this specification as they are obtained from a much wider band. Intuitively, the two groups of features have slightly different meaning. According to [18], the narrowband group helps to localize the occurring fault since it focuses on observation of characteristic frequencies related to individual rotating components. The increase of broadband features usually appears sometime after the increase of narrowband features and bring information of a severe fault. We propose that the broadband features have increased importance.

2.4Type of element
Varying attention can be given to different kinematic elements based on e.g. statistics of failures. Gathered statistical analyses are published frequently for certain types of machinery and carry information about elements of that were mostly prone to damage in the period of time the report concerns. In [19] authors presented statistics of causes of failure for offshore wind turbines. This information can result in emphasizing selected threshold violations rather than the others. Another matter is length of a downtime caused by the failure of a particular element [20] which may cause significant loss to the machine's owner and therefore should be taken into consideration.

3. Methodology
For the purpose of alarm classification, the authors propose to use the violation priority coefficient (VPC), calculated for the individual violations using to the following formula:

$$VPC_i = \prod_{j=1}^{k} A_j(s), \tag{1}$$

where i represents violation index, j – attribute index, k stands for total number of attributes used to describe violation, while s is index of jth attribute's level, also referred to as 'significance'. Values of $A_j(s)$ could be also understood as weights of particular levels and be designated based on the statistical estimations or on the personnel's experience.

Exemplary possible calculations of VPC for two attributes are presented in Table 1, where m is number of levels of the first attribute, while n stands for number of levels of the second attribute.

Table 1 Table with values of VPC for two attributes

Attribute 2 / Attribute 1	$A_2(1)$	$A_2(2)$	…	$A_2(n-1)$	$A_2(n)$
$A_1(1)$	$A_1(1) * A_2(1)$	$A_1(1) * A_2(2)$	…	$A_1(1) * A_2(n-1)$	$A_1(1) * A_2(n)$
$A_1(2)$	$A_1(2) * A_2(1)$	$A_1(2) * A_2(2)$	…	$A_1(2) * A_2(n-1)$	$A_1(2) * A_2(n)$
…	…	…	…	…	…
$A_1(m-1)$	$A_1(m-1) * A_2(1)$	$A_1(m-1) * A_2(2)$	…	$A_1(m-1) * A_2(n-1)$	$A_1(m-1) * A_2(n)$
$A_1(m)$	$A_1(m) * A_2 1$	$A_1(m) * A_2(2)$	…	$A_1(m) * A_2(n-1)$	$A_1(m) * A_2(n)$

In the final step, for the obtained sequence of VPCs: $\langle VPC_1, VPC_2, VPC_3, \dots, VPC_n \rangle$, perform permutation to obtain new vector of VPCs: $\langle VPC'_1, VPC'_2, VPC'_3, \dots, VPC'_n \rangle$, so that $VPC'_1 \geq VPC'_2 \geq VPC'_3 \geq \cdots \geq VPC'_n$. It is equivalent to sorting of violations in the descending order.

4. Case study

The authors would like to present the VPC properties via observing how the state-of-the-art CMS and proposed approach would categorize violations. The comparison is performed using simulated trend data presented in Figure 1. Although the trend plots were generated using MATLAB® software, yet it mimics diagnostic estimates behavior from a real life system – in this case from the wind turbine's front bearing on generator. For the clarity purposes authors selected two narrowband features essentially used for bearing diagnosis, i.e. fundamental train frequency (FTF) and ball-passing frequency of outer race (BPFO), while as wideband one, the vibration energy estimate calculated as root-mean-square (RMS)

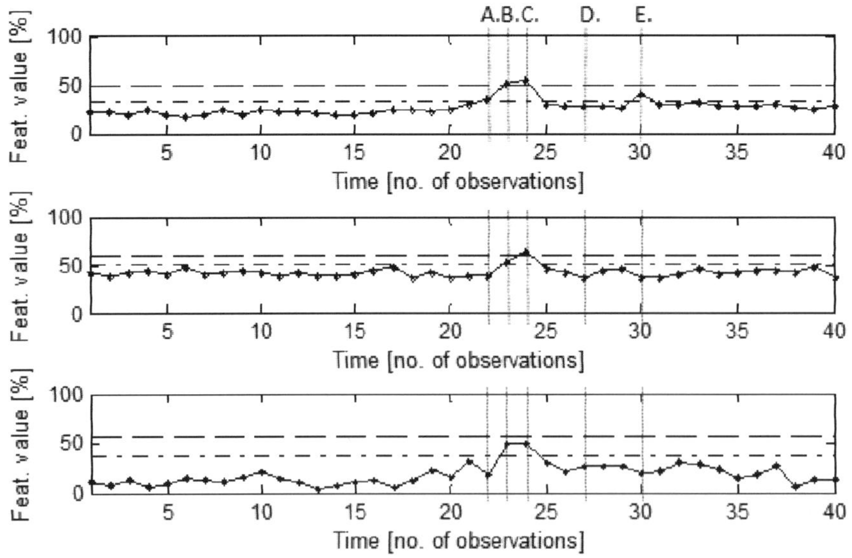

Figure 1. Trend plots of three diagnostic features: FTF (top), RMS (middle) and BPFO (bottom)

was selected. Please notice that although trends are plotted on separate figures they concern simultaneous events, what is denoted by the time axis. We did not assign any time to the observations, as the method can be adapted to different sampling frequencies. For this test we assumed that a significant change may happen between consecutive observations, so for clarity a reader can associate it with daily measurements.

The two levels of violations are set for the presentation purposes: warning (dotted line) and alarm (dashed line). They were designated using the two-step procedure presented in [5].

In order to highlight more important violations, the significance factors need to be adjusted. This can be done using either statistical approach as e.g. presented in [19-20] or from the experience of the responsible personnel. For the purpose of presentation, three attributes are distinguished, i.e. type of a feature under investigation, level of vibration understood as violation's severity and novelty factor dependent on the date of violation's occurrence. Values attached to each of the attribute levels, i.e. their 'significance', are presented in Table 2.

Table 2 Attributes of violations with it significance factors

Attribute	Type of feature		Novelty			Severity	
	Wideband	Narrowband	New	Recent	Old	Alarm	Warning
Significance	9	3	10	4	1	10	5

As one may notice the highest attention should be given to wideband feature, i.e. RMS (over FTF and BPFO) and exceeding of level identified as 'alarm' (over 'warning'). As for the novelty factor, violations occurring at the exact time should receive the highest priority, middle one should be given to events that happened one observation ago, and least importance is assigned to the older violations.

Base on this, the VPC is calculated on individual threshold violations. In the second step the violations are sorted in descending order to show the event with the greatest importance to the user.

The interesting events that are to be described in this section are ticked on Fig. 1 with letters A.-E. The first event, marked as Observation A. occurs in the moment of single warning-level violation on FTF estimate. This circumstance would be recorded similarly for the state-of-the-art approach and for the one proposed in this paper (Table 3). Instinctively, since there is only the individual violation, there is no difference between the results.

Table 3 Comparison between alarm display for two approaches for Observation A.

Event	State-of-the-art						Proposed				
	#	Status	Date	Analysis	Value	Limit	#	Status	Date	Analysis	VPC
A.	1	warning	22	FTF	0.0196	0.0179	1	warning	22	FTF	150

The first difference could be spotted for Observation B. In this case the three channels cross the threshold concurrently: RMS and BPFO their warning levels, while FTF its alarm level.

For the typical approach, all of this event are simply added to the list with no special remark (Table 4). In the contrary, the proposed technique positions the RMS on the top of the list since it might be indicating a serious threat to the objects performance. This is due to the higher VPC value related to high significance of wideband feature and its novelty factor. At the second place there is the alarm-level exceeding on FTF that should be treated with greater attention than warning-level violation of BPFO.

At Observation C. one may observe further violations on all three channels. For the state-of-the-art approach user needs to analyze the list seeking for channel that has the greatest importance for the machine's maintenance while in the proposed approach, the VPC reaches maximum for RMS's new

Table 4 Comparison between alarm display for two approaches for Observation B.

Event	State-of-the-art						Proposed				
	#	Status	Date	Analysis	Value	Limit	#	Status	Date	Analysis	VPC
B.	1	warning	22	FTF	0.0196	0.0179	1	warning	23	RMS	450
	2	alarm	23	FTF	0.0280	0.0269	2	alarm	23	FTF	350
	3	warning	23	RMS	2.1015	2.0341	3	warning	23	BPFO	300
	4	warning	23	BPFO	0.0251	0.0189	4	warning	22	FTF	60

violation of alarm level (Table 5). RMS has supremacy over also alarm level exceeding, but by the narrowband feature of FTF. Recent alarms are pushed down the list with one exception, i.e. the RMS warning information is still more important than actual BPFO one.

Table 5 Comparison between alarm display for two approaches for Observation C.

Event	State-of-the-art						Proposed				
	#	Status	Date	Analysis	Value	Limit	#	Status	Date	Analysis	VPC
C.	1	warning	22	FTF	0.0196	0.0179	1	alarm	24	RMS	900
	2	alarm	23	FTF	0.0280	0.0269	2	alarm	24	FTF	300
	3	warning	23	RMS	2.1015	2.0341	3	warning	23	RMS	180
	4	warning	23	BPFO	0.0251	0.0189	4	warning	24	BPFO	150
	5	alarm	24	FTF	0.0296	0.0269	5	alarm	23	FTF	120
	6	warning	24	BPFO	0.0288	0.0189	6	warning	23	BPFO	60
	7	alarm	24	RMS	2.5456	2.4111	7	warning	22	FTF	15

During the Observation D. there were no new events upcoming to the list, nonetheless the Reader might notice that VPCs are actualized according to the novelty factor – previously new events became recent, while recent ones – old. This affected the VPCs values (Table 6).

Table 6 Comparison between alarm display for two approaches for Observation D.

Event	State-of-the-art						Proposed				
	#	Status	Date	Analysis	Value	Limit	#	Status	Date	Analysis	VPC
D.	1	warning	22	FTF	0.0196	0.0179	1	alarm	24	RMS	360
	2	alarm	23	FTF	0.0280	0.0269	2	alarm	24	FTF	120
	3	warning	23	RMS	2.1015	2.0341	3	warning	23	RMS	60
	4	warning	23	BPFO	0.0251	0.0189	4	warning	24	BPFO	45
	5	alarm	24	FTF	0.0296	0.0269	5	alarm	23	FTF	30
	6	warning	24	BPFO	0.0288	0.0189	6	warning	23	BPFO	15
	7	alarm	24	RMS	2.5456	2.4111	7	warning	22	FTF	15

There were six time intervals that passed between Observations C and E. This allowed all of the events on the list change its VPC to the minimum according to the novelty factor, i.e. all of them became 'Old'. Therefore new violation appearing at Observation E, although coming from narrowband feature and is related 'only' to warning level of severity, for the system's user is most important and thus is positioned at the top of the list. This is the opposite to the state-of-the-art approach, as presented in Table 7.

This mimics the situation in real CMS when, old and nearly forgotten events cover up the newest ones, and therefore delays the reaction.

Table 7 Comparison between alarm display for two approaches for Observation E.

Event	State-of-the-art						Proposed				
	#	Status	Date	Analysis	Value	Limit	#	Status	Date	Analysis	VPC
E.	1	warning	22	FTF	0.0196	0.0179	1	warning	30	FTF	150
	2	alarm	23	FTF	0.0280	0.0269	2	alarm	24	RMS	90
	3	warning	23	RMS	2.1015	2.0341	3	alarm	24	FTF	45
	4	warning	23	BPFO	0.0251	0.0189	4	warning	23	RMS	30
	5	alarm	24	FTF	0.0296	0.0269	5	warning	24	BPFO	30
	6	warning	24	BPFO	0.0288	0.0189	6	alarm	23	FTF	15
	7	alarm	24	RMS	2.5456	2.4111	7	warning	23	BPFO	15
	8	warning	30	FTF	0.0220	0.0179	8	warning	22	FTF	15

5. Discussion and Conclusions

As presented in the above analysis, for CMS with multiple diagnostic estimates there is a need to introduce auxiliary method for categorizing the violations of threshold levels. Therefore the violation priority function (VPC) is proposed. Application of this approach to diagnostic reasoning may help to keep track to the new and crucial events in the monitored system, and thus enhance its performance, as presented in the discussed example. The undoubted advantages of the proposed method are its flexibility and intuitiveness. The first one is because it could be employed for monitoring of any sort of machinery or processes. The second comes from the idea that intuitively certain violations are more important to overall success of operation and maintenance process than the others. The VPC enables to reveal this fact.

The current shortcomings of the proposed method are simultaneously the challenges for nearest research. The authors note that there is a strong need for optimal weights justification in order to obtain more automatic way of violation organizing with limited human factor.

The second direction of study should be headed toward extension of violation types that might indicate fault development, e.g. the relative measure that takes into account any abrupt variation of diagnostic features.

Acknowledgement

The work presented in this paper was supported by funding from the research project number PBS3/B6/21/2015 financed by the Polish National Centre for Research and Development.

References

[1] Barszcz T, Strączkiewicz M, 2013, Novel intuitive hierarchical structure for condition monitoring system of wind turbines, *Diagnostyka*, 14/3, pp. 53-60

[2] Barszcz T, 2008, Selection of diagnostic algorithms for wind turbines, *Proc. 4th Int. Congress on Technical Diagnostics*, 9-12 September 2008, Olsztyn, Poland

[3] Cempel Cz, 1990, Limit value in practice of vibration diagnosis, *Mech. Syst. Signal Pr.*, 4/6, pp. 483-493

[4] Duda RO, Hart PE, Stork DG, 2000, Pattern Classification (2nd ed.), John Wiley & Sons, New York

[5] Jabłoński A, Barszcz T, Bielecka M, Breuhaus P, 2013, Modeling of probability distribution functions for automatic threshold calculation in condition monitoring systems, *Measurement*, 46, pp. 727-738

[6] Timusk M A, Lipsett M G, McBain J, Mechefske Ch K, 2009, Automated Operating Mode Classification for Online Monitoring Systems, *J. Vib. Acoust.*, 131, pp. 131-141

[7] Mykhaylyshyn V, Prytula A, Petrovych N, Isko V, 2010, Classification of wind turbine generator operation states for the purpose of advanced vibration monitoring, *European Wind Energy Conference and Exhibition*, 20-23 April 2010, Warsaw, Poland

[8] Jablonski A, Barszcz T, Wiciak P, 2013, Non-clustering method for automatic selection of machine operational states, *Proc. of 3rd Int. Conf. of Condition Monitoring in Non-Stationary Operations*, 8-10 May, Ferrara, Italy

[9] Andow P K, 1980, Real-time analysis of process plant alarms using a mini-computer, *Comp. Chem. Eng.*, 4, Pergamon

[10] Lihou D A, 1981, Aiding process plant operators in fault finding and corrective action, *Human Detection and Diagnosis of System Failure,* pp. 501-522

[11] Bransby M, Jenkinson J, 1998, The management of alarm systems, HSE contract research report 166/198, HSE Books

[12] ANSI/ISA-18.2, 2009, Management of Alarm Systems for the Process Industries, Research Triangle Park, NC, USA

[13] Stauffer T, VanCamp K, 2012, Make Some Alarming Moves. Tackle distractions that impair operator performance and process efficiency, *Chem. Proc.*

[14] ISO 10816, Part I, 1995, Mechanical vibration – Evaluation of machine vibration by measurements on non-rotating parts – Part I: General guidelines. International Organization for Standardization, Geneva, Switzerland

[15] VDI 3834, Part I, 2009, Measurement and evaluation of the mechanical vibration of wind energy turbines and their components, The Association of German Engineers (VDI), Düsseldorf, Germany

[16] Cempel Cz, 1987, Simple condition forecasting techniques in vibroacoustical diagnostics, *Mech. Syst. Signal Pr.*, 1/1, pp. 75-82

[17] Alt F B, 1985, Multivariate quality control, *The Encyclopedia of Statistical Sciences*, Wiley, New York, pp. 110-122

[18] Urbanek J, Barszcz T, Uhl T, 2012, Comparison of advanced signal-processing methods for roller bearing faults detection, *Metrol. Meas. Syst.*, 19/4, pp. 715-726

[19] Bussel G J W van, Zaajer M B, 2001, Reliability, availability and maintenance aspects of large-scale offshore wind farms, a concept study, *Proc. of Marine Renewable Energy Conference*, March 2001 Newcastle, UK

[20] McMillian D, Ault G W, 2007, Quantification of Condition Monitoring Benefit for Offshore Wind Turbines, *Wind Eng.*, 31/4, pp. 267-285

Autoregressive modelling for rolling element bearing fault diagnosis

H Al-Bugharbee*[1,2] and I Trendafilova[1]

[1]Department of Mechanical and Aerospace engineering

University of Strathclyde, James Weir Building, 75 Montrose Street, Glasgow G1 1XJ, United Kingdom

[2 On leave] Department of Mechanical engineering, University of Wasit, Wasit, Iraq

E-mail: hussein.al-bugharbee@strath.ac.uk

Abstract. In this study, time series analysis and pattern recognition analysis are used effectively for the purposes of rolling bearing fault diagnosis. The main part of the suggested methodology is the autoregressive (AR) modelling of the measured vibration signals. This study suggests the use of a linear AR model applied to the signals after they are stationarized. The obtained coefficients of the AR model are further used to form pattern vectors which are in turn subjected to pattern recognition for differentiating among different faults and different fault sizes. This study explores the behavior of the AR coefficients and their changes with the introduction and the growth of different faults. The idea is to gain more understanding about the process of AR modelling for roller element bearing signatures and the relation of the coefficients to the vibratory behavior of the bearings and their condition.

1. Introduction

Diagnosis of faults in rolling element bearings is a very essential task in industry. There are so many methods are used for the purpose of fault diagnosis in rolling element bearings.

Mainly, the methods focused on analyzing the vibration signals in its various domains time, frequency and time-frequency. Several papers are published on reviewing and critical evaluation of the methods used [1, 2]. The principles of time series analysis is also used by some researchers to model the vibration signal acquired from the bearing house. Linear autoregressive is a very common model used for the purpose of fault diagnosis in machinery diagnosis. It is simply related each data point of the signal to a set of past data points using some weighted values (i.e. coefficients). However one of the challenges of these with using such models is that existence of non-stationary part in the signal. One of the solutions to cover this problem is use adaptive (time variant) autoregressive in which the model coefficients are evolving with time [3-5]. However, this solution is not that direct use as it requires the assumption of a suitable initial set of model coefficients as well as the assumption of the way of coefficients model evolving. Another way is to transform the non-stationary signals to non-stationary to a stationary using a suitable stationarization technique and then to use the autoregressive of time invariant coefficients. It is important before starting of the modeling process is to test if the signal is stationary in order to decide if transformation is required or not. One of the most common tests is Kwiatkowski,Phillips, Schmidt, and Shin (KPSS) test [6].

In both cases time (variant or invariant) model, the model coefficients are very important features as they reflect the condition of the signal which provides the chance to use them as feature vectors of the signal. They can also be used to obtain the parametric spectrum of the signal.

Several papers are published in using autoregressive modeling for the machinery fault diagnosis problem [7-10]. However, according to the knowledge of the authors none of these papers analyse the model coefficients individually and that is the part what this study aims to do.

This study suggests the use of a linear AR model applied to the signals after achieving the local stationarity of the signal by testing and transformation if necessary. The obtained coefficients of the AR model are further used to form features vectors which are in turn subjected to pattern recognition for distinguishing among different faults and different fault sizes. This study also explores the behavior of the AR coefficients individually and their changes with the introduction and the growth of different faults. The goal is to gain more understanding about the relation of the of AR coefficients to the vibratory behavior of the bearings and their condition. The study is divided in three parts, namely: signal pre-treatment, diagnosis method and exploration of AR coefficients.

The paper divided in a couple of sections as follow: - section 2: signal pre-treatment. Section 3. Diagnosis method. Section 4. Exploration of AR coefficients. Section 5. Method verification. Section 6. Results and discussion. Section7. Conclusion.

2. Signal pre-treatment

The stage of signal pre-treatments includes a number of processes that starts by signal segmentation and finishes with obtaining AR model's coefficients.

2.1. Signal segmentation

Signal segmentation is essential step to provide more segments for analysis purposes especially when it is difficult to acquire more signals experimentally. In addition, it can be useful in enhance the stationarity of the signal where the whole signal can be non-stationarity but some/all of its segment are locally stationary [11].

2.2. Stationarity test and stationarization.

As we mentioned above that linear autoregressive model is suitable for only stationary signal (*i.e signals that the four statistical moments are time invariant*). It is necessary to investigate the stationarity of the signal before subjecting to modeling. This is important to decide whether the signal needs for stationarization transformation or not. The existence of stationarity can simply investigated by the Kwiatkowski, Phillips, Schmidt and Shin (KPSS) test which proposes a null hypothesis that the observed signal is stationary around its deterministic trend [6]. In case that a segment is discovered to be non-stationarity, then it can be subjected to a differencing technique which transforms it to stationarity. Differencing can be described as [12]:-

$$x_i^{dif} = x_i - x_{i-1} \quad ,where\ i = 2,3,\dots n \tag{1}$$

Where x_i^{dif} are the components of the new stationarized segment, x_i and x_{i-1} are adjacent signal data points and n is the length of the original segment.

2.3. Linear autoregressive modeling.

A linear Autoregressive (LAR) model is that representation where the output variable is predicted based on linearly depending on its own previous values. In this research, it is used to predict the stationary segmented signals. The mathematical structure of a linear AR model can be described as follow [13]

$$x_i = a_o + a_1.x_{i-1} + a_3.x_{i-2} + \cdots + a_p.x_{i-p} + \varepsilon_i \tag{2}$$

where
x_i is the predicted output value at time i which is linearly related to (p) past points,
p is the order of the model
a_i (i =0,1,2…,p) are weighting coefficients (i.e model parameters).
ε_i ,the error term, is a white noise process, which represents the difference between the actual and linearly predicted values. In this research, the model order is determined based on final prediction error (FPE). The model coefficients are estimated using the least square algorithm. The goodness of model's fit to the experimental data is evaluated by the Normalized Mean Square Error (NMSE) value which given by the equation below: -.

$$Goodness\ of\ Fit|_{NMSE} = \left(1 - \frac{\left(\sum_{i=1}^{i=n}\left(x_{pr}(i) - x_{ms}(i)\right)^2\right)^{1/2}}{\sum_{i=1}^{i=n} x_{pr}(i) - \bar{x}_{ms}}\right) * 100\% \tag{3}$$

where: -
x_{pr} is predicted signal.
x_{ms} is the real time measured signal
$\bar{x}_{ms}$ is the mean value of real time signal.
n is the number of data points (i.e segment length)

3. Diagnosis method

In the suggested methodology, pattern recognition is used for distinguishing among different classes that corresponds to different bearing conditions. The methodology is suggested to detect different fault locations and different fault severity. In the first stage of diagnosis signals are segmented and distributed into two samples training and testing sample. From training sample the four bearing signal categories are made (i.e H, IRF, BF, ORF). Every segment of the training sample is tested for stationarity using KPSS test (see &2.2) and if required the stationarization transformation is done using equation (1). Next, the segments are subjected to AR modeling equation (2) and coefficients are obtained using least square method. These model coefficients are arranged as rows to form the four signal categories mentioned above (i.e H, IRF, BF and ORF).

Up to this step there will be four matrices corresponding to different bearing condition (i.e. H, IRF, BF and ORF) and another three corresponding for fault severity (S, M, L) of each fault location.

The size of each matrix will be (N X p) dimensions. The number of rows N equals to the number of segments of the training sample while the number of column (p) equals to the number of AR model coefficients (i.e the optimum model order). In case there are different model orders (p) corresponding to the different signal categories, the minimum one is considered to be the number of columns of features vectors matrix (H_K) as in below.

$$\mathbf{H}_K = \begin{bmatrix} a_{K11} & a_{K12} & . & a_{K1p} \\ a_{K21} & a_{K22} & . & a_{K2p} \\ . & . & . & . \\ a_{KN1} & a_{KN2} & . & a_{KNp} \end{bmatrix} \tag{4}$$

Now it is required to calculate the mean of the $\mathbf{H}_K$ rows in order to be used later in Mahalanobis distance calculation.
The mean vector is calculated as follows:

$$\mathbf{E}_K = \frac{\sum_{i=1}^{N}(a_{Ki1}\ldots\ldots\ldots a_{Kip})}{N} \quad (5)$$

Eventually the Mahalanobis distances are calculated between the new tested features vectors (i.e model coefficients vectors from testing sample) and mean feature vectors as defined in Eq. (5):

$$D_K(\mathbf{x}, \mathbf{E}_K) = (\mathbf{x} - \mathbf{E}_K).\mathbf{S}^{-1}.(\mathbf{x} - \mathbf{E}_K)' \quad (6)$$

where
D_K is mahalnobis distance.
$\mathbf{x}$ a feature vector from the testing sample
$\mathbf{E}_K$ mean features vector of category K
$\mathbf{S}^{-1}$ the inverse ofcategory covarience matrix $\mathbf{H}_K$

The prime in the end of equation (6) means the transpose.
Then the new feature vector is classified based on the NN method. According to this method, each new vector is assigned to the category for which the Mahalanobis distance (see Equation (6)) has a minimum. That is **x** belongs to the category m for which $D_K(\mathbf{x}, \mathbf{E}_K)$ has minimum over all K.

4. Exploration of AR coefficients behavior.

A further exploration of the AR coefficients is also carried out for further understanding of individual coefficients sensitivity to the different bearing conditions. In this regard, reference regions including upper and lower boundaries are built from the healthy values of each coefficient. These values are taken from the features vector matrix of healthy class ($\mathbf{H}_H$). The boundaries are determined using the mean and standard deviations of the healthy values as in equations (7&8). For a coefficient number (i), the boundaries formulas will be: -

$$b_{upper} = \mu(\boldsymbol{H}_H(:,i)) + d * \sigma(\boldsymbol{H}_H(:,i)) \quad (7)$$

$$b_{lower} = \mu(\boldsymbol{H}_H(:,i)) - d * \sigma(\boldsymbol{H}_H(:,i)) \quad (8)$$

$\mu(\boldsymbol{H}_H(:,i))$ is the mean value of the column (i) of the matrix $\boldsymbol{H}_H$.

$\sigma(\boldsymbol{H}_H(:,i))$ is the standard deviation value of column (i) of the matrix $\boldsymbol{H}_H$.

d is a constant, which should be properly selected so that at least 80% of the points are within the boundaries.

The boundaries are normalized by computing their percent deviation from the $\mu(\boldsymbol{H}_H(:,i))$ as

$$bni = \frac{\mu(\boldsymbol{H}_H(:,i)) - bi}{\mu(\boldsymbol{H}_H(:,i))} * 100\% \quad (9)$$

Where
bi upper or lower boundary corresponding to coefficient i.
bni normalized upper or lower boundary corresponding to coefficient i.

Then, to investigate the sensitivity of a parameter a_i to the change of bearing condition (i.e presence or growth of the fault), the values of a_i from other than healthy signals are normalized as in equation (10) to find their percent deviation from the reference values and projected to investigate if they deviate from the reference boundaries (sensitive) or not (insensitive).

$$\% \, dev_{KHi} = \frac{\mu(\boldsymbol{H}_H(:,i)) - \mu(\boldsymbol{H}_K(j,i))}{\mu(\boldsymbol{H}_H(:,i))} * 100\% \tag{10}$$

Where $\% \, dev_{KHi}$ is the percent deviation of coefficient *i* of category *K* from the reference value. $j = 1{:}N$ is the number of segment.

In order to reduce the effect of false alarm, coefficient a_i is assigned as a fault sensitive one if the $\%dev_{KHi}$values are out the previous reference region for several successive segments. When it is so, and new reference boundaries are built from the new sensitive a_i.

Finally, the common coefficients that are sensitive to fault presence and growth in different fault locations are determined. Eventually, the first three coefficients that showing higher values of deviation among the signal categories are selected for comparison.

5. Method verification.

Bearing vibration data provided by the Case Western Reserve University (CWRU) are used for the purpose of validation[14]. The data considered in this research are shown in Table 1. The raw signal is segmented equally into 2048 points sub signals.

The segmentation process gives 232 segments (4 categories*58 segments for each category) for detection of fault location and 174 segments (3 categories * 58 segments for each category) for detection of fault severity. These are divided equally into training and testing sample. All the segments are checked to be stationary before subjected to modeling.

6. Results and Discussion.

Table 1. The List of data used for analysis. H-healthy, IRF-inner race fault, BF- ball fault, ORF-outer race fault, S-small fault 0.007inch (fault diameter), M=Medium size fault 0.014'', B-big fault 0.021''.

Motor speed(rpm)	Signal Category for fault identification (for every motor speed)	Signal Category for fault severity estimation (for every motor speed)
1797		• IRF set (S, M, B)
1772	Data mix set (Healthy, (IRF, BF,ORF)$_{\text{fault size 0.007''}}$)	• BF set (S, M, B)
1750		• ORF set (S, M, B)
1730		

Tables 2&3 show the correct classification rates of the testing segments for detection of fault locations and detection of fault severity respectively. In Table 2 all apart from those segments which recorded at speed of 1772 rpm are totally correctly classified. In Table 3, all the segments are perfectly classified for all the speeds considered in this analysis.

The first three most fault sensitive model coefficients are shown in Table 4. It is clear that these coefficients sets are not identical. However, some remarks can be highlighted. Among all the cases, the coefficient number 3 (i.e a_3) are existed within the first three higher sensitivity for the change of bearing conditions.

This can be helpful in using it as condition monitoring index. The a_4 are also present as a sensitive coefficient for fault location change or fault growth in the inner and outer races. However this is not completely true when the speed is further increased to 1797 or when the fault growth is monitored on the ball.

Table 2. Classification rate of testing samples.

Motor speed (rpm)	Signal Category used in classification	No.of Test samples	Diagnosis Accuracy %
1797	Healthy, $(IRF, BF, ORF)_{0.007''}$	116	100%
1772	Healthy, $(IRF, BF, ORF)_{0.007''}$	116	97.4%
1750	Healthy, $(IRF, BF, ORF)_{0.007''}$	116	100%
1730	Healthy, $(IRF, BF, ORF)_{0.007''}$	116	100%

Table 3. Classification rate of fault size estimation.

Motor speed (rpm)	Signal Category used in classification	No.of Test samples	IRF_{SMB}	BF_{SMB}	ORF_{SMB}
1797	**S**mall, **M**edium, **B**ig	174	100%	100%	100%
1772	**S**mall, **M**edium, **B**ig	174	100%	100%	100%
1750	**S**mall, **M**edium, **B**ig	174	100%	100%	100%
1730	**S**mall, **M**edium, **B**ig	174	100%	100%	100%

Table 4. Model coefficients of first three higher fault sensitivity ad *d=1.5 as in equation (7&8).*

Motor speed	**Sensitive coefficients set**			
(rpm)	**Data Mixed**	**IRF_{SMB}**	**BF_{SMB}**	**ORF_{SMB}**
1797	[8 3 6]	[3 8 2]	[2 4 3]	[4 3 2]
1772	[4 3 9]	[4 3 5]	[3 1 2]	[3 4 2]
1750	[4 9 3]	[4 3 5]	[3 6 5]	[4 3 2]
1730	[4 3 5]	[4 3 5]	[4 3 5]	[4 3 2]

Figure 1 shows an example of how the coefficient a_4 responses to different bearing fault location at 1730 rpm.

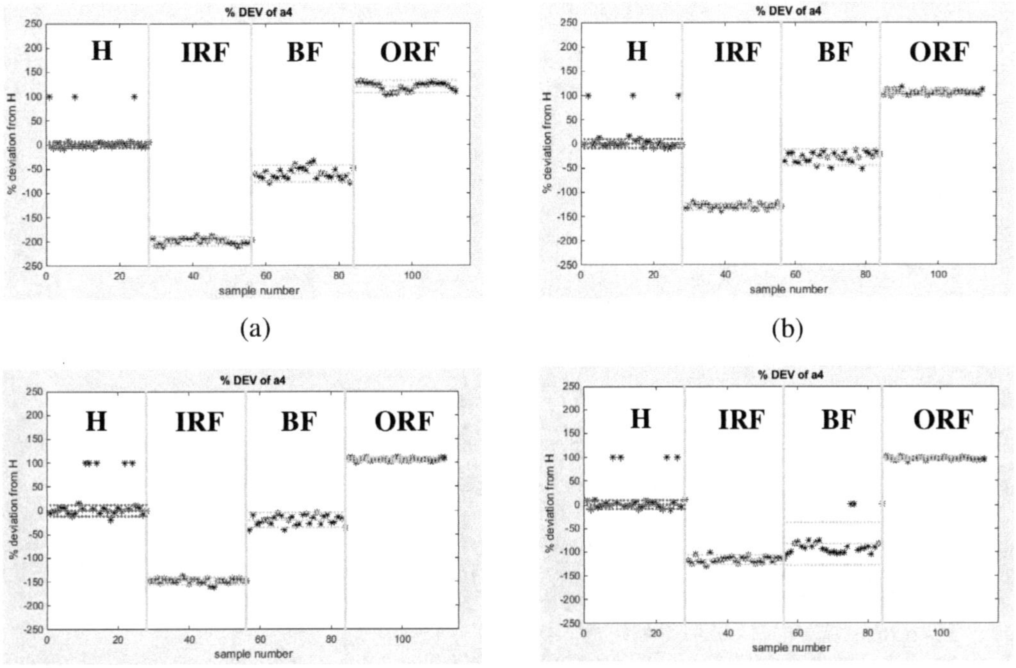

(a) (b) (c) (d)

Figure 1. The behaviour of coefficient (a_4) in correspondence to the change of bearing condition (i.e fault location). a) 1797 rpm b) 1772 rpm c) 1750 rpm d)1730 rpm.

7. Conclusion.

In this study, a methodology for fault diagnosis in rolling element bearings is introduced. It shows a very good performance in detection different types and severity of faults. The further analysis of model coefficients shows that there are certain coefficients which are more sensitive to the fault presence and growth when compared to their reference values. However, these sets of coefficients are not unique for all the cases but some of them are common for different bearing conditions. This study is useful in extraction some indices that can be used for bearing condition monitoring.

References

[1] Tandon N, Choudhury A 1999. *Tribology international.* A review of vibration and acoustic measurement methods for the detection of defects in rolling element bearings.**32**(8).469-80.

[2] Nataraj C, Kappaganthu K 2011. *13 World Congress in Mechanism and Machine Science;* Vibration on- based diagnostics of rolling element bearings: state of the art and challenges; Guanajuato, Mexico.

[3] Arnold M, Milner X, Witte H, Bauer R, Braun C 1998. *Biomedical Engineering, IEEE Transactions on.* Adaptive AR modeling of nonstationary time series by means of Kalman filtering.**45**(5).553-62.

[4] Zhan Y, Jardine A 2005. *Journal of Sound and Vibration*. Adaptive autoregressive modeling of non-stationary vibration signals under distinct gear states. Part 1: modeling.**286**(3).429-50.

[5] Zhan Y, Jardine A 2005. *Journal of sound and vibration*. Adaptive autoregressive modeling of non-stationary vibration signals under distinct gear states. Part 2: experimental analysis.**286**(3).451-76.

[6] Kwiatkowski D, Phillips P, Schmidt P, Shin Y 1992. *Journal of econometrics*. Testing the null hypothesis of stationarity against the alternative of a unit root: How sure are we that economic time series have a unit root?;**54**(1).159-78.

[7] Altmann J, Mathew J 2001. *Mechanical systems and signal processing*. Multiple band-pass autoregressive demodulation for rolling-element bearing fault diagnosis.**15**(5).963-77.

[8] Thanagasundram S, Schlindwein F 2006. *Proceedings of ISMA2006 Noise and Vibration Engineering Conference;* Autoregressive based diagnostics scheme for detection of bearing faults.

[9] Thanagasundram S, Spurgeon S, Soares Schlindwein F 2008. *Journal of Sound and Vibration*. A fault detection tool using analysis from an autoregressive model pole trajectory.**317**(3).975-93.

[10] Wang G, Luo Z, Qin X, Leng Y, Wang T 2008. *Mechanical Systems and Signal Processing*. Fault identification and classification of rolling element bearing based on time-varying autoregressive spectrum.**22**(4).934-47.

[11] Pardey J, Roberts S, Tarassenko L 1996. *Medical engineering & physics*. A review of parametric modelling techniques for EEG analysis.**18**(1).2-11.

[12] Mader H 2006. Statistics in volcanology: Geological Society of London.

[13] Box G, Jenkins G, Reinsel G 2013. *Time series analysis: forecasting and control*: John Wiley & Sons.

[14] Website CWRUBDC. [cited 2014 01-10-2014].
Available from: http://csegroups.case.edu/bearingdatacenter/pages/welcome-case-western-reserve-university-bearing-data-center-website.

A control strategy for adaptive absorber based on variable mass

Qiang Gao [1], Ning Han [1], Yanqing Zhao [1], Chendong Duan [2], Wanqin Wang [3]

[1] School of Automobile, Chang'an University, Xi'an, 710064, China
[2] School of Electronic and Control Engineering, Chang'an University, Xi'an, 710064, China
[3] School of Science, Xi'an Jiaotong University, Xi'an, 710049, China

Email: gaoqiang@chd.edu.cn

Abstract. The tuned vibration absorber (TVA) has been an effective tool for vibration control. However, the application of TVA can cause resonance of the primary system and increase its vibration when the absorber is mistuned. In this paper, a novel control strategy based on adaptive tuned vibration absorber (ATVA) of variable mass is proposed to reduce the resonance of the primary system. Unlike most ATVAs suggested by other researchers which adjust the absorber natural frequency by changing the stiffness, the variable mass ATVA varies its natural frequency by changing absorber mass to match the excitation frequency. Some simulations and experiments were conducted to test the performance of the control strategy. The results show that the proposed control plan can widen the frequency bandwidth of the absorber, as well as suppress the resonance of the primary system significantly. This implies that the work is useful for practical applications of ATVA.

1. Introduction

As an effective vibration control device, the tuned vibration absorber (TVA) has been used extensively in the vibration control fields since it was invented by Frahm in1911 [1]. When the TVA is tuned to the excitation frequency of the primary system, it can suppress the host vibration greatly. However, the effective frequency bandwidth of the conventional TVA is very narrow. Once the excitation frequency drifts slightly, the TVA will lose effectiveness and probably increase the primary system vibration. In practical applications, the vibration source is complicated and its frequency often changes with time. This greatly limits the wide applications of the TVA. In order to deal with this problem, many new tuned vibration absorbers have been proposed in recent years, such as active tuned vibration absorber, and adaptive tuned vibration absorber.

The adaptive tuned vibration absorber (ATVA) can adjust its dynamic parameters (and then its natural frequency) in real-time to match the excitation frequency and significantly improve the performance of absorber in practical engineering applications. Many ATVA configurations have been studied these years. Franchek et al. designed an ATVA with variable stiffness by changing the number of active coils of a spring [2]. Walsh et al. used a pair of leaf spring as the variable stiffness element and adjusted the stiffness by changing the gap between the two beams [3]. Smart materials such as shape memory alloy (SMA), magnetorheological elastomer (MRE), have been applied on absorber design as well. Williams et al. have made a series of studies on SMA-based ATVA in which the stiffness can be adjusted by changing the temperature of SMA [4]. Deng and Gong suggested an ATVA that used a MRE as the stiffness element. The stiffness of the absorber can be adjusted by controlling the magnet field applied on MRE [5]. To widen the effective frequency bandwidth of the absorber, the authors of this paper suggested the variable mass (VM) ATVA where a variable mass element is introduced in previous paper [6]. Different from other approaches, the natural frequency of the VM ATVA is adjusted by changing the absorber mass rather than the stiffness. Some simulation and experiments have been carried out to verify that the VM ATVA can suppress the vibration in a wider frequency range.

Besides the narrow effective bandwidth, it should be noticed that a new resonance peak is introduced into the frequency response of the primary system when the ATVA is attached to the primary system. Evidently, this will cause the serious vibration of the primary system when the excitation frequency is equal to this resonant frequency. However, to the best of authors' knowledge, few studies have been reported to reduce this resonance. A novel control strategy based on the VM ATVA is proposed in this paper to prevent this big vibration of the primary system. Following the introduction, the traditional control strategy of the TVA is introduced briefly and then the novel control strategy is proposed in section 2. Section 3 presents the vibration attenuation effectiveness of the control strategy via some experiments. And then, the conclusions are drawn in the last part of this paper.

2. Novel control strategy of the VM ATVA

2.1. Traditional control strategy

Usually, a TVA consists of a mass, a spring and a damper. In the VM ATVA, a variable mass element, i.e. a liquid tank is adopted to change the absorber mass by adding or releasing some liquid from the tank, as shown in Figure 1, where m_1, k_1, c_1 are the mass, stiffness and damping of the primary system, respectively; k_2, c_2 are stiffness, damping of the ATVA, respectively; the absorber mass includes two parts: the constant mass m_2 and the variable mass m_v, whose value ranges from 0 to $m_{v\max}$. When the variable mass changes, the natural frequency of the ATVA can be computed following the equation:

$$\omega_v = \sqrt{\frac{k_2}{m_2 + m_v}} \tag{1}$$

It is obvious that the maximal and minimal natural frequencies of the ATVA, $\omega_{v\max}$ and $\omega_{v\min}$, occur while the liquid tank is empty and filled. Hence, the effective frequency band of the adaptive absorber ranges from $\omega_{v\min}$ to $\omega_{v\max}$ by changing the variable mass. When the excitation frequency ω_e satisfies the equation of $\omega_{v\min} \le \omega_e \le \omega_{v\max}$, the variable mass m_e can be computed by the next equation to keep the VM ATVA tuned:

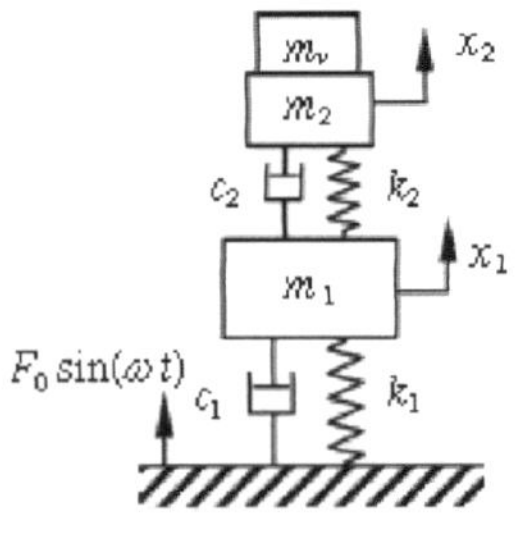

Figure 1. Model of VM ATVA attached to a primary system

$$m_e = \frac{k_2}{\omega_e^2} - m_2 \tag{2}$$

To obtain a wideband absorber, many researchers suggested a control strategy, which can be formulated as followings:

$$m_v = \begin{cases} m_{v\max} & 0 \le \omega < \omega_m \\ m_e & \omega_m \le \omega < \omega_n \\ 0 & \omega_n \le \omega \end{cases} \tag{3}$$

where ω_m and ω_n are the minimal and maximal natural frequencies of the ATVA, i.e. the frequencies of the bottom point of the frequency responses of the primary system when the absorber obtains its minimal and maximal mass, as shown in Figure 2. Point P and Q are the peaks of these two frequency responses.

The dot dash curve (red) in Figure 2 illustrates the frequency response of the primary system while the control strategy above is applied. One can see that the primary system vibration is suppressed by 12.2 and 9.1 dB at points M and N with the frequencies of 7.1 and 10 rad/s, respectively. Moreover, when the variable mass is within the rang of 0 to $m_{v\max}$, a frequency bandwidth from ω_m to ω_n is obtained.

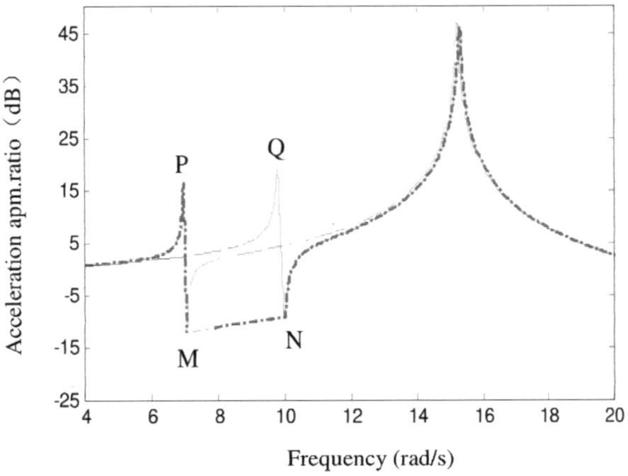

Figure 2. Frequency responses of the primary system with a VM ATVA attached. --- (green): liquid tank filled; — (blue): liquid tank empty; - · (red): old control plan used.

2.2. *New control strategy to reduce the resonance*

As discussed in the previous section, the VM ATVA can widen the effective bandwidth of the absorber by using the old control strategy. However, there is still a resonance peak at point P in the frequency response of the primary system, as shown in Figure 2. That means the primary system vibration will increase while the excitation frequency is equal to ω_p .

To suppress this big vibration, a new control strategy based on the VM ATVA is proposed, which can be expressed as follows:

$$m_v = \begin{cases} 0 & \omega < \omega_s \\ m_{v\max} & \omega_s \le \omega < \omega_m \\ m_e & \omega_m \le \omega < \omega_n \\ 0 & \omega_n \le \omega \end{cases} \tag{4}$$

where ω_s is the frequency of point S that is the intersection point of the frequency responses of the primary system while the variable mass obtains its maximal and minimal value.

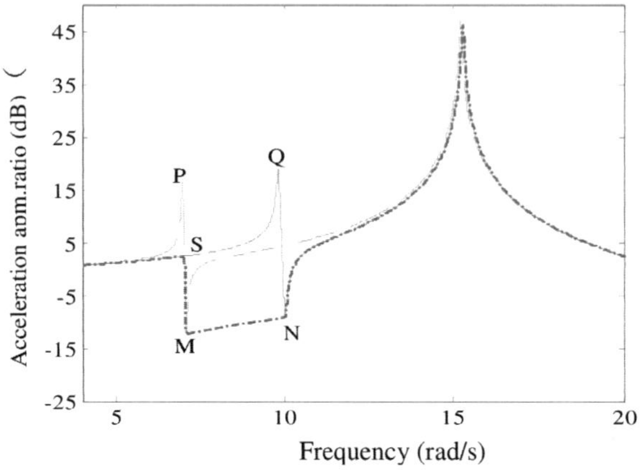

Figure 3. Frequency responses of the primary system with a VM ATVA. --- (green): liquid tank filled; — (blue): liquid tank empty; -· (red): proposed control plan used

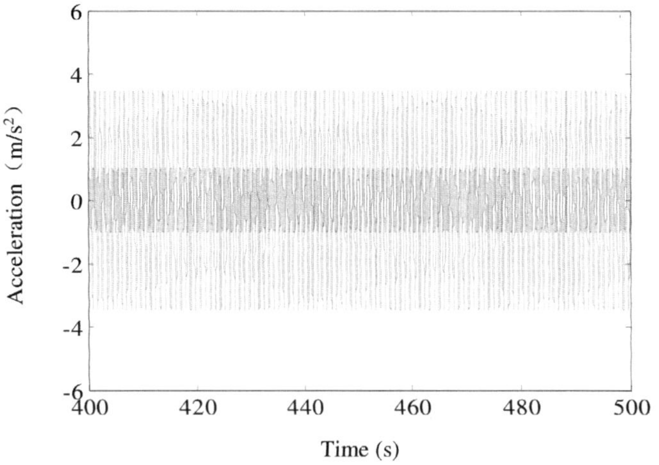

Figure 4. Vibration acceleration signal of the primary system. — (green): vibration curve at point P; ---(blue): vibration curve at point S.

Different from the previous control strategy, the liquid tank keeps empty while the excitation frequency is less than ω_s when the new control plan is used. Figure 3 shows the frequency response of the primary system. It can be seen that the vibration of the primary system is attenuated by 12.2 and 9.1dB at 7.1 and 10 HZ while the liquid tank is filled and empty, hence the bandwidth of the ATVA is from 7.1 to 10 HZ. When the excitation frequency lies within this range, the vibration of the primary system can be suppressed significantly. It is same as that when the old control plan is applied. However, the resonance at point P with amplification ratio of 16.3 dB is eliminated to approximate 2.1 dB, meanwhile a small lump about 2.3 dB at point S appears. This implies that the primary system vibration is suppressed about 14.0 dB when the new control plan is adopted. Figure 4 presents the vibration acceleration signal of the primary system while it is working at point P and S. It can be seen that the vibration of the host at point S is remarkably reduced and the amplitude is suppressed by 70.4% comparing with that at point P. Therefore, the proposed control plan can significantly reduces the resonance of the primary system caused by using the absorber, as well as widen the effective bandwidth of the ATVA.

3. Experiment tests

To verify the performance of the proposed control strategy, an experimental system was designed and some experiments, in which a glass bottle was used as variable mass element and water worked as the medium to change the absorber mass, were conducted. The frequency responses of the primary system when the glass bottle is filled and empty are shown in Figure 5. The frequency responses while the glass bottle is one-third and two-thirds water filled, and the excitation frequency varies from 10 Hz to 13.5 Hz are also conducted to obtain more information. From Figure 5 one can see that the vibration of the primary system is reduced by 27.0 and 29.0 dB at frequencies of 12.3 and 13.6 HZ while the glass bottle is filled and empty. Hence, by changing the water mass, the variable mass ATVA can effectively suppress the primary system vibration from 12.3 to 13.6 HZ.

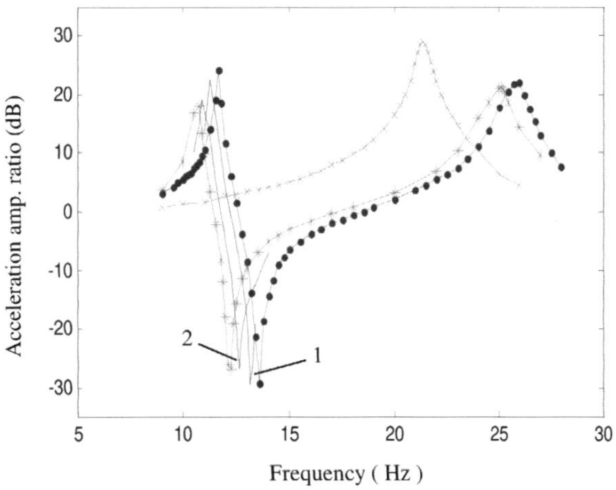

Figure 5. Frequency responses of the primary system with a VM ATVA. –•–(blue): empty bottle; –*– (green): water-filled bottle; –×–(red): with no VM ATVA mounted; 1: one-third water-filled bottle; and 2: two-thirds water-filled bottle.

In order to show the performance of the new control strategy more directly, a comparative figure of the old and new control strategies is illustrated in Figure 6. The previous resonant peak with the value of 18.1 dB at 10.8 HZ is suppressed to approximate 9.9 dB (at 11.0 Hz). This demonstrates that a resonance reduction of 8.2 dB is achieved by using the suggested control plan.

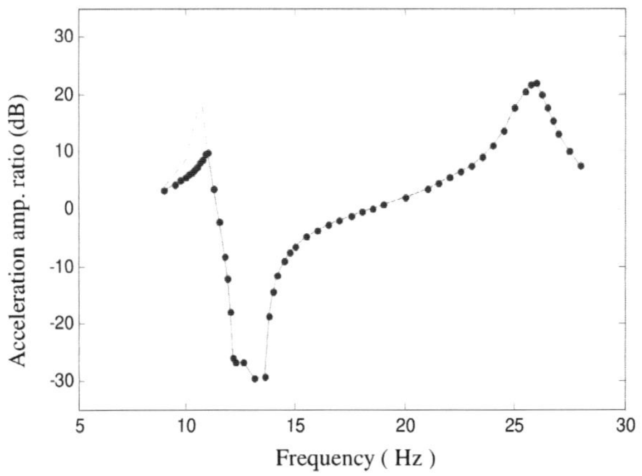

Figure 6. Frequency responses of the primary system. -•- (blue): proposed control plan used; ··· (red): old control plan used.

4. Conclusions

Although the ATVA can widen the absorber bandwidth by adjusting its natural frequency to match the excitation frequency, using the ATVA can cause the resonance of the primary system as well when it is mistuned. A new control strategy based on variable mass ATVA is proposed in this paper to suppress the resonance of the primary system. Some simulations and experiments were conducted to test the performance of the control plan. The results show that the suggested control plan can reduce the resonance of the primary system remarkably, meanwhile extend the effective frequency bandwidth of the absorber. Therefore, the performance of the ATVA can be promoted by using the proposed control strategy.

Acknowledgments

This work was supported by the National Natural Science Foundation of China (Nos. 51175049 and 51275380) and the Special Fund for Basic Scientific Research of Central Colleges, Chang'an University (CHD2012ZD007).

Reference

[1] Frahm, H, Device for damping vibrations of bodies, US Patent, 1911, 989, 958
[2] Franchek, MA., Ryan, MW. and Bernhard, R.J. Adaptive–passive vibration control, Journal of Sound and Vibration, 189, 565–585, (1995).
[3] Walsh, PL. and Lamancusa, JS. A variable stiffness vibration absorber for minimization of transient vibrations. Journal of Sound and Vibration, 158(2):195-211, (1992).
[4] Williams, K., Chiu, G. and Bernhard, R. Dynamic modeling of a shape memory alloy adaptive tuned vibration absorber. Journal of Sound and Vibration 280, 211–234, (2005).
[5] Deng, H. and Gong, X. Application of magnetorheological elastomer to vibration absorber, Communications in Nonlinear Science and Numerical Simulation, 13, 1938–1947, (2008).
[6] Gao, Q., Zhang, W. and Liu, B. An adaptive tuned vibration absorber based on variable mass. Noise Control Engineering Journal. 59,491–496, (2011)

Fuzzy Modal Control Applied to Smart Composite Structure

E H Koroishi[1], A W Faria[2], F A Lara-Molina[1], V Steffen Jr[3]

[1] Federal Technological University of Paraná, Department of Mechanical Engineering, Av. Alberto Carazzai, 1640 - Campus Cornélio Procópio - Cornélio Procópio – PR - 86300-000, Brazil.
[2] Federal University of Triangulo Mineiro, Institute of Technology and Exact Sciences, Department of Civil Engineering, Av. Dr. Randolfo Borges Junior, 1250 – Unidade I – Uberaba – MG – 38064-200, Brazil.
[3] School of Mechanical Engineering, Federal University of Uberlandia, Av. João Naves de Ávila 2121 - Campus Santa Monica - Uberlândia - MG - 38408-100, Brazil.

E-mail: edsonh@utfpr.edu.br, albert@icte.uftm.edu.br, fabianmolina@utfpr.edu.br, vsteffen@mecanica.ufu.br

Abstract. This paper proposes an active vibration control technique, which is based on Fuzzy Modal Control, as applied to a piezoelectric actuator bonded to a composite structure forming a so-called smart composite structure. Fuzzy Modal Controllers were found to be well adapted for controlling structures with nonlinear behavior, whose characteristics change considerably with respect to time. The smart composite structure was modelled by using a so called mixed theory. This theory uses a single equivalent layer for the discretization of the mechanical displacement field and a layerwise representation of the electrical field. Temperature effects are neglected. Due to numerical reasons it was necessary to reduce the size of the model of the smart composite structure so that the design of the controllers and the estimator could be performed. The role of the Kalman Estimator in the present contribution is to estimate the modal states of the system, which are used by the Fuzzy Modal controllers. Simulation results illustrate the effectiveness of the proposed vibration control methodology for composite structures.

1. Introduction

The recent years have seen the appearance of innovative materials, such as the so-called composite materials, particularly in aerospace applications. The structures constructed by this innovated arrangement are characterized by lightness, mechanical resistance, and the possibility of optimization for a specific working condition. Unlike the regular materials (steel, aluminum, etc), the composites are formed by various layers with different fiber orientations, which allows them to be adequate for particular applications [1]. Aircraft, aerospace and automotive industries are examples for which composite materials have been increasingly used.

Generally, piezoelectric layers (PZT sensor/actuator patches) are incorporated to the composite materials in order to offer potential benefits in a wide range of applications such as structural health monitoring, noise suppression, precision positioning and active vibration control [2]. Thus, the set encompassing the composite material, piezoelectric layers and monitoring and control systems, is known as Smart Composite Structure.

Currently, it is observed an increasing number of research works in engineering devoted to the development of new active vibration control techniques (AVC). Among the well-known techniques for AVC, nowadays the Active Vibration Modal Control (AVMC) is a highlight and presents successful applications in several areas [3-8].

In the present contribution the active vibration modal control is tested only numerically. The control approach now studied is the Fuzzy Logic Controller (FLC). In this case, the modal displacement and modal velocities are used by the controllers to determine the control force, and the controllers considered were the Fuzzy Modal Controllers (FMCs). The FMCs were found to be well adapted for controlling structures with nonlinear behavior, whose characteristics change considerably with respect to time. Since the modal states cannot be accessed directly during experimentation, they have to be rebuilt by using an estimator algorithm. Consequently, for determining the modal states the Kalman estimator was chosen. As an important characteristic, this estimator can rebuild the modal states from noisy signal responses. As mentioned above, numerical results are presented to illustrate the proposed methodology. Additionally, the balanced realization method is used to reduce the complexity of the computational model for only the most relevant vibration modes of the structure in order to increase the efficiency of the controller. The simulations are performed so that the procedure is as similar as possible to the experimental conditions.

2. System Modeling

The coupling between the composite structure and the piezoelectric element is made through the Hamilton's variational principle that incorporates all energy contributions presented in the structure.

According to Chee [9], the coupling elementary matrices using Mixed Theory are the following:

$$[M^e] = \int_{V_e} \rho [N_u]^T [A_u]^T [A_u][N_u] dV_e \tag{1}$$

$$[K^e_{uu}] = \sum_{k=1}^{nc} \int_{\xi=-1}^{+1} \int_{\eta}^{+1} \int_{z=z_k}^{z_{k+1}} \left([B_u]^T [c][B_u]\right) J dz d\eta d\xi \tag{2}$$

$$[K^e_{u\varphi}] = \sum_{k=1}^{nc} \int_{\xi=-1}^{+1} \int_{\eta}^{+1} \int_{z=z_k}^{z_{k+1}} \left([B_u]^T [e][B_\varphi]\right) J dz d\eta d\xi \tag{3}$$

$$[K^e_{\varphi\varphi}] = \sum_{k=1}^{nc} \int_{\xi=\ 1}^{+1} \int_{\eta}^{+1} \int_{z=z_k}^{z_{k+1}} \left([B_\varphi]^T [\chi][B_\varphi]\right) J dz d\eta d\xi \tag{4}$$

where ρ is the material density, $[M^e]$ is the elementary mass matrix and $[K^e_{uu}]$ is the elementary matrix of elastic stiffness. $[K^e_{u\varphi}]$ and $[K^e_{\varphi u}]$ are stiffness elementary matrices of electromechanical coupling and $[K^e_{\varphi\varphi}]$ is known as dielectric elementary matrix. [c], [e] and [χ] are, respectively, the elastic stiffness, piezoelectric stress and electric permissivity matrices of constant values. $[B_u]$ and $[B_\varphi]$ are input matrices. V_e is the elementary volume. J is the Jacobian of the transformation [1].

Equation (5) shows the global matrices of the model constructed through the standard procedure, in which the subscript g indicates global quantities.

$$\begin{bmatrix} [M_g] & [0] \\ [0] & [0] \end{bmatrix} \begin{Bmatrix} \{\ddot{u}_g\} \\ \{\ddot{\phi}_g\} \end{Bmatrix} + \begin{bmatrix} [K_{uu}] & [K_{u\varphi}] \\ [K_{\varphi u}] & [K_{\varphi\varphi}] \end{bmatrix} \begin{Bmatrix} \{u_g\} \\ \{\phi_g\} \end{Bmatrix} = \begin{Bmatrix} \{F_g\} \\ \{Q_g\} \end{Bmatrix} \tag{5}$$

where $\{F_g\}$ and $\{Q_g\}$ are, respectively, the generalized force and nodal charge (elementary).

2.1. Balanced Realization

Balanced realization consists in describing the model of the system in the space state form and combines the controllability and observability matrices for each state of the system by using the Gramians of controllability and observability of the system. The linear transformation that leads the system to this representation is called balanced transformation. In this method, the reduced model is obtained by neglecting the states associated with the small singular values [10-11].

A balanced realization is an asymptotically stable minimal realization in which the controllability and observability Gramians are equal and diagonal [12]. Consider a stable linear-time invariant system defined by the state space equations below:

$$\{\dot{x}(t)\} = [A]\{x(t)\} + [B]\{u(t)\} \tag{6}$$

$$\{y(t)\} = [C]\{x(t)\} \tag{7}$$

where $x(t)$ is the state vector, $[A]$ is the nxn dynamic matrix, $[B]$ is the nxm input matrix, $[C]$ is the sxn output matrix, $\{u(t)\}$ is the input force and $\{y(t)\}$ is the output vector, n is the order of the system, m is the number of inputs and s is the number of outputs. The system is called balanced if the solutions to the following Lyapunov equations (8) and (9) are $P = Q = \text{diag}\ (\sigma_1, \sigma_2 \ldots \sigma_n)$,

$$[A][P] + [P][A]^T + [B][B]^T = 0 \tag{8}$$

$$[A]^T[Q] + [Q][A] + [C]^T[C] = 0 \tag{9}$$

where $[P]$ and $[Q]$ are, respectively, the controllability and observability Gramians and σ_i, $i = 1,2,\ldots, n$, are the singular values of the system ($\sigma_1 \geq \sigma_2 \geq \ldots \geq \sigma_n \geq 0$).

Every σ_i is associated with a state x_i of the balanced system. Its value quantifies the contribution that x_i makes to the input-output behavior of the system. As $\sigma_1 \geq \sigma_2$ then x_1 affects the behavior of the system more than x_2, due to the fact that the singular values are ranked from the most important to the least important one. In the balanced realization the fidelity of the reduced model with respect to the full model (A, B, C) depends on the relation $\sigma_r \geq \sigma_{r+1}$, where r is the reduced model order. Thus, the reduced state space system is given by:

$$\{\dot{x}_r(t)\} = [A_r]\{x_r(t)\} + [B_r]\{u_r(t)\} \tag{10}$$

$$\{y_r(t)\} = [C_r]\{x_r(t)\} \tag{11}$$

where $\{x_r(t)\}$is the reduced state vector, $[A_r]$ is the rxr reduced dynamic matrix, $[B_r]$ is the rxm reduced input matrix and $[C_r]$ is the sxr reduced output matrix.

3. Control Approach

Active modal control is used as control strategy for a smart composite structure as shown in figure 1. The advantage of using active modal control is that this technique is very effective for flexible structure applications, requiring a reduced number of actuators and sensors. The estimator is responsible for determining the modal states required by the controllers. The Kalman Estimator is able to estimate the states by using noise contaminated measurement signals. More details regarding the Kalman Estimator can be found in the literature [13-15]. The states are then used by the controllers to determine the control force.

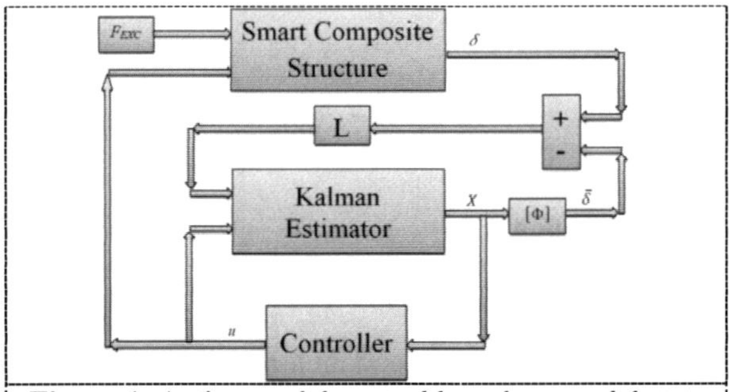

Figure 1. Active modal control based on modal state feedback control (adapted from [8]).

Figure 1 shows that in the modal state feedback control a number of controllers are necessary. The method requires the modal displacements and modal velocities to determine the control effort of the controllers. For this purpose, the Fuzzy Modal Controllers were used to determine the control effort. This method permits to take into account slight nonlinearities and uncertainties in the model.

4. Fuzzy Modal Controllers

The first step of this approach consists in the fuzzification of each controlled mode $\{X_i(t)\}$in the fuzzy input. In this work, two membership GBELL Matlab[@] functions "Positive" and "Negative" are used. For these membership functions the relevance of each variable is given by:

$$\mu_A(X_i(t)) = \frac{1}{1+\left|\frac{X_i(t)-c_f}{a_f}\right|^{b_f}} \tag{12}$$

where a_f, b_f and c_f are the parameters of the GBELL function.

The inference engine implements the "minimum" function w_j. Mahfoud et al [16] presented Fuzzy controller rules applied to flexible structures. These rules are used in the present contribution and they are presented in table 1.

Table 1. Fuzzy controller rules [16-17].

Rule	Condition	Decision
1	IF positive displacement AND positive velocity	Action
2	IF positive displacement AND negative velocity	No action
3	IF negative displacement AND positive velocity	No action
4	IF negative displacement AND negative velocity	Action

Finally, the control is obtained after defuzzification. This defuzzification requires the knowledge of the fuzzy output variables corresponding to the rules and the aggregation of these rules as well as the output membership functions. In this paper the Takagi-Sugeno method is used due to its good adaptation to controller real time computation. The membership functions for the four modes are similar and are presented in equation (13) (z_1 corresponds to "no action" and z_2 to "action"):

$$z_1 = 0 \text{ and } z_2 = \alpha_i\, q_i + \beta_i\, \dot{q}_i \tag{13}$$

where, α_i and β_i are, respectively, the weights assigned to the displacement and velocity of the mode i, respectively.

The driving force F_i, for mode i, is given by the output of the controller and can be written as:

$$F_i = \frac{\sum_{j=1}^{2} w_j z_j}{\sum_{j=1}^{2} w_j} \tag{14}$$

5. Numerical Simulations

The studied laminated composite beam, illustrated in figure 2, is such that it has 306 mm length *(L)*, 25.5 mm width *(b)* and 1 mm thickness *(h)*. A piezoelectric ceramic actuator of dimensions 45.9 x 25.5 mm^2 is bonded to the beam top surface, 1 mm away from the clamp. The composite laminate has a total of five layers made of graphite/epoxy and oriented as [45° / 0° / 45° / 0° / 45°]. The layers oriented at 0° are parallel to the x axis. The thicknesses of the layers are 0.2 mm and the thickness of the piezoceramic actuator is 1 mm.

The constants of elastic stiffness of the beam made of AS4/3501 carbon/epoxy composite are the following (given in GPa): C_{11} = 173.6; C_{22} = C_{33} = 7.61; C_{12} = C_{13} = 2.48; C_{23} = 2.31; C_{44} = 1.38; C_{55} = C_{66} = 3.45. The pzts are: C_{11} = C_{22} = C_{33} = 102.23; C_{12} = C_{13} = C_{23} = 5.035; C_{44} = C_{55} = C_{66} = 2.594. The PZT patch piezoelectric constants are (given in C/m^2) as follows: e_{31} = -18.300; e_{32} = e_{33} = -9.013. The electric permissivity are (given in F/m) given by: $\chi_{11} = \chi_{22} = \chi_{33}$ = $1800\varepsilon_0$. These constants were used in the matrices $[c]$, $[e]$ and $[\chi]$, respectively, in equations (2), (3) and (4). The mass densities, in kg/m^3, are 1578 for the composite laminated material and 7700 for the PZT patch.

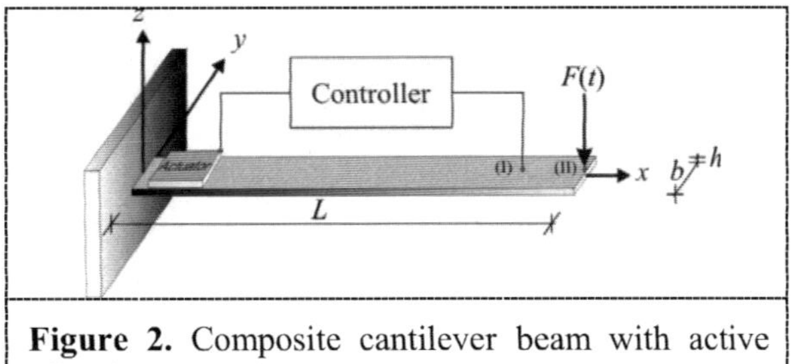

Figure 2. Composite cantilever beam with active vibration control.

The FE model has been derived by using a 10x1 uniform mesh. The excitation force (1N) was applied at point II and the time domain structure responses were captured at point I (see figure 2). The piezoelectric actuator is connected to an active control system, and the vibration amplitudes are to be minimized over time. Two cases of active vibration control considering non-parametric uncertainties were analyzed. The samples of the uncertainty term Δ are determined by using Monte Carlo simulation associated with Latin Hypercube. The number of samples was 100.

(a) Robustness analyses considering variation in the model of smart composite structure so that the uncertainties were considered in the dynamical matrix [A]+[ΔA].

(b) Robustness analyses considering variation in the dynamical matrix of Kalman Estimator $[A_r]+[\Delta A_r]$.

For the first two modes, the system was observable and controllable. The balanced realization was used to reduce the model for the first two modes of the structure. For these two modes, 4 uncertain modes result. The table 2 presents the 4 uncertain model configurations (matrix A_i). These models were used to determine the robust controllers.

Table 2: Uncertain Model (Matrix A_i).

A_i	Mode 1		Mode 2	
	-10%	+10%	-10%	+10%
1	X	--	X	--
2	X	--	--	X
3	--	X	X	--
4	--	X	--	X

6. Numerical Results

6.1. Case (a)

The figures 3, 4 and 5 present the results considering uncertainties in the dynamic matrix of the model of the smart composite structure.

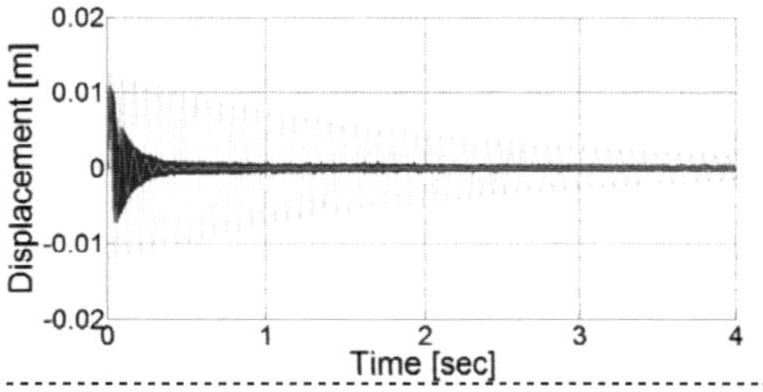

Figure 3. Impact Response (case a): deterministic value (red line) and envelope (blue lines).

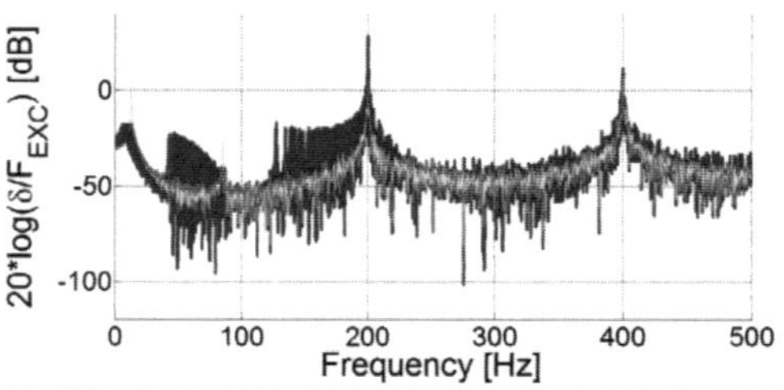

Figure 4. Frequency Response Function (case a): deterministic value (red line) and envelope (blue lines).

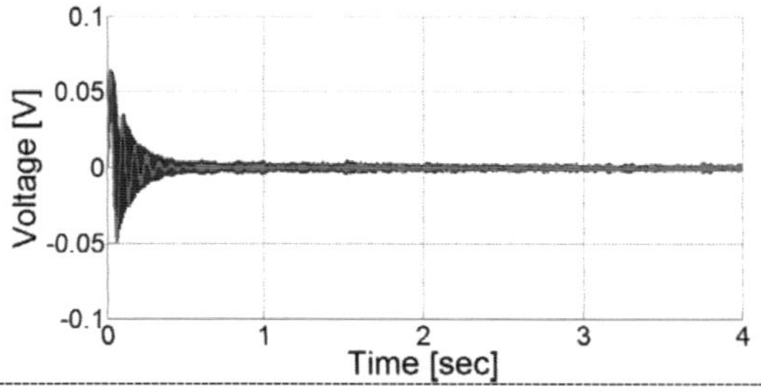

Figure 5. Control effort (case a): deterministic value (red line) and envelope (blue lines).

In terms of impact response, it is possible to see that the system´s response was completely attenuated with the use of FMCs, as seen in figure 3, where the green line represents the uncontrolled system. Another result presented in the figure 3 is the robustness of the FMCs that is demonstrated by the small envelope (blue lines). The same trends were observed in terms of the control effort (see figure 5).

Finally, the FRF shows that the two modes considered in the design of the FMCs were completely attenuated in the deterministic case (see the red line in figure 4). Therefore, in terms of robustness, the FMCs were only robust for the first mode (see the blue line in figure 4).

6.2. Case (b)

The figures 6, 7 and 8 present results for the case (b), which considers uncertainty in the dynamical matrix used by the Kalman Estimator.

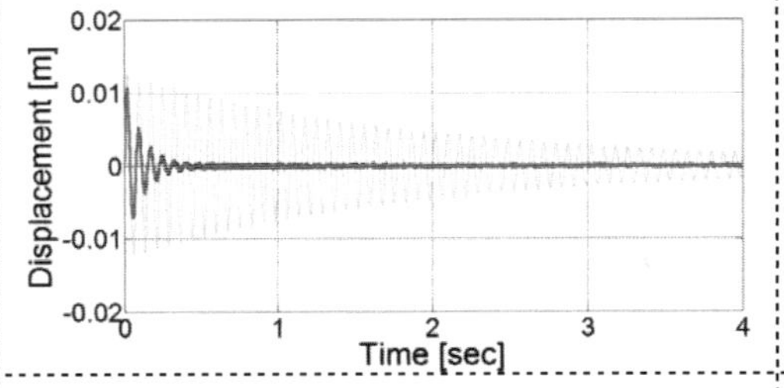

Figure 6. Impact Response (case b) – Robust Control: deterministic value (red line) and envelope (blue lines).

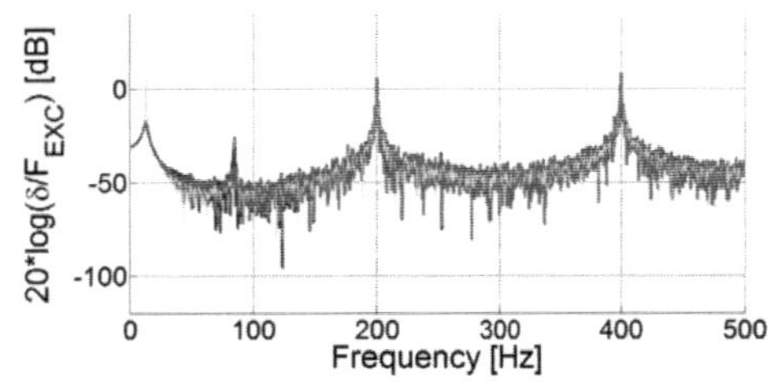

Figure 7. Frequency Response Function (case b) – Robust control: deterministic value (red line) and envelope (blue lines).

The case (b) presents envelopes smaller than those associated with the case (a) since when variation is considered in the structure's model, the natural frequency varies according to this variation, once the dynamic matrix changes with the value of Δ. The system´s response were complete attenuated as shown both in time and frequency domains. Figure 7 demonstrates that the first two modes were attenuated for the considered uncertainty models; it is shown that the FMCs were more robust in case (b).

Another important result observed in the second case is the energy consumption (figure 8). The FMCs were more robust in case (b) than in case (a), which demonstrates the same trends previously observed in the responses of the system (figures 6 and 7).

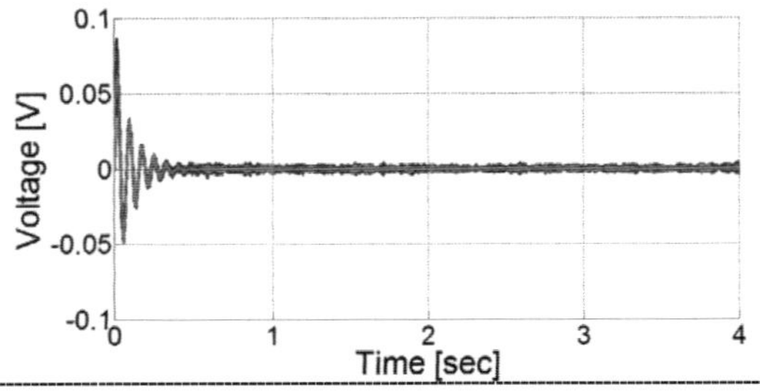

Figure 8. Control effort (case b) – Robust Control: deterministic value (red line) and envelope (blue lines).

7. Conclusions

The present contribution was dedicated to the study of active vibration control in smart composite structures by using Fuzzy Modal Controllers. The results demonstrated the efficiency of the FMCs in the AVC. The results have revealed that the number of considered modes (2 modes) was sufficient to achieve satisfactory control. It should be emphasized the importance of performing balanced realization in this stage, since this technique ranks the modes in order of relevance, regarding the dynamic behavior of the system. This means that the considered modes are the most important to the system response. The robustness analysis demonstrated that the FMCs were robust mainly for the second case presented above.

8. Acknowledgments

The authors are thankful to the Brazilian Research Agencies FAPEMIG, CNPq and CAPES for the financial support to this work through the INCT-EIE.

9. References

[1] Reddy J N 1997 *Mechanics of Laminated Composite Plates: Theory and Analysis* (Florida: CRC Press).

[2] Thinh T I and Ngoc L K 2010 Static behavior and vibration control of piezoelectric cantilever composite plates and comparison with experiments *Computacional Materials Science.*

[3] Hurbelaus S, Stöbener U and Gaul L 2008 Vibration reduction of curved panels by active modal control *Computers & Structures* 251-7.

[4] Li S 2011 Active modal control simulation of vibro-acoustic response of a fluid-loaded plate *Journal of Sound and Vibration* 5545-57.

[5] Bratland M, Haugen B and Rølvåg T 2011 Modal analysis of active flexible multibody systems *Computers & Structures* 750-61.

[6] Resta F, Ripamonti F, Cazzulani G and Ferrari M 2010 Independent modal control for nonlinear flexible structures: An experimental test rig *Journal of Sound and Vibration* 961-72.

[7] Houlston P R, Garvey S D and Popov A A 2007 Modal control of vibration in rotating machines and other generally damped systems *Journal of Sound and Vibration* 104-16.

[8] Koroishi E H, Borges A S, Cavalini Jr A A and Steffen Jr V 2014 Numerical and Experimental Modal Control of Flexible Rotor Using Electromagnetic Actuator *Mathematical Problems in Engineering (Print).*

[9] Chee C Y K 2000 *Static Shape Control of Laminated Composite Plate Smart Structure using Piezoelectric Actuators* (Sydney: Ph.D. thesis - University of Sydney).

[10] Assunção E and Hemerly E M 1992 Redução de Modelos de Sistemas Dinâmicos *9º Congresso Brasileiro de Automática* 159-64.

[11] Laub A J, Heath M T, Paige C C and Ward R C 1987 Computation of Systems Balancing Transformation and Other Applications of Simultaneous Diagonalization Algorithms *IEEE Transactions on Automatic Control* **AC-32**.

[12] Conceição S M, Bueno D D, Cavalini Jr A A, Abreu G L, Melo G P and Lopes Jr V 2009, Model Reduction Methods for Smart Truss like Structure *Proceedings of 8th Brazilian Conference on Dynamics, Control and Applications* (Bauru-SP).

[13] Welch G and Bishop G 1995 *An Introduction to Kalman Filter, Technical Report* (University of North Carolina at Chapel Hill).

[14] Anderson B D O and Moore J B 1979 *Optimal Filtering* (Prentice-Hall) 357p.

[15] Durbin J and Koopman S J 2002 *Time Series Analysis by State Space Methods* (Oxford University Press).

[16] Mahfoud J and Der Hagopian J 2011 Fuzzy Active Control of Flexible Structures by Using Electromagnetic Actuators *Journal of Aerospace Engineering* 329-37.

[17] Ayyub B M, Guran A and Haldar A 1997 *Uncertainty Modeling in Vibration, Control and Fuzzy Analysis of Structural Systems.*

Natural frequency changes due to damage in composite beams

I Negru[1], G R Gillich[1*], Z I Praisach[1], M Tufoi[1], N Gillich[2]
[1] Department of Mechanical Engineering, "Eftimie Murgu" University of Resita, P-ta Traian Vuia 1-4, 320085 Resita, Romania
[2] Department of Electrical Engineering, "Eftimie Murgu" University of Resita, P-ta Traian Vuia 1-4, 320085 Resita, Romania

E-mail: gr.gillich@uem.ro

Abstract. Transversal cracks in structures affect their stiffness as well as the natural frequency values. This paper presents a research performed to find the way how frequencies of sandwich beams change by the occurrence of damage. The influence of the locally stored energy, for ten transverse vibration modes, on the frequency shifts is derived from a study regarding the effect of stiffness decrease, realized by means of the finite element analysis. The relation between the local value of the bending moment and the frequency drop is exemplified by a concrete case. It is demonstrated that a reference curve representing the damage severity exists whence any frequency shift is derivable in respect to damage depth and location. This curve is obtained, for isotropic and multi-layer beams as well, from the stored energy (i.e. stiffness decrease), and is similar to that attained using the stress intensity factor in fracture mechanics. Also, it is proved that, for a given crack, irrespective to its depth, the frequency drop ratio of any two transverse modes is similar. This permitted separating the effect of damage location from that of its severity and to define a Damage Location Indicator as a sequence of squared of the normalized mode shape curvatures.

1. Introduction

Proper maintenance of engineering structures implies their periodic inspection and control, in order to observe the occurrence of faults and assure proper rehabilitation operations. This leads to reduced operational costs, by interventions in the incipient damage state, or can justify replacing structural elements or even the whole structure, in order to avoid collapse accidents. Numerous global evaluation methods were developed in last decades, the concern of researchers being oriented to find algorithms applicable to a large range of structures in terms of configuration and material properties.

The role of composite materials increases continuously because of their high mechanical characteristics to weight ratio. An advantage, of using these materials, consists in the possibility to elaborate them to fulfil predefined needs, permitting new approaches in structural design. Sandwich structures are a special class of composites; they are composed of sheets with diverse mechanical and physical properties, spatially distributed in a way to provide desired static and dynamical behaviour. This class of composites has been studied from the mid of the twentieth century, the main attempts being focused on bending and buckling [1-2], shock-resistance [3-4] and vibrational behaviour [5-7]. Works devoted to detection of various types of damages in sandwich structures [7-10] shown the complexity and difficulty of this performance, the authors resolving specific cases in regard to the damage type, boundary conditions and structures' mechanical and geometrical structure characteristics. The difficulty exists because actual vibration-base damage detection methods make use of the stress intensity factor to quantify the crack extend, this being difficult to be performed for layered structures. In previous researches, the authors contrived a mathematical relation that predicts the natural frequency values for isotropic beams with transversal cracks of any depth and location [11-14]. Using this relation, a two-step damage assessment algorithm was developed; in the first step the crack location is precisely found, afterwards the damage severity is evaluated. Since it is possible to separate the problem of location from that of severity evaluation, the method was tested on layered structures. In this paper, the damage detection method tailored for multi-layered beams is present.

2. Vibration of multi-layered beams

For isotropic beams the natural frequencies of bending modes are derived as:

$$f_i = \frac{(\alpha_i L)^2}{2\pi}\sqrt{\frac{EI_z}{mL^3}} \tag{1}$$

where f_i is the natural frequency and $\lambda_i = \alpha_i L$ is the wave number of mode i for a beam with length L, width B and height H. E denotes the Young's modulus, ρ the volumetric mass density, $I_z = B \cdot H^3 / 12$ is the moment of inertia regarding the weak axis of the prismatic beam and $A = B \cdot H$ is the beam cross-sectional area.

For multilayered beams, the natural frequencies of the bending modes can be derived if an appropriate expression of the rigidity EI is used. Since the interest is to attain high rigidity and low weight, the outer faces are manufactured from metals (usually steel or aluminum) and the inner layers of lightweight material; thus, an odd number of layers results. In this paper, a beam having 5 layers is considered. It has the total height of 5 mm, the layers being defined as follows:

- the outer layers 1 and 5 are made of steel (OL) of thickness $s_1 = s_5 = 1mm$;
- the intermediate layers 2 and 4 layers are of PVC, with the thickness $s_2 = s_4 = 1mm$;
- the central layer 3 is also of steel and has the thickness $s_3 = 1mm$,

the configuration being shown in Figure 1.

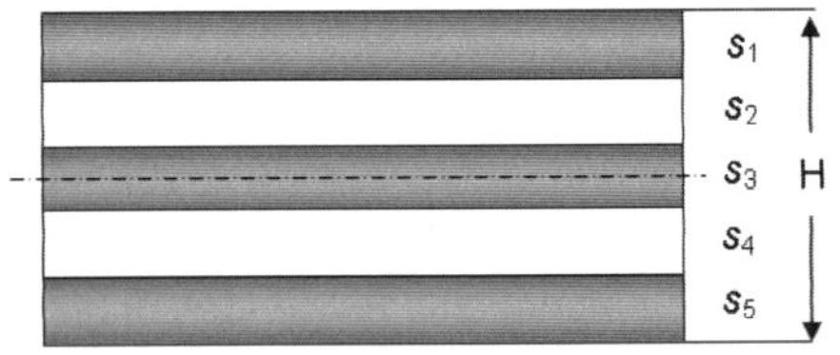

Figure 1. Cross-section in the composite beam showing the layer distribution

For this composite, the rigidity EI of the individual layers is derived, in concordance to the parallel axis theorem, as:

$$EI_{1,5} = E_{OL}\left[\frac{Bs_{1,5}^3}{12} + d_{1,5}^2(Bs_{1,5})\right] = E_{OL}\left[\frac{Bs^3}{12} + 4Bs^3\right] \tag{2}$$

$$EI_{2,4} = E_{PVC}\left[\frac{Bs_{2,4}^3}{12} + d_{2,4}^2(Bs_{2,4})\right] = E_{PVC}\left[\frac{Bs^3}{12} + Bs^3\right] \tag{3}$$

$$EI_3 = \frac{EBs_3^3}{12} = \frac{EBs^3}{12} \tag{4}$$

where, $d_{1,5}$ is the distance from the centre of gravity of the external layers (1 and 5) to the neutral axis and $d_{2,4}$ the distance from the centre of gravity of layers 2 and 4 to the neutral axis.

Because all layers have the same thickness s, and the distances between the centre of gravity and the neutral axis are $d_1 = d_5 = 2s$ and $d_2 = d_4 = s$, the bending rigidity is:

$$EI_{comp} = 2EI_1 + 2EI_2 + EI_3 = E_{OL}\left[3\frac{Bs^3}{12} + 8Bs^3\right] + E_{PVC}\left[2\frac{Bs^3}{12} + 2Bs^3\right] =$$

$$= (8.25E_{OL} + 2.167E_{PVC}) \cdot Bs^3 \tag{5}$$

Regarding the 5-layer composite structure weight, appearing in the frequency equation (1), it is derived as:

$$m_{comp} = 3\rho_{OL}BsL + 2\rho_{PVC}BsL = (3\rho_{OL} + 2\rho_{PVC})BsL \tag{6}$$

By introducing equations (5) and (7) in the equation (1), the frequency of the analysed composite beam results as:

$$f_i = \frac{(\alpha_i L)^2}{2\pi}\sqrt{\frac{(8.25E_{OL} + 2.167E_{PVC}) \cdot Bs^3}{(3\rho_{OL} + 2\rho_{PVC})BsL^4}} \tag{7}$$

Note that the modal shapes are similar with those determined for isotropic materials as long as the stiffness-mass ratio does not exceed a critical value. Else, the Euler-Bernoulli model is no longer adequate and more complex models are necessary to be used, such as Shear or Timoshenko [4-5].

In the numerical example, two beams are used. The first is a composite beam, as described above, in which the steel sheets have the Young's modulus $E_{OL} = 2.06 \cdot 10^{11}\ N/m^2$ and the mass density $\rho_{OL} = 7850\ kg/m^3$, while the PVC sheets have $E_{PVC} = 2.41 \cdot 10^9\ N/m^2$ and $\rho_{PVC} = 1300\ kg/m^3$. Each layer has the length $L = 1m$ and the cross-section edges $B = 0.02m$ and $H = 0.005m$. The results related to the stiffness and the mass, are shown in Table 1 together with the results achieved for the same beam geometry entirely made of steel. The steel beam has similar global dimensions as the composite one.

Table 1. Stiffness and mass values

Beam type	EI_{OL} (N/m^2)	EI_{PVC} (N/m^2)	EI_{total} (N/m^2)	m_{OL} (kg)	m_{PVC} (kg)	m_{total} (kg)
Steel beam	42.7	-	42.7	0.785	-	0.785
Composite beam	33.8	10.4	33.9	0.471	0.052	0.523

From Table 1 results that the stiffness of composite structures is comparable to that of steel, but the mass decreases significantly when PVC layers are introduced. It can be also noted that the central layer brings an insignificant contribution to the structure rigidity, thus it is preferable to use lightweight materials in the centre of the beam. However, in structural design, it is important to take attention of the natural frequency increase associated to the weight decrease, which can be a disadvantage, in some cases.

3. Numerical model description

The model analysed by numerical simulations was a cantilever beam with the geometry described in the previous section, for which the modal analysis has been performed using the *SolidWorks* software. For the 3D model meshing, tetrahedral finite elements with an average element size of 1 mm were used. The materials available in the software library have identical properties to that earlier described. Additionally, the structural steel **AISI 1045** has the Yield strength $5.3 \cdot 10^8$ *N/mm²*, Tensile strength $6.25 \cdot 10^8$ *N/mm²* and Poisson's ratio 0.29, while the **PVC rigid** has the Yield strength $4.5 \cdot 10^7$ *N/mm²*,

Tensile strength $4.07 \cdot 10^7$ *N/mm*^{2} and Poisson's ratio 0.3825. The ambient temperature has been set to 22 *°C*. By using the *Modal analysis* module the first 20 natural frequencies were determined; from them the first weak-axis bending vibration modes are selected and used in this study.

To get a reference, first simulation has been made for the intact beam. Afterwards, a damage placed at $x = 0.274$ *m* from the fixed end was achieved. It has the width of 1 *mm*, the depth being gradually increased from 0.2 *mm* to 4.8 *mm* by a 0.2 *mm* step.

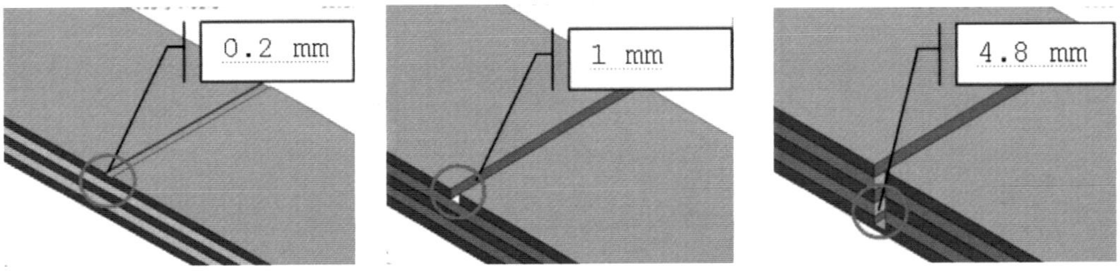

Figure 2. Beam subjected to different damage scenarios.

The beam section affected by the damage considered in the incipient stage, with dimensions 1x0.2 *mm*, is illustrated in Figure 2. This Figure shows also the damage when the outer steel sheet is completely disrupted, i.e. damage dimensions 1x1 *mm*, and the last analysed damage scenario, when damage has the dimensions 1x4.8 *mm*.

4. Results and discussions

Simulation results revealed the natural frequency values for the healthy beam and the beam with a series of damages. The first conclusion regards the availability of equation (8) to provide correctly the natural frequency values. Table 2 compare the frequencies attained by FEM simulations and analytically; one can observe that the values fit well, especially for the first five modes. This leads to the conclusion that this configuration of the composite beam is at the limit of applicability of the equation (7). For more rigid beams, i.e. increased moment of inertia or decreased length, the Timoshenko or Share model are more adequate.

Table 2. Natural frequencies of first ten bending vibration modes of the healthy beam

Mode i	Frequency - analytical f_{i-U}^{a} (Hz)	Frequency - FEM f_{i-U}^{FEM} (Hz)	Error ε (%)
1	4.48443	4.5076	-0.51668
2	28.10534	28.158	-0.18736
3	78.69577	78.439	0.326278
4	154.21230	152.578	1.05977
5	254.92397	249.857	1.987641
6	380.81227	369.034	3.092935
7	531.87831	508.711	4.355753
8	708.12200	667.375	5.754235
9	909.54337	843.447	7.266984
10	1136.14241	1035.34	8.87234

The frequency values for the beam with the damage located at $x = 0.274$ *m* from the fixed end, and having different levels of depth, are presented in Table 3. The damage depth is indicated as part of the subscript, the value 0.4 in the $f_{i-0.4}^{FEM}$ meaning the depth of 0.4 *mm*. Figure 3 shows the changes occurring for the frequency values with respect to the damage depth. Note that normalized frequencies are indicated, thus for all vibration modes the frequency for the undamaged beam is 1.

Table 3. Natural frequencies of the first ten bending vibration modes of the healthy beam and the beam with a damage at distance x = 0.247 m from the fixed end (Hz)

Mode i	f_{i-U}^{FEM}	$f_{i-0.4}^{FEM}$	$f_{i-0.6}^{FEM}$	$f_{i-0.8}^{FEM}$	f_{i-1}^{FEM}	$f_{i-1.4}^{FEM}$	f_{i-2}^{FEM}	$f_{i-2.2}^{FEM}$	$f_{i-2.4}^{FEM}$	$f_{i-2.6}^{FEM}$
1	4.5076	4.5035	4.4975	4.487	4.3908	4.3611	4.3475	4.3374	4.3155	4.2794
2	28.158	28.1562	28.152	28.145	28.073	28.050	28.040	28.033	28.018	27.992
3	78.439	78.3794	78.282	78.108	76.537	76.074	75.868	75.719	75.394	74.8667
4	152.578	152.493	152.35	152.12	150.09	149.51	149.27	149.09	148.71	148.106
5	249.857	249.857	249.85	249.86	249.85	249.84	249.83	249.83	249.83	249.831
6	369.034	368.838	368.53	368	363.27	361.89	361.28	360.84	359.88	358.351
7	508.711	508.234	507.49	506.22	495.98	493.31	492.18	491.37	489.64	486.988
8	667.375	657.463	666.90	666.41	662.63	661.67	661.26	660.97	660.37	659.444
9	843.447	843.387	843.29	843.12	841.72	841.28	841.05	840.92	840.64	840.18
10	1035.34	1034.58	1033.4	1031.3	1013.9	1009.2	1007.1	1005.5	1002.1	996.663

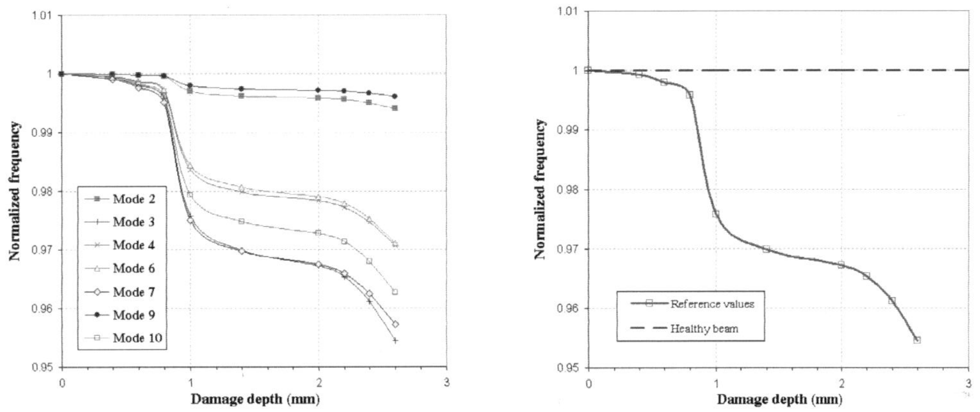

Figure 3. Normalized frequencies for the beam subjected to damages with different depths.

It is obvious that the curves depicted in left image of Figure 3 have the same allure, just the amplitude is different. Recent research demonstrated that the amplitude is totally dependent on the energy stored in the slice where the damage is located [10], i.e. propositional to the squared mode shape curvature or, similar, to the bending moment. The image on the right side of Figure 3 represents the reference for the frequency changes for a cantilever having a given damage. To obtain the normalized frequency values for the affected beam, for any vibration mode, the curve is scaled by a factor which is the value of the dimensionless squared of the mode shape curvature. Each location provides different sequences of values for the squared of curvatures, which individualize the sequence and make it suitable for the location recognition.

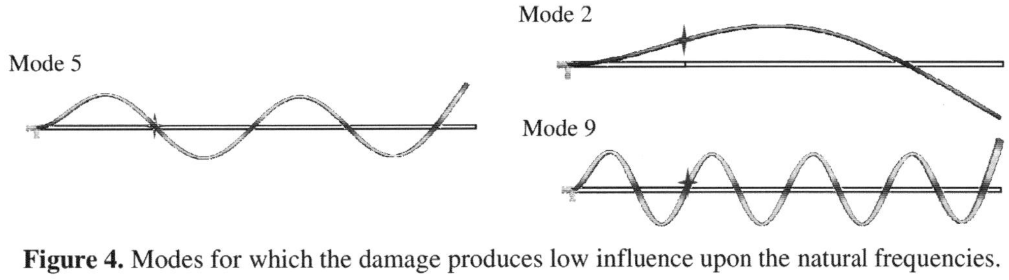

Figure 4. Modes for which the damage produces low influence upon the natural frequencies.

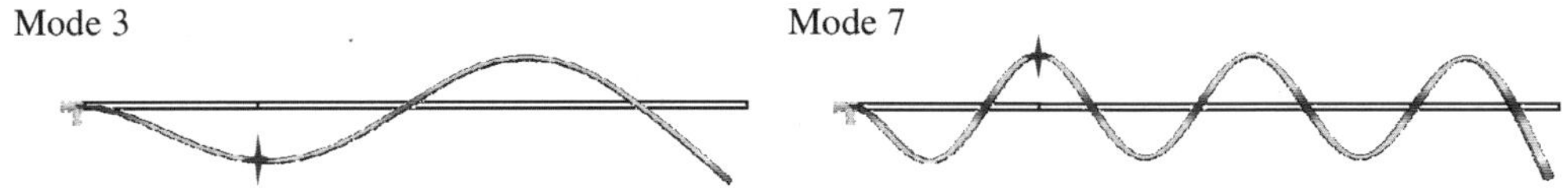

Figure 5. Modes for which the natural frequencies are mostly affected by damage.

The modes shapes for which the damage located at $x = 0.274\ m$ produce small or inexistent frequency changes, namely mode 5 and modes 2 respectively 9, are presented in Figure 4. One can observe that, for mode 5, the slice on which the damage is placed is not subjected of curving, so that the stored energy is null. In the modes 2 and 9, the damaged slices undergo low curving, thus the slice store a reduced amount of modal strain energy. Consequently, the lost energy is low as well as the frequency decreases.

Opposite, Figure 5 indicates the modes where the frequencies have most important changes, namely modes 3 and 7. It is visible that the damage is located on the slice that achieves the highs local value of curvature in these modes, storing important amount of modal strain energy. In this way, the damage causes important modal strain energy loss and relevant frequency decreasing.

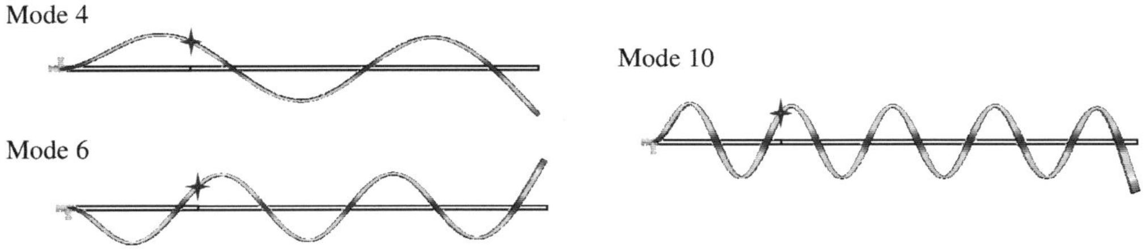

Figure 6. Modes for which the natural frequencies are subject of significant changes.

Figure 6 presents the cases in which an intermediate frequency drop occurs. This happens, for instance, at modes 4 and 6 in a moderate way, and in mode 10 more accentuated. The curves representing the frequency changes due to damages of various depths, plotted in Figure 3, confirm this finding; the stronger the curvature, the bigger the frequency drop.

A confirmation of the enunciated theory is also obtained by analysing the normalized frequency shifts, derived from the mathematical relation

$$\Delta f_i^* = \frac{f_{i-U} - f_{i-D}}{f_{i-U}} \cdot 100\ (\%) \tag{8}$$

Values of this feature, for a damage located at $x = 0.274\ m$ from the fixed end, and three damage levels (1, 2 and 2.6 *mm*), are presented in Figure 7 and Table 4. The input values for the calculus are introduced by Table 3. For the modes in which the damaged slice attains a comparable curvature, modes 4 and 6 for instance (see Figure 6), the normalized frequency shifts are similar (see Figure 7). Remarkable in Figure 7 is also the fact that the frequency shift ratio is identical for all modes of any pair of damage scenarios/depths. Additionally, this ratio is similar, for the pairs of damage depths, with that identifiable in Figure 3. This validates the existence of a unique curve describing the damage severity, as depicted in Figure 3, scalable in respect to the damage depth. This curve is, for the multi-layer structure, analogous to the curves contrived from the stress intensity factor for isotropic materials [15-17].

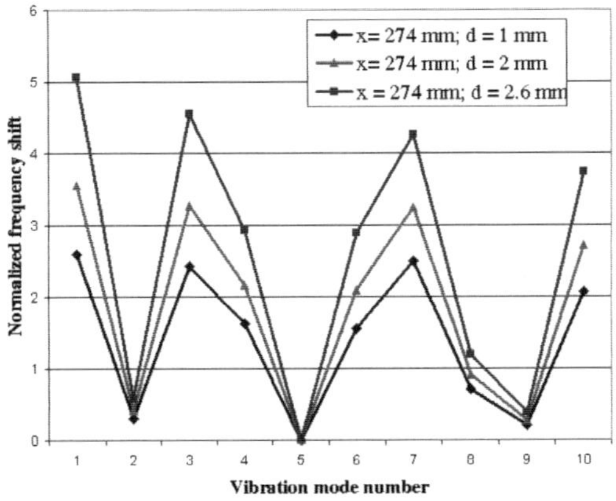

Figure 7. Normalized frequency shifts for the damage located at x = 274 *mm* and three severity scenarios: depth 1 *mm*, 2 *mm* and 2.6 *mm*.

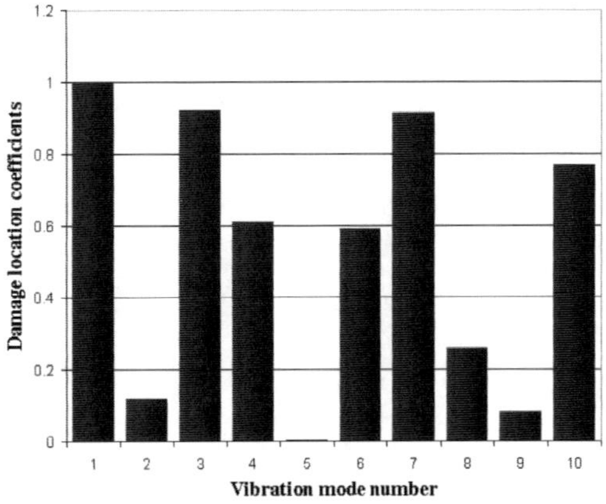

Figure 8. Damage location coefficients for location x = 274 *mm*.

Table 4. Normalized frequency shifts of the first ten bending vibration modes of the beam with the damage at distance x = 0.247 *m* from the fixed end

Mode i	Δf^*_{i-1}	Δf^*_{i-2}	$\Delta f^*_{i-2.6}$
1	4.3908	3.550226	5.062561
2	28.073	0.417643	0.589531
3	76.537	3.277069	4.55424
4	150.09	2.168043	2.932232
5	249.85	0.009605	0.011606
6	363.27	2.098745	2.893803
7	495.98	3.248216	4.270016
8	662.63	0.915234	1.187647
9	841.72	0.284664	0.387693
10	1013.9	2.722882	3.731962

The study revealed that, if the same damage is located on different slices of the beam it will produce different sets of values regarding normalized frequency shifts. For the cantilever, as an asymmetric beam, theses sets are unique for every location and represent, in fact, the damage signatures. After a new normalization, by dividing the normalized frequency shift values for a location to the highest value of the sequence, the damage location coefficients are resulted, as a new set of sub-unitary values. These values are the same for a given damage location irrespective to the damage severity, thus proper to be used as a benchmark in non-destructive vibration-based damage location.

5. Conclusions

This paper introduces a new view of interpreting frequency changes occurring due to damage in composite beams. The results highlighted that a unique curve representing the frequency change of transverse modes in respect to damage depth exists as presented in Section 4. This curve is similar for all modes, if a scaling factor is applied. This factor depends on the locally stored energy in the damaged slice, i.e. the local value of the squared of the mode shape curvature. If the slice is on an inflection point the change is null - see mode five and its normalized frequency shift. Opposite, if damaged slices achieve important curving, as mode three and seven are, the scaling factor and consequently the normalized frequency shifts are important. The ratios between the normalized frequency shifts due to damages of similar position but different depths are identical for any transverse modes. Normalizing again the set of certain normalized frequency shifts, the effect of damage depth is annulated, and a Damage Location Indicator in form of a one row vector is obtained. Thus, for location of damage in multi-layered structures the method is also applicable.

Acknowledgement
The authors gratefully acknowledge the support of the Romanian Government by co-financing the project "Doctoral and postdoctoral grants to sustain innovation and competiveness in research" (InnoRESEARCH), ID 132395.

References

[1] Plantema F J 1966 *Sandwich construction: The bending and buckling of sandwich beams, plates and shells* (New York: Wiley)
[2] Magnucka-Blandzi E and Magnucki K 2007 Effective design of a sandwich beam with a metal foam core *J. Sound Vib.* **301** 253–77
[3] Shahdin A, Morlier J, Mezeix L, Bouvet C and Gourinat Y 2011 Evaluation of the impact resistance of various composite sandwich beams by vibration tests *Shock Vib.* **18**(6) 789-805
[4] Malekzadeh K, Khalili M R and Mittal R K 2007 Response of composite sandwich panels with transversely flexible core to low-velocity transverse impact: A new dynamic model *Int. J. Impact Eng.* **34** 522–43
[5] Carrera E 2004 Assessment of theories for free vibration analysis of homogeneous and multilayered plates *Shock Vib.* **11**(3-4) 261-70
[6] Hause T and Librescu L 2006 Flexural free vibration of sandwich flat panels with laminated anisotropic face sheets *J. Sound Vib.* **297** 823–41
[7] Gillich G R, Praisach Z I, Wahab M A and Vasile O 2014 Localization of transversal cracks in sandwich beams and evaluation of their severity *Shock Vib.* **9**(4-5) 607125
[8] Moreira R A S and Dias Rodrigues 2010 J Static and dynamic analysis of soft core sandwich panels with through-thickness deformation *Compos. Struct.* **92** 201–15
[9] Schwarts-Givli H, Rabinovitch O and Frostig Y 2007 Free vibrations of delaminated unidirectional sandwich panels with a transversely flexible core - a modified Galerkin approach *J. Sound Vib.* **301** 253–77
[10] Deshpande V S and Fleck N A 2001 Collapse of truss core sandwich beams in 3-point bending *Int. J. Solids Struct.* **38**(36–37) 6275–305
[11] Gillich G R and Praisach Z I 2014 Modal identification and damage detection in beam-like structures using the power spectrum and time–frequency analysis *Signal Process.* **96**(Part A) 29–44
[12] Praisach Z I, Gillich G R and Birdeanu D E 2010 Considerations on natural frequency changes in damaged cantilever beams using FEM *Latest trends on engineering mechanics, structures, engineering geology* 214-9
[13] Gillich G R and Praisach Z I 2013 Damage-patterns based method to locate discontinuities in beams *Proceedings of SPIE (San Diego)* vol 8695 article number 869532
[14] Gillich G R, Praisach Z I and Negru I 2012 Damages influence on dynamic behaviour of composite structures reinforced with continuous fibers *Mater. Plast.* **49**(3) 186-191
[15] Liebowitz H and Claus Jr. W D S 1968 Failure of notched columns *Eng. Fract. Mech.* **1**(1) 379–383
[16] Ostachowicz W M and Krawczuk C 1991 Analysis of the effect of cracks on the natural frequencies of a cantilever beam *J. Sound Vib.* **150**(2) 191–201
[17] Chondros T J, Dimarogonas A D and Yao J 1998 A continuous cracked beam vibration theory *J. Sound Vib.* **215**(1)17–34

Ballistic damage in hybrid composite laminates

Vaibhav A Phadnis[1], Kedar S Pandya[2], Niranjan K Naik[3], Anish Roy[4], Vadim V Silberschmidt[4,*]

[1]Composite Centre, AMRC with Boeing, University of Sheffield, S60 5TZ, UK
[2]Department of Engineering, University of Cambridge, UK
[3]Aerospace Engineering Department, Indian Institute of Technology, Bombay - 400076, India
[4]Wolfson School of Mechanical and Manufacturing Engineering, Loughborough University, Loughborough, LE11 3TU, UK

[*]Email :V.Silberschmidt@lboro.ac.uk

Abstract. Ballistic damage of hybrid woven-fabric composites made of plain-weave E-glass-fabric/epoxy and 8H satin-weave T300 carbon-fabric/epoxy is studied using a combination of experimental tests, microstructural studies and finite-element (FE) analysis. Ballistic tests were conducted with a single-stage gas gun. Fibre damage and delamination were observed to be dominating failure modes. A ply-level FE model was developed, with a fabric-reinforced ply modelled as a homogeneous orthotropic material with capacity to sustain progressive stiffness degradation due to fibre/matrix cracking, fibre breaking and plastic deformation under shear loading. Simulated damage patterns on the front and back faces of fabric-reinforced composite plates provided an insight into their damage mechanisms under ballistic loading.

1. Background

Woven fabric-based polymer-matrix composites (PMCs) are finding an increased use in defence-related applications thanks to their high strength and stiffness and ability to produce structures with tailored shapes and mechanical properties. Additionally, such PMCs lead to better energy absorption in ballistic impact events, especially thanks to their balanced in-plane properties. Recent studies [1, 2] showed that a hybrid composite structure - a combination of glass- and carbon-based epoxy composites may further improve the energy absorbing capacity of such composites. Thus, it is of great interest to a scientific community to understand their mechanical behaviour in high-velocity impact events.

In ballistic impact events, PMCs absorb projectile's kinetic energy by undergoing either elastic or permanent deformation. In the latter case, PMCs often exhibit different damage modes such as delamination, punching and fibre breakage. The condition for perforation, known as the *ballistic limit velocity* (V_{50}), is one of the most important factors for design of a suitable protective structure in this regard [1-4]. It represents an average of equal numbers of highest partial-penetration velocities and lowest complete-penetration velocities of a projectile for a specific velocity range, resulting in 50% probability of partial penetration and perforation of a target. Recently, Naik and Doshi [2] reported on ballistic-impact behaviour of typical woven-fabric E-glass/epoxy thick composites employing an analytical approach and found that shear plugging was the major energy-absorbing mechanism in these laminates. The level of V_{50} was also calculated and found to vary non-linearly with a linear increase in the laminate's thickness. Zhu [3] investigated a response of woven Kevlar/polyester laminates of varying thickness to quasi-static and dynamic penetration by cylindro-conical projectiles. Damage patterns in these composites were found to differ significantly under dynamic loading from those for quasi-static penetration conditions. There are some shortcomings of experimental ballistic schemes: typical experiments involve a plethora of safety protocols and often their output is difficult to quantify in a reliable manner given a short duration of such events. Thus, computer simulations are often employed as a virtual tool to analyse performance of structures in ballistic events and support their optimisation and design.

Simulating the mechanical behaviour of a fabric-reinforced composite structure under ballistic impact is a challenging task. Unlike metallic components that can yield and dissipate energy by undergoing plastic deformation, composites can mostly dissipate energy through various damage processes that usually degrade stiffness of structural components. Hence, an advanced modelling tool that can adequately simulate such events is essential in the design process. However, due to the complexity of involved processes and mechanisms, most models attempt to provide acceptable trade-off in performance analysis [4-8]. To model a woven-fabric down to a level of crossovers of individual yarn would certainly be preferred in order to study the underlying frictional and crimping effects, but such studies are computationally impracticable for dynamic problems. In this regards, this work focuses on the development of a FE model of a ballistic-impact response of woven-fabric-reinforced composites. The experimental studies, reproduced in numerical simulations, are discussed first, followed by a brief description of the developed FE model. Its results and discussion are presented next.

2. Ballistic experiments

A ballistic-impact test apparatus operated by a single-stage gas-gun (Fig. 1) was used to carry out experimental studies. It consisted of a projectile-propelling mechanism, a chronograph for velocity measurement, a support stand for holding the specimen, a containment chamber, safety devices and a strain-measuring facility. Compressed air was used as a propellant in the system. The main components of this propelling mechanism were a cylindrical barrel to guide the projectile, a quick release valve to relieve the trapped air and a nitrogen gas-driven solenoid valve to operate this valve. The cylindrical barrel (through which a projectile was propelled) was 1.5 m long. Its inner diameter was chosen to suit a projectile used in these experiments.

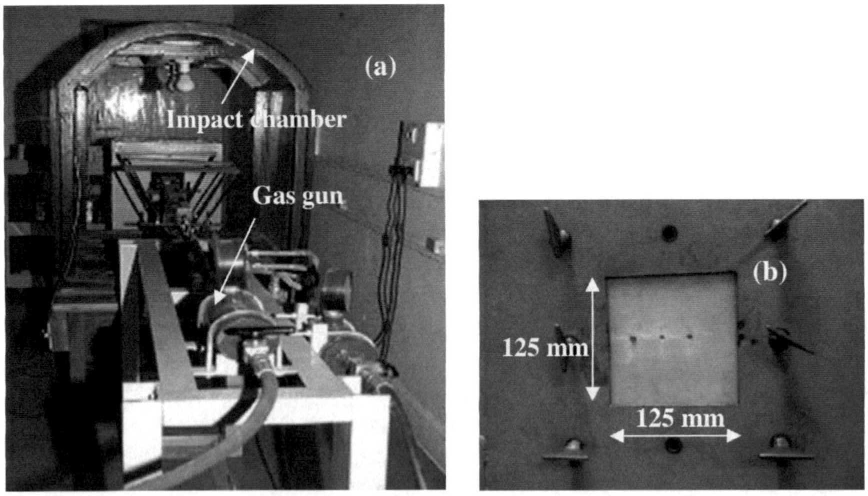

Figure 1. Ballistic test setup: (a) single-stage gas gun; (b) typical test specimen

A velocity of projectiles was varied up to 200 m/s by changing the air pressure in the cylinder. The experimental studies were carried out on flat specimens with dimensions 125 mm × 125 mm with different thicknesses (refer to Table 1). The specimen's dimensions were chosen in such a way that they can be accommodated into a fixture designed as an integral part of the ballistic-impact test apparatus. The mass of flat-end cylindrical projectiles made of hardened steel and their diameter was kept constant for all the tests. The experiments were carried out on at least four specimens for each impact condition to ensure repeatability.

Table 1. Parameters of ballistic tests

(target dimensions: 125 mm × 125 mm; projectile diameter: 6.36 mm)

Sr. No.	Projectile mass (g)	Projectile length (mm)	Target thickness (mm)
(A)	Plain-weave E-glass epoxy		
1	6.42	25.3	2.5
2	6.42	25.3	3.0
3	6.42	25.3	4.5
4	6.42	25.3	5.0
(B)	8H satin-weave T300 carbon/epoxy		
1	6.42	25.3	3.0
(C)	Hybrid (H4 and H5)		
1	6.42	25.3	3.0
2	6.42	25.3	3.0

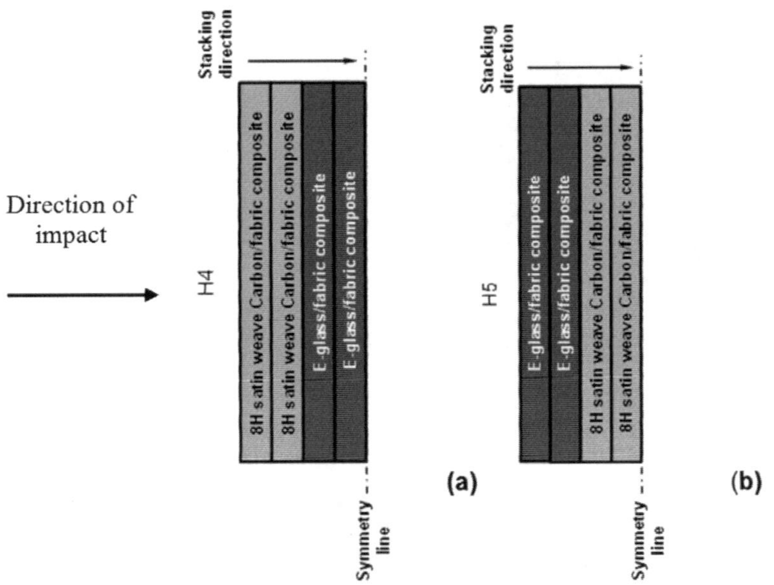

Figure 2. Architecture of hybrid composite laminates: (a) H4; (b) H5

Four symmetric cross-ply woven fabric composites were studied: plain-weave E-glass fabric/epoxy, 8H satin-weave T300 carbon fabric/epoxy and their hybrids. Specifications of tows/stands, fabrics, resin and composites for plain weave E-glass/epoxy and 8H satin-weave T300 carbon/epoxy composites can be found in [1]. For simplicity, these composite laminates are designated below as *E, C, H4* and *H5,* respectively. The ply architecture of hybrids *H4* and *H5* is shown in Figs. 2a and b, respectively.

3. Finite-element model

A ply-level constitutive model was developed to analyse the mechanical response of the fabric/epoxy composites. This model was implemented as a material subroutine VUMAT in ABAQUS [9]. Each ply was modelled as a homogeneous orthotropic material with potential to sustain progressive stiffness degradation due to fibre/matrix cracking and plastic deformation under shear loading. Delamination between the neighbouring plies was modelled using a cohesive-zone-element (CZE) technique. The setup of FE model is shown in Fig 3. Details of a constitutive model for fabric-reinforced composites and damage modelling framework used in this FE model are not explained here for the purpose of brevity; they can be found in [10].

A schematic of the developed FE model is shown in Fig. 3. Both the woven-fabric-reinforced plate and the bullet were modelled as 3D deformable solids. A dynamic explicit solver was used in simulations to account for the time-dependent loading and complex interaction between the target and the projectile.

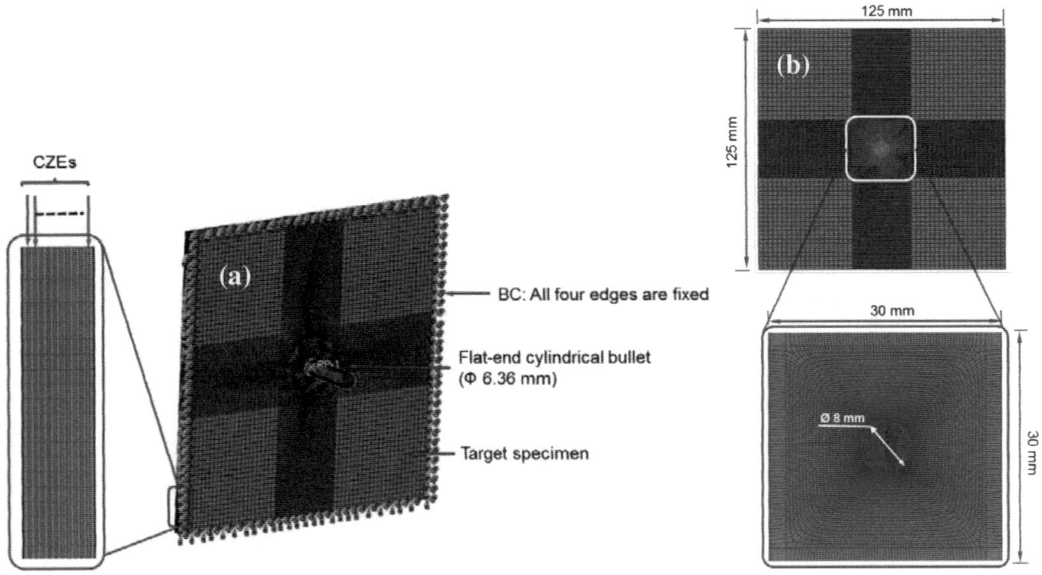

Figure 3. FE model of ballistic impact: (a) setup; (b) typical meshed specimen, t = 3 mm

3 mm-thick symmetric cross-ply hybrid laminates were modelled. The laminate consisted of 5 plies each of *C* and *E* composites with a thickness of 0.32 mm and 0.28 mm, respectively. The local co-ordinate systems were defined to account for orientations of individual plies. In experimental trials, a cylindrical projectile with mass of 6.42 g and length of 25.3 mm impacted the centre of the workpiece in the axial direction (see Fig. 3a). In simulations, this was achieved using a pre-defined velocity boundary condition. Contacts between the projectile, on the one hand, and the composite plate and all contacted plies of the laminate, on the other, were defined by a general contact algorithm available in ABAQUS/Explicit. This algorithm generated the contact forces based on a penalty-enforced contact

method. The friction coefficient μ was used to account for shear stress of the surface traction with contact pressure p and can be represented as $\tau = \mu p$. In this case, the frictional contact between a bullet and composite laminate was modelled with a constant coefficient of friction of 0.3 [7]. The models require on average 11 hours on 24 Intel quad-core processors with 48 GB RAM each to finish the analysis using the High Performance Computing (HPC) facility available at Loughborough University, UK.

4. Results and discussions

The finite-element models of ballistic-impact on studied composite laminates were validated using experimental measured ballistic limit velocity (V_{50}) and analytically calculated energy absorbed by a laminate. The ballistic limit velocity (V_{50}) was assessed for the same thickness of laminates, impactor geometry and mass to provide an assessment of their relative performance. It should be noted that, though, thickness of all the studied laminates was the same, the areal weight of two underlying fabric materials (E-glass plain-weave and carbon satin-weave) was different. Thus, a more viable comparison of their ballistic performance was provided in terms of a normalised parameter – V_{50} per unit areal weight of the target. The fracture mechanisms in these laminates were also studied, and their contribution to absorption of incident impact energy is discussed below.

(a) V_{50} for same target thickness and per unit areal density

In FE simulations, V_{50} was calculated at the reference point R_f tied to a bullet using an equation constraint to reduce the computational efforts. Figures 4a-b present the ballistic limit velocity V_{50} for *G, C, H4* and *H5* composites for the same thickness and per unit areal density, respectively.

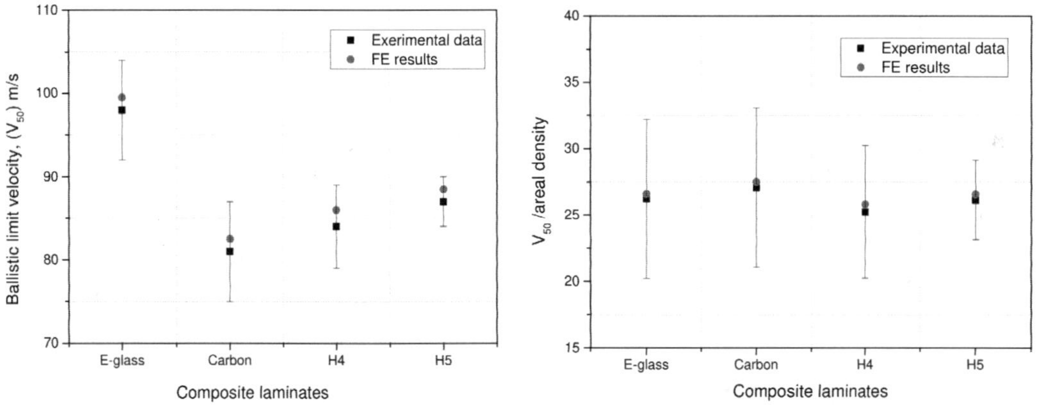

Figure 4. Ballistic-impact velocities for same thickness (a) and per unit areal density (b) for studied composite laminate (projectile mass 6.42 g, projectile length 25.3 mm)

It can be observed that the ballistic limit velocity V_{50} had considerably different hierarchies for the assessed criteria and laminates; thus, selection of appropriate composite mostly depends on the type application, in other words, it should consider the importance of higher stiffness or better specific energy absorption capacity for the working structure. In the former case, E-glass-based composite provided a better performance but at a price of a higher weight.

(b) Damage in composite panels

Damage assessment in the studied composite laminates was carried out using numerical simulations with the developed FE model. Damage patterns obtained at the front and the back face of studied composite laminates are shown in Fig 5. Impact-generated damage was visible on the both faces of the composite panels and was measured along both warp and fill directions from the centres of specimens (where they were impacted) and compared with respective magnitudes obtained with the FE analysis (Fig. 6). The damage mechanisms in *G* and *C* laminates were observed to be similar, though overall damage at the front and back faces was more pronounced for *G* plates than for *C* ones due to lower stiffness of the glass-fibre plies (see cases A and B in Fig. 5).

Hybridisation of *G* and *C* laminates could potentially give a rise to either improved or reduced material properties. In case of *H4* (hybrid laminate with exterior *G* plies), the dominant damage processes were the same as found in *C* and *G*, namely, tensile fracture of the back plies and crushing of the front plies under the impactor. The intraply cracks in the front and back plies grew upwards and downwards, respectively, leading to development of delamination cracks, and final fracture took place by formation of a crack through the laminate thickness. However, the amount of damage at the back face of *H4* laminate was far lower than in *G* and moderately less than in *C* under equivalent impact conditions (cases A and C in Fig. 5). On the other hand, in case of *H5* (the hybrid laminate with *C* plies at the exterior), apart from ply fracture at the back and crushing at the front, large deformation in the stiff carbon plies led to extensive delamination and intraply fracture that grew rapidly towards the back. Ply fracture due to crushing at the front also accelerated this process resulting in greater damage on the back face of *H5* (case D in Fig. 5).

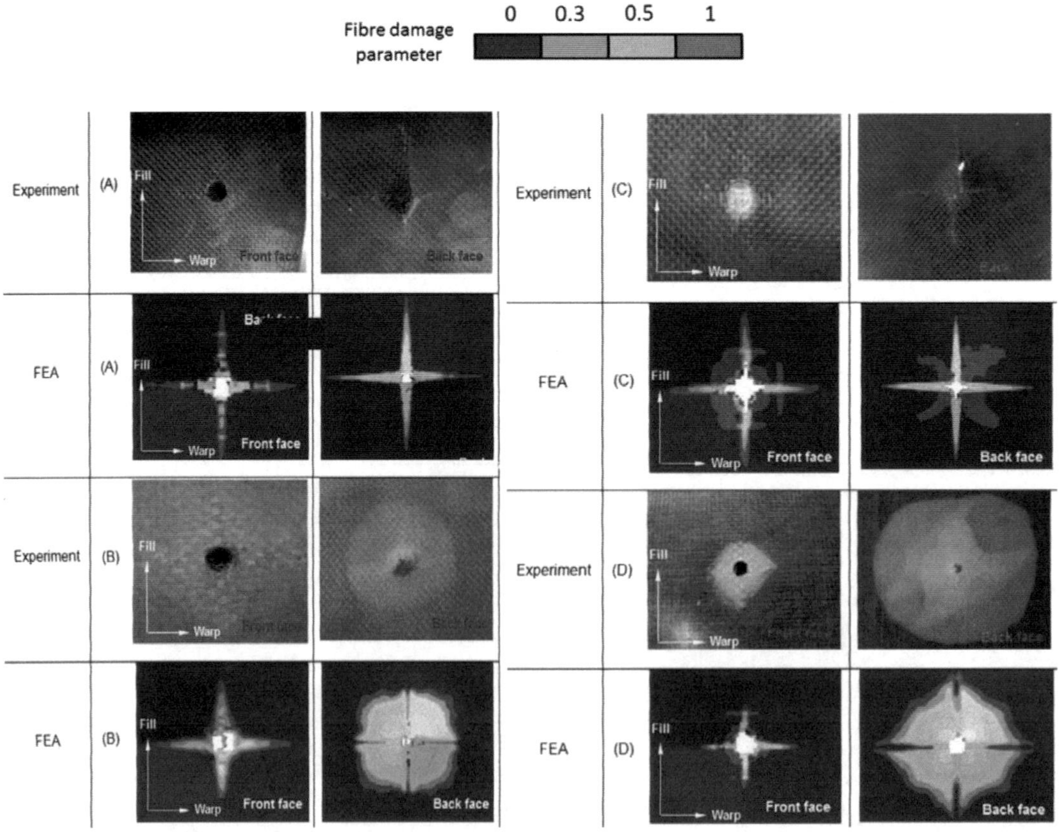

Figure 5 Damage patterns on front and back faces of composite panels: (A) 8H satin-weave T300 carbon/epoxy composite (V_{50} = 99.5 m/s); (B) plain-weave E-glass/epoxy composite (V_{50} = 82 m/s); (C) hybrid H4 (V_{50} = 86 m/s); (D) hybrid H5 (V_{50} = 88 m/s) (laminate thickness 3 mm)

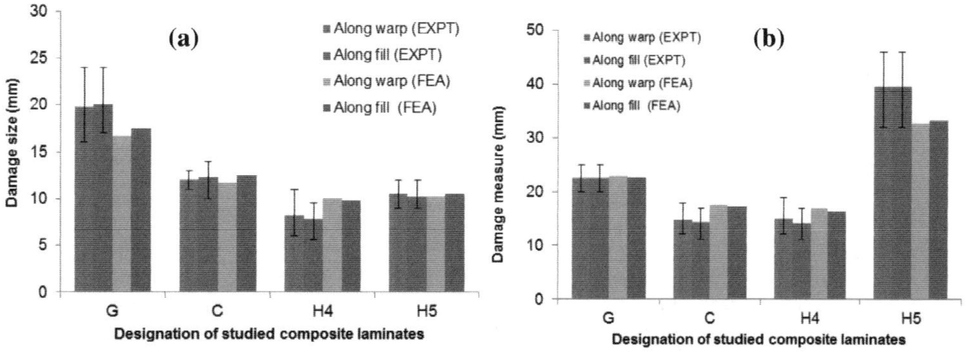

Figure 6. Measured and calculated damage size for studied composite laminates: (a) front face; (b) back face (laminate thickness 3 mm) Used notation: G - plain-weave E-glass/epoxy composites impacted at V_{50} = 99.5 m/s; C - 8H satin-weave T300 carbon/epoxy composites at V_{50} = 82 m/s; H4 - hybrid composite H4 at V_{50} = 86 m/s; H5 - hybrid composite H5 at V_{50} = 88 m/s.

(c) Contribution of damage modes to energy absorption

The trends of energy absorption mechanisms in the studied laminates were quite similar (Fig. 7), with most of incident kinetic energy absorbed due to deformation of secondary yarns, while energy absorption related to failure of primary yarn/fibre was relatively low and that due to matrix cracking and delamination was marginal.

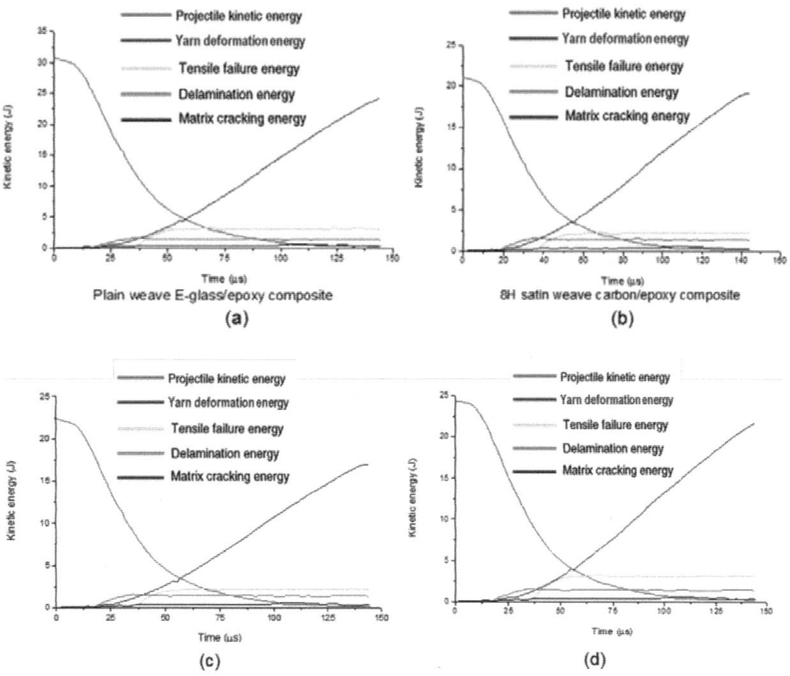

Figure 7. Contribution of damage modes to kinetic energy absorption: (a) plain-weave E-glass/epoxy composite at V_{50} = 99.5 m/s; (b) 8H satin-weave carbon/epoxy composite at V_{50} = 82 m/s; (c) hybrid composite H4 at V_{50} = 86 m/s; (d) hybrid composite H5 at V_{50} = 88 m/s (laminate thickness 3 mm)

This may be explained by a highly transient nature of impact events, with a time of contact between projectile and target of some 150 microseconds and highly localised deformation of the target, where damage initiated in the form of fibre failure directly below the projectile and had hardly any time to diffuse through the laminate before the projectile penetrates it. The contribution of damage modes to total energy absorbed for the studied composite laminates is shown in Fig. 7.

5. Conclusions

A ballistic-impact response of four woven fabric composite laminates *E, G, H4* and *H5* was studied experimentally and using the developed finite-element model. The ballistic damage on the front and back faces of the studied laminates was observed. The effect of hybridisation on damage modes and their contribution to energy-absorption capacity of the laminates was also discussed. The FE model predicted the pattern and extent of damage in the studied laminates reasonably and provided an insight into their probable damage mechanisms. Some fundamental observations based on this study are listed:

- Improvements due to hybridization in the behaviour of the laminates under impact were due to the higher strain-to-fracture ratio of the E-glass-fibre plies located near the front and back laminate surfaces. These plies were able to sustain higher deformations before fracture and hindered propagation of damage to the inner plies from the broken plies on the front and back surfaces, increasing the maximum load-bearing capability of the composite.
- The hybridisation provided a reasonable trade-off between in-plain strength and failure strain that resulted in better ballistic-impact resistance properties compared to high-modulus fibre-reinforced composites. In addition, the presence of E-glass fibres helped to sustain higher deformations before fracture of the laminate by the percolation of a through-thickness crack, significantly increasing the energy dissipated under impact. Most of the benefits of the E-glass fibre could be attributed to the plies located near the laminate surfaces. The presence of inner plies provided more limited improvements, particularly in terms of specific properties.

References

[1] Pandya KS, Pothnis JR, Ravikumar G and Naik NK 2013 *Mater. Design* **44** 128-135
[2] Naik N and Doshi A 2008 *Compos. Struct.* 82 447–64
[3] Zhu G, Goldsmith W and Dharan CKH 1992 *Int. J. Solids Struct.* **29** 399–420
[4] Cheng WL, Langlie S and Itoh S 2003 *Int. J.Impact Eng.* **29** 167–84
[5] Silva M 2005 *Int.J. Impact Eng.* **31** 289–306
[6] Pinho ST, Robinson P and Iannucci L 2006 *Compos. Sci. Tech.* **66** 13
[7] Phadnis VA, Makhdum F, Roy A and Silberschmidt VV 2013 *Compos. Part A.* **47** 41-51
[8] Ullah H, Harland AR and Silberschmidt VV 2012 *Comput. Mater Sci.* **64** 130-135
[9] Abaqus 6.11 2011 *User Manual* Dassault System Rhode Island US
[10] Phadnis VA, Pandya KS, Naik NK, Roy A and Silberschmidt VV 2013 *J Phys: Conf Ser.* **451** 012019

Damage mechanisms of random fibrous networks

E. Sozumert[1], F. Farukh[1], E. Demirci[1], M. Acar[1], B. Pourdeyhimi[2] and V. V. Silberschmidt[1]

[5]Wolfson School of Mechanical and Manufacturing Engineering, Loughborough University, Loughborough, LE11 3TU, UK
[1]Nonwovens Cooperative Research Center, North Carolina State University, Raleigh, NC, USA

E-mail: e.sozumert@lboro.ac.uk

Abstract. Fibrous networks are ubiquitous: they can be found in various engineering applications as well as in biological tissues. Due to complexity of their random microstructure, anisotropic properties and large deformation, their modelling is challenging. Though, there are numerous studies in literature focusing either on numerical simulations of fibrous networks or explaining their damage mechanisms at micro or meso-scale, the respective models usually do not include actual random microstructure and failure mechanisms. The microstructure of fibrous networks, together with highly non-linear mechanical behaviour of their fibres, is a key to initiation of damage, its spatial localization and ultimate failure [1]. Numerical models available in literature are not capable of elucidating actual microstructure of the material and, hence, its influence on damage processes in fibrous networks. To emulate a real-life microstructure in a developed finite-element model, an orientation distribution function for fibres obtained from X-ray micro computed-tomography images was considered to provide actual alignment of fibres. To validate the suggested model, notched and unnotched rectangular specimens were experimentally tested. A good correlation between the experimental data and simulation results was observed. This study revealed a significant effect of a notch on damage evolution.

1. Introduction

Fibrous materials are commonly found around us . Despite their extensive use in many products including automotive, hygiene, health applications, understanding of their mechanical properties and performance is not simple due to complexity of their microstructure. This complexity stems from manufacturing processes, non-trivial mechanical behaviour of constituent fibres and their random distribution in the microstructure of the material.

Although some of those networks are man-made, such as, cellulose, papers and most of nonwovens, some are natural. Some of these materials are composed of fibrous network layers as in elastomeric fibrous scaffolds used in engineered soft tissues [2-3]. A main source of constituent fibres is natural or synthetic polymers [3]. As a result, these fibres might show time-dependent and/or non-linear mechanical behaviour, i.e. elasto-plasticity or visco-plasticity [3].

The first attempt to model microstructured nonwoven materials was made by Liao *et al*[4] with the model based on discontinous modelling approach. The model was composed of many discrete cell elements, each representing a number of fibres. The orientation distribution of fibres in real fabrics was partially reflected in the model. Also, fibre-to-fibre interactions (friction or contacts between fibres) in the model were not included, and a partial account of fibre orientation, the model did not reproduce the actual deformation mechanisms and fracture. Similarly, other researchers simulated the mechanical behaviour of fibrous systems (paper, biomaterials, nonwovens)[5-7]. Even though a continuous modelling approach provides a relatively easy way to generate a macroscopic (global) material's structure, it is incapable of representing real microstructure and, hence, deformation and failure mechanisms.

Another type of model was developed by Kim and Pourdeyhimi in 2001 [8] and introduced to predict the behaviour of thermally point bonded nonwovens. This numerical model was based on a principle of incremental deformation with contribution of deformation of each individual fibre calculated. Then, these contributions were summed up to obtain an overall mechanical response. This model had an advantage of incorporation of an fibre orientation distribution function (ODF) to give an account for anisotropy of fibres due to different alignments. Still, it could not reflect properly important deformation and fracture mechanisms of thermally point bonded nonwovens, such as, rearrangement of bond points and changing orientation of fibres.

A model for thermally point bonded low density nonwovens was proposed by Hou *et al* [9]. It was capable of representing an anisotropic nature of that type of nonwoven materials since bond points and the fibre orientation function were introduced directly, allowing to include the actual alignment of fibres. The investigation with this model pointed out that fibre alignment was one of the major factors affecting the mechanical behaviour of nonwoven material.

Subsequently, Somuncuoglu *et al* [10] presented a new model with parametric capabilities to investigate tensile performance of thermally point-bonded nonwoven materials. This new development (its superiority compared to previous finite-element models introduced in the literature), it allows the users to study conveniently the effect of changing manufacturing parameters such as base density (number of fibres on unit area), fibre orientation, pattern and shape of bond points(and their dimensions). This parametric finite-element model has demonstrated its success and reliability in a variety of investigations of thermally point-bonded nonwovens [1, 11-14]. The model is used in this paper to model experimentally observed damage processes.

2. Experimental Studies

Properties of constituent fibres and the entire fabric play important role in damage and deformation mechanisms of fibrous materials. In fact, each fibre contributes to the overall mechanical response of the materials. For this reason, characterization of physical and mechanical properties of single fibres is vital. In experimental studies, material parameters and damage (fracture) properties of fibres were obtained by means of single-fibre tests. As for the tests on fabrics, a set of notched and virginal (un-notched) specimen was subjected to tensile loading.

.1. Material

The material in this study is a thermally bonded nonwoven with a density of 30g/m^2, made of polypropylene fibres. The manufacturing process of this material is as follow: polypropylene fibers were extruded and then stretched to increase their crystallinity, resulting in fibers of up to 40-60 μ-m in diameter. The continuous spun fibres were then laid down randomly on a flat surface, producing an isotropic fiber web sheet (before bonding) which was then bonded by a simultaneous application of pressure and heat. The fibres typically appear as isolated, although it is not infrequent that they form bundles of 2, 3 or even 4 fibres. Partial fusion between fibres at the entanglement points is normally observed. The overall view of the nonwoven fabric at a larger scale demonstrates its isotropic microstructure. The main physical properties of a typical polypropylene fibre are shown in Tab. 1.

Table 1. Physical properties of typical polypropylene fibre.

Linear density (dtex)	Tenacity (N/tex)	Ultimate tensile strength (MPa)	Initial modulus (GPa)
6.6	0.415	548	1.3

. . *Material micro tructure*
In this study, the main focus is on microstructure of fibrous networks that affects their deformation and damage behaviour. Its main feature is a presence of two distinct domains: bond points and fibrous matrix, see Fig. 1. Due to spatial distribution of those regions, the nonwoven fabric exhibits its anisotropic mechanical characteristics. Though bond points are formed by large numbers of randomly oriented fibres, they can be assumed as a solid and continuous structure, i.e. isotropic material, as a result of the melting and compaction during their manufacture. Unlike bond points, the fibrous web demonstrates a high level of porosity and, hence, compressibility as well as randomness of the orientation distribution of fibres. A fibre diameter, variations in diameter and length of fibres, frequency of a crimp wave are some of other factors defining anisotropy of the web structure [15].

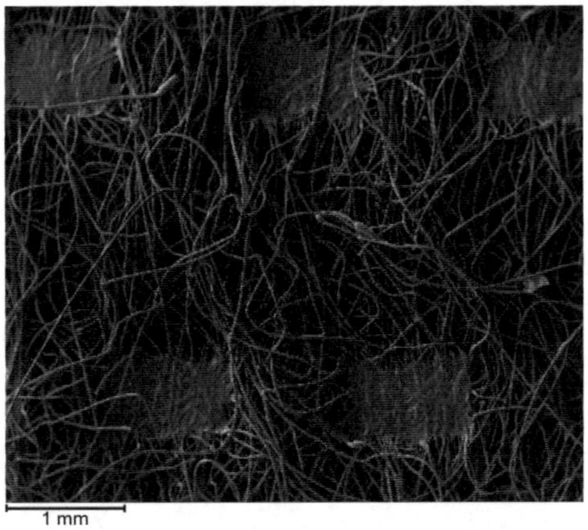

Figure 1. SEM image of studied thermally point-bonded nonwoven

.3. *Single fi re te t*
As reported, since fibres originate from different raw materials and manufacturing processes, they have different physical properties. It was also revealed [16-19] that fibres even made of the same material demonstrate distinct deformation and failure behaviours before and after their manufacturing process. Thus, individual fibres were extracted from the fabric and tested to assess their material properties. A detailed explanation of this extraction and testing processes can be found elsewhere [18].

.4. *Ten ile te t on nonwoven ecimen*
The experiments focused on rectangular samples with dimensions of 10 mm x 16.5 mm. Additionally, notched samples of nonwovens were produced with a square hole in their centre,

Table 2. Parameters of tensile test settings

Load range (N)	Speed (mm/min)	Max. extension	Preload (N)
100.0	300.0	200.0	0.5

see Fig. 3. Thus, un-notched (virginal) and notched specimens were prepared for tensile tests, with five specimens tested for each combination of parameters (see Tab. 2). The specimens were tested in their machine and cross-machine directions using AMTM D4595 (2011) [20].

3. Numerical Studies

The experimental studies were complemented by development of a new finite-element model emulating microstructure of the fibrous material to understand its fracture mechanisms under tension. The modelling approaches for fibrous networks in the literature can be divided into two basic groups: discontinuous and continuous. Apparently, the former consider a continuum formulation for the modelled material without any account for porosity inside it. They do not introduce individual fibres into the model, whereas the discontinuous numerical modelling approach aims to incorporate the real microstructure of the material [21]. To have a realistic microstructure, this modelling approach requires modelling single fibres, resulting in a natural introduction of voids and gaps observed in the network as well as another domain - bond points. To model the discontinuous microstructure of fibrous networks, several techniques can be utilized. The first technique is based on building up the microstructure with single fibres to form a complete fibrous network (macrostructure). This scheme allows researchers to perform investigations according to the purpose of their research. In the second technique, SEM or micro CT data are used for direct replication of the microstructure of the fibrous network is replicated based on direct processing of images). Despite being capable to produce similar microstructural models, this technique is cumbersome with regard to account for intersecting or overlapping regions of fibres. Also, to make any change related to other manufacturing parameters, new samples should be scanned for those parameters, so that a new microstructural model could be constructed. Considering advantages and drawbacks of these schemes, the discontinuous modelling approach was employed in this research. In the case of the studied thermally point-bonded nonwoven materials, fibres in the matrix are randomly distributed, and this randomness is affected by the manufacturing process. The fibres are aligned mostly in the direction parallel to that of the production line. This causes a dierence in the strength of fabric in the machine direction (MD) parallel to manufacturing line-and cross direction (CD) - perpendicular to MD. As a result, the fabric is stier and stronger in MD than in CD [22]. Thus, a precise assessment of fibre distribution is of great importance for elucidating of the effect of the microstructure. To express this randomness and orientation characteristics of fibrous materials, a distribution function was introduced by Cox in 1952 [23]. It is known as an *orientation i tri ution function* (ODF). The ODF was employed in this study to introduce a heterogeneous microstructure; it was determined with an Hugh-transform-based image-processing software developed at Loughborough University (for further information, see [24-25]).

3.1. FE mo elling of ecimen

A finite-element software package MSC Marc as solver and its Mentat module as pre/post-processor were used in our numerical investigations. The software provides a good support for

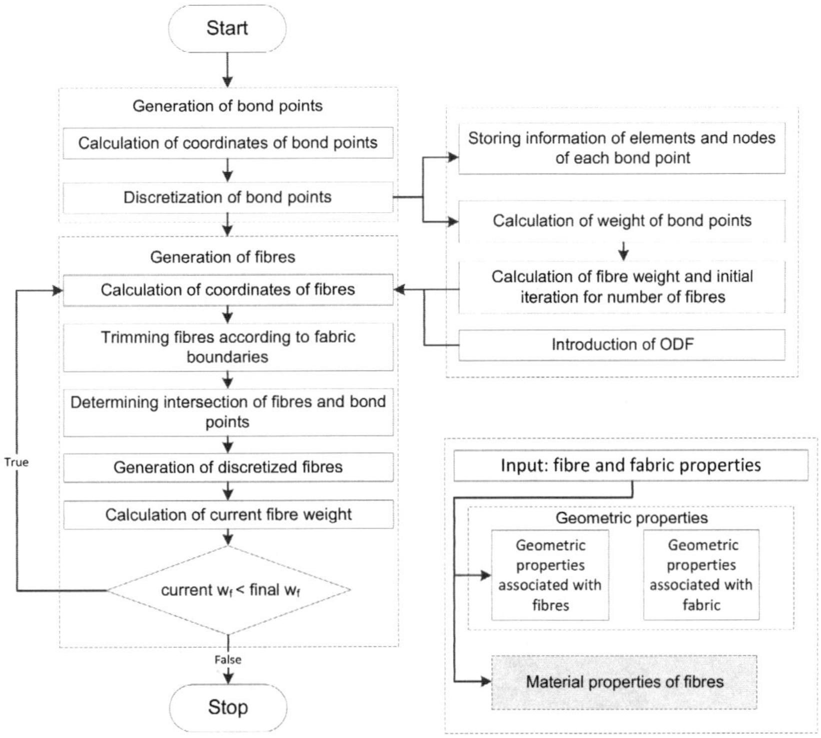

Figure 2. Flow chart of generation of nonwoven network

generation of complex geometries using a python scripting language. Figure 2 illustrates a flow chart for generation of the nonwoven network. In the python script, four important parts were considered: (i) input of fibre and fabric properties; (ii) generation of bond points; (iii) generation of fibres; (iv) storing information of nodes and elements of the finite- element model. In the input block, class geometry properties of the fibres and fabric, such as the average fibre length and cross-section and dimensions of the fabric are defined as well as the materials properties of fibres such as their flow curve and elastic modulus, obtained in the single-fibres tests. After reading this information, coordinates of the bond points were calculated depending on their shape and pattern for the modelled fabric. By the end of this calculation, bond points geometry was discretized, and the nodes and element numbers in each individual bond point were stored for the subsequent steps. At this point, according to the weight of bond points and of the total fibrous structure, the weight of fibrous matrix (i.e. without bond points) was calculated, and the coordinates and the angles of fibres with respect to MD of the web were determined based on the measured ODF. Meanwhile, the intersection points of fibres and bond points were computed, and the nearest nodes in the vicinity found with a search algorithm. Together with this, to match nodes of fibres and bond points, fibres were rotated by a small angle, and discretization of the fibre geometries took place. The generation of the fibrous matrix in the model was performed according to the condition that individual fibres were added to the model until the current weight of the web reached the prescribed magnitude. In the developed finite-element model, the microstructure of material was generated using truss elements for fibres and thin shell elements for bond points. The reason for this choice was the low density of the material used in this research. Truss elements were aligned in the same plane. The shell elements could be used to model bond points since their thickness-to-length and thickness-to-width ratios were

Journal of Physics: Conference Series **628** (2015) 012093 doi:10.1088/1742-6596/628/1/012093

low for the use of volumetric elements. The generated finite-element web with a square notch is depicted in Fig.3.

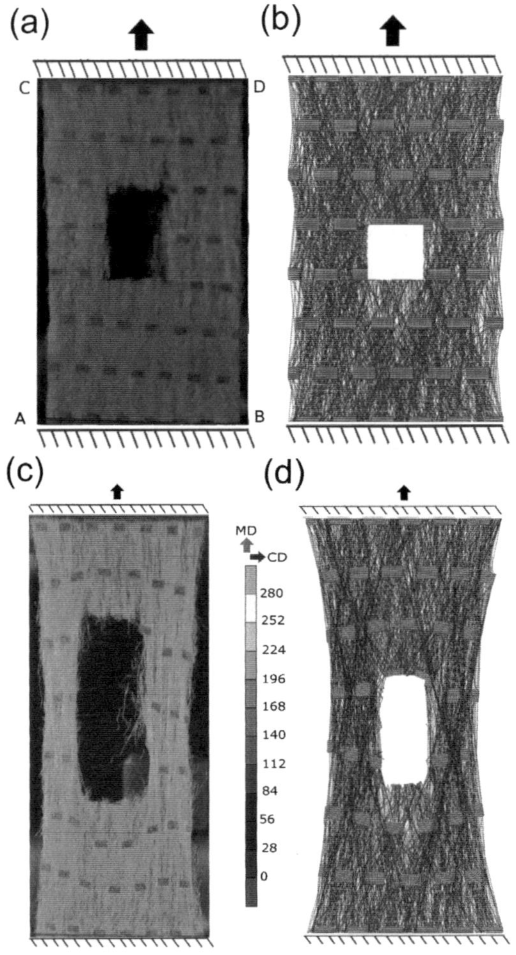

Figure 3. Undeformed specimen with square notch (dimensions 10 mm x 16.5 mm) (a) and finite element model(b). Notched specimens stretched by 30 percent demonstrate similar deformed shape in experiments and simulations c and d, showing also von Mises stress).

3. . Numerical re ult an com ari on

At micro scale, at the initial stage of deformation, reorientation of fibres defined the respective stress-strain behaviour. Then, stress concentration around the notch began to increase, generating tension in fibres surrounding the notch. Further increases in the load overcame local frictional forces between these fibres, and the size of a damaged area grew significantly, limiting contraction in the lateral direction. The stress transfer was also channelled to other fibres due to the presence of frictional fibre-to-fibre contacts, and when these inter-fibre forces were overcome, the final rupture occurred in the nonwoven material.

From the experimental tests, the shapes of notch-induced damages were found to change significantly at high levels of stain. The square hole became rectangular, stretched along

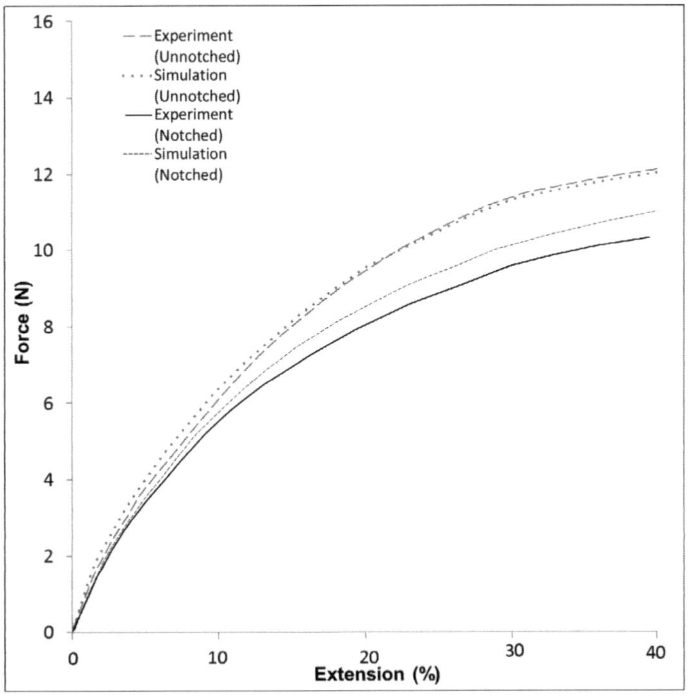

Figure 4. Force-displacement curves for notched and un-notched specimens obtained in experiments and simulations

the machine direction. In addition, force-displacement diagrams of notched and un-notched specimens obtained in experiments (mean curves) and simulations were compared, see Fig. 4; these curves show a good correlation. The trend of the curves on un-notched specimens is clearly above square notched specimens.

4. Conclusions

In this research, the effect of a central notch on nonwoven fibrous networks was investigated experimentally and numerically. To represent a realistic microstructure of the material in the finite-element model, a random distribution of fibres was introduced employing the measured orientation distribution function of fibres. Using this method, gaps and voids, typical for low-density nonwovens, were successfully represented in the model as well as the distinct character of orientation of fibres. Specimens with a square notch were used in the experiments, and the same geometry was generated in a virtual environment. The growth and final shape of the central notch in simulations are similar to those in the experiments. Similarly, the change in alignment of bond points during the simulations demonstrated behaviour comparable to that in the tests. It is concluded that experiments and simulations are in a good agreement quantitatively and qualitatively. Thus, the current model is capable of reproducing deformation until the onset of damage. Since the damage progress is determined by the fibre scale and the properties of fibres in used in the simulation were obtained from the fibres extracted from fabric rather than unprocessed fibres, the same model can be utilized to simulate fracture of the real material. From the industrial and academic points of view, this continuous modelling approach allows designers and researchers to simulate mechanical behaviour of a wide spectrum of fibrous materials. Factors affecting damage and failure mechanism can be accounted for, and, ultimately, products with optimum manufacturing and material parameters could be designed for their better performance. In the future work, larger experimental specimens will be used to investigate the effect of a notch size on damage progress, though an increase in the size of specimens results in significantly higher computation times. The effect of notch shape on damage mechanism will be also investigated. This investigation can be extended by incorporating fibre-to-fibre interactions.

5. Acknowledgements

We greatly acknowledge support of the Nonwoven Cooperation Research Center of North Carolina State University, Raleigh, USA.

References

[1] Farukh F, Demirci E, Sabuncuoglu B, Acar M, Pourdeyhimi B and Silberschmidt V V 2014 *International Journal of Solids and Structures* **51** 1670 – 1685
[2] Carleton J B, DAmore A, Feaver K R, Rodin G J and Sacks M S 2015 *Acta Biomaterialia* **12** 93 – 101
[3] Stella J A, DAmore A, Wagner W R and Sacks M S 2010 *Acta Biomaterialia* 2365 – 2381
[4] Liao T, Adanur S and Drean J Y 1997 *Textile Research Journal* **67** 753–760
[5] Liu J, Chen Z and Li K 2010 *Theoretical and Applied Fracture Mechanics* **54** 1 – 10
[6] Liu J, Chen Z, Wang H and Li K 2011 *Theoretical and Applied Fracture Mechanics* **55** 131 – 139
[7] Silberstein M N, Pai C L, Rutledge G C and Boyce M C 2012 *Journal of the Mechanics and Physics of Solids* **60** 295 – 318
[8] Kim H and Pourdeyhimi B 2001 *Journal of Textile and Apparel, Technology and Management* **1** 1–7
[9] Hou X, Acar M and Silberschmidt V V 2009 *Computational Materials Science* **46** 700 – 707
[10] Sabuncuoglu B, Acar M and Silberschmidt V V 2013 *International Journal for Numerical Methods in Engineering* **94** 441–453
[11] Farukh F, Demirci E, Sabuncuoglu B, Acar M, Pourdeyhimi B and Silberschmidt V V 2012 *Computational Materials Science* **64** 112 – 115
[12] Farukh F, Demirci E, Sabuncuolu B, Acar M, Pourdeyhimi B and Silberschmidt V V 2013 *Computational Materials Science* **71** 165 – 171
[13] Farukh F, Demirci E, Sabuncuoglu B, Acar M, Pourdeyhimi B and Silberschmidt V V 2014 *Computational Materials Science* **94** 8 – 16
[14] Farukh F, Demirci E, Sabuncuoglu B, Acar M, Pourdeyhimi B and Silberschmidt V V 2015 *Composites Part B: Engineering* **68** 327 – 335
[15] Russell S (ed) 2007 Woodhead Publishing Series in Textiles (Woodhead Publishing)
[16] Michielsen S, Pourdeyhimi B and Desai P 2006 *Journal of Applied Polymer Science* **99** 2489–2496
[17] Bhat G S, Jangala P K and Spruiell J E 2004 *Journal of Applied Polymer Science* **92** 3593–3600
[18] Farukh F, Demirci E, Acar M, Pourdeyhimi B and Silberschmidt V V 2013 *Journal of Materials Science* **48** 2334–2345
[19] Farukh F, Demirci E, Acar M, Pourdeyhimi B and Silberschmidt V V 2012 *Journal of Physics: Conference Series* **382** 012018
[20] 2003 *ASTM International USA* ASTM Standard D 4595
[21] Demirci E, Hou X N, Acar M, Pourdeyhimi B and Silberschmidt V V 2012 *Solid State Phenomena* **188** 164–169
[22] Demirci E, Acar M, Pourdeyhimi B and Silberschmidt V V 2010 ASME 2010 10th Biennial Conference on Engineering Systems Design and Analysis Vol. **4** 117-122
[23] Cox H L 1952 *British Journal of Applied Physics* **3** 72–79
[24] Demirci E, Acar M, Pourdeyhimi B and Silberschmidt V V 2012 *Computational Materials Science* **52** 157 – 163
[25] Xu B 1996 *Textile Research Journal* **66** 496–506

Dynamic response of RC beams strengthened with near surface mounted Carbon-FRP rods subjected to damage

R Capozucca[1]
Assoc. Prof., DICEA, Polytechnic University of Marche, 60100 Ancona, Italy – Ph. +39.071.2204576 Fax +39.071.2204576

E-mail: r.capozucca@univpm.it

M G Blasi
Engineering, DICEA, Polytechnic University of Marche, 60100Ancona, Italy

V Corina
PhD Student, DICEA, Polytechnic University of Marche, 60100Ancona, Italy

Abstract. Near surface mounted (NSM) technique with fiber reinforced polymer (FRP) is becoming a common method in the strengthening of concrete beams. The availability of NSM FRP technique depends on many factors linked to materials and geometry - dimensions of the rods used, type of FRP material employed, rods' surface configuration, groove size – and to adhesion between concrete and FRP rods. In this paper detection of damage is investigated measuring the natural frequency values of beam in the case of free-free ends. Damage was due both to reduction of adhesion between concrete and carbon-FRP rectangular and circular rods and cracking of concrete under static bending tests on beams. Comparison between experimental and theoretical frequency values evaluating frequency changes due to damage permits to monitor actual behaviour of RC beams strengthened by NSM CFRP rods.

1.Introduction

In the last years, the strengthening of reinforced concrete (RC) beams using fiber reinforced polymers (FRPs) has become a significantly common technique in the practice of civil engineering. In general, the use of two different strengthening techniques is adopted: FRP sheet/strips glued on the concrete surface as external bond composite materials; FRP rods with various types of sections, for example circular and rectangular, inserted into grooves on concrete covers. This last method is known as the near surface mounted (NSM) technique. The NSM FRP technique appears capable of solving a number of aspects which may threaten the strengthening with composite materials on the external surface: their susceptibility to damage deriving from collision, high temperature, and fire. The NSM method has been proven to be a promising one for increasing the capacity of RC beams [1-3] although many factors may affect the bond between FRP rod, resin and concrete: the bond length, the diameter of the rods, the type of FRP material employed, rods' surface configuration, and groove size [4-5].
NSM FRP rods are prone to exhibit greater slips than steel reinforcement due to FRP materials' potentially lower bond shear stress because of the presence of surrounding adhesive layers and local

1 Roberto Capozucca, Polytechnic University of Marche, 60100 Ancona, Italy r.capozucca@univpm.it .

cracking in the cover concrete. In this paper the assessment of strengthened RC beams with NSM CFRP rectangular and circular rods under cracking concrete has been experimentally treated by analysing dynamic response of beams. Natural frequency measures were obtained through dynamic tests on free-free beams both strengthened and un-strengthened with NSM CFRP rods. The beams were subjected to different levels of static loading with increasing damage due to cracking of concrete. The basic concept of vibration analysis is that dynamic characteristics are functions of structures' physical properties, therefore any change caused by damage results in change in dynamic response [6]. In the NSM method actual bond-slip may be influenced by the cracking of concrete and loss of adhesion of rods, which can modify frequency values and beams' modes of vibration. Over the years, many studies based on frequency measures have been developed to detect damage in uniform beams [7,8] and significant researches have been carried out in the last years with RC beams.
Below, the results of dynamic tests on beams are used to assess the availability of NSM strengthening with the reduction of bond increasing the damage due to bending loading.

2. Static and dynamic tests

2.1. Beams with CFRP circular rods

Static and dynamic tests were planned for two RC beams; one was reinforced with steel bars (Beam B0) while the B1 beam was reinforced inserted with steel bars and two CFRP circular rods into rectangular grooves of 20mm·20mm of section (Fig.1). Dimensions of beam section are 150mm·220mm and length 1700mm. Steel reinforcement was 4 bars of diameter 10mm and stirrups of diameter 6mm at interval of 60mm. The beams were characterized by concrete having, a tested, average cylinder strength equal to $f_{c,av.}$~ 44.3N/mm^2 (Young's modulus ~36.0·10^{3}N/mm^2) and steel bars with an average yielding stress equal to $f_{y,av.}$~500 N/mm^2 (Young's modulus about~2.1·10^{5}N/mm^2). CFRP rods with nominal diameter 8mm and section 65.04 mm^2 were glued to the concrete using epoxy resin. The strength of CFRP rod is $f_{av.}$~2153.27N/mm^2; Young's modulus $E_{b,av}$~142·10^{3}N/mm^2.

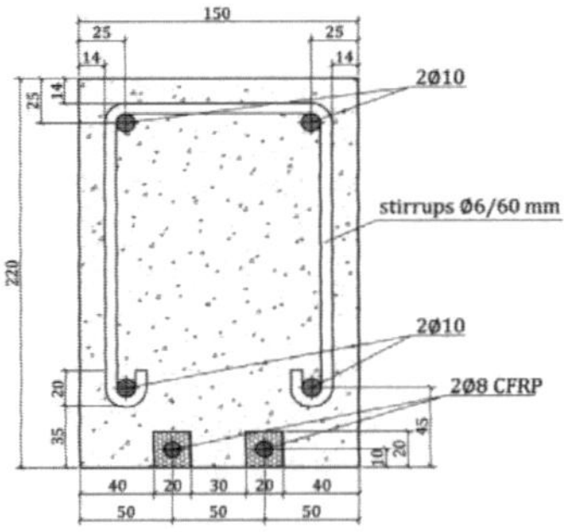

Figure 1 Section of RC beam with NSM CFRP circular rods

Table 1 – Exp. cycles of loading P_i (kN) at damage degrees D_i for tested beams.

B0		B1	
D_1	P_1= 7.2kN	D_1	P_1= 7.1kN
D_2	P_2=14.3kN	D_2	P_2=14.3kN
D_3	P_3=31.5kN	D_3	P_3=31.4kN
D_4	P_4=39.0kN	D_4	P_4=39.0kN
-	-	D_5	P_5=87.5kN
-	-	D_6	P_6=105.0kN
D_5	Failure load: P_u= 47.7kN	D_7	Failure load: P_u= 115.0kN

Static tests were carried out on beams by bending, increasing the load P applied in two points measuring 300mm from the middle of beam (Table 1).

The theoretical natural frequencies for free vibration of an undamaged beam with deflection, v, function of point, x, and time, t, may be obtained from the following equation:

$$EI \cdot \frac{\partial^4 v}{\partial x^4} + \rho \cdot A \cdot \frac{\partial^2 v}{\partial t^2} = 0 \tag{1}$$

where: ρ = density of beam material; A = the cross-sectional area.

The solution of Eq. (1) must be a harmonic function of time i.e.:

$$v(x,t) = V(x) \cdot \sin(\omega t + \alpha) \tag{2}$$

Introducing Eq. (2) in Eq. (1), assuming $\lambda^4 = \frac{\rho \cdot A \cdot \omega^2}{EI}$, applying the following boundary conditions:

$$V^{II}\Big|_{x=0} = 0; \quad V^{III}\Big|_{x=0} = 0; \quad V^{II}\Big|_{x=L} = 0; \quad V^{III}\Big|_{x=L} = 0 \tag{3}$$

an algebraic linear system in the unknown constants B_i may be obtained. A non trivial solution exists when the determinant becomes:

$$\cos\lambda L \cdot \cosh\lambda L - 1 = 0 \tag{4}$$

The expression of circular frequency for undamaged free-free beam at the r-mode is:

$$\omega = \left(a_r \cdot \frac{r\pi}{L} \right)^2 \cdot \sqrt{\frac{EI}{\rho A}} \tag{5}$$

with coefficients a_r = 1.506, 1.25, 1.167, respectively, for the first three modes (r=1, 2, 3) […].
The first three frequency values evaluated for undamaged beams B0 and B1 are shown in Table 2.

Table 2- Frequency values for free vibration of beam without cracking of concrete.

Frequency values for undamaged beam	**f_1** (Hz)	**f_2** (Hz)	**f_3** (Hz)	**f_4** (Hz)
Theor. Euler-Bernoulli beam	310	857	1680	2777
Exp. frequency value – B_0	275	706	1274	1878
Exp. frequency value - B_1	284	723	1299	1923

The experimental dynamic tests were carried out on B0, B1 hanging the beams with springs to simulate the free-free condition.

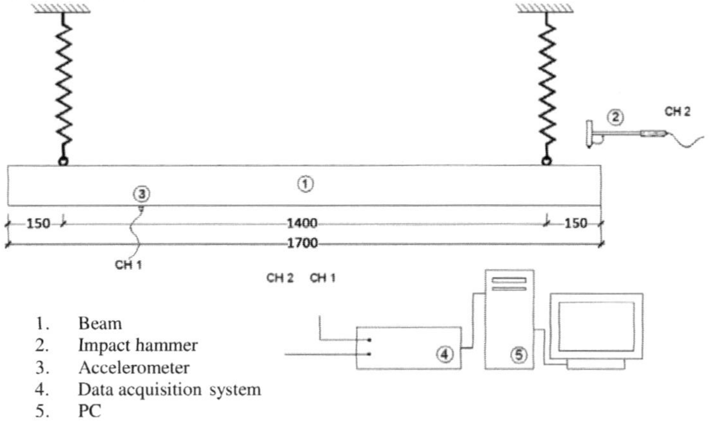

Figure 2 Set-up of free vibration text.

The free vibration tests were carried out with the impact technique using an impact hammer (*Brüel & Kjær, Type 8202*) able to transfer impulse in one point at 45mm from the edge; the measures were recorded with a mobile accelerometer (m*odel 4508 – Piezoelectric CCLD accelerometer, Brüel & Kjær*) located in 14 points at regular interval. A set of 10 hits was recorded at each of the different points and the average value was acquired. Frequency values were extracted by transformed signals in frequency domain through the fast fourier transform (FFT) technique using Pulse software. In Figure 2 set up of vibration is shown. In Figs. 3 and 4 the envelope of FRFs at damage degree Di, respectively, for B0 and strengthened beam B1 are shown. In Figs. 3 and 4 the diagrams indicate the values obtained in two positions 1 and 3 of accelerometer. Trend of diagrams is similar for other points of measure of accelerometer. It may be noted that for both beams there is a shift on the left of diagrams due to reduction of frequency values at the increasing of damage condition Di by loading.

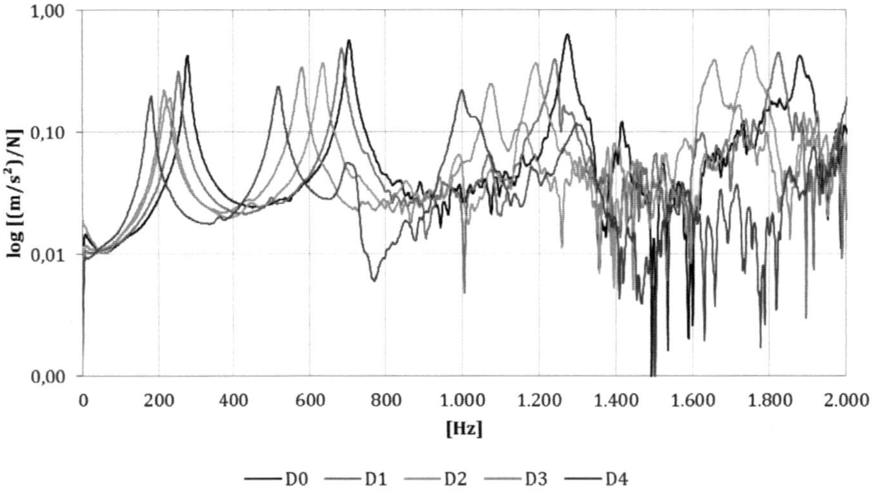

(a) Mark 1

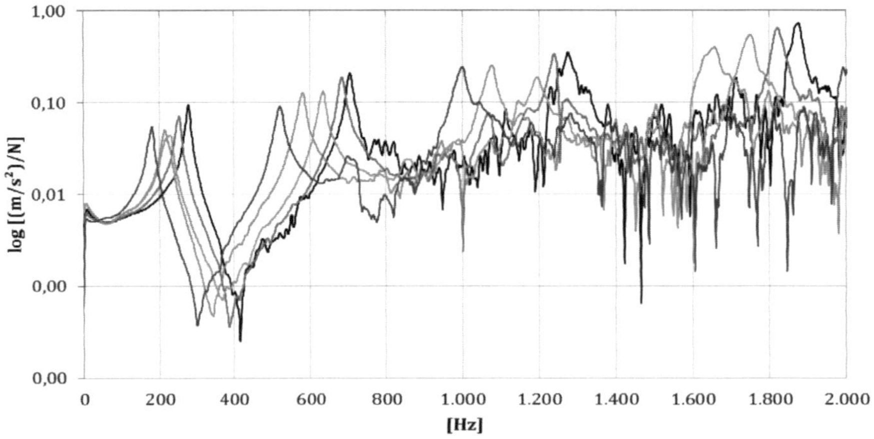

(b) Mark 3

Figure 3 Envelope of FRFs at damage degree Di, i=1,…,4 for B0 recorded with accelerometer in (a) point 1 and (b) 3.

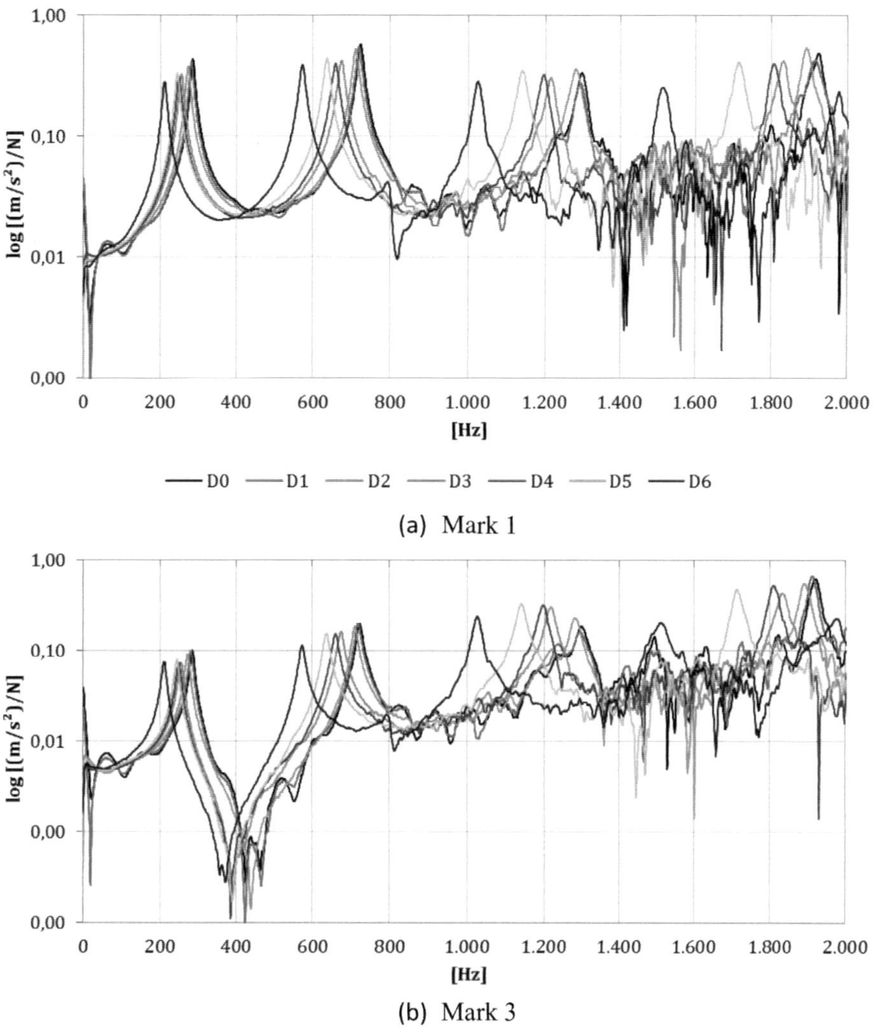

Figure 4 Envelope of FRFs at damage degree Di, i=1,...,6 for B1 recorded with accelerometer in (a) point 1 and (b) point 3.

The variations of frequency values for B0 and B1 evidence that in unstrengthened beam B0 the decrease of frequency values are greater than in the strengthened beam B1. In Figs. 5(a),(b) at the different levels of loading reached during the beams' bending tests, the variation in percent of frequency values, $\frac{\Delta f_r}{f_r^{D_0}} = 100 \cdot \frac{f_r^{D_0} - f_r^{D_i}}{f_r^{D_0}}$, for the first four modes r= 1,...4, are indicated.

For example, at the first mode, a reduction of frequency values for B0 is equal to 34% at last loading stage, while in B1 it is almost equal to 26%. In the same way at the first damage degree D1-D0 at the first mode r=1 the variation is equal to 9% for B0 and minor than 2% in B1. Same trend of results is evident in the other modes of vibrations.

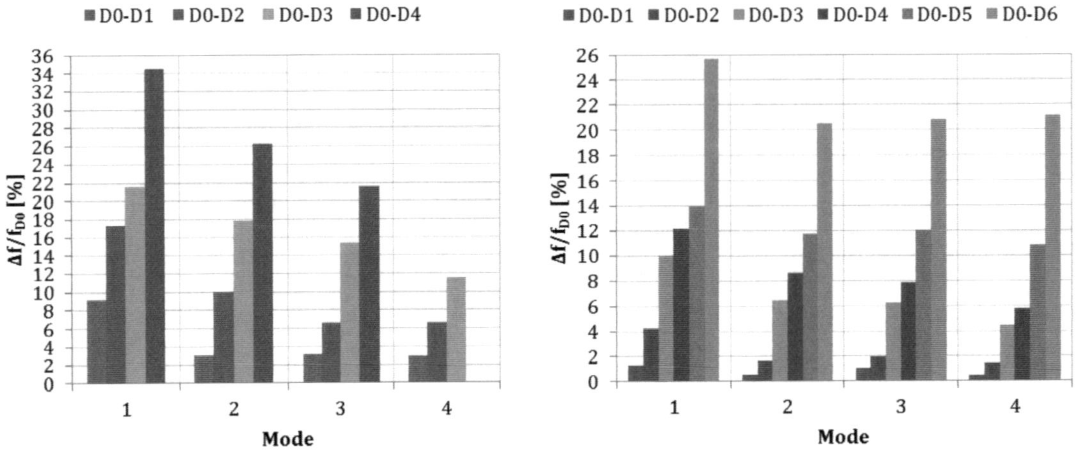

(a) (b)

Figure 5 Comparison of frequency variation (%) between different damage degrees Di for (a) B0 and (b) B1.

2.2. Beams with CFRP rectangular rods.

Two RC beams were built and subjected to testing in order to monitor static and dynamic behavior: beam B0R was not reinforced with CFRP rods; beam B1R was reinforced using three rectangular NSM CFRP rods inserted in rectangular grooves at the bottom of the beam (Fig. 6). The steel reinforcement consisted in 2 bars, measuring 10mm in diameter, placed in the lower area, and two other bars placed in the upper area, using 6mm stirrups every 80mm.

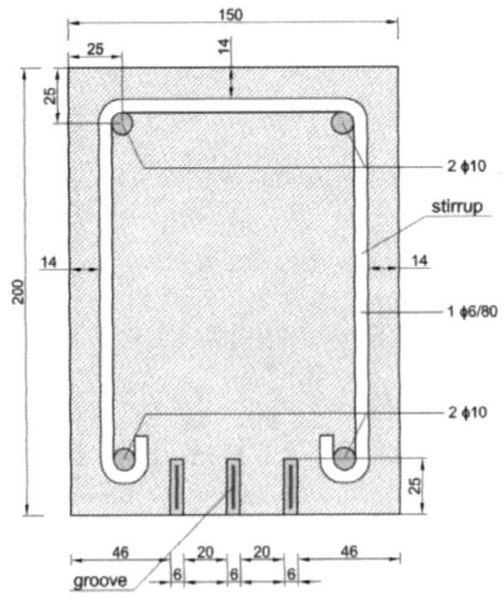

Figure 6 RC beam with NSM CFRP rectangular rods.

The bending tests for non strengthened and strengthened beams, B0R and B1R, foresaw loading cycles (damage degree D1 until P=5kN; D2 - P=20kN; D3 - P=30kN; D4 - P=40kN) with vertical load maximum equal to P= 50kN as damage degree D5 and, subsequently, the ultimate loading test until failure.

In Figs 7(b) experimental diagrams load, F, versus deflection at mid span section are shown for B0R and B1R; in the case of beam B1R, the capacity of the beam strengthened with NSM rectangular CFRP rods is higher than RC beam B0R. In reinforced beam B1R, failure is due to reinforcement delamination and not to concrete failure in the compressed zone. In particular, delamination occurred solely at the resin – concrete interface and a detachment of the concrete cover was also recorded (Fig. 7(a)). Beams B0R and B1R were analysed carrying out free vibration tests in order to obtain experimental dynamic parameters at different damage conditions due to bending loading; D0 indicates the undamaged state of the beams and with D_i with i=1,…,5 being the different levels of loading and relative damage degrees.

(a)

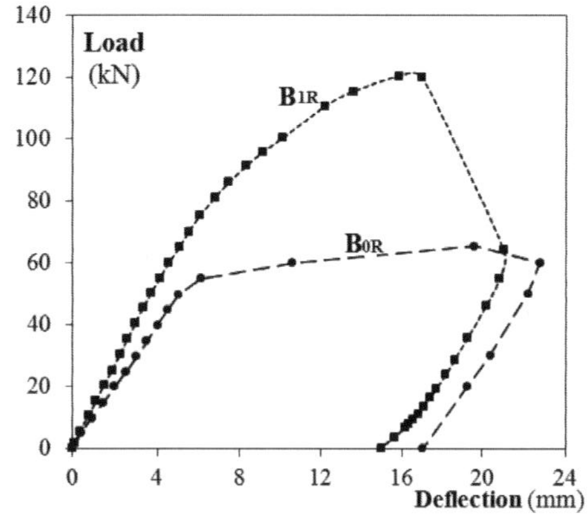

(b)

Figure 7 (a) View of B1R beam at failure; (b) comparison between exp. diagrams load, P, vs. deflection at the midspan of B0R and B1R beam reinforced with NSM CFRP rectangular rods.

The dynamic test was conducted according to the set-up in Fig. 2. The experimental free vibration test was carried out for beams after each different level of load, P_i, hanging by flexible springs to simulate free-free condition. The dynamic responses were obtained using an impact hammer and an accelerometer at the beams' intrados, 500mm from the end, connected to an acquisition system working in a range of frequencies between 0- 2000Hz. The signals were recorded and elaborated in frequency domain through the FFT technique and FRFs were obtained using Labview software.

Table 3- Experimental frequency values and variation (%) for beam B0R.

Damage degree	f_1 [Hz]	$\frac{\Delta f_1}{f_1^{D_0}}$ %	f_2 [Hz]	$\frac{\Delta f_2}{f_2^{D_0}}$ %	f_3 [Hz]	$\frac{\Delta f_3}{f_3^{D_0}}$ %
D0	233.6	0	604.8	0	1085	0
D1	225.8	3.3	599.2	0.9	1074.2	1.0
D2	176.8	24.3	492.8	18.5	939.2	13.4
D3	158.4	32.2	429.8	28.9	805	25.8
D4	166.8	28.6	439.8	27.3	796.8	26.5
D5	167.2	28.4	445.4	26.3	816.4	24.7

Beams B0R and B1R at the end of each bending load step for damage degree D_i with i=1,…,5 were subjected to free vibration; in Tables 1 and 2 experimental frequency values and variation in percent, $\frac{\Delta f_r}{f_r^{D_0}} = 100 \cdot \frac{f_r^{D0} - f_r^{D0}}{f_r^{D_0}}$, at the different levels of loading reached during the beams' bending tests, are also indicated. Regarding the dynamic response of beams BR0 and BR1, the frequency values decrease with a consequent increase of load and shift to the left in the diagrams of the FRF's maximum values, which implies an increase of damage degree. In the strengthened beam B1R, the variation of frequency values is less than the values recorded for the non-strengthened beam, thus demonstrating the validity of the NSM technique for reinforced concrete beams; in fact with the NSM technique there is a reduction of concrete crack width and more limited diffusion of cracks on the surface of beam.

Table 4 - Exp. frequency values and variation (%) for beam B1R.

Damage degree	f_1 [Hz]	$\frac{\Delta f_1}{f_1^{D_0}}$ %	f_2 [Hz]	$\frac{\Delta f_2}{f_2^{D_0}}$ %	f_3 [Hz]	$\frac{\Delta f_3}{f_3^{D_0}}$ %
D0	234.8	0	604.2	0	1098.2	0
D1	233.4	0.6	598.4	0.9	-	-
D2	203.8	13.2	553.6	8.4	1057.8	3.6
D3	200.2	14.8	532.2	12.0	1012.0	7.8
D4	200.4	14.6	527.6	12.6	988.0	9.1
D5	197.4	15.9	520.8	13.8	964.8	12.1

3. Conclusions

The experimental dynamic tests were found to be a convenient non-destructive method for verifying the safety of intact, damaged, and strengthened reinforced concrete beams. Once the dynamic parameters of these types of beams are known, monitoring their dynamic response at different load levels and assessing the validity of the strengthening becomes feasible.
In this paper the dynamic response of concrete beams with reinforced steel bars and reinforced also with NSM CFRP circular and rectangular rods is evaluated by experimental tests measuring the natural frequency values in the condition of free-free ends. Main results obtained are:

1. the strengthening of RC beams with NSM CFRP rods permits to improve the performance under static tests;
2. reduction of frequency values is greater in the unstrengthened beams than in strengthened NSM CFRP rods both for rectangular and circular rods;
3. the variation in percent of frequency values is smaller in the strengthened RC beams at every damage degree.

References

[1] El-Hacha R and Rizkalla S 2004 *ACI Structural Journal* **101**(5) 717.
[2] De Lorenzis L and Teng J G 2007 *Composites Part B: Eng.* **38** 119.
[3] Szabò Z K and Balazsw G L 2007 *Civil Engineering* **51**(1) 33.
[4] Sharaky I A, Torres L, Baena M, Miàs C. 2013 *Composite Structures* **99** 350.
[5] Sena Cruz J M, Barros J A O, Gettu R, Azevedo A F M 2006, J. of Comp. for Constr. **10** 295.
[6] Salawu O.S. 1997, *Engineering Structures* **19** 718.
[7] Capozucca R 2009 *Composite Structures* **91** 237.
[8] Capozucca R 2013 *Engineering Structures* **48** 508.

Acknowledgments
This research was supported by research funds provided by Polytechnic University of Marche. The authors would like to express their gratitude to all the technicians and students who collaborated to develop the experimental research.

Damage localization in a glass fiber reinforced composite plate via the surface interpolation method

M P Limongelli [1], **V Carvelli**[2]
[1,2]Politecnico di Milano, Department of Architecture, Built Environment and Construction Engineering, Milano, Italy.

E-mail: mariagiuseppina.limongelli@polimi.it

Abstract. This work deals with the application to composite plates of the surface interpolation method (SIM) for damage localization. The procedure, which is a generalization to the two-dimensional case of the previously published Interpolation Damage Detection Method (IDDM), locates reductions of stiffness in two-dimensional structures such as plates. The method is based on the damage sensitive of a spline function accuracy in fitting the operational displacement shapes, relies on the so-called Gibbs' phenomenon for splines. This phenomenon occurs when a spline function interpolates discontinuous functions and consists in sharp oscillations and overshoots (values higher than those of the function to be interpolated) near a discontinuous point. The operational deformed shapes are recovered from frequency response functions (FRF's) measured at different locations of the structure during vibrations. The accuracy of the spline interpolation is measured by an error function defined as the difference between the measured and interpolated operational deformed shapes. At a certain location an increase (statistically meaningful) of the interpolation error, with respect to a reference configuration, points out a localized variation of the operational shapes thus revealing the existence of damage. The accuracy of the surface interpolation method is experimentally assessed by impact hammer tests on glass fiber/vinylester composite plates progressively damaged and using finite element numerical modelling.

1. Introduction

Plates are important structural elements, widely used in several engineering applications and the monitoring of their condition is an important aspect for the assessment of the global structural condition. Damage detection techniques traditionally consist in visual inspection and/or non-destructive testing (acoustic, ultrasonic, magnetic field, thermal field methods). A different approach consists in vibration based methods detecting changes of feature related to damage. In the last twenty years several features have been proposed in literature; some of these rely on modal frequencies and/or shapes or their derivatives. In reference [1] a comprehensive state of the art of modal methods is reported.

One drawback of modal methods is related to the need of estimating modal parameters introducing error related to experimental noise that sometime hampers a correct identification of damage. This makes more feasible, approaches based on the use of operational deformed shapes [2]-[4]. The first

[1] corresponding author

author of this paper recently proposed the Interpolation Damage Detection Method (IDDM) for beam-like structures [4], [5] and [6]. The focus of this paper is on the extension of the IDDM to plates. In the following, after a brief overview of the method, the results of experimental tests carried out on a glass fiber reinforced composite plate in the undamaged and in a damaged configuration are reported.
The outline of the paper is as follows. In section 2, the interpolation via bi-cubic spline functions is briefly described and the Gibbs phenomenon for splines is presented and discussed with reference to a simple case. In section 3 the extension of the Surface Interpolation Method is presented and in section 4 the results of experimental tests carried out on a glass fiber reinforced composite plate are detailed.

2. The surface interpolation damage detection method

2.1. Bicubic spline interpolation

A bi-cubic spline interpolation can be used to interpolate an $F(x, y)$ with a smooth surface, which is a function that guarantees continuity of first and second derivatives.
A bi-cubic spline can be thought as a surface constructed from families of cubic spline functions.
Given the values of a function $F(x, y)$ defined on a rectangular grid of $n_{l+2} \times n_{p+2}$ nodes, consider the following equation:

$$s(x, y):\ s_p(x, y) = \sum_{j=0}^{3} c_{jp}(x)(y - y_p)^j \qquad y \in [y_p, y_{p+1}] \qquad \text{with } p=0, 2..., n_p \qquad (1)$$

For each value of x, equation (1) gives a spline function along y (see black curves in Figure 1). When x varies, the coefficients $c_{jp}(x)$ also vary and a family of cubic spline functions $s(x, y)$ is obtained.

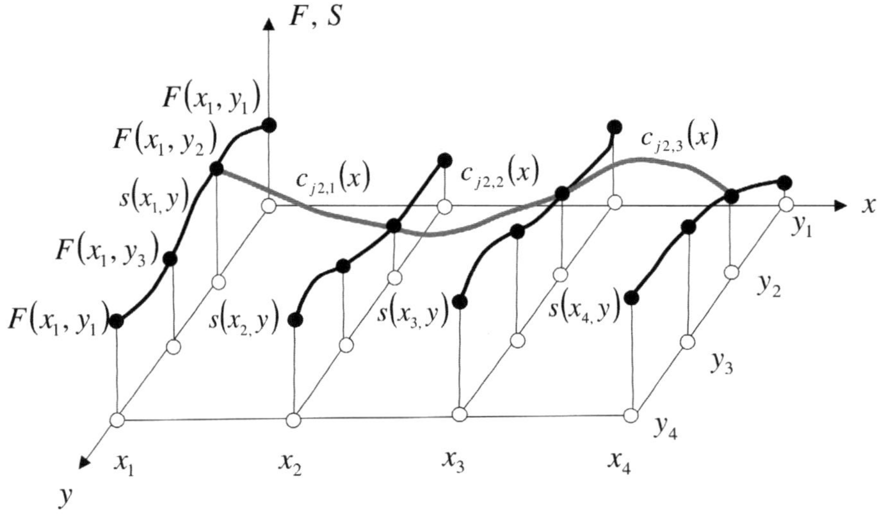

Figure 1. Bi-cubic spline interpolation.

If also the coefficients $c_{jp}(x)$ are defined by cubic spline functions (see red line in Figure 1) in each sub-area defined by $x \in [x_l, x_{l+1}];\ y \in [y_p, y_{p+1}]$ they can be written as:

$$c_{jp}(x): \qquad c_{jp,l}(x) = \sum_{i=0}^{3} a_{jp,il}(x - x_l)^i \qquad x \in [x_l, x_{l+1}];\ y \in [y_p, y_{p+1}]; \qquad (2)$$

p=0, 2..., n_p *and l*=0, 2..., n_l
In this case, the following function $S(x,y)$ is a bi-cubic spline function defined through its restrictions to sub-areas as:

$$S(x, y): \qquad S_{l,p}(x, y) = \sum_{j=0}^{3}\sum_{i=0}^{3} a_{jp,il}(x - x_l)^i (y - y_p)^j \qquad x \in [x_l, x_{l+1}];\ y \in [y_p, y_{p+1}] \qquad (3)$$

The coefficients $a_{jp,il}$ of the bi-cubic spline function are calculated knowing the function $F(x,l)$ at the nodes and, once the coefficients $a_{jp,ol}$ are known, the bi-cubic interpolation can be finally calculated using equation (3).

2.2. *The Gibbs phenomenon for cubic splines*

The Interpolation Method is based on the assumption of a beam-like behavior of the monitored structure and on the interpolation of the operational displacement shapes via a cubic spline function.
The process of interpolation consists in recovering continuous data from a discrete set of data, basing on a given model; this is a fundamental operation in several fields were resampling of a limited set of data is needed. For example, image processing may be used when there is the need to reconstruct the data lost in a previous image sampling process in order to change the size of images and/or to correct spatial distortions. Cubic Spline functions in image resampling are widespread used because they are the smoothest among all the functions that interpolate a given set of data and they are stable. Furthermore the cubic spline interpolating functions have a local character, that is a steep variation of the data in a small interval affects only locally the interpolating function. For these reasons cubic spline interpolation are capable of providing high accuracy at a low computational cost which makes them a good choice in case smooth transitions between sampled data are required.

On the other hand the high smoothness of the spline function may be a drawback for a reliable interpolation in case of non-smooth regions of curves. This is due to a phenomenon known as '*Gibbs' phenomenon of splines*' occurring when the spline interpolates functions with discontinuities: the spline interpolation oscillates near a discontinuous point and has an overshoot (values higher than those of the function to be interpolated[7]). In Figure 2 this phenomenon is illustrated with reference to a continuous (C^1) function $f(x)$ with a jump of curvature at section x=1: the cubic spline interpolation $s(x)$ of the function exhibits strong oscillation around the actual value of the function (see Figure 2a).

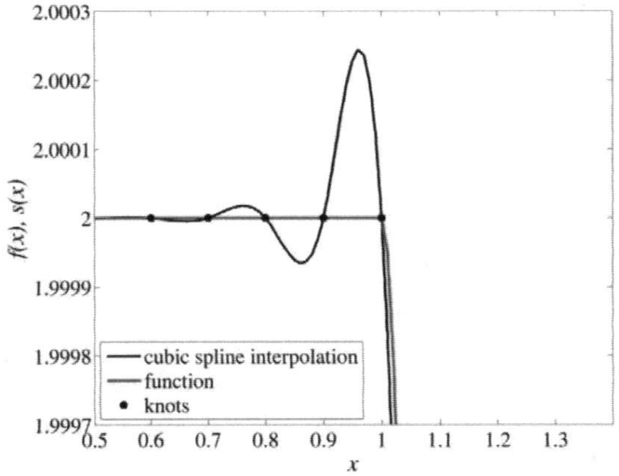

(a)

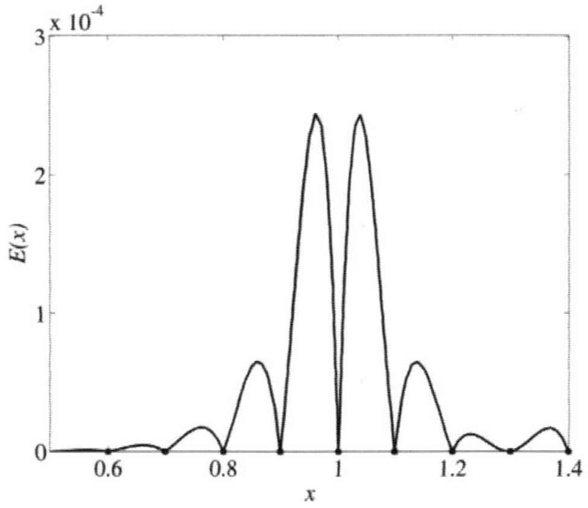

(b)

Figure 2. Gibbs phenomenon of splines.

Due to the oscillations of the interpolating spline function around the discontinuity, the interpolation error:

$$E(x) = |f(x) - s(x)| \tag{4}$$

reaches the highest values in the two intervals at the sides of the discontinuity and gradually decreases with distance (see Figure 2b).The Interpolation Method for damage localization is based on turning this 'drawback' related to the Gibbs' phenomenon into a 'useful feature' to use as a sort of 'curvature discontinuity detector'.

Basing on all these observations, the increase of the interpolation error can be reasonably used in order to detect a localized increase of curvature. As shown in reference [7] a decrease of the interpolation error may correspond to a reduction of the discontinuity of curvature or may occur in the region between two very close discontinuities. The Interpolation Damage detection Method, as will be detailed in the following section, uses as a damage feature the interpolation error calculated for the Operational Deformed Shapes (ODS).

3. The Surface Interpolation Method

The basic idea of the Surface Interpolation Method can be described with reference to Figure 3. Let (x_0, y_0),..., (x_l, y_p),..., (x_{n_l+1}, y_{n_p+1}), be instrumented location of the plate where responses in terms of acceleration have been recorded. At each frequency value, the set of Frequency Response Functions (FRFs) H_R, calculated from measured responses, gives the operational deformed shape (ODS) at that frequency.

A localized reduction of stiffness at a certain location, causes a change of the operational shapes in the region close to that location hence this variation can be used to detect the location of damage. The change of the ODS between the undamaged and the damaged configurations can be highlighted by the interpolation of the ODSs through a smooth shape function and by the comparison between the interpolation errors related to the (possibly) damaged and the undamaged configurations. In the case of plate-like (that is bi-dimensional) structures, the interpolation of the operating deflection shapes can be carried out using bi-cubic spline functions.

Assuming that sensors lye on a rectangular grid $n_{l+1} \times n_{p+1}$, for each value of frequency f_i, the actual value $H_R(x_l, y_p)$ of the FRF at a given location (x_l, y_p), and its spline interpolation $H_S(x_l, y_p)$ must be known in order to calculate the value of the interpolation error at that location: $E(x_l, y_p)$, as shown in Figure 3. To this aim, the spline interpolation $H_S(x_l, y_p)$ is calculated assuming the FRFs at all the instrumented locations except $H_R(x_l, y_p)$.

Actually, if only one response is neglected in the interpolation, a non-regular grid is obtained. In order to overcome this problem, two regular grids have been defined, by neglecting all the responses recorded on one row or on one column of the grid.

Following the two-steps procedure outlined in section 2.1, at each instrumented location the two values of the bi-cubic spline interpolation (one for each of the two grids obtained removing one row or one column) can be calculated as:

$$H_{S,x}(x, y, f_i) = \sum_{i=0}^{3}\sum_{j=0}^{3} X_{jp,i(l-1)}(f_i)(x - x_{l-1})^i (y - y_p)^j \qquad x_l \in [x_{l-1}, x_{l+1}];\; y_p \in [y_p, y_{p+1}] \tag{5}$$

$$H_{S,y}(x, y, f_i) = \sum_{i=0}^{3}\sum_{j=0}^{3} Y_{j(p-1),il}(f_i)(x - x_{l-1})^i (y - y_{p-1})^j \qquad x_l \in [x_l, x_{l+1}];\; y_p \in [y_{p-1}, y_{p+1}] \tag{6}$$

For each frequency value f_i, the coefficients, $X_{jp,i(l-1)}$ and $Y_{j(p-1),il}$ are calculated as functions of the H_R measured at all the nodes of the grid except those with $x=x_l$ (for $X_{jp,i(l-1)}$) and except those with $y=y_p$ (for $Y_{j(p-1),il}$).

Considering the two values of the interpolated FRF, at each instrumented location (x_l, y_p) and for each frequency f_i, the value of the interpolation error at the given location can be calculated as:

$$E(x_l, y_p, f_i) = E_x(x_l, y_p, f_i) + E_y(x_l, y_p, f_i) \tag{7}$$

In order to characterize each location $P(x, y)$ (in the following the suffix p and l will be dropped for clarity of notation) with a single error parameter, the norm of the error on the significant frequency range (that is the frequency range with a signal to noise ratio sufficiently high to allow a correct definition of the FRF) is calculated:

$$E(x,y)=\sqrt{\sum_{i=n_o}^{n_o+N} E(x,y,f_i)^2} \tag{8}$$

The significant frequency range is selected limiting the summation in equation (8) to the frequency range of the fundamental modes of the structure that can be tuned basing on vibration tests carried out on the undamaged structure. In equation (8) N is the number of frequency lines in the Fourier transform correspondent to the frequency range starting at line n_o, where the signal to noise ratio is high enough to allow a correct definition of the FRFs.

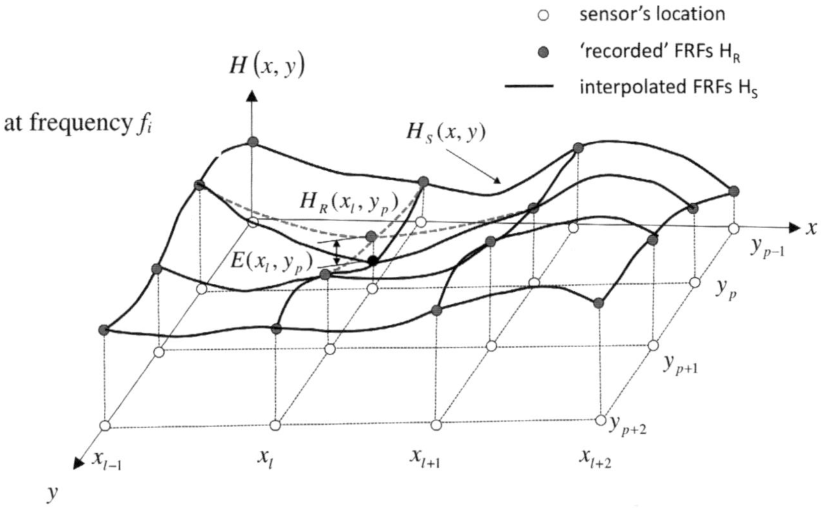

Figure 3. The interpolation error $E(x,y)$.

The values of the transfer functions H_R depend on the state of the structure. Hence, if the estimation of the error function through equation (8) is repeated in the baseline (undamaged) and in the inspection (possibly damaged) configuration, then the difference between the two values, denoted respectively by $E_o(x,y)$ and $E_d(x,y)$, can provide an indication about the existence of degradation at location (x, y):

$$\Delta E(x,y)=E_d(x,y)-E_0(x,y) \tag{9}$$

Due to the Gibbs phenomenon described in section 2.2, an increase in the interpolation error between the reference configuration and the current configuration at a station $P(x,y)$, i.e. $\Delta E(x,y)>0$, highlights a localized reduction of smoothness in the vibration amplitude profile.

A positive value of ΔE is thus assumed to be a symptom of a local decrease of stiffness at location $P(x,y)$ associated with the occurrence of damage. By this assumption, the following conditions are assumed to define the damage index $IDI(x, y)$:

$$\begin{aligned} &IDI(x,y)=\Delta E &&\text{if} &&\Delta E(x,y)\geq 0\\ &IDI(x,y)=0 &&\text{if} &&\Delta E(x,y)<0 \end{aligned} \tag{10}$$

The first condition ($\Delta E>0$) corresponds to an increase of the interpolation error, that is to a decrease of the smoothness of the amplitude profile. On the contrary a decrease of the interpolation error, as shown in section 2.2, may be associated to an increase of smoothness or may occur in the region between two damaged sections due to an insufficient spatial sampling. In both cases it is not associated to a local reduction of stiffness hence a negative variation of the interpolation error will be assumed to correspond to a null value of the damage index.

Due to several sources, positive variations of the interpolation error E occur even if no damage is present. In order to remove the effect of these random variations, the statistical variation of this parameter can be studied and a threshold can be defined considering the tolerable value of the probability of false alarm P_f (see reference [6]). The threshold of ΔE can be estimated in terms of the average $\mu_{\Delta E}$

and the variance $\sigma_{\Delta E}$ of the values of ΔE at all the instrumented locations. Assuming a Normal distribution, the threshold ΔE_t can be defined as:

$$\Delta E_t(x, y) = \mu_{\Delta E} + \nu\sigma_{\Delta E} \tag{11}$$

The threshold can be fixed setting a value of the accepted probability of false alarm assuming the parameter ν as the quantile 1-P_f of the standard normal distribution.
The interpolation damage index (*IDI*) is then calculated as:

$$\begin{cases} IDI(z) = \Delta E(z) & \text{if} \quad \Delta E(z) \geq \Delta E_t(z) \\ IDI(z) = 0 & \text{if} \quad \Delta E(z) < \Delta E_t(z) \end{cases} \tag{12}$$

A localized decrease of stiffness causes a localized increase of curvature that, having in mind the discussion reported in section 2.2, corresponds to an increase of the interpolation error.

4. Experimental tests and results

3.1. Material and specimen

The plates were produced by poltrusion technique using E-glass fibres and vinylester matrix. The Heat Deflection Temperature (HDT) of the resin, measured by the producer, was 105°C. Fibre reinforcement consisted of three non-crimp fabrics. Each fabric had plies construction 0° / ± 45° and total areal weight of 600 g/m^2. The temperature in the heated mould was set to have 95% of the complete polymerization inside the mould.

	Elastic modulus [GPa]	Strength [MPa]
Traction 0°	27.5	437
Traction 90°	9.9	76
Bending 0°	20.6	551
Bending 90°	10.8	109

Table 1. Mechanical properties of the composite material.

The fibre volume fraction was almost 60%. Some mechanical properties of the composite were measured according to international standards (ISO 527 and ISO 14125 for tensile and bending test, respectively). The average results are listed in Table 1.
Specimens for testing were square plates with average thickness of ≈8 mm and sides of ≈500 mm.

3.2. Impact hammer tests

The plate was suspended with a bungee cord to approximate free-free boundary conditions and subjected to a pulse load applied to each of the intersections of a grid depicted on the specimen (see **Figure 4**).

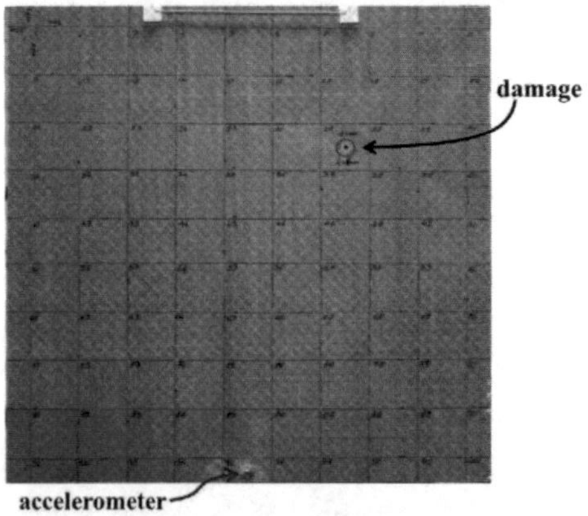

Figure 4. Grid for pulse load application and position of the accelerometer and damage.

The load is applied by an impulse hammer Dytran 5801A5 equipped with a load cell of 4.45 kN. A piezoelectric accelerometer Bruel&Kiaer 4384, fixed at the reference location shown in Figure 4 (middle of the plate close to the lower edge) was used to measure acceleration. Data were recorded with an acquisition system National Instruments 9234, 24 bit resolution and maximum sample rate of 51200 Hz. For each pulse load, 51200 measurements were stored. The Frequency Response Functions between each grid point and the reference one were determined from three impacts at each point and, in order to reduce the random fluctuations in their estimation, three FRFs were averaged at each point. The plate was tested before (as reference) and after a damage simulating a manufacturing flaw. The imparted damage was of cylindrical shape 5 mm diameter and half thickness depth. The red circle in Figure 4 is shows the position of the damage.

3.3. Results

The Operational Deformed Shapes are contained in each row or column of the FRF matrix that can be calculated in terms of receptances (ratio between acceleration and force). In this case, the receptances at each node p (p=1, 2... 100) were calculated as the ratio between $R_q(\omega)$, the FFT of the acceleration recorded at the location q (see **Figure 4**) and $F_p(\omega)$, the FFT of the force impacted by the hammer at node p:

$$H_p(\omega) = \frac{R_q(\omega)}{F_p(\omega)} \tag{13}$$

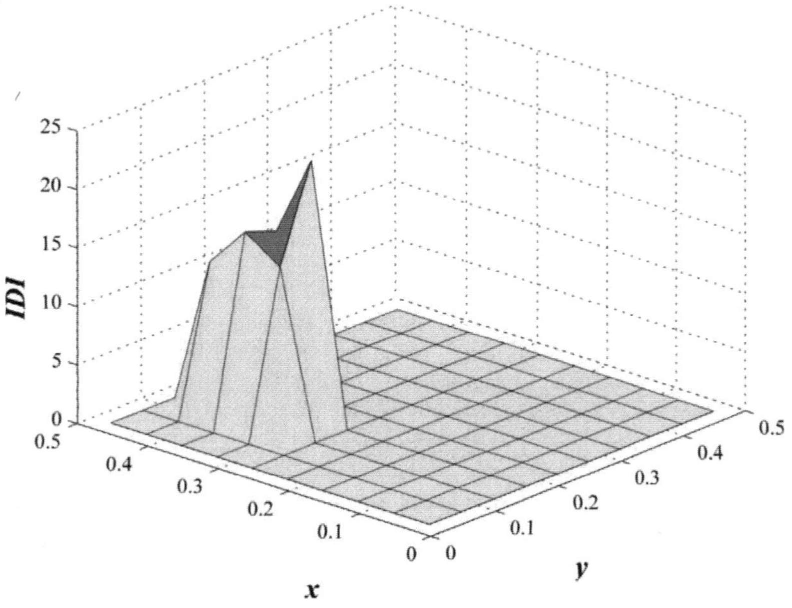

Figure 5. Results for the damage induced in the composite plate

For example, the function $H_{55}(\omega)$ was obtained by impacting the beam at point 55 and measuring the response at the location of the accelerometer.

For the application of the SIM, the probability of false alarm has been assumed equal to 2% (ν=2 in equation 11) and the frequency range for the application of the SIM, assuming the FRF measured in the undamaged configuration, was set as 40-800 Hz (n_o= 40, N=800 in equation 8).

In Figure 5 the results of the application of the SIM are reported showing that the damaged location is correctly detected by the algorithm. It is noted that the damage index exhibits the highest values at the four nodes of the damaged elements even if high values are found also at the adjacent locations. This is due, as explained in reference [5], to the interpolation process that somehow spreads the reduction of the accuracy in the interpolation to the locations close to the one where the reduction of smoothness occurs. In any case, this phenomenon occurs only in the vicinity of the effectively damaged area, even if it can reduce the precision in the identification of the contour of the damaged area, it does not hamper the correct localization.

5. Conclusions

In this paper are summarized the results of the application of the Surface Interpolation Method (SIM) to the case of a damaged glass fiber composite plate in order to experimentally verify the feasibility application of the proposed algorithm. Results indicate that the method is a viable method for damage localization providing, with respect to other method proposed in literature, the same accuracy in damage localization with at a lower computational effort. Advances in sensors technology and miniaturized signal processing platforms allow exploring the possibility of realizing an autonomous monitoring system based on responses recorded by a large number of sensors widely distributed over the structure. From this point of view the minimum interaction with a human operator required by the SIM makes it appealing for automatic damage detection through a permanent monitoring system integrated with data processing software.

Acknowledgments

The present work was partially supported by the ReLuis-DPC Executive Project 2014-2016, Special Project “Monitoring”, whose contribution is gratefully acknowledged.

References

[1] Fan W, Qiao P 2011 Vibration-based Damage Identification Methods: A Review and Comparative Study *Structural Health Monitoring* **10**(1) p. 83–29

[2] Ratcliffe CP 2000 A frequency and curvature based experimental method for locating damage in structures. *Transactions of the American Society of Mechanical Engineers. Journal of Vibration and Acoustics*, **122**: 324–329

[3] Sampaio R P C , Maia N M M Silva J M M 1999 Damage detection using the frequency response function curvature method. *Journal of Sound and Vibration*, **226**(5) p. 1029 1042

[4] Limongelli M P 2010 Frequency Response Function Interpolation for Damage Detection under Changing Environment. *Mechanical Systems and Signal Processing* **24**(8), p. 2898-2913

[5] Limongelli M P 2011 The interpolation damage detection method for frames under seismic excitation. *Journal of Sound and Vibration*, **330** p.5474–5489

[6] Limongelli M P 2014 Seismic health monitoring of an instrumented multistorey building using the Interpolation Method *Earthquake Engng. Struct. Dyn.* **43**(11) p.1581-1602

[7] Zhan Z, Martin C F 1997 Convergence and Gibbs' phenomenon in cubic spline interpolation of discontinuous functions *Journal of Computational and Applied Mathematics* **87** p. 359-371

On the monitoring and implications of growing damages caused by manufacturing defects in composite structures

M Schagerl, C Viechtbauer and S Hörrmann

Johannes Kepler University Linz, Institute of Constructional Lightweight Design,
Christian Doppler Laboratory for Structural Strength Control of Lightweight Constructions,
Altenberger Str. 69, A-4040 Linz, Austria

E-mail: `martin.schagerl@jku.at`

Abstract. Damage tolerance is a classical safety concept for the design of aircraft structures. Basically, this approach considers possible damages in the structure, predicts the damage growth under applied loading conditions and predicts the following decrease of the structural strength. As a fundamental result the damage tolerance approach yields the maximum inspection interval, which is the time a damage grows from a detectable to a critical level. The above formulation of the damage tolerance safety concept targets on metallic structures where the damage is typically a simple fatigue crack. Fiber-reinforced polymers show a much more complex damage behavior, such as delaminations in laminated composites. Moreover, progressive damage in composites is often initiated by manufacturing defects. The complex manufacturing processes for composite structures almost certainly yield parts with defects, e.g. pores in the matrix or undulations of fibers. From such defects growing damages may start after a certain time of operation.

The demand to simplify or even avoid the inspection of composite structures has therefore led to a comeback of the traditional safe-life safety concept. The aim of the so-called safe-life flaw tolerance concept is a structure that is capable of carrying the static loads during operation, despite significant damages *and* after a representative fatigue load spectrum. A structure with this property does not need to be inspected, respectively monitored at all during its service life. However, its load carrying capability is thereby not fully utilized.

This article presents the possible refinement of the state-of-the-art safe-life flaw tolerance concept for composite structures towards a damage tolerance approach considering also the influence of manufacturing defects on damage initiation and growth. Based on fundamental physical relations and experimental observations the challenges when developing damage growth and residual strength curves are discussed.

1. Introduction

Lightweight structures, such as aircrafts, spacecrafts, automobiles, wind energy plants, etc. are designed in line with safety concepts to optimally operate the structure within specified limits. The strength limits are given (i) by the static strength allowables, (ii) by the structural durability pertaining to fatigue and (iii) by the crash resistance referring to dynamic loads. Static strength includes the design against fracture, instability and detrimental deformation (e.g. not acceptable permanent deformation). Structural durability includes issues such as fatigue and damage initiation, damage growth and the residual strength of the damaged structure. This latter topic is the core of the damage tolerance concept.

The most advanced safety concepts for lightweight design can be found in civil aircraft engineering. Historically this development was often initiated by tragic accidents, followed by an immediate refinement of the certification specifications by the certification authorities in collaboration with the aircraft industry:

- The traditional safety concept is the *safe-life* approach, i.e. the design of a structure for a specific time (or for a number of events) in which there is a low probability that the strength degrades below the applied loading during operation.
- Particularly the accidents with the Comet 1 aircrafts showed that the pure safe-life concept is insufficient [1]. This simple safety concept may yield a design which has no load carrying capability at all if one structural member fails. The *fail-safe* concept introduced in 1956 requires that the structure retains a residual strength for a period of unrepaired use after failure or partial failure of a structural member. Accordingly, a fail-safe design is redundant and statically indeterminate.
- However, in 1977 the fail-safe concept failed [1]. A Boeing 707-300 lost its horizontal stabilizer in flight since cracks in the lug connections (which were designed redundantly) were not detected early enough. As a consequence the *damage tolerance* requirements were introduced in 1978. The major message of this safety concept is the acceptance that damages may always be present and need to be considered in design (until identification and repair). Figure 1 shows the classical picture illustrating this safety concept: The lower curve shows the increase of damage size due to applied loading conditions. The upper curve shows the according decrease of the structure's strength. The structure would fail if this residual strength falls below the design load (the maximum load level expected during operation [2]) in which case the damage would reach its tolerable level. This critical damage level may called fault. On the one hand, within the inspection program the damage needs to be detected in advance. On the other hand, with a specific inspection method, e.g. ultrasonic testing, it is only possible to detect a damage of a specific size. For this method the corresponding inspection interval is the time a damage grows from the detectable level to a fault. If the structure is inspected in these intervals, no damage can develop into a fault. The decision for a rougher (but more convenient and cheaper) inspection method, e.g. only visual inspection, naturally leads to a shorter inspection interval.
- In principal the above safety concepts focus on metallic structures where the damage is typically a simple crack at the surface. In contrast composite structures made from fiber-reinforced polymers show a rather complex damage behavior, for example delaminations or intra-laminar damages which are difficult to detect without sophisticated Non-Destructive Testing methods. The demand to simplify the inspection of composite structures has led to a comeback of the safe-life concept. The target of the so-called *safe-life flaw tolerance* concept is a structure that is capable of carrying the static loads during operation, despite significant defects and damages *and* after a representative fatigue load spectrum. A structure of this kind does not need to be inspected at all during its service life. A thorough formulation of this concept can be found in the certification specification CS-29 for large rotorcrafts [3].

The latter approach is pragmatic, practical, safe, thoroughly formulated and thus fulfills the current demands of industry and certification authorities. Nevertheless, it does not yield an optimized lightweight structure as the damage tolerance concept would offer. It should therefore be an aim to find an appropriate damage tolerance approach for composite structures.

However, with composites an issue arises, which complicates the introduction of a damage tolerance design concept. The complex manufacturing processes for composite structures almost certainly yield parts with defects, e.g. pores in the polymer matrix or undulations of fibers. Such defects may initiate growing damages after a certain time of operation which of course need to be considered by a refined damage tolerance approach.

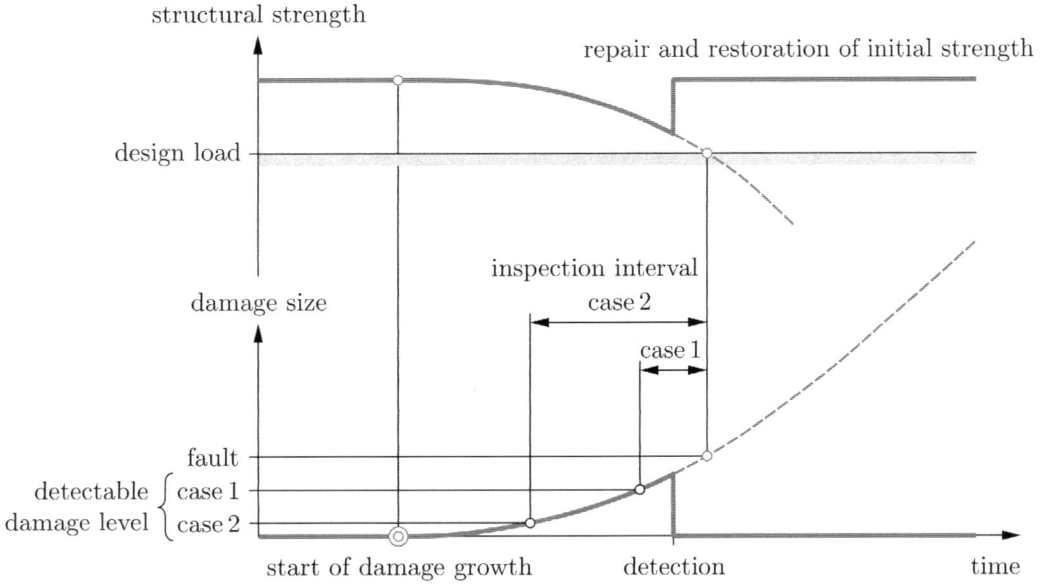

Figure 1. The concept of damage tolerance as introduced in the Federal Aviation Regulation 25 Amendment 45 by the Federal Aviation Administration in 1978.

2. Implications of manufacturing defects on the damage tolerance approach

As long as the structure is not totally damaged, its remaining life time can be predicted by damage growth considerations. As a result damage tolerance diagrams as shown in figure 1 follow. The development of such graphs is state-of-the-art in modern aircraft design for metal structures [4]. There are consequently two thrusts of current research: (i) the enhancement of the damage tolerance concept to consider also structures made from laminated composites (away from the currently accepted safe-life flaw tolerance approach described in section 1), and (ii) the enhancement of this concept for structural parts with manufacturing defects to consider not only structural damages (which occur during handling or use of the structure, as e.g. cracks, dents, or delaminations) but also defects (which are direct results of the manufacturing process, as e.g. pores or undulations).

For the latter enhancement we propose to derive modified damage tolerance diagrams as depicted in figure 2: The upper curves show the decrease of the structure's strength due to flaws, i.e. the initial reduction of the strength due to defects and the progressive degradation due to growing damages. These residual strength curves define the current allowable loading. Finding the upper curves requires methods of flaw assessment. The lower curves show the progress of damages, which are initiated by defects after a certain time of operation. These damage growth curves are derived from applied loading conditions. Finding the lower curves requires methods of flaw prognosis. The structure fails if its strength falls below the design load; a statistically determined maximum load expected in service and increased by deterministic safety factors (see [2]). The determination of the design load is thus a topic of safety requirements and legal regulations. If the decreasing strength of the structure reaches the threshold of the design load the damage reaches its tolerable level, i.e. it becomes a fault.

The idealized diagrams in figure 2 indicate that a small defect leads to a higher initial strength of the structure immediately after manufacturing which further leads to a higher maximum tolerable damage level during handling or use. The actual tolerable damage level within the structure's design life depends on the project-defined design requirements and is

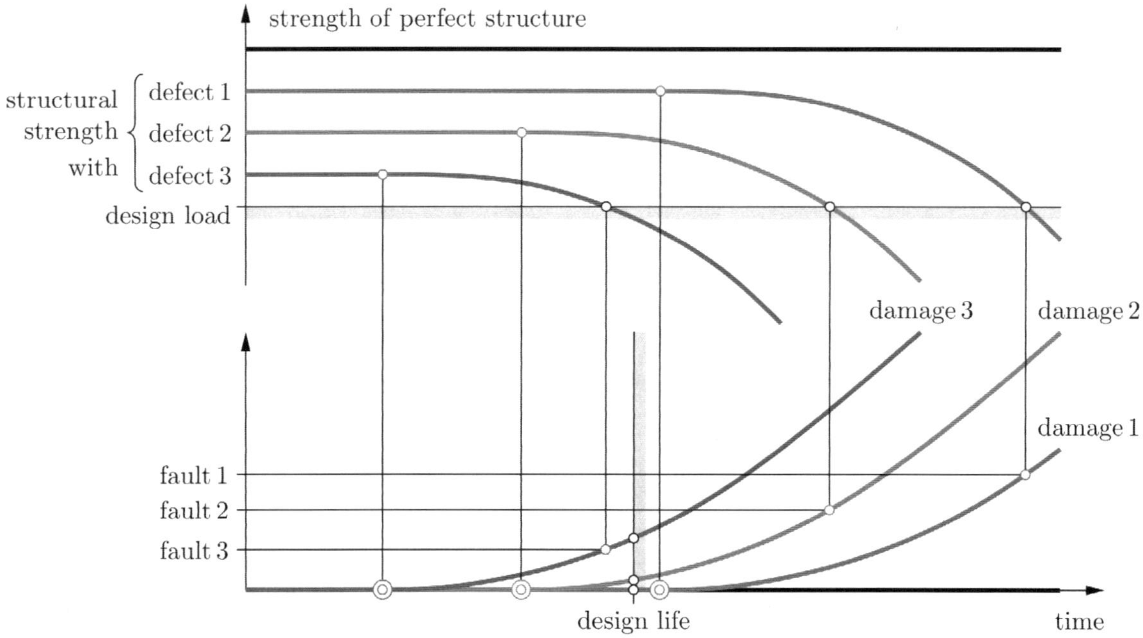

Figure 2. The effect of manufacturing defects on the concept of damage tolerance (cf. figure 1).

again a subject of safety requirements and legal regulations. For example, from the effects of the three manufacturing defects depicted in figure 2 one could draw the following conclusions:

- Defect 1 initiates damage 1 after the design life has passed. If no damage growth is allowed within the design life only such a defect level is acceptable.
- Defect 2 initiates damage 2. However, this damage does not develop into a fault within the design life. If damage growth is allowed defect 2 is therefore acceptable.
- Defect 3 initiates damage 3 which becomes a fault within the design life. This defect level is therefore not acceptable.
- Of course, there exists also a (low) defect level near a perfect structure which initiates no damage growth at all (in the sense of an endurance strength limit).

In this way design requirements yield assessment criteria to tolerate manufacturing defects. These criteria may then be listed in a concession catalog. However, developing such an extensive catalog can become rather complicated, as it must also cover the parallel occurrences of several different defects. Here a possible approach could be the introduction of a score system, assigning each defect a specific number of points. The structural part is then divided into zones within which a maximum number of points is allowed.

3. Stiffness degradation and residual strength of fiber-reinforced composites

Figures 1 and 2 hypothesize graphs on the progress of the residual strength if a structural damage grows. In the following we scrutinize this very idealized view. It is assumed that a damage is initiated and starts to grow at the manufacturing defect after a certain time of operation. This limit is called fatigue life and the according allowable load spectrum is called fatigue strength. However, it can be expected that some complex physical processes lead to this damage growth threshold if manufacturing defects are present. Actually, two reasons can be identified. At first the defect causes a finite stress concentration and the material reaches its therefore lowered fatigue strength under cyclic loading. If we think of the defect as a small hole in the structure (cf. figure 5) we formally can assign a notch factor to this stress concentration, say e.g. $\alpha = 3$. It is plausible and also indicated by the test results in figures 3 and 4 that the fatigue strength is lower than the initial static strength and that the fatigue strength decreases as the severity of a manufacturing defect increases. Damage initiation occurs if the cumulation of the far-field stress – oscillating with amplitude σ_a and concentrated by α – exceeds the endurance fatigue limit S_e of the undamaged (unnotched) material. For constant-amplitude cyclic loading the fatigue strength, i.e. the allowable stress amplitude S_a may show a linear dependency on the number N of cycles in a double-logarithmic diagram, i.e. $S_a = S_e\,(N/N_e)^{-1/k}$. This so-called line of Wöhler can be typically drawn for metals for high cycle fatigue loading with good accuracy. Also some composite materials allow this simplification by accepting a higher variance from this line [5].

If the so introduced damage is a simple crack or a delamination the second physical reason which finally causes the damage to grow is a stress intensity at the tip of this damage. Damage grow occurs if the range ΔK of the oscillating stress intensity factor is larger than a threshold value ΔK_{th}, i.e. larger than this endurance limit for crack grow. According to simple linear fracture mechanics the range of the stress intensity factor at the crack tip is formally $\Delta K = \Delta\sigma\, Y(a)\sqrt{\pi a}$ where a is the length of the crack and $Y(a)$ is a function defined by the crack geometry and the location of the crack. When exceeding the threshold value ΔK_{th} the damage will grow following some constitutive relation. This finally yields curves as sketched in the lower diagrams of figures 1 and 2. Paris' simple law $da/dN = C\,(\Delta K)^m$ may be applied for cracks in metals but is certainly insufficient for progressive damage of composite materials. The damage mechanisms in composites are typically much more complex than a simple crack or a delamination; particularly when manufacturing defects are present. For example, progressive damage can be initiated and advanced by microscopic inter-fiber matrix cracking.

The crucial point now is that the parameters of both theories, initial damage *and* initiation of damage growth are determined by different test setups and test specimens. For fatigue strength tests small coupons are used. For damage growth tests simple structures are used which are sufficiently large to observe stable damage propagation. However, to draw finally curves as in the lower diagrams of figures 1 and 2 the theories of damage initiation and of the following progressive growth of the damaged area shall match consistently.

For example, consider an oscillating stress concentration with constant amplitude $\alpha\sigma_a$ at the manufacturing defect. The fatigue strength $S_a = \alpha\sigma_a$ of the material is reached after N cycles and a significant macroscopic, i.e. measurable or even visible crack is introduced. Let a_0 be the length of this initial crack which now causes a stress intensity. The far-field stress level to consider for the crack tip is the stress influenced, i.e. concentrated by the defect. The determining stress range for crack grow is thus $\Delta\sigma = 2\alpha\sigma_a$ near the defect. For an initial crack length a_0 (and therefore a crack tip distance) which is much larger than the size of the defect the stress range reduces to $\Delta\sigma = 2\sigma_a$. A following grow requires then $\Delta K = 2\sigma_a Y(a_0)\sqrt{\pi a_0} > \Delta K_{\text{th}}$. Since $S_a > S_e$ the endurance limits for both damage growth reasons, initial stress concentration and following stress intensity, should therefore be – at least – related by

$$\Delta K = 2\,\frac{S_e}{\alpha}\,Y(a_0)\sqrt{\pi a_0} \approx \Delta K_{\text{th}} \tag{1}$$

For metals this condition is typically quite well fulfilled with reasonable values a_0 determining the transition between crack initiation and crack growth. For example, consider the aluminum alloy 2024-T3 which has the material limits $S_e = 104\,\text{N/mm}^2$ and $\Delta K_{\text{th}} = 87\,\text{N/mm}^{3/2}$ for cyclic tension-tension loading $R = 0$ [6]. Assuming the geometrical values $\alpha = 3$ and $Y(a_0) = 1$ relation (1) yields a transition crack length of at least $a_0 = 0.5\,\text{mm}$. For composite materials much more constitutive parameters are involved which will certainly complicate an according

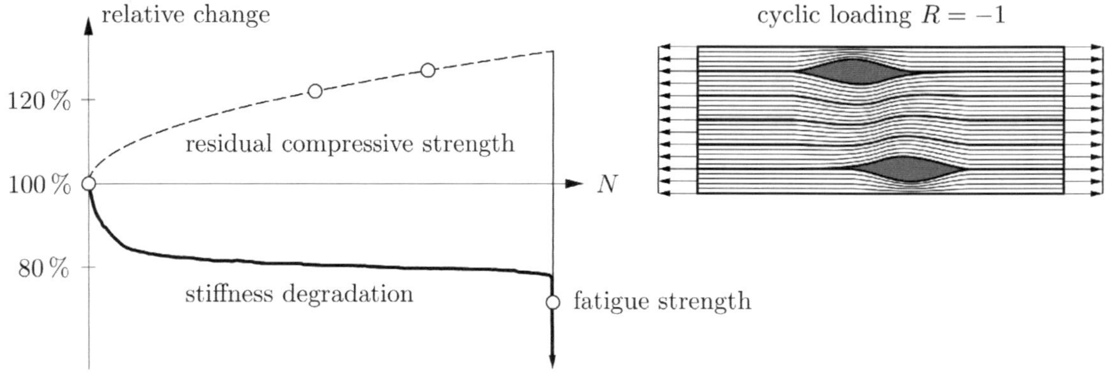

Figure 3. Stiffness degradation and residual strength of specimens with a defect (out-of-plane undulation). The stiffness degradation is measured continuously during the fatigue strength test whereas markings correspond to the single residual strength measurements.

analysis. However, equations similar to (1) may be derived to check consistency and to match the theories of damage initiation and of the following progressive growth of the damaged area.

Another aspect concerns the residual strength curves, i.e. the development of the static strength after a certain fatigue load spectrum. The upper diagrams in figures 1 and 2 indicate that the initial strength is lower for a more severe manufacturing defect and that the residual strength further decreases as the damage grows. This assumed behavior is based on the experience with metallic materials where the residual strength of a cracked structure is associated with the fracture toughness K_c. If the stress intensity factor $K = \sigma Y(a)\sqrt{\pi a}$ at the crack tip is larger than this material parameter, $K > K_c$, unstable crack grow sets in. Since the stress intensity factor K increases as the damage, i.e. the crack length a increases, the residual strength, i.e. the allowable stress level σ to reach structural failure decreases. Although this conclusion sounds plausible it is particularly for composite materials not necessarily true.

Figure 3 presents test results obtained with specimens having an artificially introduced defect. The test specimen is a standard coupon with 2 mm × 25 mm cross-section and with six uni-directional layers, i.e. all fibers are pointing in so-called 0° loading direction. Between the outer layers deliberately introduced resin deposits force an out-of-plane undulation of the fibers. Figure 3 shows the degrading stiffness of the material during a cyclic fatigue loading (in the current case a tension-compression loading $R = -1$). As the number N of cycles increases the curve passes three regions which is typical for this kind of material (cf. also figure 4). At first the curve shows a strong stiffness decrease, then a linear decrease (here on a linear scale and not on logarithmic one), and finally again a strong decrease towards total failure of the specimen. The constant load amplitude applied in this test is the fatigue strength for the gained number of cycles until total failure. In figure 3 this value is plotted in relation to the initial strength under static loading.

Figure 3 shows also a few residual strength measurements. It shall be noted that drawing a residual strength curve is a laborious procedure. After a certain number of fatigue load cycles the fatigue test is stopped and a static strength test follows. Whereas the whole stiffness degradation curve is already shown by one specimen, each measurement point along the residual strength curve needs one specimen to be spent. However, the measurements in figure 3 indicate that remarkably the residual strength increases as the number of cycles, i.e. the fatigue loading increases. An explanation for this partly surprising behavior may give the tests presented by [7] which results are summarized in figure 4. Analogously, specimens of 38 mm width but having a hole of 9.5 mm diameter undergo a cyclic fatigue loading with a following residual strength test.

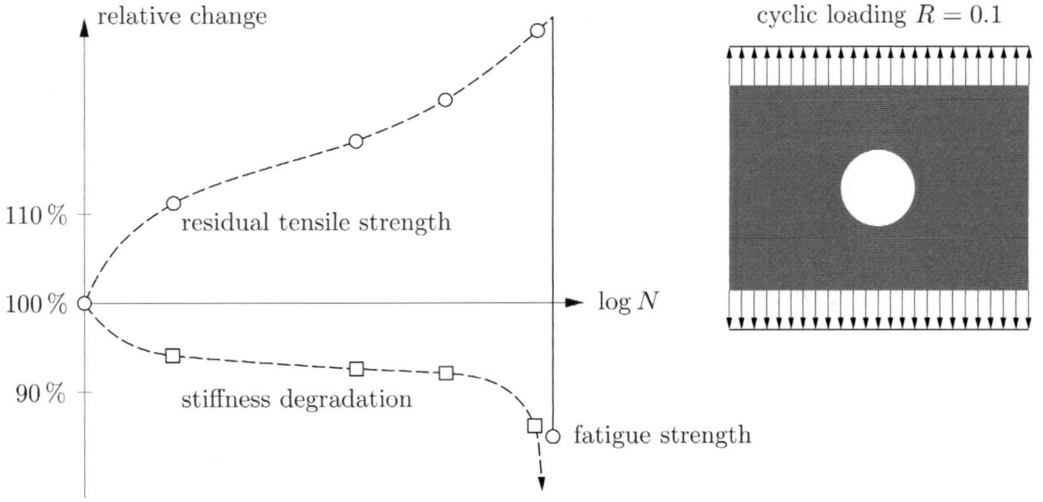

Figure 4. Stiffness degradation and residual strength of notched specimens according to the test results presented in [7]. Markings correspond to the measurements.

Both, the stiffness degradation curve and the residual strength curve show a similar behavior than in figure 3. The reason is found out by X-ray radiography in [7]. The stress concentration at the hole increasingly deteriorate the material around the hole as the fatigue loading continuous (cf. figure 5). This massively damaged area acts like a bigger hole which in turn yields a lower stress concentration during the following residual strength test. Similarly the fatigue loading at the defect depicted in figure 3 may cause a local deterioration of the material at points of stress concentration which further results in a relief of this highly stressed regions in the following residual strength test.

The results presented in the preceding paragraph reveal a diametrical different strength behavior of metals and fiber-reinforced composites under cyclic and static loading. On the one hand, as depicted in figure 5, ductile metals react on high static loading with yielding at points of stress concentration. The cross-section does not fail until all the material yields. Typically, fiber-reinforced composites do not show this damage tolerant behavior under single static loading. Particularly in fiber direction they show a nearly linear-elastic behavior until failure. Static loading causes the brittle fibers to fracture at points of stress concentration. The load carrying capacity of the whole cross-section is then suddenly reached. On the other hand, cyclic fatigue loading causes crack initiation in metals. The stress intensity at the crack tip further reduces the residual strength dramatically. Contrarily, the deterioration of composite materials in areas of stress concentration may yield an advantageous relief of the material in a following high static loading. Nevertheless the composite material is damaged after fatigue loading. The nominal increase of the residual strength in figures 3 and 4 indicates rather a certain damage tolerance than showing a potential for technical use, i.e. weight saving. Moreover, for many lightweight applications as e.g. automotive structures, the determining design criteria is the structural stiffness, which definitely decreases as the fatigue loading continues.

4. Conclusions

The brief discussion of the currently applied safety concepts for composite aircraft structures in section 1 concludes that the state-of-the-art safe-life flaw tolerance concept is not weight-optimal. It is suggested to enhance the damage tolerance approach which originally was developed for metallic materials (cf. figure 1). Thereby it is also emphasized that such a modification requires

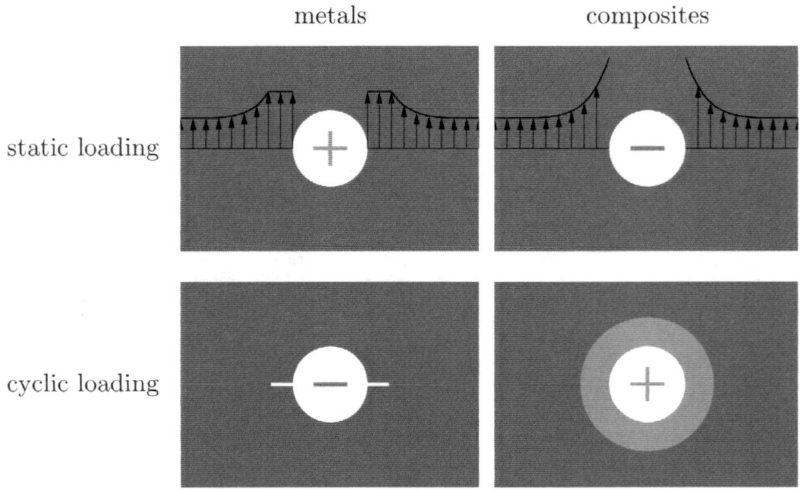

Figure 5. Notch sensitivity of metals and composites under static and cyclic loading.

definitely the consideration of manufacturing defects, which are in many cases the origin of subsequently growing damages. Section 2 presents how this modification can be introduced in principal (cf. figure 2) and how design guidelines with such an enhancement could work.

However, the presented damage tolerance diagrams show a very idealized view on the effects of manufacturing defects and growing damages. Apart from the difficulty to identify, respectively define damage initiation the discussions in section 3 reveal that the static strength and residual strength behavior of metals and fiber-reinforced composites is diametrical different (cf. figure 5). In fact, the residual strength of composite materials may increase after a certain cyclic fatigue loading (cf. figures 3 and 4). These results show that residual strength measurements need a careful interpretation, first when drawing conclusions regarding the implications of growing damages caused by manufacturing defects and second when drawing damage tolerance diagrams to develop monitoring strategies.

Acknowledgments

The financial support by the Austrian Federal Ministry of Science, Research and Economy and the National Foundation for Research, Technology and Development is gratefully acknowledged. The authors thank Erich Humer for performing the structural tests presented in figure 3.

References

[1] Wanhill R J H 2003 Milestone case histories in aircraft structural integrity *Comprehensive Structural Integrity* vol 1 Structural Integrity Assessment – Examples and Case Studies ed Milne I, Ritchie R and Karihaloo B (Elsevier Science Ltd., Kidlington) pp 61–72

[2] Cuntze R 2010 Essential topics in the determination of a reliable reserve factor *Handbook Structure Analysis* vol HSB 01100-02 A (Industry Committee for Structural Analysis IASB)

[3] European Aviation Safety Agency 14 November 2003 Certification Specifications for Large Rotorcraft CS-29

[4] European Aviation Safety Agency 17 October 2003 Certification Specifications for Large Aeroplanes CS-25

[5] Vassilopoulos A P and Keller T 2011 *Fatigue of Fiber-reinforced Composites* (Springer, London)

[6] Schwarmann L and Hoang V N 1986 Fatigue strength and crack propagation rates of aluminum alloy 2024-T3 *Handbook Structure Analysis* vol HSB 63111-01 F and HSB 63205-01 C (Industry Committee for Structural Analysis IASB)

[7] Kress G R and Stinchcomb W W 1985 Fatigue response of notched graphite/epoxy laminates *Recent Advances in Composites in the United States and Japan, ASTM STP* vol 864 pp 173–196

Nonlinear dynamic response analysis of localized damaged laminated composite structures in the context of component mode synthesis

S Mahmoudi, F Trivaudey and N Bouhaddi(*)

Applied Mechanics Department, FEMTO-ST institute - UMR 6174, University of Franche-Comté-24, chemin de l'Épitaphe, 25000 Besançon, France

E-mail: (*) noureddine.bouhaddi@univ-fcomte.fr

Abstract. The aim of this study is the prediction of the dynamic response of damaged laminated composite structures in the context of component mode synthesis. Hence, a method of damage localization of complex structures is proposed. The dynamic behavior of transversely isotropic layers is expressed through elasticity coupled with damage based on an existing macro model for cracked structures. The damage is located only in some regions of the whole structure, which is decomposed on substructures. The incremental linear dynamic governing equations are obtained by using the classical linear Kirchhoff-Love theory of plates. Then, considering the damage-induced nonlinearity, the obtained nonlinear dynamic equations are solved in time domain. However, a detailed finite element modelling of such structure on the scale of localized damage would generate very high computational costs. To reduce this cost, Component Mode Synthesis method (CMS) is used for modelling a nonlinear fine-scale substructure damaged, connected to linear dynamic models of the remaining substructures, which can be condensed and not updated at each iteration. Numerical results show that the mechanical properties of the structure highly change when damage is taken into account. Under an impact load, damage increases and reaches its highest value with the maximum of the applied load and then remains unchanged. Besides, the eigenfrequencies of the damaged structure decrease comparing with those of an undamaged one. This methodology can be used for monitoring strategies and lifetime estimations of hybrid complex structures due to the damage state is known in space and time.
Keywords: *nonlinear dynamic, damage prediction, structural monitoring, composite structures.*

1. Introduction

Over the last decades, composite materials have become increasingly used in industry, especially in high-tech products such as the defense and aerospace industries. This wide use of these materials is due to thier excellent mechanical characteristics, particularly the high values of *rigidity/weight* and *strength/weight* ratios. As an example, in the new project of Airbus, about 53% of the total weight of the A-350 aircraft is made of composite materials. Comparing to metallic materials, composite structures are characterized by a sufficient strength with low density. In addition, they have other characteristics such as their resistance to corrosion and to chemical attack. The heterogeneous nature of these composite materials can provide a source of damage mechanism growing if the composite structure is subjected to a sufficient load. Indeed, the ruin of composite materials is not due to a single mechanism but is the consequence of the accumulation of several modes of degradation (matrix cracking, delamination, fiber breakage

...). So, the knowledge and the understanding of these mechanisms become a necessity during the design phase. It is then necessary to investigate the dynamic behavior of these composite structures taking into account the damage phenomenon and the effect of this mechanism on the overall behavior of the system. Thus, detection of the location and especially the degree of damage severity is of great importance in order to ensure the reliability and safety of service structures since the static or the dynamic responses can be modified by the damage evolution. To characterize this change in mechanical properties and in the dynamic characteristics, various damage indicators are used such as natural frequencies [1]. Kulkarni *et al* [2] have studied the presence and the detection of delamination in a composite structure based on the decrease of eigenfrequencies. This idea was further investigated and deepened by Cawley *et al* [3] suggesting that the shift of eigenfrequencies can be considered as a basis for a new nondestructive control technique. Moreover, there are other indices which can be used to study this relationship between the damage and the decrease in mechanical properties, such as flexibility matrix [4] and mode shapes [5].
The main objective of this paper is the construction of a new numerical modeling in order to predict the dynamic response and the damage evolution of laminated composites structures. Hence, a method of damage localization of complex structures is proposed in context of component mode synthesis. Based on a phenomenological macro model for cracked composites structures, the dynamic behavior of transversely isotropic layers is expressed through elasticity coupled with damage. The studied structure is made of transversely isotropic layers of polymer matrix renforced with long fibers and it is decomposed in undamaged and damaged substructures. The damage is expressed by a single scalar variable and its evolution is determined by the principal of maximum dissipation. Then, using the classical Kirchhoff-Love theory, and assuming that damage induces nonlinearity, the resulting nonlinear formulation is implemented in MATLAB® software. Component Mode Synthesis (CMS) is introduced for modelling a nonlinear fine-scale damaged substructure which should be updated at each iteration due to the damage evolution. This nonlinear modelling is then connected to a linear dynamic models of the remaining substructures which can be condensed without updating in order to reduce the computational cost. Numerical results show that the mechanical properties of the structure highly change when damage is taken into account. Under an impact load, damage increases and reaches its highest value with the maximum of the applied load and then remains unchanged. Besides, the eigenfrequencies of the damaged structure decrease comparing with those of an undamaged one. This methodology can be used for monitoring strategies and lifetime estimations of hybrid complex structures.

2. Problem formulation: finite element modeling

The finite element method is currently a very reliable engineering tool as a result of its flexibility and its relative ease in numerical implementation. There are several alternative techniques of modeling by finite elements of the composites structures. In this context, several theories were extended from modeling isotropic structures towards modeling composite structures. These formulations are essentially divided into two categories; those which are based on the Equivalent Single Layer Theory and those which are based on the Layer-Wise theory . The first category includes mainly the Classical Laminate Theory, the First-order (FSDT) and High-order (HSDT) Shear Deformation Theories. While the second category includes the Independent and Dependent layers theories. These formulations are more detailed in several works including [6, 7]. The primary difference between FSDT and HSDT theories is the order the polynomial function adopted for approximating the displacement field variables of the structure. The choice of one or other of these two formulations is mainly conditioned by the thickness of the structure to be modeled. For thin composite structures, it is recommended to use the FSDT theory which has a lower computational cost comparing to the HSDT theory. So, the displacement field of a composite plate can be written as follow:

$$\{\underline{\mathbf{u}}(x,y,z,t)\} = \left\{ \begin{array}{c} u(x,y,z,t) \\ v(x,y,z,t) \\ w(x,y,z,t) \end{array} \right\} + z \left\{ \begin{array}{c} \varphi_y(x,y,t) \\ -\varphi_x(x,y,t) \\ 0 \end{array} \right\} \quad (1)$$

where $\varphi_x(x,y,t)$ and $\varphi_y(x,y,t)$ are the rotation, respectively, with respect to the $\vec{x}$ and $\vec{y}$ axes. Equation (1) shows that the finite element model should include the degrees of freedom in translations (u,v,w) as well as rotations $\varphi_x(x,y,t)$ and $\varphi_y(x,y,t)$, *i,e* five degrees of freedom per node. The finite element approximation of the global displacements can be expressed in terms of nodal displacements $\{u_n^e\}$ by introducing the matrix $[N(\xi,\eta)]$ of interpolation function such that:

$$\{u\} = [N(\xi,\eta)]\,\{u_n^e\} \quad (2)$$

The quadrangular element Serendip Q8 with eight nodes is used to mesh the structure. Using virtual work principle where E_p and E_c are respectively the total potential and kinetic energies of the structure, then:

$$\delta \int_{t_1}^{t_2} (E_p + E_c)dt = 0 \quad (3)$$

The Gauss integration method with 4 integration points and a single point in the thickness direction, is used to obtain the global mass and stiffness matrices $[M]$ and $[K]$. After admitting the assumption of proportionality of the damping expressed by $[B] = \alpha_1[K]+\alpha_2[M]$, the dynamic equation of motion is obtained as shown in (4) where D is a scalar variable representing the relative reduction of the transverse Young's modulus and it will be characterized later.

$$[M]\ddot{u} + [B]\dot{u} + [K(D)]u = F(t) \quad (4)$$

Damage plays an important role in dynamic response modification of composite materials. The elastic behavior of the matrix is modified by the formation and evolution of micro cracks and cavities during the loading F. These defects cause primarily the degradation of the structure

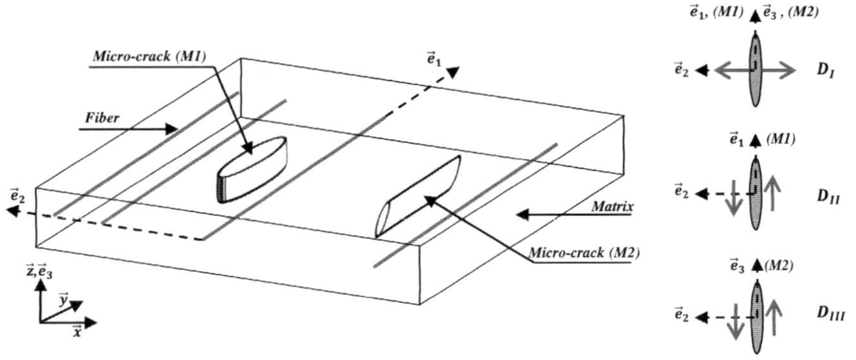

Figure 1. *Micro-cracks orientation in the matrix [8]*

stiffness through the decrease of three elastic modulus E_2, G_{12} and G_{23}, according to the stress state. In order to caracterise the reduction in the transverse Young's modulus E_2 and the shear module G_{12} and G_{23}, three parameters D_I, D_{II} and D_{III} are introduced such as:

$$D_I = \frac{\Delta E_2}{E_2} \quad ; \quad D_{II} = \frac{\Delta G_{12}}{G_{12}} \quad ; \quad D_{III} = \frac{\Delta G_{23}}{G_{23}} \tag{5}$$

Using a self-consistent method [9] which permits to express D_{II} and D_{III} as a function of D_I, a damage matrix $[H(D)]$ is fully expressed by a single scalar variable D representing the relative reduction of the transverse Young's modulus:

$$[H(D)] = \begin{bmatrix} 0 & 0 & 0 & 0 & 0 & 0 \\ 0 & H_{22} & 0 & 0 & 0 & 0 \\ 0 & 0 & 0 & 0 & 0 & 0 \\ 0 & 0 & 0 & H_{44} & 0 & 0 \\ 0 & 0 & 0 & 0 & 0 & 0 \\ 0 & 0 & 0 & 0 & 0 & H_{66} \end{bmatrix} \quad ; \quad \begin{cases} H_{22} = \dfrac{D}{1-D} S_{22} \\ H_{44} = \dfrac{D}{\sqrt{1-D}} \sqrt{S_{11} S_{22}} \\ H_{66} = \dfrac{D}{\sqrt{1-D}} S_{22} \end{cases} \tag{6}$$

Where S_{ii} depict the flexibility components of $[S]$ of the undamaged material. The changement of the mechanical properties is visible in equation (7) where the $[H(D)]$ is added to the flexibility matrix to describe the damaged elastic behavior of the composite structure.

$$\varepsilon = [S + H(D)] : \sigma \tag{7}$$

The damage evolution law is expressed trough the thermodynamics of irreversible processes. A thermodynamic force Y associated to the damage is obtained, $Y = \frac{1}{2}\sigma^T \left(\frac{\partial H}{\partial D}\right) \sigma$, where σ^T is the transpose of the stress tensor σ and $\left(\frac{\partial H}{\partial D}\right)$ is the derivative of $[H(D)]$ with respect to D. A threshold function $\bar{Y}$ is defined in order to record the damage level and adapt the threshold of damage activation to the damage level. Thereby, the following charge function is established:

$$f_d = Y - \bar{Y} = \frac{1}{2}\sigma^T \left(\frac{\partial H}{\partial D}\right) \sigma - (Y_c + qD^p) \tag{8}$$

where the constants Y_c, q and p are damage parameters.

3. Component mode synthesis (CMS): Craig-Bampton method

The numerical costs related to computing time and the important data storage for complex systems are important. So, looking for models with a small number of degrees of freedom, called Reduced Order Model, is a major issue which reduces the computation time and the storage space. CMS approch is commonly used in complex finite element analysis models whose consist of an assembly of reduced models. In this paper, the studied structure is composed of damaged and undamaged components and its nonlinear finite element modelling has a large size leading to an important computational costs. For these reasons, an approach by substructuring of the whole complex structure using the Craig-Bampton method [10] is used as a technique of model reduction where the structure is decomposed in two substructures as shown is figure 2. Indeed, this method permits to limit the impact of a modification to a considered component. The method requires first to treat each component separately and determine its reduction basis. In order to perform the Craig-Bampton method, the model is divided into two sets of degrees of freedom (DOFs). The DOFs of interface with subscript "j" and the interior DOFs with subscript "i", *i.e:* $q^k = \left\{ \begin{matrix} q_i^k \\ q_j^k \end{matrix} \right\}$ where the exposant k denotes the number of the substructure. Then, a

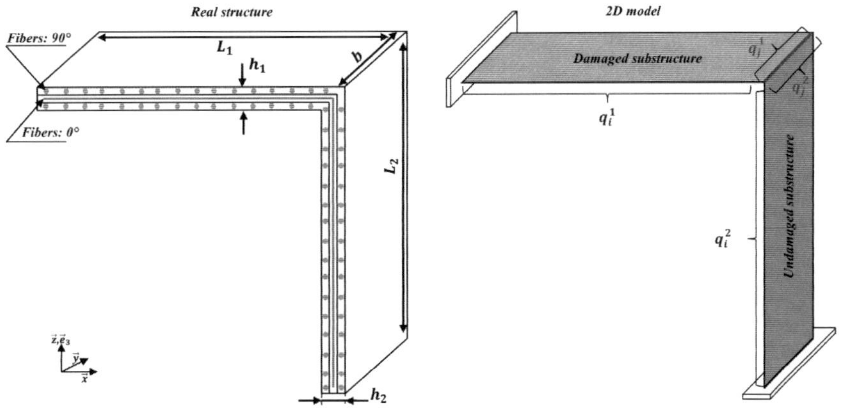

Figure 2. Real composite structure and its decomposition for the 2D finite elment reduced model

reduction basis is generated consisting of normal modes and constrained modes. Normal modes are modes of the substructure where all nodes of the interfaces with adjacent substructures are blocked. A constrained mode is defined by the static modes obtained from moving one interface DOF by one unit, while holding all other interface DOFs fixed. The mass and stiffness matrices of each substructure are partitioned as following:

$$M^k = \begin{bmatrix} M_{ii} & M_{ij} \\ M_{ji} & M_{jj} \end{bmatrix}^k \quad ; \quad K^k = \begin{bmatrix} K_{ii} & K_{ij} \\ K_{ji} & K_{jj} \end{bmatrix}^k \quad ; \quad B^k = \begin{bmatrix} B_{ii} & B_{ij} \\ B_{ji} & B_{jj} \end{bmatrix}^k \tag{9}$$

The component fixed-interface normal modes φ^k_{ii} are obtained by solving the generalized aigenvalue problem:

$$\left[K^k_{ii} - w^2 M^k_{ii}\right]\left[\varphi^k_{ii}\right] = 0 \tag{10}$$

The interface constrained modes matrix ψ^k_c is :

$$\psi^k_c = \begin{bmatrix} \psi_{ij} \\ I_{jj} \end{bmatrix}^k = \begin{bmatrix} -K^{-1}_{ii} K_{ij} \\ I_{jj} \end{bmatrix}^k \tag{11}$$

So, the Craig-Bampton transformation matrix T^k_{CB} is defined as follow where the truncated basis Φ_{ii} contains the first retained normal modes φ^k_{ii}:

$$T^k_{CB} = \begin{bmatrix} \Phi_{ii} & \psi_{ij} \\ 0 & I_{jj} \end{bmatrix}^k \tag{12}$$

Then, the reduced component mass, stiffness and damping matrices are:

$$M^k_{CB} =^T T^k_{CB} M^k T^k_{CB} \quad ; \quad K^k_{CB} =^T T^k_{CB} K^k T^k_{CB} \quad ; \quad B^k_{CB} =^T T^k_{CB} B^k T^k_{CB} \tag{13}$$

Finally, the reduced model is formed of the assembly of the components models which are condensed separately.

4. Solving

Noting that the damage phenomenon is irreversible, the damaged state is kept in memory when the structure is subjected to a load. For this reason an incremental way in solving process is recommended. So, at each time iteration, a stress increment $\Delta\sigma$ is applied. Then, the accumulation of stress continues until the function f_d becomes positive which implies the creation of a damage increment ΔD in order to kept the function $\dot{f}_d = 0$. It will be obtained by solving the consistence equation $\dot{f}_d$ using the Newton-Raphson method. Expressing $\dot{f}_d$ as a function of ΔD and knowing the damage D_i at $(i)^{th}$ iteration, the solution of $\dot{f}_d = 0$ is reduced to seek the damage increment ΔD which cancels the function $f_{d,i+1}$ for the $(i+1)^{th}$ iteration $i.e$ Eq(14):

$$f_{d,i+1} = \frac{1}{2}\left[\sigma_{i+1}\, H'(D_i + \Delta D)\, \sigma_{i+1}\right] - \left[Y_c + q(D_i + \Delta D)^p\right] \tag{14}$$

With $H'(D_i + \Delta D)$ denotes the derivative of the damage matrix $[H(D)]$ with respect to D and expressed at $(D_i + \Delta D)$.

Once the damage increment is calculated, the Craig-Bampton reduction method is applied. The undamaged substructure is condensed and keeps its initial properties. Whereas, the stiffness of the damaged substructure depends on the damage level D obtained at each time iteration. The number of retained modes of the reduced substructures is chosen so that the frequency band is twice of the frequency band of interest. The Craig-Bampton reduction is applied and only the stiffness matrix of the damaged substructure is updated to take into account of this change of material properties. The Newmark implicit predictor-corrector scheme [11] is used to solve the resulting dynamic nonlinear problem in time domain.

5. Results and discussion

Several numerical simulations have been performed in order to highlight the influence of damage on the dynamic response. The geometrical and mechanical properties of the considered laminated structure are given in table 1. The structure is made of three layers oriented as

Table 1. Geometrical and mechanical properties of the laminated beam

Elastic modulus E_1 $(\vec{e}_1)$	45680 MPa
Elastic modulus E_2 $(\vec{e}_2)$	16470 MPa
Shear modulus G_{12}	6760 MPa
Poisson ratio ν_{12}	0.34
Poisson ratio ν_{23}	0.34
Y_c	0.0027 MPa
q	1.246 MPa
p	0.816
L_1 : $(\vec{x})$	0.3 m
L_2 : $(\vec{z})$	−0.3 m
b : $(\vec{y})$	0.03 m
Thickness $(\vec{z})$:h_1	0.001 m
Thickness $(\vec{x})$:h_2	0.001 m

$(90°/0°/90°)$, it is clamped at $(x = 0, y, z = 0)$ and at $(x = 0.3, y, z = -0.3)$. It is subjected to a distributed impact in $\vec{x}$ direction and along the side of junction between the two beams with $F = 1917.5\,N$ as magnitude and $\tau = 1\,ms$ as a duration.

In order to validate the chosen CMS method, a comparison in terms of eigenfrequencies was made as depicted in Figure 3. The black bars show the eigenfrequency difference in percent between the eigenfrequencies f_{red} and f_{unred}, respectively, of the reduced and the unreduced

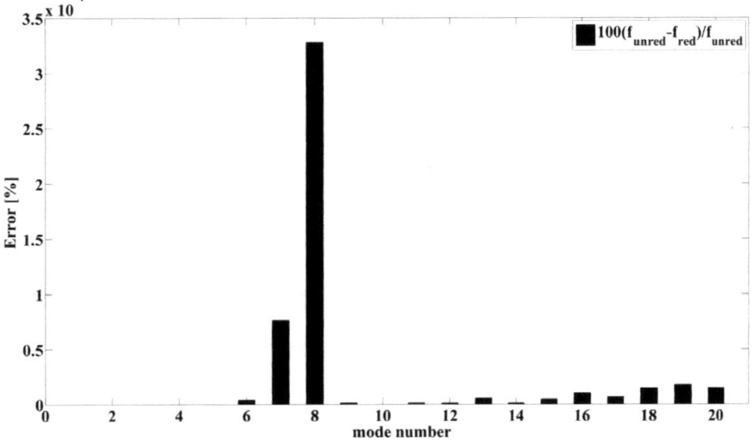

Figure 3. Error in percent of the eigenfrequencies between reduced and unreduced (full) models

(full) models. Figure 3 shows a larger eigenfrequency difference in the 7th and 8th modes which are respectively a torsion and a membrane vibration mode. The largest difference in these frequencies is lower than 3.5 10^{-4}% which permits to validate the CMS method.
The nodal displacement with respect to $\vec{x}$ direction of the junction side are depicted in figure 4 where the displacement amplitude is smaller in the undamaged case than in the damaged one. As a consequence, damage has an important impact on the dynamic response since it affects the structure stiffness. The damage state in the layer oriented as 90° of the damaged substucture is depicted in figure 5-(b), where the damage is localized and quantified according to the (x,y) coordinates . According to figure 5-(a) and (b), it can be seen that there is a spatial correlation between the stress propagation and the damage state. The stress and the damage are maximum at ($x = 0.25m$) where there is significant bending during impact load. The effect of damage is more visible in figure 6 when the eigenfrequencies f_D and f_0 respectively on the damaged and undamaged cases are compared. As a result, it can be seen that when the damage is taken into account, the eigenfrequencies have a significant decrease which reaches up to 10% for some modes.

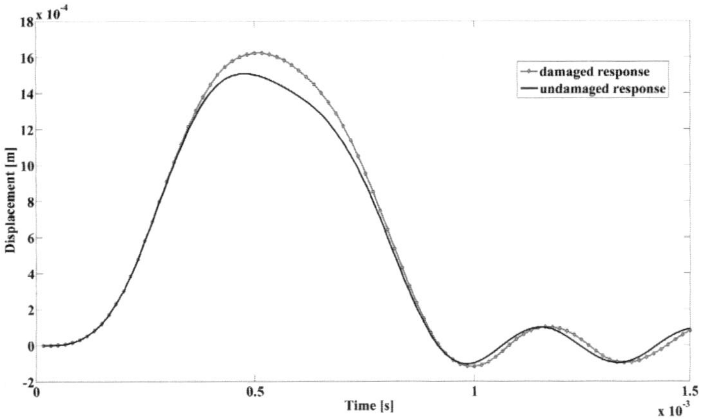

Figure 4. Dynamic impulse response in the $\vec{x}$ direction of the nodes of the junction side

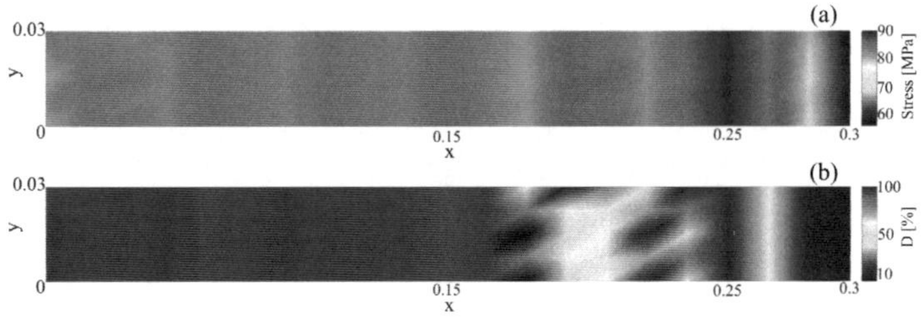

Figure 5. Stress and damage states in the first layer oriented as 90° of the damaged substructure: (a):propagation of the stress in the $\vec{x}$ direction when the load is maximum, (b):damage state

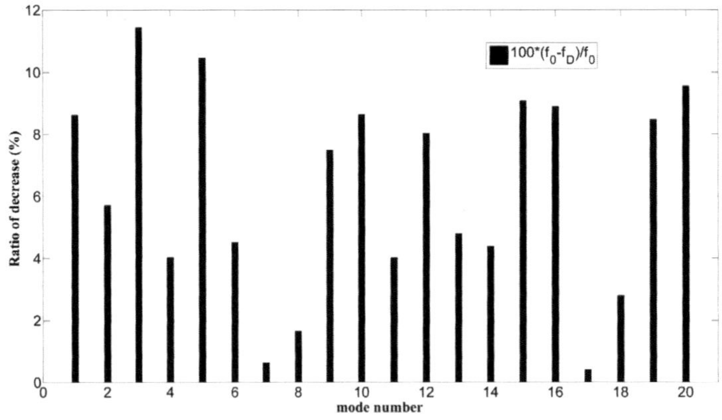

Figure 6. Ratio of decrease between the eigenfrequencies, f_D and f_0, respectively of damaged and undamaged cases

6. Conclusion

The non-linear dynamic response of a locally damaged composite structure composed of two laminated beams was investigated including the material nonlinearity generated by damage. Thus, based on a phenomenological macro model for cracked composites structures, a method of damage prediction and localization of complex structures is proposed and investigated in the context of CMS . Hence, the Craig-Bampton reduction method is introduced as a CMS technique. The resulting nonlinear modelling of the damaged component is connected to the condensed linear modelling of the remaining substructure reducing the computational cost. Several numerical simulations have been performed to highlight the effect of the damage on the dynamic behavior, particularly, the eigenfrequencies and the displacement responses. Thereby, the damage modifies significantly the dynamic response by the increasing the displacement response amplitude. Indeed, the eigenfrequencies decrease when the damage is taken into account due to the modification of the material properties of the damaged substructure. These modifications are introduced into the dynamic analysis during loading. Therefore, this methodology can be used in life time estimations and monitoring strategies of the complex structures.

[1] Vestroni F and Capecchi D 2000 *Journal of Engineering Mechanics* **126** 761-768
[2] Kulkarni S V and Frederick D 1971 *Journal of Composite Materials* **5** 112-119
[3] Cawley P and Adams R D 1979 *Journal of Composite Materials* **13** 161-175
[4] Duan Z, Yan G, Ou J and Spencer B 2005 *Journal of Sound and Vibration* **284** 455-466
[5] Au F T K, Cheng Y S, Tham L G and Bai Z Z 2003 *Journal of Sound and Vibration* **259** 1081-1094
[6] Reddy J N 1997 *Mechanics of laminated composites plate: Theory and Analysis* (2.ed. Florida: CRC Press)
[7] Berthelot J M 1992 *Matériaux composites:comportement mécanique et analyse des structures* (Masson)
[8] Boubakar M L, Trivaudey F, Perreux D and Vang L 2002 *Composite Structures* **58** 271-286
[9] Perreux D and Oytana C 1993 *Composites Engineering* **3** 115-122
[10] Bampton M C C and Craig R R 1968 *AIAA Journal* **6** 1313-1319
[11] Géradin M and Rixen D 1997 *Theory and Application to Structural Dynamics* (Wiley)

A comparative study of a self strain-monitoring carbon nanotube film and carbon fibers under flexural loading by electrical resistance changes

B Pinto[1], S Kern[1], J J Ku-Herrera[2], J Yasui[1], V La Saponara[1] and K J Loh[1]
[1]University of California, Davis, One Shields Ave, Davis, CA 95616, USA
[2]Centro de Investigación Científica de Yucatán A.C., C.P. 97200, Mérida, Yucatán, México

E-mail: vlasaponara@ucdavis.edu

Abstract. With the increased use of composites for load-carrying structures, the ability to obtain strain and damage information is critical to maintain reliable structures in the field. A promising class of multifunctional composites with the ability to self-sense strain and/or damage is possible through the use of carbon fibers and carbon nanotubes. This paper presents a comparative study of two sensors for fiber-reinforced polymer composites: the sensors are the carbon fibers themselves, and a non-structural carbon nanotube (CNT) film applied through spray deposition. The changes in resistance of the sensors are compared under monotonic and cyclic flexural loading. Within the limits of this study and the current CNT sensor configuration, the carbon fibers are shown to have a higher sensitivity for strain and damage sensing. However, the CNT film appears to track the performance of GFRPs reasonably well for tensile strains.

1. Introduction

The application of fiber-reinforced polymer (FRP) composite materials has impacted a wide variety of engineering structures, due to their low weight and tailorability to support specific loading conditions. With composites becoming increasingly common, low-cost fabrication processes have been developed in industry. However, one remaining challenge is the ability to sense and monitor damage within a composites structure. This challenge is further complicated by several complex failure mechanisms, including fiber fracture, delamination, matrix cracking and interlaminar shear failures, e.g. [1]. Unlike isotropic materials such as metals and polymers, these failures can easily go unnoticed in inspections routines, as these failure modes may not be discernable to the naked eye.

In order to address this challenge, several techniques and strategies have emerged to monitor strain and damage within a composite structure. The method of interest in this paper is to utilize electrically conductive constituent materials, with potentially piezoresistive qualities, to yield composites with multifunctional capabilities. "Multifunctional" refers to structural and non-structural functionality, such as sensing/activation, energy harvesting, and electrical conductivity, [2]. For the case of flexural loadings, electromechanical coupling between strain and surface resistance measurements can be obtained for conductive materials. Traditionally, metal foil strain gages give localized strain information. However, some of their limitations are that they can only be applied to the surface of a composite, and can only cover a limited sensing area. The application of multifunctional composites could provide strain and/or damage sensing capabilities that are not limited to the elastic range, as a composite could be monitored from its elastic regime until rupture. Through the use of carbon fibers and carbon-based nanomaterials, a much larger sensing area could be incorporated within a structure to utilize this multifunctional capability.

Carbon fibers (CF) have been widely used in the aerospace industry since the 1970s, and are slowly emerging also in wind turbines, for rotor diameters larger than 45 meters. One of the first investigations of the electrical properties of CF was completed by Owston, [3], who showed that CF reinforcements have the unique property of being intrinsically conductive. Later, studies monitoring resistance changes during monotonic tension, [4, 5], and cyclic flexure, [6, 7], were performed to demonstrate these sensing capabilities.

Carbon-based nanomaterials, in particular carbon nanotubes (CNT), have been utilized in several composite architecture configurations to create electrically conductive composites. CNT have been used to monitor damage when they are dispersed into matrix systems, [8-10], deposited onto glass fibers [11], and layer by layer fabrication [12, 13]. In this paper, a CNT-based thin film is investigated as a sensor, [14], this time under flexure loading. Films like this have been validated for uniaxial monotonic and cyclic tensile piezoresistivity, [15]. One of the main advantages of this film, as compared to other CNT composite architectures, is the ease of processing and scalability, through which it can be applied to existing structures' surfaces.

Most structures and beams are not subjected to pure uniaxial loadings. A far more common load experienced by structures and beams is bending. The work in this paper will compare resistance changes of a CNT thin film to the demonstrated self-sensing ability of carbon fibers, for FRP specimens under flexural loading. Resistance changes of both the tension and compression side of a coupon are monitored simultaneously.

2. Materials and methods

2.1. Composite specimen preparation

A vacuum-assisted resin transfer method was utilized to fabricate carbon fiber-reinforced polymer (CFRP) coupons, using carbon unidirectional fibers (Torayca T700SC 12K, Soller Composites, NH, USA). The CF have an electrical resistivity of 1.6×10^{-3} Ω cm with a filament diameter of 7 μm. In order to investigate the sensing properties of the CNT films, glass fiber-reinforced polymer (GFRP) composites were also prepared, with unidirectional E-glass fibers (Vectorply E-LR 0908, AL, USA). A LAM125/LAM237 epoxy/hardener system (Pro-set, MI, USA) was infused to obtain the CFRP and GFRP coupons, using a 100:28 resin:hardener ratio. Before infusion, the epoxy mixture was degassed for 25 min, to remove air bubbles and reduce voids within the coupons. Both composite laminates were cured at room temperature for 14 hours, and then post-cured at 82 °C for an additional 8 hours in a convection oven. Individual coupons were cut to size with a water-cooled diamond saw, to nominal dimensions of 183 mm length × 25.4 mm width. The layup orientation was $[0]_{13}$ for the CFRP samples and $[0]_{10}$ for the GFRP samples, with the zero degrees plies aligned with the span of the samples. The number of layers for both fiber architectures was chosen to maintain a 40:1 support span to thickness ratio following "Procedure B" of ASTM D7264 [16].

2.2. CNT thin film fabrication

The carbon nanotube (CNT) film applied by spray deposition in this study has been successfully employed as a sensor in GFRP samples, [14]. The constituents of this film can be categorized into three components: ink, latex, and water solution. The ink consists of 1 wt.% of multiwall carbon nanotubes (MWCNT) (SMW 200, SouthWest NanoTechnologies), 2 wt.% of poly(sodium 4-styrenesulfonate) (PSS) (~1M M_W, Sigma-Aldrich) and a small amount of N-methyl-2-pyrrolidinone (NMP). MWCNTs form an electrically conductive network within the film. The PSS helps to decrease contact resistance among adjacent MWCNTs while maintaining dispersion of the nanotubes. In lieu of the MWCNT being functionalized through a chemical treatment, the NMP solvent is used to improve dispersability of the MWCNT in the solution. This ink solution is prepared by first adding the PSS into deionized water, then adding the MWCNT, and finally adding the NMP to help aid dispersion. Then, the MWNT/PSS/NMP mixture is sonicated for 60 min using an S-450D Branson tip sonicator operating at 225 W and 20 kHz. After the MWCNT are dispersed in the solvent solution, a poly(vinylidene fluoride) (PVDF) latex-based solution is added to the mixture, to promote adhesion with the FRP coupons. Finally, the MWCNT/PSS/NMP/PVDF mixture is sprayed on the coupons using an airbrush inside a fume hood, as seen in Figure 1. For CFRP samples only, an additional epoxy layer was applied to the cured sample before the spraying of the CNT sensor, in order to ensure insulation of the electrical properties between the CNT film and the conductive carbon fibers. The film was left to dry at room temperature for one hour. Since the coupons were to be subject to flexural loading, the CNT thin film was sprayed onto both sides of the coupons to monitor their tension and compression responses.

Figure 1. Spray deposition of CNT film onto glass fibers.

2.3. Test setup and instrumentation

An MTS 810 hydraulic test frame with a 244 kN load cell was used to apply force on a four-point bending fixture (WTF-FL-52 Wyoming Test Fixtures, USA), as shown in Figure 2a. A quarter point loading was utilized with a support span of 152.4 mm and loading span of 76.4 mm. Monotonic flexural loading was applied at a displacement rate of 1 mm/min. Cyclic flexural loading was applied at 0.01 Hz with a constant deflection amplitude of ±1.5 mm.

The instruments that measured the electrical resistance from the electrodes applied to the coupons (bottom and top) consisted of two digital multimeters (DMM) (Agilent 34401A and 34410A). An internal 4-wire setup was implemented to account for contact resistance between the electrodes and the sensors and the resistance of attached wires (see Figure 2b). Each DMM collected data for a single channel using Agilent's Benchvue data acquisition software. Each DMM was allocated for sensing one of the samples' sides (one DMM for tension, one for compression). Cross-head displacement and load *versus* time were collected from the MTS, while both DMMs collected resistance values *versus* time. Post-processing of the data via linear interpolation correlated stress, strain and fractional resistance changes, meant as $(R-R_o)/R_o$, where R_o is the baseline value.

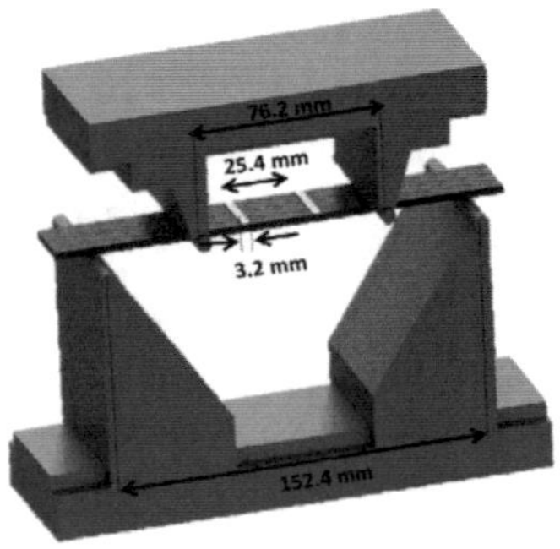

a)

b)

Figure 2. a) Sketch of 4-pt fixture, with support span, loading span, and electrode dimensions, b) instrumented CFRP sample.

2.4. Electrode application
In order to monitor changes in electrical resistance, electrodes were applied directly to both types of sensors (CF and CNT film). Surface-based measurements were utilized in order to separately monitor the strain and damage due to compression and tension loads at the edges of the samples' depth. For the CF sensor, the surface of the coupons was slightly sanded prior to application of the electrodes in order to remove the epoxy matrix and expose the carbon fibers. Electrically conductive silver paint (Ted Pella) was applied over the coupons to create a conductive path among the CF tows, while silver epoxy (MG Chemicals 8331G) was used to attach 30 AWG copper wires. For the CNT film sensor, silver epoxy alone was used to create a conductive path between the film and the copper wire electrodes.

3. Results and discussion
3.1. Baseline resistance measurements
After instrumenting the samples with electrodes, baseline resistance values were recorded using a 4-wire measurement. Baseline resistance values of carbon fibers were on the order of ohms, while those of the CNT sensor were on the order of hundreds of ohms, for an electrode distance of 25.4 mm.

3.2 Monotonic flexural loading
Stress and fractional resistance changes *versus* strain are shown in Figure 3 for a CFRP sample with CF sensor. Above and below the neutral axis of the samples are regions dominated by tensile and compressive strains. Surface resistance values are shown to increase with strain linearly up until failure on the specimen side with tensile strains, as seen in Figure 3a. At the failure point of the specimen, indicated by a sharp drop in the sample's stress, a corresponding drop in resistance can be seen in Figure 3a. Typically, fiber rupture will cause resistance to increase under tension [4]. However, under flexural loading, a CFRP composite fails in compression before tension. The side with compressive strains yields a nonlinear decrease in resistance values initially until approximately 0.3% strain, upon which the decreasing trend of resistance values changes directions. It is hypothesized that this transition, where resistance begins to increase, could be an early indicator of damage within the sample. The increase in fractional resistance in Figure 3b would indicate that the CFRP composite is capable to sense its own strain and damage.

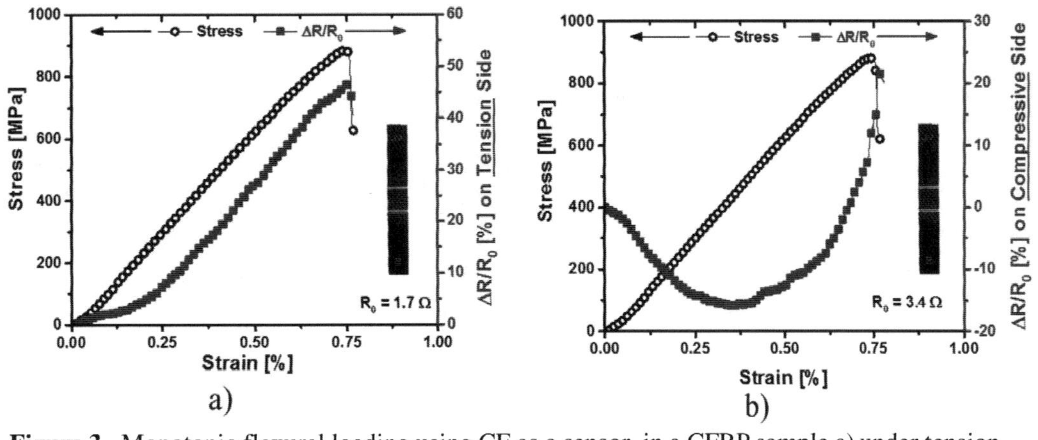

Figure 3. Monotonic flexural loading using CF as a sensor, in a CFRP sample a) under tension, b) under compression.

The gauge factor, K, i.e. the reversible change in fractional resistance changes per unit strain value, was classified as the region between 0.25% to 0.5% strain, for this study. For the few CFRP samples with CF sensors tested, K values were within reasonable agreement with the literature [17]. Figure 4 shows the response of the CNT sensor in one CFRP sample. There was noise in the tension side. The change in resistance was particularly inconsistent on the compressive side, among the samples. K was of the order of 1-3 under tensile strains, comparable to those of metal foil strain gauges, hence this sensor in its current configuration and CNT concentration exhibits a lower sensitivity to strain and damage with respect to the CF sensor.

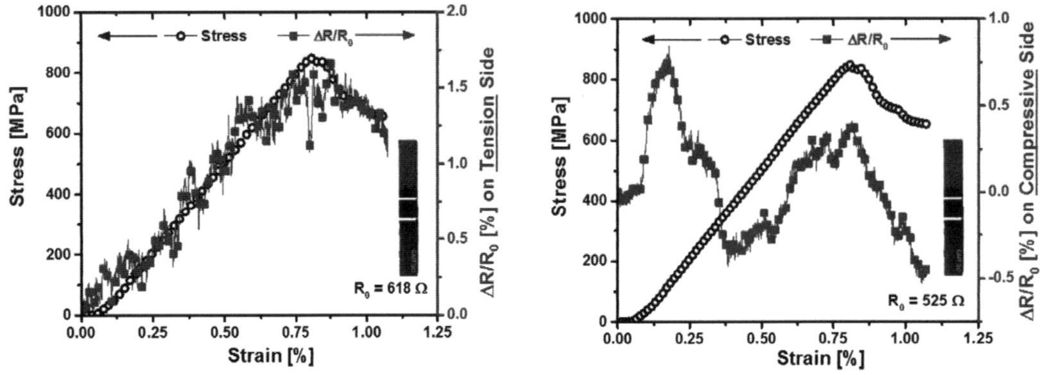

a) b)

Figure 4. Monotonic flexural loading using the CNT film as a sensor, in a CFRP sample a) under tension, b) under compression.

To further study the sensing behavior of the CNT film under flexure, the film was also applied to GFRP samples. Figure 5 show similar trends on the tensile and compression side to trends found using CF as a sensor in CFRP specimens as seen in Figure 4. However, the maximum fractional change in resistance is found to be much lower. This results in the CNT film appearing less sensitive to strain induced from flexural loading. For the compressive side with the CNT film, again there is a transition point from decreasing to increasing resistance changes. In this case, the transition point is assumed not to be caused by damage within the specimen. However, it is possible that, at high strain levels, the film could be prone to damage, causing this transition in resistance changes.

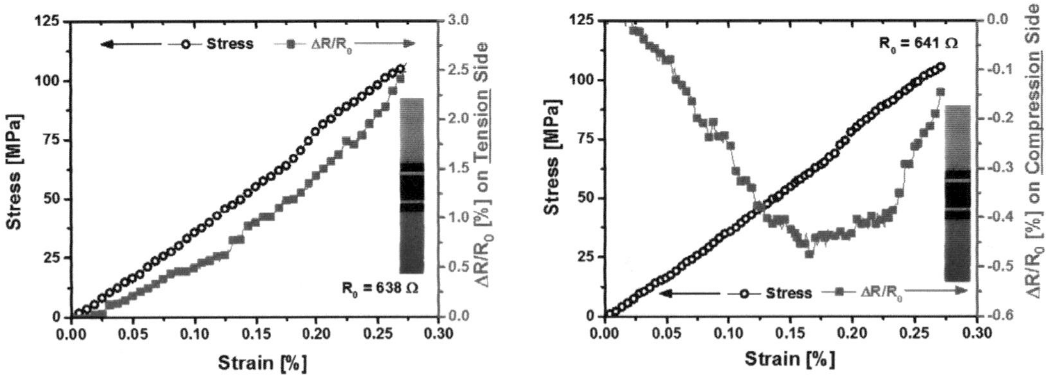

a) b)

Figure 5. Monotonic flexural loading using a CNT film as a sensor, in a GFRP sample a) under tension, b) under compression.

3.3. Cyclic flexural loading

Cyclic flexural loading plots of displacement and fractional resistance changes *versus* time are shown for the CF (Figure 6) and the CNT sensor applied to GFRP (Figure 7). The samples are loaded for 5 cycles at three different displacement amplitudes. Resistance values before loading are shown at time *t* before zero, in order to indicate the amount of noise before mechanical loading.

The CF sensor under cyclic loading follows the same trends under monotonic loading for tensile and compressive strains. While the fractional resistance peaks decrease with loading cycles, the amplitude between peaks and troughs are consistent throughout all the cycles. At the third displacement amplitude of 9 mm, the fractional resistance changes no longer follow the trends found under monotonic loading. This may be attributed to samples' damage and/or the viscoelastic behavior of the resin.

For the CNT film sensor applied to GFRP samples, it can be seen that the resistance increases with loading and decreases while unloading. The amplitude of the fractional resistance changes indicates that there are no residuals within the film after repeated cycles. Comparing Figure 6 and Figure 7, the CNT films yields lower fractional resistance changes than the CF sensor. Under cyclic flexure, the CNT film applied to GFRP samples exhibits a response similar to that of Figure 7.

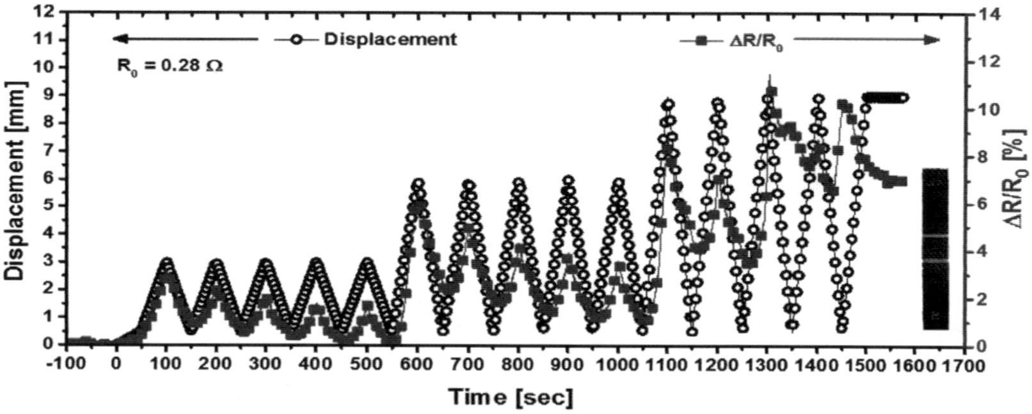

Figure 6. Cyclic flexural loading with increasing displacement amplitudes using CF as a sensor in a CFRP sample.

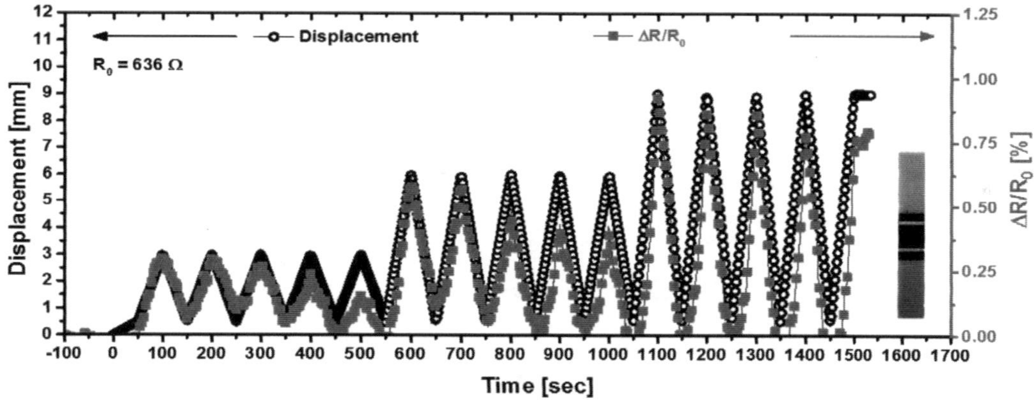

Figure 7. Cyclic flexural loading with increasing displacement amplitudes using a CNT film as a sensor in a GFRP sample.

4. Summary, conclusions and future work

CNT films and CF were investigated as resistance-based sensors under monotonic and cyclic flexural loading, with electrodes applied to the tensile and compressive surfaces in each sample. For the CNT films applied to GFRP samples and the CF in CFRP samples, resistance values increased with tensile strains and decreased with compressive strains. Under compressive strains, a transition point of decreasing-to-increasing resistance trends were found for both the CNT film on GFRP samples and the CF sensor. For CNT films applied to CFRP samples, the response exhibited a lot of scatter, and no straightforward, common trend was found among the tested samples. Overall, the CF exhibits a much higher sensitivity of resistance changes under both tensile and compressive strains. Furthermore, the CF multifunctional ability is shown to include both strain and damage sensing of the samples. However, the CNT film may allow the structural health of non-conductive GFRP to be monitored under flexure.

Future work will further investigate the performance of the CNT film in terms of its CNT concentration, to better tune the sensor. These structural health monitoring methods will also be applied to ~1.2 m long carbon/epoxy I-beams with different orientations in the flanges and web, for a proof of concept applied to an actual structure.

Acknowledgements
The authors acknowledge the support of the National Science Foundation (grant number CMMI-1200521 to KJL and VLS). VLS also acknowledges the partial support of the University of California Institute for Mexico and the United States (UC MEXUS) and CONACYT, and of the Office of Naval Research (grant N00014-13-1-0604, managed by Dr. Y. Rajapakse). Insightful technical discussions with Dr. Francis Avilés, Centro de Investigación Científica de Yucatán (Mexico), are appreciated. Finally, the authors thank Dr. Ricardo Castro (Materials Science and Engineering, UC Davis) for the loan of electrical equipment.

References
[1] Barbero E J 2010 *Introduction to Composite Materials Design* 2nd edition (CRC Press)
[2] Gibson R F 2010 *Compos Struct* **92** 2793
[3] Owston C N 1970 *J Phys D-Appl Phys* **3** 1615
[4] Schulte K and Baron C 1989 *Compos Sci Technol* **36** 63
[5] Todoroki A and Yoshida J 2004 *JSME Int J A* **47** 357
[6] Abry J C *et al* 1999 *Compos Sci Technol* **59** 925
[7] Wang S K and Chung D D L 2006 *Carbon* **44** 2739
[8] Kostopoulos V *et al* 2009 *J Intell Mater Syst Struct* **20** 1025
[9] Thostenson E T and Chou T-W 2008 *Nanotechnology* **19** 6
[10] Yesil S *et al* 2010 *Mater Sci Eng A*-Struct **527** 7340
[11] Ku-Herrera J J *et al* 2014 *Nanosci Nanotechnol Lett* **6** 932
[12] Loyola B R *et al* 2010 *J Mater Sci* **45** 6786
[13] Loh K J *et al* 2008 *J Intell Mater Syst Struct* **19** 747
[14] Loyola B R et al 2013 *IEEE Sens* J **13** 2357
[15] Thostenson E T *et al* 2001 *Compos Sci Technol* **61** 1899
[16] ASTM D7264 / D7264M-07 Standard Test Method for Flexural Properties of Polymer Matrix Composite Materials 2007 *ASTM International* (West Conshohocken, PA)
[17] Xiaojun W and Chung D D L 1996 *Smart Mater Struct* **5** 796

Impact damage resistance of thin stitched carbon/epoxy laminates

L Francesconi and F Aymerich

Department of Mechanical, Chemical and Materials Engineering
University of Cagliari. Via Marengo, 2 - 09123 Cagliari, Italy.

E-mail: lucafrancesconi@unica.it

Abstract. The study examines the influence of through-thickness stitching on the damage response of thin cross-ply carbon/epoxy laminates subjected to low-velocity impacts. Instrumented impact tests were carried out on unstitched and polyethylene stitched laminates and the resulting damage was assessed in detail by X-radiography analyses. The results of the observations carried out during the experimental analyses are illustrated and discussed to identify the mechanical role played by through-thickness reinforcement and to highlight the influence of the laminate layup on the impact resistance of stitched laminates.

1. Introduction

Delamination, i.e. separation between layers that occurs by failure of the resin rich interlaminar interface, is the most common failure mode in laminated composites, as a consequence of their inherent weakness in the thickness direction due to the lack of out-of-plane fibrous reinforcement [1]. The presence of internal delaminations may reduce considerably, especially under compressive loads, the structural integrity and the load bearing capacity of composite structures [2]. Delaminations induced by low-velocity impacts are of primary concern, since impact damage is often difficult or impossible to detect by visual inspection of the surface (Barely Visible Impact Damage – BVID) Preventing the initiation of delaminations and delaying their propagation are thus key strategies for the improvement of the reliability and lifetime of laminated composite structures.

Various approaches have been proposed in the last decades to improve delamination resistance of composite laminates [3-7]. They include the increase of interfacial strength between fibre and matrix, the adoption of high-toughness resins, the insertion of adhesive or nano-modified films between layers (interleaving), the introduction of through-thickness reinforcement by stitching or Z-pinning. Among them, the insertion of translaminar (Z direction) reinforcements by stitching has proven to be a successful strategy for the enhancement of the interlaminar properties in composite laminates.

Stitching is a technique which consists in sewing a continuous reinforcing thread through an uncured pre-preg laminate or fabric pre-form before the consolidation or impregnation phase [5]. As demonstrated by numerous studies [8,9], the bridging mechanism provided by stitches across delamination cracks reduces significantly relative opening and tangential displacements of the delaminated layers, thus limiting the interlaminar stresses at the crack tip and increasing the intrinsic resistance of the material to delamination growth. It should be remarked, however, that the stitching process inevitably perturbs the original material architecture in the vicinity of stitches, inducing localized defects such as resin rich regions, fibre misalignment and fibre breakage that may act as stress concentration and damage initiation sites and seriously degrade in-plane and flexural properties of the base laminates.

A number of studies have been conducted on the effect of stitching on the impact response of composite laminates [10-17]. However, because of the influence of many factors, such as thread and composite materials, stitching density and pattern, laminate layup and thickness, manufacturing procedures, the general role played by stitches in controlling the damage response of impacted laminates has not yet been completely characterized and described.

In this study, a series of experimental analyses have been carried out to investigate and compare the behaviour of thin unstitched and stitched cross-ply laminates under low-velocity impact. The main findings of the analyses are illustrated and discussed to assess the effect of through-thickness stitches on damage evolution, delamination resistance and structural performance of the examined laminates.

2. Experimental

Cross-ply $[0_2/90_2]$s laminated panels were manufactured using unidirectional carbon/epoxy prepreg tapes (Seal Texipreg® HS160/REM). Before curing, some of the panels were stitched using twisted polyethylene rovings (3x200 dtex Dyneema® SK60) by an industrial sewing machine. The panels were stitched perpendicularly to the 0° direction of the laminate, and both the distance between consecutive stitches along the stitching line (stitch step) and that between two adjacent stitching lines (stitch row spacing) were kept at 5 mm. A modified lock stitch, where the intersection between the upper needle thread and the lower bobbin thread occurs at the surface of laminate was adopted in order to reduce distortion within the laminate. Both unstitched (base) and stitched panels were vacuum bagged and cured in a hydraulic hot press under a pressure of 5 bar. The temperature curing cycle consisted of an initial 1.5 °C/min ramp followed by a 2 h hold at 125 °C and a ramp down to room temperature at a rate of 1 °C/min. The final thickness of cured panels was 1.2 mm. No significant difference in thickness was observed between unstitched and stitched panels after curing.

Type A and type B rectangular samples 65 mm × 87.5 mm in size were finally obtained for impact testing by cutting the consolidated plates with the orientations schematically indicated in Figure 1 with reference to stitched samples. As visible in Figure 1, type A and type B samples were cut with their long sides respectively parallel and perpendicular to the 0° direction of the laminated panel. With respect to the direction of the long side of the samples, the layups of the two classes of samples were thus $[0_2/90_2]$s (type A) and $[90_2/0_2]$s (type B).

Impact tests were conducted using an instrumented drop-weight testing machine with a 2.28 kg impactor, equipped with a hemispherically-ended rod of 12.5 mm in diameter. The velocity of the impactor immediately before and after the impact was obtained by an infra-red sensor, while the contact force between impactor and specimen was measured by means of a semiconductor strain-gauge bridge bonded to the impactor rod. The velocity and the displacement of the impactor were obtained by integration of the contact force signal.

During impact, the specimens were supported by a steel plate having a rectangular cutout 45 mm × 67.5 mm in size. The rectangular samples were always placed on the support with their long sides parallel to the long sides of the cutout.

Stitched and unstitched laminates were impacted at the central point of the sample with energies ranging between approximately 1 J and 9 J, obtained by varying the drop height of the impactor. The corresponding impact velocities ranged between 0.9 m/s and 2.8 m/s. All stitched samples were impacted on the side where the needle thread loops with the bobbin thread to form the stitch. Impact damage was characterized by penetrant enhanced X-radiography and by direct observation of the surfaces of the samples.

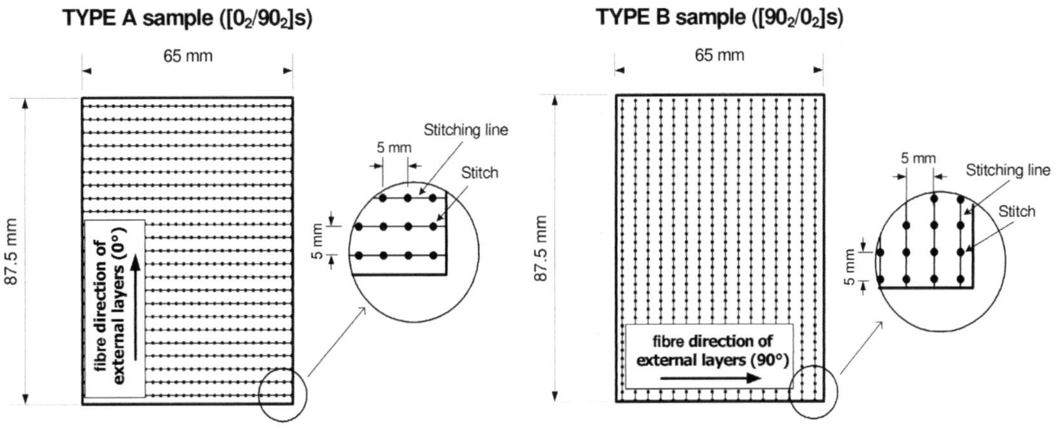

Figure 1. Stitching pattern of Type A and Type B samples

3. Experimental results and discussion

Figure 2 show force-time histories and force-displacement curves for unstitched and stitched type A samples impacted at two impact energies. It is seen that force-displacement plots are highly nonlinear, with a continuous increase in stiffness up to a deflection of about 3 mm, as a direct consequence of the low thickness of the laminate [17]. Unstitched and stitched samples exhibit analogous impact responses, with no evident irregularities in the force trace, for impacts with energies below a threshold value of about 6 J. When the impact energy exceeds this threshold, a major force drop is observed at a force level of about 3 kN in both unstitched and stitched laminates (Figure 2).

Figure 3 shows representative radiographs of damage induced by impact in unstitched and stitched type A laminates for two levels of impact energy (5 J and 7 J). Damage starts at impact energies lower than 1 J and consists of bending and shear matrix cracks that develop respectively in the bottom 0° layers and in the middle 90° layers; a delamination is then triggered at the intersection between matrix cracks on the 90°/0° interface closer to the tensile side of the sample. With increasing impact energies, the delaminated area grows in size, mainly propagating along the fibre orientation of the lower ply of the interface (0 direction) with a characteristic two-lobed shape. For impact energies higher than 6 J, the delaminated area of unstitched samples experiences a sudden and unstable growth towards the edges of the sample, finally inducing significant fibre fracture on the 0° layers. In stitched samples, major fibre breakage is again observed on the 0° layers, but the delaminated area measured after impact is significantly smaller than in unstitched samples. The sharp load drop recorded during the impact tests (see the plots of Figure 2) corresponds to the unstable growth of delamination and the onset of major fibre damage mechanisms occurring in the samples.

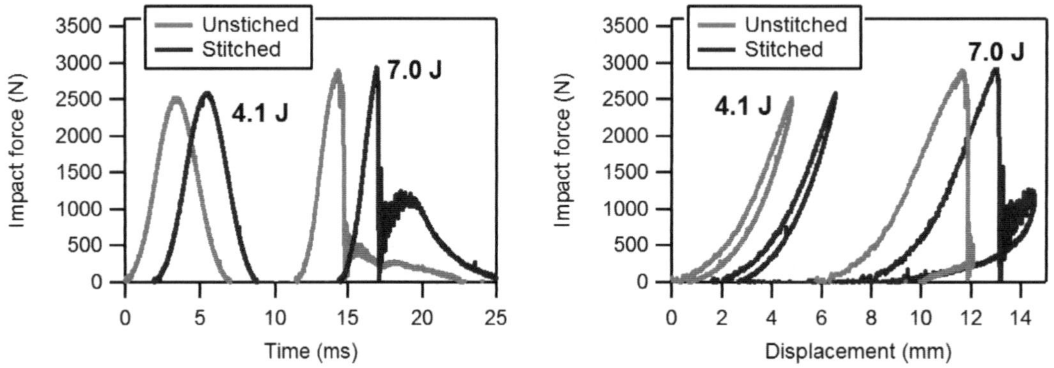

Figure 2. Force-time (left) and force-displacement (right) curves for unstitched and stitched type A samples

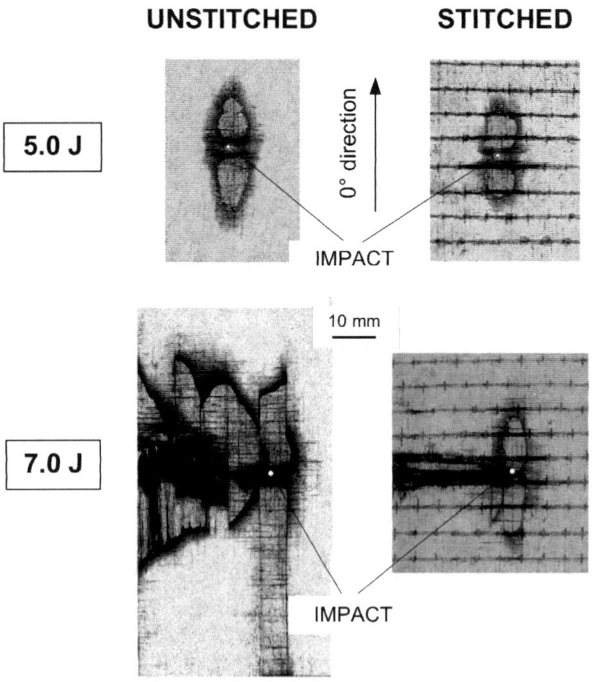

Figure 3. X-radiographs of impact damage in unstitched and stitched type A samples

The analysis of the damage patterns as obtained by X-radiography indicates that, while stitching does not modify the basic mechanisms of impact damage, it substantially reduces the area of the delamination induced by impact. The graph of Figure 4, which plots the values of the projected delaminated area measured after impact as a function of impact energy, shows that the delaminated areas of stitched samples are, on average, approximately 70% those of unstitched samples for impacts with energies up to 6 J; the reduction of the delamination size is even larger for impacts with energies higher than 6 J, with delaminated areas of stitched samples roughly one-half those of unstitched samples.

The influence of stitching on the damage resistance of type A laminates may be further examined with reference to the impact response under high-energy impacts. When the threshold 3 kN load for unstable damage propagation is exceeded, in unstitched laminates the delaminated area first propagates - in the form of a delaminated strip oriented along the 0° direction - up to the short side of the sample, and afterwards extends across the sample width along the 90° direction. When the size of the delaminated area reaches a critical extent, the load-carrying 0° plies are no longer capable of sustaining the applied load, leading to extensive fibre fracture of the top and bottom 0° layers of the sample (Figure 3, impact energy = 7 J). In stitched specimens, in contrast, the presence of stitches greatly inhibits the propagation of the delamination along both the 0° and 90° directions, thus resulting in a substantially smaller delaminated area as compared to unstitched samples (Figure 3, impact energy = 7 J). However, the stress concentration in the vicinity of stitches is sufficient to promote the onset and growth of fibre fracture, which is observed to start at stitch locations and to propagate along a path linking contiguous stitches.

The differences in the structural response and dissipative behaviour of unstitched and stitched type A laminates under high energy impacts may be also explored by the analysis of the plots of Figure 5, which show histories of the contact force and of the energy of the impactor for a 7 J impact energy. We may see that while in unstitched laminates the initial energy is almost entirely dissipated during the impact, in stitched samples a considerable portion of the initial impact energy is transferred back to the impactor after the rebound. This suggests that stitched samples may provide better post-impact properties and residual structural capacity than unstitched ones even in the presence of extensive and widespread damage as that induced by high-energy impacts.

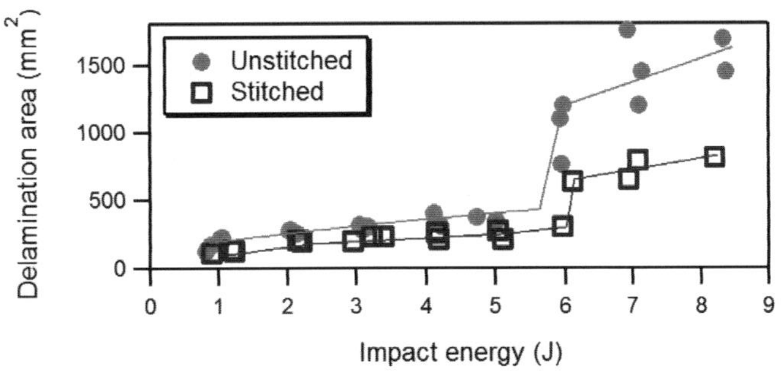

Figure 4. Projected delamination area vs impact energy for unstitched and stitched type A samples

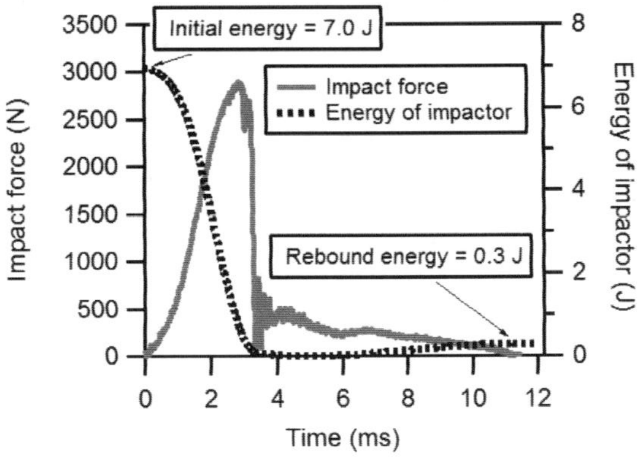

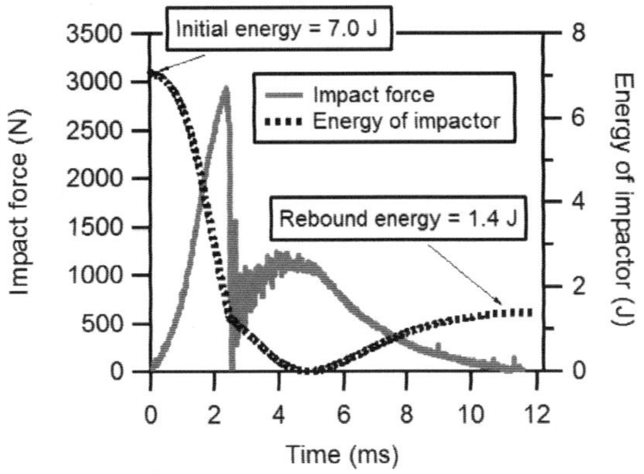

Figure 5. Histories of contact force and energy of impactor during impacts on unstitched (left) and stitched (right) type A samples

Typical force-time and force-displacement plots for unstitched and stitched type B samples impacted at approximately 3 J and 5 J energy are shown in Figure 6. Similarly to type A samples, the response of type B laminates is characterized by a continuous increase in slope with increasing out-of-plane displacement, due to membrane effects induced by large laminate deflections. Comparable force-time and force –displacement curves can be observed for unstitched and stitched samples for impacts up to about 3 J energy. Above this energy level, a sudden load decrease is experienced by unstitched samples at a load level slightly higher than 2 kN, while a comparable load drop can be observed in stitched samples only at a significantly higher load (approximately 2.5 kN).

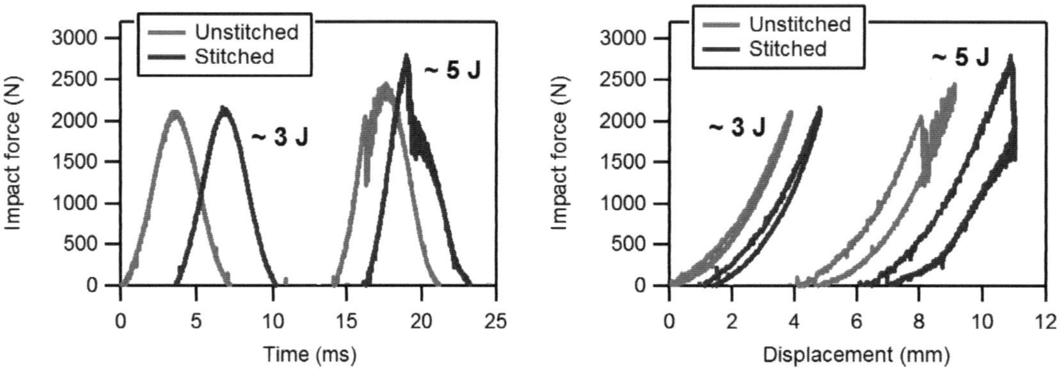

Figure 6. Force-time(left) and force-displacement (right) curves for unstitched and stitched type B samples

Representative impact damage scenarios occurring in type B samples are illustrated in the X-radiographs of Figure 7. Damage initiates at impact energies below 1 J as tensile matrix cracks in the bottom 90° layers and as shear matrix cracks around the contact area in mid 0° layers. Matrix cracking is followed by a two-lobe delamination that develops on the adjacent 0/90 interface and propagates along the fibre direction of the lower ply of the interface (90° direction). When unstitched samples are impacted with an energy exceeding a threshold value of about 3 J, the delamination exhibits an unstable growth, developing into a rectangular shape and quickly propagating to one edge of the specimen, with a damage evolution similar to that observed in unstitched type A laminates. The onset of the phase of unstable damage development is signalled by the load drop feature visible in the force-time and in the force-displacement traces.

In stitched laminates, as expected, the bridging effect of stitches hinders the growth of the two-lobe delaminated area, thus delaying the unstable propagation of the delamination and increasing the critical load at which the force signal exhibits the sudden load drop. However, the stress concentration induced by stitches promote the initiation of local delaminations at multiple stitch locations. Under increasing impact loads, local delaminations tend to coalesce and propagate following the direction of the stitching lines (Figure 7). This behaviour explains why, in type B samples, in spite of the beneficial effect of stitches, the projected delamination area of stitched laminates is only marginally smaller than that of analogous unstitched samples, as visible in the plot of Figure 8.

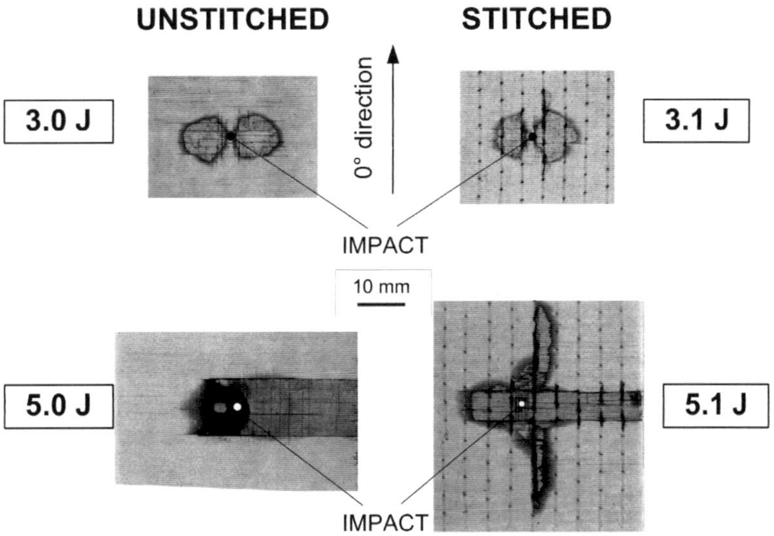

Figure 7. X-radiographs of impact damage in unstitched and stitched type B samples

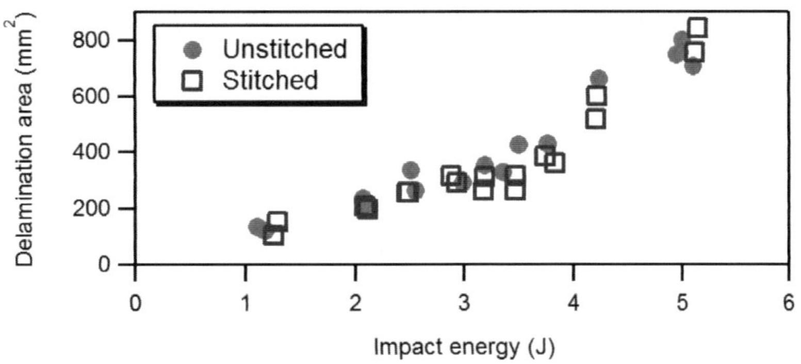

Figure 8. Projected delamination area vs impact energy for unstitched and stitched type B samples

4. Conclusion

The influence of through-thickness stitching on the structural and damage response of two classes of cross-ply carbon/epoxy laminates ($[0_2/90_2]s$ and $[90_2/0_2]s$) subjected to low-velocity impact was investigated in this study. It was found that the efficiency of stitching in improving the resistance to delamination of the laminates is affected by the orientation of the layers with respect to the loading configuration. In particular, while stitching greatly improved the delamination resistance of type A ($[0_2/90_2]s$) laminates, similar sizes of delaminated area were measured after impact in unstitched and stitched type B ($[90_2/0_2]s$) laminates, due to the growing of delaminations at multiple stitch locations.

These findings indicate that design procedures for primary structural elements based on stitched composites should also account for the sensitivity of the specific layup to individual failure modes potentially triggered by the presence of through-thickness reinforcements.

Acknowledgments

The support of the Regional Government of Sardinia through grant CRP-60404-2012 (*L.R. 7*, 2007) is gratefully acknowledged

References

[1] Sridharan S 2008 *Delamination Behaviour of Composites* (Cambridge, UK: Woodhead Publishing)
[2] Wisnom M R 2012 The role of delamination in failure of fibre-reinforced composites Philos. T. Roy. Soc. A 370 1850-70
[3] Kim J K and Mai Y W 1998 Engineered interfaces in fiber reinforced composites (Oxford, Elsevier)
[4] Dickinson LC, Farley GL and Hinders MK 1999 Translaminar reinforced composites: A review *J. Compos. Tech. Res.* 21(1) 3-15.
[5] Tong L, Mouritz A P and Bannister M K 2002 *3D fibre reinforced polymer composites* (Oxford, Elsevier)
[6] Greenhalgh E and Hiley M 2003 The Assessment of Novel Materials and Processes for the Impact Tolerant Design of Stiffened Composite Aerospace Structures *Compos. Part A-Appl. S.* 34 151-61.
[7] Zeng Y, Liu H Y, Mai Y W and Du X S 2012 Improving interlaminar fracture toughness of carbon fibre/epoxy laminates by incorporation of nano-particles *Compos. Part B-Eng.* 43(1) 90-94.
[8] Massabò R, Mumm D R and Cox B 1998 Characterizing Mode II Delamination Cracks in Stitched Composites *Int. J. Fract.* 92(1) 1-38
[9] Mouritz A P and Cox B N 2000 A mechanistic approach to the properties of stitched laminates *Compos. Part A-Appl. S.* 31(1) 1–27
[10] Wu E and Wang J 1995 Behavior of stitched laminates under in-plane tensile and transverse impact loading. *J. Compos. Mater.* 29(17) 2254–89
[11] Lopresto V, Melito V, Leone C and Caprino G 2006 Effect of stitches on the impact behaviour of graphite/epoxy composites, *Compos. Sci. Technol.* 66(2) 206-14
[12] Aymerich F, Pani C and Priolo P 2007 Damage response of stitched cross-ply laminates under impact loadings *Eng. Fract. Mech.* 74(4) 500-14
[13] Yoshimura A, Nakao T, Yashiro S and Takeda N 2008 Improvement on out-of-plane impact resistance of CFRP laminates due to through-the-thickness stitching *Compos. Part A-Appl. S.* 39(9) 1370-79
[14] Aymerich F and Priolo P 2008 Characterization of fracture modes in stitched and unstitched cross-ply laminates subjected to low-velocity impact and compression after impact loading *Int. J. Impact Eng.* 35(7) 591-608
[15] Tan K T, Watanabe N and Iwahori Y 2010 Effect of stitch density and stitch thread thickness on low-velocity impact damage of stitched composites *Compos. Part A-Appl. S.* 41(12) 1857-1868.
[16] Tan K T, Watanabe N and Iwahori Y 2012 Impact damage resistance, response, and mechanisms of laminated composites reinforced by through-thickness stitching *Int. J. Damage Mech.* 21(1) 51-80.
[17] Aymerich F and Francesconi L 2014 Damage mechanisms in thin stitched laminates subjected to low-velocity impact, *Procedia Eng.* 88 133-40

Moisture detection in composites by terahertz spectroscopy

Paweł Malinowski[1], Norbert Pałka[2], Szymon Opoka[1], Tomasz Wandowski[1], Wiesław Ostachowicz[1,3]

[1]Institute of Fluid–Flow Machinery, Polish Academy of Sciences, 14 Fiszera Street, 80–231, Gdansk, Poland

[2]Institute of Optoelectronics, Military University of Technology, 2 Kaliski Street, Warsaw, Poland

[3]Faculty of Automotive and Construction Machinery, Warsaw University of Technology, 84 Narbutta Street, 02-524 Warsaw, Poland

E-mail: pmalinowski@imp.gda.pl

Abstract. The application of Glass Fibre Reinforced Polymers (GFRP) in many branches of industry has been increasing steadily. Many research works focus on damage identification for structures made out of such materials. However, not only delaminations, cracks or other damage can have a negative influence of GFRP parts performance. Previous research proved that fluid absorption influences the mechanical performance of composites. GFRP parts can be contaminated by moisture or release agent during manufacturing, while fuel, hydraulic fluid and moisture ingression into the composite can be the in-service treats. In the reported research authors focus on moisture detection. There are numerous sources of moisture such as post manufacturing NDT inspection with ultrasonics coupled by water or exposition to moisture during transportation and in service. An NDT tool used for the research is a terahertz (THz) spectrometer. The device uses an electromagnetic radiation in the terahertz range (0.1-3 THz) and allows for reflection and transmission measurements. The spectrometer is equipped with moving table that allows for XY scanning of large objects such as GFRP panels. In the conducted research refractive indices were experimentally extracted from the materials of interest (water and GFRP). Time signals as well as C-scans were analysed for samples with moisture contamination. In order to be sure that the observed effects are related to moisture contamination reference measurements were conducted. The obtained results showed that the THz NDT technique can detect moisture hidden under a GFRP with multiple layers.

1. Introduction

The application of Glass Fibre Reinforced Polymers (GFRP) in many branches of industry has been increasing steadily. This is the main material used for wind turbine blades. Many research works focus on damage identification for structures made out of such materials. However, not only delaminations, cracks or other damage can have a negative influence of GFRP parts performance. Previous research proved that fluid absorption influences the mechanical performance of composites [1,2]. GFRP parts can be contaminated by moisture or release agent during manufacturing, while fuel, hydraulic fluid and moisture ingression into the composite can be the in-service treats. In the reported research authors focus on moisture detection. There are numerous sources of moisture such as post manufacturing non-destructive inspection with ultrasonics coupled by water or exposition to moisture during transportation and in service.

Moisture can be detected by means of destructive tests [1], [2] but the need is for a robust non-destructive testing (tool) that would allow to inspect composite elements. Such NDT tool could be used after manufacturing to check if the element is dry and ready for further processing such as adhesive bonding to another element. Such tool would be also helpful for in-service inspection. For example before applying composite repair patch. A very promising tool for moisture detection is based of electromagnetic radiation in the terahertz range [3]. In this research we used a TDS THz spectrometer [4] for inspecting GFRP contaminated with water. In the conducted research refractive indices were experimentally extracted from the materials of interest (water and GFRP). Time signals as well as C-scans were analyzed for samples with moisture contamination. In order to be sure that the observed effects are related to moisture contamination reference measurements were conducted. The obtained results showed that the THz NDT technique can detect moisture hidden under a GFRP layers.

2. Investigated samples

The object chosen for investigation was a GFRP dogbone sample with thickness 1.8 mm and 6 unidirectional layers oriented along the sample longer edge. The inspected area was limited by using a metal (aluminium) adhesive tape (Figure 1). The tape gives strong reflection in the reflection mode and blocks the THz radiation in the transmission mode. So in the THz image it gives clear borders of the inspected region and facilitates interpretation of the results.
Firstly, a sample with water layer on the surface was investigated. The estimated water layer thickness was 100-200 μm. Secondly, more sophisticated scenario was investigated. We joined two samples with moisture in their interface. There was a thin layer (50-100 μm) and later thick layer (100-300 μm) of water investigated.

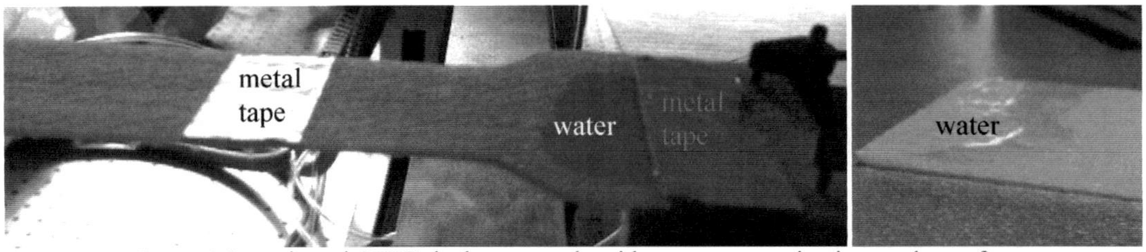

Figure 1 Investigated GFRP dogbone sample with water contamination on the surface

3. Experimental setup

Terahertz time-domain spectroscopy (TDS) in the frequency range from 100 GHz to around 4 THz has been used in recent years for the purpose of non-destructive material testing. The submillimeter-waves are able to penetrate through many nonconducting materials. At each dielectric interface, e.g. from air to the GFRP material, the electromagnetic waves experience partial transmission and reflection. In this paper, we use the THz-TDS-system as a tool to remotely measure presence of moisture in/on composite structure. The scanning heads of the THz-TDS-system worked in transmission and in reflection modes. THz TDS system TPS Spectra 3000+ by TeraView Limited generates radiation in the form of repeated, very narrow picosecond pulses which are focused and sent toward the investigated structure. Wide frequency content, small power and non-ionizing radiation are the main characteristics of the pulse. This approach is very interesting when the material allow for penetration at THz-frequencies, such as GFRP. The experimental setup using THz TDS system is shown in Figure 2. The investigated samples are supported by steel bars mounted to system responsible for the movement in the horizontal plane.

Figure 2 Terahertz spectroscopy device

4. THz transmission study

Figure 3 illustrate the transmission C-scan of the sample with 100-300 μm water layer. The image represents the maximum in the frequency content of the time signal. On the right hand side radiation attenuation due to water is visible. The amplitude of the signal reflected from water-air interface dropped twice.

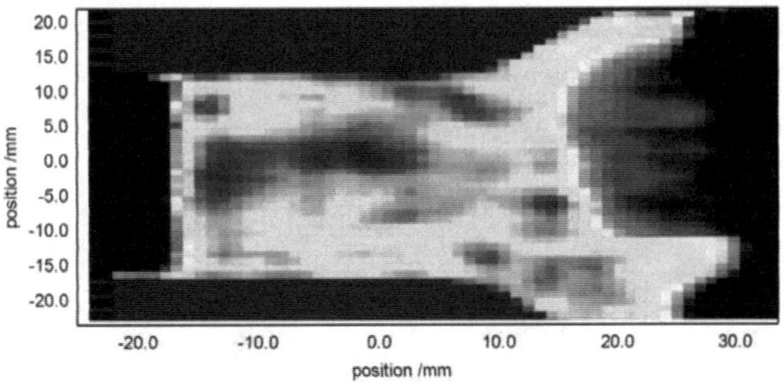

Figure 3 THz C-scan view of sample with water layer on the surface

The increase attenuation is also visible in the B-scan image (Figure 4) obtained as a horizontal slice at y=0 position of Figure 3.

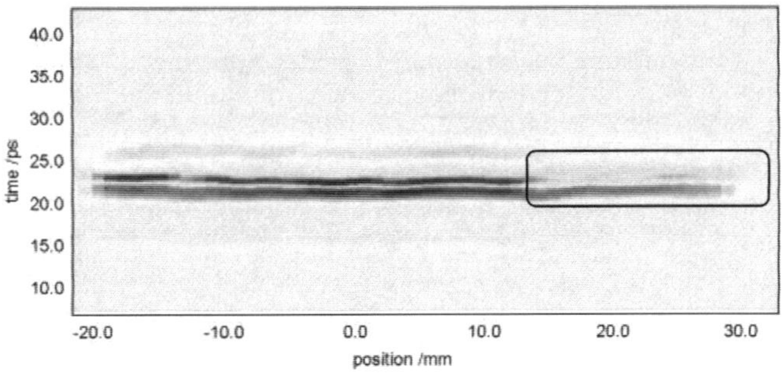

Figure 4 THz B-scan of sample with water layer on the surface. Attenuation area marked by a rectangle.

Next, the samples with moisture at the interface between samples were measured. The samples were hold together with metal clips. The thin layer (50-100 μm) of water in range x=40–50 mm caused slight attenuation in comparison to the reference measurement (Figure 5).

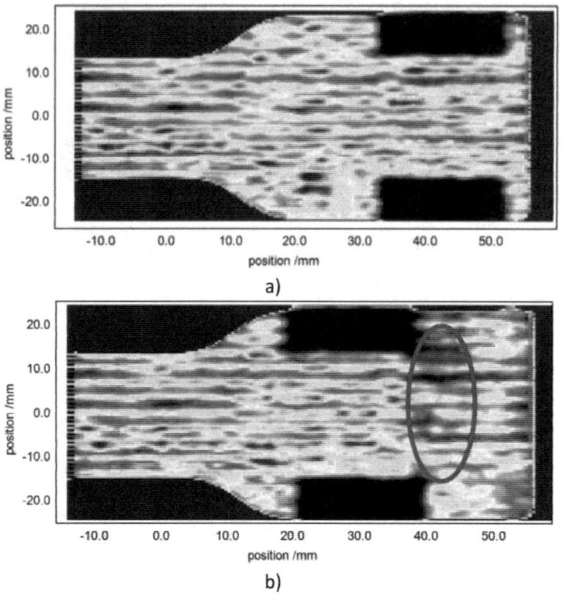

Figure 5 C-scan of reference samples (a) and samples with thin water layer in the interface (b); black rectangle denote locations of the clips

The low amount of moisture caused small change in the obtained image. The thick layer (100-300 μm) causes significant attenuation visible in time (Figure 6a) and in frequency domain (Figure 6b). Looking at the time signals the presence of moisture caused the reduction of pulse amplitude by three times (Figure 7).

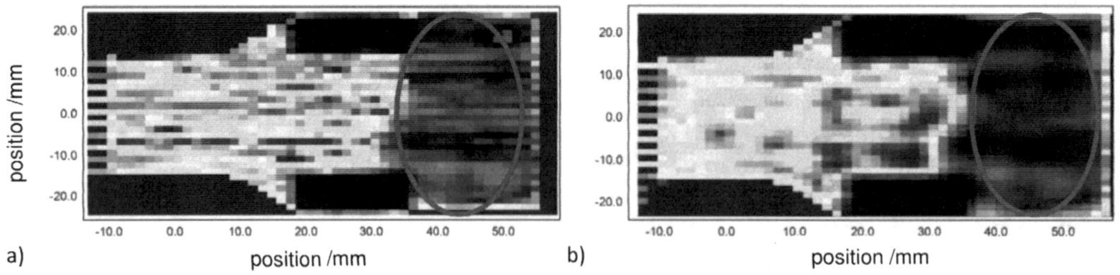

Figure 6 C-scan samples with thick water layer in the interface; a) maximum in time domain (b) maximum in frequency domain

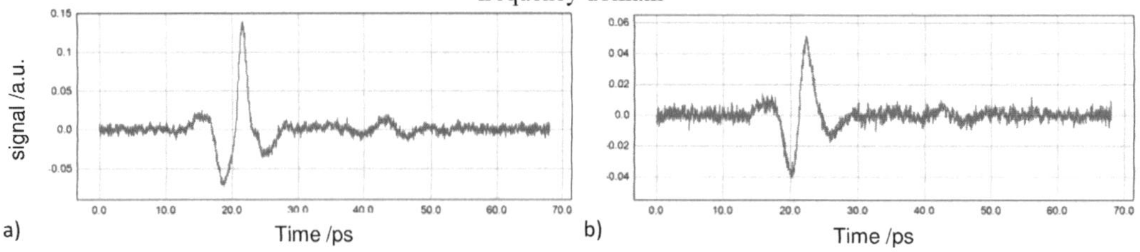

Figure 7 THz pulse transmitted through dry area (a) and water contaminated area (b).

5. Estimation of the refractive index

The investigation of the samples in reflection measurement mode was preceded by an estimation of the samples absorption and refractive index. At 0.5 THz 10^{-2} of power is transmitted while for 1 THz only 10^{-4} of power is transmitted through the material. It can be assumed that above 1.2 THz (10^{-6} of power transmitted) the sample blocks the THz radiation. The complex refractive index for GFRP was estimated from measurements. The mean value at 0.5 THz was used giving

$$\begin{aligned} N_G &= n_G + i\kappa_G \\ n_G &= 2 \\ \kappa_G &= 0.057 \end{aligned} \tag{1}$$

n_G and κ_G were calculated according to

$$n = \sqrt{\frac{\sqrt{\varepsilon_1^2 + \varepsilon_1^2} + \varepsilon_1}{2}} \tag{2}$$

$$\kappa = \sqrt{\frac{\sqrt{\varepsilon_1^2 + \varepsilon_1^2} - \varepsilon_1}{2}} \tag{3}$$

using electric permittivity ε_1 and ε_2 from the measurements.

Water absorption and refractive index can be found in [5]. It can be assumed that for 0.5 THz the complex refractive index for water is

$$\begin{aligned} N_W &= n_W + i\kappa_W \\ n_W &= 2.4 \\ \kappa_W &= 0.81 \end{aligned} \tag{4}$$

Noticable differences between n_G and n_W justify the use of THz spectroscopy which is sensitive to differences between real parts of refractive indexes of the interfaces. The comparison of imaginary parts of refractive indexes shows that THz radiation is much more attenuated in water than in GFRP sample suppressing the inspection of sample below the water interface.

6. THz reflection study

Firstly the samples without water in the interface were measured. The measured and deconvolved time signal is presented in Figure 8. The deconvolution makes the signal smoother and easier to interpret. The initial pulse is a reflection from upper surface. Then there are two wider middle pulses located at middle of the signal. First one represents the GFRP-air interface (180° phase shift). Second one corresponds to air-GFRP interface. The last pulse is the pulse reflected from the lower surface. The medium is dispersive – pulses are getting wider with the increase of propagation time through the material. Between initial and middle pulses there are very weak reflections indicating the layered structures of the sample.

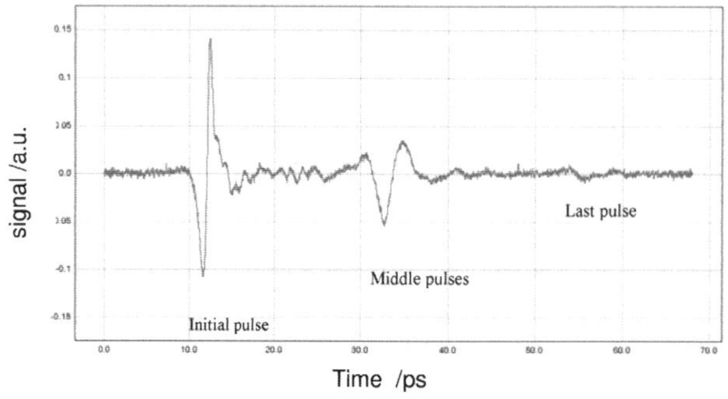

a)

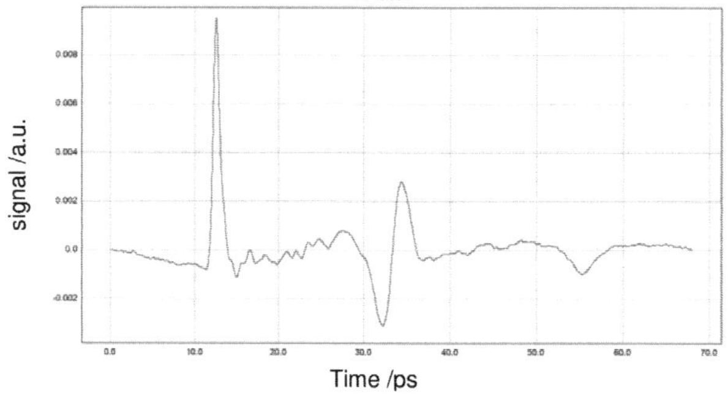

b)

Figure 8 Reflected THz time signal before (a) and after (b) deconvolution

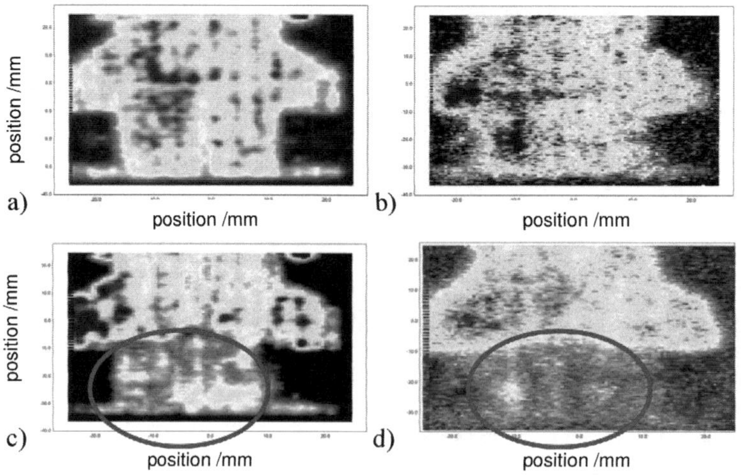

Figure 9 C-scans of the middle (a) and end (b) part of the dry samples; C-scans of the middle (c) and end (d) part of the samples with moisture contaminated interface

In the Figure 9 a,b) C-scan of the pulses reflected from middle and end part of the reference sample are depicted. Secondly, the sample with water layer between two dogbone composites were measured. Figure 9c,d) presents the sample with moisture contaminated interface. There is moisture indication in the lower part of Figure 9c). The end-reflected signal has also changed (Figure 9d), because the reflection was weakened by the presence of moisture in the interface. This effect is visible also in the B-scan (Figure 10). The moisture reduces the end reflection (left hand side), because of high value of κ_W .

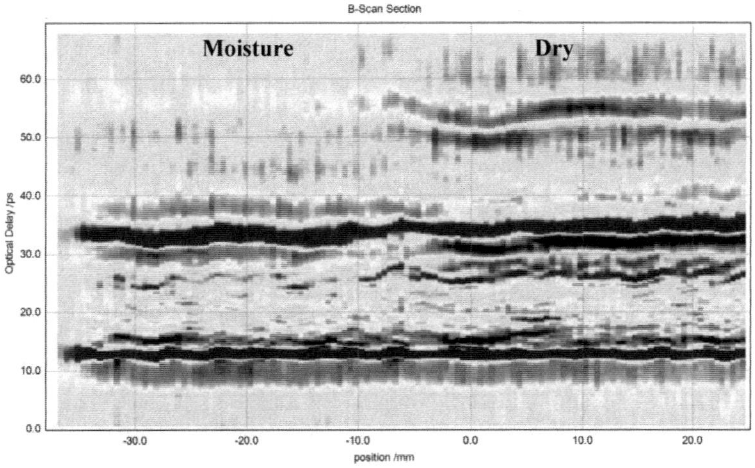

Figure 10 Horizontal B-scan image

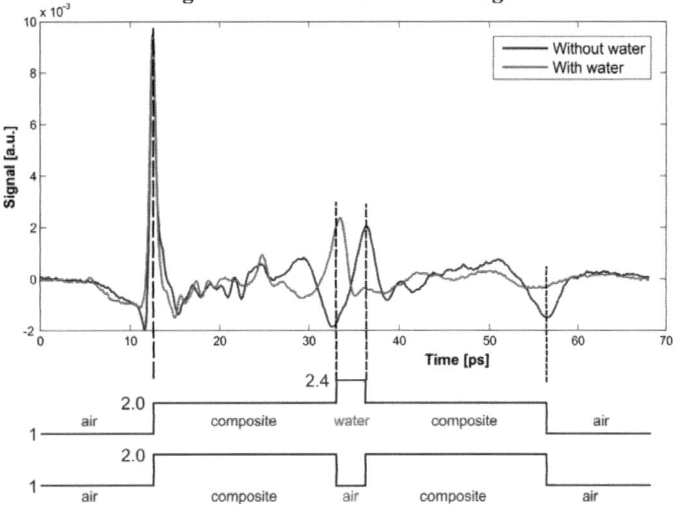

Figure 11 Signals after deconvolution and refractive profiles of both cases

Figure 11 presents the deconvolved signal with and without water at the interface of two GFRP samples. The initial pulse is identical and depends mainly on the surface morphology. The pulse reflected from the interface changes when there is moisture in the interface. The end of the upper sample is denoted by the vertical dash line. Pulses reflected from water and air differ due to relation of the refractive indexes:

$$n_{AIR} < n_G < n_W \tag{5}$$

According to Fresnel conditions a wave travelling from lower n medium to higher n medium changes its phase by 180°. While travelling in opposite direction the phase stays unchanged. In the measured case the signals are phase shifted by the used software so the interpretation will be opposite. According to the rule the refractive profiles were plotted in Figure 11.

7. Conclusions

In the conducted research water/moisture influence on terahertz signals was studied. Both transmission and reflection approach showed the ability to detect moisture presence. The moisture was hidden under a GFRP layer comprising of 6 plies. The low amount of 50-100 μm) caused small change in the obtained image and the result is ambiguous. The thick layer (100-300 μm) causes significant attenuation causing the reduction of pulse amplitude by three times.

The reflection mode measurement gives additional information. It was possible to see the moisture influence at the GFRP samples interface and also at the end of the samples.

The technique is promising for water detection and localization in the GFRP composites. The future research should focus on water detection limit of the technique.

8. Acknowledgments

This work was supported by the grant no. PBS1/B6/8/2012 (project KOMPNDT) of Polish National Centre for Research and Development (NCBR).

References

[1] Sala G 2000 Composite degradation due to fluid absorption *Compos B Eng* **31** 357–73

[2] Pantelakis Sp G, Katsiropoulos Ch V, Tserpes K I and Felgeyrolles Th 2010 Effect of water absorption on strength of the aeronautical composite material Fiberdux HTA/6376 *Key Eng Mat* **417-418** 457–60

[3] Federici J F 2012 Review of moisture and liquid detection and mapping using terhaertz imaging *J Infrared Millim Terahertz Waves* **33** 97-126

[4] Pałka N and Miedzińska D 2014 Detailed non-destructive evaluation of UHMWPE compistes in the terahertz range, *Opt Quant Electron* **46** 515–25

[5] Bertie J and Lan Z 1996 Infrared Intensities of Liquids XX: The Intensity of the OH Stretching Band of Liquid Water Revisited, and the Best Current Values of the Optical Constants of H2O(l) at 25°C between 15,000 and 1 cm-1 *Appl Spectrosc* **50** 1047–57

Assessment of Damage Detection in Composite Structures Using 3D Vibrometry

S Grigg, M Pearson, R Marks, C Featherston and R Pullin

Cardiff School of Engineering, Cardiff University, The Parade, Newport Road, Cardiff Wales, UK, CF24 3AA

griggs@cardiff.ac.uk

Abstract. Carbon fibre reinforced polymers (CFRP) have been used significantly more in recent years due to their increased specific strength over aluminium structures. One major area in which their use has grown is the aerospace industry where many now use CFRP in their construction. One major problem with CFRP's is their low resistance to impacts. Structural health monitoring (SHM) aims to continually monitor a structure throughout its entire life and can allow aircraft owners to identify impact damage as it occurs. This means that it can be repaired prior to growth, saving weight with the repair and the time that aircraft is grounded. Two areas of SHM being researched are Acoustic Emission (AE) monitoring and Acousto-Ultrasonics (AU) both based on an understanding of the propagation of ultrasonic waves. 3D Scanning laser vibrometry was used to monitor the propagation of AU waves with the aim of gaining a better understanding their interaction with delamination in carbon fibre reinforced polymers. Three frequencies were exited with a PZT transducer and the received signal analysed by a cross correlation method. The results from this and the vibrometer scans revealed 100 kHz as the most effective propagating frequency of the three. A high resolution scan was then conducted at this frequency where it could be seen that only the out of plane component of the wave interacted with the damage, in particular the A_0 mode. A 3D Fast Fourier Transform was then plotted, which identified the most effective frequency as 160 kHz.

1. Introduction

Carbon Fibre reinforced polymers (CFRP) components have many additional issues over aluminium ones, a major one being their ability to withstand impact damage and the difficulties associated with spotting said damage, due to delamination between the ply's of the material. This is becoming a serious issue in many safety critical industries, in particular the aerospace sector where they are becoming more widely used. Structural Health Monitoring (SHM) aims to continually monitor a structure throughout its entire life, both during manufacture and throughout its service life. The benefits of this over the traditional method of inspections at regular intervals is the early detection and repair of damage, leading to lighter and more efficient structures. Integration of SHM onto an aircraft structure would also lead to less down time of the aircraft, as inspection and maintenance could be conducted as and when damage was detected.

Acousto-Ultrasonics (AU) is an active form of SHM which consists of exiting ultrasonic lamb waves using a transducer bonded to a surface of the structure. This wave is then received by a sensor and by comparing the change in received signal over time using a method such as cross correlation, an indication of whether damage is present can be gained. In this work, the "xcorr" function from

MATLAB had been utilised to determine a cross correlation coefficient. This function compares two waveforms on their similarity, for example if two waveforms are exactly the same, they will have a cross correlation of 1 [1].

Zhao et al [2] utilised the cross correlation technique across a circular array of sensors bonded with a complex geometry. The sensors took turns pulsing and receiving waves and by comparing to baseline values damage could be detected and localised using a reconstruction algorithm. This test setup however required a closely spaced array with lots of sensors. Better understanding of wave propagation and interaction with damage would allow for greater optimisation of operating frequency, sensor location and a reduction in the number of sensors needed for arrays of sensors to detect damage.

The ultrasonic waves utilised for AU applications in plate like structures are known as lamb waves, which are complex elastic stress waves which propagate parallel to the surface of a solid medium throughout its thickness. These high frequency waves are able to travel long distances and are heavily influenced by damage or boundary's, making them perfect for use in damage detection [3]. Lamb waves exist in a number of modes, symmetrical and asymmetrical, existing both in and out of plane. Symmetrical modes move throughout the structure symmetrically with respect to the mid-plane whilst extending and compressing into an elliptical shape [4]. The main symmetrical mode, S_0, tends to be dominated by longitudinal vibration, i.e. in plane. S_0 also tends to be quicker and less susceptible to attenuation (the reduction in signal amplitude as the wave propagates) than its corresponding asymmetric mode, A_0 [5]. Asymmetrical modes produce more out of plane displacement due to the waves moving in the same directions on the top and bottom surfaces [6]. In theory an infinite number of S and A modes can be present within an elastic medium, as when the frequency of the wave increases more occur [4]. Dispersion curves show the modes present within a plate and their velocities at various frequencies, Figure 1 shows the dispersion curve for a 2.15mm composite panel, similar to the ones used for the testing presented in this paper. This dispersion curve shows that the S_0 and A_0 should be the only modes present and that velocity should not change significantly up to 500 kHz.

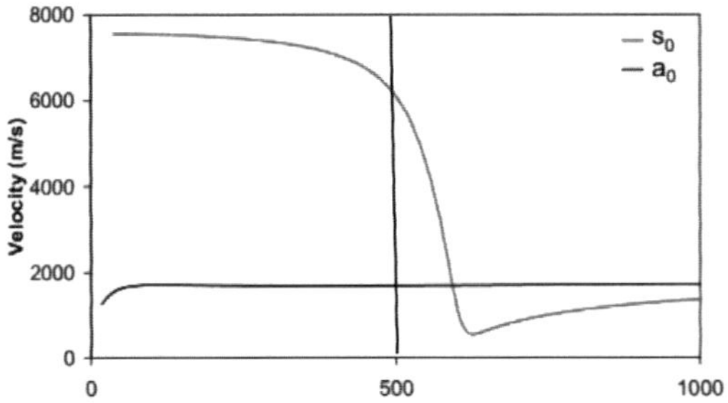

Figure 1 - Dispersion curve for a 2.15mm thick composite plate [5]

The Polytec PSV-500-3D Scanning Vibrometer is able to measure the velocity of vibration not only out of plane (Z axis), as can be measured on a standard 1D vibrometer, but also in plane (X and Y axis). Using this technology to monitor an AU system is equivalent to bonding thousands of sensors to the surface of the structure and enables a non-contact visualisation of wave propagation. The system consists of three laser heads which are alighted on the surface of the specimen and operates by analysing the change in frequency and phase caused by Doppler shift within the backscattered light from a surface. This shift is caused by changes in vibrational velocity and displacement on the surface of the structure.

Schubert et al [7] preformed a similar study using 3D laser vibrometry to analyse the interaction on lamb waves with impact damage in CFRP specimens, whilst comparing results with AU data. The results showed that only the out of plane parts of the A_0 and S_0 mode interacted with the damage and that a time delay was created when the A_0 wave front passed over the damage location.

2. Experimental Procedure

The tested specimen was manufactured from CFRP 2/2 twill with fibres running 0^0 and 90^0, eight ply's were laid in a 0/0 format. The specimen was 300mm x 300mm and 1.75 mm thick with cut edges to ensure constant reflections of the AU waves. Prior to any testing, an ultrasonic inspection was conducted with a C-Scanner, which revealed no delamination present within the specimen.

AU baseline scans were then taken using two Pancom Pico-Z piezoelectric sensors bonded to the structure, in the positions shown in Figure 2 (a), which have a high frequency response from 100-500 kHz. A five cycle square wave was used to excite the transducer at 100 kHz, 300 kHz and 500 kHz so to test the impact of the damage on a range of frequencies. A sine envelope was then applied to the generated signal as was a 5MHz filter to reduce noise. The tests were conducted 50 times and the cross correlation technique applied to these to ensure good repeatability of the received signal. Laser vibrometer scans were then conducted at each frequency to observe wave propagation in an undamaged structure. 3000 points were scanned over a 280mm x 120mm area in the centre of the plate.

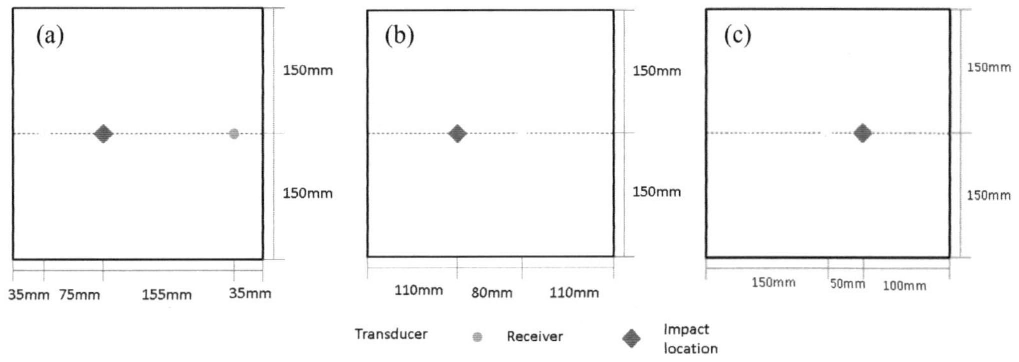

Figure 2- Layout of sensors and impact location for initial setup (a), high resolution scan (b) and second panel high resolution scan (c)

Once initial testing had been conducted, the specimen was impacted at rising energies, 4, 4, 4, 4, 5, 5, 6 and 7 joules, using an INSTRON dynatup 9250HV impact tester. The aim of these impacts was to ensure sufficient delamination was present in the structure without causing a hole to be created. The location of the impact was close to the transducing sensor, so to get a greater change in the received wave. In-between each impact AU tests were conducted, which were correlated to a baseline taken once the specimen had been clamped into the impact test rig. Though damage was visible with the naked eye a C-Scan was conducted which revealed sufficient.

After reviewing the first set of testing, it was decided that a higher resolution scan covering a smaller area would give clearer results. To reduce the reflections from the boundaries of the panel present when the wave fronts interact with the damage, the transducer was moved to the setup shown in Figure 2 (b). The same signal was used as in the first test setup, however only 100 kHz was tested due to the long period required for the scan to take place. Over 10,000 points were scanned over a 94mm × 80mm area with the damage in the center.

In order to create a 3D Matrix Fast Fourier Transform (FFT), a chirp signal was pulsed though the structure and a high resolution scan of the area around the damage was recorded. A chirp signal has a range of frequencies in this case 10-500 kHz within it; the purpose of this was to excite each frequency to see how it interacts with the damage. By taking Fourier Transforms for each frequency it was possible to see which frequencies have the most interaction with the damage. A cross section was plotted showing the magnitude of each frequency across the entire range.

The results from the 3D matrix FFT identified a frequency where wave interaction with the damage should be at its greatest. This frequency was used to produce another high resolution scan on a non-damaged panel of equal layup and size to the one used in the first set of tests. The transducer was located in the middle of the specimen, so to reduce the effect of edge reflections on the leading edge of the propagation wave as it passes over the area of interest. The impact tester applied a 9 J impact to the panel in the location shown in Figure 2 (c), which produced a delamination of similar size to the one from the initial specimen. A final high resolution scan was then conducted on the post impacted specimen.

3. Results and discussion

The initial vibrometry pre impact scans show the wave propagating through the structure with the S_0 propagation elliptically at 0° and 90°. The A_0 mode can also be seen to be propagating cylindrically at a much slower rate than the S_0 mode [5]. It was also apparent that the in plane (X and Y) plots were only observing the propagation of the S_0 mode whereas the out of plane Z axis was detecting both, but predominantly A_0. Examples of these various modes can be seen in Figure 3 where some of the 100 kHz scan plots can be seen.

By comparing the plots at each frequency it could be seen that there was little difference in wave velocity as the frequency changed, this is in line with the dispersion curve for a 2.15mm composite panel, shown in Figure 1. It was noticed however that there was far greater attenuation as the frequency increases, i.e. the rate at which amplitude drops. This is shown Table 1 which gives the amplitude of the out of plane (Y Axis) S_0 signal and the in plane (Z axis) A_0 signal for each frequency, taken from the vibrometry data in the middle of the panel.

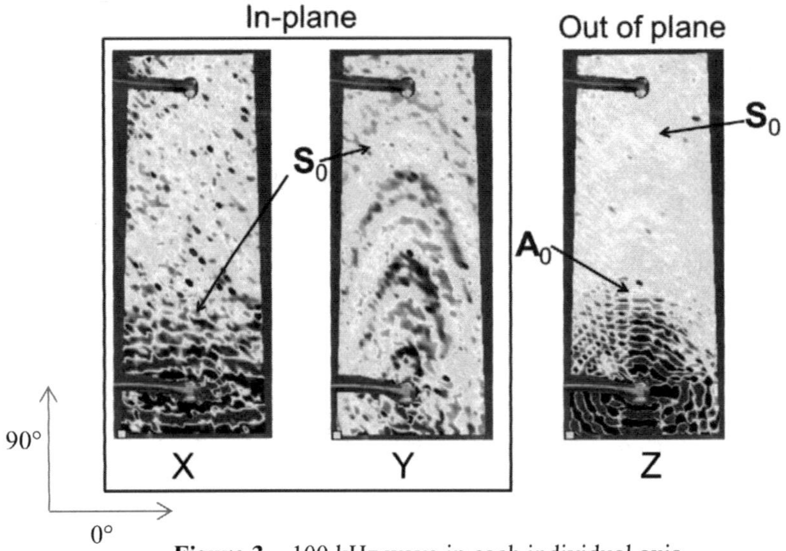

Figure 3 – 100 kHz wave in each individual axis

Table 1 – Amplitude of vibrational velocity at the centre of the specimen

Frequency (kHz)	In plane S_0 (μm/s)	Out of plane A_0 (μm/s)
100	850	540
300	800	250
500	700	200

The results of the AU tests throughout the impact testing are shown in Figure 4. It is clear from these results that 500 kHz gives the greatest change in wave correlation, however also the most un-repeatable results, due to its high standard deviation. The results show clear indication that damage is present in the structure after the 6th impact, where the correlation began to lower. The C-Scan conducted on the specimen after the impacts revealed a delamination 22mm in diameter, this compares was a visual inspection where only 13mm of damage was observed.

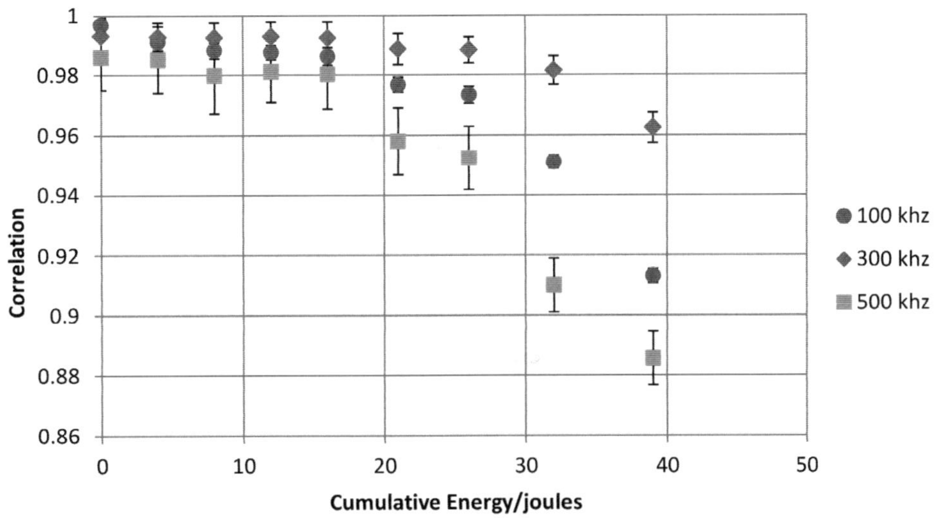

Figure 4 - AU correlation throughout impact testing with error bars showing standard deviation

The vibrometer scans conducted on the specimens after the impacts revealed that the A_0 mode could be seen to interact with the damage. Reflections were seen from the point of impact once the wave had passed over, and a changed wave front of the AU wave as it continues through the specimen was observed. This was far clearer at 100 kHz, due to the lower attenuation of the wave; however it was also seen for the other frequencies.

The higher resolution scans of the same specimen showed very clearly the A_0 interaction with the delamination shown in Figure 5. Closer inspection was conducted with the out of plane S_0 mode using a FFT of only the time period where the S_0 was passing over the area of damage. This is shown in Figure 6, and worth noting the semicircle of higher energy from the left is the approaching A_0 mode. This revealed some minor effect of the wave due to the damage; however this did not affect the continuing wave. The reason for the greater effect on the A_0 than the S_0 was probably due to its greater wavelength, meaning that it was passing over the delamination, rather than interacting with it in any significant way. It was also clear that there was no observable interaction of any in plane waves with the damage; this is likely to be because delamination occurs through the Z plane, as opposed to cracks which are predominantly in the X/Y. These results are in line with the work conducted by Zhao et al [2] where only the out of plane part of the modes were observed to interact with the delamination.

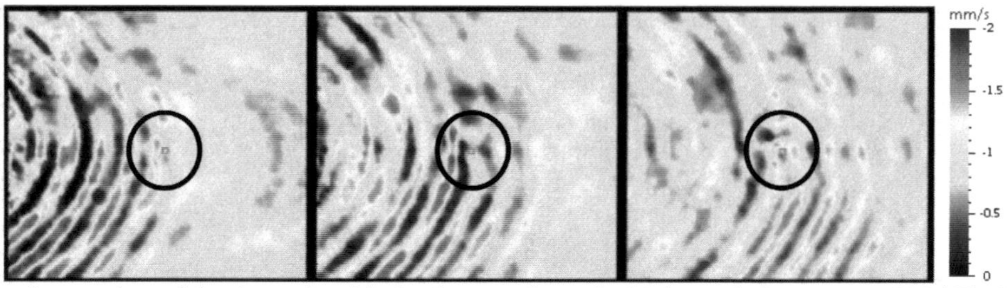

Figure 5 – Interaction of the A_0 mode with the damage from the high resolution scan at 100 kHz (205 μs, 220 μs and 235 μs) with area of damage circled

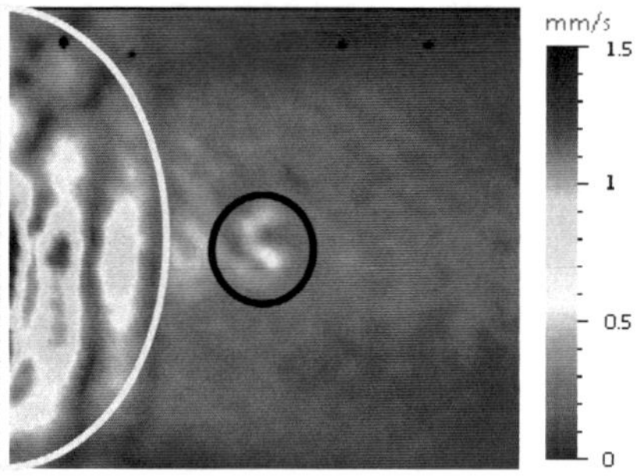

Figure 6 – FFT for the time period where the S_0 mode passes over the damage (approaching A_0 mode (white) and area of damage (black) circled)

The 3D matrix FFT plot can be seen in Figure 7. This plot shows a range of FFT results at different frequencies for a cross section of x at the point of impact. It identifies a peak at 160 kHz, meaning that this frequency has the most interaction with the delamination. This should also mean that this frequency should yield the greatest change in cross correlation.

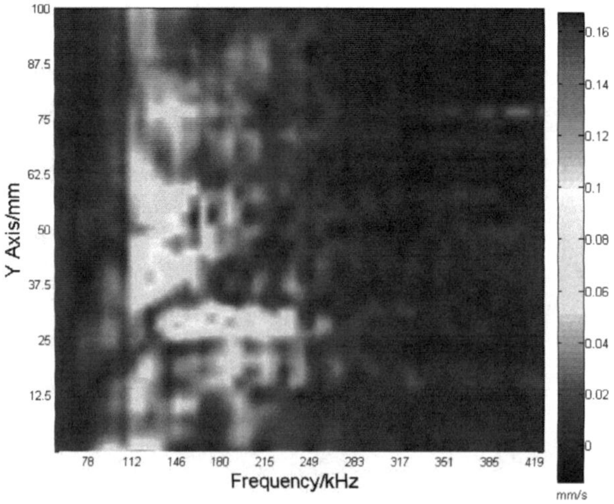

Figure 7 – 3D matrix FFT

The results from the 160 kHz scan showed that the wave at this frequency had a very strong interaction of the out of plane part of the A_0 mode with the delamination, as shown in Figure 8. Investigation into the S_0 mode did not reveal any interaction, as was present in the previous test setup at 100 kHz. The most probable reason for this is its greater wavelength causing it to fully pass over the damage, with no observable interaction. At this time it is hard to say if the 3D matrix FFT results are correctly showing the optimum frequency for transmitting in AU tests, however it gives a promising indication.

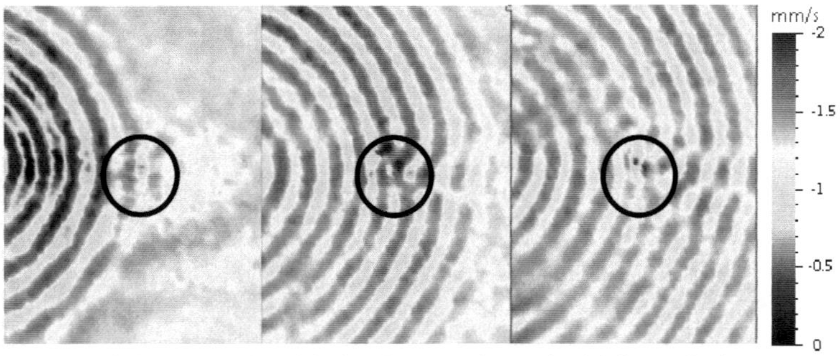

Figure 8 - Interaction of the A_0 mode with the damage from the high resolution scan at 160 kHz (205 μs, 220 μs and 235 μs) with area of damage circled

4. Conclusion

Acousto-ultrasonic wave propagation and damage interaction through a CFRP panel was recorded and analysed. Having greater knowledge of this interaction can be applied to future AU testing to better optimise the systems and tailor the sensors used to each structure. The wave propagation was monitored both in and out of plane; the latter clearly being the component of the wave interacting with the delamination. No in plane interaction was observed. Of the two modes produced by the transducer, the A_0 had far more interaction with the delamination than the S_0 and it would appear that with an increased frequency, the S_0 mode has no interaction.

The 3D matrix FFT shows that for an AU system to effectively detect delamination in a panel similar to which tested upon, transducers with a frequency response of around 160 kHz should be selected which are able to exited and receive the fundamental A_0 mode effectively. This is not by any means conclusive and testing should be conducted to further investigate this as well as the system's ability to detect smaller and different types of damage.

5. References

[1] M. R. Pearson, M. J. Eaton, C. A. Featherston, K. M. Holford and R. Pullin, "Impact Damage Detection and Assessment in Composite Panels using Macro Fibre Composites Transducers," Damas, 2011.

[2] X. Zhao, H. Gao, G. Zhang, B. Avhan, F. Yan, C. Kwan and J. Rose, "Active health monitoring of an aircraft wing with embedded piexoelectric sensor/actuator network: I. Defect detection, localization and growth monitoring," *Smart Materials and Structures,* vol. 16, pp. 1208-1217, 2007.

[3] Z. Su, L. Ye and L. Le, "Guided Lamb Waves for Identification of Damage in Composite Structures," *A review of Sound and Vibration,* vol. 295, pp. 753-780, 2006.

[4] O. Diligent, "Interaction between Fundamental Lamb Modes and Decects in Plates," University of London, London, 2003.

[5] M. J. Eaton, "Acoustic Emission (AE) Monitoring of Buckling and Failure in Carbon Fibre Composite Structures," Cardiff School of Engineering , Cardiff, 2007.

[6] W. J. Staszewshi, B. C. Lee and R. Traynor, "Fatigue Crack Detection in Metallic Structures with Lamb Waves and 3D Laser Vibrometry," *Measurement Science and Technology,* vol. 18, pp. 729-739, 2007.

[7] L. Schubert, U. Lieske, B. Köhler and B. Frankenstein, "Interaction of Lamb Waves with Impact Damaged CFRP's – Effect and Conclusions for Acousto-Ultrasonic Applications," Stanford, 2009.

Non-destructive testing on aramid fibres for the long-term assessment of interventions on heritage structures.

R Ceravolo[1], A De Marchi[2], E Pinotti[1], C Surace[1], L Zanotti Fragonara[1]
[1] Politecnico di Torino, Department of Structural, Building and Geotechnical Engineering, Corso Duca degli Abruzzi 24, Turin 10129, Italy
[2] Politecnico di Torino, Department of Electronic and Telecommunications, Corso Duca degli Abruzzi 24, Turin 10129, Italy

E-mail: rosario.ceravolo@polito.it

Abstract. High strength fibre reinforced polymers (FRPs) are composite materials made of fibres such as carbon, aramid and/or glass, and a resin matrix. FRPs are commonly used for structural repair and strengthening interventions and exhibit high potential for applications to existing constructions, including heritage buildings. In regard to aramid fibres, uncertainties about the long-term behaviour of these materials have often made the designers reluctant to use them in structural engineering. The present study describes simple and non-destructive non-linearity tests for assessing damage or degradation of structural properties in Kevlar fibres. This was obtained by using high precision measurements to detect small deviations in the dynamic response measured on fibres and ropes. The change in dynamic properties was then related to a damage produced by exposure of the sample to UV rays for a defined time period, which simulated long-term sun exposure. In order to investigate the sensitivity of such an approach to damage detection, non-linearity characterisation tests were conducted on aramid fibres in both damaged and undamaged states. With the purpose of carrying out dynamic tests on small fibre specimens, a dedicated instrumentation was designed and built in cooperation with the Metrology Laboratory of the Department of Electronics at the Politecnico di Torino.

1. Introduction

Aramid, carbon and glass fibres are typically exploited in composite materials and have been taken into account for use in structures for more than 20 years. The fibres are the primary load-carrying elements and provide the composites with their unique structural properties.

They possess a combination of high strength, high stiffness and good resistance to creep and corrosion, non-magnetic properties and they are resistant to various types of chemicals. For these reasons, they are perfect for the application in harsh environment [1, 2]. Depending on the type of application, the fibres can be oriented in different direction in order to enhance the mechanical properties of the composite in desired direction.

Areas that exploit these novel materials are, for example, external and internal pre-stressing devices, strengthening of structures through composite plates, composite bars as reinforcement, composites in the marine and railway industries and in ground engineering. The different uses of novel materials in civil engineering can be summarized in applications for new construction, repair and rehabilitation. It has been demonstrated that structures such as a bridges and columns built according to this technique show exceptional durability, and effective resistance to effects of environmental exposure. Several companies around the world begin to wrap damaged bridge piers to prevent collapse and steel-reinforced columns to improve the structural integrity and to prevent buckling of the reinforcement.

Interesting applications of FRP materials are typically found in the cultural heritage field, where invasiveness of interventions is a major issue. Carbon fibre reinforced polymers (FRPs) are commonly employed for the strengthening and seismic retrofitting of vaults and other masonry elements of heritage buildings (Figure 1). In any case, experimental studies are still required to better understand the durability and compatibility of interventions in polymeric materials and, in general, in other non-traditional materials. For instance, uncertainties about the long term behaviour of aramid fibres made often designers reluctant to their use in the structural engineering fields.

Figure 1. (a) The masonry vault and reinforcement arches dated 1600 during surface preparation for CFRP application. (b) Smart CFRP sheet placed on vault [3].

The aim of this study is to show how simple and non-destructive non-linearity tests can be used for assessing the degradation of structural characteristics in aramid fibres. The idea is to detect weak non-linearities [4, 5, 6] in the dynamic response measured on fibres and ropes and to use this non-linearity level as an indicator of the aramid element's residual life. A degradation of the material properties was produced by exposing samples to UV rays for a defined time period, so as to simulate long-term sun exposure.

2. Testing equipment

In order to analyse the dynamic behaviour and to detect the presence of non-linearities in two samples of Kevlar®, in the case of pristine and damaged conditions, respectively, a dedicated testing device was designed and built ad hoc in the Metrology Laboratory of the Department of Electronics of the Politecnico di Torino.

The testing machine frame (Figure 2.a) is composed of a rectangular crossbeam (L=45 cm, cross section: 4x4 cm) made of aluminium, transversely bounded at two cylindrical aluminium holed columns (h=95.4 cm; diameter: 5 cm, thickness: 0.5 cm) on a massive square basement (60x60 cm) made of marble with a thickness of 14.5 cm. An isolation steel table supports the marble mass, this allowing to neutralize most of environment vibrations. The materials used for the testing system were selected in order to minimise the effect of the temperature excursion in the laboratory.

The specimen, consisting of parallel fibres, were mounted vertically, with the top fixed at the horizontal crossbeam and at an aluminium block at the lower part; a special device was inserted at the bottom in order to measure the specimen displacement during the test (Figure 2.b).

To evaluate the displacement of the fibre, an optical measurement system composed by an 8-millimeter sphere made of optical glass was used. It was used to focus the beam of light inside the photodiode. The photodiode was wired to an electronic circuit, capable of turning the optical signal into an electric one.

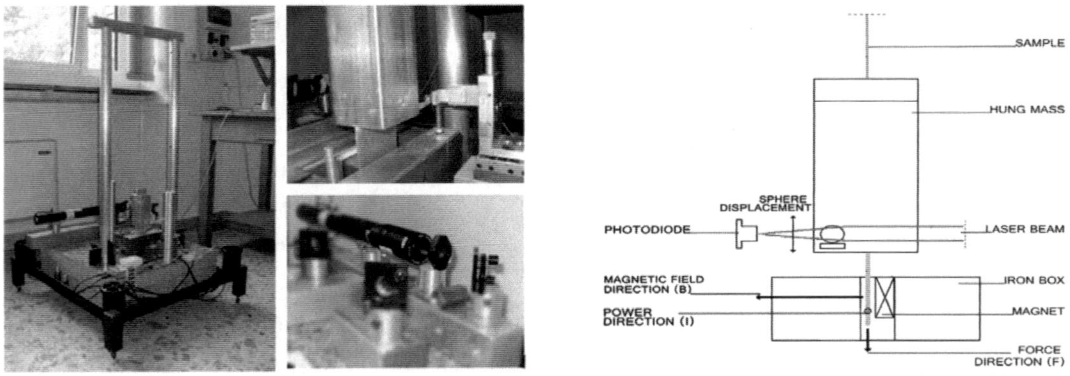

Figure 2. (a) The testing setup. (b) Micrometer slides, magnet for forced oscillation testing and hung mass. (c) He-Ne laser beam generator, attenuator and mirrors. (d) Schematic representation.

Therefore, a laser generator, an attenuator, 3 mirrors, a photodiode and a micrometre slide compose the measurement equipment.

A helium-neon generator was used for the experiment (Figure 2.c) in order to produce the laser beam. The device is able to emit a coherent, monochromatic and concentrated light beam in a highly collimated beam straight, through the process of stimulated emission. The laser generator is mounted on a suitable support, so that the laser beam was emitted in the horizontal direction with respect to the anti-vibration table.

An attenuator is an electronic device that reduces the power of a signal without appreciably distorting its waveform. The attenuator was used to prevent saturation of the photodiode, attenuating the beam light (Figure 2.c).

Metal mirrors were used to adjust the direction of the laser beam. They were mounted on supports in order to be adjusted for different inclinations. The combined movement of rotation about the horizontal and vertical axes of the two mirrors allows to determine completely the direction of the light beam (Figure 2.c).

A photodiode, which converts an optical signal to an electric signal, was mounted on three precision micrometre slides. Two of these were assembled together in reference to the X-direction (parallel axis to the plane formed by the optical bench and orthogonal to the laser beam) and Z-direction (orthogonal axis to the plane formed by the optical bench and orthogonal to the laser beam direction). The third slide is positioned in a parallel direction to the bench and the laser beam (Figure 2.b).

In Figure 2.b it is also possible to notice the hung mass of 1.375 kg which was used to form a 1-DoF system with the specimen acting as spring. A copper wire coil was placed inside a magnet and aligned so that an electromagnetic force in the vertical direction could be produced and applied to the hung mass and, consequently, to the fibres (Figure 2.b).

2.1. Principle of operation

The principle of operation of the instrumental apparatus used in the tests (Figure 3) can be summarized in the following stages. The light source produced by the laser generator was forced to follow a predetermined path by means of a prism and two lenses. These lenses, if properly set, deflected the laser beam causing it to converge towards the instrument test, where the sample was allocated. In order to allow the photodiode to read the optical signal and, consequently, the possible displacement of the fiber, a glass sphere (connected to the hung mass) with a diameter of 8 mm was used, which, working as a convex lens, focuses the beam to a very precise point of the marker located inside the photodiode (Figure 2.b). The marker was divided into four sectors, generally square. The sensor read the movement of the glass sphere and indirectly the elongation of the fiber, in relation to the quadrant where the laser beam was focused. Therefore, it was important to make a correct positioning of the photodiode using the micrometer slides. Indeed, for the system in a quiescent state, the read signal was set equal to zero. An electronic circuit was used to link the photodiode to an oscilloscope. An acquisition card and a Lab View application were used to record the collected data on csv files.

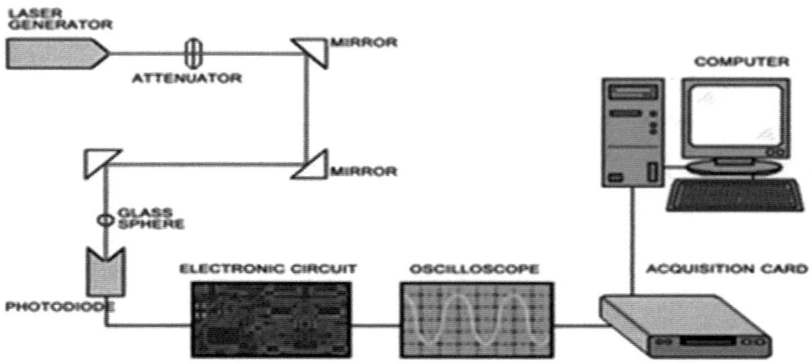

Figure 3. Principle of operation - schematic representation.

3. Samples and UV damage

The samples used for the experimental tests were provided by SIGIT s.n.c [7], which trades high-tech materials. In this specific case, the material under investigation is a para-aramid yarn of Kevlar-29 (100% purity), made with DuPont production, length 0.67 m. The bundle of fibres in Kevlar-29, used as a sample, has counted equal to 2500 dTex; this means that 10000 m of yarn (in length) weigh 2500 g. Thanks to these data, it was possible to evaluate with good accuracy the value of the cross section area of the sample that is equal to $1.77 \cdot 10^{-7}$ m^2; the diameter of the single fibre is 15 μm and thereby, the yarn is composed by about 1000 fibres. To obtain the elastic modulus, tensile tests were carried out on specimens with a length of 0.067 m, and the result is 84.5 GPa on single fibre.

The type of process chosen to cause damage has been the prolonged exposure to ultraviolet rays. Many natural and synthetic polymers are attacked by UV rays. This problem is known as UV degradation, and is especially common in products constantly exposed to sunlight. In this work, this method of damage was chosen because a continuous exposure results to be more harmful than intermittent one, since the danger depends on the intensity and the degree of exposure. The polymers possessing UV-absorbing groups such as aromatic rings can be subjected to the effect of ultraviolet radiation. In particular, aramid fibres such as Kevlar, are very sensitive to UV light and must be often protected from the deleterious effects that the sunlight might have.

In order to perform such a test, a UV lamp for industrial applications, an OSRAM Ultra Vitalux, was chosen. The composition of the radiation of this light is similar to the mixture of radiation of natural sunlight, in the alpine environment. It emits a mixed light, generated from a tungsten filament in conditions of high pressure; furthermore, the special mushroom shape, with internal reflector, ensures the passage of a high fraction of UVA and UVB, while it tends to absorb more harmful UVC radiation.

Regarding the damage test, the lamp was mounted on a support constructed ad hoc in the Metrology Laboratory of Politecnico di Torino with recycled materials, so that the distance between the edge of the lamp bulb and the sample was about 0.50 m. A cylinder made of sheet metal with a diameter of 0.13 m was used for conveying light toward the sample (which must be damaged) and there was also taken care to leave and open space below the cylinder, to ensure that natural recirculation of air during the test.

The sample was then subjected to a cycle of continuous exposure to 772 hours, turning the fibres once a week, in order to obtain a uniform damage as possible over the entire surface of the material.

4. Experimental tests and results

Once all possible uncertainties on the measurements of the testing machine were analysed and handled with, real data begun to be acquired. Through experimental measurements, it is possible to characterise the material under investigation and identify the behaviour from the dynamic point of view. It should be underlined that, in order to compare the results obtained from the pristine and damaged fibres, the experimental tests were performed at a constant temperature (21°C).

The dynamic response of a system can be analysed in different ways. Free-decay and forced oscillations at resonance tests were carried out on both pristine and damaged specimens. The mass of 1.375 kg was sufficient to let the fibers tight and it induced a tension that corresponded roughly to a 10% of the tensile strength for the Kevlar®-29 samples.

4.1. Free-decay testing

In order to characterise the dynamic behaviour of the fibres, free-decay tests were performed. The tests allowed for the identification of natural frequencies and damping ratios of both specimens. The system identification was achieved with time domain fitting techniques applied to the Hilbert transform envelops of the acquired signals (Figures 4 and Figure 5). Results were also validated by using standard ERA techniques [8].

The application of curve fitting techniques over free-decay signals sampled at 2500 Hz gave for the Kevlar-29 pristine samples, a resonance frequency of (20.30±0.03) Hz and an experimental damping ratio of (0.7±0.1) %; whilst for the Kevlar-29 damaged samples, a value for natural frequency of (19.80±0.02) Hz and experimental damping ratio equal to (0.6±0.1) % were evaluated. The spread of the results was computed using 7 free-decay measurements for both samples. The high sensitivity of the measurement device allows to quantify low damping ratio and frequencies with a high level of accuracy. These damping ratios can be directly linked to the intrinsic material dissipation factor, because all other dissipation phenomena can be actually neglected. The mean value of the natural frequency was used to estimate the elastic dynamic modulus of the fibres.

The results show that there is a relevant variation of the natural frequency correlated to the damage suffered by the continuous UV exposure for 772 hours the Kevlar-29 sample.

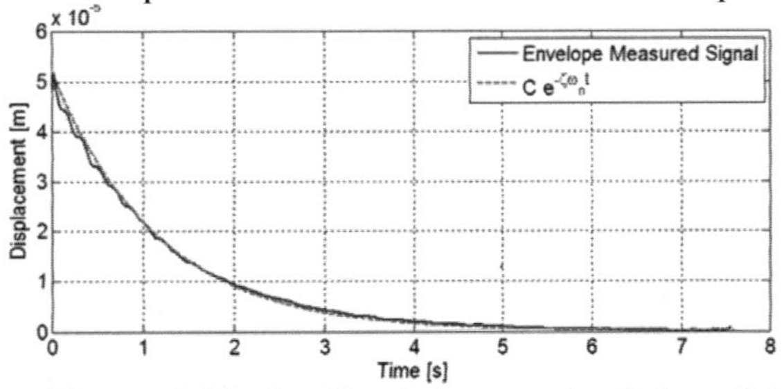

Figure 4. Kevlar-29 pristine sample: fitting of the envelope of the acquired signal.

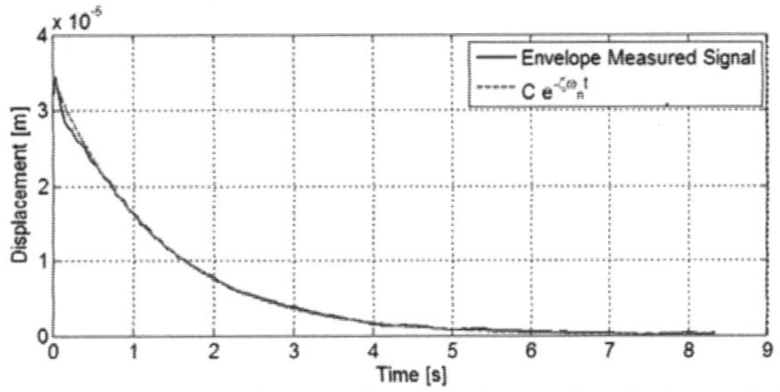

Figure 5. Kevlar-29 damaged sample: fitting of the envelope of the acquired signal.

4.2. Forced oscillations

In order to highlight possible non-linearities, campaign with harmonic forced excitation tests was set up. The frequency range investigated was of ±0.200 Hz with respect to the resonance frequency of the system, with a step excitation frequency chosen equal to 0.010 Hz while keeping the force constant. Then the test was repeated for different level of force, the amplitude being varied by adjusting the voltage applied through a function generator in steps from 3 V to 10 V, corresponding to forces induced by the electromagnetic field that were in the approximate range of 10-30 mN (see Figure 2.b and 2.d). The value of the amplitude of the static force at resonance was estimated at approximately 0.9 N at the lowest level applied, and 3.1 N at the highest.

Once the input voltage interval has been defined, changing the frequency of the driving force (via the function generator) it is possible to record the values of the output voltage, from which the displacement can be determined by using a conversion factor established during the photodiode calibration. The results are reported in Figure 6 and Figure 7.

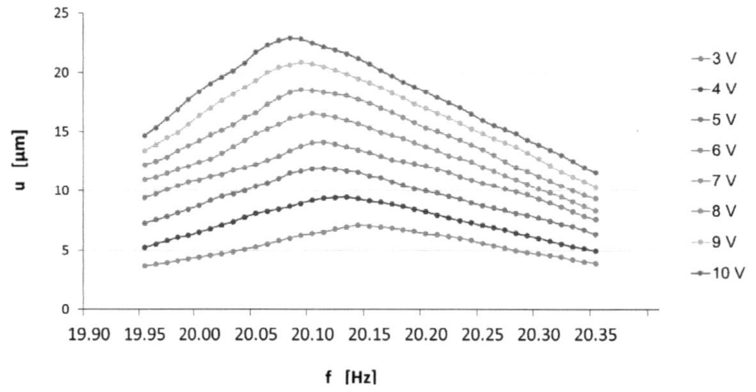

Figure 6. Response spectra for increasing values of the input force – pristine sample.

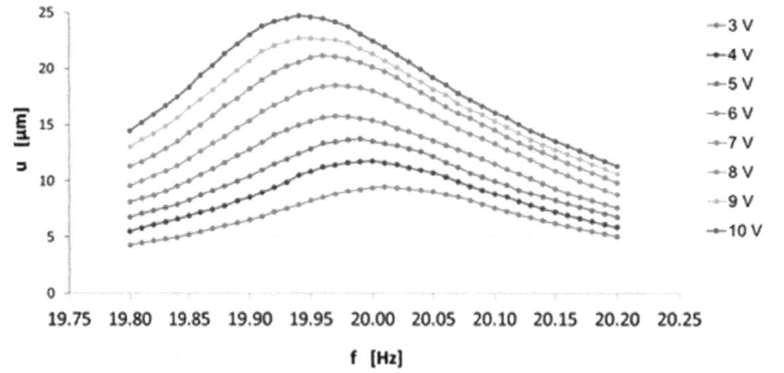

Figure 7. Response spectra for increasing values of the input force – damaged sample

4.2.1. Non-linear identification of the backbone curve. To define the dynamics of a system characterised by cubic non-linearity excited by a harmonic force, the following expression found in literature is exploited. The distortion in frequency f was quantified as:

$$f = f_n \left[1 + \frac{3}{8}\frac{\chi}{k}a^2\right] \tag{1}$$

where a is the amplitude of the response, k is the linear stiffness, f_n is the natural frequency and χ is the cubic stiffness coefficient. Equation (1) is related to the following Duffing oscillator equation:

$$m\ddot{x} + c\dot{x} + kx + \chi x^3 = F \tag{2}$$

where F is the external force, m the mass of the system and c the linear viscous damping.

In order to try to identify the non-linearity, the Krylov-Bogoliubov method was used to determine the variation of the first non-linear resonance frequency of a Duffing oscillator [6, 9].

Equation (1) was used to identify the so-called back-bone non-linear curve of the response, as it is highlighted by a red line in Figure 8 and Figure 9.

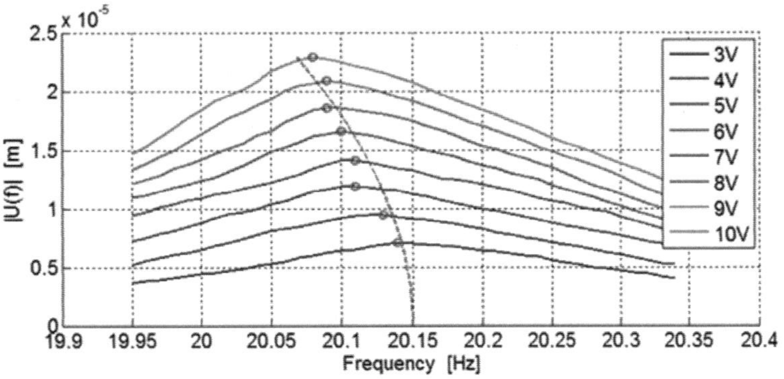

Figure 8. Back-bone nonlinear curve identification - undamaged Kevlar-29 sample.

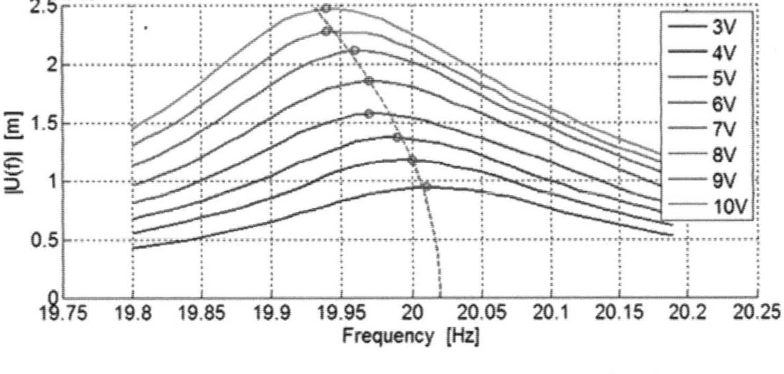

Figure 9. Back-bone nonlinear curve identification - damaged Kevlar-29 sample.

For the first Kevlar-29 (not damaged sample) the non-linear stiffness χ was identified at -3.23*10^{11} N/m^3, whilst for the second Kevlar-29 sample (after damage) it was ca -3.04*10^{11} N/m^3.

From Figures 8 and 9, where the backbone non-linear curves were represented, no significant differences can be appreciated between the pristine and damaged samples, as regards the inclination of the back-bone curve and the value of the index of non-linearity.

Anyway, this approach confirmed the high-sensitivity of the testing device also to low-levels of non-linearity.

Conclusions

The aim of this research has been to perform simple non-destructive non-linearity tests for damage assessment of aramid fibres and elements. The underlying idea was to use the existence of weak non-linearities, which can be detected in the forced dynamic response of the specimen, as indicators for safety and lifecycle assessment. To reach this goal, two samples of the same innovative material were tested: one pristine bundle of fibres in Kevlar-29 and another bundle which had been damaged.

Comparing the results obtained for the two samples, no significant differences emerged with respect to the non-linearity index under investigation ($-3.23\cdot10^{11}$ N/m^3 for the pristine sample as compared to $-3.04\cdot10^{11}$ N/m^3 for the damaged sample). Instead a weak variation was evident in the natural frequency value related to the free-decay test responses (20.30 ± 0.03) Hz in the pristine sample vs. (19.80 ± 0.02) Hz, in the UV damaged sample).

The experimental results presented in this paper will be of use in evaluating the actual feasibility of Kevlar retrofitting interventions on existing structures. In particular, the demonstrated sensitivity to UV rays, in terms of decreased stiffness and potential relaxation, may be a discriminating factor when choosing the type of aramid material to be employed for the confinement interventions on concrete or masonry columns, or vaults in historic buildings.

References

[1] C.J. Burgoyne, "Structural use of parafil ropes", Construction and Building Materials, 1(1), 3-13, 1987.

[2] L. Hollaway, C.J. Burgoyne, "Furhter applications of polymer and polymer composites", in L. Hollaway "Polymer and Polymer composites in contruction" Thomas Telford, ch. Chapter 11, 1990.

[3] F. Bastianini, M. Corradi, A. Borri, A. di Tommaso, "Retrofit and monitoring of an historical building using "Smart" CFRP with embedded fibre optic Brillouin sensors", Construction and Building Materials 19, 525-535, 2005 .

[4] G.V. Demarie, R. Ceravolo, A. De Stefano, "Instantaneous identification of polynomial nonlinearity based on Volterra series representation", Key Eng. Mater., 293-294 703-710, 2005.

[5] R. Ceravolo, S. Erlicher, L. Zanotti Fragonara, "Comparison of restoring force models for the identification of structures with hysteresis and degradation", Journal of Sound and Vibration, 332 (26), 6982-6999, 2013.

[6] A.H. Nayfeh, D.T. Mook, "Nonlinear Oscillations", John Wiley & Sons, New York, 1995.

[7] S.I.G.I.T s.n.c (Turin Company), production of textiles, http://www.sigit.to/eng/index.html

[8] J. N Juang, R. S Pappa, "An eigensystem realisation algorithm (ERA) for modal parameter identification and modal reduction." NASA/JPL Workshop on Identification and Control of Flexible Space Structures, 1984.

[9] T. Kalmar-Nagy, B. Balachandran, "Forced harmonic vibration of a Duffing oscillator with linear viscous damping", in I. Kovacic, M.J. Brennan "The Duffing equation: nonlinear

Analysis of elastic nonlinearity for impact damage detection in composite laminates

A Frau[1], L Pieczonka[2], M C Porcu[1], W J Staszewski[2] and F Aymerich[1]
[1] University of Cagliari – Department of Mechanical, Chemical and Materials Engineering, Piazza d'Armi, 09123 Cagliari, Italy
[2] AGH University of Science and Technology - Department of Robotic and Mechatronics, Al. A. Mickiewicza 30, 30-059 Krakow, Poland

E-mail: andrea.frau@unica.it, lukasz.pieczonka@agh.edu.pl, mcporcu@unica.it, w.j.staszewski@agh.edu.pl, francesco.aymerich@dimcm.unica.it

Abstract. This paper concerns the experimental analysis of nonlinear response features of a composite laminate plate for impact damage detection. The measurement procedure is based on the Scaling Subtraction Method (SSM) and consists in exciting the damaged specimen with two sinusoidal signals at different amplitude. The linearly rescaled response signal at low amplitude excitation is subtracted from the response at large amplitude excitation to extract the nonlinear signatures. The latter are analysed in the time domain to infer the presence of damage. Results are compared with frequency domain analyses using the nonlinear vibro-acoustic modulation technique (NWMS). Changes in amplitude and phase as well as modulation effects of the acquired responses are also monitored. Surface-bonded, low profile piezoceramic transducers are used for excitation and sensing. Both measurements techniques are applied to detect barely visible impact damage in laminate composite plate. Non-destructive penetrant-enhanced X-ray inspections are carried out to characterize the extent of internal damage. The behavior of the nonlinear features and the sensitivity of each technique are also investigated in the paper.

1. Introduction

Non-destructive evaluation techniques (NDE) are commonly adopted to assess structural integrity. Among them, the NDE methods based on ultrasounds are very successful in detecting structural damage, owing to the close relation between wave propagation and material properties. In recent years there has been an increasing interest in methods which investigate the nonlinear acoustic response of materials [1]. The sensitivity shown by such techniques is of particular interest when composite materials, which are typically susceptible to impact damage, are concerned. The nonlinear acoustic phenomena affecting the elastic response to dynamic excitations are related to material anharmonicity and can be radically enhanced by the presence of small defects, micro cracks or surfaces in friction. Resonance frequency shift, higher/sub harmonic generation, modulation of elastic waves are just some of the experimentally observed effects that reveal the presence of damage. Although different theoretical models (classical nonlinear theory of elasticity, contact acoustic nonlinearity models or hysteretic models) have been proposed to explain the generation of nonlinear effects, the physical mechanism behind these experimental evidences is still a matter of scientific debate [2].

In general, the nonlinear acoustic techniques take into consideration some particular features of the nonlinear elastic response such as higher harmonics or modulation sidebands, whilst they neglect other aspects such as amplitude nonlinearities occurring at the fundamental frequency. Due to the slight variations that may typically occur between intact and damaged case, especially when considering early damage stages, these methods are often effective only in laboratory environments. Other limitations may also come from intrinsic material nonlinearities, boundary conditions and inherent nonlinearities related to the acquisition instruments.

A new nonlinear ultrasound technique, relying on nonlinear effects such as amplitude attenuation and phase delay at the fundamental frequency has been recently proposed [3] as a valid alternative to traditional frequency domain techniques. Referred to as Scaling Subtraction Method (SSM), it is based on the assumption that the nonlinearities related to cracks and material defects break the proportionality between excitation and response of the structure. This nonlinear contribution is negligible for low amplitude excitations and becomes progressively more significant as the amplitude of the excitation increases. Based on this assumption, some damage indicators are introduced in [3] to identify the nonlinearities related to the damage. The SSM technique has been experimentally applied to characterize the behaviour of materials with different degree of nonlinearities (lead, steel, mortar, concrete) [4] and to describe damage evolution [5].

This paper presents an application of the SSM to detect barely visible impact damage in a composite laminate plate. The specimen was analysed in both intact and damaged conditions. Non-destructive penetrant-enhanced X-ray inspections were also performed to identify and characterize internal damage. Surface bonded, low profile piezoceramic transducers were used in the test both to excite the plate with sinusoidal signals at different amplitudes and to acquire the response in different points. The SSM damage indicators are analysed in the time domain and then compared with the results of a frequency domain analysis carried out with the Nonlinear Acoustic Wave Modulation technique (NWMS) [6,7,8]. The layout of the paper is as follows. The SSM and NWMS approaches for damage detection are briefly introduced in section 2. Section 3 describes the experimental procedures adopted in the tests. Damage detection results are provided in section 4 and the paper is concluded in section 5.

2. Scaling Subtraction Method and Nonlinear Acoustic Wave Modulation Technique

The idea behind the SSM is that the nonlinear contribution related to the damage begins to show its effects only when the amount of energy transferred to the sample reaches a certain value [4]. It follows that low amplitude excitation signals do not have sufficient energy to excite nonlinearities and the corresponding output signal is expected to be equivalent to the one obtained from an intact structure. Damage is typically defined as a change introduced into a material that adversely influences its parameters. In particular, discontinuity (like crack or delamination) introduces nonlinearity into a material, therefore the analysis of nonlinearity can be used for the purpose of damage detection. If no damage is present inside the specimen, i.e. the material can be assumed linear, the response of the structure at a higher excitation signal should be perfectly proportional (except from noise effects) to that obtained at the lowest excitation level. On the contrary, when the structure is damaged, the proportionality should no longer be valid due to the nonlinearities related to damage.

The SSM approach consists at first in exciting the damaged specimen with a sinusoidal signal at a comparatively low amplitude A_{low} and detecting its response v_{low}. The latter is thus linearly rescaled in amplitude to obtain the output signal v_{ref} that is expected to be the same as that of an intact specimen excited with a scaled amplitude $A_{\text{high}}=kA_{\text{low}}$ signal; k being the scale factor. The response of the structure v_{high} at an excitation amplitude equal to A_{high} is then acquired. The difference between the recorded signal v_{high} and the linearly rescaled signal v_{ref} is referred to as the scaled subtracted signal and can be expressed as

$$w(t)=v_{\text{high}}(t)-v_{\text{ref}}(t)=v_{\text{high}}(t)-kv_{\text{low}}(t) \qquad (1)$$

If the structure is damaged, the subtracted signal *w(t)* will contain the effects of the nonlinearities related to damage such as amplitude and phase variations at the fundamental frequency as well as at higher order harmonics. Conversely, the subtracted signal vanishes (except for noise effects) when the material is perfectly linear.

Two indicators may be adopted to assess the nonlinearity content, see also [3,4]. The first one refers to the maximum amplitude of the subtracted signal

$$\alpha = \max(w) + \left|\min(w)\right| \tag{2}$$

The second indicator is the root mean square of the subtracted signal

$$\beta = \left(\frac{1}{n} \sum_{i=1}^{n} w_i^2 \right)^{1/2} \tag{3}$$

Here n and w_i are the number of sampled points and the amplitude of the ith point of the subtracted signal, respectively. Once analysed in the time domain, the signals can be investigated in the frequency domain to put in evidence magnitude and phase differences between the linearly rescaled signal and the acquired one.

An application of the NWMS technique to the same specimen is also provided for comparison. This technique relies on the analysis of the modulation effect on a weaker high frequency probing wave by a stronger low frequency pumping wave that occurs in presence of structural nonlinearities. Exciting a damaged structure simultaneously with high and low frequency signals, the sideband components around the high frequency peak can be observed in the response signal as a result of the modulation effect. If the structure is intact this effect is less pronounced whilst it becomes more and more evident as damage increases. The frequencies at which the sidebands occur are

$$f_{S_n} = f_H \pm n f_L \tag{4}$$

where f_{Sn} is the frequency of the nth sideband, f_H is the frequency of the ultrasonic wave and f_L is the resonance low frequency. To quantify the extent of modulation it may be useful to refer to the following parameter

$$R = \frac{(A_1 + A_2)}{A_0} \tag{5}$$

relating the A_1 and A_2 of the first pair of sidebands to the amplitude A_0 of the high-frequency ultrasonic component.

3. Experimental Studies

This section presents the experimental procedure involved in the current investigations. The composite specimen together with the experimental equipment and the X-ray tests performed to assess the severity of damage are described in detail.

3.1. Composite specimen and experimental equipment

The specimen used in experimental tests was a rectangular 220 × 160 mm laminated composite plate with $[0_2/90_2]_s$ stacking sequence. The average laminate thickness was equal to 2 mm. The plate was cut from a panel made up from *Seal Texipreg® HS160/REM* carbon epoxy prepreg layers and cured in autoclave at a maximum temperature of 160 °C and pressure of 6 Bar.

The plate was instrumented with four *PI Ceramic PIC 151* low-profile piezoceramic transducers (10 mm diameter, 0.25 mm thickness) and a *PI PL055.31, PICMA* piezoceramic stack actuator positioned as shown in figure 1 and 2. The stack actuator and three piezo transducers (A, B, C) were bonded on the same surface of the impact damage whilst the remaining transducer (D) was bonded in the opposite surface in correspondence with the transducers B. The transducers were surface bonded using a two-component epoxy adhesive and wired using additionally bonded connectors. The panel was freely suspended using thin nylon cords to minimize any nonlinearity associated with structural boundaries.

The excitation signals transmitted to the piezoceramic transducers were generated using a *TTI TGA5011A* function generator and amplified twenty times with a *FLC F20A* high voltage linear amplifier. The signals exciting the stack actuator were generated by a *TTI TG2000* function generator and amplified ten times with a *PI E-508 PICA* piezo amplifier. The main equipment used in the experimental investigation is shown in figure 3. A drop-weight impact testing machine with a 2.2 kg impactor was used in order to introduce damage onto the plate. The impactor was provided with a hemispherical indentor of 12.5 mm diameter and instrumented with a semiconductor strain-gauge bridge for dynamic load acquisition. An impact of 4.6 J energy was introduced at central position of the plate in order to obtain a delamination in the area of Barely Visible Impact Damage (BVID).

Non-destructive penetrant-enhanced X-ray inspections were carried out to identify and characterize internal damage. Radiographic analyses were done with a *HP Faxitron* cabinet. An X-ray picture of internal damage after the impact is shown in figure 4. Damage induced by the impact mainly consists of a delamination developing on the 90°/0° interface farthest from the impact side of the panel, together with a major matrix crack in the bottom 0° layers and shear matrix cracks in the 90° layers.

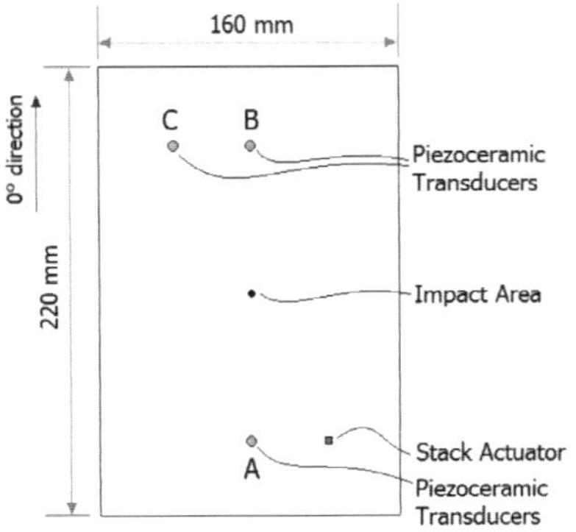

Figure 1. Representation of the composite plate.

Figure 2. Experimental set-up for the composite plate.

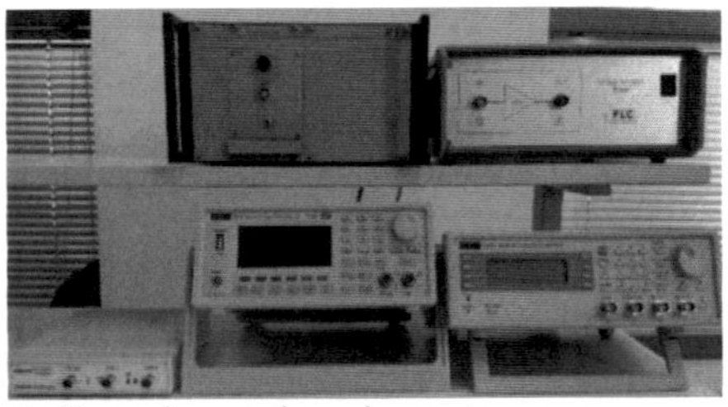

Figure 3. Experimental equipment.

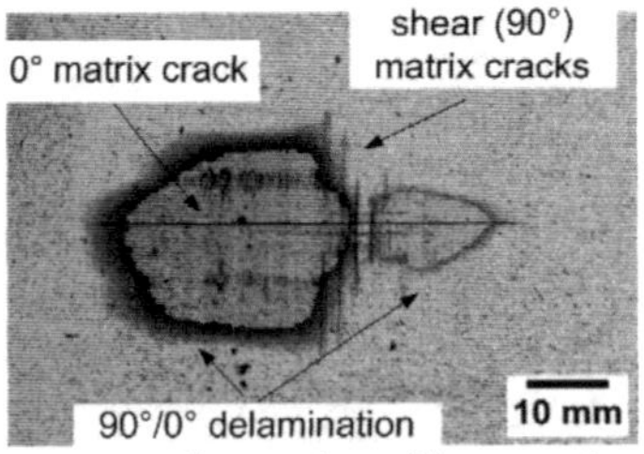

Figure 4. X-radiographs of impact damage.

3.2. Experimental procedure for damage detection – Scaling Subtraction Method

A modal analysis of the intact plate was firstly performed to find natural frequencies of the specimen. A sweep sine signal was used to excite the plate in the range between 10 and 60 kHz. Within this range, a resonant frequency of 10340 Hz was selected for further analyses as it had a clear frequency peak and good signal-to-noise ratio.

To perform the SSM test, the plate was firstly excited with a low amplitude (0.5 V peak-to-peak before amplification) sinusoidal wave applied to the piezoceramic transducer A. When stationary conditions were reached, the response signals were acquired from sensors B, C and D. A time window of 50 ms (517 cycles) and a sample frequency of about 10 MHz have been adopted in the test. The same procedure was then applied by varying the excitation amplitude between 1 and 8 V peak-to-peak (sixteen times the amplitude of the first excitation level) with a step of 1 V peak-to-peak. Each signal was averaged 256 times in order to improve the signal-to-noise ratio. Excitation and acquisition signals were triggered and synchronized using the external trigger channel of the oscilloscope. The testing procedure was applied to the panel in both intact and damaged conditions.

Through a post-processing procedure, the offset of the signals was removed and the output signal related to the 0.5 V peak-to-peak excitation was rescaled in amplitude to be compared with those acquired at higher amplitudes (figure 5). The subtracted signals for each excitation level were calculated and analysed in time and frequency domain. The nonlinear indicators α and β, defined in equations (2) and (3), were then obtained and plotted against the excitation amplitude to assess nonlinear effects in the time domain. Once transformed in the frequency domain, the amplitude/phase differences at fundamental frequency between the linearly rescaled signals and the ones acquired at higher amplitudes were analysed and plotted against the excitation amplitude.

3.3. Experimental procedure for damage detection – Nonlinear Acoustic Wave Modulation

The same specimen was also analysed through the NWMS test, by simultaneously applying two excitation signals. The first one, excited by the *PI PL055.31* stack actuator, was a low-frequency sinusoidal wave at the resonant frequency of 410 Hz. The second signal, excited by a *PI Ceramic PIC 151* piezoceramic transducer, was a high-frequency ultrasonic wave. Two different resonant ultrasonic frequencies have been considered in the test, namely 28460 Hz and 45340 Hz. Transducers A, B and C were used in turn as actuators or sensors. The amplitude of the high-frequency signal was fixed at 2 V peak-to-peak (before-amplification) for both frequencies. Various amplitude levels of the low-frequency excitation were considered, ranging from 1 to 10 V peak-to-peak (before amplification). The test was performed in both intact and damaged conditions. The amplitude of the spectral sidebands was analysed by means of the R parameter defined in equation (5). The extent of modulation effects was investigated for different values of the high frequency excitation and for different configurations of excitations and sensing.

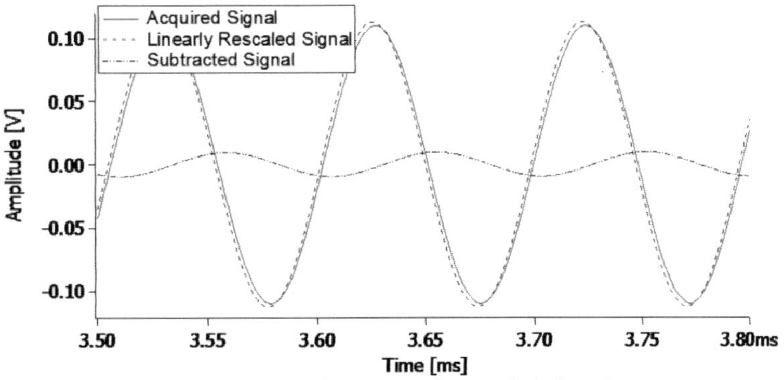

Figure 5. An instance of recorded signals.

4. Results and discussion

4.1. Scaling Subtraction Method

The nonlinear indicators α and β, are plotted in figures 6, 7 and 8 against the amplitude of the excitation signal (before amplification). Diagrams refer to the signals acquired from sensors B, C and D when the excitation is applied at sensor A. Both indicators α and β show that the difference between the linearly rescaled signals and the corresponding acquired ones increases as the input excitation increases. Although a nonlinear behaviour is detected also in the intact case, the extent of nonlinearities in the damaged case is found to be rather more marked.

To highlight the contribution to nonlinearities given by amplitude and phase variation, the recorded signals were analysed in the frequency domain. Figures 9 and 10 show the amplitude and phase differences between undamaged and damaged cases. Relevant to the fundamental frequency, the difference in amplitude between the linearly rescaled signals and the recorded ones is plotted in figure 9a as a function of the amplitude of the input excitation (in this case the signals are recorded at sensor B). In the damaged case, the peak amplitude of the rescaled signal is slightly lower than that of the recorded signal (this leads to positive values of the difference between the two signals as shown in figure 9a). A similar behaviour is found in the intact case for lower excitation values whilst an opposite trend can be seen for higher values. Phase differences are plotted in figure 9b, showing that in the damaged case the rescaled signal always exhibits a phase delay with respect to the acquired one, which actually increases as the excitation amplitude increases. On the contrary, in the intact case the phase delay is very small, at least for low amplitude excitations. Figures 9 and 10 show that both amplitude and phase differences contribute to the damage indicators, even if a slightly larger influence of phase may be noted.

4.2. Nonlinear Wave Modulation Technique

Figure 11 gives an instance of the zoomed spectra obtained for the 45340 Hz ultrasonic excitation. The peak of the fundamental high-frequency signal and the first pair of sidebands are clearly visible in both undamaged and damaged conditions. The amplitude of these sidebands was analysed through the nonlinear parameter R. Some of the results obtained from the sideband analysis are provided in figure 12, where the parameter R is plotted against the low-frequency excitation level. In both damaged and undamaged cases the amplitude of the spectral sidebands increases with the low-frequency excitation amplitude. Generally, the values of R are higher in the damaged case and the differences between damaged and undamaged case become more and more evident as the low-frequency signal amplitude increases. The results show no significant difference between the two ultrasonic frequencies used. However, a more marked difference between damaged and undamaged cases was found when the high frequency excitation was transmitted from piezoceramic transducers located in B and C (that is, when high and low frequency excitations were in opposite positions with respect to the damage).

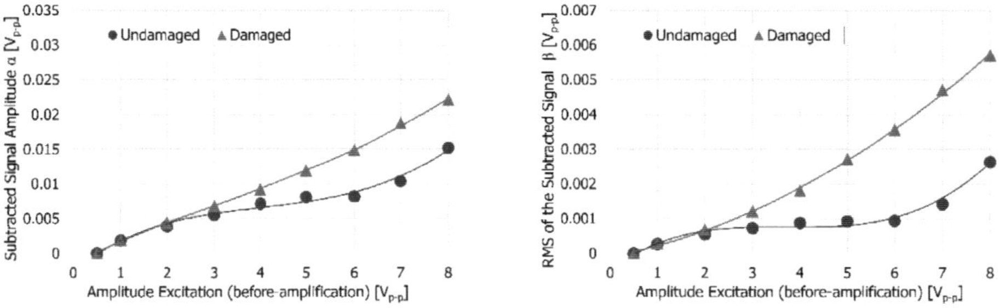

(a) **(b)**

Figure 6. Nonlinear indicators α (a) and β (b) of the subtracted signal obtained for sensor B.

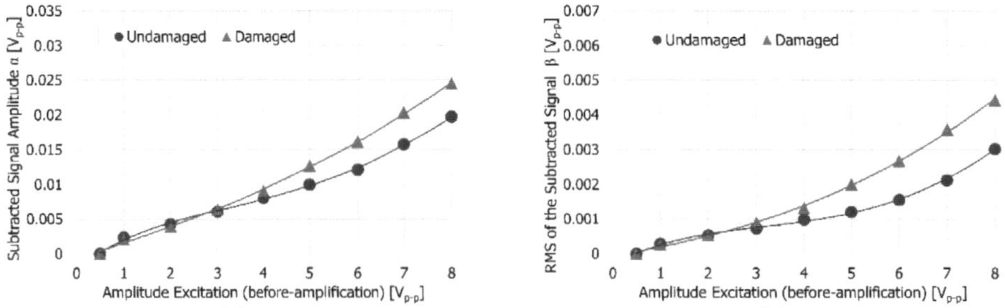

(a) **(b)**

Figure 7. Nonlinear indicators α (a) and β (b) of the subtracted signal obtained for sensor C.

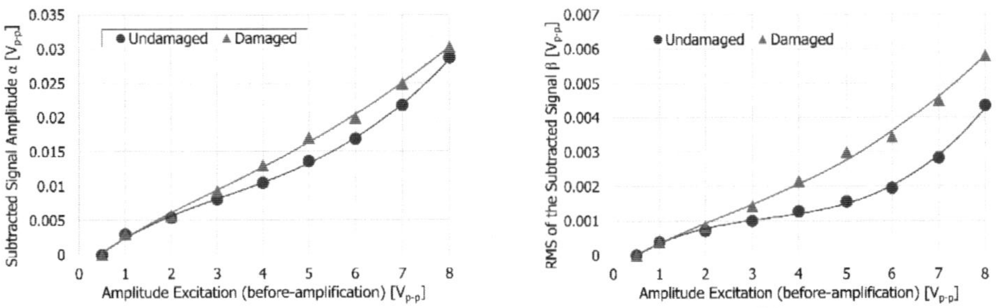

(a) **(b)**

Figure 8. Nonlinear indicators α (a) and β (b) of the subtracted signal obtained for sensor D.

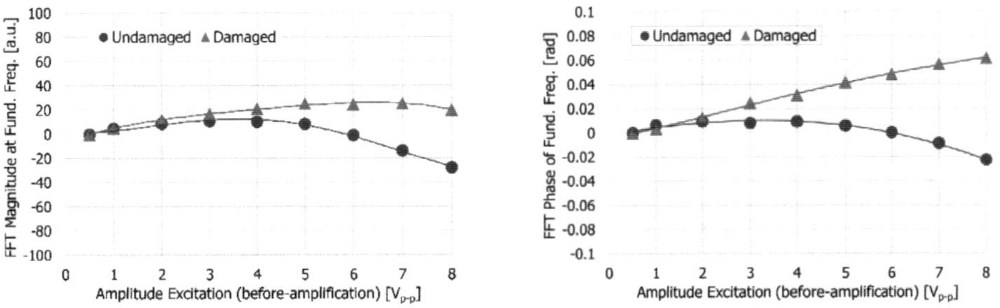

(a) **(b)**

Figure 9. Amplitude (a) and phase (b) differences obtained in correspondence of the fundamental frequency peak.

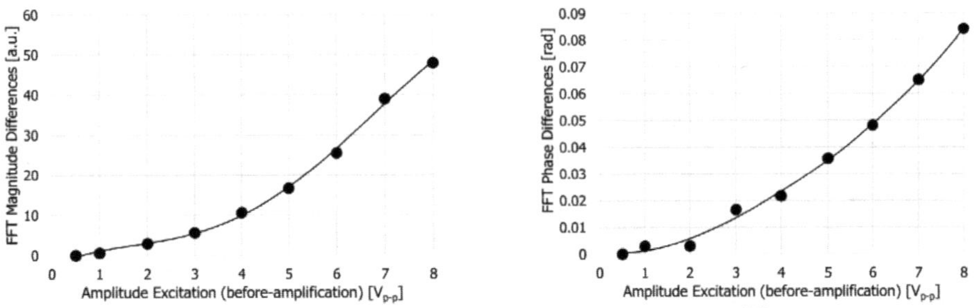

(a) **(b)**

Figure 10. Amplitude (a) and phase (b) differences trend between intact and damaged cases.

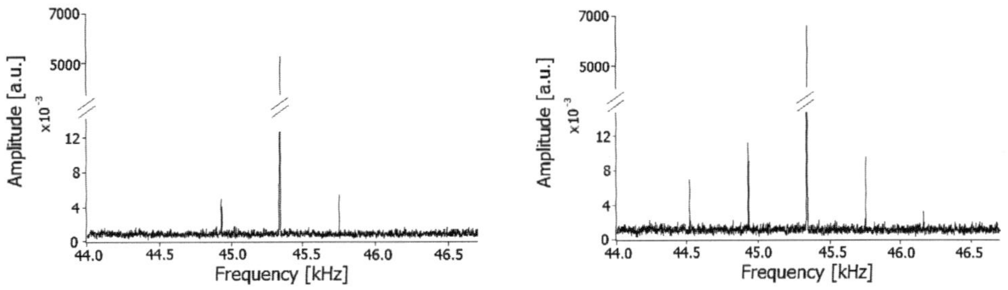

(a) **(b)**

Figure 11. Examples of the power spectra for the undamaged (a) and damaged (b) cases.

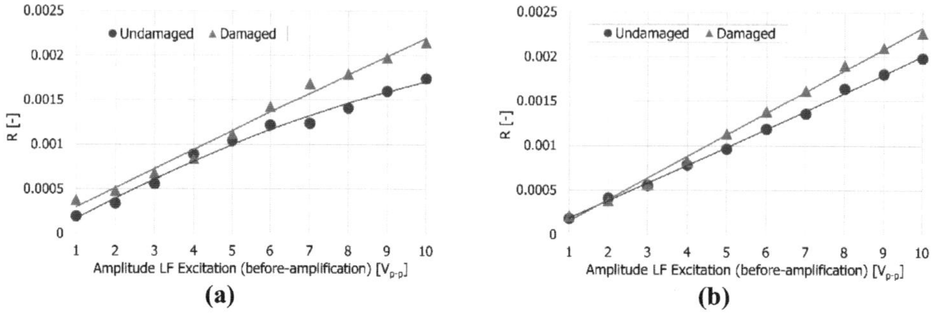

Figure 12. Intensity of modulation vs. low-frequency excitation level for 28460 Hz (a) and 45340 Hz (b) frequency excitations.

5. Conclusions

An application of Scaling Subtraction Method and Nonlinear Acoustic Wave Modulation techniques has been presented in the paper. The test specimen was a composite laminate plate with a barely visible impact damage. Surface-bonded piezoceramic transducers were used as actuators and as sensors. The response of the specimen has been analysed in time and frequency domain. Amplitude/phase nonlinearities and modulations effects have been analysed in order to discriminate between damaged and undamaged conditions. Although a nonlinear behaviour is detected also in the intact case, the extent of nonlinearities in the damaged case is found to be rather more evident for both the analysed techniques. When applied to composite specimens, the SSM may thus leads to a similar understanding of the nonlinear phenomena to those obtained for the NWMS technique. Further analyses are required to confirm these findings. Different kinds of composites and different damage severities should be investigated for monitoring the evolution of the nonlinear behaviour. Future works should also attempt to explain the obtained results with respect to the interaction mechanism between waves and damage.

6. References

[1] Jhang K Y 2009 *Int. J. Precis. Eng. Manuf.* **10** 123–35
[2] Broda D, Staszewsky W J, Martowicz A, Uhl T and Silberschmidt V V 2014 *J. Sound Vib.* **333** 1097–118
[3] Scalerandi M, Gliozzi A S, Bruno C L E, Masera D and Bocca P 2008 *Appl. Phys. Lett.* **92** 101912
[4] Bruno C L E, Gliozzi A S, Scalerandi M and Antonaci P 2009 *Phys. Rev.* B **79** 064108
[5] Antonaci P, Gliozzi A S, Griffa M and Scalerandi M 2013 *E-J. of Nondestructive Testing* **19**
[6] Aymerich F and Staszewski W J 2010 *Compos.* A **41** 1084–92
[7] Klepka A, Pieczonka L, Staszewski W J and Aymerich F 2014 *Compos.* B **65** 99–108
[8] Pieczonka L, Ukowski P, Klepka A, Staszewski W J, Uhl T and Aymerich F 2014 *Smart Mater. Struct.* **23** 105021

Acknowledgments

Andrea Frau gratefully acknowledges Sardinia Regional Government for the financial support of his PhD scholarship (P.O.R. Sardegna F.S.E. Operational Programme of the Autonomous Region of Sardinia, European Social Fund 2007-2013 - Axis IV Human Resources, Objective l.3, Line of Activity l.3.1.). The second author would like to acknowledge financial support from the Foundation for Polish Science (FNP) within the scope of the WELCOME Programme-project no. 2010-3/2.

Non-destructive visualization of linear explosive-induced Pyroshock using phase arrayed laser-induced shock in a space launcher composite

Jae kyeong Jang[1, 2] and Jung Ryul Lee[2, 3, 4]

[1]PhD student, Department of Aerospace Engineering, [2]LANL-CBNU Engineering Institute-Korea, [3]X-NDT Inc., Chonbuk National University, 567 Baekje-daero, Duckjin-gu, Jeonju, Jeonbuk, 561-756, Republic of Korea
[4]Associate Professor, Department of Aerospace Engineering, Korea Advanced Institute of Science and Technology, 291 Daehak-ro, Yuseong-gu, Daejeon, 305-338, Republic of Korea

E-mail: leejrr@kaist.ac.kr

Abstract. Separation mechanism of Space launch vehicles are used in various separation systems and pyrotechnic devices. The operation of these pyrotechnic devices generates Pyroshock that can cause failures in electronic components. The prediction of high frequency structural response, especially the shock response spectrum (SRS), is important. This paper presents a non-destructive visualization and simulation of linear explosive-induced Pyroshock using phase arrayed Laser-induced shock. The proposed method includes a laser shock test based on laser beam and filtering zone conditioning to predict the SRS of Pyroshock. A ballistic test based on linear explosive and non-contact Laser Doppler Vibrometers and a non-destructive Laser shock measurement using laser excitation and several PZT sensors, are performed using a carbon composite sandwich panel. The similarity of the SRS of the conditioned laser shock to that of the real explosive Pyroshock is evaluated with the Mean Acceleration Difference. The average of MADs over the two training points was 33.64%. And, MAD at verification point was improved to 31.99%. After that, experimentally found optimal conditions are applied to any arbitrary points in laser scanning area. Finally, it is shown that linear explosive-induced real Pyroshock wave propagation can be visualized with high similarity based on the proposed laser technology.

1. Introduction

Pyrotechnic devices have been developed for the success of the critical mechanical functions in launch vehicle programs and missile launching technology over the past decades. Majority of pyrotechnically actuated functions is achieved through piston/cylinder devices which are like linear actuators, valves, separation nuts/bolts, and cutters of guillotines. Other functions are accomplished by linear explosives such as mild detonating cord (MDC) and flexible linear shaped charge (FLSC) [1]. On the other hand, when the pyrotechnic device is operating, the structure is exposed to a shock environment because pyrotechnic devices produce a high structural shock wave, on the order of thousands of g's, which can physically damage the structure and give rise to contamination [2]. Therefore, the prediction of high shock structural response, especially the shock response spectrum (SRS), is very important for safe operation of pyrotechnic devices. Estimation and prediction of a pyroshock effects on a structure are performed using numerical simulations [3] and experimental [4]. And, many researchers have been developing a low-impact pyrotechnic mechanical device [5].

In general, an SRS is the most commonly used characteristic and test description for the pyroshock environment. The SRS acceleration is also called the maximum or peak absolute response acceleration. An SRS curve shows the maximum positive and negative absolute response experienced by a single degree-of-freedom system as a function of its natural frequency. The SRS concept has been a very useful tool in pyro-structural design, because it provides a consistent way to compare different shock time histories.

In this study, we propose a non-destructive visualization method of linear explosive-induced pyroshock wave propagation based on a linear explosive-induced pyroshock prediction algorithm. A pyrotechnic device based on a linear explosive is used in CFRP sandwich panel, for real pyroshock measurements. The linear explosive-induced pyroshock is measured by three laser Doppler vibrometers (LDVs) in a noncontact manner. On the other hand, the laser shocks generated at three points by a Q-switched laser (QL) are measured by several accelerometers installed at intervals of 23 mm to describe the linear explosive in the same target structure. In laser shock measurement, various experimental conditions are explored to find the optimal conditions associated with the optical characteristics of the laser beam and the band-pass ranges of in-line and numerical filter. Then, the similarity between the different shock waves is evaluated on the basis of the mean acceleration difference (MAD, %) in their SRS curves. In addition, the propagation of the linear explosive-induced pyroshock wave is imaged.

2. Shock measurement technique and validation method

2.1. Linear explosive-induced pyroshock wave measurement setup

Figure 1 shows the experimental setup. The test structure was a carbon fiber-reinforced plastic (CFRP) sandwich panel with an aluminum honeycomb core. The specimen dimensions were 700 mm × 1140 mm × 12.68 mm and the thickness of both face sheets was 1.34 mm. Three laser Doppler vibrometers (LDVs) were used for noncontact acceleration measurement at Pts 1, 2, and 3 as shown in Fig1 (b).

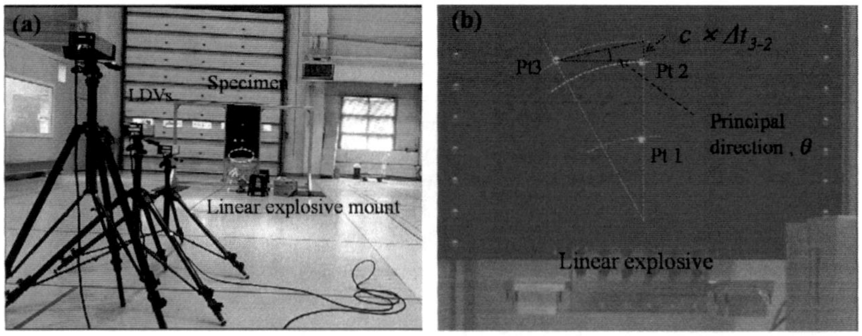

Figure 1. (a) Overall experimental setup for the pyroshock measurement, (b) location of the linear explosive and the three LDV sensing points.

Shock generation and sensing points in pyroshock and laser shock experiments are summarized in Table 1. Each LDV head was set up on a tripod, and its laser beam was focused normal to the sensing point. Retro-reflective films were used at each sensing point to improve sensitivity. A pyrotechnic device with a linear explosive in a metal tube [6], which is generally used in stage and fairing separation systems, was installed with a steel mount at the bottom edge of the specimen, as shown in the inset of Fig. 1 (b). Two steel blocks were installed at both ends of the pyrotechnic device to protect against debris.

The bandwidth of interest for the pyroshock wave is generally up to 100 kHz; hence, the signals were filtered using a low-pass filter at 100 kHz by the respective controllers of the LDVs. Next, the signals acquired by a high-speed digitizer were stored at a sampling time interval of T = 0.72 μs and a signal length of p = 900,000 samples in a LabVIEW platform. The pyroshock signals for each sensing point plotted in the time domain is shown in Fig. 2.

Table 1. Shock generation and sensing points in pyroshock and laser shock experiments.

Point	Coordinate (mm)	Pyroshock measurement	Laser shock measurement
0	(230–460, 0)	Linear explosion	Accelerometer array sensing
1	(350, 230)		
2	(350, 380)	LDV sensing	Laser shock
3	(200, 380)		

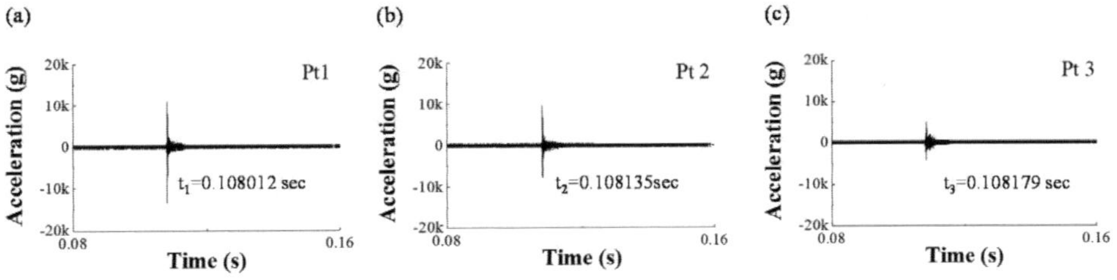

Figure 2. Pyroshock wave detected: (a) point 1, (b) point 2, and (c) point 3.

2.2. *Evaluation of the pyroshock wave propagation direction*

The propagation direction of the pyroshock wave formed by the linear explosive can be calculated by using arrival times of the maximum peaks of the measured pyroshock waves. LDV sensing points for the estimation of the speed and direction of wave propagation is shown in Fig. 1(b). The wave propagation direction can be estimated based on previous study [8] as shown in Eq. (1).

$$\theta = \tan^{-1}\left(\frac{c\Delta t}{d}\right) \tag{1}$$

Where, Δt is the difference in arrival time between Pt2 and Pt3. The arrival times of the maximum peaks of the measured pyroshock waves were 108,122, 108,135, and 108,179 μs at Pt1, Pt2, and Pt3, respectively. As a result, the wave propagation speed was calculated as 1217.53 m/s. In this case, the wave propagation direction was calculated as 19.57°. The pyroshock wave generation speed was estimated to be 3,700 m/s using the wave propagation direction and speed. Here, it turns out that the pyroshock wave generation speed is much smaller than the detonation speed of the used linear explosive, i.e., 7,030 m/s.

2.3. *Laser-induced shock measurement*

Figure 3 shows configuration of an experimental setup for laser shock wave measurement. The laser shock measurement system consist of a QL with 1064 nm wavelength, a QL controller, a galvanometric laser mirror scanner (LMS), a laser shock sensing system, and a computer with a LabVIEW platform for data acquisition and synchronization control. The laser shock sensing system consists of an amplifier-integrated PZT accelerometer and an in-line signal conditioner with filters and amplifiers.

As shown in Fig. 3, the same specimen used was used for the laser shock wave measurement. In this reciprocal setup, the LDV sensing points are now laser shock points and the linear explosive is replaced by the accelerometer array, as summarized in Table 1. Laser shock waves generated at the three points were acquired by the PZT accelerometer array at the coordinate (230–460, 0).

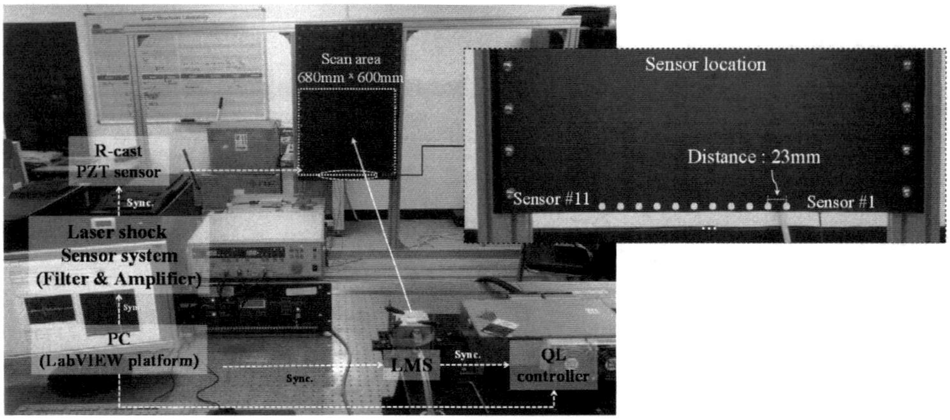

Figure 3. Experiment setup for the laser shock wave measurements

The multiple accelerometer array employed to describe the linear explosive as a set of successive point sources is shown in the inset of Fig 3. Eleven accelerometers were attached at intervals of 23 mm on the bottom of the test structure.

The number of signals for temporal running averaging at each sensing point was set as 10, and the pulse repetition rate (PRR) was selected to be 1.25 Hz. Then, the measured laser-induced shocks were band-pass filtered between 100 Hz and 100 kHz, which was the frequency band of interest to be analyzed for SRS. The time delay was determined to be 6.33 μs. The resulted laser shock wave can be represented as follows:

$$L(t) = \sum_{i=0}^{M-1} L_i(t + i\Delta\tau) \tag{2}$$

Where, the measured laser shock wave at each sensing point is denoted as Li, i=0, 1, 2, ···, M-1; L(t) is the result of the laser shock signal superposition with a time delay; and M is the number of sensors in the array.

Table 2. Optimized experimental conditions for prediction of SRS and time-domain signals.

Laser beam condition	Beam size (cm^2)	0.21
	Energy (mJ)	0.72
	Fluence (mJ/cm^2)	3.43
Laser shock measurement condition	Pulse repetition frequency (PRF)	1.25Hz
	Repetition measurement	5 times
	In-line filter	BP: 100 Hz to 100 kHz
	Numerical filter	BP: 500 Hz to 70 kHz (Butterworth)
	Time delay (μs)	6.33

2.4. Linear pyroshock reconstruction method
During the measurement of laser-induced shocks, laser beam conditions such as energy, size, and fluence were investigated based on the reconstruction process of the linear explosive-induced pyroshock. Then, the ranges of the in-line and numerical band-pass filters were explored to enhance the similarity of the laser shock wave to the real pyroshock wave. More detailed procedures about this process can be found in the previous study [7]. Optimal experimental conditions to generate laser shock waves similar to the pyroshock are summarized in Table 2.

Figure 4 shows SRS comparison between the conditioned laser SRS and the pyroshock SRS for Pt1–Pt3. The similarity between the SRS curves of the pyroshock and the conditioned laser shock is also quantitatively evaluated based on MAD [7].

The constant gain was found to be 249.5 g/mV and hence the average of MADs over the three points (Pt1–Pt3) was 33.09%. The average of MADs in training points, Pt1 and Pt2, was 33.64% and the conditioned laser SRS at verification point, Pt3, shows an MAD of 31.99%.

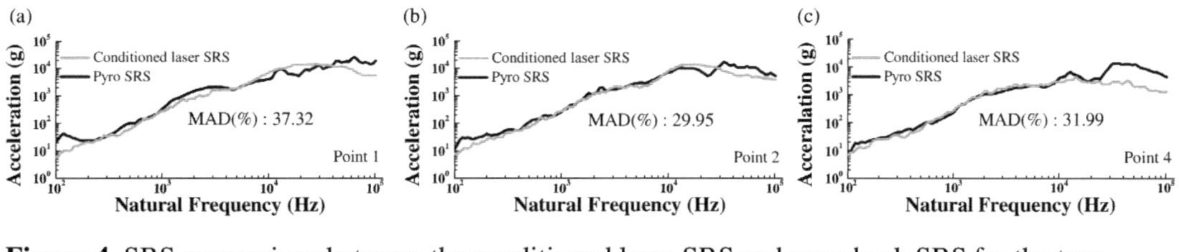

Figure 4. SRS comparison between the conditioned laser SRS and pyroshock SRS for the two training points, (a) Pt1 and (b) Pt2, and for one verification point, (c) Pt3.

3. Pyroshock wave propagation imaging results
The laser shock scanning was repeated ten times and averaged for each accelerometer position. Then, the measured laser shock signals were superimposed onto the result. As shown in the inset of Fig. 3 (a), the scan area and interval were 680 mm × 600 mm and 2.5 mm, respectively. Once these resulted signals were ready, constant gain was applied for all the scanned signals. With these converted signals, the linear explosive-induced pyroshock propagation wave can be visualized using the PWPI algorithm. Figure 5 show the freeze-frames of the PWPI generated at three different time delays, Δt = 3.2 and 6.33 μs. The wave propagation directions stand at 9.6°, and 19.6°, respectively. As a result, the more the time delay increases, i.e., when higher detonation speed explosives are used, the more the wave propagation angle increases. The wave propagation direction was determined to be 9.6° at the time delay of 3.2 μs corresponding to the pyroshock wave generation speed of 7,050 m/s, which should be a linear explosive with a higher detonation speed than the one used for the real pyroshock experiment.

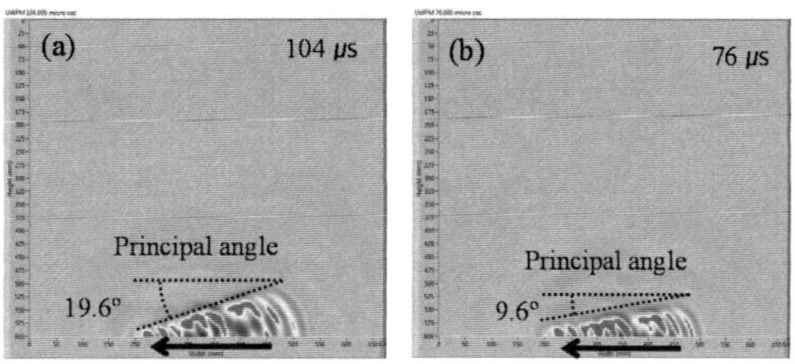

Figure 5. PWPI snapshots: (a) $\Delta t = 3.2\,\mu s$, (b) $\Delta t = 6.33\,\mu s$

4. Conclusions

This paper shows a nondestructive visualization method of linear explosive-induced pyroshock wave propagation using superposition of point source array with a time delay realized by laser shock and multiple accelerometers. For the verification of the proposed testing method, a real pyroshock wave and a laser-induced shock wave were measured in the same CFRP sandwich panel. For the generation of the linear explosive-induced pyroshock, a linear explosive was used, and the pyroshock wave was measured simultaneously by three LDVs at three sensing points. From the measured pyroshock waves, the pyroshock wave generation and propagation speeds of the linear explosive were estimated to be 3,700 m/s and 1,217.53 m/s, respectively, and the wave propagation direction was calculated to be 19.57°. The laser-induced shock wave generated at each training point by the QL was measured by 11 accelerometers installed at intervals of 23 mm on the bottom of the test panel. The waves captured by the 11 accelerometers were superimposed with a time delay of 6.33 μs.

The laser-induced shocks were trained to improve the similarity of the laser shock wave to the real pyroshock wave by using the SRS prediction process. The average of MADs over the two training points (Pt1 and Pt2) was 33.64% and the MAD at the verification point (Pt3) was 31.99%. Next, this process was also applied to the laser shock waves at arbitrary points realized by the laser shock scanning. By applying the measured time delay of 6.33 μs to the PWPI algorithm for the linear explosive, a pyroshock wave propagation video could be generated and it showed a wave propagation angle of 19.57°, which was in agreement with the results of the real pyroshock measurement.

Acknowledgments
This research was supported by the Space Core Technology Development Program (2013-042548) and Leading Foreign Research Institute Recruitment Program (2011-0330065), through the National Research Foundation of Korea, funded by the Ministry of Science, ICT and Future Planning.

References

[1] Bement L J, Schimmel M L 1995 *A manual for pyrotechnic design, development and qualification* NASA Technical Memorandum 110172
[2] Anon. 2011 *Pyroshock Test Criteria* NASA Technical Standard NASA-STD 7003A
[3] Mao Y, Li Y, Huang H and Wang J 2011 Fast Simulation of Pyroshock Responses of a Conical Structure Using Rotation-Superposition Method. *Appl. Math. & Inf. Sci.* **5(2)** 187-193
[4] Eun S W, Kong C W, Jang Y S, Yi Y M 2005 Simulation of KSLV-1 downscaled nose fairing separation motion. *Aero. Eng.and Tech.***2(1)** 8-11
[5] Pete W, Daryn E O 2006 *Proc. Of th 38th Aero. Mech. Sym. (Williamsburg, Virginia)* p149
[6] Lee J W, Kong C W, Jang Y S, Lee Y S 2013 *proc. 9th Int. conf. on Fracture & Strength of Solids (jeju, Korea)* p1
[7] Jang J K, Lee J R 2014 Nondestructive Prediction of Point Source Pyroshock Response Spectra based on Experimental Conditioning of Laser Shocks *Opt. & Laser Technol.* **61** 24-33
[8] Lee J R, Jang J K, M Choi, C W Kong 2015 Visualization and simulation of a linear explosive-induced pyroshock wave using Q-switched laser and phased array transducers in a space launcher composite structure *Opt. & Laser Technol.* **67** 12-9

Accurate Damage Location in Complex Composite Structures and Industrial Environments using Acoustic Emission

M Eaton[1,3], M Pearson[1], W Lee[2] and R Pullin[1]

[1]Cardiff School of Engineering, Cardiff University, The Parade, Newport Road, Cardiff, CF24 3AA, UK
[2]MBDA (UK) Ltd, Six Hills Way, Stevenage, Hertfordshire, SG1 2DA, UK

E-mail: EatonM@Cardiff.ac.uk

Abstract. The ability to accurately locate damage in any given structure is a highly desirable attribute for an effective structural health monitoring system and could help to reduce operating costs and improve safety. This becomes a far greater challenge in complex geometries and materials, such as modern composite airframes. The poor translation of promising laboratory based SHM demonstrators to industrial environments forms a barrier to commercial up take of technology. The acoustic emission (AE) technique is a passive NDT method that detects elastic stress waves released by the growth of damage. It offers very sensitive damage detection, using a sparse array of sensors to detect and globally locate damage within a structure. However its application to complex structures commonly yields poor accuracy due to anisotropic wave propagation and the interruption of wave propagation by structural features such as holes and thickness changes. This work adopts an empirical mapping technique for AE location, known as Delta T Mapping, which uses experimental training data to account for such structural complexities. The technique is applied to a complex geometry composite aerospace structure undergoing certification testing. The component consists of a carbon fibre composite tube with varying wall thickness and multiple holes, that was loaded under bending. The damage location was validated using X-ray CT scanning and the Delta T Mapping technique was shown to improve location accuracy when compared with commercial algorithms. The onset and progression of damage were monitored throughout the test and used to inform future design iterations.

1. Introduction

The use of damage detection tools to monitor the health of engineering structures throughout their service lives has the potential to facilitate significant cost saving by reducing maintenance and down time, whilst improving safety. For example commercial passenger aircraft spend on average 6.6 days per year undergoing inspection. A key requirement of such a structural health monitoring (SHM) system is the ability to accurately determine the position of any damage occurring within a structure. This becomes a very difficult challenge in complex materials and geometries, such as those found in modern composite aerospace structures. It is this poor translation from promising laboratory based

[3] To whom any correspondence should be addressed.

SHM demonstrators to industrial environments that forms the greatest barrier to commercial up take of technology.

The acoustic emission (AE) technique is a passive damage detection technique with a number of advantages that make it well suited to the in service monitoring required for SHM. The technique uses piezoelectric transducers to detect elastic stress waves (similar to ultrasound waves) that are released by the growth of damage itself and no external excitation is required. As such a damage mechanism must be active in order for it to be detected using AE, i.e. when a structure is under load, or in service. This means that minimum detectable defect size is very small, because a defect need only be active/growing in order for it to be detected. The most powerful attribute of the AE technique is the ability to globally locate damage within a structure using a sparse array of sensors [1]. However, as with all other damage detection approaches the accuracy of AE location calculation is seen to reduce in complex materials and structures. The observed errors results from the assumptions in current commercial algorithms that AE waves travel at a single and constant velocity in all directions and that their path from release at a damage site to detection at a transducer is direct and uninterrupted. Neither of these assumptions are true in complex geometry composite structures where anisotropic wave speed are observed and holes, thickness changes and curvatures can interrupt the propagation path [2].

A variety of approaches have been investigated to account for the anisotropic propagation behaviour experienced in composite materials. Some improvements have been demonstrated through the assumption of an elliptical wave front [3] and commercial software packages now allow the specification of an elliptical velocity profile. However, the velocity profiles in composite materials can often be more complex than this. Beamforming techniques have been used with localised arrays to determine the direction of an arriving wave [4, 5] and Aljets *et al* used the temporal separation of different wave propagation modes to determine the distance, as well as the direction, from the array[5]. Ciampa and Meo [6] used a triangular array of three closely spaced sensor pairs (six sensors in total), allowing the source position to be described by six non-linear equations. Solving the equations with an iterative Newton method provides a source position without the need for prior knowledge of the wave velocity behaviour in the material. However, processing times are high, at around 2 seconds per event, and sensor numbers required are double for similar coverage. Despite very promising results all of the above techniques are flawed by their inability account for geometric complexities that invariably exist in an industrial environment.

An alternative approach to achieving accurate AE source location in complex geometries is that of mapping. The concept of mapping requires that a relationship is formed between known physical positions of AE sources upon a structure and the resulting wave arrival times (and therefore the difference in arrival times between sensor pairs Δts) at an array of transducers from signals originating at these positions. Such a methodology has the potential to facilitate accurate source location in highly complex materials and structures, accounting for variations in wave speed, changes in thickness, holes, etc. Scholey *et al* [7] approached the mapping concept by analytically determining the expected arrival times (and therefore Δts) for an anisotropic composite panel from an array of points, or source positions. They describe the best-matched point search method which compares measured Δts with the analytically determined map to identify the array point at which the difference is minimised and hence give the location. The accuracy is affected by the resolution of the mapping array, so small spacings of 1-2mm are used, and it is also important that accurate wave velocities are known for a given material. The approach is also not well suited to dealing with complex geometries, in which the calculation of arrival times becomes far more problematic. A more robust approach, known as Delta T Mapping, was proposed by Baxter *et al* [8] for location in complex geometries. used artificial AE sources to determine Δts from known grid positions and therefore generate a map of Δts to aid source location in complex metallic structures. For Delta T Mapping H-N sources [9, 10], generated at known positions, are used to generate contour maps of constant Δt for each sensor pair, linearly interpolating between training points improves the mapping resolution. Mapped contours corresponding to measured Δts from real AE test data can then be selected for each sensor pair and overlaid to find a crossing point and hence a prediction of source location. Hensman et al [11] followed a similar methodology, but represented the relationship between the Δts and the spatial grid using Gaussian processes. Both mapping approaches were shown to improve source location in metallic structures with complex geometries and inherently compensate for variations in wave speed and any obstructions in the wave propagation path. The Delta T Mapping approach has also been demonstrated to improve location accuracy in complex geometry composite structures; detecting and locating fatigue damage [12].

In this work the authors demonstrate the use of the Delta T Mapping technique on a complex geometry composite aerospace component undergoing validation testing in an industrial environment. The tubular component is manufactured from carbon fibre composite with varying wall thicknesses and multiple holes or cut outs. The tube was loaded under bending and the location of damage was validated using X-ray CT scanning. The Delta T Mapping technique was shown to improve location accuracy when compared with commercial algorithms. The onset and progression of damage were monitored throughout the test and used to inform future design and manufacturing iterations.

2. Delta T Mapping Location

The following section details the Delta T Mapping procedure. Firstly the mechanism for wave arrival time determination is discussed, then the methodologies for map training and location calculation are described.

2.1. Arrival time determination

Traditionally the detection of arriving AE waves relies on a user defined threshold level, the crossing of which indicates the signal arrival. It is common that some signal is present prior to the first threshold crossing and the effects of attenuation can mean different phases of the wave trigger its arrival. All of which can lead to additional errors in a calculated location. In this work we adopt an Akaike Information Criterion (AIC) based method to determine wave arrival times. The approach aims to identify the first wave motion by identifying a change in entropy between the uncorrelated noise prior to signal onset and the highly correlated signal after signal arrival. It involves the minimisation of equation (1) for a time series x. The minimum of the AIC function corresponds to the signal onset time, as can be seen in Figure 1 and has been demonstrated to be a very accurate and reliable measure of AE signal arrival time for location calculation [11, 13, 14].

$$AIC(t) = t\log_{10}\big(var(x[1;t])\big) + (T-t-1)\log_{10}\big(var(x[t;T])\big) \quad (1)$$

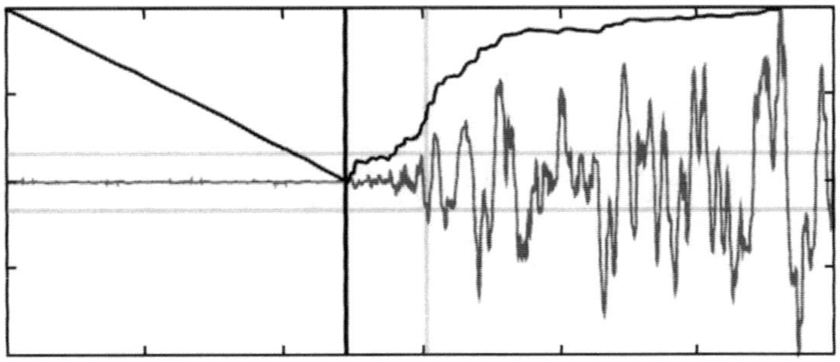

Figure 1. Arrival time estimation using AIC based approach. The vertical grey line indicates the threshold based arrival time. The black trend is the AIC function and the vertical black line indicates the estimated arrival time at its minimum.

2.2. Delta T Mapping Methodology

The Delta T Mapping procedure is described in detail by Baxter *et al* [8] and consists of 5 main steps which are outlined in brief below:

Determine area of interest - Delta-T source location can provide complete coverage of a part or structure, or it can be employed as a tool to improve source location around specific areas of expected fracture, which could potentially be identified via finite element modelling.

Construct a Map System - A grid is placed over the area of interest within which AE events will be located. It should be noted that sources are located with reference to the grid and not the sensors and it is not required that sensors be placed within the grid.

Obtain time of arrival data from an artificial source – An artificial source (nominally a H-N source [9, 10]) is generated at the nodes of the grid to provide AE data for each sensor. An average result of several sources is used for each node. Missing data points can be interpolated from surrounding nodes.

Calculate Delta T map – Each artificial source results in a difference in arrival time or Delta T for each sensor pair (an array of four sensors has six sensor pairs). The average Delta T at each node is stored in a map for each sensor pair. The resulting maps can be visualised as contours of constant Delta T.

Locating real AE data – The Delta T values from a real AE event are calculated for each sensor pair. A line of constant Delta T equivalent to that of the real AE event can then be identified on the map of each sensor pair. By overlaying the resulting contours, a convergence point can be found that indicates the source location. As with time of arrival, a minimum of three sensors is required to provide a point location and more sensors will improve the location. In theory all the lines should intersect at one location, however in practice this is not the case. Thus in order to estimate a location all convergence points are calculated and a cluster analysis provides the most likely location.

3. Test setup and configuration

The component under test takes the form of a composite tube, with dimensions in the order of 550mm in circumference and 600mm in length, shown in Figure 2. The complexity of the structure is increased by a number of holes and inserts seen as well as a significant increase in thickness under central mounting bracket. The tube was restrained at its centre using the mounting bracket observed in Figure 2 and loaded in bending. The mounting orientation and loading direction are shown schematically in Figure 3. The tube was subjected to five quasi-static loading cycles and was unloading between each cycle. The applied loads represented 11%, 48%, 58%, 77% and 100% of the final load achieved. The final failure occurred at the bracket attachment point and failure of the tube itself was not achieved.

Figure 2. Sensor layout and Delta T Mapping grid on composite component

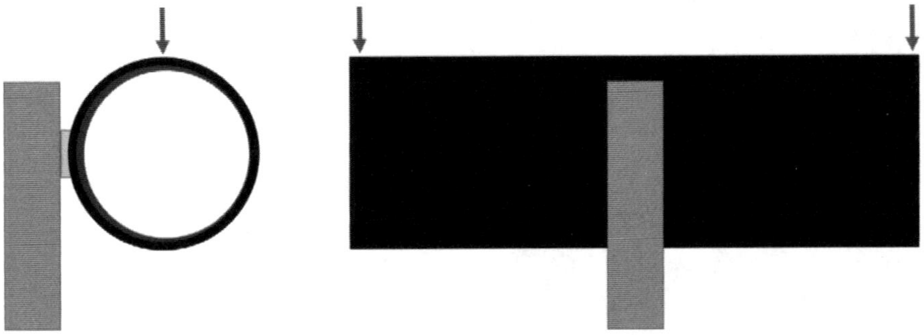

Figure 3. Schematic of composite tube loading arrangement

The tube was instrumented with eight Pancom P15 AE transducers, attached using a cyanoacrylate adhesive that provides an acoustic coupling as well as mechanical attachment. They were arranged in two rows along the length of the tube, as shown in Figure 2, placed 100mm circumferentially either side of the mounting bracket (200mm total separation). The transducer spacing along the rows was 160mm and the two rows were offset by 80mm in the axial direction. A Delta T Mapping training grid was applied to the structure using a 20mm spacing (Figure 2). The grid was an 'inverted T' shape spanning up to 380mm in the axial direction and 200mm in the circumferential direction. Five H-N sources were conducted at each available grid point and the resulting signals were recorded for the training of delta t maps. A MISTRAS PCI-2 AE system was used to record all AE data using a threshold level of 45dB. AE waveform data were recorded at 2MHz and used in post processing to correct the signal arrival times based on the AIC function in equation (1). AE monitoring was conducted throughout all five loading and unloading cycles.

4. Results and discussion

Figure 4 presents an image from the X-ray CT inspection undertaken post-test, after completion of all five load cycles. The image shown represents a slice through the thicker reinforced section beneath the central mounting bracket and reveals the cross-section of a delamination crack identified by the red ellipse. A second feature can also be identified in the CT image at the opposite side of the thicker reinforced section. It is not clear from the geometry of this feature weather or not it is a crack, it is noted that the geometry of this feature is different to that of the identified delamination crack. In particular the orientation with which it propagates through the material is different and in previous similar the observed damage in these regions is similar to that of the identified delamination crack. The image in figure 4 represents the final state of the damage following all loading cycles; however, it is cannot be determined from this image at what stage of testing this damage initiated.

During the first loading cycle only six AE events were recorded hence these results are not presented here and instead we first consider the second load cycle. Figures 5 and 6 present the traditional time-of-arrival location results and the Delta T Mapping location results, respectively, for the second loading cycle. The position of the hanger is represented in both figures by a solid rectangle and the position of the hole and insert below the hanger is identified by a solid circle. An area of damage identified by post-test X-ray CT inspection is indicated by a light grey rectangle and represents a large area of delamination. It can be seen in both cases that significant amounts of AE have been detected and located in the locality of this damage. This demonstrates the sensitivity of the AE technique to detect the onset of such damage at an early stage of the test and at a load less than 50% of maximum applied load, which did not induce ultimate failure in the composite structure. The TOA locations are located close to the identified damage region but the majority are located outside of this area. Whereas the Delta T Mapping locations are centred within the damage region; demonstrating the improvement in accuracy achieved. This equates to an improvement in location accuracy over the traditional TOA algorithm of approximately 50mm.

The presented location results suggest that the second feature identified in figure 4 did not develop during the testing undertaken. The origin of this feature is not clear and will be the subject of further investigation. It is possible that it is a manufacturing defect existing prior to testing that did not grow during these tests or that it is damage that occurred during the failure of the mounting bracket (a violent failure that resulted in the decoupling of the transducers).

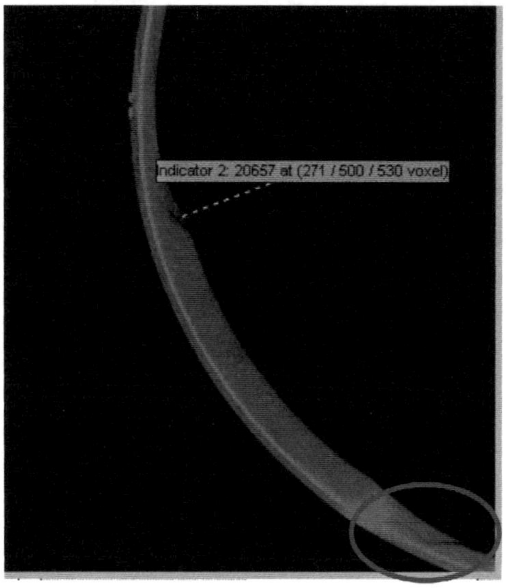

Figure 4. X-ray CT data from post-test inspection of the tube. The image shows a slice through the reinforced section of the tube below the attachment bracket. The red ellipse identifies the damage corresponding to the highlighted regions in figures 4 and 5.

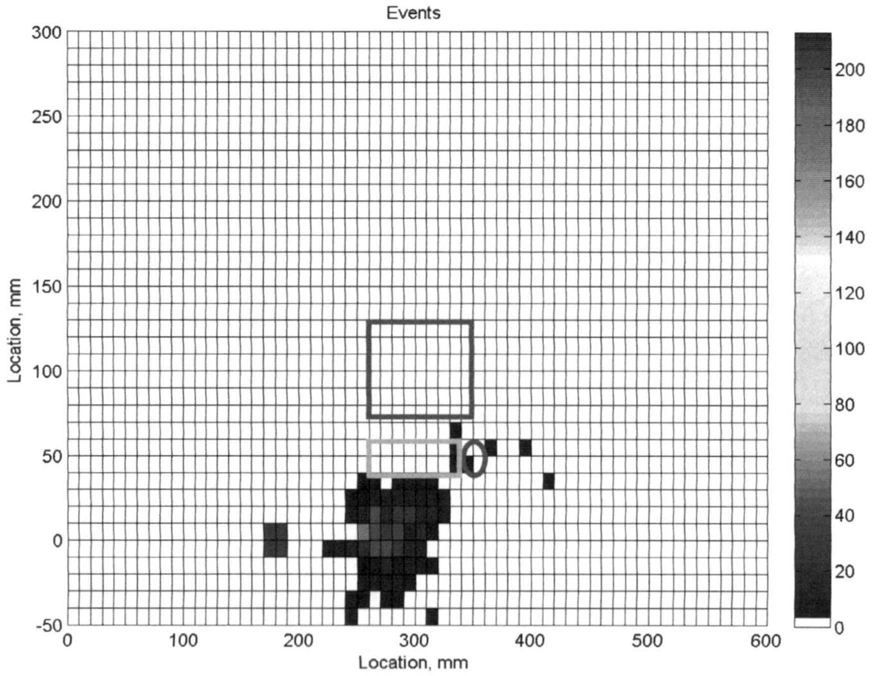

Figure 5. time-of-arrival locations for load cycle 2 – the red rectangle and circle represent the bracket location and the bolt hole seen below it, the light blue rectangle represents the region in which damage was identified by X-ray CT scanning

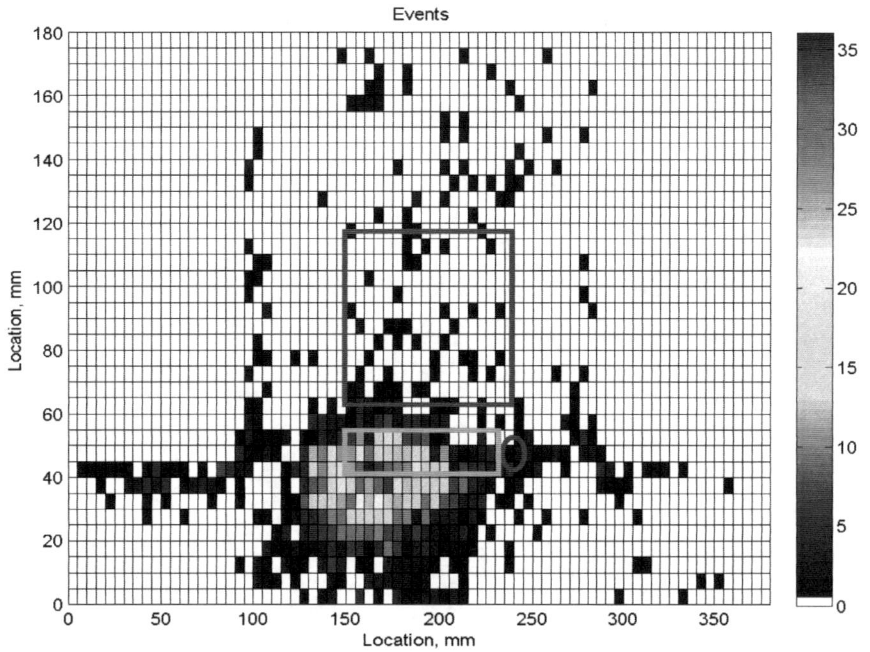

Figure 6. time-of-arrival locations for load cycle 2 – the red rectangle and circle represent the bracket location and the bolt hole seen below it, the grey rectangle represents the region in which damage was identified by X-ray CT scanning

The number of AE events located within the region of the identified damage continued to increase as the remainder of the loading cycles were completed, suggesting that the damage continued to grow throughout the testing. Figure 7 shows a plot of cumulative AE hits located in the region of identified damage versus the applied load for all five cycles (cycle 1 is not visible on this scale). Following the second load cycle the Felicity effect can be observed, whereby significant AE activity begins at a load lower than that of the previously applied maximum. The ratio of the load at which AE activity begins again and that of the previous maximum is known as the Felicity ratio and it can be used to quantify this effect [1]. Composite materials often demonstrate the Felicity effect during repeated loading, even if no significant damage has been induced. As a general rule, if the Felicity ratio is below 0.9 then it can be assumed that damage is present in a structure and the lower the Felicity ratio becomes the confidence that damage is present becomes greater and its expected severity of damage increases. Following the second load cycle Felicity ratios of 0.8, 0.8 and 0.7 were recorded for cycles 3, 4 and 5, respectively. This indicates that damage was induced in this region during the second load cycle and that reducing Felicity ratio indicates that this damage continued to grow with subsequent load applications. It should be noted that this demonstrates a very confident detection of damage using the AE technique and this was achieved at a load far below the maximum applied in a test where catastrophic failure of the composite structure did not occur. This highlights the potential of the AE technique to provide sensitive SHM with the ability to detect and accurately locate damage well before ultimate failure of a structure.

5. Conclusion

It has been demonstrated that the Delta T Mapping algorithm can improve location accuracy in complex structures compared with the traditional TOA algorithm. The ability to detect and accurately locate damage within a complex geometry composite aerospace component was demonstrated and validated using X-ray CT inspections. The presented results highlight the potential for the AE technique to bridge the gap between laboratory demonstrators and application in challenging industrial environments. This is seen as a key barrier to the uptake of SHM technology and systems by industry and these results show that the AE technique has the capability to achieve this.

The AE analysis undertaken was used to informed component designers of the onset and growth of damage occurring within the component, which occurred far below the ultimate failure load. CT data normally gives a snapshot of the final condition of a component, whereas the AE data informed designers when and where damage occurs during the test and has provided valuable understanding to support future design and manufacture iterations.

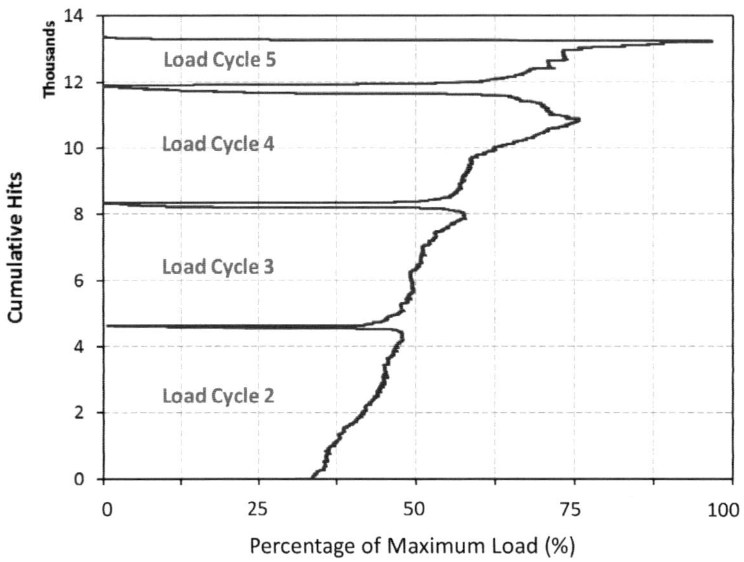

Figure 7. Cumulative AE hits versus applied load for all five load cycles

6. References

[1] Miller RK, Moore PO. 2005 *Nondestructive Testing Handbook Vol 6: Acoustic Emission Testing. Third ed.*

[2] Eaton MJ, Pullin R, Holford K. 2012 *Proc. of the Instit. of Mech. Engin. Part C.* **226** 2141

[3] Paget CA, Atherton, K., and O'Brien, E.W. 2003 *Proc. of IWSHM-4. Stanford, CA, USA*

[4] McLaskey GC, Glaser, S.D., and Grosse, C.U. 2010 *J. SoundVib.* **329** 2384

[5] Aljets D, Chong A, Wilcox S, Holford K. 2010 *J. Acoust. Emis.* **28** 85

[6] Ciampa F, and, Meo, M. 2010 *Compos. Part A* **41** 1777

[7] Scholey JJ, Wilcox, P.D., Wisnom, M.R., Friswell, M.I., Pavier, M., and Aliha, M.R. 2009 *J. Acoust. Emis.* **27** 291

[8] Baxter MG, Pullin R, Holford KM, Evans SL. 2007 *Mech. Syst. Signal Pr.* **21** 1512

[9] Hsu NN, Breckenridge FR. 1981 *Mater. Eval.***39** 60

[10] ASTM E976. 2010 *Am. Soc. for Testing and Mater.*

[11] Hensman JJ, Mills, R., Pierce, S.G., Worden, K., and Eaton, M. 2010 *Mech. Syst. Signal Pr.* **24** 211

[12] Eaton MJ, Pullin R, Holford K. 2012 *Compos. Part A* **43** 856

[13] Kurz JH, Grosse, C.U., and Reinhardt, H-W. 2005 *Ultrasonics* **43** 538

[14] Eaton MJ, Pearson MR, Holford KM, Featherston CA, Pullin R. 2012 *2nd Int. Conf. of Advanced Compos. Mater. and Tech. for Aerospace Appl. Glyndwr University, UK* 29

Finite element based damage assessment of composite tidal turbine blades

Edward M Fagan[1,3,4,a], Sean B Leen[2,3,4,c], Ciaran R Kennedy[2,3,4], Jamie Goggins[1,3,4,b]

[1] Civil Engineering, College of Engineering & Informatics, NUI Galway, Ireland

[2] Mechanical Engineering, College of Engineering & Informatics, NUI Galway.

[3] Marine Research Energy Ireland (MaREI) Research Centre, Galway, Ireland

[4] Ryan Institute for Environmental,Marine and Energy Research, NUI Galway, Ireland

[a] e.fagan1@nuigalway.ie, [b] jamie.goggins@nuigalway.ie, [c] sean.leen@nuigalway.ie

Keywords: Finite element, damage, marine current turbines, composite materials

Abstract. With significant interest growing in the ocean renewables sector, horizontal axis tidal current turbines are in a position to dominate the marketplace. The test devices that have been placed in operation so far have suffered from premature failures, caused by difficulties with structural strength prediction. The goal of this work is to develop methods of predicting the damage level in tidal turbines under their maximum operating tidal velocity. The analysis was conducted using the finite element software package Abaqus; shell models of three representative tidal turbine blades are produced. Different construction methods will affect the damage level in the blade and for this study models were developed with varying hydrofoil profiles. In order to determine the risk of failure, a user material subroutine (UMAT) was created. The UMAT uses the failure criteria designed by Alfred Puck to calculate the risk of fibre and inter-fibre failure in the blades. The results show that degradation of the stiffness is predicted for the operating conditions, having an effect on the overall tip deflection. The failure criteria applied via the UMAT form a useful tool for analysis of high risk regions within the blade designs investigated.

1. Introduction

One of the leading devices in the emerging ocean energy space is the horizontal axis tidal turbine. Its construction and development stem from decades of experience in the wind turbine industry. The world's oceans contain a vast resource of predictable and consistent renewable energy. An Irish report on tidal energy has estimated a resource of 915 GWh/yr [1] for the 10-40 m water depth range around the coast of Ireland, with practical, economic, and physical limitations accounted for. Marine current turbines operate in much the same way as their wind-based counterparts. The turbines use the pressure difference caused by a blade profile generating lift and drag. The kinetic energy of the tidal flow is converted to mechanical energy by the rotating blades and to electrical energy by the generator. The critical components in the device are the turbine blades, as the design of these face unique challenges. The industry standard for the construction of tidal turbines has not yet been identified; however, fibre-reinforced composite materials are currently the favoured design choice for the blades [2]. There are a number of advantages for their use, including corrosion resistance in a harsh environment, good specific strength and specific stiffness characteristics. In addition, construction and testing facilities can be modified from existing wind turbine operations. An analysis of possible configurations for structural designs of tidal turbine blades [3] has led to an investigation of a typical construction for a turbine blade, with spar caps and webs providing flexural stiffness and shear strength, respectively.

One of the major challenges faced by tidal turbine blade designers is due to the density of water. Comparing similarly rated wind and tidal turbines, the issue becomes apparent: a 1 MW wind turbine would require a diameter of approximately 55 m, while a similarly rated tidal turbine's diameter would be on the order of 22 m [2]. This results in considerably higher loading on a much smaller structure, and means the tidal turbine blade will require increased stiffness and strength [4]. Talreja [5] has explained how an extensive distribution of microcracks in a composite become responsible for a deterioration in stiffness. This degradation has also been attributed to composites under complex and cyclic loading, such as the type experienced by tidal turbine blades [6]. The failure theory of Puck and Schurmann [7] was chosen for assessing the damage level of the composite laminates. This theory has a proven record for providing a phenomenological basis for the analysis of fibre reinforced composites, scoring highly in the world wide failure exercise (WWFE) of leading failure criteria [8]. The method determines failure envelopes for the initiation of fibre failure and inter-fibre failure, i.e. matrix cracking. The criteria consist of a number of stress interaction equations and uses material strengths to determine the onset of failure. Its use in finite element methods is also proven [9], and it can readily be applied to the plane stress shell element case.

In this paper, the initial design process is based on a hydrodynamic model for the preliminary design of composite tidal turbine blades [10, 11]. The study presented here uses the Puck-based failure criteria in a finite element context to evaluate the potential damage due to inter-fibre failure and fibre failure in composite tidal turbine blades. Three models were constructed from a number of airfoil profiles [12-14] with a focus on thick blade sections for the inner region of the blades, in order to maximise the blades' second moment of area at the location of the highest bending moment. An overview of the hydrodynamic model used for the initial generation of the chord and twist distribution is provided, together with details of the blade models' construction and the finite element procedure employed. The results were analysed with consideration of experimental work on structural testing of full scale wind turbine blades [15-18]. The intention is to use knowledge of the failure mode in actual blades under a similar (flapwise) loading situation to inform the analysis and help identify possible high-risk regions on the blades.

2. Methods

2.1. Hydrodynamic model

The hydrodynamic model used as the preliminary design tool in this work was developed by Kennedy et al [10]. The model uses a streamtube approach and blade element momentum theory (BEMT) to balance the loads acting on a number of annular streamtube sections of the blades. The balancing procedure determines the optimum chord length for each radial blade section. A twist angle is also required, since the relative velocity of the impinging water increases along the length of the blade with increasing angular momentum. Figure 1 demonstrates the distribution of twist angle and chord length on one of the completed blades. The flap- and edgewise forces at one of the sections are also shown. These are output for each section and are applied in the finite element model for loading the blade at the maximum rated operating speed of 2.5 m/s tidal velocity.

2.2. Blade design

The three blade configurations under investigation were created using a combination of four airfoil profiles. The first 2 m of each of the three blades used one from the NACA 63-421, FFA-W3-241, and FB3500-0875 profiles (referred to as blades 1, 2, and 3 respectively hereafter). These three profiles were chosen due to their high thickness percentage (21%, 23%, and 35%, respectively), thereby maximising the second moment of area of the blades and their ability to resist the bending moment towards the root of the blade. The outboard sections of the blade use the thinner NACA 2415 profile for better hydrodynamic efficiency and power generation. The thicker sections have drawbacks with regard to increased drag and roughness sensitivity, and they can suffer from flow separation [19]. For example, studies [19-20] have demonstrated an increased lift coefficient with the flat-back profiles; however, they also show increased base drag. Methods of offsetting this with trailing edge splitter plates have proven effective as a workaround for this issue. Since approximately 80% of the power is produced by the profiles in the outer four metres of the blade [21], the reduced lift-drag ratio has minimal effect on power production while potentially improving structural performance.

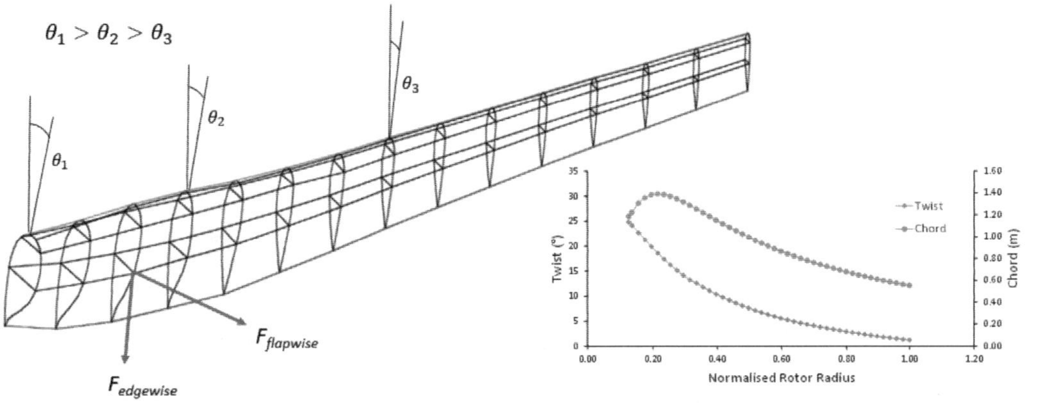

Figure 1. Wireframe of blade showing the chord and twist distribution obtained from the hydrodynamic model.

The typical construction of a wind turbine blade is used as the basis in developing the blade under consideration in this paper. The blade consists of a box spar running through its core, which acts as the main structural component of the blade. The box spar has horizontal spar caps, with a layup of $[0°/\pm45°/(0°)_N]_S$, where the outer three plies form a protective layer of GFRP over the inner unidirectional CFRP laminate. The spar caps extend from 15% of the chord length to 50%. Vertical shear webs support the spar caps and are constructed of double biased (DB) $[\pm45°]_N$ GFRP laminates. The root section requires additional reinforcement and a quasi-isotropic (QI) layup of $[0°/0°/\pm45]_{NS}$ CFRP is applied there. The remainder of the blade, including the leading and trailing edge sections, also consists of a quasi-isotropic layup, as the primary loading direction over the curved surfaces is difficult to predict as the blade performs. The QI layup is formed of $[0°/\pm45]_{NS}$ GFRP. Figure 2 and Table 1 outline the layups in the various regions, their total respective thicknesses, and the details of the materials used in the finite element models. In order to compare the three blades the same laminate thicknesses and layups were used in each of the blades.

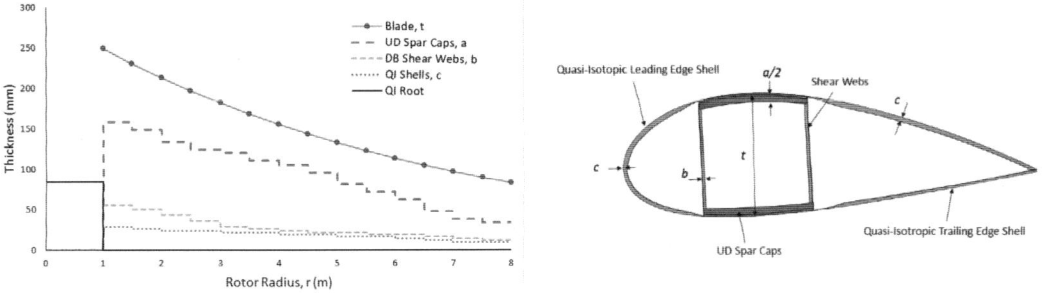

Figure 2. (a) Plot of the thickness of the different laminates in the blade, and the distribution of maximum thickness along the blade's span, r. The same thicknesses were applied to all three blades (b) diagram of the sections composing the blade, and the thickness referred to in (a).

Table 1. Material properties for Silenka E-Glass MY750 epoxy (GFRP) and AS4 Carbon-Fibre 3501-6 epoxy (CFRP) laminates [22].

Material Characteristics for a Unidirectional Layer	GFRP	CFRP
Longitudinal Modulus, E_1 (GPa)	45.6	138
Transverse Modulus, E_2 (GPa)	16.2	11
In-plane Shear Modulus, G_{12} (GPa)	5.83	5.5
Major Poisson's Ratio, ν_{12}	0.278	0.28
Longitudinal Tensile Strength, X_T (MPa)	1280	1500
Longitudinal Compressive Strength, X_C (MPa)	800	900
In-plane Shear Strength, S_{21} (MPa)	73	80
Transverse Tensile Strength, Y_T (MPa)	40	27
Transverse Compressive Strength, Y_C (MPa)	145	200
Longitudinal Tensile Failure Strain, ε_{1T} (%)	2.807	1.087
Longitudinal Compressive Failure Strain, ε_{1C} (%)	1.754	0.652

2.3. FE modelling

The loads on the blade were applied to the spar caps using the polynomial expression in Equation (1) for a distributed pressure load, *P*, varying along the blade span, *r*. The pressure load was defined using the thrust force distribution output from the streamtube model.

$$P(MPa) = 4x10^{-9}r^2 + 4x10^{-6}r + 0.0106 \qquad (1)$$

The blades were modelled from the rotor of a three-bladed tidal turbine with a theoretical output of 800 kW, a 16 m rotor diameter, and a blade length of 7.5 m. The calculated root bending moment is approximately 850 kNm, and the total thrust per blade is approximately 190 kN. The root section of an actual tidal turbine blade could be secured by a number of methods, such as a linear or circular T-bolt connection [5]. This complexity of modelling is outside the scope of this study, due to the lack of through-thickness stress components when using the plane stress shell elements. A zero-deflection boundary condition was used instead. While this is likely to be overly stiff (the pitch bearing in an actual blade for example has its own stiffness level and some small amount of deflection would be expected), it is the best method available without adding unnecessary complexity to the model. The models comprise S4R 4-noded, doubly curved, reduced integration shell elements.

Once the blade design was finalised the analysis was run using a user material subroutine (UMAT) in Abaqus. The UMAT determines the risk of failure for each element. The risk is determined by using the Puck stress interaction equations, which define a failure envelope for matrix cracking. The criteria include two equations for fibre failure (2-3) and three for inter-fibre failure (4-6), depending on which mode of IFF occurs. That is, Mode A is a tensile-shearing failure, Mode B is a compression-shearing failure, and Mode C is an extreme case of compression-shearing which results in catastrophic laminate failure. The risk of failure, or stress exposure (f_E), for each type of failure is found by calculating the left-hand side of equations (2-6). Once its value reaches one, failure has occurred. The equations used are [7],

$$\frac{1}{\varepsilon_{1T}}\left(\epsilon_1 + \frac{\nu_{f12}}{E_1} m_{\sigma f}\sigma_2\right) = 1 \qquad (2)$$

$$\frac{1}{\varepsilon_{1C}}\left|\left(\varepsilon_1 + \frac{\nu_{12}}{E_1} m_{\sigma f}\sigma_2\right)\right| + (10\gamma_{12})^2 = 1 \qquad (3)$$

$$\sqrt{\left(\frac{\tau_{12}}{S_{21}}\right)^2 + \left(1 - p_{\parallel\perp}^{(+)}\frac{Y_T}{S_{21}}\right)^2\left(\frac{\sigma_2}{Y_T}\right)^2} + p_{\parallel\perp}^{(+)}\frac{\sigma_2}{S_{21}} + \left|\frac{\sigma_1}{\sigma_{1D}}\right| = 1 \qquad (4)$$

$$\frac{1}{S_{21}}\left(\sqrt{\tau_{12}^2 + \left(p_{\perp\parallel}^{(+)}\sigma_2\right)^2} + p_{\perp\parallel}^{(-)}\sigma_2\right) + \left|\frac{\sigma_1}{\sigma_{1D}}\right| = 1 \qquad (5)$$

$$\left[\left(\frac{\tau_{12}}{2(1 + p_{\perp\perp}^{(-)})S_{21}}\right)^2 + \left(\frac{\sigma_2}{Y_C}\right)^2\right]\frac{Y_C}{(-\sigma_2)} + \left|\frac{\sigma_1}{\sigma_{1D}}\right| = 1 \qquad (6)$$

These equations contain the terms for the material properties given in Table 1. In addition, several other parameters are required: $m_{\sigma f}$ is the mean stress magnification factor for the fibres in the transverse direction, ν_{f12} is the Poisson's ratio of the fibres, $p_{\parallel\perp}^{(+)}$, $p_{\perp\parallel}^{(+)}$, $p_{\perp\parallel}^{(-)}$ and $p_{\perp\perp}^{(-)}$ are parameters that control the shape of the failure envelope around the zero transverse stress point, and σ_{1D} is used to decrease the resistance to inter-fibre failure due to early fibre breaks ($\approx 1.1X_T$).

If inter-fibre failure (IFF) occurs, the UMAT degrades the transverse and shear stiffnesses using the following equations [23], [24],

$$E_\perp = \left(\frac{1-\eta_{rE}}{1+C_E\left(f_{E(IFF)}-1\right)^{\xi_E}} + \eta_{rE}\right) E_{\perp s} \tag{7}$$

$$G_{\perp\parallel} = \left(\frac{1-\eta_{rG}}{1+C_G\left(f_{E(IFF)}-1\right)^{\xi_G}} + \eta_{rG}\right) G_{\perp\parallel s} \tag{8}$$

$$G_{\perp\parallel} = \left(\frac{\left[f_E^{(\tau_{21})} + C^{(\sigma_2)}\left(f_E - f_E^{(\tau_{21})}\right)\right] - f_{Ethr}^{(\tau_{21})}}{1-f_{Ethr}^{\tau_{21}}}\right)^{n^{(\tau_{21})}} x\,\left(G_{\perp\parallel} - G_{\perp\parallel}\big|_{f_E^{(\tau_{21})}=1}\right) \tag{9}$$

$$E_\perp = \left(\frac{\left[f_E^{(\sigma_2)} + C^{(\tau_{21})}\left(f_E - f_E^{(\sigma_2)}\right)\right] - f_{Ethr}^{(\sigma_2)}}{1-f_{Ethr}^{\sigma_2}}\right)^{n^{(\sigma_2)}} x\,\left(E_\perp - E_\perp\big|_{f_E^{(\sigma_2)}=1}\right) \tag{10}$$

where $E_{\perp s}, G_{\perp\parallel s}$ are the secant moduli at IFF-initiation. η_r, c *and* ξ are parameters that need to be calibrated by experiments [23], $f_{E(IFF)}$ is the stress exposure at IFF. $G_{\perp\parallel}$ and $E_\perp$ are the shear and secant moduli, while $G_{\perp\parallel}\big|_{f_E=1}$ is the shear modulus at IFF initiation with only shear stress applied. f_E is the stress exposure for normal and shear stresses, and f_{Ethr} is the threshold stress exposure below which no damage occurs. n and C are degradation response control factors for equations (9) and (10).

Figure 3 displays the results of validation of the plane stress shell element UMAT against experimental data from the WWFE. The model shows good agreement with the results for both carbon and glass fibre material properties. The stress exposure values can be used as indicators for the risk of damage in the laminates throughout the blade, and the level of degradation to the transverse and shear moduli is also calculated. The UMAT has been calibrated here against the experiment data of Puck [24] for GFRP and CFRP.

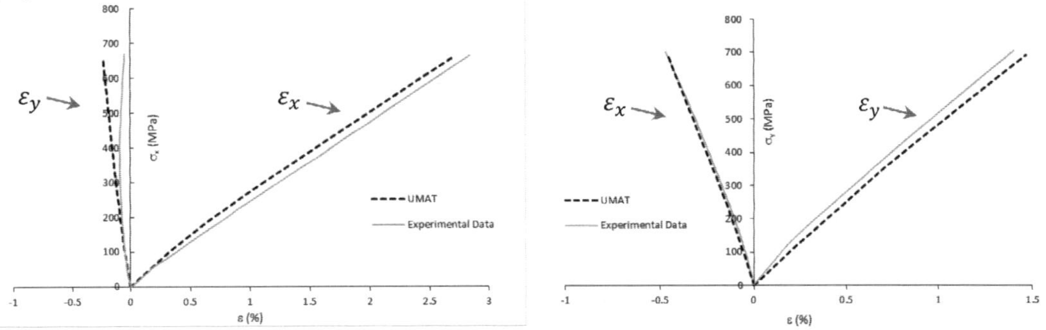

Figure 3. (a) Validation of the model for (0°/90°) GFRP laminate for applied σ_x uniaxial tension (b) validation of the model for a (0°/±45°/90°) CFRP laminate for applied σ_y uniaxial tension [7].

3. Results and discussion

Some results of the finite element analyses are illustrated in figures 4 to 9. Before the analysis commenced a mesh sensitivity study was performed; it was found there was no significant difference in the results calculated between the current mesh density (approx. 26,000) and a much finer mesh of 200,000+ elements. Hence, the lower density mesh was chosen.

Figure 6 displays the stress response of the spar caps for the three blades analysed. The plots display the values from the outermost 0° carbon fibre ply. Since the outer three layers form a glass

Journal of Physics: Conference Series **628** (2015) 012106 doi:10.1088/1742-6596/628/1/012106

fibre protective layer, these are the results for the highest loaded carbon fibre layer. The longitudinal failure strain for carbon fibre is approximately 60% lower than for glass fibres and so is more critical when considering possible failures. It is clear that the increased blade thickness due to the flat-back profile (Blade 3) is beneficial to the structural response of the blade. The stresses in the first 2 m of Blade 3 are considerably lower, almost 50% of the other two blades. A number of peaks can be seen in the resulting plots, due to the decreasing laminate thicknesses.

In conjunction with the lower predicted stresses, Figure 4 demonstrates the improved predicted stiffness of the thicker blade. Blades 1 and 2 display total flap-wise tip deflection of approximately 9% and 7.5% of the total blade length, respectively, while Blade 3 outperforms them with only a 4.5% deflection. It should be noted, however, that although the profiles for blades 1 and 2 had very similar maximum thickness values, they show a notable difference in tip deflection and stress distribution. Profile geometry should, therefore, also be considered when analysing the results. An example of where this is important is around the 0.3 mark on the normalised rotor radius. Blades 2 and 3 both exhibit a significant discontinuity in stress, which is less prominent in Blade 1 (see Fig. 6). However, Blade 1 consists of two NACA profiles, which is not the case with the other two blades, and may imply the difference in geometry between these two profiles is less severe.

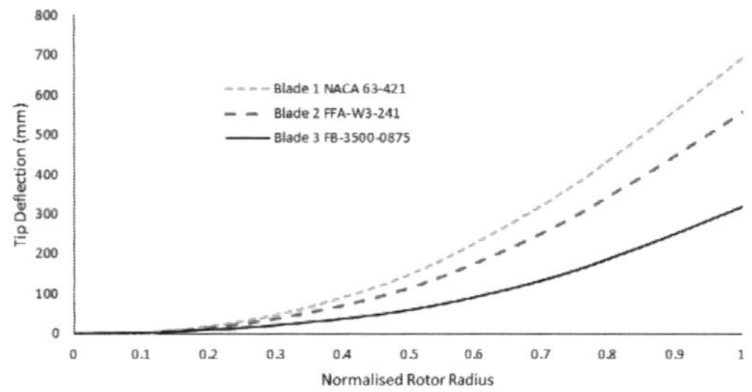

Figure 4. Deflection of the blades under max operating load, 2.5 m/s tidal velocity.

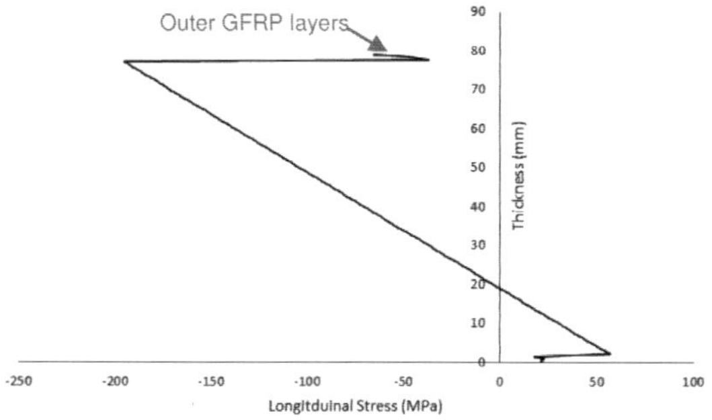

Figure 5. Longitudinal ply-orientated stresses through the compression-side spar cap thickness.

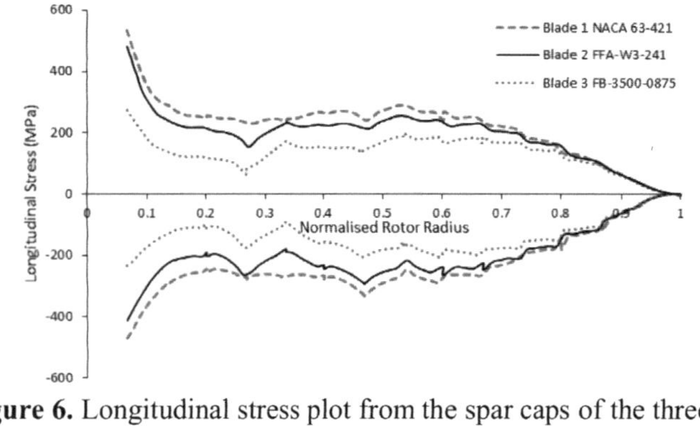

Figure 6. Longitudinal stress plot from the spar caps of the three blades.

A combination of failure modes is possible, with the faces of the blade in tension and compression. Due to the position of the neutral bending axis of the blade, some of the thick laminates near the root experience tension on one side and compression on the other side of the same laminate (Figure 5). The possibility for complex failure interaction becomes large when the compressive side has a relatively large value for Mode C stress exposure. The results are presented for the maximum operating load case, so low values of stress exposure for all modes are expected. Flood conditions are expected to be the driving factor for the failure of the blades, especially in cases where control systems fail or are slow to react and the turbines operate beyond their rated velocity.

Degradation is applied in the UMAT via a number of damage variables (valued from 0 to 1, where 1 corresponds to failure). The two damage parameters affected by equations (7) to (10) are for the transverse stiffness and shear stiffness, and are applied to the unidirectional material data before the stiffness matrix is calculated for each element. Contour plots of these variables for the three blades are shown outlined in figures 7 to 9. The changes in shear and transverse material stiffness applied in the UMAT are too low in magnitude and too localised to have a significant effect on tip deflection, increasing only from 691.3 mm for blade 1 in an elastic analysis to 693.1 mm in the non-linear UMAT analysis.

Elements at a risk of IFF in Mode A are visible on both sides of the blade due to the position of the neutral bending axis. The grey area in figures 7 and 8 near the root indicates that IFF has occurred on the tension side in the outer plies and, hence, the next design iteration should involve further strengthening the laminates in this region. From the maximum stress exposure in the contour plots it is apparent there is less chance of a compressive form of IFF occurring, with maximum values of 0.79, 0.54, and 0.42 for Mode B in the three blades. However, a higher risk of Mode C is visible from the plots, with 0.77 and 0.40 in blades 2 and 3 and 1.08 in Blade 1 implying failure. Since Mode C is a more catastrophic type of failure, care should be taken with the design. Figure 9 shows that a small amount of shear degradation has occurred throughout the blade's box-spar region, which presumably accounts for the change in overall tip deflection of the blade.

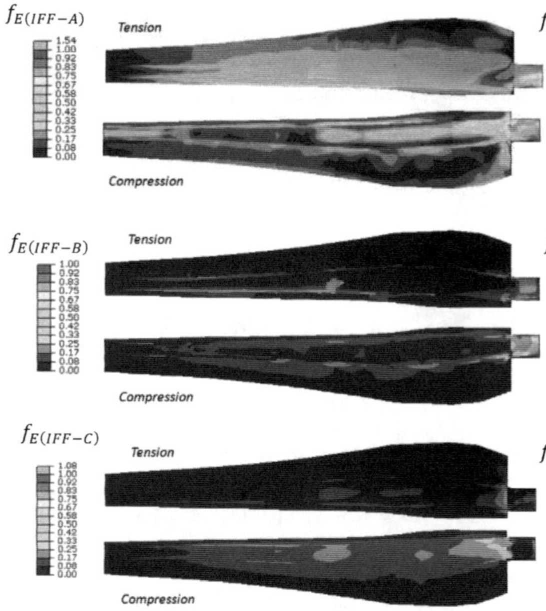

Figure 7. Contour plots of the stress exposure level for Blade 1 for Mode A (top), Mode B (middle), and Mode C (bottom).

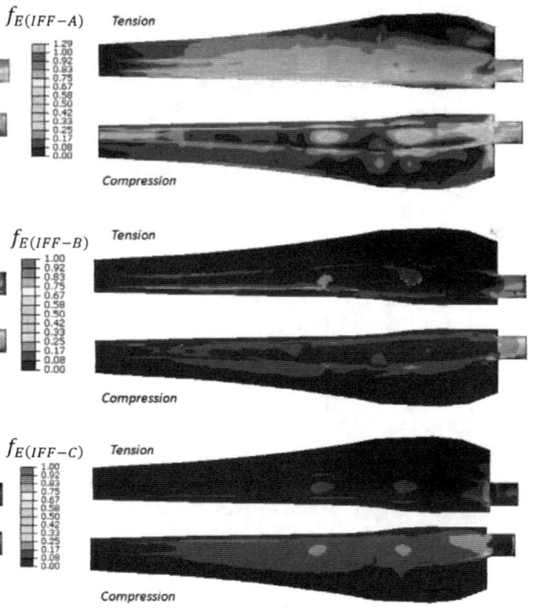

Figure 8. Contour plots of the stress exposure level for Blade 2 for Mode A (top), Mode B (middle), and Mode C (bottom).

4. Conclusions

Key results from this study showed that thicker blade sections provide a beneficial structural response, and help to reduce the risk of inter-fibre failure damage near the blade roots. The outermost highest loaded plies were predicted to have a high risk of IFF. Further work will investigate solid models for sections of the blade where through-thickness stresses are likely. The possibility of more computationally efficient modelling methods will be explored for such a model.

The CFRP and GFRP results both demonstrated a close match to the experimental data. However, issues arise with the triaxial nature of the stresses in the thick sections of the blade, especially around geometric discontinuities such as the root attachments. This is a limit to this type of shell based analysis. When used as a failure criteria, the UMAT proves to be a powerful tool for understanding the risks of inter-fibre failure for a fully loaded blade, and even modelling the reduction in stiffness in the regions that do suffer damage. Blade failures in some of the first test devices [25, 26] indicate the importance of developing better predictive techniques for assessing the damage occurring in large composites-based structures. The UMAT has added advantages in that it is not specific for turbine blade applications, but can be applied to the analysis of any glass or carbon fibre reinforced composite structure, where the plane stress assumption is valid.

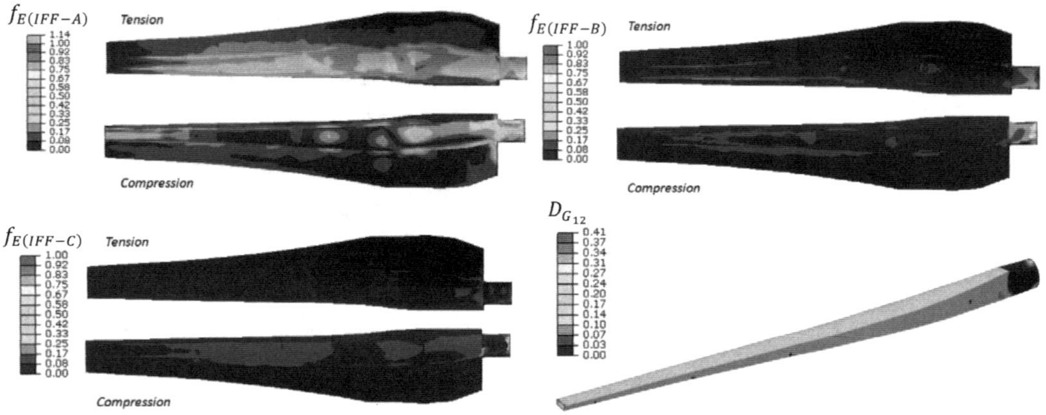

Figure 9. (Clockwise) Contour plots of the stress exposure level for Blade 3 for Mode A (top-left), Mode B (top-right), and Mode C (bottom-left). Shear damage variable for the box-spar section of the blade (bottom-right).

Acknowledgements
This material is in part based upon works supported by the Science Foundation Ireland Marine Renewable Energy Ireland (MaREI) research centre under Grant No. 12/RC/2302. It was also funded by a fellowship from the College of Engineering and Informatics, NUI Galway, and was supported by an NUI Travelling Studentship, 2014.

References

[1] Sustainable Energy Ireland 2010 Tidal & Current Energy Resources in Ireland (Dublin)
[2] Fraenkel P L 2002 Power from marine currents *Proc. Inst. Mech. Eng. Part A J. Power Energy* **216** 1–14
[3] Bir G S, Lawson M J and Li Y 2011 Structural Design of a Horizontal-Axis Tidal Current Turbine Composite Blade *ASME 30th Int. Conf. on Ocean, Offshore, and Arctic Engineering* (Rotterdam: ASME) pp 1-12
[4] Evans R, Mcadam R, Royle M, Mcewen L, Cross S, Park B and Wight I 2013 Optimum geometry for axial flow free stream tidal turbine blades *European Wave and Tidal Energy Conf. Series* ed P Fridaard, J P Kofoed, et al (Aalbord, Denmark: Technical committee EWTEC)
[5] Talreja R 2008 Damage and fatigue in composites – a personal account *Compos Sci Technol* **68** 2585–91
[6] Liu P F and Zheng J Y 2010 Recent developments on damage modeling and finite element analysis for composite laminates: A review *Mater. Des.* **31** 3825–34
[7] Puck A and Schurmann H 1998 Failure Analysis of FRP Laminates by Means of Physically

Based Phenomenological Models *Compos. Sci. Technol.* **58** 1045–67

[8] Hinton M, Kaddour A and Soden P 2002 A comparison of the predictive capabilities of current failure theories for composite laminates, judged against experimental evidence *Compos. Sci. Technol* **62** 1725–97

[9] Kennedy C R, Ó Brádaigh C M and Leen S B 2013 A multiaxial fatigue damage model for fibre reinforced polymer composites *Compos. Struct.* **106** 201–10

[10] Kennedy C R, Leen S B and Bradaigh C M 2012 A preliminary design methodology for fatigue life prediction of polymer composites for tidal turbine blades *Proc. Inst. Mech. Eng. Part L J. Mater. Des. Appl* **226** 203–18

[11] Grogan D M, Leen S B, Kennedy C R and Ó Brádaigh C M 2013 Design of composite tidal turbine blades *Renew. Energy* **57** 151–62

[12] Van Rooij R P J O M and Timmer W A 2003 Roughness Sensitivity Considerations for Thick Rotor Blade Airfoils *J. Sol. Energy Eng.* **125** 468–78

[13] Baker J P, Van Dam C P and Gilbert B L 2008 Flatback Airfoil Wind Tunnel Experiment (Albuquerque, New Mexico: Sandia National Laboratories)

[14] Bertagnolio F and Sorensen N 2001 Wind turbine airfoil catalogue (Roskilde, Denmark: Risø National Laboratory)

[15] Jensen F M, Falzon B G, Ankersen J and Stang H 2006 Structural testing and numerical simulation of a 34m composite wind turbine blade *Compos. Struct.* **76** 52–61

[16] Yang J, Peng C, Xiao J, Zeng J, Xing S, Jin J and Deng H 2013 Structural investigation of composite wind turbine blade considering structural collapse in full-scale static tests *Compos. Struct.* **97** 15–29

[17] Kong C, Bang J and Sugiyama Y 2005 Structural investigation of composite wind turbine blade considering various load cases and fatigue life *Energy* **30** 2101–14

[18] Overgaard LCT, Lund E and Thomsen O 2010 Structural collapse of a wind turbine blade. Part A: Static test and equivalent single layered models *Compos. Part A Appl. Sci. Manuf.* **41** 257-70

[19] Standish K J and van Dam C P 2003 Aerodynamic Analysis of Blunt Trailing Edge Airfoils *J. Sol. Energy Eng.* **125** 479-87

[20] Winnemöller T and Van Dam C P 2007 Design and Numerical Optimization of Thick Airfoils Including Blunt Trailing Edges *J. Aircr.* **44** 232–40

[21] Goundar J N and Ahmed M R 2013 Design of a horizontal axis tidal current turbine *Appl. Energy* **111** 161–74

[22] Soden P 1998 Lamina properties, lay-up configurations and loading conditions for a range of fibre-reinforced composite laminates *Compos. Sci. Technol.* **58** 1011–22

[23] Knops M and Bögle C 2006 Gradual failure in fibre/polymer laminates *Compos. Sci. Technol.* **66** 616–25

[24] Puck A and Mannigel M 2007 Physically based non-linear stress–strain relations for the inter-fibre fracture analysis of FRP laminates *Compos. Sci. Technol.* **67** 1955–64

[25] White D 2011 Editorial focus: tidal system challenge e blade design! *Ocean News & Technology* **17**

[26] Liu P and Veitch B 2012 Design and optimization for strength and integrity of tidal turbine rotor blades *Energy* **46** 393–404

An investigation into the impact of cryogenic environment on mechanical stresses in FRP composites

O Fifo and B Basu
Department of Civil, Structural & Environmental Engineering, Trinity College Dublin, Dublin 2, Ireland

Email: basub@tcd.ie

Abstract: Fibre reinforced polymer (FRP) composites are fast becoming a highly utilised engineering material for high performance applications due to their light weight and high strength. Carbon fibre and other high strength fibres are commonly used in design of aerospace structures, wind turbine blades, etc. and potentially for propellant tanks of launch vehicles. For the aforementioned fields of application, stability of the material is essential over a wide range of temperature particularly for structures in hostile environments. Many studies have been conducted, experimentally, over the last decade to investigate the mechanical behaviour of FRP materials at varying subzero temperature. Likewise, tests on aging and cycling effect (room to low temperature) on the mechanical response of FRP have been reported. However, a relatively lesser focused area has been the mechanical behaviour of FRP composites under cryogenic environment. This article reports a finite element method of investigating the changes in the mechanical characteristics of an FRP material when temperature based analysis falls below zero. The simulated tests are carried out using a finite element package with close material properties used in the cited literatures. Tensile test was conducted and the results indicate that the mechanical responses agree with those reported in the literature sited.

1. Introduction

It is widely accepted that fibre reinforced polymer (FRP) composites are the materials for the present and future in many engineering applications, especially for the high performance structures. FRPs are now the primary structural components in aerospace structures, found in the fuselage and wings (Boeing 787 Dreamliner); wind turbine blades; and are increasingly being considered for the propellant tanks of launch vehicles. Also, FRPs are being considered for thermal functionalities as part of an insulating system for superconducting magnet coils [1]. In the aforementioned fields, FRPs are exposed to varying degree of temperature, from room temperature to low and cryogenic temperatures. Therefore, the understanding of their mechanical response at those temperature is becoming important particularly in the event of a fracture or damage to the composite part at low temperature [2]. The behaviour of these materials at elevated temperature has been well documented over the years. However, not so much has been published on the characteristics of the FRPs at low temperatures.

In the last two decades quite a lot has been published on how the microstructure of an FRP composite changes when exposed to low temperature environment. Normally, the formation of micro-cracks on FRP composites warns of an impending failure. However, as the temperature drops so does the level of micro-crack formation in the matrix and the composite becomes more brittle due to loss of toughness in the matrix [3]. In the examination of the fracture toughness of FRPs at low temperature, Shindo et al. [4-5] have conducted tests to evaluate the fracture toughness of woven glass fibre epoxy composite at low temperatures. These involved the introduction of mixed-mode delamination to the laminate. Their results show that there is higher resistance at cryogenic temperature than at room temperature. Likewise, the behaviour of an adhesively bonded joint of an FRP material has been studied under the mode I delamination mode [6]. Research has also indicated that the interlaminar shear strength decreases with wall thickness of a carbon epoxy torus. Effect of the low temperature on the mechanical properties makes the matrix more brittle as the strain reduces with temperature [7]. Similar tests have been carried out for FRP composites use in tanks of reusable rockets as well as the numerical predictions of delamination and matrix cracks [8].

A large number of the available literature in this area of study has mainly reported experimental tests for both property characterisation and investigative fracture mechanics. Although a number of literature has been published on the thermal behaviour of FRP, some have examined the buckling analysis of FRP materials under thermal conditions [9-10]. Some models assumed temperature dependent-elastic properties in modelling the determination of critical buckling temperature for an FRP laminated composite plates. Some have investigated the influence of fibre orientation, modulus ratio, uniform and non-uniform temperature distribution [9]. Aside the investigation on thermal buckling of FRP, thermal effect on FRP material under tensile loading has also been investigated through finite element method (FEM). FEM has been used in analysing thermal residual stresses and the induced matrix failure has been conducted on a transversely tensile loaded carbon fibre reinforced polymer material[11]. The main temperature range considered for the reported FEM of analysing thermal behaviour of FRP materials has been within the room temperature and above.

Research conducted on cryogenic behaviour of FRP materials and validated using a finite element approach has been to some extent limited. This article reports the examination of FRP material at subzero and cryogenic temperatures. A simple model is developed in a commercial finite element package, ANSYS, and it is subjected to tensile loading at different subzero temperature. The results are also presented for each ply in the model.

2. Damage in Composite

At standard and elevated temperature, FRP composites have been reported to fail in various ways: under loading the matrix fails progressively in cracking; the fibre fractures; there is debonding between the fibres and the matrix; and interlamina failure between the layers. The latter is caused by out-of-plane stresses on the composite [12]. At lower temperature, the failure modes are largely similar. However, the matrix behaves differently. For a thermosetting polymer matrix with high molecular weight upon curing, forms high crosslink density, which amplifies the stiffness as the temperature drops [13]. The matrix shows high concentration of micro-cracks, which is the progressive failure of the matrix at room temperature, but at lower or cryogenic temperatures, the appearance of micro-cracks are scanty as a result of the matrix increasing in stiffness as the temperature reduces[3]. Hence, failure at low temperature is further sudden and abrupt.

3. Model Setup and Material Properties

The model was developed according to the ASTM D 3039 [14]. The material properties chosen for this study corresponds to a high strength carbon reinforced epoxy composite commonly used in structures designed for extreme environments. The IM7/8551-7 carbon-epoxy composite by Hexcel has been used in aerospace industry for decades. The properties are as shown in Table 1. The model was developed using Shell181, which can be used for analysing thin to moderately thick laminate structure. It has four nodes with six degree of freedom at each node.

The model was assumed to be transversely isotropic laminate of 2.5 mm thickness and was represented by a thin rectangular shell for tensile test. It was modelled as an eight-layer material with $[45/0/-45/90]_S$ stacking for the layers. The model was subjected to 1000N tensile load at 25, 0, -50, -100, and -150 degree Celsius.

Table 1. IM7/8551-7

IM7/8551-7		
E_{11}	158.58	GPa
E_{22}	8.34	GPa
E_{33}	8.34	GPa
G_{12}	5.86	GPa
G_{13}	3.5	GPa
G_{23}	5.86	GPa
ν_{12}	0.28	
ν_{13}	0.28	
ν_{23}	0.35	
Coefficient of Thermal Expansion		*@20 °C*
CTEx	$-4e^{-7}$	C^{-1}
CTEy	$3e^{-5}$	C^{-1}
CTEz	$3e^{-5}$	C^{-1}

4. Results and Discussion

The room temperature model is included for comparison of the changes in the material as its temperature drops. Among the main response that were evaluated are the deformation in the primary loading direction, the stress and the strain in each ply, all in the loading direction. Given the type of material and its high stiffness, it was expected that under normal conditions the deformation experienced from 1000 N would be minuscule. Under this tensile load, the extension at room temperature was 3.02 mm, Figure 1. The extension is minimal at the fixed left end and maximal at the right end where the load is applied. However, as the temperature reduces so does the tensile elongation of the material, microscopically. It appears that the material retains its behaviour even at low temperatures except for small reduction in elongation of the material in comparison to the behaviour at room temperature. The deformation was reduced by 8 μm, 24 μm, 40 μm, and 56 μm, for 0°C, -50°C, -100°C, and -150°C respectively in comparison with the room temperature (25°C) deformation.

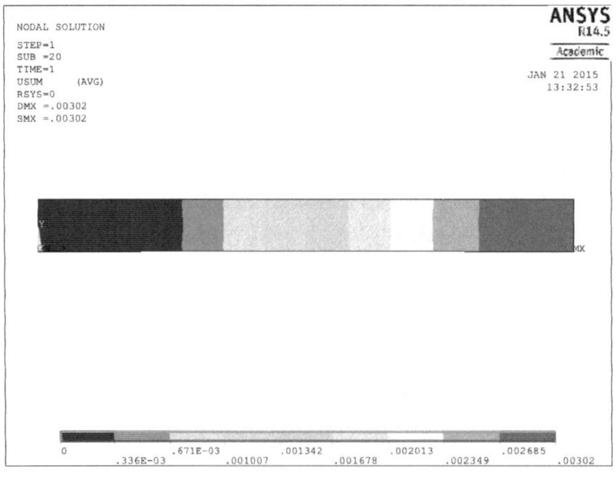

Figure 1. Deformation along the loading direction

Similarly, the stress and elastic strain developed in the material under the tensile axial loading were examined in individual layer of the composite. Since the graphic representation of the results for each temperature level bares close similarities, only the figures from the 25 °C model are presented. Also, due to the symmetric stacking sequence of the laminate, only the results for three unique layers are presented. The results for the test at room temperature are presented in the Table 2 and it is used as the standard to compare the changes in the material behaviour for the lower temperature tests.

Table 2. Stress (σ_x) and strain (ε_x) developed in the material at room temperature (25 °C), where (n_1, n_2) is the ply position

Orientation of ply (n_1, n_2) (degrees)	Stress in the loading direction, σ_x, (GPa)	Strain in the loading direction, mε
0 (2,7)	2.84	17.88
45 (1,8)	0.733	17.812
-45 (3, 6)	0.733	17.812
90 (4,5)	0.136	17.736

Figure 2 shows the elastic strain in the material. As would be expected under a normal test, the area between the two clamped tab ends illustrates the region of maximum strain for the first and third layers, the ±45° oriented plies. This is true for the other orientations (0°'s and 90°'s) and all the temperature levels considered. This demonstrates that the pattern of strain development is not significantly affected at low or at cryogenic temperatures. However, this is not true for the magnitude of the strain. As shown in Figure 3, the strain linearly increases for the diagonally and transversely oriented layers (the ±45°s and 90°s respectively) and decreases steadily for longitudinal layers, the 0°'s layers. The disparity in the strain level in the layers can be attributed to the matrix being stiffer at low temperatures and the apparent high strength of the layers oriented parallel to the loading direction. Since the former, variable of high stiffness, at cryogenic conditions applies to all layers, it can be said the latter, variable of orientation, is the defining factor in influencing the material's response to axial tensile loading.

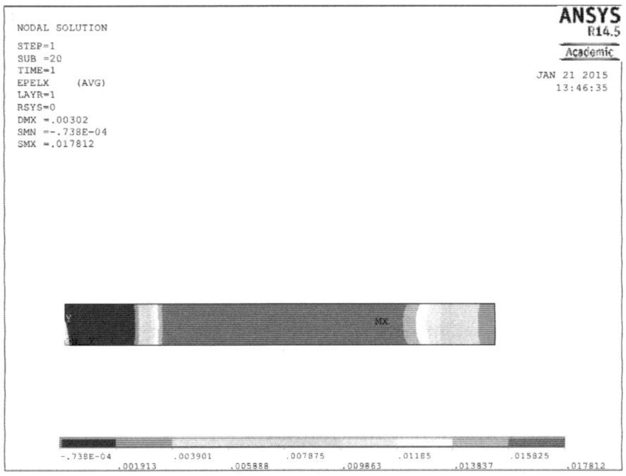

Figure 2. Plot of strain intensity at 25 degree for first layer

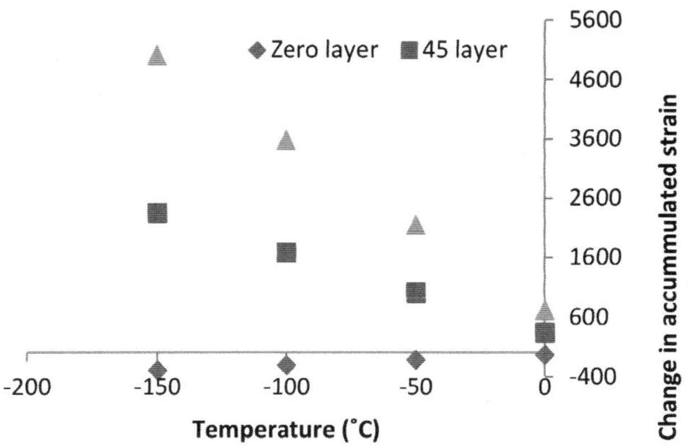

Figure 3. Change in induced strain with decreasing temperature

Figure 4 shows a graphical presentation of the change in the stress developed in the material benchmarked against those at room temperature. It shows less stress in the longitudinal fibres, which are parallel to the loading direction. This indicates that low temperature stiffens the matrix and delays the transfer of stress from the matrix to the fibres. This is particularly evident in the plies oriented perpendicularly to the loading the direction. The results reveal that the 90 degree plies were subjected to more stress than any other in the composite laminate as the temperature reduces. It appears that the stresses shed-off by the longitudinal fibres were transferred to the transverse plies. Although, the stress magnitude in the longitudinal fibres is significantly higher than those in the 90 degree plies, it however shows the vulnerabilities of the fibres oriented at large angles. The stresses in the two principal axes change within close margin of each other. Figure 5 shows that longitudinal stress in the 0° oriented ply drops by 20 percent with every degree fall in temperature and similarly, 24 percent rise in stress experienced by the transverse plies. Given that the similarity in the rate at which the longitudinal fibres and transverse fibres are experiencing reduction, and accumulation of stress respectively, it can be envisaged that in the event of increase in load and high fatigue, the transverse layers will be expected to fail first under the current loading condition. It can be said that these vulnerable transverse layers presented the least resistance path and almost no existence of fortification against the axial tensile load.

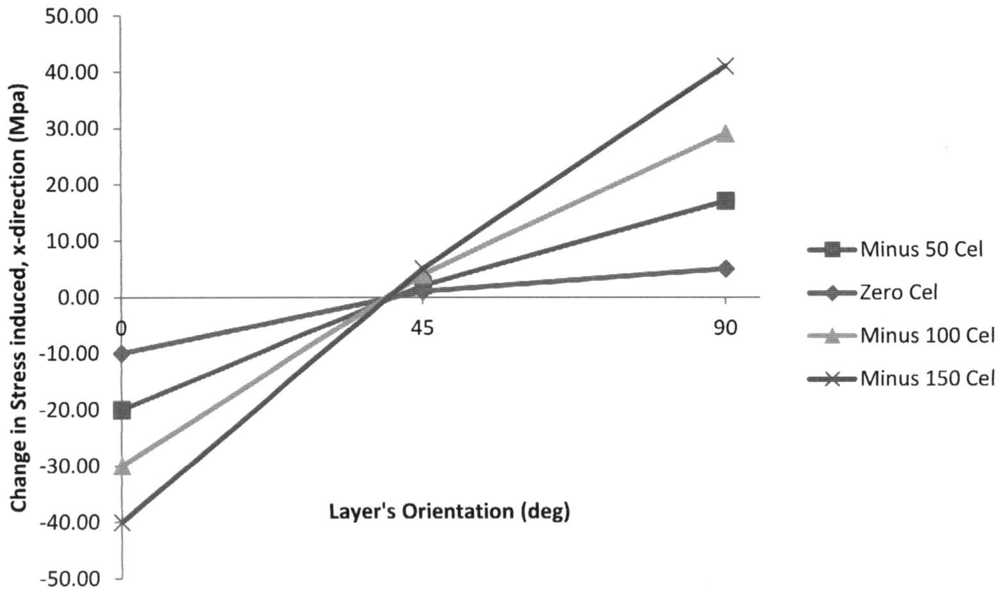

Figure 4. Change in induced stress as the temperature drops

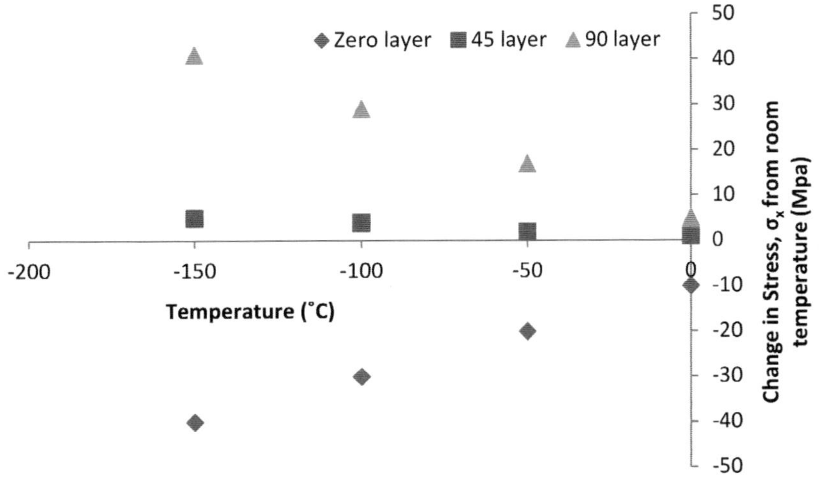

Figure 5. Relative change in stress from room temperature with falling temperature

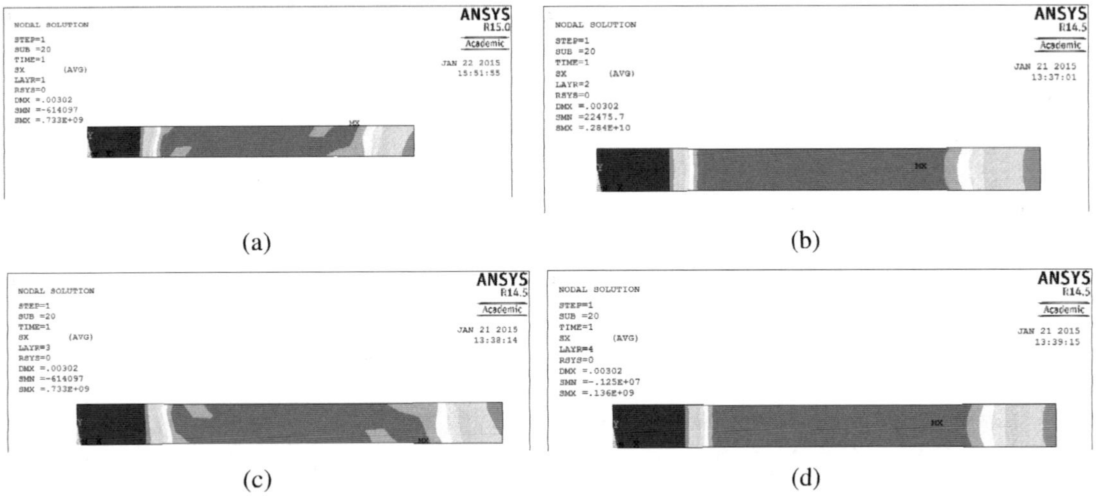

Figure 6. Illustration of the stress in the laminate at 25 for a) 45 layers; b) 0 layers; c) -45 layers; d) 90 plies

5. Conclusion

This article investigates the mechanical response of fibre reinforced polymer composite under tensile loading in a cryogenic environment using ANSYS. The results agree with some of the published experimental work in this area, in that, the FRPs do not drastically deviate from their behaviour at room temperature when exposed to cryogenic temperature. This study allows individual ply in such laminate at cryogenic temperature to be examined and analysed. It can be said that plies oriented parallel to loading direction or axis, though they account for majority of the load carrying capacity of the laminate, are the least vulnerable as strain builds up in them at a very slow rate. However, the off axis plies or the transverse plies appear to accumulate strain rapidly and therefore have the propensity to initiate the failure of the composite.

Reference:

[1] Mitchell N, Bauer P, Bessette D, Devred A, Gallix R, Jong C, Knaster J, Libeyre P, Lim B, Sahu A and Simon F 2009 Status of the ITER magnets *Fusion Engineering and Design* **84** 113-21
[2] Yokozeki T, Ogasawara T, Aoki T and Ishikawa T 2009 Experimental evaluation of gas permeability through damaged composite laminates for cryogenic tank *Compos. Sci. Technol.* **69** 1334-40
[3] Nettles A T and Biss E J 1996 Low temperature mechanical testing of carbone-fiber/epoxy composite materials *NASA Technical Paper* **3663** *George C Marshal Space Flight Center Report*
[4] Shindo Y, Miura M, Takeda T, Saito N and Narita F 2011 Cryogenic delamination growth in woven glass/epoxy composite laminates under mixed-mode I/II fatigue loading *Compos. Sci. Technol.* **71** 647-52
[5] Shindo Y, Takeda T, Narita F, Saito N, Watanabe S and Sanada KY 2009 Delamination growth mechanisms in woven glass fiber reinforced polymer composites under Mode II fatigue loading at cryogenic temperatures *Compos. Sci. Technol.* **69** 1904-11
[6] Melcher R J and Johnson W S 2007 Mode I fracture toughness of an adhesively bonded composite–composite joint in a cryogenic environment *Compos. Sci. Technol.* **67** 501-6
[7] Ahlborn K and Knaak S 1988 Cryogenic mechanical behaviour of a thick-walled carbon fibre reinforced plastic structure *Cryogenics* **28** 273-7
[8] Aoki T I T, Kumazawa H and Morino Y 2001 Cryogenic mechanical properties of CF/polymer composites for tanks of reusable rockets *Adv. Compos. Mater.* **10**
[9] Chen L W and Chen L Y 1991 Thermal postbuckling behaviors of laminated composite plates with temperature-dependent properties *Compos. Struct.* **19** 267-83
[10] Kabir H R H, Askar H and Chaudhuri R A 2003 Thermal buckling response of shear flexible laminated anisotropic plates using a three-node isoparametric element *Compos. Struct.* **59** 173-87
[11] Fiedler B, Hojo M, Ochiai S, Schulte K and Ochi M 2001 Finite-element modeling of initial matrix failure in CFRP under static transverse tensile load *Compos. Sci. Technol.* **61** 95-105
[12] Fifo O, Ryan K and Basu B 2013 Investigations on the effectiveness of cyanoacrylate adhesive for repair of fibre-reinforced polymer laminates under different environmental conditions *J. Compos. Mater.* **24**
[13] Callaghan M T 1991Use of resin composites for cryogenic tankage *Cryogenics,* **31** 282-7
[14] ASTM 2014 Standard Test Method for Tensile Properties of Polymer Matrix Composite Materials in *ASTM D3039/D3039M-14*, (ed. West Conshohocken, PA: ASTM International)

Prediction of dynamic strains on a monopile offshore wind turbine using virtual sensors

A N Iliopoulos[1], W Weijtjens[2], D Van Hemelrijck[1] and C Devriendt[2]

[1]Vrije Universiteit Brussel, Department of Mechanics of Materials and Constructions, Pleinlaan 2, B-1050,Brussels, Belgium
[2]Vrije Universiteit Brussel, Acoustic and Vibration Research Group, Pleinlaan 2, B-1050,Brussels, Belgium

Alexandros.Iliopoulos@vub.ac.be

Abstract. The monitoring of the condition of the offshore wind turbine during its operational states offers the possibility of performing accurate assessments of the remaining life-time as well as supporting maintenance decisions during its entire life. The efficacy of structural monitoring in the case of the offshore wind turbine, though, is undermined by the practical limitations connected to the measurement system in terms of cost, weight and feasibility of sensor mounting (e.g. at muddline level 30m below the water level). This limitation is overcome by reconstructing the full-field response of the structure based on the limited number of measured accelerations and a calibrated Finite Element Model of the system. A modal decomposition and expansion approach is used for reconstructing the responses at all degrees of freedom of the finite element model. The paper will demonstrate the possibility to predict dynamic strains from acceleration measurements based on the aforementioned methodology. These virtual dynamic strains will then be evaluated and validated based on actual strain measurements obtained from a monitoring campaign on an offshore Vestas V90 3 MW wind turbine on a monopile foundation.

1. Introduction

Many large-scale offshore wind farm projects use monopile foundation to obtain a cost effective design. The complexity of the offshore wind turbine (OWT) due to the added support structure as well as the unknowns in the soil-structure interaction encourage a thorough investigation of the dynamics of the structure. Furthermore, the harsh offshore conditions including combined wind and wave excitations, a corrosive environment, currents and shifts in the sea bed or soil conditions such as scour, i.e. erosion of the sea bed near the monopile, pose a threat to the structural integrity of the offshore wind turbine. Therefore, the monitoring of the condition of the OWT during its operational states offers the possibility of improving the safety of a mission. Fatigue deterioration, affecting the wind turbine, represents, in fact, one of the major issues regarding mission safety that needs to be addressed during its entire life. To assure the dynamic and fatigue strength of the structure, it is necessary to measure its dynamic strain distribution. However, strain measurements with conventional strain gauges are not always possible since the gauges are not reusable and cannot be moved from point to point when they have been attached. Moreover, to accurately assess the fatigue life consumption the strains at the mudline need to be measured. However, this implies installing sensors 30m beneath water level which is unfeasible for installation and maintenance. By contrast, a method that allows to estimate the strains at fatigue sensitive spots using accelerometers in the tower is cheaper and far more reliable.

Different response estimation techniques have been investigated by researchers to predict response time histories at various locations within structural components using the measurements available at a limited number of locations. A common approach is based on the use of Kalman filter [1] methods [2-4]. All these applications, though, are restricted to the study of linear structures and make the assumptions that the input-forces are either known or broadband, so that they can be modeled as zero mean stationary white process. Moreover, the iterative algorithm makes these approaches time consuming and computationally expensive.

Another approach includes response estimation based on the concept of modal expansion. These methods intelligently extrapolate measured data to generate virtual strain measurements. Non-contact methods like Digital Image Correlation and Dynamic Photogrammetry are known techniques used to measure sets of limited surface data that can be used in conjunction with an expansion algorithm for full field information [5-8]. However, all the aforementioned surveys correspond to laboratory scale applications and the excitation is exactly known or/and the responses can be measured in a sufficient large number of points.

This paper deals with real scale offshore wind turbine instrumented with a limited number of sensors (accelerometers, strain gauges) and demonstrates the possibility to predict dynamic strains from acceleration measurements. The full field response of the structure is reconstructed based on a calibrated finite element model and a limited number of measured accelerations. The prediction is based on a modal decomposition of the measured accelerations that results in the estimation of the modal coordinates [9, 10]. The relation between the modal coordinate and the acceleration/strain in an arbitrary point is established by making use of the corresponding numerically obtained mode shapes [9, 10].

The proposed algorithm and the virtual dynamic strains will be evaluated and validated based on actual measurements obtained from a monitoring campaign on an offshore Vestas V90 3 MW wind turbine on a monopile foundation.

2. Theory

Herein the theoretical background of the modal decomposition and expansion approach used to predict the dynamic responses is discussed in detail. The quality indicators used as correlation tools are presented too.

2.1. Modal decomposition

The displacement vector can be written as a linear combination of the mode shape vectors by making use of the modal decomposition approach [11-13]:

$$\begin{gathered} \boldsymbol{x}(\mathrm{t}) = \Phi \boldsymbol{q}(t) \\ \Phi = [\boldsymbol{\varphi_1} \quad \boldsymbol{\varphi_2} \quad \dots \quad \boldsymbol{\varphi_n}] \\ \boldsymbol{q}(t) = [q_1(t) \quad q_2(t) \quad \dots \quad q_n(t)]^T \end{gathered} \tag{1}$$

where $\boldsymbol{q}$ (t) is the vector of the modal coordinates for each time instant t, Φ is the mode shape matrix, n is the total number of modes and $[.]^{\mathrm{T}}$ denotes the transpose of a matrix. Since our interest lies within the lower frequency range, the first two or three tower modes in the Fore-Aft (FA) or Side-Side (SS) direction are the considered subset of modes.

2.2. Modal expansion

If we consider a subset of a active DOF's corresponding to the measured locations of the real experimental model and d remaining DOF's corresponding to predicted locations, then the relationship for only the measured DOF's is written as:

$$\boldsymbol{x_m} = \Phi_m \boldsymbol{q_m} \tag{2}$$

where subscript m corresponds to the measured DOF's.

Assuming that the number of modes, n_m , is less than the number of active DOF's, a, the modal coordinates $\boldsymbol{q_m}$ are calculated by using the pseudo inverse as follows:

$$\boldsymbol{q_m} = (\Phi_m^T \Phi_m)^{-1}\, \Phi_m^T\, \boldsymbol{x_m} = {\Phi_m}^{\dagger}\, \boldsymbol{x_m} \tag{3}$$

This result can be expanded to the predicted locations (d remaining DOF's) with the following expression:

$$\boldsymbol{x_p} = \Phi_p \boldsymbol{q_m} = \Phi_p\, {\Phi_m}^{\dagger}\, \boldsymbol{x_m} \tag{4}$$

where subscript p corresponds to the predicted DOF's.

A schematic representation of the modal decomposition and expansion approach for the case of the offshore wind turbine is presented in figure 1.

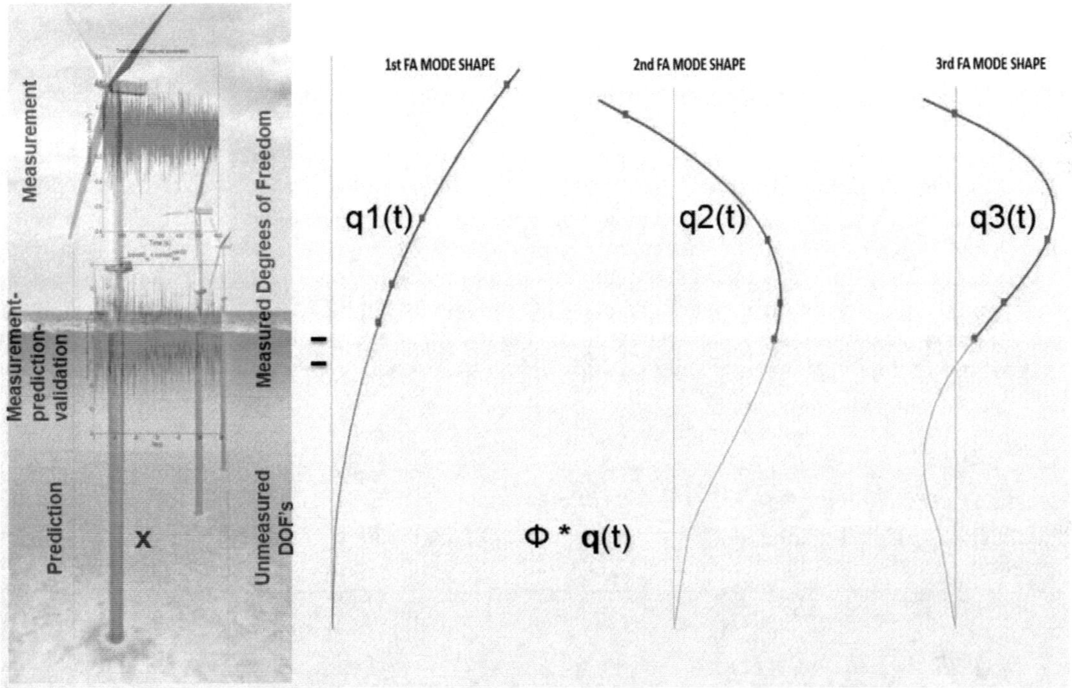

Figure 1. Modal decomposition and expansion in a schematic representation for the case of OWT. In red is the accessible area of the OWT with the corresponding measured locations and in green is the critical and inaccessible part of the OWT.

2.3. Prediction of vibrational accelerations

In order to be able to predict the acceleration signal [9, 10, 14], the acceleration mode shape matrix first needs to be constructed. This is done by making use of the numerically obtained displacement mode shape components as follows:

$$\omega_j = 2\pi f_j \tag{5}$$

$$\Phi_{a,ij} = \omega_j^2 . \Phi_{ij} \tag{6}$$

where ω_j is the angular frequency of the jth mode calculated from the corresponding natural frequency f_j derived from the Finite Element Model, Φ_{ij} is the numerically (FEM) obtained absolute displacement mode shape component of the ith degree of freedom that corresponds to the jth mode and $\Phi_{a,ij}$ is the resulting acceleration mode shape component of the ith degree of freedom that corresponds to the jth mode. The modal coordinates $\boldsymbol{q_m}(t)$ are calculated from the limited number of experimentally known acceleration time signals at measured locations $\boldsymbol{a_m}(t)$ with the following expression:

$$\boldsymbol{a_m}(t) = \Phi_{\alpha_m}\boldsymbol{q_m}(t)$$
$$\boldsymbol{q_m}(t) = (\Phi_{\alpha_m})^{-1}\boldsymbol{a_m}(t) \quad (7)$$

where superscript m indicates measured degrees of freedom (DOF's). Then, the acceleration can be predicted at any inaccessible point of the structure as follows:

$$\boldsymbol{a_p}(t) = \Phi_{\alpha_p}\boldsymbol{q_m}(t) \quad (8)$$

where superscript p indicates predicted degrees of freedom (DOF's).

2.4. Prediction of dynamic strains

The combined use of operational acceleration data and strain mode shape components derived from a well-tuned FEM [9, 10, 14] allows for successful prediction of the dynamic strain responses at different levels along the height of the structure. The prediction is based upon a modal decomposition of the measured accelerations,$\boldsymbol{a_m}(t)$, that results in the estimation of the modal coordinates, $\boldsymbol{q_m}(t)$, according to (7). The relation between the modal coordinate and the strain in any arbitrary point is established by making use of the corresponding numerically obtained strain mode shapes with the following formula:

$$\boldsymbol{\varepsilon_p}(t) = \Phi_{\varepsilon_p}\boldsymbol{q_m}(t) \quad (9)$$

where superscript p indicates predicted degrees of freedom (DOF's), Φ_{ε_p} is the strain mode shape matrix that consists of the strain mode shape components of each acting mode at the inaccessible DOF of interest.

2.5. Quality indicators

The quality indicators utilized to compare the results of the predicted and the measured signals are discussed here. First is the Time Response Assurance Criterion (TRAC) [5] which is the correlation for one DOF over all time of the predicted response time domain signal ($\boldsymbol{g_p}(t)$) with the corresponding measured response time domain signal ($\boldsymbol{g_m}(t)$).

$$TRAC = \frac{\left[\boldsymbol{g_m}(t)^T\boldsymbol{g_p}(t)\right]^2}{\left[\boldsymbol{g_m}(t)^T\boldsymbol{g_m}(t)\right]\left[\boldsymbol{g_p}(t)^T\boldsymbol{g_p}(t)\right]} \quad (10)$$

Additionally, the frequency domain equivalent to the Time Response Assurance Criterion called the Frequency Response Assurance Criterion (FRAC) [14] is introduced. This is the correlation for one DOF over all frequencies of the predicted complex frequency domain signal ($\boldsymbol{g_p}(f)$) with the measured complex frequency domain signal ($\boldsymbol{g_m}(f)$).

$$FRAC = \frac{\left[|\boldsymbol{g}_m(f)^H \boldsymbol{g}_p(f)|\right]^2}{\left[\boldsymbol{g}_m(f)^H \boldsymbol{g}_m(f)\right]\left[\boldsymbol{g}_p(f)^H \boldsymbol{g}_p(f)\right]} \tag{11}$$

where $(.)^H$ denotes the Hermitian conjugate and $(.)^T$ denotes the transpose of the vector and $\boldsymbol{g}_m$ is the generalised measured response (acceleration $\boldsymbol{a}_m$ or strain $\boldsymbol{\varepsilon}_m$) and $\boldsymbol{g}_p$ is the generalised predicted response (acceleration $\boldsymbol{a}_p$ or strain $\boldsymbol{\varepsilon}_p$) respectively. The Mean Absolute Error in the time domain and the Mean Absolute Error of the Spectral Densities in the frequency domain are utilised too.

3. Offshore field measurements

The measurement campaign is performed at the Belwind wind farm, which consists of 55 Vestas V90 3MW wind turbines. The wind farm is located in the North Sea on the Bligh Bank, 46km off the Belgian coast. Measurements are taken at 4 levels on 9 locations using a total of 10 accelerometers. Eight accelerometers capture the vibrations in the X-Y direction and the two additional accelerometers are utilised to identify torsional vibrations in the tower. The locations are chosen based on the convenience of sensor mounting, such as the vicinity of platforms. The chosen levels are 69 m, 41 m, 27 m and 19 m above Lowest Astronomical Tide Level (LAT). On the level 19m LAT, 4 optical fiber Bragg strain sensors have been installed. On the level 41m LAT, 2 fiber brag strain sensors have been installed. Figure 2 gives an overview of the instrumented OWT with the corresponding measurement locations. For more information the reader is referred to [9, 10].

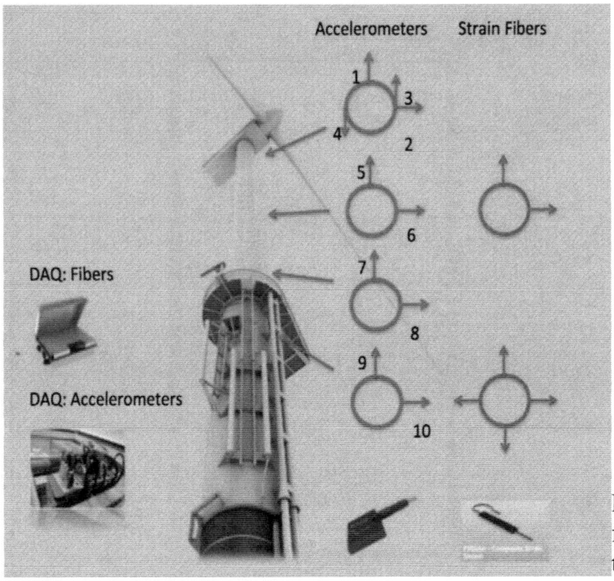

Figure 2. BBC01 OWT at Belwind farm with the instrumentation and the measurement locations.

4. Results

The acceleration time histories recorded at several locations of the structure are used in conjunction with a modal decomposition and expansion algorithm in order to compose a complete methodology for the prediction of vibrational accelerations and subsequently prediction of dynamic strains, as discussed in section 2. The results of the prediction of the dynamic strains in the time domain as well as in the frequency domain accompanied with the corresponding quality indicators are presented herein. As shown in figures 3-4, there is a great agreement between the measured and predicted dynamic strain time domain signal at both levels (+19m LAT and +41m LAT) under parked conditions of the OWT.

Journal of Physics: Conference Series **628** (2015) 012108 doi:10.1088/1742-6596/628/1/012108

Moreover there is high spectral coherence between the two signals. The qualitative good match indicated in the figures is quantified through the quality indicators (figure 6) which prove the high correlation in terms of amplitude, temporal evolution and frequency content between the measured and predicted signals (very high TRAC and FRAC indexes exceeding 0.975 for both cases and simultaneously low mean absolute errors).

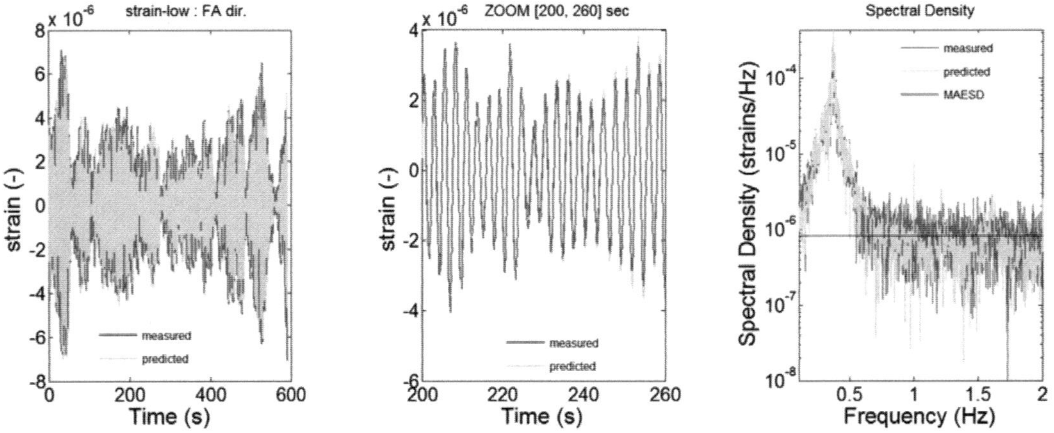

Figure 3. Comparison of the measured dynamic strain (red) at **19m LAT** and the predicted dynamic strain (green) in the FA direction under **parked** conditions of the OWT.

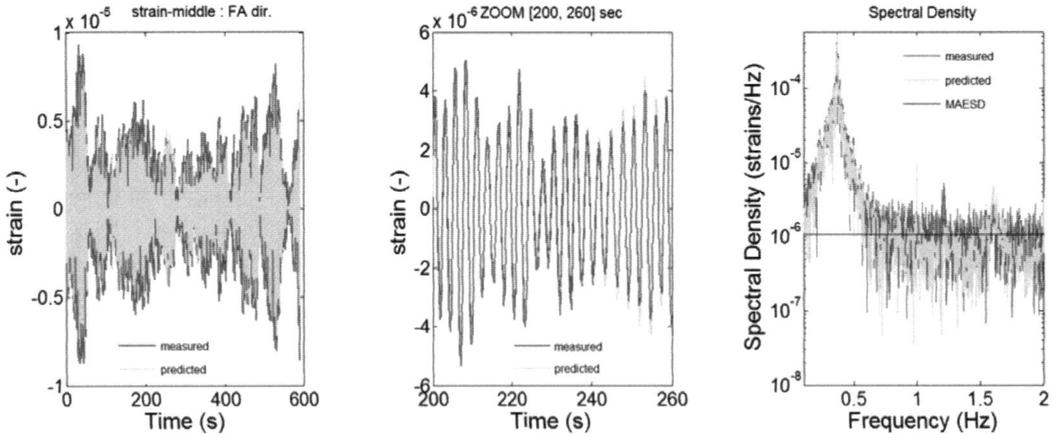

Figure 4. Comparison of the measured dynamic strain (red) at **41m LAT** and the predicted dynamic strain (green) in the FA direction under **parked** conditions of the OWT.

The prediction of dynamic strains is also shown for the case that the OWT is under normal rotating conditions in figure 5. According to this figure and the indicators in figure 6, it is worth stating that the prediction of dynamic strains under this type of loading remains a challenge with a good potential for further improvement of the quality indicators used to describe the correlation between the measured and predicted signals. The observed differences can be attributed to the non-optimally tuned Finite element model resulting in a mismatch between the experimentally obtained modal characteristics and the corresponding numerical used for the analysis. Moreover, changing operational and/or ambient conditions might be a possible source of difference.

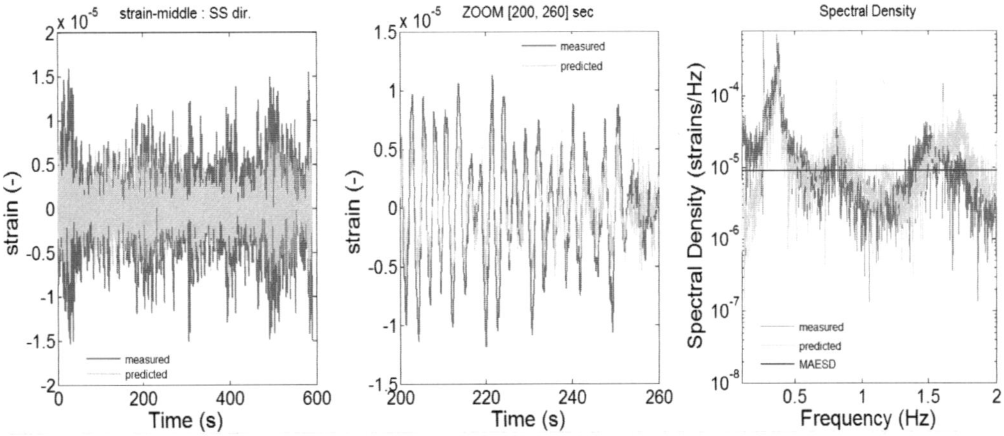

Figure 5. Comparison of the measured dynamic strain (red) at **41m LAT** and the predicted dynamic strain (green) in the SS direction under **rotating** conditions of the OWT.

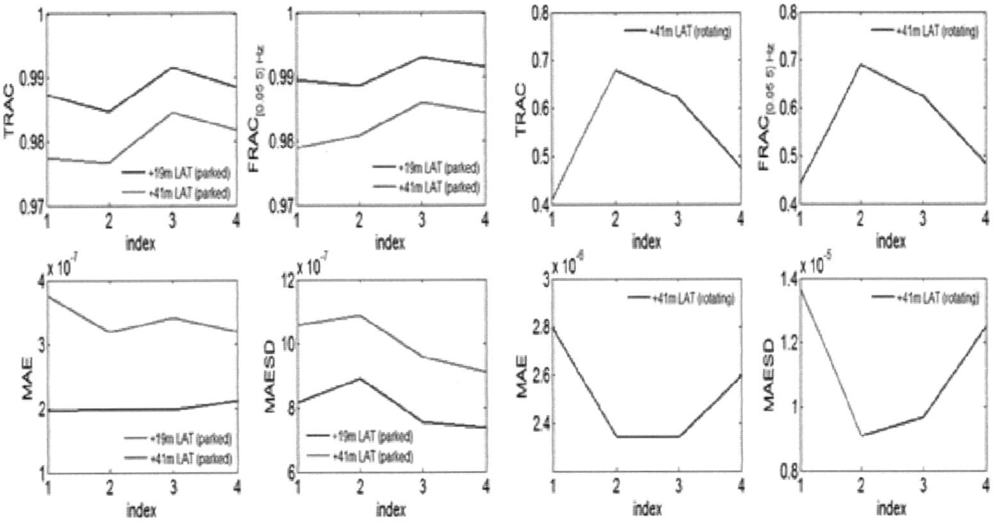

Figure 6. Quality indicators for 4 consecutive 10-min dynamic strain datasets (measured Vs predicted) for parked conditions (2 levels) and rotating conditions (1 level) of the OWT

5. Conclusion

This paper presents the results of a dynamic response prediction method applied on an offshore Vestas V90 3 MW wind turbine on a monopile foundation. The proposed method is based on a modal decomposition and expansion approach that is used to estimate the response at unmeasured locations by combining a limited set of acceleration response measurements and a Finite Element (FE) model. The method has proven to be fast, easy to implement, reliable and effective for the prediction of dynamic strain signals. The proposed algorithm and the virtual dynamic strains were evaluated and validated based on actual measurements obtained from a monitoring campaign on the examined OWT.

Acknowledgements
This research has been performed in the frame-work of the Offshore Wind Infrastructure Project (http://www.owi-lab.be) and the IWT SBO Project, OptiWind and the IWT O&O Parkwind project. The authors therefore acknowledge the financial support from the Agency for innovation by Science and Technology (IWT). The authors also gratefully thank the people of Belwind NV for their continuous support within this project.

References
[1] Kalman R E and Bucy R S 1961 New results in linear filtering and prediction theory *J. Basic Engineering Transactions on ASME, Series D* **83**(3) pp 95–108.
[2] Papadimitriou C, Fritzen C-P, Kraemer P and Ntotsios E 2011 Fatigue predictions in entire body of metallic structures from a limited number of vibration sensors using Kalman filtering *J. Struct. Control Health Monitor.* **18**(5) pp 554–73.
[3] Lourens E, Papadimitriou C, Gillijns S, Reynders E, De Roeck Gand Lombaert G 2012 Joint-input-response estimation for structural systems based on reduced-order models and vibration data from a limited number of sensors *J. Mech. Syst. Signal Process.* **29** pp 310-27
[4] Lourens E, Reynders E, De Roeck G, Degrande G and Lombaert G 2012 An augmented Kalman filter for force identification in structural dynamics, *J. Mech. Syst. Signal Process.* **27** pp 446–60
[5] Avitabile P and Pingle P 2012 Prediction of full field dynamic strain from limited sets of measured data *J. Shock and vibration* **19** pp 765-85
[6] Baqersad J, Poozesh P, Niezrecki C and Avitabile P 2015 Predicting full-field strain on a wind turbine for arbitrary excitation using displacements of optical targets measured with photogrammetry *Proc. IMAC 33 Conf. on Structural Dynamics(Orlando,FL)*
[7] Helfrick M N, Niezrecki C, Avitabile P and Schmidt T 2011 3D digital image correlation methods for full-field vibration measurement *J. Mech. Syst. Signal Process.* **25** pp 917-27. DOI:10.1016/j.ymssp.2010.08.013.
[8] Chipman C and Avitabile P 2012 Expansion of transient operating data *J. Mech. Syst. Signal Process.* **31** pp 1-12. DOI:10.1016/j.ymssp.2012.04.013.
[9] Iliopoulos A, Devriendt C, Iliopoulos S and Van Hemelrijck D 2014 Continuous fatigue assessment of offshore wind turbines using a stress prediction technique *Proc. SPIE 9064, Health Monitoring of Structural and Biological Systems, 90640S (San Diego, CA, USA,March 2014)* doi:10.1117/12.2045576.
[10] Iliopoulos A, Devriendt C, Guillaume P and D. Van Hemelrijck 2014 Continuous fatigue assessment of an offshore wind turbine using a limited number of vibration sensors *Proc. of the 7th European Workshop on Structural Health Monitoring (EWSHM14- Nantes, France, July 2014)* <hal-01020445>
[11] Heylen W, Lammens S and Sas P 1997 *Modal Analysis: Theory and Testing* Katholieke Universiteit Leuven.
[12] Maia N, Silva J, He J, Lieven N, Lin R-M, Skingle G, To W and Urgueira 1997 *Theoretical and Experimental Modal Analysis* Research Studies Press Ltd.
[13] Ewins D 2000 *Modal Testing 2, theory, practice and application* Research Studies Press Ltd second ed.
[14] Iliopoulos A, Devriendt C, Shirzadeh R, Guillaume P and Van Hemelrijck D 2015 A modal decomposition and expansion approach for prediction of dynamic responses on a monopile offshore wind turbine using a limited number of vibration sensors *J. Mech. Syst. Signal Process. (under review)*

Local interrogations for detection of multiple defects using an active sensor network

F Gao, J Lin[1], L Zeng and Z Luo

School of Mechanical Engineering, Xi'an Jiaotong University, Xi'an, China, 710049

E-mail: jinglin@mail.xjtu.edu.cn

Abstract: With the application of a clock-like sensor network, a local interrogation strategy is established to extend the reconstruction algorithm for the probabilistic inspection of damage (RAPID) into multiple defects identification. In this strategy, the monitoring area is divided into several monitoring subareas, and each subarea is covered by a particular sensor network. Local interrogations are then employed in each subarea using the associated network of sensing paths. Benefit from that, the artifacts arise from the intersection of key sensing paths associated with different defects may be absent completely from some monitoring subarea. With the application of a proper image fusion strategy, the artifacts would be reduced and the damage indications could be preserved. The algorithm is applied to an aluminum plate with two artificial through-thickness holes introduced. The results demonstrated that the RAPID algorithm with the application of local interrogations is capable of identifying multiple defects in plate-like structures.

1. Introduction

Interrogation algorithms using ultrasonic guided waves have attracted considerable attentions in the nondestructive evaluation (NDE) and structural health monitoring (SHM), because of the advantages including fast scanning capabilities, low cost, long-range inspection, and testing inaccessible or complex components [1-2]. Especially, small and conformal sensing devices, e.g. piezoelectric ceramics, have been widely studied for generating and receiving guided waves for structural integrity monitoring [3].

Theoretically, if damage arises, some changes, more or less, always occur in the captured signals. The key process in signal-based identification is to correctly tease out these changes and then establish the linkages between them and damage [4]. Because of the dispersive and the multimode character,

[1] To whom any correspondence should be addressed.

it's really hard to find out the changes caused by the defects from the Lamb wave signal. Signal difference coefficient (SDC) is a tomographic feature derived from the normalized cross covariance of the signals between the reference and present states. It captures the overall changes of guided wave signals, which are caused by the occurrence of damage. Making use of SDC as the tomographic feature, the capability of the reconstruction algorithm for the probabilistic inspection of damage (RAPID) would not be affected by the complexity of structural geometry or material property, as these influences are implicitly included in both the reference signals and present signals [3,5-7].

The objective and motivation of this study is to investigate the RAPID algorithm for damage localization of structures with multiple defects. A clock-like sensor array is used and a local interrogation strategy is established. Then, its performance in multiple defects identification is investigated. Specifically, the entire monitoring area is divided into several monitoring subareas. Since different subareas correspond to distinct sensor networks, a particular artifact may appear at different locations on individual images, or be absent completely from some images. In contrast, the dominant response from an actual defect occurs at the same location. Benefit from that, the artifacts would be reduced and the damage indications could be preserved, with the application of a proper image fusion strategy.

2. Review of RAPID Algorithm

The physical intuition behind RAPID algorithm is that a defect would cause the most significant signal changes in the direct wave path, and if the defect is away from the sensing path, the degree of signal change would decrease [3].

In practice, the severity of the signal changes could be evaluated by the tomographic feature, signal difference coefficient (SDC), which is defined as a statistical comparison between the signal in the present state and that in the reference state [8].

$$SDC = 1 - \frac{\sum_{k=1}^{K}(X_k - \mu_X)(Y_k - \mu_Y)}{\sqrt{\sum_{k=1}^{K}(X_k - \mu_X)^2}\sqrt{\sum_{k=1}^{K}(Y_k - \mu_Y)^2}} \tag{1}$$

where X is the reference and Y is the new set of data recorded with respect to current state, μ is the mean of the respective data set and K is the length of the data set. If the environmental and measurement conditions are fixed, these changes are all attributed to the presence of damage.

In the tomographic reconstruction process, besides the severity of signal changes caused by the defect, its spatial distribution also influences the estimation of probability of a defect occurrence at a certain point. For a particular transmitter-receiver pair, S_{ij} (transmitter i and receiver j), the spatial distribution function is non-negative and linearly decreasing [3].

$$w_{ij}(x,y) = \begin{cases} 1 + \frac{1}{\beta - 1}\left[1 - R_{ij}(x,y)\right] & when\ R_{ij}(x,y) < \beta \\ 0 & when\ R_{ij}(x,y) \geq \beta \end{cases} \tag{2}$$

where

$$R_{ij}(x,y)=\frac{\sqrt{(x_i-x)^2+(y_i-y)^2}+\sqrt{(x_j-x)^2+(y_j-y)^2}}{\sqrt{(x_i-x_j)^2+(y_i-y_j)^2}} \tag{3}$$

Here, (x, y) is the coordinate of any individual point within the reconstruction region, (x_i, y_i) and (x_j, y_j) are the coordinates of transmitter i and receiver j, respectively.

Assuming that there are N array elements in the active sensor network used for damage identification, the estimation of probability of the presence of damage at position (x, y) within the monitoring area can be expressed as [3],

$$P(x,y)=\sum_{i=1}^{N-1}\sum_{j=i+1}^{N}P_{ij}(x,y)=\sum_{i=1}^{N-1}\sum_{j=i+1}^{N}w_{ij}(x,y)A_{ij} \tag{4}$$

Here, $P_{ij}(x, y)$ is the defect distribution probability estimation from the transmitter i and receiver j sensor pair, S_{ij}. A_{ij} is the SDC of the sensor pair S_{ij}.

3. Local Interrogation Strategy Using Clock-like Sensor Network

In this section, an appropriate strategy is established, which enables a clock-like sensor array (as shown in figure 1a) to be employed for local interrogations. The scheme consists of three main steps: region partition, data fusion and image stitching, all of which are elaborated as follows.

3.1. Region partition

The monitoring area enclosed by the clock-like sensor network is divided into eight sectors by the straight lines which coincide with sensing paths connecting PZT A0 to other array elements, i.e. Sectors ①-⑧ in figure 1b. In this sensor network, the longest sensing paths pass through three contiguous sectors (e.g. Path 11 and Path 17 in figure 1b). Without loss of information from these sensing paths, the subarea for each local interrogation must contain no less than three adjacent sectors. In this experiment, the subarea to be interrogated consists of four contiguous sectors in the clockwise direction. For instance, the subarea enclosed by the red bold line consists of sectors ①②③④ is the first monitoring region, and the subarea enclosed by the green dashed line, which consists of sectors ②③④⑤, is the second one (figure 1b).

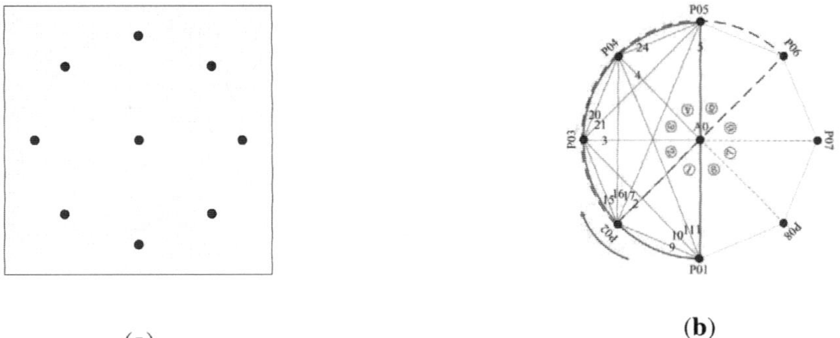

(a) (b)

Figure 1. (a) A clock-like sensor array, and (b) Sectors partition and subareas for local interrogations.

There are eight monitoring subareas in total, and each sector is covered by four of them. Figure 2 (a)-(d) give the sensor networks of sensing paths associated with Subarea 1; Subarea 2; Subarea 3 and Subarea 4, all of which comprise Sector ④. It can be seen that there are eight transmitter-receiver pairs the sensing paths of which pass through Sector ④, including Path 17, Path 22, Path 26, Path 21, Path 25, Path 24, Path 4 and Path 5. Among these, each of the longest sensing paths (i.e. Path 17, Path 22 and Path 26) only exists in two subareas; and the shortest ones (i.e. Path 24, Path 4 and Path 5) exist in all the four subareas; while the others (i.e. Path 21 and Path 25) exist in three of them.

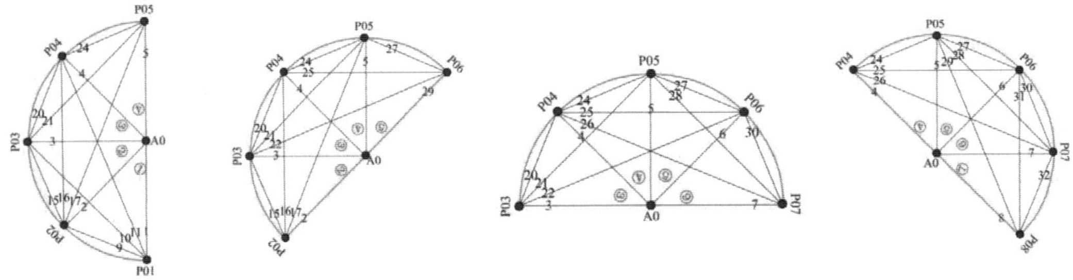

(a) **(b)** **(c)** **(d)**

Figure 2. Network of sensing paths for local interrogation of Sector ④: (a) Subarea 1; (b) Subarea 2; (c) Subarea 3; (d) Subarea 4.

Once the entire monitoring area is separated into several different monitoring subareas (e.g. Subarea 1, Subarea 2, Subarea 3 and Subarea 4), RAPID algorithm is employed for local interrogations. In each local interrogation, only one monitoring subarea is inspected. Thus in the corresponding reconstructive image, only the probability distribution in this subarea is valid.

3.2. Data fusion

In this step, data fusion is employed to combine the four input images (i.e. the imaging results of the four monitoring subareas which comprise the same sector) obtained from different sensor networks into a resulting image, so that a decision or consensus could be obtained in the sector which is covered by these sensor networks. To reduce artifacts and preserve damage indications, an image fusion scheme based on the geometric mean is adopted [9].

$$mean_{geo} = \sqrt[n]{a_1 \cdot a_2 \cdots a_n} \quad (5)$$

The equation means that the probability value of any grid of the fused image is the geometric mean of that of the same grid in the four corresponding individual images. For instance, the imaging results of Subarea 1, Subarea 2, Subarea 3 and Subarea 4 are fused to estimated the probability value of any grid in Sector ④.

3.3. Image stitching

In each individual image obtained in Step 2, the possibilities of damage occurrence are quantified at all mesh nodes of the entire monitoring area, sharing the same coordinates. However, only a sector is interrogated each time, and thus the probability distribution outside is invalid in the corresponding image. To present the prediction results in terms of the probability of damage occurrence across the entire monitoring area, the image stitching technique is employed. In practice, several strategies of image stitching are possible [10]. The algorithm implemented here is to take the maximum pixel value, *i.e.* the pixel value of the fused image is the maximum value of all the corresponding pixels of the individual images.

4. Test Specimens and Procedure

The specimens used in this study are aluminum plates with dimensions of 500×500×2 mm, and the artificial defects are introduced in the form of through-thickness holes with a diameter of 8.5 mm (figure 3). The active clock-like sensor networks are configured by 9 surface-mounted piezoelectric ceramic discs with a diameter of 8 mm and 0.5 mm in thickness. The coordinate systems are employed with the plane of the monitoring area spanned by the horizontal, x, and vertical, y, axes, where the origin of coordinate is set to be the centre of PZT A0. Two application examples are introduced, where configuration of sensors and distributing of artificial defects are different. In the first case, the PZTs (except PZT A0) are centered on a circular template 252 mm in diameter at equi-angular position, and the coordinates of the centre of actual defects are (-16, 89.2), (59, -44), in the form of (x, y). In the second case, the monitoring area enclosed by the circular array has a diameter of 300 mm, and the coordinates of the centre of actual defects are (-74, 33), (67, 83) respectively.

The PZTs are excited at 150 kHz using a 5-count sinusoid toneburst signal. Lamb wave signals captured in the pristine condition are employed as the reference signals. Then, the two defects are introduced. The present signals are captured after the introduction of the two defects.

(a) (b)

Figure 3. Configuration of sensors and distributing of artificial defects: (a) the first example, and (b) the second example.

5. Results and Discussions

The whole monitoring area (300mm×300mm) was meshed into 300×300 uniformly distributed grids to define the probability of the presence of damage. The data length, L, and the scaling parameter, β, can be determined according to [11]. In this paper, β takes 1.1. Each sensor network in a subarea provides 14 sensing paths (figure 2). Local interrogations are performed in all subareas using their associated sensor networks. For illustration, the corresponding constructed images of the first application example are displayed in figure 4. In each image, the grids indicated by a small cross are the locations of centre of the two actual defects. It can be seen that the highest probability density value always occurs at the neighborhood of the locations of actual defects.

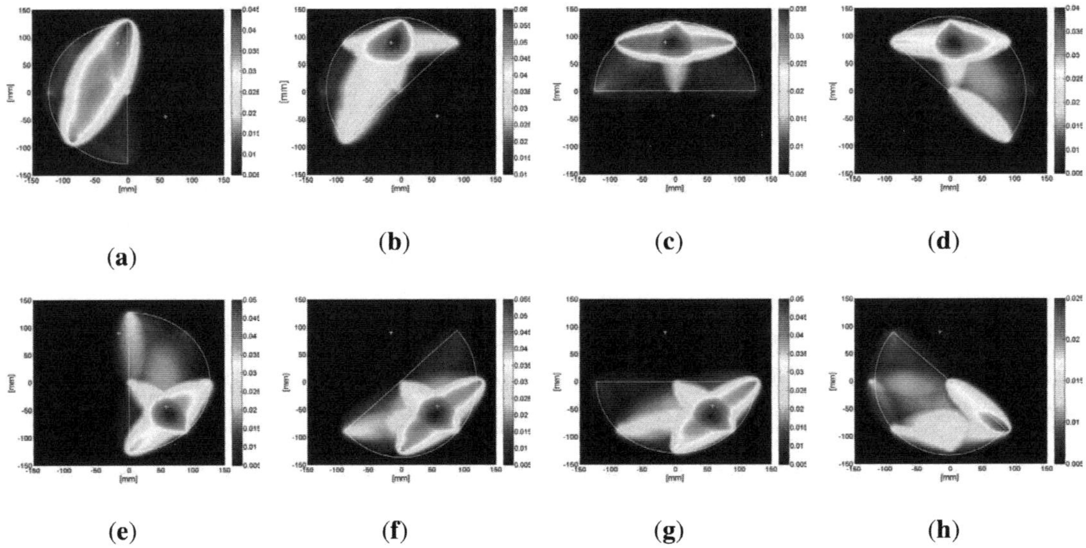

Figure 4. Probability of the presence of damage estimated using local interrogations: (a) Subarea 1; (b) Subarea 2; (c) Subarea 3; (d) Subarea 4; (e) Subarea 5; (f) Subarea 6; (g) Subarea 7; (h) Subarea 8.

The fused images corresponding to different sectors could be calculated from figure 4 by Equation (5), as shown in figure 5. It is noted that if a defect locates in or close to a sensing path connecting PZT A0 and another PZT element, the probability value would spill over into another sector in the associated fused images (e.g. figure 5d and figure 5e). The reason lies that the affected zone of this kind of sensing paths locates symmetrically in the two adjacent sectors

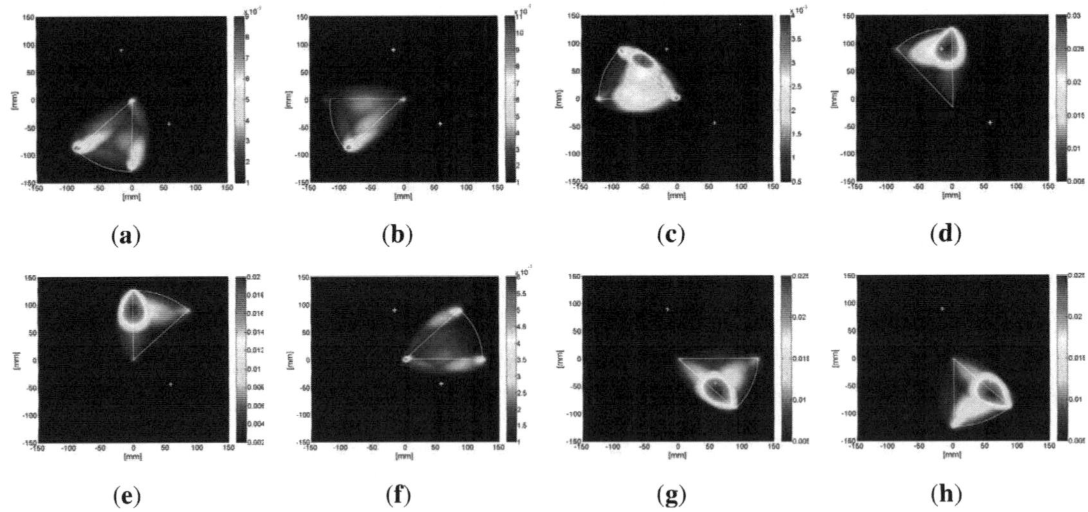

Figure 5. Fused images corresponding to different sectors: (a) Sector ①; (b) Sector ②; (c) Sector ③; (d) Sector ④; (e) Sector ⑤; (f) Sector ⑥; (g) Sector ⑦; (h) Sector ⑧.

Figure 6a shows the reconstruction image of the entire monitoring area after the eight images in figure 5 are fused. It can be seen that the confusing information due to the intersection of key sensing paths are alleviated in the image. The reason could be explained as follows. Different subareas correspond to distinct sensor networks (figure 3), and thus the intersection of key sensing paths corresponding to different defects may be absent from one or more associated subareas of a particular sector. As a result, its effects would be mitigated or eliminated in the fusion process. However, if there is an actual defect occurred at the crossover point of some key sensing paths, even the intersection of key sensing paths is eliminated, other sensing paths surrounding it would also be possible to detect it and result in a high probability value for the presence of damage. In addition, the reconstruction result of the second application example is shown in figure 6c. It can be seen that the two defects could be identified simultaneously, and further, the grids with high probability value only locate in the areas around the two defects.

For comparison, the reconstruction results using global interrogation are also shown in figure 6. From figure 6b, it can be seen that the intersection of key sensing paths (i.e. Path 10 and Path 22) brings a false alarm at the neighborhood of the crossover point. The reasons could be elaborated as follows. When multiple defects occur, if key sensing paths corresponding to different defects (e.g. Path 10 and Path 22) intersect with each other, a large spatial distribution weight would arise at the neighborhood of the crossover point. Since the probabilistic distribution maps is reconstructed as a linear summation of the product of SDC value and the spatial distribution weight of every possible sensing path, a high probability value will appear for the presence of damage at these interfered areas, and further, cause confusing information for different instances of damage.

In the constructed images of the second application using global interrogation (*i.e.* figure 6d), there is a particular sensing path which locates close to more than one defect and thus brings an extremely high SDC feature value. In this case, the damage area consisting of the grids with high probability values for the presence of damage would locate in its affected zone. As a result, the identification of the defects becomes confusing.

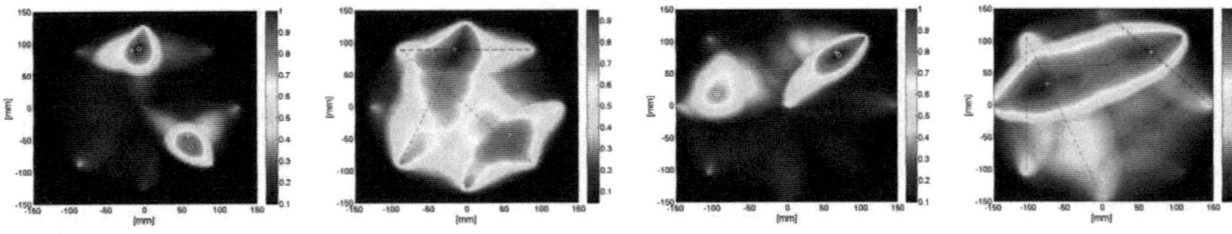

(a) **(b)** **(c)** **(d)**

Figure 6. Reconstruction images of two application defects using local and global interrogation: (a) the first application using local interrogation, (b) the first application using global interrogation; (c) the second application using local interrogation, (d) the second application using global interrogation.

6. Conclusion

In this paper, a clock-like array network and a local interrogation strategy are investigated. In addition, the local interrogation reconstruction results and global interrogation reconstruction results have been compared. Some conclusions are obtained as follows.

(i) The RAPID algorithm using global interrogations is of a potential risk that the intersection of key sensing paths corresponding to different defects may lead to an incorrect identification of multiple defects.

(ii) With the application of local interrogations via an clock-like sensor network, a particular artifact results from the intersection of key sensing paths may be absent from some images, and its effects would be alleviated or eliminated in the image fusion process.

Acknowledgments

The work is supported by the National Natural Science Foundation of China (Grant No. 51421004, 51125022), and the Ph.D. Programs Foundation of Ministry of Education of China (Grant No. 20110201110025), which are highly appreciated by the authors.

References

[1] Valdes S H D and Soutis C 2002 Real-time nondestructive evaluation of fiber composite laminates using low-frequency Lamb waves *J. Acoust. Soc. Am.* **111** 2026-2033.

[2] Alleyne D N and Cawley P 1992 The Interaction of Lamb Waves with Defects *IEEE Trans. Ultrason. Ferroelectr. Freq. Control* **39** 381-397

[3] Zhao X, Gao H, Zhang G, Ayhan B, Yan F, Kwan C and Rose J L 2007 Active health monitoring of an aircraft wing with embedded piezoelectric sensor/actuator network: I. Defect detection, localization and growth monitoring *Smart Mater. Struct.* **16** 1208-1217

[4] Su Z and Ye L 2009 Identification of Damage Using Lamb Waves: From Fundamentals to Applications *Springer Press, Berlin* Ch4 & Ch5

[5] Wang D, Ye L and Lu Y 2009 A probabilistic diagnostic algorithm for identification of multiple notches using digital damage fingerprints (DDFs) *J. Intel. Mat. Syst. Str.* **20** 1439-1450

[6] Gao H, Shi Y and Rose J L 2005 Guided wave tomography on an aircraft wing with leave in place sensors *Rev. Prog. QNDE* **24** 1788-1794.

[7] Wang D, Ye L, Su Z, Lu Y, Li F and Meng G 2010 Probabilistic damage identification based on correlation analysis using guided wave signals in aluminum plates *Struct. Health Monit.* **9** 133-144

[8] Hay T R, Royer R L, Gao H, Zhao X and Rose J L 2006 A comparison of embedded sensor Lamb wave ultrasonic tomography approaches for material loss detection *Smart Mater. Struct.* **15** 946-951

[9] Su Z, Cheng L, Wang X, Yu L and Zhou C 2009 Predicting delamination of composite laminates using an imaging approach *Smart Mater. Struct.* **18** 074002 (8pp)

[10] Su Z, Wang X, Cheng L, Yu L and Chen Z 2009 On selection of data fusion schemes for structural damage evaluation *Struct. Health Monit.* **8** 223-241

[11] Zeng L, Lin J, Hua J and Shi W 2013 Interference resisting design for guided wave tomography *Smart Mater. Struct.* **22** 055017

Damage detection monitoring applications in self-healing concrete structures using embedded piezoelectric transducers and recovery

G Karaiskos[1,4], E Tsangouri[1,2], D G Aggelis[1], A Deraemaeker[3] and D Van Hemelrijck[1]

[1] Vrije Universiteit Brussels (VUB), Department Mechanics of Materials and Constructions (MeMC), Pleinlaan 2, B-1050, Brussels, Belgium
[2] SIM vzw-program SHE, Technologiepark 935, B-9052, Zwijnaarde, Belgium
[3] Université Libre de Bruxelles (ULB), Building, Architecture & Town Planning (BATir), 50 av F.D. Roosevelt, CP 194/02, B-1050, Brussels, Belgium

E-mail: gkaraisk@vub.ac.be

Abstract. The ageing, operational and ambient loadings have a great impact in the operational and maintenance cost of concrete structures. Their service life prolongation is of utmost importance and this can be efficiently achieved by using reliable and low-cost monitoring and self-healing techniques. In the present study, the ultrasonic pulse velocity (UPV) method using embedded small-size and low-cost piezoelectric PZT (lead zirconate titanate) ceramic transducers in concrete with self-healing properties is implemented for monitoring not only the setting and hardening phases of concrete since casting time, but also for the detection of damage initiation, propagation and recovery of integrity after healing. A couple of small-scale notched unreinforced concrete beams are subjected to mode-I fracture through three-point bending tests. After a 24-hour healing agent curing period, the beams are reloaded using the same loading scenario. The results demonstrate the excellent performance of the proposed monitoring technique during the hydration, damage generation and recovery periods.

1. Introduction

Concrete is still the dominating structural material mainly due to its low production cost and the great structural design flexibility. Except the ageing, the concrete structures are vulnerable to a series of loadings such as ambient and operational. The monitoring, maintenance and renewal costs of the continuous ageing current concrete infrastructure are a major issue concerning a non-compromised prolongation of their operational service life. Reliable tools are required for error-free concrete strength and integrity performance evaluation, so that the most cost effective strategy of rehabilitation can be adopted. The main goal is to reduce the inspection costs and risks of unexpected failure not only in hardened state, but also during construction (setting and hardening phases) using real-time online monitoring systems [1].

[4] Author to whom any correspondence should be addressed.

Depending on the damage level caused, the monitoring techniques used for testing the structural integrity of concrete can be either destructive or non-destructive. Although the destructive methods provide generally direct, accurate and reliable structural assessment, these methods are harmful for the concrete structures. On the other hand, the non-destructive testing (NDT) methods indirectly evaluate the integrity of the structures. They mainly measure some other properties which can be correlated to compressive strength using an established relationship between the measured property and concrete strength. Moreover, the methods of concrete testing can be also divided into laboratory and in-situ applications. Until today, the mainstream way to inspect and test the integrity of concrete infrastructure with a high precision is a combination of destructive strength testing in a few samples coming from the existing structure as well as a thorough visual inspection of the whole structure. By definition, the aforementioned combination of monitoring techniques are extremely costly and time demanding, especially in the case of large-scale concrete facilities such as dams and tunnels, where a dense grid of measurement points is necessary. In the recent years, there is an increasing interest in NDT methods for both early-age and hardened concrete not only in the laboratory, but also in in-situ applications [2-5]. From cost and time points of view, the manual NDT inspection techniques are inefficient. Since the early 80's, the radical development in electronics and computer science have contributed a lot in the development of automated NDT structural health monitoring (SHM) for in-situ applications. The current trend is to use very large sensor networks which could be easily integrated in the design of the structures [6].

The ultrasonic pulse velocity (UPV) method has a major role among the various NDT methods in the field of SHM. It is one of the most reliable NDT methods for concrete early-age monitoring, strength estimation and damage detection. The majority of the commercially developed UPV systems are based on using bulky external piezoelectric transducers firmly held in contact with the surface of concrete and placed on two opposite faces of the tested structure (figure 1a). The testing system consists of a high voltage pulser, an amplifier and the main control unit. Because of the piezoelectric effect, the transducers can be used either as actuators (i.e. transmitters) or as sensors (i.e. receivers). A short-duration high-voltage pulse (500 to 1000 V) is introduced into the concrete through the transmitter. Then, longitudinal ultrasonic stress wave (i.e. P-wave) is generated and travels through the concrete. Finally, the P-wave is detected by the receiver which is placed at an opposite surface of the concrete at a distance L. A synchronized electronic timer, which is embedded in the main control unit, is responsible for measuring the transit time T of the P-wave traveling from the transmitter to the receiver. The transmitted pulses are subjected to loss and scattering because of the reflection, refraction and mode conversion caused by the aggregates, the possible voids and steel reinforcement possibly found along the travel path.

The longitudinal pulse velocity (in km/s or m/s) is simply calculated by dividing L by T and it could be correlated to the condition of the concrete [7].

$$V = \frac{L}{T} \tag{1}$$

The calculated velocity is directly related to the dynamic modulus of elasticity E, dynamic Poisson's ratio μ and mass density ρ through the following equation [8].

$$V = \left(\frac{E(1-\mu)}{\rho(1+\mu)(1-2\mu)} \right)^{1/2} \tag{2}$$

It is obvious that the quality of the received signal depends on a series of parameters such as the accuracy of the measurement of the distance L as well as the quality of the coupling of the transducers to the concrete surfaces. Additionally, the necessity of placing the transducers in two opposite faces of the concrete structure, the lack of flexibility in the transducers arrangement as well as the lack of real-time online application make prohibitive the practical use of the technique.

One possible alternative of the bulky external transducers is to replace them by low-cost piezoelectric transducers which could be directly embedded inside the concrete (figure 1b). These transducers were initially developed and applied in the University of Houston [9, 10]. Because of the stable coupling between the transducer and the concrete matrix the level of signal to noise ratio is quite higher than the one using the external transducers. They can be equally used for monitoring the hydration process of fresh concrete [11] as well as for SHM applications in hardened concrete [12].

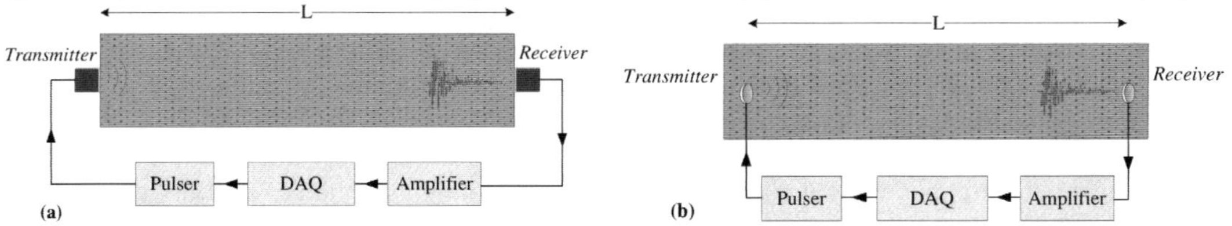

Figure 1. (a) Ordinary UPV test system and (b) UPV system based on small-size and low-cost embedded piezoelectric transducers.

Following the concept of 'smart aggregates' [9] (SMAG), a few transducers used in the present study were designed and fabricated in BATir at ULB. Each transducer consists of a low-cost flat piezoelectric PZT (lead zirconate titanate) small-dimension (12 mm x 12 mm x 0.2 mm) ceramic patch. The manufacturing steps of these transducers are shown in figure 2. Initially the cables are conductively glued with the patch (figure 2a-b). In order to avoid capacitive coupling interference between the transducers embedded in the fresh concrete (due to the presence of water) and to also mechanically protect it, the patch is properly insulated using a waterproof epoxy coating (figure 2c) and an extra electromagnetic protection is provided by applying a thin layer of conductive paint (figure 2d).

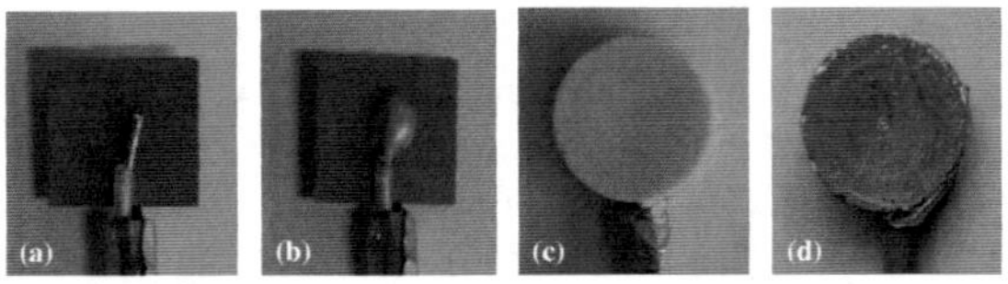

Figure 2. Manufacturing steps of the embedded transducers.

The current study is focused on the implementation of UPV method based on embedded low-cost piezoceramic transducers in a couple of small-scale notched unreinforced concrete beams with self-healing properties. Traditionally, the cracked concrete is manually healed by injecting epoxy and cement based agents into the gap of the cracks [13]. Here, the healing agent, which is filled within fragile small-dimension glass capsules, looks forward to self-healing the cracked concrete. One of the tested beams was monitored during setting and hardening period (hydration process) and both of them, after the hardening of concrete, were subjected to mode-I fracture through three-point bending tests. The beams were loaded until a wide crack opening and rupture of the capsules and release of the healing agent in the gap of the generated cracks were achieved. After a 24-hour curing period of the healing agent, the beams were reloaded using the same loading scenario. The performance of the proposed NDT technique for monitoring the concrete during the hydration period, the fracture process and fracture recovery is evaluated.

2. Experimental setup

A couple of small-scale (840 mm x 100 mm x 100 mm) notched unreinforced concrete beams with embedded self-healing system and low-cost piezoceramic transducers (SMAG–smart aggregate) were designed and manufactured in MeMC at VUB. The setup used in the experiments follows the Rilem Technical Report FMS-50 specifications [14]. The two-component expansive polyurethane-based healing agent used in the present study was encapsulated in fragile small dimensioned glass capsules which were attached on very low stiffness plastic cord tightly fixed on the wooden mold (figure 3) [15]. In beam 1, two SMAGs (SMAG 1a and 1b) were embedded with a distance between them equal to 80 mm and at a height of 40 mm from the bottom of the beam. Additionally, four couples of short glass capsules (50 mm length, 3 mm diameter) filled with the two-component healing agent were symmetrically suspended at heights 35 mm and 60 mm from the bottom of the beam and a between distance of 50 mm on the same level. In beam 2, the monitoring system consisted of two couples of SMAGs which were embedded with a distance between them equal to 100 mm and at height levels equal to 38 mm (SMAGs 2a and 2b) and 63 mm (SMAGs 2c and 2d) from the bottom of the beam. Here, four couples of long glass capsules (75 mm length, 3 mm diameter) filled with the two-component healing agent were symmetrically suspended at heights 25 mm and 50 mm from the bottom of the beam and a distance between them equal to 50 mm on the same level.

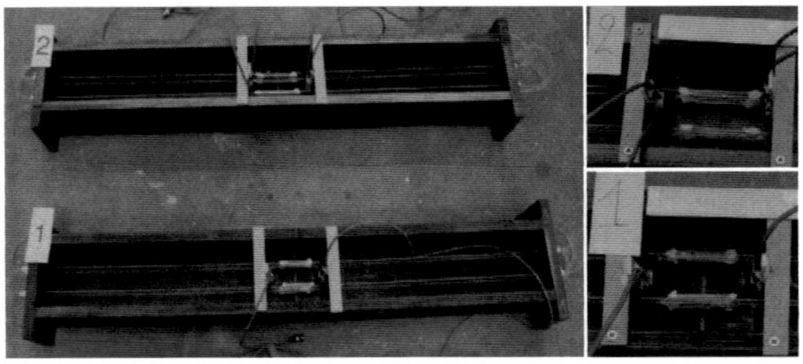

Figure 3. Preparation of the beams with the self-healing system (glass capsules filled with healing agent) and the piezoceramic transducers (SMAGs).

After the concrete casting in the molds and until their demolding (three days later) both beams were subjected to membrane curing in order to prevent moisture loss. During that (early-age) period, the beam 1 was monitored by the UPV method based on the embedded piezoceramic transducers. Monitoring the transit time T needed by the transmitted P-waves to reach the receiver, it is feasible to estimate the mechanical properties of the concrete which can be related to parameters such as the percolation of solid particles and the degree of hydration. During that task, the SMAG-1a was the transmitter and the SMAG-1b was the receiver of the monitoring system. After demolding the beams and until their loading test (eleven days later), both of them were subjected to water curing in order to improve the setting and hardening procedure. Two weeks after the casting of concrete, both beams were subjected to mode-I fracture through three-point bending tests until a wide crack opening, rupture of the capsules and filling of the cracked area with the healing agent were achieved. Then the beams were unloaded and left in rest during the 24-hour curing period. Finally, the beams were subjected to the same loading tests in order to assess the monitoring and healing performance used in the current study. During loading and reloading tests, in beam 1 the SMAG-1a was the transmitter and the SMAG-1b was the receiver, as well as in beam 2 the SMAG-2a was the transmitter and the opposite placed SMAGs-2b, 2d were the receivers of the monitoring system (figure 6).

2.1. Early-age monitoring

In beam 1, after the concrete casting, high-voltage short-duration pulses (800 V / 2.5 μs) were generated by the high voltage pulser and introduced into the concrete specimen through the transmitter (SMAG-1a). After traveling through the concrete, the generated P-waves were picked up by the receiver (SMAG-1b) whose signal was amplified. Initially and for the first 24 hours of the monitored early-age period, five tests every five minutes were performed and afterwards and until demolding the beam 1, five tests every twenty minutes were performed. By calculating the transit time T the wave takes to travel from the transmitter to the receiver, the wave velocity can be computed. In figure 4a, the evolution of the wave arrival time as a function of time during the monitored early-age period of beam 1 is shown. In the very early-age period (0-8 h) the signal to noise ratio of the received signals is very low and it was impossible to reliably compute the respective wave arrival time values. Using equation (2) and assuming a dynamic Poisson's ratio equal to 0.3, the dynamic modulus of elasticity of the concrete during the same early-age is computed and its progression is shown in figure 4b.

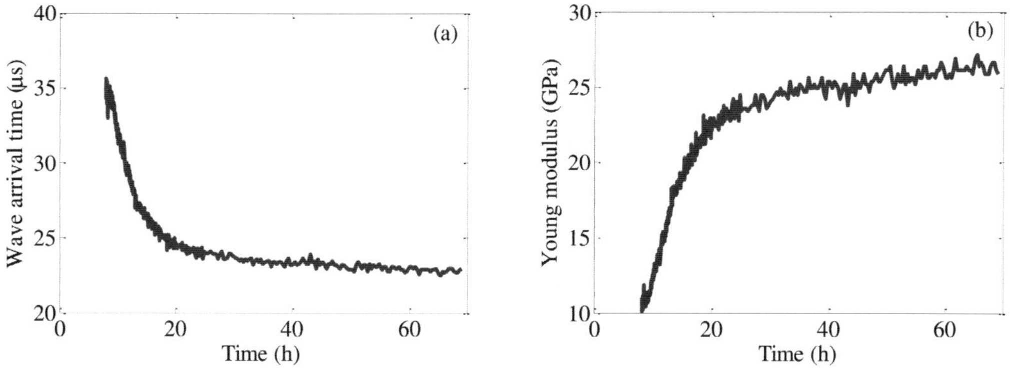

Figure 4. Evolution of the (a) wave arrival time and (b) dynamic modulus of elasticity as a function of time for beam 1.

By taking into account the time delay (2 μs) of the transducers, computed through a calibration test before embedding them in the concrete and the distance between them (80 mm), the evolution of the P-wave velocity as a function of time for the monitored early-age period (figure 5) is easily calculated. A few recorded signals at different times are also shown in figure 5. It is obvious that the signal to noise ratio and the maximum amplitude of the recorded signals gradually increase and the transit time T is decreased during the transition from fluid to solid. The maximum amplitude of the recorded signals at very early-age period is however about one hundred times smaller than the one at an age of 60 h onwards.

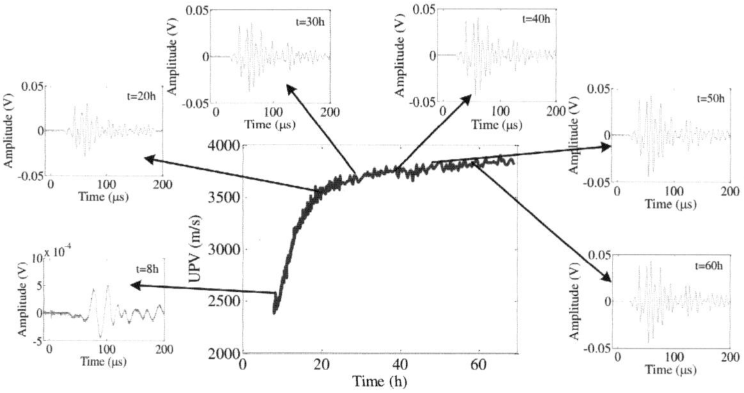

Figure 5. Evolution of the P-wave velocity as a function of time for beam 1.

2.2. *Damage and healing monitoring*

After a three-day membrane curing, both beams were demolded and subjected to water curing (submerged in a water bath) in order to improve the setting and hardening procedure until their loading test (eleven days later). Then, they were subjected to mode-I fracture through three-point bending tests under slow speed (0.04 mm/min) of the crosshead which was set in displacement control mode (figure 6). Following the loading rate, ten tests every fifteen seconds were performed and the mean value of each received group of results is used in the post processing of the results that follows. The loading is applied until a wide crack opening (0.3 mm), the rupture of the capsules and the filling of the cracked area with the healing agent were achieved. After unloading the beams, a 24-hour curing period of the healing agent followed. Then, the beams were reloaded, following the same loading pattern, in order to assess the monitoring and healing performance used in the current study.

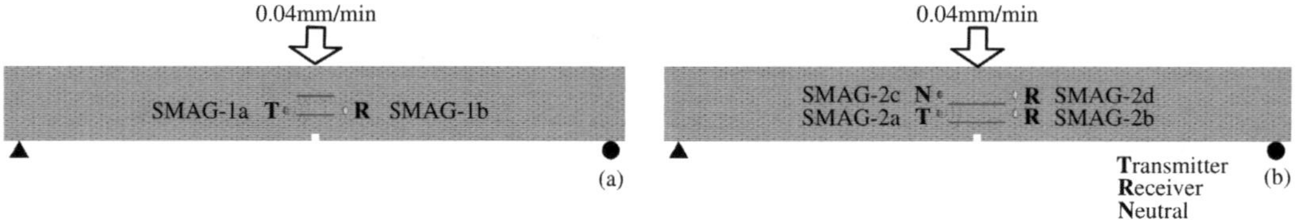

Figure 6. Experimental setup of the (a) beam 1 and (b) beam 2.

Due to the presence of aggregates and voids in the wave path between transmitter and receiver, the P-wave reaching the receiver was transformed into a complex waveform. The damage index used in the present study is based on the early part of that waveform and it considered both the shift of the arrival time as well as the amplitude of the received signals [12]. In figure 7a, the applied load (kN) vs strain (mm/mm) curves corresponding to the loading and reloading tests of beam 1 are shown. Load is the force applied by the crosshead of the testing machine used and strain is the normalized data (mm/mm) received by the crack mouth opening displacement (CMOD) gauge which was placed on the notch limits of the beam. Before testing, the initial opening of the gauge was fixed at 10 mm. Additionally, the gradual evolution of the early part of the recorded signals with increasing load corresponding to the loading test of beam 1 is shown in figure 7b.

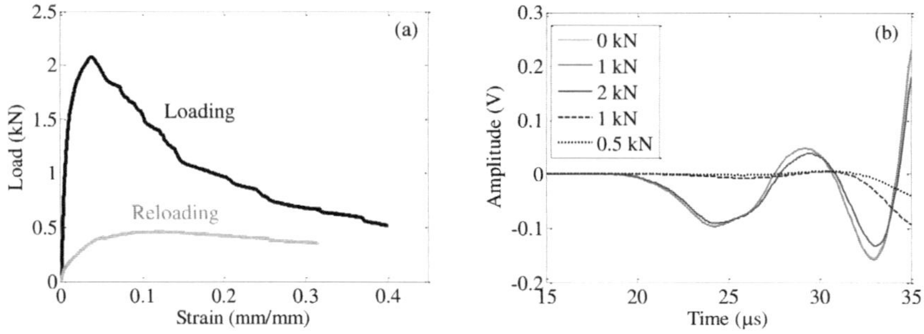

Figure 7. (a) Load vs strain (loading and reloading tests) and (b) evolution of the early part of the measured signals for beam 1 (loading test).

The change of the signals is well captured by the damage index for both beams in both tests as shown in figure 8, where its evolution is plotted as a function of the applied load.

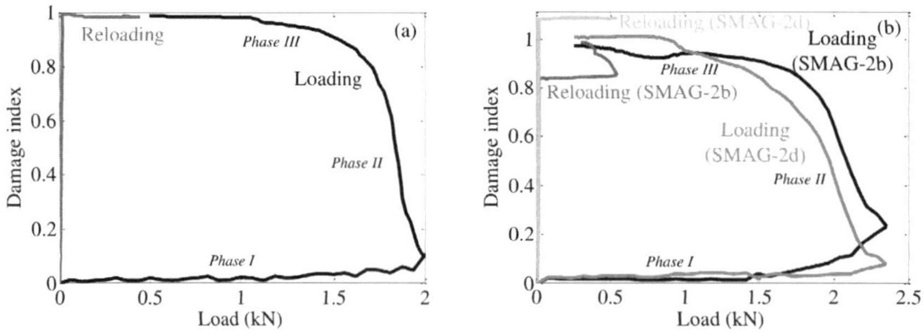

Figure 8. Damage index vs load for loading and reloading tests in (a) beam 1 and (b) beam 2.

In the loading tests, the damage index evolution graphs consist of three phases. The initiation (phase I) and progression (phase II) of the damage as well as the total failure (phase III) of the beams due to loading are clearly detected. In the first phase of loading tests in both beams, the damage index is very low but non-zero and this could be attributed to a certain level of noise in the received signals. The second phase corresponds to the initiation and evolution of damage (crack) in the concrete until the severe failure (third phase) where the damage index is almost unitary due to the fact that the wave cannot reach the receiver anymore. Additionally, in beam 2 and in the area between 1.5 kN and 2 kN, the damage index calculated by the results received by the lower embedded transducer (SMAG-2b) grows faster than the one by the higher placed transducer (SMAG-2d). This makes great sense as the crack initiates at the bottom of the beam 2, where the SMAG-2b is closely embedded. This is a solid indicator that the present UPV method cannot only efficiently detect the initiation of the damage, but also can track its evolution which makes it suitable for real-time online in-situ monitoring applications without giving a quantitative evaluation of the damage (size of crack). The damage index evolution graphs of the reloading tests, for both beams, consist of two parts. Initially, it is shown how the damage index evolves due to the curing process which is followed by the complete failure of the concrete. Unfortunately, in all the cases there is not any significant healing (recovery of mechanical properties) of the concrete due to the presence of the healing agent.

The recorded signals of beam 2 (SMAG-2b) in the end of the loading test and just before the reloading one are shown in figure 9. The presence of the dry healing agent has been well detected as in the signals before the reloading test, the arrival time has been decreased and the amplitude of the received signals has been increased.

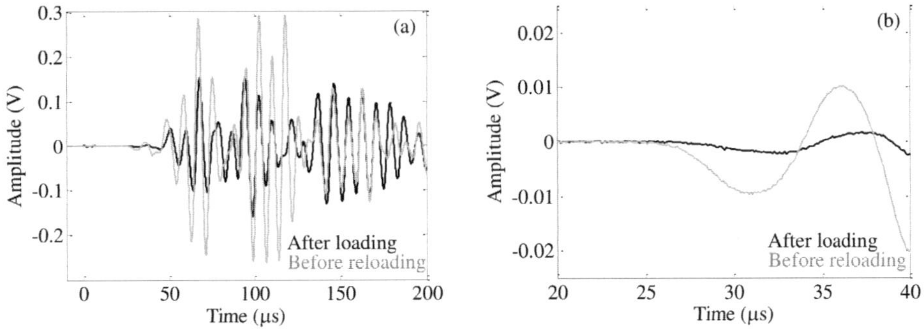

Figure 9. (a) Received signals by SMAG-2b (beam 2) after beam loading and before beam reloading and (b) the respective zooms in the early part of the graphs.

3. Conclusions and perspectives

Embedded piezoelectric transducers have been used for the monitoring of concrete with self-healing properties. Two small-scale notched unreinforced concrete beams with embedded self-healing system and low-cost piezoceramic transducers have been designed and manufactured. In one of the beams, the *P*-wave velocity has been monitored during the first three days after the concrete casting. After a two-week curing period, two three-point bending tests have been carried out on. The present monitoring technique is able not only to monitor the concrete structures during the setting and hardening period, but also to detect the initiation and follow the evolution of the cracking until complete failure of the structure. Even though the resulted performance of the self-healing technique used in the present study was not the desired one, the UPV method using embedded piezoelectric transducers was able to detect the presence of the healing agent. Next steps on the topic should be the quantitative assessment of damage using the received signals as well as the improvement of quality of the healing agent. The optimal placement of the capsules filled with the healing agent is a further topic that should be studied.

Acknowledgements

The research described in this paper has been performed in the frame of the SIM program on Engineered Self-Healing Materials (SHE). The authors are grateful to SIM (Strategic Initiative Materials Flanders) and FNRS (Fonds de la Research Scientifique) for providing financial support.

References

[1] Bungey J H, Millard S G and Grantham M G 2006 *Testing of Concrete in Structures* (London and New York: Taylor & Francis)

[2] ACI 228.2R-98 1998 Nondestructive test methods for evaluation of concrete in structures *American Concrete Institute*

[3] ACI 228.1R-03 2003 In-place methods to estimate concrete strength *American Concrete Institute*

[4] Popovics S 1998 *Strength and Related Properties of Concrete: A Quantitative Approach* (New York: John Wiley & Sons)

[5] Malhorta V M and Carino N J 2004 *Handbook on Nondestructive Testing of Concrete* (Boca Raton: CRC Press)

[6] Farrar C R and Worden K 2012 *Structural Health Monitoring: A Machine Learning Perspective* (New York: John Wiley & Sons)

[7] ASTM C 597 – 67 1967 Standard test method for pulse velocity through concrete *American Society for Testing and Materials*

[8] Krautkrämer J and Krautkrämer H 1969 *Ultrasonic Testing of Materials* (G. Allen & Unwin)

[9] Gu H, Song G, Dhonde H, Mo Y L and Yan S 2006 Concrete early-age strength monitoring using embedded piezoelectric transducers *Smart Mater. Struct.* **15** 1837-45

[10] Song G, Gu H, Mo Y L, Hsu T T C and Dhonde H 2007 Concrete structural health monitoring using embedded piezoceramic transducers *Smart Mater. Struct.* **16** 959-68

[11] Dumoulin C, Karaiskos G, Carette J, Staquet S and Deraemaeker A 2012 Monitoring of the ultrasonic P-wave velocity in early-age concrete with embedded piezoelectric transducers, *Smart Mater. Struct.* **21** 047001

[12] Karaiskos G, Flawinne S, Sener J Y and Deraemaeker A 2013 Design and validation of embedded piezoelectric transducers for damage detection applications in concrete structures *Proc. 10th Int. Conf. on Damage Assessment of Structures* (Trinity College Dublin, Ireland, 8-10 July 2013)

[13] Smoak G 2002 *Guide to Concrete Repair* (New York: Books for Business)

[14] RILEM TC-50-FCM 1985 Determination of fracture energy of mortar and concrete by means of three-point bend tests on notched beams *Mater. Struct.* **18** 285-90

[15] Van Tittelboom K, De Belie N, Van Loo D, and Jacobs P 2011 Self-healing efficiency of cementitious materials containing tubular capsules filled with healing agent *Cement Concrete Comp.* **33** 497-505

Frequency Domain Decomposition performed on the strain data obtained from the aluminium model of an offshore support structure

M Mieloszyk[1], **S Opoka**[1] **and W Ostachowicz**[1,2]

[1] Mechanics of Intelligent Structures Department, Institute of Fluid-Flow Machinery, Polish Academy of Sciences, 14 Fiszera Street, 80-231, Gdansk, POLAND.
[2] Faculty of Automotive and Construction Machinery, Warsaw University of Technology, 84 Narbutta Street, 02-524, Warsaw, POLAND

E-mail: mmieloszyk@imp.gda.pl

Abstract. This paper presents an application of Fibre Bragg Grating (FBG) sensors for Structural Health Monitoring (SHM) of offshore wind energy support structure model. The analysed structure is a tripod equipped with 16 FBG sensors.
From a wide variety of Operational Modal Analysis (OMA) methods Frequency Domain Decomposition (FDD) technique is used in this paper under assumption that the input loading is similar to a white noise excitation. The FDD method can be applied using different sets of sensors, i.e. the one which contains all FBG sensors and the other set of sensors localised only on a particular tripod's leg. The cases considered during investigation were as follows: damaged and undamaged scenarios, different support conditions. The damage was simulated as an dismantled flange on an upper brace in one of the tripod legs. First the model was fixed to an antishaker table and investigated in the air under impulse excitations. Next the tripod was submerged into water basin in order to check the quality of the measurement set-up in different environmental condition. In this case the model was excited by regular waves.

1. Introduction

This paper presents an application of Fibre Bragg Grating (FBG) sensors for Structural Health Monitoring (SHM) of offshore support structure. FBG sensors have small size and weight, multiplexing capabilities, immunity to electromagnetic fields, high corrosion resistance and require no calibration [5]. These sensors can be used for measuring strain on structures and, in particular, for damage detection and localization. The advantages of FBG sensors make them a very interesting tool in SHM systems used for civil engineering structures (bridges, [3], buildings [4]) and marine structures (marine vessels [12], offshore structures [9]).
In the paper the results of an operational modal analysis using Frequency Domain Decomposition (FDD) method applied to strain measurements on tripod model are presented. The FDD method has been applied for different sets of sensors, i.e. the one which contains all FBG sensors and the other sets of sensors localised only on a particular leg of the tripod. The analysis was performed on thermo-mechanical strain ε_{cs} calculated from single axis FBG sensors measurements. All signals before FDD decomposition were standardised to have zero mean and unit sample standard deviation. The strain change due to the bonding procedure and the influence of change in the support conditions on static

strain readings of selected sensors for the same tripod model was earlier presented in [7]. The FDD method allows to determine differences in natural frequencies for undamaged and damaged tripod structure in the air as well as to distinguish excited wave frequency during measurements performed in the water environment.

2. FDD method applied to scale model of tripod

Modal analysis methods can be divided into two categories: experimental modal analysis (EMA) and Operational Modal Analysis (OMA). In EMA both input (excitation of structure) and output (structural response) signals are used while OMA is based only on structural response [11]. OMA can determine modal properties by the structure provided that the excitation is broadband and stochastic (e.g. white noise). In such case excitations are from environmental loadings like wind or waves. OMA methods can be divided into parametric in time domain and non-parametric in frequency domain [2]. In parametric methods, a certain numerical model of the system is assumed, then its selected uncertain parameters are optimizing (updating) based on the measured structural response signals. Non-parametric methods extract the dynamic parameters of a structure by performing set of mathematical operations using measurement data [10]. Parametric methods are preferred in determination of damping parameters while non-parametric ones are used for determination of frequencies and mode shapes [2].

2.1. FDD method description

FDD method is one of non-parametric OMA methods applied especially for analysing of large civil engineering structures like bridges or buildings [1], [11]. This method allows extraction of modal parameters (mode shapes with their frequencies and damping ratios) under ambient excitation caused by wind [6]. The main advantage of FDD method is possibility of discrimination of close mode shapes and its ability to analyze signals with high noise to signal ratio.

The FDD method uses the power spectral density (PSD) of the signals instead of the Fourier transform of the impulse response function (i.e. frequency response function) [10].

The relationship between unknown input signal vector $x(t)$ and measured response signal vector $y(t)$ can be described as

$$G_{yy}(j\omega) = \bar{H}(j\omega)G_{xx}(j\omega)H(j\omega)^T \tag{1}$$

where matrix $G_{xx}(j\omega)$ with a size $(r \ x \ r)$ is the Power Spectral Density matrix (PSD) of input signals, r – number of input signals, matrix $G_{yy}(j\omega)$ with a size $(m \ x \ m)$ - the Power Spectral Density matrix of the system response, m - number of output signals, $H(j\omega)$ with a size $(m \ x \ r)$ is Frequency Response Function (FRF) of the system. The dash over the letter H denotes complex conjugate while the index T – transposition [1].

According to (1) when the PSD matrix of input signals is constant matrix ($G_{xx}(j\omega)=C$) the system response matrix can be determined based on measured output signals. This assumption is satisfied for broadband stochastic excitation like white noise [1]. In special circumstances other environmental excitations (e.g. wind) can be also interpreted as broadband stochastic excitation.

FDD algorithm can be divided into several steps. First, PSD matrix of output signals $G_{yy}(j\omega)$ for discrete frequencies ($\omega=\omega_i$) is calculated. The power spectral density matrix was calculated using cpsd procedure in Matlab environment with Welch's averaged, modified periodogram method for spectral estimation:

$$G_{xy}(\omega) = \sum_{m=-\infty}^{\infty} R_{xy}(m)e^{-j\omega m}, \qquad R_{xy}(m) = E\{x_{n+m}y*_n\} = E\{x_n y*_{n-m}\}, \tag{2}$$

where R_{xy} means cross-correlation sequence and x_n and y_n are jointly stationary random processes, $-\infty<n<\infty$, and E is the expected value operator. During the procedure the measurement signal is divided into sections with a length of 4096 points. A Hann window is applied to each measurement section before spectral density estimation. In next step singular value decomposition (SVD) method is performed on this matrix [1]:

$$G_{yy}(\omega_i) = U_i S_i U_i^H \tag{3}$$

where matrix $U_i = [u_{i1}, u_{i2}, \dots, u_{im}]$ is an unitary matrix including singular vectors u_{ij} and S_i is a diagonal matrix including scalar singular values s_i, index H denotes conjugate transpose.
The outcome of FDD method is a decomposition of PSD matrix to a set of PSD functions. Each PSD function is related to single degree of freedom (SDOF) of the system [1]. Peaks of the first singular $s_1(\omega)$ are equal to the natural frequency of the system and singular vectors $u_{i1}(\omega)$ corresponding to the peaks of the first singular values approximate mode shape vectors [10].
The FDD method performs best if the structure is excited by white noise, the damping is relatively small and the natural mode shapes are geometrically orthogonal. In other cases the decomposition process into the set SDOF systems is approximate, but still yields better results than classical analysis using Fast Fourier Transformation (FFT) [1].

2.2. Experimental investigation
The analysed offshore support structure is a 2 m height tripod model equipped with 15 single os3120 Micron Optics FBG sensors together with os4100 Micron Optics temperature compensation sensor, see Figure 1(a). The si425-500 Micron Optics interrogator with sampling frequency 250 Hz was used in experiments.
FDD method was applied to measurements performed on tripod structure for two experimental cases. In the first case, the structure was investigated in the air, under impulse excitations (repeated hammer impacts in chosen points on the tripod structure). In the second case the investigation was performed in the water basin (Figure 1(b)) in the Ship Design and Research Centre (CTO S.A.) laboratory and the structure fixed to the support base and partially submerged up to the water level about 150 cm from tripod's bottom flanges. The partially submerged model was excited by artificial waves originating from wave generator installed on one side of the basin.
For every case FDD method was applied for four different set of sensors. The first one denoted as CC contains all FBG sensors, while the others (S1, S2, S3) contains only sensors located on a particular side of the leg, SX denotes set of sensors ubX 1, ubX 2, pgX, tsX, ts Xa for X=1,2,3, compare Figure 1(a).

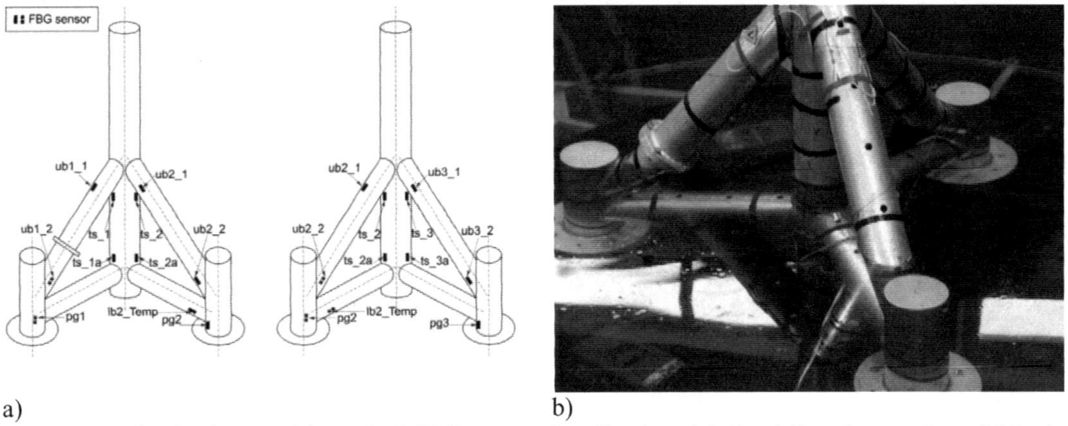

a) b)
Figure 1. Tripod scheme with marked FBG sensors localisations (a). Partially submerged model in the water basin in CTO S.A. laboratory (b).

For the tripod standing on the antishaker table different structural states were considered: the tripod with fixed flange on an upper brace in one of the legs was considered as intact one (A1), while unbolted flange (visible on the left in Figure 1b) on the upper brace (A2) simulated occurrence of damage in the model. The structure was fixed directly to the table (B3) or relatively thick foams were inserted under the tripod legs which were further fixed by bolts to the table (B4) in order to simulate different cases of boundary conditions. Additional masses 54 kg (C1) and 35 kg (C2) were put on main tube of the model. The case C3 denotes tripod with no additional mass on the top of the structure.

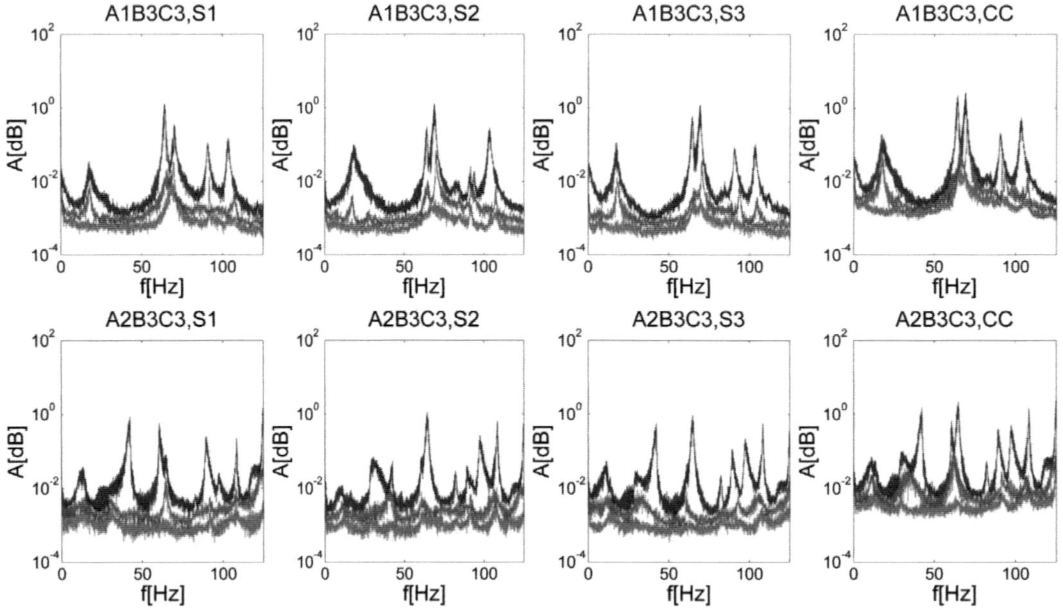

Figure 2. Comparison of first three spectral eigenvalues calculated using FDD decomposition (CC, S1, S2, S3) for tests A1B3C3 and A2B3C3.

In Figure 2 the comparison of spectral eigenvalues received after decomposition using FDD method (CC, S1, S2, S3) for tests denoted as A1B3C3 (intact structure, fixed to the table, without additional mass) and A2B3C3 (damaged structure, fixed to the table, without additional mass) are presented. All curves denoted by blue colour (first eigenvalues of decompositions) are similar in shape. Not all peaks in Figure 2 correspond to vibration modes of the tripod. Numerical simulations [8] performed for intact structure revealed the following eigenfrequencies of tripod structure: 61.75 Hz, 61.99 Hz, 94.28 Hz, 95.36 Hz. The first numerical frequency value is 61.75 Hz, so any frequency peak below such value cannot be treated as indication of vibration mode of the structure. For a structure without any damage (A1) the results for set of sensors S2 differs the most from the others. In such case instead of one frequency peak (91 Hz) two smaller frequency peaks (91 Hz and 94 Hz) are visible. Also small differences in amplitudes between frequency peaks (64 Hz and 70 Hz) can be noticed for SX and CC sets of sensors. The occurrence of damage (A2) results in appearance of additional peaks (82.4 Hz and 97.5 Hz). Moreover for set of sensors S2 and S3 instead of two frequencies (60 Hz and 64 Hz) as in A1 case only one frequency is visible (64 Hz). For a set of sensors S1 calculated for sensors from leg with flange in a range of frequencies above 80 Hz only three peaks are visible, instead of four like for the other sensor sets. Comparing the above results with the numerical natural frequencies [8] the main differences in the number of frequency peaks are visible in A2 case (about 90 Hz) where only three natural frequencies are visible. For all sensors sets (S1, S2, S3, CC) the differences between A1 and A2 case are noticeable in frequency domain especially in the overall number of frequency peaks. This feature can be useful in the future for damage detection and localisation.

The energy distribution of the measurement results presented above were compared with Le & Tamura [2] analysis performed for a building with velocity sensors. In their investigation the energy distribution for the first four eigenvalues and eigenvectors was as follow: 99,9%, 0,07%, 0,01%, 0,00%. That means that the first three eigenvalues contain almost whole information about the excited dynamic behaviour of the structure. For tripod structure the energy in first eigenvalue strongly depends on number and localisation of sensors as well as on structural state (Table 2). The highest energy value (91,5%) is obtained for S3 sensor set for an intact structure, while the lowest (81,0%) for CC sensor set case for a damaged structure. For every case the first three eigenvalues contains more than 90% of whole vibration energy. The first-order spectral eigenvalues contain most of the energy and they possess almost all information about dominating natural frequencies (see Figure 2).

Table 1. Comparison of percentage energy distribution among three first spectral eigenvalues after FDD decomposition.

	sensor set for A1B3C3				sensor set for A2B3C3			
Eigenvalue	CC	S1	S2	S3	CC	S1	S2	S3
1	83,7%	88,1%	89,7%	91,5%	81,0%	86,1%	87,8%	87,7%
2	8,5%	7,0%	6,3%	5,2%	9,4%	9,2%	7,5%	7,3%
3	2,8%	2,7%	2,1%	2,0%	3,4%	3,0%	2,7%	2,9%

The application of FDD method for partially submerged model (up to 1.5 m) are presented in Figure 3. The results are presented for intact structure. The leg with flange is parallel to the direction of travelling waves in the basin. The experimental investigation was performed for regular wave excitation with height (peak to peak amplitude) *H* and frequency *f*. The measured signal length was long enough to register the strain values on the structure for undisturbed water, influence of regular wave on the tripod structure and the waves after the wave exciter was turned off. In this case tripod vibration amplitude was about 4 με, while the interrogator strain resolution is 1-2 με. For tripod in the water the highest energy values were obtained for sensor set S1, while the lowest – for CC.

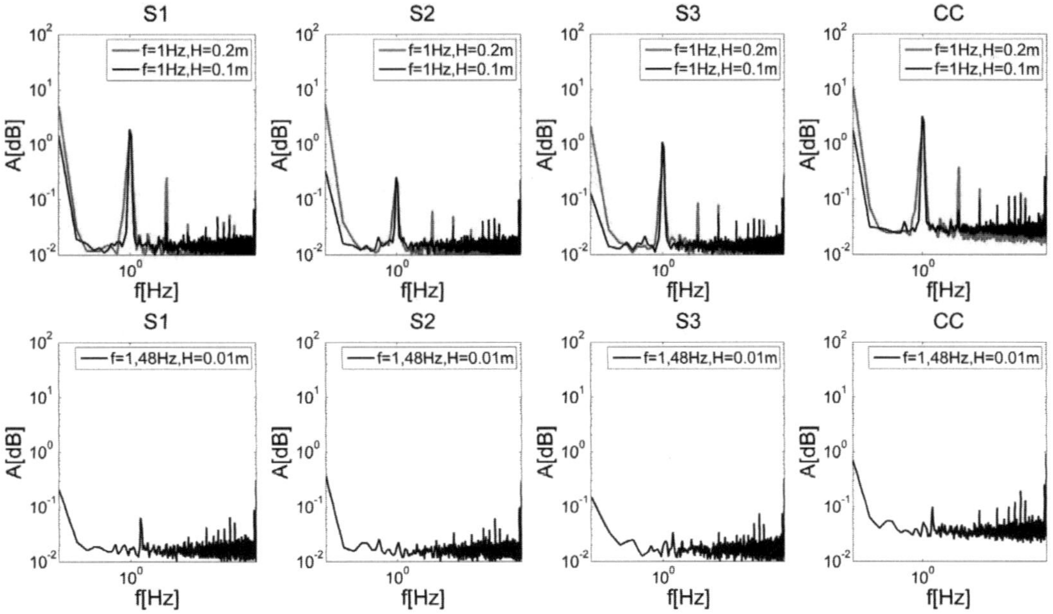

Figure 3. Comparison of first spectral eigenvalues received after FDD decomposition (CC, S1, S2, S3) for different wave excitation.

The preliminary purpose of tripod submersion was experimental verification of underwater strain measurements performed by FBG sensors. The measurement system survived mounting activities performed on the tripod in the basin as well as no deterioration of the quality of submerged sensors was observed. The results for FDD decomposition of strain signals under different wave excitations are presented in Figure 3. The final spectrum of the regular waves is too narrow to excite natural frequencies of the structure due to frequency mismatch between excitation and first natural frequencies of the model, but the sensitivity of the sensors was strong enough to allow determination of excited waves frequencies. For the wave height 0.2 m and 0.1 m the excited wave frequency 1 Hz is visible for all sensor sets while for wave height 0.01 m the excited wave frequency 1.48 Hz is observable only for sensors set S1, S3 and CC.

3. Conclusions

In this paper an application of FDD decomposition method for strain measurements using FBG sensors on offshore support structure are presented and discussed. The results allow to confirm the usefulness of FDD method for determination of natural frequencies of an offshore structure in the air for different structural conditions. In particular the frequency content of the set of measurement signals varies depending on case considered but not all results are fully consistent for the number of peaks in considered frequency interval. These results could be used in the future for damage detection procedures but more sophisticated analysis is needed. The investigation performed in the basin verify the usefulness of FBG sensors as a part of SHM system permanently installed on underwater elements of offshore structure. They also show the possibility of the FDD method to determination of exciting waves frequencies even when the wave height is very small (0.01 m) but the discrimination between origins of different peaks (response or excitation) need to be developed if possible.

ACKNOWLEDGMENTS

The research devoted to tripod structure was supported by the project entitled: Development of the selection method of the offshore wind turbine support structure for Polish maritime areas (PBS1/A6/8/2012) granted by National Center for Research and Development in Poland.

The authors direct the thanks to the Ship Design and Research Centre (CTO S.A.) especially to Sebastian Kowalczyk and Marek Kraskowski for help during research.

References

[1] Brincker R, Zhang L and Andersen P 2001 Modal identification of output-only systems using frequency domain decomposition *Smart Mater. Struct.* **10** 441

[2] Le TH and Tamura Y 2009 Modal identification of ambient vibration structure using frequency domain decomposition and wavelet transform *Proc. of the 7th Asia-Pacific Conference on Wind Engineering (APCWEVII)* Taipei Taiwan

[3] Li D Zhou Z and Ou J 2012 Dynamic behavior monitoring and damage evaluation for arch bridge suspender using GFRP optical fiber Bragg grating sensors *Optics & Laser Technology* **44** 1031

[4] Lima HF, da Silva Vicente R, Nogueira RN, Abe I, de Brito Andre PS, Fernandes C, Rodrigues H, Varum H, Kalinowski HJ, Costa A and de Lemos Pinto J. 2008, Structural health monitoring of the Church of Santa Casa da Misericordia of Aveiro using FBG sensors, *IEEE Sensors J.* **8** 1236

[5] Majumder M, Gangopaadhyay TK, Chakraborty AK, Dasgupta K and Bhattacharya DK 2008 Fibre Bragg gratings in structural health monitoring-present status and applications *Sens Actuators A Phys* **147** 150

[6] Mollasalehi E, Sun Q and Wood D 2013 Contribution of Small Wind Turbine Structural Vibration to Noise Emission *Energies* **6** 3669

[7] Opoka S, Mieloszyk M, Derra S and Ostachowicz W 2014 Comments on structural condition analysis based on strain measurements on tripod model *Proc. of the 7th EWSHM 2014* Nantes France
[8] Opoka S and Soman R 2014 Detection of structural states of tripod structure using strain data (in Polish) *internal report IMP PAN*
[9] Ren L, Li HN, Zhou J, Li DS and Sun L 2006 Health monitoring system for offshore platform with fiber Bragg grating sensors *Opt Eng* **45** 352.
[10] Tarinejad R and Damadipour M 2014 Modal identification of structures by a novel approach based on FDD-wavelet method *Journal of Sound and Vibration* **333** 1024
[11] Tong W, Lingmi Z and Yukio T 2006 An Operational Modal Analysis Method in Frequency and Spatial Domain *Earthquake Engineering and Engineering Vbration* **4** 295
[12] Torkildsen HE, Grovlen A, Skaugen A, Wang G, Jensen AE, Pran K and Sagvolden G 2005 Development and applications of full-scale ship hull health monitoring systems for the Royal Norwegian Navy *Proc of RTO-MP-AVT-124* Neuilly-sur-Seine, France

Influence of support conditions and temperature on the EMI characteristics

A Stancalie[1,2], D Sporea[1], P Malinowski[3], M Mieloszyk[3], S Opoka[3], T Wandowski[3], W Ostachowicz[3]

[1] Laser Metrology and Standardization Laboratory, National Institute for Laser, Plasma & Radiation Physics, Magurele -Ilfov, ROMANIA
[2] University 'Politehnica' of Bucharest, Faculty of Applied Sciences, Splaiul Independentei 303, 060042 ROMANIA
[3] Mechanics of Intelligent Structures Department, Institute of Fluid-Flow Machinery, Polish Academy of Sciences, Gdansk, POLAND

E-mail: andrei.stancalie@inflpr.ro

Abstract. The present work reports results from an extensive set of measurements which has been performed in order to formulate standard method to remove temperature changes from impedance measurements. The other issue addressed here is the investigation of influence of boundary condition and temperature changes on electromechanical impedance measurement for structural member. Due to electromechanical coupling, changes in dynamic characteristics can be seen in electrical impedance of piezoelectric transducer. Two different systems have been used during this measurement process. System based on FBG sensors has been used for temperature changes measurement while PZT transducers mounted on the structure with impedance analyser were used for electrical parameters measurement. During research electrical impedance and resistance of piezoelectric transducer were measured in order to analyze changes in amplitude of peaks and the frequency shift of peaks due to temperature variations and different configurations of the beam.

1. Introduction

Fiber Bragg gratings (FBG) are widely used in–fiber sensors [1-5] based on their core-refractive index modulation. This modulation along the fiber axis creates periodic structures of refractive index change, δn. In most of the cases, these refractive index changes are produced upon the optical fiber exposure to UV radiation. Various techniques to increase fiber photosensitivity have been reported [6, 7]. The theory behind the operation of FBGs is that periodically modified refractive index induces interference effect for some light's wavelength in the form of a reflected wave from the grating. For other wavelengths the light is transmitted through the grating. The condition for the light to be reflected is that its wavelength of the incident optical signal equals to the Bragg resonance wavelength, λ_{Bragg}. If the grating period, Λ, and the refractive index change,δn, are constant over whole length of grating, then the fiber is uniform. Investigations and practical applications show that FBG sensors are becoming one of the key elements in long-term temperature and strain monitoring. Technologies in this domain are widely used in applications from bio-medicine to Structural Health Monitoring (SHM) for buildings, vehicles, planes, bridges, trains or wind turbines. Due to their properties such as: low mass, low dimensions as well as the immunity to electromagnetic interference, FBG are being found in various applications in space systems or Earth Monitoring. Two important characteristics define Bragg sensors applications domain: their sensitivity to temperature, pressure and mechanical strain.

Sensors included in a Structural Health Monitoring system are often susceptible to damage themselves. Many damage detection systems require a large number of sensors distributed over a structure for effective coverage. Various studies have been focused on the influence of the bonding layer on impedance measurements [8-10]. The authors of these studies investigated the electromechanical impedance models of piezoelectric sensors incorporating the effects of the boundary layer in the measured electrical admittance signals. The sensitivity of the susceptance, or imaginary part of the admittance measurement, to changes in the bonding layer has also been pointed out in these works.

An impedance-based sensor diagnostics model uses a mass-spring-damper system to come up with an expression revealing that the electrical admittance of a PZT patch Y(ω) is a combination of the structure's mechanical impedance $Z_s(\omega)$ and the electrical impedance of the piezoelectric $Z_a(\omega)$, as follows [11]:

$$Y(\omega) = i\omega\frac{wl}{t}\left[\varepsilon_{33}^{T}(1-i\delta) - d_{31}^{2}Y_{p}^{E} + \frac{Z_a(\omega)}{Z_a(\omega)+Z_s(\omega)}d_{31}^{2}\hat{Y}_{p}^{E}\left(\frac{\tan \kappa l}{\kappa l}\right)\right] \quad (1)$$

where *w*, *l*, and *t* are the width, length, and thickness of the PZT, ε the dielectric constant of the PZT, δ the dielectric loss factor for PZT, *d* the piezoelectric coupling constant, and Y_p^E is the complex Young's modulus of the PZT at zero electric field. The κ is the wavenumber of the PZT patch described by:

$$\kappa = \omega\sqrt{\frac{\rho}{Y_p^E}} \quad (2)$$

where ρ is the material density of the PZT patch. The magnitude of the mechanical impedance of the PZT is generally several orders of magnitude lower than that of the structure they are bonded to. The PZT mass and stiffness are nearly negligible, especially at lower frequencies. The problem with the PZT sensor is that any temperature fluctuations can change the slope of the susceptance just as much as a damaged sensor. For this reason, a technique needs to be developed to compensate any temperature variation. This compensation could be used to correct the real part of impedance or even the imaginary part of the admittance [12].

The work here is an initial attempt to generate a standard method to remove temperature changes from impedance measurements in Structural Health Monitoring. Authors of this work evaluate the change in the temperature sensitivity as well as the reproducibility of temperature measurements using selected commercial Bragg grating sensors. The influence of temperature and boundary conditions on electromechanical impedance characteristics of piezoelectric transducer mounted on the structural elements is investigated. Due to electromechanical coupling, changes in dynamic characteristics can be seen in electrical impedance of piezoelectric transducer. Section 2 presents method and experimental details. Results from data analysis are also presented here. Section 3 gives concluding remarks and further possible directions of research.

2. Method and experimental procedure. Results.

The experiments were conducted in order to: *a*) compare the stability of electric characteristics of five PZT transducers and to chose the suitable one for further investigations, *b*) obtain the influence of different boundary conditions on electromechanical impedance characteristics of the piezoelectric transducer mounted on the structural elements; *c*) obtain the influence of temperature changes on electromechanical impedance characteristics of the piezoelectric transducer mounted on the structural elements. In these experiments the FBG temperature probe was used for temperature monitoring during impedance measurements in temperature chamber.

Two important characteristics define Bragg sensors applications domain: the sensitivity to temperature and strain, both being affected by the optical fiber types that we use, and the performance of their parameters. Before using the FBG were subjected to the calibration procedure aiming to: a) study their parameters at constant temperature and b) study the parameters during the heating - cooling cycles. In the first case, the Bragg grating sensors parameters were stored in databases being simultaneously recorded with respect to the Bragg Peak Shift. The measurements have been made in the presence of a thermocouple in order to compare, in real time, the temperature values as given by sensor and thermocouple itself. These measurements have been conducted in the presence of a thermocouple of high sensitivity (below 0.1°C).

Five Sonox P5 PZT transducers were utilized for measurement of real part of electrical impedance or resistance characteristics. These measurements were performed at different temperatures starting from the room temperature up to 50°C. Measurements of electric quantities for freely hanging PZT transducers in temperature chamber were made regarding resistance shift with temperature. Once determined the most stable electric characteristics, the corresponding PZT sensor was mounted on an aluminium alloy beam of 18.5cm length and 2cm width using cyanoacrylate adhesive. Next, one end of the beam was clamped in the vice and this configuration was inserted into a temperature chamber. This setup was subjected to controlled heating from 25°C to 50°C with 5°C step. With an impedance analyser (Hioki IM3570) the data was collected for each temperature step, and analysed from the point of view of induced changes observed in resistance Rs (serial resistance) due to the temperature shift. The temperature has been measured with a FBG sensor and corrections have been made in the final calculated data results.

The similar procedure was applied to the same beam fixed to two vices. The beam with PZT transducer has been clamped on both ends using vices, and the set of measurements has been repeated. The setup was heated from 25°C to 50°C. During each step the impedance measurements on PZT sensor have been recorded until the temperature was stable in the chamber. The temperature has been monitored using the FBG probe. From this research it was concluded that the shift of impedance curves with temperature for the 5 PZT sensors under test, is constant for the considered narrow frequency interval of the excitation signal.

The setup consisting of aluminum beam fixed by only one vice and PZT transducer bonded to the beam was monitored at room temperature. After that it was subjected to heating cycles up to 50 degrees. Resistance peaks changes (R_s) have been determined at different temperature steps and different frequencies. Following these measurements authors conclude that the impedance shift with temperature for the 5 PZT sensors under test is related to frequency of the excitation signal. In other set of experiments, authors have conducted similar test with two vices on the beam. Tests were run in order to compare and find differences between the two set-ups, containing one and two vices. The set-up with the beam plus one vice inside the heating chamber and the configuration of the beam with two vices are presented in Fig. 1. Fig. 2 and Fig. 3 present the changes in impedance shift at different values of temperature and frequency, for given experimental configuration: beam fixed by only one vice, and with two vices on the beam, respectively. These figures show the influence of temperature on impedance shift for wide frequency interval where the sensitivity of electrical impedance to temperature is also dependent on frequency range considered.

Figure 1. The PZT mounted on aluminium beam: (left) the beam with one vice inside the heating chamber, (right) the beam with two vices

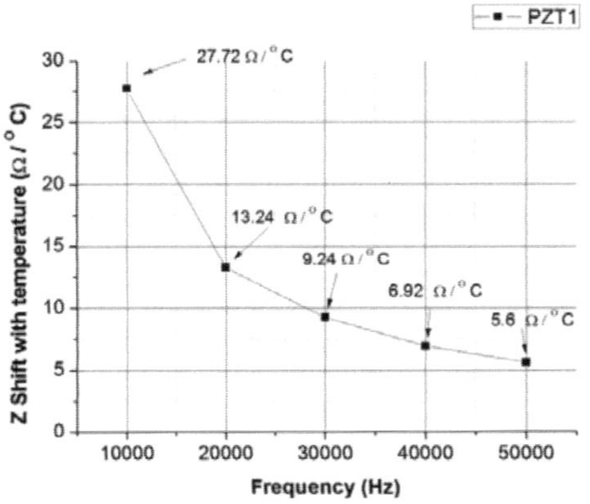

Figure 2. The impedance shift related to temperature for wide frequency interval (beam mounted on one vice)

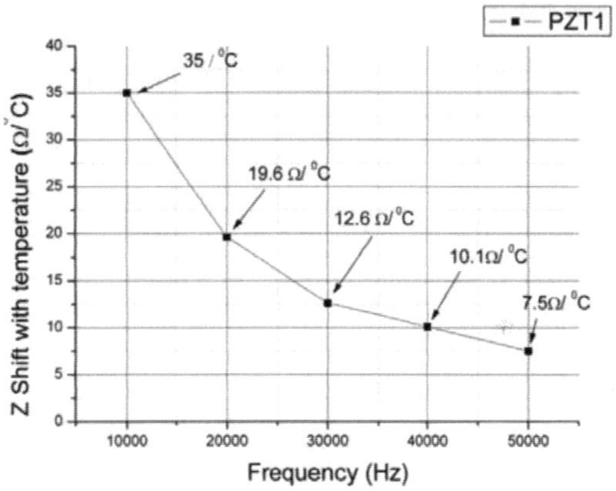

Figure 3. The impedance shift related to temperature for wide frequency interval (beam with two vices)

The piezoelectric sensors used in the EMI method are also significantly pyroelectric. Therefore, the temperature can be determined by measuring the electrical impedance of the piezoelectric sensors. If we consider variations in the amplitude of the electrical impedance resulting from temperature changes, the natural frequencies (*i.e.*, the resonance peaks in the impedance signatures) of the structure are temperature dependent. The natural frequencies decrease as the temperature increases. These variations in the natural frequencies cause frequency shifts in the electrical impedance signatures, as shown in Fig. 2 and Fig. 3.

Several tests over amplitude and frequency shift of resistance peaks Rs caused by temperature have been performed for beam with one vice and with two vices, respectively. Selected results are plotted in Fig. 4 and Fig. 5, respectively, for chosen peaks. For the beam with one vice (Fig. 4) it can be observed that resistance peak is shifted to the left with increasing temperature. The same situation can be observed for beam with two vices, see Fig. 5. Moreover in Fig. 5 characteristic monotonic amplitude reduction of resistance with increasing temperature can be noticed. Similar situation was noticed in paper [13]. This monotonic behavior does not exist for all peaks which were measured here; see for example Ref. [13]. In all experiments the temperature measurements have been done using one FBG temperature sensor. As discussed before, the FBG sensors have been monitored comparing their response with that offered by a thermocouple under same experimental condition.

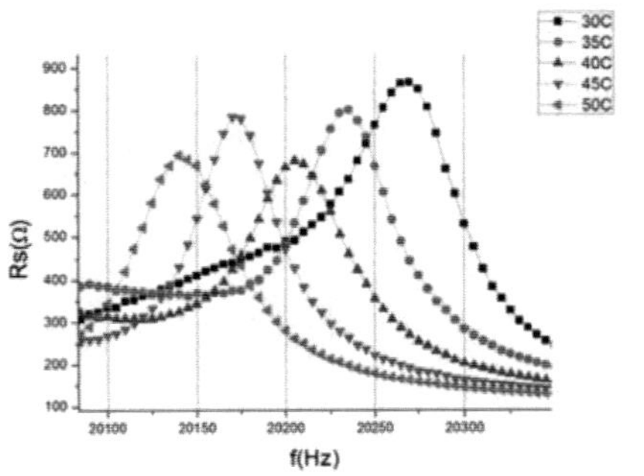

Figure 4. The influence of temperature on frequency shift of chosen peak for beam with one vice.

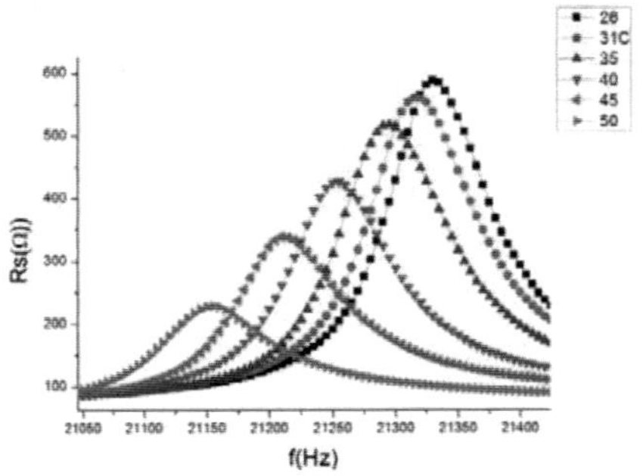

Figure 5. The influence of temperature on frequency shift of chosen peak for beam with two vices.

In Fig.6 and Fig. 7 resistance characteristics are presented for the two cases of applied boundary conditions: one vice and two vices. This simulates condition of beam fixed from one and both sides.

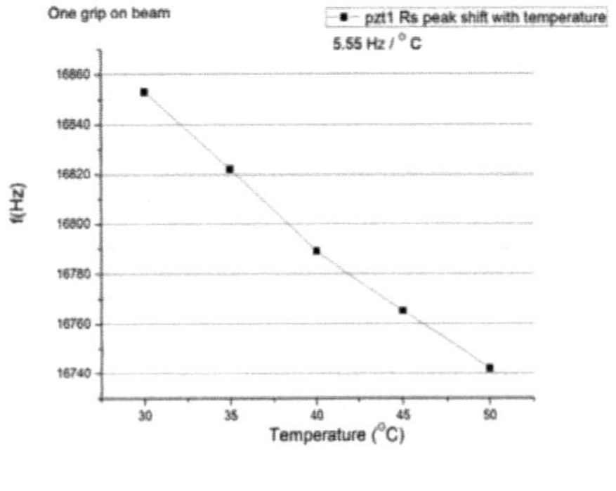

Figure 6. The influence of temperature on frequency shift of chosen peak for beam with one vice: the Rs peak shift measured is equal to 5.55Hz/°C

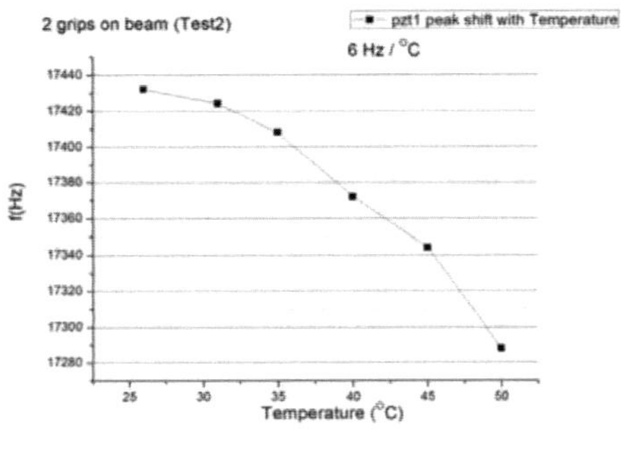

Figure 7.The influence of temperature on frequency shift of chosen peak for beam with two vices: the Rs peak shift is equal to 6Hz/°C

In the impedance approach [12] to model PZT-structure electromechanical interaction, the analytical model is bilaterally symmetric in both the longitudinal and the width directions. The strains of the PZT are uniform distributed. The selected resonance peaks plotted in Figures 4 and 5 clearly exhibited different behaviour as temperature increases. Using one vice (Fig.4) can produce erroneous or misleading results, such as over estimation of peak amplitudes. The results presented in Fig. 2 to Fig. 5 show that the width of the frequency band is a critical issue for compensation techniques based on the frequency shift. From the resistance characteristics presented in Figures 4 and 5, and the corresponding shifts presented in Figures 6 and 7 authors conclude that in one case, namely the beam with one vice, the shifts comprise almost linear line but change in amplitude of peaks is not consistent with temperature variation, while for the second configuration set-up, the shifts do not comprise linear line but amplitude decreases monotonically with increase in temperature.

Fig. 8 gives EMI results from measurements taken in the temperature 25°C from different beam configurations. Comparing these results it can be concluded that support conditions (boundary condition) has very large influence on vibration characteristic and in turn on resistance characteristic of piezoelectric transducer mounted on this structure.

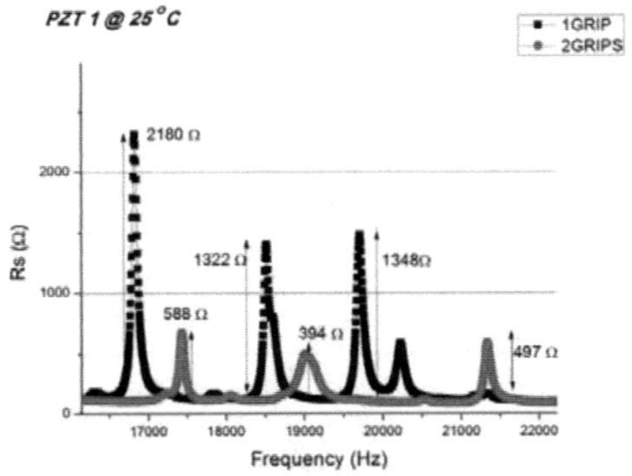

Figure 8. Comparison of the results for the beam with the two experimental configurations.

The temperature variation and strain change due to additional mass hanged on the beam is shown in Fig. 9. The graph shows that FBG strain sensor is influenced by both the tension and temperature changes. The difference between total strain (ε_c) and mechanical strain (ε_m) is well visible in Fig. 9 on the right.

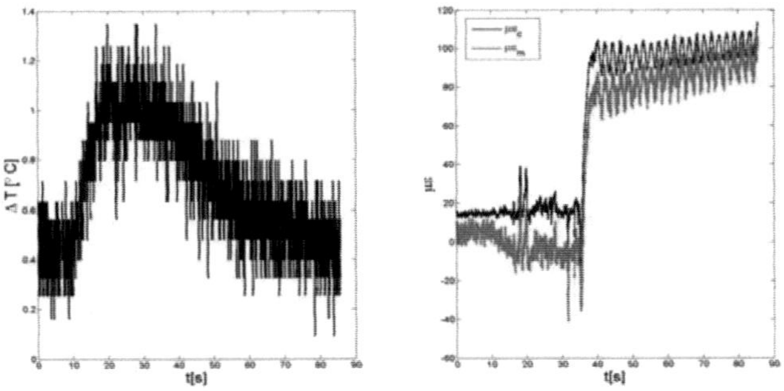

Figure 9. The FBG sensor response to the temperature changes (left, Δ T in °C) and to the strain (right, ε in μm/m); with ε_c - the total strain and ε_m - mechanical strain.

3. Conclusion

The paper presents results from an extensive set of measurements on EMI using combined methods with PZT transducers and FBG sensors in order to provide more insight in Structural Health Monitoring. Authors studied the influence of boundary condition and temperature changes on electromechanical impedance characteristics of the piezoelectric transducer mounted on the structural elements. Due to electromechanical coupling changes in dynamic characteristics can be seen in electrical impedance characteristics of piezoelectric transducer. Two conclusions follow to the present work: a) consistent decrease of frequency of the peak with increase of temperature (fig 6, 7) in the frequency interval considered regardless the type of boundary conditions used; b) the appearance of different resonance peaks for different boundary conditions of the beam.

Further research work is focused on dedicated experiments to determine the influence of strain on EMI characteristics.

Acknowledgments

The authors acknowledge the financial support of the COST action TD1001. A. Stancalie acknowledges the partial support by a grant from “Politehnica” University of Bucharest, under the project POSDRU cod 132397. A Stancalie and D Sporea acknowledge the support by the Romanian Executive Agency for Higher Education, Research, Development and Innovation Funding (UEFISCDI), under Grant 8/2012, the project “Sensor Systems for Secure Operation of Critical Installations”.

4. References

[1] Culshaw B 2000 IEEE *Journal of selected topics in quantum electronics* **6** 1014
[2] Afshar S, Chen L, Bao X 2004 *SPIE* **5567** 22
[3] Afsar S, Ferrier A, Bao X, Chen L 2003 *Optics Letters* **28** 1418
[4] Grobnic D, Smelser CW, Mihailov SJ, Walker RB 2006 *Meas. Sci. Technol.* **17** 1009
[5] Giurgiutiu V, Zagral AN, Bao J 2002 *Structural Health Monitoring* **1** 41
[6] Park G, Farrar CR, Lanza F, Scalea D, Coccia S 2006 *Smart Materials and Structures* **15** 1673
[7] Sun D, Tong L 2003 *International Journal of Solids and Structures* **40** 2449
[8] Park G, Farrar CR, Rutherford AC, Robertson AN 2005 *Journal of Vibration and Acoustics* **128** 459
[9] Simmers Jr GE, Hodgkins JR, Mascarenas DD, Park G, Sohn H 2004 *Journal of Intelligent Material Systems and Structures* **15** 941
[10] Grisso BL, Inman DJ 2008 *Smart Structures and Systems* **4** 305
[11] Grisso BL, Inman DJ 2010 *Journal of Sound and Vibration* **329** 2323
[12] Liang C, Sun FP, Rogers CA 1994 *Journal of Vibration and Acoustics* **115** 121
[13] Siebel T, Lilov M 2013 *Key engineering materials* **5** 69

Collecting data from a sensor network in a single-board computer

F Casciati[1], S Casciati[2], Z-C. Chen[3], L Faravelli[1], M Vece[1]

[1] DICAR, University of Pavia, Via Ferrata 3, 27100 Pavia, Italy
[2] DICAr, School of Architecture, University of Catania at Siracusa, P.za Federico di Svevia, 96100 Siracusa, Italy
[3] Institute of Micro/Nano Devices and Solar Cell, College of Physics and Information Engineering, Fuzhou University, QiShan Campus, Fuzhou 350108, China

E-mail: miche.vece@gmail.com

Abstract. The EU-FP7 project SPARTACUS, currently in progress, sees the international cooperation of several partners toward the design and implementation of a satellite based asset tracking for supporting emergency management in crisis operations. Due to the emergency environment, one has to rely on a low power consumption wireless communication. Therefore, the communication hardware and software must be designed to match requirements which can only be foreseen at the level of more or less likely scenarios. The latter aspect suggests a deep use of a simulator (instead of a real network of sensors) to cover extreme situations. The former power consumption remark suggests the use of a minimal computer (Raspberry Pi) as data collector. In this paper, the results of a broad simulation campaign are reported in order to investigate the accuracy of the received data and the global power consumption for each of the considered scenarios.

1. Introduction

The European Union FP7 project named "Satellite Based Asset Tracking for Supporting Emergency Management in Crisis Operations (Spartacus)" is currently in progress. It is devoted to the implementation of a local low power wireless communication network. The Spartacus project aims to develop Galileo-ready satellite-based applications in order to: (i) track, trace and localize critical transport assets especially in case of major failure of existing networks, (ii) track the flow of relief goods from the sending side to the receiving ones and (iii) support coordination and ensure the safety of first responders in disaster management operations [1]. The aspect of this project, of interest for the authors, is to develop a mobile communication infrastructure on site by integrating terrestrial wireless access technologies with satellite backhauling ones in which the low power local communication network is required to connect GNSS based tracking units and a centralized collecting unit. Indeed, since many GNSS based tracking units are expected to be placed in any considered transportation vehicle, the conventional cabling method suffers problems due to inflexible installation and negative impact on the vehicle structure [2-4]. Moreover, the local wireless communication network should be low power in order to operate long enough to support the first 24 hours after an emergency and it should have encryption function to ensure a high security. This paper details the way to achieve a link to the positioning data transfer, with a NMEA protocol, from a GNSS sensor to a credit-card sized computer (Raspberry Pi) used as data collector of a reliable and independent wireless communication network.

2. State of the art

At present, most investigation efforts aim at improving the current framework related to the exploitation of the potential offered by the satellite network. Indeed, the coordination of operations in emergencies, based on precise positioning and timing system, can find help in the transfer of information coming from a network of units reading the Global Navigation Satellite System (GNSS). Commercially, most GNSS signal generators and record/playback systems are designed specifically for high volume production test applications for devices that use commercial GPS, SBAS, GLONASS, Galileo and other GNSS receivers [5]. The quality of these commercial systems continues also to increase with the addition of new GNSS signals and features [6], but despite all the flexibility provided by these commercial systems, there is still a need for full access to GNSS signal simulation environments for some GNSS developers to generate their own customized simulation scenarios [7].

Figure 1. Focus on satellite navigation system orbits

As well known in literature [8-10], global positioning systems (GPS) offer a valid solution because this technology covers a large area and it is suitable for the monitoring of constructions requiring accuracy of centimetres and which are characterized by a large own period. Indeed, the main feature of GPS units is to rely on internet for transferring the local information to a recording nucleus, however in any disaster areas, it is quite likely that the existing terrestrial communication network and power grid could be cut off after the catastrophic event. Therefore, a reliable asset tracking system requires being independent of those terrestrial networks, especially in terms of communication.

3. System architecture

The developed system architecture consists of an inertial sensor with internal GNSS receiver and a ZigBee wireless data transmit device both connected to a hardware interface, as shown in Figure 2.

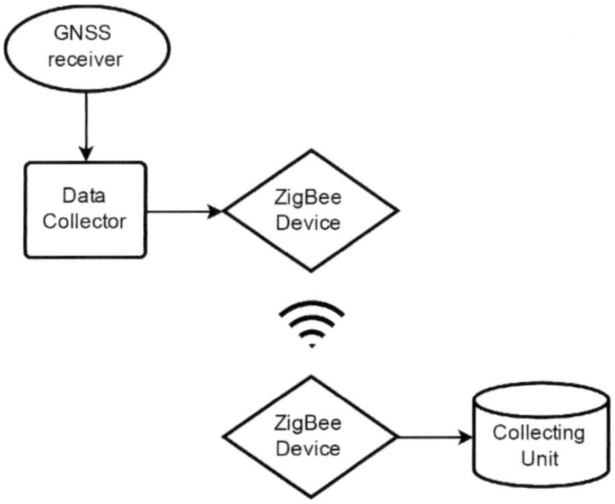

Figure 2. System Architecture

In order to satisfy all the requirement of the project, selected devices are a sensor provided by SBG Systems (Ellipse N) [11], a credit-card sized computer (Raspberry Pi Model B) [13] and GBAN ZigBee Wireless Data Transmission Devices.

4. The output code

These NMEA-0183 messages [12] allow external devices to use selected data collected or computed by the GNSS receiver. All messages conformed to the above format begin with $ and end with a carriage return and a line feed. Data fields follow comma delimiters and are variable in length; null fields still follow comma delimiters, although they do not contain information, and an asterisk delimiter with a checksum value follow the last field of data. The checksum is the 8-bit exclusive of all characters in the message, including the commas between fields, but not including the $ and asterisk delimiters. The hexadecimal result is converted to two ASCII characters and the most significant character appears first [12]. The NMEA messages handled by the sensor to input aiding data are the follow:

- GGA, for position and altitude;
- RMC, for horizontal, velocity and course;
- HDT, for true heading;
- ZDA, for UTC time data.

In the case under study, the required messages (RMC and GGA) are achieved using the configuration settings of the sensor within the proprietary software. An example of a RMC message string is $GPRMC,123519,A,4807.038,N,01131.000,E,022.4,084.4,230394,003.1,W*6A and in Table 1 the message fields are reported in detail.

Table 1. GPRMC message fields.

Field	Meaning
0	Message ID $GPRMC
1	UTC of position fix
2	Status A=active or V=void
3	Latitude
4	Longitude
5	Speed over the ground in knots
6	Track angle in degrees (True)
7	Date
8	Magnetic variation in degrees
9	The checksum data, always begins with *

Field	Meaning
0	Message ID $GPRMC
1	UTC of position fix
2	Status A=active or V=void
3	Latitude
4	Longitude
5	Speed over the ground in knots
6	Track angle in degrees (True)
7	Date
8	Magnetic variation in degrees
9	The checksum data, always begins with *

When NMEA-0183 output is enabled, a subset of NMEA-0183 messages (version 3.01) can be sent to external instruments and equipment connected to the receiver serial ports. The output data transmission is provided by an USB 2.0 interface, which is compatible with the data collector (Raspberry Pi Model B) input ports.

5. The data collector

The Raspberry Pi is a credit card-sized single-board computer developed in the UK by the Raspberry Pi Foundation [13] with the intention of promoting the teaching of basic computer science in schools. The Raspberry Pi is based on the Broadcom BCM2835 system on a chip (SoC), which includes an ARM1176JZF-S 700 MHz processor, VideoCore IV GPU and was originally shipped with 256 megabytes of RAM, later upgraded to 512 MB. The system has Secure Digital (SD) or Micro SD (Model A+ and B+) sockets for boot media and persistent storage [13]. Figure 4 shows the chosen model with a view of its components: a micro USB power supply, an SD card slot, an RCA video and audio jack output, two USB 2.0 and Ethernet ports.

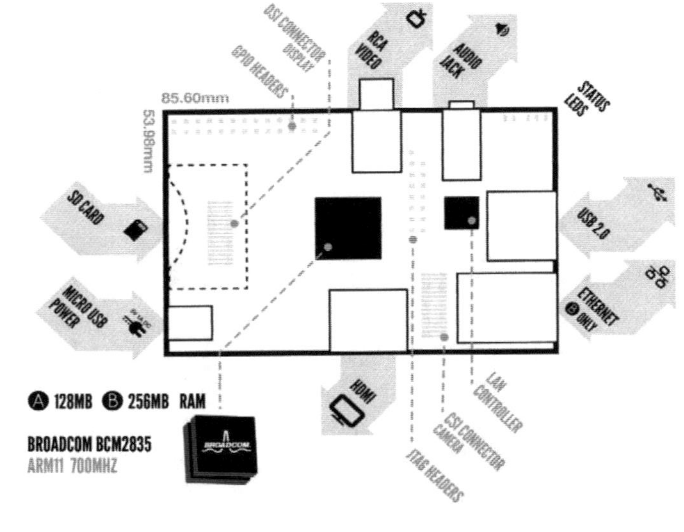

Figure 3. Raspberry Pi Model B (elaborated from [13])

The Foundation also provides Debian and Arch Linux ARM distributions for download and several tools are available for Python as the main programming language, with support for BBC BASIC (via the RISC OS image or the Brandy Basic clone for Linux), C, C++, Java, Perl and Ruby. The RasPi can support USB mouse and keyboard and for USB devices drawing more than 100 mA a powered USB hub is strongly recommended. The most relevant pins for the project, in order to process the data are the GPIO pins, are shown in Figure 5.

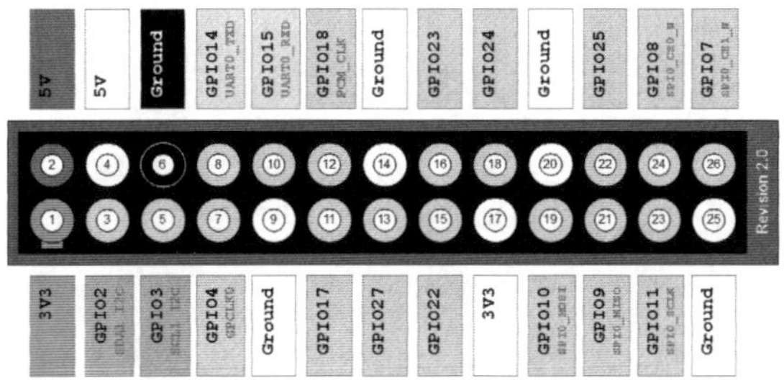

Figure 4. RasPi Model B GPIO layout

The header provides 17 Pins that can be configured as inputs and outputs. By default they are all configured as inputs except GPIO 14 and 15.

6. The wireless data transmission devices

The ZigBee standard network is a low-cost, low-power, wireless mesh network. ZigBee operates in the industrial, scientific and medical (ISM) license-free radio bands: 868 MHz in Europe, 915 MHz in the USA and Australia and 2.4 GHz in most jurisdictions worldwide. Data transmission rates vary from 20 kilobits/second in the 868 MHz frequency band to 250 kilobits/second in the 2.4 GHz frequency band. ZigBee networks are secured by 128 bit symmetric encryption keys. Its transmission distances range from 10 to 100 meters line-of-sight, depending on power output and environmental characteristics. In order to have a longer range, an external radio power amplifier can be adopted with

ZigBee compliant transceiver. There are many ZigBee compliant transceivers and protocol stacks offered by different companies, which can implement the ZigBee network. The chosen device for the project is the GBAN GB-RFTO (Figure 6), based on CC2530 SoC.

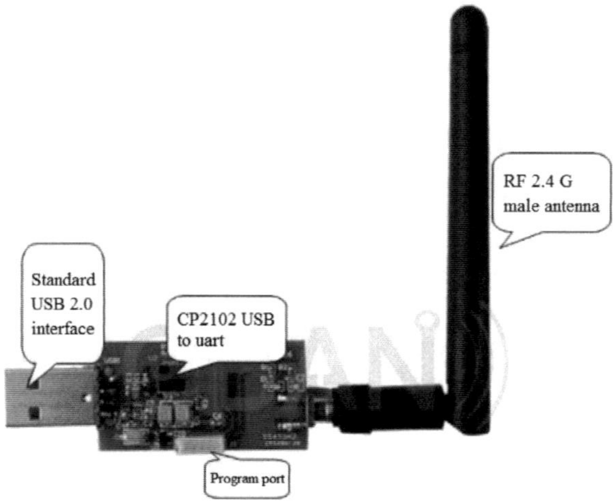

Figure 5. GBAN GB-RFTO ZigBee device

GB-RFTO [14] series ZigBee wireless data transmit device is a reinforced ZigBee module, integrated with an RF transceiver and a microprocessor based on ZigBee standard protocol. It can achieve long distance communication, strong anti-jamming capability, flexible networking. It can also realize a point-to-multipoint and multipoint-to-multipoint transfer of data. Its equipment can support the transmission of the serial port RS232, USB, Wi-Fi and NET data. The layout is reported in Figure 7.

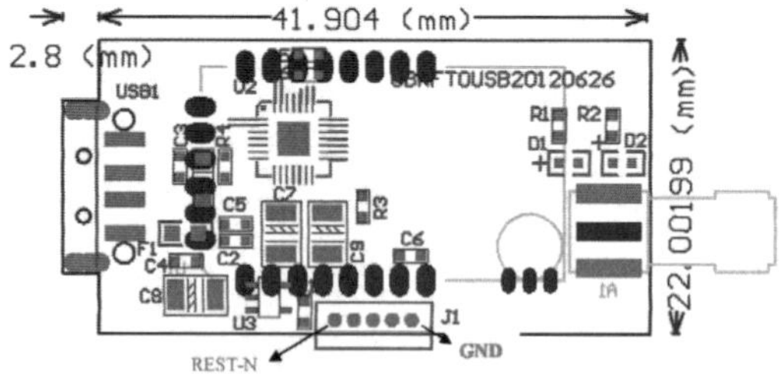

Figure 6. GBAN GB-RFTO ZigBee layout (elaborated from [14])

The CC2530 [15] is a true System on Chip (SoC) solution for IEEE 802.15.4, ZigBee and RF4CE applications. It enables robust network nodes to be built with very low total bill-of-material costs. The CC2530 combines the excellent performance of a leading RF transceiver with an industry-standard enhanced 8051 MCU, in-system programmable flash memory, 8-KB RAM. The CC2530 has various operating modes, making it highly suited for systems where ultralow power consumption is required. Short transition times between operating modes further ensure low energy consumption. Combined with the industry-leading and golden-unit-status ZigBee protocol stack (Z-Stack™) from Texas Instruments, the CC2530F256 provides a robust and complete ZigBee solution. An integrated development environment (IDE) is needed to program and compile the firmware code for the 8051 microcontroller core integrated in the CC2530 transceiver. The adopted and compatible IDE software is IAR Embedded Workbench 8051, provided by Texas Instruments [15].

It is also possible to program the boards with the Texas Instruments Flash Programmer [16] as an alternative to IAR.

A ZigBee protocol stack is needed, which is provided free by Texas Instruments and is named Z-Stack™ [17]. It is a complete protocol stack and application development solution, conformed to ZigBee Alliance standards. The downloaded Z-Stack installation package contains all of the documentation and software required to install, configure, and develop applications using Z-Stack. The package employs a Microsoft Windows-based installation application, which guides the installation process. Texas Instrument also provides two other software tools for free to facilitate the debug and testing of ZigBee transceiver and wireless network: SmartRF Studio 7 [18] and Packet Sniffer [19]. SmartRF Studio 7 is a PC application that can be used in combination with CC2530. It runs on Windows XP, Windows Vista and Windows 7 (32 and 64 bit) and uses USB (or parallel port for legacy boards) to communicate with the evaluation board (EB) which has an evaluation module (EM) with the RF chip mounted. The program provides an easy-to- operate PC interface to all of the chip's radio configuration registers, and it is very helpful for functional testing and for finding the appropriate radio settings. SmartRF Studio 7 can also be used without any hardware, but then only to generate, edit and export radio register values. The main feature of SmartRF Studio 7 is its control panel for direct access to the RF-IC's chip registers and radio-related features. It gives an overview of the many device specific features and provides full read and write access to the chip's radio registers [18]. The Device Control Panel expert mode screenshot is reported in Figure 7.

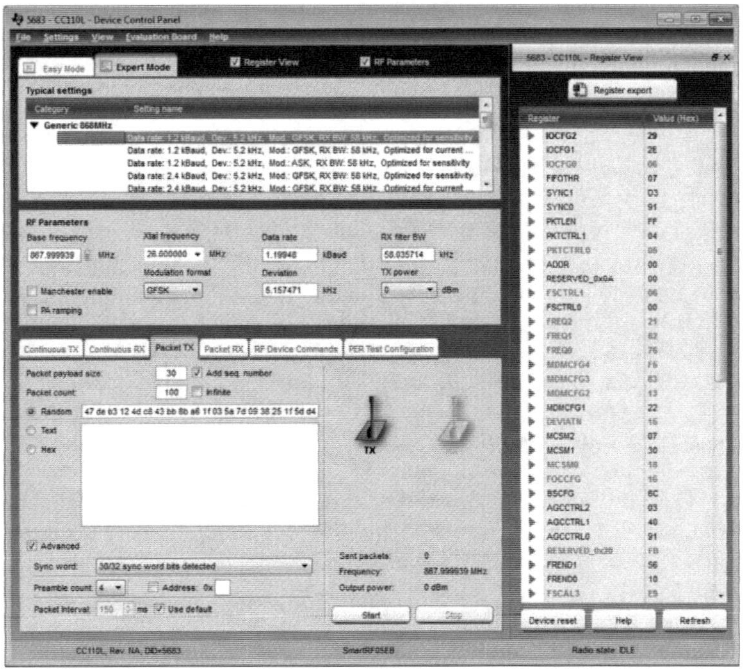

Figure 7. SmartRF Studio 7 Device Control Panel

SmartRF Studio 7 communicates with the Evaluation Board over the USB interface via a library called CEBAL (Chipcon Evaluation Board Access Layer). This is a SW library developed to interface the USB driver and firmware running on the evaluation board, containing all the functions required to read/write data over the SPI interface between the USB MCU and the Transceiver or the debug interface in case of a System-on-Chip. The SmartRF Packet Sniffer is a PC software application used to display and store RF packets captured with a listening RF HW node [19].

The Packet Sniffer filters and decodes packets and displays them in a convenient way, with options for filtering and storage to a binary file format. The supported protocols can be seen in the Launch window when starting the packet sniffer and for IEEE 802.15.4 devices, the required channel must be selected.

7. Conclusions

This contribution drives the reader across the selection path followed by the author to achieve a preliminary solution for a low-power wireless communication network. The task was carried out within the European FP7 Spartacus project. Mainly one adopts the ZigBee-based technology, i.e., CC2530 SoC transceiver and Z-Stack. More or less likely scenarios are addressed through the detailed study of communications protocols and reception systems to implement several functions, such as saving collected data that allow (one to set up) the data post-processing.

References

[1] F. Casciati, S. Casciati, Z-C. Chen, L. Faravelli, M. Vece. Development of a Reliable Wireless GNSS Sensor Network. EWSHM 2014, 7th European Workshop on Structural Health Monitoring, Nantes, France, 8-11 July 2014.
[2] F. Casciati, L-J. Wu. Wireless links for global positioning system receivers. Smart Structures and Systems, vol.10, n.1, pp 1-14, 2012.
[3] S. Casciati, L. Faravelli, Z-C. Chen. Energy harvesting and power management of wireless sensors for structural control applications in civil engineering. Smart Structures and Systems, vol.10, n. 3, pp 299-312, 2012.
[4] S. Casciati, Z-C. Chen, A multi-channel wireless connection system for structural health monitoring applications. Structural Control & Health Monitoring, vol. 18, n. 5, pp 588-600, 2011.
[5] A. Brown, R. Tredway, R. Taylor. GPS Signal Simulation using Open Source GPS Receiver Platform. Proceedings of the 21st Virginia Tech Symposium on Wireless Personal Communications, Blacksburg, Virginia, June 2011.
[6] 2013 Simulator Buyers Guide. GPS World, May 1, 2013.
[7] F. Casciati, C. Fuggini. Engineering Vibration Monitoring by GPS: Long Duration Records. Earthquake Engineering and Engineering Vibration, vol. 8, no. 3, pp. 459-467, 2009.
[8] F. Casciati, C. Fuggini. Monitoring a steel building using GPS sensors, Smart Structures and Systems, 7(5):349–363, 2011.
[9] L-J. Wu, F. Casciati. Local positioning systems versus structural monitoring: a review. Structural Control and Health Monitoring, DOI: 10.1002/stc.1643, 2014.
[10] Alison Brown and Neil Gerein. Advanced GPS Hybrid Simulator Architecture.
[11] Ellipse AHRS & INS - High Performance, Miniature Inertial Sensors User Manual. Document ELLIPSEUM.1. Revision 1 - Oct 7, 2014.
[12] http://www.trimble.com/OEM_ReceiverHelp/V4.44/en/NMEA-0183messages_MessageOverview.html
[13] http://en.wikipedia.org/wiki/Raspberry_Pi
[14] http://www.gban.cn/en/product_show.asp?action=p297&id=43
[15] http://www.ti.com/product/cc2530
[16] CC2530 Software Examples User's Guide. Copyright © 2009, Texas Instruments Incorporated.
[17] http://vast.uccs.edu/files/projects/traumagps_files/docs/chipcon/Z-Stack.pdf
[18] SmartRF Studio 7 Tutorial Product. Copyright © 2011, Texas Instruments Incorporated
[19] SmartRF™ Packet Sniffer User Manual. Copyright © 2010, Texas Instruments Incorporated

Assessment of piezoelectric sensor adhesive bonding

T Wandowski[1], J Moll[2], P Malinowski[1], S Opoka[1], W Ostachowicz[1,3]

[1] Institute of Fluid–Flow Machinery of Polish Academy of Sciences,
Fiszera 14 St., 80–952, Gdansk, Poland
Tel.: +48 (58) 341-1271

[2] Goethe University of Frankfurt am Main, Terahertz Photonics,
Max-von-Laue-Strasse, 160438, Frankfurt am Main, Germany
Tel.: +49 (0) 69 798-47208

[3] Warsaw University of Technology, Faculty of Automotive and Construction Machinery,
Narbutta 84 St., 02-524, Warsaw, Poland
Tel.: +48 (22) 849-0534

E-mail: tomaszw@imp.gda.pl

Abstract. Piezoelectric transducers are widely utilized in Structural Health Monitoring (SHM). They are used both in guided wave-based and electromechanical impedance-based methods. Transducer debonding or unevenly distributed glue underneath the transducer reduce the performance and reliability of the SHM system. Therefore, quality assessment methods for glue layer need to be developed. In this paper, the authors present results obtained from two methods that allow the quality assessment of adhesive bonds of piezoelectric transducers.

The electromechanical impedance method is utilized to analyze transducer adhesive bonding. An improperly prepared bonding layer is a source for changes in the electromechanical impedance characteristics in comparison to a perfectly bonded transducer. In the resistance characteristics of the properly bonded transducer the resonance peaks of the structure were clearly visible. In the case when adhesive layer is not equally distributed under sensor, the amplitudes of structural resonance peaks are reduced. In the case of completely detached transducer, the structural resonance peaks disappear and only resonance peaks of the transducer itself are visible. These peaks (peaks of free transducer hanging on wires) are significantly larger than the resonance peaks of the investigated structure in the considered frequency interval.

The bonding layer shape is also analyzed using time-domain terahertz spectroscopy in reflection mode. This method allows to visualize the adhesive layer distribution based on C-scan analysis. C-scans of signals or envelope-detected signals can be used to estimate the area of proper adhesion between bonding agent and transducer and hence provides a more quantitative approach towards transducer inspection.

1. Introduction

Structural Health Monitoring (SHM) is a set of techniques that allow monitoring the mechanical condition of structures in real-time by utilization a permanent transducer network distributed over the structure. Two approaches can be distinguished: passive and active measurement methods. In a passive system, the sensor network (strain gages, fiber optic strain sensors, accelerometers) is utilized for "listening" of signals coming from evaluating damage. In active systems, the transducer network (piezoelectric transducers) is utilized for generation and sensing of signals. Very often in SHM systems, especially active, piezoelectric transducers are utilized. These transducers can generate and sense diagnostic signals based on inverse and simple piezoelectric effect. Based on registered signals diagnostic process is conducted. In such a piezoelectric based SHM system two very popular diagnostic methods are utilized which are based on elastic wave propagation phenomenon and electromechanical-impedance measurements. First method is based on the fact that damage in the structure is a source of changes in elastic wave propagation. In second method due to electromechanical coupling of piezoelectric transducer and structure all structural changes can be noticed in measurements of electrical impedance of transducers. This method is very analogous to conventional modal analysis (vibration based method). However, here much higher frequencies are utilized, very often up to few hundred kHz.

Transducer diagnostics in SHM is a very important topic. Degradation of sensors can lead to reduction of damage detection ability or can generate false diagnostic reports. In this topic two subtopics can be distinguished: first related to fault or degradation of sensor itself and second related to low performance or degradation of sensor bonding layer. This paper deals with the second problem – diagnostic of transducer bond quality assessment.

Many authors reported possibility of detection piezoelectric transducer debonding by utilization electromechanical impedance method. In [1] authors have analyzed changes in imaginary part of admittance (susceptance) for free and surface bonded transducers. Authors have observed decrease in the slope in the susceptance characteristics caused by the existence of bonding layer. In relation to the bonded transducer its debonding causes increase in the slope of the susceptance characteristics. This problem is related to changes in capacitance of the transducer after its bonding to the structure. Moreover, authors have observed that degradation of transducer performance (fracture in the sensor) causes downward shift of susceptance (both transducers are bonded). It allows distinguishing the problem of bonding performance from the transducer's performance itself. In [2] authors have found that the bond layer properties can significantly modify the measured admittance and lead to incorrect identification of the frequencies peaks. They also conclude that there is no universal method or metric available to assess the quality of the bonding condition. In [3] the author has utilized the imaginary part of impedance (reactance) as a parameter allowing to detect transducer debonding, where real part of impedance (resistance) was utilized for assessment of the host`s structural state. The author concluded that in the case of debonding transducer's free vibrations are observed and the structural resonances disappear. In [4] resistance and susceptance measurements were used to detect sensor faults. These measurements were focused on regions with resonant frequencies.

In this paper, the authors propose the combination of two methods: Time-domain Terahertz spectroscopy [5] and electromechanical impedance testing. The aim of the first method is to determine the distribution of the bonding layer between piezoelectric transducer and host structure. The second method was utilized for verification of results obtained by THz spectroscopy. This THz technique is utilized in nondestructive testing [6]. In [7] THz spectroscopy was utilized for damage detection in fiber reinforced composite parts of wind turbine. Defects were in the form of saw-cut delaminations. This technique was also utilized for heat induced damages in porous materials [8]. THz spectroscopy is also a very promising tool for moisture detection in materials [9].

2. Experimental set-up

Two experimental set-ups are used here. The first one is Terahertz spectrometer Teraview TPS Spectra 3000 which generate impulses in frequency range from 0.1 up to 3 THz which are sent repeatedly and interact with the investigated material. This non-contact measurement system allows to perform measurements in reflection and transmission modes but the results presented in this paper were based exclusively on measurements done for the scanning heads working in reflection mode. Reflection mode is more feasible for analyzing real structures where access to the structure is very often limited to only one side. The spectrometer is equipped with moving table that allows for XY scanning of large objects.

The second experimental set-up is electrical impedance analyzer HIOKI IM3570 connected to the piezoelectric transducer. This equipment was utilized for measurements of electrical characteristics of piezoelectric transducers bonded to the investigated sample.

Investigated sample was in the form of panel with dimensions 500 mm x 200 mm x 2 mm made out of glass fiber reinforced polymer (GFRP) with fiber layer orientation [0/90/0/0/90/0]. Composite sample was clamped on one edge during electromechanical impedance measurements.

This sample was equipped with three piezoelectric transducers (SONOX P5 discs with diameter 10 mm and thickness 0.5 mm), compare Figure 1. One transducer was perfectly bonded (marked as #1 in the Figure 1) whereas two of them were partially debonded (#2 and #3 in the Figure 1). Bonding agent was two part epoxy glue PRO WELD® QUICK Pro Seal which is suitable for real environmental condition and for temperatures up to 150 ^{0}C.

3. Results for Terahertz spectroscopy

In the case of THz spectroscopy the scanning heads were on the opposite side of the sample with respect to the side with bonded piezoelectric transducers. The aim of measurements was to determine bonding agent planar distribution between sample and transducers. It should be mentioned that in the case of GFRP sample, THz radiation penetrates through the thickness and next reflects from bonding agent and piezoelectric transducer. This approach is not suitable for carbon fibers reinforcement polymers (CFRP) because conducting fibers will reflect or absorb the incoming signal either immediately at the surface or within a few sample layers, depending on the polarization of the incoming waves [10]. Hence, the deep penetration of the electromagnetic-radiation into the material is currently not possible.

The analyzed sample with bonded transducers is presented in Figure 1. Results of THz spectroscopy in the form of C-scans were presented in Figure 2. In the Figure 2a) slices of a C-scan based on raw radio-frequency signals was presented, whereas in Figure 2b) envelope-detected signals were utilized. The signal processing was performed using envelope notion of time-domain signal used in C-scan. Analyzing results presented in the Figure 2a) and comparing it with bonding agent distribution presented in Figure 1 it is hard to determine this distribution from THz measurements. Additional signal processing applied to raw signals improved contrast in the new C-scan. Analyzing that result (Figure 2b) it is much easier to determine how the bonding agent is distributed. Moreover circular shape of transducer is much easier visible for perfectly bonded transducer (marked by circles in Figure 1).

In the Figure 3 glue layer distribution after post-processing of THz measurements is presented. The signal processing procedure was as follows: a) the non-equidistant measurement points from the continuous measurement were firstly interpolated to a regular grid; b) isosurfaces were calculated to extract regions of constant THz-waveform amplitudes. This technique is the straightforward extension of the planar slices approach shown in Figure 2 and allows description of oblique effects. This figure shows distribution of bonding agent between composite sample and piezoelectric transducers. This should be kept in mind that proposed approach could not be suitable for other bonding agents. Authors noticed problems with very popular cyanoacrylate Super Glue bonding agent. However, results obtained for THz spectroscopy for PRO WELD bonding agent can be utilized in order to validate the electromechanical impedance method used for bond state assessment.

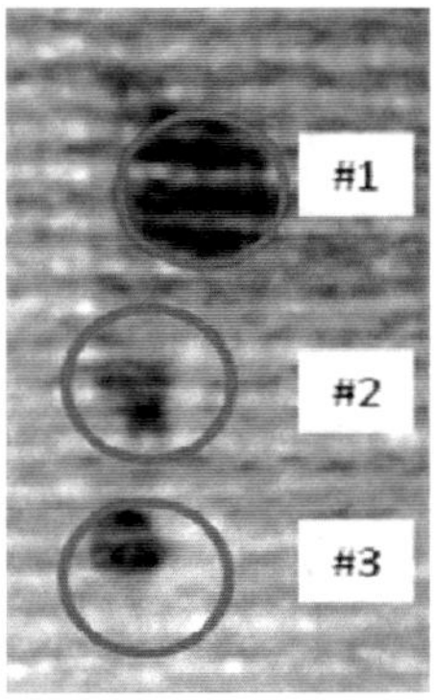

Figure 1. Photo of GFRP sample equipped with three piezoelectric transducers: #1 perfectly bonded, #2, #3 partially debonded.

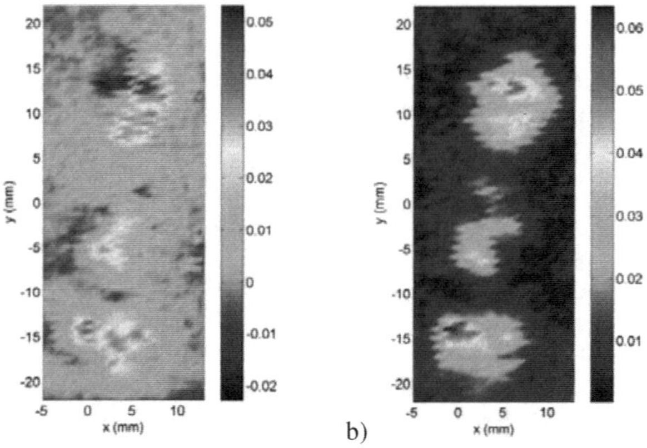

a) b)

Figure 2. Terahertz C-scan for: a) raw measurements, b) measurements after signal processing

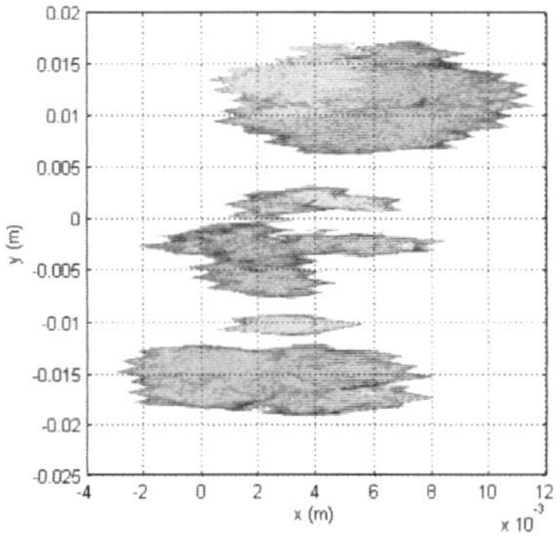

Figure 3. Glue layer distribution obtained from THz measurements after signal processing.

4. Results for electromechanical impedance method

In this section the results for electromechanical impedance method were presented. This method was utilized for quality assessment of bonding layer which is located between piezoelectric transducer and composite sample. The aim of this method was to validate results obtained for THz spectroscopy. Despite of the name of method (electromechanical impedance) very often instead of electrical impedance, such parameters like its real part (resistance) or imaginary part (reactance) are measured and analysed. Very often parameters of admittance (inversion of impedance): its real part (conductance) and imaginary part susceptance are utilized. Generally real part of electrical impedance/admittance of piezoelectric transducer coupled with the host structure is utilized for assessment of structural state of the structure. On the other hand imaginary parts of electrical impedance/admittance are related to the piezoelectric transducer bonding layer assessment are widely used in literature [1], [3].

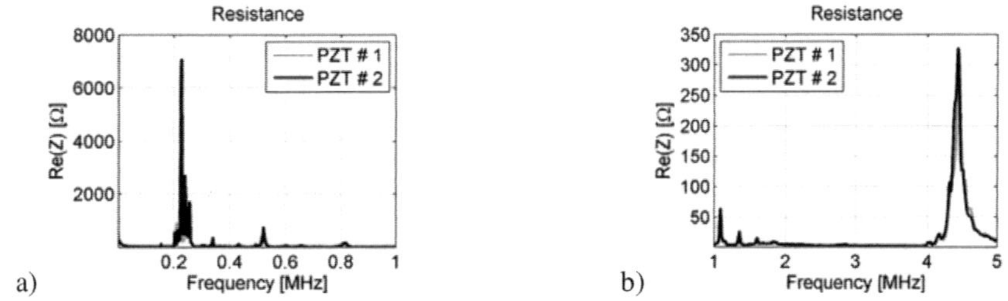

Figure 4. Resistance plot for two free (unbonded) piezoelectric transducers, frequency ranges: a) 1 kHz–1 MHz, b) 1–5 MHz.

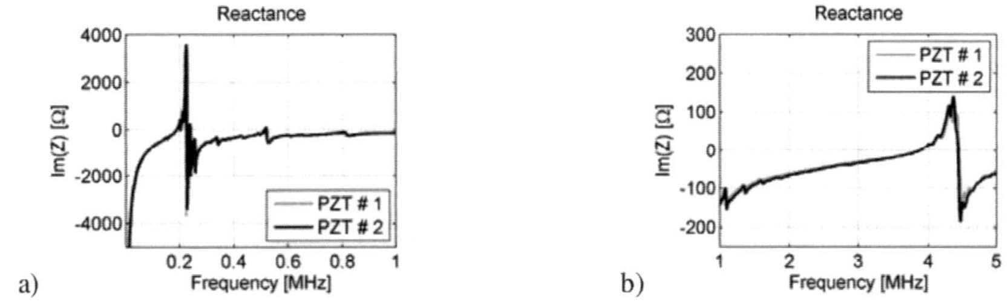

Figure 5. Reactance plot for two free (unbonded) piezoelectric transducers, frequency ranges: a) 1 kHz–1 MHz, b) 1–5 MHz.

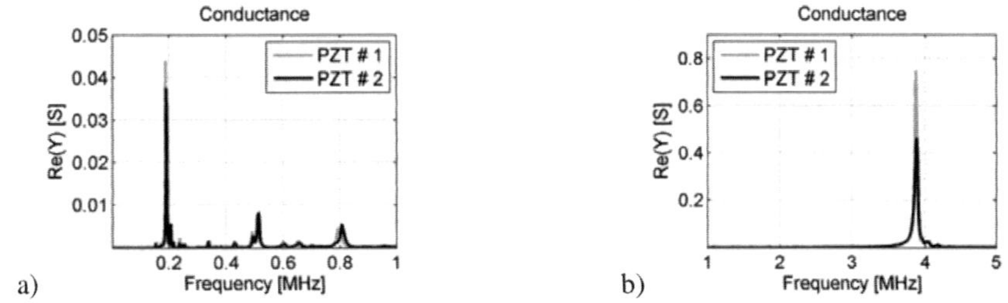

Figure 6. Conductance plot for two free (unbonded) piezoelectric transducers, frequency ranges: a) 1 kHz–1 MHz, b) 1–5 MHz.

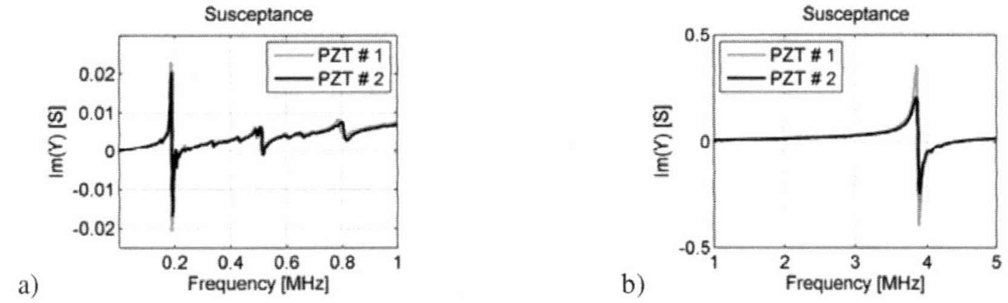

Figure 7. Susceptance plot for two free (unbonded) piezoelectric transducers, frequency ranges: a) 1 kHz–1 MHz, b) 1–5 MHz.

Conducted research has shown that real part of impedance/admittance can also be utilized for performance assessment of piezoelectric transducer bonding layer. However it is important to distinguish the resonances of piezoelectric transducer from the resonances of host structure.

In Figure 4–Figure **7** the plots of resistance, reactance conductance and suscepatance for frequency ranges 1 kHz–1 MHz and 1–5 MHz were presented. Measured parameters were: resistance and reactance (real and imaginary parts of electrical impedance), and conductance and suscepatance (real and imaginary parts of electrical admittance). Analysing plot of parameters related to electrical impedance measured for frequency range 1 kHz–1 MHz (Figure 4a and Figure 5a) group of resonances can be observed, where the one is dominating (around 230 kHz). Similar situation can be observed for frequency range 1–5 MHz, (Figure 4b and Figure 5b) where group of resonances exist but one peak (around 4.4 MHz) has largest value. Analysing parameters plots related to the electrical admittance for lower frequency band (Figure 6a and Figure 7a) and higher frequency band (Figure 6b and Figure 7b) similar behavior can be noticed, but frequencies of resonances are slightly different. For lower frequency range largest peak is located around frequency 190 kHz (Figure 6a and Figure 7a) and for higher frequency range around 3.9 MHz (Figure 6b and Figure 7b). Differences in peak frequencies related to components of impedance and admittance results from serial and parallel resonant frequencies of piezoelectric transducer. Generally peaks visible in Figure 4–Figure **7** are related to transducer resonant frequencies.

The situation changes when piezoelectric transducer is bonded to host structure. Due to mechanical coupling, peaks related to the resonant frequencies of host structure are observed in the measurements. Peaks related to resonant frequencies of piezoelectric transducers can be still visible when the transducer is improperly bonded or partially debonded. When the stiffness of the bonding layer is decreased, then the resonances of free piezoelectric transducer are more and more visible.

Next part of results is related to the transducers bonded to the GFRP sample. Three frequency ranges: 1–10 kHz (host structure resonances), 180–280 kHz (free transducer resonances – low frequency range), 3–5 MHz (free transducer resonances – high frequency range) were analysed. Moreover four transducer bonding cases were analysed: perfectly bonded, two cases of partially debonded, completely debonded (free).

Starting from frequency range 1–10 kHz it can be noticed that peaks related to the resonant frequencies of composite/host structure are clearly visible for resistance (Figure 8a) and conductance (Figure 9a) only for the case of perfectly bonded transducer (marked by #1 in the figures). In the case of imaginary parameters like reactance (Figure 10a) and susceptance (Figure 11a) instead of peaks a zero–crossing can be noticed. However they are not so clear for perfectly bonded transducer (#1) in comparison with real parameters like resistance and conductance. Analyzing all parameters for the cases with partially debonded transducers (#2 and #3), peaks have very small amplitudes and completely disappear for unbounded transducer (#4). Case #4 corresponds to the PZT #1. Its debonding causes characteristics upward shift is observed in susceptance curve (Figure 11a) like in [1].

Debonding causes also significant reduction of peak amplitudes in resistance (Figure 8a) and conductance (Figure 9a). In the next step, analysis was related to frequency region equal 180–280 kHz. Debonding of piezoelectric transducer causes increase of peak values (real components) and maximum/minimum value near zero-crossing (imaginary components) for all parameters (Figure 8b, Figure 9b, Figure 10b, Figure 11b). However, the location of these peak values (zero crossing for imaginary parts) are related to resonance of free piezoelectric transducer. They are located around frequency 230 kHz in resistance (Figure 8b, peak) and reactance (Figure 10b, zero crossing). For conductance and susceptance curves the peak and the zero-crossing are located around frequency 190 kHz (Figure 9b, Figure 11b, respectively). It is clearly visible that characteristic values (peak amplitude and maximum/minimum values near zero-crossing) of all parameters are lowest for perfectly bonded transducer (resonant frequencies of free transducer disappear). For partially debonded transducers values of all parameters are located between two mentioned cases. Moreover some frequency shifts of peaks and shape modifications of the curves are observed.

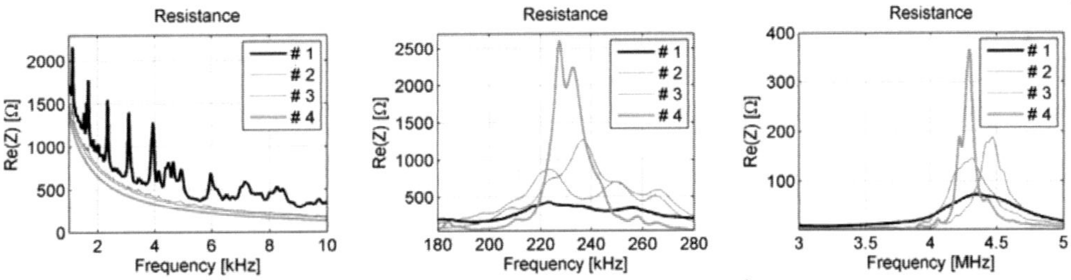

a) b) c)

Figure 8. Electrical resistance for different piezoelectric transducer bonding states: a) 1–10 kHz, b) 180–280 kHz, c) 3–5 MHz; #1 – bonded, #2, #3 – partially debonded, #4 – free (PZT #1).

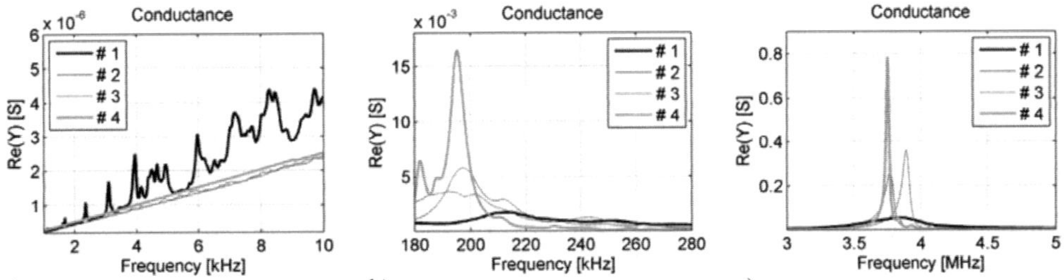

a) b) c)

Figure 9. Electrical conductance for different piezoelectric transducer bonding states: a) 1–10 kHz, b) 180–280 kHz, c) 3–5 MHz; #1 – bonded, #2, #3 – partially debonded, #4 – free (PZT #1).

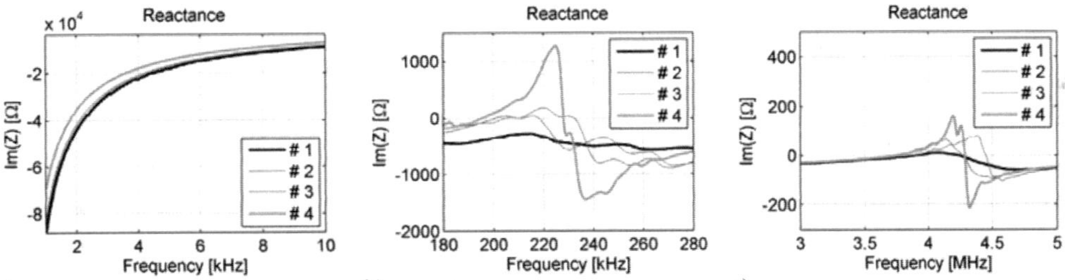

a) b) c)

Figure 10. Electrical reactance for different piezoelectric transducer bonding states: a) 1–10 kHz, b) 180–280 kHz, c) 3–5 MHz; #1 – bonded, #2, #3 – partially debonded, #4 – free (PZT #1).

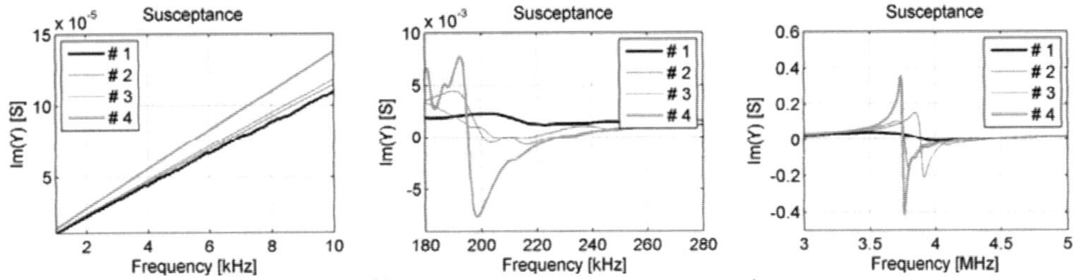

a) b) c)

Figure 11. Electrical susceptance for different piezoelectric transducer bonding states: a) 1–10 kHz, b) 180–280 kHz, c) 3–5 MHz; #1 – bonded, #2, #3 – partially debonded, #4 – free (PZT #1).

The same situation can be noticed for analysis of all plots (Figure 8c, Figure 9c, Figure 10c and Figure 11c) for highest frequency range (3–5 MHz). Different frequencies for resonances in resistance/reactance (4–4.5 MHz) and conductance/susceptance (3.5–4 MHz). Largest amplitude of all parameters were achieved for unbonded transducers and lowest for perfectly bonded.

For partially debonded transducers (#2, #3) parameters have values located between the ones for two extreme states (#1 – perfectly bonded and #2 – free). Frequency shift for different bonding states (#1–#3) and free transducer (#4) are also observed for each parameter.

5. Conclusions

Results presented in the paper have proved that both proposed measurement methods can be utilized for quality assessment of bonding layer.

Terahertz spectroscopy allows imaging of bonding layer distribution for specified types of adhesive. The considered adhesive must produce high contrast in THz image scans. Additionally, this technique is limited rather to GFRP samples. It can be utilized also for CFRP however it should be taken into account that in this case polarized THz radiation can penetrate only through the single layer. This method can be used as non-destructive testing (NDT) of bonding layer.

Electromechanical impedance method was utilized in order to verify results obtained for THz spectroscopy. This verification was successful and shown that for performance of bonding layer real and imaginary components of electrical impedance/admittance can be utilized. However, frequencies of unbounded (free) piezoelectric have to be known. This method can be utilized as NDT technique as well as in continuous mode in SHM systems. The electromechanical method can be used for quality assessment of glue of any type. The resonant frequencies of free transducer posses higher amplitudes of the vibration than amplitudes of the structural peaks visible in electrical characteristics of the transducer. This feature is responsible for the success of transducer's debonding detection by the use of electromechanical impedance in the frequency region where resonances of free transducer exist.

Acknowledgements

This research was supported by the project entitled: Non-invasive Methods for Assessment of Physicochemical and Mechanical Degradation (PBS1/B6/8/2012) granted by National Centre for Research and Development in Poland.

References

[1] G. Park, C.R. Farrar, F. L. di Scalea, S. Coccia: Performance Assessment and Validation of Piezoelectric Active-Sensors in Structural Health Monitoring. Smart Materials and Structures, 15(6), pp. 1673-1683, 2006.

[2] S. Bhalla, C.K. Soh.: Electromechanical Impedance Modeling for Adhesively Bonded Piezo-transducers. Journal of Intelligent Material Systems and Structures, Vol. 15, 2004.

[3] V. Giurgiutiu: Structural Health Monitoring with piezoelectric wafer active sensors. Elsevier, pp. 747, 2008.

[4] I. Buethe, M. Moix-Bonet, P. Wierach, C.-P. Fritzen: Check of Piezoelectric Transducers Using the Electro-Mechanical Impedance. 7th European Workshop on Structural Health Monitoring, Nantes, France, 2014.

[5] N. Palka, D. Miedzinska: Detailed non-destructive evaluation of UHMWPE compistes in the terahertz range, Optical and Quantum Electronics, 46(4), 515-525, 2014.

[6] I. Amenabar, F. Lopez, A. Mendikute; In Introductory Review to THz Non-Destructive Testing of Composite Mater. Journal of Infrared, Millimeter and Terahertz Waves, 34, 152–169, 2013.

[7] J.W. Park et al.: Terahertz Radiation NDE of Composite Materials for Wind Turbine Applications. International Journal of Precision Engineering and Manufacturing, 15(6), pp. 1247-1254, 2014.

[8] E. K. Rahani, T. Kundu, Z. Wu, H.Xin: Heat Induced Damage Detection by Terahertz (THz) Radiation. Journal of Infrared, Millimeter and Terahertz Waves, 32, 848-856, 2011.

[9] J.F. Federici, Review of moisture and liquid detection and mapping using terhaertz imaging.Journal of Infrared, Millimeter and Terahertz Waves, 33, 97-126, 2012.
[10] N. Karpowicz, D. Dawes, M.J. Perry, X.-C. Zhan: Fire damage on carbon fiber materials characterized by THz waves. Proc. of SPIE Vol. 6212, 62120G, 2006.

11th International Conference on Damage Assessment of Structures (DAMAS 2015) IOP Publishing
Journal of Physics: Conference Series **628** (2015) 012115 doi:10.1088/1742-6596/628/1/012115

Embedded Fibre Bragg Grating Sensor Response Model: Crack Growing Detection in Fibre Reinforced Plastic Materials

G.Pereira[1], L.P.Mikkelsen[1], M.McGugan[1]
[1]Technical University of Denmark, Department of Wind Energy, Frederiksborgvej 399, 4000 Roskilde
E-mail: gfpe@dtu.dk

Abstract. This article presents a novel method to simulate the sensor output response of a Fibre Bragg Grating (FBG) sensor when embedded in a host material (Composite material or adhesive), during a crack growing/damage event. A finite element model of the crack growth mechanisms was developed, and different fracture modes were addressed. Then an output algorithm was developed to predict the sensor spectrum change during the different stages of the crack growing. Thus, it is possible to identify specific phenomenon that will only happen within the proximity of a crack, as compression field ahead the crack or non-uniform strain, and then identify the presence of such damage in the structure. Experimental tests were conducted in order to validate this concept and support the model. The FBG sensor response model was applied in a delamination of a Wind Turbine trailing edge, to demonstrate the applicability of this technique to more complicated structures, and to be used as a structural health monitoring design tool.

1. Introduction

Fibre Reinforced Plastic materials (FRP or as often called composite materials) have been extensively used in aerospace, automotive, naval, civil engineering and wind energy applications. These FRP materials consist of two macroscopic phases, a stiff fibre phase, usually glass or carbon fibre, and polymeric matrix, usually polyester or epoxy. One of the advantages of fibre reinforced polymer (FRP) composites is that the alignment of the fibres can be arranged to suit the required properties of the intended structure. Taking a wind turbine blade as an example, the requirement for a stiff (but light-weight) structure means fibre orientation primarily along the length of the blade, however in the axial direction this stiff requirement is much lower, meaning that less material is required and used. In this way FRP has the capability to be tailored for a specific application, and a high level of customization of mechanical properties, such as weight-stiffness ration, thermal expansion, chemical/corrosion resistance, fatigue behaviour etc.[1].

However, the increased demand for more cost-effective lightweight composite materials is pushing for advances in material technology, design philosophies and monitoring techniques. An emerging design concept uses the damage tolerance property of FRP materials, where it's capability to hold damage is taken into consideration during the design process. This is achieved by pushing the structure operation limits, and developing models capable of simulating the behaviour of the damaged structure to give a prognosis of the remaining operational life of the structure. However, accepting the damage and it's unpredictability, requires monitoring technology to track it, by integrated sensors that give information about the presence of damage in an accurate way, it's location and the type of damage [2].

Delamination is one of the most important failure mechanisms in FRP materials and one of the most widespread cause of FRP structures life reduction, this article will focus on this failure mechanism and fibre Bragg gratings as a damage monitoring technique.

2. Delamination in Fibre Reinforced Plastic Materials

Interface fracture by crack growth along interfaces in laminated structures is called delamination, and often is accompanied by the formation of a crack bridging zone, where intact fibres connect the crack faces behind the crack tip, as showed in figure 1. This fibre bridging phenomenon creates an extra obstacle that the crack has to overcome to grow, meaning that the energy required for the crack to grow is higher than that required to initiate it.

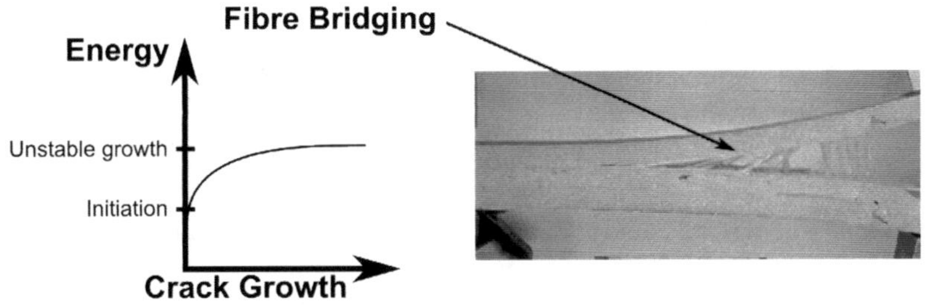

Figure 1. Fibre bridging phenomenon during delamination in a glass fibre specimen.

This large-scale crack bridging zone, can't be handled by linear elastic fracture mechanics. Instead, a cohesive model can be used to describe the fracture process zone. The cohesive law $\sigma_n(\delta_n)$, can be briefly described as a normal traction, σ_n, as function of the normal opening, δ_n in the active cohesive zone [3].

3. Strain Distribution During Crack Growth

To track delamination/crack in FRP materials the sensor/monitoring system selected should be able to track specific fracture features, which only happen in the vicinities of the crack, independent of geometry or loading conditions. To link these fracture features with the measured parameters, the strain distribution around a crack tip during delamination was analysed using Digital Image Correlation (DIC). DIC technique is a non-contact optical method that, by tracking changes in a random pattern on the specimen, can correlate it with deformation/strain of the material.

A pattern was painted on the side surface of the specimen, and ARAMIS^{TM} V6.3 software was used to calculated the strains in each measurement. To perform the measurements, ARAMIS recognizes the surface pattern in the unloaded specimen, and allocates it coordinates in to the image pixels. Then, ARAMIS compares the pattern in the pictures of the loaded specimen, and by tracking the changes calculates the displacement and consequently the strain distribution in the specimen face [4].

To simulate the delamination in a FRP structure, Double Cantilever Beam (DCB) specimens were tested using a fracture testing procedure developed by *Sørensen* [5]. In this testing apparatus the loading is applied through moments and is displacement controlled, which gives a stable crack growth. The DCB specimens were loaded in different conditions at 1mm/min,

*** Formatting Issue - Best Available Paper/Graphic ***

ranging pure Mode I to pure Mode II, in order to simulate different crack/delamination situations.

The DCB specimens were manufactured using two FRP material arms, made of unidirectional and triaxial glass fibre layers (SAERTEX UD and TRIAX), with a layup stacking of : $[90/+45/-45/0_4/0_4/+45/-45/90]$, glued by a commercial epoxy structural adhesive (Epikote MGS BPR 135G/Epikote MGS BPH137G). A thin slip foil was placed in the edge of the structural adhesive, to act as a pre-crack and ease crack initiation.

In figure 2 the DCB specimen geometry is presented.

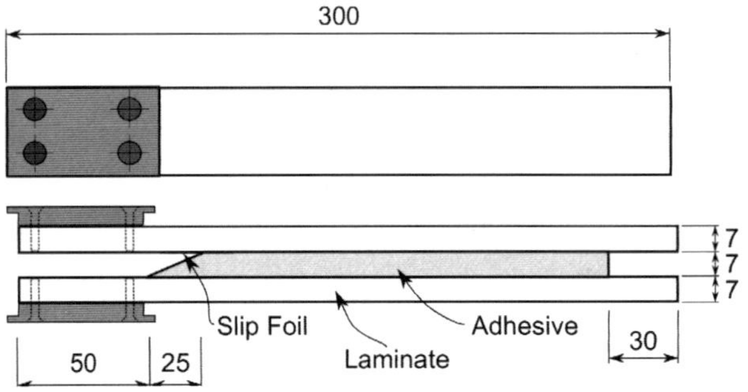

Figure 2. DCB specimen components and geometry.

In figure 3 the strain state at the crack tip during Mode II fracture is shown. We can divide the strain distribution in two distinct contributions: crack tip singularity/material damage and fibre bridging. Near the crack tip, the stress field closely approaches the singular stress field of linear elastic fracture mechanics, meaning that the stress tends to infinity and has a fast variation (high gradient). Also, with the progression of the crack the material/structure losses stiffness increasing the strain, as showed in figure 3a), here it is possible to observe higher values of strain ε_{11} at the crack faces. In the fibre bridging zone ($L < x < 0$), a positive strain ε_{22} was observed, as shown in figure 3c), due to the forces transferred by the fibre that are connecting the two crack faces. These forces are balanced by a compression stress that appears ahead of the crack tip ($x > 0$), which creates a negative strain ε_{22} shown in figure 3c) as a blue area.

4. Crack/Delamination Detection by Embedded Fibre Bragg Gratings

Different sensing technology has been implemented in order to track delamination in FRP materials, such as acoustic emission, vibration/modal analysis, piezo-electric actuators/sensor, strain gages. However, these measurement systems have several limitation in terms of cost, the need of qualified operators and impracticability to be used under operation.

Fibre Bragg Gratings (FBG) are a very a promising technology to track the presence of delamination/cracks in an operational FRP structure, due to their capability to be embedded in the material, without compromising the structural resistance. The FBG small size, 125 μm of diameter, makes it virtually non-intrusive to the material. FBG sensors, also have other interesting features, such high resolution, multiplexing capability, immunity to electromagnetic fields, chemical inertness and long term stability (fatigue behaviour). A Fibre Bragg Grating is formed when a permanent periodic modulation of the refractive index is induced along a section of an optical fibre, by exposing the optical fibre to an interference pattern of intense ultra-violet light [2]. The photosensitivity of the silica exposed to the ultra-violet light is increased, so when the optical fibre is illuminated by a broadband light source, the grating diffractive properties are such that only a very narrow wavelength band is reflected back. When any external phenomenon

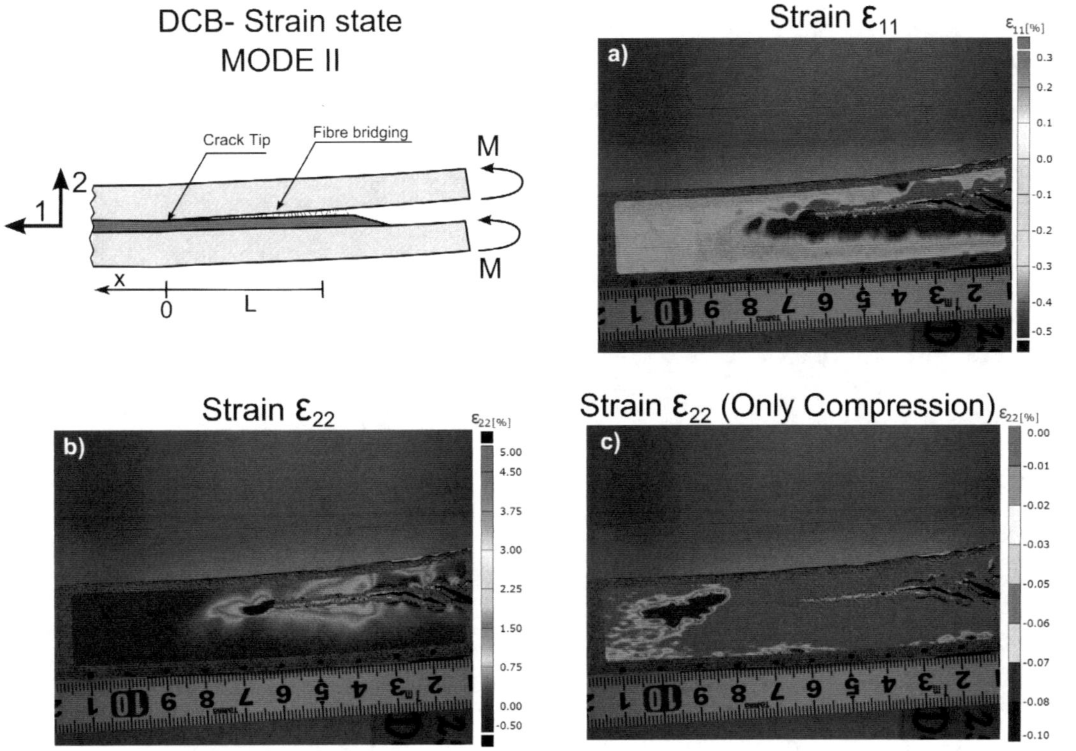

Figure 3. Strain distribution at the crack tip during Mode II fracture.

creates a change in the grating, like temperature, strain, compression, non-uniform strain fields, etc. this will create a change in the reflected light.

As discussed in section 3, during a crack/delamination event different fracture features will be present near the crack tip. Being able to identify and measure this specific phenomena with a FBG sensor is the key factor to correctly determining the presence of damage and it's growth.

In figure 4, the different stages of the FBG responses under a crack growth event are presented. First, before the crack reaches the proximity of the grating, figure 4a), the material will build up uniform strain, that will make a uniform wavelength shift in the FBG reflected peak. Next, a compression field is formed ahead of the crack tip due to the formation of a crack bridging zone, as showed in figure 3c), when it reaches the grating area it creates a peak splitting of the FGB response, figure 4b). Then, when the grating is near the influence of the crack singularity, as discused in section 3 and presented in figure 3a), a non-uniform strain field will also modify the shape of the reflected peak, as showed in figure 4c). Finally, after the crack passes the FBG sensor, the shape of the reflected peak will go back to the original shape, and the sensor response will again be a simple wavelength shift, because at this stage only uniform strains will be present at the FBG, figure 4d).

4.1. Response to Uniform Strain

The wavelength shift $\Delta\lambda_b$ represented in figure 4a), of an embedded FBG under a uniform variation of strain ϵ_{xx}, along the fibre direction, and considering no temperature variation (during crack growth temperature variation is neglected), is given by the equation 1,

*** Formatting Issue - Best Available Paper/Graphic ***

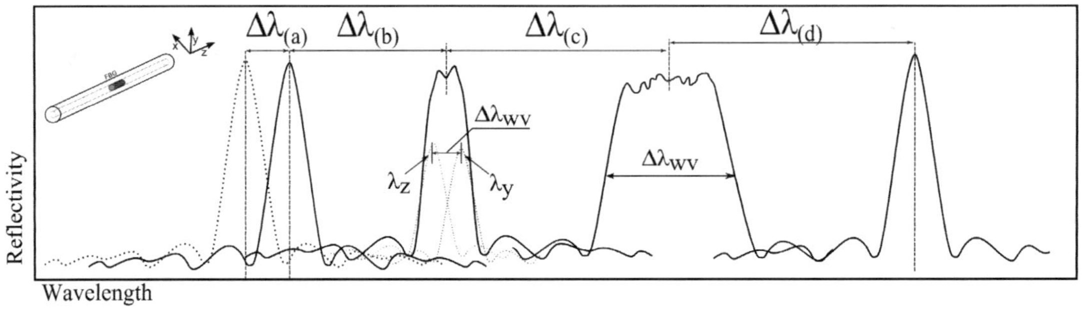

Figure 4. Different stages of the FBG response under a crack growth event.

$$\frac{\Delta\lambda_b}{\lambda_b} = (1 - p_e)\varepsilon_{xx} \tag{1}$$

where p_e is a photoelastic coefficient [6].

4.2. Response to Transverse Force

An optical fibre can present a birefringent behaviour, defined by the change of the refractive index n_{eff} in the two directions n_{effy} and n_{effz}, when the grating is subjected to a transverse force [7, 8, 9, 10]. The increase in the width of the reflected peak represented in figure 4b), is given by the equation 2,

$$\Delta\lambda_{WV} = 2\Lambda|\Delta n_{effz} - \Delta n_{effy}| = \frac{\Lambda n_o^3}{E_f}[(1+\nu_f)p_{12} - (1+\nu_f)p_{11}]|\sigma_z - \sigma_y| \tag{2}$$

where $\sigma_{y,x}$ is the transverse stress, E_f is the elastic modulus of the optical fibre,ν_f is the Poisson's ration, n_0 is the initial refractive index, p_{11} and p_{12} are the photo-elastic coefficients of the optical fibre.

4.3. Response to Non-Uniform Strain

As demonstrated by Peters [11], in a uniform grating the applied strain will induce a change in both grating period and the mean index. These two effects can be superimposed by applying an effective strain of "$(1 - p_e)\varepsilon_{xx}(x)$". Then it is possible to rewrite the grating period as:

$$\Lambda(z) = \Lambda_0[1 + (1 - pe) \times \varepsilon_{xx}(x)] \tag{3}$$

where Λ_0 is the grating period with zero strain. The non-uniform strain effect can be approximated by using the maximum and minimum strain values along the grating. So, the maximum grating period Λ_{max} and minimum Λ_{min} can be calculated using the equation (3). Thus, an approximated increase of the width of the reflected peak due to a non-uniform strain $\Delta\lambda_{IW}$ is given by the equation 4:

$$\Delta\lambda_{WV} = 2n_{eff}\Lambda_{max} - 2n_{eff}\Lambda_{min} \tag{4}$$

5. Finite Element Method Model of the FBG response: Crack Detection Model

The major goal of this study was to developed a numerical model to predict the FBG output in a general crack growth situation. This sensor-damage-structure model provides a tool to study the application of this monitoring technique in other locations, by predicting the sensor

11th International Conference on Damage Assessment of Structures (DAMAS 2015) IOP Publishing
Journal of Physics: Conference Series **628** (2015) 012115 doi:10.1088/1742-6596/628/1/012115

output and deciding, based on this, the best sensor-structure configuration. To accomplish this, an algorithm was developed, using the measurement principals/variables and equations developed in section 3 and 4. This algorithm was incorporated in to the FEM Model, using *Python* programming language.

Table 1. Material Properties used in the DCB Model

Composite Material		Interface (Cohesive Law)			Adhesive
Triaxial Fabric (Composite)	Uniaxial Fabric (Composite)	Elastic	Damage (Quadratic Stress	Damage Evolution	Elastic
$E_1 = 44.3$ GPa; $E_2 = E_3 = 12.9$ GPa; $\nu_{12} = \nu_{13} = \nu_{13} = 0.23$; $G_{12} = G_{13} = G_{23} = 4.4$ GPa	$E_1 = 23.8$ GPa; $E_2 = E_3 = 15.05$ GPa; $\nu_{12} = \nu_{13} = \nu_{13} = 0.513$; $G_{12} = G_{13} = G_{23} = 4.4$ GPa	K= 4.2 E12 Pa;	$\sigma_n = 2.64$ MPa; $\sigma_t = 22.15$MPa	δ_n= 1.4; δ_t= 0.37;	E= 4.56 GPa; ν= 0.35

As a study case, the delamination in the trailing edge of a wind turbine blade was addressed, based on a real trailing edge configuration used by the *DTU 10 MW Reference Wind Turbine* [12]. The trailing edge was modelled as a 2D and 3D Double Cantilever Beam, where different loading conditions (ranging pure Mode I to pure Mode II fracture modes) simulate the different fracture cases that can occur in this structure. The crack growth in the DCB specimens was modeled using cohesive elements, which describe the cohesive law that governs the crack growth mechanism. The geometry of the DCB specimen is the same as used in figure 2 and section 3, and the material properties used in the model are presented in the table 1.

Figure 5. FEM model Sensor output for Mode I loading case.

5.1. Model Experimental Validation

DCB specimens with the same material-sensor configuration as the model, were tested following the procedure described in section 3. The DCB specimens were instrumented with an array of FBG sensors embedded in the host material. A good agreement between the finite element model and the experiments was found. The FEM was able to represent the crack growth under different loading cases, and the sensor output model matched the experiments, showing the expected sensor response to fracture phenomena, already measured by DIC technique and described in section 3.

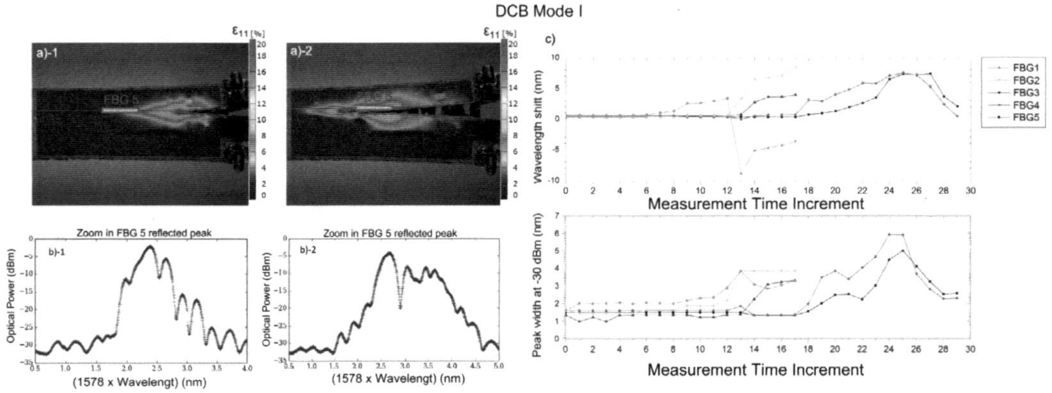

Figure 6. Model Experimental Validation for a Mode I loading case. a) Digital Image Correlation technique: Longitudinal strain ε_{11}, b) FBG 5 reflected peak, c) FBG sensor response during crack growth.

6. Application of the Model

Figure 7. FBG sensor output simulation: Crack detection in Wind turbine trailing edge.

The FBG response model was applied in to a "full" Wind turbine blade model to predict the sensor output during failure of the trailing edge, as schematic represented in figure 7. To address this, a sub-model of the blade containing a detailed trailing edge geometry was modelled using the loads given by the global model. Then the FBG-Output model was applied and the results extrapolated to the full structure model.

7. Conclusion

In this article the capability of Fibre Bragg Gratings embedded in composite material to detect and track cracks/delamination was demonstrated. The use of the digital image correlation technique proved that specific fracture features near the crack can create a change in the shape of the reflected peak. Thus, it is possible to extract information from the sensor that is independent of the loading type, geometry and boundary conditions, and depends only on the proximity of the crack. This fact allows the application of this technique in general composite material structures.

The prediction of the sensor response by the FEM model makes it possible to study the application of this monitoring technique in other locations, predict the sensor output, and decide the best sensor-structure configuration. This *"FBG output model"* concept can have an impact in condition monitoring methodology and maintenance, by optimising the measurement by sensors, enabling more diverse and accurate damage detection, giving better characterization of the damage (type and size), and giving information for prediction of the residual structural life, and if possible, repair.

8. Acknowledgement

The author acknowledges the Seventh Framework Programme (FP7) for funding the project MareWint (Project reference: 309395) as Marie-Curie Initial Training Network, Fibersensing for providing the FBG sensors and hardware, and SSP-Technology for providing the material tested.

References

[1] Jones R 1999 *Mechanics of composite materials* (Taylor & Francis) ISBN 156032712x, 9781560327127
[2] McGugan M, Pereira G, Srensen B, Toftegaard H and Branner K 2015 *Royal Society of London. Philosophical Transactions A. Mathematical, Physical and Engineering Sciences* **373** ISSN 1364-503X
[3] Sørensen B 2010 *Cohesive laws for assessment of materials failure:Theory, experimental methods and application* Ph.D. thesis
[4] ARAMIS^{TM} 2011 *User Manual. Version 6.3, GOM optical measuring Techniques, Mittelweg 7-8, D-38106 Braunschweig, Germany* (Germany)
[5] Sørensen B F, Jørgensen K, Jacobsen T K and Østergaard R C 2006 *International Journal of Fracture* **141** 163–176 ISSN 0376-9429
[6] Morey W, Meltz G and Glenn W 1990 *OE/FIBERS'89*
[7] Sørensen L, Botsis J, Gmür T and Humbert L 2008 *Composites Science and Technology* **68** 2350–2358 ISSN 02663538
[8] Zhang W, Chen W, Shu Y, Lei X and Liu X 2014 *Applied Optics* **53** 885 ISSN 0003-6935
[9] Jülich F and Roths J 2010 **7726** 77261N–77261N–9
[10] Bosia F, Giaccari P, Botsis J, Facchini M and Limberger H G 2003 *Smart Materials and Structures* **925** 925–934
[11] Peters K, Studer M, Botsis J, Iocco A, Limberger H and Salath R 19–28
[12] Bak C, Zahle F, Bitsche R, Kim T, Yde A, Henriksen L, Hansen M, Blasques J, Gaunaa M and Natarajan A 2013 The dtu 10-mw reference wind turbine URL `http://dtu-10mw-rwt.vindenergi.dtu.dk/`

Using genetic algorithms to optimize an active sensor network on a stiffened aerospace panel with 3D scanning laser vibrometry data

R.Marks[1], A. Clarke[1], C. Featherston[1], L. Kawashita[1],C.Paget[2], R. Pullin[1]

1. Cardiff School of Engineering, Queen's Buildings, The Parade, Cardiff, United Kingdom. CF24 3AA
2. Airbus UK Ltd, Filton, Bristol, United Kingdom. BS99 7AR

marksra@cardifff.ac.uk

Abstract. With the increasing complexity of aircraft structures and materials there is an essential need to continually monitor the structure for damage. This also drives the requirement for optimizing the location of sensors for damage detection to ensure full damage detection coverage of the structure whilst minimizing the number of sensors required, hence reducing costs, weight and data processing.

An experiment was carried out to investigate the optimal sensor locations of an active sensor network for detecting adhesive disbonds of a stiffened panel. A piezoelectric transducer was coupled to two different stiffened aluminium panels; one healthy and one with a 25.4mm long disbond. The transducer was positioned at five individual locations to assess the effectiveness of damage detection at different transmission locations. One excitation frequency of 100kHz was used for this study.

The panels were scanned with a 3D scanning laser vibrometer which represented a network of 'ideal' receiving transducers. The responses measured on the disbonded panel were cross-correlated with those measured on the healthy panel at a large number of potential sensor locations. This generated a cost surface which a genetic algorithm could interrogate in order to find the optimal sensor locations for a given size of sensor network. Probabilistic techniques were used to consider multiple disbond location scenarios, in order to optimise the sensor network for maximum probability of detection across a range of disbond locations.

1. Introduction

Acousto-ultrasonic induced Lamb waves are a long-established technique for detecting damage in structures. The principle involves exciting a piezoelectric transducer mounted to the structure's surface which induces a Lamb wave that is then detected by another transducer mounted at a different location on the structure. If damage occurs within the field between the two sensors, the signal propagation is altered resulting in a quantifiable difference in the signal received.

Adhesively bonded metallic joints have been used in the primary structure of aircraft for the past fifty years [1]. Bonding as opposed to mechanically fastening stiffeners to the aircraft's skin is advantageous as it can reduce the mass of the joints, distribute the stress loading more evenly and be more resistant to environmental effects which an aircraft experiences [2]. Adhesive bonding is also less labour intensive and more cost effective than using mechanical fasteners [3].

Though adhesives have been used in the secondary structure without arrestment fasteners they are rarely used in the primary structure (i.e. the structure in which the occurrence of failure would be catastrophic). Without arrestment fasteners there is a lack of confidence in the reliability of the joint [4]. From the experience of the Royal Australian Air Force (RAAF) 53% of defects detected on aircraft structures such as the F-111 were found to be bond failures [5]. Many cases of adhesive bond failure have been found to be the result of improper application of the adhesive or insufficient surface preparation [4] i.e. human error in the manufacture of the joint.

Though it has been proven that bonded joints are more resistant to fatigue loading as well as being able to withstand higher peak loading [5], the degradation of the adhesive layer over time is not fully understood [6]. This uncertainty in the condition of the bond throughout the aircraft's in-service life requires monitoring through non-destructive techniques to ensure airworthiness. If a sensor network could be installed to monitor damage and degradation of the bonded joint an "as required" inspection program could be utilised. However sensor networks need a high probability of detection whilst minimizing the weight penalty they add to the overall structure. Hence, understanding the optimal location of sensors for this monitoring is of the utmost importance.

2. Experimental Setup

Two stiffened panels of dimensions 1250mm x 1250mm were constructed from 3mm thick 6082-T6 aluminium plate. An unequal angle stiffener made from 6082-T6 aluminium was bonded to each plate using commercially available Araldite® 420 structural adhesive. The film thickness of the adhesive was regulated using 0.1mm copper wire gauges to achieve the optimum shear strength [7].

On one of the panels an intentional disbond was induced at the centre of the stiffener of 25.4mm in length which covered the width of the stiffener. This was induced by installing PTFE tape over this region of the bond which was then removed once the adhesive had cured.

A square region of 625mm x 625mm in the centre of each panel, on the face with the stiffener was designated the area of investigation in an attempt to reduce reflections from the edges of the panels.

Five excitation sites were designated along the left hand boundary of the investigation area. By having multiple excitation sites the effectiveness of the transducer-disbond path could be investigated, in effect simulating different disbond positions relative to a single transducer site.

Vibrometer measurement points were set up within the investigation region. No measurement points were positioned within a 25mm wide region on either side of the stiffener. This was due to the positioning of the laser heads which resulted in the stiffener casting a 'shadow' where it was impossible for all three lasers to align.

There also were no measurement points positioned in a 78mm wide region to the right of the excitation sites in order to save acquisition time as preliminary tests had demonstrated insignificant findings in this region. The dimensions of the panel are shown in Figure 1.

A PANCOM Pico-Z transducer was excited with a 10-cycle sine burst generated by Physical Acoustics WaveGen function generator software connected to Physical Acoustics μdisp/NB-8 hardware. The peak-to-peak amplitude of the excitation signal was 160 V. An excitation frequency of 100kHz was selected for this experiment. A 10 V peak-to-peak wave was also generated and used as a reference signal for triggering the acquisition of the vibrometer. A repetitive trigger rate of 20 Hz was used as this gave sufficient time for the induced wave energy to fully dissipate before taking the next measurement. Velocity amplitudes were investigated at 825 points with two hundred measurements being recorded at each point. These measurements were then averaged in order to improve the signal to noise ratios.

The vibrometer used for this experiment was the Polytec PSV-500-3D-M. This vibrometer uses three laser heads to measure each scan point. The measurements recorded by each laser head can then be used to calculate, using trigonometry, both the in-plane and out-of-plane modes. Due to taking measurements from three heads it is less important for the laser heads to be perpendicular to the structure under test with a 3D scanning system when compared to a 1D system. The investigation region was coated with retro reflective glass spheres which improves the signal quality by increasing the amount of back scattered light. During testing each panel was laid on 'bubble wrap' packaging material to acoustically uncouple it from the floor. A series of markers were laid within the laboratory and on the panels to ensure that each panel was positioned in the same place relative to the laser heads. A low pass front end filter was applied to the measurements with the filter frequency set 20kHz above the excitation frequency to filter out high frequency noise.

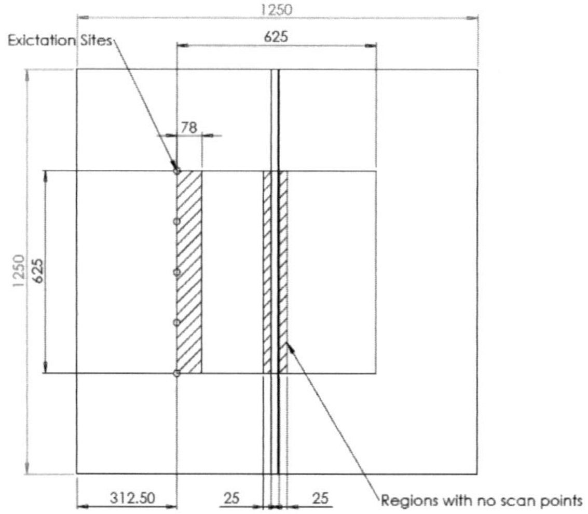

Figure 1: Dimensions of the Stiffened panel, excitations sites and scan area

3. Optimisation problem

The number of possible sensor network combinations increases with the number of sensors by a multiple of the respective binomial coefficient. In this experimental study, the number of possible sensor locations is 825.

It is massively computationally expensive to evaluate the effectiveness of every possible solution as the number of sensors in the network increases, ruling out simple optimisation techniques such as an exhaustive search. A list of the maximum number of sensor combinations is shown in Table 1 for networks of up to 5 sensors.

Table 1: The number of sensor network combinations for a given number of sensors in the network

Number of Sensors in network, N_s	Number of possible sensor network combinations
1	825
2	339900
3	93245900
4	$1.92x10^{10}$
5	$3.15x10^{12}$

3.1. Genetic Algorithms

Genetic Algorithms (GAs) are an optimisation technique developed by *Holland* [8] that are based on the theory of evolution to find the optimal solution. Each 'generation' produces more suitable solutions and less suitable solutions are discarded in a manner analogous to Darwinian evolution.

Genetic algorithms have advantages over other optimisation techniques as they have the ability to 'mutate', allowing the algorithm to interrogate the whole solution space to find the global minimum or maximum rather than converging on local minima or maxima. The way that algorithm interrogates a relatively low proportion of the total solution space is also not as computationally expensive as other techniques[9]. Figure 2 shows the schematic representation of a GA [10]. A brief overview of a genetic algorithm is given here. For further explanation the reader is referred to *Haupt and Haupt* [9].

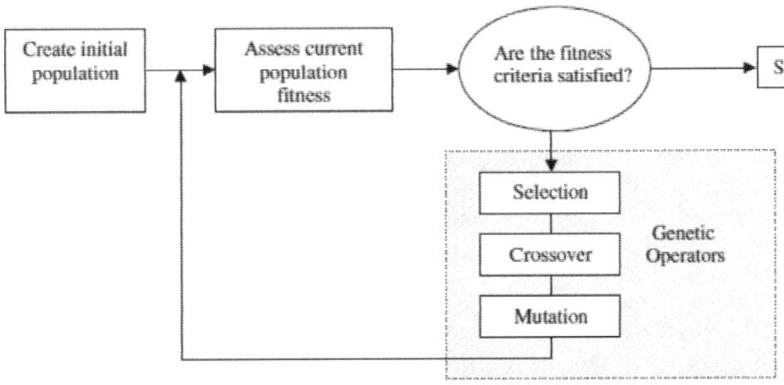

Figure 2: Schematic representation of a GA [10]

There have been many studies conducted that have used genetic algorithms to optimise sensor networks for many active and passive sensing problems [11-13]. One of the biggest issues however, as with any optimisation problem, is defining the cost or fitness function for the GA to interrogate.

3.2. Time Window

The data sets (healthy and delaminated) were first correlated for the entire sample length of the measurements (1.6ms). This however produced very low cross-correlation coefficient values due to the reflections and refraction patterns caused by the disbond resulting in significant differences in the edge reflections.

To improve the correlation between the two data sets, a 200μs time window containing only the incoming wave was used for the cross-correlation. The time window was decided by taking the maximum velocity value in the first 700μs of the entire sample length for the healthy data set. A 40% threshold value was then calculated and the arrival of the wave was determined when the threshold was first crossed. A pre-trigger of 14% was also applied to ensure that the wave front was captured.

3.3. Cross Correlation

The cross-correlation technique has been proven to be a sufficient means of identifying the presence of damage in an acousto-ultrasonic (AU) system [14] where waves received are correlated with each other to identify any differences. Typically a value of unity for the cross-correlation coefficient indicates that the received waves are identical. A value less than unity identifies that the waves are different and hence typically the presence of damage.

By calculating the cross-correlation coefficient for each measurement pair it is possible to produce five data sets with an assigned cross-correlation coefficient for each point. These five data sets serve as a fitness for sensor locations for each excitation site. The aim of having numerous excitation sites was to produce results that were representative of having the disbond in different locations relative to the excitation site. A plot of the cross-correlation data set for excitation site 3 is shown in Figure 3. For this study, only the out-of-plane wave modes have been considered.

The low correlation behind the disbond observed in Figure 3 is the result of a refraction pattern produced by the Lamb wave's interaction with the disbond. This is also observed on the excitation side of the stiffener as diagonal refraction fringes produce areas of low cross correlation values. This identifies the presence of disbond.

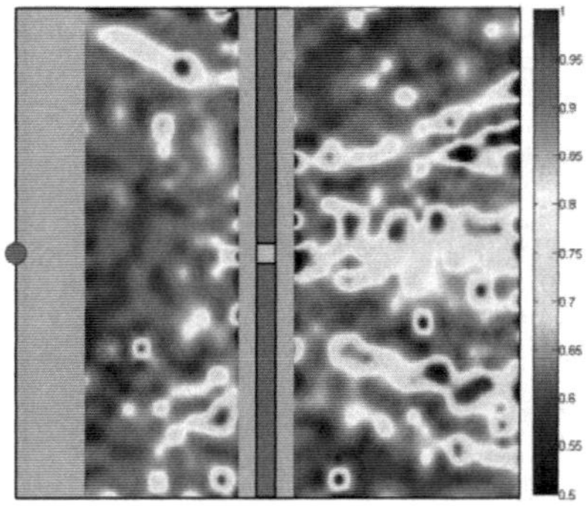

Figure 3: Cross correlation coefficient plot for out-of-plane modes for excitation site 3. The grey areas denote the regions with no measurement points.

3.4. Cross-Correlation Error

In this study there was a degree of error observed in the cross-correlation coefficient as only values below unity were observed. The cause of this was the fact that two separate stiffened panels were used giving rise to errors due to slight manufacturing differences, minor alignment errors between panels and, most likely, due to inconsistencies in the acoustic coupling of the excitation transducers on each panel.

To quantify the cross-correlation error, a time window of 80ms from the arrival of the wave at nearest measurement to each excitation site was taken so as to minimise any reflection effects. Samples of 5 x 5 measurement points nearest to excitation site were cross-correlated for each excitation study and the cross-correlation coefficients were averaged. Over the five sets of data the average cross-correlation coefficient was calculated to be 0.91 giving a 9% error.

3.5. Cost Function

In order to determine the optimal sensor locations for a sensor network that would detect the disbond regardless of excitation site, a statistical measure was needed in order to determine the fitness of the location across the five data sets – i.e. how well would the proposed sensor network allow the detection of damage for any of the five potential excitation locations?

Each sensor location had five cross-correlation coefficients (one for each excitation site). For each sensor network, the lowest cross-correlation coefficient was selected for each excitation site data set. These values were than used for the cost calculation as demonstrated for a two sensor network in Table 2.

Table 2: Cross-Correlation coefficient for a two sensor network

	Excitation Sites				
	1	2	3	4	5
Sensor A	0.8087	***0.6007***	0.7355	***0.6502***	0.8791
Sensor B	***0.6412***	0.9576	***0.6435***	0.8222	***0.5681***
Best Cross-Correlation Coefficient	0.6412	0.6007	0.6435	0.6502	0.5681

The cost was calculated by adding the standard deviation, σ, and the square root of the mean squared error of the five cross-correlation values as shown in Equation (1). The objecting of the GA was then to seek the minimum cost for a given sensor network.

$$cost = \sigma + \sqrt{\frac{1}{5}\sum_{i=1}^{5} x_i^2} \qquad \textbf{(1)}$$

Where σ is the standard deviation, n is the number of datasets and x is the cross – correlation coeffeicent

If sensors in the network were found to be redundant (i.e none of their cross-correlation coefficients were selected for calculating the cost) then a cost penalty was assigned to that sensor network. A schematic for calculating the cost is outlined in Figure 4.

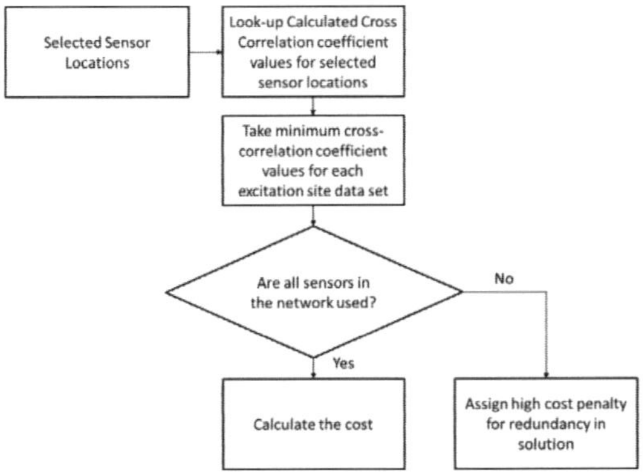

Figure 4: Schematic of the process for calculating the cost

4. Results

4.1. Genetic Algorithm performance

Binary encoding was used by the genetic algorithm in this experimental study. Each potential sensor location (i.e. locations 1-825) was described by a binary string (known as a 'gene'). The number of sensors in the in the network defined how many genes made up the full binary string (known as a 'chromosome'). Each chromosome described one candidate sensor network within the population of solutions. These chromosomes were used by the algorithm to mate and mutate.

As the total number of measurement points was 825, which requires a 10 digit binary string to resolve, artificially high cross-correlation penalties were assigned to binary values representing locations greater than 825 hence giving solutions containing these genes poor fitness. The initial population used was 40 x the number sensors in the network, N_s. The size of the initial population was increased as the size of the search space increased as to ensure sufficient sampling of the search space.

The selection method used for mating the chromosomes was a simple pairing technique where the best two solutions in the population are selected for mating, then the next best two until the whole mating pool has been paired up. A randomly assigned cross-over point was used for each mating pair. This gave the algorithm a good ability to carry forward the best attributes of the sol ution without continuously producing poor solutions which can be a drawback of fixed point cross-over schemes.

An allele mutation allowed the algorithm to explore more of the search space and hence prevented the algorithm from seeking local minima by continually exploring new regions of the solution space. A probability of mutation of 0.1 was used for this problem. The fittest 10% of solutions were made immune from allele mutation. This prevented good solutions being mutated into poorer solutions, and hence being discarded.

As with all genetic algorithms, convergence can prove to be an issue as it is not always clear that the optimal solution has been reached. This is a drawback of GAs as every possible solution is not exhaustively considered. In this study, a solution was deemed to be optimal when the minimum cost had not been improved upon for 2000 generations. This approach has previously be found to be successful by *Clarke* [10]. The performance of the GA is displayed in Figure 5.

4.2. Sensor locations

Having run the GA for different sizes of sensor network it is possible to make a direct comparison of the performance of the proposed solutions as shown in Figure 6.

The proposed solutions show a 22% reduction in cost as the number of sensors in the network increases from one to two sensors. Improved sensitivity can be achieved as more sensors are added to the network with the cost reducing by 50% with a five sensor network. The best network solution entirely depends on the requirements and constraints of the network. However these results suggest that a three sensor solution offers an acceptable compromise between performance and mass / expense penalty.

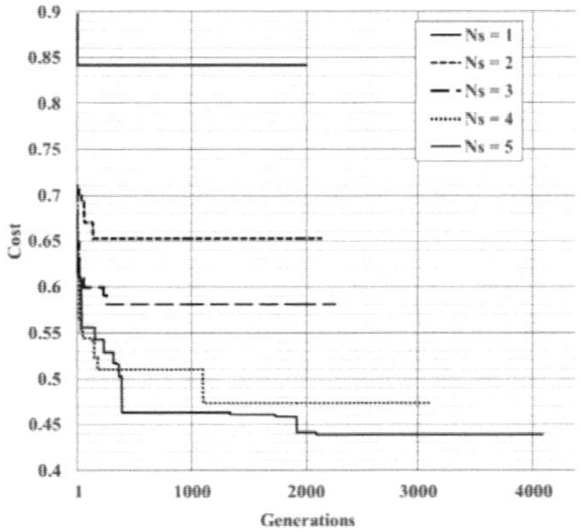

Figure 5: Convergence of the minimum cost for different values of N_s

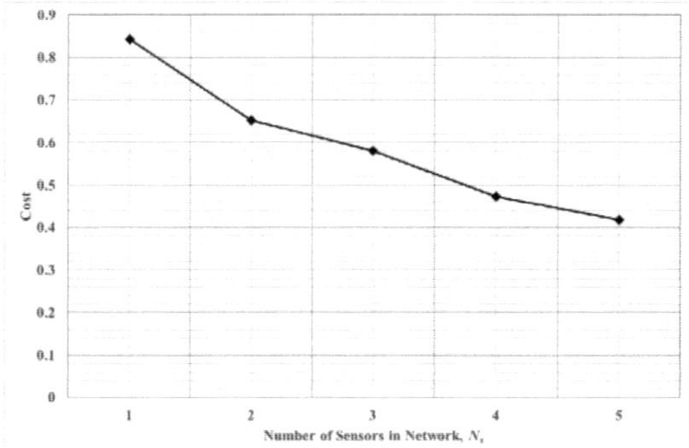

Figure 6: Calculated costs for each number of sensors in a sensor network

GAs are renowned for producing 'less-intuitive' solutions to problems. The locations of the sensors produced by the GA are shown in Figure 7. Some of the sensor locations are reused in more than one sensor network demonstrating that many of their cross-correlation coefficients are low.

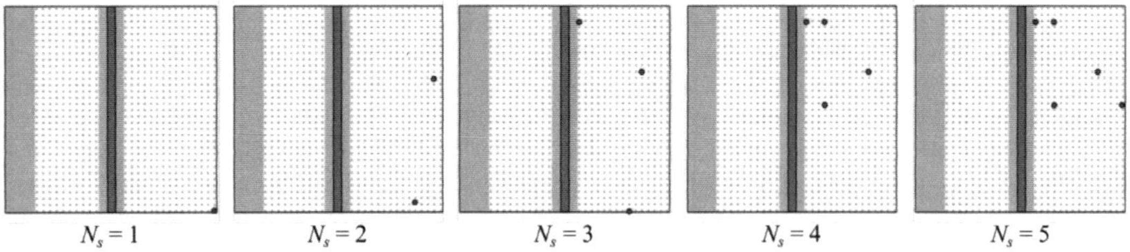

Figure 7: Optimal sensor locations for number of sensors, N_s, for five disbond locations

5. Discussion and Conclusions

The results presented demonstrate a methodology for optimising sensor locations with the use of scanning laser vibrometery. The ability to autonomously collect data from several potential locations dramatically decreases experimental time allowing a more thorough study to be conducted.

The locations that have been selected by the algorithm at first may not seem entirely intuitive such as selected location on the upper right on of the stiffener in the three sensor network. This particular location detected low correlation for one disbond location with the other sensor in the network detecting low correlation for the other disbond locations.

The work presented here is part of a more extensive study in which the effects of different excitation frequencies will be investigated together with the influence of in-plane wave modes on sensor locations.

Conclusions that can drawn be from this work are:

- Sensor location for active structural health monitoring is a complex problem and the positioning of the sensors is important in order to achieve the greatest likelihood of detecting the damage.
- Laser vibrometery is a powerful tool in producing data for a vast number of potential sensor locations.
- Genetic algorithms offer many advantages in producing solutions that offer acceptable fitness
- Though optimisation studies are of great importance, it is up to the sensor network designer to ultimately set the objective of what the network is trying to achieve. Hence, it is the responsibility of the designer to assess the best solution, using optimisation techniques such as this as a decision making aid rather than blindly relying on computational optimisation techniques.
- This study only considers centrally located disbonds. The study could be expanded to investigate disbonds at stiffener edges.

6. Acknowledgements

The authors would like to the support of EPSRC DTG together with Airbus UK Ltd. for the funding of this project as well as Dr Mark Eaton and Dr Mathew Pearson for their advice during this project.

References

1. Higgins, A., *Adhesive bonding of aircraft structures.* International Journal of Adhesion and Adhesives, 2000. **20**(5): p. 367-376.
2. Arenas, J.M., et al., *Considerations for the industrial application of structural adhesive joints in the aluminium-composite material bonding.* Composites Part B: Engineering, 2013. **44**(1): p. 417-423.
3. Petrie, E.M., *Adhesives for the assembly of aircraft structures and components: Decades of performance improvement, with the new applications of the horizon.* Metal Finishing, 2008. **106**(2): p. 26-31.
4. Bossi, R. and M.J. Piehl, *BONDING PRIMARY AIRCRAFT STRUCTURE: THE ISSUES.* Manufacturing Engineering, 2011. **146**(3): p. 101-102,104-109.
5. Davis, M. and D. Bond, *Principles and practices of adhesive bonded structural joints and repairs.* International journal of adhesion and adhesives, 1999. **19**(2): p. 91-105.
6. N Quaegebeur, P.M., P Masson, M Castaings, *Methodology for optimal configuration in structural health monitoring of composite bonded joints.* Smart Materials and Structures, 2012. **21**(10): p. 11.
7. Huntsman, *Aerospace Adhesives: Araldite 420 A/B Two component epoxy adhesive*. 2000.
8. Holland, J.H., *Adaptation in natural and artificial systems: An introductory analysis with applications to biology, control, and artificial intelligence.* 1975, Oxford, England: U Michigan Press. viii, 183.
9. Haupt, R.L. and S.E. Haupt, *Practical Genetic Algorithms*. 1998, Canada: John Wiley & Sons.
10. Clarke, A. and J.C. Miles, *Strategic Fire and Rescue Service decision making using evolutionary algorithms.* Advances in Engineering Software, 2012. **50**(1): p. 29-36.
11. Kammer, D.C. *Sensor placement for on-orbit modal identification and correlation of large space structures.* in *American Control Conference*. 1990. San Diego, CA, USA.
12. Worden, K. and A.P. Burrows, *Optimal sensor placement for fault detection.* Engineering Structures, 2001. **23**(8): p. 885-901.
13. Worden, K. and W.J. Staszewski, *Impact location and quantification on a composite panel using neural networks and a genetic algorithm.* Strain, 2000. **36**(2): p. 61-70.
14. Pullin, R., et al., *On the Development of a Damage Detection System using Macro-fibre Composite Sensors.* Journal of Physics: Conference Series, 2012. **382**(1): p. 012049.

Sensor Measurement Strategies for Monitoring Offshore Wind and Wave Energy Devices

Deirdre O'Donnell[1,2], Bruno Srbinovsky[2,3], Jimmy Murphy[2], Emanuel Popovici[3] and Vikram Pakrashi[1,2]

[1] Dynamical Systems and Risk Laboratory, Civil and Environmental Engineering, School of Engineering, University College Cork, Ireland
[2] Beaufort, Environmental Research Institute, University College Cork, Ireland
[3]Electrical and Electronics Engineering, School of Engineering, Universtity College Cork, Ireland

v.pakrashi@ucc.ie Ph: 00353214903862 Fax: 00353214276648

Abstract. While the potential of offshore wind and wave energy devices is well established and accepted, operations and maintenance issues are still not very well researched or understood. In this regard, scaled model testing has gained popularity over time for such devices at various technological readiness levels. The dynamic response of these devices are typically measured by different instruments during such scaled tests but agreed sensor choice, measurement and placement guidelines are still not in place. This paper compared the dynamic responses of some of these sensors from a scaled ocean wave testing to highlight the importance of sensor measurement strategies. The possibility of using multiple, cheaper sensors of seemingly inferior performance as opposed to the deployment of a small number of expensive and accurate sensors are also explored. An energy aware adaptive sampling theory is applied to highlight the possibility of more efficient computing when large volumes of data are available from the tested structures. Efficient sensor measurement strategies are expected to have a positive impact on the development of an device at different technological readiness levels while it is expected to be helpful in reducing operation and maintenance costs if such an approach is considered for the devices when they are in operation.

1. Introduction

Offshore wind and wave energy technologies have seen major advances in recent years. Wave energy in particular is growing in popularity [1,2]. Operations and maintenance (O&M) costs are a highly relevant factor in the overall financial assessment of such projects, all the more so in offshore projects due to lower availability of the device [3]. This has pushed the need for reliable structural monitoring systems for accurate and reliable information about the health of these energy conversion devices. With a move in recent times towards offshore energy solutions, loss in ease of accessibility may lead to damage going undetected, and the increased risk of catastrophic failure [4].

There is clear financial benefit to optimizing time between inspections and scheduled maintenance work, which affects the uptime of systems while also coming with their own costs- an unscheduled maintenance event is five times more costly than one that is scheduled [5]. However, high costs related to some sensing systems outweigh the benefits to O&M cost savings so the value of expensive sensing systems must be evaluated, usually at the scaled testing stage. There are many forms of sensing systems, based on various technologies. Accelerometers have been successfully applied to identifying and locating the presence of structural damage in offshore structures [6], as well as motion cameras and load cells [7,8] and Fiber Bragg Grating (FBG) to measure strain. Cameras can even be employed in underwater situations to detect damage [9] where marine growth exasperates fatigue damage. However, little is known of the relative merits of these technologies.

Wireless sensor networks (WSN) are a promising technology which have in recent years gained much attention from academia and industry alike. The application of WSN technology to structural health monitoring (SHM) has the potential to provide a substantial and quantifiable improvement to existing monitoring solutions for civil infrastructure [10].While wired SHM systems would require more maintenance and more frequent site visits as wires can be damaged over time, wireless SHM systems offer flexibility, even on difficult to access structures, and significantly reduced costs of installation and maintenance. However, some of the existing wireless systems for SHM still have high power consumption. The high power consumption and the limited power budget make these systems unsuitable for long-term installation on a structure and require frequent site visits for system maintenance. WSN nodes are battery powered and because of their limited energy source they are not suitable for long-term structural health monitoring applications. With the focus on enhancing the life time of a wireless sensor node, a popular is by complementing an energy harvesting technique with an efficient energy management algorithm [11]. This approach has the potential to achieve self-sustainability of the node with harvesting energy from the environment and effectively managing the node activity (i.e. the sampling rate of the sensors) according to the energy levels and the dynamics of the phenomenon observed.

While possibilities related to the deployment of sensors remain relevant and topical, there is little experimental information present from wave basin testing that compares the measured data from different sensors and highlights different issues related to the choice, positioning and sampling aspects of such sensors. This paper attempts to address this gap by reporting a wave basin testing for a Tension Leg Platform (TLP) where a number of different sensors where deployed to measure and monitor the dynamic responses of these structures.

2. Experimental Model

2.1. Scaled model

A scaled Tension Leg Platform (TLP), a truss like structure with a hexagonal base, was tested in this study. This device consists of a gravity base connected by six mooring tethers to the Buoyancy Ring and the Upper Structure and the Tower and Nacelle. The mooring lines are designed to reduce heave motion and increase horizontal stability, but the lines become a weak point against surge motions. A photograph of the tested device is presented as Figure 1.

2.2. Instrumentation and testing

The model was instrumented with 6 Tedea-Huntleigh stainless steel single ended bending beam load cells which were attached to the six mooring line cables and bolted to the gravity base. These measured the cable tension in Newtons (N). The instantaneous positions of 3 reflective markers, which were attached to the six corners of the hexagonal base, were monitored by 4 Qualisys 3-Series Oqus Marker Tracking Cameras with a sampling frequency of 32Hz. A Laser Doppler Vibrometer (LDV) was also employed during testing to record the velocity of the TLP. This high resolution technology samples at a rate of 480 Hz. Displacements and velocities were recorded in the wave direction, as this was considered the most critical plane. The model was tested at the Hydraulics and Maritime Research Centre (HMRC), University College Cork (UCC), Ireland in its Ocean Wave Basin. A variety of periods and wave amplitudes were used and the Bretschneider wave spectrum was chosen, to represent a more realistic sea state.

Figure 1. Photograph of the Tension Leg Platform tested in the ocean wave basin.

3. Results

3.1. Dynamic Displacement Response

The camera recorded the position of the TLP at 3 different locations; the Inner Ring, the Outer Ring and the Middle Mast. The dynamic response was recorded in 3 perpendicular directions but only the surge motion of the device has been used in this paper since this was the motion that was most relevant. The velocity of the structure as recorded by the LDV was used to find displacement values and consequently, the displacement obtained by the LDV is a derived quantity, while that measured by the camera is not. Figure 2(a) shows the peak displacements recorded by the camera at the 3 tracked positions. Due to the far larger amplitude of displacement at the mid mast position, due to the flexible nature of the mast and its sensitivity, these readings were omitted from the average value shown in Figure 2 (c), as they were viewed to be skewing the data (see Figure 2 (b)).

The observations indicate that there is an uncertainty and mismatch in the measurement from camera and from LDV. These mismatches or variability in results are dependent on the wave periods apart from the wave heights, which in turn guide the levels of displacement. Consequently, more than one source of measurement can be appropriate for such measurements to ensure that there is an agreement of the results. However, it also leads to the debate on the necessity of a very highly sophisticated or calibrated instrument, when a large number of relatively lower quality instruments are measuring data in a higher number of locations. When no benchmark is available, at least some part of data matching with a high quality instrument is recommended, though based on current observations and the investigations carried out next, deploying a large number of inexpensive sensors can then be very beneficial.

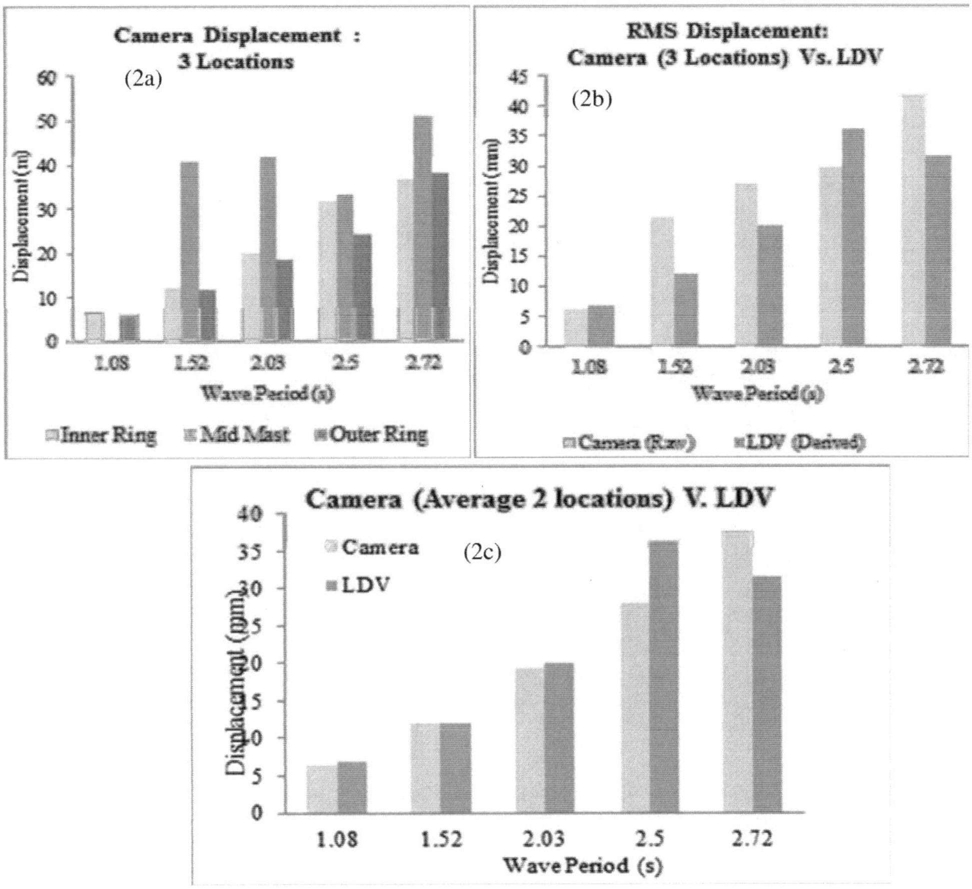

Figure 2. Variation in measured dynamic displacement responses using different devices.

3.2 Variation of Peak Frequency in Measured Responses

The displacement time series for the LDV and the motion camera were converted into the frequency domain using the Fast Fourier Transform (FFT) and the dominant peaks were picked.

In Figure 3, the response of frequency of the output for the two different instruments is compared to the known frequency of the wave input to the system. The peak frequency of the velocity output of the LDV is, on average, 18.7% lower than the wave frequency of each particular test. However, the peak frequency of the camera's displacement is an average of 31.9% higher than the same inputs. The same comparison, but for the frequency of the LDV's velocity output yielding a difference of only 7%, on average, from the wave input.

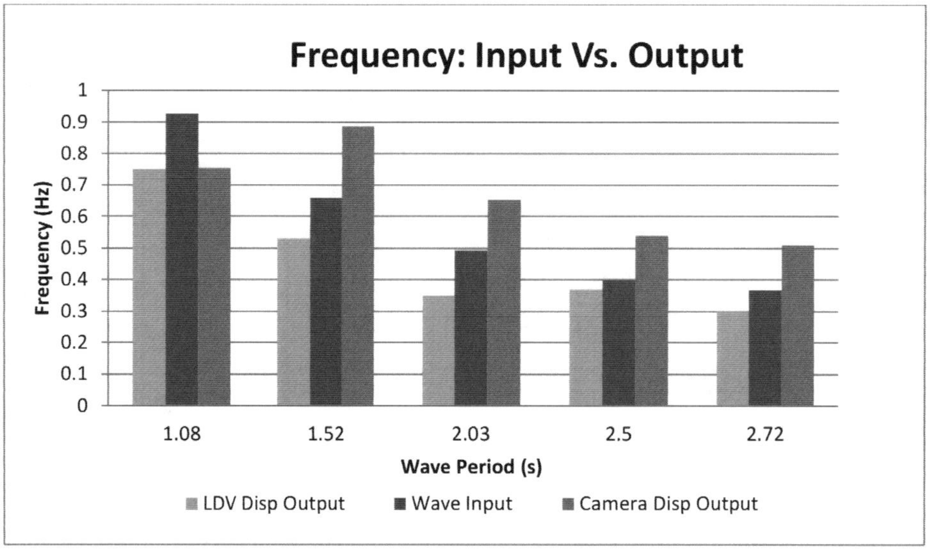

Figure 3 Frequency comparison of input wave to frequency of LDV and camera displacement outputs

3.4 Variation in Load Cell Data

Load cells were attached to tendons at Bow Port, Bow Starboard, Mid Starboard, Stern Starboard, Stern Port, Mid Port and were labelled White, Red, Yellow, Green, Brown and Blue respectively. The average peak and RMS load values of tension in the tendons for each load cell for 20 different tests are represented in Figure 4.

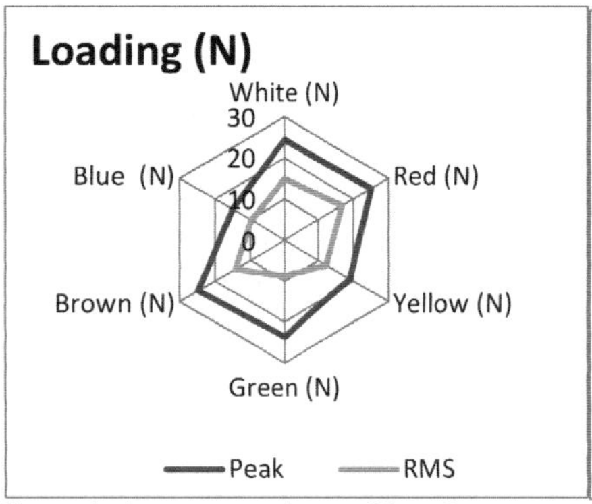

Figure 4. Average tension at 6 locations

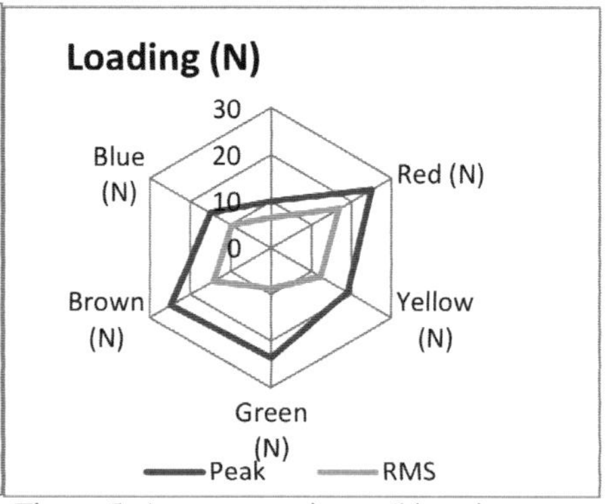

Figure 5. Average tension at 5 locations

The values were recorded in the direction of the wave, at the bow and at the stern of the structure. As expected, the tendons at the front and back of the structure, as the structre moves in the wave condition, are under the most tension. Analysis of the effect of removing different load cells to the overall data was carried out, a sample of which is shown in Figure 5. Data obtained from the white cell at Bow Port was removed from the global analysis, and the estimate shown from remaining data shows a loss in accuracy of the loading on the structure.

Such an accuracy loss may relate to one of the sensors becoming defunct during its operational regime of monitoring and how such events affect the estimated averages of loading on the structures tendons. The four load cells attached to the bow tendons and the stern tendons are shown to record the highest tension values, so should be the focus of analysis.

4 Energy Aware Adaptive Sampling Algorithm for Energy Harvesting WSNs

The development of WSN technology is hindered by their limited energy supply. In the case of monitoring applications, sensors can be expensive with respect to energy requirements. It is desirable to develop protocols that effectively manage the sensor power consumption while still meeting the requirements of the application. Adaptive sampling algorithms (ASA) are often used as a tool to minimize the communication between the sensor nodes within the network and at the same time to minimize the power consumed by the sensors by reducing the sampling rate according to the needs of the phenomenon observed. An ASA presented in [12] was implemented and evaluated using data collected with sensor for displacement response recorded by the motion cameras.

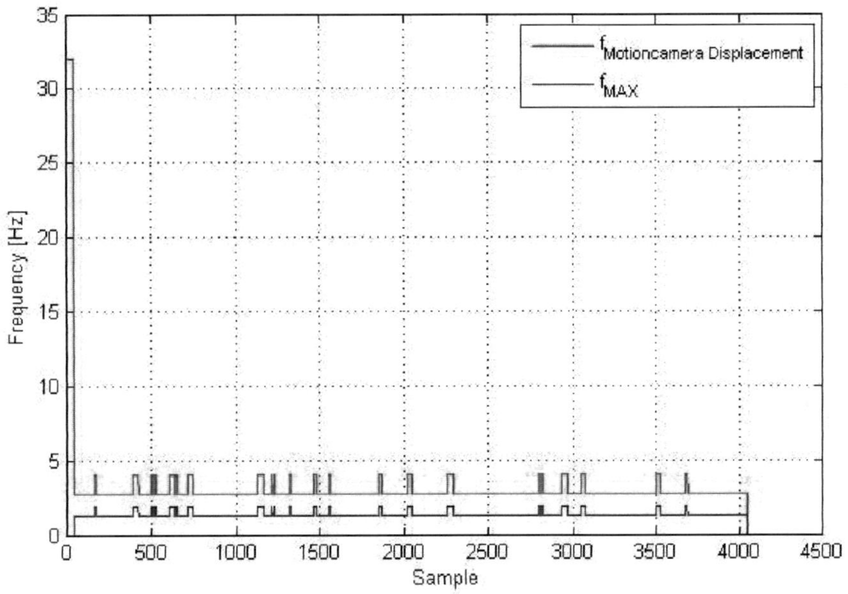

Figure 6 ASA Implementation of Camera Displacement Data

The algorithm used evaluates the maximum frequency of the signal using FFT and then decides the sampling frequency by multiplying the maximum frequency with a constant which is ≥2 satisfying the Nyquist criterion. A detailed description of the implemented algorithm with all relevant parameters explained can be found in [12].

In Figure 6, the sampling frequency according to the ASA and the maximum frequency of the signal are presented. The following values were chosen for the relevant parameters: $c = 2.1$, $h = 5$, $W = 50$, $\delta = 0.1\%$. Details for each of these parameters are explicitly given in [12]. The time between successive frames was 0.3125, thus the starting sampling frequency was 32Hz. ASA reduces the number of acquired samples with respect to the traditional fixed sampling rate approach. The approach highlights the effectiveness of using efficient algorithms to handle large volumes of monitoring data.

5 Conclusions

This paper highlights the importance of considering sensor choice, placement and measurement for monitoring offshore renewable energy devices at a scaled testing stage. A comparison was made between high quality LDV data and lower quality motion camera data which recorded 3 different locations on the structure. Additionally, load cells were used to assess the wave load that was generated for each test. It can be observed from the test that there should be an awareness of the variation of measured or estimated values of dynamic responses in the time and the frequency domain based on what instrument is being used, its location and whether the response is directly measured or derived. Additionally, there seems to value in addressing the issue of whether a large number of relatively inexpensive sensors can provide similar or better information regarding the dynamic response as opposed to a single or a small number of expensive sensors. Malfunctioning of sensors appear to have a significant effect on the estimated average dynamic responses and this leads to the importance of investigating and quantifying the sensor redundancy that is required to offset such issues, including strategies related to their placement to minimize such redundancy. More studies quantifying these issues are expected to improve wave basin testing at different technological readiness levels [13] and possible intervention schemes [14] for controlling dynamic responses. Efficient and optimized sampling of measured data is observed to be beneficial in terms of energy demand of the sensors network. In applications where a battery powered system is used to interface a power hungry sensor, reducing the sampling rate when possible will extend the life of the battery while still maintaining the application data requirements. Dynamically changing the sampling frequency according to the needs of the phenomenon under observation can also improve the data quality.

Acknowledgements. The authors acknowledge Marine Research Energy Ireland (MaREI), grant no. 12/RC/2302, a Science Foundation Ireland (SFI) supported project

References

[1] Falcão AFDO 2010 Wave energy utilization: A review of the technologies. *Renewable and Sustainable Energy Reviews* **14**(3) 899–918

[2] Clément et al. 2002 Wave energy in Europe : current status and perspectives *Renewable and Sustainable Energy Reviews* **6**(5) 405–431

[3] O'Connor M, Lewis T and Dalton G 2013 Operational expenditure costs for wave energy projects and impacts on financial returns *Renewable Energy* **50** 1119–1131

[4] Swartz RA, Lynch JP, Zerbst S, Sweetman B and Rolfes R 2010 Structural Monitoring of Wind Turbines using Wireless Sensor Networks *Smart Structures and Systems* **6**(3) 1-14.

[5] Adams D, White J, Rumsey M and Farrar C 2011 Structural health monitoring of wind turbines : method and application to a HAWT *Wind Energy* **14**(4) 603–623

[6] Mangal L, Idiichandy VG and Ganapathy C 2001 Structural monitoring of offshore platforms using impulse and relaxation response, *Ocean* Engineering, **28**(6) 689–705

[7] Jaksic V, O'Shea R, Cahill P, Murphy J, Mandic DP and Pakrashi V 2015 Dynamic Response Signatures of a Scaled Model Platform for Floating Wind Turbines in an Ocean Wave Basin. *Philosophical Transactions of the Royal Society A: Mathematical, Physical and Engineering Sciences* **373**(2035) 20140078:1-20140078:18

[8] Murphy J, O'Shea R, O'Sullivan K and Pakrashi V 2013 Dynamic Responses of a Scaled Tension Leg Platform, Wind Turbine Support Structure in a Wave Tank, Key Engineering Materials **569-570**, 563-570,

[9] O'Byrne M, Schoefs F, Pakrashi V and Ghosh B 2014 A Comparison of Image based 3D Recovery Methods for Underwater Inspections.7th European Workshop on Structural Health Monitoring EWSHM 2014, Nantes, France.

[10] Pakzad SN, Fenves GL, Kim S and Culler DE 2008 Design and Implementation of Scalable Wireless Sensor Network for Structural Monitoring *ASCE Journal of Infrastructure Systems.*

14(1) 89-101

[11] Sharma V, Mukherji U, Joseph V and Gupta S 2010 Optimal energy management policies for energy harvesting sensor nodes *Wireless Communications, IEEE Trans* **9**(4) 1326-1336

[12] Alippi C, Anastasi G, Di Francesco M. and Roveri M 2010 An Adaptive Sampling Algorithm for Effective Energy Management in Wireless Sensor Networks with Energy-Hungry Sensors. *Instrumentation and Measurement, IEEE Trans.***59**(2) 335-344.

[13] Holmes B 2009 *Tank Testing of Wave Energy Conversion Systems: Marine Renewable Energy Guides* The European Marine Energy Centre Ltd, UK

[14] Jaksic V, Wright CS, Murphy J, Afeef C, Ali SF and Pakrashi V 2015 Dynamic response mitigation of floating wind turbine platforms using tuned liquid column dampers *Philosophical Transactions of the Royal Society A: Mathematical, Physical and Engineering Sciences* **373**(2035) 20140079:1-20140079:9.

Novel FBG rosette for determining impact location in thin plate-like structure

K Majewska[♦,1], **S Opoka**[1], **P Kudela**[1], **W Ostachowicz**[1,2]
[1]Institute of Fluid Flow Machinery of PASci, Fiszera 14, 80-952 Gdansk, Poland
[2]Warsaw University of Technology, Narbutta 84, 02-524 Warsaw, Poland

[♦]E-mail: k.majewska@imp.gda.pl

Abstract. The paper consists of two interdependent parts. The first part presents numerical simulations of output response of single Fibre Bragg Grating (FBG) sensor which is driven by homogeneous deformation along sensor length. The example of such sensor is FBG sensor glued at only two points. The grating length and modulation depth of the refractive index are the critical parameters contributing to performance of the FBG sensors. Numerical analysis allowed to select an appropriate FBG sensors which will be used in the impact detection problem. In the second part of the paper a novel strain rosette with specific sensor array used in impact localisation problem is discussed and presented. The experiment was carried out on thin composite plate with the use of pulse force excitation. The method is based on estimation of principal direction of perturbed travelling wave initiated at impact point and does not use any information about wave propagation velocity.

Introduction

The ability to locate damage, impact or any other discontinuities in structural materials such as plates, rods, pipes, etc is one of the objectives of structural health monitoring (SHM) systems. The SHM systems are based on various types of sensors and their configurations i.e. PZT and fibre optic sensors or a combination of them. In addition non-contact measurement techniques like laser ultrasound, optical coherence tomography and infrared thermography could be employed. This paper is focused on the application of Fibre Bragg Grating (FBG) sensors for impact localization.

The basic principles and characteristics of FBG sensors with respect to different distribution of parameters along the grating length or under uniform and non-uniform strain fields have been analyzed in: [14], [15], [19], [21], [23]. The outcome of these works is used here to specify appropriate parameters of the FBG sensor (especially minimal grating length) in order to obtain good spectral content of the light reflected from the grating.

The technical progress in the construction engineering requires more advanced materials. The desire to reduce the weight of structure and reduce their manufacturing costs may affect the safety and reliability of designed structure. On the other hand, the structural damage can occur due to many factors which are difficult to predict in advance, e.g.: sudden impact loads. A number of researchers propose the use of FBG sensors for detecting impact loads. For this purpose two methods are used. In the first method impact location is determined by measuring the differences in time-of-flight of

Lamb waves [7], [11]. In the second method, also exploited in this paper, amplitudes of dynamic strain signals are analysed in order to find direction of the impact event [2], [3].

The paper is organised as follows: in Section 2, the parameters describing the grating influence on the reflection spectrum of FBG sensor are analyzed, in Section 3 the strain rosette bonded on the composite plate is used to calculate the angular direction of the impact. Finally, in Section 4 the conclusions are indicated.

The selection of appropriate FBG sensors

1.1. Fibre Bragg Grating sensor

The FBG sensor is an optical device with a permanent periodic modulation of the refractive index in the core of a single-mode fibre optic. The grating pattern is made by exposing the core of the optical fibre to an interference pattern of intense UV laser light [6], [15]. The change in the core refraction index is between10^{-5} and 10^{-2} [6], [13]. The periodic perturbation in the core refractive index allows coherent scattering to occur for a narrow wavelength band of incident light travelling within the fibre core. A strong narrowband back reflection of light is generated, centred around the maximum reflecting wavelength λ_B when the resonance condition, or phase match, is satisfied:

$$\lambda_B = n_{eff}\Lambda \tag{1}$$

where n_{eff} is the effective refractive index and Λ is the grating period [20]. The scheme of an FBG sensor is shown in Figure 1.

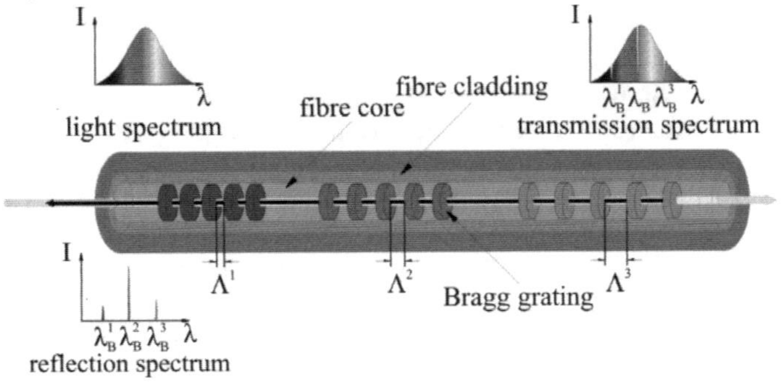

Figure 1. A scheme of a Fibre Bragg Grating sensor idea.

The maximum reflecting wavelength value λ_B is dependent on the geometrical and physical properties of both the grating and the optical fibre. Longitudinal strains ε act on the periodicity of the perturbation Δ (photoelastic effect), while temperature affects both the effective refractive index n_{eff} (thermo-optic effect) and the periodicity of the perturbation Λ (thermoelastic effect). The key point of FBG sensors is their wavelength-encoded nature, which is the absolute parameter providing repeatable measurements [20]. The FBG sensors have many advantages in comparison with electric strain sensors and are promising sensing alternatives for SHM systems. The FBG sensors have small size and weight, multiplexing capabilities, immunity to electromagnetic fields, high corrosion resistance and require no calibration [6], [10], [13]. They can also be mounted onto a structure [1] or can be embedded [3], [18] into the material of a structural element during the manufacturing process.

1.2. The basic equations for numerical simulations of FBG reflection spectrum

There are many parameters which affect the reflection spectrum of the FBG sensor, e.g. the effective refractive index, fibre bending, grating, a temperature, or sensor length [9], [10], [16], [21].

The characteristics of the response from Bragg grating are fully described by [19], [21]: the central Bragg wavelength λ_B, maximal peak reflectivity R_{max} of grating which occur at λ_B, the grating length L, the refractive index of the core of optical fibre n_0 and modulation depth of refractive index Δn_{eff}.

Using the coupled-mode theory analytical description of the reflection properties of Bragg gratings may be obtained. The reflection spectrum of a grating R is given by the following expression:

$$R(L,\lambda) = \frac{\kappa^2 \sinh(\gamma L)^2}{\Delta\beta^2 \sinh(\gamma L)^2 + \gamma^2 \cosh(\gamma L)^2} \qquad (2)$$

where $\kappa = \frac{\pi \Delta n_{eff}}{\lambda}$, $\Delta\beta = \beta - \frac{\pi}{\Lambda}$, $\beta = \frac{2\pi n_0}{\lambda}$, $\gamma = \sqrt{\kappa^2 - \Delta\beta^2}$, λ is wavelength of the reflected light from grating of the sensor and Λ is a grating period.

For light at the central Bragg wavelength, λ_B, there is no wave vector detuning and so $\Delta\beta = 0$. Then, the reflectivity function becomes

$$R(L,\lambda) = \tanh(\kappa L)^2 \qquad (3)$$

A number of numerical simulations shown below have been performed using equation (2) with tacit assumption of no amplitude losses of the signal in the fiber and the sensor is driven by homogeneous deformation along its length. The FBG sensor glued at only two points experiences such deformation.

1.2.1. The influence of the modulation depth of the refractive index and the grating length on the sensor spectrum

The parameters used for the simulation were: n_0=1.47, λ_B=1550 nm, Λ=527.21 nm. The four values for Δn_{eff} = 10^{-5}, 10^{-4}, 10^{-3}, 10^{-2} were considered. The values of R_{max} for grating length varying from L=0.25 mm to 40 mm is shown in Figure 2a. The FBG sensor must provide sufficient reflectivity for most possible variations of Δn_{eff} because the exact values for Δn_{eff} are not known for real sensors (manufactures do not provide such data). The longer the sensor is the higher value of R_{max} is up to saturation level. The results show that acceptable minimal length of the Bragg grating is about 5 mm provided Δn_{eff} is not less than 10^{-4}. For more stable results the length of FGB sensor was chosen to be 10 mm for further experimental investigations. The shapes of reflected spectrum for grating length L=10 mm for different Δn_{eff} are shown in Figure 2b. It is clearly visible that R_{max} increases with increasing Δn_{eff}. The smaller Δn_{eff}, the worse band-pass filter properties of the sensor.

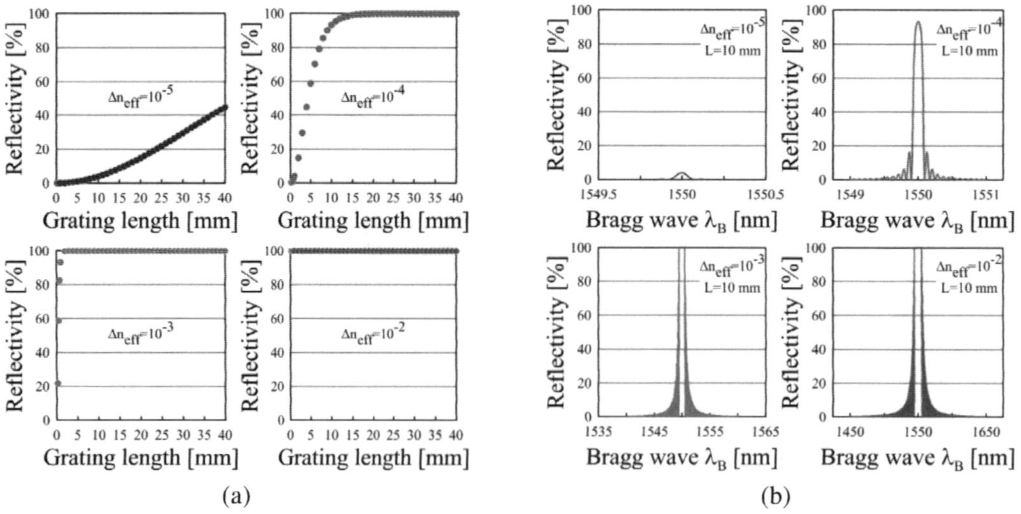

Figure 2. The influence of the modulation depth of the refractive index Δn_{eff} on sensor characteristics: (a) reflectivity R as a function of L for λ_B=1550 nm, (b) reflectivity R as a function of λ_B for L=10 mm.

1.2.2. The influence of the refractive index of the fibre core on the sensor spectrum

The parameters used for this simulation were: λ_B=1550 nm, Δn_{eff} =10^{-4}. The three values for n_0= 1.4, 1.45, 1.5 have been considered which give Λ=553.57 nm, 534.48 nm, 516.67 nm, respectively. The grating length has been varied from L=0.25 mm to 40 mm.

Analysing Figure 3a it is clearly visible that a small change of n_0 has no influence on the maximum peak reflectivity R_{max}. Only the width of the spectrum changes slightly, but this has no practical implications. In Figure 3b the influence of modulation depth of the refractive index Δn_{eff} on the reflection spectrum for particular grating length (L=10mm) is presented.

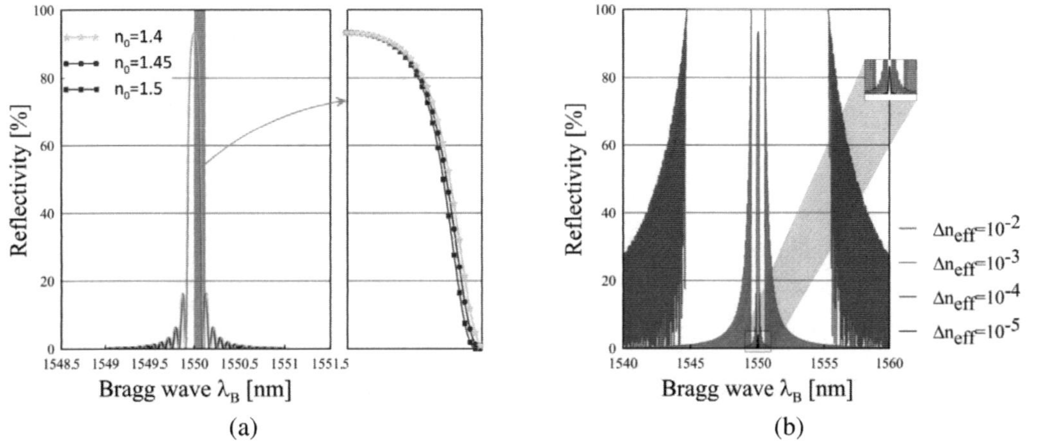

Figure 3. The influence of refractive index of the fibre core n_0: (a) reflection spectrum R as a function of λ_B for L=10 mm, (b) influence of Δn_{eff} on FWHM reflectivity peak for grating length L=10 mm.

Summarizing this part of the ideal case study the following conclusions can be highlited:

- The reflection spectrum R strongly depends on Δn_{eff} and L;
- The rate of R_{max} is strongly dependent on the value of Δn_{eff};
- The changes in n_0 have no influence on R_{max} and slighty affects the full width at half maximum (FWHM) of the sensor;
- For experimental application modulation depth of refractive index Δn_{eff} should be not smaller than 10^{-4} and the grating length should be sufficiently long in order to obtain R_{max} at least at the level of 50%. Moreover, the grating with saturated reflection spectrum should be also avoided.

Novel FGB rosettes for the impact direction

1.3. Angular localization of the impact using rosette's principal direction

Strain and the corresponding stress state for elastic solid body are described by symmetric, three-dimensional second-order tensors [22]. If the components of these tensors are independent of one of the Cartesian coordinates (usually it is z coordinate) the general three-dimensional description can be reduced to two-dimensional counterparts: plane strain state or plane stress state. Plane strain state can be described by two-dimensional symmetric second-order strain tensor $\boldsymbol{\varepsilon}$ whereas plane stress state is defined by two-dimensional symmetric second-order stress tensor $\boldsymbol{\sigma}$. The Cartesian strain tensor $\boldsymbol{\varepsilon}$ calculated from rosette measurements are then used to define line of perturbed wave propagation from the impact location. For full impact localization at least two rosettes are needed to determine two principal directions of perturbed waves coming from the impact location. The intersection of the lines determined by principal directions determines locus of mechanical impact, see Figure 4. The idea of the method comes from [6] where Macro Fibre Composite (MFC) rosettes bonded to metal plate were used for the same purpose. The formulas used in the rosette analysis have been derived using infinitesimal strain theory. The assumptions of this theory are fulfilled if the induced strain perturbation is measured at some distance from the impact location.

For general two-dimensional strain state at the surface of the structural element, at least three strain sensors are needed to calculate principal strains and directions which represent the simplest description of that state. For the purpose of the paper only the principal direction must be calculated. The principal direction is calculated as follows:

- The geometrical configuration of the sensors in rosette must be chosen: the angles β, γ must be determined, see Figure 4 (centre);

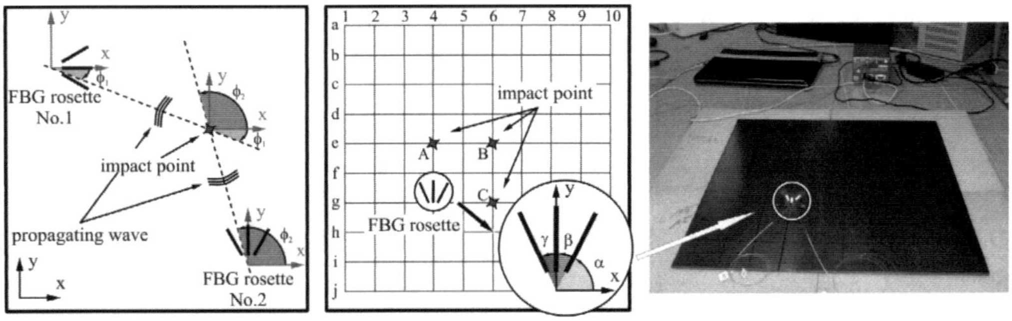

Figure 4. Two FBG sensor rosettes for the impact point localization in thin plate-like structures (left). The schema with zoomed FBG rosette and marked impact points A, B, C (centre) and the investigated CFRP plate (right).

- The plane cartesian coordinate system xy need to be oriented with respect to chosen sensor from the rosette: the angle α must be determined;
- The measurement strains ε_a, ε_b, ε_c from the sensors in rosette configuration must be converted to components ε_x, ε_y, ε_{xy} of strain tensor $\boldsymbol{\varepsilon}$ defined in cartesian coordinate system Oxy. This step can be described by the solution of the corresponding system of equations:

$$\varepsilon_a = \frac{1}{2}(\varepsilon_x - \varepsilon_y)\cos(2\alpha) + \frac{1}{2}(\varepsilon_x + \varepsilon_y) + \varepsilon_{xy}\sin(2\alpha)$$

$$\varepsilon_b = \frac{1}{2}(\varepsilon_x - \varepsilon_y)\cos[2(\alpha+\beta)] + \frac{1}{2}(\varepsilon_x + \varepsilon_y) + \varepsilon_{xy}\sin[2(\alpha+\beta)]$$

$$\varepsilon_c = \frac{1}{2}(\varepsilon_x - \varepsilon_y)\cos[2(\alpha+\beta+\gamma)] + \frac{1}{2}(\varepsilon_x + \varepsilon_y) + \varepsilon_{xy}\sin[2(\alpha+\beta+\gamma)]$$

- The rotation angle ϕ which diagonalizes tensor $\mathbf{R}^T\boldsymbol{\varepsilon}\mathbf{R}$ using definition of rotation matrix $\mathbf{R}$ determines the first principal direction associated with larger principal strain. The second angle is equal to ϕ+90° and is associated with the second principal strain. This step can be described by the solution of the corresponding equation:

$$(\varepsilon_y - \varepsilon_x)\,sin(2\phi) + 2\varepsilon_{xy}\,cos(2\phi) = 0$$

The angle ϕ is positive if it increases in counterclockwise direction from x axis. This quantity will be used for angular localization of the impact.

1.4. The experiment for determination of the direction of travelling wave initiated at impact point

The experimental set up consisted of: Carbon Fibre Reinforced Polymer (CFRP) plate (toughened epoxy matrix with carbon fibre Plain 3K fabric), the three FBG sensors, an interrogator and a laptop. The CFRP plate with a geometry as follow: 500 mm x 500 mm x 5 mm was considered. Three FBG sensors (grating length L=10 mm), arranged as it is show in Figure 4 were glued to the surface using two-point technique. The FBG rosette was placed in the point marked 4g on the plate (see Figure 4). The particular angles in rosette were: α=63.1°, β=26.6°, γ=25.5°. The distance between point 4e (point A in Figure 4) and 4g (FBG rosette) was the same as between 4e (point A) and 6e (point B), and the same as 6e (point B) and 6g (point C) and was equal 100 mm. In points marked A, B, C the hammer impacts were executed.

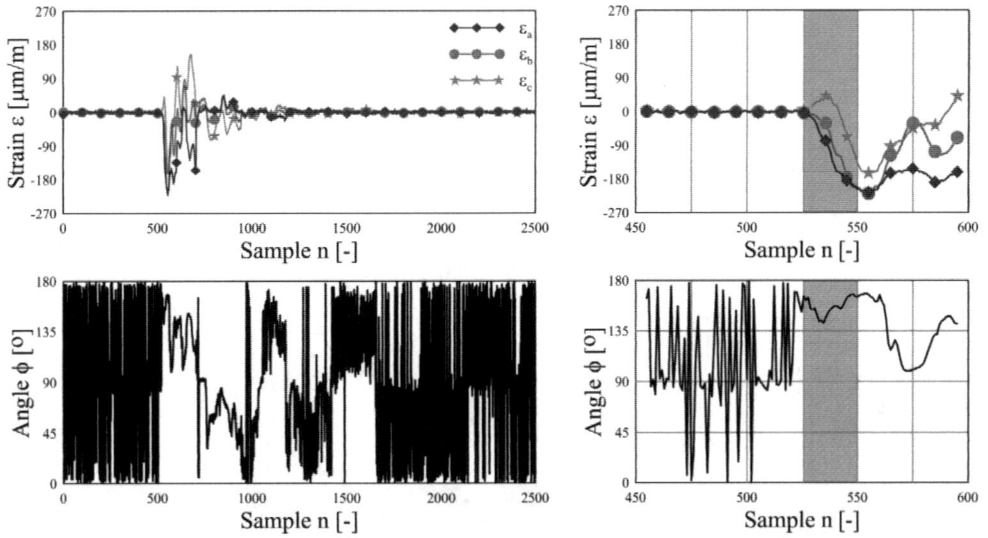

Figure 5. Rosette strain measurements (upper row) and changes in principal direction (bottom row) for impact from 0° direction (impact in point C). Shaded areas in the right column show changes caused by impact.

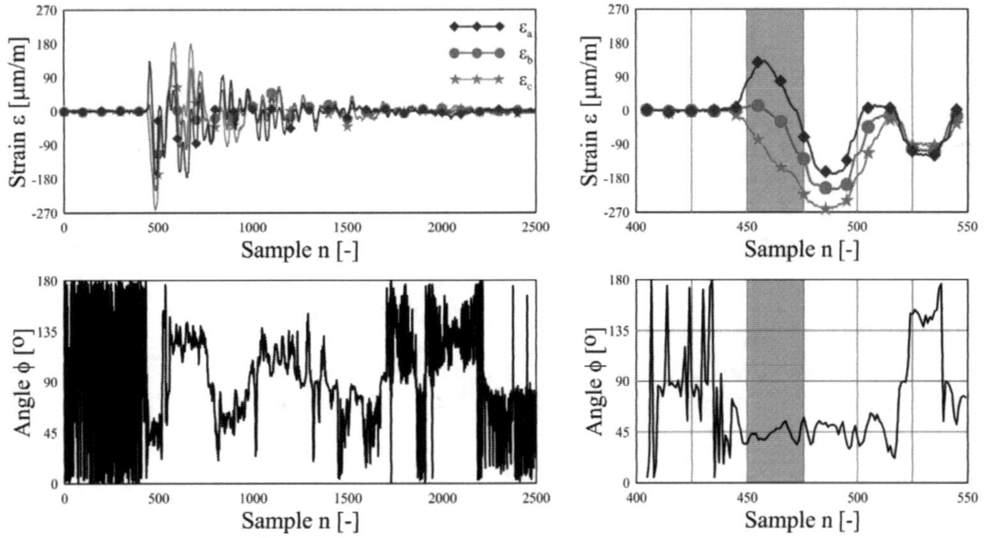

Figure 6. Rosette strain measurements (upper row) and changes in principal direction (bottom row) for impact from 45° direction (impact in point B). Shaded areas in the right column show changes caused by impact.

Rosette strain measurements and calculated principal direction from rosette analysis before, during and after hammer impact executed in three locations are shown in the Figures 5-7. The common feature of principal direction curves for these three impact cases is their stabilization after hammer impact for some time contrary to the regions before impact and after the decay of strain signals. For impacts executed in A and B the line of propagating disturbance is roughly parallel to the sensors and the proposed procedure for angular localization of the impact works very good. For case C (impact from 0° direction) the calculated angle was between 170° (-10°) and 140° (-40°). The less effective procedure for this impact direction can be partially explained by angular measurement sensitivity of FBG sensors with respect to the incoming wave which is worst when the wave is propagating in perpendicular direction to the sensor.

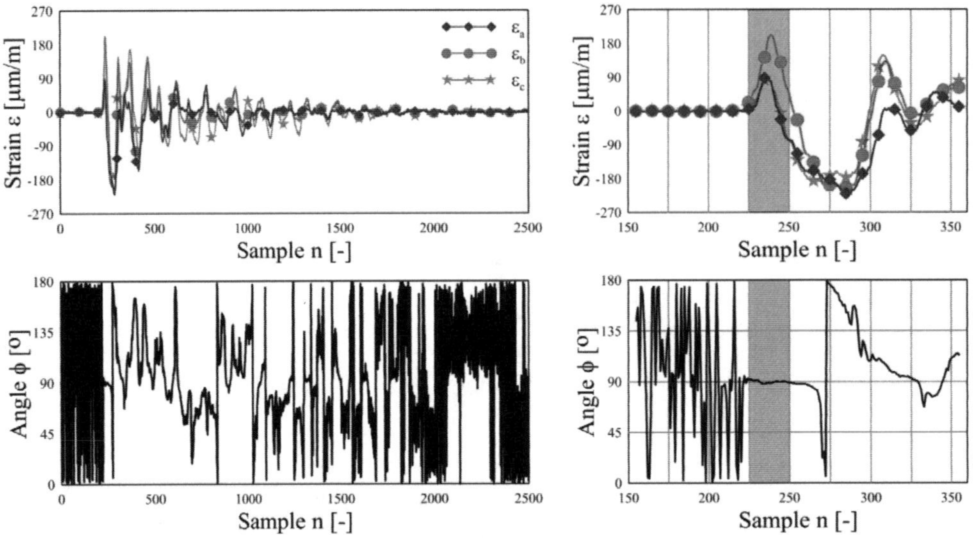

Figure 7. Rosette strain measurements (upper row) and changes in principal direction (bottom row) for impact from 90° direction (impact in point A). Shaded areas in the right column show changes caused by impact.

Discussion and conclusion

The novel FGB rosette for the impact direction as well as the proposed method was tested in the experiment.

The reflection spectrum of the FBG sensor strongly depends on Δn_{eff} and L. For experimental applications modulation depth of refractive index Δn_{eff} should not be smaller than 10^{-4} and the grating length should be sufficiently long in order to obtain maximal peak reflectivity at least at the level of 50%.

Angular localization case of the impact on composite plate by the use of the notion of principal direction from rosette analysis presented here shows very good results for two impact locations, namely for those with the line of propagating disturbance more or less parallel to sensors. The greater error in calculating line of propagating disturbance is obtained for perpendicularly oriented sensors but even in this case the result is acceptable.

Acknowledgments
The authors would like to gratefully acknowledge the support for this research provided by the National Centre for Research and Development under the Applied Research Programme via KOMPNDT Project (No. PBS1/B6/8/2012, Non-invasive Methods for Assessment of Physicochemical and Mechanical Degradation).

References:

[1] Chan THT, Yu L, Tam HY, Ni YQ, Liu SY, Chung WH and Cheng LK 2006 Fiber Bragg grating sensors for structural health monitoring of Tsing Ma bridge. Background and experimental observation *Eng. Struct.* **28** 648

[2] Chambers A, Mowlem M and Dokos L 2007 Evaluating impact damage in CFRP using fibre optic sensors *Compos. Sci. Technol.* **67** 1235

[3] Frieden J, Cugnoni J, Botsis J and Gmur T 2012 Low energy impact damage monitoring of composites using dynamic strain signals from FBG sensors – Part I: Impact detection and localization *Compos. Struct.* **94** 438

[4] Frieden J, Cugnoni J, Botsis J, Gmur T 2012 Low energy impact damage monitoring of composites using dynamic strain signals from FBG sensors – Part II: Damage identification. *Compos. Struct.* **94** 593

[5] Gebremichael YM, et al. 2005 Integration and assessment of fibre Bragg grating sensors in an all-fibre reinforced polymer composite road bridge *Sens. Actuators A Phys.* **118** 78

[6] Lanza di Scalea F and Matt HM 2007 Macro-fiber composite piezoelectric rosettes for acoustic source location in complex structures. *Smart Mater. Struct.* **16** 1489

[7] Lee j, Tsuda H and Toyama N 2007 Impact wave and damage detections using a strain-free fiber Bragg grating ultrasonic receiver *NDT&E. Int.* **40** 85–93

[8] Li HN, Li DS and Song GB 2004 Recent applications of fiber optic sensors to health monitoring in civil engineering *Eng. Struct.* **26** 1647

[9] Ikhlef A, Hedara R and Chikh-Bled M 2012 Uniform Fiber Bragg Grating modeling and simulation used matrix transfer method, *Int. J. Comp. Sci. Issues* **9(1)** 1694

[10] Kaur R and Bhamrah MS 2011 Effect of Grating Length on Reflection Spectra of Uniform Fiber Bragg Gratings *Int. J. Inform. Tech.* **3(2)**

[11] Kirkby E, de Oliveira R, Michaud V and Manson JA 2011 Impact localisation with FBG for a self-healing carbon fibre composite structure *Compos. Struct.* **94** 8

[12] Kutaran S and Kilickaya MS 2007 The modeling of fiber Bragg grating *Opt. Quant. Electron.* **39** 643.

[13] Majumder M, Gangopaadhyay TK, Chakraborty AK, Dasgupta K and Bhattacharya DK 2008 Fibre Bragg gratings in structural health monitoring-present status and applications *Sens. Actuators A Phys.* **147** 150

[14] Minardo A, Berini R, Zeni L and Giordano M 2005 Response of fibre Bragg Gratings to longitudinal ultrasonic waves *IEEE Ultrason. Ferr.* **52(2)** 304

[15] Othonos A 1997 Fiber Bragg gratings *Rev. Sci. Instrum.* **68(12)** 4309

[16] Phing H, Ali J, Rahman RA and Tahir BA 2007 Fiber Bragg grating modeling, simulation and characteristics with different grating lengths *Journal of Fundamental Science* **3** 167

[17] Rao YJ 1999 Recent progress in applications of in-fibre Bragg grating sensors *Opt. Lasers Eng.* **31** 297

[18] Shin CS and Chiang CC 2006 Fatigue damage monitoring in polymeric composites using multiple fiber Bragg gratings *Int. J. Fatigue* **28** 1315

[19] Sipe JE, Poladian L and Martijn de Sterke C 1994 Propagation through nonuniform grating structures *J. Opt. Soc. AM A* **11(4)** 1307

[20] Udd E 2006 *Fiber optic sensors: an introduction for engineers and scientists* (New York: John Wiley & Sons) ISBN 978-0-470-06810-6

[21] Ugale S and Mishra V 2010 Fiber Bragg Grating Modeling, Characterization and Optimization with different index profiles *Int. J. Eng. Sci. Technol* **2(9)** 4463

[22] Ugural AC and Fenster SK 1998 *Advanced Strength and Applied Elasticity* (New Jersey: Prentice Hall PTR) ISBN 13978-0131375895

[23] Uribe J 2012 Chapter *Optical Fiber Sensors: An Overview* [In] *Fiber Optic Sensors* InTech, ISBN: 978-953-307-922-6

The use of strain gauges in vibration-based damage detection

Fabio Luis Marques dos Santos[1,2,3], Bart Peeters[1], Jenny Lau[1], Wim Desmet[2], Luiz Carlos Sandoval Goes[3]

[1]Siemens Industry Software NV, Interleuvenlaan 68, 3001 Leuven, Belgium
[2]KU Leuven, Division PMA, Celestijnenlaan 300B, 3001 Heverlee, Belgium
[3]Instituto Tecnológico de Aeronáutica (ITA), Praça Marechal Eduardo Gomes, 50 - Vila das Acácias CEP 12.228-900 São José dos Campos SP, Brazil

E-mail: fabio.m.santos@siemens.com

Abstract. Strain gauges and strain measurements have been widely used in structural health monitoring (SHM) systems as a means of detecting and localizing damage, due to their higher sensitivity to local damage. These damage identification techniques normally use strain related measurements such as the mode curvature, strain frequency response function or strain energy as the main parameter to detect damage. However, damage detection techniques based on acceleration measurements have also been investigated in the past, using modal parameter comparison and other methodologies. In this paper, the use of vibration-based strain measurements for use in SHM systems will be evaluated, with the purpose of characterizing their higher sensitivity in damage detection, when compared to other vibration measurements, such as acceleration-based measurements. Since the choice and use of the most damage sensitive parameter can lead to a more sensitive and robust system, the assessment of the more suitable sensor and processing of information is very important. For this purpose, numerical and experimental examples will be discussed to evaluate the higher performance of the strain gauges.

1. Introduction

One of the objectives of structural health monitoring (SHM) systems is to continuously monitor a structure during its complete lifetime, being able to detect, locate and even diagnose damage present in multiple locations. In this way, many challenges lie in designing a robust and efficient monitoring system of this kind - how to detect changes, how to identify them and which types of damage or problems are to be detected - all of these play an important role in determining the sensor network and system architecture to be used, as well as the damage detection methodology.

Vibration-based and modal based methods [1, 2] use the vibration response of a structure to determine their characteristics and to be able to pinpoint some types of damage in the system. However, methods that directly use acceleration measurements are not so sensitive to damage [3]. On the other hand, there are many studies that use strain gauges and strain modes to detect damage [4, 5], or that calculate the strain modes from the displacement modes [6, 7], which can make strain modal analysis an attractive option for vibration-based damage detection methods.

2. Modal Analysis and Strain

Experimental modal analysis methodologies can have as sensors the more commonly used accelerometers, but can also be carried out using strain gauges. The use of accelerometers will lead to the identification of the displacement mode shapes, while with the strain gauges, the strain mode shapes are identified [8]. There are some small differences on the modal formulation for the use of strain gauges[9, 10], but the same modal identification techniques can be used to identify the strain mode shapes [11].

The visual interpretation of the strain modes, however, is something less commonly carried out and not as straight forward. While the displacement modes can be easily visualized and interpreted, the strain modes are harder to be interpreted and understood. As an example, Figure 1 shows the first displacement mode and strain modes (planar x and y and shear) from the finite element model of a rectangular carbon fiber composite plate. While it is easy to interpret the displacement mode, it is harder to arrive to the same conclusions when looking at the strain modes.

(a) composite plate displacement mode (b) composite plate strain mode in x (c) composite plate strain mode in y

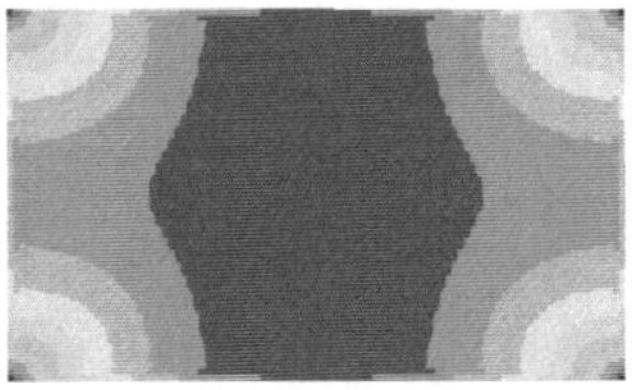

(d) composite plate strain mode in shear

Figure 1. Carbon fiber composite plate (free-free) first vibration mode (0,0): (a) displacement mode; (b) strain mode in x direction; (c) strain mode in y direction; (d) shear strain mode

Strain and displacement modes are nonetheless directly related to the resonances of the structure, and therefore they also have some degree of relation with respect to each other. These relations can come from the mode curvature (for planar strain) and twist (for shear strain) of a structure [12]. The mode curvature relations are a very well known method used for estimating the strain mode shapes from the displacement mode shapes [13], something commonly done in SHM techniques that use the modal strain energy formulation [14, 15].

For the bending modes of a beam, this relation between displacement and strain can be easily formulated - equation (1) shows the relation between the displacement mode ϕ and the mode curvature κ of a beam (approximated for small displacements), while (2) shows the relation between the strain mode ψ and the mode curvature (where h is the distance between the neutral line and the surface of the beam).

$$\kappa \approx \phi'' \tag{1}$$

$$\psi = h \cdot \kappa \tag{2}$$

By using (1) into (2), we obtain the relation between strain and displacement modes:

$$\psi = h \cdot \phi''_{beam} \tag{3}$$

The above equation is valid for the condition of small strains and displacements, which can be satisfied in modal analysis. There are similar relations for plates or plate-like structures, but then the relation is extended to strains in both normal directions.

This characteristic makes strain more sensitive to changes in the system. As an example, damage was simulated on the plate model shown in the previous figure. Stiffness was decreased by 15% in a small region in one of the plies, on the top left part of the plate. Figure 2 shows the first modes of the damaged system. As it can be seen, the damage is easily visible on the strain modes, but not noticeable at all on the displacement mode.

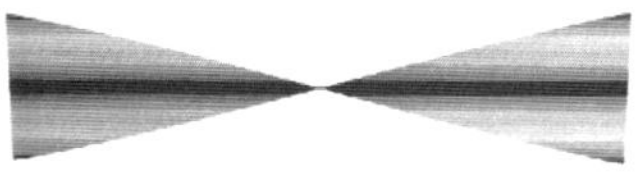

(a) composite plate displacement mode with damage on ply

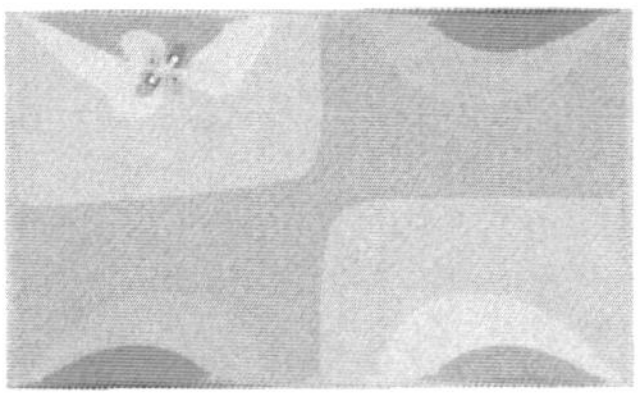

(b) composite plate strain mode in x with damage on ply

(c) composite plate strain mode in y with damage on ply

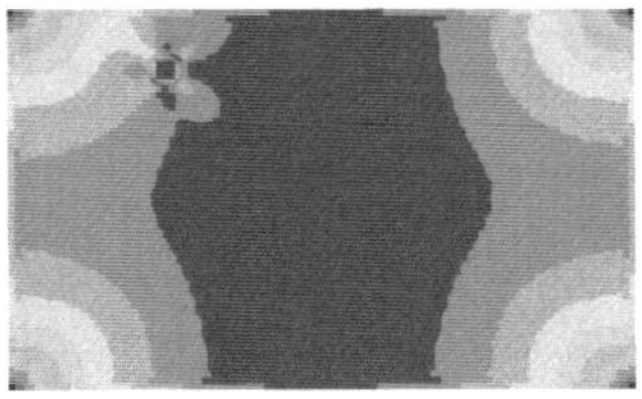

(d) composite plate strain mode in xy with damage on ply

Figure 2. Carbon fiber composite plate (free-free) with damage - first vibration mode (0,0) : (a) displacement mode; (b) strain mode in x direction; (c) strain mode in y direction; (d) shear strain mode

The example shows that if a very fine grid of strain sensors would be present on the structure (in this case, as fine as the finite element mesh), damage would be easier to detect. However, this is not the case in real life structures - there are usually few sensors and locations available on the structure, and therefore the measurement results are much more limited.

To evaluate how sensitive (and effective) strain gauges and strain modes can be when only a coarse sensor grid is available, some experimental cases will be evaluated in the next section.

3. Experimental Cases

In this section, some experimental strain modal analysis cases will be shown, with the objectives of understanding how to properly measure strain with a limited number of sensors. The first experimental case shown will be a composite carbon fiber plate, instrumented with strain gauge rosettes, where the objective was to try to characterize the full (modal) state of strain of the plate, compare it with a finite element model and come up with some conclusions with respect to how to properly interpret and use the measured data. Then, a second experimental case will be shown - the ground vibration testing of an F-16 aircraft. In this case, one of the wings of the aircraft was instrumented with piezo strain sensors to obtain the strain modes of the structure, while the full aircraft was instrumented with the commonly used accelerometers.

3.1. Carbon Fiber Composite Plate

A composite carbon fiber plate was manufactured to be used in the experiments. The composite plate is composed of 7 plies, with a stacking sequence of [0,90,0,90,0,90,0]. Each ply is made of unidirectional carbon fiber prepreg (Hexply M10R 38% UD300 CHS) and the plate was cured using an out-of-autoclave process. The dimensions of the plate are 500mm x 300mm with a thickness of 2mm.

To investigate the full state of strain of the vibration modes, the plate was instrumented with 15 strain gauge rosettes (Micro-Measurements CEA-06-250UR-350), forming a 5x3 grid of sensors. Figure 3 shows the strain gauge rosette locations on the plate.

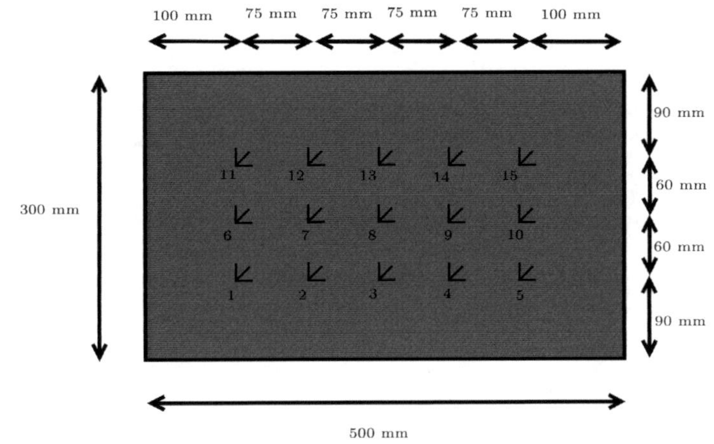

Figure 3. Composite plate: strain gauge rosette locations

Each individual strain gauge from the rosette was connected using a quarter-bridge configuration (resulting in 45 total measurement sensors). For the strain modal analysis, the plate was supported on polyurethane foam, to achieve a free-free boundary condition. Figure 4 shows the strain rosette sensor grid bonded on the plate, as well as the test set-up with the boundary conditions.

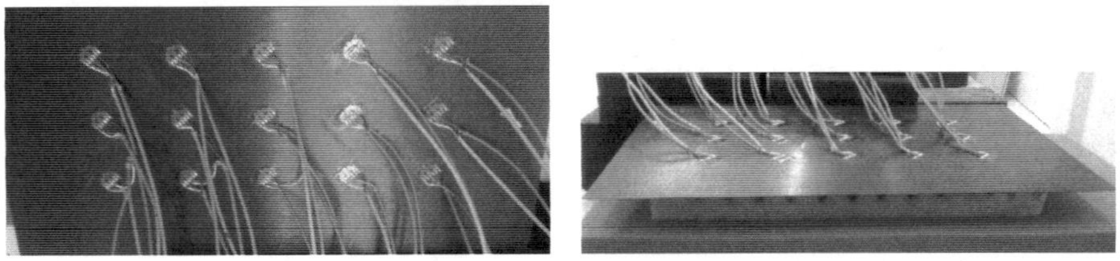

(a) Composite plate with sensors (b) Composite plate boundary conditions

Figure 4. Composite plate used for tests: strain gauge rosettes and boundary conditions

The modal analysis was carried out using impact tests. For this purpose, a modal impact hammer (PCB 086C03) was used to excite the structure. To acquire the data, two LMS SCADAS mobile acquisition units were used. Then, impact test data can be processed in LMS Test.Lab software, using the PolyMAX least squares technique [16].

Before the identification of the strain modes, a transformation has to be carried out to be able to calculate the shear strain modes from the measurements. The middle strain gauge (45 degrees) from the rosette does not directly measure shear strain, but its measurements (plus the measurements in the other 2 directions) can be used in the constitutive equations to calculate the shear strain [17]. This procedure is carried out to generate a new set of SFRFs that are representative of the shear strain, and then the strain modes in x, y and shear(xy) can be identified.

To compare and better understand the identified strain modes, a comparison and correlation with a finite element model was carried out. The model was created in LMS Samcef Composites, using the ply material properties provided by the prepreg manufacturer. As expected, the plate has bending and torsional modes, as well as modes that are a combination of bending in the x and y directions. Table 1 shows the mode pairing between simulation and test, for the first 9 modes.

Table 1. Strain modes comparison - natural frequencies (ω_n) from simulation and test

Mode number	ω_n [Hz] simulation	ω_n [Hz] test	Difference %	Mode number	ω_n [Hz] simulation	ω_n [Hz] test	Difference %
1	29.05	33.41	13	6	171.07	168.92	1.2
2	62.07	61.63	0.7	7	174.46	164.82	5.8
3	85.41	82.60	3.4	8	191.63	185.17	3.5
4	112.52	116.03	3.0	9	270.08	254.12	6.3
5	126.46	125.31	0.9				

Moreover, a good way to compare the modes is to carry out the modal assurance criterion (MAC). In this case, there are many possibilities of comparison - each direction (x,y and shear xy) can be compared individually, or all the directions can be put together in a vector for the MAC calculation. Figure 5 shows these comparisons.

It can be seen from the Figure that whenever only one direction is taken into account, the MAC is degraded significantly - the comparison between the shear shows better quality, representing the shear strain from torsion, that is always present, except in the pure bending cases. Once all the directions are taken into account, the MAC is drastically improved (modes 6 and 7, which are very close in frequency, are switched). This shows how to be able to fully represent the plate strain modes, strain measurements are required in all directions. However, the MAC improvement in quality might also be related to the increased number of points used - when the directions are compared separately, 15 measurement points are used, while when they are put together, 45 measurement points are used. To understand whether the improvement comes from the increased number of points or not, a MAC with a reduced number of measurement rosettes was calculated - rosettes 1,5,8,11 and 15 were use for the calculation, with a total of 15 measurement points (3 per rosette). Figure 6 shows the comparison between the MAC using all 15 sensors (45 measurements), and the MAC with a reduced sensor grid of 5 sensors (15 measurements).

Comparing the results in Figure 6, it can be seen that even though the quality of the correlation decreases slightly for some modes, the overall quality is very good and superior to the correlation using the individual directions. This proves that using a coarser grid of strain gauges in all directions yields better results than using a finer sensor grid, but measuring in only one direction.

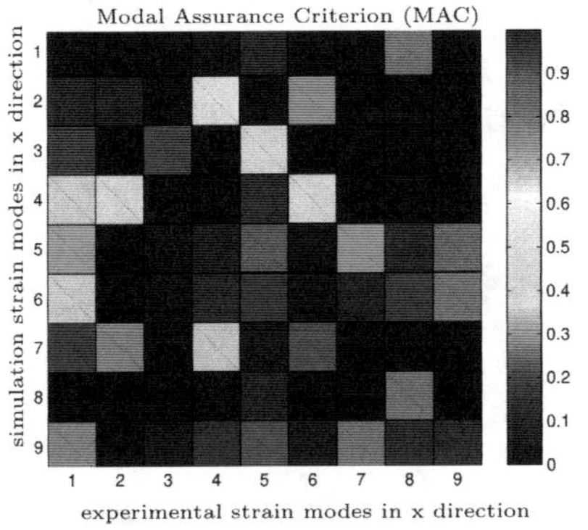

(a) Test and simulation correlation: strain measured in x direction

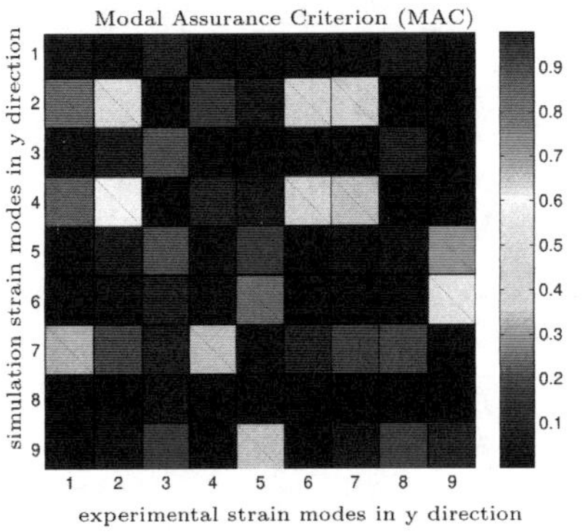

(b) Test and simulation correlation: strain measured in y direction

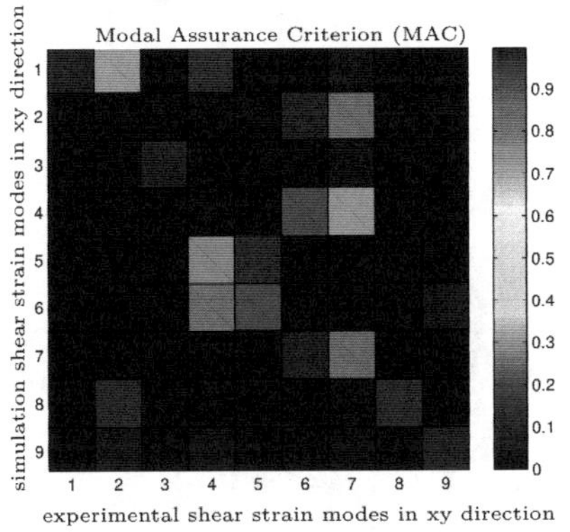

(c) Test and simulation correlation: strain measured in xy (shear) direction

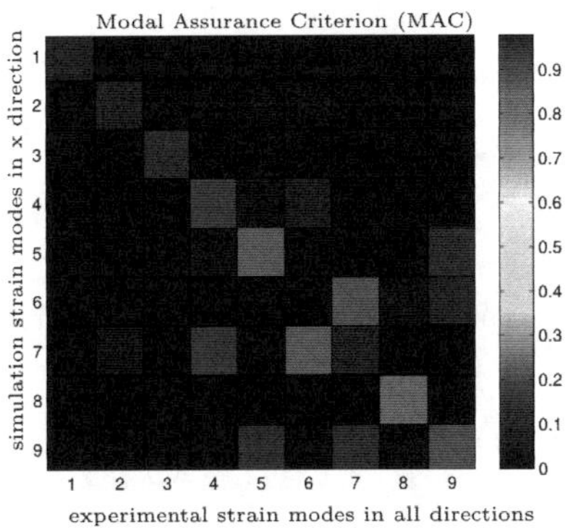

(d) Test and simulation correlation: all directions

Figure 5. Carbon fiber composite plate test correlation with strain modes: (a) strain modes only in x direction; (b) strain modes only in y direction; (c) strain modes only in xy (shear); (d) strain modes in all directions;

3.2. Strain Modal Analysis Applied on the Ground Vibration Testing of an F-16

On this second experimental example, the use of dynamic piezo strain sensors on the ground vibration testing (GVT) of an F-16 aircraft will be shown. This test campaign was part of the LMS GVT Master Class [18]. In total, 149 sensors were used to instrument the whole aircraft - of these, 20 were strain sensors instrumented on the left wing. An LMS SCADAS Lab and a SCADAS mobile were used for the data acquisition, and 2 shakers were used to excite the structure, near the tip of each wing. Figure 7 shows part of the test set-up and the left wing with collocated strain sensor and accelerometer. The objective was to visualize the similarities and differences between the displacement and strain modes on the wing.

The shakers excited the structure with a burst random signal and the modes were estimated using the PolyMAX method. The first two modes of the aircraft are shown in Figure 8. The modes show that the strain pattern from the wing cannot always be directly inferred or understood from the displacement modes. In the first case (Figure 8(a)), for the wing bending, there is a high concentration of strain in the middle of the wing, possibly because of the complex internal structure of the wing. In the second case (Figure 8(b)), the strain measurements clearly

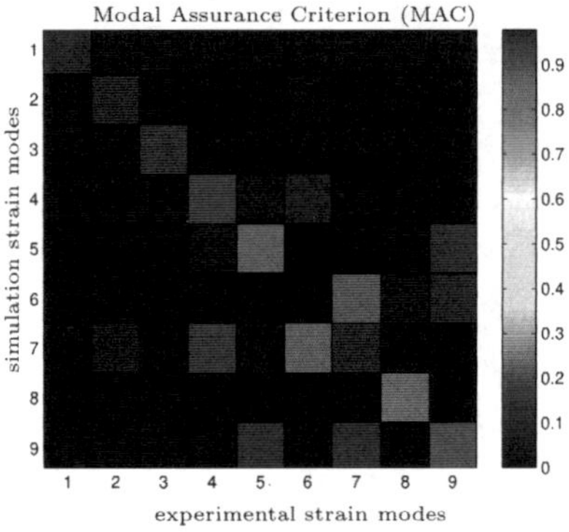

(a) Test and simulation correlation: MAC using all 15 strain gauge rosettes

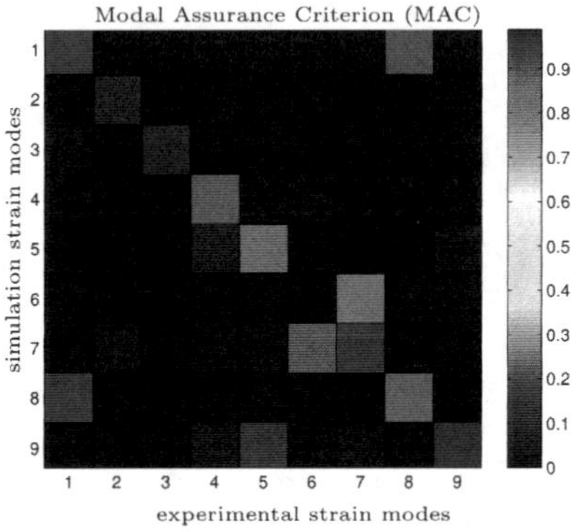

(b) Test and simulation correlation: MAC using 5 strain gauge rosettes

Figure 6. Carbon fiber composite plate MAC with strain modes: (a) MAC using all 15 strain gauge rosettes; (b) MAC using 5 strain gauge rosettes

(a) F-16 Test set-up

(b) F-16 left wing - collocated strain sensor and accelerometer

Figure 7. F-16 Test set-up and close-up on left wing with strain sensor and accelerometer

help visualizing the high amount of stress incurred on the tip of the wing, where the attachment to the bomb is.

4. Conclusions

In this paper, some advantages and difficulties of using strain gauges for modal analysis were presented. It was shown how the visual interpretation of the strain modes is harder to be carried out, when in comparison with the displacement modes. Additionally, the extra sensitivity to damage that the strain modes have when compared with displacement modes was also discussed.

Then, some experimental cases were shown - first, a modal analysis of a composite carbon fiber plate was carried out, using strain gauge rosettes to measure the full state of strain. It was shown how it is important to measure the strain in all directions, even if it means having less sensor locations. Moreover, some experimental results of the ground vibration testing of an F-16 aircraft were shown. In this case, the extra sensitivity of the strain modes was put into evidence, as they were able to properly show regions with high strain concentration, something that is not clear with the displacement modes.

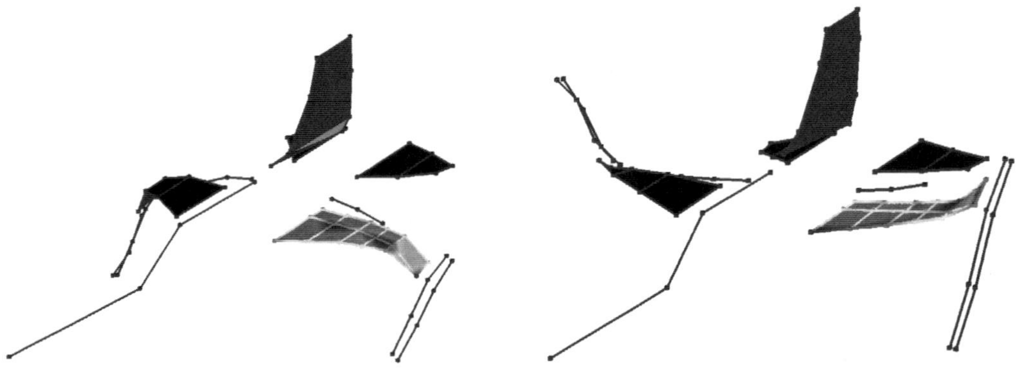

(a) F-16 first wing symmetric bending mode at 5.19 Hz (b) F-16 first wing torsion mode at 6.63 Hz

Figure 8. F-16 displacement and strain modes on left wing

References

[1] Herman A, Orifici A and Mouritz A 2013 *Composite Structures* **104** 34–42
[2] Hu H, Wang B T, Lee C H and Su J S 2006 *Composite structures* **74** 399–405
[3] Jassim Z, Ali N, Mustapha F and Jalil N A 2013 *Engineering Failure Analysis* **31** 442–461
[4] Yam L H, Li D B, Leung T P and Xue K Z 1994 Experimental study on modal strain analysis of rectangular thin plates with small holes *Proceedings of the 12th International Modal Analysis Conference, Honolulu, Hawaii* pp 1415–1421
[5] Li Y Y, Cheng L, Yam L H and Wong W O 2002 *Computers & structures* **80** 1881–1894
[6] Sehlstedt N 2001 *Journal of sound and vibration* **244** 407–430
[7] Karczub D and Norton M 1999 *Journal of sound and vibration* **226** 675–700
[8] Kranjc T, Slavič J and Boltežar M 2014 *Journal of Vibration and Control*
[9] Bernasconi O and Ewins D 1989 *International Journal of Analytical and Experimental Modal Analysis* **4** 68–76
[10] Vári L M and Heyns P S 1994 Using strain modal testing *Proceedings of the 12th International Conference on Modal Analysis* vol 2251 p 1264
[11] dos Santos F L M, Peeters B, Menchicchi M, Lau J, Gielen L, Desmet W and Góes L C S 2014 Strain-based dynamic measurements and modal testing *Topics in Modal Analysis II, Volume 8* (Springer) pp 233–242
[12] Park C, Walz C and Chopra I 1996 *Smart Materials and Structures* **5** 98–113
[13] Cornwell P, Doebling S W and Farrar C R 1999 *Journal of Sound and Vibration* **224** 359–374
[14] Dessi D and Camerlengo G 2015 *Mechanical Systems and Signal Processing* **52** 181–205
[15] Liu F, Li H, Li W and Wang B 2014 *Renewable Energy* **72** 174–181
[16] Peeters B, Van der Auweraer H, Guillaume P and Leuridan J 2004 *Shock and Vibration* **11** 395–409
[17] Sciammarella C A and Sciammarella F M 2012 111–121 URL `http://dx.doi.org/10.1002/9781119994091.ch5`
[18] Lau J, Peeters B, Debille J, Guzek Q, Flynn W, Lange D S and Kahlmann T 2011 Ground vibration testing master class: modern testing and analysis concepts applied to an f-16 aircraft *Advanced Aerospace Applications, Volume 1* (Springer) pp 221–228

Identification of faulty sensor using relative partial decomposition via independent component analysis

Z Wang and S T Quek
Department of Civil & Environmental Engineering, National University of Singapore, 1 Engineering Drive 2, #E1A-07-03, Singapore 117576, Tel: 65 6516 6283

E-mail: a0092164@u.nus.edu; ceeqst@nus.edu.sg

Abstract. Performance of any structural health monitoring algorithm relies heavily on good measurement data. Hence, it is necessary to employ robust faulty sensor detection approaches to isolate sensors with abnormal behaviour and exclude the highly inaccurate data in the subsequent analysis. The independent component analysis (ICA) is implemented to detect the presence of sensors showing abnormal behaviour. A normalized form of the relative partial decomposition contribution (rPDC) is proposed to identify the faulty sensor. Both additive and multiplicative types of faults are addressed and the detectability illustrated using a numerical and an experimental example. An empirical method to establish control limits for detecting and identifying the type of fault is also proposed. The results show the effectiveness of the ICA and rPDC method in identifying faulty sensor assuming that baseline cases are available.

1. Introduction

Various structural health monitoring techniques have been developed during the past decades such as vibration-based global health monitoring methods focusing on finding shifts in resonant frequencies as well as changes in structural mode shapes [1]. These techniques hinge on the availability of reliable data, which remains an issue to be addressed. Methods for detecting sensors with abnormal behavior in a sensor network have been proposed, such as latent variable techniques within the multivariate statistical framework [2]. Principal component analysis (PCA) and independent component analysis (ICA) are two commonly employed latent variable techniques. ICA has proven to be more efficient than PCA especially when the measured variables exhibit non-Gaussian distributions [3]. In this study, the ICA-based fault detection method is adopted, following which the cause of the anomaly needs to be identified using an appropriate fault diagnosis method. Partial decomposition contribution (PDC) has been used to isolate the group of faulty sensors [4, 5]. However, using only the magnitudes of PDCs may lead to misleading conclusions since faulty sensors with small contributions may not be properly diagnosed [5]. The current ICA-based method is unable to detect multiplicative type of fault with scaling factor smaller than one and hard-fault-type faults (such as constant responses or complete loss of signal) [5]. In order to enhance the performance of the ICA-based method, the relative partial decomposition contributions (rPDCs) and its normalized form is proposed herein for faulty sensor detection and isolation. An empirical method to establish control limits for detecting and identifying the type of fault is also proposed. The ability of the modified ICA-based method in detecting two basic categories of faults, namely, additive-type and multiplicative-type, is illustrated through both numerical and experimental examples.

2. Independent component analysis

Under ICA, a set of measured process variables, $\boldsymbol{x}$, may be decomposed as a linear combination of statistically independent non-Gaussian latent variables, $\boldsymbol{s}$, called independent components, of smaller or equal dimension compared to $\boldsymbol{x}$ [6]. This model can be expressed as

$$\boldsymbol{x} = \boldsymbol{A}\boldsymbol{s} + \boldsymbol{e} \tag{1}$$

where $\boldsymbol{A}$ is a mixing matrix and $\boldsymbol{e}$ is a residual matrix to account for imperfection in the model, known as reconstruction error. A demixing matrix $\boldsymbol{W}$ may thus be defined as

$$\boldsymbol{s} = \boldsymbol{W}\boldsymbol{x} \tag{2}$$

where $\boldsymbol{WA} = \mathbf{I}$, the identity matrix.

To determine $\boldsymbol{A}$ and $\boldsymbol{W}$, a fixed-point iteration approach called FastICA algorithm is adopted [7], where the number of independent components is chosen based on the eigenvalues of the covariance matrix $E\{\boldsymbol{x}\boldsymbol{x}^T\}$. Components corresponding to larger eigenvalues are deemed to contain more information while those with small eigenvalues are usually discarded to reduce the dimension of $\boldsymbol{s}$ [7].

In equation (1), $\boldsymbol{s}$ can be interpreted geometrically as points on the hyper-plane defined by $\boldsymbol{A}$, and coincide with $\boldsymbol{x}$ if the linear combination model is perfect, whereas $\boldsymbol{e}$ represents the perpendicular distance of the actual point $\boldsymbol{x}$ to the hyper-plane. $\boldsymbol{s}$ can be interpreted as the systematic component of $\boldsymbol{x}$ and $\boldsymbol{e}$ the residual based on the linear combination model [8]. For the purpose of monitoring the out-of-control condition in the systematic component, the D-statistic, denoted as $\boldsymbol{I}^2$, is adopted, defined as

$$\boldsymbol{I}^2 = \boldsymbol{s}^T\boldsymbol{s} = \boldsymbol{x}^T\boldsymbol{W}^T\boldsymbol{W}\boldsymbol{x} \tag{3}$$

Similarly, the Q-statistic, denoted as $\boldsymbol{SPE}$, is defined to monitor the out-of-control condition in the residual component as

$$\boldsymbol{SPE} = \boldsymbol{e}^T\boldsymbol{e} = \boldsymbol{x}^T(\boldsymbol{I} - \boldsymbol{W}^T\boldsymbol{A}^T)(\boldsymbol{I} - \boldsymbol{AW})\boldsymbol{x} \tag{4}$$

To establish the confidence limits for the D- and Q-statistics, kernel density estimation is used to extract the percentile points from the underlying probability density function [3, 9]. For the data-driven process monitoring technique adopted for detection, the baselines for normal operating data are required. If bias or drift type of fault is present, at least one of the two statistics should exceed the control limits [5]. However, this may not be the case if a multiplicative type of fault is present depending on whether the scaling factor is smaller or greater than 1.

3. Sensor fault diagnosis

3.1. Partial decomposition contributions

Once a faulty sensor is detected in the system, a diagnosis method should be used to identify the location and nature of the fault. One basis is to look at the contribution of individual variables to the D- or Q-statistics known as partial decomposition contribution (PDC) [4]. The PDCs are defined as

$$\boldsymbol{PDC}_j^{I^2} = \boldsymbol{x}^T\boldsymbol{M}^{I^2}\zeta_j\zeta_j^T\boldsymbol{x} \tag{5}$$

$$\boldsymbol{PDC}_j^{SPE} = \boldsymbol{x}^T\boldsymbol{M}^{SPE}\zeta_j\zeta_j^T\boldsymbol{x} \tag{6}$$

where $\boldsymbol{PDC}_j^{I^2}$ and $\boldsymbol{PDC}_j^{SPE}$ are the contributions of $\boldsymbol{x}_j$ to $\boldsymbol{I}^2$ and $\boldsymbol{SPE}$ respectively, $\boldsymbol{M}^{I^2} = \boldsymbol{W}^T\boldsymbol{W}$, $\boldsymbol{M}^{SPE} = (\boldsymbol{I} - \boldsymbol{W}^T\boldsymbol{A}^T)(\boldsymbol{I} - \boldsymbol{AW})$ and ζ_j is the jth column of the identity matrix.

The basic assumption is that the variable with the largest component is the primary contributor of the fault [4]. Some researchers evaluated the PDCs at one point in time [5] whereas others summed up each PDC over a certain time period [8] to determine the out-of-control situation. However, this approach is not suitable for multiplicative type of fault with scaling factor smaller than one since such fault results in a decrease in PDC. Moreover, when an additive type of fault (or a multiplicative type of fault with scaling factor larger than one) is mainly attributed to a variable with relatively small absolute contributions, as the faulty sensor may not exhibit a much larger contribution than others.

3.2. Relative partial decomposition contribution

As an improvement, the relative partial decomposition contribution (rPDC) is proposed, given by

$$\boldsymbol{rPDC_i^{Index}} = \frac{\boldsymbol{PDC_i^{Index}}}{E\{\boldsymbol{PDC_i^{Index}}\}} \quad \textbf{(7)}$$

where the superscript ***index*** refers to either to $\boldsymbol{I^2}$ and ***SPE***, and $E\{\}$ is the expectation over the same realization. The assumption is that the contribution of each variable should be statistically unchanged throughout the realization if there is no change in the condition of the sensors or structure [4]. However, a fault-free variable may exhibit a relative large change in contribution at different point in time due to the input excitation. Hence, a baseline set of representative realizations is used for the normalization, giving:

$$\boldsymbol{rPDC_i^{Index}} = \frac{\boldsymbol{PDCm_i^{Index}}}{E\{\boldsymbol{PDCb_i^{Index}}\}} \quad \textbf{(8)}$$

where $\boldsymbol{PDCb_i^{Index}}$ are the PDCs of the variables in the baseline and $\boldsymbol{PDCm_i^{Index}}$ are the PDCs of the variables in a particular set of the measurement. Instead of using the ***rPDC***s at individual point in time, the expectation over the measured realization is proposed, given by

$$\mathbf{E}\{\boldsymbol{rPDC_i^{Index}}\} = \mathbf{E}\left[\frac{\boldsymbol{PDCm_i^{Index}}}{E\{\boldsymbol{PDCb_i^{Index}}\}}\right] = \frac{E\{\boldsymbol{PDCm_i^{Index}}\}}{E\{\boldsymbol{PDCb_i^{Index}}\}} \quad \textbf{(9)}$$

When no fault is present, $E\{\boldsymbol{rPDC_i^{Index}}\}$ for all variables should be close to 1. If a fault is mainly caused by one variable, the contributions of that variable is expected to experience the largest change compared to other variables, its $E\{\boldsymbol{rPDC_i^{Index}}\}$ will change according to the type of fault. It will increase if an additive type or multiplicative type of fault with scaling factor larger than one is present.

4. Application to case studies

To illustrate the performance of the ICA-based detection and diagnosis method using the proposed rPDC technique, a numerically simulated two-dimensional truss and an actual steel cantilever three-dimensional frame were used.

4.1. Numerical simulation using a two-dimensional truss

A seven degrees-of-freedom (d.o.f.) two-dimensional truss, with seven accelerometers monitoring each d.o.f. as shown in figure 1, was analysed. All members of the structure are 2 m long hot-finished circular hollow sections with outer diameter of 88.9 mm and thickness of 5 mm. Damping ratio of 5% was assumed for all modes. The natural frequencies of the structure ranges from 4.56 to 19.19 Hz.

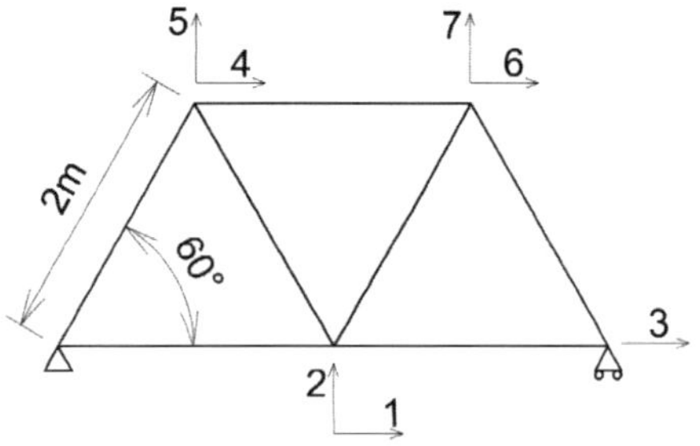

Figure 1. Two-dimensional simulated truss.

A band-limited (4 to 20 Hz) white-noise excitation was applied at d.o.f. 5. The responses were sampled at a frequency of 1000 Hz and 2 sets of data both consisting of 20001 points (duration of 20 s) were simulated for each type of fault. One data set under normal condition was used as the baseline to build the latent-variable model. Different types of faults were introduced to sensor 2 from time of 5 to 20 s (i.e. points 5001 to 20001), four examples are shown in figure 2.

4.1.1. Additive-type fault

After the ICA model is built, the 99% confidence control limits for D- and Q-statistics were obtained. For the different amount of bias introduced to sensor 2 data, the detection rates (defined as percentage of samples outside the control limits) were computed. Figure 3 shows the rates for both statistics increased with amount of shift in the mean. The mean of rPDCs for sensor 2 changed significantly with bias compared with other sensors, which had values close to 1 as shown in figure 4, indicating that sensor 2 was faulty. The negative values of rPDCs for SPE was because initially the mean of PDC for sensor 2 was negative and it increased to a positive value after an additive-type fault was introduced.

Journal of Physics: Conference Series **628** (2015) 012120 doi:10.1088/1742-6596/628/1/012120

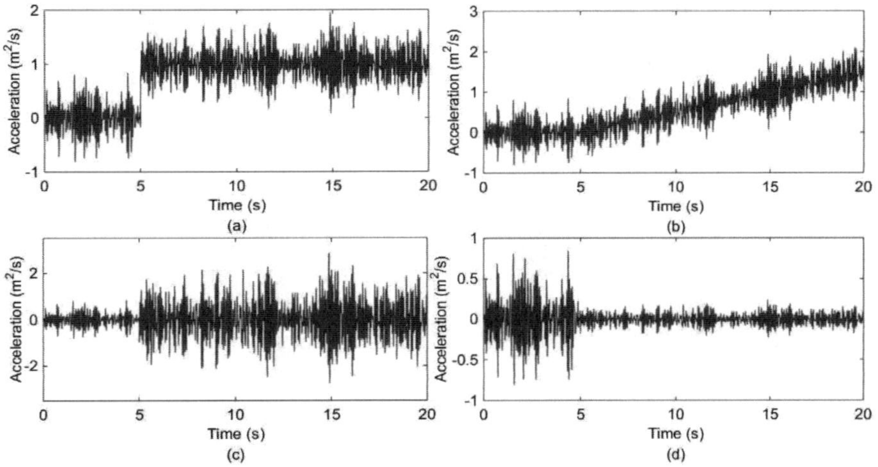

Figure 2. Simulated fault: (a) bias, (b) drift, (c) multiplicative (SF >1) and (d) multiplicative (SF < 1)

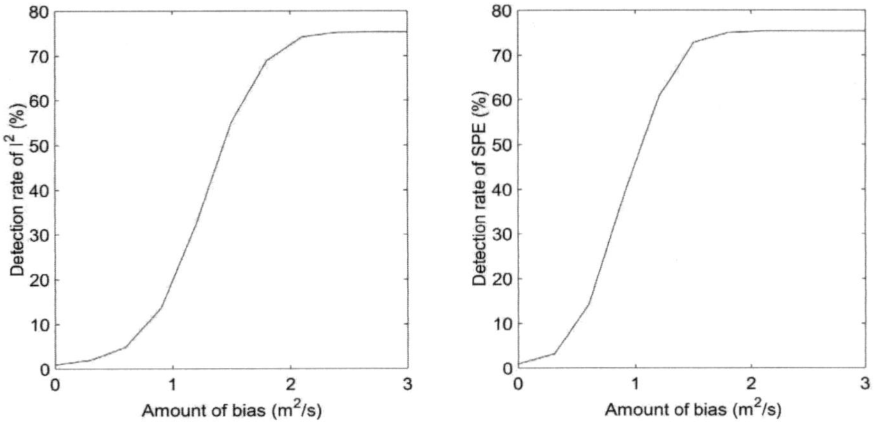

Figure 3. Detection rates for $\boldsymbol{I^2}$ and ***SPE*** for different amount of bias.

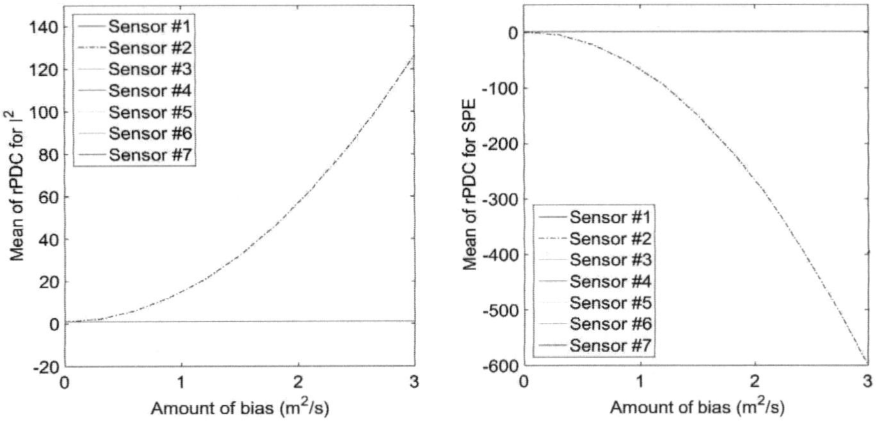

Figure 4. Variation in mean of ***rPDC*** for $\boldsymbol{I^2}$ and ***SPE*** with bias for each sensor. Dashed lines correspond to faulty sensor.

To illustrate the advantages of using rPDC instead of PDC values at a point in time, the PDC values for $\boldsymbol{I}^2$ at 13 s (data point 13001) are shown in figure 5 when a relative small amount of bias (0.5 m^2/s) was introduced. Although sensor 2 was the only faulty sensor, it did not have a much larger PDC. Instead, sensor 6 had the largest PDC. This might lead to wrong identification of the faulty sensor. In contrast, rPDC values, based on relative changes, correctly identified the faculty sensor.

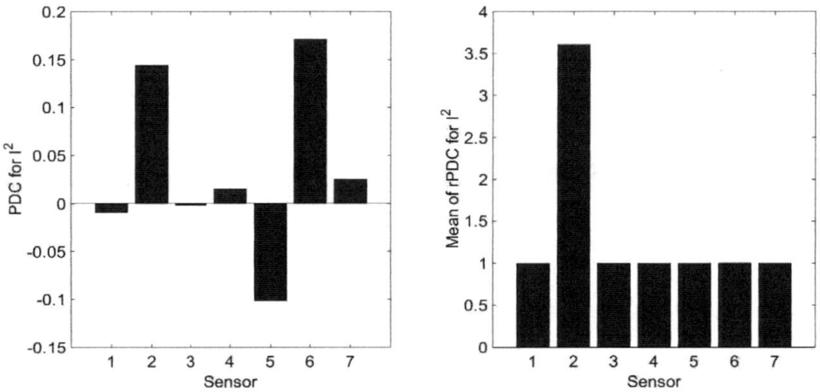

Figure 5. PDCs for $\boldsymbol{I}^2$ at 13 s and means of rPDC for $\boldsymbol{I}^2$.

Similar results were observed when a drift type of fault was introduced. The increased in both D- and Q-statistics were exhibited and by examining the rPDCs, sensor 2 was identified (not presented).

4.1.2. Multiplicative-type fault
Two cases of multiplicative-type fault were analysed. For scaling factor (SF) > 1, the rPDC results are shown in figure 6. Sensor 2, with the largest mean of rPDC, was diagnosed as faulty. The detection results were similar to the bias case and not presented here.

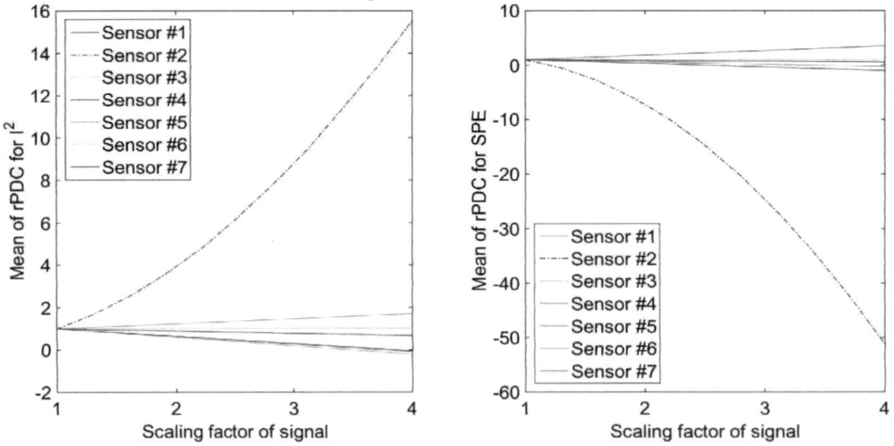

Figure 6. Mean of rPDC for $\boldsymbol{I}^2$ and ***SPE*** for different scaling factors (> 1). Dashed lines correspond to faulty sensor.

For SF < 1, the distance of each point along the model to the origin decreased. Hence, the number of points crossing the upper control limit for D-statistic will decrease, leading to decrease in detection rate. As the Q-statistic measures the distance perpendicular to the ICA model, it is hard to predict whether the detection rate will increase or decreases as SF decreases. An empirical method to help distinguish this type of fault from other types will be suggested in the experimental example.

For multiplicative-type fault with SF < 1, the rPDC for D-statistic is more useful. Due to the decrease in I^2 values, the faulty sensor should experience the largest reduction in contribution, resulting in a smaller mean of rPDC compared to other sensors. Alternatively, the reciprocal of the mean of rPDCs can be used to identify the faulty sensor. Figure 7 shows that the faulty sensor had the smallest mean of rPDC (and the largest reciprocal of mean).

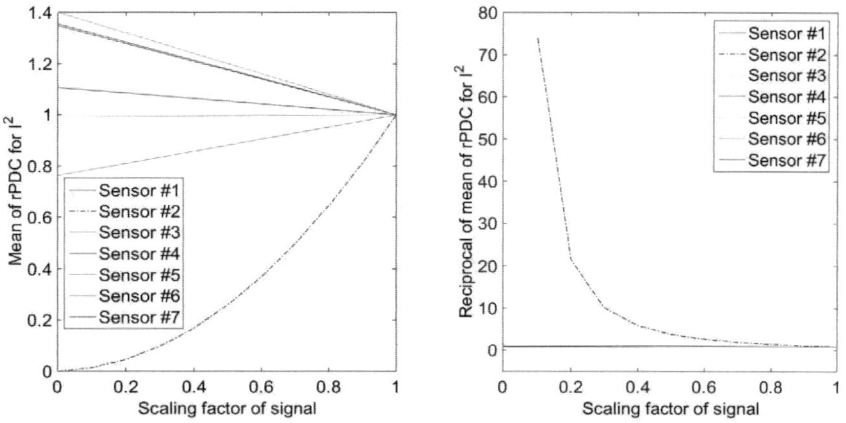

Figure 7. Mean of rPDC for I^2 and its reciprocal for different SF (< 1). Dashed lines correspond to faulty sensor.

4.2. Experiment using a steel cantilever frame

A steel cantilever frame shown in figure 8 was used to illustrate the performance of the faulty sensor identification method. Its first 15 natural frequencies ranged from 6 to 40 Hz. A shaker attached at the free end imposed a vertical white-noise excitation with a bandwidth of 0 to 50 Hz. Three wireless accelerometers, each measuring both horizontal and vertical accelerations, were placed as indicated.

Figure 8. Location of accelerometers on steel cantilever frame.

An empirical method is proposed to distinguish different types of fault. Twelve sets of responses with duration of 20s operating under normal conditions were sampled. One set of response (set 1) was used to build the ICA model while another set of response (set 2) was used to generate various abnormal conditions. The remaining sets (3 to 12) were used to generate the statistical bounds. Specifically, detection rates for D-statistics and Q-statistics tested against set 1 were computed. These detection rates will not be identical due to variation in ambient temperature and environmental noise, amongst other factors. The mean (μ) and standard deviation (σ) of these rates were used to establish upper and lower bounds to aid monitoring, typically set at $\mu +3\sigma$ and $\mu -3\sigma$, respectively, as illustrated in Table 1. If the detection rates for both statistics exceed $\mu +3\sigma$, an additive type of fault or a multiplicative type of fault with SF > 1 may be present in the system. If the detection rate is below $\mu - 3\sigma$ for $\boldsymbol{I}^2$, a multiplicative type of fault with SF < 1 may probably be present in the system.

Table 1. Empirical upper and lower bounds of detection rates.

Baseline set no.	3	4	5	…	Mean (μ)	Std deviation (σ)	μ+3σ	μ-3σ
D-statistic detection rate (%)	1.42	1.20	1.25	…	1.30	0.13	1.70	0.90
Q-statistic detection rate (%)	0.97	0.95	1.10	…	1.00	0.09	1.26	0.74

Different types of fault were introduced to sensor 2 by modifying the data in set 2 from 5 s to 20 s. The results for different magnitude of additive bias are shown in figures 9 and 10. The bias fault can be effectively detected by both D- and Q-statistics, but the latter is less sensitive due to the much smaller contributions of sensor 2 to ***SPE***. However, the faulty sensor can still be identified from the rPDC plots of both statistics. The results for drift type of fault were similar and presented here.

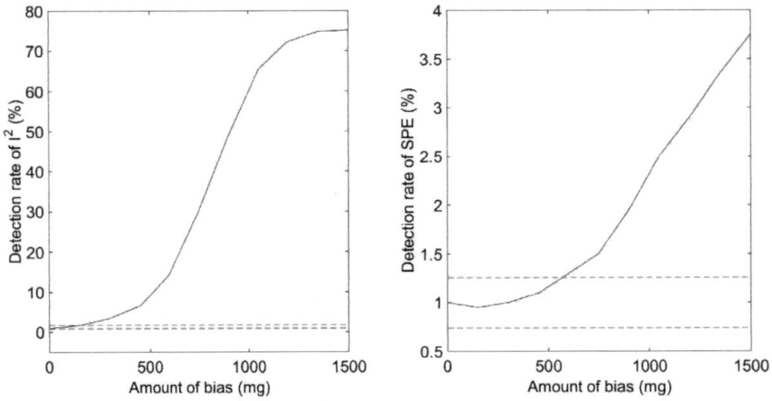

Figure 9. Detection rates for $\boldsymbol{I}^2$ and ***SPE*** for different amount of bias. Dashed lines are empirical upper and lower bounds.

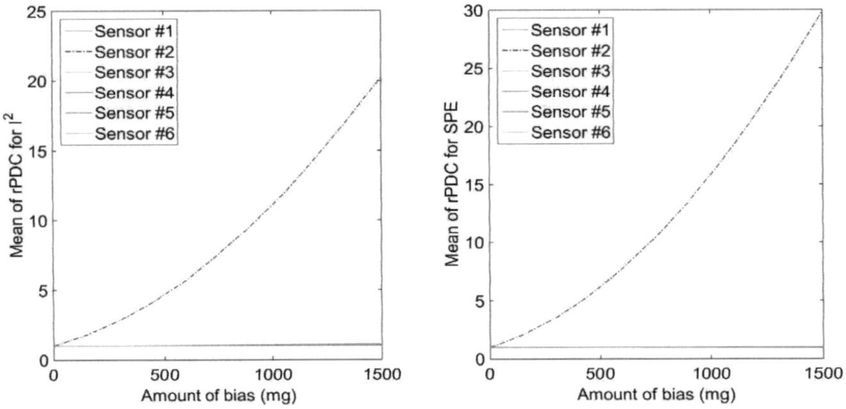

Figure 10. Mean of rPDC for $\boldsymbol{I}^2$ and ***SPE*** for different amount of bias.

The detection rates and rPDC for multiplicative type of fault with SF > 1 were investigated and the performance was similar to the additive type of fault. Although the detection rates of ***SPE*** were relatively small, the rapid change in contribution can be seen in both rPDC plots (not presented). The detection rates for multiplicative type of fault with SF < 1 are shown in figure 17. The Q-statistic was unable to detect the presence of fault while the abnormal reduction in detection rate of D-statistic can be observed for SF < 0.9 using the empirical lower bound. By examining the rPDC (or its inverse) plot for I^2, the faulty sensor can be effectively identified.

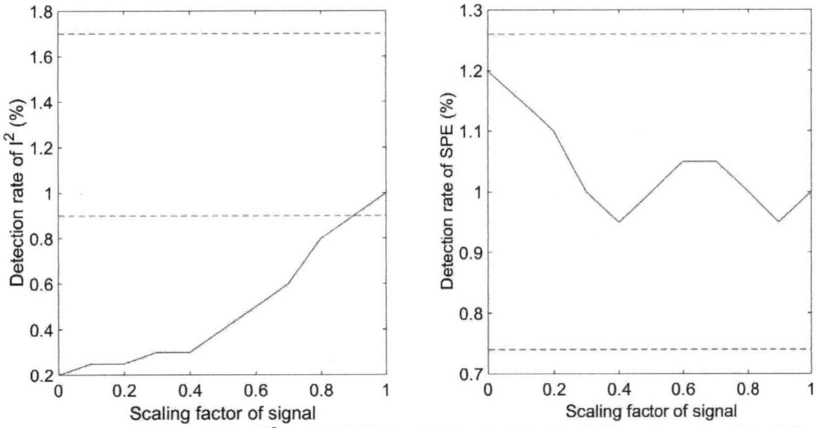

Figure 17. Detection rates for I^2 and ***SPE*** for SF (<1). Dashed lines are empirical bounds.

5. Conclusions

A normalized relative partial decomposition contribution (rPDC) method in conjunction is proposed for sensor fault detection and identification. Both numerical simulation and experimental examples suggested that for additive type of fault and multiplicative type of fault with scaling factor (SF) larger than one, the rPDC provides a more definite and accurate identification compared to using the magnitude of PDC. For a multiplicative type of fault with SF < 1, the fault causes a decrease in the detection rate of I^2 and the sensor is identified through the smallest mean of rPDC for I^2. It is proposed that upper and lower bounds for detection rates be established empirically to better account for the likely operating environmental conditions.

References

[1] Chang P, Flatau A and Liu S 2003 Review Paper: Health Monitoring of Civil Infrastructure. *Struct. Health Monit.* **2** 257-267

[2] Kourti T 2005 Application of latent variable methods to process control and multivariate statistical process control in industry *Int. J. Adapt. Control* **19** 213–246

[3] Lee J-M, Yoo C K and Lee I-B 2004 Statistical process monitoring with independent component analysis *J. Process Contr.* **14** 467–485

[4] Alcala C F and Qin S J 2011 Analysis and generalization of fault diagnosis methods for process monitoring *J. Process Contr.* **21** 322–330

[5] Hernandez-Garcia M and Masri S 2014 Application of statistical monitoring using latent-variable techniques for detection of faults in sensor networks *J. Intel. Mat. Syst. Str.* **25** 121–136

[6] Hyvärinen A, Karhunen J and Oja E 2001 *Independent Component Analysis* (New York: John Wiley & Sons)

[7] Hyvärinen A and Oja E 2000 Independent component analysis: algorithms and applications *Neural Networks* 13 411–430

[8] Westerhuis J, Gurden S and Smilde A 2000 Generalized contribution plots in multivariate statistical process monitoring *Chemometr. Intell. Lab.* **51** 95–114

[9] Bowman A W 1984 An alternative method of cross-validation for the smoothing of density estimates *Biometrika* **71** 353-360

Damage localization by statistical evaluation of signal-processed mode shapes

M.D. Ulriksen and L. Damkilde

Dept. of Civil Engineering, Aalborg University, Niels Bohrs Vej 8, Esbjerg, Denmark

E-mail: mdu@civil.aau.dk, lda@civil.aau.dk

Abstract. Due to their inherent ability to provide structural information on a local level, mode shapes and their derivatives are utilized extensively for structural damage identification. Typically, more or less advanced mathematical methods are implemented to identify damage-induced discontinuities in the spatial mode shape signals, hereby potentially facilitating damage detection and/or localization. However, by being based on distinguishing damage-induced discontinuities from other signal irregularities, an intrinsic deficiency in these methods is the high sensitivity towards measurement noise. The present article introduces a damage localization method which, compared to the conventional mode shape-based methods, has greatly enhanced robustness towards measurement noise. The method is based on signal processing of spatial mode shapes by means of continuous wavelet transformation (CWT) and subsequent application of a generalized discrete Teager-Kaiser energy operator (GDTKEO) to identify damage-induced mode shape discontinuities. In order to evaluate whether the identified discontinuities are in fact damage-induced, outlier analysis of principal components of the signal-processed mode shapes is conducted on the basis of T^2-statistics. The proposed method is demonstrated in the context of analytical work with a free-vibrating Euler-Bernoulli beam under noisy conditions.

1. Introduction

Numerous vibration-based structural health monitoring (SHM) methods have been proposed for damage detection and/or localization of structural damages in aerospace, civil, and mechanical systems, see, e.g., [1, 2, 3, 4]. The applicability of the proposed methods has primarily been tested on the basis of analytical models, finite element (FE) simulations, and controlled laboratory tests; all in which significant simplifications and idealizations are made with regard to, among other things, environmental effects. In this context, it is has been found that the simple methods directly comparing pre- and post-damage vibration-based quantities, such as modal parameters, exhibit only limited potential for damage detection and/or localization, see, e.g., [1, 5]. Under more realistic conditions, these methods turn out to be entirely inapplicable because the direct changes of the aforementioned quantities typically will be concealed by environmental effects and noise contamination. As an example, it is documented in [6, 7] how environmental effects and general operational conditions can account for up to at least 5 % shifts in eigenfrequencies, which, as documented in, e.g., [7, 8], hardly will be exceeded by damage-induced eigenfrequency changes.

As a consequence of the general inadequacy of the simple methods, current research within the field of vibration-based SHM is mainly leveled at developing more sophisticated and robust methods. Some of these methods are based on signal processing of spatial mode shapes signals, which seems auspicious as this modal parameter, contrary to global modal parameters such as eigenfrequencies, inherently provides structural information on a local level. A common approach is to exploit that a damage will introduce mode shape discontinuities, albeit not always directly visible ones, which can be captured by use of signal processing techniques, e.g. continuous wavelet transformation (CWT). Namely CWT has been employed extensively for localizing structural damages in both simple beam- and plate-like systems, see, e.g., [9, 10, 11], and more complex structures such as wind turbine blades [8, 12]. However, by being based on distinguishing damage-induced discontinuities from other signal irregularities, an intrinsic deficiency of the aforementioned CWT-based methods is the rather low robustness towards measurement noise. To treat this issue, a method in which wavelet-transformed mode shapes are processed by use of the discrete Teager-Kaiser energy operator (DTKEO) is proposed in [13]. Here, it is shown, on the basis of analytical and experimental work with different beams, that addition of the DTKEO significantly reduces the sensitivity towards noise.

Compared to the method documented in [13], the method proposed in the present paper constitutes a generalization of the applied signal processing plus an extension by means of outlier analysis to yield the final discrimination between undamaged and damaged areas. As such, the proposed method is composed of three steps; first, a filtered mode shape derivative is obtained of a spatial mode shape signal through CWT, and subsequently a generalized discrete Teager-Kaiser energy operator (GDTKEO) is applied to this derivative to form an energy-processed signal in which the damage-induced discontinuities are magnified. Finally, a statistical evaluation scheme based on T^2-statistics is applied to principal components of these energy-processed, filtered mode shape derivatives in order to label the structures as healthy or damaged at each sensor location/measurement point. To test the applicability of the method, and the potential degree of improvement compared to the method proposed in [13], a free-vibrating, analytical Euler-Bernoulli beam under noisy conditions is analysed.

2. Damage localization method

2.1. CWT

The spatial mode shape signal, $f(x) \in \mathbf{L}^2(\mathbb{R})$, is obtained as a signal-processed version of the original mode shape, $\mathbf{u}$. The signal processing consists of signal expansion in order to remove boundary distortions (resulting in $\mathbf{u} \to f$). Typically, when dealing with a sparse number of measurement points in, e.g., experiments, an interpolation scheme should also be applied to smooth the mode shape signal and hereby remove artificial discontinuities. However, since analytically derived mode shapes are utilized in this article, the smoothing scheme is omitted.

Based on the spatial mode shape signal, $f(x) \in \mathbf{L}^2(\mathbb{R})$, and the wavelet function, $\gamma(x) \in \mathbf{L}^2(\mathbb{R})$, the CWT is defined as

$$Wf(a,b) = a^{-\frac{1}{2}} \int_{\mathbb{R}} f(x)\gamma^*\Big(a^{-1}(x-b)\Big)\mathrm{d}x, \tag{1}$$

where $a > 0$ and $b \in \mathbb{R}$ are wavelet scales and positions, respectively. Evidently, the CWT is obtained as the inner product of $f(x)$ and the complex conjugated, indicated with the superscript *, of the so-called wavelet family, which consists of functions constructed from dilations and translations of γ.

The effectiveness of the CWT as a signal discontinuity scanner highly depends on the employed wavelet type; in particular, the amount of vanishing moments, m. Assume $k \in \mathbb{N}^+$, then

$$\forall k < m : \int_{\mathbb{R}} \gamma(x)x^k \mathrm{d}x = 0, \tag{2}$$

which states that a wavelet with m vanishing moments is blind to polynomial trends up to degree $m-1$. Equivalently, when analyzing a spatial mode shape signal with a wavelet with m vanishing moments, a filtered m-derivative of the mode shape is obtained. This becomes evident by realizing that the expression in equation (1) also can be seen as the convolution of f with a scaled, flipped, and conjugated wavelet, thus

$$\exists\theta(x) = \int_{\mathbb{R}} \gamma(x)\mathrm{d}^m x \ni Wf(a,b) = \int_{\mathbb{R}} f(x)\gamma_a^\times(b-x)\mathrm{d}x = \Big(f * \frac{\mathrm{d}^m}{\mathrm{d}x^m}\theta_a^\times\Big)\Big(b\Big) = \Big(\frac{\mathrm{d}^m}{\mathrm{d}x^m} f * \theta_a^\times\Big)\Big(b\Big). \tag{3}$$

This emphasises the importance of choosing a wavelet with a certain minimum of vanishing moments when searching for discontinuities in a signal. Correspondingly, there is also a certain maximum for the proper amount of vanishing moments due to, i.a., magnification of signal noise and adverse border distortions. Clearly, the optimal choice of wavelet strongly depends on signal characteristics such as mode shape type and noise conditions and should therefore be drawn based on these. This is treated further in subsection 3.2.

As stated in section 1, CWTs have previously exhibited damage localization potential. However, as noise is added to the signal, this potential is significantly reduced since the discrimination between noise- and damage-induced CWT peaks becomes troublesome, especially at the lower wavelet scales. For the higher scales, at which noise is better filtered, the low spatial frequency reduces the potential for capturing damage-induced signal discontinuities, thus emphasising the need for further enhancement of these.

2.2. GDTKEO
Since the DTKEO was proposed, originally as a signal energy estimator [14], it has been utilized extensively in speech processing [15], image processing [16], and pattern recognition [17]. Recently, the DTKEO has been adopted to the field of damage identification; as in [18], where it is included in a general damage detection algorithm, and, as mentioned in section 1, in [13], where it is applied to wavelet-transformed mode shapes for damage localization in beams, without any statistical procedure for discriminating between undamaged and damaged areas.

With the notation utilized in the present paper, the DTKEO of a wavelet-transformed mode shape signal at scale a and spatial location i can be found as

$$\psi_{a,i} = Wf_{a,i}^2 - Wf_{a,i-1}Wf_{a,i+1}. \tag{4}$$

In order to enhance the performance of the DTKEO as a signal discontinuity magnifier, it is expanded intuitively by means of a lag parameter, κ, hence yielding the GDTKEO

$$\Psi_{a,i} = Wf_{a,i}^2 - Wf_{a,i-\kappa}Wf_{a,i+\kappa}, \tag{5}$$

i.e. $\Psi_a = \psi_a$ if $\kappa = 1$. As can be realized from equation (5), the lag parameter serves directly to alleviate the adverse noise effects, in analogy to the approach explored for deriving pseudo-modal curvatures through Laplacian operators in [19]. Generally, κ should be chosen on the basis of signal characteristics such as number of measurement points, i.e. signal length, and noise conditions. In subsection 3.2, a simple, yet very effective, method for selecting κ is proposed.

2.3. Outlier analysis
In order to evaluate whether the localized mode shape discontinuities are in fact damage-induced, outlier analysis is conducted by calculating the statistical distance, based on the T^2-statistic, between the signal-processed mode shapes from the present state and similar ones from the healthy state.

Outlier analysis is conducted for each signal band located at a sensor/measurement point, j, such

$$T_j^2 = |\Psi_j|^T P_j \Sigma_j^{-1} P_j^T |\Psi_j| \tag{6}$$

where P_j and Σ_j contain, respectively, the selected principal components (i.e., eigenvectors) and eigenvalues of the covariance matrix of

$$Z_j = \begin{bmatrix} \Psi_{j,\text{train}(1)}^T \\ \vdots \\ \Psi_{j,\text{train}(S)}^T \end{bmatrix}. \tag{7}$$

Evidently, Z_j is a training matrix based on S signal bands at sensor/measurement point j from the baseline, i.e. healthy, state. A signal band at sensor/measurement point j for the current state is then tested on the basis of the following hypothesis:

$$\left.\begin{array}{l} H_{0,j} : T_j^2 \leq \vartheta_j \\ H_{1,j} : T_j^2 > \vartheta_j \end{array}\right\} \tag{8}$$

in which the null hypothesis, $H_{0,j}$, designates that no anomalies are present at the jth sensor/measurement point, whereas the alternative hypothesis, $H_{1,j}$, declares the structure damaged at sensor/measurement point j. The threshold, ϑ_j, is found through the training data. Here, a T^2-statistic between each row in Z_j and the remaining $S-1$ rows is calculated, and subsequently the S T^2-statistics are sorted in an ascending order. Finally, ϑ_j is then chosen as the value exceeded by 1 %.

3. Application example: Analytical beam model

3.1. Cracked Euler-Bernoulli beam

The application example is based on the beam model depicted in figure 1a, which is treated analytically. The beam has a length of L, cross-sectional dimensions of $v = d = 0.03L$, and is assigned an isotropic material model with a Young's modulus, E, corresponding to construction steel, i.e. $E = 200$ GPa. A structural damage with height $s = 0.15d$ is introduced as a crack at location $x/L = 0.4$.

The beam is treated analytically through Euler-Bernoulli beam theory, such transverse vibrations are governed only by the bending deformations. Assuming that the effect of the crack is apparent only in its area, the crack can be modelled as a linear rotational spring with stiffness [20]

$$K_R = \frac{vEI}{6\pi d F(\zeta)}, \tag{9}$$

where EI is the flexural stiffness, while

$$F(\zeta) = 1.86\zeta^2 - 3.95\zeta^3 + 16.37\zeta^4 + 37.22\zeta^5 + 76.81\zeta^6 + 126.9\zeta^7 + 172.5\zeta^8 - 144\zeta^9 + 66.6\zeta^{10} \tag{10}$$

is a dimensionless local compliance function, see e.g. [20], with the independent variable $\zeta = s/d$.

As seen in figure 1b, the spring divides the beam into two separate segments. The transverse motion of each beam segment, $w_i(x)$, is described spatially by the dimensionless governing equation

$$w_i''''(X) - \lambda^4 w_i(X) = 0, \quad i = 1{,}2, \tag{11}$$

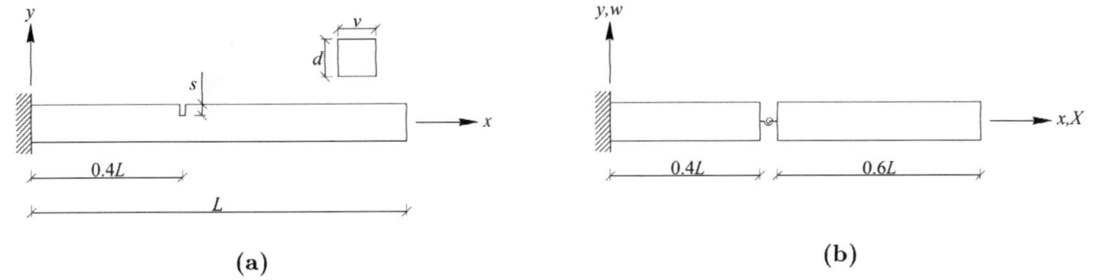

Figure 1. Cracked cantilevered beam. (a) General beam model. (b) Equivalent beam-spring model.

in which $X = x/L$, $w_i''''(X) = \mathrm{d}^4 w_i(X)/\mathrm{d}X^4$, and $\lambda^4 = \omega^2 A L^4 \rho / EI$, with ω, A, I, and ρ being the eigenfrequency, the cross-sectional area, the cross-sectional moment of inertia, and the mass density.

Solving equation (11) yields

$$w_i(X) = c_{1,i}\cos(\lambda X) + c_{2,i}\sin(\lambda X) + c_{3,i}\cosh(\lambda X) + c_{4,i}\sinh(\lambda X), \quad i = 1{,}2, \tag{12}$$

where the constants of integration are found by applying the boundary conditions of the cantilevered beam, i.e.

$$w_1(0) = 0 \quad \wedge \quad w_1'(0) = 0 \quad \wedge \quad w_2''(1) = 0 \quad \wedge \quad w_2'''(1) = 0 = 0, \tag{13}$$

plus the following compatibility conditions in the crack area ($X = X_c$):

$$w_1(X_c) = w_2(X_c) \wedge K_R\big(w_1'(X_c) - w_2'(X_c)\big) = -EIw_1''(X_c) \wedge w_1''(X_c) = w_2''(X_c) \wedge w_1'''(X_c) = w_2'''(X_c). \tag{14}$$

When substituting equation (12) into equations (13) and (14), the characteristic equation is obtained. From here, the eigenfrequencies and subsequently the constants of integration are derived and substituted into equation (12) to yield the mode shapes.

3.2. Damage localization

To demonstrate the applicability and robustness of the proposed method, 300 noise-contaminated (corresponding to a signal-to-noise ratio (SNR) of 65 dB) editions of both the first undamaged (i.e., $K_r = \infty$) bending mode shape and the first damaged bending mode shape are analyzed. 200 of the signal-processed mode shapes from the undamaged state have been randomly chosen to constitute the trained statistical baseline model against which the remaining 100 undamaged signal-processed mode shapes and the 300 damaged signal-processed mode shapes are tested.

3.2.1. CWT settings. When referring to equation (1), it is evident that the CWT is an infinite integral transformation, thus boundary distortions will be introduced when analysing the discrete mode shape signal. In order to reduce these adverse distortions, the mode shape signal is expanded, i.e. $\mathbf{u} \rightarrow f$, in accordance with the isomorphism approach [21].

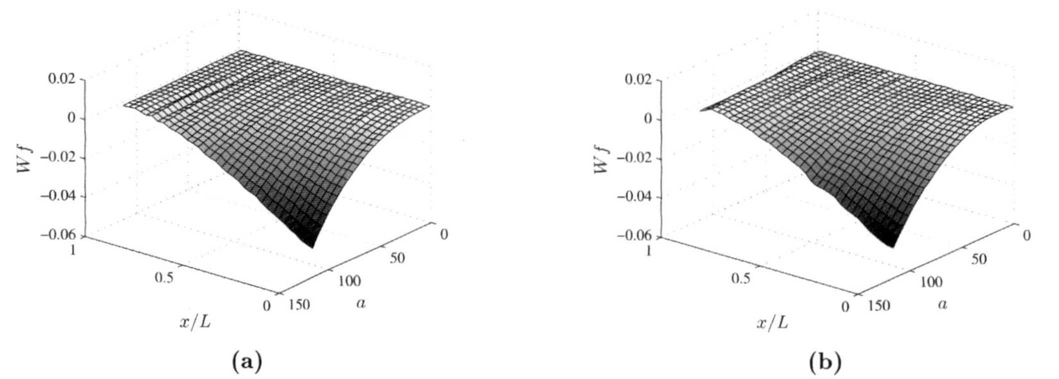

Figure 2. CWT of noise-contaminated (65 dB) first bending mode shape. (a) Undamaged state. (b) Damaged state (crack at $x/L = 0.4$).

In the choice of wavelet type, a crucial parameter is the number of vanishing moments, cf. equation (2). A wavelet with m vanishing moments is blind to polynomials up to order $m-1$, hence implying that a wavelet with a minimum of four vanishing moments would, theoretically, provide the best results when analysing the first bending mode shape. However, extensive amount of differentiation, recall that the CWT yields pseudo-derivatives, would yield detrimental magnification of the border distortions, even when applying the isomorphism approach. Consequently, it is chosen to use a second-order Gaussian wavelet as this has been found to provide a suitable tradeoff between discontinuity enhancement and border distortion magnification. In figure 2, CWTs of noise-contaminated editions of the first bending mode shape of the undamaged beam and the damaged one are presented for the wavelet scale interval $a \in [1{,}120]$.

3.2.2. GDTKEO settings. As briefly mentioned in subsection 2.2, the lag parameter, κ, serves to alleviate noise contributions. To determine an optimal value of κ, a simple approach based on the root-mean-square (RMS) is proposed. The plot in figure 3a illustrates the principle; the mean of the CWTs (at maximum wavelet scale) from the training state is signal-processed by means of the GDTKEO with different κ values. Then, the RMS is calculated for each of the derived $\Psi_a|_{a=120}$ signals, as plotted in figure 3a, and $\kappa = 342$ is chosen as the value at which the local minimum occurs. By using this setting for the GDTKEO and applying it to the wavelet transforms seen in figure 2, the results presented in figures 3b and 3c are obtained. Clearly, the damaged-induced discontinuity is located along with other irregularities, hence emphasising the need for statistical discrimination.

3.2.3. Outlier analysis. It is chosen to present outlier analysis results for four measurement points/sensor locations, namely $x/L = 0.04$, $x/L = 0.4$ (damage location), $x/L = 0.5$, and $x/L = 0.8$. Here, the statistical model for each location, composed of P_j and Σ_j, have been trained by means of

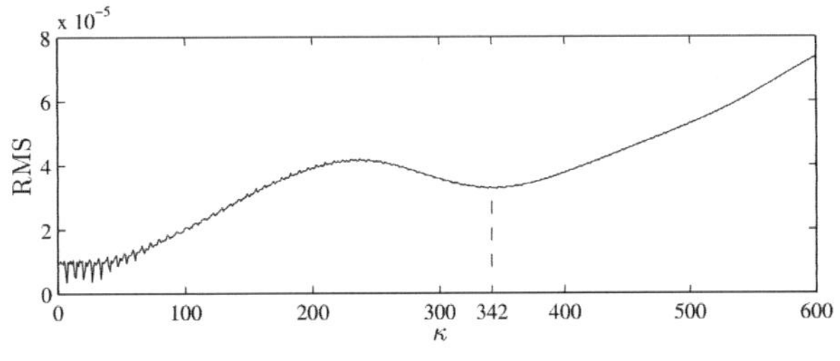

(a)

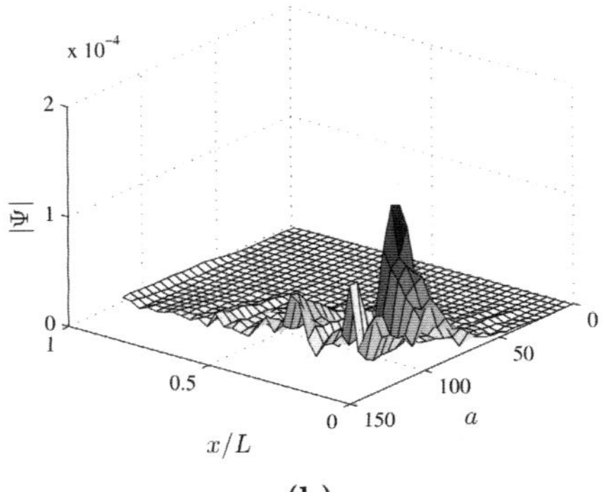

(b)

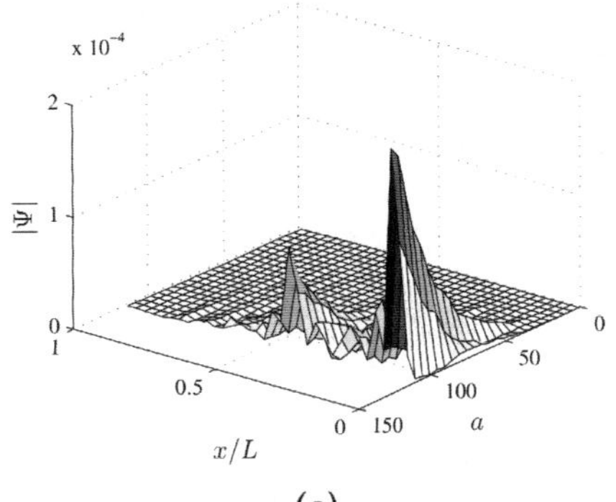

(c)

Figure 3. GDTKEO applied to CWT of noise-contaminated (65 dB) first bending mode shape. (a) Selection of κ. (b) Undamaged state for $\kappa = 342$. (c) Damaged state for $\kappa = 342$ (crack at $x/L = 0.4$).

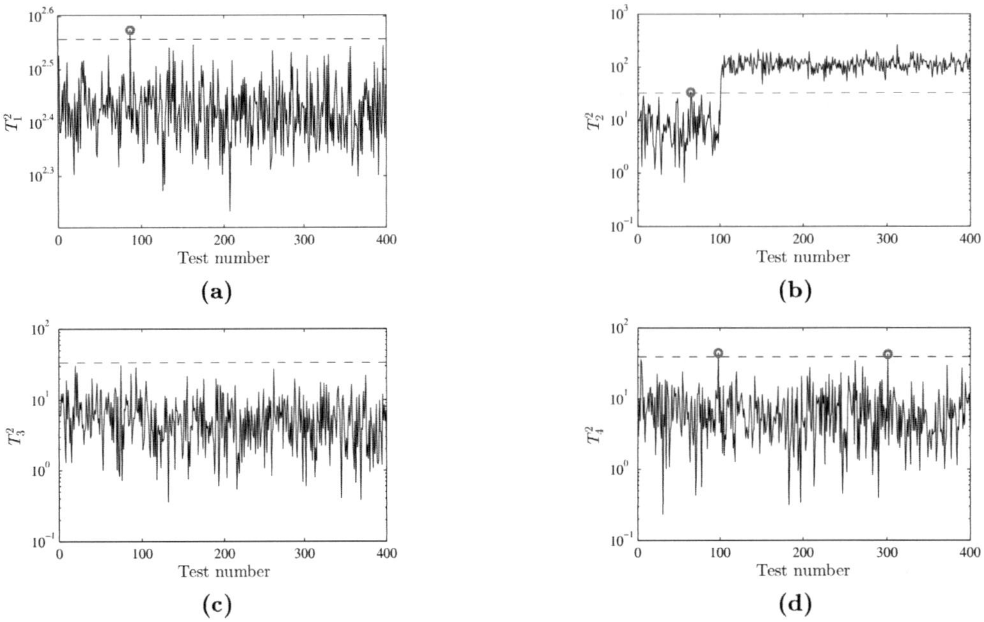

Figure 4. T_j^2-statistics on the basis of $|\Psi|$ ($\kappa = 342$) for undamaged (test number 1-100) and damaged (test number 101-400) states. (a) $x/L = 0.04$. (b) $x/L = 0.4$ (damage location). (c) $x/L = 0.5$. (d) $x/L = 0.8$. The dashed, horizontal lines mark the thresholds and the red circles designate type I errors.

the relevant signal bands from the chosen 200 undamaged signals. It is found, for all four locations, that the two first principal directions contain above 90 % of the variance, and therefore it is chosen only to use these two, thus $P_j \in \mathbb{R}^{200\times 2}$ and $\Sigma_j \in \mathbb{R}^{2\times 2}$. The corresponding thresholds are computed on the basis of the relevant bands from the 200 randomly chosen signal-processed mode shapes from the undamaged state, i.e. $Z_j \in \mathbb{R}^{200\times 120}$.

With the training and threshold computation completed, the testing is conducted. First, the relevant bands from the remaining 100 undamaged signal-processed mode shapes are tested, followed by the corresponding bands from 300 damaged counterparts. In figure 4, the results are presented. Evidently, the data from the healthy configuration, i.e. test number 1-100, are generally classified as inliers. Only a few type I errors, i.e. false outlier classification, are obtained. This is, however, to some extent expectable due to the probabilistic nature of the threshold computation. For test number 101-400, where the beam is damaged at $x/L = 0.4$, the signal-processed mode shape bands at $x/L = 0.4$ are all classified correctly as outliers, see figure 4b. For the three remaining locations, the signal-processed mode shape bands are generally classified as inliers, meaning that the damage localization method yields unambiguous localization of the crack.

As stated previously, the GDTKEO used to obtain the results presented in figure 4 is assigned a lag parameter of $\kappa = 342$. If $\kappa = 1$, the GDTKEO reduces to the DTKEO implemented in the method proposed in [13]. To demonstrate the improvement obtained by using the GDTKEO with a proper selection of κ, instead of the DTKEO, a damage localization analysis analogous to that conducted in this paper (finalised with the results presented in figure 4) is conducted for $\kappa = 1$. The final outlier analysis results, which are seen in figure 5, clearly illustrate the significant improvement obtained by implementing the GDTKEO with a proper κ.

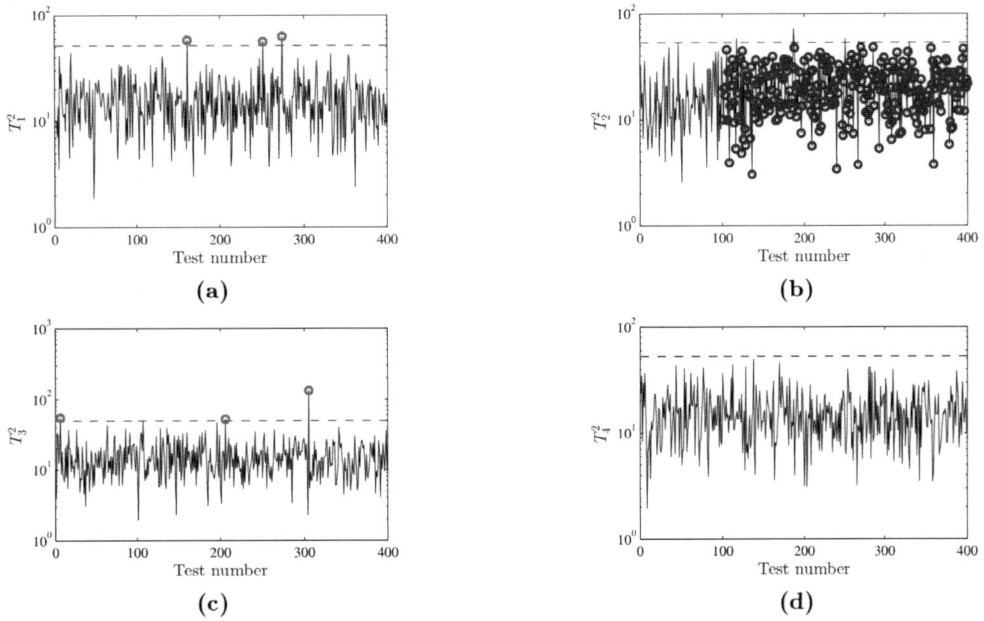

Figure 5. T_j^2-statistics on the basis of $|\psi|$ for undamaged (test number 1-100) and damaged (test number 101-400) states. (a) $x/L = 0.04$. (b) $x/L = 0.4$ (damage location). (c) $x/L = 0.5$. (d) $x/L = 0.8$. The dashed, horizontal lines mark the thresholds and the red and blue circles designate type I and II errors.

4. Conclusion

The present paper documents a proposed damage localization method whose principle is to statistically interrogate changes in signal-processed mode shapes with respect to localization of a damage. As such, the method lacks the need for measuring the excitation input and is completely independent of numerical models.

The applicability of the method has been demonstrated in the context of analytical work with a free-vibrating Euler-Bernoulli beam. Here, 300 noise-contaminated (SNR = 65 dB) editions of the first bending mode shape of the undamaged state and 300 noise-contaminated editions of a damaged state have been signal-processed by means of CWT and subsequent application of a GDTKEO. 200 of the signal-processed mode shapes from the undamaged state have been randomly chosen for training of the statistical baseline model, against which the remaining 100 signal-processed mode shapes from the undamaged state and the 300 signal-processed mode shapes from the damaged state have been tested. Hereby, it is found for the damaged state that the T^2-statistics derived for the signal bands at the damage location are consistently labeled as outliers, whereas the T^2-statistics derived for the remaining locations are classified correctly as inliers (except for a single case). For the undamaged state, only three T^2-statistics are falsely classified as outliers, thus the crack can be unambiguously localized.

References

[1] Salawu O S 1997 *Engineering Structures* **19**(9) 718–723
[2] Doebling S W, Farrar C R, Prime M B and Shevitz D W 1998 *The Shock Vibration Digest* **30** 91–105
[3] Farrar C R, Doebling S W and Nix D A 2001 *Philosophical Transactions of the Royal Society of London A: Mathematical, Physical and Engineering Sciences* **359**(1778) 131–149
[4] Fan W and Qiao P 2011 *Structural Health Monitoring* **10**(1) 83–111
[5] Larsen G C, Berring P, Tcherniak D, Nielsen P H and Branner K 2014 *Proceedings of 7th European Workshop on Structural Health Monitoring*

[6] Creed S G 1987 Assessment of large engineering structures using data collected during in-service loading *Structural Assessment: The Use of Full and Large Scale Testing* ed Garas F K, Clarke J L and Armer G S T (Butterworth-Heinemann) pp 55–62
[7] Aktan A E, Lee K L, Chuntavan C and Aksel T 2001 *Philosophical Transactions of the Royal Society of London A: Mathematical, Physical and Engineering Sciences* **359**(1778) 131–149
[8] Ulriksen M D, Tcherniak D, Kirkegaard P H and Damkilde L 2014 *Proceedings of 7th European Workshop on Structural Health Monitoring*
[9] Douka E, Loutridis S and Trochidis A 2003 *International Journal of Solids and Structures* **40**(13-14) 3557–3569
[10] Loutridis S, Douka E, Hadjileontiadis L J and Trochidis A 2005 *Engineering Structures* **27**(9) 1327–1338
[11] Rucka M and Wilde K 2006 *Journal of Sound and Vibration* **297**(3-5) 536–550
[12] Doliński L and Krawczuk M 2009 *Journal of Physics: Conference Series* **181**(1) 012086
[13] Cao M, Xu W, Ostachowicz W and Su Z 2014 *Journal of Sound and Vibration* **333**(6) 1543–1553
[14] Kaiser J F 1990 *Proceedings of 1990 International Conference on Acoustics, Speech, and Signal Processing* 381–384
[15] Maragos P, Quatieri T F and Kaiser J F 1991 *Proceedings of 1991 International Conference on Acoustics, Speech, and Signal Processing* 421–424
[16] Maragos P and Bovik A C 1995 *Journal of the Optical Society of America A* **12**(9) 1867–1876
[17] Cexus J C and Boudraa A O 2007 *International Journal of Electrical, Robotics, Electronics and Communications Engineering* **1**(2) 321–324
[18] Antoniadou I and Worden K 2014 *Proceedings of 7th European Workshop on Structural Health Monitoring*
[19] Cao M and Qiao P 2009 *Mechanical Systems and Signal Processing* **23**(4) 1223–1242
[20] Dimarogonas A D 1996 *Engineering Fracture Mechanics* **55**(5) 831–857
[21] Messina A 2008 *International Journal of Solids and Structures* **45**(14-15) 4068–4097

Pulse compression and dispersion compensation for high-resolution Lamb wave inspection

J Hua, J Lin*, L Zeng

School of Mechanical Engineering, Xi'an Jiaotong University, Xi'an, China, 710049

jinglin@mail.xjtu.edu.cn

Abstract. The dispersion of ultrasonic guided waves causes the energy of a signal to spread out in space and time as it propagates, which decreases the performance for damage detection significantly. A lot of signal processing methods have been proposed to reduce the effect of dispersion for this reason. In this paper, with the aim of developing an efficient methodology for high resolution Lamb wave inspection, a pulse compression and dispersion compensation method is established. In this method, broadband excitation and pulse compression technique are introduced to reconstruct the transform function with a high SNR. Subsequently, a scheme is established to alleviate the dispersion effects by performing compensation on the original narrowband excitation signals, and thus the time duration of received wave packet can be compressed during the extracting process. Finally, Numerical simulation and experiment are carried on aluminum specimens to investigate the behavior of the proposed method.

1. Introduction

Lamb waves are guided elastic waves that propagate in solid plates or layers with free boundaries, which are also called plate waves [1]. As a promising method for structural damage detection and evaluation, Lamb wave method has attracted considerable attentions.

Dispersion is a big challenge encountered when working with guided waves. It is manifested as an increase in received signal duration and a decrease in amplitude [2]. The former effect is especially undesirable for situations where a defect is located in close proximity to another structural feature. Since both the defect and the structural feature cause the incident Lamb wave to be reflected, the wave packets may overlap with each other at the received position. This interference worsens resolution and makes experimental data hard to interpret. A lot of methods have been developed to avoid the effect of dispersion: optimal design of the excitation waveform [3], time reversal [4], time-frequency representation [5] and dispersion removal [6,7]. Using these methods, the effect of dispersion could be reduced or even eliminated completely.

Different from above methods, the aim of this paper is to develop a rapid methodology for high resolution Lamb wave inspection. For this purpose, broadband excitation is applied to enable simultaneous acquisition of information of multiple distinct frequency ranges, and further, with the application of pulse compression technique, the transform function of the whole system could be reconstructed with a high SNR. On this basis, a new dispersion compensation algorithm is established. This algorithm is performed by applying time-reversed dispersion effects on the original excitation

* Corresponding author

signal for a given compensation distance. As a result, the response extraction and the dispersion compensation can be made simultaneously. Numerical simulation is then conducted to illustrate the influence of the compensation distance to the resolvable resolution. Finally, the effectiveness of the proposed algorithm is verified experimentally, in which the results show that the resolvable resolution is improved obviously.

2. Scheme of pulse compression and dispersion compensation

2.1 Transform function estimation via broadband excitation and pulse compression

Considering that two guided wave transducers are positioned on a plate (or layer), which act as the transmitter and receiver respectively, the entire system comprising the instrumentation, transducers and structures may be well-modeled as a linear system [8]. If the guided wave is excited with a broadband and long-duration temporal signal (*e.g.* chirp signal),

$$s(t)=\int S(\omega)e^{j\omega t}d\omega, \tag{1}$$

the response captured by the receiver satisfies,

$$y(t)=\int H(\omega)S(\omega)e^{j\omega t}d\omega. \tag{2}$$

Here, the corresponding transfer function $H(\omega)$ includes the transfer functions of the transmitter and receiver, all instrumentation effects, and the Green's function(s) needed to describe wave propagation between transmitter and receiver [8].

To estimate this transform function, the pulse compression technique is introduced. In the frequency domain, it is defined as,

$$\begin{aligned}R(\omega)&=S^*(\omega)Y(\omega)=S^*(\omega)S(\omega)H(\omega)\\&=|S(\omega)|^2H(\omega)\end{aligned}, \tag{3}$$

where the mark '*' means the complex conjugate. This formula can also be expressed in the time domain as,

$$r(t)=h(t)\,conv.\{s(t)\,corr.\,s(t)\}, \tag{4}$$

where 'corr.' and 'conv.' mean the correlation and the convolution process, respectively. The advantages of pulse-compression method are mainly twofold. Firstly, it improves the signal to noise ratios (SNRs) of the raw received signals. Secondly, it reconstructs the accurate phase of the transform function, and further, since the frequency range of the excitation signal is broad, the auto-correlation function, '$s(t)$ *corr.* $s(t)$', become pulse-like, and the pulse compression signal, $r(t)$, become closer to the pulse response, $h(t)$.

To recover the amplitude of the transform function, we divide the term $|S(\omega)|^2$ at both sides of equation (3), and thus

$$H(\omega)=R(\omega)/|S(\omega)|^2. \tag{5}$$

The broadband excitation Lamb wave signal carries abundant and comprehensive information associating with the structure. Combined with the pulse compression technique, the transform function over a wideband frequency range could be reconstructed with a high SNR. On this basis, multiple narrowband responses could be readily computed provided that the frequency ranges of the corresponding excitations fall within that of the broadband excitation. For instance, the response to a narrowband excitation, $s_d(t)$, can be obtained as,

$$Y_d(\omega)=H(\omega)S_d(\omega). \tag{6}$$

Hence, it can be concluded that the broadband excitation and the pulse compression constitute a methodology for efficiently optimizing a guided wave inspection.

2.2. Dispersion compensation for resolution improvement

In this transform function, the effect of dispersion manifests itself as a phase adjustment term, $D(x, \omega)$ [2],

$$D(x,\omega)=e^{ik(\omega)x}, \tag{7}$$

where x represents the distance measured from the actuator in the direction of propagation, $k(\omega)$ is the circular wave number of the guided wave mode, which may be obtained from the phase velocity $v_{ph}(\omega)$ dispersion curve data by

$$k(\omega)=\frac{\omega}{v_{ph}(\omega)}. \tag{8}$$

If the actuator is located at the position of x=0 while the sensor is located at the position of x=L, dispersion compensation can be accomplished by multiplying $Y_d(\omega)$, the spectrum of $y_d(t)$, with a phase correction term $D^{-1}(L, \omega)$,

$$F(\omega)=H(\omega)S_d(\omega)D^{-1}(L,\omega) \tag{9}$$

where,

$$D^{-1}(L,\omega)=e^{-ik(\omega)L} \tag{10}$$

Here, the correction term covers the same frequency range as the excitation $S_d(\omega)$. Hence, a new excitation signal, $S_n(\omega)$, is gained, with its dispersion effects are pre-compensated for a special propagating distance, L.

$$S_n(\omega)=S_d(\omega)\cdot D^{-1}(L,\omega), \tag{11}$$

Accordingly, equation (9) could be rewritten as,

$$F(\omega)=H(\omega)S_n(\omega)=H(\omega)\left[S_d(\omega)D^{-1}(L,\omega)\right]. \tag{12}$$

Similar to equation (6), the new response could also be easily computed from the transform function. More importantly, for $x\in[L-\delta, L+\delta]$, where $\delta\leqslant L/2$ is a threshold controlling the range, it may be easily concluded from equation (12) that the dispersion will be partially compensated. As a result, the resolution of the wave packets will be improved. This method is especially desirable for long range inspections.

3. The vital role of compensation distance

For a wave packet propagating through a particular distance, x, the resolvable resolution is dependent on the parameters of the new excitation waveform, $s_n(t)$. As shown in equation (11), this excitation consists of two terms, i.e. the original narrowband excitation, $S_d(\omega)$ and the phase correction term, $D^{-1}(L, \omega)$. As a routine, the former one is a windowed tone burst with a precise centre frequency and a limited bandwidth, and much research has been published on the optimal design of its waveform [3]. In this paper, our study focuses on the latter one, *i.e.* the compensation distance, L.

In practice, the compensation distance L dominates the dispersion compensation degree of the output signal. If $x<L$, the residual distance $\Delta=-(x-L)$ suffers from time reversed dispersion, as a result, the frequency components with slower group velocity appear at the front of the receive signal. Under completely compensation condition ($x=L$), all the frequency components arrive in phase at the receiver position, and the best resolution is achieved. When $x>L$, dispersion still acts on the residual distance $\Delta=x-L$, which makes the frequency component corresponding to the maximum group velocity lead the received wave, and the slower components trail.

With the application of broadband excitation and pulse compression, we can select it efficiently for high resolution inspections. For illustration, a numerical simulation is considered. Figure 1 shows the geometry of the aluminum plate used in the investigation. The damage is simulated as a rectangular slot close to the edge of the plate. A surface bonded PZT is used as both the actuator and the sensor. The broadband excitation signal is a chirp with its frequency sweeps from 4 kHz to 152 kHz over a 1*ms* window. It is assumed that the transducer is ideal and excites only the guided wave mode of interest. Referring to [9], if there is no mode conversion when the outgoing wave is reflected at a feature, the received time-trace,

$$g(t)=\sum_{j}\int_{-\infty}^{\infty}A_j(\omega)F(\omega)e^{i(k(\omega)d_j-\omega t)}d\omega \tag{13}$$

where d_j is the round trip propagation distance to each reflector, and the reflection coefficient of each reflector, $A_j(\omega)$, is taken to be a constant (*i.e.* $A_j(\omega)\equiv 0.5$). Figure 2(a) shows the response as the A0 Lamb wave mode propagating in this specimen, and Figure 2(b) shows the corresponding pulse compression signal.

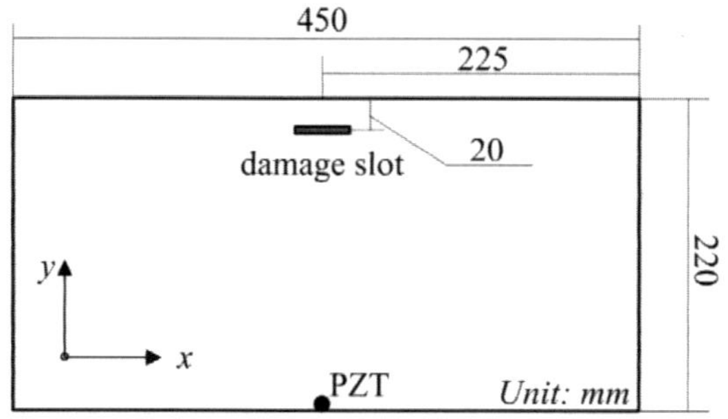

Figure 1. Schematic diagram of the aluminum plate (1-mm thick) with a damage slot.

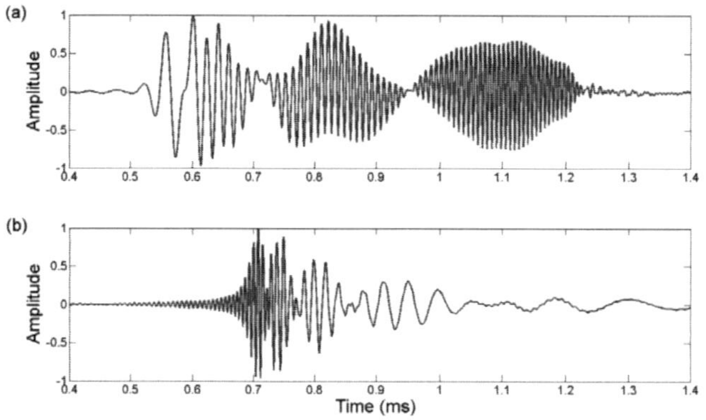

Figure 2. (a) Simulated chirp response; (b) the corresponding pulse compression signal.

It is noted that the quality of the results obtained from the proposed dispersion pre-compensation algorithm is dependent on the accuracy with which the estimated dispersion curves represent the actual characteristics of the system (see equation (10)). Referring to Wilcox's study [9], the effects of two different types of inaccuracies in the dispersion curves must be considered. As a result, besides the ideal case where the computed dispersion characteristics exactly match the actual ones (*i.e.* Case I), another two cases for the existence of these two kinds of inaccuracies (*i.e.* Case II and Case III) are also discussed in this section. For comparison, a 2 cycle Hanning windowed tone burst centered at 65kHz is used as the original narrowband excitation signal, $s_d(t)$, in these cases. In addition, all responses corresponding to different compensation distances are extracted from the broadband response shown in figure 2(a) through the transform function.

Case I

In this ideal condition, the dispersion pre-compensation algorithm can restore the reflected wave packet to the exact shape of the original narrowband excitation signal, $s_d(t)$, when the compensation distance coincides with the corresponding propagating distance. Hence, the dispersion distance is always selected as follows. ① In a single actuator-sensor pair, it is often conducted as the path length between the actuator and the sensor. ② In a transducers array, it is often considered as the average of the largest path length and the least path length between the actuator and the sensor elements or the path length of certain actuator-sensor pair which has the greatest demand of resolution. ③ If the size of the array is too large for the resolvable resolution to accommodate all the paths, the array may be divided into a few segments.

In the numerical simulation, the associated output signals as the compensation distance changes from 0.28m to 0.52m with a step of 0.04m are shown in figure 3. According to the schematic diagram shown in figure 1, the first two reflected wave packets are caused by reflections of energy in the incident guided wave from the damage and the surrounding edge respectively. Hence, their propagation distances are 0.4m and 0.44m. Applying the former wave packet for example, the time duration decreases as the increase of the compensation distance first, till the optimal point where the compensation distance is identical to the propagation distance (*i.e.* L = 0.4m), the minimum value is obtained, and then it increases as the further increase of the compensation distance.

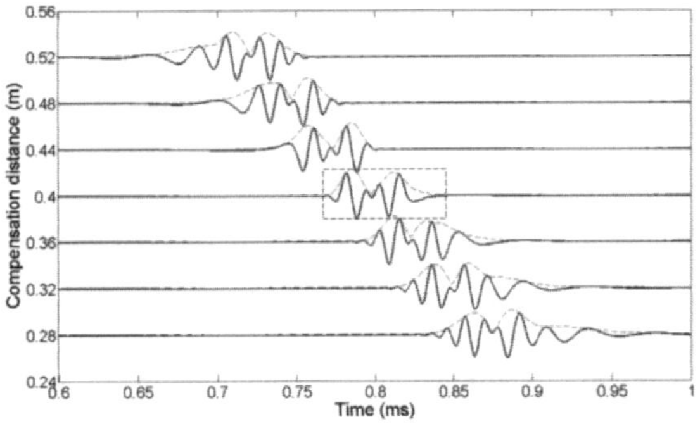

Figure 3. Responses to the new excitation signals compensated by different distances (from 0.28m to 0.52m with a step of 0.04m) in the ideal condition.

Case II

$$v_{ph}^{(0)}(f)=(1+\alpha)v_{ph}(f) \tag{14}$$

where $v_{ph}^{(0)}(f)$ is the actual phase velocity, $v_{ph}(f)$ is the phase velocity in dispersion curves, and α is the perturbation. This is representative of the cases as incorrect material properties were used to compute the dispersion curve [9].

In this condition, if the propagation distance is identical to the compensation distance (*e.g.* the first wave packet), the dispersion effects of the Lamb wave propagating are over compensated, completely compensated and under compensated corresponding to $\alpha<0$, $\alpha=0$, $\alpha>0$, respectively. Since the calculated phase velocity is exactly proportional to the actual one, the temporal duration of the received signals could be scaled by the same amount in the time domain though optimal design of compensation distance. Figure 4 shows a series of new excitation responses computed from the broadband response by varying the compensation distance from 0.28mm to 0.52m with the increment of 0.04m when the perturbation α=-0.2. The wave packets as the compensation distance consists with the propagating distance of the first wave packet (*i.e.* L=0.4m) are enclosed by a rectangle in this figure. It can be seen that the perturbation worsens the resolution of these wave packets. Fortunately, with a proper choice of the compensation distance, the resolution will be improved. For example, when the compensation distance L=0.32m, the wave packets (enclosed by the circle in figure 4) have the same shape as the ones in the ideal conditions when the compensation distance L=0.4m (enclosed by a rectangle in figure 3). It means that the first wave packet also meets its best resolution.

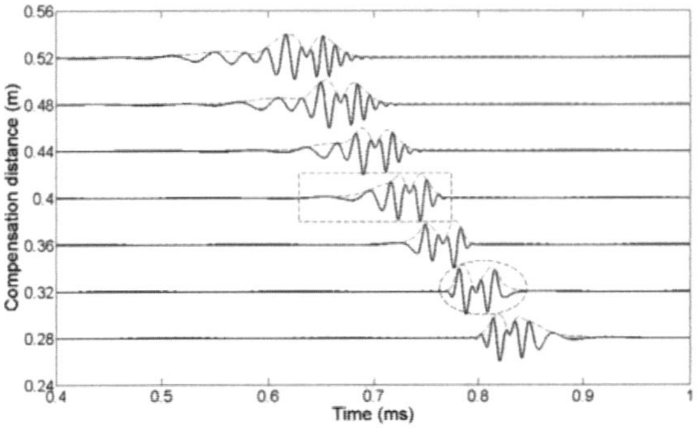

Figure 4. The results obtained from dispersion pre-compensation proposed in this paper with perturbation parameter α equal to -0.2 and the compensation distance changes from 0.28m to 0.52m.

Case III

$$v_{ph}^{(0)}(f)=v_{ph}(f(1+\beta)) \tag{15}$$

where β is the perturbation. This perturbation corresponds to the physical thickness of the plate not being the same as the thickness used in the dispersion curves calculation [9].

Different with type I, there is not an exact proportion relation between the calculated dispersion data and the actual one. As a result, the shape of the received signals cannot be compressed to the ideal one through optimal design of the compensation distance. However, the optimization of the compensation distance can also alleviate the effects of this perturbation. Figure 5 shows a series of new excitation responses computed from the broadband response by varying the compensation distance from 0.28mm to 0.52m with the increment of 0.04m when the perturbation β=0.2. It can be seen that the two wave packets are completely separated in time when the compensation distance L=0.48m (enclosed by the circle), though 20% errors are contained in the dispersion data.

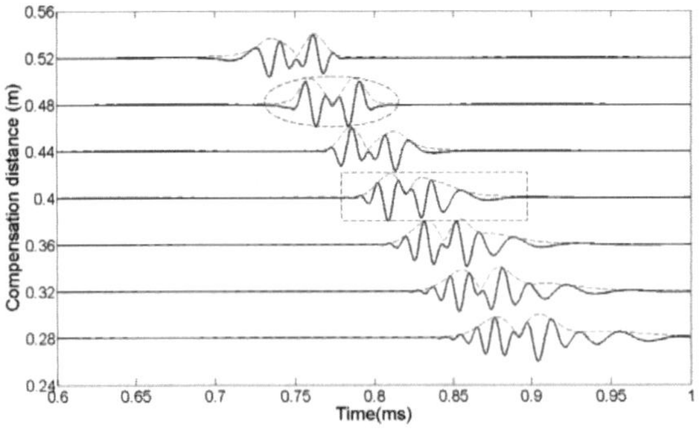

Figure 5. The results obtained from dispersion pre-compensation proposed in this paper with perturbation parameter β equal to 0.2 and the compensation distance changes from 0.36m to 0.48m.

In summary, the proposed dispersion pre-compensation algorithm is a stable technique in general. It is of good robustness to inaccuracies in calculated dispersion data. More importantly, there is a great choice of waveform parameters, with which the received wave packets are completely separated in time and thus the structural features can be separately identified (as shown in figures 3-5). This makes the proposed method efficient and effective for high-resolution Lamb wave inspection.

4. Experimental investigation

In this experiment, the excitation signal is generated by an Agilent 33220A function/arbitrary waveform generator, after being amplified with the peak-to-peak voltage of 50 V (by a Piezo Systems

EPA-104 voltage amplifier), which is applied to the actuator (PZT) of the sensor network. The Lamb wave signals pass through the monitoring area and are captured by the sensors (PZT). After being amplified by the AVANT NI-2000 conditioning amplifiers, they are acquired at a sampling rate of 10MHz by a NI PXIe-1082 data acquisition. Figure 6 shows the schematic diagram of the aluminum specimen (1200*mm*x500*mm* x1.39*mm*) and the PZT configuration. An artificial defect was introduced in the form of through-thickness notch with dimensions 50*mm*x5*mm*x1.39*mm* in the specimen.

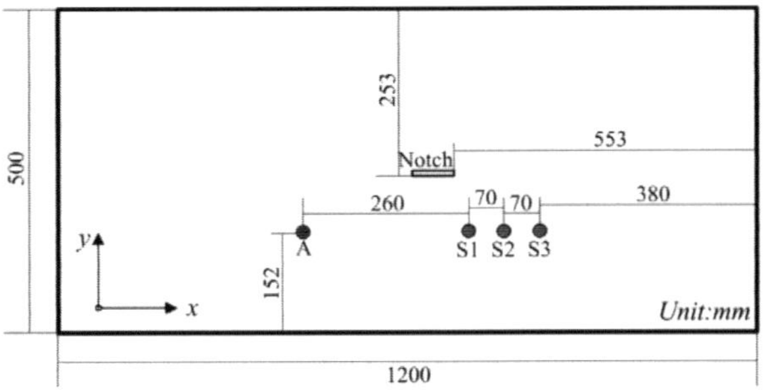

Figure 6. Schematic diagram of the PZT sensors placement in the aluminum plate.

A broadband chirp signal, sweeping from 4 kHz to 152 kHz over a 500μs window is applied to actuate Sensor A. Similar to the numerical simulation, the choice of the proper waveform parameters is performed through the broadband response. Here a 2 cycle Hanning windowed tone burst centered at 80 kHz with the dispersive effects being compensated by a distance of 400mm (equal to the distance between Actuator A and Sensor S3) is an appropriate excitation signal. Figure 7 shows the responses of the sensor array, which are calculated from the broadband response. It can be seen that the first three wave packets, *i.e.* the incident A0 mode, the damage scattered A0 mode and the edge reflected A0 mode in turn, are completely separated in time domain, which means the three structural features could be unambiguously identified.

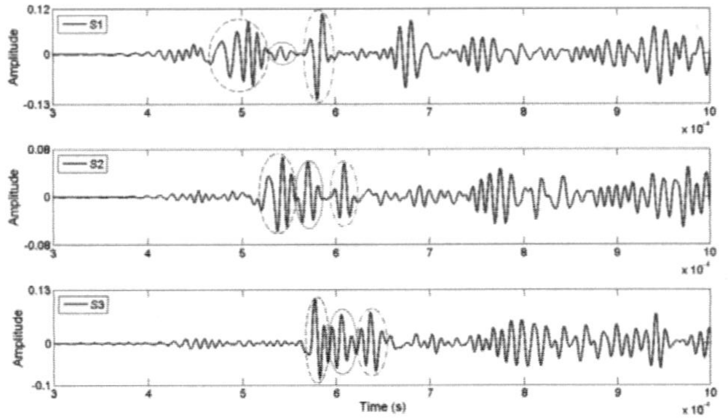

Figure 7. Responses to the new excitation signals, which are derived from 2 cycle 80 kHz tone burst being compensated by a distance of 0.4m, as generated from the measured chirp response.

Without dispersion compensation, the responses of the sensor array corresponding to the original tone burst are also extracted from the same broadband response. As shown in Figure 8, the received signals are much more complicated because all the wave packets resulting from different structural features overlap with each other.

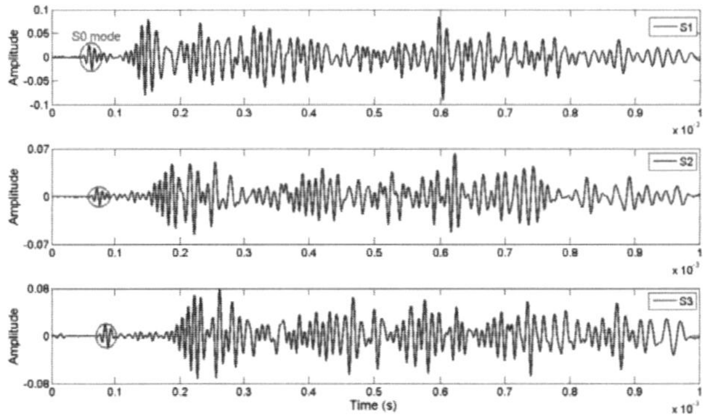

Figure 8. Responses to 2 cycle Hanning windowed tone burst centered at 80kHz as computed from the measured chirp responses.

5. Conclusions

A pulse compression and dispersion compensation method is established for high-resolution Lamb wave inspection. Some conclusions are obtained as follows.

(i) Combined with broadband excitation and pulse compression, the proposed dispersion compensation algorithm is efficient, since it can be performed during the process of the narrowband response extracting.

(ii) The effects of inaccuracies in calculated dispersion curves would possible to be alleviated by optimal design of the compensation distance.

6. Acknowledgments

The work is supported by the National Natural Science Foundation of China (Grant No. 51421004, 51125022), and the Ph.D. Programs Foundation of Ministry of Education of China (Grant No. 20110201110025), which are highly appreciated by the authors.

References

[1] Rose J L 1999 Ultrasonic Waves in Solid Media (Cambridge: Cambridge University Press)

[2] Wilcox P D, Lowe M J S and Cawley P 2001 The effect of dispersion on long-range inspection using ultrasonic guided waves *NDT & E Int.* 34 1-9

[3] Ghosh T, Kundu T and Karpur P 1998 Efficient use of Lamb modes for detecting defects in large plates *Ultrasonics* 36 791-801

[4] Qu W Z, Xiao L, Zhou Y G and Inman D J 2013 Lamb wave damage detection using time reversal DORT method *Smart Materials and Structures* 22 045014

[5] Li F, Meng G, Ye L, Lu Y and Kageyama K 2009 Dispersion analysis of Lamb waves and damage detection for aluminum structures using ridge in the time-scale domain *Meas. Sci. Technol.* 20: 095704.

[6] Cai J, Shi L H and Qing X L P 2013 A time-distance domain transform method for Lamb wave dispersion compensation considering signal waveform correction *Smart Materials and Structures* 22 105024 (13 pp.)

[7] Liu L and Yuan F G 2010 A linear mapping technique for dispersion removal of Lamb waves *Struct. Health Monit.* 9 75-86

[8] Michaels J E, Lee S J, Croxford A J and Wilcox P D 2013 Chirp excitation of ultrasonic guided waves *Ultrasonics* 53 265-270

[9] Wilcox P D 2003 A rapid signal processing technique to remove the effect of dispersion from guided wave signals *IEEE Trans Ultrason Ferroelectr Freq Control* 50 419-427

Mode purification for ultrasonic guided waves under pseudo-pulse excitation

Z Luo, J Lin[1], L Zeng and F Gao

School of Mechanical Engineering, Xi'an Jiaotong University, Xi'an, China, 710049

jinglin@mail.xjtu.edu.cn

Abstract. Under broadband excitation, the captured Lamb wave signals contain rich information of the structural properties. However, since multiple modes are highly dispersive and overlapped with each other, Lamb wave signals are especially complicated to be interpreted. To overcome this problem, a mode purification strategy is established under pseudo-pulse excitation. In this strategy, a pseudo-pulse excitation technique is introduced to obtain a high resolution for the inspection firstly. Dispersion compensation method is applied to remove the dispersion of the received signal subsequently, through which all the wave packets associated with the interferential mode propagating through different paths could be compressed into the same shape as the excitation. Benefiting from that, the energy of the wave packets corresponding to the interference mode could be concentrated in time domain as individual temporal pulses, which thereby could be eliminated by zero-amplitude rectangular time windows without affecting the desired mode much. After that, the inverse dispersion compensation is applied to the residual signals to restore the original waveform of the desired mode. Finally, experiments are introduced to validate the availability and robustness of the proposed strategy.

1. Introduction

Lamb wave techniques allow long-range inspections for large-scale plate-like waveguides (e.g., airplanes, pipelines and oil tanks) rapidly and nondestructively, because of its little energy loss and high sensitivity to both the surface and internal defects [1-4]. Actually, under broadband excitation, rich information of the structural properties could be obtained in the received Lamb wave signals [5]. Especially, the detection of incipient damage in the structure may be possible [6, 7]. However, for broadband excitations, the Lamb waves are especially complicated and thus cannot be directly interpreted in time domain, because multiple modes are highly dispersive and overlapped with each other.

One candidate way to solve this problem is the time-frequency representation (TFR), because it provides a clear illustration for the temporal variation modal energy stream in the time-frequency domain [8]. Unfortunately, with distinct dispersive effects and propagating through different distances, there are some frequency components corresponding to different modes arrive in phase at the receiver position, and thus these modes will interact at the neighborhood of those frequencies. The energy of those interacted modes, which covers the same time-frequency tiling in the time-frequency representation, will be processed together and cannot be separated by current methods [9]. Xu et al.

[1] To whom any correspondence should be addressed.

developed a mode separation method on the basis of dispersion compensation method which could compress the dispersive signal into the initial shape of the excitation with priori-knowledge of the dispersion curves [10]. However, the energy of the short spiky pulse excitation is the main limitation because guided wave measurements are frequently performed using low voltage excitations [11, 12].

In this paper, a pseudo-pulse excitation technique is introduced to obtain a high resolution for the inspection. The pseudo-pulse excitation lowers the probability of mode overlapping, while achieves a high signal-to-noise ratio. On this basis, a windowed function is employed to purify the Lamb modes in the received signals, and thus the wave-packets reflected from the structural features (defects or edges) would be much easier to be identified.

2. Methods

2.1. Pseudo-pulse excitation

Limited by the low generating voltage, when a spiky pulse acts as the excitation, it is hard to acquired response signal with high signal-to-noise ratio. Exchanging spiky pulse with the long duration excitation could significantly enhance the signal-to-noise ratio of the response signal. However, with the increase of duration, the probability of mode overlapping would rise, which complicates the interpretation of the response signal. Therefore, a pseudo-pulse excitation, with short duration and high signal-to-noise ratio, is introduced in this section.

Golay complementary code (GCC) is one type of the binary sequences, which contains two binary sequences [13-15], 8-bit for example:

$$\begin{aligned} \boldsymbol{A}[N] &= [+1,+1,+1,-1,+1,+1,-1,+1] \\ \boldsymbol{B}[N] &= [+1,+1,+1,-1,-1,-1,+1,-1] \end{aligned} \tag{1}$$

The auto-correlation functions of $\boldsymbol{A}$[N] and $\boldsymbol{B}$[N], $\boldsymbol{\Psi}_A[n]$ and $\boldsymbol{\Psi}_B[n]$, have side-lobes with equivalent magnitude but opposite sign. Therefore, the sum of the two auto-correlation functions has a main peak and zero side-lobes, which can be represented as

$$\Psi_A[n]+\Psi_B[n]=\begin{cases}16, & n=0\\ 0, & n\neq 0\end{cases} \tag{2}$$

The single-frequency sinusoidal signal with the frequency f=150 kHz and the cycle k=1 is used for GCC modulation. $a(t)$ and $b(t)$ are the $\boldsymbol{A}[N]$ and $\boldsymbol{B}[N]$ modulation, respectively, shown in figure 1(a) and 1(b). $\psi_a(t)$ and $\psi_b(t)$ are the auto-correlation of $a(t)$ and $b(t)$, respectively. As shown in figure 1(c) and 1(d), $\psi_a(t)$ and $\psi_b(t)$ have side-lobes with equivalent magnitude but opposite sign. After summing $\psi_a(t)$ and $\psi_b(t)$, as shown in figure 1(e), the sum, $s(t)$, has a main peak and zero side-lobes, which is the pseudo-pulse signal.

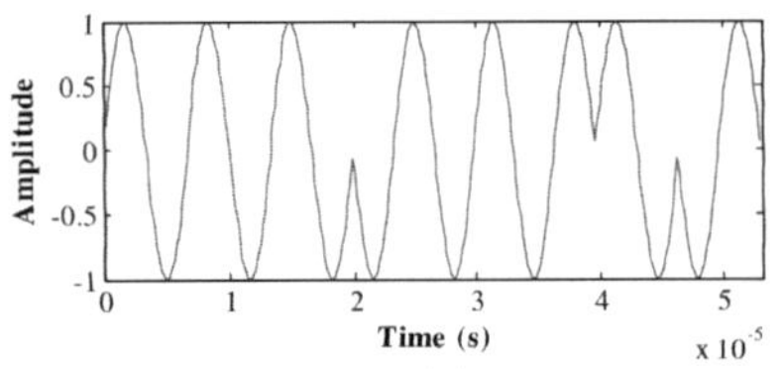

(a)

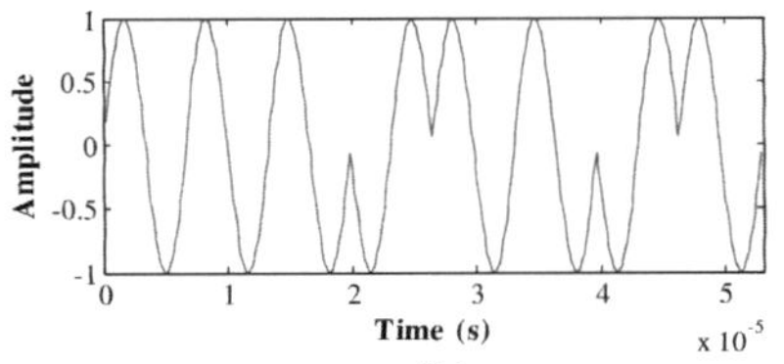

(b)

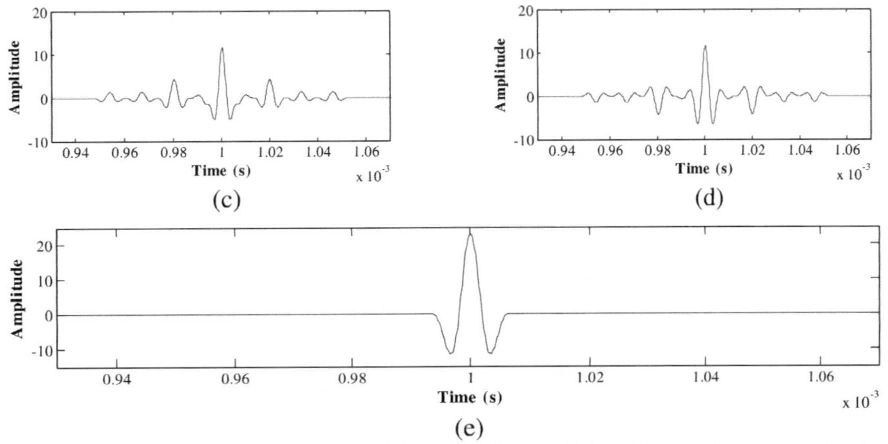

Figure 1 Cancellation property of side-lobes for an 8-bit GCC: (a) and (b): 8-bit GCC modulation; (c) and (d): auto-correlation functions of (a) and (b), respectively; (e): sum of (c) and (d).

Consider that two transducers are mounted on a plate, acting as the transmitter and receiver, respectively. The entire system, consisting of the instrumentations, transducers and structures, can be considered as a linear system [16]. If an excitation signal f(t) is generated by the transmitter, the response can be denoted as

$$r(t) = \int F(\omega)H(\omega)\exp(-i\omega t)d\omega \tag{3}$$

where $F(\omega)$ is the Fourier transform of $f(t)$, $H(\omega)$ is the transfer function of the system.

To gain the pulse compression signal $c(t)$, the response signal $g(t)$ is cross-correlated with the excitation signal $f(t)$. The pulse compression signal is represented as

$$\begin{aligned} c(t) &= \int F(\omega)H(\omega)F(\omega)^{*}\exp(-i\omega t)d\omega \\ &= \int |F(\omega)|^{2} H(\omega)\exp(-i\omega t)d\omega. \end{aligned} \tag{4}$$

where the superscript * represents the complex conjugate, $|F(\omega)|^2$ represents the Fourier transform of the auto-correlation of the excitation signal $f(t)$. Equation (4) means that the pulse compression of the response is equal to adopting this auto-correlation curve as the new excitation signal [17-19].

When $a(t)$ and $b(t)$ are acted as excitation, cross-correlating the response signals with $a(t)$ and $b(t)$, respectively, the sum of the cross-correlated signals is the response signal of pseudo-pulse excitation $s(t)$.

2.2. Dispersion compensation algorithm

The dispersion phenomenon in Lamb wave propagation has been acknowledged. Considering that the transducers generate/receive only a Lamb wave mode and the propagation distance is d_j, the amplitude attenuation is $A_j(\omega)$, the received signal can be predicted as [20]

$$g(t) = \sum_{j} \int_{-\infty}^{\infty} A_j(\omega)S(\omega)\exp[-ik(\omega)d_j]\exp(i\omega t)d\omega \tag{5}$$

where $k(\omega)$ is the wavenumber of the guided wave mode, which is a function of angular frequency ω, and $S(\omega)$ is the Fourier transform of excitation $s(t)$.

In equation (5), because the wavenumber $k(\omega)$ is a non-linear function with respect to the angular frequency ω, the non-linear phase adjustment term $\exp[-ik(\omega)d_j]$ will change the time delay of harmonic of $F(\omega)$, according to the time shift property of Fourier transform [21]. It causes the received signal $g(t)$ spreading out in time domain and deforming comparing with the original excitation, which is the so-called dispersion. Therefore, when $s(t)$ is used as excitation, the received signal will not still maintain pulse-shape.

However, if the wavenumber $k(\omega)$ is in a linear relation with respect to the angular frequency ω, $k(\omega)$ will not cause phase distortions, and there will be no dispersion in a propagated waveform [22]. On the basis of this concept, the dispersion compensation algorithm maps the dispersion relation $k(\omega)$ to the linear $k_{lin}(\omega)$ to remove the dispersion effect [23]. The linear wavenumber is approximated by using the Taylor expansion of the wavenumber $k(\omega)$ at the excitation central frequency ω_c up to the first order as

$$k(\omega) \approx k_{\text{lin}}(\omega) = k(\omega_c) + k'(\omega_c)(\omega - \omega_c) = k(\omega_c) + \frac{(\omega - \omega_c)}{c_{gr}(\omega_c)} \tag{6}$$

where $k'(\omega_c)$ is the first-order derivative of $k(\omega)$ at ω_c, and c_{gr} is the group velocity of the guided wave mode.

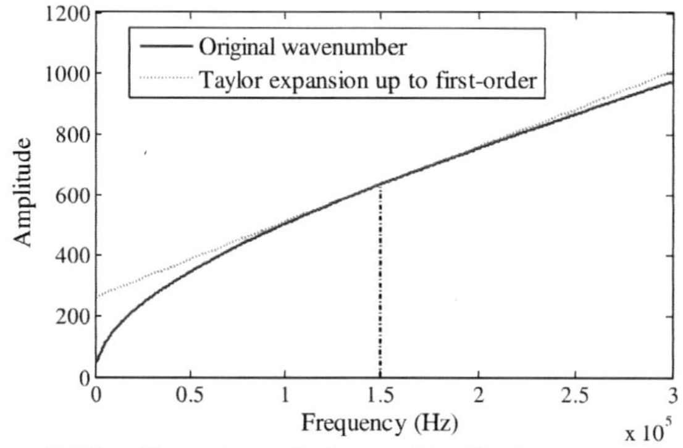

Figure 2. The dispersion relation and its Taylor expansion

The procedure of dispersion compensation can be summarized as follows:

- (i) Apply Fourier transform to the received signal $g(t)$ to get $G(\omega)$. Since k is a function of ω, $G(\omega)$ can be rewritten as $G(k)$;
- (ii) Use equation (6) to calculate the linear wavenumber $k_{\text{lin}}(\omega)$;
- (iii) Interpolate $[k, G(k)]$ pair at $k_{\text{lin}}(\omega)$ to get $G[k_{\text{lin}}(\omega)]$;
- (iv) Apply inverse Fourier transform to $G[k_{\text{lin}}(\omega)]$ to obtain a wave with the dispersion removed.

After dispersion compensation, all the wave packets associated with the compensated mode propagating through different paths could be compressed into the same shape as the excitation.

2.3. Mode elimination

When $s(t)$ is generated by transmitter, assuming that the received modes (A0 and S0, for instance) are overlopped with each other in time domain, the received signal can be represented as

$$G(\omega) = S(\omega)H_{A0}(\omega) + S(\omega)H_{S0}(\omega) = G_{A0}(\omega) + G_{S0}(\omega) \tag{7}$$

where $S(\omega)$ is the Fourier transform of the excitation, $H_{A0}(\omega)$ and $H_{S0}(\omega)$ are the transfer function of A0 and S0, repectively, $G_{A0}(\omega)$ and $G_{S0}(\omega)$ are the Fourier transform of the receive signals of the A0 and S0, repectively.

After applying dispersion compensation to the received signal with knowledge of the dispersion characteristics of A0, the processed signal can be denoted as

$$\begin{aligned} Q_{dc}(\omega) &= G_{\mathrm{A0}}(k)\circ K_{\mathrm{A0,lin}}(\omega)+G_{\mathrm{S0}}(k)\circ K_{\mathrm{A0,lin}}(\omega) \\ &= S(k)H_{\mathrm{A0}}(k)\circ K_{\mathrm{A0,lin}}(\omega)+G_{\mathrm{S0}}(k)\circ K_{\mathrm{A0,lin}}(\omega) \\ &= S_{dc}(\omega)+G_{S0}(k)\circ K_{\mathrm{A0,lin}}(\omega) \end{aligned} \tag{8}$$

where $\circ$ is composition operator, and $H_{\mathrm{A0}}(k)$ and$H_{\mathrm{S0}}(k)$ are implicit functions of k. $K_{\mathrm{A0,lin}}(\omega)$ is the Taylor expansion of the wavenumber $k(\omega)$ of A0 mode. $H_{\mathrm{A0}}(k)\circ K_{\mathrm{A0,lin}}(\omega)$ is in a linear relation with ω. $Q_{dc}(\omega)$ is the Foutier transform of the compensated signal and $S_{dc}(\omega)$ is the compensated A0 mode.

In equation (8), $S_{dc}(\omega)$ shows that the energy of the wave packets corresponding to the compensated mode would be concentrated in time domain as individual temporal pulses. $G_{\mathrm{S0}}(k)\circ K_{\mathrm{A0,lin}}(\omega)$ means that the energy of the other mode (S0 mode) will spread out further in time domain. Therefore, the main lobe of the compensated pulses could be eliminated by zero-amplitude rectangular time windows without affecting other modes much, because of short duration of the excitation.

$$Q_{dc}(\omega)\text{-}S_{dcm}(\omega)=S_{dc}(\omega)\text{-}S_{dcm}(\omega)+G_{\mathrm{S0}}(k)\circ K_{\mathrm{A0,lin}}(\omega)\approx G_{\mathrm{S0}}(k)\circ K_{\mathrm{A0,lin}}(\omega) \tag{9}$$

where $S_{dcm}(\omega)$ is the main lobe of $s(t)$, and $S_{dcm}(\omega)$ contains most energy of the compensated mode $S_{dc}(\omega)$.

After eliminating the A0 mode, the inverse dispersion compensation is applied to the residual signals to restore the original waveform of the S0 mode.

$$\left[Q_{dc}(\omega)\text{-}S_{dcm}(\omega)\right]\circ K_{\mathrm{A0,lin}}^{-1}(\omega)\approx G_{\mathrm{S0}}(\omega)\circ K_{\mathrm{A0,lin}}(\omega)\circ K_{\mathrm{A0,lin}}^{-1}(\omega)\approx G_{\mathrm{S0}}(\omega) \tag{7}$$

where $\circ K_{\mathrm{A0,lin}}^{-1}(\omega)$ is the inverse dispersion compensation.

3. Experimental Example

The experiment was carried out on an aluminium plate with dimensions 2000 mm×1000 mm×2 mm. Two piezoelectric ceramic discs with a diameter of 8 mm and 0.5 mm in thickness were mounted on the plate, acting as the transmitter and receiver, respectively, with 340 mm between them. An artificial damage is 170 mm away from the receiver and the two PZTs are 660mm away from the right boundary, shown in figure 3.

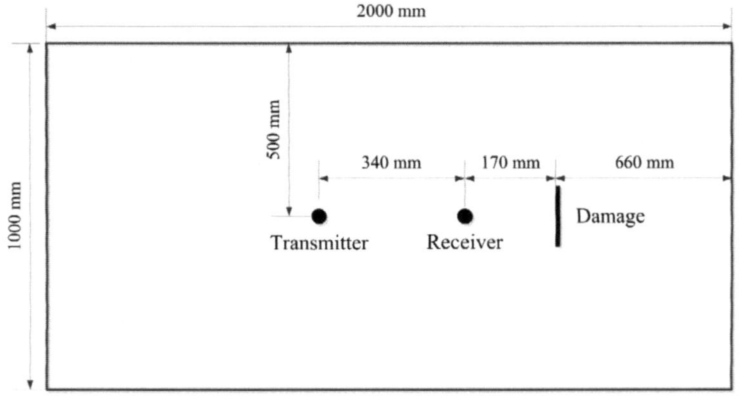

Figure 3. Experimental setup for measuring reflections of Lamb waves in an aluminium plate

In this experiment, $a(t)$ and $b(t)$ (shown in figure 1(a) and 1(b)) are generated by an Agilent 33220A function/arbitrary waveform generator, respectively, whose the centre frequency is 150 kHz. After being amplified with the peak-to-peak voltage of 5 V by a Piezo Systems EPA-104 voltage amplifier, they are applied to the actuator. The Lamb wave signals pass through the monitoring area and are captured by the receivers. After being amplified by the AVANT NI-2000 conditioning amplifiers, they are acquired at a sampling rate of 5 MHz by a NI PXIe-1082 data acquisition.

The received signals are cross-correlated with the corresponding excitations and the sum of the cross-correlated signals is the equivalent received signal, shown in figure 4(a). In this experiment, only the two fundamental A0 and S0 modes exist due to the frequency range is with [0-300] kHz. As it can be seen, A0 mode is highly dispersive and the direct arrival of A0 last a long duration, which may cover some information of other modes.

After applying dispersion compensation based on A0 mode dispersion relation, the A0 mode wave packets are compressed into pseudo-pulses and the S0 mode packets are spread out further, shown in figure 4(b).

The compensated A0 mode packets are pulse-like and can be eliminated by short zero-amplitude rectangular time windows, shown in figure 4(c). After elimination, the residual signal is restored by the inverse dispersion compensation, with the reflection of S0 mode from the damage being obvious, which could be applied to evaluate the health of the structure, as shown in figure 4(d).

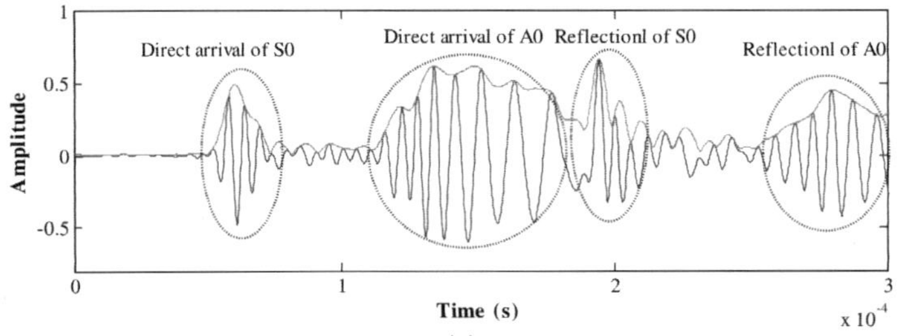

(a)

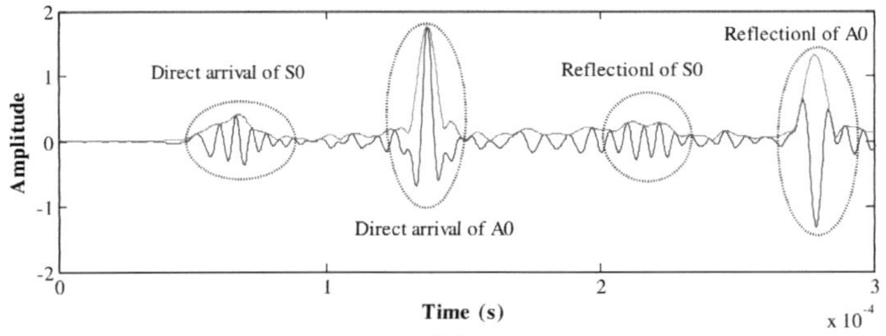

(b)

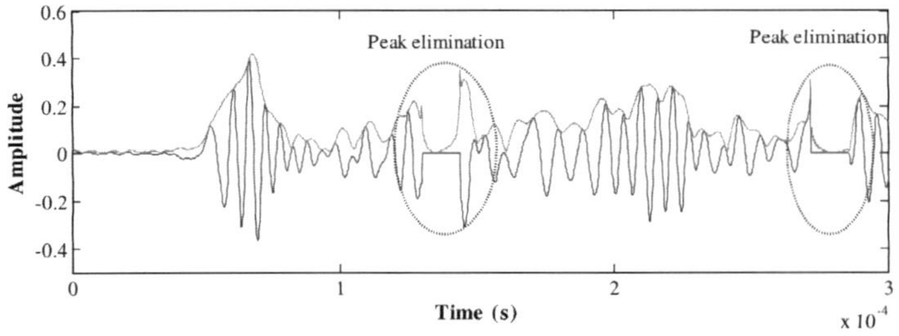

(c)

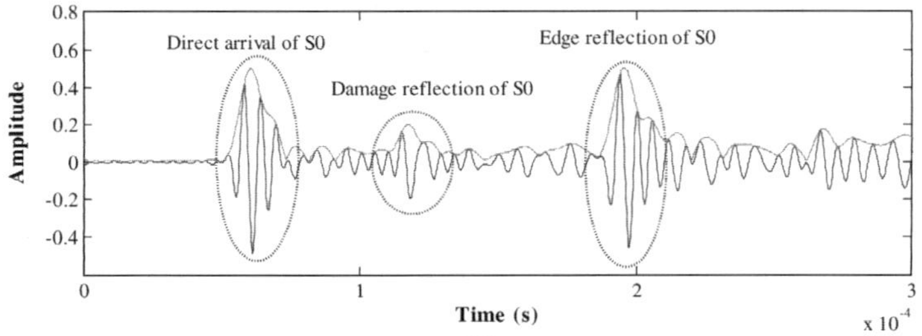

(d)

Figure 4. Experiment results of mode purification: (a) received signal; (b) compensated signal; (c) elimination of A0 mode; (d) restored signal

4. Conclusion

In this study, a mode purification strategy is established to solve mode overlapping. Some conclusions are obtained as follows.

(i) The pseudo-pulse excitation could be generated easily by PZTs with a short duration and a high signal-to-noise ratio.

(ii) Since different modes are of distinct dispersion characteristics, when the compensated mode is compressed with the application of dispersion compensation, the energy of the other modes may further spread out.

(iii) Through mode purification strategy, the energy of interferential modes could be obviously alleviated and thus wave packets corresponding to the interest mode would be much easy to be identified.

5. Acknowledgments

The work is supported by the National Natural Science Foundation of China (Grant No. 51421004, 51125022), and the Ph.D. Programs Foundation of Ministry of Education of China (Grant No. 20110201110025), which are highly appreciated by the authors.

6. References

[1] Prosser W H, Seale M D and Smith B T 1999 Time-frequency analysis of the dispersion of Lamb modes *J. Acoust. Soc. Am.* **105** 2669-76

[2] Wilcox P D, Lowe M J S and Cawley P 2001 The effect of dispersion on long-range inspection using ultrasonic guided waves *NDT & E Int.* **34** 1-9

[3] Alleyne D N, Pialucha T P and Cawley P 1993 A signal regeneration technique for long-range propagation of dispersive Lamb waves *Ultrasonics* **31** 201-4

[4] Zeng L, Lin J, Lei Y and Xie H 2013 Waveform design for high-resolution damage detection using Lamb waves *IEEE. Trans. Ultrason. Ferroelectr. Freq. Control.* **60** 1025-9

[5] Michaels J E, Lee S J, Croxford A J and Wilcox P D 2013Chirp excitation of ultrasonic guided waves *Ultrasonics* **53** 265-70

[6] Pierce S G and Culshaw B 1998 Laser generation of ultrasonic Lamb waves using low power optical sources *IEE Proc-21Meas. Technol* **145** 5 244-49

[7] Fasel T R, Olson C C and Todd M D. 2008 Optimized guided wave excitations for health monitoring of a bolted joint *The 15th International Symposium on: Smart Structures and Materials & Nondestructive Evaluation and Health Monitoring. International Society for Optics and Photonics* 69351N-69351N-8

[8] Xu K, Ta D and Wang W 2010 Multiridge-based analysis for separating individual modes from multimodal guided wave signals in long bones *IEEE Trans. Ultrason. Ferroelectr. Freq. Control* **57**112480-90

[9] Zhao M, Zeng L, Lin J and Wu W T 2014 Mode identification and extraction of broadband ultrasonic guided waves *Meas. Sci.Technol.* **25**(11): 11505

[10] Xu K, Ta D, Moilanen P and Wang W Q 2012 Mode separation of Lamb waves based on dispersion compensation method *J.Acoust. Soc. Am.* **131**(4) 2714-22

[11] Pertsch A, Kin J, Wang Y and Jacobs L 2011 An intelligent stand-alone ultrasonic device for monitoring local structural damage: implementation and preliminary experiments *Smart Mater. Struct.* **20** 015022

[12] Marchi L D, Perelli A and Marzani A 2013A signal processing approach to exploit chirp excitation in Lamb wave defect detection and localization procedures *Mech. Syst. Signal Process.* **39**(1-2) 20-31

[13] Shen C C and Shi T Y 2011 Golay-encoded excitation for dual-frequency harmonic detection of ultrasonic contrast agents *IEEE Trans. UFFC* **58** 349-56

[14] Garcia-Rodriguez M, Yanez Y, Garcia-Hernandez M J, Salazar J, Turo A and Chavez J A 2010 Application of Golay codes to improve the dynamic range in ultrasonic Lamb waves air-coupled systems *NDT & E Int.* **43** 677-86

[15] Veres I A, Cleary A, Thursby G, Mckee C, Armstrong I, Pierce G and Culshaw B 2013 Golay code modulation in low-power laser-ultrasonic *Ultrasonics* **53** 122-9

[16] Michaels J E, Lee S J, Croxford A J and Wilcox P D 2013 Chirp excitation of ultrasonic guided waves *Ultrasonics* **53** 265-70

[17] Gan T H, Hutchins D A, Billson D R and Schindel D W 2001 The use of broadband acoustic transducers and pulse-compression techniques for air-coupled ultrasonic imaging Ultrasonics **39** 181-94

[18] Toiyama K 2004 High S/N ratio guided wave inspection of pipe using chirp pulse compression *In. Proc. ASME/JSME Press, PVP* **484** 41-5

[19] Toiyama K and Hayashi T 2008 Pulse compression technique considering velocity dispersion of guided wave AIP Conf. Proc. **975** 587-93

[20] Wilcox P D 2003 A rapid signal processing technique to remove the effect of dispersion from guided wave signals *IEEE. Trans. Ultrason. Ferroelectr. Freq. Control.* **50** 419-27

[21] Liu L and Yuan F G 2010 A linear mapping technique for dispersion removal of Lamb waves *Struct. Health Monit.* **9** 75-86

[22] Zeng L and Lin J 2014 Chirp-based dispersion pre-compensation for high resolution Lamb wave inspection *NDT & E Int.* **61** 35-44

[23] Xu B, Yu L and Giurgiutiu V 2009 Lamb wave dispersion compensation in piezoelectric wafer active sensor phased-array application *Proc. SPIE, Health Monit. Struct. Biol. Syst.* **7295** 729516

Machinery vibration signal denoising based on learned dictionary and sparse representation

Liang Guo, Hongli Gao, Jun Li, Haifeng Huang and Xiaochen Zhang
School of Mechanical Engineering, Southwest Jiaotong University, ChengDu 610031 China
E-mail: guoliang2248@my.swjtu.edu.cn

Abstract. Mechanical vibration signal denoising has been an import problem for machine damage assessment and health monitoring. Wavelet transfer and sparse reconstruction are the powerful and practical methods. However, those methods are based on the fixed basis functions or atoms. In this paper, a novel method is presented. The atoms used to represent signals are learned from the raw signal. And in order to satisfy the requirements of real-time signal processing, an online dictionary learning algorithm is adopted. Orthogonal matching pursuit is applied to extract the most pursuit column in the dictionary. At last, denoised signal is calculated with the sparse vector and learned dictionary. A simulation signal and real bearing fault signal are utilized to evaluate the improved performance of the proposed method through the comparison with kinds of denoising algorithms. Then Its computing efficiency is demonstrated by an illustrative runtime example. The results show that the proposed method outperforms current algorithms with efficiency calculation.

1. Introduction

To make machines run in good operating condition and reduce the maintenance costs, health conditions needs to be monitored. Vibration signals collected from a machine contain rich information[1]. Hence, By inspecting the physical characteristics of the vibration signals, one is able to detect the presence of a fault in an operation machine, to localise the position of damaged parts, etc.

However, noise is the major barrier in defect detection for the mechanical vibration signal[2]. At present, in the field of eliminating noise, many technologies have been developed such as Moving Average algorithm[3], Fast Fourier Transform(FFT)-based Hilbert transform[4] and so on. Since the beginning of the 90's there has been considerably interest in the use of wavelet transforms for the removal of noise from signals. Mallat et.zz[5] describes the singularity of signal and noise based on multiscale, then raise the denoising algorithm based on maximum the norm of wavelet transform. Donoho develops many correlative algorithms based on wavelet transform. in[6], RiskShrink with the minimax threshold and VisuShrink with the universal threhold are introduced with solid mathematicians properties for signal denoising. At the same time, methods based on redundant representation and sparsity has drawn a lot of research attentions. Sparsity of the unitary wavelet coefficients was considered, leading to the celebrated shrinkage algorithm[7]. Wei et al.[8] combines the Morlet wavelet transform and sparse code shrinkage (SCS) to enhance the bearing vibration impulsive features and suppress residual noise. Those methods seek the maximum correlation between the fixed wavelet basic and raw signal.

In recent years, dictionary learning has emerged in signal filter and compression, due to its capability in adaptive for the raw signal. K-SVD method is proposed in [9] that generalize the K-Means clustering process. Based on the redundant dictionary trained by K-SVD, Chen et al.[10] extracts the impulse components of vibration signal and reduces the noise. And comparing with the mature shrinkage/thresholding algorithm, performances based on dictionary learning is much better. Because of the calculation of large matrix, the K-SVD algorithm has the problem of low efficiency in computing.

In this paper, a vibration signal denoising based on dictionary learning and sparse approximation (ODL-SAD) is proposed. The K-SVD algorithm and modified Online dictionary learning(ODL) algorithm are compared based on simulation and experiment mechanical vibration signal. The comparison results reveal that the modified ODL algorithm is more suitable to real calculation. Orthogonal matching pursuit is applied to sparse represent. At last, the denoised siganl can be got by reconstructing. The signal denoising results of different methods are shown and those support the proposed algorithm.

2. The adaptively vibration signal denoising

The noise reducing technique based on ODL-SAD is mathematical described as:

$$\begin{aligned} \overline{x} = \min_{x} \|x\|_0 \ s.t. \ \|y - \overline{D}x\|_2^2 \leq \gamma \\ \overline{y} = \overline{D}\overline{x}. \end{aligned} \tag{1}$$

where $\|.\|_0$ is the zero norm. $\overline{D}$ is a redundant dictionary which can be found at the *Section*2.1. $\overline{x}$ is the sparse coding of vector x. γ is the noise level. $\overline{y}$ is the reconstruction of raw signal y.

2.1. Online Dictionary learning

A dictionary $D \in R^{N \times K}$ is a collection of K atoms that are column vectors of length N. Now there are two different methods of constructing dictionary. The first one is solid dictionary, such as Fourier Transformation matrix[11], Wavelet Transformation matrix[12] and so on. The other way of constructing dictionary is dictionary learning. Dictionary learning aims at finding the adaptive dictionary that yields sparse representation for the training signals. Let X be the sparse signal, corresponding to the sparse representation of training signals Y. Mathematically, the dictionary learning can be expressed as followed:

$$\{\overline{D}, \overline{X}\} = \arg\min_{D,X} \sum_{l=1}^{L} \|x_l\|_2 + \lambda \|y_l - Dx_l\|^2, \ \ \|d_i\| = 1 \tag{2}$$

where $\overline{D}$ and $\overline{X}$ are the optimized dictionary and sparse vectors. $\|d_i\|$ is the column vector of $\overline{D}$.

The classical dictionary learning algorithm optimizes the cost funciton:

$$D_n = \arg\min_{D} \sum_{i=1}^{I} \|y_i - Dx_i\|_2^2 + \lambda \|x_i\|_p \tag{3}$$

where D_n indicates the n_{th} interated dictionary D. The sparse coefficient X_{n+1} can be calculated through the dictionary D_n:

$$X_{n+1} = \arg\min_{X} \|x_i\|_1 s.t. \ \|y_i - Dx_i\|_2^2 \leq \gamma \tag{4}$$

where X_{n+1} indicate the $n + 1_{th}$ sparse coefficients X. γ is the threshold value. In this stage, Orthogonal matching pursuit (OMP) is often applied because of its rate of convergence .

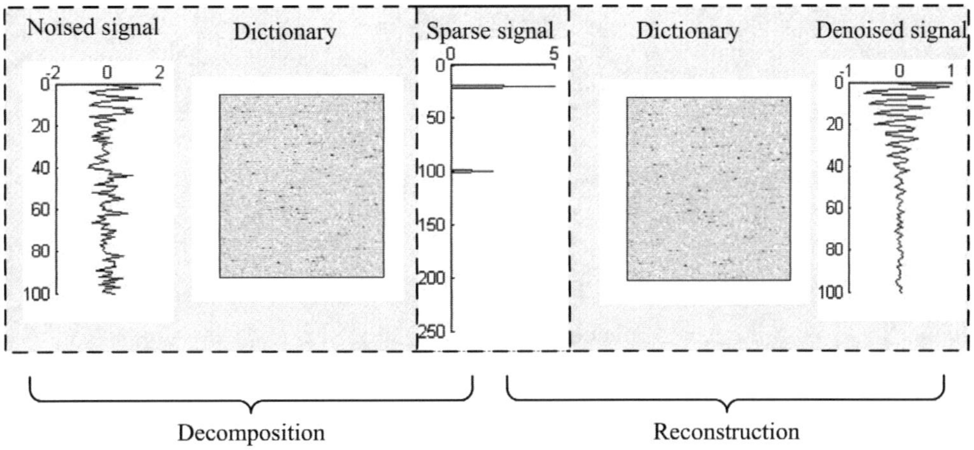

Figure 1: Two-stage dictionary sparse decomposition and reconstruction

In this paper, we use online dictionary learning to optimazation D. In online dictionary learning, Let $A_n \leftarrow A_{n-1} + x_n x_n^T, B_n \leftarrow B_{n-1} + y_n x_n^T$. With the matrix characteristics: $\|A\|_F = tr(A^H A)$, $tr(AB) = tr(AB)$ and $tr(A) = tr(A^T)$, where tr is matrix trace. Then we can derive $Eq.(3)$ to:

$$D_n = \underset{D \in C}{argmin}(\frac{1}{2}Tr(D^T D A_n) - Tr(D^T B_n)) \tag{5}$$

where C is the criteria that dictionary need to meet. Then take a derivate with Eq (6) to D:

$$\frac{1}{2}D(A_n I + (A_n I)^T) - B_n = 0 \tag{6}$$

where I is unit matrix. Then the Newton iteration method is applied to Eq (7).Let $A = [a_1, ..., a_k] \in R^{k \times k} = \sum_{i=1}^t x_i x_i^T$, $B = [b_1, ..., b_k] \in R^{m \times k} = \sum_{i=1}^t y_i x_i^T$. A parameter u can be calculated by:

$$u_j = \frac{1}{A_{jj}}(b_j - Da_j) + d_j \tag{7}$$

where Ajj is the mian value of diagonal matrix. At last, the value of D is limited by:

$$d_j = \frac{1}{max(\|u_j\|_2, 1)}u_j. \tag{8}$$

where d_j is the $j - th$ column of dictionary D. And at every iteration, the algorithm uses the value of D_{n-1} as a warm restart for cumputing D_n that speeds up convergence.

2.2. Denoising based on learned dictioanry and sparse approximate

Online learning algorithm is adopted to form the adaptive dictionary. Coefficients of the most correlation between signals and atoms are calculated at each iteration of orthogonal matching pursuit(OMP). Once the termination condition of OMP is met, the sparsity vector is obtained. As shown in $Fig.$1, reconstruction is the dual process of decomposition. Only the atoms from the certain non-zeros in sparse signal are reconstructed, the noise components are removed. The above algorithm is called Online dictionary learning and sparse approximation denoising(ODL-SAD). The detailed steps of ODL-SAD are as follows:

(1): Prepare the training sets. Truncate signals Y into segments $[y_1, ..., y_n]$.
(2): Set dictionary learning parameters. Initial dictionary D_0, iteration number K.
(3): Update the dictionary by Eqs.(7) and (8) to form the signal adaptive dictionary $\overline{D}$.
(4): Set parameters of OMP. Matrix $\overline{D}$, vector Y and the error threshold γ.
(5): Obtain the sparse signal $\overline{X}$ through OMP by Eq.(1).
(6): Compute the reconstruction signal $\overline{Y}$
(7): Calculate the envelop spectrum F

3. Simulation

In this section, a simulation experiment is designed to illustrate the denoising ability of ODL-SAD. Generally, when rotating machinery with localized faults is in operation, a serial of impulse signals will be generated. And those signal are always embeded in heavy background noise[13]. The simulation signal is designed as follows:

$$x(t) = \sum_{i=0}^{M} A_i e^{-\beta_i(t/f_s - i/f_m)} sin(2\pi f_i(t/f_s - i/f_m) + \tau_i) + v(t) \qquad (9)$$

where A_i is the amplitude, set to 1. The damping factor β_i is equalled to 500. The sampling frequency f_s is 20000Hz. f_m is the impulse frequency (100Hz). f_i is the oscillating frequency, set to 2000Hz, τ_i is the phase of the carrier wave setted random. The simulation signal waveform can be found in Fig 2(a), Then the Gaussian white noise $v(t)$ is added to obtain a noise-contaminated signal with a signal-to-noise ratio -2dB. The noise signal is shown in Fig 2(a). As you can see, the signal feature is almost swallowed by noise signal.

Table 1: Parameters of On-line dictionary learning denoising

Segment size	Dictionary size	Sparse coding	Initial dictionary
100	100*200	OMP	DCT

ODL-SAD is applied to analyse the noise-contained simulation signal. The parameters such as the Segment size, Dictionary size and so on can be found in $Table$ 1. In order to match the atom length with impulse duration and identify all the impulse components, this paper chooses the segment size with 100. The question of dictionary size is an open class, we just simply choose a value we find empirically to perform well as $K/N = 2$[9]. Over-complete DCT is chosen as the initial fixed dictionary. Its update iterations is 20. With above parameters, the resulting signal is depicted in Fig 2(d). The noise is almost wiped off while the waveform stay the same.

For comparison, the wavelet denoising and K-SVD dictioanry learning denoising are employed to analyse the same signal in Fig 2(a). The K-SVD dictionary learning method has the same parameters as the proposed method, the only difference is the K-SVD instead of On-line learning algorithm. The filtered signal shows in Fig 2(b). The signal shape is almost but large of noise is still kept. The wavelet method uses the minimax threshold selection rules and the Morlet wavelet is selected as the mother wavelet. The impulse component is clear obtained in Fig 2(c), but attenuation component of signal is significant decreased. To make accurate comparison, Room mean suqare deviation ($RMSE$) and signal to noise (SNR) are adopted. Those results show that ODL-SAD has the highest $SNR = 3.3009$ and the lowest $RMSE = 0.15276$.

For future declare the ability of ODL to real-time computation, we compare the training time between ODL and K-SVD. Fig 3 shows test performance when training on the datasets of

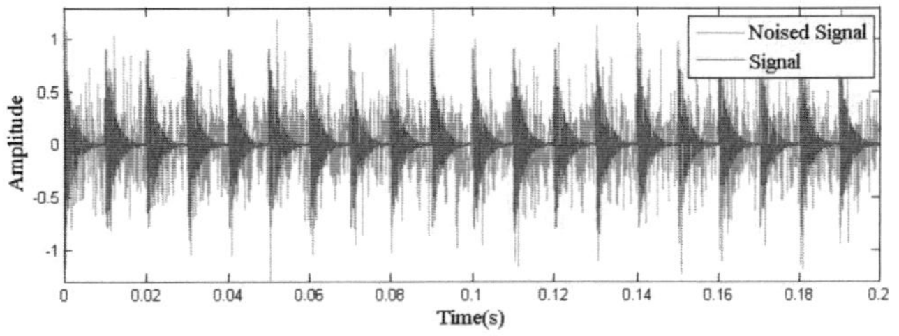

(a) The raw signal and noised signal

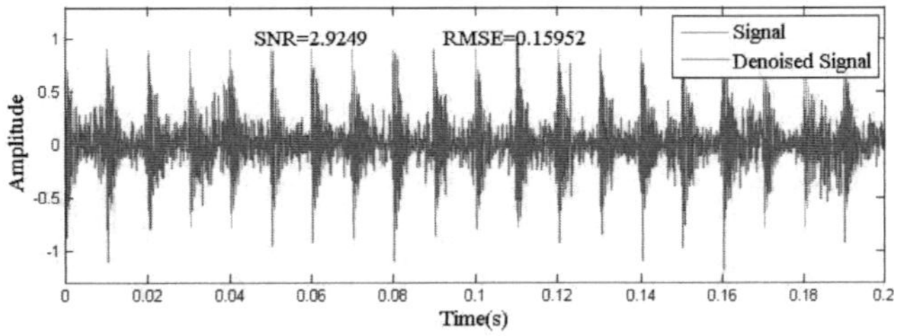

(b) The K-SVD dictioanry learning denoising

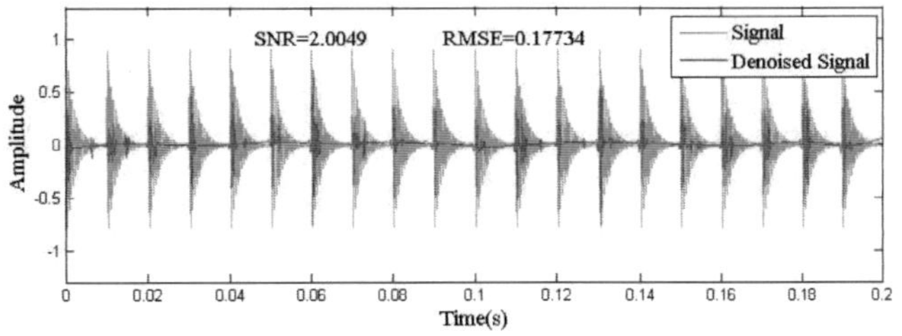

(c) The Wavelet thresholding denoising

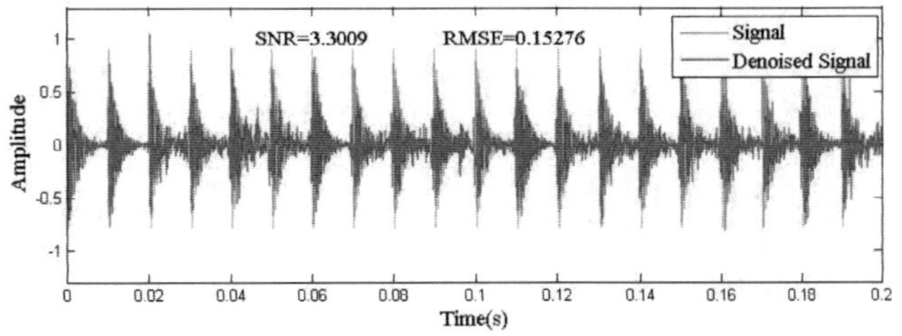

(d) The On-line dictionary learning denoising

Figure 2: The results obtained by different kinds of denoising algorithms

$2 \sim 22$. As the datasets increase, the training time of K-SVD goes up linearly. However, ODL remains about the same.

4. Experiment

To demonstrate the effectiveness of ODL-SAD in actual signal detection, an experiment of detecting the bearing outer race defect is considered.

Bearing run-to-failure tests were performed under normal load conditions on a specially designed test rig. The test rig is shown in *Fig.* 4. This test rig hosts four test bearings on

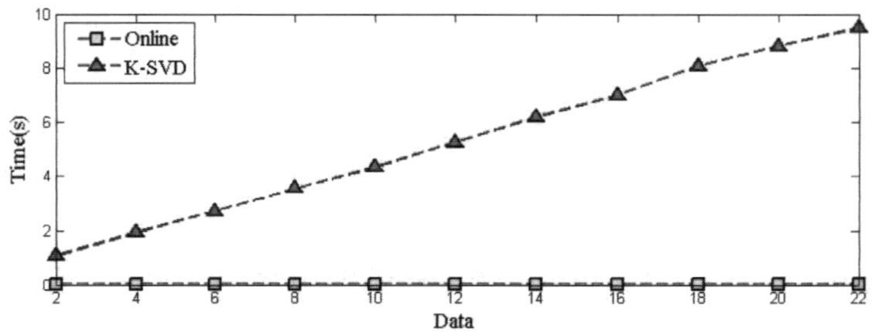

Figure 3: The runtime of K-SVD and Online dictionary learning

one shaft driven by an AC motor and coupled by rub belts. The rotation speed is kept 2000rpm. A radial load of 6000lbs is added to the shaft and bearing by a spring mechanism. All the bearings are forced lubricated.

Four Rexnord ZA-2115 double row bearings were installed on one shaft as shown in *Fig.* 4. The test bearings have 16 rollers in each row, a pitch diameter of 2.815in, roller diameter of 0.331in, and a tapered contact angle of 15.17°. Vibration data was collected every 20 minutes by a National Instruments DAQCard-6062E data acquisition card. The data sampling rate is 20kHz and the data length is 20480 points. Depending on paper[14], the outer-race fault frequency f_0 of bearing is:

$$f_0 = \frac{n}{2}(1 - \frac{d}{D_p}\cos\theta)f_r \tag{10}$$

where n is number of balls, d is the diameter of the rolling element, D_p is the groove section size, θ is the contact angle, f_r is the shafe frequency.

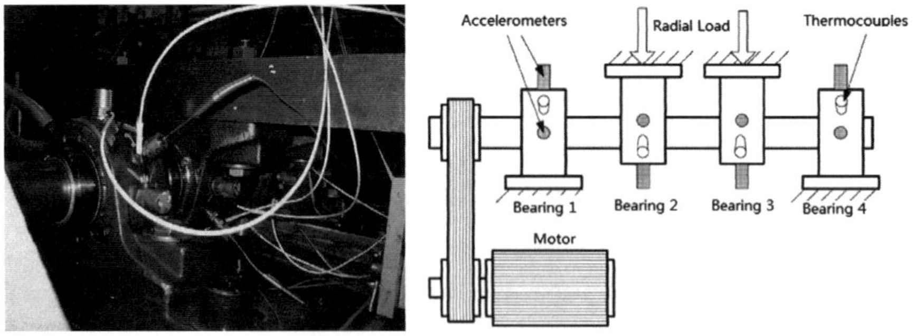

Figure 4: Bearing test rig and sensor placement illustration

Fig 5(*a*) presents the impulse periodicity vibration signal of bearing 1 with the outer race defect. By *Eq* (10), the out race band pass frequency is 263.4Hz. The waveform verifies the calculation. ODL-SAD is applied to enhance the signal shown above. The parameters are shown in *Table* 1, while the segment size is 80 and the dictionary size is $80 * 160$. *Fig*.5(*a*) shows that the noise component is eliminated. To further discover the frequency characteristic, envelop spectrum is illustrated in *Fig*5(*b*) and *Fig*5(*c*). In the envelop spectrum obtained by ODL-SAD presented in *Fig*5(*c*), even the higher order harmonic (*e.g.*, $4f_0, 5f_0$) are evident.

The computing efficiency of ODL and K-SVD based on actual signal is shown in *Fig* 6. That indicts the run times of ODL algorithm is irrelevant to the size of training datasets. Nevertheless, K-SVD is proportional to data volume.

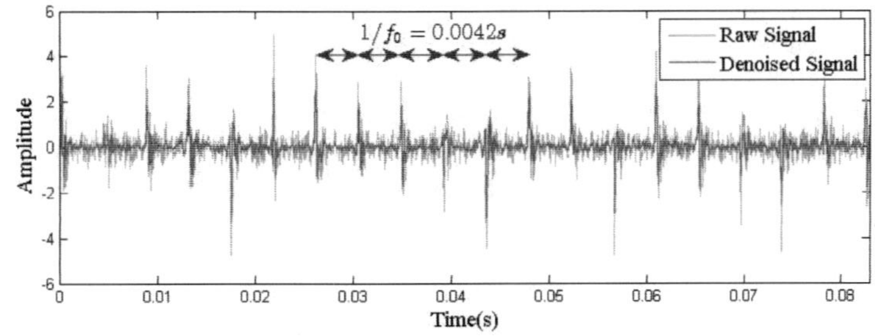

(a) The raw signal and denoised signal

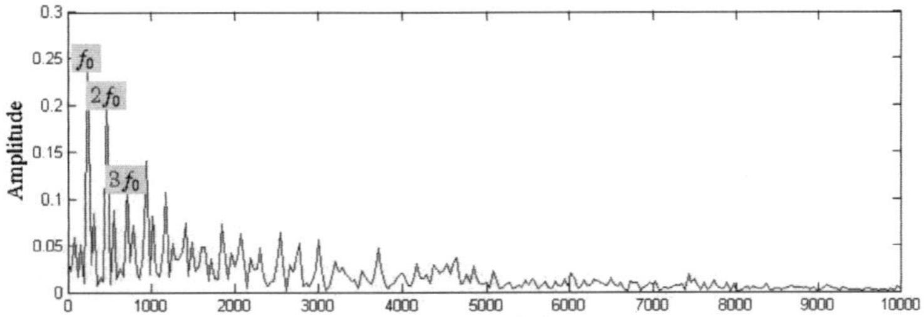

(b) The envelop spectrum of raw signal

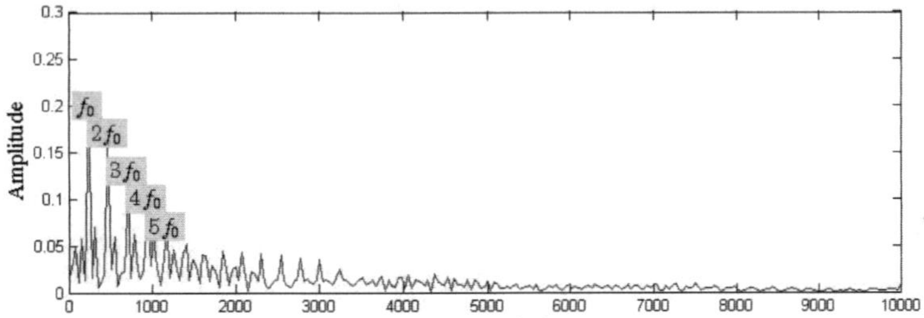

(c) The envelop spectrum of denoised signal

Figure 5: The results of ODL-SAD for bearing vibration signal

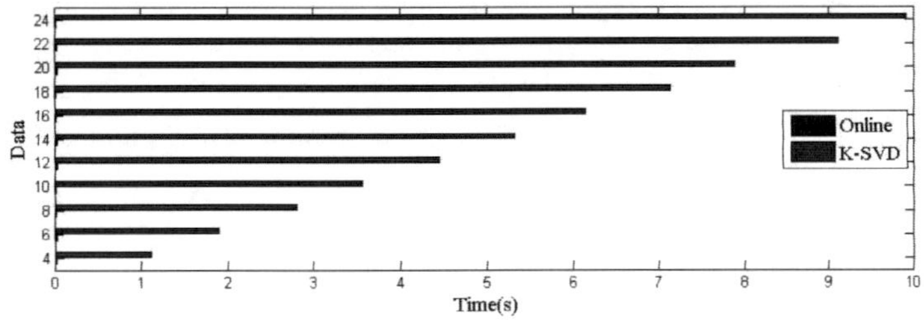

Figure 6: The runtime of K-SVD and Online dictionary learning for bearing vibration signal

5. Conclusion and future work

Signal denoising is crucial to machine health monitoring and damage assessment. The performance of traditional wavelet transform-based methods is greatly impacted by the mother wavelet. Dictionary learning is an alternative method to construct the transfer matrix. However, it is much more challenging to computing the large matrix. This paper presents a On-line dictionary learning and sparse approximation denoising algorithm(ODL-SAD). The transfer matrix of sparse signal is established by dictionary learning. The modified On-line algorithm runs the training processing real-time. In the simulation analysis experiment verification, the RMSE and SNR are compared between denoised signal and raw signal with different algorithms. Results show that the proposed algorithm capture the efficiency results with less deformation. And it is significantly faster than traditional dictionary learning algorithm. Multi-Scale dictionary learning for sparse and redundant signal representations is considering to improve the flexibility.

Acknowledgments

This research was funded by the National Science Foundation (NSF) under Grant No.51275426

References

[1] Y. Lei, J. Lin, Z. He, Y. Zi, Application of an improved kurtogram method for fault diagnosis of rolling element bearings, Mechanical Systems and Signal Processing 25 (5) (2011) 1738–1749.
[2] H. Qiu, J. Lee, J. Lin, G. Yu, Wavelet filter-based weak signature detection method and its application on rolling element bearing prognostics, Journal of sound and vibration 289 (4) (2006) 1066–1090.
[3] S. Jha, O. Singh, R. K. Sunkaria, Modified approach for ecg signal denoising based on empirical mode decomposition and moving average filter, International Journal of Medical Engineering and Informatics 6 (2) (2014) 165–188.
[4] R. B. Randall, Frequency analysis, Brül & Kjor, 1987.
[5] S. Mallat, W. L. Hwang, Singularity detection and processing with wavelets, Information Theory, IEEE Transactions on 38 (2) (1992) 617–643.
[6] D. L. Donoho, J. M. Johnstone, Ideal spatial adaptation by wavelet shrinkage, Biometrika 81 (3) (1994) 425–455.
[7] E. J. Candes, D. L. Donoho, Recovering edges in ill-posed inverse problems: Optimality of curvelet frames, Annals of statistics (2002) 784–842.
[8] W. He, Z.-N. Jiang, K. Feng, Bearing fault detection based on optimal wavelet filter and sparse code shrinkage, Measurement 42 (7) (2009) 1092–1102.
[9] M. Aharon, M. Elad, A. Bruckstein, -svd: An algorithm for designing overcomplete dictionaries for sparse representation, Signal Processing, IEEE Transactions on 54 (11) (2006) 4311–4322.
[10] J. Mairal, M. Elad, G. Sapiro, Sparse representation for color image restoration, Image Processing, IEEE Transactions on 17 (1) (2008) 53–69.
[11] M. Goodwin, Matching pursuit with damped sinusoids, in: Acoustics, Speech, and Signal Processing, 1997. ICASSP-97., 1997 IEEE International Conference on, Vol. 3, IEEE, 1997, pp. 2037–2040.
[12] P. Vera-Candeas, N. Ruiz-Reyes, M. Rosa-Zurera, D. Martinez-Munoz, F. Lopez-Ferreras, Transient modeling by matching pursuits with a wavelet dictionary for parametric audio coding, Signal Processing Letters, IEEE 11 (3) (2004) 349–352.
[13] X. Chen, Z. Du, J. Li, X. Li, H. Zhang, Compressed sensing based on dictionary learning for extracting impulse components, Signal Processing 96 (2014) 94–109.
[14] R. B. Randall, J. Antoni, Rolling element bearing diagnosticsa tutorial, Mechanical Systems and Signal Processing 25 (2) (2011) 485–520.

Identification and stochastic modelling of sources in copper ore crusher vibrations

Agnieszka Wylomanska

Department of Mathematics, Wroclaw University of Technology, Wybrzeze Wyspianskiego 27, 50-370 Wroclaw, Poland

E-mail: agnieszka.wylomanska@pwr.edu.pl

Radoslaw Zimroz

Diagnostics and Vibro-Acoustics Science Laboratory, Wroclaw University of Technology, Na Grobli 15, 50-421 Wroclaw, Poland; KGHM CUPRUM Ltd CBR Sikorskiego 2-8 , 53-659 Wroclaw, Poland

E-mail: rzimroz@cuprum.wroc.pl

Joanna Janczura

Department of Mathematics, Wroclaw University of Technology, Wybrzeze Wyspianskiego 27, 50-370 Wroclaw, Poland

E-mail: joanna.janczura@pwr.edu.pl

Abstract. A problem of rolling element bearings diagnostics for different machines is widely discussed in the literature. Most of the methods are based on the vibration signal analysis. However for some real signals the classical methods of damage detection are insufficient because of the specific nature of examined data. This specific nature is very often manifested through overlapping, mixing or interleaving of processes with different statistical properties and may be the result of different sources that have influence on the analysed signal. The problem of different sources identification and parametrisation of processes which are related to them is very challenging and requires advanced techniques. There are many methods which can be useful in this context however each signal should be analysed separately and there is no universal technique adequate to all possible time series. In this paper we propose a method of sources identification for vibration signal from the heavy duty crusher used in mineral processing plant. A crusher is a kind of machine which use a metal surface to crumble materials into small fractional pieces. During this process, as well as during entering material stream into the crusher, a lot of impacts appear. They are present in vibration signal acquired from bearings housing. Moreover, for some cases we also observe cyclic impulses which may be related to damage of rolling element bearings in the machine. The proposed sources identification method, especially useful for crushers vibrations, is based on the statistical analysis of examined data. Moreover by using advanced techniques of time series theory we propose a stochastic model that exhibits similar statistical properties as analysed signals. The introduced technique can be a starting point to damage detection of rolling element bearings of copper ore crushers.

1. Introduction
One of the main problems of rolling element bearings diagnostics is to extract information about damage. There are many methods that can be useful in this context. Most of them are based on the analysis of the vibration signal of a given machine in the time and/or frequency domain. Extraction of the information about damage is very often related to the detection of informative signal [12, 22, 23] and most of the methods are based on the impulsiveness of given time series [1, 2, 3, 5, 18, 25, 27]. However, for many real vibration signals the classical methods seem to be inadequate because of the very specific nature of examined data. One of the examples is frequency and amplitude modulated signal obtained from machine working in varying operating conditions [9] or non-Gausian character of external noise (a critical case is when noise is impulsive) [4, 22, 24]. Therefore there is need to use more advanced techniques that can be useful in the problem of fault detection for such cases. The mentioned specific nature is often manifested through overlapping, mixing or interleaving of processes with different statistical properties and may be the result of different sources that have influence on the analysed signal. Therefore in the group of damage detection methods there are segmentation techniques which lead to identification of subsignals with homogeneous structure. In the literature one can find various methods of segmentation which can be used to different signals [13, 19, 30]. Some of them are based on the statistical properties of examined time series, [14]. As it was mentioned, the segmentation problem is strongly related to identification of different sources and parametrisation of processes which are related to them. This parametrisation can be expressed by the statistical or stochastic properties of extracted segments therefore the stochastic analysis and modelling of appropriate vibration signals occur more frequently in the literature [25, 20, 21, 31].

In this paper we propose a novel segmentation method applied to vibration signal from the heavy duty crusher used in mineral processing plant. The analysed vibration signal is a mixture of different processes therefore in the first step of fault detection it is necessary to identify the appropriate subsignals and analyse them by using stochastic methods. This can lead to sources identification and finally to damage detection. The proposed method consists of two steps. First, the primary segmentation is made in order to identify distributions of selected subsignals. In the second step we make a proper segmentation in order to identify the processes which have influence on the given signal. This method is based on the two regimes switching model for which we assume that the examined time series is a mixture of two processes with different distributions. The regime switching models (called also Hidden Markov models) were initially used as a segmentation method in speech recognition [26], but they have been also exploited in other contexts, including as diverse areas as software reliability [28], electromagnetic field measurements [6] or electricity markets [15].

The rest of the paper is structured as follows: in Section 2 we describe the investigated machine. In Section 3 we present the methodology, i.e. the two steps segmentation procedure. Next, we analyse the real copper ore crushers vibration signal. Last section contains conclusions.

2. Experiment description
The machine discussed here is a heavy duty crusher used in a mineral processing plant. The crusher is a kind of machine which use a metal surface to crumble materials into small fractional pieces. During this process, as well as during entering material stream into the crusher, a lot of impacts/shocks appear. They are present in vibration signal acquired from the bearings housing, see Figure 3. Several crushers were investigated: they were in different technical state and operated under different loading conditions. Vibration from bearings housing in horizontal and vertical direction and tacho signal have been measured using Endevco accelerometers BruelKjaer Laser probe, NI DAQ card and Labview Signal Express. Signals were acquired for idle mode as well as for different values of load (volume and granulation of material stream) with approx

120s duration each, however, for further investigation shorter signals were selected. Sampling frequency was set to 25kHz, [24]. In Figure 1 we present the machine.

Figure 1. A crusher general view (note bearings with yellow housing).

3. Methodology

In this section we present the new technique of identification of different processes related to sources that have influence on copper ore crusher vibration signal. This method is based on a two step procedure: the primary segmentation and stochastic analysis of separated parts and the main method based on the two regimes switching model. In the presented methodology we assume that the analysed signal is a mixture of processes with different statistical properties which are represented by different distributions. We intuitively assume that impacts related to crushing process are rather strongly non-Gaussian, while "proper bearing vibration" will be rather Gaussian. Obviously, we will validate this assumptions. In the primary segmentation method we identify the observations adequate to different regimes and conclude on their distribution. This is a starting point to the main segmentation method for which the distribution of processes in separate regimes should be properly defined.

3.1. The primary segmentation method

The primary segmentation method of the signal $x_1, x_2, ..., x_n$ is based on positive and negative peaks finding in a fixed neighbourhood. More precisely, if in the copper ore crushers vibration signal we observe peaks, then they are treated as observations from different regime than the other data. We should mention here that the peaks we find in our procedure are identified in a given neighbourhood by the assumption that they exceed a given threshold, which we denote as th1. In order to find the peaks, we use the procedure 'findpeaks' in MATLAB. However, in the vibration signal of copper ore crushers we observe behaviour adequate to three regimes, i.e. positive peaks (regime R1), negative peaks (regime R3) and signal that arises after their removing (regime R2). Therefore the mentioned two steps procedure of peaks finding is also applied in order to find observations corresponding to regime R3. More precisely, we apply the mentioned technique to a signal which is equal to the examined dataset with negative values. In this case the thresholds used in the procedure are denoted by th3.

After three regimes identification, next we analyse them separately by using statistical methods, in order to identify the appropriate distributions. In order to identify the distribution of R2, we propose to use the advanced method presented in [8] which is based on the graphical interpretation of stability index of a given dataset. The complete description of this procedure can be found in [8].

The analysis of regimes R1 and R3 is based on the observation of the behaviour of their empirical right tails. More precisely, for observations $x_1, x_2, , x_n$ we first calculate the empirical right tail, which is equal to $1 - F_n(t)$, where $F_n(t)$ is the empirical cumulative distribution function:

$$F_n(t) = \frac{1}{n}\sum_{i=1}^{n} 1\{x_i \leq t\}, \tag{1}$$

where $1\{A\}$ is the indicator of the set A. Next, we compare the empirical tail with the power function $t^{-\gamma}$. On the basis of this we can conclude on the distribution of a given vector of observations corresponding to regimes R1 and R3. The primary analysis shows that the fitted power function of empirical right tails of observations from regimes R1 and R3 is adequate to shifted Pareto distribution, [10], which is defined through its probability density function in the following way:

$$f(t) = \frac{\gamma\lambda^{\gamma}}{t^{\gamma+1}}, \quad t > \lambda, \tag{2}$$

where γ is a scale parameter and λ - a shift parameter. For the Pareto distribution the theoretical right tail is given by:

$$1 - F(t) = \left(\frac{\lambda}{t}\right)^{\gamma}. \tag{3}$$

The Pareto distribution was used in various different applications. We mention here the insurance [10], but also the machine diagnostics to description of diagnostic features [29]. At the end, we estimate the parameters of Pareto distribution. The shift parameter is equal to the minimum of a given vector of observations, while the scale parameter is estimated by fitting the theoretical tail (3) to the empirical one (1) by using least squares method. The analysis of the vibration signal of copper ore crushers indicates that the observations from regimes R1 and R3 can be treated as Pareto distributed with the similar parameters, while for regime R2 we have the Gaussian distribution. Therefore in the main segmentation method, presented in the next part of this section, we propose to use the regime switching model with two regimes: one represented by the peaks (both negative and positive) with Pareto distribution and second represented by signal after removing the peaks, which can be modeled by a Gaussian distribution.

3.2. The two regimes switching modelling

The main segmentation method is based on the assumption that the observed crusher signal can be modelled by a Markov regime switching model with two regimes. The intuitive idea behind such a modelling approach is to describe the observed signal by separate regimes with different statistical properties of the underlying processes. Since the observed crusher signal can be treated either as a peak, being an effect of the crushing process, or a "normal" behaviour resulting from "proper bearing vibration", it seems to be natural to use such a model.

Based on the results of the primary segmentation method, the values of the observed signal, x_i are assumed to be independent, identically distributed random variables following Gaussian distribution in the "normal" regime (denoted by R2 in Section 3.1) and symmetric Pareto in the peak regime (denoted by R1 and R3 in Section 3.1, note that here we call it regime R1). Precisely:

$$x_i = \begin{cases} x_{i,1} \sim \mathrm{sP}(\gamma, x_m) & \text{if } R_i = 1, \\ x_{i,2} \sim \mathrm{N}(\mu, \sigma^2) & \text{if } R_i = 2, \end{cases} \tag{4}$$

Journal of Physics: Conference Series **628** (2015) 012125 doi:10.1088/1742-6596/628/1/012125

where $sP(\gamma, x_m)$ is the symmetric Pareto distribution with a probability density function

$$f(t) = \frac{1}{2} \frac{\gamma \lambda^{\gamma}}{|t|^{\gamma+1}}, \quad |t| > \lambda, \tag{5}$$

and R_i is the state (regime) variable for the i-th signal. Definition (4) implies that the statistical properties of the observed signal differ depending on the actual state R_i. We assume that the state process R_i is an unobserved (latent) Markov chain. As a consequence, switching from one state to another is governed by a transition matrix $\mathbf{P}$ which contains the probabilities $p_{lj} = P(R_{i+1} = j | R_i = l)$ of switching from regime l at time i to regime j at time $i+1$. The switching mechanism is not observed, meaning that we do not have the information, if the recorded signal was driven the peak or the "normal" regime. However, as a by-product of the estimation algorithm of the assumed model, we obtain probabilities $P(R_i = 1 | x_1, x_2, ..., x_n)$ and $P(R_i = 2 | x_1, x_2, ..., x_n)$ that the i-th signal came from the peak (R1) or the "normal" (R2) regime, respectively. Hence, based on these probabilities, we, in fact, get a segmentation method.

The estimation of the model parameters is based on the expectation-maximization algorithm, initially proposed in [11] and later refined for the regime-switching models in [17]. It is a two-step – expectation (E-step) and maximisation (M-step)– iterative procedure. In each iteration, first, the inferences about the state values are made based on the previously calculated parameters, second, the new set of parameters is estimated based on these inferences. The iteration is repeated until the (local) maximum of the likelihood function is reached.

The procedure of the segmentation method presented in this section is illustrated in Figure 2.

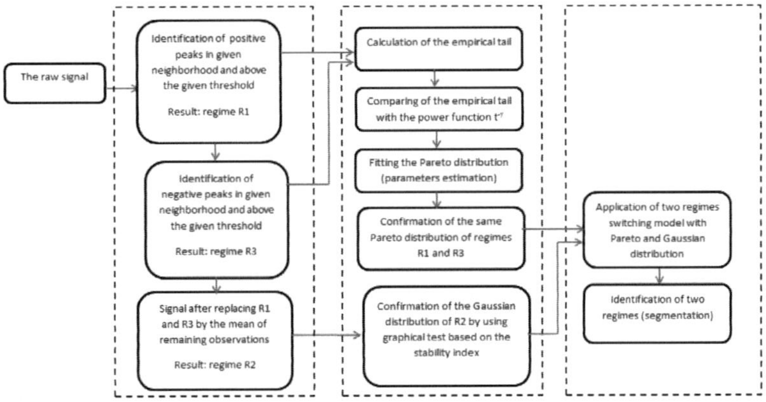

Figure 2. The scheme of the segmentation method.

4. Application to real signal

In this section we present the results of the segmentation methods (primary and main) described in Section 3. In Figure 3 (left panel) we present the analysed vibration signal of copper ore crushers. According to the presented procedure, first the primary segmentation method is made. It is worth mentioning, that the presented methodology leads to the identification of large peaks which are related to crushing process.The small peaks (related to the bearing's work) hidden in the noise are not detectable. We start with identifying the positive peaks (regime R1) for the analysed signal and the signal with negative values (negative peaks, regime R3), as well as signal that arises after replacing the observations corresponding to R1 and R3 by the mean of the remaining observations. In our analysis we select the following thresholds $th1 = 0.6$ and $th3 = 0.6$. The procedure of estimating thresholds is based on the assumption of the Gaussian distribution of the R2 regime. The result of the primary segmentation is presented in Figure 3 (right panel).

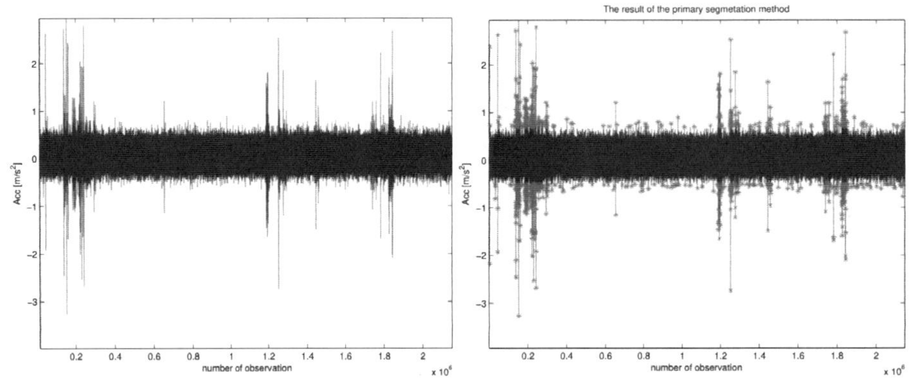

Figure 3. Left panel: the examined vibration signal of copper ore crusher. Right panel: the result of the primary segmentation method applied to vibration signal of copper ore crusher.

In the next step we examine separately the signals corresponding to the three identified regimes. First, we analyse the regimes R1 and R3. In Figure 4 we present the empirical tails corresponding to those regimes and the fitted power functions (in log-log scale). As we observe, we have a good agreement between theoretical and empirical distributions.

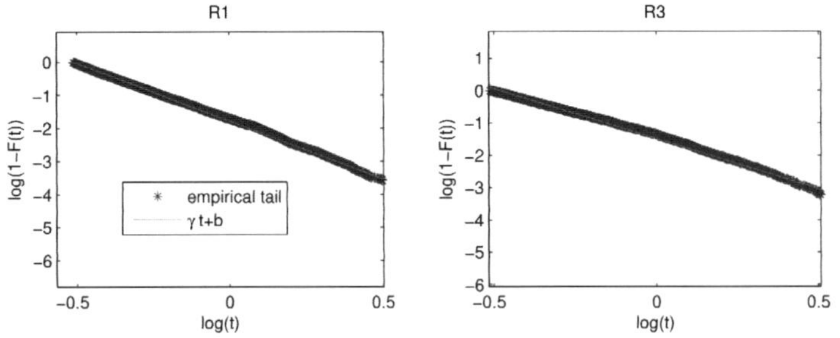

Figure 4. The empirical right tails of the observations corresponding to R1 and R3 and the fitted appropriate power functions.

The estimated parameters of the Pareto distribution for the subsignals of the regimes R1 and R3 are as follows:

$$R1: \ \gamma = 3.23, \ \lambda = 0.6, \ R3: \ \gamma = 3.21, \ \lambda = 0.6.$$

As we observe, the fitted Pareto distributions have similar parameters. Therefore, as it was mentioned earlier, we can treat the signals corresponding to those regimes like signals from the same distribution. At the end of the primary segmentation method we confirm the assumption that the signal corresponding to the regime R2 has a Gaussian distribution. Gaussianity is confirmed by using the graphical test presented in [8].

As a summary of the results of the primary segmentation method we can conclude that the examined vibration signal is a mixture of two processes. The first one, represented by positive and negative peaks has a Pareto distribution, while the second one is Gaussian. The primary results are the starting point for the main segmentation method based on the regime switching model, in which we assume two regimes – Pareto and Gaussian distributed. After applying the methodology presented in Section 3.2, we obtain the identification of the mentioned regimes. The result of the segmentation method based on the regime switching model is presented in Figure 5.

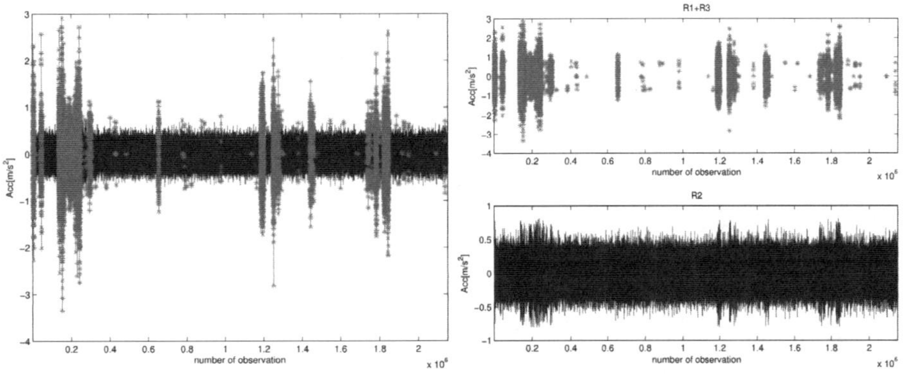

Figure 5. The result of the segmentation method based on the regime switching model with two regimes (left panel) and identified regimes R1+R3 and R2 (right panel).

5. Conclusions

In this paper we have proposed the segmentation method that can be applied to vibration signals with a specific nature. The proposed technique consists of two steps: the primary segmentation and the main one. By using the presented procedures we can identify the processes comprising the vibration signal and finally the sources that affect the signal. The proposed method is based on the characteristics of the signal in time domain and stochastic analysis of the appropriate subsignals. The main result of the primary segmentation method is the primary identification of the regimes and their distributions. This method is a starting point for the main technique based on the regime switching model with two regimes. This method is universal and can be applied to various signals with a complicated structure, like the vibration signal from copper ore crushers, where we observe a mixture of processes with different stochastic properties.

References

[1] J. Antoni, Fast computation of the kurtogram for the detection of transient faults, Mechanical Sys. and Signal Proc. 21, 108-124, 2007.
[2] J. Antoni, R. B. Randall, The spectral kurtosis: application to the vibratory surveillance and diagnostics of rotating machines, Mechanical Sys. and Signal Proc. 20, 308-331, 2006.
[3] T. Barszcz, A. Jablonski, A novel method for the optimal band selection for vibration signal demodulation and comparison with the Kurtogram, Mech. Syst and Sig Proc. 25/1, 431-451, 2011.
[4] T. Barszcz, J. Urbanek, L. Szumilas, Selection of efficient monitoring methods for machinery generating high vibration signal disturbance, Diagnostyka 4, 55-58, 2010.
[5] T. Barszcz, A. Jablonski, Analysis of Kurtogram performance in case of high level non- Gaussian noise, Proc. of ICSV, Krakw, Poland, 2009.
[6] P.Bienkowski, K. Burnecki, J.Janczura, R.Weron, B.Zubrzak, A new method for automated noise cancellation in electromagnetic field measurement, J. of Electromagn. Waves and Appl. 26, 12261236, 2012
[7] K.Burnecki, A.Wylomanska, A.Beletskii, V.Gonchar, A.Chechkin, Recognition of stable distribution with Levy index α close to 2, Phys. Rev. E 85, 056711, 2012.
[8] K. Burnecki, A. Wylomanska, A. Chechkin, Discriminating between light- and heavy-tailed distributions with limit theorem, preprint, 2014.

[9] F. Chaari, W. Bartelmus, R. Zimroz, T. Fakhfakh, M. Haddar, M. Gearbox vibration signal amplitude and frequency modulation, Shock and Vibration 19 (4), 635-652, 2012.
[10] P.Cizek, W.Hrdle, R.Weron, Statistical Tools for Finance and Insurance, Springer, Berlin, 2005.
[11] A. Dempster, N. Laird, D.B. Rubin, Maximum likelihood from incomplete data via the EM algorithm, Journal of the Royal Statistical Society 39, 1-38, 1977.
[12] J. Dybala, R. Zimroz, Rolling bearing diagnosing method based on empirical mode decomposition of machine vibration signal, Applied Acoustics, 77, 195-203, 2014.
[13] J. Gajda G. Sikora, A. Wylomanska, Regime variance testing - a quantile approach, Acta Phys. Polon B 44(5), 1015-1035, 2013.
[14] R. Granat, Detecting regional events via statistical analysis of geodetic networks, Pure and Applied Geophysics 163, 2497-2512, 2006.
[15] J.Janczura, Pricing electricity derivatives within a Markov regime-switching model: a risk premium approach, Mathematical Methods of Operations Research 79(1),1-30, 2013.
[16] J.Janczura, R.Weron, Efficient estimation of Markov regime-switching models: An application to electricity wholesale market prices, AStA - Advances in Statistical Analysis 96(3), 385-407, 2012.
[17] C.-J. Kim, Dynamic linear models with Markov-switching, Journal of Econometrics 60, 1-22, 1994.
[18] J. Lin J, M. Zuo, Gearbox fault diagnosis using adaptive wavelet filter, Mechanical Sys. and Signal Proc. 17 (6), 1259-1269, 2003.
[19] R. Makowski, R. Hossa, Automatic speech signal segmentation based on the innovation adaptive filter, International Journal of Applied Mathematics and Computer Science, 24, 259-270, 2014.
[20] R. Makowski, R. Zimroz, Adaptive bearings vibration modelling for diagnosis, Lecture Notes in Computer Science (including subseries Lecture Notes in Artificial Intelligence and Lecture Notes in Bioinformatics) 6943 LNAI, 248-259. 2011.
[21] R. Makowski, R. Zimroz, A procedure for weighted summation of the derivatives of reflection coefficients in adaptive Schur filter with application to fault detection in rolling element bearings, Mechanical Sys. and Signal Proc. 38 (1), 65-77, 2013.
[22] J. Obuchowski, A. Wylomanska, R. Zimroz, Selection of informative frequency band in local damage detection in rotating machinery, Mechanical Syst and Signal Proc 48 (1-2), 138-152, 2014.
[23] J. Obuchowski, A. Wylomanska, R. Zimroz, The local maxima method for enhancement of time-frequency map and its application to local damage detection in rotating machines, Mechanical Syst and Signal Proc 46 (2), 389-405, 2014.
[24] J. Obuchowski, A. Wylomanska, R. Zimroz, Vibration Analysis of copper ore crushers used in Mineral Processing Plant problem of bearings damage detection in presence of heavy impulsive noise, Proceedings of International conference on Condition Monitoring of Machinery in Non-Stationary Operations, 15-16.12.2014.
[25] J. Obuchowski, A. Wylomanska, R. Zimroz, Stochastic modeling of time series with application to local damage detection in rotating machinery, Key Eng. Mater. 569-570, 441-448, 2013.
[26] L.R. Rabiner, A tutorial on hidden Markov models and selected applications in speech recognition, Proceedings of the IEEE 77(2), 1989.
[27] R.B. Randall, J. Antoni, Rolling element bearing diagnostics-A tutorial, Mechanical Sys. and Signal Proc. 25 (2), 485-520, 2011.
[28] N. Ravishanker, Z. Liu, B.K. Ray, NHPP models with Markov switching for software reliability, Computational Statistics and Data Analysis 52, 3988-3999, 2008.
[29] P. Stefaniak, A. Wylomanska, J. Obuchowski, R. Zimroz, Procedures for decision thresholds finding in maintenance management of belt conveyor system statistical modeling of diagnostic data, in: Lecture Notes in Production Engineering, Christian Niemann-Delius ed., 391-402, Springer, 2015.
[30] A. Wylomanska, R. Zimroz, Signal segmentation for operational regimes detection of heavy duty mining mobile machines- a statistical approach, Diagnostyka 15(2), 33-42, 2014.
[31] G. Zak, J. Obuchowski, A. Wylomanska, R. Zimroz, Application of ARMA modelling and alpha-stable distribution for local damage detection in bearings, Diagnostyka 15 (3), 3-10, 2014.

Detection of structural damage using novelty detection algorithm under variational environmental and operational conditions

M. El Mountassir, S. Yaacoubi and F. Dahmene
Institut de Soudure, R&D department, 4 Bvd Henri Becquerel, 57970 Yutz, France

E-mail: m.elmountassir@institutdesoudure.com

Abstract. Novelty detection is a widely used algorithm in different fields of study due to its capabilities to recognize any kind of abnormalities in a specific process in order to ensure better working in normal conditions. In the context of Structural Health Monitoring (SHM), this method is utilized as damage detection technique because the presence of defects can be considered as abnormal to the structure. Nevertheless, the performance of such a method could be jeopardized if the structure is operating in harsh environmental and operational conditions (EOCs). In this paper, novelty detection statistical technique is used to investigate the detection of damages under various EOCs. Experiments were conducted with different scenarios: damage sizes and shapes. EOCs effects were simulated by adding stochastic noise to the collected experimental data. Different levels of noise were studied to determine the accuracy and the performance of the proposed method.

1. Introduction

SHM of structures using mainly ultrasonic piezoelectric sensors is a promising technique that has been developed in the recent years. The main advantage of this technique over conventional non-destructive evaluation is that the sensors are permanently mounted on the structure to be monitored [1]. Therefore, damage detection is an automatically process which does not need human intervention. It is based on several algorithms and signal processing methods that are capable of recognizing early stage structural damage. Nevertheless, sensors measurements, which should be basically sensitive to the presence of damage, are also sensitive to some environmental and operational conditions (EOCs) [2]. In other words, the effects of these EOCs could be the same as those produced by the damage. This would result in false warnings. Different methods of compensation were developed to overcome this problem, but they have been only designed to perform with an analytical damage detection method, namely subtraction method [3]. On the contrary, statistical methods are less sensitive to the EOCs and can be easily implemented to the monitoring system [4]. They fall into two categories: supervised and unsupervised. When it is a priori possible to retrieve data from damage state, damage identification is done by a supervised learning algorithm such as Artificial Neural Network (ANN) [4]. However, it is generally difficult to provide damaged data information; in this case the damage detection is achieved by an unsupervised learning algorithm based on novelty detection also known as outlier analysis.
This paper is organised as follows. First, a theoritical background wich includes outlier analysis and the Discrete Wavelt Transform (DWT) will be presented. Afterward, the experimental setup that has

Published under licence by IOP Publishing Ltd

been used to simulate different damage scenarios, will be discussed in detail as well as the results. Finally, we conclude this paper and indicate the future work.

2. Theoretical background

2.1. Novelty detection

The problem of novelty detection is to identify if the measured signal has deviated from the normal conditions while in-situ monitoring of engineering structures [5]. Obviously, two aspects can be derived from this problem, firstly how can we define the normal conditions and what is the efficient mean of calculating the deviance from these normal conditions. Since, the structure is exposed to different EOCs; the normal conditions should include all factors that can affect somehow the guided wave propagation (e.g. temperature, humidity, load vibration, etc.). In this study, to account for some variation in these EOCs, a digital stochastic noise was added to ultrasonic measured signals in form of signal-to-noise ratio. The second aspect has been inspired from statistics where the novelty detection is generally based on outlier analysis which is a well-studied field that has been recently utilized for damage detection purposes. In the data mining and statistic literature, outliers are also referred to as anomalies, abnormalities, deviants. It follows that an outlier is defined as a data which is significantly different form a baseline dataset [6]. Depending on this baseline data form, the outlier analysis can be calculated through two methods: univariate or multivariate.

2.1.1. Univariate analysis

Given a set of variables representing data which can be considered as an outlier candidate, the univariate analysis explores each variable separately. The discordancy of this outlier can be measured with different tests. The most common test is the deviation statistics, which can be defined as:

$$\mathbf{z} = \frac{|\mathbf{x}-\bar{\mathbf{x}}|}{\boldsymbol{\sigma}} \tag{1}$$

where x is the potential outlier candidate, $\bar{x}$ and σ, the main and the standard deviation of the baseline data respectively. This discordancy value is then compared to a threshold which will be discussed in detail later to determine whether the data is an outlier or not.

2.1.2. Multivariate analysis

In the multivariate analysis, all variables must be combined in a multi-dimensional vector. In this case, the deviation statistics used in the univariate analysis for discordancy test is replaced by the Mahalanobis Square Distance (MSD) given by [7]:

$$\mathbf{D} = (\mathbf{x} - \bar{\mathbf{x}})^{\mathbf{t}}.\mathbf{K}^{-\mathbf{1}}.(\mathbf{x} - \bar{\mathbf{x}}) \tag{2}$$

where x is the potentiel outlier damage index vector, $\bar{x}$, the mean vector of the baseline, K, the covariance matrix of the baseline and t, matrix transpose index. As in the univariate analysis, the MSD value must be compared to a threshold.

2.1.3. Threshold computation

The threshold value is necessary to determine whether a candidate is an outlier or inlier. Its computation depends on the baseline data distribution. When the data are supposed to be non-parametric, the threshold is calculated using the extreme value theorem (based on Monte Carlo simulation) [8]. Otherwise, following a parametric approach, more particularly, if the data is assumed to represent a Gaussian distribution, the threshold value is taken as 99.73% of the Gaussian limit confidence. Therefore, if a candidate is classified as an outlier, there is only 0.27% chance of false classification. Notice that this threshold is a choice, and hence the lower the threshold, the higher the false calls probability.

2.2. Discrete Wavelet transform (DWT)

The DWT is a discrete implementation of the wavelet transform which is a time frequency representation, well designed for the analysis of non-stationary signals (e.g. ultrasonic guided waves) [9]. The DWT decomposes the original signal by computing its correlation with a short-duration wave called the mother wavelet into approximation and detail coefficients as it is shown in Figure 1. Two major applications of this signal processing method can be exploited in this study: denoising and compression [10]. This can be achieved if only a few wavelet coefficients containing the essential information of interest are retained and the remaining coefficients, related to unnecessary information such as noise, are eliminated.

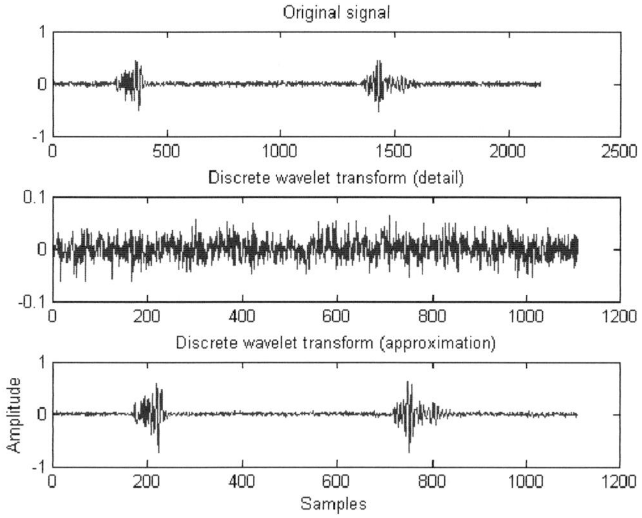

Figure 1. Discrete wavelet transform of an experimental ultrasonic signal (top) showing the detail (medium) and the approximation coefficients (down)

3. Case of study

This section aims to apply the theory of novelty detection to detect and characterise damages severity (if possible) in pipeline structure during in-situ monitoring via ultrasonic guided waves (UGW). Both the univariate and the multivariate techniques will be investigated to determine which one is the most efficient for damage detection.

3.1. Experimental setup

Data acquisition (emission and reception of UGW) was performed using MsS System, designed for pipeline non-destructive testing via ultrasonic guided waves. In order to test the performance of defect detection method, damages were created artificially by adding a magnets cluster on the surface of the structure as it is illustrated in Figure 2.

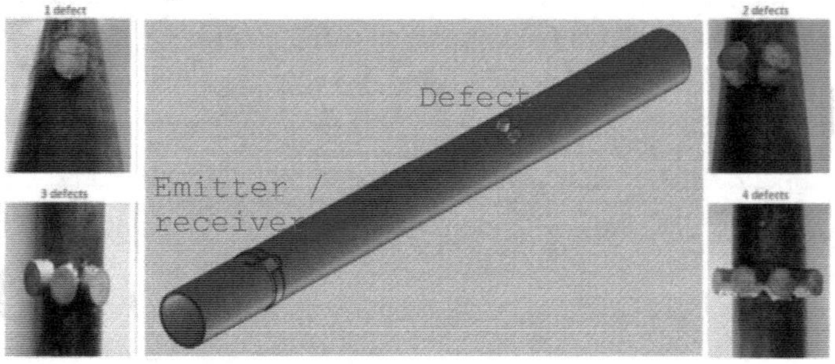

Figure 2. Artificially simulated damages in a pipeline structure

A single measurement of each damage system states (baseline, 1 defect, 2 defects, 3 defects, 4 defects) was taken. Afterward, a digital random noise was added in order to create the statistical population (database) and to simulate some variation of the signal-to-noise ratio which can be induced by several factors namely, EOCs effect, sensors ageing etc. The noise was created using randn function predefined in MATLAB. This function generates an arrays of random numbers normally distributed with zero mean and a standard deviation equal to 1. It was multiplied by a factor determining the noise level (0.01 low level noise and 0.05 for a high level noise). For each level noise, a total of 200 samples were created for all damages states. Once the database is created (Figure 3), UGW signals must first be gated to remove all redundant information as well as significantly reduce the data size. The critical information of the pipe's condition was stored in the time frame between the excitation of the tone burst pulse and the arrival of the end of the pipe (EOP). The gating technique first detects the excitation pulse and the arrival time of the EOP, which can be easily determined by knowing the signal velocity and the length of the pipe. After gating, the DWT was applied to compress and to denoise the data. The wavelets coefficients were used to calculate the damage-sensitive features. RMS, variance, peak to peak and maximum amplitude were chosen as damage index (DI).

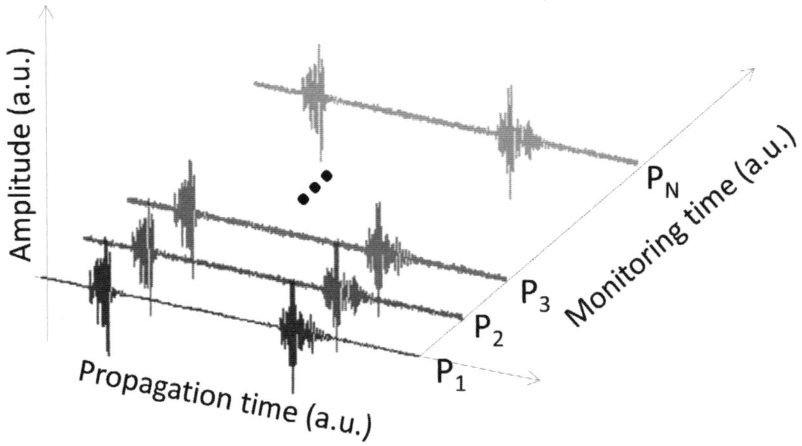

Figure 3. Building the database for the baseline population

3.2. Results and discussion

3.2.1. Univariate analysis (low noise level)

As it was described in section 2.1, in the univariate analysis all damage sensitives features (RMS, Variance, Peak to peak amplitude, maximum amplitude) were considered separately. Figure 4 illustrates the results of discordancy test for all damages types including the baseline and for all damage-sensitives features in the case of a low level of noise. Each value was obtained using equation 1 where the mean and the standard deviation were calculated from the baseline data. Since the baseline data are normally distributed, the threshold was representing 99.73% Gaussian confidence limit. Each damage type has been properly detected even in the case of one defect. Besides, the differentiation between all damage types has been clearly noticed by the increasing steps at 201, 401, 601 and 801. Interestingly, for all damage-sensitives features, there is no indication of false alarm, because all samples have been correctly classified. Despite these satisfactory results, the use of univariate analysis doesn't guarantee that all the damages-sensitive features will behave in same manner. In other words, some features could for example indicate the presence of a defect and others not. This fact reduces the applicability of such a method.

3.2.2. Univariate analysis (high noise level)

When using a high level of noise, results shown in Figure 5 and Figure 6 demonstrate that the threshold increases greatly, therefore, the sensitivity of detection decreases. In this case, all damages type's samples are correctly classified except for the type of one defect where almost all samples are below the threshold which indicates negative false alarms.

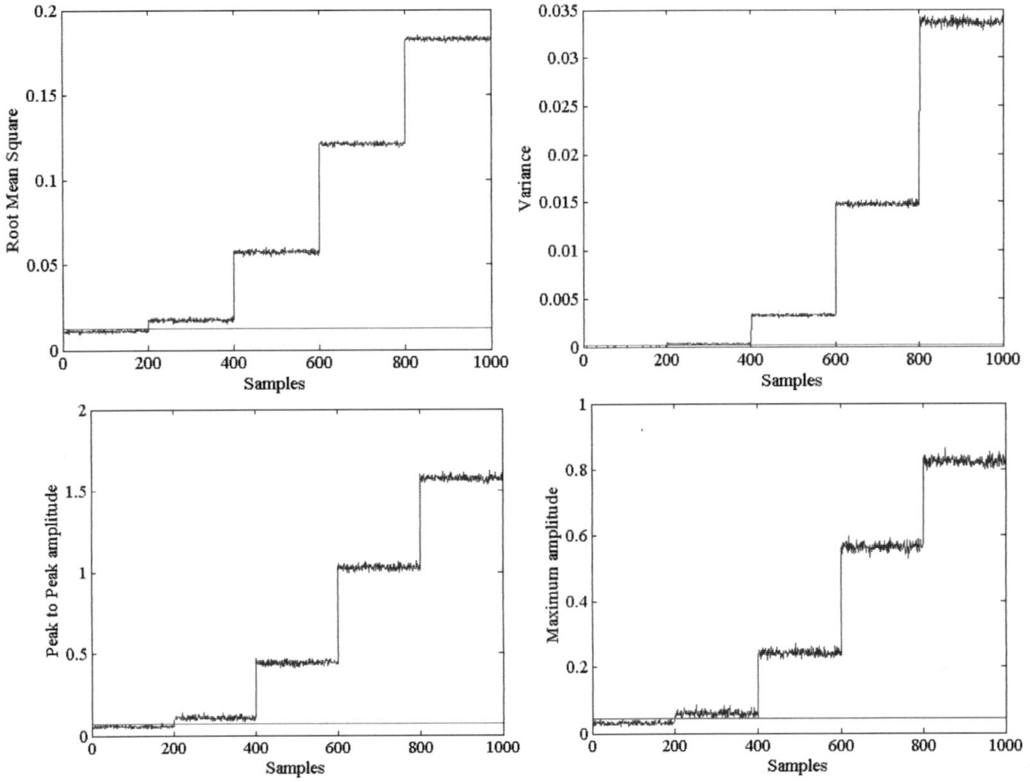

Figure 4. Univariate analysis for a low noise level and for four damage sensitive features (RMS, Variance, Peak to peak amplitude and Maximum amplitude). The red line corresponds to the threshold

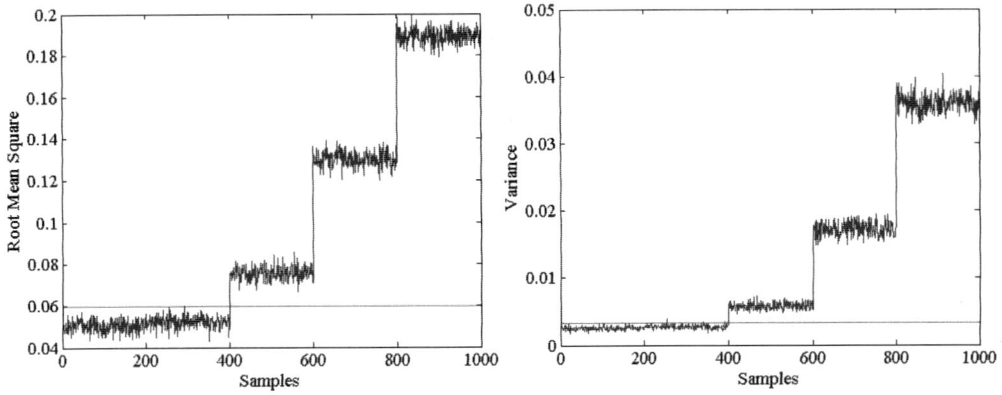

Figure 5. Univariate analysis for high level noise (left: RMS, right: Variance). The red line corresponds to the threshold

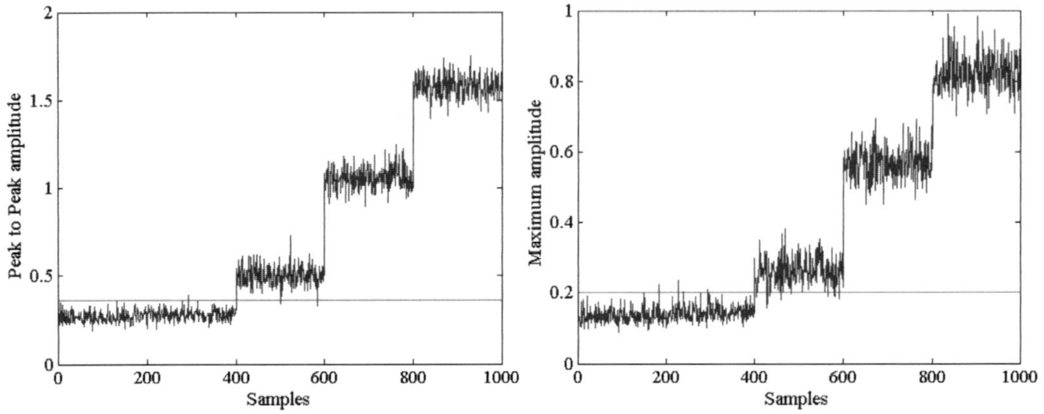

Figure 6. Univariate analysis for high level noise (left: Peak to peak amplitude, right: maximum amplitude). The red line corresponds to the threshold

3.2.3. Multivariate analysis

As it was described in section 2.1.2, in the multivariate analysis the four damages sensitive features are used simultaneously to construct a four dimensional vector. The MSD calculated from the baseline all damages conditions for data corrupted by a low noise level is illustrated in Figure 7 (left). It can be clearly noticed that the multivariate performs well than the univariate analysis. Firstly, the values of the MSDs are much greater the corresponding values of the univariate analysis. As a consequence, the sensitivity to damage is enormously improved. Secondly, there is a good discrimination between all damage types. For high noise level, results shown in Figure 7 (right), are excepted because the sensitivity to damage was decreased and hence one type defect has not been identified.

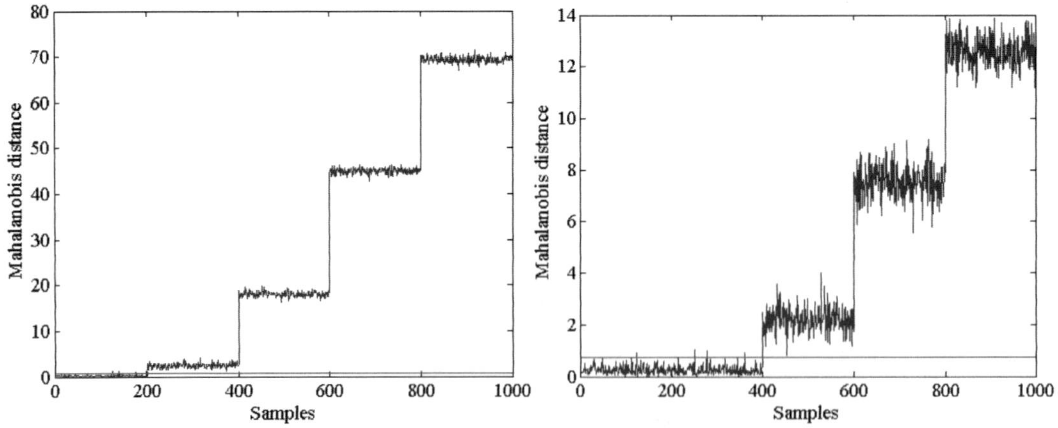

Figure 7. Multivariate analysis for low noise level (left) and for high noise level (right)

4. Conclusion

In this paper, the application of novelty detection as damage identification technique was investigated under varying EOCs. Features extraction were performed using the coefficients of DWT. The use of DWT is essential for noise suppression, and signal compression if there is need. Results have shown that the multivariate analysis is much better than the univariate analysis in terms of sensitivity to damage and discrimination between all damages types. It has been also shown that the level of noise compromises greatly the sensitivity of detection either in univariate or multivariate analysis. In the present work, RMS, Variance, Peak to peak amplitude and Maximum amplitude are chosen to be investigated. Some other features could be used and might be more sensitive that the current tried ones. Further studies concerning this item are to be performed in the near future.

5. References

[1] Farrar CR, Worden K. An introduction to structural health monitoring. Philos Trans A Math Phys Eng Sci 2007;365:303–15. doi:10.1098/rsta.2006.1928.

[2] Sohn H. Effects of environmental and operational variability on structural health monitoring. Philos Trans A Math Phys Eng Sci 2007;365:539–60. doi:10.1098/rsta.2006.1935.

[3] Croxford AJ, Moll J, Wilcox PD, Michaels JE. Efficient temperature compensation strategies for guided wave structural health monitoring. Ultrasonics 2010;50:517–28. doi:10.1016/j.ultras.2009.11.002.

[4] Sohn H, Farrar CR, Hunter NF, Worden K. Structural Health Monitoring Using Statistical Pattern Recognition Techniques. J Dyn Syst Meas Control 2001;123:706. doi:10.1115/1.1410933.

[5] Worden K, Manson G, Fieller NRJ. Damage Detection Using Outlier Analysis. J Sound Vib 2000;229:647–67. doi:10.1006/jsvi.1999.2514.

[6] Pimentel M a. F, Clifton D a., Clifton L, Tarassenko L. A review of novelty detection. Signal Processing 2014;99:215–49. doi:10.1016/j.sigpro.2013.12.026.

[7] Rizzo P, Sorrivi E, Lanza di Scalea F, Viola E. Wavelet-based outlier analysis for guided wave structural monitoring: Application to multi-wire strands. J Sound Vib 2007;307:52–68. doi:10.1016/j.jsv.2007.06.058.

[8] Odening M, Hinrichs J. Using Extreme Value Theory to Estimate Value-at-Risk. Agricultural Finance Review, 2003.

[9] Adeli H, Zhou Z, Dadmehr N. Analysis of EEG records in an epileptic patient using wavelet transform. Journal of Neuroscience Methods Volume 123, Issue 1, 15 February 2003, Pages 69–87

[10] Brussel VU. Noise suppression and signal compression using the wavelet packet transform. Chemometrics and Intelligent Laboratory Systems. Volume 36, Issue 2, April 1997, Pages 81–94

Investigation on the use of artificial neural networks to overcome the effects of environmental and operational changes on guided waves monitoring

M El Mountassir, S Yaacoubi and F Dahmene
Institut de Soudure, plateforme RDI CND, 4 Bvd Henri Becquerel, 57970 Yutz, France

E-mail: m.elmountassir@institutdesoudure.com

Abstract. Intelligent feature extraction and advanced signal processing techniques are necessary for a better interpretation of ultrasonic guided waves signals either in structural health monitoring (SHM) or in nondestructive testing (NDT). Such signals are characterized by at least multi-modal and dispersive components. In addition, in SHM, these signals are closely vulnerable to environmental and operational conditions (EOCs), and can be severely affected. In this paper we investigate the use of Artificial Neural Network (ANN) to overcome these effects and to provide a reliable damage detection method with a minimal of false indications. An experimental case of study (full scale pipe) is presented. Damages sizes have been increased and their shapes modified in different steps. Various parameters such as the number of inputs and the number of hidden neurons were studied to find the optimal configuration of the neural network.

1. Introduction

The aim of SHM is to detect, localize and identify early stage damages in operating structures in order to increase their life service, reduce maintenance cost and avoid catastrophic failure [1]. Ultrasonic guided waves are considered as a good candidate for such an application because they can travel over long distances with relatively small attenuation. Hence, they can cover large areas using only small number of permanently mounted sensors. The acquisition system continuously (or periodically, where the acquisition period is related to different factors which are not detailed in the present paper) stores data from *in-situ* structures. Consequently, signal processing and analysis are necessary to convert the acquired data into information reflecting the state of the structure. However, the collected data may contain not only information about damage but also the effect of EOCs [2]. Thus, to ensure a reliable damage detection approaches, these effects must be eliminated or at least reduced. In literature, issues (analytical and statistical) relating to the effects of EOCs have been addressed by many researchers [3,4]. By following an analytical approach, it is difficult to provide a global and reliable solution for this problem. Indeed, each environmental and operational factor should be studied carefully to demystify its effects. Once effects are known, solutions can be provided (example temperature effect in reference [7]). Since these effects are various, it becomes a challenging task to implement a solution per parameter. An alternative approach can be found in statistical methods, specifically in pattern recognition paradigm [5].

This paper starts by describing the effects of EOCs on ultrasonic guided waves monitoring technique. Time-shift, amplitude drift, distortion and dilation were determined as drastic effects of these EOCs.

Afterward, a damage detection technique based on a supervised learning algorithm, namely Artificial Neural Networks (ANNs) will be investigated. The last section will be devoted to the conclusion of the paper.

2. Environmental and operational effects

The propagation of guided ultrasonic waves can be affected by some environmental and operational factors (e.g. temperature, humidity, vibration loads, etc.). For instance, temperature variation induces generally a thermal expansion or compression of the structure which impacts the wave propagation. In the case where only single non-dispersive mode propagates through the medium, the effect of temperature has been demystified by mathematical formulations [3]. The assumption of single-mode excitation reduces the complexity of the received signals and thus makes their interpretation much easier. However, its generation depends highly on the type of transducers and the geometry of the host structures [6]. Theoretically, it has been shown that change in wave velocity, caused by temperature variation, is the main cause of time shift in signals [3]. This time shift will have an impact on the detection system because it will result in an unacceptable high residual signal if two baseline signals are subtracted. This is due to the fact that even if there is no damage, the baseline and the current signal do not match. Besides, literature review, performed recently by the authors, showed that time shift is not the only effect of temperature. In addition, the assumption of single mode propagation is often idealistic because the generation and propagation of a unique mode is not always straightforward especially in in-situ structures. Guided waves are characterized by multi-modal, dispersive and multipath components; hence it is difficult to establish a mathematical relationship between EOCs effects and wave propagation. To overcome this problem, researchers have performed empirical studies by exposing the structure to different temperature variation in an environmental chamber [7]. Results shown in Figure 1 (left) confirmed that the effect of temperature is not only a change in the arrival time but also in the amplitude of signal response. It has been also found that changes in signal shape namely distortion and dilation can be eventually noticed when dealing with diffuse ultrasonic signals [8]. These results are quiet controversial because damages might be detected via algorithms which can extract information by comparing exactly the amplitude, the shape and the arrival time of signals before and after damage. In other words, the effects which result on signals, and can have a drastic consequence on the damage detection system are: time shift, amplitude drift, distortion and dilation. Even if the other EOC factors, such as liquid loading, are not studied in this paper, the effects might be the same. As a consequence, techniques that can remove temperature effect could possibly be utilized for other EOCs factors. Therefore, the damage detection algorithm should be able to recognize if the monitored structure is damaged or not even under harsh EOCs. That's to say the algorithm should make difference between damage presence and EOCs since both impact the signal. To this end, some compensation method have been developed, but they are limited to some specific cases, and had been only designed to compensate the effect of time shift as it is illustrated in Figure 1 (right).

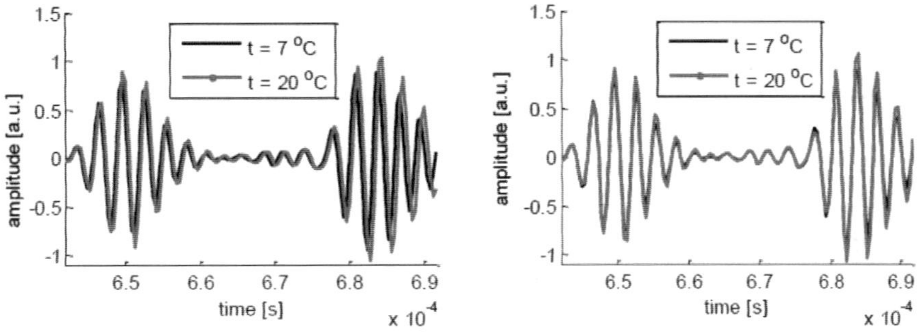

Figure 1. Example of temperature effect on sensor signals (left), and temperature compensation (right) [7]

The solution to the EOCs problem is inspired from pattern recognition paradigm (e.g. face and speech recognition) where the algorithm is capable of detecting a pattern even it includes some changes. The most utilized algorithm is the Artificial Neural Network (ANN) [9,10]. The next section will provide an analysis of this algorithm and how it can be utilized in the context of SHM applications.

3. Artificial Neural Network (ANN)

ANN is an algorithm inspired from biological neural networks. It has been initially designed for solving problems of pattern recognition, and it has been extended then to other artificial intelligence applications such as prediction, control and optimization [11,12]. In the context of SHM and non-destructive evaluations, this method has been successfully utilized for damage detection and classification [9]. It is considered as a supervised learning algorithm, because data from both system states (damaged and undamaged) are needed and should be available for classification tasks. The basic architecture of an ANN consists of three components: inputs, hidden layers and outputs. All these components must be interconnected and process information in parallel. In this study, inputs are the damage-sensitive features calculated from ultrasonic guided waves signals. It's a measure that quantifies the presence and severity of damages. The algorithm of ANN is implemented by two elementary steps: training and validation. The first step also known as learning provides the network with the desired output for specific types of inputs. Since the damage-sensitive features are dependent on damage type, the objective of this step is to learn to the detection system the relation between the features and the type of damage. The second step tests the ability of the network to recognize the type of damage whenever a new feature value is presented. Figure 2 shows the ANN architecture that has been adopted, which is the Feed Forward Back Propagation (FFBP) [13].

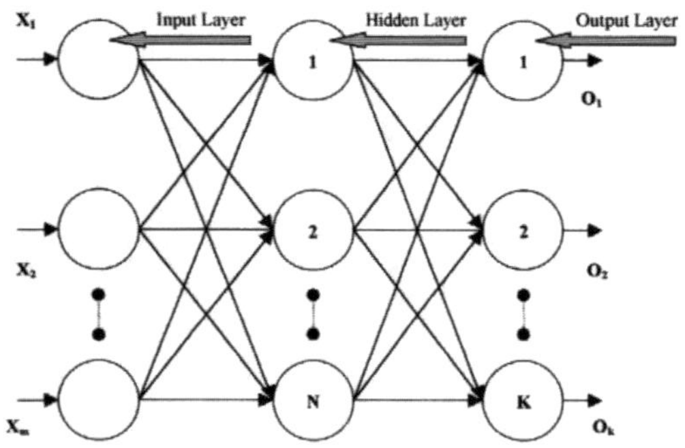

Figure 2. Feed Forward Back Propagation network.

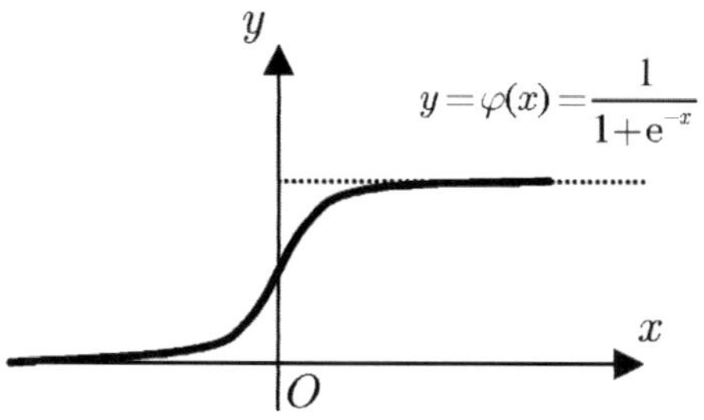

Figure 3. Sigmoid transfer function.

The input layer receives the inputs as damage-sensitive features. The hidden layer processes the data by multiplying the input vector by weights and adding biases. The results are set as input parameter of a transfer function (Figure 3). Since the output targets are generally coded with binary digit, this function returns a float between 0 and 1. The output layer provides the network with the outputs and compares them with targets. The error is calculated as:

$$E = \frac{1}{N}\sum_{i=1}^{N}\sum_{j=1}^{m}(y_{ij} - \widehat{y_{ij}})^2 \quad (1)$$

where N is the number of training samples, m is the number of output nodes, y_{ij} is the desired target and $\widehat{y_{ij}}$ is the network output. If the error E is bigger than a certain value, the training is continued by transmitting the errors backwards from the output layer, and adjusting the weight and biases. Otherwise, the learning process is stopped. Finally, it is worth noting that the performance of the network depends greatly on several parameters such as: dimension of the input vector, type of damage sensitive features and number of hidden layers. Therefore, it is recommended to test multiple combinations of these parameters in order to find the optimal configuration of the network.

4. Case of study

Experiments were conducted on a pipeline structure using ultrasonic guided waves. Damage was simulated physically by adding a mass of magnets to the surface of the pipe, as it is shown in Figure 4. This technique has been found in the literature as an acceptable way to simulate damage non-permanently in a structure [14–16]. In fact, damage is normally detected in guided waves-based systems due to the local change in the density ratio of the structure causing a wave reflexion. By adding the mass to the structure, a simulated damaged region is created. Besides, the reflexion coefficient of the incident wave is more sophisticated than the case of a realistic damage.

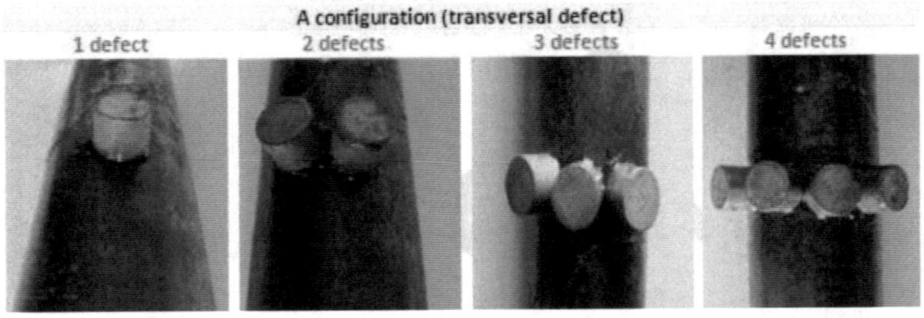

Figure 4. Different types of defects (circumferential configuration)

A single acquisition of the baseline and all damage types was acquired. To account for environmental and operational conditions changes (EOC), a stochastic noise was added to each type of damage including the baseline. For each case (for example the baseline), 199 signals is constructed from the original signal (which has not been contaminated with noise) as it is shown in Figure 5. Four damages-sensitive features were extracted: RMS, Variance, Peak to Peak amplitude and Maximum amplitude.

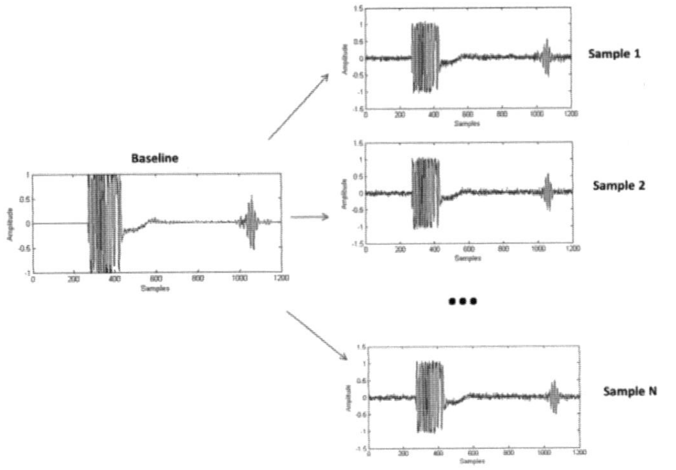

Figure 5. Database building

Table 1. Classes with codifications

Classes	Binary number
1 defect	00
2 defects	01
3 defects	10
4 defects	11

Table 1 illustrates the classes that have been considered in this algorithm. Each class of damage type was coded with a 2 digit binary numbers. Fifty signals were used as training data and 150 for validation. The network performance is quantified by the number of correctly classified samples. In order to find best network configuration, a huge and extensive number of tests of different parameters combination were accomplished. For each input vector, the number of hidden layers was changed from 1 to 40. Results illustrated in Figure 6 have shown that the best network configuration was found with twenty nine hidden layers and the four damages-sensitive features, because the accuracy of classification is 100%, it follows that all samples from different types of damages have been correctly classified. An example of a poor result has been noticed when only six hidden layers were utilized (Figure 7).

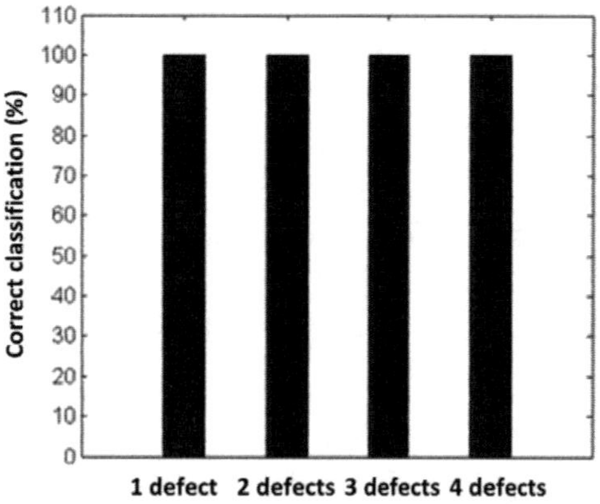

Figure 6. Accuracy of classification with the best network configuration.

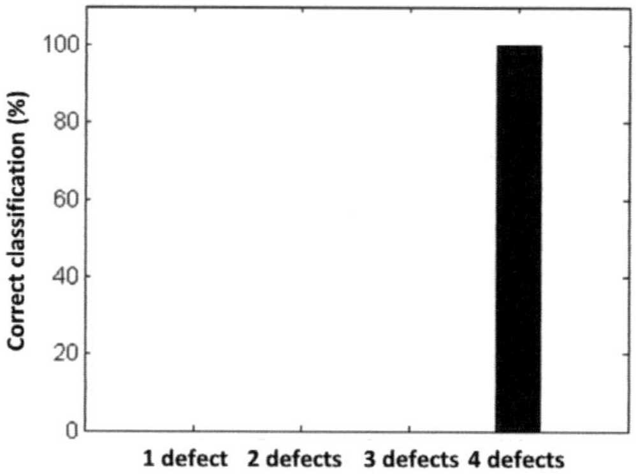

Figure 7. Accuracy of classification with 6 hidden layers.

5. Conclusion

ANN algorithm has proven its efficiency as a promising algorithm for damage detection and classification even in the presence of EOCs changes, which were simulated by a random and stochastic noise. However, there are many issues making its real-world application very limited. These issues can be summarized as follows:

- A huge and extensive set of trials must be verified to find the optimal network configuration which has the best classification performance in terms of number, type of input parameters and the number of hidden layers.
- ANN data learning requires information from damage state, which is impractical because it is generally difficult to acquire damaged data during in-situ monitoring of structures.

Finally, it must be noted that, the aim of this study was not to apply the ANN to detect and classify defects but to verify if this method can resist to the EOCs effects. Indeed, the added noise could simulate some effects of EOCs but not all of them. However, the results have shown that this approach has found to be robust. Hence, even if the other EOCs effects have not been investigated, the results would be the same.

6. References

[1] Farrar CR, Worden K. An introduction to structural health monitoring. Philos Trans A Math Phys Eng Sci 2007;365:303–15. doi:10.1098/rsta.2006.1928.

[2] Sohn H. Effects of environmental and operational variability on structural health monitoring. Philos Trans A Math Phys Eng Sci 2007;365:539–60. doi:10.1098/rsta.2006.1935.

[3] Croxford AJ, Moll J, Wilcox PD, Michaels JE. Efficient temperature compensation strategies for guided wave structural health monitoring. Ultrasonics 2010;50:517–28. doi:10.1016/j.ultras.2009.11.002.

[4] Michaels JE, Michaels TE. Detection of Structural Damage from the Local Temporal Coherence of Diffuse Ultrasonic Signals. IEEE transactions on Ultrasonics, ferroelectrics, and frequency control, vol. 52, no. 10, October 2005

[5] Sohn H, Farrar CR, Hunter NF, Worden K. Structural Health Monitoring Using Statistical Pattern Recognition Techniques. J Dyn Syst Meas Control 2001;123:706. doi:10.1115/1.1410933.

[6] Moll J, Fritzen C. Advanced Aspects of Mode-Selective Excitation of Ultrasonic Guided Waves. International Conference on Noise and Vibration Engineering. Leuven, Belgium. September 2016.

[7] Ha S, Lonkar K, Mittal a., Chang FK. Adhesive Layer Effects on PZT-induced Lamb Waves at Elevated Temperatures. Struct Heal Monit 2010;9:247–56. doi:10.1177/1475921710365267.

[8] Weaver RL, Lobkis OI. Temperature dependence of diffuse field phase. Ultrasonics 2000; 38:491–494.

[9] Nazarko P, Ziemiański L. Application of artificial neural networks in the damage identification of structural elements. Computer Assisted Mechanics and Engineering Sciences, 18: 175–189, 2011.

[10] Shi A, Yu X-H. Structural damage detection using artificial neural networks and wavelet transform. 2012 IEEE Int Conf Comput Intell Meas Syst Appl Proc 2012:7–11. doi:10.1109/CIMSA.2012.6269593.

[11] Morrist AJ. Artificial Neural Networks in Process Estimation and Control. Automatica, Vol. 28, No. 6, pp. 1181-1187, 1992

[12] Shen C, Wang L, Li Q. Optimization of injection molding process parameters using combination of artificial neural network and genetic algorithm method. J Mater Process Technol 2007;183:412–8. doi:10.1016/j.jmatprotec.2006.10.036.

[13] Hagan MT, Menhaj MB. Training feedforward networks with the Marquardt algorithm. IEEE Trans Neural Networks 1994;5:989–93. doi:10.1109/72.329697.

[14] Pitchford CW, Peairs DM. Impedance-Based Structural Health Monitoring of Wind Turbine Blades. Renew Energy 2007.

[15] Behmanesh I, Moaveni B. Probabilistic damage identification of the dowling hall footbridge using Bayesian FE model updating. Conf Proc Soc Exp Mech Ser 2013;5:43–51. doi:10.1007/978-1-4614-6564-5_5.

[16] Tarazaga P a, Inman DJ, Drive OG. Structural Health Monitoring of Space Rigidizable-Inflatable Booms. 2006. Conference & Exposition on Structural Dynamics

Pipe Leaks Classification by Using a Data-driven Approach Based on Features from Cross-Correlated Piezo-vibration Signals.

Jhonatan Camacho-Navarro[1-2], Magda Ruiz[1], Oscar Perez[2], Rodolfo Villamizar[2], Luis Mujica[1].
[1]Department of Applied Mathematics III, Escola Universitària d'Enginyeria Tècnica Industrial de Barcelona (EUETIB), Universitat Politécnica de Catalunya (UPC) BARCELONATECH, Comte d'Urgell, 187, E-08036, Barcelona, Spain.
[2]Escuela de Ingenierías Eléctrica, Electrónica y de Telecomunicaciones (E3T), Universidad Industrial de Santander (UIS). Grupo de Control Electrónica Modelado y Simulación (CEMOS). Santander, Colombia.

jhonatan.camacho@estudiant.upc.edu, magda.ruiz@upc.edu, oscar.perez.gamboa@gmail.com, rovillam@uis.edu.co, luis.eduardo.mujica@upc.edu

Abstract. This work presents a data driven approach for pipe leaks classification, validated on a steel carbon pipe section conditioned with leaks of different sizes and locations in order to emulate abnormal conditions. The tested structure was instrumented with piezoelectric devices attached at different locations over the surface, in order to induce guided waves and to record its behaviour along the structure. For each experiment, one piezo device is excited by means of a high frequency burst type signal and the other ones are used as sensors. A blind bridle is connected to one of the extremes and an air source is coupled to the other extreme to emulate operational conditions. Statistical indices of correlated piezoelectric signals are obtained by using principal component analysis to distinguish different leak scenarios. Next, a self-organizing map is used to classify them. The experimental results show an improvement of the classification-learning rate when correlated signals are used instead of uncorrelated ones

1. Introduction

If pipeline damages could be early detected, adverse environmental and economic effects can be avoided improving human safety. In this sense, pipe leaks is one of the most important type of damage with great economical and environmental impact [1]. In this point, the formulation of methodologies for pipeline leakage identification has been widely documented in the state of the art [2]. Common methods for pipelines damage identification are based on physical models and analysis of measurements obtained from available instrumentation in the structure [3], [4]. However, the big amount of false alarms is one of the drawbacks of these methods, which is a not desirable condition for continuous monitoring systems.

In recent years, the use of guided waves for damage assessment in pipeline structures has been reported as successful [5]. Thus, by taking advantage of piezoelectric properties and guided waves, piezoelectric devices are a cheap technology with promising results for pipeline damage assessment. Experimental results for detecting cuts with different depths in a pipe-like structure using piezoelectric technology are discussed in Gharibnezhad et al. [6]. Other application involves the analysis of piezo-electrical measurements from macro fiber-composite as part of guided wave methods to detect partial circumferential crack and corrosion on the pipe's surface of a pipeline structure located at Los Alamos

National Laboratory [7]. Also, piezo-ceramic transducer has been implemented to identify the position and severity of leaks in the pipeline network at Sheffield University [8]. In addition, the effectiveness of principal component analysis and self-organizing maps for structural damage detection algorithms have been widely demonstrated (for example in pipeline crack detection algorithms [9], [10]).

In this paper, a non-intrusive pipeline leaks detection and classification approach is experimentally validated. Which consists of processing induced guided waves, measured along the structure, by using cross-correlation functions and principal component analysis for detection purposes, and self-organized maps for classification task. The main contribution of this approach is the analysis of the positive influence of cross-correlation analysis as pre-processing tool in a data driven approach used for pipeline leaks detection and classification.

2. Structural damage detection methodology

Figure 1 summarizes the procedure for pipelines leaks detection and classification. The approach involves three main modules: 1. Piezo-electric instrumentation; 2. Statistical processing; and 3. Supervised classification.

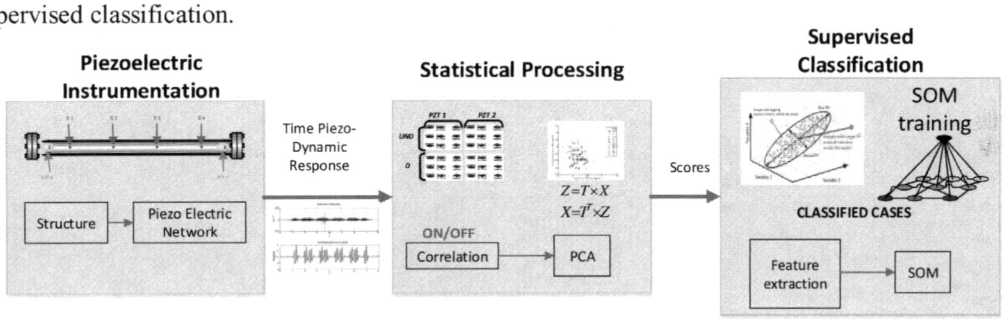

Figure 1 Damage assessment methodology

2.1. Piezoelectric instrumentation

The approach for damage identification developed in this work is based on piezo-diagnostics principle, which refers to damage identification based on the phenomenon of elastic waves propagation generated by piezo-actuators [11]. For instance, in [12] implements impedance-based structural health monitoring methods by means of piezoelectric materials analysis. Similarly, in [13] takes advantage of lamb waves in order to find patterns with high sensitivity to structural damages.

In this way, a piezoelectric active scheme is used, where several piezo devices are installed in the structure to induce and record a guided wave response at different locations of the structure. One of the piezoelectric devices is excited with a periodic high frequency burst type signal in order to induce a guide wave and the remaining piezo-devices are used as sensors. The instrumentation consists of fine-tuning filters, high wide-band amplifiers and acquisition systems, among others (Figure 2).

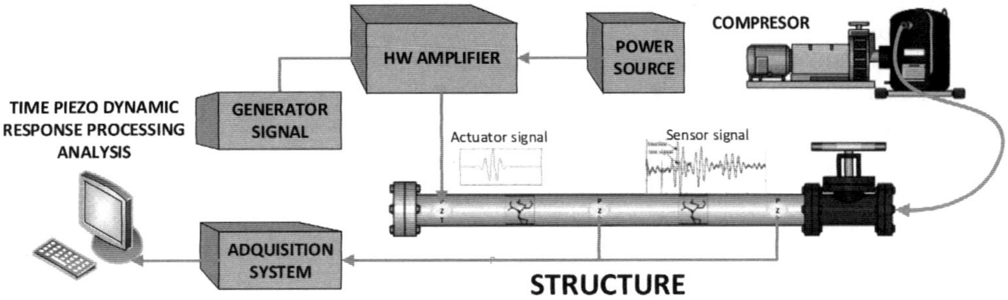

Figure 2 General components for damage identification based on piezo dynamic response analysis.

2.2. Statistical processing

Three steps are necessary for statistical processing: *i.)* Uncorrelated data matrix building; *ii.)*Cross-correlated data matrix building; and *iii.)* Model building. An important piece of the statistical processing module is the data matrix, which contains information about different types of leaks to be used for validating the classification process. The data matrix is organized as follows: types of damages studied and the number of experiments for each damage; number of piezo device sensors used and finally, number of sample times recorded from each piezo sensor. This matrix has concatenated row vectors representing several experiment repetitions for undamaged and damage conditions (**Figure 3**). The cross-correlated and uncorrelated data matrices are intended to facilitate results comparison regarding to the cross-correlation performance in the piezo-diagnostic approach. Thus, two distinct statistical models are built with both data matrices.

2.2.1. Uncorrelated data matrix building.

A first data matrix of recorded signals from piezo sensors are obtained. It contains time measurements as response to the propagation of guided waves along the structure. Thus, time piezo-electrical response for damage and undamaged conditions are folded in the data matrix (**Figure 3**).

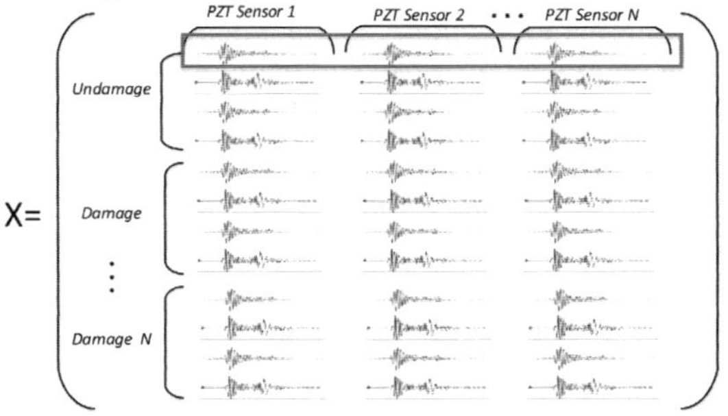

Figure 3 Data matrix

2.2.2. Cross-correlated data matrix building.

A second data matrix is built using cross-correlated functions between piezo actuation and sensing signals, since several applications for structural damage assessment have demonstrated the effectiveness of using cross-correlation functions [14]. For example, in [15] damage identification methods based on natural excitation Technique (NeXT) employs cross-correlation functions for modal analysis. It had been useful for damage identification in civil structures. Other proposal include the estimation of the time of flight of wave packages by means of cross correlation functions determining the location of the defects within a large area of a thin-plate specimen [16].

2.2.3. Model building.

The last step is modelling process. This applies Principal Component Analysis (PCA) in the data matrices. PCA is a statistical tool widely used in structural damage detection algorithms, for example in applications related to delamination detection in a composite beam [17]. The objective in PCA is to obtain a reduced space representation for multidimensional data. As is illustrated in **Figure 4**, the original data matrix **X** with ***n*** samples of ***m*** statistical variables is projected onto the orthogonal space, defined by the principal components. The new variables **Z** consists of uncorrelated data organized by down variance, but preserving the cumulative original variance.

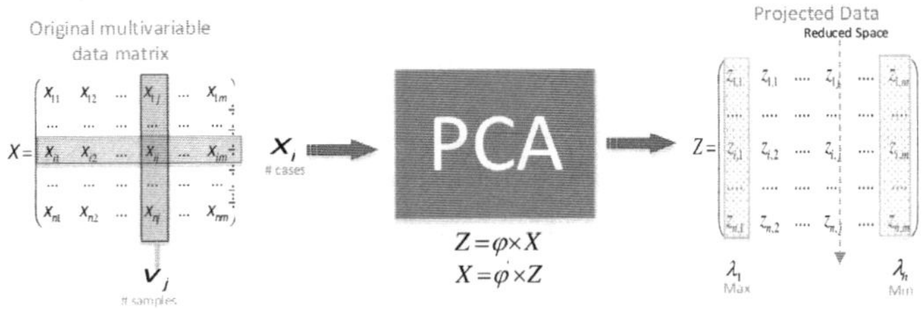

Figure 4 Principal component analysis scheme

The general procedure used to obtain the reduced space is the next [18]:

i. Normalize the original variables by means of standard deviations ($\hat{\sigma}_j$) and mean values ($\hat{\mu}_j$):

$$\hat{X} = \frac{X-\hat{\mu}_j}{\hat{\sigma}_j} \qquad (1)$$

Where $\hat{X}$ correspond to normalized data.

ii. Compute the covariance matrix of centered data matrix:

$$C_{\hat{X}} = \frac{1}{m-1}\hat{X}'\hat{X} \qquad (2)$$

iii. Estimate the singular value decomposition for covariance matrix:

$$C_{\hat{X}} = \varphi\Sigma\varphi' \qquad (3)$$

iv. Transforming the original variables onto the orthogonal space defined by the Eigenvectors ($\boldsymbol{\varphi}$) of the covariance matriz:

$$Z = \varphi \cdot \hat{X} \qquad (4)$$

v. Keep only the first $\boldsymbol{r}$ components in order to obtain a reduced representation for original variables. The variance for each new variable is equal to its respective $\boldsymbol{r}$ eigenvalues ($\boldsymbol{\Sigma}$):

$$Z_r = \varphi_r \cdot \hat{X} \qquad (5)$$

Despite, the above procedure is a common strategy for data reduction, in this approach it is used as statistical model by using the inverse transformation:

$$X = \varphi_r' \cdot Z_r + \hat{X} \qquad (6)$$

Some considerations should be take into account to build a statistical base-line model of the structure [19]:

a.) Training stage

I. Only the undamaged records of the data matrix is processed.

II. The Group-Scaling (GS) normalization method [20] is applied instead of common normalization methods. GS procedure considers the nature of data by estimating standard deviation for each block of piezo measurements. Thus, a standard deviation per piezo sensor is obtained.

III. The $\boldsymbol{r}$ principal components are estimated using an iterative algorithm as NIPALS [21], because the size of the data is big.

IV. The statistical model is obtained only from undamaged records and it consists of four elements: mean values ($\hat{\mu}_j$), standard deviations ($\hat{\sigma}_j$), $\boldsymbol{r}$ eigenvectors ($\varphi_{und,r}$) and eigenvalues ($\boldsymbol{\Sigma_r}$).

b.) Validation stage

I. Damage experiments are normalized using the mean values ($\hat{\mu}_j$) and standard deviations ($\hat{\sigma}_j$) from the statistical model according to GS method.

II. Projections onto the reduced space of Undamaged and Damage experiments are computed using the statistical model in order to differentiate damaged from undamaged states.

$$Z_{damage} = \varphi_{und} * \hat{X}_{Damage} \qquad (7)$$

2.3. Supervised Classification

Damage detection is achieved by means of statistical indexes with high sensitivity to model deviations. In this case, two widely used indexes (T^2 *and* $Q - statistic$) are computed. T^2 statistic is a variation mesurement of each experiment within the statistical PCA model (undamaged model) and Q-index is the squared 2-norm that measures deviations of the experiment respect to the lower-dimensional PCA representation [19]. Thus, T^2 and Q statistics are used to measure deviations of each experiment respect to the PCA model:

$$T^2 = X^T \varphi_{und} \left(\Sigma^T \Sigma\right)^{-1} \left(\varphi_{und,}\right)^T X \qquad (8)$$

$$Q_i = \tilde{x}_i{}^T \tilde{x}_i, \quad \tilde{x}_i = [I - \varphi_{und} * (\varphi_{und})^T] X_i \qquad (9)$$

where $\tilde{x}_i$ is the residual projection for each experiment.

In addition, the above statistical indexes are used as inputs for a self-organizing map (SOM) for cluster cases in similar damage types. This clustering is achieved by means of competitive learning and preserving topology. Accordingly, nearby data in the input space are mapped into neighbor clusters [22]. Thus, SOM network facilitates classification tasks and graphical interpretation. **Figure 5** deploys how SOM network operates over the input space, specified by T^2 and Q-indexes.

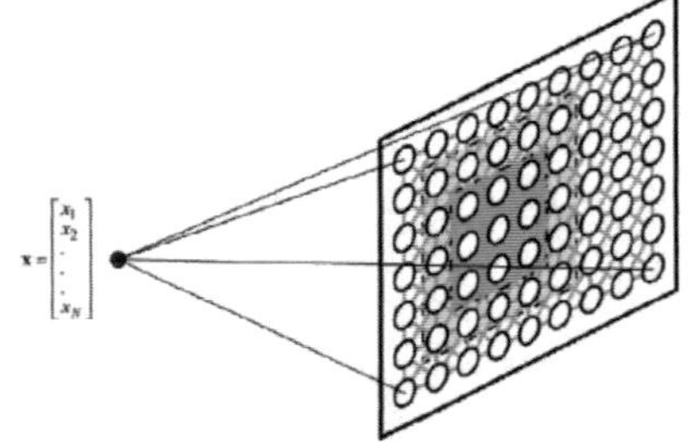

Figure 5 SOM clustering for damage classification.

From **Figure 5**, the SOM consists of **N** clusters, characterized by a prototype vector (Codebook) or cluster center, and grouping several labelled cases. Then similar cases are labelled in clusters, where each label keeps only one instance and the number of stored cases. Similarly, the validation cases are ticked assigning the label with most instances and with the most similar cluster to find the best matching units (BMU). In consequence, the classification error can be estimated by majority voting.

Finally, the SOM quality is evaluated with the quantization and topographic errors. The first one is the average distance between each experiment and its BMU. The second one corresponds to the proportion between data vectors whose first and second BMUs are not adjacent clusters and the total number of experiments [23].

3. Experimental setup

The methodology detailed in this paper was validated for leak detection in a pipeline as part of a non-intrusive damage monitoring system. The specimen used as test structure is a carbon-steel pipe section of dimensions 1 m x 0,0254 m x 0,003 m (length, diameter, thickness). The pipe section has bridles at the ends and a valve sets the air pressure from a compressor in 80 psi at one of the ends (**Figure 6**).

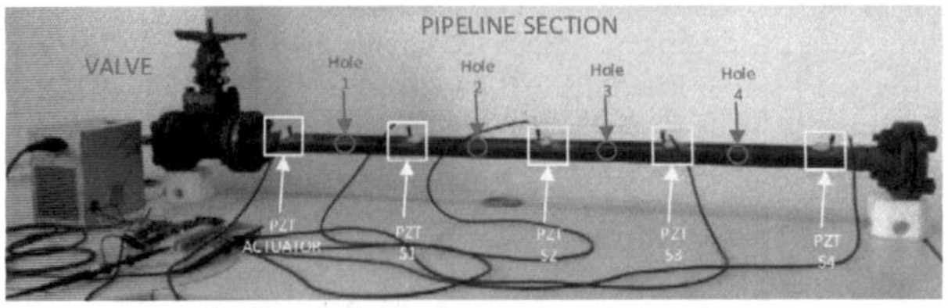

Figure 6 Experiment configuration

In order to induce leaks in the test structure, four ¼-inch holes were drilled along the pipe section wall. Graduable screws are used to control where the leak is produced. **Table 1** details the damages studied in this work. The proposed damage configuration allows concluding if classification and location of different sizes leaks is possible. For each damage type, 100 experiment repetitions were conducted with 1-second periodic excitation signal, where undamaged experiments are tagged with label 'UN'.

Table 1 Leak Damage specification

Label	Open Hole	**Label**	Open Hole
1	H1	**5**	H4 ,H3
2	H2	**6**	H4, H3, H2
3	H3	**7**	H4, H3, H2, H1
4	H4		

Five piezoelectric devices (PZT) were attached along the structure, (previously explained in **Figure 1**). One of the PZT is used as actuator and the remaining devices as sensors. The PZT actuator was excited by an 80 KHz burst signal generated by an AWG PicoScope series 2000 and then amplified to ±10 V. The piezo electrical response is recorded with a picoscope and multiplexor board. **Figure 7** deploys the time piezo electrical response for each piezo device, where it is possible to visualize time of flights for one of the experiments under undamaged state.

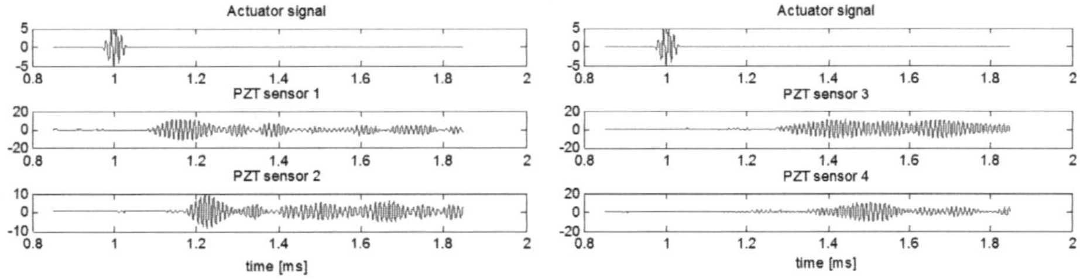

Figure 7 Time piezo dynamic structural response

The **Figure 8** presents the respective cross-correlation functions.

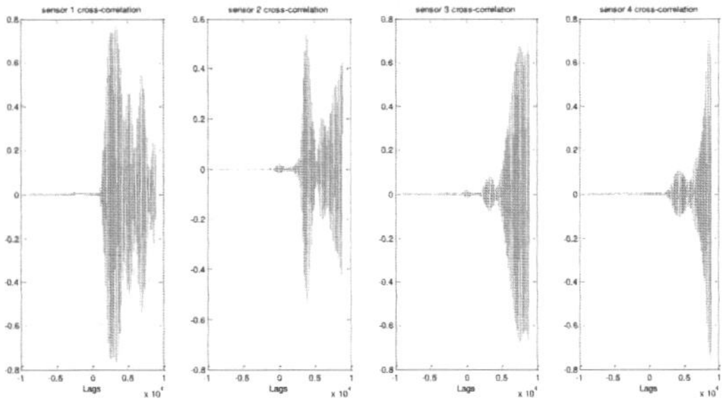

Figure 8. Evolution of the Cross-correlation function for each PZT sensor

4. Results and Discussion

Figure 9 presents the statistical indexes for the studied experiments, which were estimated preserving 15 principal components. In addition, the SOM codebooks location and their respective labels (assigned by majority voting) are shown in the right scattered plot. Empty SOM codebooks are necessary to describe the data distribution. It is observed that by the cross-correlation, better groups for studied damages are obtained.

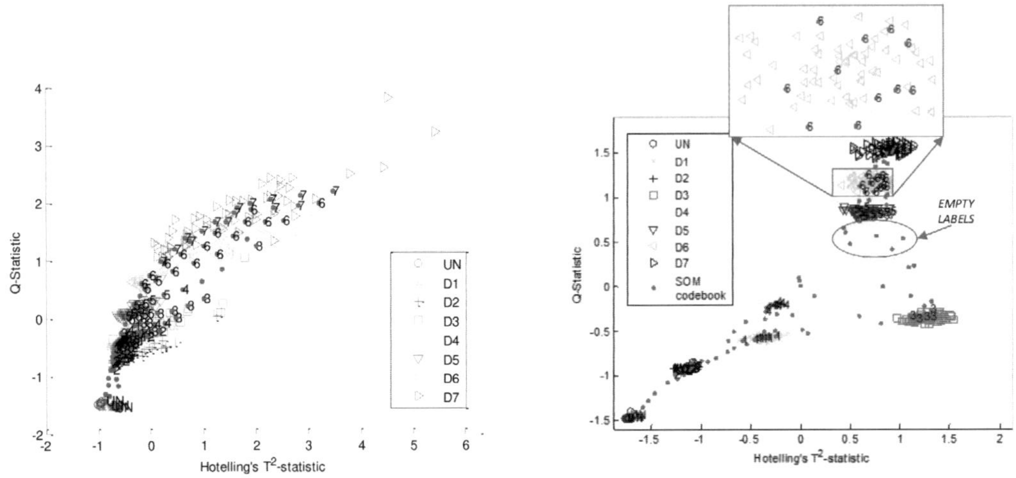

Figure 9 Distribution of damage indexes. *Left:* Without cross-correlation analysis. *Rigth:* By including cross-correlation processing.

The clusters obtained in the Self Organizing Map are depicted in **Figure 10**. A major differentiation between different damage types with boundaries clearly defined by empty clusters and BMU distance matrix (U-matrix) is observed when the cross-correlation as preprocessing stage is applied. In addition, the cases distribution avoids damages combination in one similar cluster, which allows a better classification.

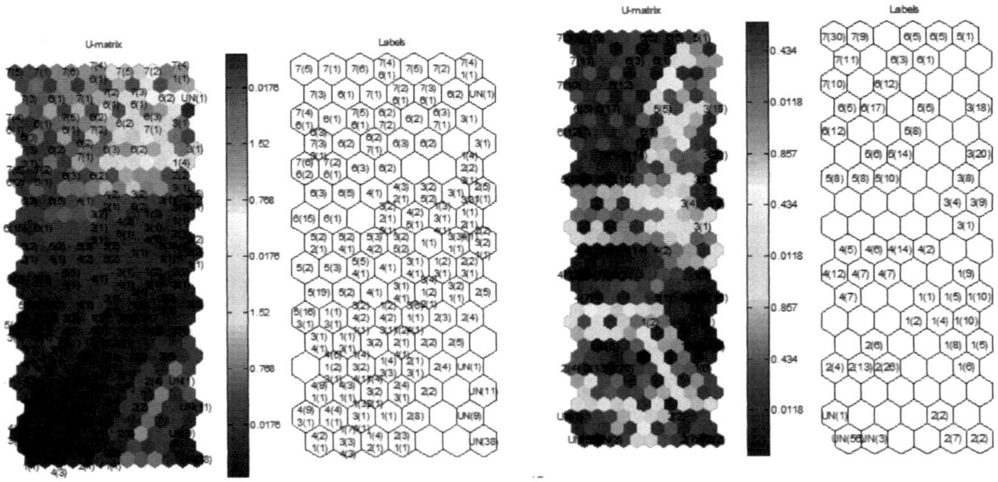

Figure 10. Training SOM network. *Left:* Without cross-correlation analysis. *Right:* By including cross-correlation processing.

The SOM quality indexes are summarized in **Table 2,** which are calculated by using 70% of cases for training purposes and 30% of cases for validation.

Table 2 . SOM quality indexes

Index	Uncorrelated Signals	Cross-correlated signals
Quantization error	0,1083	0,0336
Topographical error	0,0083	0,0833
Distortion measure	1,3828	0,4327

Training Error	18,5417	0
Empty labels in Training data	0	0
Empty Clusters	14	44
Validation Error	36,5625	1,5625
Empty labels in validation data	17	5

5. Conclusion

In this paper is shown, by means of experimental results, that damage classification is highly dependent of preprocessing stage. Thus, by using correlation of piezoelectric signals a better behavior is obtained. However, for future works normalization strategy could be considered as a preprocessing stage, also influence of correlation and normalization on damage location and quantification should be studied. Further studies are required to evaluate the influence of using cross-correlated features from piezo-electric measurements for damage detection in buried pipes.

6. Acknowledgments

This work has been developed as part of the research project: "Monitorización y Detección de Defectos en Estructuras usando Algoritmos Expertos Embebidos", financed by the Departamento Administrativo de Ciencia y Tecnología Francisco José de Caldas – COLCIENCIAS and Banca Mundial in the Universidad Industrial de Santander, Colombia. Agosto 2012-Junio 2015.

References

[1] G. S. Mavi. Feasibility of using ultrasonic guided waves for underwater pipe inspection. Thapar University, Patiala, 201.

[2] Billmann, L., & Isermann, R. (1987). Leak detection methods for pipelines. Automatica, 23(3), 381-385.

[3] Black, P. (1992). A review of pipeline leak detection technology. In Pipeline systems (pp. 287-298). Springer Netherlands

[4] Geiger, G. (2006). State-of-the-art in leak detection and localization. Oil Gas European Magazine, 32(4), 193.

[5] Lowe, M. J., Alleyne, D. N., & Cawley, P. (1998). Defect detection in pipes using guided waves. Ultrasonics, 36(1), 147-154.

[6] Gharibnezhad, F., Mujica Delgado, L. E., Rodellar Benedé, J., & Fritzen, C. P. (2012). Damage detection using robust fuzzy principal component analysis.

[7] Thien, A. B. (2006). Pipeline structural health monitoring using macro-fiber composite active sensors (Doctoral dissertation, University of Cincinnati).

[8] Taghvaei M, Beck SBM & Staszewski WJ (2007) Leak detection in pipeline networks using low-profile piezoceramic transducers. STRUCT CONTROL HLTH, 14(8), 1063-1082

[9] Meléndez, J., Colomer, J., López, B., Vehí, J., Pous, C., Ruiz, M. L., & Mújica, L. E. Experiences in exploiting data at the Universitat de Girona: Selected problems in application domains.

[10] Buethe, I., Torres-Arredondo, M. A., Mujica Delgado, L. E., Rodellar Benedé, J., & Fritzen, C. P. (2012). Damage detection in piping systems using pattern recognition techniques.

[11] Kolakowski, P. Damage Identification in Beams via Piezodiagnostics. 2nd European Workshop on Structural Health Monitoring July 7-9, 2004. Amazeum Conference Centre at Deutsches Museum Munich, Germany.

[12] Park, G., Sohn, H., Farrar, C. R., & Inman, D. J. (2003). Overview of piezoelectric impedance-based health monitoring and path forward. Shock and Vibration Digest, 35(6), 451-464.

[13] Ong, W. H., & Chiu, W. K. (2012). Redirection of Lamb waves for structural health monitoring. Smart Materials Research, 2012.

[14] Muyu Zhang, Rüdiger Schmidt, A Comparative Study of the correlation function based structural damage detection methods under sinusoidal excitation.
[15] Caicedo, J. M. (2011). Practical guidelines for the natural excitation technique (NexT) and the eigensystem realization algorithm (ERA) for modal identification using ambient vibration. Experimental Techniques, 35(4), 52-58.
[16] Yu, L., Bao, J., & Giurgiutiu, V. (2004, July). Signal processing techniques for damage detection with piezoelectric wafer active sensors and embedded ultrasonic structural radar. In Smart Structures and Materials (pp. 492-503). International Society for Optics and Photonics.
[17] Garcia, D., & Trendafilova, I. (2014). A multivariate data analysis approach towards vibration analysis and vibration-based damage assessment: Application for delamination detection in a composite beam. Journal of Sound and Vibration, 333(25), 7036-7050.
[18] Jolliffe, I. (2005). Principal component analysis. John Wiley & Sons, Ltd.
[19] Mujica, L. E., Rodellar, J., Fernandez, A., & Guemes, A. (2010). Q-statistic and T2-statistic PCA-based measures for damage assessment in structures. Structural Health Monitoring, 1475921710388972.
[20] Mujica, L. E., Vehi, J., Ruiz, M., Verleysen, M., Staszewski, W., & Worden, K. (2008). Multivariate statistics process control for dimensionality reduction in structural assessment. Mechanical Systems and Signal Processing, 22(1), 155-171.
[21] Nomikos, P., & MacGregor, J. F. (1994). Monitoring batch processes using multiway principal component analysis. AIChE Journal, 40(8), 1361-1375.
[22] Hyötyniemi, H. (2001). Multivariate regression: Techniques and tools. Helsinki University of Technology.
[23] Kohonen, T., & Maps, S. O. (1995). Springer series in information sciences. Self-organizing maps, 30.

Study of cross-correlation signals in a data-driven approach for damage classification in aircraft wings

Jhonatan Camacho-Navarro[1-2], Magda Ruiz[1], Rodolfo Villamizar[2], Luis Mujica[1], Alfredo Güemes[3], Ignacio González-Requema[3]

[1]Department of Applied Mathematics III, Escola Universitària d'Enginyeria Tècnica Industrial de Barcelona (EUETIB), Universitat Politécnica de Catalunya (UPC) BARCELONATECH, Comte d'Urgell, 187, E-08036, Barcelona, Spain.

[2]Escuela de Ingenierías Eléctrica, Electrónica y de Telecomunicaciones (E3T), Universidad Industrial de Santander (UIS). Grupo de Control Electrónica Modelado y Simulación (CEMOS). Santander, Colombia.

[3]Department of aerospace materials and processes, Universidad Politécnica de Madrid, Madrid, Spain.

jhonatan.camacho@estudiant.upc.edu, magda.ruiz@upc.edu, rovillam@uis.edu.co, luis.eduardo.mujica@upc.edu, aguemes@aero.upm.es, ignaciof.gonzalez@upm.es

Abstract. This paper discusses, experimental results of classifying several mass adding in a wing aircraft structure, using cross-correlated piezoelectric signals, represented by principal components. Piezoelectric signals are applied and recorded at specific points of the structure under analysis. Then, statistical features are obtained by means of principal component analysis to the correlation between excitation and response signals. Unsupervised learning is implemented to the reduced feature space, in order to identify clusters of damaged cases. The main result of this paper is the advantage resulting from using cross-correlated signals, evaluated through the performance of clustering indexes. Experimental data are collected from two test structures: i.) A turbine blade of a commercial aircraft and ii.) The skin panel of the torsion box of a wing. Damages are induced adding masses at different locations of the wing section surface. The results obtained show the effectiveness of the methodology to distinguish between different damage cases.

1. Introduction

An important preoccupation in the aerospace industry is the continuous monitoring of aircraft components to avoid catastrophic failures. In recent years, methods for damage identification in aerospace structures have been focused on analyzing measurements from Fiber Bragg Grating sensors and piezoelectric devices [1], [2], where effectiveness of damage monitoring systems for aircraft structures, using a piezo-diagnostics approach, has been demonstrated. For example in [3], experimental crack extending on a wing panel using vibration deflection shapes can be detected by computing changes in the response from piezo-ceramic actuator patches.

Thus, this paper deals with a previously proposed piezo-diagnostic methodology, for damage detection in aircraft structures [4]. The benefits by including a preprocessing stage based on correlated piezoelectric signals are discussed, based on experimental results. It is shown that it is possible to obtain a better classification of different damages, when cross-correlation analysis is considered.

2. Aircraft wing damage detection based on Piezo-diagnostic approach

Piezo-diagnostic principle takes advantage of the elastic wave propagation phenomenon to examine the structural signature. Where, piezoelectric devices are used both as actuators and as sensors to detect structural changes. Monitoring technologies using piezo-ceramic devices has been presented by Staszewski et al with beneficial for military aircrafts [5]. Also, some piezo-diagnostic applications include: detection of damage in aircraft joints by processing measurements from piezoelectric patches [6], inspection of rivet cracks or corrosion in aircraft wing structures [7] and, the design of a monitoring scanning system for a carbon fiber composite wing box [8]. Further application for aerospace industry can be found at Adams [9] and Stepinski et al. [10].

In piezo-diagnostics, one of the procedure used to distinguish structural damages, is the application of principal component analysis (**Figure** *1*). It consists of computing statistical indices from piezoelectric measurements recorded from several piezo devices attached to the structure surface. Typically, one of the piezoelectric devices (PZT) is excited to induce along the structure a guided wave and the others PZTs are used as sensors to capture it at different locations of the structure. Data from undamaged and damage experimental cases are projected onto the principal components space, and statistical indexes are computed to differentiate several conditions of the structure [11].

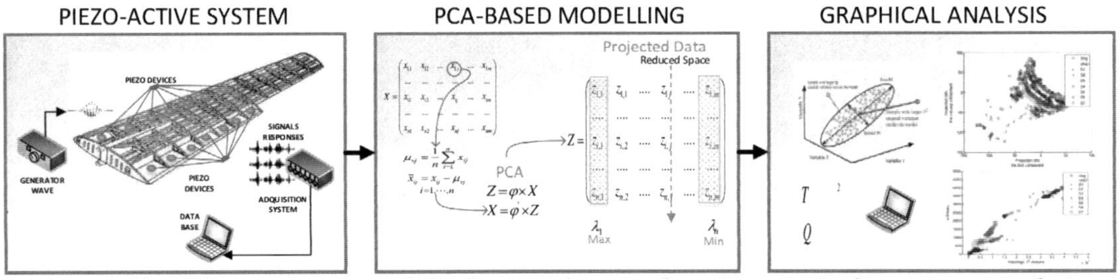

Figure 1. *Piezo-diagnostic approach for damage detection by using principal component analysis*

The methodology depicted in **Figure** *1* has been previously tested and validated using different structures: an aircraft turbine blade, an aircraft wing and an aircraft fuselage [12], [13]. However, the novelty presented in this paper with respect to previous works, is the inclusion of cross correlation analysis as a tool for improving separation boundaries for damage conditions. Thus, cross-correlation between actuation and sensing piezo-signals is estimated previous to the principal component analysis (**Figure 2**). The cross-correlation function between two signals $X(t)$ and $Y(t)$ is defined as:

$$R_{XY}(t, t+\tau) = \lim_{N\to\infty} \frac{1}{N} \sum_{k=1}^{N} X_k(t) Y_k(t+\tau) \tag{1}$$

Where N is the number of signal samples and τ is the lag time interval used to compute the cross-correlation function.

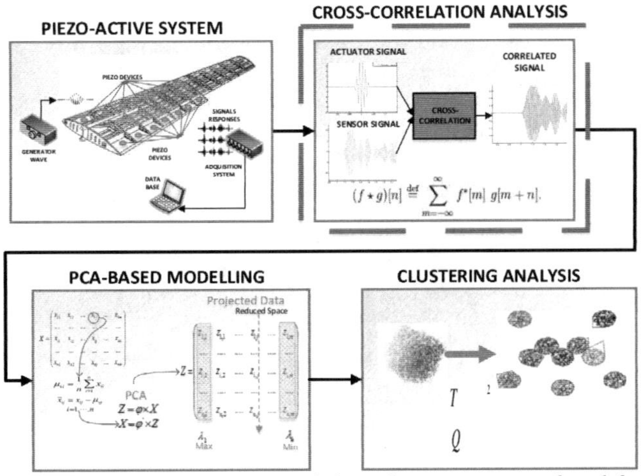

Figure 2. *Piezo-diagnostic approach based on cross-correlated features*

The K-means algorithm is proposed to organize damage cases into groups and then to evaluate the effectiveness of cross-correlation as preprocessing stage in the damage detection. This algorithm is one of the most commonly used optimization-based unsupervised learning methods. The goal of K-means clustering is to organize the data into k groups, such that the within-group sum-of-squares be minimized [14].

$$min\left[\sum_{g=1}^{k}\sum_{i=1}^{n_g}(x_{ig}-\bar{x}_g)'(x_{ig}-\bar{x}_g)\right] \quad (2)$$

Each g_{th}-cluster in the partition is defined by n_g datacase members and by its centroid $\bar{x}_g$, or center. The centroid for each cluster is the point to which the sum of distances from all members in that cluster is minimized. Gap statistic and Davies-Bouldin index are computed to evaluate the data clustering analysis. Therefore, the optimal clustering solution occurs with a largest gap statistic and smallest Davies-Bouldin index values. The K-mean centers represent adequately the data in a cluster if the observations within a group are more similar to each other.
The gap value is defined as [15]:

$$Gap_n(k) = E_n^*\{log(W_k)\} - log(W_k) \quad (3)$$

where n is the sample size, k is the number of clusters, and W_k is the cluster dispersion, defined by:

$$W_k = \sum_{r=1}^{k} 2n_r D_r \quad (4)$$

where n_r is the number of data points in the cluster r and D_r is the sum of the pairwise distances for all points in cluster r. E_n^* is the expected value determined by a reference distribution and $log(W_k)$ is computed from the sample data.

The Davies-Bouldin criterion is based on a ratio of within-cluster and between-cluster distances, defined as [16]:

$$DB = \frac{1}{k}\sum_{i=1}^{k}\max_{j\neq i}\{D_{i,j}\} \quad (5)$$

where $D_{i,j}$ is the distance ratio between the i_{th} and j_{th} clusters.

In addition to the Gap statistic and Davies-Bouldin index, the quantization error (squared sum of errors) and the dispersion in each cluster (standard deviation) are computed as quality indices to evaluate the clustering process in order to compare the piezo-diagnostic approach when cross-correlation analysis is included.

3. Experimental procedure

Two test structures were used to validate the damage detection approach (**Figure 3**). The specimens are hosted in the "Universidad Politécnica de Madrid" (UPM – Spain). The first structure belongs to an aircraft wing, which is divided by stringers and ribs and the second one to an aircraft turbine blade, which has an irregular form and includes stringers in both faces.

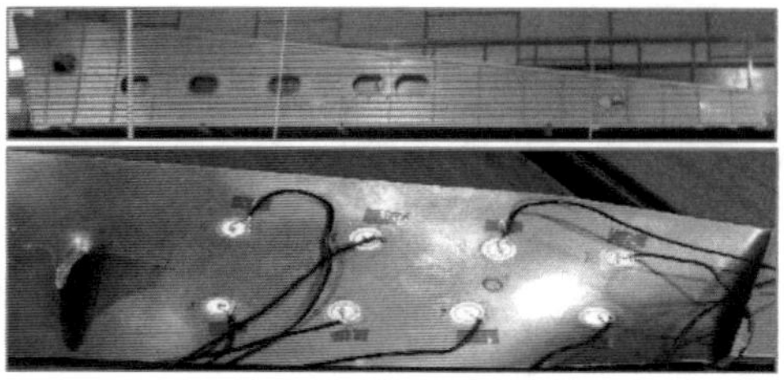

Figure 3. *Aircraft section structures used to validate the methodology.* ***Above:*** *skin panel of aircraft wing,* ***Below:*** *turbine blade*

The skin panel was instrumented with 10 PZTs, while the turbine blade with 8 PZT's. An 80 KHz burst type signal was configured to produce guided waves along the surface structures. Reversible damages were induced in both structures by adding masses in different positions, four damages for skin panel experiment (D1,..., D4) and five damages for turbine blade case (D1, ..., D5). For both specimens 150 experiment repetitions for each damage were recorded.

4. Results and discussion

Results for each experiment consider a number of clusters for the K-means algorithm equal to the number of damages in the respective structure. Therefore, each damage can be grouped in an individual cluster. In addition, 50 replicates of the K-means algorithm are executed to avoid local minima. Data normalization by means of variance values are used before K-means algorithm to minimize the within-cluster dispersion.

The steps for applying PCA in practice regarding to data collection and organization, baseline model building, scaling preprocessing, statistical indices formulation, model testing, and iterative procedure to estimate principal components, can be found with most detail in previous works [11], [17], and [18]. Thus, data normalization by means of variance values are used before K-means algorithm to minimize the within-cluster dispersion. In addition, NIPALS algorithm is configured to estimate a number of principal components equal to the number of experiments induced for each damage case, i.e. 150.

4.1. Skin Panel Experiment

Variance of the principal components are depicted in **Figure 4**. In which can be observed that the variance evolution is similar for both cases, when cross-correlated functions are used and without it. According to this, 100 components capture most of the principal components variability. In consequence, only 100 components for both uncorrelated and cross-correlated approaches are selected.

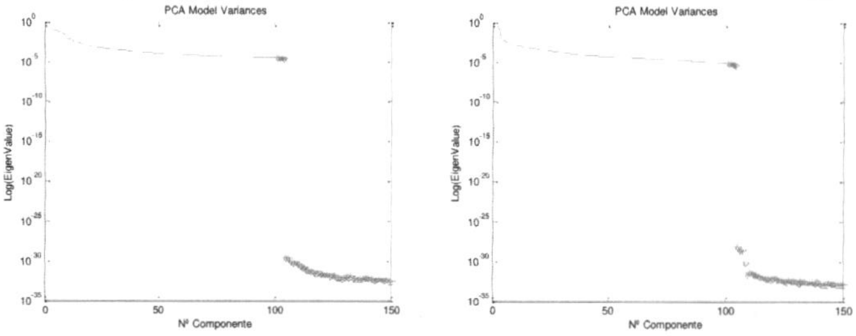

Figure 4. Principal component variances for skin panel experiment. Left: unmodified piezo-diagnostic scheme. Right: cross-correlated piezo-diagnostic approach

The **Figure 5** presents statistical indices and K-means clusters computed for uncorrelated signals. From **Figure 5** can be concluded that atypical data dominate one of the clusters. Besides, damage 2 ('D2') and undamaged cases ('UND') are closer when cross correlation are not estimated. In this case, K-means assigns a cluster for atypical data.

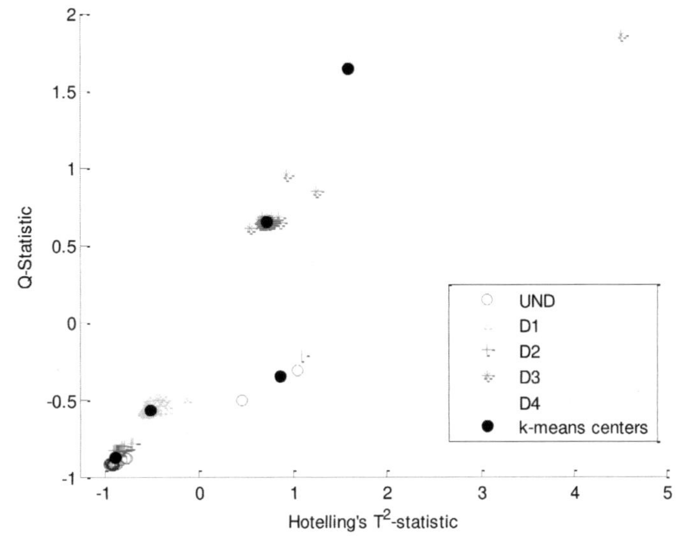

Figure 5. K-means centroids with atypical data with uncorrelated signals for skin panel experiment.

When atypical data are removed from original data matrix, the cluster centers for each damage type are located as it is shown in **Figure 6**.

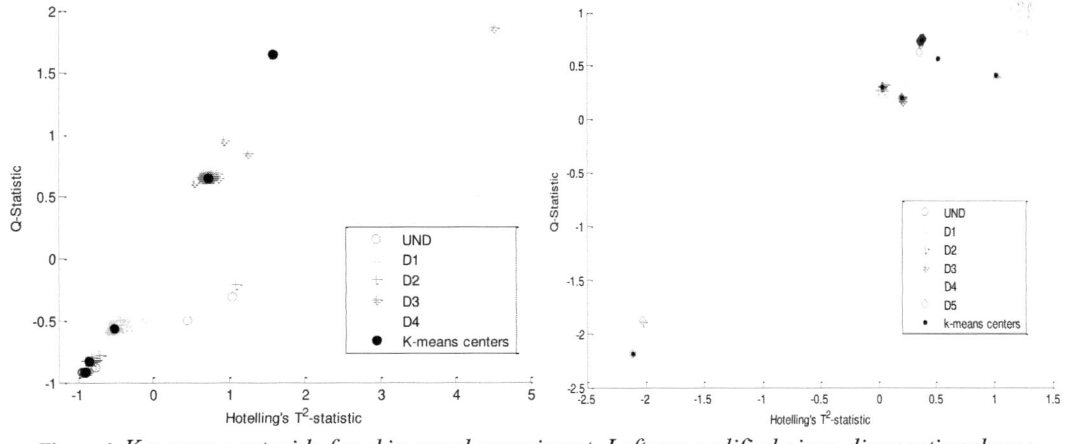

Figure 6. K-means centroids for skin panel experiment. Left: unmodified piezo-diagnostic scheme. Right: cross-correlated piezo-diagnostic approach

In **Figure 6**, it can be observed that more dispersion appears without correlation analysis. Additionally, if correlation analysis is included in the damage classification approach, then the atypical data-cases are filtered. **Table 1** summarizes the quality indexes of the clusters obtained by K-means algorithm.

Table 1. *Quality clustering indexes for skin panel experiment.*

		Approach		
		Uncorrelated with original data	**Uncorrelated eliminating atypical data**	**Cross-correlated with original data**
Index	**Gap statistic**	3,595	3,504	5,885
	Davies-Bouldin	0,241	0,275	0,089
	Quantization Error	11,798	13,615	0,439
	Dispersion for each cluster	[0.3178, 0.0655, 0.0684, 0.0570, 0.2467]	[0.0335, 0.1315, 0.0147, 0.1032, 0.2467]	[2.2434e-04, 0.0032, 0.0404, 0.0350, 0.0078]

According to **Table 1**, a better dispersion in all clusters and improved clustering indexes are obtained when cross-correlated signals are used instead of raw data. In addition, the uncorrelated approach has comparable clustering indexes values even when atypical cases are removed.

4.2. Turbine Blade Experiment
Similarly than the previous experiment, most of the principal components variability is captured with 100 components as shown in **Figure 7**.

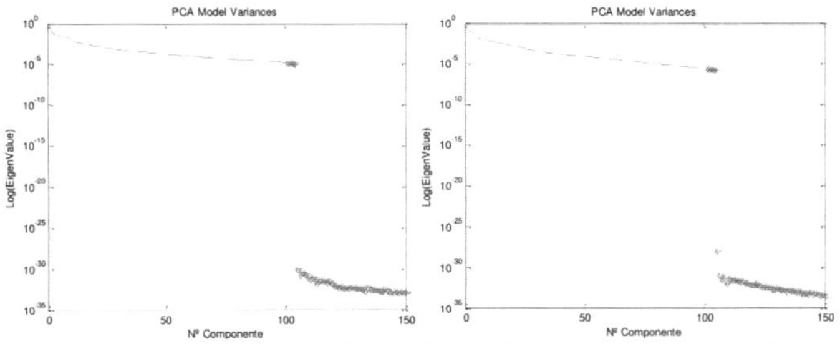

Figure 7. *Principal component variances for turbine blade experiment. Left: unmodified piezo-diagnostic scheme. Right: cross-correlated piezo-diagnostic approach*

Immediately after, the cluster procedure is performed. The final clusters for damages in turbine blade experiments are depicted in **Figure 8**. Here it is possible to highlight a clear separation between different types of damage when cross-correlation analysis are included in the piezo-diagnostic approach.

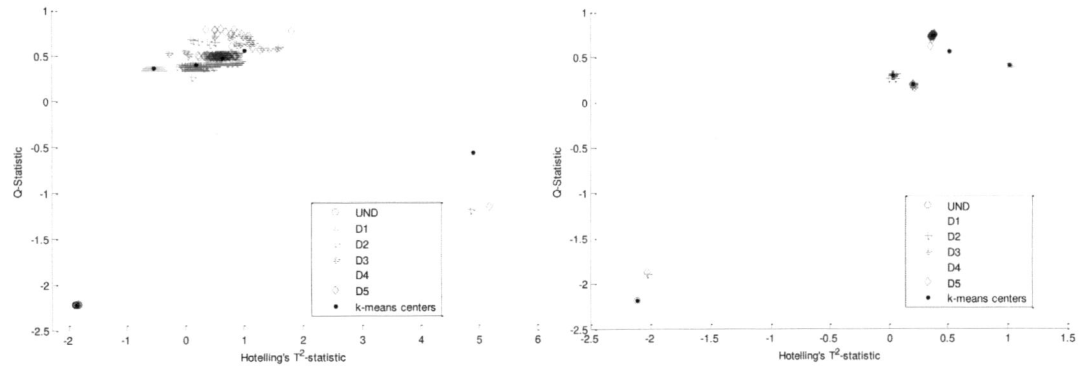

Figure 8. *K-means centers for turbine blade experiment.* Left: *uncorrelated piezo-diagnostic scheme.* Right: *cross-correlated piezo-diagnostic approach*

The respective clustering errors are summarized in **Table 2**. According to these results, the clustering indexes confirm the best performance for cross-correlated analysis.

Table 2. *Quality clustering indexes for turbine blade experiment.*

		Approach	
		Uncorrelated	**Cross-correlated**
Index	**Gap statistic**	3,681	6,865
	Davies-Bouldin	0.4054	0.0618
	Quantization Error	16,787	0.2909
	Dispersion for each cluster	[0.0092, 0.1236, 0.1362, 0.0828, 0.8833, 0.2380]	[0.0357, 0.0069, 0.0090, 0.0168, 0.0147, 0.0050]

5. Conclusion

In this paper is shown that by using common clustering techniques is possible to distinguish damages in a simple way. However, a better damage differentiation can be obtained if cross-correlation technique is used as preprocessing technique. This effectiveness was validated by computing the error clustering.

In order to avoid abnormal data, dispersion preprocessing was achieved filtering typical cases by using cross-correlated piezoelectric signals, where an adequate rejection of abnormal data was obtained.

Related to evolution of variances for principal components it can be concluded that no differences are presented for both structures of different properties. Also, since, K-means centers are assigned to damage types when cross-correlation analysis is included, classification can be achieved representing damage cases with this optimal value.

Effectiveness of cross-correlation was validated with the error clustering calculation where improvements were obtained for the two experiments.

6. Acknowledgments

This work has been developed as part of the research project: "Monitorización y Detección de Defectos en Estructuras usando Algoritmos Expertos Embebidos", financed by the Departamento Administrativo de Ciencia y Tecnología Francisco José de Caldas – COLCIENCIAS and Banca Mundial in the Universidad Industrial de Santander, Colombia. Agosto 2012-Junio 2015.

References

[1] A Sierra Pérez, J., Güemes Gordo, J. A., & Mujica, L. E. (2013). Damage detection by using FBGs and strain field pattern recognition techniques. Smart Materials and Structures, 22(2), 25011-25021.

[2] Ogisu, T., Shimanuki, M., Kiyoshima, S., Okabe, Y., & Takeda, N. (2004, July). Development of damage monitoring system for aircraft structure using a PZT actuator/FBG sensor hybrid system. In Smart Structures and Materials (pp. 425-436). International Society for Optics and Photonics.

[3] Sundaresan, M. J., Ghoshal, A., Li, J., Schulz, M. J., Pai, P. F., & Chung, J. H. (2003). Experimental damage detection on a wing panel using vibration deflection shapes. Structural Health Monitoring, 2(3), 243-256.

[4] Tibaduiza Burgos, D. A., Mujica Delgado, L. E., Anaya, M., Rodellar Benedé, J., & Güemes, A. Principal component analysis vs. independent component analysis for damage detection.

[5] Staszewski, W.J., Boller, C. and Tomlinson, G.R. (2004) – Health Monitoring of Aerospace Structures, ISBN 978-0-470-84340-3.

[6] Mickens, T., Schulz, M., Sundaresan, M., Ghoshal, A., Naser, A. S., & Reichmeider, R. (2003). Structural health monitoring of an aircraft joint. Mechanical Systems and Signal Processing, 17(2), 285-303.

[7] Zhao, X., Gao, H., Zhang, G., Ayhan, B., Yan, F., Kwan, C., & Rose, J. L. (2007). Active health monitoring of an aircraft wing with embedded piezoelectric sensor/actuator network: I. Defect detection, localization and growth monitoring. Smart materials and structures, 16(4), 1208.

[8] Lei, Q., Shenfang, Y., Qiang, W., Yajie, S., & Weiwei, Y. (2009). Design and experiment of PZT network-based structural health monitoring scanning system. Chinese Journal of Aeronautics, 22(5), 505-512.

[9] Adams, D. (2007). Health monitoring of structural materials and components: methods with applications. John Wiley & Sons.

[10] Stepinski, T., and Engholm, M. Design of piezoelectric transducers for health monitoring of composite aircraft structures, Proc. SPIE 6531, Nondestructive Characterization for Composite Materials, Aerospace Engineering, Civil Infrastructure, and Homeland Security 2007, 65310G (April 19, 2007); http://dx.doi.org/10.1117/12.715151

[11] Mujica, L. E., Rodellar, J., Fernandez, A., & Guemes, A. (2010). Q-statistic and T2-statistic PCA-based measures for damage assessment in structures. Structural Health Monitoring, 1475921710388972.

[12] Tibaduiza Burgos, D. A. (2013). Design and validation of a structural health monitoring system for aeronautical structures.

[13] Gharibnezhad, F., Rodellar, J., & Mujica Delgado, L. E. (2014). Robust damage detection in smart structures. Materia (s), 7, 04-2014.

[14] Martinez, W. L., Martinez, A., & Solka, J. (2004). Exploratory data analysis with MATLAB. CRC Press.

[15] Tibshirani, R., G. Walther, and T. Hastie. "Estimating the number of clusters in a data set via the gap statistic." Journal of the Royal Statistical Society: Series B. Vol. 63, Part 2, 2001, pp. 411–423.

[16] Davies, D. L., and D. W. Bouldin. "A Cluster Separation Measure." IEEE Transactions on Pattern Analysis and Machine Intelligence. Vol. PAMI-1, No. 2, 1979, pp. 224–227.

[17] Villamizar, R., Quiroga, J. L., Camacho, J., Mujica, L. E., & Ruiz, M. L. (2014). Structural Damage Detection Algorithm Based on Principal Component Indexes and Embedded on a Real Time Platform. In EWSHM-7th European Workshop on Structural Health Monitoring.

[18] M Ruiz, L E Mujica, X Berjaga, and J Rodellar. Partial least square/projection to latent structures (PLS) regression to estimate impact localization in structures. Smart Materials and Structures, 22(2):025028, 2013.

Protocols for Image Processing based Underwater Inspection of Infrastructure Elements

Michael O'Byrne[1,4], Bidisha Ghosh[2], Franck Schoefs[3] and Vikram Pakrashi[1,4]

[1] Beaufort, Environmental Research Institute, University College Cork, Ireland
[2]Department of Civil, Structural and Environmental Engineering, Trinity College Dublin, Ireland
[3]LUNAM Universite´ , Universite´ de Nantes-Ecole Centrale de Nantes, CNRS, GeM, Institute for Research in Civil and Mechanical Engineering, University of Nantes, Nantes, France; and IXEAD/CAPACITES Society, Nantes, France
[4]Dynamical Systems and Risk Laboratory, Civil and Environmental Engineering, School of Engineering, University College Cork, Ireland

v.pakrashi@ucc.ie Ph: 00353214903862 Fax: 00353214276648

Abstract. Image processing can be an important tool for inspecting underwater infrastructure elements like bridge piers and pile wharves. Underwater inspection often relies on visual descriptions of divers who are not necessarily trained in specifics of structural degradation and the information may often be vague, prone to error or open to significant variation of interpretation. Underwater vehicles, on the other hand can be quite expensive to deal with for such inspections. Additionally, there is now significant encouragement globally towards the deployment of more offshore renewable wind turbines and wave devices and the requirement for underwater inspection can be expected to increase significantly in the coming years. While the merit of image processing based assessment of the condition of underwater structures is understood to a certain degree, there is no existing protocol on such image based methods. This paper discusses and describes an image processing protocol for underwater inspection of structures. A stereo imaging image processing method is considered in this regard and protocols are suggested for image storage, imaging, diving, and inspection. A combined underwater imaging protocol is finally presented which can be used for a variety of situations within a range of image scenes and environmental conditions affecting the imaging conditions. An example of detecting marine growth is presented of a structure in Cork Harbour, Ireland.

1. Introduction

The monitoring of marine structures is often beset by reduced visibility, limited access, and high costs. Despite these deterrents, the need for underwater inspections is set to intensify globally in the coming decades as the offshore renewable energy sector continues to grow and major oil and gas companies move into deeper and more hostile environments [1]. Underwater inspections are mainly carried out by divers or Remotely Operated Vehicles (ROVs) [2]. ROVs can be deployed to much greater depths and for longer periods than divers, however, they are considerably more expensive, inflexible, and vulnerable to failure. Consequently, their safe operation requires great attention to detail. Underwater inspections performed by divers represent the current state-of-the-art. Divers are capable of covering a wide area and can inspect at a number of scales, from a detailed close-up assessments with the help of specialised equipment if necessary, to broader-scale visual assessments.

The quality of visual inspections largely depends on the ability of divers to observe and objectively record details of defects. This is prone to considerations such as lapses in concentration, level of knowledge, subjectivity, and fatigue, which contribute to the variability and reduced accuracy of visual inspections [3]. Additionally, a diver can expect some loss of judgement in cold conditions and due to inert gas narcosis at lower depths if using air as a breathing medium [4]. To increase reliability of visual inspections, photographs are almost always captured to include in the inspection report. However, these photographs are rarely exploited to their fullest potential in either a qualitative or a quantitative fashion. Moreover, despite having an established role in the inspection process, there has no agreed protocol for image collection and subsequent interpretation [5]. This paper aims to fill that void by specifying the technical requirements for capturing imagery in an underwater infrastructural setting, as well as outlining a set of best practice guidelines for ensuring that the acquired imagery is suited for quantitative analysis. Extracting quantitative information from images using computational techniques has received significant interest in recent years [6-7].

The development of such a protocol is especially useful for maintaining the long-term integrity of an Infrastructure Management System (IMS). A well calibrated IMS is dependent on the quality and consistency of the input information. As offshore structures are being designed with service lives of up to 35 years [8], and possibly longer if they are eligible for requalification programmes at the end of their service lives, it is of great value for the inspection data acquired during the early life stages to be accessible and relatable to the inspection data acquired decades later. This can be difficult to achieve if inspections are carried out in an ad-hoc manner, whereby divers rely on their own intuition when collecting and cataloguing imagery. By adhering to a protocol, the output from inspections is more standardised, and as a result, it becomes easier to compare and track the progression of damage.

The proposed protocol addresses each stage of the imaging pipeline from equipment set-up right through to image archiving. It is developed based on existing literature and the experience of the authors. Special attention is given to stereo based 3D shape recovery systems, which offer significant potential as an inspection tool as they are capable of recovering full 3D shape information. Stereo systems utilise a dual camera set-up to simultaneously photograph a specimen of interest from slightly different viewpoints. They have greater operational complexity than single camera systems as two cameras must be configured and also synchronised.

The following section describes the protocol. Section 3 presents an example of the protocol being implemented as part of an assessment of a submerged pile in Cork Harbour, Ireland. Finally, Section 4 concludes the paper.

2. Protocol

This protocol describes the procedural method for acquiring underwater images. The protocol addresses: 1) archiving, 2) imaging, 3) diving, and 4) a combined protocol for underwater inspections.

2.1. Image archiving

The creation of an image library requires a set of predefined guidelines to ensure that all contributions are consistent. This forms the basis for an organised and manageable library. The proposed image archiving protocol discusses the file format, metadata and information to be recorded, and file naming convention and cataloguing.

There are a wide variety of image formats available. Typically there is a trade-off between the image file size and the amount of information retained in it. As storage space is inexpensive nowadays, this protocol recommends capturing images in both JPEG and RAW format, if applicable. JPEG images, which have a ".jpg" extension are one of the most popular image formats. JPEG uses a method of lossy compression which means that the image quality degrades slightly after the image has been saved, however this is rarely perceivable. Conventional tasks, such as image processing and analysis are performed on the more manageable and light-weight JPEG images. The RAW image format should be retained for archival purposes. RAW images contain minimally processed data from the image sensor. They are capable of storing a greater level of information from a scene (i.e. wider dynamic range and colour gamut) than other image formats. The RAW format differs depending on the camera model and camera manufacturer, although each format contains essentially the same data and metadata. An additional step, if desired, would be to convert the original RAW formats to an open standard and well supported format, namely the Digital Negative (DNG) format, which is a popular and freely available format developed by Adobe Systems™. Since there is a wide range of proprietary RAW formats, it is hard for applications and programs to guarantee future compatibility with them all, especially for some of the lesser known and lesser used RAW formats. Thus, the additional step of converting to the DNG format from the original RAW format would be of particular value to users that have, or expect to have, imagery acquired from a number of devices, and would like to unify the RAW formats into a common format that retains all of the original information.

In the case of video, still frames should be extracted at relevant intervals and saved as JPEG files as analysing every frame is often unnecessary. The intervals are primarily determined by the speed of the camera relative to the subject. If a recording device is moving quite quickly then more frames should be extracted. Generally, extracting three frames per second should be sufficient for most cases. The original video should be stored in its native format.

Photographs and video have accompanying metadata which contain useful information about the content and context of the file. Metadata is automatically embedded into each digital file. It provides information such as the time and date of capture, camera model, exposure information etc. It is vital that the time and date of all contributing cameras are precisely set as this is provides a convenient way for identifying the synchronised stereo image pairs. Some metadata must be manually added such as the baseline distance for stereo imaging, or the camera calibration data. The specific nature of this metadata will vary according to the task. In the case of stereo imaging, the imagery should be marked as coming from the left or right camera.

2.2. Imaging Protocol

The first objective of establishing imaging protocols involves specifying the technical requirements of capturing imagery in an underwater setting. These technical requirements deal with underwater conditions, setting up underwater housings, and the choice of camera settings.

Underwater imaging must try to overcome the challenging environmental conditions. Even in perfectly clear water, there is a loss of colour and contrast when the structure is separated from the camera by any significant distance. Water absorbs the red component of light to a greater extent which results in underwater subjects having a blue-green tinge. Colour diminishes with distance so subjects further away will appear indistinct and devoid of colour detail. This issue can be partially offset by following one of two approaches. The first approach involves photographing the damaged region from as close up as possible. Wide-angle lenses facilitate short focusing distances, allowing the damaged region to be in close proximity to the camera(s), at the expense of some radial distortion. The second approach entails the use of artificial lighting in the form of strobes to restore some lost colour. An added complication in this case is the phenomenon of backscatter, where light reflects off particles in the water. Even seemingly clear water is affected by this. The best method for limiting backscatter is positioning the lights away from the axis of the camera lens.

There are refraction issues associated with using cameras enclosed an underwater housing. This will lead to slightly distorted images. This issue can be mitigated in the case of stereo imaging by calibrating the cameras, which accounts for any distortion.

Many modern cameras have simplified the process of choosing the optimum aperture, ISO, and shutter speed settings for a given scene through various automatic exposure modes and the use of through-the-lens (TTL) metering. ISO measures the sensitivity of the image sensor. Higher ISO settings tend to be used in darker situations to amplify the available light, however this comes at a cost of increased noise in the images. The aperture size controls how much light reaches the sensor. A wide aperture allows more light to reach the sensor at the expense of a narrow depth of field, meaning only objects within a confined range will be in focus. A small aperture will have a greater depth of field, however, the resulting image will be dark/under-exposed unless the shutter stays open for an extended period of time to let enough light to reach the sensor. Keeping the shutter open for too long (i.e. having a slow shutter speed) presents other problems, mainly a high degree of blur. The camera will automatically attempt to counteract the dimly lit underwater conditions by combining a small aperture with a low shutter speed and a high ISO. It is important that the settings remain within certain limits to minimise the impact of these problems. Recommendations of the limits are summarised in Table 1.

It should be noted that the depth of field is also affected by the focal length of the lens. Closer focusing distances will produce a shallower depth of field. Guidelines for choosing the focal length are not discussed as a good choice of focal length is best adjudged by the diver onsite. A small focal length provides a wide field of view, which can be useful in cases where there is a short distance between the camera and subject (possibly due to poor underwater visibility) and where it still wished to photograph a relatively large area of the structure, while long focal lengths give a smaller field of view [9].

Table 1. Acceptable ranges for camera settings.

Camera Setting	Limit
Aperture	f/8 - f/16
Shutter Speed	As a rule of thumb, the minimum shutter speed is the inverse of the focal length, i.e. minimum Shutter Speed (secs) = 1/Focal Length (mm).
ISO	An ISO value of 800 is a good compromise for photographing in dimly-lit conditions while still controlling noise/graininess.

The shutter priority mode is recommended as diver-held cameras employed in underwater inspections are prone to shaking, which introduces motion blur. This is a semi-automatic shooting mode that allows the user to specify the shutter speed. The camera then automatically decides the best aperture and ISO sensitivity for the specified shutter speed to get the correct exposure. The shutter speed should not be any slower than 1/15 seconds. The imagery obtained from the cameras should be reviewed at regular intervals. If it is apparent there is too much motion blur present in the images, the shutter speed should be adjusted to a faster setting.

Additional artificial lighting will be required if the camera settings exceed any of these ranges. Irrespective of these requirements, artificial lighting will be necessary at greater depths where ambient lighting is not sufficient.

2.3. Diving Protocol

The diving protocol addresses the logistical considerations (testing equipment prior to usage, route planning etc), as well as handling lighting and turbidity conditions, which will affect the optimum distance between camera(s) and subject. It is vital that any unnecessary time spent underwater by the diver is kept to a minimum. With this in mind, the diver should be presented with a clear and concise brief outlining the task at hand. A rough idea of the turbidity conditions should be known beforehand. Both lighting and turbidity are crucial factors which affect the underwater visibility and consequently the image quality [10]. Artificial lighting in the form of underwater strobe lights are required in dim lighting conditions, especially if the subject is bumpy/rugged in which case the lights would assist with revealing details that would otherwise remain in shadow.

Water is seldom optimally clear, and the dissolved and suspended matter can reduce visibility by both absorption and scattering of light. While turbidity may not be easily reduced, there are some precautions which can be taken to offset the deleterious effect in relatively high turbid waters. Firstly, caution should be taken in shallow waters to avoid disturbance with the sea/river bed which may unsettle fine sediments through either direct contact or from turbulence created from the ship.

Anchoring the vessel downstream away from the inspection site can also prevent additional sediments that would reduce the water clarity. Secondly, the distance between the cameras and the subject under consideration should be reduced. In seas and oceans, the water is generally clear so measures to counteract poor visibility need not be considered.

Choosing the distance between the subject and the diver is a trade off between a number of factors. It will vary depending on the level of visibility, the size of the subject in the scene, and level of detail required. When the underwater visibility is poor, the diver can photograph in close proximity to the subject (up to 30 cm). In the case of stereo imagery, going any closer than 30 cm leads to large perspective differences between the stereo image pairs which can hamper the 3D shape recovery process and result in a myriad of occluded regions [11]. In clear underwater conditions, the diver can photograph from a distance up to 2.5 m before stereo imaging breaks down and the error tolerance becomes unacceptably high [12]. If the object under inspection is quite large (e.g. a wide diameter pile) and it is wished to include the whole structure within an image then photographing from further back will provide better context.

Finally, the baseline distance for stereo imagery will influence the choice of subject-camera distance. While the baseline shift is sometimes fixed at a certain distance according the constraints of the available equipment, it is important to note the effect it has on the accuracy. Theoretically, a wider separation between the cameras results in a lower percentage error in the depth estimation. However, the advantages of having a wide separation are offset by the creation of large perspective differences. The baseline shift should be in the range 10 cm to 30 cm. Additionally, the cameras should be aimed inwards at an angle θ (known as the vergence angle) such that their centrelines intersect approximately at the face of the subject as shown in Figure 1. This is to ensure that the cameras capture as many of the same points in both images as possible. The vergence angle has previously been used to model the error in depth [13], where values in the range 5° - 10° were found to provide the lowest errors.

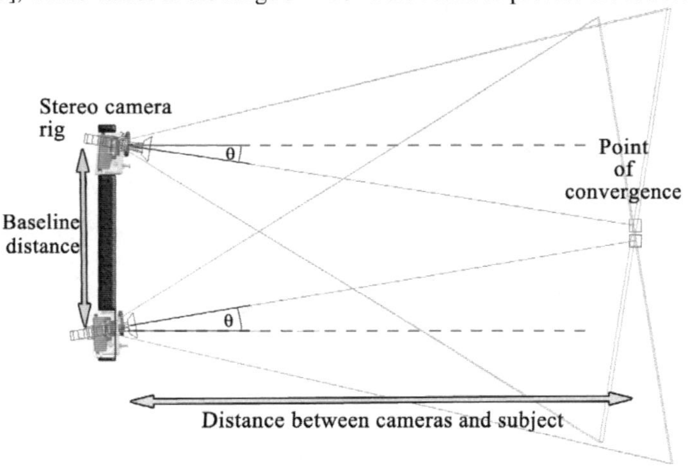

Figure 1. Stereo rig configuration.

All equipment should be checked above water prior to each inspection. The underwater housings should be checked for any signs of leakage by firstly submerging them without their contents. If necessary, the O-rings should be re-lubricated/replaced as per manual instructions. Care should be given to ensure that the time and date of the cameras are precisely set, that there is enough storage capacity in the SD cards and the battery is sufficiently charged. Appropriate settings should be configured for each camera/video recorder ensuring that the shooting modes in both are identical. It is advised to initiate filming immediately before the diver submerges as it is easier to control the simultaneous triggering of both cameras when above water. The captured imagery should be reviewed at regular intervals during the inspection and any adjustments should be made accordingly. Dives which produce sub-standard imagery should be repeated. Necessary props should be prepared such as lighting equipment or calibration tools. Finally, the diver should be familiarised with the blueprint of the structure and identify any components that are of particular interest. A suitable route should then be planned based on this. In cases where the diver cannot photograph a particular component from all sides due to restricted access, he/she should endeavour to photograph as much of it as possible.

2.4. Combined Underwater Protocol

Greater care and attention is required for stereo imaging as two cameras must act in unison. A flowchart showing the combined underwater protocol for stereo imaging is shown in Figure 2.

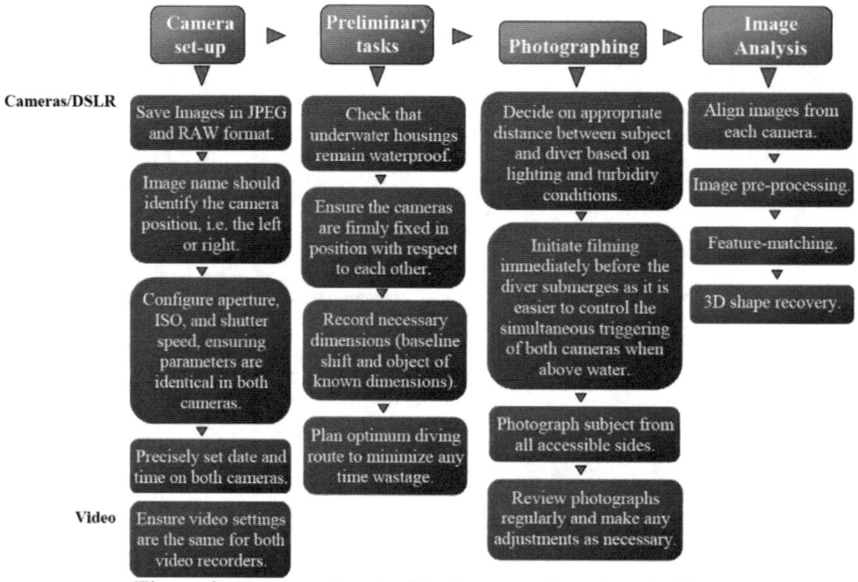

Figure 2. Key steps involved in the stereo imaging pipeline.

3. Demonstration of protocol on real world structure

This section applies the protocol during an underwater assessment of a marine growth affected pier located in Cork Harbour, Ireland. The following sub-sections provide an overview of the structure under consideration, the underwater conditions at the site, and the image acquisition procedures.

3.1. Structure and Conditions at the Test Site.

The structure under consideration is a long-serving pier in Cork Harbour that is mostly used for recreational and leisure purposes (Figure 3). The structure is affected by marine growth, which is undesirable as it changes the thickness and roughness characteristics of structural members leading to increased hydrodynamic forces. Tracking marine growth thickness is often carried out during underwater inspections so that improved estimates of the hydrodynamic forces acting on the structure can be obtained.

Figure 3. View of testing zone in Cork Harbour. Source: Google Maps (retrieved 15/09/2014).

Figure 4. The stereo system used to capture the imagery.

It was a clear and sunny day when the testing was carried out so there was no requirement for additional artificial light sources as ambient light was sufficient in the shallow waters. The turbidity was quite high so the imagery had to be acquired at a close range from the subject.

The stereo system is shown in Figure 4. This system consist of two Canon 600D DSLR cameras, enclosed in underwater cases, which are securely attached to a graduated stereo bar. The graduated stereo bar allows the baseline distance to be easily measured. In this case, the centres of the cameras were separated by 15 cm, which was duly recorded. Knowledge of the baseline is necessary for obtaining the scale factor that converts the dimensions of the reconstructed 3D shape to real world units. The Canon 600D cameras were both configured to simultaneously capture high resolution 18 MP images in RAW and JPEG format at a time interval of two seconds. The cameras were in shutter priority mode in accordance with the protocol and had a shutter speed of 1/20 seconds. The minimum focal length of 18 mm was selected so that as much of the scene could be captured as possible.

3.2. Acquired imagery
An image of the marine growth affected pile under consideration is shown in Figure 5.

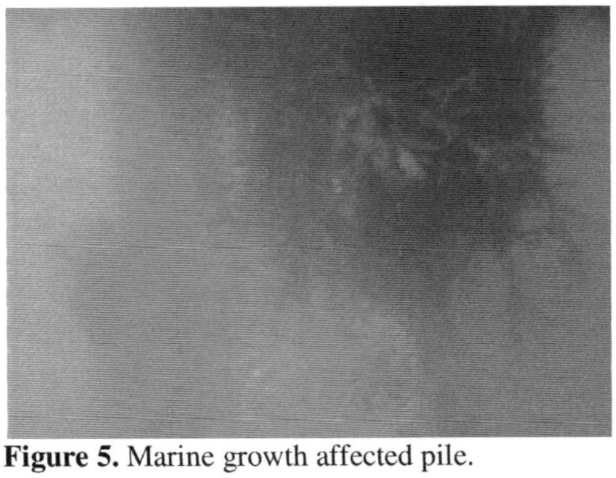

Figure 5. Marine growth affected pile.

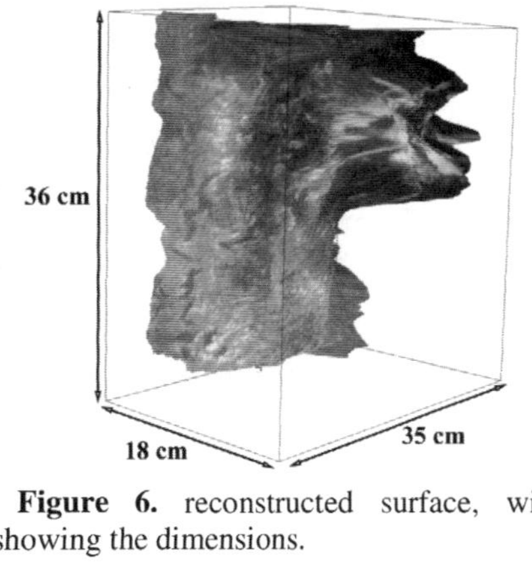

Figure 6. reconstructed surface, with (b) showing the dimensions.

It may be observed that the high turbidity creates an image lacking in contrast. In this case, the imagery greatly benefits from image pre-processing such as contrast enhancement. The associated reconstruction is shown in Figure 6, which is based on an auto-calibration technique [14]. The shape reconstruction is dependent on the quality of the imagery, which in turn is dependent on the image acquisition practices. As such, the successful results achieved here provide validation for the protocol.

4. Conclusion

The effort and expense associated with undertaking underwater inspections warrants significant forethought and planning to ensure the every technical aspect of data acquisition is covered. To address this, a comprehensive protocol is developed for the first time that sets out best practices for underwater image acquisition. The protocol provides guidance on archiving and cataloguing the imagery, overcoming challenges such as high turbidity, equipment set-up, and defining appropriate limits for camera parameters. Selecting the most suitable combination of these camera parameters for a given environment involves a trade off between minimising negative image quality factors such as blur and noise, whilst retaining sufficient brightness and ensuring enough of the subject is in focus. Ineffective image acquisition practices directly affect the performance of image algorithms. The protocol was demonstrated on a real world structure in Cork Harbour, Ireland. In spite of the high turbidity, a successful assessment was achieved through careful adherence to the protocol.

Acknowledgements

The authors wish to thank the Irish Research Council for Science, Engineering and Technology (IRCSET) and CAPACITES/IXEAD society for providing financial and practical support. The authors acknowledge Marine Research Energy Ireland (MaREI), grant no. 12/RC/2302, a Science Foundation Ireland (SFI) supported project.

References

[1] El-Reedy M A 2012 *Offshore Structures: Design, Construction and Maintenance* (Gulf Publishing)
[2] Goldberg L 1996 Diversity in underwater inspection *Materials Evaluation.* **54** 401-03
[3] Agin G 1980 Computer vision systems for industrial inspection and assembly *Computer* **13** 11-20
[4] Busby F R 1979 Underwater inspection/testing/monitoring of offshore structures *Ocean Engineering* **6** 355-491
[5] Phares B, Washer G, Rolander D, Graybeal B and Moore M 2004 Routine highway bridge inspection condition documentation accuracy and reliability *Bridge Engineering* **9** 403-13
[6] O'Byrne M., Ghosh B, Schoefs F and Pakrashi V 2014 Regionally enhanced multiphase segmentation technique for damaged surfaces *Computer Aided Civil and Infrastructure Engineering* **29** 644-58
[7] Sinha S K and Fieguth P W 2006 Automated detection of cracks in buried concrete pipe images *Automation in Construction* **15** 58-72
[8] Müller H S and Curbach M 2008 *Constitutive modelling of high strength/high performance concrete* State-of-art report International Federation for Structural Concrete Lausanne
[9] Stroebel L D and Zakia R D 1993 *The Focal encyclopedia of photography* (Boston: Focal Press)
[10] Mahiddine A, Seinturier J, Boi D P J, Drap P, Merad D and Luc L 2012 Underwater Image Preprocessing for Automated Photogrammetry in High Turbidity Water: An Application on the Arles-Rhone XIII Roman Wreck in the Rhodano River, France *Virtual Systems and Multimedia (VSMM), 2012 18th International Conference* Milan 189-194
[11] Matthies L and Shafer S 1987 Error modeling in stereo navigation *J. of Robotics and Automation* **3** 239-248
[12] Olofsson A 2010 *Modern Stereo Correspondence Algorithms: Investigation and Evaluation*, Thesis from Department of Electrical Engineering, Linköping University, Linköping
[13] Sahabi H and Basu A 1996 Analysis of error in depth perception with vergence and spatially varying sensing *Computer Vision and Image Understanding* **63** 447-461
[14] Faugeras O D, Luong Q T and Maybank S J 1992 Camera self-calibration: theory and experiments Computer Vision — ECCV'92 Springer Berlin Heidelberg 321-334

AUTHOR INDEX

Abusaad, S 633
Acar, M 748
Adesta, Erry Yullian T 266
Aggelis, D G 122, 882
Al-Bugharbee, H 710
Alegre, D M 171
Algaba, Mario 106
Ali, Shaikh Faruque 463
Ambrozinski, Lukasz 219
Anduaga, J 98
Arizmendi, C J 664
Assaeh, M 633
Avendaño-Valencia, L D 594
Aydin, E 530
Aymerich, F 796, 828
Baba, Ndao 242
Ball, A D 633
Barszcz, T 702
Basu, B 250, 543, 859
Belaidi, I 273
Benaissa, B 273
Benjumea, Antonio J 106
Besset, Sebastien 82, 610
Blasi, M G 756
Bouhaddi, N 780
Boulpaep, F 122
Bovsunovsky, A P 1
Bradshaw, T 602
Brethee, K 633
Broderick, B M 250
Bui, Truong-Son 242
Camacho-Navarro, Jhonatan 1023, 1032
Capozucca, R 756
Carvelli, V 764
Casciati, F 904
Casciati, S 904
Ceravolo, R 820
Chanayil, Afeef 463
Chang, Chia-Ming 74
Chen, Hua-Peng 400, 408
Chen, Jun 578
Chen, S F 360
Chen, Z-C. 904
Cheng, Weidong 679
Chioncel, C P 155
Chu, F L 673
Chu, Fulei 618, 679
Clarke, A 602, 929
Cockerill, A 602
Cole, P 602
Corina, V 756
Cui, Liming 641
Da Silva, Samuel 90
Dahmene, F 1010, 1017
Damkilde, L 15, 559, 969
Daoud, O 522
De Marchi, A 820
Demirci, E 748
Deraemaeker, A 882
Desmet, W 656
Desmet, Wim 953
Devriendt, C 866
Di Lorenzo, E 656
Do, Duc-Phi 242
Dollevoet, Rolf 347
Dong, Dawei 578
Dos Santos, Fabio Luis Marques 953
Duan, Chendong 718
Dworakowski, Ziemowit 219
Eaton, M 842
El Mountassir, M 1010, 1017
El-Hussein, Salisu 313
Emoto, Hisao 291
Fabbrocino, G 139
Fagan, Edward M 850
Faravelli, L 904
Faria, A W 724
Farukh, F 748
Fassois, S D 594
Fassois, Spilios D 58, 66
Featherston, C 812, 929
Fifo, O 859
Figueiredo, E 281
Firdaus, Tasnim 266
Florez, J 211
Fragonara, L Zanotti 820
Francesconi, L 796
Frau, A 828
Fritzen, C P 234
Furdui, H 163
Gachagan, Anthony 179
Galbraith, Walter 179
Galvín, Pedro 106
Gao, F 874, 986
Gao, Hai-Yang 23
Gao, Hongli 567, 994
Gao, Qiang 718
Garambois, Pierre 82
García, David 694
Garcia, W L 664
Gargaro, D 139
Ghosh, Bidisha 431, 1040
Ghouri, A A 179
Gillich, G R 31, 732
Gillich, N 155, 163, 732
Gillot, Frédéric 610
Ginsberg, D 234
Goes, Luiz Carlos Sandoval 953
Goggins, J 503
Goggins, Jamie 850

AUTHOR INDEX

Gómez, C 376
González, Arturo 513
Gonzalez, Jesus 339
González-Requema, Ignacio 1032
Grigg, S 812
Gu, F 633
Güemes, Alfredo 1032
Güney, D 530
Guo, Liang 994
Guo, Xing-Lin 23
Han, Lu 648
Han, Ning 718
Han, Q K 673
Han, Qing 648
Hanley, Ciarán 480
Hansen, L M 15, 559
Harrigan, John J 313
Hassan, M S 503
Hilmy, Irfan 266
Hoell, S 551
Holford, K M 602
Homer, Rachel M 305
Hörrmann, S 772
Hoxha, Dashnor 242
Hsu, W T 360
Hua, Chunrong 578
Hua, J 978
Huang, Haifeng 567, 994
Hung, T Y 360
Iancu, V 31
Iezzi, F 147
Iliopoulos, A N 866
Iturrospe, A 98
Izman, Sudin 258
Jablonski, A 702
Jaksic, V 463, 471
Jaksic, Vesna 455
Janczura, Joanna 1002
Jang, Jae Kyeong 836
Jankowski, R 322, 416
Jara, J M 376, 392, 423
Jara, M 376, 392, 423
Jézéquel, Louis 82
Jia, Jing 495
Jiang, M 332
Johansen, R J 15, 559
Karaiskos, G 882
Kassim, Abdullah Abdul Rahman 258
Kawashita, L 929
Kedra, R 195
Kelliher, Denis 480
Kenna, A 543
Kennedy, Ciaran R 850
Kern, S 789
Khatir, S 273
Kim, Chul-Woo 439
Konneh, Mohamed 258
Koroishi, E H 171, 724
Kosova, G 656
Kourehli, S S 9
Kudela, P 945
Ku-Herrera, J J 789
Kwiecien, A 416
La Saponara, V 789
Lara-Molina, F A 724
Lasowicz, N 416
Lau, Jenny 953
Law, David W 305
Lee, J R 114
Lee, Jung Ryul 836
Lee, W 842
Leen, Sean B 850
Li, Jianvong 679
Li, Jun 994
Li, Zili 347
Limongelli, M P 368, 764
Lin, J 874, 978, 986
Liu, J 688
Liu, Jingming 578
Loh, C H 360
Loh, Chin-Hsiung 74
Loh, K J 789
Loh, Kenneth J 339
Louis, O 122
Lúcio, Válter J G 291
Luo, Z 874, 986
Mahdavi, S H 187
Mahmoudi, S 780
Maia, N 281
Majewska, K 945
Malinowski, P 129, 897, 912
Malinowski, Pawel 804
Mandic, D P 471
Manzato, S 656
Markert, Bernd 226
Marks, R 812, 929
Marra, A 139
Martínez, F 98
Marulo, F 656
McGugan, M 921
Mehrabani, Mehrdad Bahari 400
Melo, G P 171
Mendrok, Krzysztof 219
Mieloszyk, M 890, 897
Mikkelsen, L P 921
Minda, A A 163
Mituletu, I C 31, 155
Miyamoto, Ayaho 291
Mohammed, Omar 513
Mølgaard, L L 586

AUTHOR INDEX

Moll, J 912
Molyneaux, Thomas C K 305
Montanari, L 250
Montes, O 392
Morassi, A 368
Morita, Tomoaki 439
Mortezaei, Alireza 488
Mujica, L E 211
Mujica, Luis 1023, 1032
Muntean, F 163
Murphy, J 447, 471
Murphy, Jimmy 455, 463, 937
Musella, U 656
Naeimi, Meysam 347
Naik, Niranjan K 740
Nechak, Lyes 610
Negru, I 732
Nepal, Jaya 408
Nitescu, C N 155
Noble, D 384
Nogal, M 384
O'Byrne, Michael 431, 1040
O'Connor, A J 384
O'Donnell, Deirdre 431, 937
Olmos, B A 376, 392
Olmos, Ba 423
Olmos, L 423
Omenzetter, P 551
Opoka, S 890, 897, 912, 945
Opoka, Szymon 804
O'Shea, Richard 455
Ostachowicz, W 129, 890, 897, 912, 945
Ostachowicz, Wieslaw 804
O'Sullivan, K 447
Ouyang, Huajiang 23, 567
Öztürk, B 530
Paget, C 929
Pakrashi, V 384, 447, 471
Pakrashi, Vikram 431, 455, 463, 480, 937, 1040
Palka, Norbert 804
Pandya, Kedar S 740
Pearson, M 812, 842
Peeters, B 656
Peeters, Bart 953
Pepe, M 139
Pereira, G 921
Perera, R 281
Perez, Oscar 1023
Phadnis, Vaibhav A 740
Pickwell, Andy 179
Pieczonka, L 828
Pierce, S Gareth 179
Pinotti, E 820
Pinto, B 789
Polyzos, D 122
Popovici, Emanuel 937
Porcu, M C 828
Pourdeyhimi, B 748
Praisach, Z I 163, 732
Preisler, A 39
Puga, H 423
Pullin, R 602, 812, 842, 929
Qin, Zhaoye 618
Qiu, W L 332
Quek, S T 961
Quintero, M 211
Quintero, M A 664
Quintero, S 211
Rafferty, Steven 179
Rainieri, C 139
Rainieri, G M 139
Razak, H A 187
Roy, Anish 740
Ruby, M 234
Rucka, M 195, 203
Rudawska, A 624
Ruiz, M 211
Ruiz, Magda 1023, 1032
Saada, A Ait 273
Sakaris, Christos S 58
Sakellariou, John S 58, 66
Salawdeh, S 503
Sampaio, R 281
Schagerl, M 772
Schmidt, Rüdiger 226
Schoefs, Franck 431, 1040
Schröder, K-U 39
Serra, R 273
Shao, Y M 688
Shi, Weizhen 578
Shiki, Sidney Bruce 90
Shuaib, M 522
Silberschmidt, V V 748
Silberschmidt, Vadim V 740
Sinou, Jean-Jacques 610
Solís, Mario 106
Soltysik, B 322
Soman, R 129
Sozumert, E 748
Spagnoli, A 250
Spina, D 147
Sporea, D 897
Srbinovsky, B 937
Stancalie, A 897
Starkey, Andrew 313
Staszewski, W J 828
Steenbock, C 39
Steffen Jr, V 724
Stevenson, Morris W 400
Straczkiewicz, M 702

AUTHOR INDEX

Strantza, M 122
Sugiura, Kunitomo 439
Sun, Jian 641
Surace, C 820
Tabrizi, Mohsen Mohajer 488
Tanaka, Hideaki 291
Tcherniak, D 15, 559, 586
Tcherniak, Dmitri 694
Trendafilova, I 710
Trendafilova, Irina 694
Trivaudey, F 780
Truong, T C 114
Tsangouri, E 882
Tufoi, M 31, 732
Ugalde, U 98
Ulriksen, M D 15, 559, 969
Valente, C 147
Vamvoudakis-Stefanou, Kyriakos J 66
Van Hemelrijck, D 122, 866, 882
Vece, M 904
Viechtbauer, C 772
Villamizar, Rodolfo 1023, 1032
Wahab, M Abdel 281
Wahab, M M Abdel 266
Wandowski, T 897, 912
Wandowski, Tomasz 804
Wang, Changlin 648
Wang, Fengtao 641
Wang, Tianyang 679
Wang, Wanqin 718
Wang, Z 961
Wang, Ziran 439
Weijtjens, W 866
Wright, C 447, 463, 471
Wright, Robert 431
Wu, C W 47
Wylomanska, Agnieszka 1002
Xu, Yong 641
Yaacoubi, S 1010, 1017
Yan, Bing 578
Yan, Dawen 641
Yan, Jinghong 495
Yang, Xiu-Ming 23
Yasui, J 789
Yin, L 688
Zeng, L 874, 978, 986
Zhang, Muyu 226
Zhang, R 633
Zhang, Shenghua 641
Zhang, W 47
Zhang, Xiaochen 994
Zhang, Z 332
Zhao, Yanqing 718
Zheng, H L 688
Zhou, Yun-Lai 281
Zhu, Z D 47
Zima, B 203
Zimroz, Radoslaw 1002

Institute of Physics
Dirac House, Temple Back
Bristol BS1 6BE UK